EXPONENTS AND RADICALS

$x^m x^n = x^{m+n}$

$\dfrac{x^m}{x^n} = x^{m-n}$

$(x^m)^n = x^{mn}$

$x^{-n} = \dfrac{1}{x^n}$

$(xy)^n = x^n y^n$

$\left(\dfrac{x}{y}\right)^n = \dfrac{x^n}{y^n}$

$x^{1/n} = \sqrt[n]{x}$

$x^{m/n} = \sqrt[n]{x^m} = (\sqrt[n]{x})^m$

$\sqrt[n]{xy} = \sqrt[n]{x}\,\sqrt[n]{y}$

$\sqrt[n]{\dfrac{x}{y}} = \dfrac{\sqrt[n]{x}}{\sqrt[n]{y}}$

$\sqrt[m]{\sqrt[n]{x}} = \sqrt[n]{\sqrt[m]{x}} = \sqrt[mn]{x}$

SPECIAL PRODUCTS

$(x + y)^2 = x^2 + 2xy + y^2$

$(x - y)^2 = x^2 - 2xy + y^2$

$(x + y)^3 = x^3 + 3x^2y + 3xy^2 + y^3$

$(x - y)^3 = x^3 - 3x^2y + 3xy^2 - y^3$

FACTORING FORMULAS

$x^2 - y^2 = (x + y)(x - y)$

$x^2 + 2xy + y^2 = (x + y)^2$

$x^2 - 2xy + y^2 = (x - y)^2$

$x^3 + y^3 = (x + y)(x^2 - xy + y^2)$

$x^3 - y^3 = (x - y)(x^2 + xy + y^2)$

QUADRATIC FORMULA

If $ax^2 + bx + c = 0$, then

$$x = \frac{-b \pm \sqrt{b^2 - 4ac}}{2a}$$

INEQUALITIES AND ABSOLUTE VALUE

If $a < b$ and $b < c$, then $a < c$.

If $a < b$, then $a + c < b + c$.

If $a < b$ and $c > 0$, then $ca < cb$.

If $a < b$ and $c < 0$, then $ca > cb$.

If $a > 0$, then

$|x| = a$ means $x = a$ or $x = -a$.

$|x| < a$ means $-a < x < a$.

$|x| > a$ means $x > a$ or $x < -a$.

GEOMETRIC FORMULAS

Formulas for area A, perimeter P, circumference C, volume V:

Rectangle

$A = lw$

$P = 2l + 2w$

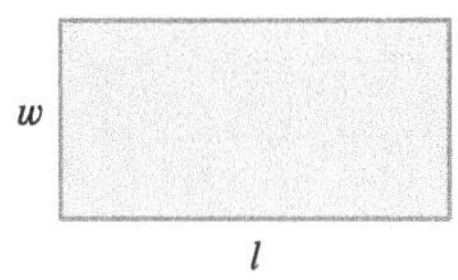

Box

$V = lwh$

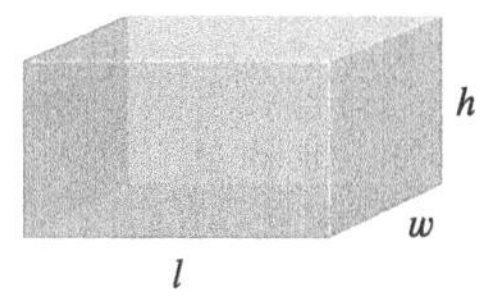

Triangle

$A = \frac{1}{2}bh$

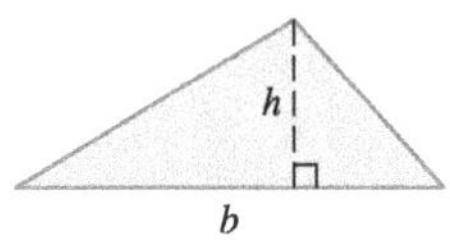

Pyramid

$V = \frac{1}{3}ha^2$

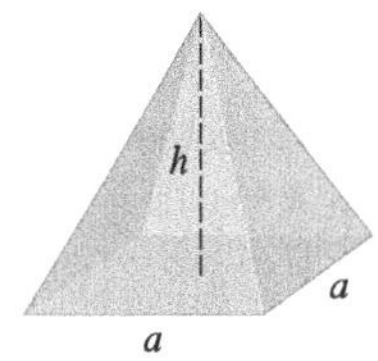

Circle

$A = \pi r^2$

$C = 2\pi r$

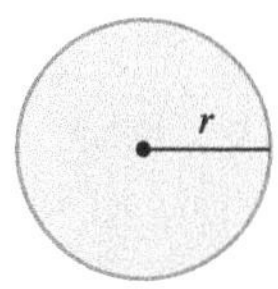

Sphere

$V = \frac{4}{3}\pi r^3$

$A = 4\pi r^2$

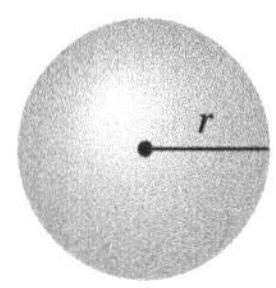

Cylinder

$V = \pi r^2 h$

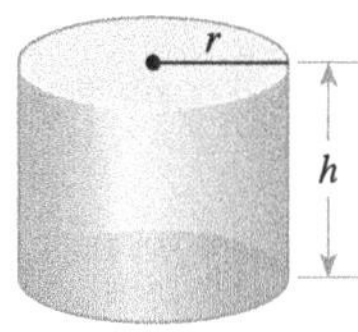

Cone

$V = \frac{1}{3}\pi r^2 h$

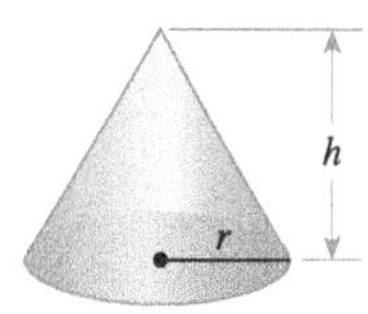

HERON'S FORMULA

$\text{Area} = \sqrt{s(s - a)(s - b)(s - c)}$

where $s = \dfrac{a + b + c}{2}$

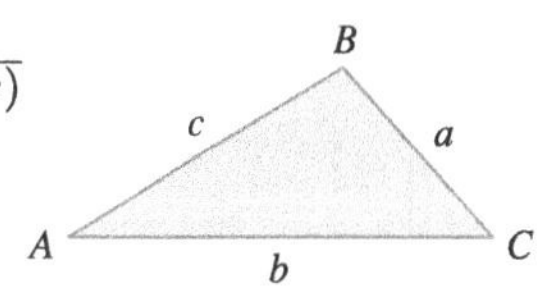

DISTANCE AND MIDPOINT FORMULAS

Distance between $P_1(x_1, y_1)$ and $P_2(x_2, y_2)$:

$$d = \sqrt{(x_2 - x_1)^2 + (y_2 - y_1)^2}$$

Midpoint of $\overline{P_1P_2}$: $\left(\dfrac{x_1 + x_2}{2}, \dfrac{y_1 + y_2}{2}\right)$

LINES

Slope of line through $P_1(x_1, y_1)$ and $P_2(x_2, y_2)$ $\qquad m = \dfrac{y_2 - y_1}{x_2 - x_1}$

Point-slope equation of line through $P_1(x_1, y_1)$ with slope m $\qquad y - y_1 = m(x - x_1)$

Slope-intercept equation of line with slope m and y-intercept b $\qquad y = mx + b$

Two-intercept equation of line with x-intercept a and y-intercept b $\qquad \dfrac{x}{a} + \dfrac{y}{b} = 1$

LOGARITHMS

$y = \log_a x$ means $a^y = x$

$\log_a a^x = x$	$a^{\log_a x} = x$
$\log_a 1 = 0$	$\log_a a = 1$
$\log x = \log_{10} x$	$\ln x = \log_e x$
$\log_a xy = \log_a x + \log_a y$	$\log_a\left(\dfrac{x}{y}\right) = \log_a x - \log_a y$
$\log_a x^b = b \log_a x$	$\log_b x = \dfrac{\log_a x}{\log_a b}$

EXPONENTIAL AND LOGARITHMIC FUNCTIONS

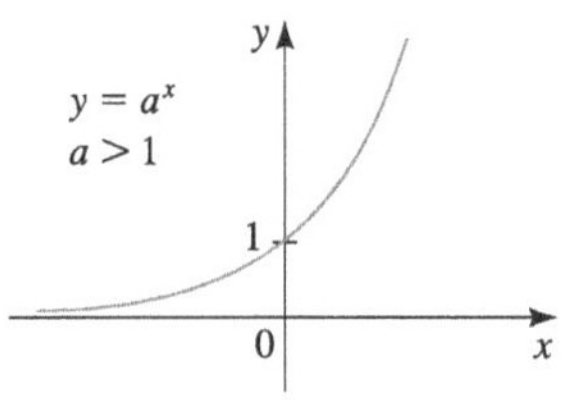

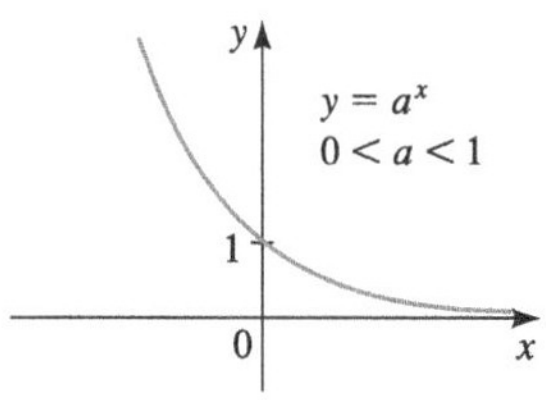

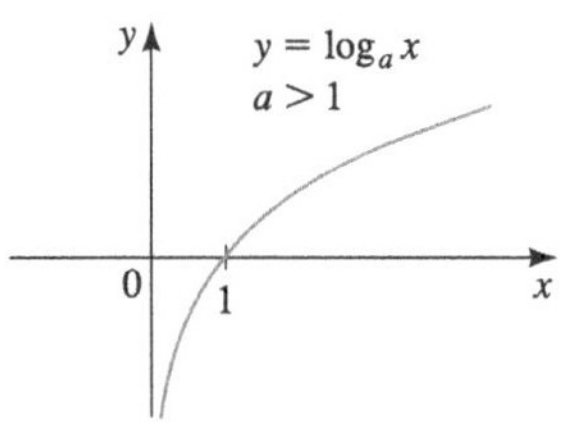

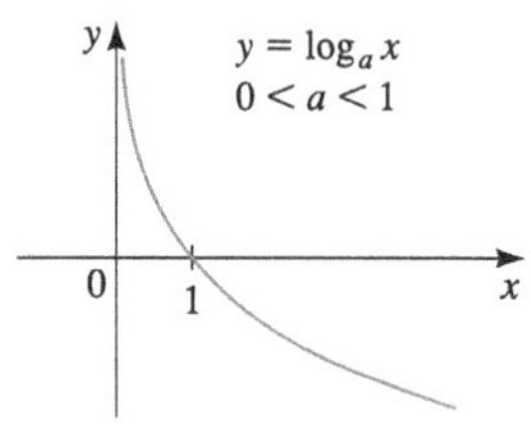

GRAPHS OF FUNCTIONS

Linear functions: $f(x) = mx + b$

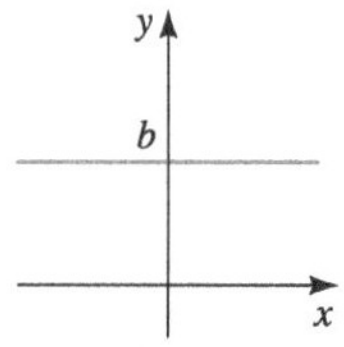

$f(x) = b$

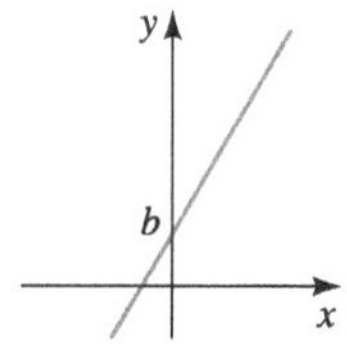

$f(x) = mx + b$

Power functions: $f(x) = x^n$

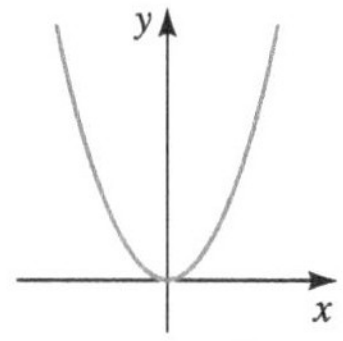

$f(x) = x^2$

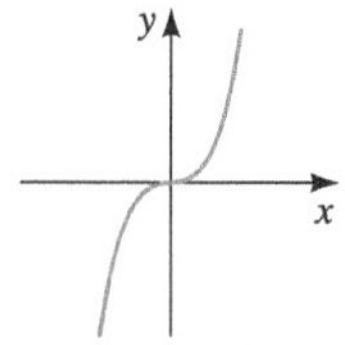

$f(x) = x^3$

Root functions: $f(x) = \sqrt[n]{x}$

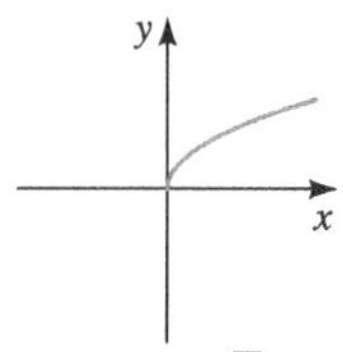

$f(x) = \sqrt{x}$

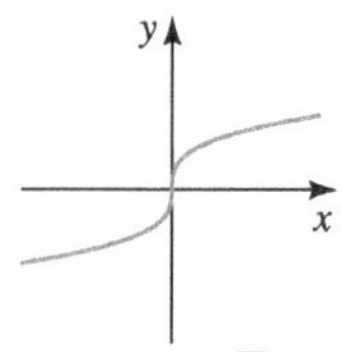

$f(x) = \sqrt[3]{x}$

Reciprocal functions: $f(x) = 1/x^n$

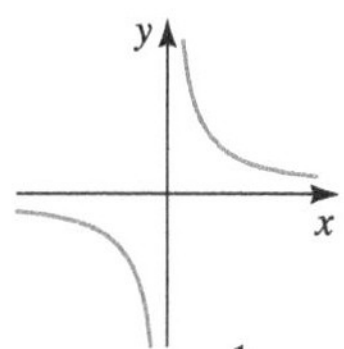

$f(x) = \dfrac{1}{x}$

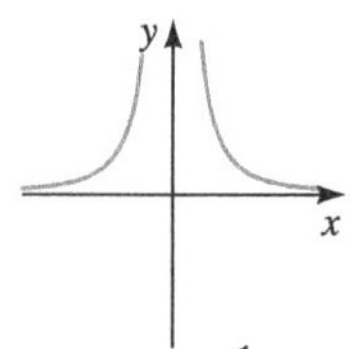

$f(x) = \dfrac{1}{x^2}$

Absolute value function

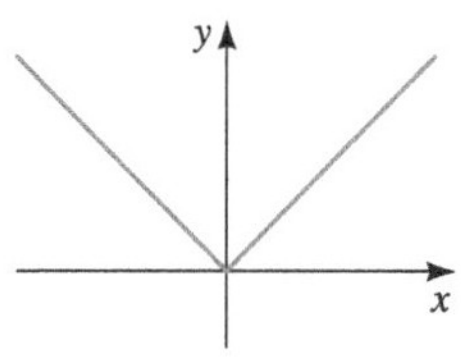

$f(x) = |x|$

Greatest integer function

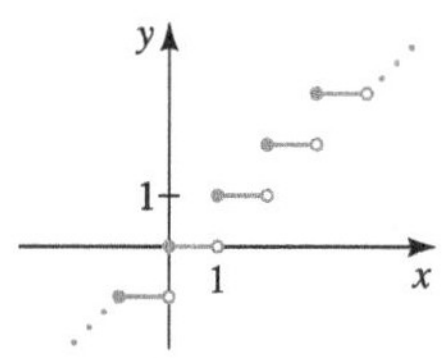

$f(x) = [\![x]\!]$

COMPLEX NUMBERS

For the complex number $z = a + bi$

the **conjugate** is $\bar{z} = a - bi$

the **modulus** is $|z| = \sqrt{a^2 + b^2}$

the **argument** is θ, where $\tan\theta = b/a$

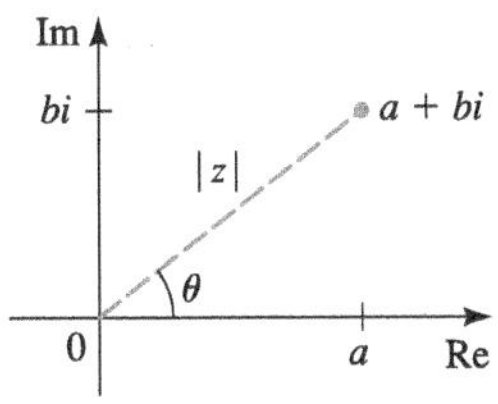

Polar form of a complex number

For $z = a + bi$, the **polar form** is

$$z = r(\cos\theta + i\sin\theta)$$

where $r = |z|$ is the modulus of z and θ is the argument of z

De Moivre's Theorem

$$z^n = [r(\cos\theta + i\sin\theta)]^n = r^n(\cos n\theta + i\sin n\theta)$$

$$\sqrt[n]{z} = [r(\cos\theta + i\sin\theta)]^{1/n}$$

$$= r^{1/n}\left(\cos\frac{\theta + 2k\pi}{n} + i\sin\frac{\theta + 2k\pi}{n}\right)$$

where $k = 0, 1, 2, \ldots, n - 1$

ROTATION OF AXES

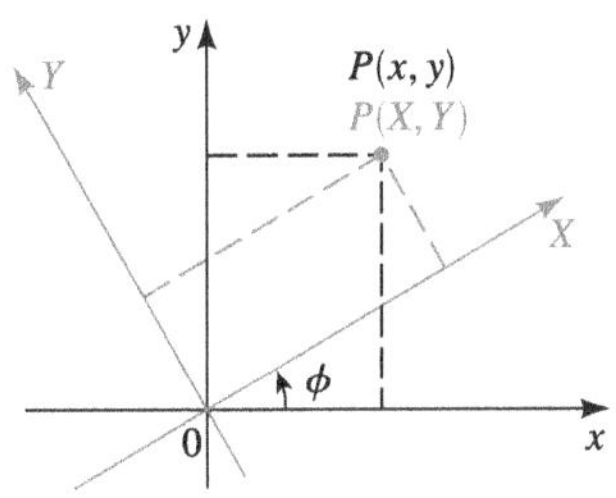

Rotation of axes formulas

$$x = X\cos\phi - Y\sin\phi$$
$$y = X\sin\phi + Y\cos\phi$$

Angle-of-rotation formula for conic sections

To eliminate the xy-term in the equation

$$Ax^2 + Bxy + Cy^2 + Dx + Ey + F = 0$$

rotate the axis by the angle ϕ that satisfies

$$\cot 2\phi = \frac{A - C}{B}$$

POLAR COORDINATES

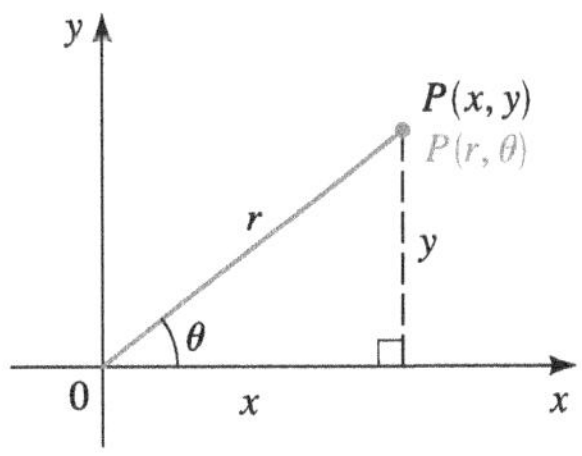

$$x = r\cos\theta$$
$$y = r\sin\theta$$
$$r^2 = x^2 + y^2$$
$$\tan\theta = \frac{y}{x}$$

SUMS OF POWERS OF INTEGERS

$$\sum_{k=1}^{n} 1 = n \qquad \sum_{k=1}^{n} k = \frac{n(n+1)}{2}$$

$$\sum_{k=1}^{n} k^2 = \frac{n(n+1)(2n+1)}{6} \qquad \sum_{k=1}^{n} k^3 = \frac{n^2(n+1)^2}{4}$$

THE DERIVATIVE

The **average rate of change** of f between a and b is

$$\frac{f(b) - f(a)}{b - a}$$

The **derivative** of f at a is

$$f'(a) = \lim_{x \to a}\frac{f(x) - f(a)}{x - a}$$

$$f'(a) = \lim_{h \to 0}\frac{f(a + h) - f(a)}{h}$$

AREA UNDER THE GRAPH OF *f*

The **area under the graph of *f*** on the interval $[a, b]$ is the limit of the sum of the areas of approximating rectangles

$$A = \lim_{n \to \infty}\sum_{k=1}^{n} f(x_k)\,\Delta x$$

where

$$\Delta x = \frac{b - a}{n}$$

$$x_k = a + k\,\Delta x$$

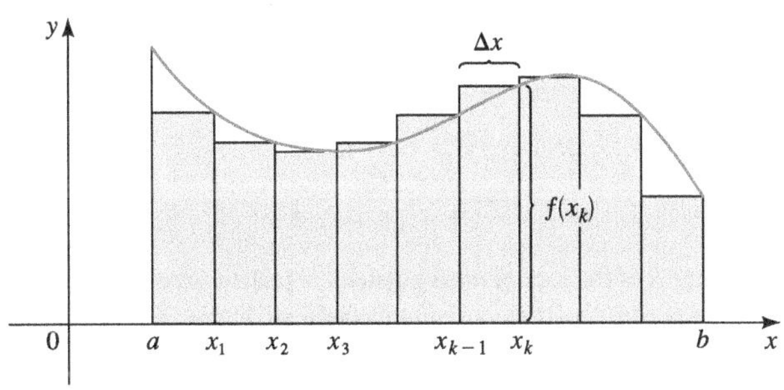

SEQUENCES AND SERIES

Arithmetic

$$a,\ a + d,\ a + 2d,\ a + 3d,\ a + 4d,\ \ldots$$

$$a_n = a + (n - 1)d$$

$$S_n = \sum_{k=1}^{n} a_k = \frac{n}{2}[2a + (n - 1)d] = n\left(\frac{a + a_n}{2}\right)$$

Geometric

$$a,\ ar,\ ar^2,\ ar^3,\ ar^4,\ \ldots$$

$$a_n = ar^{n-1}$$

$$S_n = \sum_{k=1}^{n} a_k = a\frac{1 - r^n}{1 - r}$$

If $|r| < 1$, then the sum of an infinite geometric series is

$$S = \frac{a}{1 - r}$$

THE BINOMIAL THEOREM

$$(a + b)^n = \binom{n}{0}a^n + \binom{n}{1}a^{n-1}b + \cdots + \binom{n}{n-1}ab^{n-1} + \binom{n}{n}b^n$$

FINANCE

Compound interest

$$A = P\left(1 + \frac{r}{n}\right)^{nt}$$

where A is the amount after t years, P is the principal, r is the interest rate, and the interest is compounded n times per year.

Amount of an annuity

$$A_f = R\frac{(1 + i)^n - 1}{i}$$

where A_f is the final amount, i is the interest rate per time period, and there are n payments of size R.

Present value of an annuity

$$A_p = R\frac{1 - (1 + i)^{-n}}{i}$$

where A_p is the present value, i is the interest rate per time period, and there are n payments of size R.

Installment buying

$$R = \frac{iA_p}{1 - (1 + i)^{-n}}$$

where R is the size of each payment, i is the interest rate per time period, A_p is the amount of the loan, and n is the number of payments.

CONIC SECTIONS

Circles

$$(x - h)^2 + (y - k)^2 = r^2$$

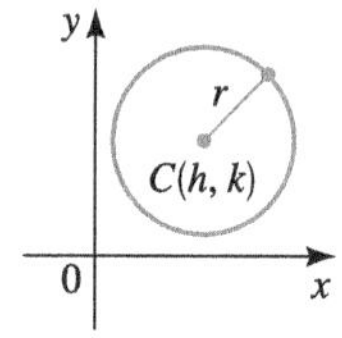

Parabolas

$$x^2 = 4py$$

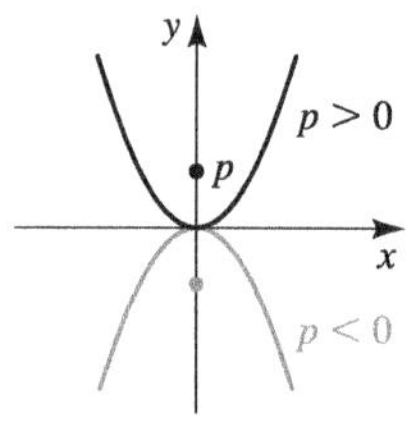

Focus $(0, p)$, directrix $y = -p$

$$y^2 = 4px$$

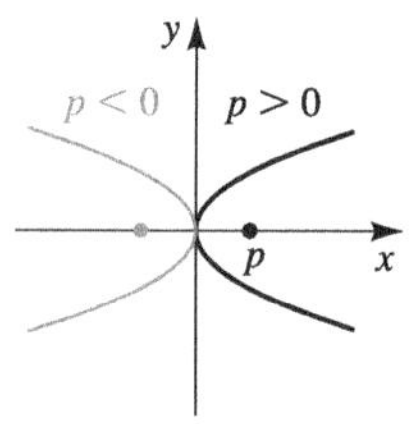

Focus $(p, 0)$, directrix $x = -p$

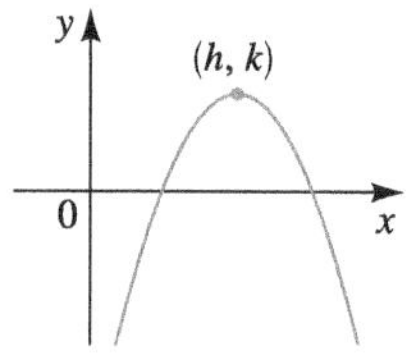

$y = a(x - h)^2 + k,$
$a < 0,\quad h > 0,\quad k > 0$

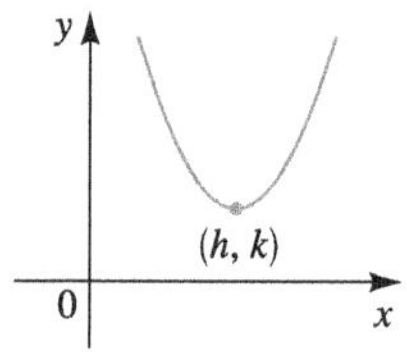

$y = a(x - h)^2 + k,$
$a > 0,\quad h > 0,\quad k > 0$

Ellipses

$$\frac{x^2}{a^2} + \frac{y^2}{b^2} = 1$$

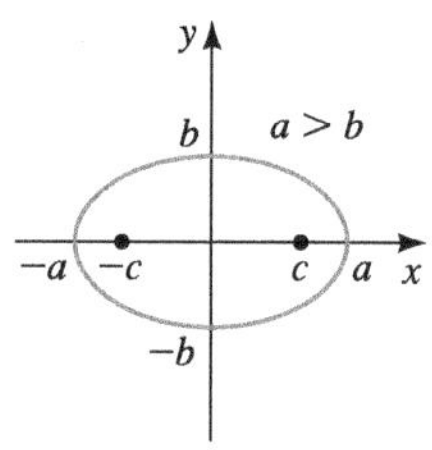

Foci $(\pm c, 0)$, $c^2 = a^2 - b^2$

$$\frac{x^2}{b^2} + \frac{y^2}{a^2} = 1$$

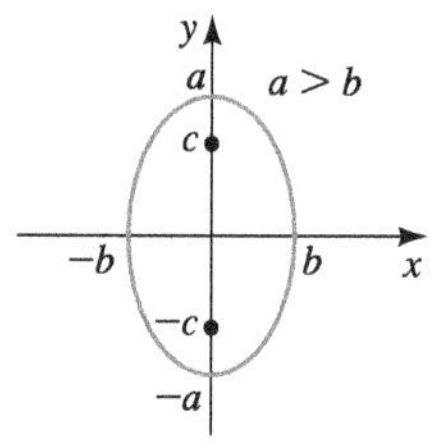

Foci $(0, \pm c)$, $c^2 = a^2 - b^2$

Hyperbolas

$$\frac{x^2}{a^2} - \frac{y^2}{b^2} = 1$$

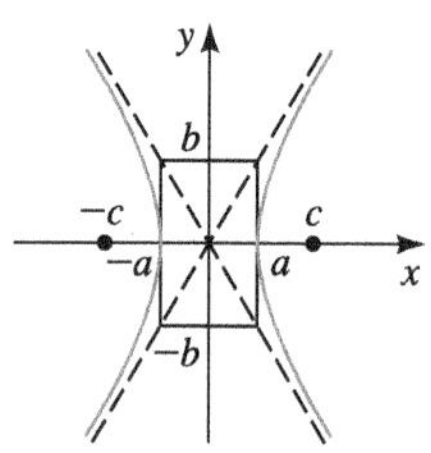

Foci $(\pm c, 0)$, $c^2 = a^2 + b^2$

$$-\frac{x^2}{b^2} + \frac{y^2}{a^2} = 1$$

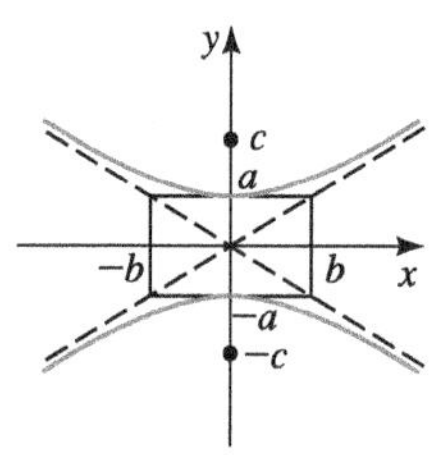

Foci $(0, \pm c)$, $c^2 = a^2 + b^2$

ANGLE MEASUREMENT

π radians $= 180°$

$1° = \dfrac{\pi}{180}$ rad $\qquad 1 \text{ rad} = \dfrac{180°}{\pi}$

$s = r\theta \qquad A = \frac{1}{2}r^2\theta \quad (\theta \text{ in radians})$

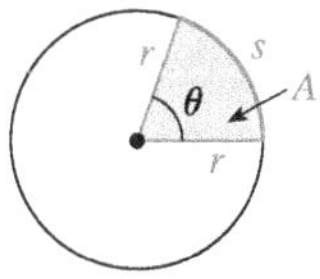

To convert from degrees to radians, multiply by $\dfrac{\pi}{180}$.

To convert from radians to degrees, multiply by $\dfrac{180}{\pi}$.

TRIGONOMETRIC FUNCTIONS OF REAL NUMBERS

$\sin t = y \qquad \csc t = \dfrac{1}{y}$

$\cos t = x \qquad \sec t = \dfrac{1}{x}$

$\tan t = \dfrac{y}{x} \qquad \cot t = \dfrac{x}{y}$

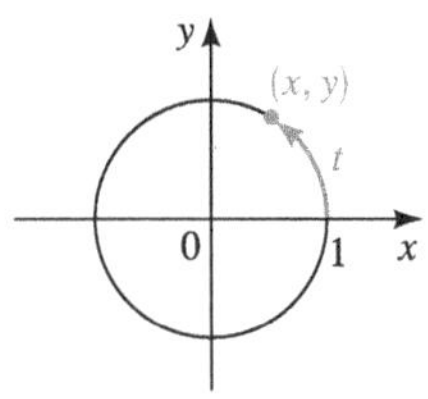

TRIGONOMETRIC FUNCTIONS OF ANGLES

$\sin \theta = \dfrac{y}{r} \qquad \csc \theta = \dfrac{r}{y}$

$\cos \theta = \dfrac{x}{r} \qquad \sec \theta = \dfrac{r}{x}$

$\tan \theta = \dfrac{y}{x} \qquad \cot \theta = \dfrac{x}{y}$

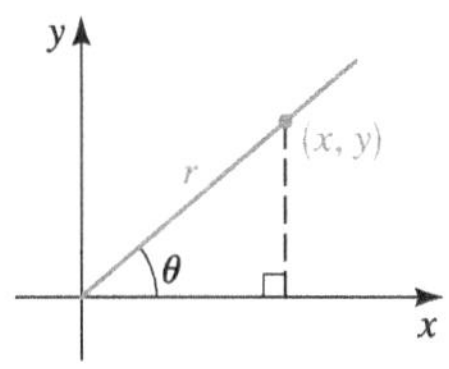

RIGHT ANGLE TRIGONOMETRY

$\sin \theta = \dfrac{\text{opp}}{\text{hyp}} \qquad \csc \theta = \dfrac{\text{hyp}}{\text{opp}}$

$\cos \theta = \dfrac{\text{adj}}{\text{hyp}} \qquad \sec \theta = \dfrac{\text{hyp}}{\text{adj}}$

$\tan \theta = \dfrac{\text{opp}}{\text{adj}} \qquad \cot \theta = \dfrac{\text{adj}}{\text{opp}}$

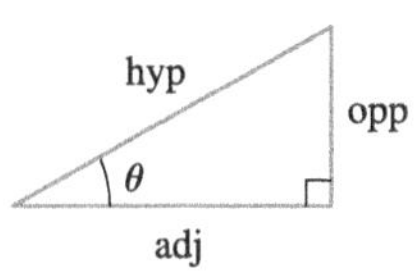

SPECIAL VALUES OF THE TRIGONOMETRIC FUNCTIONS

θ	radians	$\sin \theta$	$\cos \theta$	$\tan \theta$
0°	0	0	1	0
30°	$\pi/6$	$1/2$	$\sqrt{3}/2$	$\sqrt{3}/3$
45°	$\pi/4$	$\sqrt{2}/2$	$\sqrt{2}/2$	1
60°	$\pi/3$	$\sqrt{3}/2$	$1/2$	$\sqrt{3}$
90°	$\pi/2$	1	0	—
180°	π	0	-1	0
270°	$3\pi/2$	-1	0	—

SPECIAL TRIANGLES

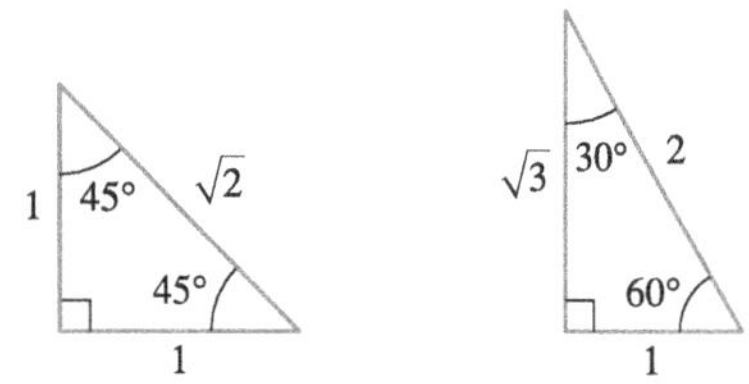

GRAPHS OF THE TRIGONOMETRIC FUNCTIONS

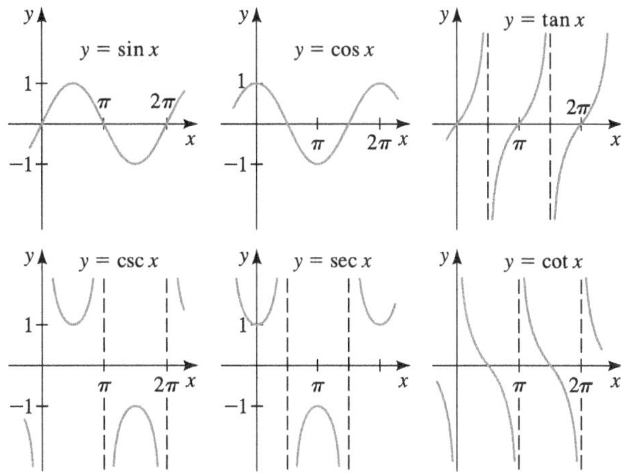

SINE AND COSINE CURVES

$y = a \sin k(x - b) \quad (k > 0) \qquad y = a \cos k(x - b) \quad (k > 0)$

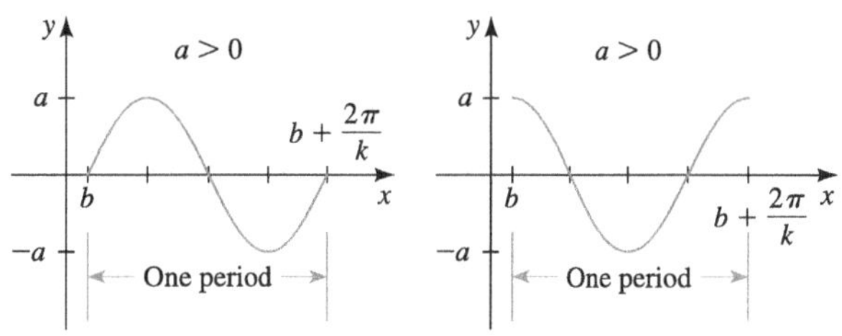

amplitude: $|a|$ period: $2\pi/k$ phase shift: b

GRAPHS OF THE INVERSE TRIGONOMETRIC FUNCTIONS

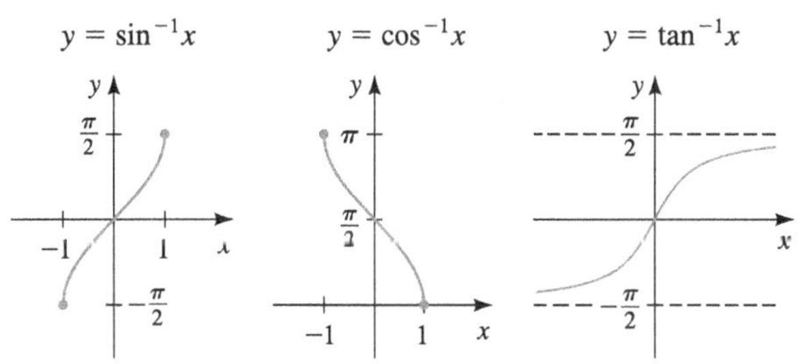

FUNDAMENTAL IDENTITIES

$$\sec x = \frac{1}{\cos x} \qquad \csc x = \frac{1}{\sin x}$$

$$\tan x = \frac{\sin x}{\cos x} \qquad \cot x = \frac{1}{\tan x}$$

$$\sin^2 x + \cos^2 x = 1 \qquad 1 + \tan^2 x = \sec^2 x \qquad 1 + \cot^2 x = \csc^2 x$$

$$\sin(-x) = -\sin x \qquad \cos(-x) = \cos x \qquad \tan(-x) = -\tan x$$

COFUNCTION IDENTITIES

$$\sin\left(\frac{\pi}{2} - x\right) = \cos x \qquad \cos\left(\frac{\pi}{2} - x\right) = \sin x$$

$$\tan\left(\frac{\pi}{2} - x\right) = \cot x \qquad \cot\left(\frac{\pi}{2} - x\right) = \tan x$$

$$\sec\left(\frac{\pi}{2} - x\right) = \csc x \qquad \csc\left(\frac{\pi}{2} - x\right) = \sec x$$

REDUCTION IDENTITIES

$$\sin(x + \pi) = -\sin x \qquad \sin\left(x + \frac{\pi}{2}\right) = \cos x$$

$$\cos(x + \pi) = -\cos x \qquad \cos\left(x + \frac{\pi}{2}\right) = -\sin x$$

$$\tan(x + \pi) = \tan x \qquad \tan\left(x + \frac{\pi}{2}\right) = -\cot x$$

ADDITION AND SUBTRACTION FORMULAS

$$\sin(x + y) = \sin x \cos y + \cos x \sin y$$

$$\sin(x - y) = \sin x \cos y - \cos x \sin y$$

$$\cos(x + y) = \cos x \cos y - \sin x \sin y$$

$$\cos(x - y) = \cos x \cos y + \sin x \sin y$$

$$\tan(x + y) = \frac{\tan x + \tan y}{1 - \tan x \tan y} \qquad \tan(x - y) = \frac{\tan x - \tan y}{1 + \tan x \tan y}$$

DOUBLE-ANGLE FORMULAS

$$\sin 2x = 2 \sin x \cos x \qquad \cos 2x = \cos^2 x - \sin^2 x$$

$$= 2\cos^2 x - 1$$

$$\tan 2x = \frac{2 \tan x}{1 - \tan^2 x} \qquad = 1 - 2\sin^2 x$$

FORMULAS FOR REDUCING POWERS

$$\sin^2 x = \frac{1 - \cos 2x}{2} \qquad \cos^2 x = \frac{1 + \cos 2x}{2}$$

$$\tan^2 x = \frac{1 - \cos 2x}{1 + \cos 2x}$$

HALF-ANGLE FORMULAS

$$\sin \frac{u}{2} = \pm\sqrt{\frac{1 - \cos u}{2}} \qquad \cos \frac{u}{2} = \pm\sqrt{\frac{1 + \cos u}{2}}$$

$$\tan \frac{u}{2} = \frac{1 - \cos u}{\sin u} = \frac{\sin u}{1 + \cos u}$$

PRODUCT-TO-SUM AND SUM-TO-PRODUCT IDENTITIES

$$\sin u \cos v = \tfrac{1}{2}[\sin(u + v) + \sin(u - v)]$$

$$\cos u \sin v = \tfrac{1}{2}[\sin(u + v) - \sin(u - v)]$$

$$\cos u \cos v = \tfrac{1}{2}[\cos(u + v) + \cos(u - v)]$$

$$\sin u \sin v = \tfrac{1}{2}[\cos(u - v) - \cos(u + v)]$$

$$\sin x + \sin y = 2 \sin \frac{x + y}{2} \cos \frac{x - y}{2}$$

$$\sin x - \sin y = 2 \cos \frac{x + y}{2} \sin \frac{x - y}{2}$$

$$\cos x + \cos y = 2 \cos \frac{x + y}{2} \cos \frac{x - y}{2}$$

$$\cos x - \cos y = -2 \sin \frac{x + y}{2} \sin \frac{x - y}{2}$$

THE LAWS OF SINES AND COSINES

The Law of Sines

$$\frac{\sin A}{a} = \frac{\sin B}{b} = \frac{\sin C}{c}$$

The Law of Cosines

$$a^2 = b^2 + c^2 - 2bc \cos A$$

$$b^2 = a^2 + c^2 - 2ac \cos B$$

$$c^2 = a^2 + b^2 - 2ab \cos C$$

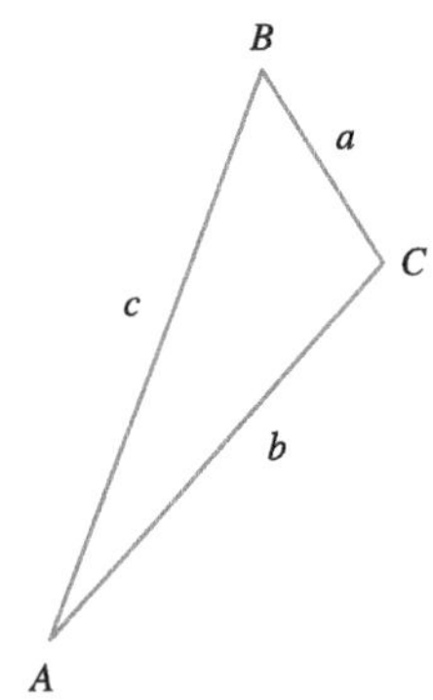

Precalculus

Mathematics for Calculus

Seventh Edition

James Stewart | Lothar Redlin | Saleem Watson

CENGAGE Learning®

Australia • Brazil • Japan • Korea • Mexico • Singapore • Spain • United Kingdom • United States

CENGAGE Learning

Precalculus: Mathematics for Calculus, Seventh Edition

Printed at CLDPC, USA, 07-18

Precalculus: Mathematics for Calculus, 7th Edition
James Stewart | Lothar Redlin | Saleem Watson

ISBN: 978-1-305-75046-3

WCN: 01-100-101

Cengage Learning
20 Channel Center Street
Boston, MA 02210
USA

Cengage Learning is a leading provider of customized learning solutions with office locations around the globe, including Singapore, the United Kingdom, Australia, Mexico, Brazil, and Japan. Locate your local office at: **www.international.cengage.com/region.**

Cengage Learning products are represented in Canada by Nelson Education, Ltd.

For your lifelong learning solutions, visit www.**cengage.com/custom.**

Visit our corporate website at www.**cengage.com.**

PREFACE

What do students really need to know to be prepared for calculus? What tools do instructors really need to assist their students in preparing for calculus? These two questions have motivated the writing of this book.

To be prepared for calculus a student needs not only technical skill but also a clear understanding of concepts. Indeed, *conceptual understanding* and *technical skill* go hand in hand, each reinforcing the other. A student also needs to gain an appreciation for the power and utility of mathematics in *modeling* the real world. Every feature of this textbook is devoted to fostering these goals.

In this Seventh Edition our objective is to further enhance the effectiveness of the book as an instructional tool for teachers and as a learning tool for students. Many of the changes in this edition are a result of suggestions we received from instructors and students who are using the current edition; others are a result of insights we have gained from our own teaching. Some chapters have been reorganized and rewritten, new sections have been added (as described below), the review material at the end of each chapter has been substantially expanded, and exercise sets have been enhanced to further focus on the main concepts of precalculus. In all these changes and numerous others (small and large) we have retained the main features that have contributed to the success of this book.

New to the Seventh Edition

- **Exercises** More than 20% of the exercises are new, and groups of exercises now have headings that identify the type of exercise. New *Skills Plus* exercises in most sections contain more challenging exercises that require students to extend and synthesize concepts.
- **Review Material** The review material at the end of each chapter now includes a summary of *Properties and Formulas* and a new *Concept Check*. Each *Concept Check* provides a step-by-step review of all the main concepts and applications of the chapter. Answers to the *Concept Check* questions are on tear-out sheets at the back of the book.
- **Discovery Projects** References to *Discovery Projects,* including brief descriptions of the content of each project, are located in boxes where appropriate in each chapter. These boxes highlight the applications of precalculus in many different real-world contexts. (The projects are located at the book companion website: **www.stewartmath.com**.)
- **Geometry Review** A new Appendix A contains a review of the main concepts of geometry used in this book, including similarity and the Pythagorean Theorem.
- **CHAPTER 1 Fundamentals** This chapter now contains two new sections. Section 1.6, "Complex Numbers" (formerly in Chapter 3), has been moved here. Section 1.12, "Modeling Variation," is now also in this chapter.
- **CHAPTER 2 Functions** This chapter now includes the new Section 2.5, "Linear Functions and Models." This section highlights the connection between the slope of a line and the rate of change of a linear function. These two interpretations of slope help prepare students for the concept of the derivative in calculus.
- **CHAPTER 3 Polynomial and Rational Functions** This chapter now includes the new Section 3.7, "Polynomial and Rational Inequalities." Section 3.6, "Rational Functions," has a new subsection on rational functions with "holes." The sections on complex numbers and on variation have been moved to Chapter 1.

- **CHAPTER 4 Exponential and Logarithmic Functions** The chapter now includes two sections on the applications of these functions. Section 4.6, "Modeling with Exponential Functions," focuses on modeling growth and decay, Newton's Law of Cooling, and other such applications. Section 4.7, "Logarithmic Scales," covers the concept of a logarithmic scale with applications involving the pH, Richter, and decibel scales.
- **CHAPTER 5 Trigonometric Functions: Unit Circle Approach** This chapter includes a new subsection on the concept of phase shift as used in modeling harmonic motion.
- **CHAPTER 10 Systems of Equations and Inequalities** The material on systems of inequalities has been rewritten to emphasize the steps used in graphing the solution of a system of inequalities.

Teaching with the Help of This Book

We are keenly aware that good teaching comes in many forms and that there are many different approaches to teaching and learning the concepts and skills of precalculus. The organization and exposition of the topics in this book are designed to accommodate different teaching and learning styles. In particular, each topic is presented algebraically, graphically, numerically, and verbally, with emphasis on the relationships between these different representations. The following are some special features that can be used to complement different teaching and learning styles:

Exercise Sets The most important way to foster conceptual understanding and hone technical skill is through the problems that the instructor assigns. To that end we have provided a wide selection of exercises.

- **Concept Exercises** These exercises ask students to use mathematical language to state fundamental facts about the topics of each section.
- **Skills Exercises** These exercises reinforce and provide practice with all the learning objectives of each section. They comprise the core of each exercise set.
- **Skills Plus Exercises** The *Skills Plus* exercises contain challenging problems that often require the synthesis of previously learned material with new concepts.
- **Applications Exercises** We have included substantial applied problems from many different real-world contexts. We believe that these exercises will capture students' interest.
- **Discovery, Writing, and Group Learning** Each exercise set ends with a block of exercises labeled *Discuss* ■ *Discover* ■ *Prove* ■ *Write.* These exercises are designed to encourage students to experiment, preferably in groups, with the concepts developed in the section and then to write about what they have learned rather than simply looking for the answer. New *Prove* exercises highlight the importance of deriving a formula.
- **Now Try Exercise . . .** At the end of each example in the text the student is directed to one or more similar exercises in the section that help to reinforce the concepts and skills developed in that example.
- **Check Your Answer** Students are encouraged to check whether an answer they obtained is reasonable. This is emphasized throughout the text in numerous *Check Your Answer* sidebars that accompany the examples (see, for instance, pages 54 and 71).

A Complete Review Chapter We have included an extensive review chapter primarily as a handy reference for the basic concepts that are preliminary to this course.

- **CHAPTER 1 Fundamentals** This is the review chapter; it contains the fundamental concepts from algebra and analytic geometry that a student needs in order to begin a precalculus course. As much or as little of this chapter can be covered in class as needed, depending on the background of the students.
- **CHAPTER 1 Test** The test at the end of Chapter 1 is designed as a diagnostic test for determining what parts of this review chapter need to be taught. It also serves to help students gauge exactly what topics they need to review.

Flexible Approach to Trigonometry The trigonometry chapters of this text have been written so that either the right triangle approach or the unit circle approach may be taught first. Putting these two approaches in different chapters, each with its relevant applications, helps to clarify the purpose of each approach. The chapters introducing trigonometry are as follows.

- **CHAPTER 5 Trigonometric Functions: Unit Circle Approach** This chapter introduces trigonometry through the unit circle approach. This approach emphasizes that the trigonometric functions are functions of real numbers, just like the polynomial and exponential functions with which students are already familiar.
- **CHAPTER 6 Trigonometric Functions: Right Triangle Approach** This chapter introduces trigonometry through the right triangle approach. This approach builds on the foundation of a conventional high-school course in trigonometry.

Another way to teach trigonometry is to intertwine the two approaches. Some instructors teach this material in the following order: Sections 5.1, 5.2, 6.1, 6.2, 6.3, 5.3, 5.4, 5.5, 5.6, 6.4, 6.5, and 6.6. Our organization makes it easy to do this without obscuring the fact that the two approaches involve distinct representations of the same functions.

Graphing Calculators and Computers We make use of graphing calculators and computers in examples and exercises throughout the book. Our calculator-oriented examples are always preceded by examples in which students must graph or calculate by hand so that they can understand precisely what the calculator is doing when they later use it to simplify the routine, mechanical part of their work. The graphing calculator sections, subsections, examples, and exercises, all marked with the special symbol , are optional and may be omitted without loss of continuity.

- **Using a Graphing Calculator** General guidelines on using graphing calculators and a quick reference guide to using TI-83/84 calculators are available at the book companion website: **www.stewartmath.com**.
- **Graphing, Regression, Matrix Algebra** Graphing calculators are used throughout the text to graph and analyze functions, families of functions, and sequences; to calculate and graph regression curves; to perform matrix algebra; to graph linear inequalities; and other powerful uses.
- **Simple Programs** We exploit the programming capabilities of a graphing calculator to simulate real-life situations, to sum series, or to compute the terms of a recursive sequence (see, for instance, pages 628, 896, and 939).

Focus on Modeling The theme of modeling has been used throughout to unify and clarify the many applications of precalculus. We have made a special effort to clarify the essential process of translating problems from English into the language of mathematics (see pages 238 and 686).

- **Constructing Models** There are many applied problems throughout the book in which students are given a model to analyze (see, for instance, page 250). But the material on modeling, in which students are required to *construct* mathematical models, has been organized into clearly defined sections and subsections (see, for instance, pages 370, 445, and 685).

- **Focus on Modeling** Each chapter concludes with a *Focus on Modeling* section. The first such section, after Chapter 1, introduces the basic idea of modeling a real-life situation by fitting lines to data (linear regression). Other sections present ways in which polynomial, exponential, logarithmic, and trigonometric functions, and systems of inequalities can all be used to model familiar phenomena from the sciences and from everyday life (see, for instance, pages 325, 392, and 466).

Review Sections and Chapter Tests Each chapter ends with an extensive review section that includes the following.

- **Properties and Formulas** The *Properties and Formulas* at the end of each chapter contains a summary of the main formulas and procedures of the chapter (see, for instance, pages 386 and 460).
- **Concept Check and Concept Check Answers** The *Concept Check* at the end of each chapter is designed to get the students to think about and explain each concept presented in the chapter and then to use the concept in a given problem. This provides a step-by-step review of all the main concepts in a chapter (see, for instance, pages 230, 319, and 769). Answers to the *Concept Check* questions are on tear-out sheets at the back of the book.
- **Review Exercises** The *Review Exercises* at the end of each chapter recapitulate the basic concepts and skills of the chapter and include exercises that combine the different ideas learned in the chapter.
- **Chapter Test** Each review section concludes with a Chapter Test designed to help students gauge their progress.
- **Cumulative Review Tests** *Cumulative Review Tests* following selected chapters are available at the book companion website. These tests contain problems that combine skills and concepts from the preceding chapters. The problems are designed to highlight the connections between the topics in these related chapters.
- **Answers** Brief answers to odd-numbered exercises in each section (including the review exercises) and to all questions in the *Concepts* exercises and *Chapter Tests,* are given in the back of the book.

Mathematical Vignettes Throughout the book we make use of the margins to provide historical notes, key insights, or applications of mathematics in the modern world. These serve to enliven the material and show that mathematics is an important, vital activity and that even at this elementary level it is fundamental to everyday life.

- **Mathematical Vignettes** These vignettes include biographies of interesting mathematicians and often include a key insight that the mathematician discovered (see, for instance, the vignettes on Viète, page 50; Salt Lake City, page 93; and radiocarbon dating, page 367).
- **Mathematics in the Modern World** This is a series of vignettes that emphasize the central role of mathematics in current advances in technology and the sciences (see, for instance, pages 302, 753, and 784).

Book Companion Website A website that accompanies this book is located at **www.stewartmath.com**. The site includes many useful resources for teaching precalculus, including the following.

- **Discovery Projects** *Discovery Projects* for each chapter are available at the book companion website. The projects are referenced in the text in the appropriate sections. Each project provides a challenging yet accessible set of activities that enable students (perhaps working in groups) to explore in greater depth an interesting aspect of the topic they have just learned (see, for instance, the Discovery Projects *Visualizing a Formula, Relations and Functions, Will the Species Survive?,* and *Computer Graphics I* and *II*, referenced on pages 29, 163, 719, 738, and 820).

- **Focus on Problem Solving** Several *Focus on Problem Solving* sections are available on the website. Each such section highlights one of the problem-solving principles introduced in the Prologue and includes several challenging problems (see for instance *Recognizing Patterns, Using Analogy, Introducing Something Extra, Taking Cases,* and *Working Backward*).
- **Cumulative Review Tests** *Cumulative Review Tests* following Chapters 4, 7, 9, 11, and 13 are available on the website.
- **Appendix B: Calculations and Significant Figures** This appendix, available at the book companion website, contains guidelines for rounding when working with approximate values.
- **Appendix C: Graphing with a Graphing Calculator** This appendix, available at the book companion website, includes general guidelines on graphing with a graphing calculator as well as guidelines on how to avoid common graphing pitfalls.
- **Appendix D: Using the TI-83/84 Graphing Calculator** In this appendix, available at the book companion website, we provide simple, easy-to-follow, step-by-step instructions for using the TI-83/84 graphing calculators.

Acknowledgments

We feel fortunate that all those involved in the production of this book have worked with exceptional energy, intense dedication, and passionate interest. It is surprising how many people are essential in the production of a mathematics textbook, including content editors, reviewers, faculty colleagues, production editors, copy editors, permissions editors, solutions and accuracy checkers, artists, photo researchers, text designers, typesetters, compositors, proofreaders, printers, and many more. We thank them all. We particularly mention the following.

Reviewers for the Sixth Edition Raji Baradwaj, UMBC; Chris Herman, Lorain County Community College; Irina Kloumova, Sacramento City College; Jim McCleery, Skagit Valley College, Whidbey Island Campus; Sally S. Shao, Cleveland State University; David Slutzky, Gainesville State College; Edward Stumpf, Central Carolina Community College; Ricardo Teixeira, University of Texas at Austin; Taixi Xu, Southern Polytechnic State University; and Anna Wlodarczyk, Florida International University.

Reviewers for the Seventh Edition Mary Ann Teel, University of North Texas; Natalia Kravtsova, The Ohio State University; Belle Sigal, Wake Technical Community College; Charity S. Turner, The Ohio State University; Yu-ing Hargett, Jefferson State Community College–Alabama; Alicia Serfaty de Markus, Miami Dade College; Cathleen Zucco-Teveloff, Rider University; Minal Vora, East Georgia State College; Sutandra Sarkar, Georgia State University; Jennifer Denson, Hillsborough Community College; Candice L. Ridlon, University of Maryland Eastern Shore; Alin Stancu, Columbus State University; Frances Tishkevich, Massachusetts Maritime Academy; Phil Veer, Johnson County Community College; Cathleen Zucco-Teveloff, Rider University; Phillip Miller, Indiana University–Southeast; Mildred Vernia, Indiana University–Southeast; Thurai Kugan, John Jay College–CUNY.

We are grateful to our colleagues who continually share with us their insights into teaching mathematics. We especially thank Robert Mena at California State University, Long Beach; we benefited from his many insights into mathematics and its history. We thank Cecilia McVoy at Penn State Abington for her helpful suggestions. We thank Andrew Bulman-Fleming for writing the Solutions Manual and Doug Shaw at the University of Northern Iowa for writing the Instructor Guide and the Study Guide. We are very grateful to Frances Gulick at the University of Maryland for checking the accuracy of the entire manuscript and doing each and every exercise; her many suggestions and corrections have contributed greatly to the accuracy and consistency of the contents of this book.

We thank Martha Emry, our production service and art editor; her energy, devotion, and experience are essential components in the creation of this book. We are grateful for her remarkable ability to instantly recall, when needed, any detail of the entire manuscript as well as her extraordinary ability to simultaneously manage several interdependent editing tracks. We thank Barbara Willette, our copy editor, for her attention to every detail in the manuscript and for ensuring a consistent, appropriate style throughout the book. We thank our designer, Diane Beasley, for the elegant and appropriate design for the interior of the book. We thank Graphic World for their attractive and accurate graphs and Precision Graphics for bringing many of our illustrations to life. We thank our compositors at Graphic World for ensuring a balanced and coherent look for each page of the book.

At Cengage Learning we thank Jennifer Risden, content project manager, for her professional management of the production of the book. We thank Lynh Pham, media developer, for his expert handling of many technical issues, including the creation of the book companion website. We thank Vernon Boes, art director, for his capable administration of the design of the book. We thank Mark Linton, marketing manager, for helping bring the book to the attention of those who may wish to use it in their classes.

We particularly thank our developmental editor, Stacy Green, for skillfully guiding and facilitating every aspect of the creation of this book. Her interest in the book, her familiarity with the entire manuscript, and her almost instant responses to our many queries have made the writing of the book an even more enjoyable experience for us.

Above all we thank our acquisitions editor, Gary Whalen. His vast editorial experience, his extensive knowledge of current issues in the teaching of mathematics, his skill in managing the resources needed to enhance this book, and his deep interest in mathematics textbooks have been invaluable assets in the creation of this book.

Ancillaries

Instructor Resources

Instructor Companion Site
Everything you need for your course in one place! This collection of book-specific lecture and class tools is available online via **www.cengage.com/login**. Access and download PowerPoint presentations, images, instructor's manual, and more.

Complete Solutions Manual
The Complete Solutions Manual provides worked-out solutions to all of the problems in the text. Located on the companion website.

Test Bank
The Test Bank provides chapter tests and final exams, along with answer keys. Located on the companion website.

Instructor's Guide
The Instructor's Guide contains points to stress, suggested time to allot, text discussion topics, core materials for lecture, workshop/discussion suggestions, group work exercises in a form suitable for handout, and suggested homework problems. Located on the companion website.

Lesson Plans
The Lesson Plans provides suggestions for activities and lessons with notes on time allotment in order to ensure timeliness and efficiency during class. Located on the companion website.

Cengage Learning Testing Powered by Cognero
(ISBN-10: 1-305-25853-3; ISBN-13: 978-1-305-25853-2)
CLT is a flexible online system that allows you to author, edit, and manage test bank content; create multiple test versions in an instant; and deliver tests from your LMS, your classroom or wherever you want. This is available online via **www.cengage.com/login**.

Enhanced WebAssign
Printed Access Card: 978-1-285-85833-3
Instant Access Code: 978-1-285-85831-9
Enhanced WebAssign combines exceptional mathematics content with the most powerful online homework solution, WebAssign®. Enhanced WebAssign engages students with immediate feedback, rich tutorial content, and an interactive, fully customizable eBook, Cengage YouBook, to help students to develop a deeper conceptual understanding of their subject matter.

Student Resources

Student Solutions Manual (ISBN-10: 1-305-25361-2; ISBN-13: 978-1-305-25361-2)
The Student Solutions Manual contains fully worked-out solutions to all of the odd-numbered exercises in the text, giving students a way to check their answers and ensure that they took the correct steps to arrive at an answer.

Study Guide (ISBN-10: 1-305-25363-9; ISBN-13: 978-1-305-25363-6)
The Study Guide reinforces student understanding with detailed explanations, worked-out examples, and practice problems. It also lists key ideas to master and builds problem-solving skills. There is a section in the Study Guide corresponding to each section in the text.

Note-Taking Guide (ISBN-10: 1-305-25383-3; ISBN-13: 978-1-305-25383-4)
The Note-Taking Guide is an innovative study aid that helps students develop a section-by-section summary of key concepts.

Text-Specific DVDs (ISBN-10: 1-305-25400-7; ISBN-13: 978-1-305-25400-8)
The Text-Specific DVDs include new learning objective–based lecture videos. These DVDs provide comprehensive coverage of the course—along with additional explanations of concepts, sample problems, and applications—to help students review essential topics.

CengageBrain.com
To access additional course materials, please visit **www.cengagebrain.com**. At the CengageBrain.com home page, search for the ISBN of your title (from the back cover of your book) using the search box at the top of the page. This will take you to the product page where these resources can be found.

Enhanced WebAssign
Printed Access Card: 978-1-285-85833-3
Instant Access Code: 978-1-285-85831-9
Enhanced WebAssign combines exceptional mathematics content with the most powerful online homework solution, WebAssign. Enhanced WebAssign engages students with immediate feedback, rich tutorial content, and an interactive, fully customizable eBook, Cengage YouBook, helping students to develop a deeper conceptual understanding of the subject matter.

TO THE STUDENT

This textbook was written for you to use as a guide to mastering precalculus mathematics. Here are some suggestions to help you get the most out of your course.

First of all, you should read the appropriate section of text *before* you attempt your homework problems. Reading a mathematics text is quite different from reading a novel, a newspaper, or even another textbook. You may find that you have to reread a passage several times before you understand it. Pay special attention to the examples, and work them out yourself with pencil and paper as you read. Then do the linked exercises referred to in *"Now Try Exercise . . ."* at the end of each example. With this kind of preparation you will be able to do your homework much more quickly and with more understanding.

Don't make the mistake of trying to memorize every single rule or fact you may come across. Mathematics doesn't consist simply of memorization. Mathematics is a *problem-solving art*, not just a collection of facts. To master the subject you must solve problems—lots of problems. Do as many of the exercises as you can. Be sure to write your solutions in a logical, step-by-step fashion. Don't give up on a problem if you can't solve it right away. Try to understand the problem more clearly—reread it thoughtfully and relate it to what you have learned from your teacher and from the examples in the text. Struggle with it until you solve it. Once you have done this a few times you will begin to understand what mathematics is really all about.

Answers to the odd-numbered exercises, as well as all the answers (even and odd) to the concept exercises and chapter tests, appear at the back of the book. If your answer differs from the one given, don't immediately assume that you are wrong. There may be a calculation that connects the two answers and makes both correct. For example, if you get $1/(\sqrt{2} - 1)$ but the answer given is $1 + \sqrt{2}$, your answer *is* correct, because you can multiply both numerator and denominator of your answer by $\sqrt{2} + 1$ to change it to the given answer. In rounding approximate answers, follow the guidelines in Appendix B: *Calculations and Significant Figures.*

The symbol is used to warn against committing an error. We have placed this symbol in the margin to point out situations where we have found that many of our students make the same mistake.

Abbreviations

The following abbreviations are used throughout the text.

cm	centimeter	**kPa**	kilopascal	**N**	Newton
dB	decibel	**L**	liter	**qt**	quart
F	farad	**lb**	pound	**oz**	ounce
ft	foot	**lm**	lumen	**s**	second
g	gram	**M**	mole of solute per liter of solution	**Ω**	ohm
gal	gallon			**V**	volt
h	hour			**W**	watt
H	henry	**m**	meter	**yd**	yard
Hz	Hertz	**mg**	milligram	**yr**	year
in.	inch	**MHz**	megahertz	**°C**	degree Celsius
J	Joule	**mi**	mile	**°F**	degree Fahrenheit
kcal	kilocalorie	**min**	minute	**K**	Kelvin
kg	kilogram	**mL**	milliliter	$\Rightarrow$	implies
km	kilometer	**mm**	millimeter	$\Leftrightarrow$	is equivalent to

PROLOGUE PRINCIPLES OF PROBLEM SOLVING

AP Images

GEORGE POLYA (1887–1985) is famous among mathematicians for his ideas on problem solving. His lectures on problem solving at Stanford University attracted overflow crowds whom he held on the edges of their seats, leading them to discover solutions for themselves. He was able to do this because of his deep insight into the psychology of problem solving. His well-known book *How To Solve It* has been translated into 15 languages. He said that Euler (see page 63) was unique among great mathematicians because he explained how he found his results. Polya often said to his students and colleagues, "Yes, I see that your proof is correct, but how did you discover it?" In the preface to *How To Solve It,* Polya writes, "A great discovery solves a great problem but there is a grain of discovery in the solution of any problem. Your problem may be modest; but if it challenges your curiosity and brings into play your inventive faculties, and if you solve it by your own means, you may experience the tension and enjoy the triumph of discovery."

The ability to solve problems is a highly prized skill in many aspects of our lives; it is certainly an important part of any mathematics course. There are no hard and fast rules that will ensure success in solving problems. However, in this Prologue we outline some general steps in the problem-solving process and we give principles that are useful in solving certain types of problems. These steps and principles are just common sense made explicit. They have been adapted from George Polya's insightful book *How To Solve It.*

1. Understand the Problem

The first step is to read the problem and make sure that you understand it. Ask yourself the following questions:

What is the unknown?

What are the given quantities?

What are the given conditions?

For many problems it is useful to

draw a diagram

and identify the given and required quantities on the diagram. Usually, it is necessary to

introduce suitable notation

In choosing symbols for the unknown quantities, we often use letters such as a, b, c, m, n, x, and y, but in some cases it helps to use initials as suggestive symbols, for instance, V for volume or t for time.

2. Think of a Plan

Find a connection between the given information and the unknown that enables you to calculate the unknown. It often helps to ask yourself explicitly: "How can I relate the given to the unknown?" If you don't see a connection immediately, the following ideas may be helpful in devising a plan.

■ Try to Recognize Something Familiar

Relate the given situation to previous knowledge. Look at the unknown and try to recall a more familiar problem that has a similar unknown.

■ Try to Recognize Patterns

Certain problems are solved by recognizing that some kind of pattern is occurring. The pattern could be geometric, numerical, or algebraic. If you can see regularity or repetition in a problem, then you might be able to guess what the pattern is and then prove it.

■ Use Analogy

Try to think of an analogous problem, that is, a similar or related problem but one that is easier than the original. If you can solve the similar, simpler problem, then it might give you the clues you need to solve the original, more difficult one. For instance, if a problem involves very large numbers, you could first try a similar problem with smaller numbers. Or if the problem is in three-dimensional geometry, you could look for something similar in two-dimensional geometry. Or if the problem you start with is a general one, you could first try a special case.

Introduce Something Extra

You might sometimes need to introduce something new—an auxiliary aid—to make the connection between the given and the unknown. For instance, in a problem for which a diagram is useful, the auxiliary aid could be a new line drawn in the diagram. In a more algebraic problem the aid could be a new unknown that relates to the original unknown.

Take Cases

You might sometimes have to split a problem into several cases and give a different argument for each case. For instance, we often have to use this strategy in dealing with absolute value.

Work Backward

Sometimes it is useful to imagine that your problem is solved and work backward, step by step, until you arrive at the given data. Then you might be able to reverse your steps and thereby construct a solution to the original problem. This procedure is commonly used in solving equations. For instance, in solving the equation $3x - 5 = 7$, we suppose that x is a number that satisfies $3x - 5 = 7$ and work backward. We add 5 to each side of the equation and then divide each side by 3 to get $x = 4$. Since each of these steps can be reversed, we have solved the problem.

Establish Subgoals

In a complex problem it is often useful to set subgoals (in which the desired situation is only partially fulfilled). If you can attain or accomplish these subgoals, then you might be able to build on them to reach your final goal.

Indirect Reasoning

Sometimes it is appropriate to attack a problem indirectly. In using **proof by contradiction** to prove that P implies Q, we assume that P is true and Q is false and try to see why this cannot happen. Somehow we have to use this information and arrive at a contradiction to what we absolutely know is true.

Mathematical Induction

In proving statements that involve a positive integer n, it is frequently helpful to use the Principle of Mathematical Induction, which is discussed in Section 12.5.

3. Carry Out the Plan

In Step 2, a plan was devised. In carrying out that plan, you must check each stage of the plan and write the details that prove that each stage is correct.

4. Look Back

Having completed your solution, it is wise to look back over it, partly to see whether any errors have been made and partly to see whether you can discover an easier way to solve the problem. Looking back also familiarizes you with the method of solution, which may be useful for solving a future problem. Descartes said, "Every problem that I solved became a rule which served afterwards to solve other problems."

We illustrate some of these principles of problem solving with an example.

PROBLEM ■ Average Speed

A driver sets out on a journey. For the first half of the distance, she drives at the leisurely pace of 30 mi/h; during the second half she drives 60 mi/h. What is her average speed on this trip?

THINKING ABOUT THE PROBLEM

It is tempting to take the average of the speeds and say that the average speed for the entire trip is

$$\frac{30 + 60}{2} = 45 \text{ mi/h}$$

But is this simple-minded approach really correct?

Try a special case. ▶

Let's look at an easily calculated special case. Suppose that the total distance traveled is 120 mi. Since the first 60 mi is traveled at 30 mi/h, it takes 2 h. The second 60 mi is traveled at 60 mi/h, so it takes one hour. Thus, the total time is $2 + 1 = 3$ hours and the average speed is

$$\frac{120}{3} = 40 \text{ mi/h}$$

So our guess of 45 mi/h was wrong.

SOLUTION

Understand the problem. ▶

We need to look more carefully at the meaning of average speed. It is defined as

$$\text{average speed} = \frac{\text{distance traveled}}{\text{time elapsed}}$$

Introduce notation. ▶

State what is given. ▶

Let d be the distance traveled on each half of the trip. Let t_1 and t_2 be the times taken for the first and second halves of the trip. Now we can write down the information we have been given. For the first half of the trip we have

$$30 = \frac{d}{t_1}$$

and for the second half we have

$$60 = \frac{d}{t_2}$$

Identify the unknown. ▶

Now we identify the quantity that we are asked to find:

$$\text{average speed for entire trip} = \frac{\text{total distance}}{\text{total time}} = \frac{2d}{t_1 + t_2}$$

Connect the given with the unknown. ▶

To calculate this quantity, we need to know t_1 and t_2, so we solve the above equations for these times:

$$t_1 = \frac{d}{30} \qquad t_2 = \frac{d}{60}$$

Now we have the ingredients needed to calculate the desired quantity:

$$\text{average speed} = \frac{2d}{t_1 + t_2} = \frac{2d}{\dfrac{d}{30} + \dfrac{d}{60}}$$

$$= \frac{60(2d)}{60\left(\dfrac{d}{30} + \dfrac{d}{60}\right)} \qquad \text{Multiply numerator and denominator by 60}$$

$$= \frac{120d}{2d + d} = \frac{120d}{3d} = 40$$

So the average speed for the entire trip is 40 mi/h. ■

Bettmann/Corbis

Don't feel bad if you can't solve these problems right away. Problems 1 and 4 were sent to Albert Einstein by his friend Wertheimer. Einstein (and his friend Bucky) enjoyed the problems and wrote back to Wertheimer. Here is part of his reply:

> Your letter gave us a lot of amusement. The first intelligence test fooled both of us (Bucky and me). Only on working it out did I notice that no time is available for the downhill run! Mr. Bucky was also taken in by the second example, but I was not. Such drolleries show us how stupid we are!

(See *Mathematical Intelligencer*, Spring 1990, page 41.)

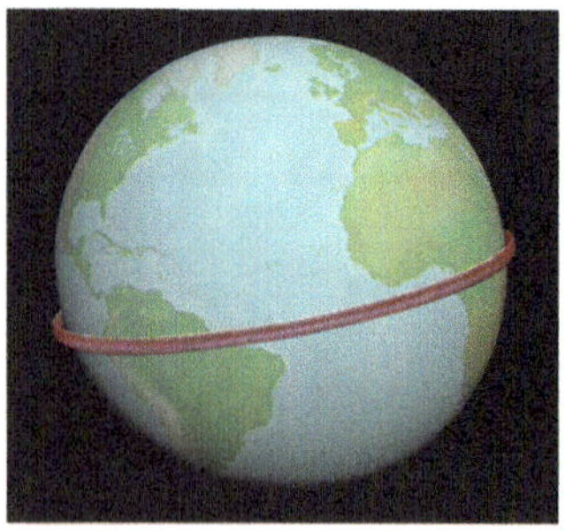

PROBLEMS

1. **Distance, Time, and Speed** An old car has to travel a 2-mile route, uphill and down. Because it is so old, the car can climb the first mile—the ascent—no faster than an average speed of 15 mi/h. How fast does the car have to travel the second mile—on the descent it can go faster, of course—to achieve an average speed of 30 mi/h for the trip?

2. **Comparing Discounts** Which price is better for the buyer, a 40% discount or two successive discounts of 20%?

3. **Cutting up a Wire** A piece of wire is bent as shown in the figure. You can see that one cut through the wire produces four pieces and two parallel cuts produce seven pieces. How many pieces will be produced by 142 parallel cuts? Write a formula for the number of pieces produced by n parallel cuts.

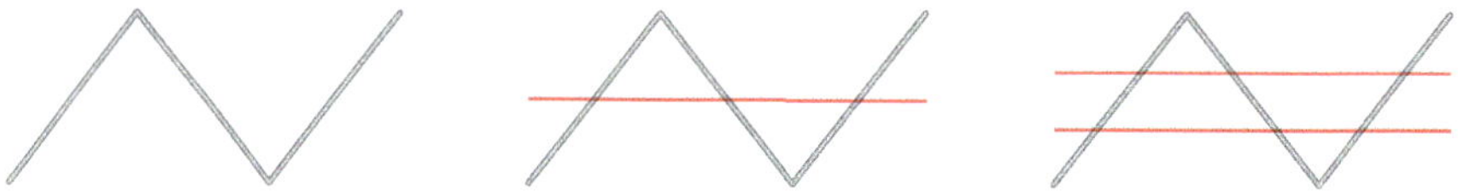

4. **Amoeba Propagation** An amoeba propagates by simple division; each split takes 3 minutes to complete. When such an amoeba is put into a glass container with a nutrient fluid, the container is full of amoebas in one hour. How long would it take for the container to be filled if we start with not one amoeba, but two?

5. **Batting Averages** Player A has a higher batting average than player B for the first half of the baseball season. Player A also has a higher batting average than player B for the second half of the season. Is it necessarily true that player A has a higher batting average than player B for the entire season?

6. **Coffee and Cream** A spoonful of cream is taken from a pitcher of cream and put into a cup of coffee. The coffee is stirred. Then a spoonful of this mixture is put into the pitcher of cream. Is there now more cream in the coffee cup or more coffee in the pitcher of cream?

7. **Wrapping the World** A ribbon is tied tightly around the earth at the equator. How much more ribbon would you need if you raised the ribbon 1 ft above the equator everywhere? (You don't need to know the radius of the earth to solve this problem.)

8. **Ending Up Where You Started** A woman starts at a point P on the earth's surface and walks 1 mi south, then 1 mi east, then 1 mi north, and finds herself back at P, the starting point. Describe all points P for which this is possible. [*Hint:* There are infinitely many such points, all but one of which lie in Antarctica.]

Many more problems and examples that highlight different problem-solving principles are available at the book companion website: www.stewartmath.com. You can try them as you progress through the book.

Blend Images/Alamy

1 Fundamentals

In this first chapter we review the real numbers, equations, and the coordinate plane. You are probably already familiar with these concepts, but it is helpful to get a fresh look at how these ideas work together to solve problems and model (or describe) real-world situations.

In the *Focus on Modeling* at the end of the chapter we learn how to find linear trends in data and how to use these trends to make predictions about the future.

1.1 REAL NUMBERS

■ Real Numbers ■ Properties of Real Numbers ■ Addition and Subtraction ■ Multiplication and Division ■ The Real Line ■ Sets and Intervals ■ Absolute Value and Distance

In the real world we use numbers to measure and compare different quantities. For example, we measure temperature, length, height, weight, blood pressure, distance, speed, acceleration, energy, force, angles, age, cost, and so on. Figure 1 illustrates some situations in which numbers are used. Numbers also allow us to express relationships between different quantities—for example, relationships between the radius and volume of a ball, between miles driven and gas used, or between education level and starting salary.

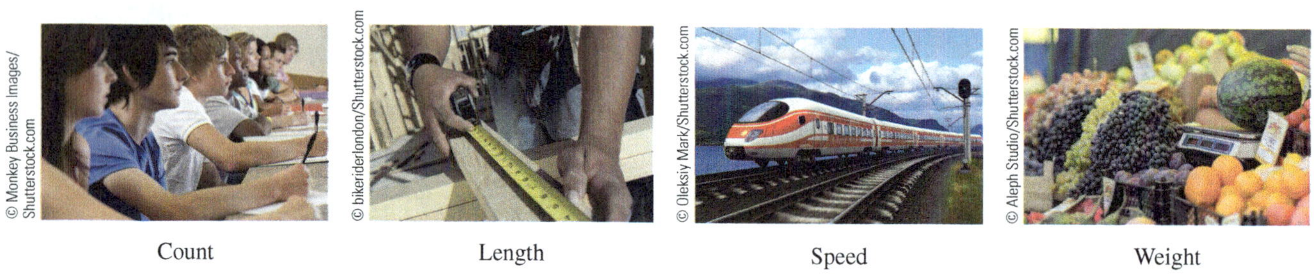

FIGURE 1 Measuring with real numbers

■ Real Numbers

Let's review the types of numbers that make up the real number system. We start with the **natural numbers**:

$$1, 2, 3, 4, \ldots$$

The different types of real numbers were invented to meet specific needs. For example, natural numbers are needed for counting, negative numbers for describing debt or below-zero temperatures, rational numbers for concepts like "half a gallon of milk," and irrational numbers for measuring certain distances, like the diagonal of a square.

The **integers** consist of the natural numbers together with their negatives and 0:

$$\ldots, -3, -2, -1, 0, 1, 2, 3, 4, \ldots$$

We construct the **rational numbers** by taking ratios of integers. Thus any rational number r can be expressed as

$$r = \frac{m}{n}$$

where m and n are integers and $n \neq 0$. Examples are

$$\tfrac{1}{2} \qquad -\tfrac{3}{7} \qquad 46 = \tfrac{46}{1} \qquad 0.17 = \tfrac{17}{100}$$

(Recall that division by 0 is always ruled out, so expressions like $\frac{3}{0}$ and $\frac{0}{0}$ are undefined.) There are also real numbers, such as $\sqrt{2}$, that cannot be expressed as a ratio of integers and are therefore called **irrational numbers**. It can be shown, with varying degrees of difficulty, that these numbers are also irrational:

$$\sqrt{3} \qquad \sqrt{5} \qquad \sqrt[3]{2} \qquad \pi \qquad \frac{3}{\pi^2}$$

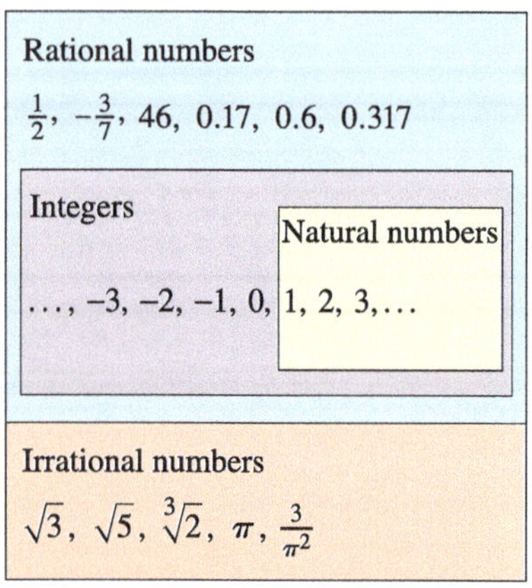

FIGURE 2 The real number system

The set of all real numbers is usually denoted by the symbol $\mathbb{R}$. When we use the word *number* without qualification, we will mean "real number." Figure 2 is a diagram of the types of real numbers that we work with in this book.

Every real number has a decimal representation. If the number is rational, then its corresponding decimal is repeating. For example,

$$\tfrac{1}{2} = 0.5000\ldots = 0.5\overline{0} \qquad \tfrac{2}{3} = 0.66666\ldots = 0.\overline{6}$$

$$\tfrac{157}{495} = 0.3171717\ldots = 0.3\overline{17} \qquad \tfrac{9}{7} = 1.285714285714\ldots = 1.\overline{285714}$$

A repeating decimal such as

$$x = 3.5474747...$$

is a rational number. To convert it to a ratio of two integers, we write

$$\begin{aligned} 1000x &= 3547.47474747... \\ 10x &= 35.47474747... \\ \hline 990x &= 3512.0 \end{aligned}$$

Thus $x = \frac{3512}{990}$. (The idea is to multiply x by appropriate powers of 10 and then subtract to eliminate the repeating part.)

(The bar indicates that the sequence of digits repeats forever.) If the number is irrational, the decimal representation is nonrepeating:

$$\sqrt{2} = 1.414213562373095... \qquad \pi = 3.141592653589793...$$

If we stop the decimal expansion of any number at a certain place, we get an approximation to the number. For instance, we can write

$$\pi \approx 3.14159265$$

where the symbol $\approx$ is read "is approximately equal to." The more decimal places we retain, the better our approximation.

Properties of Real Numbers

We all know that $2 + 3 = 3 + 2$, and $5 + 7 = 7 + 5$, and $513 + 87 = 87 + 513$, and so on. In algebra we express all these (infinitely many) facts by writing

$$a + b = b + a$$

where a and b stand for any two numbers. In other words, "$a + b = b + a$" is a concise way of saying that "when we add two numbers, the order of addition doesn't matter." This fact is called the *Commutative Property* of addition. From our experience with numbers we know that the properties in the following box are also valid.

PROPERTIES OF REAL NUMBERS

Property	Example	Description
Commutative Properties		
$a + b = b + a$	$7 + 3 = 3 + 7$	When we add two numbers, order doesn't matter.
$ab = ba$	$3 \cdot 5 = 5 \cdot 3$	When we multiply two numbers, order doesn't matter.
Associative Properties		
$(a + b) + c = a + (b + c)$	$(2 + 4) + 7 = 2 + (4 + 7)$	When we add three numbers, it doesn't matter which two we add first.
$(ab)c = a(bc)$	$(3 \cdot 7) \cdot 5 = 3 \cdot (7 \cdot 5)$	When we multiply three numbers, it doesn't matter which two we multiply first.
Distributive Property		
$a(b + c) = ab + ac$ $(b + c)a = ab + ac$	$2 \cdot (3 + 5) = 2 \cdot 3 + 2 \cdot 5$ $(3 + 5) \cdot 2 = 2 \cdot 3 + 2 \cdot 5$	When we multiply a number by a sum of two numbers, we get the same result as we get if we multiply the number by each of the terms and then add the results.

The Distributive Property applies whenever we multiply a number by a sum. Figure 3 explains why this property works for the case in which all the numbers are positive integers, but the property is true for any real numbers a, b, and c.

The Distributive Property is crucial because it describes the way addition and multiplication interact with each other.

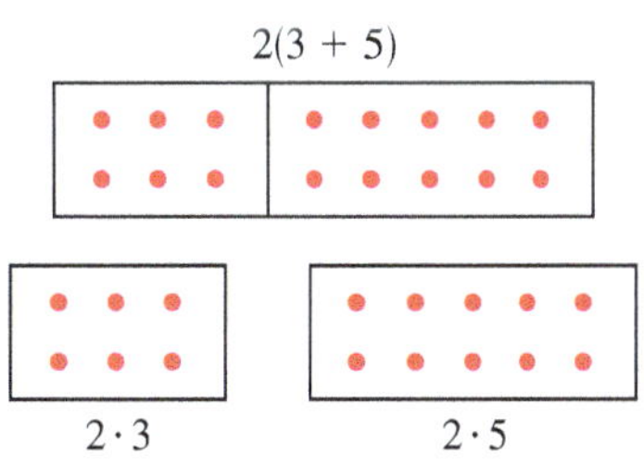

FIGURE 3 The Distributive Property

EXAMPLE 1 ■ Using the Distributive Property

(a) $2(x + 3) = 2 \cdot x + 2 \cdot 3$ — Distributive Property

$= 2x + 6$ — Simplify

(b) $(a + b)(x + y) = (a + b)x + (a + b)y$ — Distributive Property

$= (ax + bx) + (ay + by)$ — Distributive Property

$= ax + bx + ay + by$ — Associative Property of Addition

In the last step we removed the parentheses because, according to the Associative Property, the order of addition doesn't matter.

Now Try Exercise 15 ■

■ Addition and Subtraction

Don't assume that $-a$ is a negative number. Whether $-a$ is negative or positive depends on the value of a. For example, if $a = 5$, then $-a = -5$, a negative number, but if $a = -5$, then $-a = -(-5) = 5$ (Property 2), a positive number.

The number 0 is special for addition; it is called the **additive identity** because $a + 0 = a$ for any real number a. Every real number a has a **negative**, $-a$, that satisfies $a + (-a) = 0$. **Subtraction** is the operation that undoes addition; to subtract a number from another, we simply add the negative of that number. By definition

$$a - b = a + (-b)$$

To combine real numbers involving negatives, we use the following properties.

PROPERTIES OF NEGATIVES

Property	Example
1. $(-1)a = -a$	$(-1)5 = -5$
2. $-(-a) = a$	$-(-5) = 5$
3. $(-a)b = a(-b) = -(ab)$	$(-5)7 = 5(-7) = -(5 \cdot 7)$
4. $(-a)(-b) = ab$	$(-4)(-3) = 4 \cdot 3$
5. $-(a + b) = -a - b$	$-(3 + 5) = -3 - 5$
6. $-(a - b) = b - a$	$-(5 - 8) = 8 - 5$

Property 6 states the intuitive fact that $a - b$ and $b - a$ are negatives of each other. Property 5 is often used with more than two terms:

$$-(a + b + c) = -a - b - c$$

EXAMPLE 2 ■ Using Properties of Negatives

Let x, y, and z be real numbers.

(a) $-(x + 2) = -x - 2$ — Property 5: $-(a + b) = -a - b$

(b) $-(x + y - z) = -x - y - (-z)$ — Property 5: $-(a + b) = -a - b$

$= -x - y + z$ — Property 2: $-(-a) = a$

Now Try Exercise 23 ■

Multiplication and Division

The number 1 is special for multiplication; it is called the **multiplicative identity** because $a \cdot 1 = a$ for any real number a. Every nonzero real number a has an **inverse**, $1/a$, that satisfies $a \cdot (1/a) = 1$. **Division** is the operation that undoes multiplication; to divide by a number, we multiply by the inverse of that number. If $b \neq 0$, then, by definition,

$$a \div b = a \cdot \frac{1}{b}$$

We write $a \cdot (1/b)$ as simply a/b. We refer to a/b as the **quotient** of a and b or as the **fraction** a over b; a is the **numerator** and b is the **denominator** (or **divisor**). To combine real numbers using the operation of division, we use the following properties.

PROPERTIES OF FRACTIONS

Property	Example	Description
1. $\frac{a}{b} \cdot \frac{c}{d} = \frac{ac}{bd}$	$\frac{2}{3} \cdot \frac{5}{7} = \frac{2 \cdot 5}{3 \cdot 7} = \frac{10}{21}$	When **multiplying fractions**, multiply numerators and denominators.
2. $\frac{a}{b} \div \frac{c}{d} = \frac{a}{b} \cdot \frac{d}{c}$	$\frac{2}{3} \div \frac{5}{7} = \frac{2}{3} \cdot \frac{7}{5} = \frac{14}{15}$	When **dividing fractions**, invert the divisor and multiply.
3. $\frac{a}{c} + \frac{b}{c} = \frac{a+b}{c}$	$\frac{2}{5} + \frac{7}{5} = \frac{2+7}{5} = \frac{9}{5}$	When **adding fractions** with the **same denominator**, add the numerators.
4. $\frac{a}{b} + \frac{c}{d} = \frac{ad+bc}{bd}$	$\frac{2}{5} + \frac{3}{7} = \frac{2 \cdot 7 + 3 \cdot 5}{35} = \frac{29}{35}$	When **adding fractions** with **different denominators**, find a common denominator. Then add the numerators.
5. $\frac{ac}{bc} = \frac{a}{b}$	$\frac{2 \cdot 5}{3 \cdot 5} = \frac{2}{3}$	**Cancel** numbers that are **common factors** in numerator and denominator.
6. If $\frac{a}{b} = \frac{c}{d}$, then $ad = bc$	$\frac{2}{3} = \frac{6}{9}$, so $2 \cdot 9 = 3 \cdot 6$	**Cross-multiply**.

When adding fractions with different denominators, we don't usually use Property 4. Instead we rewrite the fractions so that they have the smallest possible common denominator (often smaller than the product of the denominators), and then we use Property 3. This denominator is the **L**east **C**ommon **D**enominator (LCD) described in the next example.

EXAMPLE 3 ■ Using the LCD to Add Fractions

Evaluate: $\frac{5}{36} + \frac{7}{120}$

SOLUTION Factoring each denominator into prime factors gives

$$36 = 2^2 \cdot 3^2 \qquad \text{and} \qquad 120 = 2^3 \cdot 3 \cdot 5$$

We find the least common denominator (LCD) by forming the product of all the prime factors that occur in these factorizations, using the highest power of each prime factor. Thus the LCD is $2^3 \cdot 3^2 \cdot 5 = 360$. So

$$\frac{5}{36} + \frac{7}{120} = \frac{5 \cdot 10}{36 \cdot 10} + \frac{7 \cdot 3}{120 \cdot 3} \qquad \text{Use common denominator}$$

$$= \frac{50}{360} + \frac{21}{360} = \frac{71}{360} \qquad \text{Property 3: Adding fractions with the same denominator}$$

Now Try Exercise 29 ■

The Real Line

The real numbers can be represented by points on a line, as shown in Figure 4. The positive direction (toward the right) is indicated by an arrow. We choose an arbitrary reference point O, called the **origin**, which corresponds to the real number 0. Given any convenient unit of measurement, each positive number x is represented by the point on the line a distance of x units to the right of the origin, and each negative number $-x$ is represented by the point x units to the left of the origin. The number associated with the point P is called the coordinate of P, and the line is then called a **coordinate line**, or a **real number line**, or simply a **real line**. Often we identify the point with its coordinate and think of a number as being a point on the real line.

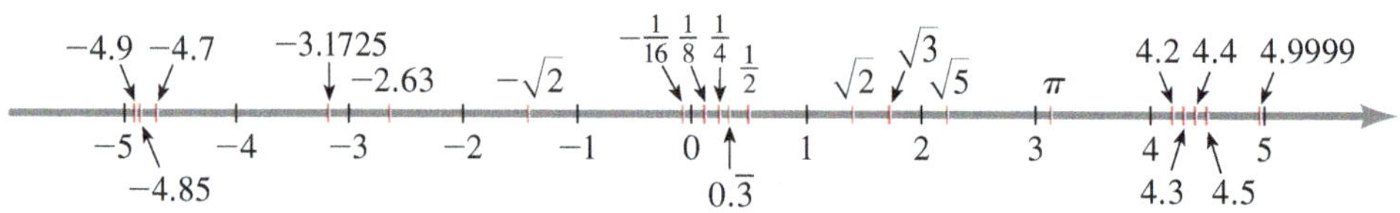

FIGURE 4 The real line

The real numbers are *ordered*. We say that ***a* is less than *b*** and write $a < b$ if $b - a$ is a positive number. Geometrically, this means that a lies to the left of b on the number line. Equivalently, we can say that ***b* is greater than *a*** and write $b > a$. The symbol $a \le b$ (or $b \ge a$) means that either $a < b$ or $a = b$ and is read "a is less than or equal to b." For instance, the following are true inequalities (see Figure 5):

$$7 < 7.4 < 7.5 \qquad -\pi < -3 \qquad \sqrt{2} < 2 \qquad 2 \le 2$$

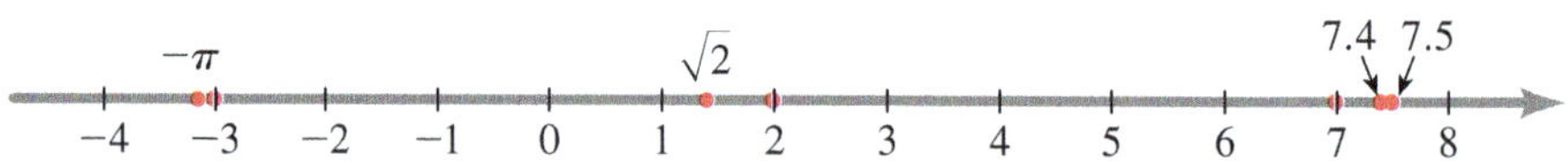

FIGURE 5

Sets and Intervals

A **set** is a collection of objects, and these objects are called the **elements** of the set. If S is a set, the notation $a \in S$ means that a is an element of S, and $b \notin S$ means that b is not an element of S. For example, if Z represents the set of integers, then $-3 \in Z$ but $\pi \notin Z$.

Some sets can be described by listing their elements within braces. For instance, the set A that consists of all positive integers less than 7 can be written as

$$A = \{1, 2, 3, 4, 5, 6\}$$

We could also write A in **set-builder notation** as

$$A = \{x \mid x \text{ is an integer and } 0 < x < 7\}$$

which is read "A is the set of all x such that x is an integer and $0 < x < 7$."

DISCOVERY PROJECT

Real Numbers in the Real World

Real-world measurements always involve units. For example, we usually measure distance in feet, miles, centimeters, or kilometers. Some measurements involve different types of units. For example, speed is measured in miles per hour or meters per second. We often need to convert a measurement from one type of unit to another. In this project we explore different types of units used for different purposes and how to convert from one type of unit to another. You can find the project at **www.stewartmath.com**.

If S and T are sets, then their **union** $S \cup T$ is the set that consists of all elements that are in S *or* T (or in both). The **intersection** of S and T is the set $S \cap T$ consisting of all elements that are in both S *and* T. In other words, $S \cap T$ is the common part of S and T. The **empty set**, denoted by $\varnothing$, is the set that contains no element.

EXAMPLE 4 ■ Union and Intersection of Sets

If $S = \{1, 2, 3, 4, 5\}$, $T = \{4, 5, 6, 7\}$, and $V = \{6, 7, 8\}$, find the sets $S \cup T$, $S \cap T$, and $S \cap V$.

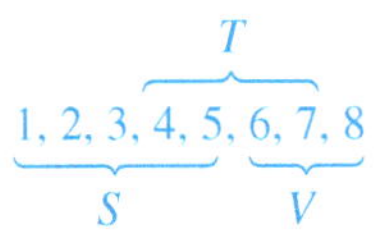

SOLUTION

$S \cup T = \{1, 2, 3, 4, 5, 6, 7\}$ — All elements in S or T

$S \cap T = \{4, 5\}$ — Elements common to both S and T

$S \cap V = \varnothing$ — S and V have no element in common

Now Try Exercise 41 ■

Certain sets of real numbers, called **intervals**, occur frequently in calculus and correspond geometrically to line segments. If $a < b$, then the **open interval** from a to b consists of all numbers between a and b and is denoted (a, b). The **closed interval** from a to b includes the endpoints and is denoted $[a, b]$. Using set-builder notation, we can write

$$(a, b) = \{x \mid a < x < b\} \qquad [a, b] = \{x \mid a \le x \le b\}$$

FIGURE 6 The open interval (a, b)

FIGURE 7 The closed interval $[a, b]$

Note that parentheses ( ) in the interval notation and open circles on the graph in Figure 6 indicate that endpoints are *excluded* from the interval, whereas square brackets [ ] and solid circles in Figure 7 indicate that the endpoints are *included*. Intervals may also include one endpoint but not the other, or they may extend infinitely far in one direction or both. The following table lists the possible types of intervals.

Notation	Set description	Graph
(a, b)	$\{x \mid a < x < b\}$	
$[a, b]$	$\{x \mid a \le x \le b\}$	
$[a, b)$	$\{x \mid a \le x < b\}$	
$(a, b]$	$\{x \mid a < x \le b\}$	
(a, ∞)	$\{x \mid a < x\}$	
$[a, \infty)$	$\{x \mid a \le x\}$	
$(-\infty, b)$	$\{x \mid x < b\}$	
$(-\infty, b]$	$\{x \mid x \le b\}$	
$(-\infty, \infty)$	$\mathbb{R}$ (set of all real numbers)	

The symbol ∞ ("infinity") does not stand for a number. The notation (a, ∞), for instance, simply indicates that the interval has no endpoint on the right but extends infinitely far in the positive direction.

EXAMPLE 5 ■ Graphing Intervals

Express each interval in terms of inequalities, and then graph the interval.

(a) $[-1, 2) = \{x \mid -1 \le x < 2\}$

(b) $[1.5, 4] = \{x \mid 1.5 \le x \le 4\}$

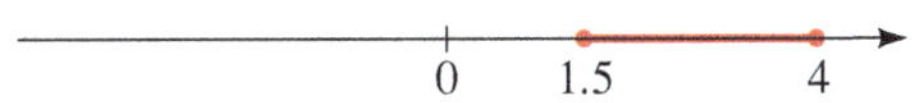

(c) $(-3, \infty) = \{x \mid -3 < x\}$

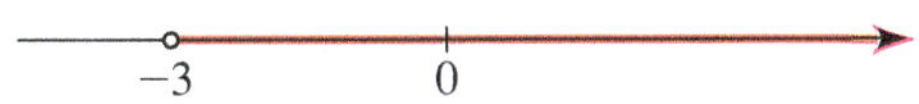

Now Try Exercise 47 ■

No Smallest or Largest Number in an Open Interval

Any interval contains infinitely many numbers—every point on the graph of an interval corresponds to a real number. In the closed interval $[0, 1]$, the smallest number is 0 and the largest is 1, but the open interval $(0, 1)$ contains no smallest or largest number. To see this, note that 0.01 is close to zero, but 0.001 is closer, 0.0001 is closer yet, and so on. We can always find a number in the interval $(0, 1)$ closer to zero than any given number. Since 0 itself is not in the interval, the interval contains no smallest number. Similarly, 0.99 is close to 1, but 0.999 is closer, 0.9999 closer yet, and so on. Since 1 itself is not in the interval, the interval has no largest number.

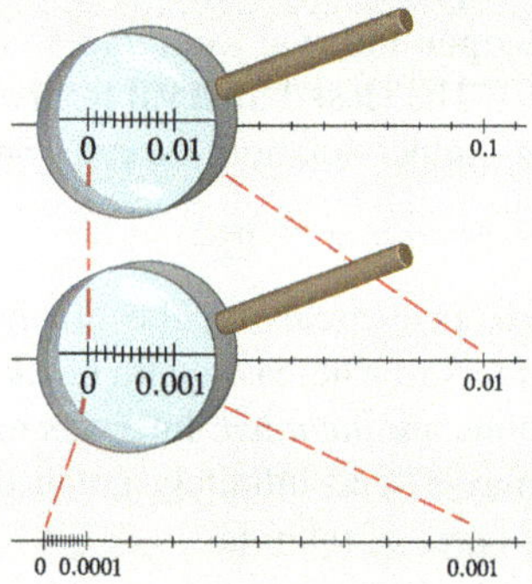

EXAMPLE 6 ■ Finding Unions and Intersections of Intervals

Graph each set.

(a) $(1, 3) \cap [2, 7]$ **(b)** $(1, 3) \cup [2, 7]$

SOLUTION

(a) The intersection of two intervals consists of the numbers that are in both intervals. Therefore

$$\begin{aligned}(1, 3) \cap [2, 7] &= \{x \mid 1 < x < 3 \text{ and } 2 \le x \le 7\} \\ &= \{x \mid 2 \le x < 3\} = [2, 3)\end{aligned}$$

This set is illustrated in Figure 8.

(b) The union of two intervals consists of the numbers that are in either one interval or the other (or both). Therefore

$$\begin{aligned}(1, 3) \cup [2, 7] &= \{x \mid 1 < x < 3 \text{ or } 2 \le x \le 7\} \\ &= \{x \mid 1 < x \le 7\} = (1, 7]\end{aligned}$$

This set is illustrated in Figure 9.

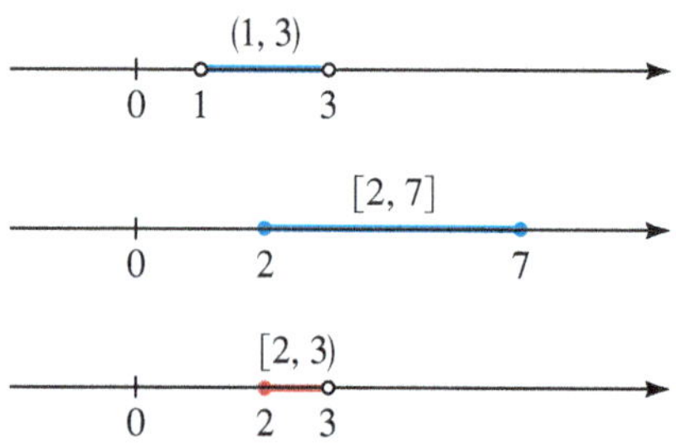

FIGURE 8 $(1, 3) \cap [2, 7] = [2, 3)$

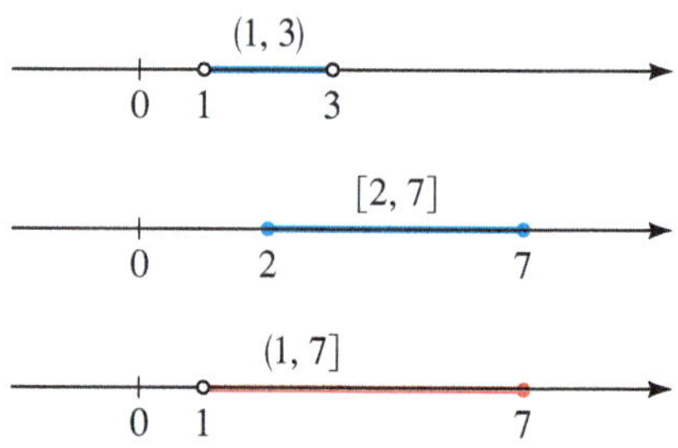

FIGURE 9 $(1, 3) \cup [2, 7] = (1, 7]$

Now Try Exercise 61 ■

Absolute Value and Distance

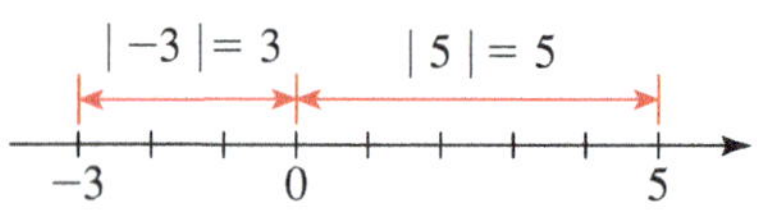

FIGURE 10

The **absolute value** of a number a, denoted by $|a|$, is the distance from a to 0 on the real number line (see Figure 10). Distance is always positive or zero, so we have $|a| \ge 0$ for every number a. Remembering that $-a$ is positive when a is negative, we have the following definition.

DEFINITION OF ABSOLUTE VALUE

If a is a real number, then the **absolute value** of a is

$$|a| = \begin{cases} a & \text{if } a \ge 0 \\ -a & \text{if } a < 0 \end{cases}$$

EXAMPLE 7 ■ Evaluating Absolute Values of Numbers

(a) $|3| = 3$

(b) $|-3| = -(-3) = 3$

(c) $|0| = 0$

(d) $|3 - \pi| = -(3 - \pi) = \pi - 3 \qquad (\text{since } 3 < \pi \Rightarrow 3 - \pi < 0)$

Now Try Exercise 67 ■

When working with absolute values, we use the following properties.

PROPERTIES OF ABSOLUTE VALUE

Property	Example	Description
1. $\lvert a \rvert \geq 0$	$\lvert -3 \rvert = 3 \geq 0$	The absolute value of a number is always positive or zero.
2. $\lvert a \rvert = \lvert -a \rvert$	$\lvert 5 \rvert = \lvert -5 \rvert$	A number and its negative have the same absolute value.
3. $\lvert ab \rvert = \lvert a \rvert \lvert b \rvert$	$\lvert -2 \cdot 5 \rvert = \lvert -2 \rvert \lvert 5 \rvert$	The absolute value of a product is the product of the absolute values.
4. $\left\lvert \dfrac{a}{b} \right\rvert = \dfrac{\lvert a \rvert}{\lvert b \rvert}$	$\left\lvert \dfrac{12}{-3} \right\rvert = \dfrac{\lvert 12 \rvert}{\lvert -3 \rvert}$	The absolute value of a quotient is the quotient of the absolute values.
5. $\lvert a + b \rvert \leq \lvert a \rvert + \lvert b \rvert$	$\lvert -3 + 5 \rvert \leq \lvert -3 \rvert + \lvert 5 \rvert$	Triangle Inequality

What is the distance on the real line between the numbers -2 and 11? From Figure 11 we see that the distance is 13. We arrive at this by finding either $\lvert 11 - (-2) \rvert = 13$ or $\lvert (-2) - 11 \rvert = 13$. From this observation we make the following definition (see Figure 12).

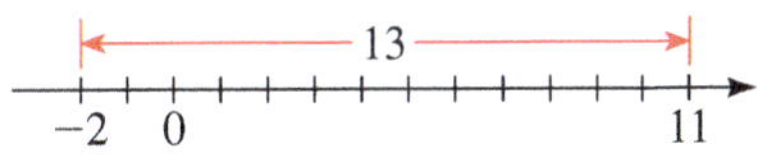

FIGURE 11

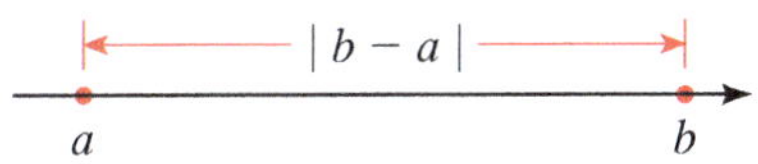

FIGURE 12 Length of a line segment is $\lvert b - a \rvert$

DISTANCE BETWEEN POINTS ON THE REAL LINE

If a and b are real numbers, then the **distance** between the points a and b on the real line is

$$d(a, b) = \lvert b - a \rvert$$

From Property 6 of negatives it follows that

$$\lvert b - a \rvert = \lvert a - b \rvert$$

This confirms that, as we would expect, the distance from a to b is the same as the distance from b to a.

EXAMPLE 8 ■ Distance Between Points on the Real Line

The distance between the numbers -8 and 2 is

$$d(a, b) = \lvert 2 - (-8) \rvert = \lvert -10 \rvert = 10$$

We can check this calculation geometrically, as shown in Figure 13.

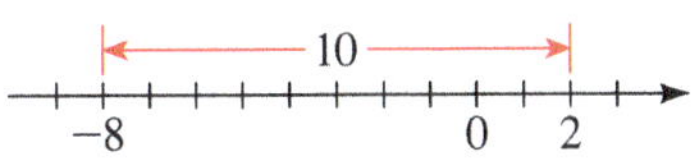

FIGURE 13

Now Try Exercise 75 ■

1.1 EXERCISES

CONCEPTS

1. Give an example of each of the following:

(a) A natural number

(b) An integer that is not a natural number

(c) A rational number that is not an integer

(d) An irrational number

2. Complete each statement and name the property of real numbers you have used.

(a) $ab =$ ______; ______ Property

(b) $a + (b + c) =$ ______; ______ Property

(c) $a(b + c) =$ ______; ______ Property

3. Express the set of real numbers between but not including 2 and 7 as follows.

(a) In set-builder notation: ______________

(b) In interval notation: ______________

4. The symbol $|x|$ stands for the ______ of the number x. If x is not 0, then the sign of $|x|$ is always ______.

5. The distance between a and b on the real line is $d(a, b) =$ ______. So the distance between -5 and 2 is ______.

6–8 ■ *Yes or No*? If *No*, give a reason. Assume that a and b are nonzero real numbers.

6. (a) Is the sum of two rational numbers always a rational number?

(b) Is the sum of two irrational numbers always an irrational number?

7. (a) Is $a - b$ equal to $b - a$?

(b) Is $-2(a - 5)$ equal to $-2a - 10$?

8. (a) Is the distance between any two different real numbers always positive?

(b) Is the distance between a and b the same as the distance between b and a?

SKILLS

9–10 ■ Real Numbers List the elements of the given set that are

(a) natural numbers

(b) integers

(c) rational numbers

(d) irrational numbers

9. $\{-1.5, 0, \frac{5}{2}, \sqrt{7}, 2.71, -\pi, 3.1\overline{4}, 100, -8\}$

10. $\{1.3, 1.3333\ldots, \sqrt{5}, 5.34, -500, 1\frac{2}{3}, \sqrt{16}, \frac{246}{579}, -\frac{20}{5}\}$

11–18 ■ Properties of Real Numbers State the property of real numbers being used.

11. $3 + 7 = 7 + 3$

12. $4(2 + 3) = (2 + 3)4$

13. $(x + 2y) + 3z = x + (2y + 3z)$

14. $2(A + B) = 2A + 2B$

15. $(5x + 1)3 = 15x + 3$

16. $(x + a)(x + b) = (x + a)x + (x + a)b$

17. $2x(3 + y) = (3 + y)2x$

18. $7(a + b + c) = 7(a + b) + 7c$

19–22 ■ Properties of Real Numbers Rewrite the expression using the given property of real numbers.

19. Commutative Property of Addition, $x + 3 =$ ______

20. Associative Property of Multiplication, $7(3x) =$ ______

21. Distributive Property, $4(A + B) =$ ______

22. Distributive Property, $5x + 5y =$ ______

23–28 ■ Properties of Real Numbers Use properties of real numbers to write the expression without parentheses.

23. $3(x + y)$ **24.** $(a - b)8$

25. $4(2m)$ **26.** $\frac{4}{3}(-6y)$

27. $-\frac{5}{2}(2x - 4y)$ **28.** $(3a)(b + c - 2d)$

29–32 ■ Arithmetic Operations Perform the indicated operations.

29. (a) $\frac{3}{10} + \frac{4}{15}$ **(b)** $\frac{1}{4} + \frac{1}{5}$

30. (a) $\frac{2}{3} - \frac{3}{5}$ **(b)** $1 + \frac{5}{8} - \frac{1}{6}$

31. (a) $\frac{2}{3}(6 - \frac{3}{2})$ **(b)** $(3 + \frac{1}{4})(1 - \frac{4}{5})$

32. (a) $\dfrac{2}{\frac{2}{3}} - \dfrac{\frac{2}{3}}{2}$ **(b)** $\dfrac{\frac{2}{5} + \frac{1}{2}}{\frac{1}{10} + \frac{3}{15}}$

33–34 ■ Inequalities Place the correct symbol ($<$, $>$, or $=$) in the space.

33. (a) 3 ___ $\frac{7}{2}$ **(b)** -3 ___ $-\frac{7}{2}$ **(c)** 3.5 ___ $\frac{7}{2}$

34. (a) $\frac{2}{3}$ ___ 0.67 **(b)** $\frac{2}{3}$ ___ -0.67

(c) $|0.67|$ ___ $|-0.67|$

35–38 ■ Inequalities State whether each inequality is true or false.

35. (a) $-3 < -4$ **(b)** $3 < 4$

36. (a) $\sqrt{3} > 1.7325$ **(b)** $1.732 \geq \sqrt{3}$

37. (a) $\frac{10}{2} \geq 5$ **(b)** $\frac{6}{10} \geq \frac{5}{6}$

38. (a) $\frac{7}{11} \geq \frac{8}{13}$ **(b)** $-\frac{3}{5} > -\frac{3}{4}$

39–40 ■ Inequalities Write each statement in terms of inequalities.

39. (a) x is positive.

(b) t is less than 4.

(c) a is greater than or equal to π.

(d) x is less than $\frac{1}{3}$ and is greater than -5.

(e) The distance from p to 3 is at most 5.

40. **(a)** y is negative.
(b) z is greater than 1.
(c) b is at most 8.
(d) w is positive and is less than or equal to 17.
(e) y is at least 2 units from π.

41–44 ■ Sets Find the indicated set if

$$A = \{1, 2, 3, 4, 5, 6, 7\} \qquad B = \{2, 4, 6, 8\}$$
$$C = \{7, 8, 9, 10\}$$

41. **(a)** $A \cup B$ **(b)** $A \cap B$

42. **(a)** $B \cup C$ **(b)** $B \cap C$

43. **(a)** $A \cup C$ **(b)** $A \cap C$

44. **(a)** $A \cup B \cup C$ **(b)** $A \cap B \cap C$

45–46 ■ Sets Find the indicated set if

$$A = \{x \mid x \geq -2\} \qquad B = \{x \mid x < 4\}$$
$$C = \{x \mid -1 < x \leq 5\}$$

45. **(a)** $B \cup C$ **(b)** $B \cap C$

46. **(a)** $A \cap C$ **(b)** $A \cap B$

47–52 ■ Intervals Express the interval in terms of inequalities, and then graph the interval.

47. $(-3, 0)$ **48.** $(2, 8]$

49. $[2, 8)$ **50.** $[-6, -\frac{1}{2}]$

51. $[2, \infty)$ **52.** $(-\infty, 1)$

53–58 ■ Intervals Express the inequality in interval notation, and then graph the corresponding interval.

53. $x \leq 1$ **54.** $1 \leq x \leq 2$

55. $-2 < x \leq 1$ **56.** $x \geq -5$

57. $x > -1$ **58.** $-5 < x < 2$

59–60 ■ Intervals Express each set in interval notation.

59. **(a)** 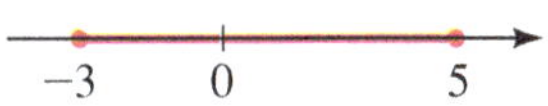

(b) 

60. **(a)** 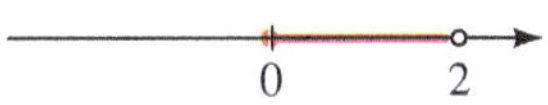

(b) 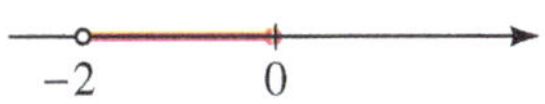

61–66 ■ Intervals Graph the set.

61. $(-2, 0) \cup (-1, 1)$ **62.** $(-2, 0) \cap (-1, 1)$

63. $[-4, 6] \cap [0, 8)$ **64.** $[-4, 6) \cup [0, 8)$

65. $(-\infty, -4) \cup (4, \infty)$ **66.** $(-\infty, 6] \cap (2, 10)$

67–72 ■ Absolute Value Evaluate each expression.

67. **(a)** $|100|$ **(b)** $|-73|$

68. **(a)** $|\sqrt{5} - 5|$ **(b)** $|10 - \pi|$

69. **(a)** $\big||6| - |-4|\big|$ **(b)** $\dfrac{-1}{|-1|}$

70. **(a)** $\big|2 - |-12|\big|$ **(b)** $-1 - \big|1 - |-1|\big|$

71. **(a)** $|(-2) \cdot 6|$ **(b)** $|(-\frac{1}{3})(-15)|$

72. **(a)** $\left|\dfrac{-6}{24}\right|$ **(b)** $\left|\dfrac{7 - 12}{12 - 7}\right|$

73–76 ■ Distance Find the distance between the given numbers.

73. 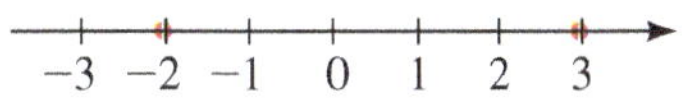

74. 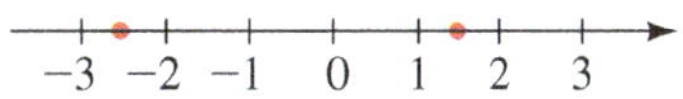

75. **(a)** 2 and 17 **(b)** −3 and 21 **(c)** $\frac{11}{8}$ and $-\frac{3}{10}$

76. **(a)** $\frac{7}{15}$ and $-\frac{1}{21}$ **(b)** −38 and −57 **(c)** −2.6 and −1.8

SKILLS Plus

77–78 ■ Repeating Decimal Express each repeating decimal as a fraction. (See the margin note on page 3.)

77. **(a)** $0.\overline{7}$ **(b)** $0.2\overline{8}$ **(c)** $0.\overline{57}$

78. **(a)** $5.\overline{23}$ **(b)** $1.3\overline{7}$ **(c)** $2.\overline{135}$

79–82 ■ Simplifying Absolute Value Express the quantity without using absolute value.

79. $|\pi - 3|$ **80.** $|1 - \sqrt{2}|$

81. $|a - b|$, where $a < b$

82. $a + b + |a - b|$, where $a < b$

83–84 ■ Signs of Numbers Let a, b, and c be real numbers such that $a > 0$, $b < 0$, and $c < 0$. Find the sign of each expression.

83. **(a)** $-a$ **(b)** bc **(c)** $a - b$ **(d)** $ab + ac$

84. **(a)** $-b$ **(b)** $a + bc$ **(c)** $c - a$ **(d)** ab^2

APPLICATIONS

85. Area of a Garden Mary's backyard vegetable garden measures 20 ft by 30 ft, so its area is $20 \times 30 = 600$ ft^2. She decides to make it longer, as shown in the figure, so that the area increases to $A = 20(30 + x)$. Which property of real numbers tells us that the new area can also be written $A = 600 + 20x$?

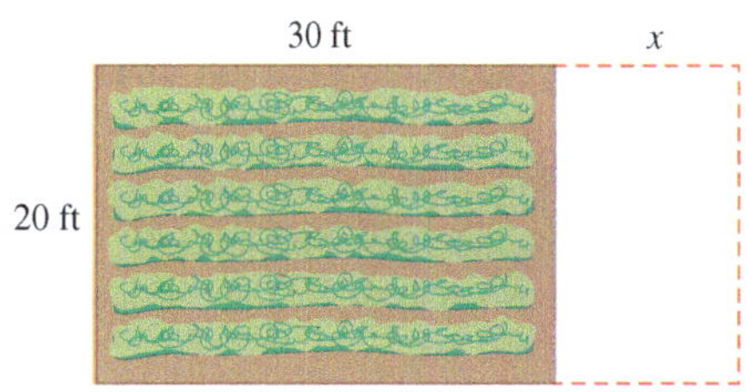

86. Temperature Variation The bar graph shows the daily high temperatures for Omak, Washington, and Geneseo, New York, during a certain week in June. Let T_O represent the temperature in Omak and T_G the temperature in Geneseo. Calculate $T_O - T_G$ and $|T_O - T_G|$ for each day shown. Which of these two values gives more information?

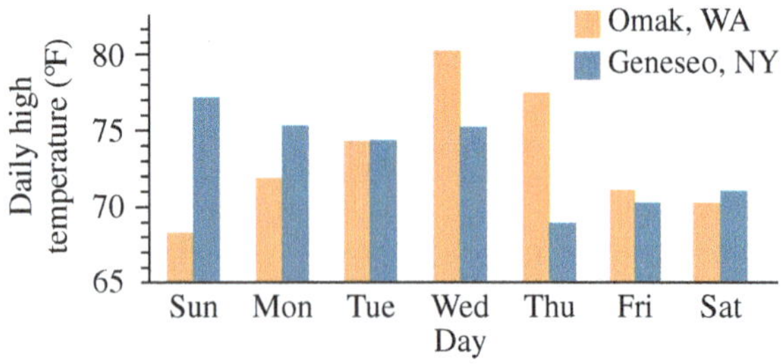

87. Mailing a Package The post office will accept only packages for which the length plus the "girth" (distance around) is no more than 108 in. Thus for the package in the figure, we must have

$$L + 2(x + y) \le 108$$

(a) Will the post office accept a package that is 6 in. wide, 8 in. deep, and 5 ft long? What about a package that measures 2 ft by 2 ft by 4 ft?

(b) What is the greatest acceptable length for a package that has a square base measuring 9 in. by 9 in.?

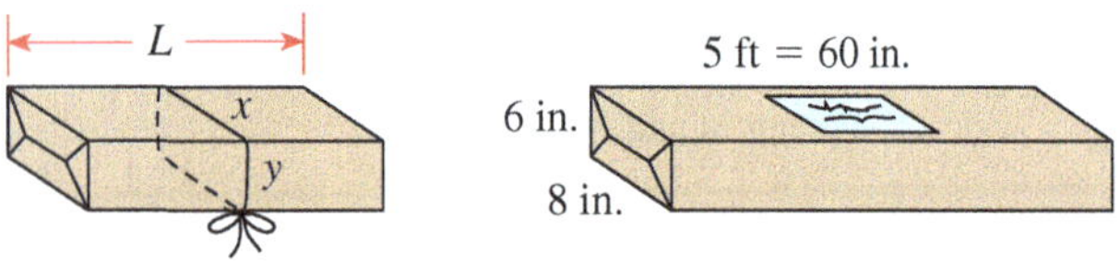

DISCUSS ■ DISCOVER ■ PROVE ■ WRITE

88. DISCUSS: Sums and Products of Rational and Irrational Numbers Explain why the sum, the difference, and the product of two rational numbers are rational numbers. Is the product of two irrational numbers necessarily irrational? What about the sum?

89. DISCOVER ■ PROVE: Combining Rational and Irrational Numbers Is $\frac{1}{2} + \sqrt{2}$ rational or irrational? Is $\frac{1}{2} \cdot \sqrt{2}$ rational or irrational? Experiment with sums and products of other rational and irrational numbers. Prove the following.

(a) The sum of a rational number r and an irrational number t is irrational.

(b) The product of a rational number r and an irrational number t is irrational.

[*Hint:* For part (a), suppose that $r + t$ is a rational number q, that is, $r + t = q$. Show that this leads to a contradiction. Use similar reasoning for part (b).]

90. DISCOVER: Limiting Behavior of Reciprocals Complete the tables. What happens to the size of the fraction $1/x$ as x gets large? As x gets small?

x	$1/x$
1	
2	
10	
100	
1000	

x	$1/x$
1.0	
0.5	
0.1	
0.01	
0.001	

91. DISCOVER: Locating Irrational Numbers on the Real Line Using the figures below, explain how to locate the point $\sqrt{2}$ on a number line. Can you locate $\sqrt{5}$ by a similar method? How can the circle shown in the figure help us to locate π on a number line? List some other irrational numbers that you can locate on a number line.

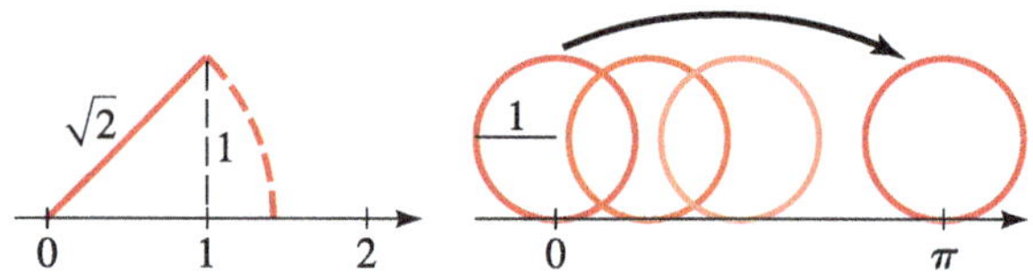

92. PROVE: Maximum and Minimum Formulas Let $\max(a, b)$ denote the maximum and $\min(a, b)$ denote the minimum of the real numbers a and b. For example, $\max(2, 5) = 5$ and $\min(-1, -2) = -2$.

(a) Prove that $\max(a, b) = \dfrac{a + b + |a - b|}{2}$.

(b) Prove that $\min(a, b) = \dfrac{a + b - |a - b|}{2}$.

[*Hint:* Take cases and write these expressions without absolute values. See Exercises 81 and 82.]

93. WRITE: Real Numbers in the Real World Write a paragraph describing different real-world situations in which you would use natural numbers, integers, rational numbers, and irrational numbers. Give examples for each type of situation.

94. DISCUSS: Commutative and Noncommutative Operations We have learned that addition and multiplication are both commutative operations.

(a) Is subtraction commutative?

(b) Is division of nonzero real numbers commutative?

(c) Are the actions of putting on your socks and putting on your shoes commutative?

(d) Are the actions of putting on your hat and putting on your coat commutative?

(e) Are the actions of washing laundry and drying it commutative?

95. PROVE: Triangle Inequality We prove Property 5 of absolute values, the Triangle Inequality:

$$|x + y| \le |x| + |y|$$

(a) Verify that the Triangle Inequality holds for $x = 2$ and $y = 3$, for $x = -2$ and $y = -3$, and for $x = -2$ and $y = 3$.

(b) Prove that the Triangle Inequality is true for all real numbers x and y. [*Hint:* Take cases.]

1.2 EXPONENTS AND RADICALS

■ Integer Exponents ■ Rules for Working with Exponents ■ Scientific Notation
■ Radicals ■ Rational Exponents ■ Rationalizing the Denominator; Standard Form

In this section we give meaning to expressions such as $a^{m/n}$ in which the exponent m/n is a rational number. To do this, we need to recall some facts about integer exponents, radicals, and nth roots.

■ Integer Exponents

A product of identical numbers is usually written in exponential notation. For example, $5 \cdot 5 \cdot 5$ is written as 5^3. In general, we have the following definition.

EXPONENTIAL NOTATION

If a is any real number and n is a positive integer, then the ***n*th power** of a is

$$a^n = \underbrace{a \cdot a \cdot \cdots \cdot a}_{n \text{ factors}}$$

The number a is called the **base**, and n is called the **exponent**.

EXAMPLE 1 ■ Exponential Notation

(a) $\left(\frac{1}{2}\right)^5 = \left(\frac{1}{2}\right)\left(\frac{1}{2}\right)\left(\frac{1}{2}\right)\left(\frac{1}{2}\right)\left(\frac{1}{2}\right) = \frac{1}{32}$

(b) $(-3)^4 = (-3) \cdot (-3) \cdot (-3) \cdot (-3) = 81$

(c) $-3^4 = -(3 \cdot 3 \cdot 3 \cdot 3) = -81$

Note the distinction between $(-3)^4$ and -3^4. In $(-3)^4$ the exponent applies to -3, but in -3^4 the exponent applies only to 3.

Now Try Exercise 17 ■

We can state several useful rules for working with exponential notation. To discover the rule for multiplication, we multiply 5^4 by 5^2:

$$5^4 \cdot 5^2 = \underbrace{(5 \cdot 5 \cdot 5 \cdot 5)}_{4 \text{ factors}}\underbrace{(5 \cdot 5)}_{2 \text{ factors}} = \underbrace{5 \cdot 5 \cdot 5 \cdot 5 \cdot 5 \cdot 5}_{6 \text{ factors}} = 5^6 = 5^{4+2}$$

It appears that *to multiply two powers of the same base, we add their exponents*. In general, for any real number a and any positive integers m and n, we have

$$a^m a^n = \underbrace{(a \cdot a \cdot \cdots \cdot a)}_{m \text{ factors}}\underbrace{(a \cdot a \cdot \cdots \cdot a)}_{n \text{ factors}} = \underbrace{a \cdot a \cdot a \cdot \cdots \cdot a}_{m+n \text{ factors}} = a^{m+n}$$

Thus $a^m a^n = a^{m+n}$.

We would like this rule to be true even when m and n are 0 or negative integers. For instance, we must have

$$2^0 \cdot 2^3 = 2^{0+3} = 2^3$$

But this can happen only if $2^0 = 1$. Likewise, we want to have

$$5^4 \cdot 5^{-4} = 5^{4+(-4)} = 5^{4-4} = 5^0 = 1$$

and this will be true if $5^{-4} = 1/5^4$. These observations lead to the following definition.

ZERO AND NEGATIVE EXPONENTS

If $a \neq 0$ is a real number and n is a positive integer, then

$$a^0 = 1 \qquad \text{and} \qquad a^{-n} = \frac{1}{a^n}$$

EXAMPLE 2 ■ Zero and Negative Exponents

(a) $\left(\frac{4}{7}\right)^0 = 1$

(b) $x^{-1} = \dfrac{1}{x^1} = \dfrac{1}{x}$

(c) $(-2)^{-3} = \dfrac{1}{(-2)^3} = \dfrac{1}{-8} = -\dfrac{1}{8}$

Now Try Exercise 19 ■

■ Rules for Working with Exponents

Familiarity with the following rules is essential for our work with exponents and bases. In the table the bases a and b are real numbers, and the exponents m and n are integers.

LAWS OF EXPONENTS

Law	Example	Description
1. $a^m a^n = a^{m+n}$	$3^2 \cdot 3^5 = 3^{2+5} = 3^7$	To multiply two powers of the same number, add the exponents.
2. $\dfrac{a^m}{a^n} = a^{m-n}$	$\dfrac{3^5}{3^2} = 3^{5-2} = 3^3$	To divide two powers of the same number, subtract the exponents.
3. $(a^m)^n = a^{mn}$	$(3^2)^5 = 3^{2\cdot 5} = 3^{10}$	To raise a power to a new power, multiply the exponents.
4. $(ab)^n = a^n b^n$	$(3\cdot 4)^2 = 3^2 \cdot 4^2$	To raise a product to a power, raise each factor to the power.
5. $\left(\dfrac{a}{b}\right)^n = \dfrac{a^n}{b^n}$	$\left(\dfrac{3}{4}\right)^2 = \dfrac{3^2}{4^2}$	To raise a quotient to a power, raise both numerator and denominator to the power.
6. $\left(\dfrac{a}{b}\right)^{-n} = \left(\dfrac{b}{a}\right)^n$	$\left(\dfrac{3}{4}\right)^{-2} = \left(\dfrac{4}{3}\right)^2$	To raise a fraction to a negative power, invert the fraction and change the sign of the exponent.
7. $\dfrac{a^{-n}}{b^{-m}} = \dfrac{b^m}{a^n}$	$\dfrac{3^{-2}}{4^{-5}} = \dfrac{4^5}{3^2}$	To move a number raised to a power from numerator to denominator or from denominator to numerator, change the sign of the exponent.

Proof of Law 3 If m and n are positive integers, we have

$$(a^m)^n = (\underbrace{a \cdot a \cdot \cdots \cdot a}_{m \text{ factors}})^n$$

$$= \underbrace{(\underbrace{a \cdot a \cdot \cdots \cdot a}_{m \text{ factors}})(\underbrace{a \cdot a \cdot \cdots \cdot a}_{m \text{ factors}}) \cdots (\underbrace{a \cdot a \cdot \cdots \cdot a}_{m \text{ factors}})}_{n \text{ groups of factors}}$$

$$= \underbrace{a \cdot a \cdot \cdots \cdot a}_{mn \text{ factors}} = a^{mn}$$

The cases for which $m \le 0$ or $n \le 0$ can be proved by using the definition of negative exponents. ■

Proof of Law 4 If n is a positive integer, we have

$$(ab)^n = \underbrace{(ab)(ab)\cdots(ab)}_{n \text{ factors}} = \underbrace{(a\cdot a\cdot\cdots\cdot a)}_{n \text{ factors}}\cdot\underbrace{(b\cdot b\cdot\cdots\cdot b)}_{n \text{ factors}} = a^n b^n$$

Here we have used the Commutative and Associative Properties repeatedly. If $n \le 0$, Law 4 can be proved by using the definition of negative exponents. ■

You are asked to prove Laws 2, 5, 6, and 7 in Exercises 108 and 109.

EXAMPLE 3 ■ Using Laws of Exponents

(a) $x^4x^7 = x^{4+7} = x^{11}$ Law 1: $a^m a^n = a^{m+n}$

(b) $y^4y^{-7} = y^{4-7} = y^{-3} = \dfrac{1}{y^3}$ Law 1: $a^m a^n = a^{m+n}$

(c) $\dfrac{c^9}{c^5} = c^{9-5} = c^4$ Law 2: $\dfrac{a^m}{a^n} = a^{m-n}$

(d) $(b^4)^5 = b^{4\cdot 5} = b^{20}$ Law 3: $(a^m)^n = a^{mn}$

(e) $(3x)^3 = 3^3x^3 = 27x^3$ Law 4: $(ab)^n = a^n b^n$

(f) $\left(\dfrac{x}{2}\right)^5 = \dfrac{x^5}{2^5} = \dfrac{x^5}{32}$ Law 5: $\left(\dfrac{a}{b}\right)^n = \dfrac{a^n}{b^n}$

Now Try Exercises 29, 31, and 33 ■

EXAMPLE 4 ■ Simplifying Expressions with Exponents

Simplify:

(a) $(2a^3b^2)(3ab^4)^3$ **(b)** $\left(\dfrac{x}{y}\right)^3\left(\dfrac{y^2x}{z}\right)^4$

SOLUTION

(a)
$$\begin{aligned}(2a^3b^2)(3ab^4)^3 &= (2a^3b^2)[3^3a^3(b^4)^3] && \text{Law 4: } (ab)^n = a^n b^n\\ &= (2a^3b^2)(27a^3b^{12}) && \text{Law 3: } (a^m)^n = a^{mn}\\ &= (2)(27)a^3a^3b^2b^{12} && \text{Group factors with the same base}\\ &= 54a^6b^{14} && \text{Law 1: } a^m a^n = a^{m+n}\end{aligned}$$

(b)
$$\begin{aligned}\left(\frac{x}{y}\right)^3\left(\frac{y^2x}{z}\right)^4 &= \frac{x^3}{y^3}\frac{(y^2)^4x^4}{z^4} && \text{Laws 5 and 4}\\ &= \frac{x^3}{y^3}\frac{y^8x^4}{z^4} && \text{Law 3}\\ &= (x^3x^4)\left(\frac{y^8}{y^3}\right)\frac{1}{z^4} && \text{Group factors with the same base}\\ &= \frac{x^7y^5}{z^4} && \text{Laws 1 and 2}\end{aligned}$$

Now Try Exercises 35 and 39 ■

When simplifying an expression, you will find that many different methods will lead to the same result; you should feel free to use any of the rules of exponents to arrive at your own method. In the next example we see how to simplify expressions with negative exponents.

Mathematics in the Modern World

Although we are often unaware of its presence, mathematics permeates nearly every aspect of life in the modern world. With the advent of modern technology, mathematics plays an ever greater role in our lives. Today you were probably awakened by a digital alarm clock, sent a text, surfed the Internet, watched HDTV or a streaming video, listened to music on your cell phone, drove a car with digitally controlled fuel injection, then fell asleep in a room whose temperature is controlled by a digital thermostat. In each of these activities mathematics is crucially involved. In general, a property such as the intensity or frequency of sound, the oxygen level in the exhaust emission from a car, the colors in an image, or the temperature in your bedroom is transformed into sequences of numbers by sophisticated mathematical algorithms. These numerical data, which usually consist of many millions of bits (the digits 0 and 1), are then transmitted and reinterpreted. Dealing with such huge amounts of data was not feasible until the invention of computers, machines whose logical processes were invented by mathematicians.

The contributions of mathematics in the modern world are not limited to technological advances. The logical processes of mathematics are now used to analyze complex problems in the social, political, and life sciences in new and surprising ways. Advances in mathematics continue to be made, some of the most exciting of these just within the past decade.

In other *Mathematics in the Modern World*, we will describe in more detail how mathematics affects all of us in our everyday activities.

EXAMPLE 5 ■ Simplifying Expressions with Negative Exponents

Eliminate negative exponents, and simplify each expression.

(a) $\dfrac{6st^{-4}}{2s^{-2}t^2}$ **(b)** $\left(\dfrac{y}{3z^3}\right)^{-2}$

SOLUTION

(a) We use Law 7, which allows us to move a number raised to a power from the numerator to the denominator (or vice versa) by changing the sign of the exponent.

t^{-4} moves to denominator and becomes t^4

s^{-2} moves to numerator and becomes s^2

$$\frac{6st^{-4}}{2s^{-2}t^2} = \frac{6ss^2}{2t^2t^4} \qquad \text{Law 7}$$

$$= \frac{3s^3}{t^6} \qquad \text{Law 1}$$

(b) We use Law 6, which allows us to change the sign of the exponent of a fraction by inverting the fraction.

$$\left(\frac{y}{3z^3}\right)^{-2} = \left(\frac{3z^3}{y}\right)^2 \qquad \text{Law 6}$$

$$= \frac{9z^6}{y^2} \qquad \text{Laws 5 and 4}$$

Now Try Exercise 41 ■

■ Scientific Notation

Scientists use exponential notation as a compact way of writing very large numbers and very small numbers. For example, the nearest star beyond the sun, Proxima Centauri, is approximately 40,000,000,000,000 km away. The mass of a hydrogen atom is about 0.00000000000000000000000166 g. Such numbers are difficult to read and to write, so scientists usually express them in *scientific notation*.

SCIENTIFIC NOTATION

A positive number x is said to be written in **scientific notation** if it is expressed as follows:

$$x = a \times 10^n \qquad \text{where } 1 \le a < 10 \text{ and } n \text{ is an integer}$$

For instance, when we state that the distance to the star Proxima Centauri is 4×10^{13} km, the positive exponent 13 indicates that the decimal point should be moved 13 places to the *right*:

$$4 \times 10^{13} = 40{,}000{,}000{,}000{,}000$$

Move decimal point 13 places to the right

When we state that the mass of a hydrogen atom is 1.66×10^{-24} g, the exponent -24 indicates that the decimal point should be moved 24 places to the *left*:

$$1.66 \times 10^{-24} = 0.00000000000000000000000166$$

Move decimal point 24 places to the left

EXAMPLE 6 ■ Changing from Decimal to Scientific Notation

Write each number in scientific notation.

(a) 56,920 **(b)** 0.000093

SOLUTION

(a) $\underbrace{56{,}920}_{\text{4 places}} = 5.692 \times 10^4$ **(b)** $\underbrace{0.000093}_{\text{5 places}} = 9.3 \times 10^{-5}$

Now Try Exercise 83 ■

EXAMPLE 7 ■ Changing from Scientific Notation to Decimal Notation

Write each number in decimal notation.

(a) 6.97×10^9 **(b)** 4.6271×10^{-6}

SOLUTION

(a) $6.97 \times 10^9 = \underbrace{6{,}970{,}000{,}000}_{\text{9 places}}$ Move decimal 9 places to the right

(b) $4.6271 \times 10^{-6} = \underbrace{0.000004}_{\text{6 places}}6271$ Move decimal 6 places to the left

Now Try Exercise 85 ■

To use scientific notation on a calculator, press the key labeled EE or EXP or EEX to enter the exponent. For example, to enter the number 3.629×10^{15} on a TI-83 or TI-84 calculator, we enter

3.629 2ND EE 15

and the display reads

3.629E15

Scientific notation is often used on a calculator to display a very large or very small number. For instance, if we use a calculator to square the number 1,111,111, the display panel may show (depending on the calculator model) the approximation

1.234568 12 or 1.234568 E12

Here the final digits indicate the power of 10, and we interpret the result as

$$1.234568 \times 10^{12}$$

EXAMPLE 8 ■ Calculating with Scientific Notation

If $a \approx 0.00046$, $b \approx 1.697 \times 10^{22}$, and $c \approx 2.91 \times 10^{-18}$, use a calculator to approximate the quotient ab/c.

SOLUTION We could enter the data using scientific notation, or we could use laws of exponents as follows:

$$\begin{aligned}\frac{ab}{c} &\approx \frac{(4.6 \times 10^{-4})(1.697 \times 10^{22})}{2.91 \times 10^{-18}} \\ &= \frac{(4.6)(1.697)}{2.91} \times 10^{-4+22+18} \\ &\approx 2.7 \times 10^{36}\end{aligned}$$

We state the answer rounded to two significant figures because the least accurate of the given numbers is stated to two significant figures.

For guidelines on working with significant figures, see Appendix B, *Calculations and Significant Figures*. Go to **www.stewartmath.com**.

Now Try Exercises 89 and 91 ■

■ Radicals

We know what 2^n means whenever n is an integer. To give meaning to a power, such as $2^{4/5}$, whose exponent is a rational number, we need to discuss radicals.

It is true that the number 9 has two square roots, 3 and -3, but the notation $\sqrt{9}$ is reserved for the *positive* square root of 9 (sometimes called the *principal square root* of 9). If we want the negative root, we must write $-\sqrt{9}$, which is -3.

The symbol $\sqrt{\ }$ means "the positive square root of." Thus

$$\sqrt{a} = b \qquad \text{means} \qquad b^2 = a \qquad \text{and} \qquad b \geq 0$$

Since $a = b^2 \geq 0$, the symbol $\sqrt{a}$ makes sense only when $a \geq 0$. For instance,

$$\sqrt{9} = 3 \qquad \text{because} \qquad 3^2 = 9 \qquad \text{and} \qquad 3 \geq 0$$

Square roots are special cases of nth roots. The nth root of x is the number that, when raised to the nth power, gives x.

DEFINITION OF nth ROOT

If n is any positive integer, then the **principal nth root** of a is defined as follows:

$$\sqrt[n]{a} = b \qquad \text{means} \qquad b^n = a$$

If n is even, we must have $a \geq 0$ and $b \geq 0$.

For example,

$$\sqrt[4]{81} = 3 \qquad \text{because} \qquad 3^4 = 81 \qquad \text{and} \qquad 3 \geq 0$$

$$\sqrt[3]{-8} = -2 \qquad \text{because} \qquad (-2)^3 = -8$$

But $\sqrt{-8}$, $\sqrt[4]{-8}$, and $\sqrt[6]{-8}$ are not defined. (For instance, $\sqrt{-8}$ is not defined because the square of every real number is nonnegative.)

Notice that

$$\sqrt{4^2} = \sqrt{16} = 4 \qquad \text{but} \qquad \sqrt{(-4)^2} = \sqrt{16} = 4 = |-4|$$

So the equation $\sqrt{a^2} = a$ is not always true; it is true only when $a \geq 0$. However, we can always write $\sqrt{a^2} = |a|$. This last equation is true not only for square roots, but for any even root. This and other rules used in working with nth roots are listed in the following box. In each property we assume that all the given roots exist.

PROPERTIES OF nth ROOTS

Property	Example
1. $\sqrt[n]{ab} = \sqrt[n]{a}\sqrt[n]{b}$	$\sqrt[3]{-8 \cdot 27} = \sqrt[3]{-8}\sqrt[3]{27} = (-2)(3) = -6$
2. $\sqrt[n]{\dfrac{a}{b}} = \dfrac{\sqrt[n]{a}}{\sqrt[n]{b}}$	$\sqrt[4]{\dfrac{16}{81}} = \dfrac{\sqrt[4]{16}}{\sqrt[4]{81}} = \dfrac{2}{3}$
3. $\sqrt[m]{\sqrt[n]{a}} = \sqrt[mn]{a}$	$\sqrt{\sqrt[3]{729}} = \sqrt[6]{729} = 3$
4. $\sqrt[n]{a^n} = a$ if n is odd	$\sqrt[3]{(-5)^3} = -5$, $\sqrt[5]{2^5} = 2$
5. $\sqrt[n]{a^n} = \|a\|$ if n is even	$\sqrt[4]{(-3)^4} = \|-3\| = 3$

EXAMPLE 9 ■ Simplifying Expressions Involving nth Roots

(a) $\sqrt[3]{x^4} = \sqrt[3]{x^3 x}$ — Factor out the largest cube

$= \sqrt[3]{x^3}\sqrt[3]{x}$ — Property 1: $\sqrt[3]{ab} = \sqrt[3]{a}\sqrt[3]{b}$

$= x\sqrt[3]{x}$ — Property 4: $\sqrt[3]{a^3} = a$

(b) $\sqrt[4]{81x^8y^4} = \sqrt[4]{81}\sqrt[4]{x^8}\sqrt[4]{y^4}$ Property 1: $\sqrt[n]{abc} = \sqrt[n]{a}\sqrt[n]{b}\sqrt[n]{c}$

$= 3\sqrt[4]{(x^2)^4}|y|$ Property 5: $\sqrt[4]{a^4} = |a|$

$= 3x^2|y|$ Property 5: $\sqrt[4]{a^4} = |a|$, $|x^2| = x^2$

Now Try Exercises 45 and 47 ■

It is frequently useful to combine like radicals in an expression such as $2\sqrt{3} + 5\sqrt{3}$. This can be done by using the Distributive Property. For example,

$$2\sqrt{3} + 5\sqrt{3} = (2 + 5)\sqrt{3} = 7\sqrt{3}$$

The next example further illustrates this process.

Avoid making the following error:

$\sqrt{a+b} \;\cancel{=}\; \sqrt{a} + \sqrt{b}$

For instance, if we let $a = 9$ and $b = 16$, then we see the error:

$\sqrt{9+16} \stackrel{?}{=} \sqrt{9} + \sqrt{16}$

$\sqrt{25} \stackrel{?}{=} 3 + 4$

$5 \stackrel{?}{=} 7$ Wrong!

EXAMPLE 10 ■ Combining Radicals

(a) $\sqrt{32} + \sqrt{200} = \sqrt{16 \cdot 2} + \sqrt{100 \cdot 2}$ Factor out the largest squares

$= \sqrt{16}\sqrt{2} + \sqrt{100}\sqrt{2}$ Property 1: $\sqrt{ab} = \sqrt{a}\sqrt{b}$

$= 4\sqrt{2} + 10\sqrt{2} = 14\sqrt{2}$ Distributive Property

(b) If $b > 0$, then

$\sqrt{25b} - \sqrt{b^3} = \sqrt{25}\sqrt{b} - \sqrt{b^2}\sqrt{b}$ Property 1: $\sqrt{ab} = \sqrt{a}\sqrt{b}$

$= 5\sqrt{b} - b\sqrt{b}$ Property 5, $b > 0$

$= (5 - b)\sqrt{b}$ Distributive Property

(c) $\sqrt{49x^2 + 49} = \sqrt{49(x^2 + 1)}$ Factor out the perfect square

$= 7\sqrt{x^2 + 1}$ Property 1: $\sqrt{ab} = \sqrt{a}\sqrt{b}$

Now Try Exercises 49, 51, and 53 ■

Rational Exponents

To define what is meant by a *rational exponent* or, equivalently, a *fractional exponent* such as $a^{1/3}$, we need to use radicals. To give meaning to the symbol $a^{1/n}$ in a way that is consistent with the Laws of Exponents, we would have to have

$$(a^{1/n})^n = a^{(1/n)n} = a^1 = a$$

So by the definition of nth root,

$$a^{1/n} = \sqrt[n]{a}$$

In general, we define rational exponents as follows.

DEFINITION OF RATIONAL EXPONENTS

For any rational exponent m/n in lowest terms, where m and n are integers and $n > 0$, we define

$$a^{m/n} = (\sqrt[n]{a})^m \qquad \text{or equivalently} \qquad a^{m/n} = \sqrt[n]{a^m}$$

If n is even, then we require that $a \geq 0$.

With this definition it can be proved that *the Laws of Exponents also hold for rational exponents*.

DIOPHANTUS lived in Alexandria about 250 A.D. His book *Arithmetica* is considered the first book on algebra. In it he gives methods for finding integer solutions of algebraic equations. *Arithmetica* was read and studied for more than a thousand years. Fermat (see page 117) made some of his most important discoveries while studying this book. Diophantus' major contribution is the use of symbols to stand for the unknowns in a problem. Although his symbolism is not as simple as what we use today, it was a major advance over writing everything in words. In Diophantus' notation the equation

$$x^5 - 7x^2 + 8x - 5 = 24$$

is written

$$\Delta K^{\gamma}\alpha\varsigma\eta ⋔ \Delta^{\gamma}\zeta \mathring{M}\varepsilon\iota^{\sigma}\kappa\delta$$

Our modern algebraic notation did not come into common use until the 17th century.

EXAMPLE 11 ■ Using the Definition of Rational Exponents

(a) $4^{1/2} = \sqrt{4} = 2$

(b) $8^{2/3} = (\sqrt[3]{8})^2 = 2^2 = 4$ Alternative solution: $8^{2/3} = \sqrt[3]{8^2} = \sqrt[3]{64} = 4$

(c) $125^{-1/3} = \dfrac{1}{125^{1/3}} = \dfrac{1}{\sqrt[3]{125}} = \dfrac{1}{5}$

Now Try Exercises 55 and 57 ■

EXAMPLE 12 ■ Using the Laws of Exponents with Rational Exponents

(a) $a^{1/3}a^{7/3} = a^{8/3}$ Law 1: $a^m a^n = a^{m+n}$

(b) $\dfrac{a^{2/5}a^{7/5}}{a^{3/5}} = a^{2/5+7/5-3/5} = a^{6/5}$ Law 1, Law 2: $\dfrac{a^m}{a^n} = a^{m-n}$

(c)
$$\begin{aligned}(2a^3b^4)^{3/2} &= 2^{3/2}(a^3)^{3/2}(b^4)^{3/2} && \text{Law 4: } (abc)^n = a^nb^nc^n\\ &= (\sqrt{2})^3 a^{3(3/2)}b^{4(3/2)} && \text{Law 3: } (a^m)^n = a^{mn}\\ &= 2\sqrt{2}a^{9/2}b^6\end{aligned}$$

(d)
$$\begin{aligned}\left(\frac{2x^{3/4}}{y^{1/3}}\right)^3\left(\frac{y^4}{x^{-1/2}}\right) &= \frac{2^3(x^{3/4})^3}{(y^{1/3})^3}\cdot(y^4x^{1/2}) && \text{Laws 5, 4, and 7}\\ &= \frac{8x^{9/4}}{y}\cdot y^4x^{1/2} && \text{Law 3}\\ &= 8x^{11/4}y^3 && \text{Laws 1 and 2}\end{aligned}$$

Now Try Exercises 61, 63, 67, and 69 ■

EXAMPLE 13 ■ Simplifying by Writing Radicals as Rational Exponents

(a) $\dfrac{1}{\sqrt[3]{x^4}} = \dfrac{1}{x^{4/3}} = x^{-4/3}$ Definition of rational and negative exponents

(b)
$$\begin{aligned}(2\sqrt{x})(3\sqrt[3]{x}) &= (2x^{1/2})(3x^{1/3}) && \text{Definition of rational exponents}\\ &= 6x^{1/2+1/3} = 6x^{5/6} && \text{Law 1}\end{aligned}$$

(c)
$$\begin{aligned}\sqrt{x\sqrt{x}} &= (xx^{1/2})^{1/2} && \text{Definition of rational exponents}\\ &= (x^{3/2})^{1/2} && \text{Law 1}\\ &= x^{3/4} && \text{Law 3}\end{aligned}$$

Now Try Exercises 73 and 77 ■

■ Rationalizing the Denominator; Standard Form

It is often useful to eliminate the radical in a denominator by multiplying both numerator and denominator by an appropriate expression. This procedure is called **rationalizing the denominator**. If the denominator is of the form $\sqrt{a}$, we multiply numerator and denominator by $\sqrt{a}$. In doing this we multiply the given quantity by 1, so we do not change its value. For instance,

$$\frac{1}{\sqrt{a}} = \frac{1}{\sqrt{a}}\cdot 1 = \frac{1}{\sqrt{a}}\cdot\frac{\sqrt{a}}{\sqrt{a}} = \frac{\sqrt{a}}{a}$$

Note that the denominator in the last fraction contains no radical. In general, if the denominator is of the form $\sqrt[n]{a^m}$ with $m < n$, then multiplying the numerator and denominator by $\sqrt[n]{a^{n-m}}$ will rationalize the denominator, because (for $a > 0$)

$$\sqrt[n]{a^m}\sqrt[n]{a^{n-m}} = \sqrt[n]{a^{m+n-m}} = \sqrt[n]{a^n} = a$$

A fractional expression whose denominator contains no radicals is said to be in **standard form**.

EXAMPLE 14 ■ Rationalizing Denominators

Put each fractional expression into standard form by rationalizing the denominator.

(a) $\dfrac{2}{\sqrt{3}}$ **(b)** $\dfrac{1}{\sqrt[3]{5}}$ **(c)** $\sqrt[7]{\dfrac{1}{a^2}}$

SOLUTION

(a) $\dfrac{2}{\sqrt{3}} = \dfrac{2}{\sqrt{3}} \cdot \dfrac{\sqrt{3}}{\sqrt{3}}$ This equals 1. Multiply by $\dfrac{\sqrt{3}}{\sqrt{3}}$

$= \dfrac{2\sqrt{3}}{3}$ $\sqrt{3} \cdot \sqrt{3} = 3$

(b) $\dfrac{1}{\sqrt[3]{5}} = \dfrac{1}{\sqrt[3]{5}} \cdot \dfrac{\sqrt[3]{5^2}}{\sqrt[3]{5^2}}$ Multiply by $\dfrac{\sqrt[3]{5^2}}{\sqrt[3]{5^2}}$

$= \dfrac{\sqrt[3]{25}}{5}$ $\sqrt[3]{5} \cdot \sqrt[3]{5^2} = \sqrt[3]{5^3} = 5$

(c) $\sqrt[7]{\dfrac{1}{a^2}} = \dfrac{1}{\sqrt[7]{a^2}}$ Property 2: $\sqrt[n]{\dfrac{a}{b}} = \dfrac{\sqrt[n]{a}}{\sqrt[n]{b}}$

$= \dfrac{1}{\sqrt[7]{a^2}} \cdot \dfrac{\sqrt[7]{a^5}}{\sqrt[7]{a^5}}$ Multiply by $\dfrac{\sqrt[7]{a^5}}{\sqrt[7]{a^5}}$

$= \dfrac{\sqrt[7]{a^5}}{a}$ $\sqrt[7]{a^2} \cdot \sqrt[7]{a^5} = a$

Now Try Exercises 79 and 81 ■

1.2 EXERCISES

CONCEPTS

1. (a) Using exponential notation, we can write the product $5 \cdot 5 \cdot 5 \cdot 5 \cdot 5 \cdot 5$ as ________.

(b) In the expression 3^4 the number 3 is called the ________, and the number 4 is called the ________.

2. (a) When we multiply two powers with the same base, we ________ the exponents. So $3^4 \cdot 3^5 =$ ________.

(b) When we divide two powers with the same base, we ________ the exponents. So $\dfrac{3^5}{3^2} =$ ________.

3. (a) Using exponential notation, we can write $\sqrt[3]{5}$ as ________.

(b) Using radicals, we can write $5^{1/2}$ as ________.

(c) Is there a difference between $\sqrt{5^2}$ and $(\sqrt{5})^2$? Explain.

4. Explain what $4^{3/2}$ means, then calculate $4^{3/2}$ in two different ways:

$(4^{1/2})^{\square} =$ ________ or $(4^3)^{\square} =$ ________

5. Explain how we rationalize a denominator, then complete the following steps to rationalize $\dfrac{1}{\sqrt{3}}$:

$$\frac{1}{\sqrt{3}} = \frac{1}{\sqrt{3}} \cdot \frac{\square}{\square} = \frac{\square}{\square}$$

6. Find the missing power in the following calculation: $5^{1/3} \cdot 5^{\square} = 5$.

7–8 ■ *Yes or No*? If *No*, give a reason.

7. (a) Is the expression $\left(\frac{2}{3}\right)^{-2}$ equal to $\frac{3}{4}$?

(b) Is there a difference between $(-5)^4$ and -5^4?

8. (a) Is the expression $(x^2)^3$ equal to x^5?

(b) Is the expression $(2x^4)^3$ equal to $2x^{12}$?

(c) Is the expression $\sqrt{4a^2}$ equal to $2a$?

(d) Is the expression $\sqrt{a^2 + 4}$ equal to $a + 2$?

SKILLS

9–16 ■ Radicals and Exponents Write each radical expression using exponents, and each exponential expression using radicals.

	Radical expression	Exponential expression
9.	$\frac{1}{\sqrt{3}}$	
10.	$\sqrt[3]{7^2}$	
11.		$4^{2/3}$
12.		$10^{-3/2}$
13.	$\sqrt[5]{5^3}$	
14.		$2^{-1.5}$
15.		$a^{2/5}$
16.	$\frac{1}{\sqrt{x^5}}$	

17–28 ■ Radicals and Exponents Evaluate each expression.

17. **(a)** -2^6 **(b)** $(-2)^6$ **(c)** $(\frac{1}{5})^2\cdot(-3)^3$

18. **(a)** $(-5)^3$ **(b)** -5^3 **(c)** $(-5)^2\cdot(\frac{2}{5})^2$

19. **(a)** $(\frac{5}{3})^0\cdot 2^{-1}$ **(b)** $\frac{2^{-3}}{3^0}$ **(c)** $(\frac{2}{3})^{-2}$

20. **(a)** $-2^3\cdot(-2)^0$ **(b)** $-2^{-3}\cdot(-2)^0$ **(c)** $(\frac{-3}{5})^{-3}$

21. **(a)** $5^3\cdot 5$ **(b)** $5^4\cdot 5^{-2}$ **(c)** $(2^2)^3$

22. **(a)** $3^8\cdot 3^5$ **(b)** $\frac{10^7}{10^4}$ **(c)** $(3^5)^4$

23. **(a)** $3\sqrt[3]{16}$ **(b)** $\frac{\sqrt{18}}{\sqrt{81}}$ **(c)** $\sqrt{\frac{27}{4}}$

24. **(a)** $2\sqrt[3]{81}$ **(b)** $\frac{\sqrt{18}}{\sqrt{25}}$ **(c)** $\sqrt{\frac{12}{49}}$

25. **(a)** $\sqrt{3}\sqrt{15}$ **(b)** $\frac{\sqrt{48}}{\sqrt{3}}$ **(c)** $\sqrt[3]{24}\sqrt[3]{18}$

26. **(a)** $\sqrt{10}\sqrt{32}$ **(b)** $\frac{\sqrt{54}}{\sqrt{6}}$ **(c)** $\sqrt[3]{15}\sqrt[3]{75}$

27. **(a)** $\frac{\sqrt{132}}{\sqrt{3}}$ **(b)** $\sqrt[3]{2}\sqrt[3]{32}$ **(c)** $\sqrt[4]{\frac{1}{4}}\sqrt[4]{\frac{1}{64}}$

28. **(a)** $\sqrt[5]{\frac{1}{8}}\sqrt[5]{\frac{1}{4}}$ **(b)** $\sqrt[6]{\frac{1}{2}}\sqrt[6]{128}$ **(c)** $\frac{\sqrt[3]{4}}{\sqrt[3]{108}}$

29–34 ■ Exponents Simplify each expression, and eliminate any negative exponents.

29. **(a)** $x^3\cdot x^4$ **(b)** $(2y^2)^3$ **(c)** $y^{-2}y^7$

30. **(a)** $y^5\cdot y^2$ **(b)** $(8x)^2$ **(c)** x^4x^{-3}

31. **(a)** $x^{-5}\cdot x^3$ **(b)** $w^{-2}w^{-4}w^5$ **(c)** $\frac{x^{16}}{x^{10}}$

32. **(a)** $y^2\cdot y^{-5}$ **(b)** $z^5z^{-3}z^{-4}$ **(c)** $\frac{y^7y^0}{y^{10}}$

33. **(a)** $\frac{a^9a^{-2}}{a}$ **(b)** $(a^2a^4)^3$ **(c)** $\left(\frac{x}{2}\right)^3(5x^6)$

34. **(a)** $\frac{z^2z^4}{z^3z^{-1}}$ **(b)** $(2a^3a^2)^4$ **(c)** $(-3z^2)^3(2z^3)$

35–44 ■ Exponents Simplify each expression, and eliminate any negative exponents.

35. **(a)** $(3x^3y^2)(2y^3)$ **(b)** $(5w^2z^{-2})^2(z^3)$

36. **(a)** $(8m^{-2}n^4)(\frac{1}{2}n^{-2})$ **(b)** $(3a^4b^{-2})^3(a^2b^{-1})$

37. **(a)** $\frac{x^2y^{-1}}{x^{-5}}$ **(b)** $\left(\frac{a^3}{2b^2}\right)^3$

38. **(a)** $\frac{y^{-2}z^{-3}}{y^{-1}}$ **(b)** $\left(\frac{x^3y^{-2}}{x^{-3}y^2}\right)^{-2}$

39. **(a)** $\left(\frac{a^2}{b}\right)^5\left(\frac{a^3b^2}{c^3}\right)^3$ **(b)** $\frac{(u^{-1}v^2)^2}{(u^3v^{-2})^3}$

40. **(a)** $\left(\frac{x^4z^2}{4y^5}\right)\left(\frac{2x^3y^2}{z^3}\right)^2$ **(b)** $\frac{(rs^2)^3}{(r^{-3}s^2)^2}$

41. **(a)** $\frac{8a^3b^{-4}}{2a^{-5}b^5}$ **(b)** $\left(\frac{y}{5x^{-2}}\right)^{-3}$

42. **(a)** $\frac{5xy^{-2}}{x^{-1}y^{-3}}$ **(b)** $\left(\frac{2a^{-1}b}{a^2b^{-3}}\right)^{-3}$

43. **(a)** $\left(\frac{3a}{b^3}\right)^{-1}$ **(b)** $\left(\frac{q^{-1}r^{-1}s^{-2}}{r^{-5}sq^{-8}}\right)^{-1}$

44. **(a)** $\left(\frac{s^2t^{-4}}{5s^{-1}t}\right)^{-2}$ **(b)** $\left(\frac{xy^{-2}z^{-3}}{x^2y^3z^{-4}}\right)^{-3}$

45–48 ■ Radicals Simplify the expression. Assume that the letters denote any positive real numbers.

45. **(a)** $\sqrt[4]{x^4}$ **(b)** $\sqrt[4]{16x^8}$

46. **(a)** $\sqrt[5]{x^{10}}$ **(b)** $\sqrt[3]{x^3y^6}$

47. **(a)** $\sqrt[6]{64a^6b^7}$ **(b)** $\sqrt[3]{a^2b}\sqrt[3]{64a^4b}$

48. **(a)** $\sqrt[4]{x^4y^2z^2}$ **(b)** $\sqrt[3]{\sqrt{64x^6}}$

49–54 ■ Radical Expressions Simplify the expression.

49. **(a)** $\sqrt{32}+\sqrt{18}$ **(b)** $\sqrt{75}+\sqrt{48}$

50. **(a)** $\sqrt{125}+\sqrt{45}$ **(b)** $\sqrt[3]{54}-\sqrt[3]{16}$

51. **(a)** $\sqrt{9a^3}+\sqrt{a}$ **(b)** $\sqrt{16x}+\sqrt{x^5}$

52. **(a)** $\sqrt[3]{x^4}+\sqrt[3]{8x}$ **(b)** $4\sqrt{18rt^3}+5\sqrt{32r^3t^5}$

53. **(a)** $\sqrt{81x^2+81}$ **(b)** $\sqrt{36x^2+36y^2}$

54. **(a)** $\sqrt{27a^2+63a}$ **(b)** $\sqrt{75t+100t^2}$

55–60 ■ Rational Exponents Evaluate each expression.

55. **(a)** $16^{1/4}$ **(b)** $-8^{1/3}$ **(c)** $9^{-1/2}$

56. **(a)** $27^{1/3}$ **(b)** $(-8)^{1/3}$ **(c)** $-(\frac{1}{8})^{1/3}$

57. **(a)** $32^{2/5}$ **(b)** $(\frac{4}{9})^{-1/2}$ **(c)** $(\frac{16}{81})^{3/4}$

58. **(a)** $125^{2/3}$ **(b)** $(\frac{25}{64})^{3/2}$ **(c)** $27^{-4/3}$

59. (a) $5^{2/3} \cdot 5^{1/3}$ (b) $\dfrac{3^{3/5}}{3^{2/5}}$ (c) $(\sqrt[3]{4})^3$

60. (a) $3^{2/7} \cdot 3^{12/7}$ (b) $\dfrac{7^{2/3}}{7^{5/3}}$ (c) $(\sqrt[5]{6})^{-10}$

61–70 ■ Rational Exponents Simplify the expression and eliminate any negative exponent(s). Assume that all letters denote positive numbers.

61. (a) $x^{3/4}x^{5/4}$ (b) $y^{2/3}y^{4/3}$

62. (a) $(4b)^{1/2}(8b^{1/4})$ (b) $(3a^{3/4})^2(5a^{1/2})$

63. (a) $\dfrac{w^{4/3}w^{2/3}}{w^{1/3}}$ (b) $\dfrac{a^{5/4}(2a^{3/4})^3}{a^{1/4}}$

64. (a) $(8y^3)^{-2/3}$ (b) $(u^4v^6)^{-1/3}$

65. (a) $(8a^6b^{3/2})^{2/3}$ (b) $(4a^6b^8)^{3/2}$

66. (a) $(x^{-5}y^{1/3})^{-3/5}$ (b) $(4r^8s^{-1/2})^{1/2}(32s^{-5/4})^{-1/5}$

67. (a) $\dfrac{(8s^3t^3)^{2/3}}{(s^4t^{-8})^{1/4}}$ (b) $\dfrac{(32x^5y^{-3/2})^{2/5}}{(x^{5/3}y^{2/3})^{3/5}}$

68. (a) $\left(\dfrac{x^8y^{-4}}{16y^{4/3}}\right)^{-1/4}$ (b) $\left(\dfrac{4s^3t^4}{s^2t^{9/2}}\right)^{-1/2}$

69. (a) $\left(\dfrac{x^{3/2}}{y^{-1/2}}\right)^4\left(\dfrac{x^{-2}}{y^3}\right)$ (b) $\left(\dfrac{4y^3z^{2/3}}{x^{1/2}}\right)^2\left(\dfrac{x^{-3}y^6}{8z^4}\right)^{1/3}$

70. (a) $\left(\dfrac{a^{1/6}b^{-3}}{x^{-1}y}\right)^3\left(\dfrac{x^{-2}b^{-1}}{a^{3/2}y^{1/3}}\right)$ (b) $\dfrac{(9st)^{3/2}}{(27s^3t^{-4})^{2/3}}\left(\dfrac{3s^{-2}}{4t^{1/3}}\right)^{-1}$

71–78 ■ Radicals Simplify the expression, and eliminate any negative exponents(s). Assume that all letters denote positive numbers.

71. (a) $\sqrt{x^3}$ (b) $\sqrt[5]{x^6}$

72. (a) $\sqrt{x^5}$ (b) $\sqrt[4]{x^6}$

73. (a) $\sqrt[6]{y^5}\sqrt[3]{y^2}$ (b) $(5\sqrt[3]{x})(2\sqrt[4]{x})$

74. (a) $\sqrt[4]{b^3}\sqrt{b}$ (b) $(2\sqrt{a})(\sqrt[3]{a^2})$

75. (a) $\sqrt{4st^3}\sqrt[6]{s^3t^2}$ (b) $\dfrac{\sqrt[4]{x^7}}{\sqrt[4]{x^3}}$

76. (a) $\sqrt[5]{x^3y^2}\sqrt[10]{x^4y^{16}}$ (b) $\dfrac{\sqrt[3]{8x^2}}{\sqrt{x}}$

77. (a) $\sqrt[3]{y\sqrt{y}}$ (b) $\sqrt{\dfrac{16u^3v}{uv^5}}$

78. (a) $\sqrt{s\sqrt{s^3}}$ (b) $\sqrt[3]{\dfrac{54x^2y^4}{2x^5y}}$

79–82 ■ Rationalize Put each fractional expression into standard form by rationalizing the denominator.

79. (a) $\dfrac{1}{\sqrt{6}}$ (b) $\sqrt{\frac{3}{2}}$ (c) $\dfrac{9}{\sqrt[4]{2}}$

80. (a) $\dfrac{12}{\sqrt{3}}$ (b) $\sqrt{\frac{12}{5}}$ (c) $\dfrac{8}{\sqrt[3]{5^2}}$

81. (a) $\dfrac{1}{\sqrt{5x}}$ (b) $\sqrt{\dfrac{x}{5}}$ (c) $\sqrt[5]{\dfrac{1}{x^3}}$

82. (a) $\sqrt{\dfrac{s}{3t}}$ (b) $\dfrac{a}{\sqrt[6]{b^2}}$ (c) $\dfrac{1}{c^{3/5}}$

83–84 ■ Scientific Notation Write each number in scientific notation.

83. (a) 69,300,000 (b) 7,200,000,000,000
(c) 0.000028536 (d) 0.0001213

84. (a) 129,540,000 (b) 7,259,000,000
(c) 0.0000000014 (d) 0.0007029

85–86 ■ Decimal Notation Write each number in decimal notation.

85. (a) 3.19×10^5 (b) 2.721×10^8
(c) 2.670×10^{-8} (d) 9.999×10^{-9}

86. (a) 7.1×10^{14} (b) 6×10^{12}
(c) 8.55×10^{-3} (d) 6.257×10^{-10}

87–88 ■ Scientific Notation Write the number indicated in each statement in scientific notation.

87. (a) A light-year, the distance that light travels in one year, is about 5,900,000,000,000 mi.
(b) The diameter of an electron is about 0.0000000000004 cm.
(c) A drop of water contains more than 33 billion billion molecules.

88. (a) The distance from the earth to the sun is about 93 million miles.
(b) The mass of an oxygen molecule is about 0.000000000000000000000053 g.
(c) The mass of the earth is about 5,970,000,000,000,000,000,000,000 kg.

89–94 ■ Scientific Notation Use scientific notation, the Laws of Exponents, and a calculator to perform the indicated operations. State your answer rounded to the number of significant digits indicated by the given data.

89. $(7.2 \times 10^{-9})(1.806 \times 10^{-12})$

90. $(1.062 \times 10^{24})(8.61 \times 10^{19})$

91. $\dfrac{1.295643 \times 10^9}{(3.610 \times 10^{-17})(2.511 \times 10^6)}$

92. $\dfrac{(73.1)(1.6341 \times 10^{28})}{0.0000000019}$

93. $\dfrac{(0.0000162)(0.01582)}{(594{,}621{,}000)(0.0058)}$ **94.** $\dfrac{(3.542 \times 10^{-6})^9}{(5.05 \times 10^4)^{12}}$

SKILLS Plus

95. Let a, b, and c be real numbers with $a > 0$, $b < 0$, and $c < 0$. Determine the sign of each expression.
(a) b^5 (b) b^{10} (c) ab^2c^3
(d) $(b - a)^3$ (e) $(b - a)^4$ (f) $\dfrac{a^3c^3}{b^6c^6}$

96. Comparing Roots Without using a calculator, determine which number is larger in each pair.
(a) $2^{1/2}$ or $2^{1/3}$ (b) $\left(\frac{1}{2}\right)^{1/2}$ or $\left(\frac{1}{2}\right)^{1/3}$
(c) $7^{1/4}$ or $4^{1/3}$ (d) $\sqrt[3]{5}$ or $\sqrt{3}$

APPLICATIONS

97. Distance to the Nearest Star Proxima Centauri, the star nearest to our solar system, is 4.3 light-years away. Use the information in Exercise 87(a) to express this distance in miles.

98. Speed of Light The speed of light is about 186,000 mi/s. Use the information in Exercise 88(a) to find how long it takes for a light ray from the sun to reach the earth.

99. Volume of the Oceans The average ocean depth is 3.7×10^3 m, and the area of the oceans is 3.6×10^{14} m^2. What is the total volume of the ocean in liters? (One cubic meter contains 1000 liters.)

100. National Debt As of July 2013, the population of the United States was 3.164×10^8, and the national debt was 1.674×10^{13} dollars. How much was each person's share of the debt?
[*Source:* U.S. Census Bureau and U.S. Department of Treasury]

101. Number of Molecules A sealed room in a hospital, measuring 5 m wide, 10 m long, and 3 m high, is filled with pure oxygen. One cubic meter contains 1000 L, and 22.4 L of any gas contains 6.02×10^{23} molecules (Avogadro's number). How many molecules of oxygen are there in the room?

102. How Far Can You See? Because of the curvature of the earth, the maximum distance D that you can see from the top of a tall building of height h is estimated by the formula

$$D = \sqrt{2rh + h^2}$$

where $r = 3960$ mi is the radius of the earth and D and h are also measured in miles. How far can you see from the observation deck of the Toronto CN Tower, 1135 ft above the ground?

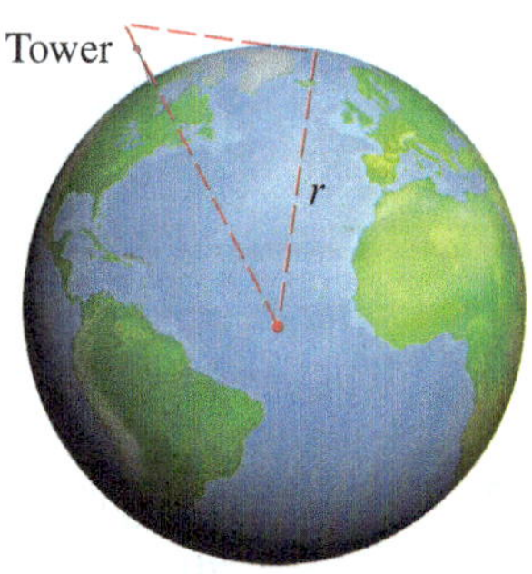

103. Speed of a Skidding Car Police use the formula $s = \sqrt{30fd}$ to estimate the speed s (in mi/h) at which a car is traveling if it skids d feet after the brakes are applied suddenly. The number f is the coefficient of friction of the road, which is a measure of the "slipperiness" of the road. The table gives some typical estimates for f.

	Tar	Concrete	Gravel
Dry	1.0	0.8	0.2
Wet	0.5	0.4	0.1

(a) If a car skids 65 ft on wet concrete, how fast was it moving when the brakes were applied?

(b) If a car is traveling at 50 mi/h, how far will it skid on wet tar?

104. Distance from the Earth to the Sun It follows from **Kepler's Third Law** of planetary motion that the average distance from a planet to the sun (in meters) is

$$d = \left(\frac{GM}{4\pi^2}\right)^{1/3} T^{2/3}$$

where $M = 1.99 \times 10^{30}$ kg is the mass of the sun, $G = 6.67 \times 10^{-11}$ N·m^2/kg^2 is the gravitational constant, and T is the period of the planet's orbit (in seconds). Use the fact that the period of the earth's orbit is about 365.25 days to find the distance from the earth to the sun.

DISCUSS ■ DISCOVER ■ PROVE ■ WRITE

105. DISCUSS: How Big is a Billion? If you had a million (10^6) dollars in a suitcase, and you spent a thousand (10^3) dollars each day, how many years would it take you to use all the money? Spending at the same rate, how many years would it take you to empty a suitcase filled with a *billion* (10^9) dollars?

106. DISCUSS: Easy Powers that Look Hard Calculate these expressions in your head. Use the Laws of Exponents to help you.

(a) $\dfrac{18^5}{9^5}$ **(b)** $20^6 \cdot (0.5)^6$

107. DISCOVER: Limiting Behavior of Powers Complete the following tables. What happens to the nth root of 2 as n gets large? What about the nth root of $\frac{1}{2}$?

n	$2^{1/n}$
1	
2	
5	
10	
100	

n	$\left(\frac{1}{2}\right)^{1/n}$
1	
2	
5	
10	
100	

Construct a similar table for $n^{1/n}$. What happens to the nth root of n as n gets large?

108. PROVE: Laws of Exponents Prove the following Laws of Exponents for the case in which m and n are positive integers and $m > n$.

(a) Law 2: $\dfrac{a^m}{a^n} = a^{m-n}$ **(b)** Law 5: $\left(\dfrac{a}{b}\right)^n = \dfrac{a^n}{b^n}$

109. PROVE: Laws of Exponents Prove the following Laws of Exponents.

(a) Law 6: $\left(\dfrac{a}{b}\right)^{-n} = \dfrac{b^n}{a^n}$ **(b)** Law 7: $\dfrac{a^{-n}}{b^{-m}} = \dfrac{b^m}{a^n}$

1.3 ALGEBRAIC EXPRESSIONS

■ Adding and Subtracting Polynomials ■ Multiplying Algebraic Expressions ■ Special Product Formulas ■ Factoring Common Factors ■ Factoring Trinomials ■ Special Factoring Formulas ■ Factoring by Grouping Terms

A **variable** is a letter that can represent any number from a given set of numbers. If we start with variables, such as x, y, and z, and some real numbers and combine them using addition, subtraction, multiplication, division, powers, and roots, we obtain an **algebraic expression**. Here are some examples:

$$2x^2 - 3x + 4 \qquad \sqrt{x} + 10 \qquad \frac{y - 2z}{y^2 + 4}$$

A **monomial** is an expression of the form ax^k, where a is a real number and k is a nonnegative integer. A **binomial** is a sum of two monomials and a **trinomial** is a sum of three monomials. In general, a sum of monomials is called a *polynomial*. For example, the first expression listed above is a polynomial, but the other two are not.

POLYNOMIALS

A **polynomial** in the variable x is an expression of the form

$$a_nx^n + a_{n-1}x^{n-1} + \cdots + a_1x + a_0$$

where $a_0, a_1, \ldots, a_n$ are real numbers, and n is a nonnegative integer. If $a_n \neq 0$, then the polynomial has **degree *n***. The monomials a_kx^k that make up the polynomial are called the **terms** of the polynomial.

Note that the degree of a polynomial is the highest power of the variable that appears in the polynomial.

Polynomial	Type	Terms	Degree
$2x^2 - 3x + 4$	trinomial	$2x^2, -3x, 4$	2
$x^8 + 5x$	binomial	$x^8, 5x$	8
$8 - x + x^2 - \frac{1}{2}x^3$	four terms	$-\frac{1}{2}x^3, x^2, -x, 3$	3
$5x + 1$	binomial	$5x, 1$	1
$9x^5$	monomial	$9x^5$	5
6	monomial	6	0

Adding and Subtracting Polynomials

Distributive Property

$ac + bc = (a + b)c$

We **add** and **subtract** polynomials using the properties of real numbers that were discussed in Section 1.1. The idea is to combine **like terms** (that is, terms with the same variables raised to the same powers) using the Distributive Property. For instance,

$$5x^7 + 3x^7 = (5 + 3)x^7 = 8x^7$$

In subtracting polynomials, we have to remember that if a minus sign precedes an expression in parentheses, then the sign of every term within the parentheses is changed when we remove the parentheses:

$$-(b + c) = -b - c$$

[This is simply a case of the Distributive Property, $a(b + c) = ab + ac$, with $a = -1$.]

EXAMPLE 1 ■ Adding and Subtracting Polynomials

(a) Find the sum $(x^3 - 6x^2 + 2x + 4) + (x^3 + 5x^2 - 7x)$.

(b) Find the difference $(x^3 - 6x^2 + 2x + 4) - (x^3 + 5x^2 - 7x)$.

SOLUTION

(a) $(x^3 - 6x^2 + 2x + 4) + (x^3 + 5x^2 - 7x)$

$= (x^3 + x^3) + (-6x^2 + 5x^2) + (2x - 7x) + 4$ Group like terms

$= 2x^3 - x^2 - 5x + 4$ Combine like terms

(b) $(x^3 - 6x^2 + 2x + 4) - (x^3 + 5x^2 - 7x)$

$= x^3 - 6x^2 + 2x + 4 - x^3 - 5x^2 + 7x$ Distributive Property

$= (x^3 - x^3) + (-6x^2 - 5x^2) + (2x + 7x) + 4$ Group like terms

$= -11x^2 + 9x + 4$ Combine like terms

Now Try Exercises 17 and 19 ■

Multiplying Algebraic Expressions

To find the **product** of polynomials or other algebraic expressions, we need to use the Distributive Property repeatedly. In particular, using it three times on the product of two binomials, we get

$$(a + b)(c + d) = a(c + d) + b(c + d) = ac + ad + bc + bd$$

This says that we multiply the two factors by multiplying each term in one factor by each term in the other factor and adding these products. Schematically, we have

$$(a + b)(c + d) = \underset{\text{F}}{ac} + \underset{\text{O}}{ad} + \underset{\text{I}}{bc} + \underset{\text{L}}{bd}$$

The acronym **FOIL** helps us remember that the product of two binomials is the sum of the products of the **F**irst terms, the **O**uter terms, the **I**nner terms, and the **L**ast terms.

In general, we can multiply two algebraic expressions by using the Distributive Property and the Laws of Exponents.

EXAMPLE 2 ■ Multiplying Binomials Using FOIL

$(2x + 1)(3x - 5) = \underset{\text{F}}{6x^2} - \underset{\text{O}}{10x} + \underset{\text{I}}{3x} - \underset{\text{L}}{5}$ Distributive Property

$= 6x^2 - 7x - 5$ Combine like terms

Now Try Exercise 25 ■

When we multiply trinomials or other polynomials with more terms, we use the Distributive Property. It is also helpful to arrange our work in table form. The next example illustrates both methods.

EXAMPLE 3 ■ Multiplying Polynomials

Find the product: $(2x + 3)(x^2 - 5x + 4)$

SOLUTION 1: Using the Distributive Property

$$\begin{aligned}(2x + 3)(x^2 - 5x + 4) &= 2x(x^2 - 5x + 4) + 3(x^2 - 5x + 4) && \text{Distributive Property}\\ &= (2x \cdot x^2 - 2x \cdot 5x + 2x \cdot 4) + (3 \cdot x^2 - 3 \cdot 5x + 3 \cdot 4) && \text{Distributive Property}\\ &= (2x^3 - 10x^2 + 8x) + (3x^2 - 15x + 12) && \text{Laws of Exponents}\\ &= 2x^3 - 7x^2 - 7x + 12 && \text{Combine like terms}\end{aligned}$$

SOLUTION 2: Using Table Form

$$\begin{array}{rrrrl} & x^2 & -5x & +4 & \\ & & 2x & +3 & \\ \hline & 3x^2 & -15x & +12 & \text{Multiply } x^2 - 5x + 4 \text{ by } 3\\ 2x^3 & -10x^2 & +8x & & \text{Multiply } x^2 - 5x + 4 \text{ by } 2x\\ \hline 2x^3 & -7x^2 & -7x & +12 & \text{Add like terms}\end{array}$$

Now Try Exercise 47 ■

■ Special Product Formulas

Certain types of products occur so frequently that you should memorize them. You can verify the following formulas by performing the multiplications.

SPECIAL PRODUCT FORMULAS

If A and B are any real numbers or algebraic expressions, then

1. $(A + B)(A - B) = A^2 - B^2$ — Sum and difference of same terms
2. $(A + B)^2 = A^2 + 2AB + B^2$ — Square of a sum
3. $(A - B)^2 = A^2 - 2AB + B^2$ — Square of a difference
4. $(A + B)^3 = A^3 + 3A^2B + 3AB^2 + B^3$ — Cube of a sum
5. $(A - B)^3 = A^3 - 3A^2B + 3AB^2 - B^3$ — Cube of a difference

The key idea in using these formulas (or any other formula in algebra) is the **Principle of Substitution**: We may substitute any algebraic expression for any letter in a formula. For example, to find $(x^2 + y^3)^2$ we use Product Formula 2, substituting x^2 for A and y^3 for B, to get

$$(x^2 + y^3)^2 = (x^2)^2 + 2(x^2)(y^3) + (y^3)^2$$

$$(A + B)^2 = A^2 + 2AB + B^2$$

Mathematics in the Modern World

Changing Words, Sound, and Pictures into Numbers

Pictures, sound, and text are routinely transmitted from one place to another via the Internet, fax machines, or modems. How can such things be transmitted through telephone wires? The key to doing this is to change them into numbers or bits (the digits 0 or 1). It's easy to see how to change text to numbers. For example, we could use the correspondence A = 00000001, B = 00000010, C = 00000011, D = 00000100, E = 00000101, and so on. The word "BED" then becomes 000000100000010100000100. By reading the digits in groups of eight, it is possible to translate this number back to the word "BED."

Changing sound to bits is more complicated. A sound wave can be graphed on an oscilloscope or a computer. The graph is then broken down mathematically into simpler components corresponding to the different frequencies of the original sound. (A branch of mathematics called Fourier analysis is used here.) The intensity of each component is a number, and the original sound can be reconstructed from these numbers. For example, music is stored on a CD as a sequence of bits; it may look like 101010001010010100101010 10000010 11110101000101011. . . . (One second of music requires 1.5 million bits!) The CD player reconstructs the music from the numbers on the CD.

Changing pictures into numbers involves expressing the color and brightness of each dot (or pixel) into a number. This is done very efficiently using a branch of mathematics called wavelet theory. The FBI uses wavelets as a compact way to store the millions of fingerprints they need on file.

EXAMPLE 4 ■ Using the Special Product Formulas

Use the Special Product Formulas to find each product.

(a) $(3x + 5)^2$ **(b)** $(x^2 - 2)^3$

SOLUTION

(a) Substituting $A = 3x$ and $B = 5$ in Product Formula 2, we get

$$(3x + 5)^2 = (3x)^2 + 2(3x)(5) + 5^2 = 9x^2 + 30x + 25$$

(b) Substituting $A = x^2$ and $B = 2$ in Product Formula 5, we get

$$\begin{aligned}(x^2 - 2)^3 &= (x^2)^3 - 3(x^2)^2(2) + 3(x^2)(2)^2 - 2^3 \\ &= x^6 - 6x^4 + 12x^2 - 8\end{aligned}$$

Now Try Exercises 31 and 43 ■

EXAMPLE 5 ■ Using the Special Product Formulas

Find each product.

(a) $(2x - \sqrt{y})(2x + \sqrt{y})$ **(b)** $(x + y - 1)(x + y + 1)$

SOLUTION

(a) Substituting $A = 2x$ and $B = \sqrt{y}$ in Product Formula 1, we get

$$(2x - \sqrt{y})(2x + \sqrt{y}) = (2x)^2 - (\sqrt{y})^2 = 4x^2 - y$$

(b) If we group $x + y$ together and think of this as one algebraic expression, we can use Product Formula 1 with $A = x + y$ and $B = 1$.

$$\begin{aligned}(x + y - 1)(x + y + 1) &= [(x + y) - 1][(x + y) + 1] \\ &= (x + y)^2 - 1^2 && \text{Product Formula 1} \\ &= x^2 + 2xy + y^2 - 1 && \text{Product Formula 2}\end{aligned}$$

Now Try Exercises 57 and 61 ■

■ Factoring Common Factors

We use the Distributive Property to expand algebraic expressions. We sometimes need to reverse this process (again using the Distributive Property) by **factoring** an expression as a product of simpler ones. For example, we can write

FACTORING →

$$x^2 - 4 = (x - 2)(x + 2)$$

← EXPANDING

We say that $x - 2$ and $x + 2$ are **factors** of $x^2 - 4$.

The easiest type of factoring occurs when the terms have a common factor.

EXAMPLE 6 ■ Factoring Out Common Factors

Factor each expression.

(a) $3x^2 - 6x$ **(b)** $8x^4y^2 + 6x^3y^3 - 2xy^4$ **(c)** $(2x + 4)(x - 3) - 5(x - 3)$

SOLUTION

(a) The greatest common factor of the terms $3x^2$ and $-6x$ is $3x$, so we have

$$3x^2 - 6x = 3x(x - 2)$$

CHECK YOUR ANSWER

Multiplying gives

$$3x(x - 2) = 3x^2 - 6x \checkmark$$

(b) We note that

$$8,\ 6,\ \text{and } -2 \text{ have the greatest common factor } 2$$
$$x^4,\ x^3,\ \text{and } x \text{ have the greatest common factor } x$$
$$y^2,\ y^3,\ \text{and } y^4 \text{ have the greatest common factor } y^2$$

So the greatest common factor of the three terms in the polynomial is $2xy^2$, and we have

$$\begin{aligned} 8x^4y^2 + 6x^3y^3 - 2xy^4 &= (2xy^2)(4x^3) + (2xy^2)(3x^2y) + (2xy^2)(-y^2) \\ &= 2xy^2(4x^3 + 3x^2y - y^2) \end{aligned}$$

CHECK YOUR ANSWER

Multiplying gives

$$2xy^2(4x^3 + 3x^2y - y^2) = 8x^4y^2 + 6x^3y^3 - 2xy^4 \ ✓$$

(c) The two terms have the common factor $x - 3$.

$$\begin{aligned} (2x + 4)(x - 3) - 5(x - 3) &= [(2x + 4) - 5](x - 3) && \text{Distributive Property} \\ &= (2x - 1)(x - 3) && \text{Simplify} \end{aligned}$$

Now Try Exercises 63, 65, and 67 ■

Factoring Trinomials

To factor a trinomial of the form $x^2 + bx + c$, we note that

$$(x + r)(x + s) = x^2 + (r + s)x + rs$$

so we need to choose numbers r and s so that $r + s = b$ and $rs = c$.

EXAMPLE 7 ■ Factoring $x^2 + bx + c$ by Trial and Error

Factor: $x^2 + 7x + 12$

SOLUTION We need to find two integers whose product is 12 and whose sum is 7. By trial and error we find that the two integers are 3 and 4. Thus the factorization is

$$x^2 + 7x + 12 = (x + 3)(x + 4)$$

factors of 12

CHECK YOUR ANSWER

Multiplying gives

$$(x + 3)(x + 4) = x^2 + 7x + 12 \ ✓$$

Now Try Exercise 69 ■

To factor a trinomial of the form $ax^2 + bx + c$ with $a \neq 1$, we look for factors of the form $px + r$ and $qx + s$:

$$ax^2 + bx + c = (px + r)(qx + s) = pqx^2 + (ps + qr)x + rs$$

factors of a

$$ax^2 + bx + c = (px + r)(qx + s)$$

factors of c

Therefore we try to find numbers p, q, r, and s such that $pq = a$, $rs = c$, $ps + qr = b$. If these numbers are all integers, then we will have a limited number of possibilities to try for p, q, r, and s.

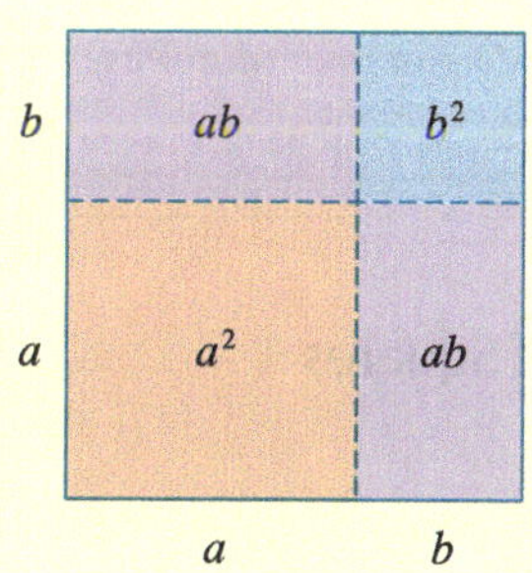

DISCOVERY PROJECT

Visualizing a Formula

Many of the Special Product Formulas in this section can be "seen" as geometrical facts about length, area, and volume. For example, the formula about the square of a sum can be interpreted to be about areas of squares and rectangles. The ancient Greeks always interpreted algebraic formulas in terms of geometric figures. Such figures give us special insight into how these formulas work. You can find the project at **www.stewartmath.com**.

EXAMPLE 8 ■ Factoring $ax^2 + bx + c$ by Trial and Error

Factor: $6x^2 + 7x - 5$

SOLUTION We can factor 6 as $6 \cdot 1$ or $3 \cdot 2$, and -5 as $-5 \cdot 1$ or $5 \cdot (-1)$. By trying these possibilities, we arrive at the factorization

factors of 6

$$6x^2 + 7x - 5 = (3x + 5)(2x - 1)$$

factors of -5

CHECK YOUR ANSWER

Multiplying gives

$(3x + 5)(2x - 1) = 6x^2 + 7x - 5$ ✓

Now Try Exercise 71 ■

EXAMPLE 9 ■ Recognizing the Form of an Expression

Factor each expression.

(a) $x^2 - 2x - 3$ **(b)** $(5a + 1)^2 - 2(5a + 1) - 3$

SOLUTION

(a) $x^2 - 2x - 3 = (x - 3)(x + 1)$ Trial and error

(b) This expression is of the form

$$\square^2 - 2\square - 3$$

where $\square$ represents $5a + 1$. This is the same form as the expression in part (a), so it will factor as $(\square - 3)(\square + 1)$.

$$(5a + 1)^2 - 2(5a + 1) - 3 = [(5a + 1) - 3][(5a + 1) + 1]$$
$$= (5a - 2)(5a + 2)$$

Now Try Exercise 75 ■

■ Special Factoring Formulas

Some special algebraic expressions can be factored by using the following formulas. The first three are simply Special Product Formulas written backward.

SPECIAL FACTORING FORMULAS

Formula	Name
1. $A^2 - B^2 = (A - B)(A + B)$	Difference of squares
2. $A^2 + 2AB + B^2 = (A + B)^2$	Perfect square
3. $A^2 - 2AB + B^2 = (A - B)^2$	Perfect square
4. $A^3 - B^3 = (A - B)(A^2 + AB + B^2)$	Difference of cubes
5. $A^3 + B^3 = (A + B)(A^2 - AB + B^2)$	Sum of cubes

EXAMPLE 10 ■ Factoring Differences of Squares

Factor each expression.

(a) $4x^2 - 25$ **(b)** $(x + y)^2 - z^2$

SOLUTION

(a) Using the Difference of Squares Formula with $A = 2x$ and $B = 5$, we have

$$4x^2 - 25 = (2x)^2 - 5^2 = (2x - 5)(2x + 5)$$

$$A^2 - B^2 = (A - B)(A + B)$$

(b) We use the Difference of Squares Formula with $A = x + y$ and $B = z$.

$$(x + y)^2 - z^2 = (x + y - z)(x + y + z)$$

Now Try Exercises 77 and 111 ■

Terms and Factors

When we multiply two numbers together, each of the numbers is called a **factor** of the product. When we add two numbers together, each number is called a **term** of the sum.

2×3	$2 + 3$
Factors	Terms

If a factor is common to each term of an expression we can factor it out. The following expression has two terms.

$$ax + 2ay$$

a is a factor of each term

Each term contains the factor a, so we can factor a out and write the expression as

$$ax + 2ay = a(x + 2y)$$

A trinomial is a perfect square if it is of the form

$$A^2 + 2AB + B^2 \qquad \text{or} \qquad A^2 - 2AB + B^2$$

So we **recognize a perfect square** if the middle term ($2AB$ or $-2AB$) is plus or minus twice the product of the square roots of the outer two terms.

EXAMPLE 11 ■ Recognizing Perfect Squares

Factor each trinomial.

(a) $x^2 + 6x + 9$ **(b)** $4x^2 - 4xy + y^2$

SOLUTION

(a) Here $A = x$ and $B = 3$, so $2AB = 2 \cdot x \cdot 3 = 6x$. Since the middle term is $6x$, the trinomial is a perfect square. By the Perfect Square Formula we have

$$x^2 + 6x + 9 = (x + 3)^2$$

(b) Here $A = 2x$ and $B = y$, so $2AB = 2 \cdot 2x \cdot y = 4xy$. Since the middle term is $-4xy$, the trinomial is a perfect square. By the Perfect Square Formula we have

$$4x^2 - 4xy + y^2 = (2x - y)^2$$

Now Try Exercises 107 and 109 ■

EXAMPLE 12 ■ Factoring Differences and Sums of Cubes

Factor each polynomial.

(a) $27x^3 - 1$ **(b)** $x^6 + 8$

SOLUTION

(a) Using the Difference of Cubes Formula with $A = 3x$ and $B = 1$, we get

$$\begin{aligned} 27x^3 - 1 &= (3x)^3 - 1^3 = (3x - 1)[(3x)^2 + (3x)(1) + 1^2] \\ &= (3x - 1)(9x^2 + 3x + 1) \end{aligned}$$

(b) Using the Sum of Cubes Formula with $A = x^2$ and $B = 2$, we have

$$x^6 + 8 = (x^2)^3 + 2^3 = (x^2 + 2)(x^4 - 2x^2 + 4)$$

Now Try Exercises 79 and 81 ■

When we factor an expression, the result can sometimes be factored further. In general, *we first factor out common factors*, then inspect the result to see whether it can be factored by any of the other methods of this section. We repeat this process until we have factored the expression completely.

EXAMPLE 13 ■ Factoring an Expression Completely

Factor each expression completely.

(a) $2x^4 - 8x^2$ **(b)** $x^5y^2 - xy^6$

SOLUTION

(a) We first factor out the power of x with the smallest exponent.

$$\begin{aligned} 2x^4 - 8x^2 &= 2x^2(x^2 - 4) && \text{Common factor is } 2x^2 \\ &= 2x^2(x - 2)(x + 2) && \text{Factor } x^2 - 4 \text{ as a difference of squares} \end{aligned}$$

(b) We first factor out the powers of x and y with the smallest exponents.

$$\begin{aligned} x^5y^2 - xy^6 &= xy^2(x^4 - y^4) && \text{Common factor is } xy^2 \\ &= xy^2(x^2 + y^2)(x^2 - y^2) && \text{Factor } x^4 - y^4 \text{ as a difference of squares} \\ &= xy^2(x^2 + y^2)(x + y)(x - y) && \text{Factor } x^2 - y^2 \text{ as a difference of squares} \end{aligned}$$

Now Try Exercises 117 and 119 ■

In the next example we factor out variables with fractional exponents. This type of factoring occurs in calculus.

EXAMPLE 14 ■ Factoring Expressions with Fractional Exponents

Factor each expression.

(a) $3x^{3/2} - 9x^{1/2} + 6x^{-1/2}$ **(b)** $(2 + x)^{-2/3}x + (2 + x)^{1/3}$

SOLUTION

(a) Factor out the power of x with the *smallest exponent*, that is, $x^{-1/2}$.

$$\begin{aligned} 3x^{3/2} - 9x^{1/2} + 6x^{-1/2} &= 3x^{-1/2}(x^2 - 3x + 2) && \text{Factor out } 3x^{-1/2} \\ &= 3x^{-1/2}(x - 1)(x - 2) && \text{Factor the quadratic } x^2 - 3x + 2 \end{aligned}$$

To factor out $x^{-1/2}$ from $x^{3/2}$, we *subtract* exponents:

$$\begin{aligned} x^{3/2} &= x^{-1/2}(x^{3/2-(-1/2)}) \\ &= x^{-1/2}(x^{3/2+1/2}) \\ &= x^{-1/2}(x^2) \end{aligned}$$

(b) Factor out the power of $2 + x$ with the *smallest exponent*, that is, $(2 + x)^{-2/3}$.

$$\begin{aligned} (2 + x)^{-2/3}x + (2 + x)^{1/3} &= (2 + x)^{-2/3}[x + (2 + x)] && \text{Factor out } (2 + x)^{-2/3} \\ &= (2 + x)^{-2/3}(2 + 2x) && \text{Simplify} \\ &= 2(2 + x)^{-2/3}(1 + x) && \text{Factor out 2} \end{aligned}$$

CHECK YOUR ANSWERS

To see that you have factored correctly, multiply using the Laws of Exponents.

(a) $3x^{-1/2}(x^2 - 3x + 2)$
$= 3x^{3/2} - 9x^{1/2} + 6x^{-1/2}$ ✓

(b) $(2 + x)^{-2/3}[x + (2 + x)]$
$= (2 + x)^{-2/3}x + (2 + x)^{1/3}$ ✓

Now Try Exercises 93 and 95 ■

■ Factoring by Grouping Terms

Polynomials with at least four terms can sometimes be factored by grouping terms. The following example illustrates the idea.

EXAMPLE 15 ■ Factoring by Grouping

Factor each polynomial.

(a) $x^3 + x^2 + 4x + 4$ **(b)** $x^3 - 2x^2 - 9x + 18$

SOLUTION

(a) $x^3 + x^2 + 4x + 4 = (x^3 + x^2) + (4x + 4)$ Group terms

$= x^2(x + 1) + 4(x + 1)$ Factor out common factors

$= (x^2 + 4)(x + 1)$ Factor $x + 1$ from each term

(b) $x^3 - 2x^2 - 9x + 18 = (x^3 - 2x^2) - (9x - 18)$ Group terms

$= x^2(x - 2) - 9(x - 2)$ Factor common factors

$= (x^2 - 9)(x - 2)$ Factor $(x - 2)$ from each term

$= (x - 3)(x + 3)(x - 2)$ Factor completely

Now Try Exercises 85 and 121

1.3 EXERCISES

CONCEPTS

1. Consider the polynomial $2x^5 + 6x^4 + 4x^3$.

(a) How many terms does this polynomial have? ______

List the terms: ____________.

(b) What factor is common to each term? ______

Factor the polynomial: $2x^5 + 6x^4 + 4x^3 =$ ______.

2. To factor the trinomial $x^2 + 7x + 10$, we look for two integers whose product is ______ and whose sum is ______. These integers are ______ and ______, so the trinomial factors as ____________.

3. The greatest common factor in the expression $3x^3 + x^2$ is ______, and the expression factors as ___ (___ + ___).

4. The Special Product Formula for the "square of a sum" is $(A + B)^2 =$ ____________.

So $(2x + 3)^2 =$ ____________.

5. The Special Product Formula for the "product of the sum and difference of terms" is $(A + B)(A - B) =$ ____________.

So $(5 + x)(5 - x) =$ ____________.

6. The Special Factoring Formula for the "difference of squares" is $A^2 - B^2 =$ ____________. So $4x^2 - 25$ factors as ____________.

7. The Special Factoring Formula for a "perfect square" is $A^2 + 2AB + B^2 =$ ____________. So $x^2 + 10x + 25$ factors as ____________.

8. *Yes or No?* If *No*, give a reason.

(a) Is the expression $(x + 5)^2$ equal to $x^2 + 25$?

(b) When you expand $(x + a)^2$, where $a \neq 0$, do you get three terms?

(c) Is the expression $(x + 5)(x - 5)$ equal to $x^2 - 25$?

(d) When you expand $(x + a)(x - a)$, where $a \neq 0$, do you get two terms?

SKILLS

9–14 ■ Polynomials Complete the following table by stating whether the polynomial is a monomial, binomial, or trinomial; then list its terms and state its degree.

Polynomial	Type	Terms	Degree
9. $5x^3 + 6$			
10. $-2x^2 + 5x - 3$			
11. -8			
12. $\frac{1}{2}x^7$			
13. $x - x^2 + x^3 - x^4$			
14. $\sqrt{2}x - \sqrt{3}$			

15–24 ■ Polynomials Find the sum, difference, or product.

15. $(12x - 7) - (5x - 12)$

16. $(5 - 3x) + (2x - 8)$

17. $(-2x^2 - 3x + 1) + (3x^2 + 5x - 4)$

18. $(3x^2 + x + 1) - (2x^2 - 3x - 5)$

19. $(5x^3 + 4x^2 - 3x) - (x^2 + 7x + 2)$

20. $3(x - 1) + 4(x + 2)$

21. $8(2x + 5) - 7(x - 9)$

22. $4(x^2 - 3x + 5) - 3(x^2 - 2x + 1)$

23. $2(2 - 5t) + t^2(t - 1) - (t^4 - 1)$

24. $5(3t - 4) - (t^2 + 2) - 2t(t - 3)$

25–30 ■ Using FOIL Multiply the algebraic expressions using the FOIL method and simplify.

25. $(3t - 2)(7t - 4)$

26. $(4s - 1)(2s + 5)$

27. $(3x + 5)(2x - 1)$

28. $(7y - 3)(2y - 1)$

29. $(x + 3y)(2x - y)$

30. $(4x - 5y)(3x - y)$

31–46 ■ Using Special Product Formulas Multiply the algebraic expressions using a Special Product Formula and simplify.

31. $(5x + 1)^2$
32. $(2 - 7y)^2$
33. $(2u + v)^2$
34. $(x - 3y)^2$
35. $(2x + 3y)^2$
36. $(r - 2s)^2$
37. $(x + 6)(x - 6)$
38. $(5 - y)(5 + y)$
39. $(3x - 4)(3x + 4)$
40. $(2y + 5)(2y - 5)$
41. $(\sqrt{x} + 2)(\sqrt{x} - 2)$
42. $(\sqrt{y} + \sqrt{2})(\sqrt{y} - \sqrt{2})$
43. $(y + 2)^3$
44. $(x - 3)^3$
45. $(1 - 2r)^3$
46. $(3 + 2y)^3$

47–62 ■ Multiplying Algebraic Expressions Perform the indicated operations and simplify.

47. $(x + 2)(x^2 + 2x + 3)$
48. $(x + 1)(2x^2 - x + 1)$
49. $(2x - 5)(x^2 - x + 1)$
50. $(1 + 2x)(x^2 - 3x + 1)$
51. $\sqrt{x}(x - \sqrt{x})$
52. $x^{3/2}(\sqrt{x} - 1/\sqrt{x})$
53. $y^{1/3}(y^{2/3} + y^{5/3})$
54. $x^{1/4}(2x^{3/4} - x^{1/4})$
55. $(x^2 - a^2)(x^2 + a^2)$
56. $(x^{1/2} + y^{1/2})(x^{1/2} - y^{1/2})$
57. $(\sqrt{a} - b)(\sqrt{a} + b)$
58. $(\sqrt{h^2 + 1} + 1)(\sqrt{h^2 + 1} - 1)$
59. $((x - 1) + x^2)((x - 1) - x^2)$
60. $(x + (2 + x^2))(x - (2 + x^2))$
61. $(2x + y - 3)(2x + y + 3)$
62. $(x + y + z)(x - y - z)$

63–68 ■ Factoring Common Factor Factor out the common factor.

63. $-2x^3 + x$
64. $3x^4 - 6x^3 - x^2$
65. $y(y - 6) + 9(y - 6)$
66. $(z + 2)^2 - 5(z + 2)$
67. $2x^2y - 6xy^2 + 3xy$
68. $-7x^4y^2 + 14xy^3 + 21xy^4$

69–76 ■ Factoring Trinomials Factor the trinomial.

69. $x^2 + 8x + 7$
70. $x^2 + 4x - 5$
71. $8x^2 - 14x - 15$
72. $6y^2 + 11y - 21$
73. $3x^2 - 16x + 5$
74. $5x^2 - 7x - 6$
75. $(3x + 2)^2 + 8(3x + 2) + 12$
76. $2(a + b)^2 + 5(a + b) - 3$

77–84 ■ Using Special Factoring Formulas Use a Special Factoring Formula to factor the expression.

77. $9a^2 - 16$
78. $(x + 3)^2 - 4$
79. $27x^3 + y^3$
80. $a^3 - b^6$
81. $8s^3 - 125t^3$
82. $1 + 1000y^3$
83. $x^2 + 12x + 36$
84. $16z^2 - 24z + 9$

85–90 ■ Factoring by Grouping Factor the expression by grouping terms.

85. $x^3 + 4x^2 + x + 4$
86. $3x^3 - x^2 + 6x - 2$
87. $5x^3 + x^2 + 5x + 1$
88. $18x^3 + 9x^2 + 2x + 1$
89. $x^3 + x^2 + x + 1$
90. $x^5 + x^4 + x + 1$

91–96 ■ Fractional Exponents Factor the expression completely. Begin by factoring out the lowest power of each common factor.

91. $x^{5/2} - x^{1/2}$
92. $3x^{-1/2} + 4x^{1/2} + x^{3/2}$
93. $x^{-3/2} + 2x^{-1/2} + x^{1/2}$
94. $(x - 1)^{7/2} - (x - 1)^{3/2}$
95. $(x^2 + 1)^{1/2} + 2(x^2 + 1)^{-1/2}$
96. $x^{-1/2}(x + 1)^{1/2} + x^{1/2}(x + 1)^{-1/2}$

97–126 ■ Factoring Completely Factor the expression completely.

97. $12x^3 + 18x$
98. $30x^3 + 15x^4$
99. $x^2 - 2x - 8$
100. $x^2 - 14x + 48$
101. $2x^2 + 5x + 3$
102. $2x^2 + 7x - 4$
103. $9x^2 - 36x - 45$
104. $8x^2 + 10x + 3$
105. $49 - 4y^2$
106. $4t^2 - 9s^2$
107. $t^2 - 6t + 9$
108. $x^2 + 10x + 25$
109. $4x^2 + 4xy + y^2$
110. $r^2 - 6rs + 9s^2$
111. $(a + b)^2 - (a - b)^2$
112. $\left(1 + \frac{1}{x}\right)^2 - \left(1 - \frac{1}{x}\right)^2$
113. $x^2(x^2 - 1) - 9(x^2 - 1)$
114. $(a^2 - 1)b^2 - 4(a^2 - 1)$
115. $8x^3 - 125$
116. $x^6 + 64$
117. $x^3 + 2x^2 + x$
118. $3x^3 - 27x$
119. $x^4y^3 - x^2y^5$
120. $18y^3x^2 - 2xy^4$
121. $3x^3 - x^2 - 12x + 4$
122. $9x^3 + 18x^2 - x - 2$
123. $(x - 1)(x + 2)^2 - (x - 1)^2(x + 2)$
124. $y^4(y + 2)^3 + y^5(y + 2)^4$
125. $(a^2 + 1)^2 - 7(a^2 + 1) + 10$
126. $(a^2 + 2a)^2 - 2(a^2 + 2a) - 3$

127–130 ■ Factoring Completely Factor the expression completely. (This type of expression arises in calculus when using the "Product Rule.")

127. $5(x^2 + 4)^4(2x)(x - 2)^4 + (x^2 + 4)^5(4)(x - 2)^3$
128. $3(2x - 1)^2(2)(x + 3)^{1/2} + (2x - 1)^3(\frac{1}{2})(x + 3)^{-1/2}$
129. $(x^2 + 3)^{-1/3} - \frac{2}{3}x^2(x^2 + 3)^{-4/3}$
130. $\frac{1}{2}x^{-1/2}(3x + 4)^{1/2} - \frac{3}{2}x^{1/2}(3x + 4)^{-1/2}$

SKILLS Plus

131–132 ■ Verifying Identities Show that the following identities hold.

131. (a) $ab = \frac{1}{2}[(a + b)^2 - (a^2 + b^2)]$
(b) $(a^2 + b^2)^2 - (a^2 - b^2)^2 = 4a^2b^2$

132. $(a^2 + b^2)(c^2 + d^2) = (ac + bd)^2 + (ad - bc)^2$

133. Factoring Completely Factor the following expression completely: $4a^2c^2 - (a^2 - b^2 + c^2)^2$.

134. Factoring $x^4 + ax^2 + b$ A trinomial of the form $x^4 + ax^2 + b$ can sometimes be factored easily. For example,

$$x^4 + 3x^2 - 4 = (x^2 + 4)(x^2 - 1)$$

But $x^4 + 3x^2 + 4$ cannot be factored in this way. Instead, we can use the following method.

$$x^4 + 3x^2 + 4 = (x^4 + 4x^2 + 4) - x^2 \quad \text{Add and subtract } x^2$$
$$= (x^2 + 2)^2 - x^2 \quad \text{Factor perfect square}$$
$$= [(x^2 + 2) - x][(x^2 + 2) + x] \quad \text{Difference of squares}$$
$$= (x^2 - x + 2)(x^2 + x + 2)$$

Factor the following, using whichever method is appropriate.

(a) $x^4 + x^2 - 2$ **(b)** $x^4 + 2x^2 + 9$

(c) $x^4 + 4x^2 + 16$ **(d)** $x^4 + 2x^2 + 1$

APPLICATIONS

135. Volume of Concrete A culvert is constructed out of large cylindrical shells cast in concrete, as shown in the figure. Using the formula for the volume of a cylinder given on the inside front cover of this book, explain why the volume of the cylindrical shell is

$$V = \pi R^2 h - \pi r^2 h$$

Factor to show that

$$V = 2\pi \cdot \text{average radius} \cdot \text{height} \cdot \text{thickness}$$

Use the "unrolled" diagram to explain why this makes sense geometrically.

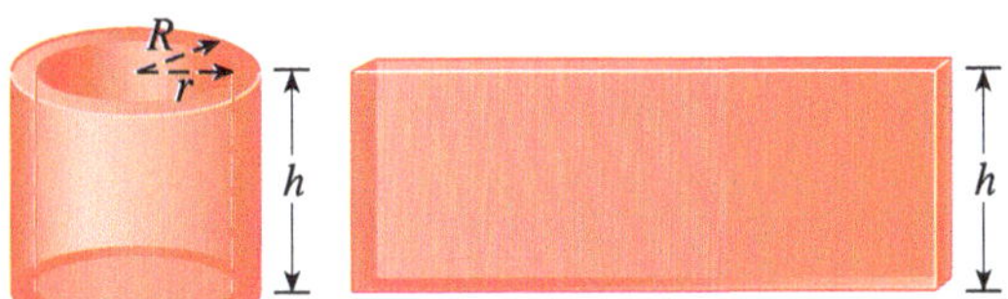

136. Mowing a Field A square field in a certain state park is mowed around the edges every week. The rest of the field is kept unmowed to serve as a habitat for birds and small animals (see the figure). The field measures b feet by b feet, and the mowed strip is x feet wide.

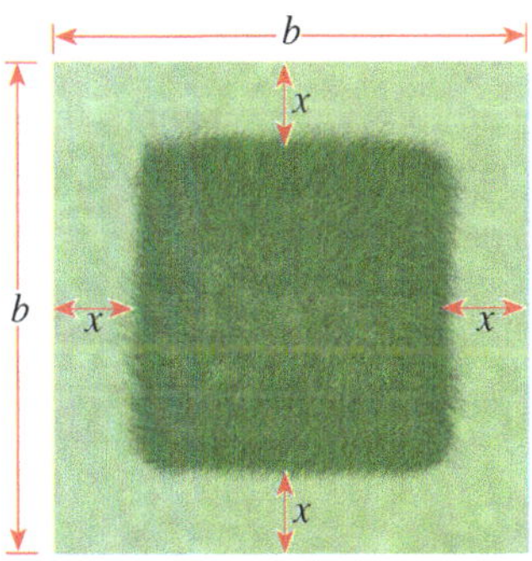

(a) Explain why the area of the mowed portion is $b^2 - (b - 2x)^2$.

(b) Factor the expression in part (a) to show that the area of the mowed portion is also $4x(b - x)$.

DISCUSS ■ DISCOVER ■ PROVE ■ WRITE

137. DISCOVER: Degree of a Sum or Product of Polynomials Make up several pairs of polynomials, then calculate the sum and product of each pair. On the basis of your experiments and observations, answer the following questions.

(a) How is the degree of the product related to the degrees of the original polynomials?

(b) How is the degree of the sum related to the degrees of the original polynomials?

138. DISCUSS: The Power of Algebraic Formulas Use the Difference of Squares Formula $A^2 - B^2 = (A + B)(A - B)$ to evaluate the following differences of squares in your head. Make up more such expressions that you can do in your head.

(a) $528^2 - 527^2$

(b) $122^2 - 120^2$

(c) $1020^2 - 1010^2$

139. DISCUSS: The Power of Algebraic Formulas Use the Special Product Formula $(A + B)(A - B) = A^2 - B^2$ to evaluate the following products of numbers in your head. Make up more such products that you can do in your head.

(a) $501 \cdot 499$

(b) $79 \cdot 61$

(c) $2007 \cdot 1993$

140. DISCOVER: Differences of Even Powers

(a) Factor the expressions completely: $A^4 - B^4$ and $A^6 - B^6$.

(b) Verify that $18{,}335 = 12^4 - 7^4$ and that $2{,}868{,}335 = 12^6 - 7^6$.

(c) Use the results of parts (a) and (b) to factor the integers 18,335 and 2,868,335. Then show that in both of these factorizations, all the factors are prime numbers.

141. DISCOVER: Factoring $A^n - 1$

(a) Verify the following formulas by expanding and simplifying the right-hand side.

$$A^2 - 1 = (A - 1)(A + 1)$$
$$A^3 - 1 = (A - 1)(A^2 + A + 1)$$
$$A^4 - 1 = (A - 1)(A^3 + A^2 + A + 1)$$

(b) On the basis of the pattern displayed in this list, how do you think $A^5 - 1$ would factor? Verify your conjecture. Now generalize the pattern you have observed to obtain a factoring formula for $A^n - 1$, where n is a positive integer.

142. PROVE: Special Factoring Formulas Prove the following formulas by expanding the right-hand side.

(a) Difference of Cubes:
$A^3 - B^3 = (A - B)(A^2 + AB + B^2)$

(b) Sum of Cubes:
$A^3 + B^3 = (A + B)(A^2 - AB + B^2)$

1.4 RATIONAL EXPRESSIONS

■ The Domain of an Algebraic Expression ■ Simplifying Rational Expressions ■ Multiplying and Dividing Rational Expressions ■ Adding and Subtracting Rational Expressions ■ Compound Fractions ■ Rationalizing the Denominator or the Numerator ■ Avoiding Common Errors

A quotient of two algebraic expressions is called a **fractional expression**. Here are some examples:

$$\frac{2x}{x-1} \qquad \frac{y-2}{y^2+4} \qquad \frac{x^3-x}{x^2-5x+6} \qquad \frac{x}{\sqrt{x^2+1}}$$

A **rational expression** is a fractional expression in which both the numerator and the denominator are polynomials. For example, the first three expressions in the above list are rational expressions, but the fourth is not, since its denominator contains a radical. In this section we learn how to perform algebraic operations on rational expressions.

■ The Domain of an Algebraic Expression

Expression	Domain
$\frac{1}{x}$	$\{x \mid x \neq 0\}$
$\sqrt{x}$	$\{x \mid x \geq 0\}$
$\frac{1}{\sqrt{x}}$	$\{x \mid x > 0\}$

In general, an algebraic expression may not be defined for all values of the variable. The **domain** of an algebraic expression is the set of real numbers that the variable is permitted to have. The table in the margin gives some basic expressions and their domains.

EXAMPLE 1 ■ Finding the Domain of an Expression

Find the domains of the following expressions.

(a) $2x^2 + 3x - 1$ **(b)** $\dfrac{x}{x^2 - 5x + 6}$ **(c)** $\dfrac{\sqrt{x}}{x-5}$

SOLUTION

(a) This polynomial is defined for every x. Thus the domain is the set $\mathbb{R}$ of real numbers.

(b) We first factor the denominator.

$$\frac{x}{x^2-5x+6} = \frac{x}{(x-2)(x-3)}$$

Denominator would be 0 if $x = 2$ or $x = 3$

Since the denominator is zero when $x = 2$ or 3, the expression is not defined for these numbers. The domain is $\{x \mid x \neq 2 \text{ and } x \neq 3\}$.

(c) For the numerator to be defined, we must have $x \geq 0$. Also, we cannot divide by zero, so $x \neq 5$.

Must have $x \geq 0$ to take square root

$$\frac{\sqrt{x}}{x-5}$$

Denominator would be 0 if $x = 5$

Thus the domain is $\{x \mid x \geq 0 \text{ and } x \neq 5\}$.

Now Try Exercise 13 ■

Simplifying Rational Expressions

To **simplify rational expressions**, we factor both numerator and denominator and use the following property of fractions:

$$\frac{AC}{BC} = \frac{A}{B}$$

This allows us to **cancel** common factors from the numerator and denominator.

EXAMPLE 2 ■ Simplifying Rational Expressions by Cancellation

Simplify: $\dfrac{x^2 - 1}{x^2 + x - 2}$

SOLUTION

$$\frac{x^2 - 1}{x^2 + x - 2} = \frac{(x - 1)(x + 1)}{(x - 1)(x + 2)} \qquad \text{Factor}$$

$$= \frac{x + 1}{x + 2} \qquad \text{Cancel common factors}$$

We can't cancel the x^2's in $\dfrac{x^2 - 1}{x^2 + x - 2}$ because x^2 is not a factor.

Now Try Exercise 19 ■

Multiplying and Dividing Rational Expressions

To **multiply rational expressions**, we use the following property of fractions:

$$\frac{A}{B} \cdot \frac{C}{D} = \frac{AC}{BD}$$

This says that to multiply two fractions, we multiply their numerators and multiply their denominators.

EXAMPLE 3 ■ Multiplying Rational Expressions

Perform the indicated multiplication and simplify: $\dfrac{x^2 + 2x - 3}{x^2 + 8x + 16} \cdot \dfrac{3x + 12}{x - 1}$

SOLUTION We first factor.

$$\frac{x^2 + 2x - 3}{x^2 + 8x + 16} \cdot \frac{3x + 12}{x - 1} = \frac{(x - 1)(x + 3)}{(x + 4)^2} \cdot \frac{3(x + 4)}{x - 1} \qquad \text{Factor}$$

$$= \frac{3(x - 1)(x + 3)(x + 4)}{(x - 1)(x + 4)^2} \qquad \text{Property of fractions}$$

$$= \frac{3(x + 3)}{x + 4} \qquad \text{Cancel common factors}$$

Now Try Exercise 27 ■

To **divide rational expressions**, we use the following property of fractions:

$$\frac{A}{B} \div \frac{C}{D} = \frac{A}{B} \cdot \frac{D}{C}$$

This says that to divide a fraction by another fraction, we invert the divisor and multiply.

EXAMPLE 4 ■ Dividing Rational Expressions

Perform the indicated division and simplify: $\dfrac{x-4}{x^2-4} \div \dfrac{x^2-3x-4}{x^2+5x+6}$

SOLUTION

$$\begin{aligned}
\frac{x-4}{x^2-4} \div \frac{x^2-3x-4}{x^2+5x+6} &= \frac{x-4}{x^2-4} \cdot \frac{x^2+5x+6}{x^2-3x-4} && \text{Invert and multiply}\\
&= \frac{(x-4)(x+2)(x+3)}{(x-2)(x+2)(x-4)(x+1)} && \text{Factor}\\
&= \frac{x+3}{(x-2)(x+1)} && \text{Cancel common factors}
\end{aligned}$$

Now Try Exercise 33 ■

■ Adding and Subtracting Rational Expressions

Avoid making the following error:

$$\frac{A}{B+C} \neq \frac{A}{B} + \frac{A}{C}$$

For instance, if we let $A = 2$, $B = 1$, and $C = 1$, then we see the error:

$$\frac{2}{1+1} \stackrel{?}{=} \frac{2}{1} + \frac{2}{1}$$

$$\frac{2}{2} \stackrel{?}{=} 2 + 2$$

$$1 \stackrel{?}{=} 4 \quad \text{Wrong!}$$

To **add or subtract rational expressions**, we first find a common denominator and then use the following property of fractions:

$$\frac{A}{C} + \frac{B}{C} = \frac{A+B}{C}$$

Although any common denominator will work, it is best to use the **least common denominator** (LCD) as explained in Section 1.1. The LCD is found by factoring each denominator and taking the product of the distinct factors, using the highest power that appears in any of the factors.

EXAMPLE 5 ■ Adding and Subtracting Rational Expressions

Perform the indicated operations and simplify.

(a) $\dfrac{3}{x-1} + \dfrac{x}{x+2}$ **(b)** $\dfrac{1}{x^2-1} - \dfrac{2}{(x+1)^2}$

SOLUTION

(a) Here the LCD is simply the product $(x-1)(x+2)$.

$$\begin{aligned}
\frac{3}{x-1} + \frac{x}{x+2} &= \frac{3(x+2)}{(x-1)(x+2)} + \frac{x(x-1)}{(x-1)(x+2)} && \text{Write fractions using LCD}\\
&= \frac{3x+6+x^2-x}{(x-1)(x+2)} && \text{Add fractions}\\
&= \frac{x^2+2x+6}{(x-1)(x+2)} && \text{Combine terms in numerator}
\end{aligned}$$

(b) The LCD of $x^2 - 1 = (x - 1)(x + 1)$ and $(x + 1)^2$ is $(x - 1)(x + 1)^2$.

$$\frac{1}{x^2 - 1} - \frac{2}{(x + 1)^2} = \frac{1}{(x - 1)(x + 1)} - \frac{2}{(x + 1)^2} \qquad \text{Factor}$$

$$= \frac{(x + 1) - 2(x - 1)}{(x - 1)(x + 1)^2} \qquad \text{Combine fractions using LCD}$$

$$= \frac{x + 1 - 2x + 2}{(x - 1)(x + 1)^2} \qquad \text{Distributive Property}$$

$$= \frac{3 - x}{(x - 1)(x + 1)^2} \qquad \text{Combine terms in numerator}$$

Now Try Exercises 43 and 45 ■

Compound Fractions

A **compound fraction** is a fraction in which the numerator, the denominator, or both, are themselves fractional expressions.

EXAMPLE 6 ■ Simplifying a Compound Fraction

Simplify: $\dfrac{\dfrac{x}{y} + 1}{1 - \dfrac{y}{x}}$

SOLUTION 1 We combine the terms in the numerator into a single fraction. We do the same in the denominator. Then we invert and multiply.

$$\frac{\dfrac{x}{y} + 1}{1 - \dfrac{y}{x}} = \frac{\dfrac{x + y}{y}}{\dfrac{x - y}{x}} = \frac{x + y}{y} \cdot \frac{x}{x - y}$$

$$= \frac{x(x + y)}{y(x - y)}$$

Mathematics in the Modern World

Courtesy of NASA

Error-Correcting Codes

The pictures sent back by the *Pathfinder* spacecraft from the surface of Mars on July 4, 1997, were astoundingly clear. But few viewing these pictures were aware of the complex mathematics used to accomplish that feat. The distance to Mars is enormous, and the background noise (or static) is many times stronger than the original signal emitted by the spacecraft. So when scientists receive the signal, it is full of errors. To get a clear picture, the errors must be found and corrected. This same problem of errors is routinely encountered in transmitting bank records when you use an ATM machine or voice when you are talking on the telephone.

To understand how errors are found and corrected, we must first understand that to transmit pictures, sound, or text, we transform them into bits (the digits 0 or 1; see page 28). To help the receiver recognize errors, the message is "coded" by inserting additional bits. For example, suppose you want to transmit the message "10100." A very simple-minded code is as follows: Send each digit a million times. The person receiving the message reads it in blocks of a million digits. If the first block is mostly 1's, he concludes that you are probably trying to transmit a 1, and so on. To say that this code is not efficient is a bit of an understatement; it requires sending a million times more data than the original message. Another method inserts "check digits." For example, for each block of eight digits insert a ninth digit; the inserted digit is 0 if there is an even number of 1's in the block and 1 if there is an odd number. So if a single digit is wrong (a 0 changed to a 1 or vice versa), the check digits allow us to recognize that an error has occurred. This method does not tell us where the error is, so we can't correct it. Modern error-correcting codes use interesting mathematical algorithms that require inserting relatively few digits but that allow the receiver to not only recognize, but also correct, errors. The first error-correcting code was developed in the 1940s by Richard Hamming at MIT. It is interesting to note that the English language has a built-in error correcting mechanism; to test it, try reading this error-laden sentence: Gve mo libty ox giv ne deth.

SOLUTION 2 We find the LCD of all the fractions in the expression, then multiply numerator and denominator by it. In this example the LCD of all the fractions is xy. Thus

$$\frac{\frac{x}{y}+1}{1-\frac{y}{x}} = \frac{\frac{x}{y}+1}{1-\frac{y}{x}} \cdot \frac{xy}{xy} \qquad \text{Multiply numerator and denominator by } xy$$

$$= \frac{x^2+xy}{xy-y^2} \qquad \text{Simplify}$$

$$= \frac{x(x+y)}{y(x-y)} \qquad \text{Factor}$$

Now Try Exercises 59 and 65 ■

The next two examples show situations in calculus that require the ability to work with fractional expressions.

EXAMPLE 7 ■ Simplifying a Compound Fraction

Simplify: $\dfrac{\frac{1}{a+h}-\frac{1}{a}}{h}$

SOLUTION We begin by combining the fractions in the numerator using a common denominator.

We can also simplify by multiplying the numerator and the denominator by $a(a+h)$.

$$\frac{\frac{1}{a+h}-\frac{1}{a}}{h} = \frac{\frac{a-(a+h)}{a(a+h)}}{h} \qquad \text{Combine fractions in the numerator}$$

$$= \frac{a-(a+h)}{a(a+h)} \cdot \frac{1}{h} \qquad \text{Property 2 of fractions (invert divisor and multiply)}$$

$$= \frac{a-a-h}{a(a+h)} \cdot \frac{1}{h} \qquad \text{Distributive Property}$$

$$= \frac{-h}{a(a+h)} \cdot \frac{1}{h} \qquad \text{Simplify}$$

$$= \frac{-1}{a(a+h)} \qquad \text{Property 5 of fractions (cancel common factors)}$$

Now Try Exercise 73 ■

EXAMPLE 8 ■ Simplifying a Compound Fraction

Simplify: $\dfrac{(1+x^2)^{1/2}-x^2(1+x^2)^{-1/2}}{1+x^2}$

SOLUTION 1 Factor $(1+x^2)^{-1/2}$ from the numerator.

Factor out the power of $1+x^2$ with the *smallest* exponent, in this case $(1+x^2)^{-1/2}$.

$$\frac{(1+x^2)^{1/2}-x^2(1+x^2)^{-1/2}}{1+x^2} = \frac{(1+x^2)^{-1/2}[(1+x^2)-x^2]}{1+x^2}$$

$$= \frac{(1+x^2)^{-1/2}}{1+x^2} = \frac{1}{(1+x^2)^{3/2}}$$

SOLUTION 2 Since $(1 + x^2)^{-1/2} = 1/(1 + x^2)^{1/2}$ is a fraction, we can clear all fractions by multiplying numerator and denominator by $(1 + x^2)^{1/2}$.

$$\frac{(1 + x^2)^{1/2} - x^2(1 + x^2)^{-1/2}}{1 + x^2} = \frac{(1 + x^2)^{1/2} - x^2(1 + x^2)^{-1/2}}{1 + x^2} \cdot \frac{(1 + x^2)^{1/2}}{(1 + x^2)^{1/2}}$$

$$= \frac{(1 + x^2) - x^2}{(1 + x^2)^{3/2}} = \frac{1}{(1 + x^2)^{3/2}}$$

Now Try Exercise 81 ■

■ Rationalizing the Denominator or the Numerator

If a fraction has a denominator of the form $A + B\sqrt{C}$, we can rationalize the denominator by multiplying numerator and denominator by the **conjugate radical** $A - B\sqrt{C}$. This works because, by Special Product Formula 1 in Section 1.3, the product of the denominator and its conjugate radical does not contain a radical:

$$(A + B\sqrt{C})(A - B\sqrt{C}) = A^2 - B^2C$$

EXAMPLE 9 ■ Rationalizing the Denominator

Rationalize the denominator: $\dfrac{1}{1 + \sqrt{2}}$

SOLUTION We multiply both the numerator and the denominator by the conjugate radical of $1 + \sqrt{2}$, which is $1 - \sqrt{2}$.

Special Product Formula 1
$(A + B)(A - B) = A^2 - B^2$

$$\frac{1}{1 + \sqrt{2}} = \frac{1}{1 + \sqrt{2}} \cdot \frac{1 - \sqrt{2}}{1 - \sqrt{2}}$$ Multiply numerator and denominator by the conjugate radical

$$= \frac{1 - \sqrt{2}}{1^2 - (\sqrt{2})^2}$$ Special Product Formula 1

$$= \frac{1 - \sqrt{2}}{1 - 2} = \frac{1 - \sqrt{2}}{-1} = \sqrt{2} - 1$$

Now Try Exercise 85 ■

EXAMPLE 10 ■ Rationalizing the Numerator

Rationalize the numerator: $\dfrac{\sqrt{4 + h} - 2}{h}$

SOLUTION We multiply numerator and denominator by the conjugate radical $\sqrt{4 + h} + 2$.

Special Product Formula 1
$(A + B)(A - B) = A^2 - B^2$

$$\frac{\sqrt{4 + h} - 2}{h} = \frac{\sqrt{4 + h} - 2}{h} \cdot \frac{\sqrt{4 + h} + 2}{\sqrt{4 + h} + 2}$$ Multiply numerator and denominator by the conjugate radical

$$= \frac{(\sqrt{4 + h})^2 - 2^2}{h(\sqrt{4 + h} + 2)}$$ Special Product Formula 1

$$= \frac{4 + h - 4}{h(\sqrt{4 + h} + 2)}$$

$$= \frac{h}{h(\sqrt{4 + h} + 2)} = \frac{1}{\sqrt{4 + h} + 2}$$ Property 5 of fractions (cancel common factors)

Now Try Exercise 91 ■

Avoiding Common Errors

Don't make the mistake of applying properties of multiplication to the operation of addition. Many of the common errors in algebra involve doing just that. The following table states several properties of multiplication and illustrates the error in applying them to addition.

Correct multiplication property	Common error with addition
$(a \cdot b)^2 = a^2 \cdot b^2$	$(a + b)^2 \neq a^2 + b^2$
$\sqrt{a \cdot b} = \sqrt{a}\sqrt{b} \quad (a, b \geq 0)$	$\sqrt{a + b} \neq \sqrt{a} + \sqrt{b}$
$\sqrt{a^2 \cdot b^2} = a \cdot b \quad (a, b \geq 0)$	$\sqrt{a^2 + b^2} \neq a + b$
$\frac{1}{a} \cdot \frac{1}{b} = \frac{1}{a \cdot b}$	$\frac{1}{a} + \frac{1}{b} \neq \frac{1}{a + b}$
$\frac{ab}{a} = b$	$\frac{a + b}{a} \neq b$
$a^{-1} \cdot b^{-1} = (a \cdot b)^{-1}$	$a^{-1} + b^{-1} \neq (a + b)^{-1}$

To verify that the equations in the right-hand column are wrong, simply substitute numbers for a and b and calculate each side. For example, if we take $a = 2$ and $b = 2$ in the fourth error, we get different values for the left- and right-hand sides:

$$\frac{1}{a} + \frac{1}{b} = \frac{1}{2} + \frac{1}{2} = 1 \qquad \frac{1}{a + b} = \frac{1}{2 + 2} = \frac{1}{4}$$

Left-hand side   Right-hand side

Since $1 \neq \frac{1}{4}$, the stated equation is wrong. You should similarly convince yourself of the error in each of the other equations. (See Exercises 101 and 102.)

1.4 EXERCISES

CONCEPTS

1. Which of the following are rational expressions?

(a) $\frac{3x}{x^2 - 1}$ **(b)** $\frac{\sqrt{x + 1}}{2x + 3}$ **(c)** $\frac{x(x^2 - 1)}{x + 3}$

2. To simplify a rational expression, we cancel *factors* that are common to the ______ and ______. So the expression

$$\frac{(x + 1)(x + 2)}{(x + 3)(x + 2)}$$

simplifies to ______.

3. To multiply two rational expressions, we multiply their ______ together and multiply their ______ together.

So $\frac{2}{x + 1} \cdot \frac{x}{x + 3}$ is the same as ______.

4. Consider the expression $\frac{1}{x} - \frac{2}{x + 1} - \frac{x}{(x + 1)^2}$.

(a) How many terms does this expression have?

(b) Find the least common denominator of all the terms.

(c) Perform the addition and simplify.

5–6 ■ *Yes or No*? If *No*, give a reason. (Disregard any value that makes a denominator zero.)

5. (a) Is the expression $\frac{x(x + 1)}{(x + 1)^2}$ equal to $\frac{x}{x + 1}$?

(b) Is the expression $\sqrt{x^2 + 25}$ equal to $x + 5$?

6. (a) Is the expression $\frac{3 + a}{3}$ equal to $1 + \frac{a}{3}$?

(b) Is the expression $\frac{2}{4 + x}$ equal to $\frac{1}{2} + \frac{2}{x}$?

SKILLS

7–14 ■ **Domain** Find the domain of the expression.

7. $4x^2 - 10x + 3$

8. $-x^4 + x^3 + 9x$

9. $\frac{x^2 - 1}{x - 3}$

10. $\frac{2t^2 - 5}{3t + 6}$

11. $\sqrt{x + 3}$

12. $\frac{1}{\sqrt{x - 1}}$

13. $\frac{x^2 + 1}{x^2 - x - 2}$

14. $\frac{\sqrt{2x}}{x + 1}$

15–24 ■ Simplify Simplify the rational expression.

15. $\dfrac{5(x-3)(2x+1)}{10(x-3)^2}$

16. $\dfrac{4(x^2-1)}{12(x+2)(x-1)}$

17. $\dfrac{x-2}{x^2-4}$

18. $\dfrac{x^2-x-2}{x^2-1}$

19. $\dfrac{x^2+5x+6}{x^2+8x+15}$

20. $\dfrac{x^2-x-12}{x^2+5x+6}$

21. $\dfrac{y^2+y}{y^2-1}$

22. $\dfrac{y^2-3y-18}{2y^2+7y+3}$

23. $\dfrac{2x^3-x^2-6x}{2x^2-7x+6}$

24. $\dfrac{1-x^2}{x^3-1}$

25–38 ■ Multiply or Divide Perform the multiplication or division and simplify.

25. $\dfrac{4x}{x^2-4}\cdot\dfrac{x+2}{16x}$

26. $\dfrac{x^2-25}{x^2-16}\cdot\dfrac{x+4}{x+5}$

27. $\dfrac{x^2+2x-15}{x^2-25}\cdot\dfrac{x-5}{x+2}$

28. $\dfrac{x^2+2x-3}{x^2-2x-3}\cdot\dfrac{3-x}{3+x}$

29. $\dfrac{t-3}{t^2+9}\cdot\dfrac{t+3}{t^2-9}$

30. $\dfrac{x^2-x-6}{x^2+2x}\cdot\dfrac{x^3+x^2}{x^2-2x-3}$

31. $\dfrac{x^2+7x+12}{x^2+3x+2}\cdot\dfrac{x^2+5x+6}{x^2+6x+9}$

32. $\dfrac{x^2+2xy+y^2}{x^2-y^2}\cdot\dfrac{2x^2-xy-y^2}{x^2-xy-2y^2}$

33. $\dfrac{x+3}{4x^2-9}\div\dfrac{x^2+7x+12}{2x^2+7x-15}$

34. $\dfrac{2x+1}{2x^2+x-15}\div\dfrac{6x^2-x-2}{x+3}$

35. $\dfrac{\dfrac{x^3}{x+1}}{\dfrac{x}{x^2+2x+1}}$

36. $\dfrac{\dfrac{2x^2-3x-2}{x^2-1}}{\dfrac{2x^2+5x+2}{x^2+x-2}}$

37. $\dfrac{x/y}{z}$

38. $\dfrac{x}{y/z}$

39–58 ■ Add or Subtract Perform the addition or subtraction and simplify.

39. $1+\dfrac{1}{x+3}$

40. $\dfrac{3x-2}{x+1}-2$

41. $\dfrac{1}{x+5}+\dfrac{2}{x-3}$

42. $\dfrac{1}{x+1}+\dfrac{1}{x-1}$

43. $\dfrac{3}{x+1}-\dfrac{1}{x+2}$

44. $\dfrac{x}{x-4}-\dfrac{3}{x+6}$

45. $\dfrac{5}{2x-3}-\dfrac{3}{(2x-3)^2}$

46. $\dfrac{x}{(x+1)^2}+\dfrac{2}{x+1}$

47. $u+1+\dfrac{u}{u+1}$

48. $\dfrac{2}{a^2}-\dfrac{3}{ab}+\dfrac{4}{b^2}$

49. $\dfrac{1}{x^2}+\dfrac{1}{x^2+x}$

50. $\dfrac{1}{x}+\dfrac{1}{x^2}+\dfrac{1}{x^3}$

51. $\dfrac{2}{x+3}-\dfrac{1}{x^2+7x+12}$

52. $\dfrac{x}{x^2-4}+\dfrac{1}{x-2}$

53. $\dfrac{1}{x+3}+\dfrac{1}{x^2-9}$

54. $\dfrac{x}{x^2+x-2}-\dfrac{2}{x^2-5x+4}$

55. $\dfrac{2}{x}+\dfrac{3}{x-1}-\dfrac{4}{x^2-x}$

56. $\dfrac{x}{x^2-x-6}-\dfrac{1}{x+2}-\dfrac{2}{x-3}$

57. $\dfrac{1}{x^2+3x+2}-\dfrac{1}{x^2-2x-3}$

58. $\dfrac{1}{x+1}-\dfrac{2}{(x+1)^2}+\dfrac{3}{x^2-1}$

59–72 ■ Compound Fractions Simplify the compound fractional expression.

59. $\dfrac{1+\dfrac{1}{x}}{\dfrac{1}{x}-2}$

60. $\dfrac{1-\dfrac{2}{y}}{\dfrac{3}{y}-1}$

61. $\dfrac{1+\dfrac{1}{x+2}}{1-\dfrac{1}{x+2}}$

62. $\dfrac{1+\dfrac{1}{c-1}}{1-\dfrac{1}{c-1}}$

63. $\dfrac{\dfrac{1}{x-1}+\dfrac{1}{x+3}}{x+1}$

64. $\dfrac{\dfrac{x-3}{x-4}-\dfrac{x+2}{x+1}}{x+3}$

65. $\dfrac{x-\dfrac{x}{y}}{y-\dfrac{y}{x}}$

66. $\dfrac{x+\dfrac{y}{x}}{y+\dfrac{x}{y}}$

67. $\dfrac{\dfrac{x}{y}-\dfrac{y}{x}}{\dfrac{1}{x^2}-\dfrac{1}{y^2}}$

68. $x-\dfrac{y}{\dfrac{x}{y}+\dfrac{y}{x}}$

69. $\dfrac{x^{-2}-y^{-2}}{x^{-1}+y^{-1}}$

70. $\dfrac{x^{-1}+y^{-1}}{(x+y)^{-1}}$

71. $1-\dfrac{1}{1-\dfrac{1}{x}}$

72. $1+\dfrac{1}{1+\dfrac{1}{1+x}}$

73–78 ■ Expressions Found in Calculus Simplify the fractional expression. (Expressions like these arise in calculus.)

73. $\dfrac{\dfrac{1}{1+x+h}-\dfrac{1}{1+x}}{h}$

74. $\dfrac{\dfrac{1}{\sqrt{x+h}}-\dfrac{1}{\sqrt{x}}}{h}$

75. $\dfrac{\dfrac{1}{(x+h)^2}-\dfrac{1}{x^2}}{h}$

76. $\dfrac{(x+h)^3-7(x+h)-(x^3-7x)}{h}$

77. $\sqrt{1+\left(\dfrac{x}{\sqrt{1-x^2}}\right)^2}$ 78. $\sqrt{1+\left(x^3-\dfrac{1}{4x^3}\right)^2}$

79–84 ■ Expressions Found in Calculus Simplify the expression. (This type of expression arises in calculus when using the "quotient rule.")

79. $\dfrac{3(x+2)^2(x-3)^2-(x+2)^3(2)(x-3)}{(x-3)^4}$

80. $\dfrac{2x(x+6)^4-x^2(4)(x+6)^3}{(x+6)^8}$

81. $\dfrac{2(1+x)^{1/2}-x(1+x)^{-1/2}}{x+1}$

82. $\dfrac{(1-x^2)^{1/2}+x^2(1-x^2)^{-1/2}}{1-x^2}$

83. $\dfrac{3(1+x)^{1/3}-x(1+x)^{-2/3}}{(1+x)^{2/3}}$

84. $\dfrac{(7-3x)^{1/2}+\frac{3}{2}x(7-3x)^{-1/2}}{7-3x}$

85–90 ■ Rationalize Denominator Rationalize the denominator.

85. $\dfrac{1}{5-\sqrt{3}}$ 86. $\dfrac{3}{2-\sqrt{5}}$

87. $\dfrac{2}{\sqrt{2}+\sqrt{7}}$ 88. $\dfrac{1}{\sqrt{x}+1}$

89. $\dfrac{y}{\sqrt{3}+\sqrt{y}}$ 90. $\dfrac{2(x-y)}{\sqrt{x}-\sqrt{y}}$

91–96 ■ Rationalize Numerator Rationalize the numerator.

91. $\dfrac{1-\sqrt{5}}{3}$ 92. $\dfrac{\sqrt{3}+\sqrt{5}}{2}$

93. $\dfrac{\sqrt{r}+\sqrt{2}}{5}$ 94. $\dfrac{\sqrt{x}-\sqrt{x+h}}{h\sqrt{x}\sqrt{x+h}}$

95. $\sqrt{x^2+1}-x$ 96. $\sqrt{x+1}-\sqrt{x}$

APPLICATIONS

97. **Electrical Resistance** If two electrical resistors with resistances R_1 and R_2 are connected in parallel (see the figure), then the total resistance R is given by

$$R=\frac{1}{\dfrac{1}{R_1}+\dfrac{1}{R_2}}$$

(a) Simplify the expression for R.

(b) If $R_1 = 10$ ohms and $R_2 = 20$ ohms, what is the total resistance R?

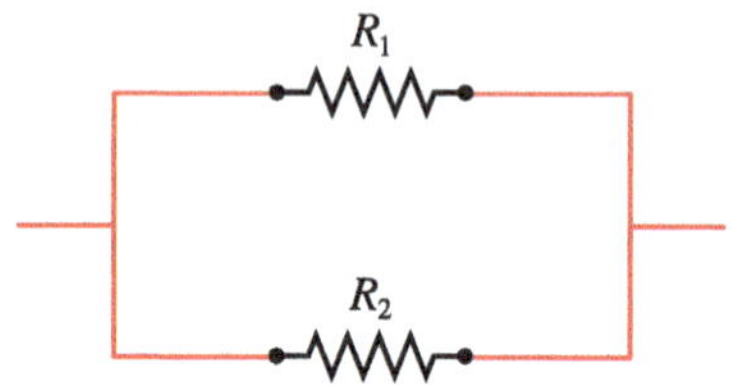

98. **Average Cost** A clothing manufacturer finds that the cost of producing x shirts is $500 + 6x + 0.01x^2$ dollars.

(a) Explain why the average cost per shirt is given by the rational expression

$$A=\frac{500+6x+0.01x^2}{x}$$

(b) Complete the table by calculating the average cost per shirt for the given values of x.

x	Average cost
10	
20	
50	
100	
200	
500	
1000	

DISCUSS ■ DISCOVER ■ PROVE ■ WRITE

99. **DISCOVER: Limiting Behavior of a Rational Expression** The rational expression

$$\frac{x^2-9}{x-3}$$

is not defined for $x = 3$. Complete the tables, and determine what value the expression approaches as x gets closer and closer to 3. Why is this reasonable? Factor the numerator of the expression and simplify to see why.

x	$\dfrac{x^2-9}{x-3}$
2.80	
2.90	
2.95	
2.99	
2.999	

x	$\dfrac{x^2-9}{x-3}$
3.20	
3.10	
3.05	
3.01	
3.001	

100. **DISCUSS ■ WRITE: Is This Rationalization?** In the expression $2/\sqrt{x}$ we would eliminate the radical if we were to square both numerator and denominator. Is this the same thing as rationalizing the denominator? Explain.

101. DISCUSS: Algebraic Errors The left-hand column of the table lists some common algebraic errors. In each case, give an example using numbers that shows that the formula is not valid. An example of this type, which shows that a statement is false, is called a *counterexample*.

Algebraic errors	Counterexample
$\dfrac{1}{a} + \dfrac{1}{b} \neq \dfrac{1}{a+b}$	$\dfrac{1}{2} + \dfrac{1}{2} \neq \dfrac{1}{2+2}$
$(a+b)^2 \neq a^2 + b^2$	
$\sqrt{a^2 + b^2} \neq a + b$	
$\dfrac{a+b}{a} \neq b$	
$\dfrac{a}{a+b} \neq \dfrac{1}{b}$	
$\dfrac{a^m}{a^n} \neq a^{m/n}$	

102. DISCUSS: Algebraic Errors Determine whether the given equation is true for all values of the variables. If not, give a counterexample. (Disregard any value that makes a denominator zero.)

(a) $\dfrac{5+a}{5} = 1 + \dfrac{a}{5}$ **(b)** $\dfrac{x+1}{y+1} = \dfrac{x}{y}$

(c) $\dfrac{x}{x+y} = \dfrac{1}{1+y}$ **(d)** $2\left(\dfrac{a}{b}\right) = \dfrac{2a}{2b}$

(e) $\dfrac{-a}{b} = -\dfrac{a}{b}$ **(f)** $\dfrac{1+x+x^2}{x} = \dfrac{1}{x} + 1 + x$

103. DISCOVER ■ PROVE: Values of a Rational Expression Consider the expression

$$x + \frac{1}{x}$$

for $x > 0$.

(a) Fill in the table, and try other values for x. What do you think is the smallest possible value for this expression?

x	1	3	$\frac{1}{2}$	$\frac{9}{10}$	$\frac{99}{100}$	
$x + \dfrac{1}{x}$						

(b) Prove that for $x > 0$,

$$x + \frac{1}{x} \geq 2$$

[*Hint:* Multiply by x, move terms to one side, and then factor to arrive at a true statement. Note that each step you made is reversible.]

1.5 EQUATIONS

■ Solving Linear Equations ■ Solving Quadratic Equations ■ Other Types of Equations

An equation is a statement that two mathematical expressions are equal. For example,

$$3 + 5 = 8$$

is an equation. Most equations that we study in algebra contain variables, which are symbols (usually letters) that stand for numbers. In the equation

$$4x + 7 = 19$$

$x = 3$ is a solution of the equation $4x + 7 = 19$, because substituting $x = 3$ makes the equation true:

$$x = 3$$
$$4(3) + 7 = 19 \quad ✓$$

the letter x is the variable. We think of x as the "unknown" in the equation, and our goal is to find the value of x that makes the equation true. The values of the unknown that make the equation true are called the **solutions** or **roots** of the equation, and the process of finding the solutions is called **solving the equation**.

Two equations with exactly the same solutions are called **equivalent equations**. To solve an equation, we try to find a simpler, equivalent equation in which the variable stands alone on one side of the "equal" sign. Here are the properties that we use to solve an equation. (In these properties, A, B, and C stand for any algebraic expressions, and the symbol $\Leftrightarrow$ means "is equivalent to.")

PROPERTIES OF EQUALITY

Property	Description
1. $A = B \quad \Leftrightarrow \quad A + C = B + C$	Adding the same quantity to both sides of an equation gives an equivalent equation.
2. $A = B \quad \Leftrightarrow \quad CA = CB \quad (C \neq 0)$	Multiplying both sides of an equation by the same nonzero quantity gives an equivalent equation.

These properties require that you *perform the same operation on both sides of an equation* when solving it. Thus if we say "*add* -7" when solving an equation, that is just a short way of saying "*add* -7 to each side of the equation."

Solving Linear Equations

The simplest type of equation is a *linear equation*, or first-degree equation, which is an equation in which each term is either a constant or a nonzero multiple of the variable.

LINEAR EQUATIONS

A **linear equation** in one variable is an equation equivalent to one of the form

$$ax + b = 0$$

where a and b are real numbers and x is the variable.

Here are some examples that illustrate the difference between linear and nonlinear equations.

Linear equations	Nonlinear equations	
$4x - 5 = 3$	$x^2 + 2x = 8$	Not linear; contains the square of the variable
$2x = \frac{1}{2}x - 7$	$\sqrt{x} - 6x = 0$	Not linear; contains the square root of the variable
$x - 6 = \dfrac{x}{3}$	$\dfrac{3}{x} - 2x = 1$	Not linear; contains the reciprocal of the variable

EXAMPLE 1 ■ Solving a Linear Equation

Solve the equation $7x - 4 = 3x + 8$.

SOLUTION We solve this by changing it to an equivalent equation with all terms that have the variable x on one side and all constant terms on the other.

$7x - 4 = 3x + 8$	Given equation
$(7x - 4) + 4 = (3x + 8) + 4$	Add 4
$7x = 3x + 12$	Simplify
$7x - 3x = (3x + 12) - 3x$	Subtract $3x$
$4x = 12$	Simplify
$\frac{1}{4} \cdot 4x = \frac{1}{4} \cdot 12$	Multiply by $\frac{1}{4}$
$x = 3$	Simplify

Because it is important to CHECK YOUR ANSWER, we do this in many of our examples. In these checks, LHS stands for "left-hand side" and RHS stands for "right-hand side" of the original equation.

CHECK YOUR ANSWER

	$x = 3$	$x = 3$
$x = 3$:	$\text{LHS} = 7(3) - 4$	$\text{RHS} = 3(3) + 8$
	$= 17$	$= 17$

LHS = RHS ✓

Now Try Exercise 17 ■

Many formulas in the sciences involve several variables, and it is often necessary to express one of the variables in terms of the others. In the next example we solve for a variable in Newton's Law of Gravity.

EXAMPLE 2 ■ Solving for One Variable in Terms of Others

This is Newton's Law of Gravity. It gives the gravitational force F between two masses m and M that are a distance r apart. The constant G is the universal gravitational constant.

Solve for the variable M in the equation

$$F = G\frac{mM}{r^2}$$

SOLUTION Although this equation involves more than one variable, we solve it as usual by isolating M on one side and treating the other variables as we would numbers.

$$F = \left(\frac{Gm}{r^2}\right)M \qquad \text{Factor } M \text{ from RHS}$$

$$\left(\frac{r^2}{Gm}\right)F = \left(\frac{r^2}{Gm}\right)\left(\frac{Gm}{r^2}\right)M \qquad \text{Multiply by reciprocal of } \frac{Gm}{r^2}$$

$$\frac{r^2F}{Gm} = M \qquad \text{Simplify}$$

The solution is $M = \dfrac{r^2F}{Gm}$.

Now Try Exercise 31 ■

EXAMPLE 3 ■ Solving for One Variable in Terms of Others

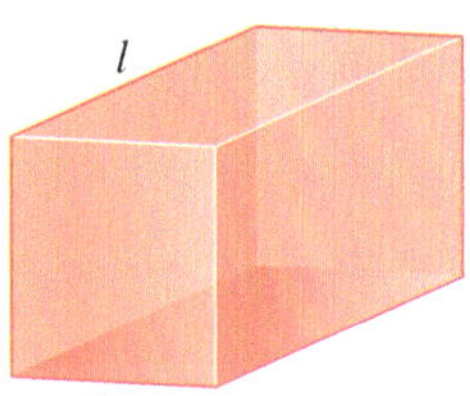

FIGURE 1 A closed rectangular box

The surface area A of the closed rectangular box shown in Figure 1 can be calculated from the length l, the width w, and the height h according to the formula

$$A = 2lw + 2wh + 2lh$$

Solve for w in terms of the other variables in this equation.

SOLUTION Although this equation involves more than one variable, we solve it as usual by isolating w on one side, treating the other variables as we would numbers.

$$A = (2lw + 2wh) + 2lh \qquad \text{Collect terms involving } w$$

$$A - 2lh = 2lw + 2wh \qquad \text{Subtract } 2lh$$

$$A - 2lh = (2l + 2h)w \qquad \text{Factor } w \text{ from RHS}$$

$$\frac{A - 2lh}{2l + 2h} = w \qquad \text{Divide by } 2l + 2h$$

The solution is $w = \dfrac{A - 2lh}{2l + 2h}$.

Now Try Exercise 33 ■

Solving Quadratic Equations

Linear equations are first-degree equations like $2x + 1 = 5$ or $4 - 3x = 2$. Quadratic equations are second-degree equations like $x^2 + 2x - 3 = 0$ or $2x^2 + 3 = 5x$.

Quadratic Equations

$x^2 - 2x - 8 = 0$

$3x + 10 = 4x^2$

$\frac{1}{2}x^2 + \frac{1}{3}x - \frac{1}{6} = 0$

QUADRATIC EQUATIONS

A **quadratic equation** is an equation of the form

$$ax^2 + bx + c = 0$$

where a, b, and c are real numbers with $a \neq 0$.

Some quadratic equations can be solved by factoring and using the following basic property of real numbers.

ZERO-PRODUCT PROPERTY

$$AB = 0 \quad \text{if and only if} \quad A = 0 \quad \text{or} \quad B = 0$$

This means that if we can factor the left-hand side of a quadratic (or other) equation, then we can solve it by setting each factor equal to 0 in turn. This method works only when the right-hand side of the equation is 0.

EXAMPLE 4 ■ Solving a Quadratic Equation by Factoring

Find all real solutions of the equation $x^2 + 5x = 24$.

SOLUTION We must first rewrite the equation so that the right-hand side is 0.

$$x^2 + 5x = 24$$

$$x^2 + 5x - 24 = 0 \qquad \text{Subtract 24}$$

$$(x - 3)(x + 8) = 0 \qquad \text{Factor}$$

$$x - 3 = 0 \quad \text{or} \quad x + 8 = 0 \qquad \text{Zero-Product Property}$$

$$x = 3 \qquad x = -8 \qquad \text{Solve}$$

The solutions are $x = 3$ and $x = -8$.

CHECK YOUR ANSWERS

$x = 3$:

$(3)^2 + 5(3) = 9 + 15 = 24$ ✓

$x = -8$:

$(-8)^2 + 5(-8) = 64 - 40 = 24$ ✓

Now Try Exercise 45 ■

Do you see why one side of the equation must be 0 in Example 4? Factoring the equation as $x(x + 5) = 24$ does not help us find the solutions, since 24 can be factored in infinitely many ways, such as $6 \cdot 4$, $\frac{1}{2} \cdot 48$, $\left(-\frac{2}{5}\right) \cdot (-60)$, and so on.

A quadratic equation of the form $x^2 - c = 0$, where c is a positive constant, factors as $(x - \sqrt{c})(x + \sqrt{c}) = 0$, so the solutions are $x = \sqrt{c}$ and $x = -\sqrt{c}$. We often abbreviate this as $x = \pm\sqrt{c}$.

SOLVING A SIMPLE QUADRATIC EQUATION

The solutions of the equation $x^2 = c$ are $x = \sqrt{c}$ and $x = -\sqrt{c}$.

EXAMPLE 5 ■ Solving Simple Quadratics

Find all real solutions of each equation.

(a) $x^2 = 5$ **(b)** $(x - 4)^2 = 5$

SOLUTION

(a) From the principle in the preceding box we get $x = \pm\sqrt{5}$.

(b) We can take the square root of each side of this equation as well.

$$
\begin{aligned}
(x-4)^2 &= 5 \\
x - 4 &= \pm\sqrt{5} && \text{Take the square root} \\
x &= 4 \pm \sqrt{5} && \text{Add 4}
\end{aligned}
$$

The solutions are $x = 4 + \sqrt{5}$ and $x = 4 - \sqrt{5}$.

Now Try Exercises 53 and 55 ■

See page 31 for how to recognize when a quadratic expression is a perfect square.

As we saw in Example 5, if a quadratic equation is of the form $(x \pm a)^2 = c$, then we can solve it by taking the square root of each side. In an equation of this form, the left-hand side is a *perfect square*: the square of a linear expression in x. So if a quadratic equation does not factor readily, then we can solve it using the technique of **completing the square**. This means that we add a constant to an expression to make it a perfect square. For example, to make $x^2 - 6x$ a perfect square, we must add 9, since $x^2 - 6x + 9 = (x - 3)^2$.

Completing the Square

The area of the blue region is

$$x^2 + 2\left(\frac{b}{2}\right)x = x^2 + bx$$

Add a small square of area $(b/2)^2$ to "complete" the square.

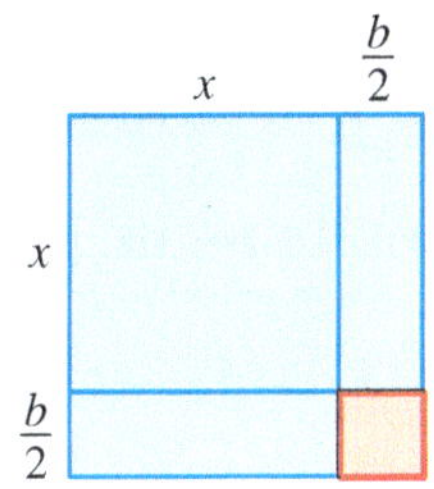

COMPLETING THE SQUARE

To make $x^2 + bx$ a perfect square, add $\left(\frac{b}{2}\right)^2$, the square of half the coefficient of x. This gives the perfect square

$$x^2 + bx + \left(\frac{b}{2}\right)^2 = \left(x + \frac{b}{2}\right)^2$$

EXAMPLE 6 ■ Solving Quadratic Equations by Completing the Square

Find all real solutions of each equation.

(a) $x^2 - 8x + 13 = 0$ **(b)** $3x^2 - 12x + 6 = 0$

SOLUTION

(a)

$$
\begin{aligned}
x^2 - 8x + 13 &= 0 && \text{Given equation} \\
x^2 - 8x &= -13 && \text{Subtract 13} \\
x^2 - 8x + 16 &= -13 + 16 && \text{Complete the square: add } \left(\frac{-8}{2}\right)^2 = 16 \\
(x-4)^2 &= 3 && \text{Perfect square} \\
x - 4 &= \pm\sqrt{3} && \text{Take square root} \\
x &= 4 \pm \sqrt{3} && \text{Add 4}
\end{aligned}
$$

When completing the square, make sure the coefficient of x^2 is 1. If it isn't, you must factor this coefficient from both terms that contain x:

$$ax^2 + bx = a\left(x^2 + \frac{b}{a}x\right)$$

Then complete the square inside the parentheses. Remember that the term added inside the parentheses is multiplied by a.

(b) After subtracting 6 from each side of the equation, we must factor the coefficient of x^2 (the 3) from the left side to put the equation in the correct form for completing the square.

$$
\begin{aligned}
3x^2 - 12x + 6 &= 0 && \text{Given equation} \\
3x^2 - 12x &= -6 && \text{Subtract 6} \\
3(x^2 - 4x) &= -6 && \text{Factor 3 from LHS}
\end{aligned}
$$

Now we complete the square by adding $(-2)^2 = 4$ *inside* the parentheses. Since everything inside the parentheses is multiplied by 3, this means that we are actually adding $3 \cdot 4 = 12$ to the left side of the equation. Thus we must add 12 to the right side as well.

$$
\begin{aligned}
3(x^2 - 4x + 4) &= -6 + 3 \cdot 4 && \text{Complete the square: add 4} \\
3(x - 2)^2 &= 6 && \text{Perfect square} \\
(x - 2)^2 &= 2 && \text{Divide by 3} \\
x - 2 &= \pm\sqrt{2} && \text{Take square root} \\
x &= 2 \pm \sqrt{2} && \text{Add 2}
\end{aligned}
$$

Now Try Exercises 57 and 61 ■

We can use the technique of completing the square to derive a formula for the roots of the general quadratic equation $ax^2 + bx + c = 0$.

THE QUADRATIC FORMULA

The roots of the quadratic equation $ax^2 + bx + c = 0$, where $a \neq 0$, are

$$x = \frac{-b \pm \sqrt{b^2 - 4ac}}{2a}$$

Proof First, we divide each side of the equation by a and move the constant to the right side, giving

$$x^2 + \frac{b}{a}x = -\frac{c}{a} \qquad \text{Divide by } a$$

We now complete the square by adding $(b/2a)^2$ to each side of the equation:

$$
\begin{aligned}
x^2 + \frac{b}{a}x + \left(\frac{b}{2a}\right)^2 &= -\frac{c}{a} + \left(\frac{b}{2a}\right)^2 && \text{Complete the square: Add } \left(\frac{b}{2a}\right)^2 \\
\left(x + \frac{b}{2a}\right)^2 &= \frac{-4ac + b^2}{4a^2} && \text{Perfect square} \\
x + \frac{b}{2a} &= \pm\frac{\sqrt{b^2 - 4ac}}{2a} && \text{Take square root} \\
x &= \frac{-b \pm \sqrt{b^2 - 4ac}}{2a} && \text{Subtract } \frac{b}{2a}
\end{aligned}
$$

■

The Quadratic Formula could be used to solve the equations in Examples 4 and 6. You should carry out the details of these calculations.

Library of Congress Prints and Photographs Division [LC-USZ62-62123]

FRANÇOIS VIÈTE (1540–1603) had a successful political career before taking up mathematics late in life. He became one of the most famous French mathematicians of the 16th century. Viète introduced a new level of abstraction in algebra by using letters to stand for *known* quantities in an equation. Before Viète's time, each equation had to be solved on its own. For instance, the quadratic equations

$$3x^2 + 2x + 8 = 0$$
$$5x^2 - 6x + 4 = 0$$

had to be solved separately by completing the square. Viète's idea was to consider all quadratic equations at once by writing

$$ax^2 + bx + c = 0$$

where a, b, and c are known quantities. Thus he made it possible to write a *formula* (in this case the quadratic formula) involving a, b, and c that can be used to solve all such equations in one fell swoop.

Viète's mathematical genius proved quite valuable during a war between France and Spain. To communicate with their troops, the Spaniards used a complicated code that Viète managed to decipher. Unaware of Viète's accomplishment, the Spanish king, Philip II, protested to the Pope, claiming that the French were using witchcraft to read his messages.

EXAMPLE 7 ■ Using the Quadratic Formula

Find all real solutions of each equation.

(a) $3x^2 - 5x - 1 = 0$ **(b)** $4x^2 + 12x + 9 = 0$ **(c)** $x^2 + 2x + 2 = 0$

SOLUTION

(a) In this quadratic equation $a = 3$, $b = -5$, and $c = -1$.

$$b = -5$$

$$3x^2 - 5x - 1 = 0$$

$$a = 3 \qquad c = -1$$

By the Quadratic Formula,

$$x = \frac{-(-5) \pm \sqrt{(-5)^2 - 4(3)(-1)}}{2(3)} = \frac{5 \pm \sqrt{37}}{6}$$

If approximations are desired, we can use a calculator to obtain

$$x = \frac{5 + \sqrt{37}}{6} \approx 1.8471 \qquad \text{and} \qquad x = \frac{5 - \sqrt{37}}{6} \approx -0.1805$$

Another Method

$$4x^2 + 12x + 9 = 0$$
$$(2x + 3)^2 = 0$$
$$2x + 3 = 0$$
$$x = -\tfrac{3}{2}$$

(b) Using the Quadratic Formula with $a = 4$, $b = 12$, and $c = 9$ gives

$$x = \frac{-12 \pm \sqrt{(12)^2 - 4 \cdot 4 \cdot 9}}{2 \cdot 4} = \frac{-12 \pm 0}{8} = -\frac{3}{2}$$

This equation has only one solution, $x = -\frac{3}{2}$.

(c) Using the Quadratic Formula with $a = 1$, $b = 2$, and $c = 2$ gives

$$x = \frac{-2 \pm \sqrt{2^2 - 4 \cdot 2}}{2} = \frac{-2 \pm \sqrt{-4}}{2} = \frac{-2 \pm 2\sqrt{-1}}{2} = -1 \pm \sqrt{-1}$$

Since the square of any real number is nonnegative, $\sqrt{-1}$ is undefined in the real number system. The equation has no real solution.

Now Try Exercises 67, 73, and 77 ■

In the next section we study the complex number system, in which the square roots of negative numbers do exist. The equation in Example 7(c) does have solutions in the complex number system.

The quantity $b^2 - 4ac$ that appears under the square root sign in the quadratic formula is called the *discriminant* of the equation $ax^2 + bx + c = 0$ and is given the symbol D. If $D < 0$, then $\sqrt{b^2 - 4ac}$ is undefined, and the quadratic equation has no real solution, as in Example 7(c). If $D = 0$, then the equation has only one real solution, as in Example 7(b). Finally, if $D > 0$, then the equation has two distinct real solutions, as in Example 7(a). The following box summarizes these observations.

THE DISCRIMINANT

The **discriminant** of the general quadratic equation $ax^2 + bx + c = 0$ $(a \neq 0)$ is $D = b^2 - 4ac$.

1. If $D > 0$, then the equation has two distinct real solutions.
2. If $D = 0$, then the equation has exactly one real solution.
3. If $D < 0$, then the equation has no real solution.

EXAMPLE 8 ■ Using the Discriminant

Use the discriminant to determine how many real solutions each equation has.

(a) $x^2 + 4x - 1 = 0$ **(b)** $4x^2 - 12x + 9 = 0$ **(c)** $\frac{1}{3}x^2 - 2x + 4 = 0$

SOLUTION

(a) The discriminant is $D = 4^2 - 4(1)(-1) = 20 > 0$, so the equation has two distinct real solutions.

(b) The discriminant is $D = (-12)^2 - 4 \cdot 4 \cdot 9 = 0$, so the equation has exactly one real solution.

(c) The discriminant is $D = (-2)^2 - 4(\frac{1}{3})4 = -\frac{4}{3} < 0$, so the equation has no real solution.

Now Try Exercises 81, 83, and 85 ■

This formula depends on the fact that acceleration due to gravity is constant near the earth's surface. Here we neglect the effect of air resistance.

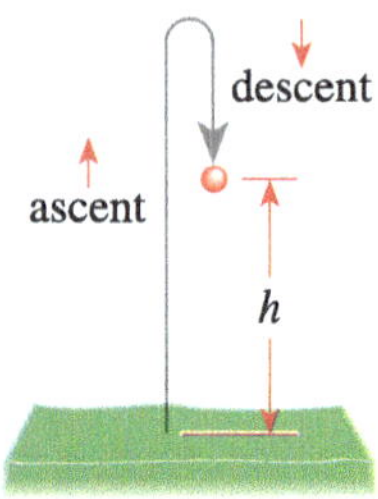

FIGURE 2

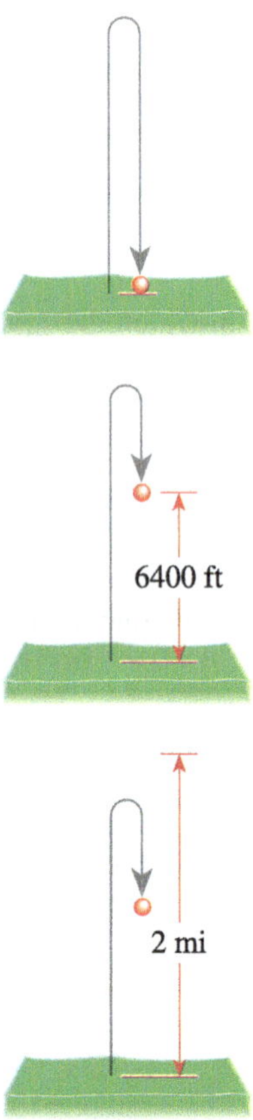

Now let's consider a real-life situation that can be modeled by a quadratic equation.

EXAMPLE 9 ■ The Path of a Projectile

An object thrown or fired straight upward at an initial speed of v_0 ft/s will reach a height of h feet after t seconds, where h and t are related by the formula

$$h = -16t^2 + v_0 t$$

Suppose that a bullet is shot straight upward with an initial speed of 800 ft/s. Its path is shown in Figure 2.

(a) When does the bullet fall back to ground level?

(b) When does it reach a height of 6400 ft?

(c) When does it reach a height of 2 mi?

(d) How high is the highest point the bullet reaches?

SOLUTION Since the initial speed in this case is $v_0 = 800$ ft/s, the formula is

$$h = -16t^2 + 800t$$

(a) Ground level corresponds to $h = 0$, so we must solve the equation

$$\begin{aligned} 0 &= -16t^2 + 800t && \text{Set } h = 0 \\ 0 &= -16t(t - 50) && \text{Factor} \end{aligned}$$

Thus $t = 0$ or $t = 50$. This means the bullet starts $(t = 0)$ at ground level and returns to ground level after 50 s.

(b) Setting $h = 6400$ gives the equation

$$\begin{aligned} 6400 &= -16t^2 + 800t && \text{Set } h = 6400 \\ 16t^2 - 800t + 6400 &= 0 && \text{All terms to LHS} \\ t^2 - 50t + 400 &= 0 && \text{Divide by 16} \\ (t - 10)(t - 40) &= 0 && \text{Factor} \\ t = 10 \quad &\text{or} \quad t = 40 && \text{Solve} \end{aligned}$$

The bullet reaches 6400 ft after 10 s (on its ascent) and again after 40 s (on its descent to earth).

(c) Two miles is $2 \times 5280 = 10{,}560$ ft.

$$\begin{aligned} 10{,}560 &= -16t^2 + 800t && \text{Set } h = 10{,}560 \\ 16t^2 - 800t + 10{,}560 &= 0 && \text{All terms to LHS} \\ t^2 - 50t + 660 &= 0 && \text{Divide by 16} \end{aligned}$$

The discriminant of this equation is $D = (-50)^2 - 4(660) = -140$, which is negative. Thus the equation has no real solution. The bullet never reaches a height of 2 mi.

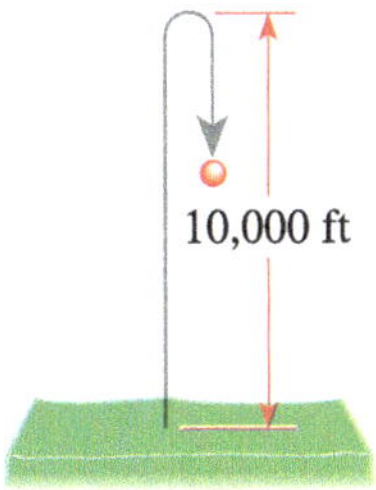

(d) Each height the bullet reaches is attained twice, once on its ascent and once on its descent. The only exception is the highest point of its path, which is reached only once. This means that for the highest value of h, the following equation has only one solution for t:

$$h = -16t^2 + 800t$$

$$16t^2 - 800t + h = 0 \qquad \text{All terms to LHS}$$

This in turn means that the discriminant D of the equation is 0, so

$$D = (-800)^2 - 4(16)h = 0$$

$$640{,}000 - 64h = 0$$

$$h = 10{,}000$$

The maximum height reached is 10,000 ft.

Now Try Exercise 129 ■

■ Other Types of Equations

So far we have learned how to solve linear and quadratic equations. Now we study other types of equations, including those that involve higher powers, fractional expressions, and radicals.

When we solve an equation that involves fractional expressions or radicals, we must be especially careful to check our final answers. The next two examples demonstrate why.

EXAMPLE 10 ■ An Equation Involving Fractional Expressions

Solve the equation $\dfrac{3}{x} - \dfrac{2}{x-3} = \dfrac{-12}{x^2 - 9}$.

SOLUTION We eliminate the denominators by multiplying each side by the lowest common denominator.

$$\left(\frac{3}{x} - \frac{2}{x-3}\right)x(x^2 - 9) = \frac{-12}{x^2-9}x(x^2 - 9) \qquad \text{Multiply by LCD, } x(x^2 - 9)$$

$$3(x^2 - 9) - 2x(x + 3) = -12x \qquad \text{Expand}$$

$$3x^2 - 27 - 2x^2 - 6x = -12x \qquad \text{Expand LHS}$$

$$x^2 - 6x - 27 = -12x \qquad \text{Add like terms on LHS}$$

$$x^2 + 6x - 27 = 0 \qquad \text{Add } 12x$$

$$(x - 3)(x + 9) = 0 \qquad \text{Factor}$$

$$x - 3 = 0 \quad \text{or} \quad x + 9 = 0 \qquad \text{Zero-Product Property}$$

$$x = 3 \qquad x = -9 \qquad \text{Solve}$$

CHECK YOUR ANSWERS

$x = 3$:

$$\text{LHS} = \frac{3}{3} - \frac{2}{3-3} \text{ undefined}$$

$$\text{RHS} = \frac{-12}{3^2 - 9} \text{ undefined} \quad ✗$$

$x = -9$:

$$\text{LHS} = \frac{3}{-9} - \frac{2}{-9-3} = -\frac{1}{6}$$

$$\text{RHS} = \frac{-12}{(-9)^2 - 9} = -\frac{1}{6}$$

LHS = RHS ✓

We must check our answer because multiplying by an expression that contains the variable can introduce extraneous solutions. From *Check Your Answers* we see that the only solution is $x = -9$.

Now Try Exercise 89 ■

EXAMPLE 11 ■ An Equation Involving a Radical

Solve the equation $2x = 1 - \sqrt{2 - x}$.

SOLUTION To eliminate the square root, we first isolate it on one side of the equal sign, then square.

$$2x - 1 = -\sqrt{2 - x} \qquad \text{Subtract 1}$$

$$(2x - 1)^2 = 2 - x \qquad \text{Square each side}$$

$$4x^2 - 4x + 1 = 2 - x \qquad \text{Expand LHS}$$

$$4x^2 - 3x - 1 = 0 \qquad \text{Add } -2 + x$$

$$(4x + 1)(x - 1) = 0 \qquad \text{Factor}$$

$$4x + 1 = 0 \quad \text{or} \quad x - 1 = 0 \qquad \text{Zero-Product Property}$$

$$x = -\tfrac{1}{4} \qquad x = 1 \qquad \text{Solve}$$

The values $x = -\frac{1}{4}$ and $x = 1$ are only potential solutions. We must check them to see whether they satisfy the original equation. From *Check Your Answers* we see that $x = -\frac{1}{4}$ is a solution but $x = 1$ is not. The only solution is $x = -\frac{1}{4}$.

CHECK YOUR ANSWERS

$x = -\frac{1}{4}$:

$$\text{LHS} = 2\left(-\tfrac{1}{4}\right) = -\tfrac{1}{2}$$

$$\text{RHS} = 1 - \sqrt{2 - \left(-\tfrac{1}{4}\right)} = 1 - \sqrt{\tfrac{9}{4}} = 1 - \tfrac{3}{2} = -\tfrac{1}{2}$$

LHS = RHS ✓

$x = 1$:

$$\text{LHS} = 2(1) = 2$$

$$\text{RHS} = 1 - \sqrt{2 - 1} = 1 - 1 = 0$$

LHS ≠ RHS ✗

Now Try Exercise 97 ■

When we solve an equation, we may end up with one or more **extraneous solutions**, that is, potential solutions that do not satisfy the original equation. In Example 10 the value $x = 3$ is an extraneous solution, and in Example 11 the value $x = 1$ is an extraneous solution. In the case of equations involving fractional expressions, potential solutions may be undefined in the original equation and hence become extraneous solutions. In the case of equations involving radicals, extraneous solutions may be introduced when we square each side of an equation because the operation of squaring can turn a false equation into a true one. For example, $-1 \neq 1$, but $(-1)^2 = 1^2$. Thus the squared equation may be true for more values of the variable than the original equation. That is why you must always check your answers to make sure that each satisfies the original equation.

An equation of the form $aW^2 + bW + c = 0$, where W is an algebraic expression, is an equation of **quadratic type**. We solve equations of quadratic type by substituting for the algebraic expression, as we see in the next two examples.

EXAMPLE 12 ■ A Fourth-Degree Equation of Quadratic Type

Find all solutions of the equation $x^4 - 8x^2 + 8 = 0$.

SOLUTION If we set $W = x^2$, then we get a quadratic equation in the new variable W.

$$(x^2)^2 - 8x^2 + 8 = 0 \qquad \text{Write } x^4 \text{ as } (x^2)^2$$

$$W^2 - 8W + 8 = 0 \qquad \text{Let } W = x^2$$

$$W = \frac{-(-8) \pm \sqrt{(-8)^2 - 4 \cdot 8}}{2} = 4 \pm 2\sqrt{2} \qquad \text{Quadratic Formula}$$

$$x^2 = 4 \pm 2\sqrt{2} \qquad W = x^2$$

$$x = \pm\sqrt{4 \pm 2\sqrt{2}} \qquad \text{Take square roots}$$

So there are four solutions:

$$\sqrt{4 + 2\sqrt{2}} \qquad \sqrt{4 - 2\sqrt{2}} \qquad -\sqrt{4 + 2\sqrt{2}} \qquad -\sqrt{4 - 2\sqrt{2}}$$

Using a calculator, we obtain the approximations $x \approx 2.61, 1.08, -2.61, -1.08$.

Now Try Exercise 103 ■

EXAMPLE 13 ■ An Equation Involving Fractional Powers

Find all solutions of the equation $x^{1/3} + x^{1/6} - 2 = 0$.

SOLUTION This equation is of quadratic type because if we let $W = x^{1/6}$, then $W^2 = (x^{1/6})^2 = x^{1/3}$.

$$x^{1/3} + x^{1/6} - 2 = 0$$

$$W^2 + W - 2 = 0 \qquad \text{Let } W = x^{1/6}$$

$$(W - 1)(W + 2) = 0 \qquad \text{Factor}$$

$$W - 1 = 0 \quad \text{or} \quad W + 2 = 0 \qquad \text{Zero-Product Property}$$

$$W = 1 \qquad W = -2 \qquad \text{Solve}$$

$$x^{1/6} = 1 \qquad x^{1/6} = -2 \qquad W = x^{1/6}$$

$$x = 1^6 = 1 \qquad x = (-2)^6 = 64 \qquad \text{Take the 6th power}$$

From *Check Your Answers* we see that $x = 1$ is a solution but $x = 64$ is not. The only solution is $x = 1$.

CHECK YOUR ANSWERS

$x = 1$:

$\text{LHS} = 1^{1/3} + 1^{1/6} - 2 = 0$

$\text{RHS} = 0$

$\text{LHS} = \text{RHS}$ ✓

$x = 64$:

$\text{LHS} = 64^{1/3} + 64^{1/6} - 2$

$= 4 + 2 - 2 = 4$

$\text{RHS} = 0$

$\text{LHS} \neq \text{RHS}$ ✗

Now Try Exercise 107 ■

When solving an absolute value equation, we use the following property

$$|X| = C \quad \text{is equivalent to} \quad X = C \quad \text{or} \quad X = -C$$

where X is any algebraic expression. This property says that to solve an absolute value equation, we must solve two separate equations.

EXAMPLE 14 ■ An Absolute Value Equation

Solve the equation $|2x - 5| = 3$.

SOLUTION By the definition of absolute value, $|2x - 5| = 3$ is equivalent to

$$2x - 5 = 3 \quad \text{or} \quad 2x - 5 = -3$$

$$2x = 8 \qquad 2x = 2$$

$$x = 4 \qquad x = 1$$

The solutions are $x = 1$, $x = 4$.

Now Try Exercise 113 ■

1.5 EXERCISES

CONCEPTS

1. *Yes or No*? If *No*, give a reason.

   (a) When you add the same number to each side of an equation, do you always get an equivalent equation?

   (b) When you multiply each side of an equation by the same nonzero number, do you always get an equivalent equation?

   (c) When you square each side of an equation, do you always get an equivalent equation?

2. What is a logical first step in solving the equation?
(a) $(x + 5)^2 = 64$ (b) $(x + 5)^2 + 5 = 64$
(c) $x^2 + x = 2$

3. Explain how you would use each method to solve the equation $x^2 - 4x - 5 = 0$.
(a) By factoring: ________
(b) By completing the square: ________
(c) By using the Quadratic Formula: ________

4. (a) The solutions of the equation $x^2(x - 4) = 0$ are ________.
(b) To solve the equation $x^3 - 4x^2 = 0$, we ________ the left-hand side.

5. Solve the equation $\sqrt{2x} + x = 0$ by doing the following steps.
(a) Isolate the radical: ________________.
(b) Square both sides: ________________.
(c) The solutions of the resulting quadratic equation are ________.
(d) The solution(s) that satisfy the original equation are ________.

6. The equation $(x + 1)^2 - 5(x + 1) + 6 = 0$ is of ________ type. To solve the equation, we set $W =$ ________. The resulting quadratic equation is ________.

7. To eliminate the denominators in the equation $\frac{3}{x} + \frac{5}{x+2} = 2$, we multiply each side by the lowest common denominator ________________ to get the equivalent equation ________________.

8. To eliminate the square root in the equation $2x + 1 = \sqrt{x + 1}$, we ________________ each side to get the equation ________________.

SKILLS

9–12 ■ Solution? Determine whether the given value is a solution of the equation.

9. $4x + 7 = 9x - 3$
(a) $x = -2$ (b) $x = 2$

10. $1 - [2 - (3 - x)] = 4x - (6 + x)$
(a) $x = 2$ (b) $x = 4$

11. $\frac{1}{x} - \frac{1}{x - 4} = 1$
(a) $x = 2$ (b) $x = 4$

12. $\frac{x^{3/2}}{x - 6} = x - 8$
(a) $x = 4$ (b) $x = 8$

13–30 ■ Linear Equations The given equation is either linear or equivalent to a linear equation. Solve the equation.

13. $5x - 6 = 14$
14. $3x + 4 = 7$
15. $\frac{1}{2}x - 8 = 1$
16. $3 + \frac{1}{3}x = 5$
17. $-x + 3 = 4x$
18. $2x + 3 = 7 - 3x$
19. $\frac{x}{3} - 2 = \frac{5}{3}x + 7$
20. $\frac{2}{5}x - 1 = \frac{3}{10}x + 3$
21. $2(1 - x) = 3(1 + 2x) + 5$
22. $\frac{2}{3}y + \frac{1}{2}(y - 3) = \frac{y + 1}{4}$
23. $x - \frac{1}{3}x - \frac{1}{2}x - 5 = 0$
24. $2x - \frac{x}{2} + \frac{x + 1}{4} = 6x$
25. $\frac{1}{x} = \frac{4}{3x} + 1$
26. $\frac{2x - 1}{x + 2} = \frac{4}{5}$
27. $\frac{3}{x + 1} - \frac{1}{2} = \frac{1}{3x + 3}$
28. $\frac{4}{x - 1} + \frac{2}{x + 1} = \frac{35}{x^2 - 1}$
29. $(t - 4)^2 = (t + 4)^2 + 32$
30. $\sqrt{3}x + \sqrt{12} = \frac{x + 5}{\sqrt{3}}$

31–44 ■ Solving for a Variable Solve the equation for the indicated variable.

31. $PV = nRT$; for R
32. $F = G\frac{mM}{r^2}$; for m
33. $P = 2l + 2w$; for w
34. $\frac{1}{R} = \frac{1}{R_1} + \frac{1}{R_2}$; for R_1
35. $\frac{ax + b}{cx + d} = 2$; for x
36. $a - 2[b - 3(c - x)] = 6$; for x
37. $a^2x + (a - 1) = (a + 1)x$; for x
38. $\frac{a + 1}{b} = \frac{a - 1}{b} + \frac{b + 1}{a}$; for a
39. $V = \frac{1}{3}\pi r^2 h$; for r
40. $F = G\frac{mM}{r^2}$; for r
41. $a^2 + b^2 = c^2$; for b
42. $A = P\left(1 + \frac{i}{100}\right)^2$; for i
43. $h = \frac{1}{2}gt^2 + v_0 t$; for t
44. $S = \frac{n(n + 1)}{2}$; for n

45–56 ■ Solving by Factoring Find all real solutions of the equation by factoring.

45. $x^2 + x - 12 = 0$
46. $x^2 + 3x - 4 = 0$
47. $x^2 - 7x + 12 = 0$
48. $x^2 + 8x + 12 = 0$
49. $4x^2 - 4x - 15 = 0$
50. $2y^2 + 7y + 3 = 0$
51. $3x^2 + 5x = 2$
52. $6x(x - 1) = 21 - x$
53. $2x^2 = 8$
54. $3x^2 - 27 = 0$
55. $(2x - 5)^2 = 81$
56. $(5x + 1)^2 + 3 = 10$

57–64 ■ Completing the Square Find all real solutions of the equation by completing the square.

57. $x^2 + 2x - 5 = 0$
58. $x^2 - 4x + 2 = 0$
59. $x^2 - 6x - 11 = 0$
60. $x^2 + 3x - \frac{7}{4} = 0$
61. $2x^2 + 8x + 1 = 0$
62. $3x^2 - 6x - 1 = 0$
63. $4x^2 - x = 0$
64. $x^2 = \frac{3}{4}x - \frac{1}{8}$

65–80 ■ Quadratic Equations Find all real solutions of the quadratic equation.

65. $x^2 - 2x - 15 = 0$ **66.** $x^2 + 5x - 6 = 0$

67. $x^2 - 13x + 42 = 0$ **68.** $x^2 + 10x - 600 = 0$

69. $2x^2 + x - 3 = 0$ **70.** $3x^2 + 7x + 4 = 0$

71. $3x^2 + 6x - 5 = 0$ **72.** $x^2 - 6x + 1 = 0$

73. $9x^2 + 12x + 4 = 0$ **74.** $4x^2 - 4x + 1 = 0$

75. $4x^2 + 16x - 9 = 0$ **76.** $0 = x^2 - 4x + 1$

77. $7x^2 - 2x + 4 = 0$ **78.** $w^2 = 3(w - 1)$

79. $10y^2 - 16y + 5 = 0$ **80.** $25x^2 + 70x + 49 = 0$

81–86 ■ Discriminant Use the discriminant to determine the number of real solutions of the equation. Do not solve the equation.

81. $x^2 - 6x + 1 = 0$ **82.** $3x^2 = 6x - 9$

83. $x^2 + 2.20x + 1.21 = 0$ **84.** $x^2 + 2.21x + 1.21 = 0$

85. $4x^2 + 5x + \frac{13}{8} = 0$ **86.** $x^2 + rx - s = 0 \quad (s > 0)$

87–116 ■ Other Equations Find all real solutions of the equation.

87. $\dfrac{x^2}{x + 100} = 50$ **88.** $\dfrac{1}{x - 1} - \dfrac{2}{x^2} = 0$

89. $\dfrac{1}{x - 1} + \dfrac{1}{x + 2} = \dfrac{5}{4}$ **90.** $\dfrac{x + 5}{x - 2} = \dfrac{5}{x + 2} + \dfrac{28}{x^2 - 4}$

91. $\dfrac{10}{x} - \dfrac{12}{x - 3} + 4 = 0$ **92.** $\dfrac{x}{2x + 7} - \dfrac{x + 1}{x + 3} = 1$

93. $5 = \sqrt{4x - 3}$ **94.** $\sqrt{8x - 1} = 3$

95. $\sqrt{2x - 1} = \sqrt{3x - 5}$ **96.** $\sqrt{3 + x} = \sqrt{x^2 + 1}$

97. $\sqrt{2x + 1} + 1 = x$ **98.** $\sqrt{5 - x} + 1 = x - 2$

99. $2x + \sqrt{x + 1} = 8$ **100.** $x - \sqrt{9 - 3x} = 0$

101. $\sqrt{3x + 1} = 2 + \sqrt{x + 1}$ **102.** $\sqrt{1 + x} + \sqrt{1 - x} = 2$

103. $x^4 - 13x^2 + 40 = 0$ **104.** $x^4 - 5x^2 + 4 = 0$

105. $2x^4 + 4x^2 + 1 = 0$ **106.** $x^6 - 2x^3 - 3 = 0$

107. $x^{4/3} - 5x^{2/3} + 6 = 0$ **108.** $\sqrt{x} - 3\sqrt[4]{x} - 4 = 0$

109. $4(x + 1)^{1/2} - 5(x + 1)^{3/2} + (x + 1)^{5/2} = 0$

110. $x^{1/2} + 3x^{-1/2} = 10x^{-3/2}$

111. $x^{1/2} - 3x^{1/3} = 3x^{1/6} - 9$ **112.** $x - 5\sqrt{x} + 6 = 0$

113. $|3x + 5| = 1$ **114.** $|2x| = 3$

115. $|x - 4| = 0.01$ **116.** $|x - 6| = -1$

SKILLS Plus

117–122 ■ More on Solving Equations Find all real solutions of the equation.

117. $\dfrac{1}{x^3} + \dfrac{4}{x^2} + \dfrac{4}{x} = 0$ **118.** $4x^{-4} - 16x^{-2} + 4 = 0$

119. $\sqrt{\sqrt{x + 5} + x} = 5$ **120.** $\sqrt[3]{4x^2 - 4x} = x$

121. $x^2\sqrt{x + 3} = (x + 3)^{3/2}$ **122.** $\sqrt{11 - x^2} - \dfrac{2}{\sqrt{11 - x^2}} = 1$

123–126 ■ More on Solving Equations Solve the equation for the variable x. The constants a and b represent positive real numbers.

123. $x^4 - 5ax^2 + 4a^2 = 0$ **124.** $a^3x^3 + b^3 = 0$

125. $\sqrt{x + a} + \sqrt{x - a} = \sqrt{2}\sqrt{x + 6}$

126. $\sqrt{x} - a\sqrt[3]{x} + b\sqrt[6]{x} - ab = 0$

APPLICATIONS

127–128 ■ Falling-Body Problems Suppose an object is dropped from a height h_0 above the ground. Then its height after t seconds is given by $h = -16t^2 + h_0$, where h is measured in feet. Use this information to solve the problem.

127. If a ball is dropped from 288 ft above the ground, how long does it take to reach ground level?

128. A ball is dropped from the top of a building 96 ft tall.
(a) How long will it take to fall half the distance to ground level?
(b) How long will it take to fall to ground level?

129–130 ■ Falling-Body Problems Use the formula $h = -16t^2 + v_0t$ discussed in Example 9.

129. A ball is thrown straight upward at an initial speed of $v_0 = 40$ ft/s.
(a) When does the ball reach a height of 24 ft?
(b) When does it reach a height of 48 ft?
(c) What is the greatest height reached by the ball?
(d) When does the ball reach the highest point of its path?
(e) When does the ball hit the ground?

130. How fast would a ball have to be thrown upward to reach a maximum height of 100 ft? [*Hint:* Use the discriminant of the equation $16t^2 - v_0t + h = 0$.]

131. Shrinkage in Concrete Beams As concrete dries, it shrinks—the higher the water content, the greater the shrinkage. If a concrete beam has a water content of w kg/m^3, then it will shrink by a factor

$$S = \frac{0.032w - 2.5}{10{,}000}$$

where S is the fraction of the original beam length that disappears due to shrinkage.
(a) A beam 12.025 m long is cast in concrete that contains 250 kg/m^3 water. What is the shrinkage factor S? How long will the beam be when it has dried?
(b) A beam is 10.014 m long when wet. We want it to shrink to 10.009 m, so the shrinkage factor should be $S = 0.00050$. What water content will provide this amount of shrinkage?

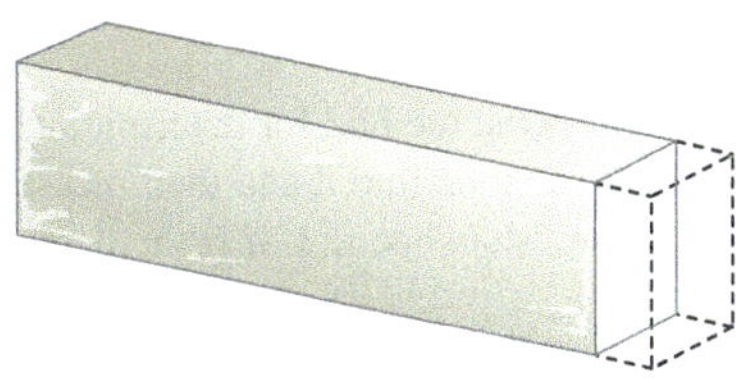

132. The Lens Equation If F is the focal length of a convex lens and an object is placed at a distance x from the lens, then its image will be at a distance y from the lens, where F, x, and y are related by the *lens equation*

$$\frac{1}{F} = \frac{1}{x} + \frac{1}{y}$$

Suppose that a lens has a focal length of 4.8 cm and that the image of an object is 4 cm closer to the lens than the object itself. How far from the lens is the object?

133. Fish Population The fish population in a certain lake rises and falls according to the formula

$$F = 1000(30 + 17t - t^2)$$

Here F is the number of fish at time t, where t is measured in years since January 1, 2002, when the fish population was first estimated.

(a) On what date will the fish population again be the same as it was on January 1, 2002?

(b) By what date will all the fish in the lake have died?

134. Fish Population A large pond is stocked with fish. The fish population P is modeled by the formula $P = 3t + 10\sqrt{t} + 140$, where t is the number of days since the fish were first introduced into the pond. How many days will it take for the fish population to reach 500?

135. Profit A small-appliance manufacturer finds that the profit P (in dollars) generated by producing x microwave ovens per week is given by the formula $P = \frac{1}{10}x(300 - x)$, provided that $0 \le x \le 200$. How many ovens must be manufactured in a given week to generate a profit of \$1250?

136. Gravity If an imaginary line segment is drawn between the centers of the earth and the moon, then the net gravitational force F acting on an object situated on this line segment is

$$F = \frac{-K}{x^2} + \frac{0.012K}{(239 - x)^2}$$

where $K > 0$ is a constant and x is the distance of the object from the center of the earth, measured in thousands of miles. How far from the center of the earth is the "dead spot" where no net gravitational force acts upon the object? (Express your answer to the nearest thousand miles.)

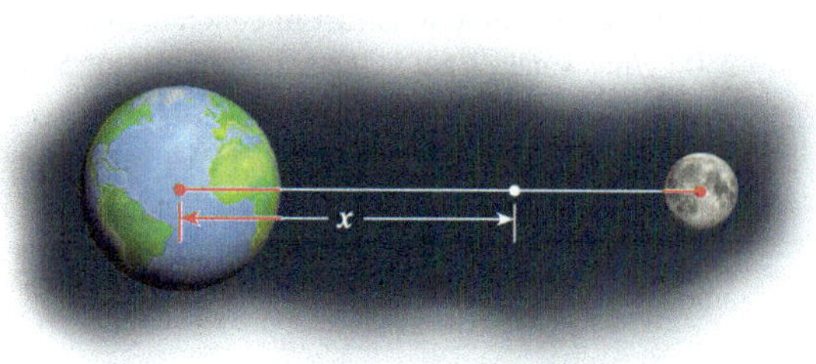

137. Depth of a Well One method for determining the depth of a well is to drop a stone into it and then measure the time it takes until the splash is heard. If d is the depth of the well (in feet) and t_1 the time (in seconds) it takes for the stone to fall, then $d = 16t_1^2$, so $t_1 = \sqrt{d}/4$. Now if t_2 is the time it takes for the sound to travel back up, then $d = 1090t_2$ because the speed of sound is 1090 ft/s. So $t_2 = d/1090$. Thus the total time elapsed between dropping the stone and hearing the splash is

$$t_1 + t_2 = \frac{\sqrt{d}}{4} + \frac{d}{1090}$$

How deep is the well if this total time is 3 s?

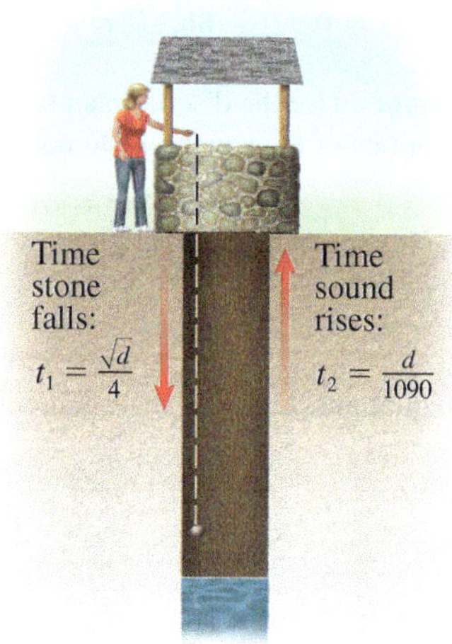

DISCUSS ■ DISCOVER ■ PROVE ■ WRITE

138. DISCUSS: A Family of Equations The equation

$$3x + k - 5 = kx - k + 1$$

is really a **family of equations**, because for each value of k, we get a different equation with the unknown x. The letter k is called a **parameter** for this family. What value should we pick for k to make the given value of x a solution of the resulting equation?

(a) $x = 0$ **(b)** $x = 1$ **(c)** $x = 2$

139. DISCUSS: Proof That 0 = 1? The following steps appear to give equivalent equations, which seem to prove that $1 = 0$. Find the error.

$x = 1$	Given
$x^2 = x$	Multiply by x
$x^2 - x = 0$	Subtract x
$x(x - 1) = 0$	Factor
$\frac{x(x-1)}{x-1} = \frac{0}{x-1}$	Divide by $x - 1$
$x = 0$	Simplify
$1 = 0$	Given $x = 1$

140. DISCOVER ■ PROVE: Relationship Between Solutions and Coefficients The Quadratic Formula gives us the solutions of a quadratic equation from its coefficients. We can also obtain the coefficients from the solutions.

(a) Find the solutions of the equation $x^2 - 9x + 20 = 0$, and show that the product of the solutions is the constant term 20 and the sum of the solutions is 9, the negative of the coefficient of x.

(b) Show that the same relationship between solutions and coefficients holds for the following equations:

$$x^2 - 2x - 8 = 0$$

$$x^2 + 4x + 2 = 0$$

(c) Use the Quadratic Formula to prove that in general, if the equation $x^2 + bx + c = 0$ has solutions r_1 and r_2, then $c = r_1 r_2$ and $b = -(r_1 + r_2)$.

141. DISCUSS: Solving an Equation in Different Ways We have learned several different ways to solve an equation in this section. Some equations can be tackled by more than one method. For example, the equation $x - \sqrt{x} - 2 = 0$ is of quadratic type. We can solve it by letting $\sqrt{x} = u$ and $x = u^2$, and factoring. Or we could solve for $\sqrt{x}$, square each side, and then solve the resulting quadratic equation. Solve the following equations using both methods indicated, and show that you get the same final answers.

(a) $x - \sqrt{x} - 2 = 0$ quadratic type; solve for the radical, and square

(b) $\dfrac{12}{(x-3)^2} + \dfrac{10}{x-3} + 1 = 0$ quadratic type; multiply by LCD

1.6 COMPLEX NUMBERS

■ Arithmetic Operations on Complex Numbers ■ Square Roots of Negative Numbers ■ Complex Solutions of Quadratic Equations

In Section 1.5 we saw that if the discriminant of a quadratic equation is negative, the equation has no real solution. For example, the equation

$$x^2 + 4 = 0$$

has no real solution. If we try to solve this equation, we get $x^2 = -4$, so

$$x = \pm\sqrt{-4}$$

But this is impossible, since the square of any real number is positive. [For example, $(-2)^2 = 4$, a positive number.] Thus negative numbers don't have real square roots.

See the note on Cardano (page 292) for an example of how complex numbers are used to find real solutions of polynomial equations.

To make it possible to solve *all* quadratic equations, mathematicians invented an expanded number system, called the *complex number system*. First they defined the new number

$$i = \sqrt{-1}$$

This means that $i^2 = -1$. A complex number is then a number of the form $a + bi$, where a and b are real numbers.

DEFINITION OF COMPLEX NUMBERS

A **complex number** is an expression of the form

$$a + bi$$

where a and b are real numbers and $i^2 = -1$. The **real part** of this complex number is a, and the **imaginary part** is b. Two complex numbers are **equal** if and only if their real parts are equal and their imaginary parts are equal.

Note that both the real and imaginary parts of a complex number are real numbers.

EXAMPLE 1 ■ Complex Numbers

The following are examples of complex numbers.

$3 + 4i$	Real part 3, imaginary part 4
$\frac{1}{2} - \frac{2}{3}i$	Real part $\frac{1}{2}$, imaginary part $-\frac{2}{3}$
$6i$	Real part 0, imaginary part 6
-7	Real part -7, imaginary part 0

Now Try Exercises 7 and 11 ■

A number such as $6i$, which has real part 0, is called a **pure imaginary number**. A real number such as -7 can be thought of as a complex number with imaginary part 0.

In the complex number system every quadratic equation has solutions. The numbers $2i$ and $-2i$ are solutions of $x^2 = -4$ because

$$(2i)^2 = 2^2i^2 = 4(-1) = -4 \quad \text{and} \quad (-2i)^2 = (-2)^2i^2 = 4(-1) = -4$$

Although we use the term *imaginary* in this context, imaginary numbers should not be thought of as any less "real" (in the ordinary rather than the mathematical sense of that word) than negative numbers or irrational numbers. All numbers (except possibly the positive integers) are creations of the human mind—the numbers -1 and $\sqrt{2}$ as well as the number i. We study complex numbers because they complete, in a useful and elegant fashion, our study of the solutions of equations. In fact, imaginary numbers are useful not only in algebra and mathematics, but in the other sciences as well. To give just one example, in electrical theory the *reactance* of a circuit is a quantity whose measure is an imaginary number.

Arithmetic Operations on Complex Numbers

Complex numbers are added, subtracted, multiplied, and divided just as we would any number of the form $a + b\sqrt{c}$. The only difference that we need to keep in mind is that $i^2 = -1$. Thus the following calculations are valid.

$$\begin{aligned}(a + bi)(c + di) &= ac + (ad + bc)i + bdi^2 && \text{Multiply and collect like terms} \\ &= ac + (ad + bc)i + bd(-1) && i^2 = -1 \\ &= (ac - bd) + (ad + bc)i && \text{Combine real and imaginary parts}\end{aligned}$$

We therefore define the sum, difference, and product of complex numbers as follows.

ADDING, SUBTRACTING, AND MULTIPLYING COMPLEX NUMBERS

Definition	Description
Addition $(a + bi) + (c + di) = (a + c) + (b + d)i$	To add complex numbers, add the real parts and the imaginary parts.
Subtraction $(a + bi) - (c + di) = (a - c) + (b - d)i$	To subtract complex numbers, subtract the real parts and the imaginary parts.
Multiplication $(a + bi) \cdot (c + di) = (ac - bd) + (ad + bc)i$	Multiply complex numbers like binomials, using $i^2 = -1$.

EXAMPLE 2 ■ Adding, Subtracting, and Multiplying Complex Numbers

Graphing calculators can perform arithmetic operations on complex numbers.

```
(3+5i)+(4-2i)
                7+3i
(3+5i)*(4-2i)
              22+14i
```

Express the following in the form $a + bi$.

(a) $(3 + 5i) + (4 - 2i)$ **(b)** $(3 + 5i) - (4 - 2i)$

(c) $(3 + 5i)(4 - 2i)$ **(d)** i^{23}

SOLUTION

(a) According to the definition, we add the real parts and we add the imaginary parts:

$$(3 + 5i) + (4 - 2i) = (3 + 4) + (5 - 2)i = 7 + 3i$$

(b) $(3 + 5i) - (4 - 2i) = (3 - 4) + [5 - (-2)]i = -1 + 7i$

(c) $(3 + 5i)(4 - 2i) = [3 \cdot 4 - 5(-2)] + [3(-2) + 5 \cdot 4]i = 22 + 14i$

(d) $i^{23} = i^{22+1} = (i^2)^{11}i = (-1)^{11}i = (-1)i = -i$

Now Try Exercises 19, 23, 29, and 47 ■

Complex Conjugates

Number	Conjugate
$3 + 2i$	$3 - 2i$
$1 - i$	$1 + i$
$4i$	$-4i$
5	5

Division of complex numbers is much like rationalizing the denominator of a radical expression, which we considered in Section 1.2. For the complex number $z = a + bi$ we define its **complex conjugate** to be $\bar{z} = a - bi$. Note that

$$z \cdot \bar{z} = (a + bi)(a - bi) = a^2 + b^2$$

So the product of a complex number and its conjugate is always a nonnegative real number. We use this property to divide complex numbers.

DIVIDING COMPLEX NUMBERS

To simplify the quotient $\dfrac{a + bi}{c + di}$, multiply the numerator and the denominator by the complex conjugate of the denominator:

$$\frac{a + bi}{c + di} = \left(\frac{a + bi}{c + di}\right)\left(\frac{c - di}{c - di}\right) = \frac{(ac + bd) + (bc - ad)i}{c^2 + d^2}$$

Rather than memorizing this entire formula, it is easier to just remember the first step and then multiply out the numerator and the denominator as usual.

EXAMPLE 3 ■ Dividing Complex Numbers

Express the following in the form $a + bi$.

(a) $\dfrac{3 + 5i}{1 - 2i}$ **(b)** $\dfrac{7 + 3i}{4i}$

SOLUTION We multiply both the numerator and denominator by the complex conjugate of the denominator to make the new denominator a real number.

(a) The complex conjugate of $1 - 2i$ is $\overline{1 - 2i} = 1 + 2i$. Therefore

$$\frac{3 + 5i}{1 - 2i} = \left(\frac{3 + 5i}{1 - 2i}\right)\left(\frac{1 + 2i}{1 + 2i}\right) = \frac{-7 + 11i}{5} = -\frac{7}{5} + \frac{11}{5}i$$

(b) The complex conjugate of $4i$ is $-4i$. Therefore

$$\frac{7 + 3i}{4i} = \left(\frac{7 + 3i}{4i}\right)\left(\frac{-4i}{-4i}\right) = \frac{12 - 28i}{16} = \frac{3}{4} - \frac{7}{4}i$$

Now Try Exercises 39 and 43 ■

Square Roots of Negative Numbers

Just as every positive real number r has two square roots ($\sqrt{r}$ and $-\sqrt{r}$), every negative number has two square roots as well. If $-r$ is a negative number, then its square roots are $\pm i\sqrt{r}$, because $(i\sqrt{r})^2 = i^2 r = -r$ and $(-i\sqrt{r})^2 = (-1)^2 i^2 r = -r$.

SQUARE ROOTS OF NEGATIVE NUMBERS

If $-r$ is negative, then the **principal square root** of $-r$ is

$$\sqrt{-r} = i\sqrt{r}$$

The two square roots of $-r$ are $i\sqrt{r}$ and $-i\sqrt{r}$.

We usually write $i\sqrt{b}$ instead of $\sqrt{b}\,i$ to avoid confusion with $\sqrt{bi}$.

EXAMPLE 4 ■ Square Roots of Negative Numbers

(a) $\sqrt{-1} = i\sqrt{1} = i$ **(b)** $\sqrt{-16} = i\sqrt{16} = 4i$ **(c)** $\sqrt{-3} = i\sqrt{3}$

Now Try Exercises 53 and 55 ■

Special care must be taken in performing calculations that involve square roots of negative numbers. Although $\sqrt{a}\cdot\sqrt{b} = \sqrt{ab}$ when a and b are positive, this is *not* true when both are negative. For example,

$$\sqrt{-2}\cdot\sqrt{-3} = i\sqrt{2}\cdot i\sqrt{3} = i^2\sqrt{6} = -\sqrt{6}$$

but $$\sqrt{(-2)(-3)} = \sqrt{6}$$

so $$\sqrt{-2}\cdot\sqrt{-3} \neq \sqrt{(-2)(-3)}$$

When multiplying radicals of negative numbers, express them first in the form $i\sqrt{r}$ (where $r > 0$) to avoid possible errors of this type.

EXAMPLE 5 ■ Using Square Roots of Negative Numbers

Evaluate $(\sqrt{12} - \sqrt{-3})(3 + \sqrt{-4})$, and express the result in the form $a + bi$.

SOLUTION

$$\begin{aligned}(\sqrt{12} - \sqrt{-3})(3 + \sqrt{-4}) &= (\sqrt{12} - i\sqrt{3})(3 + i\sqrt{4})\\ &= (2\sqrt{3} - i\sqrt{3})(3 + 2i)\\ &= (6\sqrt{3} + 2\sqrt{3}) + i(2\cdot 2\sqrt{3} - 3\sqrt{3})\\ &= 8\sqrt{3} + i\sqrt{3}\end{aligned}$$

Now Try Exercise 57 ■

Complex Solutions of Quadratic Equations

We have already seen that if $a \neq 0$, then the solutions of the quadratic equation $ax^2 + bx + c = 0$ are

$$x = \frac{-b \pm \sqrt{b^2 - 4ac}}{2a}$$

If $b^2 - 4ac < 0$, then the equation has no real solution. But in the complex number system this equation will always have solutions, because negative numbers have square roots in this expanded setting.

EXAMPLE 6 ■ Quadratic Equations with Complex Solutions

Solve each equation.

(a) $x^2 + 9 = 0$ **(b)** $x^2 + 4x + 5 = 0$

SOLUTION

(a) The equation $x^2 + 9 = 0$ means $x^2 = -9$, so

$$x = \pm\sqrt{-9} = \pm i\sqrt{9} = \pm 3i$$

The solutions are therefore $3i$ and $-3i$.

(b) By the Quadratic Formula we have

$$x = \frac{-4 \pm \sqrt{4^2 - 4 \cdot 5}}{2} = \frac{-4 \pm \sqrt{-4}}{2} = \frac{-4 \pm 2i}{2} = -2 \pm i$$

So the solutions are $-2 + i$ and $-2 - i$.

Now Try Exercises 61 and 63 ■

We see from Example 6 that if a quadratic equation with real coefficients has complex solutions, then these solutions are complex conjugates of each other. So if $a + bi$ is a solution, then $a - bi$ is also a solution.

EXAMPLE 7 ■ Complex Conjugates as Solutions of a Quadratic

Show that the solutions of the equation

$$4x^2 - 24x + 37 = 0$$

are complex conjugates of each other.

SOLUTION We use the Quadratic Formula to get

$$x = \frac{24 \pm \sqrt{(24)^2 - 4(4)(37)}}{2(4)} = \frac{24 \pm \sqrt{-16}}{8} = \frac{24 \pm 4i}{8} = 3 \pm \frac{1}{2}i$$

So the solutions are $3 + \frac{1}{2}i$ and $3 - \frac{1}{2}i$, and these are complex conjugates.

Now Try Exercise 69 ■

Library of Congress Prints and Photographs Division

LEONHARD EULER (1707–1783) was born in Basel, Switzerland, the son of a pastor. When Euler was 13, his father sent him to the University at Basel to study theology, but Euler soon decided to devote himself to the sciences. Besides theology he studied mathematics, medicine, astronomy, physics, and Asian languages. It is said that Euler could calculate as effortlessly as "men breathe or as eagles fly." One hundred years before Euler, Fermat (see page 117) had conjectured that $2^{2^n} + 1$ is a prime number for all n. The first five of these numbers are 5, 17, 257, 65537, and 4,294,967,297. It is easy to show that the first four are prime. The fifth was also thought to be prime until Euler, with his phenomenal calculating ability, showed that it is the product $641 \times 6{,}700{,}417$ and so is not prime. Euler published more than any other mathematician in history. His collected works comprise 75 large volumes. Although he was blind for the last 17 years of his life, he continued to work and publish. In his writings he popularized the use of the symbols π, e, and i, which you will find in this textbook. One of Euler's most lasting contributions is his development of complex numbers.

1.6 EXERCISES

CONCEPTS

1. The imaginary number i has the property that $i^2 =$ ______.
2. For the complex number $3 + 4i$ the real part is ______, and the imaginary part is ______.
3. **(a)** The complex conjugate of $3 + 4i$ is $\overline{3 + 4i} =$ ______.
   (b) $(3 + 4i)(\overline{3 + 4i}) =$ ______.
4. If $3 + 4i$ is a solution of a quadratic equation with real coefficients, then ______ is also a solution of the equation.

5–6 ■ *Yes or No?* If *No*, give a reason.

5. Is every real number also a complex number?
6. Is the sum of a complex number and its complex conjugate a real number?

SKILLS

7–16 ■ Real and Imaginary Parts Find the real and imaginary parts of the complex number.

7. $5 - 7i$
8. $-6 + 4i$
9. $\dfrac{-2 - 5i}{3}$
10. $\dfrac{4 + 7i}{2}$
11. 3
12. $-\frac{1}{2}$
13. $-\frac{2}{3}i$
14. $i\sqrt{3}$
15. $\sqrt{3} + \sqrt{-4}$
16. $2 - \sqrt{-5}$

17–26 ■ Sums and Differences Evaluate the sum or difference, and write the result in the form $a + bi$.

17. $(3 + 2i) + 5i$
18. $3i - (2 - 3i)$
19. $(5 - 3i) + (-4 - 7i)$
20. $(-3 + 4i) - (2 - 5i)$
21. $(-6 + 6i) + (9 - i)$
22. $(3 - 2i) + (-5 - \frac{1}{3}i)$
23. $(7 - \frac{1}{2}i) - (5 + \frac{3}{2}i)$
24. $(-4 + i) - (2 - 5i)$
25. $(-12 + 8i) - (7 + 4i)$
26. $6i - (4 - i)$

27–36 ■ Products Evaluate the product, and write the result in the form $a + bi$.

27. $4(-1 + 2i)$
28. $-2(3 - 4i)$
29. $(7 - i)(4 + 2i)$
30. $(5 - 3i)(1 + i)$
31. $(6 + 5i)(2 - 3i)$
32. $(-2 + i)(3 - 7i)$
33. $(2 + 5i)(2 - 5i)$
34. $(3 - 7i)(3 + 7i)$
35. $(2 + 5i)^2$
36. $(3 - 7i)^2$

37–46 ■ Quotients Evaluate the quotient, and write the result in the form $a + bi$.

37. $\dfrac{1}{i}$
38. $\dfrac{1}{1 + i}$
39. $\dfrac{2 - 3i}{1 - 2i}$
40. $\dfrac{5 - i}{3 + 4i}$
41. $\dfrac{10i}{1 - 2i}$
42. $(2 - 3i)^{-1}$
43. $\dfrac{4 + 6i}{3i}$
44. $\dfrac{-3 + 5i}{15i}$
45. $\dfrac{1}{1 + i} - \dfrac{1}{1 - i}$
46. $\dfrac{(1 + 2i)(3 - i)}{2 + i}$

47–52 ■ Powers Evaluate the power, and write the result in the form $a + bi$.

47. i^3
48. i^{10}
49. $(3i)^5$
50. $(2i)^4$
51. i^{1000}
52. i^{1002}

53–60 ■ Radical Expressions Evaluate the radical expression, and express the result in the form $a + bi$.

53. $\sqrt{-49}$
54. $\sqrt{\dfrac{-81}{16}}$
55. $\sqrt{-3}\sqrt{-12}$
56. $\sqrt{\frac{1}{3}}\sqrt{-27}$
57. $(3 - \sqrt{-5})(1 + \sqrt{-1})$
58. $(\sqrt{3} - \sqrt{-4})(\sqrt{6} - \sqrt{-8})$
59. $\dfrac{2 + \sqrt{-8}}{1 + \sqrt{-2}}$
60. $\dfrac{\sqrt{-36}}{\sqrt{-2}\sqrt{-9}}$

61–72 ■ Quadratic Equations Find all solutions of the equation and express them in the form $a + bi$.

61. $x^2 + 49 = 0$
62. $3x^2 + 1 = 0$
63. $x^2 - x + 2 = 0$
64. $x^2 + 2x + 2 = 0$
65. $x^2 + 3x + 7 = 0$
66. $x^2 - 6x + 10 = 0$
67. $x^2 + x + 1 = 0$
68. $x^2 - 3x + 3 = 0$
69. $2x^2 - 2x + 1 = 0$
70. $t + 3 + \dfrac{3}{t} = 0$
71. $6x^2 + 12x + 7 = 0$
72. $x^2 + \frac{1}{2}x + 1 = 0$

SKILLS Plus

73–76 ■ Conjugates Evaluate the given expression for $z = 3 - 4i$ and $w = 5 + 2i$.

73. $\overline{z} + \overline{w}$
74. $\overline{z + w}$
75. $z \cdot \overline{z}$
76. $\overline{z} \cdot \overline{w}$

77–84 ■ Conjugates Recall that the symbol $\overline{z}$ represents the complex conjugate of z. If $z = a + bi$ and $w = c + di$, show that each statement is true.

77. $\overline{z} + \overline{w} = \overline{z + w}$
78. $\overline{zw} = \overline{z} \cdot \overline{w}$
79. $(\overline{z})^2 = \overline{z^2}$
80. $\overline{\overline{z}} = z$
81. $z + \overline{z}$ is a real number.
82. $z - \overline{z}$ is a pure imaginary number.
83. $z \cdot \overline{z}$ is a real number.
84. $z = \overline{z}$ if and only if z is real.

DISCUSS ■ DISCOVER ■ PROVE ■ WRITE

85. **PROVE: Complex Conjugate Roots** Suppose that the equation $ax^2 + bx + c = 0$ has real coefficients and complex roots. Why must the roots be complex conjugates of each other? [*Hint:* Think about how you would find the roots using the Quadratic Formula.]

86. **DISCUSS: Powers of *i*** Calculate the first 12 powers of i, that is, $i, i^2, i^3, \ldots, i^{12}$. Do you notice a pattern? Explain how you would calculate any whole number power of i, using the pattern that you have discovered. Use this procedure to calculate i^{4446}.

1.7 MODELING WITH EQUATIONS

■ Making and Using Models ■ Problems About Interest ■ Problems About Area or Length ■ Problems About Mixtures ■ Problems About the Time Needed to Do a Job ■ Problems About Distance, Rate, and Time

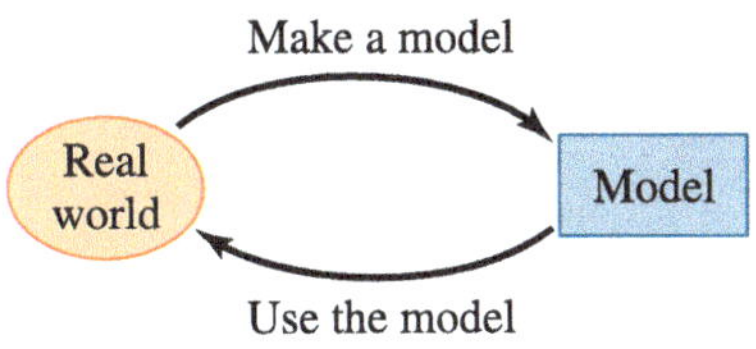

In this section a **mathematical model** is an equation that describes a real-world object or process. Modeling is the process of finding such equations. Once the model or equation has been found, it is then used to obtain information about the thing being modeled. The process is described in the diagram in the margin. In this section we learn how to make and use models to solve real-world problems.

■ Making and Using Models

We will use the following guidelines to help us set up equations that model situations described in words. To show how the guidelines can help you to set up equations, we note them as we work each example in this section.

GUIDELINES FOR MODELING WITH EQUATIONS

1. **Identify the Variable.** Identify the quantity that the problem asks you to find. This quantity can usually be determined by a careful reading of the question that is posed at the end of the problem. Then **introduce notation** for the variable (call it x or some other letter).
2. **Translate from Words to Algebra.** Read each sentence in the problem again, and express all the quantities mentioned in the problem in terms of the variable you defined in Step 1. To organize this information, it is sometimes helpful to **draw a diagram** or **make a table**.
3. **Set Up the Model.** Find the crucial fact in the problem that gives a relationship between the expressions you listed in Step 2. **Set up an equation** (or **model**) that expresses this relationship.
4. **Solve the Equation and Check Your Answer.** Solve the equation, check your answer, and express it as a sentence that answers the question posed in the problem.

The following example illustrates how these guidelines are used to translate a "word problem" into the language of algebra.

EXAMPLE 1 ■ Renting a Car

A car rental company charges \$30 a day and 15¢ a mile for renting a car. Helen rents a car for two days, and her bill comes to \$108. How many miles did she drive?

SOLUTION **Identify the variable.** We are asked to find the number of miles Helen has driven. So we let

$$x = \text{number of miles driven}$$

Translate from words to algebra. Now we translate all the information given in the problem into the language of algebra.

In Words	In Algebra
Number of miles driven	x
Mileage cost (at \$0.15 per mile)	$0.15x$
Daily cost (at \$30 per day)	$2(30)$

Set up the model. Now we set up the model.

$$\text{mileage cost} + \text{daily cost} = \text{total cost}$$

$$0.15x + 2(30) = 108$$

Solve. Now we solve for x.

$$0.15x = 48 \quad \text{Subtract 60}$$

$$x = \frac{48}{0.15} \quad \text{Divide by 0.15}$$

$$x = 320 \quad \text{Calculator}$$

Helen drove her rental car 320 miles.

Now Try Exercise 21 ■

CHECK YOUR ANSWER

total cost = mileage cost + daily cost

$= 0.15(320) + 2(30)$

$= 108$ ✓

In the examples and exercises that follow, we construct equations that model problems in many different real-life situations.

Problems About Interest

When you borrow money from a bank or when a bank "borrows" your money by keeping it for you in a savings account, the borrower in each case must pay for the privilege of using the money. The fee that is paid is called **interest**. The most basic type of interest is **simple interest**, which is just an annual percentage of the total amount borrowed or deposited. The amount of a loan or deposit is called the **principal** P. The annual percentage paid for the use of this money is the **interest rate** r. We will use the variable t to stand for the number of years that the money is on deposit and the variable I to stand for the total interest earned. The following **simple interest formula** gives the amount of interest I earned when a principal P is deposited for t years at an interest rate r.

$$\boxed{I = Prt}$$

When using this formula, remember to convert r from a percentage to a decimal. For example, in decimal form, 5% is 0.05. So at an interest rate of 5%, the interest paid on a \$1000 deposit over a 3-year period is $I = Prt = 1000(0.05)(3) = \$150$.

EXAMPLE 2 ■ Interest on an Investment

Mary inherits \$100,000 and invests it in two certificates of deposit. One certificate pays 6% and the other pays $4\frac{1}{2}$% simple interest annually. If Mary's total interest is \$5025 per year, how much money is invested at each rate?

DISCOVERY PROJECT

Equations Through the Ages

Equations have always been important in solving real-world problems. Very old manuscripts from Babylon, Egypt, India, and China show that ancient peoples used equations to solve real-world problems that they encountered. In this project we discover that they also solved equations just for fun or for practice. You can find the project at **www.stewartmath.com**.

SOLUTION **Identify the variable.** The problem asks for the amount she has invested at each rate. So we let

$$x = \text{the amount invested at } 6\%$$

Translate from words to algebra. Since Mary's total inheritance is \$100,000, it follows that she invested $100{,}000 - x$ at $4\frac{1}{2}\%$. We translate all the information given into the language of algebra.

In Words	In Algebra
Amount invested at 6%	x
Amount invested at $4\frac{1}{2}\%$	$100{,}000 - x$
Interest earned at 6%	$0.06x$
Interest earned at $4\frac{1}{2}\%$	$0.045(100{,}000 - x)$

Set up the model. We use the fact that Mary's total interest is \$5025 to set up the model.

$$\text{interest at } 6\% \quad + \quad \text{interest at } 4\tfrac{1}{2}\% \quad = \quad \text{total interest}$$

$$0.06x + 0.045(100{,}000 - x) = 5025$$

Solve. Now we solve for x.

$$0.06x + 4500 - 0.045x = 5025 \qquad \text{Distributive Property}$$

$$0.015x + 4500 = 5025 \qquad \text{Combine the } x\text{-terms}$$

$$0.015x = 525 \qquad \text{Subtract 4500}$$

$$x = \frac{525}{0.015} = 35{,}000 \qquad \text{Divide by 0.015}$$

So Mary has invested \$35,000 at 6% and the remaining \$65,000 at $4\frac{1}{2}\%$.

CHECK YOUR ANSWER

$$\text{total interest} = 6\% \text{ of } \$35{,}000 + 4\tfrac{1}{2}\% \text{ of } \$65{,}000$$

$$= \$2100 + \$2925 = \$5025 \quad ✓$$

Now Try Exercise 25 ■

Problems About Area or Length

When we use algebra to model a physical situation, we must sometimes use basic formulas from geometry. For example, we may need a formula for an area or a perimeter, or the formula that relates the sides of similar triangles, or the Pythagorean Theorem. Most of these formulas are listed in the front endpapers of this book. The next two examples use these geometric formulas to solve some real-world problems.

EXAMPLE 3 ■ Dimensions of a Garden

A square garden has a walkway 3 ft wide around its outer edge, as shown in Figure 1. If the area of the entire garden, including the walkway, is 18,000 ft^2, what are the dimensions of the planted area?

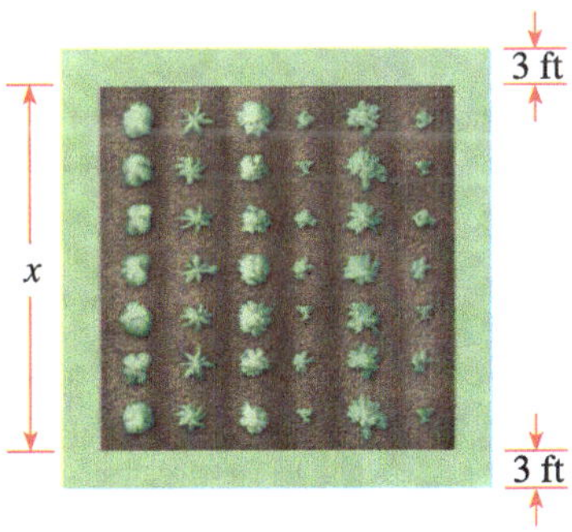

FIGURE 1

SOLUTION **Identify the variable.** We are asked to find the length and width of the planted area. So we let

$$x = \text{the length of the planted area}$$

Translate from words to algebra. Next, translate the information from Figure 1 into the language of algebra.

In Words	In Algebra
Length of planted area	x
Length of entire garden	$x + 6$
Area of entire garden	$(x + 6)^2$

Set up the model. We now set up the model.

$$\text{area of entire garden} = 18{,}000 \text{ ft}^2$$

$$(x + 6)^2 = 18{,}000$$

Solve. Now we solve for x.

$$x + 6 = \sqrt{18{,}000} \qquad \text{Take square roots}$$

$$x = \sqrt{18{,}000} - 6 \qquad \text{Subtract 6}$$

$$x \approx 128$$

The planted area of the garden is about 128 ft by 128 ft.

Now Try Exercise 49 ■

EXAMPLE 4 ■ Dimensions of a Building Lot

A rectangular building lot is 8 ft longer than it is wide and has an area of 2900 ft^2. Find the dimensions of the lot.

SOLUTION **Identify the variable.** We are asked to find the width and length of the lot. So let

$$w = \text{width of lot}$$

FIGURE 2

Translate from words to algebra. Then we translate the information given in the problem into the language of algebra (see Figure 2).

In Words	In Algebra
Width of lot	w
Length of Lot	$w + 8$

Set up the model. Now we set up the model.

$$\text{width of lot} \cdot \text{length of lot} = \text{area of lot}$$

$$w(w + 8) = 2900$$

Solve. Now we solve for w.

$$w^2 + 8w = 2900 \qquad \text{Expand}$$

$$w^2 + 8w - 2900 = 0 \qquad \text{Subtract 2900}$$

$$(w - 50)(w + 58) = 0 \qquad \text{Factor}$$

$$w = 50 \quad \text{or} \quad w = -58 \qquad \text{Zero-Product Property}$$

Since the width of the lot must be a positive number, we conclude that $w = 50$ ft. The length of the lot is $w + 8 = 50 + 8 = 58$ ft.

Now Try Exercise 41 ■

EXAMPLE 5 ■ Determining the Height of a Building Using Similar Triangles

A man who is 6 ft tall wishes to find the height of a certain four-story building. He measures its shadow and finds it to be 28 ft long, while his own shadow is $3\frac{1}{2}$ ft long. How tall is the building?

SOLUTION **Identify the variable.** The problem asks for the height of the building. So let

$$h = \text{the height of the building}$$

Translate from words to algebra. We use the fact that the triangles in Figure 3 are similar. Recall that for any pair of similar triangles the ratios of corresponding sides are equal. Now we translate these observations into the language of algebra.

In Words	In Algebra
Height of building	h
Ratio of height to base in large triangle	$\frac{h}{28}$
Ratio of height to base in small triangle	$\frac{6}{3.5}$

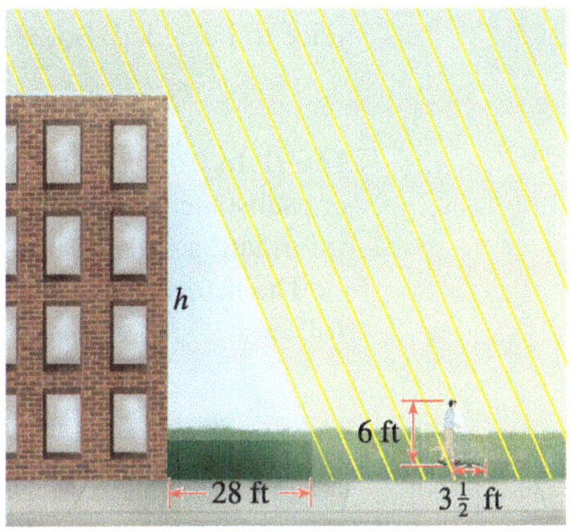

FIGURE 3

Set up the model. Since the large and small triangles are similar, we get the equation

$$\text{ratio of height to base in large triangle} = \text{ratio of height to base in small triangle}$$

$$\frac{h}{28} = \frac{6}{3.5}$$

Solve. Now we solve for h.

$$h = \frac{6 \cdot 28}{3.5} = 48 \qquad \text{Multiply by 28}$$

So the building is 48 ft tall.

Now Try Exercise 53 ■

■ Problems About Mixtures

Many real-world problems involve mixing different types of substances. For example, construction workers may mix cement, gravel, and sand; fruit juice from concentrate may involve mixing different types of juices. Problems involving mixtures and concentrations make use of the fact that if an amount x of a substance is dissolved in a solution with volume V, then the concentration C of the substance is given by

$$C = \frac{x}{V}$$

So if 10 g of sugar is dissolved in 5 L of water, then the sugar concentration is $C = 10/5 = 2$ g/L. Solving a mixture problem usually requires us to analyze the amount x of the substance that is in the solution. When we solve for x in this equation, we see that $x = CV$. Note that in many mixture problems the concentration C is expressed as a percentage, as in the next example.

EXAMPLE 6 ■ Mixtures and Concentration

A manufacturer of soft drinks advertises their orange soda as "naturally flavored," although it contains only 5% orange juice. A new federal regulation stipulates that to be called "natural," a drink must contain at least 10% fruit juice. How much pure orange juice must this manufacturer add to 900 gal of orange soda to conform to the new regulation?

SOLUTION **Identify the variable.** The problem asks for the amount of pure orange juice to be added. So let

$x =$ the amount (in gallons) of pure orange juice to be added

Translate from words to algebra. In any problem of this type—in which two different substances are to be mixed—drawing a diagram helps us to organize the given information (see Figure 4).

The information in the figure can be translated into the language of algebra, as follows.

In Words	In Algebra
Amount of orange juice to be added	x
Amount of the mixture	$900 + x$
Amount of orange juice in the first vat	$0.05(900) = 45$
Amount of orange juice in the second vat	$1 \cdot x = x$
Amount of orange juice in the mixture	$0.10(900 + x)$

FIGURE 4

Set up the model. To set up the model, we use the fact that the total amount of orange juice in the mixture is equal to the orange juice in the first two vats.

amount of orange juice in first vat	+	amount of orange juice in second vat	=	amount of orange juice in mixture

$$45 + x = 0.1(900 + x) \quad \text{From Figure 4}$$

Solve. Now we solve for x.

$$45 + x = 90 + 0.1x \quad \text{Distributive Property}$$
$$0.9x = 45 \quad \text{Subtract } 0.1x \text{ and } 45$$
$$x = \frac{45}{0.9} = 50 \quad \text{Divide by } 0.9$$

The manufacturer should add 50 gal of pure orange juice to the soda.

CHECK YOUR ANSWER

$$\text{amount of juice before mixing} = 5\% \text{ of } 900 \text{ gal} + 50 \text{ gal pure juice}$$
$$= 45 \text{ gal} + 50 \text{ gal} = 95 \text{ gal}$$
$$\text{amount of juice after mixing} = 10\% \text{ of } 950 \text{ gal} = 95 \text{ gal}$$

Amounts are equal. ✓

Now Try Exercise 55 ■

Problems About the Time Needed to Do a Job

When solving a problem that involves determining how long it takes several workers to complete a job, we use the fact that if a person or machine takes H time units to complete the task, then in one time unit the fraction of the task that has been completed is $1/H$. For example, if a worker takes 5 hours to mow a lawn, then in 1 hour the worker will mow $1/5$ of the lawn.

EXAMPLE 7 ■ Time Needed to Do a Job

Because of an anticipated heavy rainstorm, the water level in a reservoir must be lowered by 1 ft. Opening spillway A lowers the level by this amount in 4 hours, whereas opening the smaller spillway B does the job in 6 hours. How long will it take to lower the water level by 1 ft if both spillways are opened?

SOLUTION **Identify the variable.** We are asked to find the time needed to lower the level by 1 ft if both spillways are open. So let

x = the time (in hours) it takes to lower the water level by 1 ft if both spillways are open

Translate from words to algebra. Finding an equation relating x to the other quantities in this problem is not easy. Certainly x is not simply $4 + 6$, because that would mean that together the two spillways require longer to lower the water level than either spillway alone. Instead, *we look at the fraction of the job that can be done in 1 hour by each spillway.*

In Words	In Algebra
Time it takes to lower level 1 ft with A and B together	x h
Distance A lowers level in 1 h	$\frac{1}{4}$ ft
Distance B lowers level in 1 h	$\frac{1}{6}$ ft
Distance A and B together lower levels in 1 h	$\frac{1}{x}$ ft

Set up the model. Now we set up the model.

$$\text{fraction done by A} + \text{fraction done by B} = \text{fraction done by both}$$

$$\frac{1}{4} + \frac{1}{6} = \frac{1}{x}$$

Solve. Now we solve for x.

$$3x + 2x = 12 \qquad \text{Multiply by the LCD, } 12x$$

$$5x = 12 \qquad \text{Add}$$

$$x = \frac{12}{5} \qquad \text{Divide by 5}$$

It will take $2\frac{2}{5}$ hours, or 2 h 24 min, to lower the water level by 1 ft if both spillways are open.

Now Try Exercise 63 ■

Problems About Distance, Rate, and Time

The next example deals with distance, rate (speed), and time. The formula to keep in mind here is

$$\text{distance} = \text{rate} \times \text{time}$$

where the rate is either the constant speed or average speed of a moving object. For example, driving at 60 mi/h for 4 hours takes you a distance of $60 \cdot 4 = 240$ mi.

EXAMPLE 8 ■ A Distance-Speed-Time Problem

A jet flew from New York to Los Angeles, a distance of 4200 km. The speed for the return trip was 100 km/h faster than the outbound speed. If the total trip took 13 hours of flying time, what was the jet's speed from New York to Los Angeles?

SOLUTION **Identify the variable.** We are asked for the speed of the jet from New York to Los Angeles. So let

$$s = \text{speed from New York to Los Angeles}$$

Then $$s + 100 = \text{speed from Los Angeles to New York}$$

Translate from words to algebra. Now we organize the information in a table. We fill in the "Distance" column first, since we know that the cities are 4200 km apart. Then we fill in the "Speed" column, since we have expressed both speeds (rates) in terms of the variable s. Finally, we calculate the entries for the "Time" column, using

$$\text{time} = \frac{\text{distance}}{\text{rate}}$$

	Distance (km)	Speed (km/h)	Time (h)
N.Y. to L.A.	4200	s	$\frac{4200}{s}$
L.A. to N.Y.	4200	$s + 100$	$\frac{4200}{s+100}$

Set up the model. The total trip took 13 hours, so we have the model

$$\text{time from N.Y. to L.A.} + \text{time from L.A. to N.Y.} = \text{total time}$$

$$\frac{4200}{s} + \frac{4200}{s+100} = 13$$

Solve. Multiplying by the common denominator, $s(s + 100)$, we get

$$4200(s + 100) + 4200s = 13s(s + 100)$$

$$8400s + 420{,}000 = 13s^2 + 1300s$$

$$0 = 13s^2 - 7100s - 420{,}000$$

Although this equation does factor, with numbers this large it is probably quicker to use the Quadratic Formula and a calculator.

$$s = \frac{7100 \pm \sqrt{(-7100)^2 - 4(13)(-420{,}000)}}{2(13)}$$

$$= \frac{7100 \pm 8500}{26}$$

$$s = 600 \quad \text{or} \quad s = \frac{-1400}{26} \approx -53.8$$

Since s represents speed, we reject the negative answer and conclude that the jet's speed from New York to Los Angeles was 600 km/h.

Now Try Exercise 69 ■

EXAMPLE 9 ■ Energy Expended in Bird Flight

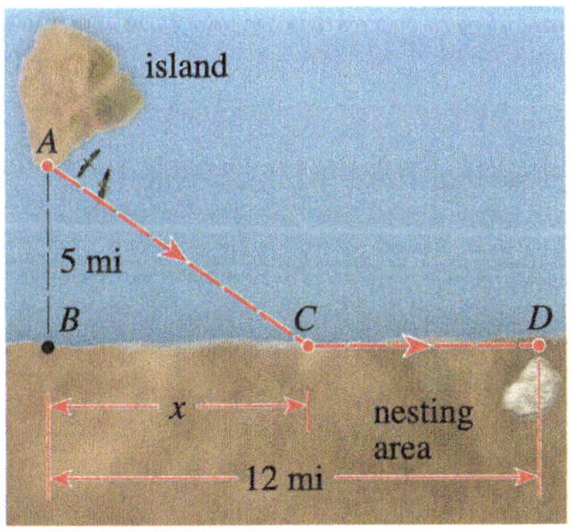

FIGURE 5

Ornithologists have determined that some species of birds tend to avoid flights over large bodies of water during daylight hours, because air generally rises over land and falls over water in the daytime, so flying over water requires more energy. A bird is released from point A on an island, 5 mi from B, the nearest point on a straight shoreline. The bird flies to a point C on the shoreline and then flies along the shoreline to its nesting area D, as shown in Figure 5. Suppose the bird has 170 kcal of energy reserves. It uses 10 kcal/mi flying over land and 14 kcal/mi flying over water.

(a) Where should the point C be located so that the bird uses exactly 170 kcal of energy during its flight?

(b) Does the bird have enough energy reserves to fly directly from A to D?

BHASKARA (born 1114) was an Indian mathematician, astronomer, and astrologer. Among his many accomplishments was an ingenious proof of the Pythagorean Theorem. (See *Focus on Problem Solving* 5, Problem 12, at the book companion website **www.stewartmath.com**.) His important mathematical book *Lilavati* [*The Beautiful*] consists of algebra problems posed in the form of stories to his daughter Lilavati. Many of the problems begin "Oh beautiful maiden, suppose . . ." The story is told that using astrology, Bhaskara had determined that great misfortune would befall his daughter if she married at any time other than at a certain hour of a certain day. On her wedding day, as she was anxiously watching the water clock, a pearl fell unnoticed from her headdress. It stopped the flow of water in the clock, causing her to miss the opportune moment for marriage. Bhaskara's *Lilavati* was written to console her.

SOLUTION

(a) **Identify the variable.** We are asked to find the location of C. So let

$$x = \text{distance from } B \text{ to } C$$

Translate from words to algebra. From the figure, and from the fact that

$$\text{energy used} = \text{energy per mile} \times \text{miles flown}$$

we determine the following:

In Words	In Algebra	
Distance from B to C	x	
Distance flown over water (from A to C)	$\sqrt{x^2 + 25}$	Pythagorean Theorem
Distance flown over land (from C to D)	$12 - x$	
Energy used over water	$14\sqrt{x^2 + 25}$	
Energy used over land	$10(12 - x)$	

Set up the model. Now we set up the model.

$$\text{total energy used} = \text{energy used over water} + \text{energy used over land}$$

$$170 = 14\sqrt{x^2 + 25} + 10(12 - x)$$

Solve. To solve this equation, we eliminate the square root by first bringing all other terms to the left of the equal sign and then squaring each side.

$$170 - 10(12 - x) = 14\sqrt{x^2 + 25} \quad \text{Isolate square-root term on RHS}$$

$$50 + 10x = 14\sqrt{x^2 + 25} \quad \text{Simplify LHS}$$

$$(50 + 10x)^2 = (14)^2(x^2 + 25) \quad \text{Square each side}$$

$$2500 + 1000x + 100x^2 = 196x^2 + 4900 \quad \text{Expand}$$

$$0 = 96x^2 - 1000x + 2400 \quad \text{Move all terms to RHS}$$

This equation could be factored, but because the numbers are so large, it is easier to use the Quadratic Formula and a calculator.

$$x = \frac{1000 \pm \sqrt{(-1000)^2 - 4(96)(2400)}}{2(96)}$$

$$= \frac{1000 \pm 280}{192} = 6\tfrac{2}{3} \quad \text{or} \quad 3\tfrac{3}{4}$$

Point C should be either $6\frac{2}{3}$ mi or $3\frac{3}{4}$ mi from B so that the bird uses exactly 170 kcal of energy during its flight.

See Appendix A, *Geometry Review*, for the Pythagorean Theorem.

(b) By the Pythagorean Theorem the length of the route directly from A to D is $\sqrt{5^2 + 12^2} = 13$ mi, so the energy the bird requires for that route is $14 \times 13 = 182$ kcal. This is more energy than the bird has available, so it can't use this route.

Now Try Exercise 85 ■

1.7 EXERCISES

CONCEPTS

1. Explain in your own words what it means for an equation to model a real-world situation, and give an example.

2. In the formula $I = Prt$ for simple interest, P stands for ________, r for ________, and t for ________.

3. Give a formula for the area of the geometric figure.

   (a) A square of side x: $A =$ ________.

   (b) A rectangle of length l and width w: $A =$ ________.

   (c) A circle of radius r: $A =$ ________.

4. Balsamic vinegar contains 5% acetic acid, so a 32-oz bottle of balsamic vinegar contains ________ ounces of acetic acid.

5. A painter paints a wall in x hours, so the fraction of the wall that she paints in 1 hour is ________.

6. The formula $d = rt$ models the distance d traveled by an object moving at the constant rate r in time t. Find formulas for the following quantities.

   $r =$ ________ $t =$ ________

SKILLS

7–20 ■ Using Variables Express the given quantity in terms of the indicated variable.

7. The sum of three consecutive integers; $n =$ first integer of the three

8. The sum of three consecutive integers; $n =$ middle integer of the three

9. The sum of three consecutive even integers; $n =$ first integer of the three

10. The sum of the squares of two consecutive integers; $n =$ first integer of the two

11. The average of three test scores if the first two scores are 78 and 82; $s =$ third test score

12. The average of four quiz scores if each of the first three scores is 8; $q =$ fourth quiz score

13. The interest obtained after 1 year on an investment at $2\frac{1}{2}\%$ simple interest per year; $x =$ number of dollars invested

14. The total rent paid for an apartment if the rent is \$795 a month; $n =$ number of months

15. The area (in ft^2) of a rectangle that is four times as long as it is wide; $w =$ width of the rectangle (in ft)

16. The perimeter (in cm) of a rectangle that is 6 cm longer than it is wide; $w =$ width of the rectangle (in cm)

17. The time (in hours) it takes to travel a given distance at 55 mi/h; $d =$ given distance (in mi)

18. The distance (in mi) that a car travels in 45 min; $s =$ speed of the car (in mi/h)

19. The concentration (in oz/gal) of salt in a mixture of 3 gal of brine containing 25 oz of salt to which some pure water has been added; $x =$ volume of pure water added (in gal)

20. The value (in cents) of the change in a purse that contains twice as many nickels as pennies, four more dimes than nickels, and as many quarters as dimes and nickels combined; $p =$ number of pennies

APPLICATIONS

21. **Renting a Truck** A rental company charges \$65 a day and 20 cents a mile for renting a truck. Michael rented a truck for 3 days, and his bill came to \$275. How many miles did he drive?

22. **Cell Phone Costs** A cell phone company charges a monthly fee of \$10 for the first 1000 text messages and 10 cents for each additional text message. Miriam's bill for text messages for the month of June is \$38.50. How many text messages did she send that month?

23. **Average** Linh has obtained scores of 82, 75, and 71 on her midterm algebra exams. If the final exam counts twice as much as a midterm, what score must she make on her final exam to get an average score of 80? (Assume that the maximum possible score on each test is 100.)

24. **Average** In a class of 25 students, the average score is 84. Six students in the class each received a maximum score of 100, and three students each received a score of 60. What is the average score of the remaining students?

25. **Investments** Phyllis invested \$12,000, a portion earning a simple interest rate of $4\frac{1}{2}\%$ per year and the rest earning a rate of 4% per year. After 1 year the total interest earned on these investments was \$525. How much money did she invest at each rate?

26. **Investments** If Ben invests \$4000 at 4% interest per year, how much additional money must he invest at $5\frac{1}{2}\%$ annual interest to ensure that the interest he receives each year is $4\frac{1}{2}\%$ of the total amount invested?

27. **Investments** What annual rate of interest would you have to earn on an investment of \$3500 to ensure receiving \$262.50 interest after 1 year?

28. **Investments** Jack invests \$1000 at a certain annual interest rate, and he invests another \$2000 at an annual rate that is one-half percent higher. If he receives a total of \$190 interest in 1 year, at what rate is the \$1000 invested?

29. **Salaries** An executive in an engineering firm earns a monthly salary plus a Christmas bonus of \$8500. If she earns a total of \$97,300 per year, what is her monthly salary?

30. **Salaries** A woman earns 15% more than her husband. Together they make \$69,875 per year. What is the husband's annual salary?

31. **Overtime Pay** Helen earns \$7.50 an hour at her job, but if she works more than 35 hours in a week, she is paid $1\frac{1}{2}$ times her regular salary for the overtime hours worked. One week her gross pay was \$352.50. How many overtime hours did she work that week?

32. Labor Costs A plumber and his assistant work together to replace the pipes in an old house. The plumber charges \$45 an hour for his own labor and \$25 an hour for his assistant's labor. The plumber works twice as long as his assistant on this job, and the labor charge on the final bill is \$4025. How long did the plumber and his assistant work on this job?

33. A Riddle A movie star, unwilling to give his age, posed the following riddle to a gossip columnist: "Seven years ago, I was eleven times as old as my daughter. Now I am four times as old as she is." How old is the movie star?

34. Career Home Runs During his major league career, Hank Aaron hit 41 more home runs than Babe Ruth hit during his career. Together they hit 1469 home runs. How many home runs did Babe Ruth hit?

35. Value of Coins A change purse contains an equal number of pennies, nickels, and dimes. The total value of the coins is \$1.44. How many coins of each type does the purse contain?

36. Value of Coins Mary has \$3.00 in nickels, dimes, and quarters. If she has twice as many dimes as quarters and five more nickels than dimes, how many coins of each type does she have?

37. Length of a Garden A rectangular garden is 25 ft wide. If its area is 1125 ft^2, what is the length of the garden?

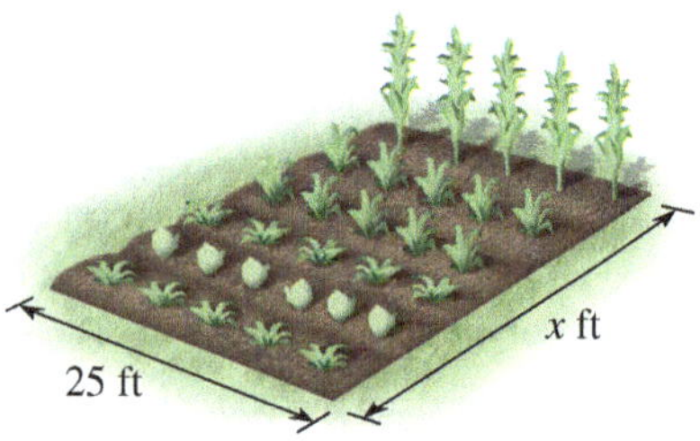

38. Width of a Pasture A pasture is twice as long as it is wide. Its area is 115,200 ft^2. How wide is the pasture?

39. Dimensions of a Lot A square plot of land has a building 60 ft long and 40 ft wide at one corner. The rest of the land outside the building forms a parking lot. If the parking lot has area 12,000 ft^2, what are the dimensions of the entire plot of land?

40. Dimensions of a Lot A half-acre building lot is five times as long as it is wide. What are its dimensions? [*Note:* 1 acre = 43,560 ft^2.]

41. Dimensions of a Garden A rectangular garden is 10 ft longer than it is wide. Its area is 875 ft^2. What are its dimensions?

42. Dimensions of a Room A rectangular bedroom is 7 ft longer than it is wide. Its area is 228 ft^2. What is the width of the room?

43. Dimensions of a Garden A farmer has a rectangular garden plot surrounded by 200 ft of fence. Find the length and width of the garden if its area is 2400 ft^2.

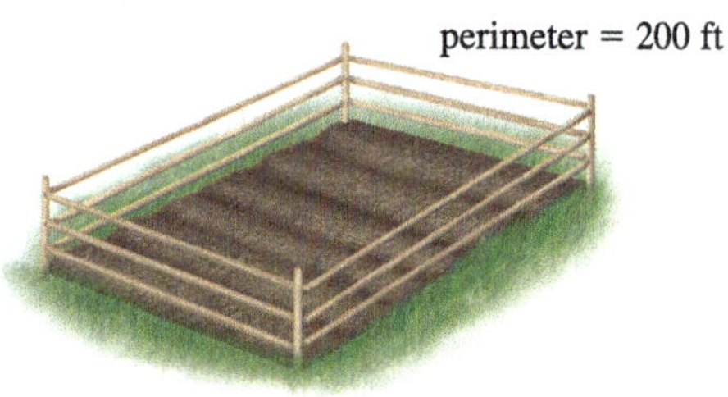

44. Dimensions of a Lot A parcel of land is 6 ft longer than it is wide. Each diagonal from one corner to the opposite corner is 174 ft long. What are the dimensions of the parcel?

45. Dimensions of a Lot A rectangular parcel of land is 50 ft wide. The length of a diagonal between opposite corners is 10 ft more than the length of the parcel. What is the length of the parcel?

46. Dimensions of a Track A running track has the shape shown in the figure, with straight sides and semicircular ends. If the length of the track is 440 yd and the two straight parts are each 110 yd long, what is the radius of the semicircular parts (to the nearest yard)?

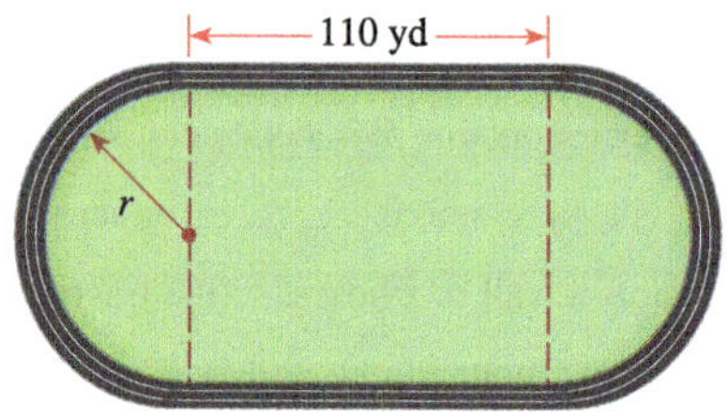

47. Length and Area Find the length x in the figure. The area of the shaded region is given.

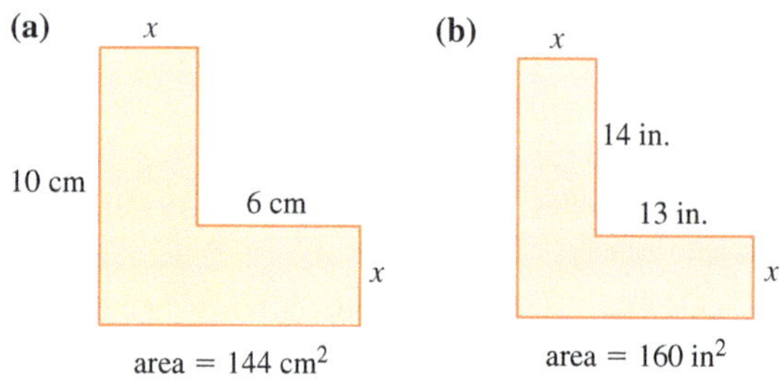

48. Length and Area Find the length y in the figure. The area of the shaded region is given.

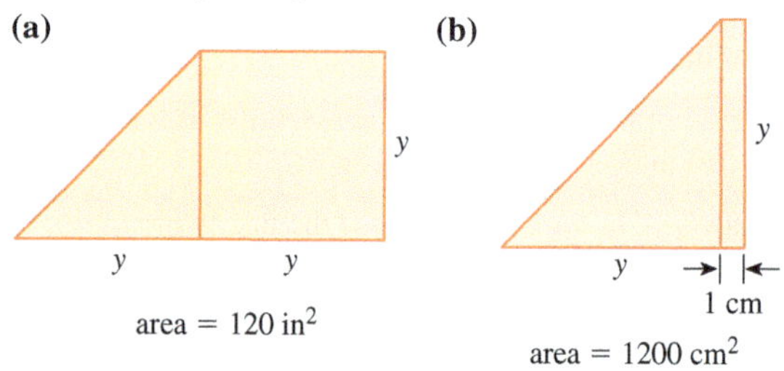

49. Framing a Painting Ali paints with watercolors on a sheet of paper 20 in. wide by 15 in. high. He then places this sheet on a mat so that a uniformly wide strip of the mat shows all around the picture. The perimeter of the mat is 102 in. How wide is the strip of the mat showing around the picture?

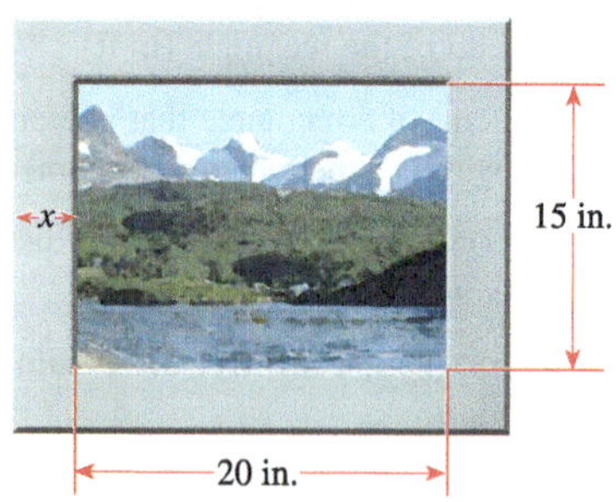

50. Dimensions of a Poster A poster has a rectangular printed area 100 cm by 140 cm and a blank strip of uniform width around the edges. The perimeter of the poster is $1\frac{1}{2}$ times the perimeter of the printed area. What is the width of the blank strip?

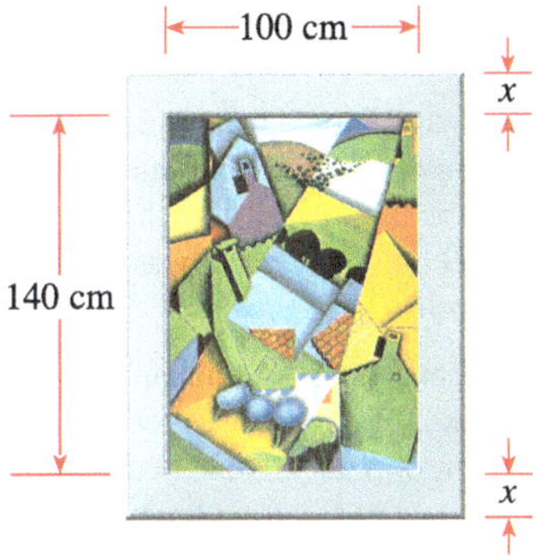

51. Reach of a Ladder A $19\frac{1}{2}$-foot ladder leans against a building. The base of the ladder is $7\frac{1}{2}$ ft from the building. How high up the building does the ladder reach?

52. Height of a Flagpole A flagpole is secured on opposite sides by two guy wires, each of which is 5 ft longer than the pole. The distance between the points where the wires are fixed to the ground is equal to the length of one guy wire. How tall is the flagpole (to the nearest inch)?

53. Length of a Shadow A man is walking away from a lamppost with a light source 6 m above the ground. The man is 2 m tall. How long is the man's shadow when he is 10 m from the lamppost? [*Hint:* Use similar triangles.]

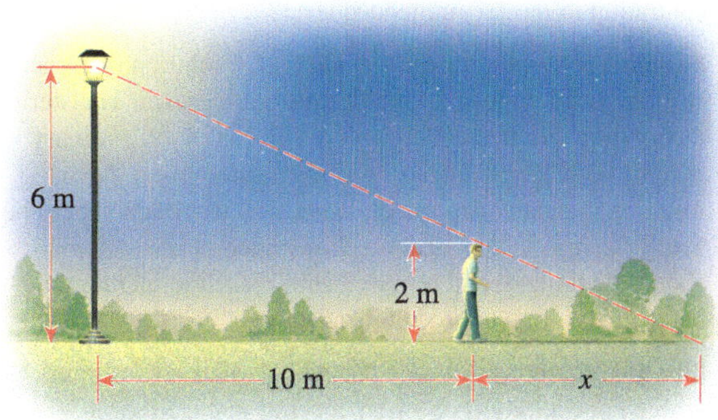

54. Height of a Tree A woodcutter determines the height of a tall tree by first measuring a smaller one 125 ft away, then moving so that his eyes are in the line of sight along the tops of the trees and measuring how far he is standing from the small tree (see the figure). Suppose the small tree is 20 ft tall, the man is 25 ft from the small tree, and his eye level is 5 ft above the ground. How tall is the taller tree?

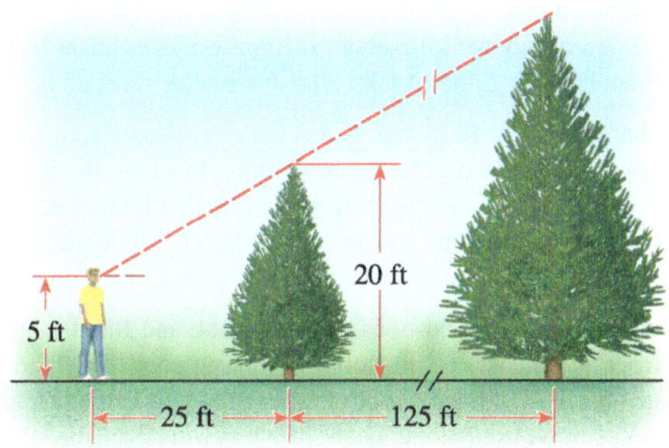

55. Mixture Problem What amount of a 60% acid solution must be mixed with a 30% solution to produce 300 mL of a 50% solution?

56. Mixture Problem What amount of pure acid must be added to 300 mL of a 50% acid solution to produce a 60% acid solution?

57. Mixture Problem A jeweler has five rings, each weighing 18 g, made of an alloy of 10% silver and 90% gold. She decides to melt down the rings and add enough silver to reduce the gold content to 75%. How much silver should she add?

58. Mixture Problem A pot contains 6 L of brine at a concentration of 120 g/L. How much of the water should be boiled off to increase the concentration to 200 g/L?

59. Mixture Problem The radiator in a car is filled with a solution of 60% antifreeze and 40% water. The manufacturer of the antifreeze suggests that for summer driving, optimal cooling of the engine is obtained with only 50% antifreeze. If the capacity of the radiator is 3.6 L, how much coolant should be drained and replaced with water to reduce the antifreeze concentration to the recommended level?

60. Mixture Problem A health clinic uses a solution of bleach to sterilize petri dishes in which cultures are grown. The sterilization tank contains 100 gal of a solution of 2% ordinary household bleach mixed with pure distilled water. New research indicates that the concentration of bleach should be 5% for complete sterilization. How much of the solution should be drained and replaced with bleach to increase the bleach content to the recommended level?

61. Mixture Problem A bottle contains 750 mL of fruit punch with a concentration of 50% pure fruit juice. Jill drinks 100 mL of the punch and then refills the bottle with an equal amount of a cheaper brand of punch. If the concentration of juice in the bottle is now reduced to 48%, what was the concentration in the punch that Jill added?

62. Mixture Problem A merchant blends tea that sells for $3.00 an ounce with tea that sells for $2.75 an ounce to produce 80 oz of a mixture that sells for $2.90 an ounce. How many ounces of each type of tea does the merchant use in the blend?

63. Sharing a Job Candy and Tim share a paper route. It takes Candy 70 min to deliver all the papers, and it takes Tim 80 min. How long does it take the two when they work together?

64. Sharing a Job Stan and Hilda can mow the lawn in 40 min if they work together. If Hilda works twice as fast as Stan, how long does it take Stan to mow the lawn alone?

65. Sharing a Job Betty and Karen have been hired to paint the houses in a new development. Working together, the women can paint a house in two-thirds the time that it takes Karen working alone. Betty takes 6 h to paint a house alone. How long does it take Karen to paint a house working alone?

66. Sharing a Job Next-door neighbors Bob and Jim use hoses from both houses to fill Bob's swimming pool. They know that it takes 18 h using both hoses. They also know that Bob's hose, used alone, takes 20% less time than Jim's hose alone. How much time is required to fill the pool by each hose alone?

67. Sharing a Job Henry and Irene working together can wash all the windows of their house in 1 h 48 min. Working alone, it takes Henry $1\frac{1}{2}$ h more than Irene to do the job. How long does it take each person working alone to wash all the windows?

68. Sharing a Job Jack, Kay, and Lynn deliver advertising flyers in a small town. If each person works alone, it takes Jack 4 h to deliver all the flyers, and it takes Lynn 1 h longer than it takes Kay. Working together, they can deliver all the flyers in 40% of the time it takes Kay working alone. How long does it take Kay to deliver all the flyers alone?

69. Distance, Speed, and Time Wendy took a trip from Davenport to Omaha, a distance of 300 mi. She traveled part of the way by bus, which arrived at the train station just in time for Wendy to complete her journey by train. The bus averaged 40 mi/h, and the train averaged 60 mi/h. The entire trip took $5\frac{1}{2}$ h. How long did Wendy spend on the train?

70. Distance, Speed, and Time Two cyclists, 90 mi apart, start riding toward each other at the same time. One cycles twice as fast as the other. If they meet 2 h later, at what average speed is each cyclist traveling?

71. Distance, Speed, and Time A pilot flew a jet from Montreal to Los Angeles, a distance of 2500 mi. On the return trip, the average speed was 20% faster than the outbound speed. The round-trip took 9 h 10 min. What was the speed from Montreal to Los Angeles?

72. Distance, Speed, and Time A woman driving a car 14 ft long is passing a truck 30 ft long. The truck is traveling at 50 mi/h. How fast must the woman drive her car so that she can pass the truck completely in 6 s, from the position shown in figure (a) to the position shown in figure (b)? [*Hint:* Use feet and seconds instead of miles and hours.]

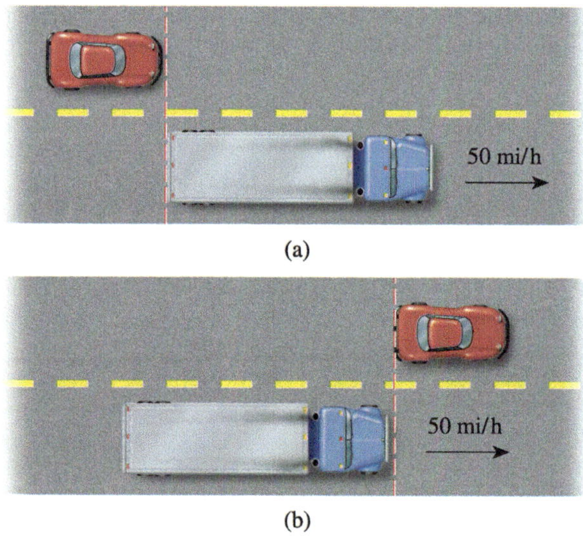

73. Distance, Speed, and Time A salesman drives from Ajax to Barrington, a distance of 120 mi, at a steady speed. He then increases his speed by 10 mi/h to drive the 150 mi from Barrington to Collins. If the second leg of his trip took 6 min more time than the first leg, how fast was he driving between Ajax and Barrington?

74. Distance, Speed, and Time Kiran drove from Tortula to Cactus, a distance of 250 mi. She increased her speed by 10 mi/h for the 360-mi trip from Cactus to Dry Junction. If the total trip took 11 h, what was her speed from Tortula to Cactus?

75. Distance, Speed, and Time It took a crew 2 h 40 min to row 6 km upstream and back again. If the rate of flow of the stream was 3 km/h, what was the rowing speed of the crew in still water?

76. Speed of a Boat Two fishing boats depart a harbor at the same time, one traveling east, the other south. The eastbound boat travels at a speed 3 mi/h faster than the southbound boat. After 2 h the boats are 30 mi apart. Find the speed of the southbound boat.

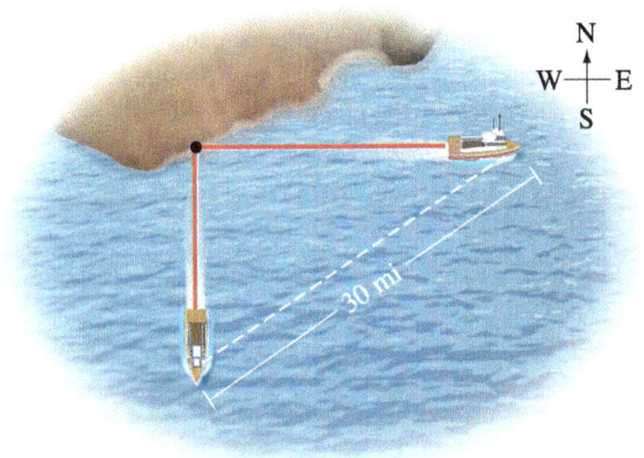

77. Law of the Lever The figure shows a lever system, similar to a seesaw that you might find in a children's playground. For the system to balance, the product of the weight and its distance from the fulcrum must be the same on each side; that is,

$$w_1x_1 = w_2x_2$$

This equation is called the **law of the lever** and was first discovered by Archimedes (see page 787).

A woman and her son are playing on a seesaw. The boy is at one end, 8 ft from the fulcrum. If the son weighs 100 lb and the mother weighs 125 lb, where should the woman sit so that the seesaw is balanced?

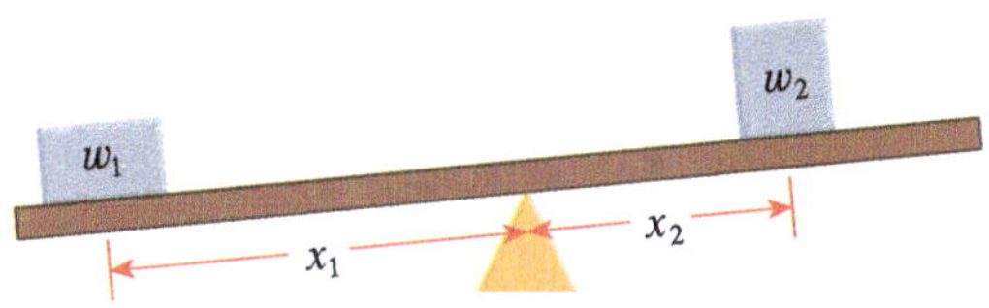

78. Law of the Lever A plank 30 ft long rests on top of a flat-roofed building, with 5 ft of the plank projecting over the edge, as shown in the figure. A worker weighing 240 lb sits on one end of the plank. What is the largest weight that can be hung on the projecting end of the plank if it is to remain in balance? (Use the law of the lever stated in Exercise 77.)

79. Dimensions of a Box A large plywood box has a volume of 180 ft^3. Its length is 9 ft greater than its height, and its width is 4 ft less than its height. What are the dimensions of the box?

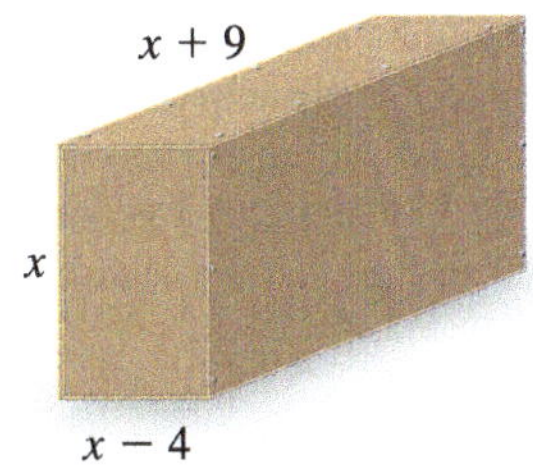

80. Radius of a Sphere A jeweler has three small solid spheres made of gold, of radius 2 mm, 3 mm, and 4 mm. He decides to melt these down and make just one sphere out of them. What will the radius of this larger sphere be?

81. Dimensions of a Box A box with a square base and no top is to be made from a square piece of cardboard by cutting 4-in. squares from each corner and folding up the sides, as shown in the figure. The box is to hold 100 in^3. How big a piece of cardboard is needed?

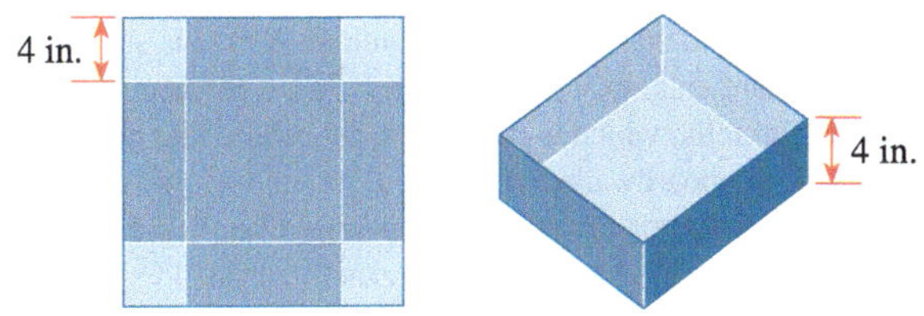

82. Dimensions of a Can A cylindrical can has a volume of 40π cm^3 and is 10 cm tall. What is its diameter? [*Hint:* Use the volume formula listed on the inside front cover of this book.]

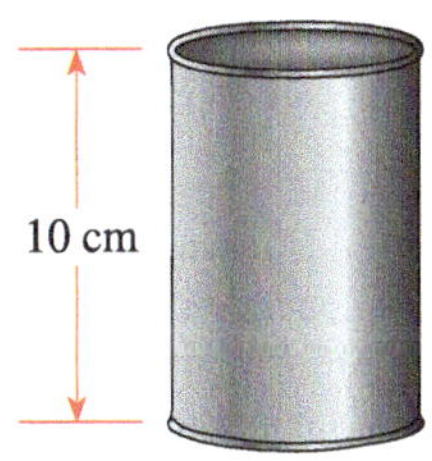

83. Radius of a Tank A spherical tank has a capacity of 750 gallons. Using the fact that one gallon is about 0.1337 ft^3, find the radius of the tank (to the nearest hundredth of a foot).

84. Dimensions of a Lot A city lot has the shape of a right triangle whose hypotenuse is 7 ft longer than one of the other sides. The perimeter of the lot is 392 ft. How long is each side of the lot?

85. Construction Costs The town of Foxton lies 10 mi north of an abandoned east-west road that runs through Grimley, as shown in the figure. The point on the abandoned road closest to Foxton is 40 mi from Grimley. County officials are about to build a new road connecting the two towns. They have determined that restoring the old road would cost \$100,000 per mile, whereas building a new road would cost \$200,000 per mile. How much of the abandoned road should be used (as indicated in the figure) if the officials intend to spend exactly \$6.8 million? Would it cost less than this amount to build a new road connecting the towns directly?

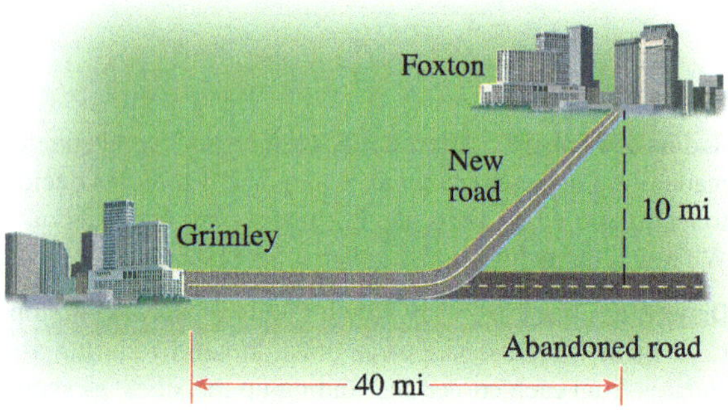

86. Distance, Speed, and Time A boardwalk is parallel to and 210 ft inland from a straight shoreline. A sandy beach lies between the boardwalk and the shoreline. A man is standing on the boardwalk, exactly 750 ft across the sand from his beach umbrella, which is right at the shoreline. The man walks 4 ft/s on the boardwalk and 2 ft/s on the sand. How far should he walk on the boardwalk before veering off onto the sand if he wishes to reach his umbrella in exactly 4 min 45 s?

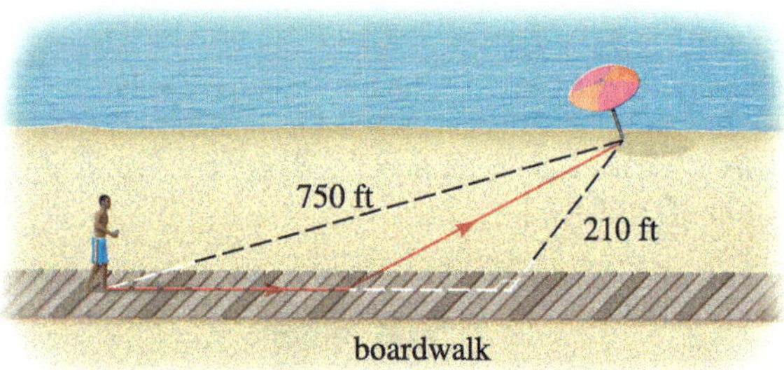

87. Volume of Grain Grain is falling from a chute onto the ground, forming a conical pile whose diameter is always three times its height. How high is the pile (to the nearest hundredth of a foot) when it contains 1000 ft^3 of grain?

88. Computer Monitors Two computer monitors sitting side by side on a shelf in an appliance store have the same screen height. One has a screen that is 7 in. wider than it is high. The other has a wider screen that is 1.8 times as wide as it is high. The diagonal measure of the wider screen is 3 in. more than the diagonal measure of the smaller screen. What is the height of the screens, correct to the nearest 0.1 in.?

89. Dimensions of a Structure A storage bin for corn consists of a cylindrical section made of wire mesh, surmounted by a conical tin roof, as shown in the figure. The height of the roof is one-third the height of the entire structure. If the total volume of the structure is 1400π ft^3 and its radius is 10 ft, what is its height? [*Hint:* Use the volume formulas listed on the inside front cover of this book.]

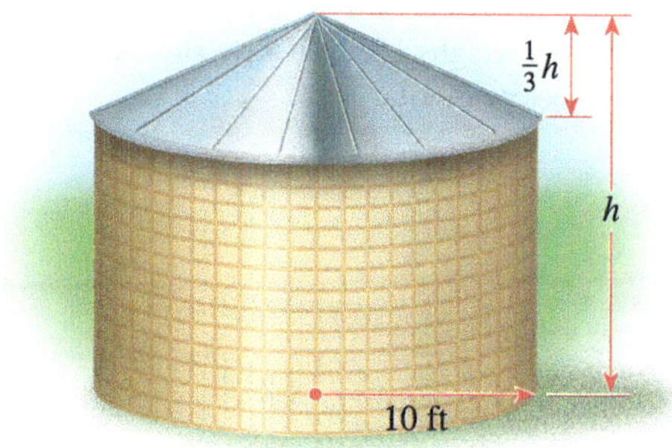

90. Comparing Areas A wire 360 in. long is cut into two pieces. One piece is formed into a square, and the other is formed into a circle. If the two figures have the same area, what are the lengths of the two pieces of wire (to the nearest tenth of an inch)?

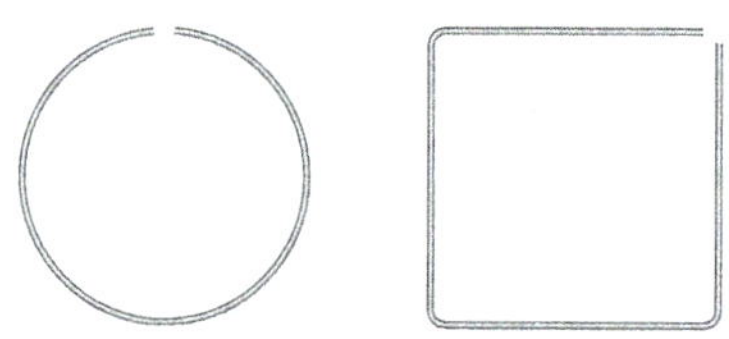

91. An Ancient Chinese Problem This problem is taken from a Chinese mathematics textbook called *Chui-chang suan-shu*, or *Nine Chapters on the Mathematical Art*, which was written about 250 B.C.

A 10-ft-long stem of bamboo is broken in such a way that its tip touches the ground 3 ft from the base of the stem, as shown in the figure. What is the height of the break?

[*Hint:* Use the Pythagorean Theorem.]

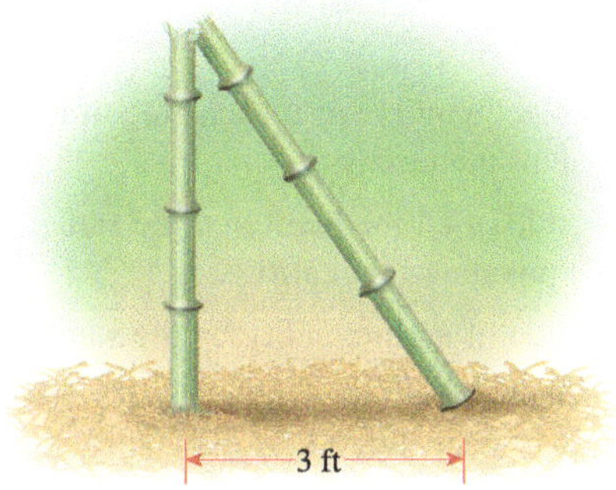

DISCUSS ■ DISCOVER ■ PROVE ■ WRITE

92. WRITE: Historical Research Read the biographical notes on Pythagoras (page 241), Euclid (page 542), and Archimedes (page 787). Choose one of these mathematicians, and find out more about him from the library or on the Internet. Write a short essay on your findings. Include both biographical information and a description of the mathematics for which he is famous.

93. WRITE: Real-world Equations In this section we learned how to translate words into algebra. In this exercise we try to find real-world situations that could correspond to an algebraic equation. For instance, the equation $A = (x + y)/2$ could model the average amount of money in two bank accounts, where x represents the amount in one account and y the amount in the other. Write a story that could correspond to the given equation, stating what the variables represent.

(a) $C = 20{,}000 + 4.50x$

(b) $A = w(w + 10)$

(c) $C = 10.50x + 11.75y$

94. DISCUSS: A Babylonian Quadratic Equation The ancient Babylonians knew how to solve quadratic equations. Here is a problem from a cuneiform tablet found in a Babylonian school dating back to about 2000 B.C.

> I have a reed, I know not its length. I broke from it one cubit, and it fit 60 times along the length of my field. I restored to the reed what I had broken off, and it fit 30 times along the width of my field. The area of my field is 375 square nindas. What was the original length of the reed?

Solve this problem. Use the fact that 1 ninda = 12 cubits.

1.8 INEQUALITIES

■ Solving Linear Inequalities ■ Solving Nonlinear Inequalities ■ Absolute Value Inequalities ■ Modeling with Inequalities

Some problems in algebra lead to **inequalities** instead of equations. An inequality looks just like an equation, except that in the place of the equal sign is one of the symbols, $<$, $>$, $\le$, or $\ge$. Here is an example of an inequality:

$$4x + 7 \le 19$$

x	$4x + 7 \le 19$
1	$11 \le 19$ ✓
2	$15 \le 19$ ✓
3	$19 \le 19$ ✓
4	$23 \le 19$ ✗
5	$27 \le 19$ ✗

The table in the margin shows that some numbers satisfy the inequality and some numbers don't.

To **solve** an inequality that contains a variable means to find all values of the variable that make the inequality true. Unlike an equation, an inequality generally has infinitely many solutions, which form an interval or a union of intervals on the real line. The following illustration shows how an inequality differs from its corresponding equation:

		Solution	**Graph**
Equation:	$4x + 7 = 19$	$x = 3$	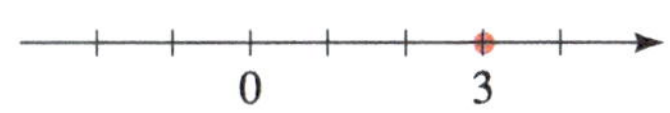
Inequality:	$4x + 7 \le 19$	$x \le 3$	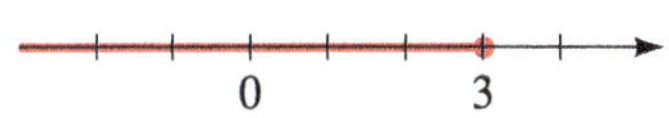

To solve inequalities, we use the following rules to isolate the variable on one side of the inequality sign. These rules tell us when two inequalities are *equivalent* (the symbol $\Leftrightarrow$ means "is equivalent to"). In these rules the symbols A, B, and C stand for real numbers or algebraic expressions. Here we state the rules for inequalities involving the symbol $\le$, but they apply to all four inequality symbols.

RULES FOR INEQUALITIES

Rule	Description
1. $A \le B \quad \Leftrightarrow \quad A + C \le B + C$	**Adding** the same quantity to each side of an inequality gives an equivalent inequality.
2. $A \le B \quad \Leftrightarrow \quad A - C \le B - C$	**Subtracting** the same quantity from each side of an inequality gives an equivalent inequality.
3. If $C > 0$, then $A \le B \quad \Leftrightarrow \quad CA \le CB$	**Multiplying** each side of an inequality by the same *positive* quantity gives an equivalent inequality.
4. If $C < 0$, then $A \le B \quad \Leftrightarrow \quad CA \ge CB$	**Multiplying** each side of an inequality by the same *negative* quantity *reverses the direction* of the inequality.
5. If $A > 0$ and $B > 0$, then $A \le B \quad \Leftrightarrow \quad \dfrac{1}{A} \ge \dfrac{1}{B}$	**Taking reciprocals** of each side of an inequality involving *positive* quantities *reverses the direction* of the inequality.
6. If $A \le B$ and $C \le D$, then $A + C \le B + D$	Inequalities can be added.
7. If $A \le B$ and $B \le C$, then $A \le C$	Inequality is transitive.

Pay special attention to Rules 3 and 4. Rule 3 says that we can multiply (or divide) each side of an inequality by a *positive* number, but Rule 4 says that if we multiply each side of an inequality by a *negative* number, then we reverse the direction of the inequality. For example, if we start with the inequality

$$3 < 5$$

and multiply by 2, we get

$$6 < 10$$

but if we multiply by -2, we get

$$-6 > -10$$

Solving Linear Inequalities

An inequality is **linear** if each term is constant or a multiple of the variable. To solve a linear inequality, we isolate the variable on one side of the inequality sign.

EXAMPLE 1 ■ Solving a Linear Inequality

Solve the inequality $3x < 9x + 4$, and sketch the solution set.

SOLUTION

$3x < 9x + 4$	Given inequality
$3x - 9x < 9x + 4 - 9x$	Subtract $9x$
$-6x < 4$	Simplify
$\left(-\frac{1}{6}\right)(-6x) > \left(-\frac{1}{6}\right)(4)$	Multiply by $-\frac{1}{6}$ and reverse inequality
$x > -\frac{2}{3}$	Simplify

Multiplying by the negative number $-\frac{1}{6}$ *reverses* the direction of the inequality.

The solution set consists of all numbers greater than $-\frac{2}{3}$. In other words the solution of the inequality is the interval $\left(-\frac{2}{3}, \infty\right)$. It is graphed in Figure 1.

FIGURE 1

Now Try Exercise 21 ■

EXAMPLE 2 ■ Solving a Pair of Simultaneous Inequalities

Solve the inequalities $4 \le 3x - 2 < 13$.

SOLUTION The solution set consists of all values of x that satisfy both of the inequalities $4 \le 3x - 2$ and $3x - 2 < 13$. Using Rules 1 and 3, we see that the following inequalities are equivalent:

$$4 \le 3x - 2 < 13 \qquad \text{Given inequality}$$

$$6 \le 3x < 15 \qquad \text{Add 2}$$

$$2 \le x < 5 \qquad \text{Divide by 3}$$

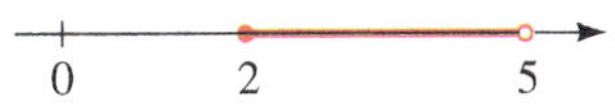

FIGURE 2

Therefore the solution set is $[2, 5)$, as shown in Figure 2.

Now Try Exercise 33 ■

■ Solving Nonlinear Inequalities

To solve inequalities involving squares and other powers of the variable, we use factoring, together with the following principle.

THE SIGN OF A PRODUCT OR QUOTIENT

- If a product or a quotient has an *even* number of *negative* factors, then its value is *positive*.
- If a product or a quotient has an *odd* number of *negative* factors, then its value is *negative*.

For example, to solve the inequality $x^2 - 5x \le -6$, we first move all terms to the left-hand side and factor to get

$$(x - 2)(x - 3) \le 0$$

This form of the inequality says that the product $(x - 2)(x - 3)$ must be negative or zero, so to solve the inequality, we must determine where each factor is negative or positive (because the sign of a product depends on the sign of the factors). The details are explained in Example 3, in which we use the following guidelines.

GUIDELINES FOR SOLVING NONLINEAR INEQUALITIES

1. **Move All Terms to One Side.** If necessary, rewrite the inequality so that all nonzero terms appear on one side of the inequality sign. If the nonzero side of the inequality involves quotients, bring them to a common denominator.
2. **Factor.** Factor the nonzero side of the inequality.
3. **Find the Intervals.** Determine the values for which each factor is zero. These numbers will divide the real line into intervals. List the intervals that are determined by these numbers.
4. **Make a Table or Diagram.** Use **test values** to make a table or diagram of the signs of each factor on each interval. In the last row of the table determine the sign of the product (or quotient) of these factors.
5. **Solve.** Use the sign table to find the intervals on which the inequality is satisfied. Check whether the **endpoints** of these intervals satisfy the inequality. (This may happen if the inequality involves $\le$ or $\ge$.)

The factoring technique that is described in these guidelines works only if all nonzero terms appear on one side of the inequality symbol. If the inequality is not written in this form, first rewrite it, as indicated in Step 1.

EXAMPLE 3 ■ Solving a Quadratic Inequality

Solve the inequality $x^2 \le 5x - 6$.

SOLUTION We will follow the guidelines given above.

Move all terms to one side. We move all the terms to the left-hand side.

$$x^2 \le 5x - 6 \qquad \text{Given inequality}$$

$$x^2 - 5x + 6 \le 0 \qquad \text{Subtract } 5x\text{, add } 6$$

Factor. Factoring the left-hand side of the inequality, we get

$$(x - 2)(x - 3) \le 0 \qquad \text{Factor}$$

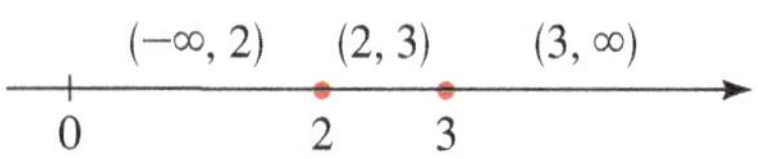

FIGURE 3

Find the intervals. The factors of the left-hand side are $x - 2$ and $x - 3$. These factors are zero when x is 2 and 3, respectively. As shown in Figure 3, the numbers 2 and 3 divide the real line into the three intervals

$$(-\infty, 2), (2, 3), (3, \infty)$$

The factors $x - 2$ and $x - 3$ change sign only at 2 and 3, respectively. So these factors maintain their sign on each of these three intervals.

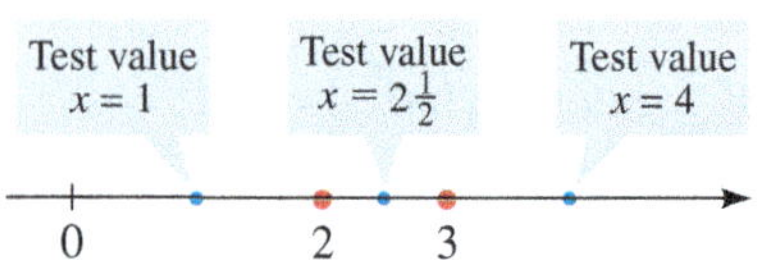

FIGURE 4

Make a table or diagram. To determine the sign of each factor on each of the intervals that we found, we use test values. We choose a number inside each interval and check the sign of the factors $x - 2$ and $x - 3$ at the number we chose. For the interval $(-\infty, 2)$, let's choose the test value 1 (see Figure 4). Substituting 1 for x in the factors $x - 2$ and $x - 3$, we get

$$x - 2 = 1 - 2 = -1 < 0$$

$$x - 3 = 1 - 3 = -2 < 0$$

So both factors are negative on this interval. Notice that we need to check only one test value for each interval because the factors $x - 2$ and $x - 3$ do not change sign on any of the three intervals we found.

Using the test values $x = 2\frac{1}{2}$ and $x = 4$ for the intervals $(2, 3)$ and $(3, \infty)$ (see Figure 4), respectively, we construct the following sign table. The final row of the table is obtained from the fact that the expression in the last row is the product of the two factors.

Interval	$(-\infty, 2)$	$(2, 3)$	$(3, \infty)$
Sign of $x - 2$	−	+	+
Sign of $x - 3$	−	−	+
Sign of $(x - 2)(x - 3)$	+	−	+

If you prefer, you can represent this information on a real line, as in the following sign diagram. The vertical lines indicate the points at which the real line is divided into intervals:

		2		3	
Sign of $x - 2$	−		+		+
Sign of $x - 3$	−		−		+
Sign of $(x - 2)(x - 3)$	+		−		+

Solve. We read from the table or the diagram that $(x - 2)(x - 3)$ is negative on the interval $(2, 3)$. You can check that the endpoints 2 and 3 satisfy the inequality, so the solution is

$$\{x \mid 2 \leq x \leq 3\} = [2, 3]$$

The solution is illustrated in Figure 5.

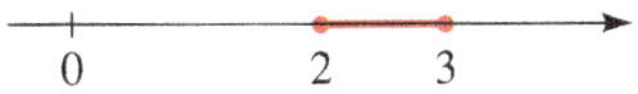

FIGURE 5

Now Try Exercise 43 ■

EXAMPLE 4 ■ Solving an Inequality with Repeated Factors

Solve the inequality $x(x - 1)^2(x - 3) < 0$.

SOLUTION All nonzero terms are already on one side of the inequality, and the nonzero side of the inequality is already factored. So we begin by finding the intervals for this inequality.

Find the intervals. The factors of the left-hand side are x, $(x - 1)^2$, and $x - 3$. These are zero when $x = 0, 1, 3$. These numbers divide the real line into the intervals

$$(-\infty, 0), (0, 1), (1, 3), (3, \infty)$$

Make a diagram. We make the following diagram, using test points to determine the sign of each factor in each interval.

		0		1		3	
Sign of x	$-$		$+$		$+$		$+$
Sign of $(x-1)^2$	$+$		$+$		$+$		$+$
Sign of $(x-3)$	$-$		$-$		$-$		$+$
Sign of $x(x-1)^2(x-3)$	$+$		$-$		$-$		$+$

Solve. From the diagram we see that the inequality is satisfied on the intervals $(0, 1)$ and $(1, 3)$. Since this inequality involves $<$, the endpoints of the intervals do not satisfy the inequality. So the solution set is the union of these two intervals:

$$(0, 1) \cup (1, 3)$$

The solution set is graphed in Figure 6.

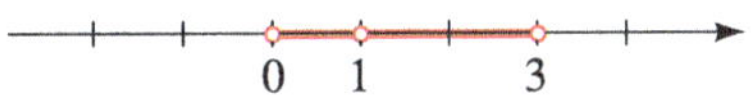

FIGURE 6

Now Try Exercise 55 ■

EXAMPLE 5 ■ Solving an Inequality Involving a Quotient

Solve the inequality $\dfrac{1 + x}{1 - x} \geq 1$.

SOLUTION **Move all terms to one side.** We move the terms to the left-hand side and simplify using a common denominator.

$$\frac{1 + x}{1 - x} \geq 1 \qquad \text{Given inequality}$$

$$\frac{1 + x}{1 - x} - 1 \geq 0 \qquad \text{Subtract 1}$$

$$\frac{1 + x}{1 - x} - \frac{1 - x}{1 - x} \geq 0 \qquad \text{Common denominator } 1 - x$$

$$\frac{1 + x - 1 + x}{1 - x} \geq 0 \qquad \text{Combine the fractions}$$

$$\frac{2x}{1 - x} \geq 0 \qquad \text{Simplify}$$

It is tempting to simply multiply both sides of the inequality by $1 - x$ (as you would if this were an *equation*). But this doesn't work because we don't know whether $1 - x$ is positive or negative, so we can't tell whether the inequality needs to be reversed. (See Exercise 127.)

Find the intervals. The factors of the left-hand side are $2x$ and $1 - x$. These are zero when x is 0 and 1. These numbers divide the real line into the intervals

$$(-\infty, 0), (0, 1), (1, \infty)$$

Make a diagram. We make the following diagram using test points to determine the sign of each factor in each interval.

		0		1	
Sign of $2x$	$-$		$+$		$+$
Sign of $1 - x$	$+$		$+$		$-$
Sign of $\dfrac{2x}{1-x}$	$-$		$+$		$-$

Solve. From the diagram we see that the inequality is satisfied on the interval $(0, 1)$. Checking the endpoints, we see that 0 satisfies the inequality but 1 does not (because the quotient in the inequality is not defined at 1). So the solution set is the interval

$$[0, 1)$$

The solution set is graphed in Figure 7.

FIGURE 7

Now Try Exercise 61 ■

Example 5 shows that we should always check the endpoints of the solution set to see whether they satisfy the original inequality.

Absolute Value Inequalities

We use the following properties to solve inequalities that involve absolute value.

These properties hold when x is replaced by any algebraic expression. (In the graphs we assume that $c > 0$.)

PROPERTIES OF ABSOLUTE VALUE INEQUALITIES

Inequality	Equivalent form	Graph
1. $\lvert x \rvert < c$	$-c < x < c$	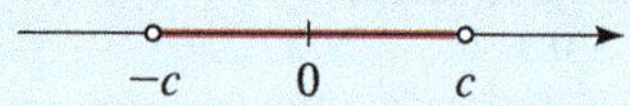
2. $\lvert x \rvert \le c$	$-c \le x \le c$	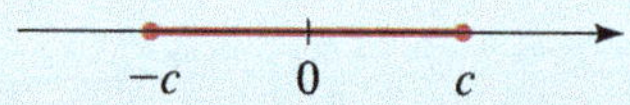
3. $\lvert x \rvert > c$	$x < -c$ or $c < x$	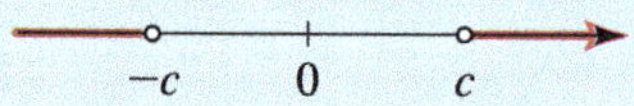
4. $\lvert x \rvert \ge c$	$x \le -c$ or $c \le x$	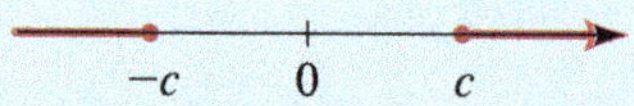

These properties can be proved using the definition of absolute value. To prove Property 1, for example, note that the inequality $\lvert x \rvert < c$ says that the distance from x to 0 is less than c, and from Figure 8 you can see that this is true if and only if x is between $-c$ and c.

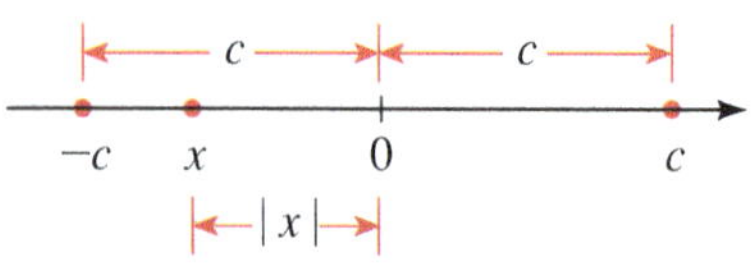

FIGURE 8

EXAMPLE 6 ■ Solving an Absolute Value Inequality

Solve the inequality $\lvert x - 5 \rvert < 2$.

SOLUTION 1 The inequality $\lvert x - 5 \rvert < 2$ is equivalent to

$$-2 < x - 5 < 2 \qquad \text{Property 1}$$

$$3 < x < 7 \qquad \text{Add 5}$$

The solution set is the open interval $(3, 7)$.

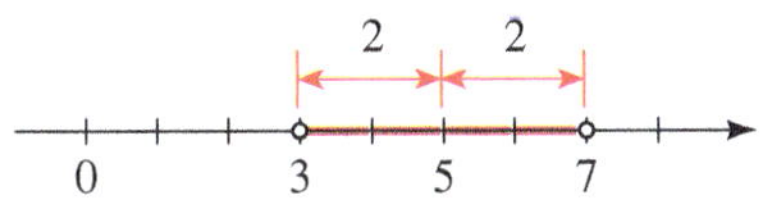

FIGURE 9

SOLUTION 2 Geometrically, the solution set consists of all numbers x whose distance from 5 is less than 2. From Figure 9 we see that this is the interval $(3, 7)$.

Now Try Exercise 81 ■

EXAMPLE 7 ■ Solving an Absolute Value Inequality

Solve the inequality $|3x + 2| \geq 4$.

SOLUTION By Property 4 the inequality $|3x + 2| \geq 4$ is equivalent to

$$3x + 2 \geq 4 \quad \text{or} \quad 3x + 2 \leq -4$$

$$3x \geq 2 \qquad 3x \leq -6 \qquad \text{Subtract 2}$$

$$x \geq \tfrac{2}{3} \qquad x \leq -2 \qquad \text{Divide by 3}$$

So the solution set is

$$\{x \mid x \leq -2 \quad \text{or} \quad x \geq \tfrac{2}{3}\} = (-\infty, -2] \cup [\tfrac{2}{3}, \infty)$$

The set is graphed in Figure 10.

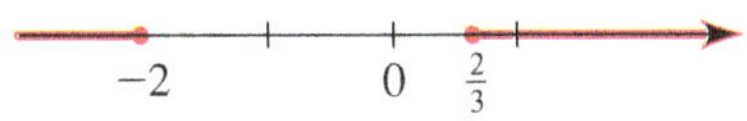

FIGURE 10

Now Try Exercise 83 ■

■ Modeling with Inequalities

Modeling real-life problems frequently leads to inequalities because we are often interested in determining when one quantity is more (or less) than another.

EXAMPLE 8 ■ Carnival Tickets

A carnival has two plans for tickets.

Plan A: \$5 entrance fee and 25¢ each ride

Plan B: \$2 entrance fee and 50¢ each ride

How many rides would you have to take for Plan A to be less expensive than Plan B?

SOLUTION **Identify the variable.** We are asked for the number of rides for which Plan A is less expensive than Plan B. So let

$$x = \text{number of rides}$$

Translate from words to algebra. The information in the problem may be organized as follows.

In Words	In Algebra
Number of rides	x
Cost with Plan A	$5 + 0.25x$
Cost with Plan B	$2 + 0.50x$

Set up the model. Now we set up the model.

$$\text{cost with Plan A} < \text{cost with Plan B}$$

$$5 + 0.25x < 2 + 0.50x$$

Solve. Now we solve for x.

$$3 + 0.25x < 0.50x \qquad \text{Subtract 2}$$
$$3 < 0.25x \qquad \text{Subtract } 0.25x$$
$$12 < x \qquad \text{Divide by 0.25}$$

So if you plan to take *more than* 12 rides, Plan A is less expensive.

Now Try Exercise 111 ■

EXAMPLE 9 ■ Relationship Between Fahrenheit and Celsius Scales

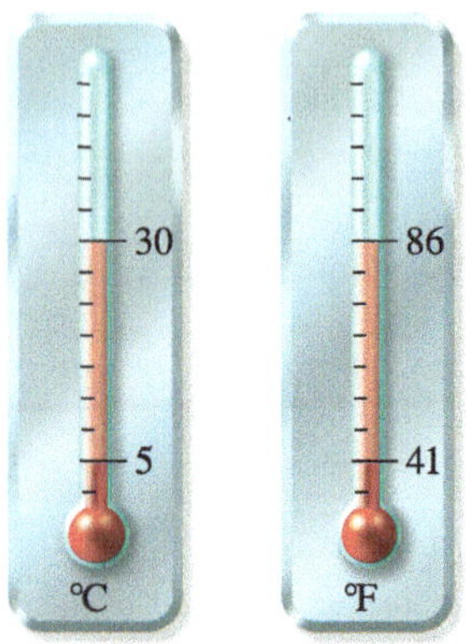

The instructions on a bottle of medicine indicate that the bottle should be stored at a temperature between 5°C and 30°C. What range of temperatures does this correspond to on the Fahrenheit scale?

SOLUTION The relationship between degrees Celsius (C) and degrees Fahrenheit (F) is given by the equation $C = \frac{5}{9}(F - 32)$. Expressing the statement on the bottle in terms of inequalities, we have

$$5 < C < 30$$

So the corresponding Fahrenheit temperatures satisfy the inequalities

$$5 < \tfrac{5}{9}(F - 32) < 30 \qquad \text{Substitute } C = \tfrac{5}{9}(F - 32)$$
$$\tfrac{9}{5} \cdot 5 < F - 32 < \tfrac{9}{5} \cdot 30 \qquad \text{Multiply by } \tfrac{9}{5}$$
$$9 < F - 32 < 54 \qquad \text{Simplify}$$
$$9 + 32 < F < 54 + 32 \qquad \text{Add 32}$$
$$41 < F < 86 \qquad \text{Simplify}$$

The medicine should be stored at a temperature between 41°F and 86°F.

Now Try Exercise 109 ■

1.8 EXERCISES

CONCEPTS

1. Fill in the blank with an appropriate inequality sign.

(a) If $x < 5$, then $x - 3$ ______ 2.

(b) If $x \le 5$, then $3x$ ______ 15.

(c) If $x \ge 2$, then $-3x$ ______ -6.

(d) If $x < -2$, then $-x$ ______ 2.

2. To solve the nonlinear inequality $\dfrac{x + 1}{x - 2} \le 0$, we first observe that the numbers ____ and ____ are zeros of the numerator and denominator. These numbers divide the real line into three intervals. Complete the table.

Interval			
Sign of $x + 1$ **Sign of $x - 2$**			
Sign of $(x + 1)/(x - 2)$			

Do any of the endpoints fail to satisfy the inequality? If so, which one(s)? ______. The solution of the inequality is ____________.

3. (a) The solution of the inequality $|x| \le 3$ is the interval ______.

(b) The solution of the inequality $|x| \ge 3$ is a union of two intervals ______ $\cup$ ______.

4. (a) The set of all points on the real line whose distance from zero is less than 3 can be described by the absolute value inequality $|x|$ ______.

(b) The set of all points on the real line whose distance from zero is greater than 3 can be described by the absolute value inequality $|x|$ ______.

5. *Yes or No?* If *No*, give an example.

(a) If $x(x + 1) > 0$, does it follow that x is positive?

(b) If $x(x + 1) > 5$, does it follow that $x > 5$?

6. What is a logical first step in solving the inequality?

(a) $3x \le 7$ **(b)** $5x - 2 \ge 1$ **(c)** $|3x + 2| \le 8$

SKILLS

7–12 ■ Solutions? Let $S = \{-5, -1, 0, \frac{2}{3}, \frac{5}{6}, 1, \sqrt{5}, 3, 5\}$. Determine which elements of S satisfy the inequality.

7. $-2 + 3x \ge \frac{1}{3}$

8. $1 - 2x \ge 5x$

9. $1 < 2x - 4 \le 7$

10. $-2 \le 3 - x < 2$

11. $\frac{1}{x} \le \frac{1}{2}$

12. $x^2 + 2 < 4$

13–36 ■ Linear Inequalities Solve the linear inequality. Express the solution using interval notation and graph the solution set.

13. $2x \le 7$

14. $-4x \ge 10$

15. $2x - 5 > 3$

16. $3x + 11 < 5$

17. $7 - x \ge 5$

18. $5 - 3x \le -16$

19. $2x + 1 < 0$

20. $0 < 5 - 2x$

21. $4x - 7 < 8 + 9x$

22. $5 - 3x \ge 8x - 7$

23. $\frac{1}{2}x - \frac{2}{3} > 2$

24. $\frac{2}{5}x + 1 < \frac{1}{5} - 2x$

25. $\frac{1}{3}x + 2 < \frac{1}{6}x - 1$

26. $\frac{2}{3} - \frac{1}{2}x \ge \frac{1}{6} + x$

27. $4 - 3x \le -(1 + 8x)$

28. $2(7x - 3) \le 12x + 16$

29. $2 \le x + 5 < 4$

30. $5 \le 3x - 4 \le 14$

31. $-1 < 2x - 5 < 7$

32. $1 < 3x + 4 \le 16$

33. $-2 < 8 - 2x \le -1$

34. $-3 \le 3x + 7 \le \frac{1}{2}$

35. $\frac{1}{6} < \frac{2x - 13}{12} \le \frac{2}{3}$

36. $-\frac{1}{2} \le \frac{4 - 3x}{5} \le \frac{1}{4}$

37–58 ■ Nonlinear Inequalities Solve the nonlinear inequality. Express the solution using interval notation and graph the solution set.

37. $(x + 2)(x - 3) < 0$

38. $(x - 5)(x + 4) \ge 0$

39. $x(2x + 7) \ge 0$

40. $x(2 - 3x) \le 0$

41. $x^2 - 3x - 18 \le 0$

42. $x^2 + 5x + 6 > 0$

43. $2x^2 + x \ge 1$

44. $x^2 < x + 2$

45. $3x^2 - 3x < 2x^2 + 4$

46. $5x^2 + 3x \ge 3x^2 + 2$

47. $x^2 > 3(x + 6)$

48. $x^2 + 2x > 3$

49. $x^2 < 4$

50. $x^2 \ge 9$

51. $(x + 2)(x - 1)(x - 3) \le 0$

52. $(x - 5)(x - 2)(x + 1) > 0$

53. $(x - 4)(x + 2)^2 < 0$

54. $(x + 3)^2(x + 1) > 0$

55. $(x + 3)^2(x - 2)(x + 5) \ge 0$

56. $4x^2(x^2 - 9) \le 0$

57. $x^3 - 4x > 0$

58. $16x \le x^3$

59–74 ■ Inequalities Involving Quotients Solve the nonlinear inequality. Express the solution using interval notation, and graph the solution set.

59. $\frac{x - 3}{x + 1} \ge 0$

60. $\frac{2x + 6}{x - 2} < 0$

61. $\frac{x}{x + 1} > 3$

62. $\frac{x - 4}{2x + 1} < 5$

63. $\frac{2x + 1}{x - 5} \le 3$

64. $\frac{3 + x}{3 - x} \ge 1$

65. $\frac{4}{x} < x$

66. $\frac{x}{x + 1} > 3x$

67. $1 + \frac{2}{x + 1} \le \frac{2}{x}$

68. $\frac{3}{x - 1} - \frac{4}{x} \ge 1$

69. $\frac{6}{x - 1} - \frac{6}{x} \ge 1$

70. $\frac{x}{2} \ge \frac{5}{x + 1} + 4$

71. $\frac{x + 2}{x + 3} < \frac{x - 1}{x - 2}$

72. $\frac{1}{x + 1} + \frac{1}{x + 2} \le 0$

73. $x^4 > x^2$

74. $x^5 > x^2$

75–90 ■ Absolute Value Inequalities Solve the absolute value inequality. Express the answer using interval notation and graph the solution set.

75. $|5x| < 20$

76. $|16x| \le 8$

77. $|2x| > 7$

78. $\frac{1}{2}|x| \ge 1$

79. $|x - 5| \le 3$

80. $|x + 1| \ge 1$

81. $|3x + 2| < 4$

82. $|5x - 2| < 8$

83. $|3x - 2| \ge 5$

84. $|8x + 3| > 12$

85. $\left|\frac{x - 2}{3}\right| < 2$

86. $\left|\frac{x + 1}{2}\right| \ge 4$

87. $|x + 6| < 0.001$

88. $3 - |2x + 4| \le 1$

89. $8 - |2x - 1| \ge 6$

90. $7|x + 2| + 5 > 4$

91–94 ■ Absolute Value Inequalities A phrase describing a set of real numbers is given. Express the phrase as an inequality involving an absolute value.

91. All real numbers x less than 3 units from 0

92. All real numbers x more than 2 units from 0

93. All real numbers x at least 5 units from 7

94. All real numbers x at most 4 units from 2

95–100 ■ Absolute Value Inequalities A set of real numbers is graphed. Find an inequality involving an absolute value that describes the set.

95. 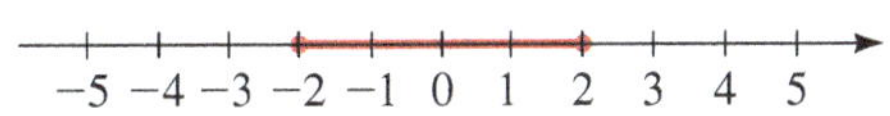

96. 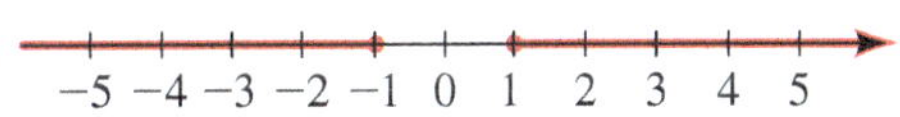

97. 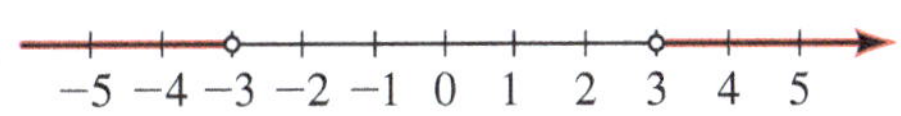

98. 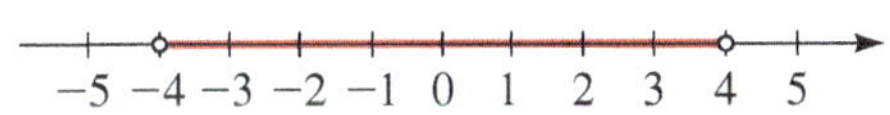

99. 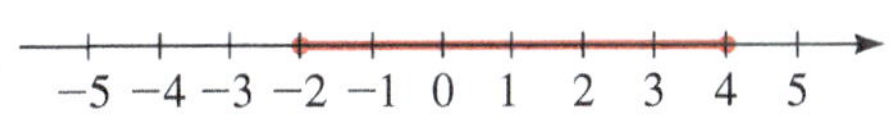

100. 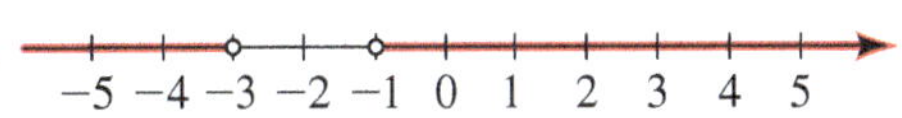

101–104 ■ Domain Determine the values of the variable for which the expression is defined as a real number.

101. $\sqrt{x^2 - 9}$

102. $\sqrt{x^2 - 5x - 50}$

103. $\left(\dfrac{1}{x^2 - 3x - 10}\right)^{1/2}$ **104.** $\sqrt[4]{\dfrac{1 - x}{2 + x}}$

SKILLS Plus

105–108 ■ Inequalities Solve the inequality for x. Assume that a, b, and c are positive constants.

105. $a(bx - c) \geq bc$ **106.** $a \leq bx + c < 2a$

107. $a|bx - c| + d \geq 4a$ **108.** $\left|\dfrac{bx + c}{a}\right| > 5a$

APPLICATIONS

109. Temperature Scales Use the relationship between C and F given in Example 9 to find the interval on the Fahrenheit scale corresponding to the temperature range $20 \leq C \leq 30$.

110. Temperature Scales What interval on the Celsius scale corresponds to the temperature range $50 \leq F \leq 95$?

111. Car Rental Cost A car rental company offers two plans for renting a car.

Plan A: $30 per day and 10¢ per mile
Plan B: $50 per day with free unlimited mileage

For what range of miles will Plan B save you money?

112. International Plans A phone service provider offers two international plans.

Plan A: $25 per month and 5¢ per minute
Plan B: $5 per month and 12¢ per minute

For what range of minutes of international calls would Plan B be financially advantageous?

113. Driving Cost It is estimated that the annual cost of driving a certain new car is given by the formula

$$C = 0.35m + 2200$$

where m represents the number of miles driven per year and C is the cost in dollars. Jane has purchased such a car and decides to budget between $6400 and $7100 for next year's driving costs. What is the corresponding range of miles that she can drive her new car?

114. Air Temperature As dry air moves upward, it expands and, in so doing, cools at a rate of about 1°C for each 100-m rise, up to about 12 km.

(a) If the ground temperature is 20°C, write a formula for the temperature at height h.

(b) What range of temperatures can be expected if a plane takes off and reaches a maximum height of 5 km?

115. Airline Ticket Price A charter airline finds that on its Saturday flights from Philadelphia to London all 120 seats will be sold if the ticket price is $200. However, for each $3 increase in ticket price, the number of seats sold decreases by one.

(a) Find a formula for the number of seats sold if the ticket price is P dollars.

(b) Over a certain period the number of seats sold for this flight ranged between 90 and 115. What was the corresponding range of ticket prices?

116. Accuracy of a Scale A coffee merchant sells a customer 3 lb of Hawaiian Kona at $6.50 per pound. The merchant's scale is accurate to within ± 0.03 lb. By how much could the customer have been overcharged or undercharged because of possible inaccuracy in the scale?

117. Gravity The gravitational force F exerted by the earth on an object having a mass of 100 kg is given by the equation

$$F = \frac{4{,}000{,}000}{d^2}$$

where d is the distance (in km) of the object from the center of the earth, and the force F is measured in newtons (N). For what distances will the gravitational force exerted by the earth on this object be between 0.0004 N and 0.01 N?

118. Bonfire Temperature In the vicinity of a bonfire the temperature T in °C at a distance of x meters from the center of the fire was given by

$$T = \frac{600{,}000}{x^2 + 300}$$

At what range of distances from the fire's center was the temperature less than 500 °C?

119. Falling Ball Using calculus, it can be shown that if a ball is thrown upward with an initial velocity of 16 ft/s from the top of a building 128 ft high, then its height h above the ground t seconds later will be

$$h = 128 + 16t - 16t^2$$

During what time interval will the ball be at least 32 ft above the ground?

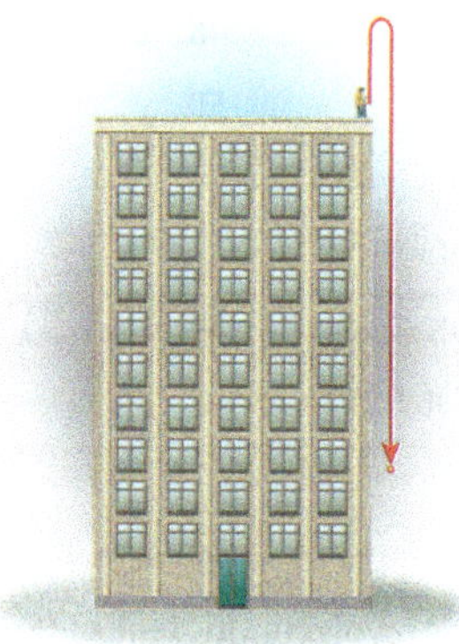

120. Gas Mileage The gas mileage g (measured in mi/gal) for a particular vehicle, driven at v mi/h, is given by the formula $g = 10 + 0.9v - 0.01v^2$, as long as v is between 10 mi/h and 75 mi/h. For what range of speeds is the vehicle's mileage 30 mi/gal or better?

121. Stopping Distance For a certain model of car the distance d required to stop the vehicle if it is traveling at v mi/h is given by the formula

$$d = v + \frac{v^2}{20}$$

where d is measured in feet. Kerry wants her stopping distance not to exceed 240 ft. At what range of speeds can she travel?

122. Manufacturer's Profit If a manufacturer sells x units of a certain product, revenue R and cost C (in dollars) are given by

$$R = 20x$$
$$C = 2000 + 8x + 0.0025x^2$$

Use the fact that

$$\text{profit} = \text{revenue} - \text{cost}$$

to determine how many units the manufacturer should sell to enjoy a profit of at least \$2400.

123. Fencing a Garden A determined gardener has 120 ft of deer-resistant fence. She wants to enclose a rectangular vegetable garden in her backyard, and she wants the area that is enclosed to be at least 800 ft^2. What range of values is possible for the length of her garden?

124. Thickness of a Laminate A company manufactures industrial laminates (thin nylon-based sheets) of thickness 0.020 in., with a tolerance of 0.003 in.

(a) Find an inequality involving absolute values that describes the range of possible thickness for the laminate.

(b) Solve the inequality you found in part (a).

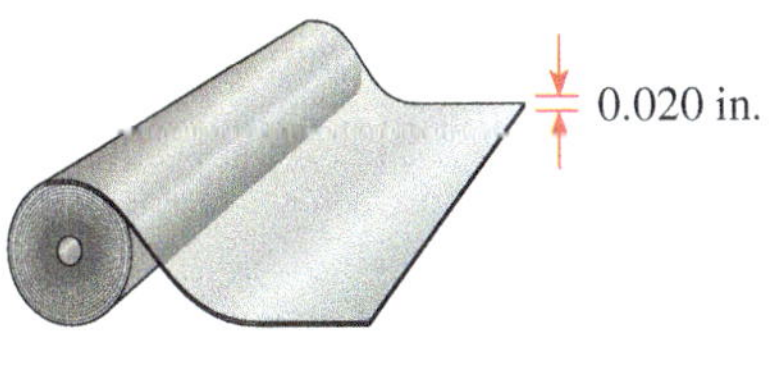

125. Range of Height The average height of adult males is 68.2 in., and 95% of adult males have height h that satisfies the inequality

$$\left| \frac{h - 68.2}{2.9} \right| \le 2$$

Solve the inequality to find the range of heights.

DISCUSS ■ DISCOVER ■ PROVE ■ WRITE

126. DISCUSS ■ DISCOVER: Do Powers Preserve Order? If $a < b$, is $a^2 < b^2$? (Check both positive and negative values for a and b.) If $a < b$, is $a^3 < b^3$? On the basis of your observations, state a general rule about the relationship between a^n and b^n when $a < b$ and n is a positive integer.

127. DISCUSS ■ DISCOVER: What's Wrong Here? It is tempting to try to solve an inequality like an equation. For instance, we might try to solve $1 < 3/x$ by multiplying both sides by x, to get $x < 3$, so the solution would be $(-\infty, 3)$. But that's wrong; for example, $x = -1$ lies in this interval but does not satisfy the original inequality. Explain why this method doesn't work (think about the *sign* of x). Then solve the inequality correctly.

128. DISCUSS ■ DISCOVER: Using Distances to Solve Absolute Value Inequalities Recall that $|a - b|$ is the distance between a and b on the number line. For any number x, what do $|x - 1|$ and $|x - 3|$ represent? Use this interpretation to solve the inequality $|x - 1| < |x - 3|$ geometrically. In general, if $a < b$, what is the solution of the inequality $|x - a| < |x - b|$?

129–130 ■ PROVE: Inequalities Use the properties of inequalities to prove the following inequalities.

129. Rule 6 for Inequalities: If a, b, c, and d are any real numbers such that $a < b$ and $c < d$, then $a + c < b + d$. [*Hint:* Use Rule 1 to show that $a + c < b + c$ and $b + c < b + d$. Use Rule 7.]

130. If a, b, c, and d are positive numbers such that $\frac{a}{b} < \frac{c}{d}$, then $\frac{a}{b} < \frac{a + c}{b + d} < \frac{c}{d}$. [*Hint:* Show that $\frac{ad}{b} + a < c + a$ and $a + c < \frac{cb}{d} + c$.]

131. PROVE: Arithmetic-Geometric Mean Inequality If $a_1, a_2, \ldots, a_n$ are nonnegative numbers, then their arithmetic mean is $\frac{a_1 + a_2 + \cdots + a_n}{n}$, and their geometric mean is $\sqrt[n]{a_1 a_2 \ldots a_n}$. The arithmetic-geometric mean inequality states that the geometric mean is always less than or equal to the arithmetic mean. In this problem we prove this in the case of two numbers x and y.

(a) If x and y are nonnegative and $x \le y$, then $x^2 \le y^2$. [*Hint:* First use Rule 3 of Inequalities to show that $x^2 \le xy$ and $xy \le y^2$.]

(b) Prove the arithmetic-geometric mean inequality

$$\sqrt{xy} \le \frac{x + y}{2}$$

1.9 THE COORDINATE PLANE; GRAPHS OF EQUATIONS; CIRCLES

■ The Coordinate Plane ■ The Distance and Midpoint Formulas ■ Graphs of Equations in Two Variables ■ Intercepts ■ Circles ■ Symmetry

The *coordinate plane* is the link between algebra and geometry. In the coordinate plane we can draw graphs of algebraic equations. The graphs, in turn, allow us to "see" the relationship between the variables in the equation. In this section we study the coordinate plane.

■ The Coordinate Plane

The Cartesian plane is named in honor of the French mathematician René Descartes (1596–1650), although another Frenchman, Pierre Fermat (1601–1665), also invented the principles of coordinate geometry at the same time. (See their biographies on pages 201 and 117.)

Just as points on a line can be identified with real numbers to form the coordinate line, points in a plane can be identified with ordered pairs of numbers to form the **coordinate plane** or **Cartesian plane**. To do this, we draw two perpendicular real lines that intersect at 0 on each line. Usually, one line is horizontal with positive direction to the right and is called the ***x*-axis**; the other line is vertical with positive direction upward and is called the ***y*-axis**. The point of intersection of the x-axis and the y-axis is the **origin *O***, and the two axes divide the plane into four **quadrants**, labeled I, II, III, and IV in Figure 1. (The points *on* the coordinate axes are not assigned to any quadrant.)

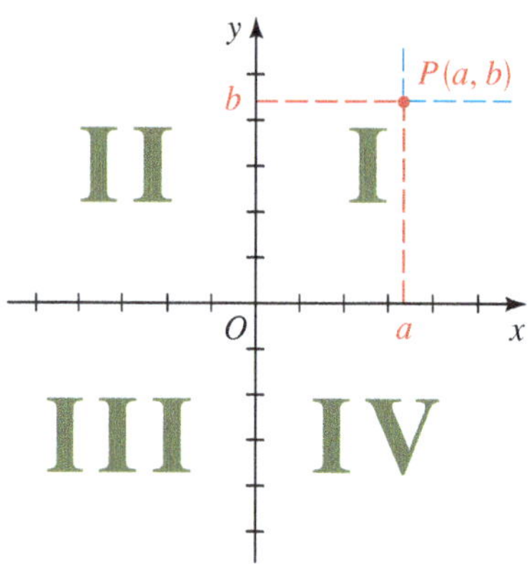

FIGURE 1

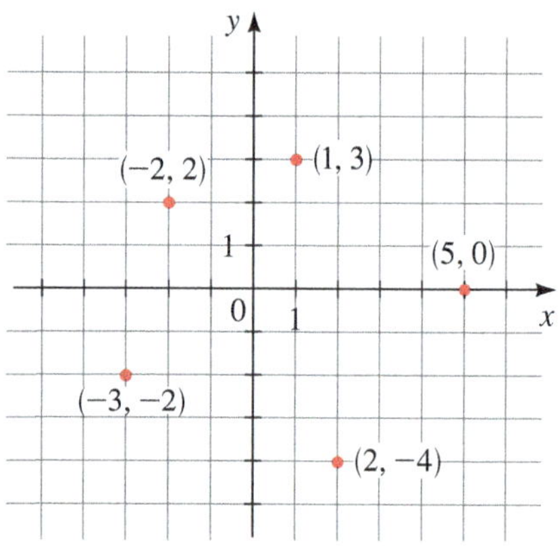

FIGURE 2

Any point P in the coordinate plane can be located by a unique **ordered pair** of numbers (a, b), as shown in Figure 1. The first number a is called the ***x*-coordinate** of P; the second number b is called the ***y*-coordinate** of P. We can think of the coordinates of P as its "address," because they specify its location in the plane. Several points are labeled with their coordinates in Figure 2.

Although the notation for a point (a, b) is the same as the notation for an open interval (a, b), the context should make clear which meaning is intended.

EXAMPLE 1 ■ Graphing Regions in the Coordinate Plane

Describe and sketch the regions given by each set.

(a) $\{(x, y) \mid x \geq 0\}$ **(b)** $\{(x, y) \mid y = 1\}$ **(c)** $\{(x, y) \mid |y| < 1\}$

SOLUTION

(a) The points whose x-coordinates are 0 or positive lie on the y-axis or to the right of it, as shown in Figure 3(a).

(b) The set of all points with y-coordinate 1 is a horizontal line one unit above the x-axis, as shown in Figure 3(b).

Coordinates as Addresses

The coordinates of a point in the *xy*-plane uniquely determine its location. We can think of the coordinates as the "address" of the point. In Salt Lake City, Utah, the addresses of most buildings are in fact expressed as coordinates. The city is divided into quadrants with Main Street as the vertical (North-South) axis and S. Temple Street as the horizontal (East-West) axis. An address such as

1760 W 2100 S

indicates a location 17.6 blocks west of Main Street and 21 blocks south of S. Temple Street. (This is the address of the main post office in Salt Lake City.) With this logical system it is possible for someone unfamiliar with the city to locate any address immediately, as easily as one locates a point in the coordinate plane.

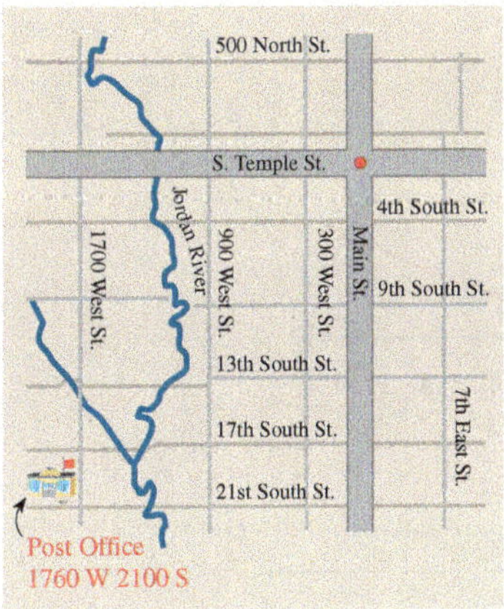

(c) Recall from Section 1.8 that

$$|y| < 1 \qquad \text{if and only if} \qquad -1 < y < 1$$

So the given region consists of those points in the plane whose y-coordinates lie between -1 and 1. Thus the region consists of all points that lie between (but not on) the horizontal lines $y = 1$ and $y = -1$. These lines are shown as broken lines in Figure 3(c) to indicate that the points on these lines are not in the set.

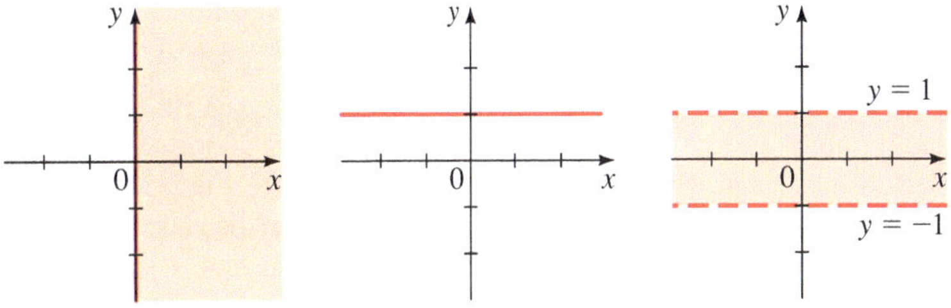

(a) $x \geq 0$ (b) $y = 1$ (c) $|y| < 1$

FIGURE 3

Now Try Exercises 15 and 17 ■

The Distance and Midpoint Formulas

We now find a formula for the distance $d(A, B)$ between two points $A(x_1, y_1)$ and $B(x_2, y_2)$ in the plane. Recall from Section 1.1 that the distance between points a and b on a number line is $d(a, b) = |b - a|$. So from Figure 4 we see that the distance between the points $A(x_1, y_1)$ and $C(x_2, y_1)$ on a horizontal line must be $|x_2 - x_1|$, and the distance between $B(x_2, y_2)$ and $C(x_2, y_1)$ on a vertical line must be $|y_2 - y_1|$.

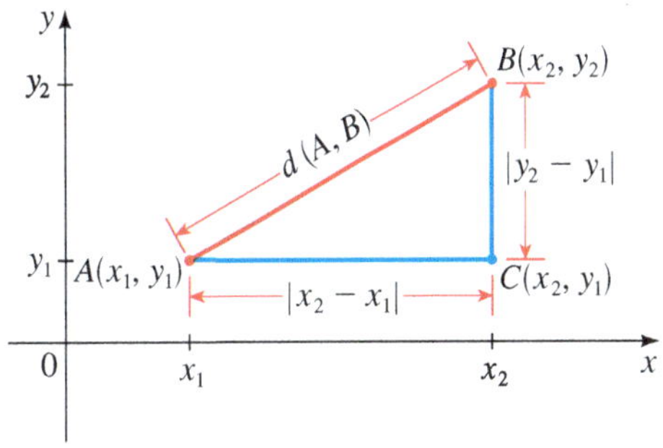

FIGURE 4

Since triangle ABC is a right triangle, the Pythagorean Theorem gives

$$d(A, B) = \sqrt{|x_2 - x_1|^2 + |y_2 - y_1|^2} = \sqrt{(x_2 - x_1)^2 + (y_2 - y_1)^2}$$

DISTANCE FORMULA

The distance between the points $A(x_1, y_1)$ and $B(x_2, y_2)$ in the plane is

$$d(A, B) = \sqrt{(x_2 - x_1)^2 + (y_2 - y_1)^2}$$

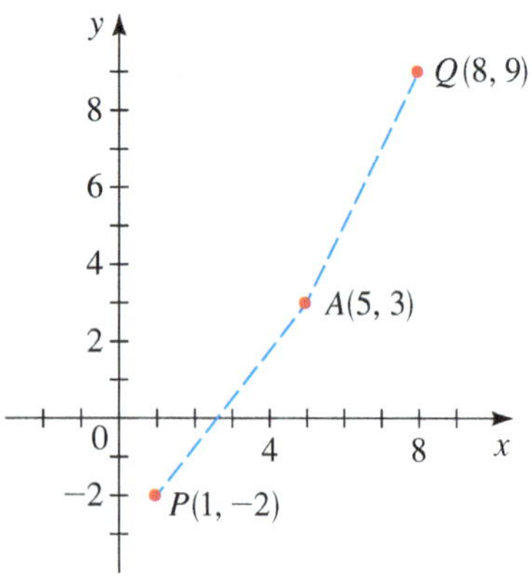

FIGURE 5

EXAMPLE 2 ■ Applying the Distance Formula

Which of the points $P(1, -2)$ or $Q(8, 9)$ is closer to the point $A(5, 3)$?

SOLUTION By the Distance Formula we have

$$d(P, A) = \sqrt{(5-1)^2 + [3-(-2)]^2} = \sqrt{4^2 + 5^2} = \sqrt{41}$$

$$d(Q, A) = \sqrt{(5-8)^2 + (3-9)^2} = \sqrt{(-3)^2 + (-6)^2} = \sqrt{45}$$

This shows that $d(P, A) < d(Q, A)$, so P is closer to A (see Figure 5).

Now Try Exercise 35 ■

Now let's find the coordinates (x, y) of the midpoint M of the line segment that joins the point $A(x_1, y_1)$ to the point $B(x_2, y_2)$. In Figure 6 notice that triangles APM and MQB are congruent because $d(A, M) = d(M, B)$ and the corresponding angles are equal. It follows that $d(A, P) = d(M, Q)$, so

$$x - x_1 = x_2 - x$$

Solving this equation for x, we get $2x = x_1 + x_2$, so $x = \frac{x_1 + x_2}{2}$. Similarly, $y = \frac{y_1 + y_2}{2}$.

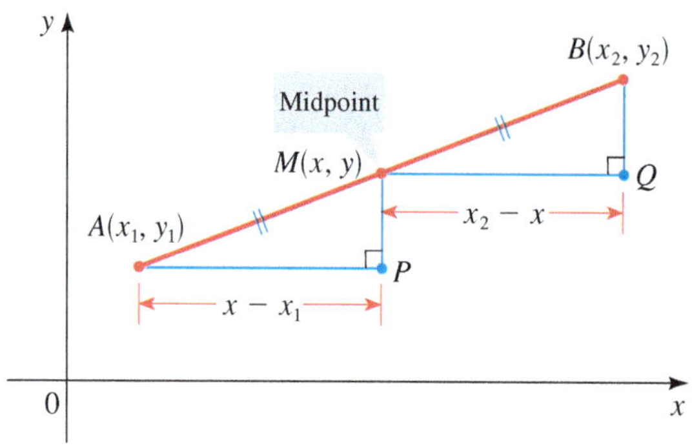

FIGURE 6

MIDPOINT FORMULA

The midpoint of the line segment from $A(x_1, y_1)$ to $B(x_2, y_2)$ is

$$\left(\frac{x_1 + x_2}{2}, \frac{y_1 + y_2}{2}\right)$$

EXAMPLE 3 ■ Applying the Midpoint Formula

Show that the quadrilateral with vertices $P(1, 2)$, $Q(4, 4)$, $R(5, 9)$, and $S(2, 7)$ is a parallelogram by proving that its two diagonals bisect each other.

SOLUTION If the two diagonals have the same midpoint, then they must bisect each other. The midpoint of the diagonal PR is

$$\left(\frac{1+5}{2}, \frac{2+9}{2}\right) = \left(3, \frac{11}{2}\right)$$

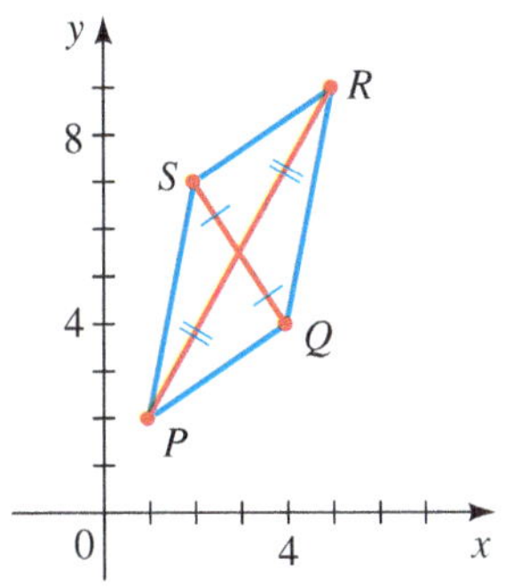

FIGURE 7

Fundamental Principle of Analytic Geometry
A point (x, y) lies on the graph of an equation if and only if its coordinates satisfy the equation.

and the midpoint of the diagonal QS is

$$\left(\frac{4+2}{2}, \frac{4+7}{2}\right) = \left(3, \frac{11}{2}\right)$$

so each diagonal bisects the other, as shown in Figure 7. (A theorem from elementary geometry states that the quadrilateral is therefore a parallelogram.)

Now Try Exercise 49 ■

Graphs of Equations in Two Variables

An **equation in two variables**, such as $y = x^2 + 1$, expresses a relationship between two quantities. A point (x, y) **satisfies** the equation if it makes the equation true when the values for x and y are substituted into the equation. For example, the point $(3, 10)$ satisfies the equation $y = x^2 + 1$ because $10 = 3^2 + 1$, but the point $(1, 3)$ does not, because $3 \neq 1^2 + 1$.

THE GRAPH OF AN EQUATION

The **graph** of an equation in x and y is the set of all points (x, y) in the coordinate plane that satisfy the equation.

The graph of an equation is a curve, so to graph an equation, we plot as many points as we can, then connect them by a smooth curve.

EXAMPLE 4 ■ Sketching a Graph by Plotting Points

Sketch the graph of the equation $2x - y = 3$.

SOLUTION We first solve the given equation for y to get

$$y = 2x - 3$$

This helps us calculate the y-coordinates in the following table.

x	$y = 2x - 3$	(x, y)
-1	-5	$(-1, -5)$
0	-3	$(0, -3)$
1	-1	$(1, -1)$
2	1	$(2, 1)$
3	3	$(3, 3)$
4	5	$(4, 5)$

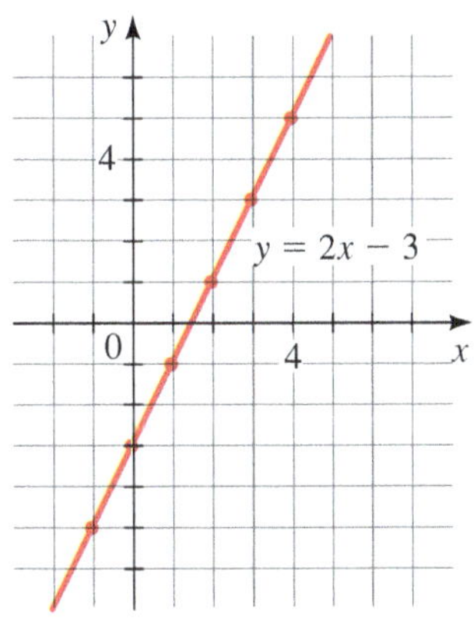

FIGURE 8

Of course, there are infinitely many points on the graph, and it is impossible to plot all of them. But the more points we plot, the better we can imagine what the graph represented by the equation looks like. We plot the points we found in Figure 8; they appear to lie on a line. So we complete the graph by joining the points by a line. (In Section 1.10 we verify that the graph of an equation of this type is indeed a line.)

Now Try Exercise 55 ■

A detailed discussion of parabolas and their geometric properties is presented in Sections 3.1 and 11.1.

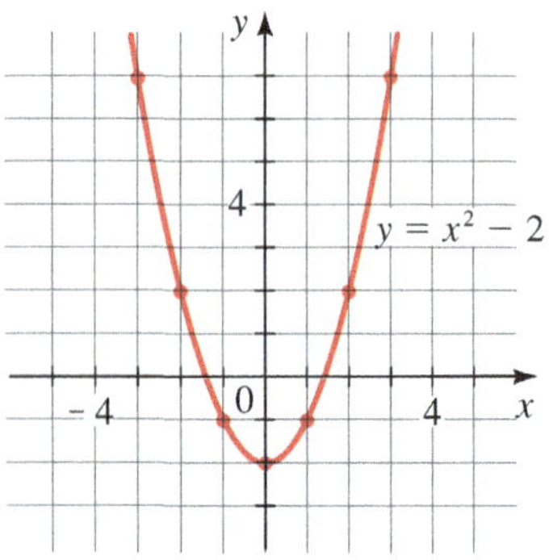

FIGURE 9

EXAMPLE 5 ■ Sketching a Graph by Plotting Points

Sketch the graph of the equation $y = x^2 - 2$.

SOLUTION We find some of the points that satisfy the equation in the following table. In Figure 9 we plot these points and then connect them by a smooth curve. A curve with this shape is called a *parabola*.

x	$y = x^2 - 2$	(x, y)
-3	7	$(-3, 7)$
-2	2	$(-2, 2)$
-1	-1	$(-1, -1)$
0	-2	$(0, -2)$
1	-1	$(1, -1)$
2	2	$(2, 2)$
3	7	$(3, 7)$

Now Try Exercise 57 ■

EXAMPLE 6 ■ Graphing an Absolute Value Equation

Sketch the graph of the equation $y = |x|$.

SOLUTION We make a table of values:

x	$y = \|x\|$	(x, y)
-3	3	$(-3, 3)$
-2	2	$(-2, 2)$
-1	1	$(-1, 1)$
0	0	$(0, 0)$
1	1	$(1, 1)$
2	2	$(2, 2)$
3	3	$(3, 3)$

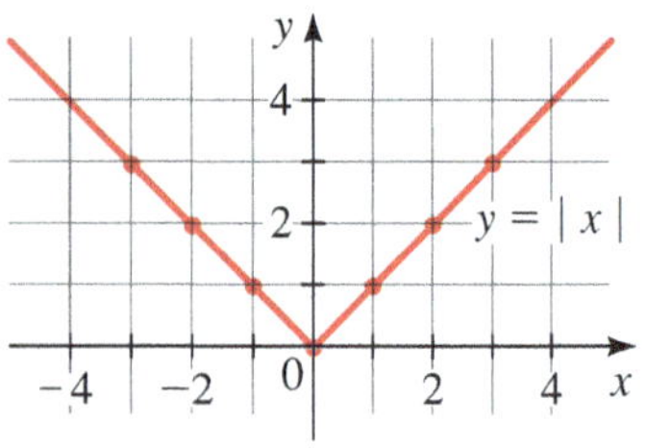

FIGURE 10

In Figure 10 we plot these points and use them to sketch the graph of the equation.

Now Try Exercise 59 ■

See Appendix C, *Graphing with a Graphing Calculator*, for general guidelines on using a graphing calculator. See Appendix D, *Using the TI-83/84 Graphing Calculator*, for specific graphing instructions. Go to **www.stewartmath.com**.

We can use a graphing calculator to graph equations. A graphing calculator draws the graph of an equation by plotting points, just as we would do by hand.

EXAMPLE 7 ■ Graphing an Equation with a Graphing Calculator

Use a graphing calculator to graph the following equation in the viewing rectangle $[-5, 5]$ by $[-1, 2]$.

$$y = \frac{1}{1 + x^2}$$

SOLUTION The graph is shown in Figure 11.

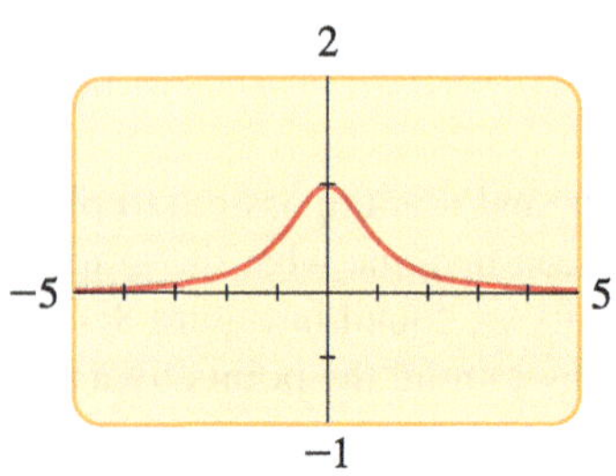

FIGURE 11 Graph of $y = \dfrac{1}{1 + x^2}$

Now Try Exercise 63 ■

■ Intercepts

The x-coordinates of the points where a graph intersects the x-axis are called the **x-intercepts** of the graph and are obtained by setting $y = 0$ in the equation of the graph. The y-coordinates of the points where a graph intersects the y-axis are called the **y-intercepts** of the graph and are obtained by setting $x = 0$ in the equation of the graph.

DEFINITION OF INTERCEPTS

Intercepts	How to find them	Where they are on the graph
x-intercepts: The x-coordinates of points where the graph of an equation intersects the x-axis	Set $y = 0$ and solve for x	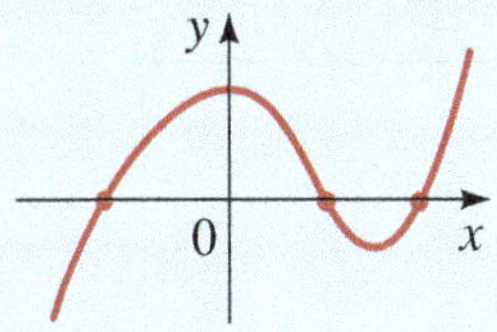
y-intercepts: The y-coordinates of points where the graph of an equation intersects the y-axis	Set $x = 0$ and solve for y	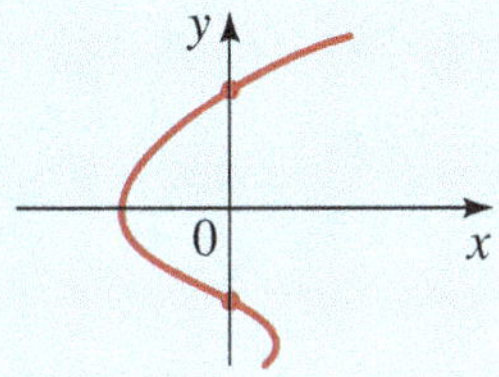

EXAMPLE 8 ■ Finding Intercepts

Find the x- and y-intercepts of the graph of the equation $y = x^2 - 2$.

SOLUTION To find the x-intercepts, we set $y = 0$ and solve for x. Thus

$$0 = x^2 - 2 \qquad \text{Set } y = 0$$

$$x^2 = 2 \qquad \text{Add 2 to each side}$$

$$x = \pm\sqrt{2} \qquad \text{Take the square root}$$

The x-intercepts are $\sqrt{2}$ and $-\sqrt{2}$.

To find the y-intercepts, we set $x = 0$ and solve for y. Thus

$$y = 0^2 - 2 \qquad \text{Set } x = 0$$

$$y = -2$$

The y-intercept is -2.

The graph of this equation was sketched in Example 5. It is repeated in Figure 12 with the x- and y-intercepts labeled.

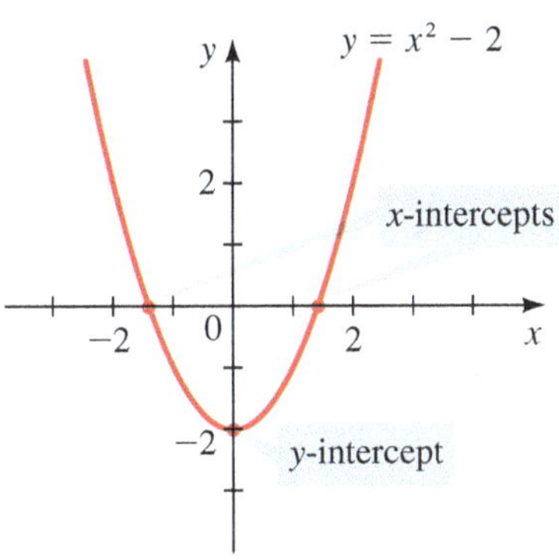

FIGURE 12

Now Try Exercise 71 ■

■ Circles

So far, we have discussed how to find the graph of an equation in x and y. The converse problem is to find an equation of a graph, that is, an equation that represents a given curve in the xy-plane. Such an equation is satisfied by the coordinates of the points on the curve and by no other point. This is the other half of the fundamental principle of analytic geometry as formulated by Descartes and Fermat. The idea is that if a geometric curve can be represented by an algebraic equation, then the rules of algebra can be used to analyze the curve.

As an example of this type of problem, let's find the equation of a circle with radius r and center (h, k). By definition the circle is the set of all points $P(x, y)$ whose

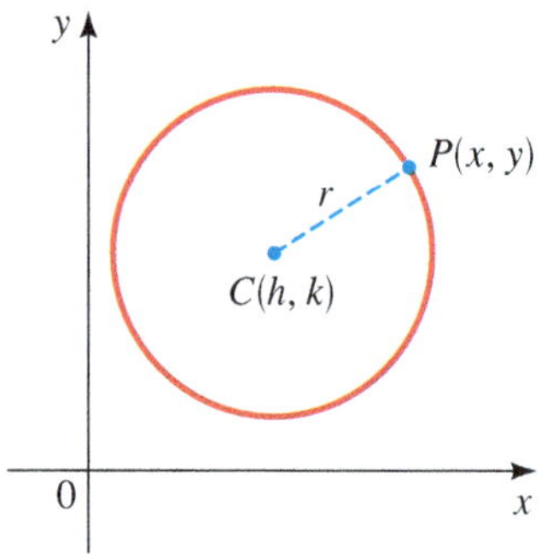

FIGURE 13

distance from the center $C(h, k)$ is r (see Figure 13). Thus P is on the circle if and only if $d(P, C) = r$. From the distance formula we have

$$\sqrt{(x-h)^2 + (y-k)^2} = r$$

$$(x-h)^2 + (y-k)^2 = r^2 \qquad \text{Square each side}$$

This is the desired equation.

> **EQUATION OF A CIRCLE**
>
> An equation of the circle with center (h, k) and radius r is
>
> $$(x-h)^2 + (y-k)^2 = r^2$$
>
> This is called the **standard form** for the equation of the circle. If the center of the circle is the origin $(0, 0)$, then the equation is
>
> $$x^2 + y^2 = r^2$$

EXAMPLE 9 ■ Graphing a Circle

Graph each equation.

(a) $x^2 + y^2 = 25$ **(b)** $(x-2)^2 + (y+1)^2 = 25$

SOLUTION

(a) Rewriting the equation as $x^2 + y^2 = 5^2$, we see that this is an equation of the circle of radius 5 centered at the origin. Its graph is shown in Figure 14.

(b) Rewriting the equation as $(x-2)^2 + (y+1)^2 = 5^2$, we see that this is an equation of the circle of radius 5 centered at $(2, -1)$. Its graph is shown in Figure 15.

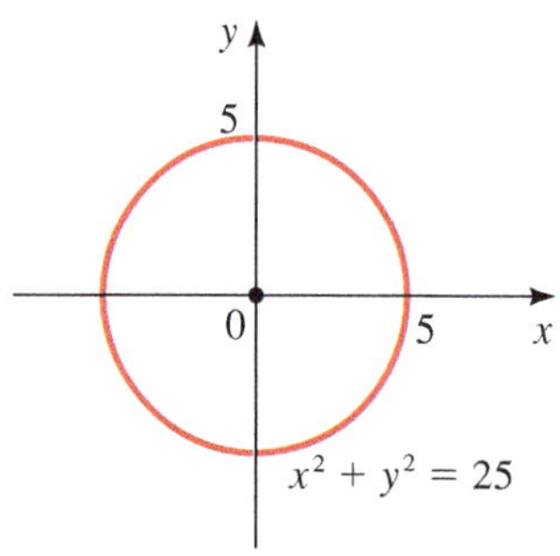

FIGURE 14

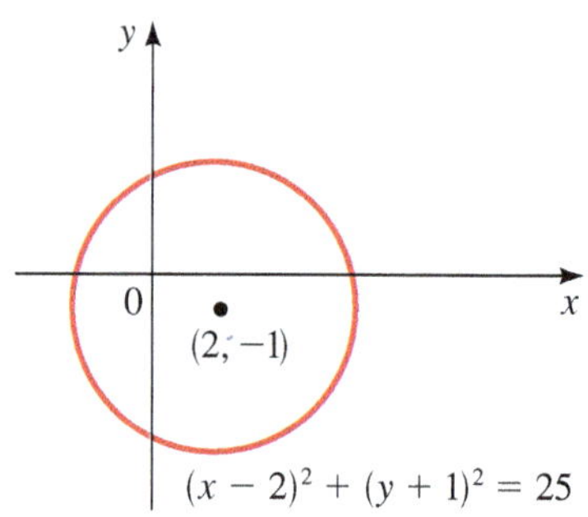

FIGURE 15

Now Try Exercises 83 and 85 ■

EXAMPLE 10 ■ Finding an Equation of a Circle

(a) Find an equation of the circle with radius 3 and center $(2, -5)$.

(b) Find an equation of the circle that has the points $P(1, 8)$ and $Q(5, -6)$ as the endpoints of a diameter.

SOLUTION

(a) Using the equation of a circle with $r = 3$, $h = 2$, and $k = -5$, we obtain

$$(x-2)^2 + (y+5)^2 = 9$$

The graph is shown in Figure 16.

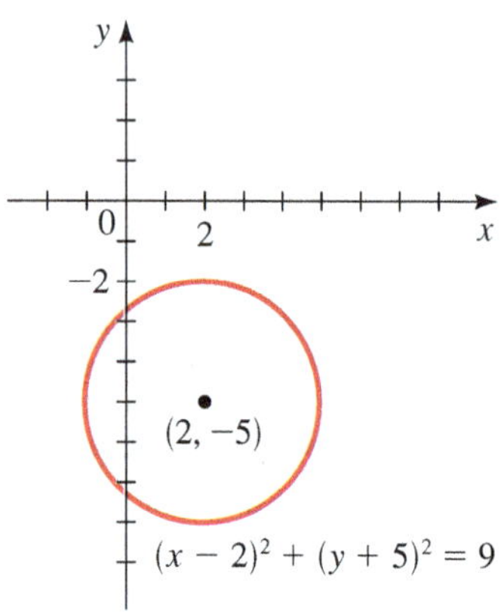

FIGURE 16

(b) We first observe that the center is the midpoint of the diameter PQ, so by the Midpoint Formula the center is

$$\left(\frac{1+5}{2}, \frac{8-6}{2}\right) = (3, 1)$$

The radius r is the distance from P to the center, so by the Distance Formula

$$r^2 = (3-1)^2 + (1-8)^2 = 2^2 + (-7)^2 = 53$$

Therefore the equation of the circle is

$$(x-3)^2 + (y-1)^2 = 53$$

The graph is shown in Figure 17.

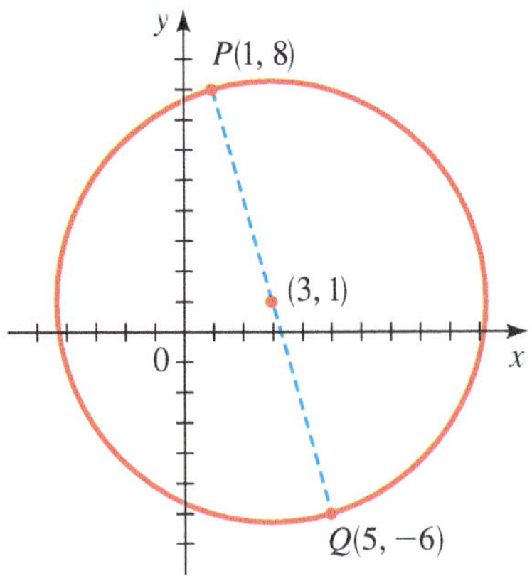

$(x-3)^2 + (y-1)^2 = 53$

FIGURE 17

Now Try Exercises 89 and 93

Let's expand the equation of the circle in the preceding example.

$$(x-3)^2 + (y-1)^2 = 53 \qquad \text{Standard form}$$
$$x^2 - 6x + 9 + y^2 - 2y + 1 = 53 \qquad \text{Expand the squares}$$
$$x^2 - 6x + y^2 - 2y = 43 \qquad \text{Subtract 10 to get expanded form}$$

Completing the square is used in many contexts in algebra. In Section 1.5 we used completing the square to solve quadratic equations.

Suppose we are given the equation of a circle in expanded form. Then to find its center and radius, we must put the equation back in standard form. That means that we must reverse the steps in the preceding calculation, and to do that, we need to know what to add to an expression like $x^2 - 6x$ to make it a perfect square—that is, we need to complete the square, as in the next example.

EXAMPLE 11 ■ Identifying an Equation of a Circle

Show that the equation $x^2 + y^2 + 2x - 6y + 7 = 0$ represents a circle, and find the center and radius of the circle.

SOLUTION We first group the x-terms and y-terms. Then we complete the square within each grouping. That is, we complete the square for $x^2 + 2x$ by adding $\left(\frac{1}{2}\cdot 2\right)^2 = 1$, and we complete the square for $y^2 - 6y$ by adding $\left[\frac{1}{2}\cdot(-6)\right]^2 = 9$.

We must add the same numbers to *each side* to maintain equality.

$$(x^2 + 2x \quad\;) + (y^2 - 6y \quad\;) = -7 \qquad \text{Group terms}$$
$$(x^2 + 2x + 1) + (y^2 - 6y + 9) = -7 + 1 + 9 \qquad \text{Complete the square by adding 1 and 9 to each side}$$
$$(x+1)^2 + (y-3)^2 = 3 \qquad \text{Factor and simplify}$$

Comparing this equation with the standard equation of a circle, we see that $h = -1$, $k = 3$, and $r = \sqrt{3}$, so the given equation represents a circle with center $(-1, 3)$ and radius $\sqrt{3}$.

Now Try Exercise 99

■ Symmetry

Figure 18 shows the graph of $y = x^2$. Notice that the part of the graph to the left of the y-axis is the mirror image of the part to the right of the y-axis. The reason is that if the point (x, y) is on the graph, then so is $(-x, y)$, and these points are reflections of each other about the y-axis. In this situation we say that the graph is **symmetric with respect to the y-axis**. Similarly, we say that a graph is **symmetric with respect to the x-axis** if whenever the point (x, y) is on the graph, then so is $(x, -y)$. A graph is **symmetric with respect to the origin** if whenever (x, y) is on the graph, so is $(-x, -y)$. (We often say symmetric "about" instead of "with respect to.")

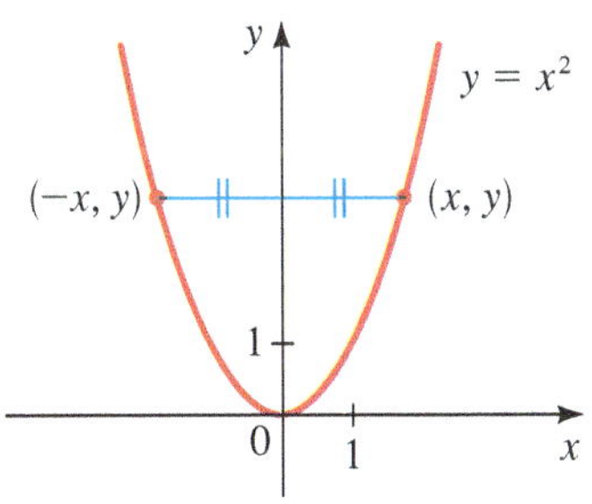

FIGURE 18

TYPES OF SYMMETRY

<table>
<tr><th>Symmetry</th><th>Test</th><th>Graph</th><th>Property of Graph</th></tr>
<tr><td>With respect to the x-axis</td><td>Replace y by $-y$. The resulting equation is equivalent to the original one.</td><td>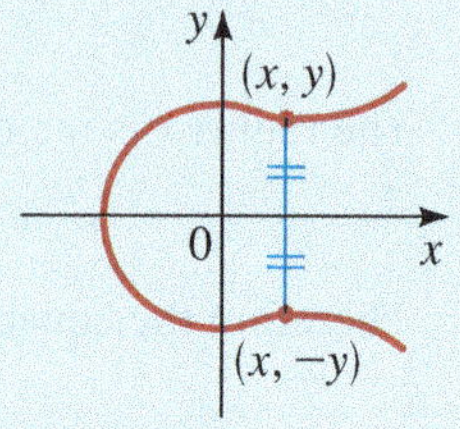
</td><td>Graph is unchanged when reflected about the x-axis. See Figures 14 and 19.</td></tr>
<tr><td>With respect to the y-axis</td><td>Replace x by $-x$. The resulting equation is equivalent to the original one.</td><td>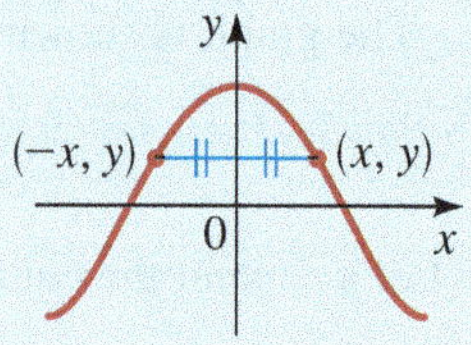
</td><td>Graph is unchanged when reflected about the y-axis. See Figures 9, 10, 11, 12, 14, and 18.</td></tr>
<tr><td>With respect to the origin</td><td>Replace x by $-x$ and y by $-y$. The resulting equation is equivalent to the original one.</td><td>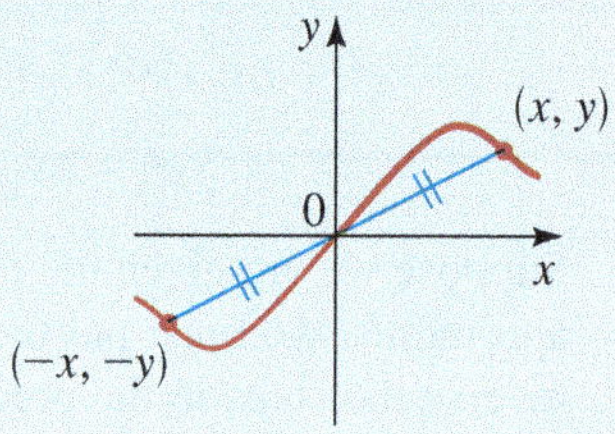
</td><td>Graph is unchanged when rotated 180° about the origin. See Figures 14 and 20.</td></tr>
</table>

The remaining examples in this section show how symmetry helps us to sketch the graphs of equations.

EXAMPLE 12 ■ Using Symmetry to Sketch a Graph

Test the equation $x = y^2$ for symmetry and sketch the graph.

SOLUTION If y is replaced by $-y$ in the equation $x = y^2$, we get

$$x = (-y)^2 \quad \text{Replace } y \text{ by } -y$$

$$x = y^2 \quad \text{Simplify}$$

and so the equation is equivalent to the original one. Therefore the graph is symmetric about the x-axis. But changing x to $-x$ gives the equation $-x = y^2$, which is not equivalent to the original equation, so the graph is not symmetric about the y-axis.

We use the symmetry about the x-axis to sketch the graph by first plotting points just for $y > 0$ and then reflecting the graph about the x-axis, as shown in Figure 19.

y	$x = y^2$	(x, y)
0	0	(0, 0)
1	1	(1, 1)
2	4	(4, 2)
3	9	(9, 3)

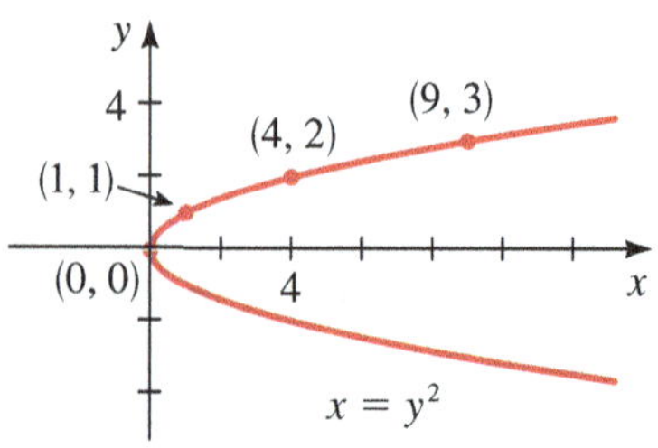

FIGURE 19

Now Try Exercises 105 and 111 ■

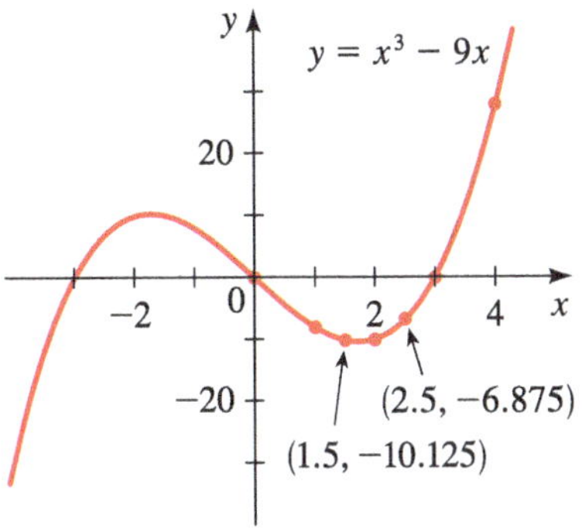

FIGURE 20

EXAMPLE 13 ■ Testing an Equation for Symmetry

Test the equation $y = x^3 - 9x$ for symmetry.

SOLUTION If we replace x by $-x$ and y by $-y$ in the equation, we get

$$-y = (-x)^3 - 9(-x) \quad \text{Replace } x \text{ by } -x \text{ and } y \text{ by } -y$$

$$-y = -x^3 + 9x \quad \text{Simplify}$$

$$y = x^3 - 9x \quad \text{Multiply by } -1$$

and so the equation is equivalent to the original one. This means that the graph is symmetric with respect to the origin, as shown in Figure 20.

Now Try Exercise 107 ■

1.9 EXERCISES

CONCEPTS

1. The point that is 3 units to the right of the y-axis and 5 units below the x-axis has coordinates (____, ____).

2. The distance between the points (a, b) and (c, d) is __________. So the distance between (1, 2) and (7, 10) is ______.

3. The point midway between (a, b) and (c, d) is ______. So the point midway between (1, 2) and (7, 10) is ______.

4. If the point (2, 3) is on the graph of an equation in x and y, then the equation is satisfied when we replace x by ______ and y by ______. Is the point (2, 3) on the graph of the equation $2y = x + 1$? Complete the table, and sketch a graph.

x	y	(x, y)
−2		
−1		
0		
1		
2		

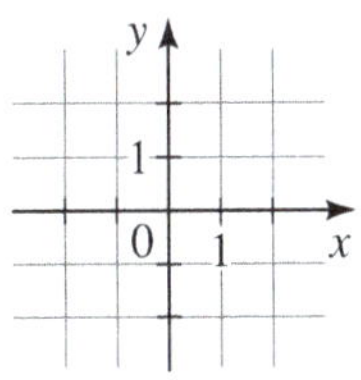

5. (a) To find the x-intercept(s) of the graph of an equation, we set ______ equal to 0 and solve for ______. So the x-intercept of $2y = x + 1$ is ______.

(b) To find the y-intercept(s) of the graph of an equation, we set ______ equal to 0 and solve for ______. So the y-intercept of $2y = x + 1$ is ______.

6. The graph of the equation $(x - 1)^2 + (y - 2)^2 = 9$ is a circle with center (____, ____) and radius ______.

7. (a) If a graph is symmetric with respect to the x-axis and (a, b) is on the graph, then (____, ____) is also on the graph.

(b) If a graph is symmetric with respect to the y-axis and (a, b) is on the graph, then (____, ____) is also on the graph.

(c) If a graph is symmetric about the origin and (a, b) is on the graph, then (____, ____) is also on the graph.

8. The graph of an equation is shown below.

(a) The x-intercept(s) are ______, and the y-intercept(s) are ______.

(b) The graph is symmetric about the ______ (x-axis/y-axis/origin).

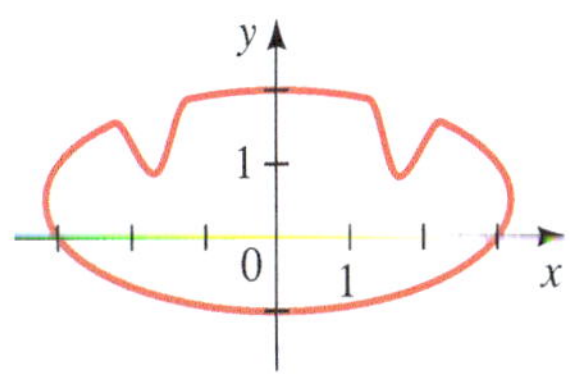

9–10 ■ *Yes or No*? If *No*, give a reason.

9. If the graph of an equation is symmetric with respect to both the x- and y-axes, is it necessarily symmetric with respect to the origin?

10. If the graph of an equation is symmetric with respect to the origin, is it necessarily symmetric with respect to the x- or y-axes?

SKILLS

11–12 ■ Points in a Coordinate Plane Refer to the figure below.

11. Find the coordinates of the points shown.

12. List the points that lie in Quadrants I and III.

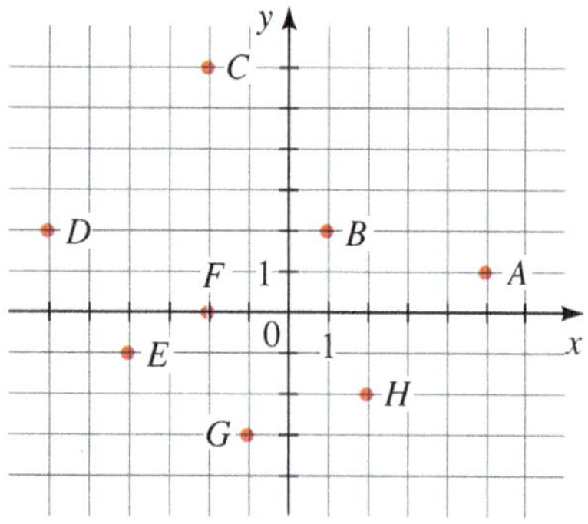

13–14 ■ Points in a Coordinate Plane Plot the given points in a coordinate plane.

13. $(0, 5), (-1, 0), (-1, -2), (\frac{1}{2}, \frac{2}{3})$

14. $(-5, 0), (2, 0), (2.6, -1.3), (-2.5, 3.5)$

15–20 ■ Sketching Regions Sketch the region given by the set.

15. (a) $\{(x, y) \mid x \geq 2\}$ **(b)** $\{(x, y) \mid y = 2\}$

16. (a) $\{(x, y) \mid y < 3\}$ **(b)** $\{(x, y) \mid x = -4\}$

17. (a) $\{(x, y) \mid -3 < x < 3\}$ **(b)** $\{(x, y) \mid |x| \leq 2\}$

18. (a) $\{(x, y) \mid 0 \leq y \leq 2\}$ **(b)** $\{(x, y) \mid |y| > 2\}$

19. (a) $\{(x, y) \mid -2 < x < 2 \text{ and } y \geq 1\}$

(b) $\{(x, y) \mid xy < 0\}$

20. (a) $\{(x, y) \mid |x| \leq 1 \text{ and } |y| \leq 3\}$

(b) $\{(x, y) \mid xy > 0\}$

21–24 ■ Distance and Midpoint A pair of points is graphed. **(a)** Find the distance between them. **(b)** Find the midpoint of the segment that joins them.

21.

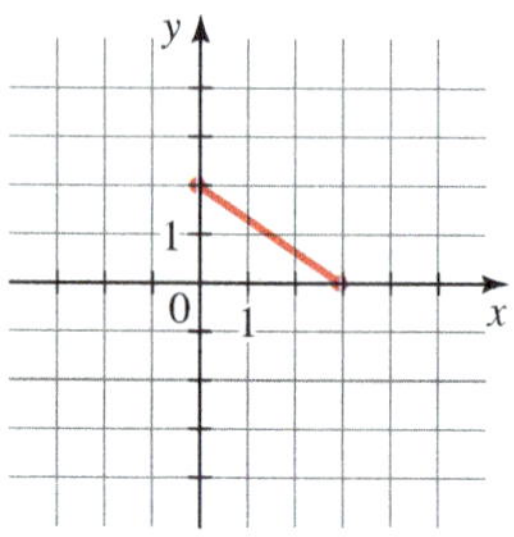

22.

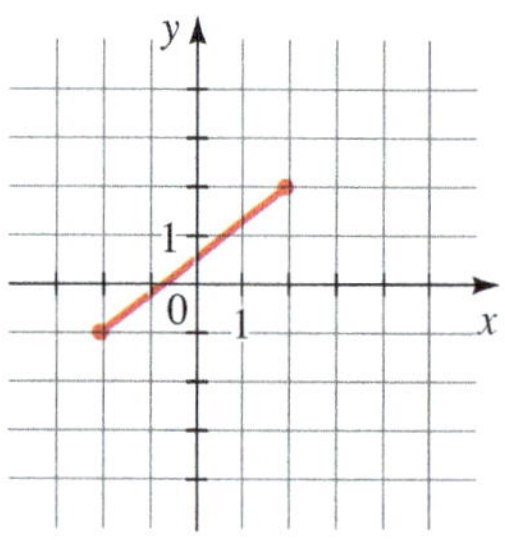

23.

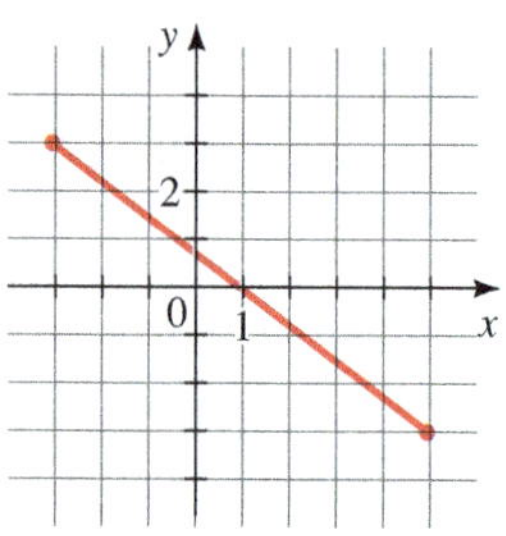

24.

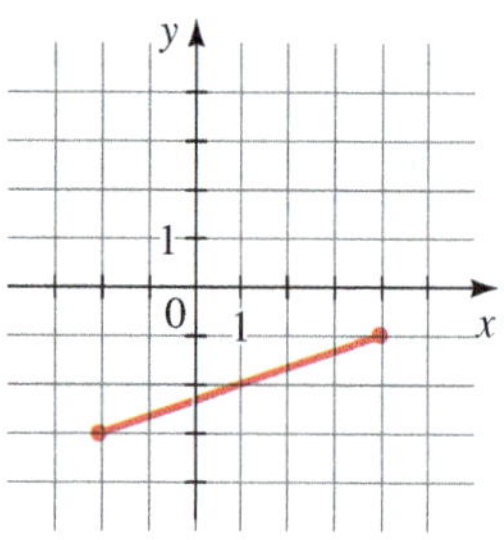

25–30 ■ Distance and Midpoint A pair of points is given. **(a)** Plot the points in a coordinate plane. **(b)** Find the distance between them. **(c)** Find the midpoint of the segment that joins them.

25. $(0, 8), (6, 16)$ **26.** $(-2, 5), (10, 0)$

27. $(3, -2), (-4, 5)$ **28.** $(-1, 1), (-6, -3)$

29. $(6, -2), (-6, 2)$ **30.** $(0, -6), (5, 0)$

31–34 ■ Area In these exercises we find the areas of plane figures.

31. Draw the rectangle with vertices $A(1, 3)$, $B(5, 3)$, $C(1, -3)$, and $D(5, -3)$ on a coordinate plane. Find the area of the rectangle.

32. Draw the parallelogram with vertices $A(1, 2)$, $B(5, 2)$, $C(3, 6)$, and $D(7, 6)$ on a coordinate plane. Find the area of the parallelogram.

33. Plot the points $A(1, 0)$, $B(5, 0)$, $C(4, 3)$, and $D(2, 3)$ on a coordinate plane. Draw the segments AB, BC, CD, and DA. What kind of quadrilateral is $ABCD$, and what is its area?

34. Plot the points $P(5, 1)$, $Q(0, 6)$, and $R(-5, 1)$ on a coordinate plane. Where must the point S be located so that the quadrilateral $PQRS$ is a square? Find the area of this square.

35–39 ■ Distance Formula In these exercises we use the Distance Formula.

35. Which of the points $A(6, 7)$ or $B(-5, 8)$ is closer to the origin?

36. Which of the points $C(-6, 3)$ or $D(3, 0)$ is closer to the point $E(-2, 1)$?

37. Which of the points $P(3, 1)$ or $Q(-1, 3)$ is closer to the point $R(-1, -1)$?

38. (a) Show that the points $(7, 3)$ and $(3, 7)$ are the same distance from the origin.

(b) Show that the points (a, b) and (b, a) are the same distance from the origin.

39. Show that the triangle with vertices $A(0, 2)$, $B(-3, -1)$, and $C(-4, 3)$ is isosceles.

40. Area of Triangle Find the area of the triangle shown in the figure.

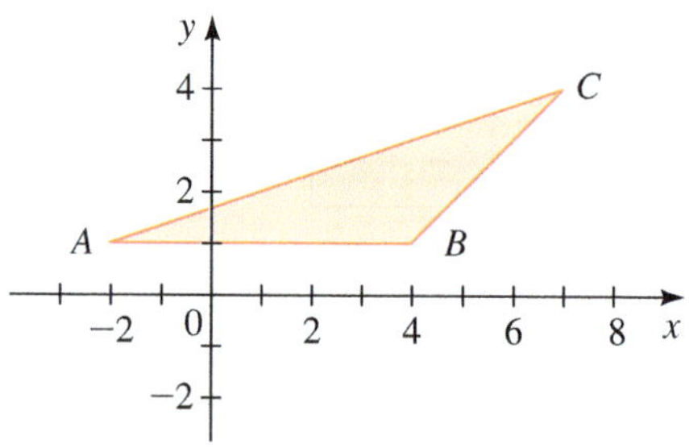

41–42 ■ Pythagorean Theorem In these exercises we use the converse of the Pythagorean Theorem (Appendix A) to show that the given triangle is a right triangle.

41. Refer to triangle ABC in the figure below.

(a) Show that triangle ABC is a right triangle by using the converse of the Pythagorean Theorem.

(b) Find the area of triangle ABC.

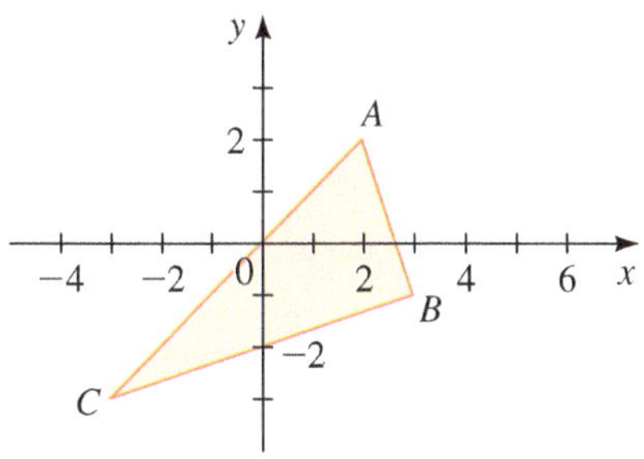

42. Show that the triangle with vertices $A(6, -7)$, $B(11, -3)$, and $C(2, -2)$ is a right triangle by using the converse of the Pythagorean Theorem. Find the area of the triangle.

43–45 ■ Distance Formula In these exercises we use the Distance Formula.

43. Show that the points $A(-2, 9)$, $B(4, 6)$, $C(1, 0)$, and $D(-5, 3)$ are the vertices of a square.

44. Show that the points $A(-1, 3)$, $B(3, 11)$, and $C(5, 15)$ are collinear by showing that $d(A, B) + d(B, C) = d(A, C)$.

45. Find a point on the y-axis that is equidistant from the points $(5, -5)$ and $(1, 1)$.

46–50 ■ Distance and Midpoint Formulas In these exercises we use the Distance Formula and the Midpoint Formula.

46. Find the lengths of the medians of the triangle with vertices $A(1, 0)$, $B(3, 6)$, and $C(8, 2)$. (A *median* is a line segment from a vertex to the midpoint of the opposite side.)

47. Plot the points $P(-1, -4)$, $Q(1, 1)$, and $R(4, 2)$ on a coordinate plane. Where should the point S be located so that the figure $PQRS$ is a parallelogram?

48. If $M(6, 8)$ is the midpoint of the line segment AB and if A has coordinates $(2, 3)$, find the coordinates of B.

49. (a) Sketch the parallelogram with vertices $A(-2, -1)$, $B(4, 2)$, $C(7, 7)$, and $D(1, 4)$.

(b) Find the midpoints of the diagonals of this parallelogram.

(c) From part (b) show that the diagonals bisect each other.

50. The point M in the figure is the midpoint of the line segment AB. Show that M is equidistant from the vertices of triangle ABC.

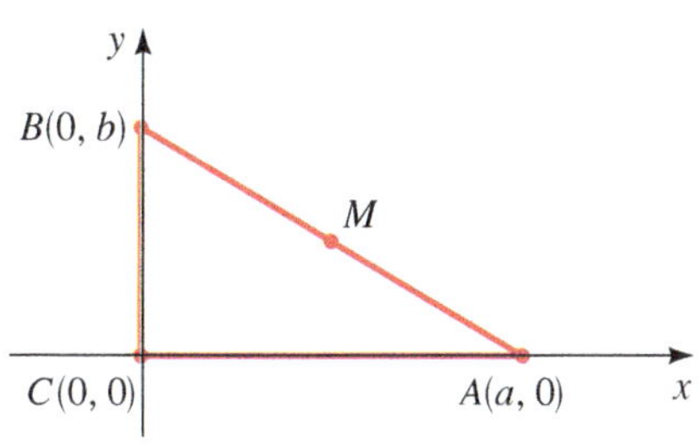

51–54 ■ Points on a Graph? Determine whether the given points are on the graph of the equation.

51. $x - 2y - 1 = 0$; $(0, 0), (1, 0), (-1, -1)$

52. $y(x^2 + 1) = 1$; $(1, 1), (1, \frac{1}{2}), (-1, \frac{1}{2})$

53. $x^2 + xy + y^2 = 4$; $(0, -2), (1, -2), (2, -2)$

54. $x^2 + y^2 = 1$; $(0, 1), \left(\frac{1}{\sqrt{2}}, \frac{1}{\sqrt{2}}\right), \left(\frac{\sqrt{3}}{2}, \frac{1}{2}\right)$

55–60 ■ Graphing Equations Make a table of values, and sketch the graph of the equation.

55. $4x + 5y = 40$ **56.** $3x - 5y = 30$

57. $y = x^2 + 4$ **58.** $y = 3 - x^2$

59. $y = |x| - 1$ **60.** $y = |x + 1|$

61–64 ■ Graphing Equations Use a graphing calculator to graph the equation in the given viewing rectangle.

61. $y = 0.01x^3 - x^2 + 5$; $[-100, 150]$ by $[-2000, 2000]$

62. $y = \sqrt{12x - 17}$; $[0, 10]$ by $[0, 20]$

63. $y = \dfrac{x}{x^2 + 25}$; $[-50, 50]$ by $[-0.2, 0.2]$

64. $y = x^4 - 4x^3$; $[-4, 6]$ by $[-50, 100]$

65–70 ■ Graphing Equations Make a table of values, and sketch the graph of the equation. Find the x- and y-intercepts, and test for symmetry.

65. (a) $2x - y = 6$ (b) $y = -(x + 1)^2$

66. (a) $x - 4y = 8$ (b) $y = -x^2 + 3$

67. (a) $y = \sqrt{x + 1}$ (b) $y = -|x|$

68. (a) $y = 3 - \sqrt{x}$ (b) $x = |y|$

69. (a) $y = \sqrt{4 - x^2}$ (b) $x = y^3 + 2y$

70. (a) $y = -\sqrt{4 - x^2}$ (b) $x = y^3$

71–74 ■ Intercepts Find the x- and y-intercepts of the graph of the equation.

71. (a) $y = x + 6$ (b) $y = x^2 - 5$

72. (a) $4x^2 + 25y^2 = 100$ (b) $x^2 - xy + 3y = 1$

73. (a) $9x^2 - 4y^2 = 36$ (b) $y - 2xy + 4x = 1$

74. (a) $y = \sqrt{x^2 - 16}$ (b) $y = \sqrt{64 - x^3}$

75–78 ■ Intercepts An equation and its graph are given. Find the x- and y-intercepts.

75. $y = 4x - x^2$

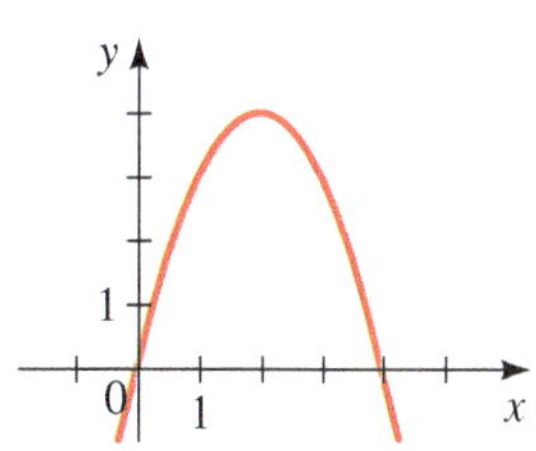

76. $\dfrac{x^2}{9} + \dfrac{y^2}{4} = 1$

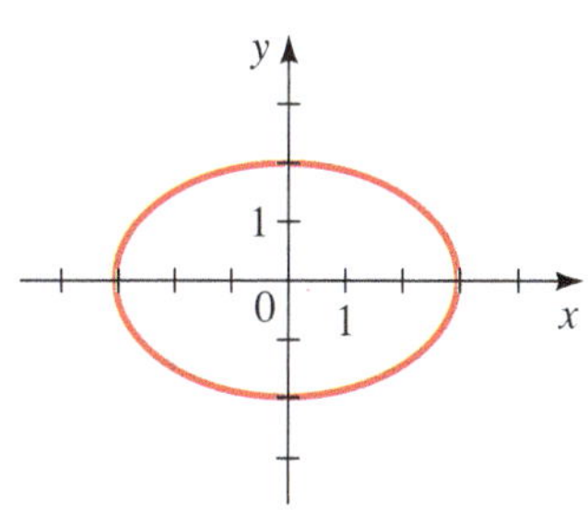

77. $x^4 + y^2 - xy = 16$

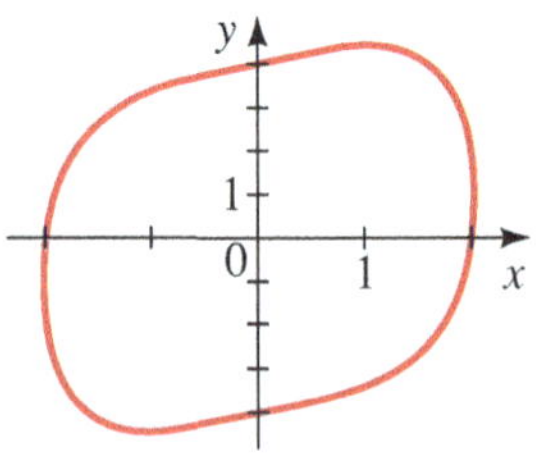

78. $x^2 + y^3 - x^2y^2 = 64$

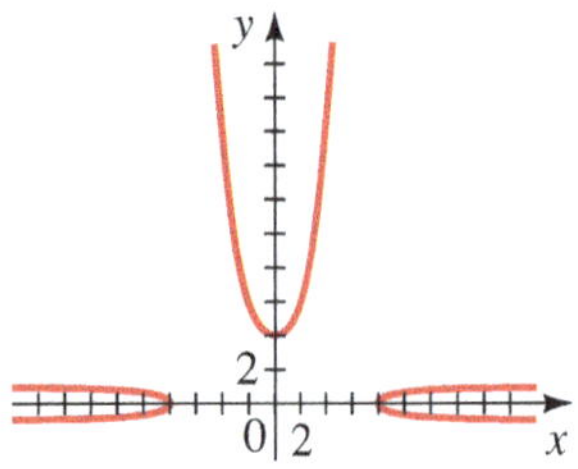

79–82 ■ Graphing Equations An equation is given. **(a)** Use a graphing calculator to graph the equation in the given viewing rectangle. **(b)** Find the x- and y-intercepts from the graph. **(c)** Verify your answers to part (b) algebraically (from the equation).

79. $y = x^3 - x^2$; $[-2, 2]$ by $[-1, 1]$

80. $y = x^4 - 2x^3$; $[-2, 3]$ by $[-3, 3]$

81. $y = -\dfrac{2}{x^2 + 1}$; $[-5, 5]$ by $[-3, 1]$

82. $y = \sqrt[3]{1 - x^2}$; $[-5, 5]$ by $[-5, 3]$

83–88 ■ Graphing Circles Find the center and radius of the circle, and sketch its graph.

83. $x^2 + y^2 = 9$

84. $x^2 + y^2 = 5$

85. $x^2 + (y - 4)^2 = 1$

86. $(x + 1)^2 + y^2 = 9$

87. $(x + 3)^2 + (y - 4)^2 = 25$

88. $(x + 1)^2 + (y + 2)^2 = 36$

89–96 ■ Equations of Circles Find an equation of the circle that satisfies the given conditions.

89. Center $(2, -1)$; radius 3

90. Center $(-1, -4)$; radius 8

91. Center at the origin; passes through $(4, 7)$

92. Center $(-1, 5)$; passes through $(-4, -6)$

93. Endpoints of a diameter are $P(-1, 1)$ and $Q(5, 9)$

94. Endpoints of a diameter are $P(-1, 3)$ and $Q(7, -5)$

95. Center $(7, -3)$; tangent to the x-axis

96. Circle lies in the first quadrant, tangent to both x- and y-axes; radius 5

97–98 ■ Equations of Circles Find the equation of the circle shown in the figure.

97.

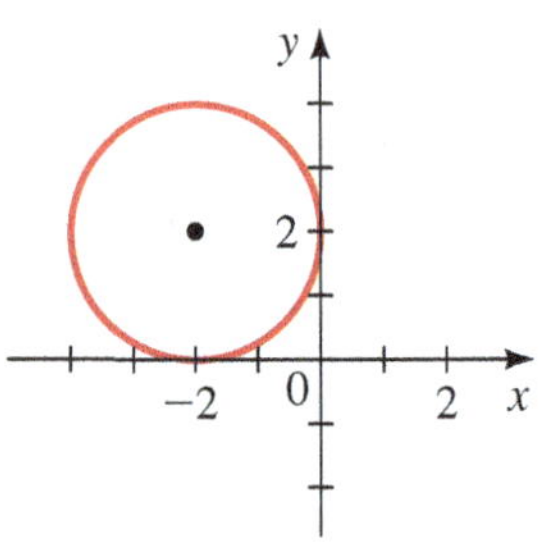

98.

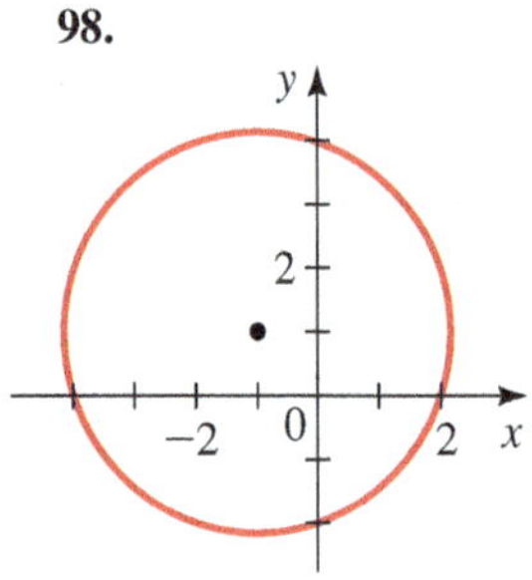

99–104 ■ Equations of Circles Show that the equation represents a circle, and find the center and radius of the circle.

99. $x^2 + y^2 + 4x - 6y + 12 = 0$

100. $x^2 + y^2 + 6y + 2 = 0$

101. $x^2 + y^2 - \frac{1}{2}x + \frac{1}{2}y = \frac{1}{8}$

102. $x^2 + y^2 + \frac{1}{2}x + 2y + \frac{1}{16} = 0$

103. $2x^2 + 2y^2 - 3x = 0$

104. $3x^2 + 3y^2 + 6x - y = 0$

105–110 ■ Symmetry Test the equation for symmetry.

105. $y = x^4 + x^2$

106. $x = y^4 - y^2$

107. $x^2y^2 + xy = 1$

108. $x^4y^4 + x^2y^2 = 1$

109. $y = x^3 + 10x$

110. $y = x^2 + |x|$

111–114 ■ Symmetry Complete the graph using the given symmetry property.

111. Symmetric with respect to the y-axis

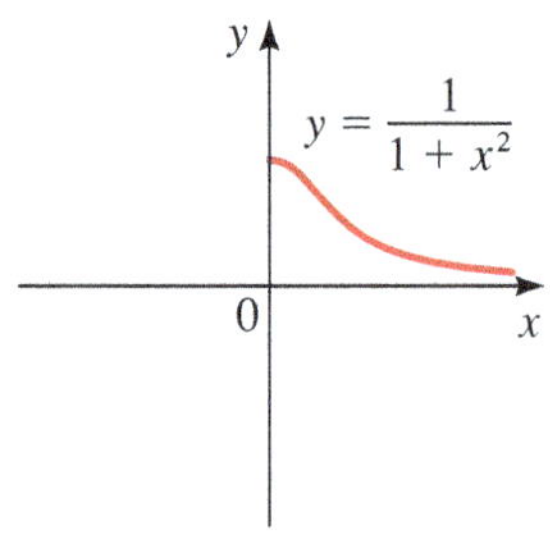

112. Symmetric with respect to the x-axis

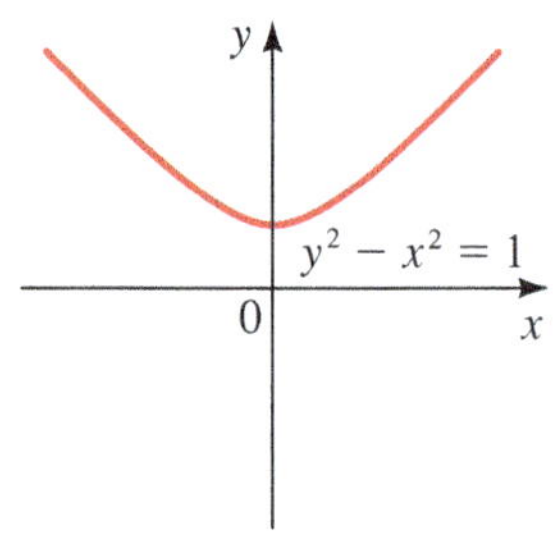

113. Symmetric with respect to the origin

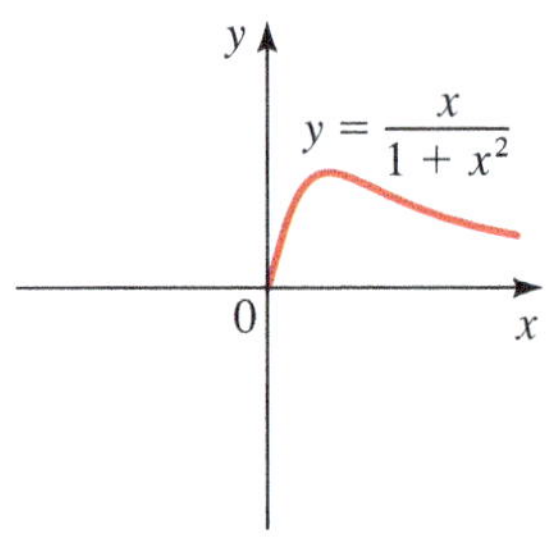

114. Symmetric with respect to the origin

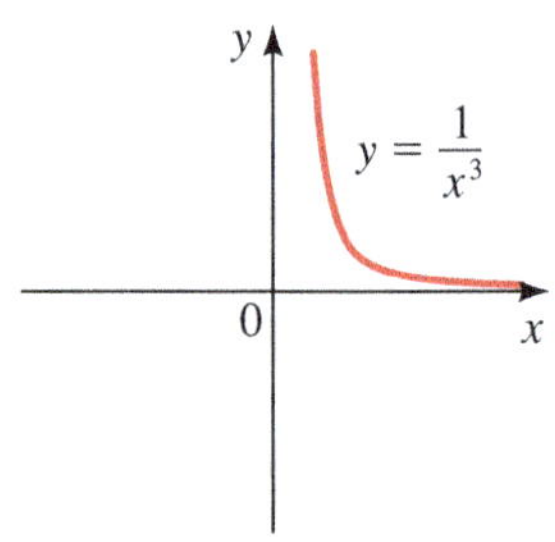

SKILLS Plus

115–116 ■ Graphing Regions Sketch the region given by the set.

115. $\{(x, y) \mid x^2 + y^2 \le 1\}$

116. $\{(x, y) \mid x^2 + y^2 > 4\}$

117. Area of a Region Find the area of the region that lies outside the circle $x^2 + y^2 = 4$ but inside the circle

$$x^2 + y^2 - 4y - 12 = 0$$

118. Area of a Region Sketch the region in the coordinate plane that satisfies both the inequalities $x^2 + y^2 \le 9$ and $y \ge |x|$. What is the area of this region?

119. Shifting the Coordinate Plane Suppose that each point in the coordinate plane is shifted 3 units to the right and 2 units upward.

(a) The point $(5, 3)$ is shifted to what new point?

(b) The point (a, b) is shifted to what new point?

(c) What point is shifted to $(3, 4)$?

(d) Triangle ABC in the figure has been shifted to triangle $A'B'C'$. Find the coordinates of the points A', B', and C'.

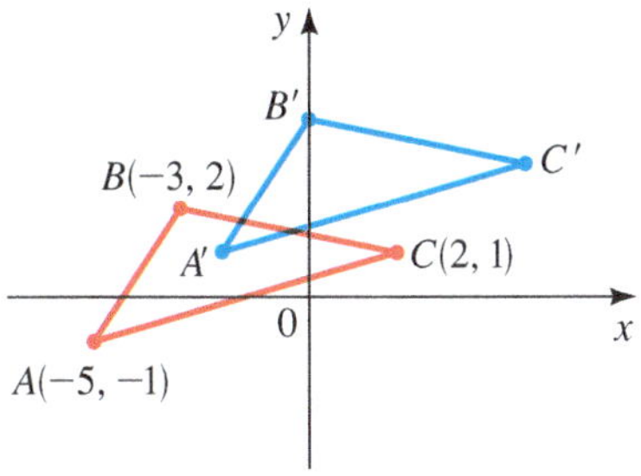

120. Reflecting in the Coordinate Plane Suppose that the y-axis acts as a mirror that reflects each point to the right of it into a point to the left of it.

(a) The point $(3, 7)$ is reflected to what point?

(b) The point (a, b) is reflected to what point?

(c) What point is reflected to $(-4, -1)$?

(d) Triangle ABC in the figure is reflected to triangle $A'B'C'$. Find the coordinates of the points A', B', and C'.

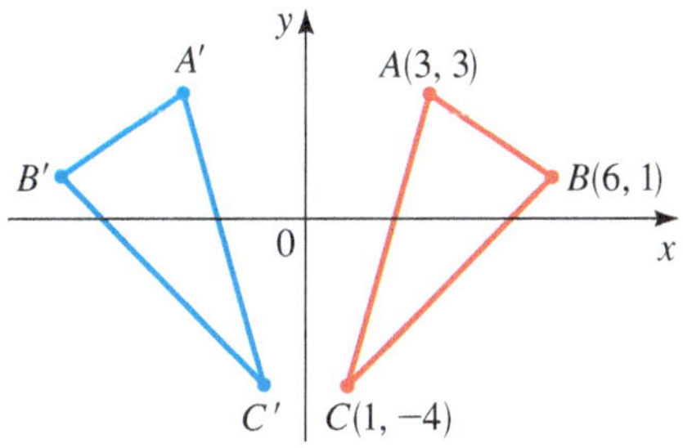

121. Making a Graph Symmetric The graph shown in the figure is not symmetric about the x-axis, the y-axis, or the origin. Add more line segments to the graph so that it exhibits the indicated symmetry. In each case, add as little as possible.

(a) Symmetry about the x-axis

(b) Symmetry about the y-axis

(c) Symmetry about the origin

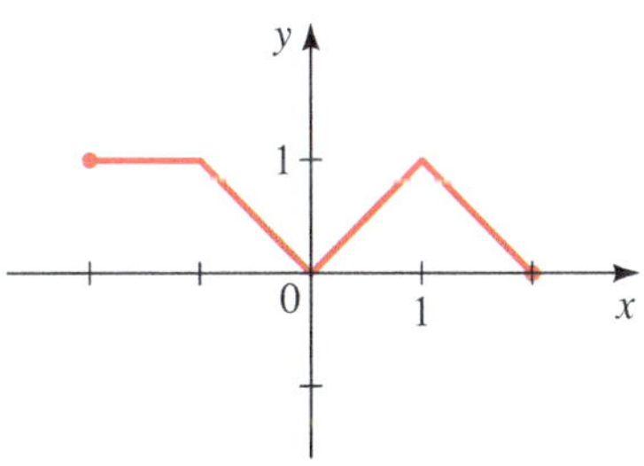

APPLICATIONS

122. Distances in a City A city has streets that run north and south and avenues that run east and west, all equally spaced. Streets and avenues are numbered sequentially, as shown in the figure. The *walking* distance between points A and B is 7 blocks—that is, 3 blocks east and 4 blocks north. To find the *straight-line* distance d, we must use the Distance Formula.

(a) Find the straight-line distance (in blocks) between A and B.

(b) Find the walking distance and the straight-line distance between the corner of 4th St. and 2nd Ave. and the corner of 11th St. and 26th Ave.

(c) What must be true about the points P and Q if the walking distance between P and Q equals the straight-line distance between P and Q?

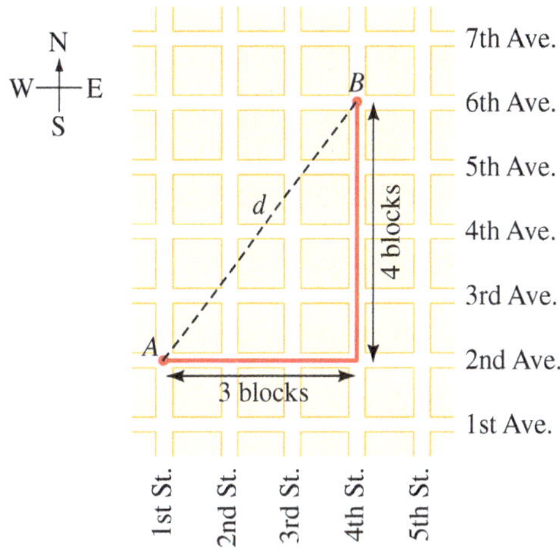

123. Halfway Point Two friends live in the city described in Exercise 122, one at the corner of 3rd St. and 7th Ave., the other at the corner of 27th St. and 17th Ave. They frequently meet at a coffee shop halfway between their homes.

(a) At what intersection is the coffee shop located?

(b) How far must each of them walk to get to the coffee shop?

124. Orbit of a Satellite A satellite is in orbit around the moon. A coordinate plane containing the orbit is set up with the center of the moon at the origin, as shown in the graph, with distances measured in megameters (Mm). The equation of the satellite's orbit is

$$\frac{(x-3)^2}{25} + \frac{y^2}{16} = 1$$

(a) From the graph, determine the closest and the farthest that the satellite gets to the center of the moon.

(b) There are two points in the orbit with y-coordinates 2. Find the x-coordinates of these points, and determine their distances to the center of the moon.

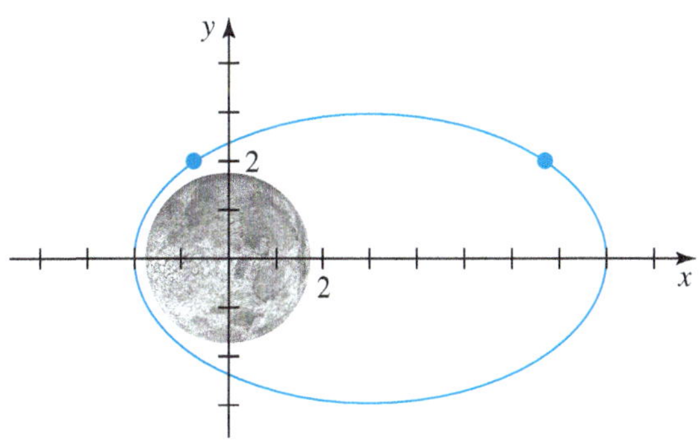

DISCUSS ■ DISCOVER ■ PROVE ■ WRITE

125. WRITE: Completing a Line Segment Plot the points $M(6, 8)$ and $A(2, 3)$ on a coordinate plane. If M is the midpoint of the line segment AB, find the coordinates of B. Write a brief description of the steps you took to find B and your reasons for taking them.

126. WRITE: Completing a Parallelogram Plot the points $P(0, 3)$, $Q(2, 2)$, and $R(5, 3)$ on a coordinate plane. Where should the point S be located so that the figure $PQRS$ is a parallelogram? Write a brief description of the steps you took and your reasons for taking them.

127. DISCOVER: Circle, Point, or Empty Set? Complete the squares in the general equation $x^2 + ax + y^2 + by + c = 0$, and simplify the result as much as possible. Under what conditions on the coefficients a, b, and c does this equation represent a circle? A single point? The empty set? In the case in which the equation does represent a circle, find its center and radius.

1.10 LINES

■ The Slope of a Line ■ Point-Slope Form of the Equation of a Line ■ Slope-Intercept Form of the Equation of a Line ■ Vertical and Horizontal Lines ■ General Equation of a Line ■ Parallel and Perpendicular Lines

In this section we find equations for straight lines lying in a coordinate plane. The equations will depend on how the line is inclined, so we begin by discussing the concept of slope.

■ The Slope of a Line

We first need a way to measure the "steepness" of a line, or how quickly it rises (or falls) as we move from left to right. We define *run* to be the distance we move to the right and *rise* to be the corresponding distance that the line rises (or falls). The *slope* of a line is the ratio of rise to run:

$$\text{slope} = \frac{\text{rise}}{\text{run}}$$

Figure 1 shows situations in which slope is important. Carpenters use the term *pitch* for the slope of a roof or a staircase; the term *grade* is used for the slope of a road.

Slope of a ramp
Slope $= \frac{1}{12}$

Pitch of a roof
Slope $= \frac{1}{3}$

Grade of a road
Slope $= \frac{8}{100}$

FIGURE 1

If a line lies in a coordinate plane, then the **run** is the change in the x-coordinate and the **rise** is the corresponding change in the y-coordinate between any two points on the line (see Figure 2). This gives us the following definition of slope.

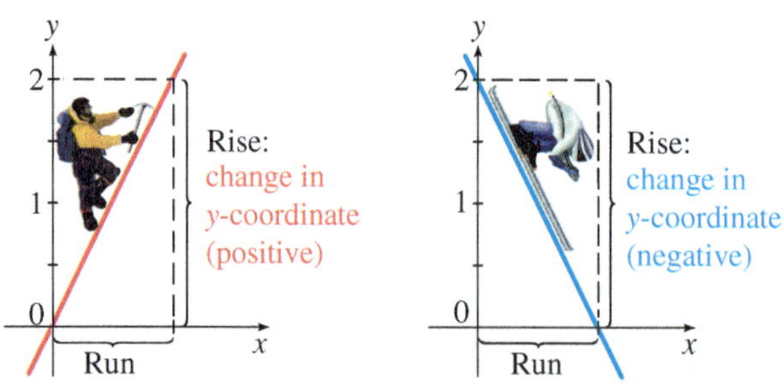

FIGURE 2

SLOPE OF A LINE

The **slope** m of a nonvertical line that passes through the points $A(x_1, y_1)$ and $B(x_2, y_2)$ is

$$m = \frac{\text{rise}}{\text{run}} = \frac{y_2 - y_1}{x_2 - x_1}$$

The slope of a vertical line is not defined.

The slope is independent of which two points are chosen on the line. We can see that this is true from the similar triangles in Figure 3.

$$\frac{y_2 - y_1}{x_2 - x_1} = \frac{y_2' - y_1'}{x_2' - x_1'}$$

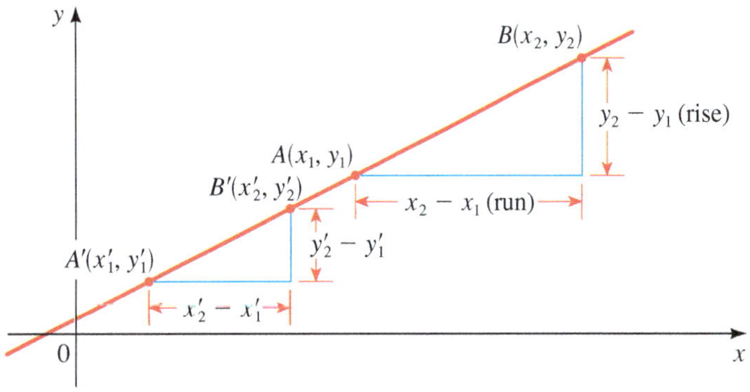

FIGURE 3

The figures in the box below show several lines labeled with their slopes. Notice that lines with positive slope slant upward to the right, whereas lines with negative slope slant downward to the right. The steepest lines are those for which the absolute value of the slope is the largest; a horizontal line has slope 0. The slope of a vertical line is undefined (it has a 0 denominator), so we say that a vertical line has no slope.

SLOPE OF A LINE

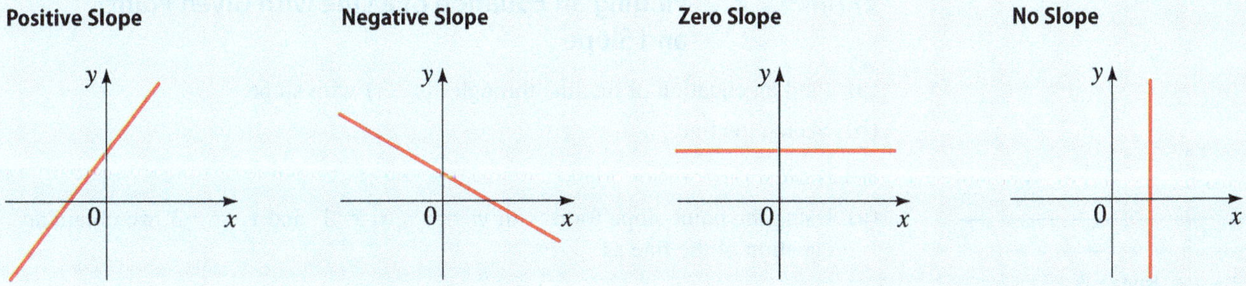

EXAMPLE 1 ■ Finding the Slope of a Line Through Two Points

Find the slope of the line that passes through the points $P(2, 1)$ and $Q(8, 5)$.

SOLUTION Since any two different points determine a line, only one line passes through these two points. From the definition the slope is

$$m = \frac{y_2 - y_1}{x_2 - x_1} = \frac{5 - 1}{8 - 2} = \frac{4}{6} = \frac{2}{3}$$

This says that for every 3 units we move to the right, the line rises 2 units. The line is drawn in Figure 4.

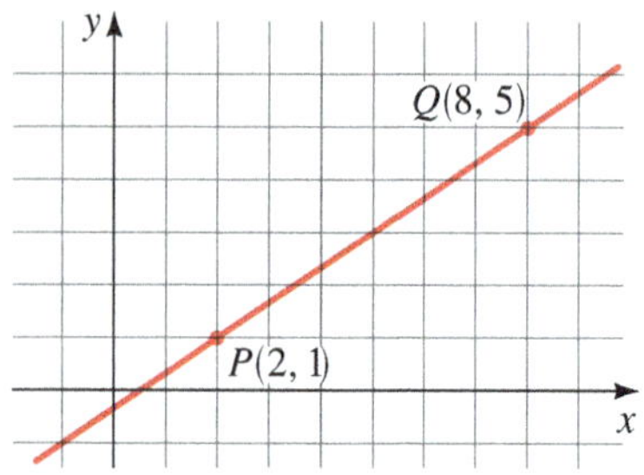

FIGURE 4

Now Try Exercise 9 ■

Point-Slope Form of the Equation of a Line

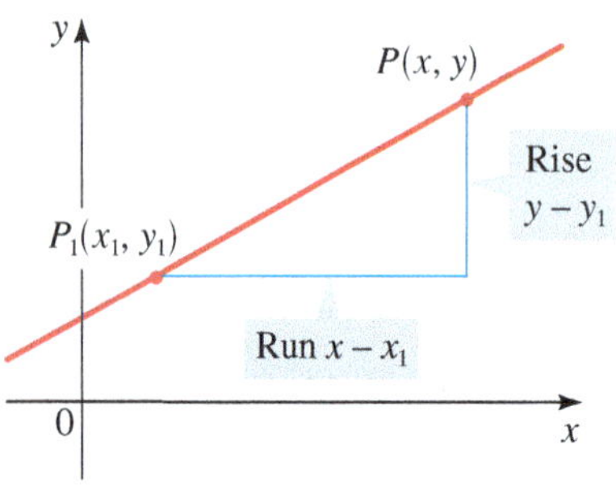

FIGURE 5

Now let's find the equation of the line that passes through a given point $P(x_1, y_1)$ and has slope m. A point $P(x, y)$ with $x \neq x_1$ lies on this line if and only if the slope of the line through P_1 and P is equal to m (see Figure 5), that is,

$$\frac{y - y_1}{x - x_1} = m$$

This equation can be rewritten in the form $y - y_1 = m(x - x_1)$; note that the equation is also satisfied when $x = x_1$ and $y = y_1$. Therefore it is an equation of the given line.

POINT-SLOPE FORM OF THE EQUATION OF A LINE

An equation of the line that passes through the point (x_1, y_1) and has slope m is

$$y - y_1 = m(x - x_1)$$

EXAMPLE 2 ■ Finding an Equation of a Line with Given Point and Slope

(a) Find an equation of the line through $(1, -3)$ with slope $-\frac{1}{2}$.

(b) Sketch the line.

SOLUTION

(a) Using the point-slope form with $m = -\frac{1}{2}$, $x_1 = 1$, and $y_1 = -3$, we obtain an equation of the line as

$$y + 3 = -\tfrac{1}{2}(x - 1) \qquad \text{Slope } m = -\tfrac{1}{2}, \text{ point } (1, -3)$$

$$2y + 6 = -x + 1 \qquad \text{Multiply by 2}$$

$$x + 2y + 5 = 0 \qquad \text{Rearrange}$$

(b) The fact that the slope is $-\frac{1}{2}$ tells us that when we move to the right 2 units, the line drops 1 unit. This enables us to sketch the line in Figure 6.

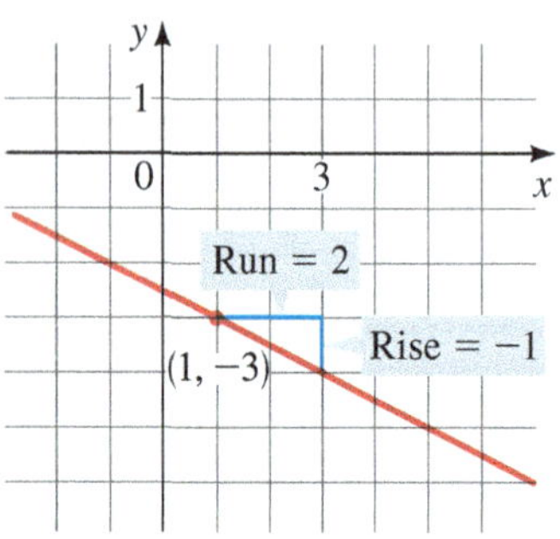

FIGURE 6

Now Try Exercise 25 ■

EXAMPLE 3 ■ Finding an Equation of a Line Through Two Given Points

Find an equation of the line through the points $(-1, 2)$ and $(3, -4)$.

SOLUTION The slope of the line is

$$m = \frac{-4 - 2}{3 - (-1)} = -\frac{6}{4} = -\frac{3}{2}$$

We can use *either* point, $(-1, 2)$ *or* $(3, -4)$, in the point-slope equation. We will end up with the same final answer.

Using the point-slope form with $x_1 = -1$ and $y_1 = 2$, we obtain

$$y - 2 = -\tfrac{3}{2}(x + 1) \qquad \text{Slope } m = -\tfrac{3}{2}, \text{ point } (-1, 2)$$

$$2y - 4 = -3x - 3 \qquad \text{Multiply by 2}$$

$$3x + 2y - 1 = 0 \qquad \text{Rearrange}$$

Now Try Exercise 29 ■

■ Slope-Intercept Form of the Equation of a Line

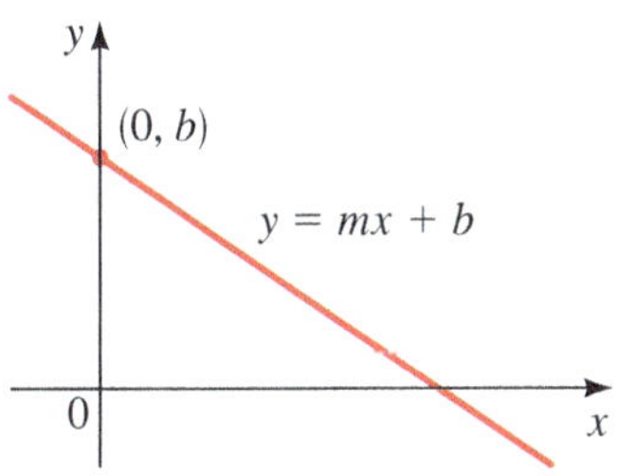

FIGURE 7

Suppose a nonvertical line has slope m and y-intercept b (see Figure 7). This means that the line intersects the y-axis at the point $(0, b)$, so the point-slope form of the equation of the line, with $x = 0$ and $y = b$, becomes

$$y - b = m(x - 0)$$

This simplifies to $y = mx + b$, which is called the **slope-intercept form** of the equation of a line.

SLOPE-INTERCEPT FORM OF THE EQUATION OF A LINE

An equation of the line that has slope m and y-intercept b is

$$y = mx + b$$

EXAMPLE 4 ■ Lines in Slope-Intercept Form

(a) Find an equation of the line with slope 3 and y-intercept -2.

(b) Find the slope and y-intercept of the line $3y - 2x = 1$.

SOLUTION

(a) Since $m = 3$ and $b = -2$, from the slope-intercept form of the equation of a line we get

$$y = 3x - 2$$

(b) We first write the equation in the form $y = mx + b$.

$$3y - 2x = 1$$

$$3y = 2x + 1 \qquad \text{Add } 2x$$

$$y = \tfrac{2}{3}x + \tfrac{1}{3} \qquad \text{Divide by 3}$$

Slope y-intercept

$y = \frac{2}{3}x + \frac{1}{3}$

From the slope-intercept form of the equation of a line, we see that the slope is $m = \frac{2}{3}$ and the y-intercept is $b = \frac{1}{3}$.

Now Try Exercises 23 and 61 ■

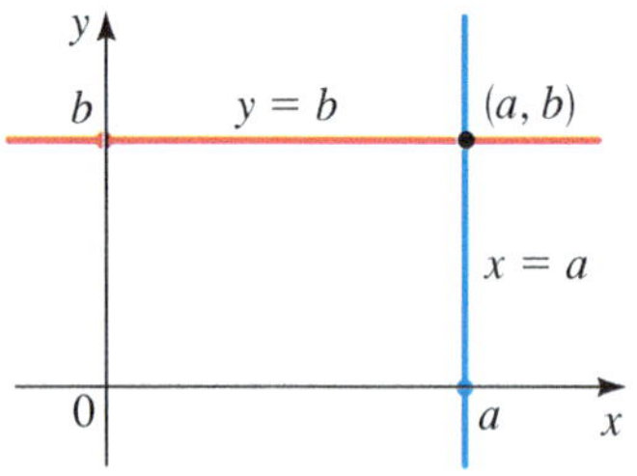

FIGURE 8

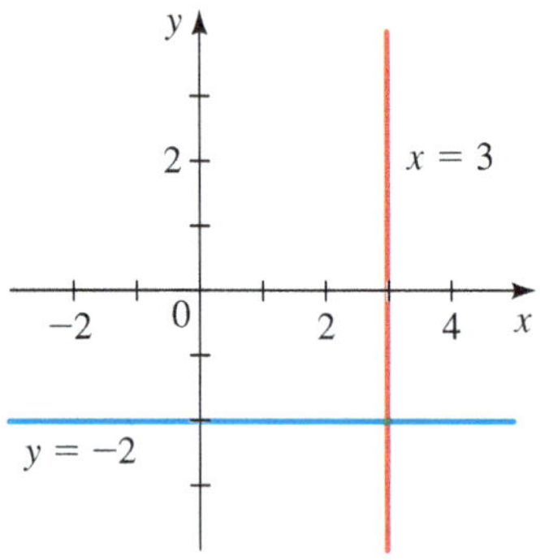

FIGURE 9

Vertical and Horizontal Lines

If a line is horizontal, its slope is $m = 0$, so its equation is $y = b$, where b is the y-intercept (see Figure 8). A vertical line does not have a slope, but we can write its equation as $x = a$, where a is the x-intercept, because the x-coordinate of every point on the line is a.

> **VERTICAL AND HORIZONTAL LINES**
>
> - An equation of the vertical line through (a, b) is $x = a$.
> - An equation of the horizontal line through (a, b) is $y = b$.

EXAMPLE 5 ■ Vertical and Horizontal Lines

(a) An equation for the vertical line through $(3, 5)$ is $x = 3$.

(b) The graph of the equation $x = 3$ is a vertical line with x-intercept 3.

(c) An equation for the horizontal line through $(8, -2)$ is $y = -2$.

(d) The graph of the equation $y = -2$ is a horizontal line with y-intercept -2.

The lines are graphed in Figure 9.

Now Try Exercises 35, 37, 63, and 65 ■

General Equation of a Line

A **linear equation** in the variables x and y is an equation of the form

$$Ax + By + C = 0$$

where A, B, and C are constants and A and B are not both 0. An equation of a line is a linear equation:

- A nonvertical line has the equation $y = mx + b$ or $-mx + y - b = 0$, which is a linear equation with $A = -m$, $B = 1$, and $C = -b$.
- A vertical line has the equation $x = a$ or $x - a = 0$, which is a linear equation with $A = 1$, $B = 0$, and $C = -a$.

  Conversely, the graph of a linear equation is a line.
- If $B \neq 0$, the equation becomes

  $$y = -\frac{A}{B}x - \frac{C}{B} \qquad \text{Divide by } B$$

  and this is the slope-intercept form of the equation of a line (with $m = -A/B$ and $b = -C/B$).
- If $B = 0$, the equation becomes

  $$Ax + C = 0 \qquad \text{Set } B = 0$$

  or $x = -C/A$, which represents a vertical line.

We have proved the following.

> **GENERAL EQUATION OF A LINE**
>
> The graph of every **linear equation**
>
> $$Ax + By + C = 0 \qquad (A, B \text{ not both zero})$$
>
> is a line. Conversely, every line is the graph of a linear equation.

EXAMPLE 6 ■ Graphing a Linear Equation

Sketch the graph of the equation $2x - 3y - 12 = 0$.

SOLUTION 1 Since the equation is linear, its graph is a line. To draw the graph, it is enough to find any two points on the line. The intercepts are the easiest points to find.

x-intercept: Substitute $y = 0$, to get $2x - 12 = 0$, so $x = 6$

y-intercept: Substitute $x = 0$, to get $-3y - 12 = 0$, so $y = -4$

With these points we can sketch the graph in Figure 10.

SOLUTION 2 We write the equation in slope-intercept form.

$$\begin{aligned} 2x - 3y - 12 &= 0 \\ 2x - 3y &= 12 && \text{Add 12} \\ -3y &= -2x + 12 && \text{Subtract } 2x \\ y &= \tfrac{2}{3}x - 4 && \text{Divide by } -3 \end{aligned}$$

This equation is in the form $y = mx + b$, so the slope is $m = \frac{2}{3}$ and the y-intercept is $b = -4$. To sketch the graph, we plot the y-intercept and then move 3 units to the right and 2 units up as shown in Figure 11.

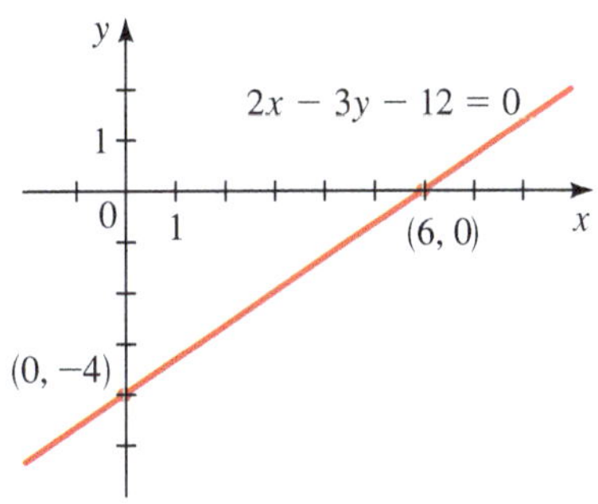

FIGURE 10

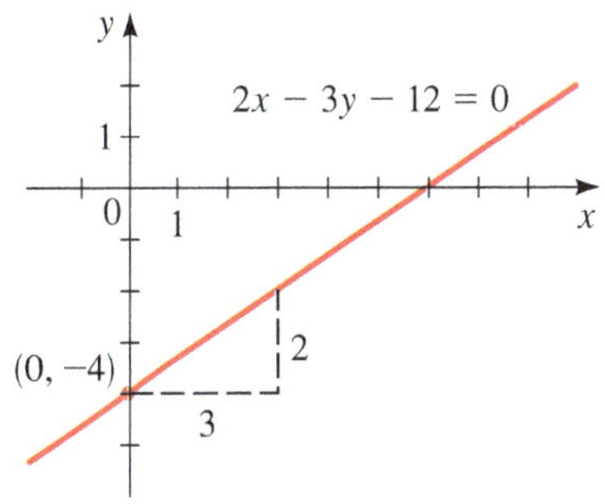

FIGURE 11

Now Try Exercise 67 ■

■ Parallel and Perpendicular Lines

Since slope measures the steepness of a line, it seems reasonable that parallel lines should have the same slope. In fact, we can prove this.

> **PARALLEL LINES**
>
> Two nonvertical lines are parallel if and only if they have the same slope.

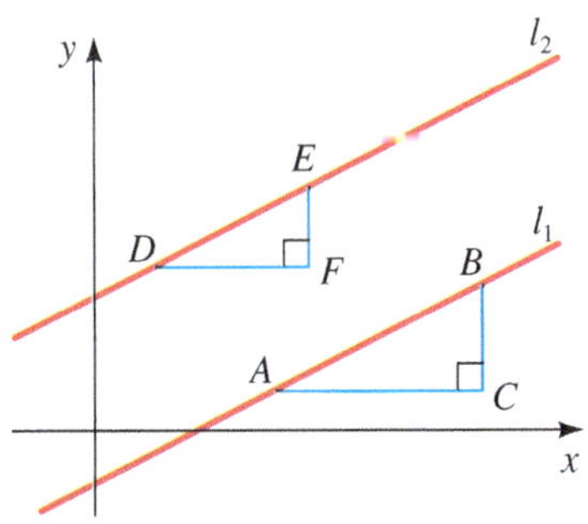

FIGURE 12

Proof Let the lines l_1 and l_2 in Figure 12 have slopes m_1 and m_2. If the lines are parallel, then the right triangles ABC and DEF are similar, so

$$m_1 = \frac{d(B, C)}{d(A, C)} = \frac{d(E, F)}{d(D, F)} = m_2$$

Conversely, if the slopes are equal, then the triangles will be similar, so $\angle BAC = \angle EDF$ and the lines are parallel. ■

EXAMPLE 7 ■ Finding an Equation of a Line Parallel to a Given Line

Find an equation of the line through the point $(5, 2)$ that is parallel to the line $4x + 6y + 5 = 0$.

SOLUTION First we write the equation of the given line in slope-intercept form.

$$4x + 6y + 5 = 0$$

$$6y = -4x - 5 \qquad \text{Subtract } 4x + 5$$

$$y = -\tfrac{2}{3}x - \tfrac{5}{6} \qquad \text{Divide by 6}$$

So the line has slope $m = -\frac{2}{3}$. Since the required line is parallel to the given line, it also has slope $m = -\frac{2}{3}$. From the point-slope form of the equation of a line we get

$$y - 2 = -\tfrac{2}{3}(x - 5) \qquad \text{Slope } m = -\tfrac{2}{3}, \text{ point } (5, 2)$$

$$3y - 6 = -2x + 10 \qquad \text{Multiply by 3}$$

$$2x + 3y - 16 = 0 \qquad \text{Rearrange}$$

Thus an equation of the required line is $2x + 3y - 16 = 0$.

Now Try Exercise 43 ■

The condition for perpendicular lines is not as obvious as that for parallel lines.

PERPENDICULAR LINES

Two lines with slopes m_1 and m_2 are perpendicular if and only if $m_1 m_2 = -1$, that is, their slopes are negative reciprocals:

$$m_2 = -\frac{1}{m_1}$$

Also, a horizontal line (slope 0) is perpendicular to a vertical line (no slope).

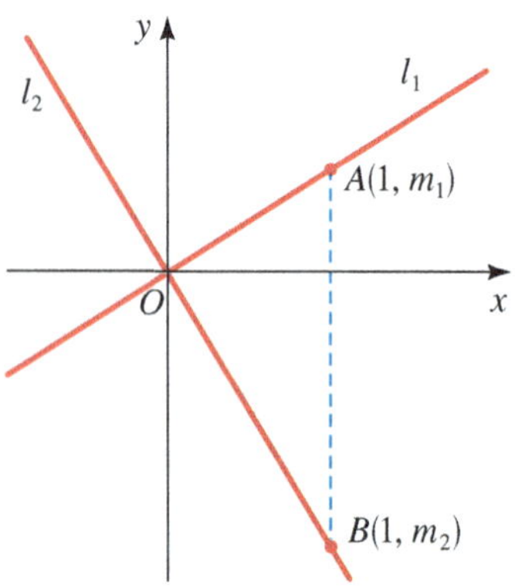

FIGURE 13

Proof In Figure 13 we show two lines intersecting at the origin. (If the lines intersect at some other point, we consider lines parallel to these that intersect at the origin. These lines have the same slopes as the original lines.)

If the lines l_1 and l_2 have slopes m_1 and m_2, then their equations are $y = m_1 x$ and $y = m_2 x$. Notice that $A(1, m_1)$ lies on l_1 and $B(1, m_2)$ lies on l_2. By the Pythagorean Theorem and its converse (see Appendix A) $OA \perp OB$ if and only if

$$[d(O, A)]^2 + [d(O, B)]^2 = [d(A, B)]^2$$

By the Distance Formula this becomes

$$(1^2 + m_1^2) + (1^2 + m_2^2) = (1 - 1)^2 + (m_2 - m_1)^2$$

$$2 + m_1^2 + m_2^2 = m_2^2 - 2m_1 m_2 + m_1^2$$

$$2 = -2m_1 m_2$$

$$m_1 m_2 = -1$$

■

EXAMPLE 8 ■ Perpendicular Lines

Show that the points $P(3, 3)$, $Q(8, 17)$, and $R(11, 5)$ are the vertices of a right triangle.

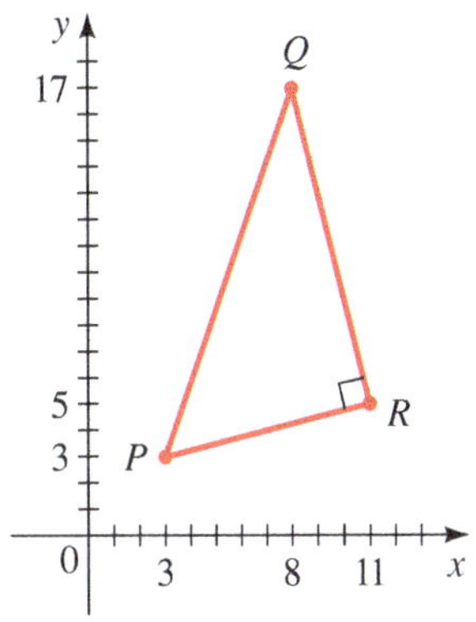

FIGURE 14

SOLUTION The slopes of the lines containing PR and QR are, respectively,

$$m_1 = \frac{5-3}{11-3} = \frac{1}{4} \qquad \text{and} \qquad m_2 = \frac{5-17}{11-8} = -4$$

Since $m_1 m_2 = -1$, these lines are perpendicular, so PQR is a right triangle. It is sketched in Figure 14.

Now Try Exercise 81 ■

EXAMPLE 9 ■ Finding an Equation of a Line Perpendicular to a Given Line

Find an equation of the line that is perpendicular to the line $4x + 6y + 5 = 0$ and passes through the origin.

SOLUTION In Example 7 we found that the slope of the line $4x + 6y + 5 = 0$ is $-\frac{2}{3}$. Thus the slope of a perpendicular line is the negative reciprocal, that is, $\frac{3}{2}$. Since the required line passes through $(0, 0)$, the point-slope form gives

$$y - 0 = \tfrac{3}{2}(x - 0) \qquad \text{Slope } m = \tfrac{3}{2}\text{, point } (0, 0)$$

$$y = \tfrac{3}{2}x \qquad \text{Simplify}$$

Now Try Exercise 47 ■

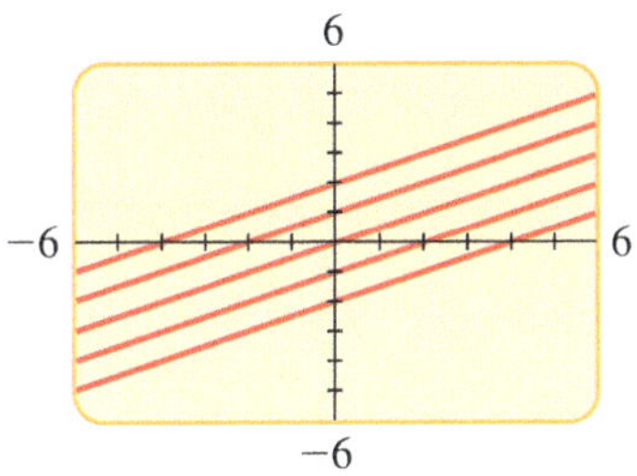

FIGURE 15 $y = 0.5x + b$

EXAMPLE 10 ■ Graphing a Family of Lines

Use a graphing calculator to graph the family of lines

$$y = 0.5x + b$$

for $b = -2, -1, 0, 1, 2$. What property do the lines share?

SOLUTION We use a graphing calculator to graph the lines in the viewing rectangle $[-6, 6]$ by $[-6, 6]$. The graphs are shown in Figure 15. The lines all have the same slope, so they are parallel.

Now Try Exercise 53 ■

EXAMPLE 11 ■ Application: Interpreting Slope

A swimming pool is being filled with a hose. The water depth y (in feet) in the pool t hours after the hose is turned on is given by

$$y = 1.5t + 2$$

(a) Find the slope and y-intercept of the graph of this equation.

(b) What do the slope and y-intercept represent?

SOLUTION

(a) This is the equation of a line with slope 1.5 and y-intercept 2.

(b) The slope represents an increase of 1.5 ft. in water depth for every hour. The y-intercept indicates that the water depth was 2 ft. at the time the hose was turned on.

Now Try Exercise 87 ■

1.10 EXERCISES

CONCEPTS

1. We find the "steepness," or slope, of a line passing through two points by dividing the difference in the _____-coordinates of these points by the difference in the _____-coordinates. So the line passing through the points (0, 1) and (2, 5) has slope _________.

2. A line has the equation $y = 3x + 2$.

(a) This line has slope _________.

(b) Any line parallel to this line has slope _________.

(c) Any line perpendicular to this line has slope _________.

3. The point-slope form of the equation of the line with slope 3 passing through the point (1, 2) is _________.

4. For the linear equation $2x + 3y - 12 = 0$, the x-intercept is _________ and the y-intercept is _________. The equation in slope-intercept form is $y =$ _________________. The slope of the graph of this equation is _________.

5. The slope of a horizontal line is _________. The equation of the horizontal line passing through (2, 3) is _________.

6. The slope of a vertical line is _________. The equation of the vertical line passing through (2, 3) is _________.

7. *Yes or No*? If *No*, give a reason.

(a) Is the graph of $y = -3$ a horizontal line?

(b) Is the graph of $x = -3$ a vertical line?

(c) Does a line perpendicular to a horizontal line have slope 0?

(d) Does a line perpendicular to a vertical line have slope 0?

8. Sketch a graph of the lines $y = -3$ and $x = -3$. Are the lines perpendicular?

SKILLS

9–16 ■ Slope Find the slope of the line through P and Q.

9. $P(-1, 2), Q(0, 0)$

10. $P(0, 0), Q(3, -1)$

11. $P(2, -2), Q(7, -1)$

12. $P(-5, 1), Q(3, -2)$

13. $P(5, 4), Q(0, 4)$

14. $P(4, 3), Q(1, -1)$

15. $P(10, -2), Q(6, -5)$

16. $P(3, -2), Q(6, -2)$

17. Slope Find the slopes of the lines l_1, l_2, l_3, and l_4 in the figure below.

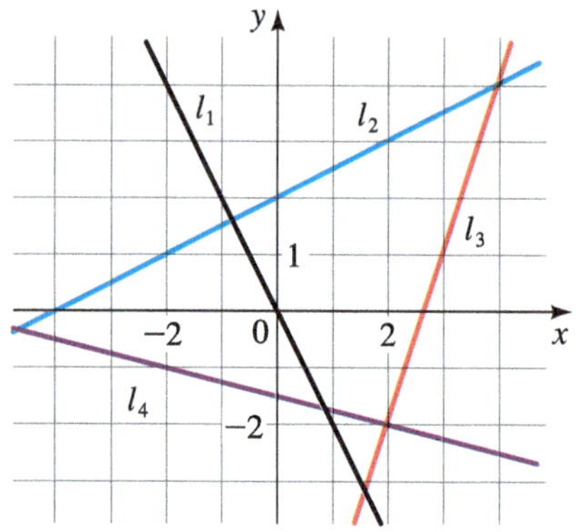

18. Slope

(a) Sketch lines through (0, 0) with slopes 1, 0, $\frac{1}{2}$, 2, and −1.

(b) Sketch lines through (0, 0) with slopes $\frac{1}{3}$, $\frac{1}{2}$, $-\frac{1}{3}$, and 3.

19–22 ■ Equations of Lines Find an equation for the line whose graph is sketched.

19.

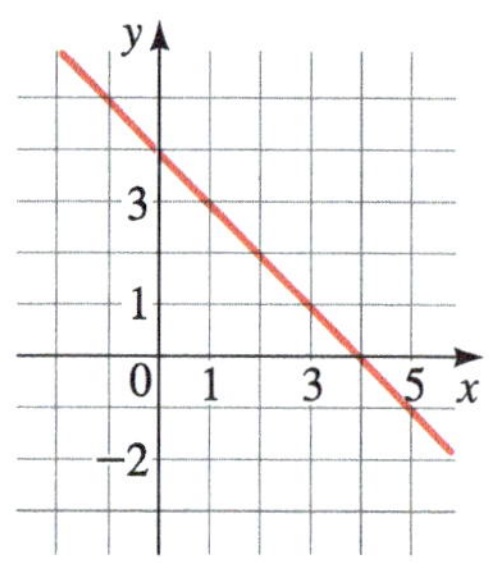

20.

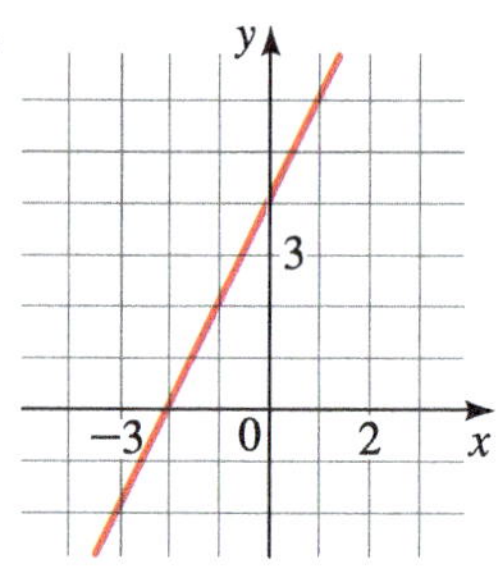

21.

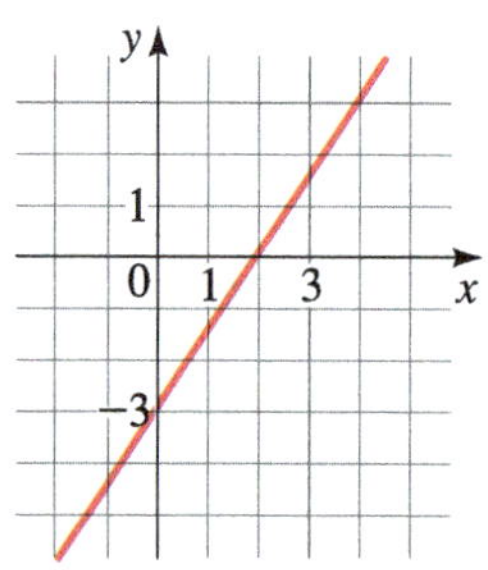

22.

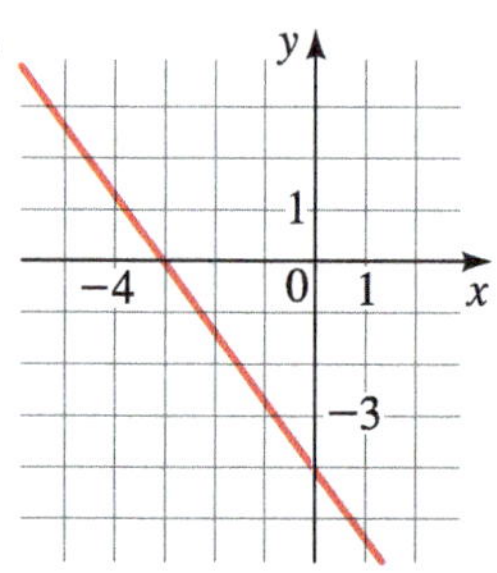

23–50 ■ Finding Equations of Lines Find an equation of the line that satisfies the given conditions.

23. Slope 3; y-intercept −2

24. Slope $\frac{2}{5}$; y-intercept 4

25. Through (2, 3); slope 5

26. Through (−2, 4); slope −1

27. Through (1, 7); slope $\frac{2}{3}$

28. Through (−3, −5); slope $-\frac{7}{2}$

29. Through (2, 1) and (1, 6)

30. Through $(-1, -2)$ and $(4, 3)$

31. Through $(-2, 5)$ and $(-1, -3)$

32. Through $(1, 7)$ and $(4, 7)$

33. x-intercept 1; y-intercept -3

34. x-intercept -8; y-intercept 6

35. Through $(1, 3)$; slope 0

36. Through $(-1, 4)$; slope undefined

37. Through $(2, -1)$; slope undefined

38. Through $(5, 1)$; slope 0

39. Through $(1, 2)$; parallel to the line $y = 3x - 5$

40. Through $(-3, 2)$; perpendicular to the line $y = -\frac{1}{2}x + 7$

41. Through $(4, 5)$; parallel to the x-axis

42. Through $(4, 5)$; parallel to the y-axis

43. Through $(1, -6)$; parallel to the line $x + 2y = 6$

44. y-intercept 6; parallel to the line $2x + 3y + 4 = 0$

45. Through $(-1, 2)$; parallel to the line $x = 5$

46. Through $(2, 6)$; perpendicular to the line $y = 1$

47. Through $(-1, -2)$; perpendicular to the line $2x + 5y + 8 = 0$

48. Through $(\frac{1}{2}, -\frac{2}{3})$; perpendicular to the line $4x - 8y = 1$

49. Through $(1, 7)$; parallel to the line passing through $(2, 5)$ and $(-2, 1)$

50. Through $(-2, -11)$; perpendicular to the line passing through $(1, 1)$ and $(5, -1)$

51. **Finding Equations of Lines and Graphing**
   (a) Sketch the line with slope $\frac{3}{2}$ that passes through the point $(-2, 1)$.
   (b) Find an equation for this line.

52. **Finding Equations of Lines and Graphing**
   (a) Sketch the line with slope -2 that passes through the point $(4, -1)$.
   (b) Find an equation for this line.

53–56 ■ Families of Lines Use a graphing device to graph the given family of lines in the same viewing rectangle. What do the lines have in common?

53. $y = -2x + b$ for $b = 0, \pm 1, \pm 3, \pm 6$

54. $y = mx - 3$ for $m = 0, \pm 0.25, \pm 0.75, \pm 1.5$

55. $y = m(x - 3)$ for $m = 0, \pm 0.25, \pm 0.75, \pm 1.5$

56. $y = 2 + m(x + 3)$ for $m = 0, \pm 0.5, \pm 1, \pm 2, \pm 6$

57–66 ■ Using Slopes and y-Intercepts to Graph Lines Find the slope and y-intercept of the line, and draw its graph.

57. $y = 3 - x$
58. $y = \frac{2}{3}x - 2$
59. $-2x + y = 7$
60. $2x - 5y = 0$
61. $4x + 5y = 10$
62. $3x - 4y = 12$
63. $y = 4$
64. $x = 5$
65. $x = 3$
66. $y = -2$

67–72 ■ Using x- and y-Intercepts to Graph Lines Find the x- and y-intercepts of the line, and draw its graph.

67. $5x + 2y - 10 = 0$
68. $6x - 7y - 42 = 0$
69. $\frac{1}{2}x - \frac{1}{3}y + 1 = 0$
70. $\frac{1}{3}x - \frac{1}{5}y - 2 = 0$
71. $y = 6x + 4$
72. $y = -4x - 10$

73–78 ■ Parallel and Perpendicular Lines The equations of two lines are given. Determine whether the lines are parallel, perpendicular, or neither.

73. $y = 2x + 3$; $2y - 4x - 5 = 0$

74. $y = \frac{1}{2}x + 4$; $2x + 4y = 1$

75. $-3x + 4y = 4$; $4x + 3y = 5$

76. $2x - 3y = 10$; $3y - 2x - 7 = 0$

77. $7x - 3y = 2$; $9y + 21x = 1$

78. $6y - 2x = 5$; $2y + 6x = 1$

SKILLS Plus

79–82 ■ Using Slopes Verify the given geometric property.

79. Use slopes to show that $A(1, 1)$, $B(7, 4)$, $C(5, 10)$, and $D(-1, 7)$ are vertices of a parallelogram.

80. Use slopes to show that $A(-3, -1)$, $B(3, 3)$, and $C(-9, 8)$ are vertices of a right triangle.

81. Use slopes to show that $A(1, 1)$, $B(11, 3)$, $C(10, 8)$, and $D(0, 6)$ are vertices of a rectangle.

82. Use slopes to determine whether the given points are collinear (lie on a line).
   (a) $(1, 1), (3, 9), (6, 21)$
   (b) $(-1, 3), (1, 7), (4, 15)$

83. **Perpendicular Bisector** Find an equation of the perpendicular bisector of the line segment joining the points $A(1, 4)$ and $B(7, -2)$.

84. **Area of a Triangle** Find the area of the triangle formed by the coordinate axes and the line

$$2y + 3x - 6 = 0$$

85. **Two-Intercept Form**
   (a) Show that if the x- and y-intercepts of a line are nonzero numbers a and b, then the equation of the line can be written in the form

$$\frac{x}{a} + \frac{y}{b} = 1$$

   This is called the **two-intercept form** of the equation of a line.
   (b) Use part (a) to find an equation of the line whose x-intercept is 6 and whose y-intercept is -8.

86. Tangent Line to a Circle

(a) Find an equation for the line tangent to the circle $x^2 + y^2 = 25$ at the point $(3, -4)$. (See the figure.)

(b) At what other point on the circle will a tangent line be parallel to the tangent line in part (a)?

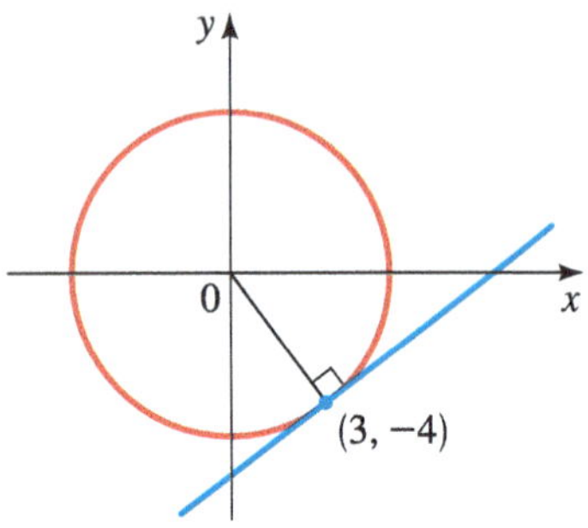

APPLICATIONS

87. Global Warming Some scientists believe that the average surface temperature of the world has been rising steadily. The average surface temperature can be modeled by

$$T = 0.02t + 15.0$$

where T is temperature in °C and t is years since 1950.

(a) What do the slope and T-intercept represent?

(b) Use the equation to predict the average global surface temperature in 2050.

88. Drug Dosages If the recommended adult dosage for a drug is D (in mg), then to determine the appropriate dosage c for a child of age a, pharmacists use the equation

$$c = 0.0417D(a + 1)$$

Suppose the dosage for an adult is 200 mg.

(a) Find the slope. What does it represent?

(b) What is the dosage for a newborn?

89. Flea Market The manager of a weekend flea market knows from past experience that if she charges x dollars for a rental space at the flea market, then the number y of spaces she can rent is given by the equation $y = 200 - 4x$.

(a) Sketch a graph of this linear equation. (Remember that the rental charge per space and the number of spaces rented must both be nonnegative quantities.)

(b) What do the slope, the y-intercept, and the x-intercept of the graph represent?

90. Production Cost A small-appliance manufacturer finds that if he produces x toaster ovens in a month, his production cost is given by the equation

$$y = 6x + 3000$$

(where y is measured in dollars).

(a) Sketch a graph of this linear equation.

(b) What do the slope and y-intercept of the graph represent?

91. Temperature Scales The relationship between the Fahrenheit (F) and Celsius (C) temperature scales is given by the equation $F = \frac{9}{5}C + 32$.

(a) Complete the table to compare the two scales at the given values.

(b) Find the temperature at which the scales agree. [*Hint:* Suppose that a is the temperature at which the scales agree. Set $F = a$ and $C = a$. Then solve for a.]

C	F
−30°	
−20°	
−10°	
0°	
	50°
	68°
	86°

92. Crickets and Temperature Biologists have observed that the chirping rate of crickets of a certain species is related to temperature, and the relationship appears to be very nearly linear. A cricket produces 120 chirps per minute at 70°F and 168 chirps per minute at 80°F.

(a) Find the linear equation that relates the temperature t and the number of chirps per minute n.

(b) If the crickets are chirping at 150 chirps per minute, estimate the temperature.

93. Depreciation A small business buys a computer for \$4000. After 4 years the value of the computer is expected to be \$200. For accounting purposes the business uses *linear depreciation* to assess the value of the computer at a given time. This means that if V is the value of the computer at time t, then a linear equation is used to relate V and t.

(a) Find a linear equation that relates V and t.

(b) Sketch a graph of this linear equation.

(c) What do the slope and V-intercept of the graph represent?

(d) Find the depreciated value of the computer 3 years from the date of purchase.

94. Pressure and Depth At the surface of the ocean the water pressure is the same as the air pressure above the water, 15 lb/in^2. Below the surface the water pressure increases by 4.34 lb/in^2 for every 10 ft of descent.

(a) Find an equation for the relationship between pressure and depth below the ocean surface.

(b) Sketch a graph of this linear equation.

(c) What do the slope and y-intercept of the graph represent?

(d) At what depth is the pressure 100 lb/in^2?

DISCUSS ■ DISCOVER ■ PROVE ■ WRITE

95. DISCUSS: What Does the Slope Mean? Suppose that the graph of the outdoor temperature over a certain period of time is a line. How is the weather changing if the slope of the line is positive? If it is negative? If it is zero?

96. DISCUSS: Collinear Points Suppose that you are given the coordinates of three points in the plane and you want to see whether they lie on the same line. How can you do this using slopes? Using the Distance Formula? Can you think of another method?

1.11 SOLVING EQUATIONS AND INEQUALITIES GRAPHICALLY

■ Solving Equations Graphically ■ Solving Inequalities Graphically

"Algebra is a merry science," Uncle Jakob would say. "We go hunting for a little animal whose name we don't know, so we call it x. When we bag our game we pounce on it and give it its right name."

ALBERT EINSTEIN

In Section 1.5 we learned how to solve equations by the **algebraic method**. In this method we view x as an *unknown* and then use the rules of algebra to "hunt it down," by isolating it on one side of the equation. In Section 1.8 we solved inequalities by this same method.

Sometimes an equation or inequality may be difficult or impossible to solve algebraically. In this case we use the **graphical method**. In this method we view x as a *variable* and sketch an appropriate graph. We can then obtain an approximate solution from the graph.

■ Solving Equations Graphically

To solve a one-variable equation such as $3x - 5 = 0$ graphically, we first draw a graph of the two-variable equation $y = 3x - 5$ obtained by setting the nonzero side of the equation equal to a variable y. The solutions of the given equation are the values of x for which y is equal to zero. That is, the solutions are the x-intercepts of the graph. The following describes the method.

SOLVING AN EQUATION

Algebraic Method

Use the rules of algebra to isolate the unknown x on one side of the equation.

Example: $3x - 4 = 1$

$3x = 5$ Add 4

$x = \frac{5}{3}$ Divide by 3

The solution is $x = \frac{5}{3}$.

Graphical Method

Move all terms to one side, and set equal to y. Graph the resulting equation, and find the x-intercepts.

Example: $3x - 4 = 1$

$3x - 5 = 0$

Set $y = 3x - 5$ and graph. From the graph we see that the solution is $x \approx 1.7$

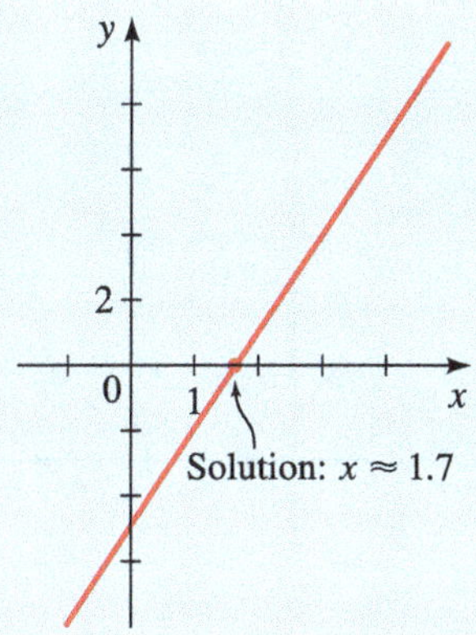

The advantage of the algebraic method is that it gives exact answers. Also, the process of unraveling the equation to arrive at the answer helps us to understand the algebraic structure of the equation. On the other hand, for many equations it is difficult or impossible to isolate x.

Bettmann/Corbis

PIERRE DE FERMAT (1601–1665) was a French lawyer who became interested in mathematics at the age of 30. Because of his job as a magistrate, Fermat had little time to write complete proofs of his discoveries and often wrote them in the margin of whatever book he was reading at the time. After his death his copy of Diophantus' *Arithmetica* (see page 20) was found to contain a particularly tantalizing comment. Where Diophantus discusses the solutions of $x^2 + y^2 = z^2$ (for example, $x = 3, y = 4$, and $z = 5$), Fermat states in the margin that for $n \geq 3$ there are no natural number solutions to the equation $x^n + y^n = z^n$. In other words, it's impossible for a cube to equal the sum of two cubes, a fourth power to equal the sum of two fourth powers, and so on. Fermat writes, "I have discovered a truly wonderful proof for this but the margin is too small to contain it." All the other margin comments in Fermat's copy of *Arithmetica* have been proved. This one, however, remained unproved, and it came to be known as "Fermat's Last Theorem."

In 1994, Andrew Wiles of Princeton University announced a proof of Fermat's Last Theorem, an astounding 350 years after it was conjectured. His proof is one of the most widely reported mathematical results in the popular press.

The graphical method gives a numerical approximation to the answer. This is an advantage when a numerical answer is desired. (For example, an engineer might find an answer expressed as $x \approx 2.6$ more immediately useful than $x = \sqrt{7}$.) Also, graphing an equation helps us to visualize how the solution is related to other values of the variable.

The *Discovery Project* referenced on page 276 describes a numerical method for solving equations.

EXAMPLE 1 ■ Solving a Quadratic Equation Algebraically and Graphically

Find all real solutions of the quadratic equation. Use the algebraic method and the graphical method.

(a) $x^2 - 4x + 2 = 0$ **(b)** $x^2 - 4x + 4 = 0$ **(c)** $x^2 - 4x + 6 = 0$

SOLUTION 1: Algebraic

The Quadratic Formula is discussed on page 50.

You can check that the Quadratic Formula gives the following solutions.

(a) There are two real solutions, $x = 2 + \sqrt{2}$ and $x = 2 - \sqrt{2}$.

(b) There is one real solution, $x = 2$.

(c) There is no real solution. (The two complex solutions are $x = 2 + \sqrt{2}i$ and $x = 2 - \sqrt{2}i$.)

SOLUTION 2: Graphical

We use a graphing calculator to graph the equations $y = x^2 - 4x + 2$, $y = x^2 - 4x + 4$, and $y = x^2 - 4x + 6$ in Figure 1. By determining the x-intercepts of the graphs, we find the following solutions.

(a) The two x-intercepts give the two solutions $x \approx 0.6$ and $x \approx 3.4$.

(b) The one x-intercept gives the one solution $x = 2$.

(c) There is no x-intercept, so the equation has no real solutions.

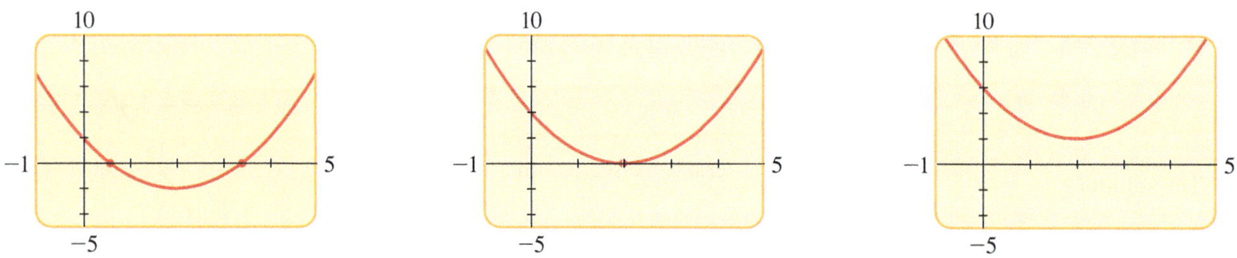

(a) $y = x^2 - 4x + 2$ (b) $y = x^2 - 4x + 4$ (c) $y = x^2 - 4x + 6$

FIGURE 1

Now Try Exercises 9, 11, and 15 ■

The graphs in Figure 1 show visually why a quadratic equation may have two solutions, one solution, or no real solution. We proved this fact algebraically in Section 1.5 when we studied the discriminant.

Science Source

ALAN TURING (1912–1954) was at the center of two pivotal events of the 20th century: World War II and the invention of computers. At the age of 23 Turing made his mark on mathematics by solving an important problem in the foundations of mathematics that had been posed by David Hilbert at the 1928 International Congress of Mathematicians (see page 735). In this research he invented a theoretical machine, now called a Turing machine, which was the inspiration for modern digital computers. During World War II Turing was in charge of the British effort to decipher secret German codes. His complete success in this endeavor played a decisive role in the Allies' victory. To carry out the numerous logical steps that are required to break a coded message, Turing developed decision procedures similar to modern computer programs. After the war he helped to develop the first electronic computers in Britain. He also did pioneering work on artificial intelligence and computer models of biological processes. At the age of 42 Turing died of poisoning after eating an apple that had mysteriously been laced with cyanide.

EXAMPLE 2 ■ Another Graphical Method

Solve the equation algebraically and graphically: $5 - 3x = 8x - 20$

SOLUTION 1: Algebraic

$5 - 3x = 8x - 20$	Given equation
$-3x = 8x - 25$	Subtract 5
$-11x = -25$	Subtract $8x$
$x = \dfrac{-25}{-11} = 2\frac{3}{11}$	Divide by –11 and simplify

SOLUTION 2: Graphical

We could move all terms to one side of the equal sign, set the result equal to y, and graph the resulting equation. But to avoid all this algebra, we use a graphing calculator to graph the two equations instead:

$$y_1 = 5 - 3x \qquad \text{and} \qquad y_2 = 8x - 20$$

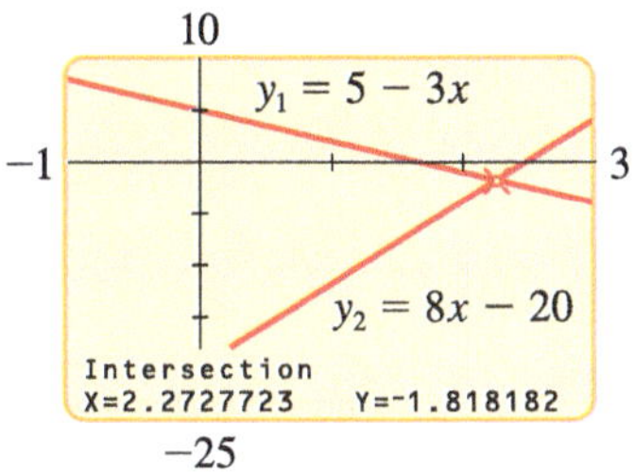

FIGURE 2

The solution of the original equation will be the value of x that makes y_1 equal to y_2; that is, the solution is the x-coordinate of the intersection point of the two graphs. Using the TRACE feature or the `intersect` command on a graphing calculator, we see from Figure 2 that the solution is $x \approx 2.27$.

Now Try Exercise 5 ■

In the next example we use the graphical method to solve an equation that is extremely difficult to solve algebraically.

EXAMPLE 3 ■ Solving an Equation in an Interval

Solve the equation

$$x^3 - 6x^2 + 9x = \sqrt{x}$$

in the interval $[1, 6]$.

SOLUTION We are asked to find all solutions x that satisfy $1 \le x \le 6$, so we use a graphing calculator to graph the equation in a viewing rectangle for which the x-values are restricted to this interval.

$x^3 - 6x^2 + 9x = \sqrt{x}$	Given equation
$x^3 - 6x^2 + 9x - \sqrt{x} = 0$	Subtract $\sqrt{x}$

We can also use the `zero` command to find the solutions, as shown in Figures 3(a) and 3(b).

Figure 3 shows the graph of the equation $y = x^3 - 6x^2 + 9x - \sqrt{x}$ in the viewing rectangle $[1, 6]$ by $[-5, 5]$. There are two x-intercepts in this viewing rectangle; zooming in, we see that the solutions are $x \approx 2.18$ and $x \approx 3.72$.

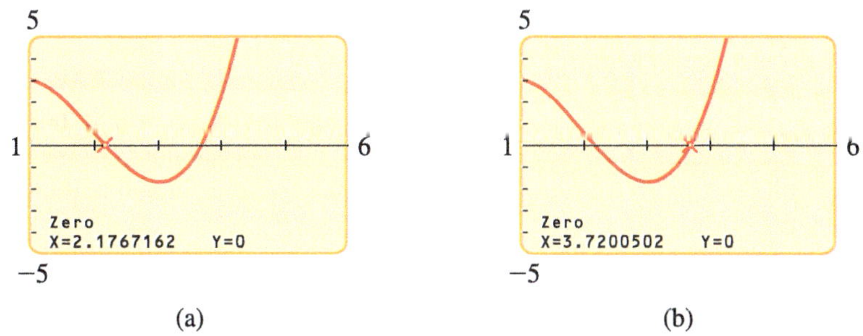

FIGURE 3

Now Try Exercise 17 ■

The equation in Example 3 actually has four solutions. You are asked to find the other two in Exercise 46.

Solving Inequalities Graphically

To solve a one-variable inequality such as $3x - 5 \geq 0$ graphically, we first draw a graph of the two-variable equation $y = 3x - 5$ obtained by setting the nonzero side of the inequality equal to a variable y. The solutions of the given inequality are the values of x for which y is greater than or equal to 0. That is, the solutions are the values of x for which the graph is above the x-axis.

SOLVING AN INEQUALITY

Algebraic Method

Use the rules of algebra to isolate the unknown x on one side of the inequality.

Example: $3x - 4 \geq 1$

$3x \geq 5$ — Add 4

$x \geq \frac{5}{3}$ — Divide by 3

The solution is $\left[\frac{5}{3}, \infty\right)$.

Graphical Method

Move all terms to one side, and set equal to y. Graph the resulting equation, and find the values of x where the graph is above or on the x-axis.

Example: $3x - 4 \geq 1$

$3x - 5 \geq 0$

Set $y = 3x - 5$ and graph. From the graph we see that the solution is $[1.7, \infty)$.

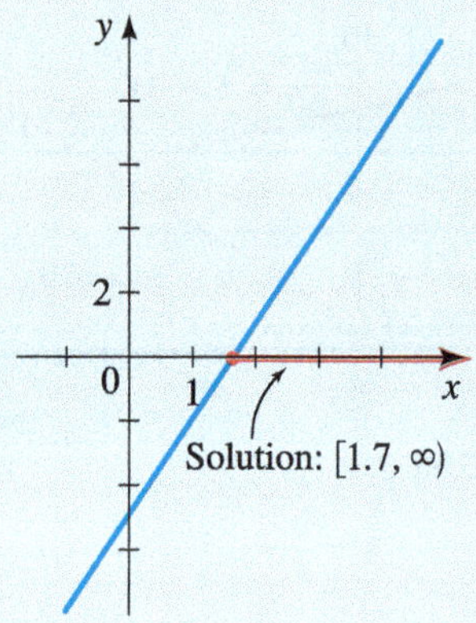

EXAMPLE 4 ■ Solving an Inequality Graphically

Solve the inequality $x^2 - 5x + 6 \leq 0$ graphically.

SOLUTION This inequality was solved algebraically in Example 3 of Section 1.8. To solve the inequality graphically, we use a graphing calculator to draw the graph of

$$y = x^2 - 5x + 6$$

Our goal is to find those values of x for which $y \leq 0$. These are simply the x-values for which the graph lies below the x-axis. From the graph in Figure 4 we see that the solution of the inequality is the interval $[2, 3]$.

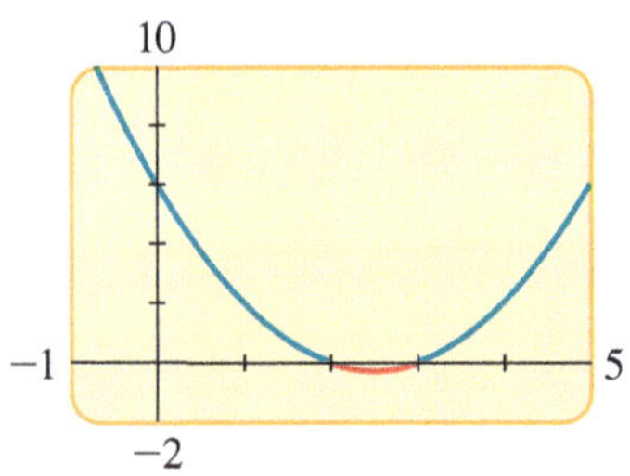

FIGURE 4

Now Try Exercise 33 ■

EXAMPLE 5 ■ Solving an Inequality Graphically

Solve the inequality $3.7x^2 + 1.3x - 1.9 \leq 2.0 - 1.4x$.

SOLUTION We use a graphing calculator to graph the equations

$$y_1 = 3.7x^2 + 1.3x - 1.9 \quad \text{and} \quad y_2 = 2.0 - 1.4x$$

The graphs are shown in Figure 5. We are interested in those values of x for which $y_1 \leq y_2$; these are points for which the graph of y_2 lies on or above the graph of y_1. To determine the appropriate interval, we look for the x-coordinates of points where the graphs intersect. We conclude that the solution is (approximately) the interval $[-1.45, 0.72]$.

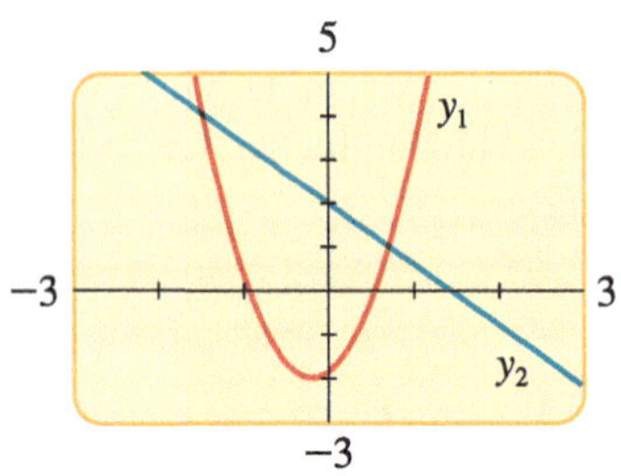

FIGURE 5
$y_1 = 3.7x^2 + 1.3x - 1.9$
$y_2 = 2.0 - 1.4x$

Now Try Exercise 35 ■

EXAMPLE 6 ■ Solving an Inequality Graphically

Solve the inequality $x^3 - 5x^2 \geq -8$.

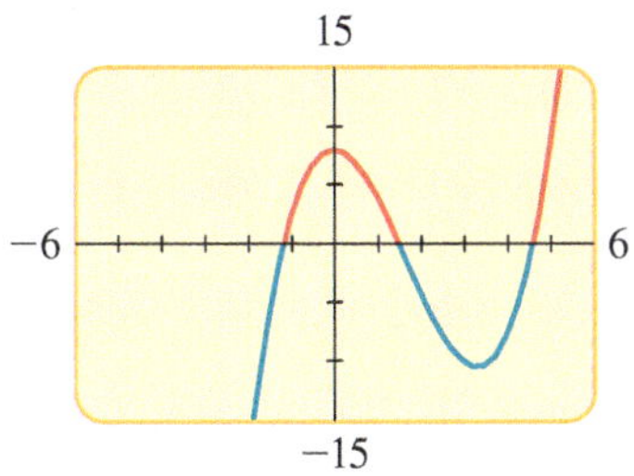

FIGURE 6 $x^3 - 5x^2 + 8 \geq 0$

SOLUTION We write the inequality as

$$x^3 - 5x^2 + 8 \geq 0$$

and then graph the equation

$$y = x^3 - 5x^2 + 8$$

in the viewing rectangle $[-6, 6]$ by $[-15, 15]$, as shown in Figure 6. The solution of the inequality consists of those intervals on which the graph lies on or above the x-axis. By moving the cursor to the x-intercepts, we find that, rounded to one decimal place, the solution is $[-1.1, 1.5] \cup [4.6, \infty)$.

Now Try Exercise 37

1.11 EXERCISES

CONCEPTS

1. The solutions of the equation $x^2 - 2x - 3 = 0$ are the ____-intercepts of the graph of $y = x^2 - 2x - 3$.

2. The solutions of the inequality $x^2 - 2x - 3 > 0$ are the x-coordinates of the points on the graph of $y = x^2 - 2x - 3$ that lie ________ the x-axis.

3. The figure shows a graph of $y = x^4 - 3x^3 - x^2 + 3x$. Use the graph to do the following.
   (a) Find the solutions of the equation $x^4 - 3x^3 - x^2 + 3x = 0$.
   (b) Find the solutions of the inequality $x^4 - 3x^3 - x^2 + 3x \leq 0$.

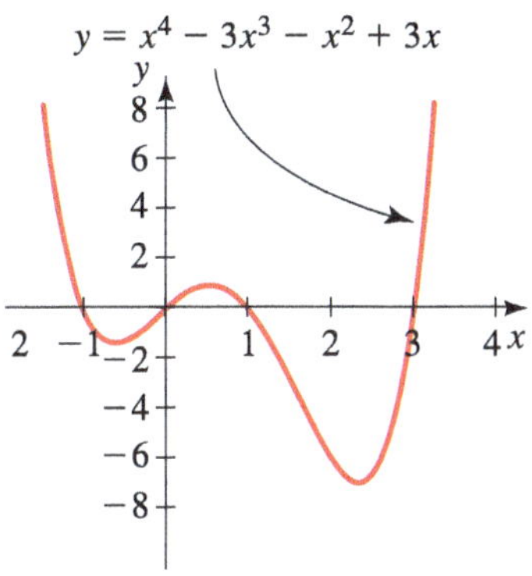

4. The figure shows the graphs of $y = 5x - x^2$ and $y = 4$. Use the graphs to do the following.
   (a) Find the solutions of the equation $5x - x^2 = 4$.
   (b) Find the solutions of the inequality $5x - x^2 > 4$.

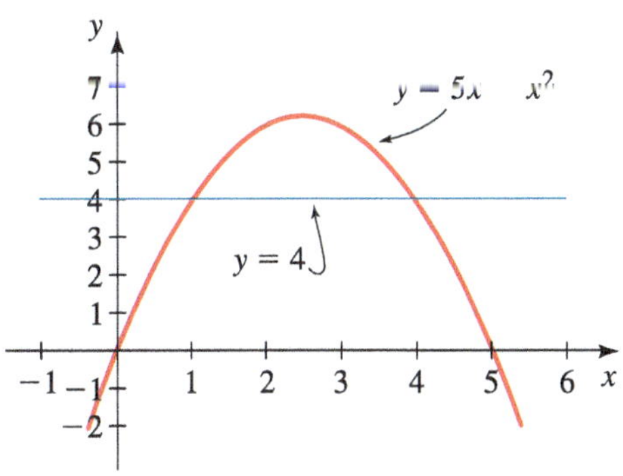

SKILLS

5–16 ■ Equations Solve the equation both algebraically and graphically.

5. $x - 4 = 5x + 12$
6. $\frac{1}{2}x - 3 = 6 + 2x$
7. $\dfrac{2}{x} + \dfrac{1}{2x} = 7$
8. $\dfrac{4}{x+2} - \dfrac{6}{2x} = \dfrac{5}{2x+4}$
9. $x^2 - 32 = 0$
10. $x^3 + 16 = 0$
11. $x^2 + 9 = 0$
12. $x^2 + 3 = 2x$
13. $16x^4 = 625$
14. $2x^5 - 243 = 0$
15. $(x - 5)^4 - 80 = 0$
16. $6(x + 2)^5 = 64$

17–24 ■ Equations Solve the equation graphically in the given interval. State each answer rounded to two decimals.

17. $x^2 - 7x + 12 = 0$; $[0, 6]$
18. $x^2 - 0.75x + 0.125 = 0$; $[-2, 2]$
19. $x^3 - 6x^2 + 11x - 6 = 0$; $[-1, 4]$
20. $16x^3 + 16x^2 = x + 1$; $[-2, 2]$
21. $x - \sqrt{x + 1} = 0$; $[-1, 5]$
22. $1 + \sqrt{x} = \sqrt{1 + x^2}$; $[-1, 5]$
23. $x^{1/3} - x = 0$; $[-3, 3]$
24. $x^{1/2} + x^{1/3} - x = 0$; $[-1, 5]$

25–28 ■ Equations Use the graphical method to solve the equation in the indicated exercise from Section 1.5.

25. Exercise 97.
26. Exercise 98.
27. Exercise 105.
28. Exercise 106.

29–32 ■ Equations Find all real solutions of the equation, rounded to two decimals.

29. $x^3 - 2x^2 - x - 1 = 0$
30. $x^4 - 8x^2 + 2 = 0$
31. $x(x - 1)(x + 2) = \frac{1}{6}x$
32. $x^4 = 16 - x^3$

33–40 ■ Inequalities Find the solutions of the inequality by drawing appropriate graphs. State each answer rounded to two decimals.

33. $x^2 \leq 3x + 10$
34. $0.5x^2 + 0.875x \leq 0.25$
35. $x^3 + 11x \leq 6x^2 + 6$
36. $16x^3 + 24x^2 > -9x - 1$

37. $x^{1/3} < x$

38. $\sqrt{0.5x^2 + 1} \le 2|x|$

39. $(x + 1)^2 < (x - 1)^2$

40. $(x + 1)^2 \le x^3$

41–44 ■ Inequalities Use the graphical method to solve the inequality in the indicated exercise from Section 1.8.

41. Exercise 45.

42. Exercise 46.

43. Exercise 55.

44. Exercise 56.

SKILLS Plus

45. Another Graphical Method In Example 2 we solved the equation $5 - 3x = 8x - 20$ by drawing graphs of two equations. Solve the equation by drawing a graph of only one equation. Compare your answer to the one obtained in Example 2.

46. Finding More Solutions In Example 3 we found two solutions of the equation $x^3 - 6x^2 + 9x = \sqrt{x}$ in the interval $[1, 6]$. Find two more solutions, rounded to two decimals.

APPLICATIONS

47. Estimating Profit An appliance manufacturer estimates that the profit y (in dollars) generated by producing x cooktops per month is given by the equation

$$y = 10x + 0.5x^2 - 0.001x^3 - 5000$$

where $0 \le x \le 450$.

(a) Graph the equation.

(b) How many cooktops must be produced to begin generating a profit?

(c) For what range of values of x is the company's profit greater than $15,000?

48. How Far Can You See? If you stand on a ship in a calm sea, then your height x (in ft) above sea level is related to the farthest distance y (in mi) that you can see by the equation

$$y = \sqrt{1.5x + \left(\frac{x}{5280}\right)^2}$$

(a) Graph the equation for $0 \le x \le 100$.

(b) How high up do you have to be to be able to see 10 mi?

DISCUSS ■ DISCOVER ■ PROVE ■ WRITE

49. WRITE: Algebraic and Graphical Solution Methods Write a short essay comparing the algebraic and graphical methods for solving equations. Make up your own examples to illustrate the advantages and disadvantages of each method.

50. DISCUSS: Enter Equations Carefully A student wishes to graph the equations

$$y = x^{1/3} \qquad \text{and} \qquad y = \frac{x}{x + 4}$$

on the same screen, so he enters the following information into his calculator:

```
Y1 = X^1/3        Y2 = X/X+4
```

The calculator graphs two lines instead of the equations he wanted. What went wrong?

1.12 MODELING VARIATION

■ Direct Variation ■ Inverse Variation ■ Combining Different Types of Variation

When scientists talk about a *mathematical model* for a real-world phenomenon, they often mean a function that describes the dependence of one physical quantity on another. For instance, the model may describe the population of an animal species as a function of time or the pressure of a gas as a function of its volume. In this section we study a kind of modeling that occurs frequently in the sciences, called *variation*.

■ Direct Variation

One type of variation is called *direct variation*; it occurs when one quantity is a constant multiple of the other. We use a function of the form $f(x) = kx$ to model this dependence.

DIRECT VARIATION

If the quantities x and y are related by an equation

$$y = kx$$

for some constant $k \neq 0$, we say that y **varies directly as** x, or y is **directly proportional to** x, or simply y **is proportional to** x. The constant k is called the **constant of proportionality**.

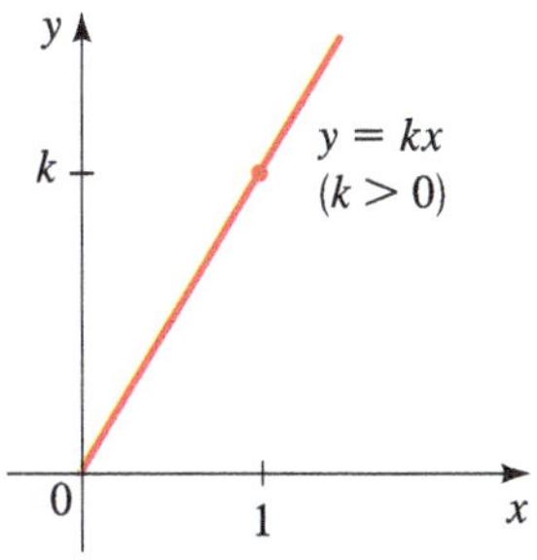

FIGURE 1

Recall that the graph of an equation of the form $y = mx + b$ is a line with slope m and y-intercept b. So the graph of an equation $y = kx$ that describes direct variation is a line with slope k and y-intercept 0 (see Figure 1).

EXAMPLE 1 ■ Direct Variation

During a thunderstorm you see the lightning before you hear the thunder because light travels much faster than sound. The distance between you and the storm varies directly as the time interval between the lightning and the thunder.

(a) Suppose that the thunder from a storm 5400 ft away takes 5 s to reach you. Determine the constant of proportionality, and write the equation for the variation.

(b) Sketch the graph of this equation. What does the constant of proportionality represent?

(c) If the time interval between the lightning and thunder is now 8 s, how far away is the storm?

SOLUTION

(a) Let d be the distance from you to the storm, and let t be the length of the time interval. We are given that d varies directly as t, so

$$d = kt$$

where k is a constant. To find k, we use the fact that $t = 5$ when $d = 5400$. Substituting these values in the equation, we get

$$5400 = k(5) \quad \text{Substitute}$$

$$k = \frac{5400}{5} = 1080 \quad \text{Solve for } k$$

Substituting this value of k in the equation for d, we obtain

$$d = 1080t$$

as the equation for d as a function of t.

(b) The graph of the equation $d = 1080t$ is a line through the origin with slope 1080 and is shown in Figure 2. The constant $k = 1080$ is the approximate speed of sound (in ft/s).

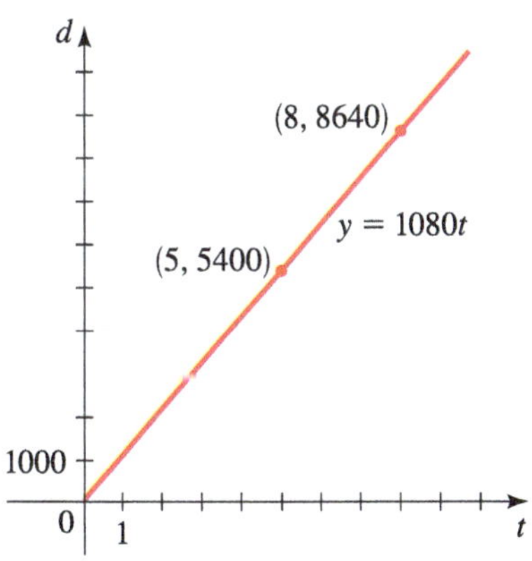

FIGURE 2

(c) When $t = 8$, we have

$$d = 1080 \cdot 8 = 8640$$

So the storm is 8640 ft $\approx$ 1.6 mi away.

Now Try Exercises 19 and 35 ■

Inverse Variation

Another function that is frequently used in mathematical modeling is $f(x) = k/x$, where k is a constant.

> **INVERSE VARIATION**
>
> If the quantities x and y are related by the equation
>
> $$y = \frac{k}{x}$$
>
> for some constant $k \neq 0$, we say that y **is inversely proportional to** x or y **varies inversely as** x. The constant k is called the **constant of proportionality**.

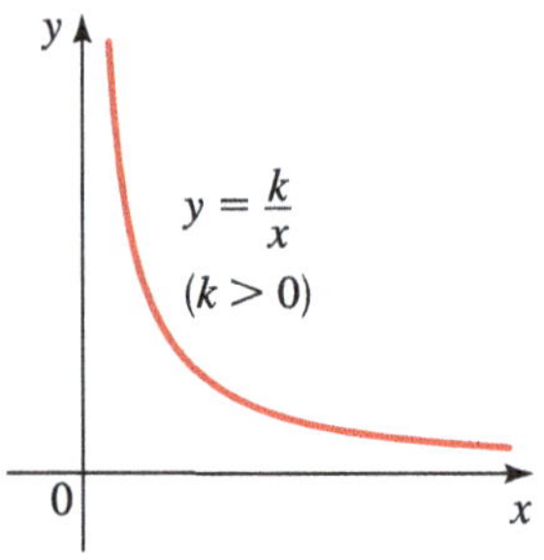

FIGURE 3 Inverse variation

The graph of $y = k/x$ for $x > 0$ is shown in Figure 3 for the case $k > 0$. It gives a picture of what happens when y is inversely proportional to x.

EXAMPLE 2 ■ Inverse Variation

Boyle's Law states that when a sample of gas is compressed at a constant temperature, the pressure of the gas is inversely proportional to the volume of the gas.

(a) Suppose the pressure of a sample of air that occupies 0.106 m^3 at 25°C is 50 kPa. Find the constant of proportionality, and write the equation that expresses the inverse proportionality. Sketch a graph of this equation.

(b) If the sample expands to a volume of 0.3 m^3, find the new pressure.

SOLUTION

(a) Let P be the pressure of the sample of gas, and let V be its volume. Then, by the definition of inverse proportionality, we have

$$P = \frac{k}{V}$$

where k is a constant. To find k, we use the fact that $P = 50$ when $V = 0.106$. Substituting these values in the equation, we get

$$50 = \frac{k}{0.106} \qquad \text{Substitute}$$

$$k = (50)(0.106) = 5.3 \qquad \text{Solve for } k$$

Putting this value of k in the equation for P, we have

$$P = \frac{5.3}{V}$$

Since V represents volume (which is never negative), we sketch the part of the graph for which $V > 0$ only. The graph is shown in Figure 4.

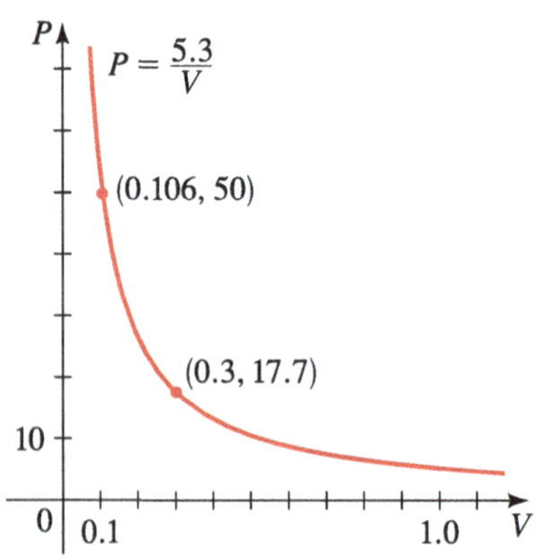

FIGURE 4

(b) When $V = 0.3$, we have

$$P = \frac{5.3}{0.3} \approx 17.7$$

So the new pressure is about 17.7 kPa.

Now Try Exercises 21 and 43 ■

Combining Different Types of Variation

In the sciences, relationships between three or more variables are common, and any combination of the different types of proportionality that we have discussed is possible. For example, if the quantities x, y, and z are related by the equation

$$z = kxy$$

then we say that z is **proportional to the product** of x and y. We can also express this relationship by saying that z **varies jointly** as x and y or that z **is jointly proportional to** x and y. If the quantities x, y, and z are related by the equation

$$z = k\frac{x}{y}$$

we say that z **is proportional to** x **and inversely proportional to** y or that z **varies directly as** x **and inversely as** y.

EXAMPLE 3 ■ Combining Variations

The apparent brightness B of a light source (measured in W/m^2) is directly proportional to the luminosity L (measured in W) of the light source and inversely proportional to the square of the distance d from the light source (measured in meters).

(a) Write an equation that expresses this variation.

(b) If the distance is doubled, by what factor will the brightness change?

(c) If the distance is cut in half and the luminosity is tripled, by what factor will the brightness change?

SOLUTION

(a) Since B is directly proportional to L and inversely proportional to d^2, we have

$$B = k\frac{L}{d^2} \qquad \text{Brightness at distance } d \text{ and luminosity } L$$

where k is a constant.

(b) To obtain the brightness at double the distance, we replace d by $2d$ in the equation we obtained in part (a).

$$B = k\frac{L}{(2d)^2} = \frac{1}{4}\left(k\frac{L}{d^2}\right) \qquad \text{Brightness at distance } 2d$$

Comparing this expression with that obtained in part (a), we see that the brightness is $\frac{1}{4}$ of the original brightness.

DISCOVERY PROJECT

Proportionality: Shape and Size

Many real-world quantities are related by proportionalities. We use the proportionality symbol $\propto$ to express proportionalities in the natural world. For example, for animals of the same shape, the skin area and volume are proportional, in different ways, to the length of the animal. In one situation we use proportionality to determine how a frog's size relates to its sensitivity to pollutants in the environment. You can find the project at **www.stewartmath.com**.

(c) To obtain the brightness at half the distance d and triple the luminosity L, we replace d by $d/2$ and L by $3L$ in the equation we obtained in part (a).

$$B = k\frac{3L}{\left(\frac{1}{2}d\right)^2} = \frac{3}{\frac{1}{4}}\left(k\frac{L}{d^2}\right) = 12\left(k\frac{L}{d^2}\right) \qquad \text{Brightness at distance } \tfrac{1}{2}d \text{ and luminosity } 3L$$

Comparing this expression with that obtained in part (a), we see that the brightness is 12 times the original brightness.

Now Try Exercises 23 and 45 ■

The relationship between apparent brightness, actual brightness (or luminosity), and distance is used in estimating distances to stars (see Exercise 56).

EXAMPLE 4 ■ Newton's Law of Gravity

Newton's Law of Gravity says that two objects with masses m_1 and m_2 attract each other with a force F that is jointly proportional to their masses and inversely proportional to the square of the distance r between the objects. Express Newton's Law of Gravity as an equation.

SOLUTION Using the definitions of joint and inverse variation and the traditional notation G for the gravitational constant of proportionality, we have

$$F = G\frac{m_1 m_2}{r^2}$$

Now Try Exercises 31 and 37 ■

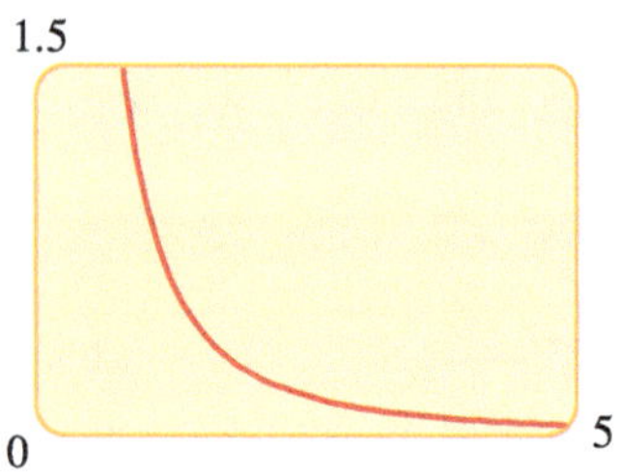

FIGURE 5 Graph of $F = \frac{1}{r^2}$

If m_1 and m_2 are fixed masses, then the gravitational force between them is $F = C/r^2$ (where $C = Gm_1m_2$ is a constant). Figure 5 shows the graph of this equation for $r > 0$ with $C = 1$. Observe how the gravitational attraction decreases with increasing distance.

Like the Law of Gravity, many laws of nature are *inverse square laws*. There is a geometric reason for this. Imagine a force or energy originating from a point source and spreading its influence equally in all directions, just like the light source in Example 3 or the gravitational force exerted by a planet in Example 4. The influence of the force or energy at a distance r from the source is spread out over the surface of a sphere of radius r, which has area $A = 4\pi r^2$ (see Figure 6). So the intensity I at a distance r from the source is the source strength S divided by the area A of the sphere:

$$I = \frac{S}{4\pi r^2} = \frac{k}{r^2}$$

where k is the constant $S/(4\pi)$. Thus point sources of light, sound, gravity, electromagnetic fields, and radiation must all obey inverse square laws, simply because of the geometry of space.

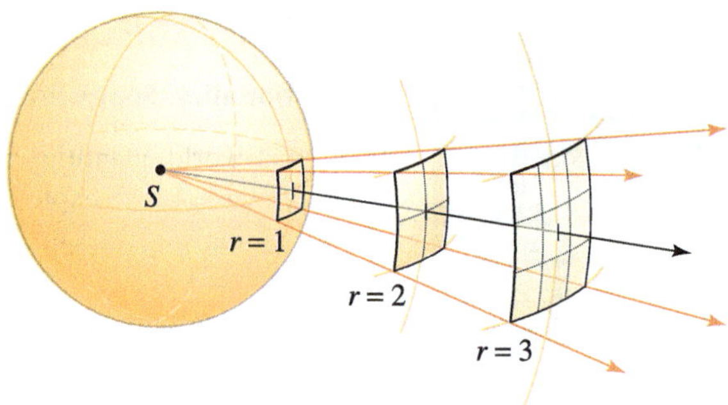

FIGURE 6 Energy from a point source S

1.12 EXERCISES

CONCEPTS

1. If the quantities x and y are related by the equation $y = 3x$, then we say that y is ______________ ______________ to x and the constant of ______________ is 3.

2. If the quantities x and y are related by the equation $y = \dfrac{3}{x}$, then we say that y is ______________ ______________ to x and the constant of ______________ is 3.

3. If the quantities x, y, and z are related by the equation $z = 3\dfrac{x}{y}$, then we say that z is ______________ ______________ to x and ______________ ______________ to y.

4. If z is directly proportional to the product of x and y and if z is 10 when x is 4 and y is 5, then x, y, and z are related by the equation $z =$ ________.

5–6 ■ In each equation, is y directly proportional, inversely proportional, or not proportional to x?

5. **(a)** $y = 3x$ **(b)** $y = 3x + 1$

6. **(a)** $y = \dfrac{3}{x + 1}$ **(b)** $y = \dfrac{3}{x}$

SKILLS

7–18 ■ **Equations of Proportionality** Write an equation that expresses the statement.

7. T varies directly as x.

8. P is directly proportional to w.

9. v is inversely proportional to z.

10. w is proportional to the product of m and n.

11. y is proportional to s and inversely proportional to t.

12. P varies inversely as T.

13. z is proportional to the square root of y.

14. A is proportional to the square of x and inversely proportional to the cube of t.

15. V is proportional to the product of l, w, and h.

16. S is proportional to the product of the squares of r and θ.

17. R is proportional to the product of the squares of P and t and inversely proportional to the cube of b.

18. A is jointly proportional to the square roots of x and y.

19–30 ■ **Constants of Proportionality** Express the statement as an equation. Use the given information to find the constant of proportionality.

19. y is directly proportional to x. If $x = 6$, then $y = 42$.

20. w is inversely proportional to t. If $t = 8$, then $w = 3$.

21. A varies inversely as r. If $r = 3$, then $A = 7$.

22. P is directly proportional to T. If $T = 300$, then $P = 20$.

23. A is directly proportional to x and inversely proportional to t. If $x = 7$ and $t = 3$, then $A = 42$.

24. S is proportional to the product of p and q. If $p = 4$ and $q = 5$, then $S = 180$.

25. W is inversely proportional to the square of r. If $r = 6$, then $W = 10$.

26. t is proportional to the product of x and y and inversely proportional to r. If $x = 2$, $y = 3$, and $r = 12$, then $t = 25$.

27. C is jointly proportional to l, w, and h. If $l = w = h = 2$, then $C = 128$.

28. H is jointly proportional to the squares of l and w. If $l = 2$ and $w = \frac{1}{3}$, then $H = 36$.

29. R is inversely proportional to the square root of x. If $x = 121$, then $R = 2.5$.

30. M is jointly proportional to a, b, and c and inversely proportional to d. If a and d have the same value and if b and c are both 2, then $M = 128$.

31–34 ■ **Proportionality** A statement describing the relationship between the variables x, y, and z is given. **(a)** Express the statement as an equation of proportionality. **(b)** If x is tripled and y is doubled, by what factor does z change? (See Example 3.)

31. z varies directly as the cube of x and inversely as the square of y.

32. z is directly proportional to the square of x and inversely proportional to the fourth power of y.

33. z is jointly proportional to the cube of x and the fifth power of y.

34. z is inversely proportional to the square of x and the cube of y.

APPLICATIONS

35. **Hooke's Law** Hooke's Law states that the force needed to keep a spring stretched x units beyond its natural length is directly proportional to x. Here the constant of proportionality is called the **spring constant**.

(a) Write Hooke's Law as an equation.

(b) If a spring has a natural length of 5 cm and a force of 30 N is required to maintain the spring stretched to a length of 9 cm, find the spring constant.

(c) What force is needed to keep the spring stretched to a length of 11 cm?

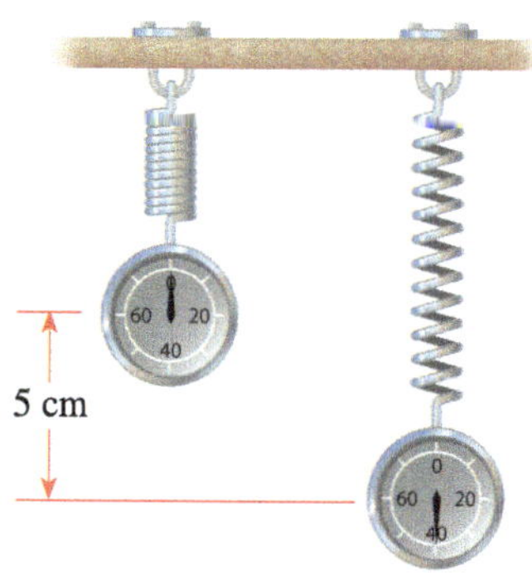

36. Printing Costs The cost C of printing a magazine is jointly proportional to the number of pages p in the magazine and the number of magazines printed m.

(a) Write an equation that expresses this joint variation.

(b) Find the constant of proportionality if the printing cost is \$60,000 for 4000 copies of a 120-page magazine.

(c) How much would the printing cost be for 5000 copies of a 92-page magazine?

37. Power from a Windmill The power P that can be obtained from a windmill is directly proportional to the cube of the wind speed s.

(a) Write an equation that expresses this variation.

(b) Find the constant of proportionality for a windmill that produces 96 watts of power when the wind is blowing at 20 mi/h.

(c) How much power will this windmill produce if the wind speed increases to 30 mi/h?

38. Power Needed to Propel a Boat The power P (measured in horsepower, hp) needed to propel a boat is directly proportional to the cube of the speed s.

(a) Write an equation that expresses this variation.

(b) Find the constant of proportionality for a boat that needs an 80-hp engine to propel the boat at 10 knots.

(c) How much power is needed to drive this boat at 15 knots?

39. Stopping Distance The stopping distance D of a car after the brakes have been applied varies directly as the square of the speed s. A certain car traveling at 40 mi/h can stop in 150 ft. What is the maximum speed it can be traveling if it needs to stop in 200 ft?

40. Aerodynamic Lift The lift L on an airplane wing at takeoff varies jointly as the square of the speed s of the plane and the area A of its wings. A plane with a wing area of 500 ft^2 traveling at 50 mi/h experiences a lift of 1700 lb. How much lift would a plane with a wing area of 600 ft^2 traveling at 40 mi/h experience?

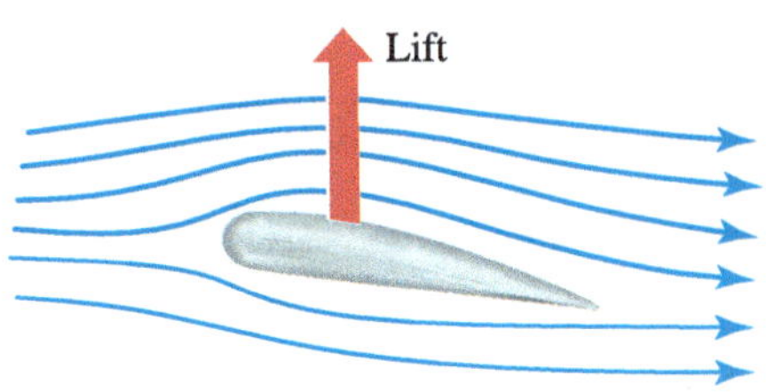

41. Drag Force on a Boat The drag force F on a boat is jointly proportional to the wetted surface area A on the hull and the square of the speed s of the boat. A boat experiences a drag force of 220 lb when traveling at 5 mi/h with a wetted surface area of 40 ft^2. How fast must a boat be traveling if it has 28 ft^2 of wetted surface area and is experiencing a drag force of 175 lb?

42. Kepler's Third Law Kepler's Third Law of planetary motion states that the square of the period T of a planet (the time it takes for the planet to make a complete revolution about the sun) is directly proportional to the cube of its average distance d from the sun.

(a) Express Kepler's Third Law as an equation.

(b) Find the constant of proportionality by using the fact that for our planet the period is about 365 days and the average distance is about 93 million miles.

(c) The planet Neptune is about 2.79×10^9 mi from the sun. Find the period of Neptune.

43. Ideal Gas Law The pressure P of a sample of gas is directly proportional to the temperature T and inversely proportional to the volume V.

(a) Write an equation that expresses this variation.

(b) Find the constant of proportionality if 100 L of gas exerts a pressure of 33.2 kPa at a temperature of 400 K (absolute temperature measured on the Kelvin scale).

(c) If the temperature is increased to 500 K and the volume is decreased to 80 L, what is the pressure of the gas?

44. Skidding in a Curve A car is traveling on a curve that forms a circular arc. The force F needed to keep the car from skidding is jointly proportional to the weight w of the car and the square of its speed s and is inversely proportional to the radius r of the curve.

(a) Write an equation that expresses this variation.

(b) A car weighing 1600 lb travels around a curve at 60 mi/h. The next car to round this curve weighs 2500 lb and requires the same force as the first car to keep from skidding. How fast is the second car traveling?

45. Loudness of Sound The loudness L of a sound (measured in decibels, dB) is inversely proportional to the square of the distance d from the source of the sound.

(a) Write an equation that expresses this variation.

(b) Find the constant of proportionality if a person 10 ft from a lawn mower experiences a sound level of 70 dB.

(c) If the distance in part (b) is doubled, by what factor is the loudness changed?

(d) If the distance in part (b) is cut in half, by what factor is the loudness changed?

46. A Jet of Water The power P of a jet of water is jointly proportional to the cross-sectional area A of the jet and to the cube of the velocity v.

(a) Write an equation that expresses this variation.

(b) If the velocity is doubled and the cross-sectional area is halved, by what factor is the power changed?

(c) If the velocity is halved and the cross-sectional area is tripled, by what factor is the power changed?

47. Electrical Resistance The resistance R of a wire varies directly as its length L and inversely as the square of its diameter d.

(a) Write an equation that expresses this joint variation.

(b) Find the constant of proportionality if a wire 1.2 m long and 0.005 m in diameter has a resistance of 140 ohms.

(c) Find the resistance of a wire made of the same material that is 3 m long and has a diameter of 0.008 m.

(d) If the diameter is doubled and the length is tripled, by what factor is the resistance changed?

48. Growing Cabbages In the short growing season of the Canadian arctic territory of Nunavut, some gardeners find it possible to grow gigantic cabbages in the midnight sun. Assume that the final size of a cabbage is proportional to the amount of nutrients it receives and inversely proportional to the number of other cabbages surrounding it. A cabbage that received 20 oz of nutrients and had 12 other cabbages around it grew to 30 lb. What size would it grow to if it received 10 oz of nutrients and had only 5 cabbage "neighbors"?

49. Radiation Energy The total radiation energy E emitted by a heated surface per unit area varies as the fourth power of its absolute temperature T. The temperature is 6000 K at the surface of the sun and 300 K at the surface of the earth.

(a) How many times more radiation energy per unit area is produced by the sun than by the earth?

(b) The radius of the earth is 3960 mi, and the radius of the sun is 435,000 mi. How many times more total radiation does the sun emit than the earth?

50. Value of a Lot The value of a building lot on Galiano Island is jointly proportional to its area and the quantity of water produced by a well on the property. A 200 ft by 300 ft lot has a well producing 10 gal of water per minute and is valued at \$48,000. What is the value of a 400 ft by 400 ft lot if the well on the lot produces 4 gal of water per minute?

51. Law of the Pendulum The period of a pendulum (the time elapsed during one complete swing of the pendulum) varies directly with the square root of the length of the pendulum.

(a) Express this relationship by writing an equation.

(b) To double the period, how would we have to change the length l?

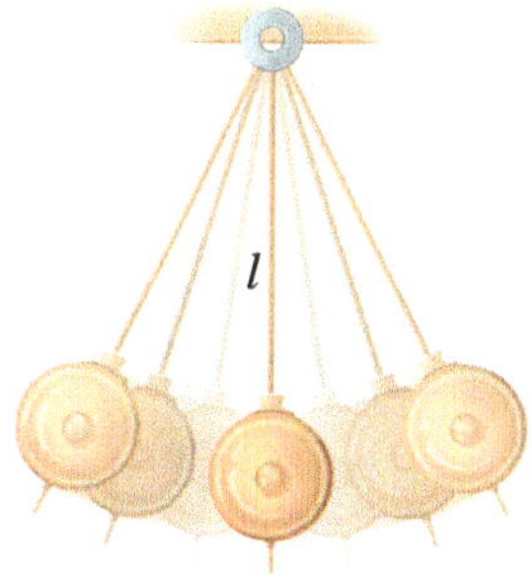

52. Heat of a Campfire The heat experienced by a hiker at a campfire is proportional to the amount of wood on the fire and inversely proportional to the cube of his distance from the fire. If the hiker is 20 ft from the fire and someone doubles the amount of wood burning, how far from the fire would he have to be so that he feels the same heat as before?

53. Frequency of Vibration The frequency f of vibration of a violin string is inversely proportional to its length L. The constant of proportionality k is positive and depends on the tension and density of the string.

(a) Write an equation that represents this variation.

(b) What effect does doubling the length of the string have on the frequency of its vibration?

54. Spread of a Disease The rate r at which a disease spreads in a population of size P is jointly proportional to the number x of infected people and the number $P - x$ who are not infected. An infection erupts in a small town that has population $P = 5000$.

(a) Write an equation that expresses r as a function of x.

(b) Compare the rate of spread of this infection when 10 people are infected to the rate of spread when 1000 people are infected. Which rate is larger? By what factor?

(c) Calculate the rate of spread when the entire population is infected. Why does this answer make intuitive sense?

55–56 ■ Combining Variations Solve the problem using the relationship between brightness B, luminosity L, and distance d derived in Example 3. The proportionality constant is $k = 0.080$.

55. Brightness of a Star The luminosity of a star is $L = 2.5 \times 10^{26}$ W, and its distance from the earth is $d = 2.4 \times 10^{19}$ m. How bright does the star appear on the earth?

56. Distance to a Star The luminosity of a star is $L = 5.8 \times 10^{30}$ W, and its brightness as viewed from the earth is $B = 8.2 \times 10^{-16}$ W/m^2. Find the distance of the star from the earth.

DISCUSS ■ DISCOVER ■ PROVE ■ WRITE

57. DISCUSS: Is Proportionality Everything? A great many laws of physics and chemistry are expressible as proportionalities. Give at least one example of a function that occurs in the sciences that is *not* a proportionality.

CHAPTER 1 ■ REVIEW

PROPERTIES AND FORMULAS

Properties of Real Numbers (p. 3)

Commutative: $a + b = b + a$

$ab = ba$

Associative: $(a + b) + c = a + (b + c)$

$(ab)c = a(bc)$

Distributive: $a(b + c) = ab + ac$

Absolute Value (pp. 8–9)

$$|a| = \begin{cases} a & \text{if } a \geq 0 \\ -a & \text{if } a < 0 \end{cases}$$

$|ab| = |a||b|$

$$\left|\frac{a}{b}\right| = \frac{|a|}{|b|}$$

Distance between a and b:

$$d(a, b) = |b - a|$$

Exponents (p. 14)

$a^m a^n = a^{m+n}$

$$\frac{a^m}{a^n} = a^{m-n}$$

$(a^m)^n = a^{mn}$

$(ab)^n = a^n b^n$

$$\left(\frac{a}{b}\right)^n = \frac{a^n}{b^n}$$

Radicals (p. 18)

$\sqrt[n]{a} = b$ means $b^n = a$

$\sqrt[n]{ab} = \sqrt[n]{a}\sqrt[n]{b}$

$$\sqrt[n]{\frac{a}{b}} = \frac{\sqrt[n]{a}}{\sqrt[n]{b}}$$

$\sqrt[m]{\sqrt[n]{a}} = \sqrt[mn]{a}$

$a^{m/n} = \sqrt[n]{a^m}$

If n is odd, then $\sqrt[n]{a^n} = a$.

If n is even, then $\sqrt[n]{a^n} = |a|$.

Special Product Formulas (p. 27)

Sum and difference of same terms:

$$(A + B)(A - B) = A^2 - B^2$$

Square of a sum or difference:

$$(A + B)^2 = A^2 + 2AB + B^2$$

$$(A - B)^2 = A^2 - 2AB + B^2$$

Cube of a sum or difference:

$$(A + B)^3 = A^3 + 3A^2B + 3AB^2 + B^3$$

$$(A - B)^3 = A^3 - 3A^2B + 3AB^2 - B^3$$

Special Factoring Formulas (p. 30)

Difference of squares:

$$A^2 - B^2 = (A + B)(A - B)$$

Perfect squares:

$$A^2 + 2AB + B^2 = (A + B)^2$$

$$A^2 - 2AB + B^2 = (A - B)^2$$

Sum or difference of cubes:

$$A^3 - B^3 = (A - B)(A^2 + AB + B^2)$$

$$A^3 + B^3 = (A + B)(A^2 - AB + B^2)$$

Rational Expressions (pp. 37–38)

We can cancel common factors:

$$\frac{AC}{BC} = \frac{A}{B}$$

To multiply two fractions, we multiply their numerators together and their denominators together:

$$\frac{A}{B} \times \frac{C}{D} = \frac{AC}{BD}$$

To divide fractions, we invert the divisor and multiply:

$$\frac{A}{B} \div \frac{C}{D} = \frac{A}{B} \times \frac{D}{C}$$

To add fractions, we find a common denominator:

$$\frac{A}{C} + \frac{B}{C} = \frac{A + B}{C}$$

Properties of Equality (p. 46)

$A = B \quad \Leftrightarrow \quad A + C = B + C$

$A = B \quad \Leftrightarrow \quad CA = CB \quad (C \neq 0)$

Linear Equations (p. 46)

A **linear equation** is an equation of the form $ax + b = 0$

Zero-Product Property (p. 48)

If $AB = 0$, then $A = 0$ or $B = 0$.

Completing the Square (p. 49)

To make $x^2 + bx$ a perfect square, add $\left(\frac{b}{2}\right)^2$. This gives the perfect square

$$x^2 + bx + \left(\frac{b}{2}\right)^2 = \left(x + \frac{b}{2}\right)^2$$

Quadratic Formula (p. 50)

A **quadratic equation** is an equation of the form

$$ax^2 + bx + c = 0$$

Its solutions are given by the **Quadratic Formula**:

$$x = \frac{-b \pm \sqrt{b^2 - 4ac}}{2a}$$

The **discriminant** is $D = b^2 - 4ac$.

If $D > 0$, the equation has two real solutions.

If $D = 0$, the equation has one solution.

If $D < 0$, the equation has two complex solutions.

Complex Numbers (pp. 59–61)

A **complex number** is a number of the form $a + bi$, where $i = \sqrt{-1}$.

The **complex conjugate** of $a + bi$ is

$$\overline{a + bi} = a - bi$$

To **multiply** complex numbers, treat them as binomials and use $i^2 = -1$ to simplify the result.

To **divide** complex numbers, multiply numerator and denominator by the complex conjugate of the denominator:

$$\frac{a + bi}{c + di} = \left(\frac{a + bi}{c + di}\right) \cdot \left(\frac{c - di}{c - di}\right) = \frac{(a + bi)(c - di)}{c^2 + d^2}$$

Inequalities (p. 82)

Adding the same quantity to each side of an inequality gives an equivalent inequality:

$$A < B \quad \Leftrightarrow \quad A + C < B + C$$

Multiplying each side of an inequality by the same *positive* quantity gives an equivalent inequality. Multiplying each side by the same *negative* quantity reverses the direction of the inequality:

If $C > 0$, then $A < B \quad \Leftrightarrow \quad CA < CB$

If $C < 0$, then $A < B \quad \Leftrightarrow \quad CA > CB$

Absolute Value Inequalities (p. 86)

To solve absolute value inequalities, we use

$$|x| < C \quad \Leftrightarrow \quad -C < x < C$$

$$|x| > C \quad \Leftrightarrow \quad x < -C \quad \text{or} \quad x > C$$

The Distance Formula (p. 93)

The distance between the points $A(x_1, y_1)$ and $B(x_2, y_2)$ is

$$d(A, B) = \sqrt{(x_1 - x_2)^2 + (y_1 - y_2)^2}$$

The Midpoint Formula (p. 94)

The midpoint of the line segment from $A(x_1, y_1)$ to $B(x_2, y_2)$ is

$$\left(\frac{x_1 + x_2}{2}, \frac{y_1 + y_2}{2}\right)$$

Intercepts (p. 97)

To find the ***x*-intercepts** of the graph of an equation, set $y = 0$ and solve for x.

To find the ***y*-intercepts** of the graph of an equation, set $x = 0$ and solve for y.

Circles (p. 98)

The circle with center $(0, 0)$ and radius r has equation

$$x^2 + y^2 = r^2$$

The circle with center (h, k) and radius r has equation

$$(x - h)^2 + (y - k)^2 = r^2$$

Symmetry (p. 100)

The graph of an equation is **symmetric with respect to the *x*-axis** if the equation remains unchanged when y is replaced by $-y$.

The graph of an equation is **symmetric with respect to the *y*-axis** if the equation remains unchanged when x is replaced by $-x$.

The graph of an equation is **symmetric with respect to the origin** if the equation remains unchanged when x is replaced by $-x$ and y by $-y$.

Slope of a Line (p. 107)

The slope of the nonvertical line that contains the points $A(x_1, y_1)$ and $B(x_2, y_2)$ is

$$m = \frac{\text{rise}}{\text{run}} = \frac{y_2 - y_1}{x_2 - x_1}$$

Equations of Lines (pp. 108–110)

If a line has slope m, has y-intercept b, and contains the point (x_1, y_1), then:

the **point-slope form** of its equation is

$$y - y_1 = m(x - x_1)$$

the **slope-intercept form** of its equation is

$$y = mx + b$$

The equation of any line can be expressed in the **general form**

$$Ax + By + C = 0$$

(where A and B can't both be 0).

Vertical and Horizontal Lines (p. 110)

The **vertical** line containing the point (a, b) has the equation $x = a$.

The **horizontal** line containing the point (a, b) has the equation $y = b$.

Parallel and Perpendicular Lines (pp. 111–112)

Two lines with slopes m_1 and m_2 are

parallel if and only if $m_1 = m_2$

perpendicular if and only if $m_1 m_2 = -1$

Variation (pp. 123–124)

If y is **directly proportional** to x, then

$$y = kx$$

If y is **inversely proportional** to x, then

$$y = \frac{k}{x}$$

CONCEPT CHECK

1. **(a)** What does the set of natural numbers consist of? What does the set of integers consist of? Give an example of an integer that is not a natural number.
   (b) What does the set of rational numbers consist of? Give an example of a rational number that is not an integer.
   (c) What does the set of irrational numbers consist of? Give an example of an irrational number.
   (d) What does the set of real numbers consist of?
2. A property of real numbers is given. State the property and give an example in which the property is used.
   (i) Commutative Property
   (ii) Associative Property
   (iii) Distributive Property
3. Explain the difference between the open interval (a, b) and the closed interval $[a, b]$. Give an example of an interval that is neither open nor closed.
4. Give the formula for finding the distance between two real numbers a and b. Use the formula to find the distance between 103 and -52.
5. Suppose $a \neq 0$ is any real number.
   (a) In the expression a^n, which is the base and which is the exponent?
   (b) What does a^n mean if n is a positive integer? What does 6^5 mean?
   (c) What does a^{-n} mean if n is a positive integer? What does 3^{-2} mean?
   (d) What does a^n mean if n is zero?
   (e) If m and n are positive integers, what does $a^{m/n}$ mean? What does $4^{3/2}$ mean?
6. State the first five Laws of Exponents. Give examples in which you would use each law.
7. When you multiply two powers of the same number, what should you do with the exponents? When you raise a power to a new power, what should you do with the exponents?
8. **(a)** What does $\sqrt[n]{a} = b$ mean?
   (b) Is it true that $\sqrt{a^2}$ is equal to $|a|$? Try values for a that are positive and negative.
   (c) How many real nth roots does a positive real number have if n is even? If n is odd?
   (d) Is $\sqrt[4]{-2}$ a real number? Is $\sqrt[3]{-2}$ a real number? Explain why or why not.
9. Explain the steps involved in rationalizing a denominator. What is the logical first step in rationalizing the denominator of the expression $\frac{5}{\sqrt{3}}$?
10. Explain the difference between expanding an expression and factoring an expression.
11. State the Special Product Formulas used for expanding the given expression.
    (i) $(a + b)^2$ **(ii)** $(a - b)^2$ **(iii)** $(a + b)^3$
    (iv) $(a - b)^3$ **(v)** $(a + b)(a - b)$
    Use the appropriate formula to expand $(x + 5)^2$ and $(x + 5)(x - 5)$.
12. State the following Special Factoring Formulas.
    (i) Difference of Squares
    (ii) Perfect Square
    (iii) Sum of Cubes
    Use the appropriate formula to factor $x^2 - 9$.
13. If the numerator and the denominator of a rational expression have a common factor, how would you simplify the expression? Simplify the expression $\frac{x^2 + x}{x + 1}$.
14. Explain the following.
    (a) How to multiply and divide rational expressions.
    (b) How to add and subtract rational expressions.
    (c) What LCD do we use to perform the addition in the expression $\frac{3}{x - 1} + \frac{5}{x + 2}$?
15. What is the logical first step in rationalizing the denominator of $\frac{3}{1 + \sqrt{x}}$?
16. What is the difference between an algebraic expression and an equation? Give examples.
17. Write the general form of each type of equation.
    (i) Linear equation
    (ii) Quadratic equation

18. What are the three ways to solve a quadratic equation?

19. State the Zero-Product Property. Use the property to solve the equation $x(x - 1) = 0$.

20. What do you need to add to $ax^2 + bx$ to complete the square? Complete the square for the expression $x^2 + 6x$.

21. State the Quadratic Formula for the quadratic equation $ax^2 + bx + c = 0$, and use it to solve the equation $x^2 + 6x - 1 = 0$.

22. What is the discriminant of the quadratic equation $ax^2 + bx + c = 0$? Find the discriminant of $2x^2 - 3x + 5 = 0$. How many real solutions does this equation have?

23. What is the logical first step in solving the equation $\sqrt{x - 1} = x - 3$? Why is it important to check your answers when solving equations of this type?

24. What is a complex number? Give an example of a complex number, and identify the real and imaginary parts.

25. What is the complex conjugate of a complex number $a + bi$?

26. (a) How do you add complex numbers?
    (b) How do you multiply $(3 + 5i)(2 - i)$?
    (c) Is $(3 - i)(3 + i)$ a real number?
    (d) How do you simplify the quotient $(3 + 5i)/(3 - i)$?

27. State the guidelines for modeling with equations.

28. Explain how to solve the given type of problem.
    (a) Linear inequality: $2x \geq 1$
    (b) Nonlinear inequality: $(x - 1)(x - 4) < 0$
    (c) Absolute value equation: $|2x - 5| = 7$
    (d) Absolute value inequality: $|2x - 5| \leq 7$

29. (a) In the coordinate plane, what is the horizontal axis called and what is the vertical axis called?
    (b) To graph an ordered pair of numbers (x, y), you need the coordinate plane. For the point $(2, 3)$, which is the x-coordinate and which is the y-coordinate?
    (c) For an equation in the variables x and y, how do you determine whether a given point is on the graph? Is the point $(5, 3)$ on the graph of the equation $y = 2x - 1$?

30. (a) What is the formula for finding the distance between the points (x_1, y_1) and (x_2, y_2)?
    (b) What is the formula to finding the midpoint between (x_1, y_1) and (x_2, y_2)?

31. How do you find x-intercepts and y-intercepts of a graph of an equation?

32. (a) Write an equation of the circle with center (h, k) and radius r.
    (b) Find the equation of the circle with center $(2, -1)$ and radius 3.

33. (a) How do you test whether the graph of an equation is symmetric with respect to the (i) x-axis, (ii) y-axis, and (iii) origin?
    (b) What type of symmetry does the graph of the equation $xy^2 + y^2x^2 = 3x$ have?

34. (a) What is the slope of a line? How do you compute the slope of the line through the points $(-1, 4)$ and $(1, -2)$?
    (b) How do you find the slope and y-intercept of the line $6x + 3y = 12$?
    (c) How do you write the equation for a line that has slope 3 and passes through the point $(1, 2)$?

35. Give an equation of a vertical line and of a horizontal line that passes through the point $(2, 3)$.

36. State the general equation of a line.

37. Given lines with slopes m_1 and m_2, explain how you can tell whether the lines are (i) parallel, (ii) perpendicular.

38. How do you solve an equation (i) algebraically? (ii) graphically?

39. How do you solve an inequality (i) algebraically? (ii) graphically?

40. Write an equation that expresses each relationship.
    (a) y is directly proportional to x.
    (b) y is inversely proportional to x.
    (c) z is jointly proportional to x and y.

ANSWERS TO THE CONCEPT CHECK CAN BE FOUND AT THE BACK OF THE BOOK.

EXERCISES

1–4 ■ Properties of Real Numbers State the property of real numbers being used.

1. $3x + 2y = 2y + 3x$
2. $(a + b)(a - b) = (a - b)(a + b)$
3. $4(a + b) = 4a + 4b$
4. $(A + 1)(x + y) = (A + 1)x + (A + 1)y$

5–6 ■ Intervals Express the interval in terms of inequalities, and then graph the interval.

5. $[-2, 6)$
6. $(-\infty, 4]$

7–8 ■ Intervals Express the inequality in interval notation, and then graph the corresponding interval.

7. $x \geq 5$
8. $-1 < x \leq 5$

9–16 ■ Evaluate Evaluate the expression.

9. $|1 - |-4||$

10. $5 - |10 - |-4||$

11. $2^{1/2}8^{1/2}$

12. $2^{-3} - 3^{-2}$

13. $216^{-1/3}$

14. $64^{2/3}$

15. $\dfrac{\sqrt{242}}{\sqrt{2}}$

16. $\sqrt{2}\sqrt{50}$

17–20 ■ Radicals and Exponents Simplify the expression.

17. (a) $(a^2)^{-3}(a^3b)^2(b^3)^4$ **(b)** $(3xy^2)^3(\frac{2}{3}x^{-1}y)^2$

18. (a) $\dfrac{x^2(2x)^4}{x^3}$ **(b)** $\left(\dfrac{r^2s^{4/3}}{r^{1/3}s}\right)^6$

19. (a) $\sqrt[3]{(x^3y)^2y^4}$ **(b)** $\sqrt{x^2y^4}$

20. (a) $\dfrac{8r^{1/2}s^{-3}}{2r^{-2}s^4}$ **(b)** $\left(\dfrac{ab^2c^{-3}}{2a^3b^{-4}}\right)^{-2}$

21–24 ■ Scientific Notation These exercises involve scientific notation.

21. Write the number 78,250,000,000 in scientific notation.

22. Write the number 2.08×10^{-8} in ordinary decimal notation.

23. If $a \approx 0.00000293$, $b \approx 1.582 \times 10^{-14}$, and $c \approx 2.8064 \times 10^{12}$, use a calculator to approximate the number ab/c.

24. If your heart beats 80 times per minute and you live to be 90 years old, estimate the number of times your heart beats during your lifetime. State your answer in scientific notation.

25–38 ■ Factoring Factor the expression completely.

25. $x^2 + 5x - 14$

26. $12x^2 + 10x - 8$

27. $x^4 - 2x^2 + 1$

28. $12x^2y^4 - 3xy^5 + 9x^3y^2$

29. $16 - 4t^2$

30. $2y^6 - 32y^2$

31. $x^6 - 1$

32. $16a^4b^2 + 2ab^5$

33. $-3x^{-1/2} + 2x^{1/2} + 5x^{3/2}$

34. $7x^{-3/2} - 8x^{-1/2} + x^{1/2}$

35. $4x^3 - 8x^2 + 3x - 6$

36. $w^3 - 3w^2 - 4w + 12$

37. $(a + b)^2 - 3(a + b) - 10$

38. $(x + 2)^2 - 7(x + 2) + 6$

39–50 ■ Operations with Algebraic Expressions Perform the indicated operations and simplify.

39. $(2y - 7)(2y + 7)$

40. $(1 + x)(2 - x) - (3 - x)(3 + x)$

41. $x^2(x - 2) + x(x - 2)^2$

42. $\sqrt{x}(\sqrt{x} + 1)(2\sqrt{x} - 1)$

43. $\dfrac{x^2 + 2x - 3}{x^2 + 8x + 16} \cdot \dfrac{3x + 12}{x - 1}$

44. $\dfrac{x^2 - 2x - 15}{x^2 - 6x + 5} \div \dfrac{x^2 - x - 12}{x^2 - 1}$

45. $\dfrac{2}{x} + \dfrac{1}{x - 2} + \dfrac{3}{(x - 2)^2}$

46. $\dfrac{1}{x + 2} + \dfrac{1}{x^2 - 4} - \dfrac{2}{x^2 - x - 2}$

47. $\dfrac{\dfrac{1}{x} - \dfrac{1}{2}}{x - 2}$

48. $\dfrac{\dfrac{1}{x} - \dfrac{1}{x + 1}}{\dfrac{1}{x} + \dfrac{1}{x + 1}}$

49. $\dfrac{\sqrt{6}}{\sqrt{3} + \sqrt{2}}$ (rationalize the denominator)

50. $\dfrac{\sqrt{x + h} - \sqrt{x}}{h}$ (rationalize the numerator)

51–54 ■ Rationalizing Rationalize the denominator and simplify.

51. $\dfrac{1}{\sqrt{11}}$

52. $\dfrac{3}{\sqrt{6}}$

53. $\dfrac{10}{\sqrt{2} - 1}$

54. $\dfrac{\sqrt{x} - 2}{\sqrt{x} + 2}$

55–70 ■ Solving Equations Find all real solutions of the equation.

55. $7x - 6 = 4x + 9$

56. $8 - 2x = 14 + x$

57. $\dfrac{x + 1}{x - 1} = \dfrac{3x}{3x - 6}$

58. $(x + 2)^2 = (x - 4)^2$

59. $x^2 - 9x + 14 = 0$

60. $x^2 + 24x + 144 = 0$

61. $2x^2 + x = 1$

62. $3x^2 + 5x - 2 = 0$

63. $4x^3 - 25x = 0$

64. $x^3 - 2x^2 - 5x + 10 = 0$

65. $3x^2 + 4x - 1 = 0$

66. $\dfrac{1}{x} + \dfrac{2}{x - 1} = 3$

67. $\dfrac{x}{x - 2} + \dfrac{1}{x + 2} = \dfrac{8}{x^2 - 4}$

68. $x^4 - 8x^2 - 9 = 0$

69. $|x - 7| = 4$

70. $|2x - 5| = 9$

71–74 ■ Complex Numbers Evaluate the expression and write in the form $a + bi$.

71. (a) $(2 - 3i) + (1 + 4i)$ **(b)** $(2 + i)(3 - 2i)$

72. (a) $(3 - 6i) - (6 - 4i)$ **(b)** $4i(2 - \frac{1}{2}i)$

73. (a) $\dfrac{4 + 2i}{2 - i}$ **(b)** $(1 - \sqrt{-1})(1 + \sqrt{-1})$

74. (a) $\dfrac{8 + 3i}{4 + 3i}$ **(b)** $\sqrt{-10} \cdot \sqrt{-40}$

75–80 ■ Real and Complex Solutions Find all real and complex solutions of the equation.

75. $x^2 + 16 = 0$

76. $x^2 = -12$

77. $x^2 + 6x + 10 = 0$

78. $2x^2 - 3x + 2 = 0$

79. $x^4 - 256 = 0$

80. $x^3 - 2x^2 + 4x - 8 = 0$

81. Mixtures The owner of a store sells raisins for \$3.20 per pound and nuts for \$2.40 per pound. He decides to mix the raisins and nuts and sell 50 lb of the mixture for \$2.72 per pound. What quantities of raisins and nuts should he use?

82. Distance and Time Anthony leaves Kingstown at 2:00 P.M. and drives to Queensville, 160 mi distant, at 45 mi/h. At 2:15 P.M. Helen leaves Queensville and drives to Kingstown at 40 mi/h. At what time do they pass each other on the road?

83. Distance and Time A woman cycles 8 mi/h faster than she runs. Every morning she cycles 4 mi and runs $2\frac{1}{2}$ mi, for a total of one hour of exercise. How fast does she run?

84. Geometry The hypotenuse of a right triangle has length 20 cm. The sum of the lengths of the other two sides is 28 cm. Find the lengths of the other two sides of the triangle.

85. Doing the Job Abbie paints twice as fast as Beth and three times as fast as Cathie. If it takes them 60 min to paint a living room with all three working together, how long would it take Abbie if she worked alone?

86. Dimensions of a Garden A homeowner wishes to fence in three adjoining garden plots, one for each of her children, as shown in the figure. If each plot is to be 80 ft^2 in area and she has 88 ft of fencing material at hand, what dimensions should each plot have?

87–94 ■ Inequalities Solve the inequality. Express the solution using interval notation and graph the solution set on the real number line.

87. $3x - 2 > -11$

88. $-1 < 2x + 5 \leq 3$

89. $x^2 + 4x - 12 > 0$

90. $x^2 \leq 1$

91. $\dfrac{x - 4}{x^2 - 4} \leq 0$

92. $\dfrac{5}{x^3 - x^2 - 4x + 4} < 0$

93. $|x - 5| \leq 3$

94. $|x - 4| < 0.02$

95–96 ■ Coordinate Plane Two points P and Q are given. **(a)** Plot P and Q on a coordinate plane. **(b)** Find the distance from P to Q. **(c)** Find the midpoint of the segment PQ. **(d)** Sketch the line determined by P and Q, and find its equation in slope-intercept form. **(e)** Sketch the circle that passes through Q and has center P, and find the equation of this circle.

95. $P(2, 0)$, $Q(-5, 12)$

96. $P(7, -1)$, $Q(2, -11)$

97–98 ■ Graphing Regions Sketch the region given by the set.

97. $\{(x, y) \mid -4 < x < 4 \text{ and } -2 < y < 2\}$

98. $\{(x, y) \mid x \geq 4 \text{ or } y \geq 2\}$

99. Distance Formula Which of the points $A(4, 4)$ or $B(5, 3)$ is closer to the point $C(-1, -3)$?

100–102 ■ Circles In these exercises we find equations of circles.

100. Find an equation of the circle that has center $(2, -5)$ and radius $\sqrt{2}$.

101. Find an equation of the circle that has center $(-5, -1)$ and passes through the origin.

102. Find an equation of the circle that contains the points $P(2, 3)$ and $Q(-1, 8)$ and has the midpoint of the segment PQ as its center.

103–106 ■ Circles **(a)** Complete the square to determine whether the equation represents a circle or a point or has no graph. **(b)** If the equation is that of a circle, find its center and radius, and sketch its graph.

103. $x^2 + y^2 + 2x - 6y + 9 = 0$

104. $2x^2 + 2y^2 - 2x + 8y = \frac{1}{2}$

105. $x^2 + y^2 + 72 = 12x$

106. $x^2 + y^2 - 6x - 10y + 34 = 0$

107–112 ■ Graphing Equations Sketch the graph of the equation by making a table and plotting points.

107. $y = 2 - 3x$

108. $2x - y + 1 = 0$

109. $y = 16 - x^2$

110. $8x + y^2 = 0$

111. $x = \sqrt{y}$

112. $y = -\sqrt{1 - x^2}$

113–118 ■ Symmetry and Intercepts **(a)** Test the equation for symmetry with respect to the x-axis, the y-axis, and the origin. **(b)** Find the x- and y-intercepts of the graph of the equation.

113. $y = 9 - x^2$

114. $6x + y^2 = 36$

115. $x^2 + (y - 1)^2 = 1$

116. $9x^2 - 16y^2 = 144$

117. $x^2 + 4xy + y^2 = 1$

118. $x^3 + xy^2 = 5$

119–122 ■ Graphing Equations **(a)** Use a graphing device to graph the equation in an appropriate viewing rectangle. **(b)** Use the graph to find the x- and y-intercepts.

119. $y = x^2 - 6x$

120. $y = \sqrt{5 - x}$

121. $y = x^3 - 4x^2 - 5x$

122. $\dfrac{x^2}{4} + y^2 = 1$

123–130 ■ Lines A description of a line is given. **(a)** Find an equation for the line in slope-intercept form. **(b)** Find an equation for the line in general form. **(c)** Graph the line.

123. The line that has slope 2 and y-intercept 6

124. The line that has slope $-\frac{1}{2}$ and passes through the point $(6, -3)$

125. The line that passes through the points $(-1, -6)$ and $(2, -4)$

126. The line that has x-intercept 4 and y-intercept 12

127. The vertical line that passes through the point $(3, -2)$

128. The horizontal line with y-intercept 5

129. The line that passes through the origin and is parallel to the line containing $(2, 4)$ and $(4, -4)$

130. The line that passes through the point $(1, 7)$ and is perpendicular to the line $x - 3y + 16 = 0$

131. Stretching a Spring Hooke's Law states that if a weight w is attached to a hanging spring, then the stretched length s of the spring is linearly related to w. For a particular spring we have

$$s = 0.3w + 2.5$$

where s is measured in inches and w in pounds.

(a) What do the slope and s-intercept in this equation represent?

(b) How long is the spring when a 5-lb weight is attached?

132. Annual Salary Margarita is hired by an accounting firm at a salary of \$60,000 per year. Three years later her annual salary has increased to \$70,500. Assume that her salary increases linearly.

(a) Find an equation that relates her annual salary S and the number of years t that she has worked for the firm.

(b) What do the slope and S-intercept of her salary equation represent?

(c) What will her salary be after 12 years with the firm?

133–138 ■ Equations and Inequalities Graphs of the equations $y = x^2 - 4x$ and $y = x + 6$ are given. Use the graphs to solve the equation or inequality.

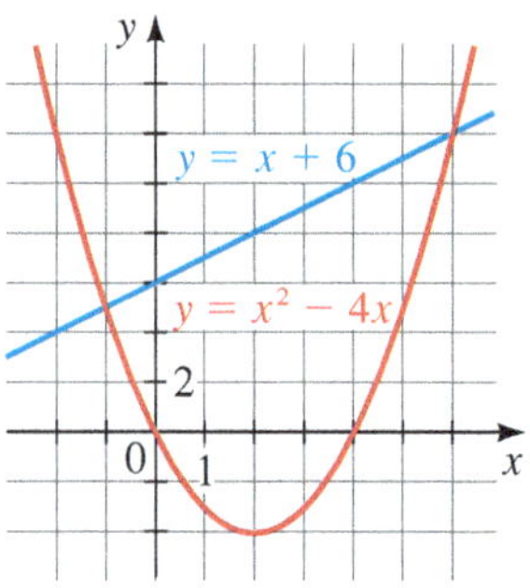

133. $x^2 - 4x = x + 6$

134. $x^2 - 4x = 0$

135. $x^2 - 4x \le x + 6$

136. $x^2 - 4x \ge x + 6$

137. $x^2 - 4x \ge 0$

138. $x^2 - 4x \le 0$

139–142 ■ Equations Solve the equation graphically.

139. $x^2 - 4x = 2x + 7$

140. $\sqrt{x + 4} = x^2 - 5$

141. $x^4 - 9x^2 = x - 9$

142. $||x + 3| - 5| = 2$

143–146 ■ Inequalities Solve the inequality graphically.

143. $4x - 3 \ge x^2$

144. $x^3 - 4x^2 - 5x > 2$

145. $x^4 - 4x^2 < \frac{1}{2}x - 1$

146. $|x^2 - 16| - 10 \ge 0$

147–148 ■ Circles and Lines Find equations for the circle and the line in the figure.

147.

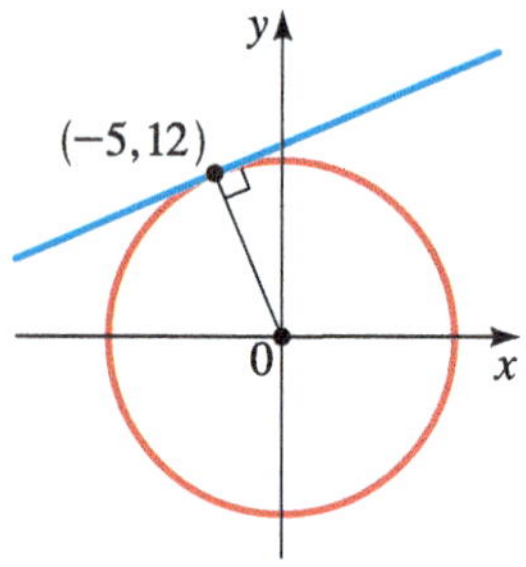

148.

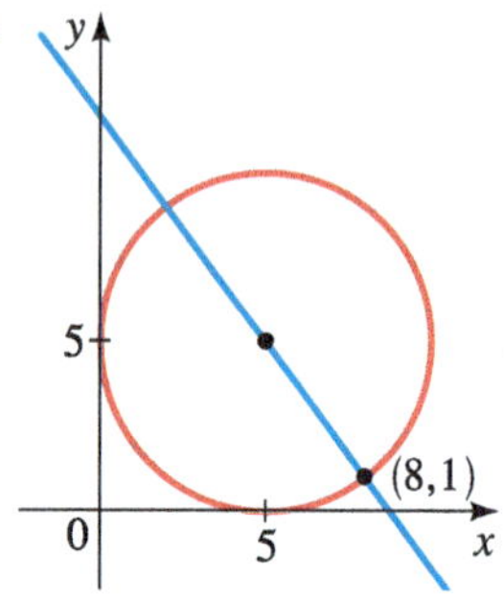

149. Variation Suppose that M varies directly as z, and $M = 120$ when $z = 15$. Write an equation that expresses this variation.

150. Variation Suppose that z is inversely proportional to y, and that $z = 12$ when $y = 16$. Write an equation that expresses z in terms of y.

151. Light Intensity The intensity of illumination I from a light varies inversely as the square of the distance d from the light.

(a) Write this statement as an equation.

(b) Determine the constant of proportionality if it is known that a lamp has an intensity of 1000 candles at a distance of 8 m.

(c) What is the intensity of this lamp at a distance of 20 m?

152. Vibrating String The frequency of a vibrating string under constant tension is inversely proportional to its length. If a violin string 12 inches long vibrates 440 times per second, to what length must it be shortened to vibrate 660 times per second?

153. Terminal Velocity The terminal velocity of a parachutist is directly proportional to the square root of his weight. A 160-lb parachutist attains a terminal velocity of 9 mi/h. What is the terminal velocity for a parachutist weighing 240 lb?

154. Range of a Projectile The maximum range of a projectile is directly proportional to the square of its velocity. A baseball pitcher throws a ball at 60 mi/h, with a maximum range of 242 ft. What is his maximum range if he throws the ball at 70 mi/h?

CHAPTER 1 TEST

1. (a) Graph the intervals $(-5, 3]$ and $(2, \infty)$ on the real number line.
   (b) Express the inequalities $x \le 3$ and $-1 \le x < 4$ in interval notation.
   (c) Find the distance between -7 and 9 on the real number line.

2. Evaluate each expression.
   (a) $(-3)^4$ (b) -3^4 (c) 3^{-4} (d) $\dfrac{5^{23}}{5^{21}}$ (e) $\left(\dfrac{2}{3}\right)^{-2}$ (f) $16^{-3/4}$

3. Write each number in scientific notation.
   (a) 186,000,000,000 (b) 0.0000003965

4. Simplify each expression. Write your final answer without negative exponents.
   (a) $\sqrt{200} - \sqrt{32}$ (b) $(3a^3b^3)(4ab^2)^2$ (c) $\left(\dfrac{3x^{3/2}y^3}{x^2y^{-1/2}}\right)^{-2}$
   (d) $\dfrac{x^2 + 3x + 2}{x^2 - x - 2}$ (e) $\dfrac{x^2}{x^2 - 4} - \dfrac{x + 1}{x + 2}$ (f) $\dfrac{\dfrac{y}{x} - \dfrac{x}{y}}{\dfrac{1}{y} - \dfrac{1}{x}}$

5. Rationalize the denominator and simplify: $\dfrac{\sqrt{10}}{\sqrt{5} - 2}$

6. Perform the indicated operations and simplify.
   (a) $3(x + 6) + 4(2x - 5)$ (b) $(x + 3)(4x - 5)$ (c) $(\sqrt{a} + \sqrt{b})(\sqrt{a} - \sqrt{b})$
   (d) $(2x + 3)^2$ (e) $(x + 2)^3$

7. Factor each expression completely.
   (a) $4x^2 - 25$ (b) $2x^2 + 5x - 12$ (c) $x^3 - 3x^2 - 4x + 12$
   (d) $x^4 + 27x$ (e) $3x^{3/2} - 9x^{1/2} + 6x^{-1/2}$ (f) $x^3y - 4xy$

8. Find all real solutions.
   (a) $x + 5 = 14 - \frac{1}{2}x$ (b) $\dfrac{2x}{x + 1} = \dfrac{2x - 1}{x}$ (c) $x^2 - x - 12 = 0$
   (d) $2x^2 + 4x + 1 = 0$ (e) $\sqrt{3 - \sqrt{x + 5}} = 2$ (f) $x^4 - 3x^2 + 2 = 0$
   (g) $3|x - 4| = 10$

9. Perform the indicated operations, and write the result in the form $a + bi$.
   (a) $(3 - 2i) + (4 + 3i)$ (b) $(3 - 2i) - (4 + 3i)$
   (c) $(3 - 2i)(4 + 3i)$ (d) $\dfrac{3 - 2i}{4 + 3i}$
   (e) i^{48} (f) $(\sqrt{2} - \sqrt{-2})(\sqrt{8} + \sqrt{-2})$

10. Find all real and complex solutions of the equation $2x^2 + 4x + 3 = 0$.

11. Mary drove from Amity to Belleville at a speed of 50 mi/h. On the way back, she drove at 60 mi/h. The total trip took $4\frac{2}{5}$ h of driving time. Find the distance between these two cities.

12. A rectangular parcel of land is 70 ft longer than it is wide. Each diagonal between opposite corners is 130 ft. What are the dimensions of the parcel?

13. Solve each inequality. Write the answer using interval notation, and sketch the solution on the real number line.
   (a) $-4 < 5 - 3x \le 17$ (b) $x(x - 1)(x + 2) > 0$
   (c) $|x - 4| < 3$ (d) $\dfrac{2x - 3}{x + 1} \le 1$

14. A bottle of medicine is to be stored at a temperature between 5°C and 10°C. What range does this correspond to on the Fahrenheit scale? [*Note:* Fahrenheit (F) and Celsius (C) temperatures satisfy the relation $C = \frac{5}{9}(F - 32)$.]

15. For what values of x is the expression $\sqrt{6x - x^2}$ defined as a real number?

16. **(a)** Plot the points $P(0, 3)$, $Q(3, 0)$, and $R(6, 3)$ in the coordinate plane. Where must the point S be located so that $PQRS$ is a square?

(b) Find the area of $PQRS$.

17. **(a)** Sketch the graph of $y = x^2 - 4$.

(b) Find the x- and y-intercepts of the graph.

(c) Is the graph symmetric about the x-axis, the y-axis, or the origin?

18. Let $P(-3,1)$ and $Q(5,6)$ be two points in the coordinate plane.

(a) Plot P and Q in the coordinate plane.

(b) Find the distance between P and Q.

(c) Find the midpoint of the segment PQ.

(d) Find the slope of the line that contains P and Q.

(e) Find the perpendicular bisector of the line that contains P and Q.

(f) Find an equation for the circle for which the segment PQ is a diameter.

19. Find the center and radius of each circle, and sketch its graph.

(a) $x^2 + y^2 = 25$ **(b)** $(x - 2)^2 + (y + 1)^2 = 9$ **(c)** $x^2 + 6x + y^2 - 2y + 6 = 0$

20. Write the linear equation $2x - 3y = 15$ in slope-intercept form, and sketch its graph. What are the slope and y-intercept?

21. Find an equation for the line with the given property.

(a) It passes through the point $(3, -6)$ and is parallel to the line $3x + y - 10 = 0$.

(b) It has x-intercept 6 and y-intercept 4.

22. A geologist measures the temperature T (in °C) of the soil at various depths below the surface and finds that at a depth of x cm, the temperature is given by $T = 0.08x - 4$.

(a) What is the temperature at a depth of 1 m (100 cm)?

(b) Sketch a graph of the linear equation.

(c) What do the slope, the x-intercept, and T-intercept of the graph represent?

23. Solve the equation and the inequality graphically.

(a) $x^3 - 9x - 1 = 0$ **(b)** $x^2 - 1 \leq |x + 1|$

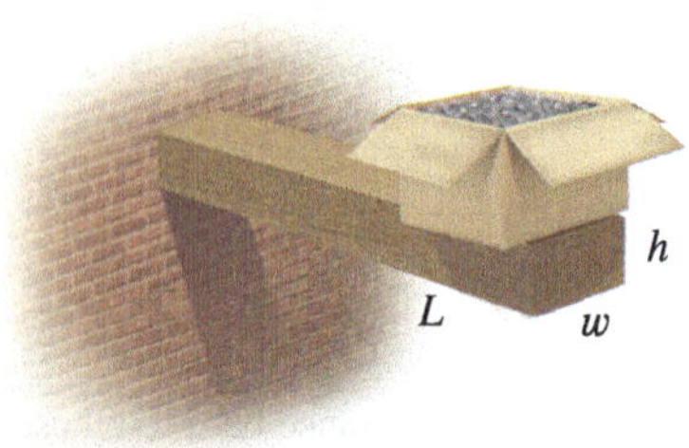

24. The maximum weight M that can be supported by a beam is jointly proportional to its width w and the square of its height h and inversely proportional to its length L.

(a) Write an equation that expresses this proportionality.

(b) Determine the constant of proportionality if a beam 4 in. wide, 6 in. high, and 12 ft long can support a weight of 4800 lb.

(c) If a 10-ft beam made of the same material is 3 in. wide and 10 in. high, what is the maximum weight it can support?

If you had difficulty with any of these problems, you may wish to review the section of this chapter indicated below.

Problem	Section	Problem	Section
1	Section 1.1	13, 14, 15	Section 1.8
2, 3, 4(a), 4(b), 4(c)	Section 1.2	23	Section 1.11
4(d), 4(e), 4(f), 5	Section 1.4	16, 17, 18(a), 18(b)	Section 1.9
6, 7	Section 1.3	18(c), 18(d)	Section 1.10
8	Section 1.5	18(e), 18(f), 19	Section 1.9
9, 10	Section 1.6	20, 21, 22	Section 1.10
11, 12	Section 1.7	24	Section 1.12

Fitting Lines to Data

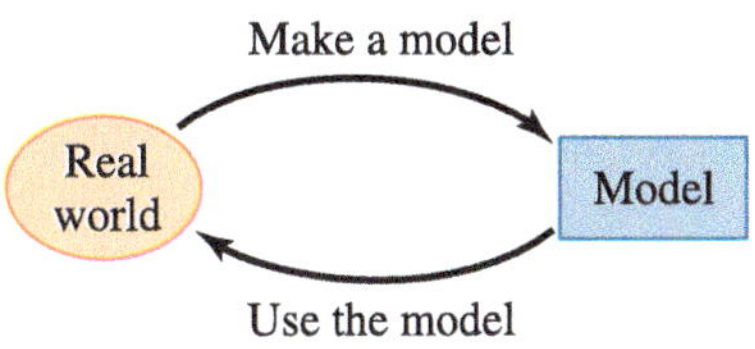

A model is a representation of an object or process. For example, a toy Ferrari is a model of the actual car; a road map is a model of the streets in a city. A **mathematical model** is a mathematical representation (usually an equation) of an object or process. Once a mathematical model has been made, it can be used to obtain useful information or make predictions about the thing being modeled. The process is described in the diagram in the margin. In these *Focus on Modeling* sections we explore different ways in which mathematics is used to model real-world phenomena.

The Line That Best Fits the Data

In Section 1.10 we used linear equations to model relationships between varying quantities. In practice, such relationships are discovered by collecting data. But real-world data seldom fall into a precise line. The **scatter plot** in Figure 1(a) shows the result of a study on childhood obesity. The graph plots the body mass index (BMI) versus the number of hours of television watched per day for 25 adolescent subjects. Of course, we would not expect the data to be exactly linear as in Figure 1(b). But there is a linear *trend* indicated by the blue line in Figure 1(a): The more hours a subject watches TV, the higher the BMI. In this section we learn how to find the line that best fits the data.

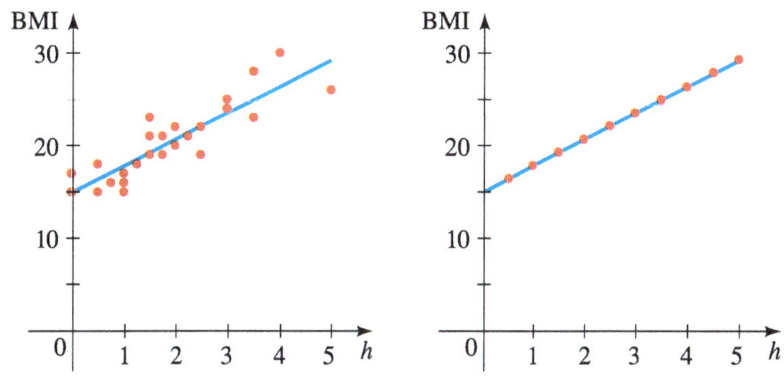

(a) Line of best fit

(b) Line fits data exactly

FIGURE 1

Table 1 gives the nationwide infant mortality rate for the period from 1950 to 2000. The *rate* is the number of infants who die before reaching their first birthday, out of every 1000 live births.

TABLE 1
U.S. Infant Mortality

Year	Rate
1950	29.2
1960	26.0
1970	20.0
1980	12.6
1990	9.2
2000	6.9

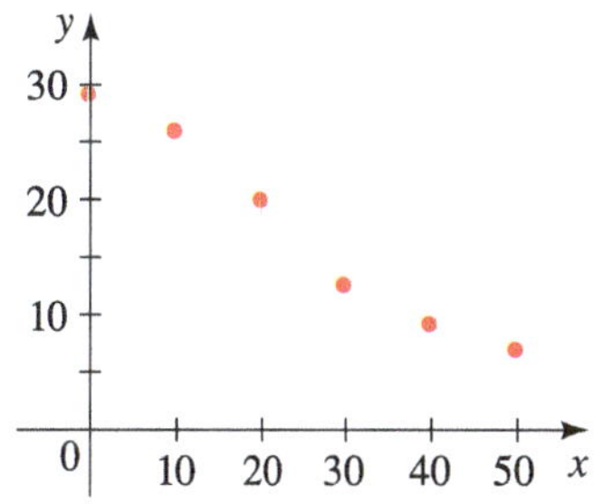

FIGURE 2 U.S. infant mortality rate

The scatter plot in Figure 2 shows that the data lie roughly on a straight line. We can try to fit a line visually to approximate the data points, but since the data aren't *exactly* linear, there are many lines that might seem to work. Figure 3 shows two attempts at "eyeballing" a line to fit the data.

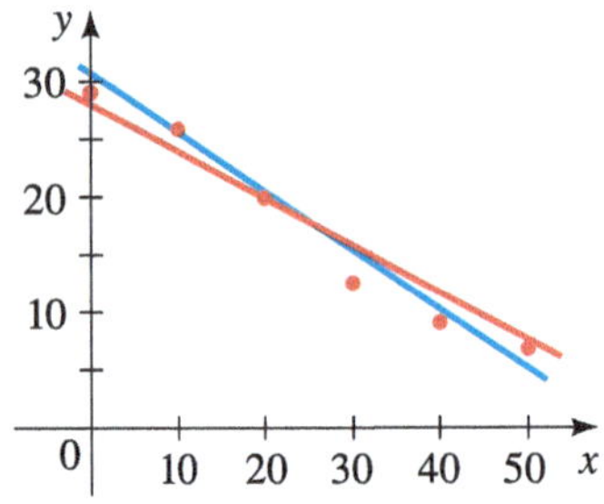

FIGURE 3 Visual attempts to fit line to data

Of all the lines that run through these data points, there is one that "best" fits the data, in the sense that it provides the most accurate linear model for the data. We now describe how to find this line.

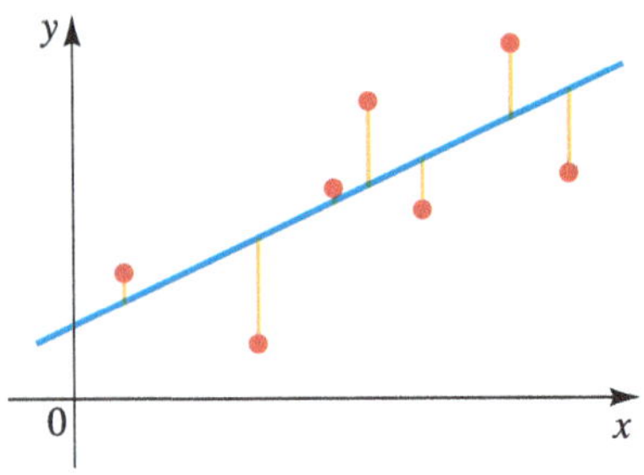

FIGURE 4 Distance from the data points to the line

It seems reasonable that the line of best fit is the line that is as close as possible to all the data points. This is the line for which the sum of the vertical distances from the data points to the line is as small as possible (see Figure 4). For technical reasons it is better to use the line where the sum of the squares of these distances is smallest. This is called the **regression line**. The formula for the regression line is found by using calculus, but fortunately, the formula is programmed into most graphing calculators. In Example 1 we see how to use a TI-83 calculator to find the regression line for the infant mortality data described above. (The process for other calculators is similar.)

EXAMPLE 1 ■ Regression Line for U.S. Infant Mortality Rates

(a) Find the regression line for the infant mortality data in Table 1.

(b) Graph the regression line on a scatter plot of the data.

(c) Use the regression line to estimate the infant mortality rates in 1995 and 2006.

SOLUTION

(a) To find the regression line using a TI-83 calculator, we must first enter the data into the lists L_1 and L_2, which are accessed by pressing the STAT key and selecting `Edit`. Figure 5 shows the calculator screen after the data have been entered. (Note that we are letting $x = 0$ correspond to the year 1950 so that $x = 50$ corresponds to 2000. This makes the equations easier to work with.) We then press the STAT key again and select `Calc`, then `4:LinReg(ax+b)`, which provides the output shown in Figure 6(a). This tells us that the regression line is

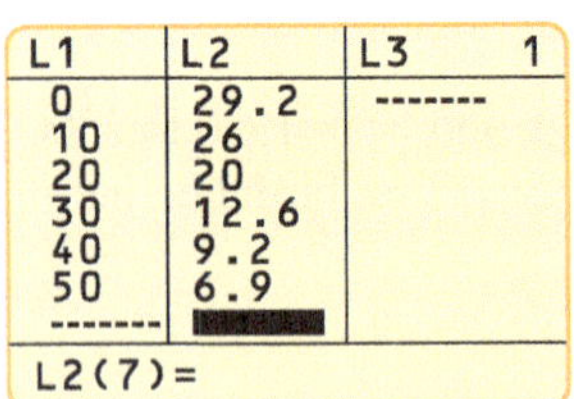

FIGURE 5 Entering the data

$$y = -0.48x + 29.4$$

Here x represents the number of years since 1950, and y represents the corresponding infant mortality rate.

(b) The scatter plot and the regression line have been plotted on a graphing calculator screen in Figure 6(b).

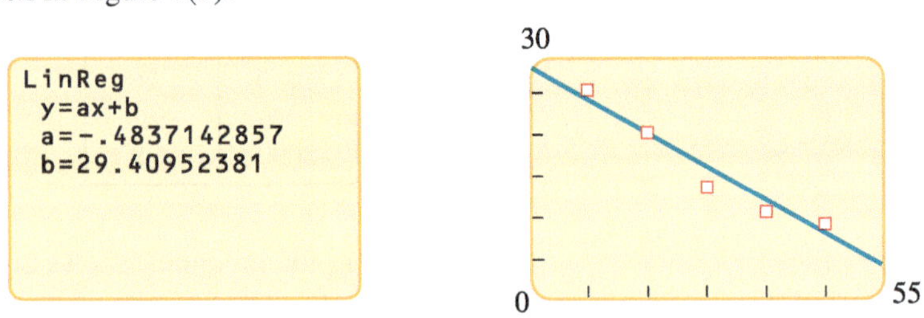

FIGURE 6

(a) Output of the `LinReg` command

(b) Scatter plot and regression line

(c) The year 1995 is 45 years after 1950, so substituting 45 for x, we find that $y = -0.48(45) + 29.4 = 7.8$. So the infant mortality rate in 1995 was about 7.8. Similarly, substituting 56 for x, we find that the infant mortality rate predicted for 2006 was about $-0.48(56) + 29.4 \approx 2.5$. ■

An Internet search shows that the actual infant mortality rate was 7.6 in 1995 and 6.4 in 2006. So the regression line is fairly accurate for 1995 (the actual rate was slightly lower than the predicted rate), but it is considerably off for 2006 (the actual rate was more than twice the predicted rate). The reason is that infant mortality in the United States stopped declining and actually started rising in 2002, for the first time in more than a century. This shows that we have to be very careful about extrapolating linear models outside the domain over which the data are spread.

Examples of Regression Analysis

Leo Mason sports photos/Alamy

Renaud Lavillenie, 2012 Olympic gold medal winner, men's pole vault

Since the modern Olympic Games began in 1896, achievements in track and field events have been improving steadily. One example in which the winning records have shown an upward linear trend is the pole vault. Pole vaulting began in the northern Netherlands as a practical activity: When traveling from village to village, people would vault across the many canals that crisscrossed the area to avoid having to go out of their way to find a bridge. Households maintained a supply of wooden poles of lengths appropriate for each member of the family. Pole vaulting for height rather than distance became a collegiate track and field event in the mid-1800s and was one of the events in the first modern Olympics. In the next example we find a linear model for the gold-medal-winning records in the men's Olympic pole vault.

EXAMPLE 2 ■ Regression Line for Olympic Pole Vault Records

Table 2 gives the men's Olympic pole vault records up to 2008.

(a) Find the regression line for the data.

(b) Make a scatter plot of the data, and graph the regression line. Does the regression line appear to be a suitable model for the data?

(c) What does the slope of the regression line represent?

(d) Use the model to predict the winning pole vault height for the 2012 Olympics.

TABLE 2
Men's Olympic Pole Vault Records

Year	x	Gold medalist	Height (m)	Year	x	Gold medalist	Height (m)
1896	−4	William Hoyt, USA	3.30	1960	60	Don Bragg, USA	4.70
1900	0	Irving Baxter, USA	3.30	1964	64	Fred Hansen, USA	5.10
1904	4	Charles Dvorak, USA	3.50	1968	68	Bob Seagren, USA	5.40
1906	6	Fernand Gonder, France	3.50	1972	72	W. Nordwig, E. Germany	5.64
1908	8	A. Gilbert, E. Cook, USA	3.71	1976	76	Tadeusz Slusarski, Poland	5.64
1912	12	Harry Babcock, USA	3.95	1980	80	W. Kozakiewicz, Poland	5.78
1920	20	Frank Foss, USA	4.09	1984	84	Pierre Quinon, France	5.75
1924	24	Lee Barnes, USA	3.95	1988	88	Sergei Bubka, USSR	5.90
1928	28	Sabin Can, USA	4.20	1992	92	M. Tarassob, Unified Team	5.87
1932	32	William Miller, USA	4.31	1996	96	Jean Jaffione, France	5.92
1936	36	Earle Meadows, USA	4.35	2000	100	Nick Hysong, USA	5.90
1948	48	Guinn Smith, USA	4.30	2004	104	Timothy Mack, USA	5.95
1952	52	Robert Richards, USA	4.55	2008	108	Steven Hooker, Australia	5.96
1956	56	Robert Richards, USA	4.56				

```
LinReg
 y=ax+b
 a=.0265652857
 b=3.400989881
```

Output of the LinReg function on the TI-83

SOLUTION

(a) Let x = year − 1900, so 1896 corresponds to $x = -4$, 1900 to $x = 0$, and so on. Using a calculator, we find the following regression line:

$$y = 0.0260x + 3.42$$

(b) The scatter plot and the regression line are shown in Figure 7. The regression line appears to be a good model for the data.

(c) The slope is the average rate of increase in the pole vault record per year. So on average, the pole vault record increased by 0.0266 m/year.

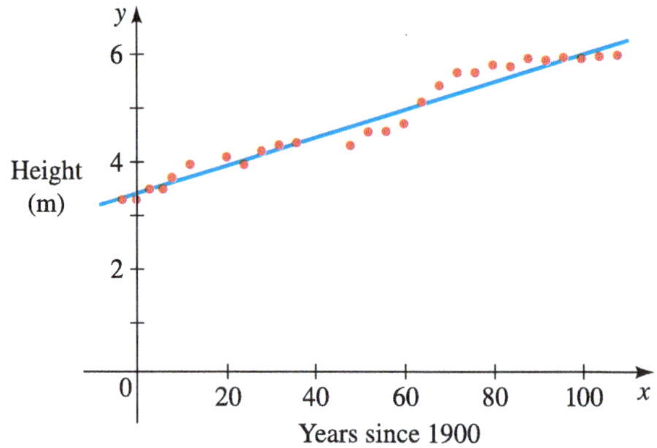

FIGURE 7 Scatter plot and regression line for pole vault data

(d) The year 2012 corresponds to $x = 112$ in our model. The model gives

$$y = 0.0260(112) + 3.42$$
$$\approx 6.33$$

So the model predicts that in 2012 the winning pole vault would be 6.33 m. ■

At the 2012 Olympics in London, England, the men's Olympic gold medal in the pole vault was won by Renaud Lavillenie of France, with a vault of 5.97 m. Although this height set an Olympic record, it was considerably lower than the 6.33 m predicted by the model of Example 2. In Problem 10 we find a regression line for the pole vault data from 1972 to 2008. Do the problem to see whether this restricted set of more recent data provides a better predictor for the 2012 record.

Is a linear model really appropriate for the data of Example 2? In subsequent *Focus on Modeling* sections we study regression models that use other types of functions, and we learn how to choose the best model for a given set of data.

In the next example we see how linear regression is used in medical research to investigate potential causes of diseases such as cancer.

TABLE 3
Asbestos–Tumor Data

Asbestos exposure (fibers/mL)	Percent that develop lung tumors
50	2
400	6
500	5
900	10
1100	26
1600	42
1800	37
2000	28
3000	50

EXAMPLE 3 ■ Regression Line for Links Between Asbestos and Cancer

When laboratory rats are exposed to asbestos fibers, some of the rats develop lung tumors. Table 3 lists the results of several experiments by different scientists.

(a) Find the regression line for the data.

(b) Make a scatter plot and graph the regression line. Does the regression line appear to be a suitable model for the data?

(c) What does the y-intercept of the regression line represent?

Eric and David Hosking/Terra/Corbis

SOLUTION

(a) Using a calculator, we find the following regression line (see Figure 8(a)):

$$y = 0.0177x + 0.5405$$

(b) The scatter plot and regression line are graphed in Figure 8(b). The regression line appears to be a reasonable model for the data.

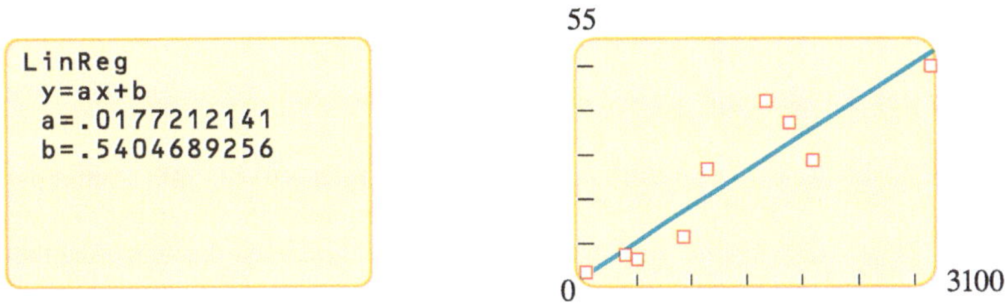

(a) Output of the `LinReg` command

(b) Scatter plot and regression line

FIGURE 8 Linear regression for the asbestos–tumor data

(c) The y-intercept is the percentage of rats that develop tumors when no asbestos fibers are present. In other words, this is the percentage that normally develop lung tumors (for reasons other than asbestos). ■

How Good Is the Fit? The Correlation Coefficient

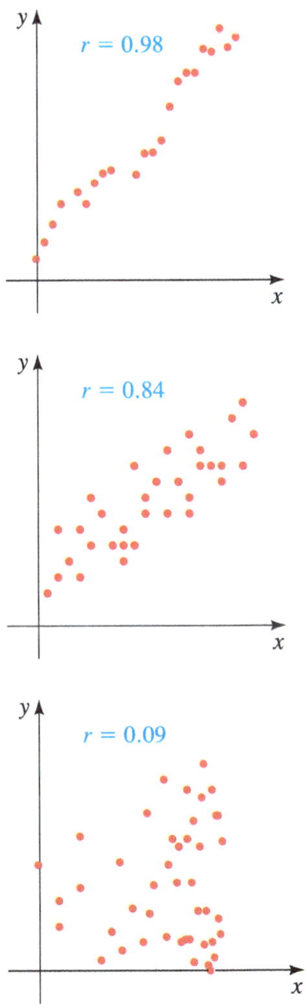

FIGURE 9

For any given set of two-variable data it is always possible to find a regression line, even if the data points do not tend to lie on a line and even if the variables don't seem to be related at all. Look at the three scatter plots in Figure 9. In the first scatter plot, the data points lie close to a line. In the second plot, there is still a linear trend but the points are more scattered. In the third plot there doesn't seem to be any trend at all, linear or otherwise.

A graphing calculator can give us a regression line for each of these scatter plots. But how well do these lines represent or "fit" the data? To answer this question, statisticians have invented the **correlation coefficient**, usually denoted r. The correlation coefficient is a number between -1 and 1 that measures how closely the data follow the regression line—or, in other words, how strongly the variables are **correlated**. Many graphing calculators give the value of r when they compute a regression line. If r is close to -1 or 1, then the variables are strongly correlated—that is, the scatter plot follows the regression line closely. If r is close to 0, then the variables are weakly correlated or not correlated at all. (The sign of r depends on the slope of the regression line.) The correlation coefficients of the scatter plots in Figure 9 are indicated on the graphs. For the first plot, r is close to 1 because the data are very close to linear. The second plot also has a relatively large r, but it is not as large as the first, because the data, while fairly linear, are more diffuse. The third plot has an r close to 0, since there is virtually no linear trend in the data.

There are no hard and fast rules for deciding what values of r are sufficient for deciding that a linear correlation is "significant." The correlation coefficient is only a rough guide in helping us decide how much faith to put into a given regression line. In Example 1 the correlation coefficient is -0.99, indicating a very high level of correlation, so we can safely say that the drop in infant mortality rates from 1950 to 2000 was strongly linear. (The value of r is negative, since infant mortality trended *down* over this period.) In Example 3 the correlation coefficient is 0.92, which also indicates a strong correlation between the variables. So exposure to asbestos is clearly associated with the growth of lung tumors in rats. Does this mean that asbestos *causes* lung cancer?

If two variables are correlated, it does not necessarily mean that a change in one variable *causes* a change in the other. For example, the mathematician John Allen Paulos points out that shoe size is strongly correlated to mathematics scores among schoolchildren. Does this mean that big feet cause high math scores? Certainly not—

both shoe size and math skills increase independently as children get older. So it is important not to jump to conclusions: Correlation and causation are not the same thing. You can explore this topic further in *Discovery Project: Correlation and Causation* at **www.stewartmath.com**. Correlation is a useful tool in bringing important cause-and-effect relationships to light; but to prove causation, we must explain the mechanism by which one variable affects the other. For example, the link between smoking and lung cancer was observed as a correlation long before science found the mechanism through which smoking causes lung cancer.

PROBLEMS

1. **Femur Length and Height** Anthropologists use a linear model that relates femur length to height. The model allows an anthropologist to determine the height of an individual when only a partial skeleton (including the femur) is found. In this problem we find the model by analyzing the data on femur length and height for the eight males given in the table.
   (a) Make a scatter plot of the data.
   (b) Find and graph a linear function that models the data.
   (c) An anthropologist finds a femur of length 58 cm. How tall was the person?

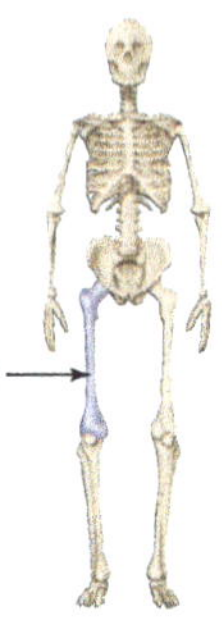

Femur length (cm)	Height (cm)
50.1	178.5
48.3	173.6
45.2	164.8
44.7	163.7
44.5	168.3
42.7	165.0
39.5	155.4
38.0	155.8

2. **Demand for Soft Drinks** A convenience store manager notices that sales of soft drinks are higher on hotter days, so he assembles the data in the table.
   (a) Make a scatter plot of the data.
   (b) Find and graph a linear function that models the data.
   (c) Use the model to predict soft drink sales if the temperature is 95°F.

High temperature (°F)	Number of cans sold
55	340
58	335
64	410
68	460
70	450
75	610
80	735
84	780

Diameter (in.)	Age (years)
2.5	15
4.0	24
6.0	32
8.0	56
9.0	49
9.5	76
12.5	90
15.5	89

3. **Tree Diameter and Age** To estimate ages of trees, forest rangers use a linear model that relates tree diameter to age. The model is useful because tree diameter is much easier to measure than tree age (which requires special tools for extracting a representative cross section of the tree and counting the rings). To find the model, use the data in the table, which were collected for a certain variety of oaks.
   (a) Make a scatter plot of the data.
   (b) Find and graph a linear function that models the data.
   (c) Use the model to estimate the age of an oak whose diameter is 18 in.

4. Carbon Dioxide Levels The Mauna Loa Observatory, located on the island of Hawaii, has been monitoring carbon dioxide (CO_2) levels in the atmosphere since 1958. The table lists the average annual CO_2 levels measured in parts per million (ppm) from 1990 to 2012.

(a) Make a scatter plot of the data.

(b) Find and graph the regression line.

(c) Use the linear model in part (b) to estimate the CO_2 level in the atmosphere in 2011. Compare your answer with the actual CO_2 level of 391.6 that was measured in 2011.

Year	CO_2 level (ppm)
1990	354.4
1992	356.4
1994	358.8
1996	362.6
1998	366.7
2000	369.5
2002	373.2
2004	377.5
2006	381.9
2008	385.6
2010	389.9
2012	393.8

Source: Mauna Loa Observatory

Temperature (°F)	Chirping rate (chirps/min)
50	20
55	46
60	79
65	91
70	113
75	140
80	173
85	198
90	211

5. Temperature and Chirping Crickets Biologists have observed that the chirping rate of crickets of a certain species appears to be related to temperature. The table in the margin shows the chirping rates for various temperatures.

(a) Make a scatter plot of the data.

(b) Find and graph the regression line.

(c) Use the linear model in part (b) to estimate the chirping rate at 100°F.

6. Extent of Arctic Sea Ice The National Snow and Ice Data Center monitors the amount of ice in the Arctic year round. The table below gives approximate values for the sea ice extent in millions of square kilometers from 1986 to 2012, in two-year intervals.

(a) Make a scatter plot of the data.

(b) Find and graph the regression line.

(c) Use the linear model in part (b) to estimate the ice extent in the year 2016.

Year	Ice extent (million km^2)	Year	Ice extent (million km^2)
1986	7.5	2000	6.3
1988	7.5	2002	6.0
1990	6.2	2004	6.0
1992	7.5	2006	5.9
1994	7.2	2008	4.7
1996	7.9	2010	4.9
1998	6.6	2012	3.6

Source: National Snow and Ice Data Center

Flow rate (%)	Mosquito positive rate (%)
0	22
10	16
40	12
60	11
90	6
100	2

7. Mosquito Prevalence The table in the margin lists the relative abundance of mosquitoes (as measured by the mosquito positive rate) versus the flow rate (measured as a percentage of maximum flow) of canal networks in Saga City, Japan.

(a) Make a scatter plot of the data.

(b) Find and graph the regression line.

(c) Use the linear model in part (b) to estimate the mosquito positive rate if the canal flow is 70% of maximum.

Noise level (dB)	MRT score (%)
80	99
84	91
88	84
92	70
96	47
100	23
104	11

8. Noise and Intelligibility Audiologists study the intelligibility of spoken sentences under different noise levels. Intelligibility, the MRT score, is measured as the percent of a spoken sentence that the listener can decipher at a certain noise level in decibels (dB). The table shows the results of one such test.

(a) Make a scatter plot of the data.

(b) Find and graph the regression line.

(c) Find the correlation coefficient. Is a linear model appropriate?

(d) Use the linear model in part (b) to estimate the intelligibility of a sentence at a 94-dB noise level.

Year	Life expectancy
1920	54.1
1930	59.7
1940	62.9
1950	68.2
1960	69.7
1970	70.8
1980	73.7
1990	75.4
2000	76.9

9. Life Expectancy The average life expectancy in the United States has been rising steadily over the past few decades, as shown in the table.

(a) Make a scatter plot of the data.

(b) Find and graph the regression line.

(c) Use the linear model you found in part (b) to predict the life expectancy in the year 2006.

(d) Search the Internet or your campus library to find the actual 2006 average life expectancy. Compare to your answer in part (c).

Year	x	Height (m)
1972	0	5.64
1976	4	
1980	8	
1984		
1988		
1992		
1996		
2000		
2004		
2008		

10. Olympic Pole Vault The graph in Figure 7 indicates that in recent years the winning Olympic men's pole vault height has fallen below the value predicted by the regression line in Example 2. This might have occurred because when the pole vault was a new event, there was much room for improvement in vaulters' performances, whereas now even the best training can produce only incremental advances. Let's see whether concentrating on more recent results gives a better predictor of future records.

(a) Use the data in Table 2 (page 141) to complete the table of winning pole vault heights shown in the margin. (Note that we are using $x = 0$ to correspond to the year 1972, where this restricted data set begins.)

(b) Find the regression line for the data in part (a).

(c) Plot the data and the regression line on the same axes. Does the regression line seem to provide a good model for the data?

(d) What does the regression line predict as the winning pole vault height for the 2012 Olympics? Compare this predicted value to the actual 2012 winning height of 5.97 m, as described on page 141. Has this new regression line provided a better prediction than the line in Example 2?

11. Shoe Size and Height Do you think that shoe size and height are correlated? Find out by surveying the shoe sizes and heights of people in your class. (Of course, the data for men and women should be separate.) Find the correlation coefficient.

Would you buy a candy bar from the vending machine in the hallway if the price is as indicated?

Price	*Yes or No*
50¢	
75¢	
$1.00	
$1.25	
$1.50	
$1.75	
$2.00	

12. Demand for Candy Bars In this problem you will determine a linear demand equation that describes the demand for candy bars in your class. Survey your classmates to determine what price they would be willing to pay for a candy bar. Your survey form might look like the sample to the left.

(a) Make a table of the number of respondents who answered "yes" at each price level.

(b) Make a scatter plot of your data.

(c) Find and graph the regression line $y = mp + b$, which gives the number of responents y who would buy a candy bar if the price were p cents. This is the *demand equation.* Why is the slope m negative?

(d) What is the p-intercept of the demand equation? What does this intercept tell you about pricing candy bars?

© iStockphoto.com/Jeff McDonald

2 Functions

A ***function* is a rule** that describes how one quantity depends on another. Many real-world situations follow precise rules, so they can be modeled by functions. For example, there is a rule that relates the distance a skydiver falls to the time he or she has been falling. So the distance traveled by the skydiver is a *function* of time. Knowing this function model allows skydivers to determine when to open their parachute. In this chapter we study functions and their graphs, as well as many real-world applications of functions. In the *Focus on Modeling* at the end of the chapter we explore different real-world situations that can be modeled by functions.

2.1 FUNCTIONS

■ Functions All Around Us ■ Definition of Function ■ Evaluating a Function
■ The Domain of a Function ■ Four Ways to Represent a Function

In this section we explore the idea of a function and then give the mathematical definition of function.

■ Functions All Around Us

In nearly every physical phenomenon we observe that one quantity depends on another. For example, your height depends on your age, the temperature depends on the date, the cost of mailing a package depends on its weight (see Figure 1). We use the term *function* to describe this dependence of one quantity on another. That is, we say the following:

- Height is a function of age.
- Temperature is a function of date.
- Cost of mailing a package is a function of weight.

The U.S. Post Office uses a simple rule to determine the cost of mailing a first-class parcel on the basis of its weight. But it's not so easy to describe the rule that relates height to age or the rule that relates temperature to date.

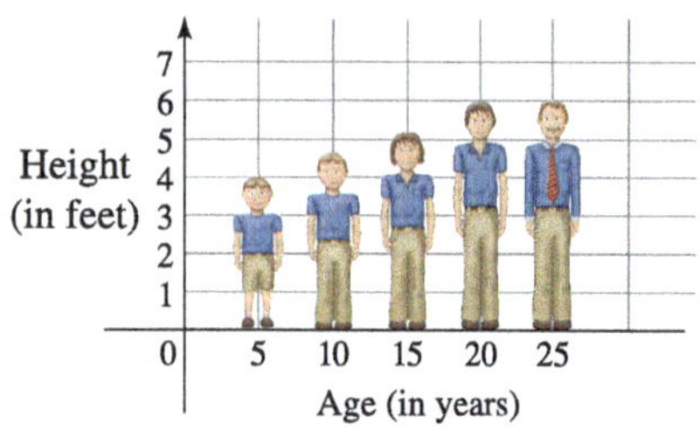

Height is a function of age.

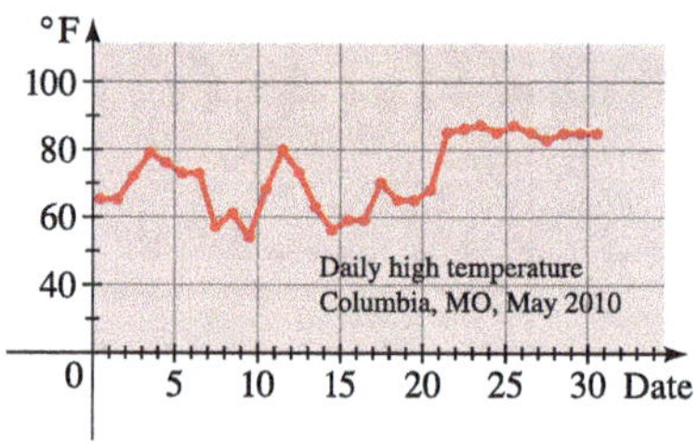

Temperature is a function of date.

w (ounces)	2014 Postage (dollars)
$0 < w \le 1$	0.98
$1 < w \le 2$	1.19
$2 < w \le 3$	1.40
$3 < w \le 4$	1.61
$4 < w \le 5$	1.82
$5 < w \le 6$	2.03

Postage is a function of weight.

FIGURE 1

Can you think of other functions? Here are some more examples:

- The area of a circle is a function of its radius.
- The number of bacteria in a culture is a function of time.
- The weight of an astronaut is a function of her elevation.
- The price of a commodity is a function of the demand for that commodity.

The rule that describes how the area A of a circle depends on its radius r is given by the formula $A = \pi r^2$. Even when a precise rule or formula describing a function is not available, we can still describe the function by a graph. For example, when you turn on a hot water faucet, the temperature of the water depends on how long the water has been running. So we can say:

- The temperature of water from the faucet is a function of time.

Figure 2 shows a rough graph of the temperature T of the water as a function of the time t that has elapsed since the faucet was turned on. The graph shows that the initial temperature of the water is close to room temperature. When the water from the hot water tank reaches the faucet, the water's temperature T increases quickly. In the next phase,

T is constant at the temperature of the water in the tank. When the tank is drained, T decreases to the temperature of the cold water supply.

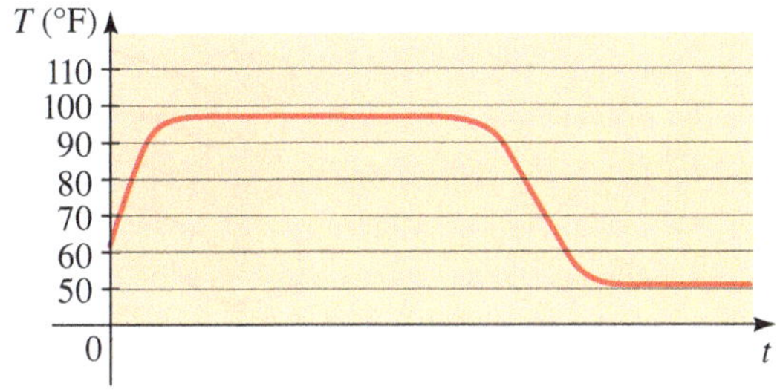

FIGURE 2 Graph of water temperature T as a function of time t

Definition of Function

We have previously used letters to stand for numbers. Here we do something quite different: We use letters to represent *rules*.

A function is a rule. To talk about a function, we need to give it a name. We will use letters such as $f, g, h, \ldots$ to represent functions. For example, we can use the letter f to represent a rule as follows:

$$\text{“}f\text{”} \quad \text{is the rule} \quad \text{“square the number”}$$

When we write $f(2)$, we mean "apply the rule f to the number 2." Applying the rule gives $f(2) = 2^2 = 4$. Similarly, $f(3) = 3^2 = 9$, $f(4) = 4^2 = 16$, and in general $f(x) = x^2$.

DEFINITION OF A FUNCTION

A **function** f is a rule that assigns to each element x in a set A exactly one element, called $f(x)$, in a set B.

The $\boxed{\sqrt{\ }}$ key on your calculator is a good example of a function as a machine. First you input x into the display. Then you press the key labeled $\boxed{\sqrt{\ }}$. (On most *graphing* calculators the order of these operations is reversed.) If $x < 0$, then x is not in the domain of this function; that is, x is not an acceptable input, and the calculator will indicate an error. If $x \geq 0$, then an approximation to $\sqrt{x}$ appears in the display, correct to a certain number of decimal places. (Thus the $\boxed{\sqrt{\ }}$ key on your calculator is not quite the same as the exact mathematical function f defined by $f(x) = \sqrt{x}$.)

We usually consider functions for which the sets A and B are sets of real numbers. The symbol $f(x)$ is read "f of x" or "f at x" and is called the **value of f at x**, or the **image of x under f**. The set A is called the **domain** of the function. The **range** of f is the set of all possible values of $f(x)$ as x varies throughout the domain, that is,

$$\text{range of } f = \{f(x) \mid x \in A\}$$

The symbol that represents an arbitrary number in the domain of a function f is called an **independent variable**. The symbol that represents a number in the range of f is called a **dependent variable**. So if we write $y = f(x)$, then x is the independent variable and y is the dependent variable.

It is helpful to think of a function as a **machine** (see Figure 3). If x is in the domain of the function f, then when x enters the machine, it is accepted as an **input** and the machine produces an **output** $f(x)$ according to the rule of the function. Thus we can think of the domain as the set of all possible inputs and the range as the set of all possible outputs.

FIGURE 3 Machine diagram of f

Another way to picture a function f is by an **arrow diagram** as in Figure 4(a). Each arrow associates an input from A to the corresponding output in B. Since a function associates *exactly* one output to each input, the diagram in Figure 4(a) represents a function but the diagram in Figure 4(b) does *not* represent a function.

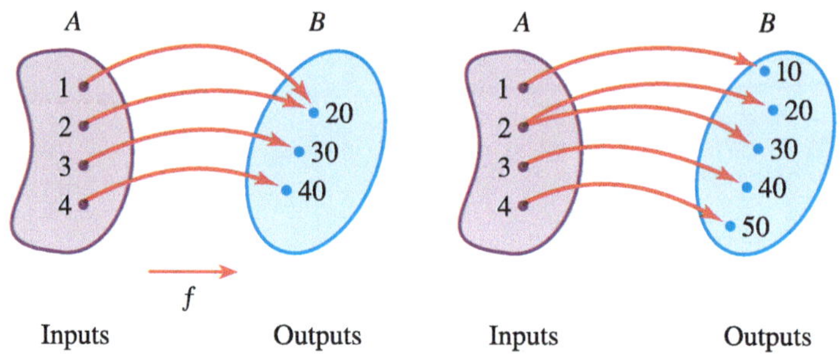

(a) Function (b) Not a function

FIGURE 4 Arrow diagrams

EXAMPLE 1 ■ Analyzing a Function

A function f is defined by the formula

$$f(x) = x^2 + 4$$

(a) Express in words how f acts on the input x to produce the output $f(x)$.

(b) Evaluate $f(3)$, $f(-2)$, and $f(\sqrt{5})$.

(c) Find the domain and range of f.

(d) Draw a machine diagram for f.

SOLUTION

(a) The formula tells us that f first squares the input x and then adds 4 to the result. So f is the function

"square, then add 4"

(b) The values of f are found by substituting for x in the formula $f(x) = x^2 + 4$.

$$f(3) = 3^2 + 4 = 13 \qquad \text{Replace } x \text{ by } 3$$

$$f(-2) = (-2)^2 + 4 = 8 \qquad \text{Replace } x \text{ by } -2$$

$$f(\sqrt{5}) = (\sqrt{5})^2 + 4 = 9 \qquad \text{Replace } x \text{ by } \sqrt{5}$$

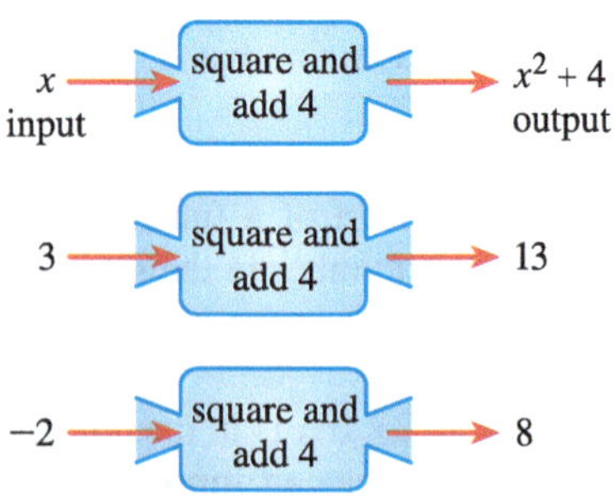

FIGURE 5 Machine diagram

(c) The domain of f consists of all possible inputs for f. Since we can evaluate the formula $f(x) = x^2 + 4$ for every real number x, the domain of f is the set $\mathbb{R}$ of all real numbers.

The range of f consists of all possible outputs of f. Because $x^2 \geq 0$ for all real numbers x, we have $x^2 + 4 \geq 4$, so for every output of f we have $f(x) \geq 4$. Thus the range of f is $\{y \mid y \geq 4\} = [4, \infty)$.

(d) A machine diagram for f is shown in Figure 5.

Now Try Exercises 11, 15, 19, and 51 ■

■ Evaluating a Function

In the definition of a function the independent variable x plays the role of a placeholder. For example, the function $f(x) = 3x^2 + x - 5$ can be thought of as

$$f(\square) = 3 \cdot \square^2 + \square - 5$$

To evaluate f at a number, we substitute the number for the placeholder.

EXAMPLE 2 ■ Evaluating a Function

Let $f(x) = 3x^2 + x - 5$. Evaluate each function value.

(a) $f(-2)$ **(b)** $f(0)$ **(c)** $f(4)$ **(d)** $f(\frac{1}{2})$

SOLUTION To evaluate f at a number, we substitute the number for x in the definition of f.

(a) $f(-2) = 3 \cdot (-2)^2 + (-2) - 5 = 5$

(b) $f(0) = 3 \cdot 0^2 + 0 - 5 = -5$

(c) $f(4) = 3 \cdot (4)^2 + 4 - 5 = 47$

(d) $f(\frac{1}{2}) = 3 \cdot (\frac{1}{2})^2 + \frac{1}{2} - 5 = -\frac{15}{4}$

Now Try Exercise 21 ■

EXAMPLE 3 ■ A Piecewise Defined Function

A cell phone plan costs \$39 a month. The plan includes 2 gigabytes (GB) of free data and charges \$15 per gigabyte for any additional data used. The monthly charges are a function of the number of gigabytes of data used, given by

$$C(x) = \begin{cases} 39 & \text{if } 0 \le x \le 2 \\ 39 + 15(x - 2) & \text{if } x > 2 \end{cases}$$

Find $C(0.5)$, $C(2)$, and $C(4)$.

SOLUTION Remember that a function is a rule. Here is how we apply the rule for this function. First we look at the value of the input, x. If $0 \le x \le 2$, then the value of $C(x)$ is 39. On the other hand, if $x > 2$, then the value of $C(x)$ is $39 + 15(x - 2)$.

Since $0.5 \le 2$, we have $C(0.5) = 39$.

Since $2 \le 2$, we have $C(2) = 39$.

Since $4 > 2$, we have $C(4) = 39 + 15(4 - 2) = 69$.

Thus the plan charges \$39 for 0.5 GB, \$39 for 2 GB, and \$69 for 4 GB.

A **piecewise defined function** is defined by different formulas on different parts of its domain. The function C of Example 3 is piecewise defined.

Now Try Exercises 31 and 85 ■

From Examples 2 and 3 we see that the values of a function can change from one input to another. The **net change** in the value of a function f as the input changes from a to b (where $a \le b$) is given by

$$\boxed{f(b) - f(a)}$$

The next example illustrates this concept.

EXAMPLE 4 ■ Finding Net Change

Let $f(x) = x^2$. Find the net change in the value of f between the given inputs.

(a) From 1 to 3 **(b)** From -2 to 2

SOLUTION

(a) The net change is $f(3) - f(1) = 9 - 1 = 8$.

(b) The net change is $f(2) - f(-2) = 4 - 4 = 0$.

The values of the function in Example 4 decrease and then increase between -2 and 2, but the net change from -2 to 2 is 0 because $f(-2)$ and $f(2)$ have the same value.

Now Try Exercise 39 ■

EXAMPLE 5 ■ Evaluating a Function

If $f(x) = 2x^2 + 3x - 1$, evaluate the following.

Expressions like the one in part (d) of Example 5 occur frequently in calculus; they are called *difference quotients*, and they represent the average change in the value of f between $x = a$ and $x = a + h$.

(a) $f(a)$ **(b)** $f(-a)$ **(c)** $f(a + h)$ **(d)** $\dfrac{f(a+h) - f(a)}{h}, \quad h \neq 0$

SOLUTION

(a) $f(a) = 2a^2 + 3a - 1$

(b) $f(-a) = 2(-a)^2 + 3(-a) - 1 = 2a^2 - 3a - 1$

(c)
$$\begin{aligned} f(a+h) &= 2(a+h)^2 + 3(a+h) - 1 \\ &= 2(a^2 + 2ah + h^2) + 3(a+h) - 1 \\ &= 2a^2 + 4ah + 2h^2 + 3a + 3h - 1 \end{aligned}$$

(d) Using the results from parts (c) and (a), we have

$$\begin{aligned} \frac{f(a+h) - f(a)}{h} &= \frac{(2a^2 + 4ah + 2h^2 + 3a + 3h - 1) - (2a^2 + 3a - 1)}{h} \\ &= \frac{4ah + 2h^2 + 3h}{h} = 4a + 2h + 3 \end{aligned}$$

Now Try Exercise 43 ■

A **table of values** for a function is a table with two headings, one for inputs and one for the corresponding outputs. A table of values helps us to analyze a function numerically, as in the next example.

EXAMPLE 6 ■ The Weight of an Astronaut

The weight of an object on or near the earth is the gravitational force that the earth exerts on it. When in orbit around the earth, an astronaut experiences the sensation of "weightlessness" because the centripetal force that keeps her in orbit is exactly the same as the gravitational pull of the earth.

If an astronaut weighs 130 lb on the surface of the earth, then her weight when she is h miles above the earth is given by the function

$$w(h) = 130\left(\frac{3960}{3960 + h}\right)^2$$

(a) What is her weight when she is 100 mi above the earth?

(b) Construct a table of values for the function w that gives her weight at heights from 0 to 500 mi. What do you conclude from the table?

(c) Find the net change in the astronaut's weight from ground level to a height of 500 mi.

SOLUTION

(a) We want the value of the function w when $h = 100$; that is, we must calculate $w(100)$:

$$w(100) = 130\left(\frac{3960}{3960 + 100}\right)^2 \approx 123.67$$

So at a height of 100 mi she weighs about 124 lb.

(b) The table gives the astronaut's weight, rounded to the nearest pound, at 100-mi increments. The values in the table are calculated as in part (a).

h	$w(h)$
0	130
100	124
200	118
300	112
400	107
500	102

The table indicates that the higher the astronaut travels, the less she weighs.

(c) The net change in the astronaut's weight from $h = 0$ to $h = 500$ is

$$w(500) - w(0) = 102 - 130 = -28$$

The negative sign indicates that the astronaut's weight *decreased* by about 28 lb.

Now Try Exercise 79 ■

The Domain of a Function

Recall that the *domain* of a function is the set of all inputs for the function. The domain of a function may be stated explicitly. For example, if we write

$$f(x) = x^2 \qquad 0 \le x \le 5$$

then the domain is the set of all real numbers x for which $0 \le x \le 5$. If the function is given by an algebraic expression and the domain is not stated explicitly, then by convention *the domain of the function is the domain of the algebraic expression—that is, the set of all real numbers for which the expression is defined as a real number*. For example, consider the functions

Domains of algebraic expressions are discussed on page 36.

$$f(x) = \frac{1}{x - 4} \qquad g(x) = \sqrt{x}$$

The function f is not defined at $x = 4$, so its domain is $\{x \mid x \neq 4\}$. The function g is not defined for negative x, so its domain is $\{x \mid x \ge 0\}$.

EXAMPLE 7 ■ Finding Domains of Functions

Find the domain of each function.

(a) $f(x) = \dfrac{1}{x^2 - x}$ **(b)** $g(x) = \sqrt{9 - x^2}$ **(c)** $h(t) = \dfrac{t}{\sqrt{t + 1}}$

SOLUTION

(a) A rational expression is not defined when the denominator is 0. Since

$$f(x) = \frac{1}{x^2 - x} = \frac{1}{x(x - 1)}$$

we see that $f(x)$ is not defined when $x = 0$ or $x = 1$. Thus the domain of f is

$$\{x \mid x \neq 0, x \neq 1\}$$

The domain may also be written in interval notation as

$$(\infty, 0) \cup (0, 1) \cup (1, \infty)$$

(b) We can't take the square root of a negative number, so we must have $9 - x^2 \ge 0$. Using the methods of Section 1.8, we can solve this inequality to find that $-3 \le x \le 3$. Thus the domain of g is

$$\{x \mid -3 \le x \le 3\} = [-3, 3]$$

(c) We can't take the square root of a negative number, and we can't divide by 0, so we must have $t + 1 > 0$, that is, $t > -1$. So the domain of h is

$$\{t \mid t > -1\} = (-1, \infty)$$

Now Try Exercises 55, 59, and 69 ■

Four Ways to Represent a Function

To help us understand what a function is, we have used machine and arrow diagrams. We can describe a specific function in the following four ways:

- verbally (by a description in words)
- algebraically (by an explicit formula)
- visually (by a graph)
- numerically (by a table of values)

A single function may be represented in all four ways, and it is often useful to go from one representation to another to gain insight into the function. However, certain functions are described more naturally by one method than by the others. An example of a verbal description is the following rule for converting between temperature scales:

> "To find the Fahrenheit equivalent of a Celsius temperature, multiply the Celsius temperature by $\frac{9}{5}$, then add 32."

In Example 8 we see how to describe this verbal rule or function algebraically, graphically, and numerically. A useful representation of the area of a circle as a function of its radius is the algebraic formula

$$A(r) = \pi r^2$$

The graph produced by a seismograph (see the box below) is a visual representation of the vertical acceleration function $a(t)$ of the ground during an earthquake. As a final example, consider the function $C(w)$, which is described verbally as "the cost of mailing a large first-class letter with weight w." The most convenient way of describing this function is numerically—that is, using a table of values.

We will be using all four representations of functions throughout this book. We summarize them in the following box.

FOUR WAYS TO REPRESENT A FUNCTION

Verbal Using words:

"To convert from Celsius to Fahrenheit, multiply the Celsius temperature by $\frac{9}{5}$, then add 32."

Relation between Celsius and Fahrenheit temperature scales

Algebraic Using a formula:

$$A(r) = \pi r^2$$

Area of a circle

Visual Using a graph:

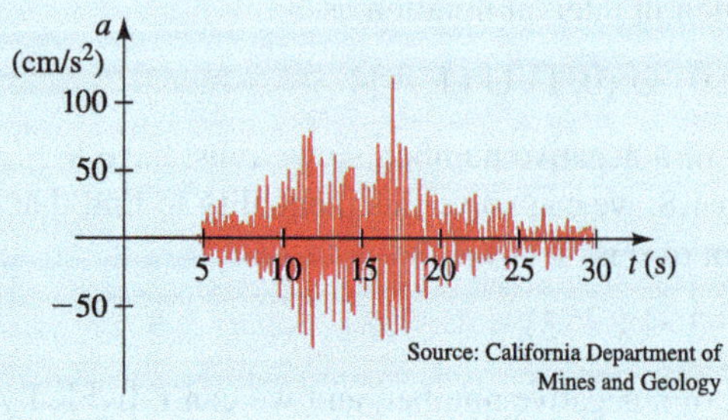

Vertical acceleration during an earthquake

Numerical Using a table of values:

w (ounces)	$C(w)$ (dollars)
$0 < w \le 1$	\$0.98
$1 < w \le 2$	\$1.19
$2 < w \le 3$	\$1.40
$3 < w \le 4$	\$1.61
$4 < w \le 5$	\$1.82
⋮	⋮

Cost of mailing a large first-class envelope

EXAMPLE 8 ■ Representing a Function Verbally, Algebraically, Numerically, and Graphically

Let $F(C)$ be the Fahrenheit temperature corresponding to the Celsius temperature C. (Thus F is the function that converts Celsius inputs to Fahrenheit outputs.) The box on page 154 gives a verbal description of this function. Find ways to represent this function

(a) Algebraically (using a formula)

(b) Numerically (using a table of values)

(c) Visually (using a graph)

SOLUTION

(a) The verbal description tells us that we should first multiply the input C by $\frac{9}{5}$ and then add 32 to the result. So we get

$$F(C) = \tfrac{9}{5}C + 32$$

(b) We use the algebraic formula for F that we found in part (a) to construct a table of values:

C (Celsius)	F (Fahrenheit)
−10	14
0	32
10	50
20	68
30	86
40	104

(c) We use the points tabulated in part (b) to help us draw the graph of this function in Figure 6.

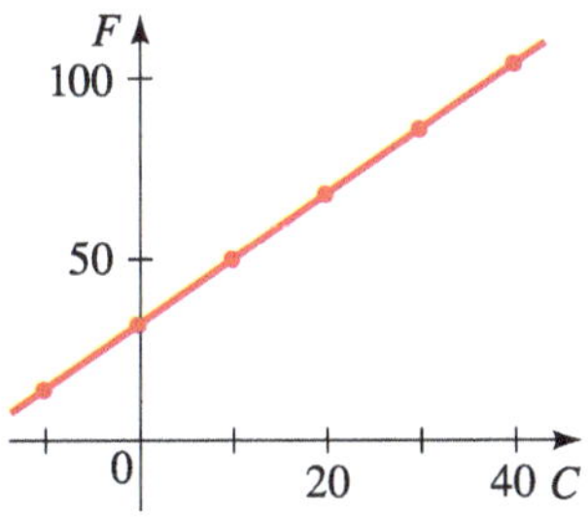

FIGURE 6 Celsius and Fahrenheit

Now Try Exercise 73 ■

2.1 EXERCISES

CONCEPTS

1. If $f(x) = x^3 + 1$, then

(a) the value of f at $x = -1$ is $f(\_\_\_) = \_\_\_\_\_$.

(b) the value of f at $x = 2$ is $f(\_\_\_) = \_\_\_\_\_$.

(c) the net change in the value of f between $x = -1$ and $x = 2$ is $f(\_\_\_) - f(\_\_\_) = \_\_\_\_\_$.

2. For a function f, the set of all possible inputs is called the _____ of f, and the set of all possible outputs is called the _____ of f.

3. (a) Which of the following functions have 5 in their domain?

$$f(x) = x^2 - 3x \qquad g(x) = \frac{x-5}{x} \qquad h(x) = \sqrt{x - 10}$$

(b) For the functions from part (a) that *do* have 5 in their domain, find the value of the function at 5.

4. A function is given algebraically by the formula $f(x) = (x-4)^2 + 3$. Complete these other ways to represent f:

(a) *Verbal:* "Subtract 4, then _____ and _____."

(b) *Numerical:*

x	$f(x)$
0	19
2	
4	
6	

5. A function f is a rule that assigns to each element x in a set A exactly ________ element(s) called $f(x)$ in a set B. Which of the following tables defines y as a function of x?

(i)

x	y
1	5
2	7
3	6
4	8

(ii)

x	y
1	5
1	7
2	6
3	8

6. *Yes or No*? If *No*, give a reason. Let f be a function.

(a) Is it possible that $f(1) = 5$ and $f(2) = 5$?

(b) Is it possible that $f(1) = 5$ and $f(1) = 6$?

SKILLS

7–10 ■ Function Notation Express the rule in function notation. (For example, the rule "square, then subtract 5" is expressed as the function $f(x) = x^2 - 5$.)

7. Multiply by 3, then subtract 5

8. Square, then add 2

9. Subtract 1, then square

10. Add 1, take the square root, then divide by 6

11–14 ■ Functions in Words Express the function (or rule) in words.

11. $f(x) = 2x + 3$

12. $g(x) = \dfrac{x + 2}{3}$

13. $h(x) = 5(x + 1)$

14. $k(x) = \dfrac{x^2 - 4}{3}$

15–16 ■ Machine Diagram Draw a machine diagram for the function.

15. $f(x) = \sqrt{x - 1}$

16. $f(x) = \dfrac{3}{x - 2}$

17–18 ■ Table of Values Complete the table.

17. $f(x) = 2(x - 1)^2$

x	$f(x)$
−1	
0	
1	
2	
3	

18. $g(x) = |2x + 3|$

x	$g(x)$
−3	
−2	
0	
1	
3	

19–30 ■ Evaluating Functions Evaluate the function at the indicated values.

19. $f(x) = x^2 - 6$; $f(-3), f(3), f(0), f(\frac{1}{2})$

20. $f(x) = x^3 + 2x$; $f(-2), f(-1), f(0), f(\frac{1}{2})$

21. $f(x) = \dfrac{1 - 2x}{3}$;

$f(2), f(-2), f(\frac{1}{2}), f(a), f(-a), f(a - 1)$

22. $h(x) = \dfrac{x^2 + 4}{5}$;

$h(2), h(-2), h(a), h(-x), h(a - 2), h(\sqrt{x})$

23. $f(x) = x^2 + 2x$;

$f(0), f(3), f(-3), f(a), f(-x), f\left(\dfrac{1}{a}\right)$

24. $h(t) = t + \dfrac{1}{t}$;

$h(-1), h(2), h(\frac{1}{2}), h(x - 1), h\left(\dfrac{1}{x}\right)$

25. $g(x) = \dfrac{1 - x}{1 + x}$;

$g(2), g(-1), g(\frac{1}{2}), g(a), g(a - 1), g(x^2 - 1)$

26. $g(t) = \dfrac{t + 2}{t - 2}$;

$g(-2), g(2), g(0), g(a), g(a^2 - 2), g(a + 1)$

27. $k(x) = -x^2 - 2x + 3$;

$k(0), k(2), k(-2), k(\sqrt{2}), k(a + 2), k(-x), k(x^2)$

28. $k(x) = 2x^3 - 3x^2$;

$k(0), k(3), k(-3), k(\frac{1}{2}), k(\frac{a}{2}), k(-x), k(x^3)$

29. $f(x) = 2|x - 1|$;

$f(-2), f(0), f(\frac{1}{2}), f(2), f(x + 1), f(x^2 + 2)$

30. $f(x) = \dfrac{|x|}{x}$;

$f(-2), f(-1), f(0), f(5), f(x^2), f\left(\dfrac{1}{x}\right)$

31–34 ■ Piecewise Defined Functions Evaluate the piecewise defined function at the indicated values.

31. $f(x) = \begin{cases} x^2 & \text{if } x < 0 \\ x + 1 & \text{if } x \geq 0 \end{cases}$

$f(-2), f(-1), f(0), f(1), f(2)$

32. $f(x) = \begin{cases} 5 & \text{if } x \leq 2 \\ 2x - 3 & \text{if } x > 2 \end{cases}$

$f(-3), f(0), f(2), f(3), f(5)$

33. $f(x) = \begin{cases} x^2 + 2x & \text{if } x \leq -1 \\ x & \text{if } -1 < x \leq 1 \\ -1 & \text{if } x > 1 \end{cases}$

$f(-4), f(-\frac{3}{2}), f(-1), f(0), f(25)$

34. $f(x) = \begin{cases} 3x & \text{if } x < 0 \\ x + 1 & \text{if } 0 \leq x \leq 2 \\ (x - 2)^2 & \text{if } x > 2 \end{cases}$

$f(-5), f(0), f(1), f(2), f(5)$

35–38 ■ Evaluating Functions Use the function to evaluate the indicated expressions and simplify.

35. $f(x) = x^2 + 1;\quad f(x + 2), f(x) + f(2)$

36. $f(x) = 3x - 1;\quad f(2x), 2f(x)$

37. $f(x) = x + 4;\quad f(x^2), (f(x))^2$

38. $f(x) = 6x - 18;\quad f\left(\dfrac{x}{3}\right), \dfrac{f(x)}{3}$

39–42 ■ Net Change Find the net change in the value of the function between the given inputs.

39. $f(x) = 3x - 2;$ from 1 to 5

40. $f(x) = 4 - 5x;$ from 3 to 5

41. $g(t) = 1 - t^2;$ from -2 to 5

42. $h(t) = t^2 + 5;$ from -3 to 6

43–50 ■ Difference Quotient Find $f(a)$, $f(a + h)$, and the difference quotient $\dfrac{f(a + h) - f(a)}{h}$, where $h \neq 0$.

43. $f(x) = 5 - 2x$

44. $f(x) = 3x^2 + 2$

45. $f(x) = 5$

46. $f(x) = \dfrac{1}{x + 1}$

47. $f(x) = \dfrac{x}{x + 1}$

48. $f(x) = \dfrac{2x}{x - 1}$

49. $f(x) = 3 - 5x + 4x^2$

50. $f(x) = x^3$

51–54 ■ Domain and Range Find the domain and range of the function.

51. $f(x) = 3x$

52. $f(x) = 5x^2 + 4$

53. $f(x) = 3x, \quad -2 \le x \le 6$

54. $f(x) = 5x^2 + 4, \quad 0 \le x \le 2$

55–72 ■ Domain Find the domain of the function.

55. $f(x) = \dfrac{1}{x - 3}$

56. $f(x) = \dfrac{1}{3x - 6}$

57. $f(x) = \dfrac{x + 2}{x^2 - 1}$

58. $f(x) = \dfrac{x^4}{x^2 + x - 6}$

59. $f(t) = \sqrt{t + 1}$

60. $g(t) = \sqrt{t^2 + 9}$

61. $f(t) = \sqrt[3]{t - 1}$

62. $g(x) = \sqrt{7 - 3x}$

63. $f(x) = \sqrt{1 - 2x}$

64. $g(x) = \sqrt{x^2 - 4}$

65. $g(x) = \dfrac{\sqrt{2 + x}}{3 - x}$

66. $g(x) = \dfrac{\sqrt{x}}{2x^2 + x - 1}$

67. $g(x) = \sqrt[4]{x^2 - 6x}$

68. $g(x) = \sqrt{x^2 - 2x - 8}$

69. $f(x) = \dfrac{3}{\sqrt{x - 4}}$

70. $f(x) = \dfrac{x^2}{\sqrt{6 - x}}$

71. $f(x) = \dfrac{(x + 1)^2}{\sqrt{2x - 1}}$

72. $f(x) = \dfrac{x}{\sqrt[4]{9 - x^2}}$

73–76 ■ Four Ways to Represent a Function A verbal description of a function is given. Find **(a)** algebraic, **(b)** numerical, and **(c)** graphical representations for the function.

73. To evaluate $f(x)$, divide the input by 3 and add $\frac{2}{3}$ to the result.

74. To evaluate $g(x)$, subtract 4 from the input and multiply the result by $\frac{3}{4}$.

75. Let $T(x)$ be the amount of sales tax charged in Lemon County on a purchase of x dollars. To find the tax, take 8% of the purchase price.

76. Let $V(d)$ be the volume of a sphere of diameter d. To find the volume, take the cube of the diameter, then multiply by π and divide by 6.

SKILLS Plus

77–78 ■ Domain and Range Find the domain and range of f.

77. $f(x) = \begin{cases} 1 & \text{if } x \text{ is rational} \\ 5 & \text{if } x \text{ is irrational} \end{cases}$

78. $f(x) = \begin{cases} 1 & \text{if } x \text{ is rational} \\ 5x & \text{if } x \text{ is irrational} \end{cases}$

APPLICATIONS

79. Torricelli's Law A tank holds 50 gal of water, which drains from a leak at the bottom, causing the tank to empty in 20 min. The tank drains faster when it is nearly full because the pressure on the leak is greater. **Torricelli's Law** gives the volume of water remaining in the tank after t minutes as

$$V(t) = 50\left(1 - \frac{t}{20}\right)^2 \qquad 0 \le t \le 20$$

(a) Find $V(0)$ and $V(20)$.

(b) What do your answers to part (a) represent?

(c) Make a table of values of $V(t)$ for $t = 0, 5, 10, 15, 20$.

(d) Find the net change in the volume V as t changes from 0 min to 20 min.

80. Area of a Sphere The surface area S of a sphere is a function of its radius r given by

$$S(r) = 4\pi r^2$$

(a) Find $S(2)$ and $S(3)$.

(b) What do your answers in part (a) represent?

81. Relativity According to the Theory of Relativity, the length L of an object is a function of its velocity v with respect to an observer. For an object whose length at rest is 10 m, the function is given by

$$L(v) = 10\sqrt{1 - \frac{v^2}{c^2}}$$

where c is the speed of light (300,000 km/s).

(a) Find $L(0.5c)$, $L(0.75c)$, and $L(0.9c)$.

(b) How does the length of an object change as its velocity increases?

82. Pupil Size When the brightness x of a light source is increased, the eye reacts by decreasing the radius R of the pupil. The dependence of R on x is given by the function

$$R(x) = \sqrt{\frac{13 + 7x^{0.4}}{1 + 4x^{0.4}}}$$

where R is measured in millimeters and x is measured in appropriate units of brightness.

(a) Find $R(1)$, $R(10)$, and $R(100)$.

(b) Make a table of values of $R(x)$.

(c) Find the net change in the radius R as x changes from 10 to 100.

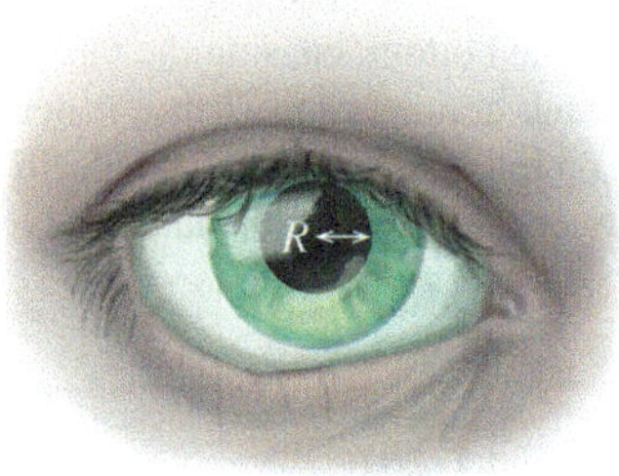

83. Blood Flow As blood moves through a vein or an artery, its velocity v is greatest along the central axis and decreases as the distance r from the central axis increases (see the figure). The formula that gives v as a function of r is called the **law of laminar flow**. For an artery with radius 0.5 cm, the relationship between v (in cm/s) and r (in cm) is given by the function

$$v(r) = 18{,}500(0.25 - r^2) \qquad 0 \le r \le 0.5$$

(a) Find $v(0.1)$ and $v(0.4)$.

(b) What do your answers to part (a) tell you about the flow of blood in this artery?

(c) Make a table of values of $v(r)$ for $r = 0, 0.1, 0.2, 0.3, 0.4, 0.5$.

(d) Find the net change in the velocity v as r changes from 0.1 cm to 0.5 cm.

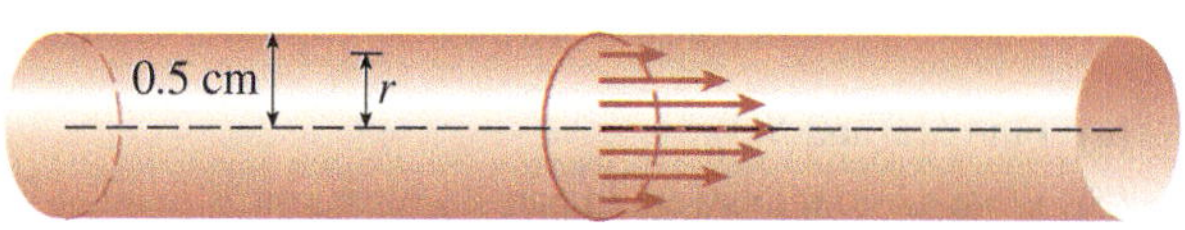

84. How Far Can You See? Because of the curvature of the earth, the maximum distance D that you can see from the top of a tall building or from an airplane at height h is given by the function

$$D(h) = \sqrt{2rh + h^2}$$

where $r = 3960$ mi is the radius of the earth and D and h are measured in miles.

(a) Find $D(0.1)$ and $D(0.2)$.

(b) How far can you see from the observation deck of Toronto's CN Tower, 1135 ft above the ground?

(c) Commercial aircraft fly at an altitude of about 7 mi. How far can the pilot see?

(d) Find the net change in the value of distance D as h changes from 1135 ft to 7 mi.

85. Income Tax In a certain country, income tax T is assessed according to the following function of income x:

$$T(x) = \begin{cases} 0 & \text{if } 0 \le x \le 10{,}000 \\ 0.08x & \text{if } 10{,}000 < x \le 20{,}000 \\ 1600 + 0.15x & \text{if } 20{,}000 < x \end{cases}$$

(a) Find $T(5{,}000)$, $T(12{,}000)$, and $T(25{,}000)$.

(b) What do your answers in part (a) represent?

86. Internet Purchases An Internet bookstore charges \$15 shipping for orders under \$100 but provides free shipping for orders of \$100 or more. The cost C of an order is a function of the total price x of the books purchased, given by

$$C(x) = \begin{cases} x + 15 & \text{if } x < 100 \\ x & \text{if } x \ge 100 \end{cases}$$

(a) Find $C(75)$, $C(90)$, $C(100)$, and $C(105)$.

(b) What do your answers in part (a) represent?

87. Cost of a Hotel Stay A hotel chain charges \$75 each night for the first two nights and \$50 for each additional night's stay. The total cost T is a function of the number of nights x that a guest stays.

(a) Complete the expressions in the following piecewise defined function.

$$T(x) = \begin{cases} \quad\quad & \text{if } 0 \le x \le 2 \\ \quad\quad & \text{if } x > 2 \end{cases}$$

(b) Find $T(2)$, $T(3)$, and $T(5)$.

(c) What do your answers in part (b) represent?

88. Speeding Tickets In a certain state the maximum speed permitted on freeways is 65 mi/h, and the minimum is 40 mi/h. The fine F for violating these limits is \$15 for every mile above the maximum or below the minimum.

(a) Complete the expressions in the following piecewise defined function, where x is the speed at which you are driving.

$$F(x) = \begin{cases} \quad\quad & \text{if } 0 < x < 40 \\ \quad\quad & \text{if } 40 \le x \le 65 \\ \quad\quad & \text{if } x > 65 \end{cases}$$

(b) Find $F(30)$, $F(50)$, and $F(75)$.

(c) What do your answers in part (b) represent?

89. Height of Grass A home owner mows the lawn every Wednesday afternoon. Sketch a rough graph of the height of the grass as a function of time over the course of a four-week period beginning on a Sunday.

90. Temperature Change You place a frozen pie in an oven and bake it for an hour. Then you take the pie out and let it cool before eating it. Sketch a rough graph of the temperature of the pie as a function of time.

91. Daily Temperature Change Temperature readings T (in °F) were recorded every 2 hours from midnight to noon in Atlanta, Georgia, on March 18, 2014. The time t was measured in hours from midnight. Sketch a rough graph of T as a function of t.

t	0	2	4	6	8	10	12
T	58	57	53	50	51	57	61

92. Population Growth The population P (in thousands) of San Jose, California, from 1980 to 2010 is shown in the table. (Midyear estimates are given.) Draw a rough graph of P as a function of time t.

t	1980	1985	1990	1995	2000	2005	2010
P	629	714	782	825	895	901	946

Source: U.S. Census Bureau

DISCUSS ■ DISCOVER ■ PROVE ■ WRITE

93. DISCUSS: Examples of Functions At the beginning of this section we discussed three examples of everyday, ordinary functions: Height is a function of age, temperature is a function of date, and postage cost is a function of weight. Give three other examples of functions from everyday life.

94. DISCUSS: Four Ways to Represent a Function In the box on page 154 we represented four different functions verbally, algebraically, visually, and numerically. Think of a function that can be represented in all four ways, and give the four representations.

95. DISCUSS: Piecewise Defined Functions In Exercises 85–88 we worked with real-world situations modeled by piecewise defined functions. Find other examples of real-world situations that can be modeled by piecewise defined functions, and express the models in function notation.

2.2 GRAPHS OF FUNCTIONS

■ Graphing Functions by Plotting Points ■ Graphing Functions with a Graphing Calculator ■ Graphing Piecewise Defined Functions ■ The Vertical Line Test: Which Graphs Represent Functions? ■ Which Equations Represent Functions?

The most important way to visualize a function is through its graph. In this section we investigate in more detail the concept of graphing functions.

■ Graphing Functions by Plotting Points

To graph a function f, we plot the points $(x, f(x))$ in a coordinate plane. In other words, we plot the points (x, y) whose x-coordinate is an input and whose y-coordinate is the corresponding output of the function.

THE GRAPH OF A FUNCTION

If f is a function with domain A, then the **graph** of f is the set of ordered pairs

$$\{(x, f(x)) \mid x \in A\}$$

plotted in a coordinate plane. In other words, the graph of f is the set of all points (x, y) such that $y = f(x)$; that is, the graph of f is the graph of the equation $y = f(x)$.

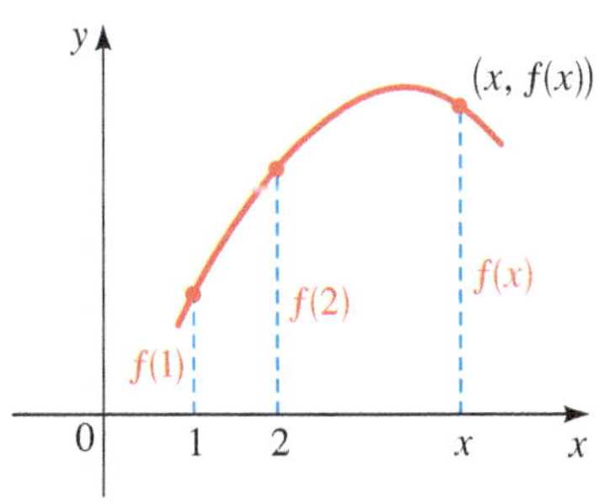

FIGURE 1 The height of the graph above the point x is the value of $f(x)$.

The graph of a function f gives a picture of the behavior or "life history" of the function. We can read the value of $f(x)$ from the graph as being the height of the graph above the point x (see Figure 1).

A function f of the form $f(x) = mx + b$ is called a **linear function** because its graph is the graph of the equation $y = mx + b$, which represents a line with slope m and y-intercept b. A special case of a linear function occurs when the slope is $m = 0$. The function $f(x) = b$, where b is a given number, is called a **constant function** because all its values are the same number, namely, b. Its graph is the horizontal line $y = b$. Figure 2 shows the graphs of the constant function $f(x) = 3$ and the linear function $f(x) = 2x + 1$.

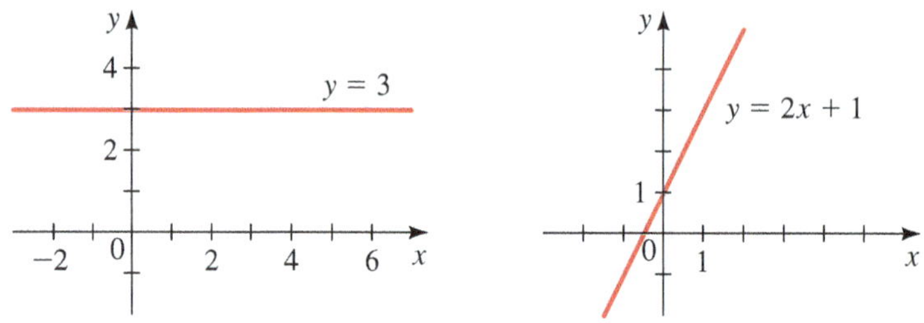

FIGURE 2 The constant function $f(x) = 3$ The linear function $f(x) = 2x + 1$

Functions of the form $f(x) = x^n$ are called **power functions**, and functions of the form $f(x) = x^{1/n}$ are called **root functions**. In the next example we graph two power functions and a root function.

EXAMPLE 1 ■ Graphing Functions by Plotting Points

Sketch graphs of the following functions.

(a) $f(x) = x^2$ **(b)** $g(x) = x^3$ **(c)** $h(x) = \sqrt{x}$

SOLUTION We first make a table of values. Then we plot the points given by the table and join them by a smooth curve to obtain the graph. The graphs are sketched in Figure 3.

x	$f(x) = x^2$
0	0
$\pm\frac{1}{2}$	$\frac{1}{4}$
± 1	1
± 2	4
± 3	9

x	$g(x) = x^3$
0	0
$\frac{1}{2}$	$\frac{1}{8}$
1	1
2	8
$-\frac{1}{2}$	$-\frac{1}{8}$
-1	-1
-2	-8

x	$h(x) = \sqrt{x}$
0	0
1	1
2	$\sqrt{2}$
3	$\sqrt{3}$
4	2
5	$\sqrt{5}$

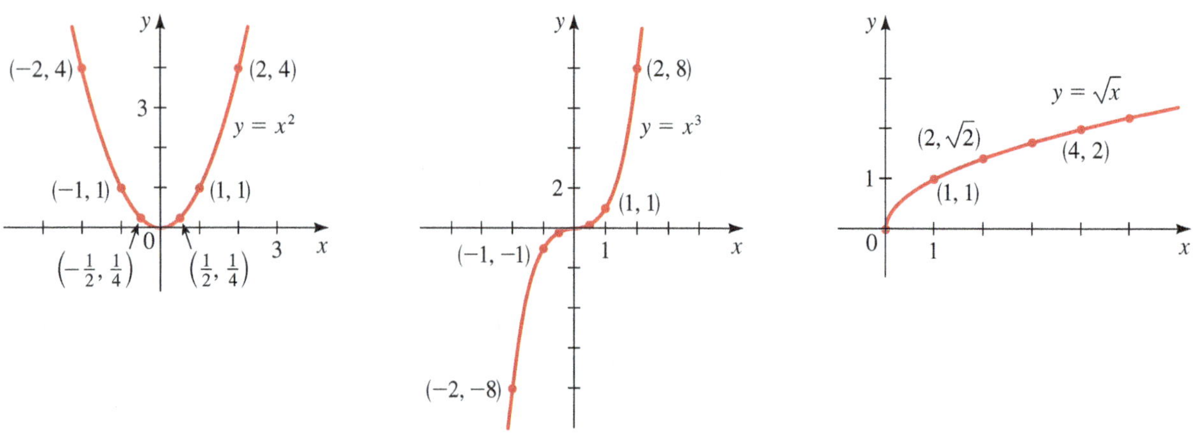

(a) $f(x) = x^2$ (b) $g(x) = x^3$ (c) $h(x) = \sqrt{x}$

FIGURE 3

Now Try Exercises 9, 15, and 19 ■

Graphing Functions with a Graphing Calculator

See Appendix C, *Graphing with a Graphing Calculator,* for general guidelines on using a graphing calculator. See Appendix D, *Using the TI-83/84 Graphing Calculator,* for specific instructions. Go to **www.stewartmath.com**.

A convenient way to graph a function is to use a graphing calculator. To graph the function f, we use a calculator to graph the equation $y = f(x)$.

EXAMPLE 2 ■ Graphing a Function with a Graphing Calculator

Use a graphing calculator to graph the function $f(x) = x^3 - 8x^2$ in an appropriate viewing rectangle.

SOLUTION To graph the function $f(x) = x^3 - 8x^2$, we must graph the equation $y = x^3 - 8x^2$. On the TI-83 graphing calculator the default viewing rectangle gives the graph in Figure 4(a). But this graph appears to spill over the top and bottom of the screen. We need to expand the vertical axis to get a better representation of the graph. The viewing rectangle $[-4, 10]$ by $[-100, 100]$ gives a more complete picture of the graph, as shown in Figure 4(b).

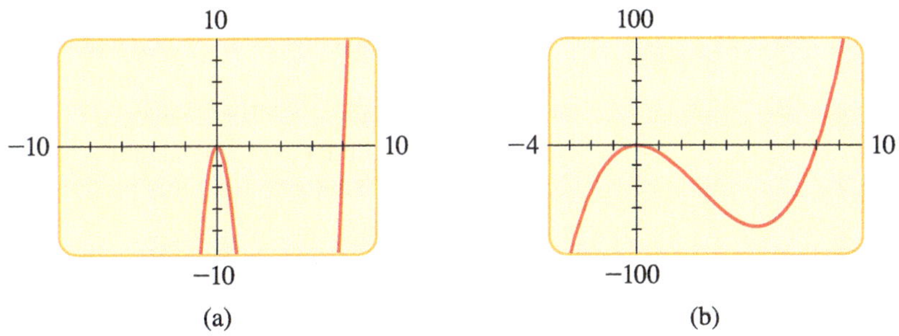

FIGURE 4 Graphing the function $f(x) = x^3 - 8x^2$

Now Try Exercise 29 ■

EXAMPLE 3 ■ A Family of Power Functions

(a) Graph the functions $f(x) = x^n$ for $n = 2$, 4, and 6 in the viewing rectangle $[-2, 2]$ by $[-1, 3]$.

(b) Graph the functions $f(x) = x^n$ for $n = 1$, 3, and 5 in the viewing rectangle $[-2, 2]$ by $[-2, 2]$.

(c) What conclusions can you draw from these graphs?

SOLUTION To graph the function $f(x) = x^n$, we graph the equation $y = x^n$. The graphs for parts (a) and (b) are shown in Figure 5.

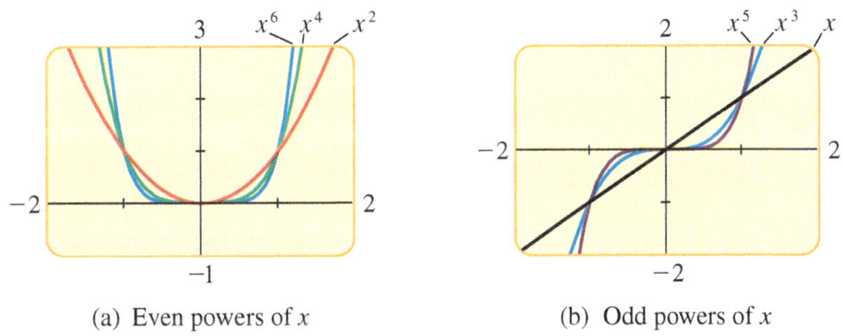

FIGURE 5 A family of power functions: $f(x) = x^n$

(c) We see that the general shape of the graph of $f(x) = x^n$ depends on whether n is even or odd.

If n is even, the graph of $f(x) = x^n$ is similar to the parabola $y = x^2$.
If n is odd, the graph of $f(x) = x^n$ is similar to that of $y = x^3$.

Now Try Exercise 69 ■

Notice from Figure 5 that as n increases, the graph of $y = x^n$ becomes flatter near 0 and steeper when $x > 1$. When $0 < x < 1$, the lower powers of x are the "bigger" functions. But when $x > 1$, the higher powers of x are the dominant functions.

Graphing Piecewise Defined Functions

A piecewise defined function is defined by different formulas on different parts of its domain. As you might expect, the graph of such a function consists of separate pieces.

EXAMPLE 4 ■ Graph of a Piecewise Defined Function

Sketch the graph of the function

$$f(x) = \begin{cases} x^2 & \text{if } x \le 1 \\ 2x + 1 & \text{if } x > 1 \end{cases}$$

On many graphing calculators the graph in Figure 6 can be produced by using the logical functions in the calculator. For example, on the TI-83 the following equation gives the required graph:

```
Y1=(X≤1)X²+(X>1)(2X+1)
```

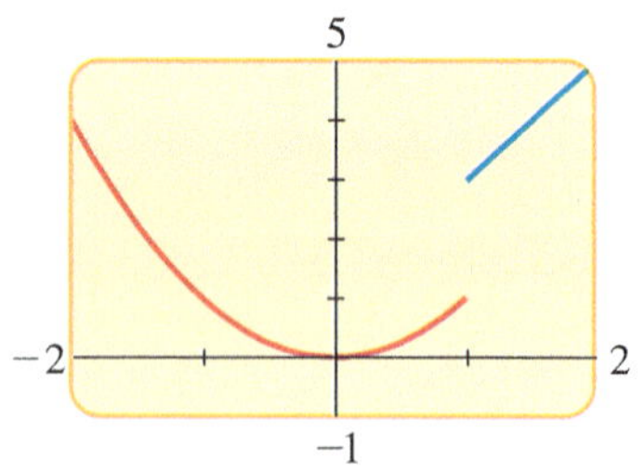

(To avoid the extraneous vertical line between the two parts of the graph, put the calculator in `Dot` mode.)

SOLUTION If $x \le 1$, then $f(x) = x^2$, so the part of the graph to the left of $x = 1$ coincides with the graph of $y = x^2$, which we sketched in Figure 3. If $x > 1$, then $f(x) = 2x + 1$, so the part of the graph to the right of $x = 1$ coincides with the line $y = 2x + 1$, which we graphed in Figure 2. This enables us to sketch the graph in Figure 6.

The solid dot at $(1, 1)$ indicates that this point is included in the graph; the open dot at $(1, 3)$ indicates that this point is excluded from the graph.

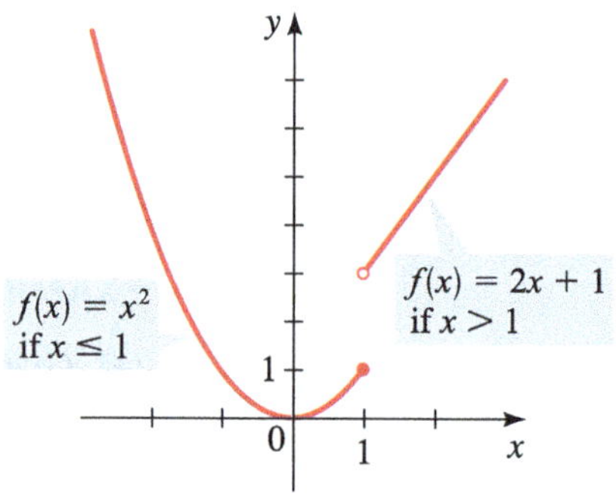

FIGURE 6

$$f(x) = \begin{cases} x^2 & \text{if } x \le 1 \\ 2x + 1 & \text{if } x > 1 \end{cases}$$

Now Try Exercise 35 ■

EXAMPLE 5 ■ Graph of the Absolute Value Function

Sketch a graph of the absolute value function $f(x) = |x|$.

SOLUTION Recall that

$$|x| = \begin{cases} x & \text{if } x \ge 0 \\ -x & \text{if } x < 0 \end{cases}$$

Using the same method as in Example 4, we note that the graph of f coincides with the line $y = x$ to the right of the y-axis and coincides with the line $y = -x$ to the left of the y-axis (see Figure 7).

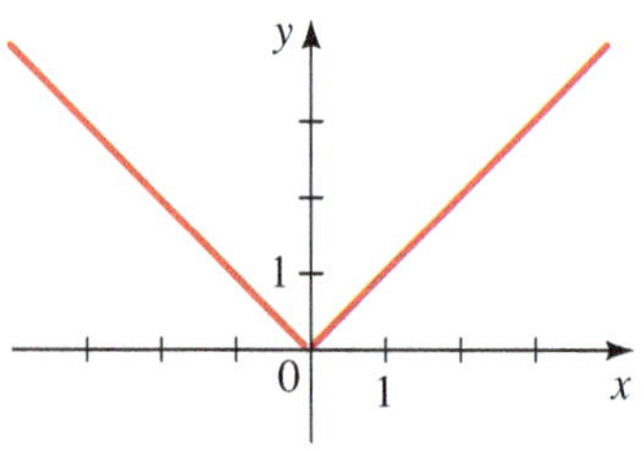

FIGURE 7 Graph of $f(x) = |x|$

Now Try Exercise 23 ■

The **greatest integer function** is defined by

$$[\![x]\!] = \text{greatest integer less than or equal to } x$$

For example, $[\![2]\!] = 2$, $[\![2.3]\!] = 2$, $[\![1.999]\!] = 1$, $[\![0.002]\!] = 0$, $[\![-3.5]\!] = -4$, and $[\![-0.5]\!] = -1$.

EXAMPLE 6 ■ Graph of the Greatest Integer Function

Sketch a graph of $f(x) = [\![x]\!]$.

SOLUTION The table shows the values of f for some values of x. Note that $f(x)$ is constant between consecutive integers, so the graph between integers is a horizontal line segment, as shown in Figure 8.

x	$[\![x]\!]$
⋮	⋮
$-2 \le x < -1$	-2
$-1 \le x < 0$	-1
$0 \le x < 1$	0
$1 \le x < 2$	1
$2 \le x < 3$	2
⋮	⋮

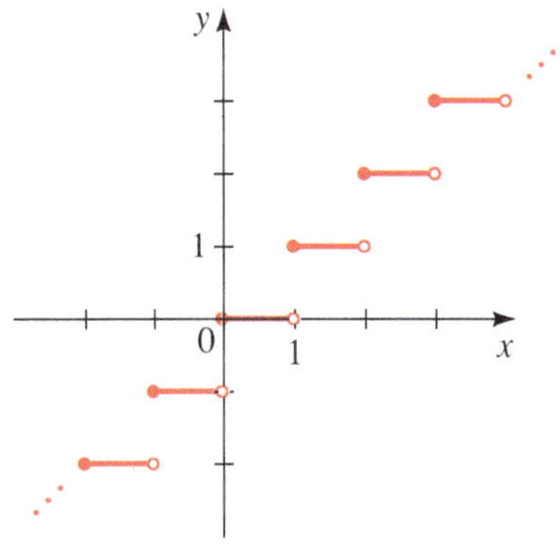

FIGURE 8 The greatest integer function, $y = [\![x]\!]$

The greatest integer function is an example of a **step function**. The next example gives a real-world example of a step function.

EXAMPLE 7 ■ The Cost Function for a Global Data Plan

A global data plan costs \$25 a month for the first 100 megabytes and \$20 for each additional 100 megabytes (or portion thereof). Draw a graph of the cost C (in dollars) as a function of the number of megabytes x used per month.

SOLUTION Let $C(x)$ be the cost of using x megabytes of data in a month. Since $x \ge 0$, the domain of the function is $[0, \infty)$. From the given information we have

$$\begin{aligned} C(x) &= 25 & &\text{if } 0 < x \le 100 \\ C(x) &= 25 + 20 = 45 & &\text{if } 100 < x \le 200 \\ C(x) &= 25 + 2(20) = 65 & &\text{if } 200 < x \le 300 \\ C(x) &= 25 + 3(20) = 85 & &\text{if } 300 < x \le 400 \\ &\vdots & &\vdots \end{aligned}$$

The graph is shown in Figure 9.

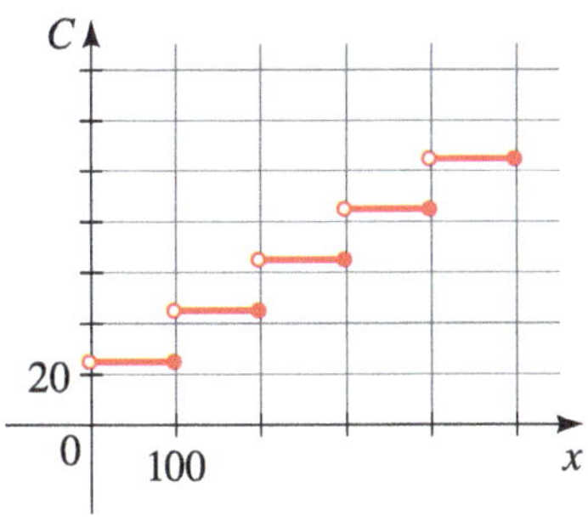

FIGURE 9 Cost of data usage

Now Try Exercise 83

DISCOVERY PROJECT

Relations and Functions

Many real-world relationships are functions, but many are not. For example, the rule that assigns to each student his or her school ID number is a function. But what about the rule that assigns to each date those persons born in Chicago on that date? Do you see why this "relation" is not a function? A set of ordered pairs is called a *relation*. In this project we explore the question of which relations are functions. You can find the project at **www.stewartmath.com**.

A function is called **continuous** if its graph has no "breaks" or "holes." The functions in Examples 1, 2, 3, and 5 are continuous; the functions in Examples 4, 6, and 7 are not continuous.

The Vertical Line Test: Which Graphs Represent Functions?

The graph of a function is a curve in the xy-plane. But the question arises: Which curves in the xy-plane are graphs of functions? This is answered by the following test.

> **THE VERTICAL LINE TEST**
>
> A curve in the coordinate plane is the graph of a function if and only if no vertical line intersects the curve more than once.

We can see from Figure 10 why the Vertical Line Test is true. If each vertical line $x = a$ intersects a curve only once at (a, b), then exactly one functional value is defined by $f(a) = b$. But if a line $x = a$ intersects the curve twice, at (a, b) and at (a, c), then the curve cannot represent a function because a function cannot assign two different values to a.

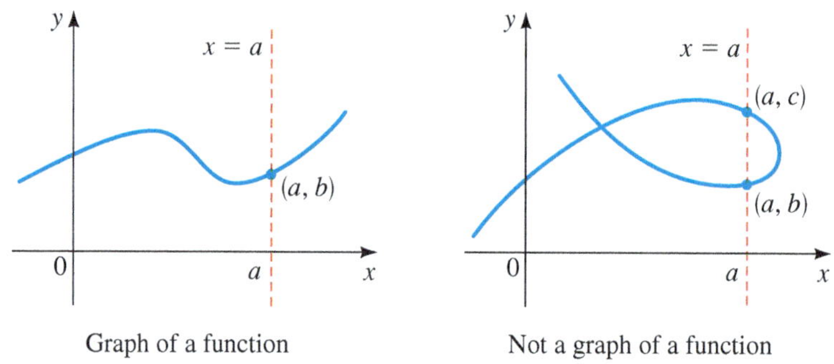

FIGURE 10 Vertical Line Test

EXAMPLE 8 ■ Using the Vertical Line Test

Using the Vertical Line Test, we see that the curves in parts (b) and (c) of Figure 11 represent functions, whereas those in parts (a) and (d) do not.

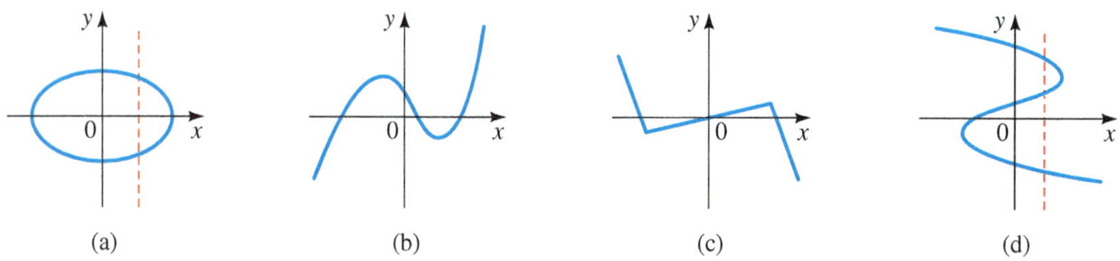

FIGURE 11

Now Try Exercise 51 ■

Which Equations Represent Functions?

Any equation in the variables x and y defines a relationship between these variables. For example, the equation

$$y - x^2 = 0$$

BBVA/NOTIMEX/Newscom

DONALD KNUTH was born in Milwaukee in 1938 and is Professor Emeritus of Computer Science at Stanford University. When Knuth was a high school student, he became fascinated with graphs of functions and laboriously drew many hundreds of them because he wanted to see the behavior of a great variety of functions. (Today, of course, it is far easier to use computers and graphing calculators to do this.) While still a graduate student at Caltech, he started writing a monumental series of books entitled *The Art of Computer Programming.*

Knuth is famous for his invention of TEX, a system of computer-assisted typesetting. This system was used in the preparation of the manuscript for this textbook.

Knuth has received numerous honors, among them election as an associate of the French Academy of Sciences, and as a Fellow of the Royal Society. President Carter awarded him the National Medal of Science in 1979.

defines a relationship between y and x. Does this equation define y as a *function* of x? To find out, we solve for y and get

$$y = x^2 \qquad \text{Equation form}$$

We see that the equation defines a rule, or function, that gives one value of y for each value of x. We can express this rule in function notation as

$$f(x) = x^2 \qquad \text{Function form}$$

But not every equation defines y as a function of x, as the following example shows.

EXAMPLE 9 ■ Equations That Define Functions

Does the equation define y as a function of x?

(a) $y - x^2 = 2$ **(b)** $x^2 + y^2 = 4$

SOLUTION

(a) Solving for y in terms of x gives

$$y - x^2 = 2$$
$$y = x^2 + 2 \qquad \text{Add } x^2$$

The last equation is a rule that gives one value of y for each value of x, so it defines y as a function of x. We can write the function as $f(x) = x^2 + 2$.

(b) We try to solve for y in terms of x.

$$x^2 + y^2 = 4$$
$$y^2 = 4 - x^2 \qquad \text{Subtract } x^2$$
$$y = \pm\sqrt{4 - x^2} \qquad \text{Take square roots}$$

The last equation gives two values of y for a given value of x. Thus the equation does not define y as a function of x.

Now Try Exercises 57 and 61 ■

The graphs of the equations in Example 9 are shown in Figure 12. The Vertical Line Test shows graphically that the equation in Example 9(a) defines a function but the equation in Example 9(b) does not.

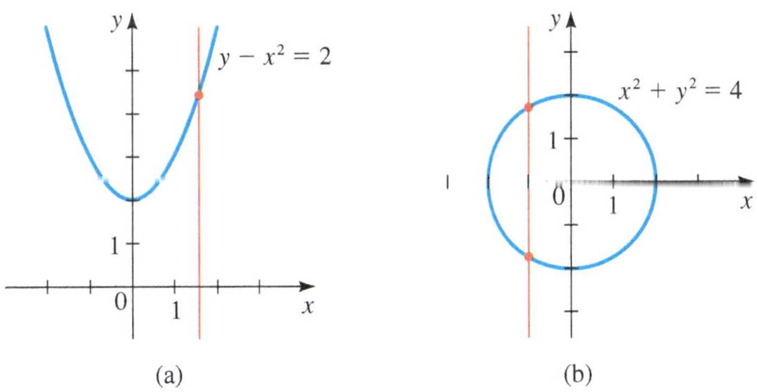

FIGURE 12

The following box shows the graphs of some functions that you will see frequently in this book.

SOME FUNCTIONS AND THEIR GRAPHS

Linear functions

$f(x) = mx + b$

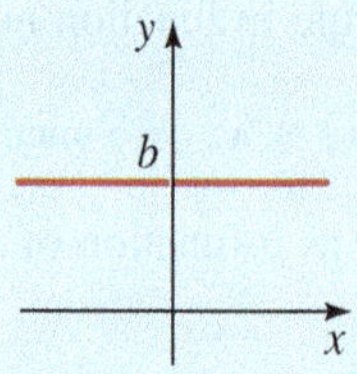

$f(x) = b$

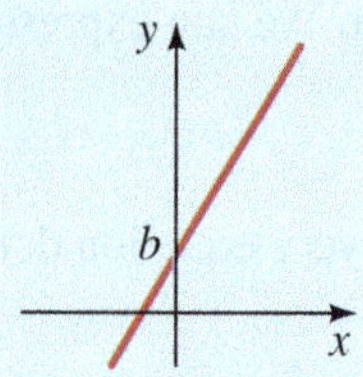

$f(x) = mx + b$

Power functions

$f(x) = x^n$

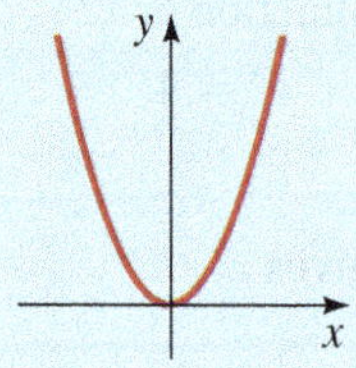

$f(x) = x^2$

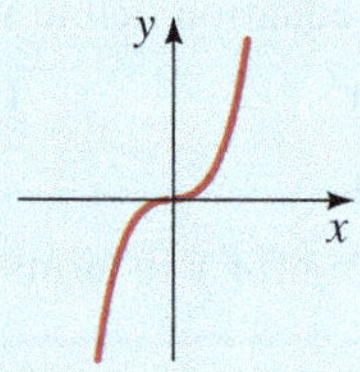

$f(x) = x^3$

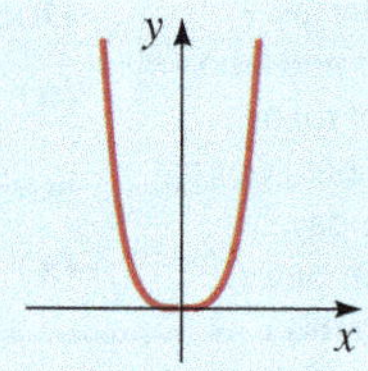

$f(x) = x^4$

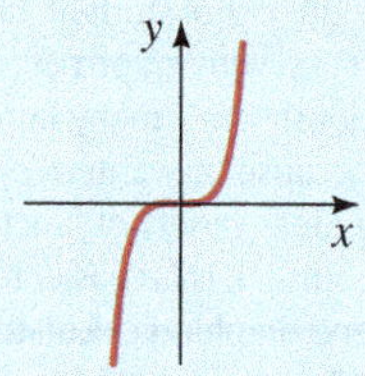

$f(x) = x^5$

Root functions

$f(x) = \sqrt[n]{x}$

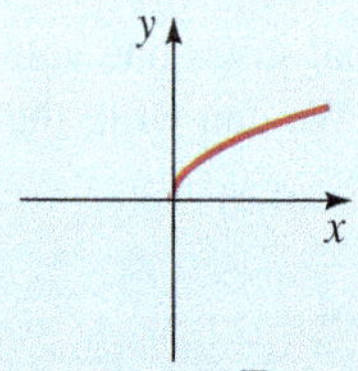

$f(x) = \sqrt{x}$

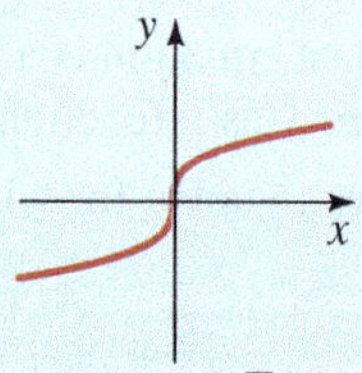

$f(x) = \sqrt[3]{x}$

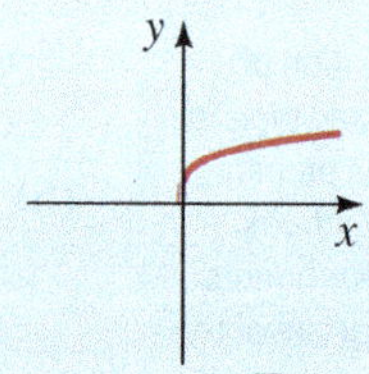

$f(x) = \sqrt[4]{x}$

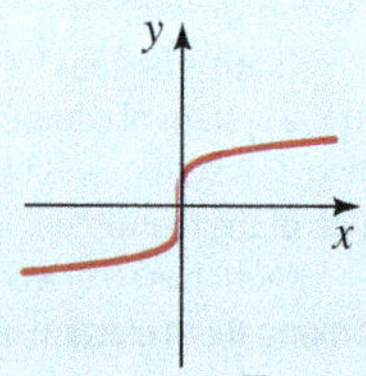

$f(x) = \sqrt[5]{x}$

Reciprocal functions

$f(x) = \dfrac{1}{x^n}$

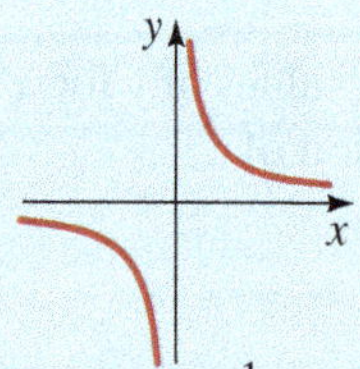

$f(x) = \dfrac{1}{x}$

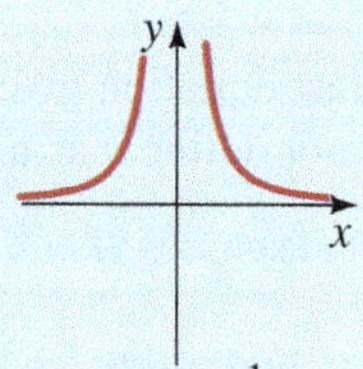

$f(x) = \dfrac{1}{x^2}$

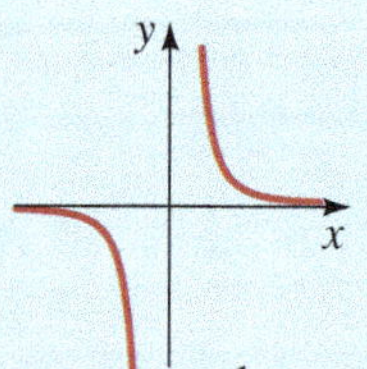

$f(x) = \dfrac{1}{x^3}$

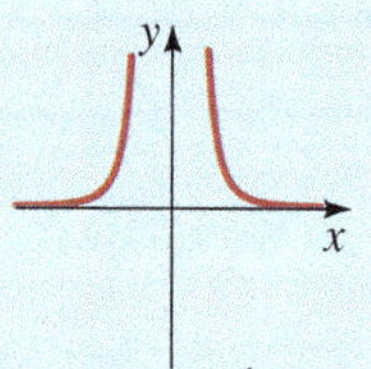

$f(x) = \dfrac{1}{x^4}$

Absolute value function

$f(x) = |x|$

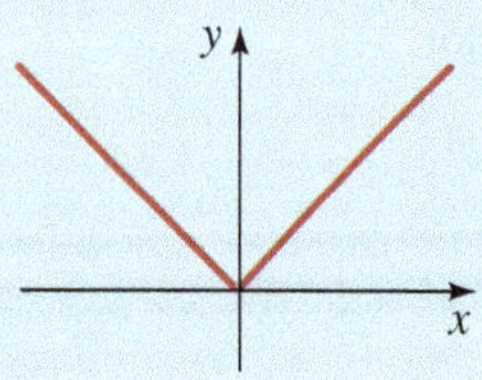

$f(x) = |x|$

Greatest integer function

$f(x) = [\![x]\!]$

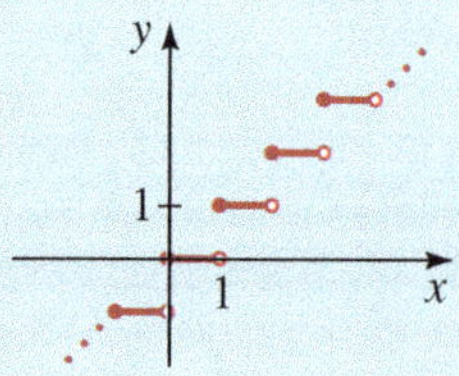

$f(x) = [\![x]\!]$

2.2 EXERCISES

CONCEPTS

1. To graph the function f, we plot the points (x, ________) in a coordinate plane. To graph $f(x) = x^2 - 2$, we plot the points (x, ________). So the point (3, ________) is on the graph of f. The height of the graph of f above the x-axis when $x = 3$ is ______. Complete the table, and sketch a graph of f.

x	$f(x)$	(x, y)
-2		
-1		
0		
1		
2		

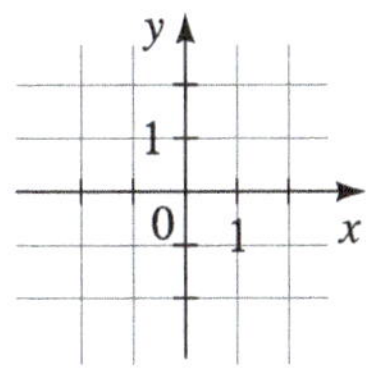

2. If $f(4) = 10$ then the point $(4, \_\_\_)$ is on the graph of f.

3. If the point $(3, 7)$ is on the graph of f, then $f(3) =$ ______.

4. Match the function with its graph.

(a) $f(x) = x^2$ **(b)** $f(x) = x^3$

(c) $f(x) = \sqrt{x}$ **(d)** $f(x) = |x|$

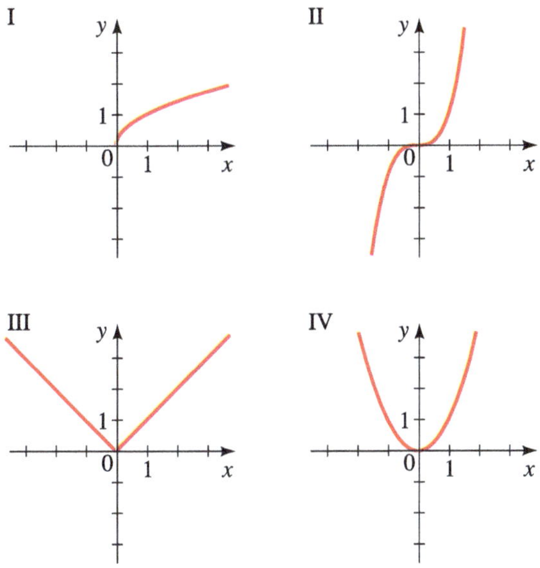

SKILLS

5–28 ■ Graphing Functions Sketch a graph of the function by first making a table of values.

5. $f(x) = x + 2$ **6.** $f(x) = 4 - 2x$

7. $f(x) = -x + 3, \quad -3 \le x \le 3$

8. $f(x) = \dfrac{x - 3}{2}, \quad 0 \le x \le 5$

9. $f(x) = -x^2$ **10.** $f(x) = x^2 - 4$

11. $g(x) = -(x + 1)^2$ **12.** $g(x) = x^2 + 2x + 1$

13. $r(x) = 3x^4$ **14.** $r(x) = 1 - x^4$

15. $g(x) = x^3 - 8$ **16.** $g(x) = (x - 1)^3$

17. $k(x) = \sqrt[3]{-x}$ **18.** $k(x) = -\sqrt[3]{x}$

19. $f(x) = 1 + \sqrt{x}$ **20.** $f(x) = \sqrt{x - 2}$

21. $C(t) = \dfrac{1}{t^2}$ **22.** $C(t) = -\dfrac{1}{t + 1}$

23. $H(x) = |2x|$ **24.** $H(x) = |x + 1|$

25. $G(x) = |x| + x$ **26.** $G(x) = |x| - x$

27. $f(x) = |2x - 2|$ **28.** $f(x) = \dfrac{x}{|x|}$

29–32 ■ Graphing Functions Graph the function in each of the given viewing rectangles, and select the one that produces the most appropriate graph of the function.

29. $f(x) = 8x - x^2$

(a) $[-5, 5]$ by $[-5, 5]$

(b) $[-10, 10]$ by $[-10, 10]$

(c) $[-2, 10]$ by $[-5, 20]$

(d) $[-10, 10]$ by $[-100, 100]$

30. $g(x) = x^2 - x - 20$

(a) $[-2, 2]$ by $[-5, 5]$

(b) $[-10, 10]$ by $[-10, 10]$

(c) $[-7, 7]$ by $[-25, 20]$

(d) $[-10, 10]$ by $[-100, 100]$

31. $h(x) = x^3 - 5x - 4$

(a) $[-2, 2]$ by $[-2, 2]$

(b) $[-3, 3]$ by $[-10, 10]$

(c) $[-3, 3]$ by $[-10, 5]$

(d) $[-10, 10]$ by $[-10, 10]$

32. $k(x) = \frac{1}{32}x^4 - x^2 + 2$

(a) $[-1, 1]$ by $[-1, 1]$

(b) $[-2, 2]$ by $[-2, 2]$

(c) $[-5, 5]$ by $[-5, 5]$

(d) $[-10, 10]$ by $[-10, 10]$

33–46 ■ Graphing Piecewise Defined Functions Sketch a graph of the piecewise defined function.

33. $f(x) = \begin{cases} 0 & \text{if } x < 2 \\ 1 & \text{if } x \ge 2 \end{cases}$

34. $f(x) = \begin{cases} 1 & \text{if } x \le 1 \\ x + 1 & \text{if } x > 1 \end{cases}$

35. $f(x) = \begin{cases} 3 & \text{if } x < 2 \\ x - 1 & \text{if } x \ge 2 \end{cases}$

36. $f(x) = \begin{cases} 1 - x & \text{if } x < -2 \\ 5 & \text{if } x \ge -2 \end{cases}$

37. $f(x) = \begin{cases} x & \text{if } x \le 0 \\ x + 1 & \text{if } x > 0 \end{cases}$

38. $f(x) = \begin{cases} 2x + 3 & \text{if } x < -1 \\ 3 - x & \text{if } x \ge -1 \end{cases}$

39. $f(x) = \begin{cases} -1 & \text{if } x < -1 \\ 1 & \text{if } -1 \le x \le 1 \\ -1 & \text{if } x > 1 \end{cases}$

40. $f(x) = \begin{cases} -1 & \text{if } x < -1 \\ x & \text{if } -1 \le x \le 1 \\ 1 & \text{if } x > 1 \end{cases}$

41. $f(x) = \begin{cases} 2 & \text{if } x \le -1 \\ x^2 & \text{if } x > -1 \end{cases}$

42. $f(x) = \begin{cases} 1 - x^2 & \text{if } x \le 2 \\ x & \text{if } x > 2 \end{cases}$

43. $f(x) = \begin{cases} 0 & \text{if } |x| \le 2 \\ 3 & \text{if } |x| > 2 \end{cases}$

44. $f(x) = \begin{cases} x^2 & \text{if } |x| \le 1 \\ 1 & \text{if } |x| > 1 \end{cases}$

45. $f(x) = \begin{cases} 4 & \text{if } x < -2 \\ x^2 & \text{if } -2 \le x \le 2 \\ -x + 6 & \text{if } x > 2 \end{cases}$

46. $f(x) = \begin{cases} -x & \text{if } x \le 0 \\ 9 - x^2 & \text{if } 0 < x \le 3 \\ x - 3 & \text{if } x > 3 \end{cases}$

47–48 ■ Graphing Piecewise Defined Functions Use a graphing device to draw a graph of the piecewise defined function. (See the margin note on page 162.)

47. $f(x) = \begin{cases} x + 2 & \text{if } x \le -1 \\ x^2 & \text{if } x > -1 \end{cases}$

48. $f(x) = \begin{cases} 2x - x^2 & \text{if } x > 1 \\ (x - 1)^3 & \text{if } x \le 1 \end{cases}$

49–50 ■ Finding Piecewise Defined Functions A graph of a piecewise defined function is given. Find a formula for the function in the indicated form.

49.

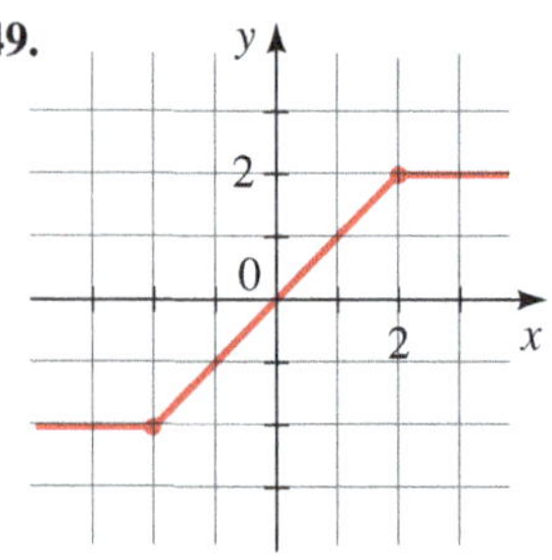

$f(x) = \begin{cases} \_\_\_ & \text{if } x < -2 \\ \_\_\_ & \text{if } -2 \le x \le 2 \\ \_\_\_ & \text{if } x > 2 \end{cases}$

50.

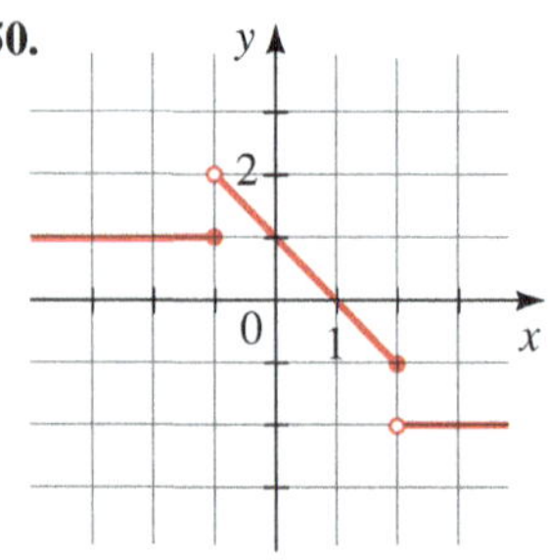

$f(x) = \begin{cases} \_\_\_ & \text{if } x \le -1 \\ \_\_\_ & \text{if } -1 < x \le 2 \\ \_\_\_ & \text{if } x > 2 \end{cases}$

51–52 ■ Vertical Line Test Use the Vertical Line Test to determine whether the curve is a graph of a function of x.

51. (a)

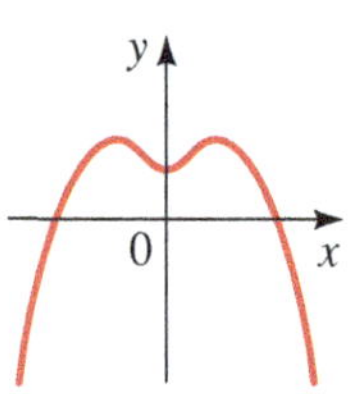

(b)

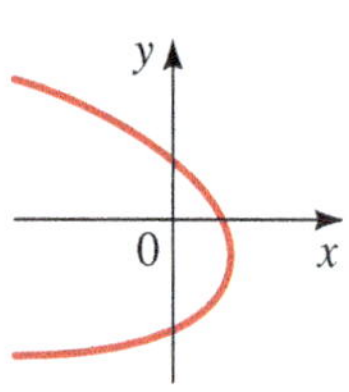

(c)

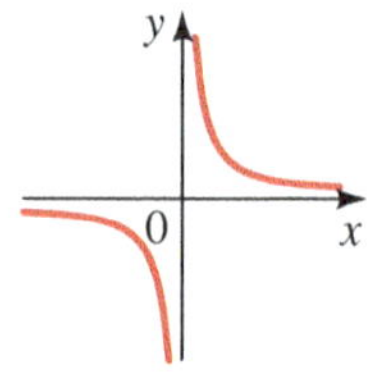

(d)

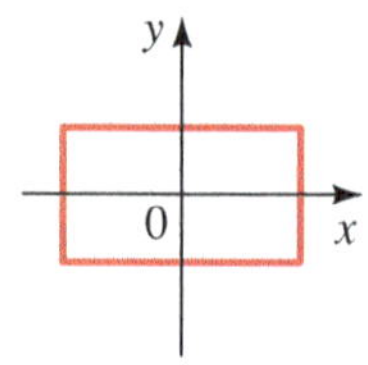

52. (a)

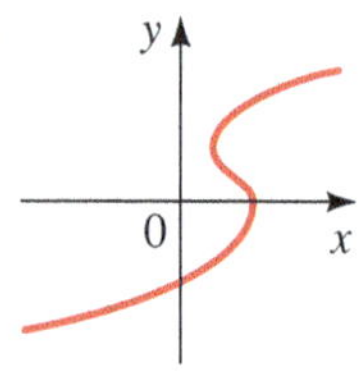

(b)

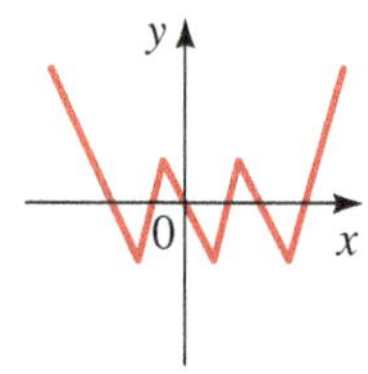

(c)

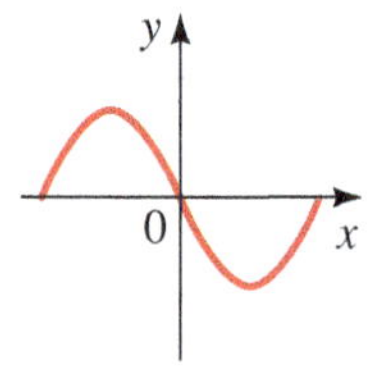

(d)

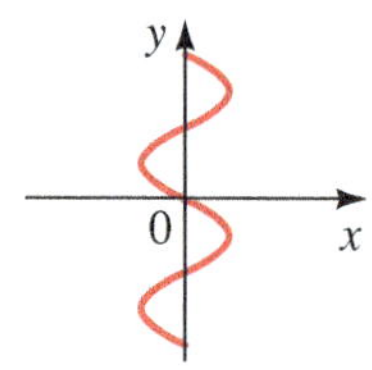

53–56 ■ Vertical Line Test: Domain and Range Use the Vertical Line Test to determine whether the curve is a graph of a function of x. If it is, state the domain and range of the function.

53.

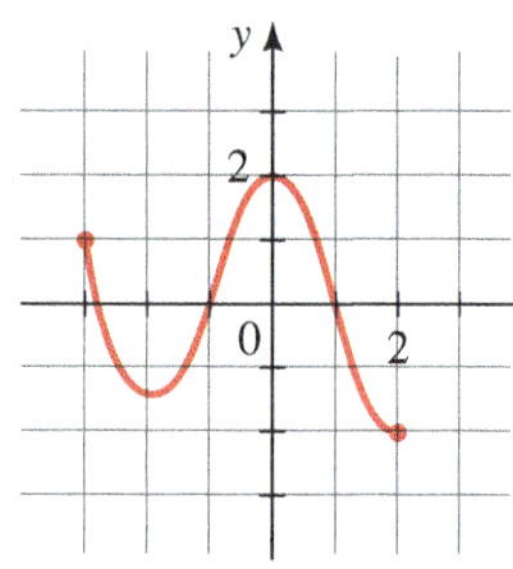

54.

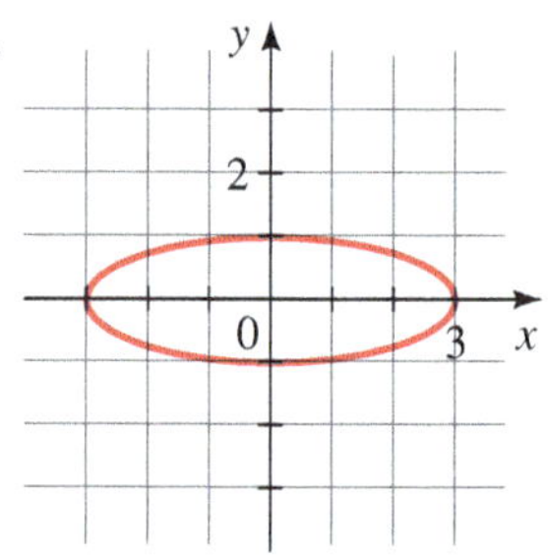

55.

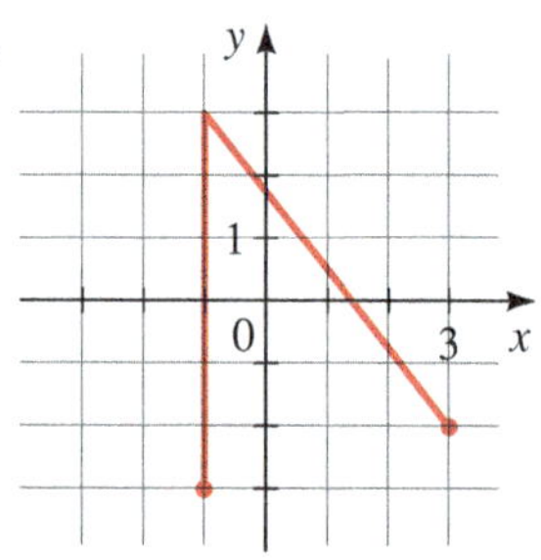

56.

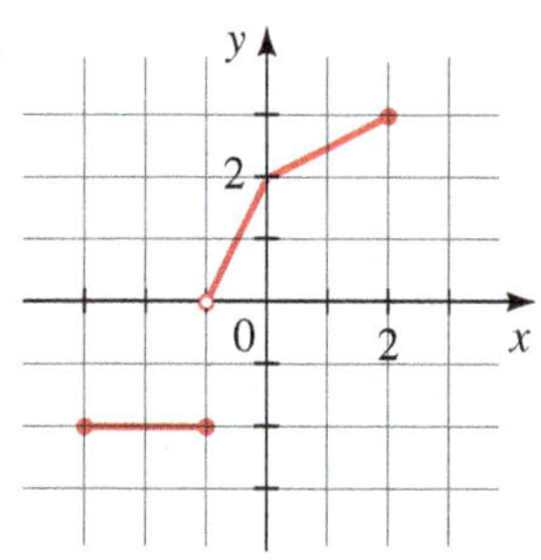

57–68 ■ Equations That Define Functions Determine whether the equation defines y as a function of x. (See Example 9.)

57. $3x - 5y = 7$

58. $3x^2 - y = 5$

59. $x = y^2$

60. $x^2 + (y - 1)^2 = 4$

61. $2x - 4y^2 = 3$

62. $2x^2 - 4y^2 = 3$

63. $2xy - 5y^2 = 4$

64. $\sqrt{y} - x = 5$

65. $2|x| + y = 0$

66. $2x + |y| = 0$

67. $x = y^3$

68. $x = y^4$

69–74 ■ Families of Functions A family of functions is given. In parts (a) and (b) graph all the given members of the family in the viewing rectangle indicated. In part (c) state the conclusions that you can make from your graphs.

69. $f(x) = x^2 + c$

(a) $c = 0, 2, 4, 6;\ [-5, 5]$ by $[-10, 10]$

(b) $c = 0, -2, -4, -6;\ [-5, 5]$ by $[-10, 10]$

(c) How does the value of c affect the graph?

70. $f(x) = (x - c)^2$

(a) $c = 0, 1, 2, 3;\ [-5, 5]$ by $[-10, 10]$

(b) $c = 0, -1, -2, -3;\ [-5, 5]$ by $[-10, 10]$

(c) How does the value of c affect the graph?

71. $f(x) = (x - c)^3$

(a) $c = 0, 2, 4, 6;\ [-10, 10]$ by $[-10, 10]$

(b) $c = 0, -2, -4, -6;\ [-10, 10]$ by $[-10, 10]$

(c) How does the value of c affect the graph?

72. $f(x) = cx^2$

(a) $c = 1, \frac{1}{2}, 2, 4;\ [-5, 5]$ by $[-10, 10]$

(b) $c = 1, -1, -\frac{1}{2}, -2;\ [-5, 5]$ by $[-10, 10]$

(c) How does the value of c affect the graph?

73. $f(x) = x^c$

(a) $c = \frac{1}{2}, \frac{1}{4}, \frac{1}{6};\ [-1, 4]$ by $[-1, 3]$

(b) $c = 1, \frac{1}{3}, \frac{1}{5};\ [-3, 3]$ by $[-2, 2]$

(c) How does the value of c affect the graph?

74. $f(x) = \dfrac{1}{x^n}$

(a) $n = 1, 3;\ [-3, 3]$ by $[-3, 3]$

(b) $n = 2, 4;\ [-3, 3]$ by $[-3, 3]$

(c) How does the value of n affect the graph?

SKILLS Plus

75–78 ■ Finding Functions for Certain Curves Find a function whose graph is the given curve.

75. The line segment joining the points $(-2, 1)$ and $(4, -6)$

76. The line segment joining the points $(-3, -2)$ and $(6, 3)$

77. The top half of the circle $x^2 + y^2 = 9$

78. The bottom half of the circle $x^2 + y^2 = 9$

APPLICATIONS

79. Weather Balloon As a weather balloon is inflated, the thickness T of its rubber skin is related to the radius of the balloon by

$$T(r) = \frac{0.5}{r^2}$$

where T and r are measured in centimeters. Graph the function T for values of r between 10 and 100.

80. Power from a Wind Turbine The power produced by a wind turbine depends on the speed of the wind. If a windmill has blades 3 meters long, then the power P produced by the turbine is modeled by

$$P(v) = 14.1v^3$$

where P is measured in watts (W) and v is measured in meters per second (m/s). Graph the function P for wind speeds between 1 m/s and 10 m/s.

81. Utility Rates Westside Energy charges its electric customers a base rate of \$6.00 per month, plus 10¢ per kilowatt-hour (kWh) for the first 300 kWh used and 6¢ per kWh for all usage over 300 kWh. Suppose a customer uses x kWh of electricity in one month.

(a) Express the monthly cost E as a piecewise defined function of x.

(b) Graph the function E for $0 \le x \le 600$.

82. Taxicab Function A taxi company charges \$2.00 for the first mile (or part of a mile) and 20 cents for each succeeding tenth of a mile (or part). Express the cost C (in dollars) of a ride as a piecewise defined function of the distance x traveled (in miles) for $0 < x < 2$, and sketch a graph of this function.

83. Postage Rates The 2014 domestic postage rate for first-class letters weighing 3.5 oz or less is 49 cents for the first ounce (or less), plus 21 cents for each additional ounce (or part of an ounce). Express the postage P as a piecewise defined function of the weight x of a letter, with $0 < x \le 3.5$, and sketch a graph of this function.

DISCUSS ■ DISCOVER ■ PROVE ■ WRITE

84. DISCOVER: When Does a Graph Represent a Function? For every integer n, the graph of the equation $y = x^n$ is the graph of a function, namely $f(x) = x^n$. Explain why the graph of $x = y^2$ is *not* the graph of a function of x. Is the graph of $x = y^3$ the graph of a function of x? If so, of what function of x is it the graph? Determine for what integers n the graph of $x = y^n$ is a graph of a function of x.

85. DISCUSS: Step Functions In Example 7 and Exercises 82 and 83 we are given functions whose graphs consist of horizontal line segments. Such functions are often called *step functions*, because their graphs look like stairs. Give some other examples of step functions that arise in everyday life.

86. DISCOVER: Stretched Step Functions Sketch graphs of the functions $f(x) = [\![x]\!]$, $g(x) = [\![2x]\!]$, and $h(x) = [\![3x]\!]$ on separate graphs. How are the graphs related? If n is a positive integer, what does a graph of $k(x) = [\![nx]\!]$ look like?

87. DISCOVER: Graph of the Absolute Value of a Function

(a) Draw graphs of the functions

$$f(x) = x^2 + x - 6$$

and $$g(x) = |x^2 + x - 6|$$

How are the graphs of f and g related?

(b) Draw graphs of the functions $f(x) = x^4 - 6x^2$ and $g(x) = |x^4 - 6x^2|$. How are the graphs of f and g related?

(c) In general, if $g(x) = |f(x)|$, how are the graphs of f and g related? Draw graphs to illustrate your answer.

2.3 GETTING INFORMATION FROM THE GRAPH OF A FUNCTION

■ Values of a Function; Domain and Range ■ Comparing Function Values: Solving Equations and Inequalities Graphically ■ Increasing and Decreasing Functions ■ Local Maximum and Minimum Values of a Function

Many properties of a function are more easily obtained from a graph than from the rule that describes the function. We will see in this section how a graph tells us whether the values of a function are increasing or decreasing and also where the maximum and minimum values of a function are.

■ Values of a Function; Domain and Range

A complete graph of a function contains all the information about a function, because the graph tells us which input values correspond to which output values. To analyze the graph of a function, we must keep in mind that *the height of the graph is the value of the function*. So we can read off the values of a function from its graph.

EXAMPLE 1 ■ Finding the Values of a Function from a Graph

The function T graphed in Figure 1 gives the temperature between noon and 6:00 P.M. at a certain weather station.

(a) Find $T(1)$, $T(3)$, and $T(5)$.

(b) Which is larger, $T(2)$ or $T(4)$?

(c) Find the value(s) of x for which $T(x) = 25$.

(d) Find the value(s) of x for which $T(x) \geq 25$.

(e) Find the net change in temperature from 1 P.M. to 3 P.M.

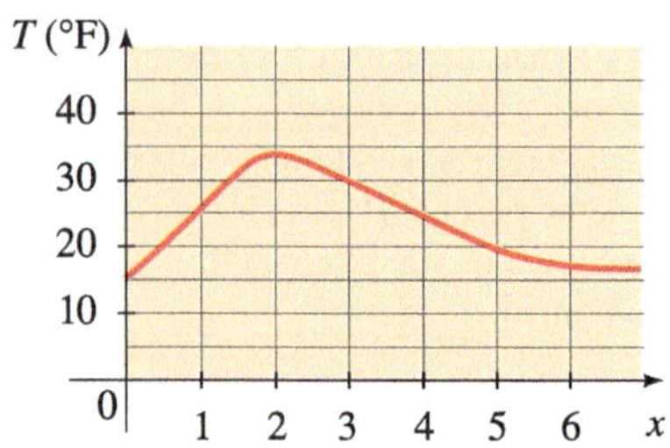

FIGURE 1 Temperature function

SOLUTION

(a) $T(1)$ is the temperature at 1:00 P.M. It is represented by the height of the graph above the x-axis at $x = 1$. Thus $T(1) = 25$. Similarly, $T(3) = 30$ and $T(5) = 20$.

(b) Since the graph is higher at $x = 2$ than at $x = 4$, it follows that $T(2)$ is larger than $T(4)$.

(c) The height of the graph is 25 when x is 1 and when x is 4. In other words, the temperature is 25 at 1:00 P.M. and 4:00 P.M.

(d) The graph is higher than 25 for x between 1 and 4. In other words, the temperature was 25 or greater between 1:00 P.M. and 4:00 P.M.

Net change is defined on page 151.

(e) The net change in temperature is

$$T(3) - T(1) = 30 - 25 = 5$$

So there was a net increase of 5°F from 1 P.M. to 3 P.M.

Now Try Exercises 7 and 55 ■

The graph of a function helps us to picture the domain and range of the function on the x-axis and y-axis, as shown in the box below.

DOMAIN AND RANGE FROM A GRAPH

The **domain** and **range** of a function $y = f(x)$ can be obtained from a graph of f as shown in the figure. The domain is the set of all x-values for which f is defined, and the range is all the corresponding y-values.

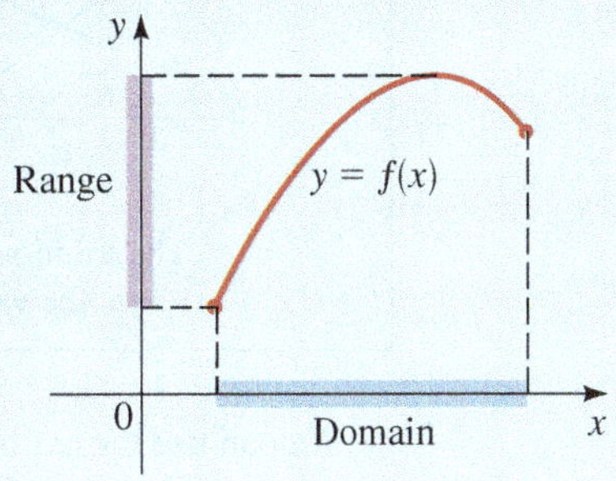

EXAMPLE 2 ■ Finding the Domain and Range from a Graph

See Appendix C, *Graphing with a Graphing Calculator*, for guidelines on using a graphing calculator. See Appendix D, *Using the TI-83/84 Graphing Calculator*, for specific graphing instructions. Go to **www.stewartmath.com**.

(a) Use a graphing calculator to draw the graph of $f(x) = \sqrt{4 - x^2}$.

(b) Find the domain and range of f.

SOLUTION

(a) The graph is shown in Figure 2.

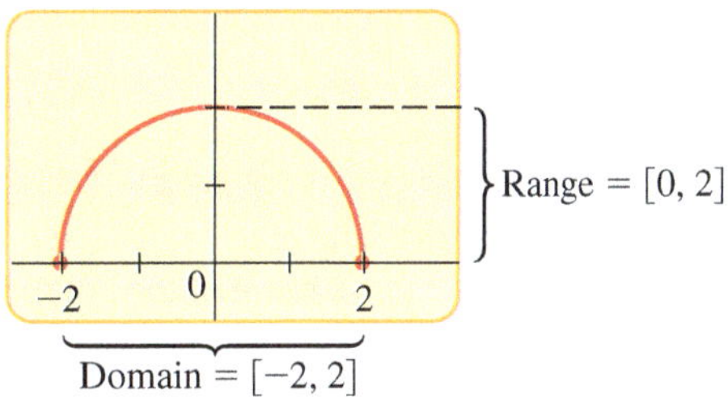

FIGURE 2 Graph of $f(x) = \sqrt{4 - x^2}$

(b) From the graph in Figure 2 we see that the domain is $[-2, 2]$ and the range is $[0, 2]$.

Now Try Exercise 21 ■

■ Comparing Function Values: Solving Equations and Inequalities Graphically

We can compare the values of two functions f and g visually by drawing their graphs. The points at which the graphs intersect are the points where the values of the two functions are equal. So the solutions of the equation $f(x) = g(x)$ are the values of x at which the two graphs intersect. The points at which the graph of g is higher than the graph of f are the points where the values of g are greater than the values of f. So the solutions of the inequality $f(x) < g(x)$ are the values of x at which the graph of g is *higher than* the graph of f.

SOLVING EQUATIONS AND INEQUALITIES GRAPHICALLY

The **solution(s) of the equation** $f(x) = g(x)$ are the values of x where the graphs of f and g intersect.

The **solution(s) of the inequality** $f(x) < g(x)$ are the values of x where the graph of g is higher than the graph of f.

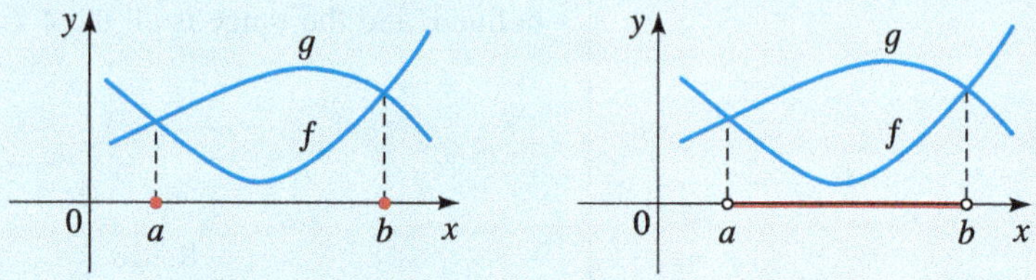

The solutions of $f(x) = g(x)$ are the values a and b.

The solution of $f(x) < g(x)$ is the interval (a, b).

We can use these observations to solve equations and inequalities graphically, as the next example illustrates.

EXAMPLE 3 ■ Solving Graphically

Solve the given equation or inequality graphically.

(a) $2x^2 + 3 = 5x + 6$

(b) $2x^2 + 3 \le 5x + 6$

(c) $2x^2 + 3 > 5x + 6$

You can also solve the equations and inequalities algebraically. Check that your solutions match the solutions we obtained graphically.

SOLUTION We first define functions f and g that correspond to the left-hand side and to the right-hand side of the equation or inequality. So we define

$$f(x) = 2x^2 + 3 \qquad \text{and} \qquad g(x) = 5x + 6$$

Next, we sketch graphs of f and g on the same set of axes.

(a) The given equation is equivalent to $f(x) = g(x)$. From the graph in Figure 3(a) we see that the solutions of the equation are $x = -0.5$ and $x = 3$.

(b) The given inequality is equivalent to $f(x) \le g(x)$. From the graph in Figure 3(b) we see that the solution is the interval $[-0.5, 3]$.

(c) The given inequality is equivalent to $f(x) > g(x)$. From the graph in Figure 3(c) we see that the solution is $(-\infty, -0.5) \cup (3, \infty)$.

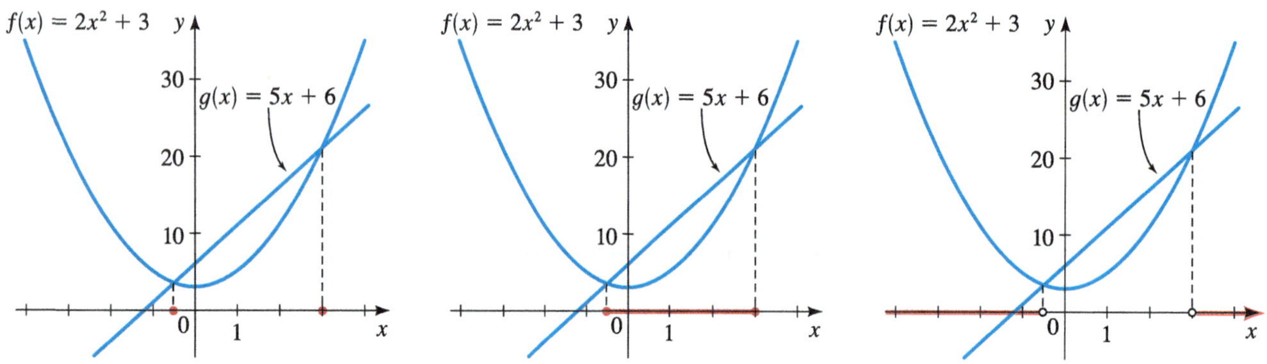

(a) Solution: $x = -0.5, 3$ (b) Solution: $[-0.5, 3]$ (c) Solution: $(-\infty, -0.5) \cup (3, \infty)$

FIGURE 3 Graphs of $f(x) = 2x^2 + 3$ and $g(x) = 5x + 6$

Now Try Exercises 9 and 23 ■

To solve an equation graphically, we can first move all terms to one side of the equation and then graph the function that corresponds to the nonzero side of the equation. In this case the solutions of the equation are the x-intercepts of the graph. We can use this same method to solve inequalities graphically, as the following example shows.

EXAMPLE 4 ■ Solving Graphically

Solve the given equation or inequality graphically.

(a) $x^3 + 6 = 2x^2 + 5x$

(b) $x^3 + 6 \geq 2x^2 + 5x$

SOLUTION We first move all terms to one side to obtain an equivalent equation (or inequality). For the equation in part (a) we obtain

$$x^3 - 2x^2 - 5x + 6 = 0 \qquad \text{Move terms to LHS}$$

Then we define a function f by

$$f(x) = x^3 - 2x^2 - 5x + 6 \qquad \text{Define } f$$

Next, we use a graphing calculator to graph f, as shown in Figure 4.

(a) The given equation is the same as $f(x) = 0$, so the solutions are the x-intercepts of the graph. From Figure 4(a) we see that the solutions are $x = -2$, $x = 1$, and $x = 3$.

(b) The given inequality is the same as $f(x) \geq 0$, so the solutions are the x-values at which the graph of f is on or above the x-axis. From Figure 4(b) we see the solution is $[-2, 1] \cup [3, \infty]$.

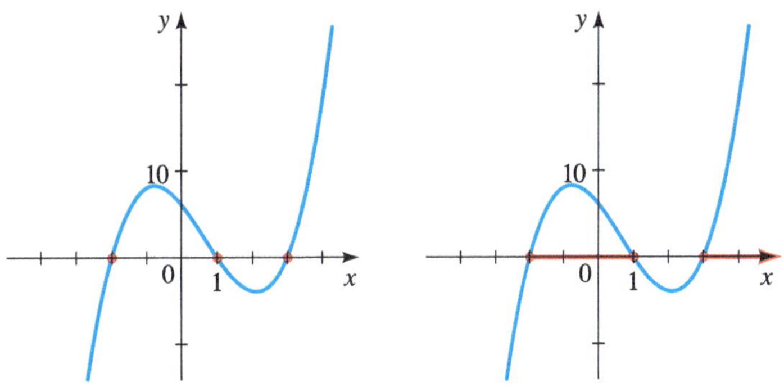

(a) Solution: $x = -2, 1, 3$

(b) Solution: $[-2, 1] \cup [3, \infty]$

FIGURE 4 Graphs of $f(x) = x^3 - 2x^2 - 5x + 6$

Now Try Exercise 27 ■

■ Increasing and Decreasing Functions

It is very useful to know where the graph of a function rises and where it falls. The graph shown in Figure 5 rises, falls, then rises again as we move from left to right: It rises from A to B, falls from B to C, and rises again from C to D. The function f is said to be *increasing* when its graph rises and *decreasing* when its graph falls.

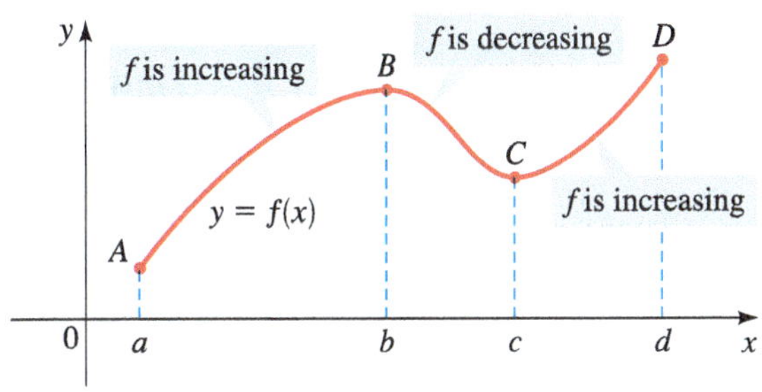

FIGURE 5 f is increasing on (a, b) and (c, d); f is decreasing on (b, c)

We have the following definition.

From the definition we see that a function increases or decreases *on an interval*. It does not make sense to apply these definitions at a single point.

DEFINITION OF INCREASING AND DECREASING FUNCTIONS

f is **increasing** on an interval I if $f(x_1) < f(x_2)$ whenever $x_1 < x_2$ in I.
f is **decreasing** on an interval I if $f(x_1) > f(x_2)$ whenever $x_1 < x_2$ in I.

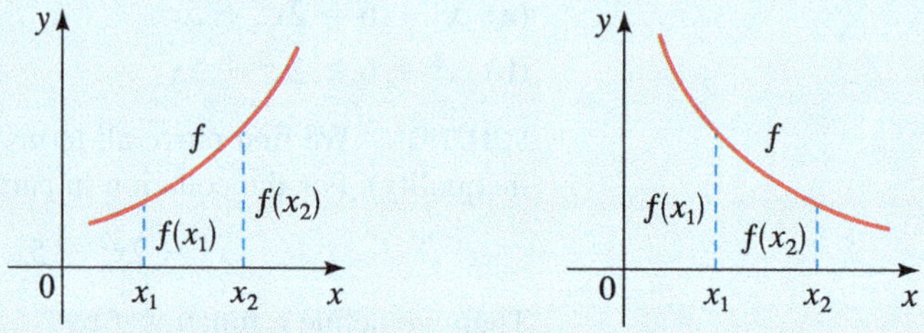

f is increasing $\qquad$ f is decreasing

EXAMPLE 5 ■ Intervals on Which a Function Increases or Decreases

The graph in Figure 6 gives the weight W of a person at age x. Determine the intervals on which the function W is increasing and on which it is decreasing.

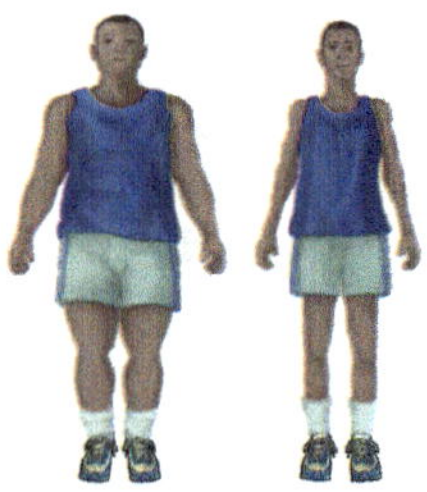

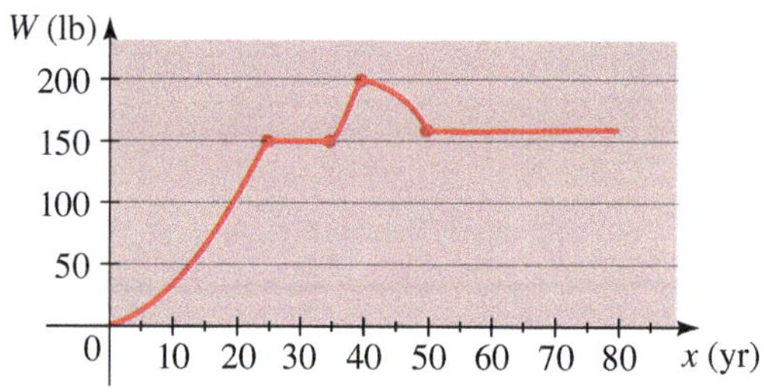

FIGURE 6 Weight as a function of age

SOLUTION The function W is increasing on $(0, 25)$ and $(35, 40)$. It is decreasing on $(40, 50)$. The function W is constant (neither increasing nor decreasing) on $(25, 35)$ and $(50, 80)$. This means that the person gained weight until age 25, then gained weight again between ages 35 and 40. He lost weight between ages 40 and 50.

Now Try Exercise 57 ■

By convention we write the intervals on which a function is increasing or decreasing as open intervals. (It would also be true to say that the function is increasing or decreasing on the corresponding closed interval. So for instance, it is also correct to say that the function W in Example 5 is decreasing on $[40, 50]$.)

EXAMPLE 6 ■ Finding Intervals on Which a Function Increases or Decreases

(a) Sketch a graph of the function $f(x) = 12x^2 + 4x^3 - 3x^4$.

(b) Find the domain and range of f.

(c) Find the intervals on which f is increasing and on which f is decreasing.

SOLUTION

(a) We use a graphing calculator to sketch the graph in Figure 7.

(b) The domain of f is $\mathbb{R}$ because f is defined for all real numbers. Using the TRACE feature on the calculator, we find that the highest value is $f(2) = 32$. So the range of f is $(-\infty, 32]$.

(c) From the graph we see that f is increasing on the intervals $(-\infty, -1)$ and $(0, 2)$ and is decreasing on $(-1, 0)$ and $(2, \infty)$.

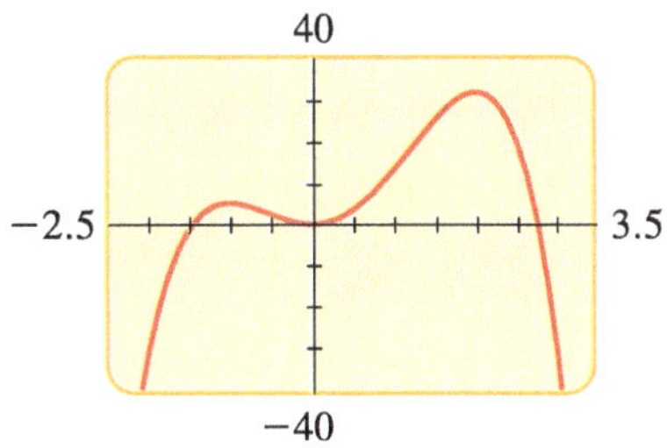

FIGURE 7 Graph of $f(x) = 12x^2 + 4x^3 - 3x^4$

Now Try Exercise 35

EXAMPLE 7 ■ Finding Intervals Where a Function Increases and Decreases

(a) Sketch the graph of the function $f(x) = x^{2/3}$.

(b) Find the domain and range of the function.

(c) Find the intervals on which f is increasing and on which f is decreasing.

SOLUTION

(a) We use a graphing calculator to sketch the graph in Figure 8.

(b) From the graph we observe that the domain of f is $\mathbb{R}$ and the range is $[0, \infty)$.

(c) From the graph we see that f is decreasing on $(-\infty, 0)$ and increasing on $(0, \infty)$.

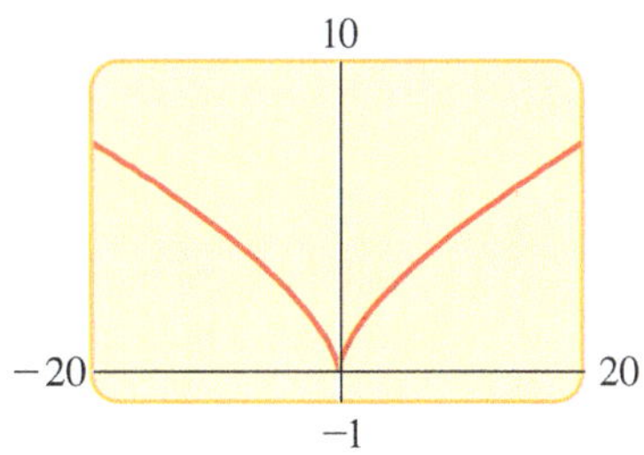

FIGURE 8 Graph of $f(x) = x^{2/3}$

Now Try Exercise 41

■ Local Maximum and Minimum Values of a Function

Finding the largest or smallest values of a function is important in many applications. For example, if a function represents revenue or profit, then we are interested in its maximum value. For a function that represents cost, we would want to find its minimum value. (See *Focus on Modeling: Modeling with Functions* on pages 237–244 for many such examples.) We can easily find these values from the graph of a function. We first define what we mean by a local maximum or minimum.

LOCAL MAXIMA AND MINIMA OF A FUNCTION

1. The function value $f(a)$ is a **local maximum value** of f if

$$f(a) \geq f(x) \quad \text{when } x \text{ is near } a$$

(This means that $f(a) \geq f(x)$ for all x in some open interval containing a.) In this case we say that f has a **local maximum** at $x = a$.

2. The function value $f(a)$ is a **local minimum value** of f if

$$f(a) \leq f(x) \quad \text{when } x \text{ is near } a$$

(This means that $f(a) \leq f(x)$ for all x in some open interval containing a.) In this case we say that f has a **local minimum** at $x = a$.

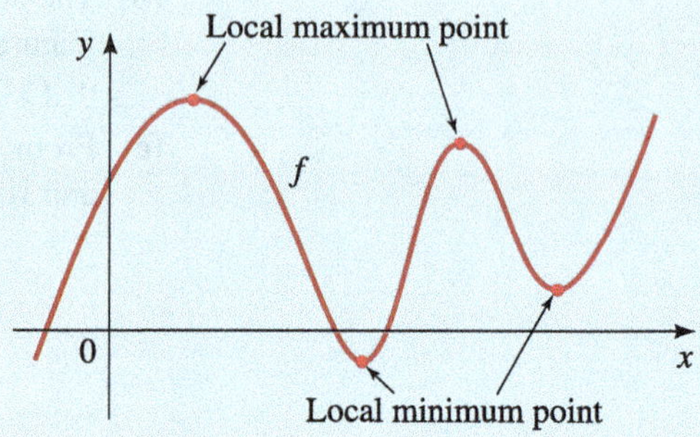

We can find the local maximum and minimum values of a function using a graphing calculator. If there is a viewing rectangle such that the point $(a, f(a))$ is the highest point on the graph of f *within* the viewing rectangle (not on the edge), then the number $f(a)$ is a local maximum value of f (see Figure 9). Notice that $f(a) \geq f(x)$ for all numbers x that are close to a.

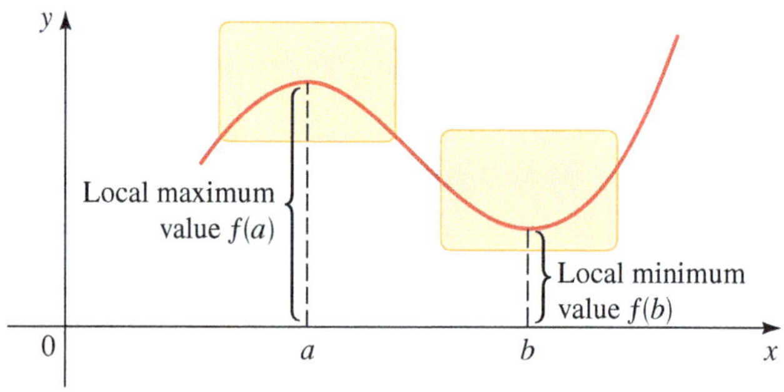

FIGURE 9

Similarly, if there is a viewing rectangle such that the point $(b, f(b))$ is the lowest point on the graph of f within the viewing rectangle, then the number $f(b)$ is a local minimum value of f. In this case $f(b) \leq f(x)$ for all numbers x that are close to b.

EXAMPLE 8 ■ Finding Local Maxima and Minima from a Graph

Find the local maximum and minimum values of the function $f(x) = x^3 - 8x + 1$, rounded to three decimal places.

SOLUTION The graph of f is shown in Figure 10. There appears to be one local maximum between $x = -2$ and $x = -1$, and one local minimum between $x = 1$ and $x = 2$.

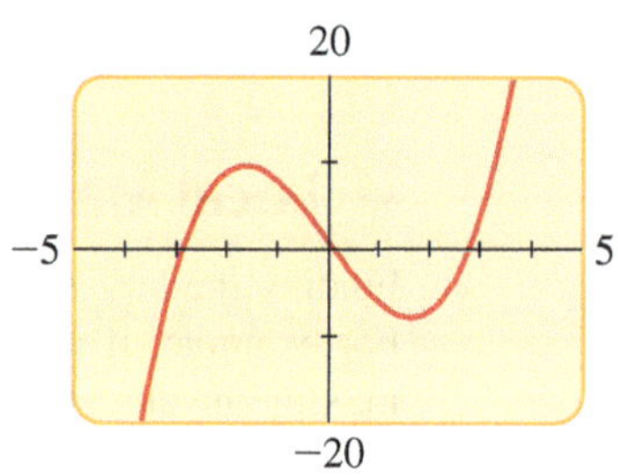

FIGURE 10 Graph of $f(x) = x^3 - 8x + 1$

Let's find the coordinates of the local maximum point first. We zoom in to enlarge the area near this point, as shown in Figure 11. Using the TRACE feature on the graphing device, we move the cursor along the curve and observe how the y-coordinates change. The local maximum value of y is 9.709, and this value occurs when x is -1.633, correct to three decimal places.

We locate the minimum value in a similar fashion. By zooming in to the viewing rectangle shown in Figure 12, we find that the local minimum value is about -7.709, and this value occurs when $x \approx 1.633$.

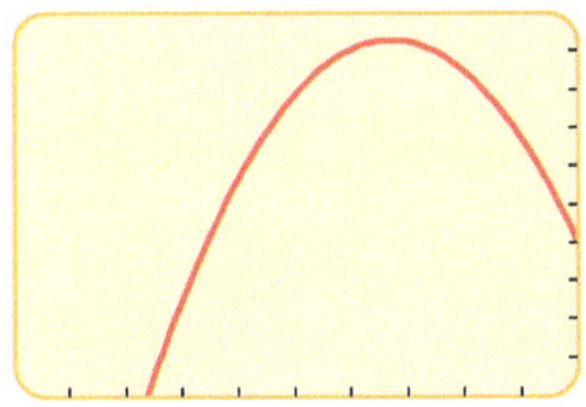

FIGURE 11

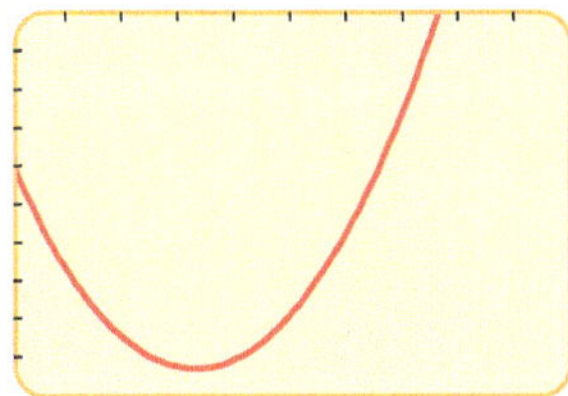

FIGURE 12

Now Try Exercise 47 ■

The `maximum` and `minimum` commands on a TI-83 or TI-84 calculator provide another method for finding extreme values of functions. We use this method in the next example.

EXAMPLE 9 ■ A Model for Managing Traffic

See the *Discovery Project* referenced in Chapter 3, on page 295, for how this model is obtained.

A highway engineer develops a formula to estimate the number of cars that can safely travel a particular highway at a given speed. She assumes that each car is 17 ft long, travels at a speed of x mi/h, and follows the car in front of it at the safe following distance for that speed. She finds that the number N of cars that can pass a given point per minute is modeled by the function

$$N(x) = \frac{88x}{17 + 17\left(\dfrac{x}{20}\right)^2}$$

Graph the function in the viewing rectangle $[0, 100]$ by $[0, 60]$.

(a) Find the intervals on which the function N is increasing and on which it is decreasing.

(b) Find the maximum value of N. What is the maximum carrying capacity of the road, and at what speed is it achieved?

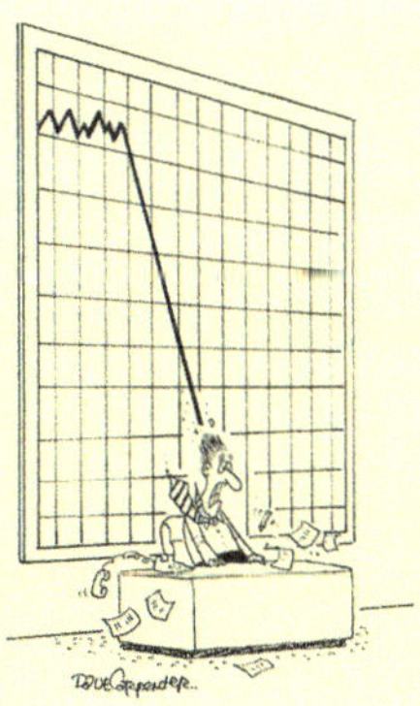

Dave Carpenter/www.CartoonStock.com

DISCOVERY PROJECT

Every Graph Tells a Story

A graph can often describe a real-world "story" much more quickly and effectively than many words. For example, the stock market crash of 1929 is effectively described by a graph of the Dow Jones Industrial Average. No words are needed to convey the message in the cartoon shown here. In this project we describe, or tell the story that corresponds to, a given graph as well as make graphs that correspond to a real-world "story." You can find the project at **www.stewartmath.com**.

SOLUTION The graph is shown in Figure 13(a).

(a) From the graph we see that the function N is increasing on $(0, 20)$ and decreasing on $(20, \infty)$.

(b) There appears to be a maximum between $x = 19$ and $x = 21$. Using the `maximum` command, as shown in Figure 13(b), we see that the maximum value of N is about 51.78, and it occurs when x is 20. So the maximum carrying capacity is about 52 cars per minute at a speed of 20 mi/h.

See Appendix D, *Using the TI-83/84 Graphing Calculator,* for specific instructions on using the `maximum` command. Go to **www.stewartmath.com.**

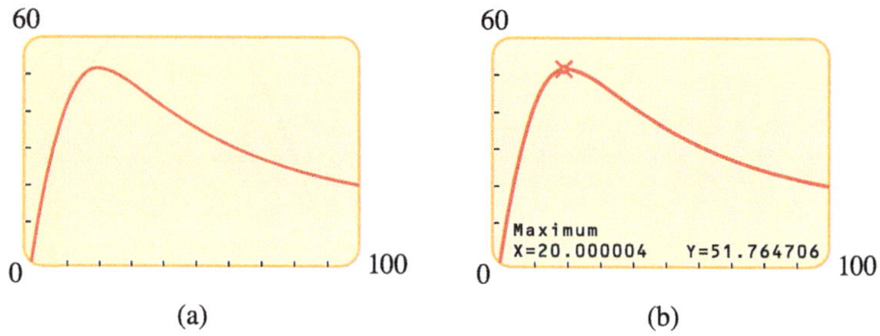

FIGURE 13 Highway capacity at speed x

Now Try Exercise 65 ■

2.3 EXERCISES

CONCEPTS

1–5 ■ The function f graphed below is defined by a polynomial expression of degree 4. Use the graph to solve the exercises.

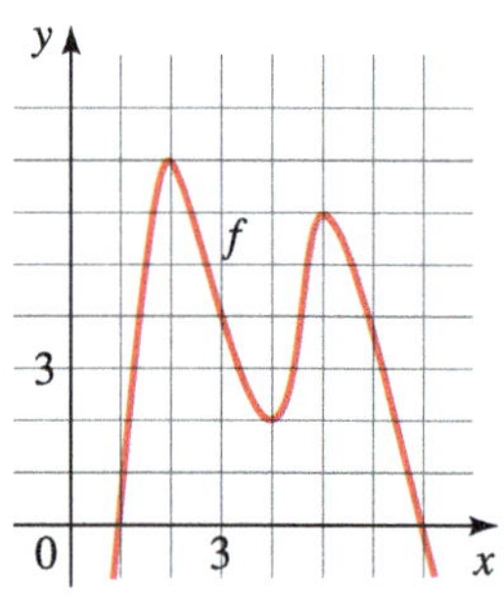

1. To find a function value $f(a)$ from the graph of f, we find the height of the graph above the x-axis at $x =$ ________. From the graph of f we see that $f(3) =$ ________ and $f(1) =$ ____. The net change in f between $x = 1$ and $x = 3$ is $f($____$) - f($____$) =$ ____.

2. The domain of the function f is all the ____-values of the points on the graph, and the range is all the corresponding ____-values. From the graph of f we see that the domain of f is the interval ________ and the range of f is the interval ________.

3. (a) If f is increasing on an interval, then the y-values of the points on the graph ________ as the x-values increase. From the graph of f we see that f is increasing on the intervals ________ and ________.

(b) If f is decreasing on an interval, then the y-values of the points on the graph ________ as the x-values increase. From the graph of f we see that f is decreasing on the intervals ________ and ________.

4. (a) A function value $f(a)$ is a local maximum value of f if $f(a)$ is the ________ value of f on some open interval containing a. From the graph of f we see that there are two local maximum values of f: One local maximum is ________, and it occurs when $x = 2$; the other local maximum is ________, and it occurs when $x =$ ________.

(b) The function value $f(a)$ is a local minimum value of f if $f(a)$ is the ________ value of f on some open interval containing a. From the graph of f we see that there is one local minimum value of f. The local minimum value is ________, and it occurs when $x =$ ________.

5. The solutions of the equation $f(x) = 0$ are the ____-intercepts of the graph of f. The solution of the inequality $f(x) \geq 0$ is the set of x-values at which the graph of f is on or above the ____-axis. From the graph of f we find that the solutions of the equation $f(x) = 0$ are $x =$ ____ and $x =$ ____, and the solution of the inequality $f(x) \geq 0$ is ______________.

6. (a) To solve the equation $2x + 1 = -x + 4$ graphically, we graph the functions $f(x) =$ ______________ and $g(x) =$ ______________ on the same set of axes and determine the values of x at which the graphs of f and g intersect. Graph f and g below, and use the graphs to solve the equation. The solution is $x =$ ________.

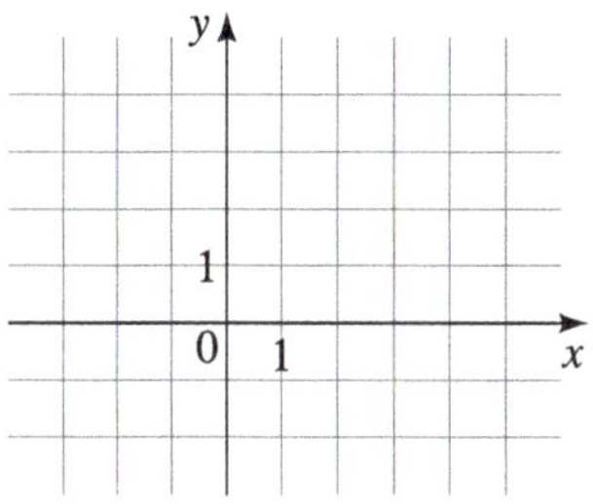

(b) To solve the inequality $2x + 1 < -x + 4$ graphically, we graph the functions $f(x) =$ ______________ and $g(x) =$ ______________ on the same set of axes and find the values of x at which the graph of g is ______________ (higher/lower) than the graph of f. From the graphs in part (a) we see that the solution of the inequality is the interval (____, ____).

SKILLS

7. Values of a Function The graph of a function h is given.

(a) Find $h(-2)$, $h(0)$, $h(2)$, and $h(3)$.
(b) Find the domain and range of h.
(c) Find the values of x for which $h(x) = 3$.
(d) Find the values of x for which $h(x) \leq 3$.
(e) Find the net change in h between $x = -3$ and $x = 3$.

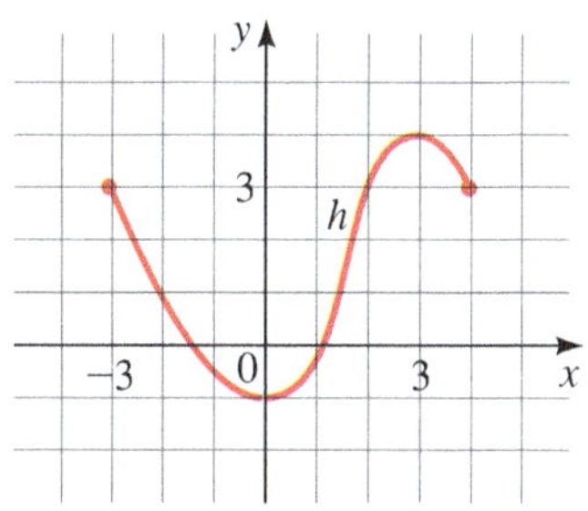

8. Values of a Function The graph of a function g is given.

(a) Find $g(-4)$, $g(-2)$, $g(0)$, $g(2)$, and $g(4)$.
(b) Find the domain and range of g.
(c) Find the values of x for which $g(x) = 3$.
(d) Estimate the values of x for which $g(x) \leq 0$.
(e) Find the net change in g between $x = -1$ and $x = 2$.

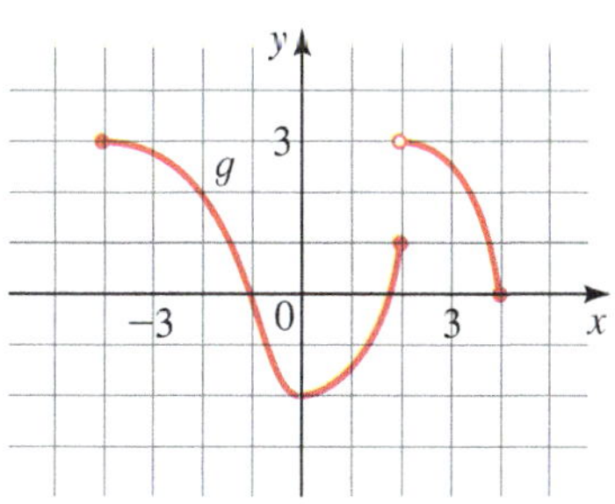

9. Solving Equations and Inequalities Graphically Graphs of the functions f and g are given.

(a) Which is larger, $f(0)$ or $g(0)$?
(b) Which is larger, $f(-3)$ or $g(-3)$?
(c) For which values of x is $f(x) = g(x)$?
(d) Find the values of x for which $f(x) \leq g(x)$.
(e) Find the values of x for which $f(x) > g(x)$.

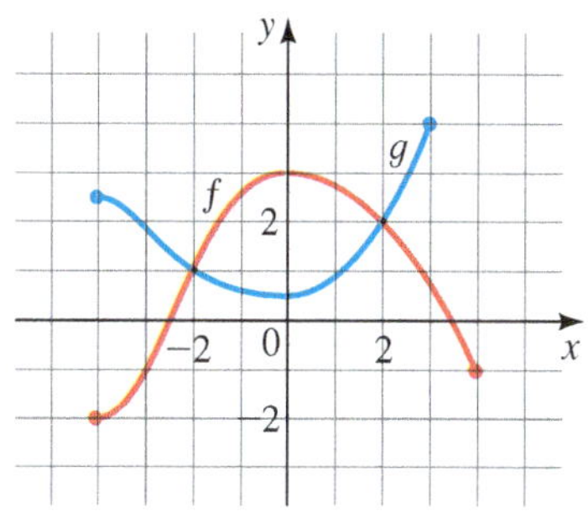

10. Solving Equations and Inequalities Graphically Graphs of the functions f and g are given.

(a) Which is larger, $f(6)$ or $g(6)$?
(b) Which is larger, $f(3)$ or $g(3)$?
(c) Find the values of x for which $f(x) = g(x)$.
(d) Find the values of x for which $f(x) \leq g(x)$.
(e) Find the values of x for which $f(x) > g(x)$.

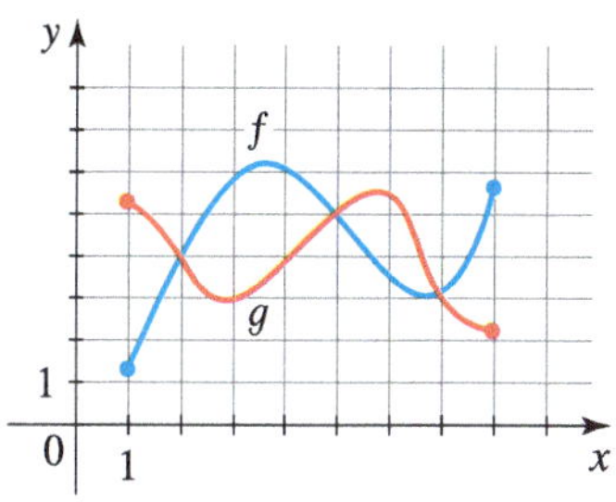

11–16 ■ Domain and Range from a Graph A function f is given. **(a)** Sketch a graph of f. **(b)** Use the graph to find the domain and range of f.

11. $f(x) = 2x + 3$
12. $f(x) = 3x - 2$
13. $f(x) = x - 2, \quad -2 \leq x \leq 5$
14. $f(x) = 4 - 2x, \quad 1 < x < 4$
15. $f(x) = x^2 - 1, \quad -3 \leq x \leq 3$
16. $f(x) = 3 - x^2, \quad -3 \leq x \leq 3$

17–22 ■ Finding Domain and Range Graphically A function f is given. **(a)** Use a graphing calculator to draw the graph of f. **(b)** Find the domain and range of f from the graph.

17. $f(x) = x^2 + 4x + 3$
18. $f(x) = -x^2 + 2x + 1$
19. $f(x) = \sqrt{x - 1}$
20. $f(x) = \sqrt{x + 2}$
21. $f(x) = \sqrt{16 - x^2}$
22. $f(x) = -\sqrt{25 - x^2}$

23–26 ■ Solving Equations and Inequalities Graphically Solve the given equation or inequality graphically.

23. **(a)** $x - 2 = 4 - x$ **(b)** $x - 2 > 4 - x$

24. **(a)** $-2x + 3 = 3x - 7$ **(b)** $-2x + 3 \leq 3x - 7$

25. **(a)** $x^2 = 2 - x$ **(b)** $x^2 \leq 2 - x$

26. **(a)** $-x^2 = 3 - 4x$ **(b)** $-x^2 \geq 3 - 4x$

27–30 ■ Solving Equations and Inequalities Graphically Solve the given equation or inequality graphically. State your answers rounded to two decimals.

27. **(a)** $x^3 + 3x^2 = -x^2 + 3x + 7$
(b) $x^3 + 3x^2 \geq -x^2 + 3x + 7$

28. **(a)** $5x^2 - x^3 = -x^2 + 3x + 4$
(b) $5x^2 - x^3 \leq -x^2 + 3x + 4$

29. **(a)** $16x^3 + 16x^2 = x + 1$
(b) $16x^3 + 16x^2 \geq x + 1$

30. **(a)** $1 + \sqrt{x} = \sqrt{x^2 + 1}$
(b) $1 + \sqrt{x} > \sqrt{x^2 + 1}$

31–34 ■ Increasing and Decreasing The graph of a function f is given. Use the graph to estimate the following. **(a)** The domain and range of f. **(b)** The intervals on which f is increasing and on which f is decreasing.

31.

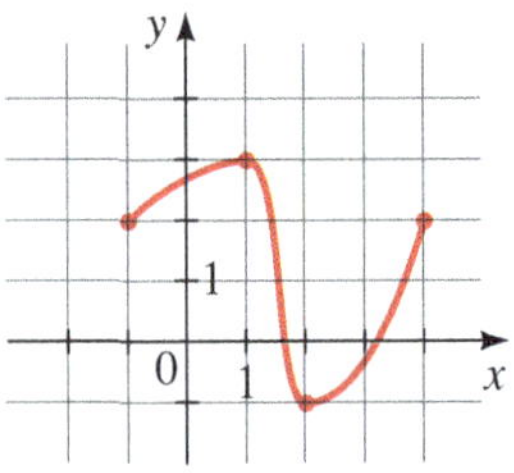

32.

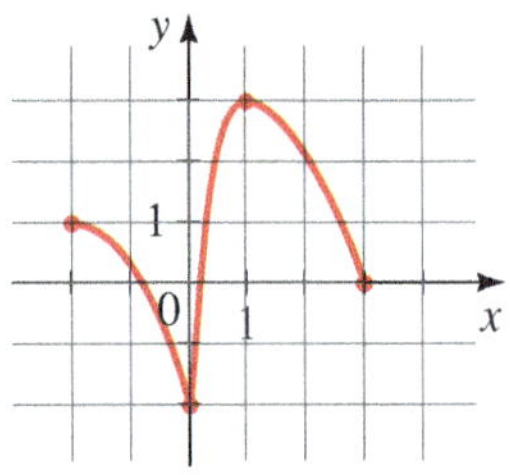

33.

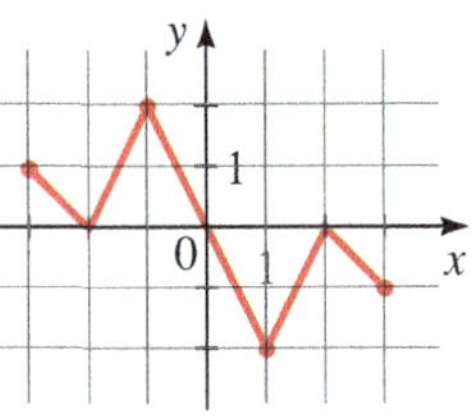

34.

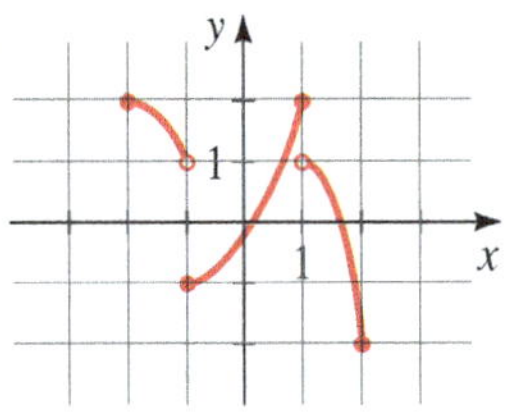

35–42 ■ Increasing and Decreasing A function f is given. **(a)** Use a graphing calculator to draw the graph of f. **(b)** Find the domain and range of f. **(c)** State approximately the intervals on which f is increasing and on which f is decreasing.

35. $f(x) = x^2 - 5x$

36. $f(x) = x^3 - 4x$

37. $f(x) = 2x^3 - 3x^2 - 12x$

38. $f(x) = x^4 - 16x^2$

39. $f(x) = x^3 + 2x^2 - x - 2$

40. $f(x) = x^4 - 4x^3 + 2x^2 + 4x - 3$

41. $f(x) = x^{2/5}$

42. $f(x) = 4 - x^{2/3}$

43–46 ■ Local Maximum and Minimum Values The graph of a function f is given. Use the graph to estimate the following. **(a)** All the local maximum and minimum values of the function and the value of x at which each occurs. **(b)** The intervals on which the function is increasing and on which the function is decreasing.

43.

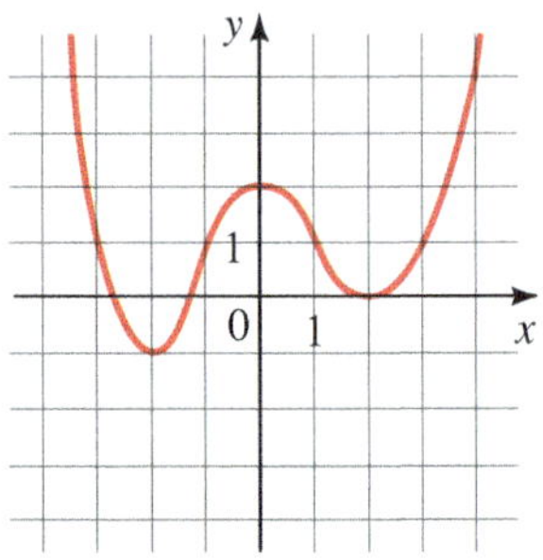

44.

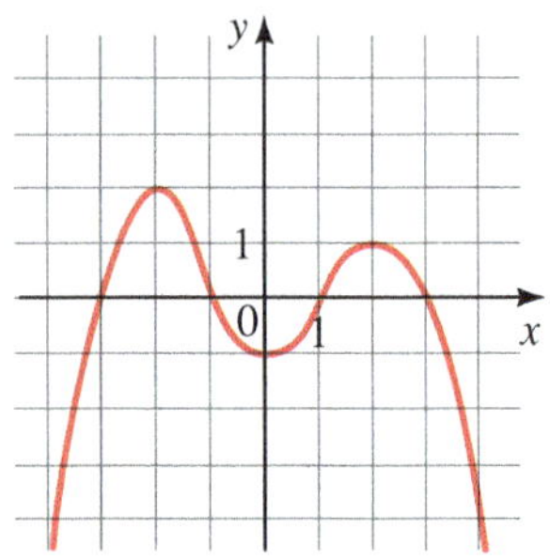

45.

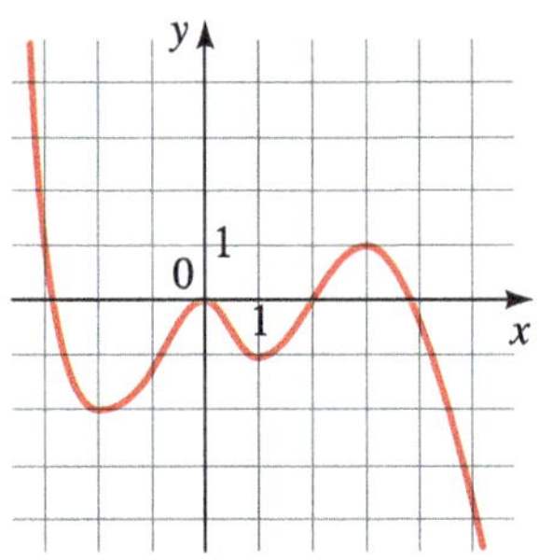

46.

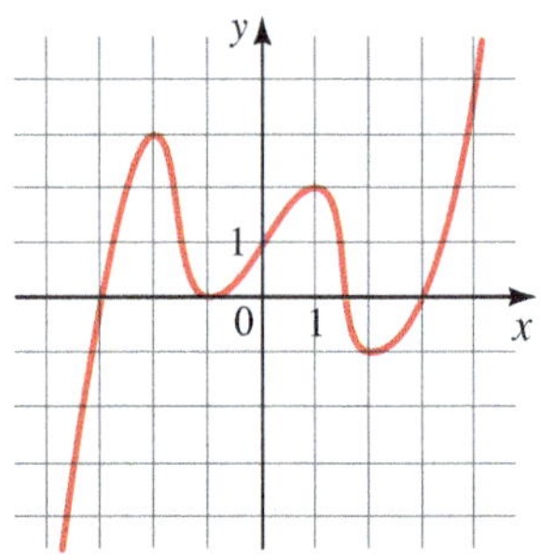

47–54 ■ Local Maximum and Minimum Values A function is given. **(a)** Find all the local maximum and minimum values of the function and the value of x at which each occurs. State each answer rounded to two decimal places. **(b)** Find the intervals on which the function is increasing and on which the function is decreasing. State each answer rounded to two decimal places.

47. $f(x) = x^3 - x$

48. $f(x) = 3 + x + x^2 - x^3$

49. $g(x) = x^4 - 2x^3 - 11x^2$

50. $g(x) = x^5 - 8x^3 + 20x$

51. $U(x) = x\sqrt{6 - x}$

52. $U(x) = x\sqrt{x - x^2}$

53. $V(x) = \dfrac{1 - x^2}{x^3}$

54. $V(x) = \dfrac{1}{x^2 + x + 1}$

APPLICATIONS

55. Power Consumption The figure shows the power consumption in San Francisco for a day in September (P is measured in megawatts; t is measured in hours starting at midnight).

(a) What was the power consumption at 6:00 A.M.? At 6:00 P.M.?

(b) When was the power consumption the lowest?

(c) When was the power consumption the highest?

(d) Find the net change in the power consumption from 9:00 A.M. to 7:00 P.M.

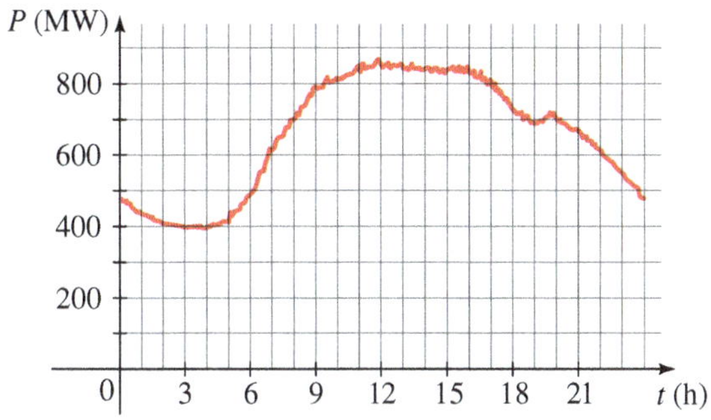

Source: Pacific Gas & Electric

56. Earthquake The graph shows the vertical acceleration of the ground from the 1994 Northridge earthquake in Los Angeles, as measured by a seismograph. (Here t represents the time in seconds.)

(a) At what time t did the earthquake first make noticeable movements of the earth?

(b) At what time t did the earthquake seem to end?

(c) At what time t was the maximum intensity of the earthquake reached?

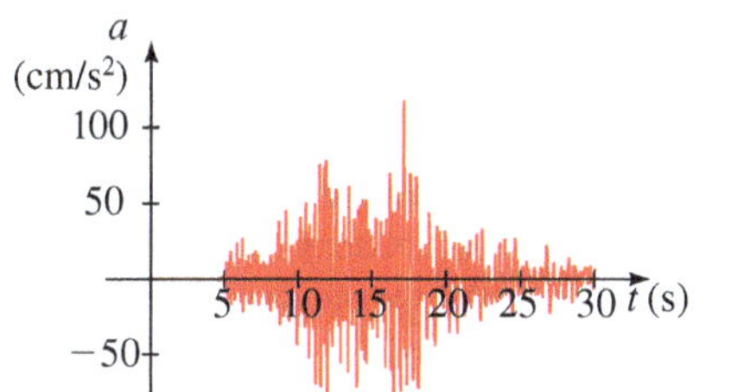

Source: California Department of Mines and Geology

57. Weight Function The graph gives the weight W of a person at age x.

(a) Determine the intervals on which the function W is increasing and those on which it is decreasing.

(b) What do you think happened when this person was 30 years old?

(c) Find the net change in the person's weight W from age 10 to age 20.

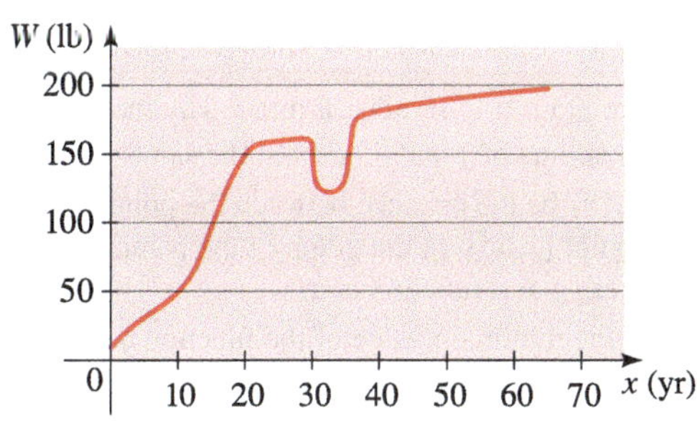

58. Distance Function The graph gives a sales representative's distance from his home as a function of time on a certain day.

(a) Determine the time intervals on which his distance from home was increasing and those on which it was decreasing.

(b) Describe in words what the graph indicates about his travels on this day.

(c) Find the net change in his distance from home between noon and 1:00 P.M.

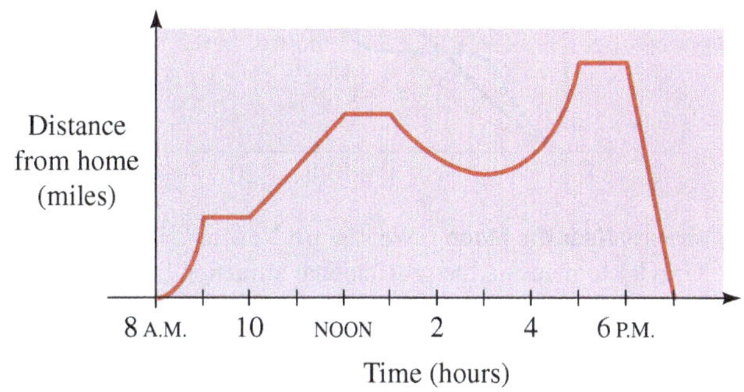

59. Changing Water Levels The graph shows the depth of water W in a reservoir over a one-year period as a function of the number of days x since the beginning of the year.

(a) Determine the intervals on which the function W is increasing and on which it is decreasing.

(b) At what value of x does W achieve a local maximum? A local minimum?

(c) Find the net change in the depth W from 100 days to 300 days.

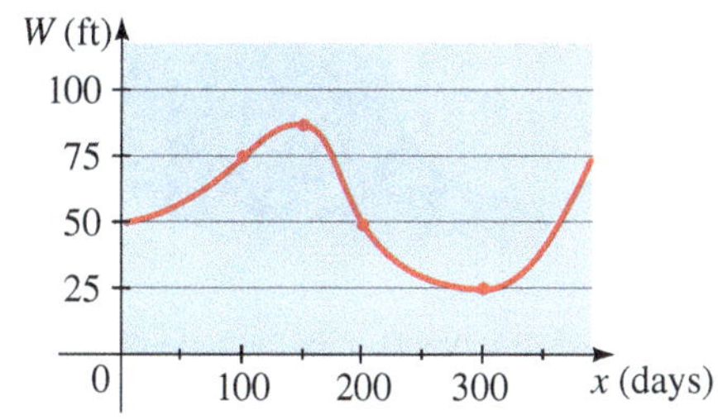

60. Population Growth and Decline The graph shows the population P in a small industrial city from 1950 to 2000. The variable x represents the number of years since 1950.

(a) Determine the intervals on which the function P is increasing and on which it is decreasing.

(b) What was the maximum population, and in what year was it attained?

(c) Find the net change in the population P from 1970 to 1990.

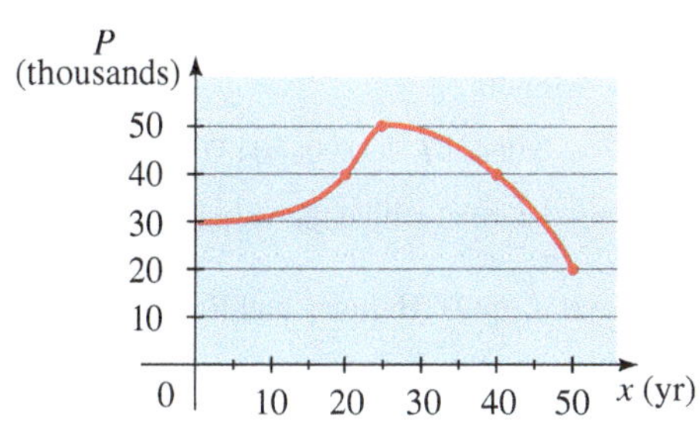

61. **Hurdle Race** Three runners compete in a 100-meter hurdle race. The graph depicts the distance run as a function of time for each runner. Describe in words what the graph tells you about this race. Who won the race? Did each runner finish the race? What do you think happened to Runner B?

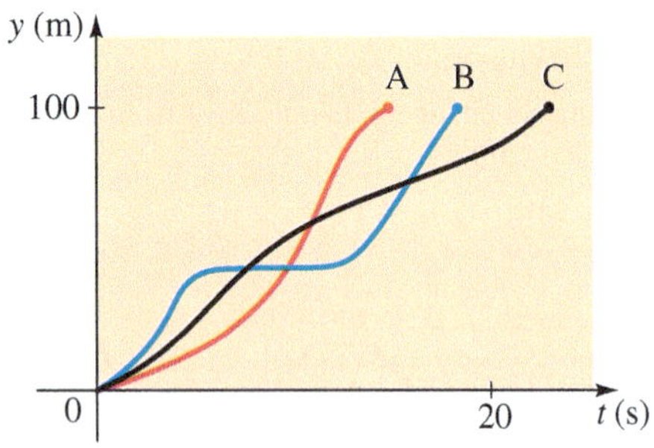

62. **Gravity Near the Moon** We can use Newton's Law of Gravity to measure the gravitational attraction between the moon and an algebra student in a spaceship located a distance x above the moon's surface:

$$F(x) = \frac{350}{x^2}$$

Here F is measured in newtons (N), and x is measured in millions of meters.

(a) Graph the function F for values of x between 0 and 10.

(b) Use the graph to describe the behavior of the gravitational attraction F as the distance x increases.

63. **Radii of Stars** Astronomers infer the radii of stars using the Stefan Boltzmann Law:

$$E(T) = (5.67 \times 10^{-8})T^4$$

where E is the energy radiated per unit of surface area measured in watts (W) and T is the absolute temperature measured in kelvins (K).

(a) Graph the function E for temperatures T between 100 K and 300 K.

(b) Use the graph to describe the change in energy E as the temperature T increases.

64. **Volume of Water** Between 0°C and 30°C, the volume V (in cubic centimeters) of 1 kg of water at a temperature T is given by the formula

$$V = 999.87 - 0.06426T + 0.0085043T^2 - 0.0000679T^3$$

Find the temperature at which the volume of 1 kg of water is a minimum.
[*Source: Physics,* by D. Halliday and R. Resnick]

65. **Migrating Fish** A fish swims at a speed v relative to the water, against a current of 5 mi/h. Using a mathematical model of energy expenditure, it can be shown that the total energy E required to swim a distance of 10 mi is given by

$$E(v) = 2.73v^3\frac{10}{v - 5}$$

Biologists believe that migrating fish try to minimize the total energy required to swim a fixed distance. Find the value of v that minimizes energy required.
[*Note:* This result has been verified; migrating fish swim against a current at a speed 50% greater than the speed of the current.]

66. **Coughing** When a foreign object that is lodged in the trachea (windpipe) forces a person to cough, the diaphragm thrusts upward, causing an increase in pressure in the lungs. At the same time, the trachea contracts, causing the expelled air to move faster and increasing the pressure on the foreign object. According to a mathematical model of coughing, the velocity v (in cm/s) of the airstream through an average-sized person's trachea is related to the radius r of the trachea (in cm) by the function

$$v(r) = 3.2(1 - r)r^2 \qquad \tfrac{1}{2} \le r \le 1$$

Determine the value of r for which v is a maximum.

DISCUSS ■ DISCOVER ■ PROVE ■ WRITE

67. **DISCUSS: Functions That Are Always Increasing or Decreasing** Sketch rough graphs of functions that are defined for all real numbers and that exhibit the indicated behavior (or explain why the behavior is impossible).

(a) f is always increasing, and $f(x) > 0$ for all x

(b) f is always decreasing, and $f(x) > 0$ for all x

(c) f is always increasing, and $f(x) < 0$ for all x

(d) f is always decreasing, and $f(x) < 0$ for all x

68. **DISCUSS: Maximum and Minimum Values** In Example 9 we saw a real-world situation in which the maximum value of a function is important. Name several other everyday situations in which a maximum or minimum value is important.

69. **DISCUSS ■ DISCOVER: Minimizing a Distance** When we seek a minimum or maximum value of a function, it is sometimes easier to work with a simpler function instead.

(a) Suppose

$$g(x) = \sqrt{f(x)}$$

where $f(x) \ge 0$ for all x. Explain why the local minima and maxima of f and g occur at the same values of x.

(b) Let $g(x)$ be the distance between the point $(3, 0)$ and the point (x, x^2) on the graph of the parabola $y = x^2$. Express g as a function of x.

(c) Find the minimum value of the function g that you found in part (b). Use the principle described in part (a) to simplify your work.

2.4 AVERAGE RATE OF CHANGE OF A FUNCTION

■ Average Rate of Change ■ Linear Functions Have Constant Rate of Change

Functions are often used to model changing quantities. In this section we learn how to find the rate at which the values of a function change as the input variable changes.

■ Average Rate of Change

We are all familiar with the concept of speed: If you drive a distance of 120 miles in 2 hours, then your average speed, or rate of travel, is $\frac{120 \text{ mi}}{2 \text{ h}} = 60$ mi/h. Now suppose you take a car trip and record the distance that you travel every few minutes. The distance s you have traveled is a function of the time t:

$$s(t) = \text{total distance traveled at time } t$$

We graph the function s as shown in Figure 1. The graph shows that you have traveled a total of 50 miles after 1 hour, 75 miles after 2 hours, 140 miles after 3 hours, and so on. To find your *average* speed between any two points on the trip, we divide the distance traveled by the time elapsed.

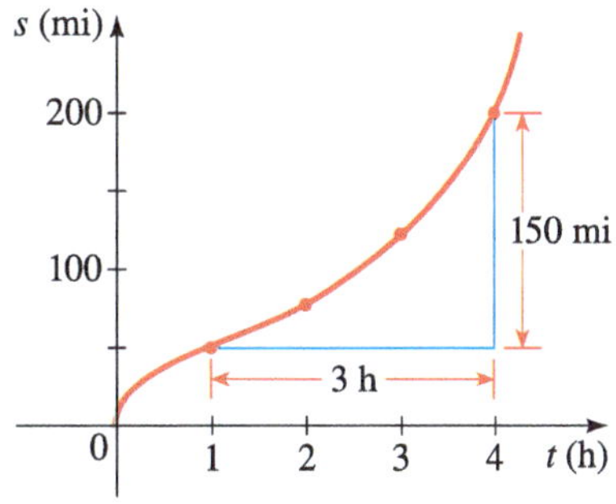

FIGURE 1 Average speed

Let's calculate your average speed between 1:00 P.M. and 4:00 P.M. The time elapsed is $4 - 1 = 3$ hours. To find the distance you traveled, we subtract the distance at 1:00 P.M. from the distance at 4:00 P.M., that is, $200 - 50 = 150$ mi. Thus your average speed is

$$\text{average speed} = \frac{\text{distance traveled}}{\text{time elapsed}} = \frac{150 \text{ mi}}{3 \text{ h}} = 50 \text{ mi/h}$$

The average speed that we have just calculated can be expressed by using function notation:

$$\text{average speed} = \frac{s(4) - s(1)}{4 - 1} = \frac{200 - 50}{3} = 50 \text{ mi/h}$$

Note that the average speed is different over different time intervals. For example, between 2:00 P.M. and 3:00 P.M. we find that

$$\text{average speed} = \frac{s(3) - s(2)}{3 - 2} = \frac{140 - 75}{1} = 65 \text{ mi/h}$$

Finding average rates of change is important in many contexts. For instance, we might be interested in knowing how quickly the air temperature is dropping as a storm approaches or how fast revenues are increasing from the sale of a new product. So we need to know how to determine the average rate of change of the functions that model these quantities. In fact, the concept of average rate of change can be defined for any function.

AVERAGE RATE OF CHANGE

The **average rate of change** of the function $y = f(x)$ between $x = a$ and $x = b$ is

$$\text{average rate of change} = \frac{\text{change in } y}{\text{change in } x} = \frac{f(b) - f(a)}{b - a}$$

The average rate of change is the slope of the **secant line** between $x = a$ and $x = b$ on the graph of f, that is, the line that passes through $(a, f(a))$ and $(b, f(b))$.

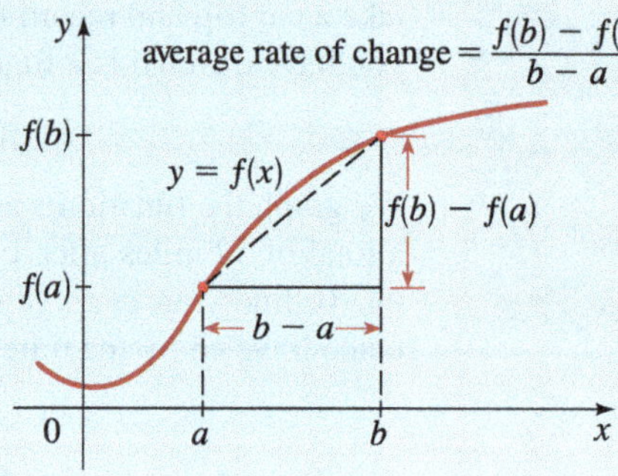

In the expression for average rate of change, the numerator $f(b) - f(a)$ is the net change in the value of f between $x = a$ and $x = b$ (see page 151).

EXAMPLE 1 ■ Calculating the Average Rate of Change

For the function $f(x) = (x - 3)^2$, whose graph is shown in Figure 2, find the net change and the average rate of change between the following points:

(a) $x = 1$ and $x = 3$ **(b)** $x = 4$ and $x = 7$

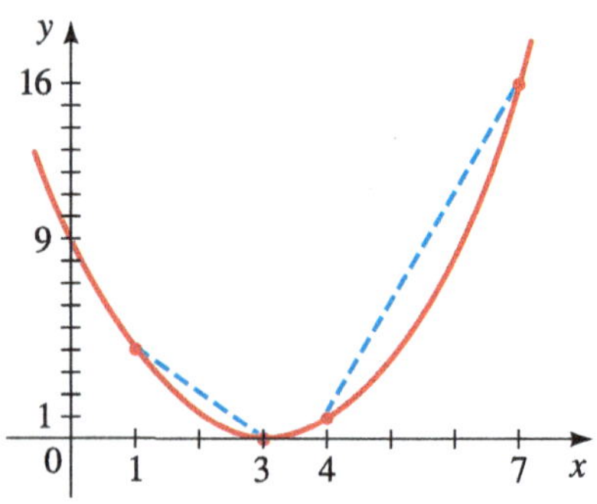

FIGURE 2 $f(x) = (x - 3)^2$

SOLUTION

(a) Net change $= f(3) - f(1)$ Definition

$= (3 - 3)^2 - (1 - 3)^2$ Use $f(x) = (x - 3)^2$

$= -4$ Calculate

$$\text{Average rate of change} = \frac{f(3) - f(1)}{3 - 1}$$ Definition

$$= \frac{-4}{2} = -2$$ Calculate

(b) Net change $= f(7) - f(4)$ Definition

$= (7 - 3)^2 - (4 - 3)^2$ Use $f(x) = (x - 3)^2$

$= 15$ Calculate

$$\text{Average rate of change} = \frac{f(7) - f(4)}{7 - 4}$$ Definition

$$= \frac{15}{3} = 5$$ Calculate

Now Try Exercise 15 ■

EXAMPLE 2 ■ Average Speed of a Falling Object

If an object is dropped from a high cliff or a tall building, then the distance it has fallen after t seconds is given by the function $d(t) = 16t^2$. Find its average speed (average rate of change) over the following intervals:

(a) Between 1 s and 5 s **(b)** Between $t = a$ and $t = a + h$

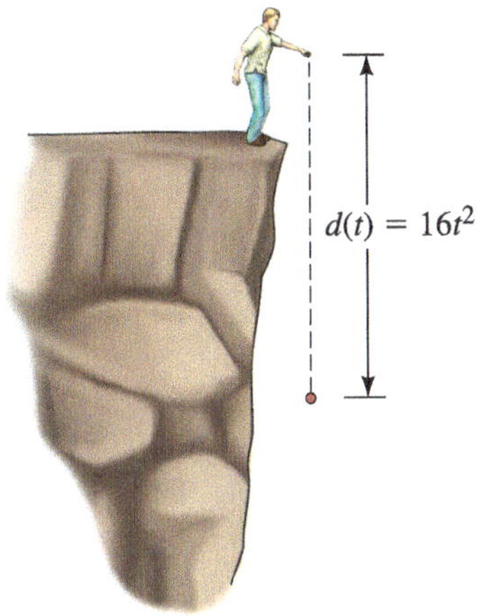

Function: In t seconds the stone falls $16t^2$ ft.

SOLUTION

(a)

$$\begin{aligned}\text{Average rate of change} &= \frac{d(5) - d(1)}{5 - 1} && \text{Definition}\\ &= \frac{16(5)^2 - 16(1)^2}{5 - 1} && \text{Use } d(t) = 16t^2\\ &= \frac{400 - 16}{4} && \text{Calculate}\\ &= 96 \text{ ft/s} && \text{Calculate}\end{aligned}$$

(b)

$$\begin{aligned}\text{Average rate of change} &= \frac{d(a + h) - d(a)}{(a + h) - a} && \text{Definition}\\ &= \frac{16(a + h)^2 - 16(a)^2}{(a + h) - a} && \text{Use } d(t) = 16t^2\\ &= \frac{16(a^2 + 2ah + h^2 - a^2)}{h} && \text{Expand and factor 16}\\ &= \frac{16(2ah + h^2)}{h} && \text{Simplify numerator}\\ &= \frac{16h(2a + h)}{h} && \text{Factor } h\\ &= 16(2a + h) && \text{Simplify}\end{aligned}$$

Now Try Exercise 19 ■

The average rate of change calculated in Example 2(b) is known as a *difference quotient*. In calculus we use difference quotients to calculate *instantaneous* rates of change. An example of an instantaneous rate of change is the speed shown on the speedometer of your car. This changes from one instant to the next as your car's speed changes.

The graphs in Figure 3 show that if a function is increasing on an interval, then the average rate of change between any two points is positive, whereas if a function is decreasing on an interval, then the average rate of change between any two points is negative.

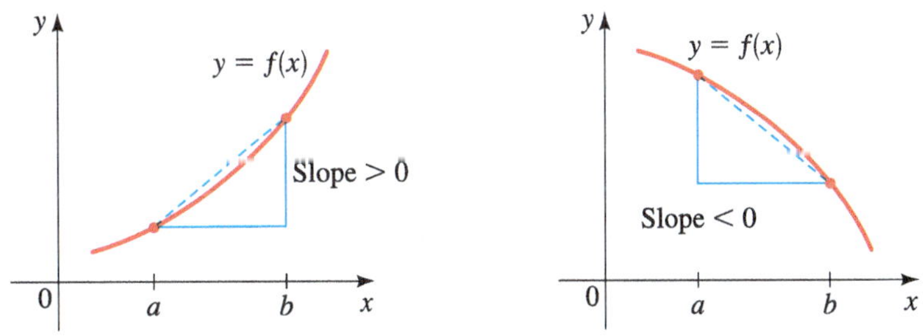

f increasing
Average rate of change positive

f decreasing
Average rate of change negative

FIGURE 3

EXAMPLE 3 ■ Average Rate of Temperature Change

Time	Temperature (°F)
8:00 A.M.	38
9:00 A.M.	40
10:00 A.M.	44
11:00 A.M.	50
12:00 NOON	56
1:00 P.M.	62
2:00 P.M.	66
3:00 P.M.	67
4:00 P.M.	64
5:00 P.M.	58
6:00 P.M.	55
7:00 P.M.	51

The table in the margin gives the outdoor temperatures observed by a science student on a spring day. Draw a graph of the data, and find the average rate of change of temperature between the following times:

(a) 8:00 A.M. and 9:00 A.M.

(b) 1:00 P.M. and 3:00 P.M.

(c) 4:00 P.M. and 7:00 P.M.

SOLUTION A graph of the temperature data is shown in Figure 4. Let t represent time, measured in hours since midnight (so, for example, 2:00 P.M. corresponds to $t = 14$). Define the function F by

$$F(t) = \text{temperature at time } t$$

Temperature at 9:00 A.M. Temperature at 8:00 A.M.

(a) $$\text{Average rate of change} = \frac{F(9) - F(8)}{9 - 8} = \frac{40 - 38}{9 - 8} = 2$$

The average rate of change was 2°F per hour.

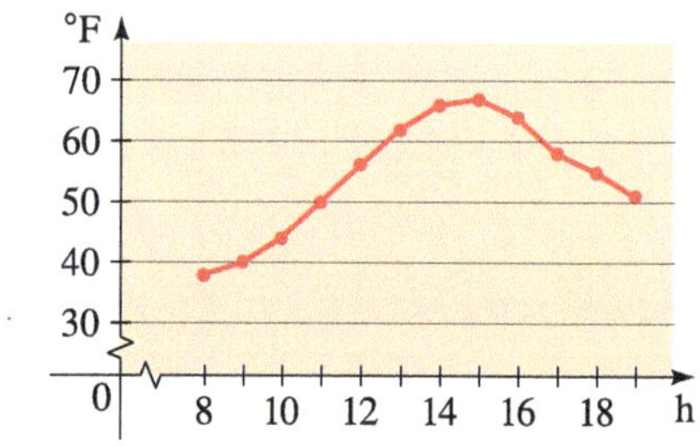

FIGURE 4

(b) $$\text{Average rate of change} = \frac{F(15) - F(13)}{15 - 13} = \frac{67 - 62}{2} = 2.5$$

The average rate of change was 2.5°F per hour.

(c) $$\text{Average rate of change} = \frac{F(19) - F(16)}{19 - 16} = \frac{51 - 64}{3} \approx -4.3$$

The average rate of change was about −4.3°F per hour during this time interval. The negative sign indicates that the temperature was dropping.

Now Try Exercise 31 ■

DISCOVERY PROJECT

When Rates of Change Change

In the real world, rates of change often themselves change. A statement like "inflation is rising, but at a slower rate" involves a change of a rate of change. When you drive your car, your speed (rate of change of distance) increases when you accelerate and decreases when you decelerate. From Example 4 we see that functions whose graph is a line (linear functions) have constant rate of change. In this project we explore how the shape of a graph corresponds to a changing rate of change. You can find the project at **www.stewartmath.com**.

Linear Functions Have Constant Rate of Change

Recall that a function of the form $f(x) = mx + b$ is a linear function (see page 160). Its graph is a line with slope m. On the other hand, if a function f has constant rate of change, then it must be a linear function. (You are asked to prove these facts in Exercises 51 and 52 in Section 2.5.) In general, the average rate of change of a linear function between any two points is the constant m. In the next example we find the average rate of change for a particular linear function.

EXAMPLE 4 ■ Linear Functions Have Constant Rate of Change

Let $f(x) = 3x - 5$. Find the average rate of change of f between the following points.

(a) $x = 0$ and $x = 1$

(b) $x = 3$ and $x = 7$

(c) $x = a$ and $x = a + h$

What conclusion can you draw from your answers?

SOLUTION

(a)
$$\text{Average rate of change} = \frac{f(1) - f(0)}{1 - 0} = \frac{(3 \cdot 1 - 5) - (3 \cdot 0 - 5)}{1}$$
$$= \frac{(-2) - (-5)}{1} = 3$$

(b)
$$\text{Average rate of change} = \frac{f(7) - f(3)}{7 - 3} = \frac{(3 \cdot 7 - 5) - (3 \cdot 3 - 5)}{4}$$
$$= \frac{16 - 4}{4} = 3$$

(c)
$$\text{Average rate of change} = \frac{f(a + h) - f(a)}{(a + h) - a} = \frac{[3(a + h) - 5] - [3a - 5]}{h}$$
$$= \frac{3a + 3h - 5 - 3a + 5}{h} = \frac{3h}{h} = 3$$

It appears that the average rate of change is always 3 for this function. In fact, part (c) proves that the rate of change between any two arbitrary points $x = a$ and $x = a + h$ is 3.

Now Try Exercise 25

2.4 EXERCISES

CONCEPTS

1. If you travel 100 miles in two hours, then your average speed for the trip is

   average speed = ——— = ______

2. The average rate of change of a function f between $x = a$ and $x = b$ is

   average rate of change = ______

3. The average rate of change of the function $f(x) = x^2$ between $x = 1$ and $x = 5$ is

   average rate of change = ______ = ______

4. **(a)** The average rate of change of a function f between $x = a$ and $x = b$ is the slope of the ______ line between $(a, f(a))$ and $(b, f(b))$.

   (b) The average rate of change of the linear function $f(x) = 3x + 5$ between any two points is ______.

5–6 ■ *Yes or No*? If *No*, give a reason.

5. (a) Is the average rate of change of a function between $x = a$ and $x = b$ the slope of the secant line through $(a, f(a))$ and $(b, f(b))$?

(b) Is the average rate of change of a linear function the same for all intervals?

6. (a) Can the average rate of change of an increasing function ever be negative?

(b) If the average rate of change of a function between $x = a$ and $x = b$ is negative, then is the function necessarily decreasing on the interval (a, b)?

SKILLS

7–10 ■ Net Change and Average Rate of Change The graph of a function is given. Determine **(a)** the net change and **(b)** the average rate of change between the indicated points on the graph.

7.

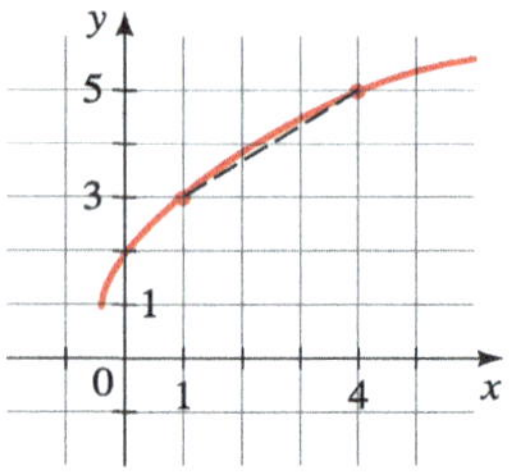

8.

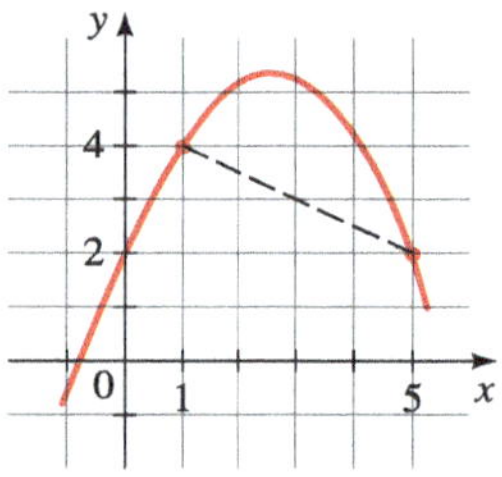

9.

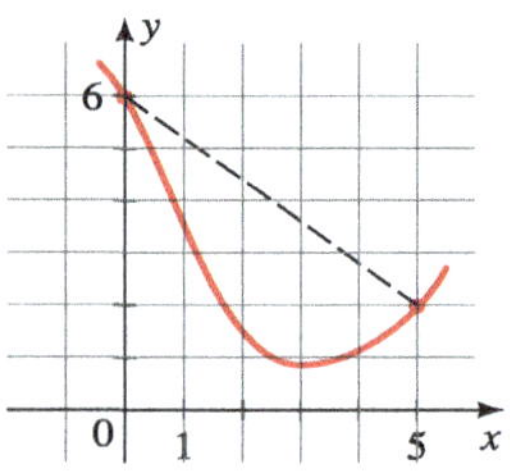

10.

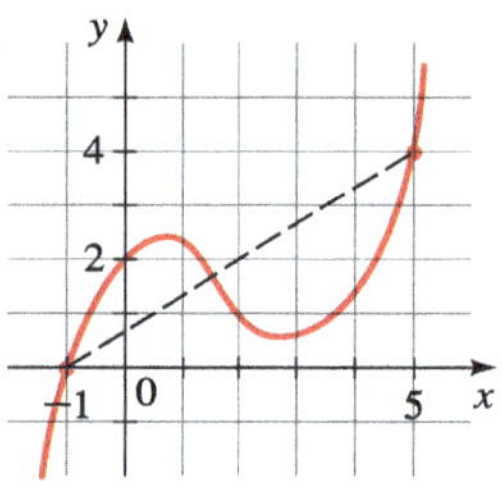

11–24 ■ Net Change and Average Rate of Change A function is given. Determine **(a)** the net change and **(b)** the average rate of change between the given values of the variable.

11. $f(x) = 3x - 2;\quad x = 2, x = 3$

12. $r(t) = 3 - \frac{1}{3}t;\quad t = 3, t = 6$

13. $h(t) = -t + \frac{3}{2};\quad t = -4, t = 1$

14. $g(x) = 2 - \frac{2}{3}x;\quad x = -3, x = 2$

15. $h(t) = 2t^2 - t;\quad t = 3, t = 6$

16. $f(z) = 1 - 3z^2;\quad z = -2, z = 0$

17. $f(x) = x^3 - 4x^2;\quad x = 0, x = 10$

18. $g(t) = t^4 - t^3 + t^2;\quad t = -2, t = 2$

19. $f(t) = 5t^2;\quad t = 3, t = 3 + h$

20. $f(x) = 1 - 3x^2;\quad x = 2, x = 2 + h$

21. $g(x) = \dfrac{1}{x};\quad x = 1, x = a$

22. $g(x) = \dfrac{2}{x + 1};\quad x = 0, x = h$

23. $f(t) = \dfrac{2}{t};\quad t = a, t = a + h$

24. $f(t) = \sqrt{t};\quad t = a, t = a + h$

25–26 ■ Average Rate of Change of a Linear Function A linear function is given. **(a)** Find the average rate of change of the function between $x = a$ and $x = a + h$. **(b)** Show that the average rate of change is the same as the slope of the line.

25. $f(x) = \frac{1}{2}x + 3$

26. $g(x) = -4x + 2$

SKILLS Plus

27. Average Rate of Change The graphs of the functions f and g are shown. The function ____ (f or g) has a greater average rate of change between $x = 0$ and $x = 1$. The function ____ (f or g) has a greater average rate of change between $x = 1$ and $x = 2$. The functions f and g have the same average rate of change between $x =$ ____ and $x =$ ____.

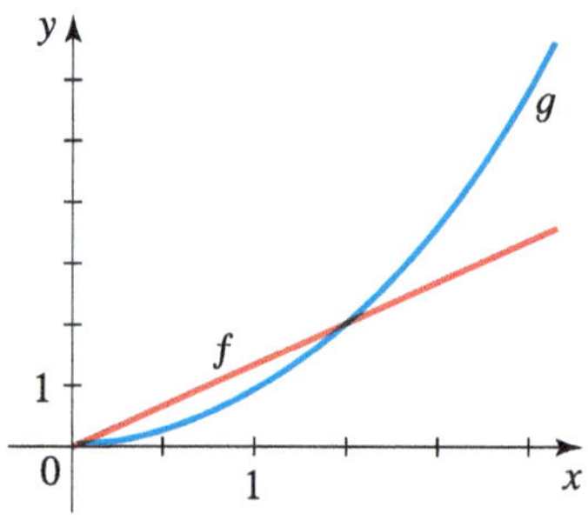

28. Average Rate of Change Graphs of the functions f, g, and h are shown below. What can you say about the average rate of change of each function on the successive intervals $[0, 1], [1, 2], [2, 3], \ldots$?

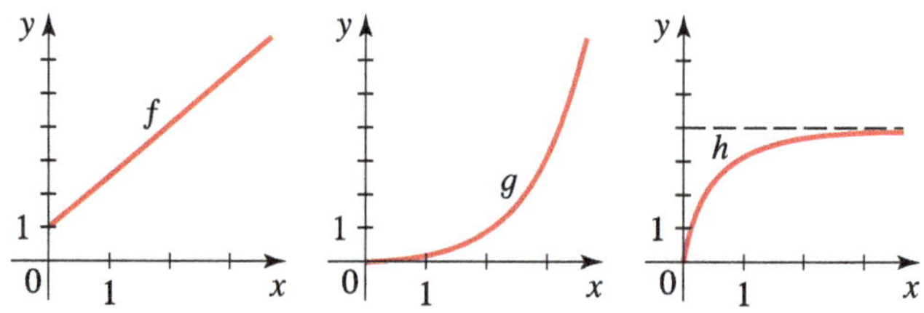

APPLICATIONS

29. Changing Water Levels The graph shows the depth of water W in a reservoir over a one-year period as a function of the number of days x since the beginning of the year. What was the average rate of change of W between $x = 100$ and $x = 200$?

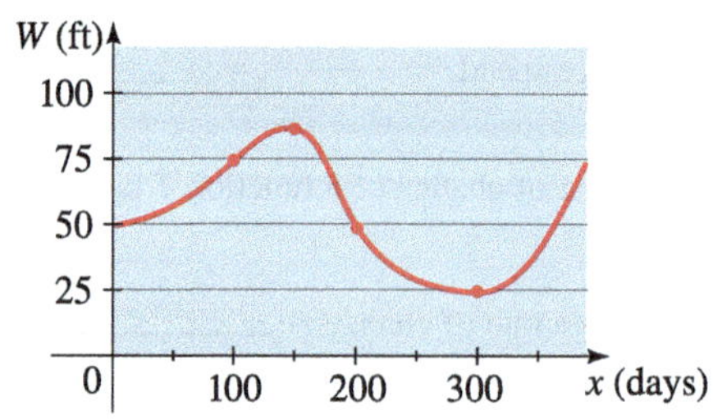

30. Population Growth and Decline The graph shows the population P in a small industrial city from 1950 to 2000. The variable x represents the number of years since 1950.

(a) What was the average rate of change of P between $x = 20$ and $x = 40$?

(b) Interpret the value of the average rate of change that you found in part (a).

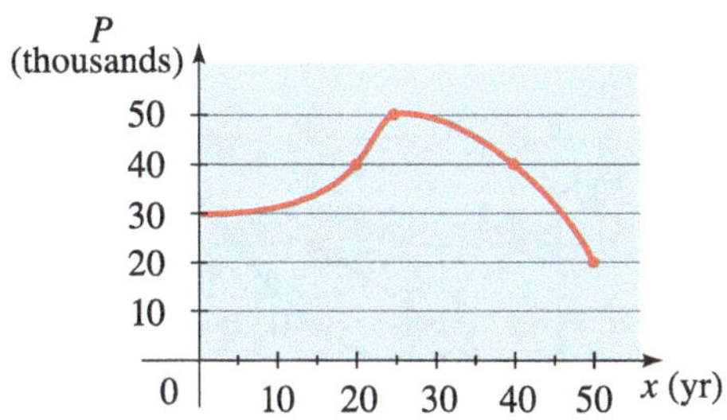

31. Population Growth and Decline The table gives the population in a small coastal community for the period 1997–2006. Figures shown are for January 1 in each year.

(a) What was the average rate of change of population between 1998 and 2001?

(b) What was the average rate of change of population between 2002 and 2004?

(c) For what period of time was the population increasing?

(d) For what period of time was the population decreasing?

Year	Population
1997	624
1998	856
1999	1,336
2000	1,578
2001	1,591
2002	1,483
2003	994
2004	826
2005	801
2006	745

32. Running Speed A man is running around a circular track that is 200 m in circumference. An observer uses a stopwatch to record the runner's time at the end of each lap, obtaining the data in the following table.

(a) What was the man's average speed (rate) between 68 s and 152 s?

(b) What was the man's average speed between 263 s and 412 s?

(c) Calculate the man's speed for each lap. Is he slowing down, speeding up, or neither?

Time (s)	Distance (m)
32	200
68	400
108	600
152	800
203	1000
263	1200
335	1400
412	1600

33. DVD Player Sales The table shows the number of DVD players sold in a small electronics store in the years 2003–2013.

Year	DVD players sold
2003	495
2004	513
2005	410
2006	402
2007	520
2008	580
2009	631
2010	719
2011	624
2012	582
2013	635

(a) What was the average rate of change of sales between 2003 and 2013?

(b) What was the average rate of change of sales between 2003 and 2004?

(c) What was the average rate of change of sales between 2004 and 2005?

(d) Between which two successive years did DVD player sales *increase* most quickly? *Decrease* most quickly?

34. Book Collection Between 1980 and 2000 a rare book collector purchased books for his collection at the rate of 40 books per year. Use this information to complete the following table. (Note that not every year is given in the table.)

Year	Number of books	Year	Number of books
1980	420	1995	
1981	460	1997	
1982		1998	
1985		1999	
1990		2000	1220
1992			

35. Cooling Soup When a bowl of hot soup is left in a room, the soup eventually cools down to room temperature. The temperature T of the soup is a function of time t. The table below gives the temperature (in °F) of a bowl of soup t minutes after it was set on the table. Find the average rate of change of the temperature of the soup over the first 20 minutes and over the next 20 minutes. During which interval did the soup cool off more quickly?

t (min)	T (°F)	t (min)	T (°F)
0	200	35	94
5	172	40	89
10	150	50	81
15	133	60	77
20	119	90	72
25	108	120	70
30	100	150	70

36. Farms in the United States The graph gives the number of farms in the United States from 1850 to 2000.

(a) Estimate the average rate of change in the number of farms between (i) 1860 and 1890 and (ii) 1950 and 1970.

(b) In which decade did the number of farms experience the greatest average rate of decline?

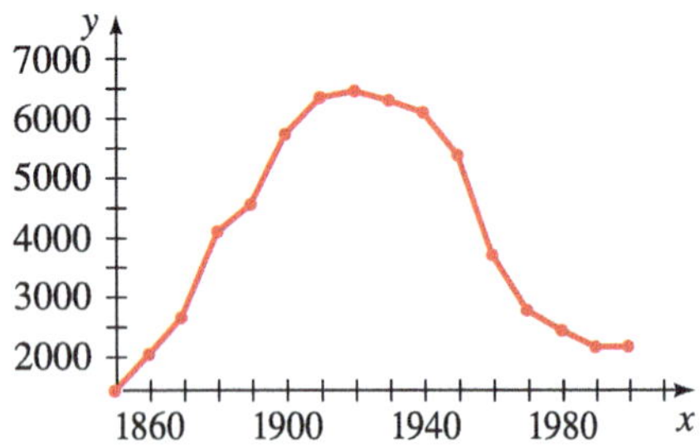

37. Three-Way Tie A downhill skiing race ends in a three-way tie for first place. The graph shows distance as a function of time for each of the three winners, A, B, and C.

(a) Find the average speed for each skier

(b) Describe the differences between the ways in which the three participants skied the race.

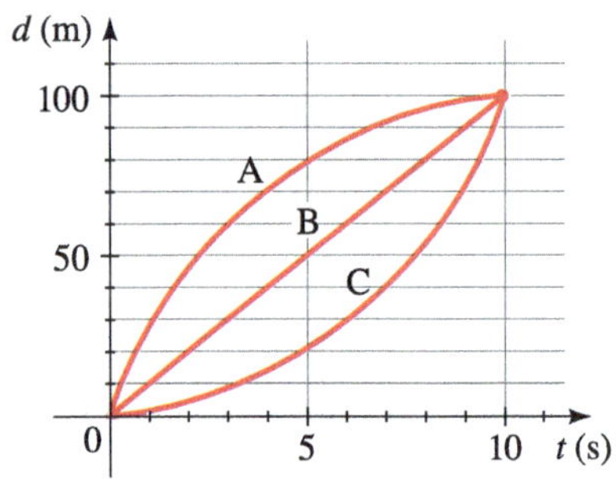

38. Speed Skating Two speed skaters, A and B, are racing in a 500-m event. The graph shows the distance they have traveled as a function of the time from the start of the race.

(a) Who won the race?

(b) Find the average speed during the first 10 s for each skater.

(c) Find the average speed during the last 15 s for each skater.

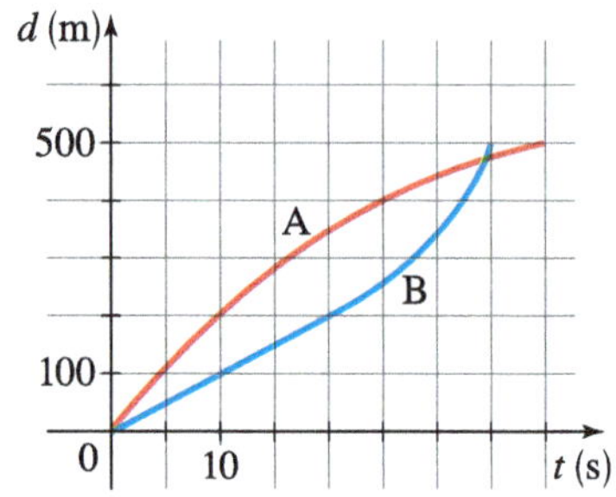

DISCUSS ■ DISCOVER ■ PROVE ■ WRITE

39. DISCOVER: Limiting Behavior of Average Speed An object is dropped from a high cliff, and the distance (in feet) it has fallen after t seconds is given by the function $d(t) = 16t^2$. Complete the table to find the average speed during the given time intervals. Use the table to determine what value the average speed approaches as the time intervals get smaller and smaller. Is it reasonable to say that this value is the speed of the object at the instant $t = 3$? Explain.

$t = a$	$t = b$	Average speed $= \dfrac{d(b) - d(a)}{b - a}$
3	3.5	
3	3.1	
3	3.01	
3	3.001	
3	3.0001	

2.5 LINEAR FUNCTIONS AND MODELS

■ Linear Functions ■ Slope and Rate of Change ■ Making and Using Linear Models

In this section we study the simplest functions that can be expressed by an algebraic expression: linear functions.

■ Linear Functions

Recall that a *linear function* is a function of the form $f(x) = ax + b$. So in the expression defining a linear function the variable occurs to the first power only. We can also express a linear function in equation form as $y = ax + b$. From Section 1.10 we know that the graph of this equation is a line with slope a and y-intercept b.

LINEAR FUNCTIONS

A **linear function** is a function of the form $f(x) = ax + b$.

The graph of a linear function is a line with slope a and y-intercept b.

EXAMPLE 1 ■ Identifying Linear Functions

Determine whether the given function is linear. If the function is linear, express the function in the form $f(x) = ax + b$.

(a) $f(x) = 2 + 3x$ **(b)** $g(x) = 3(1 - 2x)$

(c) $h(x) = x(4 + 3x)$ **(d)** $k(x) = \dfrac{1 - 5x}{4}$

SOLUTION

(a) We have $f(x) = 2 + 3x = 3x + 2$. So f is a linear function in which a is 3 and b is 2.

(b) We have $g(x) = 3(1 - 2x) = -6x + 3$. So g is a linear function in which a is -6 and b is 3.

(c) We have $h(x) = x(4 + 3x) = 4x + 3x^2$, which is not a linear function because the variable x is squared in the second term of the expression for h.

(d) We have $k(x) = \dfrac{1 - 5x}{4} = -\dfrac{5}{4}x + \dfrac{1}{4}$. So k is a linear function in which a is $-\frac{5}{4}$ and b is $\frac{1}{4}$.

Now Try Exercise 7 ■

EXAMPLE 2 ■ Graphing a Linear Function

Let f be the linear function defined by $f(x) = 3x + 2$.

(a) Make a table of values, and sketch a graph.

(b) What is the slope of the graph of f?

SOLUTION

x	$f(x)$
−2	−4
−1	−1
0	2
1	5
2	8
3	11
4	14
5	17

(a) A table of values is shown in the margin. Since f is a linear function, its graph is a line. So to obtain the graph of f, we plot any two points from the table and draw the straight line that contains the points. We use the points $(1, 5)$ and $(4, 14)$. The graph is the line shown in Figure 1. You can check that the other points in the table of values also lie on the line.

(b) Using the points given in Figure 1, we see that the slope is

$$\text{slope} = \frac{14 - 5}{4 - 1} = 3$$

So the slope is 3.

From the box at the top of this page, you can see that the slope of the graph of $f(x) = 3x + 2$ is 3.

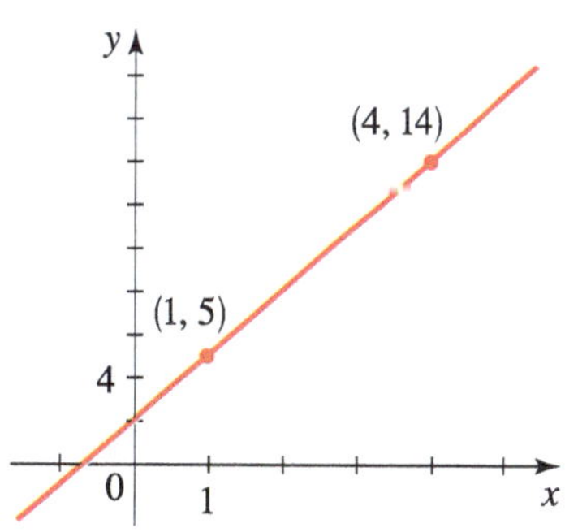

FIGURE 1 Graph of the linear function $f(x) = 3x + 2$

Now Try Exercise 15 ■

Slope and Rate of Change

In Exercise 52 we prove that all functions with constant rate of change are linear.

Let $f(x) = ax + b$ be a linear function. If x_1 and x_2 are two different values for x and if $y_1 = f(x_1)$ and $y_2 = f(x_2)$, then the points (x_1, y_1) and (x_2, y_2) lie on the graph of f. From the definitions of slope and average rate of change we have

$$\text{slope} = \frac{y_2 - y_1}{x_2 - x_1} = \frac{f(x_2) - f(x_1)}{x_2 - x_1} = \text{average rate of change}$$

From Section 1.10 we know that the *slope* of a linear function is the same between any two points. From the above equation we conclude that the *average rate of change* of a linear function is the same between any two points. Moreover, the average rate of change is equal to the slope (see Exercise 51). Since the average rate of change of a linear function is the same between any two points, it is simply called the **rate of change**.

SLOPE AND RATE OF CHANGE

For the linear function $f(x) = ax + b$, the slope of the graph of f and the rate of change of f are both equal to a, the coefficient of x.

$$a = \text{slope of graph of } f = \text{rate of change of } f$$

The difference between "slope" and "rate of change" is simply a difference in point of view. For example, to describe how a reservoir fills up over time, it is natural to talk about the rate at which the water level is rising, but we can also think of the slope of the graph of the water level (see Example 3). To describe the steepness of a staircase, it is natural to talk about the slope of the trim board of the staircase, but we can also think of the rate at which the stairs rise (see Example 5).

EXAMPLE 3 ■ Slope and Rate of Change

A dam is built on a river to create a reservoir. The water level $f(t)$ in the reservoir at time t is given by

$$f(t) = 4.5t + 28$$

where t is the number of years since the dam was constructed and $f(t)$ is measured in feet.

(a) Sketch a graph of f.

(b) What is the slope of the graph?

(c) At what rate is the water level in the reservoir changing?

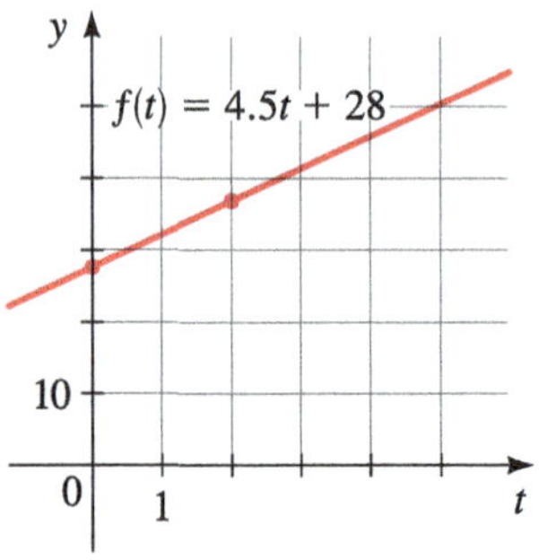

FIGURE 2 Water level as a function of time

SOLUTION

(a) A graph of f is shown in Figure 2.

(b) The graph is a line with slope 4.5, the coefficient of t.

(c) The rate of change of f is 4.5, the coefficient of t. Since time t is measured in years and the water level $f(t)$ is measured in feet, the water level in the reservoir is changing at the rate of 4.5 ft per year. Since this rate of change is positive, the water level is rising.

Now Try Exercises 19 and 39 ■

Making and Using Linear Models

When a linear function is used to model the relationship between two quantities, the slope of the graph of the function is the rate of change of the one quantity with respect to the other. For example, the graph in Figure 3(a) gives the amount of gas in a tank that is being filled. The slope between the indicated points is

$$a = \frac{6 \text{ gal}}{3 \text{ min}} = 2 \text{ gal/min}$$

The slope is the rate at which the tank is being filled, 2 gal per minute. In Figure 3(b) the tank is being drained at the rate of 0.03 gal per minute, and the slope is -0.03.

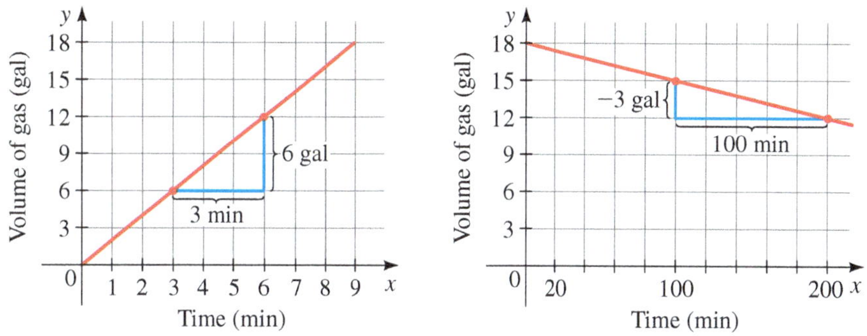

(a) Tank filled at 2 gal/min
Slope of line is 2

(b) Tank drained at 0.03 gal/min
Slope of line is −0.03

FIGURE 3 Amount of gas as a function of time

In the following examples we model real-world situations using linear functions. In each of these examples the model involves a constant rate of change (or a constant slope).

EXAMPLE 4 ■ Making a Linear Model from a Rate of Change

There are 200 gallons of water in the pool at time $t = 0$.

Water is being pumped into a swimming pool at the rate of 5 gal per min. Initially, the pool contains 200 gal of water.

(a) Find a linear function V that models the volume of water in the pool at any time t.

(b) If the pool has a capacity of 600 gal, how long does it take to completely fill the pool?

SOLUTION

(a) We need to find a linear function

$$V(t) = at + b$$

that models the volume $V(t)$ of water in the pool after t minutes. The rate of change of volume is 5 gal per min, so $a = 5$. Since the pool contains 200 gal to begin with, we have $V(0) = a \cdot 0 + b = 200$, so $b = 200$. Now that we know a and b, we get the model

$$V(t) = 5t + 200$$

(b) We want to find the time t at which $V(t) = 600$. So we need to solve the equation

$$600 = 5t + 200$$

Solving for t, we get $t = 80$. So it takes 80 min to fill the pool.

Now Try Exercise 41 ■

EXAMPLE 5 ■ Making a Linear Model from a Slope

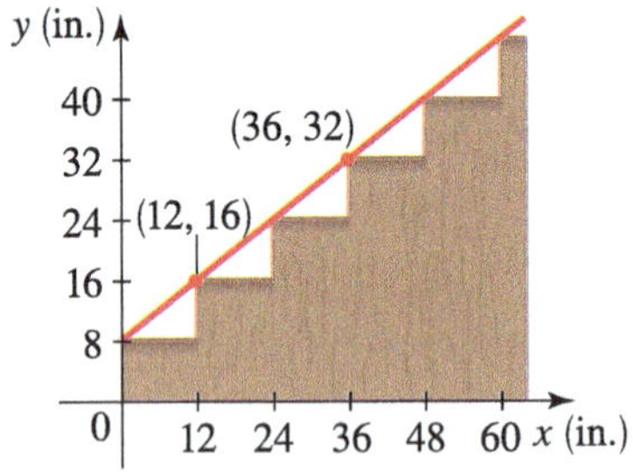

FIGURE 4 Slope of a staircase

In Figure 4 we have placed a staircase in a coordinate plane, with the origin at the bottom left corner. The red line in the figure is the edge of the trim board of the staircase.

(a) Find a linear function H that models the height of the trim board above the floor.

(b) If the space available to build a staircase is 11 ft wide, how high does the staircase reach?

SOLUTION

(a) We need to find a function

$$H(x) = ax + b$$

that models the red line in the figure. First we find the value of a, the slope of the line. From Figure 4 we see that two points on the line are $(12, 16)$ and $(36, 32)$, so the slope is

$$a = \frac{32 - 16}{36 - 12} = \frac{2}{3}$$

Another way to find the slope is to observe that each of the steps is 8 in. high (the rise) and 12 in. deep (the run), so the slope of the line is $\frac{8}{12} = \frac{2}{3}$. From Figure 4 we see that the y-intercept is 8, so $b = 8$. So the model we want is

$$H(x) = \tfrac{2}{3}x + 8$$

(b) Since 11 ft is 132 in., we need to evaluate the function H when x is 132. We have

$$H(132) = \tfrac{2}{3}(132) + 8 = 96$$

So the staircase reaches a height of 96 in., or 8 ft.

Now Try Exercise 43 ■

EXAMPLE 6 ■ Making Linear Models Involving Speed

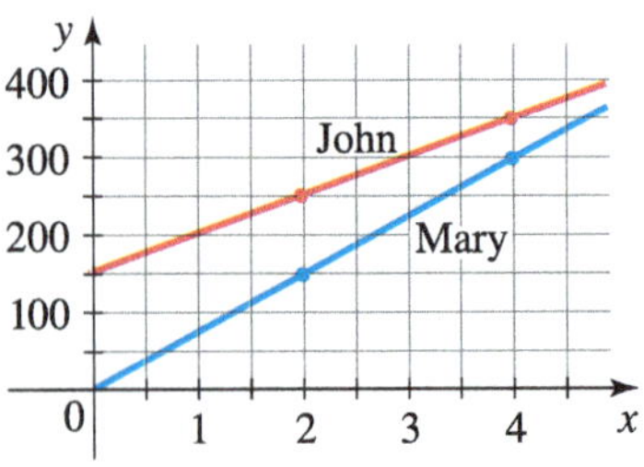

FIGURE 5 John and Mary's trips

John and Mary are driving westward along I-76 at constant speeds. The graphs in Figure 5 show the distance y (in miles) that they have traveled from Philadelphia at time x (in hours), where $x = 0$ corresponds to noon. (Note that at noon John has already traveled 150 mi.)

(a) At what speeds are John and Mary traveling? Who is traveling faster, and how does this show up in the graph?

(b) Find functions that model the distances that John and Mary have traveled as functions of x.

(c) How far will John and Mary have traveled at 5:00 P.M.?

(d) For what time period is Mary behind John? Will Mary overtake John? If so, at what time?

SOLUTION

(a) From the graph we see that John has traveled 250 mi at 2:00 P.M. and 350 mi at 4:00 P.M. The speed is the rate of change of distance with respect to time. So the speed is the slope of the graph. Therefore John's speed is

$$\frac{350 \text{ mi} - 250 \text{ mi}}{4 \text{ h} - 2 \text{ h}} = 50 \text{ mi/h} \qquad \text{John's speed}$$

Mary has traveled 150 mi at 2:00 P.M. and 300 mi at 4:00 P.M., so we calculate Mary's speed to be

$$\frac{300 \text{ mi} - 150 \text{ mi}}{4 \text{ h} - 2 \text{ h}} = 75 \text{ mi/h} \qquad \text{Mary's speed}$$

Mary is traveling faster than John. We can see this from the graph because Mary's line is steeper (has a greater slope) than John's line.

(b) Let $f(x)$ be the distance John has traveled at time x. Since the speed (average rate of change) is constant, it follows that f is a linear function. Thus we can write f in the form $f(x) = ax + b$. From part (a) we know that the slope a is 50, and from the graph we see that the y-intercept b is 150. Thus the distance that John has traveled at time x is modeled by the linear function

$$f(x) = 50x + 150 \qquad \text{Model for John's distance}$$

Similarly, Mary is traveling at 75 mi/h, and the y-intercept of her graph is 0. Thus the distance she has traveled at time x is modeled by the linear function

$$g(x) = 75x \qquad \text{Model for Mary's distance}$$

(c) Replacing x by 5 in the models that we obtained in part (b), we find that at 5:00 P.M. John has traveled $f(5) = 50(5) + 150 = 400$ mi and Mary has traveled $g(5) = 75(5) = 375$ mi.

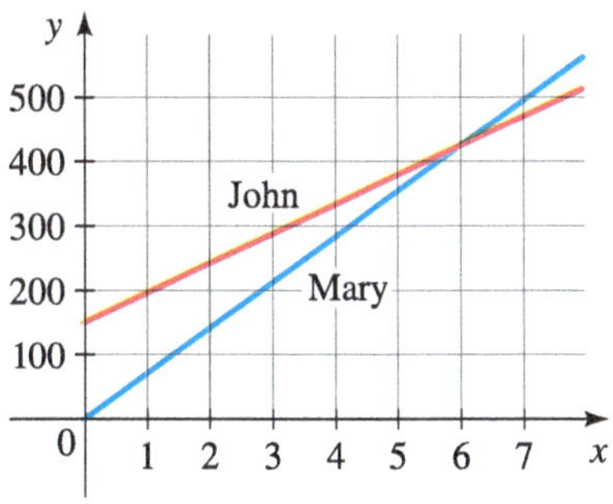

FIGURE 6 John and Mary's trips

(d) Mary overtakes John at the time when each has traveled the same distance, that is, at the time x when $f(x) = g(x)$. So we must solve the equation

$$50x + 150 = 75x \qquad \text{John's distance = Mary's distance}$$

Solving this equation, we get $x = 6$. So Mary overtakes John after 6 h, that is, at 6:00 P.M. We can confirm our solution graphically by drawing the graphs of f and g on a larger domain as shown in Figure 6. The graphs intersect when $x = 6$. From the graph we see that the graph of Mary's trip is below the graph of John's trip from $x = 0$ to $x = 6$, so Mary is behind John from noon until 6:00 P.M.

Now Try Exercise 45 ■

2.5 EXERCISES

CONCEPTS

1. Let f be a function with constant rate of change. Then
   (a) f is a ________ function and f is of the form $f(x) =$ ____ $x +$ ____.
   (b) The graph of f is a ________.

2. Let f be the linear function $f(x) = -5x + 7$.
   (a) The rate of change of f is ________.
   (b) The graph of f is a ________ with slope ________ and y-intercept ________.

3–4 ■ A swimming pool is being filled. The graph shows the number of gallons y in the pool after x minutes.

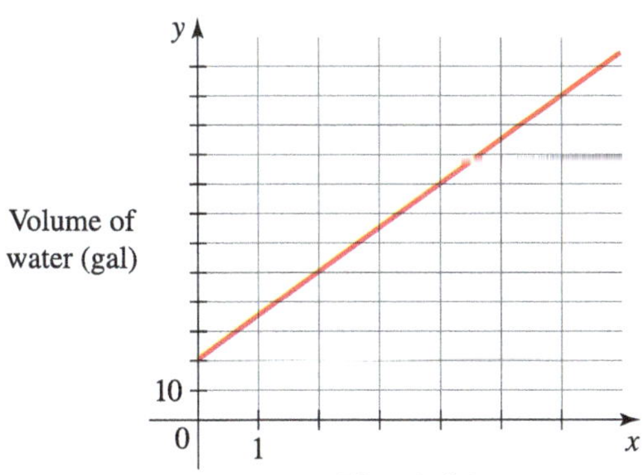

3. What is the slope of the graph?
4. At what rate is the pool being filled?
5. If a linear function has positive rate of change, does its graph slope upward or downward?
6. Is $f(x) = 3$ a linear function? If so, what are the slope and the rate of change?

SKILLS

7–14 ■ Identifying Linear Functions Determine whether the given function is linear. If the function is linear, express the function in the form $f(x) = ax + b$.

7. $f(x) = 3 + \frac{1}{3}x$
8. $f(x) = 2 - 4x$
9. $f(x) = x(4 - x)$
10. $f(x) = \sqrt{x} + 1$
11. $f(x) = \dfrac{x+1}{5}$
12. $f(x) = \dfrac{2x-3}{x}$
13. $f(x) = (x + 1)^2$
14. $f(x) = \frac{1}{2}(3x - 1)$

15–18 ■ Graphing Linear Functions For the given linear function, make a table of values and sketch its graph. What is the slope of the graph?

15. $f(x) = 2x - 5$
16. $g(x) = 4 - 2x$
17. $r(t) = -\frac{2}{3}t + 2$
18. $h(t) = \frac{1}{2} - \frac{3}{4}t$

19–26 ■ Slope and Rate of Change A linear function is given. **(a)** Sketch the graph. **(b)** Find the slope of the graph. **(c)** Find the rate of change of the function.

19. $f(x) = 2x - 6$ **20.** $g(z) = -3z - 9$

21. $h(t) = -0.5t - 2$ **22.** $s(w) = -0.2w - 6$

23. $v(t) = -\frac{10}{3}t - 20$ **24.** $A(r) = -\frac{2}{3}r - 1$

25. $f(t) = -\frac{3}{2}t + 2$ **26.** $g(x) = \frac{5}{4}x - 10$

27–30 ■ Linear Functions Given Verbally A verbal description of a linear function f is given. Express the function f in the form $f(x) = ax + b$.

27. The linear function f has rate of change 3 and initial value -1.

28. The linear function g has rate of change -12 and initial value 100.

29. The graph of the linear function h has slope $\frac{1}{2}$ and y-intercept 3.

30. The graph of the linear function k has slope $-\frac{4}{5}$ and y-intercept -2.

31–32 ■ Linear Functions Given Numerically A table of values for a linear function f is given. **(a)** Find the rate of change of f. **(b)** Express f in the form $f(x) = ax + b$

31.

x	$f(x)$
0	7
2	10
4	13
6	16
8	19

32.

x	$f(x)$
-3	11
0	2
2	-4
5	-13
7	-19

33–36 ■ Linear Functions Given Graphically The graph of a linear function f is given. **(a)** Find the rate of change of f. **(b)** Express f in the form $f(x) = ax + b$.

33.

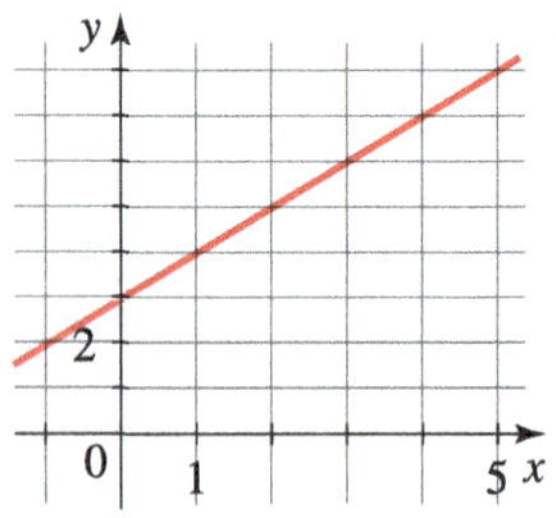

34.

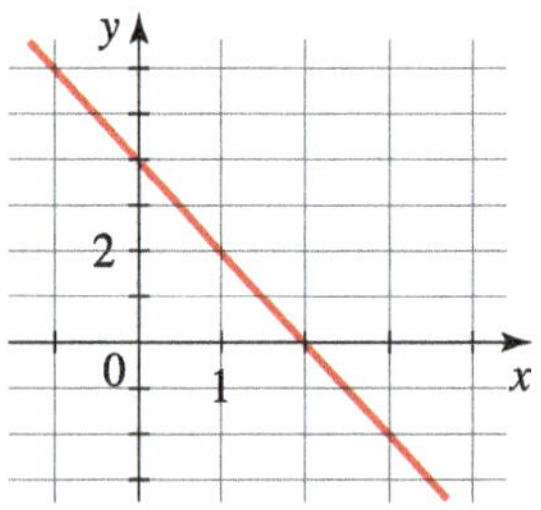

35.

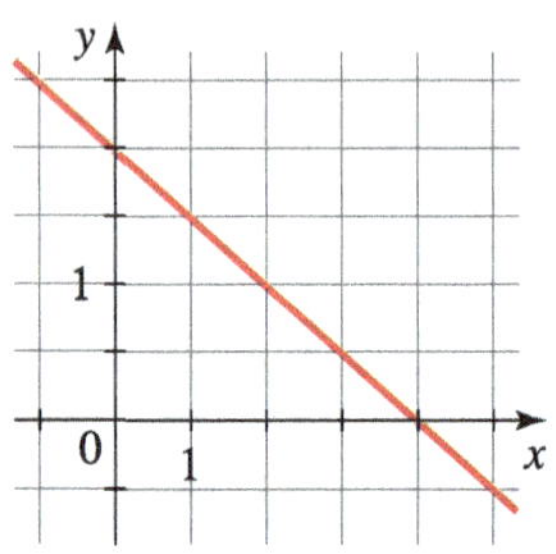

36.

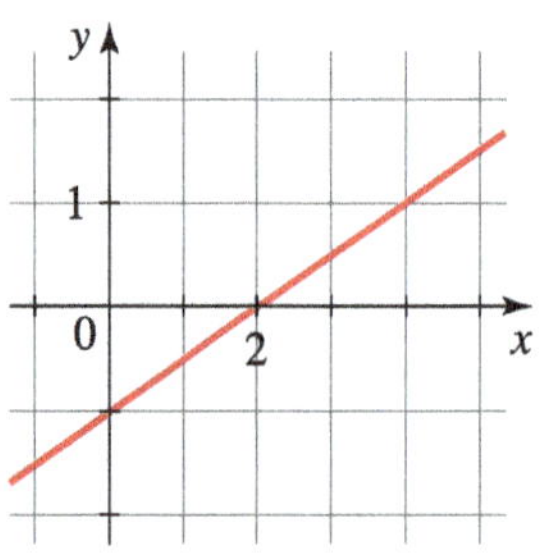

SKILLS Plus

37. Families of Linear Functions Graph $f(x) = ax$ for $a = \frac{1}{2}$, $a = 1$, and $a = 2$, all on the same set of axes. How does increasing the value of a affect the graph of f? What about the rate of change of f?

38. Families of Linear Functions Graph $f(x) = x + b$ for $b = \frac{1}{2}$, $b = 1$, and $b = 2$, all on the same set of axes. How does increasing the value of b affect the graph of f? What about the rate of change of f?

APPLICATIONS

39. Landfill The amount of trash in a county landfill is modeled by the function

$$T(x) = 150x + 32{,}000$$

where x is the number of years since 1996 and $T(x)$ is measured in thousands of tons.

(a) Sketch a graph of T.

(b) What is the slope of the graph?

(c) At what rate is the amount of trash in the landfill increasing per year?

40. Copper Mining The amount of copper ore produced from a copper mine in Arizona is modeled by the function

$$f(x) = 200 + 32x$$

where x is the number of years since 2005 and $f(x)$ is measured in thousands of tons.

(a) Sketch a graph of f.

(b) What is the slope of the graph?

(c) At what rate is the amount of ore produced changing?

41. Weather Balloon Weather balloons are filled with hydrogen and released at various sites to measure and transmit data about conditions such as air pressure and temperature. A weather balloon is filled with hydrogen at the rate of 0.5 ft^3/s. Initially, the balloon contains 2 ft^3 of hydrogen.

(a) Find a linear function V that models the volume of hydrogen in the balloon at any time t.

(b) If the balloon has a capacity of 15 ft^3, how long does it take to completely fill the balloon?

42. Filling a Pond A large koi pond is filled from a garden hose at the rate of 10 gal/min. Initially, the pond contains 300 gal of water.

(a) Find a linear function V that models the volume of water in the pond at any time t.

(b) If the pond has a capacity of 1300 gal, how long does it take to completely fill the pond?

43. Wheelchair Ramp A local diner must build a wheelchair ramp to provide handicap access to the restaurant. Federal building codes require that a wheelchair ramp must have a maximum rise of 1 in. for every horizontal distance of 12 in.

(a) What is the maximum allowable slope for a wheelchair ramp? Assuming that the ramp has maximum rise, find a linear function H that models the height of the ramp above the ground as a function of the horizontal distance x.

(b) If the space available to build a ramp is 150 in. wide, how high does the ramp reach?

44. Mountain Biking Meilin and Brianna are avid mountain bikers. On a spring day they cycle down straight roads with steep grades. The graphs give a representation of the elevation of the road on which each of them cycles. Find the grade of each road.

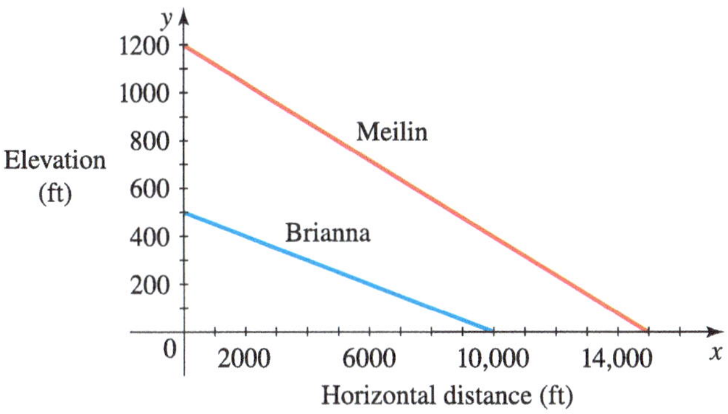

45. Commute to Work Jade and her roommate Jari commute to work each morning, traveling west on I-10. One morning Jade left for work at 6:50 A.M., but Jari left 10 minutes later. Both drove at a constant speed. The following graphs show the distance (in miles) each of them has traveled on I-10 at time t (in minutes), where $t = 0$ is 7:00 A.M.

(a) Use the graph to decide which of them is traveling faster.

(b) Find the speed (in mi/h) at which each of them is driving.

(c) Find linear functions f and g that model the distances that Jade and Jari travel as functions of t (in minutes).

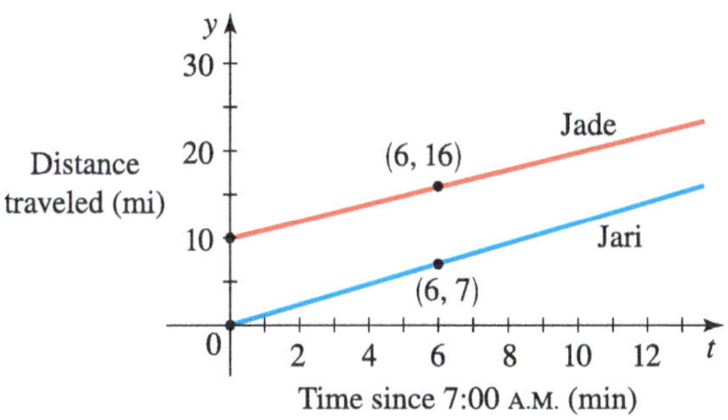

46. Distance, Speed, and Time Jacqueline leaves Detroit at 2:00 P.M. and drives at a constant speed, traveling west on I-90. She passes Ann Arbor, 40 mi from Detroit, at 2:50 P.M.

(a) Find a linear function d that models the distance (in mi) she has traveled after t min.

(b) Draw a graph of d. What is the slope of this line?

(c) At what speed (in mi/h) is Jacqueline traveling?

47. Grade of Road West of Albuquerque, New Mexico, Route 40 eastbound is straight and makes a steep descent toward the city. The highway has a 6% grade, which means that its slope is $-\frac{6}{100}$. Driving on this road, you notice from elevation signs that you have descended a distance of 1000 ft. What is the change in your horizontal distance in miles?

48. Sedimentation Devils Lake, North Dakota, has a layer of sedimentation at the bottom of the lake that increases every year. The depth of the sediment layer is modeled by the function

$$D(x) = 20 + 0.24x$$

where x is the number of years since 1980 and $D(x)$ is measured in centimeters.

(a) Sketch a graph of D.

(b) What is the slope of the graph?

(c) At what rate (in cm) is the sediment layer increasing per year?

49. Cost of Driving The monthly cost of driving a car depends on the number of miles driven. Lynn found that in May her driving cost was \$380 for 480 mi and in June her cost was \$460 for 800 mi. Assume that there is a linear relationship between the monthly cost C of driving a car and the distance x driven.

(a) Find a linear function C that models the cost of driving x miles per month.

(b) Draw a graph of C. What is the slope of this line?

(c) At what rate does Lynn's cost increase for every additional mile she drives?

50. Manufacturing Cost The manager of a furniture factory finds that it costs \$2200 to produce 100 chairs in one day and \$4800 to produce 300 chairs in one day.

(a) Assuming that the relationship between cost and the number of chairs produced is linear, find a linear function C that models the cost of producing x chairs in one day.

(b) Draw a graph of C. What is the slope of this line?

(c) At what rate does the factory's cost increase for every additional chair produced?

DISCUSS ■ DISCOVER ■ PROVE ■ WRITE

51. PROVE: Linear Functions Have Constant Rate of Change Suppose that $f(x) = ax + b$ is a linear function.

(a) Use the definition of the average rate of change of a function to calculate the average rate of change of f between any two real numbers x_1 and x_2.

(b) Use your calculation in part (a) to show that the average rate of change of f is the same as the slope a.

52. PROVE: Functions with Constant Rate of Change Are Linear Suppose that the function f has the same average rate of change c between any two points.

(a) Find the average rate of change of f between the points a and x to show that

$$c = \frac{f(x) - f(a)}{x - a}$$

(b) Rearrange the equation in part (a) to show that

$$f(x) = cx + (f(a) - ca)$$

How does this show that f is a linear function? What is the slope, and what is the y-intercept?

2.6 TRANSFORMATIONS OF FUNCTIONS

■ Vertical Shifting ■ Horizontal Shifting ■ Reflecting Graphs ■ Vertical Stretching and Shrinking ■ Horizontal Stretching and Shrinking ■ Even and Odd Functions

In this section we study how certain transformations of a function affect its graph. This will give us a better understanding of how to graph functions. The transformations that we study are shifting, reflecting, and stretching.

■ Vertical Shifting

Adding a constant to a function shifts its graph vertically: upward if the constant is positive and downward if it is negative.

In general, suppose we know the graph of $y = f(x)$. How do we obtain from it the graphs of

Recall that the graph of the function f is the same as the graph of the equation $y = f(x)$.

$$y = f(x) + c \qquad \text{and} \qquad y = f(x) - c \qquad (c > 0)$$

The y-coordinate of each point on the graph of $y = f(x) + c$ is c units above the y-coordinate of the corresponding point on the graph of $y = f(x)$. So we obtain the graph of $y = f(x) + c$ simply by shifting the graph of $y = f(x)$ upward c units. Similarly, we obtain the graph of $y = f(x) - c$ by shifting the graph of $y = f(x)$ downward c units.

VERTICAL SHIFTS OF GRAPHS

Suppose $c > 0$.

To graph $y = f(x) + c$, shift the graph of $y = f(x)$ upward c units.

To graph $y = f(x) - c$, shift the graph of $y = f(x)$ downward c units.

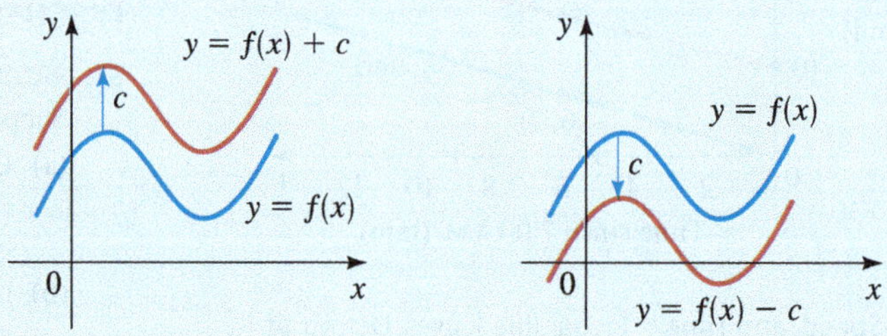

EXAMPLE 1 ■ Vertical Shifts of Graphs

Use the graph of $f(x) = x^2$ to sketch the graph of each function.

(a) $g(x) = x^2 + 3$ **(b)** $h(x) = x^2 - 2$

SOLUTION The function $f(x) = x^2$ was graphed in Example 1(a), Section 2.2. It is sketched again in Figure 1.

(a) Observe that

$$g(x) = x^2 + 3 = f(x) + 3$$

So the y-coordinate of each point on the graph of g is 3 units above the corresponding point on the graph of f. This means that to graph g, we shift the graph of f upward 3 units, as in Figure 1.

(b) Similarly, to graph h we shift the graph of f downward 2 units, as shown in Figure 1.

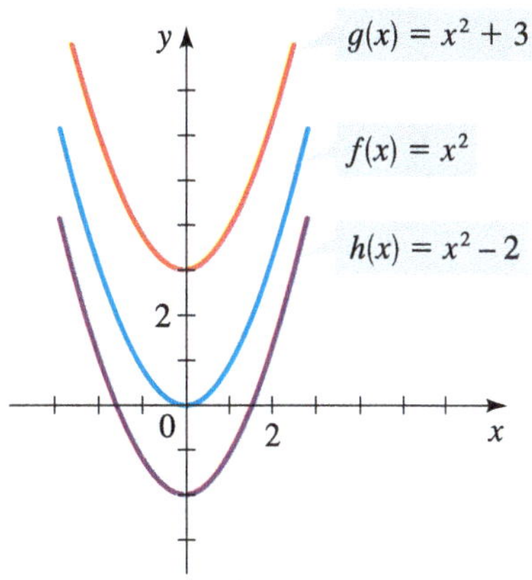

FIGURE 1

Now Try Exercises 29 and 31 ■

Horizontal Shifting

Suppose that we know the graph of $y = f(x)$. How do we use it to obtain the graphs of

$$y = f(x + c) \quad \text{and} \quad y = f(x - c) \quad (c > 0)$$

The value of $f(x - c)$ at x is the same as the value of $f(x)$ at $x - c$. Since $x - c$ is c units to the left of x, it follows that the graph of $y = f(x - c)$ is just the graph of $y = f(x)$ shifted to the right c units. Similar reasoning shows that the graph of $y = f(x + c)$ is the graph of $y = f(x)$ shifted to the left c units. The following box summarizes these facts.

HORIZONTAL SHIFTS OF GRAPHS

Suppose $c > 0$.

To graph $y = f(x - c)$, shift the graph of $y = f(x)$ to the right c units.

To graph $y = f(x + c)$, shift the graph of $y = f(x)$ to the left c units.

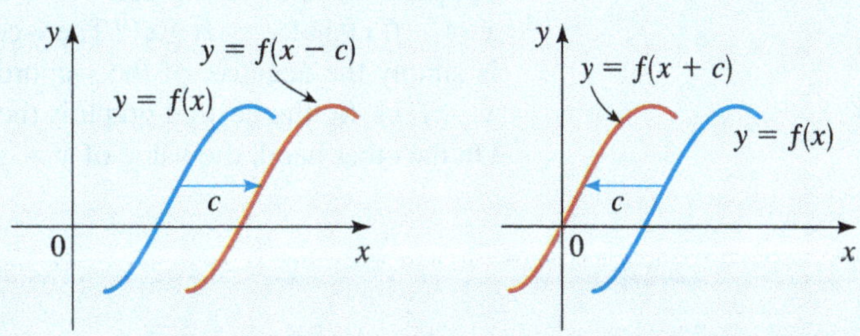

EXAMPLE 2 ■ Horizontal Shifts of Graphs

Use the graph of $f(x) = x^2$ to sketch the graph of each function.

(a) $g(x) = (x + 4)^2$ **(b)** $h(x) = (x - 2)^2$

SOLUTION

(a) To graph g, we shift the graph of f to the left 4 units.

(b) To graph h, we shift the graph of f to the right 2 units.

The graphs of g and h are sketched in Figure 2.

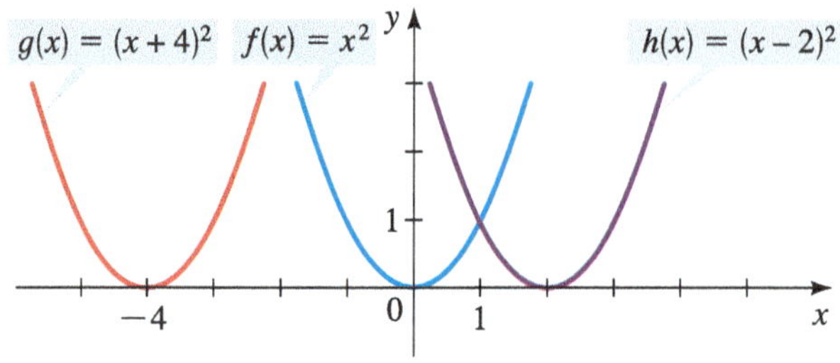

FIGURE 2

Now Try Exercises 33 and 35 ■

EXAMPLE 3 ■ Combining Horizontal and Vertical Shifts

Sketch the graph of $f(x) = \sqrt{x-3} + 4$.

SOLUTION We start with the graph of $y = \sqrt{x}$ (Example 1(c), Section 2.2) and shift it to the right 3 units to obtain the graph of $y = \sqrt{x-3}$. Then we shift the resulting graph upward 4 units to obtain the graph of $f(x) = \sqrt{x-3} + 4$ shown in Figure 3.

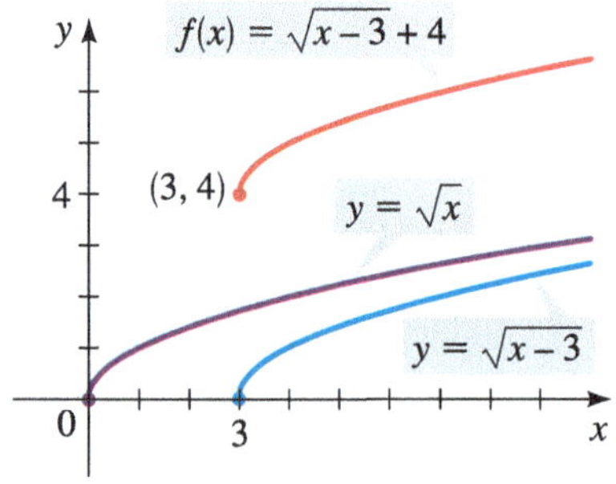

FIGURE 3

Now Try Exercise 45 ■

■ Reflecting Graphs

Suppose we know the graph of $y = f(x)$. How do we use it to obtain the graphs of $y = -f(x)$ and $y = f(-x)$? The y-coordinate of each point on the graph of $y = -f(x)$ is simply the negative of the y-coordinate of the corresponding point on the graph of $y = f(x)$. So the desired graph is the reflection of the graph of $y = f(x)$ in the x-axis. On the other hand, the value of $y = f(-x)$ at x is the same as the value of $y = f(x)$ at

DISCOVERY PROJECT

Transformation Stories

If a real-world situation, or "story," is modeled by a function, how does transforming the function change the story? For example, if the distance traveled on a road trip is modeled by a function, then how does shifting or stretching the function change the story of the trip? How does changing the story of the trip transform the function that models the trip? In this project we explore some real-world stories and transformations of these stories. You can find the project at **www.stewartmath.com**.

$-x$, so the desired graph here is the reflection of the graph of $y = f(x)$ in the y-axis. The following box summarizes these observations.

REFLECTING GRAPHS

To graph $y = -f(x)$, reflect the graph of $y = f(x)$ in the x-axis.

To graph $y = f(-x)$, reflect the graph of $y = f(x)$ in the y-axis.

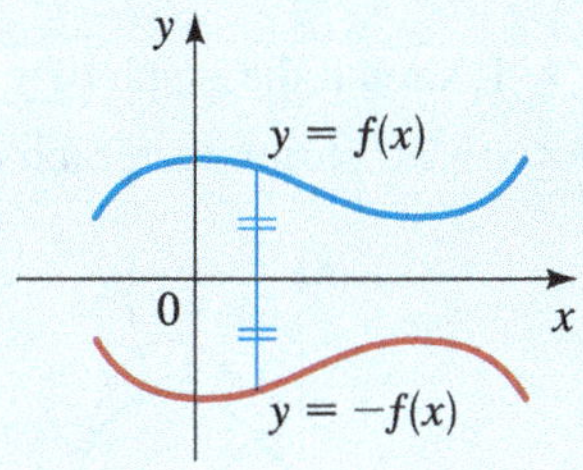

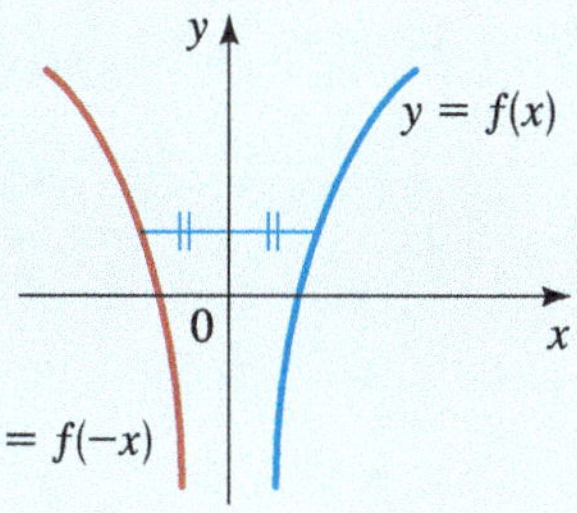

EXAMPLE 4 ■ Reflecting Graphs

Sketch the graph of each function.

(a) $f(x) = -x^2$ **(b)** $g(x) = \sqrt{-x}$

SOLUTION

(a) We start with the graph of $y = x^2$. The graph of $f(x) = -x^2$ is the graph of $y = x^2$ reflected in the x-axis (see Figure 4).

(b) We start with the graph of $y = \sqrt{x}$ (Example 1(c) in Section 2.2). The graph of $g(x) = \sqrt{-x}$ is the graph of $y = \sqrt{x}$ reflected in the y-axis (see Figure 5). Note that the domain of the function $g(x) = \sqrt{-x}$ is $\{x \mid x \le 0\}$.

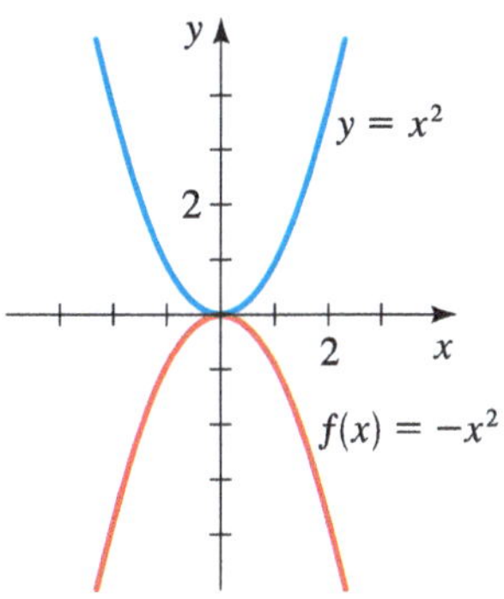

FIGURE 4

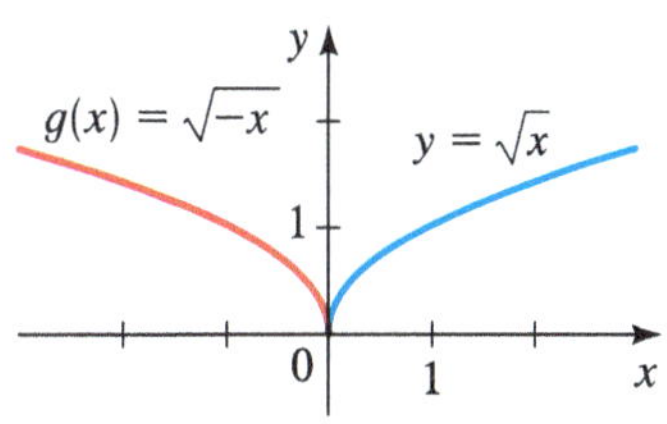

FIGURE 5

Now Try Exercises 37 and 39 ■

Library of Congress Prints and Photographs Division

RENÉ DESCARTES (1596–1650) was born in the town of La Haye in southern France. From an early age Descartes liked mathematics because of "the certainty of its results and the clarity of its reasoning." He believed that to arrive at truth, one must begin by doubting everything, including one's own existence; this led him to formulate perhaps the best-known sentence in all of philosophy: "I think, therefore I am." In his book *Discourse on Method* he described what is now called the Cartesian plane. This idea of combining algebra and geometry enabled mathematicians for the first time to graph functions and thus "see" the equations they were studying. The philosopher John Stuart Mill called this invention "the greatest single step ever made in the progress of the exact sciences." Descartes liked to get up late and spend the morning in bed thinking and writing. He invented the coordinate plane while lying in bed watching a fly crawl on the ceiling, reasoning that he could describe the exact location of the fly by knowing its distance from two perpendicular walls. In 1649 Descartes became the tutor of Queen Christina of Sweden. She liked her lessons at 5 o'clock in the morning, when, she said, her mind was sharpest. However, the change from his usual habits and the ice-cold library where they studied proved too much for Descartes. In February 1650, after just two months of this, he caught pneumonia and died.

Vertical Stretching and Shrinking

Suppose we know the graph of $y = f(x)$. How do we use it to obtain the graph of $y = cf(x)$? The y-coordinate of $y = cf(x)$ at x is the same as the corresponding y-coordinate of $y = f(x)$ multiplied by c. Multiplying the y-coordinates by c has the effect of vertically stretching or shrinking the graph by a factor of c (if $c > 0$).

VERTICAL STRETCHING AND SHRINKING OF GRAPHS

To graph $y = cf(x)$:

If $c > 1$, stretch the graph of $y = f(x)$ vertically by a factor of c.

If $0 < c < 1$, shrink the graph of $y = f(x)$ vertically by a factor of c.

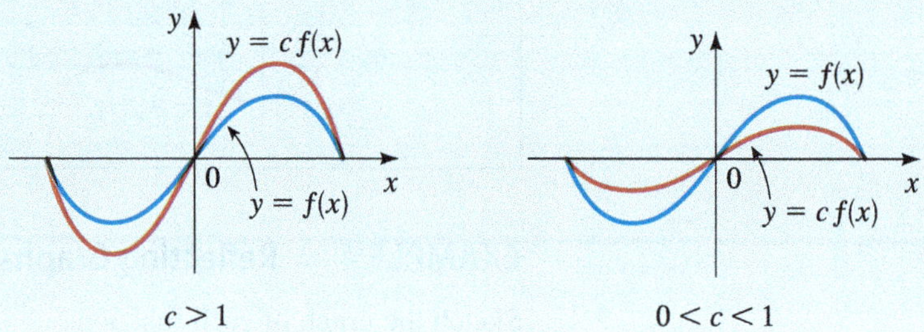

EXAMPLE 5 ■ Vertical Stretching and Shrinking of Graphs

Use the graph of $f(x) = x^2$ to sketch the graph of each function.

(a) $g(x) = 3x^2$ **(b)** $h(x) = \frac{1}{3}x^2$

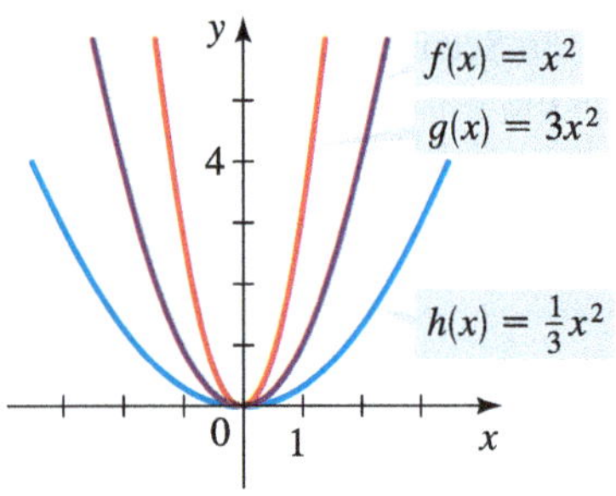

FIGURE 6

SOLUTION

(a) The graph of g is obtained by multiplying the y-coordinate of each point on the graph of f by 3. That is, to obtain the graph of g, we stretch the graph of f vertically by a factor of 3. The result is the narrowest parabola in Figure 6.

(b) The graph of h is obtained by multiplying the y-coordinate of each point on the graph of f by $\frac{1}{3}$. That is, to obtain the graph of h, we shrink the graph of f vertically by a factor of $\frac{1}{3}$. The result is the widest parabola in Figure 6.

Now Try Exercises 41 and 43 ■

We illustrate the effect of combining shifts, reflections, and stretching in the following example.

Mathematics in the Modern World

© Peter Gudella/Shutterstock.com

Computers

For centuries machines have been designed to perform specific tasks. For example, a washing machine washes clothes, a weaving machine weaves cloth, an adding machine adds numbers, and so on. The computer has changed all that.

The computer is a machine that does nothing—until it is given instructions on what to do. So your computer can play games, draw pictures, or calculate π to a million decimal places; it all depends on what program (or instructions) you give the computer. The computer can do all this because it is able to accept instructions and logically change those instructions based on incoming data. This versatility makes computers useful in nearly every aspect of human endeavor.

The idea of a computer was described theoretically in the 1940s by the mathematician Allan Turing (see page 118) in what he called a *universal machine*. In 1945 the mathematician John Von Neumann, extending Turing's ideas, built one of the first electronic computers.

Mathematicians continue to develop new theoretical bases for the design of computers. The heart of the computer is the "chip," which is capable of processing logical instructions. To get an idea of the chip's complexity, consider that the Pentium chip has over 3.5 million logic circuits!

EXAMPLE 6 ■ Combining Shifting, Stretching, and Reflecting

Sketch the graph of the function $f(x) = 1 - 2(x - 3)^2$.

SOLUTION Starting with the graph of $y = x^2$, we first shift to the right 3 units to get the graph of $y = (x - 3)^2$. Then we reflect in the x-axis and stretch by a factor of 2 to get the graph of $y = -2(x - 3)^2$. Finally, we shift upward 1 unit to get the graph of $f(x) = 1 - 2(x - 3)^2$ shown in Figure 7.

Note that the shifts and stretches follow the normal order of operations when evaluating the function. In particular, the upward shift must be performed *last*.

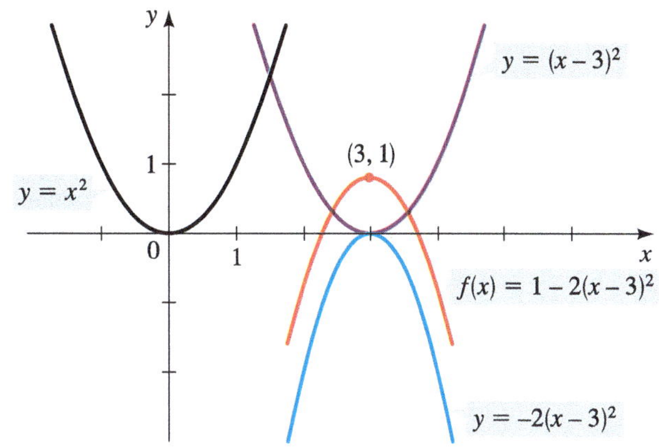

FIGURE 7

Now Try Exercise 47 ■

Horizontal Stretching and Shrinking

Now we consider horizontal shrinking and stretching of graphs. If we know the graph of $y = f(x)$, then how is the graph of $y = f(cx)$ related to it? The y-coordinate of $y = f(cx)$ at x is the same as the y-coordinate of $y = f(x)$ at cx. Thus the x-coordinates in the graph of $y = f(x)$ correspond to the x-coordinates in the graph of $y = f(cx)$ multiplied by c. Looking at this the other way around, we see that the x-coordinates in the graph of $y = f(cx)$ are the x-coordinates in the graph of $y = f(x)$ multiplied by $1/c$. In other words, to change the graph of $y = f(x)$ to the graph of $y = f(cx)$, we must shrink (or stretch) the graph horizontally by a factor of $1/c$ (if $c > 0$), as summarized in the following box.

HORIZONTAL SHRINKING AND STRETCHING OF GRAPHS

To graph $y = f(cx)$:

If $c > 1$, shrink the graph of $y = f(x)$ horizontally by a factor of $1/c$.

If $0 < c < 1$, stretch the graph of $y = f(x)$ horizontally by a factor of $1/c$.

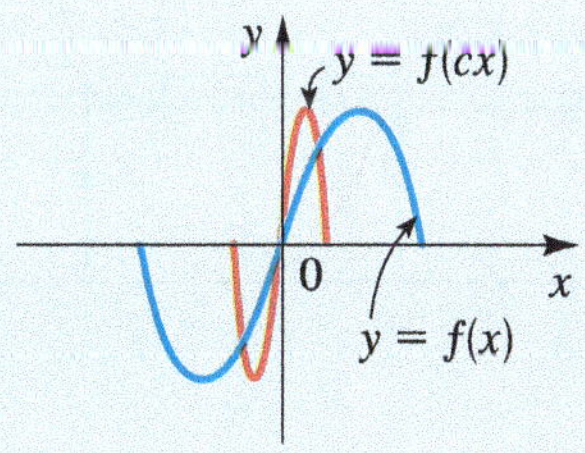

$c > 1$

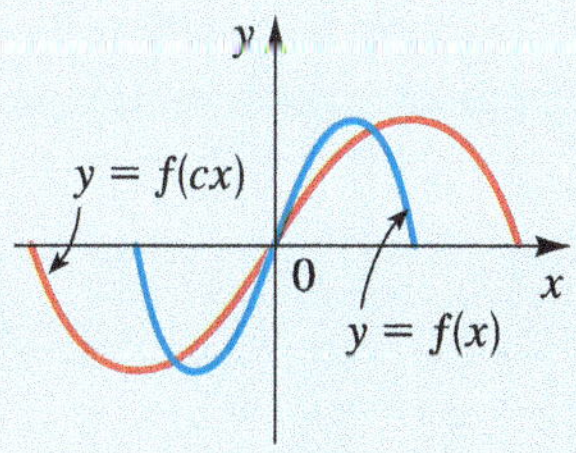

$0 < c < 1$

EXAMPLE 7 ■ Horizontal Stretching and Shrinking of Graphs

The graph of $y = f(x)$ is shown in Figure 8. Sketch the graph of each function.

(a) $y = f(2x)$ **(b)** $y = f(\frac{1}{2}x)$

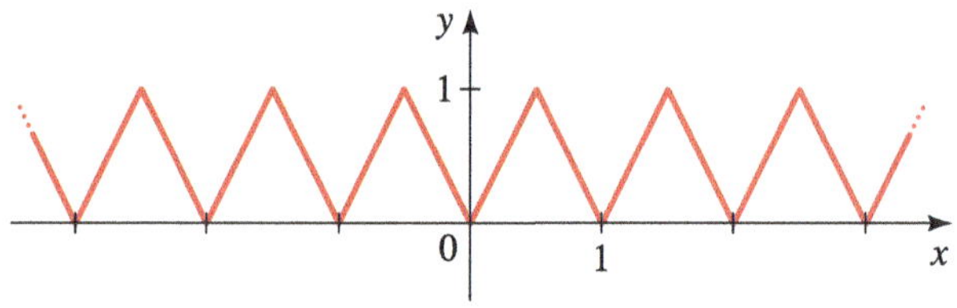

FIGURE 8 $y = f(x)$

SOLUTION Using the principles described on page 203, we **(a)** *shrink* the graph horizontally by the factor $\frac{1}{2}$ to obtain the graph in Figure 9, and **(b)** *stretch* the graph horizontally by the factor 2 to obtain the graph in Figure 10.

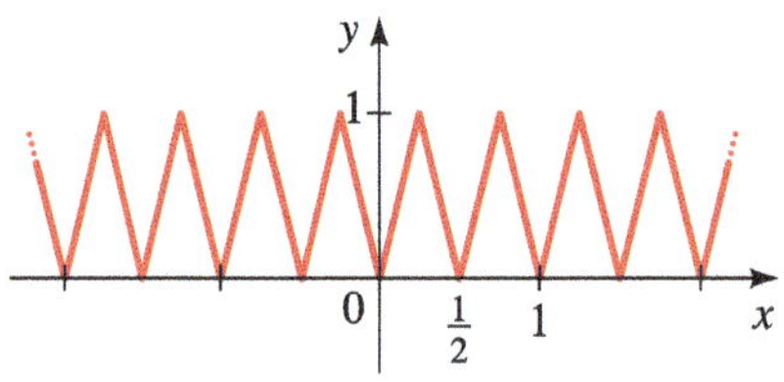

FIGURE 9 $y = f(2x)$

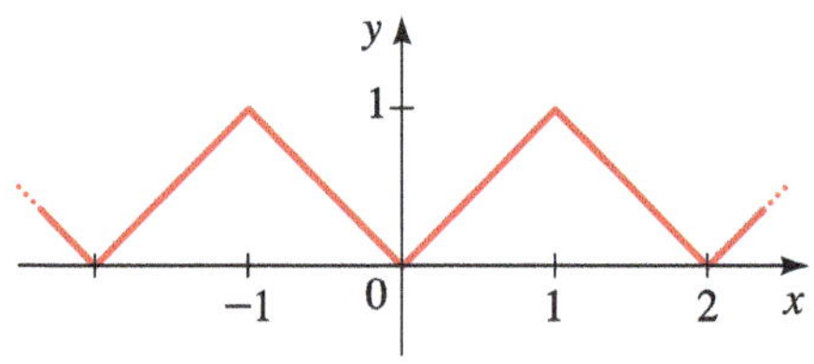

FIGURE 10 $y = f(\frac{1}{2}x)$

Now Try Exercise 71 ■

■ Even and Odd Functions

If a function f satisfies $f(-x) = f(x)$ for every number x in its domain, then f is called an **even function**. For instance, the function $f(x) = x^2$ is even because

$$f(-x) = (-x)^2 = (-1)^2x^2 = x^2 = f(x)$$

The graph of an even function is symmetric with respect to the y-axis (see Figure 11). This means that if we have plotted the graph of f for $x \geq 0$, then we can obtain the entire graph simply by reflecting this portion in the y-axis.

If f satisfies $f(-x) = -f(x)$ for every number x in its domain, then f is called an **odd function**. For example, the function $f(x) = x^3$ is odd because

$$f(-x) = (-x)^3 = (-1)^3x^3 = -x^3 = -f(x)$$

The graph of an odd function is symmetric about the origin (see Figure 12). If we have plotted the graph of f for $x \geq 0$, then we can obtain the entire graph by rotating this portion through 180° about the origin. (This is equivalent to reflecting first in the x-axis and then in the y-axis.)

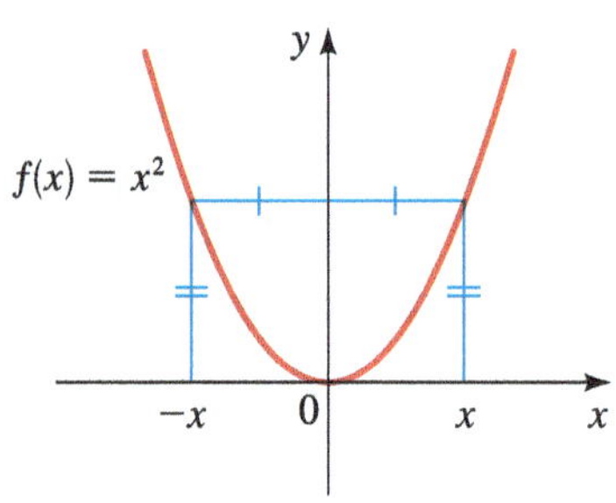

FIGURE 11 $f(x) = x^2$ is an even function.

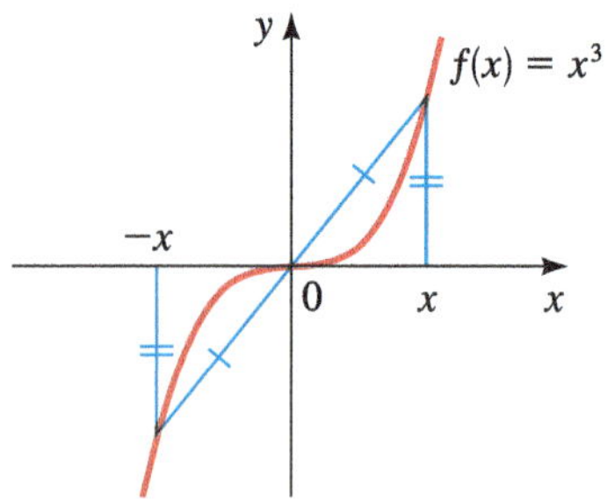

FIGURE 12 $f(x) = x^3$ is an odd function.

SONYA KOVALEVSKY (1850–1891) is considered the most important woman mathematician of the 19th century. She was born in Moscow to an aristocratic family. While a child, she was exposed to the principles of calculus in a very unusual fashion: Her bedroom was temporarily wallpapered with the pages of a calculus book. She later wrote that she "spent many hours in front of that wall, trying to understand it." Since Russian law forbade women from studying in universities, she entered a marriage of convenience, which allowed her to travel to Germany and obtain a doctorate in mathematics from the University of Göttingen. She eventually was awarded a full professorship at the University of Stockholm, where she taught for eight years before dying in an influenza epidemic at the age of 41. Her research was instrumental in helping to put the ideas and applications of functions and calculus on a sound and logical foundation. She received many accolades and prizes for her research work.

EVEN AND ODD FUNCTIONS

Let f be a function.

f is **even** if $f(-x) = f(x)$ for all x in the domain of f.

f is **odd** if $f(-x) = -f(x)$ for all x in the domain of f.

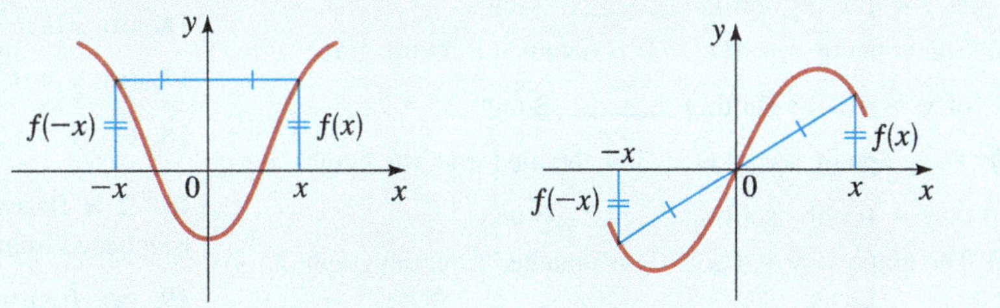

The graph of an even function is symmetric with respect to the y-axis.

The graph of an odd function is symmetric with respect to the origin.

EXAMPLE 8 ■ Even and Odd Functions

Determine whether the functions are even, odd, or neither even nor odd.

(a) $f(x) = x^5 + x$

(b) $g(x) = 1 - x^4$

(c) $h(x) = 2x - x^2$

SOLUTION

(a) $f(-x) = (-x)^5 + (-x)$

$= -x^5 - x = -(x^5 + x)$

$= -f(x)$

Therefore f is an odd function.

(b) $g(-x) = 1 - (-x)^4 = 1 - x^4 = g(x)$

So g is even.

(c) $h(-x) = 2(-x) - (-x)^2 = -2x - x^2$

Since $h(-x) \neq h(x)$ and $h(-x) \neq -h(x)$, we conclude that h is neither even nor odd.

Now Try Exercises 83, 85, and 87 ■

The graphs of the functions in Example 8 are shown in Figure 13. The graph of f is symmetric about the origin, and the graph of g is symmetric about the y-axis. The graph of h is not symmetric about either the y-axis or the origin.

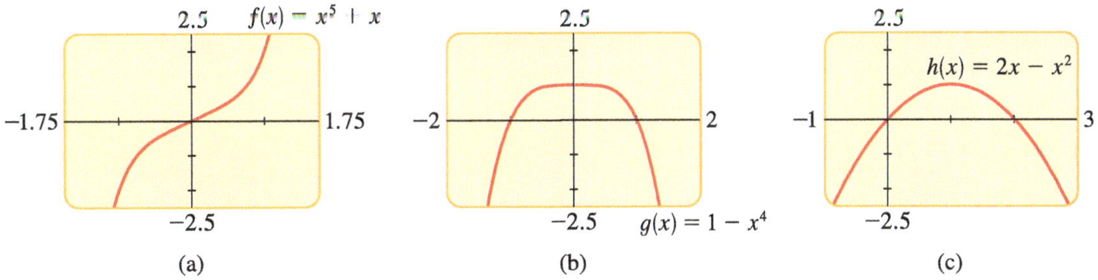

FIGURE 13

2.6 EXERCISES

CONCEPTS

1–2 ■ Fill in the blank with the appropriate direction (left, right, up, or down).

1. **(a)** The graph of $y = f(x) + 3$ is obtained from the graph of $y = f(x)$ by shifting ________ 3 units.

(b) The graph of $y = f(x + 3)$ is obtained from the graph of $y = f(x)$ by shifting ________ 3 units.

2. **(a)** The graph of $y = f(x) - 3$ is obtained from the graph of $y = f(x)$ by shifting ________ 3 units.

(b) The graph of $y = f(x - 3)$ is obtained from the graph of $y = f(x)$ by shifting ________ 3 units.

3. Fill in the blank with the appropriate axis (x-axis or y-axis).

(a) The graph of $y = -f(x)$ is obtained from the graph of $y = f(x)$ by reflecting in the ________.

(b) The graph of $y = f(-x)$ is obtained from the graph of $y = f(x)$ by reflecting in the ________.

4. A graph of a function f is given. Match each equation with one of the graphs labeled I–IV.

(a) $f(x) + 2$ **(b)** $f(x + 3)$

(c) $f(x - 2)$ **(d)** $f(x) - 4$

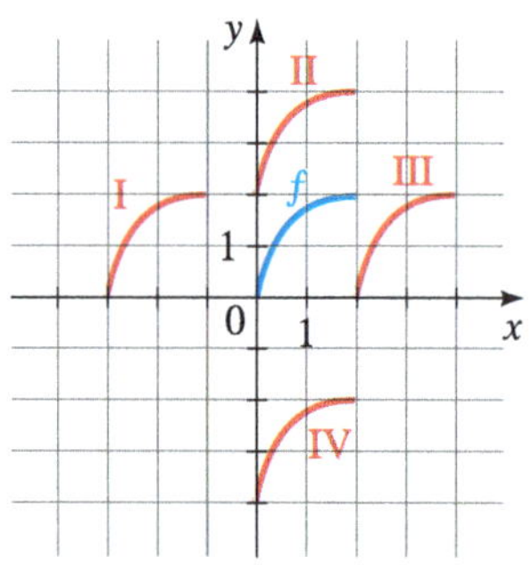

5. If a function f is an even function, then what type of symmetry does the graph of f have?

6. If a function f is an odd function, then what type of symmetry does the graph of f have?

SKILLS

7–18 ■ **Describing Transformations** Suppose the graph of f is given. Describe how the graph of each function can be obtained from the graph of f.

7. **(a)** $f(x) - 1$ **(b)** $f(x - 2)$

8. **(a)** $f(x + 5)$ **(b)** $f(x) + 4$

9. **(a)** $f(-x)$ **(b)** $3f(x)$

10. **(a)** $-f(x)$ **(b)** $\frac{1}{3}f(x)$

11. **(a)** $y = f(x - 5) + 2$ **(b)** $y = f(x + 1) - 1$

12. **(a)** $y = f(x + 3) + 2$ **(b)** $y = f(x - 7) - 3$

13. **(a)** $y = -f(x) + 5$ **(b)** $y = 3f(x) - 5$

14. **(a)** $1 - f(-x)$ **(b)** $2 - \frac{1}{5}f(x)$

15. **(a)** $2f(x + 5) - 1$ **(b)** $\frac{1}{4}f(x - 3) + 5$

16. **(a)** $\frac{1}{3}f(x - 2) + 5$ **(b)** $4f(x + 1) + 3$

17. **(a)** $y = f(4x)$ **(b)** $y = f(\frac{1}{4}x)$

18. **(a)** $y = f(2x) - 1$ **(b)** $y = 2f(\frac{1}{2}x)$

19–22 ■ **Describing Transformations** Explain how the graph of g is obtained from the graph of f.

19. **(a)** $f(x) = x^2, \quad g(x) = (x + 2)^2$

(b) $f(x) = x^2, \quad g(x) = x^2 + 2$

20. **(a)** $f(x) = x^3, \quad g(x) = (x - 4)^3$

(b) $f(x) = x^3, \quad g(x) = x^3 - 4$

21. **(a)** $f(x) = |x|, \quad g(x) = |x + 2| - 2$

(b) $f(x) = |x|, \quad g(x) = |x - 2| + 2$

22. **(a)** $f(x) = \sqrt{x}, \quad g(x) = -\sqrt{x} + 1$

(b) $f(x) = \sqrt{x}, \quad g(x) = \sqrt{-x} + 1$

23. **Graphing Transformations** Use the graph of $y = x^2$ in Figure 4 to graph the following.

(a) $g(x) = x^2 + 1$ **(b)** $g(x) = (x - 1)^2$

(c) $g(x) = -x^2$ **(d)** $g(x) = (x - 1)^2 + 3$

24. **Graphing Transformations** Use the graph of $y = \sqrt{x}$ in Figure 5 to graph the following.

(a) $g(x) = \sqrt{x - 2}$ **(b)** $g(x) = \sqrt{x} + 1$

(c) $g(x) = \sqrt{x + 2} + 2$ **(d)** $g(x) = -\sqrt{x} + 1$

25–28 ■ **Identifying Transformations** Match the graph with the function. (See the graph of $y = |x|$ on page 96.)

25. $y = |x + 1|$ **26.** $y = |x - 1|$

27. $y = |x| - 1$ **28.** $y = -|x|$

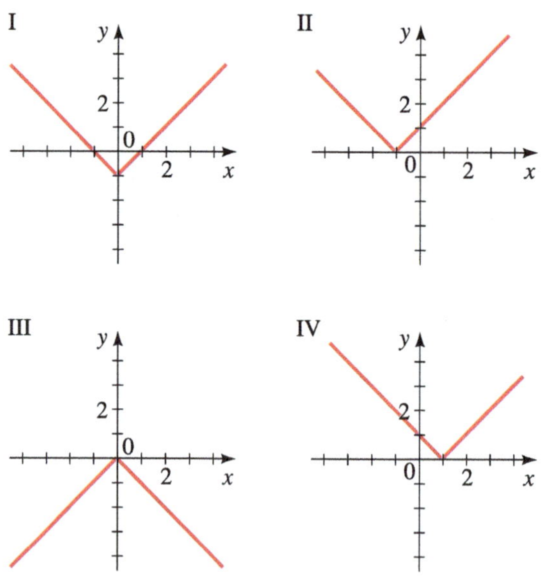

29–52 ■ Graphing Transformations Sketch the graph of the function, not by plotting points, but by starting with the graph of a standard function and applying transformations.

29. $f(x) = x^2 + 3$

30. $f(x) = x^2 - 4$

31. $f(x) = |x| - 1$

32. $f(x) = \sqrt{x} + 1$

33. $f(x) = (x - 5)^2$

34. $f(x) = (x + 1)^2$

35. $f(x) = |x + 2|$

36. $f(x) = \sqrt{x - 4}$

37. $f(x) = -x^3$

38. $f(x) = -|x|$

39. $y = \sqrt[4]{-x}$

40. $y = \sqrt[3]{-x}$

41. $y = \frac{1}{4}x^2$

42. $y = -5\sqrt{x}$

43. $y = 3|x|$

44. $y = \frac{1}{2}|x|$

45. $y = (x - 3)^2 + 5$

46. $y = \sqrt{x + 4} - 3$

47. $y = 3 - \frac{1}{2}(x - 1)^2$

48. $y = 2 - \sqrt{x + 1}$

49. $y = |x + 2| + 2$

50. $y = 2 - |x|$

51. $y = \frac{1}{2}\sqrt{x + 4} - 3$

52. $y = 3 - 2(x - 1)^2$

53–62 ■ Finding Equations for Transformations A function f is given, and the indicated transformations are applied to its graph (in the given order). Write an equation for the final transformed graph.

53. $f(x) = x^2$; shift downward 3 units

54. $f(x) = x^3$; shift upward 5 units

55. $f(x) = \sqrt{x}$; shift 2 units to the left

56. $f(x) = \sqrt[3]{x}$; shift 1 unit to the right

57. $f(x) = |x|$; shift 2 units to the left and shift downward 5 units

58. $f(x) = |x|$; reflect in the x-axis, shift 4 units to the right, and shift upward 3 units.

59. $f(x) = \sqrt[4]{x}$; reflect in the y-axis and shift upward 1 unit

60. $f(x) = x^2$; shift 2 units to the left and reflect in the x-axis

61. $f(x) = x^2$; stretch vertically by a factor of 2, shift downward 2 units, and shift 3 units to the right

62. $f(x) = |x|$; shrink vertically by a factor of $\frac{1}{2}$, shift to the left 1 unit, and shift upward 3 units

63–68 ■ Finding Formulas for Transformations The graphs of f and g are given. Find a formula for the function g.

63.

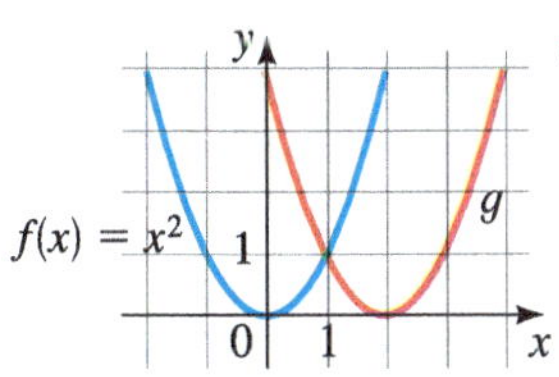

64.

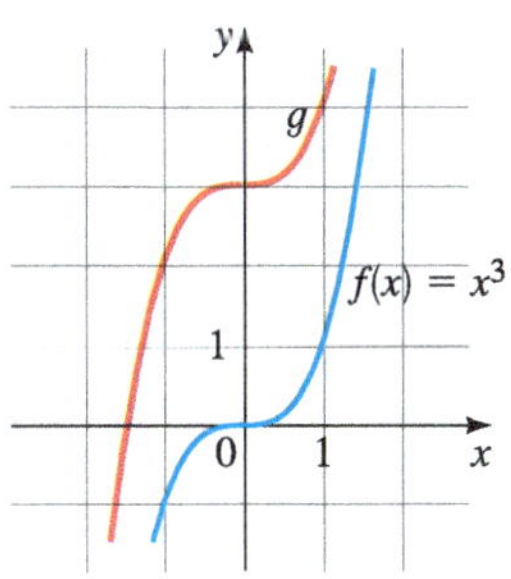

65.

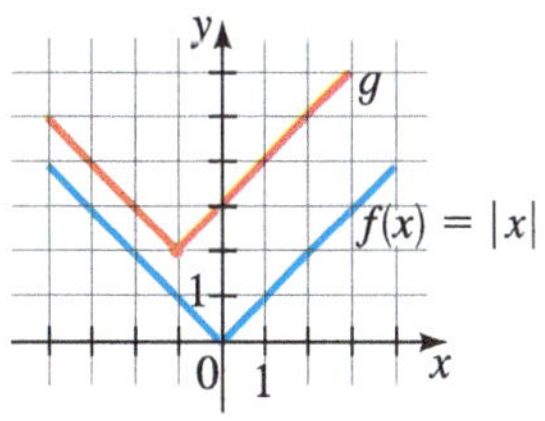

66.

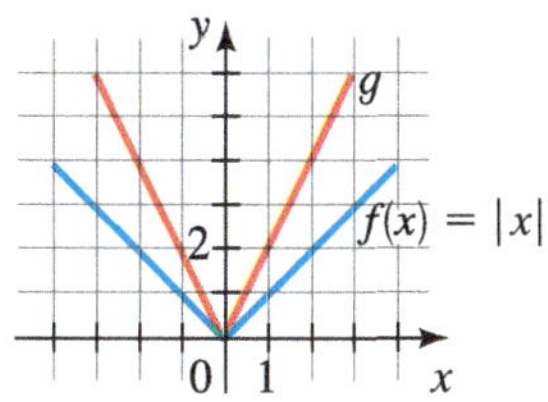

67.

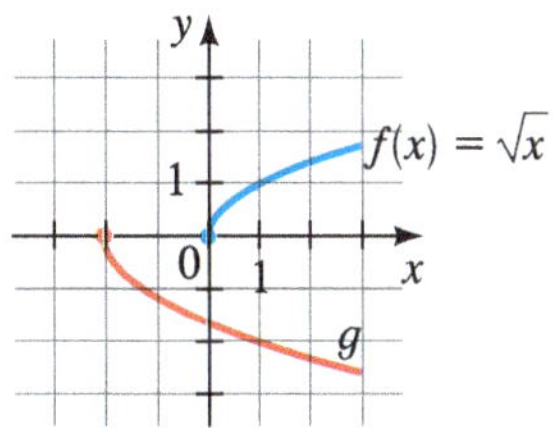

68.

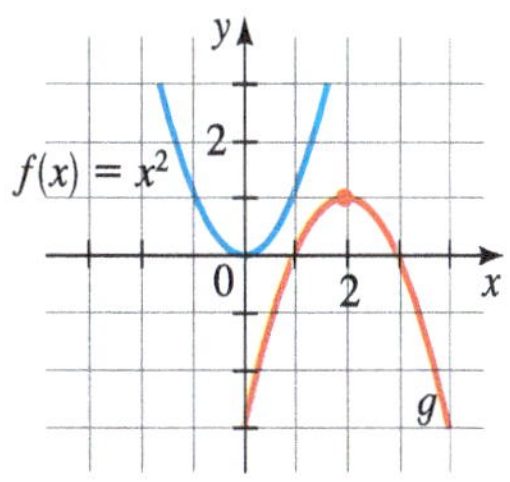

69–70 ■ Identifying Transformations The graph of $y = f(x)$ is given. Match each equation with its graph.

69. **(a)** $y = f(x - 4)$ **(b)** $y = f(x) + 3$
(c) $y = 2f(x + 6)$ **(d)** $y = -f(2x)$

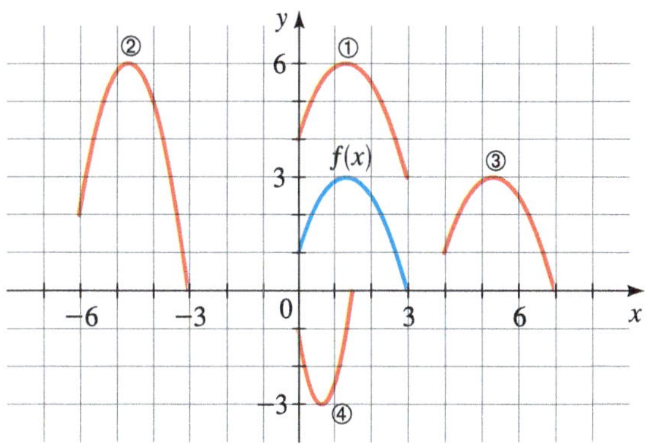

70. **(a)** $y = \frac{1}{3}f(x)$ **(b)** $y = -f(x + 4)$
(c) $y = f(x - 4) + 3$ **(d)** $y = f(-x)$

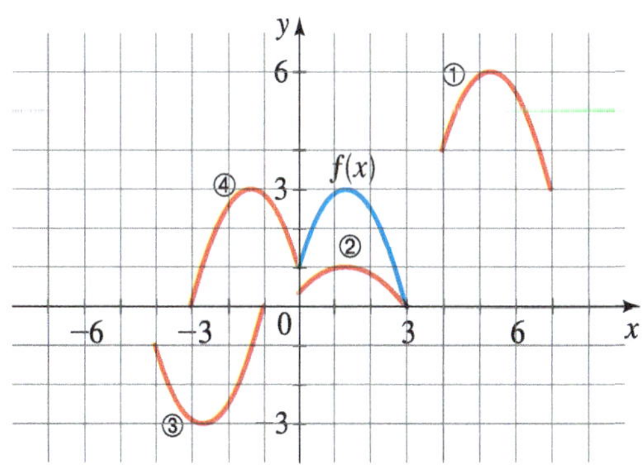

71–74 ■ Graphing Transformations The graph of a function f is given. Sketch the graphs of the following transformations of f.

71. **(a)** $y = f(x - 2)$ **(b)** $y = f(x) - 2$
(c) $y = 2f(x)$ **(d)** $y = -f(x) + 3$
(e) $y = f(-x)$ **(f)** $y = \frac{1}{2}f(x - 1)$

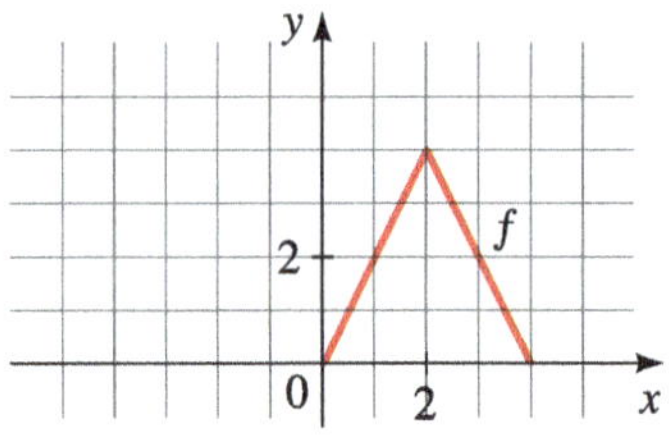

72. **(a)** $y = f(x + 1)$ **(b)** $y = f(-x)$
(c) $y = f(x - 2)$ **(d)** $y = f(x) - 2$
(e) $y = -f(x)$ **(f)** $y = 2f(x)$

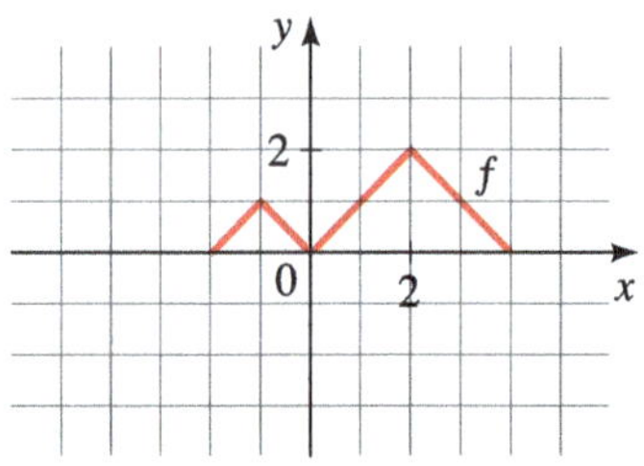

73. **(a)** $y = f(2x)$ **(b)** $y = f(\frac{1}{2}x)$

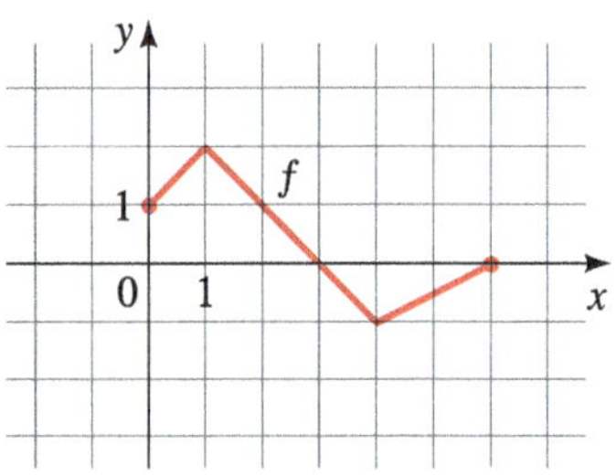

74. **(a)** $y = f(3x)$ **(b)** $y = f(\frac{1}{3}x)$

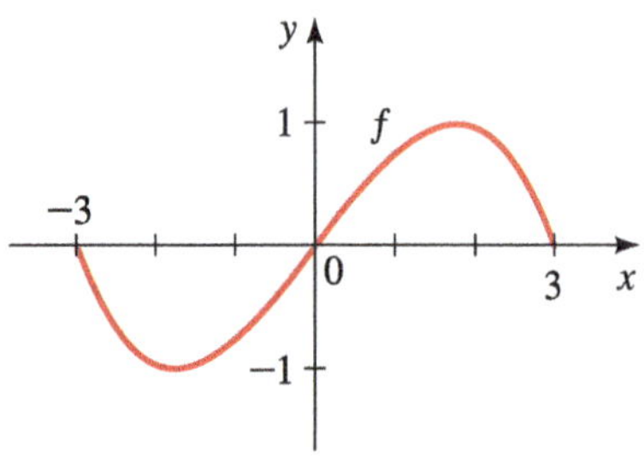

75–76 ■ Graphing Transformations Use the graph of $f(x) = [\![x]\!]$ described on page 163 to graph the indicated function.

75. $y = [\![2x]\!]$ **76.** $y = [\![\frac{1}{4}x]\!]$

77–80 ■ Graphing Transformations Graph the functions on the same screen using the given viewing rectangle. How is each graph related to the graph in part (a)?

77. Viewing rectangle $[-8, 8]$ by $[-2, 8]$
(a) $y = \sqrt[4]{x}$ **(b)** $y = \sqrt[4]{x + 5}$
(c) $y = 2\sqrt[4]{x + 5}$ **(d)** $y = 4 + 2\sqrt[4]{x + 5}$

78. Viewing rectangle $[-8, 8]$ by $[-6, 6]$
(a) $y = |x|$ **(b)** $y = -|x|$
(c) $y = -3|x|$ **(d)** $y = -3|x - 5|$

79. Viewing rectangle $[-4, 6]$ by $[-4, 4]$
(a) $y = x^6$ **(b)** $y = \frac{1}{3}x^6$
(c) $y = -\frac{1}{3}x^6$ **(d)** $y = -\frac{1}{3}(x - 4)^6$

80. Viewing rectangle $[-6, 6]$ by $[-4, 4]$
(a) $y = \dfrac{1}{\sqrt{x}}$ **(b)** $y = \dfrac{1}{\sqrt{x + 3}}$
(c) $y = \dfrac{1}{2\sqrt{x + 3}}$ **(d)** $y = \dfrac{1}{2\sqrt{x + 3}} - 3$

81–82 ■ Graphing Transformations If $f(x) = \sqrt{2x - x^2}$, graph the following functions in the viewing rectangle $[-5, 5]$ by $[-4, 4]$. How is each graph related to the graph in part (a)?

81. **(a)** $y = f(x)$ **(b)** $y = f(2x)$ **(c)** $y = f(\frac{1}{2}x)$

82. **(a)** $y = f(x)$ **(b)** $y = f(-x)$
(c) $y = -f(-x)$ **(d)** $y = f(-2x)$
(e) $y = f(-\frac{1}{2}x)$

83–90 ■ Even and Odd Functions Determine whether the function f is even, odd, or neither. If f is even or odd, use symmetry to sketch its graph.

83. $f(x) = x^4$ **84.** $f(x) = x^3$
85. $f(x) = x^2 + x$ **86.** $f(x) = x^4 - 4x^2$
87. $f(x) = x^3 - x$ **88.** $f(x) = 3x^3 + 2x^2 + 1$
89. $f(x) = 1 - \sqrt[3]{x}$ **90.** $f(x) = x + \dfrac{1}{x}$

SKILLS Plus

91–92 ■ Graphing Even and Odd Functions The graph of a function defined for $x \geq 0$ is given. Complete the graph for $x < 0$ to make **(a)** an even function and **(b)** an odd function.

91.

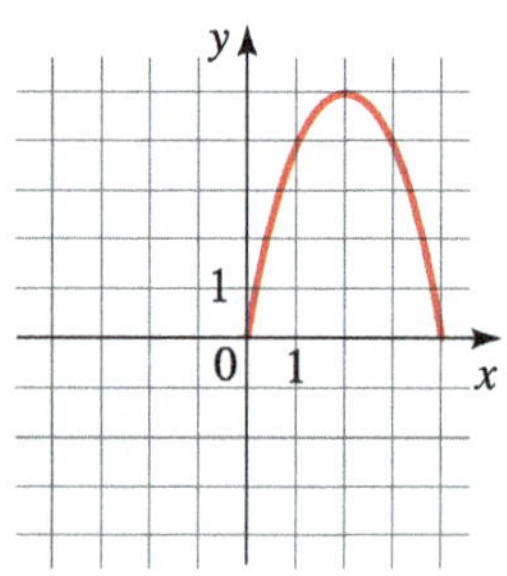

92.

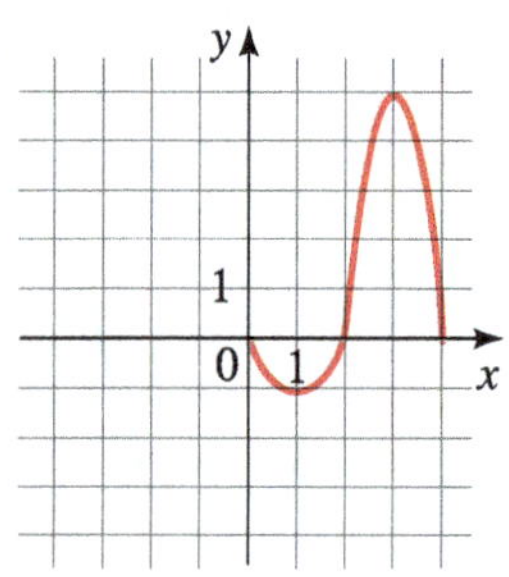

93–94 ■ Graphing the Absolute Value of a Function These exercises show how the graph of $y = |f(x)|$ is obtained from the graph of $y = f(x)$.

93. The graphs of $f(x) = x^2 - 4$ and $g(x) = |x^2 - 4|$ are shown. Explain how the graph of g is obtained from the graph of f.

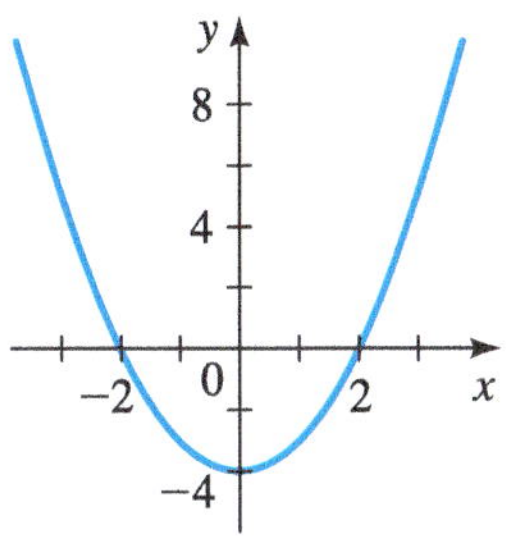

$f(x) = x^2 - 4$

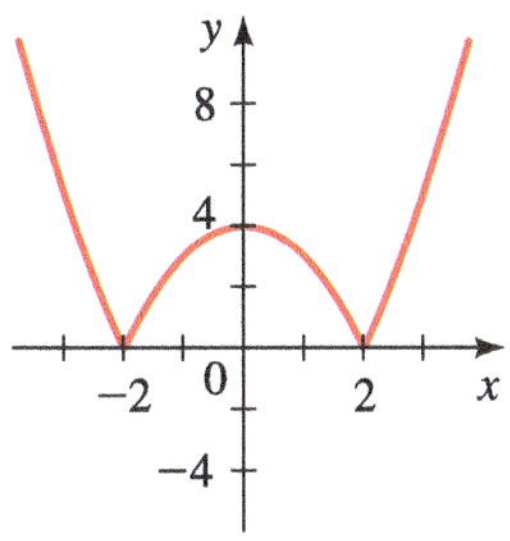

$g(x) = |x^2 - 4|$

94. The graph of $f(x) = x^4 - 4x^2$ is shown. Use this graph to sketch the graph of $g(x) = |x^4 - 4x^2|$.

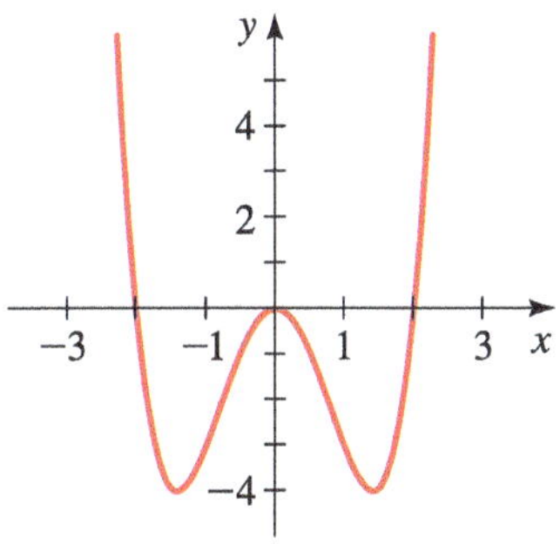

95–96 ■ Graphing the Absolute Value of a Function Sketch the graph of each function.

95. (a) $f(x) = 4x - x^2$ **(b)** $g(x) = |4x - x^2|$

96. (a) $f(x) = x^3$ **(b)** $g(x) = |x^3|$

APPLICATIONS

97. Bungee Jumping Luisa goes bungee jumping from a 500-ft-high bridge. The graph shows Luisa's height $h(t)$ (in ft) after t seconds.

(a) Describe in words what the graph indicates about Luisa's bungee jump.

(b) Suppose Luisa goes bungee jumping from a 400-ft-high bridge. Sketch a new graph that shows Luisa's height $H(t)$ after t seconds.

(c) What transformation must be performed on the function h to obtain the function H? Express the function H in terms of h.

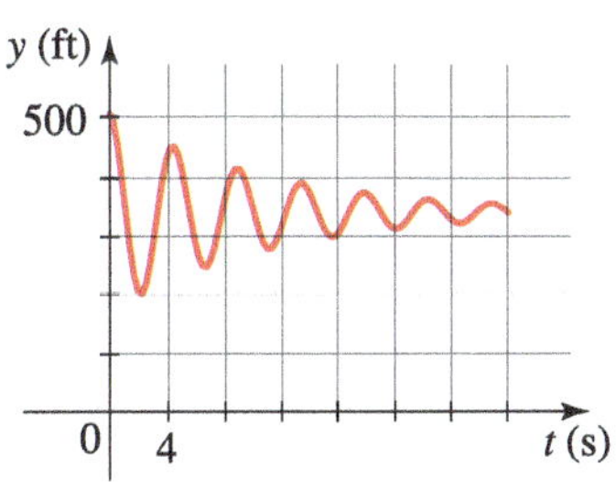

98. Swimming Laps Miyuki practices swimming laps with her team. The function $y = f(t)$ graphed below gives her distance (in meters) from the starting edge of the pool t seconds after she starts her laps.

(a) Describe in words Miyuki's swim practice. What is her average speed for the first 30 s?

(b) Graph the function $y = 1.2f(t)$. How is the graph of the new function related to the graph of the original function?

(c) What is Miyuki's new average speed for the first 30 s?

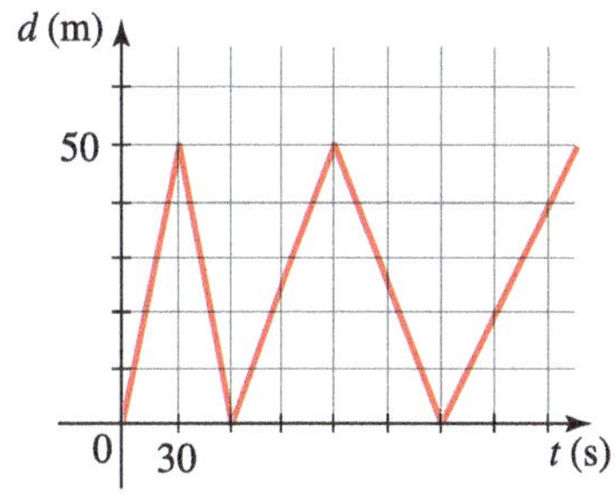

99. Field Trip A class of fourth graders walks to a park on a field trip. The function $y = f(t)$ graphed below gives their distance from school (in ft) t minutes after they left school.

(a) What is the average speed going to the park? How long was the class at the park? How far away is the park?

(b) Graph the function $y = 0.5f(t)$. How is the graph of the new function related to the graph of the original function? What is the average speed going to the new park? How far away is the new park?

(c) Graph the function $y = f(t - 10)$. How is the graph of the new function related to the graph of the original function? How does the field trip descibed by this function differ from the original trip?

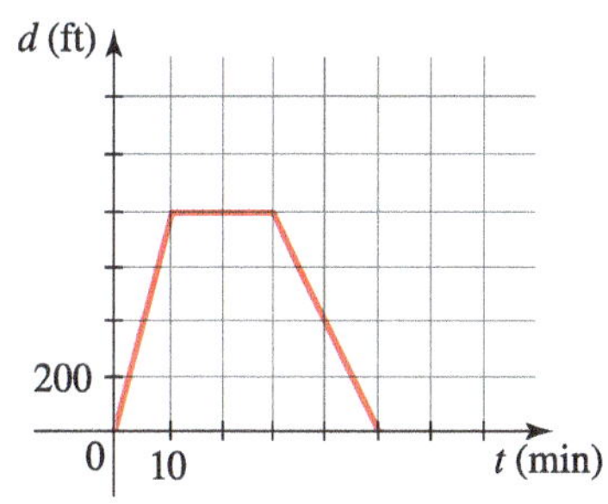

DISCUSS ■ DISCOVER ■ PROVE ■ WRITE

100–101 ■ DISCUSS: Obtaining Transformations Can the function g be obtained from f by transformations? If so, describe the transformations needed.

100. The functions f and g are described algebraically as follows:

$$f(x) = (x + 2)^2 \qquad g(x) = (x - 2)^2 + 5$$

101. The functions f and g are described graphically in the figure.

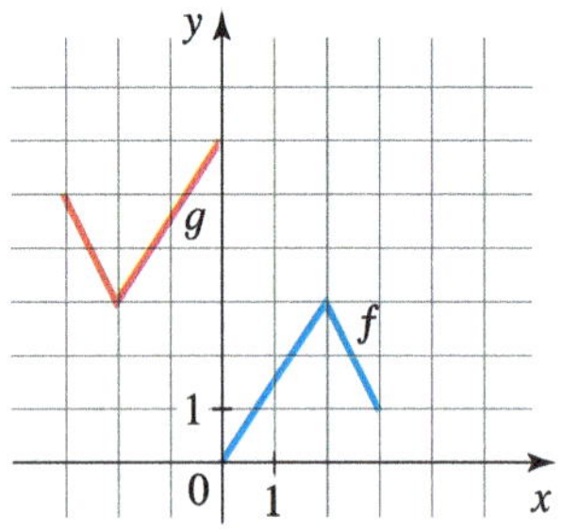

102. DISCUSS: Sums of Even and Odd Functions If f and g are both even functions, is $f + g$ necessarily even? If both are odd, is their sum necessarily odd? What can you say about the sum if one is odd and one is even? In each case, prove your answer.

103. DISCUSS: Products of Even and Odd Functions Answer the same questions as in Exercise 102, except this time consider the product of f and g instead of the sum.

104. DISCUSS: Even and Odd Power Functions What must be true about the integer n if the function

$$f(x) = x^n$$

is an even function? If it is an odd function? Why do you think the names "even" and "odd" were chosen for these function properties?

2.7 COMBINING FUNCTIONS

■ Sums, Differences, Products, and Quotients ■ Composition of Functions
■ Applications of Composition

In this section we study different ways to combine functions to make new functions.

■ Sums, Differences, Products, and Quotients

The sum of f and g is defined by

$$(f + g)(x) = f(x) + g(x)$$

The name of the new function is "$f + g$." So this + sign stands for the operation of addition of *functions*. The + sign on the right side, however, stands for addition of the *numbers* $f(x)$ and $g(x)$.

Two functions f and g can be combined to form new functions $f + g$, $f - g$, fg, and f/g in a manner similar to the way we add, subtract, multiply, and divide real numbers. For example, we define the function $f + g$ by

$$(f + g)(x) = f(x) + g(x)$$

The new function $f + g$ is called the **sum** of the functions f and g; its value at x is $f(x) + g(x)$. Of course, the sum on the right-hand side makes sense only if both $f(x)$ and $g(x)$ are defined, that is, if x belongs to the domain of f and also to the domain of g. So if the domain of f is A and the domain of g is B, then the domain of $f + g$ is the intersection of these domains, that is, $A \cap B$. Similarly, we can define the **difference** $f - g$, the **product** fg, and the **quotient** f/g of the functions f and g. Their domains are $A \cap B$, but in the case of the quotient we must remember not to divide by 0.

ALGEBRA OF FUNCTIONS

Let f and g be functions with domains A and B. Then the functions $f + g$, $f - g$, fg, and f/g are defined as follows.

$(f + g)(x) = f(x) + g(x)$	Domain $A \cap B$
$(f - g)(x) = f(x) - g(x)$	Domain $A \cap B$
$(fg)(x) = f(x)g(x)$	Domain $A \cap B$
$\left(\dfrac{f}{g}\right)(x) = \dfrac{f(x)}{g(x)}$	Domain $\{x \in A \cap B \mid g(x) \neq 0\}$

EXAMPLE 1 ■ Combinations of Functions and Their Domains

Let $f(x) = \dfrac{1}{x-2}$ and $g(x) = \sqrt{x}$.

(a) Find the functions $f+g$, $f-g$, fg, and f/g and their domains.

(b) Find $(f+g)(4)$, $(f-g)(4)$, $(fg)(4)$, and $(f/g)(4)$.

SOLUTION

(a) The domain of f is $\{x \mid x \neq 2\}$, and the domain of g is $\{x \mid x \geq 0\}$. The intersection of the domains of f and g is

$$\{x \mid x \geq 0 \text{ and } x \neq 2\} = [0, 2) \cup (2, \infty)$$

Thus we have

$$(f+g)(x) = f(x) + g(x) = \frac{1}{x-2} + \sqrt{x} \qquad \text{Domain } \{x \mid x \geq 0 \text{ and } x \neq 2\}$$

$$(f-g)(x) = f(x) - g(x) = \frac{1}{x-2} - \sqrt{x} \qquad \text{Domain } \{x \mid x \geq 0 \text{ and } x \neq 2\}$$

$$(fg)(x) = f(x)g(x) = \frac{\sqrt{x}}{x-2} \qquad \text{Domain } \{x \mid x \geq 0 \text{ and } x \neq 2\}$$

$$\left(\frac{f}{g}\right)(x) = \frac{f(x)}{g(x)} = \frac{1}{(x-2)\sqrt{x}} \qquad \text{Domain } \{x \mid x > 0 \text{ and } x \neq 2\}$$

To divide fractions, invert the denominator and multiply:

$$\frac{1/(x-2)}{\sqrt{x}} = \frac{1/(x-2)}{\sqrt{x}/1} = \frac{1}{x-2} \cdot \frac{1}{\sqrt{x}} = \frac{1}{(x-2)\sqrt{x}}$$

Note that in the domain of f/g we exclude 0 because $g(0) = 0$.

(b) Each of these values exist because $x = 4$ is in the domain of each function:

$$(f+g)(4) = f(4) + g(4) = \frac{1}{4-2} + \sqrt{4} = \frac{5}{2}$$

$$(f-g)(4) = f(4) - g(4) = \frac{1}{4-2} - \sqrt{4} = -\frac{3}{2}$$

$$(fg)(4) = f(4)g(4) = \left(\frac{1}{4-2}\right)\sqrt{4} = 1$$

$$\left(\frac{f}{g}\right)(4) = \frac{f(4)}{g(4)} = \frac{1}{(4-2)\sqrt{4}} = \frac{1}{4}$$

Now Try Exercise 9 ■

DISCOVERY PROJECT

Iteration and Chaos

The *iterates* of a function f at a point x are the numbers $f(x)$, $f(f(x))$, $f(f(f(x)))$, and so on. We examine iterates of the *logistic function,* which models the population of a species with limited potential for growth (such as lizards on an island or fish in a pond). Iterates of the model can help us to predict whether the population will eventually stabilize or whether it will fluctuate chaotically. You can find the project at **www.stewartmath.com**.

The graph of the function $f + g$ can be obtained from the graphs of f and g by **graphical addition**. This means that we add corresponding y-coordinates, as illustrated in the next example.

EXAMPLE 2 ■ Using Graphical Addition

The graphs of f and g are shown in Figure 1. Use graphical addition to graph the function $f + g$.

SOLUTION We obtain the graph of $f + g$ by "graphically adding" the value of $f(x)$ to $g(x)$ as shown in Figure 2. This is implemented by copying the line segment PQ on top of PR to obtain the point S on the graph of $f + g$.

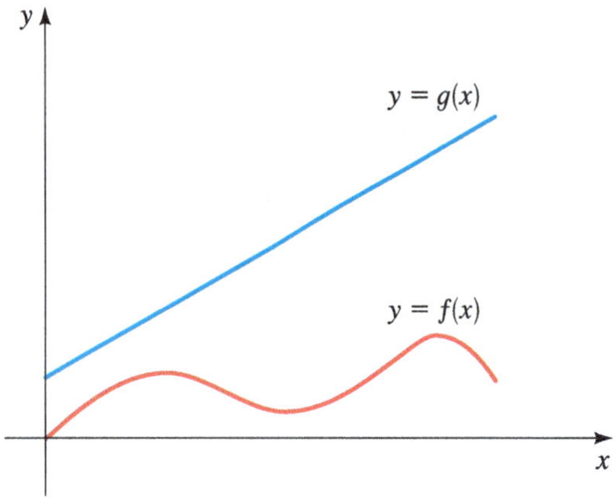

FIGURE 1

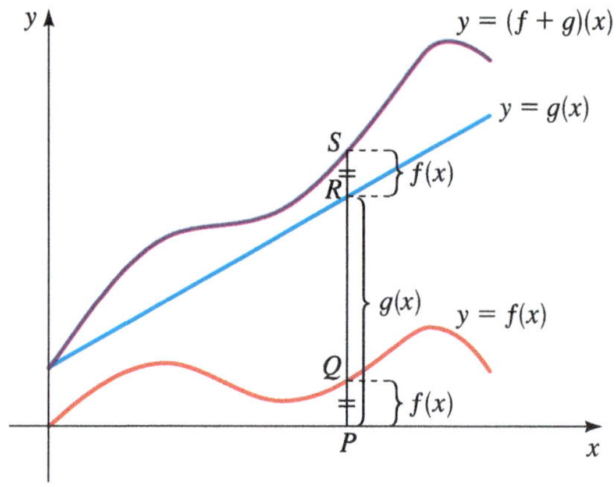

FIGURE 2 Graphical addition

Now Try Exercise 21 ■

■ Composition of Functions

Now let's consider a very important way of combining two functions to get a new function. Suppose $f(x) = \sqrt{x}$ and $g(x) = x^2 + 1$. We may define a new function h as

$$h(x) = f(g(x)) = f(x^2 + 1) = \sqrt{x^2 + 1}$$

The function h is made up of the functions f and g in an interesting way: Given a number x, we first apply the function g to it, then apply f to the result. In this case, f is the rule "take the square root," g is the rule "square, then add 1," and h is the rule "square, then add 1, then take the square root." In other words, we get the rule h by applying the rule g and then the rule f. Figure 3 shows a machine diagram for h.

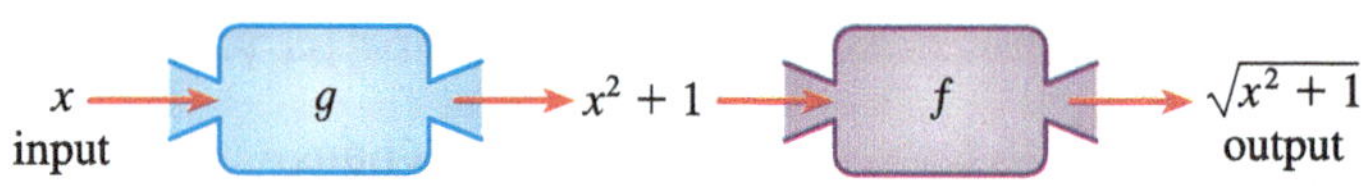

FIGURE 3 The h machine is composed of the g machine (first) and then the f machine.

In general, given any two functions f and g, we start with a number x in the domain of g and find its image $g(x)$. If this number $g(x)$ is in the domain of f, we can then calculate the value of $f(g(x))$. The result is a new function $h(x) = f(g(x))$ that is obtained by substituting g into f. It is called the *composition* (or *composite*) of f and g and is denoted by $f \circ g$ ("f composed with g").

COMPOSITION OF FUNCTIONS

Given two functions f and g, the **composite function** $f \circ g$ (also called the **composition** of f and g) is defined by

$$(f \circ g)(x) = f(g(x))$$

The domain of $f \circ g$ is the set of all x in the domain of g such that $g(x)$ is in the domain of f. In other words, $(f \circ g)(x)$ is defined whenever both $g(x)$ and $f(g(x))$ are defined. We can picture $f \circ g$ using an arrow diagram (Figure 4).

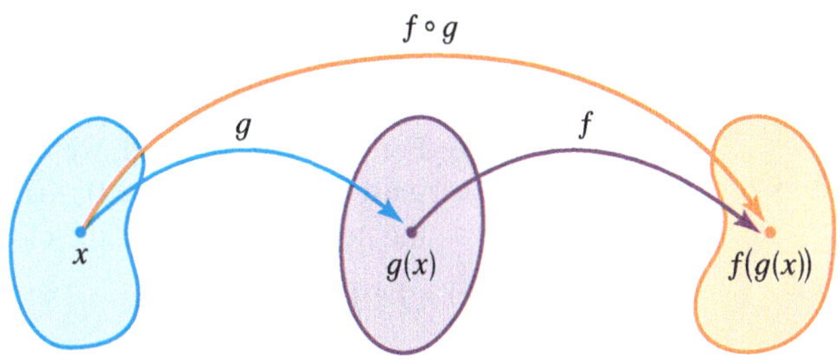

FIGURE 4 Arrow diagram for $f \circ g$

EXAMPLE 3 ■ Finding the Composition of Functions

Let $f(x) = x^2$ and $g(x) = x - 3$.

(a) Find the functions $f \circ g$ and $g \circ f$ and their domains.

(b) Find $(f \circ g)(5)$ and $(g \circ f)(7)$.

SOLUTION

In Example 3, f is the rule "square," and g is the rule "subtract 3." The function $f \circ g$ *first* subtracts 3 and *then* squares; the function $g \circ f$ *first* squares and *then* subtracts 3.

(a) We have

$$\begin{aligned} (f \circ g)(x) &= f(g(x)) && \text{Definition of } f \circ g \\ &= f(x - 3) && \text{Definition of } g \\ &= (x - 3)^2 && \text{Definition of } f \end{aligned}$$

and

$$\begin{aligned} (g \circ f)(x) &= g(f(x)) && \text{Definition of } g \circ f \\ &= g(x^2) && \text{Definition of } f \\ &= x^2 - 3 && \text{Definition of } g \end{aligned}$$

The domains of both $f \circ g$ and $g \circ f$ are $\mathbb{R}$.

(b) We have

$$(f \circ g)(5) = f(g(5)) = f(2) = 2^2 = 4$$

$$(g \circ f)(7) = g(f(7)) = g(49) = 49 - 3 = 46$$

Now Try Exercises 27 and 49 ■

You can see from Example 3 that, in general, $f \circ g \neq g \circ f$. Remember that the notation $f \circ g$ means that the function g is applied first and then f is applied second.

The graphs of f and g of Example 4, as well as those of $f \circ g$, $g \circ f$, $f \circ f$, and $g \circ g$, are shown below. These graphs indicate that the operation of composition can produce functions that are quite different from the original functions.

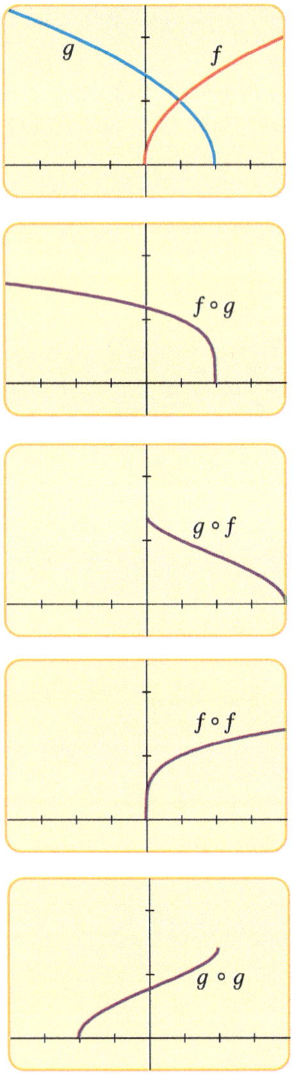

EXAMPLE 4 ■ Finding the Composition of Functions

If $f(x) = \sqrt{x}$ and $g(x) = \sqrt{2 - x}$, find the following functions and their domains.

(a) $f \circ g$ **(b)** $g \circ f$ **(c)** $f \circ f$ **(d)** $g \circ g$

SOLUTION

(a)
$$\begin{aligned}(f \circ g)(x) &= f(g(x)) && \text{Definition of } f \circ g\\ &= f(\sqrt{2 - x}) && \text{Definition of } g\\ &= \sqrt{\sqrt{2 - x}} && \text{Definition of } f\\ &= \sqrt[4]{2 - x}\end{aligned}$$

The domain of $f \circ g$ is $\{x \mid 2 - x \geq 0\} = \{x \mid x \leq 2\} = (-\infty, 2]$.

(b)
$$\begin{aligned}(g \circ f)(x) &= g(f(x)) && \text{Definition of } g \circ f\\ &= g(\sqrt{x}) && \text{Definition of } f\\ &= \sqrt{2 - \sqrt{x}} && \text{Definition of } g\end{aligned}$$

For $\sqrt{x}$ to be defined, we must have $x \geq 0$. For $\sqrt{2 - \sqrt{x}}$ to be defined, we must have $2 - \sqrt{x} \geq 0$, that is, $\sqrt{x} \leq 2$, or $x \leq 4$. Thus we have $0 \leq x \leq 4$, so the domain of $g \circ f$ is the closed interval $[0, 4]$.

(c)
$$\begin{aligned}(f \circ f)(x) &= f(f(x)) && \text{Definition of } f \circ f\\ &= f(\sqrt{x}) && \text{Definition of } f\\ &= \sqrt{\sqrt{x}} && \text{Definition of } f\\ &= \sqrt[4]{x}\end{aligned}$$

The domain of $f \circ f$ is $[0, \infty)$.

(d)
$$\begin{aligned}(g \circ g)(x) &= g(g(x)) && \text{Definition of } g \circ g\\ &= g(\sqrt{2 - x}) && \text{Definition of } g\\ &= \sqrt{2 - \sqrt{2 - x}} && \text{Definition of } g\end{aligned}$$

This expression is defined when both $2 - x \geq 0$ and $2 - \sqrt{2 - x} \geq 0$. The first inequality means $x \leq 2$, and the second is equivalent to $\sqrt{2 - x} \leq 2$, or $2 - x \leq 4$, or $x \geq -2$. Thus $-2 \leq x \leq 2$, so the domain of $g \circ g$ is $[-2, 2]$.

Now Try Exercise 55 ■

It is possible to take the composition of three or more functions. For instance, the composite function $f \circ g \circ h$ is found by first applying h, then g, and then f as follows:

$$(f \circ g \circ h)(x) = f(g(h(x)))$$

EXAMPLE 5 ■ A Composition of Three Functions

Find $f \circ g \circ h$ if $f(x) = x/(x + 1)$, $g(x) = x^{10}$, and $h(x) = x + 3$.

SOLUTION

$$\begin{aligned}(f \circ g \circ h)(x) &= f(g(h(x))) && \text{Definition of } f \circ g \circ h\\ &= f(g(x + 3)) && \text{Definition of } h\\ &= f((x + 3)^{10}) && \text{Definition of } g\\ &= \frac{(x + 3)^{10}}{(x + 3)^{10} + 1} && \text{Definition of } f\end{aligned}$$

Now Try Exercise 59 ■

So far, we have used composition to build complicated functions from simpler ones. But in calculus it is useful to be able to "decompose" a complicated function into simpler ones, as shown in the following example.

EXAMPLE 6 ■ Recognizing a Composition of Functions

Given $F(x) = \sqrt[4]{x + 9}$, find functions f and g such that $F = f \circ g$.

SOLUTION Since the formula for F says to first add 9 and then take the fourth root, we let

$$g(x) = x + 9 \qquad \text{and} \qquad f(x) = \sqrt[4]{x}$$

Then

$$\begin{aligned} (f \circ g)(x) &= f(g(x)) && \text{Definition of } f \circ g \\ &= f(x + 9) && \text{Definition of } g \\ &= \sqrt[4]{x + 9} && \text{Definition of } f \\ &= F(x) \end{aligned}$$

Now Try Exercise 63 ■

■ Applications of Composition

When working with functions that model real-world situations, we name the variables using letters that suggest the quantity being modeled. We may use t for time, d for distance, V for volume, and so on. For example, if air is being pumped into a balloon, then the radius R of the balloon is a function of the volume V of air pumped into the balloon, say, $R = f(V)$. Also the volume V is a function of the time t that the pump has been working, say, $V = g(t)$. It follows that the radius R is a function of the time t given by $R = f(g(t))$.

EXAMPLE 7 ■ An Application of Composition of Functions

A ship is traveling at 20 mi/h parallel to a straight shoreline. The ship is 5 mi from shore. It passes a lighthouse at noon.

(a) Express the distance s between the lighthouse and the ship as a function of d, the distance the ship has traveled since noon; that is, find f so that $s = f(d)$.

(b) Express d as a function of t, the time elapsed since noon; that is, find g so that $d = g(t)$.

(c) Find $f \circ g$. What does this function represent?

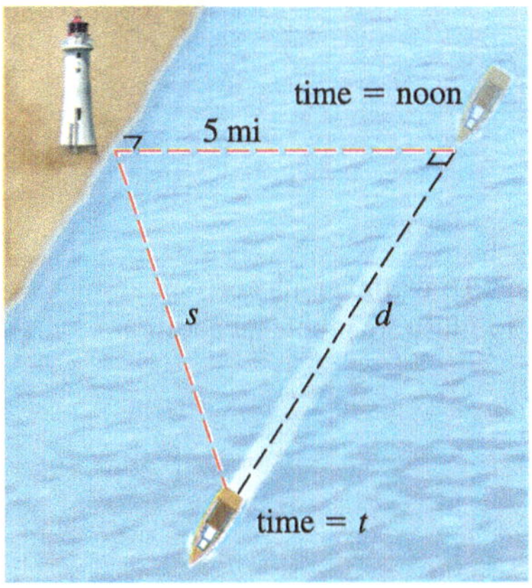

FIGURE 5

SOLUTION We first draw a diagram as in Figure 5.

(a) We can relate the distances s and d by the Pythagorean Theorem. Thus s can be expressed as a function of d by

$$s = f(d) = \sqrt{25 + d^2}$$

(b) Since the ship is traveling at 20 mi/h, the distance d it has traveled is a function of t as follows:

distance = rate × time

$$d = g(t) = 20t$$

(c) We have

$$\begin{aligned} (f \circ g)(t) &= f(g(t)) && \text{Definition of } f \circ g \\ &= f(20t) && \text{Definition of } g \\ &= \sqrt{25 + (20t)^2} && \text{Definition of } f \end{aligned}$$

The function $f \circ g$ gives the distance of the ship from the lighthouse as a function of time.

Now Try Exercise 77 ■

2.7 EXERCISES

CONCEPTS

1. From the graphs of f and g in the figure, we find

$(f+g)(2) =$ ________ $\qquad (f-g)(2) =$ ________

$(fg)(2) =$ ________ $\qquad \left(\dfrac{f}{g}\right)(2) =$ ________

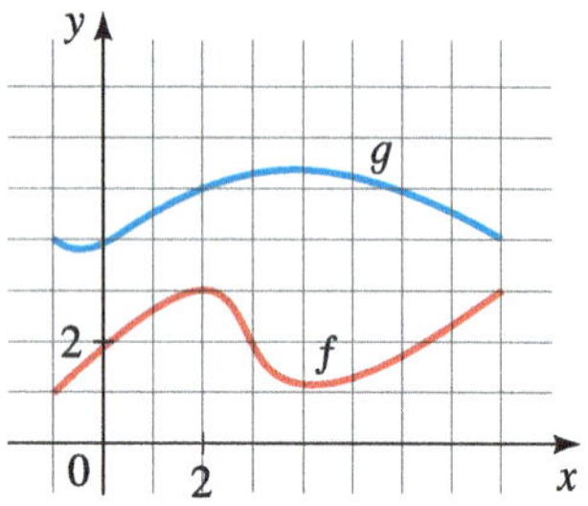

2. By definition, $(f \circ g)(x) =$ ________. So if $g(2) = 5$ and $f(5) = 12$, then $(f \circ g)(2) =$ ________.

3. If the rule of the function f is "add one" and the rule of the function g is "multiply by 2," then the rule of $f \circ g$ is

"________________________________."

and the rule of $g \circ f$ is

"________________________________."

4. We can express the functions in Exercise 3 algebraically as

$f(x) =$ ________ $\qquad g(x) =$ ________

$(f \circ g)(x) =$ ________ $\qquad (g \circ f)(x) =$ ________

5–6 ■ Let f and g be functions.

5. (a) The function $(f+g)(x)$ is defined for all values of x that are in the domains of both ________ and ________.

(b) The function $(fg)(x)$ is defined for all values of x that are in the domains of both ________ and ________.

(c) The function $(f/g)(x)$ is defined for all values of x that are in the domains of both ________ and ________, and $g(x)$ is not equal to ________.

6. The composition $(f \circ g)(x)$ is defined for all values of x for which x is in the domain of ________ and $g(x)$ is in the domain of ________.

SKILLS

7–16 ■ **Combining Functions** Find $f+g$, $f-g$, fg, and f/g and their domains.

7. $f(x) = x, \quad g(x) = 2x$

8. $f(x) = x, \quad g(x) = \sqrt{x}$

9. $f(x) = x^2 + x, \quad g(x) = x^2$

10. $f(x) = 3 - x^2, \quad g(x) = x^2 - 4$

11. $f(x) = 5 - x, \quad g(x) = x^2 - 3x$

12. $f(x) = x^2 + 2x, \quad g(x) = 3x^2 - 1$

13. $f(x) = \sqrt{25 - x^2}, \quad g(x) = \sqrt{x+3}$

14. $f(x) = \sqrt{16 - x^2}, \quad g(x) = \sqrt{x^2 - 1}$

15. $f(x) = \dfrac{2}{x}, \quad g(x) = \dfrac{4}{x+4}$

16. $f(x) = \dfrac{2}{x+1}, \quad g(x) = \dfrac{x}{x+1}$

17–20 ■ **Domain** Find the domain of the function.

17. $f(x) = \sqrt{x} + \sqrt{3 - x}$

18. $f(x) = \sqrt{x+4} - \dfrac{\sqrt{1-x}}{x}$

19. $h(x) = (x-3)^{-1/4}$

20. $k(x) = \dfrac{\sqrt{x+3}}{x-1}$

21–22 ■ **Graphical Addition** Use graphical addition to sketch the graph of $f + g$.

21.

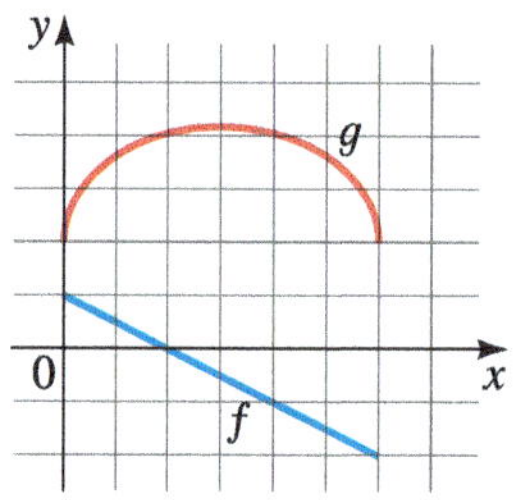

22.

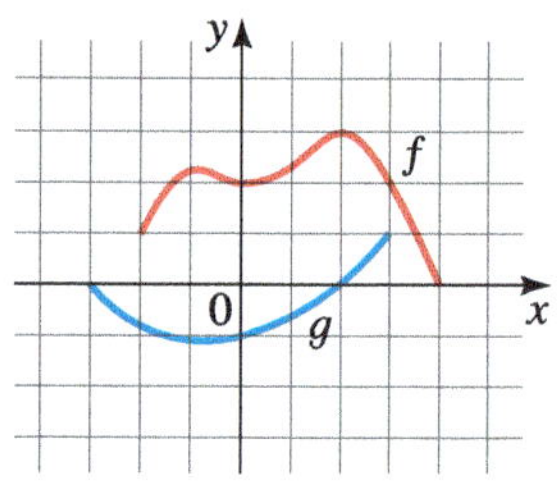

23–26 ■ **Graphical Addition** Draw the graphs of f, g, and $f + g$ on a common screen to illustrate graphical addition.

23. $f(x) = \sqrt{1+x}, \quad g(x) = \sqrt{1-x}$

24. $f(x) = x^2, \quad g(x) = \sqrt{x}$

25. $f(x) = x^2, \quad g(x) = \frac{1}{3}x^3$

26. $f(x) = \sqrt[4]{1-x}, \quad g(x) = \sqrt{1 - \dfrac{x^2}{9}}$

27–32 ■ **Evaluating Composition of Functions** Use $f(x) = 2x - 3$ and $g(x) = 4 - x^2$ to evaluate the expression.

27. (a) $f(g(0))$ **(b)** $g(f(0))$

28. (a) $f(f(2))$ **(b)** $g(g(3))$

29. (a) $(f \circ g)(-2)$ **(b)** $(g \circ f)(-2)$

30. (a) $(f \circ f)(-1)$ **(b)** $(g \circ g)(-1)$

31. (a) $(f \circ g)(x)$ **(b)** $(g \circ f)(x)$

32. (a) $(f \circ f)(x)$ **(b)** $(g \circ g)(x)$

33–38 ■ Composition Using a Graph Use the given graphs of f and g to evaluate the expression.

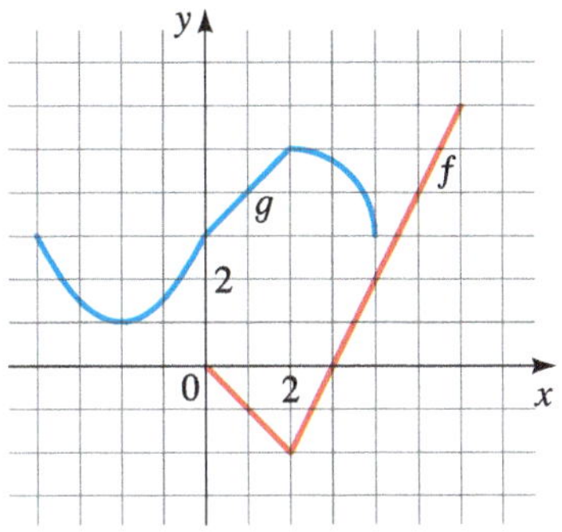

33. $f(g(2))$ **34.** $g(f(0))$

35. $(g \circ f)(4)$ **36.** $(f \circ g)(0)$

37. $(g \circ g)(-2)$ **38.** $(f \circ f)(4)$

39–46 ■ Composition Using a Table Use the table to evaluate the expression.

x	1	2	3	4	5	6
$f(x)$	2	3	5	1	6	3
$g(x)$	3	5	6	2	1	4

39. $f(g(2))$ **40.** $g(f(2))$

41. $f(f(1))$ **42.** $g(g(2))$

43. $(f \circ g)(6)$ **44.** $(g \circ f)(2)$

45. $(f \circ f)(5)$ **46.** $(g \circ g)(2)$

47–58 ■ Composition of Functions Find the functions $f \circ g$, $g \circ f$, $f \circ f$, and $g \circ g$ and their domains.

47. $f(x) = 2x + 3, \quad g(x) = 4x - 1$

48. $f(x) = 6x - 5, \quad g(x) = \dfrac{x}{2}$

49. $f(x) = x^2, \quad g(x) = x + 1$

50. $f(x) = x^3 + 2, \quad g(x) = \sqrt[3]{x}$

51. $f(x) = \dfrac{1}{x}, \quad g(x) = 2x + 4$

52. $f(x) = x^2, \quad g(x) = \sqrt{x - 3}$

53. $f(x) = |x|, \quad g(x) = 2x + 3$

54. $f(x) = x - 4, \quad g(x) = |x + 4|$

55. $f(x) = \dfrac{x}{x + 1}, \quad g(x) = 2x - 1$

56. $f(x) = \dfrac{1}{\sqrt{x}}, \quad g(x) = x^2 - 4x$

57. $f(x) = \dfrac{x}{x + 1}, \quad g(x) = \dfrac{1}{x}$

58. $f(x) = \dfrac{2}{x}, \quad g(x) = \dfrac{x}{x + 2}$

59–62 ■ Composition of Three Functions Find $f \circ g \circ h$.

59. $f(x) = x - 1, \quad g(x) = \sqrt{x}, \quad h(x) = x - 1$

60. $f(x) = \dfrac{1}{x}, \quad g(x) = x^3, \quad h(x) = x^2 + 2$

61. $f(x) = x^4 + 1, \quad g(x) = x - 5, \quad h(x) = \sqrt{x}$

62. $f(x) = \sqrt{x}, \quad g(x) = \dfrac{x}{x - 1}, \quad h(x) = \sqrt[3]{x}$

63–68 ■ Expressing a Function as a Composition Express the function in the form $f \circ g$.

63. $F(x) = (x - 9)^5$ **64.** $F(x) = \sqrt{x} + 1$

65. $G(x) = \dfrac{x^2}{x^2 + 4}$ **66.** $G(x) = \dfrac{1}{x + 3}$

67. $H(x) = |1 - x^3|$ **68.** $H(x) = \sqrt{1 + \sqrt{x}}$

69–72 ■ Expressing a Function as a Composition Express the function in the form $f \circ g \circ h$.

69. $F(x) = \dfrac{1}{x^2 + 1}$ **70.** $F(x) = \sqrt[3]{\sqrt{x} - 1}$

71. $G(x) = (4 + \sqrt[3]{x})^9$ **72.** $G(x) = \dfrac{2}{(3 + \sqrt{x})^2}$

SKILLS Plus

73. Composing Linear Functions The graphs of the functions

$$f(x) = m_1 x + b_1$$
$$g(x) = m_2 x + b_2$$

are lines with slopes m_1 and m_2, respectively. Is the graph of $f \circ g$ a line? If so, what is its slope?

74. Solving an Equation for an Unknown Function Suppose that

$$g(x) = 2x + 1$$
$$h(x) = 4x^2 + 4x + 7$$

Find a function f such that $f \circ g = h$. (Think about what operations you would have to perform on the formula for g to end up with the formula for h.) Now suppose that

$$f(x) = 3x + 5$$
$$h(x) = 3x^2 + 3x + 2$$

Use the same sort of reasoning to find a function g such that $f \circ g = h$.

APPLICATIONS

75–76 ■ Revenue, Cost, and Profit A print shop makes bumper stickers for election campaigns. If x stickers are ordered (where $x < 10{,}000$), then the price per bumper sticker is $0.15 - 0.000002x$ dollars, and the total cost of producing the order is $0.095x - 0.0000005x^2$ dollars.

75. Use the fact that

$$\text{revenue} = \text{price per item} \times \text{number of items sold}$$

to express $R(x)$, the revenue from an order of x stickers, as a product of two functions of x.

76. Use the fact that

$$\text{profit} = \text{revenue} - \text{cost}$$

to express $P(x)$, the profit on an order of x stickers, as a difference of two functions of x.

77. Area of a Ripple A stone is dropped in a lake, creating a circular ripple that travels outward at a speed of 60 cm/s.

(a) Find a function g that models the radius as a function of time.

(b) Find a function f that models the area of the circle as a function of the radius.

(c) Find $f \circ g$. What does this function represent?

78. Inflating a Balloon A spherical balloon is being inflated. The radius of the balloon is increasing at the rate of 1 cm/s.

(a) Find a function f that models the radius as a function of time.

(b) Find a function g that models the volume as a function of the radius.

(c) Find $g \circ f$. What does this function represent?

79. Area of a Balloon A spherical weather balloon is being inflated. The radius of the balloon is increasing at the rate of 2 cm/s. Express the surface area of the balloon as a function of time t (in seconds).

80. Multiple Discounts You have a \$50 coupon from the manufacturer that is good for the purchase of a cell phone. The store where you are purchasing your cell phone is offering a 20% discount on all cell phones. Let x represent the regular price of the cell phone.

(a) Suppose only the 20% discount applies. Find a function f that models the purchase price of the cell phone as a function of the regular price x.

(b) Suppose only the \$50 coupon applies. Find a function g that models the purchase price of the cell phone as a function of the sticker price x.

(c) If you can use the coupon and the discount, then the purchase price is either $(f \circ g)(x)$ or $(g \circ f)(x)$, depending on the order in which they are applied to the price. Find both $(f \circ g)(x)$ and $(g \circ f)(x)$. Which composition gives the lower price?

81. Multiple Discounts An appliance dealer advertises a 10% discount on all his washing machines. In addition, the manufacturer offers a \$100 rebate on the purchase of a washing machine. Let x represent the sticker price of the washing machine.

(a) Suppose only the 10% discount applies. Find a function f that models the purchase price of the washer as a function of the sticker price x.

(b) Suppose only the \$100 rebate applies. Find a function g that models the purchase price of the washer as a function of the sticker price x.

(c) Find $f \circ g$ and $g \circ f$. What do these functions represent? Which is the better deal?

82. Airplane Trajectory An airplane is flying at a speed of 350 mi/h at an altitude of one mile. The plane passes directly above a radar station at time $t = 0$.

(a) Express the distance s (in miles) between the plane and the radar station as a function of the horizontal distance d (in miles) that the plane has flown.

(b) Express d as a function of the time t (in hours) that the plane has flown.

(c) Use composition to express s as a function of t.

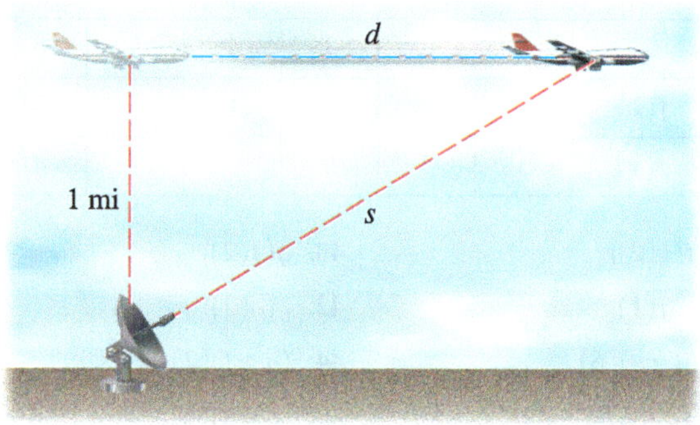

DISCUSS ■ DISCOVER ■ PROVE ■ WRITE

83. DISCOVER: Compound Interest A savings account earns 5% interest compounded annually. If you invest x dollars in such an account, then the amount $A(x)$ of the investment after one year is the initial investment plus 5%; that is,

$$A(x) = x + 0.05x = 1.05x$$

Find

$$A \circ A$$
$$A \circ A \circ A$$
$$A \circ A \circ A \circ A$$

What do these compositions represent? Find a formula for what you get when you compose n copies of A.

84. DISCUSS: Compositions of Odd and Even Functions Suppose that

$$h = f \circ g$$

If g is an even function, is h necessarily even? If g is odd, is h odd? What if g is odd and f is odd? What if g is odd and f is even?

2.8 ONE-TO-ONE FUNCTIONS AND THEIR INVERSES

■ One-to-One Functions ■ The Inverse of a Function ■ Finding the Inverse of a Function ■ Graphing the Inverse of a Function ■ Applications of Inverse Functions

The *inverse* of a function is a rule that acts on the output of the function and produces the corresponding input. So the inverse "undoes" or reverses what the function has done. Not all functions have inverses; those that do are called *one-to-one*.

One-to-One Functions

Let's compare the functions f and g whose arrow diagrams are shown in Figure 1. Note that f never takes on the same value twice (any two numbers in A have different images), whereas g does take on the same value twice (both 2 and 3 have the same image, 4). In symbols, $g(2) = g(3)$ but $f(x_1) \neq f(x_2)$ whenever $x_1 \neq x_2$. Functions that have this latter property are called *one-to-one*.

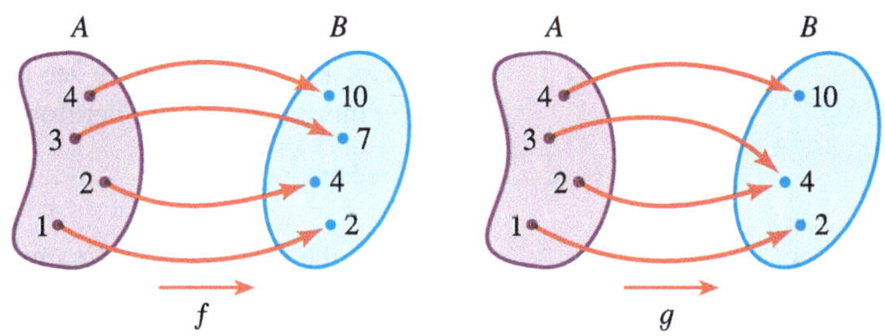

f is one-to-one $\qquad$ g is not one-to-one

FIGURE 1

DEFINITION OF A ONE-TO-ONE FUNCTION

A function with domain A is called a **one-to-one function** if no two elements of A have the same image, that is,

$$f(x_1) \neq f(x_2) \quad \text{whenever } x_1 \neq x_2$$

An equivalent way of writing the condition for a one-to-one function is this:

$$\text{If } f(x_1) = f(x_2), \text{ then } x_1 = x_2.$$

If a horizontal line intersects the graph of f at more than one point, then we see from Figure 2 that there are numbers $x_1 \neq x_2$ such that $f(x_1) = f(x_2)$. This means that f is not one-to-one. Therefore we have the following geometric method for determining whether a function is one-to-one.

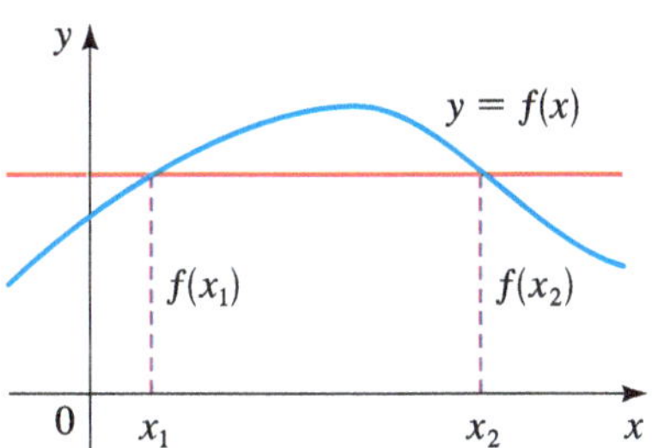

FIGURE 2 This function is not one-to-one because $f(x_1) = f(x_2)$.

HORIZONTAL LINE TEST

A function is one-to-one if and only if no horizontal line intersects its graph more than once.

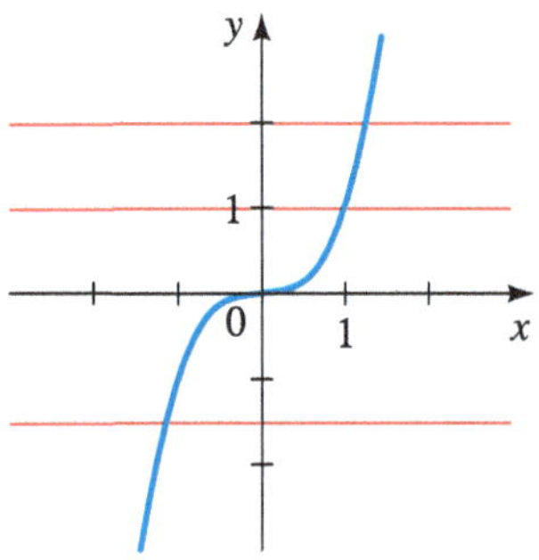

FIGURE 3 $f(x) = x^3$ is one-to-one.

EXAMPLE 1 ■ Deciding Whether a Function Is One-to-One

Is the function $f(x) = x^3$ one-to-one?

SOLUTION 1 If $x_1 \neq x_2$, then $x_1^3 \neq x_2^3$ (two different numbers cannot have the same cube). Therefore $f(x) = x^3$ is one-to-one.

SOLUTION 2 From Figure 3 we see that no horizontal line intersects the graph of $f(x) = x^3$ more than once. Therefore by the Horizontal Line Test, f is one-to-one.

Now Try Exercise 15 ■

Notice that the function f of Example 1 is increasing and is also one-to-one. In fact, it can be proved that *every increasing function and every decreasing function is one-to-one.*

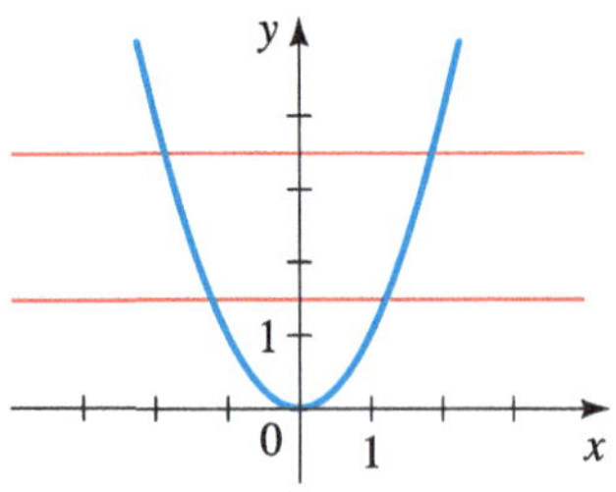

FIGURE 4 $g(x) = x^2$ is not one-to-one.

EXAMPLE 2 ■ Deciding Whether a Function Is One-to-One

Is the function $g(x) = x^2$ one-to-one?

SOLUTION 1 This function is not one-to-one because, for instance,

$$g(1) = 1 \qquad \text{and} \qquad g(-1) = 1$$

so 1 and -1 have the same image.

SOLUTION 2 From Figure 4 we see that there are horizontal lines that intersect the graph of g more than once. Therefore by the Horizontal Line Test, g is not one-to-one.

Now Try Exercise 17 ■

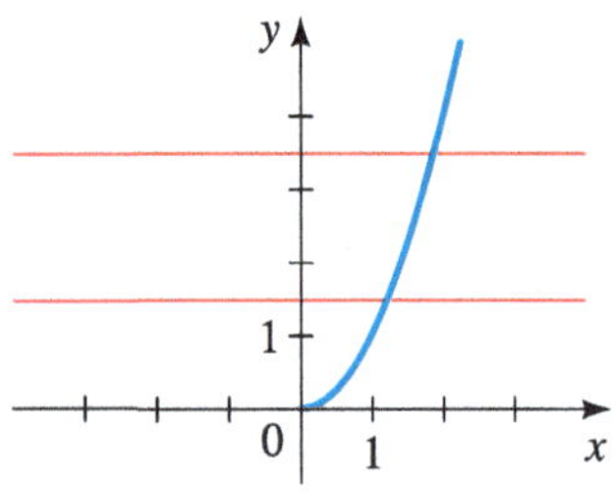

FIGURE 5 $h(x) = x^2$ $(x \geq 0)$ is one-to-one.

Although the function g in Example 2 is not one-to-one, it is possible to restrict its domain so that the resulting function is one-to-one. In fact, if we define

$$h(x) = x^2 \qquad x \geq 0$$

then h is one-to-one, as you can see from Figure 5 and the Horizontal Line Test.

EXAMPLE 3 ■ Showing That a Function Is One-to-One

Show that the function $f(x) = 3x + 4$ is one-to-one.

SOLUTION Suppose there are numbers x_1 and x_2 such that $f(x_1) = f(x_2)$. Then

$$\begin{aligned} 3x_1 + 4 &= 3x_2 + 4 && \text{Suppose } f(x_1) = f(x_2) \\ 3x_1 &= 3x_2 && \text{Subtract 4} \\ x_1 &= x_2 && \text{Divide by 3} \end{aligned}$$

Therefore f is one-to-one.

Now Try Exercise 13 ■

■ The Inverse of a Function

One-to-one functions are important because they are precisely the functions that possess inverse functions according to the following definition.

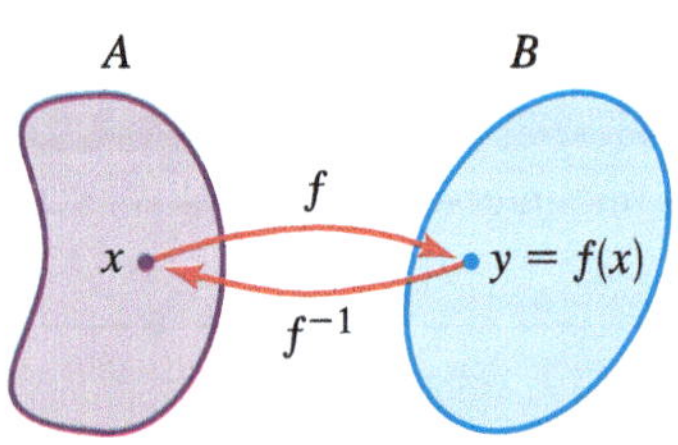

FIGURE 6

DEFINITION OF THE INVERSE OF A FUNCTION

Let f be a one-to-one function with domain A and range B. Then its **inverse function** f^{-1} has domain B and range A and is defined by

$$f^{-1}(y) = x \iff f(x) = y$$

for any y in B.

This definition says that if f takes x to y, then f^{-1} takes y back to x. (If f were not one-to-one, then f^{-1} would not be defined uniquely.) The arrow diagram in Figure 6 indicates that f^{-1} reverses the effect of f. From the definition we have

$$\text{domain of } f^{-1} = \text{range of } f$$
$$\text{range of } f^{-1} = \text{domain of } f$$

EXAMPLE 4 ■ Finding f^{-1} for Specific Values

Don't mistake the -1 in f^{-1} for an exponent.

$$f^{-1}(x) \quad \textit{does not mean} \quad \frac{1}{f(x)}$$

The reciprocal $1/f(x)$ is written as $(f(x))^{-1}$.

If $f(1) = 5$, $f(3) = 7$, and $f(8) = -10$, find $f^{-1}(5)$, $f^{-1}(7)$, and $f^{-1}(-10)$.

SOLUTION From the definition of f^{-1} we have

$$f^{-1}(5) = 1 \quad \text{because} \quad f(1) = 5$$
$$f^{-1}(7) = 3 \quad \text{because} \quad f(3) = 7$$
$$f^{-1}(-10) = 8 \quad \text{because} \quad f(8) = -10$$

Figure 7 shows how f^{-1} reverses the effect of f in this case.

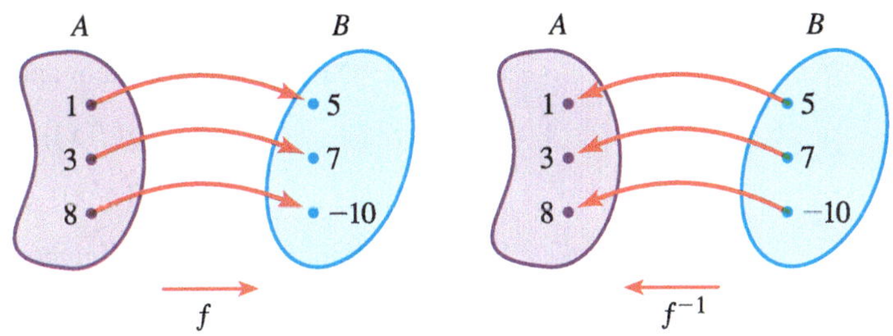

FIGURE 7

Now Try Exercise 25 ■

EXAMPLE 5 ■ Finding Values of an Inverse Function

We can find specific values of an inverse function from a table or graph of the function itself.

(a) The table below gives values of a function h. From the table we see that $h^{-1}(8) = 3$, $h^{-1}(12) = 4$, and $h^{-1}(3) = 6$.

(b) A graph of a function f is shown in Figure 8. From the graph we see that $f^{-1}(5) = 7$ and $f^{-1}(3) = 4$.

x		$h(x)$
2		5
3	←	8
4	←	12
5		1
6	←	3
7		15

Finding values of h^{-1} from a table of h

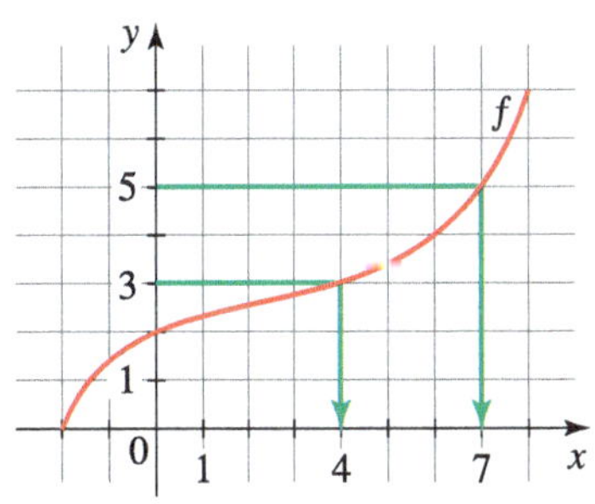

FIGURE 8 Finding values of f^{-1} from a graph of f

Now Try Exercises 29 and 31 ■

By definition the inverse function f^{-1} undoes what f does: If we start with x, apply f, and then apply f^{-1}, we arrive back at x, where we started. Similarly, f undoes what f^{-1} does. In general, any function that reverses the effect of f in this way must be the inverse of f. These observations are expressed precisely as follows.

INVERSE FUNCTION PROPERTY

Let f be a one-to-one function with domain A and range B. The inverse function f^{-1} satisfies the following cancellation properties:

$$f^{-1}(f(x)) = x \quad \text{for every } x \text{ in } A$$

$$f(f^{-1}(x)) = x \quad \text{for every } x \text{ in } B$$

Conversely, any function f^{-1} satisfying these equations is the inverse of f.

These properties indicate that f is the inverse function of f^{-1}, so we say that f and f^{-1} are *inverses of each other*.

EXAMPLE 6 ■ Verifying That Two Functions Are Inverses

Show that $f(x) = x^3$ and $g(x) = x^{1/3}$ are inverses of each other.

SOLUTION Note that the domain and range of both f and g are $\mathbb{R}$. We have

$$g(f(x)) = g(x^3) = (x^3)^{1/3} = x$$

$$f(g(x)) = f(x^{1/3}) = (x^{1/3})^3 = x$$

So by the Property of Inverse Functions, f and g are inverses of each other. These equations simply say that the cube function and the cube root function, when composed, cancel each other.

Now Try Exercise 39 ■

■ Finding the Inverse of a Function

Now let's examine how we compute inverse functions. We first observe from the definition of f^{-1} that

$$y = f(x) \quad \Leftrightarrow \quad f^{-1}(y) = x$$

So if $y = f(x)$ and if we are able to solve this equation for x in terms of y, then we must have $x = f^{-1}(y)$. If we then interchange x and y, we have $y = f^{-1}(x)$, which is the desired equation.

HOW TO FIND THE INVERSE OF A ONE-TO-ONE FUNCTION

1. Write $y = f(x)$.
2. Solve this equation for x in terms of y (if possible).
3. Interchange x and y. The resulting equation is $y = f^{-1}(x)$.

Note that Steps 2 and 3 can be reversed. In other words, we can interchange x and y first and then solve for y in terms of x.

In Example 7 note how f^{-1} reverses the effect of f. The function f is the rule "Multiply by 3, then subtract 2," whereas f^{-1} is the rule "Add 2, then divide by 3."

EXAMPLE 7 ■ Finding the Inverse of a Function

Find the inverse of the function $f(x) = 3x - 2$.

SOLUTION First we write $y = f(x)$.

$$y = 3x - 2$$

CHECK YOUR ANSWER

We use the Inverse Function Property:

$$\begin{aligned} f^{-1}(f(x)) &= f^{-1}(3x-2) \\ &= \frac{(3x-2)+2}{3} \\ &= \frac{3x}{3} = x \\ f(f^{-1}(x)) &= f\left(\frac{x+2}{3}\right) \\ &= 3\left(\frac{x+2}{3}\right) - 2 \\ &= x + 2 - 2 = x \quad ✓ \end{aligned}$$

Then we solve this equation for x:

$$\begin{aligned} 3x &= y + 2 && \text{Add 2} \\ x &= \frac{y+2}{3} && \text{Divide by 3} \end{aligned}$$

Finally, we interchange x and y:

$$y = \frac{x+2}{3}$$

Therefore, the inverse function is $f^{-1}(x) = \dfrac{x+2}{3}$.

Now Try Exercise 49 ■

In Example 8 note how f^{-1} reverses the effect of f. The function f is the rule "Take the fifth power, subtract 3, then divide by 2," whereas f^{-1} is the rule "Multiply by 2, add 3, then take the fifth root."

EXAMPLE 8 ■ Finding the Inverse of a Function

Find the inverse of the function $f(x) = \dfrac{x^5 - 3}{2}$.

SOLUTION We first write $y = (x^5 - 3)/2$ and solve for x.

$$\begin{aligned} y &= \frac{x^5-3}{2} && \text{Equation defining function} \\ 2y &= x^5 - 3 && \text{Multiply by 2} \\ x^5 &= 2y + 3 && \text{Add 3 (and switch sides)} \\ x &= (2y+3)^{1/5} && \text{Take fifth root of each side} \end{aligned}$$

Then we interchange x and y to get $y = (2x + 3)^{1/5}$. Therefore the inverse function is $f^{-1}(x) = (2x + 3)^{1/5}$.

Now Try Exercise 61 ■

CHECK YOUR ANSWER

We use the Inverse Function Property:

$$\begin{aligned} f^{-1}(f(x)) &= f^{-1}\left(\frac{x^5-3}{2}\right) \\ &= \left[2\left(\frac{x^5-3}{2}\right) + 3\right]^{1/5} \\ &= (x^5 - 3 + 3)^{1/5} \\ &= (x^5)^{1/5} = x \\ f(f^{-1}(x)) &= f((2x+3)^{1/5}) \\ &= \frac{[(2x+3)^{1/5}]^5 - 3}{2} \\ &= \frac{2x+3-3}{2} \\ &= \frac{2x}{2} = x \quad ✓ \end{aligned}$$

A **rational function** is a function defined by a rational expression. In the next example we find the inverse of a rational function.

Rational functions are studied in Section 3.6.

EXAMPLE 9 ■ Finding the Inverse of a Rational Function

Find the inverse of the function $f(x) = \dfrac{2x+3}{x-1}$.

SOLUTION We first write $y = (2x + 3)/(x - 1)$ and solve for x.

$$\begin{aligned} y &= \frac{2x+3}{x-1} && \text{Equation defining function} \\ y(x-1) &= 2x + 3 && \text{Multiply by } x - 1 \\ yx - y &= 2x + 3 && \text{Expand} \\ yx - 2x &= y + 3 && \text{Bring } x\text{-terms to LHS} \\ x(y-2) &= y + 3 && \text{Factor } x \\ x &= \frac{y+3}{y-2} && \text{Divide by } y - 2 \end{aligned}$$

Therefore the inverse function is $f^{-1}(x) = \dfrac{x+3}{x-2}$.

Now Try Exercise 55 ■

Graphing the Inverse of a Function

The principle of interchanging x and y to find the inverse function also gives us a method for obtaining the graph of f^{-1} from the graph of f. If $f(a) = b$, then $f^{-1}(b) = a$. Thus the point (a, b) is on the graph of f if and only if the point (b, a) is on the graph of f^{-1}. But we get the point (b, a) from the point (a, b) by reflecting in the line $y = x$ (see Figure 9). Therefore, as Figure 10 illustrates, the following is true.

> The graph of f^{-1} is obtained by reflecting the graph of f in the line $y = x$.

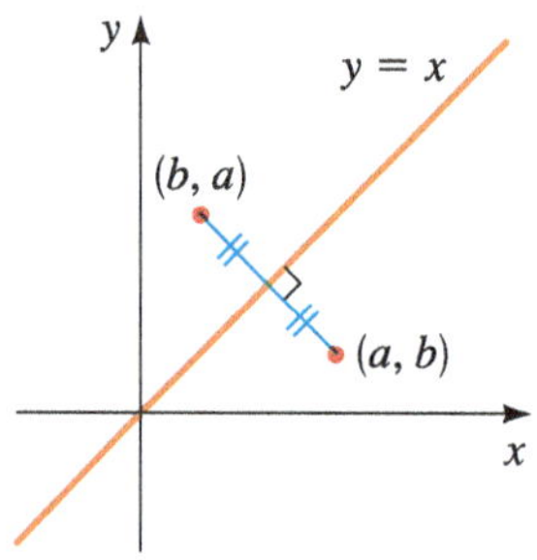

FIGURE 9

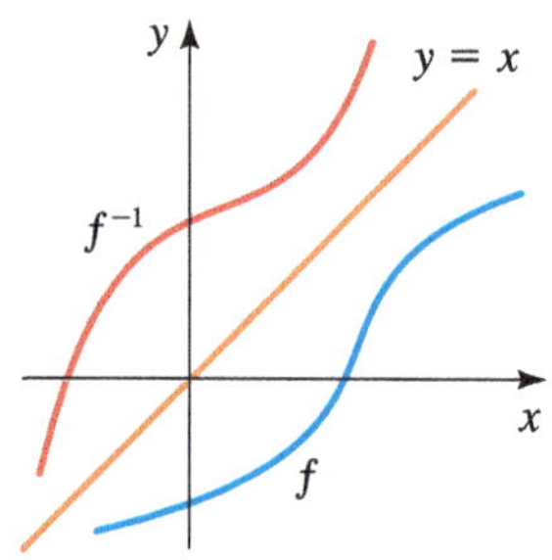

FIGURE 10

EXAMPLE 10 ■ Graphing the Inverse of a Function

(a) Sketch the graph of $f(x) = \sqrt{x - 2}$.

(b) Use the graph of f to sketch the graph of f^{-1}.

(c) Find an equation for f^{-1}.

SOLUTION

(a) Using the transformations from Section 2.6, we sketch the graph of $y = \sqrt{x - 2}$ by plotting the graph of the function $y = \sqrt{x}$ (Example 1(c) in Section 2.2) and shifting it to the right 2 units.

(b) The graph of f^{-1} is obtained from the graph of f in part (a) by reflecting it in the line $y = x$, as shown in Figure 11.

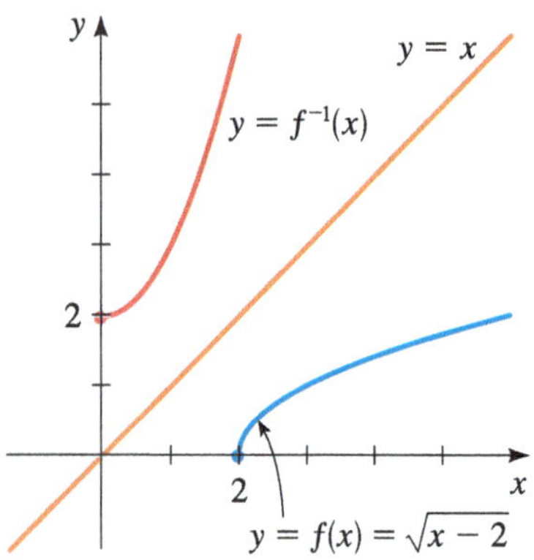

FIGURE 11

(c) Solve $y = \sqrt{x - 2}$ for x, noting that $y \geq 0$.

$$\sqrt{x - 2} = y$$

$$x - 2 = y^2 \qquad \text{Square each side}$$

$$x = y^2 + 2 \quad y \geq 0 \qquad \text{Add 2}$$

Interchange x and y, as follows:

$$y = x^2 + 2 \quad x \geq 0$$

Thus $$f^{-1}(x) = x^2 + 2 \quad x \geq 0$$

This expression shows that the graph of f^{-1} is the right half of the parabola $y = x^2 + 2$, and from the graph shown in Figure 11 this seems reasonable.

In Example 10 note how f^{-1} reverses the effect of f. The function f is the rule "Subtract 2, then take the square root," whereas f^{-1} is the rule "Square, then add 2."

Now Try Exercise 73 ■

Applications of Inverse Functions

When working with functions that model real-world situations, we name the variables using letters that suggest the quantity being modeled. For instance we may use t for time, d for distance, V for volume, and so on. When using inverse functions, we follow this convention. For example, suppose that the variable R is a function of the variable N, say, $R = f(N)$. Then $f^{-1}(R) = N$. So the function f^{-1} defines N as a function of R.

EXAMPLE 11 ■ An Inverse Function

At a local pizza parlor the daily special is \$12 for a plain cheese pizza plus \$2 for each additional topping.

(a) Find a function f that models the price of a pizza with n toppings.

(b) Find the inverse of the function f. What does f^{-1} represent?

(c) If a pizza costs \$22, how many toppings does it have?

SOLUTION Note that the price p of a pizza is a function of the number n of toppings.

(a) The price of a pizza with n toppings is given by the function

$$f(n) = 12 + 2n$$

(b) To find the inverse function, we first write $p = f(n)$, where we use the letter p instead of our usual y because $f(n)$ is the price of the pizza. We have

$$p = 12 + 2n$$

Next we solve for n:

$$\begin{aligned} p &= 12 + 2n \\ p - 12 &= 2n \\ n &= \frac{p - 12}{2} \end{aligned}$$

So $n = f^{-1}(p) = \dfrac{p - 12}{2}$. The function f^{-1} gives the number n of toppings for a pizza with price p.

(c) We have $n = f^{-1}(22) = (22 - 12)/2 = 5$. So the pizza has five toppings.

Now Try Exercise 93 ■

2.8 EXERCISES

CONCEPTS

1. A function f is one-to-one if different inputs produce ________ outputs. You can tell from the graph that a function is one-to-one by using the ________________ Test.

2. (a) For a function to have an inverse, it must be ________. So which one of the following functions has an inverse?

$$f(x) = x^2 \qquad g(x) = x^3$$

(b) What is the inverse of the function that you chose in part (a)?

3. A function f has the following verbal description: "Multiply by 3, add 5, and then take the third power of the result."

(a) Write a verbal description for f^{-1}.

(b) Find algebraic formulas that express f and f^{-1} in terms of the input x.

4. A graph of a function f is given. Does f have an inverse? If so, find $f^{-1}(1) =$ ________ and $f^{-1}(3) =$ ________.

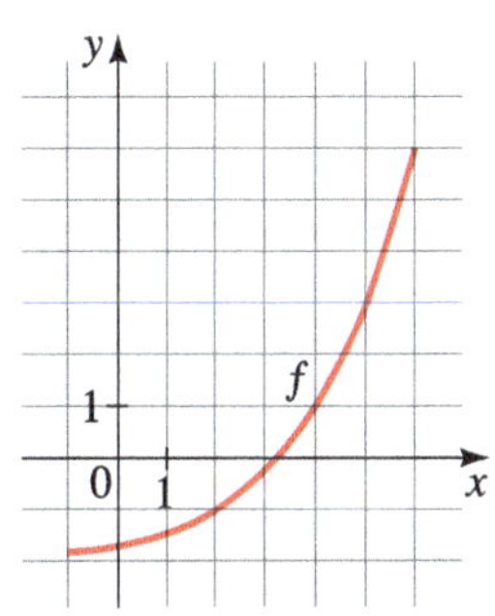

5. If the point $(3, 4)$ is on the graph of the function f, then the point (____, ____) is on the graph of f^{-1}.

6. *True or false?*

(a) If f has an inverse, then $f^{-1}(x)$ is always the same as $\dfrac{1}{f(x)}$.

(b) If f has an inverse, then $f^{-1}(f(x)) = x$.

SKILLS

7–12 ■ One-to-One Function? A graph of a function f is given. Determine whether f is one-to-one.

7.

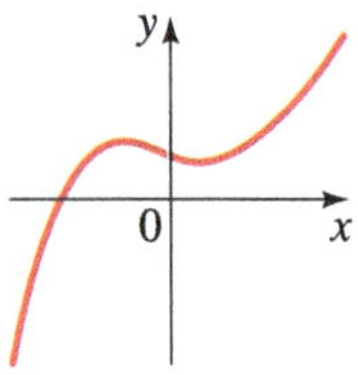

8.

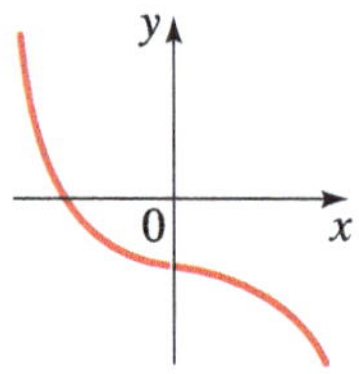

9.

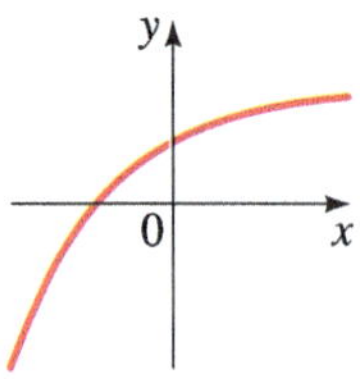

10.

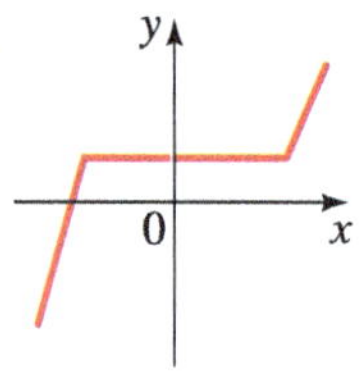

11.

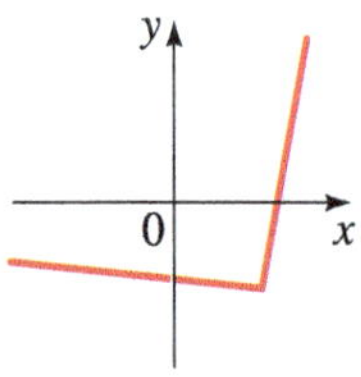

12.

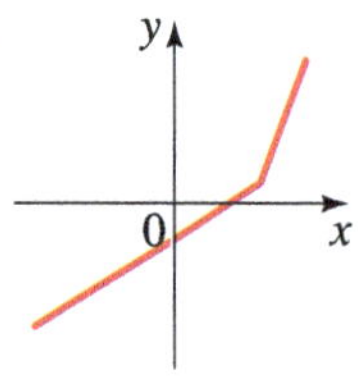

13–24 ■ One-to-One Function? Determine whether the function is one-to-one.

13. $f(x) = -2x + 4$ **14.** $f(x) = 3x - 2$

15. $g(x) = \sqrt{x}$ **16.** $g(x) = |x|$

17. $h(x) = x^2 - 2x$ **18.** $h(x) = x^3 + 8$

19. $f(x) = x^4 + 5$

20. $f(x) = x^4 + 5, \quad 0 \le x \le 2$

21. $r(t) = t^6 - 3, \quad 0 \le t \le 5$

22. $r(t) = t^4 - 1$

23. $f(x) = \dfrac{1}{x^2}$ **24.** $f(x) = \dfrac{1}{x}$

25–28 ■ Finding Values of an Inverse Function Assume that f is a one-to-one function.

25. (a) If $f(2) = 7$, find $f^{-1}(7)$.
(b) If $f^{-1}(3) = -1$, find $f(-1)$.

26. (a) If $f(5) = 18$, find $f^{-1}(18)$.
(b) If $f^{-1}(4) = 2$, find $f(2)$.

27. If $f(x) = 5 - 2x$, find $f^{-1}(3)$.

28. If $g(x) = x^2 + 4x$ with $x \ge -2$, find $g^{-1}(5)$.

29–30 ■ Finding Values of an Inverse from a Graph A graph of a function is given. Use the graph to find the indicated values.

29. (a) $f^{-1}(2)$ **(b)** $f^{-1}(5)$ **(c)** $f^{-1}(6)$

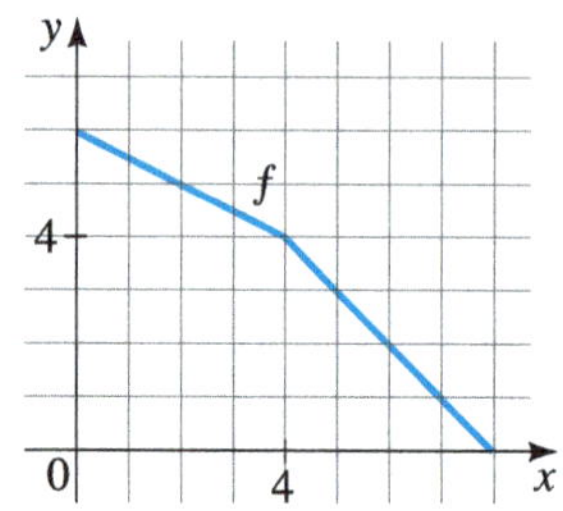

30. (a) $g^{-1}(2)$ **(b)** $g^{-1}(5)$ **(c)** $g^{-1}(6)$

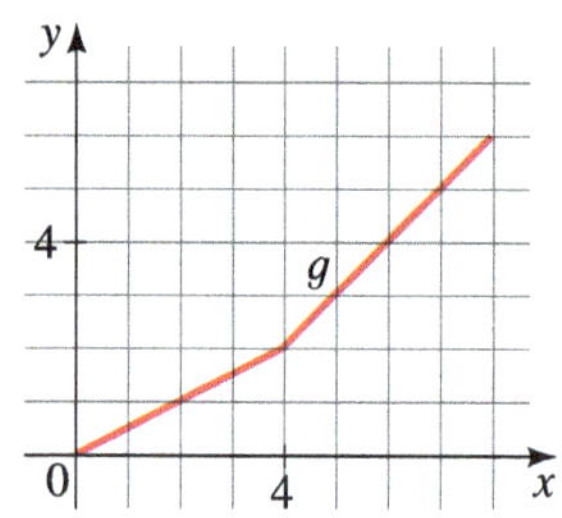

31–36 ■ Finding Values of an Inverse Using a Table A table of values for a one-to-one function is given. Find the indicated values.

31. $f^{-1}(5)$ **32.** $f^{-1}(0)$

33. $f^{-1}(f(1))$ **34.** $f(f^{-1}(6))$

35. $f^{-1}(f^{-1}(1))$ **36.** $f^{-1}(f^{-1}(0))$

x	1	2	3	4	5	6
$f(x)$	4	6	2	5	0	1

37–48 ■ Inverse Function Property Use the Inverse Function Property to show that f and g are inverses of each other.

37. $f(x) = x - 6; \quad g(x) = x + 6$

38. $f(x) = 3x; \quad g(x) = \dfrac{x}{3}$

39. $f(x) = 3x + 4; \quad g(x) = \dfrac{x - 4}{3}$

40. $f(x) = 2 - 5x; \quad g(x) = \dfrac{2 - x}{5}$

41. $f(x) = \dfrac{1}{x}; \quad g(x) = \dfrac{1}{x}$ **42.** $f(x) = x^5; \quad g(x) = \sqrt[5]{x}$

43. $f(x) = x^2 - 9, \quad x \ge 0; \quad g(x) = \sqrt{x + 9}, \quad x \ge -9$

44. $f(x) = x^3 + 1; \quad g(x) = (x - 1)^{1/3}$

45. $f(x) = \dfrac{1}{x - 1}; \quad g(x) = \dfrac{1}{x} + 1$

46. $f(x) = \sqrt{4 - x^2}, \quad 0 \le x \le 2;$
$g(x) = \sqrt{4 - x^2}, \quad 0 \le x \le 2$

47. $f(x) = \dfrac{x + 2}{x - 2}; \quad g(x) = \dfrac{2x + 2}{x - 1}$

48. $f(x) = \dfrac{x - 5}{3x + 4}; \quad g(x) = \dfrac{5 + 4x}{1 - 3x}$

49–70 ■ Finding Inverse Functions Find the inverse function of f.

49. $f(x) = 3x + 5$ **50.** $f(x) = 7 - 5x$

51. $f(x) = 5 - 4x^3$ **52.** $f(x) = 3x^3 + 8$

53. $f(x) = \dfrac{1}{x + 2}$ **54.** $f(x) = \dfrac{x - 2}{x + 2}$

55. $f(x) = \dfrac{x}{x + 4}$ **56.** $f(x) = \dfrac{3x}{x - 2}$

57. $f(x) = \dfrac{2x + 5}{x - 7}$ **58.** $f(x) = \dfrac{4x - 2}{3x + 1}$

59. $f(x) = \dfrac{2x + 3}{1 - 5x}$ **60.** $f(x) = \dfrac{3 - 4x}{8x - 1}$

61. $f(x) = 4 - x^2, \quad x \ge 0$ **62.** $f(x) = x^2 + x, \quad x \ge -\frac{1}{2}$

63. $f(x) = x^6, \quad x \ge 0$ **64.** $f(x) = \dfrac{1}{x^2}, \quad x > 0$

65. $f(x) = \dfrac{2 - x^3}{5}$ **66.** $f(x) = (x^5 - 6)^7$

67. $f(x) = \sqrt{5 + 8x}$ **68.** $f(x) = 2 + \sqrt{3 + x}$

69. $f(x) = 2 + \sqrt[3]{x}$

70. $f(x) = \sqrt{4 - x^2}, \quad 0 \le x \le 2$

71–74 ■ Graph of an Inverse Function A function f is given. **(a)** Sketch the graph of f. **(b)** Use the graph of f to sketch the graph of f^{-1}. **(c)** Find f^{-1}.

71. $f(x) = 3x - 6$ **72.** $f(x) = 16 - x^2, \quad x \ge 0$

73. $f(x) = \sqrt{x + 1}$ **74.** $f(x) = x^3 - 1$

75–80 ■ One-to-One Functions from a Graph Draw the graph of f, and use it to determine whether the function is one-to-one.

75. $f(x) = x^3 - x$ **76.** $f(x) = x^3 + x$

77. $f(x) = \dfrac{x + 12}{x - 6}$ **78.** $f(x) = \sqrt{x^3 - 4x + 1}$

79. $f(x) = |x| - |x - 6|$ **80.** $f(x) = x \cdot |x|$

81–84 ■ Finding Inverse Functions A one-to-one function is given. **(a)** Find the inverse of the function. **(b)** Graph both the function and its inverse on the same screen to verify that the graphs are reflections of each other in the line $y = x$.

81. $f(x) = 2 + x$ **82.** $f(x) = 2 - \frac{1}{2}x$

83. $g(x) = \sqrt{x + 3}$ **84.** $g(x) = x^2 + 1, \quad x \ge 0$

85–88 ■ Restricting the Domain The given function is not one-to-one. Restrict its domain so that the resulting function *is* one-to-one. Find the inverse of the function with the restricted domain. (There is more than one correct answer.)

85. $f(x) = 4 - x^2$

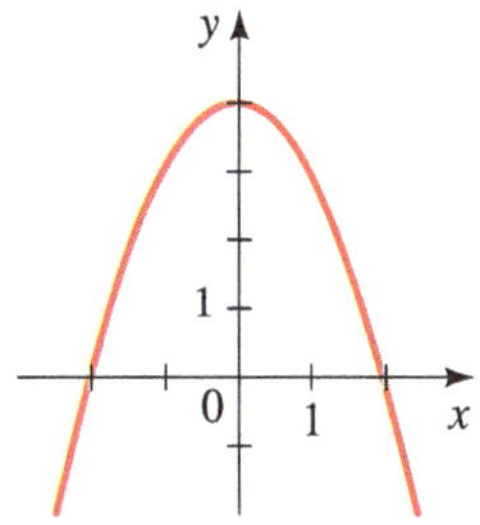

86. $g(x) = (x - 1)^2$

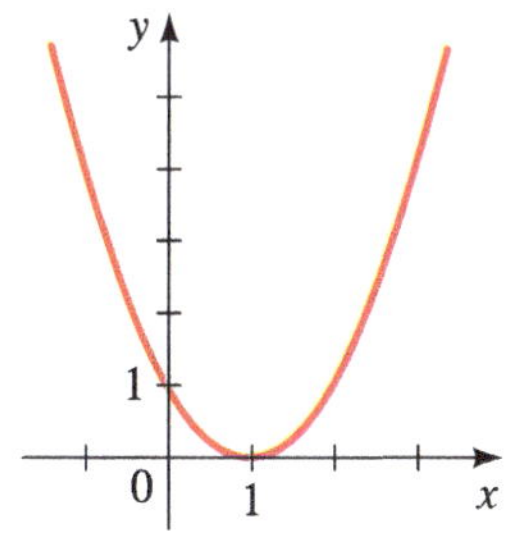

87. $h(x) = (x + 2)^2$

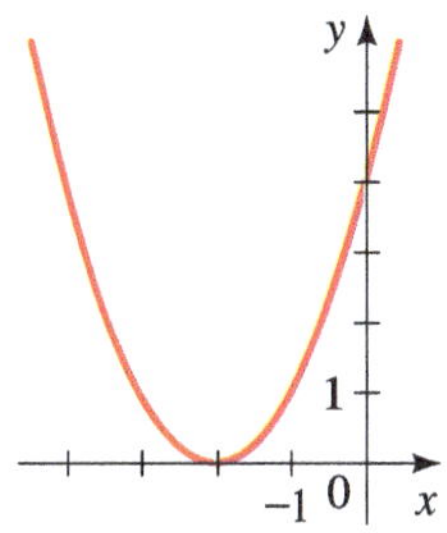

88. $k(x) = |x - 3|$

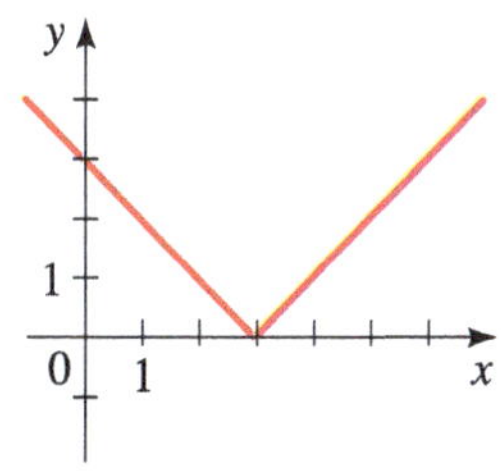

89–90 ■ Graph of an Inverse Function Use the graph of f to sketch the graph of f^{-1}.

89.

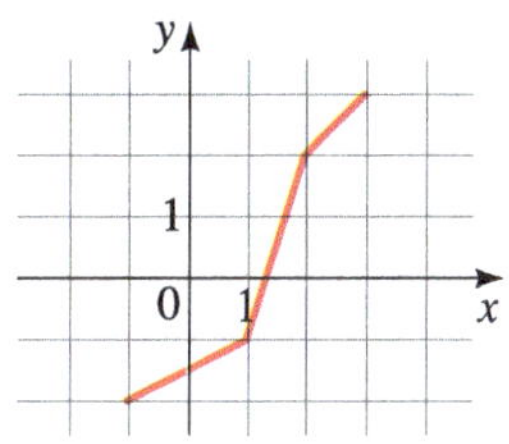

90.

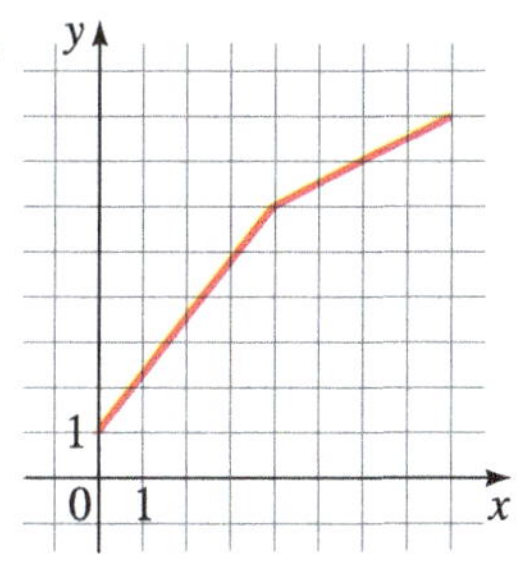

SKILLS Plus

91–92 ■ Functions That Are Their Own Inverse If a function f is its own inverse, then the graph of f is symmetric about the line $y = x$. **(a)** Graph the given function. **(b)** Does the graph indicate that f and f^{-1} are the same function? **(c)** Find the function f^{-1}. Use your result to verify your answer to part (b).

91. $f(x) = \dfrac{1}{x}$ **92.** $f(x) = \dfrac{x + 3}{x - 1}$

APPLICATIONS

93. Pizza Cost Marcello's Pizza charges a base price of $16 for a large pizza plus $1.50 for each additional topping.

(a) Find a function f that models the price of a pizza with n toppings.

(b) Find the inverse of the function f. What does f^{-1} represent?

(c) If a pizza costs $25, how many toppings does it have?

94. Fee for Service For his services, a private investigator requires a \$500 retainer fee plus \$80 per hour. Let x represent the number of hours the investigator spends working on a case.

(a) Find a function f that models the investigator's fee as a function of x.

(b) Find f^{-1}. What does f^{-1} represent?

(c) Find $f^{-1}(1220)$. What does your answer represent?

95. Torricelli's Law A tank holds 100 gallons of water, which drains from a leak at the bottom, causing the tank to empty in 40 minutes. According to Torricelli's Law, the volume V of water remaining in the tank after t min is given by the function

$$V = f(t) = 100\left(1 - \frac{t}{40}\right)^2$$

(a) Find f^{-1}. What does f^{-1} represent?

(b) Find $f^{-1}(15)$. What does your answer represent?

96. Blood Flow As blood moves through a vein or artery, its velocity v is greatest along the central axis and decreases as the distance r from the central axis increases (see the figure below). For an artery with radius 0.5 cm, v (in cm/s) is given as a function of r (in cm) by

$$v = g(r) = 18{,}500(0.25 - r^2)$$

(a) Find g^{-1}. What does g^{-1} represent?

(b) Find $g^{-1}(30)$. What does your answer represent?

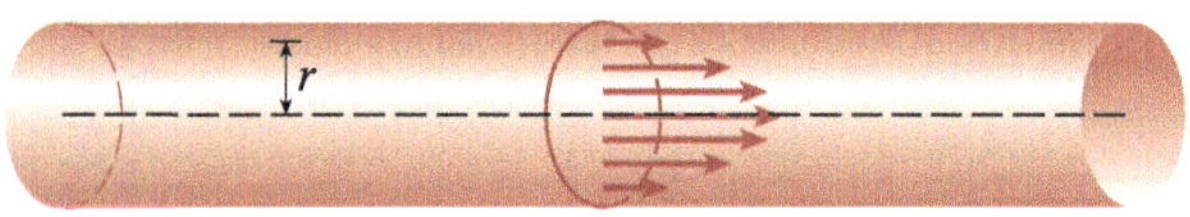

97. Demand Function The amount of a commodity that is sold is called the *demand* for the commodity. The demand D for a certain commodity is a function of the price given by

$$D = f(p) = -3p + 150$$

(a) Find f^{-1}. What does f^{-1} represent?

(b) Find $f^{-1}(30)$. What does your answer represent?

98. Temperature Scales The relationship between the Fahrenheit (F) and Celsius (C) scales is given by

$$F = g(C) = \tfrac{9}{5}C + 32$$

(a) Find g^{-1}. What does g^{-1} represent?

(b) Find $g^{-1}(86)$. What does your answer represent?

99. Exchange Rates The relative value of currencies fluctuates every day. When this problem was written, one Canadian dollar was worth 0.9766 U.S. dollars.

(a) Find a function f that gives the U.S. dollar value $f(x)$ of x Canadian dollars.

(b) Find f^{-1}. What does f^{-1} represent?

(c) How much Canadian money would \$12,250 in U.S. currency be worth?

100. Income Tax In a certain country the tax on incomes less than or equal to €20,000 is 10%. For incomes that are more than €20,000 the tax is €2000 plus 20% of the amount over €20,000.

(a) Find a function f that gives the income tax on an income x. Express f as a piecewise defined function.

(b) Find f^{-1}. What does f^{-1} represent?

(c) How much income would require paying a tax of €10,000?

101. Multiple Discounts A car dealership advertises a 15% discount on all its new cars. In addition, the manufacturer offers a \$1000 rebate on the purchase of a new car. Let x represent the sticker price of the car.

(a) Suppose that only the 15% discount applies. Find a function f that models the purchase price of the car as a function of the sticker price x.

(b) Suppose that only the \$1000 rebate applies. Find a function g that models the purchase price of the car as a function of the sticker price x.

(c) Find a formula for $H = f \circ g$.

(d) Find H^{-1}. What does H^{-1} represent?

(e) Find $H^{-1}(13{,}000)$. What does your answer represent?

DISCUSS ■ DISCOVER ■ PROVE ■ WRITE

102. DISCUSS: Determining When a Linear Function Has an Inverse For the linear function $f(x) = mx + b$ to be one-to-one, what must be true about its slope? If it is one-to-one, find its inverse. Is the inverse linear? If so, what is its slope?

103. DISCUSS: Finding an Inverse "in Your Head" In the margin notes in this section we pointed out that the inverse of a function can be found by simply reversing the operations that make up the function. For instance, in Example 7 we saw that the inverse of

$$f(x) = 3x - 2 \quad \text{is} \quad f^{-1}(x) = \frac{x+2}{3}$$

because the "reverse" of "Multiply by 3 and subtract 2" is "Add 2 and divide by 3." Use the same procedure to find the inverse of the following functions.

(a) $f(x) = \dfrac{2x+1}{5}$ (b) $f(x) = 3 - \dfrac{1}{x}$

(c) $f(x) = \sqrt{x^3 + 2}$ (d) $f(x) = (2x - 5)^3$

Now consider another function:

$$f(x) = x^3 + 2x + 6$$

Is it possible to use the same sort of simple reversal of operations to find the inverse of this function? If so, do it. If not, explain what is different about this function that makes this task difficult.

104. PROVE: The Identity Function The function $I(x) = x$ is called the **identity function**. Show that for any function f we have $f \circ I = f$, $I \circ f = f$, and $f \circ f^{-1} = f^{-1} \circ f = I$. (This means that the identity function I behaves for functions and composition just the way the number 1 behaves for real numbers and multiplication.)

105. DISCUSS: Solving an Equation for an Unknown Function In Exercises 69–72 of Section 2.7 you were asked to solve equations in which the unknowns are functions. Now that we know about inverses and the identity function (see Exercise 104), we can use algebra to solve such equations. For instance, to solve $f \circ g = h$ for the unknown function f, we perform the following steps:

$f \circ g = h$	Problem: Solve for f
$f \circ g \circ g^{-1} = h \circ g^{-1}$	Compose with g^{-1} on the right
$f \circ I = h \circ g^{-1}$	Because $g \circ g^{-1} = I$
$f = h \circ g^{-1}$	Because $f \circ I = f$

So the solution is $f = h \circ g^{-1}$. Use this technique to solve the equation $f \circ g = h$ for the indicated unknown function.

(a) Solve for f, where $g(x) = 2x + 1$ and $h(x) = 4x^2 + 4x + 7$.

(b) Solve for g, where $f(x) = 3x + 5$ and $h(x) = 3x^2 + 3x + 2$.

CHAPTER 2 ■ REVIEW

PROPERTIES AND FORMULAS

Function Notation (p. 149)

If a function is given by the formula $y = f(x)$, then x is the independent variable and denotes the **input**; y is the dependent variable and denotes the **output**; the **domain** is the set of all possible inputs x; the **range** is the set of all possible outputs y.

Net Change (p. 151)

The **net change** in the value of the function f between $x = a$ and $x = b$ is

$$\text{net change} = f(b) - f(a)$$

The Graph of a Function (p. 159)

The graph of a function f is the graph of the equation $y = f(x)$ that defines f.

The Vertical Line Test (p. 164)

A curve in the coordinate plane is the graph of a function if and only if no vertical line intersects the graph more than once.

Increasing and Decreasing Functions (p. 174)

A function f is **increasing** on an interval if $f(x_1) < f(x_2)$ whenever $x_1 < x_2$ in the interval.

A function f is **decreasing** on an interval if $f(x_1) > f(x_2)$ whenever $x_1 < x_2$ in the interval.

Local Maximum and Minimum Values (p. 176)

The function value $f(a)$ is a **local maximum value** of the function f if $f(a) \geq f(x)$ for all x near a. In this case we also say that f has a **local maximum** at $x = a$.

The function value $f(b)$ is a **local minimum value** of the function f if $f(b) \leq f(x)$ for all x near b. In this case we also say that f has a **local minimum** at $x = b$.

Average Rate of Change (p. 184)

The **average rate of change** of the function f between $x = a$ and $x = b$ is the slope of the **secant** line between $(a, f(a))$ and $(b, f(b))$:

$$\text{average rate of change} = \frac{f(b) - f(a)}{b - a}$$

Linear Functions (pp. 191–192)

A **linear function** is a function of the form $f(x) = ax + b$. The graph of f is a line with slope a and y-intercept b. The average rate of change of f has the constant value a between any two points.

$$a = \text{slope of graph of } f = \text{rate of change of } f$$

Vertical and Horizontal Shifts of Graphs (pp. 198–199)

Let c be a positive constant.

To graph $y = f(x) + c$, shift the graph of $y = f(x)$ **upward** by c units.

To graph $y = f(x) - c$, shift the graph of $y = f(x)$ **downward** by c units.

To graph $y = f(x - c)$, shift the graph of $y = f(x)$ **to the right** by c units.

To graph $y = f(x + c)$, shift the graph of $y = f(x)$ **to the left** by c units.

Reflecting Graphs (p. 201)

To graph $y = -f(x)$, **reflect** the graph of $y = f(x)$ in the **x-axis**.

To graph $y = f(-x)$, **reflect** the graph of $y = f(x)$ in the **y-axis**.

Vertical and Horizontal Stretching and Shrinking of Graphs (pp. 202, 203)

If $c > 1$, then to graph $y = cf(x)$, **stretch** the graph of $y = f(x)$ **vertically** by a factor of c.

If $0 < c < 1$, then to graph $y = cf(x)$, **shrink** the graph of $y = f(x)$ **vertically** by a factor of c.

If $c > 1$, then to graph $y = f(cx)$, **shrink** the graph of $y = f(x)$ **horizontally** by a factor of $1/c$.

If $0 < c < 1$, then to graph $y = f(cx)$, **stretch** the graph of $y = f(x)$ **horizontally** by a factor of $1/c$.

Even and Odd Functions (p. 204)

A function f is

even if $f(-x) = f(x)$

odd if $f(-x) = -f(x)$

for every x in the domain of f.

Composition of Functions (p. 213)

Given two functions f and g, the **composition** of f and g is the function $f \circ g$ defined by

$$(f \circ g)(x) = f(g(x))$$

The **domain** of $f \circ g$ is the set of all x for which both $g(x)$ and $f(g(x))$ are defined.

One-to-One Functions (p. 219)

A function f is **one-to-one** if $f(x_1) \neq f(x_2)$ whenever x_1 and x_2 are *different* elements of the domain of f.

Horizontal Line Test (p. 219)

A function is one-to-one if and only if no horizontal line intersects its graph more than once.

Inverse of a Function (p. 220)

Let f be a one-to-one function with domain A and range B.

The **inverse** of f is the function f^{-1} defined by

$$f^{-1}(y) = x \quad \Leftrightarrow \quad f(x) = y$$

The inverse function f^{-1} has domain B and range A.

The functions f and f^{-1} satisfy the following **cancellation properties**:

$$f^{-1}(f(x)) = x \quad \text{for every } x \text{ in } A$$

$$f(f^{-1}(x)) = x \quad \text{for every } x \text{ in } B$$

■ CONCEPT CHECK

1. Define each concept.

(a) Function

(b) Domain and range of a function

(c) Graph of a function

(d) Independent and dependent variables

2. Describe the four ways of representing a function.

3. Sketch graphs of the following functions by hand.

(a) $f(x) = x^2$ **(b)** $g(x) = x^3$

(c) $h(x) = |x|$ **(d)** $k(x) = \sqrt{x}$

4. What is a piecewise defined function? Give an example.

5. (a) What is the Vertical Line Test, and what is it used for?

(b) What is the Horizontal Line Test, and what is it used for?

6. Define each concept, and give an example of each.

(a) Increasing function

(b) Decreasing function

(c) Constant function

7. Suppose we know that the point $(3, 5)$ is a point on the graph of a function f. Explain how to find $f(3)$ and $f^{-1}(5)$.

8. What does it mean to say that $f(4)$ is a local maximum value of f?

9. Explain how to find the average rate of change of a function f between $x = a$ and $x = b$.

10. (a) What is the slope of a linear function? How do you find it? What is the rate of change of a linear function?

(b) Is the rate of change of a linear function constant? Explain.

(c) Give an example of a linear function, and sketch its graph.

11. Suppose the graph of a function f is given. Write an equation for each of the graphs that are obtained from the graph of f as follows.

(a) Shift upward 3 units

(b) Shift downward 3 units

(c) Shift 3 units to the right

(d) Shift 3 units to the left

(e) Reflect in the x-axis

(f) Reflect in the y-axis

(g) Stretch vertically by a factor of 3

(h) Shrink vertically by a factor of $\frac{1}{3}$

(i) Shrink horizontally by a factor of $\frac{1}{3}$

(j) Stretch horizontally by a factor of 3

12. (a) What is an even function? How can you tell that a function is even by looking at its graph? Give an example of an even function.

(b) What is an odd function? How can you tell that a function is odd by looking at its graph? Give an example of an odd function.

13. Suppose that f has domain A and g has domain B. What are the domains of the following functions?

(a) Domain of $f + g$

(b) Domain of fg

(c) Domain of f/g

14. (a) How is the composition function $f \circ g$ defined? What is its domain?

(b) If $g(a) = b$ and $f(b) = c$, then explain how to find $(f \circ g)(a)$.

15. **(a)** What is a one-to-one function?

(b) How can you tell from the graph of a function whether it is one-to-one?

(c) Suppose that f is a one-to-one function with domain A and range B. How is the inverse function f^{-1} defined? What are the domain and range of f^{-1}?

(d) If you are given a formula for f, how do you find a formula for f^{-1}? Find the inverse of the function $f(x) = 2x$.

(e) If you are given a graph of f, how do you find a graph of the inverse function f^{-1}?

ANSWERS TO THE CONCEPT CHECK CAN BE FOUND AT THE BACK OF THE BOOK.

EXERCISES

1–2 ■ Function Notation A verbal description of a function f is given. Find a formula that expresses f in function notation.

1. "Square, then subtract 5."

2. "Divide by 2, then add 9."

3–4 ■ Function in Words A formula for a function f is given. Give a verbal description of the function.

3. $f(x) = 3(x + 10)$ **4.** $f(x) = \sqrt{6x - 10}$

5–6 ■ Table of Values Complete the table of values for the given function.

5. $g(x) = x^2 - 4x$ **6.** $h(x) = 3x^2 + 2x - 5$

x	$g(x)$
-1	
0	
1	
2	
3	

x	$h(x)$
-2	
-1	
0	
1	
2	

7. Printing Cost A publisher estimates that the cost $C(x)$ of printing a run of x copies of a certain mathematics textbook is given by the function $C(x) = 5000 + 30x - 0.001x^2$.

(a) Find $C(1000)$ and $C(10{,}000)$.

(b) What do your answers in part (a) represent?

(c) Find $C(0)$. What does this number represent?

(d) Find the net change and the average rate of change of the cost C between $x = 1000$ and $x = 10{,}000$.

8. Earnings Reynalda works as a salesperson in the electronics division of a department store. She earns a base weekly salary plus a commission based on the retail price of the goods she has sold. If she sells x dollars worth of goods in a week, her earnings for that week are given by the function $E(x) = 400 + 0.03x$.

(a) Find $E(2000)$ and $E(15{,}000)$.

(b) What do your answers in part (a) represent?

(c) Find $E(0)$. What does this number represent?

(d) Find the net change and the average rate of change of her earnings E between $x = 2000$ and $x = 15{,}000$.

(e) From the formula for E, determine what percentage Reynalda earns on the goods that she sells.

9–10 ■ Evaluating Functions Evaluate the function at the indicated values.

9. $f(x) = x^2 - 4x + 6$; $f(0), f(2), f(-2), f(a), f(-a), f(x + 1), f(2x)$

10. $f(x) = 4 - \sqrt{3x - 6}$; $f(5), f(9), f(a + 2), f(-x), f(x^2)$

11. Functions Given by a Graph Which of the following figures are graphs of functions? Which of the functions are one-to-one?

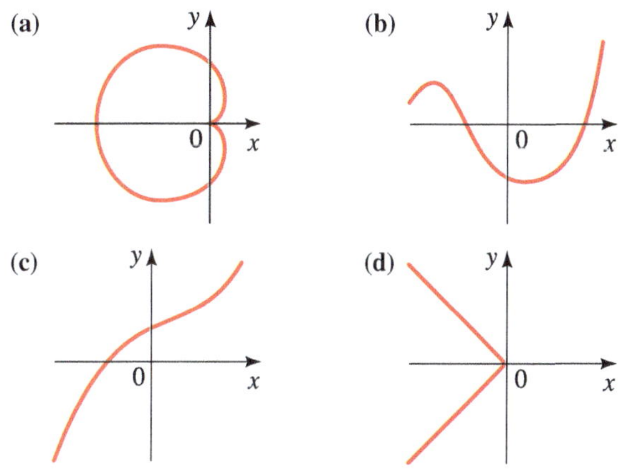

12. Getting Information from a Graph A graph of a function f is given.

(a) Find $f(-2)$ and $f(2)$.

(b) Find the net change and the average rate of change of f between $x = -2$ and $x = 2$.

(c) Find the domain and range of f.

(d) On what intervals is f increasing? On what intervals is f decreasing?

(e) What are the local maximum values of f?

(f) Is f one-to-one?

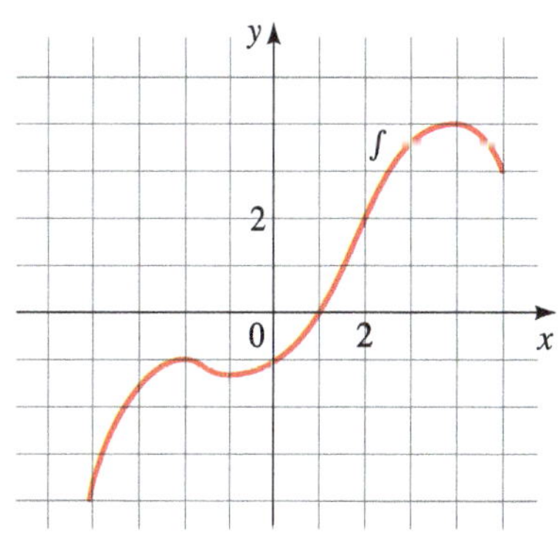

13–14 ■ Domain and Range Find the domain and range of the function.

13. $f(x) = \sqrt{x + 3}$

14. $F(t) = t^2 + 2t + 5$

15–22 ■ Domain Find the domain of the function.

15. $f(x) = 7x + 15$

16. $f(x) = \dfrac{2x + 1}{2x - 1}$

17. $f(x) = \sqrt{x + 4}$

18. $f(x) = 3x - \dfrac{2}{\sqrt{x + 1}}$

19. $f(x) = \dfrac{1}{x} + \dfrac{1}{x + 1} + \dfrac{1}{x + 2}$

20. $g(x) = \dfrac{2x^2 + 5x + 3}{2x^2 - 5x - 3}$

21. $h(x) = \sqrt{4 - x} + \sqrt{x^2 - 1}$

22. $f(x) = \dfrac{\sqrt[3]{2x + 1}}{\sqrt[3]{2x} + 2}$

23–38 ■ Graphing Functions Sketch a graph of the function. Use transformations of functions whenever possible.

23. $f(x) = 1 - 2x$

24. $f(x) = \frac{1}{3}(x - 5), \quad 2 \le x \le 8$

25. $f(x) = 3x^2$

26. $f(x) = -\frac{1}{4}x^2$

27. $f(x) = 2x^2 - 1$

28. $f(x) = -(x - 1)^4$

29. $f(x) = 1 + \sqrt{x}$

30. $f(x) = 1 - \sqrt{x + 2}$

31. $f(x) = \frac{1}{2}x^3$

32. $f(x) = \sqrt[3]{-x}$

33. $f(x) = -|x|$

34. $f(x) = |x + 1|$

35. $f(x) = -\dfrac{1}{x^2}$

36. $f(x) = \dfrac{1}{(x - 1)^3}$

37. $f(x) = \begin{cases} 1 - x & \text{if } x < 0 \\ 1 & \text{if } x \ge 0 \end{cases}$

38. $f(x) = \begin{cases} -x & \text{if } x < 0 \\ x^2 & \text{if } 0 \le x < 2 \\ 1 & \text{if } x \ge 2 \end{cases}$

39–42 ■ Equations That Represent Functions Determine whether the equation defines y as a function of x.

39. $x + y^2 = 14$

40. $3x - \sqrt{y} = 8$

41. $x^3 - y^3 = 27$

42. $2x = y^4 - 16$

43–44 ■ Graphing Functions Determine which viewing rectangle produces the most appropriate graph of the function.

43. $f(x) = 6x^3 - 15x^2 + 4x - 1$

(i) $[-2, 2]$ by $[-2, 2]$

(ii) $[-8, 8]$ by $[-8, 8]$

(iii) $[-4, 4]$ by $[-12, 12]$

(iv) $[-100, 100]$ by $[-100, 100]$

44. $f(x) = \sqrt{100 - x^3}$.

(i) $[-4, 4]$ by $[-4, 4]$

(ii) $[-10, 10]$ by $[-10, 10]$

(iii) $[-10, 10]$ by $[-10, 40]$

(iv) $[-100, 100]$ by $[-100, 100]$

45–48 ■ Domain and Range from a Graph A function f is given. **(a)** Use a graphing calculator to draw the graph of f. **(b)** Find the domain and range of f from the graph.

45. $f(x) = \sqrt{9 - x^2}$

46. $f(x) = -\sqrt{x^2 - 3}$

47. $f(x) = \sqrt{x^3 - 4x + 1}$

48. $f(x) = x^4 - x^3 + x^2 + 3x - 6$

49–50 ■ Getting Information from a Graph Draw a graph of the function f, and determine the intervals on which f is increasing and on which f is decreasing.

49. $f(x) = x^3 - 4x^2$

50. $f(x) = |x^4 - 16|$

51–56 ■ Net Change and Average Rate of Change A function is given (either numerically, graphically, or algebraically). Find the net change and the average rate of change of the function between the indicated values.

51. Between $x = 4$ and $x = 8$

x	$f(x)$
2	14
4	12
6	12
8	8
10	6

52. Between $x = 10$ and $x = 30$

x	$g(x)$
0	25
10	−5
20	−2
30	30
40	0

53. Between $x = -1$ and $x = 2$

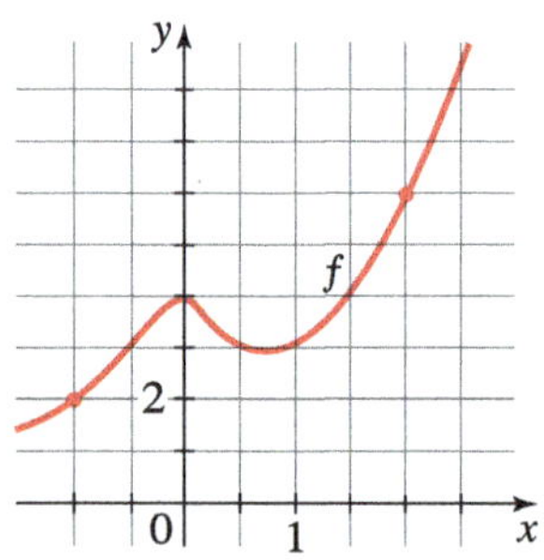

54. Between $x = 1$ and $x = 3$

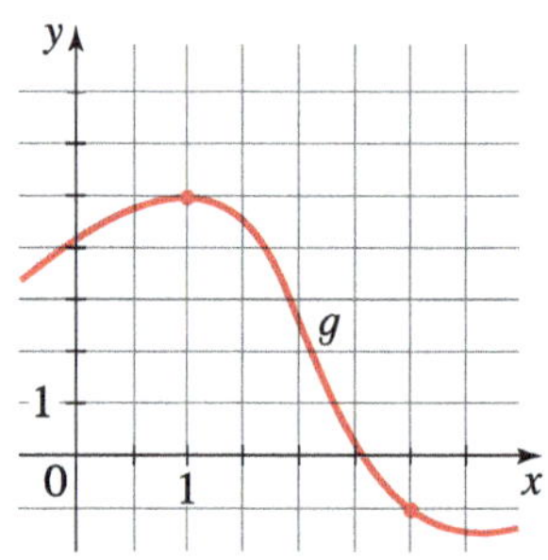

55. $f(x) = x^2 - 2x$; between $x = 1$ and $x = 4$

56. $g(x) = (x + 1)^2$; between $x = a$ and $x = a + h$

57–58 ■ Linear? Determine whether the given function is linear.

57. $f(x) = (2 + 3x)^2$

58. $g(x) = \dfrac{x + 3}{5}$

59–60 ■ Linear Functions A linear function is given. **(a)** Sketch a graph of the function. **(b)** What is the slope of the graph? **(c)** What is the rate of change of the function?

59. $f(x) = 3x + 2$

60. $g(x) = 3 - \frac{1}{2}x$

61–66 ■ Linear Functions A linear function is described either verbally, numerically, or graphically. Express f in the form $f(x) = ax + b$.

61. The function has rate of change -2 and initial value 3.

62. The graph of the function has slope $\frac{1}{2}$ and y-intercept -1.

63.

x	$f(x)$
0	3
1	5
2	7
3	9
4	11

64.

x	$f(x)$
0	6
2	5.5
4	5
6	4.5
8	4

65.

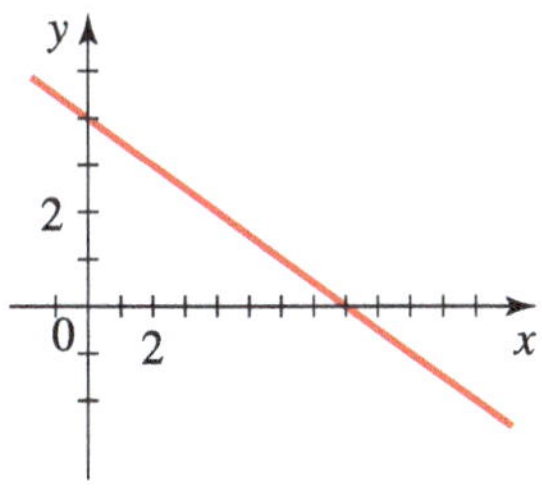

66.

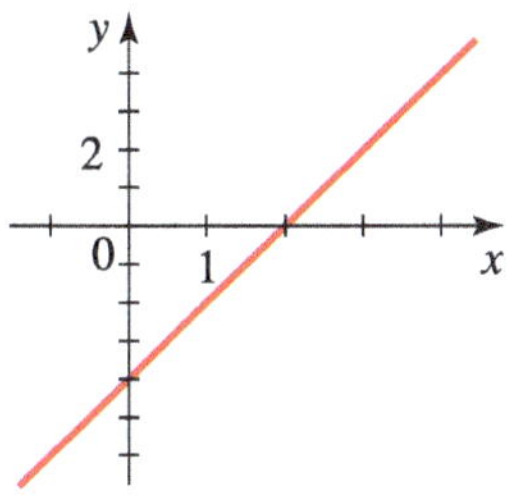

67. Population The population of a planned seaside community in Florida is given by the function $P(t) = 3000 + 200t + 0.1t^2$, where t represents the number of years since the community was incorporated in 1985.

(a) Find $P(10)$ and $P(20)$. What do these values represent?

(b) Find the average rate of change of P between $t = 10$ and $t = 20$. What does this number represent?

68. Retirement Savings Ella is saving for her retirement by making regular deposits into a 401(k) plan. As her salary rises, she finds that she can deposit increasing amounts each year. Between 1995 and 2008 the annual amount (in dollars) that she deposited was given by the function $D(t) = 3500 + 15t^2$, where t represents the year of the deposit measured from the start of the plan (so 1995 corresponds to $t = 0$, 1996 corresponds to $t = 1$, and so on).

(a) Find $D(0)$ and $D(15)$. What do these values represent?

(b) Assuming that her deposits continue to be modeled by the function D, in what year will she deposit \$17,000?

(c) Find the average rate of change of D between $t = 0$ and $t = 15$. What does this number represent?

69–70 ■ Average Rate of Change A function f is given. **(a)** Find the average rate of change of f between $x = 0$ and $x = 2$, and the average rate of change of f between $x = 15$ and $x = 50$. **(b)** Were the two average rates of change that you found in part (a) the same? **(c)** Is the function linear? If so, what is its rate of change?

69. $f(x) = \frac{1}{2}x - 6$ **70.** $f(x) = 8 - 3x$

71. Transformations Suppose the graph of f is given. Describe how the graphs of the following functions can be obtained from the graph of f.

(a) $y = f(x) + 8$ **(b)** $y = f(x + 8)$

(c) $y = 1 + 2f(x)$ **(d)** $y = f(x - 2) - 2$

(e) $y = f(-x)$ **(f)** $y = -f(-x)$

(g) $y = -f(x)$ **(h)** $y = f^{-1}(x)$

72. Transformations The graph of f is given. Draw the graphs of the following functions.

(a) $y = f(x - 2)$ **(b)** $y = -f(x)$

(c) $y = 3 - f(x)$ **(d)** $y = \frac{1}{2}f(x) - 1$

(e) $y = f^{-1}(x)$ **(f)** $y = f(-x)$

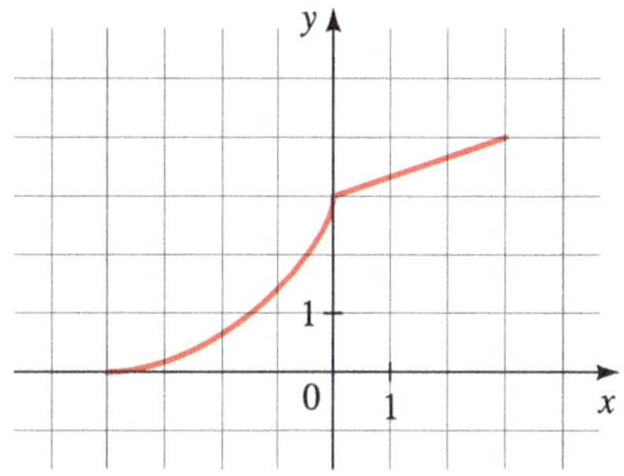

73. Even and Odd Functions Determine whether f is even, odd, or neither.

(a) $f(x) = 2x^5 - 3x^2 + 2$ **(b)** $f(x) = x^3 - x^7$

(c) $f(x) = \dfrac{1 - x^2}{1 + x^2}$ **(d)** $f(x) = \dfrac{1}{x + 2}$

74. Even and Odd Functions Determine whether the function in the figure is even, odd, or neither.

(a)

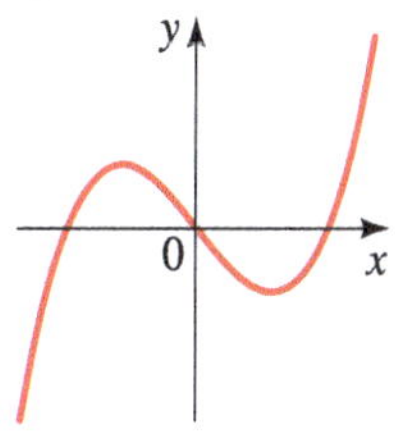

(b)

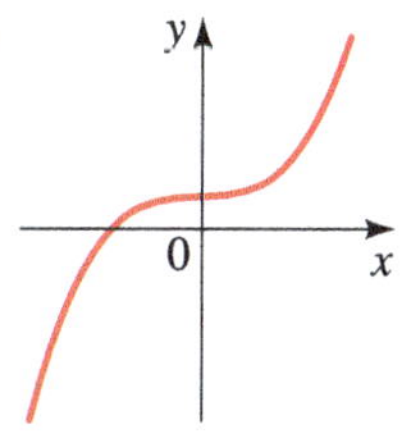

(c)

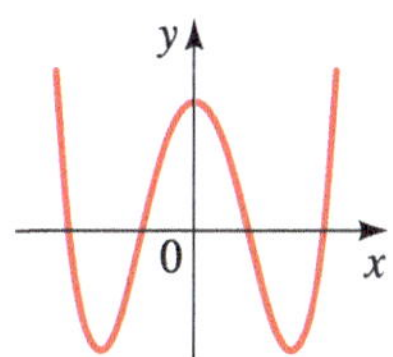

(d)

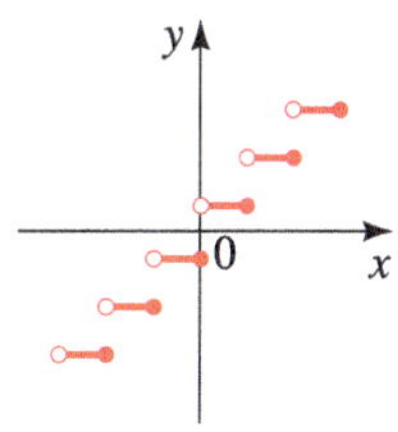

75–78 ■ Local Maxima and Minima Find the local maximum and minimum values of the function and the values of x at which they occur. State each answer rounded to two decimal places.

75. $g(x) = 2x^2 + 4x - 5$

76. $f(x) = 1 - x - x^2$

77. $f(x) = 3.3 + 1.6x - 2.5x^3$

78. $f(x) = x^{2/3}(6 - x)^{1/3}$

79. Maximum Height of Projectile A stone is thrown upward from the top of a building. Its height (in feet) above the ground after t seconds is given by

$$h(t) = -16t^2 + 48t + 32$$

What maximum height does it reach?

80. Maximum Profit The profit P (in dollars) generated by selling x units of a certain commodity is given by

$$P(x) = -1500 + 12x - 0.0004x^2$$

What is the maximum profit, and how many units must be sold to generate it?

81–82 ■ Graphical Addition Two functions, f and g, are given. Draw graphs of f, g, and $f + g$ on the same graphing calculator screen to illustrate the concept of graphical addition.

81. $f(x) = x + 2, \quad g(x) = x^2$

82. $f(x) = x^2 + 1, \quad g(x) = 3 - x^2$

83. Combining Functions If $f(x) = x^2 - 3x + 2$ and $g(x) = 4 - 3x$, find the following functions.

(a) $f + g$ **(b)** $f - g$ **(c)** fg

(d) f/g **(e)** $f \circ g$ **(f)** $g \circ f$

84. If $f(x) = 1 + x^2$ and $g(x) = \sqrt{x - 1}$, find the following.

(a) $f \circ g$ **(b)** $g \circ f$ **(c)** $(f \circ g)(2)$

(d) $(f \circ f)(2)$ **(e)** $f \circ g \circ f$ **(f)** $g \circ f \circ g$

85–86 ■ Composition of Functions Find the functions $f \circ g$, $g \circ f$, $f \circ f$, and $g \circ g$ and their domains.

85. $f(x) = 3x - 1, \quad g(x) = 2x - x^2$

86. $f(x) = \sqrt{x}, \quad g(x) = \dfrac{2}{x - 4}$

87. Finding a Composition Find $f \circ g \circ h$, where $f(x) = \sqrt{1 - x}$, $g(x) = 1 - x^2$, and $h(x) = 1 + \sqrt{x}$.

88. Finding a Composition If $T(x) = \dfrac{1}{\sqrt{1 + \sqrt{x}}}$, find functions f, g, and h such that $f \circ g \circ h = T$.

89–94 ■ One-to-One Functions Determine whether the function is one-to-one.

89. $f(x) = 3 + x^3$

90. $g(x) = 2 - 2x + x^2$

91. $h(x) = \dfrac{1}{x^4}$

92. $r(x) = 2 + \sqrt{x + 3}$

93. $p(x) = 3.3 + 1.6x - 2.5x^3$

94. $q(x) = 3.3 + 1.6x + 2.5x^3$

95–98 ■ Finding Inverse Functions Find the inverse of the function.

95. $f(x) = 3x - 2$

96. $f(x) = \dfrac{2x + 1}{3}$

97. $f(x) = (x + 1)^3$

98. $f(x) = 1 + \sqrt[5]{x - 2}$

99–100 ■ Inverse Functions from a Graph A graph of a function f is given. Does f have an inverse? If so, find $f^{-1}(0)$ and $f^{-1}(4)$.

99.

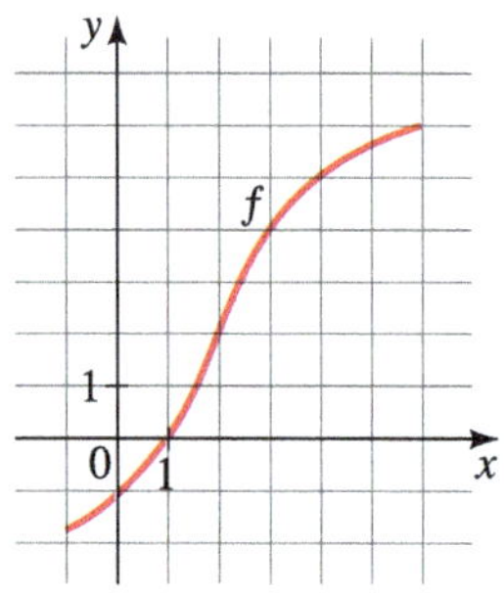

100.

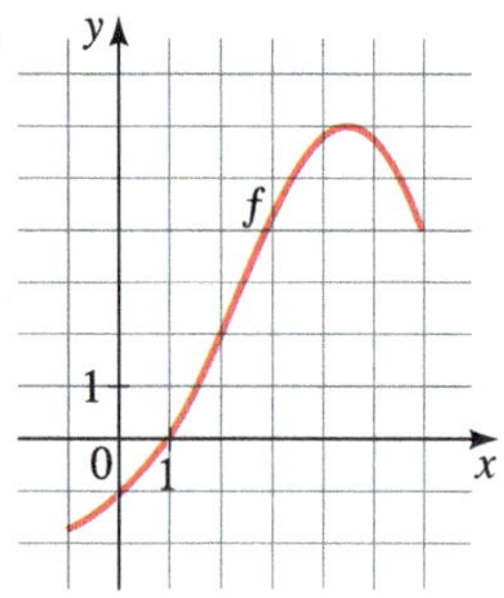

101. Graphing Inverse Functions

(a) Sketch a graph of the function

$$f(x) = x^2 - 4 \qquad x \geq 0$$

(b) Use part (a) to sketch the graph of f^{-1}.

(c) Find an equation for f^{-1}.

102. Graphing Inverse Functions

(a) Show that the function $f(x) = 1 + \sqrt[4]{x}$ is one-to-one.

(b) Sketch the graph of f.

(c) Use part (b) to sketch the graph of f^{-1}.

(d) Find an equation for f^{-1}.

CHAPTER 2 TEST

1. Which of the following are graphs of functions? If the graph is that of a function, is it one-to-one?

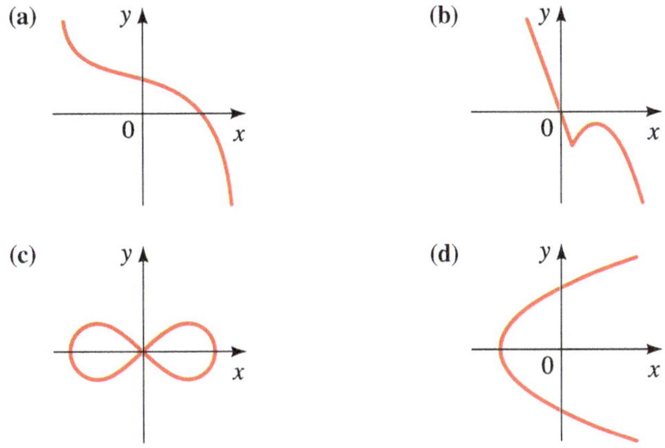

2. Let $f(x) = \dfrac{\sqrt{x}}{x + 1}$.
   (a) Evaluate $f(0)$, $f(2)$, and $f(a + 2)$.
   (b) Find the domain of f.
   (c) What is the average rate of change of f between $x = 2$ and $x = 10$?

3. A function f has the following verbal description: "Subtract 2, then cube the result."
   (a) Find a formula that expresses f algebraically.
   (b) Make a table of values of f, for the inputs $-1, 0, 1, 2, 3$, and 4.
   (c) Sketch a graph of f, using the table of values from part (b) to help you.
   (d) How do we know that f has an inverse? Give a verbal description for f^{-1}.
   (e) Find a formula that expresses f^{-1} algebraically.

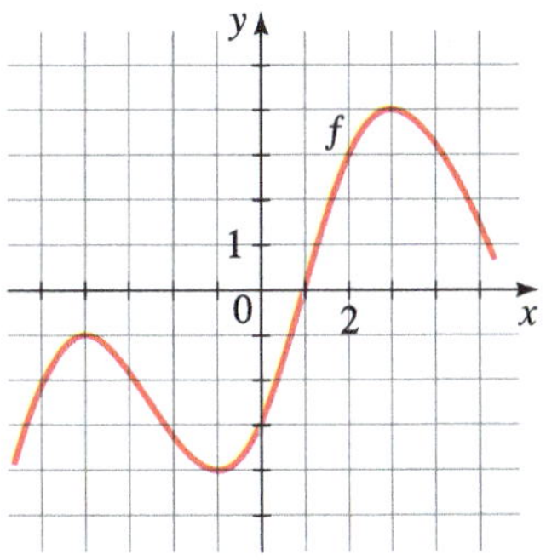

4. A graph of a function f is given in the margin.
   (a) Find the local minimum and maximum values of f and the values of x at which they occur.
   (b) Find the intervals on which f is increasing and on which f is decreasing.

5. A school fund-raising group sells chocolate bars to help finance a swimming pool for their physical education program. The group finds that when they set their price at x dollars per bar (where $0 < x \le 5$), their total sales revenue (in dollars) is given by the function $R(x) = -500x^2 + 3000x$.
   (a) Evaluate $R(2)$ and $R(4)$. What do these values represent?
   (b) Use a graphing calculator to draw a graph of R. What does the graph tell us about what happens to revenue as the price increases from 0 to 5 dollars?
   (c) What is the maximum revenue, and at what price is it achieved?

6. Determine the net change and the average rate of change for the function $f(t) = t^2 - 2t$ between $t = 2$ and $t = 2 + h$.

7. Let $f(x) = (x + 5)^2$ and $g(x) = 1 - 5x$.
   (a) Only one of the two functions f and g is linear. Which one is linear, and why is the other one not linear?
   (b) Sketch a graph of each function.
   (c) What is the rate of change of the linear function?

8. (a) Sketch the graph of the function $f(x) = x^3$.
   (b) Use part (a) to graph the function $g(x) = (x - 1)^3 - 2$.

9. (a) How is the graph of $y = f(x - 3) + 2$ obtained from the graph of f?
   (b) How is the graph of $y = f(-x)$ obtained from the graph of f?

10. Let $f(x) = \begin{cases} 1 - x & \text{if } x \le 1 \\ 2x + 1 & \text{if } x > 1 \end{cases}$

(a) Evaluate $f(-2)$ and $f(1)$.

(b) Sketch the graph of f.

11. If $f(x) = x^2 + x + 1$ and $g(x) = x - 3$, find the following.

(a) $f + g$ **(b)** $f - g$ **(c)** $f \circ g$ **(d)** $g \circ f$

(e) $f(g(2))$ **(f)** $g(f(2))$ **(g)** $g \circ g \circ g$

12. Determine whether the function is one-to-one.

(a) $f(x) = x^3 + 1$ **(b)** $g(x) = |x + 1|$

13. Use the Inverse Function Property to show that $f(x) = \dfrac{1}{x - 2}$ is the inverse of $g(x) = \dfrac{1}{x} + 2$.

14. Find the inverse function of $f(x) = \dfrac{x - 3}{2x + 5}$.

15. (a) If $f(x) = \sqrt{3 - x}$, find the inverse function f^{-1}.

(b) Sketch the graphs of f and f^{-1} on the same coordinate axes.

16–21 ■ A graph of a function f is given below.

16. Find the domain and range of f.

17. Find $f(0)$ and $f(4)$.

18. Graph $f(x - 2)$ and $f(x) + 2$.

19. Find the net change and the average rate of change of f between $x = 2$ and $x = 6$.

20. Find $f^{-1}(1)$ and $f^{-1}(3)$.

21. Sketch the graph of f^{-1}.

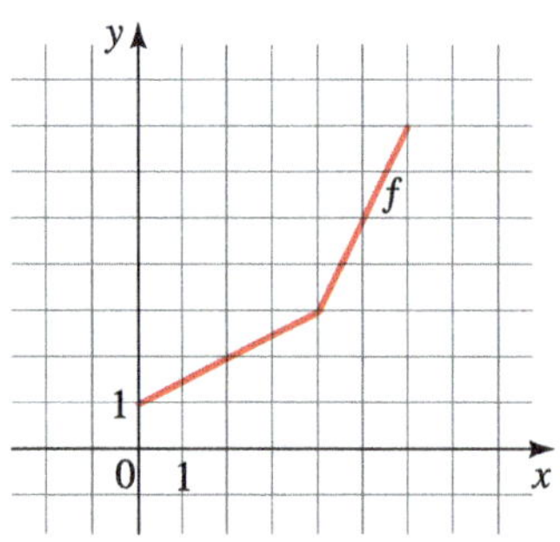

22. Let $f(x) = 3x^4 - 14x^2 + 5x - 3$.

(a) Draw the graph of f in an appropriate viewing rectangle.

(b) Is f one-to-one?

(c) Find the local maximum and minimum values of f and the values of x at which they occur. State each answer correct to two decimal places.

(d) Use the graph to determine the range of f.

(e) Find the intervals on which f is increasing and on which f is decreasing.

Many of the processes that are studied in the physical and social sciences involve understanding how one quantity varies with respect to another. Finding a function that describes the dependence of one quantity on another is called *modeling*. For example, a biologist observes that the number of bacteria in a certain culture increases with time. He tries to model this phenomenon by finding the precise function (or rule) that relates the bacteria population to the elapsed time.

In this *Focus on Modeling* we will learn how to find models that can be constructed using geometric or algebraic properties of the object under study. Once the model is found, we use it to analyze and predict properties of the object or process being studied.

Modeling with Functions

We begin by giving some general guidelines for making a function model.

GUIDELINES FOR MODELING WITH FUNCTIONS

1. **Express the Model in Words.** Identify the quantity you want to model, and express it, in words, as a function of the other quantities in the problem.
2. **Choose the Variable.** Identify all the variables that are used to express the function in Step 1. Assign a symbol, such as x, to one variable, and express the other variables in terms of this symbol.
3. **Set up the Model.** Express the function in the language of algebra by writing it as a function of the single variable chosen in Step 2.
4. **Use the Model.** Use the function to answer the questions posed in the problem. (To find a maximum or a minimum, use the methods described in Section 2.3.)

EXAMPLE 1 ■ Fencing a Garden

A gardener has 140 feet of fencing to fence in a rectangular vegetable garden.

(a) Find a function that models the area of the garden she can fence.

(b) For what range of widths is the area greater than 825 ft^2?

(c) Can she fence a garden with area 1250 ft^2?

(d) Find the dimensions of the largest area she can fence.

THINKING ABOUT THE PROBLEM

If the gardener fences a plot with width 10 ft, then the length must be 60 ft, because $10 + 10 + 60 + 60 = 140$. So the area is

$$A = \text{width} \times \text{length} = 10 \cdot 60 = 600 \text{ ft}^2$$

The table shows various choices for fencing the garden. We see that as the width increases, the fenced area increases, then decreases.

Width	Length	Area
10	60	600
20	50	1000
30	40	1200
40	30	1200
50	20	1000
60	10	600

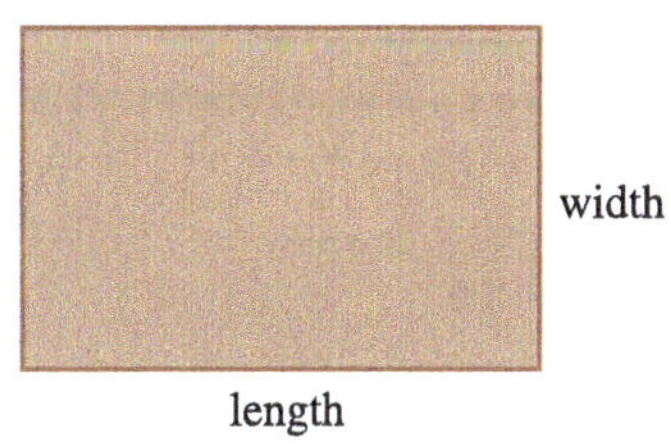

SOLUTION

(a) The model that we want is a function that gives the area she can fence.

Express the model in words. We know that the area of a rectangular garden is

$$\text{area} = \text{width} \times \text{length}$$

Choose the variable. There are two varying quantities: width and length. Because the function we want depends on only one variable, we let

$$x = \text{width of the garden}$$

FIGURE 1

Then we must express the length in terms of x. The perimeter is fixed at 140 ft, so the length is determined once we choose the width. If we let the length be l, as in Figure 1, then $2x + 2l = 140$, so $l = 70 - x$. We summarize these facts:

In Words	In Algebra
Width	x
Length	$70 - x$

Set up the model. The model is the function A that gives the area of the garden for any width x.

$$\text{area} = \text{width} \times \text{length}$$

$$A(x) = x(70 - x)$$

$$A(x) = 70x - x^2$$

The area that she can fence is modeled by the function $A(x) = 70x - x^2$.

Use the model. We use the model to answer the questions in parts (b)–(d).

(b) We need to solve the inequality $A(x) \geq 825$. To solve graphically, we graph $y = 70x - x^2$ and $y = 825$ in the same viewing rectangle (see Figure 2). We see that $15 \leq x \leq 55$.

(c) From Figure 3 we see that the graph of $A(x)$ always lies below the line $y = 1250$, so an area of $1250\ \text{ft}^2$ is never attained.

Maximum values of functions are discussed on page 176.

(d) We need to find where the maximum value of the function $A(x) = 70x - x^2$ occurs. The function is graphed in Figure 4. Using the TRACE feature on a graphing calculator, we find that the function achieves its maximum value at $x = 35$. So the maximum area that she can fence is that when the garden's width is 35 ft and its length is $70 - 35 = 35$ ft. The maximum area then is $35 \times 35 = 1225\ \text{ft}^2$.

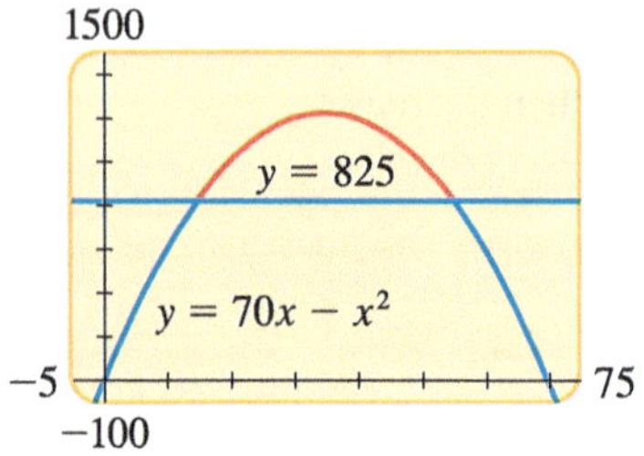

FIGURE 2

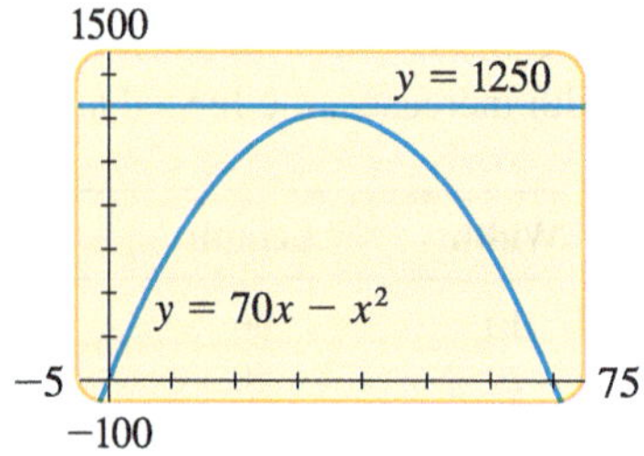

FIGURE 3

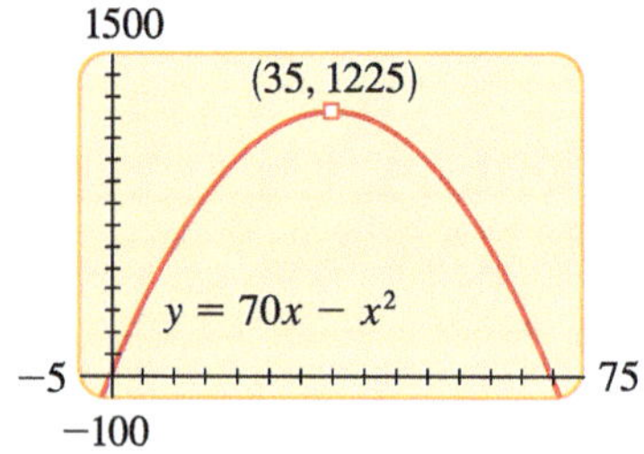

FIGURE 4

EXAMPLE 2 ■ Minimizing the Metal in a Can

A manufacturer makes a metal can that holds 1 L (liter) of oil. What radius minimizes the amount of metal in the can?

THINKING ABOUT THE PROBLEM

To use the least amount of metal, we must minimize the surface area of the can, that is, the area of the top, bottom, and the sides. The area of the top and bottom is $2\pi r^2$ and the area of the sides is $2\pi rh$ (see Figure 5), so the surface area of the can is

$$S = 2\pi r^2 + 2\pi rh$$

The radius and height of the can must be chosen so that the volume is exactly 1 L, or 1000 cm^3. If we want a small radius, say, $r = 3$, then the height must be just tall enough to make the total volume 1000 cm^3. In other words, we must have

$$\pi(3)^2h = 1000 \qquad \text{Volume of the can is } \pi r^2 h$$

$$h = \frac{1000}{9\pi} \approx 35.37 \text{ cm} \qquad \text{Solve for } h$$

Now that we know the radius and height, we can find the surface area of the can:

$$\text{surface area} = 2\pi(3)^2 + 2\pi(3)(35.4) \approx 723.2 \text{ cm}^3$$

If we want a different radius, we can find the corresponding height and surface area in a similar fashion.

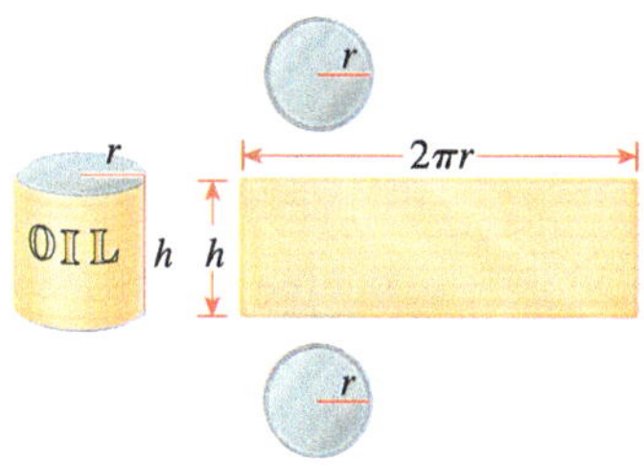

FIGURE 5

SOLUTION The model that we want is a function that gives the surface area of the can.

Express the model in words. We know that for a cylindrical can

surface area = area of top and bottom + area of sides

Choose the variable. There are two varying quantities: radius and height. Because the function we want depends on the radius, we let

$$r = \text{radius of can}$$

Next, we must express the height in terms of the radius r. Because the volume of a cylindrical can is $V = \pi r^2 h$ and the volume must be 1000 cm^3, we have

$$\pi r^2 h = 1000 \qquad \text{Volume of can is 1000 cm}^3$$

$$h = \frac{1000}{\pi r^2} \qquad \text{Solve for } h$$

We can now express the areas of the top, bottom, and sides in terms of r only:

In Words	In Algebra
Radius of can	r
Height of can	$\dfrac{1000}{\pi r^2}$
Area of top and bottom	$2\pi r^2$
Area of sides ($2\pi rh$)	$2\pi r\left(\dfrac{1000}{\pi r^2}\right)$

Set up the model. The model is the function S that gives the surface area of the can as a function of the radius r.

$$\text{surface area} = \text{area of top and bottom} + \text{area of sides}$$

$$S(r) = 2\pi r^2 + 2\pi r\left(\frac{1000}{\pi r^2}\right)$$

$$S(r) = 2\pi r^2 + \frac{2000}{r}$$

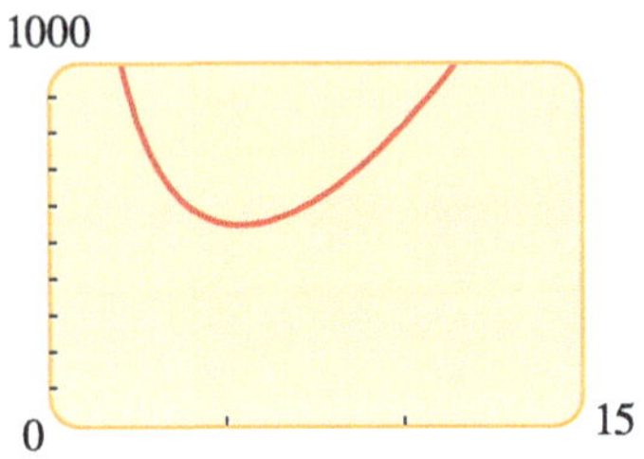

FIGURE 6 $S(r) = 2\pi r^2 + \dfrac{2000}{r}$

Use the model. We use the model to find the minimum surface area of the can. We graph S in Figure 6 and zoom in on the minimum point to find that the minimum value of S is about 554 cm^2 and occurs when the radius is about 5.4 cm. ■

PROBLEMS

1–18 ■ In these problems you are asked to find a function that models a real-life situation. Use the principles of modeling described in this Focus to help you.

1. **Area** A rectangular building lot is three times as long as it is wide. Find a function that models its area A in terms of its width w.

2. **Area** A poster is 10 in. longer than it is wide. Find a function that models its area A in terms of its width w.

3. **Volume** A rectangular box has a square base. Its height is half the width of the base. Find a function that models its volume V in terms of its width w.

4. **Volume** The height of a cylinder is four times its radius. Find a function that models the volume V of the cylinder in terms of its radius r.

5. **Area** A rectangle has a perimeter of 20 ft. Find a function that models its area A in terms of the length x of one of its sides.

6. **Perimeter** A rectangle has an area of 16 m^2. Find a function that models its perimeter P in terms of the length x of one of its sides.

7. **Area** Find a function that models the area A of an equilateral triangle in terms of the length x of one of its sides.

8. **Area** Find a function that models the surface area S of a cube in terms of its volume V.

9. **Radius** Find a function that models the radius r of a circle in terms of its area A.

10. **Area** Find a function that models the area A of a circle in terms of its circumference C.

11. **Area** A rectangular box with a volume of 60 ft^3 has a square base. Find a function that models its surface area S in terms of the length x of one side of its base.

PYTHAGORAS (circa 580–500 B.C.) founded a school in Croton in southern Italy, devoted to the study of arithmetic, geometry, music, and astronomy. The Pythagoreans, as they were called, were a secret society with peculiar rules and initiation rites. They wrote nothing down and were not to reveal to anyone what they had learned from the Master. Although women were barred by law from attending public meetings, Pythagoras allowed women in his school, and his most famous student was Theano (whom he later married).

According to Aristotle, the Pythagoreans were convinced that "the principles of mathematics are the principles of all things." Their motto was "Everything is Number," by which they meant *whole* numbers. The outstanding contribution of Pythagoras is the theorem that bears his name: In a right triangle the area of the square on the hypotenuse is equal to the sum of the areas of the square on the other two sides.

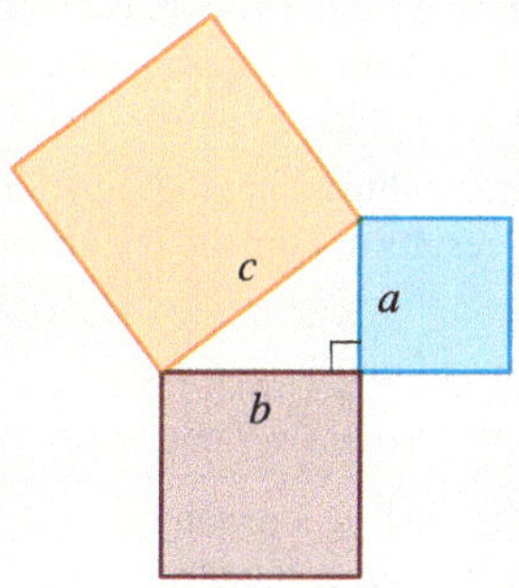

$$c^2 = a^2 + b^2$$

The converse of Pythagoras's Theorem is also true; that is, a triangle whose sides a, b, and c satisfy $a^2 + b^2 = c^2$ is a right triangle.

12. Length A woman 5 ft tall is standing near a street lamp that is 12 ft tall, as shown in the figure. Find a function that models the length L of her shadow in terms of her distance d from the base of the lamp.

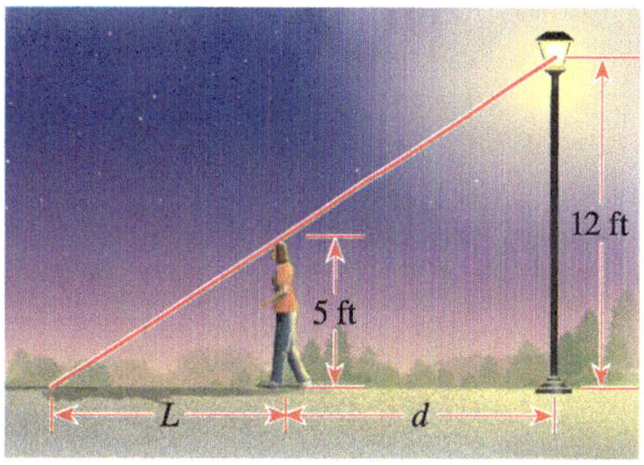

13. Distance Two ships leave port at the same time. One sails south at 15 mi/h, and the other sails east at 20 mi/h. Find a function that models the distance D between the ships in terms of the time t (in hours) elapsed since their departure.

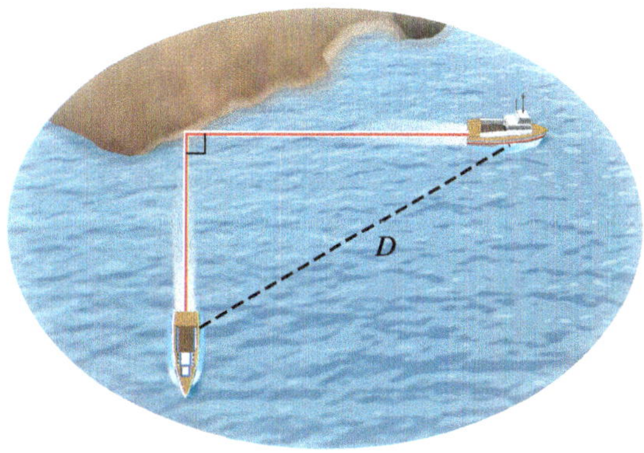

14. Product The sum of two positive numbers is 60. Find a function that models their product P in terms of x, one of the numbers.

15. Area An isosceles triangle has a perimeter of 8 cm. Find a function that models its area A in terms of the length of its base b.

16. Perimeter A right triangle has one leg twice as long as the other. Find a function that models its perimeter P in terms of the length x of the shorter leg.

17. Area A rectangle is inscribed in a semicircle of radius 10, as shown in the figure. Find a function that models the area A of the rectangle in terms of its height h.

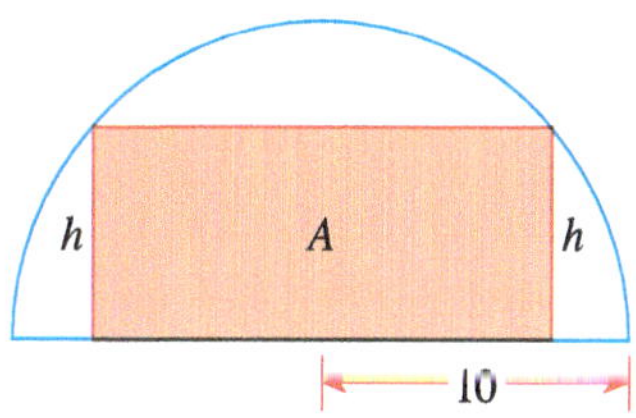

18. Height The volume of a cone is 100 in^3. Find a function that models the height h of the cone in terms of its radius r.

19–32 ■ In these problems you are asked to find a function that models a real-life situation and then use the model to answer questions about the situation. Use the guidelines on page 237 to help you.

19. Maximizing a Product Consider the following problem: Find two numbers whose sum is 19 and whose product is as large as possible.

(a) Experiment with the problem by making a table like the one following, showing the product of different pairs of numbers that add up to 19. On the basis of the evidence in your table, estimate the answer to the problem.

First number	Second number	Product
1	18	18
2	17	34
3	16	48
⋮	⋮	⋮

(b) Find a function that models the product in terms of one of the two numbers.

(c) Use your model to solve the problem, and compare with your answer to part (a).

20. Minimizing a Sum Find two positive numbers whose sum is 100 and the sum of whose squares is a minimum.

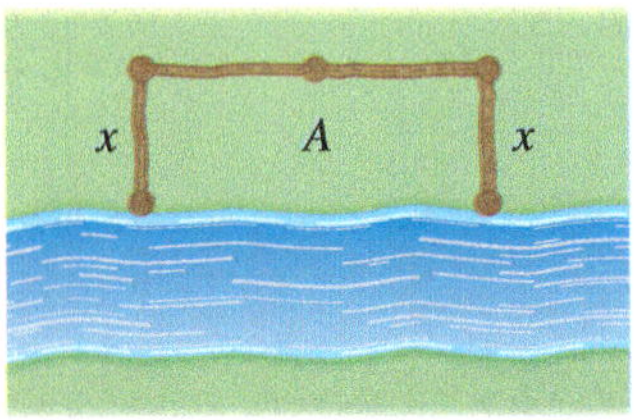

21. Fencing a Field Consider the following problem: A farmer has 2400 ft of fencing and wants to fence off a rectangular field that borders a straight river. He does not need a fence along the river (see the figure). What are the dimensions of the field of largest area that he can fence?

(a) Experiment with the problem by drawing several diagrams illustrating the situation. Calculate the area of each configuration, and use your results to estimate the dimensions of the largest possible field.

(b) Find a function that models the area of the field in terms of one of its sides.

(c) Use your model to solve the problem, and compare with your answer to part (a).

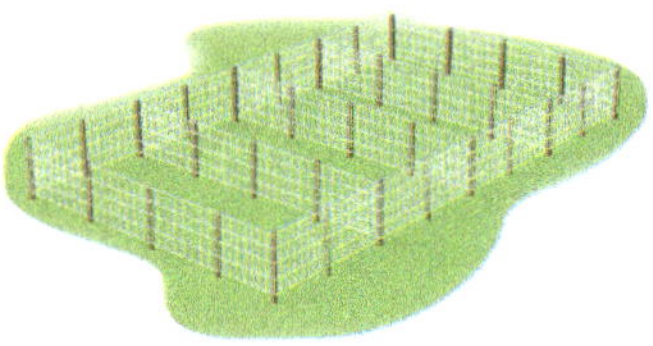

22. Dividing a Pen A rancher with 750 ft of fencing wants to enclose a rectangular area and then divide it into four pens with fencing parallel to one side of the rectangle (see the figure).

(a) Find a function that models the total area of the four pens.

(b) Find the largest possible total area of the four pens.

23. Fencing a Garden Plot A property owner wants to fence a garden plot adjacent to a road, as shown in the figure. The fencing next to the road must be sturdier and costs \$5 per foot, but the other fencing costs just \$3 per foot. The garden is to have an area of 1200 ft^2.

(a) Find a function that models the cost of fencing the garden.

(b) Find the garden dimensions that minimize the cost of fencing.

(c) If the owner has at most \$600 to spend on fencing, find the range of lengths he can fence along the road.

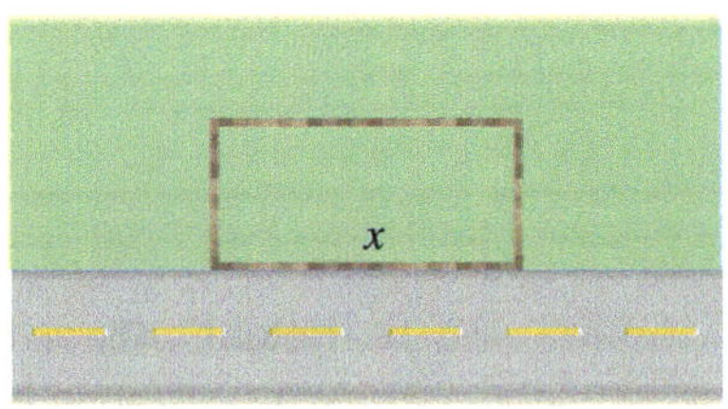

24. Maximizing Area A wire 10 cm long is cut into two pieces, one of length x and the other of length $10 - x$, as shown in the figure. Each piece is bent into the shape of a square.

(a) Find a function that models the total area enclosed by the two squares.

(b) Find the value of x that minimizes the total area of the two squares.

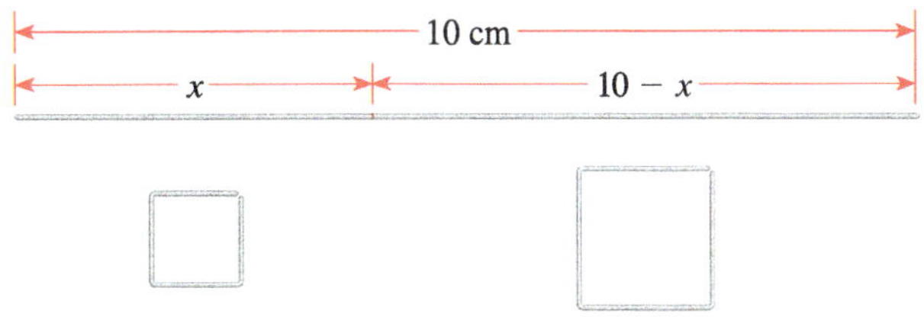

25. Light from a Window A Norman window has the shape of a rectangle surmounted by a semicircle, as shown in the figure to the left. A Norman window with perimeter 30 ft is to be constructed.

(a) Find a function that models the area of the window.

(b) Find the dimensions of the window that admits the greatest amount of light.

26. Volume of a Box A box with an open top is to be constructed from a rectangular piece of cardboard with dimensions 12 in. by 20 in. by cutting out equal squares of side x at each corner and then folding up the sides (see the figure).

(a) Find a function that models the volume of the box.

(b) Find the values of x for which the volume is greater than 200 in^3.

(c) Find the largest volume that such a box can have.

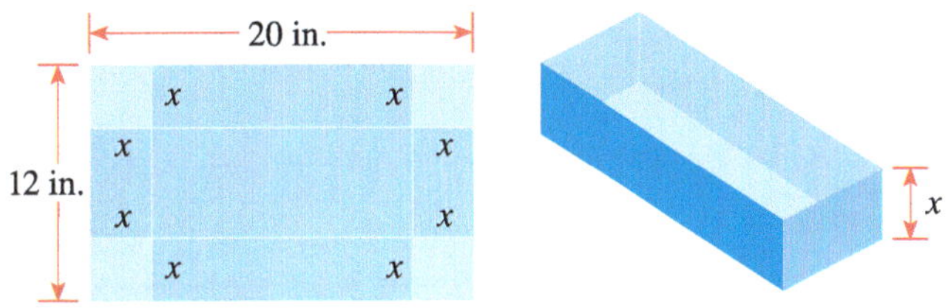

27. Area of a Box An open box with a square base is to have a volume of 12 ft^3.

(a) Find a function that models the surface area of the box.

(b) Find the box dimensions that minimize the amount of material used.

28. Inscribed Rectangle Find the dimensions that give the largest area for the rectangle shown in the figure. Its base is on the x-axis, and its other two vertices are above the x-axis, lying on the parabola $y = 8 - x^2$.

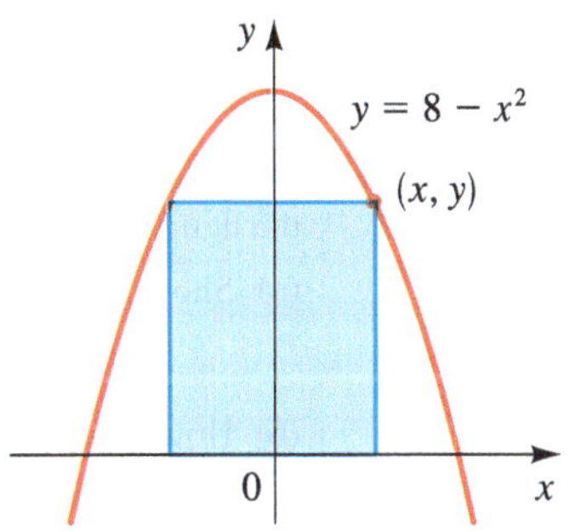

29. Minimizing Costs A rancher wants to build a rectangular pen with an area of 100 m^2.

(a) Find a function that models the length of fencing required.

(b) Find the pen dimensions that require the minimum amount of fencing.

30. Minimizing Time A man stands at a point A on the bank of a straight river, 2 mi wide. To reach point B, 7 mi downstream on the opposite bank, he first rows his boat to point P on the opposite bank and then walks the remaining distance x to B, as shown in the figure. He can row at a speed of 2 mi/h and walk at a speed of 5 mi/h.

(a) Find a function that models the time needed for the trip.

(b) Where should he land so that he reaches B as soon as possible?

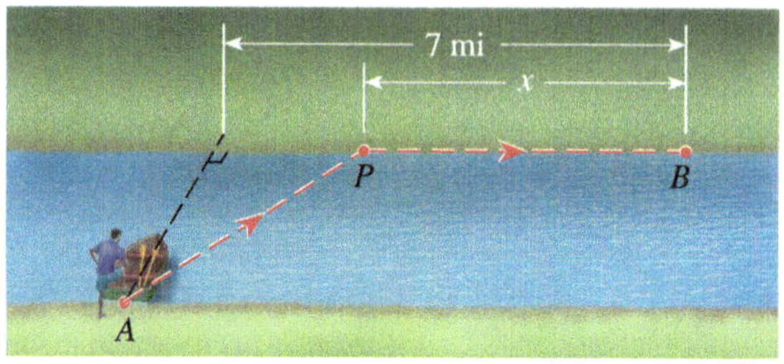

31. Bird Flight A bird is released from point A on an island, 5 mi from the nearest point B on a straight shoreline. The bird flies to a point C on the shoreline and then flies along the shoreline to its nesting area D (see the figure). Suppose the bird requires 10 kcal/mi of energy to fly over land and 14 kcal/mi to fly over water.

(a) Use the fact that

$$\text{energy used} = \text{energy per mile} \times \text{miles flown}$$

to show that the total energy used by the bird is modeled by the function

$$E(x) = 14\sqrt{x^2 + 25} + 10(12 - x)$$

(b) If the bird instinctively chooses a path that minimizes its energy expenditure, to what point does it fly?

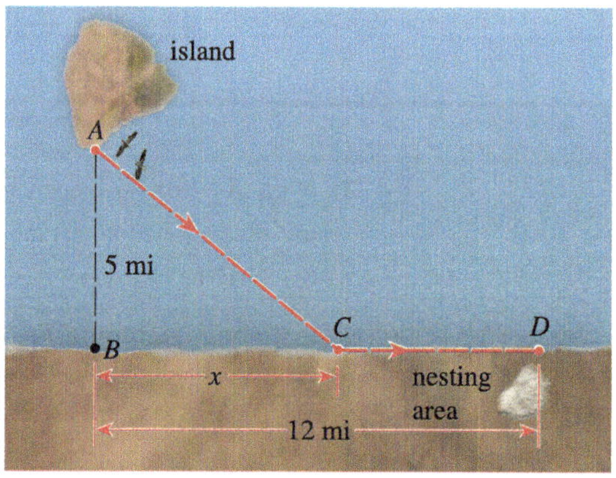

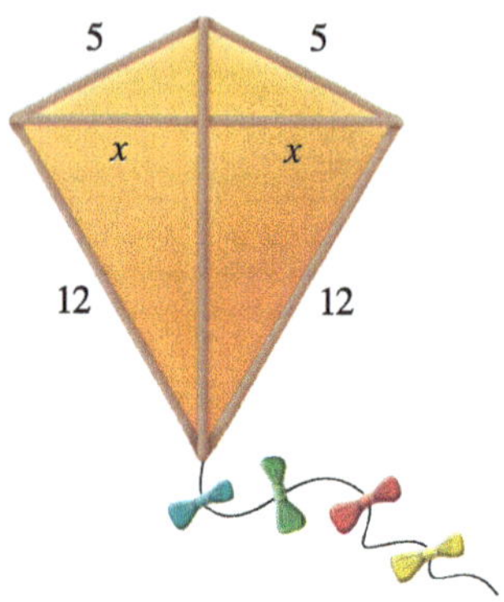

32. Area of a Kite A kite frame is to be made from six pieces of wood. The four pieces that form its border have been cut to the lengths indicated in the figure. Let x be as shown in the figure.

(a) Show that the area of the kite is given by the function

$$A(x) = x\left(\sqrt{25 - x^2} + \sqrt{144 - x^2}\right)$$

(b) How long should each of the two crosspieces be to maximize the area of the kite?

3 Polynomial and Rational Functions

Functions defined by polynomial expressions are called *polynomial functions.* The graphs of polynomial functions can have many peaks and valleys. This property makes them suitable models for many real-world situations. For example, a factory owner notices that if she increases the number of workers, productivity increases, but if there are too many workers, productivity begins to decrease. This situation is modeled by a polynomial function of degree 2 (a quadratic function). The growth of many animal species follows a predictable pattern, beginning with a period of rapid growth, followed by a period of slow growth and then a final growth spurt. This variability in growth is modeled by a polynomial of degree 3.

In the *Focus on Modeling* at the end of this chapter we explore different ways of using polynomial functions to model real-world situations.

3.1 QUADRATIC FUNCTIONS AND MODELS

■ Graphing Quadratic Functions Using the Standard Form ■ Maximum and Minimum Values of Quadratic Functions ■ Modeling with Quadratic Functions

A polynomial function is a function that is defined by a polynomial expression. So a **polynomial function of degree *n*** is a function of the form

$$P(x) = a_n x^n + a_{n-1}x^{n-1} + \cdots + a_1 x + a_0 \qquad a_n \neq 0$$

Polynomial expressions are defined in Section 1.3.

We have already studied polynomial functions of degree 0 and 1. These are functions of the form $P(x) = a_0$ and $P(x) = a_1 x + a_0$, respectively, whose graphs are lines. In this section we study polynomial functions of degree 2. These are called quadratic functions.

QUADRATIC FUNCTIONS

A **quadratic function** is a polynomial function of degree 2. So a quadratic function is a function of the form

$$f(x) = ax^2 + bx + c \qquad a \neq 0$$

We see in this section how quadratic functions model many real-world phenomena. We begin by analyzing the graphs of quadratic functions.

■ Graphing Quadratic Functions Using the Standard Form

For a geometric definition of parabolas, see Section 11.1.

If we take $a = 1$ and $b = c = 0$ in the quadratic function $f(x) = ax^2 + bx + c$, we get the quadratic function $f(x) = x^2$, whose graph is the parabola graphed in Example 1 of Section 2.2. In fact, the graph of any quadratic function is a **parabola**; it can be obtained from the graph of $f(x) = x^2$ by the transformations given in Section 2.6.

STANDARD FORM OF A QUADRATIC FUNCTION

A quadratic function $f(x) = ax^2 + bx + c$ can be expressed in the **standard form**

$$f(x) = a(x - h)^2 + k$$

by completing the square. The graph of f is a parabola with **vertex** (h, k); the parabola opens upward if $a > 0$ or downward if $a < 0$.

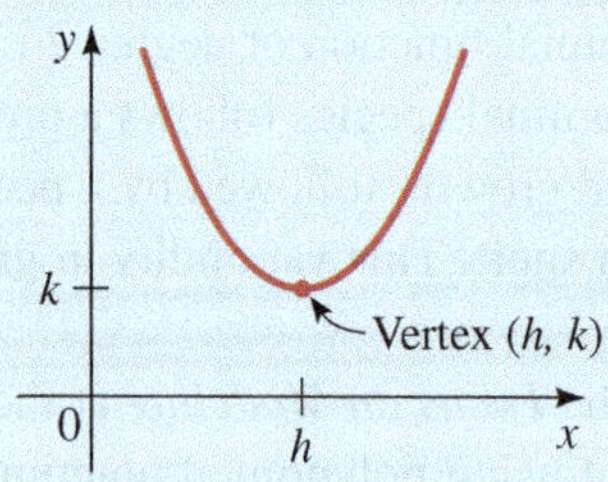

$f(x) = a(x - h)^2 + k,\ a > 0$

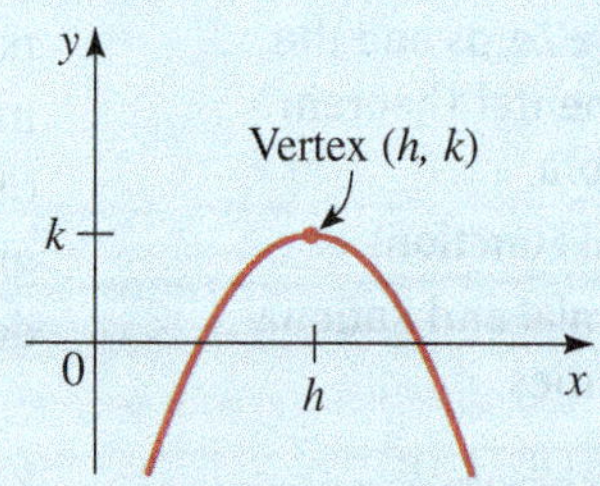

$f(x) = a(x - h)^2 + k,\ a < 0$

EXAMPLE 1 ■ Standard Form of a Quadratic Function

Let $f(x) = 2x^2 - 12x + 13$.

(a) Express f in standard form.

(b) Find the vertex and x- and y-intercepts of f.

(c) Sketch a graph of f.

(d) Find the domain and range of f.

SOLUTION

(a) Since the coefficient of x^2 is not 1, we must factor this coefficient from the terms involving x before we complete the square.

Completing the square is discussed in Section 1.5.

$$\begin{aligned} f(x) &= 2x^2 - 12x + 13 & \\ &= 2(x^2 - 6x) + 13 & \text{Factor 2 from the } x\text{-terms} \\ &= 2(x^2 - 6x + 9) + 13 - 2\cdot 9 & \text{Complete the square: Add 9 inside parentheses, subtract } 2\cdot 9 \text{ outside} \\ &= 2(x - 3)^2 - 5 & \text{Factor and simplify} \end{aligned}$$

The standard form is $f(x) = 2(x - 3)^2 - 5$.

(b) From the standard form of f we can see that the vertex of f is $(3, -5)$. The y-intercept is $f(0) = 13$. To find the x-intercepts, we set $f(x) = 0$ and solve the resulting equation. We can solve a quadratic equation by any of the methods we studied in Section 1.5. In this case we solve the equation by using the Quadratic Formula.

$$\begin{aligned} 0 &= 2x^2 - 12x + 13 & \text{Set } f(x) = 0 \\ x &= \frac{12 \pm \sqrt{144 - 4\cdot 2\cdot 13}}{4} & \text{Solve for } x \text{ using the Quadratic Formula} \\ x &= \frac{6 \pm \sqrt{10}}{2} & \text{Simplify} \end{aligned}$$

Thus the x-intercepts are $x = (6 \pm \sqrt{10})/2$. So the intercepts are approximately 1.42 and 4.58.

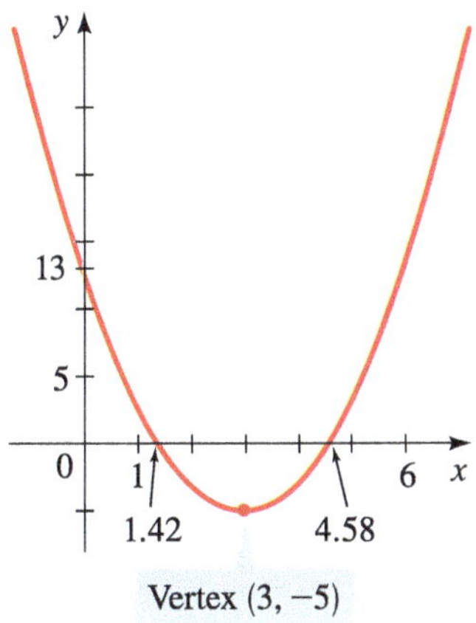

FIGURE 1 $f(x) = 2x^2 - 12x + 13$

(c) The standard form tells us that we get the graph of f by taking the parabola $y = x^2$, shifting it to the right 3 units, stretching it vertically by a factor of 2, and moving it downward 5 units. We sketch a graph of f in Figure 1, including the x- and y-intercepts found in part (b).

(d) The domain of f is the set of all real numbers $(-\infty, \infty)$. From the graph we see that the range of f is $[-5, \infty)$.

Now Try Exercise 15 ■

Maximum and Minimum Values of Quadratic Functions

If a quadratic function has vertex (h, k), then the function has a minimum value at the vertex if its graph opens upward and a maximum value at the vertex if its graph opens downward. For example, the function graphed in Figure 1 has minimum value 5 when $x = 3$, since the vertex $(3, 5)$ is the lowest point on the graph.

MAXIMUM OR MINIMUM VALUE OF A QUADRATIC FUNCTION

Let f be a quadratic function with standard form $f(x) = a(x - h)^2 + k$. The maximum or minimum value of f occurs at $x = h$.

If $a > 0$, then the **minimum value** of f is $f(h) = k$.

If $a < 0$, then the **maximum value** of f is $f(h) = k$.

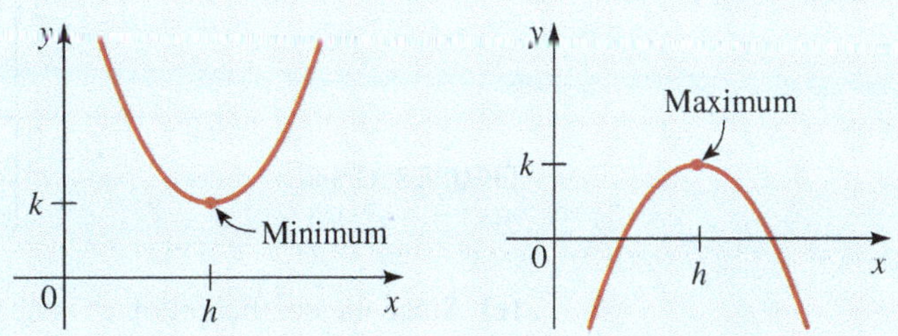

$f(x) = a(x - h)^2 + k,\ a > 0$ $f(x) = a(x - h)^2 + k,\ a < 0$

EXAMPLE 2 ■ Minimum Value of a Quadratic Function

Consider the quadratic function $f(x) = 5x^2 - 30x + 49$.

(a) Express f in standard form.

(b) Sketch a graph of f.

(c) Find the minimum value of f.

SOLUTION

(a) To express this quadratic function in standard form, we complete the square.

$$\begin{aligned} f(x) &= 5x^2 - 30x + 49 \\ &= 5(x^2 - 6x) + 49 && \text{Factor 5 from the } x\text{-terms} \\ &= 5(x^2 - 6x + 9) + 49 - 5 \cdot 9 && \text{Complete the square: Add 9 inside parentheses, subtract } 5 \cdot 9 \text{ outside} \\ &= 5(x - 3)^2 + 4 && \text{Factor and simplify} \end{aligned}$$

(b) The graph is a parabola that has its vertex at $(3, 4)$ and opens upward, as sketched in Figure 2.

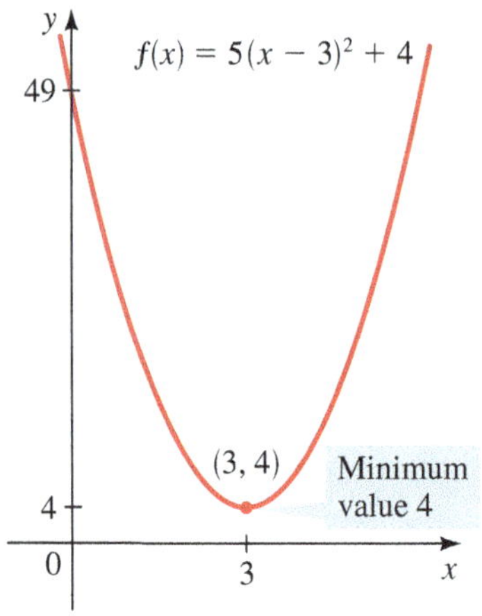

FIGURE 2

(c) Since the coefficient of x^2 is positive, f has a minimum value. The minimum value is $f(3) = 4$.

Now Try Exercise 27 ■

EXAMPLE 3 ■ Maximum Value of a Quadratic Function

Consider the quadratic function $f(x) = -x^2 + x + 2$.

(a) Express f in standard form.

(b) Sketch a graph of f.

(c) Find the maximum value of f.

SOLUTION

(a) To express this quadratic function in standard form, we complete the square.

$$\begin{aligned} f(x) &= -x^2 + x + 2 \\ &= -(x^2 - x) + 2 && \text{Factor } -1 \text{ from the } x\text{-terms} \\ &= -\left(x^2 - x + \tfrac{1}{4}\right) + 2 - (-1)\tfrac{1}{4} && \text{Complete the square: Add } \tfrac{1}{4} \text{ inside parentheses, subtract } (-1)\tfrac{1}{4} \text{ outside} \\ &= -\left(x - \tfrac{1}{2}\right)^2 + \tfrac{9}{4} && \text{Factor and simplify} \end{aligned}$$

In Example 3 you can check that the x-intercepts of the parabola are -1 and 2. These are obtained by solving the equation $f(x) = 0$.

(b) From the standard form we see that the graph is a parabola that opens downward and has vertex $\left(\frac{1}{2}, \frac{9}{4}\right)$. The graph of f is sketched in Figure 3.

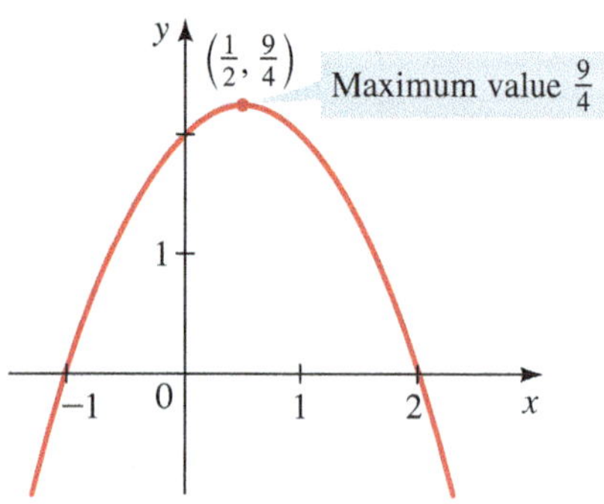

FIGURE 3 Graph of $f(x) = -x^2 + x + 2$

(c) Since the coefficient of x^2 is negative, f has a maximum value, which is $f\left(\frac{1}{2}\right) = \frac{9}{4}$.

Now Try Exercise 29 ■

Expressing a quadratic function in standard form helps us to sketch its graph as well as to find its maximum or minimum value. If we are interested only in finding the maximum or minimum value, then a formula is available for doing so. This formula is obtained by completing the square for the general quadratic function as follows.

$$\begin{aligned} f(x) &= ax^2 + bx + c \\ &= a\left(x^2 + \frac{b}{a}x\right) + c && \text{Factor } a \text{ from the } x\text{-terms} \\ &= a\left(x^2 + \frac{b}{a}x + \frac{b^2}{4a^2}\right) + c - a\left(\frac{b^2}{4a^2}\right) && \text{Complete the square: Add } \frac{b^2}{4a^2} \text{ inside parentheses, subtract } a\left(\frac{b^2}{4a^2}\right) \text{ outside} \\ &= a\left(x + \frac{b}{2a}\right)^2 + c - \frac{b^2}{4a} && \text{Factor} \end{aligned}$$

This equation is in standard form with $h = -b/(2a)$ and $k = c - b^2/(4a)$. Since the maximum or minimum value occurs at $x = h$, we have the following result.

MAXIMUM OR MINIMUM VALUE OF A QUADRATIC FUNCTION

The maximum or minimum value of a quadratic function $f(x) = ax^2 + bx + c$ occurs at

$$x = -\frac{b}{2a}$$

If $a > 0$, then the **minimum value** is $f\left(-\frac{b}{2a}\right)$.

If $a < 0$, then the **maximum value** is $f\left(-\frac{b}{2a}\right)$.

EXAMPLE 4 ■ Finding Maximum and Minimum Values of Quadratic Functions

Find the maximum or minimum value of each quadratic function.

(a) $f(x) = x^2 + 4x$

(b) $g(x) = -2x^2 + 4x - 5$

SOLUTION

(a) This is a quadratic function with $a = 1$ and $b = 4$. Thus the maximum or minimum value occurs at

$$x = -\frac{b}{2a} = -\frac{4}{2 \cdot 1} = -2$$

Since $a > 0$, the function has the *minimum* value

$$f(-2) = (-2)^2 + 4(-2) = -4$$

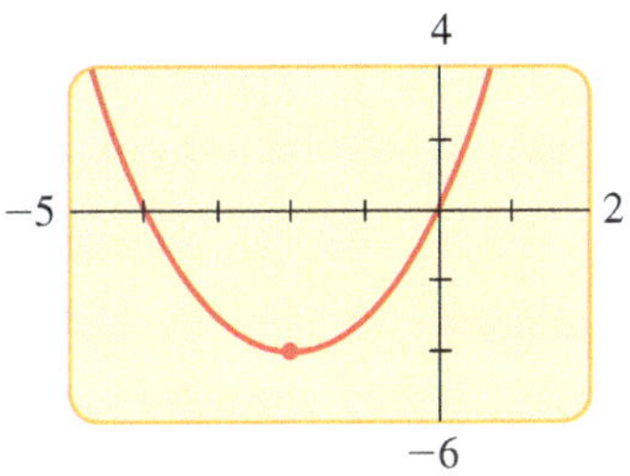

The minimum value occurs at $x = -2$.

(b) This is a quadratic function with $a = -2$ and $b = 4$. Thus the maximum or minimum value occurs at

$$x = -\frac{b}{2a} = -\frac{4}{2 \cdot (-2)} = 1$$

Since $a < 0$, the function has the *maximum* value

$$f(1) = -2(1)^2 + 4(1) - 5 = -3$$

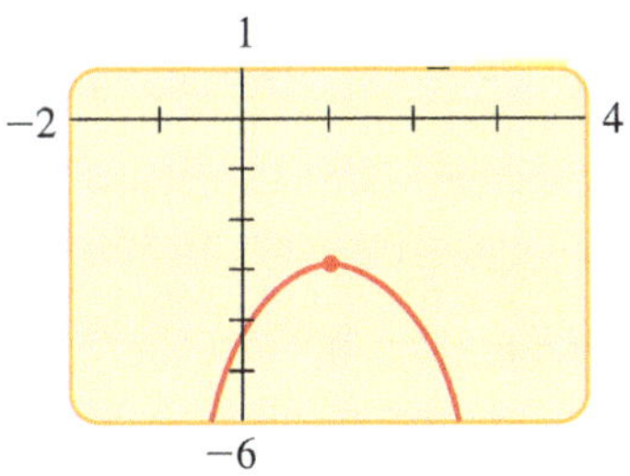

The maximum value occurs at $x = 1$.

Now Try Exercises 35 and 37 ■

■ Modeling with Quadratic Functions

We study some examples of real-world phenomena that are modeled by quadratic functions. These examples and the *Applications* exercises for this section show some of the variety of situations that are naturally modeled by quadratic functions.

EXAMPLE 5 ■ Maximum Gas Mileage for a Car

Most cars get their best gas mileage when traveling at a relatively modest speed. The gas mileage M for a certain new car is modeled by the function

$$M(s) = -\frac{1}{28}s^2 + 3s - 31 \qquad 15 \le s \le 70$$

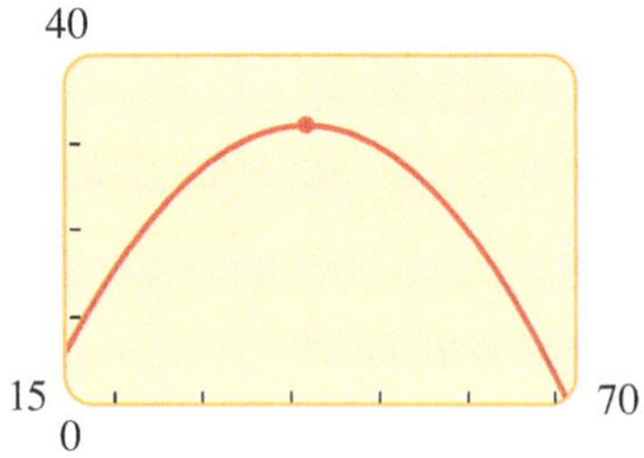

The maximum gas mileage occurs at 42 mi/h.

where s is the speed in mi/h and M is measured in mi/gal. What is the car's best gas mileage, and at what speed is it attained?

SOLUTION The function M is a quadratic function with $a = -\frac{1}{28}$ and $b = 3$. Thus its maximum value occurs when

$$s = -\frac{b}{2a} = -\frac{3}{2\left(-\frac{1}{28}\right)} = 42$$

The maximum value is $M(42) = -\frac{1}{28}(42)^2 + 3(42) - 31 = 32$. So the car's best gas mileage is 32 mi/gal when it is traveling at 42 mi/h.

Now Try Exercise 55 ■

EXAMPLE 6 ■ Maximizing Revenue from Ticket Sales

A hockey team plays in an arena that has a seating capacity of 15,000 spectators. With the ticket price set at \$14, average attendance at recent games has been 9500. A market survey indicates that for each dollar the ticket price is lowered, the average attendance increases by 1000.

(a) Find a function that models the revenue in terms of ticket price.

(b) Find the price that maximizes revenue from ticket sales.

(c) What ticket price is so high that no one attends and so no revenue is generated?

SOLUTION

(a) **Express the model in words.** The model that we want is a function that gives the revenue for any ticket price:

revenue = ticket price × attendance

DISCOVERY PROJECT

Torricelli's Law

Evangelista Torricelli (1608–1647) is best known for his invention of the barometer. He also discovered that the speed at which a fluid leaks from the bottom of a tank is related to the height of the fluid in the tank (a principle now called Torricelli's Law). In this project we conduct a simple experiment to collect data on the speed of water leaking through a hole in the bottom of a large soft-drink bottle. We then find an algebraic expression for Torricelli's Law by fitting a quadratic function to the data we obtained. You can find the project at **www.stewartmath.com**.

Choose the variable. There are two varying quantities: ticket price and attendance. Since the function we want depends on price, we let

$$x = \text{ticket price}$$

Next, we express attendance in terms of x.

In Words	In Algebra
Ticket price	x
Amount ticket price is lowered	$14 - x$
Increase in attendance	$1000(14 - x)$
Attendance	$9500 + 1000(14 - x)$

Set up the model. The model that we want is the function R that gives the revenue for a given ticket price x.

$$\text{revenue} = \text{ticket price} \times \text{attendance}$$

$$R(x) = x \times [9500 + 1000(14 - x)]$$

$$R(x) = x(23{,}500 - 1000x)$$

$$R(x) = 23{,}500x - 1000x^2$$

(b) **Use the model.** Since R is a quadratic function with $a = -1000$ and $b = 23{,}500$, the maximum occurs at

$$x = -\frac{b}{2a} = -\frac{23{,}500}{2(-1000)} = 11.75$$

So a ticket price of \$11.75 gives the maximum revenue.

(c) **Use the model.** We want to find the ticket price for which $R(x) = 0$.

$$23{,}500x - 1000x^2 = 0 \quad \text{Set } R(x) = 0$$

$$23.5x - x^2 = 0 \quad \text{Divide by 1000}$$

$$x(23.5 - x) = 0 \quad \text{Factor}$$

$$x = 0 \quad \text{or} \quad x = 23.5 \quad \text{Solve for } x$$

So according to this model, a ticket price of \$23.50 is just too high; at that price no one attends to watch this team play. (Of course, revenue is also zero if the ticket price is zero.)

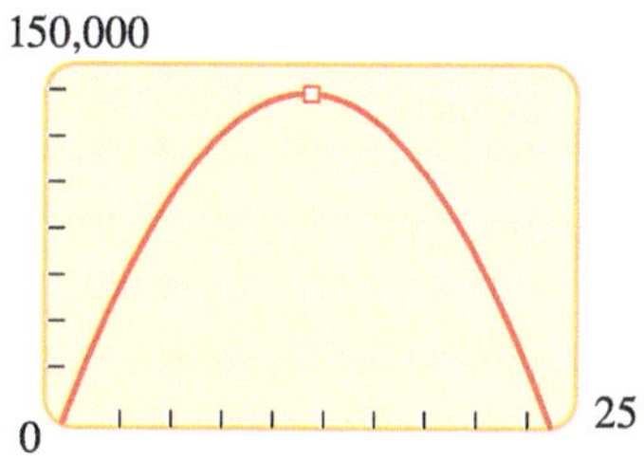

Maximum attendance occurs when ticket price is \$11.75.

Now Try Exercise 65 ■

3.1 EXERCISES

CONCEPTS

1. To put the quadratic function $f(x) = ax^2 + bx + c$ in standard form, we complete the ________.

2. The quadratic function $f(x) = a(x - h)^2 + k$ is in standard form.

(a) The graph of f is a parabola with vertex (____, ____).

(b) If $a > 0$, the graph of f opens ________. In this case $f(h) = k$ is the ________ value of f.

(c) If $a < 0$, the graph of f opens ________. In this case $f(h) = k$ is the ________ value of f.

3. The graph of $f(x) = 3(x - 2)^2 - 6$ is a parabola that opens ________, with its vertex at (____, ____), and $f(2) =$ ________ is the (minimum/maximum) ________ value of f.

4. The graph of $f(x) = -3(x - 2)^2 - 6$ is a parabola that opens ________, with its vertex at (____, ____), and $f(2) =$ ________ is the (minimum/maximum) ________ value of f.

SKILLS

5–8 ■ Graphs of Quadratic Functions The graph of a quadratic function f is given. **(a)** Find the coordinates of the vertex and the x- and y-intercepts. **(b)** Find the maximum or minimum value of f. **(c)** Find the domain and range of f.

5. $f(x) = -x^2 + 6x - 5$

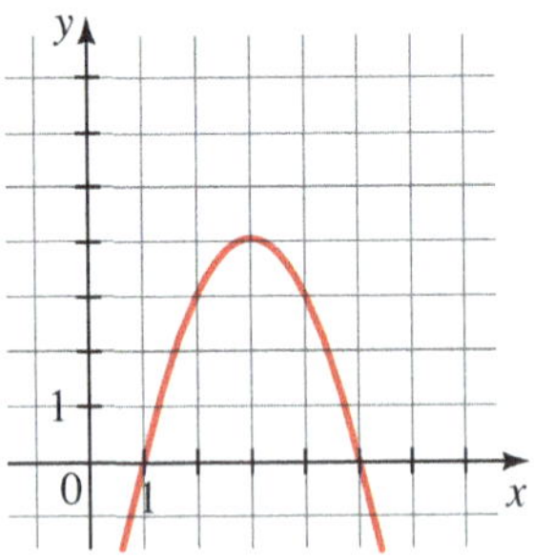

6. $f(x) = -\frac{1}{2}x^2 - 2x + 6$

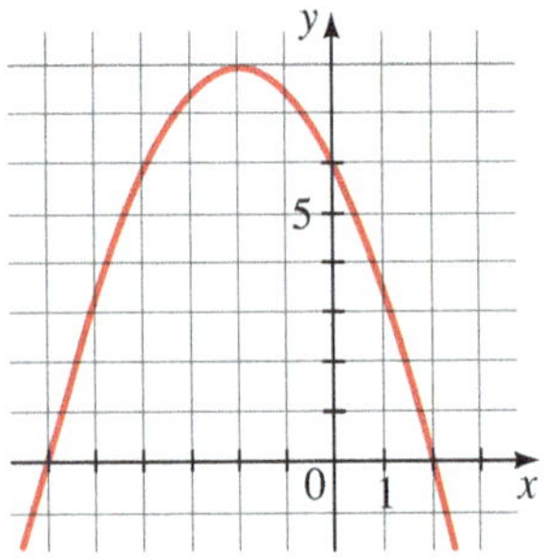

7. $f(x) = 2x^2 - 4x - 1$

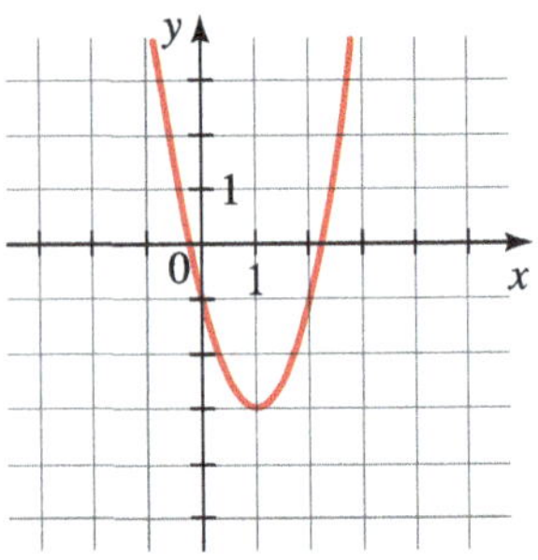

8. $f(x) = 3x^2 + 6x - 1$

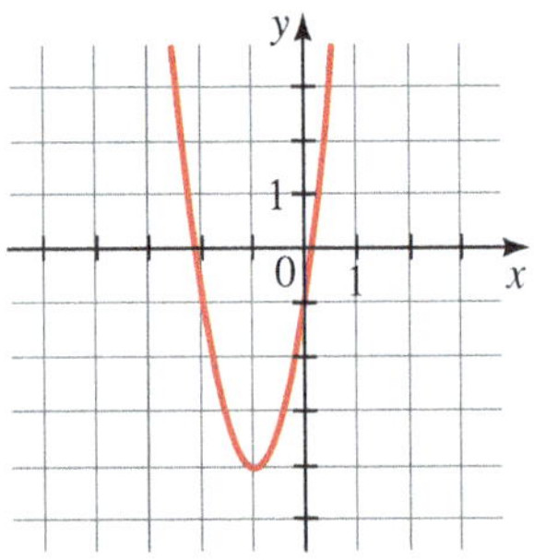

9–24 ■ Graphing Quadratic Functions A quadratic function f is given. **(a)** Express f in standard form. **(b)** Find the vertex and x- and y-intercepts of f. **(c)** Sketch a graph of f. **(d)** Find the domain and range of f.

9. $f(x) = x^2 - 2x + 3$ **10.** $f(x) = x^2 + 4x - 1$

11. $f(x) = x^2 - 6x$ **12.** $f(x) = x^2 + 8x$

13. $f(x) = 3x^2 + 6x$ **14.** $f(x) = -x^2 + 10x$

15. $f(x) = x^2 + 4x + 3$ **16.** $f(x) = x^2 - 2x + 2$

17. $f(x) = -x^2 + 6x + 4$ **18.** $f(x) = -x^2 - 4x + 4$

19. $f(x) = 2x^2 + 4x + 3$ **20.** $f(x) = -3x^2 + 6x - 2$

21. $f(x) = 2x^2 - 20x + 57$ **22.** $f(x) = 2x^2 + 12x + 10$

23. $f(x) = -4x^2 - 12x + 1$ **24.** $f(x) = 3x^2 + 2x - 2$

25–34 ■ Maximum and Minimum Values A quadratic function f is given. **(a)** Express f in standard form. **(b)** Sketch a graph of f. **(c)** Find the maximum or minimum value of f.

25. $f(x) = x^2 + 2x - 1$ **26.** $f(x) = x^2 - 8x + 8$

27. $f(x) = 3x^2 - 6x + 1$ **28.** $f(x) = 5x^2 + 30x + 4$

29. $f(x) = -x^2 - 3x + 3$ **30.** $f(x) = 1 - 6x - x^2$

31. $g(x) = 3x^2 - 12x + 13$ **32.** $g(x) = 2x^2 + 8x + 11$

33. $h(x) = 1 - x - x^2$ **34.** $h(x) = 3 - 4x - 4x^2$

35–44 ■ Formula for Maximum and Minimum Values Find the maximum or minimum value of the function.

35. $f(x) = 2x^2 + 4x - 1$ **36.** $f(x) = 3 - 4x - x^2$

37. $f(t) = -3 + 80t - 20t^2$ **38.** $f(x) = 6x^2 - 24x - 100$

39. $f(s) = s^2 - 1.2s + 16$ **40.** $g(x) = 100x^2 - 1500x$

41. $h(x) = \frac{1}{2}x^2 + 2x - 6$ **42.** $f(x) = -\dfrac{x^2}{3} + 2x + 7$

43. $f(x) = 3 - x - \frac{1}{2}x^2$ **44.** $g(x) = 2x(x - 4) + 7$

45–46 ■ Maximum and Minimum Values A quadratic function is given. **(a)** Use a graphing device to find the maximum or minimum value of the quadratic function f, rounded to two decimal places. **(b)** Find the exact maximum or minimum value of f, and compare it with your answer to part (a).

45. $f(x) = x^2 + 1.79x - 3.21$

46. $f(x) = 1 + x - \sqrt{2}x^2$

SKILLS Plus

47–48 ■ Finding Quadratic Functions Find a function f whose graph is a parabola with the given vertex and that passes through the given point.

47. Vertex $(2, -3)$; point $(3, 1)$

48. Vertex $(-1, 5)$; point $(-3, -7)$

49. Maximum of a Fourth-Degree Polynomial Find the maximum value of the function

$$f(x) = 3 + 4x^2 - x^4$$

[*Hint:* Let $t = x^2$.]

50. Minimum of a Sixth-Degree Polynomial Find the minimum value of the function

$$f(x) = 2 + 16x^3 + 4x^6$$

[*Hint:* Let $t = x^3$.]

APPLICATIONS

51. Height of a Ball If a ball is thrown directly upward with a velocity of 40 ft/s, its height (in feet) after t seconds is given by $y = 40t - 16t^2$. What is the maximum height attained by the ball?

52. Path of a Ball A ball is thrown across a playing field from a height of 5 ft above the ground at an angle of 45° to the horizontal at a speed of 20 ft/s. It can be deduced from physical principles that the path of the ball is modeled by the function

$$y = -\frac{32}{(20)^2}x^2 + x + 5$$

where x is the distance in feet that the ball has traveled horizontally.

(a) Find the maximum height attained by the ball.

(b) Find the horizontal distance the ball has traveled when it hits the ground.

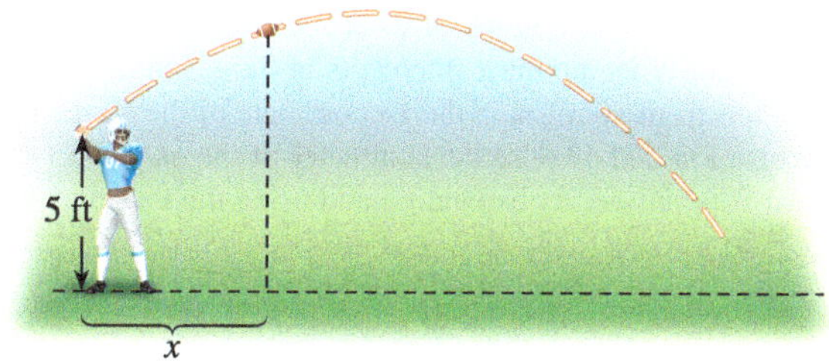

53. Revenue A manufacturer finds that the revenue generated by selling x units of a certain commodity is given by the function $R(x) = 80x - 0.4x^2$, where the revenue $R(x)$ is measured in dollars. What is the maximum revenue, and how many units should be manufactured to obtain this maximum?

54. Sales A soft-drink vendor at a popular beach analyzes his sales records and finds that if he sells x cans of soda pop in one day, his profit (in dollars) is given by

$$P(x) = -0.001x^2 + 3x - 1800$$

What is his maximum profit per day, and how many cans must he sell for maximum profit?

55. Advertising The effectiveness of a television commercial depends on how many times a viewer watches it. After some experiments an advertising agency found that if the effectiveness E is measured on a scale of 0 to 10, then

$$E(n) = \tfrac{2}{3}n - \tfrac{1}{90}n^2$$

where n is the number of times a viewer watches a given commercial. For a commercial to have maximum effectiveness, how many times should a viewer watch it?

56. Pharmaceuticals When a certain drug is taken orally, the concentration of the drug in the patient's bloodstream after t minutes is given by $C(t) = 0.06t - 0.0002t^2$, where $0 \le t \le 240$ and the concentration is measured in mg/L. When is the maximum serum concentration reached, and what is that maximum concentration?

57. Agriculture The number of apples produced by each tree in an apple orchard depends on how densely the trees are planted. If n trees are planted on an acre of land, then each tree produces $900 - 9n$ apples. So the number of apples produced per acre is

$$A(n) = n(900 - 9n)$$

How many trees should be planted per acre to obtain the maximum yield of apples?

58. Agriculture At a certain vineyard it is found that each grape vine produces about 10 lb of grapes in a season when about 700 vines are planted per acre. For each additional vine that is planted, the production of each vine decreases by about 1 percent. So the number of pounds of grapes produced per acre is modeled by

$$A(n) = (700 + n)(10 - 0.01n)$$

where n is the number of additional vines planted. Find the number of vines that should be planted to maximize grape production.

59–62 ■ Maxima and Minima Use the formulas of this section to give an alternative solution to the indicated problem in *Focus on Modeling: Modeling with Functions* on pages 237–244.

59. Problem 21 **60.** Problem 22

61. Problem 25 **62.** Problem 24

63. Fencing a Horse Corral Carol has 2400 ft of fencing to fence in a rectangular horse corral.

(a) Find a function that models the area of the corral in terms of the width x of the corral.

(b) Find the dimensions of the rectangle that maximize the area of the corral.

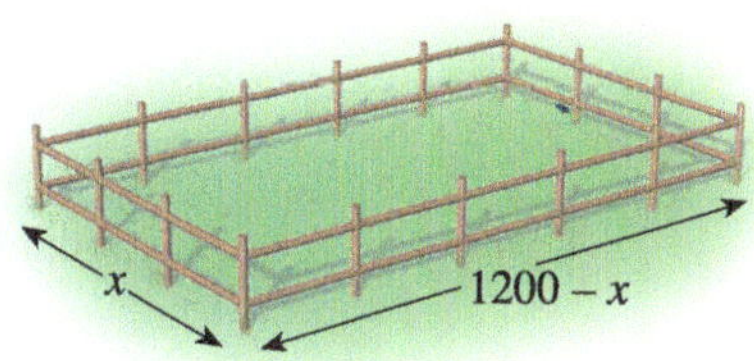

64. Making a Rain Gutter A rain gutter is formed by bending up the sides of a 30-in.-wide rectangular metal sheet as shown in the figure.

(a) Find a function that models the cross-sectional area of the gutter in terms of x.

(b) Find the value of x that maximizes the cross-sectional area of the gutter.

(c) What is the maximum cross-sectional area for the gutter?

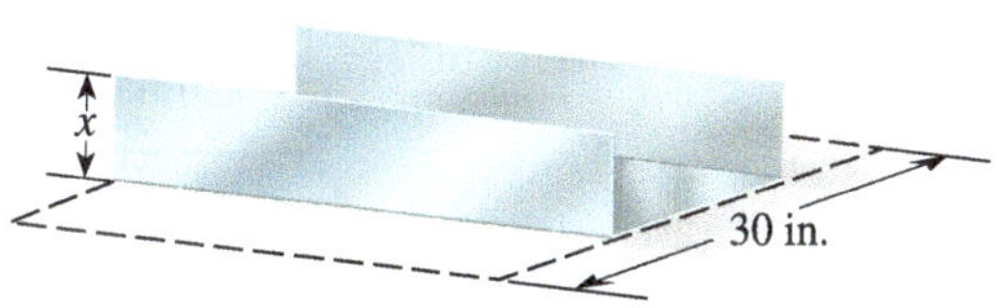

65. Stadium Revenue A baseball team plays in a stadium that holds 55,000 spectators. With the ticket price at \$10, the average attendance at recent games has been 27,000. A market survey indicates that for every dollar the ticket price is lowered, attendance increases by 3000.

(a) Find a function that models the revenue in terms of ticket price.

(b) Find the price that maximizes revenue from ticket sales.

(c) What ticket price is so high that no revenue is generated?

66. Maximizing Profit A community bird-watching society makes and sells simple bird feeders to raise money for its conservation activities. The materials for each feeder cost \$6, and the society sells an average of 20 per week at a price of \$10 each. The society has been considering raising the price, so it conducts a survey and finds that for every dollar increase, it will lose 2 sales per week.

(a) Find a function that models weekly profit in terms of price per feeder.

(b) What price should the society charge for each feeder to maximize profits? What is the maximum weekly profit?

DISCUSS ■ DISCOVER ■ PROVE ■ WRITE

67. DISCOVER: Vertex and *x*-Intercepts We know that the graph of the quadratic function $f(x) = (x - m)(x - n)$ is a parabola. Sketch a rough graph of what such a parabola would look like. What are the x-intercepts of the graph of f? Can you tell from your graph the x-coordinate of the vertex in terms of m and n? (Use the symmetry of the parabola.) Confirm your answer by expanding and using the formulas of this section.

3.2 POLYNOMIAL FUNCTIONS AND THEIR GRAPHS

■ Polynomial Functions ■ Graphing Basic Polynomial Functions ■ Graphs of Polynomial Functions: End Behavior ■ Using Zeros to Graph Polynomials ■ Shape of the Graph Near a Zero ■ Local Maxima and Minima of Polynomials

■ Polynomial Functions

In this section we study polynomial functions of any degree. But before we work with polynomial functions, we must agree on some terminology.

POLYNOMIAL FUNCTIONS

A **polynomial function of degree n** is a function of the form

$$P(x) = a_n x^n + a_{n-1}x^{n-1} + \cdots + a_1 x + a_0$$

where n is a nonnegative integer and $a_n \neq 0$.

The numbers $a_0, a_1, a_2, \ldots, a_n$ are called the **coefficients** of the polynomial.

The number a_0 is the **constant coefficient** or **constant term**.

The number a_n, the coefficient of the highest power, is the **leading coefficient**, and the term $a_n x^n$ is the **leading term**.

We often refer to polynomial functions simply as *polynomials*. The following polynomial has degree 5, leading coefficient 3, and constant term −6.

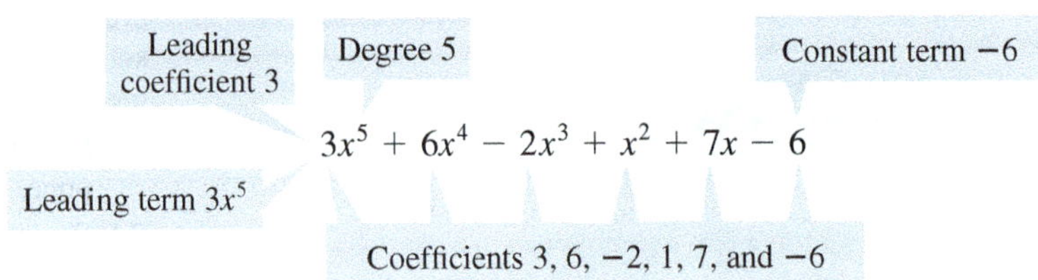

The table lists some more examples of polynomials.

Polynomial	Degree	Leading term	Constant term
$P(x) = 4x - 7$	1	$4x$	-7
$P(x) = x^2 + x$	2	x^2	0
$P(x) = 2x^3 - 6x^2 + 10$	3	$2x^3$	10
$P(x) = -5x^4 + x - 2$	4	$-5x^4$	-2

If a polynomial consists of just a single term, then it is called a **monomial**. For example, $P(x) = x^3$ and $Q(x) = -6x^5$ are monomials.

Graphing Basic Polynomial Functions

The simplest polynomial functions are the monomials $P(x) = x^n$, whose graphs are shown in Figure 1. As the figure suggests, the graph of $P(x) = x^n$ has the same general shape as the graph of $y = x^2$ when n is even and the same general shape as the graph of $y = x^3$ when n is odd. However, as the degree n becomes larger, the graphs become flatter around the origin and steeper elsewhere.

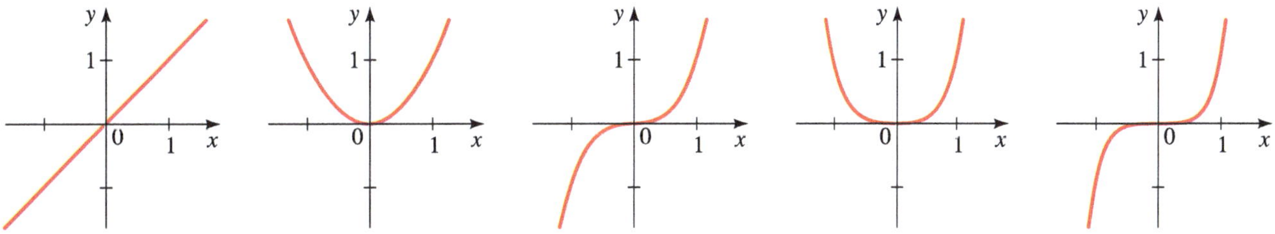

(a) $y = x$ (b) $y = x^2$ (c) $y = x^3$ (d) $y = x^4$ (e) $y = x^5$

FIGURE 1 Graphs of monomials

EXAMPLE 1 ■ Transformations of Monomials

Sketch graphs of the following functions.

(a) $P(x) = -x^3$ **(b)** $Q(x) = (x - 2)^4$

(c) $R(x) = -2x^5 + 4$

Mathematics in the Modern World

Splines

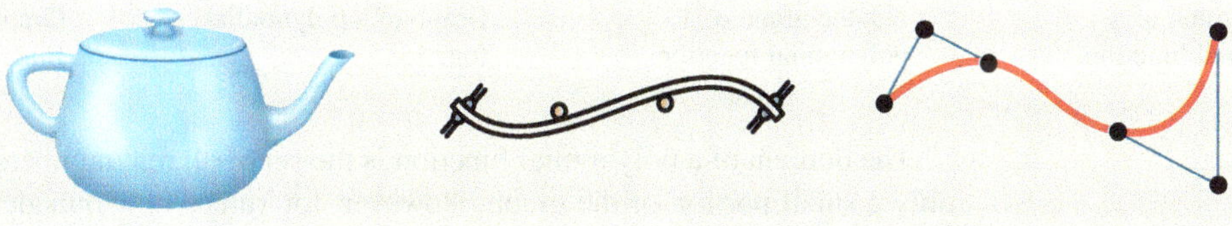

A spline is a long strip of wood that is curved while held fixed at certain points. In the old days shipbuilders used splines to create the curved shape of a boat's hull. Splines are also used to make the curves of a piano, a violin, or the spout of a teapot.

Mathematicians discovered that the shapes of splines can be obtained by piecing together parts of polynomials. For example, the graph of a cubic polynomial can be made to fit specified points by adjusting the coefficients of the polynomial (see Example 10, page 265).

Curves obtained in this way are called cubic splines. In modern computer design programs, such as Adobe Illustrator or Microsoft Paint, a curve can be drawn by fixing two points, then using the mouse to drag one or more anchor points. Moving the anchor points amounts to adjusting the coefficients of a cubic polynomial.

SOLUTION We use the graphs in Figure 1 and transform them using the techniques of Section 2.6.

(a) The graph of $P(x) = -x^3$ is the reflection of the graph of $y = x^3$ in the x-axis, as shown in Figure 2(a) below.

(b) The graph of $Q(x) = (x - 2)^4$ is the graph of $y = x^4$ shifted to the right 2 units, as shown in Figure 2(b).

(c) We begin with the graph of $y = x^5$. The graph of $y = -2x^5$ is obtained by stretching the graph vertically and reflecting it in the x-axis (see the dashed blue graph in Figure 2(c)). Finally, the graph of $R(x) = -2x^5 + 4$ is obtained by shifting upward 4 units (see the red graph in Figure 2(c)).

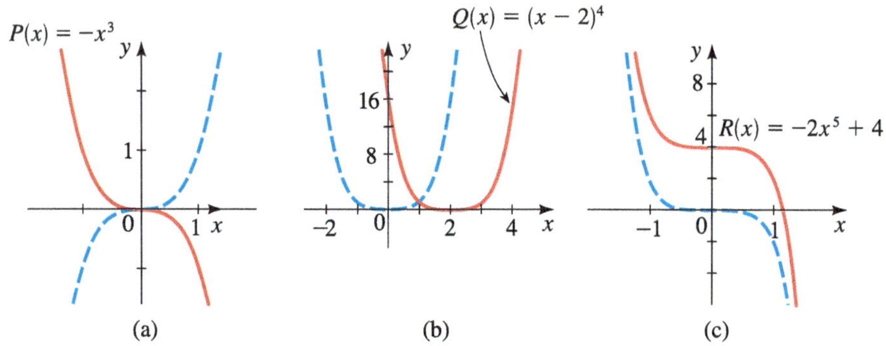

FIGURE 2

Now Try Exercise 5 ■

Graphs of Polynomial Functions: End Behavior

The graphs of polynomials of degree 0 or 1 are lines (Sections 1.10 and 2.5), and the graphs of polynomials of degree 2 are parabolas (Section 3.1). The greater the degree of a polynomial, the more complicated its graph can be. However, the graph of a polynomial function is **continuous**. This means that the graph has no breaks or holes (see Figure 3). Moreover, the graph of a polynomial function is a smooth curve; that is, it has no corners or sharp points (cusps) as shown in Figure 3.

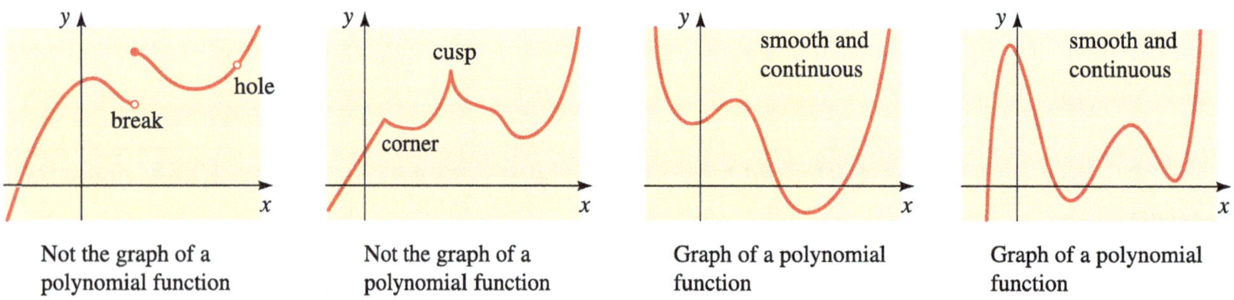

FIGURE 3

The domain of a polynomial function is the set of all real numbers, so we can sketch only a small portion of the graph. However, for values of x outside the portion of the graph we have drawn, we can describe the behavior of the graph.

The **end behavior** of a polynomial is a description of what happens as x becomes large in the positive or negative direction. To describe end behavior, we use the following **arrow notation**.

Symbol	Meaning
$x \to \infty$	x goes to infinity; that is, x increases without bound
$x \to -\infty$	x goes to negative infinity; that is, x decreases without bound

For example, the monomial $y = x^2$ in Figure 1(b) has the following end behavior.

$$y \to \infty \quad \text{as} \quad x \to \infty \qquad \text{and} \qquad y \to \infty \quad \text{as} \quad x \to -\infty$$

The monomial $y = x^3$ in Figure 1(c) has the following end behavior.

$$y \to \infty \quad \text{as} \quad x \to \infty \qquad \text{and} \qquad y \to -\infty \quad \text{as} \quad x \to -\infty$$

For any polynomial *the end behavior is determined by the term that contains the highest power of x*, because when x is large, the other terms are relatively insignificant in size. The following box shows the four possible types of end behavior, based on the highest power and the sign of its coefficient.

END BEHAVIOR OF POLYNOMIALS

The end behavior of the polynomial $P(x) = a_n x^n + a_{n-1}x^{n-1} + \cdots + a_1 x + a_0$ is determined by the degree n and the sign of the leading coefficient a_n, as indicated in the following graphs.

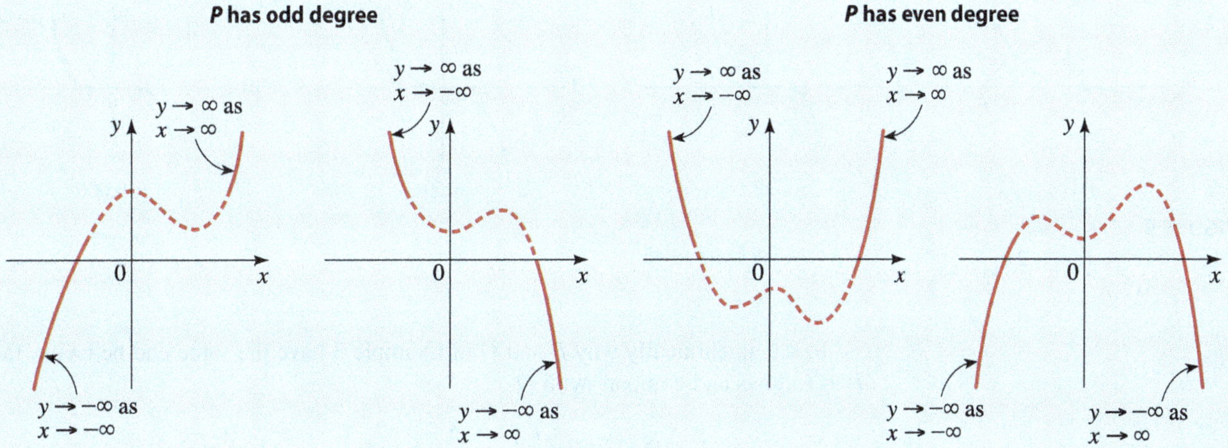

Leading coefficient positive — Leading coefficient negative — Leading coefficient positive — Leading coefficient negative

EXAMPLE 2 ■ End Behavior of a Polynomial

Determine the end behavior of the polynomial

$$P(x) = -2x^4 + 5x^3 + 4x - 7$$

SOLUTION The polynomial P has degree 4 and leading coefficient -2. Thus P has *even* degree and *negative* leading coefficient, so it has the following end behavior.

$$y \to -\infty \quad \text{as} \quad x \to \infty \qquad \text{and} \qquad y \to -\infty \quad \text{as} \quad x \to -\infty$$

The graph in Figure 4 illustrates the end behavior of P.

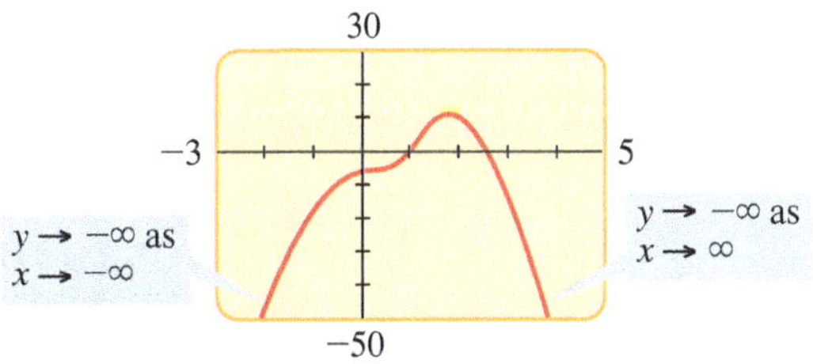

FIGURE 4 $P(x) = -2x^4 + 5x^3 + 4x - 7$

Now Try Exercise 11 ■

EXAMPLE 3 ■ End Behavior of a Polynomial

(a) Determine the end behavior of the polynomial $P(x) = 3x^5 - 5x^3 + 2x$.

(b) Confirm that P and its leading term $Q(x) = 3x^5$ have the same end behavior by graphing them together.

SOLUTION

(a) Since P has odd degree and positive leading coefficient, it has the following end behavior.

$$y \to \infty \quad \text{as} \quad x \to \infty \qquad \text{and} \qquad y \to -\infty \quad \text{as} \quad x \to -\infty$$

(b) Figure 5 shows the graphs of P and Q in progressively larger viewing rectangles. The larger the viewing rectangle, the more the graphs look alike. This confirms that they have the same end behavior.

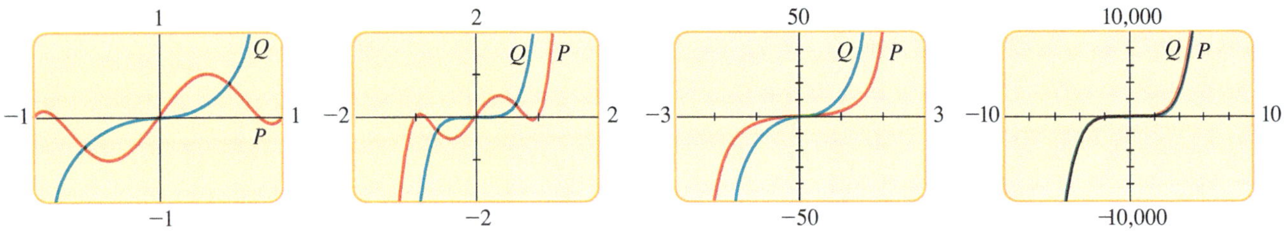

FIGURE 5
$P(x) = 3x^5 - 5x^3 + 2x$
$Q(x) = 3x^5$

Now Try Exercise 45 ■

To see algebraically why P and Q in Example 3 have the same end behavior, factor P as follows and compare with Q.

$$P(x) = 3x^5\left(1 - \frac{5}{3x^2} + \frac{2}{3x^4}\right) \qquad Q(x) = 3x^5$$

When x is large, the terms $5/(3x^2)$ and $2/(3x^4)$ are close to 0 (see Exercise 90 on page 12). So for large x we have

$$P(x) \approx 3x^5(1 - 0 - 0) = 3x^5 = Q(x)$$

So when x is large, P and Q have approximately the same values. We can also see this numerically by making a table like the one shown below.

x	$P(x)$	$Q(x)$
15	2,261,280	2,278,125
30	72,765,060	72,900,000
50	936,875,100	937,500,000

By the same reasoning we can show that the end behavior of *any* polynomial is determined by its leading term.

■ Using Zeros to Graph Polynomials

If P is a polynomial function, then c is called a **zero** of P if $P(c) = 0$. In other words, the zeros of P are the solutions of the polynomial equation $P(x) = 0$. Note that if $P(c) = 0$, then the graph of P has an x-intercept at $x = c$, so the x-intercepts of the graph are the zeros of the function.

REAL ZEROS OF POLYNOMIALS

If P is a polynomial and c is a real number, then the following are equivalent:

1. c is a zero of P.
2. $x = c$ is a solution of the equation $P(x) = 0$.
3. $x - c$ is a factor of $P(x)$.
4. c is an x-intercept of the graph of P.

To find the zeros of a polynomial P, we factor and then use the Zero-Product Property (see page 48). For example, to find the zeros of $P(x) = x^2 + x - 6$, we factor P to get

$$P(x) = (x - 2)(x + 3)$$

From this factored form we easily see that

1. 2 is a zero of P.
2. $x = 2$ is a solution of the equation $x^2 + x - 6 = 0$.
3. $x - 2$ is a factor of $x^2 + x - 6$.
4. 2 is an x-intercept of the graph of P.

The same facts are true for the other zero, -3.

The following theorem has many important consequences. (See, for instance, the *Discovery Project* referenced on page 276.) Here we use it to help us graph polynomial functions.

INTERMEDIATE VALUE THEOREM FOR POLYNOMIALS

If P is a polynomial function and $P(a)$ and $P(b)$ have opposite signs, then there exists at least one value c between a and b for which $P(c) = 0$.

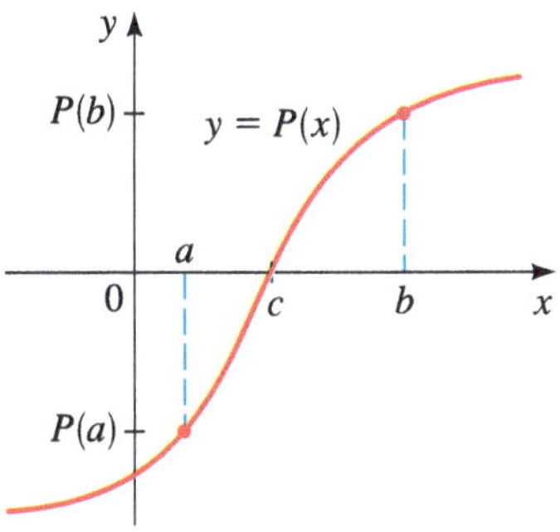

FIGURE 6

We will not prove this theorem, but Figure 6 shows why it is intuitively plausible.

One important consequence of this theorem is that between any two successive zeros the values of a polynomial are either all positive or all negative. That is, between two successive zeros the graph of a polynomial lies *entirely above* or *entirely below* the x-axis. To see why, suppose c_1 and c_2 are successive zeros of P. If P has both positive and negative values between c_1 and c_2, then by the Intermediate Value Theorem, P must have another zero between c_1 and c_2. But that's not possible because c_1 and c_2 are successive zeros. This observation allows us to use the following guidelines to graph polynomial functions.

GUIDELINES FOR GRAPHING POLYNOMIAL FUNCTIONS

1. **Zeros.** Factor the polynomial to find all its real zeros; these are the x-intercepts of the graph.
2. **Test Points.** Make a table of values for the polynomial. Include test points to determine whether the graph of the polynomial lies above or below the x-axis on the intervals determined by the zeros. Include the y-intercept in the table.
3. **End Behavior.** Determine the end behavior of the polynomial.
4. **Graph.** Plot the intercepts and other points you found in the table. Sketch a smooth curve that passes through these points and exhibits the required end behavior.

Mathematics in the Modern World

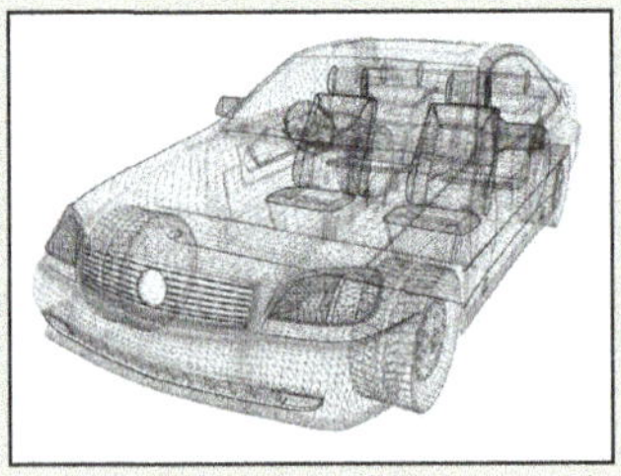

© 3DProfi/Shutterstock.com

Automotive Design

Computer-aided design (CAD) has completely changed the way in which car companies design and manufacture cars. Before the 1980s automotive engineers would build a full-scale "nuts and bolts" model of a proposed new car; this was really the only way to tell whether the design was feasible. Today automotive engineers build a mathematical model, one that exists only in the memory of a computer. The model incorporates all the main design features of the car. Certain polynomial curves, called *splines* (see page 255), are used in shaping the body of the car. The resulting "mathematical car" can be tested for structural stability, handling, aerodynamics, suspension response, and more. All this testing is done before a prototype is built. As you can imagine, CAD saves car manufacturers millions of dollars each year. More importantly, CAD gives automotive engineers far more flexibility in design; desired changes can be created and tested within seconds. With the help of computer graphics, designers can see how good the "mathematical car" looks before they build the real one. Moreover, the mathematical car can be viewed from any perspective; it can be moved, rotated, or seen from the inside. These manipulations of the car on the computer monitor translate mathematically into solving large systems of linear equations.

EXAMPLE 4 ■ Using Zeros to Graph a Polynomial Function

Sketch the graph of the polynomial function $P(x) = (x + 2)(x - 1)(x - 3)$.

SOLUTION The zeros are $x = -2$, 1, and 3. These determine the intervals $(-\infty, -2)$, $(-2, 1)$, $(1, 3)$, and $(3, \infty)$. Using test points in these intervals, we get the information in the following sign diagram (see Section 1.8).

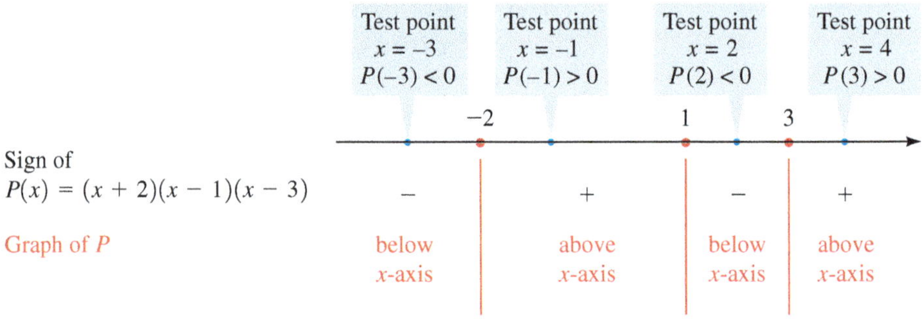

Plotting a few additional points and connecting them with a smooth curve helps us to complete the graph in Figure 7.

	x	$P(x)$
Test point →	−3	−24
	−2	0
Test point →	−1	8
	0	6
	1	0
Test point →	2	−4
	3	0
Test point →	4	18

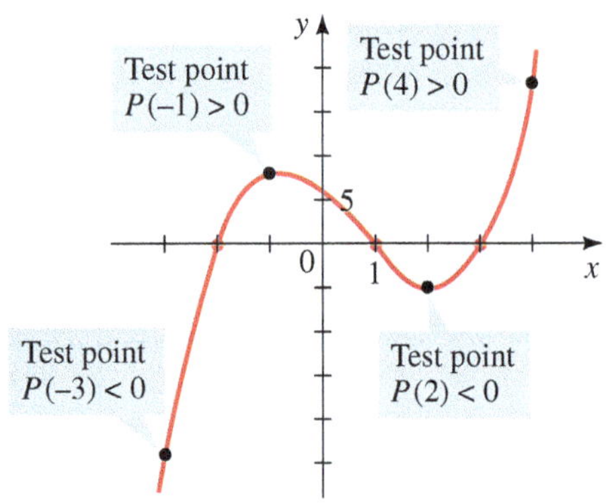

FIGURE 7 $P(x) = (x + 2)(x - 1)(x - 3)$

Now Try Exercise 17 ■

EXAMPLE 5 ■ Finding Zeros and Graphing a Polynomial Function

Let $P(x) = x^3 - 2x^2 - 3x$.

(a) Find the zeros of P. **(b)** Sketch a graph of P.

SOLUTION

(a) To find the zeros, we factor completely.

$$\begin{aligned} P(x) &= x^3 - 2x^2 - 3x \\ &= x(x^2 - 2x - 3) && \text{Factor } x \\ &= x(x - 3)(x + 1) && \text{Factor quadratic} \end{aligned}$$

Thus the zeros are $x = 0$, $x = 3$, and $x = -1$.

(b) The x-intercepts are $x = 0$, $x = 3$, and $x = -1$. The y-intercept is $P(0) = 0$. We make a table of values of $P(x)$, making sure that we choose test points between (and to the right and left of) successive zeros.

Since P is of odd degree and its leading coefficient is positive, it has the following end behavior:

$$y \to \infty \quad \text{as} \quad x \to \infty \qquad \text{and} \qquad y \to -\infty \quad \text{as} \quad x \to -\infty$$

We plot the points in the table and connect them by a smooth curve to complete the graph, as shown in Figure 8.

A table of values is most easily calculated by using a programmable calculator or a graphing calculator. See Appendix D, *Using the TI-83/84 Graphing Calculator*, for specific instructions. Go to **www.stewartmath.com**.

	x	$P(x)$
Test point →	-2	-10
	-1	0
Test point →	$-\frac{1}{2}$	$\frac{7}{8}$
	0	0
Test point →	1	-4
	2	-6
	3	0
Test point →	4	20

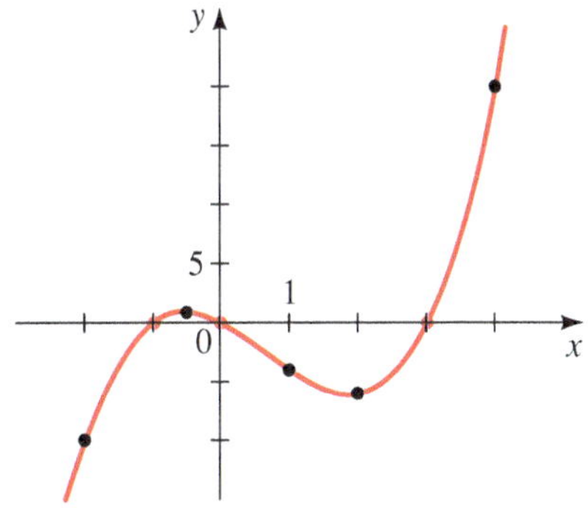

FIGURE 8 $P(x) = x^3 - 2x^2 - 3x$

Now Try Exercise 31 ■

EXAMPLE 6 ■ Finding Zeros and Graphing a Polynomial Function

Let $P(x) = -2x^4 - x^3 + 3x^2$.

(a) Find the zeros of P. **(b)** Sketch a graph of P.

SOLUTION

(a) To find the zeros, we factor completely.

$$\begin{aligned} P(x) &= -2x^4 - x^3 + 3x^2 \\ &= -x^2(2x^2 + x - 3) && \text{Factor } -x^2 \\ &= -x^2(2x + 3)(x - 1) && \text{Factor quadratic} \end{aligned}$$

Thus the zeros are $x = 0$, $x = -\frac{3}{2}$, and $x = 1$.

(b) The x-intercepts are $x = 0$, $x = -\frac{3}{2}$, and $x = 1$. The y-intercept is $P(0) = 0$. We make a table of values of $P(x)$, making sure that we choose test points between (and to the right and left of) successive zeros.

Since P is of even degree and its leading coefficient is negative, it has the following end behavior.

$$y \to -\infty \quad \text{as} \quad x \to \infty \qquad \text{and} \qquad y \to -\infty \quad \text{as} \quad x \to -\infty$$

We plot the points from the table and connect the points by a smooth curve to complete the graph in Figure 9.

x	$P(x)$
-2	-12
-1.5	0
-1	2
-0.5	0.75
0	0
0.5	0.5
1	0
1.5	-6.75

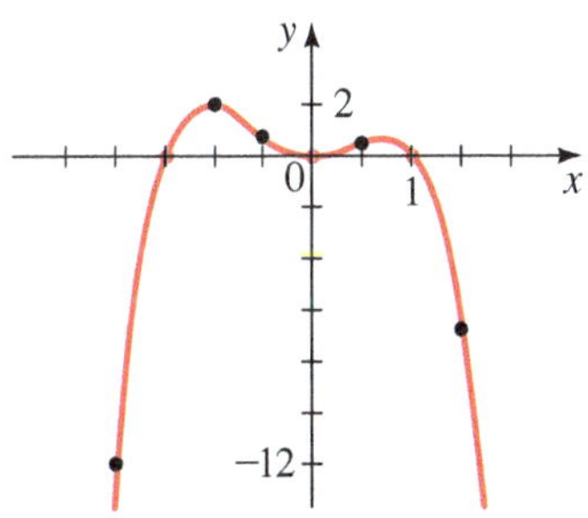

FIGURE 9 $P(x) = -2x^4 - x^3 + 3x^2$

Now Try Exercise 35 ■

EXAMPLE 7 ■ Finding Zeros and Graphing a Polynomial Function

Let $P(x) = x^3 - 2x^2 - 4x + 8$.

(a) Find the zeros of P. **(b)** Sketch a graph of P.

SOLUTION

(a) To find the zeros, we factor completely.

$$\begin{aligned} P(x) &= x^3 - 2x^2 - 4x + 8 \\ &= x^2(x-2) - 4(x-2) && \text{Group and factor} \\ &= (x^2-4)(x-2) && \text{Factor } x-2 \\ &= (x+2)(x-2)(x-2) && \text{Difference of squares} \\ &= (x+2)(x-2)^2 && \text{Simplify} \end{aligned}$$

Thus the zeros are $x = -2$ and $x = 2$.

(b) The x-intercepts are $x = -2$ and $x = 2$. The y-intercept is $P(0) = 8$. The table gives additional values of $P(x)$.

Since P is of odd degree and its leading coefficient is positive, it has the following end behavior.

$$y \to \infty \quad \text{as} \quad x \to \infty \qquad \text{and} \qquad y \to -\infty \quad \text{as} \quad x \to -\infty$$

We connect the points by a smooth curve to complete the graph in Figure 10.

x	$P(x)$
−3	−25
−2	0
−1	9
0	8
1	3
2	0
3	5

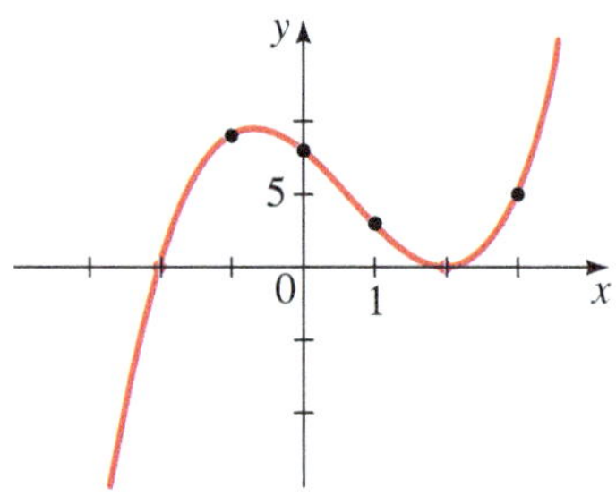

FIGURE 10
$P(x) = x^3 - 2x^2 - 4x + 8$

Now Try Exercise 37 ■

Shape of the Graph Near a Zero

Although $x = 2$ is a zero of the polynomial in Example 7, the graph does not cross the x-axis at the x-intercept 2. This is because the factor $(x - 2)^2$ corresponding to that zero is raised to an even power, so it doesn't change sign as we test points on either side of 2. In the same way the graph does not cross the x-axis at $x = 0$ in Example 6.

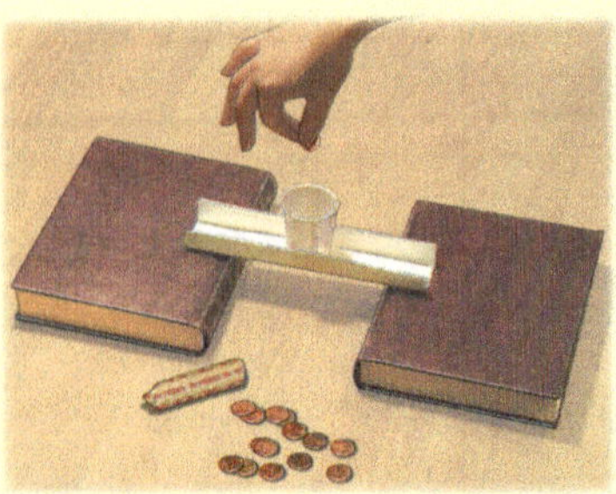

DISCOVERY PROJECT

Bridge Science

If you want to build a bridge, how can you be sure that your bridge design is strong enough to support the cars that will drive over it? In this project we perform a simple experiment using paper "bridges" to collect data on the weight our bridges can support. We model the data with linear and power functions to determine which model best fits the data. The model we obtain allows us to predict the strength of a large bridge *before* it is built. You can find the project at **www.stewartmath.com**.

In general, if c is a zero of P and the corresponding factor $x - c$ occurs exactly m times in the factorization of P, then we say that c is a **zero of multiplicity m**. By considering test points on either side of the x-intercept c, we conclude that the graph crosses the x-axis at c if the multiplicity m is odd and does not cross the x-axis if m is even. Moreover, it can be shown by using calculus that near $x = c$ the graph has the same general shape as the graph of $y = A(x - c)^m$.

SHAPE OF THE GRAPH NEAR A ZERO OF MULTIPLICITY m

If c is a zero of P of multiplicity m, then the shape of the graph of P near c is as follows.

Multiplicity of c	**Shape of the graph of P near the x-intercept c**
m odd, $m > 1$	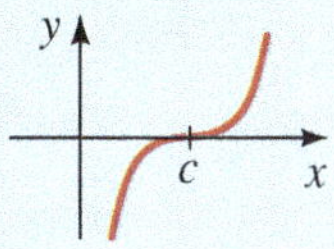 OR 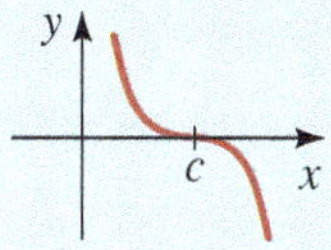
m even, $m > 1$	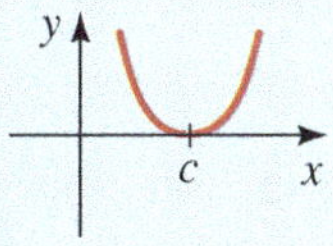 OR 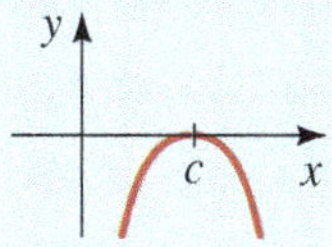

EXAMPLE 8 ■ Graphing a Polynomial Function Using Its Zeros

Graph the polynomial $P(x) = x^4(x - 2)^3(x + 1)^2$.

SOLUTION The zeros of P are -1, 0, and 2 with multiplicities 2, 4, and 3, respectively:

0 is a zero of multiplicity 4 — 2 is a zero of multiplicity 3 — –1 is a zero of multiplicity 2

$$P(x) = x^4(x - 2)^3(x + 1)^2$$

The zero 2 has *odd* multiplicity, so the graph crosses the x-axis at the x-intercept 2. But the zeros 0 and -1 have *even* multiplicity, so the graph does not cross the x-axis at the x-intercepts 0 and -1.

Since P is a polynomial of degree 9 and has positive leading coefficient, it has the following end behavior:

$$y \to \infty \quad \text{as} \quad x \to \infty \qquad \text{and} \qquad y \to -\infty \quad \text{as} \quad x \to -\infty$$

With this information and a table of values we sketch the graph in Figure 11.

x	$P(x)$
-1.3	-9.2
-1	0
-0.5	-3.9
0	0
1	-4
2	0
2.3	8.2

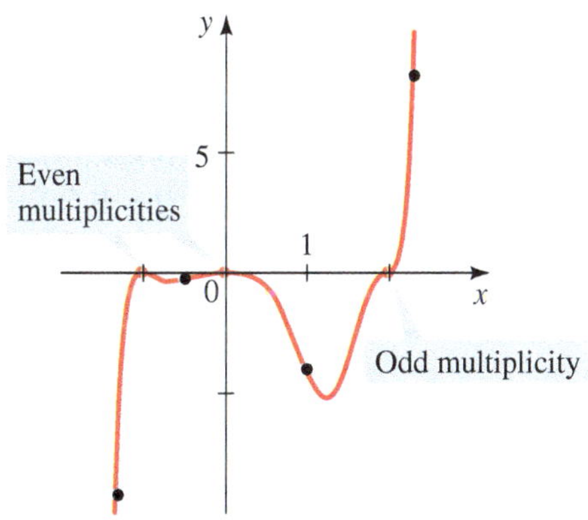

FIGURE 11 $P(x) = x^4(x - 2)^3(x + 1)^2$

Now Try Exercise 29

Local Maxima and Minima of Polynomials

Recall from Section 2.3 that if the point $(a, f(a))$ is the highest point on the graph of f within some viewing rectangle, then $f(a)$ is a local maximum value of f, and if $(b, f(b))$ is the lowest point on the graph of f within a viewing rectangle, then $f(b)$ is a local minimum value (see Figure 12). We say that such a point $(a, f(a))$ is a **local maximum point** on the graph and that $(b, f(b))$ is a **local minimum point**. The local maximum and minimum points on the graph of a function are called its **local extrema**.

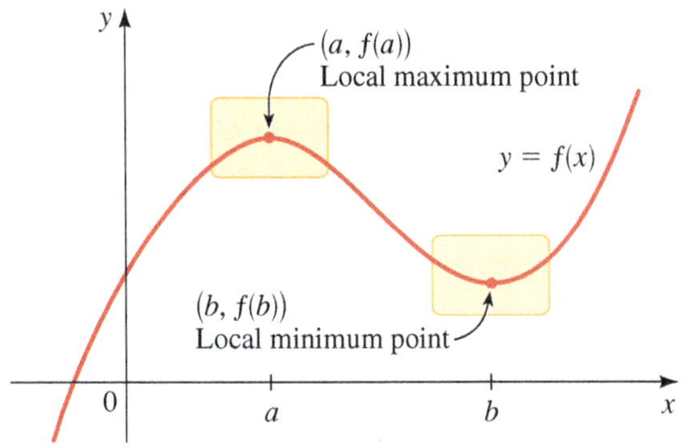

FIGURE 12

For a polynomial function the number of local extrema must be less than the degree, as the following principle indicates. (A proof of this principle requires calculus.)

LOCAL EXTREMA OF POLYNOMIALS

If $P(x) = a_nx^n + a_{n-1}x^{n-1} + \cdots + a_1x + a_0$ is a polynomial of degree n, then the graph of P has at most $n - 1$ local extrema.

A polynomial of degree n may in fact have fewer than $n - 1$ local extrema. For example, $P(x) = x^5$ (graphed in Figure 1) has *no* local extrema, even though it is of degree 5. The preceding principle tells us only that a polynomial of degree n can have no more than $n - 1$ local extrema.

EXAMPLE 9 ■ The Number of Local Extrema

Graph the polynomial and determine how many local extrema it has.

(a) $P_1(x) = x^4 + x^3 - 16x^2 - 4x + 48$

(b) $P_2(x) = x^5 + 3x^4 - 5x^3 - 15x^2 + 4x - 15$

(c) $P_3(x) = 7x^4 + 3x^2 - 10x$

SOLUTION The graphs are shown in Figure 13.

(a) P_1 has two local minimum points and one local maximum point, for a total of three local extrema.

(b) P_2 has two local minimum points and two local maximum points, for a total of four local extrema.

(c) P_3 has just one local extremum, a local minimum.

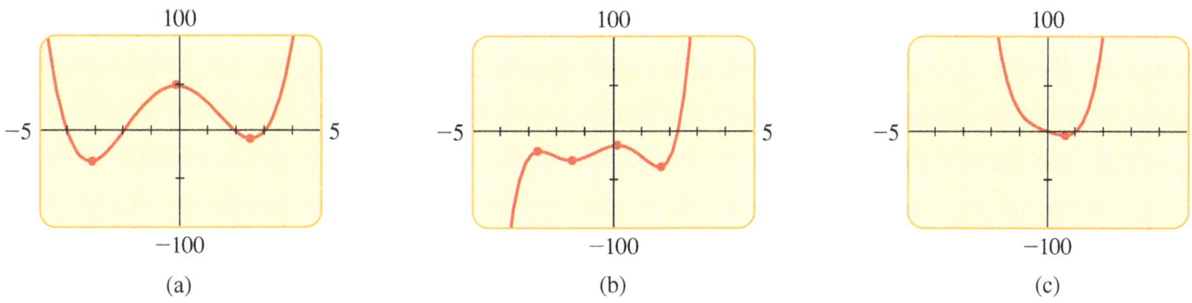

$P_1(x) = x^4 + x^3 - 16x^2 - 4x + 48$ $\quad P_2(x) = x^5 + 3x^4 - 5x^3 - 15x^2 + 4x - 15$ $\quad P_3(x) = 7x^4 + 3x^2 - 10x$

FIGURE 13

Now Try Exercises 65 and 67 ■

With a graphing calculator we can quickly draw the graphs of many functions at once, on the same viewing screen. This allows us to see how changing a value in the definition of the functions affects the shape of its graph. In the next example we apply this principle to a family of third-degree polynomials.

EXAMPLE 10 ■ A Family of Polynomials

Sketch the family of polynomials $P(x) = x^3 - cx^2$ for $c = 0, 1, 2,$ and 3. How does changing the value of c affect the graph?

SOLUTION The polynomials

$$P_0(x) = x^3 \qquad P_1(x) = x^3 - x^2$$
$$P_2(x) = x^3 - 2x^2 \qquad P_3(x) = x^3 - 3x^2$$

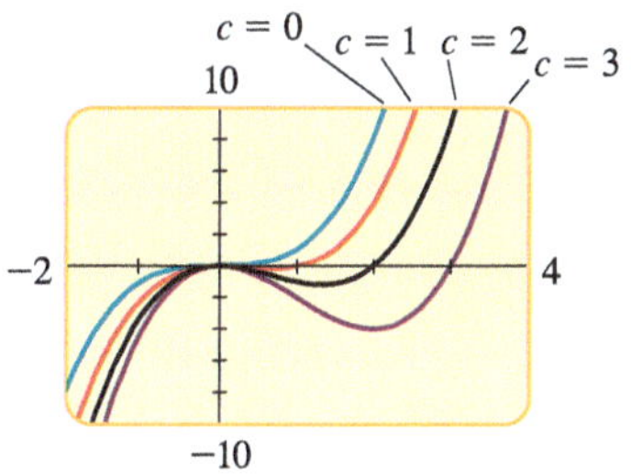

FIGURE 14 A family of polynomials $P(x) = x^3 - cx^2$

are graphed in Figure 14. We see that increasing the value of c causes the graph to develop an increasingly deep "valley" to the right of the y-axis, creating a local maximum at the origin and a local minimum at a point in Quadrant IV. This local minimum moves lower and farther to the right as c increases. To see why this happens, factor $P(x) = x^2(x - c)$. The polynomial P has zeros at 0 and c, and the larger c gets, the farther to the right the minimum between 0 and c will be.

Now Try Exercise 75 ■

3.2 EXERCISES

CONCEPTS

1. Only one of the following graphs could be the graph of a polynomial function. Which one? Why are the others not graphs of polynomials?

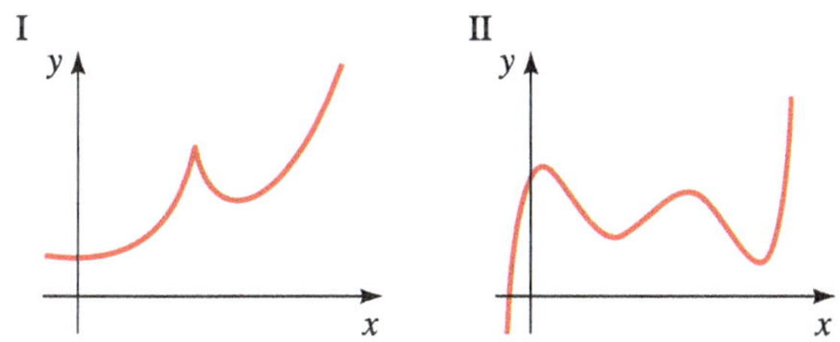

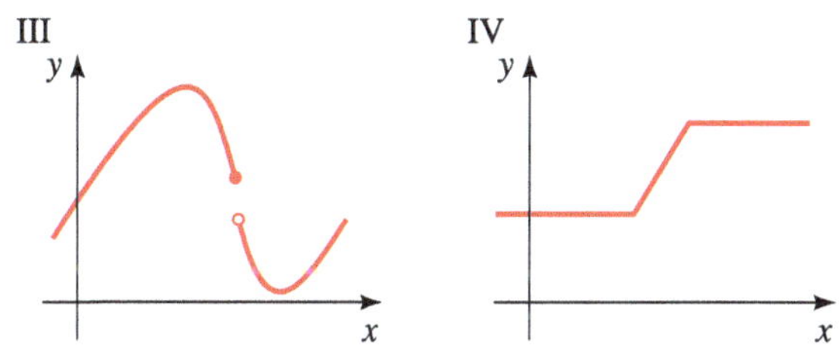

2. Describe the end behavior of each polynomial.
   (a) $y = x^3 - 8x^2 + 2x - 15$

   End behavior: $y \to$ ______ as $x \to \infty$

   $y \to$ ______ as $x \to -\infty$

(b) $y = -2x^4 + 12x + 100$

End behavior: $y \to$ ______ as $x \to \infty$

$y \to$ ______ as $x \to -\infty$

3. If c is a zero of the polynomial P, then

(a) $P(c) =$ ______.

(b) $x - c$ is a ______ of $P(x)$.

(c) c is a(n) ______ -intercept of the graph of P.

4. Which of the following statements couldn't possibly be true about the polynomial function P?

(a) P has degree 3, two local maxima, and two local minima.

(b) P has degree 3 and no local maxima or minima.

(c) P has degree 4, one local maximum, and no local minima.

SKILLS

5–8 ■ Transformations of Monomials Sketch the graph of each function by transforming the graph of an appropriate function of the form $y = x^n$ from Figure 1. Indicate all x- and y-intercepts on each graph.

5. **(a)** $P(x) = x^2 - 4$ **(b)** $Q(x) = (x - 4)^2$
(c) $P(x) = 2x^2 + 3$ **(d)** $P(x) = -(x + 2)^2$

6. **(a)** $P(x) = x^4 - 16$ **(b)** $P(x) = -(x + 5)^4$
(c) $P(x) = -5x^4 + 5$ **(d)** $P(x) = (x - 5)^4$

7. **(a)** $P(x) = x^3 - 8$ **(b)** $Q(x) = -x^3 + 27$
(c) $R(x) = -(x + 2)^3$ **(d)** $S(x) = \frac{1}{2}(x - 1)^3 + 4$

8. **(a)** $P(x) = (x + 3)^5$ **(b)** $Q(x) = 2(x + 3)^5 - 64$
(c) $R(x) = -\frac{1}{2}(x - 2)^5$ **(d)** $S(x) = -\frac{1}{2}(x - 2)^5 + 16$

9–14 ■ End Behavior A polynomial function is given.
(a) Describe the end behavior of the polynomial function.
(b) Match the polynomial function with one of the graphs I–VI.

9. $P(x) = x(x^2 - 4)$ **10.** $Q(x) = -x^2(x^2 - 4)$

11. $R(x) = -x^5 + 5x^3 - 4x$ **12.** $S(x) = \frac{1}{2}x^6 - 2x^4$

13. $T(x) = x^4 + 2x^3$ **14.** $U(x) = -x^3 + 2x^2$

I
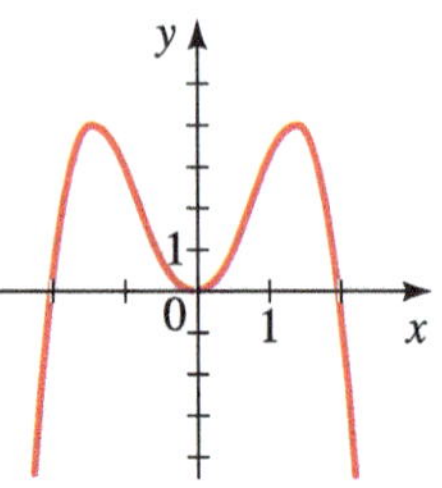

II
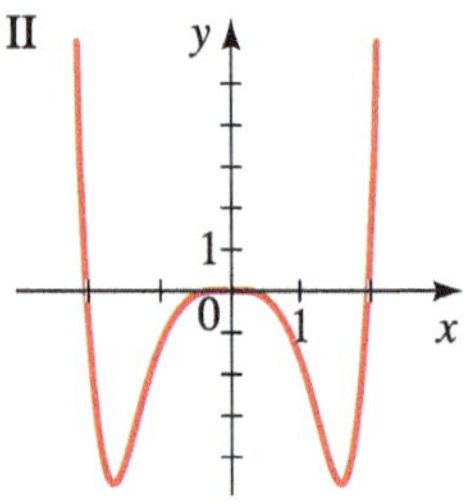

III
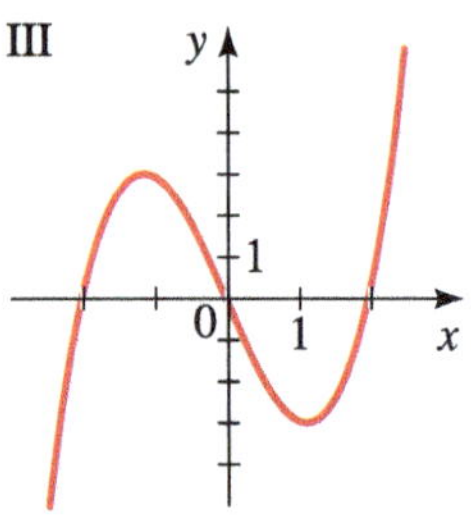

IV
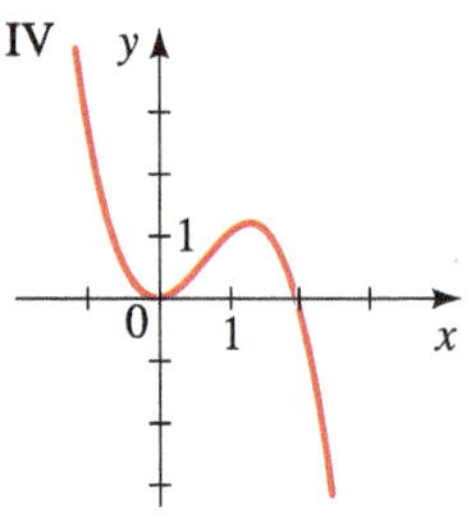

V
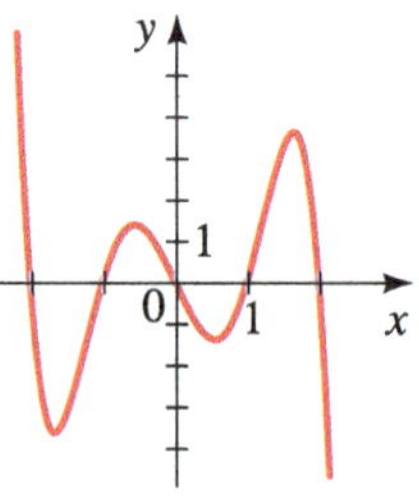

VI
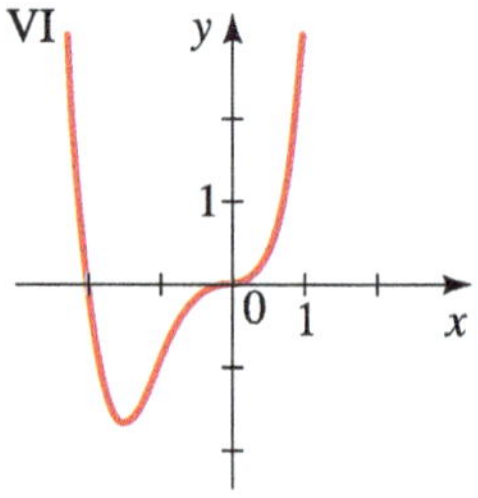

15–30 ■ Graphing Factored Polynomials Sketch the graph of the polynomial function. Make sure your graph shows all intercepts and exhibits the proper end behavior.

15. $P(x) = (x - 1)(x + 2)$

16. $P(x) = (2 - x)(x + 5)$

17. $P(x) = -x(x - 3)(x + 2)$

18. $P(x) = x(x - 3)(x + 2)$

19. $P(x) = -(2x - 1)(x + 1)(x + 3)$

20. $P(x) = (x - 3)(x + 2)(3x - 2)$

21. $P(x) = (x + 2)(x + 1)(x - 2)(x - 3)$

22. $P(x) = x(x + 1)(x - 1)(2 - x)$

23. $P(x) = -2x(x - 2)^2$

24. $P(x) = \frac{1}{5}x(x - 5)^2$

25. $P(x) = (x + 2)(x + 1)^2(2x - 3)$

26. $P(x) = -(x + 1)^2(x - 1)^3(x - 2)$

27. $P(x) = \frac{1}{12}(x + 2)^2(x - 3)^2$

28. $P(x) = (x - 1)^2(x + 2)^3$

29. $P(x) = x^3(x + 2)(x - 3)^2$

30. $P(x) = (x - 3)^2(x + 1)^2$

31–44 ■ Graphing Polynomials Factor the polynomial and use the factored form to find the zeros. Then sketch the graph.

31. $P(x) = x^3 - x^2 - 6x$ **32.** $P(x) = x^3 + 2x^2 - 8x$

33. $P(x) = -x^3 + x^2 + 12x$ **34.** $P(x) = -2x^3 - x^2 + x$

35. $P(x) = x^4 - 3x^3 + 2x^2$ **36.** $P(x) = x^5 - 9x^3$

37. $P(x) = x^3 + x^2 - x - 1$

38. $P(x) = x^3 + 3x^2 - 4x - 12$

39. $P(x) = 2x^3 - x^2 - 18x + 9$

40. $P(x) = \frac{1}{8}(2x^4 + 3x^3 - 16x - 24)^2$

41. $P(x) = x^4 - 2x^3 - 8x + 16$

42. $P(x) = x^4 - 2x^3 + 8x - 16$

43. $P(x) = x^4 - 3x^2 - 4$ **44.** $P(x) = x^6 - 2x^3 + 1$

45–50 ■ End Behavior Determine the end behavior of P. Compare the graphs of P and Q in large and small viewing rectangles, as in Example 3(b).

45. $P(x) = 3x^3 - x^2 + 5x + 1$; $Q(x) = 3x^3$

46. $P(x) = -\frac{1}{8}x^3 + \frac{1}{4}x^2 + 12x$; $Q(x) = -\frac{1}{8}x^3$

47. $P(x) = x^4 - 7x^2 + 5x + 5$; $Q(x) = x^4$

48. $P(x) = -x^5 + 2x^2 + x;\quad Q(x) = -x^5$

49. $P(x) = x^{11} - 9x^9;\quad Q(x) = x^{11}$

50. $P(x) = 2x^2 - x^{12};\quad Q(x) = -x^{12}$

51–54 ■ Local Extrema The graph of a polynomial function is given. From the graph, find **(a)** the x- and y-intercepts, and **(b)** the coordinates of all local extrema.

51. $P(x) = -x^2 + 4x$

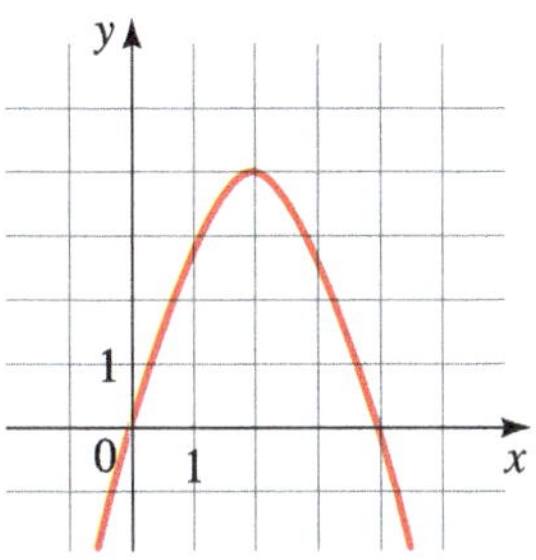

52. $P(x) = \frac{2}{9}x^3 - x^2$

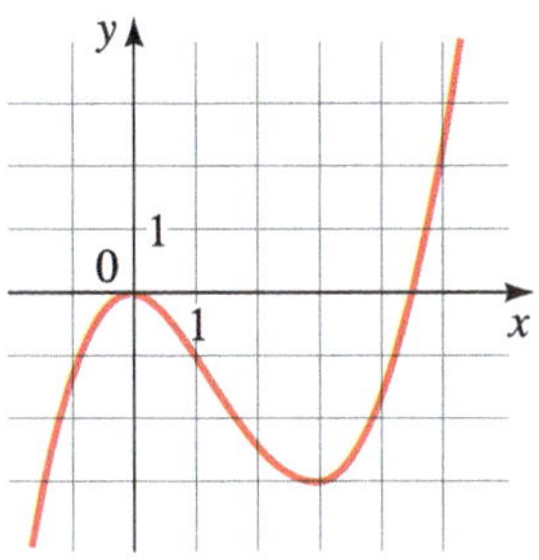

53. $P(x) = -\frac{1}{2}x^3 + \frac{3}{2}x - 1$

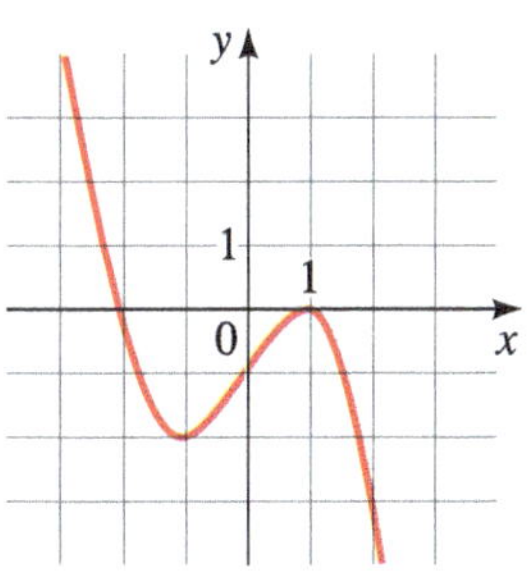

54. $P(x) = \frac{1}{9}x^4 - \frac{4}{9}x^3$

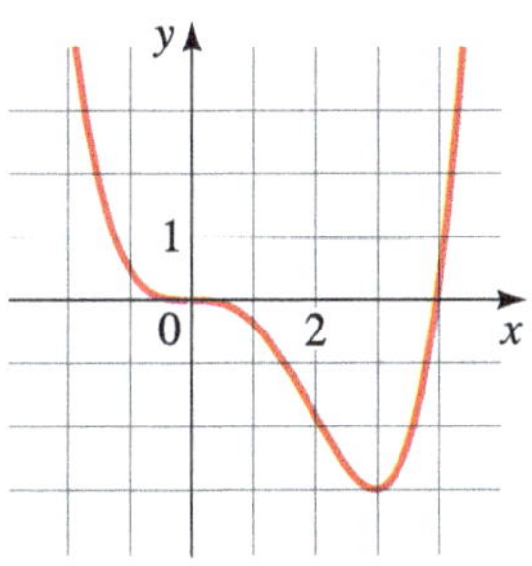

55–62 ■ Local Extrema Graph the polynomial in the given viewing rectangle. Find the coordinates of all local extrema. State each answer rounded to two decimal places. State the domain and range.

55. $y = -x^2 + 8x,\quad [-4, 12]$ by $[-50, 30]$

56. $y = x^3 - 3x^2,\quad [-2, 5]$ by $[-10, 10]$

57. $y = x^3 - 12x + 9,\quad [-5, 5]$ by $[-30, 30]$

58. $y = 2x^3 - 3x^2 - 12x - 32,\quad [-5, 5]$ by $[-60, 30]$

59. $y = x^4 + 4x^3,\quad [-5, 5]$ by $[-30, 30]$

60. $y = x^4 - 18x^2 + 32,\quad [-5, 5]$ by $[-100, 100]$

61. $y = 3x^5 - 5x^3 + 3,\quad [-3, 3]$ by $[-5, 10]$

62. $y = x^5 - 5x^2 + 6,\quad [-3, 3]$ by $[-5, 10]$

63–72 ■ Number of Local Extrema Graph the polynomial, and determine how many local maxima and minima it has.

63. $y = -2x^2 + 3x + 5$

64. $y = x^3 + 12x$

65. $y = x^3 - x^2 - x$

66. $y = 6x^3 + 3x + 1$

67. $y = x^4 - 5x^2 + 4$

68. $y = 1.2x^5 + 3.75x^4 - 7x^3 - 15x^2 + 18x$

69. $y = (x - 2)^5 + 32$

70. $y = (x^2 - 2)^3$

71. $y = x^8 - 3x^4 + x$

72. $y = \frac{1}{3}x^7 - 17x^2 + 7$

73–78 ■ Families of Polynomials Graph the family of polynomials in the same viewing rectangle, using the given values of c. Explain how changing the value of c affects the graph.

73. $P(x) = cx^3;\quad c = 1, 2, 5, \frac{1}{2}$

74. $P(x) = (x - c)^4;\quad c = -1, 0, 1, 2$

75. $P(x) = x^4 + c;\quad c = -1, 0, 1, 2$

76. $P(x) = x^3 + cx;\quad c = 2, 0, -2, -4$

77. $P(x) = x^4 - cx;\quad c = 0, 1, 8, 27$

78. $P(x) = x^c;\quad c = 1, 3, 5, 7$

SKILLS Plus

79. Intersection Points of Two Polynomials

(a) On the same coordinate axes, sketch graphs (as accurately as possible) of the functions

$$y = x^3 - 2x^2 - x + 2 \quad \text{and} \quad y = -x^2 + 5x + 2$$

(b) On the basis of your sketch in part (a), at how many points do the two graphs appear to intersect?

(c) Find the coordinates of all intersection points.

80. Power Functions Portions of the graphs of $y = x^2$, $y = x^3$, $y = x^4$, $y = x^5$, and $y = x^6$ are plotted in the figures. Determine which function belongs to each graph.

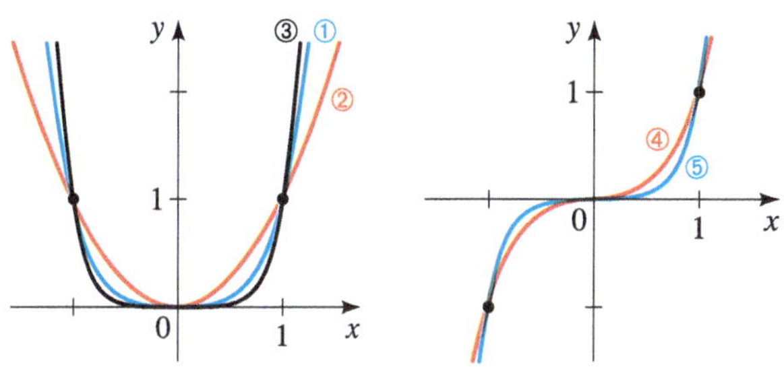

81. Odd and Even Functions Recall that a function f is *odd* if $f(-x) = -f(x)$ or *even* if $f(-x) = f(x)$ for all real x.

(a) Show that a polynomial $P(x)$ that contains only odd powers of x is an odd function.

(b) Show that a polynomial $P(x)$ that contains only even powers of x is an even function.

(c) Show that if a polynomial $P(x)$ contains both odd and even powers of x, then it is neither an odd nor an even function.

(d) Express the function

$$P(x) = x^5 + 6x^3 - x^2 - 2x + 5$$

as the sum of an odd function and an even function.

82. Number of Intercepts and Local Extrema

(a) How many x-intercepts and how many local extrema does the polynomial $P(x) = x^3 - 4x$ have?

(b) How many x-intercepts and how many local extrema does the polynomial $Q(x) = x^3 + 4x$ have?

(c) If $a > 0$, how many x-intercepts and how many local extrema does each of the polynomials $P(x) = x^3 - ax$ and $Q(x) = x^3 + ax$ have? Explain your answer.

83–86 ■ Local Extrema These exercises involve local maxima and minima of polynomial functions.

83. **(a)** Graph the function $P(x) = (x - 1)(x - 3)(x - 4)$ and find all local extrema, correct to the nearest tenth.

(b) Graph the function

$$Q(x) = (x - 1)(x - 3)(x - 4) + 5$$

and use your answers to part (a) to find all local extrema, correct to the nearest tenth.

84. **(a)** Graph the function $P(x) = (x - 2)(x - 4)(x - 5)$ and determine how many local extrema it has.

(b) If $a < b < c$, explain why the function

$$P(x) = (x - a)(x - b)(x - c)$$

must have two local extrema.

85. Maximum Number of Local Extrema What is the smallest possible degree that the polynomial whose graph is shown can have? Explain.

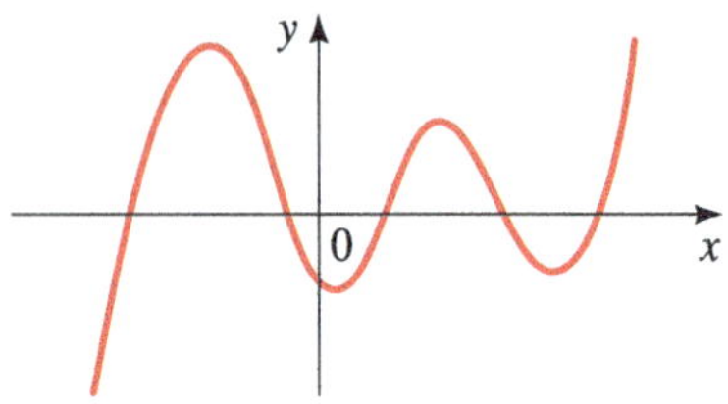

86. Impossible Situation? Is it possible for a polynomial to have two local maxima and no local minimum? Explain.

APPLICATIONS

87. Market Research A market analyst working for a small-appliance manufacturer finds that if the firm produces and sells x blenders annually, the total profit (in dollars) is

$$P(x) = 8x + 0.3x^2 - 0.0013x^3 - 372$$

Graph the function P in an appropriate viewing rectangle and use the graph to answer the following questions.

(a) When just a few blenders are manufactured, the firm loses money (profit is negative). (For example, $P(10) = -263.3$, so the firm loses \$263.30 if it produces and sells only 10 blenders.) How many blenders must the firm produce to break even?

(b) Does profit increase indefinitely as more blenders are produced and sold? If not, what is the largest possible profit the firm could have?

88. Population Change The rabbit population on a small island is observed to be given by the function

$$P(t) = 120t - 0.4t^4 + 1000$$

where t is the time (in months) since observations of the island began.

(a) When is the maximum population attained, and what is that maximum population?

(b) When does the rabbit population disappear from the island?

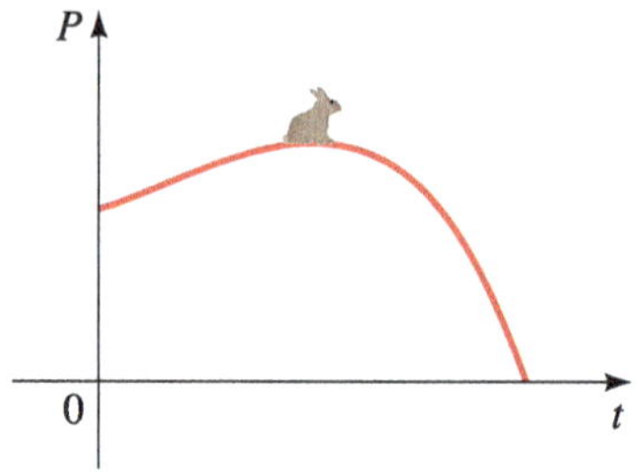

89. Volume of a Box An open box is to be constructed from a piece of cardboard 20 cm by 40 cm by cutting squares of side length x from each corner and folding up the sides, as shown in the figure.

(a) Express the volume V of the box as a function of x.

(b) What is the domain of V? (Use the fact that length and volume must be positive.)

(c) Draw a graph of the function V, and use it to estimate the maximum volume for such a box.

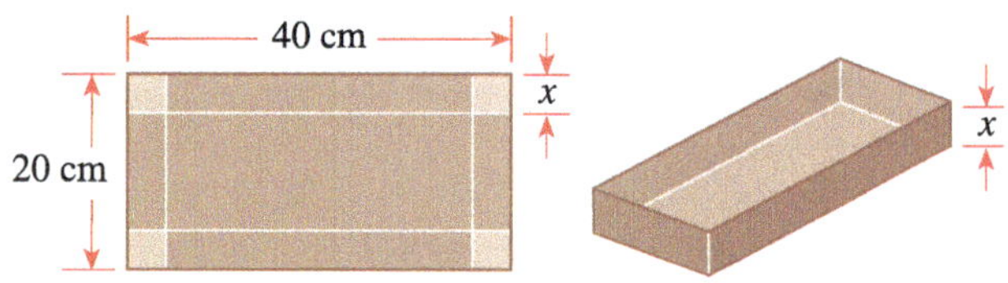

90. Volume of a Box A cardboard box has a square base, with each edge of the base having length x inches, as shown in the figure. The total length of all 12 edges of the box is 144 in.

(a) Show that the volume of the box is given by the function $V(x) = 2x^2(18 - x)$.

(b) What is the domain of V? (Use the fact that length and volume must be positive.)

(c) Draw a graph of the function V and use it to estimate the maximum volume for such a box.

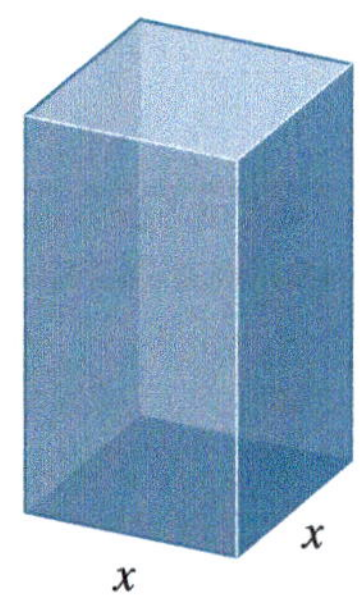

DISCUSS ■ DISCOVER ■ PROVE ■ WRITE

91. DISCOVER: Graphs of Large Powers Graph the functions $y = x^2$, $y = x^3$, $y = x^4$, and $y = x^5$, for $-1 \le x \le 1$, on the same coordinate axes. What do you think the graph of $y = x^{100}$ would look like on this same interval? What about $y = x^{101}$? Make a table of values to confirm your answers.

92. DISCUSS ■ DISCOVER: Possible Number of Local Extrema Is it possible for a third-degree polynomial to have exactly one local extremum? Can a fourth-degree polynomial have exactly two local extrema? How many local extrema can polynomials of third, fourth, fifth, and sixth degree have? (Think about the end behavior of such polynomials.) Now give an example of a polynomial that has six local extrema.

3.3 DIVIDING POLYNOMIALS

■ Long Division of Polynomials ■ Synthetic Division ■ The Remainder and Factor Theorems

So far in this chapter we have been studying polynomial functions *graphically*. In this section we begin to study polynomials *algebraically*. Most of our work will be concerned with factoring polynomials, and to factor, we need to know how to divide polynomials.

■ Long Division of Polynomials

Dividing polynomials is much like the familiar process of dividing numbers. When we divide 38 by 7, the quotient is 5 and the remainder is 3. We write

Dividend, Remainder

$$\frac{38}{7} = 5 + \frac{3}{7}$$

Divisor, Quotient

To divide polynomials, we use long division, as follows.

DIVISION ALGORITHM

If $P(x)$ and $D(x)$ are polynomials, with $D(x) \neq 0$, then there exist unique polynomials $Q(x)$ and $R(x)$, where $R(x)$ is either 0 or of degree less than the degree of $D(x)$, such that

$$\frac{P(x)}{D(x)} = Q(x) + \frac{R(x)}{D(x)} \qquad \text{or} \qquad P(x) = D(x) \cdot Q(x) + R(x)$$

Dividend, Divisor, Quotient, Remainder

The polynomials $P(x)$ and $D(x)$ are called the **dividend** and **divisor**, respectively, $Q(x)$ is the **quotient**, and $R(x)$ is the **remainder**.

EXAMPLE 1 ■ Long Division of Polynomials

Divide $6x^2 - 26x + 12$ by $x - 4$. Express the result in each of the two forms shown in the above box.

SOLUTION The *dividend* is $6x^2 - 26x + 12$, and the *divisor* is $x - 4$. We begin by arranging them as follows.

$$x - 4\overline{)6x^2 - 26x + 12}$$

Next we divide the leading term in the dividend by the leading term in the divisor to get the first term of the quotient: $6x^2/x = 6x$. Then we multiply the divisor by $6x$ and subtract the result from the dividend.

$$\begin{array}{rll} & 6x & \text{Divide leading terms: } \frac{6x^2}{x} = 6x \\ x - 4\overline{)} & 6x^2 - 26x + 12 & \\ & \underline{6x^2 - 24x} & \text{Multiply: } 6x(x - 4) = 6x^2 - 24x \\ & -2x + 12 & \text{Subtract and "bring down" 12} \end{array}$$

We repeat the process using the last line $-2x + 12$ as the dividend.

$$\begin{array}{r} 6x - 2 \\ x - 4 \overline{)6x^2 - 26x + 12} \\ \underline{6x^2 - 24x} \\ -2x + 12 \\ \underline{-2x + 8} \\ 4 \end{array}$$

Divide leading terms: $\frac{-2x}{x} = -2$

Multiply: $-2(x - 4) = -2x + 8$

Subtract

The division process ends when the last line is of lesser degree than the divisor. The last line then contains the *remainder*, and the top line contains the *quotient*. The result of the division can be interpreted in either of two ways:

$$\underbrace{\frac{6x^2 - 26x + 12}{x - 4}}_{\text{Dividend / Divisor}} = \underbrace{6x - 2}_{\text{Quotient}} + \frac{\overbrace{4}^{\text{Remainder}}}{x - 4} \quad \text{or} \quad \underbrace{6x^2 - 26x + 12}_{\text{Dividend}} = \underbrace{(x - 4)}_{\text{Divisor}}\underbrace{(6x - 2)}_{\text{Quotient}} + \overbrace{4}^{\text{Remainder}}$$

Now Try Exercises 3 and 9 ■

EXAMPLE 2 ■ Long Division of Polynomials

Let $P(x) = 8x^4 + 6x^2 - 3x + 1$ and $D(x) = 2x^2 - x + 2$. Find polynomials $Q(x)$ and $R(x)$ such that $P(x) = D(x) \cdot Q(x) + R(x)$.

SOLUTION We use long division after first inserting the term $0x^3$ into the dividend to ensure that the columns line up correctly.

$$\begin{array}{r} 4x^2 + 2x \\ 2x^2 - x + 2 \overline{)8x^4 + 0x^3 + 6x^2 - 3x + 1} \\ \underline{8x^4 - 4x^3 + 8x^2} \\ 4x^3 - 2x^2 - 3x \\ \underline{4x^3 - 2x^2 + 4x} \\ -7x + 1 \end{array}$$

Multiply divisor by $4x^2$

Subtract

Multiply divisor by $2x$

Subtract

The process is complete at this point because $-7x + 1$ is of lesser degree than the divisor $2x^2 - x + 2$. From the above long division we see that $Q(x) = 4x^2 + 2x$ and $R(x) = -7x + 1$, so

$$8x^4 + 6x^2 - 3x + 1 = (2x^2 - x + 2)(4x^2 + 2x) + (-7x + 1)$$

Now Try Exercise 19 ■

■ Synthetic Division

Synthetic division is a quick method of dividing polynomials; it can be used when the divisor is of the form $x - c$. In synthetic division we write only the essential parts of the long division. Compare the following long and synthetic divisions, in which we divide $2x^3 - 7x^2 + 5$ by $x - 3$. (We'll explain how to perform the synthetic division in Example 3.)

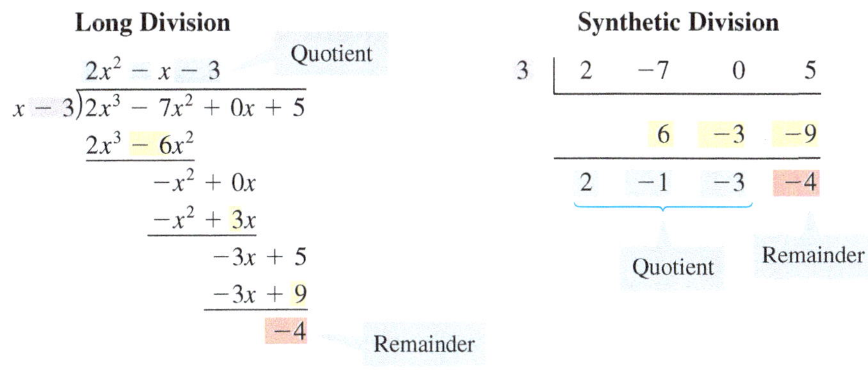

Note that in synthetic division we abbreviate $2x^3 - 7x^2 + 5$ by writing only the coefficients: 2 -7 0 5, and instead of $x - 3$, we simply write 3. (Writing 3 instead of -3 allows us to add instead of subtract, but this changes the sign of all the numbers that appear in the gold boxes.)

The next example shows how synthetic division is performed.

EXAMPLE 3 ■ Synthetic Division

Use synthetic division to divide $2x^3 - 7x^2 + 5$ by $x - 3$.

SOLUTION We begin by writing the appropriate coefficients to represent the divisor and the dividend:

Divisor $x - 3$ 3 | 2 -7 0 5 Dividend $2x^3 - 7x^2 + 0x + 5$

We bring down the 2, multiply $3 \cdot 2 = 6$, and write the result in the middle row. Then we add.

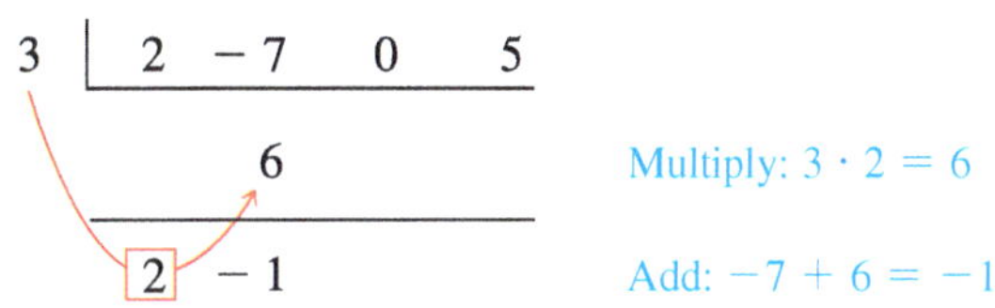

We repeat this process of multiplying and then adding until the table is complete.

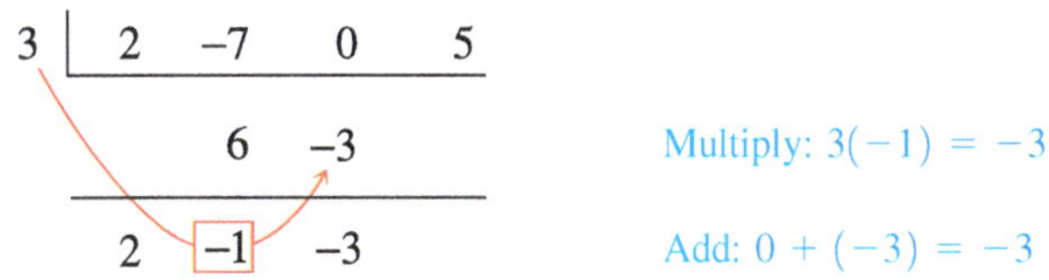

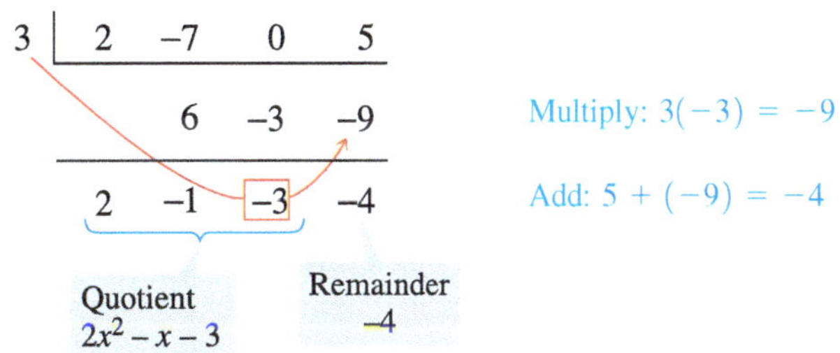

From the last line of the synthetic division we see that the quotient is $2x^2 - x - 3$ and the remainder is -4. Thus

$$2x^3 - 7x^2 + 5 = (x - 3)(2x^2 - x - 3) - 4$$

Now Try Exercise 31 ■

The Remainder and Factor Theorems

The next theorem shows how synthetic division can be used to evaluate polynomials easily.

REMAINDER THEOREM

If the polynomial $P(x)$ is divided by $x - c$, then the remainder is the value $P(c)$.

Proof If the divisor in the Division Algorithm is of the form $x - c$ for some real number c, then the remainder must be a constant (since the degree of the remainder is less than the degree of the divisor). If we call this constant r, then

$$P(x) = (x - c) \cdot Q(x) + r$$

Replacing x by c in this equation, we get $P(c) = (c - c) \cdot Q(c) + r = 0 + r = r$, that is, $P(c)$ is the remainder r. ■

EXAMPLE 4 ■ Using the Remainder Theorem to Find the Value of a Polynomial

Let $P(x) = 3x^5 + 5x^4 - 4x^3 + 7x + 3$.

(a) Find the quotient and remainder when $P(x)$ is divided by $x + 2$.

(b) Use the Remainder Theorem to find $P(-2)$.

SOLUTION

(a) Since $x + 2 = x - (-2)$, the synthetic division for this problem takes the following form:

$$\begin{array}{r|rrrrrr} -2 & 3 & 5 & -4 & 0 & 7 & 3 \\ & & -6 & 2 & 4 & -8 & 2 \\ \hline & 3 & -1 & -2 & 4 & -1 & 5 \end{array}$$

Remainder is 5, so $P(-2) = 5$

The quotient is $3x^4 - x^3 - 2x^2 + 4x - 1$, and the remainder is 5.

(b) By the Remainder Theorem, $P(-2)$ is the remainder when $P(x)$ is divided by $x - (-2) = x + 2$. From part (a) the remainder is 5, so $P(-2) = 5$.

Now Try Exercise 39 ■

The next theorem says that *zeros* of polynomials correspond to *factors*. We used this fact in Section 3.2 to graph polynomials.

FACTOR THEOREM

c is a zero of P if and only if $x - c$ is a factor of $P(x)$.

Proof If $P(x)$ factors as $P(x) = (x - c)Q(x)$, then

$$P(c) = (c - c)Q(c) = 0 \cdot Q(c) = 0$$

Conversely, if $P(c) = 0$, then by the Remainder Theorem

$$P(x) = (x - c)Q(x) + 0 = (x - c)Q(x)$$

so $x - c$ is a factor of $P(x)$. ■

EXAMPLE 5 ■ Factoring a Polynomial Using the Factor Theorem

Let $P(x) = x^3 - 7x + 6$. Show that $P(1) = 0$, and use this fact to factor $P(x)$ completely.

$$\begin{array}{r|rrrr} 1 & 1 & 0 & -7 & 6 \\ & & 1 & 1 & -6 \\ \hline & 1 & 1 & -6 & 0 \end{array}$$

$$\begin{array}{r} x^2 + x - 6 \\ x - 1\overline{)x^3 + 0x^2 - 7x + 6} \\ \underline{x^3 - x^2} \qquad\qquad \\ x^2 - 7x \qquad \\ \underline{x^2 - x} \qquad \\ -6x + 6 \\ \underline{-6x + 6} \\ 0 \end{array}$$

SOLUTION Substituting, we see that $P(1) = 1^3 - 7 \cdot 1 + 6 = 0$. By the Factor Theorem this means that $x - 1$ is a factor of $P(x)$. Using synthetic or long division (shown in the margin), we see that

$$\begin{aligned} P(x) &= x^3 - 7x + 6 && \text{Given polynomial} \\ &= (x - 1)(x^2 + x - 6) && \text{See margin} \\ &= (x - 1)(x - 2)(x + 3) && \text{Factor quadratic } x^2 + x - 6 \end{aligned}$$

Now Try Exercises 53 and 57 ■

EXAMPLE 6 ■ Finding a Polynomial with Specified Zeros

Find a polynomial of degree four that has zeros -3, 0, 1, and 5, and the coefficient of x^3 is -6.

SOLUTION By the Factor Theorem, $x - (-3)$, $x - 0$, $x - 1$, and $x - 5$ must all be factors of the desired polynomial. Let

$$\begin{aligned} P(x) &= (x + 3)(x - 0)(x - 1)(x - 5) \\ &= x^4 - 3x^3 - 13x^2 + 15x \end{aligned}$$

The polynomial $P(x)$ is of degree 4 with the desired zeros, but the coefficient of x^3 is -3, not -6. Multiplication by a nonzero constant does not change the degree, so the desired polynomial is a constant multiple of $P(x)$. If we multiply $P(x)$ by the constant 2, we get

$$Q(x) = 2x^4 - 6x^3 - 26x^2 + 30x$$

which is a polynomial with all the desired properties.The polynomial Q is graphed in Figure 1. Note that the zeros of Q correspond to the x-intercepts of the graph.

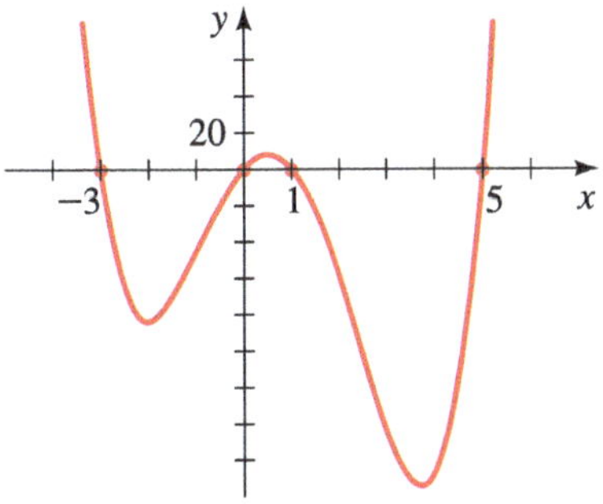

FIGURE 1
$Q(x) = 2x(x + 3)(x - 1)(x - 5)$ has zeros -3, 0, 1, and 5, and the coefficient of x^3 is -6.

Now Try Exercises 63 and 67 ■

3.3 EXERCISES

CONCEPTS

1. If we divide the polynomial P by the factor $x - c$ and we obtain the equation $P(x) = (x - c)Q(x) + R(x)$, then we say that $x - c$ is the divisor, $Q(x)$ is the ________, and $R(x)$ is the ________.

2. **(a)** If we divide the polynomial $P(x)$ by the factor $x - c$ and we obtain a remainder of 0, then we know that c is a ________ of P.

   (b) If we divide the polynomial $P(x)$ by the factor $x - c$ and we obtain a remainder of k, then we know that $P(c) =$ ________.

SKILLS

3–8 ■ Division of Polynomials Two polynomials P and D are given. Use either synthetic or long division to divide $P(x)$ by $D(x)$, and express the quotient $P(x)/D(x)$ in the form

$$\frac{P(x)}{D(x)} = Q(x) + \frac{R(x)}{D(x)}$$

3. $P(x) = 2x^2 - 5x - 7$, $D(x) = x - 2$
4. $P(x) = 3x^3 + 9x^2 - 5x - 1$, $D(x) = x + 4$
5. $P(x) = 4x^2 - 3x - 7$, $D(x) = 2x - 1$
6. $P(x) = 6x^3 + x^2 - 12x + 5$, $D(x) = 3x - 4$
7. $P(x) = 2x^4 - x^3 + 9x^2$, $D(x) = x^2 + 4$
8. $P(x) = 2x^5 + x^3 - 2x^2 + 3x - 5$, $D(x) = x^2 - 3x + 1$

9–14 ■ Division of Polynomials Two polynomials P and D are given. Use either synthetic or long division to divide $P(x)$ by $D(x)$, and express P in the form

$$P(x) = D(x) \cdot Q(x) + R(x)$$

9. $P(x) = -x^3 - 2x + 6, \quad D(x) = x + 1$

10. $P(x) = x^4 + 2x^3 - 10x, \quad D(x) = x - 3$

11. $P(x) = 2x^3 - 3x^2 - 2x, \quad D(x) = 2x - 3$

12. $P(x) = 4x^3 + 7x + 9, \quad D(x) = 2x + 1$

13. $P(x) = 8x^4 + 4x^3 + 6x^2, \quad D(x) = 2x^2 + 1$

14. $P(x) = 27x^5 - 9x^4 + 3x^2 - 3, \quad D(x) = 3x^2 - 3x + 1$

15–24 ■ Long Division of Polynomials Find the quotient and remainder using long division.

15. $\dfrac{x^2 - 3x + 7}{x - 2}$

16. $\dfrac{x^3 + 2x^2 - x + 1}{x + 3}$

17. $\dfrac{4x^3 + 2x^2 - 2x - 3}{2x + 1}$

18. $\dfrac{x^3 + 3x^2 + 4x + 3}{3x + 6}$

19. $\dfrac{x^3 + 2x + 1}{x^2 - x + 3}$

20. $\dfrac{x^4 - 3x^3 + x - 2}{x^2 - 5x + 1}$

21. $\dfrac{6x^3 + 2x^2 + 22x}{2x^2 + 5}$

22. $\dfrac{9x^2 - x + 5}{3x^2 - 7x}$

23. $\dfrac{x^6 + x^4 + x^2 + 1}{x^2 + 1}$

24. $\dfrac{2x^5 - 7x^4 - 13}{4x^2 - 6x + 8}$

25–38 ■ Synthetic Division of Polynomials Find the quotient and remainder using synthetic division.

25. $\dfrac{2x^2 - 5x + 3}{x - 3}$

26. $\dfrac{-x^2 + x - 4}{x + 1}$

27. $\dfrac{3x^2 + x}{x + 1}$

28. $\dfrac{4x^2 - 3}{x - 2}$

29. $\dfrac{x^3 + 2x^2 + 2x + 1}{x + 2}$

30. $\dfrac{3x^3 - 12x^2 - 9x + 1}{x - 5}$

31. $\dfrac{x^3 - 8x + 2}{x + 3}$

32. $\dfrac{x^4 - x^3 + x^2 - x + 2}{x - 2}$

33. $\dfrac{x^5 + 3x^3 - 6}{x - 1}$

34. $\dfrac{x^3 - 9x^2 + 27x - 27}{x - 3}$

35. $\dfrac{2x^3 + 3x^2 - 2x + 1}{x - \frac{1}{2}}$

36. $\dfrac{6x^4 + 10x^3 + 5x^2 + x + 1}{x + \frac{2}{3}}$

37. $\dfrac{x^3 - 27}{x - 3}$

38. $\dfrac{x^4 - 16}{x + 2}$

39–51 ■ Remainder Theorem Use synthetic division and the Remainder Theorem to evaluate $P(c)$.

39. $P(x) = 4x^2 + 12x + 5, \quad c = -1$

40. $P(x) = 2x^2 + 9x + 1, \quad c = \frac{1}{2}$

41. $P(x) = x^3 + 3x^2 - 7x + 6, \quad c = 2$

42. $P(x) = x^3 - x^2 + x + 5, \quad c = -1$

43. $P(x) = x^3 + 2x^2 - 7, \quad c = -2$

44. $P(x) = 2x^3 - 21x^2 + 9x - 200, \quad c = 11$

45. $P(x) = 5x^4 + 30x^3 - 40x^2 + 36x + 14, \quad c = -7$

46. $P(x) = 6x^5 + 10x^3 + x + 1, \quad c = -2$

47. $P(x) = x^7 - 3x^2 - 1, \quad c = 3$

48. $P(x) = -2x^6 + 7x^5 + 40x^4 - 7x^2 + 10x + 112, \quad c = -3$

49. $P(x) = 3x^3 + 4x^2 - 2x + 1, \quad c = \frac{2}{3}$

50. $P(x) = x^3 - x + 1, \quad c = \frac{1}{4}$

51. $P(x) = x^3 + 2x^2 - 3x - 8, \quad c = 0.1$

52. Remainder Theorem Let

$$P(x) = 6x^7 - 40x^6 + 16x^5 - 200x^4 - 60x^3 - 69x^2 + 13x - 139$$

Calculate $P(7)$ by **(a)** using synthetic division and **(b)** substituting $x = 7$ into the polynomial and evaluating directly.

53–56 ■ Factor Theorem Use the Factor Theorem to show that $x - c$ is a factor of $P(x)$ for the given value(s) of c.

53. $P(x) = x^3 - 3x^2 + 3x - 1, \quad c = 1$

54. $P(x) = x^3 + 2x^2 - 3x - 10, \quad c = 2$

55. $P(x) = 2x^3 + 7x^2 + 6x - 5, \quad c = \frac{1}{2}$

56. $P(x) = x^4 + 3x^3 - 16x^2 - 27x + 63, \quad c = 3, -3$

57–62 ■ Factor Theorem Show that the given value(s) of c are zeros of $P(x)$, and find all other zeros of $P(x)$.

57. $P(x) = x^3 + 2x^2 - 9x - 18, \quad c = -2$

58. $P(x) = x^3 - 5x^2 - 2x + 10, \quad c = 5$

59. $P(x) = x^3 - x^2 - 11x + 15, \quad c = 3$

60. $P(x) = 3x^4 - x^3 - 21x^2 - 11x + 6, \quad c = -2, \frac{1}{3}$

61. $P(x) = 3x^4 - 8x^3 - 14x^2 + 31x + 6, \quad c = -2, 3$

62. $P(x) = 2x^4 - 13x^3 + 7x^2 + 37x + 15, \quad c = -1, 3$

63–66 ■ Finding a Polynomial with Specified Zeros Find a polynomial of the specified degree that has the given zeros.

63. Degree 3; zeros $-1, 1, 3$

64. Degree 4; zeros $-2, 0, 2, 4$

65. Degree 4; zeros $-1, 1, 3, 5$

66. Degree 5; zeros $-2, -1, 0, 1, 2$

67–70 ■ Polynomials with Specified Zeros Find a polynomial of the specified degree that satisfies the given conditions.

67. Degree 4; zeros $-2, 0, 1, 3$; coefficient of x^3 is 4

68. Degree 4; zeros $-1, 0, 2, \frac{1}{2}$; coefficient of x^3 is 3

69. Degree 4; zeros $-1, 1, \sqrt{2}$; integer coefficients and constant term 6

70. Degree 5; zeros $-2, -1, 2, \sqrt{5}$; integer coefficients and constant term 40

SKILLS Plus

71–74 ■ Finding a Polynomial from a Graph Find the polynomial of the specified degree whose graph is shown.

71. Degree 3

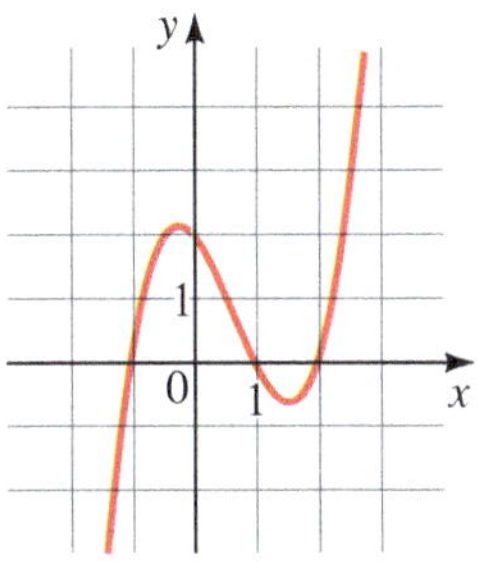

72. Degree 3

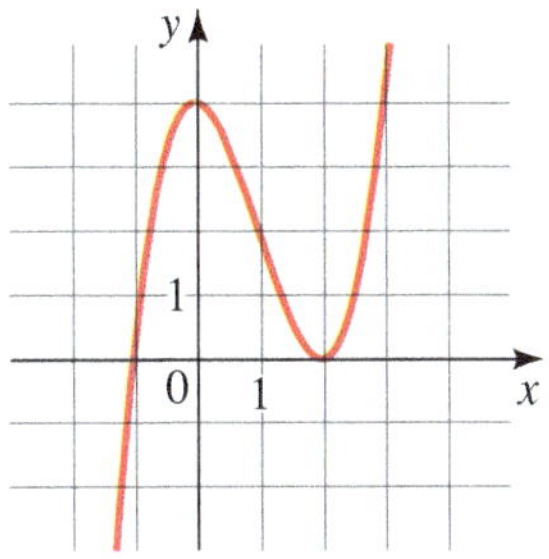

73. Degree 4

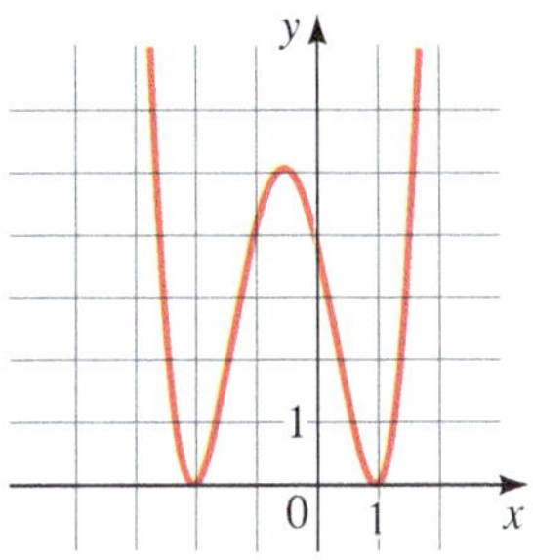

74. Degree 4

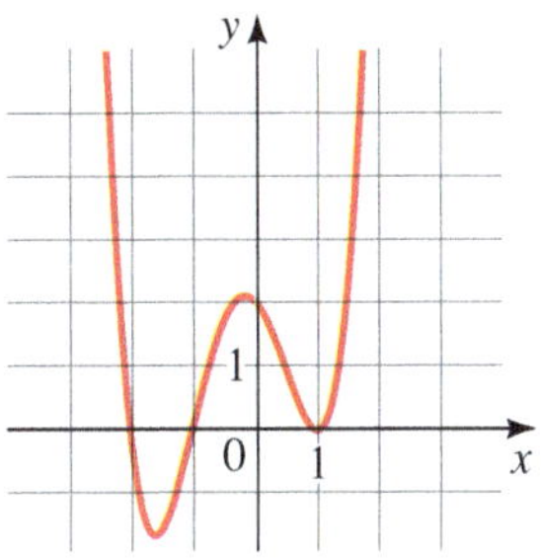

DISCUSS ■ DISCOVER ■ PROVE ■ WRITE

75. DISCUSS: Impossible Division? Suppose you were asked to solve the following two problems on a test:

A. Find the remainder when $6x^{1000} - 17x^{562} + 12x + 26$ is divided by $x + 1$.

B. Is $x - 1$ a factor of $x^{567} - 3x^{400} + x^9 + 2$?

Obviously, it's impossible to solve these problems by dividing, because the polynomials are of such large degree. Use one or more of the theorems in this section to solve these problems *without* actually dividing.

76. DISCOVER: Nested Form of a Polynomial Expand Q to prove that the polynomials P and Q are the same.

$$P(x) = 3x^4 - 5x^3 + x^2 - 3x + 5$$

$$Q(x) = (((3x - 5)x + 1)x - 3)x + 5$$

Try to evaluate $P(2)$ and $Q(2)$ in your head, using the forms given. Which is easier? Now write the polynomial $R(x) = x^5 - 2x^4 + 3x^3 - 2x^2 + 3x + 4$ in "nested" form, like the polynomial Q. Use the nested form to find $R(3)$ in your head.

Do you see how calculating with the nested form follows the same arithmetic steps as calculating the value of a polynomial using synthetic division?

3.4 REAL ZEROS OF POLYNOMIALS

■ Rational Zeros of Polynomials ■ Descartes' Rule of Signs ■ Upper and Lower Bounds Theorem ■ Using Algebra and Graphing Devices to Solve Polynomial Equations

The Factor Theorem tells us that finding the zeros of a polynomial is really the same thing as factoring it into linear factors. In this section we study some algebraic methods that help us to find the real zeros of a polynomial and thereby factor the polynomial. We begin with the *rational* zeros of a polynomial.

■ Rational Zeros of Polynomials

To help us understand the next theorem, let's consider the polynomial

$$P(x) = (x - 2)(x - 3)(x + 4) \qquad \text{Factored form}$$

$$= x^3 - x^2 - 14x + 24 \qquad \text{Expanded form}$$

From the factored form we see that the zeros of P are 2, 3, and -4. When the polynomial is expanded, the constant 24 is obtained by multiplying $(-2) \times (-3) \times 4$. This means that the zeros of the polynomial are all factors of the constant term. The following generalizes this observation.

RATIONAL ZEROS THEOREM

If the polynomial $P(x) = a_n x^n + a_{n-1}x^{n-1} + \cdots + a_1 x + a_0$ has integer coefficients (where $a_n \neq 0$ and $a_0 \neq 0$), then every rational zero of P is of the form

$$\frac{p}{q}$$

where p and q are integers and

p is a factor of the constant coefficient a_0

q is a factor of the leading coefficient a_n

Proof If p/q is a rational zero, in lowest terms, of the polynomial P, then we have

$$a_n\left(\frac{p}{q}\right)^n + a_{n-1}\left(\frac{p}{q}\right)^{n-1} + \cdots + a_1\left(\frac{p}{q}\right) + a_0 = 0$$

$$a_n p^n + a_{n-1}p^{n-1}q + \cdots + a_1 p q^{n-1} + a_0 q^n = 0 \qquad \text{Multiply by } q^n$$

$$p(a_n p^{n-1} + a_{n-1}p^{n-2}q + \cdots + a_1 q^{n-1}) = -a_0 q^n \qquad \text{Subtract } a_0q^n \text{ and factor LHS}$$

Now p is a factor of the left side, so it must be a factor of the right side as well. Since p/q is in lowest terms, p and q have no factor in common, so p must be a factor of a_0. A similar proof shows that q is a factor of a_n. ■

We see from the Rational Zeros Theorem that if the leading coefficient is 1 or -1, then the rational zeros must be factors of the constant term.

EXAMPLE 1 ■ Using the Rational Zeros Theorem

Find the rational zeros of $P(x) = x^3 - 3x + 2$.

SOLUTION Since the leading coefficient is 1, any rational zero must be a divisor of the constant term 2. So the possible rational zeros are ± 1 and ± 2. We test each of these possibilities.

$$P(1) = (1)^3 - 3(1) + 2 = 0$$
$$P(-1) = (-1)^3 - 3(-1) + 2 = 4$$
$$P(2) = (2)^3 - 3(2) + 2 = 4$$
$$P(-2) = (-2)^3 - 3(-2) + 2 = 0$$

The rational zeros of P are 1 and -2.

Now Try Exercise 15 ■

DISCOVERY PROJECT

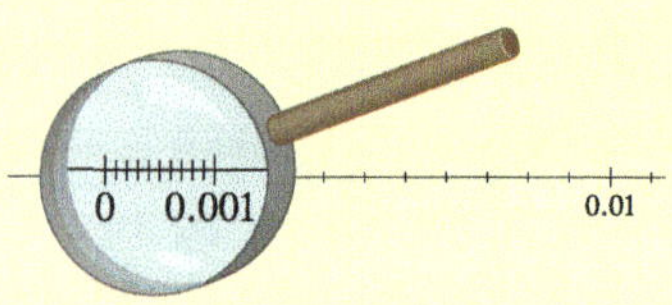

Zeroing in on a Zero

We have learned how to find the zeros of a polynomial function algebraically and graphically. In this project we investigate a *numerical* method for finding the zeros of a polynomial. With this method we can approximate the zeros of a polynomial to as many decimal places as we wish. The method involves finding smaller and smaller intervals that zoom in on a zero of a polynomial. You can find the project at **www.stewartmath.com**.

Library of Congress Prints and Photographs Division

EVARISTE GALOIS (1811–1832) is one of the very few mathematicians to have an entire theory named in his honor. Not yet 21 when he died, he completely settled the central problem in the theory of equations by describing a criterion that reveals whether a polynomial equation can be solved by algebraic operations. Galois was one of the greatest mathematicians in the world at that time, although no one knew it but him. He repeatedly sent his work to the eminent mathematicians Cauchy and Poisson, who either lost his letters or did not understand his ideas. Galois wrote in a terse style and included few details, which probably played a role in his failure to pass the entrance exams at the Ecole Polytechnique in Paris. A political radical, Galois spent several months in prison for his revolutionary activities. His brief life came to a tragic end when he was killed in a duel over a love affair. The night before his duel, fearing that he would die, Galois wrote down the essence of his ideas and entrusted them to his friend Auguste Chevalier. He concluded by writing "there will, I hope, be people who will find it to their advantage to decipher all this mess." The mathematician Camille Jordan did just that, 14 years later.

The following box explains how we use the Rational Zeros Theorem with synthetic division to factor a polynomial.

FINDING THE RATIONAL ZEROS OF A POLYNOMIAL

1. **List Possible Zeros.** List all possible rational zeros, using the Rational Zeros Theorem.
2. **Divide.** Use synthetic division to evaluate the polynomial at each of the candidates for the rational zeros that you found in Step 1. When the remainder is 0, note the quotient you have obtained.
3. **Repeat.** Repeat Steps 1 and 2 for the quotient. Stop when you reach a quotient that is quadratic or factors easily, and use the quadratic formula or factor to find the remaining zeros.

EXAMPLE 2 ■ Finding Rational Zeros

Write the polynomial $P(x) = 2x^3 + x^2 - 13x + 6$ in factored form, and find all its zeros.

SOLUTION By the Rational Zeros Theorem the rational zeros of P are of the form

$$\text{possible rational zero of } P = \frac{\text{factor of constant term}}{\text{factor of leading coefficient}}$$

The constant term is 6 and the leading coefficient is 2, so

$$\text{possible rational zero of } P = \frac{\text{factor of 6}}{\text{factor of 2}}$$

The factors of 6 are $\pm 1, \pm 2, \pm 3, \pm 6$, and the factors of 2 are $\pm 1, \pm 2$. Thus the possible rational zeros of P are

$$\pm\frac{1}{1}, \quad \pm\frac{2}{1}, \quad \pm\frac{3}{1}, \quad \pm\frac{6}{1}, \quad \pm\frac{1}{2}, \quad \pm\frac{2}{2}, \quad \pm\frac{3}{2}, \quad \pm\frac{6}{2}$$

Simplifying the fractions and eliminating duplicates, we get the following list of possible rational zeros:

$$\pm 1, \quad \pm 2, \quad \pm 3, \quad \pm 6, \quad \pm\frac{1}{2}, \quad \pm\frac{3}{2}$$

To check which of these *possible* zeros actually *are* zeros, we need to evaluate P at each of these numbers. An efficient way to do this is to use synthetic division.

Test whether 1 is a zero

1	2	1	−13	6
		2	3	−10
	2	3	−10	−4

Remainder is *not* 0, so 1 is *not* a zero

Test whether 2 is a zero

2	2	1	−13	6
		4	10	−6
	2	5	−3	0

Remainder *is* 0, so 2 *is* a zero

From the last synthetic division we see that 2 is a zero of P and that P factors as

$$\begin{aligned} P(x) &= 2x^3 + x^2 - 13x + 6 && \text{Given polynomial} \\ &= (x-2)(2x^2 + 5x - 3) && \text{From synthetic division} \\ &= (x-2)(2x-1)(x+3) && \text{Factor } 2x^2 + 5x - 3 \end{aligned}$$

From the factored form we see that the zeros of P are 2, $\frac{1}{2}$, and –3.

Now Try Exercise 29 ■

EXAMPLE 3 ■ Using the Rational Zeros Theorem and the Quadratic Formula

Let $P(x) = x^4 - 5x^3 - 5x^2 + 23x + 10$.

(a) Find the zeros of P. **(b)** Sketch a graph of P.

SOLUTION

(a) The leading coefficient of P is 1, so all the rational zeros are integers: They are divisors of the constant term 10. Thus the possible candidates are

$$\pm 1, \quad \pm 2, \quad \pm 5, \quad \pm 10$$

1	1	−5	−5	23	10
		1	−4	−9	14
	1	−4	−9	14	24

2	1	−5	−5	23	10
		2	−6	−22	2
	1	−3	−11	1	12

5	1	−5	−5	23	10
		5	0	−25	−10
	1	0	−5	−2	0

Using synthetic division (see the margin), we find that 1 and 2 are not zeros but that 5 is a zero and that P factors as

$$x^4 - 5x^3 - 5x^2 + 23x + 10 = (x-5)(x^3 - 5x - 2)$$

We now try to factor the quotient $x^3 - 5x - 2$. Its possible zeros are the divisors of -2, namely,

$$\pm 1, \quad \pm 2$$

Since we already know that 1 and 2 are not zeros of the original polynomial P, we don't need to try them again. Checking the remaining candidates, -1 and -2, we see that -2 is a zero (see the margin), and P factors as

−2	1	0	−5	−2
		−2	4	2
	1	−2	−1	0

$$\begin{aligned} x^4 - 5x^3 - 5x^2 + 23x + 10 &= (x-5)(x^3 - 5x - 2) \\ &= (x-5)(x+2)(x^2 - 2x - 1) \end{aligned}$$

Now we use the Quadratic Formula to obtain the two remaining zeros of P:

$$x = \frac{2 \pm \sqrt{(-2)^2 - 4(1)(-1)}}{2} = 1 \pm \sqrt{2}$$

The zeros of P are 5, -2, $1 + \sqrt{2}$, and $1 - \sqrt{2}$.

(b) Now that we know the zeros of P, we can use the methods of Section 3.2 to sketch the graph. If we want to use a graphing calculator instead, knowing the zeros allows us to choose an appropriate viewing rectangle—one that is wide enough to contain all the x-intercepts of P. Numerical approximations to the zeros of P are

$$5, \quad -2, \quad 2.4, \quad -0.4$$

So in this case we choose the rectangle $[-3, 6]$ by $[-50, 50]$ and draw the graph shown in Figure 1.

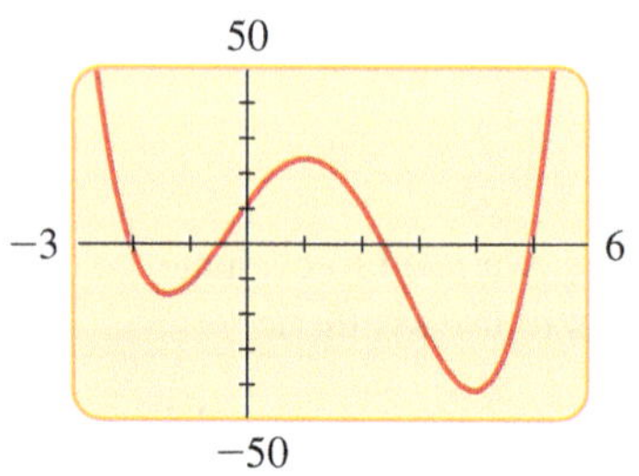

FIGURE 1
$P(x) = x^4 - 5x^3 - 5x^2 + 23x + 10$

Now Try Exercises 45 and 55 ■

■ Descartes' Rule of Signs

In some cases, the following rule—discovered by the French philosopher and mathematician René Descartes around 1637 (see page 201)—is helpful in eliminating candidates from lengthy lists of possible rational roots. To describe this rule, we need the concept

Polynomial	Variations in sign
$x^2 + 4x + 1$	0
$2x^3 + x - 6$	1
$x^4 - 3x^2 - x + 4$	2

of *variation in sign*. If $P(x)$ is a polynomial with real coefficients, written with descending powers of x (and omitting powers with coefficient 0), then a **variation in sign** occurs whenever adjacent coefficients have opposite signs. For example,

$$P(x) = 5x^7 - 3x^5 - x^4 + 2x^2 + x - 3$$

has three variations in sign.

DESCARTES' RULE OF SIGNS

Let P be a polynomial with real coefficients.

1. The number of positive real zeros of $P(x)$ either is equal to the number of variations in sign in $P(x)$ or is less than that by an even whole number.
2. The number of negative real zeros of $P(x)$ either is equal to the number of variations in sign in $P(-x)$ or is less than that by an even whole number.

Multiplicity is discussed on page 263.

In Descartes' Rule of Signs a zero with multiplicity m is counted m times. For example, the polynomial $P(x) = x^2 - 2x + 1$ has two sign changes and has the positive zero $x = 1$. But this zero is counted twice because it has multiplicity 2.

EXAMPLE 4 ■ Using Descartes' Rule

Use Descartes' Rule of Signs to determine the possible number of positive and negative real zeros of the polynomial

$$P(x) = 3x^6 + 4x^5 + 3x^3 - x - 3$$

SOLUTION The polynomial has one variation in sign, so it has one positive zero. Now

$$\begin{aligned} P(-x) &= 3(-x)^6 + 4(-x)^5 + 3(-x)^3 - (-x) - 3 \\ &= 3x^6 - 4x^5 - 3x^3 + x - 3 \end{aligned}$$

So $P(-x)$ has three variations in sign. Thus $P(x)$ has either three or one negative zero(s), making a total of either two or four real zeros.

Now Try Exercise 63 ■

Upper and Lower Bounds Theorem

We say that a is a **lower bound** and b is an **upper bound** for the zeros of a polynomial if every real zero c of the polynomial satisfies $a \le c \le b$. The next theorem helps us to find such bounds for the zeros of a polynomial.

THE UPPER AND LOWER BOUNDS THEOREM

Let P be a polynomial with real coefficients.

1. If we divide $P(x)$ by $x - b$ (with $b > 0$) using synthetic division and if the row that contains the quotient and remainder has no negative entry, then b is an upper bound for the real zeros of P.
2. If we divide $P(x)$ by $x - a$ (with $a < 0$) using synthetic division and if the row that contains the quotient and remainder has entries that are alternately nonpositive and nonnegative, then a is a lower bound for the real zeros of P.

A proof of this theorem is suggested in Exercise 109. The phrase "alternately nonpositive and nonnegative" simply means that the signs of the numbers alternate, with 0 considered to be positive or negative as required.

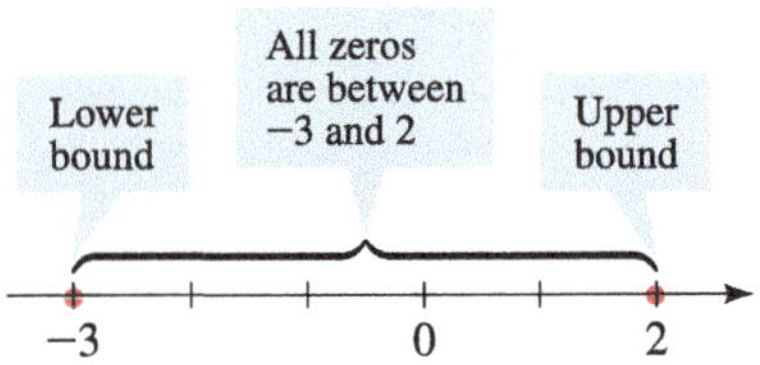

EXAMPLE 5 ■ Upper and Lower Bounds for the Zeros of a Polynomial

Show that all the real zeros of the polynomial $P(x) = x^4 - 3x^2 + 2x - 5$ lie between -3 and 2.

SOLUTION We divide $P(x)$ by $x - 2$ and $x + 3$ using synthetic division:

2	1	0	−3	2	−5
		2	4	2	8
	1	2	1	4	3

All entries nonnegative

−3	1	0	−3	2	−5
		−3	9	−18	48
	1	−3	6	−16	43

Entries alternate in sign

By the Upper and Lower Bounds Theorem -3 is a lower bound and 2 is an upper bound for the zeros. Since neither -3 nor 2 is a zero (the remainders are not 0 in the division table), all the real zeros lie between these numbers.

Now Try Exercise 69 ■

EXAMPLE 6 ■ A Lower Bound for the Zeros of a Polynomial

Show that all the real zeros of the polynomial $P(x) = x^4 + 4x^3 + 3x^2 + 7x - 5$ are greater than or equal to -4.

SOLUTION We divide $P(x)$ by $x + 4$ using synthetic division:

−4	1	4	3	7	−5
		−4	0	−12	20
	1	0	3	−5	15

Alternately nonnegative and nonpositive

Since 0 can be considered either nonnegative or nonpositive, the entries alternate in sign. So -4 is a lower bound for the real zeros of P.

Now Try Exercise 73 ■

EXAMPLE 7 ■ Factoring a Fifth-Degree Polynomial

Factor completely the polynomial

$$P(x) = 2x^5 + 5x^4 - 8x^3 - 14x^2 + 6x + 9$$

SOLUTION The possible rational zeros of P are $\pm\frac{1}{2}, \pm1, \pm\frac{3}{2}, \pm3, \pm\frac{9}{2}$, and ±9. We check the positive candidates first, beginning with the smallest:

$\frac{1}{2}$	2	5	−8	−14	6	9
		1	3	$-\frac{5}{2}$	$-\frac{33}{4}$	$-\frac{9}{8}$
	2	6	−5	$-\frac{33}{2}$	$-\frac{9}{4}$	$\frac{63}{8}$

$\frac{1}{2}$ is not a zero

1	2	5	−8	−14	6	9
		2	7	−1	−15	−9
	2	7	−1	−15	−9	0

$P(1) = 0$

So 1 is a zero, and $P(x) = (x - 1)(2x^4 + 7x^3 - x^2 - 15x - 9)$. We continue by factoring the quotient. We still have the same list of possible zeros except that $\frac{1}{2}$ has been eliminated.

$$\begin{array}{r|rrrrr} 1 & 2 & 7 & -1 & -15 & -9 \\ & & 2 & 9 & 8 & -7 \\ \hline & 2 & 9 & 8 & -7 & -16 \end{array}$$

1 is not a zero

$$\begin{array}{r|rrrrr} \frac{3}{2} & 2 & 7 & -1 & -15 & -9 \\ & & 3 & 15 & 21 & 9 \\ \hline & 2 & 10 & 14 & 6 & 0 \end{array}$$

$P(\frac{3}{2}) = 0$, all entries nonnegative

We see that $\frac{3}{2}$ is both a zero and an upper bound for the zeros of $P(x)$, so we do not need to check any further for positive zeros, because all the remaining candidates are greater than $\frac{3}{2}$.

$$\begin{aligned} P(x) &= (x - 1)(x - \tfrac{3}{2})(2x^3 + 10x^2 + 14x + 6) && \text{From synthetic division} \\ &= (x - 1)(2x - 3)(x^3 + 5x^2 + 7x + 3) && \text{Factor 2 from last factor, multiply into second factor} \end{aligned}$$

By Descartes' Rule of Signs, $x^3 + 5x^2 + 7x + 3$ has no positive zero, so its only possible rational zeros are -1 and -3:

$$\begin{array}{r|rrrr} -1 & 1 & 5 & 7 & 3 \\ & & -1 & -4 & -3 \\ \hline & 1 & 4 & 3 & 0 \end{array}$$

$P(-1) = 0$

Therefore,

$$\begin{aligned} P(x) &= (x - 1)(2x - 3)(x + 1)(x^2 + 4x + 3) && \text{From synthetic division} \\ &= (x - 1)(2x - 3)(x + 1)^2(x + 3) && \text{Factor quadratic} \end{aligned}$$

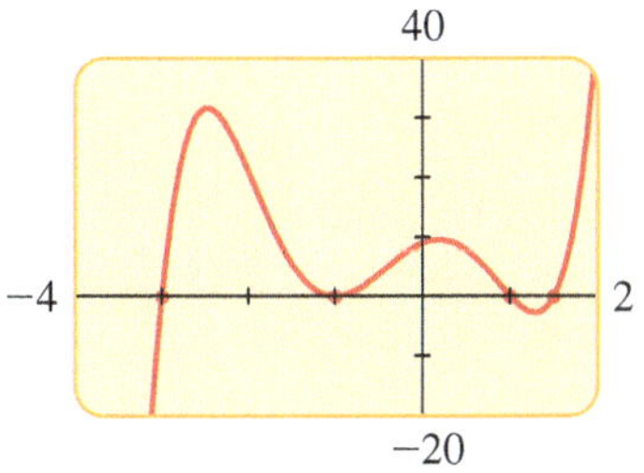

FIGURE 2
$P(x) = 2x^5 + 5x^4 - 8x^3 - 14x^2 + 6x + 9$
$= (x - 1)(2x - 3)(x + 1)^2(x + 3)$

This means that the zeros of P are 1, $\frac{3}{2}$, -1, and -3. The graph of the polynomial is shown in Figure 2.

Now Try Exercise 81 ■

■ Using Algebra and Graphing Devices to Solve Polynomial Equations

In Section 1.11 we used graphing devices to solve equations graphically. We can now use the algebraic techniques that we've learned to select an appropriate viewing rectangle when solving a polynomial equation graphically.

EXAMPLE 8 ■ Solving a Fourth-Degree Equation Graphically

Find all real solutions of the following equation, rounded to the nearest tenth:

$$3x^4 + 4x^3 - 7x^2 - 2x - 3 = 0$$

SOLUTION To solve the equation graphically, we graph

$$P(x) = 3x^4 + 4x^3 - 7x^2 - 2x - 3$$

We use the Upper and Lower Bounds Theorem to see where the solutions can be found.

First we use the Upper and Lower Bounds Theorem to find two numbers between which all the solutions must lie. This allows us to choose a viewing rectangle that is certain to contain all the x-intercepts of P. We use synthetic division and proceed by trial and error.

To find an upper bound, we try the whole numbers, 1, 2, 3, . . . , as potential candidates. We see that 2 is an upper bound for the solutions:

$$\begin{array}{r|rrrrr} 2 & 3 & 4 & -7 & -2 & -3 \\ & & 6 & 20 & 26 & 48 \\ \hline & 3 & 10 & 13 & 24 & 45 \end{array}$$

All positive

Now we look for a lower bound, trying the numbers -1, -2, and -3 as potential candidates. We see that -3 is a lower bound for the solutions:

$$\begin{array}{r|rrrrr} -3 & 3 & 4 & -7 & -2 & -3 \\ & & -9 & 15 & -24 & 78 \\ \hline & 3 & -5 & 8 & -26 & 75 \end{array}$$

Entries alternate in sign

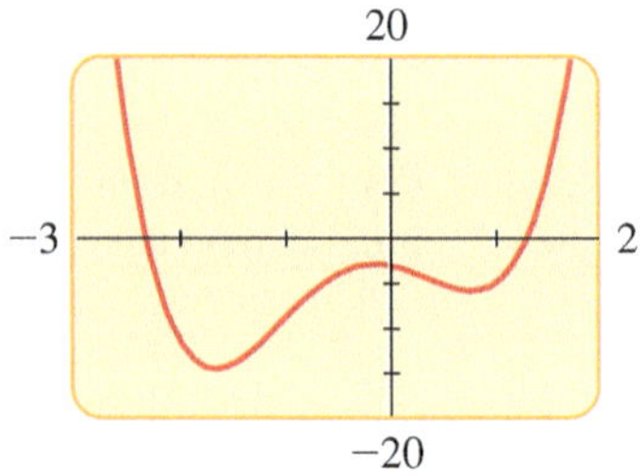

FIGURE 3

$y = 3x^4 + 4x^3 - 7x^2 - 2x - 3$

Thus all the solutions lie between -3 and 2. So the viewing rectangle $[-3, 2]$ by $[-20, 20]$ contains all the x-intercepts of P. The graph in Figure 3 has two x-intercepts, one between -3 and -2 and the other between 1 and 2. Zooming in, we find that the solutions of the equation, to the nearest tenth, are -2.3 and 1.3.

Now Try Exercise 95 ■

EXAMPLE 9 ■ Determining the Size of a Fuel Tank

A fuel tank consists of a cylindrical center section that is 4 ft long and two hemispherical end sections, as shown in Figure 4. If the tank has a volume of 100 ft^3, what is the radius r shown in the figure, rounded to the nearest hundredth of a foot?

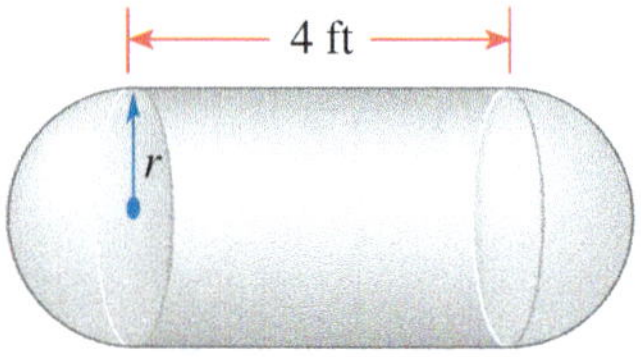

FIGURE 4

SOLUTION Using the volume formula listed on the inside front cover of this book, we see that the volume of the cylindrical section of the tank is

$$\pi \cdot r^2 \cdot 4$$

Volume of a cylinder: $V = \pi r^2 h$

The two hemispherical parts together form a complete sphere whose volume is

$$\tfrac{4}{3}\pi r^3$$

Volume of a sphere: $V = \frac{4}{3}\pi r^3$

Because the total volume of the tank is 100 ft^3, we get the following equation:

$$\tfrac{4}{3}\pi r^3 + 4\pi r^2 = 100$$

A negative solution for r would be meaningless in this physical situation, and by substitution we can verify that $r = 3$ leads to a tank that is over 226 ft^3 in volume, much larger than the required 100 ft^3. Thus we know the correct radius lies somewhere between 0 and 3 ft, so we use a viewing rectangle of $[0, 3]$ by $[50, 150]$ to graph the function $y = \frac{4}{3}\pi x^3 + 4\pi x^2$, as shown in Figure 5. Since we want the value of this function to be 100, we also graph the horizontal line $y = 100$ in the same viewing rectangle. The correct radius will be the x-coordinate of the point of intersection of the curve and the line. Using the cursor and zooming in, we see that at the point of intersection $x \approx 2.15$, rounded to two decimal places. Thus the tank has a radius of about 2.15 ft.

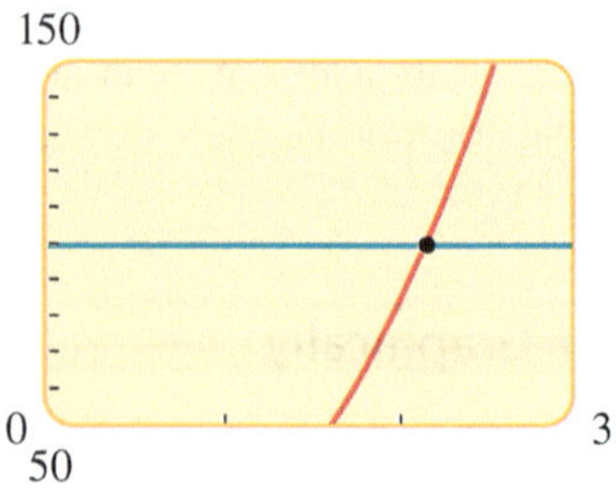

FIGURE 5

$y = \frac{4}{3}\pi x^3 + 4\pi x^2$ and $y = 100$

Now Try Exercise 99 ■

Note that we also could have solved the equation in Example 9 by first writing it as

$$\tfrac{4}{3}\pi r^3 + 4\pi r^2 - 100 = 0$$

and then finding the x-intercept of the function $y = \frac{4}{3}\pi x^3 + 4\pi x^2 - 100$.

3.4 EXERCISES

CONCEPTS

1. If the polynomial function

$$P(x) = a_n x^n + a_{n-1}x^{n-1} + \cdots + a_1 x + a_0$$

has integer coefficients, then the only numbers that could possibly be rational zeros of P are all of the form $\frac{p}{q}$, where p is a factor of ________ and q is a factor of ________. The possible rational zeros of $P(x) = 6x^3 + 5x^2 - 19x - 10$ are

____________________.

2. Using Descartes' Rule of Signs, we can tell that the polynomial $P(x) = x^5 - 3x^4 + 2x^3 - x^2 + 8x - 8$ has ________, ________, or ________ positive real zeros and ________ negative real zeros.

3. *True or False?* If c is a real zero of the polynomial P, then all the other zeros of P are zeros of $P(x)/(x - c)$.

4. *True or False?* If a is an upper bound for the real zeros of the polynomial P, then $-a$ is necessarily a lower bound for the real zeros of P.

SKILLS

5–10 ■ Possible Rational Zeros List all possible rational zeros given by the Rational Zeros Theorem (but don't check to see which actually are zeros).

5. $P(x) = x^3 - 4x^2 + 3$

6. $Q(x) = x^4 - 3x^3 - 6x + 8$

7. $R(x) = 2x^5 + 3x^3 + 4x^2 - 8$

8. $S(x) = 6x^4 - x^2 + 2x + 12$

9. $T(x) = 4x^4 - 2x^2 - 7$

10. $U(x) = 12x^5 + 6x^3 - 2x - 8$

11–14 ■ Possible Rational Zeros A polynomial function P and its graph are given. **(a)** List all possible rational zeros of P given by the Rational Zeros Theorem. **(b)** From the graph, determine which of the possible rational zeros actually turn out to be zeros.

11. $P(x) = 5x^3 - x^2 - 5x + 1$

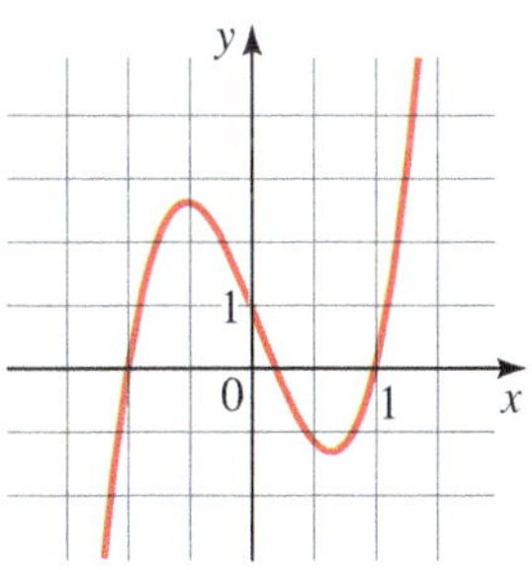

12. $P(x) = 3x^3 + 4x^2 - x - 2$

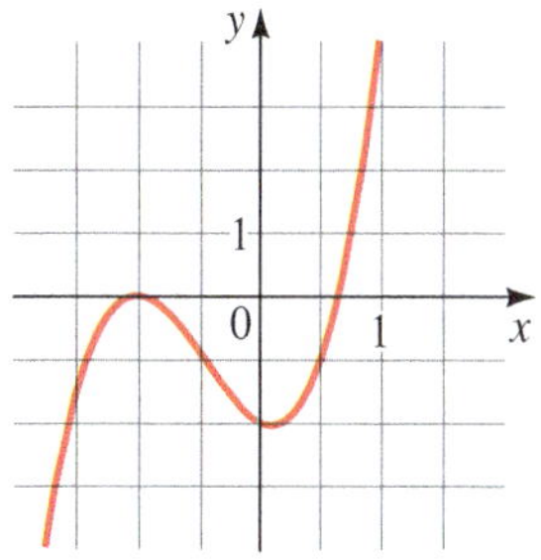

13. $P(x) = 2x^4 - 9x^3 + 9x^2 + x - 3$

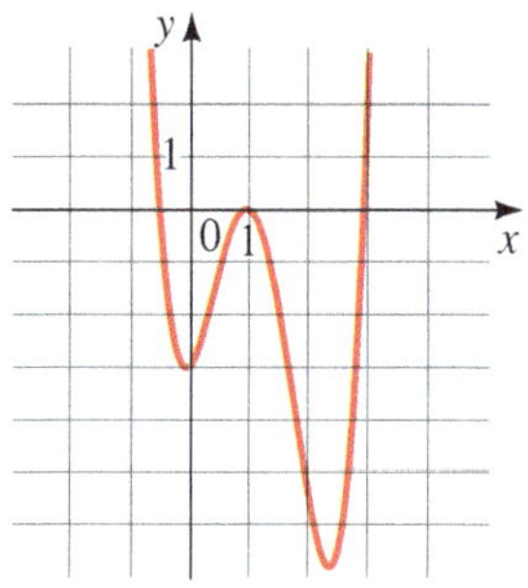

14. $P(x) = 4x^4 - x^3 - 4x + 1$

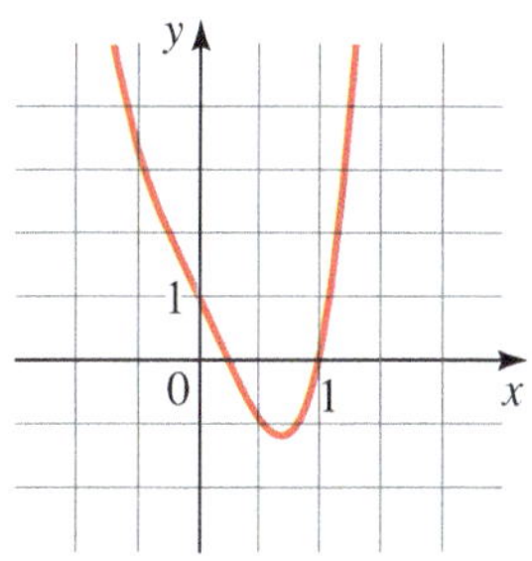

15–28 ■ Integer Zeros All the real zeros of the given polynomial are integers. Find the zeros, and write the polynomial in factored form.

15. $P(x) = x^3 + 2x^2 - 13x + 10$

16. $P(x) = x^3 - 4x^2 - 19x - 14$

17. $P(x) = x^3 + 3x^2 - 4$

18. $P(x) = x^3 - 3x - 2$

19. $P(x) = x^3 - 6x^2 + 12x - 8$

20. $P(x) = x^3 + 12x^2 + 48x + 64$

21. $P(x) = x^3 - 19x - 30$

22. $P(x) = x^3 + 11x^2 + 8x - 20$

23. $P(x) = x^3 + 3x^2 - x - 3$

24. $P(x) = x^3 - 4x^2 - 11x + 30$

25. $P(x) = x^4 - 5x^2 + 4$

26. $P(x) = x^4 - 2x^3 - 3x^2 + 8x - 4$

27. $P(x) = x^4 + 6x^3 + 7x^2 - 6x - 8$

28. $P(x) = x^4 - x^3 - 23x^2 - 3x + 90$

29–44 ■ Rational Zeros Find all rational zeros of the polynomial, and write the polynomial in factored form.

29. $P(x) = 4x^4 - 37x^2 + 9$

30. $P(x) = 6x^4 - 23x^3 - 13x^2 + 32x + 16$

31. $P(x) = 3x^4 - 10x^3 - 9x^2 + 40x - 12$

32. $P(x) = 2x^3 + 7x^2 + 4x - 4$

33. $P(x) = 4x^3 + 4x^2 - x - 1$

34. $P(x) = 2x^3 - 3x^2 - 2x + 3$

35. $P(x) = 4x^3 - 7x + 3$

36. $P(x) = 12x^3 - 25x^2 + x + 2$

37. $P(x) = 24x^3 + 10x^2 - 13x - 6$

38. $P(x) = 12x^3 - 20x^2 + x + 3$

39. $P(x) = 2x^4 - 7x^3 + 3x^2 + 8x - 4$

40. $P(x) = 6x^4 - 7x^3 - 12x^2 + 3x + 2$

41. $P(x) = x^5 + 3x^4 - 9x^3 - 31x^2 + 36$

42. $P(x) = x^5 - 4x^4 - 3x^3 + 22x^2 - 4x - 24$

43. $P(x) = 3x^5 - 14x^4 - 14x^3 + 36x^2 + 43x + 10$

44. $P(x) = 2x^6 - 3x^5 - 13x^4 + 29x^3 - 27x^2 + 32x - 12$

45–54 ■ Real Zeros of a Polynomial Find all the real zeros of the polynomial. Use the Quadratic Formula if necessary, as in Example 3(a).

45. $P(x) = 3x^3 + 5x^2 - 2x - 4$

46. $P(x) = 3x^4 - 5x^3 - 16x^2 + 7x + 15$

47. $P(x) = x^4 - 6x^3 + 4x^2 + 15x + 4$

48. $P(x) = x^4 + 2x^3 - 2x^2 - 3x + 2$

49. $P(x) = x^4 - 7x^3 + 14x^2 - 3x - 9$

50. $P(x) = x^5 - 4x^4 - x^3 + 10x^2 + 2x - 4$

51. $P(x) = 4x^3 - 6x^2 + 1$

52. $P(x) = 3x^3 - 5x^2 - 8x - 2$

53. $P(x) = 2x^4 + 15x^3 + 17x^2 + 3x - 1$

54. $P(x) = 4x^5 - 18x^4 - 6x^3 + 91x^2 - 60x + 9$

55–62 ■ Real Zeros of a Polynomial A polynomial P is given. **(a)** Find all the real zeros of P. **(b)** Sketch a graph of P.

55. $P(x) = x^3 - 3x^2 - 4x + 12$

56. $P(x) = -x^3 - 2x^2 + 5x + 6$

57. $P(x) = 2x^3 - 7x^2 + 4x + 4$

58. $P(x) = 3x^3 + 17x^2 + 21x - 9$

59. $P(x) = x^4 - 5x^3 + 6x^2 + 4x - 8$

60. $P(x) = -x^4 + 10x^2 + 8x - 8$

61. $P(x) = x^5 - x^4 - 5x^3 + x^2 + 8x + 4$

62. $P(x) = x^5 - x^4 - 6x^3 + 14x^2 - 11x + 3$

63–68 ■ Descartes' Rule of Signs Use Descartes' Rule of Signs to determine how many positive and how many negative real zeros the polynomial can have. Then determine the possible total number of real zeros.

63. $P(x) = x^3 - x^2 - x - 3$

64. $P(x) = 2x^3 - x^2 + 4x - 7$

65. $P(x) = 2x^6 + 5x^4 - x^3 - 5x - 1$

66. $P(x) = x^4 + x^3 + x^2 + x + 12$

67. $P(x) = x^5 + 4x^3 - x^2 + 6x$

68. $P(x) = x^8 - x^5 + x^4 - x^3 + x^2 - x + 1$

69–76 ■ Upper and Lower Bounds Show that the given values for a and b are lower and upper bounds for the real zeros of the polynomial.

69. $P(x) = 2x^3 + 5x^2 + x - 2;\quad a = -3, b = 1$

70. $P(x) = x^4 - 2x^3 - 9x^2 + 2x + 8;\quad a = -3, b = 5$

71. $P(x) = 8x^3 + 10x^2 - 39x + 9;\quad a = -3, b = 2$

72. $P(x) = 3x^4 - 17x^3 + 24x^2 - 9x + 1;\quad a = 0, b = 6$

73. $P(x) = x^4 + 2x^3 + 3x^2 + 5x - 1;\quad a = -2, b = 1$

74. $P(x) = x^4 + 3x^3 - 4x^2 - 2x - 7;\quad a = -4, b = 2$

75. $P(x) = 2x^4 - 6x^3 + x^2 - 2x + 3;\quad a = -1, b = 3$

76. $P(x) = 3x^4 - 5x^3 - 2x^2 + x - 1;\quad a = -1, b = 2$

77–80 ■ Upper and Lower Bounds Find integers that are upper and lower bounds for the real zeros of the polynomial.

77. $P(x) = x^3 - 3x^2 + 4$

78. $P(x) = 2x^3 - 3x^2 - 8x + 12$

79. $P(x) = x^4 - 2x^3 + x^2 - 9x + 2$

80. $P(x) = x^5 - x^4 + 1$

81–86 ■ Zeros of a Polynomial Find all rational zeros of the polynomial, and then find the irrational zeros, if any. Whenever appropriate, use the Rational Zeros Theorem, the Upper and Lower Bounds Theorem, Descartes' Rule of Signs, the Quadratic Formula, or other factoring techniques.

81. $P(x) = 2x^4 + 3x^3 - 4x^2 - 3x + 2$

82. $P(x) = 2x^4 + 15x^3 + 31x^2 + 20x + 4$

83. $P(x) = 4x^4 - 21x^2 + 5$

84. $P(x) = 6x^4 - 7x^3 - 8x^2 + 5x$

85. $P(x) = x^5 - 7x^4 + 9x^3 + 23x^2 - 50x + 24$

86. $P(x) = 8x^5 - 14x^4 - 22x^3 + 57x^2 - 35x + 6$

87–90 ■ Polynomials With No Rational Zeros Show that the polynomial does not have any rational zeros.

87. $P(x) = x^3 - x - 2$

88. $P(x) = 2x^4 - x^3 + x + 2$

89. $P(x) = 3x^3 - x^2 - 6x + 12$

90. $P(x) = x^{50} - 5x^{25} + x^2 - 1$

91–94 ■ Verifying Zeros Using a Graphing Device The real solutions of the given equation are rational. List all possible rational roots using the Rational Zeros Theorem, and then graph the polynomial in the given viewing rectangle to determine which values are actually solutions. (All solutions can be seen in the given viewing rectangle.)

91. $x^3 - 3x^2 - 4x + 12 = 0$; $[-4, 4]$ by $[-15, 15]$

92. $x^4 - 5x^2 + 4 = 0$; $[-4, 4]$ by $[-30, 30]$

93. $2x^4 - 5x^3 - 14x^2 + 5x + 12 = 0$; $[-2, 5]$ by $[-40, 40]$

94. $3x^3 + 8x^2 + 5x + 2 = 0$; $[-3, 3]$ by $[-10, 10]$

95–98 ■ Finding Zeros Using a Graphing Device Use a graphing device to find all real solutions of the equation, rounded to two decimal places.

95. $x^4 - x - 4 = 0$

96. $2x^3 - 8x^2 + 9x - 9 = 0$

97. $4.00x^4 + 4.00x^3 - 10.96x^2 - 5.88x + 9.09 = 0$

98. $x^5 + 2.00x^4 + 0.96x^3 + 5.00x^2 + 10.00x + 4.80 = 0$

APPLICATIONS

99. Volume of a Silo A grain silo consists of a cylindrical main section and a hemispherical roof. If the total volume of the silo (including the part inside the roof section) is 15,000 ft^3 and the cylindrical part is 30 ft tall, what is the radius of the silo, rounded to the nearest tenth of a foot?

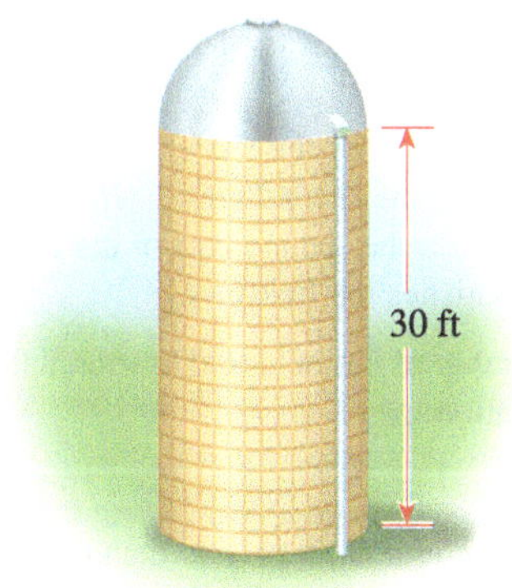

100. Dimensions of a Lot A rectangular parcel of land has an area of 5000 ft^2. A diagonal between opposite corners is measured to be 10 ft longer than one side of the parcel. What are the dimensions of the land, rounded to the nearest foot?

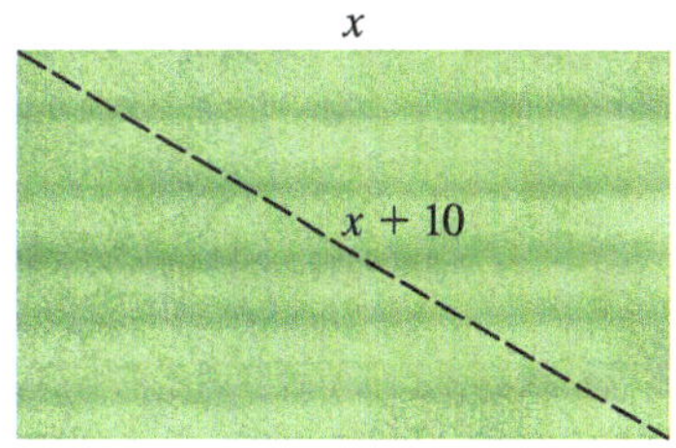

101. Depth of Snowfall Snow began falling at noon on Sunday. The amount of snow on the ground at a certain location at time t was given by the function

$$h(t) = 11.60t - 12.41t^2 + 6.20t^3 - 1.58t^4 + 0.20t^5 - 0.01t^6$$

where t is measured in days from the start of the snowfall and $h(t)$ is the depth of snow in inches. Draw a graph of this function, and use your graph to answer the following questions.

(a) What happened shortly after noon on Tuesday?

(b) Was there ever more than 5 in. of snow on the ground? If so, on what day(s)?

(c) On what day and at what time (to the nearest hour) did the snow disappear completely?

102. Volume of a Box An open box with a volume of 1500 cm^3 is to be constructed by taking a piece of cardboard 20 cm by 40 cm, cutting squares of side length x cm from each corner, and folding up the sides. Show that this can be done in two different ways, and find the exact dimensions of the box in each case.

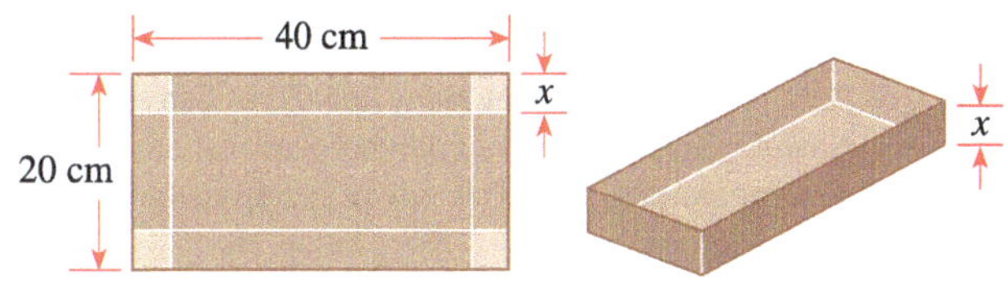

103. Volume of a Rocket A rocket consists of a right circular cylinder of height 20 m surmounted by a cone whose height and diameter are equal and whose radius is the same as that of the cylindrical section. What should this radius be (rounded to two decimal places) if the total volume is to be $500\pi/3$ m^3?

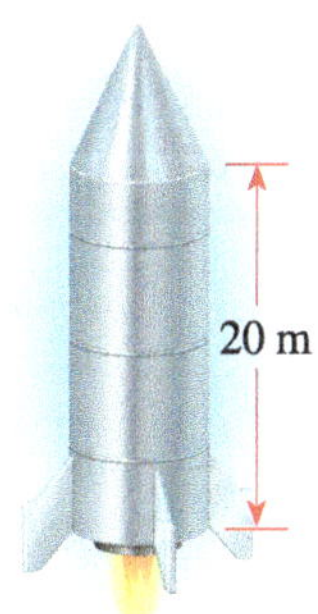

104. Volume of a Box A rectangular box with a volume of $2\sqrt{2}$ ft^3 has a square base as shown below. The diagonal of the box (between a pair of opposite corners) is 1 ft longer than each side of the base.

(a) If the base has sides of length x feet, show that

$$x^6 - 2x^5 - x^4 + 8 = 0$$

(b) Show that two different boxes satisfy the given conditions. Find the dimensions in each case, rounded to the nearest hundredth of a foot.

105. Girth of a Box A box with a square base has length plus girth of 108 in. (Girth is the distance "around" the box.) What is the length of the box if its volume is 2200 in^3?

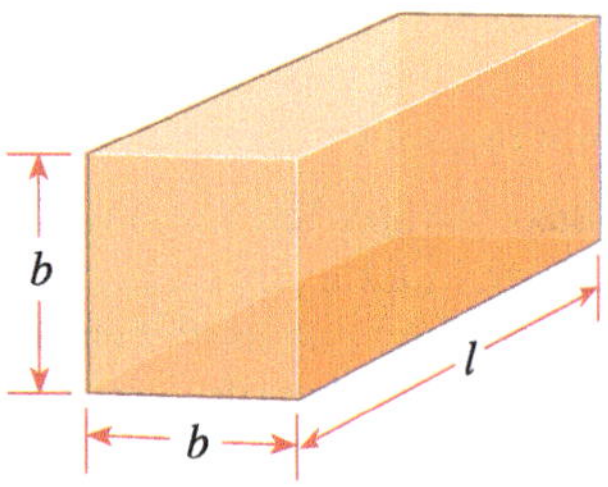

DISCUSS ■ DISCOVER ■ PROVE ■ WRITE

106. DISCUSS ■ DISCOVER: How Many Real Zeros Can a Polynomial Have? Give examples of polynomials that have the following properties, or explain why it is impossible to find such a polynomial.

(a) A polynomial of degree 3 that has no real zeros

(b) A polynomial of degree 4 that has no real zeros

(c) A polynomial of degree 3 that has three real zeros, only one of which is rational

(d) A polynomial of degree 4 that has four real zeros, none of which is rational

What must be true about the degree of a polynomial with integer coefficients if it has no real zeros?

107. DISCUSS ■ PROVE: The Depressed Cubic The most general cubic (third-degree) equation with rational coefficients can be written as

$$x^3 + ax^2 + bx + c = 0$$

(a) Prove that if we replace x by $X - a/3$ and simplify, we end up with an equation that doesn't have an X^2 term, that is, an equation of the form

$$X^3 + pX + q = 0$$

This is called a *depressed cubic*, because we have "depressed" the quadratic term.

(b) Use the procedure described in part (a) to depress the equation $x^3 + 6x^2 + 9x + 4 = 0$.

108. DISCUSS: The Cubic Formula The Quadratic Formula can be used to solve any quadratic (or second-degree) equation. You might have wondered whether similar formulas exist for cubic (third-degree), quartic (fourth-degree), and higher-degree equations. For the depressed cubic $x^3 + px + q = 0$, Cardano (page 292) found the following formula for one solution:

$$x = \sqrt[3]{\frac{-q}{2} + \sqrt{\frac{q^2}{4} + \frac{p^3}{27}}} + \sqrt[3]{\frac{-q}{2} - \sqrt{\frac{q^2}{4} + \frac{p^3}{27}}}$$

A formula for quartic equations was discovered by the Italian mathematician Ferrari in 1540. In 1824 the Norwegian mathematician Niels Henrik Abel proved that it is impossible to write a quintic formula, that is, a formula for fifth-degree equations. Finally, Galois (page 277) gave a criterion for determining which equations can be solved by a formula involving radicals.

Use the formula given above to find a solution for the following equations. Then solve the equations using the methods you learned in this section. Which method is easier?

(a) $x^3 - 3x + 2 = 0$

(b) $x^3 - 27x - 54 = 0$

(c) $x^3 + 3x + 4 = 0$

109. PROVE: Upper and Lower Bounds Theorem Let $P(x)$ be a polynomial with real coefficients, and let $b > 0$. Use the Division Algorithm to write

$$P(x) = (x - b) \cdot Q(x) + r$$

Suppose that $r \geq 0$ and that all the coefficients in $Q(x)$ are nonnegative. Let $z > b$.

(a) Show that $P(z) > 0$.

(b) Prove the first part of the Upper and Lower Bounds Theorem.

(c) Use the first part of the Upper and Lower Bounds Theorem to prove the second part. [*Hint:* Show that if $P(x)$ satisfies the second part of the theorem, then $P(-x)$ satisfies the first part.]

110. PROVE: Number of Rational and Irrational Roots Show that the equation

$$x^5 - x^4 - x^3 - 5x^2 - 12x - 6 = 0$$

has exactly one rational root, and then prove that it must have either two or four irrational roots.

3.5 COMPLEX ZEROS AND THE FUNDAMENTAL THEOREM OF ALGEBRA

■ The Fundamental Theorem of Algebra and Complete Factorization ■ Zeros and Their Multiplicities ■ Complex Zeros Come in Conjugate Pairs ■ Linear and Quadratic Factors

We have already seen that an nth-degree polynomial can have at most n real zeros. In the complex number system an nth-degree polynomial has exactly n zeros (counting multiplicity) and so can be factored into exactly n linear factors. This fact is a consequence of the Fundamental Theorem of Algebra, which was proved by the German mathematician C. F. Gauss in 1799 (see page 290).

■ The Fundamental Theorem of Algebra and Complete Factorization

The following theorem is the basis for much of our work in factoring polynomials and solving polynomial equations.

FUNDAMENTAL THEOREM OF ALGEBRA

Every polynomial

$$P(x) = a_n x^n + a_{n-1}x^{n-1} + \cdots + a_1 x + a_0 \qquad (n \geq 1, a_n \neq 0)$$

with complex coefficients has at least one complex zero.

Complex numbers are discussed in Section 1.6.

Because any real number is also a complex number, the theorem applies to polynomials with real coefficients as well.

The Fundamental Theorem of Algebra and the Factor Theorem together show that a polynomial can be factored completely into linear factors, as we now prove.

COMPLETE FACTORIZATION THEOREM

If $P(x)$ is a polynomial of degree $n \geq 1$, then there exist complex numbers $a, c_1, c_2, \ldots, c_n$ (with $a \neq 0$) such that

$$P(x) = a(x - c_1)(x - c_2)\cdots(x - c_n)$$

Proof By the Fundamental Theorem of Algebra, P has at least one zero. Let's call it c_1. By the Factor Theorem (see page 272), $P(x)$ can be factored as

$$P(x) = (x - c_1)Q_1(x)$$

where $Q_1(x)$ is of degree $n - 1$. Applying the Fundamental Theorem to the quotient $Q_1(x)$ gives us the factorization

$$P(x) = (x - c_1)(x - c_2)Q_2(x)$$

where $Q_2(x)$ is of degree $n - 2$ and c_2 is a zero of $Q_1(x)$. Continuing this process for n steps, we get a final quotient $Q_n(x)$ of degree 0, a nonzero constant that we will call a. This means that P has been factored as

$$P(x) = a(x - c_1)(x - c_2)\cdots(x - c_n)$$

■

To actually find the complex zeros of an nth-degree polynomial, we usually first factor as much as possible, then use the Quadratic Formula on parts that we can't factor further.

EXAMPLE 1 ■ Factoring a Polynomial Completely

Let $P(x) = x^3 - 3x^2 + x - 3$.

(a) Find all the zeros of P.

(b) Find the complete factorization of P.

SOLUTION

(a) We first factor P as follows.

$$\begin{aligned} P(x) &= x^3 - 3x^2 + x - 3 && \text{Given} \\ &= x^2(x - 3) + (x - 3) && \text{Group terms} \\ &= (x - 3)(x^2 + 1) && \text{Factor } x - 3 \end{aligned}$$

We find the zeros of P by setting each factor equal to 0:

$$P(x) = (x - 3)(x^2 + 1)$$

This factor is 0 when $x = 3$ | This factor is 0 when $x = i$ or $-i$

Setting $x - 3 = 0$, we see that $x = 3$ is a zero. Setting $x^2 + 1 = 0$, we get $x^2 = -1$, so $x = \pm i$. So the zeros of P are 3, i, and $-i$.

(b) Since the zeros are 3, i, and $-i$, the complete factorization of P is

$$\begin{aligned} P(x) &= (x - 3)(x - i)[x - (-i)] \\ &= (x - 3)(x - i)(x + i) \end{aligned}$$

Now Try Exercise 7 ■

EXAMPLE 2 ■ Factoring a Polynomial Completely

Let $P(x) = x^3 - 2x + 4$.

(a) Find all the zeros of P.

(b) Find the complete factorization of P.

SOLUTION

$$\begin{array}{r|rrrr} -2 & 1 & 0 & -2 & 4 \\ & & -2 & 4 & -4 \\ \hline & 1 & -2 & -2 & 0 \end{array}$$

(a) The possible rational zeros are the factors of 4, which are ± 1, ± 2, ± 4. Using synthetic division (see the margin), we find that -2 is a zero, and the polynomial factors as

$$P(x) = (x + 2)(x^2 - 2x + 2)$$

This factor is 0 when $x = -2$ | Use the Quadratic Formula to find when this factor is 0

To find the zeros, we set each factor equal to 0. Of course, $x + 2 = 0$ means that $x = -2$. We use the Quadratic Formula to find when the other factor is 0.

$$\begin{aligned} x^2 - 2x + 2 &= 0 && \text{Set factor equal to 0} \\ x &= \frac{2 \pm \sqrt{4 - 8}}{2} && \text{Quadratic Formula} \\ x &= \frac{2 \pm 2i}{2} && \text{Take square root} \\ x &= 1 \pm i && \text{Simplify} \end{aligned}$$

So the zeros of P are -2, $1 + i$, and $1 - i$.

(b) Since the zeros are -2, $1 + i$, and $1 - i$, the complete factorization of P is

$$P(x) = [x - (-2)][x - (1 + i)][x - (1 - i)]$$
$$= (x + 2)(x - 1 - i)(x - 1 + i)$$

Now Try Exercise 9 ■

■ Zeros and Their Multiplicities

In the Complete Factorization Theorem the numbers $c_1, c_2, \ldots, c_n$ are the zeros of P. These zeros need not all be different. If the factor $x - c$ appears k times in the complete factorization of $P(x)$, then we say that c is a zero of **multiplicity** k (see page 263). For example, the polynomial

$$P(x) = (x - 1)^3(x + 2)^2(x + 3)^5$$

has the following zeros:

1 (multiplicity 3) -2 (multiplicity 2) -3 (multiplicity 5)

The polynomial P has the same number of zeros as its degree: It has degree 10 and has 10 zeros, provided that we count multiplicities. This is true for all polynomials, as we prove in the following theorem.

ZEROS THEOREM

Every polynomial of degree $n \geq 1$ has exactly n zeros, provided that a zero of multiplicity k is counted k times.

Proof Let P be a polynomial of degree n. By the Complete Factorization Theorem

$$P(x) = a(x - c_1)(x - c_2) \cdots (x - c_n)$$

Now suppose that c is any given zero of P. Then

$$P(c) = a(c - c_1)(c - c_2) \cdots (c - c_n) = 0$$

Thus by the Zero-Product Property, one of the factors $c - c_i$ must be 0, so $c = c_i$ for some i. It follows that P has exactly the n zeros $c_1, c_2, \ldots, c_n$. ■

EXAMPLE 3 ■ Factoring a Polynomial with Complex Zeros

Find the complete factorization and all five zeros of the polynomial

$$P(x) = 3x^5 + 24x^3 + 48x$$

SOLUTION Since $3x$ is a common factor, we have

$$P(x) = 3x(x^4 + 8x^2 + 16)$$
$$= 3x(x^2 + 4)^2$$

This factor is 0 when $x = 0$

This factor is 0 when $x = 2i$ or $x = -2i$

CARL FRIEDRICH GAUSS (1777–1855) is considered the greatest mathematician of modern times. His contemporaries called him the "Prince of Mathematics." He was born into a poor family; his father made a living as a mason. As a very small child, Gauss found a calculation error in his father's accounts, the first of many incidents that gave evidence of his mathematical precocity. (See also page 854.) At 19, Gauss demonstrated that the regular 17-sided polygon can be constructed with straight-edge and compass alone. This was remarkable because, since the time of Euclid, it had been thought that the only regular polygons constructible in this way were the triangle and pentagon. Because of this discovery Gauss decided to pursue a career in mathematics instead of languages, his other passion. In his doctoral dissertation, written at the age of 22, Gauss proved the Fundamental Theorem of Algebra: A polynomial of degree n with complex coefficients has n roots. His other accomplishments range over every branch of mathematics, as well as physics and astronomy.

To factor $x^2 + 4$, note that $2i$ and $-2i$ are zeros of this polynomial. Thus $x^2 + 4 = (x - 2i)(x + 2i)$, so

$$P(x) = 3x[(x - 2i)(x + 2i)]^2$$
$$= 3x(x - 2i)^2(x + 2i)^2$$

0 is a zero of multiplicity 1 | $2i$ is a zero of multiplicity 2 | $-2i$ is a zero of multiplicity 2

The zeros of P are 0, $2i$, and $-2i$. Since the factors $x - 2i$ and $x + 2i$ each occur twice in the complete factorization of P, the zeros $2i$ and $-2i$ are of multiplicity 2 (or *double* zeros). Thus we have found all five zeros.

Now Try Exercise 31 ■

The following table gives further examples of polynomials with their complete factorizations and zeros.

Degree	Polynomial	Zero(s)	Number of zeros
1	$P(x) = x - 4$	4	1
2	$P(x) = x^2 - 10x + 25$ $= (x - 5)(x - 5)$	5 (multiplicity 2)	2
3	$P(x) = x^3 + x$ $= x(x - i)(x + i)$	$0, i, -i$	3
4	$P(x) = x^4 + 18x^2 + 81$ $= (x - 3i)^2(x + 3i)^2$	$3i$ (multiplicity 2), $-3i$ (multiplicity 2)	4
5	$P(x) = x^5 - 2x^4 + x^3$ $= x^3(x - 1)^2$	0 (multiplicity 3), 1 (multiplicity 2)	5

EXAMPLE 4 ■ Finding Polynomials with Specified Zeros

(a) Find a polynomial $P(x)$ of degree 4, with zeros i, $-i$, 2, and -2, and with $P(3) = 25$.

(b) Find a polynomial $Q(x)$ of degree 4, with zeros -2 and 0, where -2 is a zero of multiplicity 3.

SOLUTION

(a) The required polynomial has the form

$$P(x) = a(x - i)(x - (-i))(x - 2)(x - (-2))$$
$$= a(x^2 + 1)(x^2 - 4) \quad \text{Difference of squares}$$
$$= a(x^4 - 3x^2 - 4) \quad \text{Multiply}$$

We know that $P(3) = a(3^4 - 3 \cdot 3^2 - 4) = 50a = 25$, so $a = \frac{1}{2}$. Thus

$$P(x) = \tfrac{1}{2}x^4 - \tfrac{3}{2}x^2 - 2$$

(b) We require

$$Q(x) = a[x - (-2)]^3(x - 0)$$
$$= a(x + 2)^3x$$
$$= a(x^3 + 6x^2 + 12x + 8)x \quad \text{Special Product Formula 4 (Section 1.3)}$$
$$= a(x^4 + 6x^3 + 12x^2 + 8x)$$

Since we are given no information about Q other than its zeros and their multiplicity, we can choose any number for a. If we use $a = 1$, we get

$$Q(x) = x^4 + 6x^3 + 12x^2 + 8x$$

Now Try Exercise 37 ■

EXAMPLE 5 ■ Finding All the Zeros of a Polynomial

Find all four zeros of $P(x) = 3x^4 - 2x^3 - x^2 - 12x - 4$.

SOLUTION Using the Rational Zeros Theorem from Section 3.4, we obtain the following list of possible rational zeros: $\pm 1, \pm 2, \pm 4, \pm \frac{1}{3}, \pm \frac{2}{3}, \pm \frac{4}{3}$. Checking these using synthetic division, we find that 2 and $-\frac{1}{3}$ are zeros, and we get the following factorization.

$$\begin{aligned} P(x) &= 3x^4 - 2x^3 - x^2 - 12x - 4 \\ &= (x - 2)(3x^3 + 4x^2 + 7x + 2) && \text{Factor } x - 2 \\ &= (x - 2)(x + \tfrac{1}{3})(3x^2 + 3x + 6) && \text{Factor } x + \tfrac{1}{3} \\ &= 3(x - 2)(x + \tfrac{1}{3})(x^2 + x + 2) && \text{Factor 3} \end{aligned}$$

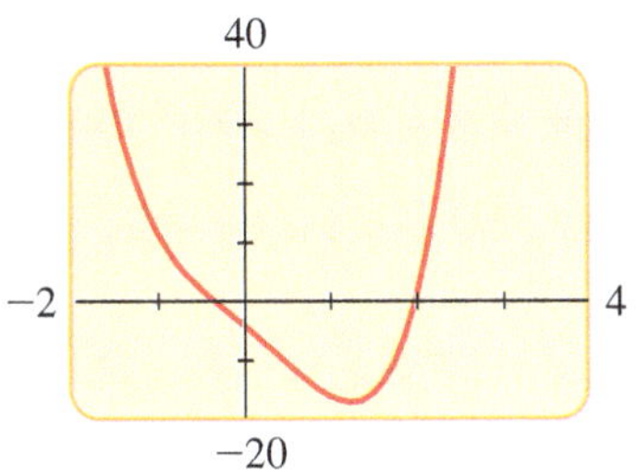

FIGURE 1
$P(x) = 3x^4 - 2x^3 - x^2 - 12x - 4$

Figure 1 shows the graph of the polynomial P in Example 5. The x-intercepts correspond to the real zeros of P. The imaginary zeros cannot be determined from the graph.

The zeros of the quadratic factor are

$$x = \frac{-1 \pm \sqrt{1 - 8}}{2} = -\frac{1}{2} \pm i\frac{\sqrt{7}}{2} \qquad \text{Quadratic Formula}$$

so the zeros of $P(x)$ are

$$2, \quad -\frac{1}{3}, \quad -\frac{1}{2} + i\frac{\sqrt{7}}{2}, \quad \text{and} \quad -\frac{1}{2} - i\frac{\sqrt{7}}{2}$$

Now Try Exercise 47 ■

Complex Zeros Come in Conjugate Pairs

As you might have noticed from the examples so far, the complex zeros of polynomials with real coefficients come in pairs. Whenever $a + bi$ is a zero, its complex conjugate $a - bi$ is also a zero.

CONJUGATE ZEROS THEOREM

If the polynomial P has real coefficients and if the complex number z is a zero of P, then its complex conjugate $\bar{z}$ is also a zero of P.

Proof Let

$$P(x) = a_n x^n + a_{n-1}x^{n-1} + \cdots + a_1 x + a_0$$

where each coefficient is real. Suppose that $P(z) = 0$. We must prove that $P(\bar{z}) = 0$. We use the facts that the complex conjugate of a sum of two complex numbers is the sum of the conjugates and that the conjugate of a product is the product of the conjugates.

$$\begin{aligned} P(\bar{z}) &= a_n(\bar{z})^n + a_{n-1}(\bar{z})^{n-1} + \cdots + a_1\bar{z} + a_0 \\ &= \overline{a_n}\,\overline{z^n} + \overline{a_{n-1}}\,\overline{z^{n-1}} + \cdots + \overline{a_1}\,\overline{z} + \overline{a_0} && \text{Because the coefficients are real} \\ &= \overline{a_n z^n} + \overline{a_{n-1}z^{n-1}} + \cdots + \overline{a_1 z} + \overline{a_0} \\ &= \overline{a_n z^n + a_{n-1}z^{n-1} + \cdots + a_1 z + a_0} \\ &= \overline{P(z)} = \overline{0} = 0 \end{aligned}$$

This shows that $\bar{z}$ is also a zero of $P(x)$, which proves the theorem. ■

GEROLAMO CARDANO (1501–1576) is certainly one of the most colorful figures in the history of mathematics. He was the best-known physician in Europe in his day, yet throughout his life he was plagued by numerous maladies, including ruptures, hemorrhoids, and an irrational fear of encountering rabid dogs. He was a doting father, but his beloved sons broke his heart—his favorite was eventually beheaded for murdering his own wife. Cardano was also a compulsive gambler; indeed, this vice might have driven him to write the *Book on Games of Chance*, the first study of probability from a mathematical point of view.

In Cardano's major mathematical work, the *Ars Magna*, he detailed the solution of the general third- and fourth-degree polynomial equations. At the time of its publication, mathematicians were uncomfortable even with negative numbers, but Cardano's formulas paved the way for the acceptance not just of negative numbers, but also of imaginary numbers, because they occurred naturally in solving polynomial equations. For example, for the cubic equation

$$x^3 - 15x - 4 = 0$$

one of his formulas gives the solution

$$x = \sqrt[3]{2 + \sqrt{-121}} + \sqrt[3]{2 - \sqrt{-121}}$$

(See page 286, Exercise 108.) This value for x actually turns out to be the *integer* 4, yet to find it, Cardano had to use the imaginary number $\sqrt{-121} = 11i$.

EXAMPLE 6 ■ A Polynomial with a Specified Complex Zero

Find a polynomial $P(x)$ of degree 3 that has integer coefficients and zeros $\frac{1}{2}$ and $3 - i$.

SOLUTION Since $3 - i$ is a zero, then so is $3 + i$ by the Conjugate Zeros Theorem. This means that $P(x)$ must have the following form.

$$\begin{aligned} P(x) &= a\left(x - \tfrac{1}{2}\right)[x - (3 - i)][x - (3 + i)] \\ &= a\left(x - \tfrac{1}{2}\right)[(x - 3) + i][(x - 3) - i] && \text{Regroup} \\ &= a\left(x - \tfrac{1}{2}\right)[(x - 3)^2 - i^2] && \text{Difference of Squares Formula} \\ &= a\left(x - \tfrac{1}{2}\right)(x^2 - 6x + 10) && \text{Expand} \\ &= a\left(x^3 - \tfrac{13}{2}x^2 + 13x - 5\right) && \text{Expand} \end{aligned}$$

To make all coefficients integers, we set $a = 2$ and get

$$P(x) = 2x^3 - 13x^2 + 26x - 10$$

Any other polynomial that satisfies the given requirements must be an integer multiple of this one.

Now Try Exercise 41 ■

■ Linear and Quadratic Factors

We have seen that a polynomial factors completely into linear factors if we use complex numbers. If we don't use complex numbers, then a polynomial with real coefficients can always be factored into linear and quadratic factors. We use this property in Section 10.7 when we study partial fractions. A quadratic polynomial with no real zeros is called **irreducible** over the real numbers. Such a polynomial cannot be factored without using complex numbers.

> **LINEAR AND QUADRATIC FACTORS THEOREM**
>
> Every polynomial with real coefficients can be factored into a product of linear and irreducible quadratic factors with real coefficients.

Proof We first observe that if $c = a + bi$ is a complex number, then

$$\begin{aligned} (x - c)(x - \bar{c}) &= [x - (a + bi)][x - (a - bi)] \\ &= [(x - a) - bi][(x - a) + bi] \\ &= (x - a)^2 - (bi)^2 \\ &= x^2 - 2ax + (a^2 + b^2) \end{aligned}$$

The last expression is a quadratic with *real* coefficients.

Now, if P is a polynomial with real coefficients, then by the Complete Factorization Theorem

$$P(x) = a(x - c_1)(x - c_2)\cdots(x - c_n)$$

Since the complex roots occur in conjugate pairs, we can multiply the factors corresponding to each such pair to get a quadratic factor with real coefficients. This results in P being factored into linear and irreducible quadratic factors. ■

EXAMPLE 7 ■ Factoring a Polynomial into Linear and Quadratic Factors

Let $P(x) = x^4 + 2x^2 - 8$.

(a) Factor P into linear and irreducible quadratic factors with real coefficients.

(b) Factor P completely into linear factors with complex coefficients.

SOLUTION

(a)

$$\begin{aligned} P(x) &= x^4 + 2x^2 - 8 \\ &= (x^2 - 2)(x^2 + 4) \\ &= (x - \sqrt{2})(x + \sqrt{2})(x^2 + 4) \end{aligned}$$

The factor $x^2 + 4$ is irreducible, since it has no real zeros.

(b) To get the complete factorization, we factor the remaining quadratic factor:

$$\begin{aligned} P(x) &= (x - \sqrt{2})(x + \sqrt{2})(x^2 + 4) \\ &= (x - \sqrt{2})(x + \sqrt{2})(x - 2i)(x + 2i) \end{aligned}$$

Now Try Exercise 67

3.5 EXERCISES

CONCEPTS

1. The polynomial $P(x) = 5x^2(x - 4)^3(x + 7)$ has degree ________. It has zeros 0, 4, and ________. The zero 0 has multiplicity ________, and the zero 4 has multiplicity ________.

2. (a) If a is a zero of the polynomial P, then ________ must be a factor of $P(x)$.
   (b) If a is a zero of multiplicity m of the polynomial P, then ________ must be a factor of $P(x)$ when we factor P completely.

3. A polynomial of degree $n \geq 1$ has exactly ________ zeros if a zero of multiplicity m is counted m times.

4. If the polynomial function P has real coefficients and if $a + bi$ is a zero of P, then ________ is also a zero of P. So if $3 + i$ is a zero of P, then ________ is also a zero of P.

5–6 ■ *True or False*? If *False*, give a reason.

5. Let $P(x) = x^4 + 1$.
   (a) The polynomial P has four complex zeros.
   (b) The polynomial P can be factored into linear factors with complex coefficients.
   (c) Some of the zeros of P are real.

6. Let $P(x) = x^3 + x$.
   (a) The polynomial P has three real zeros.
   (b) The polynomial P has at least one real zero.
   (c) The polynomial P can be factored into linear factors with real coefficients.

SKILLS

7–18 ■ Complete Factorization A polynomial P is given. **(a)** Find all zeros of P, real and complex. **(b)** Factor P completely.

7. $P(x) = x^4 + 4x^2$
8. $P(x) = x^5 + 9x^3$
9. $P(x) = x^3 - 2x^2 + 2x$
10. $P(x) = x^3 + x^2 + x$
11. $P(x) = x^4 + 2x^2 + 1$
12. $P(x) = x^4 - x^2 - 2$
13. $P(x) = x^4 - 16$
14. $P(x) = x^4 + 6x^2 + 9$
15. $P(x) = x^3 + 8$
16. $P(x) = x^3 - 8$
17. $P(x) = x^6 - 1$
18. $P(x) = x^6 - 7x^3 - 8$

19–36 ■ Complete Factorization Factor the polynomial completely, and find all its zeros. State the multiplicity of each zero.

19. $P(x) = x^2 + 25$
20. $P(x) = 4x^2 + 9$
21. $Q(x) = x^2 + 2x + 2$
22. $Q(x) = x^2 - 8x + 17$
23. $P(x) = x^3 + 4x$
24. $P(x) = x^3 - x^2 + x$
25. $Q(x) = x^4 - 1$
26. $Q(x) = x^4 - 625$
27. $P(x) = 16x^4 - 81$
28. $P(x) = x^3 - 64$
29. $P(x) = x^3 + x^2 + 9x + 9$
30. $P(x) = x^6 - 729$
31. $Q(x) = x^4 + 2x^2 + 1$
32. $Q(x) = x^4 + 10x^2 + 25$
33. $P(x) = x^4 + 3x^2 - 4$
34. $P(x) = x^5 + 7x^3$
35. $P(x) = x^5 + 6x^3 + 9x$
36. $P(x) = x^6 + 16x^3 + 64$

37–46 ■ Finding a Polynomial with Specified Zeros Find a polynomial with integer coefficients that satisfies the given conditions.

37. P has degree 2 and zeros $1 + i$ and $1 - i$.

38. P has degree 2 and zeros $1 + i\sqrt{2}$ and $1 - i\sqrt{2}$.

39. Q has degree 3 and zeros 3, $2i$, and $-2i$.

40. Q has degree 3 and zeros 0 and i.

41. P has degree 3 and zeros 2 and i.

42. Q has degree 3 and zeros -3 and $1 + i$.

43. R has degree 4 and zeros $1 - 2i$ and 1, with 1 a zero of multiplicity 2.

44. S has degree 4 and zeros $2i$ and $3i$.

45. T has degree 4, zeros i and $1 + i$, and constant term 12.

46. U has degree 5, zeros $\frac{1}{2}$, -1, and $-i$, and leading coefficient 4; the zero -1 has multiplicity 2.

47–64 ■ Finding Complex Zeros Find all zeros of the polynomial.

47. $P(x) = x^3 + 2x^2 + 4x + 8$

48. $P(x) = x^3 - 7x^2 + 17x - 15$

49. $P(x) = x^3 - 2x^2 + 2x - 1$

50. $P(x) = x^3 + 7x^2 + 18x + 18$

51. $P(x) = x^3 - 3x^2 + 3x - 2$

52. $P(x) = x^3 - x - 6$

53. $P(x) = 2x^3 + 7x^2 + 12x + 9$

54. $P(x) = 2x^3 - 8x^2 + 9x - 9$

55. $P(x) = x^4 + x^3 + 7x^2 + 9x - 18$

56. $P(x) = x^4 - 2x^3 - 2x^2 - 2x - 3$

57. $P(x) = x^5 - x^4 + 7x^3 - 7x^2 + 12x - 12$

58. $P(x) = x^5 + x^3 + 8x^2 + 8$ [*Hint:* Factor by grouping.]

59. $P(x) = x^4 - 6x^3 + 13x^2 - 24x + 36$

60. $P(x) = x^4 - x^2 + 2x + 2$

61. $P(x) = 4x^4 + 4x^3 + 5x^2 + 4x + 1$

62. $P(x) = 4x^4 + 2x^3 - 2x^2 - 3x - 1$

63. $P(x) = x^5 - 3x^4 + 12x^3 - 28x^2 + 27x - 9$

64. $P(x) = x^5 - 2x^4 + 2x^3 - 4x^2 + x - 2$

65–70 ■ Linear and Quadratic Factors A polynomial P is given. **(a)** Factor P into linear and irreducible quadratic factors with real coefficients. **(b)** Factor P completely into linear factors with complex coefficients.

65. $P(x) = x^3 - 5x^2 + 4x - 20$

66. $P(x) = x^3 - 2x - 4$

67. $P(x) = x^4 + 8x^2 - 9$

68. $P(x) = x^4 + 8x^2 + 16$

69. $P(x) = x^6 - 64$

70. $P(x) = x^5 - 16x$

SKILLS Plus

71. Number of Real and Non-Real Solutions By the Zeros Theorem, every nth-degree polynomial equation has exactly n solutions (including possibly some that are repeated). Some of these may be real, and some may be non-real. Use a graphing device to determine how many real and non-real solutions each equation has.

(a) $x^4 - 2x^3 - 11x^2 + 12x = 0$

(b) $x^4 - 2x^3 - 11x^2 + 12x - 5 = 0$

(c) $x^4 - 2x^3 - 11x^2 + 12x + 40 = 0$

72–74 ■ Real and Non-Real Coefficients So far, we have worked only with polynomials that have real coefficients. These exercises involve polynomials with real and imaginary coefficients.

72. Find all solutions of the equation.

(a) $2x + 4i = 1$

(b) $x^2 - ix = 0$

(c) $x^2 + 2ix - 1 = 0$

(d) $ix^2 - 2x + i = 0$

73. (a) Show that $2i$ and $1 - i$ are both solutions of the equation

$$x^2 - (1 + i)x + (2 + 2i) = 0$$

but that their complex conjugates $-2i$ and $1 + i$ are not.

(b) Explain why the result of part (a) does not violate the Conjugate Zeros Theorem.

74. (a) Find the polynomial with *real* coefficients of the smallest possible degree for which i and $1 + i$ are zeros and in which the coefficient of the highest power is 1.

(b) Find the polynomial with *complex* coefficients of the smallest possible degree for which i and $1 + i$ are zeros and in which the coefficient of the highest power is 1.

DISCUSS ■ DISCOVER ■ PROVE ■ WRITE

75. DISCUSS: Polynomials of Odd Degree The Conjugate Zeros Theorem says that the complex zeros of a polynomial with real coefficients occur in complex conjugate pairs. Explain how this fact proves that a polynomial with real coefficients and odd degree has at least one real zero.

76. DISCUSS ■ DISCOVER: Roots of Unity There are two square roots of 1, namely, 1 and -1. These are the solutions of $x^2 = 1$. The fourth roots of 1 are the solutions of the equation $x^4 = 1$ or $x^4 - 1 = 0$. How many fourth roots of 1 are there? Find them. The cube roots of 1 are the solutions of the equation $x^3 = 1$ or $x^3 - 1 = 0$. How many cube roots of 1 are there? Find them. How would you find the sixth roots of 1? How many are there? Make a conjecture about the number of nth roots of 1.

3.6 RATIONAL FUNCTIONS

■ Rational Functions and Asymptotes ■ Transformations of $y = 1/x$ ■ Asymptotes of Rational Functions ■ Graphing Rational Functions ■ Common Factors in Numerator and Denominator ■ Slant Asymptotes and End Behavior ■ Applications

A rational function is a function of the form

$$r(x) = \frac{P(x)}{Q(x)}$$

where P and Q are polynomials. We assume that $P(x)$ and $Q(x)$ have no factor in common. Even though rational functions are constructed from polynomials, their graphs look quite different from the graphs of polynomial functions.

■ Rational Functions and Asymptotes

Domains of rational expressions are discussed in Section 1.4.

The *domain* of a rational function consists of all real numbers x except those for which the denominator is zero. When graphing a rational function, we must pay special attention to the behavior of the graph near those x-values. We begin by graphing a very simple rational function.

EXAMPLE 1 ■ A Simple Rational Function

Graph the rational function $f(x) = 1/x$, and state the domain and range.

SOLUTION The function f is not defined for $x = 0$. The following tables show that when x is close to zero, the value of $|f(x)|$ is large, and the closer x gets to zero, the larger $|f(x)|$ gets.

For positive real numbers,

$$\frac{1}{\text{BIG NUMBER}} = \text{small number}$$

$$\frac{1}{\text{small number}} = \text{BIG NUMBER}$$

x	$f(x)$
-0.1	-10
-0.01	-100
-0.00001	$-100{,}000$
Approaching 0^-	Approaching $-\infty$

x	$f(x)$
0.1	10
0.01	100
0.00001	$100{,}000$
Approaching 0^+	Approaching ∞

We describe this behavior in words and in symbols as follows. The first table shows that as x approaches 0 from the left, the values of $y = f(x)$ decrease without bound. In symbols,

$$f(x) \to -\infty \quad \text{as} \quad x \to 0^-$$

"y approaches negative infinity as x approaches 0 from the left"

© silver-john/Shutterstock.com

DISCOVERY PROJECT

Managing Traffic

A highway engineer wants to determine the optimal safe driving speed for a road. The higher the speed limit, the more cars the road can accommodate, but safety requires a greater following distance at higher speeds. In this project we find a rational function that models the carrying capacity of a road at a given traffic speed. The model can be used to determine the speed limit at which the road has its maximum carrying capacity. You can find the project at **www.stewartmath.com**.

The second table shows that as x approaches 0 from the right, the values of $f(x)$ increase without bound. In symbols,

$$f(x) \to \infty \quad \text{as} \quad x \to 0^+$$ "y approaches infinity as x approaches 0 from the right"

The next two tables show how $f(x)$ changes as $|x|$ becomes large.

x	$f(x)$
−10	−0.1
−100	−0.01
−100,000	−0.00001
Approaching $-\infty$	Approaching 0

x	$f(x)$
10	0.1
100	0.01
100,000	0.00001
Approaching ∞	Approaching 0

These tables show that as $|x|$ becomes large, the value of $f(x)$ gets closer and closer to zero. We describe this situation in symbols by writing

$$f(x) \to 0 \quad \text{as} \quad x \to -\infty \qquad \text{and} \qquad f(x) \to 0 \quad \text{as} \quad x \to \infty$$

Using the information in these tables and plotting a few additional points, we obtain the graph shown in Figure 1.

x	$f(x) = 1/x$
−2	$-\frac{1}{2}$
−1	−1
$-\frac{1}{2}$	−2
$\frac{1}{2}$	2
1	1
2	$\frac{1}{2}$

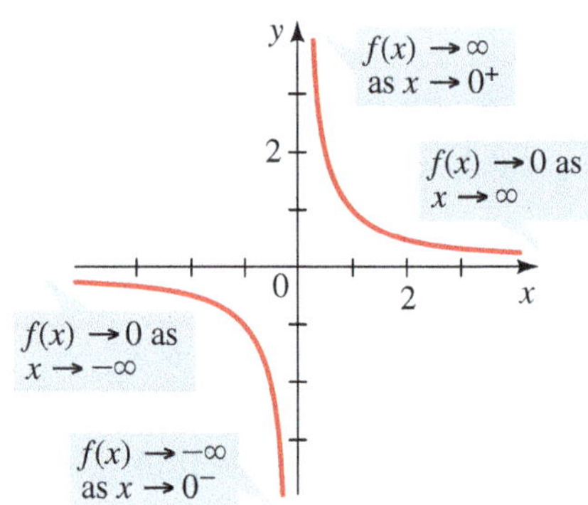

FIGURE 1
$f(x) = 1/x$

Obtaining the domain and range of a function from its graph is explained in Section 2.3, page 171.

The function f is defined for all values of x other than 0, so the domain is $\{x \mid x \neq 0\}$. From the graph we see that the range is $\{y \mid y \neq 0\}$.

Now Try Exercise 9 ■

In Example 1 we used the following **arrow notation**.

Symbol	Meaning
$x \to a^-$	x approaches a from the left
$x \to a^+$	x approaches a from the right
$x \to -\infty$	x goes to negative infinity; that is, x decreases without bound
$x \to \infty$	x goes to infinity; that is, x increases without bound

The line $x = 0$ is called a *vertical asymptote* of the graph in Figure 1, and the line $y = 0$ is a *horizontal asymptote*. Informally speaking, an asymptote of a function is a line to which the graph of the function gets closer and closer as one travels along that line.

DEFINITION OF VERTICAL AND HORIZONTAL ASYMPTOTES

1. The line $x = a$ is a **vertical asymptote** of the function $y = f(x)$ if y approaches $\pm\infty$ as x approaches a from the right or left.

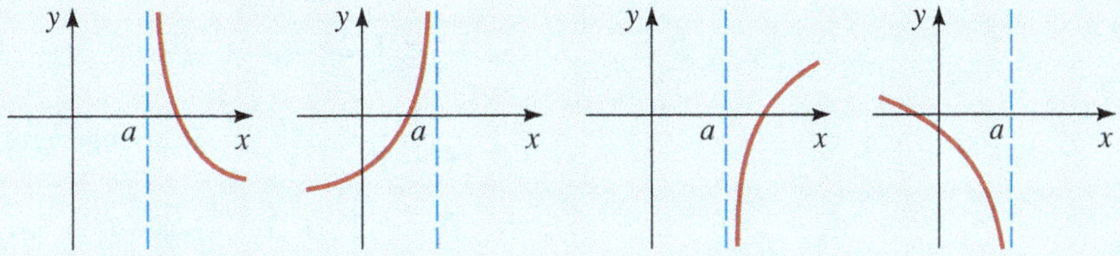

$y \to \infty$ as $x \to a^+$ $\quad$ $y \to \infty$ as $x \to a^-$ $\quad$ $y \to -\infty$ as $x \to a^+$ $\quad$ $y \to -\infty$ as $x \to a^-$

2. The line $y = b$ is a **horizontal asymptote** of the function $y = f(x)$ if y approaches b as x approaches $\pm\infty$.

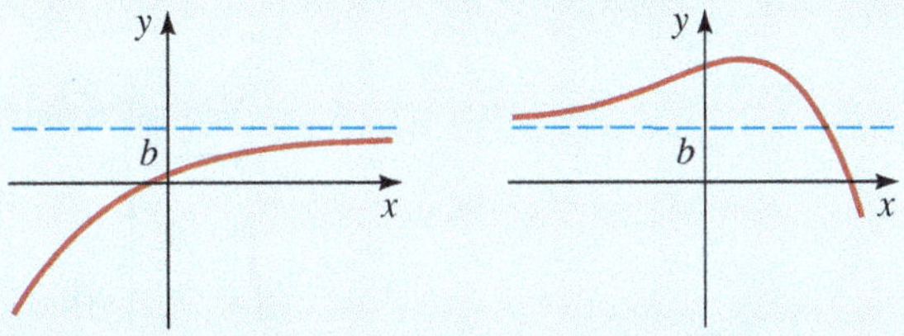

$y \to b$ as $x \to \infty$ $\quad$ $y \to b$ as $x \to -\infty$

Recall that for a rational function $R(x) = P(x)/Q(x)$, we assume that $P(x)$ and $Q(x)$ have no factor in common.

A rational function has vertical asymptotes where the function is undefined, that is, where the denominator is zero.

Transformations of $y = 1/x$

A rational function of the form

$$r(x) = \frac{ax + b}{cx + d}$$

can be graphed by shifting, stretching, and/or reflecting the graph of $f(x) = 1/x$ shown in Figure 1, using the transformations studied in Section 2.6. (Such functions are called *linear fractional transformations*.)

EXAMPLE 2 ■ Using Transformations to Graph Rational Functions

Graph each rational function, and state the domain and range.

(a) $r(x) = \dfrac{2}{x - 3}$ $\qquad$ **(b)** $s(x) = \dfrac{3x + 5}{x + 2}$

SOLUTION

(a) Let $f(x) = 1/x$. Then we can express r in terms of f as follows:

$$\begin{aligned} r(x) &= \frac{2}{x - 3} \\ &= 2\left(\frac{1}{x - 3}\right) && \text{Factor 2} \\ &= 2(f(x - 3)) && \text{Since } f(x) = 1/x \end{aligned}$$

From this form we see that the graph of r is obtained from the graph of f by shifting 3 units to the right and stretching vertically by a factor of 2. Thus r has vertical asymptote $x = 3$ and horizontal asymptote $y = 0$. The graph of r is shown in Figure 2.

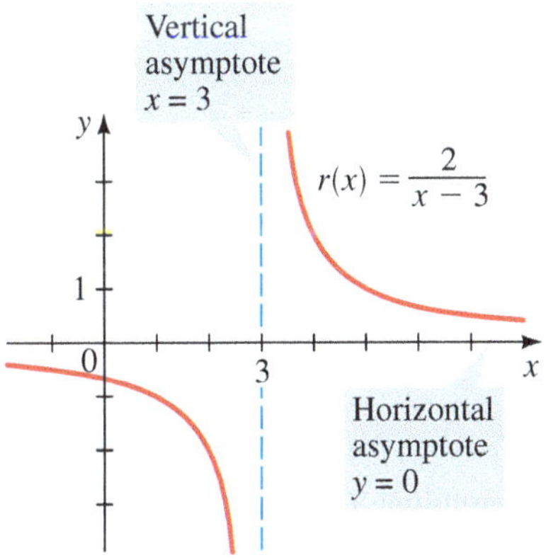

FIGURE 2

The function r is defined for all x other than 3, so the domain is $\{x \mid x \neq 3\}$. From the graph we see that the range is $\{y \mid y \neq 0\}$.

$$x + 2\overline{)3x + 5} \quad \text{quotient } 3; \quad 3x + 6; \quad -1$$

(b) Using long division (see the margin), we get $s(x) = 3 - \dfrac{1}{x+2}$. Thus we can express s in terms of f as follows.

$$\begin{aligned} s(x) &= 3 - \frac{1}{x+2} \\ &= -\frac{1}{x+2} + 3 && \text{Rearrange terms} \\ &= -f(x+2) + 3 && \text{Since } f(x) = 1/x \end{aligned}$$

From this form we see that the graph of s is obtained from the graph of f by shifting 2 units to the left, reflecting in the x-axis, and shifting upward 3 units. Thus s has vertical asymptote $x = -2$ and horizontal asymptote $y = 3$. The graph of s is shown in Figure 3.

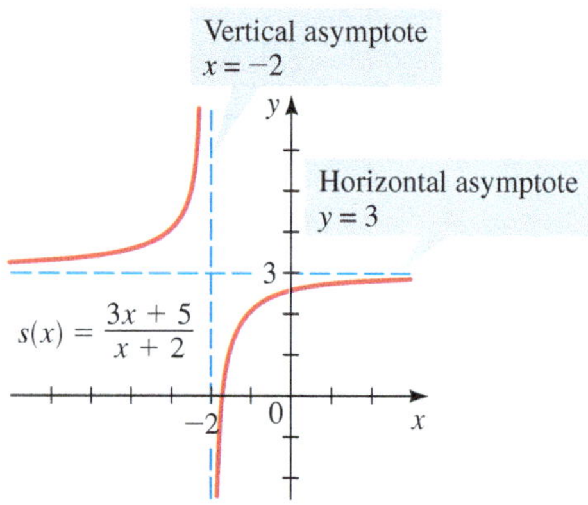

FIGURE 3

The function s is defined for all x other than -2, so the domain is $\{x \mid x \neq -2\}$. From the graph we see that the range is $\{y \mid y \neq 3\}$.

Now Try Exercises 15 and 17 ■

Asymptotes of Rational Functions

The methods of Example 2 work only for simple rational functions. To graph more complicated ones, we need to take a closer look at the behavior of a rational function near its vertical and horizontal asymptotes.

EXAMPLE 3 ■ Asymptotes of a Rational Function

Graph $r(x) = \dfrac{2x^2 - 4x + 5}{x^2 - 2x + 1}$, and state the domain and range.

SOLUTION

Vertical asymptote. We first factor the denominator

$$r(x) = \frac{2x^2 - 4x + 5}{(x-1)^2}$$

The line $x = 1$ is a vertical asymptote because the denominator of r is zero when $x = 1$.

To see what the graph of r looks like near the vertical asymptote, we make tables of values for x-values to the left and to the right of 1. From the tables shown below we see that

$$y \to \infty \quad \text{as} \quad x \to 1^- \qquad \text{and} \qquad y \to \infty \quad \text{as} \quad x \to 1^+$$

$x \to 1^-$

x	y
0	5
0.5	14
0.9	302
0.99	30,002

Approaching 1^- — Approaching ∞

$x \to 1^+$

x	y
2	5
1.5	14
1.1	302
1.01	30,002

Approaching 1^+ — Approaching ∞

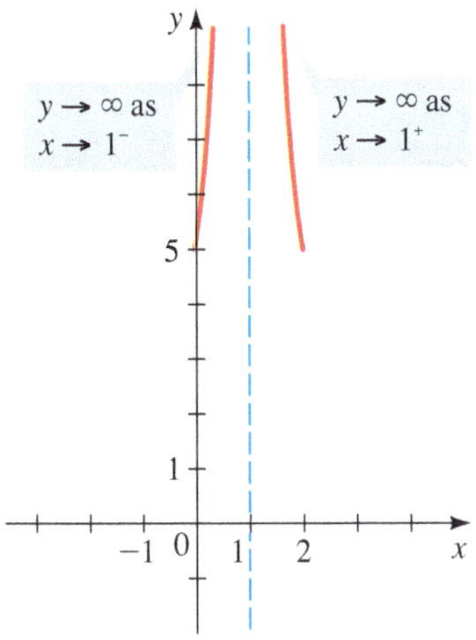

FIGURE 4

Thus near the vertical asymptote $x = 1$, the graph of r has the shape shown in Figure 4.

Horizontal asymptote. The horizontal asymptote is the value that y approaches as $x \to \pm\infty$. To help us find this value, we divide both numerator and denominator by x^2, the highest power of x that appears in the expression:

$$y = \frac{2x^2 - 4x + 5}{x^2 - 2x + 1} \cdot \frac{\frac{1}{x^2}}{\frac{1}{x^2}} = \frac{2 - \frac{4}{x} + \frac{5}{x^2}}{1 - \frac{2}{x} + \frac{1}{x^2}}$$

The fractional expressions $\frac{4}{x}$, $\frac{5}{x^2}$, $\frac{2}{x}$, and $\frac{1}{x^2}$ all approach 0 as $x \to \pm\infty$ (see Exercise 90, Section 1.1, page 12). So as $x \to \pm\infty$, we have

These terms approach 0

$$y = \frac{2 - \frac{4}{x} + \frac{5}{x^2}}{1 - \frac{2}{x} + \frac{1}{x^2}} \quad \longrightarrow \quad \frac{2 - 0 + 0}{1 - 0 + 0} = 2$$

These terms approach 0

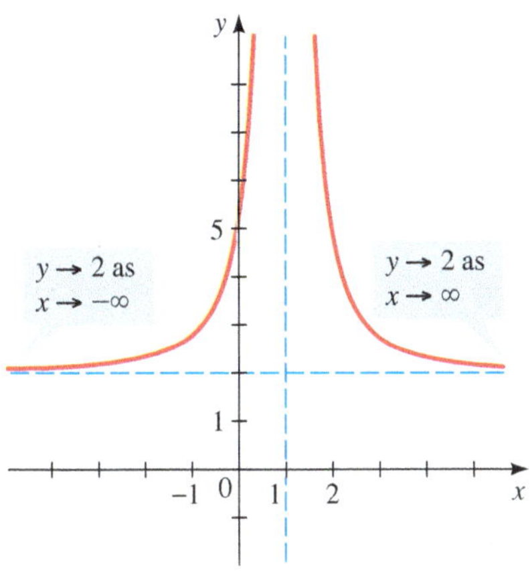

FIGURE 5

$$r(x) = \frac{2x^2 - 4x + 5}{x^2 - 2x + 1}$$

Thus the horizontal asymptote is the line $y = 2$.

Since the graph must approach the horizontal asymptote, we can complete it as in Figure 5.

Domain and range. The function r is defined for all values of x other than 1, so the domain is $\{x \mid x \neq 1\}$. From the graph we see that the range is $\{y \mid y > 2\}$.

Now Try Exercise 45 ■

From Example 3 we see that the horizontal asymptote is determined by the leading coefficients of the numerator and denominator, since after dividing through by x^2 (the highest power of x), all other terms approach zero. In general, if $r(x) = P(x)/Q(x)$ and the degrees of P and Q are the same (both n, say), then dividing both numerator and denominator by x^n shows that the horizontal asymptote is

$$y = \frac{\text{leading coefficient of } P}{\text{leading coefficient of } Q}$$

The following box summarizes the procedure for finding asymptotes.

FINDING ASYMPTOTES OF RATIONAL FUNCTIONS

Let r be the rational function

$$r(x) = \frac{a_n x^n + a_{n-1}x^{n-1} + \cdots + a_1 x + a_0}{b_m x^m + b_{m-1}x^{m-1} + \cdots + b_1 x + b_0}$$

1. The vertical asymptotes of r are the lines $x = a$, where a is a zero of the denominator.

2. **(a)** If $n < m$, then r has horizontal asymptote $y = 0$.

(b) If $n = m$, then r has horizontal asymptote $y = \dfrac{a_n}{b_m}$.

(c) If $n > m$, then r has no horizontal asymptote.

Recall that for a rational function $R(x) = P(x)/Q(x)$ we assume that $P(x)$ and $Q(x)$ have no factor in common. (See page 295.)

EXAMPLE 4 ■ Asymptotes of a Rational Function

Find the vertical and horizontal asymptotes of $r(x) = \dfrac{3x^2 - 2x - 1}{2x^2 + 3x - 2}$.

SOLUTION

Vertical asymptotes. We first factor

$$r(x) = \frac{3x^2 - 2x - 1}{(2x - 1)(x + 2)}$$

This factor is 0 when $x = \frac{1}{2}$

This factor is 0 when $x = -2$

The vertical asymptotes are the lines $x = \frac{1}{2}$ and $x = -2$.

Horizontal asymptote. The degrees of the numerator and denominator are the same, and

$$\frac{\text{leading coefficient of numerator}}{\text{leading coefficient of denominator}} = \frac{3}{2}$$

Thus the horizontal asymptote is the line $y = \frac{3}{2}$.

To confirm our results, we graph r using a graphing calculator (see Figure 6).

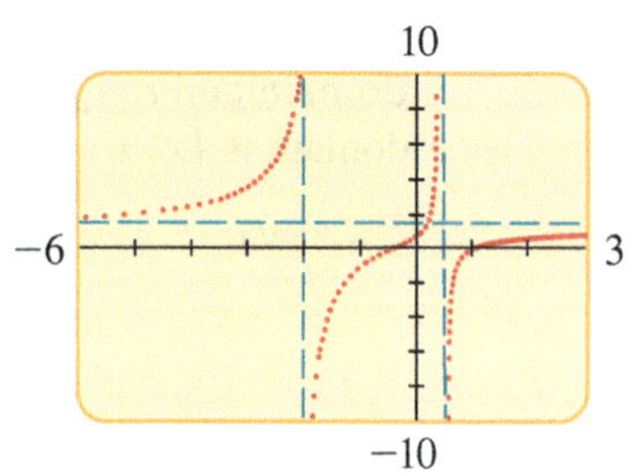

FIGURE 6

$r(x) = \dfrac{3x^2 - 2x - 1}{2x^2 + 3x - 2}$

Graph is drawn using dot mode to avoid extraneous lines.

Now Try Exercises 33 and 35 ■

Graphing Rational Functions

We have seen that asymptotes are important when graphing rational functions. In general, we use the following guidelines to graph rational functions.

A fraction is 0 only if its numerator is 0.

SKETCHING GRAPHS OF RATIONAL FUNCTIONS

1. **Factor.** Factor the numerator and denominator.
2. **Intercepts.** Find the x-intercepts by determining the zeros of the numerator and the y-intercept from the value of the function at $x = 0$.
3. **Vertical Asymptotes.** Find the vertical asymptotes by determining the zeros of the denominator, and then see whether $y \to \infty$ or $y \to -\infty$ on each side of each vertical asymptote by using test values.
4. **Horizontal Asymptote.** Find the horizontal asymptote (if any), using the procedure described in the box on page 300.
5. **Sketch the Graph.** Graph the information provided by the first four steps. Then plot as many additional points as needed to fill in the rest of the graph of the function.

EXAMPLE 5 ■ Graphing a Rational Function

Graph $r(x) = \dfrac{2x^2 + 7x - 4}{x^2 + x - 2}$, and state the domain and range.

SOLUTION We factor the numerator and denominator, find the intercepts and asymptotes, and sketch the graph.

Factor. $y = \dfrac{(2x - 1)(x + 4)}{(x - 1)(x + 2)}$

x-Intercepts. The x-intercepts are the zeros of the numerator, $x = \frac{1}{2}$ and $x = -4$.

y-Intercept. To find the y-intercept, we substitute $x = 0$ into the original form of the function.

$$r(0) = \frac{2(0)^2 + 7(0) - 4}{(0)2 + (0) - 2} = \frac{-4}{-2} = 2$$

The y-intercept is 2.

Vertical asymptotes. The vertical asymptotes occur where the denominator is 0, that is, where the function is undefined. From the factored form we see that the vertical asymptotes are the lines $x = 1$ and $x = -2$.

When choosing test values, we must make sure that there is no x-intercept between the test point and the vertical asymptote.

Behavior near vertical asymptotes. We need to know whether $y \to \infty$ or $y \to -\infty$ on each side of each vertical asymptote. To determine the sign of y for x-values near the vertical asymptotes, we use test values. For instance, as $x \to 1^-$, we use a test value close to and to the left of 1 ($x = 0.9$, say) to check whether y is positive or negative to the left of $x = 1$.

$$y = \frac{(2(0.9) - 1)((0.9) + 4)}{((0.9) - 1)((0.9) + 2)} \quad \text{whose sign is} \quad \frac{(+)(+)}{(-)(+)} \quad \text{(negative)}$$

So $y \to -\infty$ as $x \to 1^-$. On the other hand, as $x \to 1^+$, we use a test value close to and to the right of 1 ($x = 1.1$, say), to get

$$y = \frac{(2(1.1) - 1)((1.1) + 4)}{((1.1) - 1)((1.1) + 2)} \quad \text{whose sign is} \quad \frac{(+)(+)}{(+)(+)} \quad \text{(positive)}$$

Mathematics in the Modern World

Unbreakable Codes

If you read spy novels, you know about secret codes and how the hero "breaks" the code. Today secret codes have a much more common use. Most of the information that is stored on computers is coded to prevent unauthorized use. For example, your banking records, medical records, and school records are coded. Many cellular and cordless phones code the signal carrying your voice so that no one can listen in. Fortunately, because of recent advances in mathematics, today's codes are "unbreakable."

Modern codes are based on a simple principle: Factoring is much harder than multiplying. For example, try multiplying 78 and 93; now try factoring 9991. It takes a long time to factor 9991 because it is a product of two primes 97×103, so to factor it, we have to find one of these primes. Now imagine trying to factor a number N that is the product of two primes p and q, each about 200 digits long. Even the fastest computers would take many millions of years to factor such a number! But the same computer would take less than a second to multiply two such numbers. This fact was used by Ron Rivest, Adi Shamir, and Leonard Adleman in the 1970s to devise the RSA code. Their code uses an extremely large number to encode a message but requires us to know its factors to decode it. As you can see, such a code is practically unbreakable.

The RSA code is an example of a "public key encryption" code. In such codes, anyone can code a message using a publicly known procedure based on N, but to decode the message, they must know p and q, the factors of N. When the RSA code was developed, it was thought that a carefully selected 80-digit number would provide an unbreakable code. But interestingly, recent advances in the study of factoring have made much larger numbers necessary.

So $y \to \infty$ as $x \to 1^+$. The other entries in the following table are calculated similarly.

As $x \to$	-2^-	-2^+	1^-	1^+
the sign of $y = \dfrac{(2x-1)(x+4)}{(x-1)(x+2)}$ is	$\dfrac{(-)(+)}{(-)(-)}$	$\dfrac{(-)(+)}{(-)(+)}$	$\dfrac{(+)(+)}{(-)(+)}$	$\dfrac{(+)(+)}{(+)(+)}$
so $y \to$	$-\infty$	∞	$-\infty$	∞

Horizontal asymptote. The degrees of the numerator and denominator are the same, and

$$\frac{\text{leading coefficient of numerator}}{\text{leading coefficient of denominator}} = \frac{2}{1} = 2$$

Thus the horizontal asymptote is the line $y = 2$.

Graph. We use the information we have found, together with some additional values, to sketch the graph in Figure 7.

x	y
-6	0.93
-3	-1.75
-1	4.50
1.5	6.29
2	4.50
3	3.50

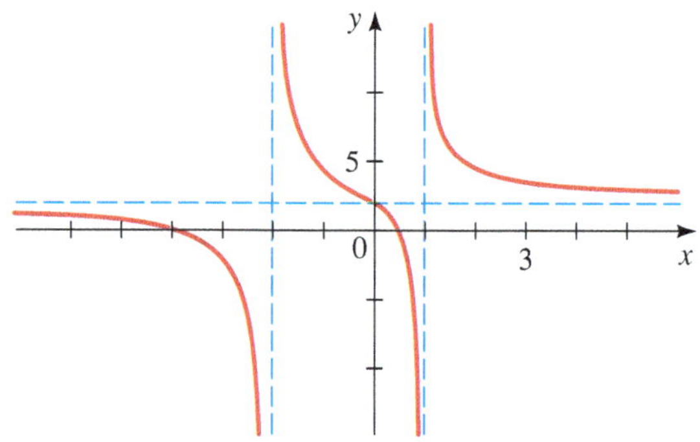

FIGURE 7

$r(x) = \dfrac{2x^2 + 7x - 4}{x^2 + x - 2}$

Domain and range. The domain is $\{x \mid x \neq 1, x \neq -2\}$. From the graph we see that the range is all real numbers.

Now Try Exercise 53 ■

EXAMPLE 6 ■ Graphing a Rational Function

Graph the rational function $r(x) = \dfrac{x^2 - 4}{2x^2 + 2x}$, and state the domain and range.

SOLUTION

Factor. $y = \dfrac{(x+2)(x-2)}{2x(x+1)}$

x-intercepts. -2 and 2, from $x + 2 = 0$ and $x - 2 = 0$

y-intercept. None, because $r(0)$ is undefined

Vertical asymptotes. $x = 0$ and $x = -1$, from the zeros of the denominator

Behavior near vertical asymptote.

As $x \to$	-1^-	-1^+	0^-	0^+
the sign of $y = \dfrac{(x+2)(x-2)}{2x(x+1)}$ is	$\dfrac{(+)(-)}{(-)(-)}$	$\dfrac{(+)(-)}{(-)(+)}$	$\dfrac{(+)(-)}{(-)(+)}$	$\dfrac{(+)(-)}{(+)(+)}$
so $y \to$	$-\infty$	∞	∞	$-\infty$

Horizontal asymptote. $y = \frac{1}{2}$, because the degree of the numerator and the degree of the denominator are the same and

$$\frac{\text{leading coefficient of numerator}}{\text{leading coefficient of denominator}} = \frac{1}{2}$$

Graph. We use the information we have found, together with some additional values, to sketch the graph in Figure 8.

x	y
-0.9	17.72
-0.5	7.50
-0.45	7.67
-0.4	8.00
-0.3	9.31
-0.1	22.17

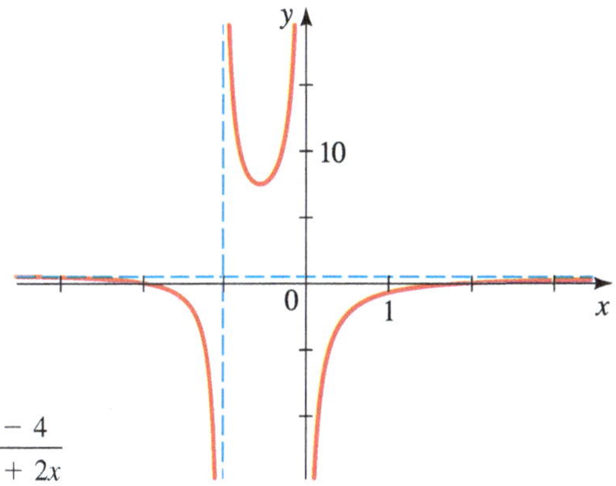

FIGURE 8

$r(x) = \dfrac{x^2 - 4}{2x^2 + 2x}$

Domain and range. The domain is $\{x \mid x \neq 0, x \neq -1\}$. From the graph we see that the range is $\{x \mid x < \frac{1}{2} \text{ or } x > 7.5\}$.

Now Try Exercise 55 ■

EXAMPLE 7 ■ Graphing a Rational Function

Graph $r(x) = \dfrac{5x + 21}{x^2 + 10x + 25}$, and state the domain and range.

SOLUTION

Factor. $y = \dfrac{5x + 21}{(x + 5)^2}$

x-Intercept. $-\dfrac{21}{5}$, from $5x + 21 = 0$

y-Intercept. $\dfrac{21}{25}$, because $r(0) = \dfrac{5 \cdot 0 + 21}{0^2 + 10 \cdot 0 + 25}$

$$= \frac{21}{25}$$

Vertical asymptote. $x = -5$, from the zeros of the denominator

Behavior near vertical asymptote.

As $x \to$	-5^-	-5^+
the sign of $y = \dfrac{5x + 21}{(x + 5)^2}$ is	$\dfrac{(-)}{(-)(-)}$	$\dfrac{(-)}{(+)(+)}$
so $y \to$	$-\infty$	$-\infty$

Horizontal asymptote. $y = 0$, because the degree of the numerator is less than the degree of the denominator

Graph. We use the information we have found, together with some additional values, to sketch the graph in Figure 9.

x	y
-15	-0.5
-10	-1.2
-3	1.5
-1	1.0
3	0.6
5	0.5
10	0.3

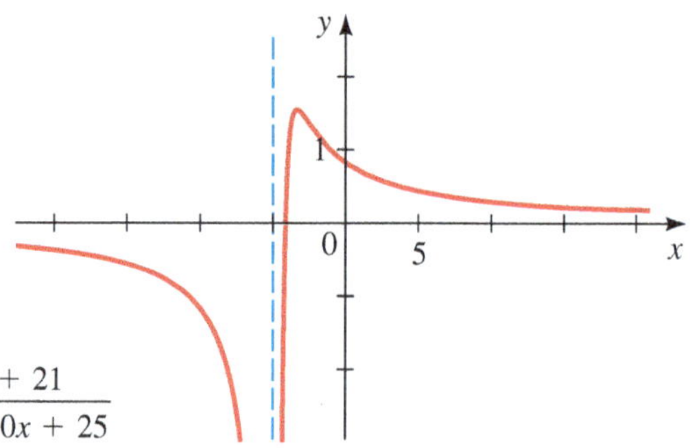

FIGURE 9

$r(x) = \dfrac{5x + 21}{x^2 + 10x + 25}$

Domain and range. The domain is $\{x \mid x \neq -5\}$. From the graph we see that the range is approximately the interval $(-\infty, 1.6]$.

Now Try Exercise 59 ■

From the graph in Figure 9 we see that, contrary to common misconception, a graph may cross a horizontal asymptote. The graph in Figure 9 crosses the x-axis (the horizontal asymptote) from below, reaches a maximum value near $x = -3$, and then approaches the x-axis from above as $x \to \infty$.

■ Common Factors in Numerator and Denominator

We have adopted the convention that the numerator and denominator of a rational function have no factor in common. If $s(x) = p(x)/q(x)$ and if p and q do have a factor in common, then we may cancel that factor, but only for those values of x for which that factor is *not zero* (because division by zero is not defined). Since s is not defined at those values of x, its graph has a "**hole**" at those points, as the following example illustrates.

EXAMPLE 8 ■ Common Factor in Numerator and Denominator

Graph the following functions.

(a) $s(x) = \dfrac{x - 3}{x^2 - 3x}$ **(b)** $t(x) = \dfrac{x^3 - 2x^2}{x - 2}$

SOLUTION

(a) We factor the numerator and denominator:

$$s(x) = \frac{x - 3}{x^2 - 3x} = \frac{(x - 3)}{x(x - 3)} = \frac{1}{x} \quad \text{for } x \neq 3$$

So s has the same graph as the rational function $r(x) = 1/x$ but with a "hole" when x is 3, as shown in Figure 10(a).

(b) We factor the numerator and denominator:

$$t(x) = \frac{x^3 - 2x^2}{x - 2} = \frac{x^2(x-2)}{x-2} = x^2 \quad \text{for } x \neq 2$$

So the graph of t is the same as the graph of $r(x) = x^2$ but with a "hole" when x is 2, as shown in Figure 10(b).

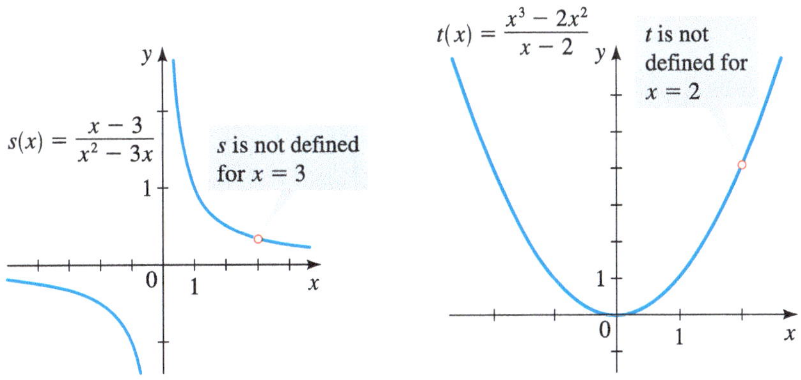

(a) $s(x) = 1/x$ for $x \neq 3$

(b) $t(x) = x^2$ for $x \neq 2$

FIGURE 10 Graphs with "holes"

Now Try Exercise 63 ■

■ Slant Asymptotes and End Behavior

If $r(x) = P(x)/Q(x)$ is a rational function in which the degree of the numerator is one more than the degree of the denominator, we can use the Division Algorithm to express the function in the form

$$r(x) = ax + b + \frac{R(x)}{Q(x)}$$

where the degree of R is less than the degree of Q and $a \neq 0$. This means that as $x \to \pm\infty$, $R(x)/Q(x) \to 0$, so for large values of $|x|$ the graph of $y = r(x)$ approaches the graph of the line $y = ax + b$. In this situation we say that $y = ax + b$ is a **slant asymptote**, or an **oblique asymptote**.

EXAMPLE 9 ■ A Rational Function with a Slant Asymptote

Graph the rational function $r(x) = \dfrac{x^2 - 4x - 5}{x - 3}$.

SOLUTION

Factor. $y = \dfrac{(x + 1)(x - 5)}{x - 3}$

x-Intercepts. -1 and 5, from $x + 1 = 0$ and $x - 5 = 0$

y-Intercept. $\frac{5}{3}$, because $r(0) = \dfrac{0^2 - 4 \cdot 0 - 5}{0 - 3} = \dfrac{5}{3}$

Vertical asymptote. $x = 3$, from the zero of the denominator

Behavior near vertical asymptote. $y \to \infty$ as $x \to 3^-$ and $y \to -\infty$ as $x \to 3^+$

Horizontal asymptote. None, because the degree of the numerator is greater than the degree of the denominator

$$\begin{array}{r} x - 1 \\ x - 3 \overline{)\, x^2 - 4x - 5} \\ \underline{x^2 - 3x} \\ -x - 5 \\ \underline{-x + 3} \\ -8 \end{array}$$

Slant asymptote. Since the degree of the numerator is one more than the degree of the denominator, the function has a slant asymptote. Dividing (see the margin), we obtain

$$r(x) = x - 1 - \frac{8}{x - 3}$$

Thus $y = x - 1$ is the slant asymptote.

Graph. We use the information we have found, together with some additional values, to sketch the graph in Figure 11.

x	y
−2	−1.4
1	4
2	9
4	−5
6	2.33

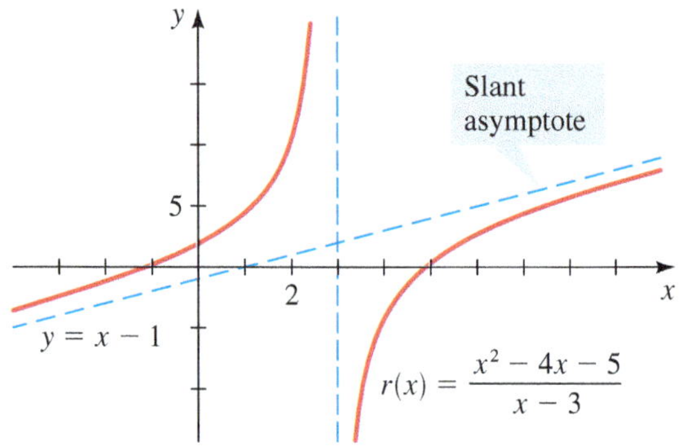

FIGURE 11

Now Try Exercise 69 ■

So far, we have considered only horizontal and slant asymptotes as end behaviors for rational functions. In the next example we graph a function whose end behavior is like that of a parabola.

EXAMPLE 10 ■ End Behavior of a Rational Function

Graph the rational function

$$r(x) = \frac{x^3 - 2x^2 + 3}{x - 2}$$

and describe its end behavior.

SOLUTION

Factor. $y = \dfrac{(x + 1)(x^2 - 3x + 3)}{x - 2}$

x-Intercept. -1, from $x + 1 = 0$ (The other factor in the numerator has no real zeros.)

y-Intercept. $-\frac{3}{2}$, because $r(0) = \dfrac{0^3 - 2 \cdot 0^2 + 3}{0 - 2} = -\dfrac{3}{2}$

Vertical asymptote. $x = 2$, from the zero of the denominator

Behavior near vertical asymptote. $y \to -\infty$ as $x \to 2^-$ and $y \to \infty$ as $x \to 2^+$

Horizontal asymptote. None, because the degree of the numerator is greater than the degree of the denominator

$$\begin{array}{r} x^2 \\ x - 2 \overline{)\, x^3 - 2x^2 + 0x + 3} \\ \underline{x^3 - 2x^2} \\ 3 \end{array}$$

End behavior. Dividing (see the margin), we get

$$r(x) = x^2 + \frac{3}{x - 2}$$

This shows that the end behavior of r is like that of the parabola $y = x^2$ because $3/(x - 2)$ is small when $|x|$ is large. That is, $3/(x - 2) \to 0$ as $x \to \pm\infty$. This means that the graph of r will be close to the graph of $y = x^2$ for large $|x|$.

Graph. In Figure 12(a) we graph r in a small viewing rectangle; we can see the intercepts, the vertical asymptotes, and the local minimum. In Figure 12(b) we graph r in a larger viewing rectangle; here the graph looks almost like the graph of a parabola. In Figure 12(c) we graph both $y = r(x)$ and $y = x^2$; these graphs are very close to each other except near the vertical asymptote.

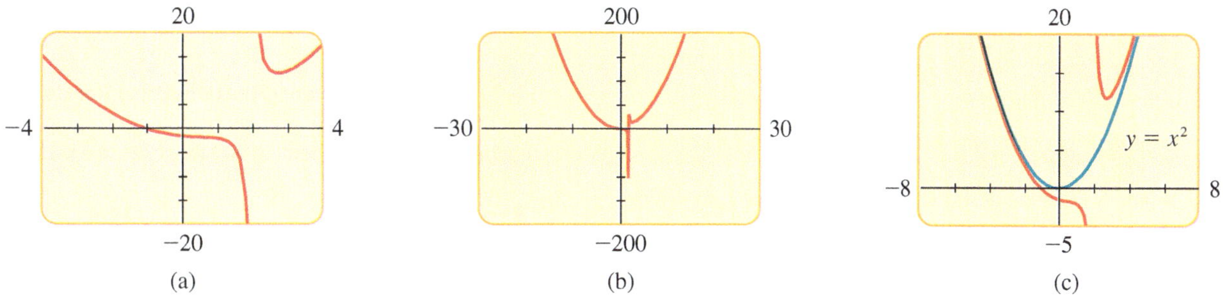

FIGURE 12
$r(x) = \dfrac{x^3 - 2x^2 + 3}{x - 2}$

Now Try Exercise 77 ■

Applications

Rational functions occur frequently in scientific applications of algebra. In the next example we analyze the graph of a function from the theory of electricity.

EXAMPLE 11 ■ Electrical Resistance

When two resistors with resistances R_1 and R_2 are connected in parallel, their combined resistance R is given by the formula

$$R = \frac{R_1 R_2}{R_1 + R_2}$$

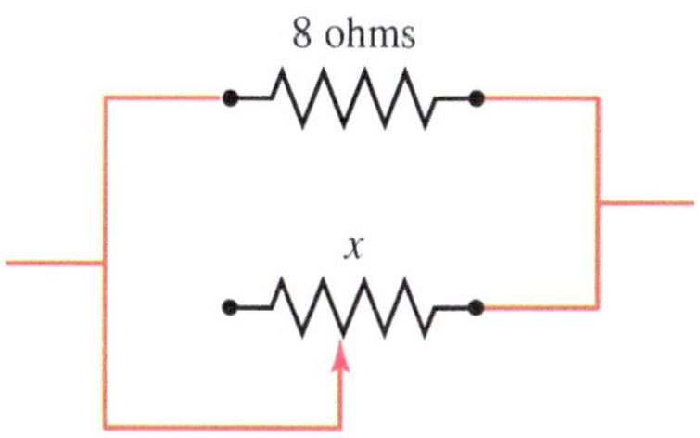

FIGURE 13

Suppose that a fixed 8-ohm resistor is connected in parallel with a variable resistor, as shown in Figure 13. If the resistance of the variable resistor is denoted by x, then the combined resistance R is a function of x. Graph R, and give a physical interpretation of the graph.

SOLUTION Substituting $R_1 = 8$ and $R_2 = x$ into the formula gives the function

$$R(x) = \frac{8x}{8 + x}$$

Since resistance cannot be negative, this function has physical meaning only when $x > 0$. The function is graphed in Figure 14(a) using the viewing rectangle $[0, 20]$ by $[0, 10]$. The function has no vertical asymptote when x is restricted to positive values. The combined resistance R increases as the variable resistance x increases. If we widen the viewing rectangle to $[0, 100]$ by $[0, 10]$, we obtain the graph in Figure 14(b). For large x the combined resistance R levels off, getting closer and closer to the horizontal asymptote $R = 8$. No matter how large the variable resistance x, the combined resistance is never greater than 8 ohms.

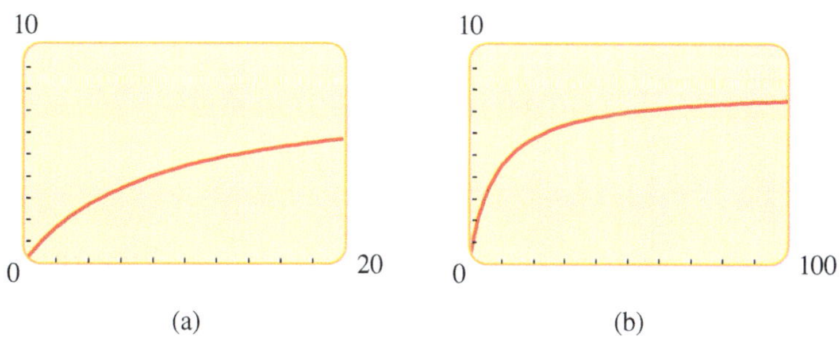

FIGURE 14
$R(x) = \dfrac{8x}{8 + x}$

Now Try Exercise 87 ■

3.6 EXERCISES

CONCEPTS

1. If the rational function $y = r(x)$ has the vertical asymptote $x = 2$, then as $x \to 2^+$, either $y \to$ ______ or $y \to$ ______.

2. If the rational function $y = r(x)$ has the horizontal asymptote $y = 2$, then $y \to$ ______ as $x \to \pm\infty$.

3–6 ■ The following questions are about the rational function

$$r(x) = \frac{(x+1)(x-2)}{(x+2)(x-3)}$$

3. The function r has x-intercepts ______ and ______.

4. The function r has y-intercept ______.

5. The function r has vertical asymptotes $x =$ ______ and $x =$ ______.

6. The function r has horizontal asymptote $y =$ ______.

7–8 ■ *True or False?*

7. Let $r(x) = \dfrac{x^2 + x}{(x+1)(2x-4)}$. The graph of r has

(a) vertical asymptote $x = -1$.

(b) vertical asymptote $x = 2$.

(c) horizontal asymptote $y = 1$.

(d) horizontal asymptote $y = \frac{1}{2}$.

8. The graph of a rational function may cross a horizontal asymptote.

SKILLS

9–12 ■ **Table of Values** A rational function is given. **(a)** Complete each table for the function. **(b)** Describe the behavior of the function near its vertical asymptote, based on Tables 1 and 2. **(c)** Determine the horizontal asymptote, based on Tables 3 and 4.

TABLE 1

x	$r(x)$
1.5	
1.9	
1.99	
1.999	

TABLE 2

x	$r(x)$
2.5	
2.1	
2.01	
2.001	

TABLE 3

x	$r(x)$
10	
50	
100	
1000	

TABLE 4

x	$r(x)$
−10	
−50	
−100	
−1000	

9. $r(x) = \dfrac{x}{x-2}$

10. $r(x) = \dfrac{4x+1}{x-2}$

11. $r(x) = \dfrac{3x-10}{(x-2)^2}$

12. $r(x) = \dfrac{3x^2+1}{(x-2)^2}$

13–20 ■ **Graphing Rational Functions Using Transformations** Use transformations of the graph of $y = 1/x$ to graph the rational function, and state the domain and range, as in Example 2.

13. $r(x) = \dfrac{1}{x-1}$

14. $r(x) = \dfrac{1}{x+4}$

15. $s(x) = \dfrac{3}{x+1}$

16. $s(x) = \dfrac{-2}{x-2}$

17. $t(x) = \dfrac{2x-3}{x-2}$

18. $t(x) = \dfrac{3x-3}{x+2}$

19. $r(x) = \dfrac{x+2}{x+3}$

20. $r(x) = \dfrac{2x-9}{x-4}$

21–26 ■ **Intercepts of Rational Functions** Find the x- and y-intercepts of the rational function.

21. $r(x) = \dfrac{x-1}{x+4}$

22. $s(x) = \dfrac{3x}{x-5}$

23. $t(x) = \dfrac{x^2-x-2}{x-6}$

24. $r(x) = \dfrac{2}{x^2+3x-4}$

25. $r(x) = \dfrac{x^2-9}{x^2}$

26. $r(x) = \dfrac{x^3+8}{x^2+4}$

27–30 ■ **Getting Information from a Graph** From the graph, determine the x- and y-intercepts and the vertical and horizontal asymptotes.

27.

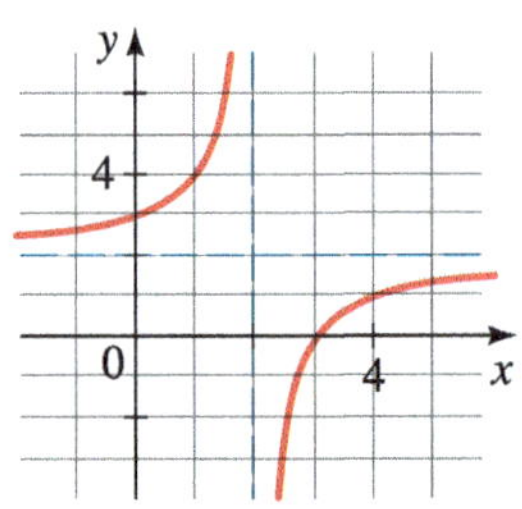

28.

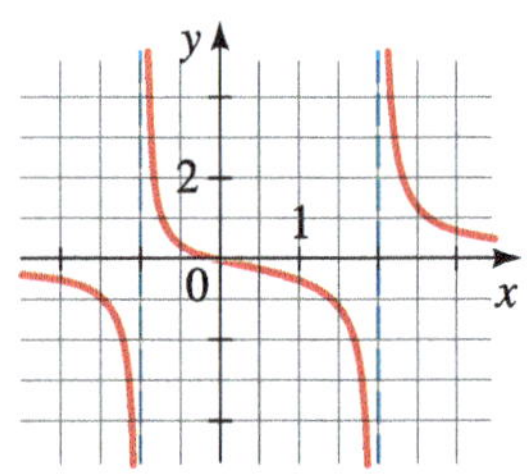

29.

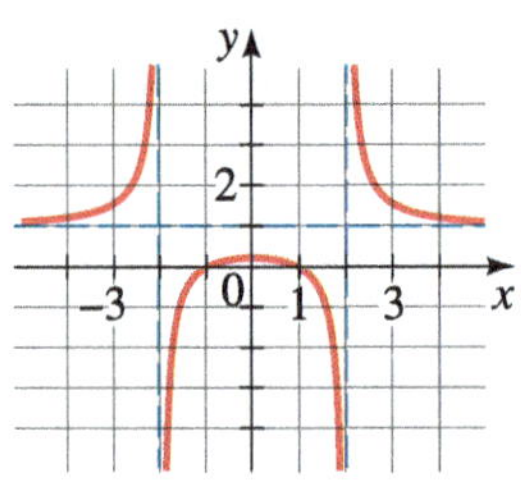

30.

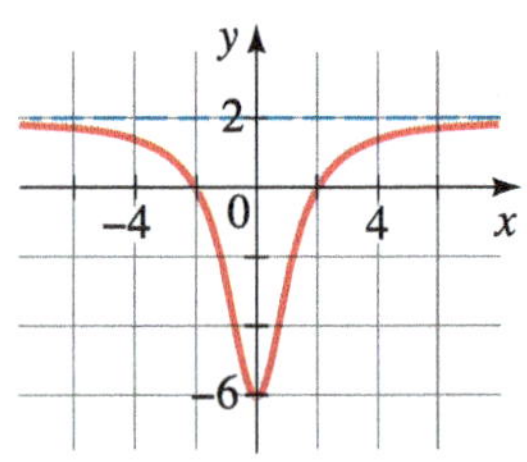

31–42 ■ **Asymptotes** Find all horizontal and vertical asymptotes (if any).

31. $r(x) = \dfrac{5}{x-2}$

32. $r(x) = \dfrac{2x-3}{x^2-1}$

33. $r(x) = \dfrac{3x + 1}{4x^2 + 1}$

34. $r(x) = \dfrac{3x^2 + 5x}{x^4 - 1}$

35. $s(x) = \dfrac{6x^2 + 1}{2x^2 + x - 1}$

36. $s(x) = \dfrac{8x^2 + 1}{4x^2 + 2x - 6}$

37. $r(x) = \dfrac{(x + 1)(2x - 3)}{(x - 2)(4x + 7)}$

38. $r(x) = \dfrac{(x - 3)(x + 2)}{(5x + 1)(2x - 3)}$

39. $r(x) = \dfrac{6x^3 - 2}{2x^3 + 5x^2 + 6x}$

40. $r(x) = \dfrac{5x^3}{x^3 + 2x^2 + 5x}$

41. $t(x) = \dfrac{x^2 + 2}{x - 1}$

42. $r(x) = \dfrac{x^3 + 3x^2}{x^2 - 4}$

43–62 ■ Graphing Rational Functions Find the intercepts and asymptotes, and then sketch a graph of the rational function and state the domain and range. Use a graphing device to confirm your answer.

43. $r(x) = \dfrac{4x - 4}{x + 2}$

44. $r(x) = \dfrac{2x + 6}{-6x + 3}$

45. $r(x) = \dfrac{3x^2 - 12x + 13}{x^2 - 4x + 4}$

46. $r(x) = \dfrac{-2x^2 - 8x - 9}{x^2 + 4x + 4}$

47. $r(x) = \dfrac{-x^2 + 8x - 18}{x^2 - 8x + 16}$

48. $r(x) = \dfrac{x^2 + 2x + 3}{2x^2 + 4x + 2}$

49. $s(x) = \dfrac{4x - 8}{(x - 4)(x + 1)}$

50. $s(x) = \dfrac{6}{x^2 - 5x - 6}$

51. $s(x) = \dfrac{2x - 4}{x^2 + x - 2}$

52. $s(x) = \dfrac{x + 2}{(x + 3)(x - 1)}$

53. $r(x) = \dfrac{(x - 1)(x + 2)}{(x + 1)(x - 3)}$

54. $r(x) = \dfrac{2x^2 + 10x - 12}{x^2 + x - 6}$

55. $r(x) = \dfrac{2x^2 + 2x - 4}{x^2 + x}$

56. $r(x) = \dfrac{3x^2 + 6}{x^2 - 2x - 3}$

57. $s(x) = \dfrac{x^2 - 2x + 1}{x^3 - 3x^2}$

58. $r(x) = \dfrac{x^2 - x - 6}{x^2 + 3x}$

59. $r(x) = \dfrac{x^2 - 2x + 1}{x^2 + 2x + 1}$

60. $r(x) = \dfrac{4x^2}{x^2 - 2x - 3}$

61. $r(x) = \dfrac{5x^2 + 5}{x^2 + 4x + 4}$

62. $t(x) = \dfrac{x^3 - x^2}{x^3 - 3x - 2}$

63–68 ■ Rational Functions with Holes Find the factors that are common in the numerator and the denominator. Then find the intercepts and asymptotes, and sketch a graph of the rational function. State the domain and range of the function.

63. $r(x) = \dfrac{x^2 + 4x - 5}{x^2 + x - 2}$

64. $r(x) = \dfrac{x^2 + 3x - 10}{(x + 1)(x - 3)(x + 5)}$

65. $r(x) = \dfrac{x^2 - 2x - 3}{x + 1}$

66. $r(x) = \dfrac{x^3 - 2x^2 - 3x}{x - 3}$

67. $r(x) = \dfrac{x^3 - 5x^2 + 3x + 9}{x + 1}$

[*Hint:* Check that $x + 1$ is a factor of the numerator.]

68. $r(x) = \dfrac{x^2 + 4x - 5}{x^3 + 7x^2 + 10x}$

69–76 ■ Slant Asymptotes Find the slant asymptote and the vertical asymptotes, and sketch a graph of the function.

69. $r(x) = \dfrac{x^2}{x - 2}$

70. $r(x) = \dfrac{x^2 + 2x}{x - 1}$

71. $r(x) = \dfrac{x^2 - 2x - 8}{x}$

72. $r(x) = \dfrac{3x - x^2}{2x - 2}$

73. $r(x) = \dfrac{x^2 + 5x + 4}{x - 3}$

74. $r(x) = \dfrac{x^3 + 4}{2x^2 + x - 1}$

75. $r(x) = \dfrac{x^3 + x^2}{x^2 - 4}$

76. $r(x) = \dfrac{2x^3 + 2x}{x^2 - 1}$

SKILLS Plus

77–80 ■ End Behavior Graph the rational function f, and determine all vertical asymptotes from your graph. Then graph f and g in a sufficiently large viewing rectangle to show that they have the same end behavior.

77. $f(x) = \dfrac{2x^2 + 6x + 6}{x + 3}$, $g(x) = 2x$

78. $f(x) = \dfrac{-x^3 + 6x^2 - 5}{x^2 - 2x}$, $g(x) = -x + 4$

79. $f(x) = \dfrac{x^3 - 2x^2 + 16}{x - 2}$, $g(x) = x^2$

80. $f(x) = \dfrac{-x^4 + 2x^3 - 2x}{(x - 1)^2}$, $g(x) = 1 - x^2$

81–86 ■ End Behavior Graph the rational function, and find all vertical asymptotes, x- and y-intercepts, and local extrema, correct to the nearest tenth. Then use long division to find a polynomial that has the same end behavior as the rational function, and graph both functions in a sufficiently large viewing rectangle to verify that the end behaviors of the polynomial and the rational function are the same.

81. $y = \dfrac{2x^2 - 5x}{2x + 3}$

82. $y = \dfrac{x^4 - 3x^3 + x^2 - 3x + 3}{x^2 - 3x}$

83. $y = \dfrac{x^5}{x^3 - 1}$

84. $y = \dfrac{x^4}{x^2 - 2}$

85. $r(x) = \dfrac{x^4 - 3x^3 + 6}{x - 3}$

86. $r(x) = \dfrac{4 + x^2 - x^4}{x^2 - 1}$

APPLICATIONS

87. Population Growth Suppose that the rabbit population on Mr. Jenkins' farm follows the formula

$$p(t) = \frac{3000t}{t+1}$$

where $t \geq 0$ is the time (in months) since the beginning of the year.

(a) Draw a graph of the rabbit population.

(b) What eventually happens to the rabbit population?

88. Drug Concentration After a certain drug is injected into a patient, the concentration c of the drug in the bloodstream is monitored. At time $t \geq 0$ (in minutes since the injection) the concentration (in mg/L) is given by

$$c(t) = \frac{30t}{t^2+2}$$

(a) Draw a graph of the drug concentration.

(b) What eventually happens to the concentration of drug in the bloodstream?

89. Drug Concentration A drug is administered to a patient, and the concentration of the drug in the bloodstream is monitored. At time $t \geq 0$ (in hours since giving the drug) the concentration (in mg/L) is given by

$$c(t) = \frac{5t}{t^2+1}$$

Graph the function c with a graphing device.

(a) What is the highest concentration of drug that is reached in the patient's bloodstream?

(b) What happens to the drug concentration after a long period of time?

(c) How long does it take for the concentration to drop below 0.3 mg/L?

90. Flight of a Rocket Suppose a rocket is fired upward from the surface of the earth with an initial velocity v (measured in meters per second). Then the maximum height h (in meters) reached by the rocket is given by the function

$$h(v) = \frac{Rv^2}{2gR - v^2}$$

where $R = 6.4 \times 10^6$ m is the radius of the earth and $g = 9.8 \text{ m/s}^2$ is the acceleration due to gravity. Use a graphing device to draw a graph of the function h. (Note that h and v must both be positive, so the viewing rectangle need not contain negative values.) What does the vertical asymptote represent physically?

91. The Doppler Effect As a train moves toward an observer (see the figure), the pitch of its whistle sounds higher to the observer than it would if the train were at rest, because the crests of the sound waves are compressed closer together. This phenomenon is called the *Doppler effect*. The observed pitch P is a function of the speed v of the train and is given by

$$P(v) = P_0\left(\frac{s_0}{s_0 - v}\right)$$

where P_0 is the actual pitch of the whistle at the source and $s_0 = 332$ m/s is the speed of sound in air. Suppose that a train has a whistle pitched at $P_0 = 440$ Hz. Graph the function $y = P(v)$ using a graphing device. How can the vertical asymptote of this function be interpreted physically?

92. Focusing Distance For a camera with a lens of fixed focal length F to focus on an object located a distance x from the lens, the film must be placed a distance y behind the lens, where F, x, and y are related by

$$\frac{1}{x} + \frac{1}{y} = \frac{1}{F}$$

(See the figure.) Suppose the camera has a 55-mm lens ($F = 55$).

(a) Express y as a function of x, and graph the function.

(b) What happens to the focusing distance y as the object moves far away from the lens?

(c) What happens to the focusing distance y as the object moves close to the lens?

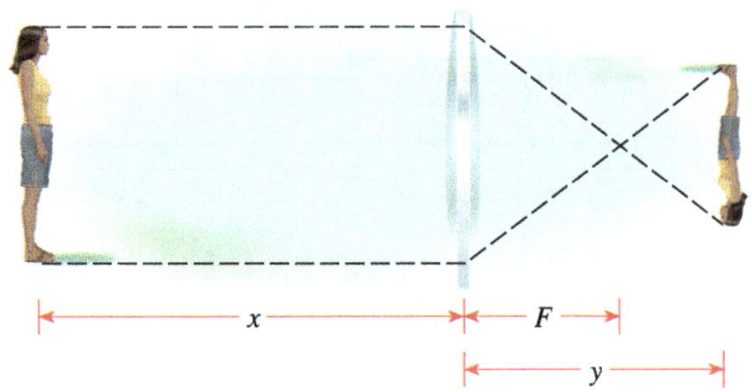

DISCUSS ■ DISCOVER ■ PROVE ■ WRITE

93. DISCUSS: Constructing a Rational Function from Its Asymptotes Give an example of a rational function that has vertical asymptote $x = 3$. Now give an example of one that has vertical asymptote $x = 3$ *and* horizontal asymptote $y = 2$. Now give an example of a rational function with vertical asymptotes $x = 1$ and $x = -1$, horizontal asymptote $y = 0$, and x-intercept 4.

94. DISCUSS: A Rational Function with No Asymptote Explain how you can tell (without graphing it) that the function

$$r(x) = \frac{x^6 + 10}{x^4 + 8x^2 + 15}$$

has no x-intercept and no horizontal, vertical, or slant asymptote. What is its end behavior?'

95. DISCOVER: Transformations of $y = 1/x^2$ In Example 2 we saw that some simple rational functions can be graphed by shifting, stretching, or reflecting the graph of $y = 1/x$. In this exercise we consider rational functions that can be graphed by transforming the graph of $y = 1/x^2$.

(a) Graph the function

$$r(x) = \frac{1}{(x-2)^2}$$

by transforming the graph of $y = 1/x^2$.

(b) Use long division and factoring to show that the function

$$s(x) = \frac{2x^2 + 4x + 5}{x^2 + 2x + 1}$$

can be written as

$$s(x) = 2 + \frac{3}{(x+1)^2}$$

Then graph s by transforming the graph of $y = 1/x^2$.

(c) One of the following functions can be graphed by transforming the graph of $y = 1/x^2$; the other cannot. Use transformations to graph the one that can be, and explain why this method doesn't work for the other one.

$$p(x) = \frac{2 - 3x^2}{x^2 - 4x + 4} \qquad q(x) = \frac{12x - 3x^2}{x^2 - 4x + 4}$$

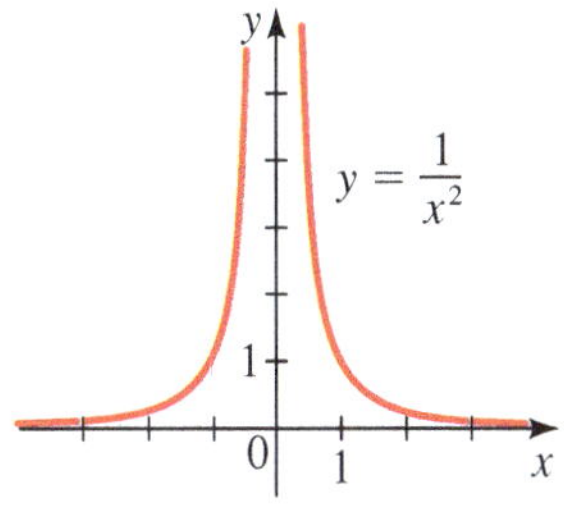

3.7 POLYNOMIAL AND RATIONAL INEQUALITIES

■ Polynomial Inequalities ■ Rational Inequalities

In Section 1.8 we solved basic inequalities. In this section we solve more advanced inequalities by using the methods we learned in Section 3.4 for factoring and graphing polynomials.

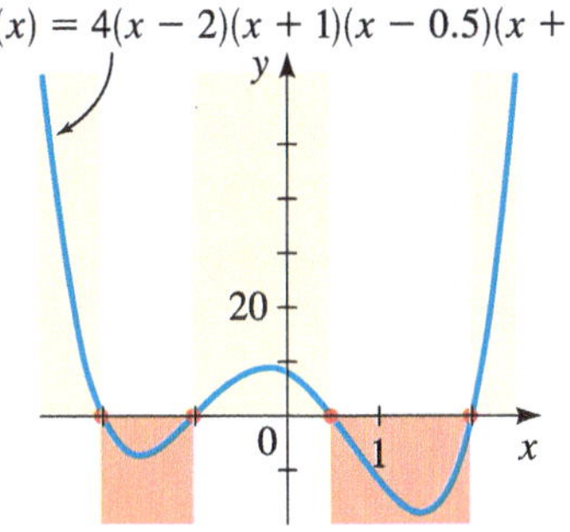

FIGURE 1 $P(x) > 0$ or $P(x) < 0$ for x between successive zeros of P

■ Polynomial Inequalities

An important consequence of the Intermediate Value Theorem (page 259) is that the values of a polynomial function P do not change sign between successive zeros. In other words, the values of P between successive zeros are either all positive or all negative. Graphically, this means that between successive x-intercepts, the graph of P is entirely above or entirely below the x-axis. Figure 1 illustrates this property of polynomials. This property of polynomials allows us to solve **polynomial inequalities** like $P(x) \geq 0$ by finding the zeros of the polynomial and using test points between successive zeros to determine the intervals that satisfy the inequality. We use the following guidelines.

SOLVING POLYNOMIAL INEQUALITIES

1. **Move All Terms to One Side.** Rewrite the inequality so that all nonzero terms appear on one side of the inequality symbol.
2. **Factor the Polynomial.** Factor the polynomial into irreducible factors, and find the **real zeros** of the polynomial.
3. **Find the Intervals.** List the intervals determined by the real zeros.
4. **Make a Table or Diagram.** Use test values to make a table or diagram of the signs of each factor in each interval. In the last row of the table determine the sign of the polynomial on that interval.
5. **Solve.** Determine the solutions of the inequality from the last row of the table. Check whether the **endpoints** of these intervals satisfy the inequality. (This may happen if the inequality involves $\leq$ or $\geq$.)

EXAMPLE 1 ■ Solving a Polynomial Inequality

Solve the inequality $2x^3 + x^2 + 6 \geq 13x$.

SOLUTION We follow the preceding guidelines.

Move all terms to one side. We move all terms to the left-hand side of the inequality to get

$$2x^3 + x^2 - 13x + 6 \geq 0$$

The left-hand side is a polynomial.

Factor the polynomial. This polynomial is factored in Example 2, Section 3.4, on page 277. We get

$$(x - 2)(2x - 1)(x + 3) \geq 0$$

The zeros of the polynomial are -3, $\frac{1}{2}$, and 2.

Find the intervals. The intervals determined by the zeros of the polynomial are

$$(-\infty, -3), \left(-3, \tfrac{1}{2}\right), \left(\tfrac{1}{2}, 2\right), (2, \infty)$$

Make a table or diagram. We make a diagram indicating the sign of each factor on each interval.

		-3		$\frac{1}{2}$		2	
Sign of $x - 2$	$-$		$-$		$-$		$+$
Sign of $2x - 1$	$-$		$-$		$+$		$+$
Sign of $x + 3$	$-$		$+$		$+$		$+$
Sign of $(x - 2)(2x - 1)(x + 3)$	$-$		$+$		$-$		$+$

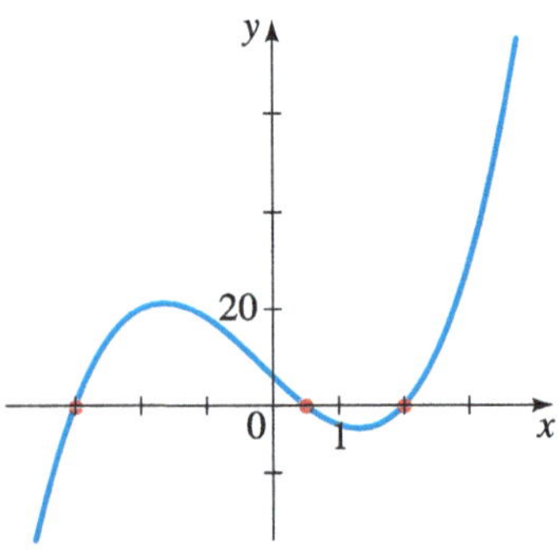

FIGURE 2

Solve. From the diagram we see that the inequality is satisfied on the intervals $\left(-3, \frac{1}{2}\right)$ and $(2, \infty)$. Checking the endpoints, we see that -3, $\frac{1}{2}$, and 2 satisfy the inequality, so the solution is $\left[-3, \frac{1}{2}\right] \cup [2, \infty)$. The graph in Figure 2 confirms our solution.

Now Try Exercise 7 ■

EXAMPLE 2 ■ Solving a Polynomial Inequality

Solve the inequality $3x^4 - x^2 - 4 < 2x^3 + 12x$.

SOLUTION We follow the above guidelines.

Move all terms to one side. We move all terms to the left-hand side of the inequality to get

$$3x^4 - 2x^3 - x^2 - 12x - 4 < 0$$

The left-hand side is a polynomial.

Factor the polynomial. This polynomial is factored into linear and irreducible quadratic factors in Example 5, Section 3.5, page 291. We get

$$(x - 2)(3x + 1)(x^2 + x + 2) < 0$$

From the first two factors we obtain the zeros 2 and $-\frac{1}{3}$. The third factor has no real zeros.

Find the intervals. The intervals determined by the zeros of the polynomial are

$$\left(-\infty, -\tfrac{1}{3}\right), \left(-\tfrac{1}{3}, 2\right), (2, \infty)$$

Make a table or diagram. We make a sign diagram.

		$-\frac{1}{3}$		2	
Sign of $x - 2$	$-$		$-$		$+$
Sign of $3x + 1$	$-$		$+$		$+$
Sign of $x^2 + x + 2$	$+$		$+$		$+$
Sign of $(x - 2)(3x + 1)(x^2 + x + 2)$	$+$		$-$		$+$

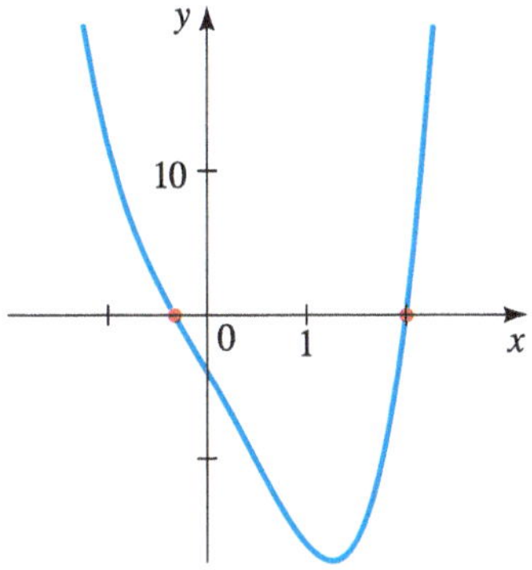

FIGURE 3

Solve. From the diagram we see that the inequality is satisfied on the interval $\left(-\frac{1}{3}, 2\right)$. You can check that the two endpoints do not satisfy the inequality, so the solution is $\left(-\frac{1}{3}, 2\right)$. The graph in Figure 3 confirms our solution.

Now Try Exercise 13 ■

■ Rational Inequalities

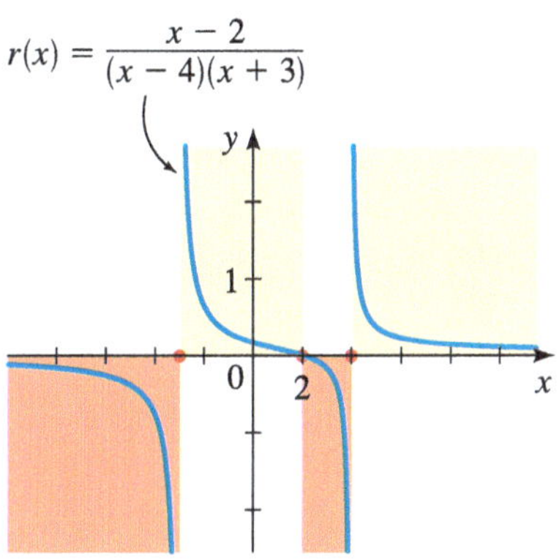

FIGURE 4 $r(x) > 0$ or $r(x) < 0$ for x between successive cut points of r

Unlike polynomial functions, rational functions are not necessarily continuous. The vertical asymptotes of a rational function r break up the graph into separate "branches." So the intervals on which r does not change sign are determined by the vertical asymptotes as well as the zeros of r. This is the reason for the following definition: If $r(x) = P(x)/Q(x)$ is a rational function, the **cut points** of r are the values of x at which either $P(x) = 0$ or $Q(x) = 0$. In other words, the cut points of r are the zeros of the numerator and the zeros of the denominator (see Figure 4). So to solve a **rational inequality** like $r(x) \geq 0$, we use test points between successive cut points to determine the intervals that satisfy the inequality. We use the following guidelines.

SOLVING RATIONAL INEQUALITIES

1. **Move All Terms to One Side.** Rewrite the inequality so that all nonzero terms appear on one side of the inequality symbol. Bring all quotients to a common denominator.
2. **Factor Numerator and Denominator.** Factor the numerator and denominator into irreducible factors, and then find the **cut points**.
3. **Find the Intervals.** List the intervals determined by the cut points.
4. **Make a Table or Diagram.** Use test values to make a table or diagram of the signs of each factor in each interval. In the last row of the table determine the sign of the rational function on that interval.
5. **Solve.** Determine the solution of the inequality from the last row of the table. Check whether the **endpoints** of these intervals satisfy the inequality. (This may happen if the inequality involves $\leq$ or $\geq$.)

EXAMPLE 3 ■ Solving a Rational Inequality

Solve the inequality

$$\frac{1 - 2x}{x^2 - 2x - 3} \geq 1$$

SOLUTION We follow the above guidelines.

Move all terms to one side. We move all terms to the left-hand side of the inequality.

$$\frac{1 - 2x}{x^2 - 2x - 3} - 1 \ge 0 \qquad \text{Move terms to LHS}$$

$$\frac{(1 - 2x) - (x^2 - 2x - 3)}{x^2 - 2x - 3} \ge 0 \qquad \text{Common denominator}$$

$$\frac{4 - x^2}{x^2 - 2x - 3} \ge 0 \qquad \text{Simplify}$$

The left-hand side of the inequality is a rational function.

Factor numerator and denominator. Factoring the numerator and denominator, we get

$$\frac{(2 - x)(2 + x)}{(x - 3)(x + 1)} \ge 0$$

The zeros of the numerator are 2 and -2, and the zeros of the denominator are -1 and 3, so the cut points are -2, -1, 2, and 3.

Find the intervals. The intervals determined by the cut points are

$$(-\infty, -2), (-2, -1), (-1, 2), (2, 3), (3, \infty)$$

Make a table or diagram. We make a sign diagram.

		-2		-1		2		3	
Sign of $2 - x$	+		+		+		−		−
Sign of $2 + x$	−		+		+		+		+
Sign of $x - 3$	−		−		−		−		+
Sign of $x + 1$	−		−		+		+		+
Sign of $\frac{(2 - x)(2 + x)}{(x - 3)(x + 1)}$	−		+		−		+		−

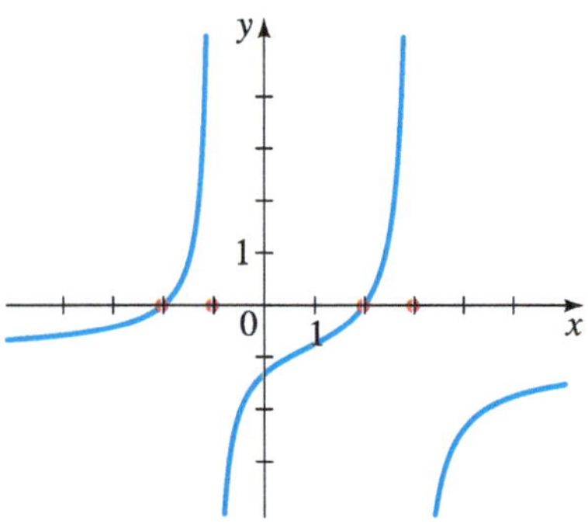

FIGURE 5

Solve. From the diagram we see that the inequality is satisfied on the intervals $(-2, -1)$ and $(2, 3)$. Checking the endpoints, we see that -2 and 2 satisfy the inequality, so the solution is $[-2, -1) \cup [2, 3)$. The graph in Figure 5 confirms our solution.

Now Try Exercise 27 ■

EXAMPLE 4 ■ Solving a Rational Inequality

Solve the inequality

$$\frac{x^2 - 4x + 3}{x^2 - 4x - 5} \ge 0$$

SOLUTION Since all nonzero terms are already on one side of the inequality symbol, we begin by factoring.

Factor numerator and denominator. Factoring the numerator and denominator, we get

$$\frac{(x - 3)(x - 1)}{(x - 5)(x + 1)} \ge 0$$

The cut points are -1, 1, 3, and 5.

Find the intervals. The intervals determined by the cut points are

$$(-\infty, -1), (-1, 1), (1, 3), (3, 5), (5, \infty)$$

Make a table or diagram. We make a sign diagram.

	$(-\infty, -1)$	$(-1, 1)$	$(1, 3)$	$(3, 5)$	$(5, \infty)$
Sign of $x - 5$	−	−	−	−	+
Sign of $x - 3$	−	−	−	+	+
Sign of $x - 1$	−	−	+	+	+
Sign of $x + 1$	−	+	+	+	+
Sign of $\dfrac{(x-3)(x-1)}{(x-5)(x+1)}$	+	−	+	−	+

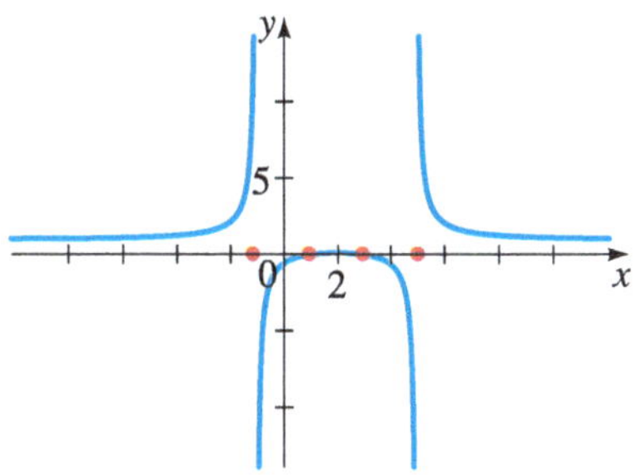

FIGURE 6

Solve. From the diagram we see that the inequality is satisfied on the intervals $(-\infty, -1)$, $(1, 3)$, and $(5, \infty)$. Checking the endpoints, we see that 1 and 3 satisfy the inequality, so the solution is $(-\infty, -1) \cup [1, 3] \cup (5, \infty)$. The graph in Figure 6 confirms our solution.

Now Try Exercise 23

We can also solve polynomial and rational inequalities graphically (see pages 120 and 172). In the next example we graph each side of the inequality and compare the values of left- and right-hand sides graphically.

EXAMPLE 5 ■ Solving a Rational Inequality Graphically

Two light sources are 10 m apart. One is three times as intense as the other. The light intensity L (in lux) at a point x meters from the weaker source is given by

$$L(x) = \frac{10}{x^2} + \frac{30}{(10 - x)^2}$$

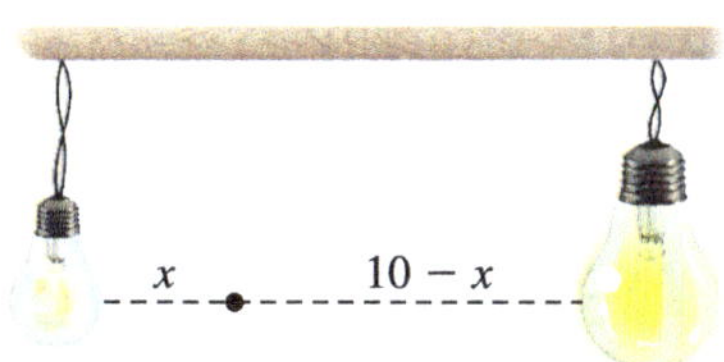

FIGURE 7

(See Figure 7.) Find the points at which the light intensity is 4 lux or less.

SOLUTION We need to solve the inequality

$$\frac{10}{x^2} + \frac{30}{(10 - x)^2} \le 4$$

We solve the inequality graphically by graphing the two functions

$$y_1 = \frac{10}{x^2} + \frac{30}{(10 - x)^2} \qquad \text{and} \qquad y_2 = 4$$

See Appendix D, *Using the TI-83/84 Graphing Calculator*, for specific instructions. Go to **www.stewartmath.com**.

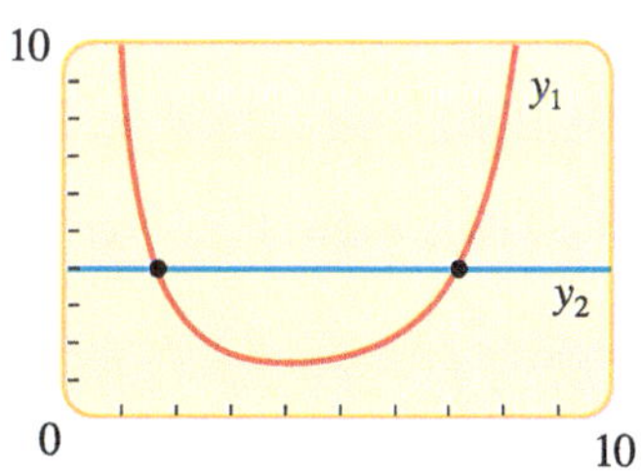

FIGURE 8

In this physical problem the possible values of x are between 0 and 10, so we graph the two functions in a viewing rectangle with x-values between 0 and 10, as shown in Figure 8. We want those values of x for which $y_1 \le y_2$. Zooming in (or using the `intersect` command), we find that the graphs intersect at $x \approx 1.67431$ and at $x \approx 7.19272$, and between these x-values the graph of y_1 lies below the graph of y_2. So the solution of the inequality is the interval $(1.67, 7.19)$, rounded to two decimal places. Thus the light intensity is less than or equal to 4 lux when the distance from the weaker source is between 1.67 m and 7.19 m.

Now Try Exercises 45 and 55

3.7 EXERCISES

CONCEPTS

1. To solve a polynomial inequality, we factor the polynomial into irreducible factors and find all the real ________ of the polynomial. Then we find the intervals determined by the real ________ and use test points in each interval to find the sign of the polynomial on that interval. Let

$$P(x) = x(x + 2)(x - 1)$$

Fill in the diagram below to find the intervals on which $P(x) \geq 0$.

Sign of		-2		0		1	
x							
$x + 2$							
$x - 1$							
$x(x + 2)(x - 1)$							

From the diagram above we see that $P(x) \geq 0$ on the intervals ________ and ________.

2. To solve a rational inequality, we factor the numerator and the denominator into irreducible factors. The cut points are the real ________ of the numerator and the real ________ denominator. Then we find the intervals determined by the ________ ________, and we use test points to find the sign of the rational function on each interval. Let

$$r(x) = \frac{(x + 2)(x - 1)}{(x - 3)(x + 4)}$$

Fill in the diagram below to find the intervals on which $r(x) \geq 0$.

Sign of		-4		-2		1		3	
$x + 2$									
$x - 1$									
$x - 3$									
$x + 4$									
$\frac{(x + 2)(x - 1)}{(x - 3)(x + 4)}$									

From the diagram we see that $r(x) \geq 0$ on the intervals ________, ________, and ________.

SKILLS

3–16 ■ Polynomial Inequalities Solve the inequality.

3. $(x - 3)(x + 5)(2x + 5) < 0$

4. $(x - 1)(x + 2)(x - 3)(x + 4) \geq 0$

5. $(x + 5)^2(x + 3)(x - 1) > 0$

6. $(2x - 7)^4(x - 1)^3(x + 1) \leq 0$

7. $x^3 + 4x^2 \geq 4x + 16$

8. $2x^3 - 18x < x^2 - 9$

9. $2x^3 - x^2 < 9 - 18x$

10. $x^4 + 3x^3 < x + 3$

11. $x^4 - 7x^2 - 18 < 0$

12. $4x^4 - 25x^2 + 36 \leq 0$

13. $x^3 + x^2 - 17x + 15 \geq 0$

14. $x^4 + 3x^3 - 3x^2 + 3x - 4 < 0$

15. $x(1 - x^2)^3 > 7(1 - x^2)^3$

16. $x^2(7 - 6x) \leq 1$

17–36 ■ Rational Inequalities Solve the inequality.

17. $\dfrac{x - 1}{x - 10} < 0$

18. $\dfrac{3x - 7}{x + 2} \leq 0$

19. $\dfrac{2x + 5}{x^2 + 2x - 35} \geq 0$

20. $\dfrac{4x^2 - 25}{x^2 - 9} \leq 0$

21. $\dfrac{x}{x^2 + 2x - 2} \leq 0$

22. $\dfrac{x + 1}{2x^2 - 4x + 1} > 0$

23. $\dfrac{x^2 + 2x - 3}{3x^2 - 7x - 6} > 0$

24. $\dfrac{x - 1}{x^3 + 1} \geq 0$

25. $\dfrac{x^3 + 3x^2 - 9x - 27}{x + 4} \leq 0$

26. $\dfrac{x^2 - 16}{x^4 - 16} < 0$

27. $\dfrac{x - 3}{2x + 5} \geq 1$

28. $\dfrac{1}{x} + \dfrac{1}{x + 1} < \dfrac{2}{x + 2}$

29. $2 + \dfrac{1}{1 - x} \leq \dfrac{3}{x}$

30. $\dfrac{1}{x - 3} + \dfrac{1}{x + 2} \geq \dfrac{2x}{x^2 + x - 2}$

31. $\dfrac{(x - 1)^2}{(x + 1)(x + 2)} > 0$

32. $\dfrac{x^2 - 2x + 1}{x^3 + 3x^2 + 3x + 1} \leq 0$

33. $\dfrac{6}{x - 1} - \dfrac{6}{x} \geq 1$

34. $\dfrac{x}{2} \geq \dfrac{5}{x + 1} + 4$

35. $\dfrac{x + 2}{x + 3} < \dfrac{x - 1}{x - 2}$

36. $\dfrac{1}{x + 1} + \dfrac{1}{x + 2} \leq \dfrac{1}{x + 3}$

37–40 ■ Graphs of Two Functions Find all values of x for which the graph of f lies above the graph of g.

37. $f(x) = x^2;\quad g(x) = 3x + 10$

38. $f(x) = \dfrac{1}{x};\quad g(x) = \dfrac{1}{x - 1}$

39. $f(x) = 4x;\quad g(x) = \dfrac{1}{x}$

40. $f(x) = x^2 + x;\quad g(x) = \dfrac{2}{x}$

41–44 ■ Domain of a Function Find the domain of the given function.

41. $f(x) = \sqrt{6 + x - x^2}$

42. $g(x) = \sqrt{\dfrac{5 + x}{5 - x}}$

43. $h(x) = \sqrt[4]{x^4 - 1}$

44. $f(x) = \dfrac{1}{\sqrt{x^4 - 5x^2 + 4}}$

45–50 ■ Solving Inequalities Graphically Use a graphing device to solve the inequality, as in Example 5. Express your answer using interval notation, with the endpoints of the intervals rounded to two decimals.

45. $x^3 - 2x^2 - 5x + 6 \geq 0$ **46.** $2x^3 + x^2 - 8x - 4 \leq 0$

47. $2x^3 - 3x + 1 < 0$ **48.** $x^4 - 4x^3 + 8x > 0$

49. $5x^4 < 8x^3$ **50.** $x^5 + x^3 \geq x^2 + 6x$

SKILLS Plus

51–52 ■ Rational Inequalities Solve the inequality. (These exercises involve expressions that arise in calculus.)

51. $\dfrac{(1-x)^2}{\sqrt{x}} \geq 4\sqrt{x}(x-1)$

52. $\frac{2}{3}x^{-1/3}(x+2)^{1/2} + \frac{1}{2}x^{2/3}(x+2)^{-1/2} < 0$

53. General Polynomial Inequality Solve the inequality

$$(x-a)(x-b)(x-c)(x-d) \geq 0$$

where $a < b < c < d$.

54. General Rational Inequality Solve the inequality

$$\frac{x^2 + (a-b)x - ab}{x+c} \leq 0$$

where $0 < a < b < c$.

APPLICATIONS

55. Bonfire Temperature In the vicinity of a bonfire the temperature T (in °C) at a distance of x meters from the center of the fire is given by

$$T(x) = \frac{500{,}000}{x^2 + 400}$$

At what range of distances from the fire's center is the temperature less than 300°C?

56. Stopping Distance For a certain model of car the distance d required to stop the vehicle if it is traveling at v mi/h is given by the function

$$d(t) = v + \frac{v^2}{25}$$

where d is measured in feet. Kerry wants her stopping distance not to exceed 175 ft. At what range of speeds can she travel?

57. Managing Traffic A highway engineer develops a formula to estimate the number of cars that can safely travel a particular highway at a given speed. She finds that the number N of cars that can pass a given point per minute is modeled by the function

$$N(x) = \frac{88x}{17 + 17\left(\dfrac{x}{20}\right)^2}$$

Graph the function in the viewing rectangle $[0, 100]$ by $[0, 60]$. If the number of cars that pass by the given point is greater than 40, at what range of speeds can the cars travel?

58. Estimating Solar Panel Profits A solar panel manufacturer estimates that the profit y (in dollars) generated by producing x solar panels per month is given by the equation

$$S(x) = 8x + 0.8x^2 - 0.002x^3 - 4000$$

Graph the function in the viewing rectangle $[0, 400]$ by $[-10{,}000, 20{,}000]$. For what range of values of x is the company's profit greater than \$12,000?

CHAPTER 3 ■ REVIEW

PROPERTIES AND FORMULAS

Quadratic Functions (pp. 246–251)

A **quadratic function** is a function of the form

$$f(x) = ax^2 + bx + c$$

It can be expressed in the **standard form**

$$f(x) = a(x-h)^2 + k$$

by completing the square.

The graph of a quadratic function in standard form is a **parabola** with **vertex** (h, k).

If $a > 0$, then the quadratic function f has the **minimum value** k at $x = h = -b/(2a)$.

If $a < 0$, then the quadratic function f has the **maximum value** k at $x = h = -b/(2a)$.

Polynomial Functions (p. 254)

A **polynomial function** of **degree** n is a function P of the form

$$P(x) = a_n x^n + a_{n-1}x^{n-1} + \cdots + a_1 x + a_0$$

(where $a_n \neq 0$). The numbers a_i are the **coefficients** of the polynomial; a_n is the **leading coefficient**, and a_0 is the **constant coefficient** (or **constant term**).

The graph of a polynomial function is a smooth, continuous curve.

Real Zeros of Polynomials (p. 259)

A **zero** of a polynomial P is a number c for which $P(c) = 0$. The following are equivalent ways of describing real zeros of polynomials:

1. c is a real zero of P.
2. $x = c$ is a solution of the equation $P(x) = 0$.
3. $x - c$ is a factor of $P(x)$.
4. c is an x-intercept of the graph of P.

Multiplicity of a Zero (pp. 262–263)

A zero c of a polynomial P has multiplicity m if m is the highest power for which $(x - c)^m$ is a factor of $P(x)$.

Local Maxima and Minima (p. 264)

A polynomial function P of degree n has $n - 1$ or fewer **local extrema** (i.e., local maxima and minima).

Division of Polynomials (p. 269)

If P and D are any polynomials (with $D(x) \neq 0$), then we can divide P by D using either **long division** or (if D is linear) **synthetic division**. The result of the division can be expressed in one of the following equivalent forms:

$$P(x) = D(x) \cdot Q(x) + R(x)$$

$$\frac{P(x)}{D(x)} = Q(x) + \frac{R(x)}{D(x)}$$

In this division, P is the **dividend**, D is the **divisor**, Q is the **quotient**, and R is the **remainder**. When the division is continued to its completion, the degree of R will be less than the degree of D (or $R(x) = 0$).

Remainder Theorem (p. 272)

When $P(x)$ is divided by the linear divisor $D(x) = x - c$, the **remainder** is the constant $P(c)$. So one way to **evaluate** a polynomial function P at c is to use synthetic division to divide $P(x)$ by $x - c$ and observe the value of the remainder.

Rational Zeros of Polynomials (pp. 275–277)

If the polynomial P given by

$$P(x) = a_n x^n + a_{n-1}x^{n-1} + \cdots + a_1 x + a_0$$

has integer coefficients, then all the **rational zeros** of P have the form

$$x = \pm\frac{p}{q}$$

where p is a divisor of the constant term a_0 and q is a divisor of the leading coefficient a_n.

So to find all the rational zeros of a polynomial, we list all the *possible* rational zeros given by this principle and then check to see which *actually* are zeros by using synthetic division.

Descartes' Rule of Signs (pp. 278–279)

Let P be a polynomial with real coefficients. Then:

The number of positive real zeros of P either is the number of **changes of sign** in the coefficients of $P(x)$ or is less than that by an even number.

The number of negative real zeros of P either is the number of **changes of sign** in the coefficients of $P(-x)$ or is less than that by an even number.

Upper and Lower Bounds Theorem (p. 279)

Suppose we divide the polynomial P by the linear expression $x - c$ and arrive at the result

$$P(x) = (x - c) \cdot Q(x) + r$$

If $c > 0$ and the coefficients of Q, followed by r, are all nonnegative, then c is an **upper bound** for the zeros of P.

If $c < 0$ and the coefficients of Q, followed by r (including zero coefficients), are alternately nonnegative and nonpositive, then c is a **lower bound** for the zeros of P.

The Fundamental Theorem of Algebra, Complete Factorization, and the Zeros Theorem (p. 287)

Every polynomial P of degree n with complex coefficients has exactly n complex zeros, provided that each zero of multiplicity m is counted m times. P factors into n linear factors as follows:

$$P(x) = a(x - c_1)(x - c_2) \cdots (x - c_n)$$

where a is the leading coefficient of P and $c_1, c_1, \ldots, c_n$ are the zeros of P.

Conjugate Zeros Theorem (p. 291)

If the polynomial P has real coefficients and if $a + bi$ is a zero of P, then its complex conjugate $a - bi$ is also a zero of P.

Linear and Quadratic Factors Theorem (p. 292)

Every polynomial with real coefficients can be factored into linear and irreducible quadratic factors with real coefficients.

Rational Functions (p. 295)

A **rational function** r is a quotient of polynomial functions:

$$r(x) = \frac{P(x)}{Q(x)}$$

We generally assume that the polynomials P and Q have no factors in common.

Asymptotes (pp. 296–297)

The line $x = a$ is a **vertical asymptote** of the function $y = f(x)$ if

$$y \to \infty \quad \text{or} \quad y \to -\infty \quad \text{as} \quad x \to a^+ \quad \text{or} \quad x \to a^-$$

The line $y = b$ is a **horizontal asymptote** of the function $y = f(x)$ if

$$y \to b \quad \text{as} \quad x \to \infty \quad \text{or} \quad x \to -\infty$$

Asymptotes of Rational Functions (pp. 298–300)

Let $r(x) = \dfrac{P(x)}{Q(x)}$ be a rational function.

The vertical asymptotes of r are the lines $x = a$ where a is a zero of Q.

If the degree of P is less than the degree of Q, then the horizontal asymptote of r is the line $y = 0$.

If the degrees of P and Q are the same, then the horizontal asymptote of r is the line $y = b$, where

$$b = \frac{\text{leading coefficient of } P}{\text{leading coefficient of } Q}$$

If the degree of P is greater than the degree of Q, then r has no horizontal asymptote.

Polynomial and Rational Inequalities (pp. 311, 313)

A **polynomial inequality** is an inequality of the form $P(x) \geq 0$, where P is a polynomial. We solve $P(x) \geq 0$ by finding the zeros of P and using test points between successive zeros to determine the intervals that satisfy the inequality.

A **rational inequality** is an inequality of the form $r(x) \geq 0$, where

$$r(x) = \frac{P(x)}{Q(x)}$$

is a rational function. The cut points of r are the values of x at which either $P(x) = 0$ or $Q(x) = 0$. We solve $r(x) \geq 0$ by using test points between successive cut points to determine the intervals that satisfy the inequality.

CONCEPT CHECK

1. **(a)** What is the degree of a quadratic function f? What is the standard form of a quadratic function? How do you put a quadratic function into standard form?

(b) The quadratic function $f(x) = a(x - h)^2 + k$ is in standard form. The graph of f is a parabola. What is the vertex of the graph of f? How do you determine whether $f(h) = k$ is a minimum or a maximum value?

(c) Express $f(x) = x^2 + 4x + 1$ in standard form. Find the vertex of the graph and the maximum or minimum value of f.

2. **(a)** Give the general form of polynomial function P of degree n.

(b) What does it mean to say that c is a zero of P? Give two equivalent conditions that tell us that c is a zero of P.

3. Sketch graphs showing the possible end behaviors of polynomials of odd degree and of even degree.

4. What steps do you follow to graph a polynomial function P?

5. **(a)** What is a local maximum point or local minimum point of a polynomial P?

(b) How many local extrema can a polynomial P of degree n have?

6. When we divide a polynomial $P(x)$ by a divisor $D(x)$, the Division Algorithm tells us that we can always obtain a quotient $Q(x)$ and a remainder $R(x)$. State the two forms in which the result of this division can be written.

7. **(a)** State the Remainder Theorem.

(b) State the Factor Theorem.

(c) State the Rational Zeros Theorem.

8. What steps would you take to find the rational zeros of a polynomial P?

9. Let $P(x) = 2x^4 - 3x^3 + x - 15$.

(a) Explain how Descartes' Rule of Signs is used to determine the possible number of positive and negative real roots of P.

(b) What does it mean to say that a is a lower bound and b is an upper bound for the zeros of a polynomial?

(c) Explain how the Upper and Lower Bounds Theorem is used to show that all the real zeros of P lie between -3 and 3.

10. **(a)** State the Fundamental Theorem of Algebra.

(b) State the Complete Factorization Theorem.

(c) State the Zeros Theorem.

(d) State the Conjugate Zeros Theorem.

11. **(a)** What is a rational function?

(b) What does it mean to say that $x = a$ is a vertical asymptote of $y = f(x)$?

(c) What does it mean to say that $y = b$ is a horizontal asymptote of $y = f(x)$?

(d) Find the vertical and horizontal asymptotes of

$$f(x) = \frac{5x^2 + 3}{x^2 - 4}$$

12. **(a)** How do you find vertical asymptotes of rational functions?

(b) Let s be the rational function

$$s(x) = \frac{a_n x^n + a_{n-1}x^{n-1} + \cdots + a_1 x + a_0}{b_m x^m + b_{m-1}x^{m-1} + \cdots + b_1 x + b_0}$$

How do you find the horizontal asymptote of s?

13. **(a)** Under what circumstances does a rational function have a slant asymptote?

(b) How do you determine the end behavior of a rational function?

14. **(a)** Explain how to solve a polynomial inequality.

(b) What are the cut points of a rational function? Explain how to solve a rational inequality.

(c) Solve the inequality $x^2 - 9 \leq 8x$.

ANSWERS TO THE CONCEPT CHECK CAN BE FOUND AT THE BACK OF THE BOOK.

EXERCISES

1–4 ■ Graphs of Quadratic Functions A quadratic function is given. **(a)** Express the function in standard form. **(b)** Graph the function.

1. $f(x) = x^2 + 6x + 2$ **2.** $f(x) = 2x^2 - 8x + 4$

3. $f(x) = 1 - 10x - x^2$ **4.** $g(x) = -2x^2 + 12x$

5–6 ■ Maximum and Minimum Values Find the maximum or minimum value of the quadratic function.

5. $f(x) = -x^2 + 3x - 1$ **6.** $f(x) = 3x^2 - 18x + 5$

7. Height of a Stone A stone is thrown upward from the top of a building. Its height (in feet) above the ground after t seconds is given by the function $h(t) = -16t^2 + 48t + 32$. What maximum height does the stone reach?

8. Profit The profit P (in dollars) generated by selling x units of a certain commodity is given by the function

$$P(x) = -1500 + 12x - 0.004x^2$$

What is the maximum profit, and how many units must be sold to generate it?

9–14 ■ Transformations of Monomials Graph the polynomial by transforming an appropriate graph of the form $y = x^n$. Show clearly all x- and y-intercepts.

9. $P(x) = -x^3 + 64$ **10.** $P(x) = 2x^3 - 16$

11. $P(x) = 2(x + 1)^4 - 32$ **12.** $P(x) = 81 - (x - 3)^4$

13. $P(x) = 32 + (x - 1)^5$ **14.** $P(x) = -3(x + 2)^5 + 96$

15–18 ■ Graphing Polynomials in Factored Form A polynomial function P is given. **(a)** Describe the end behavior. **(b)** Sketch a graph of P. Make sure your graph shows all intercepts.

15. $P(x) = (x - 3)(x + 1)(x - 5)$

16. $P(x) = -(x - 5)(x^2 - 9)(x + 2)$

17. $P(x) = -(x - 1)^2(x - 4)(x + 2)^2$

18. $P(x) = x^2(x^2 - 4)(x^2 - 9)$

19–20 ■ Graphing Polynomials A polynomial function P is given. **(a)** Determine the multiplicity of each zero of P. **(b)** Sketch a graph of P.

19. $P(x) = x^3(x - 2)^2$ **20.** $P(x) = x(x + 1)^3(x - 1)^2$

21–24 ■ Graphing Polynomials Use a graphing device to graph the polynomial. Find the x- and y-intercepts and the coordinates of all local extrema, correct to the nearest decimal. Describe the end behavior of the polynomial.

21. $P(x) = x^3 - 4x + 1$ **22.** $P(x) = -2x^3 + 6x^2 - 2$

23. $P(x) = 3x^4 - 4x^3 - 10x - 1$

24. $P(x) = x^5 + x^4 - 7x^3 - x^2 + 6x + 3$

25. Strength of a Beam The strength S of a wooden beam of width x and depth y is given by the formula $S = 13.8xy^2$. A beam is to be cut from a log of diameter 10 in., as shown in the figure.

(a) Express the strength S of this beam as a function of x only.

(b) What is the domain of the function S?

(c) Draw a graph of S.

(d) What width will make the beam the strongest?

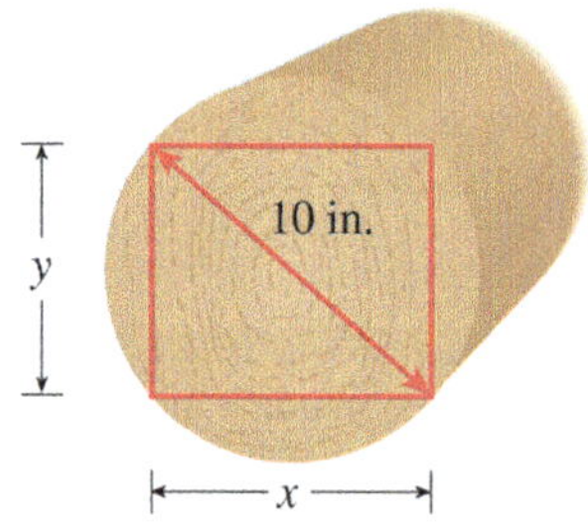

26. Volume A small shelter for delicate plants is to be constructed of thin plastic material. It will have square ends and a rectangular top and back, with an open bottom and front, as shown in the figure. The total area of the four plastic sides is to be 1200 in^2.

(a) Express the volume V of the shelter as a function of the depth x.

(b) Draw a graph of V.

(c) What dimensions will maximize the volume of the shelter?

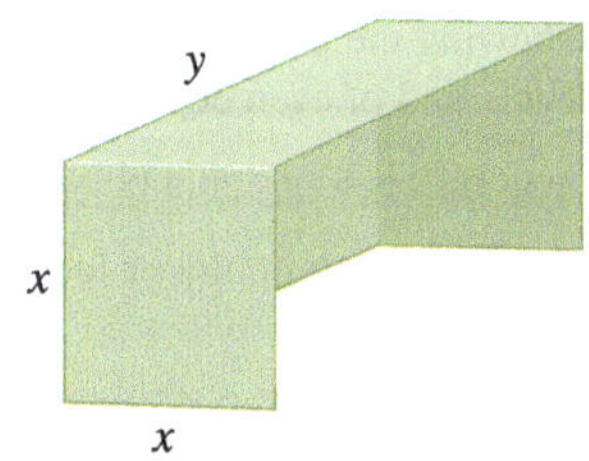

27–34 ■ Division of Polynomials Find the quotient and remainder.

27. $\dfrac{x^2 - 5x + 2}{x - 3}$ **28.** $\dfrac{3x^2 + x - 5}{x + 2}$

29. $\dfrac{2x^3 - x^2 + 3x - 4}{x + 5}$ **30.** $\dfrac{-x^3 + 2x + 4}{x - 7}$

31. $\dfrac{x^4 - 8x^2 + 2x + 7}{x + 5}$ **32.** $\dfrac{2x^4 + 3x^3 - 12}{x + 4}$

33. $\dfrac{2x^3 + x^2 - 8x + 15}{x^2 + 2x - 1}$ **34.** $\dfrac{x^4 - 2x^2 + 7x}{x^2 - x + 3}$

35–38 ■ Remainder Theorem These exercises involve the Remainder Theorem.

35. If $P(x) = 2x^3 - 9x^2 - 7x + 13$, find $P(5)$.

36. If $Q(x) = x^4 + 4x^3 + 7x^2 + 10x + 15$, find $Q(-3)$.

37. What is the remainder when the polynomial $P(x) = x^{500} + 6x^{101} - x^2 - 2x + 4$ is divided by $x - 1$?

38. What is the remainder when the polynomial $Q(x) = x^{101} - x^4 + 2$ is divided by $x + 1$?

39–40 ■ Factor Theorem Use the Factor Theorem to show that the statement in the exercise is true.

39. Show that $\frac{1}{2}$ is a zero of the polynomial

$$P(x) = 2x^4 + x^3 - 5x^2 + 10x - 4$$

40. Show that $x + 4$ is a factor of the polynomial

$$P(x) = x^5 + 4x^4 - 7x^3 - 23x^2 + 23x + 12$$

41–44 ■ Number of Possible Zeros A polynomial P is given. **(a)** List all possible rational zeros (without testing to see whether they actually are zeros). **(b)** Determine the possible number of positive and negative real zeros using Descartes' Rule of Signs.

41. $P(x) = x^5 - 6x^3 - x^2 + 2x + 18$

42. $P(x) = 6x^4 + 3x^3 + x^2 + 3x + 4$

43. $P(x) = 3x^7 - x^5 + 5x^4 + x^3 + 8$

44. $P(x) = 6x^{10} - 2x^8 - 5x^3 + 2x^2 + 12$

45–52 ■ Finding Real Zeros and Graphing Polynomials A polynomial P is given. **(a)** Find all real zeros of P, and state their multiplicities. **(b)** Sketch the graph of P.

45. $P(x) = x^3 - 16x$

46. $P(x) = x^3 - 3x^2 - 4x$

47. $P(x) = x^4 + x^3 - 2x^2$

48. $P(x) = x^4 - 5x^2 + 4$

49. $P(x) = x^4 - 2x^3 - 7x^2 + 8x + 12$

50. $P(x) = x^4 - 2x^3 - 2x^2 + 8x - 8$

51. $P(x) = 2x^4 + x^3 + 2x^2 - 3x - 2$

52. $P(x) = 9x^5 - 21x^4 + 10x^3 + 6x^2 - 3x - 1$

53–56 ■ Polynomials with Specified Zeros Find a polynomial with real coefficients of the specified degree that satisfies the given conditions.

53. Degree 3; zeros $-\frac{1}{2}, 2, 3$; constant coefficient 12

54. Degree 4; zeros 4 (multiplicity 2) and $3i$; integer coefficients; coefficient of x^2 is -25

55. **Complex Zeros of Polynomials** Does there exist a polynomial of degree 4 with integer coefficients that has zeros i, $2i$, $3i$, and $4i$? If so, find it. If not, explain why.

56. **Polynomial with no Real Roots** Prove that the equation $3x^4 + 5x^2 + 2 = 0$ has no real root.

57–68 ■ Finding Real and Complex Zeros of Polynomials Find all rational, irrational, and complex zeros (and state their multiplicities). Use Descartes' Rule of Signs, the Upper and Lower Bounds Theorem, the Quadratic Formula, or other factoring techniques to help you whenever possible.

57. $P(x) = x^3 - x^2 + x - 1$

58. $P(x) = x^3 - 8$

59. $P(x) = x^3 - 3x^2 - 13x + 15$

60. $P(x) = 2x^3 + 5x^2 - 6x - 9$

61. $P(x) = x^4 + 6x^3 + 17x^2 + 28x + 20$

62. $P(x) = x^4 + 7x^3 + 9x^2 - 17x - 20$

63. $P(x) = x^5 - 3x^4 - x^3 + 11x^2 - 12x + 4$

64. $P(x) = x^4 - 81$

65. $P(x) = x^6 - 64$

66. $P(x) = 18x^3 + 3x^2 - 4x - 1$

67. $P(x) = 6x^4 - 18x^3 + 6x^2 - 30x + 36$

68. $P(x) = x^4 + 15x^2 + 54$

69–72 ■ Solving Polynomials Graphically Use a graphing device to find all real solutions of the equation.

69. $2x^2 = 5x + 3$

70. $x^3 + x^2 - 14x - 24 = 0$

71. $x^4 - 3x^3 - 3x^2 - 9x - 2 = 0$

72. $x^5 = x + 3$

73–74 ■ Complete Factorization A polynomial function P is given. Find all the real zeros of P, and factor P completely into linear and irreducible quadratic factors with real coefficients.

73. $P(x) = x^3 - 2x - 4$

74. $P(x) = x^4 + 3x^2 - 4$

75–78 ■ Transformations of $y = 1/x$ A rational function is given. **(a)** Find all vertical and horizontal asymptotes, all x- and y-intercepts, and state the domain and range. **(b)** Use transformations of the graph of $y = 1/x$ to sketch a graph of the rational function, and state the domain and range of r.

75. $r(x) = \dfrac{3}{x + 4}$

76. $r(x) = \dfrac{-1}{x - 5}$

77. $r(x) = \dfrac{3x - 4}{x - 1}$

78. $r(x) = \dfrac{2x + 5}{x + 2}$

79–84 ■ Graphing Rational Functions Graph the rational function. Show clearly all x- and y-intercepts and asymptotes, and state the domain and range of r.

79. $r(x) = \dfrac{3x - 12}{x + 1}$

80. $r(x) = \dfrac{1}{(x + 2)^2}$

81. $r(x) = \dfrac{x - 2}{x^2 - 2x - 8}$

82. $r(x) = \dfrac{x^3 + 27}{x + 4}$

83. $r(x) = \dfrac{x^2 - 9}{2x^2 + 1}$

84. $r(x) = \dfrac{2x^2 - 6x - 7}{x - 4}$

85–88 ■ Rational Functions with Holes Find the common factors of the numerator and denominator of the rational function. Then find the intercepts and asymptotes, and sketch a graph. State the domain and range.

85. $r(x) = \dfrac{x^2 + 5x - 14}{x - 2}$

86. $r(x) = \dfrac{x^3 - 3x^2 - 10x}{x + 2}$

87. $r(x) = \dfrac{x^2 + 3x - 18}{x^2 - 8x + 15}$

88. $r(x) = \dfrac{x^2 + 2x - 15}{x^3 + 4x^2 - 7x - 10}$

89–92 ■ Graphing Rational Functions Use a graphing device to analyze the graph of the rational function. Find all x- and y-intercepts and all vertical, horizontal, and slant asymptotes. If the function has no horizontal or slant asymptote, find a polynomial that has the same end behavior as the rational function.

89. $r(x) = \dfrac{x - 3}{2x + 6}$

90. $r(x) = \dfrac{2x - 7}{x^2 + 9}$

91. $r(x) = \dfrac{x^3 + 8}{x^2 - x - 2}$

92. $r(x) = \dfrac{2x^3 - x^2}{x + 1}$

93–96 ■ Polynomial Inequalities Solve the inequality.

93. $2x^2 \geq x + 3$

94. $x^3 - 3x^2 - 4x + 12 \leq 0$

95. $x^4 - 7x^2 - 18 < 0$

96. $x^8 - 17x^4 + 16 > 0$

97–100 ■ Rational Inequalities Solve the inequality.

97. $\dfrac{5}{x^3 - x^2 - 4x + 4} < 0$

98. $\dfrac{3x + 1}{x + 2} \leq \dfrac{2}{3}$

99. $\dfrac{1}{x - 2} + \dfrac{2}{x + 3} \geq \dfrac{3}{x}$

100. $\dfrac{1}{x + 2} + \dfrac{3}{x - 3} \leq \dfrac{4}{x}$

101–102 ■ Domain of a Function Find the domain of the given function.

101. $f(x) = \sqrt{24 - x - 3x^2}$

102. $g(x) = \dfrac{1}{\sqrt[4]{x - x^4}}$

103–104 ■ Solving Inequalities Graphically Use a graphing device to solve the inequality. Express your answer using interval notation, with the endpoints of the intervals rounded to two decimals.

103. $x^4 + x^3 \leq 5x^2 + 4x - 5$

104. $x^5 - 4x^4 + 7x^3 - 12x + 2 > 0$

105. Application of Descartes' Rule of Signs We use Descartes' Rule of Signs to show that a polynomial $Q(x) = 2x^3 + 3x^2 - 3x + 4$ has no positive real zeros.

(a) Show that -1 is a zero of the polynomial $P(x) = 2x^4 + 5x^3 + x + 4$.

(b) Use the information from part (a) and Descartes' Rule of Signs to show that the polynomial $Q(x) = 2x^3 + 3x^2 - 3x + 4$ has no positive real zeros. [*Hint:* Compare the coefficients of the latter polynomial to your synthetic division table from part (a).]

106. Points of Intersection Find the coordinates of all points of intersection of the graphs of

$$y = x^4 + x^2 + 24x \quad \text{and} \quad y = 6x^3 + 20$$

CHAPTER 3 TEST

1. Express the quadratic function $f(x) = x^2 - x - 6$ in standard form, and sketch its graph.

2. Find the maximum or minimum value of the quadratic function $g(x) = 2x^2 + 6x + 3$.

3. A cannonball fired out to sea from a shore battery follows a parabolic trajectory given by the graph of the equation

$$h(x) = 10x - 0.01x^2$$

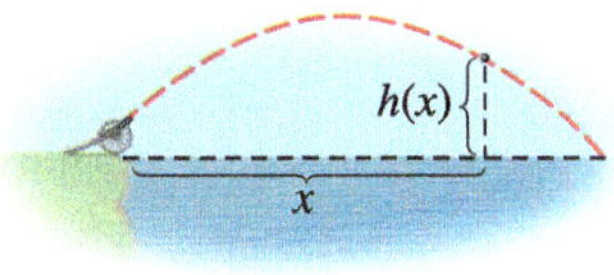

where $h(x)$ is the height of the cannonball above the water when it has traveled a horizontal distance of x feet.

   (a) What is the maximum height that the cannonball reaches?

   (b) How far does the cannonball travel horizontally before splashing into the water?

4. Graph the polynomial $P(x) = -(x + 2)^3 + 27$, showing clearly all x- and y-intercepts.

5. (a) Use synthetic division to find the quotient and remainder when $x^4 - 4x^2 + 2x + 5$ is divided by $x - 2$.

   (b) Use long division to find the quotient and remainder when $2x^5 + 4x^4 - x^3 - x^2 + 7$ is divided by $2x^2 - 1$.

6. Let $P(x) = 2x^3 - 5x^2 - 4x + 3$.

   (a) List all possible rational zeros of P.

   (b) Find the complete factorization of P.

   (c) Find the zeros of P.

   (d) Sketch the graph of P.

7. Find all real and complex zeros of $P(x) = x^3 - x^2 - 4x - 6$.

8. Find the complete factorization of $P(x) = x^4 - 2x^3 + 5x^2 - 8x + 4$.

9. Find a fourth-degree polynomial with integer coefficients that has zeros $3i$ and -1, with -1 a zero of multiplicity 2.

10. Let $P(x) = 2x^4 - 7x^3 + x^2 - 18x + 3$.

   (a) Use Descartes' Rule of Signs to determine how many positive and how many negative real zeros P can have.

   (b) Show that 4 is an upper bound and -1 is a lower bound for the real zeros of P.

   (c) Draw a graph of P, and use it to estimate the real zeros of P, rounded to two decimal places.

   (d) Find the coordinates of all local extrema of P, rounded to two decimals.

11. Consider the following rational functions:

$$r(x) = \frac{2x - 1}{x^2 - x - 2} \qquad s(x) = \frac{x^3 + 27}{x^2 + 4} \qquad t(x) = \frac{x^3 - 9x}{x + 2} \qquad u(x) = \frac{x^2 + x - 6}{x^2 - 25} \qquad w(x) = \frac{x^3 + 6x^2 + 9x}{x + 3}$$

   (a) Which of these rational functions has a horizontal asymptote?

   (b) Which of these functions has a slant asymptote?

   (c) Which of these functions has no vertical asymptote?

   (d) Which of these functions has a "hole"?

   (e) What are the asymptotes of the function $r(x)$?

   (f) Graph $y = u(x)$, showing clearly any asymptotes and x- and y-intercepts the function may have.

   (g) Use long division to find a polynomial P that has the same end behavior as t. Graph both P and t on the same screen to verify that they have the same end behavior.

12. Solve the rational inequality $x \le \dfrac{6 - x}{2x - 5}$.

13. Find the domain of the function $f(x) = \dfrac{1}{\sqrt{4 - 2x - x^2}}$.

14. **(a)** Choosing an appropriate viewing rectangle, graph the following function and find all its x-intercepts and local extrema, rounded to two decimals.

$$P(x) = x^4 - 4x^3 + 8x$$

(b) Use your graph from part (a) to solve the inequality

$$x^4 - 4x^3 + 8x \ge 0$$

Express your answer in interval form, with the endpoints rounded to two decimals.

We have learned how to fit a line to data (see *Focus on Modeling*, page 139). The line models the increasing or decreasing trend in the data. If the data exhibit more variability, such as an increase followed by a decrease, then to model the data, we need to use a curve rather than a line. Figure 1 shows a scatter plot with three possible models that appear to fit the data. Which model fits the data best?

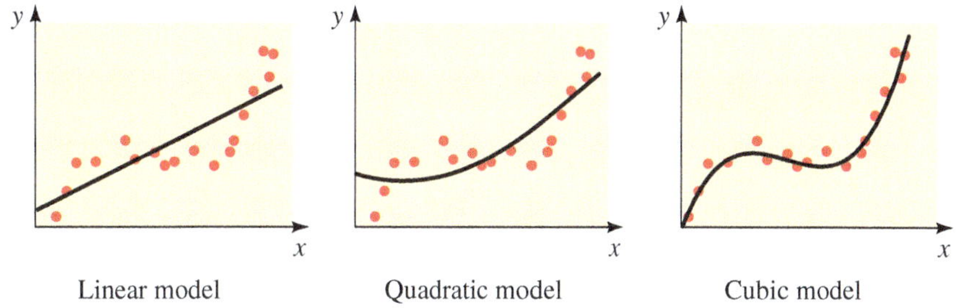

FIGURE 1

Polynomial Functions as Models

Polynomial functions are ideal for modeling data for which the scatter plot has peaks or valleys (that is, local maxima or minima). For example, if the data have a single peak as in Figure 2(a), then it may be appropriate to use a quadratic polynomial to model the data. The more peaks or valleys the data exhibit, the higher the degree of the polynomial needed to model the data (see Figure 2).

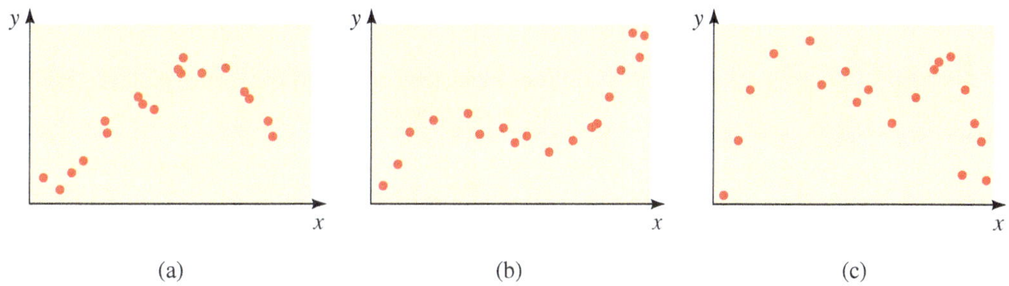

FIGURE 2

Graphing calculators are programmed to find the **polynomial of best fit** of a specified degree. As is the case for lines (see page 140), a polynomial of a given degree fits the data *best* if the sum of the squares of the distances between the graph of the polynomial and the data points is minimized.

Dennis MacDonald/Alamy

EXAMPLE 1 Rainfall and Crop Yield

Rain is essential for crops to grow, but too much rain can diminish crop yields. The data on the next page give rainfall and cotton yield per acre for several seasons in a certain county.

(a) Make a scatter plot of the data. What degree polynomial seems appropriate for modeling the data?

(b) Use a graphing calculator to find the polynomial of best fit. Graph the polynomial on the scatter plot.

(c) Use the model that you found to estimate the yield if there are 25 in. of rainfall.

Season	Rainfall (in.)	Yield (kg/acre)
1	23.3	5311
2	20.1	4382
3	18.1	3950
4	12.5	3137
5	30.9	5113
6	33.6	4814
7	35.8	3540
8	15.5	3850
9	27.6	5071
10	34.5	3881

SOLUTION

(a) The scatter plot is shown in Figure 3. The data appear to have a peak, so it is appropriate to model the data by a quadratic polynomial (degree 2).

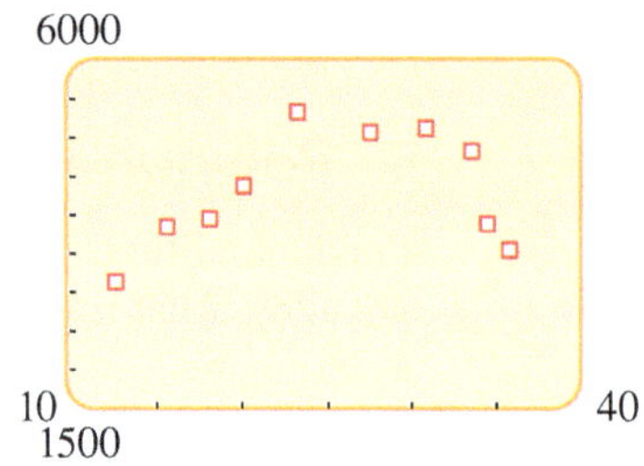

FIGURE 3 Scatter plot of yield versus rainfall data

(b) Using a graphing calculator, we find that the quadratic polynomial of best fit is

$$y = -12.6x^2 + 651.5x - 3283.2$$

The calculator output and the scatter plot, together with the graph of the quadratic model, are shown in Figure 4.

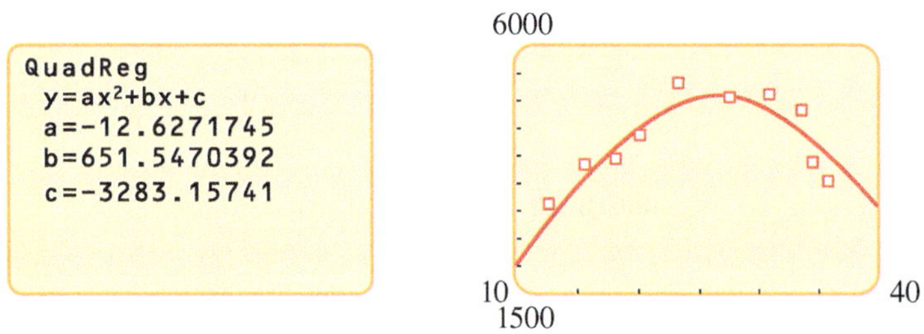

FIGURE 4 (a) (b)

(c) Using the model with $x = 25$, we get

$$y = -12.6(25)^2 + 651.5(25) - 3283.2 \approx 5129.3$$

We estimate the yield to be about 5130 kg/acre. ■

EXAMPLE 2 ■ Length-at-Age Data for Fish

Cod Redfish Hake

Otoliths for several fish species

Otoliths ("earstones") are tiny structures that are found in the heads of fish. Microscopic growth rings on the otoliths, not unlike growth rings on a tree, record the age of a fish. The following table gives the lengths of rock bass caught at different ages, as determined by the otoliths. Scientists have proposed a cubic polynomial to model this data.

(a) Use a graphing calculator to find the cubic polynomial of best fit for the data.

(b) Make a scatter plot of the data, and graph the polynomial from part (a).

(c) A fisherman catches a rock bass 20 in. long. Use the model to estimate its age.

Age (yr)	Length (in.)	Age (yr)	Length (in.)
1	4.8	9	18.2
2	8.8	9	17.1
2	8.0	10	18.8
3	7.9	10	19.5
4	11.9	11	18.9
5	14.4	12	21.7
6	14.1	12	21.9
6	15.8	13	23.8
7	15.6	14	26.9
8	17.8	14	25.1

SOLUTION

(a) Using a graphing calculator (see Figure 5(a)), we find the cubic polynomial of best fit:

$$y = 0.0155x^3 - 0.372x^2 + 3.95x + 1.21$$

(b) The scatter plot of the data and the cubic polynomial are graphed in Figure 5(b).

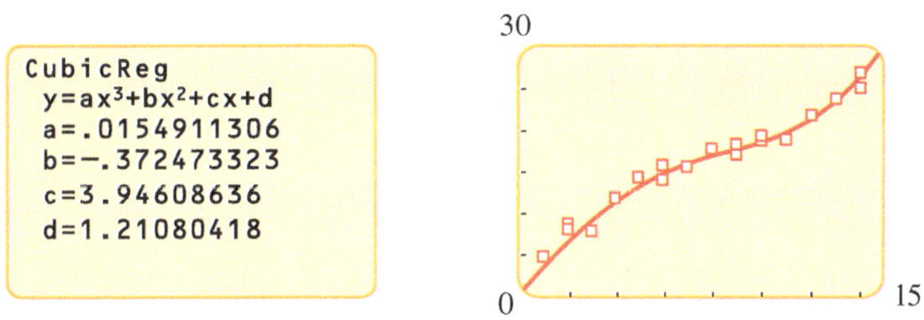

FIGURE 5 (a) (b)

(c) Moving the cursor along the graph of the polynomial, we find that $y = 20$ when $x \approx 10.8$. Thus the fish is about 11 years old. ■

PROBLEMS

Pressure (lb/in^2)	Tire life (mi)
26	50,000
28	66,000
31	78,000
35	81,000
38	74,000
42	70,000
45	59,000

1. **Tire Inflation and Treadwear** Car tires need to be inflated properly. Overinflation or underinflation can cause premature treadwear. The data in the margin show tire life for different inflation values for a certain type of tire.
   (a) Find the quadratic polynomial that best fits the data.
   (b) Draw a graph of the polynomial from part (a) together with a scatter plot of the data.
   (c) Use your result from part (b) to estimate the pressure that gives the longest tire life.

2. **Too Many Corn Plants per Acre?** The more corn a farmer plants per acre, the greater is the yield the farmer can expect, but only up to a point. Too many plants per acre can cause overcrowding and decrease yields. The data give crop yields per acre for various densities of corn plantings, as found by researchers at a university test farm.
   (a) Find the quadratic polynomial that best fits the data.
   (b) Draw a graph of the polynomial from part (a) together with a scatter plot of the data.
   (c) Use your result from part (b) to estimate the yield for 37,000 plants per acre.

Density (plants/acre)	15,000	20,000	25,000	30,000	35,000	40,000	45,000	50,000
Crop yield (bushels/acre)	43	98	118	140	142	122	93	67

3. How Fast Can You List Your Favorite Things? If you are asked to make a list of objects in a certain category, how fast you can list them follows a predictable pattern. For example, if you try to name as many vegetables as you can, you'll probably think of several right away—for example, carrots, peas, beans, corn, and so on. Then after a pause you might think of ones you eat less frequently—perhaps zucchini, eggplant, and asparagus. Finally, a few more exotic vegetables might come to mind—artichokes, jicama, bok choy, and the like. A psychologist performs this experiment on a number of subjects. The table below gives the average number of vegetables that the subjects named by a given number of seconds.

(a) Find the cubic polynomial that best fits the data.

(b) Draw a graph of the polynomial from part (a) together with a scatter plot of the data.

(c) Use your result from part (b) to estimate the number of vegetables that subjects would be able to name in 40 s.

(d) According to the model, how long (to the nearest 0.1 s) would it take a person to name five vegetables?

Seconds	Number of vegetables
1	2
2	6
5	10
10	12
15	14
20	15
25	18
30	21

Time (s)	Height (ft)
0	4.2
0.5	26.1
1.0	40.1
1.5	46.0
2.0	43.9
2.5	33.7
3.0	15.8

4. Height of a Baseball A baseball is thrown upward, and its height is measured at 0.5-s intervals using a strobe light. The resulting data are given in the table.

(a) Draw a scatter plot of the data. What degree polynomial is appropriate for modeling the data?

(b) Find a polynomial model that best fits the data, and graph it on the scatter plot.

(c) Find the times when the ball is 20 ft above the ground.

(d) What is the maximum height attained by the ball?

5. Torricelli's Law Water in a tank will flow out of a small hole in the bottom faster when the tank is nearly full than when it is nearly empty. According to Torricelli's Law, the height $h(t)$ of water remaining at time t is a quadratic function of t.

A certain tank is filled with water and allowed to drain. The height of the water is measured at different times as shown in the table.

(a) Find the quadratic polynomial that best fits the data.

(b) Draw a graph of the polynomial from part (a) together with a scatter plot of the data.

(c) Use your graph from part (b) to estimate how long it takes for the tank to drain completely.

Time (min)	Height (ft)
0	5.0
4	3.1
8	1.9
12	0.8
16	0.2

© TonyV3112/Shutterstock.com

4 Exponential and Logarithmic Functions

In this chapter we study *exponential functions*. These are functions like $f(x) = 2^x$, where the independent variable is in the exponent. Exponential functions are used in modeling many real-world phenomena, such as the growth of a population, the growth of an investment that earns compound interest, or the decay of a radioactive substance. Once an exponential model has been obtained, we can use the model to predict the size of a population, calculate the amount of an investment, or find the amount of a radioactive substance that remains. The inverse functions of exponential functions are called *logarithmic functions*. With exponential models and logarithmic functions we can answer questions such as these: When will my city be as crowded as the city street pictured here? When will my bank account have a million dollars? When will radiation from a radioactive spill decay to a safe level?

In the *Focus on Modeling* at the end of the chapter we learn how to fit exponential and power curves to data.

4.1 EXPONENTIAL FUNCTIONS

■ Exponential Functions ■ Graphs of Exponential Functions ■ Compound Interest

In this chapter we study a new class of functions called *exponential functions*. For example,

$$f(x) = 2^x$$

is an exponential function (with base 2). Notice how quickly the values of this function increase.

$$f(3) = 2^3 = 8$$
$$f(10) = 2^{10} = 1024$$
$$f(30) = 2^{30} = 1{,}073{,}741{,}824$$

Compare this with the function $g(x) = x^2$, where $g(30) = 30^2 = 900$. The point is that when the variable is in the exponent, even a small change in the variable can cause a dramatic change in the value of the function.

■ Exponential Functions

To study exponential functions, we must first define what we mean by the exponential expression a^x when x is any real number. In Section 1.2 we defined a^x for $a > 0$ and x a rational number, but we have not yet defined irrational powers. So what is meant by $5^{\sqrt{3}}$ or 2^{π}? To define a^x when x is irrational, we approximate x by rational numbers.

For example, since

$$\sqrt{3} \approx 1.73205\ldots$$

is an irrational number, we successively approximate $a^{\sqrt{3}}$ by the following rational powers:

$$a^{1.7}, a^{1.73}, a^{1.732}, a^{1.7320}, a^{1.73205}, \ldots$$

Intuitively, we can see that these rational powers of a are getting closer and closer to $a^{\sqrt{3}}$. It can be shown by using advanced mathematics that there is exactly one number that these powers approach. We define $a^{\sqrt{3}}$ to be this number.

For example, using a calculator, we find

$$5^{\sqrt{3}} \approx 5^{1.732}$$
$$\approx 16.2411\ldots$$

The more decimal places of $\sqrt{3}$ we use in our calculation, the better our approximation of $5^{\sqrt{3}}$.

The Laws of Exponents are listed on page 14.

It can be proved that the *Laws of Exponents are still true when the exponents are real numbers*.

EXPONENTIAL FUNCTIONS

The **exponential function with base *a*** is defined for all real numbers x by

$$f(x) = a^x$$

where $a > 0$ and $a \neq 1$.

We assume that $a \neq 1$ because the function $f(x) = 1^x = 1$ is just a constant function. Here are some examples of exponential functions:

$$f(x) = 2^x \qquad g(x) = 3^x \qquad h(x) = 10^x$$

Base 2 Base 3 Base 10

EXAMPLE 1 ■ Evaluating Exponential Functions

Let $f(x) = 3^x$, and evaluate the following:

(a) $f(5)$ **(b)** $f(-\frac{2}{3})$

(c) $f(\pi)$ **(d)** $f(\sqrt{2})$

SOLUTION We use a calculator to obtain the values of f.

	Calculator keystrokes	Output
(a) $f(5) = 3^5 = 243$	3 ^ 5 ENTER	243
(b) $f(-\frac{2}{3}) = 3^{-2/3} \approx 0.4807$	3 ^ ((−) 2 ÷ 3) ENTER	0.4807498
(c) $f(\pi) = 3^{\pi} \approx 31.544$	3 ^ π ENTER	31.5442807
(d) $f(\sqrt{2}) = 3^{\sqrt{2}} \approx 4.7288$	3 ^ √ 2 ENTER	4.7288043

Now Try Exercise 7 ■

■ Graphs of Exponential Functions

We first graph exponential functions by plotting points. We will see that the graphs of such functions have an easily recognizable shape.

EXAMPLE 2 ■ Graphing Exponential Functions by Plotting Points

Draw the graph of each function.

(a) $f(x) = 3^x$ **(b)** $g(x) = \left(\dfrac{1}{3}\right)^x$

SOLUTION We calculate values of $f(x)$ and $g(x)$ and plot points to sketch the graphs in Figure 1.

x	$f(x) = 3^x$	$g(x) = (\frac{1}{3})^x$
−3	$\frac{1}{27}$	27
−2	$\frac{1}{9}$	9
−1	$\frac{1}{3}$	3
0	1	1
1	3	$\frac{1}{3}$
2	9	$\frac{1}{9}$
3	27	$\frac{1}{27}$

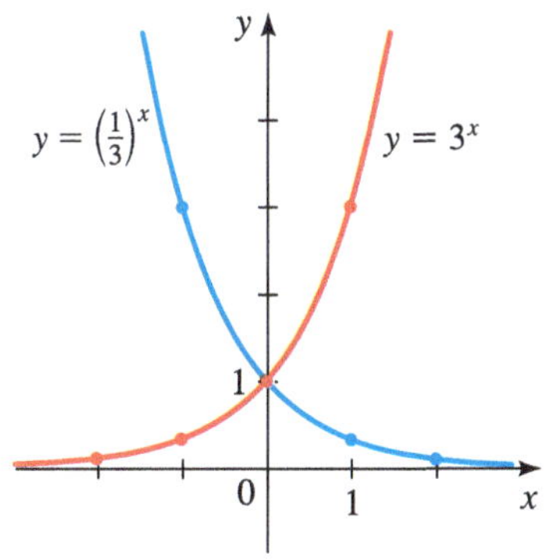

FIGURE 1

Notice that

$$g(x) = \left(\frac{1}{3}\right)^x = \frac{1}{3^x} = 3^{-x} = f(-x)$$

so we could have obtained the graph of g from the graph of f by reflecting in the y-axis.

Reflecting graphs is explained in Section 2.6.

Now Try Exercise 17 ■

Figure 2 shows the graphs of the family of exponential functions $f(x) = a^x$ for various values of the base a. All of these graphs pass through the point $(0, 1)$ because

To see just how quickly $f(x) = 2^x$ increases, let's perform the following thought experiment. Suppose we start with a piece of paper that is a thousandth of an inch thick, and we fold it in half 50 times. Each time we fold the paper, the thickness of the paper stack doubles, so the thickness of the resulting stack would be $2^{50}/1000$ inches. How thick do you think that is? It works out to be more than 17 million miles!

$a^0 = 1$ for $a \neq 0$. You can see from Figure 2 that there are two kinds of exponential functions: If $0 < a < 1$, the exponential function decreases rapidly. If $a > 1$, the function increases rapidly (see the margin note).

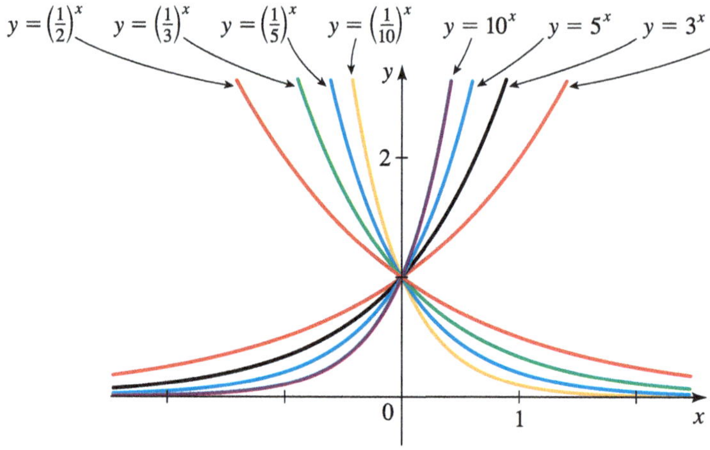

FIGURE 2 A family of exponential functions

See Section 3.6, page 295, where the arrow notation used here is explained.

The x-axis is a horizontal asymptote for the exponential function $f(x) = a^x$. This is because when $a > 1$, we have $a^x \to 0$ as $x \to -\infty$, and when $0 < a < 1$, we have $a^x \to 0$ as $x \to \infty$ (see Figure 2). Also, $a^x > 0$ for all $x \in \mathbb{R}$, so the function $f(x) = a^x$ has domain $\mathbb{R}$ and range $(0, \infty)$. These observations are summarized in the following box.

GRAPHS OF EXPONENTIAL FUNCTIONS

The exponential function

$$f(x) = a^x \qquad a > 0, a \neq 1$$

has domain $\mathbb{R}$ and range $(0, \infty)$. The line $y = 0$ (the x-axis) is a horizontal asymptote of f. The graph of f has one of the following shapes.

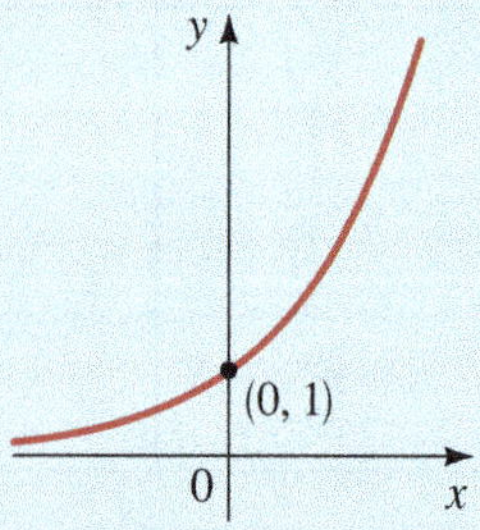

$f(x) = a^x$ for $a > 1$

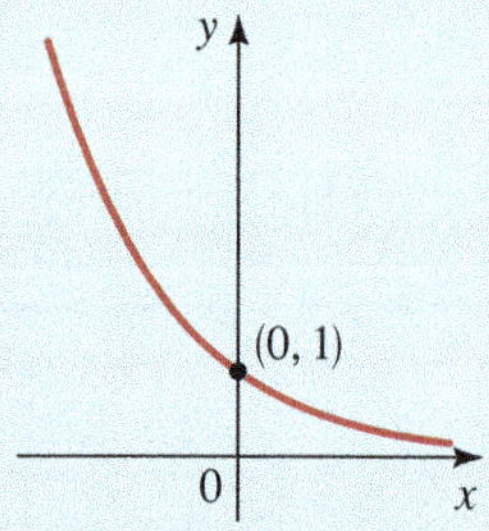

$f(x) = a^x$ for $0 < a < 1$

EXAMPLE 3 ■ Identifying Graphs of Exponential Functions

Find the exponential function $f(x) = a^x$ whose graph is given.

(a)

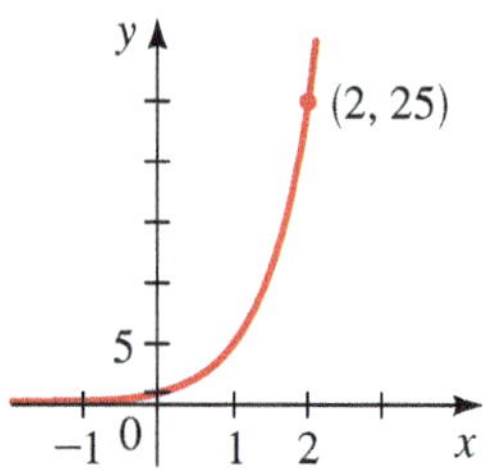

(b)

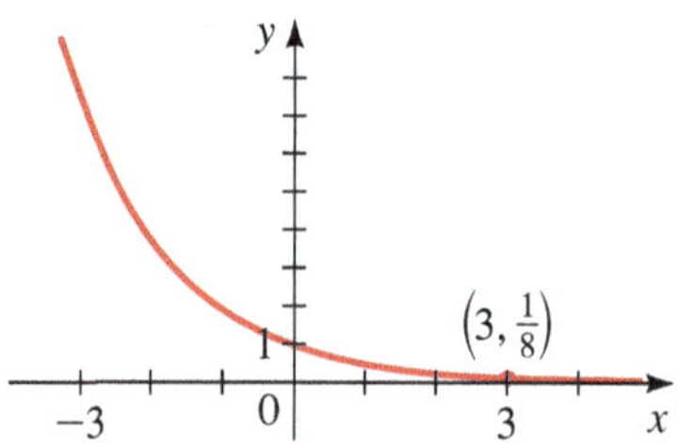

SOLUTION

(a) Since $f(2) = a^2 = 25$, we see that the base is $a = 5$. So $f(x) = 5^x$.

(b) Since $f(3) = a^3 = \frac{1}{8}$, we see that the base is $a = \frac{1}{2}$. So $f(x) = \left(\frac{1}{2}\right)^x$.

Now Try Exercise 21

In the next example we see how to graph certain functions, not by plotting points, but by taking the basic graphs of the exponential functions in Figure 2 and applying the shifting and reflecting transformations of Section 2.6.

EXAMPLE 4 ■ Transformations of Exponential Functions

Use the graph of $f(x) = 2^x$ to sketch the graph of each function. State the domain, range, and asymptote.

(a) $g(x) = 1 + 2^x$ **(b)** $h(x) = -2^x$ **(c)** $k(x) = 2^{x-1}$

SOLUTION

Shifting and reflecting of graphs are explained in Section 2.6.

(a) To obtain the graph of $g(x) = 1 + 2^x$, we start with the graph of $f(x) = 2^x$ and shift it upward 1 unit to get the graph shown in Figure 3(a). From the graph we see that the domain of g is the set $\mathbb{R}$ of real numbers, the range is the interval $(1, \infty)$, and the line $y = 1$ is a horizontal asymptote.

(b) Again we start with the graph of $f(x) = 2^x$, but here we reflect in the x-axis to get the graph of $h(x) = -2^x$ shown in Figure 3(b). From the graph we see that the domain of h is the set $\mathbb{R}$ of all real numbers, the range is the interval $(-\infty, 0)$, and the line $y = 0$ is a horizontal asymptote.

(c) This time we start with the graph of $f(x) = 2^x$ and shift it to the right by 1 unit to get the graph of $k(x) = 2^{x-1}$ shown in Figure 3(c). From the graph we see that the domain of k is the set $\mathbb{R}$ of all real numbers, the range is the interval $(0, \infty)$, and the line $y = 0$ is a horizontal asymptote.

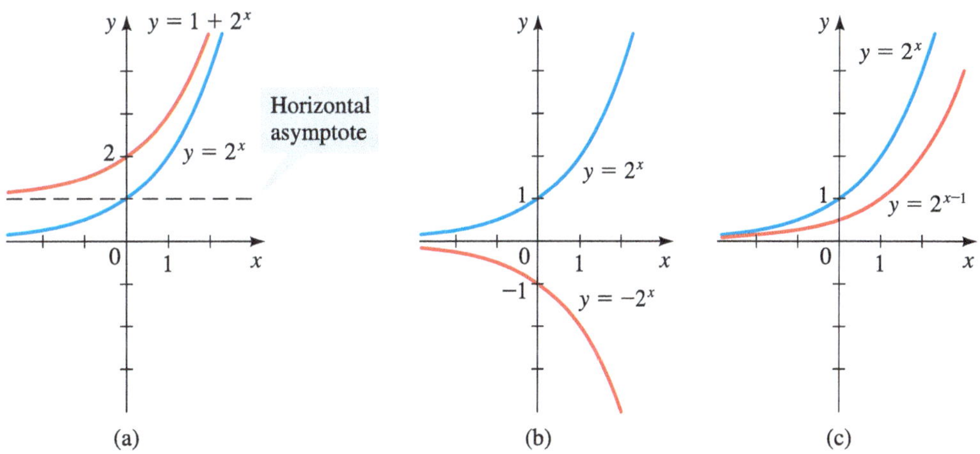

FIGURE 3

Now Try Exercises 27, 29, and 31

EXAMPLE 5 ■ Comparing Exponential and Power Functions

Compare the rates of growth of the exponential function $f(x) = 2^x$ and the power function $g(x) = x^2$ by drawing the graphs of both functions in the following viewing rectangles.

(a) $[0, 3]$ by $[0, 8]$ **(b)** $[0, 6]$ by $[0, 25]$ **(c)** $[0, 20]$ by $[0, 1000]$

SOLUTION

(a) Figure 4(a) shows that the graph of $g(x) = x^2$ catches up with, and becomes higher than, the graph of $f(x) = 2^x$ at $x = 2$.

(b) The larger viewing rectangle in Figure 4(b) shows that the graph of $f(x) = 2^x$ overtakes that of $g(x) = x^2$ when $x = 4$.

(c) Figure 4(c) gives a more global view and shows that when x is large, $f(x) = 2^x$ is much larger than $g(x) = x^2$.

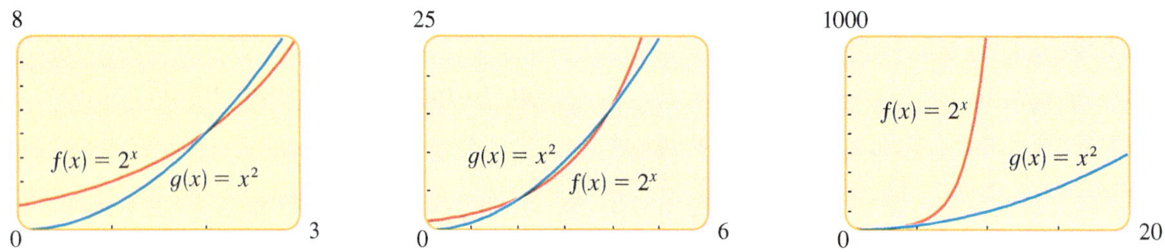

FIGURE 4 (a) (b) (c)

Now Try Exercise 45 ■

Compound Interest

Exponential functions occur in calculating compound interest. If an amount of money P, called the **principal**, is invested at an interest rate i per time period, then after one time period the interest is Pi, and the amount A of money is

$$A = P + Pi = P(1 + i)$$

If the interest is reinvested, then the new principal is $P(1 + i)$, and the amount after another time period is $A = P(1 + i)(1 + i) = P(1 + i)^2$. Similarly, after a third time period the amount is $A = P(1 + i)^3$. In general, after k periods the amount is

$$A = P(1 + i)^k$$

Notice that this is an exponential function with base $1 + i$.

If the annual interest rate is r and if interest is compounded n times per year, then in each time period the interest rate is $i = r/n$, and there are nt time periods in t years. This leads to the following formula for the amount after t years.

COMPOUND INTEREST

Compound interest is calculated by the formula

$$A(t) = P\left(1 + \frac{r}{n}\right)^{nt}$$

where $A(t)$ = amount after t years

P = principal

r = interest rate per year

n = number of times interest is compounded per year

t = number of years

r is often referred to as the nominal annual interest rate.

EXAMPLE 6 ■ Calculating Compound Interest

A sum of \$1000 is invested at an interest rate of 12% per year. Find the amounts in the account after 3 years if interest is compounded annually, semiannually, quarterly, monthly, and daily.

SOLUTION We use the compound interest formula with $P = \$1000$, $r = 0.12$, and $t = 3$.

Compounding	n	Amount after 3 years
Annual	1	$1000\left(1 + \frac{0.12}{1}\right)^{1(3)} = \$1404.93$
Semiannual	2	$1000\left(1 + \frac{0.12}{2}\right)^{2(3)} = \$1418.52$
Quarterly	4	$1000\left(1 + \frac{0.12}{4}\right)^{4(3)} = \$1425.76$
Monthly	12	$1000\left(1 + \frac{0.12}{12}\right)^{12(3)} = \$1430.77$
Daily	365	$1000\left(1 + \frac{0.12}{365}\right)^{365(3)} = \$1433.24$

Now Try Exercise 57 ■

If an investment earns compound interest, then the **annual percentage yield** (APY) is the *simple* interest rate that yields the same amount at the end of one year.

EXAMPLE 7 ■ Calculating the Annual Percentage Yield

Find the annual percentage yield for an investment that earns interest at a rate of 6% per year, compounded daily.

SOLUTION After one year, a principal P will grow to the amount

$$A = P\left(1 + \frac{0.06}{365}\right)^{365} = P(1.06183)$$

Simple interest is studied in Section 1.7.

The formula for simple interest is

$$A = P(1 + r)$$

Comparing, we see that $1 + r = 1.06183$, so $r = 0.06183$. Thus the annual percentage yield is 6.183%.

Now Try Exercise 63 ■

DISCOVERY PROJECT

So You Want to Be a Millionaire?

In this project we explore how rapidly the values of an exponential function increase by examining some real-world situations. For example, if you save a penny today, two pennies tomorrow, four pennies the next day, and so on, how long do you have to continue saving in this way before you become a millionaire? You can find out the surprising answer to this and other questions by completing this discovery project. You can find the project at **www.stewartmath.com**.

4.1 EXERCISES

CONCEPTS

1. The function $f(x) = 5^x$ is an exponential function with base ________; $f(-2) =$ ________, $f(0) =$ ________, $f(2) =$ ________, and $f(6) =$ ________.

2. Match the exponential function with one of the graphs labeled I, II, III, or IV, shown below.

(a) $f(x) = 2^x$ **(b)** $f(x) = 2^{-x}$

(c) $f(x) = -2^x$ **(d)** $f(x) = -2^{-x}$

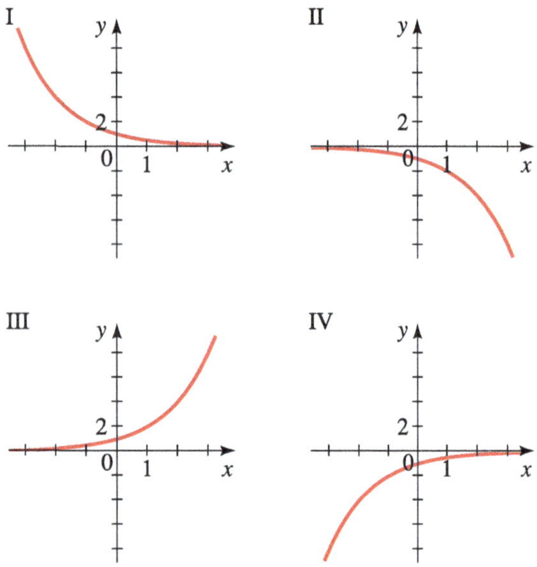

3. (a) To obtain the graph of $g(x) = 2^x - 1$, we start with the graph of $f(x) = 2^x$ and shift it ________ (upward/downward) 1 unit.

(b) To obtain the graph of $h(x) = 2^{x-1}$, we start with the graph of $f(x) = 2^x$ and shift it to the ________ (left/right) 1 unit.

4. In the formula $A(t) = P\left(1 + \frac{r}{n}\right)^{nt}$ for compound interest the letters P, r, n, and t stand for ________, ________, ________, and ________, respectively, and $A(t)$ stands for ________. So if \$100 is invested at an interest rate of 6% compounded quarterly, then the amount after 2 years is ________.

5. The exponential function $f(x) = \left(\frac{1}{2}\right)^x$ has the ________ asymptote $y =$ ________. This means that as $x \to \infty$, we have $\left(\frac{1}{2}\right)^x \to$ ________.

6. The exponential function $f(x) = \left(\frac{1}{2}\right)^x + 3$ has the ________ asymptote $y =$ ________. This means that as $x \to \infty$, we have $\left(\frac{1}{2}\right)^x + 3 \to$ ________.

SKILLS

7–10 ■ Evaluating Exponential Functions Use a calculator to evaluate the function at the indicated values. Round your answers to three decimals.

7. $f(x) = 4^x$; $f(\frac{1}{2}), f(\sqrt{5}), f(-2), f(0.3)$

8. $f(x) = 3^{x-1}$; $f(\frac{1}{2}), f(2.5), f(-1), f(\frac{1}{4})$

9. $g(x) = \left(\frac{1}{3}\right)^{x+1}$; $g(\frac{1}{2}), g(\sqrt{2}), g(-3.5), g(-1.4)$

10. $g(x) = \left(\frac{4}{3}\right)^{3x}$; $g(-\frac{1}{2}), g(\sqrt{6}), g(-3), g(\frac{4}{3})$

11–16 ■ Graphing Exponential Functions Sketch the graph of the function by making a table of values. Use a calculator if necessary.

11. $f(x) = 2^x$ **12.** $g(x) = 8^x$

13. $f(x) = \left(\frac{1}{3}\right)^x$ **14.** $h(x) = (1.1)^x$

15. $g(x) = 3(1.3)^x$ **16.** $h(x) = 2\left(\frac{1}{4}\right)^x$

17–20 ■ Graphing Exponential Functions Graph both functions on one set of axes.

17. $f(x) = 2^x$ and $g(x) = 2^{-x}$

18. $f(x) = 3^{-x}$ and $g(x) = \left(\frac{1}{3}\right)^x$

19. $f(x) = 4^x$ and $g(x) = 7^x$

20. $f(x) = \left(\frac{3}{4}\right)^x$ and $g(x) = 1.5^x$

21–24 ■ Exponential Functions from a Graph Find the exponential function $f(x) = a^x$ whose graph is given.

21.

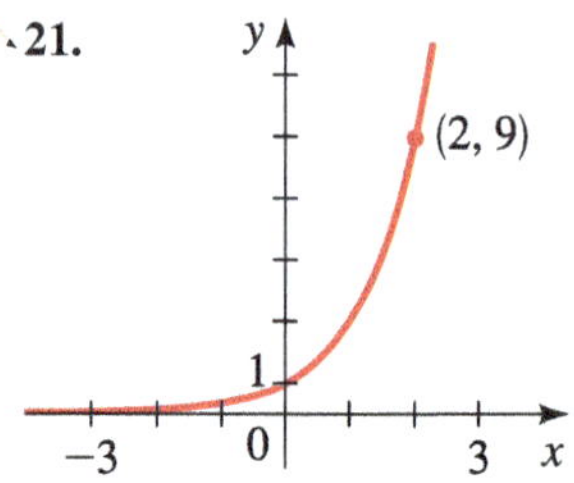

22.

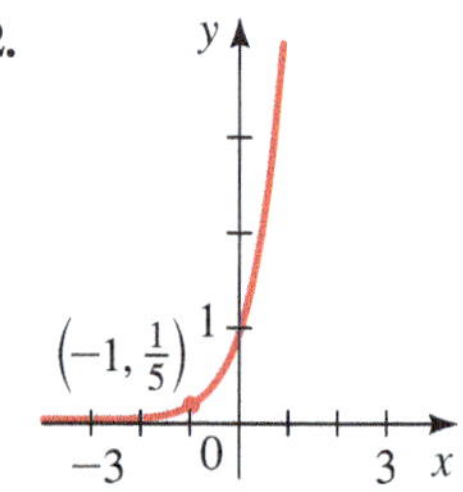

23.

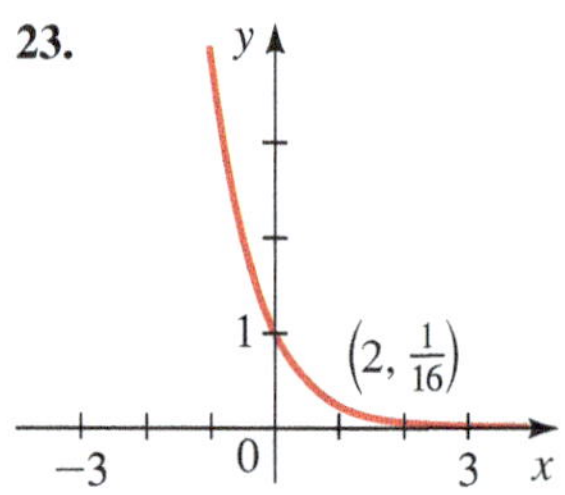

24.

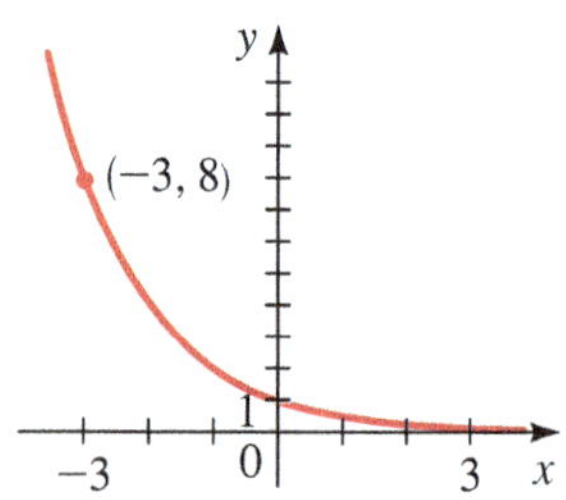

25–26 ■ Exponential Functions from a Graph Match the exponential function with one of the graphs labeled I or II.

25. $f(x) = 5^{x+1}$ **26.** $f(x) = 5^x + 1$

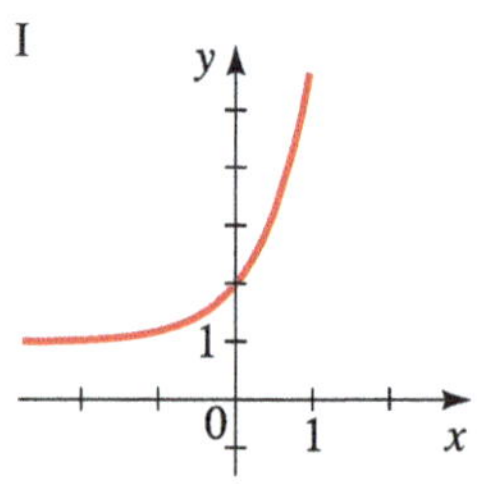

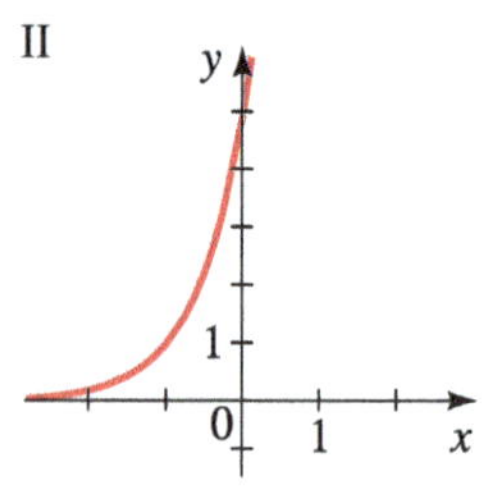

27–40 ■ Graphing Exponential Functions Graph the function, not by plotting points, but by starting from the graphs in Figure 2. State the domain, range, and asymptote.

27. $g(x) = 2^x - 3$

28. $h(x) = 4 + \left(\frac{1}{2}\right)^x$

29. $f(x) = -3^x$

30. $f(x) = 10^{-x}$

31. $f(x) = 10^{x+3}$

32. $g(x) = 2^{x-3}$

33. $y = 5^{-x} + 1$

34. $h(x) = 6 - 3^x$

35. $y = 2 - \left(\frac{1}{3}\right)^x$

36. $y = 5^{-x} - 3$

37. $h(x) = 2^{x-4} + 1$

38. $y = 3 - 10^{x-1}$

39. $g(x) = 1 - 3^{-x}$

40. $y = 3 - \left(\frac{1}{5}\right)^x$

41–42 ■ Comparing Exponential Functions In these exercises we compare the graphs of two exponential functions.

41. **(a)** Sketch the graphs of $f(x) = 2^x$ and $g(x) = 3(2^x)$.
(b) How are the graphs related?

42. **(a)** Sketch the graphs of $f(x) = 9^{x/2}$ and $g(x) = 3^x$.
(b) Use the Laws of Exponents to explain the relationship between these graphs.

43–44 ■ Comparing Exponential and Power Functions Compare the graphs of the power function f and exponential function g by evaluating both of them for $x = 0, 1, 2, 3, 4, 6, 8,$ and 10. Then draw the graphs of f and g on the same set of axes.

43. $f(x) = x^3;\quad g(x) = 3^x$

44. $f(x) = x^4;\quad g(x) = 4^x$

45–46 ■ Comparing Exponential and Power Functions In these exercises we use a graphing calculator to compare the rates of growth of the graphs of a power function and an exponential function.

45. **(a)** Compare the rates of growth of the functions $f(x) = 2^x$ and $g(x) = x^5$ by drawing the graphs of both functions in the following viewing rectangles.
(i) $[0, 5]$ by $[0, 20]$
(ii) $[0, 25]$ by $[0, 10^7]$
(iii) $[0, 50]$ by $[0, 10^8]$
(b) Find the solutions of the equation $2^x = x^5$, rounded to one decimal place.

46. **(a)** Compare the rates of growth of the functions $f(x) = 3^x$ and $g(x) = x^4$ by drawing the graphs of both functions in the following viewing rectangles:
(i) $[-4, 4]$ by $[0, 20]$
(ii) $[0, 10]$ by $[0, 5000]$
(iii) $[0, 20]$ by $[0, 10^5]$
(b) Find the solutions of the equation $3^x = x^4$, rounded to two decimal places.

SKILLS Plus

47–48 ■ Families of Functions Draw graphs of the given family of functions for $c = 0.25, 0.5, 1, 2, 4$. How are the graphs related?

47. $f(x) = c2^x$

48. $f(x) = 2^{cx}$

49–50 ■ Getting Information from a Graph Find, rounded to two decimal places, **(a)** the intervals on which the function is increasing or decreasing and **(b)** the range of the function.

49. $y = 10^{x-x^2}$

50. $y = x2^x$

51–52 ■ Difference Quotients These exercises involve a difference quotient for an exponential function.

51. If $f(x) = 10^x$, show that

$$\frac{f(x+h) - f(x)}{h} = 10^x\left(\frac{10^h - 1}{h}\right)$$

52. If $f(x) = 3^{x-1}$, show that

$$\frac{f(x+h) - f(x)}{h} = 3^{x-1}\left(\frac{3^h - 1}{h}\right)$$

APPLICATIONS

53. Bacteria Growth A bacteria culture contains 1500 bacteria initially and doubles every hour.
(a) Find a function N that models the number of bacteria after t hours.
(b) Find the number of bacteria after 24 hours.

54. Mouse Population A certain breed of mouse was introduced onto a small island with an initial population of 320 mice, and scientists estimate that the mouse population is doubling every year.
(a) Find a function N that models the number of mice after t years.
(b) Estimate the mouse population after 8 years.

55–56 ■ Compound Interest An investment of \$5000 is deposited into an account in which interest is compounded monthly. Complete the table by filling in the amounts to which the investment grows at the indicated times or interest rates.

55. $r = 4\%$

Time (years)	Amount
1	
2	
3	
4	
5	
6	

56. $t = 5$ years

Rate per year	Amount
1%	
2%	
3%	
4%	
5%	
6%	

57. Compound Interest If \$10,000 is invested at an interest rate of 3% per year, compounded semiannually, find the value of the investment after the given number of years.
(a) 5 years **(b)** 10 years **(c)** 15 years

58. Compound Interest If \$2500 is invested at an interest rate of 2.5% per year, compounded daily, find the value of the investment after the given number of years.
(a) 2 years **(b)** 3 years **(c)** 6 years

59. Compound Interest If \$500 is invested at an interest rate of 3.75% per year, compounded quarterly, find the value of the investment after the given number of years.
(a) 1 year **(b)** 2 years **(c)** 10 years

60. Compound Interest If \$4000 is borrowed at a rate of 5.75% interest per year, compounded quarterly, find the amount due at the end of the given number of years.
(a) 4 years **(b)** 6 years **(c)** 8 years

61–62 ■ Present Value The **present value** of a sum of money is the amount that must be invested now, at a given rate of interest, to produce the desired sum at a later date.

61. Find the present value of \$10,000 if interest is paid at a rate of 9% per year, compounded semiannually, for 3 years.

62. Find the present value of \$100,000 if interest is paid at a rate of 8% per year, compounded monthly, for 5 years.

63. Annual Percentage Yield Find the annual percentage yield for an investment that earns 8% per year, compounded monthly.

64. Annual Percentage Yield Find the annual percentage yield for an investment that earns $5\frac{1}{2}\%$ per year, compounded quarterly.

DISCUSS ■ DISCOVER ■ PROVE ■ WRITE

65. DISCUSS ■ DISCOVER: Growth of an Exponential Function Suppose you are offered a job that lasts one month, and you are to be very well paid. Which of the following methods of payment is more profitable for you?

(a) One million dollars at the end of the month

(b) Two cents on the first day of the month, 4 cents on the second day, 8 cents on the third day, and, in general, 2^n cents on the nth day

66. DISCUSS ■ DISCOVER: The Height of the Graph of an Exponential Function Your mathematics instructor asks you to sketch a graph of the exponential function

$$f(x) = 2^x$$

for x between 0 and 40, using a scale of 10 units to one inch. What are the dimensions of the sheet of paper you will need to sketch this graph?

4.2 THE NATURAL EXPONENTIAL FUNCTION

■ The Number *e* ■ The Natural Exponential Function ■ Continuously Compounded Interest

Any positive number can be used as a base for an exponential function. In this section we study the special base e, which is convenient for applications involving calculus.

© iStockphoto.com/Kubrak78

The **Gateway Arch** in St. Louis, Missouri, is shaped in the form of the graph of a combination of exponential functions (*not* a parabola, as it might first appear). Specifically, it is a **catenary**, which is the graph of an equation of the form

$$y = a(e^{bx} + e^{-bx})$$

(see Exercises 17 and 19). This shape was chosen because it is optimal for distributing the internal structural forces of the arch. Chains and cables suspended between two points (for example, the stretches of cable between pairs of telephone poles) hang in the shape of a catenary.

■ The Number *e*

The number e is defined as the value that $(1 + 1/n)^n$ approaches as n becomes large. (In calculus this idea is made more precise through the concept of a limit.) The table shows the values of the expression $(1 + 1/n)^n$ for increasingly large values of n.

n	$\left(1 + \dfrac{1}{n}\right)^n$
1	2.00000
5	2.48832
10	2.59374
100	2.70481
1000	2.71692
10,000	2.71815
100,000	2.71827
1,000,000	2.71828

It appears that, rounded to five decimal places, $e \approx 2.71828$; in fact, the approximate value to 20 decimal places is

$$e \approx 2.71828182845904523536$$

It can be shown that e is an irrational number, so we cannot write its exact value in decimal form.

■ The Natural Exponential Function

The notation e was chosen by Leonhard Euler (see page 63), probably because it is the first letter of the word *exponential.*

The number e is the base for the natural exponential function. Why use such a strange base for an exponential function? It might seem at first that a base such as 10 is easier to work with. We will see, however, that in certain applications the number e is the best

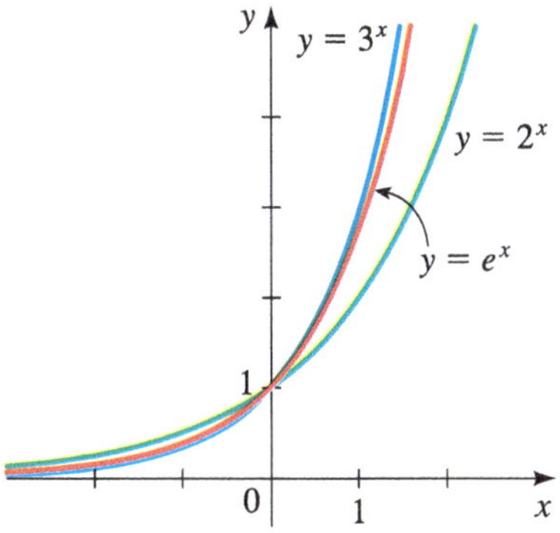

FIGURE 1 Graph of the natural exponential function

possible base. In this section we study how e occurs in the description of compound interest.

THE NATURAL EXPONENTIAL FUNCTION

The **natural exponential function** is the exponential function

$$f(x) = e^x$$

with base e. It is often referred to as *the* exponential function.

Since $2 < e < 3$, the graph of the natural exponential function lies between the graphs of $y = 2^x$ and $y = 3^x$, as shown in Figure 1.

Scientific calculators have a special key for the function $f(x) = e^x$. We use this key in the next example.

EXAMPLE 1 ■ Evaluating the Exponential Function

Evaluate each expression rounded to five decimal places.

(a) e^3 **(b)** $2e^{-0.53}$ **(c)** $e^{4.8}$

SOLUTION We use the $\boxed{e^x}$ key on a calculator to evaluate the exponential function.

(a) $e^3 \approx 20.08554$ **(b)** $2e^{-0.53} \approx 1.17721$ **(c)** $e^{4.8} \approx 121.51042$

Now Try Exercise 3 ■

EXAMPLE 2 ■ Graphing the Exponential Functions

Sketch the graph of each function. State the domain, range, and asymptote.

(a) $f(x) = e^{-x}$ **(b)** $g(x) = 3e^{0.5x}$

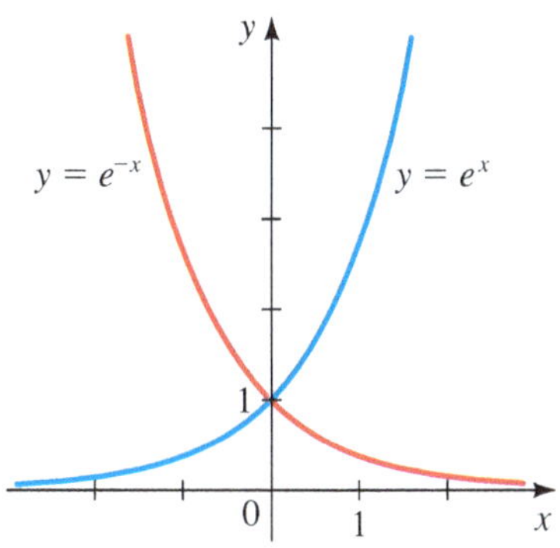

FIGURE 2

SOLUTION

(a) We start with the graph of $y = e^x$ and reflect in the y-axis to obtain the graph of $y = e^{-x}$ as in Figure 2. From the graph we see that the domain of f is the set $\mathbb{R}$ of all real numbers, the range is the interval $(0, \infty)$, and the line $y = 0$ is a horizontal asymptote.

(b) We calculate several values, plot the resulting points, then connect the points with a smooth curve. The graph is shown in Figure 3. From the graph we see that the domain of g is the set $\mathbb{R}$ of all real numbers, the range is the interval $(0, \infty)$, and the line $y = 0$ is a horizontal asymptote.

x	$f(x) = 3e^{0.5x}$
−3	0.67
−2	1.10
−1	1.82
0	3.00
1	4.95
2	8.15
3	13.45

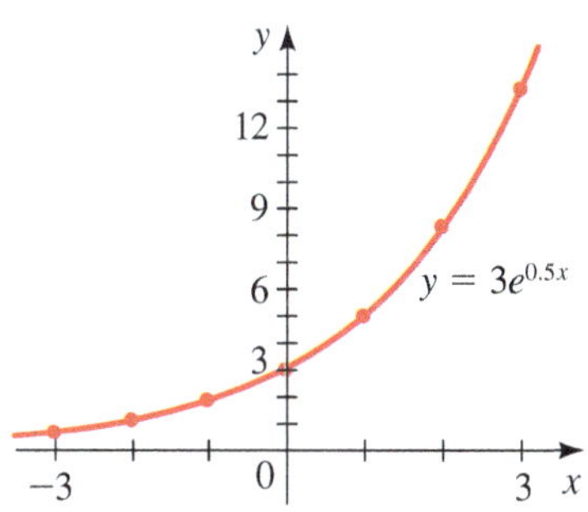

FIGURE 3

Now Try Exercises 5 and 7 ■

EXAMPLE 3 ■ An Exponential Model for the Spread of a Virus

An infectious disease begins to spread in a small city of population 10,000. After t days, the number of people who have succumbed to the virus is modeled by the function

$$v(t) = \frac{10{,}000}{5 + 1245e^{-0.97t}}$$

(a) How many infected people are there initially (at time $t = 0$)?

(b) Find the number of infected people after one day, two days, and five days.

(c) Graph the function v, and describe its behavior.

SOLUTION

(a) Since $v(0) = 10{,}000/(5 + 1245e^0) = 10{,}000/1250 = 8$, we conclude that 8 people initially have the disease.

(b) Using a calculator, we evaluate $v(1)$, $v(2)$, and $v(5)$ and then round off to obtain the following values.

Days	Infected people
1	21
2	54
5	678

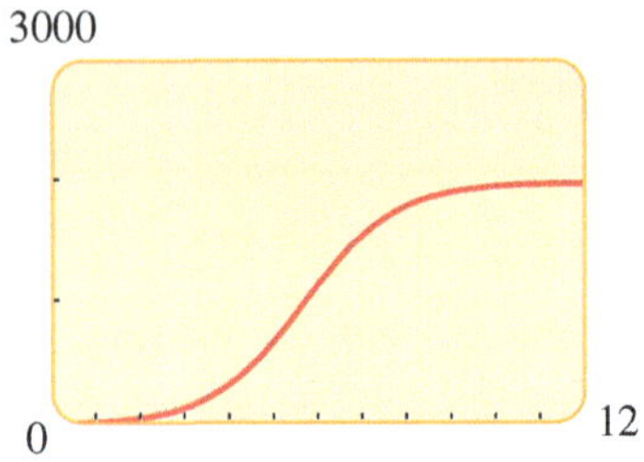

FIGURE 4

$$v(t) = \frac{10{,}000}{5 + 1245e^{-0.97t}}$$

(c) From the graph in Figure 4 we see that the number of infected people first rises slowly, then rises quickly between day 3 and day 8, and then levels off when about 2000 people are infected.

Now Try Exercise 27 ■

The graph in Figure 4 is called a *logistic curve* or a *logistic growth model*. Curves like it occur frequently in the study of population growth. (See Exercises 27–30.)

Continuously Compounded Interest

In Example 6 of Section 4.1 we saw that the interest paid increases as the number of compounding periods n increases. Let's see what happens as n increases indefinitely. If we let $m = n/r$, then

$$A(t) = P\left(1 + \frac{r}{n}\right)^{nt} = P\left[\left(1 + \frac{r}{n}\right)^{n/r}\right]^{rt} = P\left[\left(1 + \frac{1}{m}\right)^{m}\right]^{rt}$$

Recall that as m becomes large, the quantity $(1 + 1/m)^m$ approaches the number e. Thus the amount approaches $A = Pe^{rt}$. This expression gives the amount when the interest is compounded at "every instant."

CONTINUOUSLY COMPOUNDED INTEREST

Continuously compounded interest is calculated by the formula

$$A(t) = Pe^{rt}$$

where $A(t)$ = amount after t years

P = principal

r = interest rate per year

t = number of years

EXAMPLE 4 ■ Calculating Continuously Compounded Interest

Find the amount after 3 years if \$1000 is invested at an interest rate of 12% per year, compounded continuously.

SOLUTION We use the formula for continuously compounded interest with $P = \$1000$, $r = 0.12$, and $t = 3$ to get

$$A(3) = 1000e^{(0.12)3} = 1000e^{0.36} = \$1433.33$$

Compare this amount with the amounts in Example 6 of Section 4.1.

Now Try Exercise 33 ■

4.2 EXERCISES

CONCEPTS

1. The function $f(x) = e^x$ is called the ________ exponential function. The number e is approximately equal to ________.

2. In the formula $A(t) = Pe^{rt}$ for continuously compound interest, the letters P, r, and t stand for ________, ________, and ________, respectively, and $A(t)$ stands for ________. So if \$100 is invested at an interest rate of 6% compounded continuously, then the amount after 2 years is ________.

SKILLS

3–4 ■ Evaluating Exponential Functions Use a calculator to evaluate the function at the indicated values. Round your answers to three decimals.

3. $h(x) = e^x$; $h(1), h(\pi), h(-3), h(\sqrt{2})$

4. $h(x) = e^{-3x}$; $h(\frac{1}{3}), h(1.5), h(-1), h(-\pi)$

5–6 ■ Graphing Exponential Functions Complete the table of values, rounded to two decimal places, and sketch a graph of the function.

5.

x	$f(x) = 1.5e^x$
−2	
−1	
−0.5	
0	
0.5	
1	
2	

6.

x	$f(x) = 4e^{-x/3}$
−3	
−2	
−1	
0	
1	
2	
3	

7–16 ■ Graphing Exponential Functions Graph the function, not by plotting points, but by starting from the graph of $y = e^x$ in Figure 1. State the domain, range, and asymptote.

7. $g(x) = 2 + e^x$
8. $h(x) = e^{-x} - 3$
9. $f(x) = -e^x$
10. $y = 1 - e^x$
11. $y = e^{-x} - 1$
12. $f(x) = -e^{-x}$
13. $f(x) = e^{x-2}$
14. $y = e^{x-3} + 4$
15. $h(x) = e^{x+1} - 3$
16. $g(x) = -e^{x-1} - 2$

SKILLS Plus

17. **Hyperbolic Cosine Function** The *hyperbolic cosine function* is defined by

$$\cosh(x) = \frac{e^x + e^{-x}}{2}$$

(a) Sketch the graphs of the functions $y = \frac{1}{2}e^x$ and $y = \frac{1}{2}e^{-x}$ on the same axes, and use graphical addition (see Section 2.7) to sketch the graph of $y = \cosh(x)$.

(b) Use the definition to show that $\cosh(-x) = \cosh(x)$.

18. **Hyperbolic Sine Function** The *hyperbolic sine function* is defined by

$$\sinh(x) = \frac{e^x - e^{-x}}{2}$$

(a) Sketch the graph of this function using graphical addition as in Exercise 17.

(b) Use the definition to show that $\sinh(-x) = -\sinh(x)$

19. **Families of Functions**

(a) Draw the graphs of the family of functions

$$f(x) = \frac{a}{2}(e^{x/a} + e^{-x/a})$$

for $a = 0.5, 1, 1.5$, and 2.

(b) How does a larger value of a affect the graph?

20. **The Definition of *e*** Illustrate the definition of the number e by graphing the curve $y = (1 + 1/x)^x$ and the line $y = e$ on the same screen, using the viewing rectangle $[0, 40]$ by $[0, 4]$.

21–22 ■ Local Extrema Find the local maximum and minimum values of the function and the value of x at which each occurs. State each answer rounded to two decimal places.

21. $g(x) = x^x, \quad x > 0$

22. $g(x) = e^x + e^{-2x}$

APPLICATIONS

23. Medical Drugs When a certain medical drug is administered to a patient, the number of milligrams remaining in the patient's bloodstream after t hours is modeled by

$$D(t) = 50e^{-0.2t}$$

How many milligrams of the drug remain in the patient's bloodstream after 3 hours?

24. Radioactive Decay A radioactive substance decays in such a way that the amount of mass remaining after t days is given by the function

$$m(t) = 13e^{-0.015t}$$

where $m(t)$ is measured in kilograms.

(a) Find the mass at time $t = 0$.

(b) How much of the mass remains after 45 days?

25. Sky Diving A sky diver jumps from a reasonable height above the ground. The air resistance she experiences is proportional to her velocity, and the constant of proportionality is 0.2. It can be shown that the downward velocity of the sky diver at time t is given by

$$v(t) = 180(1 - e^{-0.2t})$$

where t is measured in seconds (s) and $v(t)$ is measured in feet per second (ft/s).

(a) Find the initial velocity of the sky diver.

(b) Find the velocity after 5 s and after 10 s.

(c) Draw a graph of the velocity function $v(t)$.

(d) The maximum velocity of a falling object with wind resistance is called its *terminal velocity*. From the graph in part (c) find the terminal velocity of this sky diver.

$v(t) = 180(1 - e^{-0.2t})$

26. Mixtures and Concentrations A 50-gal barrel is filled completely with pure water. Salt water with a concentration of 0.3 lb/gal is then pumped into the barrel, and the resulting mixture overflows at the same rate. The amount of salt in the barrel at time t is given by

$$Q(t) = 15(1 - e^{-0.04t})$$

where t is measured in minutes and $Q(t)$ is measured in pounds.

(a) How much salt is in the barrel after 5 min?

(b) How much salt is in the barrel after 10 min?

(c) Draw a graph of the function $Q(t)$.

(d) Use the graph in part (c) to determine the value that the amount of salt in the barrel approaches as t becomes large. Is this what you would expect?

$Q(t) = 15(1 - e^{-0.04t})$

27. Logistic Growth Animal populations are not capable of unrestricted growth because of limited habitat and food supplies. Under such conditions the population follows a *logistic growth model*:

$$P(t) = \frac{d}{1 + ke^{-ct}}$$

where c, d, and k are positive constants. For a certain fish population in a small pond $d = 1200$, $k = 11$, $c = 0.2$, and t is measured in years. The fish were introduced into the pond at time $t = 0$.

(a) How many fish were originally put in the pond?

(b) Find the population after 10, 20, and 30 years.

(c) Evaluate $P(t)$ for large values of t. What value does the population approach as $t \to \infty$? Does the graph shown confirm your calculations?

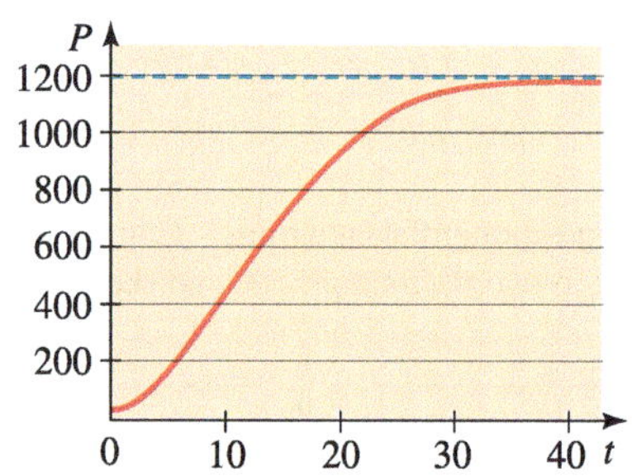

28. Bird Population The population of a certain species of bird is limited by the type of habitat required for nesting. The population behaves according to the logistic growth model

$$n(t) = \frac{5600}{0.5 + 27.5e^{-0.044t}}$$

where t is measured in years.

(a) Find the initial bird population.

(b) Draw a graph of the function $n(t)$.

(c) What size does the population approach as time goes on?

29. World Population The relative growth rate of world population has been decreasing steadily in recent years. On the basis of this, some population models predict that world population will eventually stabilize at a level that the planet can support. One such logistic model is

$$P(t) = \frac{73.2}{6.1 + 5.9e^{-0.02t}}$$

where $t = 0$ is the year 2000 and population is measured in billions.

(a) What world population does this model predict for the year 2200? For 2300?

(b) Sketch a graph of the function P for the years 2000 to 2500.

(c) According to this model, what size does the world population seem to approach as time goes on?

30. Tree Diameter For a certain type of tree the diameter D (in feet) depends on the tree's age t (in years) according to the logistic growth model

$$D(t) = \frac{5.4}{1 + 2.9e^{-0.01t}}$$

Find the diameter of a 20-year-old tree.

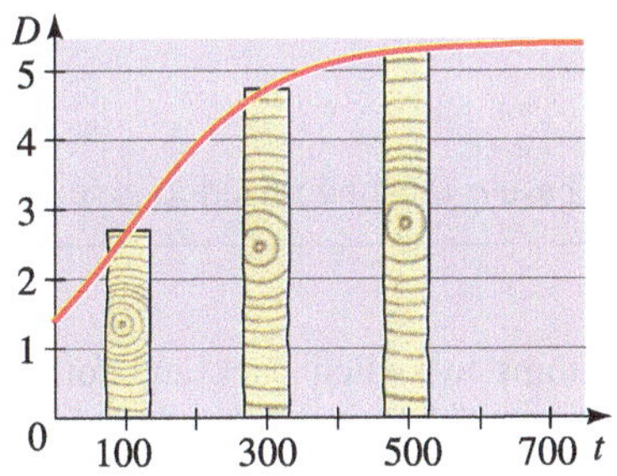

31–32 ■ Compound Interest An investment of \$7000 is deposited into an account in which interest is compounded continuously. Complete the table by filling in the amounts to which the investment grows at the indicated times or interest rates.

31. $r = 3\%$

Time (years)	Amount
1	
2	
3	
4	
5	
6	

32. $t = 10$ years

Rate per year	Amount
1%	
2%	
3%	
4%	
5%	
6%	

33. Compound Interest If \$2000 is invested at an interest rate of 3.5% per year, compounded continuously, find the value of the investment after the given number of years.

(a) 2 years

(b) 4 years

(c) 12 years

34. Compound Interest If \$3500 is invested at an interest rate of 6.25% per year, compounded continuously, find the value of the investment after the given number of years.

(a) 3 years

(b) 6 years

(c) 9 years

35. Compound Interest If \$600 is invested at an interest rate of 2.5% per year, find the amount of the investment at the end of 10 years for the following compounding methods.

(a) Annually

(b) Semiannually

(c) Quarterly

(d) Continuously

36. Compound Interest If \$8000 is invested in an account for which interest is compounded continuously, find the amount of the investment at the end of 12 years for the following interest rates.

(a) 2%

(b) 3%

(c) 4.5%

(d) 7%

37. Compound Interest Which of the given interest rates and compounding periods would provide the best investment?

(a) $2\frac{1}{2}\%$ per year, compounded semiannually

(b) $2\frac{1}{4}\%$ per year, compounded monthly

(c) 2% per year, compounded continuously

38. Compound Interest Which of the given interest rates and compounding periods would provide the better investment?

(a) $5\frac{1}{8}\%$ per year, compounded semiannually

(b) 5% per year, compounded continuously

39. Investment A sum of \$5000 is invested at an interest rate of 9% per year, compounded continuously.

(a) Find the value $A(t)$ of the investment after t years.

(b) Draw a graph of $A(t)$.

(c) Use the graph of $A(t)$ to determine when this investment will amount to \$25,000.

4.3 LOGARITHMIC FUNCTIONS

■ Logarithmic Functions ■ Graphs of Logarithmic Functions ■ Common Logarithms ■ Natural Logarithms

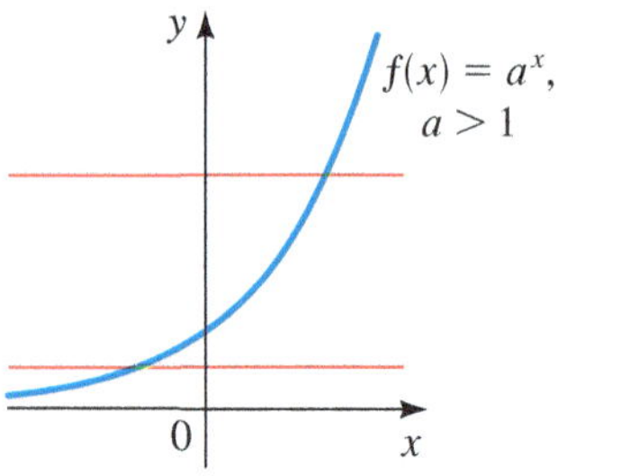

FIGURE 1 $f(x) = a^x$ is one-to-one.

In this section we study the inverses of exponential functions.

■ Logarithmic Functions

Every exponential function $f(x) = a^x$, with $a > 0$ and $a \neq 1$, is a one-to-one function by the Horizontal Line Test (see Figure 1 for the case $a > 1$) and therefore has an inverse function. The inverse function f^{-1} is called the *logarithmic function with base a* and is denoted by $\log_a$. Recall from Section 2.8 that f^{-1} is defined by

$$f^{-1}(x) = y \quad \Leftrightarrow \quad f(y) = x$$

This leads to the following definition of the logarithmic function.

We read $\log_a x = y$ as "log base a of x is y."

DEFINITION OF THE LOGARITHMIC FUNCTION

Let a be a positive number with $a \neq 1$. The **logarithmic function with base *a***, denoted by $\mathbf{log}_a$, is defined by

$$\log_a x = y \quad \Leftrightarrow \quad a^y = x$$

So $\log_a x$ is the *exponent* to which the base a must be raised to give x.

By tradition the name of the logarithmic function is $\log_a$, not just a single letter. Also, we usually omit the parentheses in the function notation and write

$$\log_a(x) = \log_a x$$

When we use the definition of logarithms to switch back and forth between the **logarithmic form** $\log_a x = y$ and the **exponential form** $a^y = x$, it is helpful to notice that, in both forms, the base is the same.

Logarithmic form	Exponential form
Exponent	Exponent
$\log_a x = y$	$a^y = x$
Base	Base

EXAMPLE 1 ■ Logarithmic and Exponential Forms

The logarithmic and exponential forms are equivalent equations: If one is true, then so is the other. So we can switch from one form to the other as in the following illustrations.

Logarithmic form	Exponential form
$\log_{10} 100{,}000 = 5$	$10^5 = 100{,}000$
$\log_2 8 = 3$	$2^3 = 8$
$\log_2\left(\frac{1}{8}\right) = -3$	$2^{-3} = \frac{1}{8}$
$\log_5 s = r$	$5^r = s$

Now Try Exercise 7 ■

x	$\log_{10} x$
10^4	4
10^3	3
10^2	2
10	1
1	0
10^{-1}	-1
10^{-2}	-2
10^{-3}	-3
10^{-4}	-4

It is important to understand that $\log_a x$ is an *exponent*. For example, the numbers in the right-hand column of the table in the margin are the logarithms (base 10) of the numbers in the left-hand column. This is the case for all bases, as the following example illustrates.

EXAMPLE 2 ■ Evaluating Logarithms

(a) $\log_{10} 1000 = 3$ because $10^3 = 1000$

(b) $\log_2 32 = 5$ because $2^5 = 32$

(c) $\log_{10} 0.1 = -1$ because $10^{-1} = 0.1$

(d) $\log_{16} 4 = \frac{1}{2}$ because $16^{1/2} = 4$

Now Try Exercises 9 and 11

Inverse Function Property:

$$f^{-1}(f(x)) = x$$

$$f(f^{-1}(x)) = x$$

When we apply the Inverse Function Property described on page 222 to $f(x) = a^x$ and $f^{-1}(x) = \log_a x$, we get

$$\log_a(a^x) = x \qquad x \in \mathbb{R}$$

$$a^{\log_a x} = x \qquad x > 0$$

We list these and other properties of logarithms discussed in this section.

PROPERTIES OF LOGARITHMS

Property	Reason
1. $\log_a 1 = 0$	We must raise a to the power 0 to get 1.
2. $\log_a a = 1$	We must raise a to the power 1 to get a.
3. $\log_a a^x = x$	We must raise a to the power x to get a^x.
4. $a^{\log_a x} = x$	$\log_a x$ is the power to which a must be raised to get x.

EXAMPLE 3 ■ Applying Properties of Logarithms

We illustrate the properties of logarithms when the base is 5.

$\log_5 1 = 0$ Property 1 $\qquad \log_5 5 = 1$ Property 2

$\log_5 5^8 = 8$ Property 3 $\qquad 5^{\log_5 12} = 12$ Property 4

Now Try Exercises 25 and 31

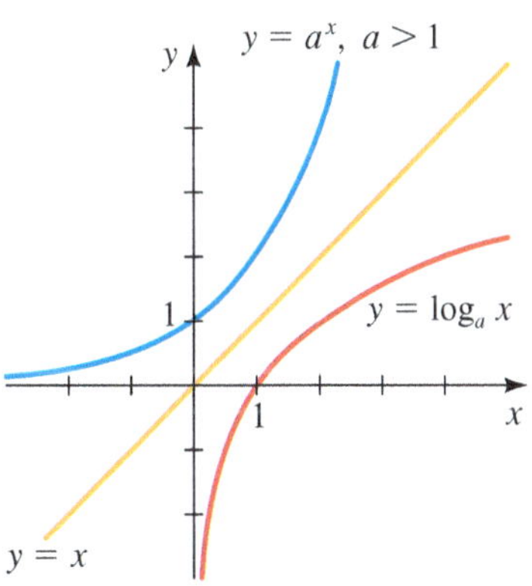

FIGURE 2 Graph of the logarithmic function $f(x) = \log_a x$

■ Graphs of Logarithmic Functions

Recall that if a one-to-one function f has domain A and range B, then its inverse function f^{-1} has domain B and range A. Since the exponential function $f(x) = a^x$ with $a \neq 1$ has domain $\mathbb{R}$ and range $(0, \infty)$, we conclude that its inverse function, $f^{-1}(x) = \log_a x$, has domain $(0, \infty)$ and range $\mathbb{R}$.

The graph of $f^{-1}(x) = \log_a x$ is obtained by reflecting the graph of $f(x) = a^x$ in the line $y = x$. Figure 2 shows the case $a > 1$. The fact that $y = a^x$ (for $a > 1$) is a very rapidly increasing function for $x > 0$ implies that $y = \log_a x$ is a very slowly increasing function for $x > 1$ (see Exercise 102).

Since $\log_a 1 = 0$, the x-intercept of the function $y = \log_a x$ is 1. The y-axis is a vertical asymptote of $y = \log_a x$ because $\log_a x \to -\infty$ as $x \to 0^+$.

EXAMPLE 4 ■ Graphing a Logarithmic Function by Plotting Points

Sketch the graph of $f(x) = \log_2 x$.

SOLUTION To make a table of values, we choose the x-values to be powers of 2 so that we can easily find their logarithms. We plot these points and connect them with a smooth curve as in Figure 3.

x	$\log_2 x$
2^3	3
2^2	2
2	1
1	0
2^{-1}	-1
2^{-2}	-2
2^{-3}	-3
2^{-4}	-4

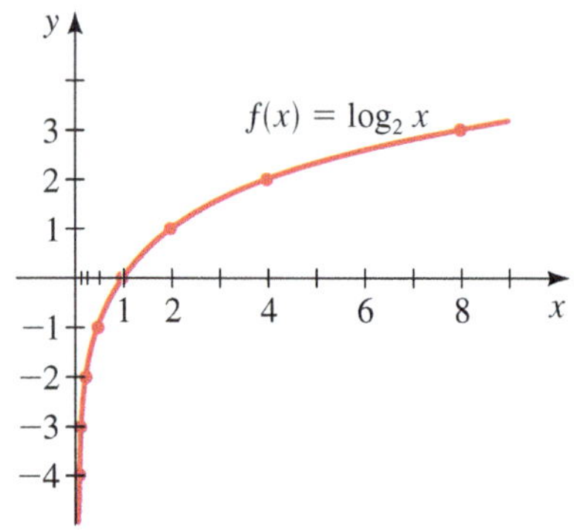

FIGURE 3

Now Try Exercise 49 ■

Figure 4 shows the graphs of the family of logarithmic functions with bases 2, 3, 5, and 10. These graphs are drawn by reflecting the graphs of $y = 2^x$, $y = 3^x$, $y = 5^x$, and $y = 10^x$ (see Figure 2 in Section 4.1) in the line $y = x$. We can also plot points as an aid to sketching these graphs, as illustrated in Example 4.

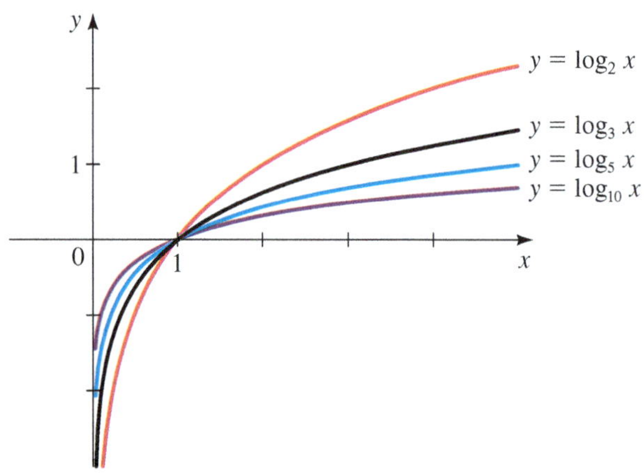

FIGURE 4 A family of logarithmic functions

In the next two examples we graph logarithmic functions by starting with the basic graphs in Figure 4 and using the transformations of Section 2.6.

EXAMPLE 5 ■ Reflecting Graphs of Logarithmic Functions

Sketch the graph of each function. State the domain, range, and asymptote.

(a) $g(x) = -\log_2 x$ **(b)** $h(x) = \log_2(-x)$

SOLUTION

(a) We start with the graph of $f(x) = \log_2 x$ and reflect in the x-axis to get the graph of $g(x) = -\log_2 x$ in Figure 5(a). From the graph we see that the domain of g is $(0, \infty)$, the range is the set $\mathbb{R}$ of all real numbers, and the line $x = 0$ is a vertical asymptote.

(b) We start with the graph of $f(x) = \log_2 x$ and reflect in the y-axis to get the graph of $h(x) = \log_2(-x)$ in Figure 5(b). From the graph we see that the domain of h is $(-\infty, 0)$, the range is the set $\mathbb{R}$ of all real numbers, and the line $x = 0$ is a vertical asymptote.

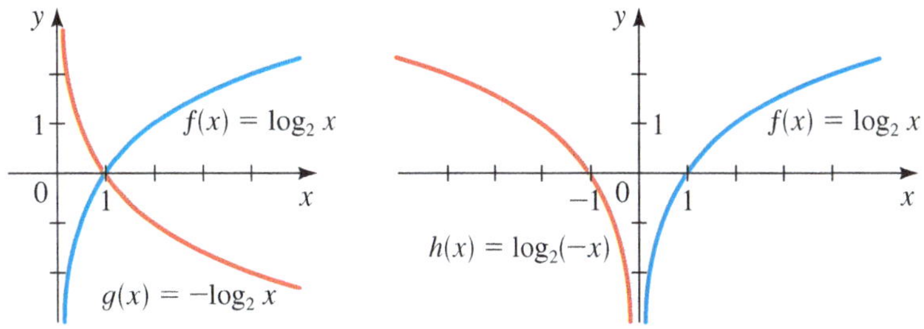

FIGURE 5

Now Try Exercise 61

EXAMPLE 6 ■ Shifting Graphs of Logarithmic Functions

Sketch the graph of each function. State the domain, range, and asymptote.

(a) $g(x) = 2 + \log_5 x$ **(b)** $h(x) = \log_{10}(x - 3)$

SOLUTION

(a) The graph of g is obtained from the graph of $f(x) = \log_5 x$ (Figure 4) by shifting upward 2 units, as shown in Figure 6. From the graph we see that the domain of g is $(0, \infty)$, the range is the set $\mathbb{R}$ of all real numbers, and the line $x = 0$ is a vertical asymptote.

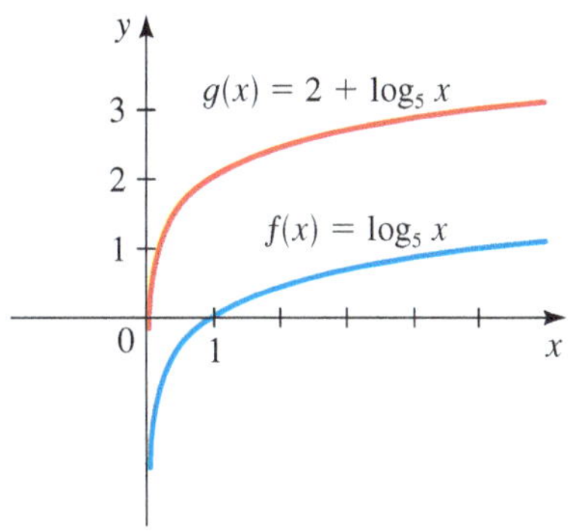

FIGURE 6

(b) The graph of h is obtained from the graph of $f(x) = \log_{10} x$ (Figure 4) by shifting to the right 3 units, as shown in Figure 7. From the graph we see that the domain of h is $(3, \infty)$, the range is the set $\mathbb{R}$ of all real numbers, and the line $x = 3$ is a vertical asymptote.

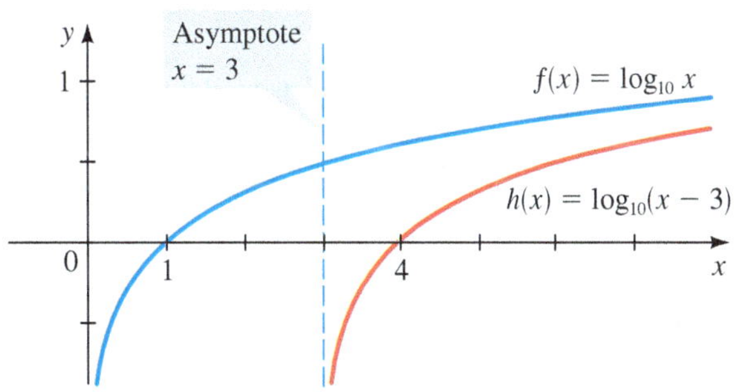

FIGURE 7

Now Try Exercises 63 and 67

Mathematics in the Modern World

Bettmann/CORBIS

Hulton-Deutsch Collection/Historical/Corbis

Law Enforcement

Mathematics aids law enforcement in numerous and surprising ways, from the reconstruction of bullet trajectories to determining the time of death to calculating the probability that a DNA sample is from a particular person. One interesting use is in the search for missing persons. A person who has been missing for several years might look quite different from his or her most recent available photograph. This is particularly true if the missing person is a child. Have you ever wondered what you will look like 5, 10, or 15 years from now?

Researchers have found that different parts of the body grow at different rates. For example, you have no doubt noticed that a baby's head is much larger relative to its body than an adult's. As another example, the ratio of arm length to height is $\frac{1}{3}$ in a child but about $\frac{2}{5}$ in an adult. By collecting data and analyzing the graphs, researchers are able to determine the functions that model growth. As in all growth phenomena, exponential and logarithmic functions play a crucial role. For instance, the formula that relates arm length l to height h is $l = ae^{kh}$ where a and k are constants. By studying various physical characteristics of a person, mathematical biologists model each characteristic by a function that describes how it changes over time. Models of facial characteristics can be programmed into a computer to give a picture of how a person's appearance changes over time. These pictures aid law enforcement agencies in locating missing persons.

Library of Congress Prints and Photographs Division

JOHN NAPIER (1550–1617) was a Scottish landowner for whom mathematics was a hobby. We know him today because of his key invention: logarithms, which he published in 1614 under the title *A Description of the Marvelous Rule of Logarithms*. In Napier's time, logarithms were used exclusively for simplifying complicated calculations. For example, to multiply two large numbers, we would write them as powers of 10. The exponents are simply the logarithms of the numbers. For instance,

$$\begin{aligned} 4532 \times 57783 &\approx 10^{3.65629} \times 10^{4.76180} \\ &= 10^{8.41809} \\ &\approx 261{,}872{,}564 \end{aligned}$$

The idea is that multiplying powers of 10 is easy (we simply add their exponents). Napier produced extensive tables giving the logarithms (or exponents) of numbers. Since the advent of calculators and computers, logarithms are no longer used for this purpose. The logarithmic functions, however, have found many applications, some of which are described in this chapter.

Napier wrote on many topics. One of his most colorful works is a book entitled *A Plaine Discovery of the Whole Revelation of Saint John*, in which he predicted that the world would end in the year 1700.

■ Common Logarithms

We now study logarithms with base 10.

> **COMMON LOGARITHM**
>
> The logarithm with base 10 is called the **common logarithm** and is denoted by omitting the base:
>
> $$\log x = \log_{10} x$$

From the definition of logarithms we can easily find that

$$\log 10 = 1 \qquad \text{and} \qquad \log 100 = 2$$

But how do we find log 50? We need to find the exponent y such that $10^y = 50$. Clearly, 1 is too small and 2 is too large. So

$$1 < \log 50 < 2$$

To get a better approximation, we can experiment to find a power of 10 closer to 50. Fortunately, scientific calculators are equipped with a [LOG] key that directly gives values of common logarithms.

EXAMPLE 7 ■ Evaluating Common Logarithms

Use a calculator to find appropriate values of $f(x) = \log x$, and use the values to sketch the graph.

SOLUTION We make a table of values, using a calculator to evaluate the function at those values of x that are not powers of 10. We plot those points and connect them by a smooth curve as in Figure 8.

x	$\log x$
0.01	−2
0.1	−1
0.5	−0.301
1	0
4	0.602
5	0.699
10	1

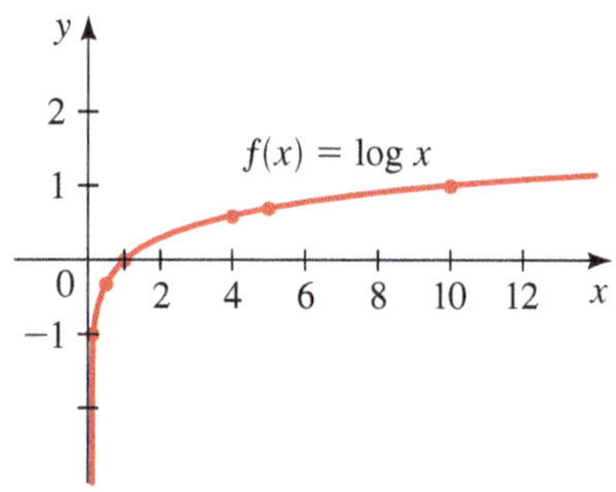

FIGURE 8

Now Try Exercise 51 ■

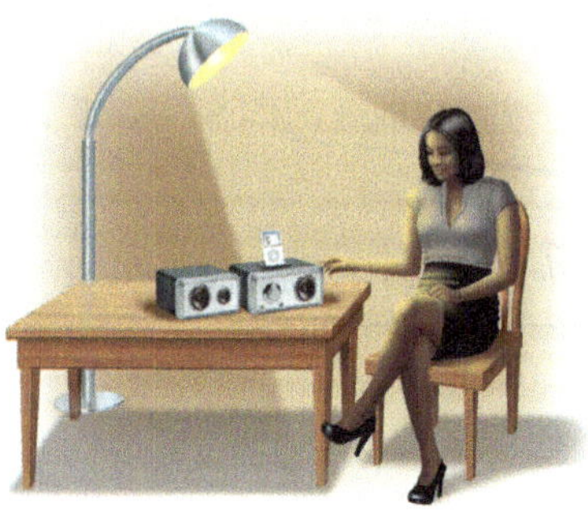

Human response to sound and light intensity is logarithmic.

Scientists model human response to stimuli (such as sound, light, or pressure) using logarithmic functions. For example, the intensity of a sound must be increased manyfold before we "feel" that the loudness has simply doubled. The psychologist Gustav Fechner formulated the law as

$$S = k \log\left(\frac{I}{I_0}\right)$$

where S is the subjective intensity of the stimulus, I is the physical intensity of the stimulus, I_0 stands for the threshold physical intensity, and k is a constant that is different for each sensory stimulus.

EXAMPLE 8 ■ Common Logarithms and Sound

We study the decibel scale in more detail in Section 4.7.

The perception of the loudness B (in decibels, dB) of a sound with physical intensity I (in W/m^2) is given by

$$B = 10 \log\left(\frac{I}{I_0}\right)$$

where I_0 is the physical intensity of a barely audible sound. Find the decibel level (loudness) of a sound whose physical intensity I is 100 times that of I_0.

SOLUTION We find the decibel level B by using the fact that $I = 100I_0$.

$$\begin{aligned} B &= 10 \log\left(\frac{I}{I_0}\right) && \text{Definition of } B \\ &= 10 \log\left(\frac{100I_0}{I_0}\right) && I = 100I_0 \\ &= 10 \log 100 && \text{Cancel } I_0 \\ &= 10 \cdot 2 = 20 && \text{Definition of log} \end{aligned}$$

The loudness of the sound is 20 dB.

Now Try Exercise 97 ■

■ Natural Logarithms

Of all possible bases a for logarithms, it turns out that the most convenient choice for the purposes of calculus is the number e, which we defined in Section 4.2.

The notation ln is an abbreviation for the Latin name *logarithmus naturalis*.

NATURAL LOGARITHM

The logarithm with base e is called the **natural logarithm** and is denoted by **ln**:

$$\ln x = \log_e x$$

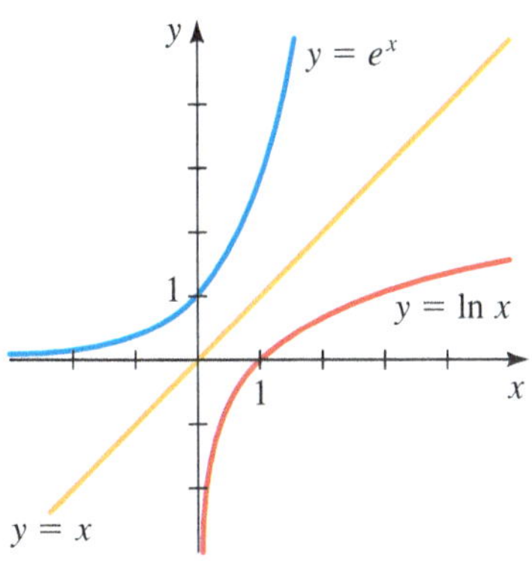

FIGURE 9 Graph of the natural logarithmic function

The natural logarithmic function $y = \ln x$ is the inverse function of the natural exponential function $y = e^x$. Both functions are graphed in Figure 9. By the definition of inverse functions we have

$$\ln x = y \quad \Leftrightarrow \quad e^y = x$$

If we substitute $a = e$ and write "ln" for "$\log_e$" in the properties of logarithms mentioned earlier, we obtain the following properties of natural logarithms.

PROPERTIES OF NATURAL LOGARITHMS

Property	Reason
1. $\ln 1 = 0$	We must raise e to the power 0 to get 1.
2. $\ln e = 1$	We must raise e to the power 1 to get e.
3. $\ln e^x = x$	We must raise e to the power x to get e^x.
4. $e^{\ln x} = x$	$\ln x$ is the power to which e must be raised to get x.

Calculators are equipped with an LN key that directly gives the values of natural logarithms.

EXAMPLE 9 ■ Evaluating the Natural Logarithm Function

(a) $\ln e^8 = 8$ — Definition of natural logarithm

(b) $\ln\left(\frac{1}{e^2}\right) = \ln e^{-2} = -2$ — Definition of natural logarithm

(c) $\ln 5 \approx 1.609$ — Use LN key on calculator

Now Try Exercise 47 ■

EXAMPLE 10 ■ Finding the Domain of a Logarithmic Function

Find the domain of the function $f(x) = \ln(4 - x^2)$.

SOLUTION As with any logarithmic function, $\ln x$ is defined when $x > 0$. Thus the domain of f is

$$\{x \mid 4 - x^2 > 0\} = \{x \mid x^2 < 4\} = \{x \mid |x| < 2\}$$
$$= \{x \mid -2 < x < 2\} = (-2, 2)$$

Now Try Exercise 73 ■

EXAMPLE 11 ■ Drawing the Graph of a Logarithmic Function

Draw the graph of the function $y = x\ln(4 - x^2)$, and use it to find the asymptotes and local maximum and minimum values.

SOLUTION As in Example 10 the domain of this function is the interval $(-2, 2)$, so we choose the viewing rectangle $[-3, 3]$ by $[-3, 3]$. The graph is shown in Figure 10, and from it we see that the lines $x = -2$ and $x = 2$ are vertical asymptotes.

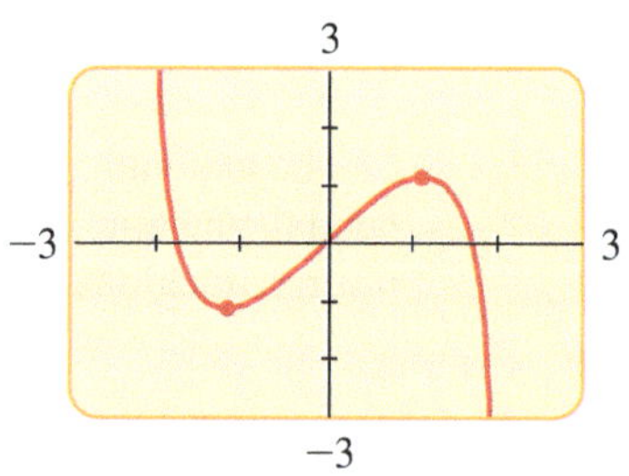

FIGURE 10
$y = x\ln(4 - x^2)$

DISCOVERY PROJECT

Orders of Magnitude

In this project we explore how to compare the sizes of real-world objects using logarithms. For example, how much bigger is an elephant than a flea? How much smaller is a man than a giant redwood? It is difficult to compare objects of such enormously varying sizes. In this project we learn how logarithms can be used to define the concept of "order of magnitude," which provides a simple and meaningful way of comparison. You can find the project at **www.stewartmath.com**.

The function has a local maximum point to the right of $x = 1$ and a local minimum point to the left of $x = -1$. By zooming in and tracing along the graph with the cursor, we find that the local maximum value is approximately 1.13 and this occurs when $x \approx 1.15$. Similarly (or by noticing that the function is odd), we find that the local minimum value is about -1.13, and it occurs when $x \approx -1.15$.

Now Try Exercise 79 ■

4.3 EXERCISES

CONCEPTS

1. $\log x$ is the exponent to which the base 10 must be raised to get ________. So we can complete the following table for $\log x$.

x	10^3	10^2	10^1	10^0	10^{-1}	10^{-2}	10^{-3}	$10^{1/2}$
$\log x$								

2. The function $f(x) = \log_9 x$ is the logarithm function with base ________. So $f(9) =$ ________, $f(1) =$ ________, $f(\frac{1}{9}) =$ ________, $f(81) =$ ________, and $f(3) =$ ________.

3. (a) $5^3 = 125$, so $\log_{\square} \square = \square$

(b) $\log_5 25 = 2$, so $\square^{\square} = \square$

4. Match the logarithmic function with its graph.

(a) $f(x) = \log_2 x$ **(b)** $f(x) = \log_2(-x)$

(c) $f(x) = -\log_2 x$ **(d)** $f(x) = -\log_2(-x)$

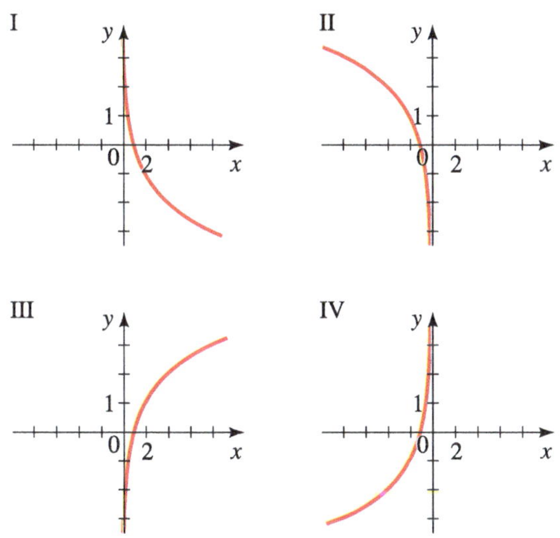

5. The natural logarithmic function $f(x) = \ln x$ has the ____________ asymptote $x =$ ____.

6. The logarithmic function $f(x) = \ln(x - 1)$ has the ____________ asymptote $x =$ ____.

SKILLS

7–8 ■ Logarithmic and Exponential Forms Complete the table by finding the appropriate logarithmic or exponential form of the equation, as in Example 1.

7.

Logarithmic form	Exponential form
$\log_8 8 = 1$	
$\log_8 64 = 2$	
	$8^{2/3} = 4$
	$8^3 = 512$
$\log_8(\frac{1}{8}) = -1$	
	$8^{-2} = \frac{1}{64}$

8.

Logarithmic form	Exponential form
	$4^3 = 64$
$\log_4 2 = \frac{1}{2}$	
	$4^{3/2} = 8$
$\log_4(\frac{1}{16}) = -2$	
$\log_4(\frac{1}{2}) = -\frac{1}{2}$	
	$4^{-5/2} = \frac{1}{32}$

9–16 ■ Exponential Form Express the equation in exponential form.

9. (a) $\log_3 81 = 4$ **(b)** $\log_3 1 = 0$

10. (a) $\log_5(\frac{1}{5}) = -1$ **(b)** $\log_4 64 = 3$

11. (a) $\log_8 2 = \frac{1}{3}$ **(b)** $\log_{10} 0.01 = -2$

12. (a) $\log_5(\frac{1}{125}) = -3$ **(b)** $\log_8 4 = \frac{2}{3}$

13. (a) $\log_3 5 = x$ **(b)** $\log_7(3y) = 2$

14. (a) $\log_6 z = 1$ **(b)** $\log_{10} 3 = 2t$

15. (a) $\ln 5 = 3y$ **(b)** $\ln(t + 1) = -1$

16. (a) $\ln(x + 1) = 2$ **(b)** $\ln(x - 1) = 4$

17–24 ■ Logarithmic Form Express the equation in logarithmic form.

17. (a) $10^4 = 10{,}000$ **(b)** $5^{-2} = \frac{1}{25}$

18. (a) $6^2 = 36$ **(b)** $10^{-1} = \frac{1}{10}$

19. **(a)** $8^{-1} = \frac{1}{8}$ **(b)** $2^{-3} = \frac{1}{8}$

20. **(a)** $4^{-3/2} = 0.125$ **(b)** $7^3 = 343$

21. **(a)** $4^x = 70$ **(b)** $3^5 = w$

22. **(a)** $3^{2x} = 10$ **(b)** $10^{-4x} = 0.1$

23. **(a)** $e^x = 2$ **(b)** $e^3 = y$

24. **(a)** $e^{x+1} = 0.5$ **(b)** $e^{0.5x} = t$

25–34 ■ Evaluating Logarithms Evaluate the expression.

25. **(a)** $\log_2 2$ **(b)** $\log_5 1$ **(c)** $\log_6 6^5$

26. **(a)** $\log_3 3^7$ **(b)** $\log_4 64$ **(c)** $\log_5 125$

27. **(a)** $\log_6 36$ **(b)** $\log_9 81$ **(c)** $\log_7 7^{10}$

28. **(a)** $\log_2 32$ **(b)** $\log_8 8^{17}$ **(c)** $\log_6 1$

29. **(a)** $\log_3(\frac{1}{27})$ **(b)** $\log_{10} \sqrt{10}$ **(c)** $\log_5 0.2$

30. **(a)** $\log_5 125$ **(b)** $\log_{49} 7$ **(c)** $\log_9 \sqrt{3}$

31. **(a)** $3^{\log_3 5}$ **(b)** $5^{\log_5 27}$ **(c)** $e^{\ln 10}$

32. **(a)** $e^{\ln \sqrt{3}}$ **(b)** $e^{\ln(1/\pi)}$ **(c)** $10^{\log 13}$

33. **(a)** $\log_8 0.25$ **(b)** $\ln e^4$ **(c)** $\ln(1/e)$

34. **(a)** $\log_4 \sqrt{2}$ **(b)** $\log_4(\frac{1}{2})$ **(c)** $\log_4 8$

35–44 ■ Logarithmic Equations Use the definition of the logarithmic function to find x.

35. **(a)** $\log_4 x = 3$ **(b)** $\log_{10} 0.01 = x$

36. **(a)** $\log_3 x = -2$ **(b)** $\log_5 125 = x$

37. **(a)** $\ln x = 3$ **(b)** $\ln e^2 = x$

38. **(a)** $\ln x = -1$ **(b)** $\ln(1/e) = x$

39. **(a)** $\log_7(\frac{1}{49}) = x$ **(b)** $\log_2 x = 5$

40. **(a)** $\log_4 2 = x$ **(b)** $\log_4 x = 2$

41. **(a)** $\log_2(\frac{1}{2}) = x$ **(b)** $\log_{10} x = -3$

42. **(a)** $\log_x 1000 = 3$ **(b)** $\log_x 25 = 2$

43. **(a)** $\log_x 16 = 4$ **(b)** $\log_x 8 = \frac{3}{2}$

44. **(a)** $\log_x 6 = \frac{1}{2}$ **(b)** $\log_x 3 = \frac{1}{3}$

45–48 ■ Evaluating Logarithms Use a calculator to evaluate the expression, correct to four decimal places.

45. **(a)** $\log 2$ **(b)** $\log 35.2$ **(c)** $\log(\frac{2}{3})$

46. **(a)** $\log 50$ **(b)** $\log \sqrt{2}$ **(c)** $\log(3\sqrt{2})$

47. **(a)** $\ln 5$ **(b)** $\ln 25.3$ **(c)** $\ln(1 + \sqrt{3})$

48. **(a)** $\ln 27$ **(b)** $\ln 7.39$ **(c)** $\ln 54.6$

49–52 ■ Graphing Logarithmic Functions Sketch the graph of the function by plotting points.

49. $f(x) = \log_3 x$ **50.** $g(x) = \log_4 x$

51. $f(x) = 2 \log x$ **52.** $g(x) = 1 + \log x$

53–56 ■ Finding Logarithmic Functions Find the function of the form $y = \log_a x$ whose graph is given.

53.

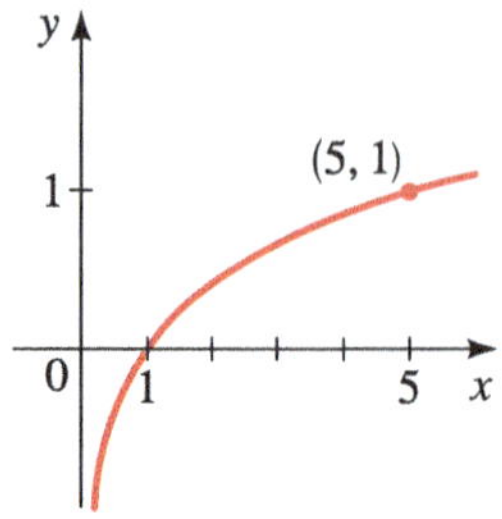

54.

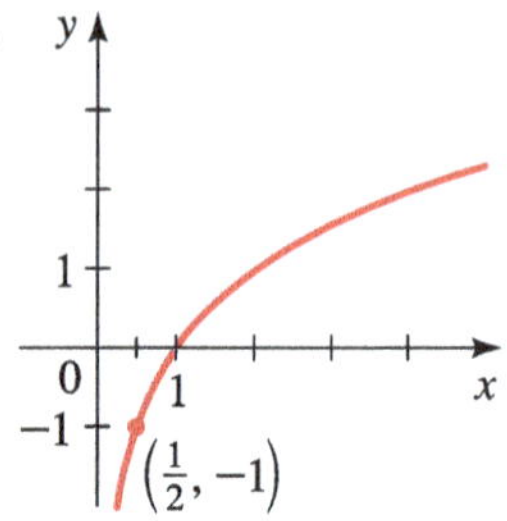

55.

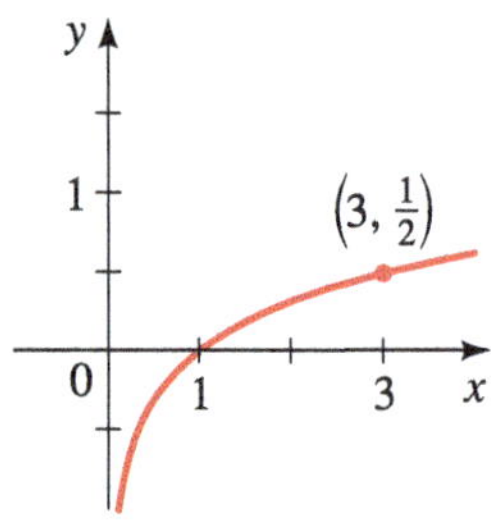

56.

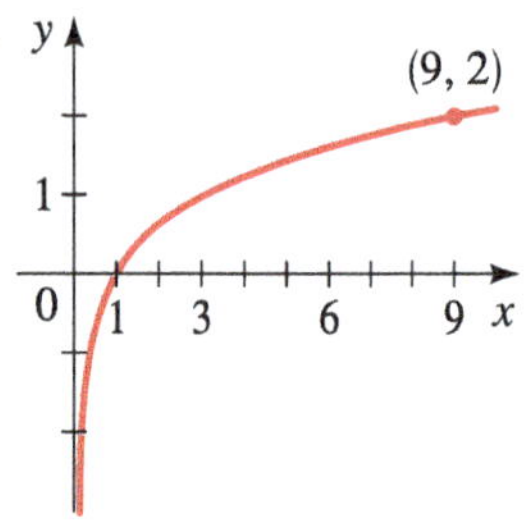

57–58 ■ Graphing Logarithmic Functions Match the logarithmic function with one of the graphs labeled I or II.

57. $f(x) = 2 + \ln x$ **58.** $f(x) = \ln(x - 2)$

I

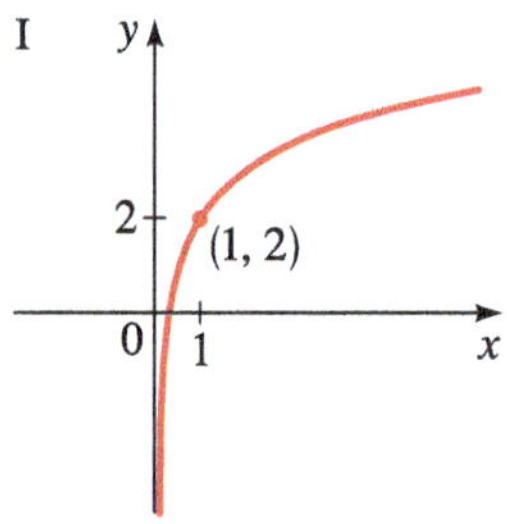

II

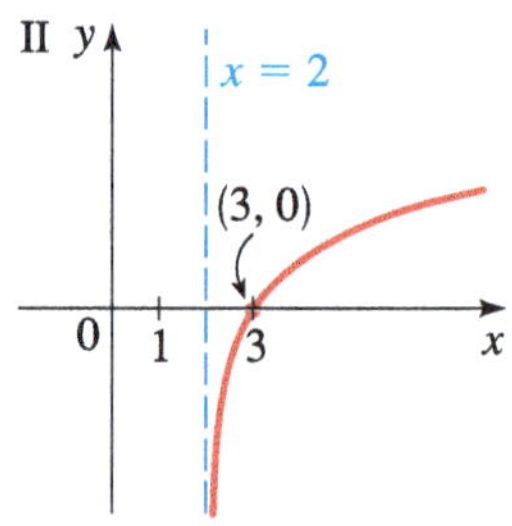

59. Graphing Draw the graph of $y = 4^x$, then use it to draw the graph of $y = \log_4 x$.

60. Graphing Draw the graph of $y = 3^x$, then use it to draw the graph of $y = \log_3 x$.

61–72 ■ Graphing Logarithmic Functions Graph the function, not by plotting points, but by starting from the graphs in Figures 4 and 9. State the domain, range, and asymptote.

61. $g(x) = \log_5(-x)$ **62.** $f(x) = -\log_{10} x$

63. $f(x) = \log_2(x - 4)$ **64.** $g(x) = \ln(x + 2)$

65. $h(x) = \ln(x + 5)$ **66.** $g(x) = \log_6(x - 3)$

67. $y = 2 + \log_3 x$ **68.** $y = 1 - \log_{10} x$

69. $y = \log_3(x - 1) - 2$ **70.** $y = 1 + \ln(-x)$

71. $y = |\ln x|$ **72.** $y = \ln|x|$

73–78 ■ Domain Find the domain of the function.

73. $f(x) = \log_{10}(x + 3)$ **74.** $f(x) = \log_5(8 - 2x)$

75. $g(x) = \log_3(x^2 - 1)$ **76.** $g(x) = \ln(x - x^2)$

77. $h(x) = \ln x + \ln(2 - x)$

78. $h(x) = \sqrt{x - 2} - \log_5(10 - x)$

79–84 ■ Graphing Logarithmic Functions Draw the graph of the function in a suitable viewing rectangle, and use it to find the domain, the asymptotes, and the local maximum and minimum values.

79. $y = \log_{10}(1 - x^2)$

80. $y = \ln(x^2 - x)$

81. $y = x + \ln x$

82. $y = x(\ln x)^2$

83. $y = \dfrac{\ln x}{x}$

84. $y = x \log_{10}(x + 10)$

SKILLS Plus

85–88 ■ Domain of a Composition Find the functions $f \circ g$ and $g \circ f$ and their domains.

85. $f(x) = 2^x, \quad g(x) = x + 1$

86. $f(x) = 3^x, \quad g(x) = x^2 + 1$

87. $f(x) = \log_2 x, \quad g(x) = x - 2$

88. $f(x) = \log x, \quad g(x) = x^2$

89. **Rates of Growth** Compare the rates of growth of the functions $f(x) = \ln x$ and $g(x) = \sqrt{x}$ by drawing their graphs on a common screen using the viewing rectangle $[-1, 30]$ by $[-1, 6]$.

90. **Rates of Growth**
   (a) By drawing the graphs of the functions
   $$f(x) = 1 + \ln(1 + x) \quad \text{and} \quad g(x) = \sqrt{x}$$
   in a suitable viewing rectangle, show that even when a logarithmic function starts out higher than a root function, it is ultimately overtaken by the root function.
   (b) Find, rounded to two decimal places, the solutions of the equation $\sqrt{x} = 1 + \ln(1 + x)$.

91–92 ■ Family of Functions A family of functions is given. **(a)** Draw graphs of the family for $c = 1, 2, 3,$ and 4. **(b)** How are the graphs in part (a) related?

91. $f(x) = \log(cx)$

92. $f(x) = c \log x$

93–94 ■ Inverse Functions A function $f(x)$ is given. **(a)** Find the domain of the function f. **(b)** Find the inverse function of f.

93. $f(x) = \log_2(\log_{10} x)$

94. $f(x) = \ln(\ln(\ln x))$

95. **Inverse Functions**
   (a) Find the inverse of the function $f(x) = \dfrac{2^x}{1 + 2^x}$.
   (b) What is the domain of the inverse function?

APPLICATIONS

96. **Absorption of Light** A spectrophotometer measures the concentration of a sample dissolved in water by shining a light through it and recording the amount of light that emerges. In other words, if we know the amount of light that is absorbed, we can calculate the concentration of the sample. For a certain substance the concentration (in moles per liter, mol/L) is found by using the formula
$$C = -2500 \ln\left(\frac{I}{I_0}\right)$$
where I_0 is the intensity of the incident light and I is the intensity of light that emerges. Find the concentration of the substance if the intensity I is 70% of I_0.

97. **Carbon Dating** The age of an ancient artifact can be determined by the amount of radioactive carbon-14 remaining in it. If D_0 is the original amount of carbon-14 and D is the amount remaining, then the artifact's age A (in years) is given by
$$A = -8267 \ln\left(\frac{D}{D_0}\right)$$
Find the age of an object if the amount D of carbon-14 that remains in the object is 73% of the original amount D_0.

98. **Bacteria Colony** A certain strain of bacteria divides every 3 hours. If a colony is started with 50 bacteria, then the time t (in hours) required for the colony to grow to N bacteria is given by
$$t = 3\frac{\log(N/50)}{\log 2}$$
Find the time required for the colony to grow to a million bacteria.

99. **Investment** The time required to double the amount of an investment at an interest rate r compounded continuously is given by
$$t = \frac{\ln 2}{r}$$
Find the time required to double an investment at 6%, 7%, and 8%.

100. **Charging a Battery** The rate at which a battery charges is slower the closer the battery is to its maximum charge C_0. The time (in hours) required to charge a fully discharged battery to a charge C is given by
$$t = -k \ln\left(1 - \frac{C}{C_0}\right)$$
where k is a positive constant that depends on the battery. For a certain battery, $k = 0.25$. If this battery is fully discharged, how long will it take to charge to 90% of its maximum charge C_0?

101. Difficulty of a Task The difficulty in "acquiring a target" (such as using your mouse to click on an icon on your computer screen) depends on the distance to the target and the size of the target. According to Fitts's Law, the index of difficulty (ID) is given by

$$\text{ID} = \frac{\log(2A/W)}{\log 2}$$

where W is the width of the target and A is the distance to the center of the target. Compare the difficulty of clicking on an icon that is 5 mm wide to clicking on one that is 10 mm wide. In each case, assume that the mouse is 100 mm from the icon.

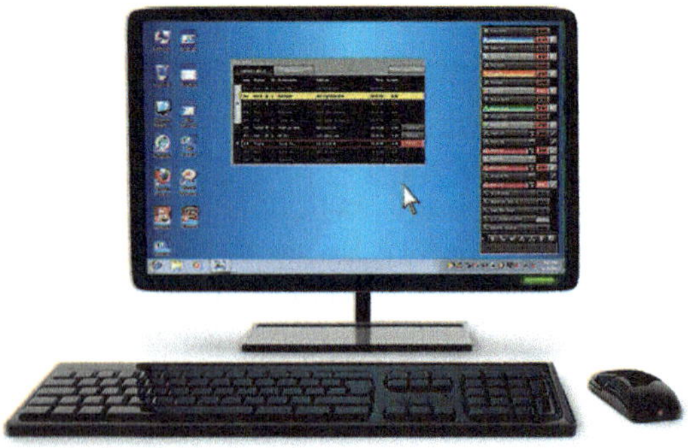

DISCUSS ■ DISCOVER ■ PROVE ■ WRITE

102. DISCUSS: The Height of the Graph of a Logarithmic Function Suppose that the graph of $y = 2^x$ is drawn on a coordinate plane where the unit of measurement is an inch.

(a) Show that at a distance 2 ft to the right of the origin the height of the graph is about 265 mi.

(b) If the graph of $y = \log_2 x$ is drawn on the same set of axes, how far to the right of the origin do we have to go before the height of the curve reaches 2 ft?

103. DISCUSS: The Googolplex A **googol** is 10^{100}, and a **googolplex** is 10^{googol}. Find

$$\log(\log(\text{googol})) \qquad \text{and} \qquad \log(\log(\log(\text{googolplex})))$$

104. DISCUSS: Comparing Logarithms Which is larger, $\log_4 17$ or $\log_5 24$? Explain your reasoning.

105. DISCUSS ■ DISCOVER: The Number of Digits in an Integer Compare log 1000 to the number of digits in 1000. Do the same for 10,000. How many digits does any number between 1000 and 10,000 have? Between what two values must the common logarithm of such a number lie? Use your observations to explain why the number of digits in any positive integer x is $[\![\log x]\!] + 1$. (The symbol $[\![n]\!]$ is the greatest integer function defined in Section 2.2.) How many digits does the number 2^{100} have?

4.4 LAWS OF LOGARITHMS

■ Laws of Logarithms ■ Expanding and Combining Logarithmic Expressions ■ Change of Base Formula

In this section we study properties of logarithms. These properties give logarithmic functions a wide range of applications, as we will see in Sections 4.6 and 4.7.

■ Laws of Logarithms

Since logarithms are exponents, the Laws of Exponents give rise to the Laws of Logarithms.

LAWS OF LOGARITHMS

Let a be a positive number, with $a \neq 1$. Let A, B, and C be any real numbers with $A > 0$ and $B > 0$.

Law	Description
1. $\log_a(AB) = \log_a A + \log_a B$	The logarithm of a product of numbers is the sum of the logarithms of the numbers.
2. $\log_a\left(\dfrac{A}{B}\right) = \log_a A - \log_a B$	The logarithm of a quotient of numbers is the difference of the logarithms of the numbers.
3. $\log_a(A^C) = C\log_a A$	The logarithm of a power of a number is the exponent times the logarithm of the number.

Proof We make use of the property $\log_a a^x = x$ from Section 4.3.

Law 1 Let $\log_a A = u$ and $\log_a B = v$. When written in exponential form, these equations become

$$a^u = A \qquad \text{and} \qquad a^v = B$$

Thus
$$\log_a(AB) = \log_a(a^u a^v) = \log_a(a^{u+v})$$
$$= u + v = \log_a A + \log_a B$$

Law 2 Using Law 1, we have

$$\log_a A = \log_a\left[\left(\frac{A}{B}\right)B\right] = \log_a\left(\frac{A}{B}\right) + \log_a B$$

so
$$\log_a\left(\frac{A}{B}\right) = \log_a A - \log_a B$$

Law 3 Let $\log_a A = u$. Then $a^u = A$, so

$$\log_a(A^C) = \log_a(a^u)^C = \log_a(a^{uC}) = uC = C\log_a A$$ ■

EXAMPLE 1 ■ Using the Laws of Logarithms to Evaluate Expressions

Evaluate each expression.

(a) $\log_4 2 + \log_4 32$

(b) $\log_2 80 - \log_2 5$

(c) $-\frac{1}{3}\log 8$

SOLUTION

(a) $\log_4 2 + \log_4 32 = \log_4(2 \cdot 32)$ Law 1

$= \log_4 64 = 3$ Because $64 = 4^3$

(b) $\log_2 80 - \log_2 5 = \log_2(\frac{80}{5})$ Law 2

$= \log_2 16 = 4$ Because $16 = 2^4$

(c) $-\frac{1}{3}\log 8 = \log 8^{-1/3}$ Law 3

$= \log(\frac{1}{2})$ Property of negative exponents

≈ -0.301 Calculator

Now Try Exercises 9, 11, and 13 ■

Expanding and Combining Logarithmic Expressions

The Laws of Logarithms allow us to write the logarithm of a product or a quotient as the sum or difference of logarithms. This process, called *expanding* a logarithmic expression, is illustrated in the next example.

EXAMPLE 2 ■ Expanding Logarithmic Expressions

Use the Laws of Logarithms to expand each expression.

(a) $\log_2(6x)$ **(b)** $\log_5(x^3y^6)$ **(c)** $\ln\left(\frac{ab}{\sqrt[3]{c}}\right)$

SOLUTION

(a) $\log_2(6x) = \log_2 6 + \log_2 x$ Law 1

(b) $\log_5(x^3y^6) = \log_5 x^3 + \log_5 y^6$ Law 1

$= 3\log_5 x + 6\log_5 y$ Law 3

(c) $\ln\left(\dfrac{ab}{\sqrt[3]{c}}\right) = \ln(ab) - \ln\sqrt[3]{c}$ Law 2

$= \ln a + \ln b - \ln c^{1/3}$ Law 1

$= \ln a + \ln b - \frac{1}{3}\ln c$ Law 3

Now Try Exercises 23, 31, and 37

The Laws of Logarithms also allow us to reverse the process of expanding that was done in Example 2. That is, we can write sums and differences of logarithms as a single logarithm. This process, called *combining* logarithmic expressions, is illustrated in the next example.

EXAMPLE 3 ■ Combining Logarithmic Expressions

Use the Laws of Logarithms to combine each expression into a single logarithm.

(a) $3\log x + \frac{1}{2}\log(x+1)$

(b) $3\ln s + \frac{1}{2}\ln t - 4\ln(t^2+1)$

SOLUTION

(a) $3\log x + \frac{1}{2}\log(x+1) = \log x^3 + \log(x+1)^{1/2}$ Law 3

$= \log(x^3(x+1)^{1/2})$ Law 1

(b) $3\ln s + \frac{1}{2}\ln t - 4\ln(t^2+1) = \ln s^3 + \ln t^{1/2} - \ln(t^2+1)^4$ Law 3

$= \ln(s^3t^{1/2}) - \ln(t^2+1)^4$ Law 1

$= \ln\left(\dfrac{s^3\sqrt{t}}{(t^2+1)^4}\right)$ Law 2

Now Try Exercises 51 and 53

Warning Although the Laws of Logarithms tell us how to compute the logarithm of a product or a quotient, *there is no corresponding rule for the logarithm of a sum or a difference*. For instance,

$$\log_a(x+y) \neq \log_a x + \log_a y$$

In fact, we know that the right side is equal to $\log_a(xy)$. Also, don't improperly simplify quotients or powers of logarithms. For instance,

$$\frac{\log 6}{\log 2} \neq \log\left(\frac{6}{2}\right) \quad \text{and} \quad (\log_2 x)^3 \neq 3\log_2 x$$

Logarithmic functions are used to model a variety of situations involving human behavior. One such behavior is how quickly we forget things we have learned. For example, if you learn algebra at a certain performance level (say, 90% on a test) and then don't use algebra for a while, how much will you retain after a week, a month, or a year? Hermann Ebbinghaus (1850–1909) studied this phenomenon and formulated the law described in the next example.

Forgetting what we've learned depends on how long ago we learned it.

EXAMPLE 4 ■ The Law of Forgetting

If a task is learned at a performance level P_0, then after a time interval t the performance level P satisfies

$$\log P = \log P_0 - c\log(t+1)$$

where c is a constant that depends on the type of task and t is measured in months.

(a) Solve for P.

(b) If your score on a history test is 90, what score would you expect to get on a similar test after two months? After a year? (Assume that $c = 0.2$.)

SOLUTION

(a) We first combine the right-hand side.

$$\log P = \log P_0 - c\log(t+1) \qquad \text{Given equation}$$

$$\log P = \log P_0 - \log(t+1)^c \qquad \text{Law 3}$$

$$\log P = \log \frac{P_0}{(t+1)^c} \qquad \text{Law 2}$$

$$P = \frac{P_0}{(t+1)^c} \qquad \text{Because log is one-to-one}$$

(b) Here $P_0 = 90$, $c = 0.2$, and t is measured in months.

In 2 months: $t = 2$ and $P = \dfrac{90}{(2+1)^{0.2}} \approx 72$

In 1 year: $t = 12$ and $P = \dfrac{90}{(12+1)^{0.2}} \approx 54$

Your expected scores after 2 months and after 1 year are 72 and 54, respectively.

Now Try Exercise 73

Change of Base Formula

For some purposes we find it useful to change from logarithms in one base to logarithms in another base. Suppose we are given $\log_a x$ and want to find $\log_b x$. Let

$$y = \log_b x$$

We write this in exponential form and take the logarithm, with base a, of each side.

$$b^y = x \qquad \text{Exponential form}$$

$$\log_a(b^y) = \log_a x \qquad \text{Take } \log_a \text{ of each side}$$

$$y \log_a b = \log_a x \qquad \text{Law 3}$$

$$y = \frac{\log_a x}{\log_a b} \qquad \text{Divide by } \log_a b$$

This proves the following formula.

We may write the Change of Base Formula as

$$\log_b x = \left(\frac{1}{\log_a b}\right)\log_a x$$

So $\log_b x$ is just a constant multiple of $\log_a x$; the constant is $\dfrac{1}{\log_a b}$.

CHANGE OF BASE FORMULA

$$\log_b x = \frac{\log_a x}{\log_a b}$$

In particular, if we put $x = a$, then $\log_a a = 1$, and this formula becomes

$$\log_b a = \frac{1}{\log_a b}$$

We can now evaluate a logarithm to *any* base by using the Change of Base Formula to express the logarithm in terms of common logarithms or natural logarithms and then using a calculator.

EXAMPLE 5 ■ Evaluating Logarithms with the Change of Base Formula

Use the Change of Base Formula and common or natural logarithms to evaluate each logarithm, rounded to five decimal places.

(a) $\log_8 5$ **(b)** $\log_9 20$

SOLUTION

(a) We use the Change of Base Formula with $b = 8$ and $a = 10$:

$$\log_8 5 = \frac{\log_{10} 5}{\log_{10} 8} \approx 0.77398$$

(b) We use the Change of Base Formula with $b = 9$ and $a = e$:

$$\log_9 20 = \frac{\ln 20}{\ln 9} \approx 1.36342$$

Now Try Exercises 59 and 61 ■

EXAMPLE 6 ■ Using the Change of Base Formula to Graph a Logarithmic Function

Use a graphing calculator to graph $f(x) = \log_6 x$.

SOLUTION Calculators don't have a key for $\log_6$, so we use the Change of Base Formula to write

$$f(x) = \log_6 x = \frac{\ln x}{\ln 6}$$

Since calculators do have an LN key, we can enter this new form of the function and graph it. The graph is shown in Figure 1.

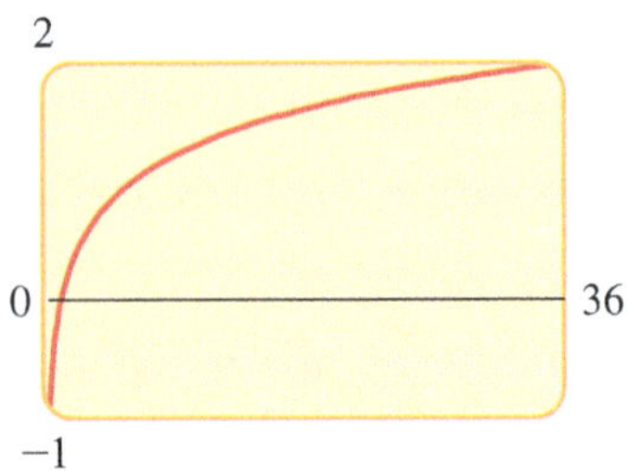

FIGURE 1

$f(x) = \log_6 x = \dfrac{\ln x}{\ln 6}$

Now Try Exercise 67 ■

4.4 EXERCISES

CONCEPTS

1. The logarithm of a product of two numbers is the same as the ________ of the logarithms of these numbers. So $\log_5(25 \cdot 125) =$ ________ + ________.

2. The logarithm of a quotient of two numbers is the same as the ________ of the logarithms of these numbers. So $\log_5(\frac{25}{125}) =$ ________ − ________.

3. The logarithm of a number raised to a power is the same as the ________ times the logarithm of the number. So $\log_5(25^{10}) =$ ________ · ________.

4. We can expand $\log\left(\dfrac{x^2 y}{z}\right)$ to get ________.

5. We can combine $2\log x + \log y - \log z$ to get ________.

6. **(a)** Most calculators can find logarithms with base ________ and base ________. To find logarithms with different bases, we use the ________ Formula. To find $\log_7 12$, we write

$$\log_7 12 = \frac{\log \square}{\log \square} \approx \text{\_\_\_\_\_\_\_\_}$$

(b) Do we get the same answer if we perform the calculation in part (a) using ln in place of log?

7–8 ■ *True or False?*

7. **(a)** $\log(A + B)$ is the same as $\log A + \log B$.
(b) $\log AB$ is the same as $\log A + \log B$.

8. **(a)** $\log \dfrac{A}{B}$ is the same as $\log A - \log B$.
(b) $\dfrac{\log A}{\log B}$ is the same as $\log A - \log B$.

SKILLS

9–22 ■ Evaluating Logarithms Use the Laws of Logarithms to evaluate the expression.

9. $\log 50 + \log 200$
10. $\log_6 9 + \log_6 24$
11. $\log_2 60 - \log_2 15$
12. $\log_3 135 - \log_3 45$
13. $\frac{1}{4}\log_3 81$
14. $-\frac{1}{3}\log_3 27$
15. $\log_5 \sqrt{5}$
16. $\log_5 \dfrac{1}{\sqrt{125}}$
17. $\log_2 6 - \log_2 15 + \log_2 20$
18. $\log_3 100 - \log_3 18 - \log_3 50$
19. $\log_4 16^{100}$
20. $\log_2 8^{33}$
21. $\log(\log 10^{10,000})$
22. $\ln(\ln e^{e^{200}})$

23–48 ■ Expanding Logarithmic Expressions Use the Laws of Logarithms to expand the expression.

23. $\log_3 8x$
24. $\log_6 7r$
25. $\log_3 2xy$
26. $\log_5 4st$
27. $\ln a^3$
28. $\log \sqrt{t^5}$
29. $\log_2(xy)^{10}$
30. $\ln \sqrt{ab}$
31. $\log_2(AB^2)$
32. $\log_3(x\sqrt{y})$
33. $\log_3 \dfrac{2x}{y}$
34. $\ln \dfrac{r}{3s}$
35. $\log_5\left(\dfrac{3x^2}{y^3}\right)$
36. $\log_2\left(\dfrac{s^5}{7t^2}\right)$
37. $\log_3 \dfrac{\sqrt{3x^5}}{y}$
38. $\log \dfrac{y^3}{\sqrt{2x}}$
39. $\log\left(\dfrac{x^3y^4}{z^6}\right)$
40. $\log_a\left(\dfrac{x^2}{yz^3}\right)$
41. $\ln \sqrt{x^4 + 2}$
42. $\log \sqrt[3]{x^2 + 4}$
43. $\ln\left(x\sqrt{\dfrac{y}{z}}\right)$
44. $\ln \dfrac{3x^2}{(x+1)^{10}}$
45. $\log \sqrt[4]{x^2 + y^2}$
46. $\log\left(\dfrac{x}{\sqrt[3]{1-x}}\right)$
47. $\log\sqrt{\dfrac{x^2+4}{(x^2+1)(x^3-7)^2}}$
48. $\log \sqrt{x\sqrt{y\sqrt{z}}}$

49–58 ■ Combining Logarithmic Expressions Use the Laws of Logarithms to combine the expression.

49. $\log_4 6 + 2\log_4 7$

50. $\frac{1}{2}\log_2 5 - 2\log_2 7$

51. $2\log x - 3\log(x+1)$

52. $3\ln 2 + 2\ln x - \frac{1}{2}\ln(x+4)$

53. $4\log x - \frac{1}{3}\log(x^2+1) + 2\log(x-1)$

54. $\log_5(x^2 - 1) - \log_5(x - 1)$

55. $\ln(a+b) + \ln(a-b) - 2\ln c$

56. $2(\log_5 x + 2\log_5 y - 3\log_5 z)$

57. $\frac{1}{3}\log(x+2)^3 + \frac{1}{2}[\log x^4 - \log(x^2 - x - 6)^2]$

58. $\log_a b + c\log_a d - r\log_a s$

59–66 ■ Change of Base Formula Use the Change of Base Formula and a calculator to evaluate the logarithm, rounded to six decimal places. Use either natural or common logarithms.

59. $\log_2 5$
60. $\log_5 2$
61. $\log_3 16$
62. $\log_6 92$
63. $\log_7 2.61$
64. $\log_6 532$
65. $\log_4 125$
66. $\log_{12} 2.5$

67. Change of Base Formula Use the Change of Base Formula to show that

$$\log_3 x = \frac{\ln x}{\ln 3}$$

Then use this fact to draw the graph of the function $f(x) = \log_3 x$.

SKILLS Plus

68. Families of Functions Draw graphs of the family of functions $y = \log_a x$ for $a = 2, e, 5$, and 10 on the same screen, using the viewing rectangle $[0, 5]$ by $[-3, 3]$. How are these graphs related?

69. Change of Base Formula Use the Change of Base Formula to show that

$$\log e = \frac{1}{\ln 10}$$

70. Change of Base Formula Simplify: $(\log_2 5)(\log_5 7)$

71. A Logarithmic Identity Show that

$$-\ln(x - \sqrt{x^2 - 1}) = \ln(x + \sqrt{x^2 - 1})$$

APPLICATIONS

72. Forgetting Use the Law of Forgetting (Example 4) to estimate a student's score on a biology test two years after he got a score of 80 on a test covering the same material. Assume that $c = 0.3$ and t is measured in months.

73. Wealth Distribution Vilfredo Pareto (1848–1923) observed that most of the wealth of a country is owned by a few members of the population. **Pareto's Principle** is

$$\log P = \log c - k\log W$$

where W is the wealth level (how much money a person has) and P is the number of people in the population having that much money.

(a) Solve the equation for P.

(b) Assume that $k = 2.1$ and $c = 8000$, and that W is measured in millions of dollars. Use part (a) to find the number of people who have \$2 million or more. How many people have \$10 million or more?

74. Biodiversity Some biologists model the number of species S in a fixed area A (such as an island) by the species-area relationship

$$\log S = \log c + k \log A$$

where c and k are positive constants that depend on the type of species and habitat.

(a) Solve the equation for S.

(b) Use part (a) to show that if $k = 3$, then doubling the area increases the number of species eightfold.

75. Magnitude of Stars The magnitude M of a star is a measure of how bright a star appears to the human eye. It is defined by

$$M = -2.5 \log\left(\frac{B}{B_0}\right)$$

where B is the actual brightness of the star and B_0 is a constant.

(a) Expand the right-hand side of the equation.

(b) Use part (a) to show that the brighter a star, the less its magnitude.

(c) Betelgeuse is about 100 times brighter than Albiero. Use part (a) to show that Betelgeuse is 5 magnitudes less bright than Albiero.

DISCUSS ■ DISCOVER ■ PROVE ■ WRITE

76. DISCUSS: True or False? Discuss each equation, and determine whether it is true for all possible values of the variables. (Ignore values of the variables for which any term is undefined.)

(a) $\log\left(\dfrac{x}{y}\right) = \dfrac{\log x}{\log y}$

(b) $\log_2(x - y) = \log_2 x - \log_2 y$

(c) $\log_5\left(\dfrac{a}{b^2}\right) = \log_5 a - 2\log_5 b$

(d) $\log 2^z = z \log 2$

(e) $(\log P)(\log Q) = \log P + \log Q$

(f) $\dfrac{\log a}{\log b} = \log a - \log b$

(g) $(\log_2 7)^x = x \log_2 7$

(h) $\log_a a^a = a$

(i) $\log(x - y) = \dfrac{\log x}{\log y}$

(j) $-\ln\left(\dfrac{1}{A}\right) = \ln A$

77. DISCUSS: Find the Error What is wrong with the following argument?

$$\begin{aligned} \log 0.1 &< 2 \log 0.1 \\ &= \log(0.1)^2 \\ &= \log 0.01 \\ \log 0.1 &< \log 0.01 \\ 0.1 &< 0.01 \end{aligned}$$

78. PROVE: Shifting, Shrinking, and Stretching Graphs of Functions Let $f(x) = x^2$. Show that $f(2x) = 4f(x)$, and explain how this shows that shrinking the graph of f horizontally has the same effect as stretching it vertically. Then use the identities $e^{2+x} = e^2e^x$ and $\ln(2x) = \ln 2 + \ln x$ to show that for $g(x) = e^x$ a horizontal shift is the same as a vertical stretch and for $h(x) = \ln x$ a horizontal shrinking is the same as a vertical shift.

4.5 EXPONENTIAL AND LOGARITHMIC EQUATIONS

■ Exponential Equations ■ Logarithmic Equations ■ Compound Interest

In this section we solve equations that involve exponential or logarithmic functions. The techniques that we develop here will be used in the next section for solving applied problems.

Exponential Equations

An *exponential equation* is one in which the variable occurs in the exponent. Some exponential equations can be solved by using the fact that exponential functions are one-to-one. This means that

$$a^x = a^y \quad \Rightarrow \quad x = y$$

We use this property in the next example.

EXAMPLE 1 ■ Exponential Equations

Solve the exponential equation.

(a) $5^x = 125$ **(b)** $5^{2x} = 5^{x+1}$

SOLUTION

(a) We first express 125 as a power of 5 and then use the fact that the exponential function $f(x) = 5^x$ is one-to-one.

$$5^x = 125 \quad \text{Given equation}$$
$$5^x = 5^3 \quad \text{Because } 125 = 5^3$$
$$x = 3 \quad \text{One-to-one property}$$

The solution is $x = 3$.

(b) We first use the fact that the function $f(x) = 5^x$ is one-to-one.

$$5^{2x} = 5^{x+1} \quad \text{Given equation}$$
$$2x = x + 1 \quad \text{One-to-one property}$$
$$x = 1 \quad \text{Solve for } x$$

The solution is $x = 1$.

Now Try Exercises 3 and 7 ■

Law 3: $\log_a A^C = C \log_a A$

The equations in Example 1 were solved by comparing exponents. This method is not suitable for solving an equation like $5^x = 160$ because 160 is not easily expressed as a power of the base 5. To solve such equations, we take the logarithm of each side and use Law 3 of logarithms to "bring down the exponent." The following guidelines describe the process.

GUIDELINES FOR SOLVING EXPONENTIAL EQUATIONS

1. Isolate the exponential expression on one side of the equation.
2. Take the logarithm of each side, then use the Laws of Logarithms to "bring down the exponent."
3. Solve for the variable.

EXAMPLE 2 ■ Solving an Exponential Equation

Consider the exponential equation $3^{x+2} = 7$.

(a) Find the exact solution of the equation expressed in terms of logarithms.

(b) Use a calculator to find an approximation to the solution rounded to six decimal places.

DISCOVERY PROJECT

Super Origami

Origami is the traditional Japanese art of folding paper to create illustrations. In this project we explore some thought experiments about folding paper. Suppose that you fold a sheet of paper in half, then fold it in half again, and continue to fold the paper in half. How many folds are needed to obtain a mile-high stack of paper? To answer this question, we need to solve an exponential equation. In this project we use logarithms to answer this and other thought questions about folding paper. You can find the project at **www.stewartmath.com**.

SOLUTION

(a) We take the common logarithm of each side and use Law 3.

$$3^{x+2} = 7 \qquad \text{Given equation}$$

$$\log(3^{x+2}) = \log 7 \qquad \text{Take log of each side}$$

$$(x + 2)\log 3 = \log 7 \qquad \text{Law 3 (bring down exponent)}$$

$$x + 2 = \frac{\log 7}{\log 3} \qquad \text{Divide by log 3}$$

$$x = \frac{\log 7}{\log 3} - 2 \qquad \text{Subtract 2}$$

The exact solution is $x = \dfrac{\log 7}{\log 3} - 2$.

(b) Using a calculator, we find the decimal approximation $x \approx -0.228756$.

We could have used natural logarithms instead of common logarithms. In fact, using the same steps, we get

$$x = \frac{\ln 7}{\ln 3} - 2 \approx -0.228756$$

CHECK YOUR ANSWER

Substituting $x = -0.228756$ into the original equation and using a calculator, we get

$$3^{(-0.228756)+2} \approx 7 \quad ✓$$

Now Try Exercise 15 ■

EXAMPLE 3 ■ Solving an Exponential Equation

Solve the equation $8e^{2x} = 20$.

SOLUTION We first divide by 8 to isolate the exponential term on one side of the equation.

$$8e^{2x} = 20 \qquad \text{Given equation}$$

$$e^{2x} = \tfrac{20}{8} \qquad \text{Divide by 8}$$

$$\ln e^{2x} = \ln 2.5 \qquad \text{Take ln of each side}$$

$$2x = \ln 2.5 \qquad \text{Property of ln}$$

$$x = \frac{\ln 2.5}{2} \qquad \text{Divide by 2 (exact solution)}$$

$$\approx 0.458 \qquad \text{Calculator (approximate solution)}$$

CHECK YOUR ANSWER

Substituting $x = 0.458$ into the original equation and using a calculator, we get

$$8e^{2(0.458)} \approx 20 \quad ✓$$

Now Try Exercise 17 ■

EXAMPLE 4 ■ Solving an Exponential Equation Algebraically and Graphically

Solve the equation $e^{3-2x} = 4$ algebraically and graphically.

SOLUTION 1: Algebraic

Since the base of the exponential term is e, we use natural logarithms to solve this equation.

$$e^{3-2x} = 4 \qquad \text{Given equation}$$

$$\ln(e^{3-2x}) = \ln 4 \qquad \text{Take ln of each side}$$

$$3 - 2x = \ln 4 \qquad \text{Property of ln}$$

$$-2x = -3 + \ln 4 \qquad \text{Subtract 3}$$

$$x = \tfrac{1}{2}(3 - \ln 4) \approx 0.807 \qquad \text{Multiply by } -\tfrac{1}{2}$$

You should check that this answer satisfies the original equation.

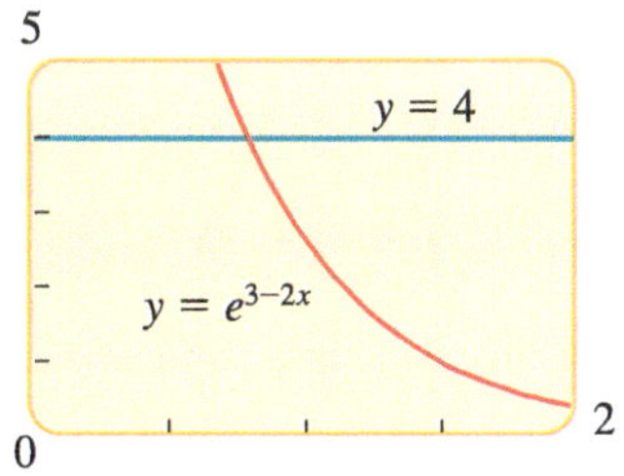

FIGURE 1

SOLUTION 2: Graphical

We graph the equations $y = e^{3-2x}$ and $y = 4$ in the same viewing rectangle as in Figure 1. The solutions occur where the graphs intersect. Zooming in on the point of intersection of the two graphs, we see that $x \approx 0.81$.

Now Try Exercise 21 ■

EXAMPLE 5 ■ An Exponential Equation of Quadratic Type

Solve the equation $e^{2x} - e^x - 6 = 0$.

SOLUTION To isolate the exponential term, we factor.

$$e^{2x} - e^x - 6 = 0 \quad \text{Given equation}$$

$$(e^x)^2 - e^x - 6 = 0 \quad \text{Law of Exponents}$$

$$(e^x - 3)(e^x + 2) = 0 \quad \text{Factor (a quadratic in } e^x\text{)}$$

$$e^x - 3 = 0 \quad \text{or} \quad e^x + 2 = 0 \quad \text{Zero-Product Property}$$

$$e^x = 3 \qquad e^x = -2$$

If we let $w = e^x$, we get the quadratic equation

$$w^2 - w - 6 = 0$$

which factors as

$$(w - 3)(w + 2) = 0$$

The equation $e^x = 3$ leads to $x = \ln 3$. But the equation $e^x = -2$ has no solution because $e^x > 0$ for all x. Thus $x = \ln 3 \approx 1.0986$ is the only solution. You should check that this answer satisfies the original equation.

Now Try Exercise 39 ■

EXAMPLE 6 ■ An Equation Involving Exponential Functions

Solve the equation $3xe^x + x^2e^x = 0$.

SOLUTION First we factor the left side of the equation.

$$3xe^x + x^2e^x = 0 \quad \text{Given equation}$$

$$x(3 + x)e^x = 0 \quad \text{Factor out common factors}$$

$$x(3 + x) = 0 \quad \text{Divide by } e^x \text{ (because } e^x \neq 0\text{)}$$

$$x = 0 \quad \text{or} \quad 3 + x = 0 \quad \text{Zero-Product Property}$$

Thus the solutions are $x = 0$ and $x = -3$.

Now Try Exercise 45 ■

CHECK YOUR ANSWER

$x = 0$:

$3(0)e^0 + 0^2e^0 = 0$ ✓

$x = -3$:

$3(-3)e^{-3} + (-3)^2e^{-3}$

$= -9e^{-3} + 9e^{-3} = 0$ ✓

Logarithmic Equations

A *logarithmic equation* is one in which a logarithm of the variable occurs. Some logarithmic equations can be solved by using the fact that logarithmic functions are one-to-one. This means that

$$\log_a x = \log_a y \quad \Rightarrow \quad x = y$$

We use this property in the next example.

EXAMPLE 7 ■ Solving a Logarithmic Equation

Solve the equation $\log(x^2 + 1) = \log(x - 2) + \log(x + 3)$.

SOLUTION First we combine the logarithms on the right-hand side, and then we use the one-to-one property of logarithms.

$$\log_5(x^2+1) = \log_5(x-2) + \log_5(x+3) \qquad \text{Given equation}$$

$$\log_5(x^2+1) = \log_5[(x-2)(x+3)] \qquad \text{Law 1: } \log_a AB = \log_a A + \log_a B$$

$$\log_5(x^2+1) = \log_5(x^2+x-6) \qquad \text{Expand}$$

$$x^2+1 = x^2+x-6 \qquad \text{log is one-to-one (or raise 5 to each side)}$$

$$x = 7 \qquad \text{Solve for } x$$

The solution is $x = 7$. (You can check that $x = 7$ satisfies the original equation.)

Now Try Exercise 49 ■

The method of Example 7 is not suitable for solving an equation like $\log_5 x = 13$ because the right-hand side is not expressed as a logarithm (base 5). To solve such equations, we use the following guidelines.

GUIDELINES FOR SOLVING LOGARITHMIC EQUATIONS

1. Isolate the logarithmic term on one side of the equation; you might first need to combine the logarithmic terms.
2. Write the equation in exponential form (or raise the base to each side of the equation).
3. Solve for the variable.

EXAMPLE 8 ■ Solving Logarithmic Equations

Solve each equation for x.

(a) $\ln x = 8$

(b) $\log_2(25 - x) = 3$

SOLUTION

(a)

$$\ln x = 8 \qquad \text{Given equation}$$

$$x = e^8 \qquad \text{Exponential form}$$

Therefore $x = e^8 \approx 2981$.

We can also solve this problem another way.

$$\ln x = 8 \qquad \text{Given equation}$$

$$e^{\ln x} = e^8 \qquad \text{Raise } e \text{ to each side}$$

$$x = e^8 \qquad \text{Property of ln}$$

(b) The first step is to rewrite the equation in exponential form.

$$\log_2(25 - x) = 3 \qquad \text{Given equation}$$

$$25 - x = 2^3 \qquad \text{Exponential form (or raise 2 to each side)}$$

$$25 - x = 8$$

$$x = 25 - 8 = 17$$

Now Try Exercises 55 and 59 ■

CHECK YOUR ANSWER

If $x = 17$, we get

$\log_2(25 - 17) = \log_2 8 = 3$ ✓

EXAMPLE 9 ■ Solving a Logarithmic Equation

Solve the equation $4 + 3\log(2x) = 16$.

SOLUTION We first isolate the logarithmic term. This allows us to write the equation in exponential form.

$$4 + 3\log(2x) = 16 \quad \text{Given equation}$$
$$3\log(2x) = 12 \quad \text{Subtract 4}$$
$$\log(2x) = 4 \quad \text{Divide by 3}$$
$$2x = 10^4 \quad \text{Exponential form (or raise 10 to each side)}$$
$$x = 5000 \quad \text{Divide by 2}$$

CHECK YOUR ANSWER

If $x = 5000$, we get

$$4 + 3\log 2(5000) = 4 + 3\log 10{,}000 = 4 + 3(4) = 16 \checkmark$$

Now Try Exercise 61 ■

EXAMPLE 10 ■ Solving a Logarithmic Equation Algebraically and Graphically

Solve the equation $\log(x + 2) + \log(x - 1) = 1$ algebraically and graphically.

SOLUTION 1: Algebraic

We first combine the logarithmic terms, using the Laws of Logarithms.

$$\log[(x + 2)(x - 1)] = 1 \quad \text{Law 1}$$
$$(x + 2)(x - 1) = 10 \quad \text{Exponential form (or raise 10 to each side)}$$
$$x^2 + x - 2 = 10 \quad \text{Expand left side}$$
$$x^2 + x - 12 = 0 \quad \text{Subtract 10}$$
$$(x + 4)(x - 3) = 0 \quad \text{Factor}$$
$$x = -4 \quad \text{or} \quad x = 3$$

We check these potential solutions in the original equation and find that $x = -4$ is not a solution (because logarithms of negative numbers are undefined), but $x = 3$ is a solution. (See *Check Your Answers*.)

CHECK YOUR ANSWERS

$x = -4$:

$$\log(-4 + 2) + \log(-4 - 1) = \log(-2) + \log(-5) \quad \text{undefined} \ \times$$

$x = 3$:

$$\log(3 + 2) + \log(3 - 1) = \log 5 + \log 2 = \log(5 \cdot 2) = \log 10 = 1 \checkmark$$

SOLUTION 2: Graphical

We first move all terms to one side of the equation:

$$\log(x + 2) + \log(x - 1) - 1 = 0$$

Then we graph

$$y = \log(x + 2) + \log(x - 1) - 1$$

as in Figure 2. The solutions are the x-intercepts of the graph. Thus the only solution is $x \approx 3$.

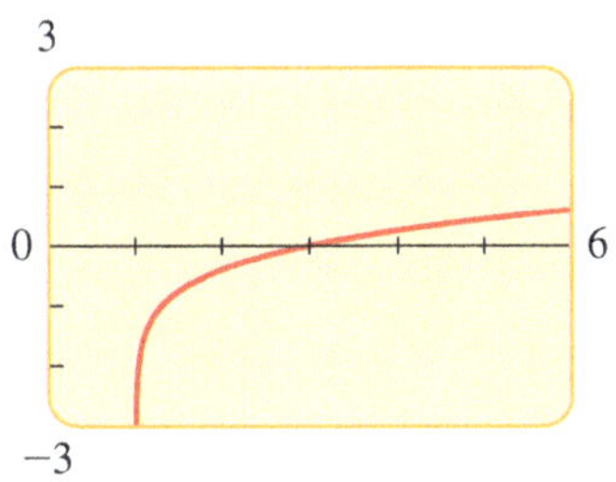

FIGURE 2

Now Try Exercise 63 ■

EXAMPLE 11 ■ Solving a Logarithmic Equation Graphically

In Example 11 it's not possible to isolate x algebraically, so we must solve the equation graphically.

Solve the equation $x^2 = 2\ln(x + 2)$.

SOLUTION We first move all terms to one side of the equation.

$$x^2 - 2\ln(x + 2) = 0$$

Then we graph

$$y = x^2 - 2\ln(x + 2)$$

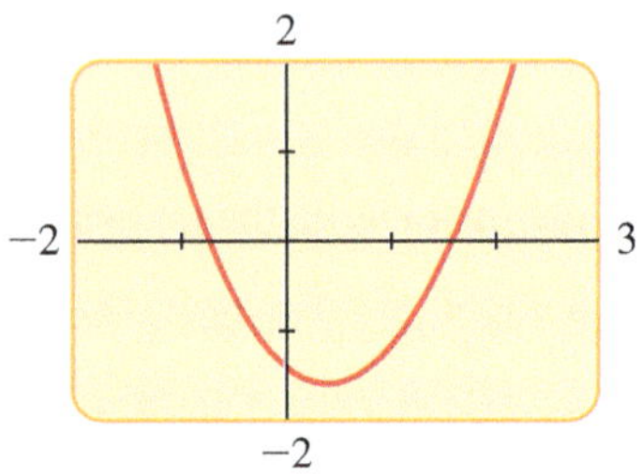

FIGURE 3

as in Figure 3. The solutions are the x-intercepts of the graph. Zooming in on the x-intercepts, we see that there are two solutions:

$$x \approx -0.71 \qquad \text{and} \qquad x \approx 1.60$$

Now Try Exercise 69 ■

Logarithmic equations are used in determining the amount of light that reaches various depths in a lake. (This information helps biologists to determine the types of life a lake can support.) As light passes through water (or other transparent materials such as glass or plastic), some of the light is absorbed. It's easy to see that the murkier the water, the more light is absorbed. The exact relationship between light absorption and the distance light travels in a material is described in the next example.

EXAMPLE 12 ■ Transparency of a Lake

The intensity of light in a lake diminishes with depth.

If I_0 and I denote the intensity of light before and after going through a material and x is the distance (in feet) the light travels in the material, then according to the **Beer-Lambert Law**,

$$-\frac{1}{k}\ln\left(\frac{I}{I_0}\right) = x$$

where k is a constant depending on the type of material.

(a) Solve the equation for I.

(b) For a certain lake $k = 0.025$, and the light intensity is $I_0 = 14$ lumens (lm). Find the light intensity at a depth of 20 ft.

SOLUTION

(a) We first isolate the logarithmic term.

$$\begin{aligned} -\frac{1}{k}\ln\left(\frac{I}{I_0}\right) &= x && \text{Given equation} \\ \ln\left(\frac{I}{I_0}\right) &= -kx && \text{Multiply by } -k \\ \frac{I}{I_0} &= e^{-kx} && \text{Exponential form} \\ I &= I_0 e^{-kx} && \text{Multiply by } I_0 \end{aligned}$$

(b) We find I using the formula from part (a).

$$\begin{aligned} I &= I_0 e^{-kx} && \text{From part (a)} \\ &= 14e^{(-0.025)(20)} && I_0 = 14,\ k = 0.025,\ x = 20 \\ &\approx 8.49 && \text{Calculator} \end{aligned}$$

The light intensity at a depth of 20 ft is about 8.5 lm.

Now Try Exercise 99 ■

■ Compound Interest

Recall the formulas for interest that we found in Section 4.1. If a principal P is invested at an interest rate r for a period of t years, then the amount A of the investment is given by

$$\begin{aligned} A &= P(1 + r) && \text{Simple interest (for one year)} \\ A(t) &= P\left(1 + \frac{r}{n}\right)^{nt} && \text{Interest compounded } n \text{ times per year} \\ A(t) &= Pe^{rt} && \text{Interest compounded continuously} \end{aligned}$$

Radiocarbon Dating is a method that archeologists use to determine the age of ancient objects. The carbon dioxide in the atmosphere always contains a fixed fraction of radioactive carbon, carbon-14 (^{14}C), with a half-life of about 5730 years. Plants absorb carbon dioxide from the atmosphere, which then makes its way to animals through the food chain. Thus, all living creatures contain the same fixed proportions of ^{14}C to nonradioactive ^{12}C as the atmosphere.

After an organism dies, it stops assimilating ^{14}C, and the amount of ^{14}C in it begins to decay exponentially. We can then determine the time that has elapsed since the death of the organism by measuring the amount of ^{14}C left in it.

For example, if a donkey bone contains 73% as much ^{14}C as a living donkey and it died t years ago, then by the formula for radioactive decay (Section 4.6),

$$0.73 = (1.00)e^{-(t \ln 2)/5730}$$

We solve this exponential equation to find $t \approx 2600$, so the bone is about 2600 years old.

We can use logarithms to determine the time it takes for the principal to increase to a given amount.

EXAMPLE 13 ■ Finding the Term for an Investment to Double

A sum of \$5000 is invested at an interest rate of 5% per year. Find the time required for the money to double if the interest is compounded according to the following methods.

(a) Semiannually **(b)** Continuously

SOLUTION

(a) We use the formula for compound interest with $P = \$5000$, $A(t) = \$10{,}000$, $r = 0.05$, and $n = 2$, and solve the resulting exponential equation for t.

$$\begin{aligned} 5000\left(1 + \frac{0.05}{2}\right)^{2t} &= 10{,}000 && P\left(1 + \frac{r}{n}\right)^{nt} = A \\ (1.025)^{2t} &= 2 && \text{Divide by 5000} \\ \log 1.025^{2t} &= \log 2 && \text{Take log of each side} \\ 2t \log 1.025 &= \log 2 && \text{Law 3 (bring down the exponent)} \\ t &= \frac{\log 2}{2 \log 1.025} && \text{Divide by 2 log 1.025} \\ t &\approx 14.04 && \text{Calculator} \end{aligned}$$

The money will double in 14.04 years.

(b) We use the formula for continuously compounded interest with $P = \$5000$, $A(t) = \$10{,}000$, and $r = 0.05$ and solve the resulting exponential equation for t.

$$\begin{aligned} 5000e^{0.05t} &= 10{,}000 && Pe^{rt} = A \\ e^{0.05t} &= 2 && \text{Divide by 5000} \\ \ln e^{0.05t} &= \ln 2 && \text{Take ln of each side} \\ 0.05t &= \ln 2 && \text{Property of ln} \\ t &= \frac{\ln 2}{0.05} && \text{Divide by 0.05} \\ t &\approx 13.86 && \text{Calculator} \end{aligned}$$

The money will double in 13.86 years.

Now Try Exercise 89 ■

EXAMPLE 14 ■ Time Required to Grow an Investment

A sum of \$1000 is invested at an interest rate of 4% per year. Find the time required for the amount to grow to \$4000 if interest is compounded continuously.

SOLUTION We use the formula for continuously compounded interest with $P = \$1000$, $A(t) = \$4000$, and $r = 0.04$ and solve the resulting exponential equation for t.

$$\begin{aligned} 1000e^{0.04t} &= 4000 && Pe^{rt} = A \\ e^{0.04t} &= 4 && \text{Divide by 1000} \\ 0.04t &= \ln 4 && \text{Take ln of each side} \\ t &= \frac{\ln 4}{0.04} && \text{Divide by 0.04} \\ t &\approx 34.66 && \text{Calculator} \end{aligned}$$

The amount will be \$4000 in about 34 years and 8 months.

Now Try Exercise 91 ■

4.5 EXERCISES

CONCEPTS

1. Let's solve the exponential equation $2e^x = 50$.
   (a) First, we isolate e^x to get the equivalent equation ________.
   (b) Next, we take ln of each side to get the equivalent equation ________________.
   (c) Now we use a calculator to find $x \approx$ ________.

2. Let's solve the logarithmic equation
$$\log 3 + \log(x - 2) = \log x$$
   (a) First, we combine the logarithms on the LHS to get the equivalent equation ________________.
   (b) Next, we use the fact that log is one-to-one to get the equivalent equation ________________.
   (c) Now we find $x =$ ________.

SKILLS

3–10 ■ Exponential Equations Find the solution of the exponential equation, as in Example 1.

3. $5^{x-1} = 125$
4. $e^{x^2} = e^9$
5. $5^{2x-3} = 1$
6. $10^{2x-3} = \frac{1}{10}$
7. $7^{2x-3} = 7^{6+5x}$
8. $e^{1-2x} = e^{3x-5}$
9. $6^{x^2-1} = 6^{1-x^2}$
10. $10^{2x^2-3} = 10^{9-x^2}$

11–38 ■ Exponential Equations (a) Find the exact solution of the exponential equation in terms of logarithms. (b) Use a calculator to find an approximation to the solution rounded to six decimal places.

11. $10^x = 25$
12. $10^{-x} = 4$
13. $e^{-5x} = 10$
14. $e^{0.4x} = 8$
15. $2^{1-x} = 3$
16. $3^{2x-1} = 5$
17. $3e^x = 10$
18. $2e^{12x} = 17$
19. $300(1.025)^{12t} = 1000$
20. $10(1.375)^{10t} = 50$
21. $e^{1-4x} = 2$
22. $e^{3-5x} = 16$
23. $2^{5-7x} = 15$
24. $2^{3x} = 34$
25. $3^{x/14} = 0.1$
26. $5^{-x/100} = 2$
27. $4(1 + 10^{5x}) = 9$
28. $2(5 + 3^{x+1}) = 100$
29. $8 + e^{1-4x} = 20$
30. $1 + e^{4x+1} = 20$
31. $4^x + 2^{1+2x} = 50$
32. $125^x + 5^{3x+1} = 200$
33. $5^x = 4^{x+1}$
34. $10^{1-x} = 6^x$
35. $2^{3x+1} = 3^{x-2}$
36. $7^{x/2} = 5^{1-x}$
37. $\dfrac{50}{1 + e^{-x}} = 4$
38. $\dfrac{10}{1 + e^{-x}} = 2$

39–44 ■ Exponential Equations of Quadratic Type Solve the equation.

39. $e^{2x} - 3e^x + 2 = 0$
40. $e^{2x} - e^x - 6 = 0$
41. $e^{4x} + 4e^{2x} - 21 = 0$
42. $3^{4x} - 3^{2x} - 6 = 0$
43. $2^x - 10(2^{-x}) + 3 = 0$
44. $e^x + 15e^{-x} - 8 = 0$

45–48 ■ Equations Involving Exponential Functions Solve the equation.

45. $x^2 2^x - 2^x = 0$
46. $x^2 10^x - x10^x = 2(10^x)$
47. $4x^3e^{-3x} - 3x^4e^{-3x} = 0$
48. $x^2e^x + xe^x - e^x = 0$

49–54 ■ Logarithmic Equations Solve the logarithmic equation for x, as in Example 7.

49. $\log x + \log(x - 1) = \log(4x)$
50. $\log_5 x + \log_5(x + 1) = \log_5 20$
51. $2 \log x = \log 2 + \log(3x - 4)$
52. $\ln(x - \frac{1}{2}) + \ln 2 = 2 \ln x$
53. $\log_2 3 + \log_2 x = \log_2 5 + \log_2(x - 2)$
54. $\log_4(x + 2) + \log_4 3 = \log_4 5 + \log_4(2x - 3)$

55–68 ■ Logarithmic Equations Solve the logarithmic equation for x.

55. $\ln x = 10$
56. $\ln(2 + x) = 1$
57. $\log x = -2$
58. $\log(x - 4) = 3$
59. $\log(3x + 5) = 2$
60. $\log_3(2 - x) = 3$
61. $4 - \log(3 - x) = 3$
62. $\log_2(x^2 - x - 2) = 2$
63. $\log_2 x + \log_2(x - 3) = 2$
64. $\log x + \log(x - 3) = 1$
65. $\log_9(x - 5) + \log_9(x + 3) = 1$
66. $\ln(x - 1) + \ln(x + 2) = 1$
67. $\log_5(x + 1) - \log_5(x - 1) = 2$
68. $\log_3(x + 15) - \log_3(x - 1) = 2$

69–76 ■ Solving Equations Graphically Use a graphing device to find all solutions of the equation, rounded to two decimal places.

69. $\ln x = 3 - x$
70. $\log x = x^2 - 2$
71. $x^3 - x = \log(x + 1)$
72. $x = \ln(4 - x^2)$
73. $e^x = -x$
74. $2^{-x} = x - 1$
75. $4^{-x} = \sqrt{x}$
76. $e^{x^2} - 2 = x^3 - x$

77–78 ■ More Exponential and Logarithmic Equations Solve the equation for x.

77. $2^{2/\log_5 x} = \frac{1}{16}$
78. $\log_2(\log_3 x) = 4$

SKILLS Plus

79–82 ■ Solving Inequalities Solve the inequality.

79. $\log(x - 2) + \log(9 - x) < 1$

80. $3 \le \log_2 x \le 4$

81. $2 < 10^x < 5$

82. $x^2 e^x - 2e^x < 0$

83–86 ■ Inverse Functions Find the inverse function of f.

83. $f(x) = 2^{2x}$

84. $f(x) = 3^{x+1}$

85. $f(x) = \log_2(x - 1)$

86. $f(x) = \log 3x$

87–88 ■ Special Logarithmic Equations Find the value(s) of x for which the equation is true.

87. $\log(x + 3) = \log x + \log 3$

88. $(\log x)^3 = 3 \log x$

APPLICATIONS

89. Compound Interest A man invests \$5000 in an account that pays 8.5% interest per year, compounded quarterly.

(a) Find the amount after 3 years.

(b) How long will it take for the investment to double?

90. Compound Interest A woman invests \$6500 in an account that pays 6% interest per year, compounded continuously.

(a) What is the amount after 2 years?

(b) How long will it take for the amount to be \$8000?

91. Compound Interest Find the time required for an investment of \$5000 to grow to \$8000 at an interest rate of 7.5% per year, compounded quarterly.

92. Compound Interest Nancy wants to invest \$4000 in saving certificates that bear an interest rate of 9.75% per year, compounded semiannually. How long a time period should she choose to save an amount of \$5000?

93. Doubling an Investment How long will it take for an investment of \$1000 to double in value if the interest rate is 8.5% per year, compounded continuously?

94. Interest Rate A sum of \$1000 was invested for 4 years, and the interest was compounded semiannually. If this sum amounted to \$1435.77 in the given time, what was the interest rate?

95. Radioactive Decay A 15-g sample of radioactive iodine decays in such a way that the mass remaining after t days is given by $m(t) = 15e^{-0.087t}$, where $m(t)$ is measured in grams. After how many days are there only 5 g remaining?

96. Sky Diving The velocity of a sky diver t seconds after jumping is given by $v(t) = 80(1 - e^{-0.2t})$. After how many seconds is the velocity 70 ft/s?

97. Fish Population A small lake is stocked with a certain species of fish. The fish population is modeled by the function

$$P = \frac{10}{1 + 4e^{-0.8t}}$$

where P is the number of fish in thousands and t is measured in years since the lake was stocked.

(a) Find the fish population after 3 years.

(b) After how many years will the fish population reach 5000 fish?

98. Transparency of a Lake Environmental scientists measure the intensity of light at various depths in a lake to find the "transparency" of the water. Certain levels of transparency are required for the biodiversity of the submerged macrophyte population. In a certain lake the intensity of light at depth x is given by

$$I = 10e^{-0.008x}$$

where I is measured in lumens and x in feet.

(a) Find the intensity I at a depth of 30 ft.

(b) At what depth has the light intensity dropped to $I = 5$?

99. Atmospheric Pressure Atmospheric pressure P (in kilopascals, kPa) at altitude h (in kilometers, km) is governed by the formula

$$\ln\left(\frac{P}{P_0}\right) = -\frac{h}{k}$$

where $k = 7$ and $P_0 = 100$ kPa are constants.

(a) Solve the equation for P.

(b) Use part (a) to find the pressure P at an altitude of 4 km.

100. Cooling an Engine Suppose you're driving your car on a cold winter day (20°F outside) and the engine overheats (at about 220°F). When you park, the engine begins to cool down. The temperature T of the engine t minutes after you park satisfies the equation

$$\ln\left(\frac{T - 20}{200}\right) = -0.11t$$

(a) Solve the equation for T.

(b) Use part (a) to find the temperature of the engine after 20 min ($t = 20$).

101. Electric Circuits An electric circuit contains a battery that produces a voltage of 60 volts (V), a resistor with a resistance of 13 ohms (Ω), and an inductor with an inductance of 5 henrys (H), as shown in the figure on the following page. Using calculus, it can be shown that the current

$I = I(t)$ (in amperes, A) t seconds after the switch is closed is $I = \frac{60}{13}(1 - e^{-13t/5})$.

(a) Use this equation to express the time t as a function of the current I.

(b) After how many seconds is the current 2 A?

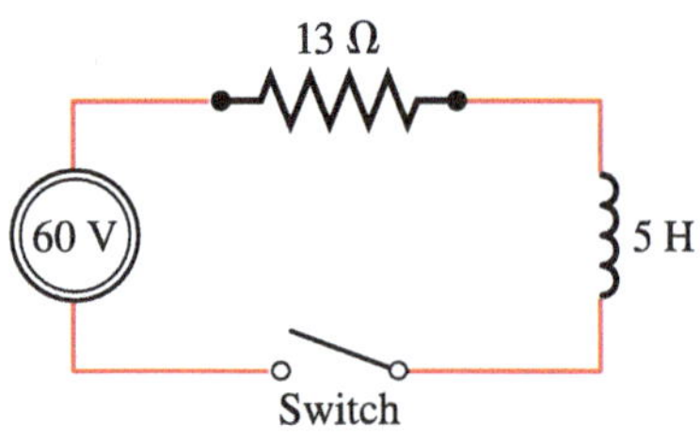

102. Learning Curve A *learning curve* is a graph of a function $P(t)$ that measures the performance of someone learning a skill as a function of the training time t. At first, the rate of learning is rapid. Then, as performance increases and approaches a maximal value M, the rate of learning decreases. It has been found that the function

$$P(t) = M - Ce^{-kt}$$

where k and C are positive constants and $C < M$ is a reasonable model for learning.

(a) Express the learning time t as a function of the performance level P.

(b) For a pole-vaulter in training, the learning curve is given by

$$P(t) = 20 - 14e^{-0.024t}$$

where $P(t)$ is the height he is able to pole-vault after t months. After how many months of training is he able to vault 12 ft?

(c) Draw a graph of the learning curve in part (b).

DISCUSS ■ DISCOVER ■ PROVE ■ WRITE

103. DISCUSS: Estimating a Solution Without actually solving the equation, find two whole numbers between which the solution of $9^x = 20$ must lie. Do the same for $9^x = 100$. Explain how you reached your conclusions.

104. DISCUSS ■ DISCOVER: A Surprising Equation Take logarithms to show that the equation

$$x^{1/\log x} = 5$$

has no solution. For what values of k does the equation

$$x^{1/\log x} = k$$

have a solution? What does this tell us about the graph of the function $f(x) = x^{1/\log x}$? Confirm your answer using a graphing device.

105. DISCUSS: Disguised Equations Each of these equations can be transformed into an equation of linear or quadratic type by applying the hint. Solve each equation.

(a) $(x - 1)^{\log(x-1)} = 100(x - 1)$
[*Hint:* Take log of each side.]

(b) $\log_2 x + \log_4 x + \log_8 x = 11$
[*Hint:* Change all logs to base 2.]

(c) $4^x - 2^{x+1} = 3$
[*Hint:* Write as a quadratic in 2^x.]

4.6 MODELING WITH EXPONENTIAL FUNCTIONS

■ Exponential Growth (Doubling Time) ■ Exponential Growth (Relative Growth Rate) ■ Radioactive Decay ■ Newton's Law of Cooling

Many processes that occur in nature, such as population growth, radioactive decay, heat diffusion, and numerous others, can be modeled by using exponential functions. In this section we study exponential models.

■ Exponential Growth (Doubling Time)

Suppose we start with a single bacterium, which divides every hour. After one hour we have 2 bacteria, after two hours we have 2^2 or 4 bacteria, after three hours we have 2^3

or 8 bacteria, and so on (see Figure 1). We see that we can model the bacteria population after t hours by $f(t) = 2^t$.

FIGURE 1 Bacteria population

If we start with 10 of these bacteria, then the population is modeled by $f(t) = 10 \cdot 2^t$. A slower-growing strain of bacteria doubles every 3 hours; in this case the population is modeled by $f(t) = 10 \cdot 2^{t/3}$. In general, we have the following.

EXPONENTIAL GROWTH (DOUBLING TIME)

If the initial size of a population is n_0 and the doubling time is a, then the size of the population at time t is

$$n(t) = n_0 2^{t/a}$$

where a and t are measured in the same time units (minutes, hours, days, years, and so on).

EXAMPLE 1 ■ Bacteria Population

Under ideal conditions a certain bacteria population doubles every three hours. Initially, there are 1000 bacteria in a colony.

(a) Find a model for the bacteria population after t hours.

(b) How many bacteria are in the colony after 15 hours?

(c) After how many hours will the bacteria count reach 100,000?

SOLUTION

(a) The population at time t is modeled by

$$n(t) = 1000 \cdot 2^{t/3}$$

where t is measured in hours.

(b) After 15 hours the number of bacteria is

$$n(15) = 1000 \cdot 2^{15/3} = 32{,}000$$

(c) We set $n(t) = 100{,}000$ in the model that we found in part (a) and solve the resulting exponential equation for t.

$$100{,}000 = 1000 \cdot 2^{t/3} \qquad n(t) = 1000 \cdot 2^{t/3}$$

$$100 = 2^{t/3} \qquad \text{Divide by 1000}$$

$$\log 100 = \log 2^{t/3} \qquad \text{Take log of each side}$$

$$2 = \frac{t}{3} \log 2 \qquad \text{Properties of log}$$

$$t = \frac{6}{\log 2} \approx 19.93 \qquad \text{Solve for } t$$

The bacteria level reaches 100,000 in about 20 hours.

Now Try Exercise 1 ■

EXAMPLE 2 ■ Rabbit Population

A certain breed of rabbit was introduced onto a small island 8 months ago. The current rabbit population on the island is estimated to be 4100 and doubling every 3 months.

(a) What was the initial size of the rabbit population?

(b) Estimate the population 1 year after the rabbits were introduced to the island.

(c) Sketch a graph of the rabbit population.

SOLUTION

(a) The doubling time is $a = 3$, so the population at time t is

$$n(t) = n_0 2^{t/3} \qquad \text{Model}$$

where n_0 is the initial population. Since the population is 4100 when t is 8 months, we have

$$n(8) = n_0 2^{8/3} \qquad \text{From model}$$

$$4100 = n_0 2^{8/3} \qquad \text{Because } n(8) = 4100$$

$$n_0 = \frac{4100}{2^{8/3}} \qquad \text{Divide by } 2^{8/3} \text{ and switch sides}$$

$$n_0 \approx 645 \qquad \text{Calculator}$$

Thus we estimate that 645 rabbits were introduced onto the island.

(b) From part (a) we know that the initial population is $n_0 = 645$, so we can model the population after t months by

$$n(t) = 645 \cdot 2^{t/3} \qquad \text{Model}$$

After 1 year $t = 12$, so

$$n(12) = 645 \cdot 2^{12/3} = 10{,}320$$

So after 1 year there would be about 10,000 rabbits.

(c) We first note that the domain is $t \geq 0$. The graph is shown in Figure 2.

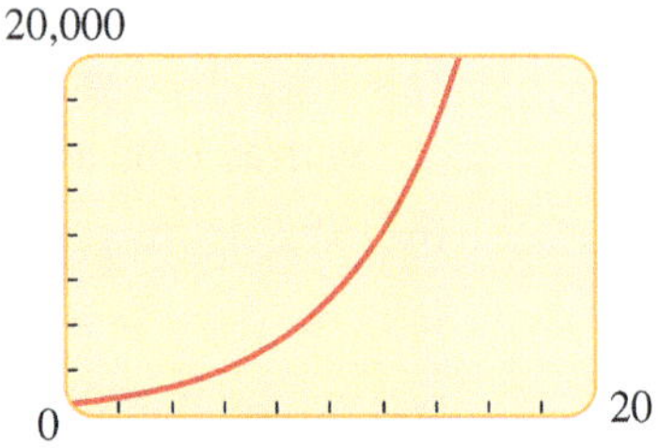

FIGURE 2 $n(t) = 645 \cdot 2^{t/3}$

Now Try Exercise 3 ■

Exponential Growth (Relative Growth Rate)

The growth of a population with relative growth rate r is analogous to the growth of an investment with continuously compounded interest rate r.

We have used an exponential function with base 2 to model population growth (in terms of the doubling time). We could also model the same population with an exponential function with base 3 (in terms of the tripling time). In fact, we can find an exponential model with any base. If we use the base e, we get a population model in terms of the **relative growth rate** r: the rate of population growth expressed as a proportion of the population at any time. In this case r is the "instantaneous" growth rate. (In calculus the concept of instantaneous rate is given a precise meaning.) For instance, if $r = 0.02$, then at any time t the growth rate is 2% of the population at time t.

EXPONENTIAL GROWTH (RELATIVE GROWTH RATE)

A population that experiences **exponential growth** increases according to the model

$$n(t) = n_0 e^{rt}$$

where $n(t)$ = population at time t

n_0 = initial size of the population

r = relative rate of growth (expressed as a proportion of the population)

t = time

Notice that the formula for population growth is the same as that for continuously compounded interest. In fact, the same principle is at work in both cases: The growth of a population (or an investment) per time period is proportional to the size of the population (or the amount of the investment). A population of 1,000,000 will increase more in one year than a population of 1000; in exactly the same way, an investment of \$1,000,000 will increase more in one year than an investment of \$1000.

In the following examples we assume that the populations grow exponentially.

EXAMPLE 3 ■ Predicting the Size of a Population

The initial bacterium count in a culture is 500. A biologist later makes a sample count of bacteria in the culture and finds that the relative rate of growth is 40% per hour.

(a) Find a function that models the number of bacteria after t hours.

(b) What is the estimated count after 10 hours?

(c) After how many hours will the bacteria count reach 80,000?

(d) Sketch a graph of the function $n(t)$.

SOLUTION

(a) We use the exponential growth model with $n_0 = 500$ and $r = 0.4$ to get

$$n(t) = 500e^{0.4t}$$

where t is measured in hours.

(b) Using the function in part (a), we find that the bacterium count after 10 hours is

$$n(10) = 500e^{0.4(10)} = 500e^4 \approx 27{,}300$$

(c) We set $n(t) = 80{,}000$ and solve the resulting exponential equation for t.

$$80{,}000 = 500 \cdot e^{0.4t} \qquad n(t) = 500 \cdot e^{0.4t}$$

$$160 = e^{0.4t} \qquad \text{Divide by 500}$$

$$\ln 160 = 0.4t \qquad \text{Take ln of each side}$$

$$t = \frac{\ln 160}{0.4} \approx 12.68 \qquad \text{Solve for } t$$

The bacteria level reaches 80,000 in about 12.7 hours.

(d) The graph is shown in Figure 3.

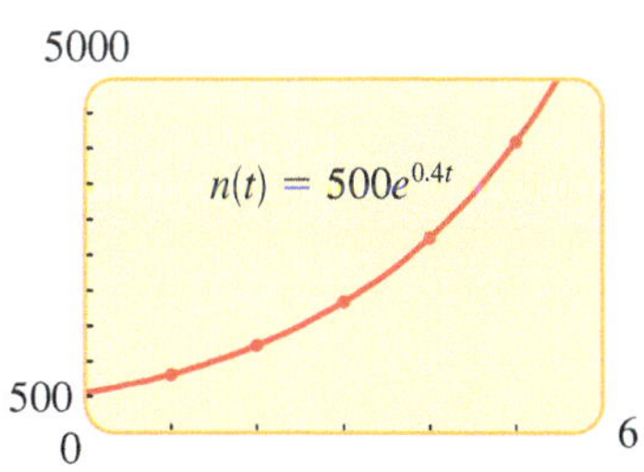

FIGURE 3

Now Try Exercise 5 ■

The relative growth of world population has been declining over the past few decades—from 2% in 1995 to 1.1% in 2013.

Standing Room Only
The population of the world was about 6.1 billion in 2000 and was increasing at 1.4% per year. Assuming that each person occupies an average of 4 ft^2 of the surface of the earth, the exponential model for population growth projects that by the year 2801 there will be standing room only! (The total land surface area of the world is about 1.8×10^{15} ft^2.)

EXAMPLE 4 ■ Comparing Different Rates of Population Growth

In 2000 the population of the world was 6.1 billion, and the relative rate of growth was 1.4% per year. It is claimed that a rate of 1.0% per year would make a significant difference in the total population in just a few decades. Test this claim by estimating the population of the world in the year 2050 using a relative rate of growth of (a) 1.4% per year and (b) 1.0% per year.

Graph the population functions for the next 100 years for the two relative growth rates in the same viewing rectangle.

SOLUTION

(a) By the exponential growth model we have

$$n(t) = 6.1e^{0.014t}$$

where $n(t)$ is measured in billions and t is measured in years since 2000. Because the year 2050 is 50 years after 2000, we find

$$n(50) = 6.1e^{0.014(50)} = 6.1e^{0.7} \approx 12.3$$

The estimated population in the year 2050 is about 12.3 billion.

(b) We use the function

$$n(t) = 6.1e^{0.010t}$$

and find

$$n(50) = 6.1e^{0.010(50)} = 6.1e^{0.50} \approx 10.1$$

The estimated population in the year 2050 is about 10.1 billion.

The graphs in Figure 4 show that a small change in the relative rate of growth will, over time, make a large difference in population size.

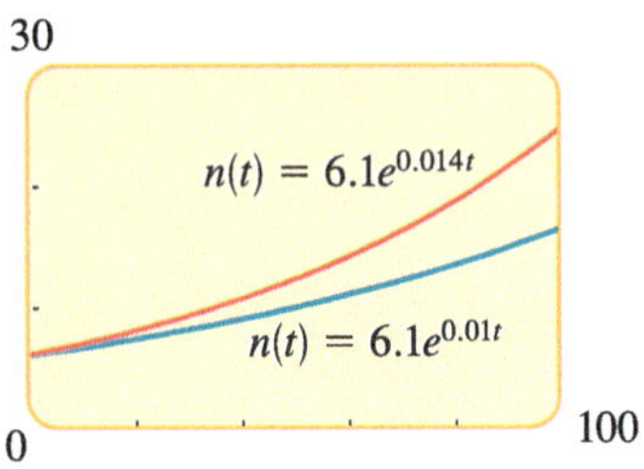

FIGURE 4

Now Try Exercise 7 ■

EXAMPLE 5 ■ Expressing a Model in Terms of *e*

A culture starts with 10,000 bacteria, and the number doubles every 40 minutes.

(a) Find a function $n(t) = n_0 2^{t/a}$ that models the number of bacteria after t hours.

(b) Find a function $n(t) = n_0 e^{rt}$ that models the number of bacteria after t hours.

(c) Sketch a graph of the number of bacteria at time t.

SOLUTION

(a) The initial population is $n_0 = 10{,}000$. The doubling time is $a = 40 \text{ min} = 2/3 \text{ h}$. Since $1/a = 3/2 = 1.5$, the model is

$$n(t) = 10{,}000 \cdot 2^{1.5t}$$

(b) The initial population is $n_0 = 10{,}000$. We need to find the relative growth rate r. Since there are 20,000 bacteria when $t = 2/3$ h, we have

$$\begin{aligned} 20{,}000 &= 10{,}000e^{r(2/3)} && n(t) = 10{,}000e^{rt} \\ 2 &= e^{r(2/3)} && \text{Divide by 10,000} \\ \ln 2 &= \ln e^{r(2/3)} && \text{Take ln of each side} \\ \ln 2 &= r(2/3) && \text{Property of ln} \\ r &= \frac{3 \ln 2}{2} \approx 1.0397 && \text{Solve for } r \end{aligned}$$

Now that we know the relative growth rate r, we can find the model:

$$n(t) = 10{,}000e^{1.0397t}$$

(c) We can graph the model in part (a) or the one in part (b). The graphs are identical. See Figure 5.

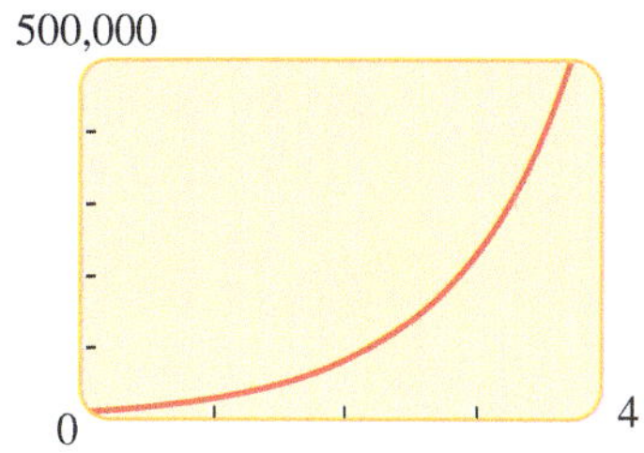

FIGURE 5 Graphs of $y = 10{,}000 \cdot 2^{1.5t}$ and $y = 10{,}000e^{1.0397t}$

Now Try Exercise 9 ■

Radioactive Decay

Radioactive substances decay by spontaneously emitting radiation. The rate of decay is proportional to the mass of the substance. This is analogous to population growth except that the mass *decreases*. Physicists express the rate of decay in terms of **half-life**, the time it takes for a sample of the substance to decay to half its original mass. For example, the half-life of radium-226 is 1600 years, so a 100-g sample decays to 50 g $\left(\text{or } \frac{1}{2} \times 100 \text{ g}\right)$ in 1600 years, then to 25 g $\left(\text{or } \frac{1}{2} \times \frac{1}{2} \times 100 \text{ g}\right)$ in 3200 years, and so on. In general, for a radioactive substance with mass m_0 and half-life h, the amount remaining at time t is modeled by

$$m(t) = m_0 2^{-t/h}$$

where h and t are measured in the same time units (minutes, hours, days, years, and so on).

To express this model in the form $m(t) = m_0 e^{rt}$, we need to find the relative decay rate r. Since h is the half-life, we have

$$m(t) = m_0 e^{-rt} \qquad \text{Model}$$

$$\frac{m_0}{2} = m_0 e^{-rh} \qquad h \text{ is the half-life}$$

$$\frac{1}{2} = e^{-rh} \qquad \text{Divide by } m_0$$

$$\ln \frac{1}{2} = -rh \qquad \text{Take ln of each side}$$

$$r = \frac{\ln 2}{h} \qquad \text{Solve for } r$$

This last equation allows us to find the relative decay rate r from the half-life h.

The half-lives of **radioactive elements** vary from very long to very short. Here are some examples.

Element	Half-life
Thorium-232	14.5 billion years
Uranium-235	4.5 billion years
Thorium-230	80,000 years
Plutonium-239	24,360 years
Carbon-14	5,730 years
Radium-226	1,600 years
Cesium-137	30 years
Strontium-90	28 years
Polonium-210	140 days
Thorium-234	25 days
Iodine-135	8 days
Radon-222	3.8 days
Lead-211	3.6 minutes
Krypton-91	10 seconds

DISCOVERY PROJECT

Modeling Radiation with Coins and Dice

Radioactive elements decay when their atoms spontaneously emit radiation and change into smaller, stable atoms. But if atoms decay randomly, how is it possible to find a function that models their behavior? We'll try to answer this question by experimenting with randomly tossing coins and rolling dice. The experiments allow us to experience how a very large number of random events can result in predictable exponential results. You can find the project at **www.stewartmath.com**.

RADIOACTIVE DECAY MODEL

If m_0 is the initial mass of a radioactive substance with half-life h, then the mass remaining at time t is modeled by the function

$$m(t) = m_0 e^{-rt}$$

where $r = \dfrac{\ln 2}{h}$ is the **relative decay rate**.

EXAMPLE 6 ■ Radioactive Decay

Polonium-210 (^{210}Po) has a half-life of 140 days. Suppose a sample of this substance has a mass of 300 mg.

(a) Find a function $m(t) = m_0 2^{-t/h}$ that models the mass remaining after t days.

(b) Find a function $m(t) = m_0 e^{-rt}$ that models the mass remaining after t days.

(c) Find the mass remaining after one year.

(d) How long will it take for the sample to decay to a mass of 200 mg?

(e) Draw a graph of the sample mass as a function of time.

SOLUTION

(a) We have $m_0 = 300$ and $h = 140$, so the amount remaining after t days is

$$m(t) = 300 \cdot 2^{-t/140}$$

(b) We have $m_0 = 300$ and $r = \ln 2/140 \approx -0.00495$, so the amount remaining after t days is

$$m(t) = 300 \cdot e^{-0.00495t}$$

In parts (c) and (d) we can also use the model found in part (a). Check that the result is the same using either model.

(c) We use the function we found in part (a) with $t = 365$ (1 year):

$$m(365) = 300e^{-0.00495(365)} \approx 49.256$$

Thus approximately 49 mg of ^{210}Po remains after 1 year.

(d) We use the function that we found in part (b) with $m(t) = 200$ and solve the resulting exponential equation for t:

$$\begin{aligned} 300e^{-0.00495t} &= 200 && m(t) = m_0 e^{-rt} \\ e^{-0.00495t} &= \tfrac{2}{3} && \text{Divide by 300} \\ \ln e^{-0.00495t} &= \ln \tfrac{2}{3} && \text{Take ln of each side} \\ -0.00495t &= \ln \tfrac{2}{3} && \text{Property of ln} \\ t &= -\frac{\ln \frac{2}{3}}{0.00495} && \text{Solve for } t \\ t &\approx 81.9 && \text{Calculator} \end{aligned}$$

The time required for the sample to decay to 200 mg is about 82 days.

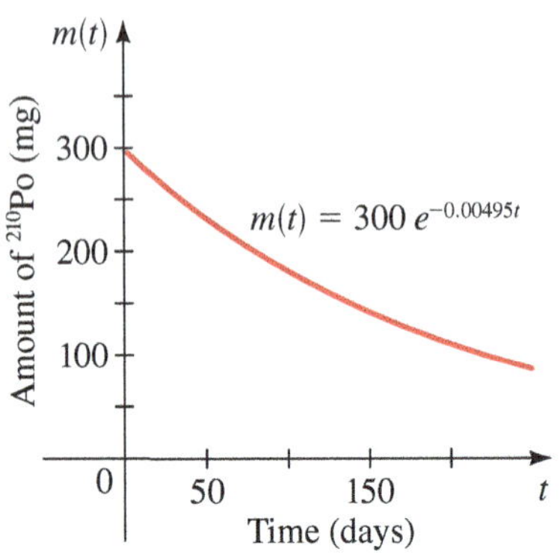

FIGURE 6

(e) We can graph the model in part (a) or the one in part (b). The graphs are identical. See Figure 6.

Now Try Exercise 17 ■

Joel W. Rogers/Encyclopedia/Corbis

Radioactive Waste

Harmful radioactive isotopes are produced whenever a nuclear reaction occurs, whether as the result of an atomic bomb test, a nuclear accident such as the one at Fukushima Daiichi in 2011, or the uneventful production of electricity at a nuclear power plant.

One radioactive material that is produced in atomic bombs is the isotope strontium-90 (^{90}Sr), with a half-life of 28 years. This is deposited like calcium in human bone tissue, where it can cause leukemia and other cancers. However, in the decades since atmospheric testing of nuclear weapons was halted, ^{90}Sr levels in the environment have fallen to a level that no longer poses a threat to health.

Nuclear power plants produce radioactive plutonium-239 (^{239}Pu), which has a half-life of 24,360 years. Because of its long half-life, ^{239}Pu could pose a threat to the environment for thousands of years. So great care must be taken to dispose of it properly. The difficulty of ensuring the safety of the disposed radioactive waste is one reason that nuclear power plants remain controversial.

Newton's Law of Cooling

Newton's Law of Cooling states that the rate at which an object cools is proportional to the temperature difference between the object and its surroundings, provided that the temperature difference is not too large. By using calculus, the following model can be deduced from this law.

NEWTON'S LAW OF COOLING

If D_0 is the initial temperature difference between an object and its surroundings, and if its surroundings have temperature T_s, then the temperature of the object at time t is modeled by the function

$$T(t) = T_s + D_0 e^{-kt}$$

where k is a positive constant that depends on the type of object.

EXAMPLE 7 ■ Newton's Law of Cooling

A cup of coffee has a temperature of 200°F and is placed in a room that has a temperature of 70°F. After 10 min the temperature of the coffee is 150°F.

(a) Find a function that models the temperature of the coffee at time t.

(b) Find the temperature of the coffee after 15 min.

(c) After how long will the coffee have cooled to 100°F?

(d) Illustrate by drawing a graph of the temperature function.

SOLUTION

(a) The temperature of the room is $T_s = 70$°F, and the initial temperature difference is

$$D_0 = 200 - 70 = 130\text{°F}$$

So by Newton's Law of Cooling, the temperature after t minutes is modeled by the function

$$T(t) = 70 + 130e^{-kt}$$

We need to find the constant k associated with this cup of coffee. To do this, we use the fact that when $t = 10$, the temperature is $T(10) = 150$. So we have

$$70 + 130e^{-10k} = 150 \qquad T_s + D_0 e^{-kt} = T(t)$$

$$130e^{-10k} = 80 \qquad \text{Subtract 70}$$

$$e^{-10k} = \tfrac{8}{13} \qquad \text{Divide by 130}$$

$$-10k = \ln \tfrac{8}{13} \qquad \text{Take ln of each side}$$

$$k = -\tfrac{1}{10} \ln \tfrac{8}{13} \qquad \text{Solve for } k$$

$$k \approx 0.04855 \qquad \text{Calculator}$$

Substituting this value of k into the expression for $T(t)$, we get

$$T(t) = 70 + 130e^{-0.04855t}$$

(b) We use the function that we found in part (a) with $t = 15$.

$$T(15) = 70 + 130e^{-0.04855(15)} \approx 133\text{°F}$$

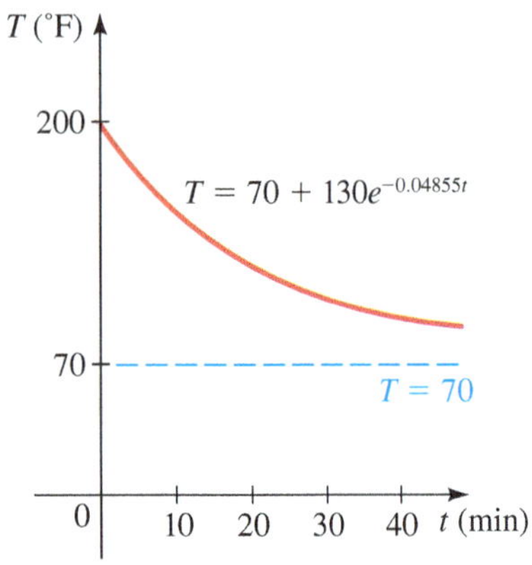

FIGURE 7 Temperature of coffee after t minutes

(c) We use the function that we found in part (a) with $T(t) = 100$ and solve the resulting exponential equation for t.

$$\begin{aligned} 70 + 130e^{-0.04855t} &= 100 && T_s + D_0e^{-kt} = T(t) \\ 130e^{-0.04855t} &= 30 && \text{Subtract 70} \\ e^{-0.04855t} &= \tfrac{3}{13} && \text{Divide by 130} \\ -0.04855t &= \ln \tfrac{3}{13} && \text{Take ln of each side} \\ t &= \frac{\ln \frac{3}{13}}{-0.04855} && \text{Solve for } t \\ t &\approx 30.2 && \text{Calculator} \end{aligned}$$

The coffee will have cooled to 100°F after about half an hour.

(d) The graph of the temperature function is sketched in Figure 7. Notice that the line $t = 70$ is a horizontal asymptote. (Why?)

Now Try Exercise 25 ■

4.6 EXERCISES

APPLICATIONS

1–16 ■ Population Growth These exercises use the population growth model.

1. Bacteria Culture A certain culture of the bacterium *Streptococcus A* initially has 10 bacteria and is observed to double every 1.5 hours.

(a) Find an exponential model $n(t) = n_0 2^{t/a}$ for the number of bacteria in the culture after t hours.

(b) Estimate the number of bacteria after 35 hours.

(c) After how many hours will the bacteria count reach 10,000?

© Sebastian Kaulitzki/Shutterstock.com

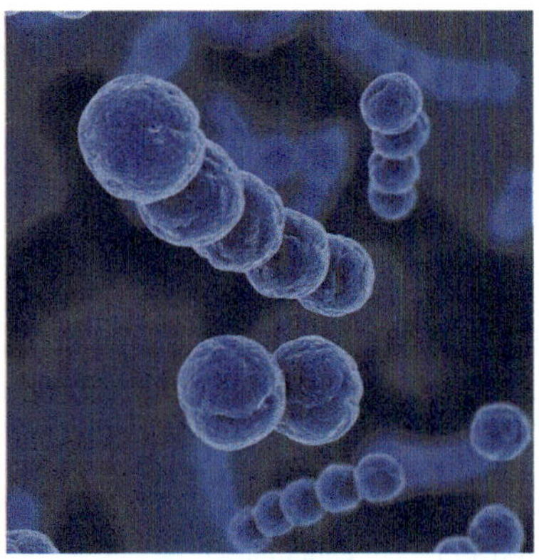

Streptococcus A
(12,000 × magnification)

2. Bacteria Culture A certain culture of the bacterium *Rhodobacter sphaeroides* initially has 25 bacteria and is observed to double every 5 hours.

(a) Find an exponential model $n(t) = n_0 2^{t/a}$ for the number of bacteria in the culture after t hours.

(b) Estimate the number of bacteria after 18 hours.

(c) After how many hours will the bacteria count reach 1 million?

3. Squirrel Population A grey squirrel population was introduced in a certain county of Great Britain 30 years ago. Biologists observe that the population doubles every 6 years, and now the population is 100,000.

(a) What was the initial size of the squirrel population?

(b) Estimate the squirrel population 10 years from now.

(c) Sketch a graph of the squirrel population.

4. Bird Population A certain species of bird was introduced in a certain county 25 years ago. Biologists observe that the population doubles every 10 years, and now the population is 13,000.

(a) What was the initial size of the bird population?

(b) Estimate the bird population 5 years from now.

(c) Sketch a graph of the bird population.

5. Fox Population The fox population in a certain region has a relative growth rate of 8% per year. It is estimated that the population in 2013 was 18,000.

(a) Find a function $n(t) = n_0e^{rt}$ that models the population t years after 2013.

(b) Use the function from part (a) to estimate the fox population in the year 2021.

(c) After how many years will the fox population reach 25,000?

(d) Sketch a graph of the fox population function for the years 2013–2021.

6. Fish Population The population of a certain species of fish has a relative growth rate of 1.2% per year. It is estimated that the population in 2010 was 12 million.

(a) Find an exponential model $n(t) = n_0e^{rt}$ for the population t years after 2010.

(b) Estimate the fish population in the year 2015.

(c) After how many years will the fish population reach 14 million?

(d) Sketch a graph of the fish population.

7. Population of a Country The population of a country has a relative growth rate of 3% per year. The government is trying to reduce the growth rate to 2%. The population in 2011 was approximately 110 million. Find the projected population for the year 2036 for the following conditions.

(a) The relative growth rate remains at 3% per year.

(b) The relative growth rate is reduced to 2% per year.

8. Bacteria Culture It is observed that a certain bacteria culture has a relative growth rate of 12% per hour, but in the presence of an antibiotic the relative growth rate is reduced to 5% per hour. The initial number of bacteria in the culture is 22. Find the projected population after 24 hours for the following conditions.

(a) No antibiotic is present, so the relative growth rate is 12%.

(b) An antibiotic is present in the culture, so the relative growth rate is reduced to 5%.

9. Population of a City The population of a certain city was 112,000 in 2014, and the observed doubling time for the population is 18 years.

(a) Find an exponential model $n(t) = n_0 2^{t/a}$ for the population t years after 2014.

(b) Find an exponential model $n(t) = n_0e^{rt}$ for the population t years after 2014.

(c) Sketch a graph of the population at time t.

(d) Estimate how long it takes the population to reach 500,000.

10. Bat Population The bat population in a certain Midwestern county was 350,000 in 2012, and the observed doubling time for the population is 25 years.

(a) Find an exponential model $n(t) = n_0 2^{t/a}$ for the population t years after 2012.

(b) Find an exponential model $n(t) = n_0e^{rt}$ for the population t years after 2012.

(c) Sketch a graph of the population at time t.

(d) Estimate how long it takes the population to reach 2 million.

11. Deer Population The graph shows the deer population in a Pennsylvania county between 2010 and 2014. Assume that the population grows exponentially.

(a) What was the deer population in 2010?

(b) Find a function that models the deer population t years after 2010.

(c) What is the projected deer population in 2018?

(d) Estimate how long it takes the population to reach 100,000.

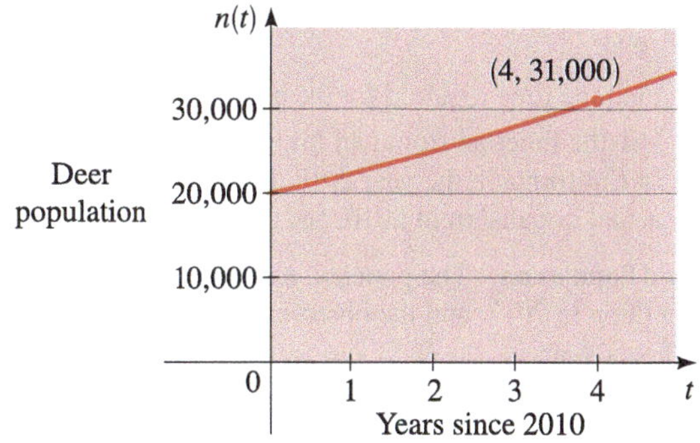

12. Frog Population Some bullfrogs were introduced into a small pond. The graph shows the bullfrog population for the next few years. Assume that the population grows exponentially.

(a) What was the initial bullfrog population?

(b) Find a function that models the bullfrog population t years since the bullfrogs were put into the pond.

(c) What is the projected bullfrog population after 15 years?

(d) Estimate how long it takes the population to reach 75,000.

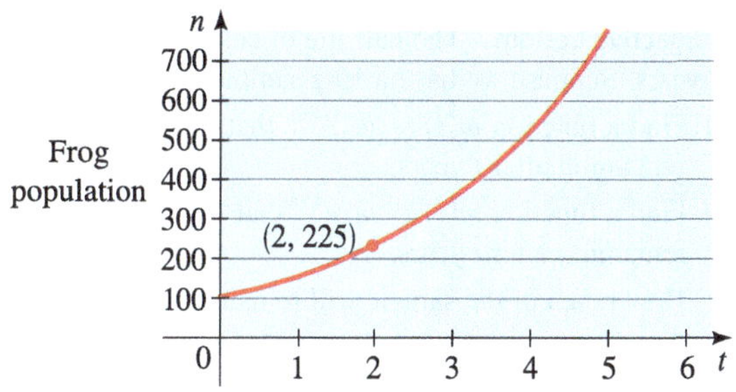

13. Bacteria Culture A culture starts with 8600 bacteria. After 1 hour the count is 10,000.

(a) Find a function that models the number of bacteria $n(t)$ after t hours.

(b) Find the number of bacteria after 2 hours.

(c) After how many hours will the number of bacteria double?

14. Bacteria Culture The count in a culture of bacteria was 400 after 2 hours and 25,600 after 6 hours.

(a) What is the relative rate of growth of the bacteria population? Express your answer as a percentage.

(b) What was the initial size of the culture?

(c) Find a function that models the number of bacteria $n(t)$ after t hours.

(d) Find the number of bacteria after 4.5 hours.

(e) After how many hours will the number of bacteria reach 50,000?

15. Population of California The population of California was 29.76 million in 1990 and 33.87 million in 2000. Assume that the population grows exponentially.

(a) Find a function that models the population t years after 1990.

(b) Find the time required for the population to double.

(c) Use the function from part (a) to predict the population of California in the year 2010. Look up California's actual population in 2010, and compare.

16. World Population The population of the world was 7.1 billion in 2013, and the observed relative growth rate was 1.1% per year.

(a) Estimate how long it takes the population to double.

(b) Estimate how long it takes the population to triple.

17–24 ■ Radioactive Decay These exercises use the radioactive decay model.

17. Radioactive Radium The half-life of radium-226 is 1600 years. Suppose we have a 22-mg sample.

(a) Find a function $m(t) = m_0 2^{-t/h}$ that models the mass remaining after t years.

(b) Find a function $m(t) = m_0 e^{-rt}$ that models the mass remaining after t years.

(c) How much of the sample will remain after 4000 years?

(d) After how many years will only 18 mg of the sample remain?

18. Radioactive Cesium The half-life of cesium-137 is 30 years. Suppose we have a 10-g sample.

(a) Find a function $m(t) = m_0 2^{-t/h}$ that models the mass remaining after t years.

(b) Find a function $m(t) = m_0 e^{-rt}$ that models the mass remaining after t years.

(c) How much of the sample will remain after 80 years?

(d) After how many years will only 2 g of the sample remain?

19. Radioactive Strontium The half-life of strontium-90 is 28 years. How long will it take a 50-mg sample to decay to a mass of 32 mg?

20. Radioactive Radium Radium-221 has a half-life of 30 s. How long will it take for 95% of a sample to decay?

21. Finding Half-Life If 250 mg of a radioactive element decays to 200 mg in 48 hours, find the half-life of the element.

22. Radioactive Radon After 3 days a sample of radon-222 has decayed to 58% of its original amount.

(a) What is the half-life of radon-222?

(b) How long will it take the sample to decay to 20% of its original amount?

23. Carbon-14 Dating A wooden artifact from an ancient tomb contains 65% of the carbon-14 that is present in living trees. How long ago was the artifact made? (The half-life of carbon-14 is 5730 years.)

24. Carbon-14 Dating The burial cloth of an Egyptian mummy is estimated to contain 59% of the carbon-14 it contained originally. How long ago was the mummy buried? (The half-life of carbon-14 is 5730 years.)

25–28 ■ Law of Cooling These exercises use Newton's Law of Cooling.

25. Cooling Soup A hot bowl of soup is served at a dinner party. It starts to cool according to Newton's Law of Cooling, so its temperature at time t is given by

$$T(t) = 65 + 145e^{-0.05t}$$

where t is measured in minutes and T is measured in °F.

(a) What is the initial temperature of the soup?

(b) What is the temperature after 10 min?

(c) After how long will the temperature be 100°F?

26. Time of Death Newton's Law of Cooling is used in homicide investigations to determine the time of death. The normal body temperature is 98.6°F. Immediately following death, the body begins to cool. It has been determined experimentally that the constant in Newton's Law of Cooling is approximately $k = 0.1947$, assuming that time is measured in hours. Suppose that the temperature of the surroundings is 60°F.

(a) Find a function $T(t)$ that models the temperature t hours after death.

(b) If the temperature of the body is now 72°F, how long ago was the time of death?

27. Cooling Turkey A roasted turkey is taken from an oven when its temperature has reached 185°F and is placed on a table in a room where the temperature is 75°F.

(a) If the temperature of the turkey is 150°F after half an hour, what is its temperature after 45 min?

(b) After how many hours will the turkey cool to 100°F?

28. Boiling Water A kettle full of water is brought to a boil in a room with temperature 20°C. After 15 min the temperature of the water has decreased from 100°C to 75°C. Find the temperature after another 10 min. Illustrate by graphing the temperature function.

4.7 LOGARITHMIC SCALES

■ The pH Scale ■ The Richter Scale ■ The Decibel Scale

Animal	W (kg)	$\log W$
Ant	0.000003	−5.5
Elephant	4000	3.6
Whale	170,000	5.2

When a physical quantity varies over a very large range, it is often convenient to take its logarithm in order to work with more manageable numbers. On a **logarithmic scale**, numbers are represented by their logarithms. For example, the table in the margin gives the weights W of some animals (in kilograms) and their logarithms ($\log W$).

The weights (W) vary enormously, but on a logarithmic scale, the weights are represented by more manageable numbers ($\log W$). Figure 1 shows that it is difficult to compare the weights W graphically but easy to compare them on a logarithmic scale.

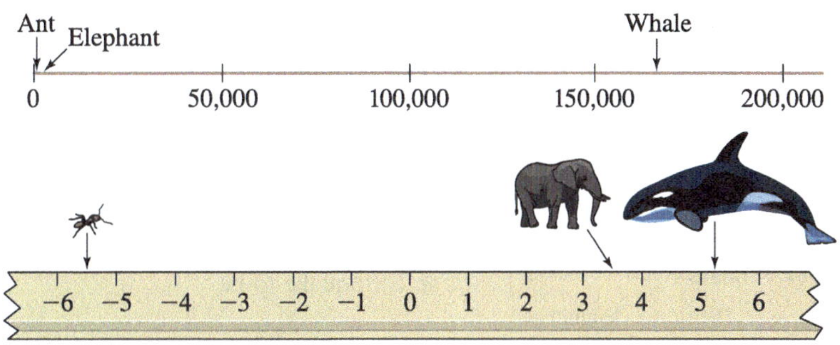

FIGURE 1 Weight graphed on the real line (top) and on a logarithmic scale (bottom)

We discuss three commonly used logarithmic scales: the pH scale, which measures acidity; the Richter scale, which measures the intensity of earthquakes; and the decibel scale, which measures the loudness of sounds. Other quantities that are measured on logarithmic scales are light intensity, information capacity, and radiation.

■ The pH Scale

Chemists measured the acidity of a solution by giving its hydrogen ion concentration until Søren Peter Lauritz Sørensen, in 1909, proposed a more convenient measure. He defined

$$\text{pH} = -\log[\text{H}^+]$$

where $[\text{H}^+]$ is the concentration of hydrogen ions measured in moles per liter (M). He did this to avoid very small numbers and negative exponents. For instance,

$$\text{if} \quad [\text{H}^+] = 10^{-4}\text{ M}, \quad \text{then} \quad \text{pH} = -\log_{10}(10^{-4}) = -(-4) = 4$$

Solutions with a pH of 7 are defined as *neutral*, those with pH < 7 are *acidic*, and those with pH > 7 are *basic*. Notice that when the pH increases by one unit, $[\text{H}^+]$ decreases by a factor of 10.

pH for Some Common Substances

Substance	pH
Milk of magnesia	10.5
Seawater	8.0–8.4
Human blood	7.3–7.5
Crackers	7.0–8.5
Hominy	6.9–7.9
Cow's milk	6.4–6.8
Spinach	5.1–5.7
Tomatoes	4.1–4.4
Oranges	3.0–4.0
Apples	2.9–3.3
Limes	1.3–2.0
Battery acid	1.0

EXAMPLE 1 ■ pH Scale and Hydrogen Ion Concentration

(a) The hydrogen ion concentration of a sample of human blood was measured to be $[\text{H}^+] = 3.16 \times 10^{-8}$ M. Find the pH, and classify the blood as acidic or basic.

(b) The most acidic rainfall ever measured occurred in Scotland in 1974; its pH was 2.4. Find the hydrogen ion concentration.

SOLUTION

(a) A calculator gives

$$\text{pH} = -\log[\text{H}^+] = -\log(3.16 \times 10^{-8}) \approx 7.5$$

Since this is greater than 7, the blood is basic.

(b) To find the hydrogen ion concentration, we need to solve for $[\text{H}^+]$ in the logarithmic equation

$$\log[\text{H}^+] = -\text{pH}$$

So we write it in exponential form:

$$[\text{H}^+] = 10^{-\text{pH}}$$

In this case pH = 2.4, so

$$[\text{H}^+] = 10^{-2.4} \approx 4.0 \times 10^{-3} \text{ M}$$

Now Try Exercises 1 and 3

The Richter Scale

In 1935 the American geologist Charles Richter (1900–1984) defined the magnitude M of an earthquake to be

$$M = \log \frac{I}{S}$$

where I is the intensity of the earthquake (measured by the amplitude of a seismograph reading taken 100 km from the epicenter of the earthquake) and S is the intensity of a "standard" earthquake (whose amplitude is 1 micron $= 10^{-4}$ cm). (In practice, seismograph stations may not be exactly 100 km from the epicenter, so appropriate adjustments are made in calculating the magnitude of an earthquake.) The magnitude of a standard earthquake is

$$M = \log \frac{S}{S} = \log 1 = 0$$

Richter studied many earthquakes that occurred between 1900 and 1950. The largest had magnitude 8.9 on the Richter scale, and the smallest had magnitude 0. This corresponds to a ratio of intensities of 800,000,000, so the Richter scale provides more

Largest Earthquakes

Location	Date	Magnitude
Chile	1960	9.5
Alaska	1964	9.2
Japan	2011	9.1
Sumatra	2004	9.1
Kamchatka	1952	9.0
Chile	2010	8.8
Ecuador	1906	8.8
Alaska	1965	8.7
Alaska	1957	8.6
Sumatra	2005	8.6
Sumatra	2012	8.6
Tibet	1950	8.6
Indonesia	1938	8.5
Kamchatka	1923	8.5

Source: U.S. Geological Society

Robert Vos/AFP/Getty Images

DISCOVERY PROJECT

The Even-Tempered Clavier

Poets, writers, philosophers, and even politicians have extolled the virtues of music—its beauty and its power to communicate emotion. But at the heart of music is a logarithmic scale. The tones that we are familiar with from our everyday listening can all be reproduced by the keys of a piano. The keys of a piano, in turn, are "evenly tempered" using a logarithmic scale. In this project we explore how exponential and logarithmic functions are used in properly tuning a piano. You can find the project at **www.stewartmath.com**.

manageable numbers to work with. For instance, an earthquake of magnitude 6 is ten times stronger than an earthquake of magnitude 5.

EXAMPLE 2 ■ Magnitude and Intensity

(a) Find the magnitude of an earthquake that has an intensity of 3.75 (that is, the amplitude of the seismograph reading is 3.75 cm).

(b) An earthquake was measured to have a magnitude of 5.1 on the Richter scale. Find the intensity of the earthquake.

SOLUTION

(a) From the definition of magnitude we see that

$$M = \log \frac{I}{S} = \log \frac{3.75}{10^{-4}} = \log 37500 \approx 4.6$$

Thus the magnitude is 4.6 on the Richter scale.

(b) To find the intensity, we need to solve for I in the logarithmic equation

$$M = \log \frac{I}{S}$$

So we write it in exponential form:

$$10^M = \frac{I}{S}$$

In this case $S = 10^{-4}$ and $M = 5.1$, so

$$10^{5.1} = \frac{I}{10^{-4}} \qquad M = 5.1, S = 10^{-4}$$

$$(10^{-4})(10^{5.1}) = I \qquad \text{Multiply by } 10^{-4}$$

$$I = 10^{1.1} \approx 12.6 \qquad \text{Add exponents}$$

Thus the intensity of the earthquake is about 12.6, which means that the amplitude of the seismograph reading is about 12.6 cm.

Now Try Exercise 9 ■

EXAMPLE 3 ■ Magnitude of Earthquakes

There are several other logarithmic scales used to calculate the magnitude of earthquakes. For instance, the U.S. Geological Survey uses the *moment magnitude scale*.

The 1906 earthquake in San Francisco had an estimated magnitude of 8.3 on the Richter scale. In the same year a powerful earthquake occurred on the Colombia-Ecuador border that was four times as intense. What was the magnitude of the Colombia-Ecuador earthquake on the Richter scale?

SOLUTION If I is the intensity of the San Francisco earthquake, then from the definition of magnitude we have

$$M = \log \frac{I}{S} = 8.3$$

The intensity of the Colombia-Ecuador earthquake was $4I$, so its magnitude was

$$M = \log \frac{4I}{S} = \log 4 + \log \frac{I}{S} = \log 4 + 8.3 \approx 8.9$$

Now Try Exercise 11 ■

EXAMPLE 4 ■ Intensity of Earthquakes

Grant Smith/Corbis

The 1989 Loma Prieta earthquake that shook San Francisco had a magnitude of 7.1 on the Richter scale. How many times more intense was the 1906 earthquake (see Example 3) than the 1989 event?

SOLUTION If I_1 and I_2 are the intensities of the 1906 and 1989 earthquakes, then we are required to find I_1/I_2. To relate this to the definition of magnitude, we divide the numerator and denominator by S.

$$\log \frac{I_1}{I_2} = \log \frac{I_1/S}{I_2/S} \qquad \text{Divide numerator and denominator by } S$$

$$= \log \frac{I_1}{S} - \log \frac{I_2}{S} \qquad \text{Law 2 of logarithms}$$

$$= 8.3 - 7.1 = 1.2 \qquad \text{Definition of earthquake magnitude}$$

Therefore

$$\frac{I_1}{I_2} = 10^{\log(I_1/I_2)} = 10^{1.2} \approx 16$$

The 1906 earthquake was about 16 times as intense as the 1989 earthquake.

Now Try Exercise 13 ■

■ The Decibel Scale

The ear is sensitive to an extremely wide range of sound intensities. We take as a reference intensity $I_0 = 10^{-12}$ W/m^2 (watts per square meter) at a frequency of 1000 hertz, which measures a sound that is just barely audible (the threshold of hearing). The psychological sensation of loudness varies with the logarithm of the intensity (the Weber-Fechner Law), so the **decibel level** B, measured in decibels (dB), is defined as

$$B = 10 \log \frac{I}{I_0}$$

The decibel level of the barely audible reference sound is

$$B = 10 \log \frac{I_0}{I_0} = 10 \log 1 = 0 \text{ dB}$$

EXAMPLE 5 ■ Decibel Level and Intensity

(a) Find the decibel level of a jet engine at takeoff if the intensity was measured at 100 W/m^2.

(b) Find the intensity level of a motorcycle engine at full throttle if the decibel level was measured at 90 dB.

SOLUTION

(a) From the definition of decibel level we see that

$$B = 10 \log \frac{I}{I_0} = 10 \log \frac{10^2}{10^{-12}} = 10 \log 10^{14} = 140 \text{ dB}$$

Thus the decibel level is 140 dB.

(b) To find the intensity, we need to solve for I in the logarithmic equation

$$B = 10 \log \frac{I}{I_0} \qquad \text{Definition of decibel level}$$

$$\frac{B}{10} = \log I - \log 10^{-12} \qquad \text{Divide by 10, } I_0 = 10^{-12}$$

$$\frac{B}{10} = \log I + 12 \qquad \text{Definition of logarithm}$$

$$\frac{B}{10} - 12 = \log I \qquad \text{Subtract 12}$$

$$\log I = \frac{90}{10} - 12 = -3 \qquad B = 90$$

$$I = 10^{-3} \qquad \text{Exponential form}$$

Thus the intensity is 10^{-3} W/m^2.

Now Try Exercises 15 and 17 ■

The **decibel levels of sounds** that we can hear vary from very loud to very soft. Here are some examples of the decibel levels of commonly heard sounds.

Source of sound	B (dB)
Jet takeoff	140
Jackhammer	130
Rock concert	120
Subway	100
Heavy traffic	80
Ordinary traffic	70
Normal conversation	50
Whisper	30
Rustling leaves	10–20
Threshold of hearing	0

The table in the margin lists decibel levels for some common sounds ranging from the threshold of human hearing to the jet takeoff of Example 5. The threshold of pain is about 120 dB.

4.7 EXERCISES

APPLICATIONS

1. **Finding pH** The hydrogen ion concentration of a sample of each substance is given. Calculate the pH of the substance.
   (a) Lemon juice: $[H^+] = 5.0 \times 10^{-3}$ M
   (b) Tomato juice: $[H^+] = 3.2 \times 10^{-4}$ M
   (c) Seawater: $[H^+] = 5.0 \times 10^{-9}$ M

2. **Finding pH** An unknown substance has a hydrogen ion concentration of $[H^+] = 3.1 \times 10^{-8}$ M. Find the pH and classify the substance as acidic or basic.

3. **Ion Concentration** The pH reading of a sample of each substance is given. Calculate the hydrogen ion concentration of the substance.
   (a) Vinegar: pH = 3.0 **(b)** Milk: pH = 6.5

4. **Ion Concentration** The pH reading of a glass of liquid is given. Find the hydrogen ion concentration of the liquid.
   (a) Beer: pH = 4.6 **(b)** Water: pH = 7.3

5. **Finding pH** The hydrogen ion concentrations in cheeses range from 4.0×10^{-7} M to 1.6×10^{-5} M. Find the corresponding range of pH readings.

6. **Ion Concentration in Wine** The pH readings for wines vary from 2.8 to 3.8. Find the corresponding range of hydrogen ion concentrations.

7. **pH of Wine** If the pH of a wine is too high, say, 4.0 or above, the wine becomes unstable and has a flat taste.
   (a) A certain California red wine has a pH of 3.2, and a certain Italian white wine has a pH of 2.9. Find the corresponding hydrogen ion concentrations of the two wines.
   (b) Which wine has the lower hydrogen ion concentration?

8. **pH of Saliva** The pH of saliva is normally in the range of 6.4 to 7.0. However, when a person is ill, the person's saliva becomes more acidic.
   (a) When Marco is sick, he tests the pH of his saliva and finds that it is 5.5. What is the hydrogen ion concentration of his saliva?
   (b) Will the hydrogen ion concentration in Marco's saliva increase or decrease as he gets better?
   (c) After Marco recovers, he tests the pH of his saliva, and it is 6.5. Was the saliva more acidic or less acidic when he was sick?

9. **Earthquake Magnitude and Intensity**
   (a) Find the magnitude of an earthquake that has an intensity that is 31.25 (that is, the amplitude of the seismograph reading is 31.25 cm).
   (b) An earthquake was measured to have a magnitude of 4.8 on the Richter scale. Find the intensity of the earthquake.

10. Earthquake Magnitude and Intensity

(a) Find the magnitude of an earthquake that has an intensity that is 72.1 (that is, the amplitude of the seismograph reading is 72.1 cm).

(b) An earthquake was measured to have a magnitude of 5.8 on the Richter scale. Find the intensity of the earthquake.

11. Earthquake Magnitudes If one earthquake is 20 times as intense as another, how much larger is its magnitude on the Richter scale?

12. Earthquake Magnitudes The 1906 earthquake in San Francisco had a magnitude of 8.3 on the Richter scale. At the same time in Japan an earthquake with magnitude 4.9 caused only minor damage. How many times more intense was the San Francisco earthquake than the Japan earthquake?

13. Earthquake Magnitudes The Japan earthquake of 2011 had a magnitude of 9.1 on the Richter scale. How many times more intense was this than the 1906 San Francisco earthquake? (See Exercise 12.)

14. Earthquake Magnitudes The Northridge, California, earthquake of 1994 had a magnitude of 6.8 on the Richter scale. A year later, a 7.2-magnitude earthquake struck Kobe, Japan. How many times more intense was the Kobe earthquake than the Northridge earthquake?

15. Traffic Noise The intensity of the sound of traffic at a busy intersection was measured at 2.0×10^{-5} W/m^2. Find the decibel level.

16. Leaf Blower The intensity of the sound from a certain leaf blower is measured at 3.2×10^{-2} W/m^2. Find the decibel level.

17. Hair Dryer The decibel level of the sound from a certain hair dryer is measured at 70 dB. Find the intensity of the sound.

18. Subway Noise The decibel level of the sound of a subway train was measured at 98 dB. Find the intensity in watts per square meter (W/m^2).

19. Hearing Loss from MP3 Players Recent research has shown that the use of earbud-style headphones packaged with MP3 players can cause permanent hearing loss.

(a) The intensity of the sound from the speakers of a certain MP3 player (without earbuds) is measured at 3.1×10^{-5} W/m^2. Find the decibel level.

(b) If earbuds are used with the MP3 player in part (a), the decibel level is 95 dB. Find the intensity.

(c) Find the ratio of the intensity of the sound from the MP3 player with earbuds to that of the sound without earbuds.

20. Comparing Decibel Levels The noise from a power mower was measured at 106 dB. The noise level at a rock concert was measured at 120 dB. Find the ratio of the intensity of the rock music to that of the power mower.

DISCUSS ■ DISCOVER ■ PROVE ■ WRITE

21. PROVE: Inverse Square Law for Sound A law of physics states that the intensity of sound is inversely proportional to the square of the distance d from the source: $I = k/d^2$.

(a) Use this model and the equation

$$B = 10 \log \frac{I}{I_0}$$

(described in this section) to show that the decibel levels B_1 and B_2 at distances d_1 and d_2 from a sound source are related by the equation

$$B_2 = B_1 + 20 \log \frac{d_1}{d_2}$$

(b) The intensity level at a rock concert is 120 dB at a distance 2 m from the speakers. Find the intensity level at a distance of 10 m.

CHAPTER 4 ■ REVIEW

■ PROPERTIES AND FORMULAS

Exponential Functions (pp. 330–332)

The **exponential function** f with base a (where $a > 0$, $a \neq 1$) is defined for all real numbers x by

$$f(x) = a^x$$

The domain of f is $\mathbb{R}$, and the range of f is $(0, \infty)$ The graph of f has one of the following shapes, depending on the value of a:

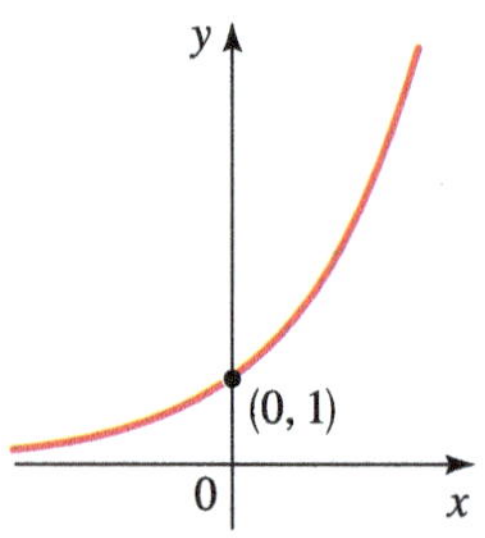

$f(x) = a^x$ for $a > 1$

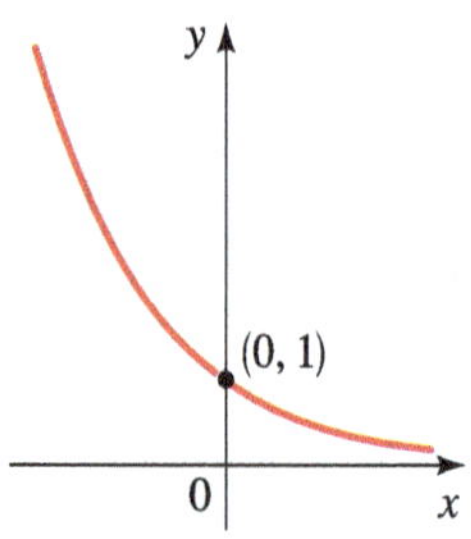

$f(x) = a^x$ for $0 < a < 1$

The Natural Exponential Function (p. 339)

The **natural exponential function** is the exponential function with base e:

$$f(x) = e^x$$

The number e is defined to be the number that the expression $(1 + 1/n)^n$ approaches as $n \to \infty$. An approximate value for the irrational number e is

$$e \approx 2.7182818284590\ldots$$

Compound Interest (pp. 334, 340)

If a principal P is invested in an account paying an annual interest rate r, compounded n times a year, then after t years the **amount** $A(t)$ in the account is

$$A(t) = P\left(1 + \frac{r}{n}\right)^{nt}$$

If the interest is compounded **continuously**, then the amount is

$$A(t) = Pe^{rt}$$

Logarithmic Functions (pp. 344–345)

The **logarithmic function** $\log_a$ with base a (where $a > 0, a \neq 1$) is defined for $x > 0$ by

$$\log_a x = y \quad \Leftrightarrow \quad a^y = x$$

So $\log_a x$ is the exponent to which the base a must be raised to give y.

The domain of $\log_a$ is $(0, \infty)$, and the range is $\mathbb{R}$. For $a > 1$, the graph of the function $\log_a$ has the following shape:

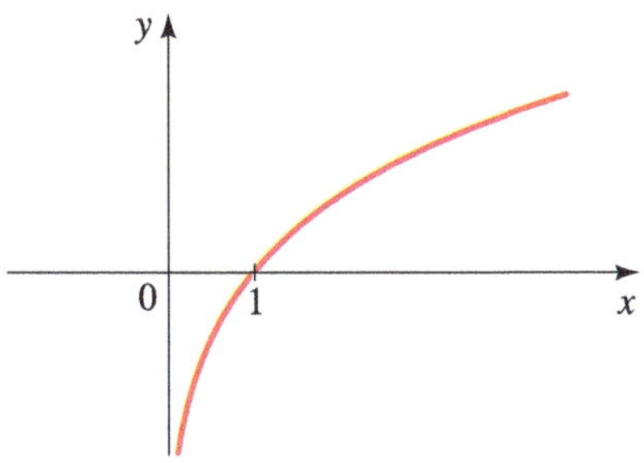

$y = \log_a x, a > 1$

Common and Natural Logarithms (pp. 348–349)

The logarithm function with base 10 is called the **common logarithm** and is denoted **log**. So

$$\log x = \log_{10} x$$

The logarithm function with base e is called the **natural logarithm** and is denoted **ln**. So

$$\ln x = \log_e x$$

Properties of Logarithms (pp. 345, 349)

1. $\log_a 1 = 0$
2. $\log_a a = 1$
3. $\log_a a^x = x$
4. $a^{\log_a x} = x$

Laws of Logarithms (p. 354)

Let a be a logarithm base $(a > 0, a \neq 1)$, and let A, B, and C be any real numbers or algebraic expressions that represent real numbers, with $A > 0$ and $B > 0$. Then:

1. $\log_a(AB) = \log_a A + \log_a B$
2. $\log_a(A/B) = \log_a A - \log_a B$
3. $\log_a(A^C) = C \log_a A$

Change of Base Formula (p. 357)

$$\log_b x = \frac{\log_a x}{\log_a b}$$

Guidelines for Solving Exponential Equations (p. 361)

1. Isolate the exponential term on one side of the equation.
2. Take the logarithm of each side, and use the Laws of Logarithms to "bring down the exponent."
3. Solve for the variable.

Guidelines for Solving Logarithmic Equations (p. 364)

1. Isolate the logarithmic term(s) on one side of the equation, and use the Laws of Logarithms to combine logarithmic terms if necessary.
2. Rewrite the equation in exponential form.
3. Solve for the variable.

Exponential Growth Model (p. 373)

A population experiences **exponential growth** if it can be modeled by the exponential function

$$n(t) = n_0 e^{rt}$$

where $n(t)$ is the population at time t, n_0 is the initial population (at time $t = 0$), and r is the relative growth rate (expressed as a proportion of the population).

Radioactive Decay Model (pp. 375–376)

If a **radioactive substance** with half-life h has initial mass m_0, then at time t the mass $m(t)$ of the substance that remains is modeled by the exponential function

$$m(t) = m_0 e^{-rt}$$

where $r = \dfrac{\ln 2}{h}$.

Newton's Law of Cooling (p. 377)

If an object has an initial temperature that is D_0 degrees warmer than the surrounding temperature T_s, then at time t the temperature $T(t)$ of the object is modeled by the function

$$T(t) = T_s + D_0 e^{-kt}$$

where the constant $k > 0$ depends on the size and type of the object.

Logarithmic Scales (pp. 381–385)

The **pH scale** measures the acidity of a solution:

$$\text{pH} = -\log[\text{H}^+]$$

The **Richter scale** measures the intensity of earthquakes:

$$M = \log \frac{I}{S}$$

The **decibel scale** measures the intensity of sound:

$$B = 10 \log \frac{I}{I_0}$$

CONCEPT CHECK

1. Let f be the exponential function with base a.
   (a) Write an equation that defines f.
   (b) Write an equation for the exponential function f with base 3.

2. Let f be the exponential function $f(x) = a^x$, where $a > 0$.
   (a) What is the domain of f?
   (b) What is the range of f?
   (c) Sketch graphs of f for the following cases.
   (i) $a > 1$ (ii) $0 < a < 1$

3. If x is large, which function grows faster, $f(x) = 2^x$ or $g(x) = x^2$?

4. (a) How is the number e defined?
   (b) Give an approximate value of e, correct to four decimal places.
   (c) What is the natural exponential function?

5. (a) How is $\log_a x$ defined?
   (b) Find $\log_3 9$.
   (c) What is the natural logarithm?
   (d) What is the common logarithm?
   (e) Write the exponential form of the equation $\log_7 49 = 2$.

6. Let f be the logarithmic function $f(x) = \log_a x$.
   (a) What is the domain of f?
   (b) What is the range of f?
   (c) Sketch a graph of the logarithmic function for the case that $a > 1$.

7. State the three Laws of Logarithms.

8. (a) State the Change of Base Formula.
   (b) Find $\log_7 30$.

9. (a) What is an exponential equation?
   (b) How do you solve an exponential equation?
   (c) Solve for x: $2^x = 19$

10. (a) What is a logarithmic equation?
    (b) How do you solve a logarithmic equation?
    (c) Solve for x: $4 \log_3 x = 7$

11. Suppose that an amount P is invested at an interest rate r and $A(t)$ is the amount of the investment after t years. Write a formula for $A(t)$ in the following cases.
    (a) Interest is compounded n times per year.
    (b) Interest is compounded continuously.

12. Suppose that the initial size of a population is n_0 and the population grows exponentially. Let $n(t)$ be the size of the population at time t.
    (a) Write a formula for $n(t)$ in terms of the doubling time a.
    (b) Write a formula for $n(t)$ in terms of the relative growth rate r.

13. Suppose that the initial mass of a radioactive substance is m_0 and the half-life of the substance is h. Let $m(t)$ be the mass remaining at time t.
    (a) What is meant by the half-life h?
    (b) Write a formula for $m(t)$ in terms of the half-life h.
    (c) Write a formula for the relative decay rate r in terms of the half-life h.
    (d) Write a formula for $m(t)$ in terms of the relative decay rate r.

14. Suppose that the initial temperature difference between an object and its surroundings is D_0 and the surroundings have temperature T_s. Let $T(t)$ be the temperature at time t. State Newton's Law of Cooling for $T(t)$.

15. What is a logarithmic scale? If we use a logarithmic scale with base 10, what do the following numbers correspond to on the logarithmic scale?
    (i) 100 (ii) 100,000 (iii) 0.0001

16. (a) What does the pH scale measure?
    (b) Define the pH of a substance with hydrogen ion concentration of $[H^+]$.

17. (a) What does the Richter scale measure?
    (b) Define the magnitude M of an earthquake in terms of the intensity I of the earthquake and the intensity S of a standard earthquake.

18. (a) What does the decibel scale measure?
    (b) Define the decibel level B of a sound in terms of the intensity I of the sound and the intensity I_0 of a barely audible sound.

ANSWERS TO THE CONCEPT CHECK CAN BE FOUND AT THE BACK OF THE BOOK.

EXERCISES

1–4 ■ Evaluating Exponential Functions Use a calculator to find the indicated values of the exponential function, rounded to three decimal places.

1. $f(x) = 5^x$; $f(-1.5), f(\sqrt{2}), f(2.5)$
2. $f(x) = 3 \cdot 2^x$; $f(-2.2), f(\sqrt{7}), f(5.5)$
3. $g(x) = 4e^{x-2}$; $g(-0.7), g(1), g(\pi)$
4. $g(x) = \frac{7}{4}e^{x+1}$; $g(-2), g(\sqrt{3}), g(3.6)$

5–16 ■ Graphing Exponential and Logarithmic Functions Sketch the graph of the function. State the domain, range, and asymptote.

5. $f(x) = 3^{x-2}$
6. $f(x) = 2^{-x+1}$
7. $g(x) = 3 + 2^x$
8. $g(x) = 5^{-x} - 5$
9. $F(x) = e^{x-1} + 1$
10. $G(x) = -e^{x+1} - 2$

11. $f(x) = \log_3(x - 1)$

12. $g(x) = \log(-x)$

13. $f(x) = 2 - \log_2 x$

14. $f(x) = 3 + \log_5(x + 4)$

15. $g(x) = 2 \ln x$

16. $g(x) = \ln(x^2)$

17–20 ■ Domain Find the domain of the function.

17. $f(x) = 10^{x^2} + \log(1 - 2x)$

18. $g(x) = \log(2 + x - x^2)$

19. $h(x) = \ln(x^2 - 4)$

20. $k(x) = \ln |x|$

21–24 ■ Exponential Form Write the equation in exponential form.

21. $\log_2 1024 = 10$

22. $\log_6 37 = x$

23. $\log x = y$

24. $\ln c = 17$

25–28 ■ Logarithmic Form Write the equation in logarithmic form.

25. $2^6 = 64$

26. $49^{-1/2} = \frac{1}{7}$

27. $10^x = 74$

28. $e^k = m$

29–44 ■ Evaluating Logarithmic Expressions Evaluate the expression without using a calculator.

29. $\log_2 128$

30. $\log_8 1$

31. $10^{\log 45}$

32. $\log 0.000001$

33. $\ln(e^6)$

34. $\log_4 8$

35. $\log_3\left(\frac{1}{27}\right)$

36. $2^{\log_2 13}$

37. $\log_5 \sqrt{5}$

38. $e^{2 \ln 7}$

39. $\log 25 + \log 4$

40. $\log_3 \sqrt{243}$

41. $\log_2 16^{23}$

42. $\log_5 250 - \log_5 2$

43. $\log_8 6 - \log_8 3 + \log_8 2$

44. $\log \log 10^{100}$

45–50 ■ Expanding Logarithmic Expressions Expand the logarithmic expression.

45. $\log(AB^2C^3)$

46. $\log_2(x\sqrt{x^2 + 1})$

47. $\ln\sqrt{\dfrac{x^2 - 1}{x^2 + 1}}$

48. $\log\left(\dfrac{4x^3}{y^2(x - 1)^5}\right)$

49. $\log_5\left(\dfrac{x^2(1 - 5x)^{3/2}}{\sqrt{x^3 - x}}\right)$

50. $\ln\left(\dfrac{\sqrt[3]{x^4 + 12}}{(x + 16)\sqrt{x - 3}}\right)$

51–56 ■ Combining Logarithmic Expressions Combine into a single logarithm.

51. $\log 6 + 4 \log 2$

52. $\log x + \log(x^2y) + 3 \log y$

53. $\frac{3}{2}\log_2(x - y) - 2\log_2(x^2 + y^2)$

54. $\log_5 2 + \log_5(x + 1) - \frac{1}{3}\log_5(3x + 7)$

55. $\log(x - 2) + \log(x + 2) - \frac{1}{2}\log(x^2 + 4)$

56. $\frac{1}{2}[\ln(x - 4) + 5\ln(x^2 + 4x)]$

57–70 ■ Exponential and Logarithmic Equations Solve the equation. Find the exact solution if possible; otherwise, use a calculator to approximate to two decimals.

57. $3^{2x-7} = 27$

58. $5^{4-x} = \frac{1}{125}$

59. $2^{3x-5} = 7$

60. $10^{6-3x} = 18$

61. $4^{1-x} = 3^{2x+5}$

62. $e^{3x/4} = 10$

63. $x^2e^{2x} + 2xe^{2x} = 8e^{2x}$

64. $3^{2x} - 3^x - 6 = 0$

65. $\log x + \log(x + 1) = \log 12$

66. $\ln(x - 2) + \ln 3 = \ln(5x - 7)$

67. $\log_2(1 - x) = 4$

68. $\ln(2x - 3) + 1 = 0$

69. $\log_3(x - 8) + \log_3 x = 2$

70. $\log_8(x + 5) - \log_8(x - 2) = 1$

71–74 ■ Exponential Equations Use a calculator to find the solution of the equation, rounded to six decimal places.

71. $5^{-2x/3} = 0.63$

72. $2^{3x-5} = 7$

73. $5^{2x+1} = 3^{4x-1}$

74. $e^{-15k} = 10{,}000$

75–78 ■ Local Extrema and Asymptotes Draw a graph of the function and use it to determine the asymptotes and the local maximum and minimum values.

75. $y = e^{x/(x+2)}$

76. $y = 10^x - 5^x$

77. $y = \log(x^3 - x)$

78. $y = 2x^2 - \ln x$

79–80 ■ Solving Equations Find the solutions of the equation, rounded to two decimal places.

79. $3 \log x = 6 - 2x$

80. $4 - x^2 = e^{-2x}$

81–82 ■ Solving Inequalities Solve the inequality graphically.

81. $\ln x > x - 2$

82. $e^x < 4x^2$

83. Increasing and Decreasing Use a graph of $f(x) = e^x - 3e^{-x} - 4x$ to find, approximately, the intervals on which f is increasing and on which f is decreasing.

84. Equation of a Line Find an equation of the line shown in the figure.

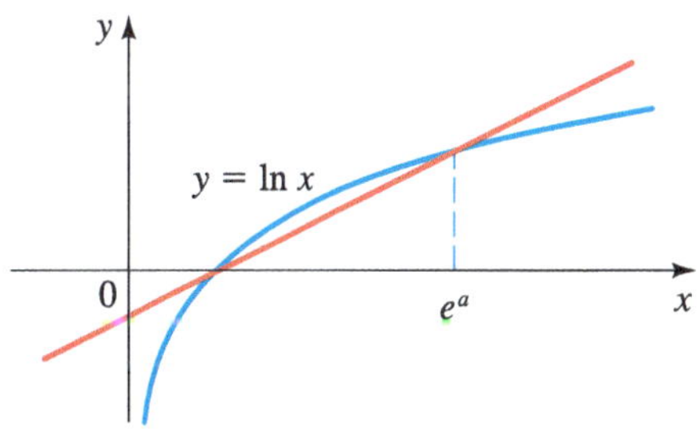

85–88 ■ Change of Base Use the Change of Base Formula to evaluate the logarithm, rounded to six decimal places.

85. $\log_4 15$

86. $\log_7\left(\frac{3}{4}\right)$

87. $\log_9 0.28$

88. $\log_{100} 250$

89. Comparing Logarithms Which is larger, $\log_4 258$ or $\log_5 620$?

90. Inverse Function Find the inverse of the function $f(x) = 2^{3^x}$, and state its domain and range.

91. Compound Interest If $12,000 is invested at an interest rate of 10% per year, find the amount of the investment at the end of 3 years for each compounding method.

(a) Semiannually **(b)** Monthly

(c) Daily **(d)** Continuously

92. Compound Interest A sum of $5000 is invested at an interest rate of $8\frac{1}{2}\%$ per year, compounded semiannually.

(a) Find the amount of the investment after $1\frac{1}{2}$ years.

(b) After what period of time will the investment amount to $7000?

(c) If interest were compounded continously instead of semiannually, how long would it take for the amount to grow to $7000?

93. Compound Interest A money market account pays 5.2% annual interest, compounded daily. If $100,000 is invested in this account, how long will it take for the account to accumulate $10,000 in interest?

94. Compound Interest A retirement savings plan pays 4.5% interest, compounded continuously. How long will it take for an investment in this plan to double?

95–96 ■ APY Determine the annual percentage yield (APY) for the given nominal annual interest rate and compounding frequency.

95. 4.25%; daily

96. 3.2%; monthly

97. Cat Population The stray-cat population in a small town grows exponentially. In 1999 the town had 30 stray cats, and the relative growth rate was 15% per year.

(a) Find a function that models the stray-cat population $n(t)$ after t years.

(b) Find the projected population after 4 years.

(c) Find the number of years required for the stray-cat population to reach 500.

98. Bacterial Growth A culture contains 10,000 bacteria initially. After 1 hour the bacteria count is 25,000.

(a) Find the doubling period.

(b) Find the number of bacteria after 3 hours.

99. Radioactive Decay Uranium-234 has a half-life of 2.7×10^5 years.

(a) Find the amount remaining from a 10-mg sample after a thousand years.

(b) How long will it take this sample to decompose until its mass is 7 mg?

100. Radioactive Decay A sample of bismuth-210 decayed to 33% of its original mass after 8 days.

(a) Find the half-life of this element.

(b) Find the mass remaining after 12 days.

101. Radioactive Decay The half-life of radium-226 is 1590 years.

(a) If a sample has a mass of 150 mg, find a function that models the mass that remains after t years.

(b) Find the mass that will remain after 1000 years.

(c) After how many years will only 50 mg remain?

102. Radioactive Decay The half-life of palladium-100 is 4 days. After 20 days a sample has been reduced to a mass of 0.375 g.

(a) What was the initial mass of the sample?

(b) Find a function that models the mass remaining after t days.

(c) What is the mass after 3 days?

(d) After how many days will only 0.15 g remain?

103. Bird Population The graph shows the population of a rare species of bird, where t represents years since 2009 and $n(t)$ is measured in thousands.

(a) Find a function that models the bird population at time t in the form $n(t) = n_0 e^{rt}$.

(b) What is the bird population expected to be in the year 2020?

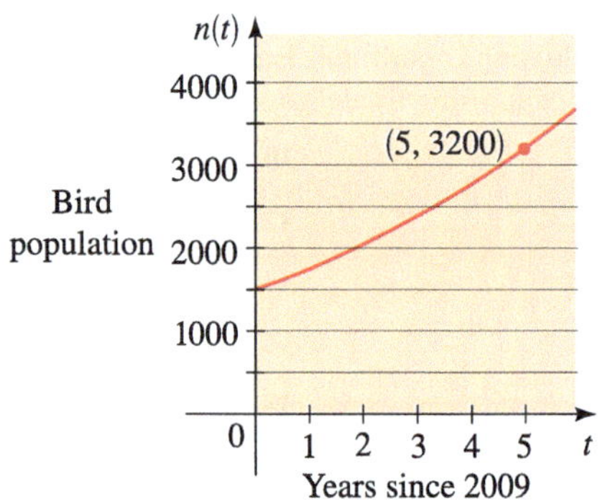

104. Law of Cooling A car engine runs at a temperature of 190°F. When the engine is turned off, it cools according to Newton's Law of Cooling with constant $k = 0.0341$, where the time is measured in minutes. Find the time needed for the engine to cool to 90°F if the surrounding temperature is 60°F.

105. pH The hydrogen ion concentration of fresh egg whites was measured as

$$[H^+] = 1.3 \times 10^{-8}\text{ M}$$

Find the pH, and classify the substance as acidic or basic.

106. pH The pH of lime juice is 1.9. Find the hydrogen ion concentration.

107. Richter Scale If one earthquake has magnitude 6.5 on the Richter scale, what is the magnitude of another quake that is 35 times as intense?

108. Decibel Scale The drilling of a jackhammer was measured at 132 dB. The sound of whispering was measured at 28 dB. Find the ratio of the intensity of the drilling to that of the whispering.

CHAPTER 4 TEST

1. Sketch the graph of each function, and state its domain, range, and asymptote. Show the x- and y-intercepts on the graph.
   (a) $f(x) = 2^{-x} + 4$ (b) $g(x) = \log_3(x + 3)$
2. Find the domain of the function.
   (a) $f(t) = \ln(2t - 3)$ (b) $g(x) = \log(x^2 - 1)$
3. (a) Write the equation $6^{2x} = 25$ in logarithmic form.
   (b) Write the equation $\ln A = 3$ in exponential form.
4. Find the exact value of the expression.
   (a) $10^{\log 36}$ (b) $\ln e^3$ (c) $\log_3 \sqrt{27}$
   (d) $\log_2 80 - \log_2 10$ (e) $\log_8 4$ (f) $\log_6 4 + \log_6 9$
5. Use the Laws of Logarithms to expand the expression.
   (a) $\log\left(\dfrac{xy^3}{z^2}\right)$ (b) $\ln\sqrt{\dfrac{x}{y}}$ (c) $\log\sqrt[3]{\dfrac{x+2}{x^4(x^2+4)}}$
6. Use the Laws of Logarithms to combine the expression into a single logarithm.
   (a) $\log a + 2\log b$ (b) $\ln(x^2 - 25) - \ln(x + 5)$ (c) $\log_2 3 - 3\log_2 x + \frac{1}{2}\log_2(x + 1)$
7. Find the solution of the exponential equation, rounded to two decimal places.
   (a) $3^{4x} = 3^{100}$ (b) $e^{3x-2} = e^{x^2}$ (c) $5^{x/10} + 1 = 7$ (d) $10^{x+3} = 6^{2x}$
8. Solve the logarithmic equation for x.
   (a) $\log(2x) = 3$ (b) $\log(x + 1) + \log 2 = \log(5x)$
   (c) $5\ln(3 - x) = 4$ (d) $\log_2(x + 2) + \log_2(x - 1) = 2$
9. Use the Change of Base Formula to evaluate $\log_{12} 27$.
10. The initial size of a culture of bacteria is 1000. After 1 hour the bacteria count is 8000.
    (a) Find a function $n(t) = n_0e^{rt}$ that models the population after t hours.
    (b) Find the population after 1.5 hours.
    (c) After how many hours will the number of bacteria reach 15,000?
    (d) Sketch the graph of the population function.
11. Suppose that \$12,000 is invested in a savings account paying 5.6% interest per year.
    (a) Write the formula for the amount in the account after t years if interest is compounded monthly.
    (b) Find the amount in the account after 3 years if interest is compounded daily.
    (c) How long will it take for the amount in the account to grow to \$20,000 if interest is compounded continuously?
12. The half-life of krypton-91 (^{91}Kr) is 10 s. At time $t = 0$ a heavy canister contains 3 g of this radioactive gas.
    (a) Find a function $m(t) = m_0 2^{-t/h}$ that models the amount of ^{91}Kr remaining in the canister after t seconds.
    (b) Find a function $m(t) = m_0 e^{-rt}$ that models the amount of ^{91}Kr remaining in the canister after t seconds.
    (c) How much ^{91}Kr remains after 1 min?
    (d) After how long will the amount of ^{91}Kr remaining be reduced to 1 μg (1 microgram, or 10^{-6} g)?
13. An earthquake measuring 6.4 on the Richter scale struck Japan in July 2007, causing extensive damage. Earlier that year, a minor earthquake measuring 3.1 on the Richter scale was felt in parts of Pennsylvania. How many times more intense was the Japanese earthquake than the Pennsylvania earthquake?

A CUMULATIVE REVIEW TEST FOR CHAPTERS 2, 3, AND 4 CAN BE FOUND AT THE BOOK COMPANION WEBSITE: **www.stewartmath.com.**

Fitting Exponential and Power Curves to Data

In a previous *Focus on Modeling* (page 325) we learned that the shape of a scatter plot helps us to choose the type of curve to use in modeling data. The first plot in Figure 1 strongly suggests that a line be fitted through it, and the second one points to a cubic polynomial. For the third plot it is tempting to fit a second-degree polynomial. But what if an exponential curve fits better? How do we decide this? In this section we learn how to fit exponential and power curves to data and how to decide which type of curve fits the data better. We also learn that for scatter plots like those in the last two plots in Figure 1, the data can be modeled by logarithmic or logistic functions.

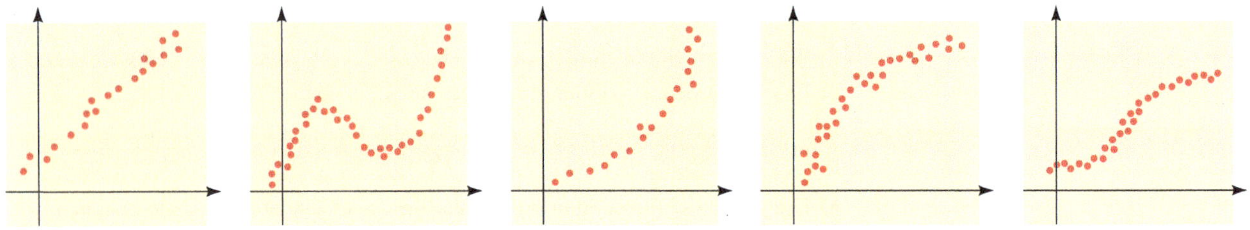

FIGURE 1

Modeling with Exponential Functions

If a scatter plot shows that the data increase rapidly, we might want to model the data using an *exponential model*, that is, a function of the form

$$f(x) = Ce^{kx}$$

where C and k are constants. In the first example we model world population by an exponential model. Recall from Section 4.6 that population tends to increase exponentially.

EXAMPLE 1 ■ An Exponential Model for World Population

Table 1 gives the population of the world in the 20th century.

(a) Draw a scatter plot, and note that a linear model is not appropriate.

(b) Find an exponential function that models population growth.

(c) Draw a graph of the function that you found together with the scatter plot. How well does the model fit the data?

(d) Use the model that you found to predict world population in the year 2020.

TABLE 1
World population

Year (t)	World population (P in millions)
1900	1650
1910	1750
1920	1860
1930	2070
1940	2300
1950	2520
1960	3020
1970	3700
1980	4450
1990	5300
2000	6060

SOLUTION

(a) The scatter plot is shown in Figure 2. The plotted points do not appear to lie along a straight line, so a linear model is not appropriate.

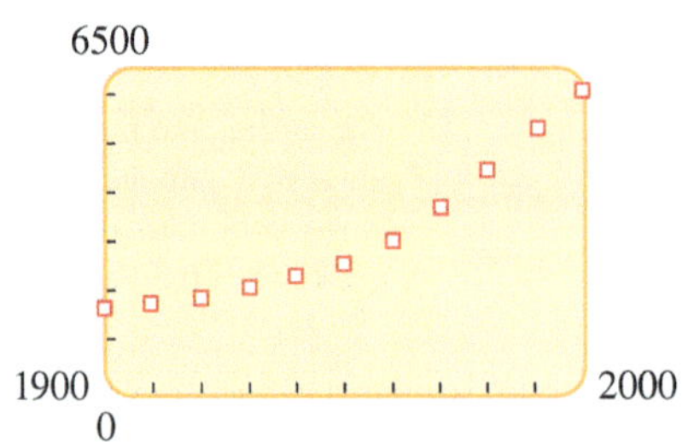

FIGURE 2 Scatter plot of world population

Chabruken/The Image Bank/Getty Images

The population of the world increases exponentially.

(b) Using a graphing calculator and the `ExpReg` command (see Figure 3(a)), we get the exponential model

$$P(t) = (0.0082543) \cdot (1.0137186)^t$$

This is a model of the form $y = Cb^t$. To convert this to the form $y = Ce^{kt}$, we use the properties of exponentials and logarithms as follows.

$$\begin{aligned} 1.0137186^t &= e^{\ln 1.0137186^t} && A = e^{\ln A} \\ &= e^{t \ln 1.0137186} && \ln A^B = B \ln A \\ &= e^{0.013625t} && \ln 1.0137186 \approx 0.013625 \end{aligned}$$

Thus we can write the model as

$$P(t) = 0.0082543e^{0.013625t}$$

(c) From the graph in Figure 3(b) we see that the model appears to fit the data fairly well. The period of relatively slow population growth is explained by the depression of the 1930s and the two world wars.

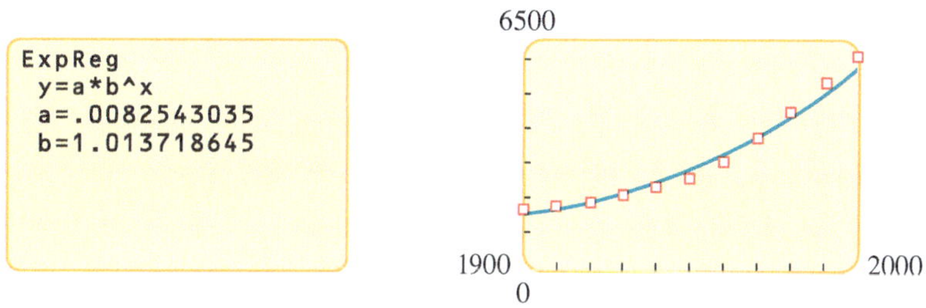

(a) (b)

FIGURE 3 Exponential model for world population

(d) The model predicts that the world population in 2020 will be

$$\begin{aligned} P(2020) &= 0.0082543e^{(0.013625)(2020)} \\ &\approx 7{,}405{,}400{,}000 \end{aligned}$$

Modeling with Power Functions

If the scatter plot of the data we are studying resembles the graph of $y = ax^2$, $y = ax^{1.32}$, or some other power function, then we seek a *power model*, that is, a function of the form

$$f(x) = ax^n$$

where a is a positive constant and n is any real number.

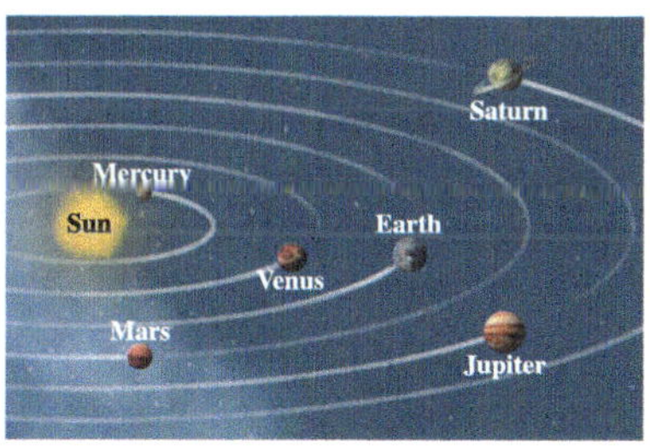

In the next example we seek a power model for some astronomical data. In astronomy, distance in the solar system is often measured in astronomical units. An *astronomical unit* (AU) is the mean distance from the earth to the sun. The *period* of a planet is the time it takes the planet to make a complete revolution around the sun (measured in earth years). In this example we derive the remarkable relationship, first discovered by Johannes Kepler (see page 808), between the mean distance of a planet from the sun and its period.

EXAMPLE 2 ■ A Power Model for Planetary Periods

Table 2 gives the mean distance d of each planet from the sun in astronomical units and its period T in years.

TABLE 2
Distances and periods of the planets

Planet	d	T
Mercury	0.387	0.241
Venus	0.723	0.615
Earth	1.000	1.000
Mars	1.523	1.881
Jupiter	5.203	11.861
Saturn	9.541	29.457
Uranus	19.190	84.008
Neptune	30.086	164.784
Pluto*	39.507	248.350

*Pluto is a "dwarf planet."

(a) Sketch a scatter plot. Is a linear model appropriate?

(b) Find a power function that models the data.

(c) Draw a graph of the function you found and the scatter plot on the same graph. How well does the model fit the data?

(d) Use the model that you found to calculate the period of an asteroid whose mean distance from the sun is 5 AU.

SOLUTION

(a) The scatter plot shown in Figure 4 indicates that the plotted points do not lie along a straight line, so a linear model is not appropriate.

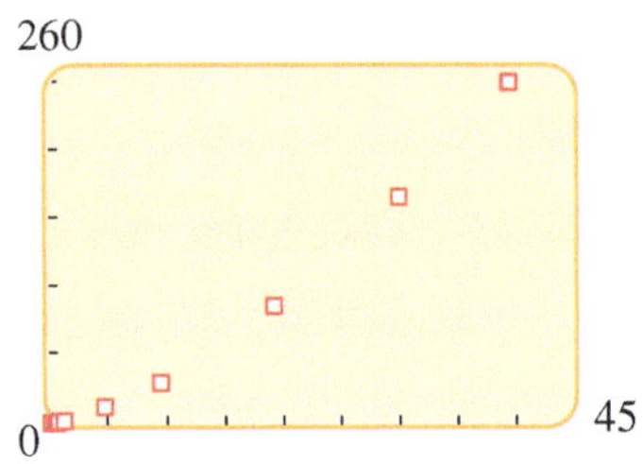

FIGURE 4 Scatter plot of planetary data

(b) Using a graphing calculator and the `PwrReg` command (see Figure 5(a)), we get the power model

$$T = 1.000396d^{1.49966}$$

If we round both the coefficient and the exponent to three significant figures, we can write the model as

$$T = d^{1.5}$$

This is the relationship discovered by Kepler (see page 808). Sir Isaac Newton (page 911) later used his Law of Gravity to derive this relationship theoretically, thereby providing strong scientific evidence that the Law of Gravity must be true.

(c) The graph is shown in Figure 5(b). The model appears to fit the data very well.

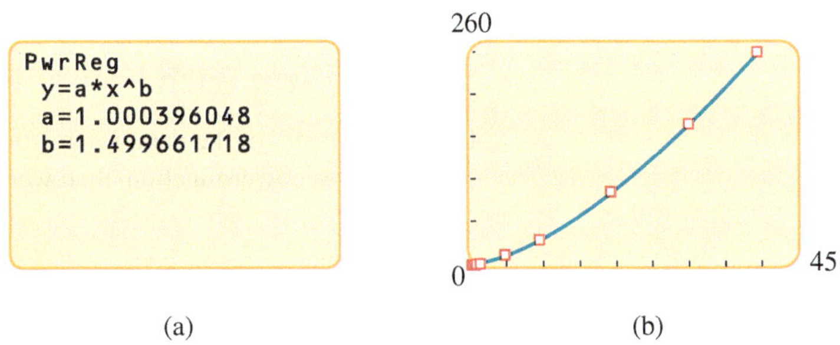

FIGURE 5 Power model for planetary data

(d) In this case $d = 5$ AU, so our model gives

$$T = 1.00039 \cdot 5^{1.49966} \approx 11.22$$

The period of the asteroid is about 11.2 years. ■

Linearizing Data

We have used the shape of a scatter plot to decide which type of model to use: linear, exponential, or power. This works well if the data points lie on a straight line. But it's difficult to distinguish a scatter plot that is exponential from one that requires a power model. So to help decide which model to use, we can *linearize* the data, that is, apply a function that "straightens" the scatter plot. The inverse of the linearizing function is then an appropriate model. We now describe how to linearize data that can be modeled by exponential or power functions.

Linearizing Exponential Data

If we suspect that the data points (x, y) lie on an exponential curve $y = Ce^{kx}$, then the points

$$(x, \ln y)$$

should lie on a straight line. We can see this from the following calculations.

$$\begin{aligned} \ln y &= \ln Ce^{kx} && \text{Assume that } y = Ce^{kx} \text{ and take ln} \\ &= \ln e^{kx} + \ln C && \text{Property of ln} \\ &= kx + \ln C && \text{Property of ln} \end{aligned}$$

To see that $\ln y$ is a linear function of x, let $Y = \ln y$ and $A = \ln C$; then

$$Y = kx + A$$

We apply this technique to the world population data (t, P) to obtain the points $(t, \ln P)$ in Table 3. The scatter plot of $(t, \ln P)$ in Figure 6, called a **semi-log plot**, shows that the linearized data lie approximately on a straight line, so an exponential model should be appropriate.

TABLE 3
World population data

t	Population P (in millions)	$\ln P$
1900	1650	21.224
1910	1750	21.283
1920	1860	21.344
1930	2070	21.451
1940	2300	21.556
1950	2520	21.648
1960	3020	21.829
1970	3700	22.032
1980	4450	22.216
1990	5300	22.391
2000	6060	22.525

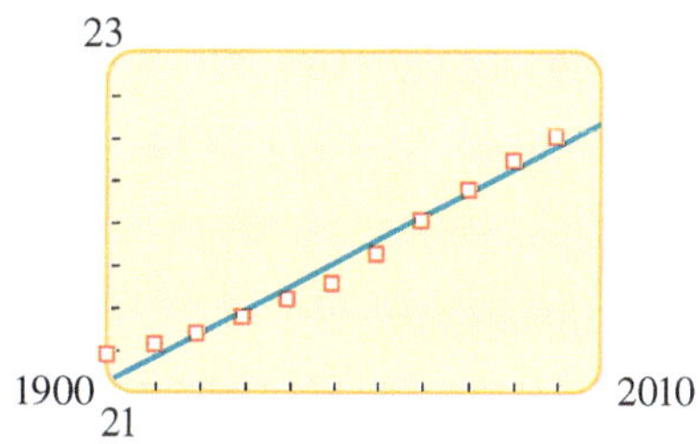

FIGURE 6 Semi-log plot of data in Table 3

Linearizing Power Data

If we suspect that the data points (x, y) lie on a power curve $y = ax^n$, then the points

$$(\ln x, \ln y)$$

should be on a straight line. We can see this from the following calculations.

$$\begin{aligned} \ln y &= \ln ax^n && \text{Assume that } y = ax^n \text{ and take ln} \\ &= \ln a + \ln x^n && \text{Property of ln} \\ &= \ln a + n \ln x && \text{Property of ln} \end{aligned}$$

To see that $\ln y$ is a linear function of $\ln x$, let $Y = \ln y$, $X = \ln x$, and $A = \ln a$; then

$$Y = nX + A$$

We apply this technique to the planetary data (d, T) in Table 2 to obtain the points $(\ln d, \ln T)$ in Table 4. The scatter plot of $(\ln d, \ln T)$ in Figure 7, called a **log-log plot**, shows that the data lie on a straight line, so a power model seems appropriate.

TABLE 4
Log-log table

$\ln d$	$\ln T$
−0.94933	−1.4230
−0.32435	−0.48613
0	0
0.42068	0.6318
1.6492	2.4733
2.2556	3.3829
2.9544	4.4309
3.4041	5.1046
3.6765	5.5148

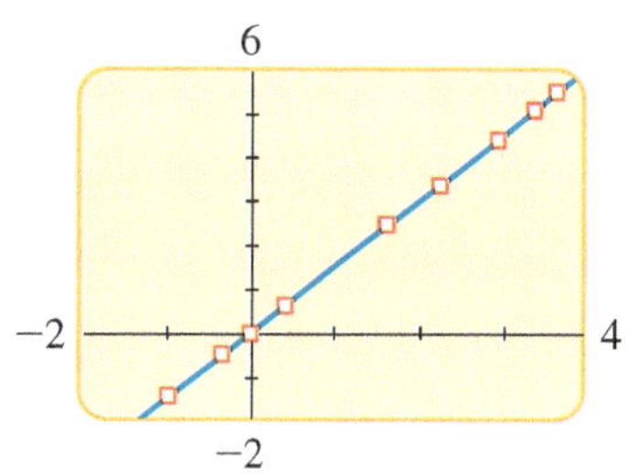

FIGURE 7 Log-log plot of data in Table 4

An Exponential or Power Model?

Suppose that a scatter plot of the data points (x, y) shows a rapid increase. Should we use an exponential function or a power function to model the data? To help us decide, we draw two scatter plots: one for the points $(x, \ln y)$ and the other for the points $(\ln x, \ln y)$. If the first scatter plot appears to lie along a line, then an exponential model is appropriate. If the second plot appears to lie along a line, then a power model is appropriate.

EXAMPLE 3 ■ An Exponential or Power Model?

TABLE 5

x	y
1	2
2	6
3	14
4	22
5	34
6	46
7	64
8	80
9	102
10	130

Data points (x, y) are shown in Table 5.

(a) Draw a scatter plot of the data.

(b) Draw scatter plots of $(x, \ln y)$ and $(\ln x, \ln y)$.

(c) Is an exponential function or a power function appropriate for modeling this data?

(d) Find an appropriate function to model the data.

SOLUTION

(a) The scatter plot of the data is shown in Figure 8.

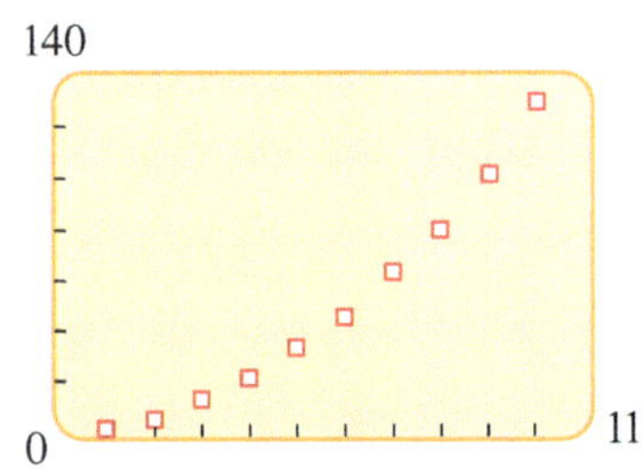

FIGURE 8

TABLE 6

x	$\ln x$	$\ln y$
1	0	0.7
2	0.7	1.8
3	1.1	2.6
4	1.4	3.1
5	1.6	3.5
6	1.8	3.8
7	1.9	4.2
8	2.1	4.4
9	2.2	4.6
10	2.3	4.9

(b) We use the values from Table 6 to graph the scatter plots in Figures 9 and 10.

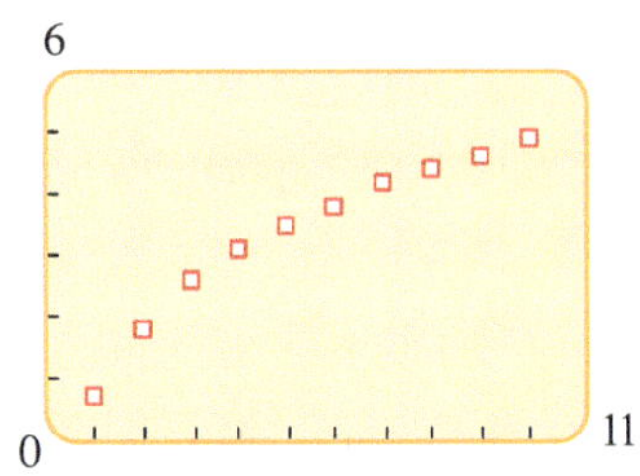

FIGURE 9 Semi-log plot

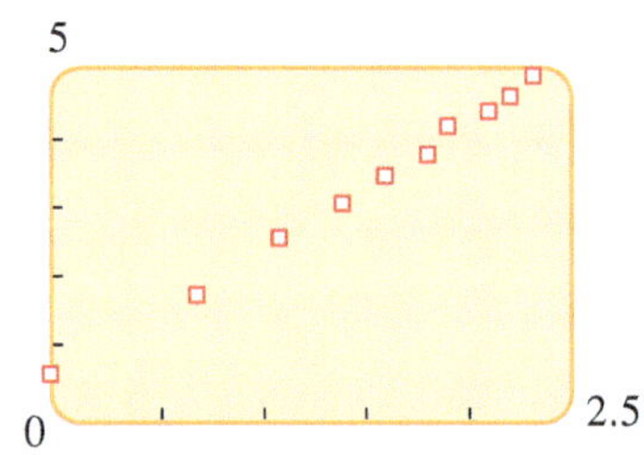

FIGURE 10 Log-log plot

(c) The scatter plot of $(x, \ln y)$ in Figure 9 does not appear to be linear, so an exponential model is not appropriate. On the other hand, the scatter plot of $(\ln x, \ln y)$ in Figure 10 is very nearly linear, so a power model is appropriate.

(d) Using the `PwrReg` command on a graphing calculator, we find that the power function that best fits the data point is

$$y = 1.85x^{1.82}$$

The graph of this function and the original data points are shown in Figure 11. ■

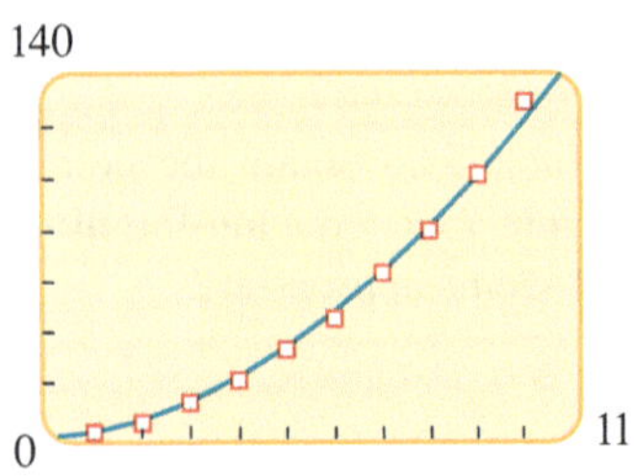

FIGURE 11

Before graphing calculators and statistical software became common, exponential and power models for data were often constructed by first finding a linear model for the linearized data. Then the model for the actual data was found by taking exponentials. For instance, if we find that $\ln y = A \ln x + B$, then by taking exponentials we get the model $y = e^B \cdot e^{A \ln x}$, or $y = Cx^A$ (where $C = e^B$). Special graphing paper called "log paper" or "log-log paper" was used to facilitate this process.

Modeling with Logistic Functions

A logistic growth model is a function of the form

$$f(t) = \frac{c}{1 + ae^{-bt}}$$

where a, b, and c are positive constants. Logistic functions are used to model populations where the growth is constrained by available resources. (See Exercises 27–30 of Section 4.2.)

EXAMPLE 4 ■ Stocking a Pond with Catfish

Much of the fish that is sold in supermarkets today is raised on commercial fish farms, not caught in the wild. A pond on one such farm is initially stocked with 1000 catfish, and the fish population is then sampled at 15-week intervals to estimate its size. The population data are given in Table 7.

TABLE 7

Week	Catfish
0	1000
15	1500
30	3300
45	4400
60	6100
75	6900
90	7100
105	7800
120	7900

(a) Find an appropriate model for the data.

(b) Make a scatter plot of the data and graph the model that you found in part (a) on the scatter plot.

(c) How does the model predict that the fish population will change with time?

SOLUTION

(a) Since the catfish population is restricted by its habitat (the pond), a logistic model is appropriate. Using the `Logistic` command on a calculator (see Figure 12(a)), we find the following model for the catfish population $P(t)$:

$$P(t) = \frac{7925}{1 + 7.7e^{-0.052t}}$$

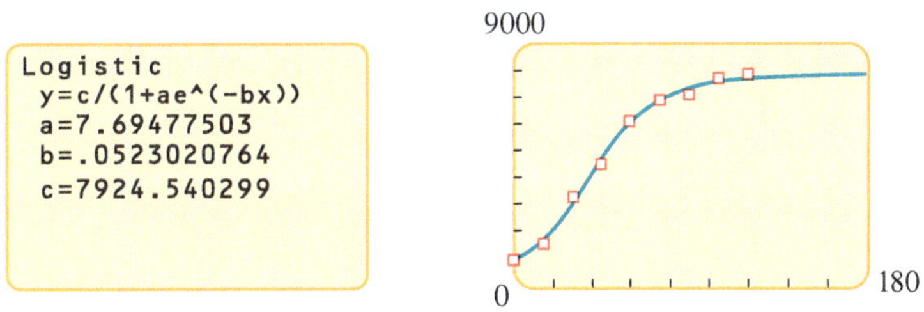

FIGURE 12 (a) (b) Catfish population $y = P(t)$

(b) The scatter plot and the logistic curve are shown in Figure 12(b).

(c) From the graph of P in Figure 12(b) we see that the catfish population increases rapidly until about $t = 80$ weeks. Then growth slows down, and at about $t = 120$ weeks the population levels off and remains more or less constant at slightly over 7900. ■

The behavior that is exhibited by the catfish population in Example 4 is typical of logistic growth. After a rapid growth phase, the population approaches a constant level called the **carrying capacity** of the environment. This occurs because as $t \to \infty$, we have $e^{-bt} \to 0$ (see Section 4.2), and so

$$P(t) = \frac{c}{1 + ae^{-bt}} \quad \longrightarrow \quad \frac{c}{1 + 0} = c$$

Thus the carrying capacity is c.

PROBLEMS

1. U.S. Population The U.S. Constitution requires a census every 10 years. The census data for 1790–2010 are given in the table.

(a) Make a scatter plot of the data.

(b) Use a calculator to find an exponential model for the data.

(c) Use your model to predict the population at the 2020 census.

(d) Use your model to estimate the population in 1965.

Year	Population (in millions)	Year	Population (in millions)	Year	Population (in millions)
1790	3.9	1870	38.6	1950	151.3
1800	5.3	1880	50.2	1960	179.3
1810	7.2	1890	63.0	1970	203.3
1820	9.6	1900	76.2	1980	226.5
1830	12.9	1910	92.2	1990	248.7
1840	17.1	1920	106.0	2000	281.4
1850	23.2	1930	123.2	2010	308.7
1860	31.4	1940	132.2		

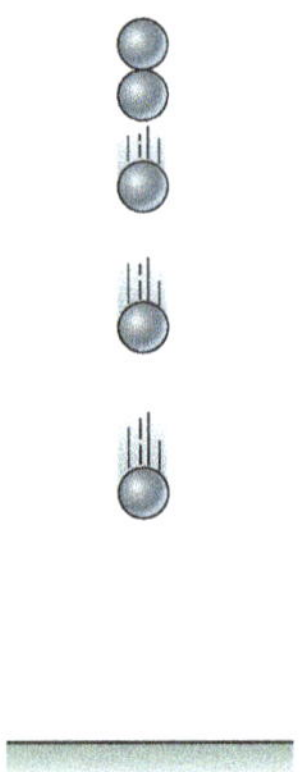

Time (s)	Distance (m)
0.1	0.048
0.2	0.197
0.3	0.441
0.4	0.882
0.5	1.227
0.6	1.765
0.7	2.401
0.8	3.136
0.9	3.969
1.0	4.902

2. A Falling Ball In a physics experiment a lead ball is dropped from a height of 5 m. The students record the distance the ball has fallen every one-tenth of a second. (This can be done by using a camera and a strobe light.) Their data are shown in the margin.

(a) Make a scatter plot of the data.

(b) Use a calculator to find a power model.

(c) Use your model to predict how far a dropped ball would fall in 3 s.

3. Half-Life of Radioactive Iodine A student is trying to determine the half-life of radioactive iodine-131. He measures the amount of iodine-131 in a sample solution every 8 hours. His data are shown in the table below.

(a) Make a scatter plot of the data.

(b) Use a calculator to find an exponential model.

(c) Use your model to find the half-life of iodine-131.

Time (h)	Amount of $^{131}\mathrm{I}$(g)
0	4.80
8	4.66
16	4.51
24	4.39
32	4.29
40	4.14
48	4.04

Light intensity decreases exponentially with depth.

4. The Beer-Lambert Law As sunlight passes through the waters of lakes and oceans, the light is absorbed, and the deeper it penetrates, the more its intensity diminishes. The light intensity I at depth x is given by the Beer-Lambert Law:

$$I = I_0 e^{-kx}$$

where I_0 is the light intensity at the surface and k is a constant that depends on the murkiness of the water (see page 366). A biologist uses a photometer to investigate light penetration in a northern lake, obtaining the data in the table.

(a) Use a graphing calculator to find an exponential function of the form given by the Beer-Lambert Law to model these data. What is the light intensity I_0 at the surface on this day, and what is the "murkiness" constant k for this lake? [*Hint:* If your calculator gives you a function of the form $I = ab^x$, convert this to the form you want using the identities $b^x = e^{\ln(b^x)} = e^{x \ln b}$. See Example 1(b).]

(b) Make a scatter plot of the data, and graph the function that you found in part (a) on your scatter plot.

(c) If the light intensity drops below 0.15 lumen (lm), a certain species of algae can't survive because photosynthesis is impossible. Use your model from part (a) to determine the depth below which there is insufficient light to support this algae.

Depth (ft)	Light intensity (lm)	Depth (ft)	Light intensity (lm)
5	13.0	25	1.8
10	7.6	30	1.1
15	4.5	35	0.5
20	2.7	40	0.3

Time	Words recalled
15 min	64.3
1 h	45.1
8 h	37.3
1 day	32.8
2 days	26.9
3 days	25.6
5 days	22.9

5. Experimenting with "Forgetting" Curves Every one of us is all too familiar with the phenomenon of forgetting. Facts that we clearly understood at the time we first learned them sometimes fade from our memory by the time the final exam rolls around. Psychologists have proposed several ways to model this process. One such model is Ebbinghaus' Law of Forgetting, described on page 356. Other models use exponential or logarithmic functions. To develop her own model, a psychologist performs an experiment on a group of volunteers by asking them to memorize a list of 100 related words. She then tests how many of these words they can recall after various periods of time. The average results for the group are shown in the table.

(a) Use a graphing calculator to find a *power* function of the form $y = at^b$ that models the average number of words y that the volunteers remember after t hours. Then find an *exponential* function of the form $y = ab^t$ to model the data.

(b) Make a scatter plot of the data, and graph both the functions that you found in part (a) on your scatter plot.

(c) Which of the two functions seems to provide the better model?

The number of different bat species in a cave is related to the size of the cave by a power function.

6. Modeling the Species-Area Relation The table gives the areas of several caves in central Mexico and the number of bat species that live in each cave.*

(a) Find a power function that models the data.

(b) Draw a graph of the function you found in part (a) and a scatter plot of the data on the same graph. Does the model fit the data well?

(c) The cave called El Sapo near Puebla, Mexico, has a surface area of $A = 205\ \text{m}^2$. Use the model to estimate the number of bat species you would expect to find in that cave.

Cave	Area (m^2)	Number of species
La Escondida	18	1
El Escorpion	19	1
El Tigre	58	1
Mision Imposible	60	2
San Martin	128	5
El Arenal	187	4
La Ciudad	344	6
Virgen	511	7

*A. K. Brunet and R. A. Medallin, "The Species-Area Relationship in Bat Assemblages of Tropical Caves." *Journal of Mammalogy, 82*(4):1114–1122, 2001.

7. Auto Exhaust Emissions A study by the U.S. Office of Science and Technology in 1972 estimated the cost of reducing automobile emissions by certain percentages. Find an exponential model that captures the "diminishing returns" trend of these data shown in the table below.

Reduction in emissions (%)	Cost per car ($)
50	45
55	55
60	62
65	70
70	80
75	90
80	100
85	200
90	375
95	600

8. Exponential or Power Model? Data points (x, y) are shown in the table.

(a) Draw a scatter plot of the data.

(b) Draw scatter plots of $(x, \ln y)$ and $(\ln x, \ln y)$.

(c) Which is more appropriate for modeling this data: an exponential function or a power function?

(d) Find an appropriate function to model the data.

x	2	4	6	8	10	12	14	16
y	0.08	0.12	0.18	0.25	0.36	0.52	0.73	1.06

9. Exponential or Power Model? Data points (x, y) are shown in the table in the margin.

(a) Draw a scatter plot of the data.

(b) Draw scatter plots of $(x, \ln y)$ and $(\ln x, \ln y)$.

(c) Which is more appropriate for modeling this data: an exponential function or a power function?

(d) Find an appropriate function to model the data.

x	10	20	30	40	50	60	70	80	90
y	29	82	151	235	330	430	546	669	797

10. Logistic Population Growth The table and scatter plot give the population of black flies in a closed laboratory container over an 18-day period.

(a) Use the `Logistic` command on your calculator to find a logistic model for these data.

(b) Use the model to estimate the time when there were 400 flies in the container.

Time (days)	Number of flies
0	10
2	25
4	66
6	144
8	262
10	374
12	446
16	492
18	498

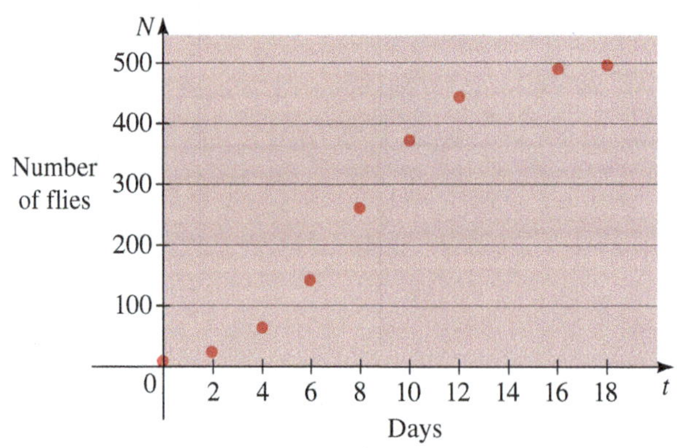

Doug Steakley/Lonely Planet Images/Getty Images

5 Trigonometric Functions: Unit Circle Approach

In this chapter and the next we introduce two different but equivalent ways of viewing the trigonometric functions. One way is to view them as *functions of real numbers* (Chapter 5); the other is to view them as *functions of angles* (Chapter 6). The two approaches to trigonometry are independent of each other, so either Chapter 5 or Chapter 6 may be studied first. The applications of trigonometry are numerous, including signal processing, digital coding of music and videos, finding distances to stars, producing CAT scans for medical imaging, and many others. These applications are very diverse, and we need to study both approaches to trigonometry because the different approaches are required for different applications.

One of the main applications of trigonometry that we study in this chapter is periodic motion. If you've ever taken a Ferris wheel ride, then you know about periodic motion—that is, motion that repeats over and over. Periodic motion occurs often in nature, as in the daily rising and setting of the sun, the daily variation in tide levels, the vibrations of a leaf in the wind, and many more. We will see in this chapter how the trigonometric functions are used to model periodic motion.

5.1 THE UNIT CIRCLE

■ The Unit Circle ■ Terminal Points on the Unit Circle ■ The Reference Number

In this section we explore some properties of the circle of radius 1 centered at the origin. These properties are used in the next section to define the trigonometric functions.

■ The Unit Circle

The set of points at a distance 1 from the origin is a circle of radius 1 (see Figure 1). In Section 1.9 we learned that the equation of this circle is $x^2 + y^2 = 1$.

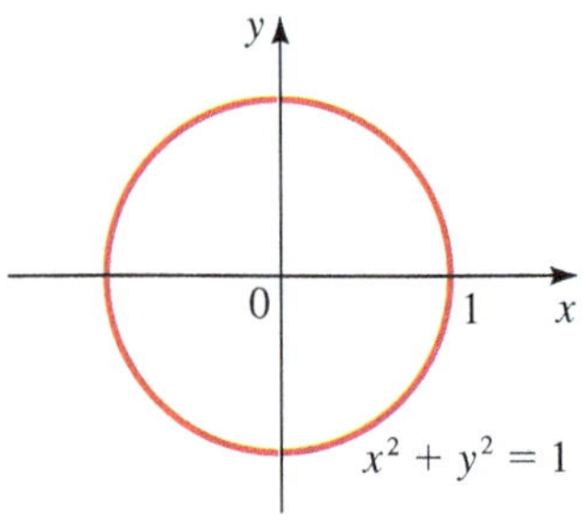

FIGURE 1 The unit circle

Circles are studied in Section 1.9, page 97.

THE UNIT CIRCLE

The **unit circle** is the circle of radius 1 centered at the origin in the xy-plane. Its equation is

$$x^2 + y^2 = 1$$

EXAMPLE 1 ■ A Point on the Unit Circle

Show that the point $P\left(\frac{\sqrt{3}}{3}, \frac{\sqrt{6}}{3}\right)$ is on the unit circle.

SOLUTION We need to show that this point satisfies the equation of the unit circle, that is, $x^2 + y^2 = 1$. Since

$$\left(\frac{\sqrt{3}}{3}\right)^2 + \left(\frac{\sqrt{6}}{3}\right)^2 = \frac{3}{9} + \frac{6}{9} = 1$$

P is on the unit circle.

Now Try Exercise 3 ■

EXAMPLE 2 ■ Locating a Point on the Unit Circle

The point $P(\sqrt{3}/2, y)$ is on the unit circle in Quadrant IV. Find its y-coordinate.

SOLUTION Since the point is on the unit circle, we have

$$\left(\frac{\sqrt{3}}{2}\right)^2 + y^2 = 1$$

$$y^2 = 1 - \frac{3}{4} = \frac{1}{4}$$

$$y = \pm\frac{1}{2}$$

Since the point is in Quadrant IV, its y-coordinate must be negative, so $y = -\frac{1}{2}$.

Now Try Exercise 9 ■

■ Terminal Points on the Unit Circle

Suppose t is a real number. If $t \geq 0$, let's mark off a distance t along the unit circle, starting at the point $(1, 0)$ and moving in a counterclockwise direction. If $t < 0$, we mark off a distance $|t|$ in a clockwise direction (Figure 2). In this way we arrive at a point $P(x, y)$ on the unit circle. The point $P(x, y)$ obtained in this way is called the **terminal point** determined by the real number t.

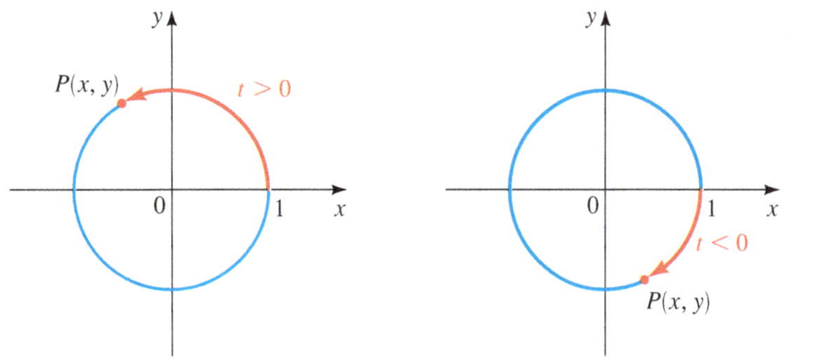

FIGURE 2 (a) Terminal point $P(x, y)$ determined by $t > 0$ (b) Terminal point $P(x, y)$ determined by $t < 0$

The circumference of the unit circle is $C = 2\pi(1) = 2\pi$. So if a point starts at $(1, 0)$ and moves counterclockwise all the way around the unit circle and returns to $(1, 0)$, it travels a distance of 2π. To move halfway around the circle, it travels a distance of $\frac{1}{2}(2\pi) = \pi$. To move a quarter of the distance around the circle, it travels a distance of $\frac{1}{4}(2\pi) = \pi/2$. Where does the point end up when it travels these distances along the circle? From Figure 3 we see, for example, that when it travels a distance of π starting at $(1, 0)$, its terminal point is $(-1, 0)$.

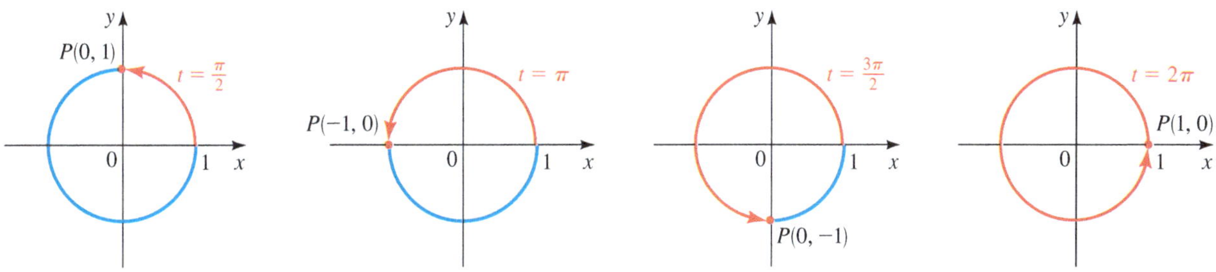

FIGURE 3 Terminal points determined by $t = \frac{\pi}{2}, \pi, \frac{3\pi}{2}$, and 2π

EXAMPLE 3 ■ Finding Terminal Points

Find the terminal point on the unit circle determined by each real number t.

(a) $t = 3\pi$ **(b)** $t = -\pi$ **(c)** $t = -\dfrac{\pi}{2}$

SOLUTION From Figure 4 we get the following:

(a) The terminal point determined by 3π is $(-1, 0)$.

(b) The terminal point determined by $-\pi$ is $(-1, 0)$.

(c) The terminal point determined by $-\pi/2$ is $(0, -1)$.

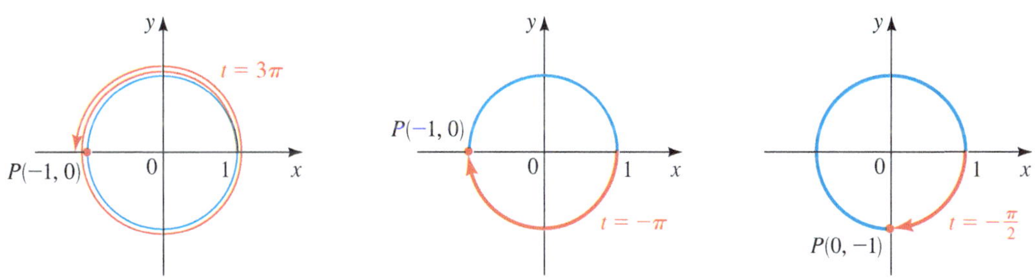

FIGURE 4

Notice that different values of t can determine the same terminal point.

Now Try Exercise 23 ■

The terminal point $P(x, y)$ determined by $t = \pi/4$ is the same distance from $(1, 0)$ as from $(0, 1)$ along the unit circle (see Figure 5).

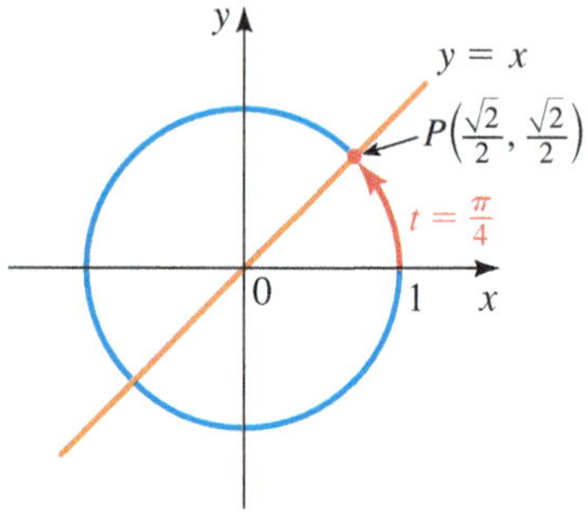

FIGURE 5

Since the unit circle is symmetric with respect to the line $y = x$, it follows that P lies on the line $y = x$. So P is the point of intersection (in the Quadrant I) of the circle $x^2 + y^2 = 1$ and the line $y = x$. Substituting x for y in the equation of the circle, we get

$$x^2 + x^2 = 1$$

$$2x^2 = 1 \qquad \text{Combine like terms}$$

$$x^2 = \frac{1}{2} \qquad \text{Divide by 2}$$

$$x = \pm\frac{1}{\sqrt{2}} \qquad \text{Take square roots}$$

Since P is in the Quadrant I, $x = 1/\sqrt{2}$ and since $y = x$, we have $y = 1/\sqrt{2}$ also. Thus the terminal point determined by $\pi/4$ is

$$P\left(\frac{1}{\sqrt{2}}, \frac{1}{\sqrt{2}}\right) = P\left(\frac{\sqrt{2}}{2}, \frac{\sqrt{2}}{2}\right)$$

Similar methods can be used to find the terminal points determined by $t = \pi/6$ and $t = \pi/3$ (see Exercises 61 and 62). Table 1 and Figure 6 give the terminal points for some special values of t.

TABLE 1

t	Terminal point determined by t
0	$(1, 0)$
$\frac{\pi}{6}$	$\left(\frac{\sqrt{3}}{2}, \frac{1}{2}\right)$
$\frac{\pi}{4}$	$\left(\frac{\sqrt{2}}{2}, \frac{\sqrt{2}}{2}\right)$
$\frac{\pi}{3}$	$\left(\frac{1}{2}, \frac{\sqrt{3}}{2}\right)$
$\frac{\pi}{2}$	$(0, 1)$

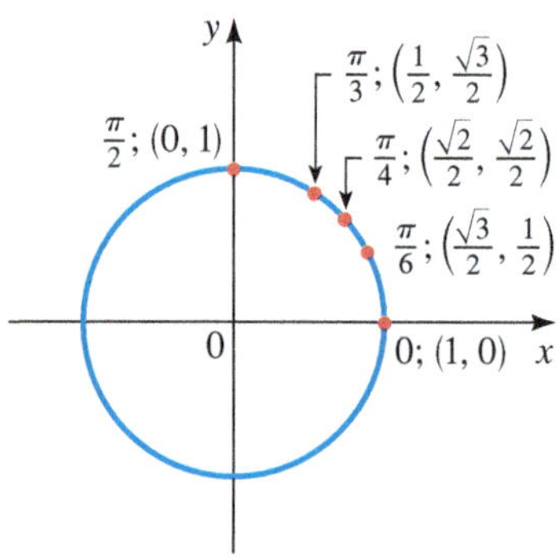

FIGURE 6

EXAMPLE 4 ■ Finding Terminal Points

Find the terminal point determined by each given real number t.

(a) $t = -\frac{\pi}{4}$ **(b)** $t = \frac{3\pi}{4}$ **(c)** $t = -\frac{5\pi}{6}$

SOLUTION

(a) Let P be the terminal point determined by $-\pi/4$, and let Q be the terminal point determined by $\pi/4$. From Figure 7(a) we see that the point P has the same coordinates as Q except for sign. Since P is in Quadrant IV, its x-coordinate is positive and its y-coordinate is negative. Thus, the terminal point is $P(\sqrt{2}/2, -\sqrt{2}/2)$.

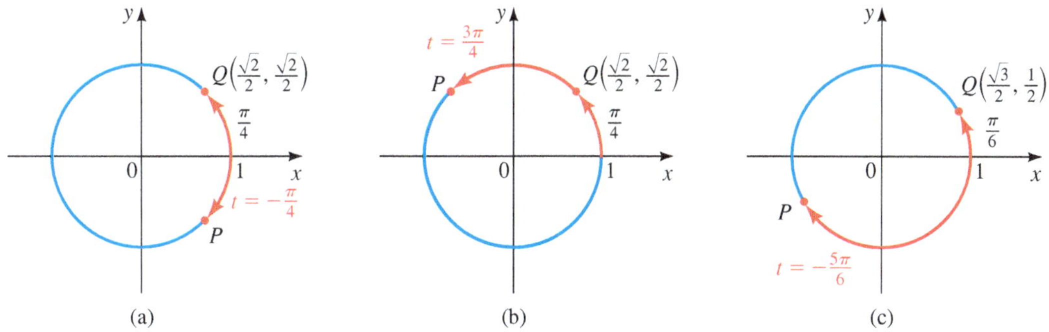

FIGURE 7

(b) Let P be the terminal point determined by $3\pi/4$, and let Q be the terminal point determined by $\pi/4$. From Figure 7(b) we see that the point P has the same coordinates as Q except for sign. Since P is in Quadrant II, its x-coordinate is negative and its y-coordinate is positive. Thus the terminal point is $P(-\sqrt{2}/2, \sqrt{2}/2)$.

(c) Let P be the terminal point determined by $-5\pi/6$, and let Q be the terminal point determined by $\pi/6$. From Figure 7(c) we see that the point P has the same coordinates as Q except for sign. Since P is in Quadrant III, its coordinates are both negative. Thus the terminal point is $P(-\sqrt{3}/2, -\frac{1}{2})$.

Now Try Exercise 27 ■

The Reference Number

From Examples 3 and 4 we see that to find a terminal point in any quadrant we need only know the "corresponding" terminal point in the first quadrant. We use the idea of the *reference number* to help us find terminal points.

> **REFERENCE NUMBER**
>
> Let t be a real number. The **reference number** $\bar{t}$ associated with t is the shortest distance along the unit circle between the terminal point determined by t and the x-axis.

Figure 8 shows that to find the reference number $\bar{t}$, it's helpful to know the quadrant in which the terminal point determined by t lies. If the terminal point lies in Quadrant I or IV, where x is positive, we find $\bar{t}$ by moving along the circle to the *positive* x-axis. If it lies in Quadrant II or III, where x is negative, we find $\bar{t}$ by moving along the circle to the *negative* x-axis.

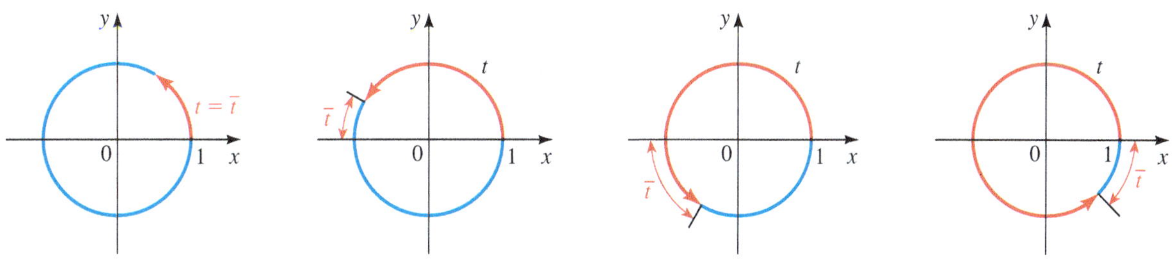

FIGURE 8 The reference number $\bar{t}$ for t

EXAMPLE 5 ■ Finding Reference Numbers

Find the reference number for each value of t.

(a) $t = \dfrac{5\pi}{6}$ **(b)** $t = \dfrac{7\pi}{4}$ **(c)** $t = -\dfrac{2\pi}{3}$ **(d)** $t = 5.80$

SOLUTION From Figure 9 we find the reference numbers as follows.

(a) $\bar{t} = \pi - \dfrac{5\pi}{6} = \dfrac{\pi}{6}$ **(b)** $\bar{t} = 2\pi - \dfrac{7\pi}{4} = \dfrac{\pi}{4}$

(c) $\bar{t} = \pi - \dfrac{2\pi}{3} = \dfrac{\pi}{3}$ **(d)** $\bar{t} = 2\pi - 5.80 \approx 0.48$

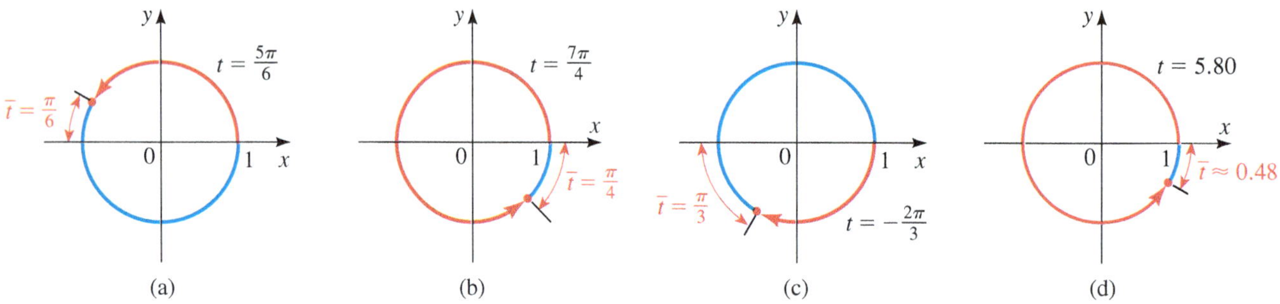

FIGURE 9

Now Try Exercise 37 ■

USING REFERENCE NUMBERS TO FIND TERMINAL POINTS

To find the terminal point P determined by any value of t, we use the following steps:

1. Find the reference number $\bar{t}$.
2. Find the terminal point $Q(a, b)$ determined by $\bar{t}$.
3. The terminal point determined by t is $P(\pm a, \pm b)$, where the signs are chosen according to the quadrant in which this terminal point lies.

EXAMPLE 6 ■ Using Reference Numbers to Find Terminal Points

Find the terminal point determined by each given real number t.

(a) $t = \dfrac{5\pi}{6}$ **(b)** $t = \dfrac{7\pi}{4}$ **(c)** $t = -\dfrac{2\pi}{3}$

SOLUTION The reference numbers associated with these values of t were found in Example 5.

(a) The reference number is $\bar{t} = \pi/6$, which determines the terminal point $\left(\sqrt{3}/2, \frac{1}{2}\right)$ from Table 1. Since the terminal point determined by t is in Quadrant II, its x-coordinate is negative and its y-coordinate is positive. Thus the desired terminal point is

$$\left(-\frac{\sqrt{3}}{2}, \frac{1}{2}\right)$$

(b) The reference number is $\bar{t} = \pi/4$, which determines the terminal point $\left(\sqrt{2}/2, \sqrt{2}/2\right)$ from Table 1. Since the terminal point is in Quadrant IV, its x-coordinate is positive and its y-coordinate is negative. Thus the desired terminal point is

$$\left(\frac{\sqrt{2}}{2}, -\frac{\sqrt{2}}{2}\right)$$

(c) The reference number is $\bar{t} = \pi/3$, which determines the terminal point $\left(\frac{1}{2}, \sqrt{3}/2\right)$ from Table 1. Since the terminal point determined by t is in Quadrant III, its coordinates are both negative. Thus the desired terminal point is

$$\left(-\frac{1}{2}, -\frac{\sqrt{3}}{2}\right)$$

Now Try Exercise 41 ■

Since the circumference of the unit circle is 2π, the terminal point determined by t is the same as that determined by $t + 2\pi$ or $t - 2\pi$. In general, we can add or subtract 2π any number of times without changing the terminal point determined by t. We use this observation in the next example to find terminal points for large t.

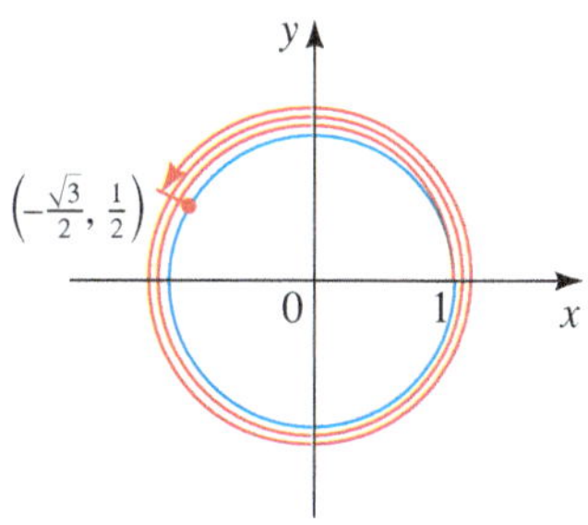

FIGURE 10

EXAMPLE 7 ■ Finding the Terminal Point for Large *t*

Find the terminal point determined by $t = \dfrac{29\pi}{6}$.

SOLUTION Since

$$t = \frac{29\pi}{6} = 4\pi + \frac{5\pi}{6}$$

we see that the terminal point of t is the same as that of $5\pi/6$ (that is, we subtract 4π). So by Example 6(a) the terminal point is $\left(-\sqrt{3}/2, \frac{1}{2}\right)$. (See Figure 10.)

Now Try Exercise 47 ■

5.1 EXERCISES

CONCEPTS

1. (a) The unit circle is the circle centered at ________ with radius ________.

(b) The equation of the unit circle is ____________.

(c) Suppose the point $P(x, y)$ is on the unit circle. Find the missing coordinate:

(i) $P(1, \quad)$ (ii) $P(\quad, 1)$

(iii) $P(-1, \quad)$ (iv) $P(\quad, -1)$

2. (a) If we mark off a distance t along the unit circle, starting at $(1, 0)$ and moving in a counterclockwise direction, we arrive at the ____________ point determined by t.

(b) The terminal points determined by $\pi/2, \pi, -\pi/2, 2\pi$ are ________, ________, ________, and ________, respectively.

SKILLS

3–8 ■ Points on the Unit Circle Show that the point is on the unit circle.

3. $\left(\frac{3}{5}, -\frac{4}{5}\right)$ **4.** $\left(-\frac{24}{25}, -\frac{7}{25}\right)$

5. $\left(\frac{3}{4}, -\frac{\sqrt{7}}{4}\right)$ **6.** $\left(-\frac{5}{7}, -\frac{2\sqrt{6}}{7}\right)$

7. $\left(-\frac{\sqrt{5}}{3}, \frac{2}{3}\right)$ **8.** $\left(\frac{\sqrt{11}}{6}, \frac{5}{6}\right)$

9–14 ■ Points on the Unit Circle Find the missing coordinate of P, using the fact that P lies on the unit circle in the given quadrant.

Coordinates	Quadrant
9. $P(-\frac{3}{5}, \quad)$	III
10. $P(\quad, -\frac{7}{25})$	IV
11. $P(\quad, \frac{1}{3})$	II

Coordinates	Quadrant
12. $P(\frac{2}{5}, \quad)$	I
13. $P(\quad, -\frac{2}{7})$	IV
14. $P(-\frac{2}{3}, \quad)$	II

15–20 ■ Points on the Unit Circle The point P is on the unit circle. Find $P(x, y)$ from the given information.

15. The x-coordinate of P is $\frac{5}{13}$, and the y-coordinate is negative.

16. The y-coordinate of P is $-\frac{3}{5}$, and the x-coordinate is positive.

17. The y-coordinate of P is $\frac{2}{3}$, and the x-coordinate is negative.

18. The x-coordinate of P is positive, and the y-coordinate of P is $-\sqrt{5}/5$.

19. The x-coordinate of P is $-\sqrt{2}/3$, and P lies below the x-axis.

20. The x-coordinate of P is $-\frac{2}{5}$, and P lies above the x-axis.

21–22 ■ Terminal Points Find t and the terminal point determined by t for each point in the figure. In Exercise 21, t increases in increments of $\pi/4$; in Exercise 22, t increases in increments of $\pi/6$.

21.

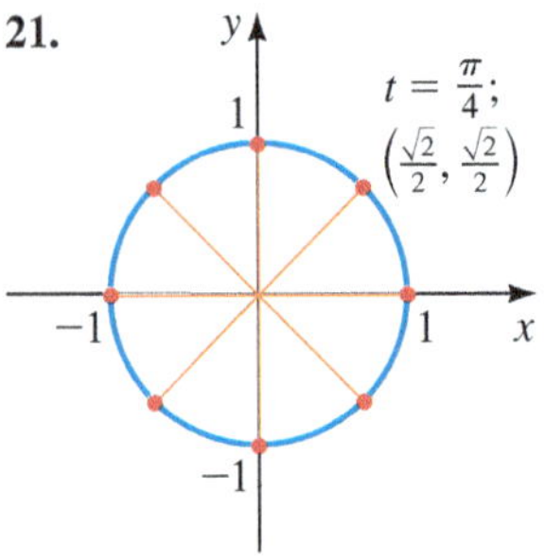

22.

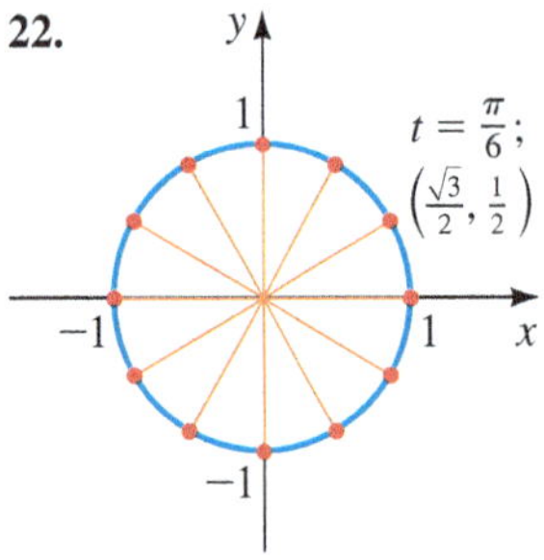

23–36 ■ Terminal Points Find the terminal point $P(x, y)$ on the unit circle determined by the given value of t.

23. $t = 4\pi$ **24.** $t = -3\pi$

25. $t = \dfrac{3\pi}{2}$ **26.** $t = \dfrac{5\pi}{2}$

27. $t = -\dfrac{\pi}{6}$ **28.** $t = \dfrac{7\pi}{6}$

29. $t = \dfrac{5\pi}{4}$ **30.** $t = \dfrac{4\pi}{3}$

31. $t = -\dfrac{7\pi}{6}$ **32.** $t = \dfrac{5\pi}{3}$

33. $t = -\dfrac{7\pi}{4}$ **34.** $t = -\dfrac{4\pi}{3}$

35. $t = -\dfrac{3\pi}{4}$ **36.** $t = \dfrac{11\pi}{6}$

37–40 ■ Reference Numbers Find the reference number for each value of t.

37. **(a)** $t = \dfrac{4\pi}{3}$ **(b)** $t = \dfrac{5\pi}{3}$

(c) $t = -\dfrac{7\pi}{6}$ **(d)** $t = 3.5$

38. **(a)** $t = 9\pi$ **(b)** $t = -\dfrac{5\pi}{4}$

(c) $t = \dfrac{25\pi}{6}$ **(d)** $t = 4$

39. **(a)** $t = \dfrac{5\pi}{7}$ **(b)** $t = -\dfrac{7\pi}{9}$

(c) $t = -3$ **(d)** $t = 5$

40. **(a)** $t = \dfrac{11\pi}{5}$ **(b)** $t = -\dfrac{9\pi}{7}$

(c) $t = 6$ **(d)** $t = -7$

41–54 ■ Terminal Points and Reference Numbers Find **(a)** the reference number for each value of t and **(b)** the terminal point determined by t.

41. $t = \dfrac{11\pi}{6}$ **42.** $t = \dfrac{2\pi}{3}$

43. $t = -\dfrac{4\pi}{3}$ **44.** $t = \dfrac{5\pi}{3}$

45. $t = -\dfrac{2\pi}{3}$ **46.** $t = -\dfrac{7\pi}{6}$

47. $t = \dfrac{13\pi}{4}$ **48.** $t = \dfrac{13\pi}{6}$

49. $t = \dfrac{41\pi}{6}$ **50.** $t = \dfrac{17\pi}{4}$

51. $t = -\dfrac{11\pi}{3}$ **52.** $t = \dfrac{31\pi}{6}$

53. $t = \dfrac{16\pi}{3}$ **54.** $t = -\dfrac{41\pi}{4}$

55–58 ■ Terminal Points The unit circle is graphed in the figure below. Use the figure to find the terminal point determined by the real number t, with coordinates rounded to one decimal place.

55. $t = 1$

56. $t = 2.5$

57. $t = -1.1$

58. $t = 4.2$

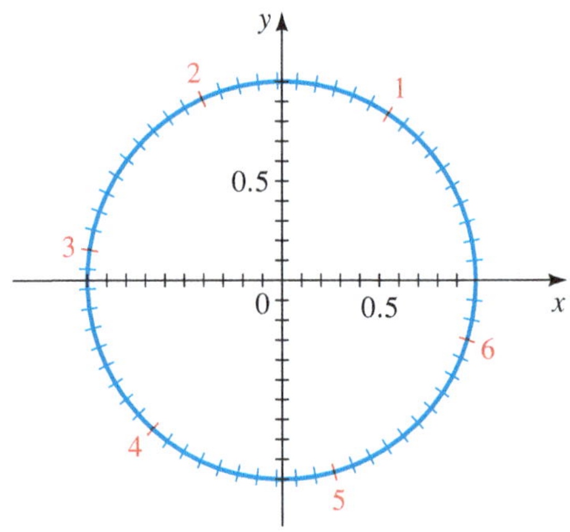

SKILLS Plus

59. Terminal Points Suppose that the terminal point determined by t is the point $\left(\frac{3}{5}, \frac{4}{5}\right)$ on the unit circle. Find the terminal point determined by each of the following.

(a) $\pi - t$ **(b)** $-t$

(c) $\pi + t$ **(d)** $2\pi + t$

60. Terminal Points Suppose that the terminal point determined by t is the point $\left(\frac{3}{4}, \sqrt{7}/4\right)$ on the unit circle. Find the terminal point determined by each of the following.

(a) $-t$ **(b)** $4\pi + t$

(c) $\pi - t$ **(d)** $t - \pi$

DISCUSS ■ DISCOVER ■ PROVE ■ WRITE

61. DISCOVER ■ PROVE: Finding the Terminal Point for $\pi/6$ Suppose the terminal point determined by $t = \pi/6$ is $P(x, y)$ and the points Q and R are as shown in the figure. Why are the distances PQ and PR the same? Use this fact, together with the Distance Formula, to show that the coordinates of P satisfy the equation $2y = \sqrt{x^2 + (y - 1)^2}$. Simplify this equation using the fact that $x^2 + y^2 = 1$. Solve the simplified equation to find $P(x, y)$.

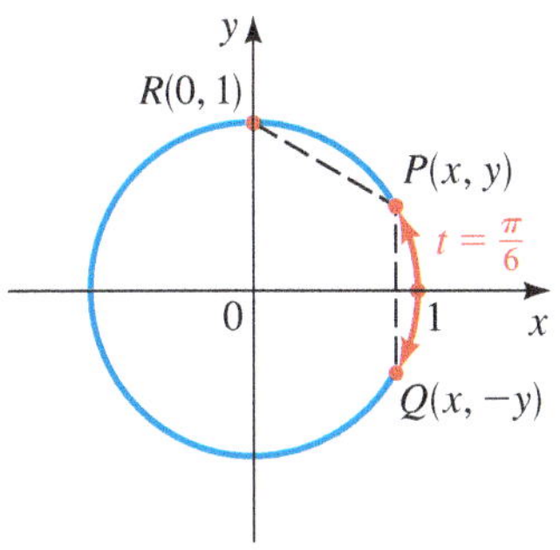

62. DISCOVER ■ PROVE: Finding the Terminal Point for $\pi/3$ Now that you know the terminal point determined by $t = \pi/6$, use symmetry to find the terminal point determined by $t = \pi/3$ (see the figure). Explain your reasoning.

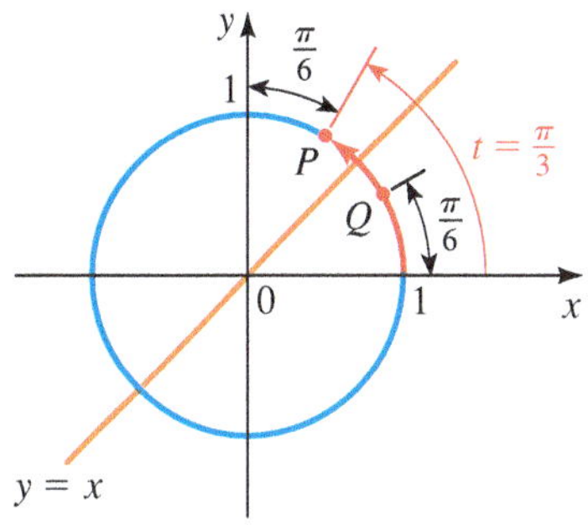

5.2 TRIGONOMETRIC FUNCTIONS OF REAL NUMBERS

■ The Trigonometric Functions ■ Values of the Trigonometric Functions
■ Fundamental Identities

A function is a rule that assigns to each real number another real number. In this section we use properties of the unit circle from the preceding section to define the trigonometric functions.

■ The Trigonometric Functions

Recall that to find the terminal point $P(x, y)$ for a given real number t, we move a distance $|t|$ along the unit circle, starting at the point $(1, 0)$. We move in a counterclockwise direction if t is positive and in a clockwise direction if t is negative (see Figure 1). We now use the x- and y-coordinates of the point $P(x, y)$ to define several functions. For instance, we define the function called *sine* by assigning to each real number t the y-coordinate of the terminal point $P(x, y)$ determined by t. The functions *cosine*, *tangent*, *cosecant*, *secant*, and *cotangent* are also defined by using the coordinates of $P(x, y)$.

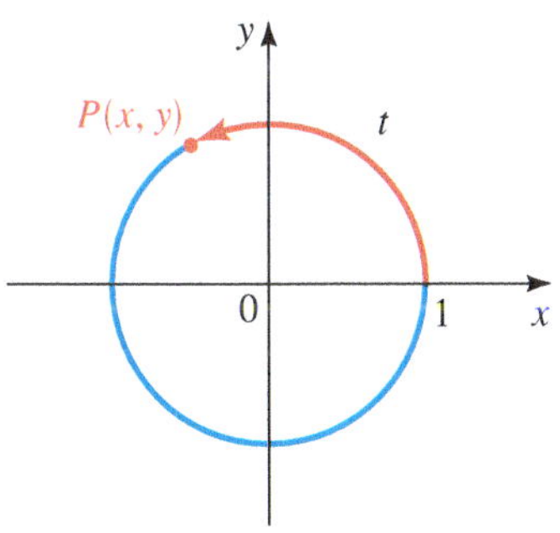

FIGURE 1

DEFINITION OF THE TRIGONOMETRIC FUNCTIONS

Let t be any real number and let $P(x, y)$ be the terminal point on the unit circle determined by t. We define

$$\sin t = y \qquad \cos t = x \qquad \tan t = \frac{y}{x} \quad (x \neq 0)$$

$$\csc t = \frac{1}{y} \quad (y \neq 0) \qquad \sec t = \frac{1}{x} \quad (x \neq 0) \qquad \cot t = \frac{x}{y} \quad (y \neq 0)$$

Because the trigonometric functions can be defined in terms of the unit circle, they are sometimes called the **circular functions**.

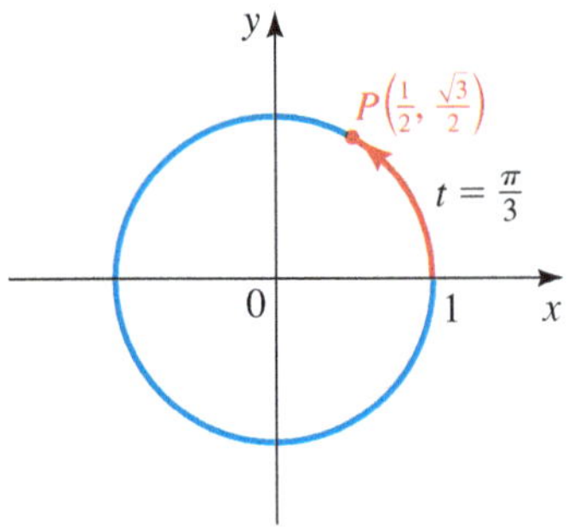

FIGURE 2

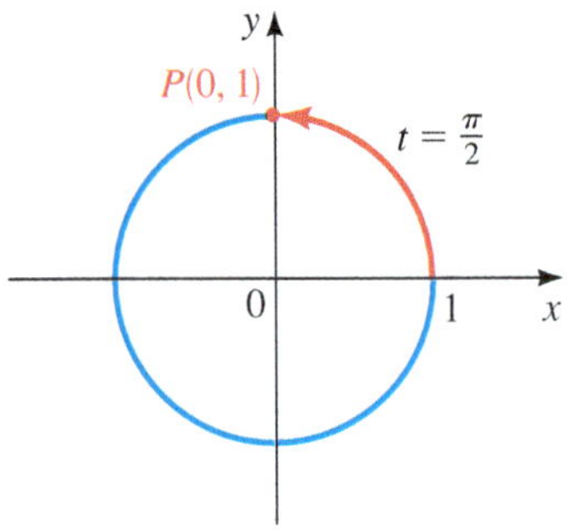

FIGURE 3

EXAMPLE 1 ■ Evaluating Trigonometric Functions

Find the six trigonometric functions of each given real number t.

(a) $t = \dfrac{\pi}{3}$ **(b)** $t = \dfrac{\pi}{2}$

SOLUTION

(a) From Table 1 on page 404, we see that the terminal point determined by $t = \pi/3$ is $P(\frac{1}{2}, \sqrt{3}/2)$. (See Figure 2.) Since the coordinates are $x = \frac{1}{2}$ and $y = \sqrt{3}/2$, we have

$$\sin\frac{\pi}{3} = \frac{\sqrt{3}}{2} \qquad \cos\frac{\pi}{3} = \frac{1}{2} \qquad \tan\frac{\pi}{3} = \frac{\sqrt{3}/2}{1/2} = \sqrt{3}$$

$$\csc\frac{\pi}{3} = \frac{2\sqrt{3}}{3} \qquad \sec\frac{\pi}{3} = 2 \qquad \cot\frac{\pi}{3} = \frac{1/2}{\sqrt{3}/2} = \frac{\sqrt{3}}{3}$$

(b) The terminal point determined by $\pi/2$ is $P(0, 1)$. (See Figure 3.) So

$$\sin\frac{\pi}{2} = 1 \qquad \cos\frac{\pi}{2} = 0 \qquad \csc\frac{\pi}{2} = \frac{1}{1} = 1 \qquad \cot\frac{\pi}{2} = \frac{0}{1} = 0$$

But $\tan \pi/2$ and $\sec \pi/2$ are undefined because $x = 0$ appears in the denominator in each of their definitions.

Now Try Exercise 3 ■

Some special values of the trigonometric functions are listed in the table below. This table is easily obtained from Table 1 of Section 5.1, together with the definitions of the trigonometric functions.

SPECIAL VALUES OF THE TRIGONOMETRIC FUNCTIONS

The following values of the trigonometric functions are obtained from the special terminal points.

TABLE 1

t	$\sin t$	$\cos t$	$\tan t$	$\csc t$	$\sec t$	$\cot t$
0	0	1	0	—	1	—
$\frac{\pi}{6}$	$\frac{1}{2}$	$\frac{\sqrt{3}}{2}$	$\frac{\sqrt{3}}{3}$	2	$\frac{2\sqrt{3}}{3}$	$\sqrt{3}$
$\frac{\pi}{4}$	$\frac{\sqrt{2}}{2}$	$\frac{\sqrt{2}}{2}$	1	$\sqrt{2}$	$\sqrt{2}$	1
$\frac{\pi}{3}$	$\frac{\sqrt{3}}{2}$	$\frac{1}{2}$	$\sqrt{3}$	$\frac{2\sqrt{3}}{3}$	2	$\frac{\sqrt{3}}{3}$
$\frac{\pi}{2}$	1	0	—	1	—	0

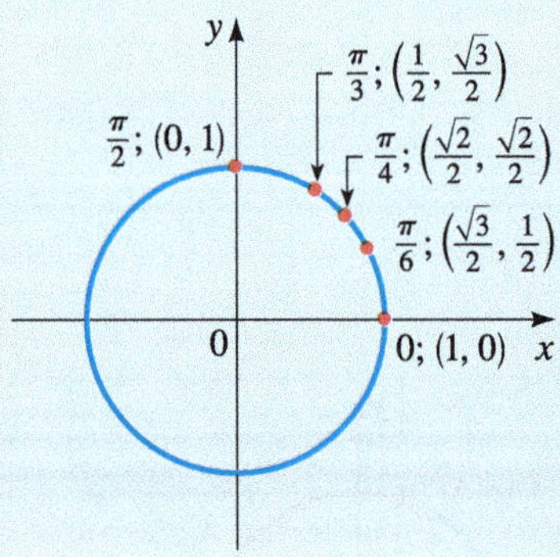

We can easily remember the sines and cosines of the basic angles by writing them in the form $\sqrt{\blacksquare}/2$:

t	$\sin t$	$\cos t$
0	$\sqrt{0}/2$	$\sqrt{4}/2$
$\pi/6$	$\sqrt{1}/2$	$\sqrt{3}/2$
$\pi/4$	$\sqrt{2}/2$	$\sqrt{2}/2$
$\pi/3$	$\sqrt{3}/2$	$\sqrt{1}/2$
$\pi/2$	$\sqrt{4}/2$	$\sqrt{0}/2$

Example 1 shows that some of the trigonometric functions fail to be defined for certain real numbers. So we need to determine their domains. The functions sine and cosine are defined for all values of t. Since the functions cotangent and cosecant have y in the denominator of their definitions, they are not defined whenever the y-coordinate of the terminal point $P(x, y)$ determined by t is 0. This happens when $t = n\pi$ for any integer n, so their domains do not include these points. The functions tangent and secant have x in the denominator in their definitions, so they are not defined whenever $x = 0$. This happens when $t = (\pi/2) + n\pi$ for any integer n.

(text continues on page 412)

Relationship to the Trigonometric Functions of Angles

If you have studied the trigonometry of right triangles in Chapter 6, you are probably wondering how the sine and cosine of an *angle* relate to those of this section. To see how, let's start with a right triangle, ΔOPQ.

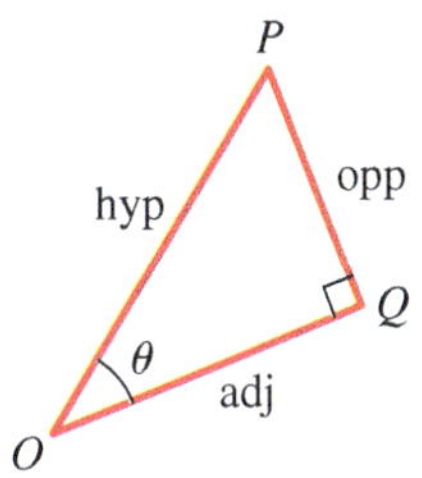

Right triangle OPQ

Place the triangle in the coordinate plane as shown, with angle θ in standard position.

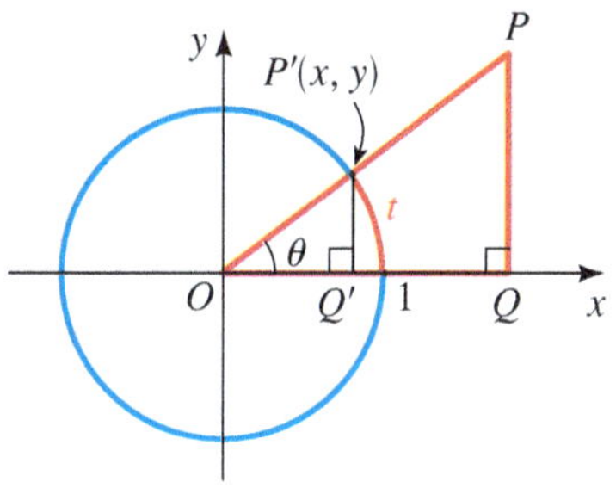

$P'(x, y)$ is the terminal point determined by t.

The point $P'(x, y)$ in the figure is the terminal point determined by t. Note that triangle OPQ is similar to the small triangle $OP'Q'$ whose legs have lengths x and y.

Now, by the definition of the trigonometric functions of the ***angle*** θ we have

$$\sin\theta = \frac{\text{opp}}{\text{hyp}} = \frac{PQ}{OP} = \frac{P'Q'}{OP'} = \frac{y}{1} = y$$

$$\cos\theta = \frac{\text{adj}}{\text{hyp}} = \frac{OQ}{OP} = \frac{OQ'}{OP'} = \frac{x}{1} = x$$

By the definition of the trigonometric functions of the ***real number t***, we have

$$\sin t = y \qquad \cos t = x$$

Now, if θ is measured in radians, then $\theta = t$ (see the figure). So the trigonometric functions of the angle with radian measure θ are exactly the same as the trigonometric functions defined in terms of the terminal point determined by the real number t.

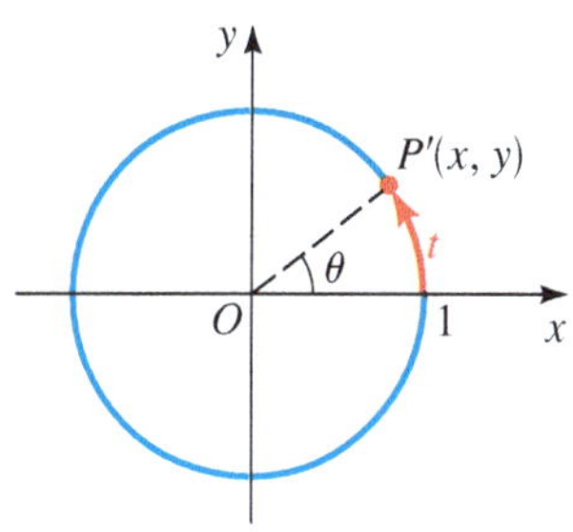

The radian measure of angle θ is t.

Why then study trigonometry in two different ways? Because different applications require that we view the trigonometric functions differently. (Compare Section 5.6 with Sections 6.2, 6.5, and 6.6.)

DOMAINS OF THE TRIGONOMETRIC FUNCTIONS

Function	Domain
sin, cos	All real numbers
tan, sec	All real numbers other than $\frac{\pi}{2} + n\pi$ for any integer n
cot, csc	All real numbers other than $n\pi$ for any integer, n

Values of the Trigonometric Functions

To compute values of the trigonometric functions for any real number t, we first determine their signs. The signs of the trigonometric functions depend on the quadrant in which the terminal point of t lies. For example, if the terminal point $P(x, y)$ determined by t lies in Quadrant III, then its coordinates are both negative. So sin t, cos t, csc t, and sec t are all negative, whereas tan t and cot t are positive. You can check the other entries in the following box.

The following mnemonic device will help you remember which trigonometric functions are positive in each quadrant: **A**ll of them, **S**ine, **T**angent, or **C**osine.

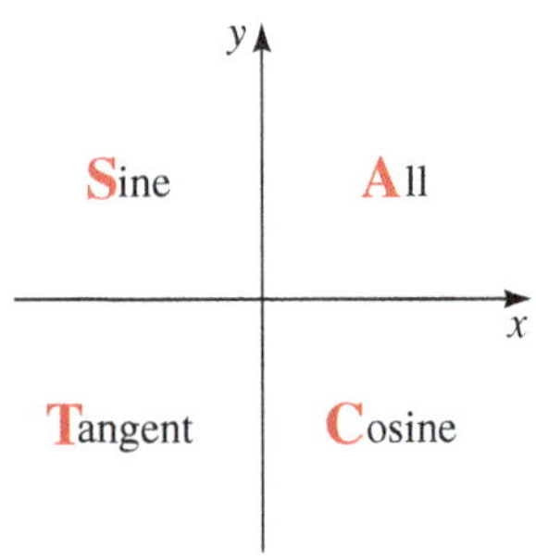

You can remember this as "All Students Take Calculus."

SIGNS OF THE TRIGONOMETRIC FUNCTIONS

Quadrant	Positive Functions	Negative Functions
I	all	none
II	sin, csc	cos, sec, tan, cot
III	tan, cot	sin, csc, cos, sec
IV	cos, sec	sin, csc, tan, cot

For example $\cos(2\pi/3) < 0$ because the terminal point of $t = 2\pi/3$ is in Quadrant II, whereas $\tan 4 > 0$ because the terminal point of $t = 4$ is in Quadrant III.

In Section 5.1 we used the reference number to find the terminal point determined by a real number t. Since the trigonometric functions are defined in terms of the coordinates of terminal points, we can use the reference number to find values of the trigonometric functions. Suppose that $\bar{t}$ is the reference number for t. Then the terminal point of $\bar{t}$ has the same coordinates, except possibly for sign, as the terminal point of t. So the value of each trigonometric function at t is the same, except possibly for sign, as its value at $\bar{t}$. We illustrate this procedure in the next example.

EVALUATING TRIGONOMETRIC FUNCTIONS FOR ANY REAL NUMBER

To find the values of the trigonometric functions for any real number t, we carry out the following steps.

1. **Find the reference number.** Find the reference number $\bar{t}$ associated with t.
2. **Find the sign.** Determine the sign of the trigonometric function of t by noting the quadrant in which the terminal point lies.
3. **Find the value.** The value of the trigonometric function of t is the same, except possibly for sign, as the value of the trigonometric function of $\bar{t}$.

EXAMPLE 2 ■ Evaluating Trigonometric Functions

Find each value.

(a) $\cos \dfrac{2\pi}{3}$ **(b)** $\tan\left(-\dfrac{\pi}{3}\right)$ **(c)** $\sin \dfrac{19\pi}{4}$

SOLUTION

(a) The reference number for $2\pi/3$ is $\pi/3$ (see Figure 4(a)). Since the terminal point of $2\pi/3$ is in Quadrant II, $\cos(2\pi/3)$ is negative. Thus

$$\cos \frac{2\pi}{3} = -\cos \frac{\pi}{3} = -\frac{1}{2}$$

Sign | Reference number | From Table 1 (page 410)

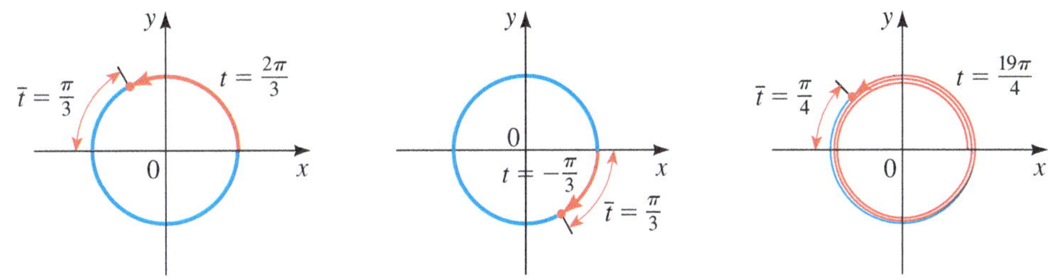

FIGURE 4 (a) (b) (c)

(b) The reference number for $-\pi/3$ is $\pi/3$ (see Figure 4(b)). Since the terminal point of $-\pi/3$ is in Quadrant IV, $\tan(-\pi/3)$ is negative. Thus

$$\tan\left(-\frac{\pi}{3}\right) = -\tan \frac{\pi}{3} = -\sqrt{3}$$

Sign | Reference number | From Table 1 (page 410)

(c) Since $(19\pi/4) - 4\pi = 3\pi/4$, the terminal points determined by $19\pi/4$ and $3\pi/4$ are the same. The reference number for $3\pi/4$ is $\pi/4$ (see Figure 4(c)). Since the terminal point of $3\pi/4$ is in Quadrant II, $\sin(3\pi/4)$ is positive. Thus

$$\sin \frac{19\pi}{4} = \sin \frac{3\pi}{4} = +\sin \frac{\pi}{4} = \frac{\sqrt{2}}{2}$$

Subtract 4π | Sign | Reference number | From Table 1 (page 410)

Now Try Exercise 5 ■

So far, we have been able to compute the values of the trigonometric functions only for certain values of t. In fact, we can compute the values of the trigonometric functions whenever t is a multiple of $\pi/6$, $\pi/4$, $\pi/3$, and $\pi/2$. How can we compute the trigonometric functions for other values of t? For example, how can we find $\sin 1.5$? One way is to carefully sketch a diagram and read the value (see Exercises 37–44); however, this method is not very accurate. Fortunately, programmed directly into scientific calculators are mathematical procedures (see the margin note on page 433) that find the values of *sine*, *cosine*, and *tangent* correct to the number of digits in the display. The calculator must be put in *radian mode* to evaluate these functions. To find values of cosecant, secant, and cotangent using a calculator, we need to use the following *reciprocal relations*:

$$\csc t = \frac{1}{\sin t} \qquad \sec t = \frac{1}{\cos t} \qquad \cot t = \frac{1}{\tan t}$$

These identities follow from the definitions of the trigonometric functions. For instance, since $\sin t = y$ and $\csc t = 1/y$, we have $\csc t = 1/y = 1/(\sin t)$. The others follow similarly.

EXAMPLE 3 ■ Using a Calculator to Evaluate Trigonometric Functions

Using a calculator, find the following.

(a) $\sin 2.2$ **(b)** $\cos 1.1$ **(c)** $\cot 28$ **(d)** $\csc 0.98$

SOLUTION Making sure our calculator is set to radian mode and rounding the results to six decimal places, we get

(a) $\sin 2.2 \approx 0.808496$

(b) $\cos 1.1 \approx 0.453596$

(c) $\cot 28 = \dfrac{1}{\tan 28} \approx -3.553286$

(d) $\csc 0.98 = \dfrac{1}{\sin 0.98} \approx 1.204098$

Now Try Exercises 39 and 41 ■

Let's consider the relationship between the trigonometric functions of t and those of $-t$. From Figure 5 we see that

$$\sin(-t) = -y = -\sin t$$

$$\cos(-t) = x = \cos t$$

$$\tan(-t) = \frac{-y}{x} = -\frac{y}{x} = -\tan t$$

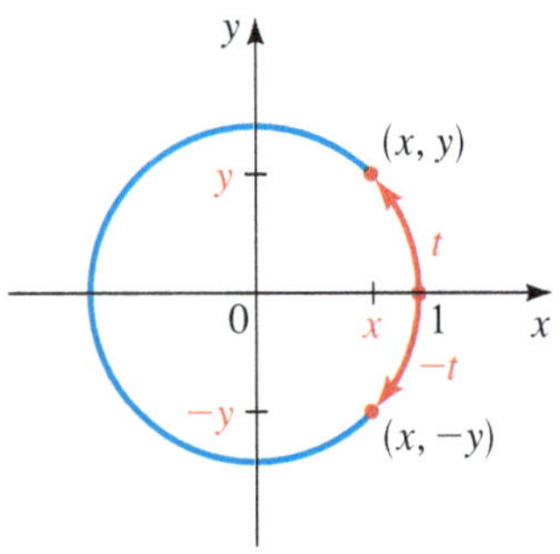

FIGURE 5

These equations show that sine and tangent are odd functions, whereas cosine is an even function. It's easy to see that the reciprocal of an even function is even and the reciprocal of an odd function is odd. This fact, together with the reciprocal relations, completes our knowledge of the even-odd properties for all the trigonometric functions.

Even and odd functions are defined in Section 2.6.

EVEN-ODD PROPERTIES

Sine, cosecant, tangent, and cotangent are odd functions; cosine and secant are even functions.

$$\sin(-t) = -\sin t \qquad \cos(-t) = \cos t \qquad \tan(-t) = -\tan t$$

$$\csc(-t) = -\csc t \qquad \sec(-t) = \sec t \qquad \cot(-t) = -\cot t$$

EXAMPLE 4 ■ Even and Odd Trigonometric Functions

Use the even-odd properties of the trigonometric functions to determine each value.

(a) $\sin\left(-\dfrac{\pi}{6}\right)$ **(b)** $\cos\left(-\dfrac{\pi}{4}\right)$

SOLUTION By the even-odd properties and Table 1 on page 410, we have

(a) $\sin\left(-\frac{\pi}{6}\right) = -\sin\frac{\pi}{6} = -\frac{1}{2}$ Sine is odd

(b) $\cos\left(-\frac{\pi}{4}\right) = \cos\frac{\pi}{4} = \frac{\sqrt{2}}{2}$ Cosine is even

Now Try Exercise 13

Fundamental Identities

The trigonometric functions are related to each other through equations called **trigonometric identities**. We give the most important ones in the following box.*

FUNDAMENTAL IDENTITIES

Reciprocal Identities

$$\csc t = \frac{1}{\sin t} \qquad \sec t = \frac{1}{\cos t} \qquad \cot t = \frac{1}{\tan t} \qquad \tan t = \frac{\sin t}{\cos t} \qquad \cot t = \frac{\cos t}{\sin t}$$

Pythagorean Identities

$$\sin^2 t + \cos^2 t = 1 \qquad \tan^2 t + 1 = \sec^2 t \qquad 1 + \cot^2 t = \csc^2 t$$

Proof The reciprocal identities follow immediately from the definitions on page 409. We now prove the Pythagorean identities. By definition $\cos t = x$ and $\sin t = y$, where x and y are the coordinates of a point $P(x, y)$ on the unit circle. Since $P(x, y)$ is on the unit circle, we have $x^2 + y^2 = 1$. Thus

$$\sin^2 t + \cos^2 t = 1$$

Dividing both sides by $\cos^2 t$ (provided that $\cos t \neq 0$), we get

$$\frac{\sin^2 t}{\cos^2 t} + \frac{\cos^2 t}{\cos^2 t} = \frac{1}{\cos^2 t}$$

$$\left(\frac{\sin t}{\cos t}\right)^2 + 1 = \left(\frac{1}{\cos t}\right)^2$$

$$\tan^2 t + 1 = \sec^2 t$$

We have used the reciprocal identities $\sin t/\cos t = \tan t$ and $1/\cos t = \sec t$. Similarly, dividing both sides of the first Pythagorean identity by $\sin^2 t$ (provided that $\sin t \neq 0$) gives us $1 + \cot^2 t = \csc^2 t$. ■

As their name indicates, the fundamental identities play a central role in trigonometry because we can use them to relate any trigonometric function to any other. So if we know the value of any one of the trigonometric functions at t, then we can find the values of all the others at t.

EXAMPLE 5 ■ Finding All Trigonometric Functions from the Value of One

If $\cos t = \frac{3}{5}$ and t is in Quadrant IV, find the values of all the trigonometric functions at t.

---

*We follow the usual convention of writing $\sin^2 t$ for $(\sin t)^2$. In general, we write $\sin^n t$ for $(\sin t)^n$ for all integers n except $n = -1$. The superscript $n = -1$ will be assigned another meaning in Section 5.5. Of course, the same convention applies to the other five trigonometric functions.

The Value of π

The number π is the ratio of the circumference of a circle to its diameter. It has been known since ancient times that this ratio is the same for all circles. The first systematic effort to find a numerical approximation for π was made by Archimedes (ca. 240 B.C.), who proved that $\frac{22}{7} < \pi < \frac{223}{71}$ by finding the perimeters of regular polygons inscribed in and circumscribed about a circle.

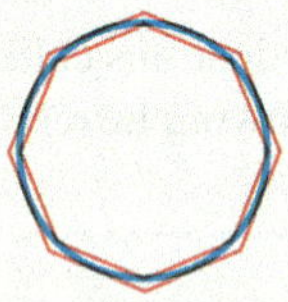

In about A.D. 480, the Chinese physicist Tsu Ch'ung-chih gave the approximation

$$\pi \approx \frac{355}{113} = 3.141592\ldots$$

which is correct to six decimals. This remained the most accurate estimation of π until the Dutch mathematician Adrianus Romanus (1593) used polygons with more than a billion sides to compute π correct to 15 decimals. In the 17th century, mathematicians began to use infinite series and trigonometric identities in the quest for π. The Englishman William Shanks spent 15 years (1858–1873) using these methods to compute π to 707 decimals, but in 1946 it was found that his figures were wrong beginning with the 528th decimal. Today, with the aid of computers, mathematicians routinely determine π correct to millions of decimals. In fact, mathematicians have recently developed new algorithms that can be programmed into computers to calculate π to many trillions of decimal places.

SOLUTION From the Pythagorean identities we have

$$\sin^2 t + \cos^2 t = 1$$

$$\sin^2 t + \left(\tfrac{3}{5}\right)^2 = 1 \qquad \text{Substitute } \cos t = \tfrac{3}{5}$$

$$\sin^2 t = 1 - \tfrac{9}{25} = \tfrac{16}{25} \qquad \text{Solve for } \sin^2 t$$

$$\sin t = \pm\tfrac{4}{5} \qquad \text{Take square roots}$$

Since this point is in Quadrant IV, $\sin t$ is negative, so $\sin t = -\frac{4}{5}$. Now that we know both $\sin t$ and $\cos t$, we can find the values of the other trigonometric functions using the reciprocal identities.

$$\sin t = -\frac{4}{5} \qquad \cos t = \frac{3}{5} \qquad \tan t = \frac{\sin t}{\cos t} = \frac{-\frac{4}{5}}{\frac{3}{5}} = -\frac{4}{3}$$

$$\csc t = \frac{1}{\sin t} = -\frac{5}{4} \qquad \sec t = \frac{1}{\cos t} = \frac{5}{3} \qquad \cot t = \frac{1}{\tan t} = -\frac{3}{4}$$

Now Try Exercise 63 ■

EXAMPLE 6 ■ Writing One Trigonometric Function in Terms of Another

Write $\tan t$ in terms of $\cos t$, where t is in Quadrant III.

SOLUTION Since $\tan t = \sin t/\cos t$, we need to write $\sin t$ in terms of $\cos t$. By the Pythagorean identities we have

$$\sin^2 t + \cos^2 t = 1$$

$$\sin^2 t = 1 - \cos^2 t \qquad \text{Solve for } \sin^2 t$$

$$\sin t = \pm\sqrt{1 - \cos^2 t} \qquad \text{Take square roots}$$

Since $\sin t$ is negative in Quadrant III, the negative sign applies here. Thus

$$\tan t = \frac{\sin t}{\cos t} = \frac{-\sqrt{1 - \cos^2 t}}{\cos t}$$

Now Try Exercise 53 ■

5.2 EXERCISES

CONCEPTS

1. Let $P(x, y)$ be the terminal point on the unit circle determined by t. Then $\sin t =$ ________, $\cos t =$ ________, and $\tan t =$ ________.

2. If $P(x, y)$ is on the unit circle, then $x^2 + y^2 =$ ________. So for all t we have $\sin^2 t + \cos^2 t =$ ________.

SKILLS

3–4 ■ Evaluating Trigonometric Functions Find $\sin t$ and $\cos t$ for the values of t whose terminal points are shown on the unit circle in the figure. In Exercise 3, t increases in increments of $\pi/4$; in Exercise 4, t increases in increments of $\pi/6$. (See Exercises 21 and 22 in Section 5.1.)

3.

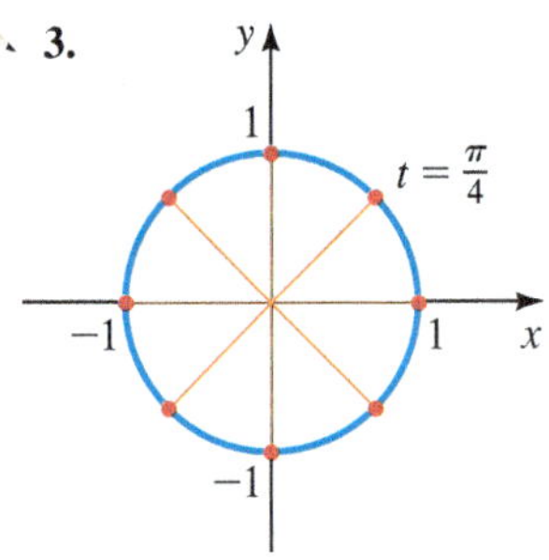

4.

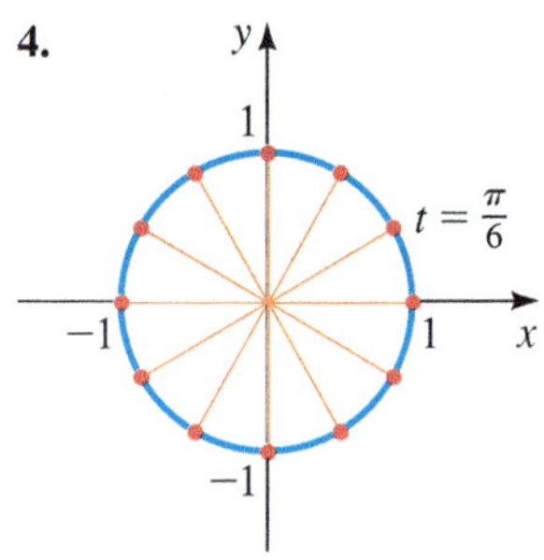

5–22 ■ Evaluating Trigonometric Functions Find the exact value of the trigonometric function at the given real number.

5. **(a)** $\sin \dfrac{7\pi}{6}$ **(b)** $\cos \dfrac{17\pi}{6}$ **(c)** $\tan \dfrac{7\pi}{6}$

6. **(a)** $\sin \dfrac{5\pi}{3}$ **(b)** $\cos \dfrac{11\pi}{3}$ **(c)** $\tan \dfrac{5\pi}{3}$

7. **(a)** $\sin \dfrac{11\pi}{4}$ **(b)** $\sin\left(-\dfrac{\pi}{4}\right)$ **(c)** $\sin \dfrac{5\pi}{4}$

8. **(a)** $\cos \dfrac{19\pi}{6}$ **(b)** $\cos\left(-\dfrac{7\pi}{6}\right)$ **(c)** $\cos\left(-\dfrac{\pi}{6}\right)$

9. **(a)** $\cos \dfrac{3\pi}{4}$ **(b)** $\cos \dfrac{5\pi}{4}$ **(c)** $\cos \dfrac{7\pi}{4}$

10. **(a)** $\sin \dfrac{3\pi}{4}$ **(b)** $\sin \dfrac{5\pi}{4}$ **(c)** $\sin \dfrac{7\pi}{4}$

11. **(a)** $\sin \dfrac{7\pi}{3}$ **(b)** $\csc \dfrac{7\pi}{3}$ **(c)** $\cot \dfrac{7\pi}{3}$

12. **(a)** $\csc \dfrac{5\pi}{4}$ **(b)** $\sec \dfrac{5\pi}{4}$ **(c)** $\tan \dfrac{5\pi}{4}$

13. **(a)** $\cos\left(-\dfrac{\pi}{3}\right)$ **(b)** $\sec\left(-\dfrac{\pi}{3}\right)$ **(c)** $\sin\left(-\dfrac{\pi}{3}\right)$

14. **(a)** $\tan\left(-\dfrac{\pi}{4}\right)$ **(b)** $\csc\left(-\dfrac{\pi}{4}\right)$ **(c)** $\cot\left(-\dfrac{\pi}{4}\right)$

15. **(a)** $\cos\left(-\dfrac{\pi}{6}\right)$ **(b)** $\csc\left(-\dfrac{\pi}{3}\right)$ **(c)** $\tan\left(-\dfrac{\pi}{6}\right)$

16. **(a)** $\sin\left(-\dfrac{\pi}{4}\right)$ **(b)** $\sec\left(-\dfrac{\pi}{4}\right)$ **(c)** $\cot\left(-\dfrac{\pi}{6}\right)$

17. **(a)** $\csc \dfrac{7\pi}{6}$ **(b)** $\sec\left(-\dfrac{\pi}{6}\right)$ **(c)** $\cot\left(-\dfrac{5\pi}{6}\right)$

18. **(a)** $\sec \dfrac{3\pi}{4}$ **(b)** $\cos\left(-\dfrac{2\pi}{3}\right)$ **(c)** $\tan\left(-\dfrac{7\pi}{6}\right)$

19. **(a)** $\sin \dfrac{4\pi}{3}$ **(b)** $\sec \dfrac{11\pi}{6}$ **(c)** $\cot\left(-\dfrac{\pi}{3}\right)$

20. **(a)** $\csc \dfrac{2\pi}{3}$ **(b)** $\sec\left(-\dfrac{5\pi}{3}\right)$ **(c)** $\cos\left(\dfrac{10\pi}{3}\right)$

21. **(a)** $\sin 13\pi$ **(b)** $\cos 14\pi$ **(c)** $\tan 15\pi$

22. **(a)** $\sin \dfrac{25\pi}{2}$ **(b)** $\cos \dfrac{25\pi}{2}$ **(c)** $\cot \dfrac{25\pi}{2}$

23–26 ■ Evaluating Trigonometric Functions Find the value of each of the six trigonometric functions (if it is defined) at the given real number t. Use your answers to complete the table.

23. $t = 0$ **24.** $t = \dfrac{\pi}{2}$

25. $t = \pi$ **26.** $t = \dfrac{3\pi}{2}$

t	$\sin t$	$\cos t$	$\tan t$	$\csc t$	$\sec t$	$\cot t$
0	0	1		undefined		
$\frac{\pi}{2}$						
π			0			undefined
$\frac{3\pi}{2}$						

27–36 ■ Evaluating Trigonometric Functions The terminal point $P(x, y)$ determined by a real number t is given. Find $\sin t$, $\cos t$, and $\tan t$.

27. $\left(-\dfrac{3}{5}, -\dfrac{4}{5}\right)$ **28.** $\left(-\dfrac{1}{2}, \dfrac{\sqrt{3}}{2}\right)$

29. $\left(-\dfrac{1}{3}, \dfrac{2\sqrt{2}}{3}\right)$ **30.** $\left(\dfrac{1}{5}, -\dfrac{2\sqrt{6}}{5}\right)$

31. $\left(-\dfrac{6}{7}, \dfrac{\sqrt{13}}{7}\right)$ **32.** $\left(\dfrac{40}{41}, \dfrac{9}{41}\right)$

33. $\left(-\dfrac{5}{13}, -\dfrac{12}{13}\right)$ **34.** $\left(\dfrac{\sqrt{5}}{5}, \dfrac{2\sqrt{5}}{5}\right)$

35. $\left(-\dfrac{20}{29}, \dfrac{21}{29}\right)$ **36.** $\left(\dfrac{24}{25}, -\dfrac{7}{25}\right)$

37–44 ■ Values of Trigonometric Functions Find an approximate value of the given trigonometric function by using **(a)** the figure and **(b)** a calculator. Compare the two values.

37. $\sin 1$

38. $\cos 0.8$

39. $\sin 1.2$

40. $\cos 5$

41. $\tan 0.8$

42. $\tan(-1.3)$

43. $\cos 4.1$

44. $\sin(-5.2)$

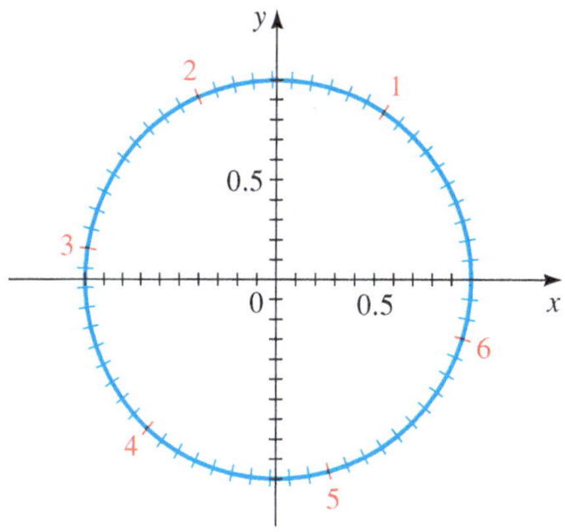

45–48 ■ Sign of a Trigonometric Expression Find the sign of the expression if the terminal point determined by t is in the given quadrant.

45. $\sin t \cos t$, Quadrant II **46.** $\tan t \sec t$, Quadrant IV

47. $\dfrac{\tan t \sin t}{\cot t}$, Quadrant III **48.** $\cos t \sec t$, any quadrant

49–52 ■ Quadrant of a Terminal Point From the information given, find the quadrant in which the terminal point determined by t lies.

49. $\sin t > 0$ and $\cos t < 0$

50. $\tan t > 0$ and $\sin t < 0$

51. $\csc t > 0$ and $\sec t < 0$

52. $\cos t < 0$ and $\cot t < 0$

53–62 ■ Writing One Trigonometric Expression in Terms of Another Write the first expression in terms of the second if the terminal point determined by t is in the given quadrant.

53. $\sin t, \cos t$; Quadrant II
54. $\cos t, \sin t$; Quadrant IV
55. $\tan t, \sin t$; Quadrant IV
56. $\tan t, \cos t$; Quadrant III
57. $\sec t, \tan t$; Quadrant II
58. $\csc t, \cot t$; Quadrant III
59. $\tan t, \sec t$; Quadrant III
60. $\sin t, \sec t$; Quadrant IV
61. $\tan^2 t, \sin t$; any quadrant
62. $\sec^2 t \sin^2 t, \cos t$; any quadrant

63–70 ■ Using the Pythagorean Identities Find the values of the trigonometric functions of t from the given information.

63. $\sin t = -\frac{4}{5}$, terminal point of t is in Quadrant IV
64. $\cos t = -\frac{7}{25}$, terminal point of t is in Quadrant III
65. $\sec t = 3$, terminal point of t is in Quadrant IV
66. $\tan t = \frac{1}{4}$, terminal point of t is in Quadrant III
67. $\tan t = -\frac{12}{5}$, $\sin t > 0$
68. $\csc t = 5$, $\cos t < 0$
69. $\sin t = -\frac{1}{4}$, $\sec t < 0$
70. $\tan t = -4$, $\csc t > 0$

SKILLS Plus

71–78 ■ Even and Odd Functions Determine whether the function is even, odd, or neither. (See page 204 for the definitions of even and odd functions.)

71. $f(x) = x^2 \sin x$
72. $f(x) = x^2 \cos 2x$
73. $f(x) = \sin x \cos x$
74. $f(x) = \sin x + \cos x$
75. $f(x) = |x| \cos x$
76. $f(x) = x \sin^3 x$
77. $f(x) = x^3 + \cos x$
78. $f(x) = \cos(\sin x)$

APPLICATIONS

79. Harmonic Motion The displacement from equilibrium of an oscillating mass attached to a spring is given by $y(t) = 4 \cos 3\pi t$ where y is measured in inches and t in seconds. Find the displacement at the times indicated in the table.

t	$y(t)$
0	
0.25	
0.50	
0.75	
1.00	
1.25	

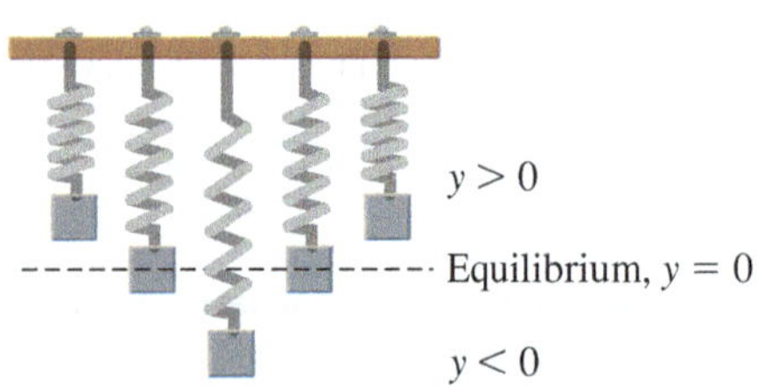

80. Circadian Rhythms Everybody's blood pressure varies over the course of the day. In a certain individual the resting diastolic blood pressure at time t is given by $B(t) = 80 + 7 \sin(\pi t/12)$, where t is measured in hours since midnight and $B(t)$ in mmHg (millimeters of mercury). Find this person's resting diastolic blood pressure at
(a) 6:00 A.M. **(b)** 10:30 A.M. **(c)** Noon **(d)** 8:00 P.M.

81. Electric Circuit After the switch is closed in the circuit shown, the current t seconds later is $I(t) = 0.8e^{-3t} \sin 10t$. Find the current at the times **(a)** $t = 0.1$ s and **(b)** $t = 0.5$ s.

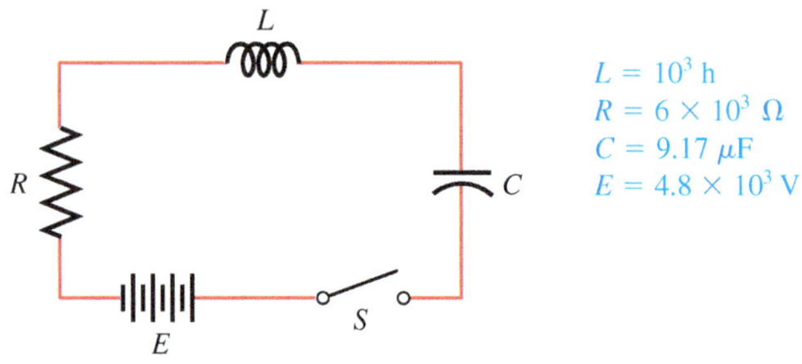

82. Bungee Jumping A bungee jumper plummets from a high bridge to the river below and then bounces back over and over again. At time t seconds after her jump, her height H (in meters) above the river is given by $H(t) = 100 + 75e^{-t/20}\cos(\frac{\pi}{4}t)$. Find her height at the times indicated in the table.

t	$H(t)$
0	
1	
2	
4	
6	
8	
12	

DISCUSS ■ DISCOVER ■ PROVE ■ WRITE

83. DISCOVER ■ PROVE: Reduction Formulas A *reduction formula* is one that can be used to "reduce" the number of terms in the input for a trigonometric function. Explain how the figure shows that the following reduction formulas are valid:

$$\sin(t + \pi) = -\sin t \qquad \cos(t + \pi) = -\cos t$$

$$\tan(t + \pi) = \tan t$$

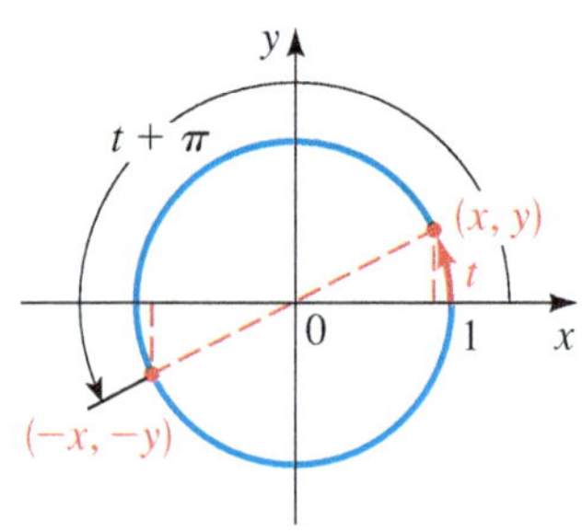

84. DISCOVER ■ PROVE: More Reduction Formulas By the Angle-Side-Angle Theorem from elementary geometry, triangles CDO and AOB in the figure to the right are congruent. Explain how this proves that if B has coordinates (x, y), then D has coordinates $(-y, x)$. Then explain how the figure shows that the following reduction formulas are valid:

$$\sin\left(t + \frac{\pi}{2}\right) = \cos t \qquad \cos\left(t + \frac{\pi}{2}\right) = -\sin t$$

$$\tan\left(t + \frac{\pi}{2}\right) = -\cot t$$

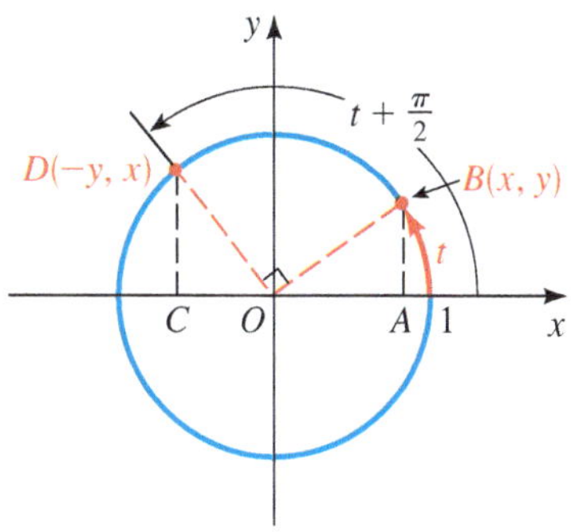

5.3 TRIGONOMETRIC GRAPHS

■ Graphs of Sine and Cosine ■ Graphs of Transformations of Sine and Cosine
■ Using Graphing Devices to Graph Trigonometric Functions

The graph of a function gives us a better idea of its behavior. So in this section we graph the sine and cosine functions and certain transformations of these functions. The other trigonometric functions are graphed in the next section.

■ Graphs of Sine and Cosine

To help us graph the sine and cosine functions, we first observe that these functions repeat their values in a regular fashion. To see exactly how this happens, recall that the circumference of the unit circle is 2π. It follows that the terminal point $P(x, y)$ determined by the real number t is the same as that determined by $t + 2\pi$. Since the sine and cosine functions are defined in terms of the coordinates of $P(x, y)$, it follows that their values are unchanged by the addition of any integer multiple of 2π. In other words,

$$\sin(t + 2n\pi) = \sin t \qquad \text{for any integer } n$$

$$\cos(t + 2n\pi) = \cos t \qquad \text{for any integer } n$$

Thus the sine and cosine functions are *periodic* according to the following definition: A function f is **periodic** if there is a positive number p such that $f(t + p) = f(t)$ for every t. The least such positive number (if it exists) is the **period** of f. If f has period p, then the graph of f on any interval of length p is called **one complete period** of f.

PERIODIC PROPERTIES OF SINE AND COSINE

The functions sine and cosine have period 2π:

$$\sin(t + 2\pi) = \sin t \qquad \cos(t + 2\pi) = \cos t$$

TABLE 1

t	$\sin t$	$\cos t$
$0 \to \frac{\pi}{2}$	$0 \to 1$	$1 \to 0$
$\frac{\pi}{2} \to \pi$	$1 \to 0$	$0 \to -1$
$\pi \to \frac{3\pi}{2}$	$0 \to -1$	$-1 \to 0$
$\frac{3\pi}{2} \to 2\pi$	$-1 \to 0$	$0 \to 1$

So the sine and cosine functions repeat their values in any interval of length 2π. To sketch their graphs, we first graph one period. To sketch the graphs on the interval $0 \le t \le 2\pi$, we could try to make a table of values and use those points to draw the graph. Since no such table can be complete, let's look more closely at the definitions of these functions.

Recall that $\sin t$ is the y-coordinate of the terminal point $P(x, y)$ on the unit circle determined by the real number t. How does the y-coordinate of this point vary as t increases? It's easy to see that the y-coordinate of $P(x, y)$ increases to 1, then decreases to -1 repeatedly as the point $P(x, y)$ travels around the unit circle. (See Figure 1.) In fact, as t increases from 0 to $\pi/2$, $y = \sin t$ increases from 0 to 1. As t increases from $\pi/2$ to π, the value of $y = \sin t$ decreases from 1 to 0. Table 1 shows the variation of the sine and cosine functions for t between 0 and 2π.

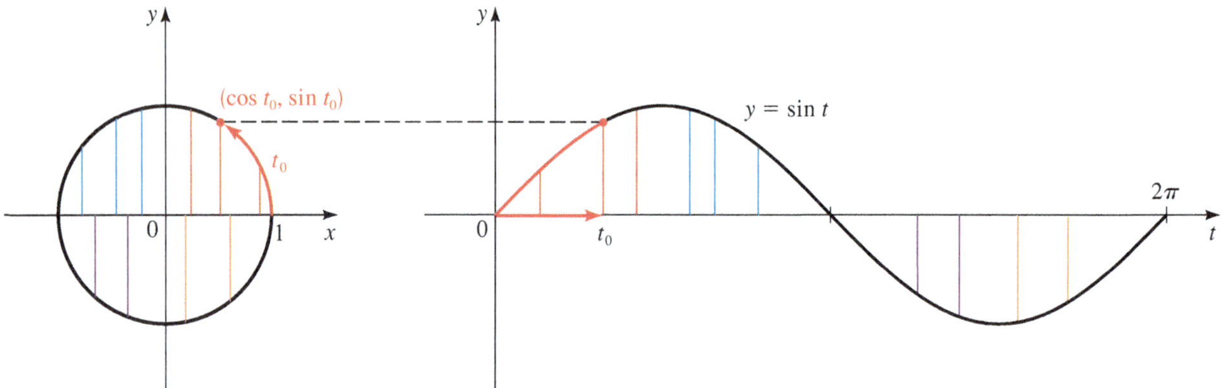

FIGURE 1

To draw the graphs more accurately, we find a few other values of $\sin t$ and $\cos t$ in Table 2. We could find still other values with the aid of a calculator.

TABLE 2

t	0	$\frac{\pi}{6}$	$\frac{\pi}{3}$	$\frac{\pi}{2}$	$\frac{2\pi}{3}$	$\frac{5\pi}{6}$	π	$\frac{7\pi}{6}$	$\frac{4\pi}{3}$	$\frac{3\pi}{2}$	$\frac{5\pi}{3}$	$\frac{11\pi}{6}$	2π
$\sin t$	0	$\frac{1}{2}$	$\frac{\sqrt{3}}{2}$	1	$\frac{\sqrt{3}}{2}$	$\frac{1}{2}$	0	$-\frac{1}{2}$	$-\frac{\sqrt{3}}{2}$	-1	$-\frac{\sqrt{3}}{2}$	$-\frac{1}{2}$	0
$\cos t$	1	$\frac{\sqrt{3}}{2}$	$\frac{1}{2}$	0	$-\frac{1}{2}$	$-\frac{\sqrt{3}}{2}$	-1	$-\frac{\sqrt{3}}{2}$	$-\frac{1}{2}$	0	$\frac{1}{2}$	$\frac{\sqrt{3}}{2}$	1

Now we use this information to graph the functions $\sin t$ and $\cos t$ for t between 0 and 2π in Figures 2 and 3. These are the graphs of one period. Using the fact that these functions are periodic with period 2π, we get their complete graphs by continuing the same pattern to the left and to the right in every successive interval of length 2π.

The graph of the sine function is symmetric with respect to the origin. This is as expected, since sine is an odd function. Since the cosine function is an even function, its graph is symmetric with respect to the y-axis.

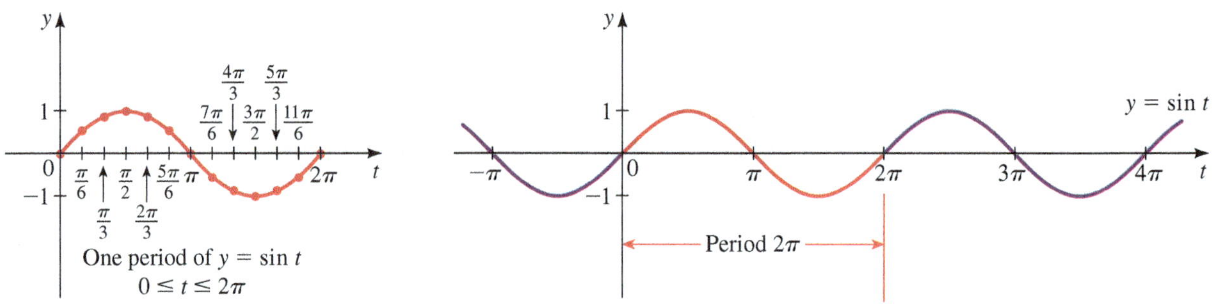

FIGURE 2 Graph of $\sin t$

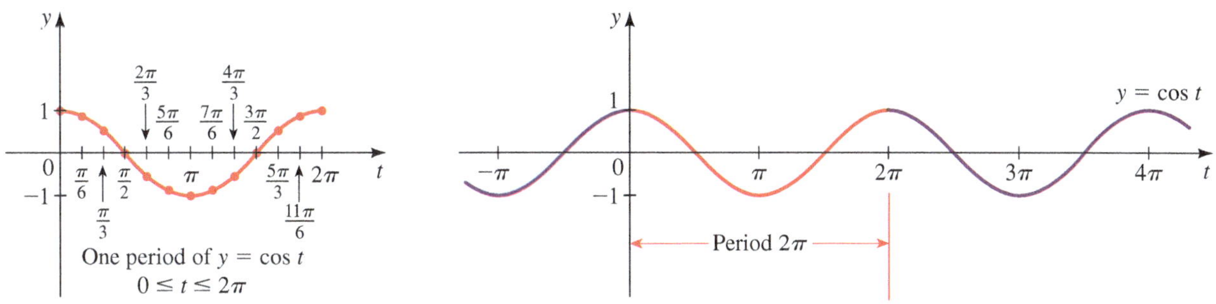

FIGURE 3 Graph of $\cos t$

■ Graphs of Transformations of Sine and Cosine

We now consider graphs of functions that are transformations of the sine and cosine functions. Thus, the graphing techniques of Section 2.6 are very useful here. The graphs we obtain are important for understanding applications to physical situations such as harmonic motion (see Section 5.6), but some of them are beautiful graphs that are interesting in their own right.

It's traditional to use the letter x to denote the variable in the domain of a function. So from here on we use the letter x and write $y = \sin x$, $y = \cos x$, $y = \tan x$, and so on to denote these functions.

EXAMPLE 1 ■ Cosine Curves

Sketch the graph of each function.

(a) $f(x) = 2 + \cos x$ **(b)** $g(x) = -\cos x$

SOLUTION

(a) The graph of $y = 2 + \cos x$ is the same as the graph of $y = \cos x$, but shifted up 2 units (see Figure 4(a)).

(b) The graph of $y = -\cos x$ in Figure 4(b) is the reflection of the graph of $y = \cos x$ in the x-axis.

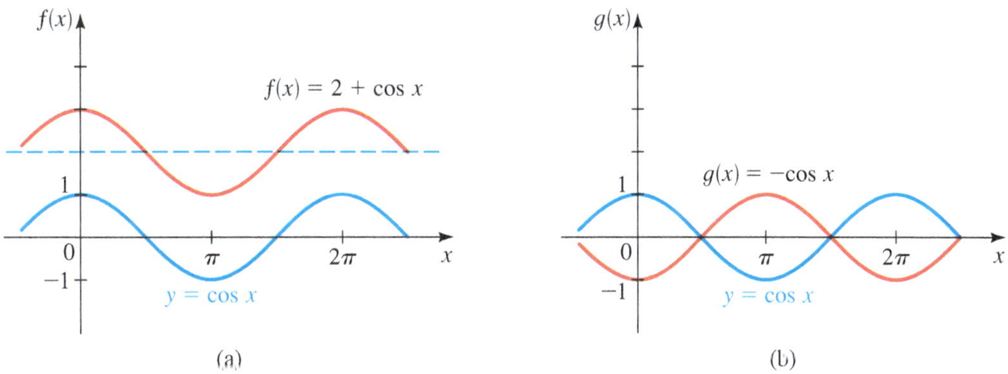

FIGURE 4

Now Try Exercises 5 and 7 ■

Let's graph $y = 2 \sin x$. We start with the graph of $y = \sin x$ and multiply the y-coordinate of each point by 2. This has the effect of stretching the graph vertically by a factor of 2. To graph $y = \frac{1}{2} \sin x$, we start with the graph of $y = \sin x$ and multiply

Vertical stretching and shrinking of graphs is discussed in Section 2.6.

the y-coordinate of each point by $\frac{1}{2}$. This has the effect of shrinking the graph vertically by a factor of $\frac{1}{2}$ (see Figure 5).

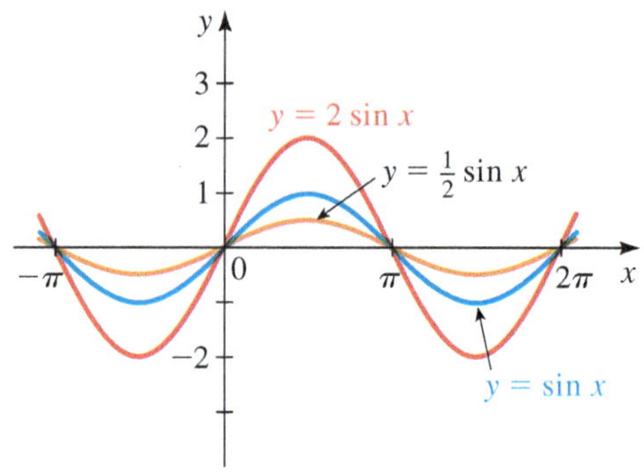

FIGURE 5

In general, for the functions

$$y = a \sin x \qquad \text{and} \qquad y = a \cos x$$

the number $|a|$ is called the **amplitude** and is the largest value these functions attain. Graphs of $y = a \sin x$ for several values of a are shown in Figure 6.

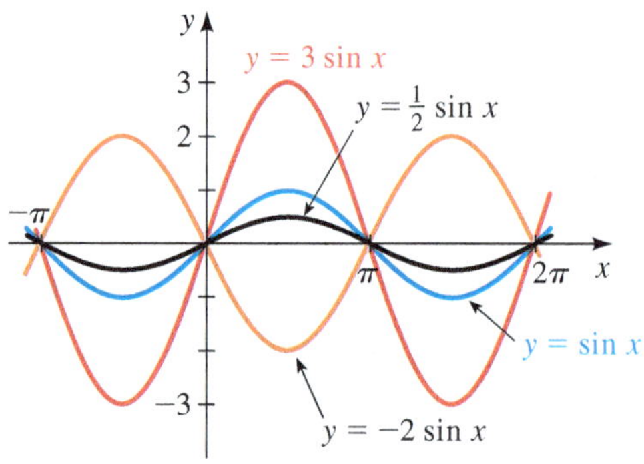

FIGURE 6

EXAMPLE 2 ■ Stretching a Cosine Curve

Find the amplitude of $y = -3 \cos x$, and sketch its graph.

SOLUTION The amplitude is $|-3| = 3$, so the largest value the graph attains is 3 and the smallest value is -3. To sketch the graph, we begin with the graph of $y = \cos x$, stretch the graph vertically by a factor of 3, and reflect in the x-axis, arriving at the graph in Figure 7.

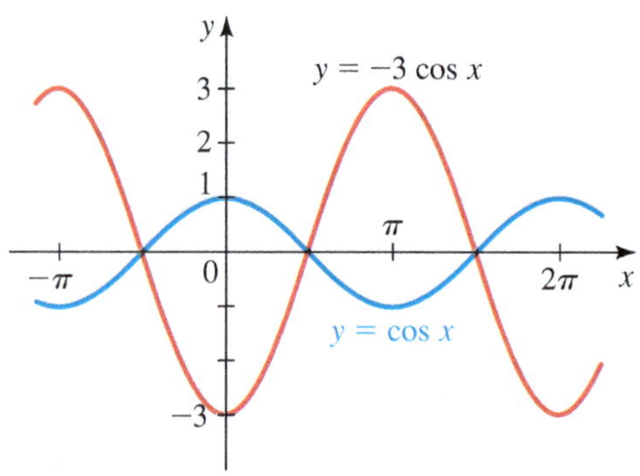

FIGURE 7

Now Try Exercise 11 ■

Since the sine and cosine functions have period 2π, the functions

$$y = a \sin kx \qquad \text{and} \qquad y = a \cos kx \qquad (k > 0)$$

complete one period as kx varies from 0 to 2π, that is, for $0 \le kx \le 2\pi$ or for $0 \le x \le 2\pi/k$. So these functions complete one period as x varies between 0 and $2\pi/k$ and thus have period $2\pi/k$. The graphs of these functions are called **sine curves** and **cosine curves**, respectively. (Collectively, sine and cosine curves are often referred to as **sinusoidal** curves.)

SINE AND COSINE CURVES

The sine and cosine curves

$$y = a \sin kx \qquad \text{and} \qquad y = a \cos kx \qquad (k > 0)$$

have **amplitude** $|a|$ and **period** $2\pi/k$.

An appropriate interval on which to graph one complete period is $[0, 2\pi/k]$.

Horizontal stretching and shrinking of graphs is discussed in Section 2.6.

To see how the value of k affects the graph of $y = \sin kx$, let's graph the sine curve $y = \sin 2x$. Since the period is $2\pi/2 = \pi$, the graph completes one period in the interval $0 \le x \le \pi$ (see Figure 8(a)). For the sine curve $y = \sin \frac{1}{2}x$ the period is $2\pi \div \frac{1}{2} = 4\pi$, so the graph completes one period in the interval $0 \le x \le 4\pi$ (see Figure 8(b)). We see that the effect is to *shrink* the graph horizontally if $k > 1$ or to *stretch* the graph horizontally if $k < 1$.

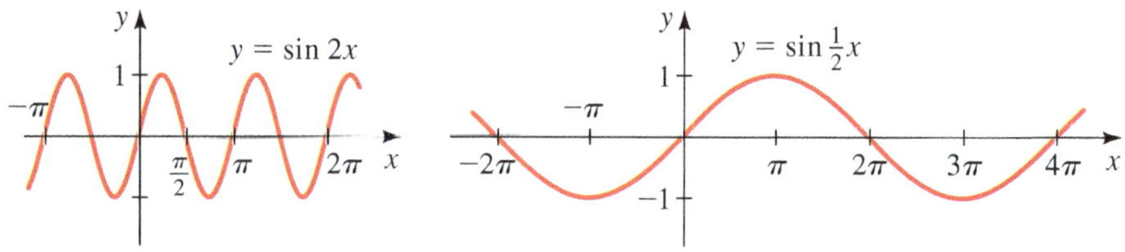

FIGURE 8

(a) (b)

For comparison, in Figure 9 we show the graphs of one period of the sine curve $y = a \sin kx$ for several values of k.

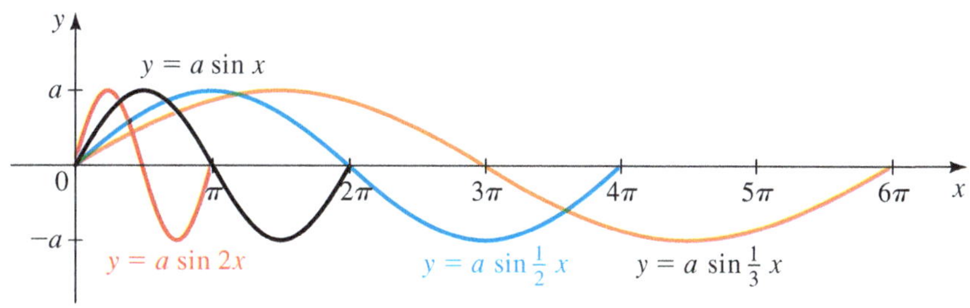

FIGURE 9

EXAMPLE 3 ■ Amplitude and Period

Find the amplitude and period of each function, and sketch its graph.

(a) $y = 4 \cos 3x$ **(b)** $y = -2 \sin \frac{1}{2}x$

SOLUTION

(a) We get the amplitude and period from the form of the function as follows.

$$\text{amplitude} = |a| = 4$$

$$y = 4 \cos 3x$$

$$\text{period} = \frac{2\pi}{k} = \frac{2\pi}{3}$$

The amplitude is 4, and the period is $2\pi/3$. The graph is shown in Figure 10.

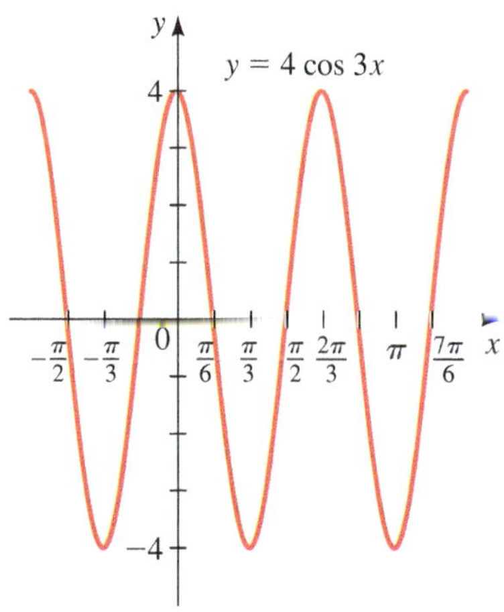

FIGURE 10

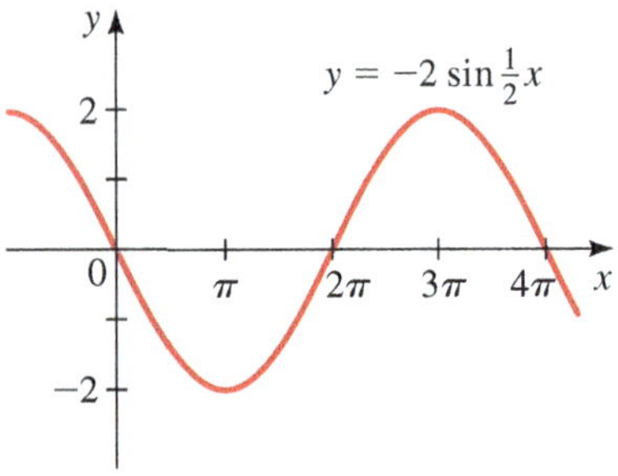

FIGURE 11

(b) For $y = -2\sin\frac{1}{2}x$,

$$\text{amplitude} = |a| = |-2| = 2$$

$$\text{period} = \frac{2\pi}{\frac{1}{2}} = 4\pi$$

The graph is shown in Figure 11.

Now Try Exercises 23 and 25 ■

The graphs of functions of the form $y = a\sin k(x - b)$ and $y = a\cos k(x - b)$ are simply sine and cosine curves shifted horizontally by an amount $|b|$. They are shifted to the right if $b > 0$ or to the left if $b < 0$. We summarize the properties of these functions in the following box.

The phase shift of a sine curve is discussed in Section 5.6.

SHIFTED SINE AND COSINE CURVES

The sine and cosine curves

$$y = a\sin k(x - b) \qquad \text{and} \qquad y = a\cos k(x - b) \qquad (k > 0)$$

have **amplitude** $|a|$, **period** $2\pi/k$, and **horizontal shift** b.

An appropriate interval on which to graph one complete period is $[b, b + (2\pi/k)]$.

The graphs of $y = \sin\left(x - \frac{\pi}{3}\right)$ and $y = \sin\left(x + \frac{\pi}{6}\right)$ are shown in Figure 12.

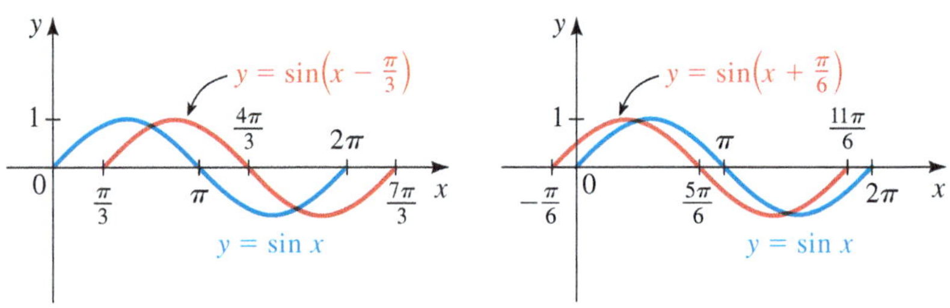

FIGURE 12 Horizontal shifts of a sine curve

EXAMPLE 4 ■ A Horizontally Shifted Sine Curve

Find the amplitude, period, and horizontal shift of $y = 3\sin 2\left(x - \frac{\pi}{4}\right)$, and graph one complete period.

SOLUTION We get the amplitude, period, and horizontal shift from the form of the function as follows:

$$\text{amplitude} = |a| = 3 \qquad \text{period} = \frac{2\pi}{k} = \frac{2\pi}{2} = \pi$$

$$y = 3\sin 2\left(x - \frac{\pi}{4}\right)$$

$$\text{horizontal shift} = \frac{\pi}{4}\ (\text{to the right})$$

Since the horizontal shift is $\pi/4$ and the period is π, one complete period occurs on the interval

$$\left[\frac{\pi}{4}, \frac{\pi}{4} + \pi\right] = \left[\frac{\pi}{4}, \frac{5\pi}{4}\right]$$

Here is another way to find an appropriate interval on which to graph one complete period. Since the period of $y = \sin x$ is 2π, the function $y = 3\sin 2(x - \frac{\pi}{4})$ will go through one complete period as $2(x - \frac{\pi}{4})$ varies from 0 to 2π.

Start of period:	End of period:
$2(x - \frac{\pi}{4}) = 0$	$2(x - \frac{\pi}{4}) = 2\pi$
$x - \frac{\pi}{4} = 0$	$x - \frac{\pi}{4} = \pi$
$x = \frac{\pi}{4}$	$x = \frac{5\pi}{4}$

So we graph one period on the interval $[\frac{\pi}{4}, \frac{5\pi}{4}]$.

As an aid in sketching the graph, we divide this interval into four equal parts, then graph a sine curve with amplitude 3 as in Figure 13.

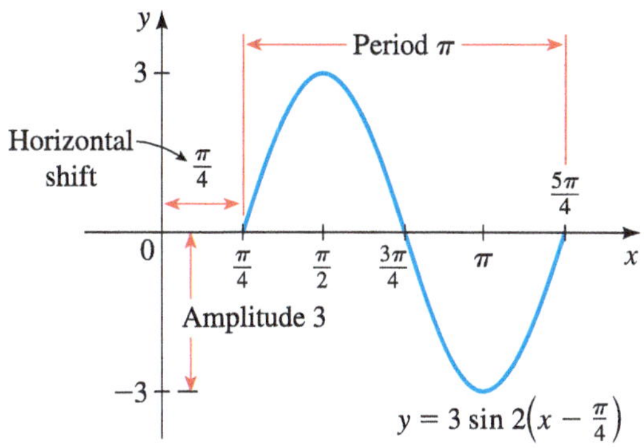

FIGURE 13

Now Try Exercise 35 ■

EXAMPLE 5 ■ A Horizontally Shifted Cosine Curve

Find the amplitude, period, and horizontal shift of $y = \frac{3}{4}\cos\left(2x + \frac{2\pi}{3}\right)$, and graph one complete period.

SOLUTION We first write this function in the form $y = a\cos k(x - b)$. To do this, we factor 2 from the expression $2x + \dfrac{2\pi}{3}$ to get

$$y = \frac{3}{4}\cos 2\left[x - \left(-\frac{\pi}{3}\right)\right]$$

Thus we have

$$\text{amplitude} = |a| = \frac{3}{4}$$

$$\text{period} = \frac{2\pi}{k} = \frac{2\pi}{2} = \pi$$

$$\text{horizontal shift} = b = -\frac{\pi}{3} \qquad \text{Shift } \frac{\pi}{3} \text{ to the } \textit{left}$$

From this information it follows that one period of this cosine curve begins at $-\pi/3$ and ends at $(-\pi/3) + \pi = 2\pi/3$. To sketch the graph over the interval $[-\pi/3, 2\pi/3]$, we divide this interval into four equal parts and graph a cosine curve with amplitude $\frac{3}{4}$ as shown in Figure 14.

We can also find one complete period as follows:

Start of period:	End of period:
$2x + \frac{2\pi}{3} = 0$	$2x + \frac{2\pi}{3} = 2\pi$
$2x = -\frac{2\pi}{3}$	$2x = \frac{4\pi}{3}$
$x = -\frac{\pi}{3}$	$x = \frac{2\pi}{3}$

So we graph one period on the interval $[-\frac{\pi}{3}, \frac{2\pi}{3}]$.

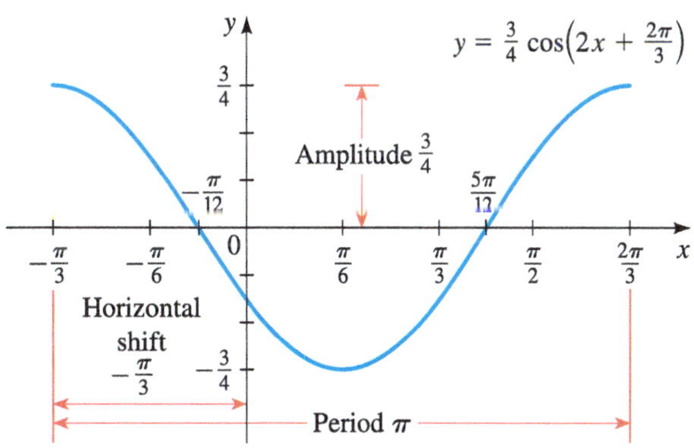

FIGURE 14

Now Try Exercise 37 ■

Using Graphing Devices to Graph Trigonometric Functions

See Appendix C, *Graphing with a Graphing Calculator,* for guidelines on choosing an appropriate viewing rectangle. Go to **www.stewartmath.com**.

When using a graphing calculator or a computer to graph a function, it is important to choose the viewing rectangle carefully in order to produce a reasonable graph of the function. This is especially true for trigonometric functions; the next example shows that, if care is not taken, it's easy to produce a very misleading graph of a trigonometric function.

EXAMPLE 6 ■ Choosing the Viewing Rectangle

Graph the function $f(x) = \sin 50x$ in an appropriate viewing rectangle.

SOLUTION Figure 15(a) shows the graph of f produced by a graphing calculator using the viewing rectangle $[-12, 12]$ by $[-1.5, 1.5]$. At first glance the graph appears to be reasonable. But if we change the viewing rectangle to the ones shown in Figure 15, the graphs look very different. Something strange is happening.

The appearance of the graphs in Figure 15 depends on the machine used. The graphs you get with your own graphing device might not look like these figures, but they will also be quite inaccurate.

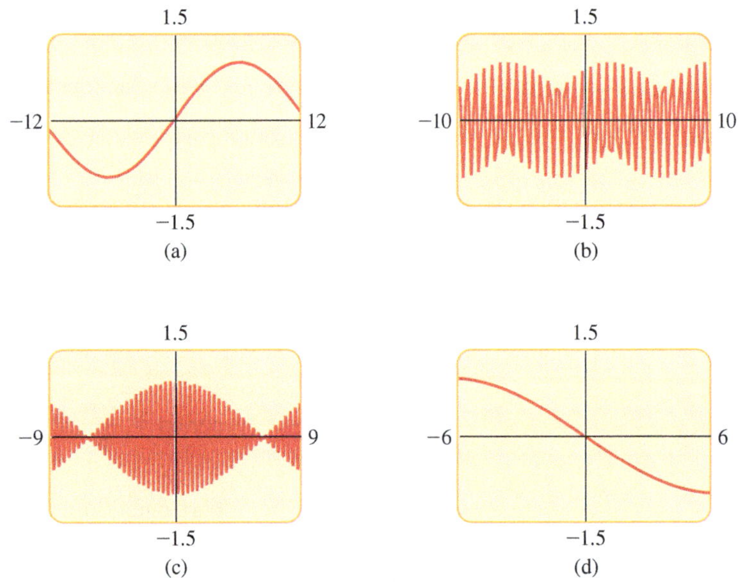

FIGURE 15 Graphs of $f(x) = \sin 50x$ in different viewing rectangles

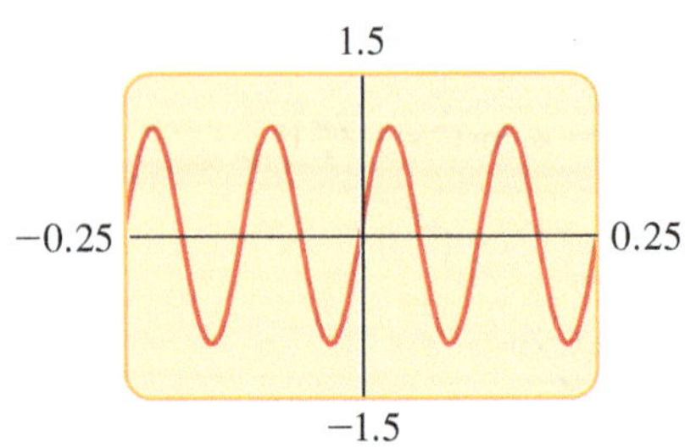

FIGURE 16 $f(x) = \sin 50x$

To explain the big differences in appearance of these graphs and to find an appropriate viewing rectangle, we need to find the period of the function $y = \sin 50x$.

$$\text{period} = \frac{2\pi}{50} = \frac{\pi}{25} \approx 0.126$$

This suggests that we should deal only with small values of x in order to show just a few oscillations of the graph. If we choose the viewing rectangle $[-0.25, 0.25]$ by $[-1.5, 1.5]$, we get the graph shown in Figure 16.

Now we see what went wrong in Figure 15. The oscillations of $y = \sin 50x$ are so rapid that when the calculator plots points and joins them, it misses most of the maximum and minimum points and therefore gives a very misleading impression of the graph.

Now Try Exercise 55 ■

The function h in Example 7 is **periodic** with period 2π. In general, functions that are sums of functions from the following list are periodic:

$$1, \cos kx, \cos 2kx, \cos 3kx, \ldots$$

$$\sin kx, \sin 2kx, \sin 3kx, \ldots$$

Although these functions appear to be special, they are actually fundamental to describing all periodic functions that arise in practice. The French mathematician J. B. J. Fourier (see page 546) discovered that nearly every periodic function can be written as a sum (usually an infinite sum) of these functions. This is remarkable because it means that any situation in which periodic variation occurs can be described mathematically using the functions sine and cosine. A modern application of Fourier's discovery is the digital encoding of sound on compact discs.

EXAMPLE 7 ■ A Sum of Sine and Cosine Curves

Graph $f(x) = 2\cos x$, $g(x) = \sin 2x$, and $h(x) = 2\cos x + \sin 2x$ on a common screen to illustrate the method of graphical addition.

SOLUTION Notice that $h = f + g$, so its graph is obtained by adding the corresponding y-coordinates of the graphs of f and g. The graphs of f, g, and h are shown in Figure 17.

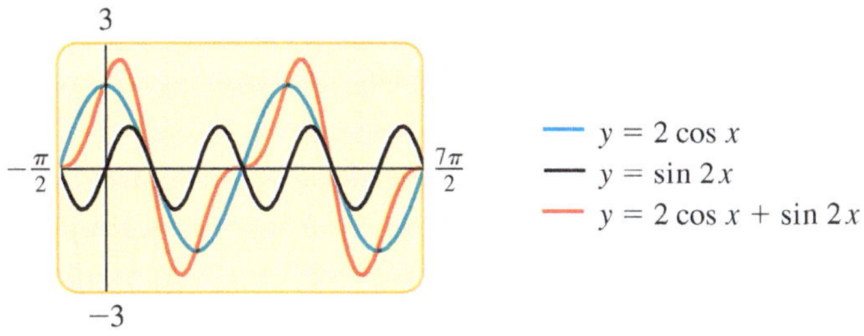

FIGURE 17

Now Try Exercise 63 ■

EXAMPLE 8 ■ A Cosine Curve with Variable Amplitude

Graph the functions $y = x^2$, $y = -x^2$, and $y = x^2 \cos 6\pi x$ on a common screen. Comment on and explain the relationship among the graphs.

SOLUTION Figure 18 shows all three graphs in the viewing rectangle $[-1.5, 1.5]$ by $[-2, 2]$. It appears that the graph of $y = x^2 \cos 6\pi x$ lies between the graphs of the functions $y = x^2$ and $y = -x^2$.

To understand this, recall that the values of $\cos 6\pi x$ lie between -1 and 1, that is,

$$-1 \le \cos 6\pi x \le 1$$

for all values of x. Multiplying the inequalities by x^2 and noting that $x^2 \ge 0$, we get

$$-x^2 \le x^2 \cos 6\pi x \le x^2$$

This explains why the functions $y = x^2$ and $y = -x^2$ form a boundary for the graph of $y = x^2 \cos 6\pi x$. (Note that the graphs touch when $\cos 6\pi x = \pm 1$.)

Now Try Exercise 69 ■

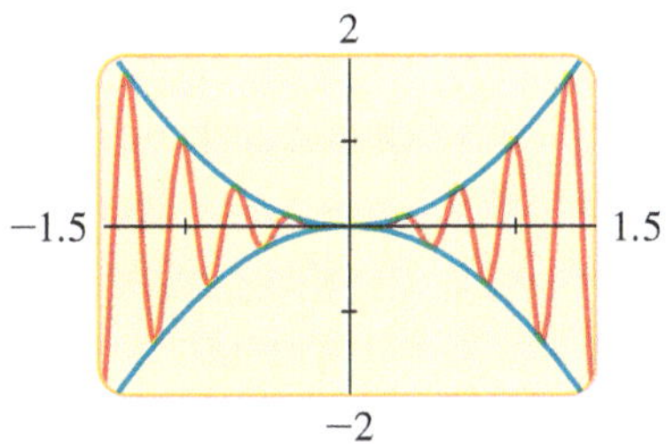

FIGURE 18 $y = x^2 \cos 6\pi x$

Example 8 shows that the function $y = x^2$ controls the amplitude of the graph of $y = x^2 \cos 6\pi x$. In general, if $f(x) = a(x) \sin kx$ or $f(x) = a(x) \cos kx$, the function a determines how the amplitude of f varies, and the graph of f lies between the graphs of $y = -a(x)$ and $y = a(x)$. Here is another example.

Jeffrey Lepore/Science Source

DISCOVERY PROJECT

Predator/Prey Models

Many animal populations fluctuate regularly in size and so can be modeled by trigonometric functions. Predicting population changes allows scientists to detect anomalies and take steps to protect a species. In this project we study the population of a predator species and the population of its prey. If the prey is abundant, the predator population grows, but too many predators tend to deplete the prey. This results in a decrease in the predator population, then the prey population increases, and so on. You can find the project at **www.stewartmath.com**.

AM and FM Radio

Radio transmissions consist of sound waves superimposed on a harmonic electromagnetic wave form called the **carrier signal**.

Sound wave

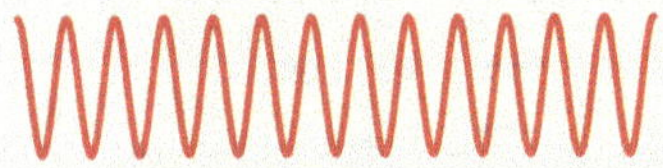

Carrier signal

There are two types of radio transmission, called **amplitude modulation (AM)** and **frequency modulation (FM)**. In AM broadcasting, the sound wave changes, or **modulates**, the amplitude of the carrier, but the frequency remains unchanged.

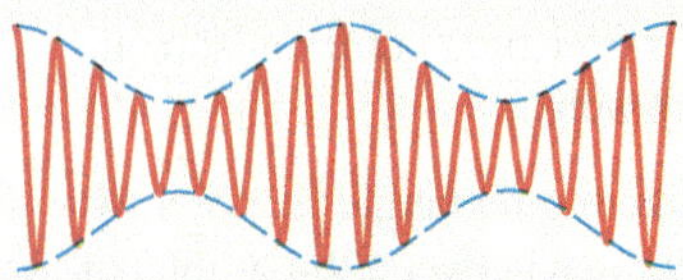

AM signal

In FM broadcasting, the sound wave modulates the frequency, but the amplitude remains the same.

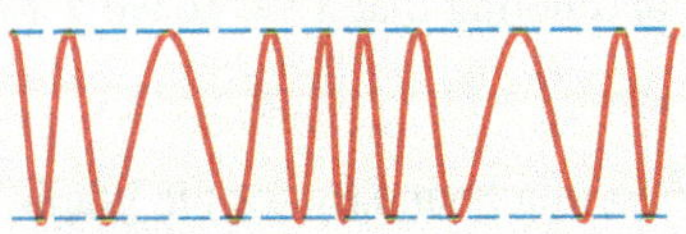

FM signal

EXAMPLE 9 ■ A Cosine Curve with Variable Amplitude

Graph the function $f(x) = \cos 2\pi x \cos 16\pi x$.

SOLUTION The graph is shown in Figure 19. Although it was drawn by a computer, we could have drawn it by hand, by first sketching the boundary curves $y = \cos 2\pi x$ and $y = -\cos 2\pi x$. The graph of f is a cosine curve that lies between the graphs of these two functions.

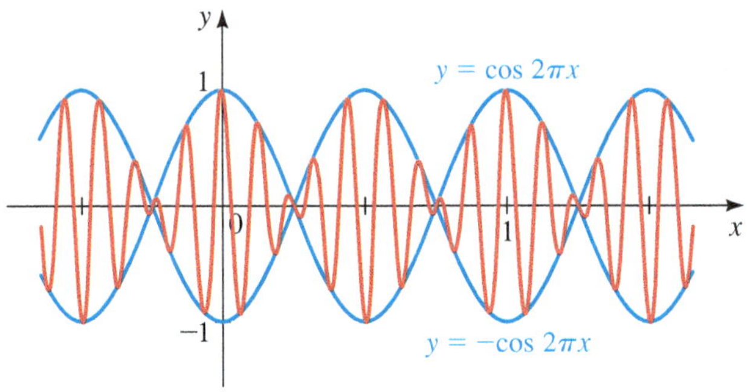

FIGURE 19 $f(x) = \cos 2\pi x \cos 16\pi x$

Now Try Exercise 71 ■

EXAMPLE 10 ■ A Sine Curve with Decaying Amplitude

The function $f(x) = \dfrac{\sin x}{x}$ is important in calculus. Graph this function, and comment on its behavior when x is close to 0.

SOLUTION The viewing rectangle $[-15, 15]$ by $[-0.5, 1.5]$ shown in Figure 20(a) gives a good global view of the graph of f. The viewing rectangle $[-1, 1]$ by $[-0.5, 1.5]$ in Figure 20(b) focuses on the behavior of f when $x \approx 0$. Notice that although $f(x)$ is not defined when $x = 0$ (in other words, 0 is not in the domain of f), the values of f seem to approach 1 when x gets close to 0. This fact is crucial in calculus.

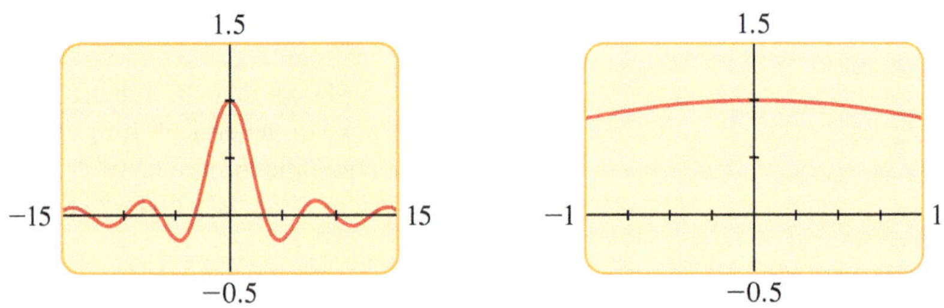

(a) (b)

FIGURE 20 $f(x) = \dfrac{\sin x}{x}$

Now Try Exercise 81 ■

The function in Example 10 can be written as

$$f(x) = \frac{1}{x} \sin x$$

and may thus be viewed as a sine function whose amplitude is controlled by the function $a(x) = 1/x$.

5.3 EXERCISES

CONCEPTS

1. If a function f is periodic with period p, then $f(t+p) =$ ________ for every t. The trigonometric functions $y = \sin x$ and $y = \cos x$ are periodic, with period ________ and amplitude ________. Sketch a graph of each function on the interval $[0, 2\pi]$.

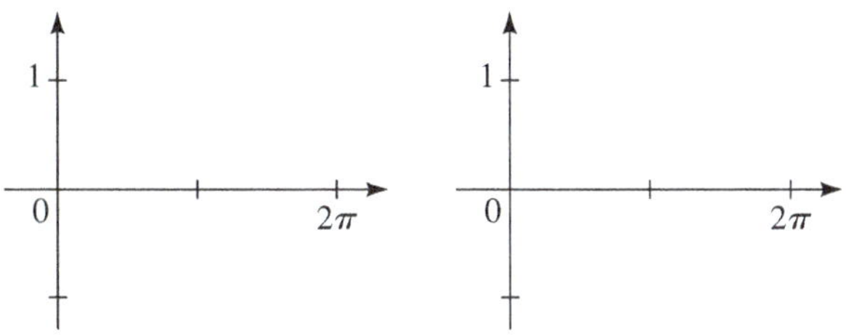

2. To obtain the graph of $y = 5 + \sin x$, we start with the graph of $y = \sin x$, then shift it 5 units ________ (upward/downward). To obtain the graph of $y = -\cos x$, we start with the graph of $y = \cos x$, then reflect it in the ____-axis.

3. The sine and cosine curves $y = a \sin kx$ and $y = a \cos kx$, $k > 0$, have amplitude ________ and period ________. The sine curve $y = 3 \sin 2x$ has amplitude ________ and period ________.

4. The sine curve $y = a \sin k(x - b)$ has amplitude ________, period ________, and horizontal shift ________. The sine curve $y = 4 \sin 3\left(x - \frac{\pi}{6}\right)$ has amplitude ________, period ________, and horizontal shift ________.

SKILLS

5–18 ■ Graphing Sine and Cosine Functions Graph the function.

5. $f(x) = 2 + \sin x$
6. $f(x) = -2 + \cos x$
7. $f(x) = -\sin x$
8. $f(x) = 2 - \cos x$
9. $f(x) = -2 + \sin x$
10. $f(x) = -1 + \cos x$
11. $g(x) = 3 \cos x$
12. $g(x) = 2 \sin x$
13. $g(x) = -\frac{1}{2} \sin x$
14. $g(x) = -\frac{2}{3} \cos x$
15. $g(x) = 3 + 3 \cos x$
16. $g(x) = 4 - 2 \sin x$
17. $h(x) = |\cos x|$
18. $h(x) = |\sin x|$

19–32 ■ Amplitude and Period Find the amplitude and period of the function, and sketch its graph.

19. $y = \cos 2x$
20. $y = -\sin 2x$
21. $y = -\sin 3x$
22. $y = \cos 4\pi x$
23. $y = -2 \cos 3\pi x$
24. $y = -3 \sin 6x$
25. $y = 10 \sin \frac{1}{2}x$
26. $y = 5 \cos \frac{1}{4}x$
27. $y = -\frac{1}{3} \cos \frac{1}{3}x$
28. $y = 4 \sin(-2x)$
29. $y = -2 \sin 2\pi x$
30. $y = -3 \sin \pi x$
31. $y = 1 + \frac{1}{2} \cos \pi x$
32. $y = -2 + \cos 4\pi x$

33–46 ■ Horizontal Shifts Find the amplitude, period, and horizontal shift of the function, and graph one complete period.

33. $y = \cos\left(x - \frac{\pi}{2}\right)$
34. $y = 2 \sin\left(x - \frac{\pi}{3}\right)$
35. $y = -2 \sin\left(x - \frac{\pi}{6}\right)$
36. $y = 3 \cos\left(x + \frac{\pi}{4}\right)$
37. $y = -4 \sin 2\left(x + \frac{\pi}{2}\right)$
38. $y = \sin \frac{1}{2}\left(x + \frac{\pi}{4}\right)$
39. $y = 5 \cos\left(3x - \frac{\pi}{4}\right)$
40. $y = 2 \sin\left(\frac{2}{3}x - \frac{\pi}{6}\right)$
41. $y = \frac{1}{2} - \frac{1}{2} \cos\left(2x - \frac{\pi}{3}\right)$
42. $y = 1 + \cos\left(3x + \frac{\pi}{2}\right)$
43. $y = 3 \cos \pi\left(x + \frac{1}{2}\right)$
44. $y = 3 + 2 \sin 3(x + 1)$
45. $y = \sin(\pi + 3x)$
46. $y = \cos\left(\frac{\pi}{2} - x\right)$

47–54 ■ Equations from a Graph The graph of one complete period of a sine or cosine curve is given.
(a) Find the amplitude, period, and horizontal shift.
(b) Write an equation that represents the curve in the form

$$y = a \sin k(x - b) \quad \text{or} \quad y = a \cos k(x - b)$$

47.
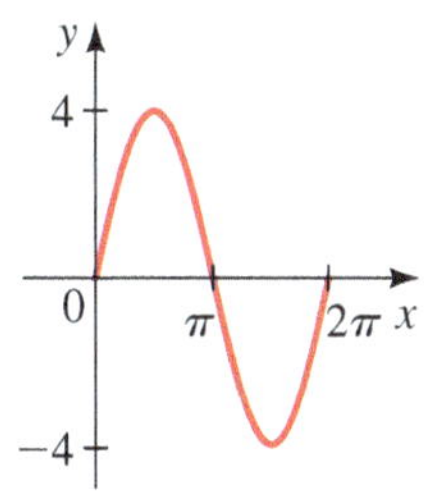

48.
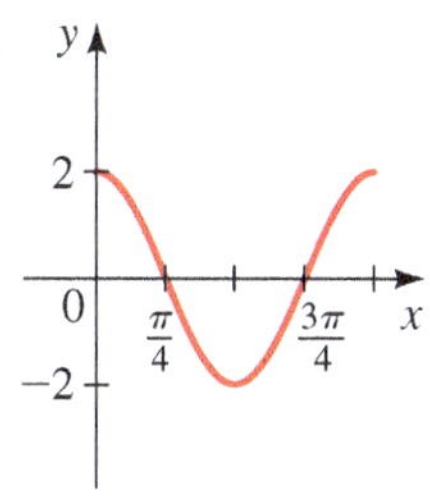

49.
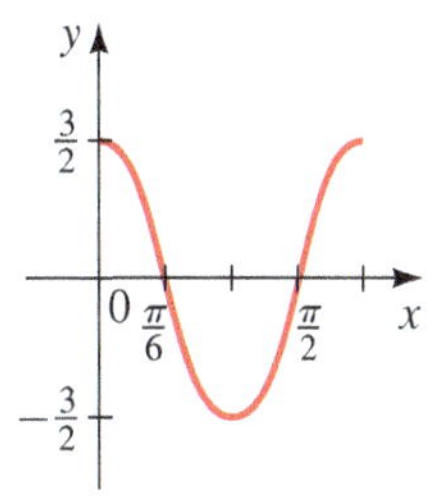

50.
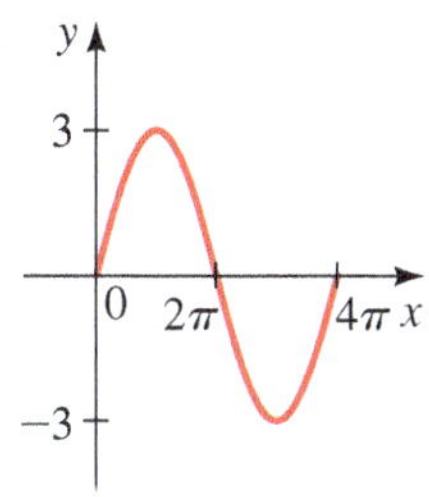

51.
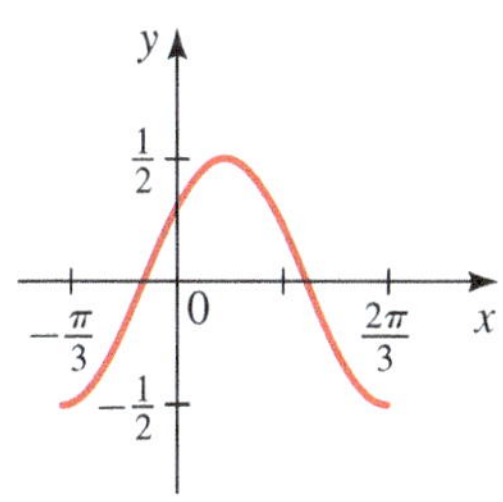

52.
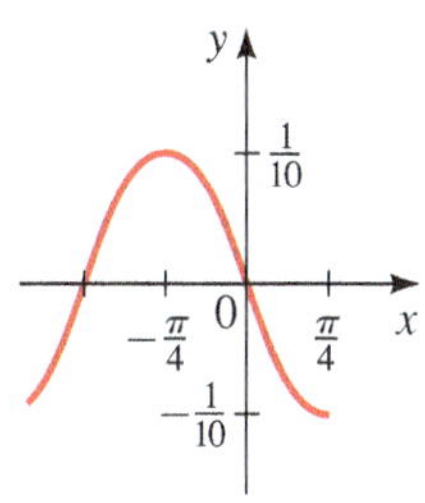

53.

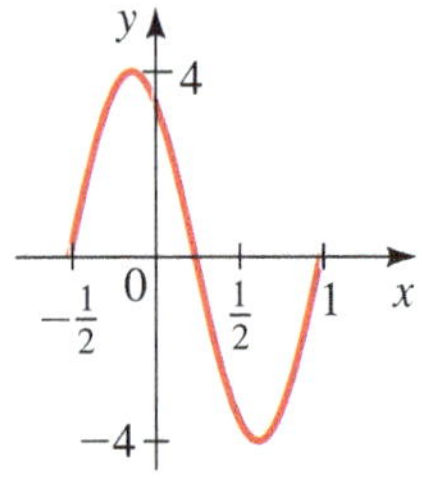

54.

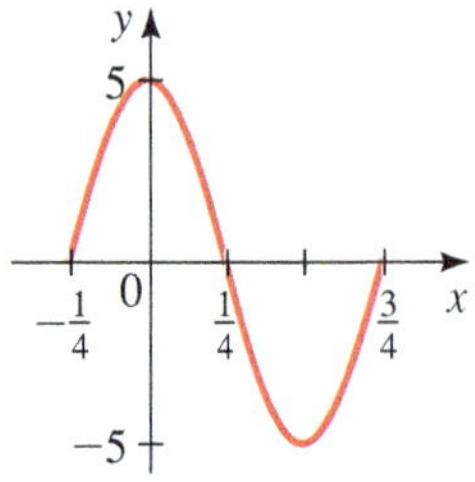

55–62 ■ Graphing Trigonometric Functions Determine an appropriate viewing rectangle for each function, and use it to draw the graph.

55. $f(x) = \cos 100x$ **56.** $f(x) = 3 \sin 120x$

57. $f(x) = \sin(x/40)$ **58.** $f(x) = \cos(x/80)$

59. $y = \tan 25x$ **60.** $y = \csc 40x$

61. $y = \sin^2 20x$ **62.** $y = \sqrt{\tan 10\pi x}$

63–66 ■ Graphical Addition Graph f, g, and $f + g$ on a common screen to illustrate graphical addition.

63. $f(x) = x, \quad g(x) = \sin x$

64. $f(x) = \sin x, \quad g(x) = \sin 2x$

65. $f(x) = \sin 3x, \quad g(x) = \cos \frac{1}{2}x$

66. $f(x) = 0.5 \sin 5x, \quad g(x) = -\cos 2x$

67–72 ■ Sine and Cosine Curves with Variable Amplitude Graph the three functions on a common screen. How are the graphs related?

67. $y = x^2, \quad y = -x^2, \quad y = x^2 \sin x$

68. $y = x, \quad y = -x, \quad y = x \cos x$

69. $y = \sqrt{x}, \quad y = -\sqrt{x}, \quad y = \sqrt{x} \sin 5\pi x$

70. $y = \dfrac{1}{1 + x^2}, \quad y = -\dfrac{1}{1 + x^2}, \quad y = \dfrac{\cos 2\pi x}{1 + x^2}$

71. $y = \cos 3\pi x, \quad y = -\cos 3\pi x, \quad y = \cos 3\pi x \cos 21\pi x$

72. $y = \sin 2\pi x, \quad y = -\sin 2\pi x, \quad y = \sin 2\pi x \sin 10\pi x$

SKILLS Plus

73–76 ■ Maxima and Minima Find the maximum and minimum values of the function.

73. $y = \sin x + \sin 2x$

74. $y = x - 2 \sin x, \quad 0 \le x \le 2\pi$

75. $y = 2 \sin x + \sin^2 x$

76. $y = \dfrac{\cos x}{2 + \sin x}$

77–80 ■ Solving Trigonometric Equations Graphically Find all solutions of the equation that lie in the interval $[0, \pi]$. State each answer rounded to two decimal places.

77. $\cos x = 0.4$ **78.** $\tan x = 2$

79. $\csc x = 3$ **80.** $\cos x = x$

81–82 ■ Limiting Behavior of Trigonometric Functions A function f is given.

(a) Is f even, odd, or neither?

(b) Find the x-intercepts of the graph of f.

(c) Graph f in an appropriate viewing rectangle.

(d) Describe the behavior of the function as $x \to \pm\infty$.

(e) Notice that $f(x)$ is not defined when $x = 0$. What happens as x approaches 0?

81. $f(x) = \dfrac{1 - \cos x}{x}$ **82.** $f(x) = \dfrac{\sin 4x}{2x}$

APPLICATIONS

83. Height of a Wave As a wave passes by an offshore piling, the height of the water is modeled by the function

$$h(t) = 3 \cos\left(\frac{\pi}{10}t\right)$$

where $h(t)$ is the height in feet above mean sea level at time t seconds.

(a) Find the period of the wave.

(b) Find the wave height, that is, the vertical distance between the trough and the crest of the wave.

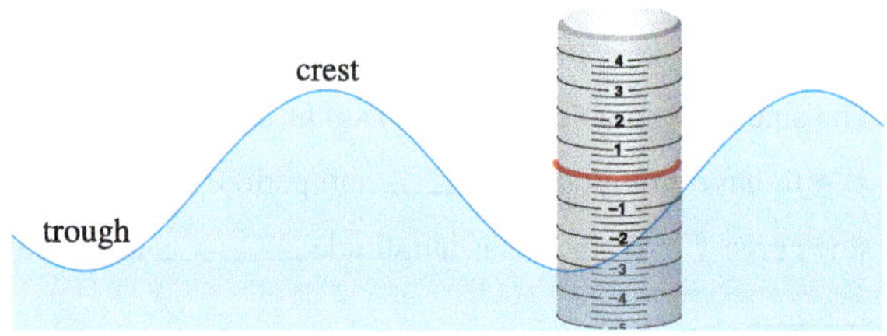

84. Sound Vibrations A tuning fork is struck, producing a pure tone as its tines vibrate. The vibrations are modeled by the function

$$v(t) = 0.7 \sin(880\pi t)$$

where $v(t)$ is the displacement of the tines in millimeters at time t seconds.

(a) Find the period of the vibration.

(b) Find the frequency of the vibration, that is, the number of times the fork vibrates per second.

(c) Graph the function v.

85. Blood Pressure Each time your heart beats, your blood pressure first increases and then decreases as the heart rests between beats. The maximum and minimum blood pressures are called the *systolic* and *diastolic* pressures, respectively. Your *blood pressure reading* is written as systolic/diastolic. A reading of 120/80 is considered normal.

A certain person's blood pressure is modeled by the function

$$p(t) = 115 + 25 \sin(160\pi t)$$

where $p(t)$ is the pressure in mmHg (millimeters of mercury), at time t measured in minutes.

(a) Find the period of p.

(b) Find the number of heartbeats per minute.

(c) Graph the function p.

(d) Find the blood pressure reading. How does this compare to normal blood pressure?

86. Variable Stars Variable stars are ones whose brightness varies periodically. One of the most visible is R Leonis; its brightness is modeled by the function

$$b(t) = 7.9 - 2.1\cos\left(\frac{\pi}{156}t\right)$$

where t is measured in days.

(a) Find the period of R Leonis.

(b) Find the maximum and minimum brightness.

(c) Graph the function b.

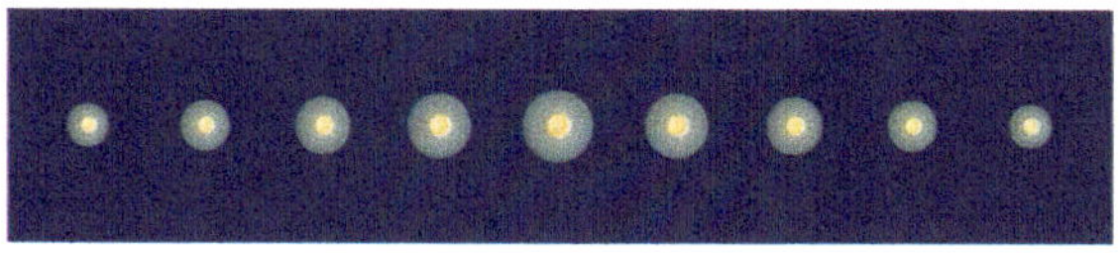

DISCUSS ■ DISCOVER ■ PROVE ■ WRITE

87. DISCUSS: Compositions Involving Trigonometric Functions This exercise explores the effect of the inner function g on a composite function $y = f(g(x))$.

(a) Graph the function $y = \sin\sqrt{x}$ using the viewing rectangle $[0, 400]$ by $[-1.5, 1.5]$. In what ways does this graph differ from the graph of the sine function?

(b) Graph the function $y = \sin(x^2)$ using the viewing rectangle $[-5, 5]$ by $[-1.5, 1.5]$. In what ways does this graph differ from the graph of the sine function?

88. DISCUSS: Periodic Functions I Recall that a function f is *periodic* if there is a positive number p such that $f(t + p) = f(t)$ for every t, and the least such p (if it exists) is the *period* of f. The graph of a function of period p looks the same on each interval of length p, so we can easily determine the period from the graph. Determine whether the function whose graph is shown is periodic; if it is periodic, find the period.

(a)

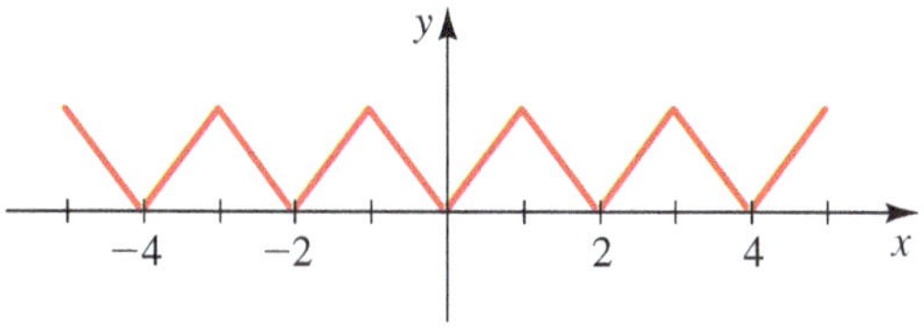

(b)

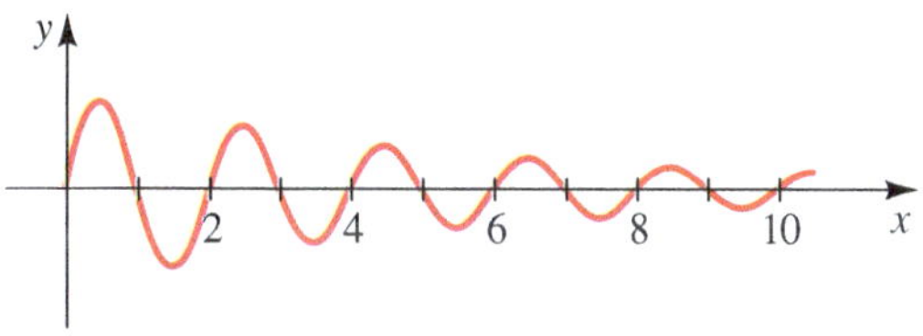

(c)

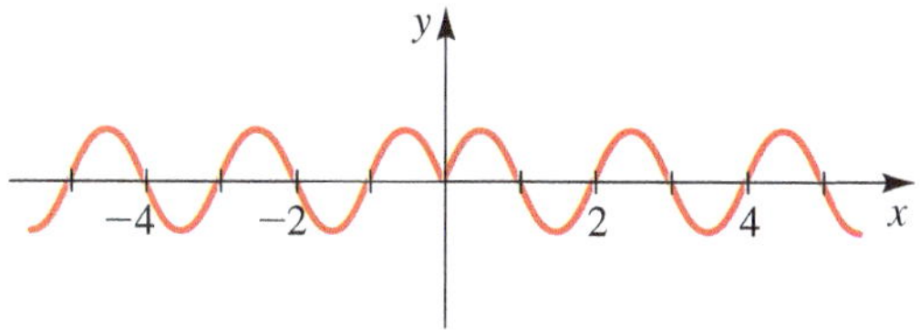

(d)

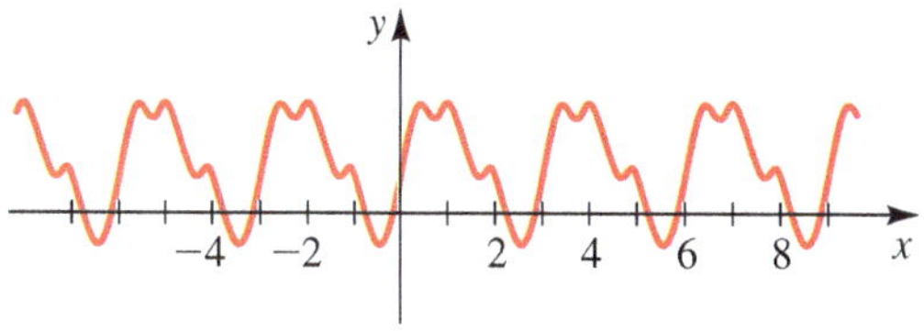

89. DISCUSS: Periodic Functions II Use a graphing device to graph the following functions. From the graph, determine whether the function is periodic; if it is periodic, find the period. (See page 163 for the definition of $[\![x]\!]$.)

(a) $y = |\sin x|$

(b) $y = \sin|x|$

(c) $y = 2^{\cos x}$

(d) $y = x - [\![x]\!]$

(e) $y = \cos(\sin x)$

(f) $y = \cos(x^2)$

90. DISCUSS: Sinusoidal Curves The graph of $y = \sin x$ is the same as the graph of $y = \cos x$ shifted to the right $\pi/2$ units. So the sine curve $y = \sin x$ is also at the same time a cosine curve: $y = \cos(x - \frac{\pi}{2})$. In fact, any sine curve is also a cosine curve with a different horizontal shift, and any cosine curve is also a sine curve. Sine and cosine curves are collectively referred to as *sinusoidal*. For the curve whose graph is shown, find all possible ways of expressing it as a sine curve $y = a\sin(x - b)$ or as a cosine curve $y = a\cos(x - b)$. Explain why you think you have found all possible choices for a and b in each case.

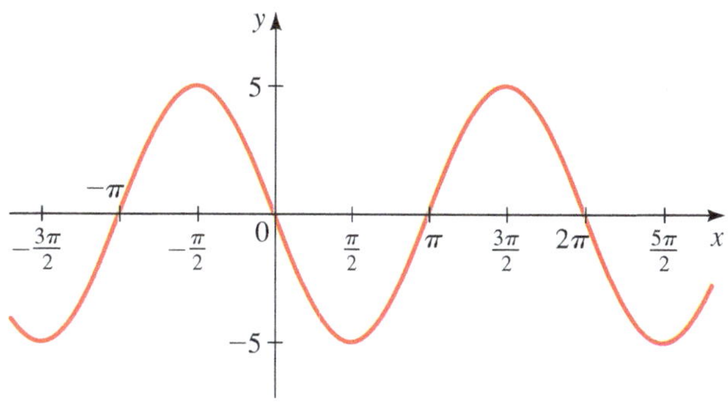

5.4 MORE TRIGONOMETRIC GRAPHS

■ Graphs of Tangent, Cotangent, Secant, and Cosecant ■ Graphs of Transformations of Tangent and Cotangent ■ Graphs of Transformations of Cosecant and Secant

In this section we graph the tangent, cotangent, secant, and cosecant functions and transformations of these functions.

■ Graphs of Tangent, Cotangent, Secant, and Cosecant

We begin by stating the periodic properties of these functions. Recall that sine and cosine have period 2π. Since cosecant and secant are the reciprocals of sine and cosine, respectively, they also have period 2π (see Exercise 63). Tangent and cotangent, however, have period π (see Exercise 83 of Section 5.2).

PERIODIC PROPERTIES

The functions tangent and cotangent have period π:

$$\tan(x + \pi) = \tan x \qquad \cot(x + \pi) = \cot x$$

The functions cosecant and secant have period 2π:

$$\csc(x + 2\pi) = \csc x \qquad \sec(x + 2\pi) = \sec x$$

x	$\tan x$
0	0
$\pi/6$	0.58
$\pi/4$	1.00
$\pi/3$	1.73
1.4	5.80
1.5	14.10
1.55	48.08
1.57	1,255.77
1.5707	10,381.33

We first sketch the graph of tangent. Since it has period π, we need only sketch the graph on any interval of length π and then repeat the pattern to the left and to the right. We sketch the graph on the interval $(-\pi/2, \pi/2)$. Since $\tan(\pi/2)$ and $\tan(-\pi/2)$ aren't defined, we need to be careful in sketching the graph at points near $\pi/2$ and $-\pi/2$. As x gets near $\pi/2$ through values less than $\pi/2$, the value of $\tan x$ becomes large. To see this, notice that as x gets close to $\pi/2$, $\cos x$ approaches 0 and $\sin x$ approaches 1 and so $\tan x = \sin x/\cos x$ is large. A table of values of $\tan x$ for x close to $\pi/2$ (≈ 1.570796) is shown in the margin.

So as x approaches $\pi/2$ from the left, the value of $\tan x$ increases without bound. We express this by writing

$$\tan x \to \infty \quad \text{as} \quad x \to \frac{\pi}{2}^-$$

This is read "tan x approaches infinity as x approaches $\pi/2$ from the left."

In a similar way, as x approaches $-\pi/2$ from the right, the value of $\tan x$ decreases without bound. We write this as

$$\tan x \to -\infty \quad \text{as} \quad x \to -\frac{\pi}{2}^+$$

Arrow notation is discussed in Section 3.6.

This is read "tan x approaches negative infinity as x approaches $-\pi/2$ from the right."

Asymptotes are discussed in Section 3.6.

Thus the graph of $y = \tan x$ approaches the vertical lines $x = \pi/2$ and $x = -\pi/2$. So these lines are **vertical asymptotes**. With the information we have so far, we sketch the graph of $y = \tan x$ for $-\pi/2 < x < \pi/2$ in Figure 1. The complete graph of tangent (see

Mathematics in the Modern World

Evaluating Functions on a Calculator

How does your calculator evaluate $\sin t$, $\cos t$, e^t, $\ln t$, $\sqrt{t}$, and other such functions? One method is to approximate these functions by polynomials because polynomials are easy to evaluate. For example,

$$\sin t = t - \frac{t^3}{3!} + \frac{t^5}{5!} - \frac{t^7}{7!} + \cdots$$

$$\cos t = 1 - \frac{t^2}{2!} + \frac{t^4}{4!} - \frac{t^6}{6!} + \cdots$$

where $n! = 1 \cdot 2 \cdot 3 \cdot \cdots \cdot n$. These remarkable formulas were found by the British mathematician Brook Taylor (1685–1731). For instance, if we use the first three terms of Taylor's series to find $\cos(0.4)$, we get

$$\cos 0.4 \approx 1 - \frac{(0.4)^2}{2!} + \frac{(0.4)^4}{4!}$$

$$\approx 0.92106667$$

(Compare this with the value you get from your calculator.) The graph shows that the more terms of the series we use, the more closely the polynomials approximate the function $\cos t$.

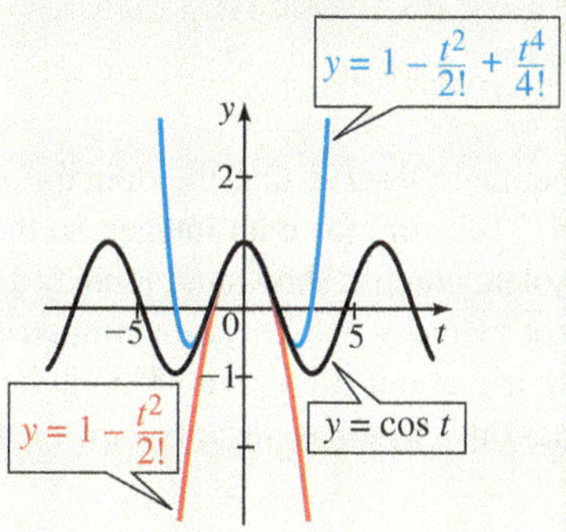

Figure 5(a) on the next page) is now obtained by using the fact that tangent is periodic with period π.

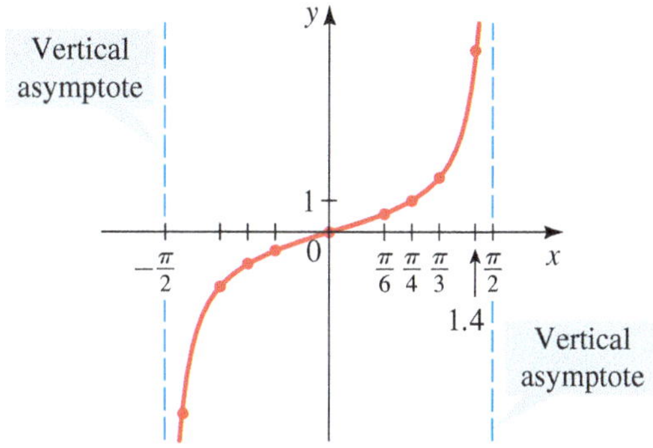

FIGURE 1 One period of $y = \tan x$

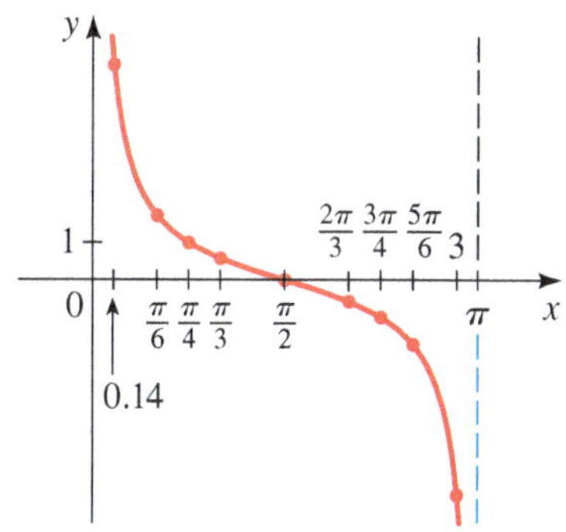

FIGURE 2 One period of $y = \cot x$

The function $y = \cot x$ is graphed on the interval $(0, \pi)$ by a similar analysis (see Figure 2). Since $\cot x$ is undefined for $x = n\pi$ with n an integer, its complete graph (in Figure 5(b) on the next page) has vertical asymptotes at these values.

To graph the cosecant and secant functions, we use the reciprocal identities

$$\csc x = \frac{1}{\sin x} \qquad \text{and} \qquad \sec x = \frac{1}{\cos x}$$

So to graph $y = \csc x$, we take the reciprocals of the y-coordinates of the points of the graph of $y = \sin x$. (See Figure 3.) Similarly, to graph $y = \sec x$, we take the reciprocals of the y-coordinates of the points of the graph of $y = \cos x$. (See Figure 4.)

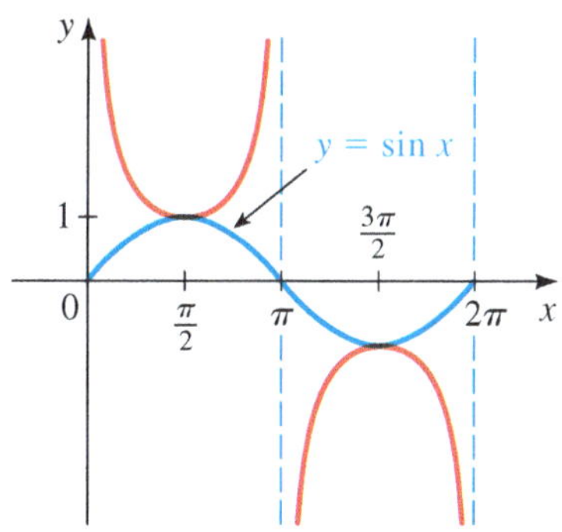

FIGURE 3 One period of $y = \csc x$

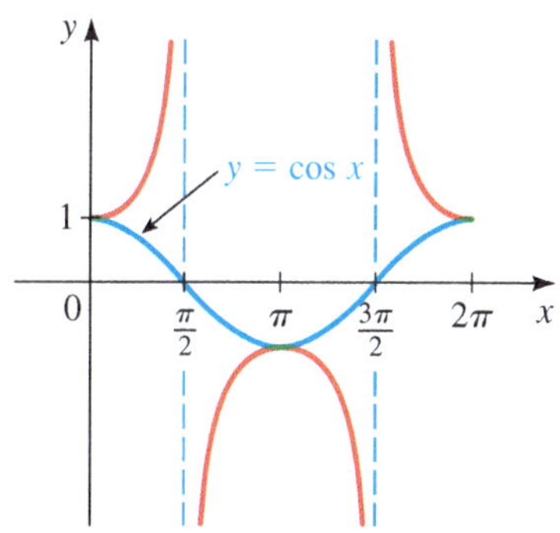

FIGURE 4 One period of $y = \sec x$

Let's consider more closely the graph of the function $y = \csc x$ on the interval $0 < x < \pi$. We need to examine the values of the function near 0 and π, since at these values $\sin x = 0$, and $\csc x$ is thus undefined. We see that

$$\csc x \to \infty \qquad \text{as} \qquad x \to 0^+$$

$$\csc x \to \infty \qquad \text{as} \qquad x \to \pi^-$$

Thus the lines $x = 0$ and $x = \pi$ are vertical asymptotes. In the interval $\pi < x < 2\pi$ the graph is sketched in the same way. The values of $\csc x$ in that interval are the same as those in the interval $0 < x < \pi$ except for sign (see Figure 3). The complete graph in Figure 5(c) is now obtained from the fact that the function cosecant is periodic with period 2π. Note that the graph has vertical asymptotes at the points where $\sin x = 0$, that is, at $x = n\pi$, for n an integer.

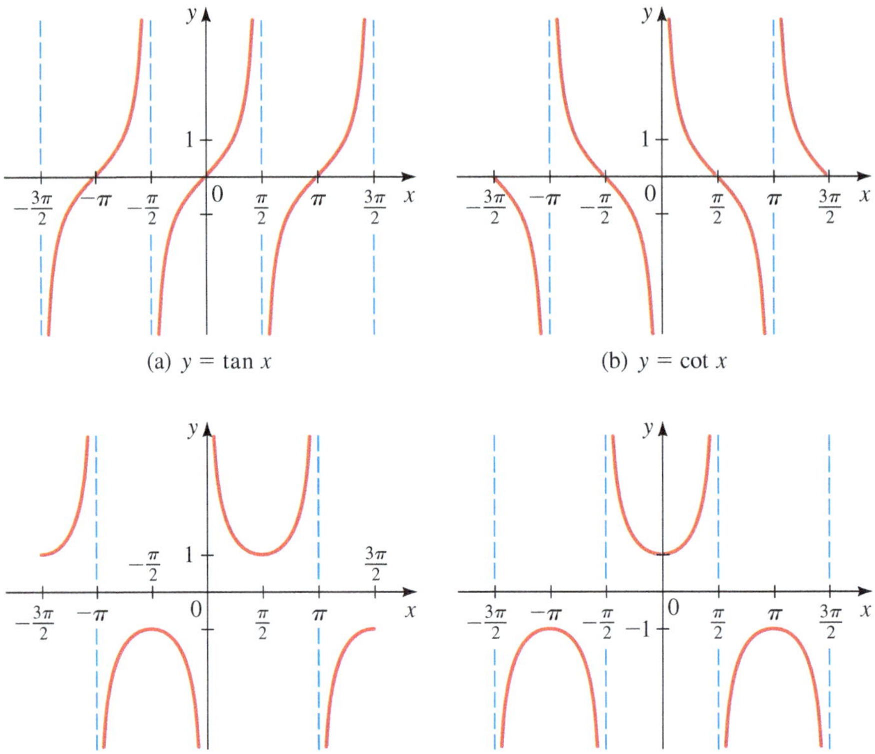

(a) $y = \tan x$ (b) $y = \cot x$ (c) $y = \csc x$ (d) $y = \sec x$

FIGURE 5

The graph of $y = \sec x$ is sketched in a similar manner. Observe that the domain of $\sec x$ is the set of all real numbers other than $x = (\pi/2) + n\pi$, for n an integer, so the graph has vertical asymptotes at those points. The complete graph is shown in Figure 5(d).

It is apparent that the graphs of $y = \tan x$, $y = \cot x$, and $y = \csc x$ are symmetric about the origin, whereas that of $y = \sec x$ is symmetric about the y-axis. This is because tangent, cotangent, and cosecant are odd functions, whereas secant is an even function.

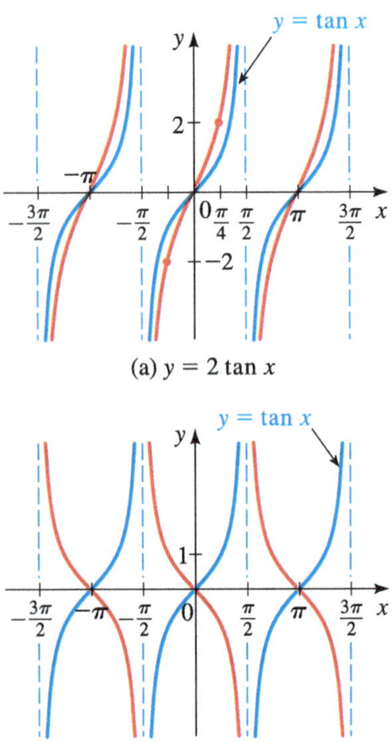

(a) $y = 2 \tan x$

(b) $y = -\tan x$

FIGURE 6

Graphs of Transformations of Tangent and Cotangent

We now consider graphs of transformations of the tangent and cotangent functions.

EXAMPLE 1 ■ Graphing Tangent Curves

Graph each function.

(a) $y = 2 \tan x$ **(b)** $y = -\tan x$

SOLUTION We first graph $y = \tan x$ and then transform it as required.

(a) To graph $y = 2 \tan x$, we multiply the y-coordinate of each point on the graph of $y = \tan x$ by 2. The resulting graph is shown in Figure 6(a).

(b) The graph of $y = -\tan x$ in Figure 6(b) is obtained from that of $y = \tan x$ by reflecting in the x-axis.

Now Try Exercises 9 and 11

Since the tangent and cotangent functions have period π, the functions

$$y = a \tan kx \qquad \text{and} \qquad y = a \cot kx \qquad (k > 0)$$

complete one period as kx varies from 0 to π, that is, for $0 \le kx \le \pi$. Solving this inequality, we get $0 \le x \le \pi/k$. So they each have period π/k.

TANGENT AND COTANGENT CURVES

The functions

$$y = a \tan kx \qquad \text{and} \qquad y = a \cot kx \qquad (k > 0)$$

have period π/k.

Thus one complete period of the graphs of these functions occurs on any interval of length π/k. To sketch a complete period of these graphs, it's convenient to select an interval between vertical asymptotes:

To graph one period of $y = a \tan kx$, an appropriate interval is $\left(-\dfrac{\pi}{2k}, \dfrac{\pi}{2k}\right)$.

To graph one period of $y = a \cot kx$, an appropriate interval is $\left(0, \dfrac{\pi}{k}\right)$.

EXAMPLE 2 ■ Graphing Tangent Curves

Graph each function.

(a) $y = \tan 2x$ **(b)** $y = \tan 2\left(x - \dfrac{\pi}{4}\right)$

SOLUTION

(a) The period is $\pi/2$ and an appropriate interval is $(-\pi/4, \pi/4)$. The endpoints $x = -\pi/4$ and $x = \pi/4$ are vertical asymptotes. Thus we graph one complete period of the function on $(-\pi/4, \pi/4)$. The graph has the same shape as that of the tangent function but is shrunk horizontally by a factor of $\frac{1}{2}$. We then repeat that portion of the graph to the left and to the right. See Figure 7(a).

(b) The graph is the same as that in part (a), but it is shifted to the right $\pi/4$, as shown in Figure 7(b).

Since $y = \tan x$ completes one period between $x = -\frac{\pi}{2}$ and $x = \frac{\pi}{2}$, the function $y = \tan 2(x - \frac{\pi}{4})$ completes one period as $2(x - \frac{\pi}{4})$ varies from $-\frac{\pi}{2}$ to $\frac{\pi}{2}$.

Start of period:	End of period:
$2(x - \frac{\pi}{4}) = -\frac{\pi}{2}$	$2(x - \frac{\pi}{4}) = \frac{\pi}{2}$
$x - \frac{\pi}{4} = -\frac{\pi}{4}$	$x - \frac{\pi}{4} = \frac{\pi}{4}$
$x = 0$	$x = \frac{\pi}{2}$

So we graph one period on the interval $(0, \frac{\pi}{2})$.

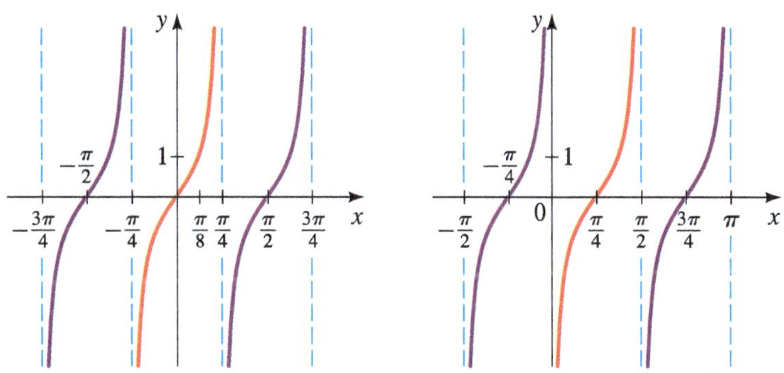

(a) $y = \tan 2x$ (b) $y = \tan 2\left(x - \frac{\pi}{4}\right)$

FIGURE 7

Now Try Exercises 19, 35, and 43 ■

EXAMPLE 3 ■ A Horizontally Shifted Cotangent Curve

Graph the function $y = 2\cot\left(3x - \dfrac{\pi}{4}\right)$.

SOLUTION We first put the equation in the form $y = a\cot k(x - b)$ by factoring 3 from the expression $3x - \dfrac{\pi}{4}$:

$$y = 2\cot\left(3x - \frac{\pi}{4}\right) = 2\cot 3\left(x - \frac{\pi}{12}\right)$$

Thus the graph is the same as that of $y = 2\cot 3x$ but is shifted to the right $\pi/12$. The period of $y = 2\cot 3x$ is $\pi/3$, and an appropriate interval for graphing one period is $(0, \pi/3)$. To get the corresponding interval for the desired graph, we shift this interval to the right $\pi/12$. So we have

$$\left(0 + \frac{\pi}{12}, \frac{\pi}{3} + \frac{\pi}{12}\right) = \left(\frac{\pi}{12}, \frac{5\pi}{12}\right)$$

Finally, we graph one period in the shape of cotangent on the interval $(\pi/12, 5\pi/12)$ and repeat that portion of the graph to the left and to the right. (See Figure 8.)

Since $y = \cot x$ completes one period between $x = 0$ and $x = \pi$, the function $y = 2\cot(3x - \frac{\pi}{4})$ completes one period as $3x - \frac{\pi}{4}$ varies from 0 to π.

Start of period:	End of period:
$3x - \frac{\pi}{4} = 0$	$3x - \frac{\pi}{4} = \pi$
$3x = \frac{\pi}{4}$	$3x = \frac{5\pi}{4}$
$x = \frac{\pi}{12}$	$x = \frac{5\pi}{12}$

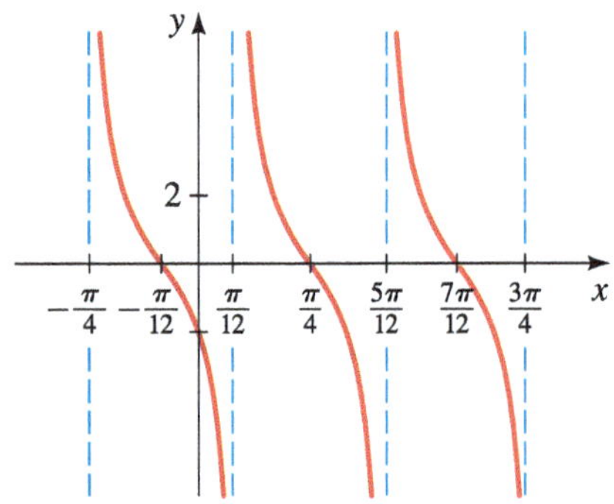

FIGURE 8
$y = 2\cot\left(3x - \dfrac{\pi}{4}\right)$

Now Try Exercises 37 and 47 ■

■ Graphs of Transformations of Cosecant and Secant

We have already observed that the cosecant and secant functions are the reciprocals of the sine and cosine functions. Thus the following result is the counterpart of the result for sine and cosine curves in Section 5.3.

> **COSECANT AND SECANT CURVES**
>
> The functions
>
> $$y = a\csc kx \quad \text{and} \quad y = a\sec kx \quad (k > 0)$$
>
> have period $2\pi/k$.

An appropriate interval on which to graph one complete period is $(0, 2\pi/k)$.

EXAMPLE 4 ■ Graphing Cosecant Curves

Graph each function.

(a) $y = \dfrac{1}{2}\csc 2x$ **(b)** $y = \dfrac{1}{2}\csc\left(2x + \dfrac{\pi}{2}\right)$

SOLUTION

(a) The period is $2\pi/2 = \pi$. An appropriate interval is $[0, \pi]$, and the asymptotes occur in this interval whenever $\sin 2x = 0$. So the asymptotes in this interval are $x = 0$, $x = \pi/2$, and $x = \pi$. With this information we sketch on the interval $[0, \pi]$ a graph with the same general shape as that of one period of the cosecant function. The complete graph in Figure 9(a) is obtained by repeating this portion of the graph to the left and to the right.

(b) We first write

$$y = \frac{1}{2}\csc\left(2x + \frac{\pi}{2}\right) = \frac{1}{2}\csc 2\left(x + \frac{\pi}{4}\right)$$

From this we see that the graph is the same as that in part (a) but shifted to the left $\pi/4$. The graph is shown in Figure 9(b).

Since $y = \csc x$ completes one period between $x = 0$ and $x = 2\pi$, the function $y = \frac{1}{2}\csc(2x + \frac{\pi}{2})$ completes one period as $2x + \frac{\pi}{2}$ varies from 0 to 2π.

Start of period:	End of period:
$2x + \frac{\pi}{2} = 0$	$2x + \frac{\pi}{2} = 2\pi$
$2x = -\frac{\pi}{2}$	$2x = \frac{3\pi}{2}$
$x = -\frac{\pi}{4}$	$x = \frac{3\pi}{4}$

So we graph one period on the interval $\left[-\frac{\pi}{4}, \frac{3\pi}{4}\right]$.

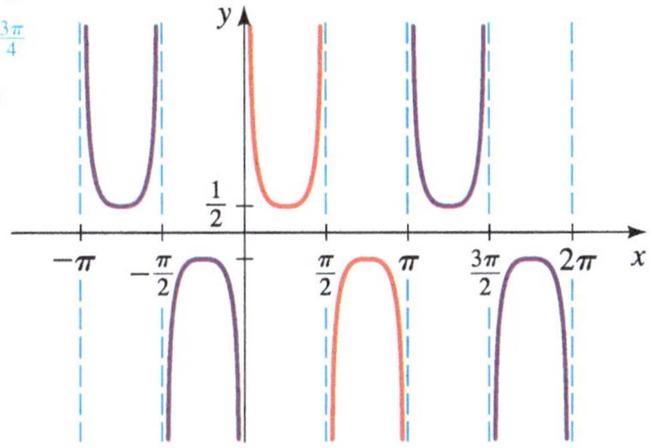

(a) $y = \frac{1}{2}\csc 2x$

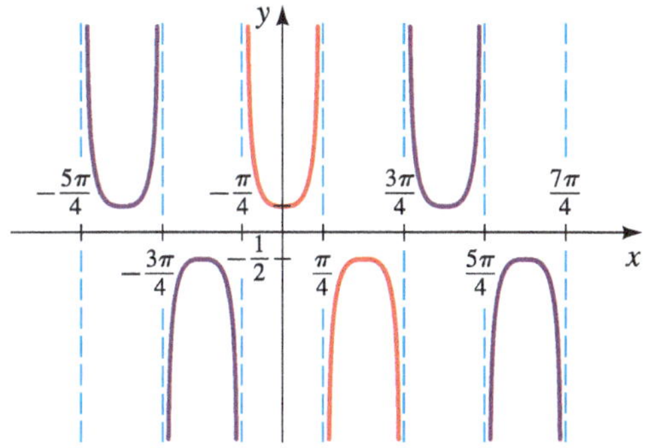

(b) $y = \frac{1}{2}\csc\left(2x + \frac{\pi}{2}\right)$

FIGURE 9

Now Try Exercises 29 and 49 ■

EXAMPLE 5 ■ Graphing a Secant Curve

Graph $y = 3\sec\frac{1}{2}x$.

SOLUTION The period is $2\pi \div \frac{1}{2} = 4\pi$. An appropriate interval is $[0, 4\pi]$, and the asymptotes occur in this interval wherever $\cos\frac{1}{2}x = 0$. Thus the asymptotes in this interval are $x = \pi$, $x = 3\pi$. With this information we sketch on the interval $[0, 4\pi]$ a graph with the same general shape as that of one period of the secant function. The complete graph in Figure 10 is obtained by repeating this portion of the graph to the left and to the right.

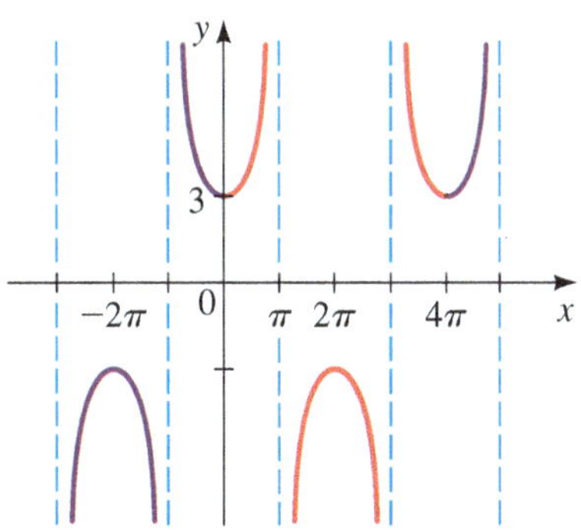

FIGURE 10
$y = 3\sec\frac{1}{2}x$

Now Try Exercises 31 and 51 ■

5.4 EXERCISES

CONCEPTS

1. The trigonometric function $y = \tan x$ has period ________ and asymptotes $x =$ ________. Sketch a graph of this function on the interval $(-\pi/2, \pi/2)$.

2. The trigonometric function $y = \csc x$ has period ________ and asymptotes $x =$ ________. Sketch a graph of this function on the interval $(-\pi, \pi)$.

SKILLS

3–8 ■ Graphs of Trigonometric Functions Match the trigonometric function with one of the graphs I–VI.

3. $f(x) = \tan\left(x + \dfrac{\pi}{4}\right)$

4. $f(x) = \sec 2x$

5. $f(x) = \cot 4x$

6. $f(x) = -\tan x$

7. $f(x) = 2 \sec x$

8. $f(x) = 1 + \csc x$

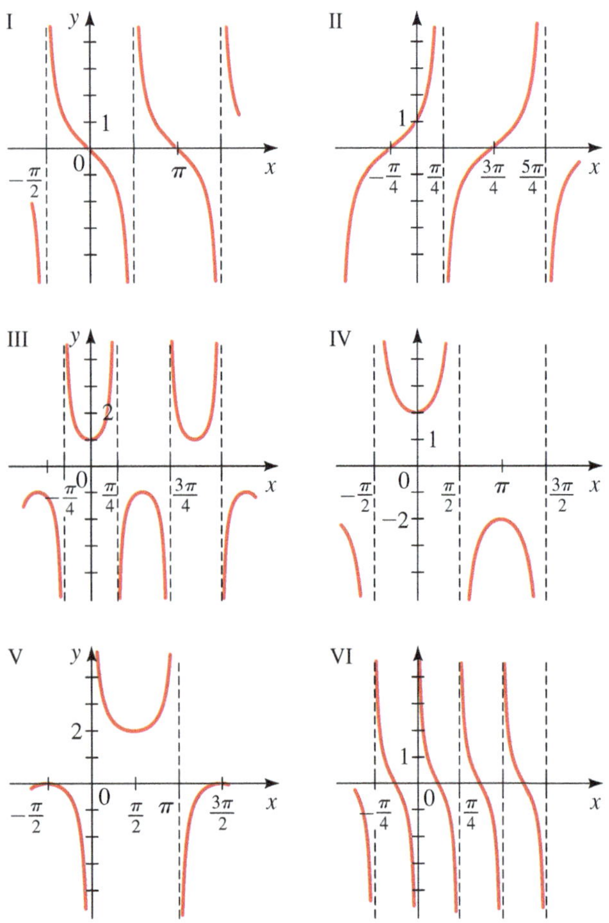

9–18 ■ Graphs of Trigonometric Functions Find the period, and graph the function.

9. $y = 3 \tan x$

10. $y = -3 \tan x$

11. $y = -\frac{3}{2} \tan x$

12. $y = \frac{3}{4} \tan x$

13. $y = -\cot x$

14. $y = 2 \cot x$

15. $y = 2 \csc x$

16. $y = \frac{1}{2} \csc x$

17. $y = 3 \sec x$

18. $y = -3 \sec x$

19–34 ■ Graphs of Trigonometric Functions with Different Periods Find the period, and graph the function.

19. $y = \tan 3x$

20. $y = \tan 4x$

21. $y = -5 \tan \pi x$

22. $y = -3 \tan 4\pi x$

23. $y = 2 \cot 3\pi x$

24. $y = 3 \cot 2\pi x$

25. $y = \tan \dfrac{\pi}{4} x$

26. $y = \cot \dfrac{\pi}{2} x$

27. $y = 2 \tan 3\pi x$

28. $y = 2 \tan \dfrac{\pi}{2} x$

29. $y = \csc 4x$

30. $y = 5 \csc 3x$

31. $y = \sec 2x$

32. $y = \frac{1}{2} \sec(4\pi x)$

33. $y = 5 \csc \dfrac{3\pi}{2} x$

34. $y = 5 \sec 2\pi x$

35–60 ■ Graphs of Trigonometric Functions with Horizontal Shifts Find the period, and graph the function.

35. $y = \tan\left(x + \dfrac{\pi}{4}\right)$

36. $y = \tan\left(x - \dfrac{\pi}{4}\right)$

37. $y = \cot\left(x + \dfrac{\pi}{4}\right)$

38. $y = 2 \cot\left(x - \dfrac{\pi}{3}\right)$

39. $y = \csc\left(x - \dfrac{\pi}{4}\right)$

40. $y = \sec\left(x + \dfrac{\pi}{4}\right)$

41. $y = \dfrac{1}{2} \sec\left(x - \dfrac{\pi}{6}\right)$

42. $y = 3 \csc\left(x + \dfrac{\pi}{2}\right)$

43. $y = \tan 2\left(x - \dfrac{\pi}{3}\right)$

44. $y = \cot\left(2x - \dfrac{\pi}{4}\right)$

45. $y = 5 \cot\left(3x + \dfrac{\pi}{2}\right)$

46. $y = 4 \tan(4x - 2\pi)$

47. $y = \cot\left(2x - \dfrac{\pi}{2}\right)$

48. $y = \frac{1}{2} \tan(\pi x - \pi)$

49. $y = 2 \csc\left(\pi x - \dfrac{\pi}{3}\right)$

50. $y = 3 \sec\left(\dfrac{1}{4} x - \dfrac{\pi}{6}\right)$

51. $y = \sec 2\left(x - \dfrac{\pi}{4}\right)$

52. $y = \csc 2\left(x + \dfrac{\pi}{2}\right)$

53. $y = 5 \sec\left(3x - \dfrac{\pi}{2}\right)$

54. $y = \frac{1}{2} \sec(2\pi x - \pi)$

55. $y = \tan\left(\dfrac{2}{3} x - \dfrac{\pi}{6}\right)$

56. $y = \tan \dfrac{1}{2}\left(x + \dfrac{\pi}{4}\right)$

57. $y = 3 \sec \pi(x + \frac{1}{2})$

58. $y = \sec\left(3x + \frac{\pi}{2}\right)$

59. $y = -2 \tan\left(2x - \frac{\pi}{3}\right)$

60. $y = 2 \cot(3\pi x + 3\pi)$

APPLICATIONS

61. Lighthouse The beam from a lighthouse completes one rotation every 2 min. At time t, the distance d shown in the figure below is

$$d(t) = 3 \tan \pi t$$

where t is measured in minutes and d in miles.

(a) Find $d(0.15)$, $d(0.25)$, and $d(0.45)$.

(b) Sketch a graph of the function d for $0 \le t < \frac{1}{2}$.

(c) What happens to the distance d as t approaches $\frac{1}{2}$?

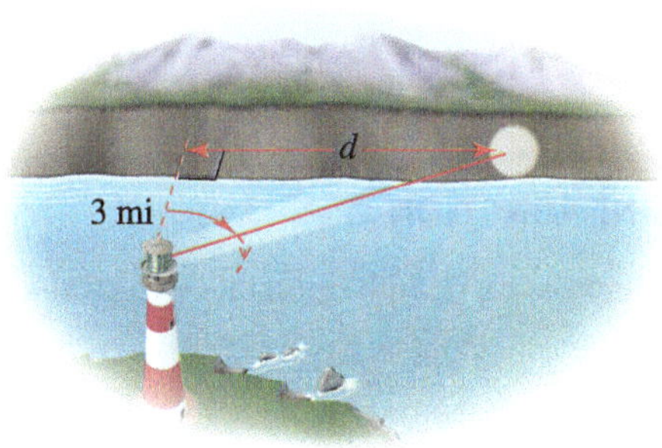

62. Length of a Shadow On a day when the sun passes directly overhead at noon, a 6-ft-tall man casts a shadow of length

$$S(t) = 6\left|\cot \frac{\pi}{12} t\right|$$

where S is measured in feet and t is the number of hours since 6 A.M.

(a) Find the length of the shadow at 8:00 A.M., noon, 2:00 P.M., and 5:45 P.M.

(b) Sketch a graph of the function S for $0 < t < 12$.

(c) From the graph, determine the values of t at which the length of the shadow equals the man's height. To what time of day does each of these values correspond?

(d) Explain what happens to the shadow as the time approaches 6 P.M. (that is, as $t \to 12^-$).

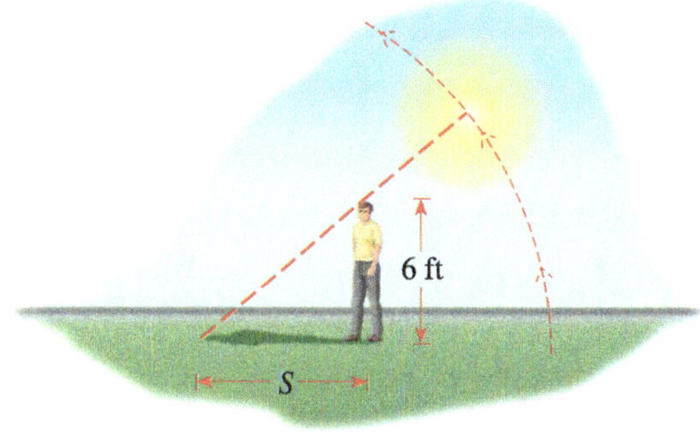

DISCUSS ■ DISCOVER ■ PROVE ■ WRITE

63. PROVE: Periodic Functions **(a)** Prove that if f is periodic with period p, then $1/f$ is also periodic with period p.

(b) Prove that cosecant and secant both have period 2π.

64. PROVE: Periodic Functions Prove that if f and g are periodic with period p, then f/g is also periodic but its period could be smaller than p.

65. PROVE: Reduction Formulas Use the graphs in Figure 5 to explain why the following formulas are true.

$$\tan\left(x - \frac{\pi}{2}\right) = -\cot x \qquad \sec\left(x - \frac{\pi}{2}\right) = \csc x$$

5.5 INVERSE TRIGONOMETRIC FUNCTIONS AND THEIR GRAPHS

■ The Inverse Sine Function ■ The Inverse Cosine Function ■ The Inverse Tangent Function ■ The Inverse Secant, Cosecant, and Cotangent Functions

We study applications of inverse trigonometric functions to triangles in Sections 6.4–6.6.

Recall from Section 2.8 that the inverse of a function f is a function f^{-1} that reverses the rule of f. For a function to have an inverse, it must be one-to-one. Since the trigonometric functions are not one-to-one, they do not have inverses. It is possible, however, to restrict the domains of the trigonometric functions in such a way that the resulting functions are one-to-one.

■ The Inverse Sine Function

Let's first consider the sine function. There are many ways to restrict the domain of sine so that the new function is one-to-one. A natural way to do this is to restrict the domain to the interval $[-\pi/2, \pi/2]$. The reason for this choice is that sine is one-to-one on this interval and moreover attains each of the values in its range on this interval. From Figure 1 we see that sine is one-to-one on this restricted domain (by the Horizontal Line Test) and so has an inverse.

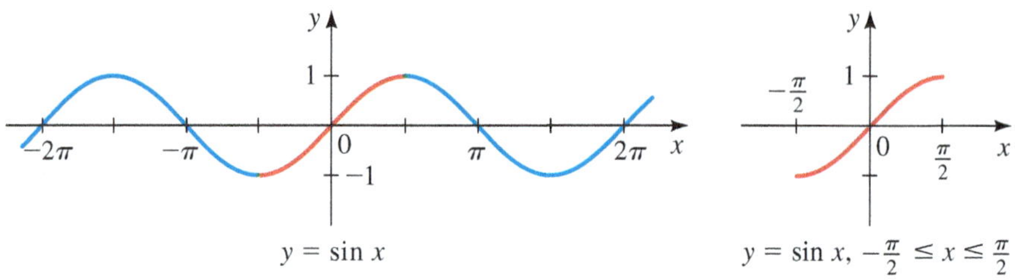

FIGURE 1 Graphs of the sine function and the restricted sine function

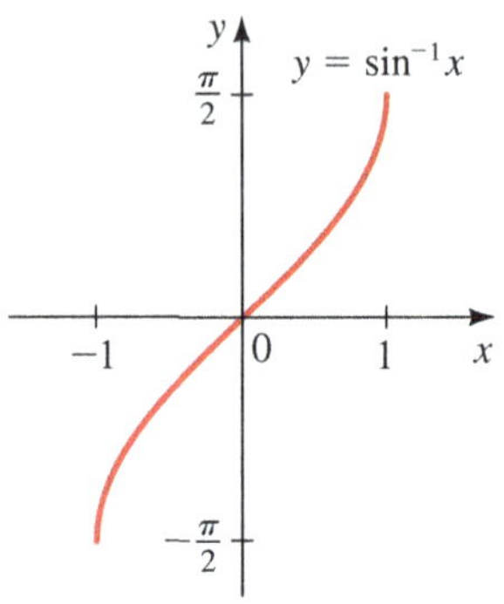

FIGURE 2 Graph of $y = \sin^{-1} x$

We can now define an inverse sine function on this restricted domain. The graph of $y = \sin^{-1} x$ is shown in Figure 2; it is obtained by reflecting the graph of $y = \sin x$, $-\pi/2 \leq x \leq \pi/2$, in the line $y = x$.

DEFINITION OF THE INVERSE SINE FUNCTION

The **inverse sine function** is the function $\sin^{-1}$ with domain $[-1, 1]$ and range $[-\pi/2, \pi/2]$ defined by

$$\sin^{-1} x = y \quad \Leftrightarrow \quad \sin y = x$$

The inverse sine function is also called **arcsine**, denoted by **arcsin**.

Thus $y = \sin^{-1} x$ *is the number in the interval* $[-\pi/2, \pi/2]$ *whose sine is* x. In other words, $\sin(\sin^{-1} x) = x$. In fact, from the general properties of inverse functions studied in Section 2.8, we have the following **cancellation properties**.

$$\sin(\sin^{-1} x) = x \quad \text{for} \quad -1 \leq x \leq 1$$

$$\sin^{-1}(\sin x) = x \quad \text{for} \quad -\frac{\pi}{2} \leq x \leq \frac{\pi}{2}$$

EXAMPLE 1 ■ Evaluating the Inverse Sine Function

Find each value.

(a) $\sin^{-1} \dfrac{1}{2}$ **(b)** $\sin^{-1}\left(-\dfrac{1}{2}\right)$ **(c)** $\sin^{-1} \dfrac{3}{2}$

SOLUTION

(a) The number in the interval $[-\pi/2, \pi/2]$ whose sine is $\frac{1}{2}$ is $\pi/6$. Thus $\sin^{-1} \frac{1}{2} = \pi/6$.

(b) The number in the interval $[-\pi/2, \pi/2]$ whose sine is $-\frac{1}{2}$ is $-\pi/6$. Thus $\sin^{-1}(-\frac{1}{2}) = -\pi/6$.

(c) Since $\frac{3}{2} > 1$, it is not in the domain of $\sin^{-1} x$, so $\sin^{-1} \frac{3}{2}$ is not defined.

Now Try Exercise 3 ■

EXAMPLE 2 ■ Using a Calculator to Evaluate Inverse Sine

Find approximate values for **(a)** $\sin^{-1}(0.82)$ and **(b)** $\sin^{-1}\frac{1}{3}$.

SOLUTION

We use a calculator to approximate these values. Using the [SIN⁻¹], or [INV] [SIN], or [ARC] [SIN] key(s) on the calculator (with the calculator in radian mode), we get

(a) $\sin^{-1}(0.82) \approx 0.96141$ **(b)** $\sin^{-1}\frac{1}{3} \approx 0.33984$

Now Try Exercises 11 and 21 ■

When evaluating expressions involving $\sin^{-1}$, we need to remember that the range of $\sin^{-1}$ is the interval $[-\pi/2, \pi/2]$.

EXAMPLE 3 ■ Evaluating Expressions with Inverse Sine

Find each value.

(a) $\sin^{-1}\left(\sin \dfrac{\pi}{3}\right)$ **(b)** $\sin^{-1}\left(\sin \dfrac{2\pi}{3}\right)$

SOLUTION

(a) Since $\pi/3$ is in the interval $[-\pi/2, \pi/2]$, we can use the cancellation properties of inverse functions (page 440):

$$\sin^{-1}\left(\sin \frac{\pi}{3}\right) = \frac{\pi}{3} \qquad \text{Cancellation property: } -\frac{\pi}{2} \le \frac{\pi}{3} \le \frac{\pi}{2}$$

(b) We first evaluate the expression in the parentheses:

$$\sin^{-1}\left(\sin \frac{2\pi}{3}\right) = \sin^{-1}\left(\frac{\sqrt{3}}{2}\right) \qquad \text{Evaluate}$$

$$= \frac{\pi}{3} \qquad \text{Because } \sin \frac{\pi}{3} = \frac{\sqrt{3}}{2}$$

Note: $\sin^{-1}(\sin x) = x$ only if $-\frac{\pi}{2} \le x \le \frac{\pi}{2}$.

Now Try Exercises 31 and 33 ■

■ The Inverse Cosine Function

If the domain of the cosine function is restricted to the interval $[0, \pi]$, the resulting function is one-to-one and so has an inverse. We choose this interval because on it, cosine attains each of its values exactly once (see Figure 3).

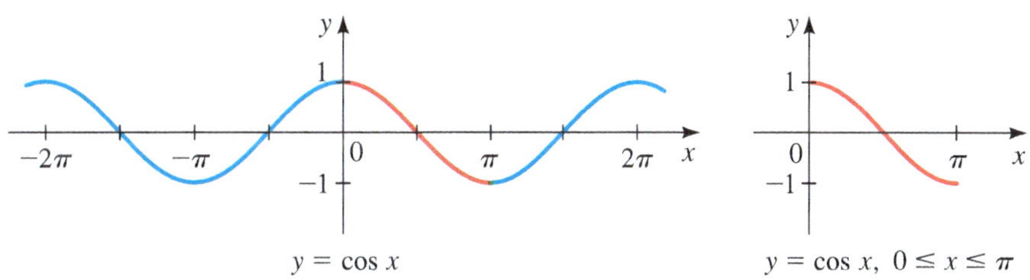

FIGURE 3 Graphs of the cosine function and the restricted cosine function

DEFINITION OF THE INVERSE COSINE FUNCTION

The **inverse cosine function** is the function $\cos^{-1}$ with domain $[-1, 1]$ and range $[0, \pi]$ defined by

$$\cos^{-1} x = y \quad \Leftrightarrow \quad \cos y = x$$

The inverse cosine function is also called **arccosine**, denoted by **arccos**.

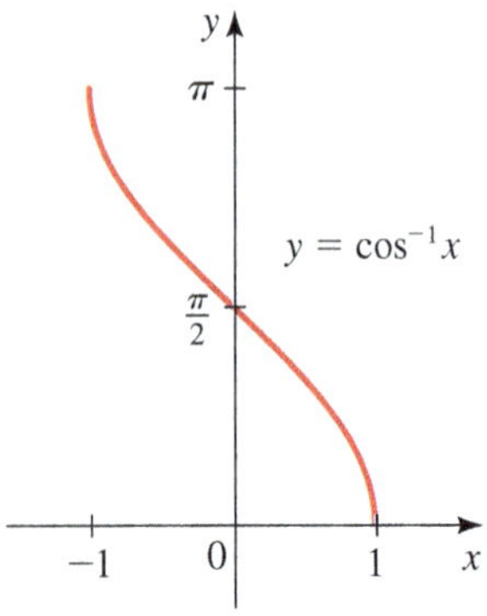

FIGURE 4 Graph of $y = \cos^{-1} x$

Thus $y = \cos^{-1} x$ *is the number in the interval* $[0, \pi]$ *whose cosine is* x. The following **cancellation properties** follow from the inverse function properties.

$$\cos(\cos^{-1} x) = x \quad \text{for} \quad -1 \le x \le 1$$
$$\cos^{-1}(\cos x) = x \quad \text{for} \quad 0 \le x \le \pi$$

The graph of $y = \cos^{-1} x$ is shown in Figure 4; it is obtained by reflecting the graph of $y = \cos x$, $0 \le x \le \pi$, in the line $y = x$.

EXAMPLE 4 ■ Evaluating the Inverse Cosine Function

Find each value.

(a) $\cos^{-1} \dfrac{\sqrt{3}}{2}$ **(b)** $\cos^{-1} 0$ **(c)** $\cos^{-1}\left(-\dfrac{1}{2}\right)$

SOLUTION

(a) The number in the interval $[0, \pi]$ whose cosine is $\sqrt{3}/2$ is $\pi/6$. Thus $\cos^{-1}(\sqrt{3}/2) = \pi/6$.

(b) The number in the interval $[0, \pi]$ whose cosine is 0 is $\pi/2$. Thus $\cos^{-1} 0 = \pi/2$.

(c) The number in the interval $[0, \pi]$ whose cosine is $-\frac{1}{2}$ is $2\pi/3$. Thus $\cos^{-1}(-\frac{1}{2}) = 2\pi/3$. (The graph in Figure 4 shows that if $-1 \le x < 0$, then $\cos^{-1} x > \pi/2$.)

Now Try Exercises 5 and 13 ■

EXAMPLE 5 ■ Evaluating Expressions with Inverse Cosine

Find each value.

(a) $\cos^{-1}\left(\cos \dfrac{2\pi}{3}\right)$ **(b)** $\cos^{-1}\left(\cos \dfrac{5\pi}{3}\right)$

SOLUTION

(a) Since $2\pi/3$ is in the interval $[0, \pi]$ we can use the above cancellation properties:

$$\cos^{-1}\left(\cos \frac{2\pi}{3}\right) = \frac{2\pi}{3} \qquad \text{Cancellation property: } 0 \le \frac{2\pi}{3} \le \pi$$

(b) We first evaluate the expression in the parentheses:

$$\cos^{-1}\left(\cos \frac{5\pi}{3}\right) = \cos^{-1}\left(\frac{1}{2}\right) \qquad \text{Evaluate}$$
$$= \frac{\pi}{3} \qquad \text{Because } \cos \frac{\pi}{3} = \frac{1}{2}$$

Note: $\cos^{-1}(\cos x) = x$ only if $0 \le x \le \pi$.

Now Try Exercises 35 and 37 ■

■ The Inverse Tangent Function

We restrict the domain of the tangent function to the interval $(-\pi/2, \pi/2)$ to obtain a one-to-one function.

DEFINITION OF THE INVERSE TANGENT FUNCTION

The **inverse tangent function** is the function $\tan^{-1}$ with domain $\mathbb{R}$ and range $(-\pi/2, \pi/2)$ defined by

$$\tan^{-1} x = y \quad \Leftrightarrow \quad \tan y = x$$

The inverse tangent function is also called **arctangent**, denoted by **arctan**.

Thus $y = \tan^{-1} x$ *is the number in the interval* $(-\pi/2, \pi/2)$ *whose tangent is* x. The following **cancellation properties** follow from the inverse function properties.

$$\tan(\tan^{-1} x) = x \quad \text{for} \quad x \in \mathbb{R}$$
$$\tan^{-1}(\tan x) = x \quad \text{for} \quad -\frac{\pi}{2} < x < \frac{\pi}{2}$$

Figure 5 shows the graph of $y = \tan x$ on the interval $(-\pi/2, \pi/2)$ and the graph of its inverse function, $y = \tan^{-1} x$.

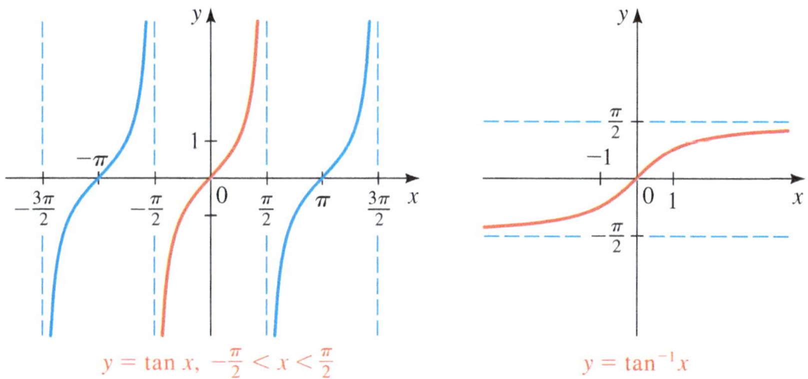

FIGURE 5 Graphs of the restricted tangent function and the inverse tangent function

EXAMPLE 6 ■ Evaluating the Inverse Tangent Function

Find each value.

(a) $\tan^{-1} 1$ **(b)** $\tan^{-1} \sqrt{3}$ **(c)** $\tan^{-1}(20)$

SOLUTION

(a) The number in the interval $(-\pi/2, \pi/2)$ with tangent 1 is $\pi/4$. Thus $\tan^{-1} 1 = \pi/4$.

(b) The number in the interval $(-\pi/2, \pi/2)$ with tangent $\sqrt{3}$ is $\pi/3$. Thus $\tan^{-1} \sqrt{3} = \pi/3$.

(c) We use a calculator (in radian mode) to find that $\tan^{-1}(20) \approx 1.52084$.

Now Try Exercises 7 and 17 ■

■ The Inverse Secant, Cosecant, and Cotangent Functions

See Exercise 46 in Section 6.4 (page 508) for a way of finding the values of these inverse trigonometric functions on a calculator.

To define the inverse functions of the secant, cosecant, and cotangent functions, we restrict the domain of each function to a set on which it is one-to-one and on which it attains all its values. Although any interval satisfying these criteria is appropriate, we choose to restrict the domains in a way that simplifies the choice of sign in computations involving inverse trigonometric functions. The choices we make are also appropriate for calculus. This explains the seemingly strange restriction for the domains of the secant and cosecant functions. We end this section by displaying the graphs of the secant, cosecant, and cotangent functions with their restricted domains and the graphs of their inverse functions (Figures 6–8).

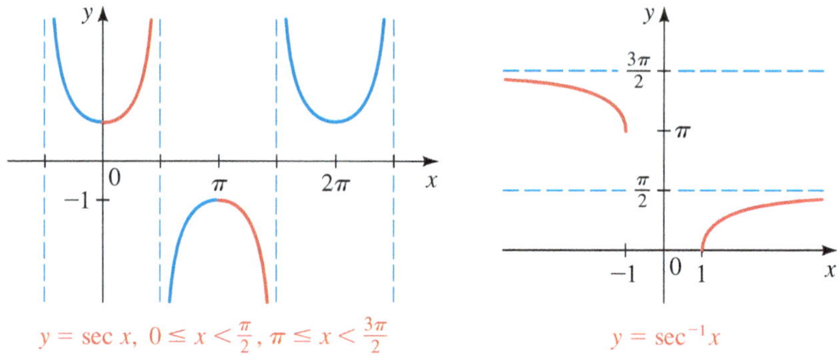

FIGURE 6 The inverse secant function

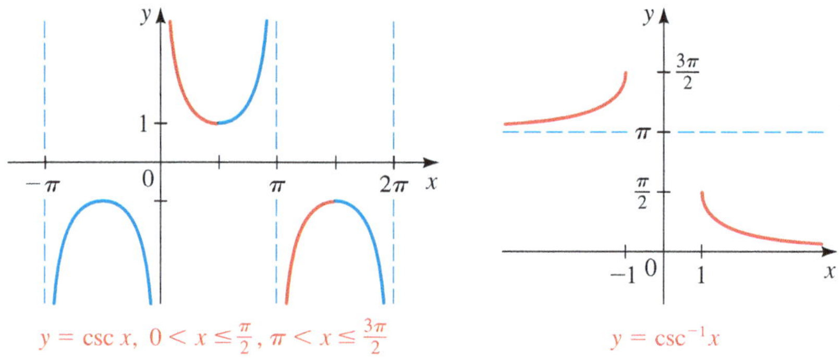

FIGURE 7 The inverse cosecant function

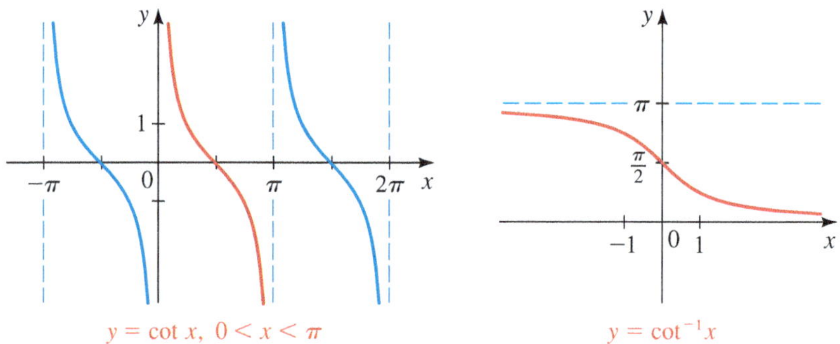

FIGURE 8 The inverse cotangent function

5.5 EXERCISES

CONCEPTS

1. (a) To define the inverse sine function, we restrict the domain of sine to the interval __________. On this interval the sine function is one-to-one, and its inverse function $\sin^{-1}$ is defined by $\sin^{-1} x = y \Leftrightarrow \sin$ _____ = _____. For example, $\sin^{-1}\frac{1}{2} =$ _____ because sin _____ = _____.

(b) To define the inverse cosine function, we restrict the domain of cosine to the interval __________. On this interval the cosine function is one-to-one and its inverse function $\cos^{-1}$ is defined by $\cos^{-1} x = y \Leftrightarrow$ cos _____ = _____. For example, $\cos^{-1}\frac{1}{2} =$ _____ because cos _____ = _____.

2. The cancellation property $\sin^{-1}(\sin x) = x$ is valid for x in the interval __________. Which of the following is not true?

(i) $\sin^{-1}\left(\sin\frac{\pi}{3}\right) = \frac{\pi}{3}$ (ii) $\sin^{-1}\left(\sin\frac{10\pi}{3}\right) = \frac{10\pi}{3}$

(iii) $\sin^{-1}\left(\sin\left(-\frac{\pi}{4}\right)\right) = -\frac{\pi}{4}$

SKILLS

3–10 ■ Evaluating Inverse Trigonometric Functions Find the exact value of each expression, if it is defined.

3. (a) $\sin^{-1} 1$ **(b)** $\sin^{-1}\frac{\sqrt{3}}{2}$ **(c)** $\sin^{-1} 2$

4. **(a)** $\sin^{-1}(-1)$ **(b)** $\sin^{-1}\dfrac{\sqrt{2}}{2}$ **(c)** $\sin^{-1}(-2)$

5. **(a)** $\cos^{-1}(-1)$ **(b)** $\cos^{-1}\frac{1}{2}$ **(c)** $\cos^{-1}\left(-\dfrac{\sqrt{3}}{2}\right)$

6. **(a)** $\cos^{-1}\dfrac{\sqrt{2}}{2}$ **(b)** $\cos^{-1}1$ **(c)** $\cos^{-1}\left(-\dfrac{\sqrt{2}}{2}\right)$

7. **(a)** $\tan^{-1}(-1)$ **(b)** $\tan^{-1}\sqrt{3}$ **(c)** $\tan^{-1}\dfrac{\sqrt{3}}{3}$

8. **(a)** $\tan^{-1}0$ **(b)** $\tan^{-1}(-\sqrt{3})$ **(c)** $\tan^{-1}\left(-\dfrac{\sqrt{3}}{3}\right)$

9. **(a)** $\cos^{-1}(-\frac{1}{2})$ **(b)** $\sin^{-1}\left(-\dfrac{\sqrt{2}}{2}\right)$ **(c)** $\tan^{-1}1$

10. **(a)** $\cos^{-1}0$ **(b)** $\sin^{-1}0$ **(c)** $\sin^{-1}(-\frac{1}{2})$

11–22 ■ **Inverse Trigonometric Functions with a Calculator** Use a calculator to find an approximate value of each expression correct to five decimal places, if it is defined.

11. $\sin^{-1}\frac{2}{3}$ **12.** $\sin^{-1}(-\frac{8}{9})$

13. $\cos^{-1}(-\frac{3}{7})$ **14.** $\cos^{-1}(\frac{4}{9})$

15. $\cos^{-1}(-0.92761)$ **16.** $\sin^{-1}(0.13844)$

17. $\tan^{-1}10$ **18.** $\tan^{-1}(-26)$

19. $\tan^{-1}(1.23456)$ **20.** $\cos^{-1}(1.23456)$

21. $\sin^{-1}(-0.25713)$ **22.** $\tan^{-1}(-0.25713)$

23–48 ■ **Simplifying Expressions Involving Trigonometric Functions** Find the exact value of the expression, if it is defined.

23. $\sin(\sin^{-1}\frac{1}{4})$ **24.** $\cos(\cos^{-1}\frac{2}{3})$

25. $\tan(\tan^{-1}5)$ **26.** $\sin(\sin^{-1}5)$

27. $\sin(\sin^{-1}\frac{3}{2})$ **28.** $\tan(\tan^{-1}\frac{3}{2})$

29. $\cos\left(\cos^{-1}\left(-\dfrac{1}{5}\right)\right)$ **30.** $\sin\left(\sin^{-1}\left(-\dfrac{3}{4}\right)\right)$

31. $\sin^{-1}\left(\sin\left(\dfrac{\pi}{4}\right)\right)$ **32.** $\cos^{-1}\left(\cos\left(\dfrac{\pi}{4}\right)\right)$

33. $\sin^{-1}\left(\sin\left(\dfrac{3\pi}{4}\right)\right)$ **34.** $\cos^{-1}\left(\cos\left(\dfrac{3\pi}{4}\right)\right)$

35. $\cos^{-1}\left(\cos\left(\dfrac{5\pi}{6}\right)\right)$ **36.** $\sin^{-1}\left(\sin\left(\dfrac{5\pi}{6}\right)\right)$

37. $\cos^{-1}\left(\cos\left(\dfrac{7\pi}{6}\right)\right)$ **38.** $\sin^{-1}\left(\sin\left(\dfrac{7\pi}{6}\right)\right)$

39. $\tan^{-1}\left(\tan\left(\dfrac{\pi}{4}\right)\right)$ **40.** $\tan^{-1}\left(\tan\left(-\dfrac{\pi}{3}\right)\right)$

41. $\tan^{-1}\left(\tan\left(\dfrac{2\pi}{3}\right)\right)$ **42.** $\sin^{-1}\left(\sin\left(\dfrac{11\pi}{4}\right)\right)$

43. $\tan(\sin^{-1}\frac{1}{2})$ **44.** $\cos(\sin^{-1}0)$

45. $\cos\left(\sin^{-1}\dfrac{\sqrt{3}}{2}\right)$ **46.** $\tan\left(\sin^{-1}\dfrac{\sqrt{2}}{2}\right)$

47. $\sin(\tan^{-1}(-1))$ **48.** $\sin(\tan^{-1}(-\sqrt{3}))$

DISCUSS ■ DISCOVER ■ PROVE ■ WRITE

49–50 ■ **PROVE: Identities Involving Inverse Trigonometric Functions** **(a)** Graph the function and make a conjecture, and **(b)** prove that your conjecture is true.

49. $y = \sin^{-1}x + \cos^{-1}x$ **50.** $y = \tan^{-1}x + \tan^{-1}\dfrac{1}{x}$

51. **DISCUSS: Two Different Compositions** Let f and g be the functions

$$f(x) = \sin(\sin^{-1}x)$$

and

$$g(x) = \sin^{-1}(\sin x)$$

By the cancellation properties, $f(x) = x$ and $g(x) = x$ for suitable values of x. But these functions are not the same for all x. Graph both f and g to show how the functions differ. (Think carefully about the domain and range of $\sin^{-1}$).

5.6 MODELING HARMONIC MOTION

■ Simple Harmonic Motion ■ Damped Harmonic Motion ■ Phase and Phase Difference

Periodic behavior—behavior that repeats over and over again—is common in nature. Perhaps the most familiar example is the daily rising and setting of the sun, which results in the repetitive pattern of day, night, day, night, ... Another example is the daily variation of tide levels at the beach, which results in the repetitive pattern of high tide, low tide, high tide, low tide, ... Certain animal populations increase and decrease in a predictable periodic pattern: A large population exhausts the food supply, which causes the population to dwindle; this in turn results in a more plentiful food supply, which makes it possible for the population to increase; and the pattern then repeats over and over (see *Discovery Project: Predator/Prey Models* referenced on page 427).

Other common examples of periodic behavior involve motion that is caused by vibration or oscillation. A mass suspended from a spring that has been compressed and then allowed to vibrate vertically is a simple example. This back-and-forth motion also occurs in such diverse phenomena as sound waves, light waves, alternating electrical current, and pulsating stars, to name a few. In this section we consider the problem of modeling periodic behavior.

Simple Harmonic Motion

The trigonometric functions are ideally suited for modeling periodic behavior. A glance at the graphs of the sine and cosine functions, for instance, tells us that these functions themselves exhibit periodic behavior. Figure 1 shows the graph of $y = \sin t$. If we think of t as time, we see that as time goes on, $y = \sin t$ increases and decreases over and over again. Figure 2 shows that the motion of a vibrating mass on a spring is modeled very accurately by $y = \sin t$.

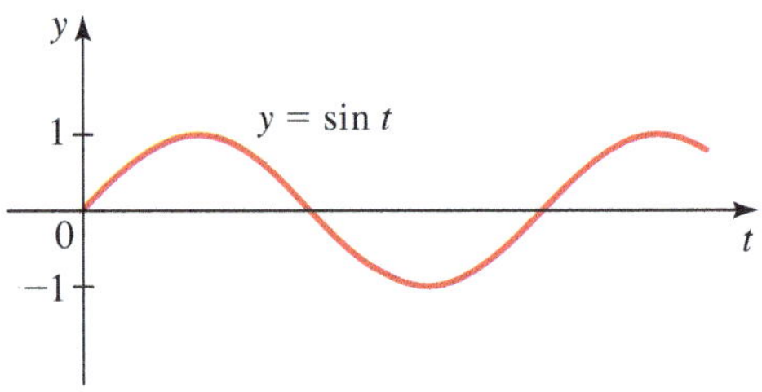

FIGURE 1 $y = \sin t$

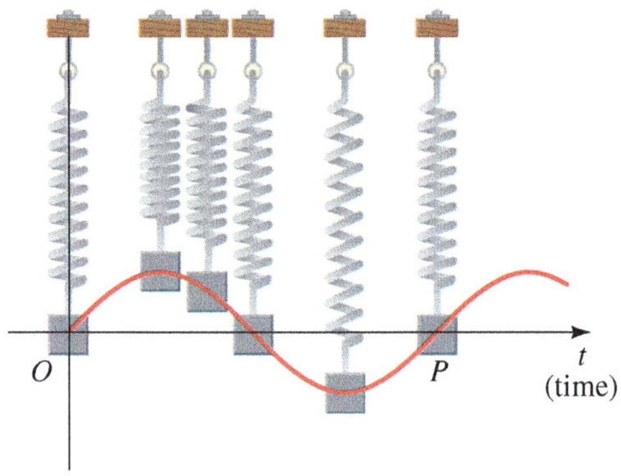

FIGURE 2 Motion of a vibrating spring is modeled by $y = \sin t$.

Notice that the mass returns to its original position over and over again. A **cycle** is one complete vibration of an object, so the mass in Figure 2 completes one cycle of its motion between O and P. Our observations about how the sine and cosine functions model periodic behavior are summarized in the following box.

The main difference between the two equations describing simple harmonic motion is the starting point. At $t = 0$ we get

$$y = a \sin \omega \cdot 0 = 0$$

$$y = a \cos \omega \cdot 0 = a$$

In the first case the motion "starts" with zero displacement, whereas in the second case the motion "starts" with the displacement at maximum (at the amplitude a).

The symbol ω is the lowercase Greek letter "omega," and ν is the letter "nu."

SIMPLE HARMONIC MOTION

If the equation describing the displacement y of an object at time t is

$$y = a \sin \omega t \qquad \text{or} \qquad y = a \cos \omega t$$

then the object is in **simple harmonic motion**. In this case,

amplitude $= |a|$ — Maximum displacement of the object

period $= \dfrac{2\pi}{\omega}$ — Time required to complete one cycle

frequency $= \dfrac{\omega}{2\pi}$ — Number of cycles per unit of time

Notice that the functions

$$y = a \sin 2\pi\nu t \qquad \text{and} \qquad y = a \cos 2\pi\nu t$$

have frequency ν, because $2\pi\nu/(2\pi) = \nu$. Since we can immediately read the frequency from these equations, we often write equations of simple harmonic motion in this form.

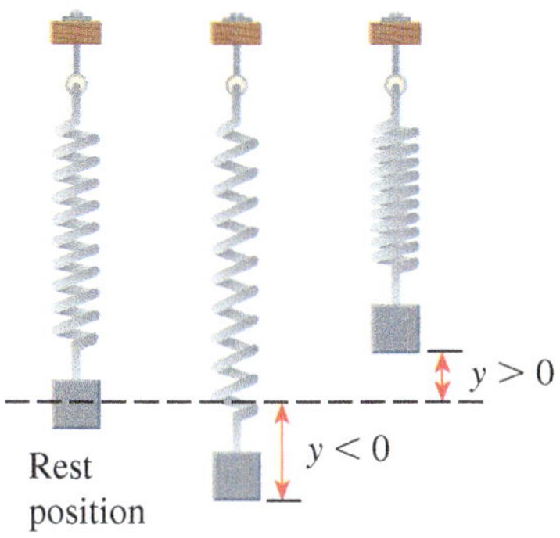

FIGURE 3

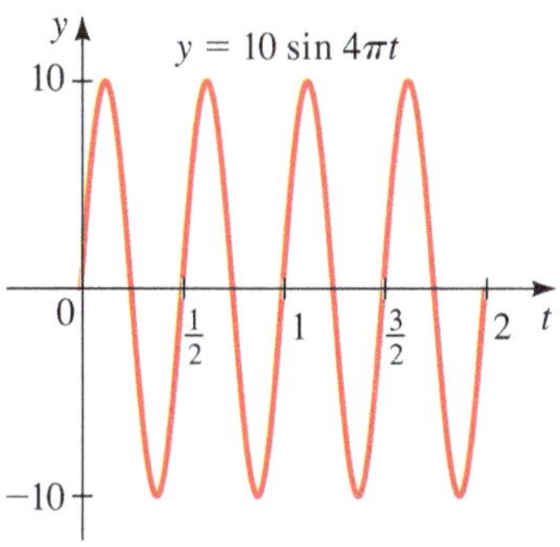

FIGURE 4

EXAMPLE 1 ■ A Vibrating Spring

The displacement of a mass suspended by a spring is modeled by the function

$$y = 10 \sin 4\pi t$$

where y is measured in inches and t in seconds (see Figure 3).

(a) Find the amplitude, period, and frequency of the motion of the mass.

(b) Sketch a graph of the displacement of the mass.

SOLUTION

(a) From the formulas for amplitude, period, and frequency we get

$$\text{amplitude} = |a| = 10 \text{ in.}$$

$$\text{period} = \frac{2\pi}{\omega} = \frac{2\pi}{4\pi} = \frac{1}{2} \text{ s}$$

$$\text{frequency} = \frac{\omega}{2\pi} = \frac{4\pi}{2\pi} = 2 \text{ cycles per second (Hz)}$$

(b) The graph of the displacement of the mass at time t is shown in Figure 4.

Now Try Exercise 5 ■

An important situation in which simple harmonic motion occurs is in the production of sound. Sound is produced by a regular variation in air pressure from the normal pressure. If the pressure varies in simple harmonic motion, then a pure sound is produced. The tone of the sound depends on the frequency, and the loudness depends on the amplitude.

EXAMPLE 2 ■ Vibrations of a Musical Note

A sousaphone player plays the note E and sustains the sound for some time. For a pure E the variation in pressure from normal air pressure is given by

$$V(t) = 0.2 \sin 80\pi t$$

where V is measured in pounds per square inch and t is measured in seconds.

(a) Find the amplitude, period, and frequency of V.

(b) Sketch a graph of V.

(c) If the player increases the loudness of the note, how does the equation for V change?

(d) If the player is playing the note incorrectly and it is a little flat, how does the equation for V change?

SOLUTION

(a) From the formulas for amplitude, period, and frequency we get

$$\text{amplitude} = |0.2| = 0.2$$

$$\text{period} = \frac{2\pi}{80\pi} = \frac{1}{40}$$

$$\text{frequency} = \frac{80\pi}{2\pi} = 40$$

(b) The graph of V is shown in Figure 5.

(c) If the player increases the loudness the amplitude increases. So the number 0.2 is replaced by a larger number.

(d) If the note is flat, then the frequency is decreased. Thus the coefficient of t is less than 80π.

Now Try Exercise 41 ■

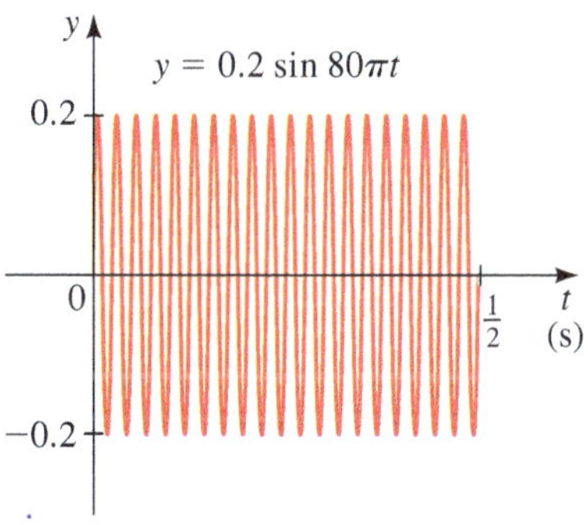

FIGURE 5

EXAMPLE 3 ■ Modeling a Vibrating Spring

A mass is suspended from a spring. The spring is compressed a distance of 4 cm and then released. It is observed that the mass returns to the compressed position after $\frac{1}{3}$ s.

(a) Find a function that models the displacement of the mass.

(b) Sketch the graph of the displacement of the mass.

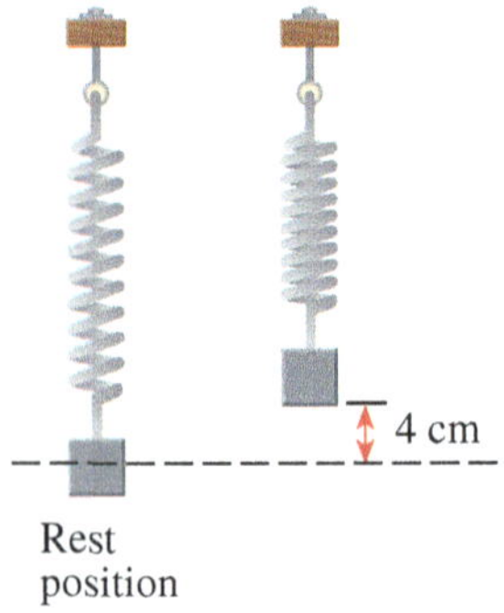

SOLUTION

(a) The motion of the mass is given by one of the equations for simple harmonic motion. The amplitude of the motion is 4 cm. Since this amplitude is reached at time $t = 0$, an appropriate function that models the displacement is of the form

$$y = a \cos \omega t$$

Since the period is $p = \frac{1}{3}$, we can find ω from the following equation:

$$\text{period} = \frac{2\pi}{\omega}$$

$$\frac{1}{3} = \frac{2\pi}{\omega} \qquad \text{Period} = \tfrac{1}{3}$$

$$\omega = 6\pi \qquad \text{Solve for } \omega$$

So the motion of the mass is modeled by the function

$$y = 4 \cos 6\pi t$$

where y is the displacement from the rest position at time t. Notice that when $t = 0$, the displacement is $y = 4$, as we expect.

(b) The graph of the displacement of the mass at time t is shown in Figure 6.

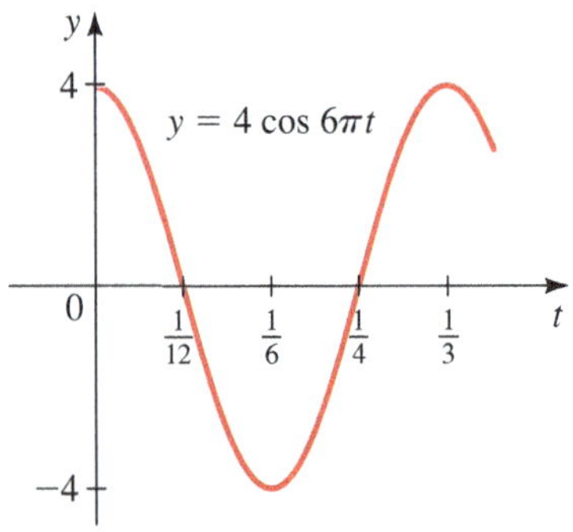

FIGURE 6

Now Try Exercises 17 and 47 ■

In general, the sine or cosine functions representing harmonic motion may be shifted horizontally or vertically. In this case the equations take the form

$$y = a \sin(\omega(t - c)) + b \qquad \text{or} \qquad y = a \cos(\omega(t - c)) + b$$

The vertical shift b indicates that the variation occurs around an average value b. The horizontal shift c indicates the position of the object at $t = 0$. (See Figure 7.)

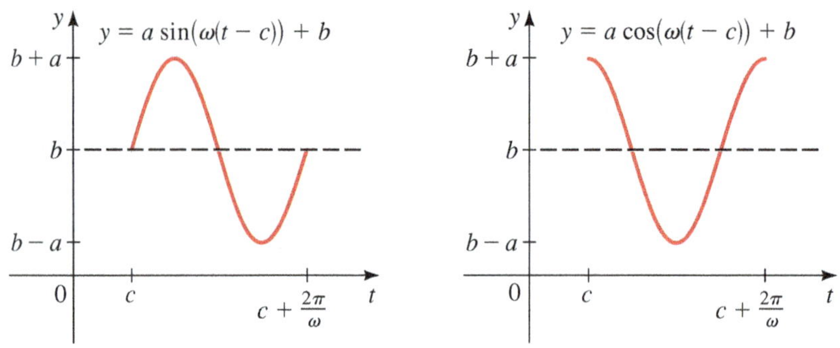

FIGURE 7

(a) (b)

EXAMPLE 4 ■ Modeling the Brightness of a Variable Star

A variable star is one whose brightness alternately increases and decreases. For the variable star Delta Cephei the time between periods of maximum brightness is 5.4 days. The average brightness (or magnitude) of the star is 4.0, and its brightness varies by ± 0.35 magnitude.

(a) Find a function that models the brightness of Delta Cephei as a function of time.

(b) Sketch a graph of the brightness of Delta Cephei as a function of time.

SOLUTION

(a) Let's find a function in the form

$$y = a\cos(\omega(t - c)) + b$$

The amplitude is the maximum variation from average brightness, so the amplitude is $a = 0.35$ magnitude. We are given that the period is 5.4 days, so

$$\omega = \frac{2\pi}{5.4} \approx 1.16$$

Since the brightness varies from an average value of 4.0 magnitudes, the graph is shifted upward by $b = 4.0$. If we take $t = 0$ to be a time when the star is at maximum brightness, there is no horizontal shift, so $c = 0$ (because a cosine curve achieves its maximum at $t = 0$). Thus the function we want is

$$y = 0.35\cos(1.16t) + 4.0$$

where t is the number of days from a time when the star is at maximum brightness.

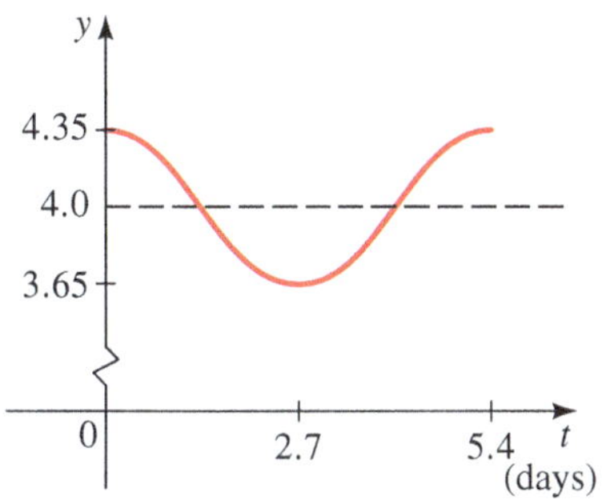

FIGURE 8

(b) The graph is sketched in Figure 8.

Now Try Exercise 51 ■

The number of hours of daylight varies throughout the course of a year. In the Northern Hemisphere the longest day is June 21, and the shortest is December 21. The average length of daylight is 12 h, and the variation from this average depends on the latitude. (For example, Fairbanks, Alaska, experiences more than 20 h of daylight on the longest day and less than 4 h on the shortest day!) The graph in Figure 9 shows the number of hours of daylight at different times of the year for various latitudes. It's apparent from the graph that the variation in hours of daylight is simple harmonic.

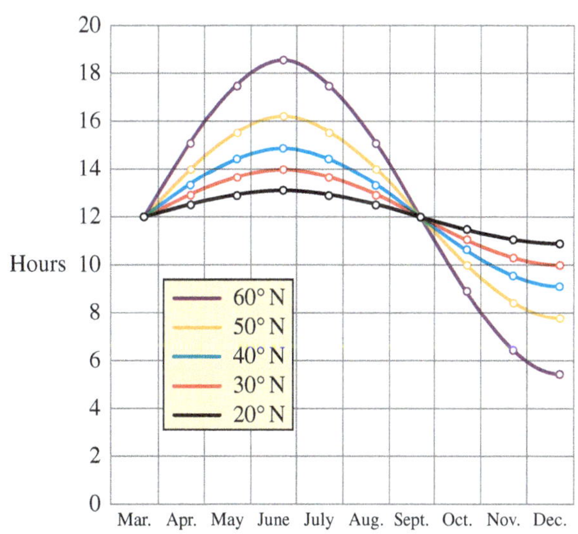

Source: Lucia C. Harrison, *Daylight, Twilight, Darkness and Time* (New York: Silver, Burdett, 1935), page 40

FIGURE 9 Graph of the length of daylight from March 21 through December 21 at various latitudes

EXAMPLE 5 ■ Modeling the Number of Hours of Daylight

In Philadelphia (40° N latitude) the longest day of the year has 14 h 50 min of daylight, and the shortest day has 9 h 10 min of daylight.

(a) Find a function L that models the length of daylight as a function of t, the number of days from January 1.

(b) An astronomer needs at least 11 hours of darkness for a long exposure astronomical photograph. On what days of the year are such long exposures possible?

SOLUTION

(a) We need to find a function in the form

$$y = a\sin(\omega(t - c)) + b$$

whose graph is the 40° N latitude curve in Figure 9. From the information given, we see that the amplitude is

$$a = \tfrac{1}{2}(14\tfrac{5}{6} - 9\tfrac{1}{6}) \approx 2.83 \text{ h}$$

Since there are 365 days in a year, the period is 365, so

$$\omega = \frac{2\pi}{365} \approx 0.0172$$

Since the average length of daylight is 12 h, the graph is shifted upward by 12, so $b = 12$. Since the curve attains the average value (12) on March 21, the 80th day of the year, the curve is shifted 80 units to the right. Thus $c = 80$. So a function that models the number of hours of daylight is

$$y = 2.83\sin(0.0172(t - 80)) + 12$$

where t is the number of days from January 1.

(b) A day has 24 h, so 11 h of night correspond to 13 h of daylight. So we need to solve the inequality $y \le 13$. To solve this inequality graphically, we graph $y = 2.83\sin 0.0172(t - 80) + 12$ and $y = 13$ on the same graph. From the graph in Figure 10 we see that there are fewer than 13 h of daylight between day 1 (January 1) and day 101 (April 11) and between day 241 (August 29) and day 365 (December 31).

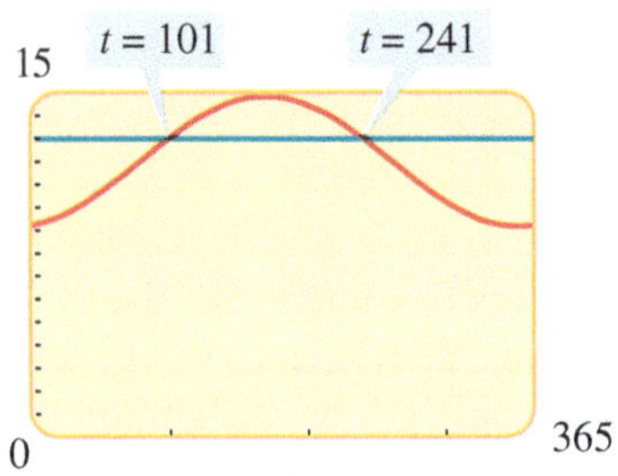

FIGURE 10

Now Try Exercise 53 ■

Another situation in which simple harmonic motion occurs is in alternating current (AC) generators. Alternating current is produced when an armature rotates about its axis in a magnetic field.

Figure 11 represents a simple version of such a generator. As the wire passes through the magnetic field, a voltage E is generated in the wire. It can be shown that the voltage generated is given by

$$E(t) = E_0 \cos \omega t$$

where E_0 is the maximum voltage produced (which depends on the strength of the magnetic field) and $\omega/(2\pi)$ is the number of revolutions per second of the armature (the frequency).

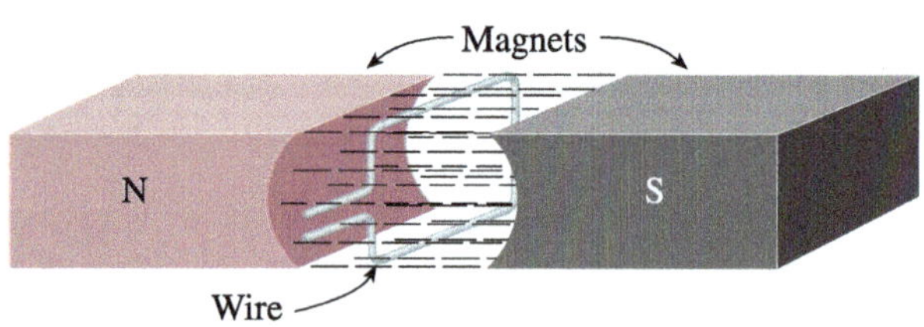

FIGURE 11

EXAMPLE 6 ■ Modeling Alternating Current

Ordinary 110-V household alternating current varies from +155 V to −155 V with a frequency of 60 Hz (cycles per second). Find an equation that describes this variation in voltage.

Why do we say that household current is 110 V when the maximum voltage produced is 155 V? From the symmetry of the cosine function we see that the average voltage produced is zero. This average value would be the same for all AC generators and so gives no information about the voltage generated. To obtain a more informative measure of voltage, engineers use the **root-mean-square** (RMS) method. It can be shown that the RMS voltage is $1/\sqrt{2}$ times the maximum voltage. So for household current the RMS voltage is

$$155 \times \frac{1}{\sqrt{2}} \approx 110 \text{ V}$$

SOLUTION The variation in voltage is simple harmonic. Since the frequency is 60 cycles per second, we have

$$\frac{\omega}{2\pi} = 60 \qquad \text{or} \qquad \omega = 120\pi$$

Let's take $t = 0$ to be a time when the voltage is +155 V. Then

$$E(t) = a \cos \omega t = 155 \cos 120\pi t$$

Now Try Exercise 55 ■

Damped Harmonic Motion

The spring in Figure 2 on page 446 is assumed to oscillate in a frictionless environment. In this hypothetical case the amplitude of the oscillation will not change. In the presence of friction, however, the motion of the spring eventually "dies down"; that is, the amplitude of the motion decreases with time. Motion of this type is called *damped harmonic motion*.

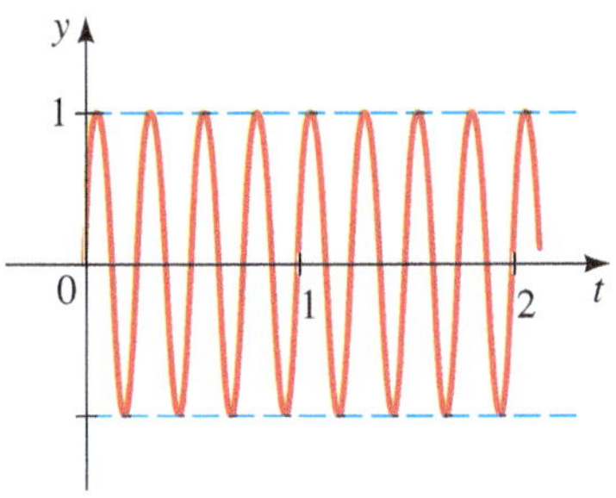

(a) Harmonic motion: $y = \sin 8\pi t$

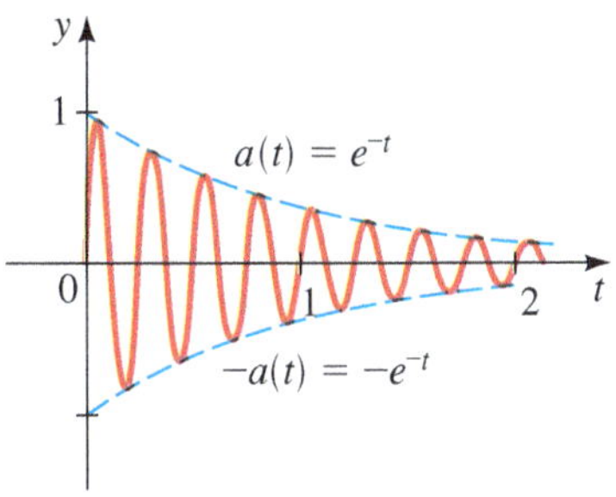

(b) Damped harmonic motion: $y = e^{-t} \sin 8\pi t$

FIGURE 12

DAMPED HARMONIC MOTION

If the equation describing the displacement y of an object at time t is

$$y = ke^{-ct} \sin \omega t \qquad \text{or} \qquad y = ke^{-ct} \cos \omega t \qquad (c > 0)$$

then the object is in **damped harmonic motion**. The constant c is the **damping constant**, k is the initial amplitude, and $2\pi/\omega$ is the period.*

Damped harmonic motion is simply harmonic motion for which the amplitude is governed by the function $a(t) = ke^{-ct}$. Figure 12 shows the difference between harmonic motion and damped harmonic motion.

EXAMPLE 7 ■ Modeling Damped Harmonic Motion

Two mass-spring systems are experiencing damped harmonic motion, both at 0.5 cycles per second and both with an initial maximum displacement of 10 cm. The first has a damping constant of 0.5, and the second has a damping constant of 0.1.

(a) Find functions of the form $g(t) = ke^{-ct} \cos \omega t$ to model the motion in each case.

(b) Graph the two functions you found in part (a). How do they differ?

SOLUTION

Hz is the abbreviation for hertz. One hertz is one cycle per second.

(a) At time $t = 0$ the displacement is 10 cm. Thus $g(0) = ke^{-c \cdot 0} \cos(\omega \cdot 0) = k$, so $k = 10$. Also, the frequency is $f = 0.5$ Hz, and since $\omega = 2\pi f$ (see page 446), we get $\omega = 2\pi(0.5) = \pi$. Using the given damping constants, we find that the motions of the two springs are given by the functions

$$g_1(t) = 10e^{-0.5t} \cos \pi t \qquad \text{and} \qquad g_2(t) = 10e^{-0.1t} \cos \pi t$$

*In the case of damped harmonic motion the term *quasi-period* is often used instead of *period* because the motion is not actually periodic—it diminishes with time. However, we will continue to use the term *period* to avoid confusion.

(b) The functions g_1 and g_2 are graphed in Figure 13. From the graphs we see that in the first case (where the damping constant is larger) the motion dies down quickly, whereas in the second case, perceptible motion continues much longer.

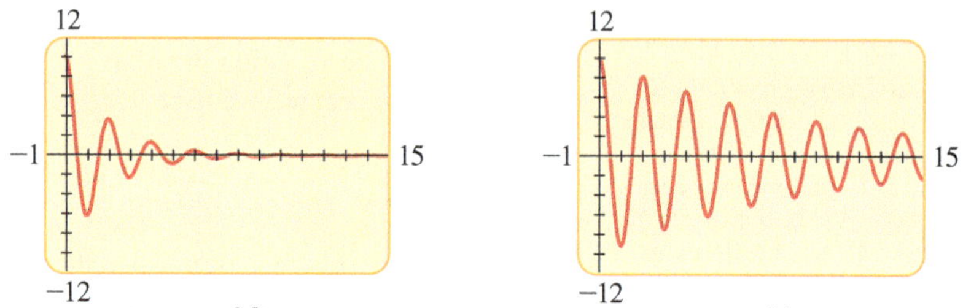

FIGURE 13

$g_1(t) = 10\,e^{-0.5t} \cos \pi t$ $\qquad$ $g_2(t) = 10\,e^{-0.1t} \cos \pi t$

Now Try Exercise 21 ■

As Example 7 indicates, the larger the damping constant c, the quicker the oscillation dies down. When a guitar string is plucked and then allowed to vibrate freely, a point on that string undergoes damped harmonic motion. We hear the damping of the motion as the sound produced by the vibration of the string fades. How fast the damping of the string occurs (as measured by the size of the constant c) is a property of the size of the string and the material it is made of. Another example of damped harmonic motion is the motion that a shock absorber on a car undergoes when the car hits a bump in the road. In this case the shock absorber is engineered to damp the motion as quickly as possible (large c) and to have the frequency as small as possible (small ω). On the other hand, the sound produced by a tuba player playing a note is undamped as long as the player can maintain the loudness of the note. The electromagnetic waves that produce light move in simple harmonic motion that is not damped.

EXAMPLE 8 ■ A Vibrating Violin String

The G-string on a violin is pulled a distance of 0.5 cm above its rest position, then released and allowed to vibrate. The damping constant c for this string is determined to be 1.4. Suppose that the note produced is a pure G (frequency = 200 Hz). Find an equation that describes the motion of the point at which the string was plucked.

SOLUTION Let P be the point at which the string was plucked. We will find a function $f(t)$ that gives the distance at time t of the point P from its original rest position. Since the maximum displacement occurs at $t = 0$, we find an equation in the form

$$y = ke^{-ct} \cos \omega t$$

From this equation we see that $f(0) = k$. But we know that the original displacement of the string is 0.5 cm. Thus $k = 0.5$. Since the frequency of the vibration is 200, we have $\omega = 2\pi f = 2\pi(200) = 400\pi$. Finally, since we know that the damping constant is 1.4, we get

$$f(t) = 0.5e^{-1.4t} \cos 400\pi t$$

Now Try Exercise 57 ■

EXAMPLE 9 ■ Ripples on a Pond

A stone is dropped in a calm lake, causing waves to form. The up-and-down motion of a point on the surface of the water is modeled by damped harmonic motion. At some time the amplitude of the wave is measured, and 20 s later it is found that the amplitude has dropped to $\frac{1}{10}$ of this value. Find the damping constant c.

SOLUTION The amplitude is governed by the coefficient ke^{-ct} in the equations for damped harmonic motion. Thus the amplitude at time t is ke^{-ct}, and 20 s later, it is $ke^{-c(t+20)}$. So because the later value is $\frac{1}{10}$ the earlier value, we have

$$ke^{-c(t+20)} = \tfrac{1}{10}ke^{-ct}$$

We now solve this equation for c. Canceling k and using the Laws of Exponents, we get

$$e^{-ct} \cdot e^{-20c} = \tfrac{1}{10}e^{-ct}$$

$$e^{-20c} = \tfrac{1}{10} \qquad \text{Cancel } e^{-ct}$$

$$e^{20c} = 10 \qquad \text{Take reciprocals}$$

Taking the natural logarithm of each side gives

$$20c = \ln(10)$$

$$c = \tfrac{1}{20}\ln(10) \approx \tfrac{1}{20}(2.30) \approx 0.12$$

Thus the damping constant is $c \approx 0.12$.

Now Try Exercise 59 ■

Phase and Phase Difference

When two objects are moving in simple harmonic motion with the same frequency, it is often important to determine whether the objects are "moving together" or by how much their motions differ. Let's consider a specific example.

Suppose that an object is rotating along the unit circle and the height y of the object at time t is given by $y = \sin(kt - b)$. When $t = 0$, the height is $y = \sin(-b)$. This means that the motion "starts" at an angle b as shown in Figure 14.

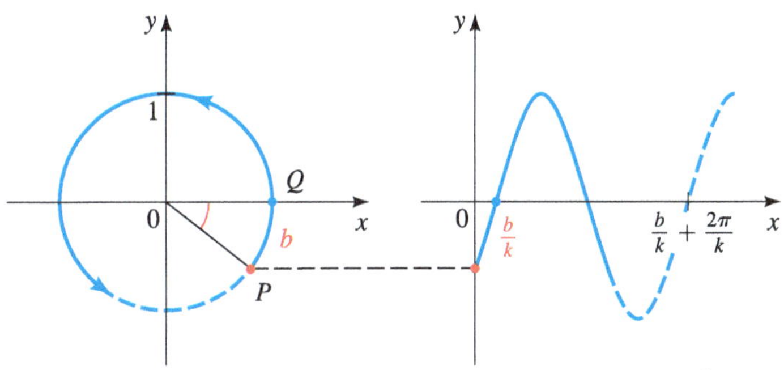

(a) Phase angle b (b) Horizontal shift $\frac{b}{k}$

FIGURE 14 Graph of $y = \sin(kt - b)$

The phase angle b depends only on the starting position of the object and not on the frequency. The lag time does depend on the frequency.

We can view the starting point in two ways: as the *angle* between P and Q on the unit circle or as the *time* required for P to "catch up" to Q. The angle b is called the **phase** (or **phase angle**). To find the time required, we factor out k:

$$y = \sin(kt - b) = \sin k\left(t - \frac{b}{k}\right)$$

We see that P "catches up" to Q (that is, $y = 0$) when $t = b/k$. This last equation also shows that the graph in Figure 14(b) is **shifted horizontally** b/k (to the right) on the t-axis. The time b/k is called the **lag time** if $b > 0$ (because P is behind, or lags, Q by b/k time units) and is called the **lead time** if $b < 0$.

PHASE

Any sine curve can be expressed in the following equivalent forms:

$y = A\sin(kt - b)$ The **phase** is b.

$y = A\sin k\left(t - \frac{b}{k}\right)$ The **horizontal shift** is $\frac{b}{k}$.

It is often important to know whether two waves with the same period (modeled by sine curves) are *in phase* or *out of phase*. For the curves

$$y_1 = A\sin(kt - b) \quad \text{and} \quad y_2 = A\sin(kt - c)$$

the **phase difference** between y_1 and y_2 is $b - c$. If the phase difference is a multiple of 2π, the waves are **in phase**; otherwise, the waves are **out of phase**. If two sine curves are in phase, then their graphs coincide.

Note that the phase difference depends on the order in which the functions are given.

EXAMPLE 10 ■ Finding Phase and Phase Difference

Objects are in harmonic motion modeled by the following curves:

$$y_1 = 10\sin\left(3t - \frac{\pi}{6}\right) \qquad y_2 = 10\sin\left(3t - \frac{\pi}{2}\right) \qquad y_3 = 10\sin\left(3t + \frac{23\pi}{6}\right)$$

(a) Find the amplitude, period, phase, and horizontal shift of the curve y_1.

(b) Find the phase difference between the curves y_1 and y_2. Are the two curves in phase?

(c) Find the phase difference between the curves y_1 and y_3. Are the two curves in phase?

(d) Sketch all three curves on the same axes.

SOLUTION

(a) The amplitude is 10, the period is $2\pi/3$, and the phase is $\pi/6$. To find the horizontal shift, we factor:

$$y_1 = 10\sin\left(3t - \frac{\pi}{6}\right) = 10\sin 3\left(t - \frac{\pi}{18}\right)$$

So the horizontal shift is $\pi/18$.

(b) The phase of y_2 is $\pi/2$. So the phase difference is

$$\frac{\pi}{2} - \frac{\pi}{6} = \frac{\pi}{3}$$

The phase difference is not a multiple of 2π, so the two curves are out of phase.

(c) The phase of y_3 is $-23\pi/6$. So the phase difference is

$$\frac{\pi}{6} - \left(-\frac{23\pi}{6}\right) = 4\pi = 2(2\pi)$$

The phase difference is a multiple of 2π, so the two curves are in phase.

(d) The graphs are shown in Figure 15. Notice that the curves y_1 and y_3 have the same graph because they are in phase.

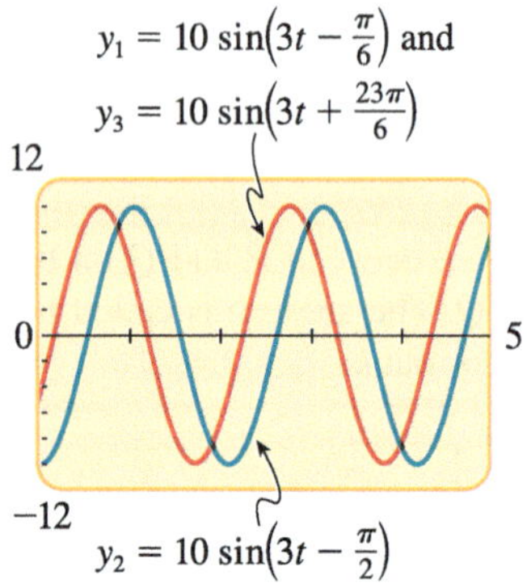

FIGURE 15

Now Try Exercises 29 and 35 ■

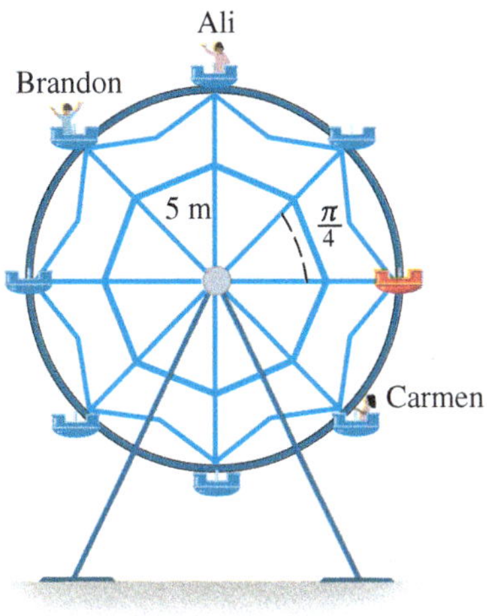

EXAMPLE 11 ■ Using Phase

Ali, Brandon, and Carmen are sitting in a stopped Ferris wheel as shown in the figure in the margin. At time $t = 0$ the Ferris wheel starts turning counterclockwise at the rate of 2 revolutions per minute.

(a) Find sine curves that model the height of each rider above the center line of the Ferris wheel at any time $t > 0$.

(b) Find the phase difference between Brandon and Ali, between Ali and Carmen, and between Brandon and Carmen.

(c) Find the horizontal shift of Ali's equation. What is Ali's lead or lag time (relative to the red seat in the figure)?

SOLUTION

(a) The motion of each rider is modeled by a function of the form $y = A\sin(kt - b)$. From the figure we see that the amplitude is $A = 5$ m. Since the Ferris wheel makes two revolutions per minute, the period is $\frac{1}{2}$ min. So

$$\text{period} = \frac{2\pi}{k} = \frac{1}{2}\text{ min}$$

It follows that $k = 4\pi$. From the figure we see that each rider starts at a different phase. Let's consider Ali and Brandon to be ahead of the red seat, and let's consider Carmen to be behind the red seat. So their phases are $-\pi/2$, $-3\pi/4$, and $\pi/4$, respectively. The equations are as follows.

Ali	Brandon	Carmen
$y_A = 5\sin\left(4\pi t + \frac{\pi}{2}\right)$	$y_B = 5\sin\left(4\pi t + \frac{3\pi}{4}\right)$	$y_C = 5\sin\left(4\pi t - \frac{\pi}{4}\right)$

(b) The phase differences are as follows.

Ali and Brandon	Ali and Carmen	Brandon and Carmen
$\frac{3\pi}{4} - \frac{\pi}{2} = \frac{\pi}{4}$	$\frac{\pi}{2} - \left(-\frac{\pi}{4}\right) = \frac{3\pi}{4}$	$\frac{3\pi}{4} - \left(-\frac{\pi}{4}\right) = \pi$

(c) The equation that models Ali's position above the center line of the Ferris wheel was found in part (b). To find the horizontal shift, we factor Ali's equation.

$$y_A = 5\sin\left(4\pi t + \frac{\pi}{2}\right) \qquad \text{Ali's equation}$$

$$y_A = 5\sin 4\pi\left(t + \frac{1}{8}\right) \qquad \text{Factor } 4\pi$$

We see that the horizontal shift is $\frac{1}{8}$ to the left. This means that Ali's lead time is $\frac{1}{8}$ of a minute (so she is $\frac{1}{8}$ of a minute ahead of the red seat).

Now Try Exercise 61 ■

5.6 EXERCISES

CONCEPTS

1. For an object in simple harmonic motion with amplitude a and period $2\pi/\omega$, find an equation that models the displacement y at time t if

(a) $y = 0$ at time $t = 0$: $y =$ ________.

(b) $y = a$ at time $t = 0$: $y =$ ________.

2. For an object in damped harmonic motion with initial amplitude a, period $2\pi/\omega$, and damping constant c, find an equation that models the displacement y at time t if

(a) $y = 0$ at time $t = 0$: $y =$ ________.

(b) $y = a$ at time $t = 0$: $y =$ ________.

3. (a) For an object in harmonic motion modeled by $y = A\sin(kt - b)$ the amplitude is ________, the period is ________, and the phase is ________. To find the horizontal shift, we factor out k to get $y =$ ________. From this form of the equation we see that the horizontal shift is ________.

(b) For an object in harmonic motion modeled by $y = 5\sin(4t - \pi)$ the amplitude is ________, the period is ________, the phase is ________, and the horizontal shift is ________.

4. Objects A and B are in harmonic motion modeled by $y = 3\sin(2t - \pi)$ and $y = 3\sin\left(2t - \dfrac{\pi}{2}\right)$. The phase of A is ________, and the phase of B is ________. The phase difference is ________, so the objects are moving ________ (in phase/out of phase).

SKILLS

5–12 ■ Simple Harmonic Motion The given function models the displacement of an object moving in simple harmonic motion.
(a) Find the amplitude, period, and frequency of the motion.
(b) Sketch a graph of the displacement of the object over one complete period.

5. $y = 2\sin 3t$ **6.** $y = 3\cos\frac{1}{2}t$

7. $y = -\cos 0.3t$ **8.** $y = 2.4\sin 3.6t$

9. $y = -0.25\cos\left(1.5t - \dfrac{\pi}{3}\right)$ **10.** $y = -\frac{3}{2}\sin(0.2t + 1.4)$

11. $y = 5\cos\left(\frac{2}{3}t + \frac{3}{4}\right)$ **12.** $y = 1.6\sin(t - 1.8)$

13–16 ■ Simple Harmonic Motion Find a function that models the simple harmonic motion having the given properties. Assume that the displacement is zero at time $t = 0$.

13. amplitude 10 cm, period 3 s

14. amplitude 24 ft, period 2 min

15. amplitude 6 in., frequency $5/\pi$ Hz

16. amplitude 1.2 m, frequency 0.5 Hz

17–20 ■ Simple Harmonic Motion Find a function that models the simple harmonic motion having the given properties. Assume that the displacement is at its maximum at time $t = 0$.

17. amplitude 60 ft, period 0.5 min

18. amplitude 35 cm, period 8 s

19. amplitude 2.4 m, frequency 750 Hz

20. amplitude 6.25 in., frequency 60 Hz

21–28 ■ Damped Harmonic Motion An initial amplitude k, damping constant c, and frequency f or period p are given. (Recall that frequency and period are related by the equation $f = 1/p$.)
(a) Find a function that models the damped harmonic motion. Use a function of the form $y = ke^{-ct}\cos\omega t$ in Exercises 21–24 and of the form $y = ke^{-ct}\sin\omega t$ in Exercises 25–28.
(b) Graph the function.

21. $k = 2,\quad c = 1.5,\quad f = 3$

22. $k = 15,\quad c = 0.25,\quad f = 0.6$

23. $k = 100,\quad c = 0.05,\quad p = 4$

24. $k = 0.75,\quad c = 3,\quad p = 3\pi$

25. $k = 7,\quad c = 10,\quad p = \pi/6$

26. $k = 1,\quad c = 1,\quad p = 1$

27. $k = 0.3,\quad c = 0.2,\quad f = 20$

28. $k = 12,\quad c = 0.01,\quad f = 8$

29–34 ■ Amplitude, Period, Phase, and Horizontal Shift For each sine curve find the amplitude, period, phase, and horizontal shift.

29. $y = 5\sin\left(2t - \dfrac{\pi}{2}\right)$ **30.** $y = 10\sin\left(t - \dfrac{\pi}{3}\right)$

31. $y = 100\sin(5t + \pi)$ **32.** $y = 50\sin\left(\dfrac{1}{2}t + \dfrac{\pi}{5}\right)$

33. $y = 20\sin 2\left(t - \dfrac{\pi}{4}\right)$ **34.** $y = 8\sin 4\left(t + \dfrac{\pi}{12}\right)$

35–38 ■ Phase and Phase Difference A pair of sine curves with the same period is given. **(a)** Find the phase of each curve. **(b)** Find the phase difference between the curves. **(c)** Determine whether the curves are in phase or out of phase. **(d)** Sketch both curves on the same axes.

35. $y_1 = 10\sin\left(3t - \dfrac{\pi}{2}\right)$; $y_2 = 10\sin\left(3t - \dfrac{5\pi}{2}\right)$

36. $y_1 = 15\sin\left(2t - \dfrac{\pi}{3}\right)$; $y_2 = 15\sin\left(2t - \dfrac{\pi}{6}\right)$

37. $y_1 = 80 \sin 5\left(t - \dfrac{\pi}{10}\right)$; $y_2 = 80 \sin\left(5t - \dfrac{\pi}{3}\right)$

38. $y_1 = 20 \sin 2\left(t - \dfrac{\pi}{2}\right)$; $y_2 = 20 \sin 2\left(t - \dfrac{3\pi}{2}\right)$

APPLICATIONS

39. A Bobbing Cork A cork floating in a lake is bobbing in simple harmonic motion. Its displacement above the bottom of the lake is modeled by

$$y = 0.2 \cos 20\pi t + 8$$

where y is measured in meters and t is measured in minutes.

(a) Find the frequency of the motion of the cork.

(b) Sketch a graph of y.

(c) Find the maximum displacement of the cork above the lake bottom.

40. FM Radio Signals The carrier wave for an FM radio signal is modeled by the function

$$y = a \sin(2\pi(9.15 \times 10^7)t)$$

where t is measured in seconds. Find the period and frequency of the carrier wave.

41. Blood Pressure Each time your heart beats, your blood pressure increases, then decreases as the heart rests between beats. A certain person's blood pressure is modeled by the function

$$p(t) = 115 + 25 \sin(160\pi t)$$

where $p(t)$ is the pressure (in mmHg) at time t, measured in minutes.

(a) Find the amplitude, period, and frequency of p.

(b) Sketch a graph of p.

(c) If a person is exercising, his or her heart beats faster. How does this affect the period and frequency of p?

42. Predator Population Model In a predator/prey model, the predator population is modeled by the function

$$y = 900 \cos 2t + 8000$$

where t is measured in years.

(a) What is the maximum population?

(b) Find the length of time between successive periods of maximum population.

43. Mass-Spring System A mass attached to a spring is moving up and down in simple harmonic motion. The graph gives its displacement $d(t)$ from equilibrium at time t. Express the function d in the form $d(t) = a \sin \omega t$.

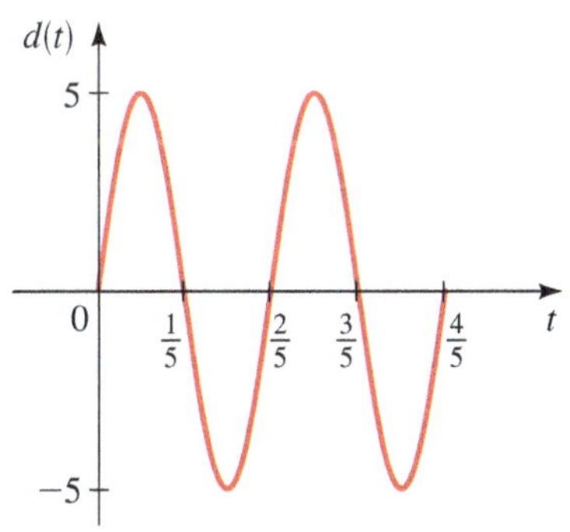

44. Tides The graph shows the variation of the water level relative to mean sea level in Commencement Bay at Tacoma, Washington, for a particular 24-h period. Assuming that this variation is modeled by simple harmonic motion, find an equation of the form $y = a \sin \omega t$ that describes the variation in water level as a function of the number of hours after midnight.

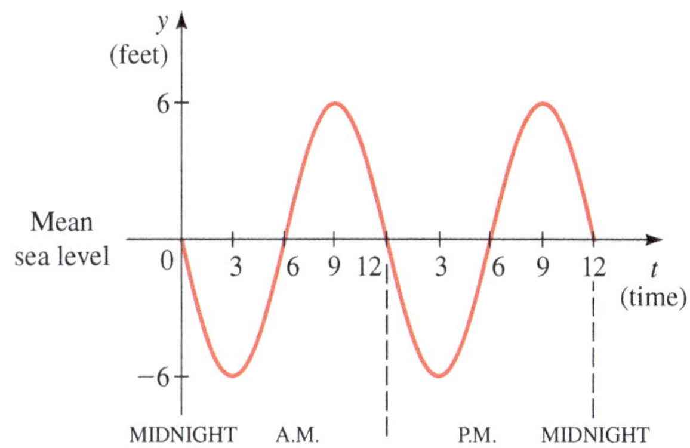

45. Tides The Bay of Fundy in Nova Scotia has the highest tides in the world. In one 12-h period the water starts at mean sea level, rises to 21 ft above, drops to 21 ft below, then returns to mean sea level. Assuming that the motion of the tides is simple harmonic, find an equation that describes the height of the tide in the Bay of Fundy above mean sea level. Sketch a graph that shows the level of the tides over a 12-h period.

46. Mass-Spring System A mass suspended from a spring is pulled down a distance of 2 ft from its rest position, as shown in the figure. The mass is released at time $t = 0$ and allowed to oscillate. If the mass returns to this position after 1 s, find an equation that describes its motion.

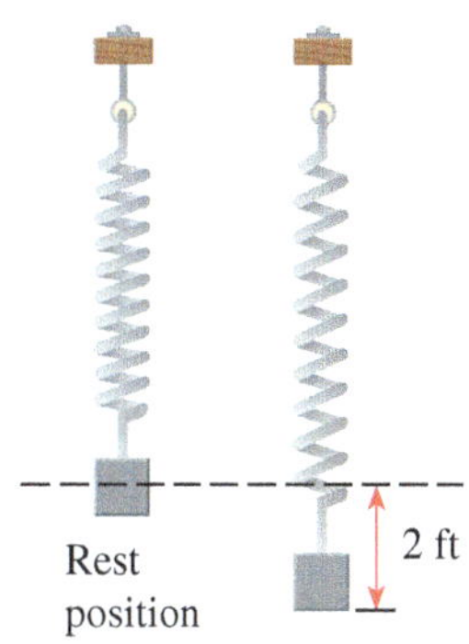

47. Mass-Spring System A mass is suspended on a spring. The spring is compressed so that the mass is located 5 cm above its rest position. The mass is released at time $t = 0$ and allowed to oscillate. It is observed that the mass reaches its lowest point $\frac{1}{2}$ s after it is released. Find an equation that describes the motion of the mass.

48. Mass-Spring System The frequency of oscillation of an object suspended on a spring depends on the stiffness k of the spring (called the *spring constant*) and the mass m of the object. If the spring is compressed a distance a and then allowed to oscillate, its displacement is given by

$$f(t) = a \cos \sqrt{k/m}\, t$$

(a) A 10-g mass is suspended from a spring with stiffness $k = 3$. If the spring is compressed a distance 5 cm and then released, find the equation that describes the oscillation of the spring.

(b) Find a general formula for the frequency (in terms of k and m).

(c) How is the frequency affected if the mass is increased? Is the oscillation faster or slower?

(d) How is the frequency affected if a stiffer spring is used (larger k)? Is the oscillation faster or slower?

49. Ferris Wheel A Ferris wheel has a radius of 10 m, and the bottom of the wheel passes 1 m above the ground. If the Ferris wheel makes one complete revolution every 20 s, find an equation that gives the height above the ground of a person on the Ferris wheel as a function of time.

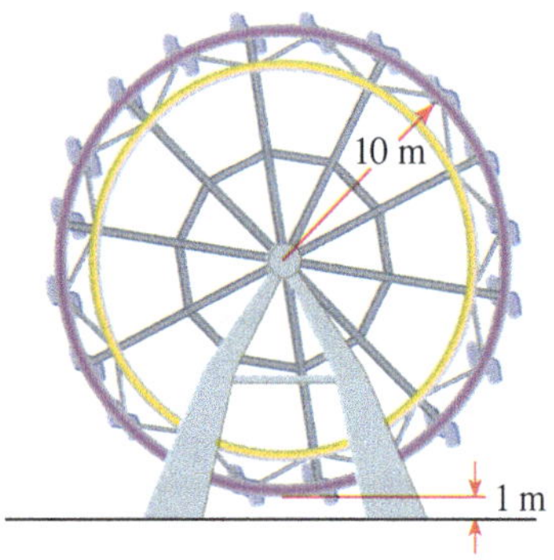

50. Clock Pendulum The pendulum in a grandfather clock makes one complete swing every 2 s. The maximum angle that the pendulum makes with respect to its rest position is 10°. We know from physical principles that the angle θ between the pendulum and its rest position changes in simple harmonic fashion. Find an equation that describes the size of the angle θ as a function of time. (Take $t = 0$ to be a time when the pendulum is vertical.)

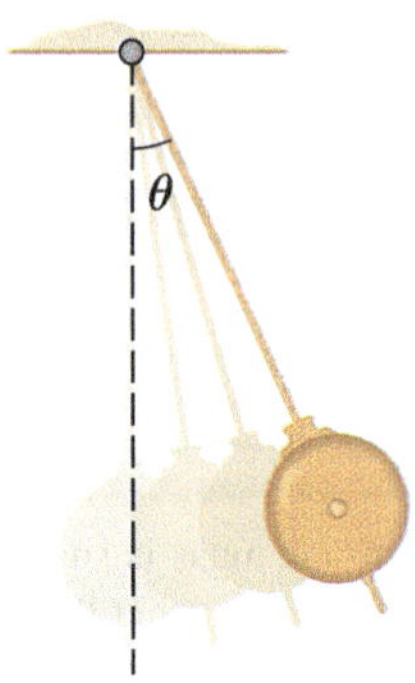

51. Variable Stars The variable star Zeta Gemini has a period of 10 days. The average brightness of the star is 3.8 magnitudes, and the maximum variation from the average is 0.2 magnitude. Assuming that the variation in brightness is simple harmonic, find an equation that gives the brightness of the star as a function of time.

52. Variable Stars Astronomers believe that the radius of a variable star increases and decreases with the brightness of the star. The variable star Delta Cephei (Example 4) has an average radius of 20 million miles and changes by a maximum of 1.5 million miles from this average during a single pulsation. Find an equation that describes the radius of this star as a function of time.

53. Biological Clocks *Circadian rhythms* are biological processes that oscillate with a period of approximately 24 h. That is, a circadian rhythm is an internal daily biological clock. Blood pressure appears to follow such a rhythm. For a certain individual the average resting blood pressure varies from a maximum of 100 mmHg at 2:00 P.M. to a minimum of 80 mmHg at 2:00 A.M. Find a sine function of the form

$$f(t) = a \sin(\omega(t - c)) + b$$

that models the blood pressure at time t, measured in hours from midnight.

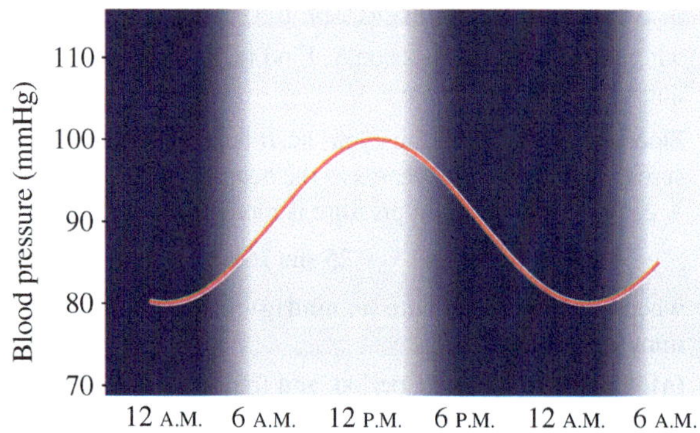

54. Electric Generator The armature in an electric generator is rotating at the rate of 100 revolutions per second (rps). If the maximum voltage produced is 310 V, find an equation that describes this variation in voltage. What is the RMS voltage? (See Example 6 and the margin note adjacent to it.)

55. Electric Generator The graph shows an oscilloscope reading of the variation in voltage of an AC current produced by a simple generator.

(a) Find the maximum voltage produced.

(b) Find the frequency (cycles per second) of the generator.

(c) How many revolutions per second does the armature in the generator make?

(d) Find a formula that describes the variation in voltage as a function of time.

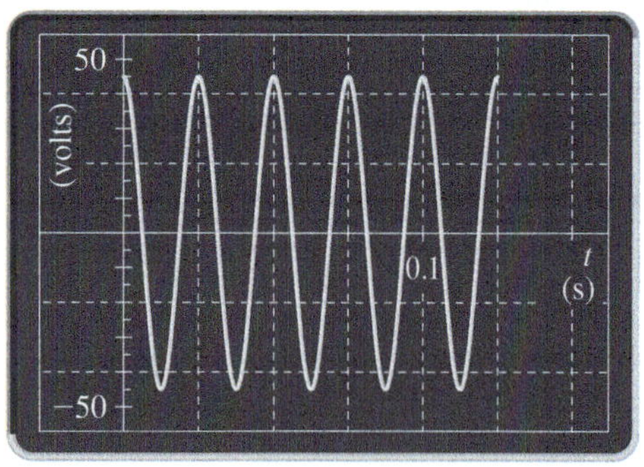

56. Doppler Effect When a car with its horn blowing drives by an observer, the pitch of the horn seems higher as it approaches and lower as it recedes (see the figure below). This phenomenon is called the **Doppler effect**. If the sound source is moving at speed v relative to the observer and if the speed of sound is v_0, then the perceived frequency f is related to the actual frequency f_0 as follows.

$$f = f_0\left(\frac{v_0}{v_0 \pm v}\right)$$

We choose the minus sign if the source is moving toward the observer and the plus sign if it is moving away.

Suppose that a car drives at 110 ft/s past a woman standing on the shoulder of a highway, blowing its horn, which has a frequency of 500 Hz. Assume that the speed of sound is 1130 ft/s. (This is the speed in dry air at 70°F.)

(a) What are the frequencies of the sounds that the woman hears as the car approaches her and as it moves away from her?

(b) Let A be the amplitude of the sound. Find functions of the form

$$y = A \sin \omega t$$

that model the perceived sound as the car approaches the woman and as it recedes.

57. Motion of a Building A strong gust of wind strikes a tall building, causing it to sway back and forth in damped harmonic motion. The frequency of the oscillation is 0.5 cycle per second, and the damping constant is $c = 0.9$. Find an equation that describes the motion of the building. (Assume that $k = 1$, and take $t = 0$ to be the instant when the gust of wind strikes the building.)

58. Shock Absorber When a car hits a certain bump on the road, a shock absorber on the car is compressed a distance of 6 in., then released (see the figure). The shock absorber vibrates in damped harmonic motion with a frequency of 2 cycles per second. The damping constant for this particular shock absorber is 2.8.

(a) Find an equation that describes the displacement of the shock absorber from its rest position as a function of time. Take $t = 0$ to be the instant that the shock absorber is released.

(b) How long does it take for the amplitude of the vibration to decrease to 0.5 in.?

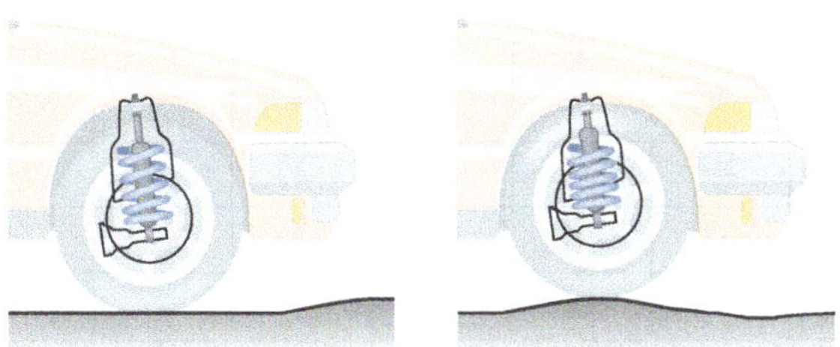

59. Tuning Fork A tuning fork is struck and oscillates in damped harmonic motion. The amplitude of the motion is measured, and 3 s later it is found that the amplitude has dropped to $\frac{1}{4}$ of this value. Find the damping constant c for this tuning fork.

60. Guitar String A guitar string is pulled at point P a distance of 3 cm above its rest position. It is then released and vibrates in damped harmonic motion with a frequency of 165 cycles per second. After 2 s, it is observed that the amplitude of the vibration at point P is 0.6 cm.

(a) Find the damping constant c.

(b) Find an equation that describes the position of point P above its rest position as a function of time. Take $t = 0$ to be the instant that the string is released.

61. Two Fans Electric fans A and B have radius 1 ft and, when switched on, rotate counterclockwise at the rate of 100 revolutions per minute. Starting with the position shown in the figure, the fans are simultaneously switched on.

(a) For each fan, find an equation that gives the height of the red dot (above the horizontal line shown) t minutes after the fans are switched on.

(b) Are the fans rotating in phase? Through what angle should fan A be rotated counterclockwise in order that the two fans rotate in phase?

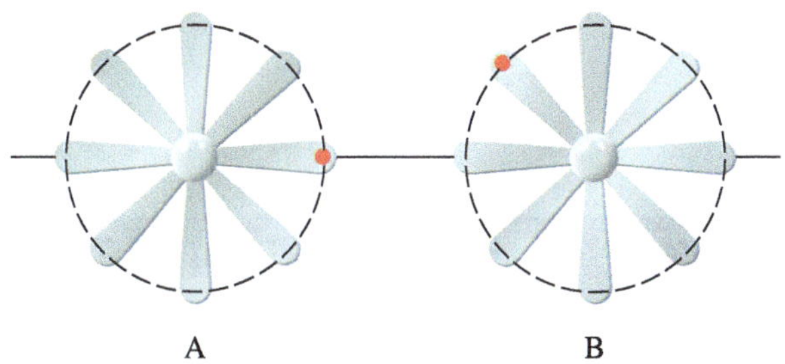

62. Alternating Current Alternating current is produced when an armature rotates about its axis in a magnetic field, as shown in the figure. Generators A and B rotate counterclockwise at 60 Hz (cycles per second) and each generator produces a maximum of 50 V. The voltage for each generator is modeled by

$$E_A = 50 \sin(120\pi t) \qquad E_B = 50 \sin\left(120\pi t - \frac{5\pi}{4}\right)$$

(a) Find the voltage phase for each generator, and find the phase difference.

(b) Are the generators producing voltage in phase? Through what angle should the armature in the second generator be rotated counterclockwise in order that the two generators produce voltage in phase?

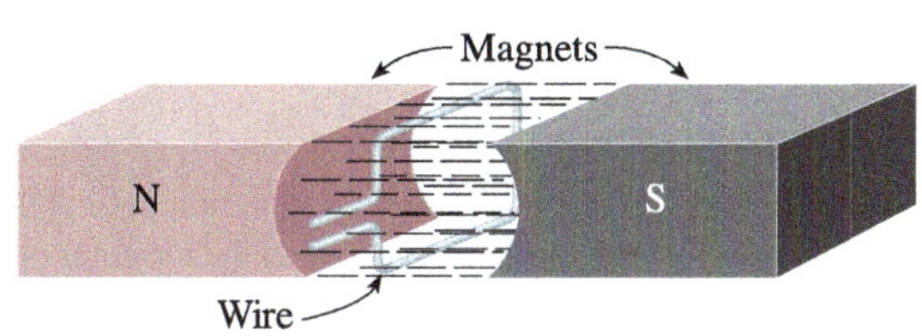

DISCUSS ■ DISCOVER ■ PROVE ■ WRITE

63. DISCUSS: Phases of Sine The phase of a sine curve $y = \sin(kt + b)$ represents a particular location on the graph of the sine function $y = \sin t$. Specifically, when $t = 0$, we have $y = \sin b$, and this corresponds to the point $(b, \sin b)$ on the graph of $y = \sin t$. Observe that each point on the graph of $y = \sin t$ has different characteristics. For example, for $t = \pi/6$, we have $\sin t = \frac{1}{2}$ and the values of sine are increasing, whereas at $t = 5\pi/6$, we also have $\sin t = \frac{1}{2}$ but the values of sine are decreasing. So each point on the graph of sine corresponds to a different "phase" of a sine curve. Complete the descriptions for each label on the graph below.

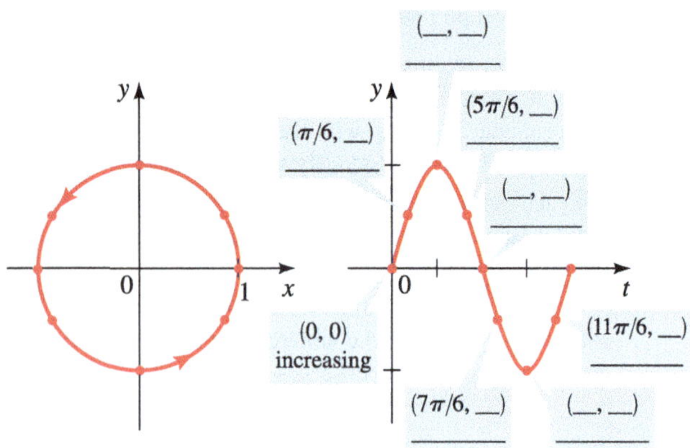

64. DISCUSS: Phases of the Moon During the course of a lunar cycle (about 1 month) the moon undergoes the familiar lunar phases. The phases of the moon are completely analogous to the phases of the sine function described in Exercise 63. The figure below shows some phases of the lunar cycle starting with a "new moon," "waxing crescent moon," "first quarter moon," and so on. The next to last phase shown is a "waning crescent moon." Give similar descriptions for the other phases of the moon shown in the figure. What are some events on the earth that follow a monthly cycle and are in phase with the lunar cycle? What are some events that are out of phase with the lunar cycle?

CHAPTER 5 ■ REVIEW

PROPERTIES AND FORMULAS

The Unit Circle (p. 402)

The **unit circle** is the circle of radius 1 centered at (0, 0). The equation of the unit circle is $x^2 + y^2 = 1$.

Terminal Points on the Unit Circle (pp. 402–404)

The **terminal point** $P(x, y)$ determined by the real number t is the point obtained by traveling counterclockwise a distance t along the unit circle, starting at $(1, 0)$.

Special terminal points are listed in Table 1 on page 404.

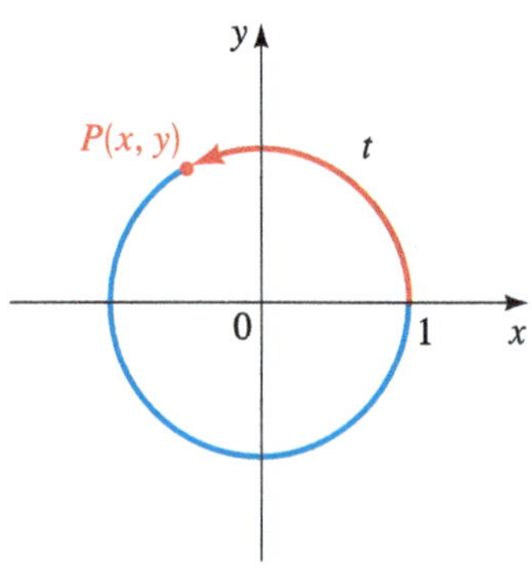

The Reference Number (pp. 405–406)

The **reference number** associated with the real number t is the shortest distance along the unit circle between the terminal point determined by t and the x-axis.

The Trigonometric Functions (p. 409)

Let $P(x, y)$ be the terminal point on the unit circle determined by the real number t. Then for nonzero values of the denominator the trigonometric functions are defined as follows.

$$\sin t = y \qquad \cos t = x \qquad \tan t = \frac{y}{x}$$

$$\csc t = \frac{1}{y} \qquad \sec t = \frac{1}{x} \qquad \cot t = \frac{x}{y}$$

Special Values of the Trigonometric Functions (p. 410)

The trigonometric functions have the following values at the special values of t.

t	$\sin t$	$\cos t$	$\tan t$	$\csc t$	$\sec t$	$\cot t$
0	0	1	0	—	1	—
$\frac{\pi}{6}$	$\frac{1}{2}$	$\frac{\sqrt{3}}{2}$	$\frac{\sqrt{3}}{3}$	2	$\frac{2\sqrt{3}}{3}$	$\sqrt{3}$
$\frac{\pi}{4}$	$\frac{\sqrt{2}}{2}$	$\frac{\sqrt{2}}{2}$	1	$\sqrt{2}$	$\sqrt{2}$	1
$\frac{\pi}{3}$	$\frac{\sqrt{3}}{2}$	$\frac{1}{2}$	$\sqrt{3}$	$\frac{2\sqrt{3}}{3}$	2	$\frac{\sqrt{3}}{3}$
$\frac{\pi}{2}$	1	0	—	1	—	0

Basic Trigonometric Identities (pp. 414–415)

An identity is an equation that is true for all values of the variable. The basic trigonometric identities are as follows.

Reciprocal Identities:

$$\csc t = \frac{1}{\sin t} \qquad \sec t = \frac{1}{\cos t} \qquad \cot t = \frac{1}{\tan t}$$

Pythagorean Identities:

$$\sin^2 t + \cos^2 t = 1$$
$$\tan^2 t + 1 = \sec^2 t$$
$$1 + \cot^2 t = \csc^2 t$$

Even-Odd Properties:

$$\sin(-t) = -\sin t \qquad \cos(-t) = \cos t \qquad \tan(-t) = -\tan t$$
$$\csc(-t) = -\csc t \qquad \sec(-t) = \sec t \qquad \cot(-t) = -\cot t$$

Periodic Properties (p. 419)

A function f is **periodic** if there is a positive number p such that $f(x + p) = f(x)$ for every x. The least such p is called the **period** of f. The sine and cosine functions have period 2π, and the tangent function has period π.

$$\sin(t + 2\pi) = \sin t$$
$$\cos(t + 2\pi) = \cos t$$
$$\tan(t + \pi) = \tan t$$

Graphs of the Sine and Cosine Functions (p. 420)

The graphs of sine and cosine have amplitude 1 and period 2π.

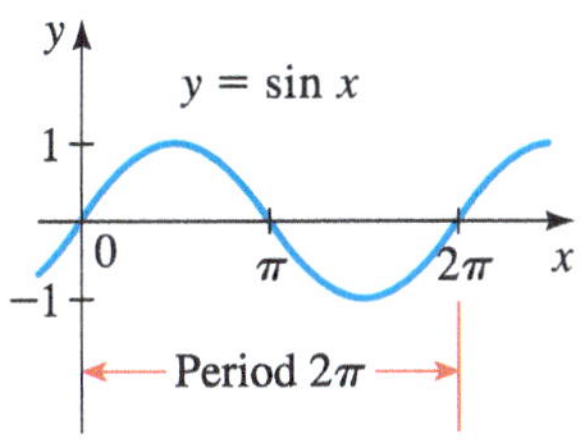

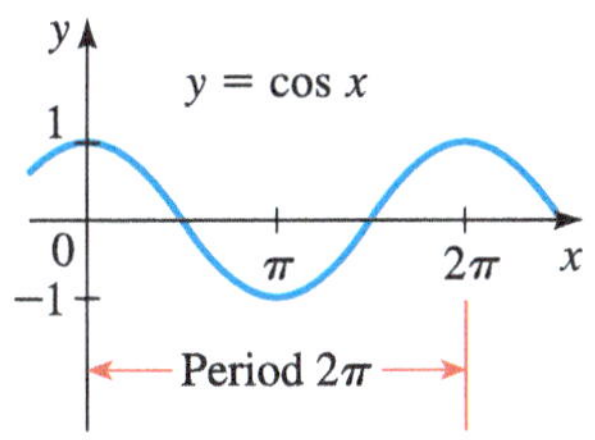

Amplitude 1, Period 2π

Graphs of Transformations of Sine and Cosine (p. 424)

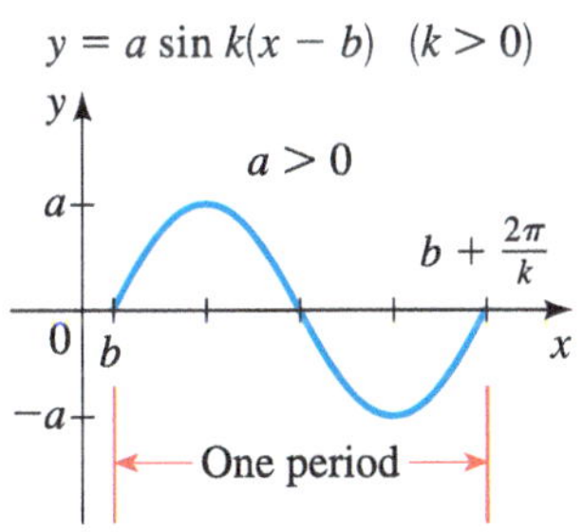

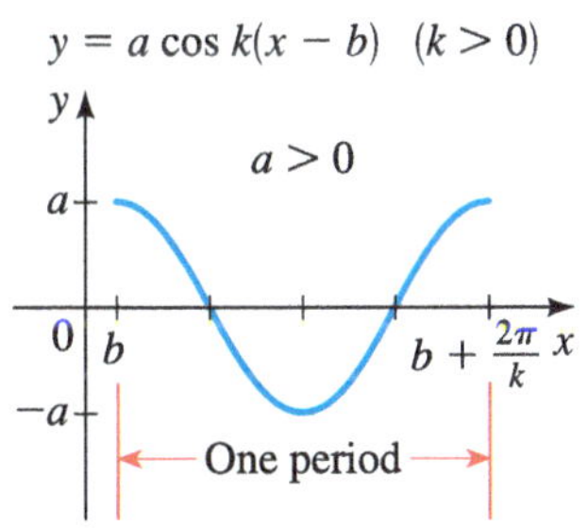

Amplitude a, Period $\frac{2\pi}{k}$, Horizontal shift b

An appropriate interval on which to graph one complete period is $[b, b + (2\pi/k)]$.

Graphs of the Tangent and Cotangent Functions (pp. 434–435)

These functions have period π.

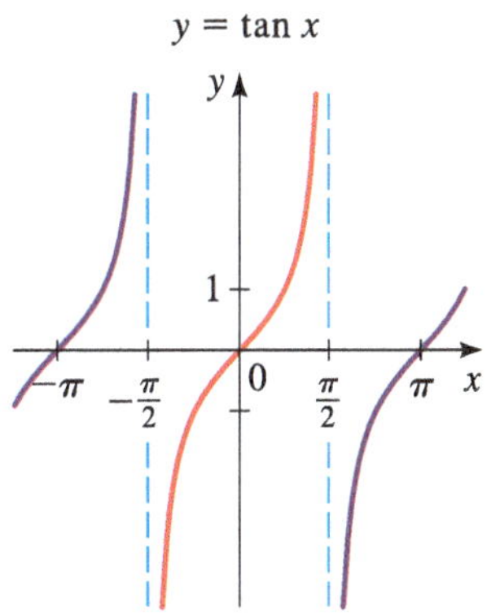

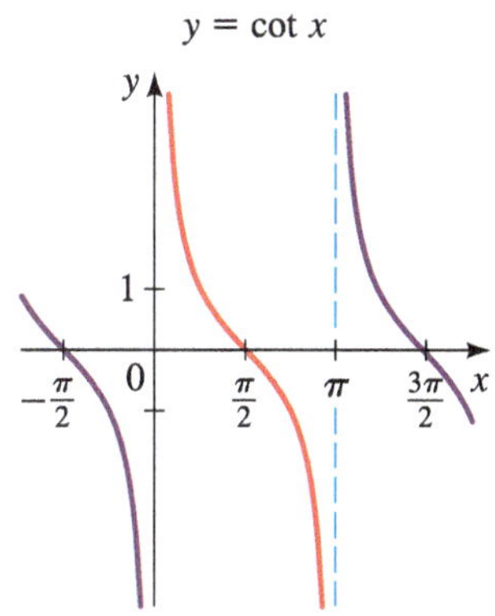

To graph one period of $y = a \tan kx$, an appropriate interval is $(-\pi/2k, \pi/2k)$.

To graph one period of $y = a \cot kx$, an appropriate interval is $(0, \pi/k)$.

Graphs of the Cosecant and Secant Functions (pp. 436–437)

These functions have period 2π.

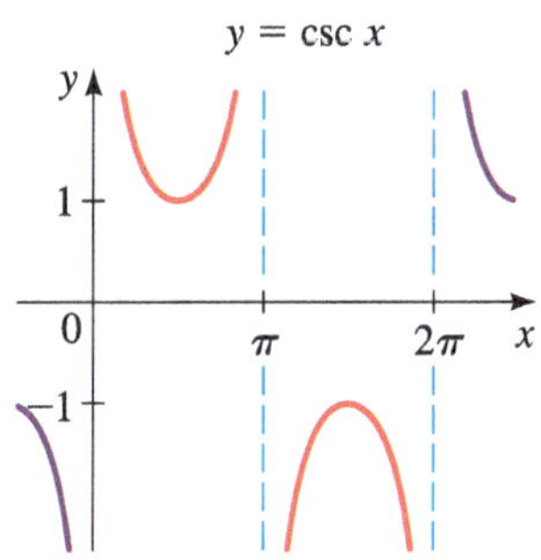

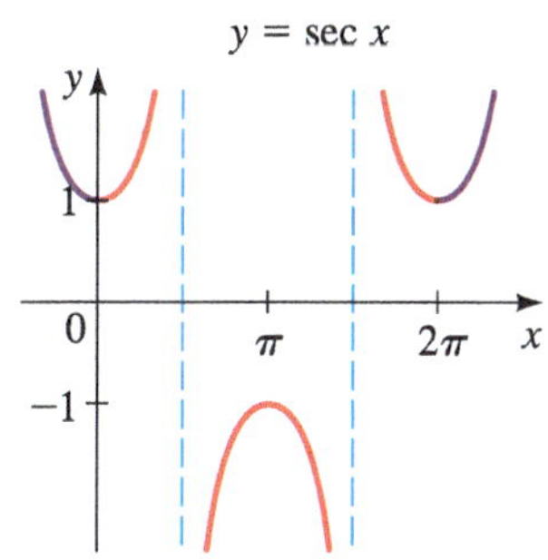

To graph one period of $y = a \csc kx$, an appropriate interval is $(0, 2\pi/k)$.

To graph one period of $y = a \sec kx$, an appropriate interval is $(0, 2\pi/k)$.

Inverse Trigonometric Functions (pp. 440–443)

Inverse functions of the trigonometric functions are defined by restricting the domains as follows.

Function	Domain	Range
$\sin^{-1}$	$[-1, 1]$	$[-\frac{\pi}{2}, \frac{\pi}{2}]$
$\cos^{-1}$	$[-1, 1]$	$[0, \pi]$
$\tan^{-1}$	$(-\infty, \infty)$	$(-\frac{\pi}{2}, \frac{\pi}{2})$

The inverse trigonometric functions are defined as follows.

$$\sin^{-1} x = y \quad \Leftrightarrow \quad \sin y = x$$
$$\cos^{-1} x = y \quad \Leftrightarrow \quad \cos y = x$$
$$\tan^{-1} x = y \quad \Leftrightarrow \quad \tan y = x$$

Graphs of these inverse functions are shown below.

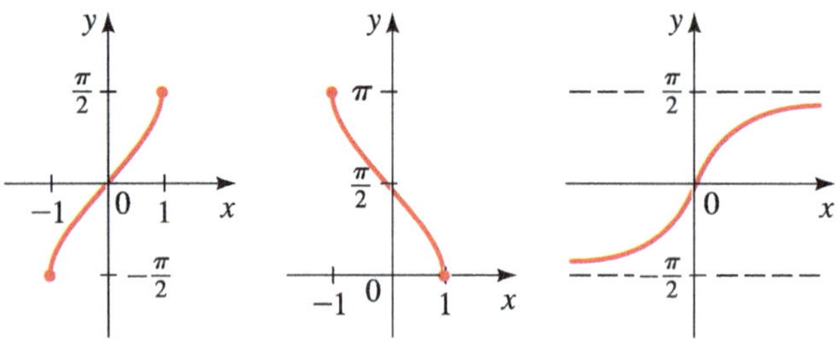

Harmonic Motion (p. 446)

An object is in **simple harmonic motion** if its displacement y at time t is modeled by $y = a \sin \omega t$ or $y = a \cos \omega t$. In this case the amplitude is $|a|$, the period is $2\pi/\omega$, and the frequency is $\omega/2\pi$.

Damped Harmonic Motion (p. 451)

An object is in **damped harmonic motion** if its displacement y at time t is modeled by $y = ke^{-ct} \sin \omega t$ or $y = ke^{-ct} \cos \omega t$, $c > 0$. In this case c is the damping constant, k is the initial amplitude, and $2\pi/\omega$ is the period.

Phase (pp. 453–454)

Any sine curve can be expressed in the following equivalent forms:

$y = A \sin(kt - b)$, the **phase** is b

$y = A \sin k\left(t - \dfrac{b}{k}\right)$, the **horizontal shift** is $\dfrac{b}{k}$

The phase (or phase angle) b is the initial angular position of the motion. The number b/k is also called the **lag time** ($b > 0$) or **lead time** ($b < 0$).

Suppose that two objects are in harmonic motion with the same period modeled by

$$y_1 = A \sin(kt - b) \qquad \text{and} \qquad y_2 = A \sin(kt - c)$$

The **phase difference** between y_1 and y_2 is $b - c$. The motions are "in phase" if the phase difference is a multiple of 2π; otherwise, the motions are "out of phase."

CONCEPT CHECK

1. (a) What is the unit circle, and what is the equation of the unit circle?
(b) Use a diagram to explain what is meant by the terminal point $P(x, y)$ determined by t.
(c) Find the terminal point for $t = \dfrac{\pi}{2}$.
(d) What is the reference number associated with t?
(e) Find the reference number and terminal point for $t = \dfrac{7\pi}{4}$.

2. Let t be a real number, and let $P(x, y)$ be the terminal point determined by t.
(a) Write equations that define $\sin t$, $\cos t$, $\tan t$, $\csc t$, $\sec t$, and $\cot t$.
(b) In each of the four quadrants, identify the trigonometric functions that are positive.
(c) List the special values of sine, cosine, and tangent.

3. (a) Describe the steps we use to find the value of a trigonometric function at a real number t.
(b) Find $\sin \dfrac{5\pi}{6}$.

4. (a) What is a periodic function?
(b) What are the periods of the six trigonometric functions?
(c) Find $\sin \dfrac{19\pi}{4}$.

5. (a) What is an even function, and what is an odd function?
(b) Which trigonometric functions are even? Which are odd?
(c) If $\sin t = 0.4$, find $\sin(-t)$.
(d) If $\cos s = 0.7$, find $\cos(-s)$.

6. (a) State the reciprocal identities.
(b) State the Pythagorean identities.

7. (a) Graph the sine and cosine functions.
(b) What are the amplitude, period, and horizontal shift for the sine curve $y = a \sin k(x - b)$ and for the cosine curve $y = a \cos k(x - b)$?
(c) Find the amplitude, period, and horizontal shift of $y = 3 \sin\left(2x - \dfrac{\pi}{6}\right)$.

8. (a) Graph the tangent and cotangent functions.
(b) For the curves $y = a \tan kx$ and $y = a \cot kx$, state appropriate intervals to graph one complete period of each curve.
(c) Find an appropriate interval to graph one complete period of $y = 5 \tan 3x$.

9. (a) Graph the cosecant and secant functions.
(b) For the curves $y = a \csc kx$ and $y = a \sec kx$, state appropriate intervals to graph one complete period of each curve.
(c) Find an appropriate interval to graph one period of $y = 3 \csc 6x$.

10. (a) Define the inverse sine function, the inverse cosine function, and the inverse tangent function.
(b) Find $\sin^{-1} \dfrac{1}{2}$, $\cos^{-1} \dfrac{\sqrt{2}}{2}$, and $\tan^{-1} 1$.
(c) For what values of x is the equation $\sin(\sin^{-1} x) = x$ true? For what values of x is the equation $\sin^{-1}(\sin x) = x$ true?

11. (a) What is simple harmonic motion?
(b) What is damped harmonic motion?
(c) Give real-world examples of harmonic motion.

12. Suppose that an object is in simple harmonic motion given by
$$y = 5\sin\left(2t - \frac{\pi}{3}\right).$$
(a) Find the amplitude, period, and frequency.
(b) Find the phase and the horizontal shift.

13. Consider the following models of harmonic motion.
$$y_1 = 5\sin(2t - 1) \qquad y_2 = 5\sin(2t - 3)$$
Do both motions have the same frequency? What is the phase for each equation? What is the phase difference? Are the objects moving in phase or out of phase?

ANSWERS TO THE CONCEPT CHECK CAN BE FOUND AT THE BACK OF THE BOOK.

EXERCISES

1–2 ■ Terminal Points A point $P(x, y)$ is given. **(a)** Show that P is on the unit circle. **(b)** Suppose that P is the terminal point determined by t. Find $\sin t$, $\cos t$, and $\tan t$.

1. $P\left(-\frac{\sqrt{3}}{2}, \frac{1}{2}\right)$ **2.** $P\left(\frac{3}{5}, -\frac{4}{5}\right)$

3–6 ■ Reference Number and Terminal Point A real number t is given. **(a)** Find the reference number for t. **(b)** Find the terminal point $P(x, y)$ on the unit circle determined by t. **(c)** Find the six trigonometric functions of t.

3. $t = \frac{2\pi}{3}$ **4.** $t = \frac{5\pi}{3}$

5. $t = -\frac{11\pi}{4}$ **6.** $t = -\frac{7\pi}{6}$

7–16 ■ Values of Trigonometric Functions Find the value of the trigonometric function. If possible, give the exact value; otherwise, use a calculator to find an approximate value rounded to five decimal places.

7. **(a)** $\sin\frac{3\pi}{4}$ **(b)** $\cos\frac{3\pi}{4}$

8. **(a)** $\tan\frac{\pi}{3}$ **(b)** $\tan\left(-\frac{\pi}{3}\right)$

9. **(a)** $\sin 1.1$ **(b)** $\cos 1.1$

10. **(a)** $\cos\frac{\pi}{5}$ **(b)** $\cos\left(-\frac{\pi}{5}\right)$

11. **(a)** $\cos\frac{9\pi}{2}$ **(b)** $\sec\frac{9\pi}{2}$

12. **(a)** $\sin\frac{\pi}{7}$ **(b)** $\csc\frac{\pi}{7}$

13. **(a)** $\tan\frac{5\pi}{2}$ **(b)** $\cot\frac{5\pi}{2}$

14. **(a)** $\sin 2\pi$ **(b)** $\csc 2\pi$

15. **(a)** $\tan\frac{5\pi}{6}$ **(b)** $\cot\frac{5\pi}{6}$

16. **(a)** $\cos\frac{\pi}{3}$ **(b)** $\sin\frac{\pi}{6}$

17–20 ■ Fundamental Identities Use the fundamental identities to write the first expression in terms of the second.

17. $\frac{\tan t}{\cos t}$, $\sin t$

18. $\tan^2 t \sec t$, $\cos t$

19. $\tan t$, $\sin t$; t in Quadrant IV

20. $\sec t$, $\sin t$; t in Quadrant II

21–24 ■ Values of Trigonometric Functions Find the values of the remaining trigonometric functions at t from the given information.

21. $\sin t = \frac{5}{13}$, $\cos t = -\frac{12}{13}$

22. $\sin t = -\frac{1}{2}$, $\cos t > 0$

23. $\cot t = -\frac{1}{2}$, $\csc t = \sqrt{5}/2$

24. $\cos t = -\frac{3}{5}$, $\tan t < 0$

25–28 ■ Values of Trigonometric Functions Find the values of the trigonometric function of t from the given information.

25. $\sec t + \cot t$; $\tan t = \frac{1}{4}$, terminal point for t in Quadrant III

26. $\csc t + \sec t$; $\sin t = -\frac{8}{17}$, terminal point for t in Quadrant IV

27. $\tan t + \sec t$; $\cos t = \frac{3}{5}$, terminal point for t in Quadrant I

28. $\sin^2 t + \cos^2 t$; $\sec t = -5$, terminal point for t in Quadrant II

29–36 ■ Horizontal Shifts A trigonometric function is given. **(a)** Find the amplitude, period, and horizontal shift of the function. **(b)** Sketch the graph.

29. $y = 10\cos\frac{1}{2}x$ **30.** $y = 4\sin 2\pi x$

31. $y = -\sin\frac{1}{2}x$ **32.** $y = 2\sin\left(x - \frac{\pi}{4}\right)$

33. $y = 3\sin(2x - 2)$ **34.** $y = \cos 2\left(x - \frac{\pi}{2}\right)$

35. $y = -\cos\left(\frac{\pi}{2}x + \frac{\pi}{6}\right)$ **36.** $y = 10\sin\left(2x - \frac{\pi}{2}\right)$

37–40 ■ Functions from a Graph The graph of one period of a function of the form $y = a \sin k(x - b)$ or $y = a \cos k(x - b)$ is shown. Determine the function.

37.

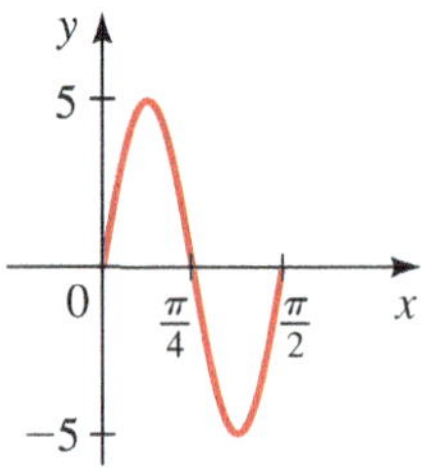

38.

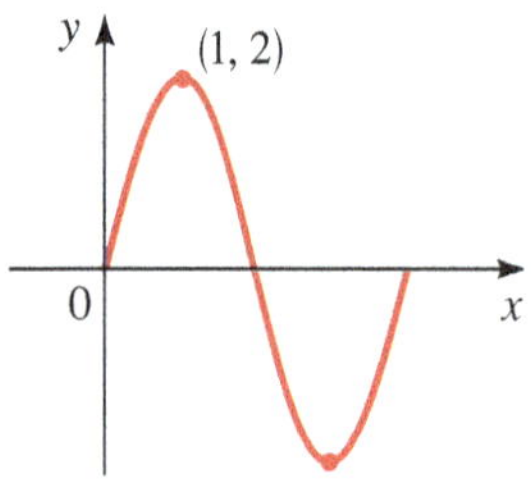

39.

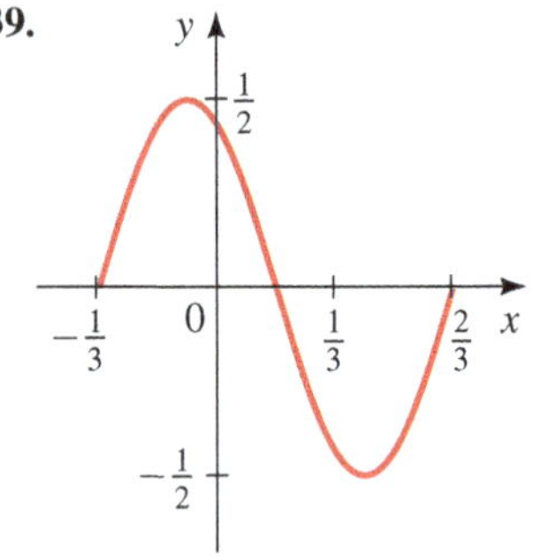

40.

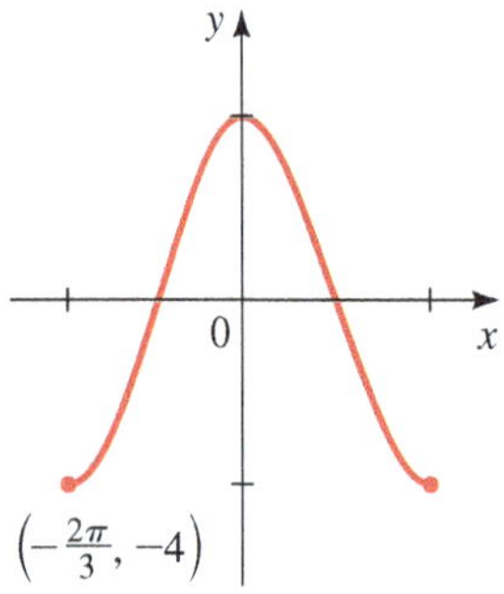

41–48 ■ Graphing Trigonometric Functions Find the period, and sketch the graph.

41. $y = 3 \tan x$

42. $y = \tan \pi x$

43. $y = 2 \cot\left(x - \frac{\pi}{2}\right)$

44. $y = \sec\left(\frac{1}{2}x - \frac{\pi}{2}\right)$

45. $y = 4 \csc(2x + \pi)$

46. $y = \tan\left(x + \frac{\pi}{6}\right)$

47. $y = \tan\left(\frac{1}{2}x - \frac{\pi}{8}\right)$

48. $y = -4 \sec 4\pi x$

49–52 ■ Evaluating Expressions Involving Inverse Trigonometric Functions Find the exact value of each expression, if it is defined.

49. $\sin^{-1} 1$

50. $\cos^{-1}\left(-\frac{1}{2}\right)$

51. $\sin^{-1}\left(\sin \frac{13\pi}{6}\right)$

52. $\tan\left(\cos^{-1}\left(\frac{1}{2}\right)\right)$

53–54 ■ Amplitude, Period, Phase, and Horizontal Shift For each sine curve find the amplitude, period, phase, and horizontal shift.

53. $y = 100 \sin 8\left(t + \frac{\pi}{16}\right)$

54. $y = 80 \sin 3\left(t - \frac{\pi}{2}\right)$

55–56 ■ Phase and Phase Difference A pair of sine curves with the same period is given. **(a)** Find the phase of each curve. **(b)** Find the phase difference between the curves. **(c)** Determine whether the curves are in phase or out of phase. **(d)** Sketch both curves on the same axes.

55. $y_1 = 25 \sin 3\left(t - \frac{\pi}{2}\right)$; $y_2 = 10 \sin\left(3t - \frac{5\pi}{2}\right)$

56. $y_1 = 50 \sin\left(10t - \frac{\pi}{2}\right)$; $y_2 = 50 \sin 10\left(t - \frac{\pi}{20}\right)$

57–62 ■ Even and Odd Functions A function is given. **(a)** Use a graphing device to graph the function. **(b)** Determine from the graph whether the function is periodic and, if so, determine the period. **(c)** Determine from the graph whether the function is odd, even, or neither.

57. $y = |\cos x|$

58. $y = \sin(\cos x)$

59. $y = \cos(2^{0.1x})$

60. $y = 1 + 2^{\cos x}$

61. $y = |x| \cos 3x$

62. $y = \sqrt{x} \sin 3x, \quad x > 0$

63–66 ■ Sine and Cosine Curves with Variable Amplitude Graph the three functions on a common screen. How are the graphs related?

63. $y = x, \quad y = -x, \quad y = x \sin x$

64. $y = 2^{-x}, \quad y = -2^{-x}, \quad y = 2^{-x} \cos 4\pi x$

65. $y = x, \quad y = \sin 4x, \quad y = x + \sin 4x$

66. $y = \sin^2 x, \quad y = \cos^2 x, \quad y = \sin^2 x + \cos^2 x$

67–68 ■ Maxima and Minima Find the maximum and minimum values of the function.

67. $y = \cos x + \sin 2x$

68. $y = \cos x + \sin^2 x$

69-70 ■ Solving Trigonometric Equations Graphically Find all solutions of the equation that lie in the given interval. State each answer rounded to two decimal places.

69. $\sin x = 0.3; \quad [0, 2\pi]$

70. $\cos 3x = x; \quad [0, \pi]$

71. Discover the Period of a Trigonometric Function Let $y_1 = \cos(\sin x)$ and $y_2 = \sin(\cos x)$.

(a) Graph y_1 and y_2 in the same viewing rectangle.

(b) Determine the period of each of these functions from its graph.

(c) Find an inequality between $\sin(\cos x)$ and $\cos(\sin x)$ that is valid for all x.

72. Simple Harmonic Motion A point P moving in simple harmonic motion completes 8 cycles every second. If the amplitude of the motion is 50 cm, find an equation that describes the motion of P as a function of time. Assume that the point P is at its maximum displacement when $t = 0$.

73. Simple Harmonic Motion A mass suspended from a spring oscillates in simple harmonic motion at a frequency of 4 cycles per second. The distance from the highest to the lowest point of the oscillation is 100 cm. Find an equation that describes the distance of the mass from its rest position as a function of time. Assume that the mass is at its lowest point when $t = 0$.

74. Damped Harmonic Motion The top floor of a building undergoes damped harmonic motion after a sudden brief earthquake. At time $t = 0$ the displacement is at a maximum, 16 cm from the normal position. The damping constant is $c = 0.72$, and the building vibrates at 1.4 cycles per second.

(a) Find a function of the form $y = ke^{-ct} \cos \omega t$ to model the motion.

(b) Graph the function you found in part (a).

(c) What is the displacement at time $t = 10$ s?

CHAPTER 5 TEST

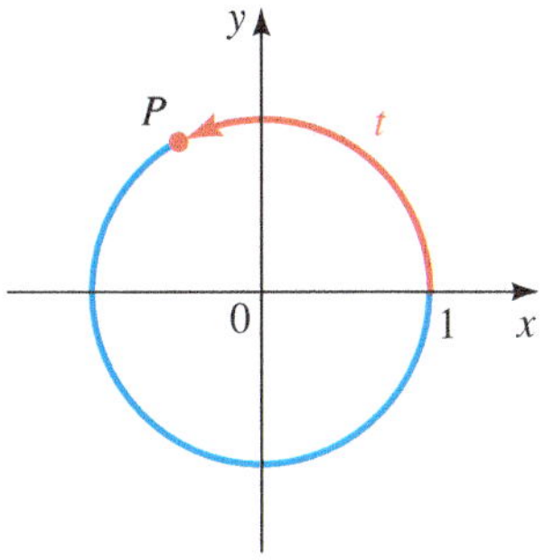

1. The point $P(x, y)$ is on the unit circle in Quadrant IV. If $x = \sqrt{11}/6$, find y.

2. The point P in the figure at the left has y-coordinate $\frac{4}{5}$. Find:

   (a) $\sin t$ (b) $\cos t$

   (c) $\tan t$ (d) $\sec t$

3. Find the exact value.

   (a) $\sin \dfrac{7\pi}{6}$ (b) $\cos \dfrac{13\pi}{4}$

   (c) $\tan\left(-\dfrac{5\pi}{3}\right)$ (d) $\csc \dfrac{3\pi}{2}$

4. Express $\tan t$ in terms of $\sin t$, if the terminal point determined by t is in Quadrant II.

5. If $\cos t = -\frac{8}{17}$ and if the terminal point determined by t is in Quadrant III, find $\tan t \cot t + \csc t$.

6–7 ■ A trigonometric function is given.

(a) Find the amplitude, period, phase, and horizontal shift of the function.

(b) Sketch the graph of one complete period.

6. $y = -5 \cos 4x$ 7. $y = 2 \sin\left(\dfrac{1}{2}x - \dfrac{\pi}{6}\right)$

8–9 ■ Find the period, and graph the function.

8. $y = -\csc 2x$ 9. $y = \tan\left(2x - \dfrac{\pi}{2}\right)$

10. Find the exact value of each expression, if it is defined.

    (a) $\tan^{-1} 1$ (b) $\cos^{-1}\left(-\dfrac{\sqrt{3}}{2}\right)$

    (c) $\tan^{-1}(\tan 3\pi)$ (d) $\cos(\tan^{-1}(-\sqrt{3}))$

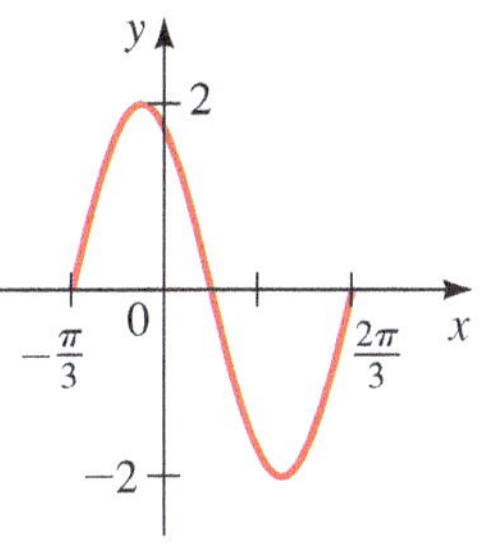

11. The graph shown at left is one period of a function of the form $y = a \sin k(x - b)$. Determine the function.

12. The sine curves $y_1 = 30 \sin\left(6t - \dfrac{\pi}{2}\right)$ and $y_2 = 30 \sin\left(6t - \dfrac{\pi}{3}\right)$ have the same period.

    (a) Find the phase of each curve.

    (b) Find the phase difference between y_1 and y_2.

    (c) Determine whether the curves are in phase or out of phase.

    (d) Sketch both curves on the same axes.

13. Let $f(x) = \dfrac{\cos x}{1 + x^2}$.

    (a) Use a graphing device to graph f in an appropriate viewing rectangle.

    (b) Determine from the graph if f is even, odd, or neither.

    (c) Find the minimum and maximum values of f.

14. A mass suspended from a spring oscillates in simple harmonic motion. The mass completes 2 cycles every second, and the distance between the highest point and the lowest point of the oscillation is 10 cm. Find an equation of the form $y = a \sin \omega t$ that gives the distance of the mass from its rest position as a function of time.

15. An object is moving up and down in damped harmonic motion. Its displacement at time $t = 0$ is 16 in.; this is its maximum displacement. The damping constant is $c = 0.1$, and the frequency is 12 Hz.

    (a) Find a function that models this motion.

    (b) Graph the function.

FOCUS ON MODELING Fitting Sinusoidal Curves to Data

In previous *Focus on Modeling* sections, we learned how to fit linear, polynomial, exponential, and power models to data. Figure 1 shows some scatter plots of data. The scatter plots can help guide us in choosing an appropriate model. (Try to determine what type of function would best model the data in each graph.) If the scatter plot indicates simple harmonic motion, then we might try to model the data with a sine or cosine function. The next example illustrates this process.

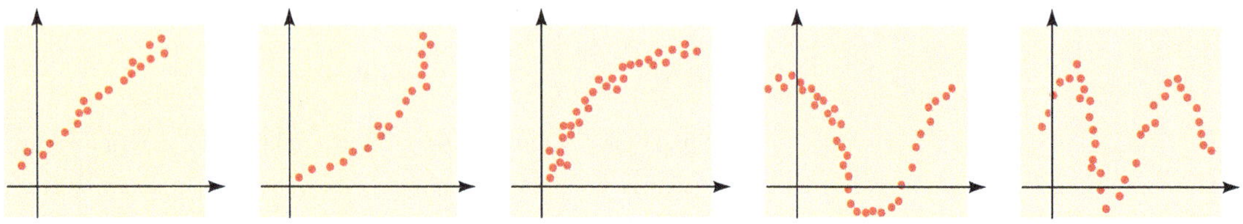

FIGURE 1

EXAMPLE 1 ■ Modeling the Height of a Tide

The water depth in a narrow channel varies with the tides. Table 1 shows the water depth over a 12-h period. A scatter plot of the data is shown in Figure 2.

(a) Find a function that models the water depth with respect to time.

(b) If a boat needs at least 11 ft of water to cross the channel, during which times can it safely do so?

TABLE 1

Time	Depth (ft)
12:00 A.M.	9.8
1:00 A.M.	11.4
2:00 A.M.	11.6
3:00 A.M.	11.2
4:00 A.M.	9.6
5:00 A.M.	8.5
6:00 A.M.	6.5
7:00 A.M.	5.7
8:00 A.M.	5.4
9:00 A.M.	6.0
10:00 A.M.	7.0
11:00 A.M.	8.6
12:00 P.M.	10.0

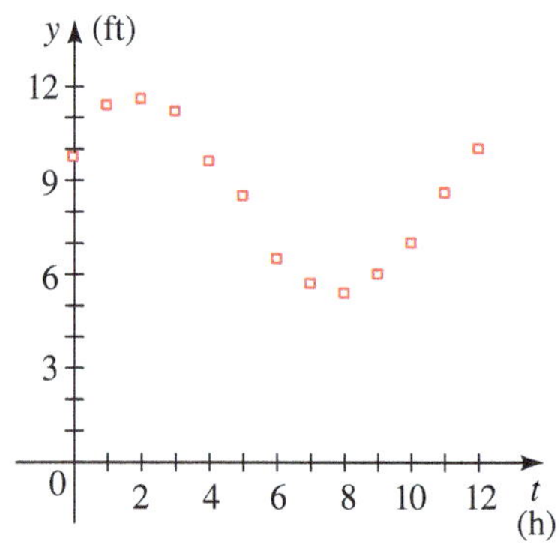

FIGURE 2

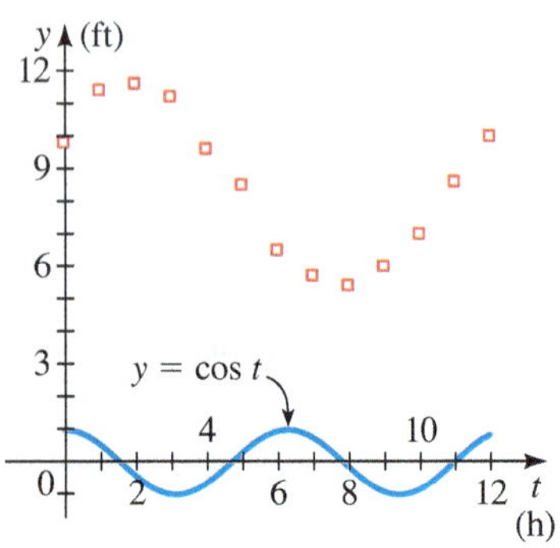

FIGURE 3

SOLUTION

(a) The data appear to lie on a cosine (or sine) curve. But if we graph $y = \cos t$ on the same graph as the scatter plot, the result in Figure 3 is not even close to the data. To fit the data, we need to adjust the vertical shift, amplitude, period, and phase shift of the cosine curve. In other words, we need to find a function of the form

$$y = a\cos(\omega(t - c)) + b$$

We use the following steps, which are illustrated by the graphs in the margin on the next page.

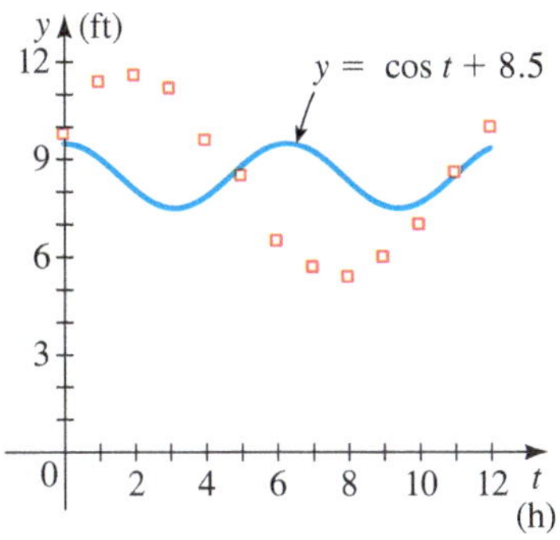

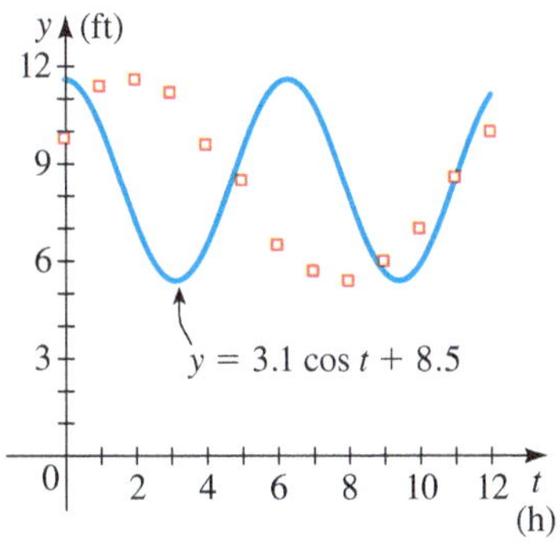

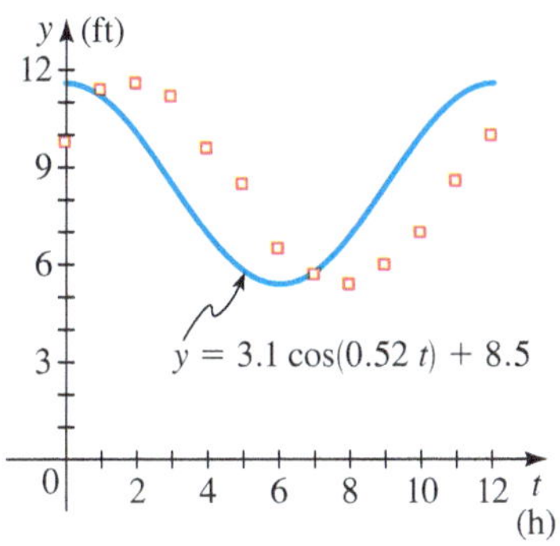

Adjust the Vertical Shift The vertical shift b is the average of the maximum and minimum values:

$$\begin{aligned} b &= \text{vertical shift} \\ &= \frac{1}{2} \cdot (\text{maximum value} + \text{minimum value}) \\ &= \frac{1}{2}(11.6 + 5.4) = 8.5 \end{aligned}$$

Adjust the Amplitude The amplitude a is half of the difference between the maximum and minimum values:

$$\begin{aligned} a &= \text{amplitude} \\ &= \frac{1}{2} \cdot (\text{maximum value} - \text{minimum value}) \\ &= \frac{1}{2}(11.6 - 5.4) = 3.1 \end{aligned}$$

Adjust the Period The time between consecutive maximum and minimum values is half of one period. Thus

$$\begin{aligned} \frac{2\pi}{\omega} &= \text{period} \\ &= 2 \cdot (\text{time of maximum value} - \text{time of minimum value}) \\ &= 2(8 - 2) = 12 \end{aligned}$$

Thus $\omega = 2\pi/12 = 0.52$.

Adjust the Horizontal Shift Since the maximum value of the data occurs at approximately $t = 2.0$, it represents a cosine curve shifted 2 h to the right. So

$$\begin{aligned} c &= \text{phase shift} \\ &= \text{time of maximum value} \\ &= 2.0 \end{aligned}$$

The Model We have shown that a function that models the tides over the given time period is given by

$$y = 3.1 \cos(0.52(t - 2.0)) + 8.5$$

A graph of the function and the scatter plot are shown in Figure 4. It appears that the model we found is a good approximation to the data.

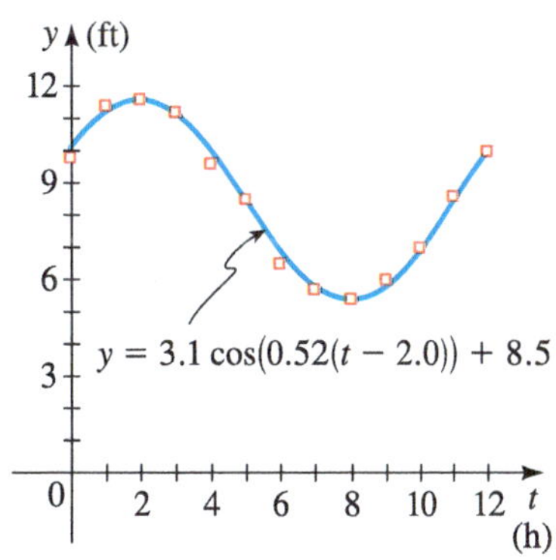

FIGURE 4

(b) We need to solve the inequality $y \geq 11$. We solve this inequality graphically by graphing $y = 3.1 \cos 0.52(t - 2.0) + 8.5$ and $y = 11$ on the same graph. From the graph in Figure 5 we see the water depth is higher than 11 ft between $t \approx 0.8$ and $t \approx 3.2$. This corresponds to the times 12:48 A.M. to 3:12 A.M.

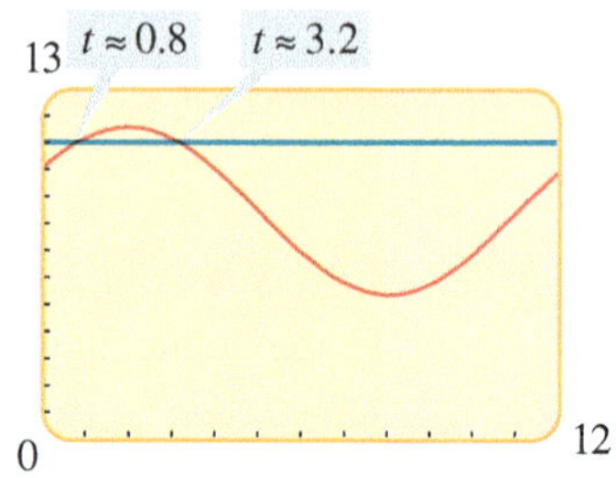

FIGURE 5

For the TI-83 and TI-84 the command `SinReg` (for sine regression) finds the sine curve that best fits the given data.

In Example 1 we used the scatter plot to guide us in finding a cosine curve that gives an approximate model of the data. Some graphing calculators are capable of finding a sine or cosine curve that best fits a given set of data points. The method these calculators use is similar to the method of finding a line of best fit, as explained on page 140.

EXAMPLE 2 ■ Fitting a Sine Curve to Data

(a) Use a graphing device to find the sine curve that best fits the depth of water data in Table 1 on page 466.

(b) Compare your result to the model found in Example 1.

SOLUTION

```
SinReg
 y=a*sin(bx+c)+d
 a=3.097877596
 b=.5268322697
 c=.5493035195
 d=8.424021899
```

Output of the `SinReg` function on the TI-83.

(a) Using the data in Table 1 and the `SinReg` command on the TI-83 calculator, we get a function of the form

$$y = a \sin(bt + c) + d$$

where

$$a = 3.1 \qquad b = 0.53$$
$$c = 0.55 \qquad d = 8.42$$

So the sine function that best fits the data is

$$y = 3.1 \sin(0.53t + 0.55) + 8.42$$

(b) To compare this with the function in Example 1, we change the sine function to a cosine function by using the reduction formula $\sin u = \cos(u - \pi/2)$.

$$\begin{aligned} y &= 3.1 \sin(0.53t + 0.55) + 8.42 \\ &= 3.1 \cos\left(0.53t + 0.55 - \frac{\pi}{2}\right) + 8.42 && \text{Reduction formula} \\ &= 3.1 \cos(0.53t - 1.02) + 8.42 \\ &= 3.1 \cos(0.53(t - 1.92)) + 8.42 && \text{Factor 0.53} \end{aligned}$$

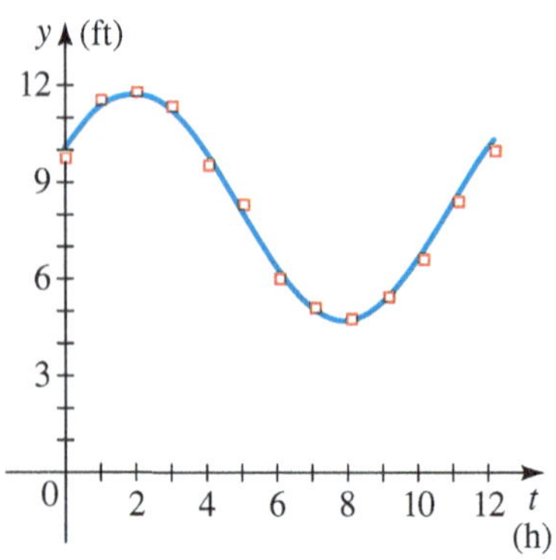

FIGURE 6

Comparing this with the function we obtained in Example 1, we see that there are small differences in the coefficients. In Figure 6 we graph a scatter plot of the data together with the sine function of best fit. ■

In Example 1 we estimated the values of the amplitude, period, and shifts from the data. In Example 2 the calculator computed the sine curve that best fits the data (that is, the curve that deviates least from the data as explained on page 140). The different ways of obtaining the model account for the differences in the functions.

PROBLEMS

1–4 ■ Modeling Periodic Data A set of data is given.

(a) Make a scatter plot of the data.

(b) Find a cosine function of the form $y = a\cos(\omega(t - c)) + b$ that models the data, as in Example 1.

(c) Graph the function you found in part (b) together with the scatter plot. How well does the curve fit the data?

(d) Use a graphing calculator to find the sine function that best fits the data, as in Example 2.

(e) Compare the functions you found in parts (b) and (d). [Use the reduction formula $\sin u = \cos(u - \pi/2)$.]

1.

t	y
0	2.1
2	1.1
4	−0.8
6	−2.1
8	−1.3
10	0.6
12	1.9
14	1.5

2.

t	y
0	190
25	175
50	155
75	125
100	110
125	95
150	105
175	120
200	140
225	165
250	185
275	200
300	195
325	185
350	165

3.

t	y
0.1	21.1
0.2	23.6
0.3	24.5
0.4	21.7
0.5	17.5
0.6	12.0
0.7	5.6
0.8	2.2
0.9	1.0
1.0	3.5
1.1	7.6
1.2	13.2
1.3	18.4
1.4	23.0
1.5	25.1

4.

t	y
0.0	0.56
0.5	0.45
1.0	0.29
1.5	0.13
2.0	0.05
2.5	−0.10
3.0	0.02
3.5	0.12
4.0	0.26
4.5	0.43
5.0	0.54
5.5	0.63
6.0	0.59

5. Circadian Rhythms Circadian rhythm (from the Latin *circa*—about, and *diem*—day) is the daily biological pattern by which body temperature, blood pressure, and other physiological variables change. The data in the table below show typical changes in human body temperature over a 24-h period ($t = 0$ corresponds to midnight).

(a) Make a scatter plot of the data.

(b) Find a cosine curve that models the data (as in Example 1).

(c) Graph the function you found in part (b) together with the scatter plot.

(d) Use a graphing calculator to find the sine curve that best fits the data (as in Example 2).

Time	Body temperature (°C)	Time	Body temperature (°C)
0	36.8	14	37.3
2	36.7	16	37.4
4	36.6	18	37.3
6	36.7	20	37.2
8	36.8	22	37.0
10	37.0	24	36.8
12	37.2		

Year	Owl population
0	50
1	62
2	73
3	80
4	71
5	60
6	51
7	43
8	29
9	20
10	28
11	41
12	49

6. Predator Population When two species interact in a predator/prey relationship, the populations of both species tend to vary in a sinusoidal fashion. (See *Discovery Project: Predator/Prey Models* referenced on page 427). In a certain midwestern county, the main food source for barn owls consists of field mice and other small mammals. The table gives the population of barn owls in this county every July 1 over a 12-year period.

(a) Make a scatter plot of the data.

(b) Find a sine curve that models the data (as in Example 1).

(c) Graph the function you found in part (b) together with the scatter plot.

(d) Use a graphing calculator to find the sine curve that best fits the data (as in Example 2). Compare to your answer from part (b).

7. Salmon Survival For reasons that are not yet fully understood, the number of fingerling salmon that survive the trip from their riverbed spawning grounds to the open ocean varies approximately sinusoidally from year to year. The table shows the number of salmon that hatch in a certain British Columbia creek and then make their way to the Strait of Georgia. The data are given in thousands of fingerlings, over a period of 16 years.

(a) Make a scatter plot of the data.

(b) Find a sine curve that models the data (as in Example 1).

(c) Graph the function you found in part (b) together with the scatter plot.

(d) Use a graphing calculator to find the sine curve that best fits the data (as in Example 2). Compare to your answer from part (b).

Year	Salmon (× 1000)	Year	Salmon (× 1000)
1985	43	1993	56
1986	36	1994	63
1987	27	1995	57
1988	23	1996	50
1989	26	1997	44
1990	33	1998	38
1991	43	1999	30
1992	50	2000	22

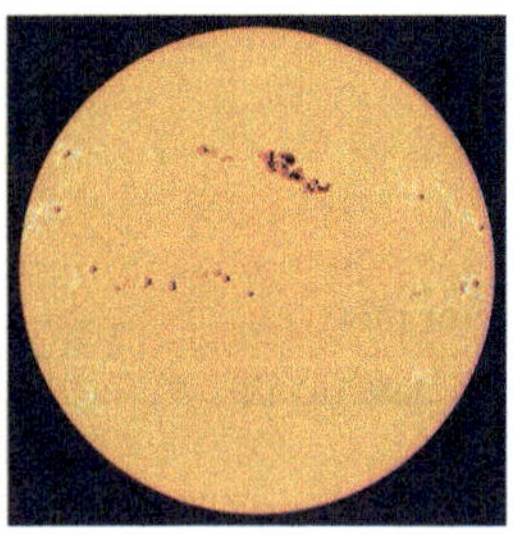

Science Source

8. Sunspot Activity Sunspots are relatively "cool" regions on the sun that appear as dark spots when observed through special solar filters. The number of sunspots varies in an 11-year cycle. The table gives the average daily sunspot count for the years 1968–2012.

(a) Make a scatter plot of the data.

(b) Find a cosine curve that models the data (as in Example 1).

(c) Graph the function you found in part (b) together with the scatter plot.

(d) Use a graphing calculator to find the sine curve that best fits the data (as in Example 2). Compare to your answer in part (b).

Year	Sunspots	Year	Sunspots	Year	Sunspots	Year	Sunspots
1968	106	1980	154	1991	145	2002	104
1969	105	1981	140	1992	94	2003	63
1970	104	1982	115	1993	54	2004	40
1971	67	1983	66	1994	29	2005	30
1972	69	1984	45	1995	17	2006	15
1973	38	1985	17	1996	8	2007	7
1974	34	1986	13	1997	21	2008	3
1975	15	1987	29	1998	64	2009	3
1976	12	1988	100	1999	93	2010	16
1977	27	1989	157	2000	119	2011	56
1978	92	1990	142	2001	111	2012	58
1979	155						

Source: Solar Influence Data Analysis Center, Belgium

john pacetti/Alamy

6 Trigonometric Functions: Right Triangle Approach

Suppose we want to find the distance from the earth to the sun. Using a tape measure is obviously impractical, so we need something other than simple measurements to tackle this problem. Angles are easier to measure than distances. For example, we can find the angle formed by the sun, earth, and moon by simply pointing to the sun with one arm and to the moon with the other and estimating the angle between them. The key idea is to find relationships between angles and distances. So if we had a way of determining distances from angles, we would be able to find the distance to the sun without having to go there. The trigonometric functions that we study in this chapter provide us with just the tools we need.

The trigonometric functions can be defined in two different but equivalent ways: as functions of real numbers (Chapter 5) or as functions of angles (Chapter 6). The two approaches are independent of each other, so either Chapter 5 or Chapter 6 may be studied first. We study both approaches because the different approaches are required for different applications.

6.1 ANGLE MEASURE

■ Angle Measure ■ Angles in Standard Position ■ Length of a Circular Arc ■ Area of a Circular Sector ■ Circular Motion

An **angle** AOB consists of two rays R_1 and R_2 with a common vertex O (see Figure 1). We often interpret an angle as a rotation of the ray R_1 onto R_2. In this case R_1 is called the **initial side**, and R_2 is called the **terminal side** of the angle. If the rotation is counterclockwise, the angle is considered **positive**, and if the rotation is clockwise, the angle is considered **negative**.

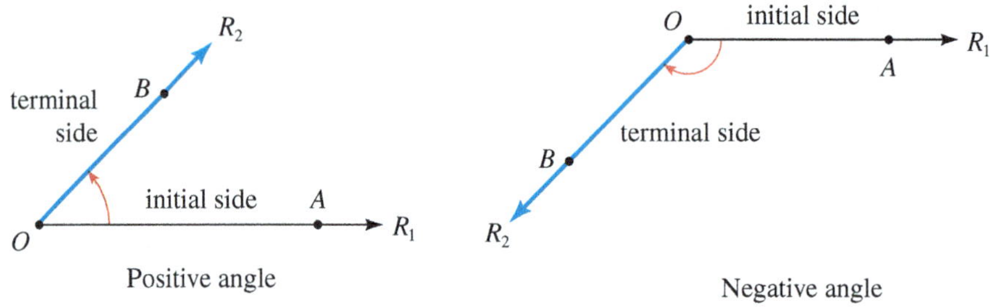

FIGURE 1

■ Angle Measure

The **measure** of an angle is the amount of rotation about the vertex required to move R_1 onto R_2. Intuitively, this is how much the angle "opens." One unit of measurement for angles is the **degree**. An angle of measure 1 degree is formed by rotating the initial side $\frac{1}{360}$ of a complete revolution. In calculus and other branches of mathematics a more natural method of measuring angles is used: *radian measure*. The amount an angle opens is measured along the arc of a circle of radius 1 with its center at the vertex of the angle.

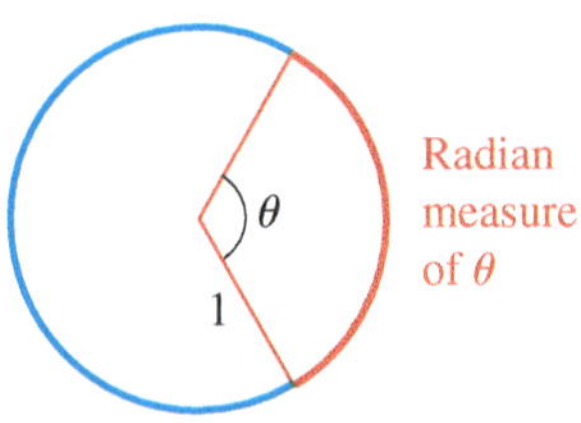

FIGURE 2

DEFINITION OF RADIAN MEASURE

If a circle of radius 1 is drawn with the vertex of an angle at its center, then the measure of this angle in **radians** (abbreviated **rad**) is the length of the arc that subtends the angle (see Figure 2).

The circumference of the circle of radius 1 is 2π, so a complete revolution has measure 2π rad, a straight angle has measure π rad, and a right angle has measure $\pi/2$ rad. An angle that is subtended by an arc of length 2 along the unit circle has radian measure 2 (see Figure 3).

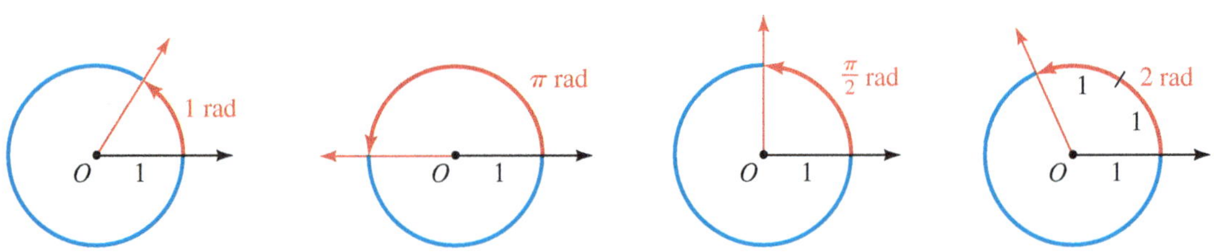

FIGURE 3 Radian measure

Since a complete revolution measured in degrees is 360° and measured in radians is 2π rad, we get the following simple relationship between these two methods of angle measurement.

RELATIONSHIP BETWEEN DEGREES AND RADIANS

$$180^\circ = \pi \text{ rad} \qquad 1 \text{ rad} = \left(\frac{180}{\pi}\right)^\circ \qquad 1^\circ = \frac{\pi}{180} \text{ rad}$$

1. To convert degrees to radians, multiply by $\dfrac{\pi}{180}$.
2. To convert radians to degrees, multiply by $\dfrac{180}{\pi}$.

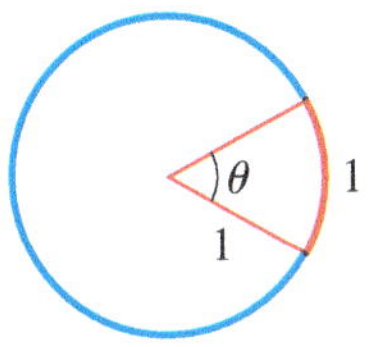

Measure of $\theta = 1$ rad
Measure of $\theta \approx 57.296^\circ$

FIGURE 4

To get some idea of the size of a radian, notice that

$$1 \text{ rad} \approx 57.296^\circ \qquad \text{and} \qquad 1^\circ \approx 0.01745 \text{ rad}$$

An angle θ of measure 1 rad is shown in Figure 4.

EXAMPLE 1 ■ Converting Between Radians and Degrees

(a) Express 60° in radians. **(b)** Express $\dfrac{\pi}{6}$ rad in degrees.

SOLUTION The relationship between degrees and radians gives

(a) $60^\circ = 60\left(\dfrac{\pi}{180}\right) \text{ rad} = \dfrac{\pi}{3} \text{ rad}$ **(b)** $\dfrac{\pi}{6} \text{ rad} = \left(\dfrac{\pi}{6}\right)\left(\dfrac{180}{\pi}\right) = 30^\circ$

Now Try Exercises 5 and 17 ■

A note on terminology: We often use a phrase such as "a 30° angle" to mean *an angle whose measure is* 30°. Also, for an angle θ we write $\theta = 30^\circ$ or $\theta = \pi/6$ to mean *the measure of* θ *is* 30° *or* $\pi/6$ *rad.* When no unit is given, the angle is assumed to be measured in radians.

Angles in Standard Position

An angle is in **standard position** if it is drawn in the xy-plane with its vertex at the origin and its initial side on the positive x-axis. Figure 5 gives examples of angles in standard position.

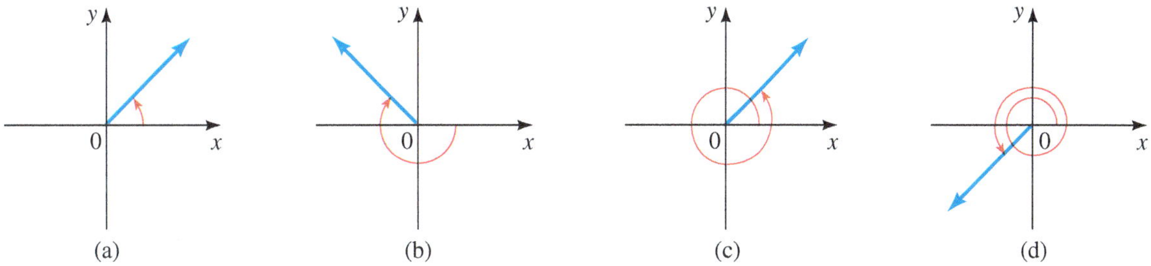

FIGURE 5 Angles in standard position

Two angles in standard position are **coterminal** if their sides coincide. In Figure 5 the angles in (a) and (c) are coterminal.

EXAMPLE 2 ■ Coterminal Angles

(a) Find angles that are coterminal with the angle $\theta = 30^\circ$ in standard position.

(b) Find angles that are coterminal with the angle $\theta = \dfrac{\pi}{3}$ in standard position.

SOLUTION

(a) To find positive angles that are coterminal with θ, we add any multiple of 360°. Thus

$$30° + 360° = 390° \quad \text{and} \quad 30° + 720° = 750°$$

are coterminal with $\theta = 30°$. To find negative angles that are coterminal with θ, we subtract any multiple of 360°. Thus

$$30° - 360° = -330° \quad \text{and} \quad 30° - 720° = -690°$$

are coterminal with θ. (See Figure 6.)

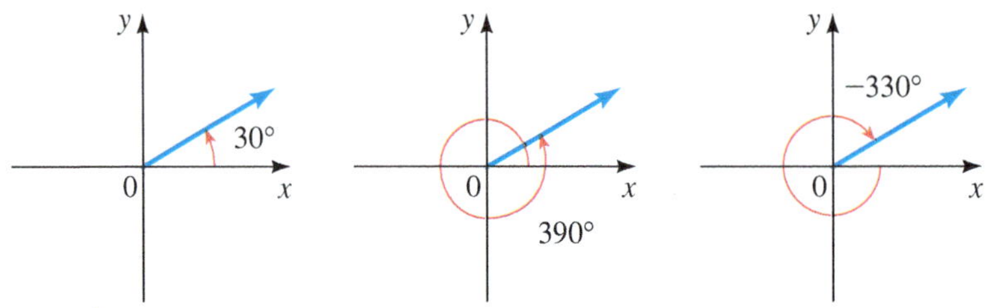

FIGURE 6

(b) To find positive angles that are coterminal with θ, we add any multiple of 2π. Thus

$$\frac{\pi}{3} + 2\pi = \frac{7\pi}{3} \quad \text{and} \quad \frac{\pi}{3} + 4\pi = \frac{13\pi}{3}$$

are coterminal with $\theta = \pi/3$. To find negative angles that are coterminal with θ, we subtract any multiple of 2π. Thus

$$\frac{\pi}{3} - 2\pi = -\frac{5\pi}{3} \quad \text{and} \quad \frac{\pi}{3} - 4\pi = -\frac{11\pi}{3}$$

are coterminal with θ. (See Figure 7.)

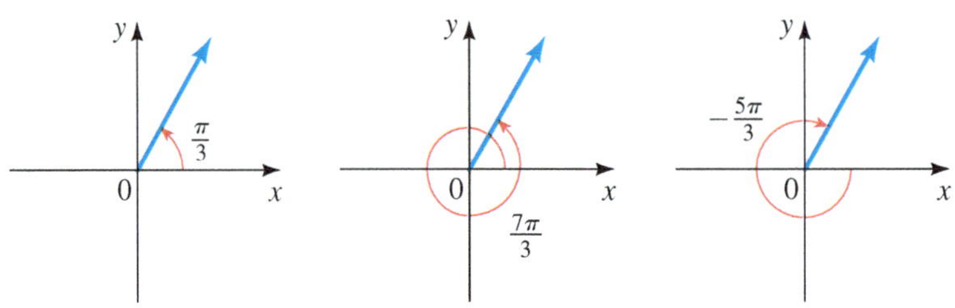

FIGURE 7

Now Try Exercises 29 and 31 ■

EXAMPLE 3 ■ Coterminal Angles

Find an angle with measure between 0° and 360° that is coterminal with the angle of measure 1290° in standard position.

SOLUTION We can subtract 360° as many times as we wish from 1290°, and the resulting angle will be coterminal with 1290°. Thus 1290° − 360° = 930° is coterminal with 1290°, and so is the angle 1290° − 2(360)° = 570°.

To find the angle we want between 0° and 360°, we subtract 360° from 1290° as many times as necessary. An efficient way to do this is to determine how many times 360° goes into 1290°, that is, divide 1290 by 360, and the remainder will be the angle we are looking for. We see that 360 goes into 1290 three times with a remainder of 210. Thus 210° is the desired angle (see Figure 8).

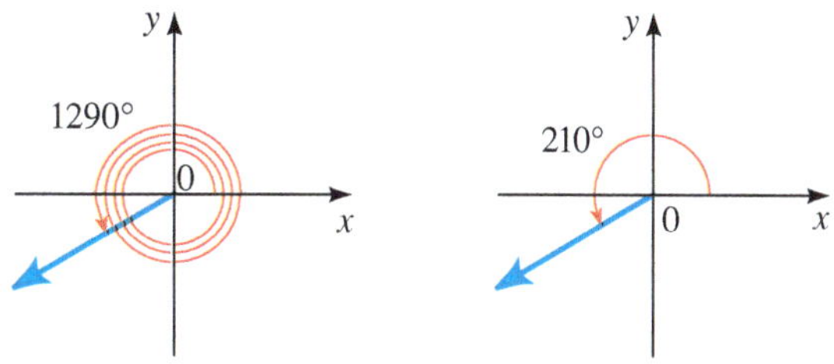

FIGURE 8

Now Try Exercise 41 ■

Length of a Circular Arc

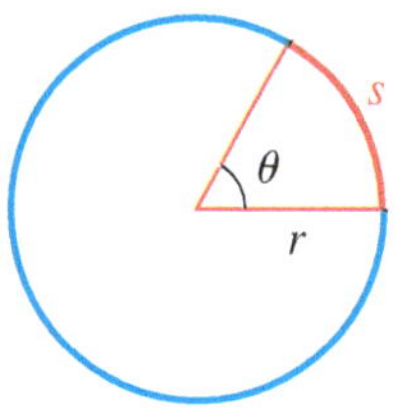

FIGURE 9 $s = \theta r$

An angle whose radian measure is θ is subtended by an arc that is the fraction $\theta/(2\pi)$ of the circumference of a circle. Thus in a circle of radius r the length s of an arc that subtends the angle θ (see Figure 9) is

$$\begin{aligned} s &= \frac{\theta}{2\pi} \times \text{circumference of circle} \\ &= \frac{\theta}{2\pi}(2\pi r) = \theta r \end{aligned}$$

LENGTH OF A CIRCULAR ARC

In a circle of radius r the length s of an arc that subtends a central angle of θ radians is

$$s = r\theta$$

Solving for θ, we get the important formula

$$\boxed{\theta = \frac{s}{r}}$$

This formula allows us to define radian measure using a circle of any radius r: The radian measure of an angle θ is s/r, where s is the length of the circular arc that subtends θ in a circle of radius r (see Figure 10).

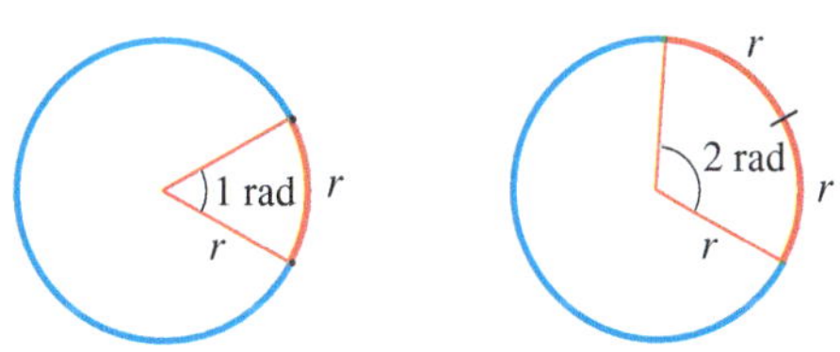

FIGURE 10 The radian measure of θ is the number of "radiuses" that can fit in the arc that subtends θ; hence the term *radian*.

EXAMPLE 4 ■ Arc Length and Angle Measure

(a) Find the length of an arc of a circle with radius 10 m that subtends a central angle of 30°.

(b) A central angle θ in a circle of radius 4 m is subtended by an arc of length 6 m. Find the measure of θ in radians.

SOLUTION

(a) From Example 1(b) we see that $30° = \pi/6$ rad. So the length of the arc is

$$s = r\theta = (10)\frac{\pi}{6} = \frac{5\pi}{3} \text{ m}$$

The formula $s = r\theta$ is true only when θ is measured in radians.

(b) By the formula $\theta = s/r$ we have

$$\theta = \frac{s}{r} = \frac{6}{4} = \frac{3}{2} \text{ rad}$$

Now Try Exercises 57 and 59

Area of a Circular Sector

The area of a circle of radius r is $A = \pi r^2$. A sector of this circle with central angle θ has an area that is the fraction $\theta/(2\pi)$ of the area of the entire circle (see Figure 11). So the area of this sector is

$$A = \frac{\theta}{2\pi} \times \text{area of circle}$$

$$= \frac{\theta}{2\pi}(\pi r^2) = \frac{1}{2}r^2\theta$$

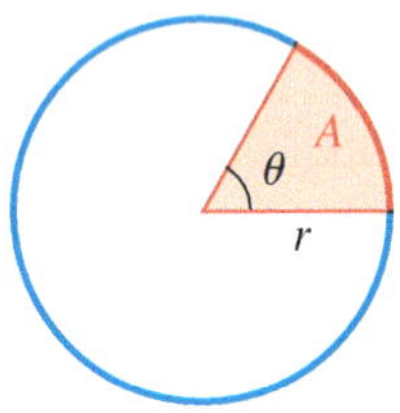

FIGURE 11
$A = \frac{1}{2}r^2\theta$

AREA OF A CIRCULAR SECTOR

In a circle of radius r the area A of a sector with a central angle of θ radians is

$$A = \frac{1}{2}r^2\theta$$

EXAMPLE 5 ■ Area of a Sector

Find the area of a sector of a circle with central angle 60° if the radius of the circle is 3 m.

SOLUTION To use the formula for the area of a circular sector, we must find the central angle of the sector in radians: $60° = 60(\pi/180)$ rad $= \pi/3$ rad. Thus the area of the sector is

$$A = \frac{1}{2}r^2\theta = \frac{1}{2}(3)^2\left(\frac{\pi}{3}\right) = \frac{3\pi}{2} \text{ m}^2$$

The formula $A = \frac{1}{2}r^2\theta$ is true only when θ is measured in radians.

Now Try Exercise 63

Circular Motion

Suppose a point moves along a circle as shown in Figure 12. There are two ways to describe the motion of the point: linear speed and angular speed. **Linear speed** is the rate at which the distance traveled is changing, so linear speed is the distance traveled divided by the time elapsed. **Angular speed** is the rate at which the central angle θ is changing, so angular speed is the number of radians this angle changes divided by the time elapsed.

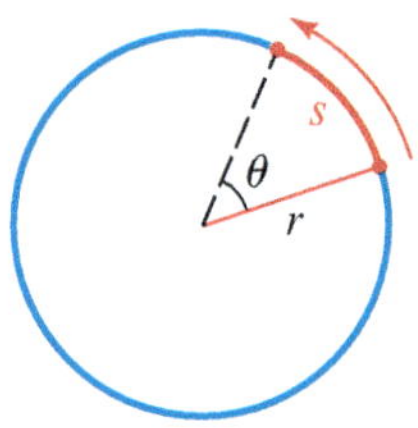

FIGURE 12

LINEAR SPEED AND ANGULAR SPEED

Suppose a point moves along a circle of radius r and the ray from the center of the circle to the point traverses θ radians in time t. Let $s = r\theta$ be the distance the point travels in time t. Then the speed of the object is given by

Angular speed $\qquad \omega = \dfrac{\theta}{t}$

Linear speed $\qquad v = \dfrac{s}{t}$

The symbol ω is the Greek letter "omega."

EXAMPLE 6 ■ Finding Linear and Angular Speed

A boy rotates a stone in a 3-ft-long sling at the rate of 15 revolutions every 10 seconds. Find the angular and linear velocities of the stone.

SOLUTION In 10 s the angle θ changes by $15 \cdot 2\pi = 30\pi$ rad. So the *angular speed* of the stone is

$$\omega = \frac{\theta}{t} = \frac{30\pi \text{ rad}}{10 \text{ s}} = 3\pi \text{ rad/s}$$

The distance traveled by the stone in 10 s is $s = 15 \cdot 2\pi r = 15 \cdot 2\pi \cdot 3 = 90\pi$ ft. So the *linear speed* of the stone is

$$v = \frac{s}{t} = \frac{90\pi \text{ ft}}{10 \text{ s}} = 9\pi \text{ ft/s}$$

Now Try Exercise 85 ■

Notice that angular speed does *not* depend on the radius of the circle; it depends only on the angle θ. However, if we know the angular speed ω and the radius r, we can find linear speed as follows: $v = s/t = r\theta/t = r(\theta/t) = r\omega$.

RELATIONSHIP BETWEEN LINEAR AND ANGULAR SPEED

If a point moves along a circle of radius r with angular speed ω, then its linear speed v is given by

$$v = r\omega$$

EXAMPLE 7 ■ Finding Linear Speed from Angular Speed

A woman is riding a bicycle whose wheels are 26 in. in diameter. If the wheels rotate at 125 revolutions per minute (rpm), find the speed (in mi/h) at which she is traveling.

SOLUTION The angular speed of the wheels is $2\pi \cdot 125 = 250\pi$ rad/min. Since the wheels have radius 13 in. (half the diameter), the linear speed is

$$v = r\omega = 13 \cdot 250\pi \approx 10{,}210.2 \text{ in./min}$$

Since there are 12 inches per foot, 5280 feet per mile, and 60 minutes per hour, her speed in miles per hour is

$$\frac{10{,}210.2 \text{ in./min} \times 60 \text{ min/h}}{12 \text{ in./ft} \times 5280 \text{ ft/mi}} = \frac{612{,}612 \text{ in./h}}{63{,}360 \text{ in./mi}}$$
$$\approx 9.7 \text{ mi/h}$$

Now Try Exercise 87 ■

6.1 EXERCISES

CONCEPTS

1. (a) The radian measure of an angle θ is the length of the ________ that subtends the angle in a circle of radius ________.

   (b) To convert degrees to radians, we multiply by ________.

   (c) To convert radians to degrees, we multiply by ________.

2. A central angle θ is drawn in a circle of radius r, as in the figure below.

   (a) The length of the arc subtended by θ is $s =$ ________.

   (b) The area of the sector with central angle θ is $A =$ ________.

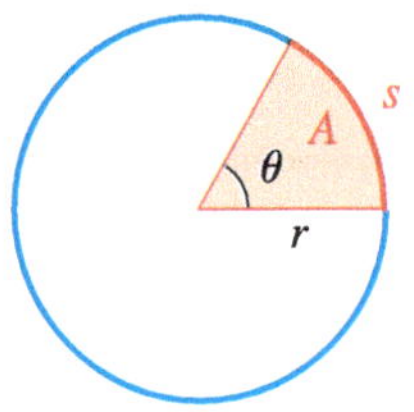

3. Suppose a point moves along a circle with radius r as shown in the figure below. The point travels a distance s along the circle in time t.

   (a) The angular speed of the point is $\omega = \dfrac{}{}$.

   (b) The linear speed of the point is $v = \dfrac{}{}$.

   (c) The linear speed v and the angular speed ω are related by the equation $v =$ ________.

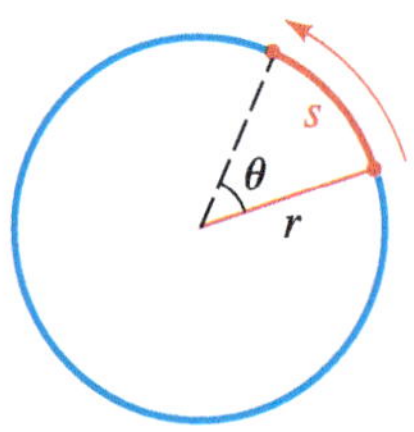

4. Object A is traveling along a circle of radius 2, and Object B is traveling along a circle of radius 5. The objects have the same angular speed. Do the objects have the same linear speed? If not, which object has the greater linear speed?

SKILLS

5–16 ■ From Degrees to Radians Find the radian measure of the angle with the given degree measure. Round your answer to three decimal places.

5. 15° 6. 36° 7. 54° 8. 75°
9. −45° 10. −30° 11. 100° 12. 200°
13. 1000° 14. 3600° 15. −70° 16. −150°

17–28 ■ From Radians to Degrees Find the degree measure of the angle with the given radian measure.

17. $\frac{5\pi}{3}$ 18. $\frac{3\pi}{4}$ 19. $\frac{5\pi}{6}$
20. $-\frac{3\pi}{2}$ 21. 3 22. −2
23. −1.2 24. 3.4 25. $\frac{\pi}{10}$
26. $\frac{5\pi}{18}$ 27. $-\frac{2\pi}{15}$ 28. $-\frac{13\pi}{12}$

29–34 ■ Coterminal Angles The measure of an angle in standard position is given. Find two positive angles and two negative angles that are coterminal with the given angle.

29. 50° 30. 135° 31. $\frac{3\pi}{4}$
32. $\frac{11\pi}{6}$ 33. $-\frac{\pi}{4}$ 34. −45°

35–40 ■ Coterminal Angles? The measures of two angles in standard position are given. Determine whether the angles are coterminal.

35. 70°, 430° 36. −30°, 330°
37. $\frac{5\pi}{6}$, $\frac{17\pi}{6}$ 38. $\frac{32\pi}{3}$, $\frac{11\pi}{3}$
39. 155°, 875° 40. 50°, 340°

41–46 ■ Finding a Coterminal Angle Find an angle between 0° and 360° that is coterminal with the given angle.

41. 400° 42. 375°
43. 780° 44. −100°
45. −800° 46. 1270°

47–52 ■ Finding a Coterminal Angle Find an angle between 0 and 2π that is coterminal with the given angle.

47. $\frac{19\pi}{6}$ 48. $-\frac{5\pi}{3}$ 49. 25π
50. 10 51. $\frac{17\pi}{4}$ 52. $\frac{51\pi}{2}$

53–62 ■ Circular Arcs Find the length s of the circular arc, the radius r of the circle, or the central angle θ, as indicated.

53.
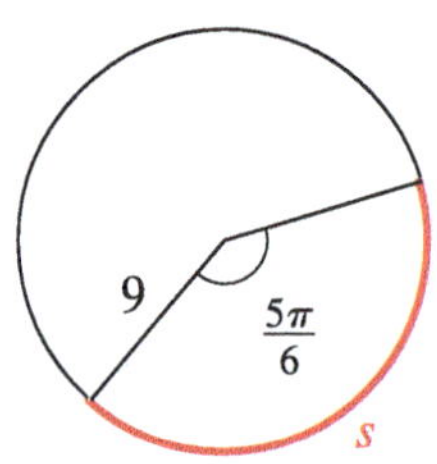

54.
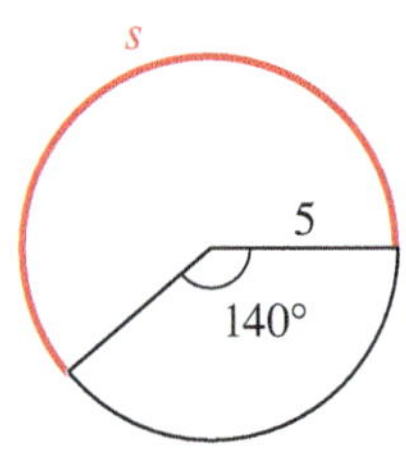

55.

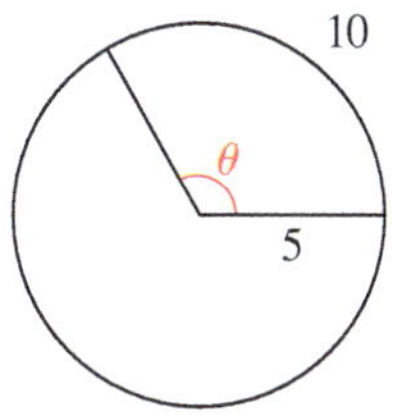

56.

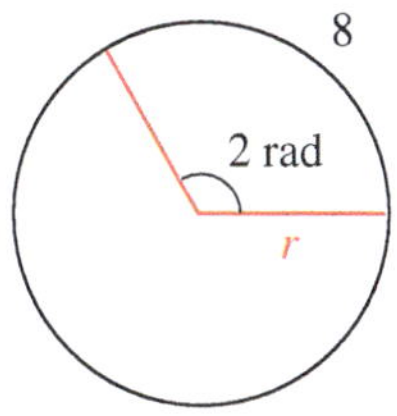

57. Find the length s of the arc that subtends a central angle of measure 3 rad in a circle of radius 5 cm.

58. Find the length s of the arc that subtends a central angle of measure 40° in a circle of radius 12 m.

59. A central angle θ in a circle of radius 9 m is subtended by an arc of length 14 m. Find the measure of θ in degrees and radians.

60. An arc of length 15 ft subtends a central angle θ in a circle of radius 9 ft. Find the measure of θ in degrees and radians.

61. Find the radius r of the circle if an arc of length 15 m on the circle subtends a central angle of $5\pi/6$.

62. Find the radius r of the circle if an arc of length 20 cm on the circle subtends a central angle of 50°.

63–70 ■ Area of a Circular Sector These exercises involve the formula for the area of a circular sector.

63. Find the area of the sector shown in each figure.

(a)

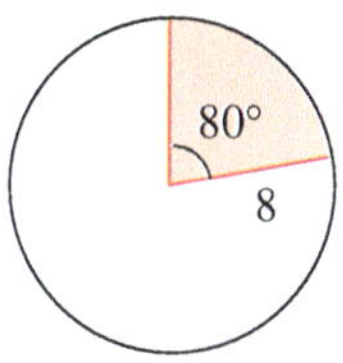

(b)

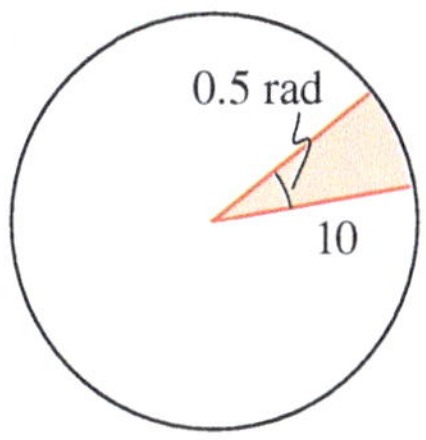

64. Find the radius of each circle if the area of the sector is 12.

(a)

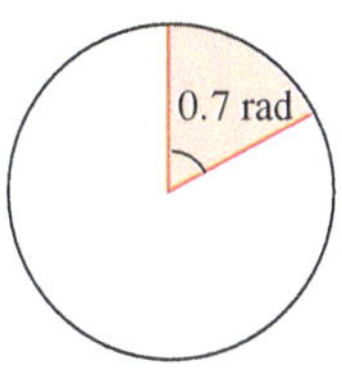

(b)

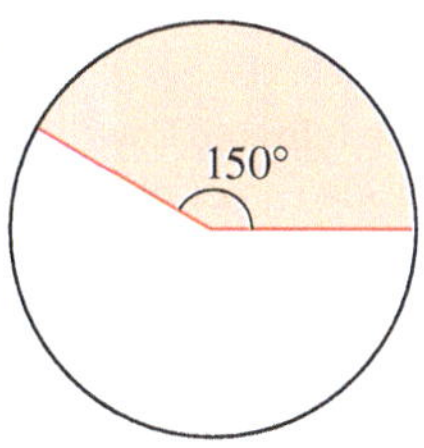

65. Find the area of a sector with central angle $2\pi/3$ rad in a circle of radius 10 m.

66. A sector of a circle has a central angle of 145°. Find the area of the sector if the radius of the circle is 6 ft.

67. The area of a sector of a circle with a central angle of 140° is 70 m^2. Find the radius of the circle.

68. The area of a sector of a circle with a central angle of $5\pi/12$ rad is 20 m^2. Find the radius of the circle.

69. A sector of a circle of radius 80 mi has an area of 1600 mi^2. Find the central angle (in radians) of the sector.

70. The area of a circle is 600 m^2. Find the area of a sector of this circle that subtends a central angle of 3 rad.

SKILLS Plus

71. Area of a Sector of a Circle Three circles with radii 1, 2, and 3 ft are externally tangent to one another, as shown in the figure. Find the area of the sector of the circle of radius 1 that is cut off by the line segments joining the center of that circle to the centers of the other two circles.

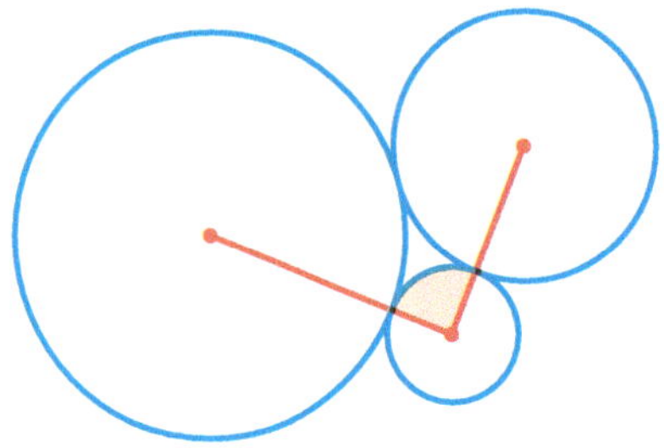

72. Comparing a Triangle and a Sector of a Circle Two wood sticks and a metal rod, each of length 1, are connected to form a triangle with angle θ_1 at the point P, as shown in the first figure below. The rod is then bent to form an arc of a circle with center P, resulting in a smaller angle θ_2 at the point P, as shown in the second figure. Find θ_1, θ_2, and $\theta_1 - \theta_2$.

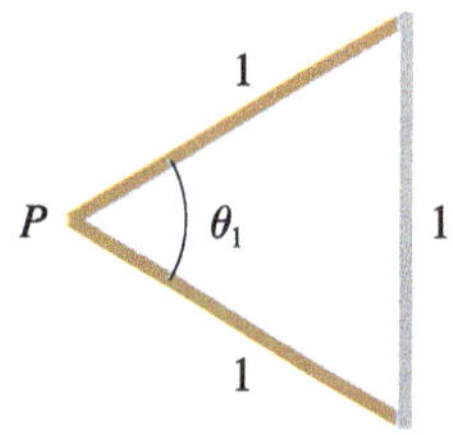

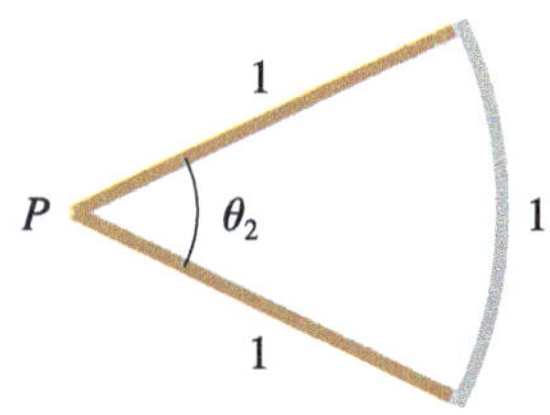

73–74 ■ Clocks and Angles In 1 h the minute hand on a clock moves through a complete circle, and the hour hand moves through $\frac{1}{12}$ of a circle.

73. Through how many radians do the minute hand and the hour hand move between 1:00 P.M. and 1:45 P.M. (on the same day)?

74. Through how many radians do the minute hand and the hour hand move between 1:00 P.M. and 6:45 P.M. (on the same day)?

APPLICATIONS

75. Travel Distance A car's wheels are 28 in. in diameter. How far (in mi.) will the car travel if its wheels revolve 10,000 times without slipping?

76. Wheel Revolutions How many revolutions will a car wheel of diameter 30 in. make as the car travels a distance of one mile?

77. Latitudes Pittsburgh, Pennsylvania, and Miami, Florida, lie approximately on the same meridian. Pittsburgh has a latitude of 40.5° N, and Miami has a latitude of 25.5° N. Find the distance between these two cities. (The radius of the earth is 3960 mi.)

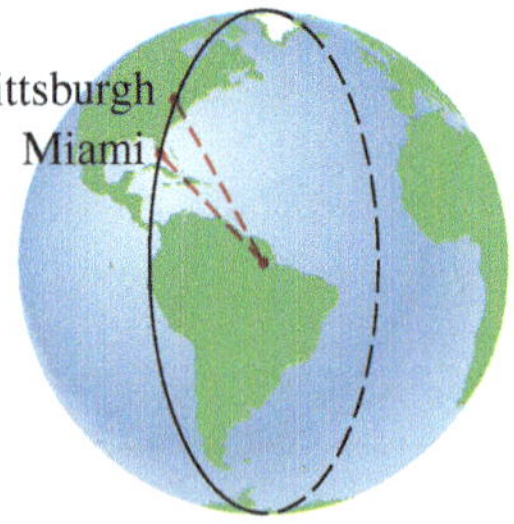

78. Latitudes Memphis, Tennessee, and New Orleans, Louisiana, lie approximately on the same meridian. Memphis has a latitude of 35° N, and New Orleans has a latitude of 30° N. Find the distance between these two cities. (The radius of the earth is 3960 mi.)

79. Orbit of the Earth Find the distance that the earth travels in one day in its path around the sun. Assume that a year has 365 days and that the path of the earth around the sun is a circle of radius 93 million miles. [*Note:* The path of the earth around the sun is actually an *ellipse* with the sun at one focus (see Section 11.2). This ellipse, however, has very small eccentricity, so it is nearly circular.]

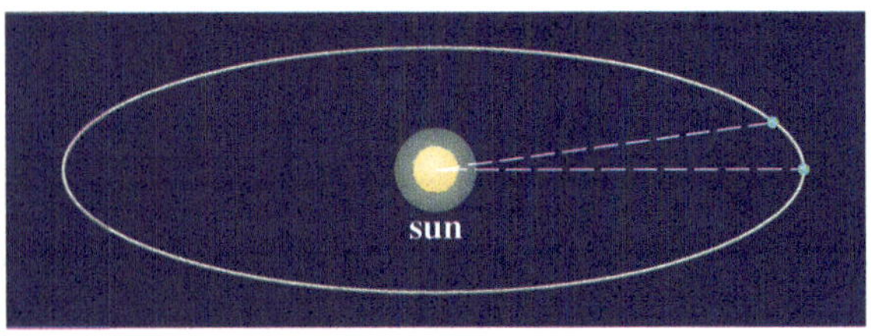

80. Circumference of the Earth The Greek mathematician Eratosthenes (ca. 276–195 B.C.) measured the circumference of the earth from the following observations. He noticed that on a certain day the sun shone directly down a deep well in Syene (modern Aswan). At the same time in Alexandria, 500 miles north (on the same meridian), the rays of the sun shone at an angle of 7.2° to the zenith. Use this information and the figure to find the radius and circumference of the earth.

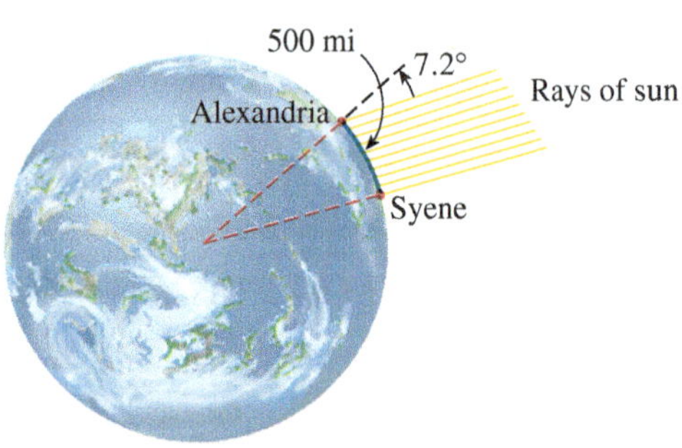

81. Nautical Miles Find the distance along an arc on the surface of the earth that subtends a central angle of 1 minute ($1 \text{ minute} = \frac{1}{60}$ degree). This distance is called a *nautical mile*. (The radius of the earth is 3960 mi.)

82. Irrigation An irrigation system uses a straight sprinkler pipe 300 ft long that pivots around a central point as shown. Because of an obstacle the pipe is allowed to pivot through 280° only. Find the area irrigated by this system.

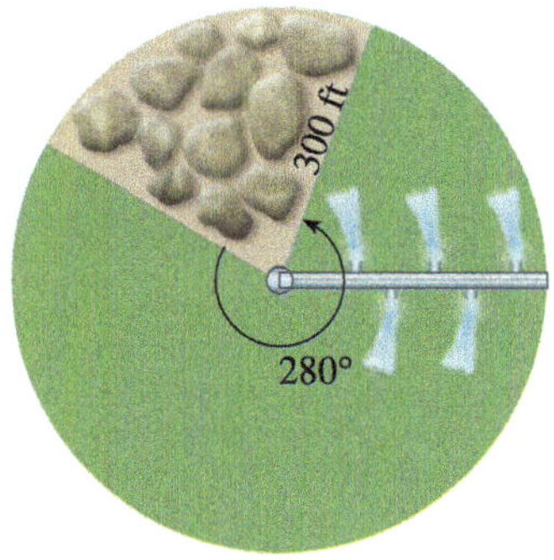

83. Windshield Wipers The top and bottom ends of a windshield wiper blade are 34 in. and 14 in., respectively, from the pivot point. While in operation, the wiper sweeps through 135°. Find the area swept by the blade.

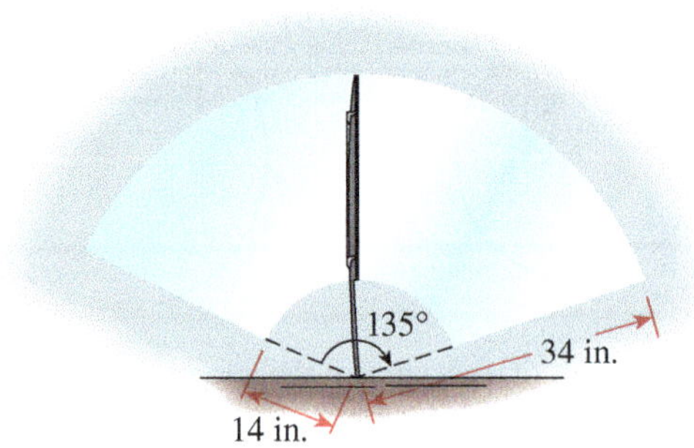

84. The Tethered Cow A cow is tethered by a 100-ft rope to the inside corner of an L-shaped building, as shown in the figure. Find the area that the cow can graze.

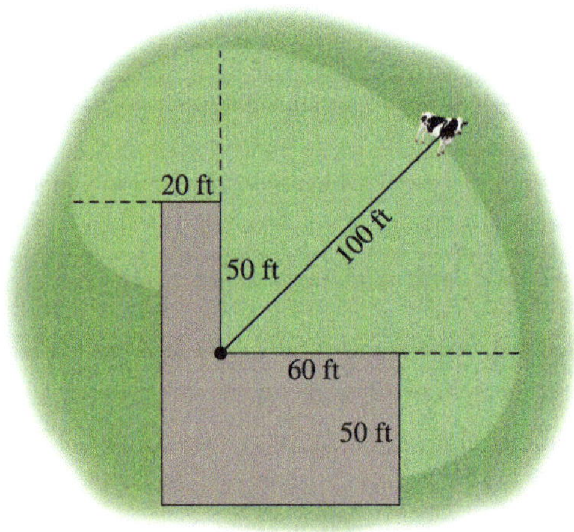

85. Fan A ceiling fan with 16-in. blades rotates at 45 rpm.

(a) Find the angular speed of the fan in rad/min.

(b) Find the linear speed of the tips of the blades in in./min.

86. Radial Saw A radial saw has a blade with a 6-in. radius. Suppose that the blade spins at 1000 rpm.

(a) Find the angular speed of the blade in rad/min.

(b) Find the linear speed of the sawteeth in ft/s.

87. Winch A winch of radius 2 ft is used to lift heavy loads. If the winch makes 8 revolutions every 15 s, find the speed at which the load is rising.

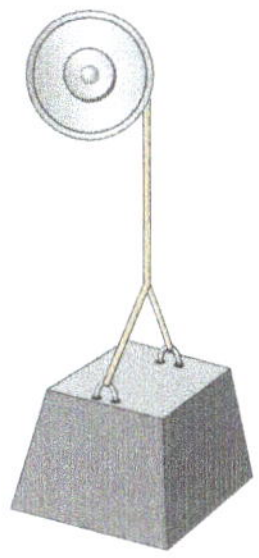

88. Speed of a Car The wheels of a car have radius 11 in. and are rotating at 600 rpm. Find the speed of the car in mi/h.

89. Speed at the Equator The earth rotates about its axis once every 23 h 56 min 4 s, and the radius of the earth is 3960 mi. Find the linear speed of a point on the equator in mi/h.

90. Truck Wheels A truck with 48-in.-diameter wheels is traveling at 50 mi/h.

(a) Find the angular speed of the wheels in rad/min.

(b) How many revolutions per minute do the wheels make?

91. Speed of a Current To measure the speed of a current, scientists place a paddle wheel in the stream and observe the rate at which it rotates. If the paddle wheel has radius 0.20 m and rotates at 100 rpm, find the speed of the current in m/s.

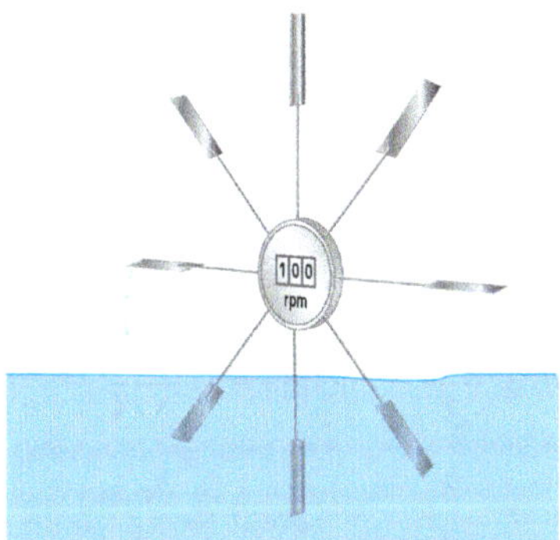

92. Bicycle Wheel The sprockets and chain of a bicycle are shown in the figure. The pedal sprocket has a radius of 4 in., the wheel sprocket a radius of 2 in., and the wheel a radius of 13 in. The cyclist pedals at 40 rpm.

(a) Find the angular speed of the wheel sprocket.

(b) Find the speed of the bicycle. (Assume that the wheel turns at the same rate as the wheel sprocket.)

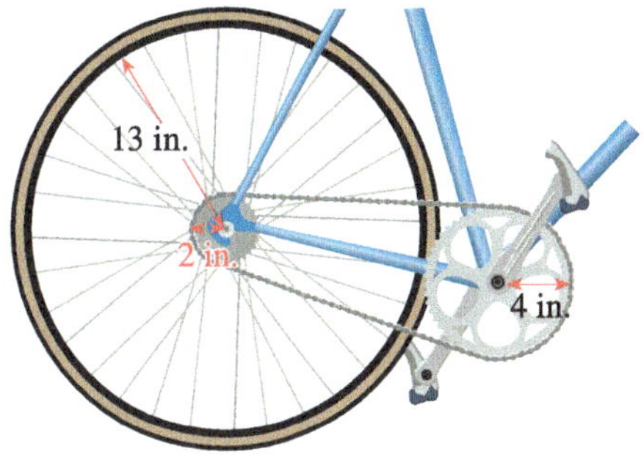

93. Conical Cup A conical cup is made from a circular piece of paper with radius 6 cm by cutting out a sector and joining the edges as shown below. Suppose $\theta = 5\pi/3$.

(a) Find the circumference C of the opening of the cup.

(b) Find the radius r of the opening of the cup. [*Hint:* Use $C = 2\pi r$.]

(c) Find the height h of the cup. [*Hint:* Use the Pythagorean Theorem.]

(d) Find the volume of the cup.

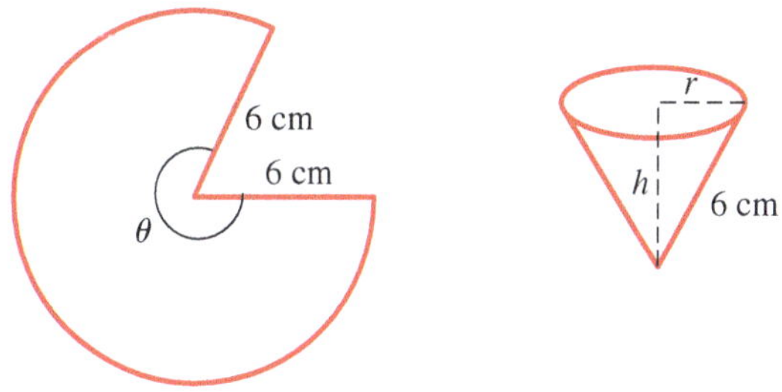

94. Conical Cup In this exercise we find the volume of the conical cup in Exercise 93 for any angle θ.

(a) Follow the steps in Exercise 93 to show that the volume of the cup as a function of θ is

$$V(\theta) = \frac{9}{\pi^2}\theta^2\sqrt{4\pi^2 - \theta^2}, \qquad 0 < \theta < 2\pi$$

(b) Graph the function V.

(c) For what angle θ is the volume of the cup a maximum?

DISCUSS ■ DISCOVER ■ PROVE ■ WRITE

95. WRITE: Different Ways of Measuring Angles The custom of measuring angles using degrees, with 360° in a circle, dates back to the ancient Babylonians, who used a number system based on groups of 60. Another system of measuring angles divides the circle into 400 units, called *grads*. In this system a right angle is 100 grad, so this fits in with our base 10 number system.

Write a short essay comparing the advantages and disadvantages of these two systems and the radian system of measuring angles. Which system do you prefer? Why?

6.2 TRIGONOMETRY OF RIGHT TRIANGLES

■ Trigonometric Ratios ■ Special Triangles; Calculators ■ Applications of Trigonometry of Right Triangles

In this section we study certain ratios of the sides of right triangles, called trigonometric ratios, and give several applications.

■ Trigonometric Ratios

Consider a right triangle with θ as one of its acute angles. The trigonometric ratios are defined as follows (see Figure 1).

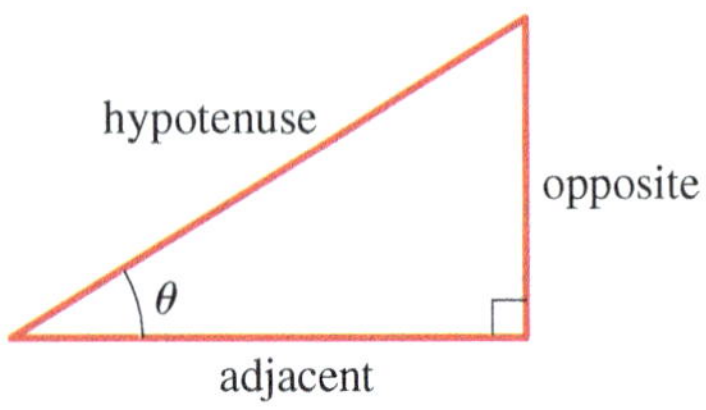

FIGURE 1

THE TRIGONOMETRIC RATIOS

$$\sin\theta = \frac{\text{opposite}}{\text{hypotenuse}} \qquad \cos\theta = \frac{\text{adjacent}}{\text{hypotenuse}} \qquad \tan\theta = \frac{\text{opposite}}{\text{adjacent}}$$

$$\csc\theta = \frac{\text{hypotenuse}}{\text{opposite}} \qquad \sec\theta = \frac{\text{hypotenuse}}{\text{adjacent}} \qquad \cot\theta = \frac{\text{adjacent}}{\text{opposite}}$$

The symbols we use for these ratios are abbreviations for their full names: **sine**, **cosine**, **tangent**, **cosecant**, **secant**, **cotangent**. Since any two right triangles with angle θ are similar, these ratios are the same, regardless of the size of the triangle; the trigonometric ratios depend only on the angle θ (see Figure 2).

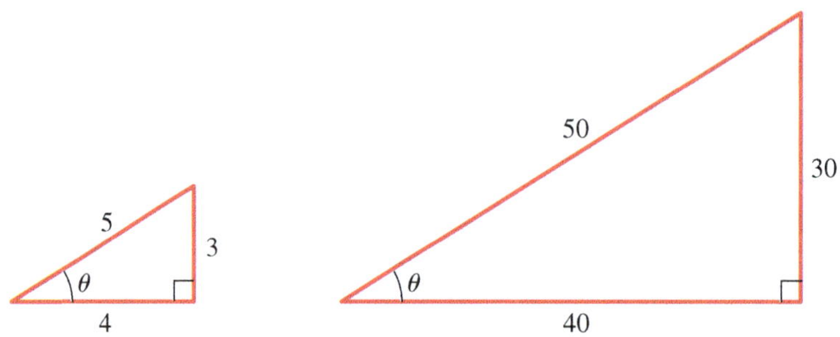

$\sin\theta = \frac{3}{5}$ $\qquad$ $\sin\theta = \frac{30}{50} = \frac{3}{5}$

FIGURE 2

EXAMPLE 1 ■ Finding Trigonometric Ratios

Find the six trigonometric ratios of the angle θ in Figure 3.

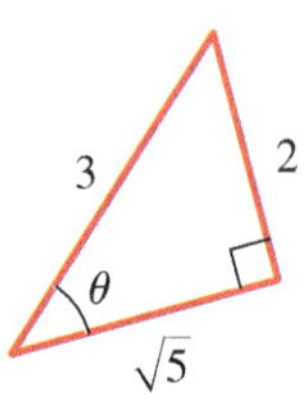

FIGURE 3

SOLUTION By the definition of trigonometric ratios, we get

$$\sin\theta = \frac{2}{3} \qquad \cos\theta = \frac{\sqrt{5}}{3} \qquad \tan\theta = \frac{2}{\sqrt{5}}$$

$$\csc\theta = \frac{3}{2} \qquad \sec\theta = \frac{3}{\sqrt{5}} \qquad \cot\theta = \frac{\sqrt{5}}{2}$$

Now Try Exercise 3 ■

EXAMPLE 2 ■ Finding Trigonometric Ratios

If $\cos\alpha = \frac{3}{4}$, sketch a right triangle with acute angle α, and find the other five trigonometric ratios of α.

SOLUTION Since $\cos \alpha$ is defined as the ratio of the adjacent side to the hypotenuse, we sketch a triangle with hypotenuse of length 4 and a side of length 3 adjacent to α. If the opposite side is x, then by the Pythagorean Theorem, $3^2 + x^2 = 4^2$ or $x^2 = 7$, so $x = \sqrt{7}$. We then use the triangle in Figure 4 to find the ratios.

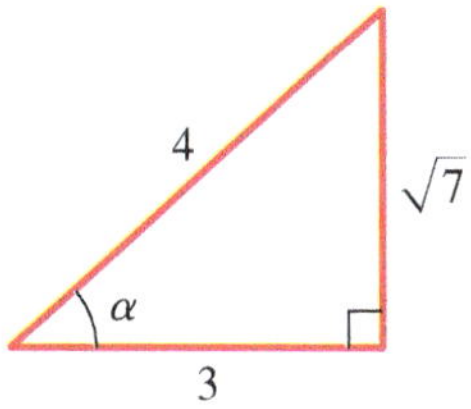

FIGURE 4

$$\sin \alpha = \frac{\sqrt{7}}{4} \qquad \cos \alpha = \frac{3}{4} \qquad \tan \alpha = \frac{\sqrt{7}}{3}$$

$$\csc \alpha = \frac{4}{\sqrt{7}} \qquad \sec \alpha = \frac{4}{3} \qquad \cot \alpha = \frac{3}{\sqrt{7}}$$

Now Try Exercise 23 ■

Special Triangles; Calculators

There are special trigonometric ratios that can be calculated from certain triangles (which we call special triangles). We can also use a calculator to find trigonometric ratios.

Special Ratios Certain right triangles have ratios that can be calculated easily from the Pythagorean Theorem. Since they are used frequently, we mention them here.

The first triangle is obtained by drawing a diagonal in a square of side 1 (see Figure 5). By the Pythagorean Theorem this diagonal has length $\sqrt{2}$. The resulting triangle has angles 45°, 45°, and 90° (or $\pi/4$, $\pi/4$, and $\pi/2$). To get the second triangle, we start with an equilateral triangle ABC of side 2 and draw the perpendicular bisector DB of the base, as in Figure 6. By the Pythagorean Theorem the length of DB is $\sqrt{3}$. Since DB bisects angle ABC, we obtain a triangle with angles 30°, 60°, and 90° (or $\pi/6$, $\pi/3$, and $\pi/2$).

HIPPARCHUS (circa 140 B.C.) is considered the founder of trigonometry. He constructed tables for a function closely related to the modern sine function and evaluated for angles at half-degree intervals. These are considered the first trigonometric tables. He used his tables mainly to calculate the paths of the planets through the heavens.

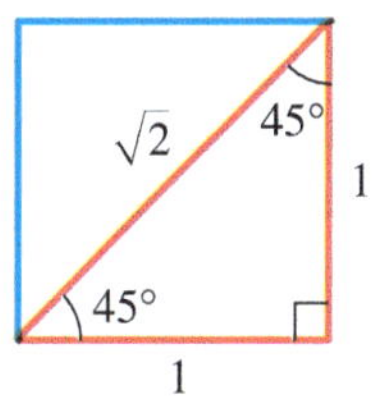

FIGURE 5

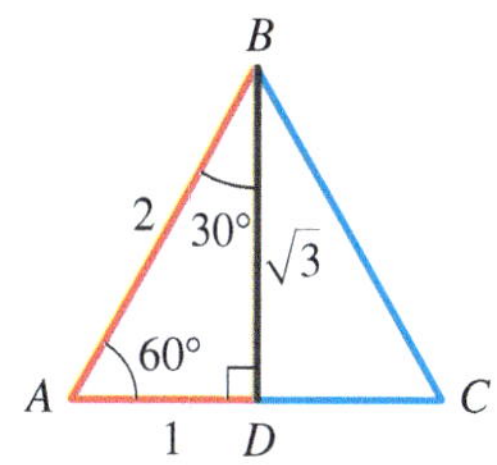

FIGURE 6

We can now use the special triangles in Figures 5 and 6 to calculate the trigonometric ratios for angles with measures 30°, 45°, and 60° (or $\pi/6$, $\pi/4$, and $\pi/3$). These are listed in the table below.

SPECIAL VALUES OF THE TRIGONOMETRIC FUNCTIONS

The following values of the trigonometric functions are obtained from the special triangles.

θ in degrees	θ in radians	$\sin \theta$	$\cos \theta$	$\tan \theta$	$\csc \theta$	$\sec \theta$	$\cot \theta$
0	0	0	1	0	—	1	—
30°	$\frac{\pi}{6}$	$\frac{1}{2}$	$\frac{\sqrt{3}}{2}$	$\frac{\sqrt{3}}{3}$	2	$\frac{2\sqrt{3}}{3}$	$\sqrt{3}$
45°	$\frac{\pi}{4}$	$\frac{\sqrt{2}}{2}$	$\frac{\sqrt{2}}{2}$	1	$\sqrt{2}$	$\sqrt{2}$	1
60°	$\frac{\pi}{3}$	$\frac{\sqrt{3}}{2}$	$\frac{1}{2}$	$\sqrt{3}$	$\frac{2\sqrt{3}}{3}$	2	$\frac{\sqrt{3}}{3}$
90°	$\frac{\pi}{2}$	1	0	—	1	—	0

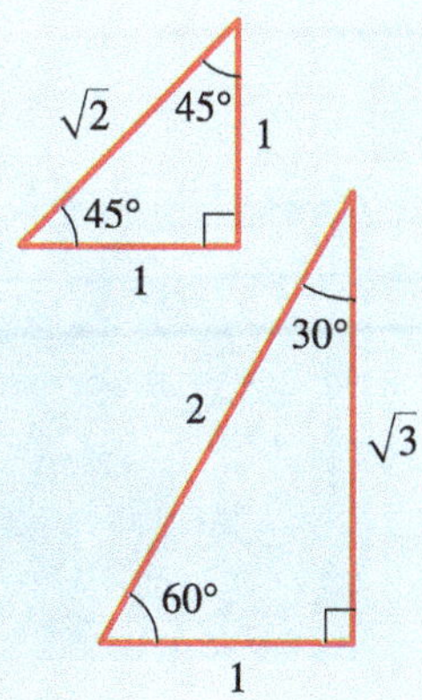

It's useful to remember these special trigonometric ratios because they occur often. Of course, they can be recalled easily if we remember the triangles from which they are obtained.

For an explanation of numerical methods, see the margin note on page 433.

Using a Calculator To find the values of the trigonometric ratios for other angles, we use a calculator. Mathematical methods (called *numerical methods*) used in finding the trigonometric ratios are programmed directly into scientific calculators. For instance, when the SIN key is pressed, the calculator computes an approximation to the value of the sine of the given angle. Calculators give the values of sine, cosine, and tangent; the other ratios can be easily calculated from these by using the following *reciprocal relations*:

$$\csc t = \frac{1}{\sin t} \qquad \sec t = \frac{1}{\cos t} \qquad \cot t = \frac{1}{\tan t}$$

You should check that these relations follow immediately from the definitions of the trigonometric ratios.

We follow the convention that when we write $\sin t$, *we mean the sine of the angle whose radian measure is* t. For instance, $\sin 1$ means the sine of the angle whose radian measure is 1. When using a calculator to find an approximate value for this number, set your calculator to radian mode; you will find that $\sin 1 \approx 0.841471$. If you want to find the sine of the angle whose measure is $1°$, set your calculator to degree mode; you will find that $\sin 1° \approx 0.0174524$.

EXAMPLE 3 ■ Using a Calculator

Using a calculator, find the following.

(a) $\tan 40°$ **(b)** $\cos 20°$ **(c)** $\cot 14°$ **(d)** $\csc 80°$

SOLUTION Making sure our calculator is set in degree mode and rounding the results to six decimal places, we get the following:

(a) $\tan 40° \approx 0.839100$ **(b)** $\cos 20° \approx 0.939693$

(c) $\cot 14° = \dfrac{1}{\tan 14°} \approx 4.010781$ **(d)** $\csc 80° = \dfrac{1}{\sin 80°} \approx 1.015427$

Now Try Exercise 11 ■

Applications of Trigonometry of Right Triangles

A triangle has six parts: three angles and three sides. To **solve a triangle** means to determine all of its parts from the information known about the triangle, that is, to determine the lengths of the three sides and the measures of the three angles.

Hulton Archive/Moviepix/Getty Images

DISCOVERY PROJECT

Similarity

Similarity of triangles is the basic concept underlying the definition of the trigonometric functions. The ratios of the sides of a triangle are the same as the corresponding ratios in any similar triangle. But the concept of similarity of figures applies to all shapes, not just triangles. In this project we explore how areas and volumes of similar figures are related. These relationships allow us to determine whether an ape the size of King Kong (that is, an ape similar to, but much larger than, a real ape) can actually exist. You can find the project at **www.stewartmath.com**.

EXAMPLE 4 ■ Solving a Right Triangle

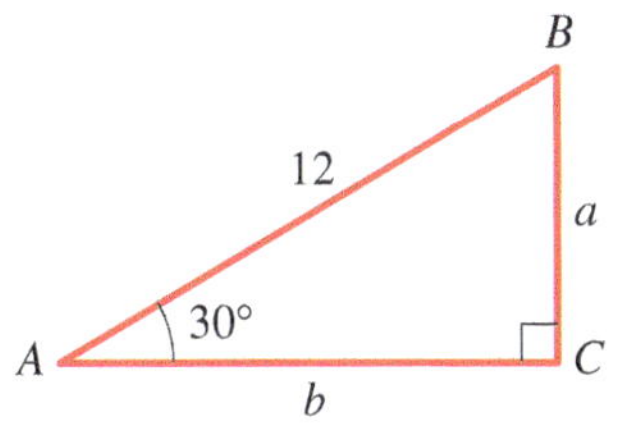

FIGURE 7

Solve triangle ABC, shown in Figure 7.

SOLUTION It's clear that $\angle B = 60°$. From Figure 7 we have

$$\sin 30° = \frac{a}{12} \qquad \text{Definition of sine}$$

$$a = 12 \sin 30° \qquad \text{Multiply by 12}$$

$$= 12\left(\tfrac{1}{2}\right) = 6 \qquad \text{Evaluate}$$

Also from Figure 7 we have

$$\cos 30° = \frac{b}{12} \qquad \text{Definition of cosine}$$

$$b = 12 \cos 30° \qquad \text{Multiply by 12}$$

$$= 12\left(\frac{\sqrt{3}}{2}\right) = 6\sqrt{3} \qquad \text{Evaluate}$$

Now Try Exercise 37 ■

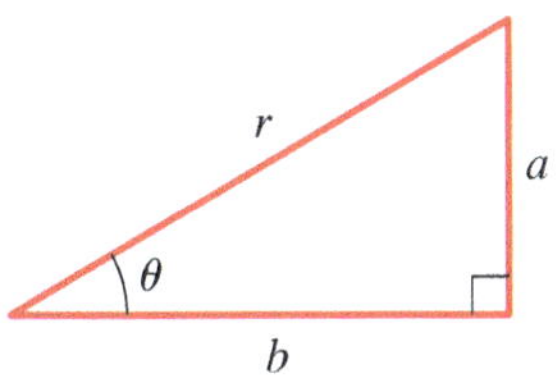

FIGURE 8
$a = r \sin\theta, \quad b = r\cos\theta$

Figure 8 shows that if we know the hypotenuse r and an acute angle θ in a right triangle, then the legs a and b are given by

$$a = r \sin\theta \qquad \text{and} \qquad b = r\cos\theta$$

The ability to solve right triangles by using the trigonometric ratios is fundamental to many problems in navigation, surveying, astronomy, and the measurement of distances. The applications we consider in this section always involve right triangles, but as we will see in the next three sections, trigonometry is also useful in solving triangles that are not right triangles.

To discuss the next examples, we need some terminology. If an observer is looking at an object, then the line from the eye of the observer to the object is called the **line of sight** (Figure 9). If the object being observed is above the horizontal, then the angle between the line of sight and the horizontal is called the **angle of elevation**. If the object is below the horizontal, then the angle between the line of sight and the horizontal is called the **angle of depression**. In many of the examples and exercises in this chapter, angles of elevation and depression will be given for a hypothetical observer at ground level. If the line of sight follows a physical object, such as an inclined plane or a hillside, we use the term **angle of inclination**.

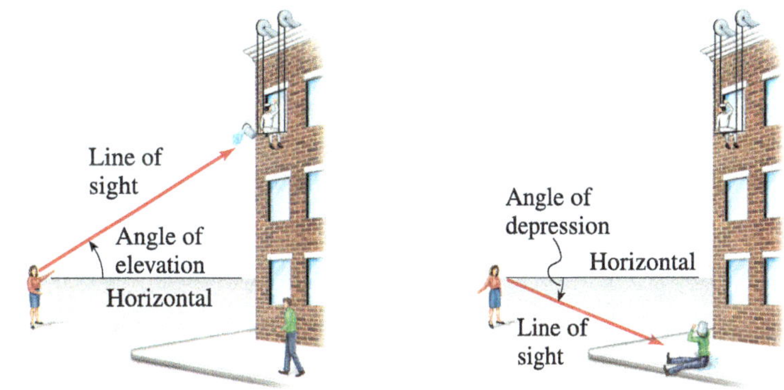

FIGURE 9

The next example gives an important application of trigonometry to the problem of measurement: We measure the height of a tall tree without having to climb it! Although the example is simple, the result is fundamental to understanding how the trigonometric ratios are applied to such problems.

ARISTARCHUS OF SAMOS (310–230 B.C.) was a famous Greek scientist, musician, astronomer, and geometer. He observed that the angle between the sun and moon can be measured directly (see the figure below). In his book *On the Sizes and Distances of the Sun and the Moon* he estimated the distance to the sun by observing that when the moon is exactly half full, the triangle formed by the sun, the moon, and the earth has a right angle at the moon. His method was similar to the one described in Exercise 67 in this section. Aristarchus was the first to advance the theory that the earth and planets move around the sun, an idea that did not gain full acceptance until after the time of Copernicus, 1800 years later. For this reason Aristarchus is often called "the Copernicus of antiquity."

THALES OF MILETUS (circa 625–547 B.C.) is the legendary founder of Greek geometry. It is said that he calculated the height of a Greek column by comparing the length of the shadow of his staff with that of the column. Using properties of similar triangles, he argued that the ratio of the height h of the column to the height h' of his staff was equal to the ratio of the length s of the column's shadow to the length s' of the staff's shadow:

$$\frac{h}{h'} = \frac{s}{s'}$$

Since three of these quantities are known, Thales was able to calculate the height of the column.

According to legend, Thales used a similar method to find the height of the Great Pyramid in Egypt, a feat that impressed Egypt's king. Plutarch wrote that "although he [the king of Egypt] admired you [Thales] for other things, yet he particularly liked the manner by which you measured the height of the pyramid without any trouble or instrument." The principle Thales used, the fact that ratios of corresponding sides of similar triangles are equal, is the foundation of the subject of trigonometry.

EXAMPLE 5 ■ Finding the Height of a Tree

A giant redwood tree casts a shadow 532 ft long. Find the height of the tree if the angle of elevation of the sun is 25.7°.

SOLUTION Let the height of the tree be h. From Figure 10 we see that

$$\frac{h}{532} = \tan 25.7^\circ \qquad \text{Definition of tangent}$$

$$h = 532 \tan 25.7^\circ \qquad \text{Multiply by 532}$$

$$\approx 532(0.48127) \approx 256 \qquad \text{Use a calculator}$$

Therefore the height of the tree is about 256 ft.

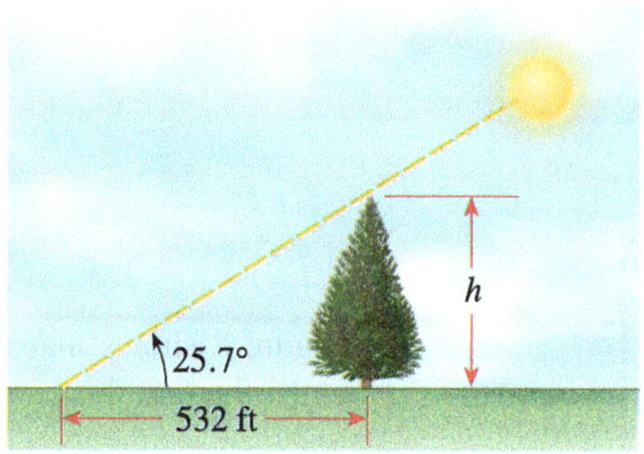

FIGURE 10

Now Try Exercise 53 ■

EXAMPLE 6 ■ A Problem Involving Right Triangles

From a point on the ground 500 ft from the base of a building, an observer finds that the angle of elevation to the top of the building is 24° and that the angle of elevation to the top of a flagpole atop the building is 27°. Find the height of the building and the length of the flagpole.

SOLUTION Figure 11 illustrates the situation. The height of the building is found in the same way that we found the height of the tree in Example 4.

$$\frac{h}{500} = \tan 24^\circ \qquad \text{Definition of tangent}$$

$$h = 500 \tan 24^\circ \qquad \text{Multiply by 500}$$

$$\approx 500(0.4452) \approx 223 \qquad \text{Use a calculator}$$

The height of the building is approximately 223 ft.

To find the length of the flagpole, let's first find the height from the ground to the top of the pole.

$$\frac{k}{500} = \tan 27^\circ \qquad \text{Definition of tangent}$$

$$k = 500 \tan 27^\circ \qquad \text{Multiply by 500}$$

$$\approx 500(0.5095) \qquad \text{Use a calculator}$$

$$\approx 255$$

To find the length of the flagpole, we subtract h from k. So the length of the pole is approximately $255 - 223 = 32$ ft.

Now Try Exercise 61 ■

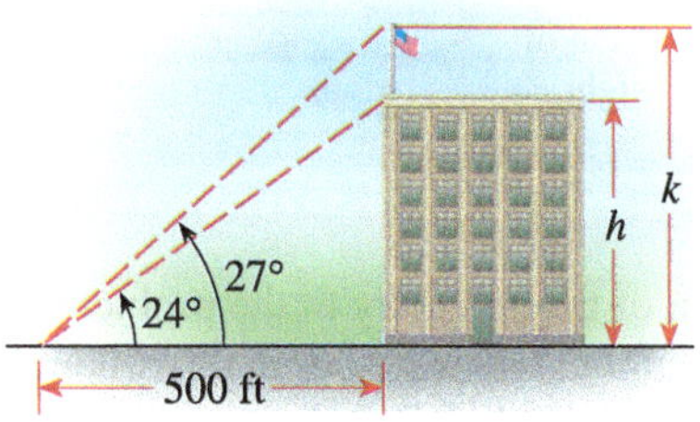

FIGURE 11

6.2 EXERCISES

CONCEPTS

1. A right triangle with an angle θ is shown in the figure.

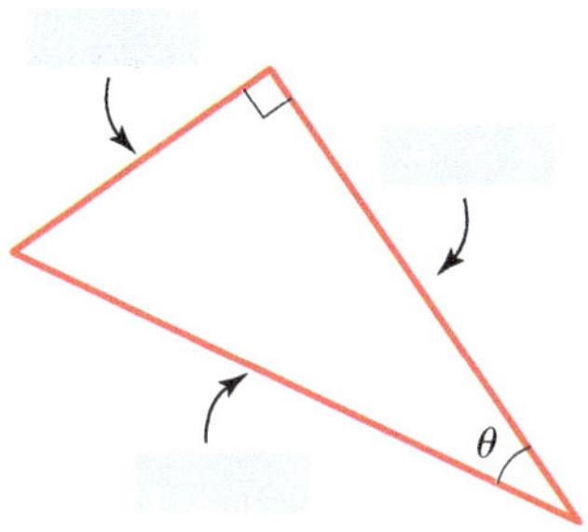

(a) Label the "opposite" and "adjacent" sides of θ and the hypotenuse of the triangle.

(b) The trigonometric functions of the angle θ are defined as follows:

$\sin\theta = \dfrac{\quad}{\quad}$ $\cos\theta = \dfrac{\quad}{\quad}$ $\tan\theta = \dfrac{\quad}{\quad}$

(c) The trigonometric ratios do not depend on the size of the triangle. This is because all right triangles with the same acute angle θ are ________________.

2. The reciprocal identities state that

$\csc\theta = \dfrac{1}{\quad}$ $\sec\theta = \dfrac{1}{\quad}$ $\cot\theta = \dfrac{1}{\quad}$

SKILLS

3–8 ■ Trigonometric Ratios Find the exact values of the six trigonometric ratios of the angle θ in the triangle.

3.

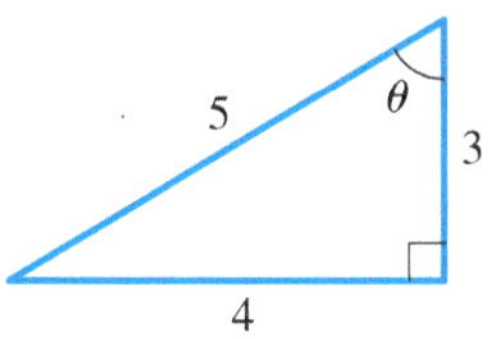

4.

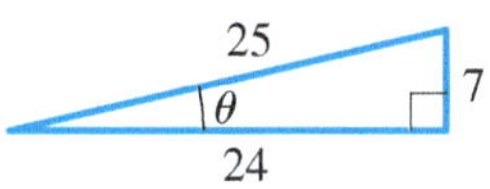

5.

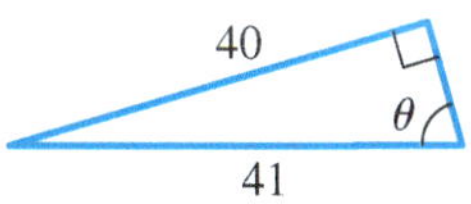

6.

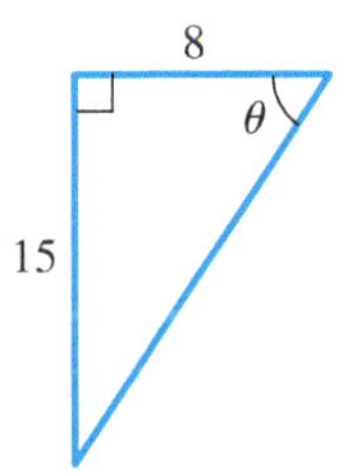

7.

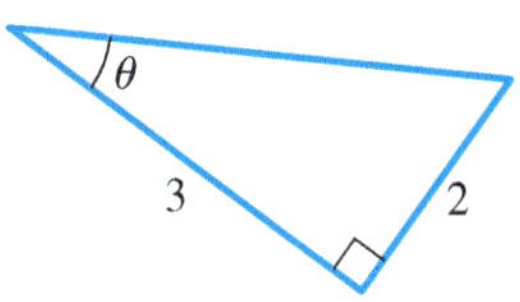

8.

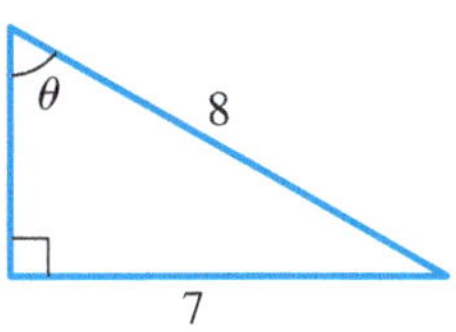

9–10 ■ Trigonometric Ratios Find **(a)** $\sin\alpha$ and $\cos\beta$, **(b)** $\tan\alpha$ and $\cot\beta$, and **(c)** $\sec\alpha$ and $\csc\beta$.

9.

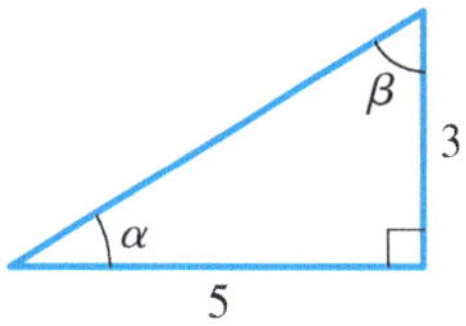

10.

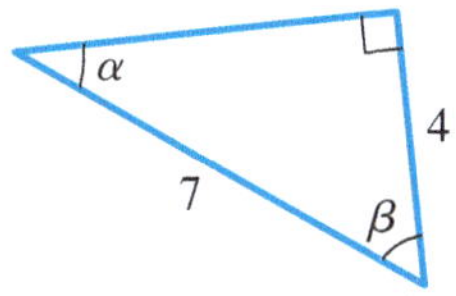

11–14 ■ Using a Calculator Use a calculator to evaluate the expression. Round your answer to five decimal places.

11. (a) $\sin 22^\circ$ **(b)** $\cot 23^\circ$

12. (a) $\cos 37^\circ$ **(b)** $\csc 48^\circ$

13. (a) $\sec 13^\circ$ **(b)** $\tan 51^\circ$

14. (a) $\csc 10^\circ$ **(b)** $\sin 46^\circ$

15–20 ■ Finding an Unknown Side Find the side labeled x. In Exercises 17 and 18 state your answer rounded to five decimal places.

15.

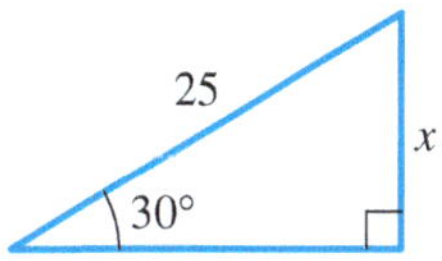

16.

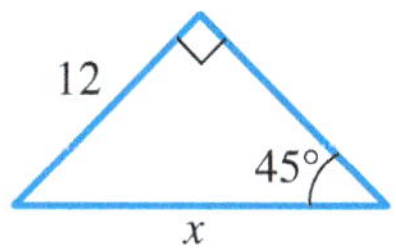

17.

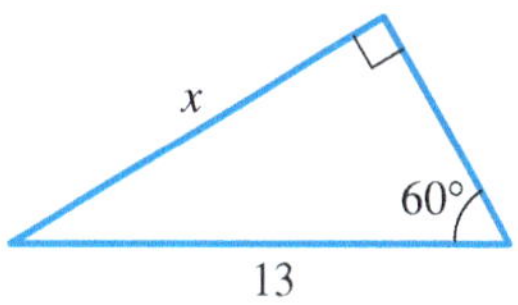

18.

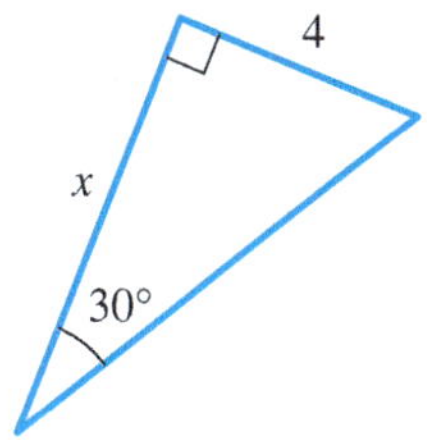

19.

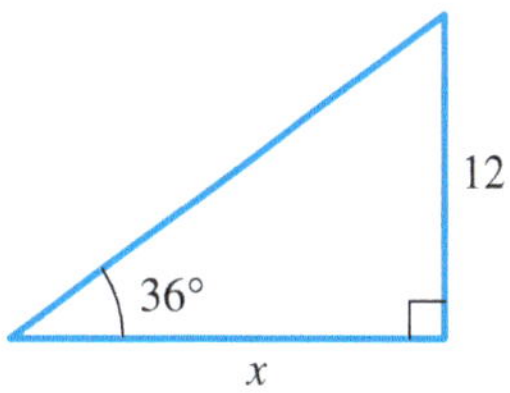

20.

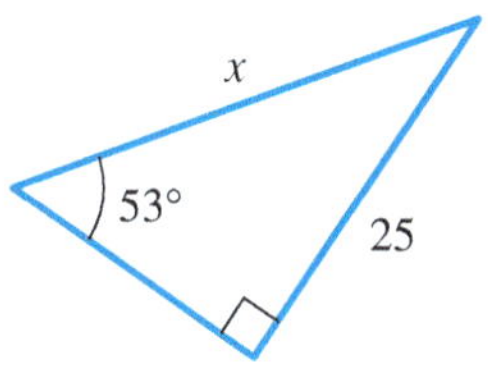

21–22 ■ Trigonometric Ratios Express x and y in terms of trigonometric ratios of θ.

21.

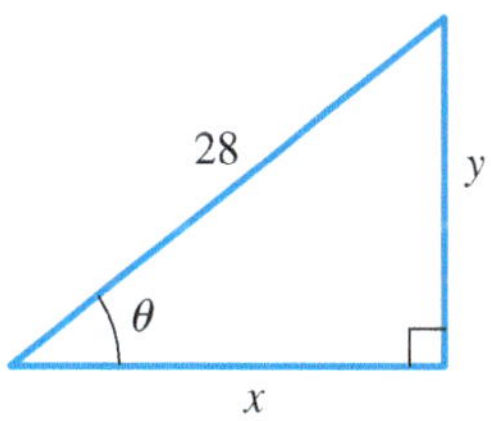

22.

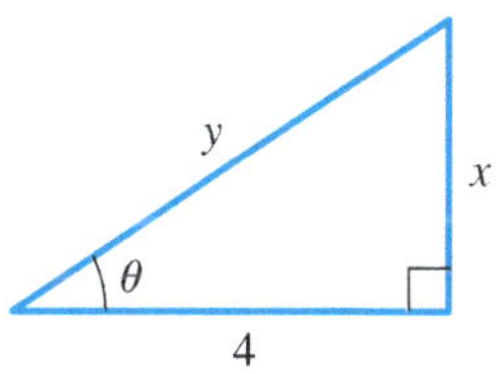

23–28 ■ Trigonometric Ratios Sketch a triangle that has acute angle θ, and find the other five trigonometric ratios of θ.

23. $\tan\theta = \frac{5}{6}$ **24.** $\cos\theta = \frac{12}{13}$ **25.** $\cot\theta = 1$

26. $\tan\theta = \sqrt{3}$ **27.** $\csc\theta = \frac{11}{6}$ **28.** $\cot\theta = \frac{5}{3}$

29–36 ■ Evaluating an Expression Evaluate the expression without using a calculator.

29. $\sin\frac{\pi}{6} + \cos\frac{\pi}{6}$

30. $\sin 30° \csc 30°$

31. $\sin 30° \cos 60° + \sin 60° \cos 30°$

32. $(\sin 60°)^2 + (\cos 60°)^2$

33. $(\cos 30°)^2 - (\sin 30°)^2$

34. $\left(\sin\frac{\pi}{3}\cos\frac{\pi}{4} - \sin\frac{\pi}{4}\cos\frac{\pi}{3}\right)^2$

35. $\left(\cos\frac{\pi}{4} + \sin\frac{\pi}{6}\right)^2$

36. $\left(\sin\frac{\pi}{3}\tan\frac{\pi}{6} + \csc\frac{\pi}{4}\right)^2$

37–44 ■ Solving a Right Triangle Solve the right triangle.

37.

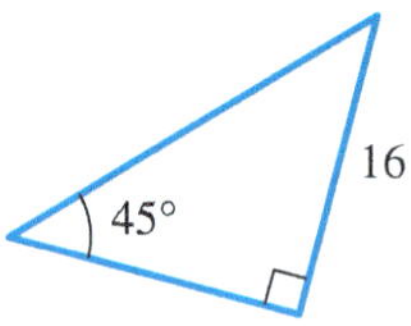

38.

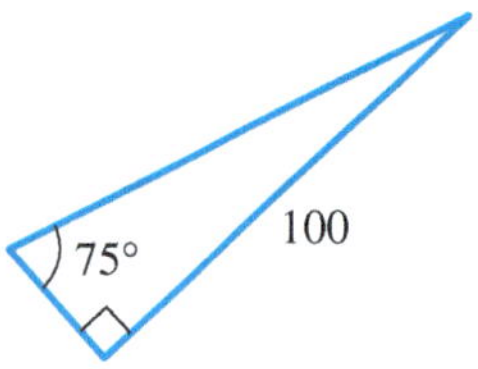

39.

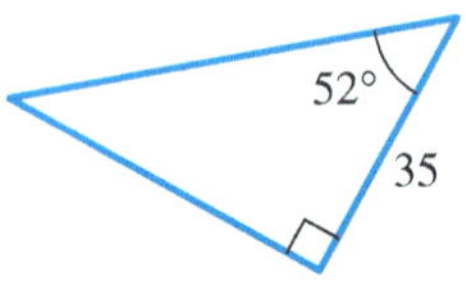

40.

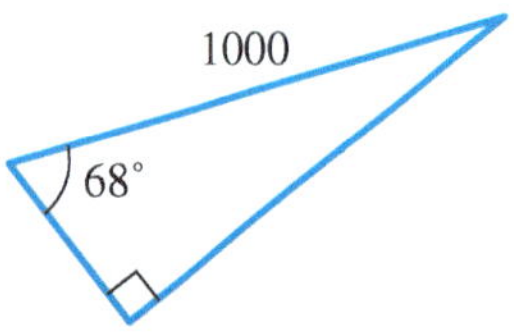

41.

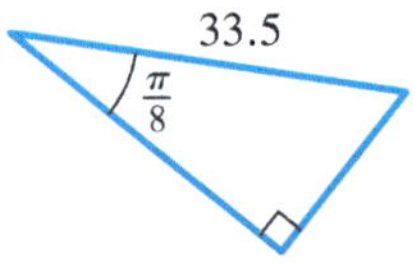

42.

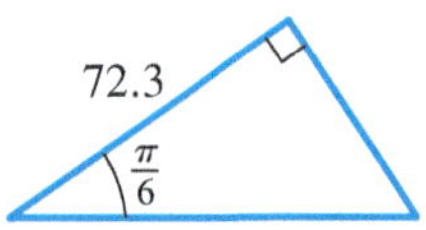

43.

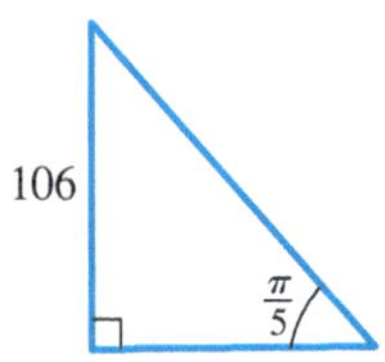

44.

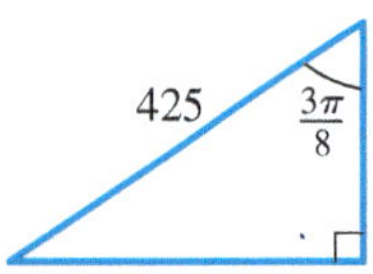

SKILLS Plus

45. Using a Ruler to Estimate Trigonometric Ratios Use a ruler to carefully measure the sides of the triangle, and then use your measurements to estimate the six trigonometric ratios of θ.

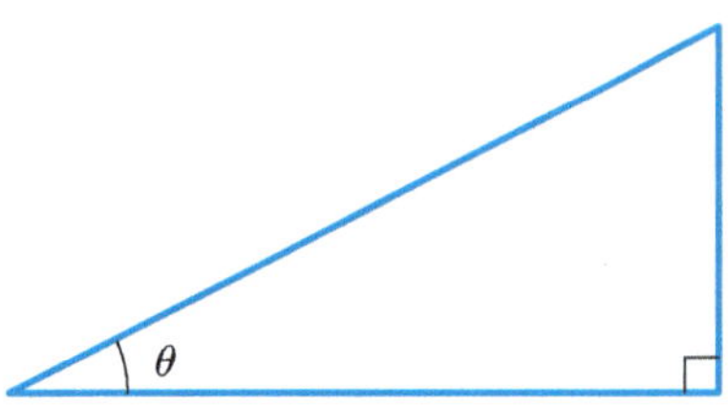

46. Using a Protractor to Estimate Trigonometric Ratios Using a protractor, sketch a right triangle that has the acute angle 40°. Measure the sides carefully, and use your results to estimate the six trigonometric ratios of 40°.

47–50 ■ Finding an Unknown Side Find x rounded to one decimal place.

47.

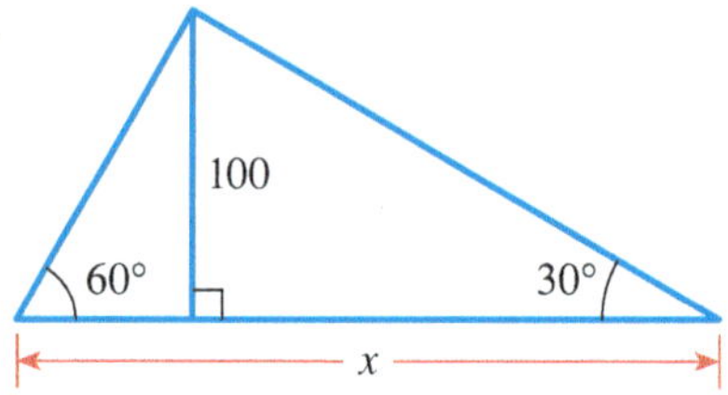

48.

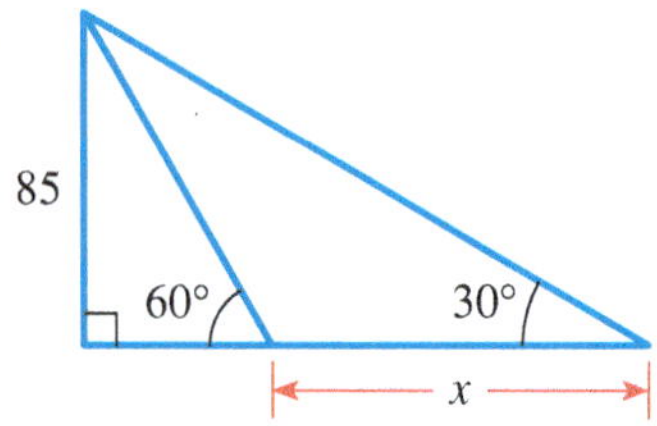

49.

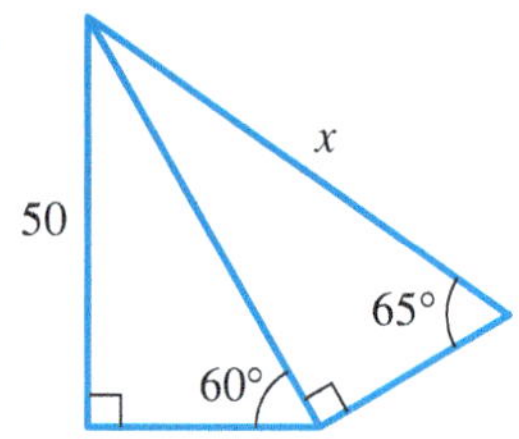

50.

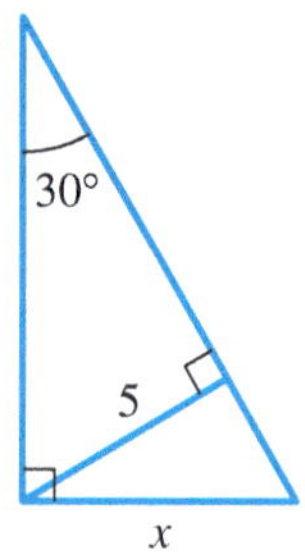

51. Trigonometric Ratios Express the length x in terms of the trigonometric ratios of θ.

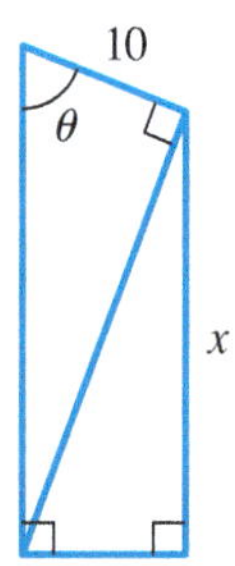

52. Trigonometric Ratios Express the lengths a, b, c, and d in the figure in terms of the trigonometric ratios of θ.

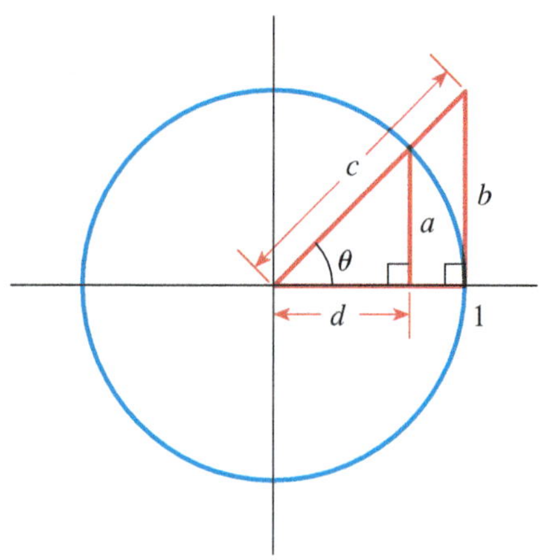

APPLICATIONS

53. Height of a Building The angle of elevation to the top of the Empire State Building in New York is found to be 11° from the ground at a distance of 1 mi from the base of the building. Using this information, find the height of the Empire State Building.

54. Gateway Arch A plane is flying within sight of the Gateway Arch in St. Louis, Missouri, at an elevation of 35,000 ft. The pilot would like to estimate her distance from the Gateway Arch. She finds that the angle of depression to a point on the ground below the arch is 22°.

(a) What is the distance between the plane and the arch?

(b) What is the distance between a point on the ground directly below the plane and the arch?

55. Deviation of a Laser Beam A laser beam is to be directed toward the center of the moon, but the beam strays 0.5° from its intended path.

(a) How far has the beam diverged from its assigned target when it reaches the moon? (The distance from the earth to the moon is 240,000 mi.)

(b) The radius of the moon is about 1000 mi. Will the beam strike the moon?

56. Distance at Sea From the top of a 200-ft lighthouse, the angle of depression to a ship in the ocean is 23°. How far is the ship from the base of the lighthouse?

57. Leaning Ladder A 20-ft ladder leans against a building so that the angle between the ground and the ladder is 72°. How high does the ladder reach on the building?

58. Height of a Tower A 600-ft guy wire is attached to the top of a communications tower. If the wire makes an angle of 65° with the ground, how tall is the communications tower?

59. Elevation of a Kite A man is lying on the beach, flying a kite. He holds the end of the kite string at ground level and estimates the angle of elevation of the kite to be 50°. If the string is 450 ft long, how high is the kite above the ground?

60. Determining a Distance A woman standing on a hill sees a flagpole that she knows is 60 ft tall. The angle of depression to the bottom of the pole is 14°, and the angle of elevation to the top of the pole is 18°. Find her distance x from the pole.

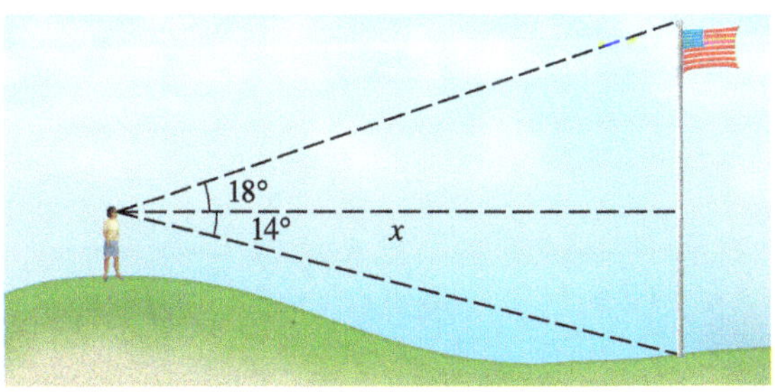

61. Height of a Tower A water tower is located 325 ft from a building (see the figure). From a window in the building, an observer notes that the angle of elevation to the top of the tower is 39° and that the angle of depression to the bottom of the tower is 25°. How tall is the tower? How high is the window?

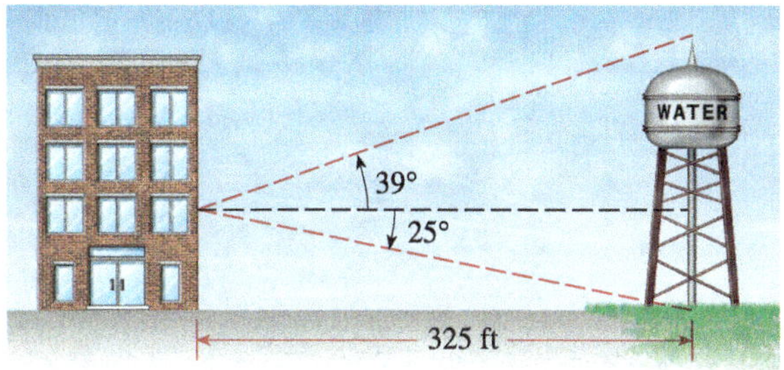

62. Determining a Distance An airplane is flying at an elevation of 5150 ft, directly above a straight highway. Two motorists are driving cars on the highway on opposite sides of the plane. The angle of depression to one car is 35°, and that to the other is 52°. How far apart are the cars?

63. Determining a Distance If both cars in Exercise 62 are on one side of the plane and if the angle of depression to one car is 38° and that to the other car is 52°, how far apart are the cars?

64. Height of a Balloon A hot-air balloon is floating above a straight road. To estimate their height above the ground, the balloonists simultaneously measure the angle of depression to two consecutive mileposts on the road on the same side of the balloon. The angles of depression are found to be 20° and 22°. How high is the balloon?

65. Height of a Mountain To estimate the height of a mountain above a level plain, the angle of elevation to the top of the mountain is measured to be 32°. One thousand feet closer to the mountain along the plain, it is found that the angle of elevation is 35°. Estimate the height of the mountain.

66. Height of Cloud Cover To measure the height of the cloud cover at an airport, a worker shines a spotlight upward at an angle 75° from the horizontal. An observer 600 m away measures the angle of elevation to the spot of light to be 45°. Find the height h of the cloud cover.

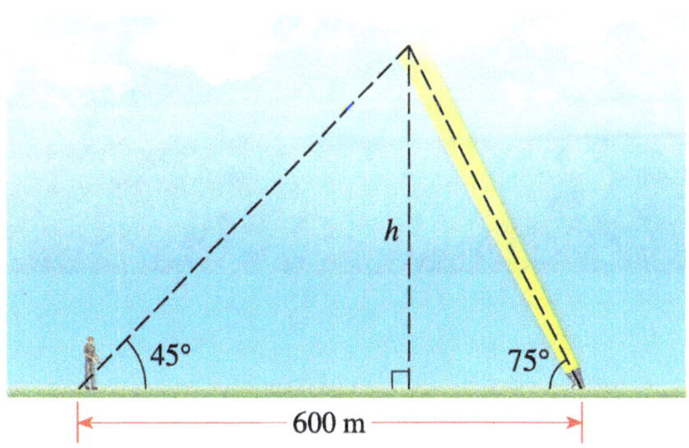

67. Distance to the Sun When the moon is exactly half full, the earth, moon, and sun form a right angle (see the figure). At that time the angle formed by the sun, earth, and moon is measured to be 89.85°. If the distance from the earth to the moon is 240,000 mi, estimate the distance from the earth to the sun.

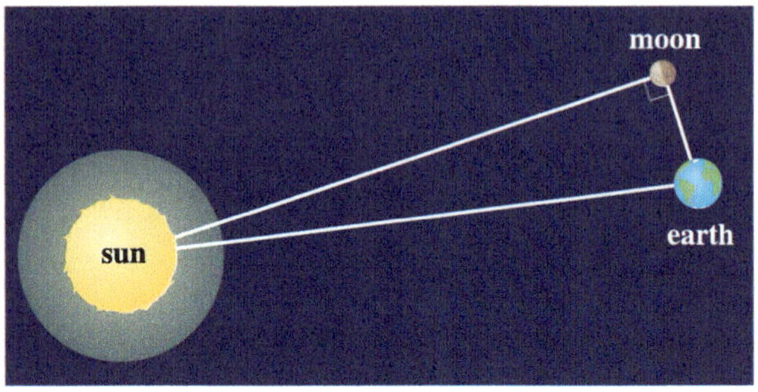

68. Distance to the Moon To find the distance to the sun as in Exercise 67, we needed to know the distance to the moon. Here is a way to estimate that distance: When the moon is seen at its zenith at a point A on the earth, it is observed to be at the horizon from point B (see the following figure). Points A and B are 6155 mi apart, and the radius of the earth is 3960 mi.

(a) Find the angle θ in degrees.

(b) Estimate the distance from point A to the moon.

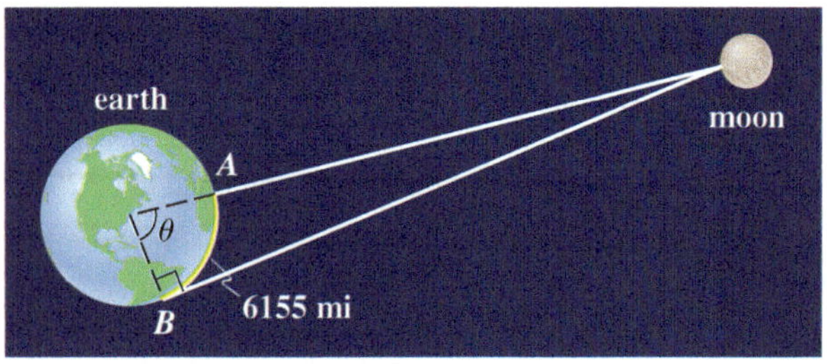

69. Radius of the Earth In Exercise 80 of Section 6.1 a method was given for finding the radius of the earth. Here is a more modern method: From a satellite 600 mi above the earth it is observed that the angle formed by the vertical and the line of sight to the horizon is 60.276°. Use this information to find the radius of the earth.

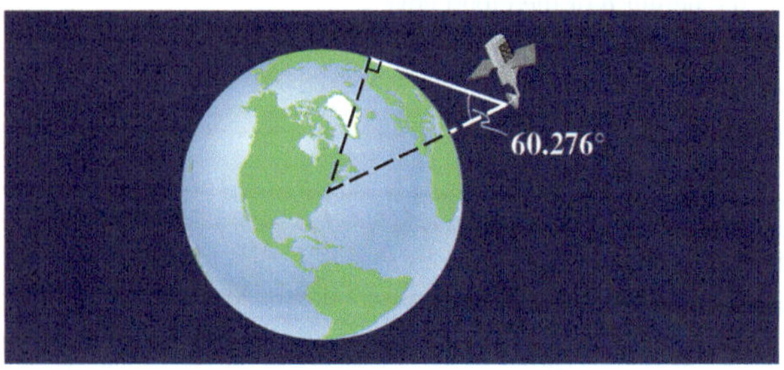

70. Parallax To find the distance to nearby stars, the method of parallax is used. The idea is to find a triangle with the star at one vertex and with a base as large as possible. To do this, the star is observed at two different times exactly 6 months apart, and its apparent change in position is recorded. From these two observations $\angle E_1SE_2$ can be calculated. (The times are chosen so that $\angle E_1SE_2$ is as large as possible, which guarantees that $\angle E_1OS$ is 90°.) The angle E_1SO is called the *parallax* of the star. Alpha Centauri, the star nearest the earth, has a parallax of 0.000211°. Estimate the distance to this star. (Take the distance from the earth to the sun to be 9.3×10^7 mi.)

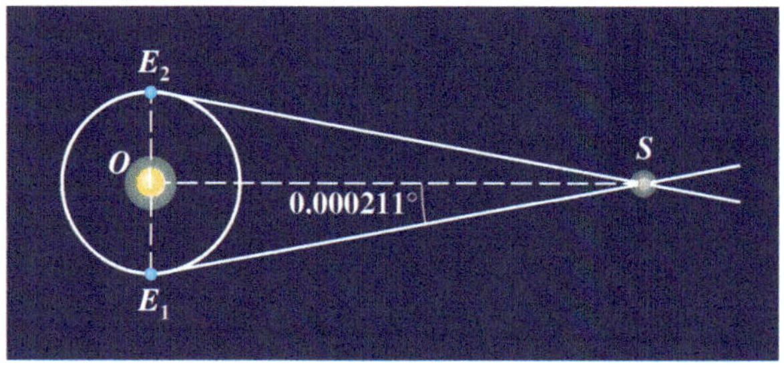

71. Distance from Venus to the Sun The **elongation** α of a planet is the angle formed by the planet, earth, and sun (see the figure). When Venus achieves its maximum elongation of 46.3°, the earth, Venus, and the sun form a triangle with a right angle at Venus. Find the distance between Venus and the sun in astronomical units (AU). (By definition the distance between the earth and the sun is 1 AU.)

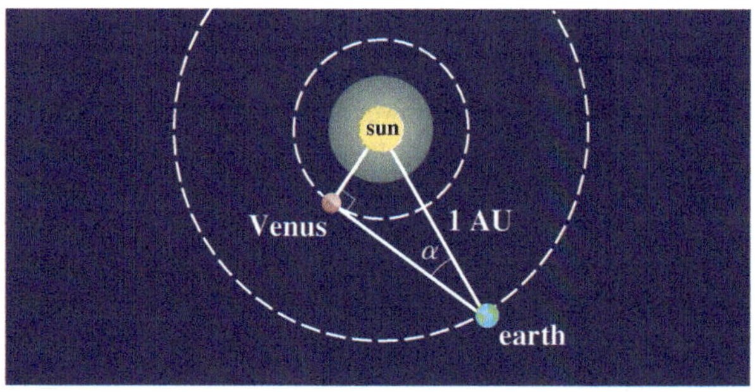

DISCUSS ■ DISCOVER ■ PROVE ■ WRITE

72. DISCUSS: Similar Triangles If two triangles are similar, what properties do they share? Explain how these properties make it possible to define the trigonometric ratios without regard to the size of the triangle.

6.3 TRIGONOMETRIC FUNCTIONS OF ANGLES

■ Trigonometric Functions of Angles ■ Evaluating Trigonometric Functions at Any Angle ■ Trigonometric Identities ■ Areas of Triangles

In Section 6.2 we defined the trigonometric ratios for acute angles. Here we extend the trigonometric ratios to all angles by defining the trigonometric functions of angles. With these functions we can solve practical problems that involve angles that are not necessarily acute.

■ Trigonometric Functions of Angles

Let POQ be a right triangle with acute angle θ as shown in Figure 1(a). Place θ in standard position as shown in Figure 1(b).

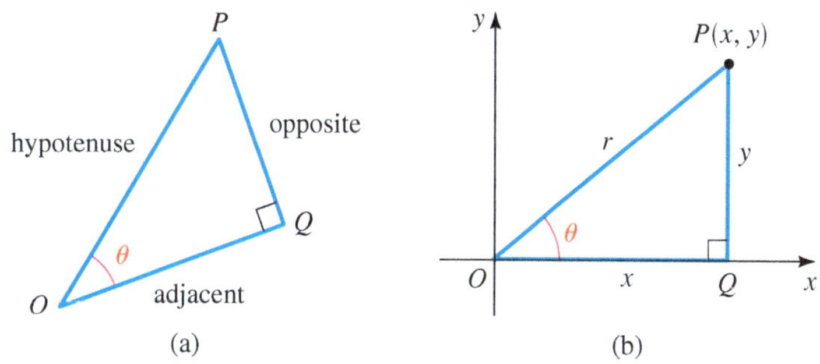

FIGURE 1

Then $P = P(x, y)$ is a point on the terminal side of θ. In triangle POQ the opposite side has length y and the adjacent side has length x. Using the Pythagorean Theorem, we see that the hypotenuse has length $r = \sqrt{x^2 + y^2}$. So

$$\sin\theta = \frac{y}{r} \qquad \cos\theta = \frac{x}{r} \qquad \tan\theta = \frac{y}{x}$$

The other trigonometric ratios can be found in the same way.

These observations allow us to extend the trigonometric ratios to any angle. We define the trigonometric functions of angles as follows (see Figure 2).

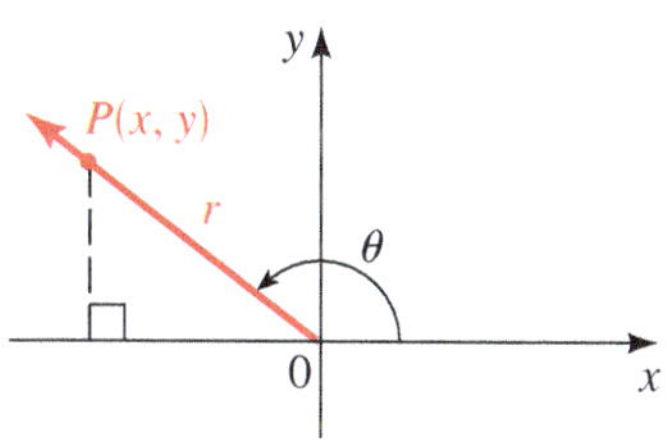

FIGURE 2

DEFINITION OF THE TRIGONOMETRIC FUNCTIONS

Let θ be an angle in standard position, and let $P(x, y)$ be a point on the terminal side. If $r = \sqrt{x^2 + y^2}$ is the distance from the origin to the point $P(x, y)$, then

$$\sin\theta = \frac{y}{r} \qquad \cos\theta = \frac{x}{r} \qquad \tan\theta = \frac{y}{x} \quad (x \neq 0)$$

$$\csc\theta = \frac{r}{y} \quad (y \neq 0) \qquad \sec\theta = \frac{r}{x} \quad (x \neq 0) \qquad \cot\theta = \frac{x}{y} \quad (y \neq 0)$$

Since division by 0 is an undefined operation, certain trigonometric functions are not defined for certain angles. For example, $\tan 90^\circ = y/x$ is undefined because $x = 0$. The angles for which the trigonometric functions may be undefined are the angles for which

Relationship to the Trigonometric Functions of Real Numbers

You may have already studied the trigonometric functions defined by using the unit circle (Chapter 5). To see how they relate to the trigonometric functions of an *angle*, let's start with the unit circle in the coordinate plane.

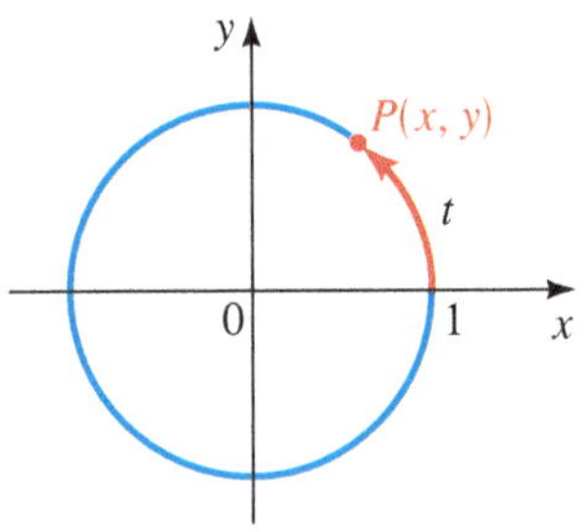

$P(x, y)$ is the terminal point determined by t.

Let $P(x, y)$ be the terminal point determined by an arc of length t on the unit circle. Then t subtends an angle θ at the center of the circle. If we drop a perpendicular from P onto the point Q on the x-axis, then triangle $\triangle OPQ$ is a right triangle with legs of length x and y, as shown in the figure.

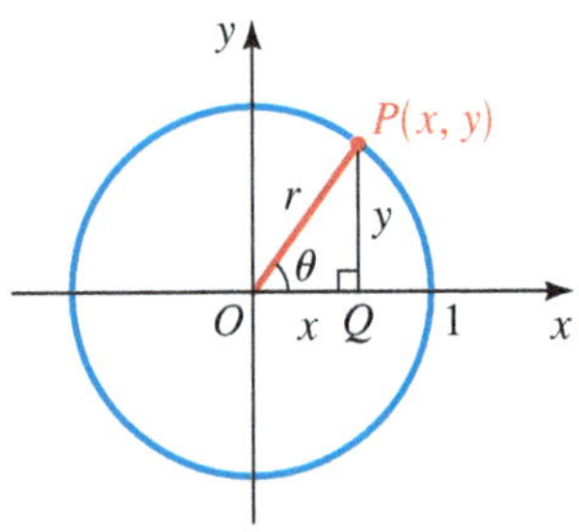

Triangle OPQ is a right triangle.

Now, by the definition of the trigonometric functions of the *real number* t we have

$$\sin t = y$$

$$\cos t = x$$

By the definition of the trigonometric functions of the *angle* θ we have

$$\sin \theta = \frac{\text{opp}}{\text{hyp}} = \frac{y}{1} = y$$

$$\cos \theta = \frac{\text{adj}}{\text{hyp}} = \frac{x}{1} = x$$

If θ is measured in radians, then $\theta = t$. (See the figure below.) Comparing the two ways of defining the trigonometric functions, we see that they are identical. In other words, as functions they assign identical values to a given real number. (The real number is the radian measure of θ in one case or the length t of an arc in the other.)

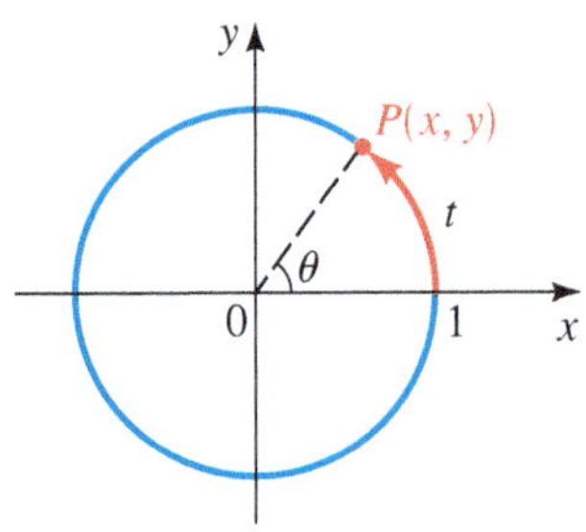

The radian measure of angle θ is t.

Why then do we study trigonometry in two different ways? Because different applications require that we view the trigonometric functions differently. (See *Focus on Modeling*, pages 466, 533, and 581, and Sections 6.2, 6.5, and 6.6.)

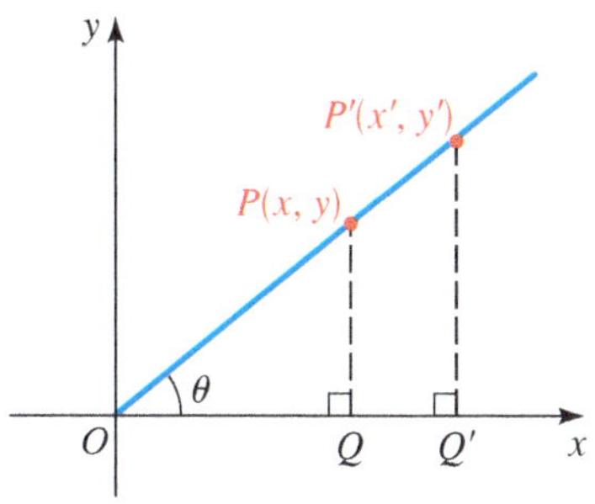

FIGURE 3

either the x- or y-coordinate of a point on the terminal side of the angle is 0. These are **quadrantal angles**—angles that are coterminal with the coordinate axes.

It is a crucial fact that the values of the trigonometric functions do *not* depend on the choice of the point $P(x, y)$. This is because if $P'(x', y')$ is any other point on the terminal side, as in Figure 3, then triangles POQ and $P'OQ'$ are similar.

Evaluating Trigonometric Functions at Any Angle

From the definition we see that the values of the trigonometric functions are all positive if the angle θ has its terminal side in Quadrant I. This is because x and y are positive in this quadrant. [Of course, r is always positive, since it is simply the distance from the origin to the point $P(x, y)$.] If the terminal side of θ is in Quadrant II, however, then x is negative and y is positive. Thus in Quadrant II the functions $\sin\theta$ and $\csc\theta$ are positive, and all the other trigonometric functions have negative values. You can check the other entries in the following table.

The following mnemonic device can be used to remember which trigonometric functions are positive in each quadrant: **A**ll of them, **S**ine, **T**angent, or **C**osine.

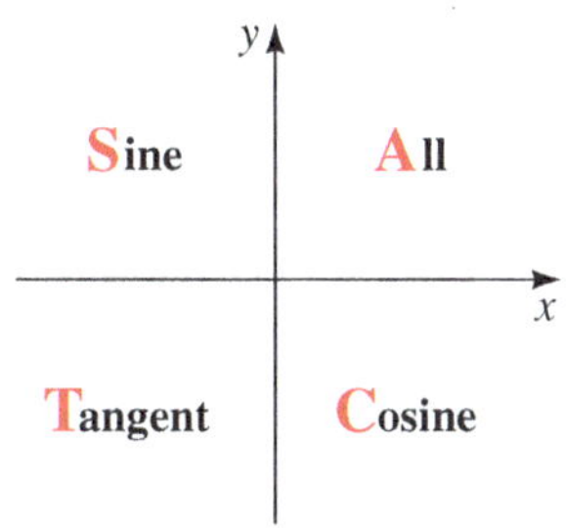

You can remember this as "All Students Take Calculus."

SIGNS OF THE TRIGONOMETRIC FUNCTIONS

Quadrant	Positive Functions	Negative Functions
I	all	none
II	sin, csc	cos, sec, tan, cot
III	tan, cot	sin, csc, cos, sec
IV	cos, sec	sin, csc, tan, cot

We now turn our attention to finding the values of the trigonometric functions for angles that are not acute.

EXAMPLE 1 ■ Finding Trigonometric Functions of Angles

Find **(a)** $\cos 135°$ and **(b)** $\tan 390°$.

SOLUTION

(a) From Figure 4 we see that $\cos 135° = -x/r$. But $\cos 45° = x/r$, and since $\cos 45° = \sqrt{2}/2$, we have

$$\cos 135° = -\frac{\sqrt{2}}{2}$$

(b) The angles 390° and 30° are coterminal. From Figure 5 it's clear that $\tan 390° = \tan 30°$, and since $\tan 30° = \sqrt{3}/3$, we have

$$\tan 390° = \frac{\sqrt{3}}{3}$$

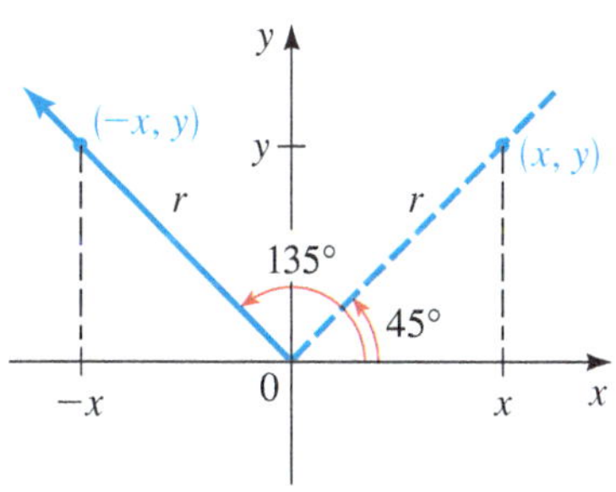

FIGURE 4

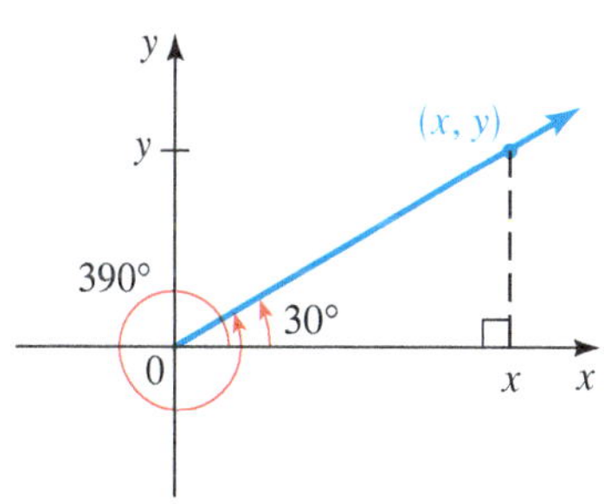

FIGURE 5

Now Try Exercises 13 and 15 ■

From Example 1 we see that the trigonometric functions for angles that aren't acute have the same value, except possibly for sign, as the corresponding trigonometric functions of an acute angle. That acute angle will be called the *reference angle*.

REFERENCE ANGLE

Let θ be an angle in standard position. The **reference angle** $\bar{\theta}$ associated with θ is the acute angle formed by the terminal side of θ and the x-axis.

Figure 6 shows that to find a reference angle $\bar{\theta}$, it's useful to know the quadrant in which the terminal side of the angle θ lies.

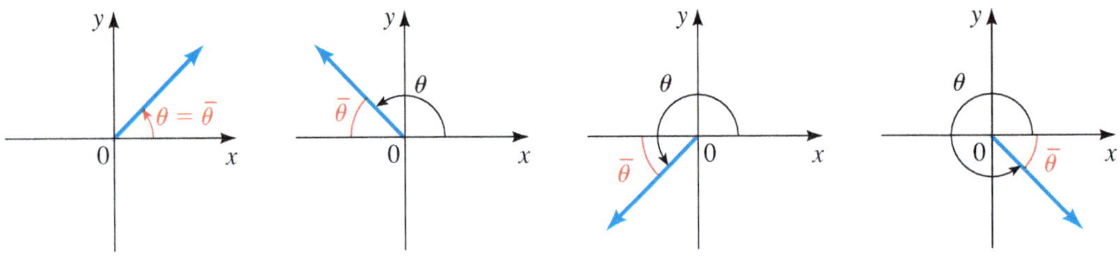

FIGURE 6 The reference angle $\bar{\theta}$ for an angle θ

EXAMPLE 2 ■ Finding Reference Angles

Find the reference angle for **(a)** $\theta = \dfrac{5\pi}{3}$ and **(b)** $\theta = 870°$.

SOLUTION

(a) The reference angle is the acute angle formed by the terminal side of the angle $5\pi/3$ and the x-axis (see Figure 7). Since the terminal side of this angle is in Quadrant IV, the reference angle is

$$\bar{\theta} = 2\pi - \frac{5\pi}{3} = \frac{\pi}{3}$$

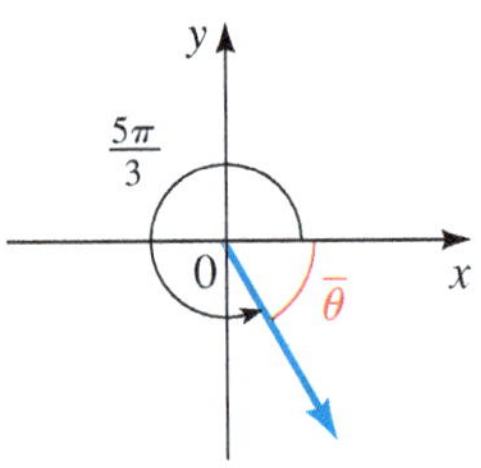

FIGURE 7

(b) The angles 870° and 150° are coterminal [because $870 - 2(360) = 150$]. Thus the terminal side of this angle is in Quadrant II (see Figure 8). So the reference angle is

$$\bar{\theta} = 180° - 150° = 30°$$

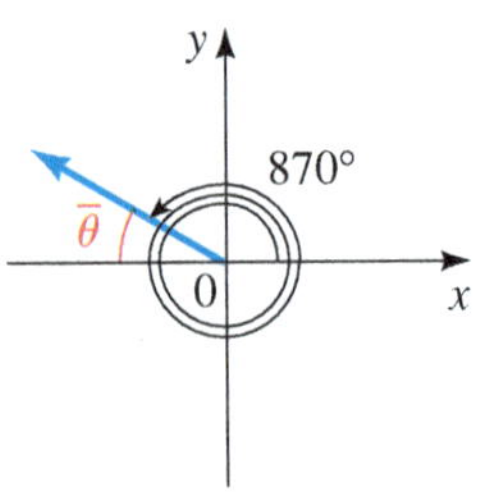

FIGURE 8

Now Try Exercises 5 and 9 ■

EVALUATING TRIGONOMETRIC FUNCTIONS FOR ANY ANGLE

To find the values of the trigonometric functions for any angle θ, we carry out the following steps.

1. Find the reference angle $\bar{\theta}$ associated with the angle θ.
2. Determine the sign of the trigonometric function of θ by noting the quadrant in which θ lies.
3. The value of the trigonometric function of θ is the same, except possibly for sign, as the value of the trigonometric function of $\bar{\theta}$.

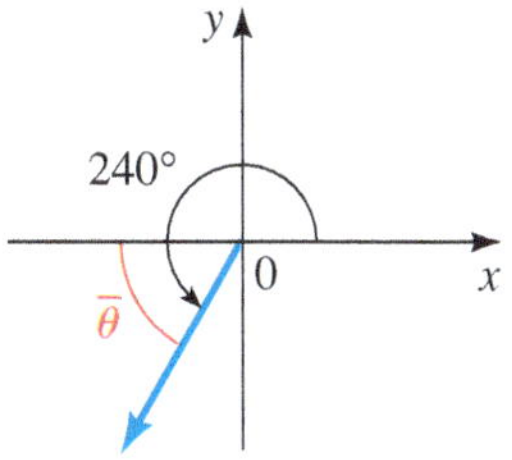

FIGURE 9

S | A
T | C

sin 240° is *negative*.

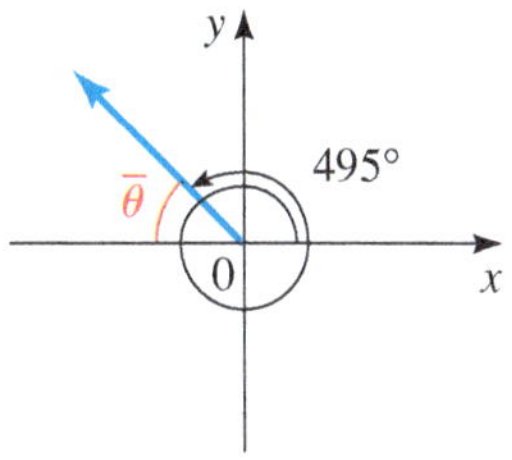

FIGURE 10

S | A
T | C

tan 495° is *negative*, so cot 495° is *negative*.

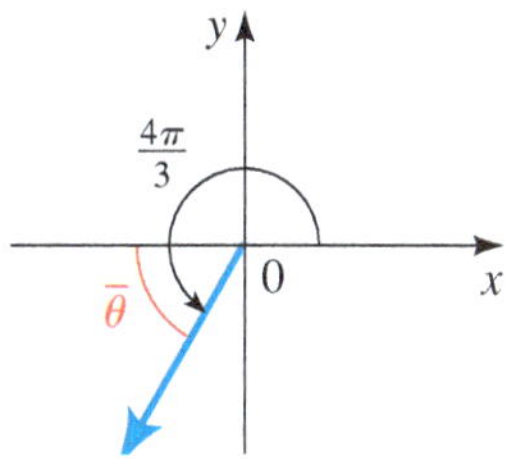

FIGURE 11

S | A
T | C

$\sin \frac{16\pi}{3}$ is *negative*.

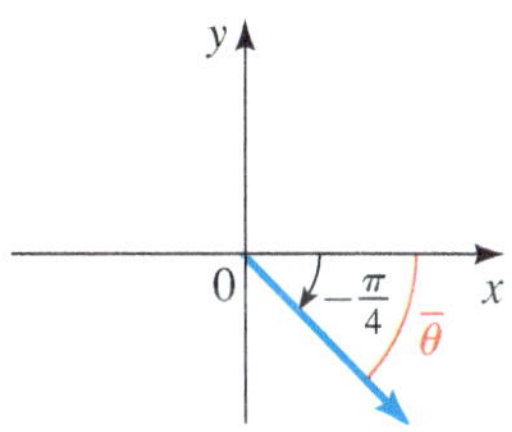

FIGURE 12

S | A
T | C

$\cos(-\frac{\pi}{4})$ is *positive*, so $\sec(-\frac{\pi}{4})$ is *positive*.

EXAMPLE 3 ■ Using the Reference Angle to Evaluate Trigonometric Functions

Find **(a)** sin 240° and **(b)** cot 495°.

SOLUTION

(a) This angle has its terminal side in Quadrant III, as shown in Figure 9. The reference angle is therefore $240° - 180° = 60°$, and the value of sin 240° is negative. Thus

$$\sin 240° = -\sin 60° = -\frac{\sqrt{3}}{2}$$

Sign Reference angle

(b) The angle 495° is coterminal with the angle 135°, and the terminal side of this angle is in Quadrant II, as shown in Figure 10. So the reference angle is $180° - 135° = 45°$, and the value of cot 495° is negative. We have

$$\cot 495° = \cot 135° = -\cot 45° = -1$$

Coterminal angles Sign Reference angle

Now Try Exercises 19 and 21 ■

EXAMPLE 4 ■ Using the Reference Angle to Evaluate Trigonometric Functions

Find **(a)** $\sin \dfrac{16\pi}{3}$ and **(b)** $\sec\left(-\dfrac{\pi}{4}\right)$.

SOLUTION

(a) The angle $16\pi/3$ is coterminal with $4\pi/3$, and these angles are in Quadrant III (see Figure 11). Thus the reference angle is $(4\pi/3) - \pi = \pi/3$. Since the value of sine is negative in Quadrant III, we have

$$\sin \frac{16\pi}{3} = \sin \frac{4\pi}{3} = -\sin \frac{\pi}{3} = -\frac{\sqrt{3}}{2}$$

Coterminal angles Sign Reference angle

(b) The angle $-\pi/4$ is in Quadrant IV, and its reference angle is $\pi/4$ (see Figure 12). Since secant is positive in this quadrant, we get

$$\sec\left(-\frac{\pi}{4}\right) = +\sec \frac{\pi}{4} = \sqrt{2}$$

Sign Reference angle

Now Try Exercises 25 and 27 ■

■ Trigonometric Identities

The trigonometric functions of angles are related to each other through several important equations called **trigonometric identities**. We've already encountered the reciprocal identities. These identities continue to hold for any angle θ, provided that both sides of the equation are defined. The Pythagorean identities are a consequence of the Pythagorean Theorem.*

FUNDAMENTAL IDENTITIES

Reciprocal Identities

$$\csc\theta = \frac{1}{\sin\theta} \qquad \sec\theta = \frac{1}{\cos\theta} \qquad \cot\theta = \frac{1}{\tan\theta}$$

$$\tan\theta = \frac{\sin\theta}{\cos\theta} \qquad \cot\theta = \frac{\cos\theta}{\sin\theta}$$

Pythagorean Identities

$$\sin^2\theta + \cos^2\theta = 1 \qquad \tan^2\theta + 1 = \sec^2\theta \qquad 1 + \cot^2\theta = \csc^2\theta$$

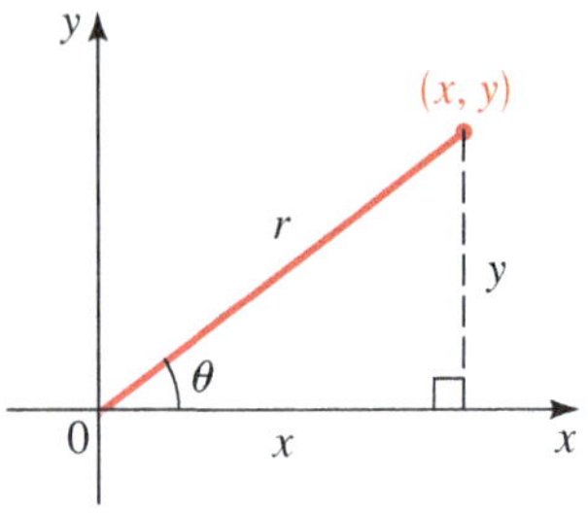

FIGURE 13

Proof Let's prove the first Pythagorean identity. Using $x^2 + y^2 = r^2$ (the Pythagorean Theorem) in Figure 13, we have

$$\sin^2\theta + \cos^2\theta = \left(\frac{y}{r}\right)^2 + \left(\frac{x}{r}\right)^2 = \frac{x^2 + y^2}{r^2} = \frac{r^2}{r^2} = 1$$

Thus $\sin^2\theta + \cos^2\theta = 1$. (Although the figure indicates an acute angle, you should check that the proof holds for all angles θ.) ■

See Exercise 76 for the proofs of the other two Pythagorean identities.

EXAMPLE 5 ■ Expressing One Trigonometric Function in Terms of Another

(a) Express $\sin\theta$ in terms of $\cos\theta$.

(b) Express $\tan\theta$ in terms of $\sin\theta$, where θ is in Quadrant II.

SOLUTION

(a) From the first Pythagorean identity we get

$$\sin\theta = \pm\sqrt{1 - \cos^2\theta}$$

where the sign depends on the quadrant. If θ is in Quadrant I or II, then $\sin\theta$ is positive, so

$$\sin\theta = \sqrt{1 - \cos^2\theta}$$

whereas if θ is in Quadrant III or IV, $\sin\theta$ is negative, so

$$\sin\theta = -\sqrt{1 - \cos^2\theta}$$

(b) Since $\tan\theta = \sin\theta/\cos\theta$, we need to write $\cos\theta$ in terms of $\sin\theta$. By part (a)

$$\cos\theta = \pm\sqrt{1 - \sin^2\theta}$$

and since $\cos\theta$ is negative in Quadrant II, the negative sign applies here. Thus

$$\tan\theta = \frac{\sin\theta}{\cos\theta} = \frac{\sin\theta}{-\sqrt{1 - \sin^2\theta}}$$

Now Try Exercise 41 ■

*We follow the usual convention of writing $\sin^2\theta$ for $(\sin\theta)^2$. In general, we write $\sin^n\theta$ for $(\sin\theta)^n$ for all integers n except $n = -1$. The superscript $n = -1$ will be assigned another meaning in Section 6.4. Of course, the same convention applies to the other five trigonometric functions.

EXAMPLE 6 ■ Evaluating a Trigonometric Function

If $\tan\theta = \frac{2}{3}$ and θ is in Quadrant III, find $\cos\theta$.

SOLUTION 1 We need to write $\cos\theta$ in terms of $\tan\theta$. From the identity $\tan^2\theta + 1 = \sec^2\theta$ we get $\sec\theta = \pm\sqrt{\tan^2\theta + 1}$. In Quadrant III, $\sec\theta$ is negative, so

$$\sec\theta = -\sqrt{\tan^2\theta + 1}$$

If you wish to rationalize the denominator, you can express $\cos\theta$ as

$$-\frac{3}{\sqrt{13}}\cdot\frac{\sqrt{13}}{\sqrt{13}} = -\frac{3\sqrt{13}}{13}$$

Thus

$$\cos\theta = \frac{1}{\sec\theta} = \frac{1}{-\sqrt{\tan^2\theta + 1}}$$

$$= \frac{1}{-\sqrt{\left(\frac{2}{3}\right)^2 + 1}} = \frac{1}{-\sqrt{\frac{13}{9}}} = -\frac{3}{\sqrt{13}}$$

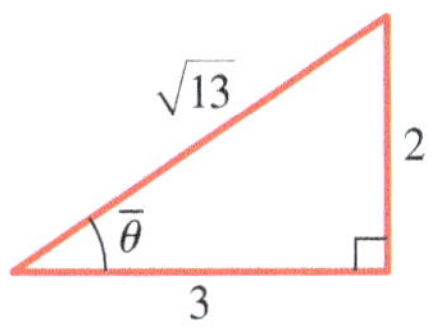

FIGURE 14

SOLUTION 2 This problem can be solved more easily by using the method of Example 2 of Section 6.2. Recall that, except for sign, the values of the trigonometric functions of any angle are the same as those of an acute angle (the reference angle). So, ignoring the sign for the moment, let's sketch a right triangle with an acute angle $\overline{\theta}$ satisfying $\tan\overline{\theta} = \frac{2}{3}$ (see Figure 14). By the Pythagorean Theorem the hypotenuse of this triangle has length $\sqrt{13}$. From the triangle in Figure 14 we immediately see that $\cos\overline{\theta} = 3/\sqrt{13}$. Since θ is in Quadrant III, $\cos\theta$ is negative, so

$$\cos\theta = -\frac{3}{\sqrt{13}}$$

Now Try Exercise 47 ■

EXAMPLE 7 ■ Evaluating Trigonometric Functions

If $\sec\theta = 2$ and θ is in Quadrant IV, find the other five trigonometric functions of θ.

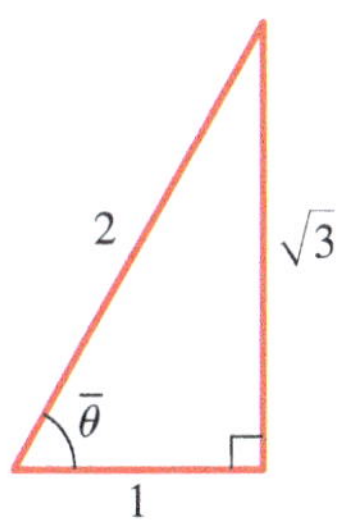

FIGURE 15

SOLUTION We sketch a triangle as in Figure 15 so that $\sec\overline{\theta} = 2$. Taking into account the fact that θ is in Quadrant IV, we get

$$\sin\theta = -\frac{\sqrt{3}}{2} \qquad \cos\theta = \frac{1}{2} \qquad \tan\theta = -\sqrt{3}$$

$$\csc\theta = -\frac{2}{\sqrt{3}} \qquad \sec\theta = 2 \qquad \cot\theta = -\frac{1}{\sqrt{3}}$$

Now Try Exercise 49 ■

Areas of Triangles

We conclude this section with an application of the trigonometric functions that involves angles that are not necessarily acute. More extensive applications appear in Sections 6.5 and 6.6.

The area of a triangle is $\mathcal{A} = \frac{1}{2} \times \text{base} \times \text{height}$. If we know two sides and the included angle of a triangle, then we can find the height using the trigonometric functions, and from this we can find the area.

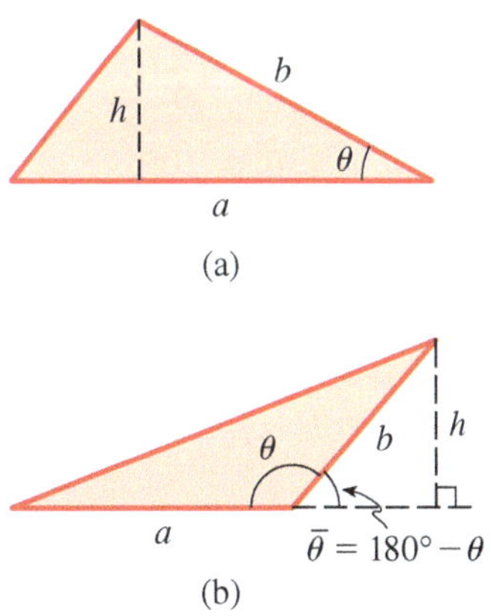

FIGURE 16

If θ is an acute angle, then the height of the triangle in Figure 16(a) is given by $h = b\sin\theta$. Thus the area is

$$\mathcal{A} = \tfrac{1}{2} \times \text{base} \times \text{height} = \tfrac{1}{2}ab\sin\theta$$

If the angle θ is not acute, then from Figure 16(b) we see that the height of the triangle is

$$h = b\sin(180° - \theta) = b\sin\theta$$

This is so because the reference angle of θ is the angle $180° - \theta$. Thus in this case also the area of the triangle is

$$\mathcal{A} = \frac{1}{2} \times \text{base} \times \text{height} = \frac{1}{2}ab \sin \theta$$

> **AREA OF A TRIANGLE**
>
> The area $\mathcal{A}$ of a triangle with sides of lengths a and b and with included angle θ is
>
> $$\mathcal{A} = \frac{1}{2}ab \sin \theta$$

EXAMPLE 8 ■ Finding the Area of a Triangle

Find the area of triangle ABC shown in Figure 17.

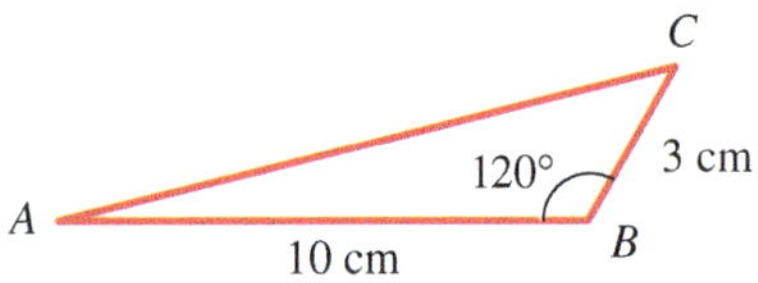

FIGURE 17

SOLUTION The triangle has sides of length 10 cm and 3 cm, with included angle 120°. Therefore

$$\begin{aligned} \mathcal{A} &= \tfrac{1}{2}ab \sin \theta \\ &= \tfrac{1}{2}(10)(3) \sin 120° \\ &= 15 \sin 60° \qquad \text{Reference angle} \\ &= 15\frac{\sqrt{3}}{2} \approx 13 \text{ cm}^2 \end{aligned}$$

Now Try Exercise 57 ■

6.3 EXERCISES

CONCEPTS

1. If the angle θ is in standard position and $P(x, y)$ is a point on the terminal side of θ, and r is the distance from the origin to P, then

$\sin \theta = \frac{\quad}{\quad}$ $\cos \theta = \frac{\quad}{\quad}$ $\tan \theta = \frac{\quad}{\quad}$

2. The sign of a trigonometric function of θ depends on the ________ in which the terminal side of the angle θ lies.

In Quadrant II, $\sin \theta$ is ________ (positive / negative).

In Quadrant III, $\cos \theta$ is ________ (positive / negative).

In Quadrant IV, $\sin \theta$ is ________ (positive / negative).

3. (a) If θ is in standard position, then the reference angle $\overline{\theta}$ is the acute angle formed by the terminal side of θ and the ________. So the reference angle for $\theta = 100°$ is $\overline{\theta} =$ ________, and that for $\theta = 190°$ is $\overline{\theta} =$ ________.

(b) If θ is any angle, the value of a trigonometric function of θ is the same, except possibly for sign, as the value of the trigonometric function of $\overline{\theta}$. So $\sin 100° = \sin$ ____, and $\sin 190° = -\sin$ ____.

4. The area $\mathcal{A}$ of a triangle with sides of lengths a and b and with included angle θ is given by the formula $\mathcal{A} =$ ________. So the area of the triangle with sides 4 and 7 and included angle $\theta = 30°$ is ________.

SKILLS

5–12 ■ Reference Angle Find the reference angle for the given angle.

5. (a) 120° **(b)** 200° **(c)** 285°

6. (a) 175° **(b)** 310° **(c)** 730°

7. (a) 225° **(b)** 810° **(c)** $-105°$

8. (a) 99° **(b)** $-199°$ **(c)** 359°

9. (a) $\frac{7\pi}{10}$ **(b)** $\frac{9\pi}{8}$ **(c)** $\frac{10\pi}{3}$

10. (a) $\frac{5\pi}{6}$ **(b)** $\frac{10\pi}{9}$ **(c)** $\frac{23\pi}{7}$

11. (a) $\frac{5\pi}{7}$ **(b)** -1.4π **(c)** 1.4

12. (a) 2.3π **(b)** 2.3 **(c)** -10π

13–36 ■ Values of Trigonometric Functions Find the exact value of the trigonometric function.

13. $\cos 150°$ **14.** $\sin 240°$ **15.** $\tan 330°$

16. $\sin(-30°)$ **17.** $\cot(-120°)$ **18.** $\csc 300°$

19. $\csc(-630°)$ **20.** $\cot 210°$ **21.** $\cos 570°$

22. $\sec 120°$ **23.** $\tan 750°$ **24.** $\cos 660°$

25. $\sin \dfrac{3\pi}{2}$ **26.** $\cos \dfrac{4\pi}{3}$ **27.** $\tan\left(-\dfrac{4\pi}{3}\right)$

28. $\cos\left(-\dfrac{11\pi}{6}\right)$ **29.** $\csc\left(-\dfrac{5\pi}{6}\right)$ **30.** $\sec \dfrac{7\pi}{6}$

31. $\sec \dfrac{17\pi}{3}$ **32.** $\csc \dfrac{5\pi}{4}$ **33.** $\cot\left(-\dfrac{\pi}{4}\right)$

34. $\cos \dfrac{7\pi}{4}$ **35.** $\tan \dfrac{5\pi}{2}$ **36.** $\sin \dfrac{11\pi}{6}$

37–40 ■ Quadrant in Which an Angle Lies Find the quadrant in which θ lies from the information given.

37. $\sin \theta < 0$ and $\cos \theta < 0$

38. $\tan \theta < 0$ and $\sin \theta < 0$

39. $\sec \theta > 0$ and $\tan \theta < 0$

40. $\csc \theta > 0$ and $\cos \theta < 0$

41–46 ■ Expressing One Trigonometric Function in Terms of Another Write the first trigonometric function in terms of the second for θ in the given quadrant.

41. $\tan \theta$, $\cos \theta$; θ in Quadrant III

42. $\cot \theta$, $\sin \theta$; θ in Quadrant II

43. $\cos \theta$, $\sin \theta$; θ in Quadrant IV

44. $\sec \theta$, $\sin \theta$; θ in Quadrant I

45. $\sec \theta$, $\tan \theta$; θ in Quadrant II

46. $\csc \theta$, $\cot \theta$; θ in Quadrant III

47–54 ■ Values of Trigonometric Functions Find the values of the trigonometric functions of θ from the information given.

47. $\sin \theta = -\frac{4}{5}$, θ in Quadrant IV

48. $\tan \theta = \frac{4}{3}$, θ in Quadrant III

49. $\cos \theta = \frac{7}{12}$, $\sin \theta < 0$

50. $\cot \theta = -\frac{8}{9}$, $\cos \theta > 0$

51. $\csc \theta = 2$, θ in Quadrant I

52. $\cot \theta = \frac{1}{4}$, $\sin \theta < 0$

53. $\cos \theta = -\frac{2}{7}$, $\tan \theta < 0$

54. $\tan \theta = -4$, $\sin \theta > 0$

55–56 ■ Values of an Expression If $\theta = \pi/3$, find the value of each expression.

55. $\sin 2\theta$, $2 \sin \theta$ **56.** $\sin^2\theta$, $\sin(\theta^2)$

57–60 ■ Area of a Triangle Find the area of the triangle with the given description.

57. A triangle with sides of length 7 and 9 and included angle 72°

58. A triangle with sides of length 10 and 22 and included angle 10°

59. An equilateral triangle with side of length 10

60. An equilateral triangle with side of length 13

61. Finding an Angle of a Triangle A triangle has an area of 16 in^2, and two of the sides have lengths 5 in. and 7 in. Find the sine of the angle included by these two sides.

62. Finding a Side of a Triangle An isosceles triangle has an area of 24 cm^2, and the angle between the two equal sides is $5\pi/6$. Find the length of the two equal sides.

SKILLS Plus

63–64 ■ Area of a Region Find the area of the shaded region in the figure.

63.

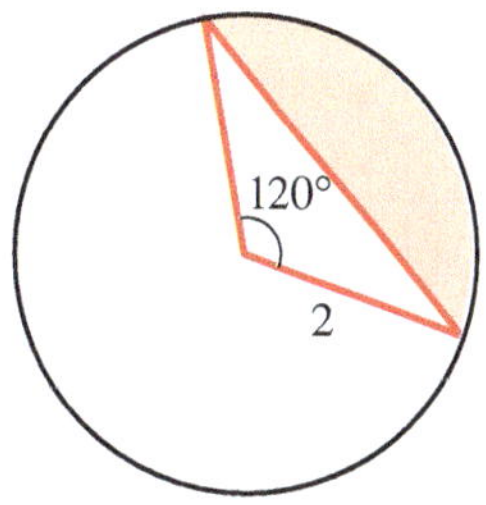

64.

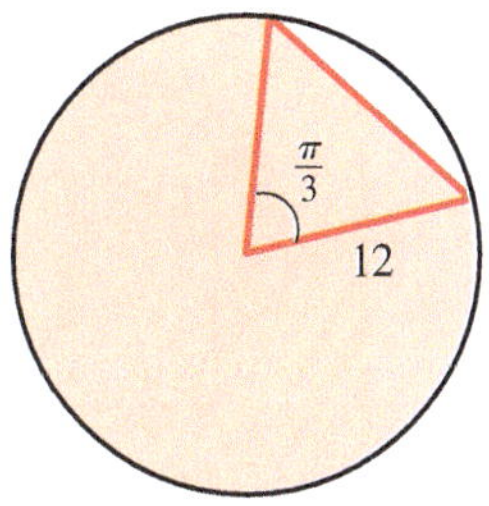

APPLICATIONS

65. Height of a Rocket A rocket fired straight up is tracked by an observer on the ground 1 mi away.

(a) Show that when the angle of elevation is θ, the height of the rocket (in ft) is $h = 5280 \tan \theta$.

(b) Complete the table to find the height of the rocket at the given angles of elevation.

θ	20°	60°	80°	85°
h				

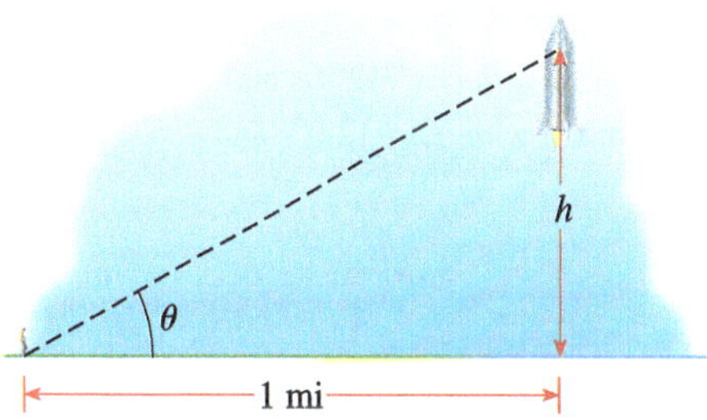

66. Rain Gutter A rain gutter is to be constructed from a metal sheet of width 30 cm by bending up one-third of the sheet on each side through an angle θ. (See the figure on the next page.)

(a) Show that the cross-sectional area of the gutter is modeled by the function

$$A(\theta) = 100 \sin \theta + 100 \sin \theta \cos \theta$$

(b) Graph the function A for $0 \le \theta \le \pi/2$.

(c) For what angle θ is the largest cross-sectional area achieved?

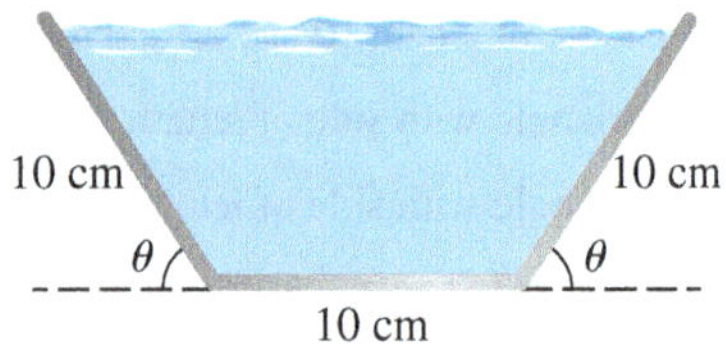

67. Wooden Beam A rectangular beam is to be cut from a cylindrical log of diameter 20 cm. The figures show different ways this can be done.

(a) Express the cross-sectional area of the beam as a function of the angle θ in the figures.

(b) Graph the function you found in part (a).

(c) Find the dimensions of the beam with largest cross-sectional area.

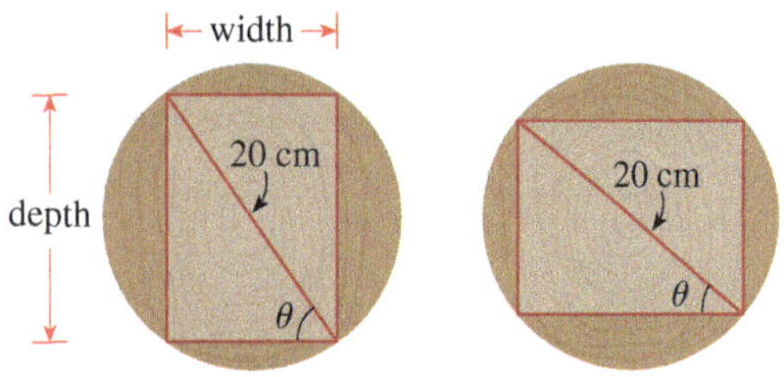

68. Strength of a Beam The strength of a beam is proportional to the width and the square of the depth. A beam is cut from a log as in Exercise 67. Express the strength of the beam as a function of the angle θ in the figures.

69. Throwing a Shot Put The range R and height H of a shot put thrown with an initial velocity of v_0 ft/s at an angle θ are given by

$$R = \frac{v_0^2 \sin(2\theta)}{g}$$

$$H = \frac{v_0^2 \sin^2\theta}{2g}$$

On the earth $g = 32$ ft/s², and on the moon $g = 5.2$ ft/s². Find the range and height of a shot put thrown under the given conditions.

(a) On the earth with $v_0 = 12$ ft/s and $\theta = \pi/6$

(b) On the moon with $v_0 = 12$ ft/s and $\theta = \pi/6$

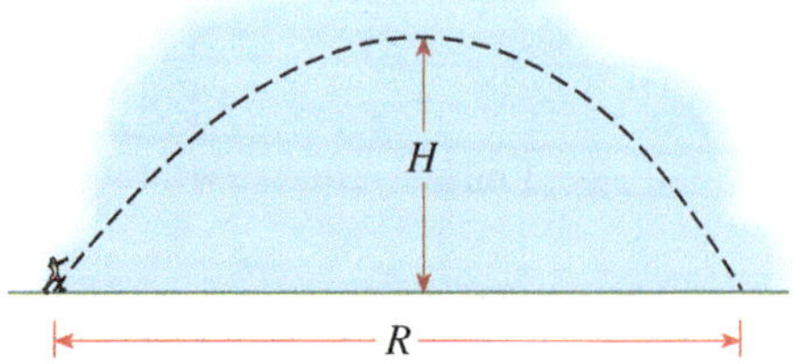

70. Sledding The time in seconds that it takes for a sled to slide down a hillside inclined at an angle θ is

$$t = \sqrt{\frac{d}{16 \sin \theta}}$$

where d is the length of the slope in feet. Find the time it takes to slide down a 2000-ft slope inclined at 30°.

71. Beehives In a beehive each cell is a regular hexagonal prism, as shown in the figure. The amount of wax W in the cell depends on the apex angle θ and is given by

$$W = 3.02 - 0.38 \cot \theta + 0.65 \csc \theta$$

Bees instinctively choose θ so as to use the least amount of wax possible.

(a) Use a graphing device to graph W as a function of θ for $0 < \theta < \pi$.

(b) For what value of θ does W have its minimum value? [*Note:* Biologists have discovered that bees rarely deviate from this value by more than a degree or two.]

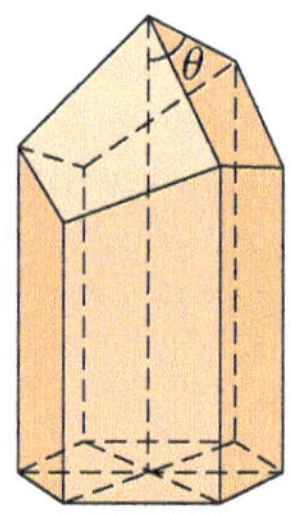

72. Turning a Corner A steel pipe is being carried down a hallway that is 9 ft wide. At the end of the hall there is a right-angled turn into a narrower hallway 6 ft wide.

(a) Show that the length of the pipe in the figure is modeled by the function

$$L(\theta) = 9 \csc \theta + 6 \sec \theta$$

(b) Graph the function L for $0 < \theta < \pi/2$.

(c) Find the minimum value of the function L.

(d) Explain why the value of L you found in part (c) is the length of the longest pipe that can be carried around the corner.

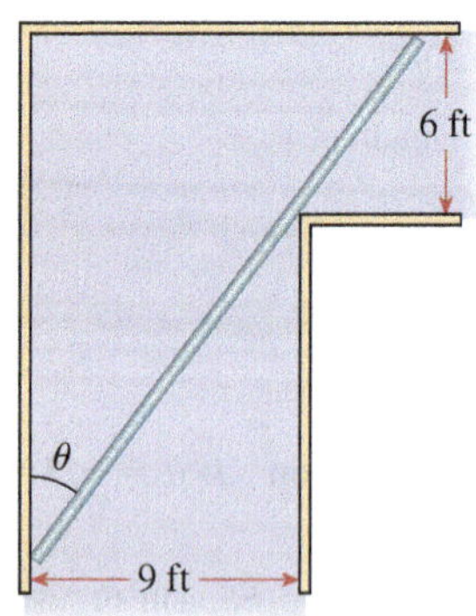

73. Rainbows Rainbows are created when sunlight of different wavelengths (colors) is refracted and reflected in raindrops. The angle of elevation θ of a rainbow is always the same. It can be shown that $\theta = 4\beta - 2\alpha$, where

$$\sin \alpha = k \sin \beta$$

and $\alpha = 59.4°$ and $k = 1.33$ is the index of refraction of water. Use the given information to find the angle of elevation θ of a rainbow. [*Hint:* Find $\sin \beta$, then use the SIN^{-1} key on your calculator to find β.] (For a mathematical explanation of rainbows see *Calculus Early Transcendentals,* 7th Edition, by James Stewart, page 282.)

DISCUSS ■ DISCOVER ■ PROVE ■ WRITE

74. DISCUSS: Using a Calculator To solve a certain problem, you need to find the sine of 4 rad. Your study partner uses his calculator and tells you that

$$\sin 4 = 0.0697564737$$

On your calculator you get

$$\sin 4 = -0.7568024953$$

What is wrong? What mistake did your partner make?

75. DISCUSS ■ DISCOVER: Viète's Trigonometric Diagram In the 16th century the French mathematician François Viète (see page 50) published the following remarkable diagram. Each of the six trigonometric functions of θ is equal to the length of a line segment in the figure. For instance, $\sin \theta = |PR|$, since from $\triangle OPR$ we see that

$$\sin \theta = \frac{\text{opp}}{\text{hyp}} = \frac{|PR|}{|OR|} = \frac{|PR|}{1} = |PR|$$

For each of the five other trigonometric functions, find a line segment in the figure whose length equals the value of the function at θ. [*Note:* The radius of the circle is 1, the center is O, segment QS is tangent to the circle at R, and $\angle SOQ$ is a right angle.]

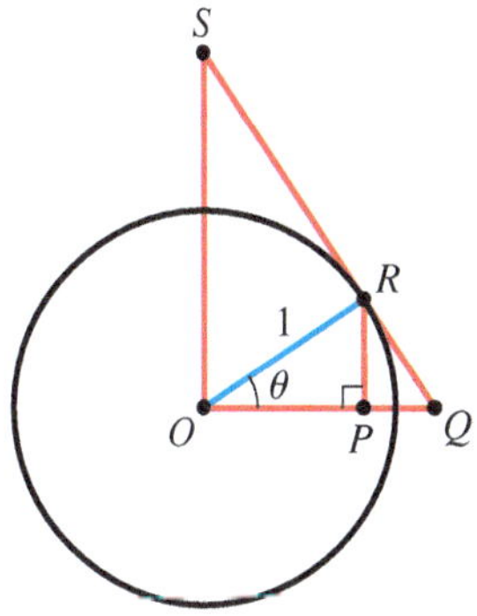

76. PROVE: Pythagorean Identities To prove the following Pythagorean identities, start with the first Pythagorean identity, $\sin^2\theta + \cos^2\theta = 1$, which was proved in the text, and then divide both sides by an appropriate trigonometric function of θ.

(a) $\tan^2\theta + 1 = \sec^2\theta$ **(b)** $1 + \cot^2\theta = \csc^2\theta$

77. DISCUSS ■ DISCOVER: Degrees and Radians What is the smallest positive real number x with the property that the sine of x degrees is equal to the sine of x radians?

6.4 INVERSE TRIGONOMETRIC FUNCTIONS AND RIGHT TRIANGLES

■ The Inverse Sine, Inverse Cosine, and Inverse Tangent Functions ■ Solving for Angles in Right Triangles ■ Evaluating Expressions Involving Inverse Trigonometric Functions

The graphs of the inverse trigonometric functions are studied in Section 5.5.

Recall that for a function to have an inverse, it must be one-to-one. Since the trigonometric functions are not one-to-one, they do not have inverses. So we restrict the domain of each of the trigonometric functions to intervals on which they attain all their values and on which they are one-to-one. The resulting functions have the same range as the original functions but are one-to-one.

■ The Inverse Sine, Inverse Cosine, and Inverse Tangent Functions

Let's first consider the sine function. We restrict the domain of the sine function to angles θ with $-\pi/2 \le \theta \le \pi/2$. From Figure 1 we see that on this domain the sine function attains each of the values in the interval $[-1, 1]$ exactly once and so is one-to-one. Similarly, we restrict the domains of cosine and tangent as shown in Figure 1.

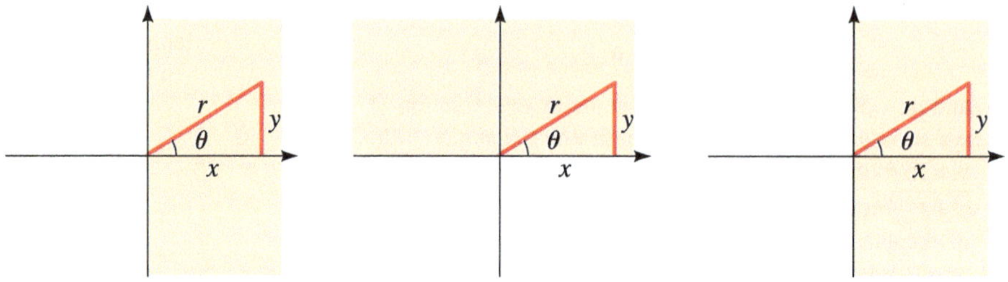

$\sin\theta = \frac{y}{r}$ $\quad -\frac{\pi}{2} \le \theta \le \frac{\pi}{2}$

$\cos\theta = \frac{x}{r}$ $\quad 0 \le \theta \le \pi$

$\tan\theta = \frac{y}{x}$ $\quad -\frac{\pi}{2} < \theta < \frac{\pi}{2}$

FIGURE 1 Restricted domains of the sine, cosine, and tangent functions

On these restricted domains we can define an inverse for each of these functions. By the definition of inverse function we have

$$\sin^{-1}x = y \quad \Leftrightarrow \quad \sin y = x$$
$$\cos^{-1}x = y \quad \Leftrightarrow \quad \cos y = x$$
$$\tan^{-1}x = y \quad \Leftrightarrow \quad \tan y = x$$

We summarize the domains and ranges of the inverse trigonometric functions in the following box.

THE INVERSE SINE, INVERSE COSINE, AND INVERSE TANGENT FUNCTIONS

The sine, cosine, and tangent functions on the restricted domains $[-\pi/2, \pi/2]$, $[0, \pi]$, and $(-\pi/2, \pi/2)$, respectively, are one-to one and so have inverses. The inverse functions have domain and range as follows.

Function	Domain	Range
$\sin^{-1}$	$[-1, 1]$	$[-\pi/2, \pi/2]$
$\cos^{-1}$	$[-1, 1]$	$[0, \pi]$
$\tan^{-1}$	$\mathbb{R}$	$(-\pi/2, \pi/2)$

The functions $\sin^{-1}$, $\cos^{-1}$, and $\tan^{-1}$ are sometimes called **arcsine**, **arccosine**, and **arctangent**, respectively.

Since these are inverse functions, they reverse the rule of the original function. For example, since $\sin \pi/6 = \frac{1}{2}$, it follows that $\sin^{-1}\frac{1}{2} = \pi/6$. The following example gives further illustrations.

EXAMPLE 1 ■ Evaluating Inverse Trigonometric Functions

Find the exact value.

(a) $\sin^{-1}\dfrac{\sqrt{3}}{2}$ **(b)** $\cos^{-1}(-\frac{1}{2})$ **(c)** $\tan^{-1} 1$

SOLUTION

(a) The angle in the interval $[-\pi/2, \pi/2]$ whose sine is $\sqrt{3}/2$ is $\pi/3$. Thus $\sin^{-1}(\sqrt{3}/2) = \pi/3$.

(b) The angle in the interval $[0, \pi]$ whose cosine is $-\frac{1}{2}$ is $2\pi/3$. Thus $\cos^{-1}(-\frac{1}{2}) = 2\pi/3$.

(c) The angle in the interval $(-\pi/2, \pi/2)$ whose tangent is 1 is $\pi/4$. Thus $\tan^{-1} 1 = \pi/4$.

Now Try Exercise 5 ■

EXAMPLE 2 ■ Evaluating Inverse Trigonometric Functions

Find approximate values for the given expression.

(a) $\sin^{-1}(0.71)$ **(b)** $\tan^{-1} 2$ **(c)** $\cos^{-1} 2$

SOLUTION We use a calculator to approximate these values.

(a) Using the [INV][SIN], or [SIN⁻¹], or [ARC][SIN] key(s) on the calculator (with the calculator in radian mode), we get

$$\sin^{-1}(0.71) \approx 0.78950$$

(b) Using the [INV][TAN], or [TAN⁻¹], or [ARC][TAN] key(s) on the calculator (with the calculator in radian mode), we get

$$\tan^{-1} 2 \approx 1.10715$$

(c) Since $2 > 1$, it is not in the domain of $\cos^{-1}$, so $\cos^{-1} 2$ is not defined.

Now Try Exercises 9, 13, and 15 ■

■ Solving for Angles in Right Triangles

In Section 6.2 we solved triangles by using the trigonometric functions to find the unknown sides. We now use the inverse trigonometric functions to solve for *angles* in a right triangle.

EXAMPLE 3 ■ Finding an Angle in a Right Triangle

FIGURE 2

Find the angle θ in the triangle shown in Figure 2.

SOLUTION Since θ is the angle opposite the side of length 10 and the hypotenuse has length 50, we have

$$\sin \theta = \frac{10}{50} = \frac{1}{5} \qquad \sin \theta = \frac{\text{opp}}{\text{hyp}}$$

Now we can use $\sin^{-1}$ to find θ.

$$\theta = \sin^{-1} \tfrac{1}{5} \qquad \text{Definition of } \sin^{-1}$$

$$\theta \approx 11.5° \qquad \text{Calculator (in degree mode)}$$

Now Try Exercise 17 ■

EXAMPLE 4 ■ Solving for an Angle in a Right Triangle

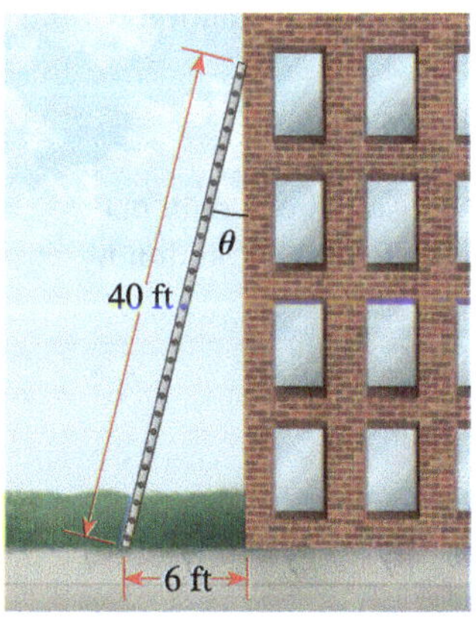

FIGURE 3

A 40-ft ladder leans against a building. If the base of the ladder is 6 ft from the base of the building, what is the angle formed by the ladder and the building?

SOLUTION First we sketch a diagram as in Figure 3. If θ is the angle between the ladder and the building, then

$$\sin \theta = \frac{6}{40} = 0.15 \qquad \sin \theta = \frac{\text{opp}}{\text{hyp}}$$

Now we use $\sin^{-1}$ to find θ.

$$\theta = \sin^{-1}(0.15) \qquad \text{Definition of } \sin^{-1}$$
$$\theta \approx 8.6° \qquad \text{Calculator (in degree mode)}$$

Now Try Exercise 39 ■

EXAMPLE 5 ■ The Angle of a Beam of Light

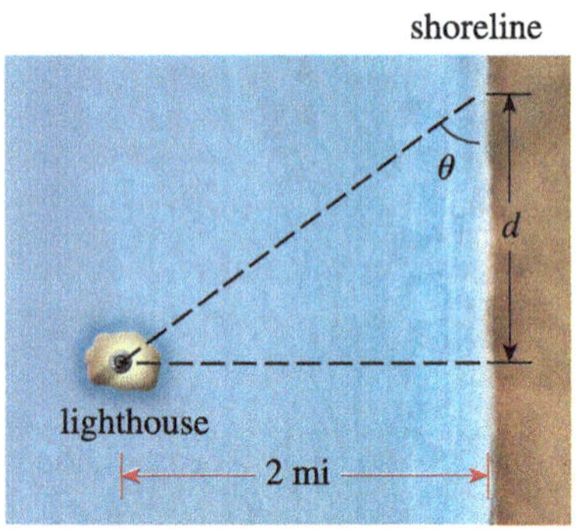

FIGURE 4

A lighthouse is located on an island that is 2 mi off a straight shoreline (see Figure 4). Express the angle formed by the beam of light and the shoreline in terms of the distance d in the figure.

SOLUTION From the figure we see that

$$\tan\theta = \frac{2}{d} \qquad \tan\theta = \frac{\text{opp}}{\text{adj}}$$

Taking the inverse tangent of both sides, we get

$$\tan^{-1}(\tan\theta) = \tan^{-1}\left(\frac{2}{d}\right) \qquad \text{Take } \tan^{-1} \text{ of both sides}$$
$$\theta = \tan^{-1}\left(\frac{2}{d}\right) \qquad \text{Property of inverse functions: } \tan^{-1}(\tan\theta) = \theta$$

Now Try Exercise 41 ■

In Sections 6.5 and 6.6 we will learn how to solve any triangle (not necessarily a right triangle). The angles in a triangle are always in the interval $(0, \pi)$ (or between 0° and 180°). We'll see that to solve such triangles, we need to find all angles in the interval $(0, \pi)$ that have a specified sine or cosine. We do this in the next example.

EXAMPLE 6 ■ Solving a Basic Trigonometric Equation on an Interval

Find all angles θ between 0° and 180° satisfying the given equation.

(a) $\sin\theta = 0.4$ **(b)** $\cos\theta = 0.4$

SOLUTION

(a) We use $\sin^{-1}$ to find one solution in the interval $[-\pi/2, \pi/2]$.

$$\sin\theta = 0.4 \qquad \text{Equation}$$
$$\theta = \sin^{-1}(0.4) \qquad \text{Take } \sin^{-1} \text{ of each side}$$
$$\theta \approx 23.6° \qquad \text{Calculator (in degree mode)}$$

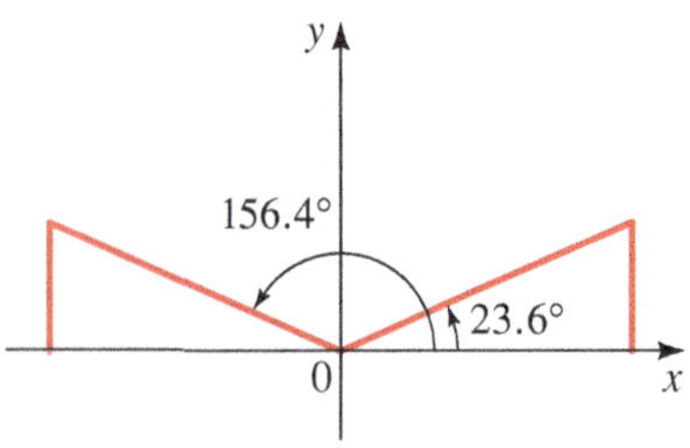

FIGURE 5

Another solution with θ between 0° and 180° is obtained by taking the supplement of the angle: $180° - 23.6° = 156.4°$ (see Figure 5). So the solutions of the equation with θ between 0° and 180° are

$$\theta \approx 23.6° \quad \text{and} \quad \theta \approx 156.4°$$

(b) The cosine function is one-to-one on the interval $[0, \pi]$, so there is only one solution of the equation with θ between 0° and 180°. We find that solution by taking $\cos^{-1}$ of each side.

$$\cos\theta = 0.4$$
$$\theta = \cos^{-1}(0.4) \qquad \text{Take } \cos^{-1} \text{ of each side}$$
$$\theta \approx 66.4° \qquad \text{Calculator (in degree mode)}$$

The solution is $\theta \approx 66.4°$

Now Try Exercises 25 and 27 ■

Evaluating Expressions Involving Inverse Trigonometric Functions

Expressions like $\cos(\sin^{-1} x)$ arise in calculus. We find exact values of such expressions using trigonometric identities or right triangles.

EXAMPLE 7 ■ Composing Trigonometric Functions and Their Inverses

Find $\cos\left(\sin^{-1} \frac{3}{5}\right)$.

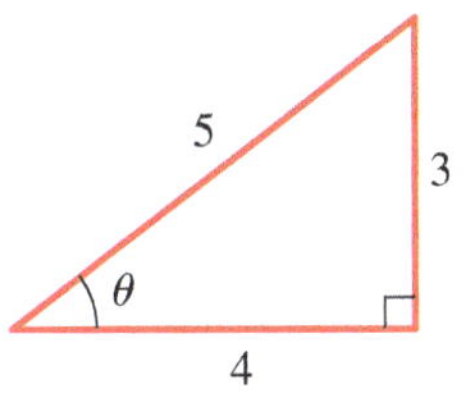

FIGURE 6

$\cos\theta = \dfrac{4}{5}$

SOLUTION 1 Let $\theta = \sin^{-1}\frac{3}{5}$. Then θ is the number in the interval $[-\pi/2, \pi/2]$ whose sine is $\frac{3}{5}$. Let's interpret θ as an angle and draw a right triangle with θ as one of its acute angles, with opposite side 3 and hypotenuse 5 (see Figure 6). The remaining leg of the triangle is found by the Pythagorean Theorem to be 4. From the figure we get

$$\begin{aligned} \cos\left(\sin^{-1}\tfrac{3}{5}\right) &= \cos\theta && \theta = \sin^{-1}\tfrac{3}{5} \\ &= \frac{4}{5} && \cos\theta = \frac{\text{adj}}{\text{hyp}} \end{aligned}$$

So $\cos\left(\sin^{-1}\frac{3}{5}\right) = \frac{4}{5}$.

SOLUTION 2 It's easy to find $\sin\left(\sin^{-1}\frac{3}{5}\right)$. In fact, by the cancellation properties of inverse functions, this value is exactly $\frac{3}{5}$. To find $\cos\left(\sin^{-1}\frac{3}{5}\right)$, we first write the cosine function in terms of the sine function. Let $u = \sin^{-1}\frac{3}{5}$. Since $-\pi/2 \le u \le \pi/2$, $\cos u$ is positive, and we can write the following:

$$\begin{aligned} \cos u &= +\sqrt{1 - \sin^2 u} && \cos^2 u + \sin^2 u = 1 \\ &= \sqrt{1 - \sin^2\left(\sin^{-1}\tfrac{3}{5}\right)} && u = \sin^{-1}\tfrac{3}{5} \\ &= \sqrt{1 - \left(\tfrac{3}{5}\right)^2} && \text{Property of inverse functions: } \sin\left(\sin^{-1}\tfrac{3}{5}\right) = \tfrac{3}{5} \\ &= \sqrt{1 - \tfrac{9}{25}} = \sqrt{\tfrac{16}{25}} = \tfrac{4}{5} && \text{Calculate} \end{aligned}$$

So $\cos\left(\sin^{-1}\frac{3}{5}\right) = \frac{4}{5}$.

Now Try Exercise 29 ■

EXAMPLE 8 ■ Composing Trigonometric Functions and Their Inverses

Write $\sin(\cos^{-1} x)$ and $\tan(\cos^{-1} x)$ as algebraic expressions in x for $-1 \le x \le 1$.

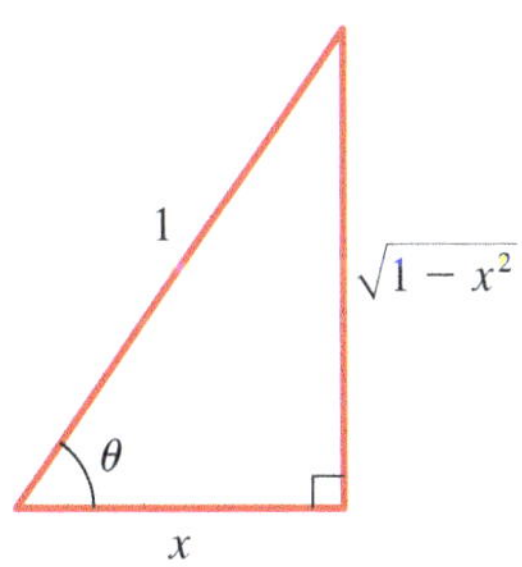

FIGURE 7

$\cos\theta = \dfrac{x}{1} = x$

SOLUTION 1 Let $\theta = \cos^{-1} x$; then $\cos\theta = x$. In Figure 7 we sketch a right triangle with an acute angle θ, adjacent side x, and hypotenuse 1. By the Pythagorean Theorem the remaining leg is $\sqrt{1 - x^2}$. From the figure we have

$$\sin(\cos^{-1} x) = \sin\theta = \sqrt{1 - x^2} \quad \text{and} \quad \tan(\cos^{-1} x) = \tan\theta = \frac{\sqrt{1 - x^2}}{x}$$

SOLUTION 2 Let $u = \cos^{-1} x$. We need to find $\sin u$ and $\tan u$ in terms of x. As in Example 7 the idea here is to write sine and tangent in terms of cosine. Note that $0 \le u \le \pi$ because $u = \cos^{-1} x$. We have

$$\sin u = \pm\sqrt{1 - \cos^2 u} \quad \text{and} \quad \tan u = \frac{\sin u}{\cos u} = \frac{\pm\sqrt{1 - \cos^2 u}}{\cos u}$$

To choose the proper signs, note that u lies in the interval $[0, \pi]$ because $u = \cos^{-1} x$. Since $\sin u$ is positive on this interval, the + sign is the correct choice. Substituting $u = \cos^{-1} x$ in the displayed equations and using the cancellation property $\cos(\cos^{-1} x) = x$, we get

$$\sin(\cos^{-1} x) = \sqrt{1 - x^2} \qquad \text{and} \qquad \tan(\cos^{-1} x) = \frac{\sqrt{1 - x^2}}{x}$$

Now Try Exercises 35 and 37 ■

Note: In Solution 1 of Example 8 it might seem that because we are sketching a triangle, the angle $\theta = \cos^{-1} x$ must be acute. But it turns out that the triangle method works for any x. The domains and ranges of all six inverse trigonometric functions have been chosen in such a way that we can always use a triangle to find $S(T^{-1}(x))$, where S and T are any trigonometric functions.

6.4 EXERCISES

CONCEPTS

1. For a function to have an inverse, it must be ________. To define the inverse sine function, we restrict the ________ of the sine function to the interval ________.

2. The inverse sine, inverse cosine, and inverse tangent functions have the following domains and ranges.

(a) The function $\sin^{-1}$ has domain ________ and range ________.

(b) The function $\cos^{-1}$ has domain ________ and range ________.

(c) The function $\tan^{-1}$ has domain ________ and range ________.

3. In the triangle shown we can find the angle θ as follows.

(a) $\theta = \sin^{-1} \frac{\square}{\square}$

(b) $\theta = \cos^{-1} \frac{\square}{\square}$

(c) $\theta = \tan^{-1} \frac{\square}{\square}$

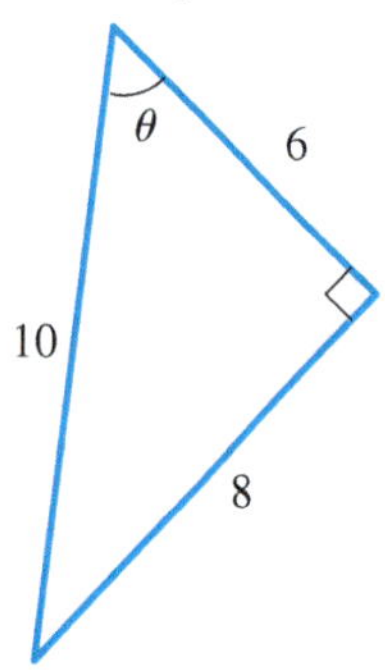

4. To find $\sin(\cos^{-1} \frac{5}{13})$, we let $\theta = \cos^{-1}(\frac{5}{13})$ and complete the right triangle at the top of the next column. We find that $\sin(\cos^{-1} \frac{5}{13}) =$ ________.

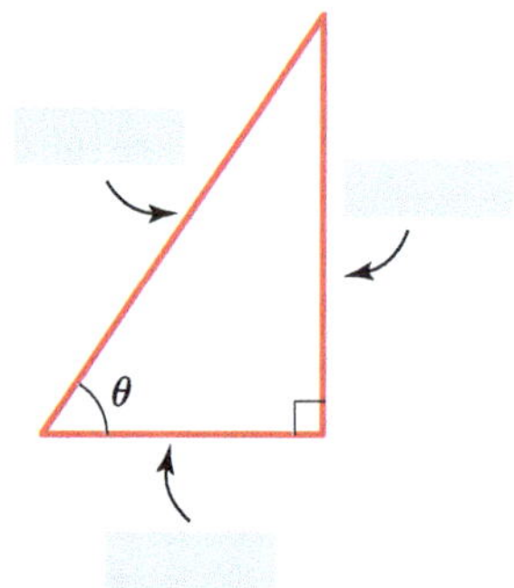

SKILLS

5–8 ■ Evaluating Inverse Trigonometric Functions Find the exact value of each expression, if it is defined. Express your answer in radians.

5. (a) $\sin^{-1} 1$ (b) $\cos^{-1} 0$ (c) $\tan^{-1} \sqrt{3}$

6. (a) $\sin^{-1} 0$ (b) $\cos^{-1}(-1)$ (c) $\tan^{-1} 0$

7. (a) $\sin^{-1}\left(-\frac{\sqrt{2}}{2}\right)$ (b) $\cos^{-1}\left(-\frac{\sqrt{2}}{2}\right)$ (c) $\tan^{-1}(-1)$

8. (a) $\sin^{-1}\left(-\frac{\sqrt{3}}{2}\right)$ (b) $\cos^{-1}(-\frac{1}{2})$ (c) $\tan^{-1}(-\sqrt{3})$

9–16 ■ Evaluating Inverse Trigonometric Functions Use a calculator to find an approximate value (in radians) of each expression rounded to five decimal places, if it is defined.

9. $\sin^{-1}(0.30)$

10. $\cos^{-1}(-0.2)$

11. $\cos^{-1} \frac{1}{3}$

12. $\sin^{-1} \frac{5}{6}$

13. $\tan^{-1} 3$

14. $\tan^{-1}(-4)$

15. $\cos^{-1} 3$

16. $\sin^{-1}(-2)$

17–22 ■ Finding Angles in Right Triangles Find the angle θ in degrees, rounded to one decimal place.

17.

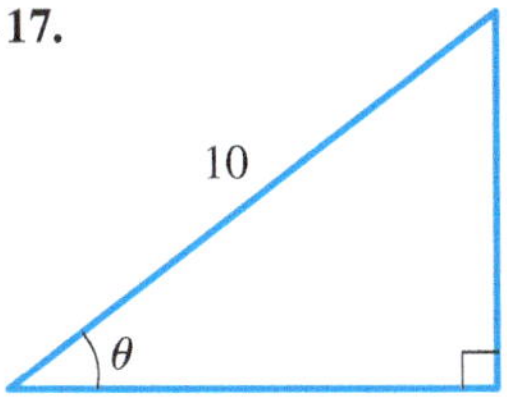

18.

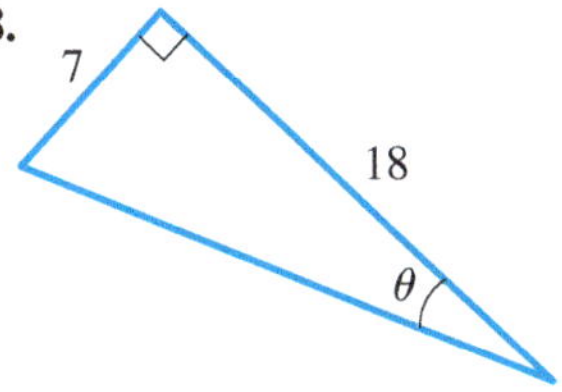

19.

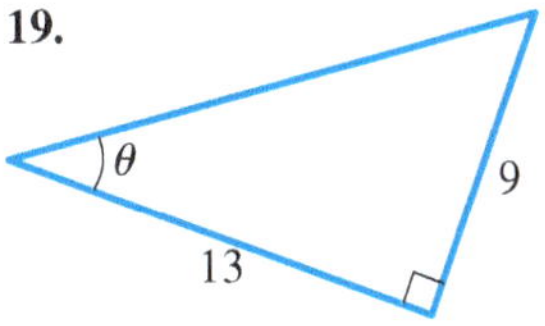

20.

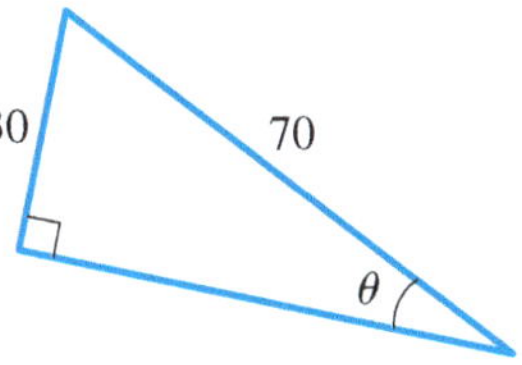

21.

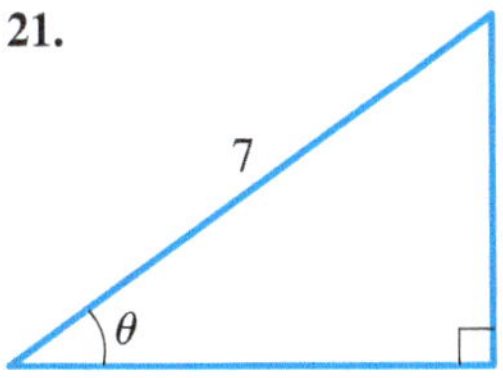

22.

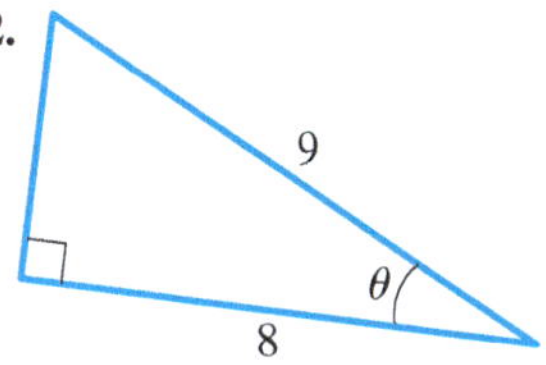

23–28 ■ Basic Trigonometric Equations Find all angles θ between 0° and 180° satisfying the given equation. Round your answer to one decimal place.

23. $\sin\theta = \frac{2}{3}$ **24.** $\cos\theta = \frac{3}{4}$

25. $\cos\theta = -\frac{2}{5}$ **26.** $\tan\theta = -20$

27. $\tan\theta = 5$ **28.** $\sin\theta = \frac{4}{5}$

29–34 ■ Value of an Expression Find the exact value of the expression.

29. $\cos(\sin^{-1}\frac{4}{5})$ **30.** $\cos(\tan^{-1}\frac{4}{3})$ **31.** $\sec(\sin^{-1}\frac{12}{13})$

32. $\csc(\cos^{-1}\frac{7}{25})$ **33.** $\tan(\sin^{-1}\frac{12}{13})$ **34.** $\cot(\sin^{-1}\frac{2}{3})$

35–38 ■ Algebraic Expressions Rewrite the expression as an algebraic expression in x.

35. $\cos(\sin^{-1}x)$ **36.** $\sin(\tan^{-1}x)$

37. $\tan(\sin^{-1}x)$ **38.** $\cos(\tan^{-1}x)$

APPLICATIONS

39. Leaning Ladder A 20-ft ladder is leaning against a building. If the base of the ladder is 6 ft from the base of the building, what is the angle of elevation of the ladder? How high does the ladder reach on the building?

40. Angle of the Sun A 96-ft tree casts a shadow that is 120 ft long. What is the angle of elevation of the sun?

41. Height of the Space Shuttle An observer views the space shuttle from a distance of 2 mi from the launch pad.

(a) Express the height of the space shuttle as a function of the angle of elevation θ.

(b) Express the angle of elevation θ as a function of the height h of the space shuttle.

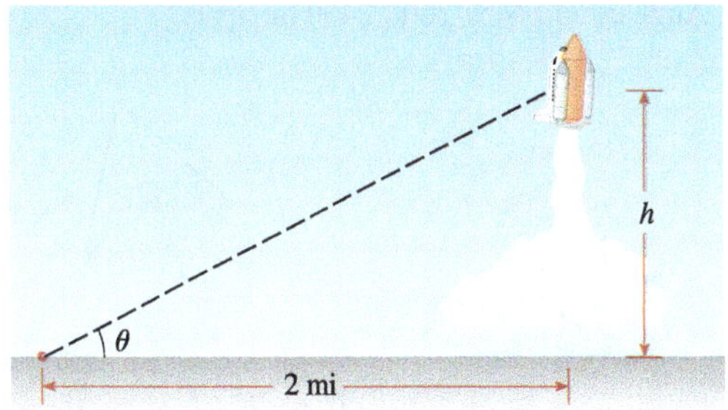

42. Height of a Pole A 50-ft pole casts a shadow as shown in the figure.

(a) Express the angle of elevation θ of the sun as a function of the length s of the shadow.

(b) Find the angle θ of elevation of the sun when the shadow is 20 ft long.

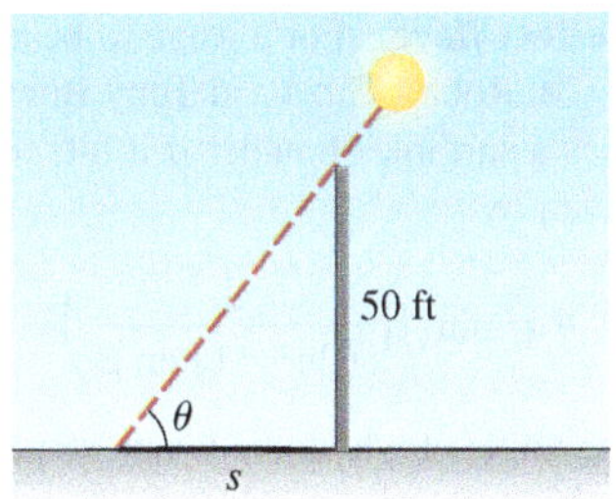

43. Height of a Balloon A 680-ft rope anchors a hot-air balloon as shown in the figure.

(a) Express the angle θ as a function of the height h of the balloon.

(b) Find the angle θ if the balloon is 500 ft high.

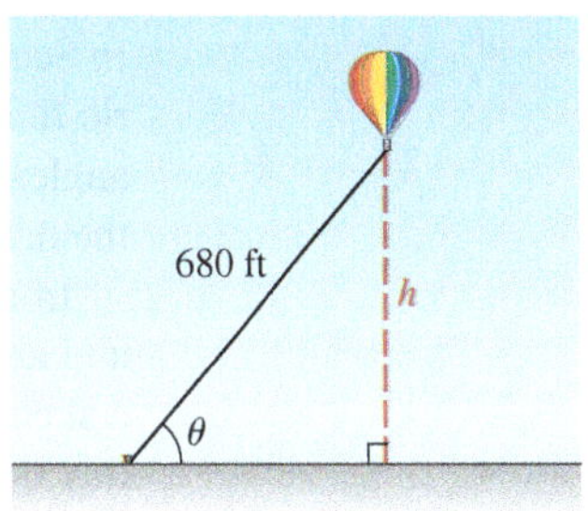

44. View from a Satellite The figures on the next page indicate that the higher the orbit of a satellite, the more of the earth the satellite can "see." Let θ, s, and h be as in the figure, and assume that the earth is a sphere of radius 3960 mi.

(a) Express the angle θ as a function of h.

(b) Express the distance s as a function of θ.

(c) Express the distance s as a function of h. [*Hint:* Find the composition of the functions in parts (a) and (b).]

(d) If the satellite is 100 mi above the earth, what is the distance s that it can see?

(e) How high does the satellite have to be to see both Los Angeles and New York, 2450 mi apart?

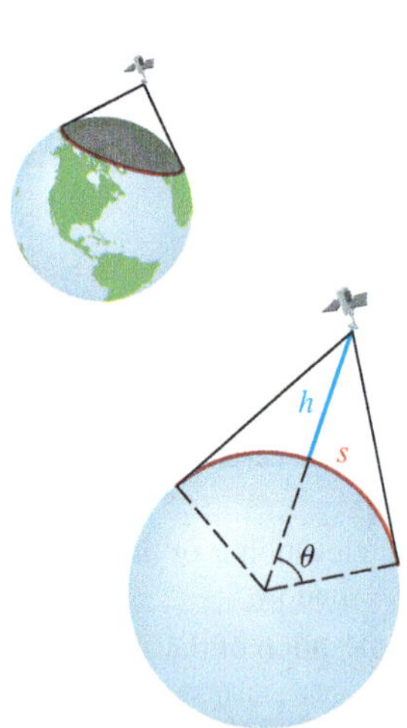

45. Surfing the Perfect Wave For a wave to be surfable, it can't break all at once. Robert Guza and Tony Bowen have shown that a wave has a surfable shoulder if it hits the shoreline at an angle θ given by

$$\theta = \sin^{-1}\left(\frac{1}{(2n + 1)\tan\beta}\right)$$

where β is the angle at which the beach slopes down and where $n = 0, 1, 2, \ldots$.

(a) For $\beta = 10°$, find θ when $n = 3$.

(b) For $\beta = 15°$, find θ when $n = 2$, 3, and 4. Explain why the formula does not give a value for θ when $n = 0$ or 1.

DISCUSS ■ DISCOVER ■ PROVE ■ WRITE

46. PROVE: Inverse Trigonometric Functions on a Calculator Most calculators do not have keys for $\sec^{-1}$, $\csc^{-1}$, or $\cot^{-1}$. Prove the following identities, and then use these identities and a calculator to find $\sec^{-1}2$, $\csc^{-1}3$, and $\cot^{-1}4$.

$$\sec^{-1}x = \cos^{-1}\left(\frac{1}{x}\right) \qquad x \ge 1$$

$$\csc^{-1}x = \sin^{-1}\left(\frac{1}{x}\right) \qquad x \ge 1$$

$$\cot^{-1}x = \tan^{-1}\left(\frac{1}{x}\right) \qquad x > 0$$

6.5 THE LAW OF SINES

■ The Law of Sines ■ The Ambiguous Case

In Section 6.2 we used the trigonometric ratios to solve right triangles. The trigonometric functions can also be used to solve *oblique triangles*, that is, triangles with no right angles. To do this, we first study the Law of Sines here and then the Law of Cosines in the next section.

In general, to solve a triangle, we need to know certain information about its sides and angles. To decide whether we have enough information, it's often helpful to make a sketch. For instance, if we are given two angles and the included side, then it's clear that one and only one triangle can be formed (see Figure 1(a)). Similarly, if two sides and the included angle are known, then a unique triangle is determined (Figure 1(c)). But if we know all three angles and no sides, we cannot uniquely determine the triangle because many triangles can have the same three angles. (All these triangles would be similar, of course.) So we won't consider this last case.

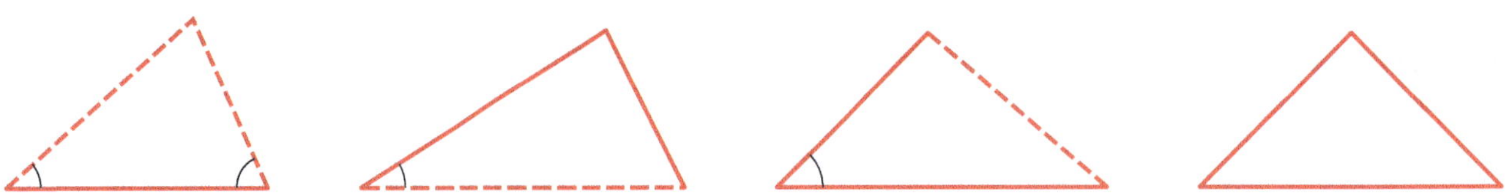

(a) ASA or SAA (b) SSA (c) SAS (d) SSS

FIGURE 1

In general, a triangle is determined by three of its six parts (angles and sides) as long as at least one of these three parts is a side. So the possibilities, illustrated in Figure 1, are as follows.

Case 1 One side and two angles (ASA or SAA)

Case 2 Two sides and the angle opposite one of those sides (SSA)

Case 3 Two sides and the included angle (SAS)

Case 4 Three sides (SSS)

Cases 1 and 2 are solved by using the Law of Sines; Cases 3 and 4 require the Law of Cosines.

The Law of Sines

The **Law of Sines** says that in any triangle the lengths of the sides are proportional to the sines of the corresponding opposite angles. To state this law (or formula) more easily, we follow the convention of labeling the angles of a triangle as A, B, and C and the lengths of the corresponding opposite sides as a, b, and c, as in Figure 2.

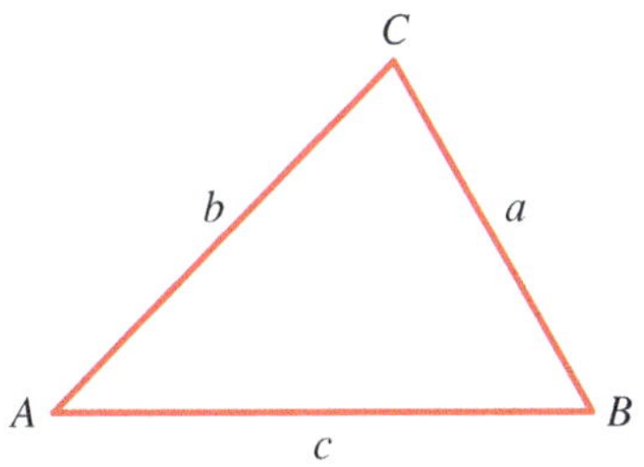

FIGURE 2

THE LAW OF SINES

In triangle ABC we have

$$\frac{\sin A}{a} = \frac{\sin B}{b} = \frac{\sin C}{c}$$

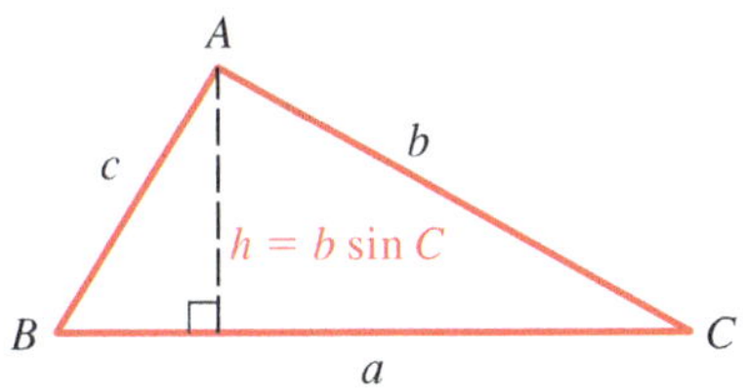

FIGURE 3

Proof To see why the Law of Sines is true, refer to Figure 3. By the formula in Section 6.3 the area of triangle ABC is $\frac{1}{2}ab\sin C$. By the same formula the area of this triangle is also $\frac{1}{2}ac\sin B$ and $\frac{1}{2}bc\sin A$. Thus

$$\tfrac{1}{2}bc\sin A = \tfrac{1}{2}ac\sin B = \tfrac{1}{2}ab\sin C$$

Multiplying by $2/(abc)$ gives the Law of Sines. ■

EXAMPLE 1 ■ Tracking a Satellite (ASA)

A satellite orbiting the earth passes directly overhead at observation stations in Phoenix and Los Angeles, 340 mi apart. At an instant when the satellite is between these two stations, its angle of elevation is simultaneously observed to be 60° at Phoenix and 75° at Los Angeles. How far is the satellite from Los Angeles?

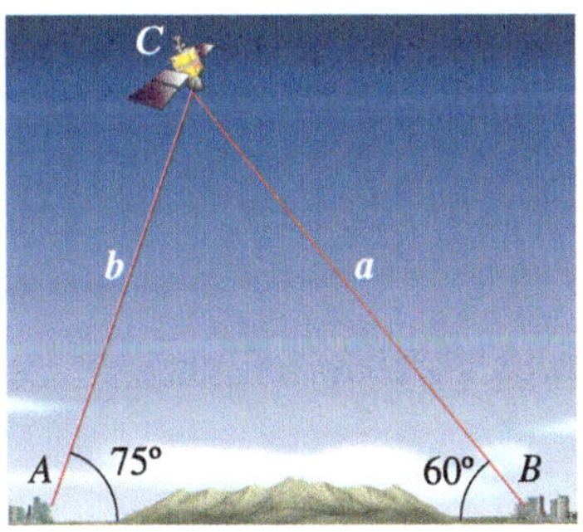

Los Angeles $c = 340$ mi Phoenix

FIGURE 4

SOLUTION We need to find the distance b in Figure 4. Since the sum of the angles in any triangle is 180°, we see that $\angle C = 180° - (75° + 60°) = 45°$ (see Figure 4), so we have

$$\frac{\sin B}{b} = \frac{\sin C}{c} \qquad \text{Law of Sines}$$

$$\frac{\sin 60°}{b} = \frac{\sin 45°}{340} \qquad \text{Substitute}$$

$$b = \frac{340 \sin 60°}{\sin 45°} \approx 416 \qquad \text{Solve for } b$$

The distance of the satellite from Los Angeles is approximately 416 mi.

Now Try Exercises 5 and 31 ■

EXAMPLE 2 ■ Solving a Triangle (SAA)

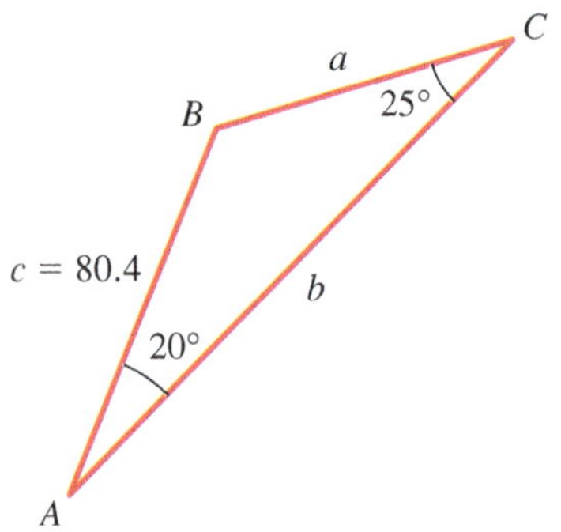

FIGURE 5

Solve the triangle in Figure 5.

SOLUTION First, $\angle B = 180° - (20° + 25°) = 135°$. Since side c is known, to find side a, we use the relation

$$\frac{\sin A}{a} = \frac{\sin C}{c} \qquad \text{Law of Sines}$$

$$a = \frac{c \sin A}{\sin C} = \frac{80.4 \sin 20°}{\sin 25°} \approx 65.1 \qquad \text{Solve for } a$$

Similarly, to find b, we use

$$\frac{\sin B}{b} = \frac{\sin C}{c} \qquad \text{Law of Sines}$$

$$b = \frac{c \sin B}{\sin C} = \frac{80.4 \sin 135°}{\sin 25°} \approx 134.5 \qquad \text{Solve for } b$$

Now Try Exercise 13 ■

The Ambiguous Case

In Examples 1 and 2 a unique triangle was determined by the information given. This is always true of Case 1 (ASA or SAA). But in Case 2 (SSA) there may be two triangles, one triangle, or no triangle with the given properties. For this reason, Case 2 is sometimes called the **ambiguous case**. To see why this is so, we show in Figure 6 the possibilities when angle A and sides a and b are given. In part (a) no solution is possible, since side a is too short to complete the triangle. In part (b) the solution is a right triangle. In part (c) two solutions are possible, and in part (d) there is a unique triangle with the given properties. We illustrate the possibilities of Case 2 in the following examples.

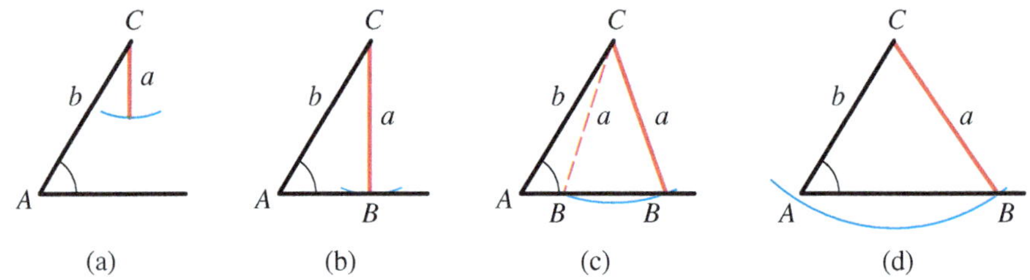

FIGURE 6 The ambiguous case

EXAMPLE 3 ■ SSA, the One-Solution Case

Solve triangle ABC, where $\angle A = 45°$, $a = 7\sqrt{2}$, and $b = 7$.

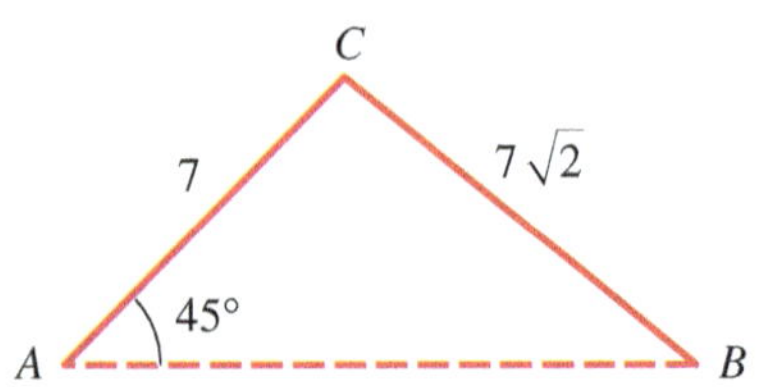

FIGURE 7

SOLUTION We first sketch the triangle with the information we have (see Figure 7). Our sketch is necessarily tentative, since we don't yet know the other angles. Nevertheless, we can now see the possibilities.

We first find $\angle B$.

$$\frac{\sin A}{a} = \frac{\sin B}{b} \qquad \text{Law of Sines}$$

$$\sin B = \frac{b \sin A}{a} = \frac{7}{7\sqrt{2}} \sin 45° = \left(\frac{1}{\sqrt{2}}\right)\left(\frac{\sqrt{2}}{2}\right) = \frac{1}{2} \qquad \text{Solve for } \sin B$$

We consider only angles smaller than 180°, since no triangle can contain an angle of 180° or larger.

Which angles B have $\sin B = \frac{1}{2}$? From the preceding section we know that there are two such angles smaller than 180° (they are 30° and 150°). Which of these angles is compatible with what we know about triangle ABC? Since $\angle A = 45°$, we cannot have $\angle B = 150°$, because $45° + 150° > 180°$. So $\angle B = 30°$, and the remaining angle is $\angle C = 180° - (30° + 45°) = 105°$.

Now we can find side c.

$$\frac{\sin B}{b} = \frac{\sin C}{c} \qquad \text{Law of Sines}$$

$$c = \frac{b \sin C}{\sin B} = \frac{7 \sin 105°}{\sin 30°} = \frac{7 \sin 105°}{\frac{1}{2}} \approx 13.5 \qquad \text{Solve for } c$$

Now Try Exercise 19 ■

In Example 3 there were two possibilities for angle B, and one of these was not compatible with the rest of the information. In general, if $\sin A < 1$, we must check the angle and its supplement as possibilities, because any angle smaller than 180° can be in the triangle. To decide whether either possibility works, we check to see whether the resulting sum of the angles exceeds 180°. It can happen, as in Figure 6(c), that both possibilities are compatible with the given information. In that case, two different triangles are solutions to the problem.

The *supplement* of an angle θ (where $0 \le \theta \le 180°$) is the angle $180° - \theta$.

EXAMPLE 4 ■ SSA, the Two-Solution Case

Solve triangle ABC if $\angle A = 43.1°$, $a = 186.2$, and $b = 248.6$.

SOLUTION From the given information we sketch the triangle shown in Figure 8. Note that side a may be drawn in two possible positions to complete the triangle. From the Law of Sines

$$\sin B = \frac{b \sin A}{a} = \frac{248.6 \sin 43.1°}{186.2} \approx 0.91225$$

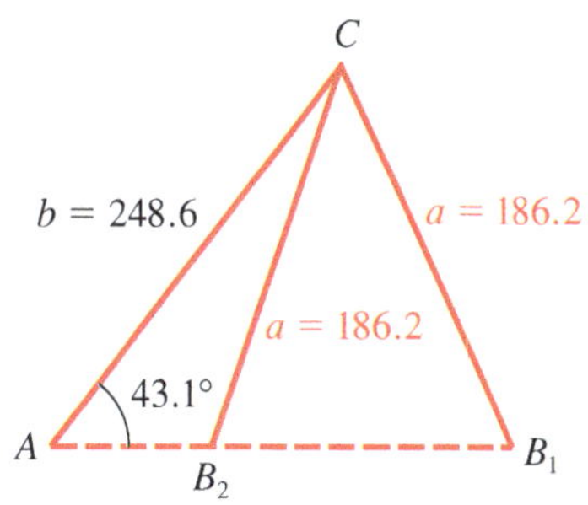

FIGURE 8

There are two possible angles B between 0° and 180° such that $\sin B = 0.91225$. Using a calculator, we find that one of the angles is

$$\sin^{-1}(0.91225) \approx 65.8°$$

The other angle is approximately $180° - 65.8° = 114.2°$. We denote these two angles by B_1 and B_2 so that

$$\angle B_1 \approx 65.8° \qquad \text{and} \qquad \angle B_2 \approx 114.2°$$

Thus two triangles satisfy the given conditions: triangle $A_1B_1C_1$ and triangle $A_2B_2C_2$.

Solve triangle $A_1B_1C_1$:

$$\angle C_1 \approx 180° - (43.1° + 65.8°) = 71.1° \qquad \text{Find } \angle C_1$$

Thus

$$c_1 = \frac{a_1 \sin C_1}{\sin A_1} \approx \frac{186.2 \sin 71.1°}{\sin 43.1°} \approx 257.8 \qquad \text{Law of Sines}$$

Surveying is a method of land measurement used for mapmaking. Surveyors use a process called *triangulation* in which a network of thousands of interlocking triangles is created on the area to be mapped. The process is started by measuring the length of a *baseline* between two surveying stations. Then, with the use of an instrument called a *theodolite*, the angles between these two stations and a third station are measured. The Law of Sines is then used to calculate the two other sides of the triangle formed by the three stations. The calculated sides are used as baselines, and the process is repeated over and over to create a network of triangles. In this method the only distance measured is the initial baseline; all other distances are calculated from the Law of Sines. This method is practical because it is much easier to measure angles than distances.

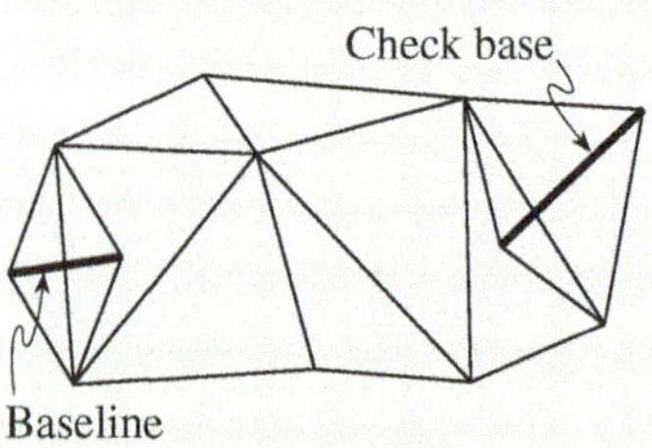

One of the most ambitious mapmaking efforts of all time was the Great Trigonometric Survey of India (see Problem 8, page 536) which required several expeditions and took over a century to complete. The famous expedition of 1823, led by **Sir George Everest**, lasted 20 years. Ranging over treacherous terrain and encountering the dreaded malaria-carrying mosquitoes, this expedition reached the foothills of the Himalayas. A later expedition, using triangulation, calculated the height of the highest peak of the Himalayas to be 29,002 ft. The peak was named in honor of Sir George Everest.

Today, with the use of satellites, the height of Mt. Everest is estimated to be 29,028 ft. The very close agreement of these two estimates shows the great accuracy of the trigonometric method.

Solve triangle $A_2B_2C_2$:

$$\angle C_2 \approx 180° - (43.1° + 114.2°) = 22.7° \qquad \text{Find } \angle C_2$$

Thus $$c_2 = \frac{a_2 \sin C_2}{\sin A_2} \approx \frac{186.2 \sin 22.7°}{\sin 43.1°} \approx 105.2 \qquad \text{Law of Sines}$$

Triangles $A_1B_1C_1$ and $A_2B_2C_2$ are shown in Figure 9.

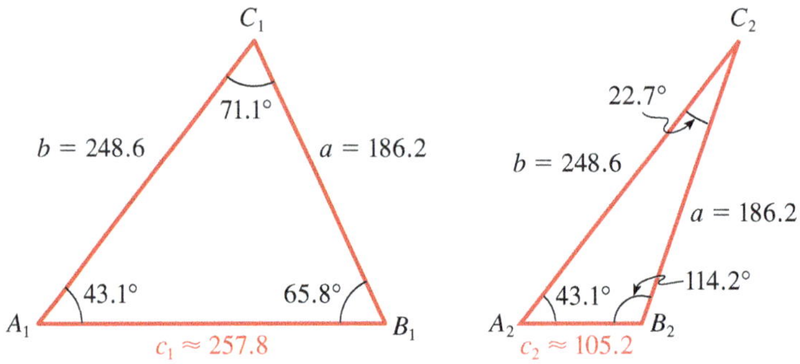

FIGURE 9

Now Try Exercise 23 ■

The next example presents a situation for which no triangle is compatible with the given data.

EXAMPLE 5 ■ SSA, the No-Solution Case

Solve triangle ABC, where $\angle A = 42°$, $a = 70$, and $b = 122$.

SOLUTION To organize the given information, we sketch the diagram in Figure 10. Let's try to find $\angle B$. We have

$$\frac{\sin A}{a} = \frac{\sin B}{b} \qquad \text{Law of Sines}$$

$$\sin B = \frac{b \sin A}{a} = \frac{122 \sin 42°}{70} \approx 1.17 \qquad \text{Solve for } \sin B$$

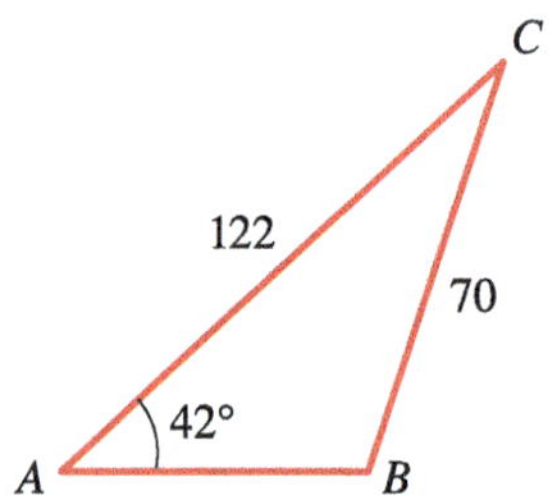

FIGURE 10

Since the sine of an angle is never greater than 1, we conclude that no triangle satisfies the conditions given in this problem.

Now Try Exercise 21 ■

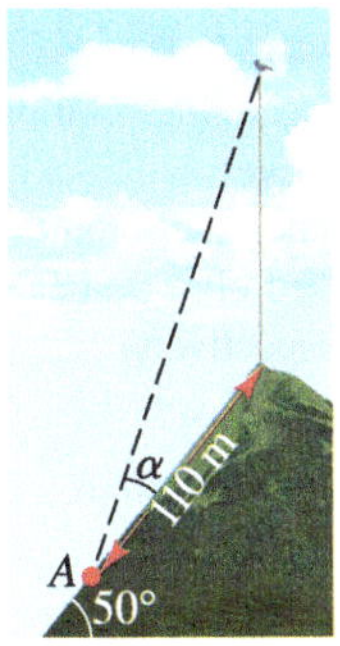

EXAMPLE 6 ■ Calculating a Distance

A bird is perched on top of a pole on a steep hill, and an observer is located at point A on the side of the hill, 110 m downhill from the base of the pole, as shown in the figure. The angle of inclination of the hill is 50°, and the angle α in the figure is 9°. Find the distance from the observer to the bird.

SOLUTION We first sketch a diagram as shown in Figure 11. We want to find the distance b in the figure. Triangle ADB is a right triangle, so $\angle DBA = 90° - 50° = 40°$. It follows that $\angle ABC = 180° - 40° = 140°$.

Now in triangle ABC we have $\angle A = 9°$ and $\angle B = 140°$, so $\angle C = 180° - 149° = 31°$. By the Law of Sines we have

$$\frac{\sin B}{b} = \frac{\sin C}{c} \qquad \text{Law of Sines}$$

Substituting $\angle B = 140°$, $\angle C = 31°$, and $c = 110$, we get

$$\frac{\sin 140°}{b} = \frac{\sin 31°}{110}$$

$$b = \frac{110 \sin 140°}{\sin 31°} \qquad \text{Solve for } b$$

$$\approx 137.3 \qquad \text{Calculator}$$

So the distance from the observer to the bird is about 137 m.

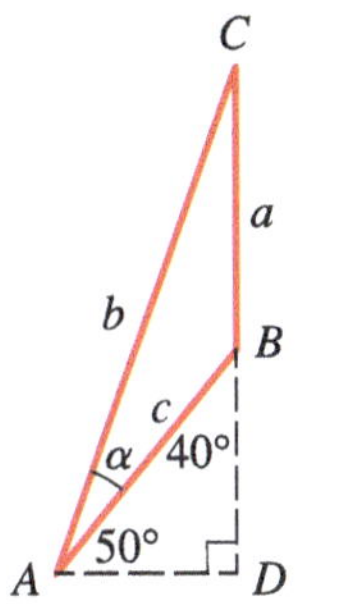

FIGURE 11

Now Try Exercise 37 ■

6.5 EXERCISES

CONCEPTS

1. In triangle ABC with sides a, b, and c the Law of Sines states that

$$\frac{\square}{\square} = \frac{\square}{\square} = \frac{\square}{\square}$$

2. The four cases in which we can solve a triangle are

ASA SSA SAS SSS

(a) In which of these cases can we use the Law of Sines to solve the triangle?

(b) Which of the cases listed can lead to more than one solution (the ambiguous case)?

SKILLS

3–8 ■ Finding an Angle or Side Use the Law of Sines to find the indicated side x or angle θ.

3. 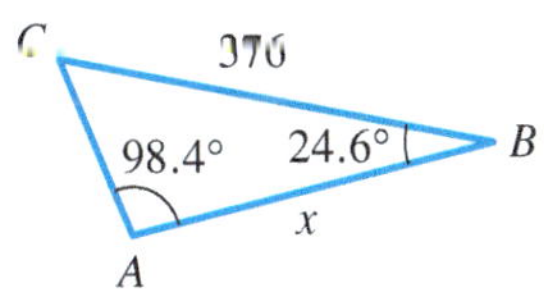

4. 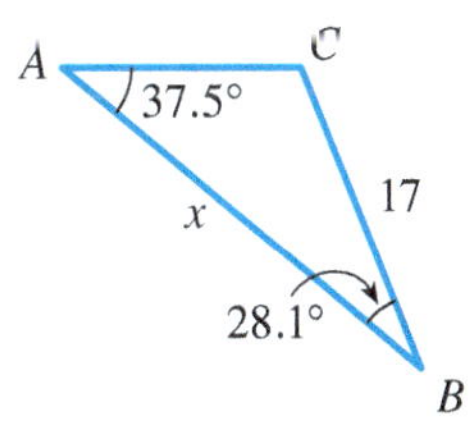

5. 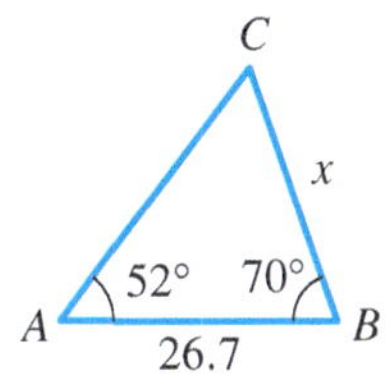

6. 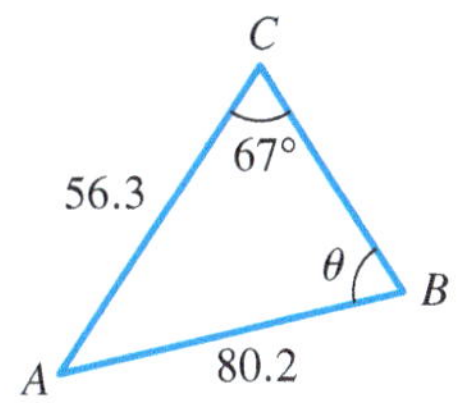

7. 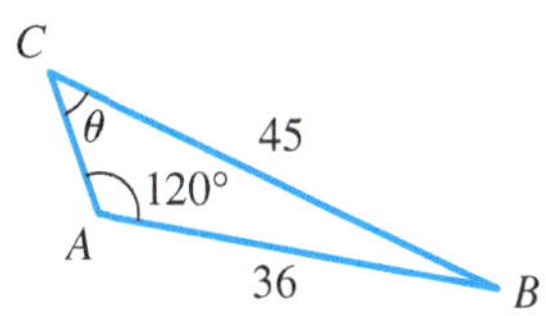

8. 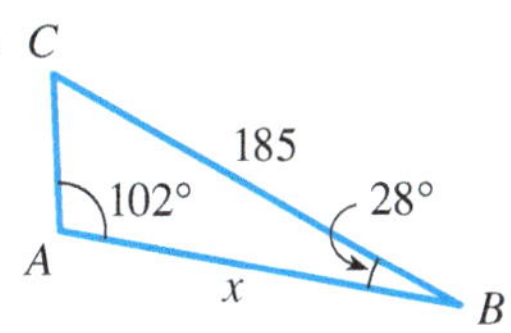

9–12 ■ Solving a Triangle Solve the triangle using the Law of Sines.

9. 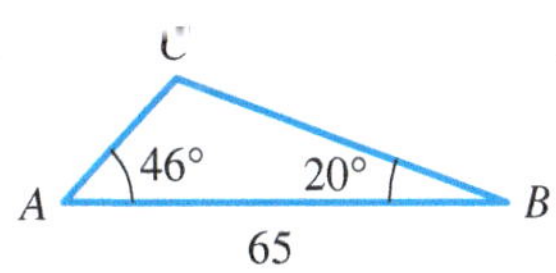

10. 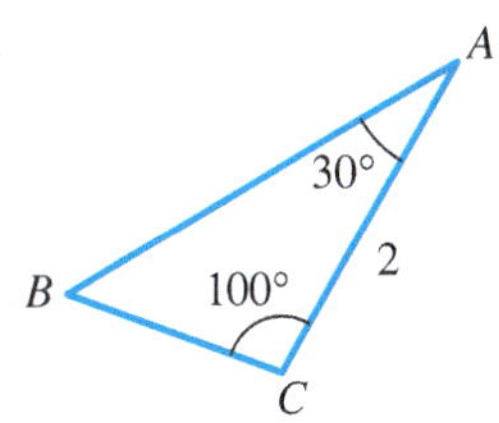

11.

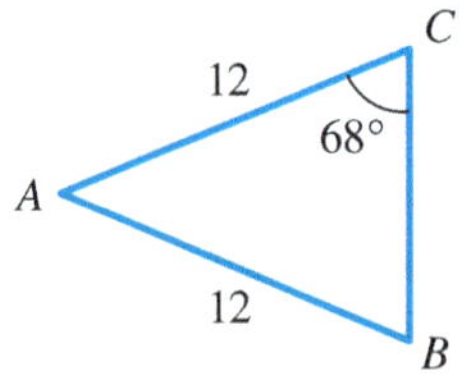

12.

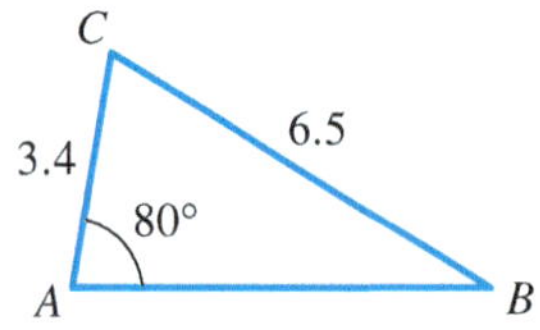

13–18 ■ Solving a Triangle Sketch each triangle, and then solve the triangle using the Law of Sines.

13. $\angle A = 50°$, $\angle B = 68°$, $c = 230$

14. $\angle A = 23°$, $\angle B = 110°$, $c = 50$

15. $\angle A = 30°$, $\angle C = 65°$, $b = 10$

16. $\angle A = 22°$, $\angle B = 95°$, $a = 420$

17. $\angle B = 29°$, $\angle C = 51°$, $b = 44$

18. $\angle B = 10°$, $\angle C = 100°$, $c = 115$

19–28 ■ Solving a Triangle Use the Law of Sines to solve for all possible triangles that satisfy the given conditions.

19. $a = 28$, $b = 15$, $\angle A = 110°$

20. $a = 30$, $c = 40$, $\angle A = 37°$

21. $a = 20$, $c = 45$, $\angle A = 125°$

22. $b = 45$, $c = 42$, $\angle C = 38°$

23. $b = 25$, $c = 30$, $\angle B = 25°$

24. $a = 75$, $b = 100$, $\angle A = 30°$

25. $a = 50$, $b = 100$, $\angle A = 50°$

26. $a = 100$, $b = 80$, $\angle A = 135°$

27. $a = 26$, $c = 15$, $\angle C = 29°$

28. $b = 73$, $c = 82$, $\angle B = 58°$

SKILLS Plus

29. Finding Angles For the triangle shown, find **(a)** $\angle BCD$ and **(b)** $\angle DCA$.

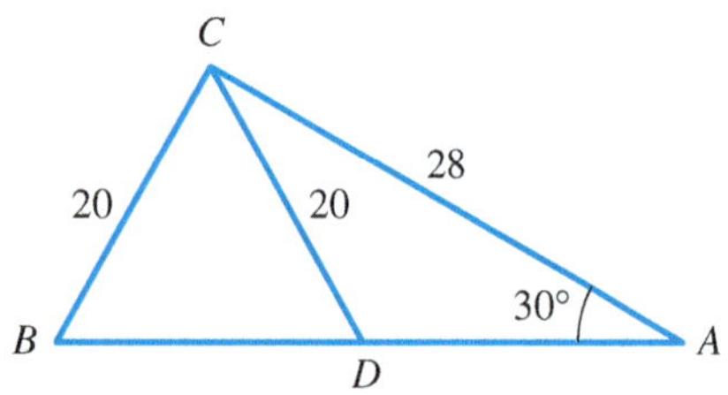

30. Finding a Side For the triangle shown, find the length AD.

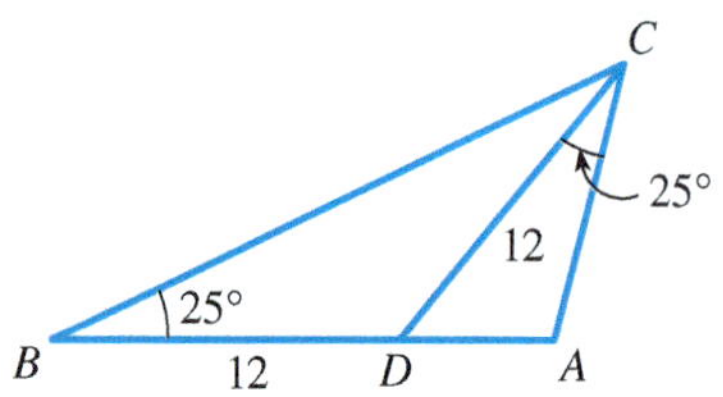

APPLICATIONS

31. Tracking a Satellite The path of a satellite orbiting the earth causes the satellite to pass directly over two tracking stations A and B, which are 50 mi apart. When the satellite is on one side of the two stations, the angles of elevation at A and B are measured to be 87.0° and 84.2°, respectively.

(a) How far is the satellite from station A?

(b) How high is the satellite above the ground?

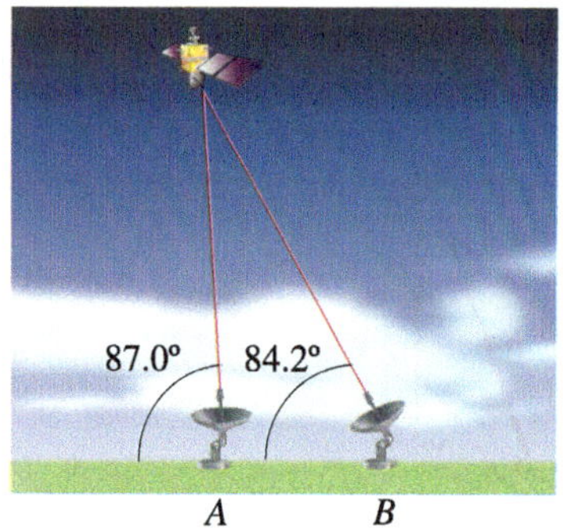

32. Flight of a Plane A pilot is flying over a straight highway. He determines the angles of depression to two mileposts, 5 mi apart, to be 32° and 48°, as shown in the figure.

(a) Find the distance of the plane from point A.

(b) Find the elevation of the plane.

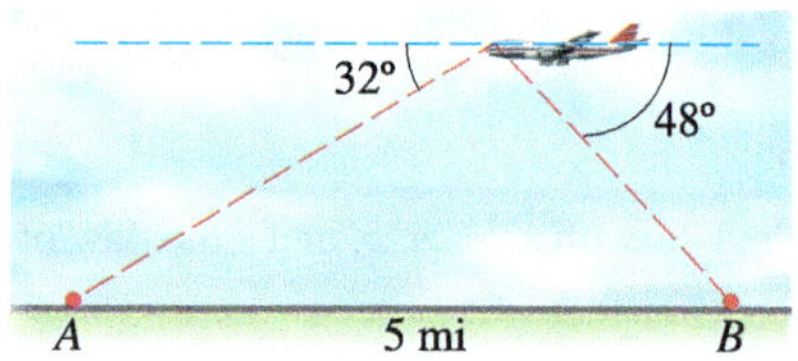

33. Distance Across a River To find the distance across a river, a surveyor chooses points A and B, which are 200 ft apart on one side of the river (see the figure). She then chooses a reference point C on the opposite side of the river and finds that $\angle BAC \approx 82°$ and $\angle ABC \approx 52°$. Approximate the distance from A to C.

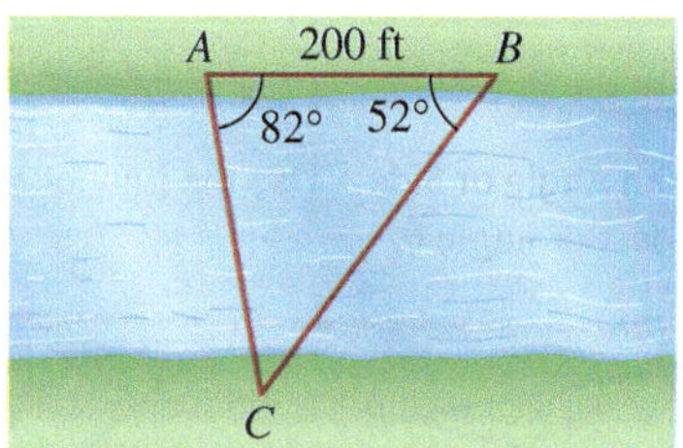

34. Distance Across a Lake Points A and B are separated by a lake. To find the distance between them, a surveyor locates a point C on land such that $\angle CAB = 48.6°$. He also measures CA as 312 ft and CB as 527 ft. Find the distance between A and B.

35. The Leaning Tower of Pisa The bell tower of the cathedral in Pisa, Italy, leans 5.6° from the vertical. A tourist stands 105 m from its base, with the tower leaning directly toward her. She measures the angle of elevation to the top of the tower to be 29.2°. Find the length of the tower to the nearest meter.

36. Radio Antenna A short-wave radio antenna is supported by two guy wires, 165 ft and 180 ft long. Each wire is attached to the top of the antenna and anchored to the ground at two anchor points on opposite sides of the antenna. The shorter wire makes an angle of 67° with the ground. How far apart are the anchor points?

37. Height of a Tree A tree on a hillside casts a shadow 215 ft down the hill. If the angle of inclination of the hillside is 22° to the horizontal and the angle of elevation of the sun is 52°, find the height of the tree.

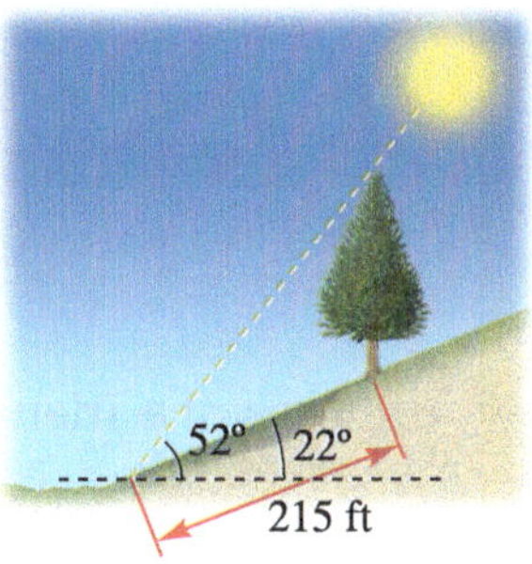

38. Length of a Guy Wire A communications tower is located at the top of a steep hill, as shown. The angle of inclination of the hill is 58°. A guy wire is to be attached to the top of the tower and to the ground, 100 m downhill from the base of the tower. The angle α in the figure is determined to be 12°. Find the length of cable required for the guy wire.

39. Calculating a Distance Observers at P and Q are located on the side of a hill that is inclined 32° to the horizontal, as shown. The observer at P determines the angle of elevation to a hot-air balloon to be 62°. At the same instant the observer at Q measures the angle of elevation to the balloon to be 71°. If P is 60 m down the hill from Q, find the distance from Q to the balloon.

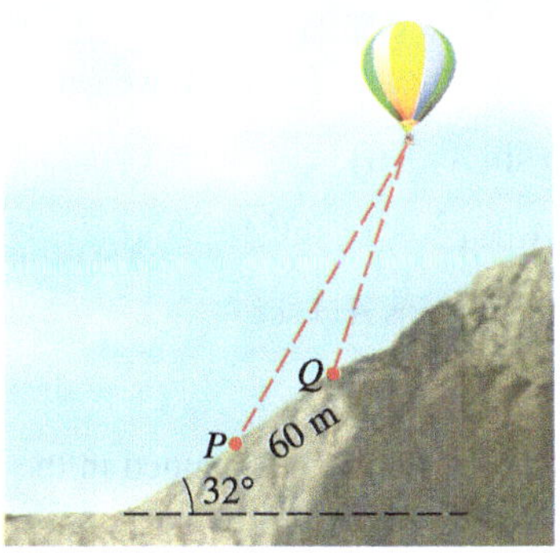

40. Calculating an Angle A water tower 30 m tall is located at the top of a hill. From a distance of 120 m down the hill it is observed that the angle formed between the top and base of the tower is 8°. Find the angle of inclination of the hill.

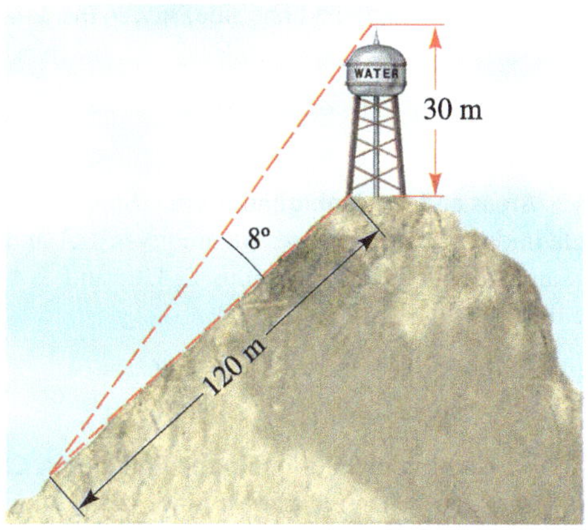

41. Distances to Venus The *elongation* α of a planet is the angle formed by the planet, earth, and sun (see the figure). It is known that the distance from the sun to Venus is 0.723 AU (see Exercise 71 in Section 6.2). At a certain time the elongation of Venus is found to be 39.4°. Find the possible distances from the earth to Venus at that time in astronomical units (AU).

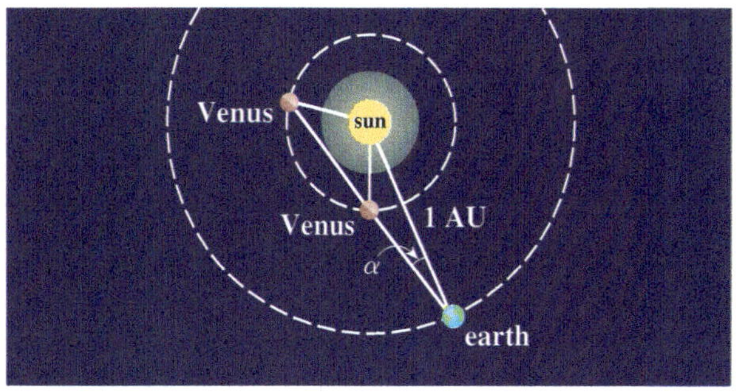

42. Soap Bubbles When two bubbles cling together in midair, their common surface is part of a sphere whose center D lies on the line passing through the centers of the bubbles (see the figure). Also, $\angle ACB$ and $\angle ACD$ each have measure 60°.

(a) Show that the radius r of the common face is given by

$$r = \frac{ab}{a - b}$$

[*Hint:* Use the Law of Sines together with the fact that an angle θ and its supplement $180° - \theta$ have the same sine.]

(b) Find the radius of the common face if the radii of the bubbles are 4 cm and 3 cm.

(c) What shape does the common face take if the two bubbles have equal radii?

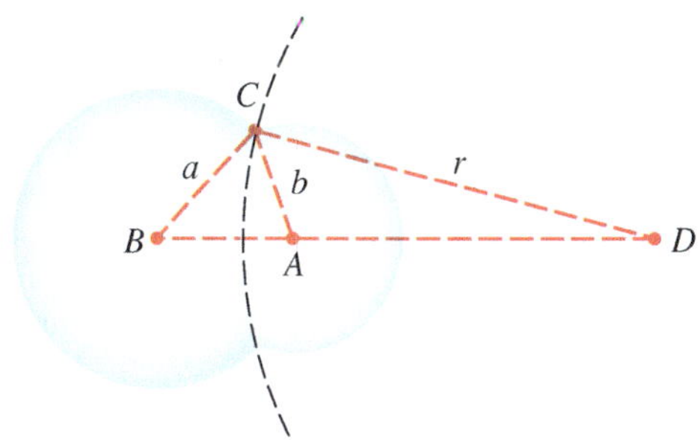

DISCUSS ■ DISCOVER ■ PROVE ■ WRITE

43. PROVE: Area of a Triangle Show that, given the three angles A, B, and C of a triangle and one side, say, a, the area of the triangle is

$$\text{area} = \frac{a^2 \sin B \sin C}{2 \sin A}$$

44. PROVE: Areas and the Ambiguous Case Suppose we solve a triangle in the ambiguous case. We are given $\angle A$ and sides a and b, and we find the two solutions $\triangle ABC$ and $\triangle A'B'C'$. Prove that

$$\frac{\text{area of } \triangle ABC}{\text{area of } \triangle A'B'C'} = \frac{\sin C}{\sin C'}$$

45. DISCOVER: Number of Solutions in the Ambiguous Case We have seen that when the Law of Sines is used to solve a triangle in the SSA case, there may be two, one, or no solution(s). Sketch triangles like those in Figure 6 to verify the criteria in the table for the number of solutions if you are given $\angle A$ and sides a and b.

Criterion	Number of solutions
$a \geq b$	1
$b > a > b \sin A$	2
$a = b \sin A$	1
$a < b \sin A$	0

If $\angle A = 30°$ and $b = 100$, use these criteria to find the range of values of a for which the triangle ABC has two solutions, one solution, or no solution.

6.6 THE LAW OF COSINES

■ The Law of Cosines ■ Navigation: Heading and Bearing ■ The Area of a Triangle

■ The Law of Cosines

The Law of Sines cannot be used directly to solve triangles if we know two sides and the angle between them or if we know all three sides (these are Cases 3 and 4 of the preceding section). In these two cases the **Law of Cosines** applies.

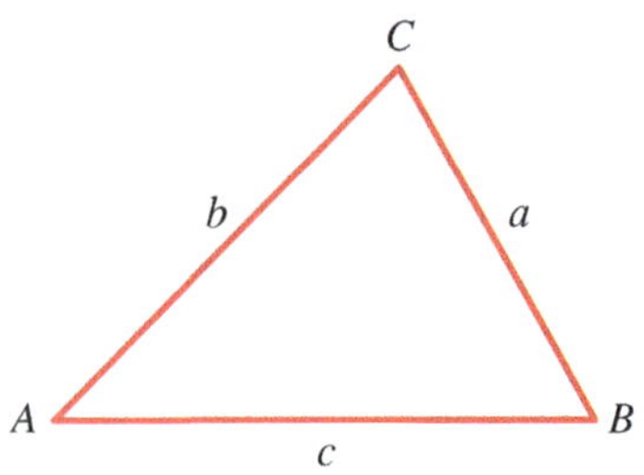

FIGURE 1

THE LAW OF COSINES

In any triangle ABC (see Figure 1) we have

$$a^2 = b^2 + c^2 - 2bc \cos A$$
$$b^2 = a^2 + c^2 - 2ac \cos B$$
$$c^2 = a^2 + b^2 - 2ab \cos C$$

Proof To prove the Law of Cosines, place triangle ABC so that $\angle A$ is at the origin, as shown in Figure 2. The coordinates of vertices B and C are $(c, 0)$ and $(b \cos A, b \sin A)$, respectively. (You should check that the coordinates of these points will be the same if we draw angle A as an acute angle.) Using the Distance Formula, we get

$$\begin{aligned} a^2 &= (b \cos A - c)^2 + (b \sin A - 0)^2 \\ &= b^2 \cos^2 A - 2bc \cos A + c^2 + b^2 \sin^2 A \\ &= b^2(\cos^2 A + \sin^2 A) - 2bc \cos A + c^2 \\ &= b^2 + c^2 - 2bc \cos A \end{aligned}$$

Because $\sin^2 A + \cos^2 A = 1$

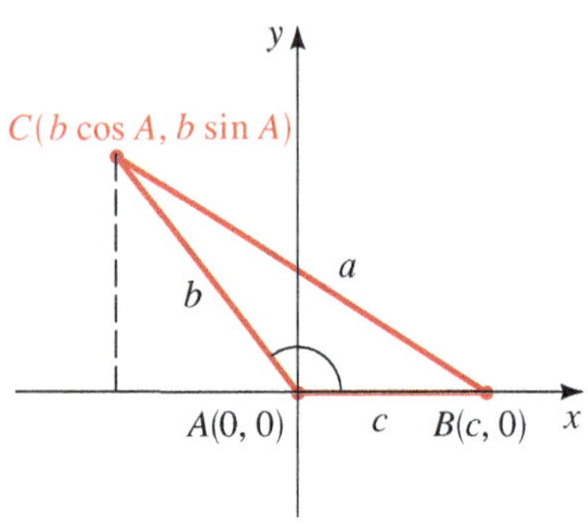

FIGURE 2

This proves the first formula. The other two formulas are obtained in the same way by placing each of the other vertices of the triangle at the origin and repeating the preceding argument. ■

In words, the Law of Cosines says that the square of any side of a triangle is equal to the sum of the squares of the other two sides minus twice the product of those two sides times the cosine of the included angle.

If one of the angles of a triangle, say, $\angle C$, is a right angle, then $\cos C = 0$, and the Law of Cosines reduces to the Pythagorean Theorem, $c^2 = a^2 + b^2$. Thus the Pythagorean Theorem is a special case of the Law of Cosines.

EXAMPLE 1 ■ Length of a Tunnel

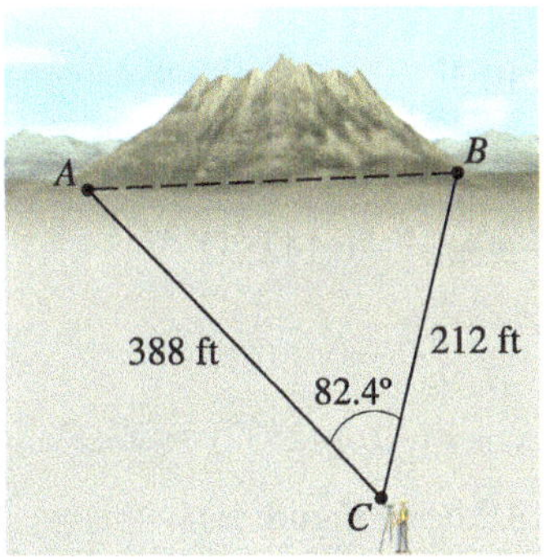

FIGURE 3

A tunnel is to be built through a mountain. To estimate the length of the tunnel, a surveyor makes the measurements shown in Figure 3. Use the surveyor's data to approximate the length of the tunnel.

SOLUTION To approximate the length c of the tunnel, we use the Law of Cosines.

$$\begin{aligned} c^2 &= a^2 + b^2 - 2ab\cos C && \text{Law of Cosines} \\ &= 212^2 + 388^2 - 2(212)(388)\cos 82.4^\circ && \text{Substitute} \\ &\approx 173730.2367 && \text{Use a calculator} \\ c &\approx \sqrt{173730.2367} \approx 416.8 && \text{Take square roots} \end{aligned}$$

Thus the tunnel will be approximately 417 ft long.

Now Try Exercises 3 and 39 ■

EXAMPLE 2 ■ SSS, the Law of Cosines

The sides of a triangle are $a = 5$, $b = 8$, and $c = 12$ (see Figure 4). Find the angles of the triangle.

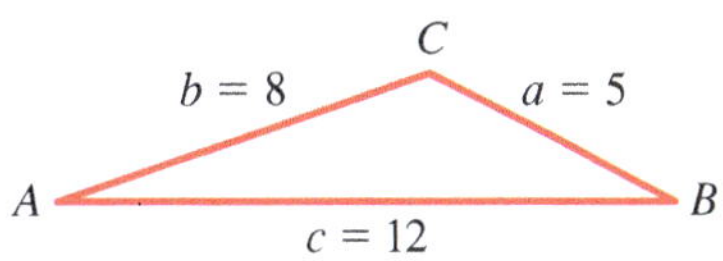

FIGURE 4

SOLUTION We first find $\angle A$. From the Law of Cosines, $a^2 = b^2 + c^2 - 2bc\cos A$. Solving for $\cos A$, we get

$$\cos A = \frac{b^2 + c^2 - a^2}{2bc} = \frac{8^2 + 12^2 - 5^2}{2(8)(12)} = \frac{183}{192} = 0.953125$$

COS⁻¹ or INV COS or ARC COS

Using a calculator, we find that $\angle A = \cos^{-1}(0.953125) \approx 18^\circ$. In the same way we get

$$\cos B = \frac{a^2 + c^2 - b^2}{2ac} = \frac{5^2 + 12^2 - 8^2}{2(5)(12)} = 0.875$$

$$\cos C = \frac{a^2 + b^2 - c^2}{2ab} = \frac{5^2 + 8^2 - 12^2}{2(5)(8)} = -0.6875$$

Using a calculator, we find that

$$\angle B = \cos^{-1}(0.875) \approx 29^\circ \quad \text{and} \quad \angle C = \cos^{-1}(-0.6875) \approx 133^\circ$$

Of course, once two angles have been calculated, the third can more easily be found from the fact that the sum of the angles of a triangle is 180°. However, it's a good idea to calculate all three angles using the Law of Cosines and add the three angles as a check on your computations.

Now Try Exercise 7 ■

EXAMPLE 3 ■ SAS, the Law of Cosines

Solve triangle ABC, where $\angle A = 46.5°$, $b = 10.5$, and $c = 18.0$.

SOLUTION We can find a using the Law of Cosines.

$$a^2 = b^2 + c^2 - 2bc \cos A$$

$$= (10.5)^2 + (18.0)^2 - 2(10.5)(18.0)(\cos 46.5°) \approx 174.05$$

Thus $a \approx \sqrt{174.05} \approx 13.2$. We also use the Law of Cosines to find $\angle B$ and $\angle C$, as in Example 2.

$$\cos B = \frac{a^2 + c^2 - b^2}{2ac} = \frac{13.2^2 + 18.0^2 - 10.5^2}{2(13.2)(18.0)} \approx 0.816477$$

$$\cos C = \frac{a^2 + b^2 - c^2}{2ab} = \frac{13.2^2 + 10.5^2 - 18.0^2}{2(13.2)(10.5)} \approx -0.142532$$

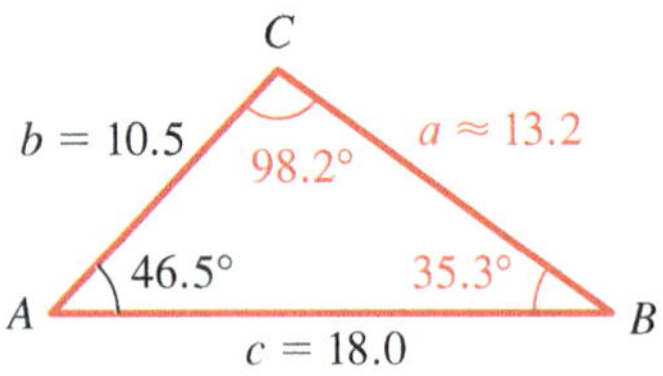

FIGURE 5

Using a calculator, we find that

$$\angle B = \cos^{-1}(0.816477) \approx 35.3° \quad \text{and} \quad \angle C = \cos^{-1}(-0.142532) \approx 98.2°$$

To summarize: $\angle B \approx 35.3°$, $\angle C \approx 98.2°$, and $a \approx 13.2$. (See Figure 5.)

Now Try Exercise 13 ■

We could have used the Law of Sines to find $\angle B$ and $\angle C$ in Example 3, since we knew all three sides and an angle in the triangle. But knowing the sine of an angle does not uniquely specify the angle, since an angle θ and its supplement $180° - \theta$ both have the same sine. Thus we would need to decide which of the two angles is the correct choice. This ambiguity does not arise when we use the Law of Cosines, because every angle between 0° and 180° has a unique cosine. So using only the Law of Cosines is preferable in problems like Example 3.

■ Navigation: Heading and Bearing

In navigation a direction is often given as a **bearing**, that is, as an acute angle measured from due north or due south. The bearing N 30° E, for example, indicates a direction that points 30° to the east of due north (see Figure 6).

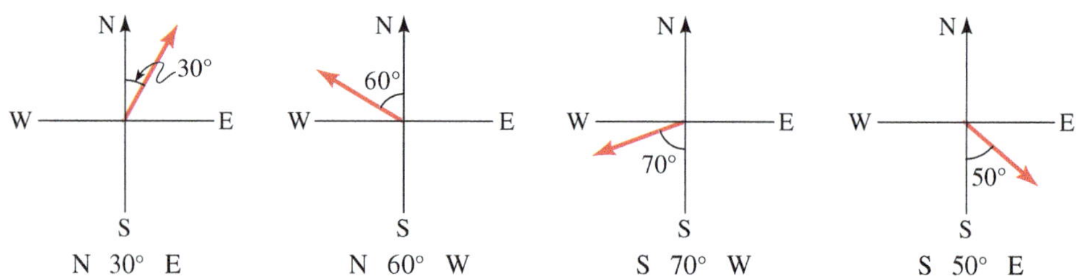

FIGURE 6

EXAMPLE 4 ■ Navigation

A pilot sets out from an airport and heads in the direction N 20° E, flying at 200 mi/h. After 1 h, he makes a course correction and heads in the direction N 40° E. Half an hour after that, engine trouble forces him to make an emergency landing.

(a) Find the distance between the airport and his final landing point.

(b) Find the bearing from the airport to his final landing point.

SOLUTION

(a) In 1 h the plane travels 200 mi, and in half an hour it travels 100 mi, so we can plot the pilot's course as in Figure 7. When he makes his course

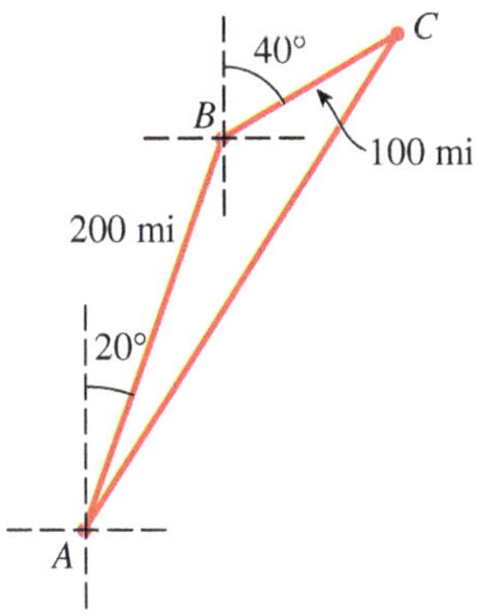

FIGURE 7

correction, he turns 20° to the right, so the angle between the two legs of his trip is $180° - 20° = 160°$. So by the Law of Cosines we have

$$b^2 = 200^2 + 100^2 - 2 \cdot 200 \cdot 100 \cos 160°$$
$$\approx 87{,}587.70$$

Thus $b \approx 295.95$. The pilot lands about 296 mi from his starting point.

(b) We first use the Law of Sines to find $\angle A$.

$$\frac{\sin A}{100} = \frac{\sin 160°}{295.95}$$
$$\sin A = 100 \cdot \frac{\sin 160°}{295.95}$$
$$\approx 0.11557$$

Another angle with sine 0.11557 is $180° - 6.636° = 173.364°$. But this is clearly too large to be $\angle A$ in $\angle ABC$.

Using the $\boxed{\text{SIN}^{-1}}$ key on a calculator, we find that $\angle A \approx 6.636°$. From Figure 7 we see that the line from the airport to the final landing site points in the direction $20° + 6.636° = 26.636°$ east of due north. Thus the bearing is about N 26.6° E.

Now Try Exercise 45 ■

The Area of a Triangle

An interesting application of the Law of Cosines involves a formula for finding the area of a triangle from the lengths of its three sides (see Figure 8).

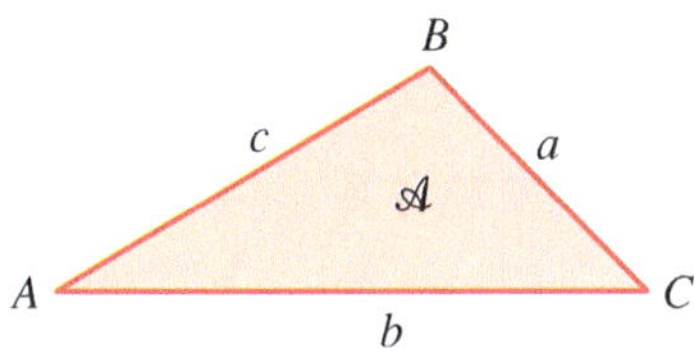

FIGURE 8

HERON'S FORMULA

The area $\mathcal{A}$ of triangle ABC is given by

$$\mathcal{A} = \sqrt{s(s-a)(s-b)(s-c)}$$

where $s = \frac{1}{2}(a + b + c)$ is the **semiperimeter** of the triangle; that is, s is half the perimeter.

Proof We start with the formula $\mathcal{A} = \frac{1}{2}ab \sin C$ from Section 6.3. Thus

$$\mathcal{A}^2 = \tfrac{1}{4}a^2b^2 \sin^2 C$$
$$= \tfrac{1}{4}a^2b^2(1 - \cos^2 C) \qquad \text{Pythagorean identity}$$
$$= \tfrac{1}{4}a^2b^2(1 - \cos C)(1 + \cos C) \qquad \text{Factor}$$

Next, we write the expressions $1 - \cos C$ and $1 + \cos C$ in terms of a, b, and c. By the Law of Cosines we have

$$\cos C = \frac{a^2 + b^2 - c^2}{2ab} \qquad \text{Law of Cosines}$$
$$1 + \cos C = 1 + \frac{a^2 + b^2 - c^2}{2ab} \qquad \text{Add 1}$$
$$= \frac{2ab + a^2 + b^2 - c^2}{2ab} \qquad \text{Common denominator}$$
$$= \frac{(a + b)^2 - c^2}{2ab} \qquad \text{Factor}$$
$$= \frac{(a + b + c)(a + b - c)}{2ab} \qquad \text{Difference of squares}$$

Similarly,

$$1 - \cos C = \frac{(c + a - b)(c - a + b)}{2ab}$$

Substituting these expressions in the formula we obtained for $\mathcal{A}^2$ gives

To see that the factors in the last two products are equal, note for example that

$$\frac{a+b-c}{2} = \frac{a+b+c}{2} - c$$
$$= s - c$$

$$\begin{aligned}\mathcal{A}^2 &= \tfrac{1}{4}a^2b^2\frac{(a+b+c)(a+b-c)}{2ab}\,\frac{(c+a-b)(c-a+b)}{2ab}\\ &= \frac{(a+b+c)}{2}\,\frac{(a+b-c)}{2}\,\frac{(c+a-b)}{2}\,\frac{(c-a+b)}{2}\\ &= s(s-c)(s-b)(s-a)\end{aligned}$$

Heron's Formula now follows from taking the square root of each side. ■

EXAMPLE 5 ■ Area of a Lot

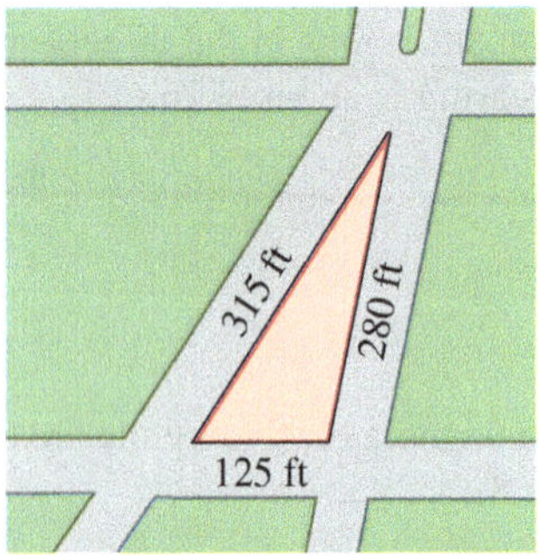

FIGURE 9

A businessman wishes to buy a triangular lot in a busy downtown location (see Figure 9). The lot frontages on the three adjacent streets are 125, 280, and 315 ft. Find the area of the lot.

SOLUTION The semiperimeter of the lot is

$$s = \frac{125 + 280 + 315}{2} = 360$$

By Heron's Formula the area is

$$\mathcal{A} = \sqrt{360(360 - 125)(360 - 280)(360 - 315)} \approx 17{,}451.6$$

Thus the area is approximately 17,452 ft^2.

Now Try Exercises 29 and 53 ■

6.6 EXERCISES

CONCEPTS

1. For triangle ABC with sides a, b, and c the Law of Cosines states

$c^2 =$ ______________________

2. In which of the following cases must the Law of Cosines be used to solve a triangle?

ASA SSS SAS SSA

SKILLS

3–10 ■ Finding an Angle or Side Use the Law of Cosines to determine the indicated side x or angle θ.

3.

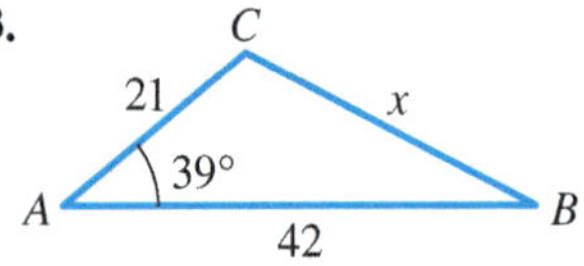

4.

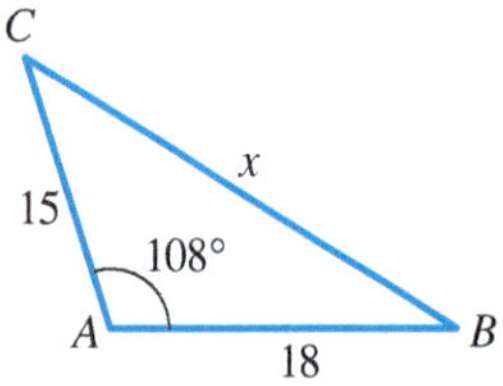

5.

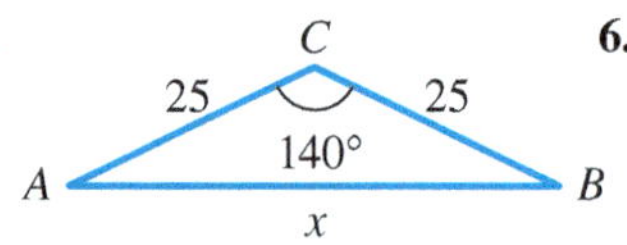

6.

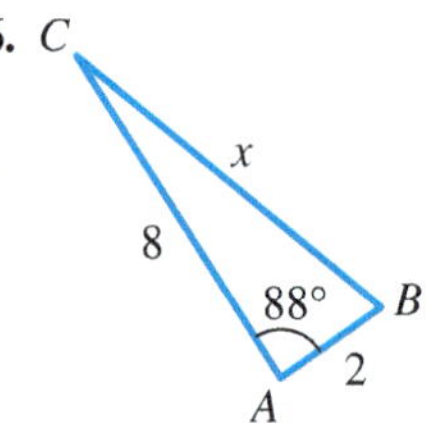

7.

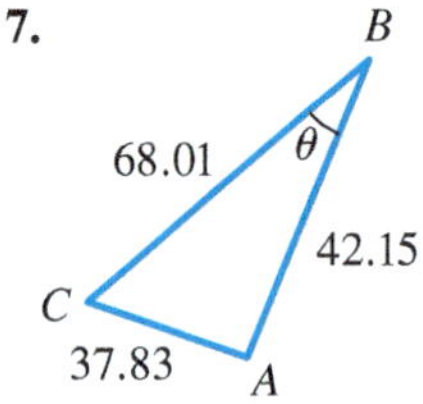

8.

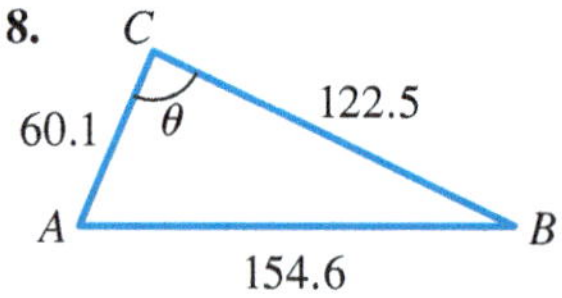

9.

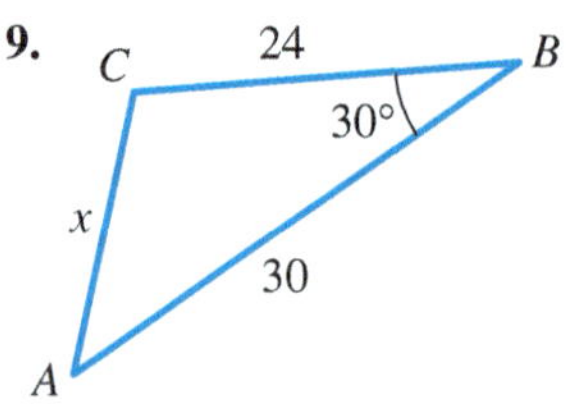

10.

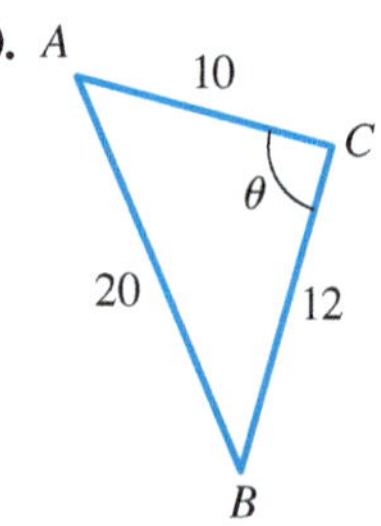

11–20 ■ Solving a Triangle Solve triangle ABC.

11.

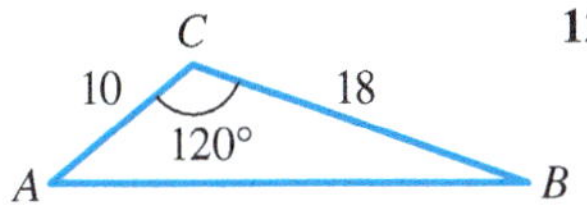

12.

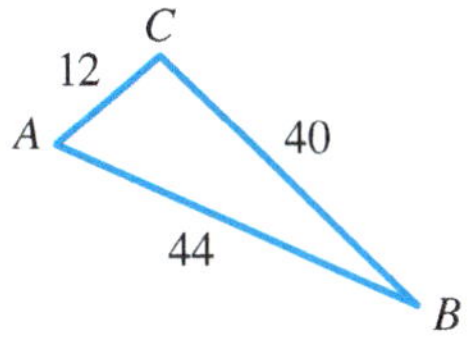

13. $a = 3.0, \quad b = 4.0, \quad \angle C = 53°$

14. $b = 60, \quad c = 30, \quad \angle A = 70°$

15. $a = 20, \quad b = 25, \quad c = 22$

16. $a = 10, \quad b = 12, \quad c = 16$

17. $b = 125, \quad c = 162, \quad \angle B = 40°$

18. $a = 65, \quad c = 50, \quad \angle C = 52°$

19. $a = 50, \quad b = 65, \quad \angle A = 55°$

20. $a = 73.5, \quad \angle B = 61°, \quad \angle C = 83°$

21–28 ■ Law of Sines or Law of Cosines? Find the indicated side x or angle θ. (Use either the Law of Sines or the Law of Cosines, as appropriate.)

21.

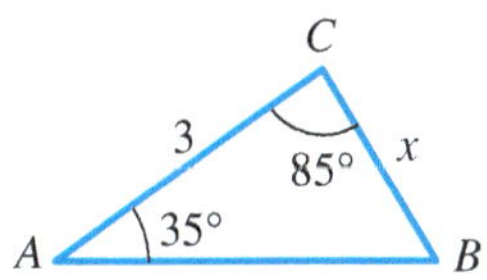

22.

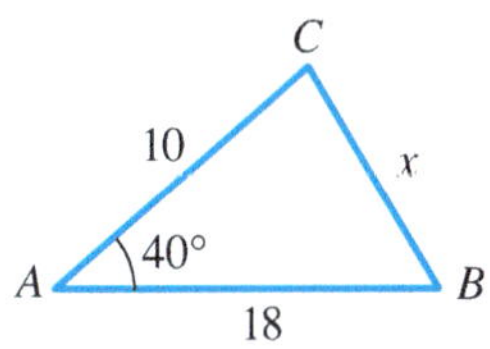

23.

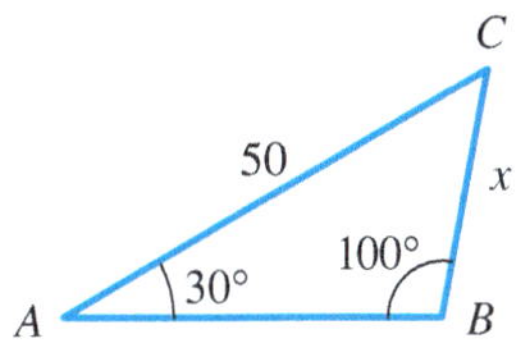

24.

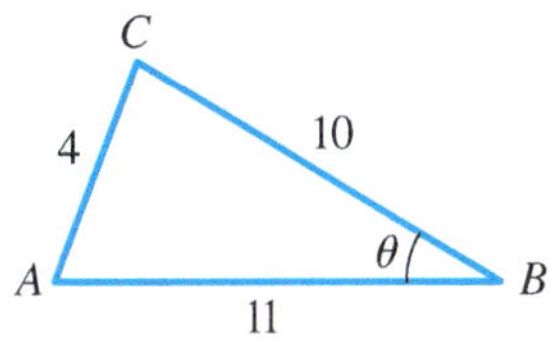

25.

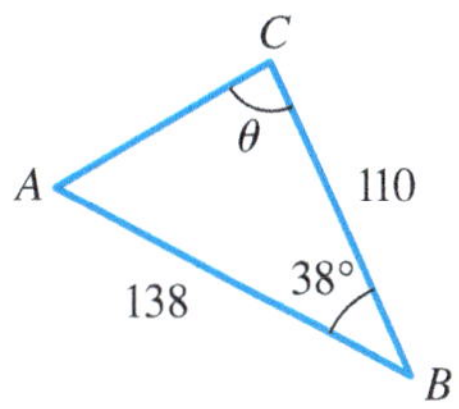

26.

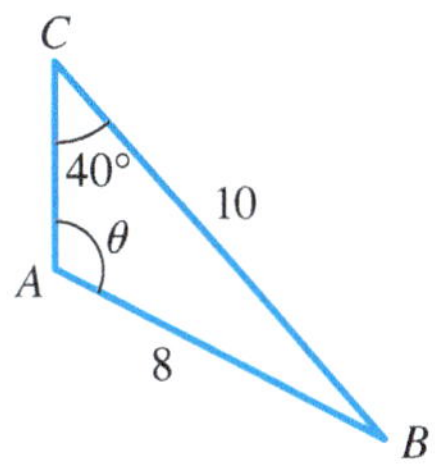

27.

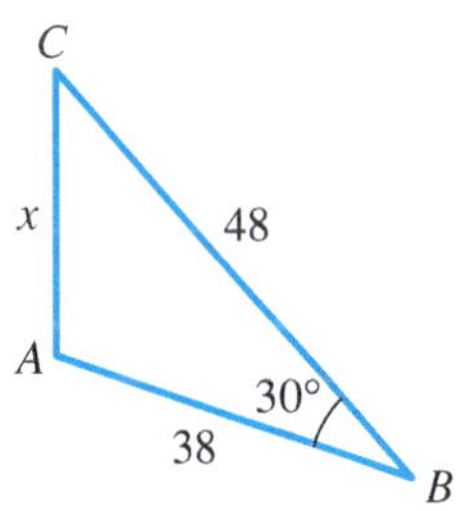

28.

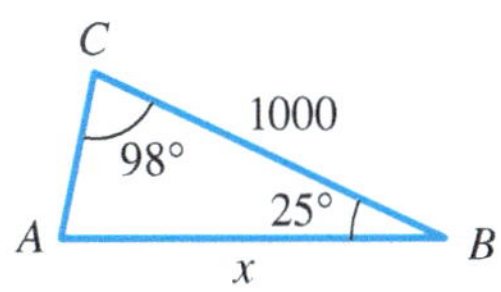

29–32 ■ Heron's Formula Find the area of the triangle whose sides have the given lengths.

29. $a = 9, \quad b = 12, \quad c = 15$

30. $a = 1, \quad b = 2, \quad c = 2$

31. $a = 7, \quad b = 8, \quad c = 9$

32. $a = 11, \quad b = 100, \quad c = 101$

SKILLS Plus

33–36 ■ Heron's Formula Find the area of the shaded figure, rounded to two decimals.

33.

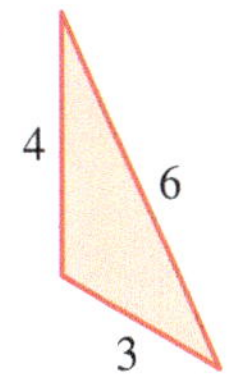

34.

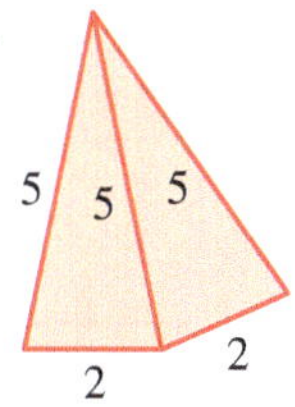

35.

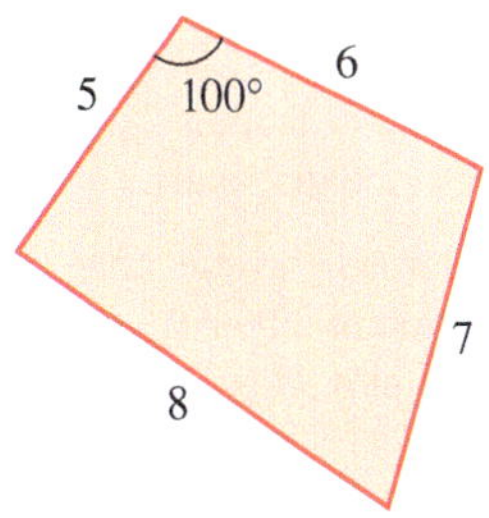

36.

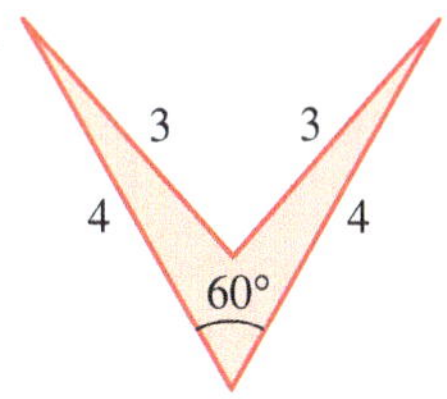

37. Area of a Region Three circles of radii 4, 5, and 6 cm are mutually tangent. Find the shaded area enclosed between the circles.

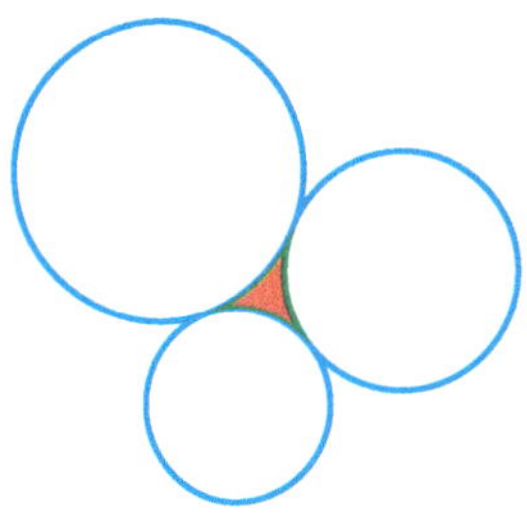

38. Finding a Length In the figure, triangle ABC is a right triangle, $CQ = 6$, and $BQ = 4$. Also, $\angle AQC = 30°$ and $\angle CQB = 45°$. Find the length of AQ. [*Hint:* First use the Law of Cosines to find expressions for a^2, b^2, and c^2.]

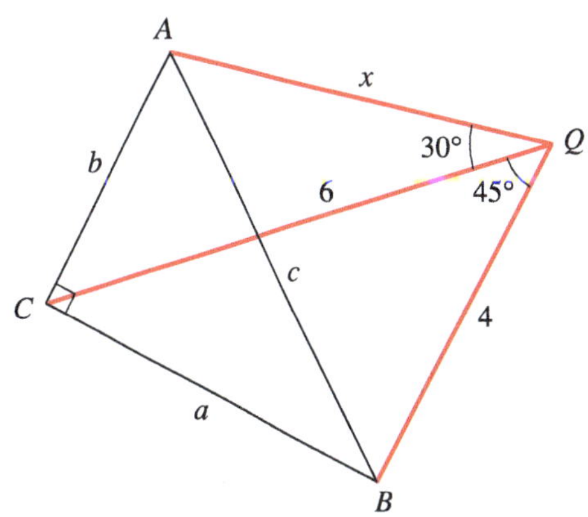

APPLICATIONS

39. Surveying To find the distance across a small lake, a surveyor has taken the measurements shown. Find the distance across the lake using this information.

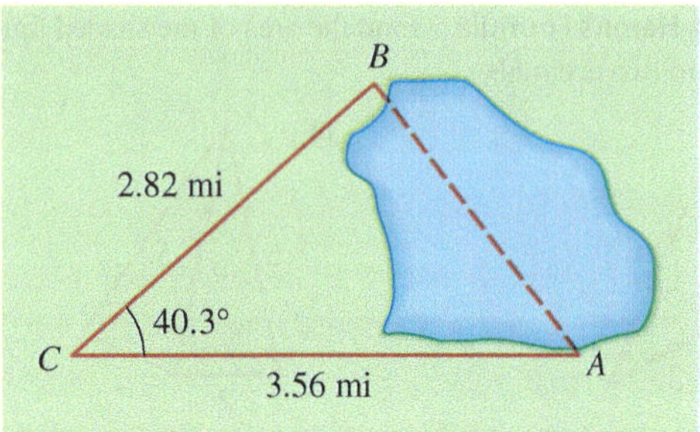

40. Geometry A parallelogram has sides of lengths 3 and 5, and one angle is 50°. Find the lengths of the diagonals.

41. Calculating Distance Two straight roads diverge at an angle of 65°. Two cars leave the intersection at 2:00 P.M., one traveling at 50 mi/h and the other at 30 mi/h. How far apart are the cars at 2:30 P.M.?

42. Calculating Distance A car travels along a straight road, heading east for 1 h, then traveling for 30 min on another road that leads northeast. If the car has maintained a constant speed of 40 mi/h, how far is it from its starting position?

43. Dead Reckoning A pilot flies in a straight path for 1 h 30 min. She then makes a course correction, heading 10° to the right of her original course, and flies 2 h in the new direction. If she maintains a constant speed of 625 mi/h, how far is she from her starting position?

44. Navigation Two boats leave the same port at the same time. One travels at a speed of 30 mi/h in the direction N 50° E, and the other travels at a speed of 26 mi/h in a direction S 70° E (see the figure). How far apart are the two boats after 1 h?

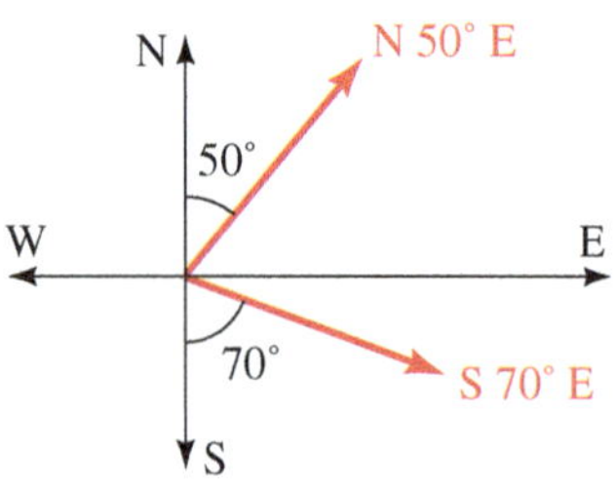

45. Navigation A fisherman leaves his home port and heads in the direction N 70° W. He travels 30 mi and reaches Egg Island. The next day he sails N 10° E for 50 mi, reaching Forrest Island.

(a) Find the distance between the fisherman's home port and Forrest Island.

(b) Find the bearing from Forrest Island back to his home port.

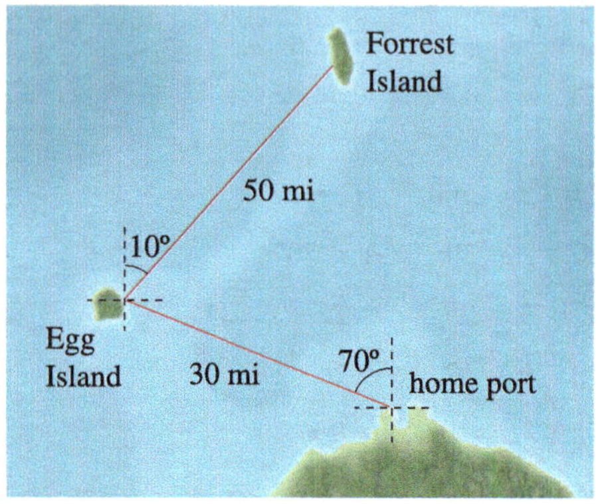

46. Navigation Airport B is 300 mi from airport A at a bearing N 50° E (see the figure). A pilot wishing to fly from A to B mistakenly flies due east at 200 mi/h for 30 min, when he notices his error.

(a) How far is the pilot from his destination at the time he notices the error?

(b) What bearing should he head his plane to arrive at airport B?

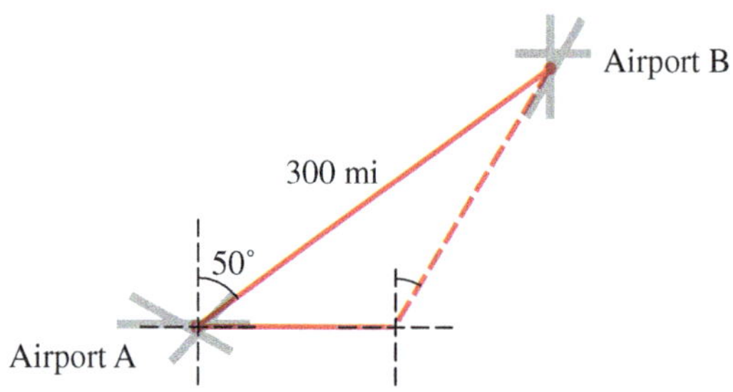

47. Triangular Field A triangular field has sides of lengths 22, 36, and 44 yd. Find the largest angle.

48. Towing a Barge Two tugboats that are 120 ft apart pull a barge, as shown. If the length of one cable is 212 ft and the length of the other is 230 ft, find the angle formed by the two cables.

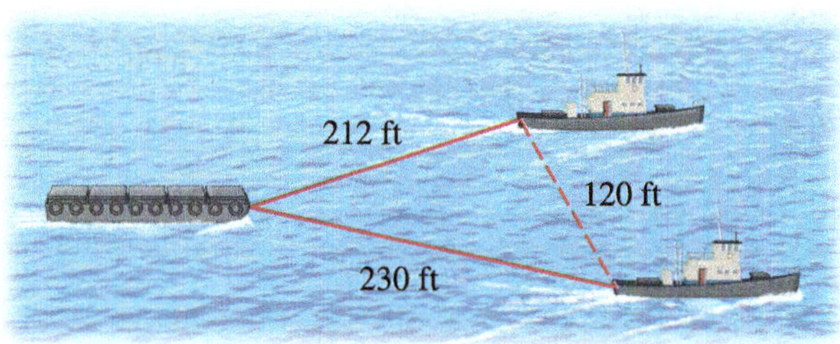

49. Flying Kites A boy is flying two kites at the same time. He has 380 ft of line out to one kite and 420 ft to the other. He estimates the angle between the two lines to be 30°. Approximate the distance between the kites.

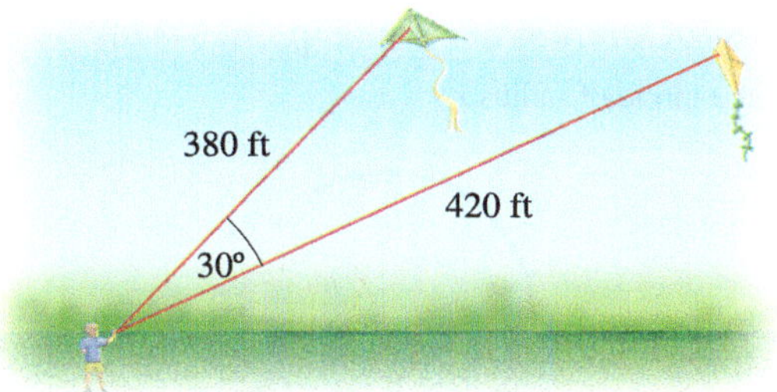

50. Securing a Tower A 125-ft tower is located on the side of a mountain that is inclined 32° to the horizontal. A guy wire is to be attached to the top of the tower and anchored at a point 55 ft downhill from the base of the tower. Find the shortest length of wire needed.

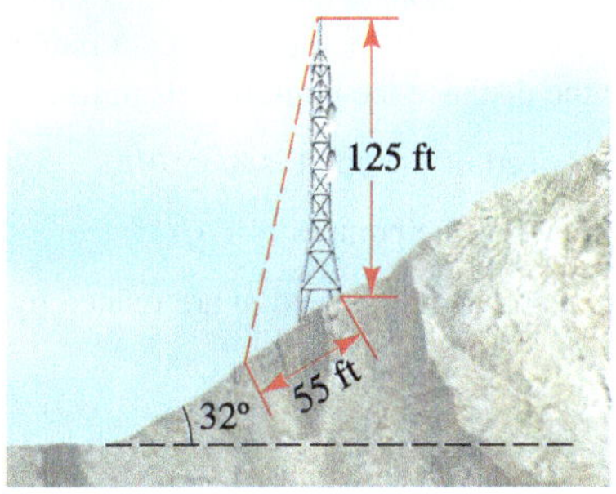

51. Cable Car A steep mountain is inclined 74° to the horizontal and rises 3400 ft above the surrounding plain. A cable car is to be installed from a point 800 ft from the base to the top of the mountain, as shown. Find the shortest length of cable needed.

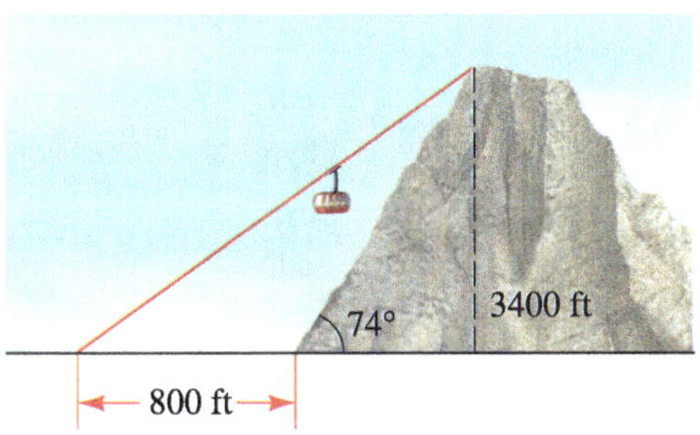

52. CN Tower The CN Tower in Toronto, Canada, is the tallest free-standing structure in North America. A woman on the observation deck, 1150 ft above the ground, wants to determine the distance between two landmarks on the ground below. She observes that the angle formed by the lines of sight to these two landmarks is 43°. She also observes that the angle between the vertical and the line of sight to one of the landmarks is 62° and that to the other landmark is 54°. Find the distance between the two landmarks.

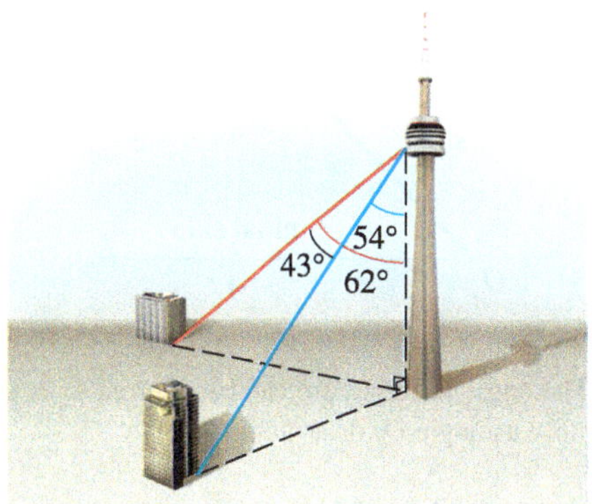

53. Land Value Land in downtown Columbia is valued at $20 a square foot. What is the value of a triangular lot with sides of lengths 112, 148, and 190 ft?

DISCUSS ■ DISCOVER ■ PROVE ■ WRITE

54. DISCUSS: Solving for the Angles in a Triangle The paragraph that follows the solution of Example 3 on page 518 explains an alternative method for finding $\angle B$ and $\angle C$, using the Law of Sines. Use this method to solve the triangle in the example, finding $\angle B$ first and then $\angle C$. Explain how you chose the appropriate value for the measure of $\angle B$. Which method do you prefer for solving an SAS triangle problem: the one explained in Example 3 or the one you used in this exercise?

55. PROVE: Projection Laws Prove that in triangle ABC

$$a = b\cos C + c\cos B$$
$$b = c\cos A + a\cos C$$
$$c = a\cos B + b\cos A$$

These are called the *Projection Laws*. [*Hint:* To get the first equation, add the second and third equations in the Law of Cosines and solve for a.]

CHAPTER 6 ■ REVIEW

PROPERTIES AND FORMULAS

Angles (p. 472)

An **angle** consists of two rays with a common vertex. One of the rays is the **initial side**, and the other the **terminal side**. An angle can be viewed as a rotation of the initial side onto the terminal side. If the rotation is counterclockwise, the angle is **positive**; if the rotation is clockwise, the angle is **negative**.

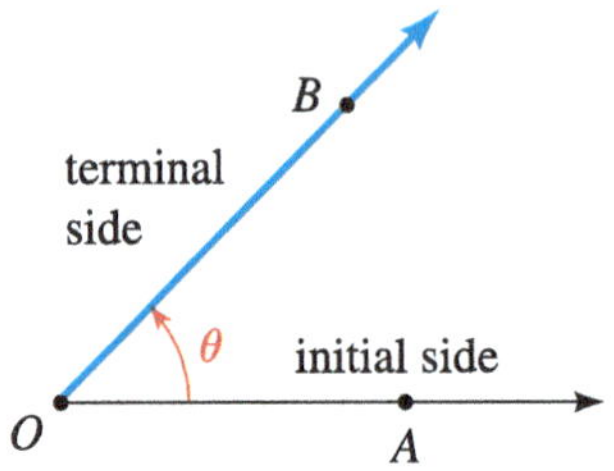

Notation: The angle in the figure can be referred to as angle AOB, or simply as angle O, or as angle θ.

Angle Measure (p. 472)

The **radian measure** of an angle (abbreviated **rad**) is the length of the arc that the angle subtends in a circle of radius 1, as shown in the figure.

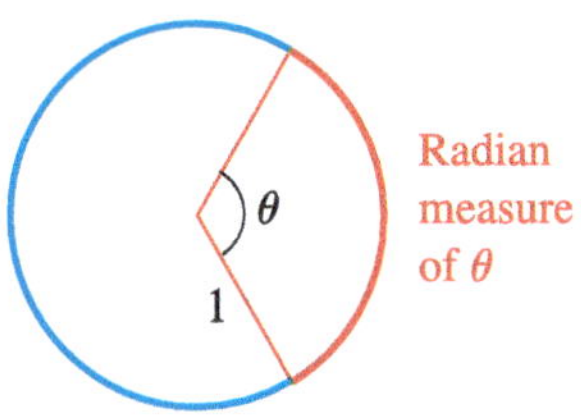

The **degree measure** of an angle is the number of degrees in the angle, where a degree is $\frac{1}{360}$ of a complete circle.

To convert degrees to radians, multiply by $\pi/180$.

To convert radians to degrees, multiply by $180/\pi$.

Angles in Standard Position (pp. 473, 494)

An angle is in **standard position** if it is drawn in the xy-plane with its vertex at the origin and its initial side on the positive x-axis.

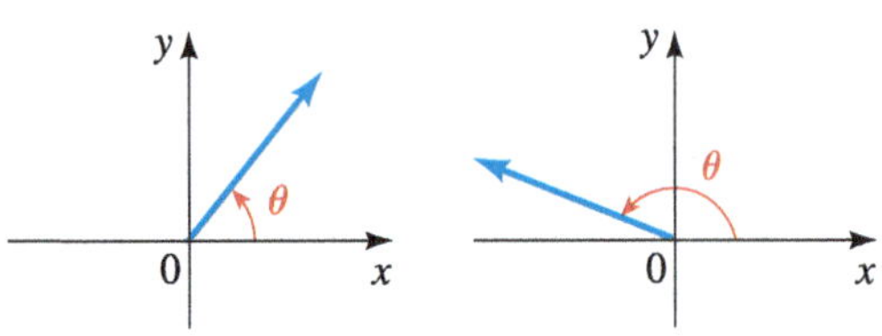

Two angles in standard position are **coterminal** if their sides coincide.

The **reference angle** $\overline{\theta}$ associated with an angle θ is the acute angle formed by the terminal side of θ and the x-axis.

Length of an Arc; Area of a Sector (pp. 475–476)

Consider a circle of radius r.

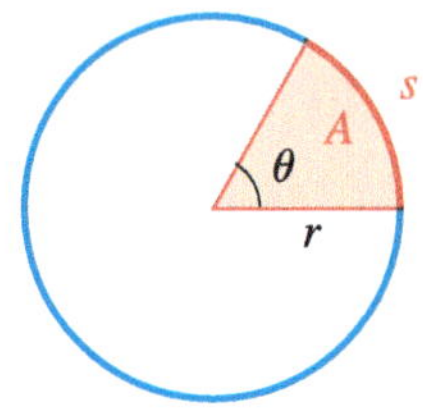

The **length s of an arc** that subtends a central angle of θ radians is $s = r\theta$.

The **area A of a sector** with central angle of θ radians is $A = \frac{1}{2}r^2\theta$.

Circular Motion (pp. 476–477)

Suppose a point moves along a circle of radius r and the ray from the center of the circle to the point traverses θ radians in time t. Let $s = r\theta$ be the distance the point travels in time t.

The **angular speed** of the point is $\omega = \theta/t$.

The **linear speed** of the point is $v = s/t$.

Linear speed v and angular speed ω are related by the formula $v = r\omega$.

Trigonometric Ratios (p. 482)

For a right triangle with an acute angle θ the trigonometric ratios are defined as follows.

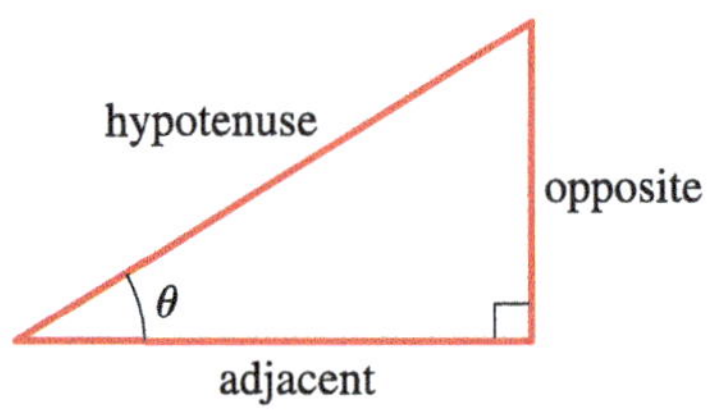

$$\sin\theta = \frac{\text{opp}}{\text{hyp}} \qquad \cos\theta = \frac{\text{adj}}{\text{hyp}} \qquad \tan\theta = \frac{\text{opp}}{\text{adj}}$$

$$\csc\theta = \frac{\text{hyp}}{\text{opp}} \qquad \sec\theta = \frac{\text{hyp}}{\text{adj}} \qquad \cot\theta = \frac{\text{adj}}{\text{opp}}$$

Special Trigonometric Ratios (p. 483)

The trigonometric functions have the following values at the special values of θ.

θ	θ	$\sin\theta$	$\cos\theta$	$\tan\theta$	$\csc\theta$	$\sec\theta$	$\cot\theta$
30°	$\frac{\pi}{6}$	$\frac{1}{2}$	$\frac{\sqrt{3}}{2}$	$\frac{\sqrt{3}}{3}$	2	$\frac{2\sqrt{3}}{3}$	$\sqrt{3}$
45°	$\frac{\pi}{4}$	$\frac{\sqrt{2}}{2}$	$\frac{\sqrt{2}}{2}$	1	$\sqrt{2}$	$\sqrt{2}$	1
60°	$\frac{\pi}{3}$	$\frac{\sqrt{3}}{2}$	$\frac{1}{2}$	$\sqrt{3}$	$\frac{2\sqrt{3}}{3}$	2	$\frac{\sqrt{3}}{3}$

Trigonometric Functions of Angles (p. 491)

Let θ be an angle in standard position, and let $P(x, y)$ be a point on the terminal side. Let $r = \sqrt{x^2 + y^2}$ be the distance from the origin to the point $P(x, y)$.

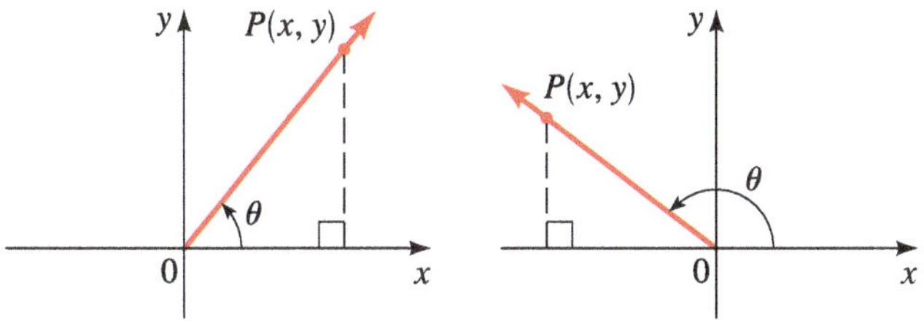

For nonzero values of the denominator the **trigonometric functions** are defined as follows.

$$\sin t = \frac{y}{r} \qquad \cos t = \frac{x}{r} \qquad \tan t = \frac{y}{x}$$

$$\csc t = \frac{r}{y} \qquad \sec t = \frac{r}{x} \qquad \cot t = \frac{x}{y}$$

Basic Trigonometric Identities (p. 496)

An identity is an equation that is true for all values of the variable. The basic trigonometric identities are as follows.

Reciprocal Identities:

$$\csc\theta = \frac{1}{\sin\theta} \qquad \sec\theta = \frac{1}{\cos\theta} \qquad \cot\theta = \frac{1}{\tan\theta}$$

Pythagorean Identities:

$$\sin^2\theta + \cos^2\theta = 1$$

$$\tan^2\theta + 1 = \sec^2\theta$$

$$1 + \cot^2\theta = \csc^2\theta$$

Area of a Triangle (p. 498)

The area $\mathcal{A}$ of a triangle with sides of lengths a and b and with included angle θ is

$$\mathcal{A} = \frac{1}{2}ab\sin\theta$$

Inverse Trigonometric Functions (p. 502)

Inverse functions of the trigonometric functions are defined by restricting the domains as follows.

Function	Domain	Range
$\sin^{-1}$	$[-1, 1]$	$[-\frac{\pi}{2}, \frac{\pi}{2}]$
$\cos^{-1}$	$[-1, 1]$	$[0, \pi]$
$\tan^{-1}$	$(-\infty, \infty)$	$(-\frac{\pi}{2}, \frac{\pi}{2})$

The inverse trigonometric functions are defined as follows.

$$\sin^{-1}x = y \quad \Leftrightarrow \quad \sin y = x$$

$$\cos^{-1}x = y \quad \Leftrightarrow \quad \cos y = x$$

$$\tan^{-1}x = y \quad \Leftrightarrow \quad \tan y = x$$

The Law of Sines and the Law of Cosines (pp. 509, 516)

We follow the convention of labeling the angles of a triangle as A, B, C and the lengths of the corresponding opposite sides as a, b, c, as in the figure.

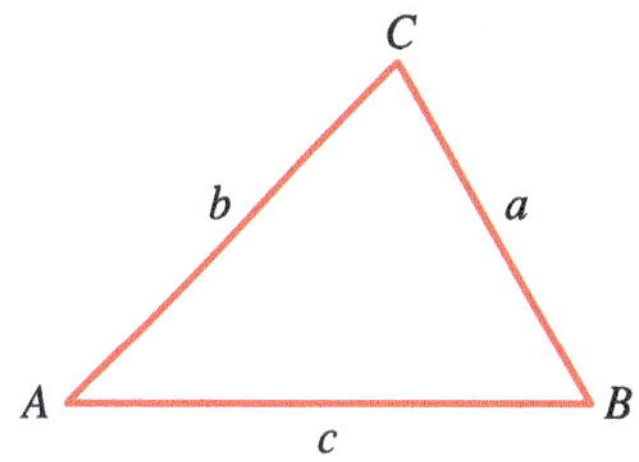

For a triangle ABC we have the following laws.

The **Law of Sines** states that

$$\frac{\sin A}{a} = \frac{\sin B}{b} = \frac{\sin C}{c}$$

The **Law of Cosines** states that

$$a^2 = b^2 + c^2 - 2bc\cos A$$

$$b^2 = a^2 + c^2 - 2ac\cos B$$

$$c^2 = a^2 + b^2 - 2ab\cos C$$

Heron's Formula (p. 519)

Let ABC be a triangle with sides a, b, and c.

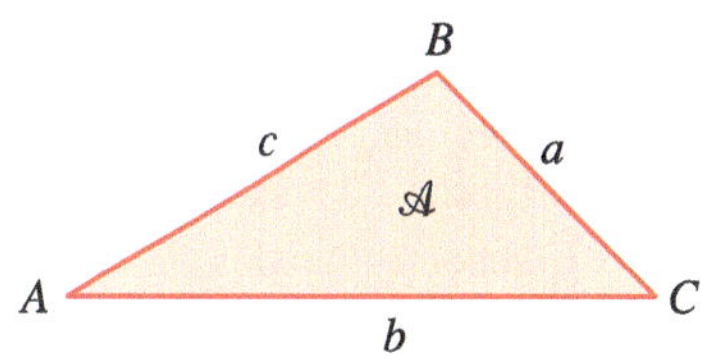

Heron's Formula states that the area $\mathcal{A}$ of triangle ABC is

$$\mathcal{A} = \sqrt{s(s-a)(s-b)(s-c)}$$

where $s = \frac{1}{2}(a + b + c)$ is the semiperimeter of the triangle.

CONCEPT CHECK

1. (a) How is the degree measure of an angle defined?
   (b) How is the radian measure of an angle defined?
   (c) How do you convert from degrees to radians? Convert 45° to radians.
   (d) How do you convert from radians to degrees? Convert 2 rad to degrees.

2. (a) When is an angle in standard position? Illustrate with a graph.
   (b) When are two angles in standard position coterminal? Illustrate with a graph.
   (c) Are the angles 25° and 745° coterminal?
   (d) How is the reference angle for an angle θ defined?
   (e) Find the reference angle for 150°.

3. (a) In a circle of radius r, what is the length s of an arc that subtends a central angle of θ radians?
   (b) In a circle of radius r, what is the area A of a sector with central angle θ radians?

4. (a) Let θ be an acute angle in a right triangle. Identify the opposite side, the adjacent side, and the hypotenuse in the figure.

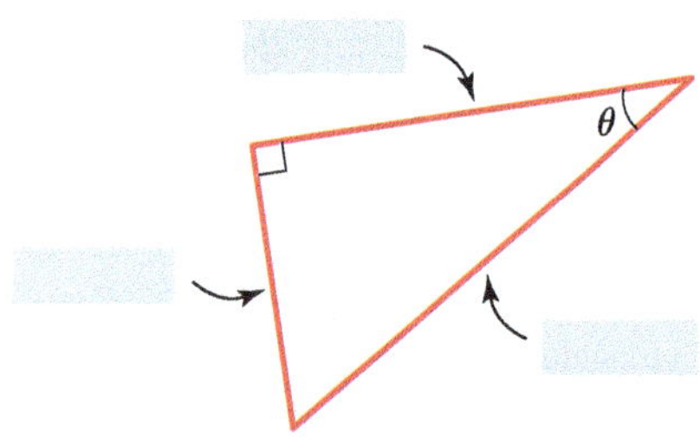

   (b) Define the six trigonometric ratios in terms of the adjacent and opposite sides and the hypotenuse.
   (c) Find the six trigonometric ratios for the angle θ shown in the figure.

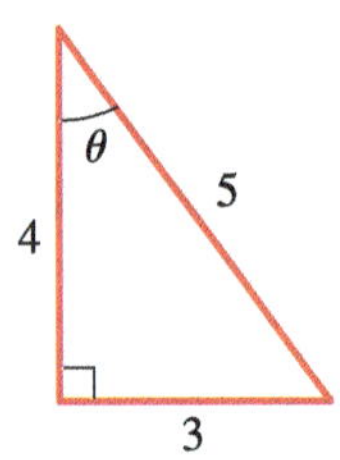

   (d) List the special values of sine, cosine, and tangent.

5. (a) What does it mean to solve a triangle?
   (b) Solve the triangle shown.

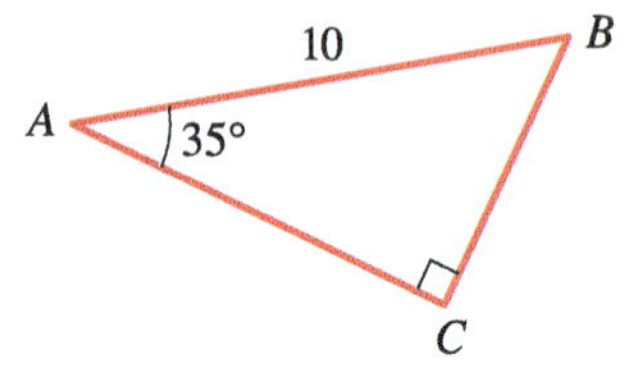

6. (a) Let θ be an angle in standard position, let $P(x, y)$ be a point on the terminal side, and let r be the distance from the origin to P, as shown in the figure. Write expressions for the six trigonometric functions of θ.

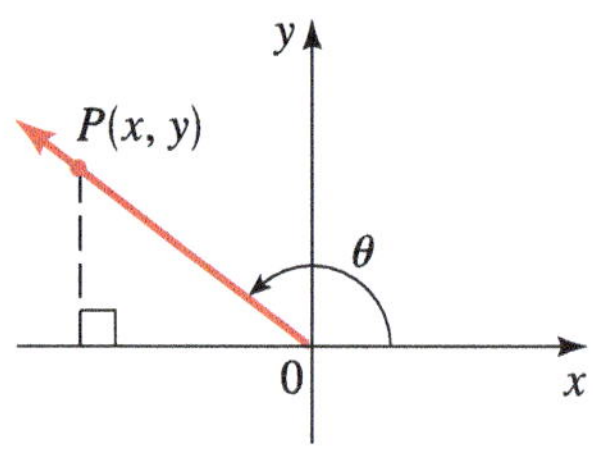

   (b) Find the sine, cosine, and tangent for the angle θ shown in the figure.

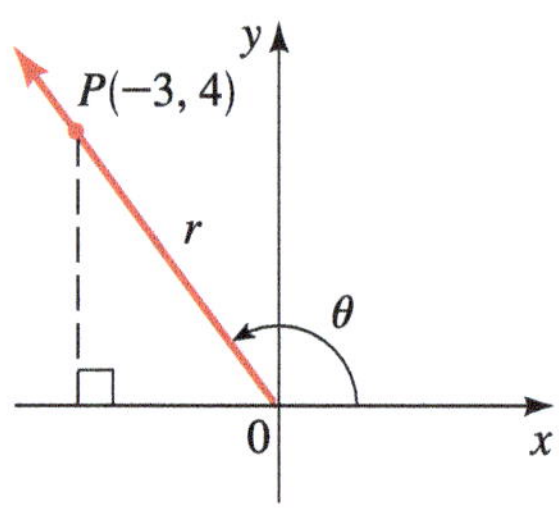

7. In each of the four quadrants, identify the trigonometric functions that are positive.

8. (a) Describe the steps we use to find the value of a trigonometric function of an angle θ.
   (b) Find $\sin 5\pi/6$.

9. (a) State the reciprocal identities.
   (b) State the Pythagorean identities.

10. (a) What is the area of a triangle with sides of length a and b and with included angle θ?
    (b) What is the area of a triangle with sides of length a, b, and c?

11. (a) Define the inverse sine function, the inverse cosine function, and the inverse tangent function.
    (b) Find $\sin^{-1}\frac{1}{2}$, $\cos^{-1}(\sqrt{2}/2)$, and $\tan^{-1}1$.
    (c) For what values of x is the equation $\sin(\sin^{-1}x) = x$ true? For what values of x is the equation $\sin^{-1}(\sin x) = x$ true?

12. (a) State the Law of Sines.
(b) Find side a in the figure.

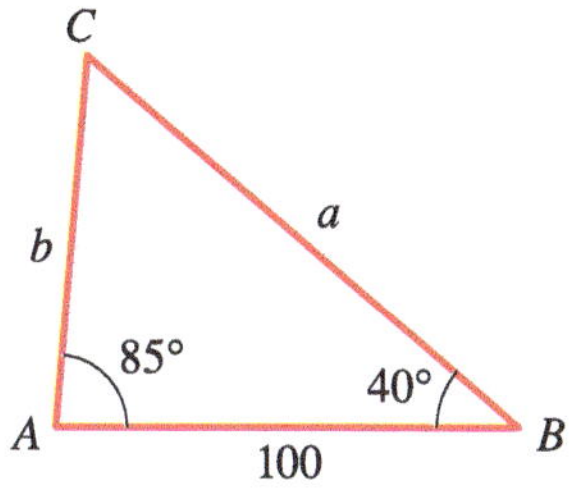

(c) Explain the ambiguous case in the Law of Sines.

13. (a) State the Law of Cosines.
(b) Find side a in the figure.

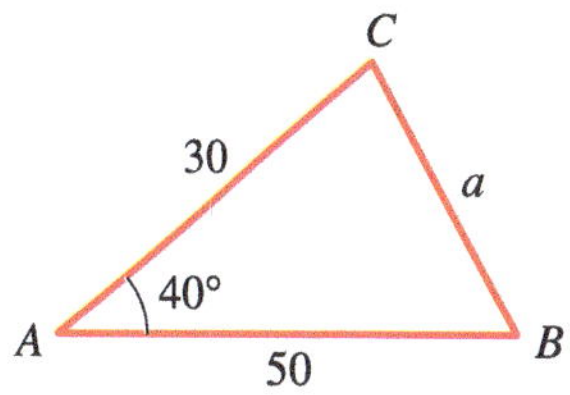

ANSWERS TO THE CONCEPT CHECK CAN BE FOUND AT THE BACK OF THE BOOK.

EXERCISES

1–2 ■ From Degrees to Radians Find the radian measure that corresponds to the given degree measure.

1. (a) 30° (b) 150° (c) −20° (d) −225°
2. (a) 105° (b) 72° (c) −405° (d) −315°

3–4 ■ From Radians to Degrees Find the degree measure that corresponds to the given radian measure.

3. (a) $\frac{5\pi}{6}$ (b) $-\frac{\pi}{9}$ (c) $-\frac{4\pi}{3}$ (d) 4
4. (a) $-\frac{5\pi}{3}$ (b) $\frac{10\pi}{9}$ (c) −5 (d) $\frac{11\pi}{3}$

5–10 ■ Length of a Circular Arc These exercises involve the formula for the length of a circular arc.

5. Find the length of an arc of a circle of radius 10 m if the arc subtends a central angle of $2\pi/5$ rad.
6. A central angle θ in a circle of radius 2.5 cm is subtended by an arc of length 7 cm. Find the measure of θ in degrees and radians.
7. A circular arc of length 25 ft subtends a central angle of 50°. Find the radius of the circle.
8. A circular arc of length 13π m subtends a central angle of 130°. Find the radius of the circle.
9. How many revolutions will a car wheel of diameter 28 in. make over a period of half an hour if the car is traveling at 60 mi/h?
10. New York and Los Angeles are 2450 mi apart. Find the angle that the arc between these two cities subtends at the center of the earth. (The radius of the earth is 3960 mi.)

11–14 ■ Area of a Circular Sector These exercises involve the formula for the area of a circular sector.

11. Find the area of a sector with central angle 2 rad in a circle of radius 5 m.
12. Find the area of a sector with central angle 52° in a circle of radius 200 ft.
13. A sector in a circle of radius 25 ft has an area of 125 ft^2. Find the central angle of the sector.
14. The area of a sector of a circle with a central angle of $11\pi/6$ radians is 50 m^2. Find the radius of the circle.

15. **Angular Speed and Linear Speed** A potter's wheel with radius 8 in. spins at 150 rpm. Find the angular and linear speeds of a point on the rim of the wheel.

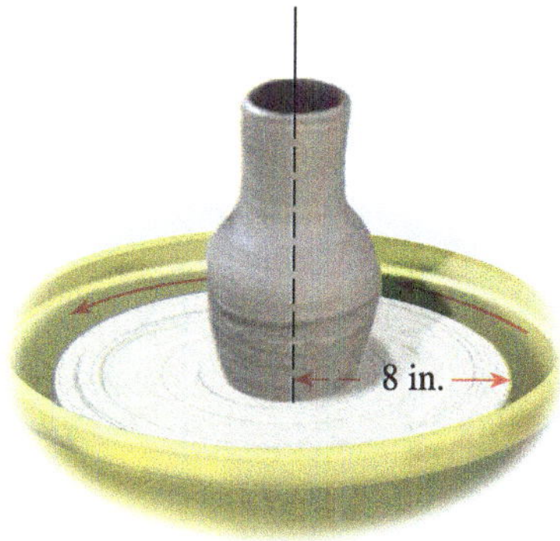

16. **Angular Speed and Linear Speed** In an automobile transmission a *gear ratio* g is the ratio

$$g = \frac{\text{angular speed of engine}}{\text{angular speed of wheels}}$$

The angular speed of the engine is shown on the tachometer (in rpm).

A certain sports car has wheels with radius 11 in. Its gear ratios are shown in the following table. Suppose the car is in fourth gear and the tachometer reads 3500 rpm.

(a) Find the angular speed of the engine.
(b) Find the angular speed of the wheels.
(c) How fast (in mi/h) is the car traveling?

Gear	Ratio
1st	4.1
2nd	3.0
3rd	1.6
4th	0.9
5th	0.7

17–18 ■ Trigonometric Ratios Find the values of the six trigonometric ratios of θ.

17.

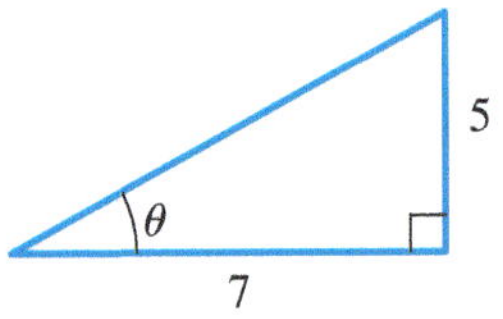

18.

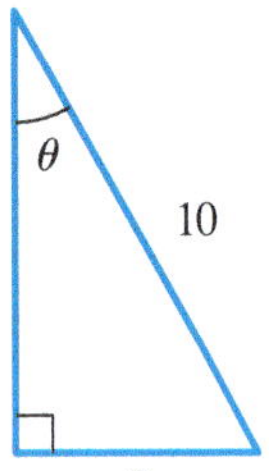

19–22 ■ Finding Sides in Right Triangles Find the sides labeled x and y, rounded to two decimal places.

19.

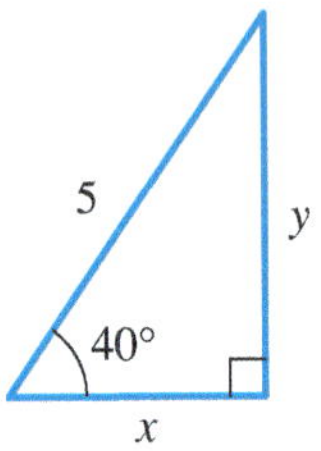

20.

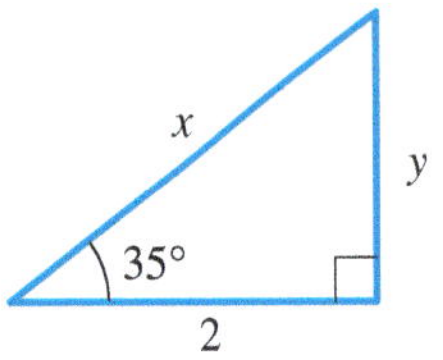

21.

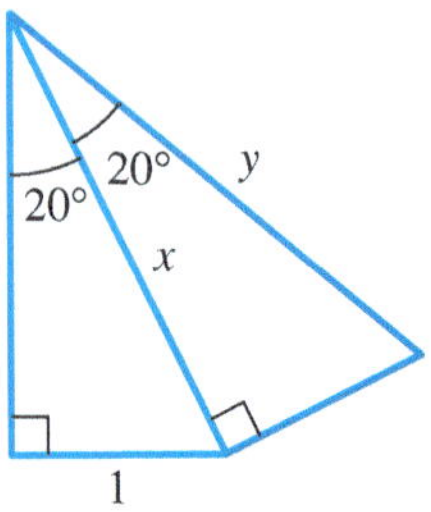

22.

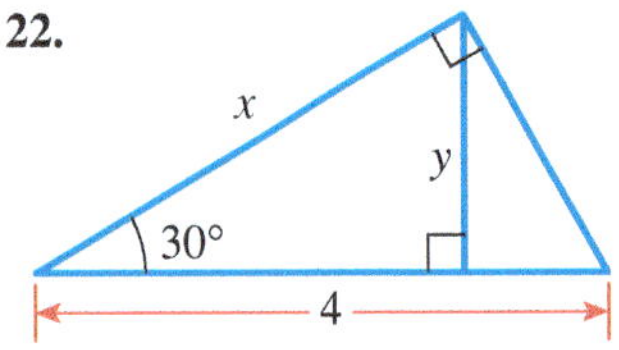

23–26 ■ Solving a Triangle Solve the triangle.

23.

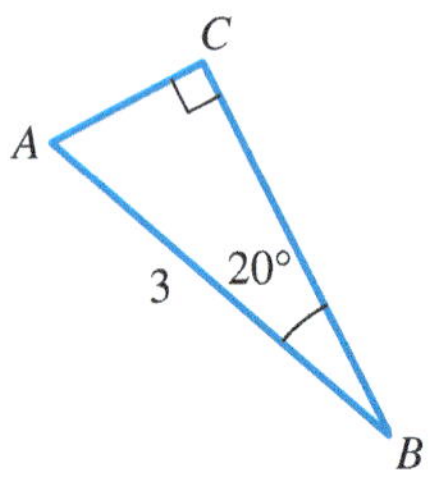

24.

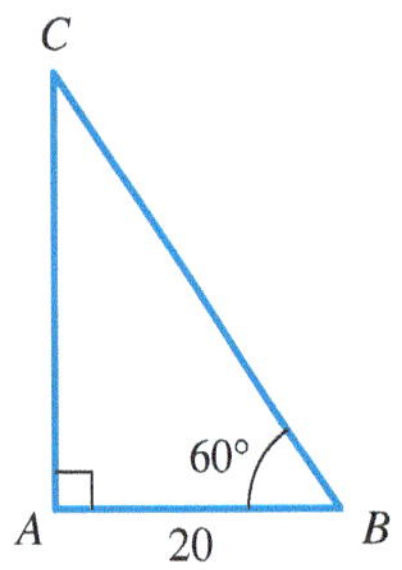

25.

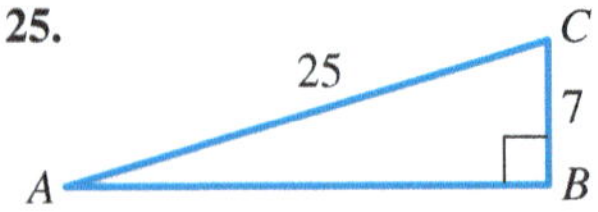

26.

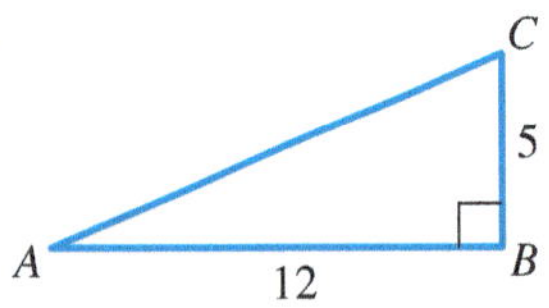

27. Trigonometric Ratios Express the lengths a and b in the figure in terms of the trigonometric ratios of θ.

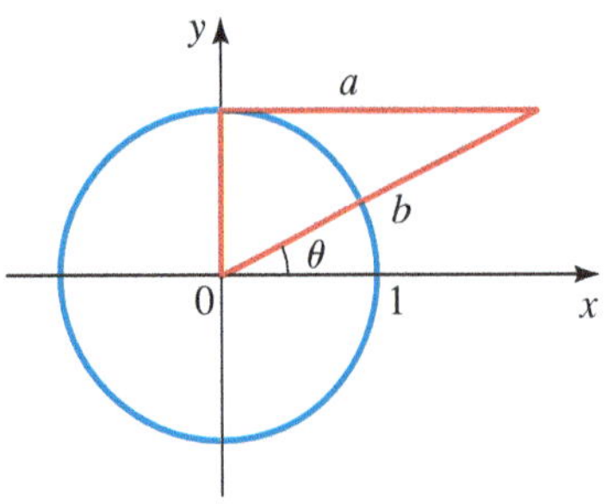

28. CN Tower The highest free-standing tower in North America is the CN Tower in Toronto, Canada. From a distance of 1 km from its base, the angle of elevation to the top of the tower is 28.81°. Find the height of the tower.

29. Perimeter of a Regular Hexagon Find the perimeter of a regular hexagon that is inscribed in a circle of radius 8 m.

30. Pistons of an Engine The pistons in a car engine move up and down repeatedly to turn the crankshaft, as shown. Find the height of the point P above the center O of the crankshaft in terms of the angle θ.

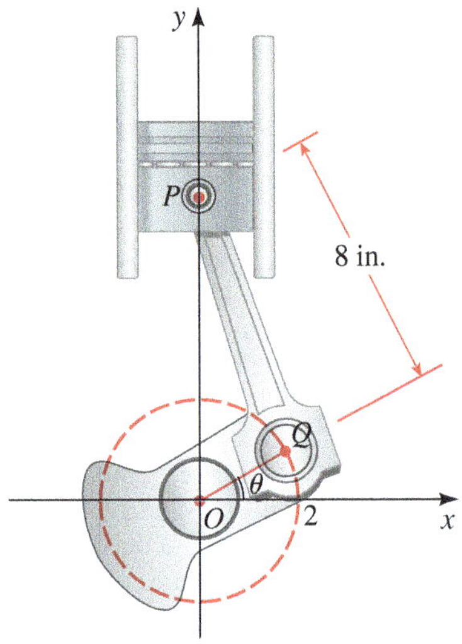

31. Radius of the Moon As viewed from the earth, the angle subtended by the full moon is 0.518°. Use this information and the fact that the distance AB from the earth to the moon is 236,900 mi to find the radius of the moon.

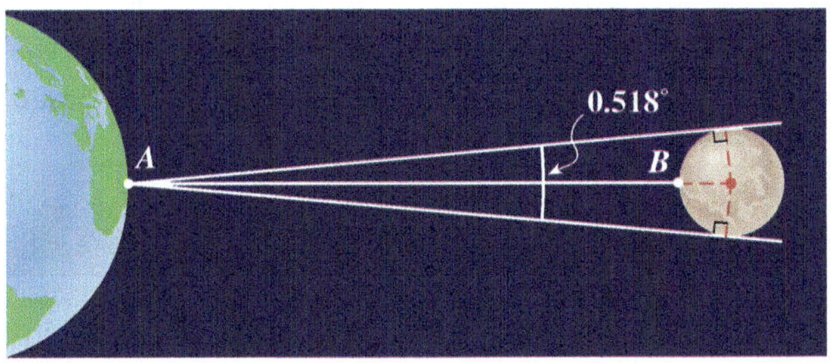

32. Distance Between Two Ships A pilot measures the angles of depression to two ships to be 40° and 52° (see the figure). If the pilot is flying at an elevation of 35,000 ft, find the distance between the two ships.

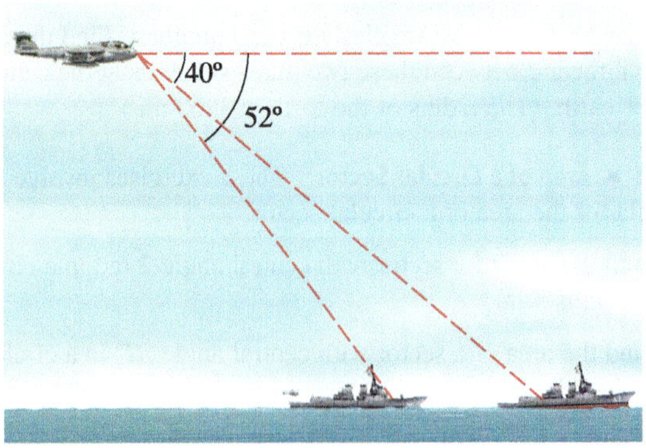

33–44 ■ Values of Trigonometric Functions Find the exact value.

33. $\sin 315°$ **34.** $\csc \dfrac{9\pi}{4}$ **35.** $\tan(-135°)$

36. $\cos \dfrac{5\pi}{6}$ **37.** $\cot\left(-\dfrac{22\pi}{3}\right)$ **38.** $\sin 405°$

39. $\cos 585°$ **40.** $\sec \dfrac{22\pi}{3}$ **41.** $\csc \dfrac{8\pi}{3}$

42. $\sec \dfrac{13\pi}{6}$ **43.** $\cot(-390°)$ **44.** $\tan \dfrac{23\pi}{4}$

45. Values of Trigonometric Functions Find the values of the six trigonometric ratios of the angle θ in standard position if the point $(-5, 12)$ is on the terminal side of θ.

46. Values of Trigonometric Functions Find $\sin\theta$ if θ is in standard position and its terminal side intersects the circle of radius 1 centered at the origin at the point $(-\sqrt{3}/2, \frac{1}{2})$.

47. Angle Formed by a Line Find the acute angle that is formed by the line $y - \sqrt{3}x + 1 = 0$ and the x-axis.

48. Values of Trigonometric Functions Find the six trigonometric ratios of the angle θ in standard position if its terminal side is in Quadrant III and is parallel to the line $4y - 2x - 1 = 0$.

49–52 ■ Expressing One Trigonometric Function in Terms of Another Write the first expression in terms of the second, for θ in the given quadrant.

49. $\tan\theta$, $\cos\theta$; θ in Quadrant II

50. $\sec\theta$, $\sin\theta$; θ in Quadrant III

51. $\tan^2\theta$, $\sin\theta$; θ in any quadrant

52. $\csc^2\theta\cos^2\theta$, $\sin\theta$; θ in any quadrant

53–56 ■ Values of Trigonometric Functions Find the values of the six trigonometric functions of θ from the information given.

53. $\tan\theta = \sqrt{7}/3$, $\sec\theta = \frac{4}{3}$

54. $\sec\theta = \frac{41}{40}$, $\csc\theta = -\frac{41}{9}$

55. $\sin\theta = \frac{3}{5}$, $\cos\theta < 0$

56. $\sec\theta = -\frac{13}{5}$, $\tan\theta > 0$

57–60 ■ Value of an Expression Find the value of the given trigonometric expression.

57. If $\tan\theta = -\frac{1}{2}$ for θ in Quadrant II, find $\sin\theta + \cos\theta$.

58. If $\sin\theta = \frac{1}{2}$ for θ in Quadrant I, find $\tan\theta + \sec\theta$.

59. If $\tan\theta = -1$, find $\sin^2\theta + \cos^2\theta$.

60. If $\cos\theta = -\sqrt{3}/2$ and $\pi/2 < \theta < \pi$, find $\sin 2\theta$.

61–64 ■ Values of Inverse Trigonometric Functions Find the exact value of the expression.

61. $\sin^{-1}(\sqrt{3}/2)$ **62.** $\tan^{-1}(\sqrt{3}/3)$

63. $\tan(\sin^{-1}\frac{2}{5})$ **64.** $\sin(\cos^{-1}\frac{3}{8})$

65–66 ■ Inverse Trigonometric Functions Rewrite the expression as an algebraic expression in x.

65. $\sin(\tan^{-1}x)$ **66.** $\sec(\sin^{-1}x)$

67–68 ■ Finding an Unknown Side Express θ in terms of x.

67.

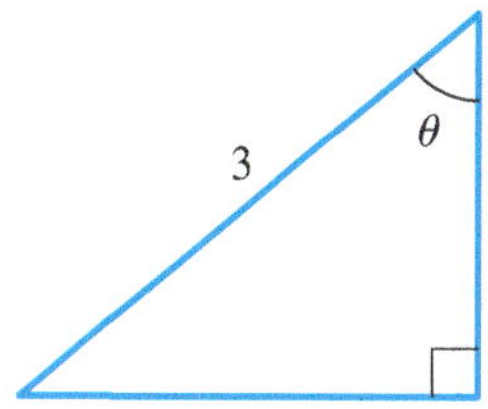

68.

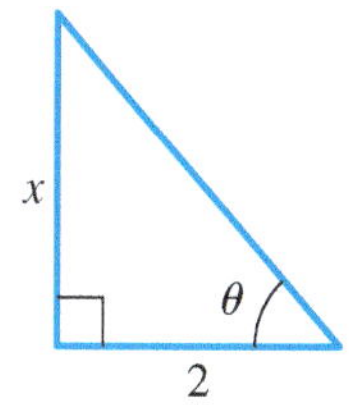

69–78 ■ Law of Sines and Law of Cosines Find the side labeled x or the angle labeled θ.

69.

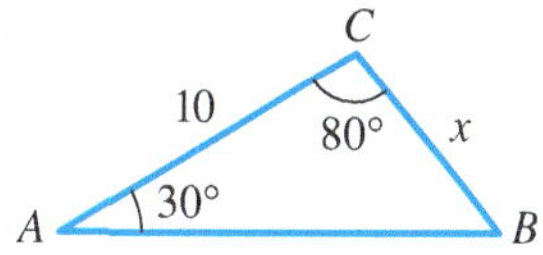

70.

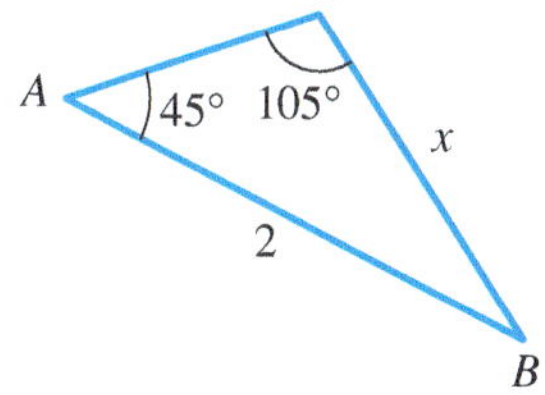

71.

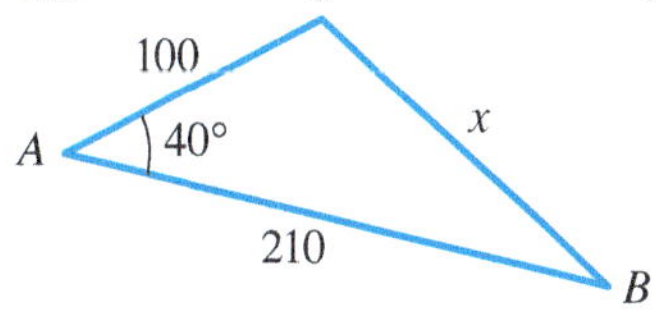

72.

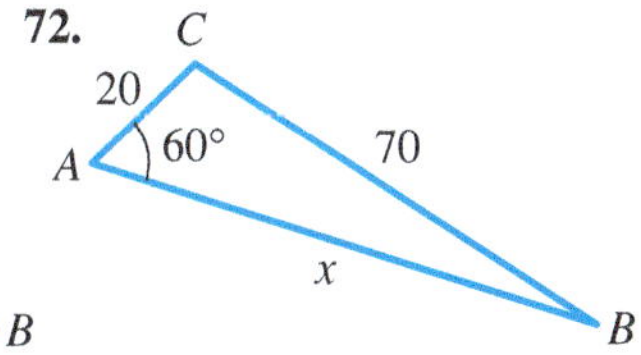

73.

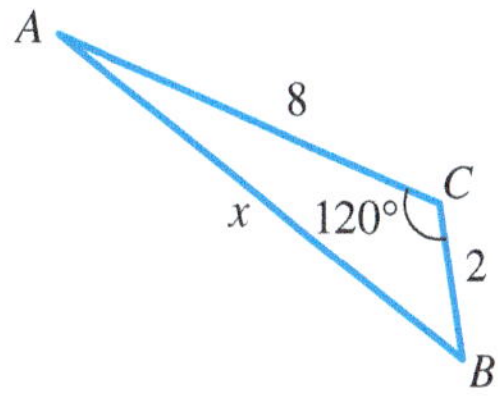

74.

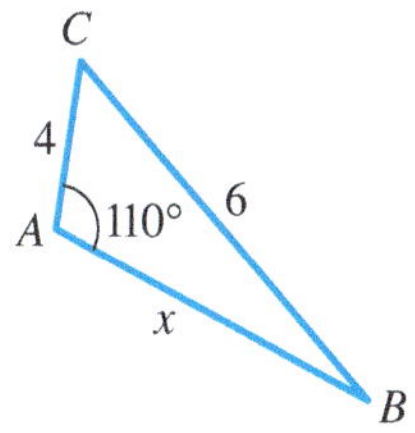

75.

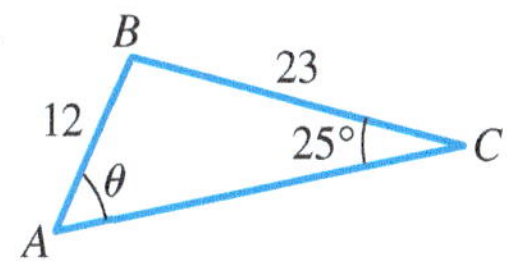

76.

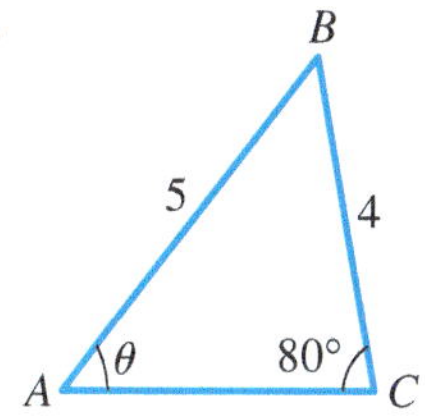

77.

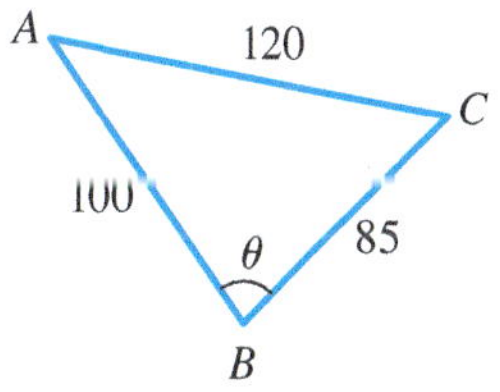

78.

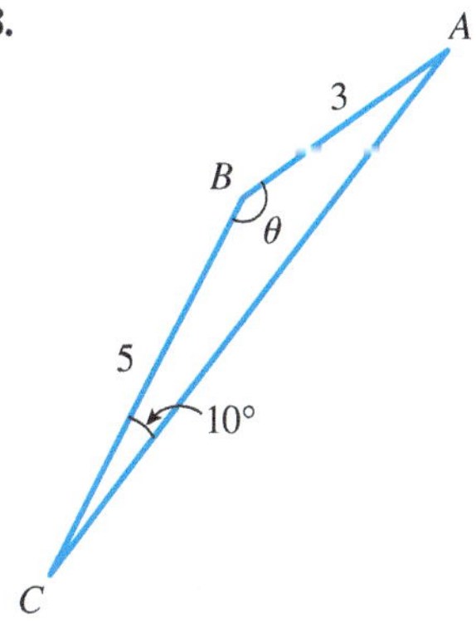

79. Distance Between Two Ships Two ships leave a port at the same time. One travels at 20 mi/h in a direction N 32° E, and the other travels at 28 mi/h in a direction S 42° E (see the figure). How far apart are the two ships after 2 h?

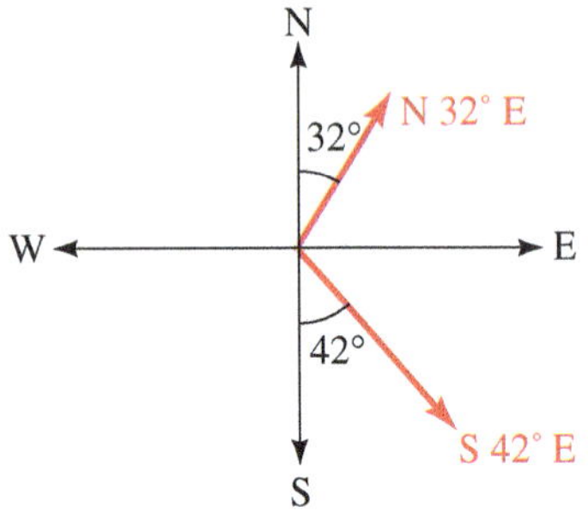

80. Height of a Building From a point A on the ground, the angle of elevation to the top of a tall building is 24.1°. From a point B, which is 600 ft closer to the building, the angle of elevation is measured to be 30.2°. Find the height of the building.

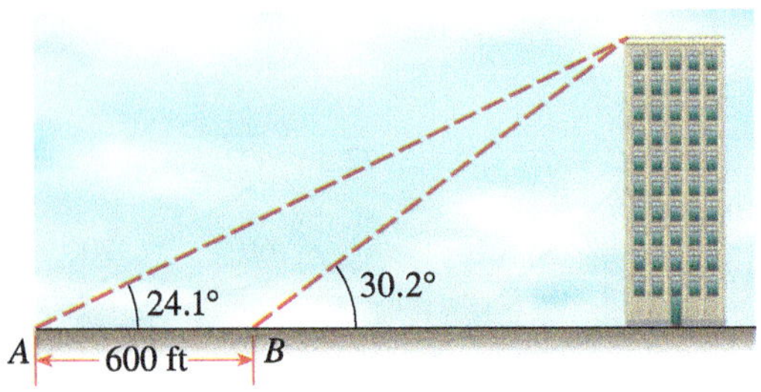

81. Distance Between Two Points Find the distance between points A and B on opposite sides of a lake from the information shown.

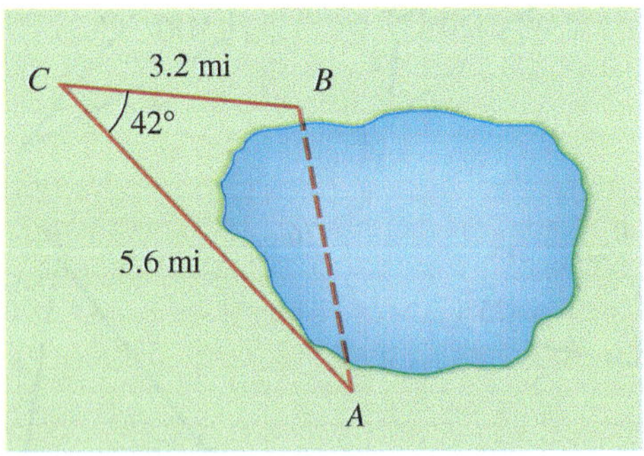

82. Distance Between a Boat and the Shore A boat is cruising the ocean off a straight shoreline. Points A and B are 120 mi apart on the shore, as shown. It is found that $\angle A = 42.3°$ and $\angle B = 68.9°$. Find the shortest distance from the boat to the shore.

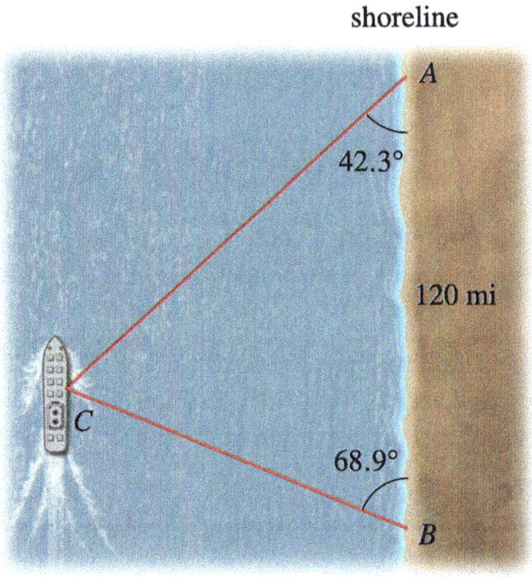

83. Area of a Triangle Find the area of a triangle with sides of length 8 and 14 and included angle 35°.

84. Heron's Formula Find the area of a triangle with sides of length 5, 6, and 8.

CHAPTER 6 TEST

1. Find the radian measures that correspond to the degree measures 330° and −135°.
2. Find the degree measures that correspond to the radian measures $4\pi/3$ and −1.3.
3. The rotor blades of a helicopter are 16 ft long and are rotating at 120 rpm.
   (a) Find the angular speed of the rotor.
   (b) Find the linear speed of a point on the tip of a blade.
4. Find the exact value of each of the following.

   (a) $\sin 405°$ **(b)** $\tan(-150°)$ **(c)** $\sec \frac{5\pi}{3}$ **(d)** $\csc \frac{5\pi}{2}$
5. Find $\tan\theta + \sin\theta$ for the angle θ shown.

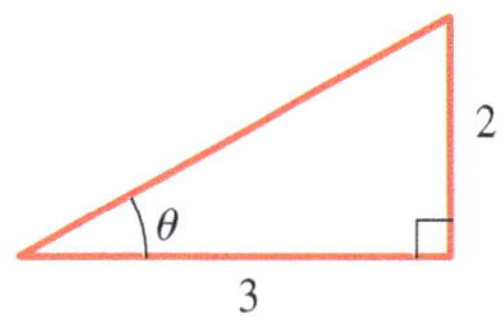

6. Express the lengths a and b shown in the figure in terms of θ.

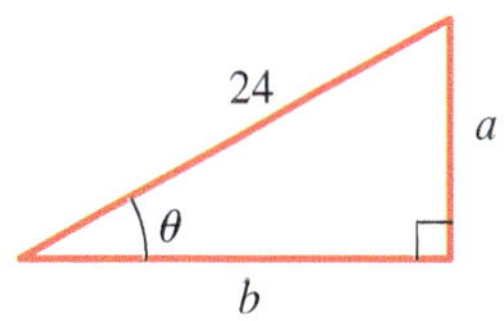

7. If $\cos\theta = -\frac{1}{3}$ and θ is in Quadrant III, find $\tan\theta \cot\theta + \csc\theta$.
8. If $\sin\theta = \frac{5}{13}$ and $\tan\theta = -\frac{5}{12}$, find $\sec\theta$.
9. Express $\tan\theta$ in terms of $\sec\theta$ for θ in Quadrant II.
10. The base of the ladder in the figure is 6 ft from the building, and the angle formed by the ladder and the ground is 73°. How high up the building does the ladder touch?

11. Express θ in each figure in terms of x.

(a)

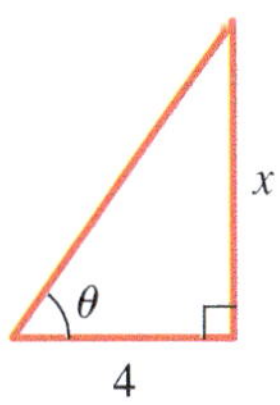

(b)

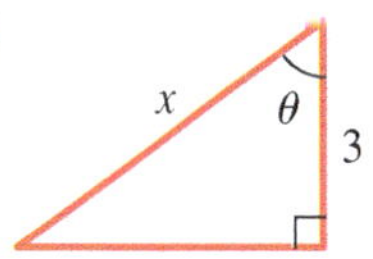

12. Find the exact value of $\cos\left(\tan^{-1}\frac{9}{40}\right)$.

13–18 ■ Find the side labeled x or the angle labeled θ.

13.

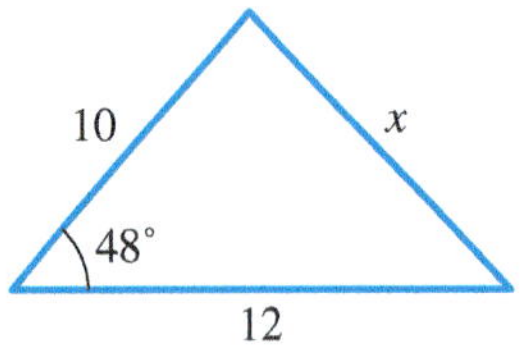

14.

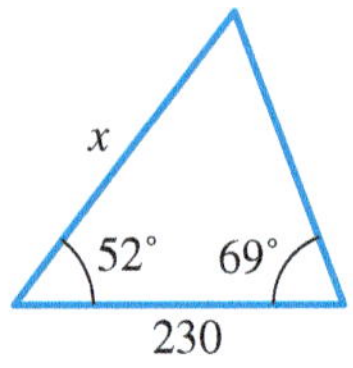

15.

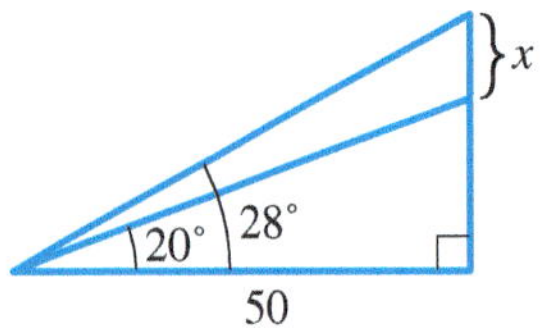

16.

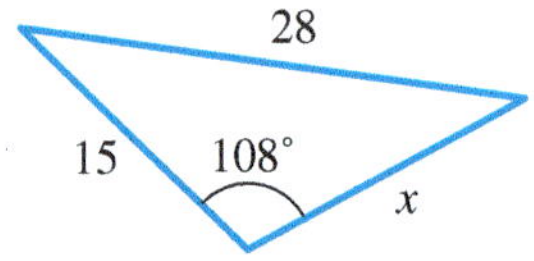

17.

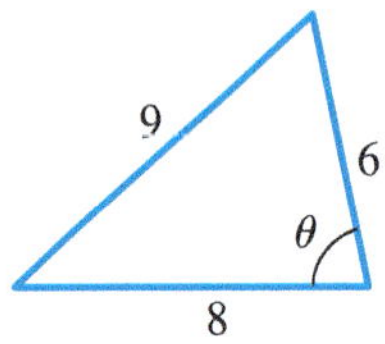

18.

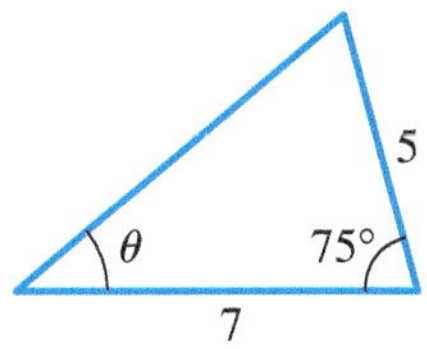

19. Refer to the figure below.

(a) Find the area of the shaded region.

(b) Find the perimeter of the shaded region.

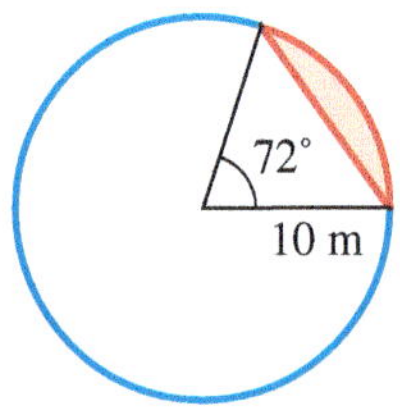

20. Refer to the figure below.

(a) Find the angle opposite the longest side.

(b) Find the area of the triangle.

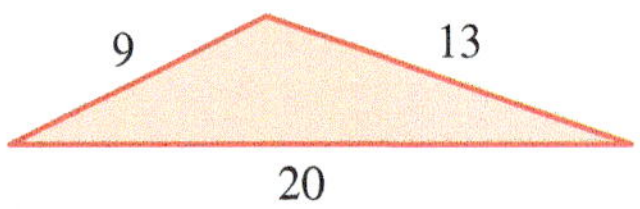

21. Two wires tether a balloon to the ground, as shown. How high is the balloon above the ground?

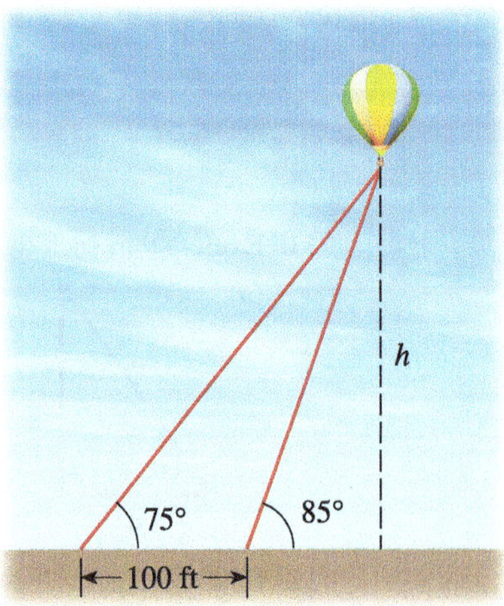

How can we measure the height of a mountain or the distance across a lake? Obviously, it may be difficult, inconvenient, or impossible to measure these distances directly (that is, by using a tape measure or a yardstick). On the other hand, it is easy to measure *angles* involving distant objects. That's where trigonometry comes in: The trigonometric ratios relate angles to distances, so they can be used to *calculate* distances from the *measured* angles. In this *Focus* we examine how trigonometry is used to map a town. Modern mapmaking methods use satellites and the Global Positioning System, but mathematics remains at the core of the process.

Mapping a Town

A student wants to draw a map of his hometown. To construct an accurate map (or scale model), he needs to find distances between various landmarks in the town. The student makes the measurements shown in Figure 1. Note that only one distance is measured: that between City Hall and the first bridge. All other measurements are angles.

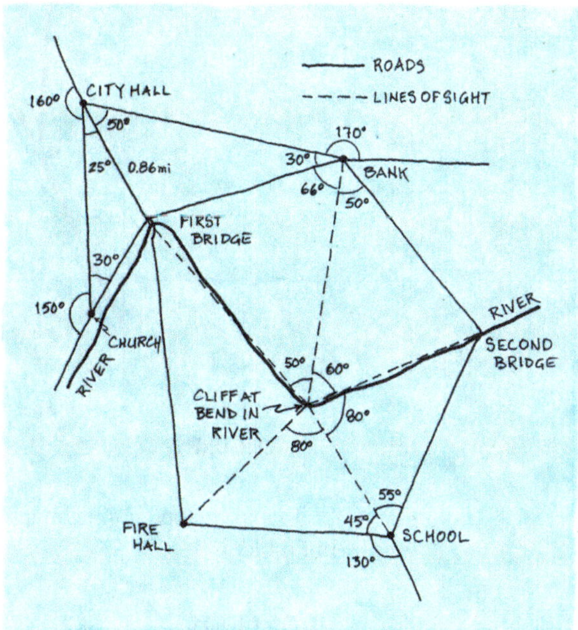

FIGURE 1

The distances between other landmarks can now be found by using the Law of Sines. For example, the distance x from the bank to the first bridge is calculated by applying the Law of Sines to the triangle with vertices at City Hall, the bank, and the first bridge.

$$\frac{x}{\sin 50^\circ} = \frac{0.86}{\sin 30^\circ} \qquad \text{Law of Sines}$$

$$x = \frac{0.86 \sin 50^\circ}{\sin 30^\circ} \qquad \text{Solve for } x$$

$$\approx 1.32 \text{ mi} \qquad \text{Calculator}$$

So the distance between the bank and the first bridge is 1.32 mi.

The distance we just found can now be used to find other distances. For instance, we find the distance y between the bank and the cliff as follows:

$$\frac{y}{\sin 64^\circ} = \frac{1.32}{\sin 50^\circ} \quad \text{Law of Sines}$$

$$y = \frac{1.32 \sin 64^\circ}{\sin 50^\circ} \quad \text{Solve for } y$$

$$\approx 1.55 \text{ mi} \quad \text{Calculator}$$

Continuing in this fashion, we can calculate all the distances between the landmarks shown in the rough sketch in Figure 1. We can use this information to draw the map shown in Figure 2.

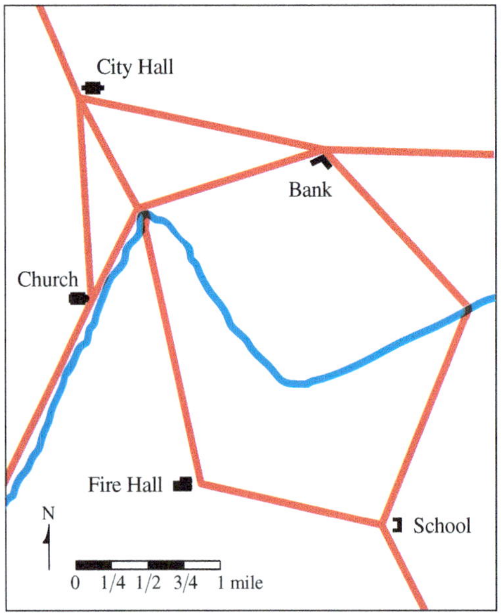

FIGURE 2

To make a topographic map, we need to measure elevation. This concept is explored in Problems 4–6.

PROBLEMS

1. **Completing the Map** Find the distance between the church and City Hall.
2. **Completing the Map** Find the distance between the fire hall and the school. [*Hint:* You will need to find other distances first.]
3. **Determining a Distance** A surveyor on one side of a river wishes to find the distance between points A and B on the opposite side of the river. On her side she chooses points C and D, which are 20 m apart, and measures the angles shown in the figure below. Find the distance between A and B.

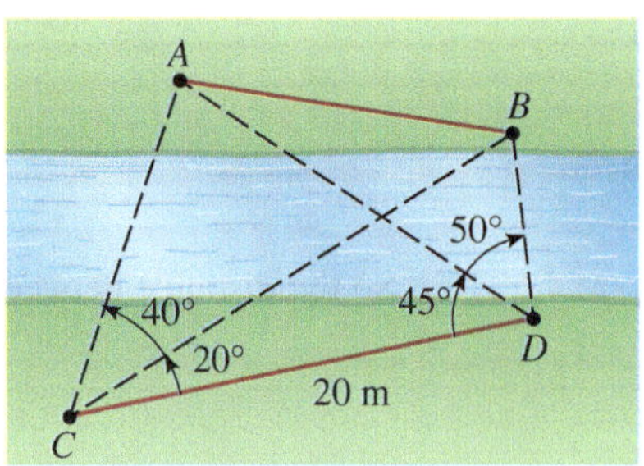

4. Height of a Cliff To measure the height of an inaccessible cliff on the opposite side of a river, a surveyor makes the measurements shown in the figure at the left. Find the height of the cliff.

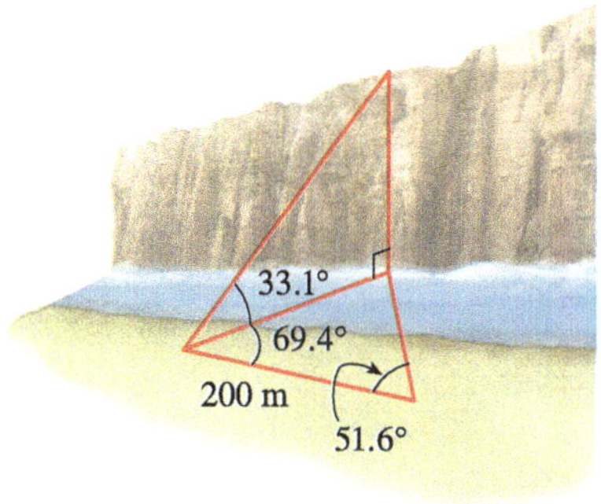

5. Height of a Mountain To calculate the height h of a mountain, angles α and β and distance d are measured, as shown in the figure below.

(a) Show that

$$h = \frac{d}{\cot\alpha - \cot\beta}$$

(b) Show that

$$h = d\frac{\sin\alpha\sin\beta}{\sin(\beta - \alpha)}$$

(c) Use the formulas from parts (a) and (b) to find the height of a mountain if $\alpha = 25°$, $\beta = 29°$, and $d = 800$ ft. Do you get the same answer from each formula?

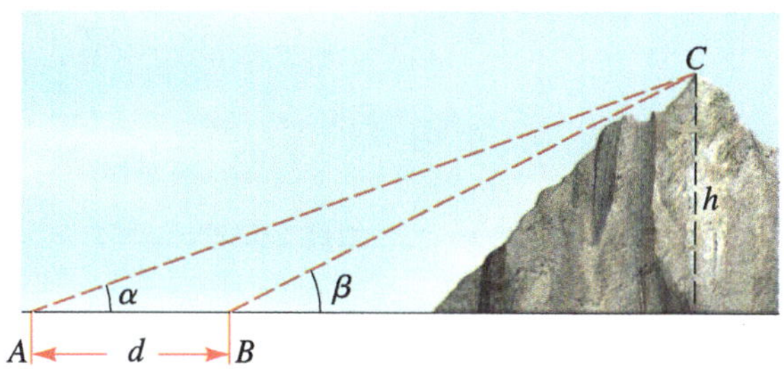

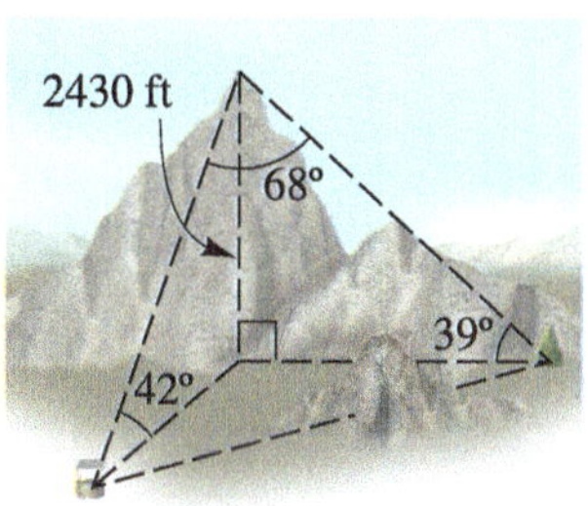

6. Determining a Distance A surveyor has determined that a mountain is 2430 ft high. From the top of the mountain he measures the angles of depression to two landmarks at the base of the mountain and finds them to be 42° and 39°. (Observe that these are the same as the angles of elevation from the landmarks as shown in the figure at the left.) The angle between the lines of sight to the landmarks is 68°. Calculate the distance between the two landmarks.

7. Surveying Building Lots A surveyor surveys two adjacent lots and makes the following rough sketch showing his measurements. Calculate all the distances shown in the figure, and use your result to draw an accurate map of the two lots.

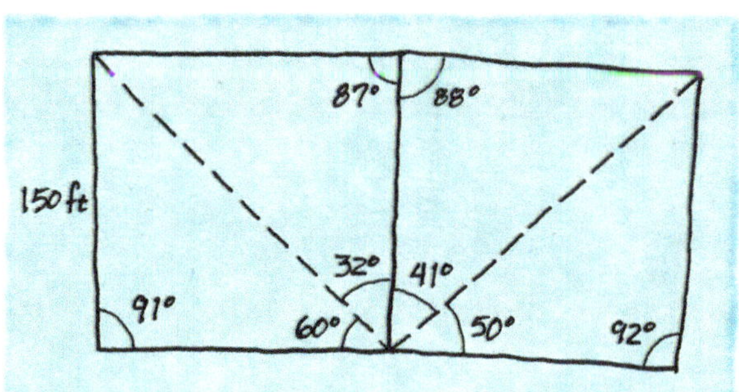

8. Great Survey of India The Great Trigonometric Survey of India was one of the most massive mapping projects ever undertaken (see the margin note on page 512). Do some research at your library or on the Internet to learn more about the Survey, and write a report on your findings.

© Jule_Berlin/Shutterstock.com

7 Analytic Trigonometry

In Chapters 5 and 6 we studied graphical and geometric properties of the trigonometric functions. In this chapter we study algebraic properties of these functions, that is, simplifying and factoring expressions and solving equations that involve trigonometric functions.

We have used the trigonometric functions to model different real-world phenomena, including periodic motion (such as the sound waves produced by a band). To obtain information from a model, we often need to solve equations. If the model involves trigonometric functions, we need to solve trigonometric equations. Solving trigonometric equations often involves using trigonometric identities. We've already encountered some basic trigonometric identities in the preceding chapters. We begin this chapter by finding many new identities.

7.1 TRIGONOMETRIC IDENTITIES

■ Simplifying Trigonometric Expressions ■ Proving Trigonometric Identities

Recall that an **equation** is a statement that two mathematical expressions are equal. For example, the following are equations:

$$x + 2 = 5$$
$$(x + 1)^2 = x^2 + 2x + 1$$
$$\sin^2 t + \cos^2 t = 1$$

An **identity** is an equation that is true for all values of the variable(s). The last two equations above are identities, but the first one is not, since it is not true for values of x other than 3.

A **trigonometric** identity is an identity involving trigonometric functions. We begin by listing some of the basic trigonometric identities. We studied most of these in Chapters 5 and 6; you are asked to prove the cofunction identities in Exercise 118.

FUNDAMENTAL TRIGONOMETRIC IDENTITIES

Reciprocal Identities

$$\csc x = \frac{1}{\sin x} \qquad \sec x = \frac{1}{\cos x} \qquad \cot x = \frac{1}{\tan x}$$

$$\tan x = \frac{\sin x}{\cos x} \qquad \cot x = \frac{\cos x}{\sin x}$$

Pythagorean Identities

$$\sin^2 x + \cos^2 x = 1 \qquad \tan^2 x + 1 = \sec^2 x \qquad 1 + \cot^2 x = \csc^2 x$$

Even-Odd Identities

$$\sin(-x) = -\sin x \qquad \cos(-x) = \cos x \qquad \tan(-x) = -\tan x$$

Cofunction Identities

$$\sin\left(\frac{\pi}{2} - x\right) = \cos x \qquad \tan\left(\frac{\pi}{2} - x\right) = \cot x \qquad \sec\left(\frac{\pi}{2} - x\right) = \csc x$$

$$\cos\left(\frac{\pi}{2} - x\right) = \sin x \qquad \cot\left(\frac{\pi}{2} - x\right) = \tan x \qquad \csc\left(\frac{\pi}{2} - x\right) = \sec x$$

■ Simplifying Trigonometric Expressions

Identities enable us to write the same expression in different ways. It is often possible to rewrite a complicated-looking expression as a much simpler one. To simplify algebraic expressions, we used factoring, common denominators, and the Special Product Formulas. To simplify trigonometric expressions, we use these same techniques together with the fundamental trigonometric identities.

EXAMPLE 1 ■ Simplifying a Trigonometric Expression

Simplify the expression $\cos t + \tan t \sin t$.

SOLUTION We start by rewriting the expression in terms of sine and cosine.

$$\begin{aligned}\cos t + \tan t \sin t &= \cos t + \left(\frac{\sin t}{\cos t}\right)\sin t && \text{Reciprocal identity}\\ &= \frac{\cos^2 t + \sin^2 t}{\cos t} && \text{Common denominator}\\ &= \frac{1}{\cos t} && \text{Pythagorean identity}\\ &= \sec t && \text{Reciprocal identity}\end{aligned}$$

Now Try Exercise 3 ■

EXAMPLE 2 ■ Simplifying by Combining Fractions

Simplify the expression $\dfrac{\sin\theta}{\cos\theta} + \dfrac{\cos\theta}{1 + \sin\theta}$.

SOLUTION We combine the fractions by using a common denominator.

$$\begin{aligned}\frac{\sin\theta}{\cos\theta} + \frac{\cos\theta}{1 + \sin\theta} &= \frac{\sin\theta\,(1 + \sin\theta) + \cos^2\theta}{\cos\theta\,(1 + \sin\theta)} && \text{Common denominator}\\ &= \frac{\sin\theta + \sin^2\theta + \cos^2\theta}{\cos\theta\,(1 + \sin\theta)} && \text{Distribute } \sin\theta\\ &= \frac{\sin\theta + 1}{\cos\theta\,(1 + \sin\theta)} && \text{Pythagorean identity}\\ &= \frac{1}{\cos\theta} = \sec\theta && \text{Cancel, and use reciprocal identity}\end{aligned}$$

Now Try Exercise 23 ■

■ Proving Trigonometric Identities

Many identities follow from the fundamental identities. In the examples that follow, we learn how to prove that a given trigonometric equation is an identity, and in the process we will see how to discover new identities.

First, it's easy to decide when a given equation is *not* an identity. All we need to do is show that the equation does not hold for some value of the variable (or variables). Thus the equation

$$\sin x + \cos x = 1$$

is not an identity, because when $x = \pi/4$, we have

$$\sin\frac{\pi}{4} + \cos\frac{\pi}{4} = \frac{\sqrt{2}}{2} + \frac{\sqrt{2}}{2} = \sqrt{2} \neq 1$$

To verify that a trigonometric equation is an identity, we transform one side of the equation into the other side by a series of steps, each of which is itself an identity.

GUIDELINES FOR PROVING TRIGONOMETRIC IDENTITIES

1. **Start with one side.** Pick one side of the equation, and write it down. Your goal is to transform it into the other side. It's usually easier to start with the more complicated side.
2. **Use known identities.** Use algebra and the identities you know to change the side you started with. Bring fractional expressions to a common denominator, factor, and use the fundamental identities to simplify expressions.
3. **Convert to sines and cosines.** If you are stuck, you may find it helpful to rewrite all functions in terms of sines and cosines.

Warning: To prove an identity, we do *not* just perform the same operations on both sides of the equation. For example, if we start with an equation that is not an identity, such as

$$\sin x = -\sin x$$

and square both sides, we get the equation

$$\sin^2 x = \sin^2 x$$

which is clearly an identity. Does this mean that the original equation is an identity? Of course not. The problem here is that the operation of squaring is not **reversible** in the sense that we cannot arrive back at the original equation by taking square roots (reversing the procedure). Only operations that are reversible will necessarily transform an identity into an identity.

EXAMPLE 3 ■ Proving an Identity by Rewriting in Terms of Sine and Cosine

Consider the equation $\cos\theta\,(\sec\theta - \cos\theta) = \sin^2\theta$.

(a) Verify algebraically that the equation is an identity.

(b) Confirm graphically that the equation is an identity.

SOLUTION

(a) The left-hand side looks more complicated, so we start with it and try to transform it into the right-hand side.

$$\begin{aligned} \text{LHS} &= \cos\theta\,(\sec\theta - \cos\theta) & \\ &= \cos\theta\left(\frac{1}{\cos\theta} - \cos\theta\right) & \text{Reciprocal identity} \\ &= 1 - \cos^2\theta & \text{Expand} \\ &= \sin^2\theta = \text{RHS} & \text{Pythagorean identity} \end{aligned}$$

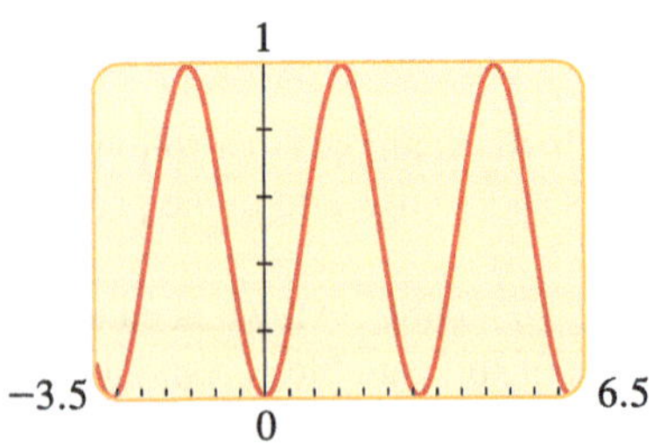

FIGURE 1

(b) We graph each side of the equation to see whether the graphs coincide. From Figure 1 we see that the graphs of $y = \cos\theta\,(\sec\theta - \cos\theta)$ and $y = \sin^2\theta$ are identical. This confirms that the equation is an identity.

Now Try Exercise 29 ■

In Example 3 it isn't easy to see how to change the right-hand side into the left-hand side, but it's definitely possible. Simply notice that each step is reversible. In other words, if we start with the last expression in the proof and work backward through the steps, the right-hand side is transformed into the left-hand side. You will probably agree, however, that it's more difficult to prove the identity this way. That's why it's often better to change the more complicated side of the identity into the simpler side.

EXAMPLE 4 ■ Proving an Identity by Combining Fractions

Verify the identity

$$2 \tan x \sec x = \frac{1}{1 - \sin x} - \frac{1}{1 + \sin x}$$

SOLUTION Finding a common denominator and combining the fractions on the right-hand side of this equation, we get

$$\begin{aligned}
\text{RHS} &= \frac{1}{1 - \sin x} - \frac{1}{1 + \sin x} \\
&= \frac{(1 + \sin x) - (1 - \sin x)}{(1 - \sin x)(1 + \sin x)} && \text{Common denominator} \\
&= \frac{2 \sin x}{1 - \sin^2 x} && \text{Simplify} \\
&= \frac{2 \sin x}{\cos^2 x} && \text{Pythagorean identity} \\
&= 2 \frac{\sin x}{\cos x}\left(\frac{1}{\cos x}\right) && \text{Factor} \\
&= 2 \tan x \sec x = \text{LHS} && \text{Reciprocal identities}
\end{aligned}$$

Now Try Exercise 65 ■

See the Prologue: *Principles of Problem Solving*, page P2

In Example 5 we introduce "something extra" to the problem by multiplying the numerator and the denominator by a trigonometric expression, chosen so that we can simplify the result.

EXAMPLE 5 ■ Proving an Identity by Introducing Something Extra

Verify the identity $\dfrac{\cos u}{1 - \sin u} = \sec u + \tan u$.

SOLUTION We start with the left-hand side and multiply the numerator and denominator by $1 + \sin u$.

We multiply by $1 + \sin u$ because we know by the difference of squares formula that

$$(1 - \sin u)(1 + \sin u) = 1 - \sin^2 u$$

and this is just $\cos^2 u$, a simpler expression.

$$\begin{aligned}
\text{LHS} &= \frac{\cos u}{1 - \sin u} \\
&= \frac{\cos u}{1 - \sin u} \cdot \frac{1 + \sin u}{1 + \sin u} && \text{Multiply numerator and denominator by } 1 + \sin u \\
&= \frac{\cos u\,(1 + \sin u)}{1 - \sin^2 u} && \text{Expand denominator} \\
&= \frac{\cos u\,(1 + \sin u)}{\cos^2 u} && \text{Pythagorean identity} \\
&= \frac{1 + \sin u}{\cos u} && \text{Cancel common factor} \\
&= \frac{1}{\cos u} + \frac{\sin u}{\cos u} && \text{Separate into two fractions} \\
&= \sec u + \tan u && \text{Reciprocal identities}
\end{aligned}$$

Now Try Exercise 77 ■

EUCLID (circa 300 B.C.) taught in Alexandria. His *Elements* is the most widely influential scientific book in history. For 2000 years it was the standard introduction to geometry in schools, and for many generations it was considered the best way to develop logical reasoning. Abraham Lincoln, for instance, studied the *Elements* as a way to sharpen his mind. The story is told that King Ptolemy once asked Euclid whether there was a faster way to learn geometry than through the *Elements*. Euclid replied that there is "no royal road to geometry"—meaning by this that mathematics does not respect wealth or social status. Euclid was revered in his own time and was referred to as "The Geometer" or "The Writer of the *Elements*." The greatness of the *Elements* stems from its precise, logical, and systematic treatment of geometry. For dealing with equality, Euclid lists the following rules, which he calls "common notions."

1. Things that are equal to the same thing are equal to each other.
2. If equals are added to equals, the sums are equal.
3. If equals are subtracted from equals, the remainders are equal.
4. Things that coincide with one another are equal.
5. The whole is greater than the part.

Here is another method for proving that an equation is an identity. If we can transform each side of the equation *separately*, by way of identities, to arrive at the same result, then the equation is an identity. Example 6 illustrates this procedure.

EXAMPLE 6 ■ Proving an Identity by Working with Both Sides Separately

Verify the identity $\dfrac{1+\cos\theta}{\cos\theta} = \dfrac{\tan^2\theta}{\sec\theta - 1}$.

SOLUTION We prove the identity by changing each side separately into the same expression. (You should supply the reasons for each step.)

$$\text{LHS} = \frac{1+\cos\theta}{\cos\theta} = \frac{1}{\cos\theta} + \frac{\cos\theta}{\cos\theta} = \sec\theta + 1$$

$$\text{RHS} = \frac{\tan^2\theta}{\sec\theta - 1} = \frac{\sec^2\theta - 1}{\sec\theta - 1} = \frac{(\sec\theta - 1)(\sec\theta + 1)}{\sec\theta - 1} = \sec\theta + 1$$

It follows that LHS = RHS, so the equation is an identity.

Now Try Exercise 83 ■

We conclude this section by describing the technique of *trigonometric substitution*, which we use to convert algebraic expressions to trigonometric ones. This is often useful in calculus, for instance, in finding the area of a circle or an ellipse.

EXAMPLE 7 ■ Trigonometric Substitution

Substitute $\sin\theta$ for x in the expression $\sqrt{1-x^2}$, and simplify. Assume that $0 \le \theta \le \pi/2$.

SOLUTION Setting $x = \sin\theta$, we have

$$\begin{aligned} \sqrt{1-x^2} &= \sqrt{1-\sin^2\theta} && \text{Substitute } x = \sin\theta \\ &= \sqrt{\cos^2\theta} && \text{Pythagorean identity} \\ &= \cos\theta && \text{Take square root} \end{aligned}$$

The last equality is true because $\cos\theta \ge 0$ for the values of θ in question.

Now Try Exercise 89 ■

7.1 EXERCISES

CONCEPTS

1. An equation is called an identity if it is valid for ________ values of the variable. The equation $2x = x + x$ is an algebraic identity, and the equation $\sin^2 x + \cos^2 x =$ ________ is a trigonometric identity.
2. For any x it is true that $\cos(-x)$ has the same value as $\cos x$. We express this fact as the identity ________________.

SKILLS

3–12 ■ Simplifying Trigonometric Expressions Write the trigonometric expression in terms of sine and cosine, and then simplify.

3. $\cos t \tan t$
4. $\cos t \csc t$
5. $\sin\theta \sec\theta$
6. $\tan\theta \csc\theta$
7. $\tan^2 x - \sec^2 x$
8. $\dfrac{\sec x}{\csc x}$

9. $\sin u + \cot u \cos u$

10. $\cos^2\theta\,(1 + \tan^2\theta)$

11. $\dfrac{\sec\theta - \cos\theta}{\sin\theta}$

12. $\dfrac{\cot\theta}{\csc\theta - \sin\theta}$

13–28 ■ Simplifying Trigonometric Expressions Simplify the trigonometric expression.

13. $\dfrac{\sin x \sec x}{\tan x}$

14. $\dfrac{\cos x \sec x}{\cot x}$

15. $\dfrac{\sin t + \tan t}{\tan t}$

16. $\dfrac{1 + \cot A}{\csc A}$

17. $\cos^3 x + \sin^2 x \cos x$

18. $\sin^4\alpha - \cos^4\alpha + \cos^2\alpha$

19. $\dfrac{\sec^2 x - 1}{\sec^2 x}$

20. $\dfrac{\sec x - \cos x}{\tan x}$

21. $\dfrac{1 + \cos y}{1 + \sec y}$

22. $\dfrac{1 + \sin y}{1 + \csc y}$

23. $\dfrac{1 + \sin u}{\cos u} + \dfrac{\cos u}{1 + \sin u}$

24. $\dfrac{\sin t}{1 - \cos t} - \csc t$

25. $\dfrac{\cos x}{\sec x + \tan x}$

26. $\dfrac{\cot A - 1}{1 + \tan(-A)}$

27. $\dfrac{1}{1 - \sin\alpha} + \dfrac{1}{1 + \sin\alpha}$

28. $\dfrac{2 + \tan^2 x}{\sec^2 x} - 1$

29–30 ■ Proving an Identity Algebraically and Graphically Consider the given equation. **(a)** Verify algebraically that the equation is an identity. **(b)** Confirm graphically that the equation is an identity.

29. $\dfrac{\cos x}{\sec x \sin x} = \csc x - \sin x$

30. $\dfrac{\tan y}{\csc y} = \sec y - \cos y$

31–88 ■ Proving Identities Verify the identity.

31. $\dfrac{\sin\theta}{\tan\theta} = \cos\theta$

32. $\dfrac{\tan x}{\sec x} = \sin x$

33. $\dfrac{\cos u \sec u}{\tan u} = \cot u$

34. $\dfrac{\cot x \sec x}{\csc x} = 1$

35. $\dfrac{\tan y}{\csc y} = \dfrac{1}{\cos y} - \dfrac{1}{\sec y}$

36. $\dfrac{\cos^2 v}{\sin v} = \csc v - \sin v$

37. $\cos(-x) - \sin(-x) = \cos x + \sin x$

38. $\cot(-\alpha)\cos(-\alpha) + \sin(-\alpha) = -\csc\alpha$

39. $\tan\theta + \cot\theta = \sec\theta\csc\theta$

40. $(\sin x + \cos x)^2 = 1 + 2\sin x\cos x$

41. $(1 - \cos\beta)(1 + \cos\beta) = \dfrac{1}{\csc^2\beta}$

42. $\dfrac{\cos x}{\sec x} + \dfrac{\sin x}{\csc x} = 1$

43. $\dfrac{1}{1 - \sin^2 y} = 1 + \tan^2 y$

44. $\csc x - \sin x = \cos x \cot x$

45. $(\tan x + \cot x)^2 = \sec^2 x + \csc^2 x$

46. $\tan^2 x - \cot^2 x = \sec^2 x - \csc^2 x$

47. $(1 - \sin^2 t + \cos^2 t)^2 + 4\sin^2 t\cos^2 t = 4\cos^2 t$

48. $\dfrac{2\sin x\cos x}{(\sin x + \cos x)^2 - 1} = 1$

49. $\csc x\cos^2 x + \sin x = \csc x$

50. $\cot^2 t - \cos^2 t = \cot^2 t\cos^2 t$

51. $\dfrac{(\sin x + \cos x)^2}{\sin^2 x - \cos^2 x} = \dfrac{\sin^2 x - \cos^2 x}{(\sin x - \cos x)^2}$

52. $(\sin x + \cos x)^4 = (1 + 2\sin x\cos x)^2$

53. $\dfrac{\sec t - \cos t}{\sec t} = \sin^2 t$

54. $(\cot x - \csc x)(\cos x + 1) = -\sin x$

55. $\cos^2 x - \sin^2 x = 2\cos^2 x - 1$

56. $2\cos^2 x - 1 = 1 - 2\sin^2 x$

57. $\sin^4\theta - \cos^4\theta = \sin^2\theta - \cos^2\theta$

58. $(1 - \cos^2 x)(1 + \cot^2 x) = 1$

59. $\dfrac{(\sin t + \cos t)^2}{\sin t\cos t} = 2 + \sec t\csc t$

60. $\sec t\csc t\,(\tan t + \cot t) = \sec^2 t + \csc^2 t$

61. $\dfrac{1 + \tan^2 u}{1 - \tan^2 u} = \dfrac{1}{\cos^2 u - \sin^2 u}$

62. $\dfrac{1 + \sec^2 x}{1 + \tan^2 x} = 1 + \cos^2 x$

63. $\dfrac{\sec x + \csc x}{\tan x + \cot x} = \sin x + \cos x$

64. $\dfrac{\sin x + \cos x}{\sec x + \csc x} = \sin x\cos x$

65. $\dfrac{1 - \cos x}{\sin x} + \dfrac{\sin x}{1 - \cos x} = 2\csc x$

66. $\dfrac{\csc x - \cot x}{\sec x - 1} = \cot x$

67. $\tan^2 u - \sin^2 u = \tan^2 u\sin^2 u$

68. $\sec^4 x - \tan^4 x = \sec^2 x + \tan^2 x$

69. $\dfrac{1 + \tan x}{1 - \tan x} = \dfrac{\cos x + \sin x}{\cos x - \sin x}$

70. $\dfrac{\cos\theta}{1 - \sin\theta} = \dfrac{\sin\theta - \csc\theta}{\cos\theta - \cot\theta}$

71. $\dfrac{1}{\sec x + \tan x} + \dfrac{1}{\sec x - \tan x} = 2\sec x$

72. $\dfrac{\cos^2 t + \tan^2 t - 1}{\sin^2 t} = \tan^2 t$

73. $\dfrac{1 + \sin x}{1 - \sin x} - \dfrac{1 - \sin x}{1 + \sin x} = 4\tan x\sec x$

74. $\dfrac{\tan x + \tan y}{\cot x + \cot y} = \tan x\tan y$

75. $\dfrac{\sin^3 x + \cos^3 x}{\sin x + \cos x} = 1 - \sin x \cos x$

76. $\dfrac{\tan v - \cot v}{\tan^2 v - \cot^2 v} = \sin v \cos v$

77. $\dfrac{1 - \cos \alpha}{\sin \alpha} = \dfrac{\sin \alpha}{1 + \cos \alpha}$

78. $\dfrac{\sin x - 1}{\sin x + 1} = \dfrac{-\cos^2 x}{(\sin x + 1)^2}$

79. $\dfrac{\sin w}{\sin w + \cos w} = \dfrac{\tan w}{1 + \tan w}$

80. $\dfrac{\sin A}{1 - \cos A} - \cot A = \csc A$

81. $\dfrac{\sec x}{\sec x - \tan x} = \sec x(\sec x + \tan x)$

82. $\sec v - \tan v = \dfrac{1}{\sec v + \tan v}$

83. $\dfrac{\cos \theta}{1 - \sin \theta} = \sec \theta + \tan \theta$

84. $\dfrac{\tan v \sin v}{\tan v + \sin v} = \dfrac{\tan v - \sin v}{\tan v \sin v}$

85. $\dfrac{1 - \sin x}{1 + \sin x} = (\sec x - \tan x)^2$

86. $\dfrac{1 + \sin x}{1 - \sin x} = (\tan x + \sec x)^2$

87. $\csc x - \cot x = \dfrac{1}{\csc x + \cot x}$

88. $\dfrac{\sec u - 1}{\sec u + 1} = \dfrac{\tan u - \sin u}{\tan u + \sin u}$

89–94 ■ Trigonometric Substitution Make the indicated trigonometric substitution in the given algebraic expression and simplify (see Example 7). Assume that $0 < \theta < \pi/2$.

89. $\dfrac{x}{\sqrt{1 - x^2}}$, $x = \sin \theta$

90. $\sqrt{1 + x^2}$, $x = \tan \theta$

91. $\sqrt{x^2 - 1}$, $x = \sec \theta$

92. $\dfrac{1}{x^2\sqrt{4 + x^2}}$, $x = 2 \tan \theta$

93. $\sqrt{9 - x^2}$, $x = 3 \sin \theta$

94. $\dfrac{\sqrt{x^2 - 25}}{x}$, $x = 5 \sec \theta$

95–98 ■ Determining Identities Graphically Graph f and g in the same viewing rectangle. Do the graphs suggest that the equation $f(x) = g(x)$ is an identity? Prove your answer.

95. $f(x) = \cos^2 x - \sin^2 x$, $g(x) = 1 - 2\sin^2 x$

96. $f(x) = \tan x(1 + \sin x)$, $g(x) = \dfrac{\sin x \cos x}{1 + \sin x}$

97. $f(x) = (\sin x + \cos x)^2$, $g(x) = 1$

98. $f(x) = \cos^4 x - \sin^4 x$, $g(x) = 2\cos^2 x - 1$

SKILLS Plus

99–104 ■ Proving More Identities Verify the identity.

99. $(\sin x \sin y - \cos x \cos y)(\sin x \sin y + \cos x \cos y) = \sin^2 y - \cos^2 x$

100. $\dfrac{1 + \cos x + \sin x}{1 + \cos x - \sin x} = \dfrac{1 + \sin x}{\cos x}$

101. $(\tan x + \cot x)^4 = \sec^4 x \csc^4 x$

102. $(\sin \alpha - \tan \alpha)(\cos \alpha - \cot \alpha) = (\cos \alpha - 1)(\sin \alpha - 1)$

103. $\dfrac{\sin^3 y - \csc^3 y}{\sin y - \csc y} = \sin^2 y + \csc^2 y + 1$

104. $\sin^6 \beta + \cos^6 \beta = 1 - 3\sin^2 \beta \cos^2 \beta$

105–108 ■ Proving Identities Involving Other Functions These identities involve trigonometric functions as well as other functions that we have studied.

105. $\ln|\tan x \sin x| = 2\ln|\sin x| + \ln|\sec x|$

106. $\ln|\tan x| + \ln|\cot x| = 0$

107. $e^{\sin^2 x}e^{\tan^2 x} = e^{\sec^2 x}e^{-\cos^2 x}$

108. $e^{x + 2\ln|\sin x|} = e^x \sin^2 x$

109–112 ■ Is the Equation an Identity? Determine whether the given equation is an identity. If the equation is not an identity, find all its solutions.

109. $e^{\sin^2 x}e^{\cos^2 x} = e$

110. $\dfrac{x}{x + 1} = 1 + x$

111. $\sqrt{\sin^2 x + 1} = \sqrt{\sin^2 x} + 1$

112. $xe^{\ln x^2} = x^3$

113. An Identity Involving Three Variables Suppose $x = R \cos \theta \sin \phi$, $y = R \sin \theta \sin \phi$, and $z = R \cos \phi$. Verify the identity $x^2 + y^2 + z^2 = R^2$.

DISCUSS ■ DISCOVER ■ PROVE ■ WRITE

114. DISCUSS: Equations That Are Identities You have encountered many identities in this course. Which of the following equations do you recognize as identities? For those that you think are identities, test several values of the variables to confirm that the equation is true for those variables.

(a) $(x + y)^2 = x^2 + 2xy + y^2$

(b) $x^2 + y^2 = 1$

(c) $x(y + z) = xy + xz$

(d) $t^2 - \cos^2 t = (t - \cos t)(t + \cos t)$

(e) $\sin t + \cos t = 1$

(f) $x^2 - \tan^2 x = 0$

115. DISCUSS: Equations That Are Not Identities How can you tell if an equation is not an identity? Show that the following equations are not identities.

(a) $\sin 2x = 2 \sin x$

(b) $\sin(x + y) = \sin x + \sin y$

(c) $\sec^2 x + \csc^2 x = 1$

(d) $\dfrac{1}{\sin x + \cos x} = \csc x + \sec x$

116. DISCUSS: Graphs and Identities Suppose you graph two functions, f and g, on a graphing device and their graphs appear identical in the viewing rectangle. Does this prove that the equation $f(x) = g(x)$ is an identity? Explain.

117. DISCOVER: Making Up Your Own Identity If you start with a trigonometric expression and rewrite it or simplify it, then setting the original expression equal to the rewritten expression yields a trigonometric identity. For instance, from Example 1 we get the identity

$$\cos t + \tan t \sin t = \sec t$$

Use this technique to make up your own identity, then give it to a classmate to verify.

118. DISCUSS: Cofunction Identities In the right triangle shown, explain why $v = (\pi/2) - u$. Explain how you can obtain all six cofunction identities from this triangle for $0 < u < \pi/2$.

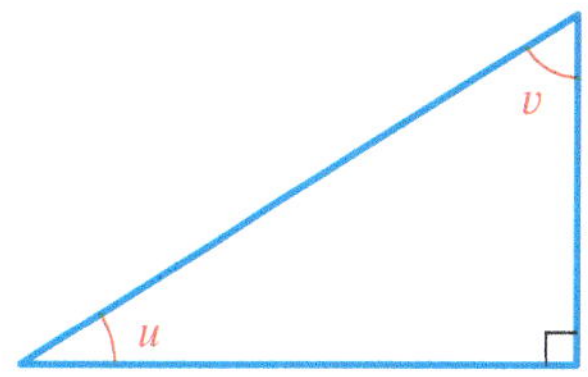

Note that u and v are complementary angles. So the cofunction identities state that "a trigonometric function of an angle u is equal to the corresponding cofunction of the complementary angle v."

7.2 ADDITION AND SUBTRACTION FORMULAS

■ Addition and Subtraction Formulas ■ Evaluating Expressions Involving Inverse Trigonometric Functions ■ Expressions of the form $A \sin x + B \cos x$

■ Addition and Subtraction Formulas

We now derive identities for trigonometric functions of sums and differences.

> **ADDITION AND SUBTRACTION FORMULAS**
>
> **Formulas for sine:**
> $$\sin(s + t) = \sin s \cos t + \cos s \sin t$$
> $$\sin(s - t) = \sin s \cos t - \cos s \sin t$$
>
> **Formulas for cosine:**
> $$\cos(s + t) = \cos s \cos t - \sin s \sin t$$
> $$\cos(s - t) = \cos s \cos t + \sin s \sin t$$
>
> **Formulas for tangent:**
> $$\tan(s + t) = \frac{\tan s + \tan t}{1 - \tan s \tan t}$$
> $$\tan(s - t) = \frac{\tan s - \tan t}{1 + \tan s \tan t}$$

Proof of Addition Formula for Cosine To prove the formula

$$\cos(s + t) = \cos s \cos t - \sin s \sin t$$

we use Figure 1. In the figure, the distances t, $s + t$, and $-s$ have been marked on the unit circle, starting at $P_0(1, 0)$ and terminating at Q_1, P_1, and Q_0, respectively. The coordinates of these points are as follows:

$$P_0(1, 0) \qquad Q_0(\cos(-s), \sin(-s))$$
$$P_1(\cos(s + t), \sin(s + t)) \qquad Q_1(\cos t, \sin t)$$

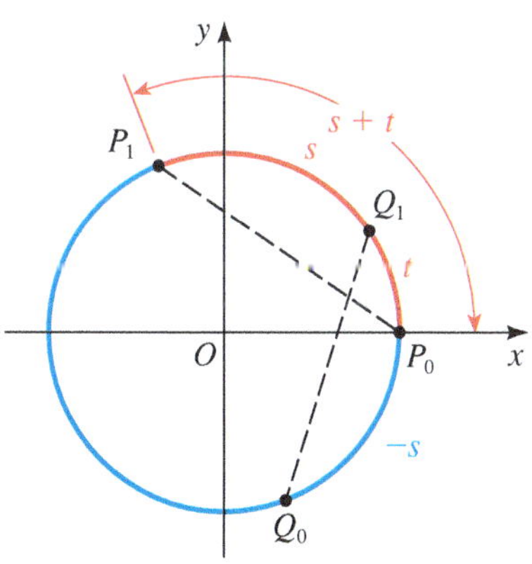

FIGURE 1

Since $\cos(-s) = \cos s$ and $\sin(-s) = -\sin s$, it follows that the point Q_0 has the coordinates $Q_0(\cos s, -\sin s)$. Notice that the distances between P_0 and P_1 and between Q_0 and Q_1 measured along the arc of the circle are equal. Since equal arcs are subtended by equal chords, it follows that $d(P_0, P_1) = d(Q_0, Q_1)$. Using the Distance Formula, we get

$$\sqrt{[\cos(s + t) - 1]^2 + [\sin(s + t) - 0]^2} = \sqrt{(\cos t - \cos s)^2 + (\sin t + \sin s)^2}$$

Stock Montage/Archive Photos/Getty Images

JEAN BAPTISTE JOSEPH FOURIER (1768–1830) is responsible for the most powerful application of the trigonometric functions (see the margin note on page 427). He used sums of these functions to describe such physical phenomena as the transmission of sound and the flow of heat.

Orphaned as a young boy, Fourier was educated in a military school, where he became a mathematics teacher at the age of 20. He was later appointed professor at the École Polytechnique but resigned this position to accompany Napoleon on his expedition to Egypt, where Fourier served as governor. After returning to France, he began conducting experiments on heat. The French Academy refused to publish his early papers on this subject because of his lack of rigor. Fourier eventually became Secretary of the Academy and in this capacity had his papers published in their original form. Probably because of his study of heat and his years in the deserts of Egypt, Fourier became obsessed with keeping himself warm—he wore several layers of clothes, even in the summer, and kept his rooms at unbearably high temperatures. Evidently, these habits overburdened his heart and contributed to his death at the age of 62.

Squaring both sides and expanding, we have

These add to 1

$$\cos^2(s + t) - 2\cos(s + t) + 1 + \sin^2(s + t)$$
$$= \cos^2 t - 2\cos s \cos t + \cos^2 s + \sin^2 t + 2\sin s \sin t + \sin^2 s$$

These add to 1 These add to 1

Using the Pythagorean identity $\sin^2\theta + \cos^2\theta = 1$ three times gives

$$2 - 2\cos(s + t) = 2 - 2\cos s \cos t + 2\sin s \sin t$$

Finally, subtracting 2 from each side and dividing both sides by -2, we get

$$\cos(s + t) = \cos s \cos t - \sin s \sin t$$

which proves the Addition Formula for Cosine. ■

Proof of Subtraction Formula for Cosine Replacing t with $-t$ in the Addition Formula for Cosine, we get

$$\cos(s - t) = \cos(s + (-t))$$
$$= \cos s \cos(-t) - \sin s \sin(-t) \qquad \text{Addition Formula for Cosine}$$
$$= \cos s \cos t + \sin s \sin t \qquad \text{Even-odd identities}$$

This proves the Subtraction Formula for Cosine. ■

See Exercises 77 and 78 for proofs of the other Addition Formulas.

EXAMPLE 1 ■ Using the Addition and Subtraction Formulas

Find the exact value of each expression.

(a) $\cos 75°$ **(b)** $\cos \dfrac{\pi}{12}$

SOLUTION

(a) Notice that $75° = 45° + 30°$. Since we know the exact values of sine and cosine at 45° and 30°, we use the Addition Formula for Cosine to get

$$\cos 75° = \cos(45° + 30°)$$
$$= \cos 45° \cos 30° - \sin 45° \sin 30°$$
$$= \frac{\sqrt{2}}{2}\frac{\sqrt{3}}{2} - \frac{\sqrt{2}}{2}\frac{1}{2} = \frac{\sqrt{2}\sqrt{3} - \sqrt{2}}{4} = \frac{\sqrt{6} - \sqrt{2}}{4}$$

(b) Since $\dfrac{\pi}{12} = \dfrac{\pi}{4} - \dfrac{\pi}{6}$, the Subtraction Formula for Cosine gives

$$\cos\frac{\pi}{12} = \cos\left(\frac{\pi}{4} - \frac{\pi}{6}\right)$$
$$= \cos\frac{\pi}{4}\cos\frac{\pi}{6} + \sin\frac{\pi}{4}\sin\frac{\pi}{6}$$
$$= \frac{\sqrt{2}}{2}\frac{\sqrt{3}}{2} + \frac{\sqrt{2}}{2}\frac{1}{2} = \frac{\sqrt{6} + \sqrt{2}}{4}$$

Now Try Exercises 3 and 9 ■

EXAMPLE 2 ■ Using the Addition Formula for Sine

Find the exact value of the expression $\sin 20° \cos 40° + \cos 20° \sin 40°$.

SOLUTION We recognize the expression as the right-hand side of the Addition Formula for Sine with $s = 20°$ and $t = 40°$. So we have

$$\sin 20° \cos 40° + \cos 20° \sin 40° = \sin(20° + 40°) = \sin 60° = \frac{\sqrt{3}}{2}$$

Now Try Exercise 15 ■

EXAMPLE 3 ■ Proving a Cofunction Identity

Prove the cofunction identity $\cos\left(\frac{\pi}{2} - u\right) = \sin u$.

SOLUTION By the Subtraction Formula for Cosine we have

$$\cos\left(\frac{\pi}{2} - u\right) = \cos\frac{\pi}{2}\cos u + \sin\frac{\pi}{2}\sin u$$
$$= 0 \cdot \cos u + 1 \cdot \sin u = \sin u$$

Now Try Exercises 21 and 25 ■

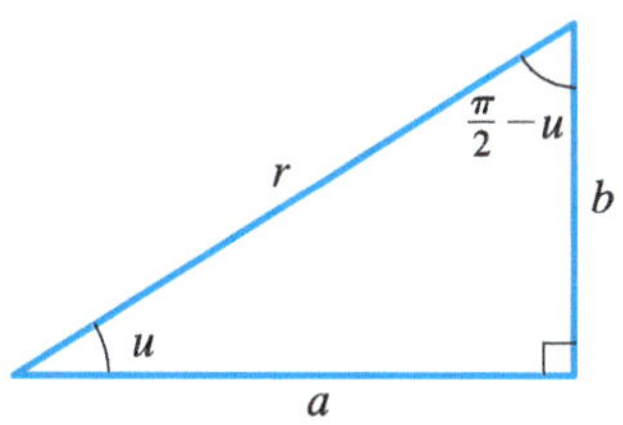

$\cos\left(\frac{\pi}{2} - u\right) = \frac{b}{r} = \sin u$

For acute angles, the cofunction identity in Example 3, as well as the other cofunction identities, can also be derived from the figure in the margin.

EXAMPLE 4 ■ Proving an Identity

Verify the identity $\dfrac{1 + \tan x}{1 - \tan x} = \tan\left(\dfrac{\pi}{4} + x\right)$.

SOLUTION Starting with the right-hand side and using the Addition Formula for Tangent, we get

$$\text{RHS} = \tan\left(\frac{\pi}{4} + x\right) = \frac{\tan\frac{\pi}{4} + \tan x}{1 - \tan\frac{\pi}{4}\tan x}$$
$$= \frac{1 + \tan x}{1 - \tan x} = \text{LHS}$$

Now Try Exercise 33 ■

The next example is a typical use of the Addition and Subtraction Formulas in calculus.

EXAMPLE 5 ■ An Identity from Calculus

If $f(x) = \sin x$, show that

$$\frac{f(x + h) - f(x)}{h} = \sin x\left(\frac{\cos h - 1}{h}\right) + \cos x\left(\frac{\sin h}{h}\right)$$

SOLUTION

$$\begin{aligned}\frac{f(x+h)-f(x)}{h} &= \frac{\sin(x+h)-\sin x}{h} && \text{Definition of } f\\ &= \frac{\sin x\cos h+\cos x\sin h-\sin x}{h} && \text{Addition Formula for Sine}\\ &= \frac{\sin x(\cos h-1)+\cos x\sin h}{h} && \text{Factor}\\ &= \sin x\left(\frac{\cos h-1}{h}\right)+\cos x\left(\frac{\sin h}{h}\right) && \text{Separate the fraction}\end{aligned}$$

Now Try Exercise 65 ■

■ Evaluating Expressions Involving Inverse Trigonometric Functions

Expressions involving trigonometric functions and their inverses arise in calculus. In the next examples we illustrate how to evaluate such expressions.

EXAMPLE 6 ■ Simplifying an Expression Involving Inverse Trigonometric Functions

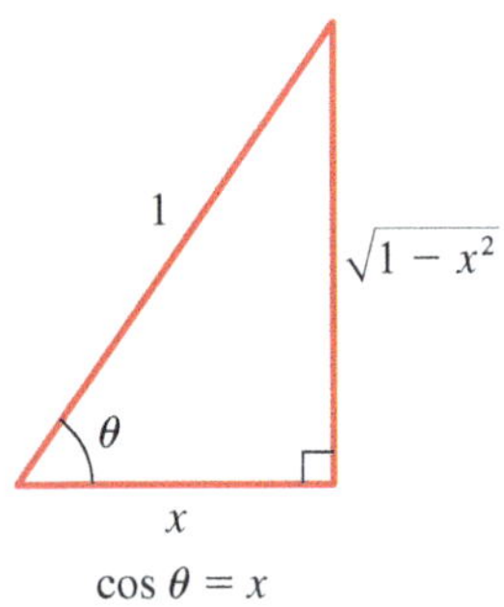

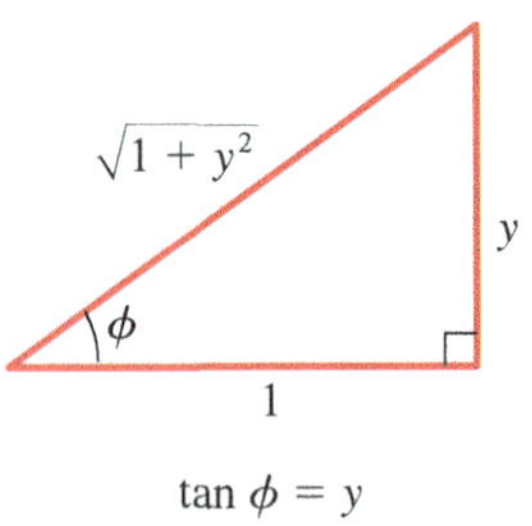

FIGURE 2

Write $\sin(\cos^{-1}x + \tan^{-1}y)$ as an algebraic expression in x and y, where $-1 \le x \le 1$ and y is any real number.

SOLUTION Let $\theta = \cos^{-1}x$ and $\phi = \tan^{-1}y$. Using the methods of Section 6.4, we sketch triangles with angles θ and ϕ such that $\cos\theta = x$ and $\tan\phi = y$ (see Figure 2). From the triangles we have

$$\sin\theta = \sqrt{1-x^2} \qquad \cos\phi = \frac{1}{\sqrt{1+y^2}} \qquad \sin\phi = \frac{y}{\sqrt{1+y^2}}$$

From the Addition Formula for Sine we have

$$\begin{aligned}\sin(\cos^{-1}x+\tan^{-1}y) &= \sin(\theta+\phi)\\ &= \sin\theta\cos\phi+\cos\theta\sin\phi && \text{Addition Formula for Sine}\\ &= \sqrt{1-x^2}\frac{1}{\sqrt{1+y^2}}+x\frac{y}{\sqrt{1+y^2}} && \text{From triangles}\\ &= \frac{1}{\sqrt{1+y^2}}\left(\sqrt{1-x^2}+xy\right) && \text{Factor } \frac{1}{\sqrt{1+y^2}}\end{aligned}$$

Now Try Exercises 47 and 51 ■

EXAMPLE 7 ■ Evaluating an Expression Involving Trigonometric Functions

Evaluate $\sin(\theta+\phi)$, where $\sin\theta = \frac{12}{13}$ with θ in Quadrant II and $\tan\phi = \frac{3}{4}$ with ϕ in Quadrant III.

SOLUTION We first sketch the angles θ and ϕ in standard position with terminal sides in the appropriate quadrants as in Figure 3. Since $\sin\theta = y/r = \frac{12}{13}$, we can label a side and the hypotenuse in the triangle in Figure 3(a). To find the remaining side, we use the Pythagorean Theorem.

$$\begin{aligned} x^2 + y^2 &= r^2 && \text{Pythagorean Theorem} \\ x^2 + 12^2 &= 13^2 && y = 12, \quad r = 13 \\ x^2 &= 25 && \text{Solve for } x^2 \\ x &= -5 && \text{Because } x < 0 \end{aligned}$$

Similarly, since $\tan \phi = y/x = \frac{3}{4}$, we can label two sides of the triangle in Figure 3(b) and then use the Pythagorean Theorem to find the hypotenuse.

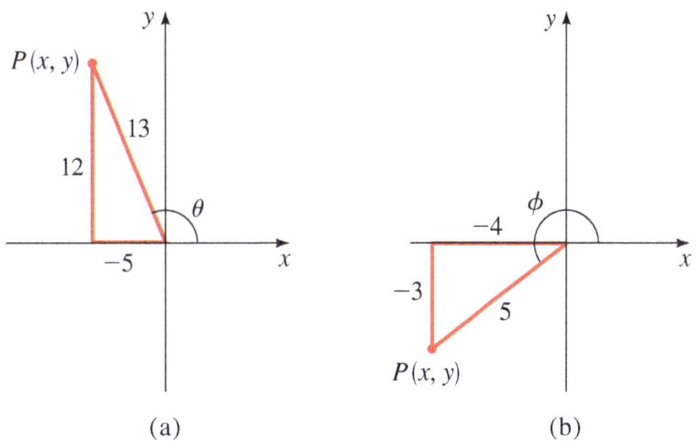

FIGURE 3

Now, to find $\sin(\theta + \phi)$, we use the Addition Formula for Sine and the triangles in Figure 3.

$$\begin{aligned} \sin(\theta + \phi) &= \sin \theta \cos \phi + \cos \theta \sin \phi && \text{Addition Formula} \\ &= \left(\tfrac{12}{13}\right)\left(-\tfrac{4}{5}\right) + \left(-\tfrac{5}{13}\right)\left(-\tfrac{3}{5}\right) && \text{From triangles} \\ &= -\tfrac{33}{65} && \text{Calculate} \end{aligned}$$

Now Try Exercise 55 ■

■ Expressions of the Form *A* sin *x* + *B* cos *x*

We can write expressions of the form $A \sin x + B \cos x$ in terms of a single trigonometric function using the Addition Formula for Sine. For example, consider the expression

$$\frac{1}{2} \sin x + \frac{\sqrt{3}}{2} \cos x$$

If we set $\phi = \pi/3$, then $\cos \phi = \frac{1}{2}$ and $\sin \phi = \sqrt{3}/2$, and we can write

$$\begin{aligned} \frac{1}{2} \sin x + \frac{\sqrt{3}}{2} \cos x &= \cos \phi \sin x + \sin \phi \cos x \\ &= \sin(x + \phi) = \sin\left(x + \frac{\pi}{3}\right) \end{aligned}$$

We are able to do this because the coefficients $\frac{1}{2}$ and $\sqrt{3}/2$ are precisely the cosine and sine of a particular number, in this case, $\pi/3$. We can use this same idea in general to write $A \sin x + B \cos x$ in the form $k \sin(x + \phi)$. We start by multiplying the numerator and denominator by $\sqrt{A^2 + B^2}$ to get

$$A \sin x + B \cos x = \sqrt{A^2 + B^2}\left(\frac{A}{\sqrt{A^2 + B^2}} \sin x + \frac{B}{\sqrt{A^2 + B^2}} \cos x\right)$$

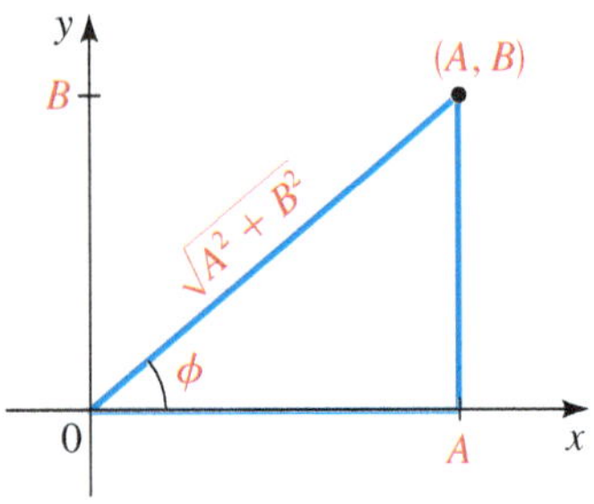

FIGURE 4

We need a number ϕ with the property that

$$\cos \phi = \frac{A}{\sqrt{A^2 + B^2}} \quad \text{and} \quad \sin \phi = \frac{B}{\sqrt{A^2 + B^2}}$$

Figure 4 shows that the point (A, B) in the plane determines a number ϕ with precisely this property. With this ϕ we have

$$\begin{aligned} A \sin x + B \cos x &= \sqrt{A^2 + B^2}(\cos \phi \sin x + \sin \phi \cos x) \\ &= \sqrt{A^2 + B^2} \sin(x + \phi) \end{aligned}$$

We have proved the following theorem.

SUMS OF SINES AND COSINES

If A and B are real numbers, then

$$A \sin x + B \cos x = k \sin(x + \phi)$$

where $k = \sqrt{A^2 + B^2}$ and ϕ satisfies

$$\cos \phi = \frac{A}{\sqrt{A^2 + B^2}} \quad \text{and} \quad \sin \phi = \frac{B}{\sqrt{A^2 + B^2}}$$

EXAMPLE 8 ■ A Sum of Sine and Cosine Terms

Express $3 \sin x + 4 \cos x$ in the form $k \sin(x + \phi)$.

SOLUTION By the preceding theorem, $k = \sqrt{A^2 + B^2} = \sqrt{3^2 + 4^2} = 5$. The angle ϕ has the property that $\sin \phi = B/k = \frac{4}{5}$ and $\cos \phi = A/k = \frac{3}{5}$, and ϕ in Quadrant I (because $\sin \phi$ and $\cos \phi$ are both positive), so $\phi = \sin^{-1} \frac{4}{5}$. Using a calculator, we get $\phi \approx 53.1°$. Thus

$$3 \sin x + 4 \cos x \approx 5 \sin(x + 53.1°)$$

Now Try Exercise 59 ■

EXAMPLE 9 ■ Graphing a Trigonometric Function

Write the function $f(x) = -\sin 2x + \sqrt{3} \cos 2x$ in the form $k \sin(2x + \phi)$, and use the new form to graph the function.

SOLUTION Since $A = -1$ and $B = \sqrt{3}$, we have $k = \sqrt{A^2 + B^2} = \sqrt{1 + 3} = 2$. The angle ϕ satisfies $\cos \phi = -\frac{1}{2}$ and $\sin \phi = \sqrt{3}/2$. From the signs of these quantities we conclude that ϕ is in Quadrant II. Thus $\phi = 2\pi/3$. By the preceding theorem we can write

$$f(x) = -\sin 2x + \sqrt{3} \cos 2x = 2 \sin\left(2x + \frac{2\pi}{3}\right)$$

Using the form

$$f(x) = 2 \sin 2\left(x + \frac{\pi}{3}\right)$$

we see that the graph is a sine curve with amplitude 2, period $2\pi/2 = \pi$, and phase shift $-\pi/3$. The graph is shown in Figure 5.

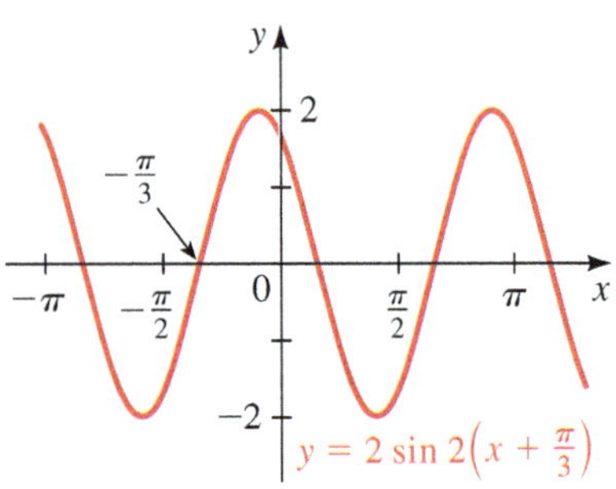

FIGURE 5

Now Try Exercise 63 ■

7.2 EXERCISES

CONCEPTS

1. If we know the values of the sine and cosine of x and y, we can find the value of $\sin(x + y)$ by using the ________ Formula for Sine. State the formula: $\sin(x + y) =$ ________________.

2. If we know the values of the sine and cosine of x and y, we can find the value of $\cos(x - y)$ by using the ________ Formula for Cosine. State the formula: $\cos(x - y) =$ ________________.

SKILLS

3–14 ■ Values of Trigonometric Functions Use an Addition or Subtraction Formula to find the exact value of the expression, as demonstrated in Example 1.

3. $\sin 75°$
4. $\sin 15°$
5. $\cos 105°$
6. $\cos 195°$
7. $\tan 15°$
8. $\tan 165°$
9. $\sin \dfrac{19\pi}{12}$
10. $\cos \dfrac{17\pi}{12}$
11. $\tan\left(-\dfrac{\pi}{12}\right)$
12. $\sin\left(-\dfrac{5\pi}{12}\right)$
13. $\cos \dfrac{11\pi}{12}$
14. $\tan \dfrac{7\pi}{12}$

15–20 ■ Values of Trigonometric Functions Use an Addition or Subtraction Formula to write the expression as a trigonometric function of one number, and then find its exact value.

15. $\sin 18° \cos 27° + \cos 18° \sin 27°$
16. $\cos 10° \cos 80° - \sin 10° \sin 80°$
17. $\cos \dfrac{3\pi}{7} \cos \dfrac{2\pi}{21} + \sin \dfrac{3\pi}{7} \sin \dfrac{2\pi}{21}$
18. $\dfrac{\tan \frac{\pi}{18} + \tan \frac{\pi}{9}}{1 - \tan \frac{\pi}{18} \tan \frac{\pi}{9}}$
19. $\dfrac{\tan 73° - \tan 13°}{1 + \tan 73° \tan 13°}$
20. $\cos \dfrac{13\pi}{15} \cos\left(-\dfrac{\pi}{5}\right) - \sin \dfrac{13\pi}{15} \sin\left(-\dfrac{\pi}{5}\right)$

21–24 ■ Cofunction Identities Prove the cofunction identity using the Addition and Subtraction Formulas.

21. $\tan\left(\dfrac{\pi}{2} - u\right) = \cot u$
22. $\cot\left(\dfrac{\pi}{2} - u\right) = \tan u$
23. $\sec\left(\dfrac{\pi}{2} - u\right) = \csc u$
24. $\csc\left(\dfrac{\pi}{2} - u\right) = \sec u$

25–46 ■ Proving Identities Prove the identity.

25. $\sin\left(x - \dfrac{\pi}{2}\right) = -\cos x$
26. $\cos\left(x - \dfrac{\pi}{2}\right) = \sin x$
27. $\sin(x - \pi) = -\sin x$
28. $\cos(x - \pi) = -\cos x$
29. $\tan(x - \pi) = \tan x$
30. $\tan\left(x - \dfrac{\pi}{2}\right) = -\cot x$
31. $\sin\left(\dfrac{\pi}{2} - x\right) = \sin\left(\dfrac{\pi}{2} + x\right)$
32. $\cos\left(x + \dfrac{\pi}{3}\right) + \sin\left(x - \dfrac{\pi}{6}\right) = 0$
33. $\tan\left(x + \dfrac{\pi}{3}\right) = \dfrac{\sqrt{3} + \tan x}{1 - \sqrt{3} \tan x}$
34. $\tan\left(x - \dfrac{\pi}{4}\right) = \dfrac{\tan x - 1}{\tan x + 1}$
35. $\sin(x + y) - \sin(x - y) = 2 \cos x \sin y$
36. $\cos(x + y) + \cos(x - y) = 2 \cos x \cos y$
37. $\cot(x - y) = \dfrac{\cot x \cot y + 1}{\cot y - \cot x}$
38. $\cot(x + y) = \dfrac{\cot x \cot y - 1}{\cot x + \cot y}$
39. $\tan x - \tan y = \dfrac{\sin(x - y)}{\cos x \cos y}$
40. $1 - \tan x \tan y = \dfrac{\cos(x + y)}{\cos x \cos y}$
41. $\dfrac{\tan x - \tan y}{1 - \tan x \tan y} = \dfrac{\sin(x - y)}{\cos(x + y)}$
42. $\dfrac{\sin(x + y) - \sin(x - y)}{\cos(x + y) + \cos(x - y)} = \tan y$
43. $\cos(x + y) \cos(x - y) = \cos^2 x - \sin^2 y$
44. $\cos(x + y) \cos y + \sin(x + y) \sin y = \cos x$
45. $\sin(x + y + z) = \sin x \cos y \cos z + \cos x \sin y \cos z + \cos x \cos y \sin z - \sin x \sin y \sin z$
46. $\tan(x - y) + \tan(y - z) + \tan(z - x) = \tan(x - y) \tan(y - z) \tan(z - x)$

47–50 ■ Expressions Involving Inverse Trigonometric Functions Write the given expression in terms of x and y only.

47. $\cos(\sin^{-1} x - \tan^{-1} y)$
48. $\tan(\sin^{-1} x + \cos^{-1} y)$
49. $\sin(\tan^{-1} x - \tan^{-1} y)$
50. $\sin(\sin^{-1} x + \cos^{-1} y)$

51–54 ■ Expressions Involving Inverse Trigonometric Functions Find the exact value of the expression.

51. $\sin\left(\cos^{-1}\frac{1}{2} + \tan^{-1} 1\right)$ **52.** $\cos\left(\sin^{-1}\frac{\sqrt{3}}{2} + \cot^{-1}\sqrt{3}\right)$

53. $\tan\left(\sin^{-1}\frac{3}{4} - \cos^{-1}\frac{1}{3}\right)$ **54.** $\sin\left(\cos^{-1}\frac{2}{3} - \tan^{-1}\frac{1}{2}\right)$

55–58 ■ Evaluating Expressions Involving Trigonometric Functions Evaluate each expression under the given conditions.

55. $\cos(\theta - \phi)$; $\cos\theta = \frac{3}{5}$, θ in Quadrant IV, $\tan\phi = -\sqrt{3}$, ϕ in Quadrant II.

56. $\sin(\theta - \phi)$; $\tan\theta = \frac{4}{3}$, θ in Quadrant III, $\sin\phi = -\sqrt{10}/10$, ϕ in Quadrant IV

57. $\sin(\theta + \phi)$; $\sin\theta = \frac{5}{13}$, θ in Quadrant I, $\cos\phi = -2\sqrt{5}/5$, ϕ in Quadrant II

58. $\tan(\theta + \phi)$; $\cos\theta = -\frac{1}{3}$, θ in Quadrant III, $\sin\phi = \frac{1}{4}$, ϕ in Quadrant II

59–62 ■ Expressions in Terms of Sine Write the expression in terms of sine only.

59. $-\sqrt{3}\sin x + \cos x$ **60.** $\sin x - \cos x$

61. $5(\sin 2x - \cos 2x)$ **62.** $3\sin\pi x + 3\sqrt{3}\cos\pi x$

63–64 ■ Graphing a Trigonometric Function **(a)** Express the function in terms of sine only. **(b)** Graph the function.

63. $g(x) = \cos 2x + \sqrt{3}\sin 2x$ **64.** $f(x) = \sin x + \cos x$

SKILLS Plus

65–66 ■ Difference Quotient Let $f(x) = \cos x$ and $g(x) = \sin x$. Use Addition or Subtraction Formulas to show the following.

65. $\dfrac{f(x+h) - f(x)}{h} = -\cos x\left(\dfrac{1 - \cos h}{h}\right) - \sin x\left(\dfrac{\sin h}{h}\right)$

66. $\dfrac{g(x+h) - g(x)}{h} = \left(\dfrac{\sin h}{h}\right)\cos x - \sin x\left(\dfrac{1 - \cos h}{h}\right)$

67–68 ■ Discovering an Identity Graphically In these exercises we discover an identity graphically and then prove the identity. **(a)** Graph the function and make a conjecture, then **(b)** prove that your conjecture is true.

67. $y = \sin^2\left(x + \dfrac{\pi}{4}\right) + \sin^2\left(x - \dfrac{\pi}{4}\right)$

68. $y = -\frac{1}{2}[\cos(x + \pi) + \cos(x - \pi)]$

69. Difference of Two Angles Show that if $\beta - \alpha = \pi/2$, then

$$\sin(x + \alpha) + \cos(x + \beta) = 0$$

70. Sum of Two Angles Refer to the figure. Show that $\alpha + \beta = \gamma$, and find $\tan\gamma$.

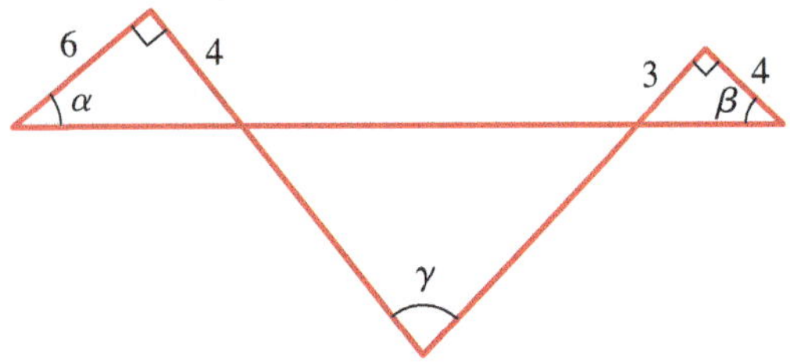

71–72 ■ Identities Involving Inverse Trigonometric Functions Prove the identity.

71. $\tan^{-1}\left(\dfrac{x + y}{1 - xy}\right) = \tan^{-1}x + \tan^{-1}y$

[*Hint:* Let $u = \tan^{-1}x$ and $v = \tan^{-1}y$, so that $x = \tan u$ and $y = \tan v$. Use an Addition Formula to find $\tan(u + v)$.]

72. $\tan^{-1}x + \tan^{-1}\left(\dfrac{1}{x}\right) = \dfrac{\pi}{2}$, $x > 0$ [*Hint:* Let $u = \tan^{-1}x$ and $v = \tan^{-1}\left(\dfrac{1}{x}\right)$, so that $x = \tan u$ and $\dfrac{1}{x} = \tan v$. Use an Addition Formula to find $\cot(u + v)$.]

73. Angle Between Two Lines In this exercise we find a formula for the angle formed by two lines in a coordinate plane.

(a) If L is a line in the plane and θ is the angle formed by the line and the x-axis as shown in the figure, show that the slope m of the line is given by

$$m = \tan\theta$$

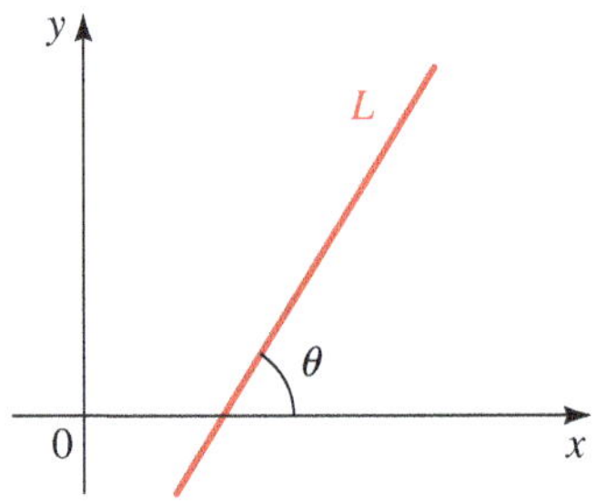

(b) Let L_1 and L_2 be two nonparallel lines in the plane with slopes m_1 and m_2, respectively. Let ψ be the acute angle formed by the two lines (see the following figure). Show that

$$\tan\psi = \frac{m_2 - m_1}{1 + m_1 m_2}$$

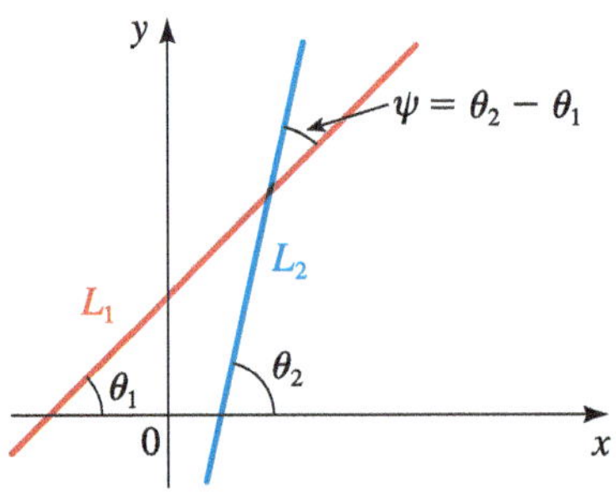

(c) Find the acute angle formed by the two lines

$$y = \tfrac{1}{3}x + 1 \quad\text{and}\quad y = \tfrac{1}{2}x - 3$$

(d) Show that if two lines are perpendicular, then the slope of one is the negative reciprocal of the slope of the other. [*Hint:* First find an expression for $\cot\psi$.]

74. Find $\angle A + \angle B + \angle C$ in the figure. [*Hint:* First use an Addition Formula to find $\tan(A + B)$.]

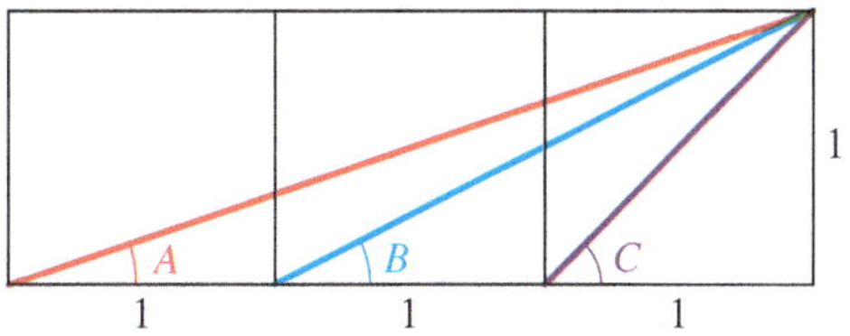

APPLICATIONS

75. Adding an Echo A digital delay device echoes an input signal by repeating it a fixed length of time after it is received. If such a device receives the pure note $f_1(t) = 5 \sin t$ and echoes the pure note $f_2(t) = 5 \cos t$, then the combined sound is $f(t) = f_1(t) + f_2(t)$.

(a) Graph $y = f(t)$, and observe that the graph has the form of a sine curve $y = k \sin(t + \phi)$.

(b) Find k and ϕ.

76. Interference Two identical tuning forks are struck, one a fraction of a second after the other. The sounds produced are modeled by $f_1(t) = C \sin \omega t$ and $f_2(t) = C \sin(\omega t + \alpha)$. The two sound waves interfere to produce a single sound modeled by the sum of these functions

$$f(t) = C \sin \omega t + C \sin(\omega t + \alpha)$$

(a) Use the Addition Formula for Sine to show that f can be written in the form $f(t) = A \sin \omega t + B \cos \omega t$, where A and B are constants that depend on α.

(b) Suppose that $C = 10$ and $\alpha = \pi/3$. Find constants k and ϕ so that $f(t) = k \sin(\omega t + \phi)$.

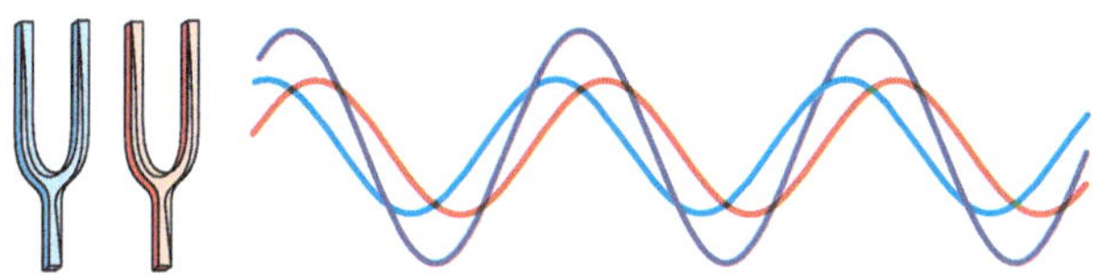

DISCUSS ■ DISCOVER ■ PROVE ■ WRITE

77. PROVE: Addition Formula for Sine In the text we proved only the Addition and Subtraction Formulas for Cosine. Use these formulas and the cofunction identities

$$\sin x = \cos\left(\frac{\pi}{2} - x\right)$$

$$\cos x = \sin\left(\frac{\pi}{2} - x\right)$$

to prove the Addition Formula for Sine. [*Hint:* To get started, use the first cofunction identity to write

$$\sin(s + t) = \cos\left(\frac{\pi}{2} - (s + t)\right)$$
$$= \cos\left(\left(\frac{\pi}{2} - s\right) - t\right)$$

and use the Subtraction Formula for Cosine.]

78. PROVE: Addition Formula for Tangent Use the Addition Formulas for Cosine and Sine to prove the Addition Formula for Tangent. [*Hint:* Use

$$\tan(s + t) = \frac{\sin(s + t)}{\cos(s + t)}$$

and divide the numerator and denominator by $\cos s \cos t$.]

7.3 DOUBLE-ANGLE, HALF-ANGLE, AND PRODUCT-SUM FORMULAS

■ Double-Angle Formulas ■ Half-Angle Formulas ■ Evaluating Expressions Involving Inverse Trigonometric Functions ■ Product-Sum Formulas

The identities we consider in this section are consequences of the addition formulas. The **Double-Angle Formulas** allow us to find the values of the trigonometric functions at $2x$ from their values at x. The **Half-Angle Formulas** relate the values of the trigonometric functions at $\frac{1}{2}x$ to their values at x. The **Product-Sum Formulas** relate products of sines and cosines to sums of sines and cosines.

■ Double-Angle Formulas

The formulas in the box on the next page are immediate consequences of the addition formulas, which we proved in Section 7.2.

DOUBLE-ANGLE FORMULAS

Formula for sine: $\sin 2x = 2 \sin x \cos x$

Formulas for cosine:

$$\begin{aligned}\cos 2x &= \cos^2 x - \sin^2 x\\ &= 1 - 2\sin^2 x\\ &= 2\cos^2 x - 1\end{aligned}$$

Formula for tangent: $\tan 2x = \dfrac{2 \tan x}{1 - \tan^2 x}$

The proofs for the formulas for cosine are given here. You are asked to prove the remaining formulas in Exercises 35 and 36.

Proof of Double-Angle Formulas for Cosine

$$\begin{aligned}\cos 2x &= \cos(x + x)\\ &= \cos x \cos x - \sin x \sin x\\ &= \cos^2 x - \sin^2 x\end{aligned}$$

The second and third formulas for $\cos 2x$ are obtained from the formula we just proved and the Pythagorean identity. Substituting $\cos^2 x = 1 - \sin^2 x$ gives

$$\begin{aligned}\cos 2x &= \cos^2 x - \sin^2 x\\ &= (1 - \sin^2 x) - \sin^2 x\\ &= 1 - 2\sin^2 x\end{aligned}$$

The third formula is obtained in the same way, by substituting $\sin^2 x = 1 - \cos^2 x$. ■

EXAMPLE 1 ■ Using the Double-Angle Formulas

If $\cos x = -\frac{2}{3}$ and x is in Quadrant II, find $\cos 2x$ and $\sin 2x$.

SOLUTION Using one of the Double-Angle Formulas for Cosine, we get

$$\begin{aligned}\cos 2x &= 2\cos^2 x - 1\\ &= 2\left(-\frac{2}{3}\right)^2 - 1 = \frac{8}{9} - 1 = -\frac{1}{9}\end{aligned}$$

To use the formula $\sin 2x = 2 \sin x \cos x$, we need to find $\sin x$ first. We have

$$\sin x = \sqrt{1 - \cos^2 x} = \sqrt{1 - \left(-\tfrac{2}{3}\right)^2} = \frac{\sqrt{5}}{3}$$

where we have used the positive square root because $\sin x$ is positive in Quadrant II. Thus

$$\begin{aligned}\sin 2x &= 2 \sin x \cos x\\ &= 2\left(\frac{\sqrt{5}}{3}\right)\left(-\frac{2}{3}\right) = -\frac{4\sqrt{5}}{9}\end{aligned}$$

Now Try Exercise 3 ■

EXAMPLE 2 ■ A Triple-Angle Formula

Write $\cos 3x$ in terms of $\cos x$.

SOLUTION

$$\begin{aligned}
\cos 3x &= \cos(2x + x) \\
&= \cos 2x \cos x - \sin 2x \sin x && \text{Addition formula} \\
&= (2\cos^2 x - 1)\cos x - (2\sin x\cos x)\sin x && \text{Double-Angle Formulas} \\
&= 2\cos^3 x - \cos x - 2\sin^2 x\cos x && \text{Expand} \\
&= 2\cos^3 x - \cos x - 2\cos x(1 - \cos^2 x) && \text{Pythagorean identity} \\
&= 2\cos^3 x - \cos x - 2\cos x + 2\cos^3 x && \text{Expand} \\
&= 4\cos^3 x - 3\cos x && \text{Simplify}
\end{aligned}$$

Now Try Exercise 109 ■

Example 2 shows that $\cos 3x$ can be written as a polynomial of degree 3 in $\cos x$. The identity $\cos 2x = 2\cos^2 x - 1$ shows that $\cos 2x$ is a polynomial of degree 2 in $\cos x$. In fact, for any natural number n we can write $\cos nx$ as a polynomial in $\cos x$ of degree n (see the note following Exercise 109). The analogous result for $\sin nx$ is not true in general.

EXAMPLE 3 ■ Proving an Identity

Prove the identity $\dfrac{\sin 3x}{\sin x\cos x} = 4\cos x - \sec x$.

SOLUTION We start with the left-hand side.

$$\begin{aligned}
\frac{\sin 3x}{\sin x\cos x} &= \frac{\sin(x + 2x)}{\sin x\cos x} \\
&= \frac{\sin x\cos 2x + \cos x\sin 2x}{\sin x\cos x} && \text{Addition Formula} \\
&= \frac{\sin x(2\cos^2 x - 1) + \cos x(2\sin x\cos x)}{\sin x\cos x} && \text{Double-Angle Formulas} \\
&= \frac{\sin x(2\cos^2 x - 1)}{\sin x\cos x} + \frac{\cos x(2\sin x\cos x)}{\sin x\cos x} && \text{Separate fraction} \\
&= \frac{2\cos^2 x - 1}{\cos x} + 2\cos x && \text{Cancel} \\
&= 2\cos x - \frac{1}{\cos x} + 2\cos x && \text{Separate fraction} \\
&= 4\cos x - \sec x && \text{Reciprocal identity}
\end{aligned}$$

Now Try Exercise 87 ■

■ Half-Angle Formulas

The following formulas allow us to write any trigonometric expression involving even powers of sine and cosine in terms of the first power of cosine only. This technique is important in calculus. The Half-Angle Formulas are immediate consequences of these formulas.

FORMULAS FOR LOWERING POWERS

$$\sin^2 x = \frac{1 - \cos 2x}{2} \qquad \cos^2 x = \frac{1 + \cos 2x}{2}$$

$$\tan^2 x = \frac{1 - \cos 2x}{1 + \cos 2x}$$

Proof The first formula is obtained by solving for $\sin^2 x$ in the Double-Angle Formula $\cos 2x = 1 - 2\sin^2 x$. Similarly, the second formula is obtained by solving for $\cos^2 x$ in the Double-Angle Formula $\cos 2x = 2\cos^2 x - 1$.

The last formula follows from the first two and the reciprocal identities:

$$\tan^2 x = \frac{\sin^2 x}{\cos^2 x} = \frac{\dfrac{1 - \cos 2x}{2}}{\dfrac{1 + \cos 2x}{2}} = \frac{1 - \cos 2x}{1 + \cos 2x}$$

■

EXAMPLE 4 ■ Lowering Powers in a Trigonometric Expression

Express $\sin^2 x \cos^2 x$ in terms of the first power of cosine.

SOLUTION We use the formulas for lowering powers repeatedly.

$$\begin{aligned}
\sin^2 x \cos^2 x &= \left(\frac{1 - \cos 2x}{2}\right)\left(\frac{1 + \cos 2x}{2}\right) \\
&= \frac{1 - \cos^2 2x}{4} = \frac{1}{4} - \frac{1}{4}\cos^2 2x \\
&= \frac{1}{4} - \frac{1}{4}\left(\frac{1 + \cos 4x}{2}\right) = \frac{1}{4} - \frac{1}{8} - \frac{\cos 4x}{8} \\
&= \frac{1}{8} - \frac{1}{8}\cos 4x = \frac{1}{8}(1 - \cos 4x)
\end{aligned}$$

Another way to obtain this identity is to use the Double-Angle Formula for Sine in the form $\sin x \cos x = \frac{1}{2}\sin 2x$. Thus

$$\begin{aligned}
\sin^2 x \cos^2 x &= \frac{1}{4}\sin^2 2x = \frac{1}{4}\left(\frac{1 - \cos 4x}{2}\right) \\
&= \frac{1}{8}(1 - \cos 4x)
\end{aligned}$$

Now Try Exercise 11 ■

HALF-ANGLE FORMULAS

$$\sin\frac{u}{2} = \pm\sqrt{\frac{1 - \cos u}{2}} \qquad \cos\frac{u}{2} = \pm\sqrt{\frac{1 + \cos u}{2}}$$

$$\tan\frac{u}{2} = \frac{1 - \cos u}{\sin u} = \frac{\sin u}{1 + \cos u}$$

The choice of the $+$ or $-$ sign depends on the quadrant in which $u/2$ lies.

Proof We substitute $x = u/2$ in the formulas for lowering powers and take the square root of each side. This gives the first two Half-Angle Formulas. In the case of the Half-Angle Formula for Tangent we get

$$\tan\frac{u}{2} = \pm\sqrt{\frac{1-\cos u}{1+\cos u}}$$

$$= \pm\sqrt{\left(\frac{1-\cos u}{1+\cos u}\right)\left(\frac{1-\cos u}{1-\cos u}\right)} \qquad \text{Multiply numerator and denominator by } 1-\cos u$$

$$= \pm\sqrt{\frac{(1-\cos u)^2}{1-\cos^2 u}} \qquad \text{Simplify}$$

$$= \pm\frac{|1-\cos u|}{|\sin u|} \qquad \sqrt{A^2} = |A| \text{ and } 1-\cos^2 u = \sin^2 u$$

Now, $1 - \cos u$ is nonnegative for all values of u. It is also true that $\sin u$ and $\tan(u/2)$ always have the same sign. (Verify this.) It follows that

$$\tan\frac{u}{2} = \frac{1-\cos u}{\sin u}$$

The other Half-Angle Formula for Tangent is derived from this by multiplying the numerator and denominator by $1 + \cos u$. ■

EXAMPLE 5 ■ Using a Half-Angle Formula

Find the exact value of $\sin 22.5°$.

SOLUTION Since 22.5° is half of 45°, we use the Half-Angle Formula for Sine with $u = 45°$. We choose the + sign because 22.5° is in the first quadrant.

$$\sin\frac{45°}{2} = \sqrt{\frac{1-\cos 45°}{2}} \qquad \text{Half-Angle Formula}$$

$$= \sqrt{\frac{1-\sqrt{2}/2}{2}} \qquad \cos 45° = \sqrt{2}/2$$

$$= \sqrt{\frac{2-\sqrt{2}}{4}} \qquad \text{Common denominator}$$

$$= \tfrac{1}{2}\sqrt{2-\sqrt{2}} \qquad \text{Simplify}$$

Now Try Exercise 17 ■

EXAMPLE 6 ■ Using a Half-Angle Formula

Find $\tan(u/2)$ if $\sin u = \frac{2}{5}$ and u is in Quadrant II.

SOLUTION To use the Half-Angle Formula for Tangent, we first need to find $\cos u$. Since cosine is negative in Quadrant II, we have

$$\cos u = -\sqrt{1-\sin^2 u}$$

$$= -\sqrt{1-\left(\tfrac{2}{5}\right)^2} = -\frac{\sqrt{21}}{5}$$

Thus

$$\tan\frac{u}{2} = \frac{1-\cos u}{\sin u}$$

$$= \frac{1+\sqrt{21}/5}{\frac{2}{5}} = \frac{5+\sqrt{21}}{2}$$

Now Try Exercise 37 ■

Evaluating Expressions Involving Inverse Trigonometric Functions

Expressions involving trigonometric functions and their inverses arise in calculus. In the next examples we illustrate how to evaluate such expressions.

EXAMPLE 7 ■ Simplifying an Expression Involving an Inverse Trigonometric Function

Write $\sin(2\cos^{-1}x)$ as an algebraic expression in x only, where $-1 \le x \le 1$.

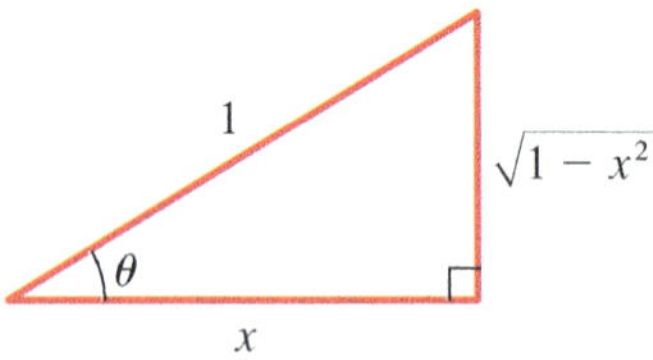

FIGURE 1

SOLUTION Let $\theta = \cos^{-1}x$, and sketch a triangle as in Figure 1. We need to find $\sin 2\theta$, but from the triangle we can find trigonometric functions of θ only, not 2θ. So we use the Double-Angle Formula for Sine.

$$\begin{aligned} \sin(2\cos^{-1}x) &= \sin 2\theta && \cos^{-1}x = \theta \\ &= 2\sin\theta\cos\theta && \text{Double-Angle Formula} \\ &= 2x\sqrt{1-x^2} && \text{From the triangle} \end{aligned}$$

Now Try Exercises 43 and 47 ■

EXAMPLE 8 ■ Evaluating an Expression Involving Trigonometric Functions

Evaluate $\sin 2\theta$, where $\cos\theta = -\frac{2}{5}$ with θ in Quadrant II.

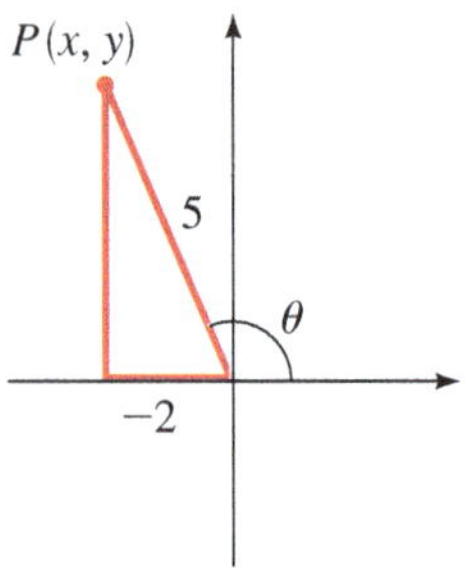

FIGURE 2

SOLUTION We first sketch the angle θ in standard position with terminal side in Quadrant II as in Figure 2. Since $\cos\theta = x/r = -\frac{2}{5}$, we can label a side and the hypotenuse of the triangle in Figure 2. To find the remaining side, we use the Pythagorean Theorem.

$$\begin{aligned} x^2 + y^2 &= r^2 && \text{Pythagorean Theorem} \\ (-2)^2 + y^2 &= 5^2 && x = -2, \quad r = 5 \\ y &= \pm\sqrt{21} && \text{Solve for } y^2 \\ y &= +\sqrt{21} && \text{Because } y > 0 \end{aligned}$$

We can now use the Double-Angle Formula for Sine.

$$\begin{aligned} \sin 2\theta &= 2\sin\theta\cos\theta && \text{Double-Angle Formula} \\ &= 2\left(\frac{\sqrt{21}}{5}\right)\left(-\frac{2}{5}\right) && \text{From the triangle} \\ &= -\frac{4\sqrt{21}}{25} && \text{Simplify} \end{aligned}$$

Now Try Exercise 51 ■

DISCOVERY PROJECT

Where to Sit at the Movies

To best view a painting or a movie requires that the viewing angle be as large as possible. If the painting or movie screen is at a height above eye level, then being too far away or too close results in a small viewing angle and hence a poor viewing experience. So what is the best distance from which to view a movie or a painting? In this project we use trigonometry to find the best location from which to view a painting or a movie. You can find the project at **www.stewartmath.com**.

Product-Sum Formulas

It is possible to write the product $\sin u \cos v$ as a sum of trigonometric functions. To see this, consider the Addition and Subtraction Formulas for Sine:

$$\sin(u + v) = \sin u \cos v + \cos u \sin v$$

$$\sin(u - v) = \sin u \cos v - \cos u \sin v$$

Adding the left- and right-hand sides of these formulas gives

$$\sin(u + v) + \sin(u - v) = 2 \sin u \cos v$$

Dividing by 2 gives the formula

$$\sin u \cos v = \tfrac{1}{2}[\sin(u + v) + \sin(u - v)]$$

The other three **Product-to-Sum Formulas** follow from the Addition Formulas in a similar way.

PRODUCT-TO-SUM FORMULAS

$$\sin u \cos v = \tfrac{1}{2}[\sin(u + v) + \sin(u - v)]$$

$$\cos u \sin v = \tfrac{1}{2}[\sin(u + v) - \sin(u - v)]$$

$$\cos u \cos v = \tfrac{1}{2}[\cos(u + v) + \cos(u - v)]$$

$$\sin u \sin v = \tfrac{1}{2}[\cos(u - v) - \cos(u + v)]$$

EXAMPLE 9 ■ Expressing a Trigonometric Product as a Sum

Express $\sin 3x \sin 5x$ as a sum of trigonometric functions.

SOLUTION Using the fourth Product-to-Sum Formula with $u = 3x$ and $v = 5x$ and the fact that cosine is an even function, we get

$$\begin{aligned} \sin 3x \sin 5x &= \tfrac{1}{2}[\cos(3x - 5x) - \cos(3x + 5x)] \\ &= \tfrac{1}{2}\cos(-2x) - \tfrac{1}{2}\cos 8x \\ &= \tfrac{1}{2}\cos 2x - \tfrac{1}{2}\cos 8x \end{aligned}$$

Now Try Exercise 55 ■

The Product-to-Sum Formulas can also be used as Sum-to-Product Formulas. This is possible because the right-hand side of each Product-to-Sum Formula is a sum and the left side is a product. For example, if we let

$$u = \frac{x + y}{2} \quad \text{and} \quad v = \frac{x - y}{2}$$

in the first Product-to-Sum Formula, we get

$$\sin \frac{x + y}{2} \cos \frac{x - y}{2} = \tfrac{1}{2}(\sin x + \sin y)$$

so

$$\sin x + \sin y = 2 \sin \frac{x + y}{2} \cos \frac{x - y}{2}$$

The remaining three of the following **Sum-to-Product Formulas** are obtained in a similar manner.

SUM-TO-PRODUCT FORMULAS

$$\sin x + \sin y = 2\sin\frac{x+y}{2}\cos\frac{x-y}{2}$$

$$\sin x - \sin y = 2\cos\frac{x+y}{2}\sin\frac{x-y}{2}$$

$$\cos x + \cos y = 2\cos\frac{x+y}{2}\cos\frac{x-y}{2}$$

$$\cos x - \cos y = -2\sin\frac{x+y}{2}\sin\frac{x-y}{2}$$

EXAMPLE 10 ■ Expressing a Trigonometric Sum as a Product

Write $\sin 7x + \sin 3x$ as a product.

SOLUTION The first Sum-to-Product Formula gives

$$\sin 7x + \sin 3x = 2\sin\frac{7x+3x}{2}\cos\frac{7x-3x}{2}$$

$$= 2\sin 5x\cos 2x$$

Now Try Exercise 61 ■

EXAMPLE 11 ■ Proving an Identity

Verify the identity $\dfrac{\sin 3x - \sin x}{\cos 3x + \cos x} = \tan x$.

SOLUTION We apply the second Sum-to-Product Formula to the numerator and the third formula to the denominator.

$$\text{LHS} = \frac{\sin 3x - \sin x}{\cos 3x + \cos x} = \frac{2\cos\frac{3x+x}{2}\sin\frac{3x-x}{2}}{2\cos\frac{3x+x}{2}\cos\frac{3x-x}{2}} \quad \text{Sum-to-Product Formulas}$$

$$= \frac{2\cos 2x\sin x}{2\cos 2x\cos x} \quad \text{Simplify}$$

$$= \frac{\sin x}{\cos x} = \tan x = \text{RHS} \quad \text{Cancel}$$

Now Try Exercise 93 ■

7.3 EXERCISES

CONCEPTS

1. If we know the values of $\sin x$ and $\cos x$, we can find the value of $\sin 2x$ by using the ________ Formula for Sine. State the formula: $\sin 2x =$ ______________.

2. If we know the value of $\cos x$ and the quadrant in which $x/2$ lies, we can find the value of $\sin(x/2)$ by using the ________ Formula for Sine. State the formula: $\sin(x/2) =$ ______________.

SKILLS

3–10 ■ Double Angle Formulas Find $\sin 2x$, $\cos 2x$, and $\tan 2x$ from the given information.

3. $\sin x = \frac{5}{13}$, x in Quadrant I

4. $\tan x = -\frac{4}{3}$, x in Quadrant II

5. $\cos x = \frac{4}{5}$, $\csc x < 0$

6. $\csc x = 4$, $\tan x < 0$

7. $\sin x = -\frac{3}{5}$, x in Quadrant III

8. $\sec x = 2$, x in Quadrant IV

9. $\tan x = -\frac{1}{3}$, $\cos x > 0$

10. $\cot x = \frac{2}{3}$, $\sin x > 0$

11–16 ■ Lowering Powers in a Trigonometric Expression Use the formulas for lowering powers to rewrite the expression in terms of the first power of cosine, as in Example 4.

11. $\sin^4 x$

12. $\cos^4 x$

13. $\cos^2 x \sin^4 x$

14. $\cos^4 x \sin^2 x$

15. $\cos^4 x \sin^4 x$

16. $\cos^6 x$

17–28 ■ Half Angle Formulas Use an appropriate Half-Angle Formula to find the exact value of the expression.

17. $\sin 15°$

18. $\tan 15°$

19. $\tan 22.5°$

20. $\sin 75°$

21. $\cos 165°$

22. $\cos 112.5°$

23. $\tan \dfrac{\pi}{8}$

24. $\cos \dfrac{3\pi}{8}$

25. $\cos \dfrac{\pi}{12}$

26. $\tan \dfrac{5\pi}{12}$

27. $\sin \dfrac{9\pi}{8}$

28. $\sin \dfrac{11\pi}{12}$

29–34 ■ Double- and Half-Angle Formulas Simplify the expression by using a Double-Angle Formula or a Half-Angle Formula.

29. (a) $2 \sin 18° \cos 18°$ **(b)** $2 \sin 3\theta \cos 3\theta$

30. (a) $\dfrac{2 \tan 7°}{1 - \tan^2 7°}$ **(b)** $\dfrac{2 \tan 7\theta}{1 - \tan^2 7\theta}$

31. (a) $\cos^2 34° - \sin^2 34°$ **(b)** $\cos^2 5\theta - \sin^2 5\theta$

32. (a) $\cos^2 \dfrac{\theta}{2} - \sin^2 \dfrac{\theta}{2}$ **(b)** $2 \sin \dfrac{\theta}{2} \cos \dfrac{\theta}{2}$

33. (a) $\dfrac{\sin 8°}{1 + \cos 8°}$ **(b)** $\dfrac{1 - \cos 4\theta}{\sin 4\theta}$

34. (a) $\sqrt{\dfrac{1 - \cos 30°}{2}}$ **(b)** $\sqrt{\dfrac{1 - \cos 8\theta}{2}}$

35. Proving a Double-Angle Formula Use the Addition Formula for Sine to prove the Double-Angle Formula for Sine.

36. Proving a Double-Angle Formula Use the Addition Formula for Tangent to prove the Double-Angle Formula for Tangent.

37–42 ■ Using a Half-Angle Formula Find $\sin \dfrac{x}{2}$, $\cos \dfrac{x}{2}$, and $\tan \dfrac{x}{2}$ from the given information.

37. $\sin x = \frac{3}{5}$, $0° < x < 90°$

38. $\cos x = -\frac{4}{5}$, $180° < x < 270°$

39. $\csc x = 3$, $90° < x < 180°$

40. $\tan x = 1$, $0° < x < 90°$

41. $\sec x = \frac{3}{2}$, $270° < x < 360°$

42. $\cot x = 5$, $180° < x < 270°$

43–46 ■ Expressions Involving Inverse Trigonometric Functions Write the given expression as an algebraic expression in x.

43. $\sin(2 \tan^{-1} x)$

44. $\tan(2 \cos^{-1} x)$

45. $\sin(\frac{1}{2} \cos^{-1} x)$

46. $\cos(2 \sin^{-1} x)$

47–50 ■ Expressions Involving Inverse Trigonometric Functions Find the exact value of the given expression.

47. $\sin\left(2 \cos^{-1} \frac{7}{25}\right)$

48. $\cos\left(2 \tan^{-1} \frac{12}{5}\right)$

49. $\sec\left(2 \sin^{-1} \frac{1}{4}\right)$

50. $\tan\left(\frac{1}{2} \cos^{-1} \frac{2}{3}\right)$

51–54 ■ Evaluating an Expression Involving Trigonometric Functions Evaluate each expression under the given conditions.

51. $\cos 2\theta$; $\sin \theta = -\frac{3}{5}$, θ in Quadrant III

52. $\sin(\theta/2)$; $\tan \theta = -\frac{5}{12}$, θ in Quadrant IV

53. $\sin 2\theta$; $\sin \theta = \frac{1}{7}$, θ in Quadrant II

54. $\tan 2\theta$; $\cos \theta = \frac{3}{5}$, θ in Quadrant I

55–60 ■ Product-to-Sum Formulas Write the product as a sum.

55. $\sin 2x \cos 3x$

56. $\sin x \sin 5x$

57. $\cos x \sin 4x$

58. $\cos 5x \cos 3x$

59. $3 \cos 4x \cos 7x$

60. $11 \sin \dfrac{x}{2} \cos \dfrac{x}{4}$

61–66 ■ Sum-to-Product Formulas Write the sum as a product.

61. $\sin 5x + \sin 3x$

62. $\sin x - \sin 4x$

63. $\cos 4x - \cos 6x$

64. $\cos 9x + \cos 2x$

65. $\sin 2x - \sin 7x$

66. $\sin 3x + \sin 4x$

67–72 ■ Value of a Product or Sum Find the value of the product or sum.

67. $2 \sin 52.5° \sin 97.5°$

68. $3 \cos 37.5° \cos 7.5°$

69. $\cos 37.5° \sin 7.5°$

70. $\sin 75° + \sin 15°$

71. $\cos 255° - \cos 195°$

72. $\cos \dfrac{\pi}{12} + \cos \dfrac{5\pi}{12}$

73–92 ■ Proving Identities Prove the identity.

73. $\cos^2 5x - \sin^2 5x = \cos 10x$

74. $\sin 8x = 2 \sin 4x \cos 4x$

75. $(\sin x + \cos x)^2 = 1 + \sin 2x$

76. $\cos^4 x - \sin^4 x = \cos 2x$

77. $\dfrac{2 \tan x}{1 + \tan^2 x} = \sin 2x$

78. $\dfrac{1 - \cos 2x}{\sin 2x} = \tan x$

79. $\tan\left(\dfrac{x}{2}\right) + \cos x \tan\left(\dfrac{x}{2}\right) = \sin x$

80. $\tan\left(\dfrac{x}{2}\right) + \csc x = \dfrac{2 - \cos x}{\sin x}$

81. $\dfrac{\sin 4x}{\sin x} = 4 \cos x \cos 2x$

82. $\dfrac{1 + \sin 2x}{\sin 2x} = 1 + \frac{1}{2} \sec x \csc x$

83. $\dfrac{2(\tan x - \cot x)}{\tan^2 x - \cot^2 x} = \sin 2x$

84. $\tan x = \dfrac{\sin 2x}{1 + \cos 2x}$

85. $\cot 2x = \dfrac{1 - \tan^2 x}{2 \tan x}$

86. $4(\sin^6 x + \cos^6 x) = 4 - 3 \sin^2 2x$

87. $\tan 3x = \dfrac{3 \tan x - \tan^3 x}{1 - 3 \tan^2 x}$

88. $\dfrac{\sin 3x + \cos 3x}{\cos x - \sin x} = 1 + 4 \sin x \cos x$

89. $\dfrac{\sin x + \sin 5x}{\cos x + \cos 5x} = \tan 3x$

90. $\dfrac{\sin 3x + \sin 7x}{\cos 3x - \cos 7x} = \cot 2x$

91. $\dfrac{\sin 10x}{\sin 9x + \sin x} = \dfrac{\cos 5x}{\cos 4x}$

92. $\dfrac{\sin x + \sin 3x + \sin 5x}{\cos x + \cos 3x + \cos 5x} = \tan 3x$

93. $\dfrac{\sin x + \sin y}{\cos x + \cos y} = \tan\left(\dfrac{x + y}{2}\right)$

94. $\tan y = \dfrac{\sin(x + y) - \sin(x - y)}{\cos(x + y) + \cos(x - y)}$

95. $\tan^2\left(\dfrac{x}{2} + \dfrac{\pi}{4}\right) = \dfrac{1 + \sin x}{1 - \sin x}$

96. $(1 - \cos 4x)(2 + \tan^2 x + \cot^2 x) = 8$

97–100 ■ Sum-to-Product Formulas Use a Sum-to-Product Formula to show the following.

97. $\sin 130° - \sin 110° = -\sin 10°$

98. $\cos 100° - \cos 200° = \sin 50°$

99. $\sin 45° + \sin 15° = \sin 75°$

100. $\cos 87° + \cos 33° = \sin 63°$

SKILLS Plus

101. Proving an Identity Prove the identity

$$\frac{\sin x + \sin 2x + \sin 3x + \sin 4x + \sin 5x}{\cos x + \cos 2x + \cos 3x + \cos 4x + \cos 5x} = \tan 3x$$

102. Proving an Identity Use the identity

$$\sin 2x = 2 \sin x \cos x$$

n times to show that

$$\sin(2^n x) = 2^n \sin x \cos x \cos 2x \cos 4x \cdots \cos 2^{n-1} x$$

103–104 ■ Identities Involving Inverse Trigonometric Functions Prove the identity.

103. $2 \sin^{-1} x = \cos^{-1}(1 - 2x^2)$, $0 \le x \le 1$ [*Hint:* Let $u = \sin^{-1} x$, so that $x = \sin u$. Use a Double-Angle Formula to show that $1 - 2x^2 = \cos 2u$.]

104. $2 \tan^{-1}\left(\dfrac{1}{x}\right) = \cos^{-1}\left(\dfrac{x^2 - 1}{x^2 + 1}\right)$

[*Hint:* Let $u = \tan^{-1}\left(\dfrac{1}{x}\right)$, so that $x = \dfrac{1}{\tan u} = \cot u$. Use a Double-Angle Formula to show that $\dfrac{x^2 - 1}{x^2 + 1} = \dfrac{\cot^2 u - 1}{\csc^2 u} = \cos 2u$.]

105–107 ■ Discovering an Identity Graphically In these problems we discover an identity graphically and then prove the identity.

105. (a) Graph $f(x) = \dfrac{\sin 3x}{\sin x} - \dfrac{\cos 3x}{\cos x}$, and make a conjecture.

(b) Prove the conjecture you made in part (a).

106. (a) Graph $f(x) = \cos 2x + 2 \sin^2 x$, and make a conjecture.

(b) Prove the conjecture you made in part (a).

107. Let $f(x) = \sin 6x + \sin 7x$.

(a) Graph $y = f(x)$.

(b) Verify that $f(x) = 2 \cos \frac{1}{2}x \sin \frac{13}{2}x$.

(c) Graph $y = 2 \cos \frac{1}{2}x$ and $y = -2 \cos \frac{1}{2}x$, together with the graph in part (a), in the same viewing rectangle. How are these graphs related to the graph of f?

108. A Cubic Equation Let $3x = \pi/3$, and let $y = \cos x$. Use the result of Example 2 to show that y satisfies the equation

$$8y^3 - 6y - 1 = 0$$

[*Note:* This equation has roots of a certain kind that are used to show that the angle $\pi/3$ cannot be trisected by using a ruler and compass only.]

109. Tchebycheff Polynomials

(a) Show that there is a polynomial $P(t)$ of degree 4 such that $\cos 4x = P(\cos x)$ (see Example 2).

(b) Show that there is a polynomial $Q(t)$ of degree 5 such that $\cos 5x = Q(\cos x)$.

[*Note:* In general, there is a polynomial $P_n(t)$ of degree n such that $\cos nx = P_n(\cos x)$. These polynomials are called *Tchebycheff polynomials*, after the Russian mathematician P. L. Tchebycheff (1821–1894).]

110. Length of a Bisector In triangle ABC (see the figure) the line segment s bisects angle C. Show that the length of s is given by

$$s = \frac{2ab\cos x}{a+b}$$

[*Hint:* Use the Law of Sines.]

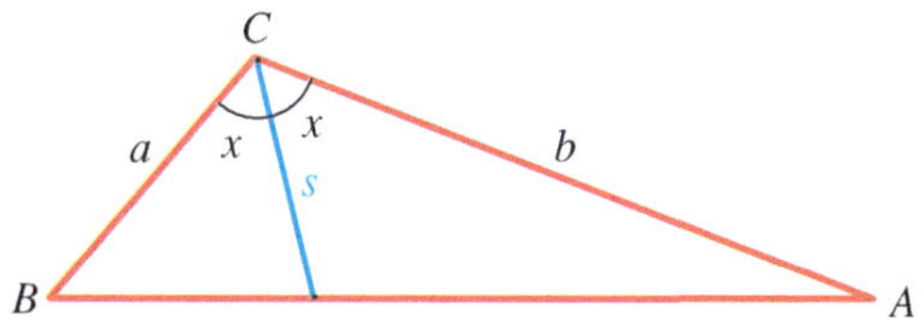

111. Angles of a Triangle If A, B, and C are the angles in a triangle, show that

$$\sin 2A + \sin 2B + \sin 2C = 4\sin A \sin B \sin C$$

112. Largest Area A rectangle is to be inscribed in a semicircle of radius 5 cm as shown in the following figure.

(a) Show that the area of the rectangle is modeled by the function

$$A(\theta) = 25\sin 2\theta$$

(b) Find the largest possible area for such an inscribed rectangle. [*Hint:* Use the fact that $\sin u$ achieves its maximum value at $u = \pi/2$.]

(c) Find the dimensions of the inscribed rectangle with the largest possible area.

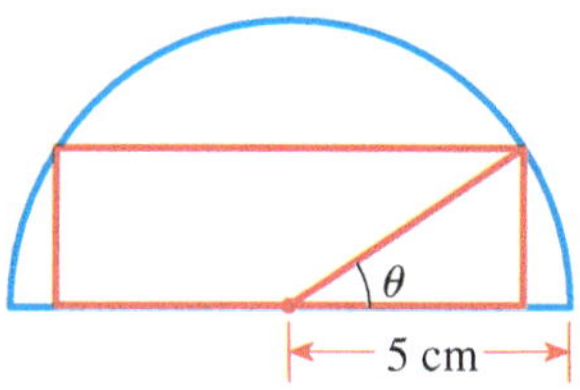

APPLICATIONS

113. Sawing a Wooden Beam A rectangular beam is to be cut from a cylindrical log of diameter 20 in.

(a) Show that the cross-sectional area of the beam is modeled by the function

$$A(\theta) = 200\sin 2\theta$$

where θ is as shown in the figure.

(b) Show that the maximum cross-sectional area of such a beam is 200 in^2. [*Hint:* Use the fact that $\sin u$ achieves its maximum value at $u = \pi/2$.]

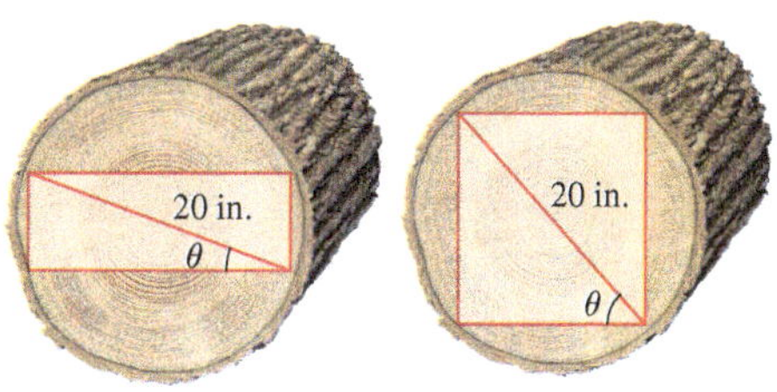

114. Length of a Fold The lower right-hand corner of a long piece of paper 6 in. wide is folded over to the left-hand edge as shown. The length L of the fold depends on the angle θ. Show that

$$L = \frac{3}{\sin\theta\cos^2\theta}$$

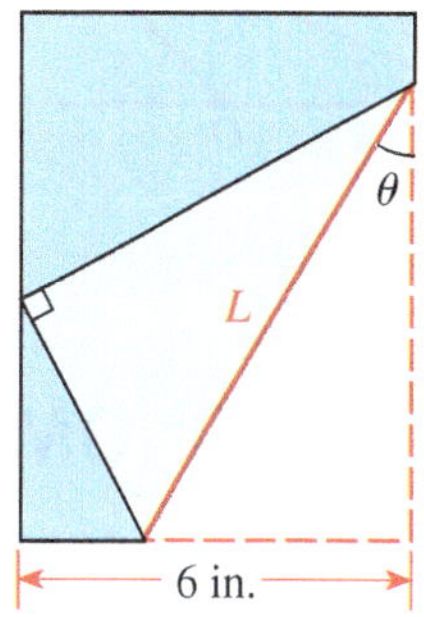

115. Sound Beats When two pure notes that are close in frequency are played together, their sounds interfere to produce *beats*; that is, the loudness (or amplitude) of the sound alternately increases and decreases. If the two notes are given by

$$f_1(t) = \cos 11t \quad \text{and} \quad f_2(t) = \cos 13t$$

the resulting sound is $f(t) = f_1(t) + f_2(t)$.

(a) Graph the function $y = f(t)$.

(b) Verify that $f(t) = 2\cos t \cos 12t$.

(c) Graph $y = 2\cos t$ and $y = -2\cos t$, together with the graph in part (a), in the same viewing rectangle. How do these graphs describe the variation in the loudness of the sound?

116. Touch-Tone Telephones When a key is pressed on a touch-tone telephone, the keypad generates two pure tones, which combine to produce a sound that uniquely identifies the key. The figure shows the low frequency f_1 and the high frequency f_2 associated with each key. Pressing a key produces the sound wave $y = \sin(2\pi f_1 t) + \sin(2\pi f_2 t)$.

(a) Find the function that models the sound produced when the 4 key is pressed.

(b) Use a Sum-to-Product Formula to express the sound generated by the 4 key as a product of a sine and a cosine function.

(c) Graph the sound wave generated by the 4 key from $t = 0$ to $t = 0.006$ s.

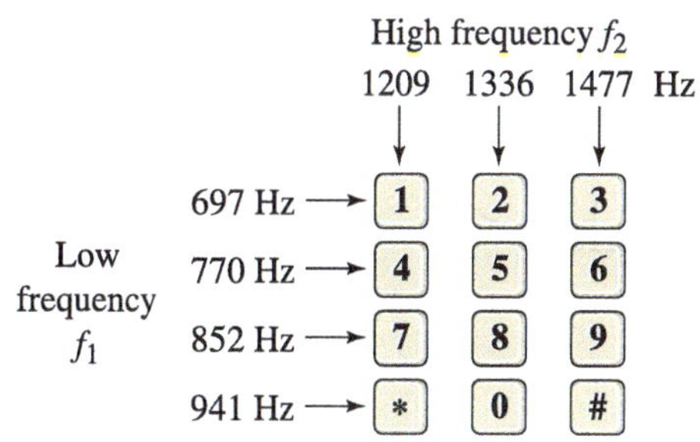

DISCUSS ■ DISCOVER ■ PROVE ■ WRITE

117. PROVE: Geometric Proof of a Double-Angle Formula Use the figure to prove that $\sin 2\theta = 2\sin\theta\cos\theta$.

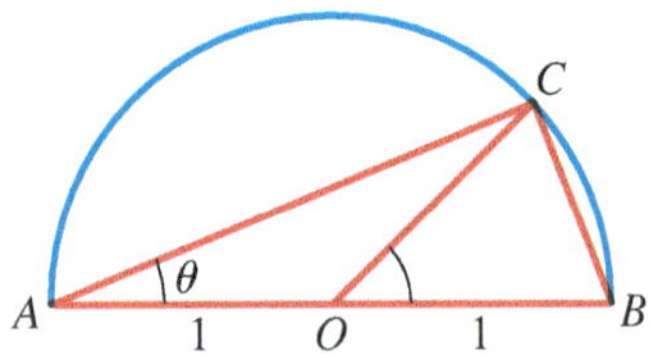

[*Hint:* Find the area of triangle ABC in two different ways. You will need the following facts from geometry:

An angle inscribed in a semicircle is a right angle, so $\angle ACB$ is a right angle.

The central angle subtended by the chord of a circle is twice the angle subtended by the chord on the circle, so $\angle BOC$ is 2θ.]

7.4 BASIC TRIGONOMETRIC EQUATIONS

■ Basic Trigonometric Equations ■ Solving Trigonometric Equations by Factoring

An equation that contains trigonometric functions is called a **trigonometric equation**. For example, the following are trigonometric equations:

$$\sin^2\theta + \cos^2\theta = 1$$

$$2\sin\theta - 1 = 0$$

$$\tan 2\theta - 1 = 0$$

The first equation is an *identity*—that is, it is true for every value of the variable θ. The other two equations are true only for certain values of θ. To solve a trigonometric equation, we find all the values of the variable that make the equation true.

■ Basic Trigonometric Equations

Solving any trigonometric equation always reduces to solving a **basic trigonometric equation**—an equation of the form $T(\theta) = c$, where T is a trigonometric function and c is a constant. In the next three examples we solve such basic equations.

EXAMPLE 1 ■ Solving a Basic Trigonometric Equation

Solve the equation $\sin\theta = \dfrac{1}{2}$.

SOLUTION **Find the solutions in one period.** Because sine has period 2π, we first find the solutions in any interval of length 2π. To find these solutions, we look at the unit circle in Figure 1. We see that $\sin\theta = \frac{1}{2}$ in Quadrants I and II, so the solutions in the interval $[0, 2\pi)$ are

$$\theta = \frac{\pi}{6} \qquad \theta = \frac{5\pi}{6}$$

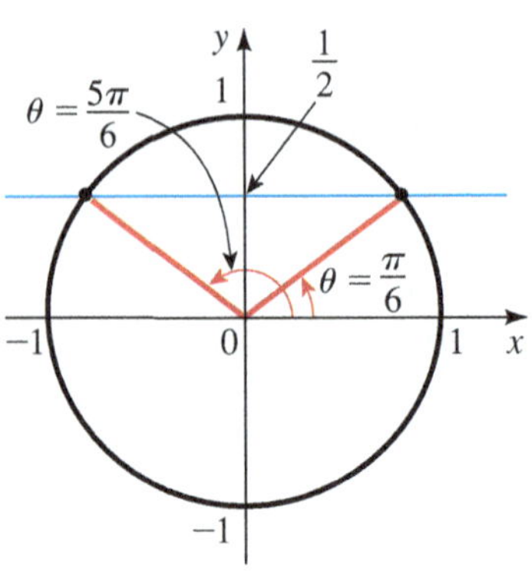

FIGURE 1

Find all solutions. Because the sine function repeats its values every 2π units, we get all solutions of the equation by adding integer multiples of 2π to these solutions:

$$\theta = \frac{\pi}{6} + 2k\pi \qquad \theta = \frac{5\pi}{6} + 2k\pi$$

where k is any integer. Figure 2 gives a graphical representation of the solutions.

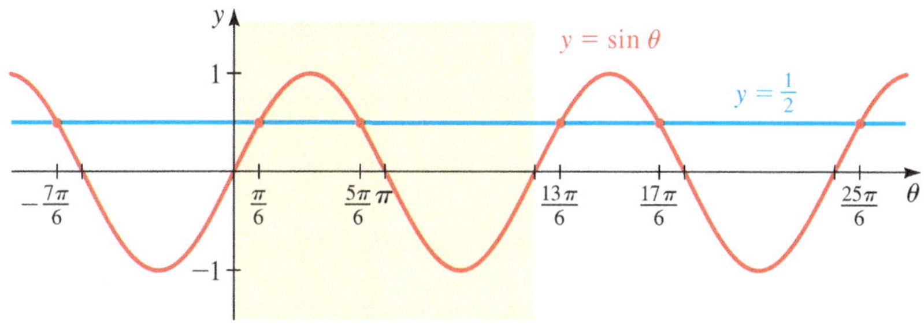

FIGURE 2

Now Try Exercise 5 ■

EXAMPLE 2 ■ Solving a Basic Trigonometric Equation

Solve the equation $\cos\theta = -\dfrac{\sqrt{2}}{2}$, and list eight specific solutions.

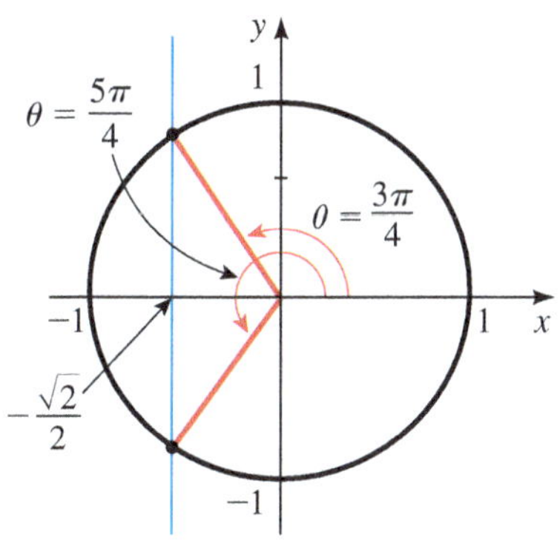

FIGURE 3

SOLUTION **Find the solutions in one period.** Because cosine has period 2π, we first find the solutions in any interval of length 2π. From the unit circle in Figure 3 we see that $\cos\theta = -\sqrt{2}/2$ in Quadrants II and III, so the solutions in the interval $[0, 2\pi)$ are

$$\theta = \frac{3\pi}{4} \qquad \theta = \frac{5\pi}{4}$$

Find all solutions. Because the cosine function repeats its values every 2π units, we get all solutions of the equation by adding integer multiples of 2π to these solutions:

$$\theta = \frac{3\pi}{4} + 2k\pi \qquad \theta = \frac{5\pi}{4} + 2k\pi$$

where k is any integer. You can check that for $k = -1, 0, 1, 2$ we get the following specific solutions:

$$\theta = \underbrace{-\frac{5\pi}{4}, -\frac{3\pi}{4}}_{k=-1}, \underbrace{\frac{3\pi}{4}, \frac{5\pi}{4}}_{k=0}, \underbrace{\frac{11\pi}{4}, \frac{13\pi}{4}}_{k=1}, \underbrace{\frac{19\pi}{4}, \frac{21\pi}{4}}_{k=2}$$

Figure 4 gives a graphical representation of the solutions.

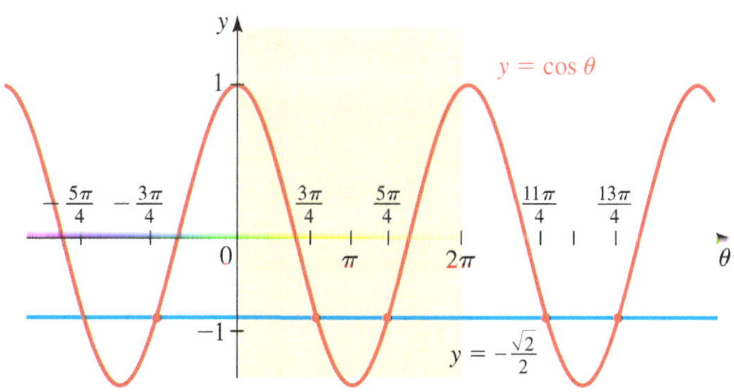

FIGURE 4

Now Try Exercise 17 ■

EXAMPLE 3 ■ Solving a Basic Trigonometric Equation

Solve the equation $\cos\theta = 0.65$.

SOLUTION **Find the solutions in one period.** We first find one solution by taking $\cos^{-1}$ of each side of the equation.

$$\cos\theta = 0.65 \qquad \text{Given equation}$$
$$\theta = \cos^{-1}(0.65) \qquad \text{Take } \cos^{-1} \text{ of each side}$$
$$\theta \approx 0.86 \qquad \text{Calculator (in radian mode)}$$

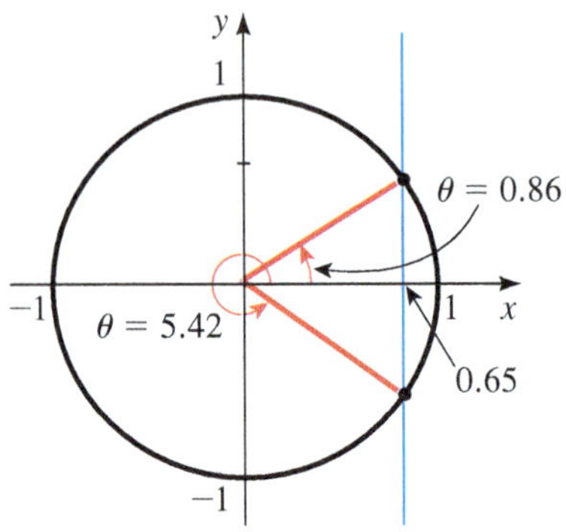

FIGURE 5

Because cosine has period 2π, we next find the solutions in any interval of length 2π. To find these solutions, we look at the unit circle in Figure 5. We see that $\cos\theta = 0.86$ in Quadrants I and IV, so the solutions are

$$\theta \approx 0.86 \qquad \theta \approx 2\pi - 0.86 \approx 5.42$$

Find all solutions. To get all solutions of the equation, we add integer multiples of 2π to these solutions:

$$\theta \approx 0.86 + 2k\pi \qquad \theta \approx 5.42 + 2k\pi$$

where k is any integer.

Now Try Exercise 21 ■

EXAMPLE 4 ■ Solving a Basic Trigonometric Equation

Solve the equation $\tan\theta = 2$.

SOLUTION **Find the solutions in one period.** We first find one solution by taking $\tan^{-1}$ of each side of the equation.

$$\tan\theta = 2 \qquad \text{Given equation}$$
$$\theta = \tan^{-1}(2) \qquad \text{Take } \tan^{-1} \text{ of each side}$$
$$\theta \approx 1.12 \qquad \text{Calculator (in radian mode)}$$

By the definition of $\tan^{-1}$ the solution that we obtained is the only solution in the interval $(-\pi/2, \pi/2)$ (which is an interval of length π).

Find all solutions. Since tangent has period π, we get all solutions of the equation by adding integer multiples of π:

$$\theta \approx 1.12 + k\pi$$

where k is any integer. A graphical representation of the solutions is shown in Figure 6. You can check that the solutions shown in the graph correspond to $k = -1, 0, 1, 2, 3$.

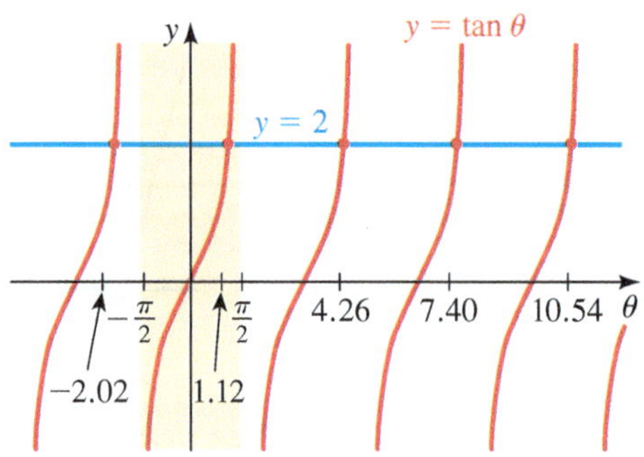

FIGURE 6

Now Try Exercise 23 ■

In the next example we solve trigonometric equations that are algebraically equivalent to basic trigonometric equations.

EXAMPLE 5 ■ Solving Trigonometric Equations

Find all solutions of the equation.

(a) $2\sin\theta - 1 = 0$ **(b)** $\tan^2\theta - 3 = 0$

SOLUTION

(a) We start by isolating $\sin\theta$.

$$2\sin\theta - 1 = 0 \quad \text{Given equation}$$
$$2\sin\theta = 1 \quad \text{Add 1}$$
$$\sin\theta = \frac{1}{2} \quad \text{Divide by 2}$$

This last equation is the same as that in Example 1. The solutions are

$$\theta = \frac{\pi}{6} + 2k\pi \qquad \theta = \frac{5\pi}{6} + 2k\pi$$

where k is any integer.

(b) We start by isolating $\tan\theta$.

$$\tan^2\theta - 3 = 0 \quad \text{Given equation}$$
$$\tan^2\theta = 3 \quad \text{Add 3}$$
$$\tan\theta = \pm\sqrt{3} \quad \text{Take the square root}$$

Because tangent has period π, we first find the solutions in any interval of length π. In the interval $(-\pi/2, \pi/2)$ the solutions are $\theta = \pi/3$ and $\theta = -\pi/3$. To get all solutions, we add integer multiples of π to these solutions:

$$\theta = \frac{\pi}{3} + k\pi \qquad \theta = -\frac{\pi}{3} + k\pi$$

where k is any integer.

Now Try Exercises 27 and 33 ■

■ Solving Trigonometric Equations by Factoring

Factoring is one of the most useful techniques for solving equations, including trigonometric equations. The idea is to move all terms to one side of the equation, factor, and then use the Zero-Product Property (see Section 1.5).

Zero-Product Property

If $AB = 0$, then $A = 0$ or $B = 0$.

EXAMPLE 6 ■ A Trigonometric Equation of Quadratic Type

Solve the equation $2\cos^2\theta - 7\cos\theta + 3 = 0$.

Equation of Quadratic Type

$$2C^2 - 7C + 3 = 0$$
$$(2C - 1)(C - 3) = 0$$

SOLUTION We factor the left-hand side of the equation.

$$2\cos^2\theta - 7\cos\theta + 3 = 0 \quad \text{Given equation}$$
$$(2\cos\theta - 1)(\cos\theta - 3) = 0 \quad \text{Factor}$$
$$2\cos\theta - 1 = 0 \quad \text{or} \quad \cos\theta - 3 = 0 \quad \text{Set each factor equal to 0}$$
$$\cos\theta = \frac{1}{2} \quad \text{or} \quad \cos\theta = 3 \quad \text{Solve for } \cos\theta$$

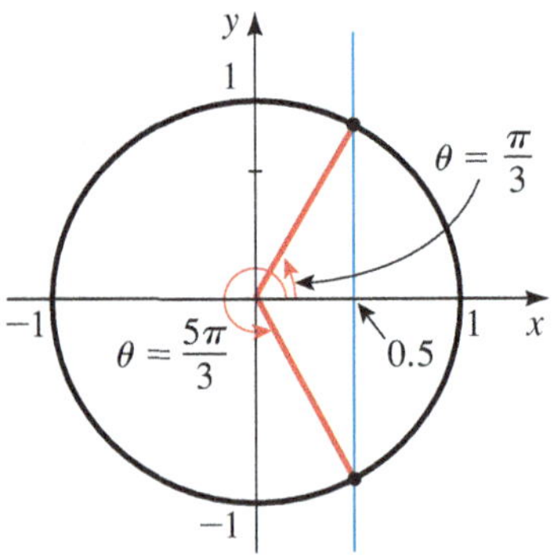

FIGURE 7

Because cosine has period 2π, we first find the solutions in the interval $[0, 2\pi)$. For the first equation the solutions are $\theta = \pi/3$ and $\theta = 5\pi/3$ (see Figure 7). The second equation has no solution because $\cos\theta$ is never greater than 1. Thus the solutions are

$$\theta = \frac{\pi}{3} + 2k\pi \qquad \theta = \frac{5\pi}{3} + 2k\pi$$

where k is any integer.

Now Try Exercise 41 ■

EXAMPLE 7 ■ Solving a Trigonometric Equation by Factoring

Solve the equation $5\sin\theta\cos\theta + 4\cos\theta = 0$.

SOLUTION We factor the left-hand side of the equation.

$$5\sin\theta\cos\theta + 2\cos\theta = 0 \qquad \text{Given equation}$$

$$\cos\theta\,(5\sin\theta + 2) = 0 \qquad \text{Factor}$$

$$\cos\theta = 0 \quad \text{or} \quad 5\sin\theta + 4 = 0 \qquad \text{Set each factor equal to 0}$$

$$\sin\theta = -0.8 \qquad \text{Solve for } \sin\theta$$

Because sine and cosine have period 2π, we first find the solutions of these equations in an interval of length 2π. For the first equation the solutions in the interval $[0, 2\pi)$ are $\theta = \pi/2$ and $\theta = 3\pi/2$. To solve the second equation, we take $\sin^{-1}$ of each side.

$$\sin\theta = -0.80 \qquad \text{Second equation}$$

$$\theta = \sin^{-1}(-0.80) \qquad \text{Take } \sin^{-1} \text{ of each side}$$

$$\theta \approx -0.93 \qquad \text{Calculator (in radian mode)}$$

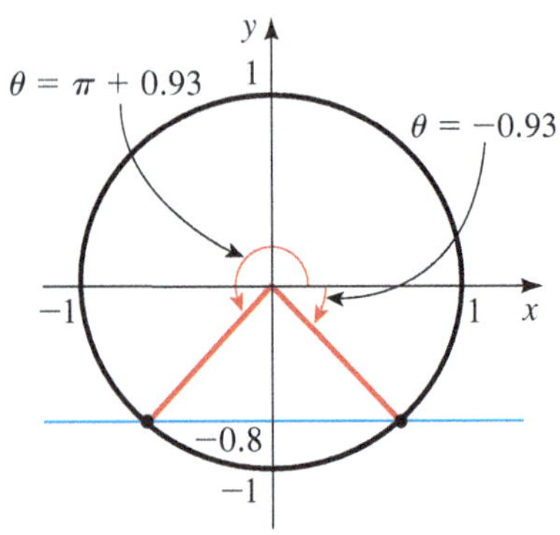

FIGURE 8

So the solutions in an interval of length 2π are $\theta = -0.93$ and $\theta = \pi + 0.93 \approx 4.07$ (see Figure 8). We get all the solutions of the equation by adding integer multiples of 2π to these solutions.

$$\theta = \frac{\pi}{2} + 2k\pi \qquad \theta = \frac{3\pi}{2} + 2k\pi \qquad \theta \approx -0.93 + 2k\pi \qquad \theta \approx 4.07 + 2k\pi$$

where k is any integer.

Now Try Exercise 53 ■

7.4 EXERCISES

CONCEPTS

1. Because the trigonometric functions are periodic, if a basic trigonometric equation has one solution, it has ________ (several/infinitely many) solutions.

2. The basic equation $\sin x = 2$ has ________ (no/one/infinitely many) solutions, whereas the basic equation $\sin x = 0.3$ has ________ (no/one/infinitely many) solutions.

3. We can find some of the solutions of $\sin x = 0.3$ graphically by graphing $y = \sin x$ and $y =$ ________. Use the graph below to estimate some of the solutions.

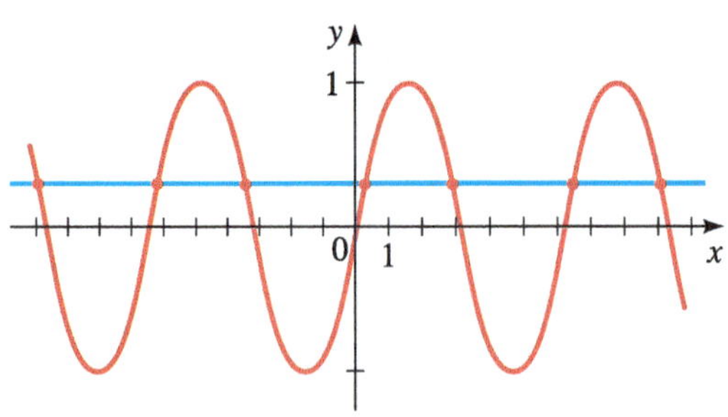

4. We can find the solutions of $\sin x = 0.3$ algebraically.
   (a) First we find the solutions in the interval $[0, 2\pi)$. We get one such solution by taking $\sin^{-1}$ to get $x \approx$ ________. The other solution in this interval is $x \approx$ ________.
   (b) We find all solutions by adding multiples of ________ to the solutions in $[0, 2\pi)$. The solutions are $x \approx$ ________ and $x \approx$ ________.

SKILLS

5–16 ■ Solving Basic Trigonometric Equations Solve the given equation.

5. $\sin\theta = \dfrac{\sqrt{3}}{2}$
6. $\sin\theta = -\dfrac{\sqrt{2}}{2}$
7. $\cos\theta = -1$
8. $\cos\theta = \dfrac{\sqrt{3}}{2}$
9. $\cos\theta = \frac{1}{4}$
10. $\sin\theta = -0.3$
11. $\sin\theta = -0.45$
12. $\cos\theta = 0.32$
13. $\tan\theta = -\sqrt{3}$
14. $\tan\theta = 1$
15. $\tan\theta = 5$
16. $\tan\theta = -\frac{1}{3}$

17–24 ■ Solving Basic Trigonometric Equations Solve the given equation, and list six specific solutions.

17. $\cos\theta = -\dfrac{\sqrt{3}}{2}$
18. $\cos\theta = \dfrac{1}{2}$
19. $\sin\theta = \dfrac{\sqrt{2}}{2}$
20. $\sin\theta = -\dfrac{\sqrt{3}}{2}$
21. $\cos\theta = 0.28$
22. $\tan\theta = 2.5$
23. $\tan\theta = -10$
24. $\sin\theta = -0.9$

25–38 ■ Solving Trigonometric Equations Find all solutions of the given equation.

25. $\cos\theta + 1 = 0$
26. $\sin\theta + 1 = 0$
27. $\sqrt{2}\sin\theta + 1 = 0$
28. $\sqrt{2}\cos\theta - 1 = 0$
29. $5\sin\theta - 1 = 0$
30. $4\cos\theta + 1 = 0$
31. $3\tan^2\theta - 1 = 0$
32. $\cot\theta + 1 = 0$
33. $2\cos^2\theta - 1 = 0$
34. $4\sin^2\theta - 3 = 0$
35. $\tan^2\theta - 4 = 0$
36. $9\sin^2\theta - 1 = 0$
37. $\sec^2\theta - 2 = 0$
38. $\csc^2\theta - 4 = 0$

39–56 ■ Solving Trigonometric Equations by Factoring Solve the given equation.

39. $(\tan^2\theta - 4)(2\cos\theta + 1) = 0$
40. $(\tan\theta - 2)(16\sin^2\theta - 1) = 0$
41. $4\cos^2\theta - 4\cos\theta + 1 = 0$
42. $2\sin^2\theta - \sin\theta - 1 = 0$
43. $3\sin^2\theta - 7\sin\theta + 2 = 0$
44. $\tan^4\theta - 13\tan^2\theta + 36 = 0$
45. $2\cos^2\theta - 7\cos\theta + 3 = 0$
46. $\sin^2\theta - \sin\theta - 2 = 0$
47. $\cos^2\theta - \cos\theta - 6 = 0$
48. $2\sin^2\theta + 5\sin\theta - 12 = 0$
49. $\sin^2\theta = 2\sin\theta + 3$
50. $3\tan^3\theta = \tan\theta$
51. $\cos\theta\,(2\sin\theta + 1) = 0$
52. $\sec\theta\,(2\cos\theta - \sqrt{2}) = 0$
53. $\cos\theta\sin\theta - 2\cos\theta = 0$
54. $\tan\theta\sin\theta + \sin\theta = 0$
55. $3\tan\theta\sin\theta - 2\tan\theta = 0$
56. $4\cos\theta\sin\theta + 3\cos\theta = 0$

APPLICATIONS

57. **Refraction of Light** It has been observed since ancient times that light refracts, or "bends," as it travels from one medium to another (from air to water, for example). If v_1 is the speed of light in one medium and v_2 is its speed in another medium, then according to **Snell's Law**,

$$\frac{\sin\theta_1}{\sin\theta_2} = \frac{v_1}{v_2}$$

where θ_1 is the *angle of incidence* and θ_2 is the *angle of refraction* (see the figure). The number v_1/v_2 is called the *index of refraction*. The index of refraction for several substances is given in the table.

If a ray of light passes through the surface of a lake at an angle of incidence of 70°, what is the angle of refraction?

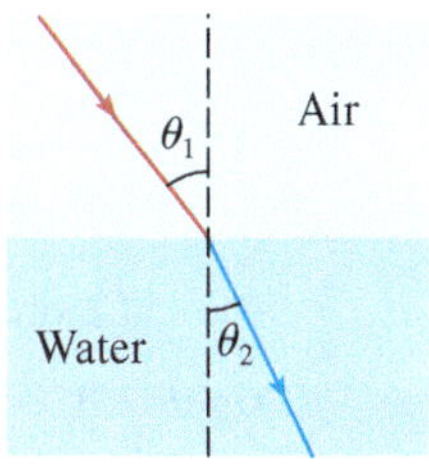

Substance	Refraction from air to substance
Water	1.33
Alcohol	1.36
Glass	1.52
Diamond	2.41

58. **Total Internal Reflection** When light passes from a more-dense to a less-dense medium—from glass to air, for example—the angle of refraction predicted by Snell's Law (see Exercise 57) can be 90° or larger. In this case the light beam is actually reflected back into the denser medium. This phenomenon, called *total internal reflection*, is the principle behind fiber optics. Set $\theta_2 = 90°$ in Snell's Law, and solve for θ_1 to determine the critical angle of incidence at which total internal reflection begins to occur when light passes from glass to air. (Note that the index of refraction from glass to air is the reciprocal of the index from air to glass.)

59. Phases of the Moon As the moon revolves around the earth, the side that faces the earth is usually just partially illuminated by the sun. The phases of the moon describe how much of the surface appears to be in sunlight. An astronomical measure of phase is given by the fraction F of the lunar disc that is lit. When the angle between the sun, earth, and moon is θ $(0 \le \theta \le 360°)$, then

$$F = \frac{1}{2}(1 - \cos\theta)$$

Determine the angles θ that correspond to the following phases:

(a) $F = 0$ (new moon)

(b) $F = 0.25$ (a crescent moon)

(c) $F = 0.5$ (first or last quarter)

(d) $F = 1$ (full moon)

DISCUSS ■ DISCOVER ■ PROVE ■ WRITE

60. DISCUSS ■ WRITE: Equations and Identities Which of the following statements is true?

A. Every identity is an equation.

B. Every equation is an identity.

Give examples to illustrate your answer. Write a short paragraph to explain the difference between an equation and an identity.

7.5 MORE TRIGONOMETRIC EQUATIONS

■ Solving Trigonometric Equations by Using Identities ■ Equations with Trigonometric Functions of Multiples of Angles

In this section we solve trigonometric equations by first using identities to simplify the equation. We also solve trigonometric equations in which the terms contain multiples of angles.

■ Solving Trigonometric Equations by Using Identities

In the next two examples we use trigonometric identities to express a trigonometric equation in a form in which it can be factored.

EXAMPLE 1 ■ Using a Trigonometric Identity

Solve the equation $1 + \sin\theta = 2\cos^2\theta$.

SOLUTION We first need to rewrite this equation so that it contains only one trigonometric function. To do this, we use a trigonometric identity.

$$1 + \sin\theta = 2\cos^2\theta \qquad \text{Given equation}$$

$$1 + \sin\theta = 2(1 - \sin^2\theta) \qquad \text{Pythagorean identity}$$

$$2\sin^2\theta + \sin\theta - 1 = 0 \qquad \text{Put all terms on one side}$$

$$(2\sin\theta - 1)(\sin\theta + 1) = 0 \qquad \text{Factor}$$

$$2\sin\theta - 1 = 0 \quad \text{or} \quad \sin\theta + 1 = 0 \qquad \text{Set each factor equal to 0}$$

$$\sin\theta = \frac{1}{2} \quad \text{or} \quad \sin\theta = -1 \qquad \text{Solve for } \sin\theta$$

$$\theta = \frac{\pi}{6}, \frac{5\pi}{6} \quad \text{or} \quad \theta = \frac{3\pi}{2} \qquad \text{Solve for } \theta \text{ in the interval } [0, 2\pi)$$

Because sine has period 2π, we get all the solutions of the equation by adding integer multiples of 2π to these solutions. Thus the solutions are

$$\theta = \frac{\pi}{6} + 2k\pi \qquad \theta = \frac{5\pi}{6} + 2k\pi \qquad \theta = \frac{3\pi}{2} + 2k\pi$$

where k is any integer.

Now Try Exercises 3 and 11 ■

EXAMPLE 2 ■ Using a Trigonometric Identity

Solve the equation $\sin 2\theta - \cos\theta = 0$.

SOLUTION The first term is a function of 2θ, and the second is a function of θ, so we begin by using a trigonometric identity to rewrite the first term as a function of θ only.

$$\sin 2\theta - \cos\theta = 0 \qquad \text{Given equation}$$

$$2\sin\theta\cos\theta - \cos\theta = 0 \qquad \text{Double-Angle Formula}$$

$$\cos\theta\,(2\sin\theta - 1) = 0 \qquad \text{Factor}$$

$$\cos\theta = 0 \quad \text{or} \quad 2\sin\theta - 1 = 0 \qquad \text{Set each factor equal to 0}$$

$$\sin\theta = \frac{1}{2} \qquad \text{Solve for } \sin\theta$$

$$\theta = \frac{\pi}{2}, \frac{3\pi}{2} \quad \text{or} \quad \theta = \frac{\pi}{6}, \frac{5\pi}{6} \qquad \text{Solve for } \theta \text{ in } [0, 2\pi)$$

Both sine and cosine have period 2π, so we get all the solutions of the equation by adding integer multiples of 2π to these solutions. Thus the solutions are

$$\theta = \frac{\pi}{2} + 2k\pi \qquad \theta = \frac{3\pi}{2} + 2k\pi \qquad \theta = \frac{\pi}{6} + 2k\pi \qquad \theta = \frac{5\pi}{6} + 2k\pi$$

where k is any integer.

Now Try Exercises 7 and 9 ■

EXAMPLE 3 ■ Squaring and Using an Identity

Solve the equation $\cos\theta + 1 = \sin\theta$ in the interval $[0, 2\pi)$.

SOLUTION To get an equation that involves either sine only or cosine only, we square both sides and use a Pythagorean identity.

$$\cos\theta + 1 = \sin\theta \qquad \text{Given equation}$$

$$\cos^2\theta + 2\cos\theta + 1 = \sin^2\theta \qquad \text{Square both sides}$$

$$\cos^2\theta + 2\cos\theta + 1 = 1 - \cos^2\theta \qquad \text{Pythagorean identity}$$

$$2\cos^2\theta + 2\cos\theta = 0 \qquad \text{Simplify}$$

$$2\cos\theta\,(\cos\theta + 1) = 0 \qquad \text{Factor}$$

$$2\cos\theta = 0 \quad \text{or} \quad \cos\theta + 1 = 0 \qquad \text{Set each factor equal to 0}$$

$$\cos\theta = 0 \quad \text{or} \quad \cos\theta = -1 \qquad \text{Solve for } \cos\theta$$

$$\theta = \frac{\pi}{2}, \frac{3\pi}{2} \quad \text{or} \quad \theta = \pi \qquad \text{Solve for } \theta \text{ in } [0, 2\pi)$$

Because we squared both sides, we need to check for extraneous solutions. From *Check Your Answers* we see that the solutions of the given equation are $\pi/2$ and π.

CHECK YOUR ANSWERS

$\theta = \frac{\pi}{2}$

$$\cos\frac{\pi}{2} + 1 = \sin\frac{\pi}{2}$$

$$0 + 1 = 1 \quad ✓$$

$\theta = \frac{3\pi}{2}$

$$\cos\frac{3\pi}{2} + 1 = \sin\frac{3\pi}{2}$$

$$0 + 1 \stackrel{?}{=} -1 \quad ✗$$

$\theta = \pi$

$$\cos\pi + 1 = \sin\pi$$

$$-1 + 1 = 0 \quad ✓$$

Now Try Exercise 13 ■

EXAMPLE 4 ■ Finding Intersection Points

Find the values of x for which the graphs of $f(x) = \sin x$ and $g(x) = \cos x$ intersect.

SOLUTION 1: Graphical

The graphs intersect where $f(x) = g(x)$. In Figure 1 we graph $y_1 = \sin x$ and $y_2 = \cos x$ on the same screen, for x between 0 and 2π. Using TRACE or the `intersect` command on the graphing calculator, we see that the two points of intersection in this interval occur where $x \approx 0.785$ and $x \approx 3.927$. Since sine and cosine are periodic with period 2π, the intersection points occur where

$$x \approx 0.785 + 2k\pi \qquad \text{and} \qquad x \approx 3.927 + 2k\pi$$

where k is any integer.

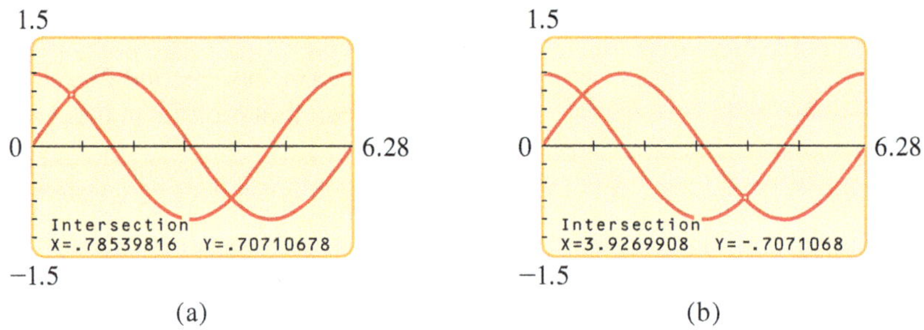

FIGURE 1

SOLUTION 2: Algebraic

To find the exact solution, we set $f(x) = g(x)$ and solve the resulting equation algebraically:

$$\sin x = \cos x \qquad \text{Equate functions}$$

Since the numbers x for which $\cos x = 0$ are not solutions of the equation, we can divide both sides by $\cos x$:

$$\frac{\sin x}{\cos x} = 1 \qquad \text{Divide by } \cos x$$

$$\tan x = 1 \qquad \text{Reciprocal identity}$$

The only solution of this equation in the interval $(-\pi/2, \pi/2)$ is $x = \pi/4$. Since tangent has period π, we get all solutions of the equation by adding integer multiples of π:

$$x = \frac{\pi}{4} + k\pi$$

where k is any integer. The graphs intersect for these values of x. You should use your calculator to check that, rounded to three decimals, these are the same values that we obtained in Solution 1.

Now Try Exercise 35 ■

■ Equations with Trigonometric Functions of Multiples of Angles

When solving trigonometric equations that involve functions of multiples of angles, we first solve for the multiple of the angle, then divide to solve for the angle.

EXAMPLE 5 ■ A Trigonometric Equation Involving a Multiple of an Angle

Consider the equation $2\sin 3\theta - 1 = 0$.

(a) Find all solutions of the equation.

(b) Find the solutions in the interval $[0, 2\pi)$.

SOLUTION

(a) We first isolate $\sin 3\theta$ and then solve for the angle 3θ.

$$2\sin 3\theta - 1 = 0 \qquad \text{Given equation}$$

$$2\sin 3\theta = 1 \qquad \text{Add 1}$$

$$\sin 3\theta = \frac{1}{2} \qquad \text{Divide by 2}$$

$$3\theta = \frac{\pi}{6}, \frac{5\pi}{6} \qquad \text{Solve for } 3\theta \text{ in the interval } [0, 2\pi) \text{ (see Figure 2)}$$

To get all solutions, we add integer multiples of 2π to these solutions. So the solutions are of the form

$$3\theta = \frac{\pi}{6} + 2k\pi \qquad 3\theta = \frac{5\pi}{6} + 2k\pi$$

To solve for θ, we divide by 3 to get the solutions

$$\theta = \frac{\pi}{18} + \frac{2k\pi}{3} \qquad \theta - \frac{5\pi}{18} + \frac{2k\pi}{3}$$

where k is any integer.

(b) The solutions from part (a) that are in the interval $[0, 2\pi)$ correspond to $k = 0$, 1, and 2. For all other values of k the corresponding values of θ lie outside this interval. So the solutions in the interval $[0, 2\pi)$ are

$$\theta = \underbrace{\frac{\pi}{18}, \frac{5\pi}{18}}_{k=0}, \underbrace{\frac{13\pi}{18}, \frac{17\pi}{18}}_{k=1}, \underbrace{\frac{25\pi}{18}, \frac{29\pi}{18}}_{k=2}$$

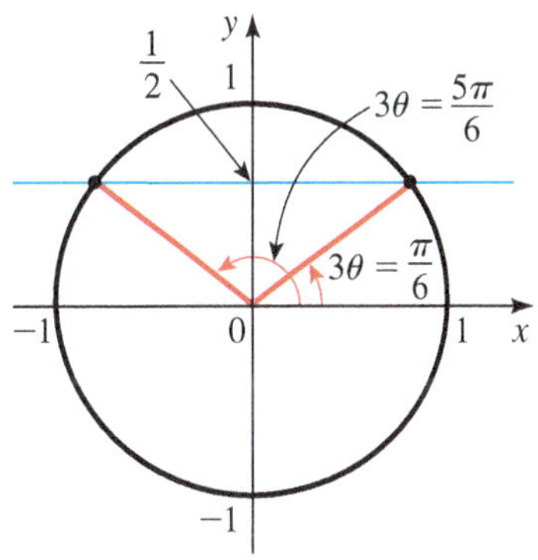

FIGURE 2

Now Try Exercise 17 ■

EXAMPLE 6 ■ A Trigonometric Equation Involving a Half Angle

Consider the equation $\sqrt{3}\tan\frac{\theta}{2} - 1 = 0$.

(a) Find all solutions of the equation.

(b) Find the solutions in the interval $[0, 4\pi)$.

SOLUTION

(a) We start by isolating $\tan \frac{\theta}{2}$.

$$\sqrt{3} \tan \frac{\theta}{2} - 1 = 0 \qquad \text{Given equation}$$

$$\sqrt{3} \tan \frac{\theta}{2} = 1 \qquad \text{Add 1}$$

$$\tan \frac{\theta}{2} = \frac{1}{\sqrt{3}} \qquad \text{Divide by } \sqrt{3}$$

$$\frac{\theta}{2} = \frac{\pi}{6} \qquad \text{Solve for } \frac{\theta}{2} \text{ in the interval } \left(-\frac{\pi}{2}, \frac{\pi}{2}\right)$$

Since tangent has period π, to get all solutions, we add integer multiples of π to this solution. So the solutions are of the form

$$\frac{\theta}{2} = \frac{\pi}{6} + k\pi$$

Multiplying by 2, we get the solutions

$$\theta = \frac{\pi}{3} + 2k\pi$$

where k is any integer.

(b) The solutions from part (a) that are in the interval $[0, 4\pi)$ correspond to $k = 0$ and $k = 1$. For all other values of k the corresponding values of x lie outside this interval. Thus the solutions in the interval $[0, 4\pi)$ are

$$x = \frac{\pi}{3}, \frac{7\pi}{3}$$

Now Try Exercise 23 ■

7.5 EXERCISES

CONCEPTS

1–2 ■ We can use identities to help us solve trigonometric equations.

1. Using a Pythagorean identity we see that the equation $\sin x + \sin^2 x + \cos^2 x = 1$ is equivalent to the basic equation ________ whose solutions are $x =$ ________.

2. Using a Double-Angle Formula we see that the equation $\sin x + \sin 2x = 0$ is equivalent to the equation ________. Factoring, we see that solving this equation is equivalent to solving the two basic equations ________ and ________.

SKILLS

3–16 ■ Solving Trigonometric Equations by Using Identities Solve the given equation.

3. $2\cos^2\theta + \sin\theta = 1$

4. $\sin^2\theta = 4 - 2\cos^2\theta$

5. $\tan^2\theta - 2\sec\theta = 2$

6. $\csc^2\theta = \cot\theta + 3$

7. $2\sin 2\theta - 3\sin\theta = 0$

8. $3\sin 2\theta - 2\sin\theta = 0$

9. $\cos 2\theta = 3\sin\theta - 1$

10. $\cos 2\theta = \cos^2\theta - \frac{1}{2}$

11. $2\sin^2\theta - \cos\theta = 1$

12. $\tan\theta - 3\cot\theta = 0$

13. $\sin\theta - 1 = \cos\theta$

14. $\cos\theta - \sin\theta = 1$

15. $\tan\theta + 1 = \sec\theta$

16. $2\tan\theta + \sec^2\theta = 4$

17–30 ■ Solving Trigonometric Equations Involving a Multiple of an Angle An equation is given. **(a)** Find all solutions of the equation. **(b)** Find the solutions in the interval $[0, 2\pi)$.

17. $2\cos 3\theta = 1$

18. $2\sin 2\theta = 1$

19. $2\cos 2\theta + 1 = 0$

20. $2\sin 3\theta + 1 = 0$

21. $\sqrt{3}\tan 3\theta + 1 = 0$

22. $\sec 4\theta - 2 = 0$

23. $\cos\frac{\theta}{2} - 1 = 0$

24. $\tan\frac{\theta}{4} + \sqrt{3} = 0$

25. $2\sin\frac{\theta}{3} + \sqrt{3} = 0$

26. $\sec\frac{\theta}{2} = \cos\frac{\theta}{2}$

27. $\sin 2\theta = 3 \cos 2\theta$

28. $\csc 3\theta = 5 \sin 3\theta$

29. $1 - 2 \sin \theta = \cos 2\theta$

30. $\tan 3\theta + 1 = \sec 3\theta$

31–34 ■ Solving Trigonometric Equations Solve the equations by factoring.

31. $3 \tan^3\theta - 3 \tan^2\theta - \tan \theta + 1 = 0$

32. $4 \sin \theta \cos \theta + 2 \sin \theta - 2 \cos \theta - 1 = 0$

33. $2 \sin \theta \tan \theta - \tan \theta = 1 - 2 \sin \theta$

34. $\sec \theta \tan \theta - \cos \theta \cot \theta = \sin \theta$

35–38 ■ Finding Intersection Points Graphically **(a)** Graph f and g in the given viewing rectangle and find the intersection points graphically, rounded to two decimal places. **(b)** Find the intersection points of f and g algebraically. Give exact answers.

35. $f(x) = 3 \cos x + 1, \quad g(x) = \cos x - 1;$
$[-2\pi, 2\pi]$ by $[-2.5, 4.5]$

36. $f(x) = \sin 2x + 1, \quad g(x) = 2 \sin 2x + 1;$
$[-2\pi, 2\pi]$ by $[-1.5, 3.5]$

37. $f(x) = \tan x, \quad g(x) = \sqrt{3};$
$[-\pi/2, \pi/2]$ by $[-10, 10]$

38. $f(x) = \sin x - 1, \quad g(x) = \cos x;$
$[-2\pi, 2\pi]$ by $[-2.5, 1.5]$

39–42 ■ Using Addition or Subtraction Formulas Use an Addition or Subtraction Formula to simplify the equation. Then find all solutions in the interval $[0, 2\pi)$.

39. $\cos \theta \cos 3\theta - \sin \theta \sin 3\theta = 0$

40. $\cos \theta \cos 2\theta + \sin \theta \sin 2\theta = \frac{1}{2}$

41. $\sin 2\theta \cos \theta - \cos 2\theta \sin \theta = \sqrt{3}/2$

42. $\sin 3\theta \cos \theta - \cos 3\theta \sin \theta = 0$

43–52 ■ Using Double- or Half-Angle Formulas Use a Double- or Half-Angle Formula to solve the equation in the interval $[0, 2\pi)$.

43. $\sin 2\theta + \cos \theta = 0$

44. $\tan \dfrac{\theta}{2} - \sin \theta = 0$

45. $\cos 2\theta + \cos \theta = 2$

46. $\tan \theta + \cot \theta = 4 \sin 2\theta$

47. $\cos 2\theta - \cos^2\theta = 0$

48. $2 \sin^2\theta = 2 + \cos 2\theta$

49. $\cos 2\theta - \cos 4\theta = 0$

50. $\sin 3\theta - \sin 6\theta = 0$

51. $\cos \theta - \sin \theta = \sqrt{2} \sin \dfrac{\theta}{2}$

52. $\sin \theta - \cos \theta = \frac{1}{2}$

53–56 ■ Using Sum-to-Product Formulas Solve the equation by first using a Sum-to-Product Formula.

53. $\sin \theta + \sin 3\theta = 0$

54. $\cos 5\theta - \cos 7\theta = 0$

55. $\cos 4\theta + \cos 2\theta = \cos \theta$

56. $\sin 5\theta - \sin 3\theta = \cos 4\theta$

57–62 ■ Solving Trigonometric Equations Graphically Use a graphing device to find the solutions of the equation, rounded to two decimal places.

57. $\sin 2x = x$

58. $\cos x = \dfrac{x}{3}$

59. $2^{\sin x} = x$

60. $\sin x = x^3$

61. $\dfrac{\cos x}{1 + x^2} = x^2$

62. $\cos x = \frac{1}{2}(e^x + e^{-x})$

SKILLS Plus

63–64 ■ Equations Involving Inverse Trigonometric Functions Solve the given equation for x.

63. $\tan^{-1}x + \tan^{-1}2x = \dfrac{\pi}{4}$ [*Hint:* Let $u = \tan^{-1}x$ and $v = \tan^{-1}2x$. Solve the equation $u + v = \dfrac{\pi}{4}$ by taking the tangent of each side.]

64. $2 \sin^{-1}x + \cos^{-1}x = \pi$ [*Hint:* Take the cosine of each side.]

APPLICATIONS

65. Range of a Projectile If a projectile is fired with velocity v_0 at an angle θ, then its *range*, the horizontal distance it travels (in ft), is modeled by the function

$$R(\theta) = \frac{v_0^2 \sin 2\theta}{32}$$

(See page 627.) If $v_0 = 2200$ ft/s, what angle (in degrees) should be chosen for the projectile to hit a target on the ground 5000 ft away?

66. Damped Vibrations The displacement of a spring vibrating in damped harmonic motion is given by

$$y = 4e^{-3t} \sin 2\pi t$$

Find the times when the spring is at its equilibrium position $(y = 0)$.

67. Hours of Daylight In Philadelphia the number of hours of daylight on day t (where t is the number of days after January 1) is modeled by the function

$$L(t) = 12 + 2.83 \sin\left(\frac{2\pi}{365}(t - 80)\right)$$

(a) Which days of the year have about 10 h of daylight?

(b) How many days of the year have more than 10 h of daylight?

68. Belts and Pulleys A thin belt of length L surrounds two pulleys of radii R and r, as shown in the figure to the right.

(a) Show that the angle θ (in rad) where the belt crosses itself satisfies the equation

$$\theta + 2\cot\frac{\theta}{2} = \frac{L}{R + r} - \pi$$

[*Hint:* Express L in terms of R, r, and θ by adding up the lengths of the curved and straight parts of the belt.]

(b) Suppose that $R = 2.42$ ft, $r = 1.21$ ft, and $L = 27.78$ ft. Find θ by solving the equation in part (a) graphically. Express your answer both in radians and in degrees.

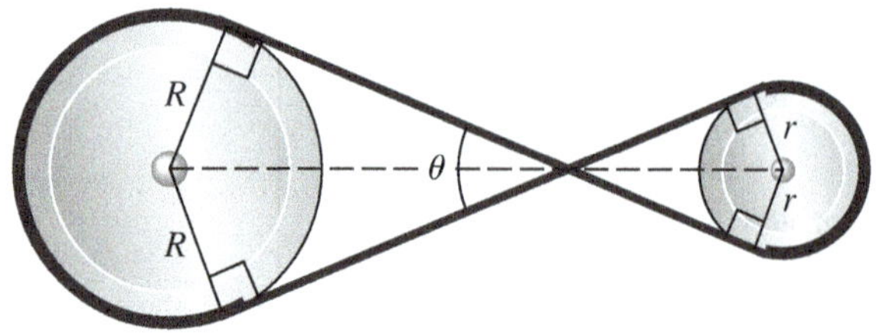

DISCUSS ■ DISCOVER ■ PROVE ■ WRITE

69. DISCUSS: A Special Trigonometric Equation What makes the equation $\sin(\cos x) = 0$ different from all the other equations we've looked at in this section? Find all solutions of this equation.

CHAPTER 7 ■ REVIEW

PROPERTIES AND FORMULAS

Fundamental Trigonometric Identities (p. 538)

An **identity** is an equation that is true for all values of the variable(s). A **trigonometric identity** is an identity that involves trigonometric functions. The fundamental trigonometric identities are as follows.

Reciprocal Identities:

$$\csc x = \frac{1}{\sin x} \qquad \sec x = \frac{1}{\cos x} \qquad \cot x = \frac{1}{\tan x}$$

$$\tan x = \frac{\sin x}{\cos x} \qquad \cot x = \frac{\cos x}{\sin x}$$

Pythagorean Identities:

$$\sin^2 x + \cos^2 x = 1$$
$$\tan^2 x + 1 = \sec^2 x$$
$$1 + \cot^2 x = \csc^2 x$$

Even-Odd Identities:

$$\sin(-x) = -\sin x$$
$$\cos(-x) = \cos x$$
$$\tan(-x) = -\tan x$$

Cofunction Identities:

$$\sin\left(\frac{\pi}{2} - x\right) = \cos x \qquad \tan\left(\frac{\pi}{2} - x\right) = \cot x$$
$$\sec\left(\frac{\pi}{2} - x\right) = \csc x$$
$$\cos\left(\frac{\pi}{2} - x\right) = \sin x \qquad \cot\left(\frac{\pi}{2} - x\right) = \tan x$$
$$\csc\left(\frac{\pi}{2} - x\right) = \sec x$$

Proving Trigonometric Identities (p. 540)

To prove that a trigonometric equation is an identity, we use the following guidelines.

1. **Start with one side.** Pick one side of the equation.
2. **Use known identities.** Use algebra and known identities to change the side you started with into the other side.
3. **Convert to sines and cosines.** Sometimes it is helpful to convert all functions in the equation to sines and cosines.

Addition and Subtraction Formulas (p. 545)

These identities involve the trigonometric functions of a sum or a difference.

Formulas for Sine:

$$\sin(s + t) = \sin s\cos t + \cos s\sin t$$
$$\sin(s - t) = \sin s\cos t - \cos s\sin t$$

Formulas for Cosine:

$$\cos(s + t) = \cos s\cos t - \sin s\sin t$$
$$\cos(s - t) = \cos s\cos t + \sin s\sin t$$

Formulas for Tangent:

$$\tan(s + t) = \frac{\tan s + \tan t}{1 - \tan s\tan t}$$
$$\tan(s - t) = \frac{\tan s - \tan t}{1 + \tan s\tan t}$$

Sums of Sines and Cosines (p. 550)

If A and B are real numbers, then

$$A\sin x + B\cos x = k\sin(x + \phi)$$

where $k = \sqrt{A^2 + B^2}$ and ϕ satisfies

$$\cos\phi = \frac{A}{\sqrt{A^2 + B^2}} \qquad \sin\phi = \frac{B}{\sqrt{A^2 + B^2}}$$

Double-Angle Formulas (p. 554)

These identities involve the trigonometric functions of twice the variable.

Formula for Sine:

$$\sin 2x = 2\sin x\cos x$$

Formulas for Cosine:

$$\begin{aligned}\cos 2x &= \cos^2 x - \sin^2 x\\ &= 1 - 2\sin^2 x\\ &= 2\cos^2 x - 1\end{aligned}$$

Formulas for Tangent:

$$\tan 2x = \frac{2\tan x}{1 - \tan^2 x}$$

Formulas for Lowering Powers (p. 556)

These formulas allow us to write a trigonometric expression involving even powers of sine and cosine in terms of the first power of cosine only.

$$\sin^2 x = \frac{1 - \cos 2x}{2} \qquad \cos^2 x = \frac{1 + \cos 2x}{2}$$

$$\tan^2 x = \frac{1 - \cos 2x}{1 + \cos 2x}$$

Half-Angle Formulas (p. 556)

These formulas involve trigonometric functions of half an angle.

$$\sin\frac{u}{2} = \pm\sqrt{\frac{1 - \cos u}{2}} \qquad \cos\frac{u}{2} = \pm\sqrt{\frac{1 + \cos u}{2}}$$

$$\tan\frac{u}{2} = \frac{1 - \cos u}{\sin u} = \frac{\sin u}{1 + \cos u}$$

Product-Sum Formulas (pp. 559–560)

These formulas involve products and sums of trigonometric functions.

Product-to-Sum Formulas:

$$\begin{aligned}\sin u\cos v &= \tfrac{1}{2}[\sin(u + v) + \sin(u - v)]\\ \cos u\sin v &= \tfrac{1}{2}[\sin(u + v) - \sin(u - v)]\\ \cos u\cos v &= \tfrac{1}{2}[\cos(u + v) + \cos(u - v)]\\ \sin u\sin v &= \tfrac{1}{2}[\cos(u - v) - \cos(u + v)]\end{aligned}$$

Sum-to-Product Formulas:

$$\begin{aligned}\sin x + \sin y &= 2\sin\frac{x + y}{2}\cos\frac{x - y}{2}\\ \sin x - \sin y &= 2\cos\frac{x + y}{2}\sin\frac{x - y}{2}\\ \cos x + \cos y &= 2\cos\frac{x + y}{2}\cos\frac{x - y}{2}\\ \cos x - \cos y &= -2\sin\frac{x + y}{2}\sin\frac{x - y}{2}\end{aligned}$$

Trigonometric Equations (p. 564)

A **trigonometric equation** is an equation that contains trigonometric functions. A basic trigonometric equation is an equation of the form $T(\theta) = c$, where T is a trigonometric function and c is a constant. For example, $\sin\theta = 0.5$ and $\tan\theta = 2$ are basic trigonometric equations. Solving any trigonometric equation involves solving a basic trigonometric equation.

If a trigonometric equation has a solution, then it has infinitely many solutions.

To find all solutions, we first find the solutions in one period and then add integer multiples of the period.

We can sometimes use trigonometric identities to simplify a trigonometric equation.

CONCEPT CHECK

1. What is an identity? What is a trigonometric identity?

2. (a) State the Pythagorean identities.

(b) Use a Pythagorean identity to express cosine in terms of sine.

3. (a) State the reciprocal identities for cosecant, secant, and cotangent.

(b) State the even-odd identities for sine and cosine.

(c) State the cofunction identities for sine, tangent, and secant.

(d) Suppose that $\cos(-x) = 0.4$; use the identities in parts (a) and (b) to find $\sec x$.

(e) Suppose that $\sin 10° = a$; use the identities in part (c) to find $\cos 80°$.

4. (a) How do you prove an identity?

(b) Prove the identity $\sin x(\csc x - \sin x) = \cos^2 x$

5. (a) State the Addition and Subtraction Formulas for Sine and Cosine.

(b) Use a formula from part (a) to find $\sin 75°$.

6. (a) State the formula for $A\sin x + B\cos x$.

(b) Express $3\sin x + 4\cos x$ as a function of sine only.

7. (a) State the Double-Angle Formula for Sine and the Double-Angle Formulas for Cosine.

(b) Prove the identity $\sec x\sin 2x = 2\sin x$.

8. (a) State the formulas for lowering powers of sine and cosine.

(b) Prove the identity $4\sin^2 x\cos^2 x = \sin^2 2x$.

9. (a) State the Half-Angle Formulas for Sine and Cosine.

(b) Find $\cos 15°$.

10. (a) State the Product-to-Sum Formula for the product $\sin u\cos v$.

(b) Express $\sin 5x\cos 3x$ as a sum of trigonometric functions.

11. **(a)** State the Sum-to-Product Formula for the sum $\sin x + \sin y$.

(b) Express $\sin 5x + \sin 7x$ as a product of trigonometric functions.

12. What is a trigonometric equation? How do we solve a trigonometric equation?

(a) Solve the equation $\cos x = \frac{1}{2}$.

(b) Solve the equation $2 \sin x \cos x = \frac{1}{2}$.

ANSWERS TO THE CONCEPT CHECK CAN BE FOUND AT THE BACK OF THE BOOK.

EXERCISES

1–22 ■ Proving Identities Verify the identity.

1. $\sin\theta\,(\cot\theta + \tan\theta) = \sec\theta$

2. $(\sec\theta - 1)(\sec\theta + 1) = \tan^2\theta$

3. $\cos^2 x \csc x - \csc x = -\sin x$

4. $\dfrac{1}{1 - \sin^2 x} = 1 + \tan^2 x$

5. $\dfrac{\cos^2 x - \tan^2 x}{\sin^2 x} = \cot^2 x - \sec^2 x$

6. $\dfrac{1 + \sec x}{\sec x} = \dfrac{\sin^2 x}{1 - \cos x}$

7. $\dfrac{\cos^2 x}{1 - \sin x} = \dfrac{\cos x}{\sec x - \tan x}$

8. $(1 - \tan x)(1 - \cot x) = 2 - \sec x \csc x$

9. $\sin^2 x \cot^2 x + \cos^2 x \tan^2 x = 1$

10. $(\tan x + \cot x)^2 = \csc^2 x \sec^2 x$

11. $\dfrac{\sin 2x}{1 + \cos 2x} = \tan x$

12. $\dfrac{\cos(x + y)}{\cos x \sin y} = \cot y - \tan x$

13. $\csc x - \tan \dfrac{x}{2} = \cot x$

14. $1 + \tan x \tan \dfrac{x}{2} = \sec x$

15. $\dfrac{\sin 2x}{\sin x} - \dfrac{\cos 2x}{\cos x} = \sec x$

16. $\tan\left(x + \dfrac{\pi}{4}\right) = \dfrac{1 + \tan x}{1 - \tan x}$

17. $\dfrac{\sec x - 1}{\sin x \sec x} = \tan \dfrac{x}{2}$

18. $(\cos x + \cos y)^2 + (\sin x - \sin y)^2 = 2 + 2\cos(x + y)$

19. $\left(\cos \dfrac{x}{2} - \sin \dfrac{x}{2}\right)^2 = 1 - \sin x$

20. $\dfrac{\cos 3x - \cos 7x}{\sin 3x + \sin 7x} = \tan 2x$

21. $\dfrac{\sin(x + y) + \sin(x - y)}{\cos(x + y) + \cos(x - y)} = \tan x$

22. $\sin(x + y)\sin(x - y) = \sin^2 x - \sin^2 y$

23–26 ■ Checking Identities Graphically **(a)** Graph f and g. **(b)** Do the graphs suggest that the equation $f(x) = g(x)$ is an identity? Prove your answer.

23. $f(x) = 1 - \left(\cos \dfrac{x}{2} - \sin \dfrac{x}{2}\right)^2, \quad g(x) = \sin x$

24. $f(x) = \sin x + \cos x, \quad g(x) = \sqrt{\sin^2 x + \cos^2 x}$

25. $f(x) = \tan x \tan \dfrac{x}{2}, \quad g(x) = \dfrac{1}{\cos x}$

26. $f(x) = 1 - 8\sin^2 x + 8\sin^4 x, \quad g(x) = \cos 4x$

27–28 ■ Determining Identities Graphically **(a)** Graph the function(s) and make a conjecture, and **(b)** prove your conjecture.

27. $f(x) = 2\sin^2 3x + \cos 6x$

28. $f(x) = \sin x \cot \dfrac{x}{2}, \quad g(x) = \cos x$

29–46 ■ Solving Trigonometric Equations Solve the equation in the interval $[0, 2\pi)$.

29. $4\sin\theta - 3 = 0$

30. $5\cos\theta + 3 = 0$

31. $\cos x \sin x - \sin x = 0$

32. $\sin x - 2\sin^2 x = 0$

33. $2\sin^2 x - 5\sin x + 2 = 0$

34. $\sin x - \cos x - \tan x = -1$

35. $2\cos^2 x - 7\cos x + 3 = 0$

36. $4\sin^2 x + 2\cos^2 x = 3$

37. $\dfrac{1 - \cos x}{1 + \cos x} = 3$

38. $\sin x = \cos 2x$

39. $\tan^3 x + \tan^2 x - 3\tan x - 3 = 0$

40. $\cos 2x \csc^2 x = 2\cos 2x$

41. $\tan \frac{1}{2}x + 2\sin 2x = \csc x$

42. $\cos 3x + \cos 2x + \cos x = 0$

43. $\tan x + \sec x = \sqrt{3}$

44. $2\cos x - 3\tan x = 0$

45. $\cos x = x^2 - 1$

46. $e^{\sin x} = x$

47. Range of a Projectile If a projectile is fired with velocity v_0 at an angle θ, then the maximum height it reaches (in ft) is modeled by the function

$$M(\theta) = \frac{v_0^2 \sin^2\theta}{64}$$

Suppose $v_0 = 400$ ft/s.

(a) At what angle θ should the projectile be fired so that the maximum height it reaches is 2000 ft?

(b) Is it possible for the projectile to reach a height of 3000 ft?

(c) Find the angle θ for which the projectile will travel highest.

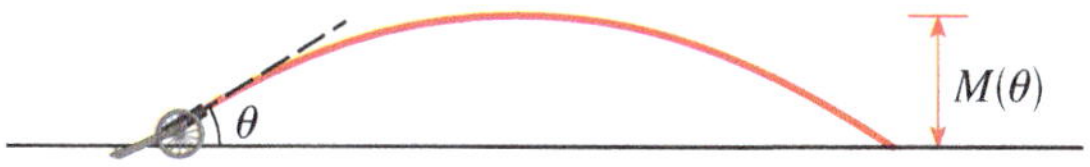

48. Displacement of a Shock Absorber The displacement of an automobile shock absorber is modeled by the function

$$f(t) = 2^{-0.2t} \sin 4\pi t$$

Find the times when the shock absorber is at its equilibrium position (that is, when $f(t) = 0$). [*Hint:* $2^x > 0$ for all real x.]

49–58 ■ Value of Expressions Find the exact value of the expression.

49. $\cos 15°$ **50.** $\sin \frac{5\pi}{12}$

51. $\tan \frac{\pi}{8}$ **52.** $2 \sin \frac{\pi}{12} \cos \frac{\pi}{12}$

53. $\sin 5° \cos 40° + \cos 5° \sin 40°$

54. $\dfrac{\tan 66° - \tan 6°}{1 + \tan 66° \tan 6°}$

55. $\cos^2 \frac{\pi}{8} - \sin^2 \frac{\pi}{8}$ **56.** $\frac{1}{2} \cos \frac{\pi}{12} + \frac{\sqrt{3}}{2} \sin \frac{\pi}{12}$

57. $\cos 37.5° \cos 7.5°$ **58.** $\cos 67.5° + \cos 22.5°$

59–64 ■ Evaluating Expressions Involving Trigonometric Functions Find the exact value of the expression given that $\sec x = \frac{3}{2}$, $\csc y = 3$, and x and y are in Quadrant I.

59. $\sin(x + y)$ **60.** $\cos(x - y)$

61. $\tan(x + y)$ **62.** $\sin 2x$

63. $\cos \frac{y}{2}$ **64.** $\tan \frac{y}{2}$

65–66 ■ Evaluating Expressions Involving Inverse Trigonometric Functions Find the exact value of the expression.

65. $\tan\left(2 \cos^{-1} \frac{3}{7}\right)$ **66.** $\sin\left(\tan^{-1} \frac{3}{4} + \cos^{-1} \frac{5}{13}\right)$

67–68 ■ Expressions Involving Inverse Trigonometric Functions Write the expression as an algebraic expression in the variable(s).

67. $\tan(2 \tan^{-1} x)$ **68.** $\cos(\sin^{-1} x + \cos^{-1} y)$

69. Viewing Angle of a Sign A 10-ft-wide highway sign is adjacent to a roadway, as shown in the figure. As a driver approaches the sign, the viewing angle θ changes.

(a) Express viewing angle θ as a function of the distance x between the driver and the sign.

(b) The sign is legible when the viewing angle is 2° or greater. At what distance x does the sign first become legible?

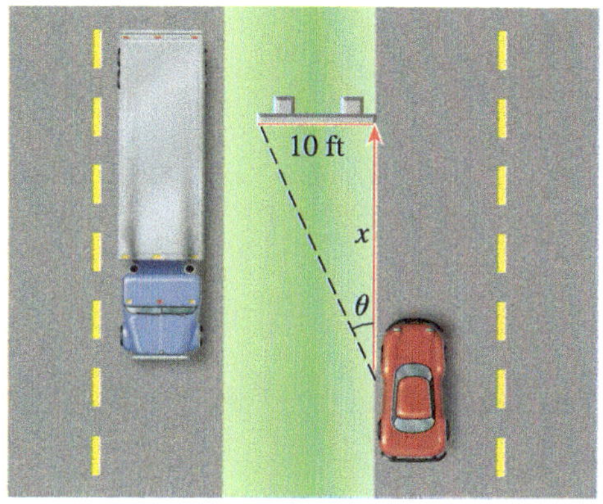

70. Viewing Angle of a Tower A 380-ft-tall building supports a 40-ft communications tower (see the figure). As a driver approaches the building, the viewing angle θ of the tower changes.

(a) Express the viewing angle θ as a function of the distance x between the driver and the building.

(b) At what distance from the building is the viewing angle θ as large as possible?

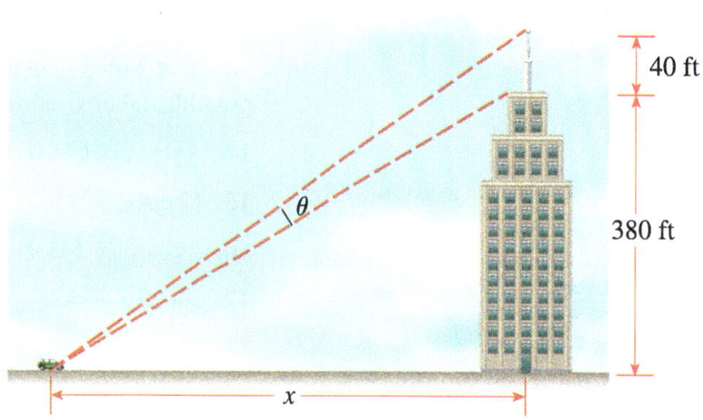

CHAPTER 7 TEST

1–8 ■ Verify each identity.

1. $\tan\theta\sin\theta + \cos\theta = \sec\theta$

2. $\dfrac{\tan x}{1 - \cos x} = \csc x\,(1 + \sec x)$

3. $\dfrac{2\tan x}{1 + \tan^2 x} = \sin 2x$

4. $\sin x \tan\left(\dfrac{x}{2}\right) = 1 - \cos x$

5. $2\sin^2(3x) = 1 - \cos(6x)$

6. $\cos 4x = 1 - 8\sin^2 x + 8\sin^4 x$

7. $\left(\sin\left(\dfrac{x}{2}\right) + \cos\left(\dfrac{x}{2}\right)\right)^2 = 1 + \sin x$

8. Let $x = 2\sin\theta$, $-\pi/2 < \theta < \pi/2$. Simplify the expression

$$\frac{x}{\sqrt{4 - x^2}}$$

9. Find the exact value of each expression.

(a) $\sin 8^\circ \cos 22^\circ + \cos 8^\circ \sin 22^\circ$ **(b)** $\sin 75^\circ$ **(c)** $\sin\dfrac{\pi}{12}$

10. For the angles α and β in the figures, find $\cos(\alpha + \beta)$.

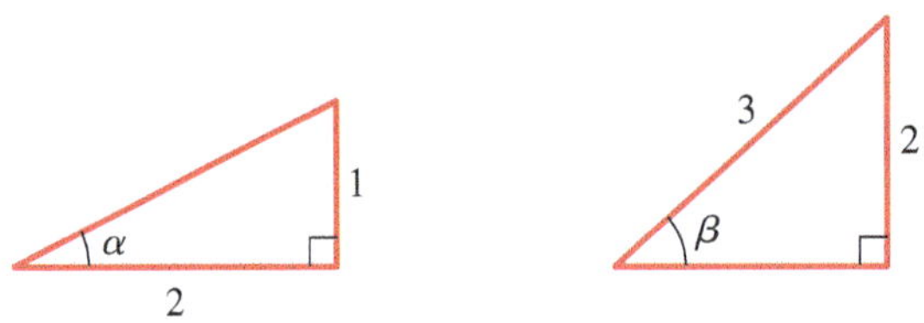

11. Write $\sin 3x \cos 5x$ as a sum of trigonometric functions.

12. Write $\sin 2x - \sin 5x$ as a product of trigonometric functions.

13. If $\sin\theta = -\frac{4}{5}$ and θ is in Quadrant III, find $\tan(\theta/2)$.

14–20 ■ Solve each trigonometric equation in the interval $[0, 2\pi)$. Give the exact value, if possible; otherwise, round your answer to two decimal places.

14. $3\sin\theta - 1 = 0$

15. $(2\cos\theta - 1)(\sin\theta - 1) = 0$

16. $2\cos^2\theta + 5\cos\theta + 2 = 0$

17. $\sin 2\theta - \cos\theta = 0$

18. $5\cos 2\theta = 2$

19. $2\cos^2 x + \cos 2x = 0$

20. $2\tan\left(\dfrac{x}{2}\right) - \csc x = 0$

21. Find the exact value of $\cos\left(2\tan^{-1}\frac{9}{40}\right)$.

22. Rewrite the expression as an algebraic function of x and y: $\sin(\cos^{-1}x - \tan^{-1}y)$.

A CUMULATIVE REVIEW TEST FOR CHAPTERS 5, 6, AND 7 CAN BE FOUND AT THE BOOK COMPANION WEBSITE: **www.stewartmath.com**.

We've learned that the position of a particle in simple harmonic motion is described by a function of the form $y = A \sin \omega t$ (see Section 5.6). For example, if a string is moved up and down as in Figure 1, then the red dot on the string moves up and down in simple harmonic motion. Of course, the same holds true for each point on the string.

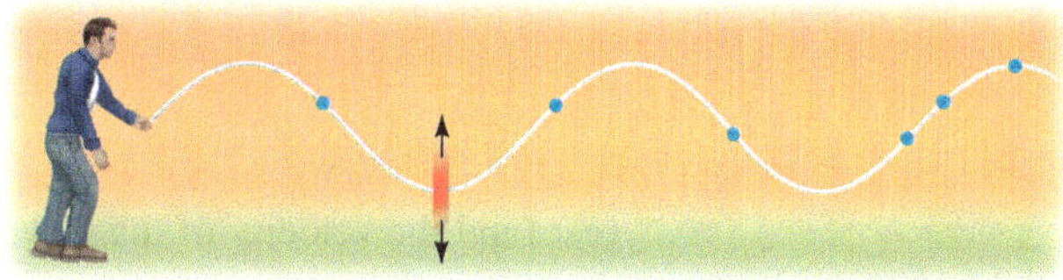

FIGURE 1

What function describes the shape of the whole string? If we fix an instant in time $(t = 0)$ and snap a photograph of the string, we get the shape in Figure 2, which is modeled by

$$y = A \sin kx$$

where y is the height of the string above the x-axis at the point x.

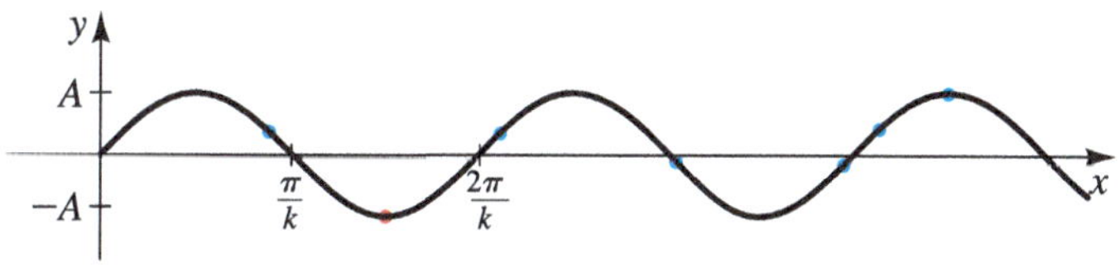

FIGURE 2 $y = A \sin kx$

Traveling Waves

If we snap photographs of the string at other instants, as in Figure 3, it appears that the waves in the string "travel" or shift to the right.

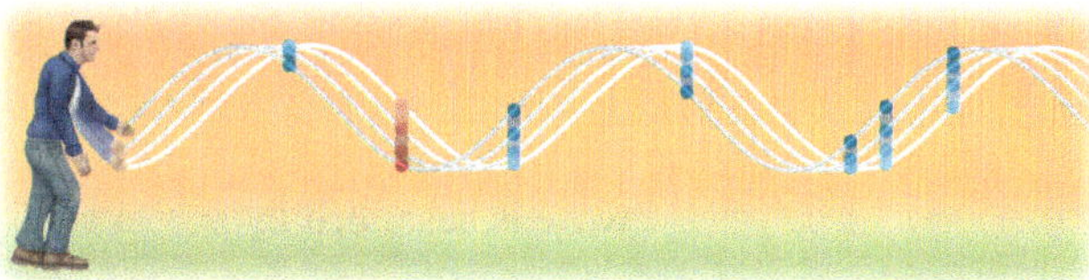

FIGURE 3

The **velocity** of the wave is the rate at which it moves to the right. If the wave has velocity v, then it moves to the right a distance vt in time t. So the graph of the shifted wave at time t is

$$y(x, t) = A \sin k(x - vt)$$

This function models the position of any point x on the string at any time t. We use the notation $y(x, t)$ to indicate that the function depends on the *two* variables x and t. Here is how this function models the motion of the string.

- **If we fix x**, then $y(x, t)$ is a function of t only, which gives the position of the fixed point x at time t.
- **If we fix t**, then $y(x, t)$ is a function of x only, whose graph is the shape of the string at the fixed time t.

EXAMPLE 1 ■ A Traveling Wave

A traveling wave is described by the function

$$y(x, t) = 3 \sin\left(2x - \frac{\pi}{2}t\right) \qquad x \geq 0$$

(a) Find the function that models the position of the point $x = \pi/6$ at any time t. Observe that the point moves in simple harmonic motion.

(b) Sketch the shape of the wave when $t = 0, 0.5, 1.0, 1.5$, and 2.0. Does the wave appear to be traveling to the right?

(c) Find the velocity of the wave.

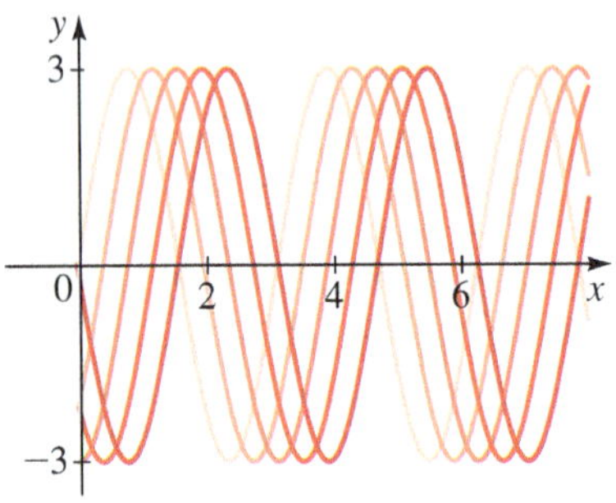

FIGURE 4 Traveling wave

SOLUTION

(a) Substituting $x = \pi/6$, we get

$$y\left(\frac{\pi}{6}, t\right) = 3 \sin\left(2 \cdot \frac{\pi}{6} - \frac{\pi}{2}t\right) = 3 \sin\left(\frac{\pi}{3} - \frac{\pi}{2}t\right)$$

The function $y = 3 \sin(\frac{\pi}{3} - \frac{\pi}{2}t)$ describes simple harmonic motion with amplitude 3 and period $2\pi/(\pi/2) = 4$.

(b) The graphs are shown in Figure 4. As t increases, the wave moves to the right.

(c) We express the given function in the standard form $y(x, t) = A \sin k(x - vt)$.

$$\begin{aligned} y(x, t) &= 3 \sin\left(2x - \frac{\pi}{2}t\right) && \text{Given} \\ &= 3 \sin 2\left(x - \frac{\pi}{4}t\right) && \text{Factor 2} \end{aligned}$$

Comparing this to the standard form, we see that the wave is moving with velocity $v = \pi/4$. ■

■ Standing Waves

If two waves are traveling along the same string, then the movement of the string is determined by the sum of the two waves. For example, if the string is attached to a wall, then the waves bounce back with the same amplitude and speed but in the opposite direction. In this case, one wave is described by $y = A \sin k(x - vt)$, and the reflected wave is described by $y = A \sin k(x + vt)$. The resulting wave is

$$\begin{aligned} y(x, t) &= A \sin k(x - vt) + A \sin k(x + vt) && \text{Add the two waves} \\ &= 2A \sin kx \cos kvt && \text{Sum-to-Product Formula} \end{aligned}$$

The points where kx is a multiple of 2π are special, because at these points $y = 0$ for any time t. In other words, these points never move. Such points are called **nodes**. Figure 5 shows the graph of the wave for several values of t. We see that the wave does not travel but simply vibrates up and down. Such a wave is called a **standing wave**.

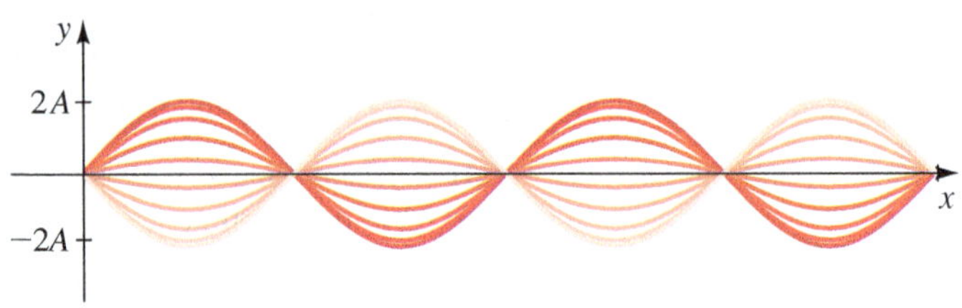

FIGURE 5 A standing wave

EXAMPLE 2 ■ A Standing Wave

Traveling waves are generated at each end of a wave tank 30 ft long, with equations

$$y = 1.5 \sin\left(\frac{\pi}{5}x - 3t\right)$$

and

$$y = 1.5 \sin\left(\frac{\pi}{5}x + 3t\right)$$

(a) Find the equation of the combined wave, and find the nodes.

(b) Sketch the graph for $t = 0, 0.17, 0.34, 0.51, 0.68, 0.85$, and 1.02. Is this a standing wave?

SOLUTION

(a) The combined wave is obtained by adding the two equations.

$$y = 1.5 \sin\left(\frac{\pi}{5}x - 3t\right) + 1.5 \sin\left(\frac{\pi}{5}x + 3t\right) \qquad \text{Add the two waves}$$

$$= 3 \sin \frac{\pi}{5}x \cos 3t \qquad \text{Sum-to-Product Formula}$$

The nodes occur at the values of x for which $\sin \frac{\pi}{5}x = 0$, that is, where $\frac{\pi}{5}x = k\pi$ (k an integer). Solving for x, we get $x = 5k$. So the nodes occur at

$$x = 0, 5, 10, 15, 20, 25, 30$$

(b) The graphs are shown in Figure 6. From the graphs we see that this is a standing wave.

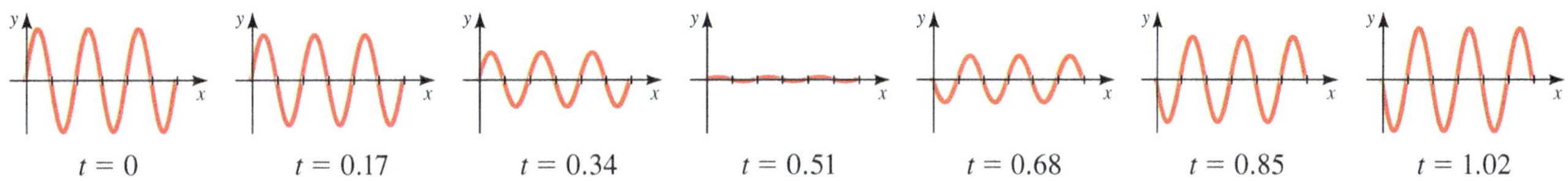

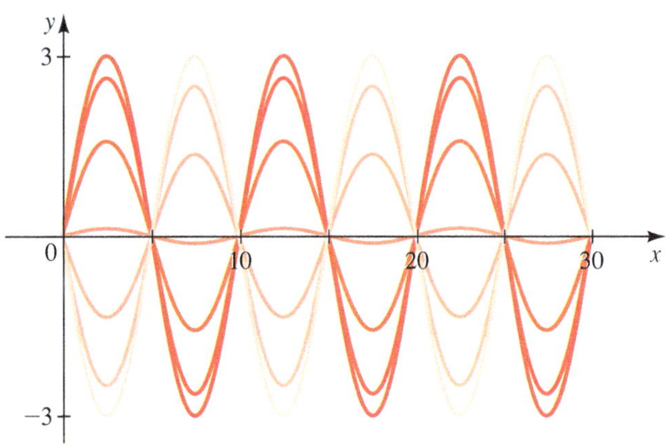

FIGURE 6
$y(x, t) = 3 \sin \frac{\pi}{5}x \cos 3t$

■

PROBLEMS

1. **Wave on a Canal** A wave on the surface of a long canal is described by the function

$$y(x, t) = 5 \sin\left(2x - \frac{\pi}{2}t\right) \qquad x \geq 0$$

   (a) Find the function that models the position of the point $x = 0$ at any time t.

   (b) Sketch the shape of the wave when $t = 0, 0.4, 0.8, 1.2$, and 1.6. Is this a traveling wave?

   (c) Find the velocity of the wave.

2. **Wave in a Rope** Traveling waves are generated at each end of a tightly stretched rope 24 ft long, with equations

$$y = 0.2 \sin(1.047x - 0.524t) \qquad \text{and} \qquad y = 0.2 \sin(1.047x + 0.524t)$$

   (a) Find the equation of the combined wave, and find the nodes.

   (b) Sketch the graph for $t = 0, 1, 2, 3, 4, 5$, and 6. Is this a standing wave?

3. **Traveling Wave** A traveling wave is graphed at the instant $t = 0$. If it is moving to the right with velocity 6, find an equation of the form $y(x, t) = A \sin(kx - kvt)$ for this wave.

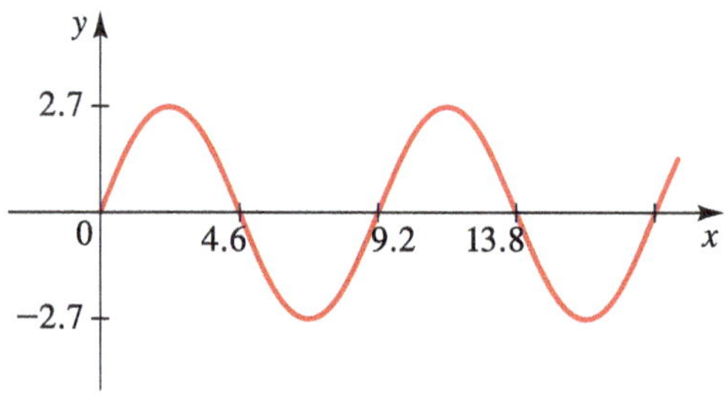

4. **Traveling Wave** A traveling wave has period $2\pi/3$, amplitude 5, and velocity 0.5.

   (a) Find the equation of the wave.

   (b) Sketch the graph for $t = 0, 0.5, 1, 1.5$, and 2.

5. **Standing Wave** A standing wave with amplitude 0.6 is graphed at several times t as shown in the figure. If the vibration has a frequency of 20 Hz, find an equation of the form $y(x, t) = A \sin \alpha x \cos \beta t$ that models this wave.

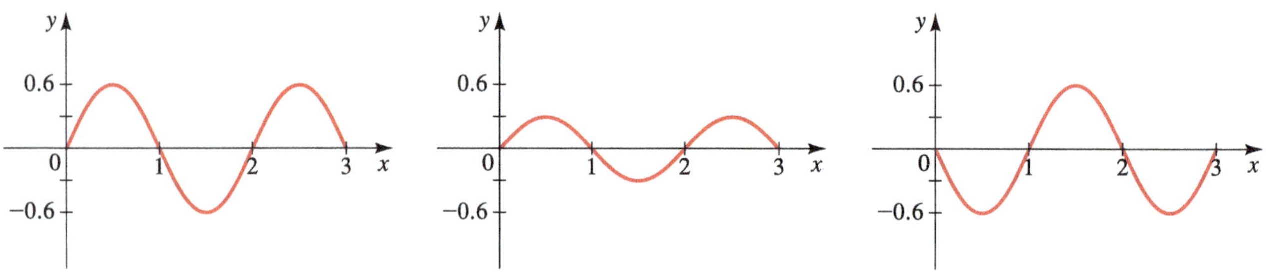

$t = 0$ s $\qquad$ $t = 0.010$ s $\qquad$ $t = 0.025$ s

6. Standing Wave A standing wave has maximum amplitude 7 and nodes at 0, $\pi/2$, π, $3\pi/2$, 2π, as shown in the figure. Each point that is not a node moves up and down with period 4π. Find a function of the form $y(x, t) = A \sin \alpha x \cos \beta t$ that models this wave.

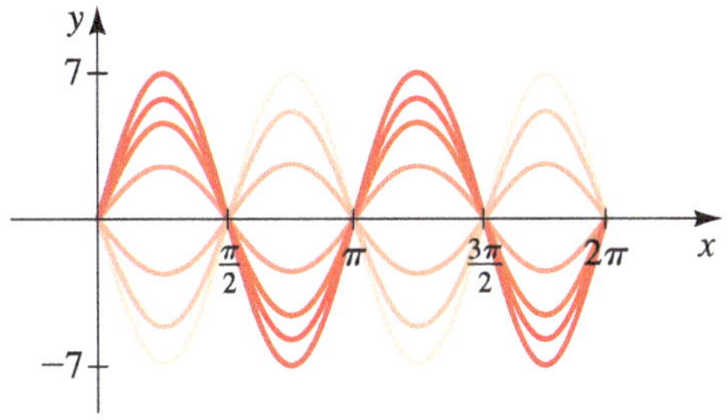

7. Vibrating String When a violin string vibrates, the sound produced results from a combination of standing waves that have evenly placed nodes. The figure illustrates some of the possible standing waves. Let's assume that the string has length π.

(a) For fixed t, the string has the shape of a sine curve $y = A \sin \alpha x$. Find the appropriate value of α for each of the illustrated standing waves.

(b) Do you notice a pattern in the values of α that you found in part (a)? What would the next two values of α be? Sketch rough graphs of the standing waves associated with these new values of α.

(c) Suppose that for fixed t, each point on the string that is not a node vibrates with frequency 440 Hz. Find the value of β for which an equation of the form $y = A \cos \beta t$ would model this motion.

(d) Combine your answers for parts (a) and (c) to find functions of the form $y(x, t) = A \sin \alpha x \cos \beta t$ that model each of the standing waves in the figure. (Assume that $A = 1$.)

8. Waves in a Tube Standing waves in a violin string must have nodes at the ends of the string because the string is fixed at its endpoints. But this need not be the case with sound waves in a tube (such as a flute or an organ pipe). The figure shows some possible standing waves in a tube.

Suppose that a standing wave in a tube 37.7 ft long is modeled by the function

$$y(x, t) = 0.3 \cos \tfrac{1}{2}x \cos 50\pi t$$

Here $y(x, t)$ represents the variation from normal air pressure at the point x feet from the end of the tube, at time t seconds.

(a) At what points x are the nodes located? Are the endpoints of the tube nodes?

(b) At what frequency does the air vibrate at points that are not nodes?

8 Polar Coordinates and Parametric Equations

In Section 1.9 we learned how to graph points in rectangular coordinates. In this chapter we study a different way of locating points in the plane, called *polar coordinates*. Using rectangular coordinates is like describing a location in a city by saying that it's at the corner of 2nd Street and 4th Avenue; these directions would help a taxi driver find the location. But we may also describe this same location "as the crow flies"; we can say that it's 1.5 miles northeast of City Hall. These directions would help an airplane or hot air balloon pilot find the location. So instead of specifying the location with respect to a grid of streets and avenues, we specify it by giving its distance and direction from a fixed reference point. That's what we do in the polar coordinate system. In polar coordinates the location of a point is given by an ordered pair of numbers: the distance of the point from the origin (or pole) and the angle from the positive x-axis.

Why do we study different coordinate systems? It's because certain curves are more naturally described in one coordinate system rather than another. For example, in rectangular coordinates lines and parabolas have simple equations, but equations of circles are rather complicated. We'll see that in polar coordinates circles have very simple equations.

8.1 POLAR COORDINATES

■ Definition of Polar Coordinates ■ Relationship Between Polar and Rectangular Coordinates ■ Polar Equations

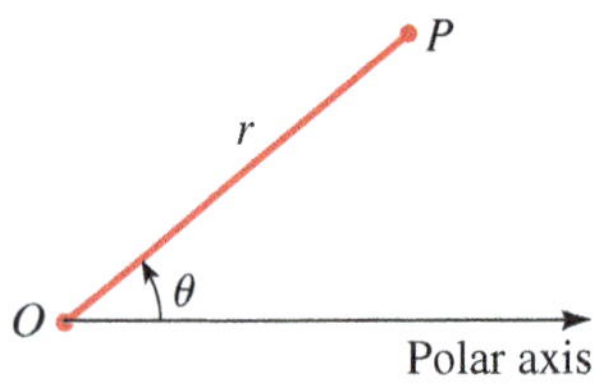

FIGURE 1

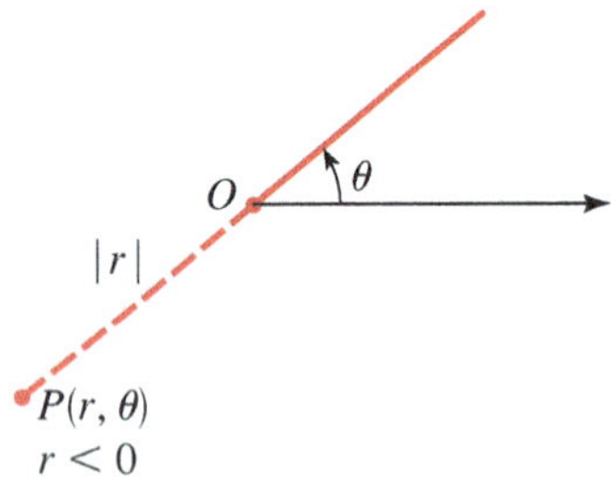

FIGURE 2

In this section we define polar coordinates, and we learn how polar coordinates are related to rectangular coordinates.

■ Definition of Polar Coordinates

The **polar coordinate system** uses distances and directions to specify the location of a point in the plane. To set up this system, we choose a fixed point O in the plane called the **pole** (or **origin**) and draw from O a ray (half-line) called the **polar axis** as in Figure 1. Then each point P can be assigned polar coordinates $P(r, \theta)$, where

r is the *distance* from O to P

θ is the *angle* between the polar axis and the segment $\overline{OP}$

We use the convention that θ is positive if measured in a counterclockwise direction from the polar axis or negative if measured in a clockwise direction. If r is negative, then $P(r, \theta)$ is defined to be the point that lies $|r|$ units from the pole in the direction opposite to that given by θ (see Figure 2).

EXAMPLE 1 ■ Plotting Points in Polar Coordinates

Plot the points whose polar coordinates are given.

(a) $(1, 3\pi/4)$ **(b)** $(3, -\pi/6)$ **(c)** $(3, 3\pi)$ **(d)** $(-4, \pi/4)$

SOLUTION The points are plotted in Figure 3. Note that the point in part (d) lies 4 units from the origin along the angle $5\pi/4$, because the given value of r is negative.

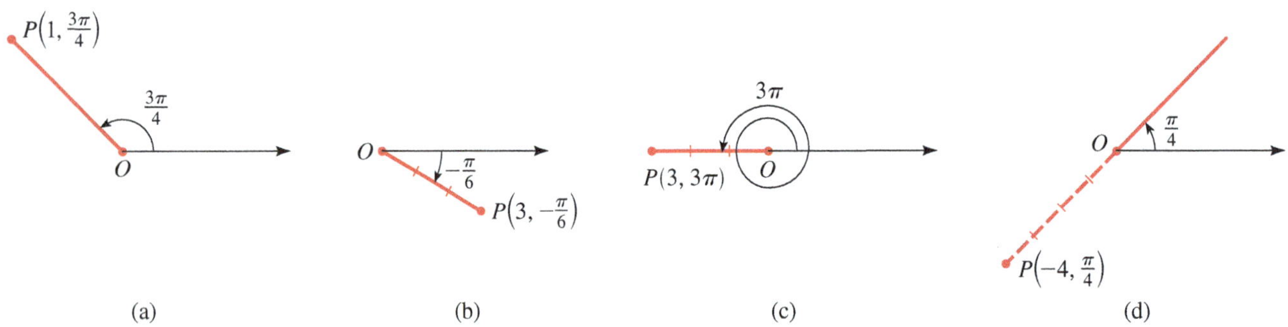

FIGURE 3

Now Try Exercises 5 and 7 ■

Note that the coordinates (r, θ) and $(-r, \theta + \pi)$ represent the same point, as shown in Figure 4. Moreover, because the angles $\theta + 2n\pi$ (where n is any integer) all

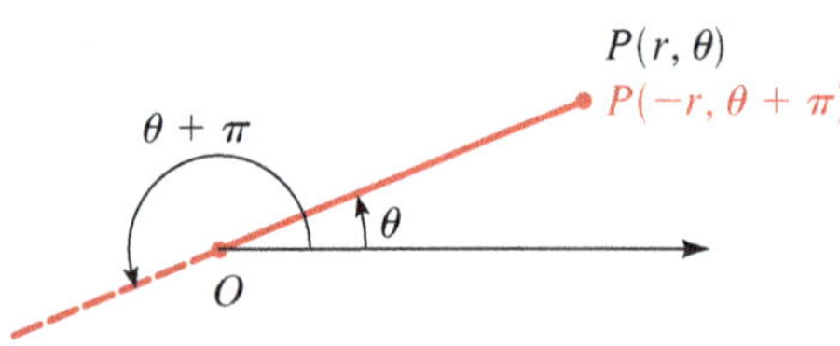

FIGURE 4

have the same terminal side as the angle θ, each point in the plane has infinitely many representations in polar coordinates. In fact, any point $P(r, \theta)$ can also be represented by

$$P(r, \theta + 2n\pi) \qquad \text{and} \qquad P(-r, \theta + (2n + 1)\pi)$$

for any integer n.

EXAMPLE 2 ■ Different Polar Coordinates for the Same Point

(a) Graph the point with polar coordinates $P(2, \pi/3)$.

(b) Find two other polar coordinate representations of P with $r > 0$ and two with $r < 0$.

SOLUTION

(a) The graph is shown in Figure 5(a).

(b) Other representations with $r > 0$ are

$$\left(2, \frac{\pi}{3} + 2\pi\right) = \left(2, \frac{7\pi}{3}\right) \qquad \text{Add } 2\pi \text{ to } \theta$$

$$\left(2, \frac{\pi}{3} - 2\pi\right) = \left(2, -\frac{5\pi}{3}\right) \qquad \text{Add } -2\pi \text{ to } \theta$$

Other representations with $r < 0$ are

$$\left(-2, \frac{\pi}{3} + \pi\right) = \left(-2, \frac{4\pi}{3}\right) \qquad \text{Replace } r \text{ by } -r \text{ and add } \pi \text{ to } \theta$$

$$\left(-2, \frac{\pi}{3} - \pi\right) = \left(-2, -\frac{2\pi}{3}\right) \qquad \text{Replace } r \text{ by } -r \text{ and add } -\pi \text{ to } \theta$$

The graphs in Figure 5 explain why these coordinates represent the same point.

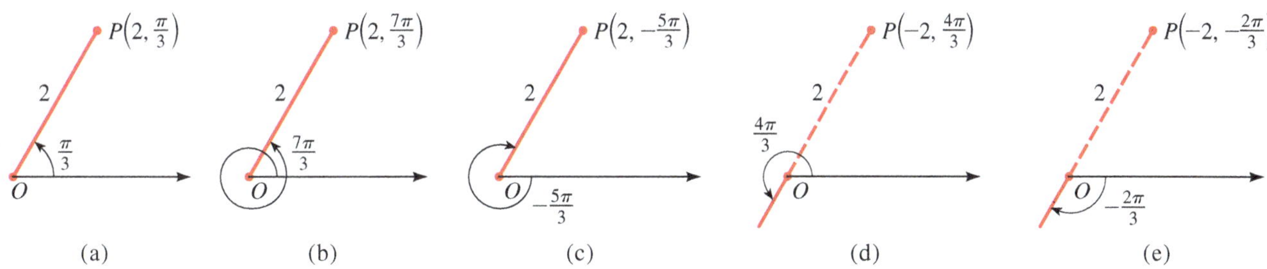

FIGURE 5

Now Try Exercise 11 ■

■ Relationship Between Polar and Rectangular Coordinates

Situations often arise in which we need to consider polar and rectangular coordinates simultaneously. The connection between the two systems is illustrated in Figure 6 (see next page), where the polar axis coincides with the positive x-axis. The formulas in the box are obtained from the figure using the definitions of the trigonometric functions and the Pythagorean Theorem. (Although we have pictured the case in which $r > 0$ and θ is acute, the formulas hold for any angle θ and for any value of r.)

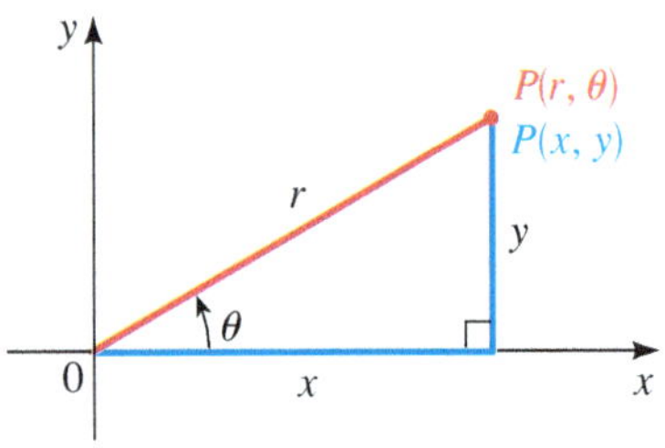

FIGURE 6

RELATIONSHIP BETWEEN POLAR AND RECTANGULAR COORDINATES

1. To change from polar to rectangular coordinates, use the formulas

$$x = r\cos\theta \qquad \text{and} \qquad y = r\sin\theta$$

2. To change from rectangular to polar coordinates, use the formulas

$$r^2 = x^2 + y^2 \qquad \text{and} \qquad \tan\theta = \frac{y}{x} \quad (x \neq 0)$$

EXAMPLE 3 ■ Converting Polar Coordinates to Rectangular Coordinates

Find rectangular coordinates for the point that has polar coordinates $(4, 2\pi/3)$.

SOLUTION Since $r = 4$ and $\theta = 2\pi/3$, we have

$$x = r\cos\theta = 4\cos\frac{2\pi}{3} = 4\cdot\left(-\frac{1}{2}\right) = -2$$

$$y = r\sin\theta = 4\sin\frac{2\pi}{3} = 4\cdot\frac{\sqrt{3}}{2} = 2\sqrt{3}$$

Thus the point has rectangular coordinates $(-2, 2\sqrt{3})$.

Now Try Exercise 29 ■

EXAMPLE 4 ■ Converting Rectangular Coordinates to Polar Coordinates

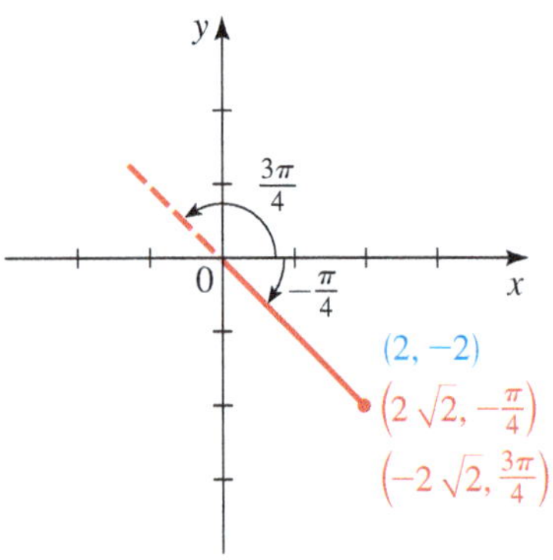

FIGURE 7

Find polar coordinates for the point that has rectangular coordinates $(2, -2)$.

SOLUTION Using $x = 2$, $y = -2$, we get

$$r^2 = x^2 + y^2 = 2^2 + (-2)^2 = 8$$

so $r = 2\sqrt{2}$ or $-2\sqrt{2}$. Also

$$\tan\theta = \frac{y}{x} = \frac{-2}{2} = -1$$

so $\theta = 3\pi/4$ or $-\pi/4$. Since the point $(2, -2)$ lies in Quadrant IV (see Figure 7), we can represent it in polar coordinates as $(2\sqrt{2}, -\pi/4)$ or $(-2\sqrt{2}, 3\pi/4)$.

Now Try Exercise 37 ■

DISCOVERY PROJECT

Mapping the World

In the *Focus on Modeling* on page 533 we learned how surveyors can make a map of a city or town. But mapping the whole world introduces a new difficulty. How is it possible to represent the *spherical* world we live on by a *flat* map? This challenge was faced by Renaissance explorers and their mapmakers, who developed several ingenious solutions. In this project we see how polar coordinates and trigonometry can help us make a map of the whole world on a flat sheet of paper. You can find the project at **www.stewartmath.com**.

Note that the equations relating polar and rectangular coordinates do not uniquely determine r or θ. When we use these equations to find the polar coordinates of a point, we must be careful that the values we choose for r and θ give us a point in the correct quadrant, as we did in Example 4.

Polar Equations

In Examples 3 and 4 we converted points from one coordinate system to the other. Now we consider the same problem for equations.

EXAMPLE 5 ■ Converting an Equation from Rectangular to Polar Coordinates

Express the equation $x^2 = 4y$ in polar coordinates.

SOLUTION We use the formulas $x = r\cos\theta$ and $y = r\sin\theta$.

$$x^2 = 4y \qquad \text{Rectangular equation}$$
$$(r\cos\theta)^2 = 4(r\sin\theta) \qquad \text{Substitute } x = r\cos\theta,\ y = r\sin\theta$$
$$r^2\cos^2\theta = 4r\sin\theta \qquad \text{Expand}$$
$$r = 4\frac{\sin\theta}{\cos^2\theta} \qquad \text{Divide by } r\cos^2\theta$$
$$r = 4\sec\theta\tan\theta \qquad \text{Simplify}$$

Now Try Exercise 47 ■

As Example 5 shows, converting from rectangular to polar coordinates is straightforward: Just replace x by $r\cos\theta$ and y by $r\sin\theta$, and then simplify. But converting polar equations to rectangular form often requires more thought.

EXAMPLE 6 ■ Converting Equations from Polar to Rectangular Coordinates

Express the polar equation in rectangular coordinates. If possible, determine the graph of the equation from its rectangular form.

(a) $r = 5\sec\theta$ **(b)** $r = 2\sin\theta$ **(c)** $r = 2 + 2\cos\theta$

SOLUTION

(a) Since $\sec\theta = 1/\cos\theta$, we multiply both sides by $\cos\theta$.

$$r = 5\sec\theta \qquad \text{Polar equation}$$
$$r\cos\theta = 5 \qquad \text{Multiply by } \cos\theta$$
$$x = 5 \qquad \text{Substitute } x = r\cos\theta$$

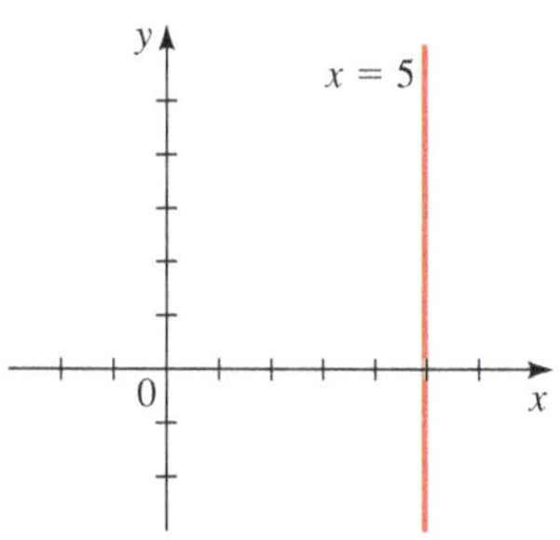

FIGURE 8

The graph of $x = 5$ is the vertical line in Figure 8.

(b) We multiply both sides of the equation by r, because then we can use the formulas $r^2 = x^2 + y^2$ and $r\sin\theta = y$.

$$r = 2\sin\theta \qquad \text{Polar equation}$$
$$r^2 = 2r\sin\theta \qquad \text{Multiply by } r$$
$$x^2 + y^2 = 2y \qquad r^2 = x^2 + y^2 \text{ and } r\sin\theta = y$$
$$x^2 + y^2 - 2y = 0 \qquad \text{Subtract } 2y$$
$$x^2 + (y-1)^2 = 1 \qquad \text{Complete the square in } y$$

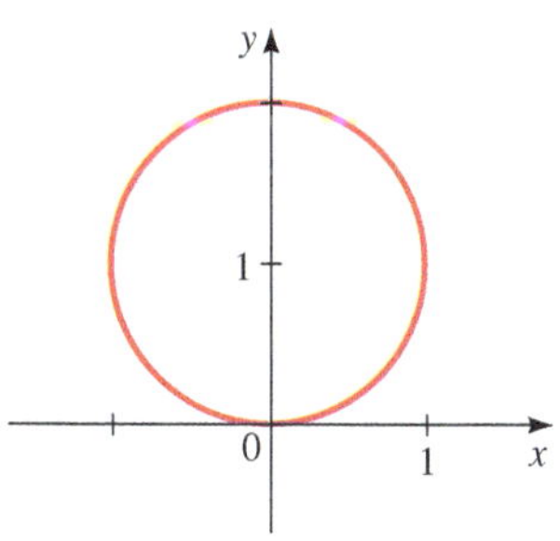

FIGURE 9

This is the equation of a circle of radius 1 centered at the point $(0, 1)$. It is graphed in Figure 9.

(c) We first multiply both sides of the equation by r:

$$r^2 = 2r + 2r\cos\theta$$

Using $r^2 = x^2 + y^2$ and $x = r\cos\theta$, we can convert two terms in the equation into rectangular coordinates, but eliminating the remaining r requires more work.

$$x^2 + y^2 = 2r + 2x \qquad r^2 = x^2 + y^2 \text{ and } r\cos\theta = x$$

$$x^2 + y^2 - 2x = 2r \qquad \text{Subtract } 2x$$

$$(x^2 + y^2 - 2x)^2 = 4r^2 \qquad \text{Square both sides}$$

$$(x^2 + y^2 - 2x)^2 = 4(x^2 + y^2) \qquad r^2 = x^2 + y^2$$

In this case the rectangular equation looks more complicated than the polar equation. Although we cannot easily determine the graph of the equation from its rectangular form, we will see in the next section how to graph it using the polar equation.

Now Try Exercises 55, 57, and 59 ■

8.1 EXERCISES

CONCEPTS

1. We can describe the location of a point in the plane using different ______ systems. The point P shown in the figure has rectangular coordinates (,) and polar coordinates (,).

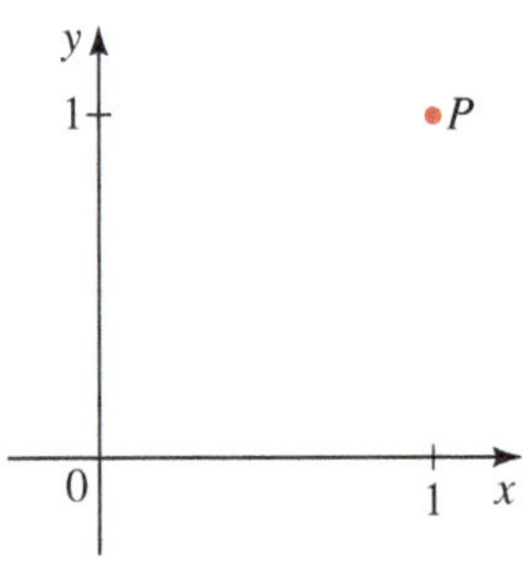

2. Let P be a point in the plane.

(a) If P has polar coordinates (r, θ) then it has rectangular coordinates (x, y) where $x =$ ______ and $y =$ ______.

(b) If P has rectangular coordinates (x, y) then it has polar coordinates (r, θ) where $r^2 =$ ______ and $\tan\theta =$ ______.

3–4 ■ *Yes or No*? If *No*, give a reason.

3. Do the polar coordinates $(2, \pi/6)$ and $(-2, 7\pi/6)$ represent the same point?

4. Do the equations relating polar and rectangular coordinates uniquely determine r and θ?

SKILLS

5–10 ■ **Plotting Points in Polar Coordinates** Plot the point that has the given polar coordinates.

5. $(4, \pi/4)$ **6.** $(1, 0)$ **7.** $(6, -7\pi/6)$

8. $(3, -2\pi/3)$ **9.** $(-2, 4\pi/3)$ **10.** $(-5, -17\pi/6)$

11–16 ■ **Different Polar Coordinates for the Same Point** Plot the point that has the given polar coordinates. Then give two other polar coordinate representations of the point, one with $r < 0$ and the other with $r > 0$.

11. $(3, \pi/2)$ **12.** $(2, 3\pi/4)$ **13.** $(-1, 7\pi/6)$

14. $(-2, -\pi/3)$ **15.** $(-5, 0)$ **16.** $(3, 1)$

17–24 ■ **Points in Polar Coordinates** Determine which point in the figure, P, Q, R, or S, has the given polar coordinates.

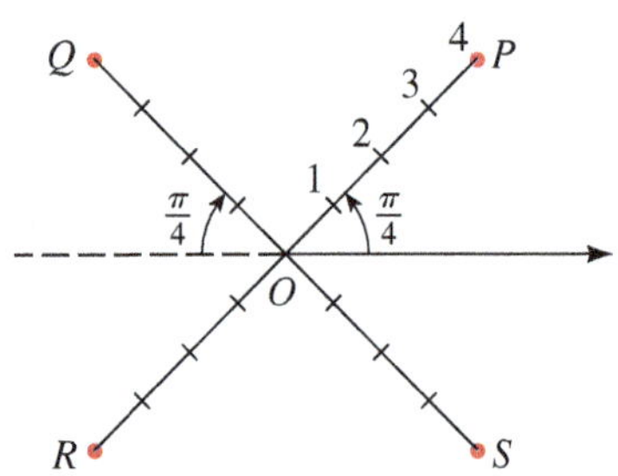

17. $(4, 3\pi/4)$ **18.** $(4, -3\pi/4)$

19. $(-4, -\pi/4)$ **20.** $(-4, 13\pi/4)$

21. $(4, -23\pi/4)$ **22.** $(-4, 23\pi/4)$

23. $(-4, 101\pi/4)$ **24.** $(4, 103\pi/4)$

25–26 ■ Rectangular Coordinates to Polar Coordinates A point is graphed in rectangular form. Find polar coordinates for the point, with $r > 0$ and $0 < \theta < 2\pi$.

25.

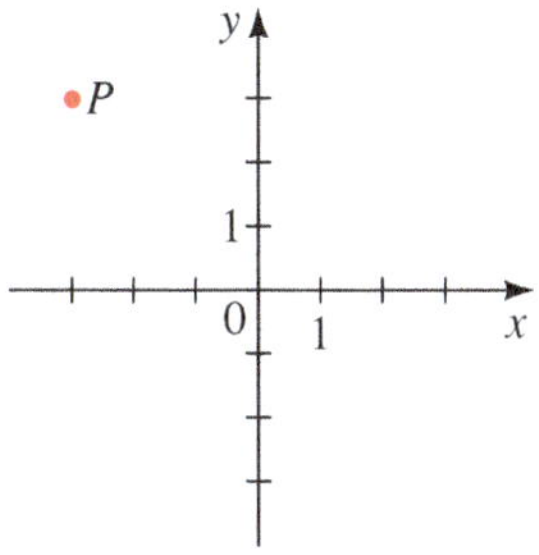

26.

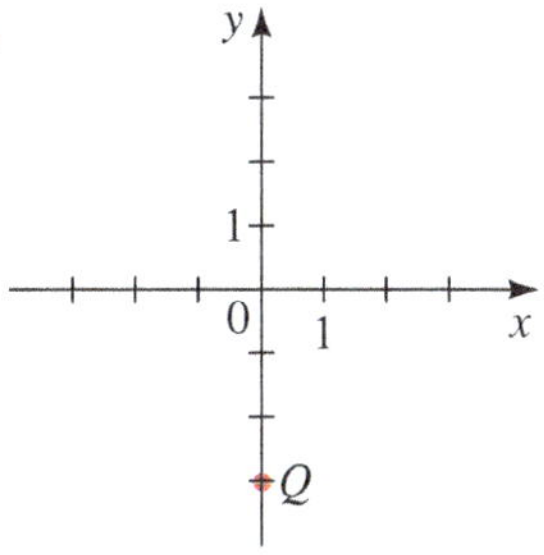

27–28 ■ Polar Coordinates to Rectangular Coordinates A point is graphed in polar form. Find its rectangular coordinates.

27.

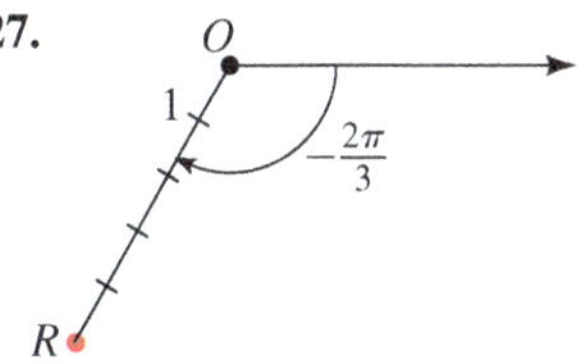

28.

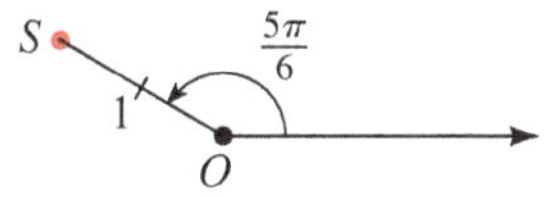

29–36 ■ Polar Coordinates to Rectangular Coordinates Find the rectangular coordinates for the point whose polar coordinates are given.

29. $(4, \pi/6)$ **30.** $(6, 2\pi/3)$

31. $(\sqrt{2}, -\pi/4)$ **32.** $(-1, 5\pi/2)$

33. $(5, 5\pi)$ **34.** $(0, 13\pi)$

35. $(6\sqrt{2}, 11\pi/6)$ **36.** $(\sqrt{3}, -5\pi/3)$

37–44 ■ Rectangular Coordinates to Polar Coordinates Convert the rectangular coordinates to polar coordinates with $r > 0$ and $0 \le \theta < 2\pi$.

37. $(-1, 1)$ **38.** $(3\sqrt{3}, -3)$

39. $(\sqrt{8}, \sqrt{8})$ **40.** $(-\sqrt{6}, -\sqrt{2})$

41. $(3, 4)$ **42.** $(1, -2)$

43. $(-6, 0)$ **44.** $(0, -\sqrt{3})$

45–50 ■ Rectangular Equations to Polar Equations Convert the equation to polar form.

45. $x = y$ **46.** $x^2 + y^2 = 9$

47. $y = x^2$ **48.** $y = 5$

49. $x = 4$ **50.** $x^2 - y^2 = 1$

51–70 ■ Polar Equations to Rectangular Equations Convert the polar equation to rectangular coordinates.

51. $r = 7$ **52.** $r = -3$

53. $\theta = -\dfrac{\pi}{2}$ **54.** $\theta = \pi$

55. $r \cos \theta = 6$ **56.** $r = 2 \csc \theta$

57. $r = 4 \sin \theta$ **58.** $r = 6 \cos \theta$

59. $r = 1 + \cos \theta$ **60.** $r = 3(1 - \sin \theta)$

61. $r = 1 + 2 \sin \theta$ **62.** $r = 2 - \cos \theta$

63. $r = \dfrac{1}{\sin \theta - \cos \theta}$ **64.** $r = \dfrac{1}{1 + \sin \theta}$

65. $r = \dfrac{4}{1 + 2 \sin \theta}$ **66.** $r = \dfrac{2}{1 - \cos \theta}$

67. $r^2 = \tan \theta$ **68.** $r^2 = \sin 2\theta$

69. $\sec \theta = 2$ **70.** $\cos 2\theta = 1$

DISCUSS ■ DISCOVER ■ PROVE ■ WRITE

71. DISCUSS ■ PROVE: The Distance Formula in Polar Coordinates

(a) Use the Law of Cosines to prove that the distance between the polar points (r_1, θ_1) and (r_2, θ_2) is

$$d = \sqrt{r_1^2 + r_2^2 - 2r_1 r_2 \cos(\theta_2 - \theta_1)}$$

(b) Find the distance between the points whose polar coordinates are $(3, 3\pi/4)$ and $(1, 7\pi/6)$, using the formula from part (a).

(c) Now convert the points in part (b) to rectangular coordinates. Find the distance between them, using the usual Distance Formula. Do you get the same answer?

72. DISCUSS: Different Coordinate Systems As was noted in the overview of the chapter, certain curves are more naturally described in one coordinate system than in another. In each of the following situations, which coordinate system would be appropriate: rectangular or polar? Give reasons to support your answer.

(a) You need to give directions to your house to a taxi driver.

(b) You need to give directions to your house to a homing pigeon.

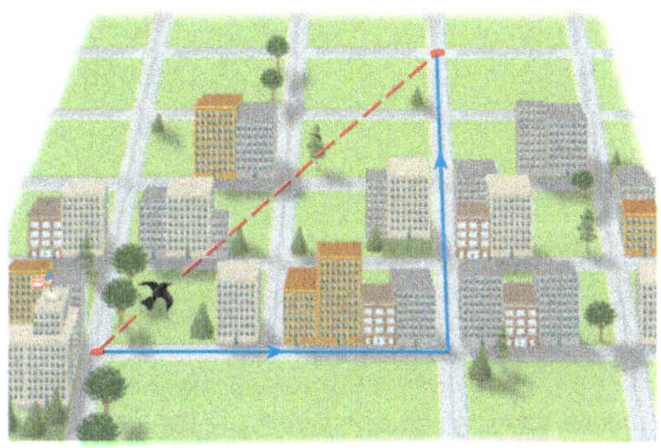

8.2 GRAPHS OF POLAR EQUATIONS

■ Graphing Polar Equations ■ Symmetry ■ Graphing Polar Equations with Graphing Devices

The **graph of a polar equation** $r = f(\theta)$ consists of all points P that have at least one polar representation (r, θ) whose coordinates satisfy the equation. Many curves that arise in mathematics and its applications are more easily and naturally represented by polar equations than by rectangular equations.

■ Graphing Polar Equations

A rectangular grid is helpful for plotting points in rectangular coordinates (see Figure 1(a)). To plot points in polar coordinates, it is convenient to use a grid consisting of circles centered at the pole and rays emanating from the pole, as in Figure 1(b). We will use such grids to help us sketch polar graphs.

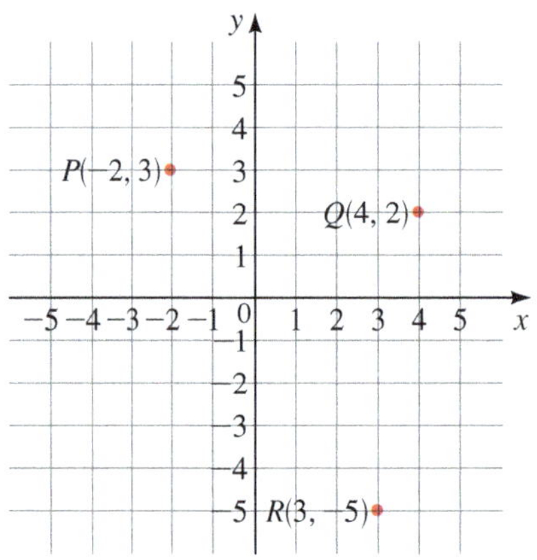

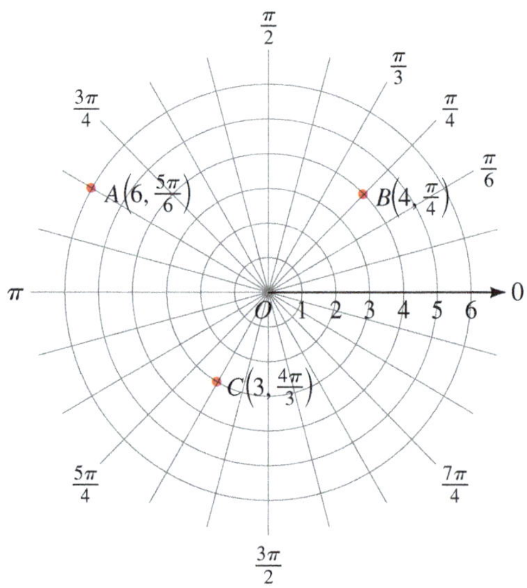

FIGURE 1 (a) Grid for rectangular coordinates (b) Grid for polar coordinates

In Examples 1 and 2 we see that circles centered at the origin and lines that pass through the origin have particularly simple equations in polar coordinates.

EXAMPLE 1 ■ Sketching the Graph of a Polar Equation

Sketch a graph of the equation $r = 3$, and express the equation in rectangular coordinates.

SOLUTION The graph consists of all points whose r-coordinate is 3, that is, all points that are 3 units away from the origin. So the graph is a circle of radius 3 centered at the origin, as shown in Figure 2.

Squaring both sides of the equation, we get

$$r^2 = 3^2 \qquad \text{Square both sides}$$

$$x^2 + y^2 = 9 \qquad \text{Substitute } r^2 = x^2 + y^2$$

So the equivalent equation in rectangular coordinates is $x^2 + y^2 = 9$.

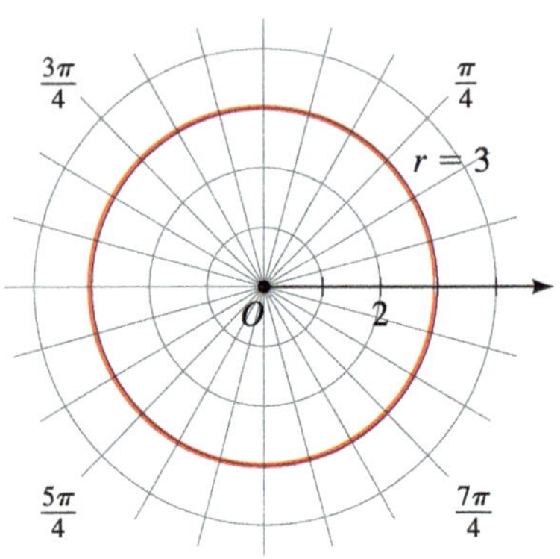

FIGURE 2

Now Try Exercise 17 ■

In general, the graph of the equation $r = a$ is a circle of radius $|a|$ centered at the origin. Squaring both sides of this equation, we see that the equivalent equation in rectangular coordinates is $x^2 + y^2 = a^2$.

EXAMPLE 2 ■ Sketching the Graph of a Polar Equation

Sketch a graph of the equation $\theta = \pi/3$, and express the equation in rectangular coordinates.

SOLUTION The graph consists of all points whose θ-coordinate is $\pi/3$. This is the straight line that passes through the origin and makes an angle of $\pi/3$ with the polar axis (see Figure 3). Note that the points $(r, \pi/3)$ on the line with $r > 0$ lie in Quadrant I, whereas those with $r < 0$ lie in Quadrant III. If the point (x, y) lies on this line, then

$$\frac{y}{x} = \tan\theta = \tan\frac{\pi}{3} = \sqrt{3}$$

Thus the rectangular equation of this line is $y = \sqrt{3}x$.

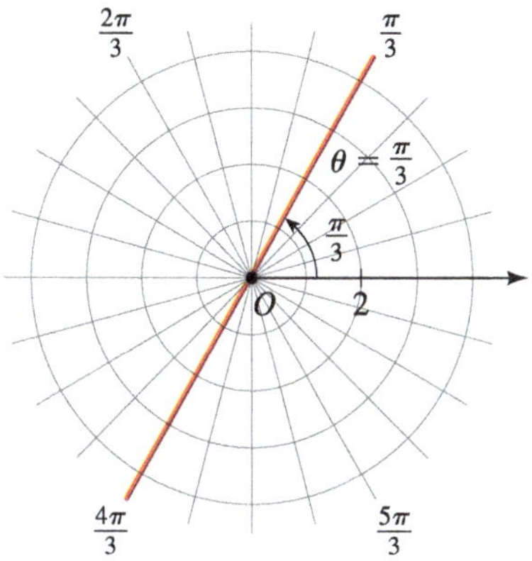

FIGURE 3

Now Try Exercise 19 ■

To sketch a polar curve whose graph isn't as obvious as the ones in the preceding examples, we plot points calculated for sufficiently many values of θ and then join them in a continuous curve. (This is what we did when we first learned to graph equations in rectangular coordinates.)

EXAMPLE 3 ■ Sketching the Graph of a Polar Equation

Sketch a graph of the polar equation $r = 2\sin\theta$.

SOLUTION We first use the equation to determine the polar coordinates of several points on the curve. The results are shown in the following table.

$\boldsymbol{\theta}$	0	$\pi/6$	$\pi/4$	$\pi/3$	$\pi/2$	$2\pi/3$	$3\pi/4$	$5\pi/6$	π
$\boldsymbol{r = 2\sin\theta}$	0	1	$\sqrt{2}$	$\sqrt{3}$	2	$\sqrt{3}$	$\sqrt{2}$	1	0

We plot these points in Figure 4 and then join them to sketch the curve. The graph appears to be a circle. We have used values of θ only between 0 and π, since the same points (this time expressed with negative r-coordinates) would be obtained if we allowed θ to range from π to 2π.

The polar equation $r = 2\sin\theta$ in rectangular coordinates is

$$x^2 + (y-1)^2 = 1$$

(see Section 8.1, Example 6(b)). From the rectangular form of the equation we see that the graph is a circle of radius 1 centered at $(0, 1)$.

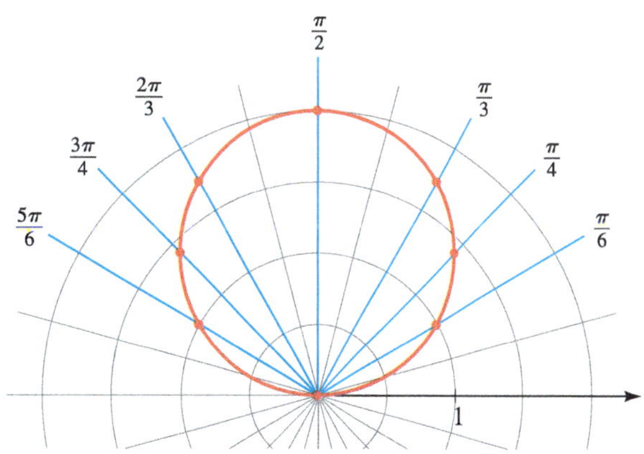

FIGURE 4 $r = 2\sin\theta$

Now Try Exercise 21 ■

In general, the graphs of equations of the form

$$r = 2a \sin \theta \qquad \text{and} \qquad r = 2a \cos \theta$$

are **circles** with radius $|a|$ centered at the points with polar coordinates $(a, \pi/2)$ and $(a, 0)$, respectively.

EXAMPLE 4 ■ Sketching the Graph of a Cardioid

Sketch a graph of $r = 2 + 2 \cos \theta$.

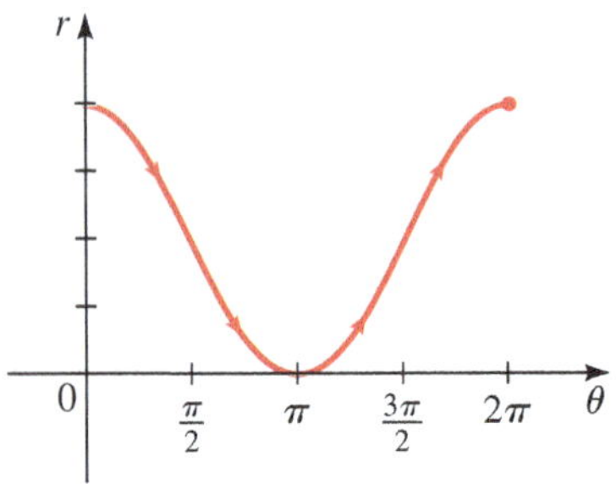

FIGURE 5 $r = 2 + 2 \cos \theta$

SOLUTION Instead of plotting points as in Example 3, we first sketch the graph of $r = 2 + 2 \cos \theta$ in *rectangular* coordinates in Figure 5. We can think of this graph as a table of values that enables us to read at a glance the values of r that correspond to increasing values of θ. For instance, we see that as θ increases from 0 to $\pi/2$, r (the distance from O) decreases from 4 to 2, so we sketch the corresponding part of the polar graph in Figure 6(a). As θ increases from $\pi/2$ to π, Figure 5 shows that r decreases from 2 to 0, so we sketch the next part of the graph as in Figure 6(b). As θ increases from π to $3\pi/2$, r increases from 0 to 2, as shown in part (c). Finally, as θ increases from $3\pi/2$ to 2π, r increases from 2 to 4, as shown in part (d). If we let θ increase beyond 2π or decrease beyond 0, we would simply retrace our path. Combining the portions of the graph from parts (a) through (d) of Figure 6, we sketch the complete graph in part (e).

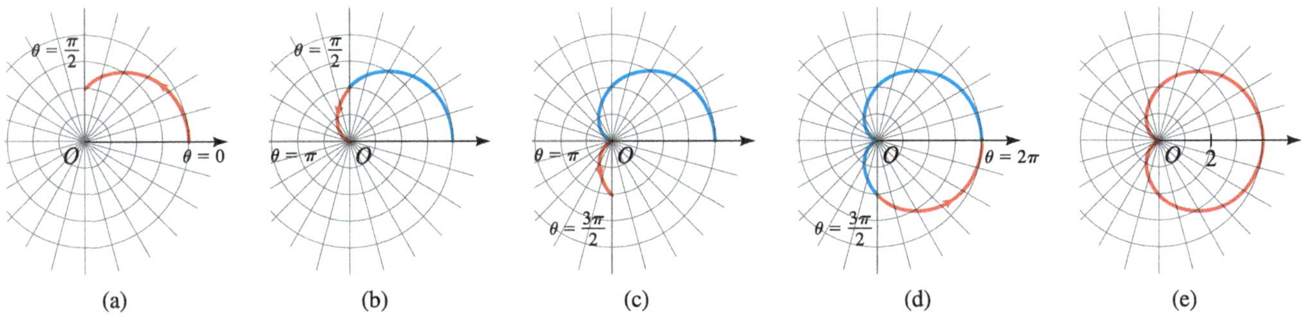

FIGURE 6 Steps in sketching $r = 2 + 2 \cos \theta$

The polar equation $r = 2 + 2 \cos \theta$ in rectangular coordinates is

$$(x^2 + y^2 - 2x)^2 = 4(x^2 + y^2)$$

(see Section 8.1, Example 6(c)). The simpler form of the polar equation shows that it is more natural to describe cardioids using polar coordinates.

Now Try Exercise 25 ■

The curve in Figure 6 is called a **cardioid** because it is heart-shaped. In general, the graph of any equation of the form

$$r = a(1 \pm \cos \theta) \qquad \text{or} \qquad r = a(1 \pm \sin \theta)$$

is a cardioid.

EXAMPLE 5 ■ Sketching the Graph of a Four-Leaved Rose

Sketch the curve $r = \cos 2\theta$.

SOLUTION As in Example 4, we first sketch the graph of $r = \cos 2\theta$ in *rectangular* coordinates, as shown in Figure 7. As θ increases from 0 to $\pi/4$, Figure 7 shows that r decreases from 1 to 0, so we draw the corresponding portion of the polar curve in Figure 8 (indicated by ①). As θ increases from $\pi/4$ to $\pi/2$, the value of r goes from 0 to -1. This means that the distance from the origin increases from 0 to 1, but instead of being in Quadrant I, this portion of the polar curve (indicated by ②) lies on the opposite side of the origin in Quadrant III. The remainder of the curve is drawn in a similar fashion, with the arrows and numbers indicating the order in which the portions are traced out. The resulting curve has four petals and is called a **four-leaved rose**.

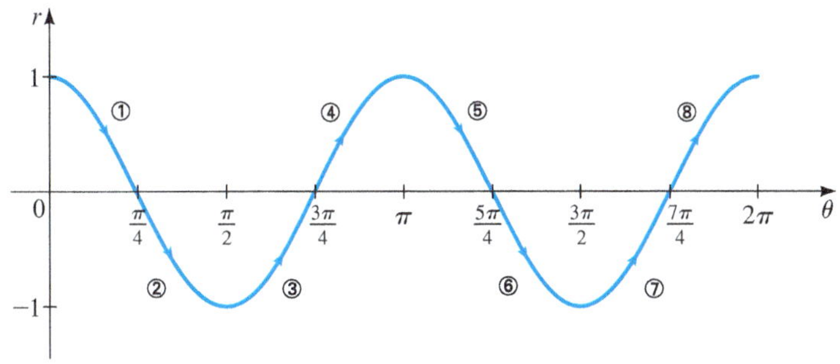

FIGURE 7 Graph of $r = \cos 2\theta$ sketched in rectangular coordinates

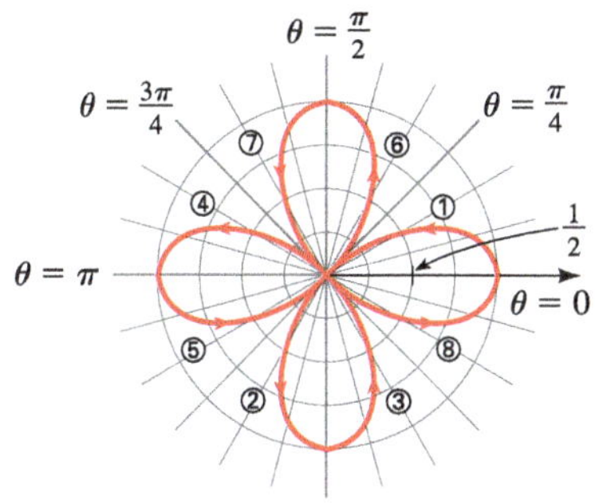

FIGURE 8 Four-leaved rose $r = \cos 2\theta$ sketched in polar coordinates

Now Try Exercise 29 ■

In general, the graph of an equation of the form

$$r = a \cos n\theta \qquad \text{or} \qquad r = a \sin n\theta$$

is an ***n*-leaved rose** if n is odd or a $2n$-leaved rose if n is even (as in Example 5).

Symmetry

In graphing a polar equation, it's often helpful to take advantage of symmetry. We list three tests for symmetry; Figure 9 shows why these tests work.

TESTS FOR SYMMETRY

1. If a polar equation is unchanged when we replace θ by $-\theta$, then the graph is symmetric about the polar axis (Figure 9(a)).
2. If the equation is unchanged when we replace r by $-r$, or θ by $\theta + \pi$, then the graph is symmetric about the pole (Figure 9(b)).
3. If the equation is unchanged when we replace θ by $\pi - \theta$, then the graph is symmetric about the vertical line $\theta = \pi/2$ (the y-axis) (Figure 9(c)).

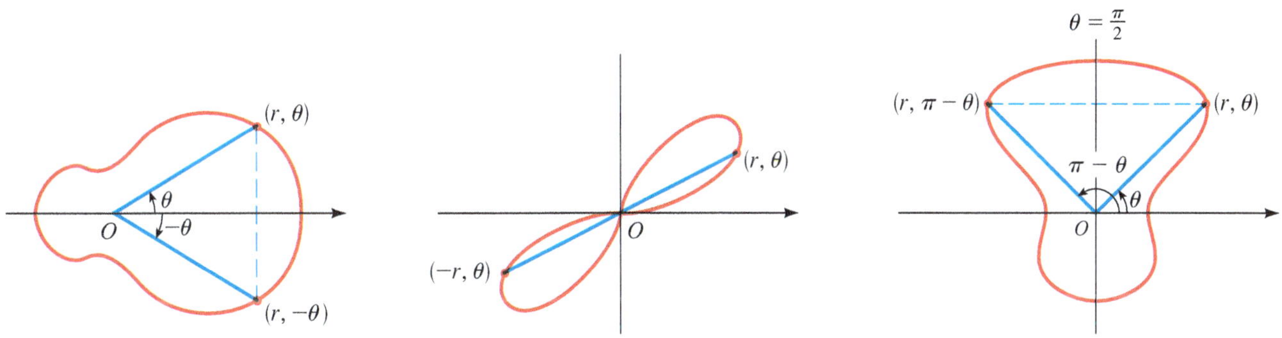

(a) Symmetry about the polar axis (b) Symmetry about the pole (c) Symmetry about the line $\theta = \frac{\pi}{2}$

FIGURE 9

The graphs in Figures 2, 6(e), and 8 are symmetric about the polar axis. The graph in Figure 8 is also symmetric about the pole. Figures 4 and 8 show graphs that are symmetric about $\theta = \pi/2$. Note that the four-leaved rose in Figure 8 meets all three tests for symmetry.

In rectangular coordinates the zeros of the function $y = f(x)$ correspond to the x-intercepts of the graph. In polar coordinates the zeros of the function $r = f(\theta)$ are the angles θ at which the curve crosses the pole. The zeros help us sketch the graph, as is illustrated in the next example.

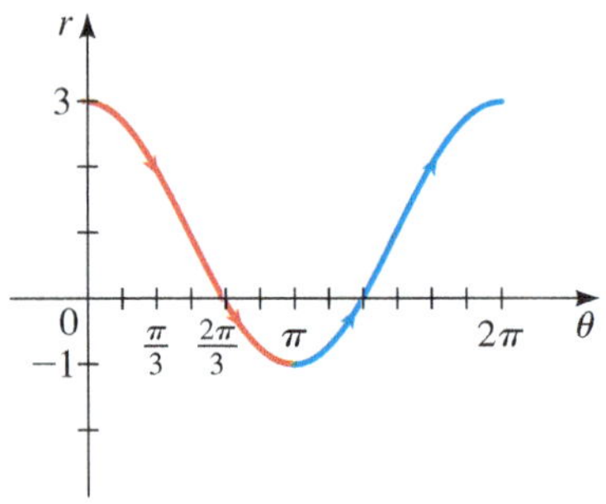

FIGURE 10

EXAMPLE 6 ■ Using Symmetry to Sketch a Limaçon

Sketch a graph of the equation $r = 1 + 2\cos\theta$.

SOLUTION We use the following as aids in sketching the graph.

Symmetry. Since the equation is unchanged when θ is replaced by $-\theta$, the graph is symmetric about the polar axis.

Zeros. To find the zeros, we solve

$$0 = 1 + 2\cos\theta$$

$$\cos\theta = -\frac{1}{2}$$

$$\theta = \frac{2\pi}{3}, \frac{4\pi}{3}$$

Table of values. As in Example 4, we sketch the graph of $r = 1 + 2\cos\theta$ in *rectangular* coordinates to serve as a table of values (Figure 10).

Now we sketch the polar graph of $r = 1 + 2\cos\theta$ from $\theta = 0$ to $\theta = \pi$ and then use symmetry to complete the graph in Figure 11.

Now Try Exercise 37 ■

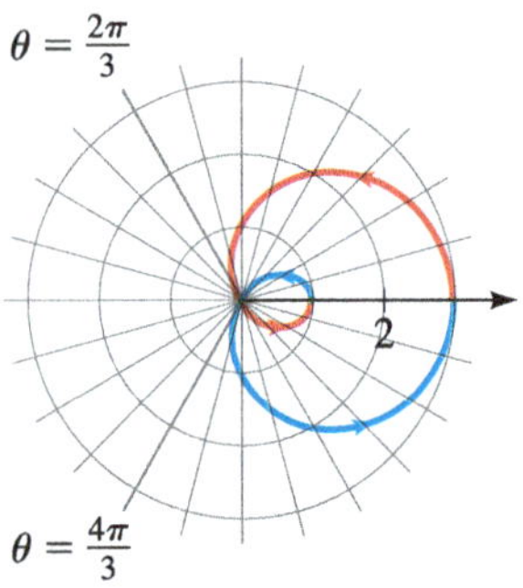

FIGURE 11 $r = 1 + 2\cos\theta$

The curve in Figure 11 is called a **limaçon**, after the Middle French word for snail. In general, the graph of an equation of the form

$$r = a \pm b\cos\theta \qquad \text{or} \qquad r = a \pm b\sin\theta$$

is a limaçon. The shape of the limaçon depends on the relative size of a and b (see the box on the next page).

Graphing Polar Equations with Graphing Devices

Although it's useful to be able to sketch simple polar graphs by hand, we need a graphing calculator or computer when the graph is as complicated as the one in Figure 12. Fortunately, most graphing calculators are capable of graphing polar equations directly.

FIGURE 12 $r = \sin\theta + \sin^3(5\theta/2)$

EXAMPLE 7 ■ Drawing the Graph of a Polar Equation

Graph the equation $r = \cos(2\theta/3)$.

SOLUTION We need to determine the domain for θ. So we ask ourselves: How many times must θ go through a complete rotation (2π radians) before the graph starts to repeat itself? The graph repeats itself when the same value of r is obtained at θ and $\theta + 2n\pi$. Thus we need to find an integer n so that

$$\cos\frac{2(\theta + 2n\pi)}{3} = \cos\frac{2\theta}{3}$$

For this equality to hold, $4n\pi/3$ must be a multiple of 2π, and this first happens when $n = 3$. Therefore we obtain the entire graph if we choose values of θ between $\theta = 0$ and $\theta = 0 + 2(3)\pi = 6\pi$. The graph is shown in Figure 13.

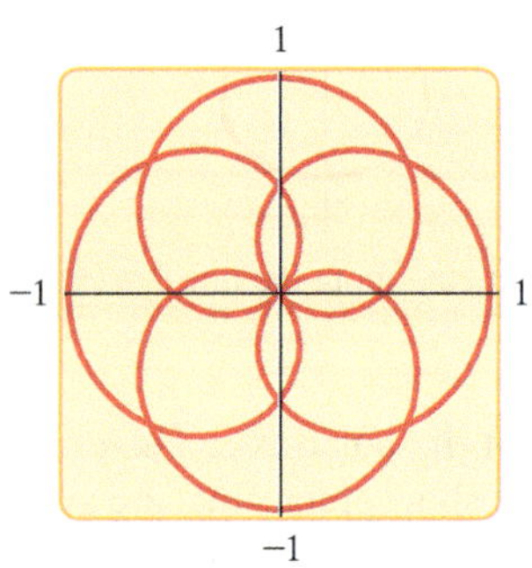

FIGURE 13 $r = \cos(2\theta/3)$

Now Try Exercise 47 ■

EXAMPLE 8 ■ A Family of Polar Equations

Graph the family of polar equations $r = 1 + c \sin \theta$ for $c = 3, 2.5, 2, 1.5, 1$. How does the shape of the graph change as c changes?

SOLUTION Figure 14 shows computer-drawn graphs for the given values of c. When $c > 1$, the graph has an inner loop; the loop decreases in size as c decreases. When $c = 1$, the loop disappears, and the graph becomes a cardioid (see Example 4).

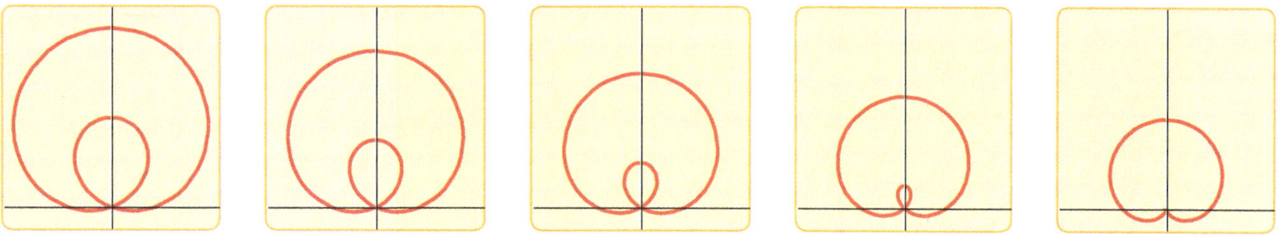

$c = 3.0$ $c = 2.5$ $c = 2.0$ $c = 1.5$ $c = 1.0$

FIGURE 14 A family of limaçons $r = 1 + c \sin \theta$ in the viewing rectangle $[-2.5, 2.5]$ by $[-0.5, 4.5]$

Now Try Exercise 51 ■

The box below gives a summary of some of the basic polar graphs used in calculus.

SOME COMMON POLAR CURVES

Circles and Spiral

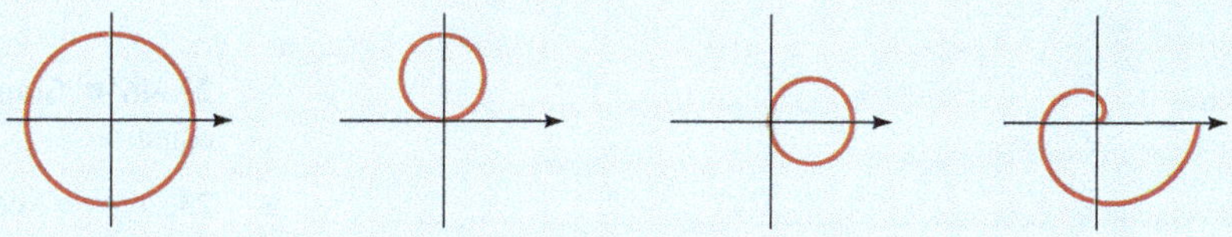

$r = a$ circle; $r = a \sin \theta$ circle; $r = a \cos \theta$ circle; $r = a\theta$ spiral

Limaçons

$r = a \pm b \sin \theta$

$r = a \pm b \cos \theta$

$(a > 0, b > 0)$

Orientation depends on the trigonometric function (sine or cosine) and the sign of b.

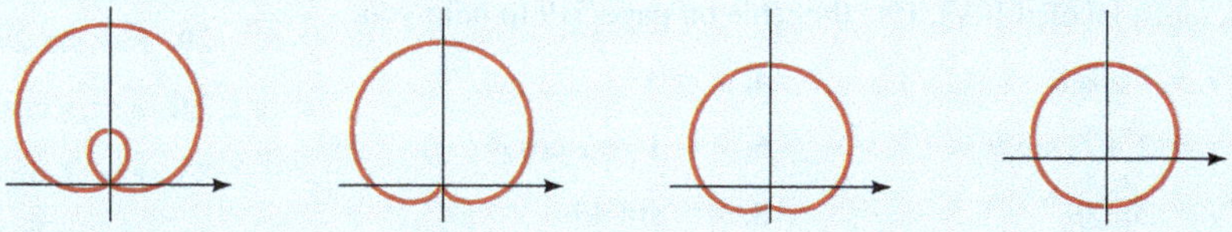

$a < b$ limaçon with inner loop; $a = b$ cardioid; $a > b$ dimpled limaçon; $a \geq 2b$ convex limaçon

Roses

$r = a \sin n\theta$

$r = a \cos n\theta$

n-leaved if n is odd

$2n$-leaved if n is even

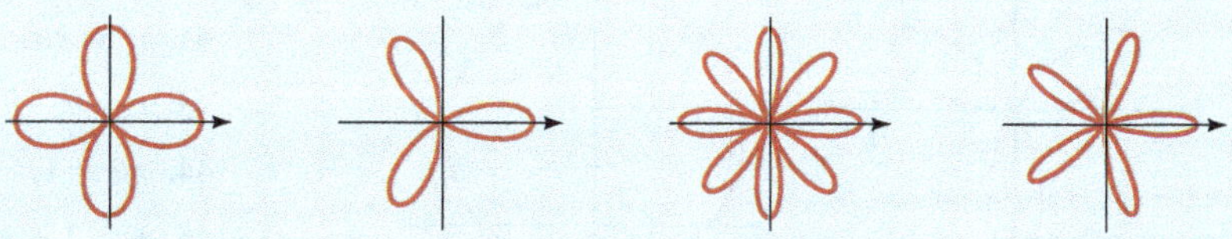

$r = a \cos 2\theta$ 4-leaved rose; $r = a \cos 3\theta$ 3-leaved rose; $r = a \cos 4\theta$ 8-leaved rose; $r = a \cos 5\theta$ 5-leaved rose

Lemniscates

Figure-eight-shaped curves

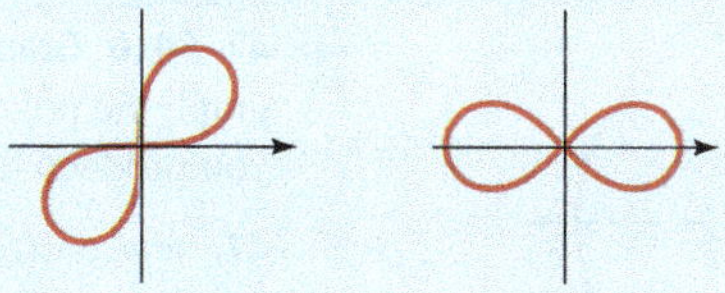

$r^2 = a^2 \sin 2\theta$ lemniscate; $r^2 = a^2 \cos 2\theta$ lemniscate

8.2 EXERCISES

CONCEPTS

1. To plot points in polar coordinates, we use a grid consisting of ________ centered at the pole and ________ emanating from the pole.

2. **(a)** To graph a polar equation $r = f(\theta)$, we plot all the points (r, θ) that ________ the equation.

(b) The simplest polar equations are obtained by setting r or θ equal to a constant. The graph of the polar equation $r = 3$ is a ________ with radius ________ centered at the ________. The graph of the polar equation $\theta = \pi/4$ is a ________ passing through the ________ with slope ________. Graph these polar equations below.

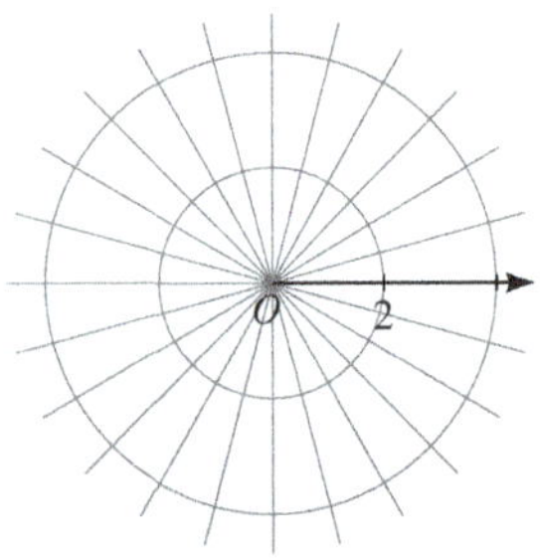

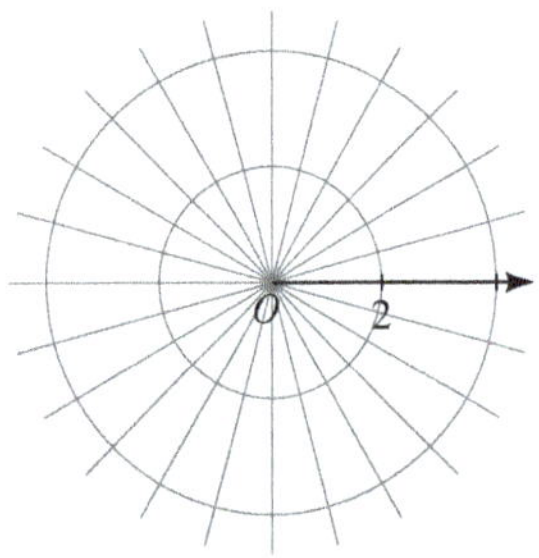

SKILLS

3–8 ■ Graphs of Polar Equations Match the polar equation with the graphs labeled I–VI. Use the table on page 519 to help you.

3. $r = 3\cos\theta$

4. $r = 3$

5. $r = 2 + 2\sin\theta$

6. $r = 1 + 2\cos\theta$

7. $r = \sin 3\theta$

8. $r = \sin 4\theta$

I

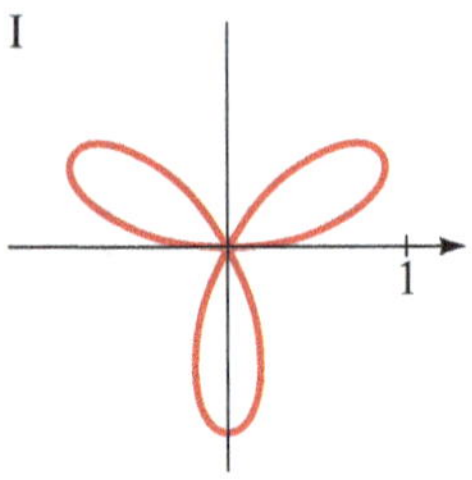

II

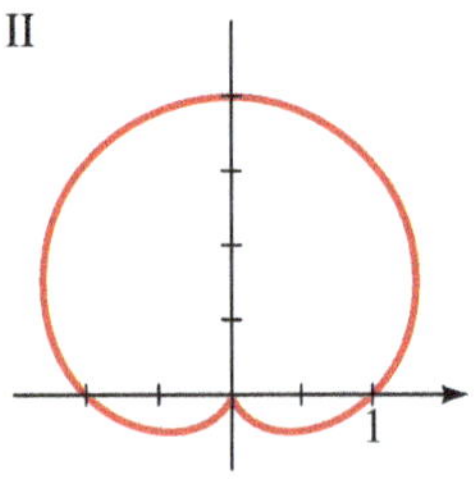

III

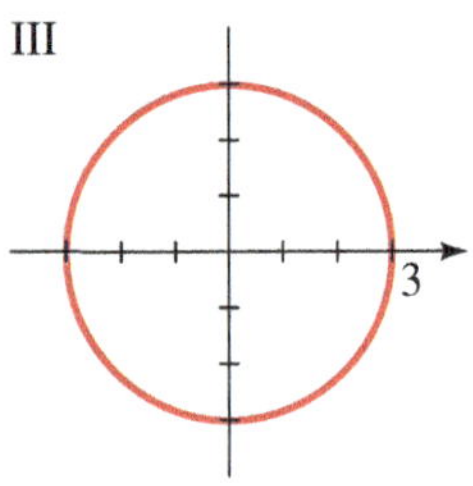

IV

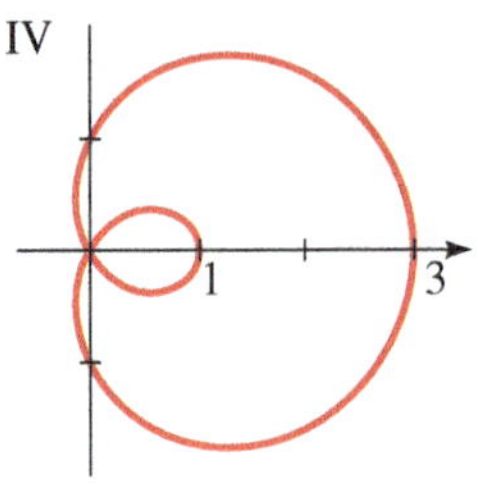

V

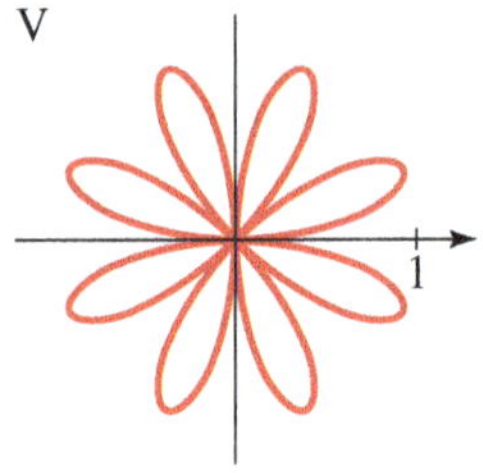

VI

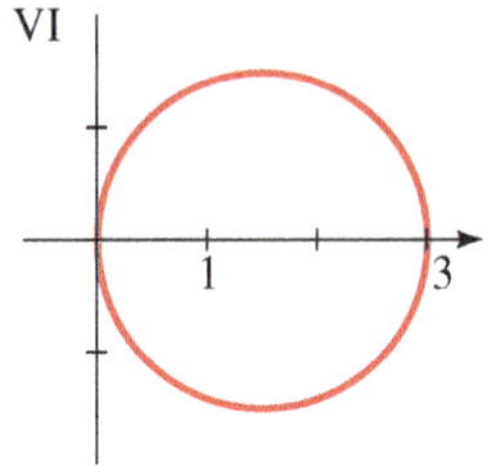

9–16 ■ Testing for Symmetry Test the polar equation for symmetry with respect to the polar axis, the pole, and the line $\theta = \pi/2$.

9. $r = 2 - \sin\theta$

10. $r = 4 + 8\cos\theta$

11. $r = 3\sec\theta$

12. $r = 5\cos\theta\csc\theta$

13. $r = \dfrac{4}{3 - 2\sin\theta}$

14. $r = \dfrac{5}{1 + 3\cos\theta}$

15. $r^2 = 4\cos 2\theta$

16. $r^2 = 9\sin\theta$

17–22 ■ Polar to Rectangular Sketch a graph of the polar equation, and express the equation in rectangular coordinates.

17. $r = 2$

18. $r = -1$

19. $\theta = -\pi/2$

20. $\theta = 5\pi/6$

21. $r = 6\sin\theta$

22. $r = \cos\theta$

23–46 ■ Graphing Polar Equations Sketch a graph of the polar equation.

23. $r = -2\cos\theta$

24. $r = 3\sin\theta$

25. $r = 2 - 2\cos\theta$

26. $r = 1 + \sin\theta$

27. $r = -3(1 + \sin\theta)$

28. $r = \cos\theta - 1$

29. $r = \sin 2\theta$

30. $r = 2\cos 3\theta$

31. $r = -\cos 5\theta$

32. $r = \sin 4\theta$

33. $r = 2\sin 5\theta$

34. $r = -3\cos 4\theta$

35. $r = \sqrt{3} - 2\sin\theta$

36. $r = 2 + \sin\theta$

37. $r = \sqrt{3} + \cos\theta$

38. $r = 1 - 2\cos\theta$

39. $r = 2 - 2\sqrt{2}\cos\theta$

40. $r = 3 + 6\sin\theta$

41. $r^2 = \cos 2\theta$

42. $r^2 = 4\sin 2\theta$

43. $r = \theta,\quad \theta \geq 0$ (spiral)

44. $r\theta = 1,\quad \theta > 0$ (reciprocal spiral)

45. $r = 2 + \sec\theta$ (conchoid)

46. $r = \sin\theta\tan\theta$ (cissoid)

47–50 ■ Graphing Polar Equations Use a graphing device to graph the polar equation. Choose the domain of θ to make sure you produce the entire graph.

47. $r = \cos(\theta/2)$

48. $r = \sin(8\theta/5)$

49. $r = 1 + 2\sin(\theta/2)$ (nephroid)

50. $r = \sqrt{1 - 0.8\sin^2\theta}$ (hippopede)

51–52 ■ Families of Polar Equations These exercises involve families of polar equations.

51. Graph the family of polar equations $r = 1 + \sin n\theta$ for $n = 1, 2, 3, 4$, and 5. How is the number of loops related to n?

52. Graph the family of polar equations $r = 1 + c \sin 2\theta$ for $c = 0.3, 0.6, 1, 1.5$, and 2. How does the graph change as c increases?

53–56 ■ Special Polar Equations Match the polar equation with the graphs labeled I–IV. Give reasons for your answers.

53. $r = \sin(\theta/2)$

54. $r = 1/\sqrt{\theta}$

55. $r = \theta \sin \theta$

56. $r = 1 + 3\cos(3\theta)$

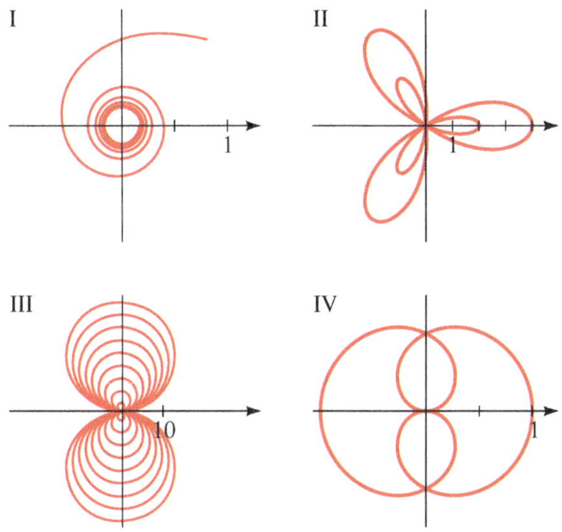

SKILLS Plus

57–60 ■ Rectangular to Polar Sketch a graph of the rectangular equation. [*Hint:* First convert the equation to polar coordinates.]

57. $(x^2 + y^2)^3 = 4x^2y^2$

58. $(x^2 + y^2)^3 = (x^2 - y^2)^2$

59. $(x^2 + y^2)^2 = x^2 - y^2$

60. $x^2 + y^2 = (x^2 + y^2 - x)^2$

61. A Circle in Polar Coordinates Consider the polar equation $r = a\cos\theta + b\sin\theta$.

(a) Express the equation in rectangular coordinates, and use this to show that the graph of the equation is a circle. What are the center and radius?

(b) Use your answer to part (a) to graph the equation $r = 2\sin\theta + 2\cos\theta$.

62. A Parabola in Polar Coordinates

(a) Graph the polar equation $r = \tan\theta \sec\theta$ in the viewing rectangle $[-3, 3]$ by $[-1, 9]$.

(b) Note that your graph in part (a) looks like a parabola (see Section 3.1). Confirm this by converting the equation to rectangular coordinates.

APPLICATIONS

63. Orbit of a Satellite Scientists and engineers often use polar equations to model the motion of satellites in earth orbit. Let's consider a satellite whose orbit is modeled by the equation $r = 22500/(4 - \cos\theta)$, where r is the distance in miles between the satellite and the center of the earth and θ is the angle shown in the following figure.

(a) On the same viewing screen, graph the circle $r = 3960$ (to represent the earth, which we will assume to be a sphere of radius 3960 mi) and the polar equation of the satellite's orbit. Describe the motion of the satellite as θ increases from 0 to 2π.

(b) For what angle θ is the satellite closest to the earth? Find the height of the satellite above the earth's surface for this value of θ.

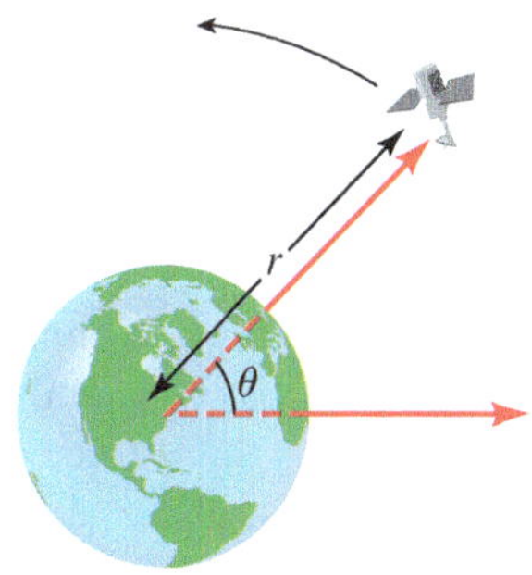

64. An Unstable Orbit The orbit described in Exercise 63 is stable because the satellite traverses the same path over and over as θ increases. Suppose that a meteor strikes the satellite and changes its orbit to

$$r = \frac{22500\left(1 - \dfrac{\theta}{40}\right)}{4 - \cos\theta}$$

(a) On the same viewing screen, graph the circle $r = 3960$ and the new orbit equation, with θ increasing from 0 to 3π. Describe the new motion of the satellite.

(b) Use the TRACE feature on your graphing calculator to find the value of θ at the moment the satellite crashes into the earth.

DISCUSS ■ DISCOVER ■ PROVE ■ WRITE

65. DISCUSS ■ DISCOVER: A Transformation of Polar Graphs How are the graphs of

$$r = 1 + \sin\left(\theta - \frac{\pi}{6}\right)$$

and

$$r = 1 + \sin\left(\theta - \frac{\pi}{3}\right)$$

related to the graph of $r = 1 + \sin\theta$? In general, how is the graph of $r = f(\theta - \alpha)$ related to the graph of $r = f(\theta)$?

66. DISCUSS: Choosing a Convenient Coordinate System Compare the polar equation of the circle $r = 2$ with its equation in rectangular coordinates. In which coordinate system is the equation simpler? Do the same for the equation of the four-leaved rose $r = \sin 2\theta$. Which coordinate system would you choose to study these curves?

67. DISCUSS: Choosing a Convenient Coordinate System Compare the rectangular equation of the line $y = 2$ with its polar equation. In which coordinate system is the equation simpler? Which coordinate system would you choose to study lines?

8.3 POLAR FORM OF COMPLEX NUMBERS; DE MOIVRE'S THEOREM

■ Graphing Complex Numbers ■ Polar Form of Complex Numbers ■ De Moivre's Theorem ■ *n*th Roots of Complex Numbers

In this section we represent complex numbers in polar (or trigonometric) form. This enables us to find the *n*th roots of complex numbers. To describe the polar form of complex numbers, we must first learn to work with complex numbers graphically.

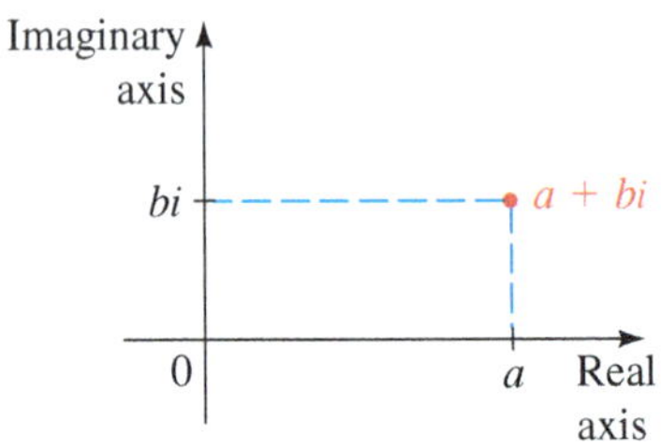

FIGURE 1

Graphing Complex Numbers

To graph real numbers or sets of real numbers, we have been using the number line, which has just one dimension. Complex numbers, however, have two components: a real part and an imaginary part. This suggests that we need two axes to graph complex numbers: one for the real part and one for the imaginary part. We call these the **real axis** and the **imaginary axis**, respectively. The plane determined by these two axes is called the **complex plane**. To graph the complex number $a + bi$, we plot the ordered pair of numbers (a, b) in this plane, as indicated in Figure 1.

EXAMPLE 1 ■ Graphing Complex Numbers

Graph the complex numbers $z_1 = 2 + 3i$, $z_2 = 3 - 2i$, and $z_1 + z_2$.

SOLUTION We have $z_1 + z_2 = (2 + 3i) + (3 - 2i) = 5 + i$. The graph is shown in Figure 2.

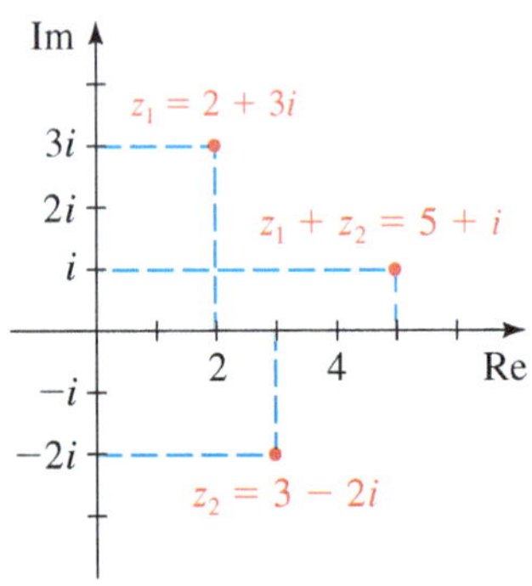

FIGURE 2

Now Try Exercise 19 ■

EXAMPLE 2 ■ Graphing Sets of Complex Numbers

Graph each set of complex numbers.

(a) $S = \{a + bi \mid a \geq 0\}$

(b) $T = \{a + bi \mid a < 1, b \geq 0\}$

SOLUTION

(a) S is the set of complex numbers whose real part is nonnegative. The graph is shown in Figure 3(a).

(b) T is the set of complex numbers for which the real part is less than 1 and the imaginary part is nonnegative. The graph is shown in Figure 3(b).

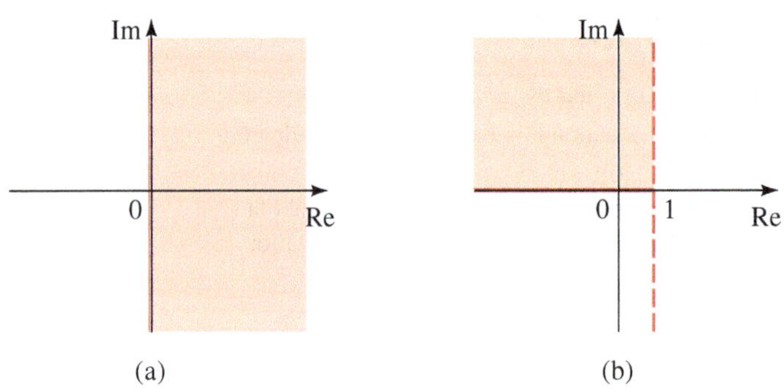

FIGURE 3

Now Try Exercise 21 ■

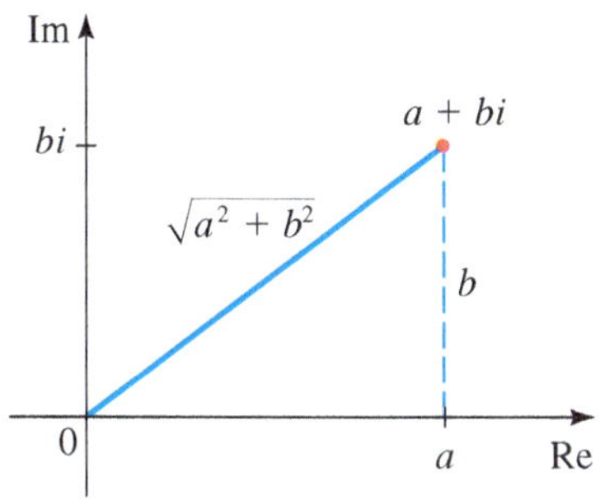

FIGURE 4

Recall that the absolute value of a real number can be thought of as its distance from the origin on the real number line (see Section 1.1). We define absolute value for complex numbers in a similar fashion. Using the Pythagorean Theorem, we can see from Figure 4 that the distance between $a + bi$ and the origin in the complex plane is $\sqrt{a^2 + b^2}$. This leads to the following definition.

MODULUS OF A COMPLEX NUMBER

The **modulus** (or **absolute value**) of the complex number $z = a + bi$ is

$$|z| = \sqrt{a^2 + b^2}$$

EXAMPLE 3 ■ Calculating the Modulus

The plural of *modulus* is *moduli*.

Find the moduli of the complex numbers $3 + 4i$ and $8 - 5i$.

SOLUTION

$$|3 + 4i| = \sqrt{3^2 + 4^2} = \sqrt{25} = 5$$

$$|8 - 5i| = \sqrt{8^2 + (-5)^2} = \sqrt{89}$$

Now Try Exercise 9 ■

EXAMPLE 4 ■ Absolute Value of Complex Numbers

Graph each set of complex numbers.

(a) $C = \{z \mid |z| = 1\}$ **(b)** $D = \{z \mid |z| \le 1\}$

SOLUTION

(a) C is the set of complex numbers whose distance from the origin is 1. Thus C is a circle of radius 1 with center at the origin, as shown in Figure 5.

(b) D is the set of complex numbers whose distance from the origin is less than or equal to 1. Thus D is the disk that consists of all complex numbers on and inside the circle C of part (a), as shown in Figure 6.

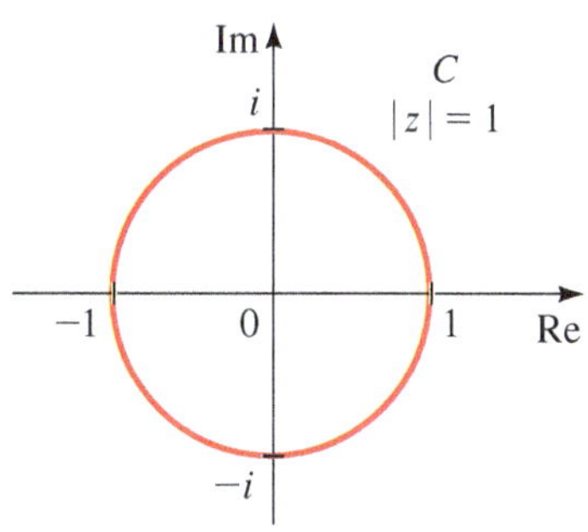

FIGURE 5

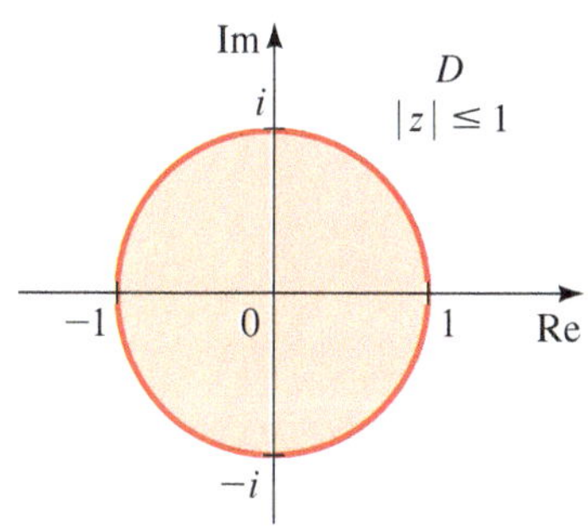

FIGURE 6

Now Try Exercises 23 and 25 ■

■ Polar Form of Complex Numbers

Let $z = a + bi$ be a complex number, and in the complex plane let's draw the line segment joining the origin to the point $a + bi$ (see Figure 7 on the next page). The length of this line segment is $r = |z| = \sqrt{a^2 + b^2}$. If θ is an angle in standard position whose terminal side coincides with this line segment, then by the definitions of sine and cosine (see Section 6.3)

$$a = r\cos\theta \quad \text{and} \quad b = r\sin\theta$$

so $z = r\cos\theta + ir\sin\theta = r(\cos\theta + i\sin\theta)$. We have shown the following.

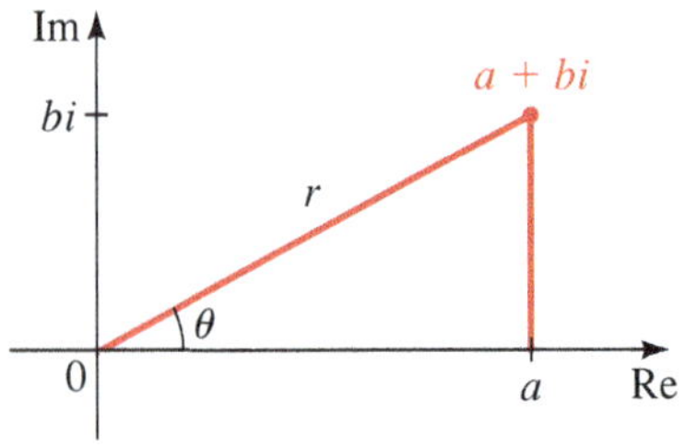

FIGURE 7

POLAR FORM OF COMPLEX NUMBERS

A complex number $z = a + bi$ has the **polar form** (or **trigonometric form**)

$$z = r(\cos\theta + i\sin\theta)$$

where $r = |z| = \sqrt{a^2 + b^2}$ and $\tan\theta = b/a$. The number r is the **modulus** of z, and θ is an **argument** of z.

The argument of z is not unique, but any two arguments of z differ by a multiple of 2π. When determining the argument, we must consider the quadrant in which z lies, as we see in the next example.

EXAMPLE 5 ■ Writing Complex Numbers in Polar Form

Write each complex number in polar form.

(a) $1 + i$ **(b)** $-1 + \sqrt{3}i$ **(c)** $-4\sqrt{3} - 4i$ **(d)** $3 + 4i$

SOLUTION These complex numbers are graphed in Figure 8, which helps us find their arguments.

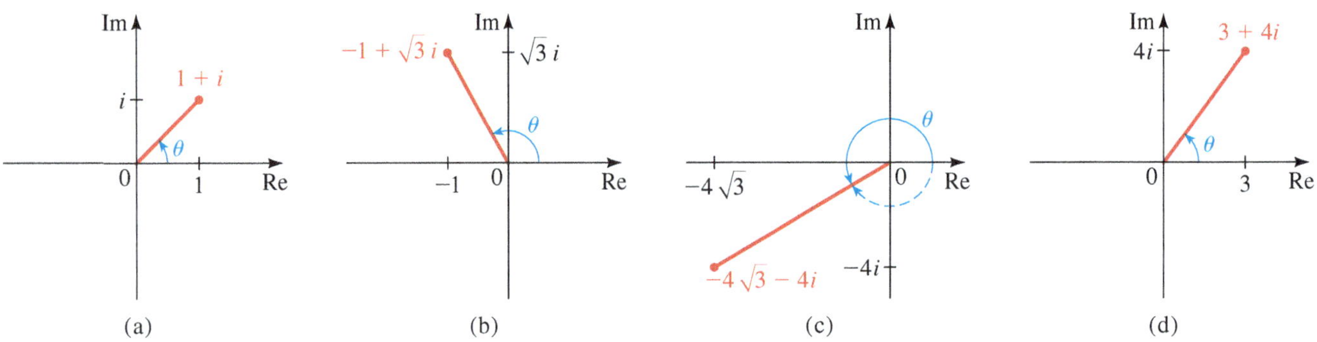

FIGURE 8

$\tan\theta = \frac{1}{1} = 1$

$\theta = \frac{\pi}{4}$

(a) An argument is $\theta = \pi/4$ and $r = \sqrt{1 + 1} = \sqrt{2}$. Thus

$$1 + i = \sqrt{2}\left(\cos\frac{\pi}{4} + i\sin\frac{\pi}{4}\right)$$

$\tan\theta = \dfrac{\sqrt{3}}{-1} = -\sqrt{3}$

$\theta = \frac{2\pi}{3}$

(b) An argument is $\theta = 2\pi/3$ and $r = \sqrt{1 + 3} = 2$. Thus

$$-1 + \sqrt{3}i = 2\left(\cos\frac{2\pi}{3} + i\sin\frac{2\pi}{3}\right)$$

$\tan\theta = \dfrac{-4}{-4\sqrt{3}} = \dfrac{1}{\sqrt{3}}$

$\theta = \frac{7\pi}{6}$

(c) An argument is $\theta = 7\pi/6$ (or we could use $\theta = -5\pi/6$), and $r = \sqrt{48 + 16} = 8$. Thus

$$-4\sqrt{3} - 4i = 8\left(\cos\frac{7\pi}{6} + i\sin\frac{7\pi}{6}\right)$$

$\tan\theta = \frac{4}{3}$

$\theta = \tan^{-1}\frac{4}{3}$

(d) An argument is $\theta = \tan^{-1}\frac{4}{3}$ and $r = \sqrt{3^2 + 4^2} = 5$. So

$$3 + 4i = 5\left[\cos\left(\tan^{-1}\tfrac{4}{3}\right) + i\sin\left(\tan^{-1}\tfrac{4}{3}\right)\right]$$

Now Try Exercises 29, 31, 33, and 43 ■

The Addition Formulas for Sine and Cosine that we discussed in Section 7.2 greatly simplify the multiplication and division of complex numbers in polar form. The following theorem shows how.

MULTIPLICATION AND DIVISION OF COMPLEX NUMBERS

If the two complex numbers z_1 and z_2 have the polar forms

$$z_1 = r_1(\cos\theta_1 + i\sin\theta_1) \qquad \text{and} \qquad z_2 = r_2(\cos\theta_2 + i\sin\theta_2)$$

then

$$z_1 z_2 = r_1 r_2[\cos(\theta_1 + \theta_2) + i\sin(\theta_1 + \theta_2)] \qquad \text{Multiplication}$$

$$\frac{z_1}{z_2} = \frac{r_1}{r_2}[\cos(\theta_1 - \theta_2) + i\sin(\theta_1 - \theta_2)] \qquad z_2 \neq 0 \qquad \text{Division}$$

This theorem says:

To multiply two complex numbers, multiply the moduli and add the arguments.

To divide two complex numbers, divide the moduli and subtract the arguments.

Proof To prove the Multiplication Formula, we simply multiply the two complex numbers:

$$\begin{aligned} z_1 z_2 &= r_1 r_2(\cos\theta_1 + i\sin\theta_1)(\cos\theta_2 + i\sin\theta_2) \\ &= r_1 r_2[\cos\theta_1\cos\theta_2 - \sin\theta_1\sin\theta_2 + i(\sin\theta_1\cos\theta_2 + \cos\theta_1\sin\theta_2)] \\ &= r_1 r_2[\cos(\theta_1 + \theta_2) + i\sin(\theta_1 + \theta_2)] \end{aligned}$$

In the last step we used the Addition Formulas for Sine and Cosine.

The proof of the Division Formula is left as an exercise. (See Exercise 101.) ■

EXAMPLE 6 ■ Multiplying and Dividing Complex Numbers

Let

$$z_1 = 2\left(\cos\frac{\pi}{4} + i\sin\frac{\pi}{4}\right) \qquad \text{and} \qquad z_2 = 5\left(\cos\frac{\pi}{3} + i\sin\frac{\pi}{3}\right)$$

Find **(a)** $z_1 z_2$ and **(b)** z_1/z_2.

SOLUTION

(a) By the Multiplication Formula

$$\begin{aligned} z_1 z_2 &= (2)(5)\left[\cos\left(\frac{\pi}{4} + \frac{\pi}{3}\right) + i\sin\left(\frac{\pi}{4} + \frac{\pi}{3}\right)\right] \\ &= 10\left(\cos\frac{7\pi}{12} + i\sin\frac{7\pi}{12}\right) \end{aligned}$$

To approximate the answer, we use a calculator in radian mode and get

$$\begin{aligned} z_1 z_2 &\approx 10(-0.2588 + 0.9659i) \\ &= -2.588 + 9.659i \end{aligned}$$

(b) By the Division Formula

$$\begin{aligned}\frac{z_1}{z_2} &= \frac{2}{5}\left[\cos\left(\frac{\pi}{4} - \frac{\pi}{3}\right) + i\sin\left(\frac{\pi}{4} - \frac{\pi}{3}\right)\right] \\ &= \frac{2}{5}\left[\cos\left(-\frac{\pi}{12}\right) + i\sin\left(-\frac{\pi}{12}\right)\right] \\ &= \frac{2}{5}\left(\cos\frac{11\pi}{12} + i\sin\frac{11\pi}{12}\right)\end{aligned}$$

Using a calculator in radian mode, we get the approximate answer:

$$\frac{z_1}{z_2} \approx \frac{2}{5}(0.9659 - 0.2588i) = 0.3864 - 0.1035i$$

Now Try Exercise 49 ■

De Moivre's Theorem

Repeated use of the Multiplication Formula gives the following useful formula for raising a complex number to a power n for any positive integer n.

DE MOIVRE'S THEOREM

If $z = r(\cos\theta + i\sin\theta)$, then for any integer n

$$z^n = r^n(\cos n\theta + i\sin n\theta)$$

This theorem says: *To take the nth power of a complex number, we take the nth power of the modulus and multiply the argument by n.*

Proof By the Multiplication Formula

$$\begin{aligned}z^2 = zz &= r^2[\cos(\theta + \theta) + i\sin(\theta + \theta)] \\ &= r^2(\cos 2\theta + i\sin 2\theta)\end{aligned}$$

Now we multiply z^2 by z to get

$$\begin{aligned}z^3 = z^2z &= r^3[\cos(2\theta + \theta) + i\sin(2\theta + \theta)] \\ &= r^3(\cos 3\theta + i\sin 3\theta)\end{aligned}$$

Repeating this argument, we see that for any positive integer n

$$z^n = r^n(\cos n\theta + i\sin n\theta)$$

A similar argument using the Division Formula shows that this also holds for negative integers. ■

EXAMPLE 7 ■ Finding a Power Using De Moivre's Theorem

Find $\left(\frac{1}{2} + \frac{1}{2}i\right)^{10}$.

SOLUTION Since $\frac{1}{2} + \frac{1}{2}i = \frac{1}{2}(1 + i)$, it follows from Example 5(a) that

$$\frac{1}{2} + \frac{1}{2}i = \frac{\sqrt{2}}{2}\left(\cos\frac{\pi}{4} + i\sin\frac{\pi}{4}\right)$$

So by De Moivre's Theorem

$$\left(\frac{1}{2}+\frac{1}{2}i\right)^{10} = \left(\frac{\sqrt{2}}{2}\right)^{10}\left(\cos\frac{10\pi}{4} + i\sin\frac{10\pi}{4}\right)$$

$$= \frac{2^5}{2^{10}}\left(\cos\frac{5\pi}{2} + i\sin\frac{5\pi}{2}\right) = \frac{1}{32}i$$

Now Try Exercise 65

nth Roots of Complex Numbers

An ***n*th root** of a complex number z is any complex number w such that $w^n = z$. De Moivre's Theorem gives us a method for calculating the nth roots of any complex number.

nth ROOTS OF COMPLEX NUMBERS

If $z = r(\cos\theta + i\sin\theta)$ and n is a positive integer, then z has the n distinct nth roots

$$w_k = r^{1/n}\left[\cos\left(\frac{\theta + 2k\pi}{n}\right) + i\sin\left(\frac{\theta + 2k\pi}{n}\right)\right]$$

for $k = 0, 1, 2, \ldots, n - 1$.

Proof To find the nth roots of z, we need to find a complex number w such that

$$w^n = z$$

Let's write z in polar form:

$$z = r(\cos\theta + i\sin\theta)$$

One nth root of z is

$$w = r^{1/n}\left(\cos\frac{\theta}{n} + i\sin\frac{\theta}{n}\right)$$

since by De Moivre's Theorem, $w^n = z$. But the argument θ of z can be replaced by $\theta + 2k\pi$ for any integer k. Since this expression gives a different value of w for $k = 0, 1, 2, \ldots, n - 1$, we have proved the formula in the theorem. ■

The following observations help us use the preceding formula.

FINDING THE nth ROOTS OF $z = r(\cos\theta + i\sin\theta)$

1. The modulus of each nth root is $r^{1/n}$.
2. The argument of the first root is θ/n.
3. We repeatedly add $2\pi/n$ to get the argument of each successive root.

These observations show that, when graphed, the nth roots of z are spaced equally on the circle of radius $r^{1/n}$.

EXAMPLE 8 ■ Finding Roots of a Complex Number

Find the six sixth roots of $z = -64$, and graph these roots in the complex plane.

SOLUTION In polar form, $z = 64(\cos \pi + i \sin \pi)$. Applying the formula for nth roots with $n = 6$, we get

$$w_k = 64^{1/6}\left[\cos\left(\frac{\pi + 2k\pi}{6}\right) + i \sin\left(\frac{\pi + 2k\pi}{6}\right)\right]$$

for $k = 0, 1, 2, 3, 4, 5$. Using $64^{1/6} = 2$, we find that the six sixth roots of -64 are

We add $2\pi/6 = \pi/3$ to each argument to get the argument of the next root.

$$w_0 = 2\left(\cos\frac{\pi}{6} + i \sin\frac{\pi}{6}\right) = \sqrt{3} + i$$

$$w_1 = 2\left(\cos\frac{\pi}{2} + i \sin\frac{\pi}{2}\right) = 2i$$

$$w_2 = 2\left(\cos\frac{5\pi}{6} + i \sin\frac{5\pi}{6}\right) = -\sqrt{3} + i$$

$$w_3 = 2\left(\cos\frac{7\pi}{6} + i \sin\frac{7\pi}{6}\right) = -\sqrt{3} - i$$

$$w_4 = 2\left(\cos\frac{3\pi}{2} + i \sin\frac{3\pi}{2}\right) = -2i$$

$$w_5 = 2\left(\cos\frac{11\pi}{6} + i \sin\frac{11\pi}{6}\right) = \sqrt{3} - i$$

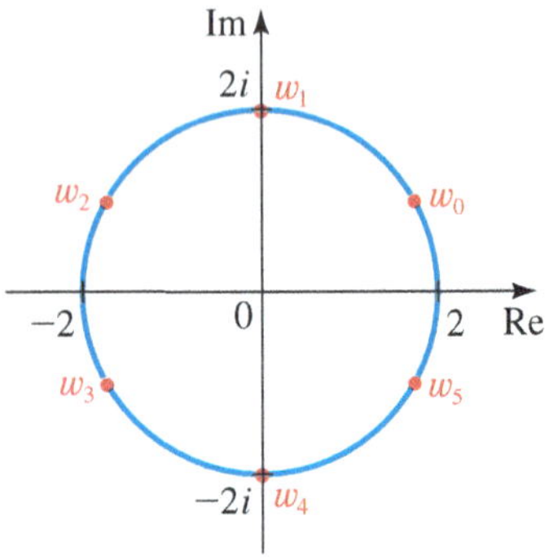

FIGURE 9 The six sixth roots of $z = -64$

All these points lie on a circle of radius 2, as shown in Figure 9.

Now Try Exercise 81

When finding roots of complex numbers, we sometimes write the argument θ of the complex number in degrees. In this case the nth roots are obtained from the formula

$$w_k = r^{1/n}\left[\cos\left(\frac{\theta + 360°k}{n}\right) + i \sin\left(\frac{\theta + 360°k}{n}\right)\right]$$

for $k = 0, 1, 2, \ldots, n - 1$.

DISCOVERY PROJECT

Fractals

Most of the things we model in this book follow regular predictable patterns. But many real-world phenomena—such as a cloud, a jagged coastline, or a flickering flame—appear to have random or even chaotic shapes. Fractals allow us to model such shapes and many others. Surprisingly, the extremely complex shapes of fractals and their infinite detail are produced by exceedingly simple rules and endless repetitions that involve iterating simple functions whose inputs and outputs are complex numbers. You can find the project at **www.stewartmath.com**.

EXAMPLE 9 ■ Finding Cube Roots of a Complex Number

Find the three cube roots of $z = 2 + 2i$, and graph these roots in the complex plane.

SOLUTION First we write z in polar form using degrees. We have $r = \sqrt{2^2 + 2^2} = 2\sqrt{2}$ and $\theta = 45°$. Thus

$$z = 2\sqrt{2}(\cos 45° + i \sin 45°)$$

We add $360°/3 = 120°$ to each argument to get the argument of the next root.

Applying the formula for nth roots (in degrees) with $n = 3$, we find that the cube roots of z are of the form

$$w_k = (2\sqrt{2})^{1/3}\left[\cos\left(\frac{45° + 360°k}{3}\right) + i \sin\left(\frac{45° + 360°k}{3}\right)\right]$$

where $k = 0, 1, 2$. Thus the three cube roots are

$$w_0 = \sqrt{2}(\cos 15° + i \sin 15°) \approx 1.366 + 0.366i$$
$$w_1 = \sqrt{2}(\cos 135° + i \sin 135°) = -1 + i$$
$$w_2 = \sqrt{2}(\cos 255° + i \sin 255°) \approx -0.366 - 1.366i$$

$(2\sqrt{2})^{1/3} = (2^{3/2})^{1/3} = 2^{1/2} = \sqrt{2}$

The three cube roots of z are graphed in Figure 10. These roots are spaced equally on a circle of radius $\sqrt{2}$.

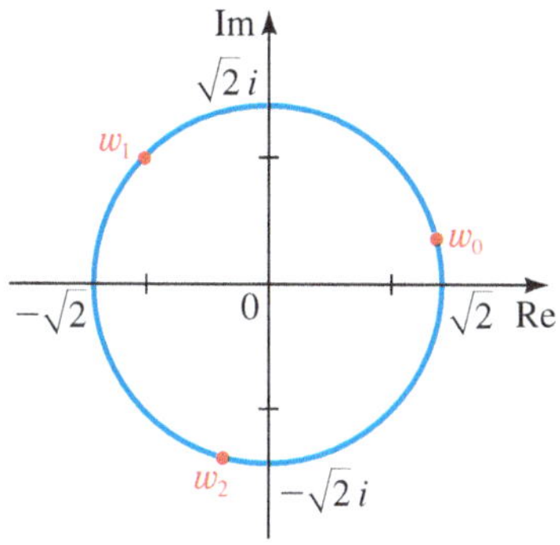

FIGURE 10 The three cube roots of $z = 2 + 2i$

Now Try Exercise 77 ■

EXAMPLE 10 ■ Solving an Equation Using the *n*th Roots Formula

Solve the equation $z^6 + 64 = 0$.

SOLUTION This equation can be written as $z^6 = -64$. Thus the solutions are the sixth roots of -64, which we found in Example 8.

Now Try Exercise 87 ■

8.3 EXERCISES

CONCEPTS

1. A complex number $z = a + bi$ has two parts: a is the ________ part, and b is the ________ part. To graph $a + bi$, we graph the ordered pair (,) in the complex plane.

2. Let $z = a + bi$.

(a) The modulus of z is $r =$ ________, and an argument of z is an angle θ satisfying $\tan\theta =$ ________.

(b) We can express z in polar form as $z =$ ____________, where r is the modulus of z and θ is the argument of z.

3. (a) The complex number $z = -1 + i$ in polar form is $z =$ ________.

(b) The complex number $z = 2\left(\cos\frac{\pi}{6} + i \sin\frac{\pi}{6}\right)$ in rectangular form is $z =$ ________.

(c) The complex number graphed below can be expressed in rectangular form as ________ or in polar form as ________.

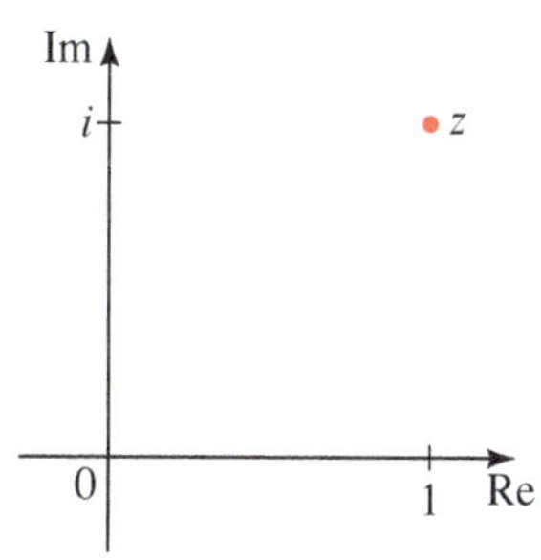

4. How many different nth roots does a nonzero complex number have? ________. The number 16 has ________ fourth roots. These roots are ________, ________, ________, and ________. In the complex plane these roots all lie on a circle of radius ________. Graph the roots on the following graph.

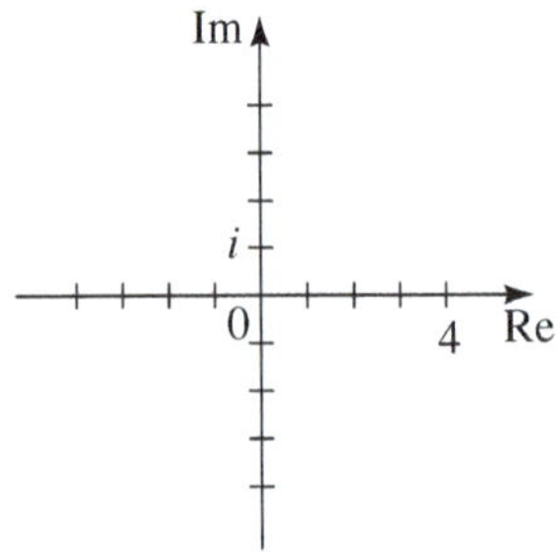

SKILLS

5–14 ■ A Complex Number and Its Modulus Graph the complex number and find its modulus.

5. $4i$ **6.** $-3i$

7. -2 **8.** 6

9. $5 + 2i$ **10.** $7 - 3i$

11. $\sqrt{3} + i$ **12.** $-1 - \dfrac{\sqrt{3}}{3}i$

13. $\dfrac{3 + 4i}{5}$ **14.** $\dfrac{-\sqrt{2} + i\sqrt{2}}{2}$

15–16 ■ Graphing Complex Numbers Sketch the complex numbers z, $2z$, $-z$, and $\frac{1}{2}z$ on the same complex plane.

15. $z = 1 + i$ **16.** $z = -1 + i\sqrt{3}$

17–18 ■ Graphing a Complex Number and Its Complex Conjugate Sketch the complex number z and its complex conjugate $\bar{z}$ on the same complex plane.

17. $z = 8 + 2i$ **18.** $z = -5 + 6i$

19–20 ■ Graphing Complex Numbers Sketch z_1, z_2, $z_1 + z_2$, and $z_1 z_2$ on the same complex plane.

19. $z_1 = 2 - i, \quad z_2 = 2 + i$

20. $z_1 = -1 + i, \quad z_2 = 2 - 3i$

21–28 ■ Graphing Sets of Complex Numbers Sketch the set in the complex plane.

21. $\{z = a + bi \mid a \le 0, b \ge 0\}$

22. $\{z = a + bi \mid a > 1, b > 1\}$

23. $\{z \mid |z| = 3\}$ **24.** $\{z \mid |z| \ge 1\}$

25. $\{z \mid |z| < 2\}$ **26.** $\{z \mid 2 \le |z| \le 5\}$

27. $\{z = a + bi \mid a + b < 2\}$

28. $\{z = a + bi \mid a \ge b\}$

29–48 ■ Polar Form of Complex Numbers Write the complex number in polar form with argument θ between 0 and 2π.

29. $1 + i$ **30.** $1 - i$ **31.** $-2 + 2i$

32. $-\sqrt{2} - \sqrt{2}i$ **33.** $-\sqrt{3} - i$ **34.** $-5 + 5\sqrt{3}i$

35. $2\sqrt{3} - 2i$ **36.** $3 + 3\sqrt{3}i$ **37.** $2i$

38. $-5i$ **39.** -3 **40.** $\sqrt{2}$

41. $-\sqrt{6} + \sqrt{2}i$ **42.** $-\sqrt{5} - \sqrt{15}i$ **43.** $4 + 3i$

44. $3 + 2i$ **45.** $4(\sqrt{3} - i)$ **46.** $i(\sqrt{2} - \sqrt{6}i)$

47. $-3(1 - i)$ **48.** $2i(1 + i)$

49–56 ■ Products and Quotients of Complex Numbers Find the product $z_1 z_2$ and the quotient z_1/z_2. Express your answer in polar form.

49. $z_1 = 3\left(\cos\dfrac{\pi}{3} + i\sin\dfrac{\pi}{3}\right), \quad z_2 = 2\left(\cos\dfrac{\pi}{6} + i\sin\dfrac{\pi}{6}\right)$

50. $z_1 = \sqrt{3}\left(\cos\dfrac{5\pi}{4} + i\sin\dfrac{5\pi}{4}\right), \quad z_2 = 2(\cos\pi + i\sin\pi)$

51. $z_1 = \sqrt{2}\left(\cos\dfrac{5\pi}{3} + i\sin\dfrac{5\pi}{3}\right)$,
$z_2 = 2\sqrt{2}\left(\cos\dfrac{3\pi}{2} + i\sin\dfrac{3\pi}{2}\right)$

52. $z_1 = \cos\dfrac{3\pi}{4} + i\sin\dfrac{3\pi}{4}, \quad z_2 = \cos\dfrac{\pi}{3} + i\sin\dfrac{\pi}{3}$

53. $z_1 = 4(\cos 120° + i\sin 120°)$,
$z_2 = 2(\cos 30° + i\sin 30°)$

54. $z_1 = \sqrt{2}(\cos 75° + i\sin 75°)$,
$z_2 = 3\sqrt{2}(\cos 60° + i\sin 60°)$

55. $z_1 = 4(\cos 200° + i\sin 200°)$,
$z_2 = 25(\cos 150° + i\sin 150°)$

56. $z_1 = \frac{4}{5}(\cos 25° + i\sin 25°)$,
$z_2 = \frac{1}{5}(\cos 155° + i\sin 155°)$

57–64 ■ Products and Quotients of Complex Numbers Write z_1 and z_2 in polar form, and then find the product $z_1 z_2$ and the quotients z_1/z_2 and $1/z_1$.

57. $z_1 = \sqrt{3} + i, \quad z_2 = 1 + \sqrt{3}i$

58. $z_1 = \sqrt{2} - \sqrt{2}i, \quad z_2 = 1 - i$

59. $z_1 = 2\sqrt{3} - 2i, \quad z_2 = -1 + i$

60. $z_1 = -\sqrt{2}i, \quad z_2 = -3 - 3\sqrt{3}i$

61. $z_1 = 5 + 5i, \quad z_2 = 4$ **62.** $z_1 = 4\sqrt{3} - 4i, \quad z_2 = 8i$

63. $z_1 = -20, \quad z_2 = \sqrt{3} + i$ **64.** $z_1 = 3 + 4i, \quad z_2 = 2 - 2i$

65–76 ■ Powers Using De Moivre's Theorem Find the indicated power using De Moivre's Theorem.

65. $(-\sqrt{3} + i)^6$ **66.** $(1 - i)^{10}$

67. $(-\sqrt{2} - \sqrt{2}i)^5$ **68.** $(1 + i)^7$

69. $\left(\dfrac{\sqrt{2}}{2} + \dfrac{\sqrt{2}}{2}i\right)^{12}$ **70.** $(\sqrt{3} - i)^{-10}$

71. $(2 - 2i)^8$

72. $\left(-\frac{1}{2} - \frac{\sqrt{3}}{2}i\right)^{15}$

73. $(-1 - i)^7$

74. $(3 + \sqrt{3}i)^4$

75. $(2\sqrt{3} + 2i)^{-5}$

76. $(1 - i)^{-8}$

77–86 ■ Roots of Complex Numbers Find the indicated roots, and graph the roots in the complex plane.

77. The square roots of $4\sqrt{3} + 4i$

78. The cube roots of $4\sqrt{3} + 4i$

79. The fourth roots of $-81i$

80. The fifth roots of 32

81. The eighth roots of 1

82. The cube roots of $1 + i$

83. The cube roots of i

84. The fifth roots of i

85. The fourth roots of -1

86. The fifth roots of $-16 - 16\sqrt{3}i$

87–92 ■ Solving Equations Using *n*th Roots Solve the equation.

87. $z^4 + 1 = 0$

88. $z^8 - i = 0$

89. $z^3 - 4\sqrt{3} - 4i = 0$

90. $z^6 \quad 1 = 0$

91. $z^3 + 1 = -i$

92. $z^3 - 1 = 0$

SKILLS Plus

93–96 ■ Complex Coefficients and the Quadratic Formula The quadratic formula works whether the coefficients of the equation are real or complex. Solve the following equations using the quadratic formula and, if necessary, De Moivre's Theorem.

93. $z^2 - iz + 1 = 0$

94. $z^2 + iz + 2 = 0$

95. $z^2 - 2iz - 2 = 0$

96. $z^2 + (1 + i)z + i = 0$

97–98 ■ Finding *n*th roots of a Complex Number Let $w = \cos(2\pi/n) + i\sin(2\pi/n)$, where n is a positive integer.

97. Show that the n distinct roots of 1 are

$$1, w, w^2, w^3, \ldots, w^{n-1}$$

98. If $z \neq 0$ and s is any nth root of z, show that the n distinct roots of z are

$$s, sw, sw^2, sw^3, \ldots, sw^{n-1}$$

DISCUSS ■ DISCOVER ■ PROVE ■ WRITE

99. **DISCUSS: Sums of Roots of Unity** Find the exact values of all three cube roots of 1 (see Exercise 97), and then add them. Do the same for the fourth, fifth, sixth, and eighth roots of 1. What do you think is the sum of the nth roots of 1 for any n?

100. **DISCUSS: Products of Roots of Unity** Find the product of the three cube roots of 1 (see Exercise 97). Do the same for the fourth, fifth, sixth, and eighth roots of 1. What do you think is the product of the nth roots of 1 for any n?

101. **PROOF: Division in Polar Form** If the two complex numbers z_1 and z_2 have the polar forms

$$z_1 = r_1(\cos\theta_1 + i\sin\theta_1)$$

and

$$z_2 = r_2(\cos\theta_2 + i\sin\theta_2)$$

show that

$$\frac{z_1}{z_2} = \frac{r_1}{r_2}[\cos(\theta_1 - \theta_2) + i\sin(\theta_1 - \theta_2)]$$

[*Hint:* Multiply numerator and denominator by the complex conjugate of z_2 and simplify.]

8.4 PLANE CURVES AND PARAMETRIC EQUATIONS

■ Plane Curves and Parametric Equations ■ Eliminating the Parameter ■ Finding Parametric Equations for a Curve ■ Using Graphing Devices to Graph Parametric Curves

So far, we have described a curve by giving an equation (in rectangular or polar coordinates) that the coordinates of all the points on the curve must satisfy. But not all curves in the plane can be described in this way. In this section we study parametric equations, which are a general method for describing any curve.

■ Plane Curves and Parametric Equations

We can think of a curve as the path of a point moving in the plane; the x- and y-coordinates of the point are then functions of time. This idea leads to the following definition.

PLANE CURVES AND PARAMETRIC EQUATIONS

If f and g are functions defined on an interval I, then the set of points $(f(t), g(t))$ is a **plane curve**. The equations

$$x = f(t) \qquad y = g(t)$$

where $t \in I$, are **parametric equations** for the curve, with **parameter** t.

EXAMPLE 1 ■ Sketching a Plane Curve

Sketch the curve defined by the parametric equations

$$x = t^2 - 3t \qquad y = t - 1$$

SOLUTION For every value of t we get a point on the curve. For example, if $t = 0$, then $x = 0$ and $y = -1$, so the corresponding point is $(0, -1)$. In Figure 1 we plot the points (x, y) determined by the values of t shown in the following table.

The arrows on the curve indicate the direction of the curve for increasing values of t.

t	x	y
−2	10	−3
−1	4	−2
0	0	−1
1	−2	0
2	−2	1
3	0	2
4	4	3
5	10	4

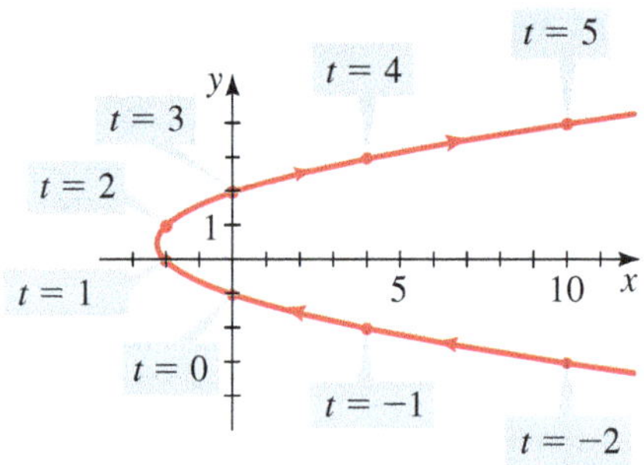

FIGURE 1

As t increases, a particle whose position is given by the parametric equations moves along the curve in the direction of the arrows.

Now Try Exercise 3 ■

If we replace t by $-t$ in Example 1, we obtain the parametric equations

$$x = t^2 + 3t \qquad y = -t - 1$$

The graph of these parametric equations (see Figure 2) is the same as the curve in Figure 1 but traced out in the opposite direction. On the other hand, if we replace t by $2t$ in Example 1, we obtain the parametric equations

$$x = 4t^2 - 6t \qquad y = 2t - 1$$

The graph of these parametric equations (see Figure 3) is again the same but is traced out "twice as fast." *Thus a parametrization contains more information than just the shape of the curve; it also indicates* how *the curve is being traced out.*

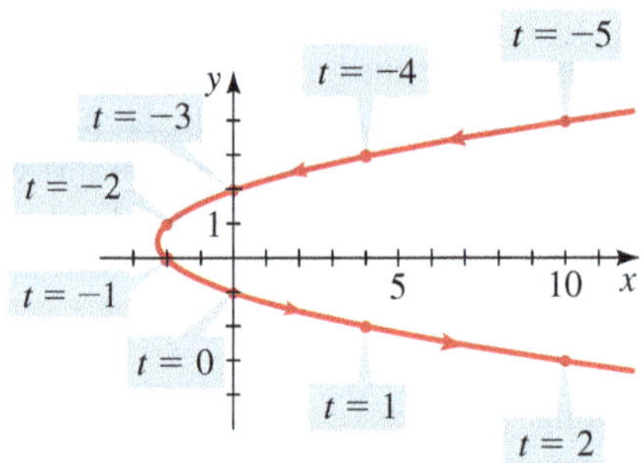

FIGURE 2 $x = t^2 + 3t$, $y = -t - 1$

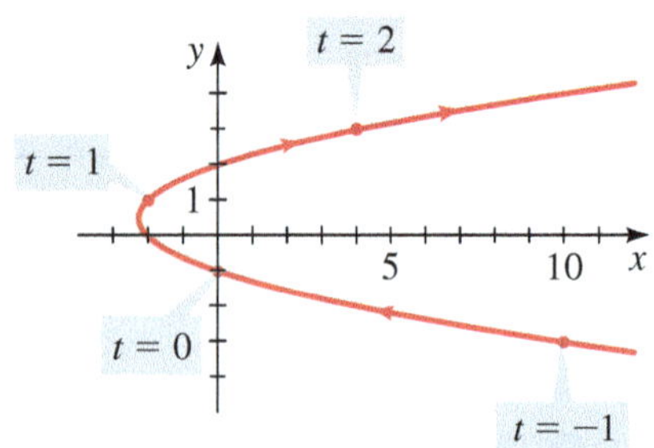

FIGURE 3 $x = 4t^2 - 6t$, $y = 2t - 1$

MARIA GAETANA AGNESI (1718–1799) is famous for having written *Instituzioni Analitiche*, one of the first calculus textbooks.

Agnesi was born into a wealthy family in Milan, Italy, the oldest of 21 children. She was a child prodigy, mastering many languages at an early age, including Latin, Greek, and Hebrew. At the age of 20 she published a series of essays on philosophy and natural science. After her mother died, Agnesi took on the task of educating her brothers. In 1748 Agnesi published her famous textbook, which she originally wrote as a text for tutoring her brothers. The book compiled and explained the mathematical knowledge of the day. It contains many carefully chosen examples, one of which is the curve now known as the "witch of Agnesi" (see Exercise 66, page 619). One review calls her book an "exposition by examples rather than by theory." The book gained Agnesi immediate recognition. Pope Benedict XIV appointed her to a position at the University of Bologna, writing, "we have had the idea that you should be awarded the well-known chair of mathematics, by which it comes of itself that you should not thank us but we you." This appointment was an extremely high honor for a woman, since very few women then were even allowed to attend university. Just two years later, Agnesi's father died, and she left mathematics completely. She became a nun and devoted the rest of her life and her wealth to caring for sick and dying women, herself dying in poverty at a poorhouse of which she had once been director.

Eliminating the Parameter

Often a curve given by parametric equations can also be represented by a single rectangular equation in x and y. The process of finding this equation is called *eliminating the parameter*. One way to do this is to solve for t in one equation, then substitute into the other.

EXAMPLE 2 ■ Eliminating the Parameter

Eliminate the parameter in the parametric equations of Example 1.

SOLUTION First we solve for t in the simpler equation, then we substitute into the other equation. From the equation $y = t - 1$ we get $t = y + 1$. Substituting into the equation for x, we get

$$x = t^2 - 3t = (y + 1)^2 - 3(y + 1) = y^2 - y - 2$$

Thus the curve in Example 1 has the rectangular equation $x = y^2 - y - 2$, so it is a parabola.

Now Try Exercise 5 ■

Eliminating the parameter often helps us identify the shape of a curve, as we see in the next two examples.

EXAMPLE 3 ■ Modeling Circular Motion

The following parametric equations model the position of a moving object at time t (in seconds):

$$x = \cos t \qquad y = \sin t \qquad t \geq 0$$

Describe and graph the path of the object.

SOLUTION To identify the curve, we eliminate the parameter. Since $\cos^2 t + \sin^2 t = 1$ and since $x = \cos t$ and $y = \sin t$ for every point (x, y) on the curve, we have

$$x^2 + y^2 = (\cos t)^2 + (\sin t)^2 = 1$$

This means that all points on the curve satisfy the equation $x^2 + y^2 = 1$, so the graph is a circle of radius 1 centered at the origin. As t increases from 0 to 2π, the point given by the parametric equations starts at $(1, 0)$ and moves counterclockwise once around the circle, as shown in Figure 4. So the object completes one revolution around the circle in 2π seconds. Notice that the parameter t can be interpreted as the angle shown in the figure.

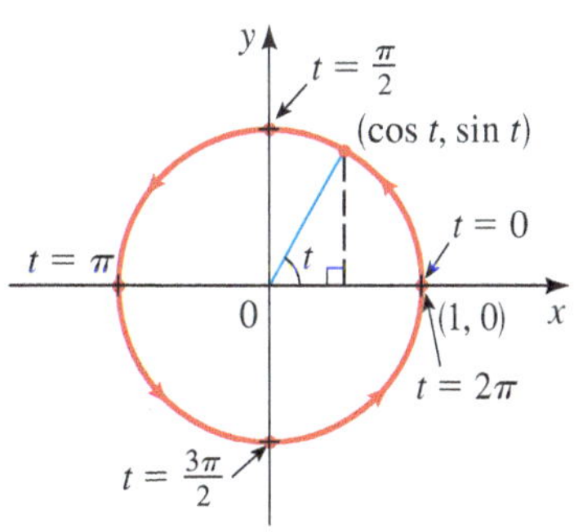

FIGURE 4

Now Try Exercise 27 ■

EXAMPLE 4 ■ Sketching a Parametric Curve

Eliminate the parameter, and sketch the graph of the parametric equations

$$x = \sin t \qquad y = 2 - \cos^2 t$$

SOLUTION To eliminate the parameter, we first use the trigonometric identity $\cos^2 t = 1 - \sin^2 t$ to change the second equation:

$$y = 2 - \cos^2 t = 2 - (1 - \sin^2 t) = 1 + \sin^2 t$$

Now we can substitute $\sin t = x$ from the first equation to get

$$y = 1 + x^2$$

so the point (x, y) moves along the parabola $y = 1 + x^2$. However, since $-1 \leq \sin t \leq 1$, we have $-1 \leq x \leq 1$, so the parametric equations represent only the part of the parabola between $x = -1$ and $x = 1$. Since $\sin t$ is periodic, the point $(x, y) = (\sin t, 2 - \cos^2 t)$ moves back and forth infinitely often along the parabola between the points $(-1, 2)$ and $(1, 2)$, as shown in Figure 5.

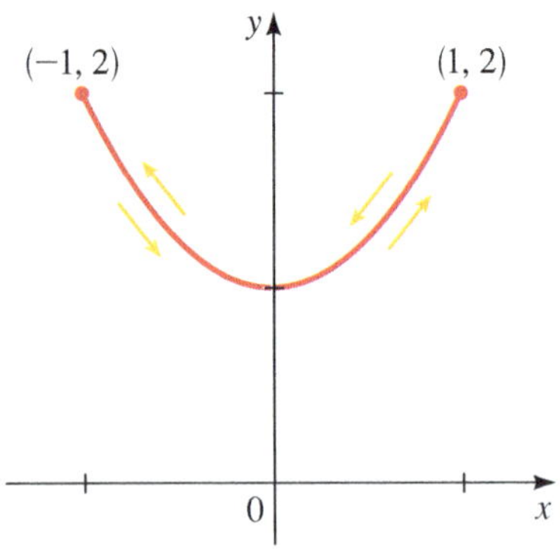

FIGURE 5

Now Try Exercise 15 ■

Finding Parametric Equations for a Curve

It is often possible to find parametric equations for a curve by using some geometric properties that define the curve, as in the next two examples.

EXAMPLE 5 ■ Finding Parametric Equations for a Graph

Find parametric equations for the line of slope 3 that passes through the point $(2, 6)$.

SOLUTION Let's start at the point $(2, 6)$ and move up and to the right along this line. Because the line has slope 3, for every 1 unit we move to the right, we must move up 3 units. In other words, if we increase the x-coordinate by t units, we must correspondingly increase the y-coordinate by $3t$ units. This leads to the parametric equations

$$x = 2 + t \qquad y = 6 + 3t$$

To confirm that these equations give the desired line, we eliminate the parameter. We solve for t in the first equation and substitute into the second to get

$$y = 6 + 3(x - 2) = 3x$$

Thus the slope-intercept form of the equation of this line is $y = 3x$, which is a line of slope 3 that does pass through $(2, 6)$ as required. The graph is shown in Figure 6.

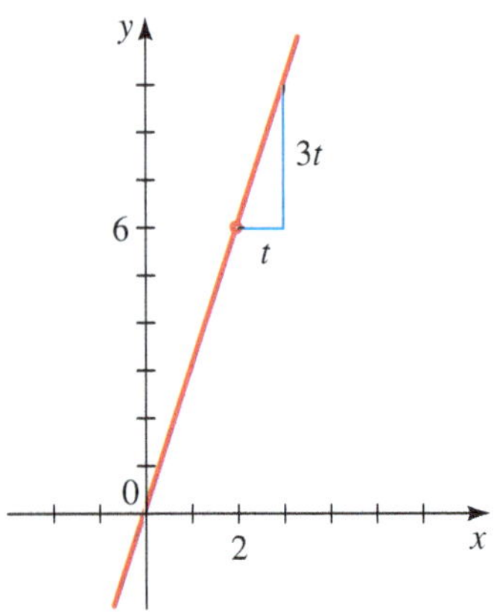

FIGURE 6

Now Try Exercise 31 ■

EXAMPLE 6 ■ Parametric Equations for the Cycloid

As a circle rolls along a straight line, the curve traced out by a fixed point P on the circumference of the circle is called a **cycloid** (see Figure 7). If the circle has radius a and rolls along the x-axis, with one position of the point P being at the origin, find parametric equations for the cycloid.

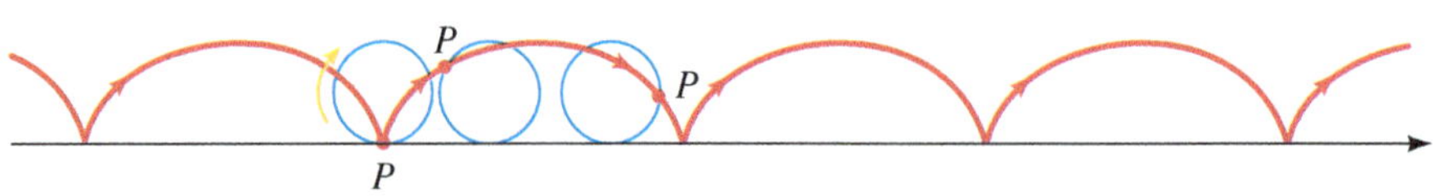

FIGURE 7

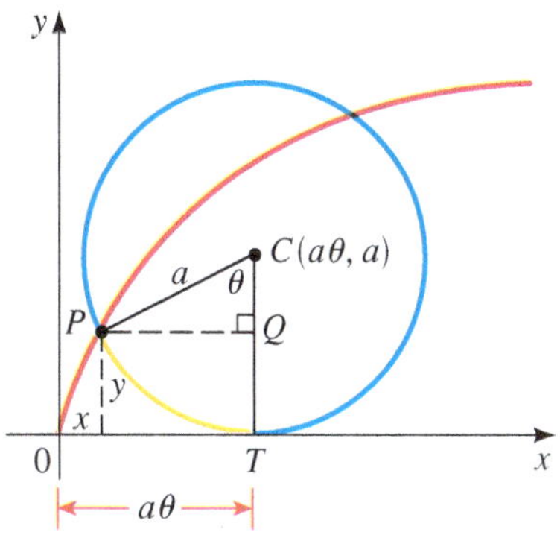

FIGURE 8

SOLUTION Figure 8 shows the circle and the point P after the circle has rolled through an angle θ (in radians). The distance $d(O, T)$ that the circle has rolled must be the same as the length of the arc PT, which, by the arc length formula, is $a\theta$ (see Section 6.1). This means that the center of the circle is $C(a\theta, a)$.

Let the coordinates of P be (x, y). Then from Figure 8 (which illustrates the case $0 < \theta < \pi/2$), we see that

$$x = d(O, T) - d(P, Q) = a\theta - a\sin\theta = a(\theta - \sin\theta)$$
$$y = d(T, C) - d(Q, C) = a - a\cos\theta = a(1 - \cos\theta)$$

so parametric equations for the cycloid are

$$x = a(\theta - \sin\theta) \qquad y = a(1 - \cos\theta)$$

Now Try Exercise 53 ■

The cycloid has a number of interesting physical properties. It is the "curve of quickest descent" in the following sense. Let's choose two points P and Q that are not directly above each other and join them with a wire. Suppose we allow a bead to slide down the wire under the influence of gravity (ignoring friction). Of all possible shapes into which the wire can be bent, the bead will slide from P to Q the fastest when the shape is half of an arch of an inverted cycloid (see Figure 9). The cycloid is also the "curve of equal descent" in the sense that no matter where we place a bead B on a cycloid-shaped wire, it takes the same time to slide to the bottom (see Figure 10). These rather surprising properties of the cycloid were proved (using calculus) in the 17th century by several mathematicians and physicists, including Johann Bernoulli, Blaise Pascal, and Christiaan Huygens.

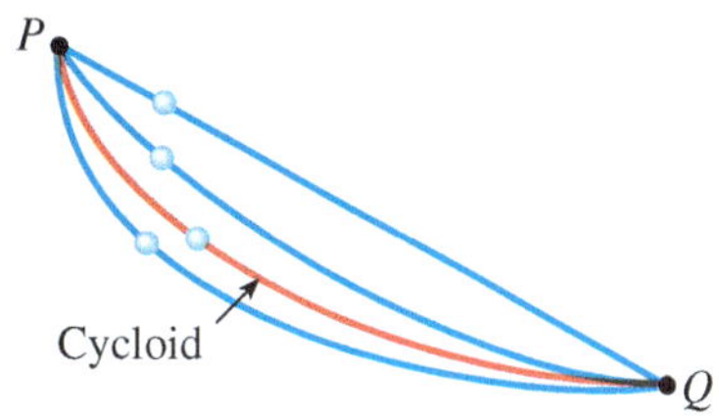

FIGURE 9

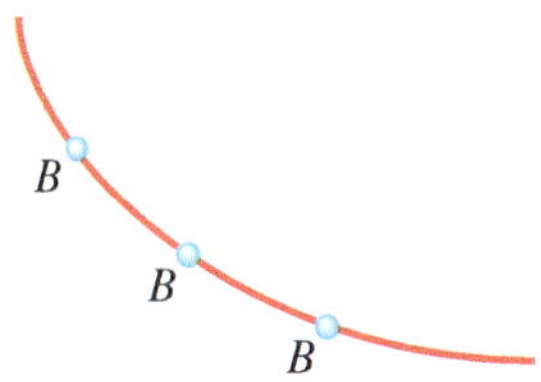

FIGURE 10

Using Graphing Devices to Graph Parametric Curves

Most graphing calculators and computer graphing programs can be used to graph parametric equations. Such devices are particularly useful in sketching complicated curves like the one shown in Figure 11.

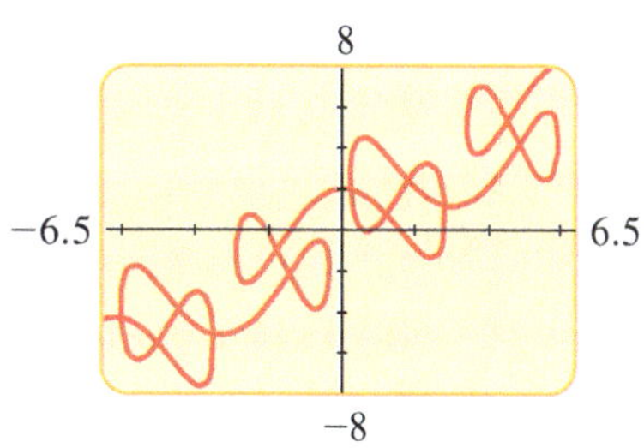

FIGURE 11
$x = t + 2\sin 2t$, $y = t + 2\cos 5t$

EXAMPLE 7 ■ Graphing Parametric Curves

Use a graphing device to draw the following parametric curves. Discuss their similarities and differences.

(a) $x = \sin 2t$
$y = 2\cos t$

(b) $x = \sin 3t$
$y = 2\cos t$

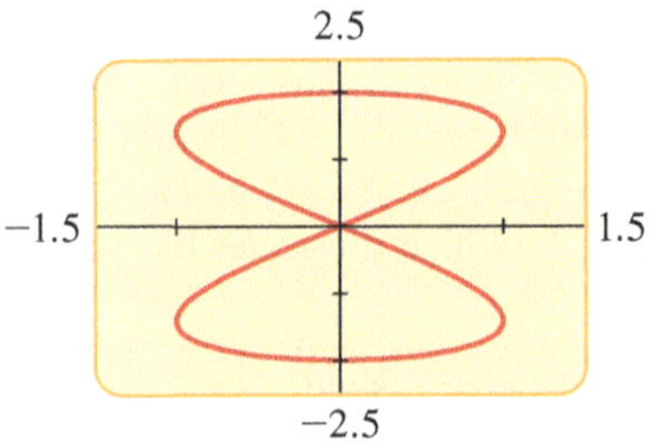

(a) $x = \sin 2t,\ y = 2\cos t$

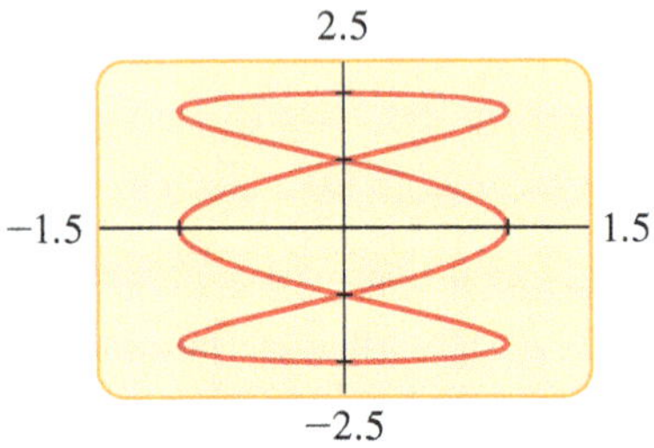

(b) $x = \sin 3t,\ y = 2\cos t$

FIGURE 12

SOLUTION In both parts (a) and (b) the graph will lie inside the rectangle given by $-1 \le x \le 1$, $-2 \le y \le 2$, since both the sine and the cosine of any number will be between -1 and 1. Thus we may use the viewing rectangle $[-1.5, 1.5]$ by $[-2.5, 2.5]$.

(a) Since $2\cos t$ is periodic with period 2π (see Section 5.3) and since $\sin 2t$ has period π, letting t vary over the interval $0 \le t \le 2\pi$ gives us the complete graph, which is shown in Figure 12(a).

(b) Again, letting t take on values between 0 and 2π gives the complete graph shown in Figure 12(b).

Both graphs are *closed curves*, which means that they form loops with the same starting and ending point; also, both graphs cross over themselves. However, the graph in Figure 12(a) has two loops, like a figure eight, whereas the graph in Figure 12(b) has three loops.

Now Try Exercise 39 ■

The curves graphed in Example 7 are called Lissajous figures. A **Lissajous figure** is the graph of a pair of parametric equations of the form

$$x = A \sin \omega_1 t \qquad y = B \cos \omega_2 t$$

where A, B, ω_1, and ω_2 are positive real constants. Since $\sin \omega_1 t$ and $\cos \omega_2 t$ are both between -1 and 1, a Lissajous figure will lie inside the rectangle determined by $-A \le x \le A$, $-B \le y \le B$. This fact can be used to choose a viewing rectangle when graphing a Lissajous figure, as in Example 7.

Recall from Section 8.1 that rectangular coordinates (x, y) and polar coordinates (r, θ) are related by the equations $x = r\cos\theta$, $y = r\sin\theta$. Thus we can graph the polar equation $r = f(\theta)$ by changing it to parametric form as follows.

$$x = r\cos\theta = f(\theta)\cos\theta \qquad \text{Since } r = f(\theta)$$

$$y = r\sin\theta = f(\theta)\sin\theta$$

Replacing θ by the standard parametric variable t, we have the following result.

POLAR EQUATIONS IN PARAMETRIC FORM

The graph of the polar equation $r = f(\theta)$ is the same as the graph of the parametric equations

$$x = f(t)\cos t \qquad y = f(t)\sin t$$

EXAMPLE 8 ■ Parametric Form of a Polar Equation

Consider the polar equation $r = \theta$, $1 \le \theta \le 10\pi$.

(a) Express the equation in parametric form.

(b) Draw a graph of the parametric equations from part (a).

SOLUTION

(a) The given polar equation is equivalent to the parametric equations

$$x = t\cos t \qquad y = t\sin t$$

(b) Since $10\pi \approx 31.42$, we use the viewing rectangle $[-32, 32]$ by $[-32, 32]$, and we let t vary from 1 to 10π. The resulting graph shown in Figure 13 is a *spiral*.

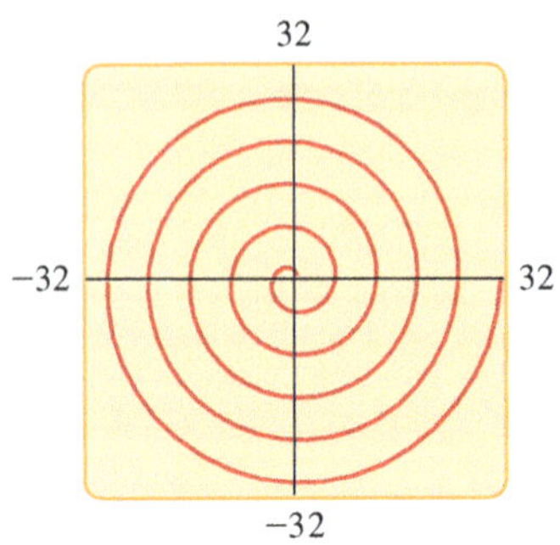

FIGURE 13 $x = t\cos t,\ y = t\sin t$

Now Try Exercise 47 ■

8.4 EXERCISES

CONCEPTS

1. (a) The parametric equations $x = f(t)$ and $y = g(t)$ give the coordinates of a point $(x, y) = (f(t), g(t))$ for appropriate values of t. The variable t is called a ________.

(b) Suppose that the parametric equations $x = t$, $y = t^2$, $t \geq 0$, model the position of a moving object at time t. When $t = 0$, the object is at (,   ), and when $t = 1$, the object is at (,   ).

(c) If we eliminate the parameter in part (b), we get the equation $y =$ ________. We see from this equation that the path of the moving object is a ________.

2. (a) *True or False?* The same curve can be described by parametric equations in many different ways.

(b) The parametric equations $x = 2t$, $y = (2t)^2$ model the position of a moving object at time t. When $t = 0$, the object is at (,   ), and when $t = 1$, the object is at (,   ).

(c) If we eliminate the parameter, we get the equation $y =$ ________, which is the same equation as in Exercise 1(c). So the objects in Exercises 1(b) and 2(b) move along the same ________ but traverse the path differently. Indicate the position of each object when $t = 0$ and when $t = 1$ on the following graph.

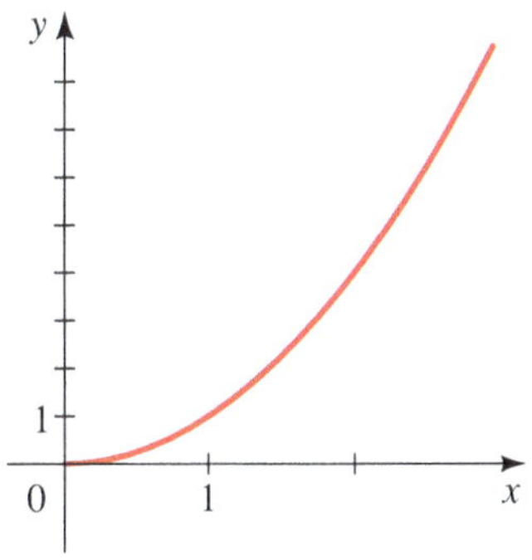

SKILLS

3–26 ■ Sketching a Curve by Eliminating the Parameter A pair of parametric equations is given. **(a)** Sketch the curve represented by the parametric equations. Use arrows to indicate the direction of the curve as t increases. **(b)** Find a rectangular-coordinate equation for the curve by eliminating the parameter.

3. $x = 2t, \quad y = t + 6$

4. $x = 6t - 4, \quad y = 3t, \quad t \geq 0$

5. $x = t^2, \quad y = t - 2, \quad 2 \leq t \leq 4$

6. $x = 2t + 1, \quad y = \left(t + \frac{1}{2}\right)^2$

7. $x = \sqrt{t}, \quad y = 1 - t$

8. $x = t^2, \quad y = t^4 + 1$

9. $x = \dfrac{1}{t}, \quad y = t + 1$

10. $x = t + 1, \quad y = \dfrac{t}{t + 1}$

11. $x = 4t^2, \quad y = 8t^3$

12. $x = |t|, \quad y = |1 - |t||$

13. $x = 2 \sin t, \quad y = 2 \cos t, \quad 0 \leq t \leq \pi$

14. $x = 2 \cos t, \quad y = 3 \sin t, \quad 0 \leq t \leq 2\pi$

15. $x = \sin^2 t, \quad y = \sin^4 t$

16. $x = \sin^2 t, \quad y = \cos t$

17. $x = \cos t, \quad y = \cos 2t$

18. $x = \cos 2t, \quad y = \sin 2t$

19. $x = \sec t, \quad y = \tan t, \quad 0 \leq t < \pi/2$

20. $x = \cot t, \quad y = \csc t, \quad 0 < t < \pi$

21. $x = \tan t, \quad y = \cot t, \quad 0 < t < \pi/2$

22. $x = e^{-t}, \quad y = e^t$

23. $x = e^{2t}, \quad y = e^t$

24. $x = \sec t, \quad y = \tan^2 t, \quad 0 \leq t < \pi/2$

25. $x = \cos^2 t, \quad y = \sin^2 t$

26. $x = \cos^3 t, \quad y = \sin^3 t, \quad 0 \leq t \leq 2\pi$

27–30 ■ Circular Motion The position of an object in circular motion is modeled by the given parametric equations. Describe the path of the object by stating the radius of the circle, the position at time $t = 0$, the orientation of the motion (clockwise or counterclockwise), and the time t that it takes to complete one revolution around the circle.

27. $x = 3 \cos t, \quad y = 3 \sin t$

28. $x = 2 \sin t, \quad y = 2 \cos t$

29. $x = \sin 2t, \quad y = \cos 2t$

30. $x = 4 \cos 3t, \quad y = 4 \sin 3t$

31–36 ■ Parametric Equations for Curves Find parametric equations for the curve with the given properties.

31. The line with slope $\frac{1}{2}$, passing through $(4, -1)$

32. The line with slope -2, passing through $(-10, -20)$

33. The line passing through $(6, 7)$ and $(7, 8)$

34. The line passing through $(12, 7)$ and the origin

35. The circle $x^2 + y^2 = a^2$.

36. The ellipse

$$\frac{x^2}{a^2} + \frac{y^2}{b^2} = 1$$

37. Path of a Projectile If a projectile is fired with an initial speed of v_0 ft/s at an angle α above the horizontal, then its position after t seconds is given by the parametric equations

$$x = (v_0 \cos \alpha)t \qquad y = (v_0 \sin \alpha)t - 16t^2$$

(where x and y are measured in feet). Show that the path of the projectile is a parabola by eliminating the parameter t.

38. Path of a Projectile Referring to Exercise 37, suppose a gun fires a bullet into the air with an initial speed of 2048 ft/s at an angle of 30° to the horizontal.

(a) After how many seconds will the bullet hit the ground?

(b) How far from the gun will the bullet hit the ground?

(c) What is the maximum height attained by the bullet?

39–44 ■ Graphs of Parametric Equations Use a graphing device to draw the curve represented by the parametric equations.

39. $x = \sin t, \quad y = 2\cos 3t$

40. $x = 2\sin t, \quad y = \cos 4t$

41. $x = 3\sin 5t, \quad y = 5\cos 3t$

42. $x = \sin 4t, \quad y = \cos 3t$

43. $x = \sin(\cos t), \quad y = \cos(t^{3/2}), \quad 0 \le t \le 2\pi$

44. $x = 2\cos t + \cos 2t, \quad y = 2\sin t - \sin 2t$

45–48 ■ Parametric Form of a Polar Equation A polar equation is given. **(a)** Express the polar equation in parametric form. **(b)** Use a graphing device to graph the parametric equations you found in part (a).

45. $r = 2^{\theta/12}, \quad 0 \le \theta \le 4\pi$

46. $r = \sin\theta + 2\cos\theta$

47. $r = \dfrac{4}{2 - \cos\theta}$

48. $r = 2^{\sin\theta}$

49–52 ■ Graphs of Parametric Equations Match the parametric equations with the graphs labeled I–IV. Give reasons for your answers.

49. $x = t^3 - 2t, \quad y = t^2 - t$

50. $x = \sin 3t, \quad y = \sin 4t$

51. $x = t + \sin 2t, \quad y = t + \sin 3t$

52. $x = \sin(t + \sin t), \quad y = \cos(t + \cos t)$

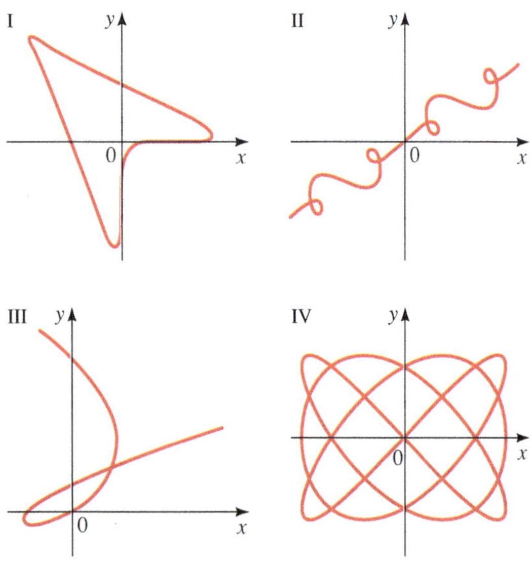

53. Finding Parametric Equations for a Curve Two circles of radius a and b are centered at the origin, as shown in the figure. As the angle θ increases, the point P traces out a curve that lies between the circles.

(a) Find parametric equations for the curve, using θ as the parameter.

(b) Graph the curve using a graphing device, with $a = 3$ and $b = 2$.

(c) Eliminate the parameter, and identify the curve.

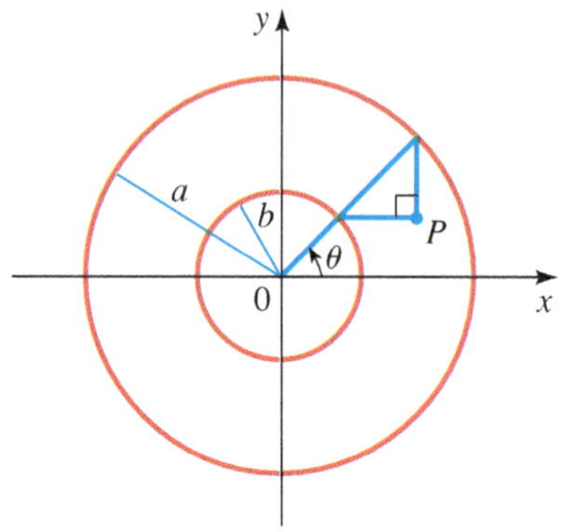

54. Finding Parametric Equations for a Curve Two circles of radius a and b are centered at the origin, as shown in the figure.

(a) Find parametric equations for the curve traced out by the point P, using the angle θ as the parameter. (Note that the line segment AB is always tangent to the larger circle.)

(b) Graph the curve using a graphing device, with $a = 3$ and $b = 2$.

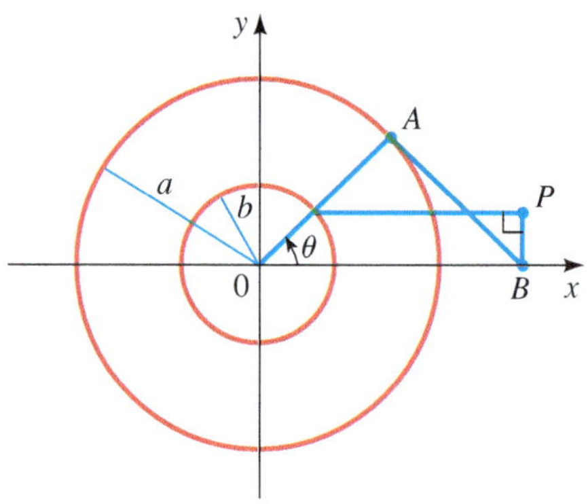

55. Curtate Cycloid

(a) In Example 6, suppose the point P that traces out the curve lies not on the edge of the circle but rather at a fixed point inside the rim, at a distance b from the center (with $b < a$). The curve traced out by P is called a **curtate cycloid** (or **trochoid**). Show that parametric equations for the curtate cycloid are

$$x = a\theta - b\sin\theta \qquad y = a - b\cos\theta$$

(b) Sketch the graph using $a = 3$ and $b = 2$.

56. Prolate Cycloid

(a) In Exercise 55 if the point P lies *outside* the circle at a distance b from the center (with $b > a$), then the curve traced out by P is called a **prolate cycloid**. Show that parametric equations for the prolate cycloid are the same as the equations for the curtate cycloid.

(b) Sketch the graph for the case in which $a = 1$ and $b = 2$.

SKILLS Plus

57. Parametric Equations of a Hyperbola Eliminate the parameter θ in the following parametric equations. (This curve is called a **hyperbola**; see page 800.)

$$x = a \tan \theta \qquad y = b \sec \theta$$

58. Parametric Equations of a Hyperbola Show that the following parametric equations represent a part of the hyperbola of Exercise 57.

$$x = a\sqrt{t} \qquad y = b\sqrt{t+1}$$

59–62 ■ Graphs of Parametric Equations Sketch the curve given by the parametric equations.

59. $x = t \cos t, \quad y = t \sin t, \quad t \geq 0$

60. $x = \sin t, \quad y = \sin 2t$

61. $x = \dfrac{3t}{1+t^3}, \quad y = \dfrac{3t^2}{1+t^3}$

62. $x = \cot t, \quad y = 2\sin^2 t, \quad 0 < t < \pi$

63. Hypocycloid A circle C of radius b rolls on the inside of a larger circle of radius a centered at the origin. Let P be a fixed point on the smaller circle, with initial position at the point $(a, 0)$ as shown in the figure. The curve traced out by P is called a **hypocycloid**.

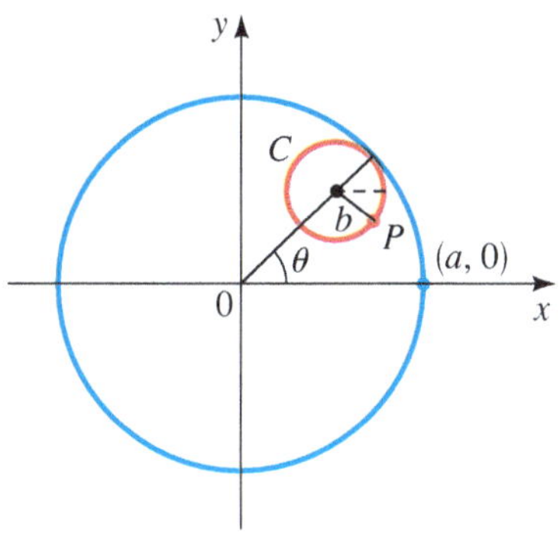

(a) Show that parametric equations for the hypocycloid are

$$x = (a-b)\cos\theta + b\cos\left(\frac{a-b}{b}\theta\right)$$

$$y = (a-b)\sin\theta - b\sin\left(\frac{a-b}{b}\theta\right)$$

(b) If $a = 4b$, the hypocycloid is called an **astroid**. Show that in this case the parametric equations can be reduced to

$$x = a\cos^3\theta \qquad y = a\sin^3\theta$$

Sketch the curve. Eliminate the parameter to obtain an equation for the astroid in rectangular coordinates.

64. Epicycloid If the circle C of Exercise 63 rolls on the *outside* of the larger circle, the curve traced out by P is called an **epicycloid**. Find parametric equations for the epicycloid.

65. Longbow Curve In the following figure, the circle of radius a is stationary, and for every θ, the point P is the midpoint of the segment QR. The curve traced out by P for $0 < \theta < \pi$ is called the **longbow curve**. Find parametric equations for this curve.

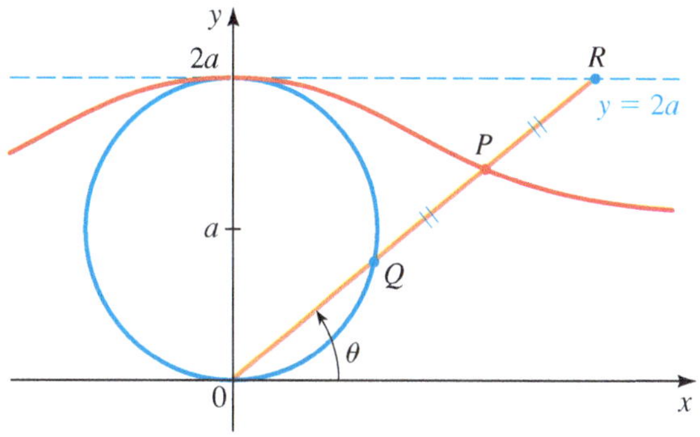

66. The Witch of Agnesi A curve, called a **witch of Agnesi**, consists of all points P determined as shown in the figure.

(a) Show that parametric equations for this curve can be written as

$$x = 2a\cot\theta \qquad y = 2a\sin^2\theta$$

(b) Graph the curve using a graphing device, with $a = 3$.

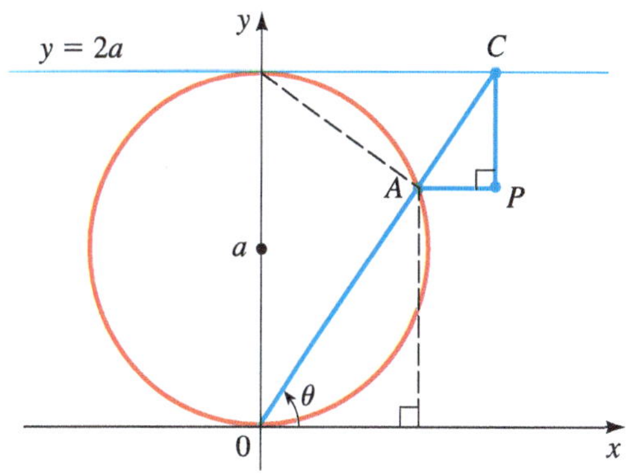

67. Eliminating the Parameter Eliminate the parameter θ in the parametric equations for the cycloid (Example 6) to obtain a rectangular coordinate equation for the section of the curve given by $0 \leq \theta \leq \pi$.

APPLICATIONS

68. The Rotary Engine The Mazda RX-8 uses an unconventional engine (invented by Felix Wankel in 1954) in which the pistons are replaced by a triangular rotor that turns in a special housing as shown in the figure on the next page. The vertices of the rotor maintain contact with the housing at all times, while the center of the triangle traces out a circle of radius r, turning the drive shaft. The shape of the housing is given by the parametric equations below (where R is the distance between the vertices and center of the rotor):

$$x = r\cos 3\theta + R\cos\theta \qquad y = r\sin 3\theta + R\sin\theta$$

(a) Suppose that the drive shaft has radius $r = 1$. Graph the curve given by the parametric equations for the following values of R: 0.5, 1, 3, 5.

(b) Which of the four values of R given in part (a) seems to best model the engine housing illustrated in the figure?

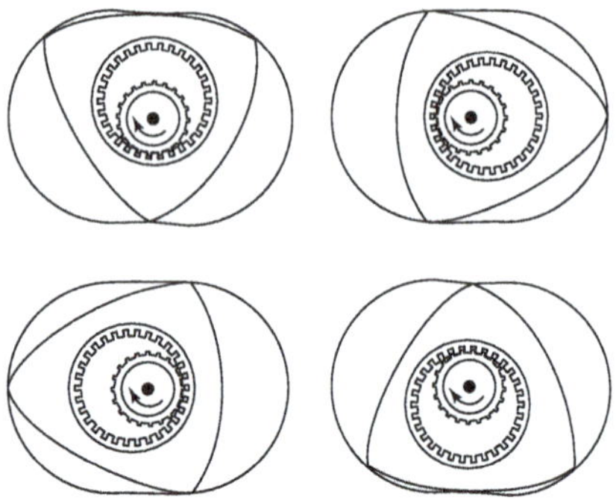

69. Spiral Path of a Dog A dog is tied to a cylindrical tree trunk of radius 1 ft by a long leash. He has managed to wrap the entire leash around the tree while playing in the yard, and he finds himself at the point $(1, 0)$ in the figure. Seeing a squirrel, he runs around the tree counterclockwise, keeping the leash taut while chasing the intruder.

(a) Show that parametric equations for the dog's path (called an **involute of a circle**) are

$$x = \cos\theta + \theta\sin\theta \qquad y = \sin\theta - \theta\cos\theta$$

[*Hint:* Note that the leash is always tangent to the tree, so OT is perpendicular to TD.]

(b) Graph the path of the dog for $0 \le \theta \le 4\pi$.

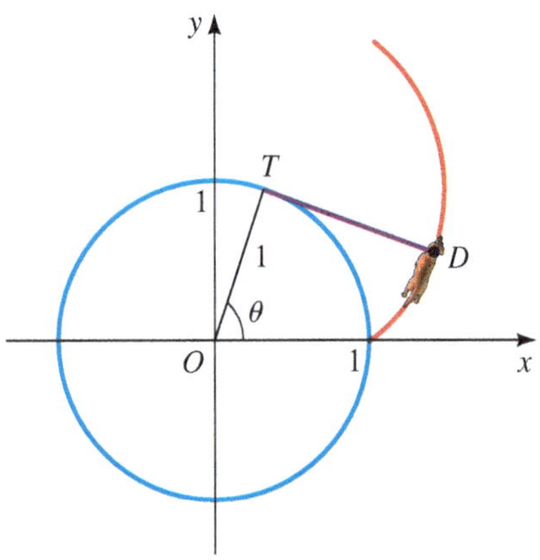

DISCUSS ■ DISCOVER ■ PROVE ■ WRITE

70. DISCOVER ■ WRITE: More Information in Parametric Equations In this section we stated that parametric equations contain more information than just the shape of a curve. Write a short paragraph explaining this statement. Use the following example and your answers to parts (a) and (b) below in your explanation.

The position of a particle is given by the parametric equations

$$x = \sin t \qquad y = \cos t$$

where t represents time. We know that the shape of the path of the particle is a circle.

(a) How long does it take the particle to go once around the circle? Find parametric equations if the particle moves twice as fast around the circle.

(b) Does the particle travel clockwise or counterclockwise around the circle? Find parametric equations if the particle moves in the opposite direction around the circle.

71. DISCUSS: Different Ways of Tracing Out a Curve The curves C, D, E, and F are defined parametrically as follows, where the parameter t takes on all real values unless otherwise stated:

$$C\colon\ x = t,\quad y = t^2$$

$$D\colon\ x = \sqrt{t},\quad y = t,\quad t \ge 0$$

$$E\colon\ x = \sin t,\quad y = \sin^2 t$$

$$F\colon\ x = 3^t,\quad y = 3^{2t}$$

(a) Show that the points on all four of these curves satisfy the same rectangular coordinate equation.

(b) Draw the graph of each curve and explain how the curves differ from one another.

CHAPTER 8 ■ REVIEW

PROPERTIES AND FORMULAS

Polar Coordinates (p. 588)

In the **polar coordinate** system the location of a point P in the plane is determined by an ordered pair (r, θ), where r is the distance from the pole O to P and θ is the angle formed by the polar axis and the ray $\overrightarrow{OP}$, as shown in the figure.

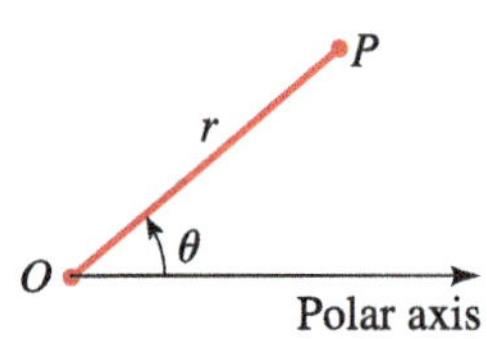

Polar and Rectangular Coordinates (p. 590)

Any point P in the plane has polar coordinates $P(r, \theta)$ and rectangular coordinates $P(x, y)$, as shown.

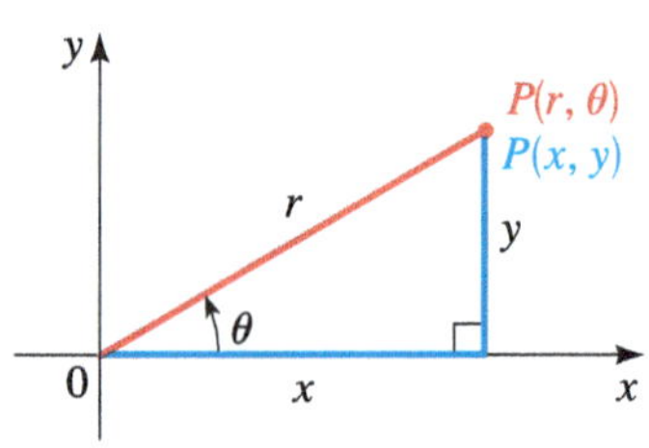

- **To change from polar to rectangular coordinates**, we use the equations

$$x = r\cos\theta \qquad \text{and} \qquad y = r\sin\theta$$

- **To change from rectangular to polar coordinates**, we use the equations

$$r^2 = x^2 + y^2 \qquad \text{and} \qquad \tan\theta = \frac{y}{x}$$

Polar Equations and Graphs (pp. 594, 599)

A **polar equation** is an equation in the variables r and θ. The **graph of a polar equation** $r = f(\theta)$ consists of all points (r, θ) whose coordinates satisfy the equation.

Symmetry in Graphs of Polar Equations (p. 597)

We can test a polar equation for symmetry as follows. The graph of a polar equation is

- **symmetric about the polar axis** if the equation is unchanged when we replace θ by $-\theta$;
- **symmetric about the pole** if the equation is unchanged when we replace r by $-r$, or θ by $\theta + \pi$.
- **symmetric about the vertical line** $\theta = \pi/2$ if the equation is unchanged when we replace θ by $\pi - \theta$.

Complex Numbers (pp. 602–603)

A **complex number** is a number of the form $a + bi$, where $i^2 = -1$ and where a and b are real numbers. For the complex number $z = a + bi$, a is called the **real part** and b is called the **imaginary part**. A complex number $a + bi$ is graphed in the complex plane as shown.

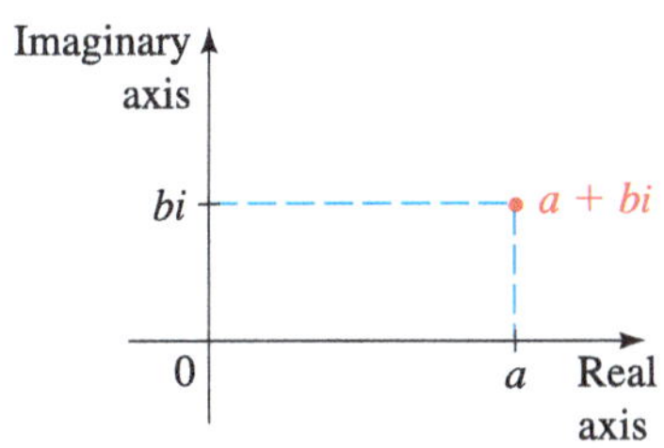

The **modulus** (or **absolute value**) of a complex number $z = a + bi$ is

$$|z| = \sqrt{a^2 + b^2}$$

Polar Form of Complex Numbers (p. 604)

A complex number $z = a + bi$ has the **polar form** (or **trigonometric form**)

$$z = r(\cos\theta + i\sin\theta)$$

where $r = |z|$ and $\tan\theta = b/a$. The number r is the modulus of z and θ is the argument of z.

Multiplication and Division of Complex Numbers in Polar Form (p. 605)

Suppose the complex numbers z_1 and z_2 have the following polar form:

$$z_1 = r_1(\cos\theta_1 + i\sin\theta_1)$$
$$z_2 = r_2(\cos\theta_2 + i\sin\theta_2)$$

Then

$$z_1z_2 = r_1r_2[\cos(\theta_1 + \theta_2) + i\sin(\theta_1 + \theta_2)]$$
$$\frac{z_1}{z_2} = \frac{r_1}{r_2}[\cos(\theta_1 - \theta_2) + i\sin(\theta_1 - \theta_2)]$$

De Moivre's Theorem (p. 606)

If $z = r(\cos\theta + i\sin\theta)$ is a complex number in polar form and n is a positive integer, then

$$z^n = r^n(\cos n\theta + i\sin n\theta)$$

*n*th Roots of Complex Numbers (p. 607)

If $z = r(\cos\theta + i\sin\theta)$ is a complex number in polar form and n is a positive integer, then z has the n distinct nth roots $w_0, w_1, \ldots, w_{n-1}$, where

$$w_k = r^{1/n}\left[\cos\left(\frac{\theta + 2k\pi}{n}\right) + i\sin\left(\frac{\theta + 2k\pi}{n}\right)\right]$$

where $k = 0, 1, 2, \ldots, n - 1$

Finding the *n*th Roots of *z* (p. 607)

To find the nth roots of $z = r(\cos\theta + i\sin\theta)$, we use the following observations:

1. The modulus of each nth root is $r^{1/n}$.
2. The argument of the first root w_0 is θ/n.
3. Repeatedly add $2\pi/n$ to get the argument of each successive root.

Parametric Equations (p. 612)

If f and g are functions defined on an interval I, then the set of points $(f(t), g(t))$ is a **plane curve**. The equations

$$x = f(t) \qquad y = g(t)$$

where $t \in I$, are **parametric equations** for the curve, with **parameter** t.

Polar Equations in Parametric Form (p. 616)

The graph of the polar equation $r = f(\theta)$ is the same as the graph of the parametric equations

$$x = f(t)\cos t \qquad y = f(t)\sin t$$

■ CONCEPT CHECK

1. **(a)** Explain the polar coordinate system.
   (b) Graph the points with polar coordinates $(2, \pi/3)$ and $(-1, 3\pi/4)$.
   (c) State the equations that relate the rectangular coordinates of a point to its polar coordinates.
   (d) Find rectangular coordinates for $(2, \pi/3)$.
   (e) Find polar coordinates for $P(-2, 2)$.

2. **(a)** What is a polar equation?

(b) Convert the polar equation $r = \sin\theta$ to an equivalent rectangular equation.

3. **(a)** How do we graph a polar equation?

(b) Sketch a graph of the polar equation $r = 4 + 4\cos\theta$. What is the graph called?

4. **(a)** What is the complex plane? How do we graph a complex number $z = a + bi$ in the complex plane?

(b) What are the modulus and argument of the complex number $z = a + bi$?

(c) Graph the point $z = \sqrt{3} - i$, and find the modulus and argument of z.

5. **(a)** How do we express the complex number z in polar form?

(b) Express $z = \sqrt{3} - i$ in polar form.

6. Let $$z_1 = 2\left(\cos\frac{\pi}{3} + i\sin\frac{\pi}{3}\right)$$

and $$z_2 = 5\left(\cos\frac{\pi}{4} + i\sin\frac{\pi}{4}\right)$$

(a) Find the product $z_1 z_2$.

(b) Find the quotient z_1/z_2.

7. **(a)** State De Moivre's Theorem.

(b) Use De Moivre's Theorem to find the fifth power of $z = 2\left(\cos\frac{\pi}{3} + i\sin\frac{\pi}{3}\right)$.

8. **(a)** State the formula for the nth roots of a complex number $z = r(\cos\theta + i\sin\theta)$.

(b) How do we find the nth roots of a complex number?

(c) Find the three third roots of $z = -8$.

9. **(a)** What are parametric equations?

(b) Sketch a graph of the following parametric equations, using arrows to indicate the direction of the curve.

$$x = t + 1 \qquad y = t^2 \qquad -2 \le t \le 2$$

(c) Eliminate the parameter to obtain an equation in x and y.

ANSWERS TO THE CONCEPT CHECK CAN BE FOUND AT THE BACK OF THE BOOK.

EXERCISES

1–6 ■ Polar Coordinates to Rectangular Coordinates A point $P(r, \theta)$ is given in polar coordinates. **(a)** Plot the point P. **(b)** Find rectangular coordinates for P.

1. $(12, \pi/6)$ **2.** $(8, -3\pi/4)$

3. $(-3, 7\pi/4)$ **4.** $(-\sqrt{3}, 2\pi/3)$

5. $(4\sqrt{3}, -5\pi/3)$ **6.** $(-6\sqrt{2}, -\pi/4)$

7–12 ■ Rectangular Coordinates to Polar Coordinates A point $P(x, y)$ is given in rectangular coordinates. **(a)** Plot the point P. **(b)** Find polar coordinates for P with $r \ge 0$. **(c)** Find polar coordinates for P with $r \le 0$.

7. $(8, 8)$ **8.** $(-\sqrt{2}, \sqrt{6})$

9. $(-6\sqrt{2}, -6\sqrt{2})$ **10.** $(3\sqrt{3}, 3)$

11. $(-3, \sqrt{3})$ **12.** $(4, -4)$

13–16 ■ Rectangular Equations to Polar Equations **(a)** Convert the equation to polar coordinates and simplify. **(b)** Graph the equation. [*Hint:* Use the form of the equation that you find easier to graph.]

13. $x + y = 4$ **14.** $xy = 1$

15. $x^2 + y^2 = 4x + 4y$ **16.** $(x^2 + y^2)^2 = 2xy$

17–24 ■ Polar Equations to Rectangular Equations **(a)** Sketch the graph of the polar equation. **(b)** Express the equation in rectangular coordinates.

17. $r = 3 + 3\cos\theta$ **18.** $r = 3\sin\theta$

19. $r = 2\sin 2\theta$ **20.** $r = 4\cos 3\theta$

21. $r^2 = \sec 2\theta$ **22.** $r^2 = 4\sin 2\theta$

23. $r = \sin\theta + \cos\theta$ **24.** $r = \dfrac{4}{2 + \cos\theta}$

25–28 ■ Graphing Polar Equations Use a graphing device to graph the polar equation. Choose the domain of θ to make sure you produce the entire graph.

25. $r = \cos(\theta/3)$

26. $r = \sin(9\theta/4)$

27. $r = 1 + 4\cos(\theta/3)$

28. $r = \theta\sin\theta, \quad -6\pi \le \theta \le 6\pi$

29–34 ■ Complex Numbers A complex number is given. **(a)** Graph the complex number in the complex plane. **(b)** Find the modulus and argument. **(c)** Write the number in polar form.

29. $4 + 4i$ **30.** $-10i$

31. $5 + 3i$ **32.** $1 + \sqrt{3}i$

33. $-1 + i$ **34.** -20

35–38 ■ Powers Using De Moivre's Theorem Use De Moivre's Theorem to find the indicated power.

35. $(1 - \sqrt{3}i)^4$ **36.** $(1 + i)^8$

37. $(\sqrt{3} + i)^{-4}$ **38.** $\left(\frac{1}{2} + \frac{\sqrt{3}}{2}i\right)^{20}$

39–42 ■ Roots of Complex Numbers Find the indicated roots.

39. The square roots of $-16i$

40. The cube roots of $4 + 4\sqrt{3}i$

41. The sixth roots of 1

42. The eighth roots of i

43–46 ■ Parametric Curves A pair of parametric equations is given. **(a)** Sketch the curve represented by the parametric equations. Use arrows to indicate the direction of the curve as t increases. **(b)** Find a rectangular-coordinate equation for the curve by eliminating the parameter.

43. $x = 1 - t^2, \quad y = 1 + t$

44. $x = t^2 - 1, \quad y = t^2 + 1$

45. $x = 1 + \cos t, \quad y = 1 - \sin t, \quad 0 \le t \le \pi/2$

46. $x = \dfrac{1}{t} + 2, \quad y = \dfrac{2}{t^2}, \quad 0 < t \le 2$

47–48 ■ Graphs of Parametric Equations Use a graphing device to draw the parametric curve.

47. $x = \cos 2t, \quad y = \sin 3t$

48. $x = \sin(t + \cos 2t), \quad y = \cos(t + \sin 3t)$

49. Finding Parametric Equations for a Curve In the figure, the point P is the midpoint of the segment QR and $0 \le \theta < \pi/2$. Using θ as the parameter, find a parametric representation for the curve traced out by P.

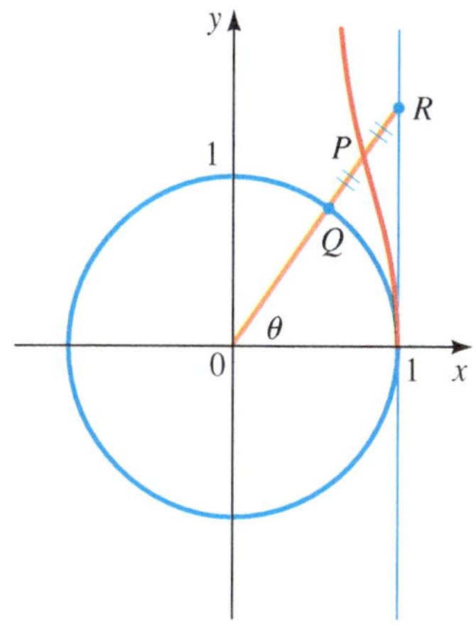

TEST

1. **(a)** Convert the point whose polar coordinates are $(8, 5\pi/4)$ to rectangular coordinates.

   (b) Find two polar coordinate representations for the rectangular coordinate point $(-6, 2\sqrt{3})$, one with $r > 0$ and one with $r < 0$ and both with $0 \le \theta < 2\pi$.

2. **(a)** Graph the polar equation $r = 8 \cos \theta$. What type of curve is this?

   (b) Convert the equation to rectangular coordinates.

3. Graph the polar equation $r = 3 + 6 \sin \theta$. What type of curve is this?

4. Let $z = 1 + \sqrt{3}i$.

   (a) Graph z in the complex plane.

   (b) Write z in polar form.

   (c) Find the complex number z^9.

5. Let $z_1 = 4\left(\cos \dfrac{7\pi}{12} + i \sin \dfrac{7\pi}{12}\right)$ and $z_2 = 2\left(\cos \dfrac{5\pi}{12} + i \sin \dfrac{5\pi}{12}\right)$.

   Find $z_1 z_2$ and $\dfrac{z_1}{z_2}$.

6. Find the cube roots of $27i$, and sketch these roots in the complex plane.

7. **(a)** Sketch the curve represented by the parametric equations below. Use arrows to indicate the direction of the curve as t increases.

   $$x = 3 \sin t + 3 \qquad y = 2 \cos t \qquad 0 \le t \le \pi$$

   (b) Eliminate the parameter t in part (a) to obtain an equation for this curve in rectangular coordinates.

8. Find parametric equations for the line of slope 2 that passes through the point $(3, 5)$.

9. The position of an object in circular motion is modeled by the parametric equations

   $$x = 3 \sin 2t \qquad y = 3 \cos 2t$$

   where t is measured in seconds.

   (a) Describe the path of the object by stating the radius of the circle, the position at time $t = 0$, the orientation of motion (clockwise or counterclockwise), and the time t it takes to complete one revolution around the circle.

   (b) Suppose the speed of the object is doubled. Find new parametric equations that model the motion of the object.

   (c) Find a rectangular-coordinate equation for the same curve by eliminating the parameter.

   (d) Find a polar equation for the same curve.

The Path of a Projectile

Modeling motion is one of the most important ideas in both classical and modern physics. Much of Isaac Newton's work dealt with creating a mathematical model for how objects move and interact—this was the main reason for his invention of calculus. Albert Einstein developed his Special Theory of Relativity in the early 1900s to refine Newton's laws of motion.

In this section we use coordinate geometry to model the motion of a projectile, such as a ball thrown upward into the air, a bullet fired from a gun, or any other sort of missile. A similar model was created by Galileo, but we have the advantage of using our modern mathematical notation to make describing the model much easier than it was for Galileo!

Parametric Equations for the Path of a Projectile

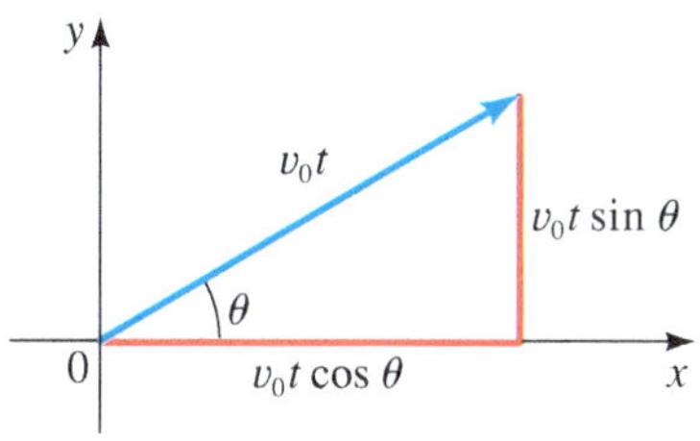

FIGURE 1

Suppose that we fire a projectile into the air from ground level, with an initial speed v_0 and at an angle θ upward from the ground. If there were no gravity (and no air resistance), the projectile would just keep moving indefinitely at the same speed and in the same direction. Since distance = speed × time, the projectile would travel a distance v_0t, so its position at time t would be given by the following parametric equations (assuming that the origin of our coordinate system is placed at the initial location of the projectile; see Figure 1):

$$x = (v_0 \cos\theta)t \qquad y = (v_0 \sin\theta)t \qquad \text{No gravity}$$

But, of course, we know that gravity will pull the projectile back to ground level. By using calculus, it can be shown that the effect of gravity can be accounted for by subtracting $\frac{1}{2}gt^2$ from the vertical position of the projectile. In this expression, g is the gravitational acceleration: $g \approx 32\ \text{ft/s}^2 \approx 9.8\ \text{m/s}^2$. Thus we have the following parametric equations for the path of the projectile:

$$x = (v_0 \cos\theta)t \qquad y = (v_0 \sin\theta)t - \tfrac{1}{2}gt^2 \qquad \text{With gravity}$$

EXAMPLE ■ The Path of a Cannonball

Find parametric equations that model the path of a cannonball fired into the air with an initial speed of 150.0 m/s at a 30° angle of elevation. Sketch the path of the cannonball.

SOLUTION Substituting the given initial speed and angle into the general parametric equations of the path of a projectile, we get

$$x = (150.0 \cos 30^\circ)t \qquad y = (150.0 \sin 30^\circ)t - \tfrac{1}{2}(9.8)t^2 \qquad \text{Substitute } v_0 = 150.0,\ \theta = 30^\circ$$

$$x = 129.9t \qquad y = 75.0t - 4.9t^2 \qquad \text{Simplify}$$

This path is graphed in Figure 2.

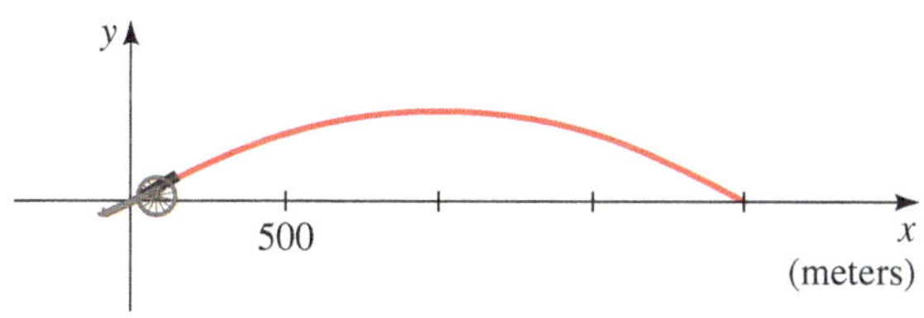

FIGURE 2 Path of a cannonball

■

Range of a Projectile

How can we tell where and when the cannonball of the above example hits the ground? Since ground level corresponds to $y = 0$, we substitute this value for y and solve for t.

$$0 = 75.0t - 4.9t^2 \qquad \text{Set } y = 0$$

$$0 = t(75.0 - 4.9t) \qquad \text{Factor}$$

$$t = 0 \quad \text{or} \quad t = \frac{75.0}{4.9} \approx 15.3 \qquad \text{Solve for } t$$

The first solution, $t = 0$, is the time when the cannon was fired; the second solution means that the cannonball hits the ground after 15.3 s of flight. To see *where* this happens, we substitute this value into the equation for x, the horizontal location of the cannonball.

$$x = 129.9(15.3) \approx 1987.5 \text{ m}$$

The cannonball travels almost 2 km before hitting the ground.

Figure 3 shows the paths of several projectiles, all fired with the same initial speed but at different angles. From the graphs we see that if the firing angle is too high or too low, the projectile doesn't travel very far.

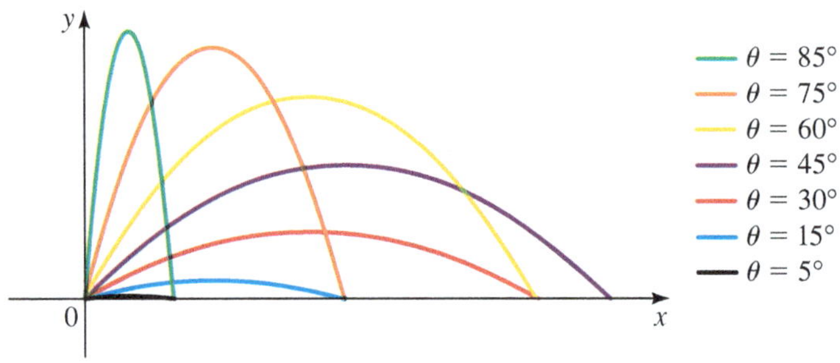

FIGURE 3 Paths of projectiles

Let's try to find the optimal firing angle—the angle that shoots the projectile as far as possible. We'll go through the same steps as we did in the preceding example, but we'll use the general parametric equations instead. First, we solve for the time when the projectile hits the ground by substituting $y = 0$.

$$0 = (v_0 \sin \theta)t - \tfrac{1}{2}gt^2 \qquad \text{Substitute } y = 0$$

$$0 = t(v_0 \sin \theta - \tfrac{1}{2}gt) \qquad \text{Factor}$$

$$0 = v_0 \sin \theta - \tfrac{1}{2}gt \qquad \text{Set second factor equal to 0}$$

$$t = \frac{2v_0 \sin \theta}{g} \qquad \text{Solve for } t$$

World History Archive/Alamy

GALILEO GALILEI (1564–1642) was born in Pisa, Italy. He studied medicine but later abandoned this in favor of science and mathematics. At the age of 25, by dropping cannonballs of various sizes from the Leaning Tower of Pisa, he demonstrated that light objects fall at the same rate as heavier ones. This contradicted the then-accepted view of Aristotle that heavier objects fall more quickly. Galileo also showed that the distance an object falls is proportional to the square of the time it has been falling, and from this he was able to prove that the path of a projectile is a parabola.

Galileo constructed the first telescope and, using it, discovered the moons of Jupiter. His advocacy of the Copernican view that the earth revolves around the sun (rather than being stationary) led to his being called before the Inquisition. By then an old man, he was forced to recant his views, but he is said to have muttered under his breath, "Nevertheless, it does move." Galileo revolutionized science by expressing scientific principles in the language of mathematics. He said, "The great book of nature is written in mathematical symbols."

Now we substitute this into the equation for x to see how far the projectile has traveled horizontally when it hits the ground.

$$\begin{aligned} x &= (v_0 \cos\theta)t && \text{Parametric equation for } x \\ &= (v_0 \cos\theta)\left(\frac{2v_0 \sin\theta}{g}\right) && \text{Substitute } t = (2v_0 \sin\theta)/g \\ &= \frac{2v_0^2 \sin\theta \cos\theta}{g} && \text{Simplify} \\ &= \frac{v_0^2 \sin 2\theta}{g} && \text{Use identity } \sin 2\theta = 2\sin\theta\cos\theta \end{aligned}$$

We want to choose θ so that x is as large as possible. The largest value that the sine of any angle can have is 1, the sine of 90°. Thus we want $2\theta = 90°$, or $\theta = 45°$. So to send the projectile as far as possible, it should be shot up at an angle of 45°. From the last equation in the preceding display, we can see that it will then travel a distance $x = v_0^2/g$.

PROBLEMS

1. **Trajectories Are Parabolas** From the graphs in Figure 3 the paths of projectiles appear to be parabolas that open downward. Eliminate the parameter t from the general parametric equations to verify that these are indeed parabolas.

2. **Path of a Baseball** Suppose a baseball is thrown at 30 ft/s at a 60° angle to the horizontal from a height of 4 ft above the ground.
   - **(a)** Find parametric equations for the path of the baseball, and sketch its graph.
   - **(b)** How far does the baseball travel, and when does it hit the ground?

3. **Path of a Rocket** Suppose that a rocket is fired at an angle of 5° from the vertical with an initial speed of 1000 ft/s.
   - **(a)** Find the length of time the rocket is in the air.
   - **(b)** Find the greatest height it reaches.
   - **(c)** Find the horizontal distance it has traveled when it hits the ground.
   - **(d)** Graph the rocket's path.

4. **Firing a Missile** The initial speed of a missile is 330 m/s.
   - **(a)** At what angle should the missile be fired so that it hits a target 10 km away? (You should find that there are two possible angles.) Graph the missile paths for both angles.
   - **(b)** For which angle is the target hit sooner?

5. **Maximum Height** Show that the maximum height reached by a projectile as a function of its initial speed v_0 and its firing angle θ is

$$y = \frac{v_0^2 \sin^2\theta}{2g}$$

6. **Shooting into the Wind** Suppose that a projectile is fired into a headwind that pushes it back so as to reduce its horizontal speed by a constant amount w. Find parametric equations for the path of the projectile.

7. **Shooting into the Wind** Using the parametric equations you derived in Problem 6, draw graphs of the path of a projectile with initial speed $v_0 = 32$ ft/s, fired into a headwind of $w = 24$ ft/s, for the angles $\theta = 5°, 15°, 30°, 40°, 45°, 55°, 60°$, and $75°$. Is it still true that the greatest range is attained when firing at 45°? Draw some more graphs for different angles, and use these graphs to estimate the optimal firing angle.

8. Simulating the Path of a Projectile The path of a projectile can be simulated on a graphing calculator. On the TI-83, use the "Path" graph style to graph the general parametric equations for the path of a projectile, and watch as the circular cursor moves, simulating the motion of the projectile. Selecting the size of the `Tstep` determines the speed of the "projectile."

(a) Simulate the path of a projectile. Experiment with various values of θ. Use $v_0 = 10$ ft/s and `Tstep` $= 0.02$. Part (a) of the figure below shows one such path.

(b) Simulate the path of two projectiles, fired simultaneously, one at $\theta = 30°$ and the other at $\theta = 60°$. This can be done on the TI-83 using `Simul` mode ("simultaneous" mode). Use $v_0 = 10$ ft/s and `Tstep` $= 0.02$. See part (b) of the figure. Where do the projectiles land? Which lands first?

(c) Simulate the path of a ball thrown straight up ($\theta = 90°$). Experiment with values of v_0 between 5 and 20 ft/s. Use the "Animate" graph style and `Tstep` $= 0.02$. Simulate the path of two balls thrown simultaneously at different speeds. To better distinguish the two balls, place them at different x-coordinates (for example, $x = 1$ and $x = 2$). See part (c) of the figure. How does doubling v_0 change the maximum height the ball reaches?

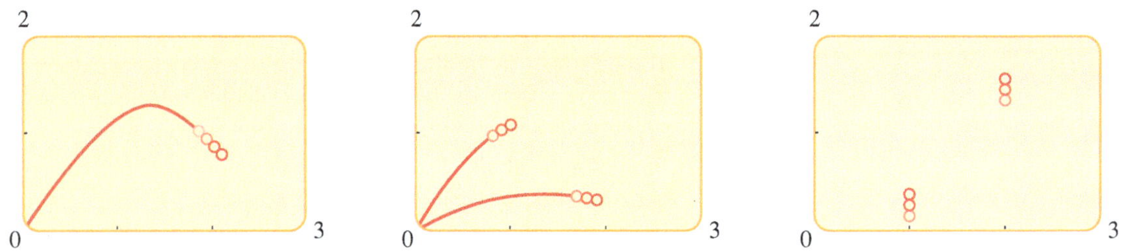

(a) (b) (c)

9 Vectors in Two and Three Dimensions

Many real-world quantities are described mathematically by just one number: their "size" or magnitude. For example, quantities such as mass, volume, distance, and temperature are described by their magnitude. But many other real-world quantities involve both magnitude *and* direction. Such quantities are described mathematically by vectors. For example, if you push a car with a certain force, the direction in which you push on the car is important; you get different results if you push the car forward, backward, or perhaps sideways. So force is a vector. The result of several forces acting on an object can be evaluated by using vectors. For example, we'll see how we can combine the vector forces of wind and water on the sails and hull of a sailboat to find the direction in which the boat will sail. Analyzing these vector forces helps sailors to sail against the wind by tacking. (See *Discovery Project: Sailing Against the Wind* referenced on page 645.)

9.1 VECTORS IN TWO DIMENSIONS

■ Geometric Description of Vectors ■ Vectors in the Coordinate Plane ■ Using Vectors to Model Velocity and Force

In applications of mathematics, certain quantities are determined completely by their magnitude—for example, length, mass, area, temperature, and energy. We speak of a length of 5 m or a mass of 3 kg; only one number is needed to describe each of these quantities. Such a quantity is called a **scalar**.

On the other hand, to describe the displacement of an object, two numbers are required: the *magnitude* and the *direction* of the displacement. To describe the velocity of a moving object, we must specify both the *speed* and the *direction* of travel. Quantities such as displacement, velocity, acceleration, and force that involve magnitude as well as direction are called *directed quantities*. One way to represent such quantities mathematically is through the use of *vectors*.

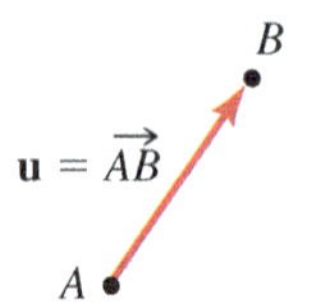

FIGURE 1

■ Geometric Description of Vectors

A **vector** in the plane is a line segment with an assigned direction. We sketch a vector as shown in Figure 1 with an arrow to specify the direction. We denote this vector by $\overrightarrow{AB}$. Point A is the **initial point**, and B is the **terminal point** of the vector $\overrightarrow{AB}$. The length of the line segment AB is called the **magnitude** or **length** of the vector and is denoted by $|\overrightarrow{AB}|$. We use boldface letters to denote vectors. Thus we write $\mathbf{u} = \overrightarrow{AB}$.

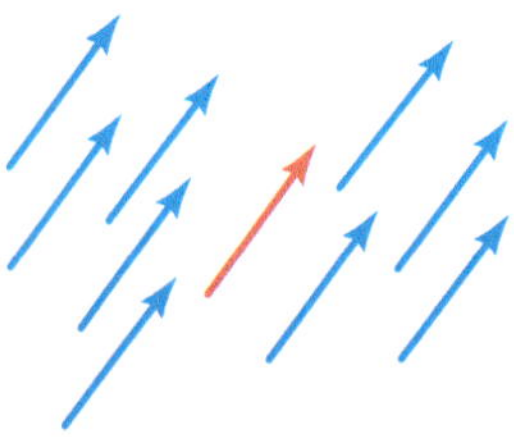

FIGURE 2

Two vectors are considered **equal** if they have equal magnitude and the same direction. Thus all the vectors in Figure 2 are equal. This definition of equality makes sense if we think of a vector as representing a displacement. Two such displacements are the same if they have equal magnitudes and the same direction. So the vectors in Figure 2 can be thought of as the *same* displacement applied to objects in different locations in the plane.

If the displacement $\mathbf{u} = \overrightarrow{AB}$ is followed by the displacement $\mathbf{v} = \overrightarrow{BC}$, then the resulting displacement is $\overrightarrow{AC}$ as shown in Figure 3. In other words, the single displacement represented by the vector $\overrightarrow{AC}$ has the same effect as the other two displacements together. We call the vector $\overrightarrow{AC}$ the **sum** of the vectors $\overrightarrow{AB}$ and $\overrightarrow{BC}$, and we write $\overrightarrow{AC} = \overrightarrow{AB} + \overrightarrow{BC}$. (The **zero vector**, denoted by $\mathbf{0}$, represents no displacement.) Thus to find the sum of any two vectors $\mathbf{u}$ and $\mathbf{v}$, we sketch vectors equal to $\mathbf{u}$ and $\mathbf{v}$ with the initial point of one at the terminal point of the other (see Figure 4(a)). If we draw $\mathbf{u}$ and $\mathbf{v}$ starting at the same point, then $\mathbf{u} + \mathbf{v}$ is the vector that is the diagonal of the parallelogram formed by $\mathbf{u}$ and $\mathbf{v}$ shown in Figure 4(b).

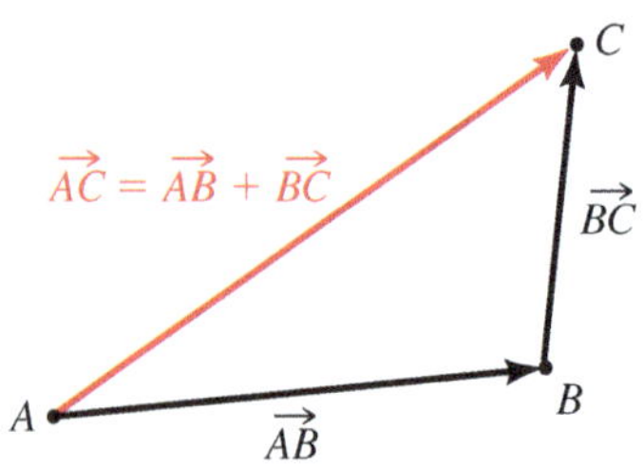

FIGURE 3

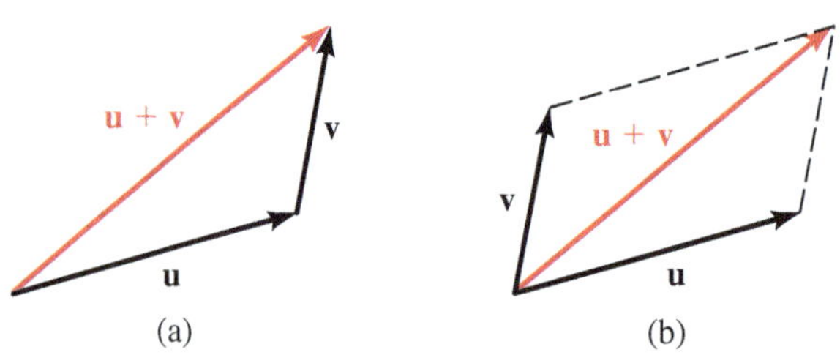

FIGURE 4 Addition of vectors

If c is a real number and $\mathbf{v}$ is a vector, we define a new vector $c\mathbf{v}$ as follows: The vector $c\mathbf{v}$ has magnitude $|c||\mathbf{v}|$ and has the same direction as $\mathbf{v}$ if $c > 0$ and the opposite direction if $c < 0$. If $c = 0$, then $c\mathbf{v} = \mathbf{0}$, the zero vector. This process is called **multiplication of a vector by a scalar**. Multiplying a vector by a scalar has the effect of stretching or shrinking the vector. Figure 5 shows graphs of the vector $c\mathbf{v}$ for different values of c. We write the vector $(-1)\mathbf{v}$ as $-\mathbf{v}$. Thus $-\mathbf{v}$ is the vector with the same length as $\mathbf{v}$ but with the opposite direction.

The **difference** of two vectors **u** and **v** is defined by $\mathbf{u} - \mathbf{v} = \mathbf{u} + (-\mathbf{v})$. Figure 6 shows that the vector $\mathbf{u} - \mathbf{v}$ is the other diagonal of the parallelogram formed by **u** and **v**.

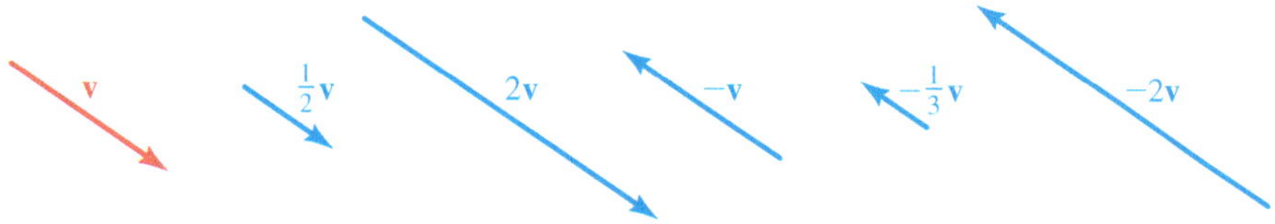

FIGURE 5 Multiplication of a vector by a scalar

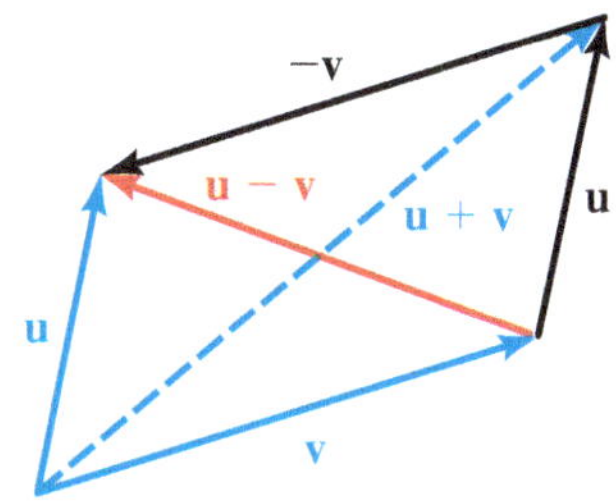

FIGURE 6 Subtraction of vectors

Vectors in the Coordinate Plane

So far, we've discussed vectors geometrically. By placing a vector in a coordinate plane, we can describe it analytically (that is, by using components). In Figure 7(a), to go from the initial point of the vector **v** to the terminal point, we move a_1 units to the right and a_2 units upward. We represent **v** as an ordered pair of real numbers.

Note the distinction between the *vector* $\langle a_1, a_2 \rangle$ and the *point* (a_1, a_2).

$$\mathbf{v} = \langle a_1, a_2 \rangle$$

where a_1 is the **horizontal component** of **v** and a_2 is the **vertical component** of **v**. Remember that a vector represents a magnitude and a direction, not a particular arrow in the plane. Thus the vector $\langle a_1, a_2 \rangle$ has many different representations, depending on its initial point (see Figure 7(b)).

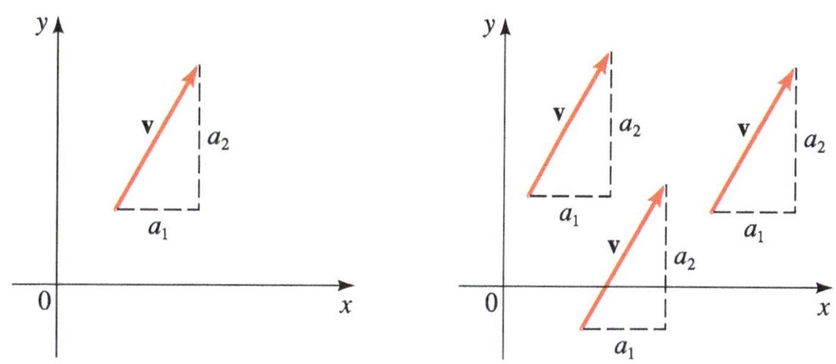

FIGURE 7 (a) (b)

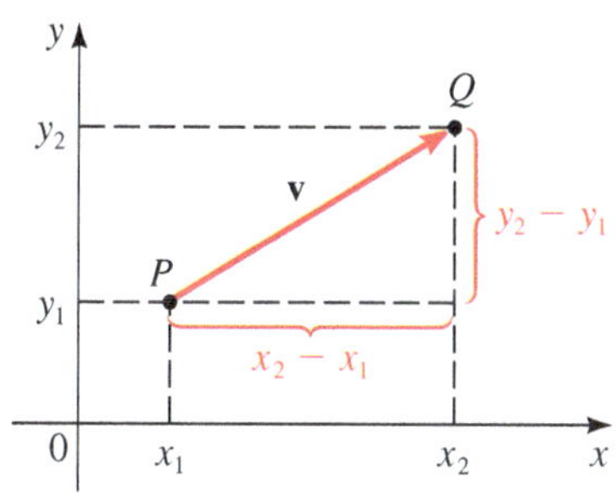

FIGURE 8

Using Figure 8, we can state the relationship between a geometric representation of a vector and the analytic one as follows.

COMPONENT FORM OF A VECTOR

If a vector **v** is represented in the plane with initial point $P(x_1, y_1)$ and terminal point $Q(x_2, y_2)$, then

$$\mathbf{v} = \langle x_2 - x_1, y_2 - y_1 \rangle$$

EXAMPLE 1 ■ Describing Vectors in Component Form

(a) Find the component form of the vector **u** with initial point $(-2, 5)$ and terminal point $(3, 7)$.

(b) If the vector $\mathbf{v} = \langle 3, 7 \rangle$ is sketched with initial point $(2, 4)$, what is its terminal point?

(c) Sketch representations of the vector $\mathbf{w} = \langle 2, 3 \rangle$ with initial points at $(0, 0)$, $(2, 2)$, $(-2, -1)$, and $(1, 4)$.

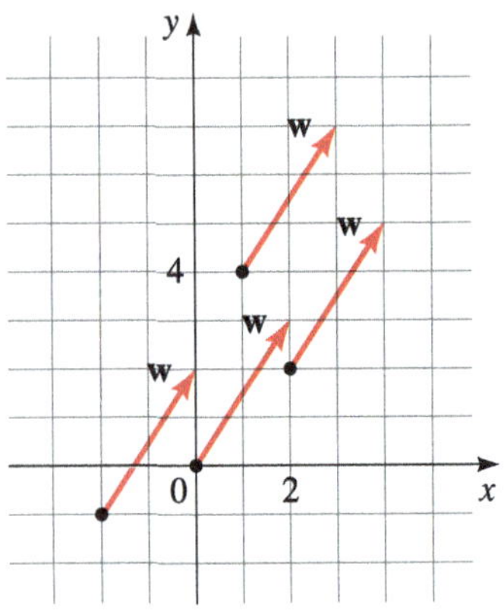

FIGURE 9

SOLUTION

(a) The desired vector is

$$\mathbf{u} = \langle 3 - (-2), 7 - 5 \rangle = \langle 5, 2 \rangle$$

(b) Let the terminal point of $\mathbf{v}$ be (x, y). Then

$$\langle x - 2, y - 4 \rangle = \langle 3, 7 \rangle$$

So $x - 2 = 3$ and $y - 4 = 7$, or $x = 5$ and $y = 11$. The terminal point is $(5, 11)$.

(c) Representations of the vector $\mathbf{w}$ are sketched in Figure 9.

Now Try Exercises 11, 19, and 23 ■

We now give analytic definitions of the various operations on vectors that we have described geometrically. Let's start with equality of vectors. We've said that two vectors are equal if they have equal magnitude and the same direction. For the vectors $\mathbf{u} = \langle a_1, a_2 \rangle$ and $\mathbf{v} = \langle b_1, b_2 \rangle$ this means that $a_1 = b_1$ and $a_2 = b_2$. In other words, two vectors are **equal** if and only if their corresponding components are equal. Thus all the arrows in Figure 7(b) represent the same vector, as do all the arrows in Figure 9.

Applying the Pythagorean Theorem to the triangle in Figure 10, we obtain the following formula for the magnitude of a vector.

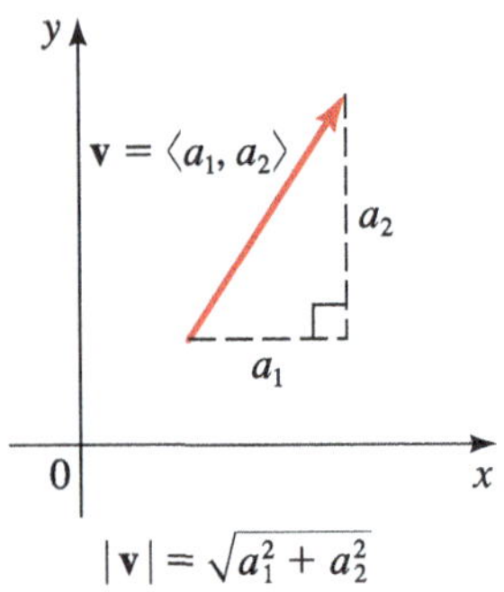

FIGURE 10

MAGNITUDE OF A VECTOR

The **magnitude** or **length** of a vector $\mathbf{v} = \langle a_1, a_2 \rangle$ is

$$|\mathbf{v}| = \sqrt{a_1^2 + a_2^2}$$

EXAMPLE 2 ■ Magnitudes of Vectors

Find the magnitude of each vector.

(a) $\mathbf{u} = \langle 2, -3 \rangle$ **(b)** $\mathbf{v} = \langle 5, 0 \rangle$ **(c)** $\mathbf{w} = \langle \frac{3}{5}, \frac{4}{5} \rangle$

SOLUTION

(a) $|\mathbf{u}| = \sqrt{2^2 + (-3)^2} = \sqrt{13}$

(b) $|\mathbf{v}| = \sqrt{5^2 + 0^2} = \sqrt{25} = 5$

(c) $|\mathbf{w}| = \sqrt{\left(\frac{3}{5}\right)^2 + \left(\frac{4}{5}\right)^2} = \sqrt{\frac{9}{25} + \frac{16}{25}} = 1$

Now Try Exercise 37 ■

The following definitions of addition, subtraction, and scalar multiplication of vectors correspond to the geometric descriptions given earlier. Figure 11 shows how the analytic definition of addition corresponds to the geometric one.

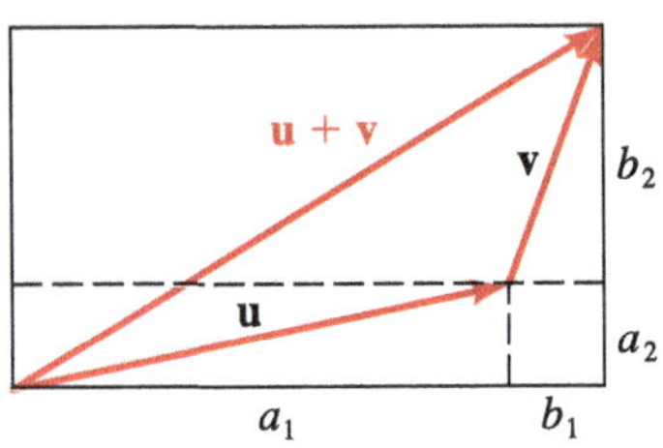

FIGURE 11

ALGEBRAIC OPERATIONS ON VECTORS

If $\mathbf{u} = \langle a_1, a_2 \rangle$ and $\mathbf{v} = \langle b_1, b_2 \rangle$, then

$$\mathbf{u} + \mathbf{v} = \langle a_1 + b_1, a_2 + b_2 \rangle$$

$$\mathbf{u} - \mathbf{v} = \langle a_1 - b_1, a_2 - b_2 \rangle$$

$$c\mathbf{u} = \langle ca_1, ca_2 \rangle \qquad c \in \mathbb{R}$$

EXAMPLE 3 ■ Operations with Vectors

If $\mathbf{u} = \langle 2, -3 \rangle$ and $\mathbf{v} = \langle -1, 2 \rangle$, find $\mathbf{u} + \mathbf{v}$, $\mathbf{u} - \mathbf{v}$, $2\mathbf{u}$, $-3\mathbf{v}$, and $2\mathbf{u} + 3\mathbf{v}$.

SOLUTION By the definitions of the vector operations we have

$$\mathbf{u} + \mathbf{v} = \langle 2, -3 \rangle + \langle -1, 2 \rangle = \langle 1, -1 \rangle$$

$$\mathbf{u} - \mathbf{v} = \langle 2, -3 \rangle - \langle -1, 2 \rangle = \langle 3, -5 \rangle$$

$$2\mathbf{u} = 2\langle 2, -3 \rangle = \langle 4, -6 \rangle$$

$$-3\mathbf{v} = -3\langle -1, 2 \rangle = \langle 3, -6 \rangle$$

$$2\mathbf{u} + 3\mathbf{v} = 2\langle 2, -3 \rangle + 3\langle -1, 2 \rangle = \langle 4, -6 \rangle + \langle -3, 6 \rangle = \langle 1, 0 \rangle$$

Now Try Exercise 31 ■

The following properties for vector operations can be easily proved from the definitions. The **zero vector** is the vector $\mathbf{0} = \langle 0, 0 \rangle$. It plays the same role for addition of vectors as the number 0 does for addition of real numbers.

PROPERTIES OF VECTORS

Vector addition	Multiplication by a scalar
$\mathbf{u} + \mathbf{v} = \mathbf{v} + \mathbf{u}$	$c(\mathbf{u} + \mathbf{v}) = c\mathbf{u} + c\mathbf{v}$
$\mathbf{u} + (\mathbf{v} + \mathbf{w}) = (\mathbf{u} + \mathbf{v}) + \mathbf{w}$	$(c + d)\mathbf{u} = c\mathbf{u} + d\mathbf{u}$
$\mathbf{u} + \mathbf{0} = \mathbf{u}$	$(cd)\mathbf{u} = c(d\mathbf{u}) = d(c\mathbf{u})$
$\mathbf{u} + (-\mathbf{u}) = \mathbf{0}$	$1\mathbf{u} = \mathbf{u}$
Length of a vector	$0\mathbf{u} = \mathbf{0}$
$\lvert c\mathbf{u} \rvert = \lvert c \rvert \lvert \mathbf{u} \rvert$	$c\mathbf{0} = \mathbf{0}$

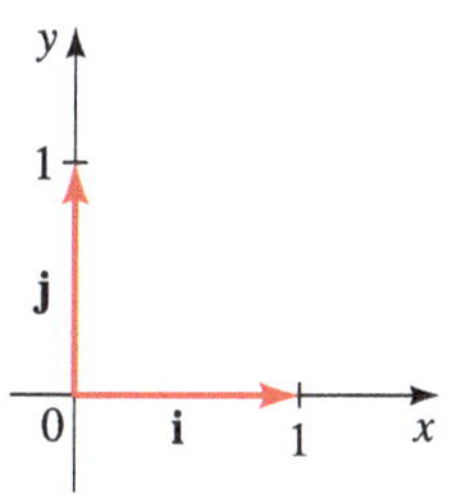

FIGURE 12

A vector of length 1 is called a **unit vector**. For instance, in Example 2(c) the vector $\mathbf{w} = \left\langle \frac{3}{5}, \frac{4}{5} \right\rangle$ is a unit vector. Two useful unit vectors are **i** and **j**, defined by

$$\mathbf{i} = \langle 1, 0 \rangle \qquad \mathbf{j} = \langle 0, 1 \rangle$$

(See Figure 12.) These vectors are special because any vector can be expressed in terms of them. (See Figure 13.)

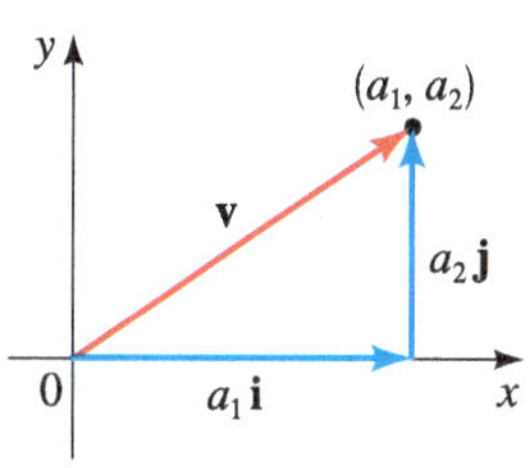

FIGURE 13

VECTORS IN TERMS OF i AND j

The vector $\mathbf{v} = \langle a_1, a_2 \rangle$ can be expressed in terms of **i** and **j** by

$$\mathbf{v} = \langle a_1, a_2 \rangle = a_1\mathbf{i} + a_2\mathbf{j}$$

EXAMPLE 4 ■ Vectors in Terms of i and j

(a) Write the vector $\mathbf{u} = \langle 5, -8 \rangle$ in terms of **i** and **j**.

(b) If $\mathbf{u} = 3\mathbf{i} + 2\mathbf{j}$ and $\mathbf{v} = -\mathbf{i} + 6\mathbf{j}$, write $2\mathbf{u} + 5\mathbf{v}$ in terms of **i** and **j**.

SOLUTION

(a) $\mathbf{u} = 5\mathbf{i} + (-8)\mathbf{j} = 5\mathbf{i} - 8\mathbf{j}$

(b) The properties of addition and scalar multiplication of vectors show that we can manipulate vectors in the same way as algebraic expressions. Thus

$$\begin{aligned} 2\mathbf{u} + 5\mathbf{v} &= 2(3\mathbf{i} + 2\mathbf{j}) + 5(-\mathbf{i} + 6\mathbf{j}) \\ &= (6\mathbf{i} + 4\mathbf{j}) + (-5\mathbf{i} + 30\mathbf{j}) \\ &= \mathbf{i} + 34\mathbf{j} \end{aligned}$$

Now Try Exercises 27 and 35 ■

Let $\mathbf{v}$ be a vector in the plane with its initial point at the origin. The **direction** of $\mathbf{v}$ is θ, the smallest positive angle in standard position formed by the positive x-axis and $\mathbf{v}$ (see Figure 14). If we know the magnitude and direction of a vector, then Figure 14 shows that we can find the horizontal and vertical components of the vector.

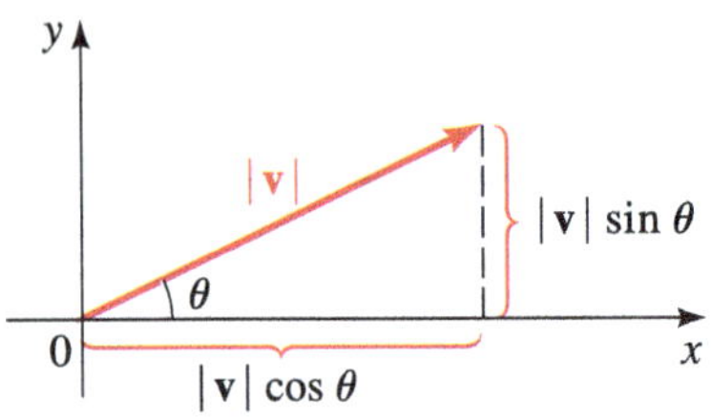

FIGURE 14

HORIZONTAL AND VERTICAL COMPONENTS OF A VECTOR

Let $\mathbf{v}$ be a vector with magnitude $|\mathbf{v}|$ and direction θ.
Then $\mathbf{v} = \langle a_1, a_2 \rangle = a_1\mathbf{i} + a_2\mathbf{j}$, where

$$a_1 = |\mathbf{v}|\cos\theta \qquad \text{and} \qquad a_2 = |\mathbf{v}|\sin\theta$$

Thus we can express $\mathbf{v}$ as

$$\mathbf{v} = |\mathbf{v}|\cos\theta\,\mathbf{i} + |\mathbf{v}|\sin\theta\,\mathbf{j}$$

EXAMPLE 5 ■ Components and Direction of a Vector

(a) A vector $\mathbf{v}$ has length 8 and direction $\pi/3$. Find the horizontal and vertical components, and write $\mathbf{v}$ in terms of $\mathbf{i}$ and $\mathbf{j}$.

(b) Find the direction of the vector $\mathbf{u} = -\sqrt{3}\mathbf{i} + \mathbf{j}$.

SOLUTION

(a) We have $\mathbf{v} = \langle a, b \rangle$, where the components are given by

$$a = 8\cos\frac{\pi}{3} = 4 \qquad \text{and} \qquad b = 8\sin\frac{\pi}{3} = 4\sqrt{3}$$

Thus $\mathbf{v} = \langle 4, 4\sqrt{3} \rangle = 4\mathbf{i} + 4\sqrt{3}\mathbf{j}$.

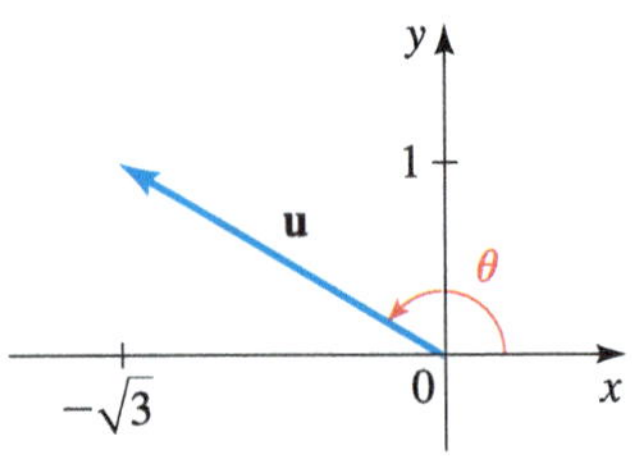

FIGURE 15

(b) From Figure 15 we see that the direction θ has the property that

$$\tan\theta = \frac{1}{-\sqrt{3}} = -\frac{\sqrt{3}}{3}$$

Thus the reference angle for θ is $\pi/6$. Since the terminal point of the vector $\mathbf{u}$ is in Quadrant II, it follows that $\theta = 5\pi/6$.

Now Try Exercises 41 and 51 ■

Using Vectors to Model Velocity and Force

The use of bearings (such as N 30° E) to describe directions is explained on page 518 in Section 6.6.

The **velocity** of a moving object is modeled by a vector whose direction is the direction of motion and whose magnitude is the speed. Figure 16 on the next page shows some vectors $\mathbf{u}$, representing the velocity of wind flowing in the direction N 30° E, and a vector $\mathbf{v}$, representing the velocity of an airplane flying through this wind at the point P. It's obvious from our experience that wind affects both the speed and the direction of an airplane. Figure 17 indicates that the true velocity of the plane (relative to the ground) is given by the vector $\mathbf{w} = \mathbf{u} + \mathbf{v}$.

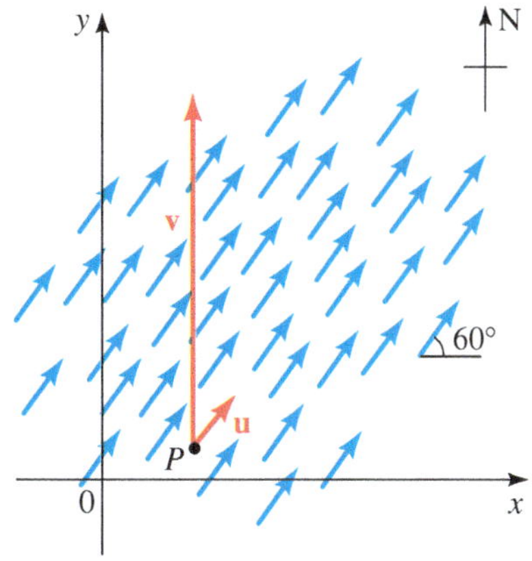

FIGURE 16

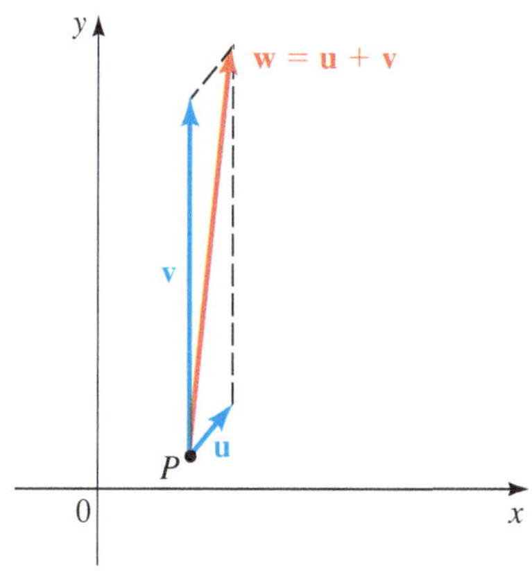

FIGURE 17

EXAMPLE 6 ■ The True Speed and Direction of an Airplane

An airplane heads due north at 300 mi/h. It experiences a 40 mi/h crosswind flowing in the direction N 30° E, as shown in Figure 16.

(a) Express the velocity $\mathbf{v}$ of the airplane relative to the air and the velocity $\mathbf{u}$ of the wind, in component form.

(b) Find the true velocity of the airplane as a vector.

(c) Find the true speed and direction of the airplane.

SOLUTION

(a) The velocity of the airplane relative to the air is $\mathbf{v} = 0\mathbf{i} + 300\mathbf{j} = 300\mathbf{j}$. By the formulas for the components of a vector we find that the velocity of the wind is

$$\begin{aligned} \mathbf{u} &= (40 \cos 60°)\mathbf{i} + (40 \sin 60°)\mathbf{j} \\ &= 20\mathbf{i} + 20\sqrt{3}\mathbf{j} \\ &\approx 20\mathbf{i} + 34.64\mathbf{j} \end{aligned}$$

The **true velocity** is the velocity relative to the ground.

(b) The true velocity of the airplane is given by the vector $\mathbf{w} = \mathbf{u} + \mathbf{v}$:

$$\begin{aligned} \mathbf{w} = \mathbf{u} + \mathbf{v} &= (20\mathbf{i} + 20\sqrt{3}\mathbf{j}) + (300\mathbf{j}) \\ &= 20\mathbf{i} + (20\sqrt{3} + 300)\mathbf{j} \\ &\approx 20\mathbf{i} + 334.64\mathbf{j} \end{aligned}$$

(c) The true speed of the airplane is given by the magnitude of $\mathbf{w}$:

$$|\mathbf{w}| \approx \sqrt{(20)^2 + (334.64)^2} \approx 335.2 \text{ mi/h}$$

The direction of the airplane is the direction θ of the vector $\mathbf{w}$. The angle θ has the property that $\tan \theta \approx 334.64/20 = 16.732$, so $\theta \approx 86.6°$. Thus the airplane is heading in the direction N 3.4° E.

Now Try Exercise 59 ■

EXAMPLE 7 ■ Calculating a Heading

A woman launches a boat from one shore of a straight river and wants to land at the point directly on the opposite shore. If the speed of the boat (relative to the water) is 10 mi/h and the river is flowing east at the rate of 5 mi/h, in what direction should she head the boat in order to arrive at the desired landing point?

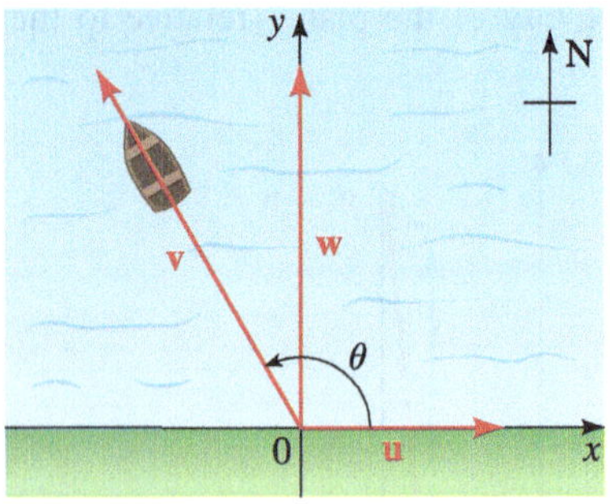

FIGURE 18

SOLUTION We choose a coordinate system with the origin at the initial position of the boat as shown in Figure 18. Let **u** and **v** represent the velocities of the river and the boat, respectively. Clearly, $\mathbf{u} = 5\mathbf{i}$, and since the speed of the boat is 10 mi/h, we have $|\mathbf{v}| = 10$, so

$$\mathbf{v} = (10 \cos \theta)\mathbf{i} + (10 \sin \theta)\mathbf{j}$$

where the angle θ is as shown in Figure 18. The true course of the boat is given by the vector $\mathbf{w} = \mathbf{u} + \mathbf{v}$. We have

$$\begin{aligned}\mathbf{w} = \mathbf{u} + \mathbf{v} &= 5\mathbf{i} + (10 \cos \theta)\mathbf{i} + (10 \sin \theta)\mathbf{j} \\ &= (5 + 10 \cos \theta)\mathbf{i} + (10 \sin \theta)\mathbf{j}\end{aligned}$$

Since the woman wants to land at a point directly across the river, her direction should have horizontal component 0. In other words, she should choose θ in such a way that

$$\begin{aligned}5 + 10 \cos \theta &= 0 \\ \cos \theta &= -\tfrac{1}{2} \\ \theta &= 120°\end{aligned}$$

Thus she should head the boat in the direction $\theta = 120°$ (or N 30° W).

Now Try Exercise 57 ■

Force is also represented by a vector. Intuitively, we can think of force as describing a push or a pull on an object, for example, a horizontal push of a book across a table or the downward pull of the earth's gravity on a ball. Force is measured in pounds (or in newtons, in the metric system). For instance, a man weighing 200 lb exerts a force of 200 lb downward on the ground. If several forces are acting on an object, the **resultant force** experienced by the object is the vector sum of these forces.

EXAMPLE 8 ■ Resultant Force

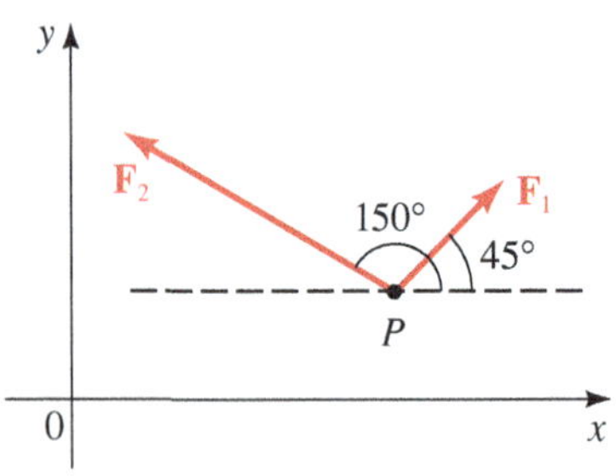

FIGURE 19

Two forces $\mathbf{F}_1$ and $\mathbf{F}_2$ with magnitudes 10 and 20 lb, respectively, act on an object at a point P as shown in Figure 19. Find the resultant force acting at P.

SOLUTION We write $\mathbf{F}_1$ and $\mathbf{F}_2$ in component form:

$$\begin{aligned}\mathbf{F}_1 &= (10 \cos 45°)\mathbf{i} + (10 \sin 45°)\mathbf{j} = 10\frac{\sqrt{2}}{2}\mathbf{i} + 10\frac{\sqrt{2}}{2}\mathbf{j} \\ &= 5\sqrt{2}\mathbf{i} + 5\sqrt{2}\mathbf{j} \\ \mathbf{F}_2 &= (20 \cos 150°)\mathbf{i} + (20 \sin 150°)\mathbf{j} = -20\frac{\sqrt{3}}{2}\mathbf{i} + 20\left(\frac{1}{2}\right)\mathbf{j} \\ &= -10\sqrt{3}\mathbf{i} + 10\mathbf{j}\end{aligned}$$

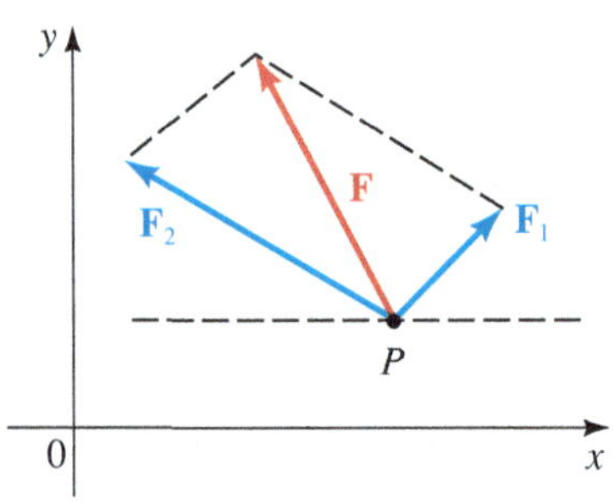

FIGURE 20

So the resultant force **F** is

$$\begin{aligned}\mathbf{F} &= \mathbf{F}_1 + \mathbf{F}_2 \\ &= (5\sqrt{2}\mathbf{i} + 5\sqrt{2}\mathbf{j}) + (-10\sqrt{3}\mathbf{i} + 10\mathbf{j}) \\ &= (5\sqrt{2} - 10\sqrt{3})\mathbf{i} + (5\sqrt{2} + 10)\mathbf{j} \\ &\approx -10\mathbf{i} + 17\mathbf{j}\end{aligned}$$

The resultant force **F** is shown in Figure 20.

Now Try Exercise 67 ■

9.1 EXERCISES

CONCEPTS

1. (a) A vector in the plane is a line segment with an assigned direction. In Figure I below, the vector **u** has initial point ________ and terminal point ________. Sketch the vectors 2**u** and **u** + **v**.

(b) A vector in a coordinate plane is expressed by using components. In Figure II below, the vector **u** has initial point (___ , ___) and terminal point (___ , ___). In component form we write **u** = ⟨ ___ , ___ ⟩, and **v** = ⟨ ___ , ___ ⟩. Then 2**u** = ⟨ ___ , ___ ⟩ and **u** + **v** = ⟨ ___ , ___ ⟩.

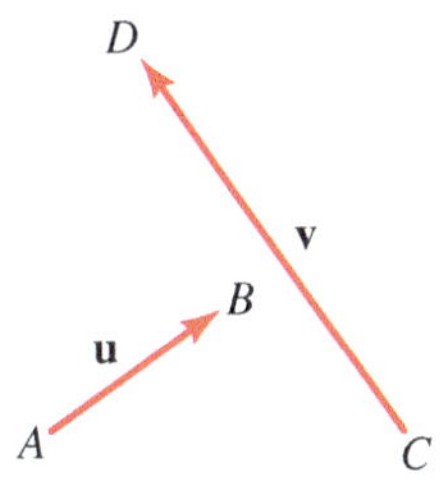

I

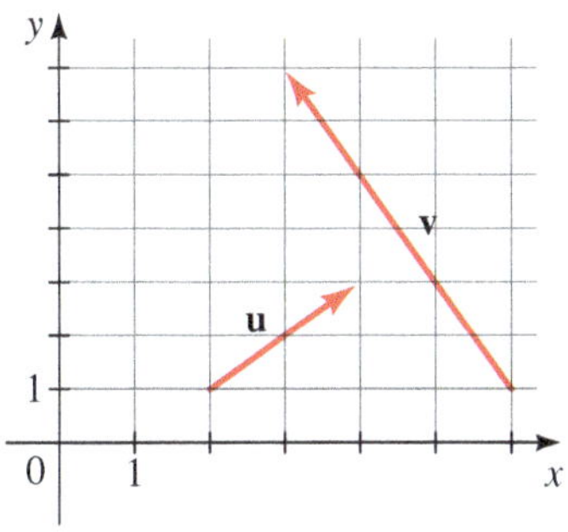

II

2. (a) The length of a vector $\mathbf{w} = \langle a_1, a_2 \rangle$ is $|\mathbf{w}|$ = ________, so the length of the vector **u** in Figure II is $|\mathbf{u}|$ = ________.

(b) If we know the length $|\mathbf{w}|$ and direction θ of a vector **w**, then we can express the vector in component form as **w** = ⟨ ________ , ________ ⟩.

SKILLS

3–8 ■ Sketching Vectors Sketch the vector indicated. (The vectors **u** and **v** are shown in the figure.)

3. 2**u**

4. −**v**

5. **u** + **v**

6. **u** − **v**

7. **v** − 2**u**

8. 2**u** + **v**

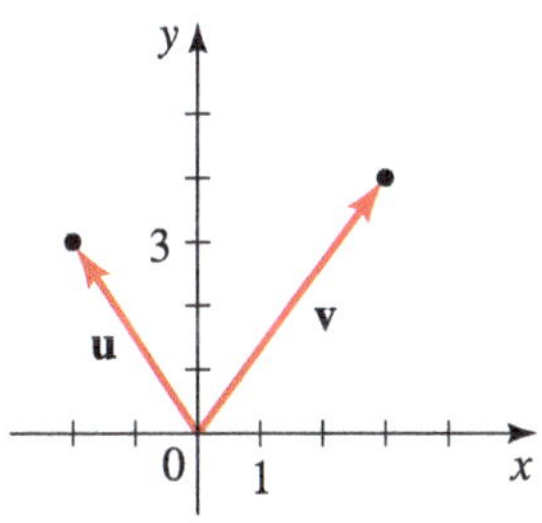

9–18 ■ Component Form of Vectors Express the vector with initial point P and terminal point Q in component form.

9.

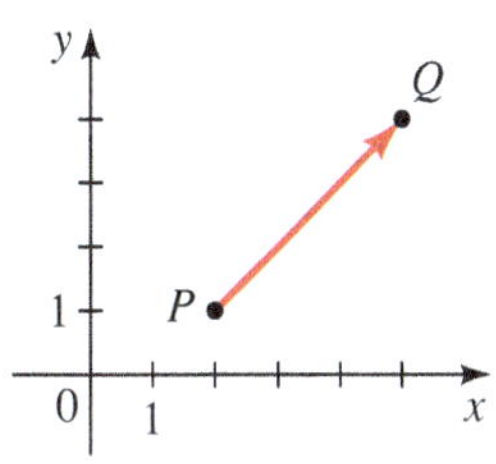

10.

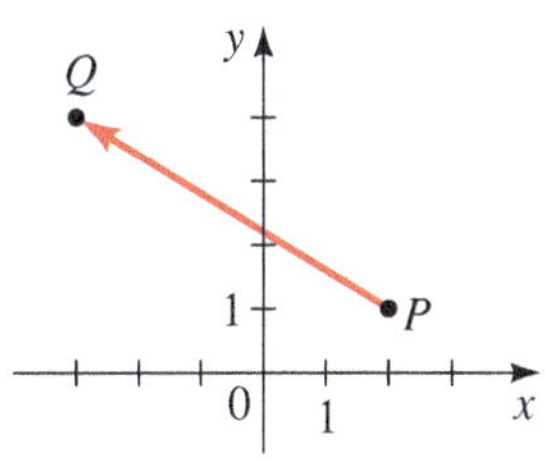

11.

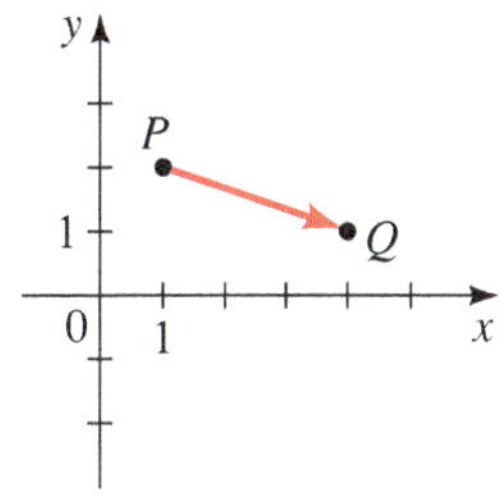

12.

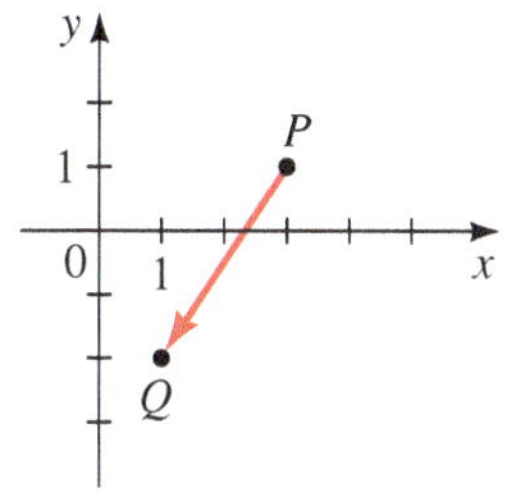

13. $P(3, 2)$, $Q(8, 9)$ **14.** $P(1, 1)$, $Q(9, 9)$

15. $P(5, 3)$, $Q(1, 0)$ **16.** $P(-1, 3)$, $Q(-6, -1)$

17. $P(-1, -1)$, $Q(-1, 1)$

18. $P(-8, -6)$, $Q(-1, -1)$

19–22 ■ Sketching Vectors Sketch the given vector with initial point (4, 3), and find the terminal point.

19. $\mathbf{u} = \langle 2, 4 \rangle$ **20.** $\mathbf{u} = \langle -1, 2 \rangle$

21. $\mathbf{u} = \langle 4, -3 \rangle$ **22.** $\mathbf{u} = \langle -8, -1 \rangle$

23–26 ■ Sketching Vectors Sketch representations of the given vector with initial points at $(0, 0)$, $(2, 3)$, and $(-3, 5)$.

23. $\mathbf{u} = \langle 3, 5 \rangle$ **24.** $\mathbf{u} = \langle 4, -6 \rangle$

25. $\mathbf{u} = \langle -7, 2 \rangle$ **26.** $\mathbf{u} = \langle 0, -9 \rangle$

27–30 ■ Writing Vectors in Terms of i and j Write the given vector in terms of **i** and **j**.

27. $\mathbf{u} = \langle 1, 4 \rangle$ **28.** $\mathbf{u} = \langle -2, 10 \rangle$

29. $\mathbf{u} = \langle 3, 0 \rangle$ **30.** $\mathbf{u} = \langle 0, -5 \rangle$

31–36 ■ Operations with Vectors Find 2**u**, −3**v**, **u** + **v**, and 3**u** − 4**v** for the given vectors **u** and **v**.

31. $\mathbf{u} = \langle 2, 7 \rangle$, $\mathbf{v} = \langle 3, 1 \rangle$ **32.** $\mathbf{u} = \langle -2, 5 \rangle$, $\mathbf{v} = \langle 2, -8 \rangle$

33. $\mathbf{u} = \langle 0, -1 \rangle$, $\mathbf{v} = \langle -2, 0 \rangle$

34. $\mathbf{u} = \mathbf{i}$, $\mathbf{v} = -2\mathbf{j}$

35. $\mathbf{u} = 2\mathbf{i}$, $\mathbf{v} = 3\mathbf{i} - 2\mathbf{j}$ **36.** $\mathbf{u} = \mathbf{i} + \mathbf{j}$, $\mathbf{v} = \mathbf{i} - \mathbf{j}$

37–40 ■ Magnitude of Vectors Find $|\mathbf{u}|$, $|\mathbf{v}|$, $|2\mathbf{u}|$, $|\frac{1}{2}\mathbf{v}|$, $|\mathbf{u} + \mathbf{v}|$, $|\mathbf{u} - \mathbf{v}|$, and $|\mathbf{u}| - |\mathbf{v}|$.

37. $\mathbf{u} = 2\mathbf{i} + \mathbf{j}$, $\mathbf{v} = 3\mathbf{i} - 2\mathbf{j}$

38. $\mathbf{u} = -2\mathbf{i} + 3\mathbf{j}$, $\mathbf{v} = \mathbf{i} - 2\mathbf{j}$

39. $\mathbf{u} = \langle 10, -1 \rangle$, $\mathbf{v} = \langle -2, -2 \rangle$

40. $\mathbf{u} = \langle -6, 6 \rangle$, $\mathbf{v} = \langle -2, -1 \rangle$

41–46 ■ Components of a Vector Find the horizontal and vertical components of the vector with given length and direction, and write the vector in terms of the vectors **i** and **j**.

41. $|\mathbf{v}| = 40$, $\theta = 30°$ **42.** $|\mathbf{v}| = 50$, $\theta = 120°$

43. $|\mathbf{v}| = 1$, $\theta = 225°$ **44.** $|\mathbf{v}| = 800$, $\theta = 125°$

45. $|\mathbf{v}| = 4$, $\theta = 10°$ **46.** $|\mathbf{v}| = \sqrt{3}$, $\theta = 300°$

47–52 ■ Magnitude and Direction of a Vector Find the magnitude and direction (in degrees) of the vector.

47. $\mathbf{v} = \langle 3, 4 \rangle$

48. $\mathbf{v} = \left\langle -\frac{\sqrt{2}}{2}, -\frac{\sqrt{2}}{2} \right\rangle$

49. $\mathbf{v} = \langle -12, 5 \rangle$

50. $\mathbf{v} = \langle 40, 9 \rangle$

51. $\mathbf{v} = \mathbf{i} + \sqrt{3}\mathbf{j}$

52. $\mathbf{v} = \mathbf{i} + \mathbf{j}$

APPLICATIONS

53. Components of a Force A man pushes a lawn mower with a force of 30 lb exerted at an angle of 30° to the ground. Find the horizontal and vertical components of the force.

54. Components of a Velocity A jet is flying in a direction N 20° E with a speed of 500 mi/h. Find the north and east components of the velocity.

55. Velocity A river flows due south at 3 mi/h. A swimmer attempting to cross the river heads due east swimming at 2 mi/h relative to the water. Find the true velocity of the swimmer as a vector.

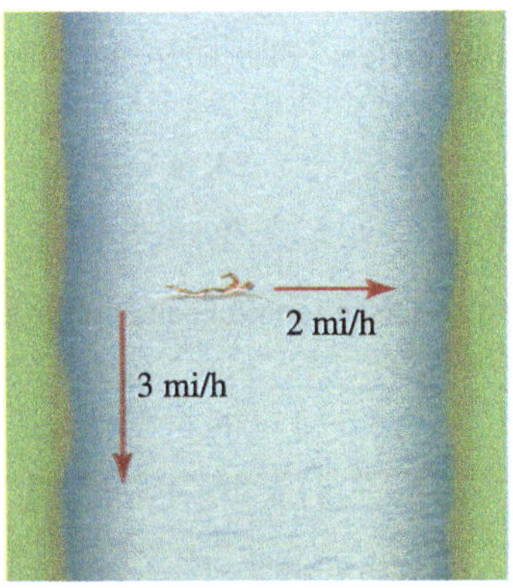

56. Velocity Suppose that in Exercise 55 the current is flowing at 1.2 mi/h due south. In what direction should the swimmer head in order to arrive at a landing point due east of his starting point?

57. Velocity The speed of an airplane is 300 mi/h relative to the air. The wind is blowing due north with a speed of 30 mi/h. In what direction should the airplane head in order to arrive at a point due west of its location?

58. Velocity A migrating salmon heads in the direction N 45° E, swimming at 5 mi/h relative to the water. The prevailing ocean currents flow due east at 3 mi/h. Find the true velocity of the fish as a vector.

59. True Velocity of a Jet A pilot heads his jet due east. The jet has a speed of 425 mi/h relative to the air. The wind is blowing due north with a speed of 40 mi/h.

(a) Express the velocity of the wind as a vector in component form.

(b) Express the velocity of the jet relative to the air as a vector in component form.

(c) Find the true velocity of the jet as a vector.

(d) Find the true speed and direction of the jet.

60. True Velocity of a Jet A jet is flying through a wind that is blowing with a speed of 55 mi/h in the direction N 30° E (see the figure). The jet has a speed of 765 mi/h relative to the air, and the pilot heads the jet in the direction N 45° E.

(a) Express the velocity of the wind as a vector in component form.

(b) Express the velocity of the jet relative to the air as a vector in component form.

(c) Find the true velocity of the jet as a vector.

(d) Find the true speed and direction of the jet.

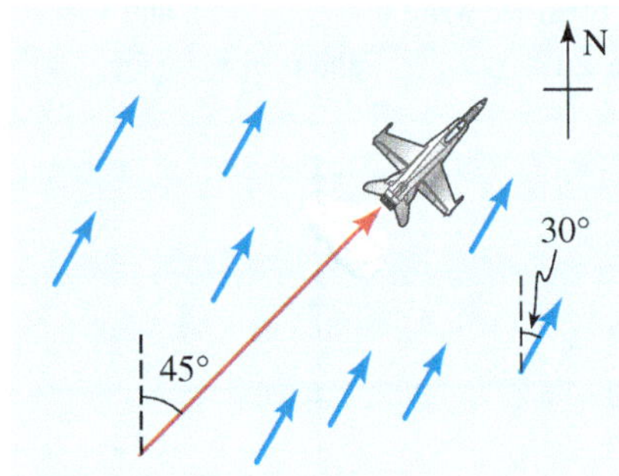

61. True Velocity of a Jet Find the true speed and direction of the jet in Exercise 60 if the pilot heads the plane in the direction N 30° W.

62. True Velocity of a Jet In what direction should the pilot in Exercise 60 head the plane for the true course to be due north?

63. Velocity of a Boat A straight river flows east at a speed of 10 mi/h. A boater starts at the south shore of the river and heads in a direction 60° from the shore (see the figure). The motorboat has a speed of 20 mi/h relative to the water.

(a) Express the velocity of the river as a vector in component form.

(b) Express the velocity of the motorboat relative to the water as a vector in component form.

(c) Find the true velocity of the motorboat.

(d) Find the true speed and direction of the motorboat.

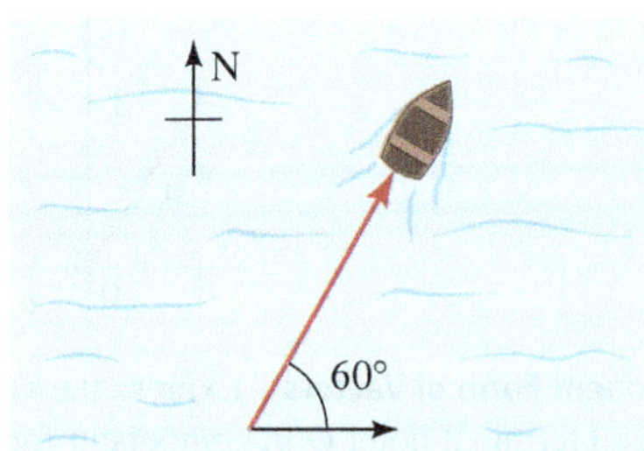

64. Velocity of a Boat The boater in Exercise 63 wants to arrive at a point on the north shore of the river directly opposite the starting point. In what direction should the boat be headed?

65. Velocity of a Boat A boat heads in the direction N 72° E. The speed of the boat relative to the water is 24 mi/h. The water is flowing directly south. It is observed that the true direction of the boat is directly east.

(a) Express the velocity of the boat relative to the water as a vector in component form.

(b) Find the speed of the water and the true speed of the boat.

66. Velocity A woman walks due west on the deck of an ocean liner at 2 mi/h. The ocean liner is moving due north at a speed of 25 mi/h. Find the speed and direction of the woman relative to the surface of the water.

67–72 ■ Equilibrium of Forces The forces $\mathbf{F}_1, \mathbf{F}_2, \ldots, \mathbf{F}_n$ acting at the same point P are said to be in equilibrium if the resultant force is zero, that is, if $\mathbf{F}_1 + \mathbf{F}_2 + \cdots + \mathbf{F}_n = \mathbf{0}$. Find **(a)** the resultant forces acting at P, and **(b)** the additional force required (if any) for the forces to be in equilibrium.

67. $\mathbf{F}_1 = \langle 2, 5 \rangle$, $\mathbf{F}_2 = \langle 3, -8 \rangle$

68. $\mathbf{F}_1 = \langle 3, -7 \rangle$, $\mathbf{F}_2 = \langle 4, -2 \rangle$, $\mathbf{F}_3 = \langle -7, 9 \rangle$

69. $\mathbf{F}_1 = 4\mathbf{i} - \mathbf{j}$, $\mathbf{F}_2 = 3\mathbf{i} - 7\mathbf{j}$, $\mathbf{F}_3 = -8\mathbf{i} + 3\mathbf{j}$, $\mathbf{F}_4 = \mathbf{i} + \mathbf{j}$

70. $\mathbf{F}_1 = \mathbf{i} - \mathbf{j}$, $\mathbf{F}_2 = \mathbf{i} + \mathbf{j}$, $\mathbf{F}_3 = -2\mathbf{i} + \mathbf{j}$

71.

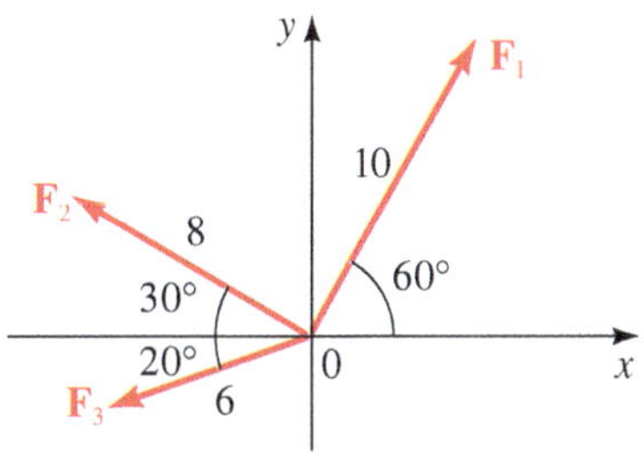

72.

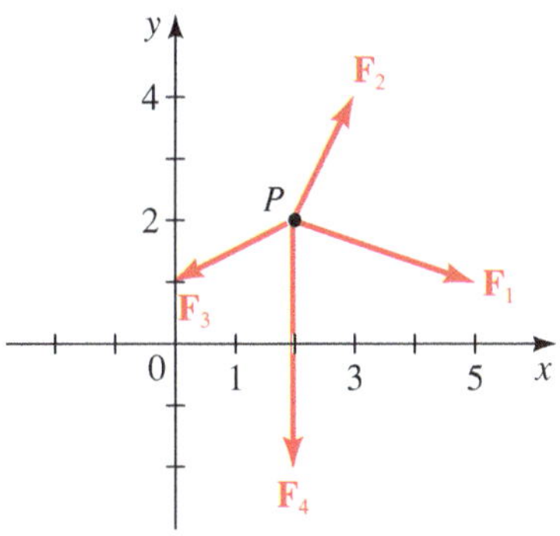

73. Equilibrium of Tensions A 100-lb weight hangs from a string as shown in the figure. Find the tensions $\mathbf{T}_1$ and $\mathbf{T}_2$ in the string.

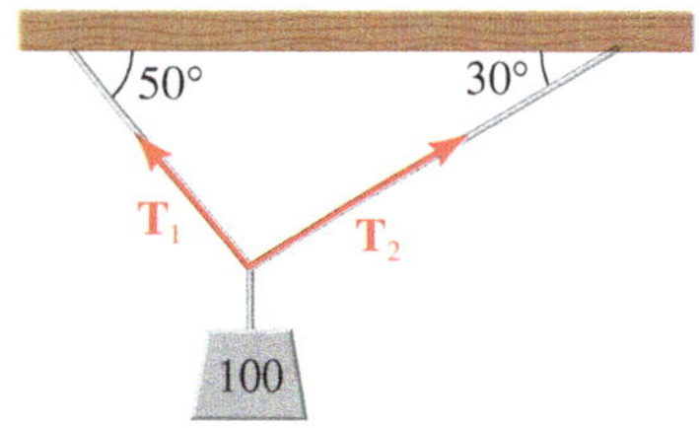

74. Equilibrium of Tensions The cranes in the figure are lifting an object that weighs 18,278 lb. Find the tensions $\mathbf{T}_1$ and $\mathbf{T}_2$.

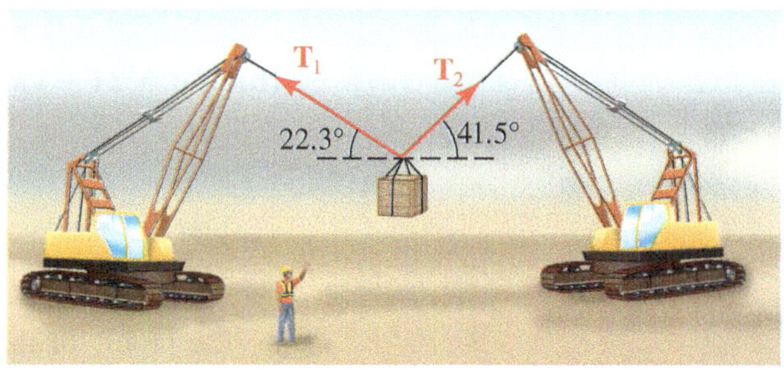

DISCUSS ■ DISCOVER ■ PROVE ■ WRITE

75. DISCUSS: Vectors That Form a Polygon Suppose that n vectors can be placed head to tail in the plane so that they form a polygon. (The figure shows the case of a hexagon.) Explain why the sum of these vectors is $\mathbf{0}$.

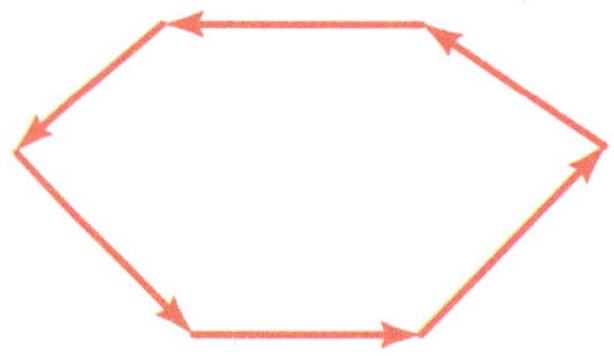

9.2 THE DOT PRODUCT

■ The Dot Product of Vectors ■ The Component of u along v ■ The Projection of u onto v ■ Work

In this section we define an operation on vectors called the dot product. This concept is especially useful in calculus and in applications of vectors to physics and engineering.

■ The Dot Product of Vectors

We begin by defining the dot product of two vectors.

DEFINITION OF THE DOT PRODUCT

If $\mathbf{u} = \langle a_1, a_2 \rangle$ and $\mathbf{v} = \langle b_1, b_2 \rangle$ are vectors, then their **dot product**, denoted by $\mathbf{u} \cdot \mathbf{v}$, is defined by

$$\mathbf{u} \cdot \mathbf{v} = a_1 b_1 + a_2 b_2$$

Thus to find the dot product of **u** and **v**, we multiply corresponding components and add. The dot product is *not* a vector; it is a real number, or scalar.

EXAMPLE 1 ■ Calculating Dot Products

(a) If $\mathbf{u} = \langle 3, -2 \rangle$ and $\mathbf{v} = \langle 4, 5 \rangle$ then

$$\mathbf{u} \cdot \mathbf{v} = (3)(4) + (-2)(5) = 2$$

(b) If $\mathbf{u} = 2\mathbf{i} + \mathbf{j}$ and $\mathbf{v} = 5\mathbf{i} - 6\mathbf{j}$, then

$$\mathbf{u} \cdot \mathbf{v} = (2)(5) + (1)(-6) = 4$$

Now Try Exercises 5(a) and 11(a) ■

The proofs of the following properties of the dot product follow easily from the definition.

PROPERTIES OF THE DOT PRODUCT

1. $\mathbf{u} \cdot \mathbf{v} = \mathbf{v} \cdot \mathbf{u}$
2. $(c\mathbf{u}) \cdot \mathbf{v} = c(\mathbf{u} \cdot \mathbf{v}) = \mathbf{u} \cdot (c\mathbf{v})$
3. $(\mathbf{u} + \mathbf{v}) \cdot \mathbf{w} = \mathbf{u} \cdot \mathbf{w} + \mathbf{v} \cdot \mathbf{w}$
4. $|\mathbf{u}|^2 = \mathbf{u} \cdot \mathbf{u}$

Proof We prove only the last property. The proofs of the others are left as exercises. Let $\mathbf{u} = \langle a_1, a_2 \rangle$. Then

$$\mathbf{u} \cdot \mathbf{u} = a_1 a_1 + a_2 a_2 = a_1^2 + a_2^2 = |\mathbf{u}|^2$$ ■

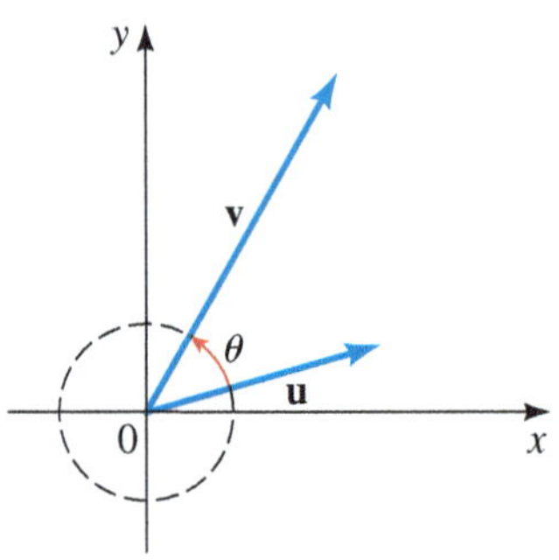

FIGURE 1

Let **u** and **v** be vectors, and sketch them with initial points at the origin. We define the **angle** θ **between u and v** to be the smaller of the angles formed by these representations of **u** and **v** (see Figure 1). Thus $0 \le \theta \le \pi$. The next theorem relates the angle between two vectors to their dot product.

THE DOT PRODUCT THEOREM

If θ is the angle between two nonzero vectors **u** and **v**, then

$$\mathbf{u} \cdot \mathbf{v} = |\mathbf{u}||\mathbf{v}| \cos \theta$$

Proof Applying the Law of Cosines to triangle AOB in Figure 2 gives

$$|\mathbf{u} - \mathbf{v}|^2 = |\mathbf{u}|^2 + |\mathbf{v}|^2 - 2|\mathbf{u}||\mathbf{v}| \cos \theta$$

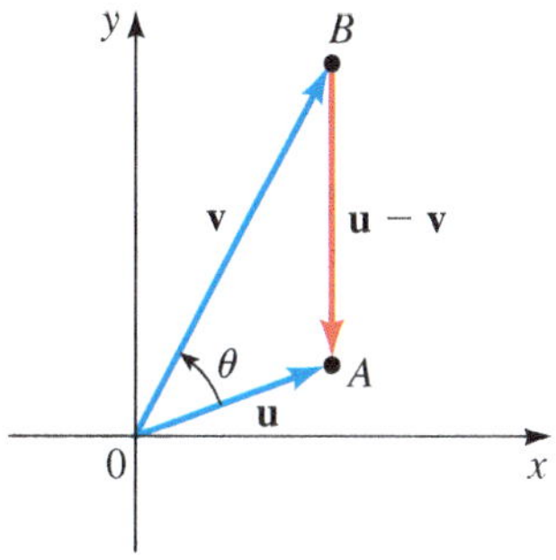

FIGURE 2

Using the properties of the dot product, we write the left-hand side as follows:

$$\begin{aligned}|\mathbf{u}-\mathbf{v}|^2 &= (\mathbf{u}-\mathbf{v})\cdot(\mathbf{u}-\mathbf{v})\\ &= \mathbf{u}\cdot\mathbf{u}-\mathbf{u}\cdot\mathbf{v}-\mathbf{v}\cdot\mathbf{u}+\mathbf{v}\cdot\mathbf{v}\\ &= |\mathbf{u}|^2-2(\mathbf{u}\cdot\mathbf{v})+|\mathbf{v}|^2\end{aligned}$$

Equating the right-hand sides of the displayed equations, we get

$$\begin{aligned}|\mathbf{u}|^2-2(\mathbf{u}\cdot\mathbf{v})+|\mathbf{v}|^2 &= |\mathbf{u}|^2+|\mathbf{v}|^2-2|\mathbf{u}||\mathbf{v}|\cos\theta\\ -2(\mathbf{u}\cdot\mathbf{v}) &= -2|\mathbf{u}||\mathbf{v}|\cos\theta\\ \mathbf{u}\cdot\mathbf{v} &= |\mathbf{u}||\mathbf{v}|\cos\theta\end{aligned}$$

This proves the theorem. ■

The Dot Product Theorem is useful because it allows us to find the angle between two vectors if we know the components of the vectors. The angle is obtained simply by solving the equation in the Dot Product Theorem for $\cos\theta$. We state this important result explicitly.

ANGLE BETWEEN TWO VECTORS

If θ is the angle between two nonzero vectors **u** and **v**, then

$$\cos\theta = \frac{\mathbf{u}\cdot\mathbf{v}}{|\mathbf{u}||\mathbf{v}|}$$

EXAMPLE 2 ■ Finding the Angle Between Two Vectors

Find the angle between the vectors $\mathbf{u} = \langle 2, 5\rangle$ and $\mathbf{v} = \langle 4, -3\rangle$.

SOLUTION By the formula for the angle between two vectors we have

$$\cos\theta = \frac{\mathbf{u}\cdot\mathbf{v}}{|\mathbf{u}||\mathbf{v}|} = \frac{(2)(4)+(5)(-3)}{\sqrt{4+25}\sqrt{16+9}} = \frac{-7}{5\sqrt{29}}$$

Thus the angle between **u** and **v** is

$$\theta = \cos^{-1}\left(\frac{-7}{5\sqrt{29}}\right) \approx 105.1°$$

Now Try Exercises 5(b) and 11(b) ■

Two nonzero vectors **u** and **v** are called **perpendicular**, or **orthogonal**, if the angle between them is $\pi/2$. The following theorem shows that we can determine whether two vectors are perpendicular by finding their dot product.

ORTHOGONAL VECTORS

Two nonzero vectors **u** and **v** are perpendicular if and only if $\mathbf{u}\cdot\mathbf{v} = 0$.

Proof If **u** and **v** are perpendicular, then the angle between them is $\pi/2$, so

$$\mathbf{u}\cdot\mathbf{v} = |\mathbf{u}||\mathbf{v}|\cos\frac{\pi}{2} = 0$$

Conversely, if $\mathbf{u}\cdot\mathbf{v} = 0$, then

$$|\mathbf{u}||\mathbf{v}|\cos\theta = 0$$

Since **u** and **v** are nonzero vectors, we conclude that $\cos\theta = 0$, so $\theta = \pi/2$. Thus **u** and **v** are orthogonal. ■

EXAMPLE 3 ■ Checking Whether Two Vectors Are Perpendicular

Determine whether the vectors in each pair are perpendicular.

(a) $\mathbf{u} = \langle 3, 5 \rangle$ and $\mathbf{v} = \langle 2, -8 \rangle$ **(b)** $\mathbf{u} = \langle 2, 1 \rangle$ and $\mathbf{v} = \langle -1, 2 \rangle$

SOLUTION

(a) $\mathbf{u} \cdot \mathbf{v} = (3)(2) + (5)(-8) = -34 \neq 0$, so **u** and **v** are not perpendicular.

(b) $\mathbf{u} \cdot \mathbf{v} = (2)(-1) + (1)(2) = 0$, so **u** and **v** are perpendicular.

Now Try Exercises 15 and 17 ■

■ The Component of u along v

The **component of u along v** (also called the **component of u in the direction of v** or the **scalar projection of u onto v**) is defined to be

Note that the component of **u** along **v** is a scalar, not a vector.

$$|\mathbf{u}| \cos \theta$$

where θ is the angle between **u** and **v**. Figure 3 gives a geometric interpretation of this concept. Intuitively, the component of **u** along **v** is the magnitude of the portion of **u** that points in the direction of **v**. Notice that the component of **u** along **v** is negative if $\pi/2 < \theta \leq \pi$.

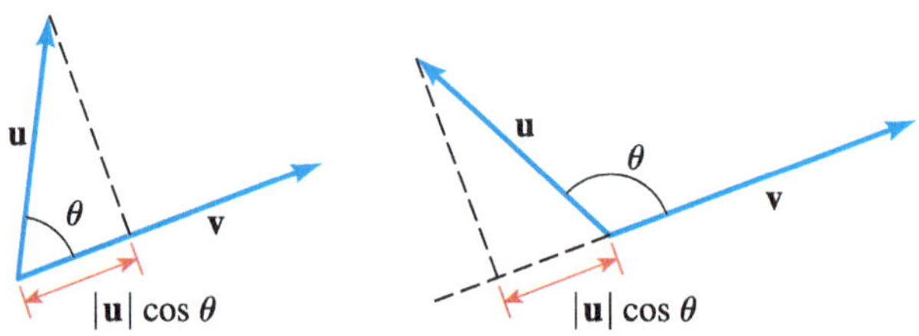

FIGURE 3

In analyzing forces in physics and engineering, it's often helpful to express a vector as a sum of two vectors lying in perpendicular directions. For example, suppose a car is parked on an inclined driveway as in Figure 4. The weight of the car is a vector **w** that points directly downward. We can write

$$\mathbf{w} = \mathbf{u} + \mathbf{v}$$

where **u** is parallel to the driveway and **v** is perpendicular to the driveway. The vector **u** is the force that tends to roll the car down the driveway, and **v** is the force experienced by the surface of the driveway. The magnitudes of these forces are the components of **w** along **u** and **v**, respectively.

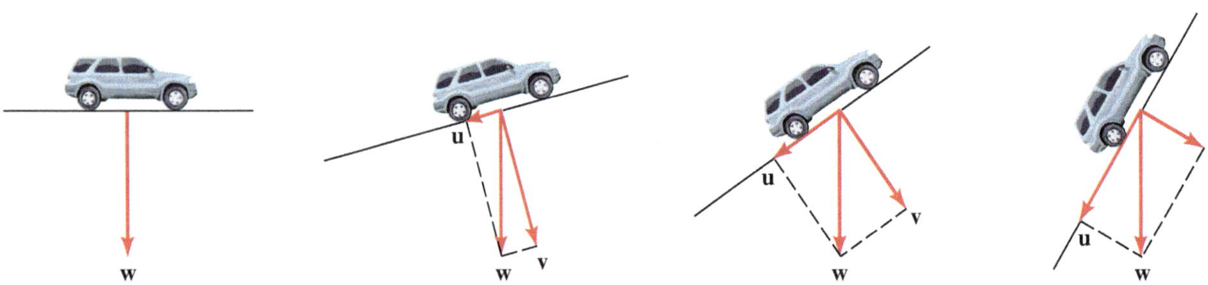

FIGURE 4

EXAMPLE 4 ■ Resolving a Force into Components

A car weighing 3000 lb is parked on a driveway that is inclined 15° to the horizontal, as shown in Figure 5.

(a) Find the magnitude of the force required to prevent the car from rolling down the driveway.

(b) Find the magnitude of the force experienced by the driveway due to the weight of the car.

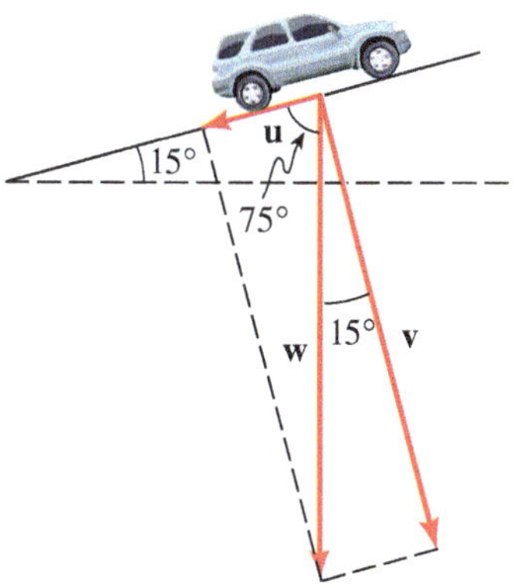

FIGURE 5

SOLUTION The car exerts a force **w** of 3000 lb directly downward. We resolve **w** into the sum of two vectors **u** and **v**, one parallel to the surface of the driveway and the other perpendicular to it, as shown in Figure 5.

(a) The magnitude of the part of the force **w** that causes the car to roll down the driveway is

$$|\mathbf{u}| = \text{component of } \mathbf{w} \text{ along } \mathbf{u} = 3000 \cos 75° \approx 776$$

Thus the force needed to prevent the car from rolling down the driveway is about 776 lb.

(b) The magnitude of the force exerted by the car on the driveway is

$$|\mathbf{v}| = \text{component of } \mathbf{w} \text{ along } \mathbf{v} = 3000 \cos 15° \approx 2898$$

The force experienced by the driveway is about 2898 lb.

Now Try Exercise 49 ■

The component of **u** along **v** can be computed by using dot products:

$$|\mathbf{u}| \cos \theta = \frac{|\mathbf{v}||\mathbf{u}| \cos \theta}{|\mathbf{v}|} = \frac{\mathbf{u} \cdot \mathbf{v}}{|\mathbf{v}|}$$

We have shown the following.

THE COMPONENT OF u ALONG v

The component of **u** along **v** (or the scalar projection of **u** onto **v**) is

$$\text{comp}_{\mathbf{v}}\, \mathbf{u} = \frac{\mathbf{u} \cdot \mathbf{v}}{|\mathbf{v}|}$$

EXAMPLE 5 ■ Finding Components

Let $\mathbf{u} = \langle 1, 4 \rangle$ and $\mathbf{v} = \langle -2, 1 \rangle$. Find the component of **u** along **v**.

SOLUTION From the formula for the component of **u** along **v** we have

$$\text{comp}_{\mathbf{v}}\, \mathbf{u} = \frac{\mathbf{u} \cdot \mathbf{v}}{|\mathbf{v}|} = \frac{(1)(-2) + (4)(1)}{\sqrt{4 + 1}} = \frac{2}{\sqrt{5}}$$

Now Try Exercise 25 ■

■ The Projection of u onto v

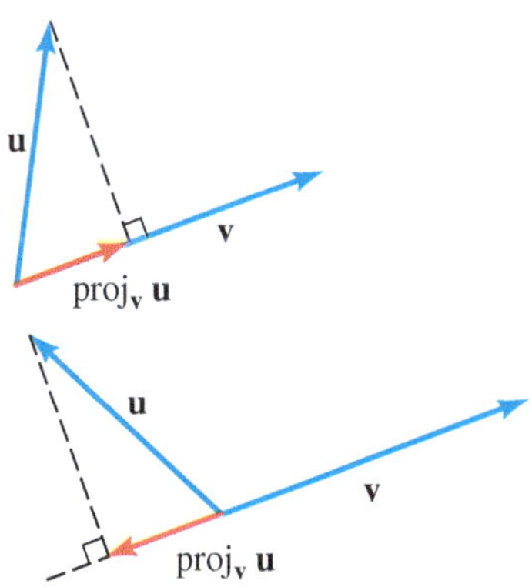

FIGURE 6

Figure 6 shows representations of the vectors **u** and **v**. The projection of **u** onto **v**, denoted by $\text{proj}_{\mathbf{v}}\, \mathbf{u}$, is the vector *parallel* to **v** and whose *length* is the component of **u** along **v** as shown in Figure 6. To find an expression for $\text{proj}_{\mathbf{v}}\, \mathbf{u}$, we first find a unit vector in the direction of **v** and then multiply it by the component of **u** along **v**:

$$\text{proj}_{\mathbf{v}}\, \mathbf{u} = (\text{component of } \mathbf{u} \text{ along } \mathbf{v})(\text{unit vector in direction of } \mathbf{v})$$

$$= \left(\frac{\mathbf{u} \cdot \mathbf{v}}{|\mathbf{v}|}\right) \frac{\mathbf{v}}{|\mathbf{v}|} = \left(\frac{\mathbf{u} \cdot \mathbf{v}}{|\mathbf{v}|^2}\right) \mathbf{v}$$

We often need to **resolve** a vector **u** into the sum of two vectors, one parallel to **v** and one orthogonal to **v**. That is, we want to write $\mathbf{u} = \mathbf{u}_1 + \mathbf{u}_2$, where $\mathbf{u}_1$ is parallel to **v** and $\mathbf{u}_2$ is orthogonal to **v**. In this case, $\mathbf{u}_1 = \text{proj}_{\mathbf{v}}\, \mathbf{u}$ and $\mathbf{u}_2 = \mathbf{u} - \text{proj}_{\mathbf{v}}\, \mathbf{u}$ (see Exercise 43).

Note that the projection of **u** onto **v** is a vector, not a scalar.

THE VECTOR PROJECTION OF u ONTO v

The **projection of u onto v** is the vector $\text{proj}_{\mathbf{v}}\,\mathbf{u}$ given by

$$\text{proj}_{\mathbf{v}}\,\mathbf{u} = \left(\frac{\mathbf{u}\cdot\mathbf{v}}{|\mathbf{v}|^2}\right)\mathbf{v}$$

If the vector **u** is **resolved** into $\mathbf{u}_1$ and $\mathbf{u}_2$, where $\mathbf{u}_1$ is parallel to **v** and $\mathbf{u}_2$ is orthogonal to **v**, then

$$\mathbf{u}_1 = \text{proj}_{\mathbf{v}}\,\mathbf{u} \qquad \text{and} \qquad \mathbf{u}_2 = \mathbf{u} - \text{proj}_{\mathbf{v}}\,\mathbf{u}$$

EXAMPLE 6 ■ Resolving a Vector into Orthogonal Vectors

Let $\mathbf{u} = \langle -2, 9\rangle$ and $\mathbf{v} = \langle -1, 2\rangle$.

(a) Find $\text{proj}_{\mathbf{v}}\,\mathbf{u}$.

(b) Resolve **u** into $\mathbf{u}_1$ and $\mathbf{u}_2$, where $\mathbf{u}_1$ is parallel to **v** and $\mathbf{u}_2$ is orthogonal to **v**.

SOLUTION

(a) By the formula for the projection of one vector onto another we have

$$\text{proj}_{\mathbf{v}}\,\mathbf{u} = \left(\frac{\mathbf{u}\cdot\mathbf{v}}{|\mathbf{v}|^2}\right)\mathbf{v} \qquad \text{Formula for projection}$$

$$= \left(\frac{\langle -2, 9\rangle\cdot\langle -1, 2\rangle}{(-1)^2 + 2^2}\right)\langle -1, 2\rangle \qquad \text{Definition of } \mathbf{u} \text{ and } \mathbf{v}$$

$$= 4\langle -1, 2\rangle = \langle -4, 8\rangle$$

(b) By the formula in the preceding box we have $\mathbf{u} = \mathbf{u}_1 + \mathbf{u}_2$, where

$$\mathbf{u}_1 = \text{proj}_{\mathbf{v}}\,\mathbf{u} = \langle -4, 8\rangle \qquad \text{From part (a)}$$

$$\mathbf{u}_2 = \mathbf{u} - \text{proj}_{\mathbf{v}}\,\mathbf{u} = \langle -2, 9\rangle - \langle -4, 8\rangle = \langle 2, 1\rangle$$

Now Try Exercise 29 ■

■ Work

One use of the dot product occurs in calculating work. In everyday use, the term *work* means the total amount of effort required to perform a task. In physics, *work* has a technical meaning that conforms to this intuitive meaning. If a constant force of magnitude F moves an object through a distance d along a straight line, then the **work** done is

$$W = Fd \qquad \text{or} \qquad \text{work} = \text{force} \times \text{distance}$$

If F is measured in pounds and d in feet, then the unit of work is a foot-pound (ft-lb). For example, how much work is done in lifting a 20-lb weight 6 ft off the ground? Since a force of 20 lb is required to lift this weight and since the weight moves through a distance of 6 ft, the amount of work done is

$$W = Fd = (20)(6) = 120 \text{ ft-lb}$$

This formula applies only when the force is directed along the direction of motion. In the general case, if the force **F** moves an object from P to Q, as in Figure 7, then only the component of the force in the direction of $\mathbf{D} = \overrightarrow{PQ}$ affects the object. Thus the effective magnitude of the force on the object is

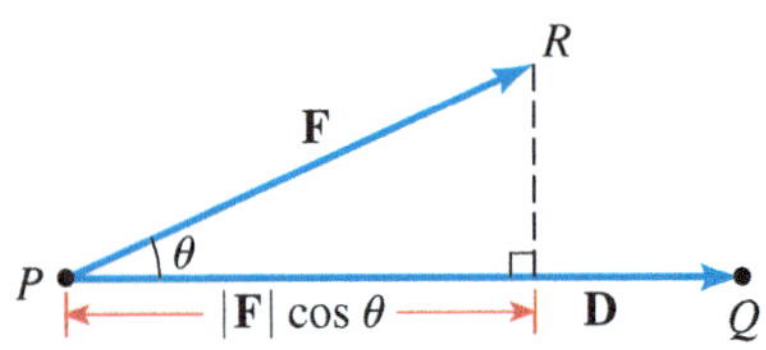

FIGURE 7

$$\text{comp}_{\mathbf{D}}\,\mathbf{F} = |\mathbf{F}|\cos\theta$$

So the work done is

$$W = \text{force} \times \text{distance} = (|\mathbf{F}|\cos\theta)|\mathbf{D}| = |\mathbf{F}||\mathbf{D}|\cos\theta = \mathbf{F}\cdot\mathbf{D}$$

We have derived the following simple formula for calculating work.

WORK

The **work** W done by a force $\mathbf{F}$ in moving along a vector $\mathbf{D}$ is

$$W = \mathbf{F} \cdot \mathbf{D}$$

EXAMPLE 7 ■ Calculating Work

A force is given by the vector $\mathbf{F} = \langle 2, 3 \rangle$ and moves an object from the point $(1, 3)$ to the point $(5, 9)$. Find the work done.

SOLUTION The displacement vector is

$$\mathbf{D} = \langle 5 - 1, 9 - 3 \rangle = \langle 4, 6 \rangle$$

So the work done is

$$W = \mathbf{F} \cdot \mathbf{D} = \langle 2, 3 \rangle \cdot \langle 4, 6 \rangle = 26$$

If the unit of force is pounds and the distance is measured in feet, then the work done is 26 ft-lb.

Now Try Exercise 35 ■

EXAMPLE 8 ■ Calculating Work

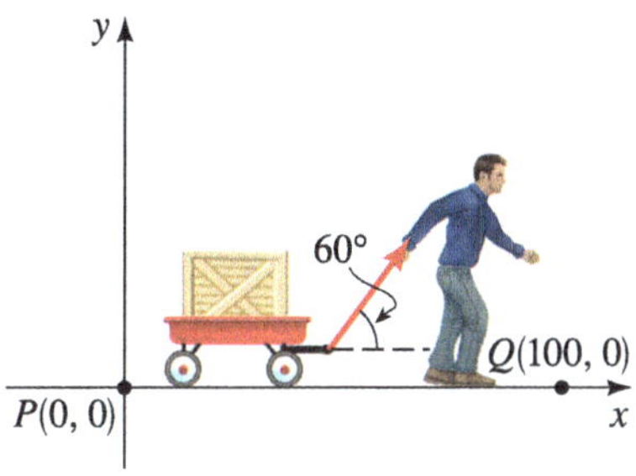

FIGURE 8

A man pulls a wagon horizontally by exerting a force of 20 lb on the handle. If the handle makes an angle of 60° with the horizontal, find the work done in moving the wagon 100 ft.

SOLUTION We choose a coordinate system with the origin at the initial position of the wagon (see Figure 8). That is, the wagon moves from the point $P(0, 0)$ to the point $Q(100, 0)$. The vector that represents this displacement is

$$\mathbf{D} = 100\mathbf{i}$$

The force on the handle can be written in terms of components (see Section 9.1) as

$$\mathbf{F} = (20 \cos 60°)\mathbf{i} + (20 \sin 60°)\mathbf{j} = 10\mathbf{i} + 10\sqrt{3}\mathbf{j}$$

Thus the work done is

$$W = \mathbf{F} \cdot \mathbf{D} = (10\mathbf{i} + 10\sqrt{3}\mathbf{j}) \cdot (100\mathbf{i}) = 1000 \text{ ft-lb}$$

Now Try Exercise 47 ■

James L. Amos/SuperStock

DISCOVERY PROJECT

Sailing Against the Wind

Sailors depend on the wind to propel their boats. But what if the wind is blowing in a direction opposite to that in which they want to travel? Although it is impossible to sail directly against the wind, it is possible to sail at an angle into the wind so that the sailboat can make headway against the wind. In this project we discover how vectors that model the sail, the keel, and the wind can be combined to find the direction in which the boat will move. You can find the project at **www.stewartmath.com**.

9.2 EXERCISES

CONCEPTS

1–2 ■ Let $\mathbf{u} = \langle a_1, a_2 \rangle$ and $\mathbf{v} = \langle b_1, b_2 \rangle$ be nonzero vectors in the plane, and let θ be the angle between them.

1. The dot product of $\mathbf{u}$ and $\mathbf{v}$ is defined by

$$\mathbf{u} \cdot \mathbf{v} = \_\_\_\_\_\_\_\_\_\_\_\_$$

The dot product of two vectors is a ________, not a vector.

2. The angle θ satisfies

$$\cos \theta = \frac{}{}$$

So if $\mathbf{u} \cdot \mathbf{v} = 0$, the vectors are ________.

3. (a) The component of $\mathbf{u}$ along $\mathbf{v}$ is the scalar $|\mathbf{u}| \cos \theta$ and can be expressed in terms of the dot product as

$\text{comp}_{\mathbf{v}} \mathbf{u} =$ ________. Sketch this component in the figure below.

(b) The projection of $\mathbf{u}$ onto $\mathbf{v}$ is the vector

$\text{proj}_{\mathbf{v}} \mathbf{u} =$ ________. Sketch this projection in the figure below.

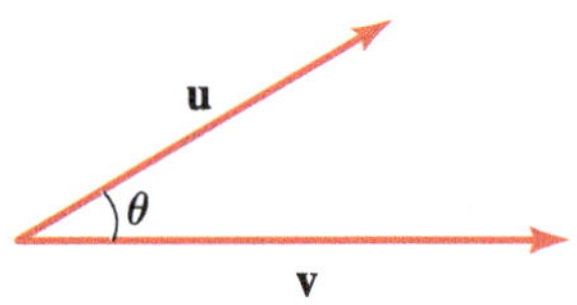

4. The work done by a force $\mathbf{F}$ in moving an object along a vector $\mathbf{D}$ is $W =$ ________.

SKILLS

5–14 ■ Dot Products and Angles Between Vectors Find **(a)** $\mathbf{u} \cdot \mathbf{v}$ and **(b)** the angle between $\mathbf{u}$ and $\mathbf{v}$ to the nearest degree.

5. $\mathbf{u} = \langle 2, 0 \rangle$, $\mathbf{v} = \langle 1, 1 \rangle$

6. $\mathbf{u} = \mathbf{i} + \sqrt{3}\mathbf{j}$, $\mathbf{v} = -\sqrt{3}\mathbf{i} + \mathbf{j}$

7. $\mathbf{u} = \langle 2, 7 \rangle$, $\mathbf{v} = \langle 3, 1 \rangle$

8. $\mathbf{u} = \langle -6, 6 \rangle$, $\mathbf{v} = \langle 1, -1 \rangle$

9. $\mathbf{u} = \langle 3, -2 \rangle$, $\mathbf{v} = \langle 1, 2 \rangle$

10. $\mathbf{u} = 2\mathbf{i} + \mathbf{j}$, $\mathbf{v} = 3\mathbf{i} - 2\mathbf{j}$

11. $\mathbf{u} = -5\mathbf{j}$, $\mathbf{v} = -\mathbf{i} - \sqrt{3}\mathbf{j}$

12. $\mathbf{u} = \mathbf{i} + \mathbf{j}$, $\mathbf{v} = \mathbf{i} - \mathbf{j}$

13. $\mathbf{u} = \mathbf{i} + 3\mathbf{j}$, $\mathbf{v} = 4\mathbf{i} - \mathbf{j}$

14. $\mathbf{u} = 3\mathbf{i} + 4\mathbf{j}$, $\mathbf{v} = -2\mathbf{i} - \mathbf{j}$

15–20 ■ Perpendicular Vectors? Determine whether the given vectors are perpendicular.

15. $\mathbf{u} = \langle 6, 4 \rangle$, $\mathbf{v} = \langle -2, 3 \rangle$

16. $\mathbf{u} = \langle 0, -5 \rangle$, $\mathbf{v} = \langle 4, 0 \rangle$

17. $\mathbf{u} = \langle -2, 6 \rangle$, $\mathbf{v} = \langle 4, 2 \rangle$

18. $\mathbf{u} = 2\mathbf{i}$, $\mathbf{v} = -7\mathbf{j}$

19. $\mathbf{u} = 2\mathbf{i} - 8\mathbf{j}$, $\mathbf{v} = -12\mathbf{i} - 3\mathbf{j}$

20. $\mathbf{u} = 4\mathbf{i}$, $\mathbf{v} = -\mathbf{i} + 3\mathbf{j}$

21–24 ■ Dot Products Find the indicated quantity, assuming that $\mathbf{u} = 2\mathbf{i} + \mathbf{j}$, $\mathbf{v} = \mathbf{i} - 3\mathbf{j}$, and $\mathbf{w} = 3\mathbf{i} + 4\mathbf{j}$.

21. $\mathbf{u} \cdot \mathbf{v} + \mathbf{u} \cdot \mathbf{w}$

22. $\mathbf{u} \cdot (\mathbf{v} + \mathbf{w})$

23. $(\mathbf{u} + \mathbf{v}) \cdot (\mathbf{u} - \mathbf{v})$

24. $(\mathbf{u} \cdot \mathbf{v})(\mathbf{u} \cdot \mathbf{w})$

25–28 ■ The Component of u along v Find the component of $\mathbf{u}$ along $\mathbf{v}$.

25. $\mathbf{u} = \langle 4, 6 \rangle$, $\mathbf{v} = \langle 3, -4 \rangle$

26. $\mathbf{u} = \langle -3, 5 \rangle$, $\mathbf{v} = \langle 1/\sqrt{2}, 1/\sqrt{2} \rangle$

27. $\mathbf{u} = 7\mathbf{i} - 24\mathbf{j}$, $\mathbf{v} = \mathbf{j}$

28. $\mathbf{u} = 7\mathbf{i}$, $\mathbf{v} = 8\mathbf{i} + 6\mathbf{j}$

29–34 ■ Vector Projection of u onto v **(a)** Calculate $\text{proj}_{\mathbf{v}} \mathbf{u}$. **(b)** Resolve $\mathbf{u}$ into $\mathbf{u}_1$ and $\mathbf{u}_2$, where $\mathbf{u}_1$ is parallel to $\mathbf{v}$ and $\mathbf{u}_2$ is orthogonal to $\mathbf{v}$.

29. $\mathbf{u} = \langle -2, 4 \rangle$, $\mathbf{v} = \langle 1, 1 \rangle$

30. $\mathbf{u} = \langle 7, -4 \rangle$, $\mathbf{v} = \langle 2, 1 \rangle$

31. $\mathbf{u} = \langle 1, 2 \rangle$, $\mathbf{v} = \langle 1, -3 \rangle$

32. $\mathbf{u} = \langle 11, 3 \rangle$, $\mathbf{v} = \langle -3, -2 \rangle$

33. $\mathbf{u} = \langle 2, 9 \rangle$, $\mathbf{v} = \langle -3, 4 \rangle$

34. $\mathbf{u} = \langle 1, 1 \rangle$, $\mathbf{v} = \langle 2, -1 \rangle$

35–38 ■ Calculating Work Find the work done by the force $\mathbf{F}$ in moving an object from P to Q.

35. $\mathbf{F} = 4\mathbf{i} - 5\mathbf{j}$; $P(0, 0)$, $Q(3, 8)$

36. $\mathbf{F} = 400\mathbf{i} + 50\mathbf{j}$; $P(-1, 1)$, $Q(200, 1)$

37. $\mathbf{F} = 10\mathbf{i} + 3\mathbf{j}$; $P(2, 3)$, $Q(6, -2)$

38. $\mathbf{F} = -4\mathbf{i} + 20\mathbf{j}$; $P(0, 10)$, $Q(5, 25)$

SKILLS Plus

39–42 ■ Properties of Vectors Let $\mathbf{u}$, $\mathbf{v}$, and $\mathbf{w}$ be vectors, and let c be a scalar. Prove the given property.

39. $\mathbf{u} \cdot \mathbf{v} = \mathbf{v} \cdot \mathbf{u}$

40. $(c\mathbf{u}) \cdot \mathbf{v} = c(\mathbf{u} \cdot \mathbf{v}) = \mathbf{u} \cdot (c\mathbf{v})$

41. $(\mathbf{u} + \mathbf{v}) \cdot \mathbf{w} = \mathbf{u} \cdot \mathbf{w} + \mathbf{v} \cdot \mathbf{w}$

42. $(\mathbf{u} - \mathbf{v}) \cdot (\mathbf{u} + \mathbf{v}) = |\mathbf{u}|^2 - |\mathbf{v}|^2$

43. Projection Show that $\text{proj}_{\mathbf{v}} \mathbf{u}$ and $\mathbf{u} - \text{proj}_{\mathbf{v}} \mathbf{u}$ are orthogonal.

44. Projection Evaluate $\mathbf{v} \cdot \text{proj}_{\mathbf{v}} \mathbf{u}$.

APPLICATIONS

45. Work The force $\mathbf{F} = 4\mathbf{i} - 7\mathbf{j}$ moves an object 4 ft along the x-axis in the positive direction. Find the work done if the unit of force is the pound.

46. Work A constant force $\mathbf{F} = \langle 2, 8 \rangle$ moves an object along a straight line from the point $(2, 5)$ to the point $(11, 13)$. Find the work done if the distance is measured in feet and the force is measured in pounds.

47. Work A lawn mower is pushed a distance of 200 ft along a horizontal path by a constant force of 50 lb. The handle of the lawn mower is held at an angle of 30° from the horizontal (see the figure). Find the work done.

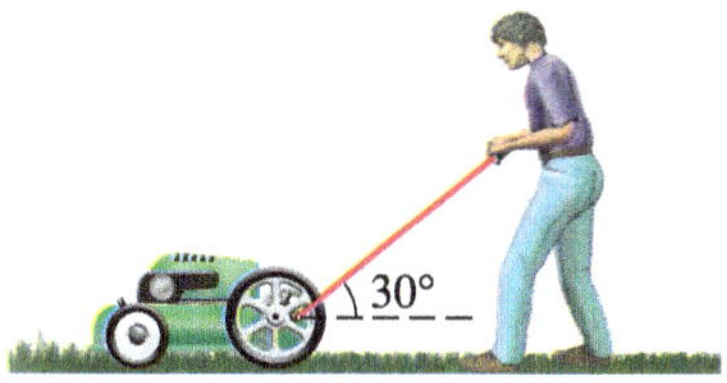

48. Work A car drives 500 ft on a road that is inclined 12° to the horizontal, as shown in the following figure. The car weighs 2500 lb. Thus gravity acts straight down on the car with a constant force $\mathbf{F} = -2500\mathbf{j}$. Find the work done by the car in overcoming gravity.

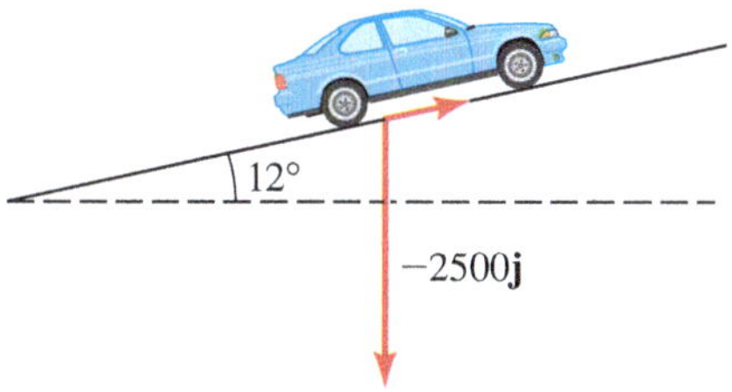

49. Force A car is on a driveway that is inclined 10° to the horizontal. A force of 490 lb is required to keep the car from rolling down the driveway.

(a) Find the weight of the car.

(b) Find the force the car exerts against the driveway.

50. Force A car is on a driveway that is inclined 25° to the horizontal. If the car weighs 2755 lb, find the force required to keep it from rolling down the driveway.

51. Force A package that weighs 200 lb is placed on an inclined plane. If a force of 80 lb is just sufficient to keep the package from sliding, find the angle of inclination of the plane. (Ignore the effects of friction.)

52. Force A cart weighing 40 lb is placed on a ramp inclined at 15° to the horizontal. The cart is held in place by a rope inclined at 60° to the horizontal, as shown in the figure. Find the force that the rope must exert on the cart to keep it from rolling down the ramp.

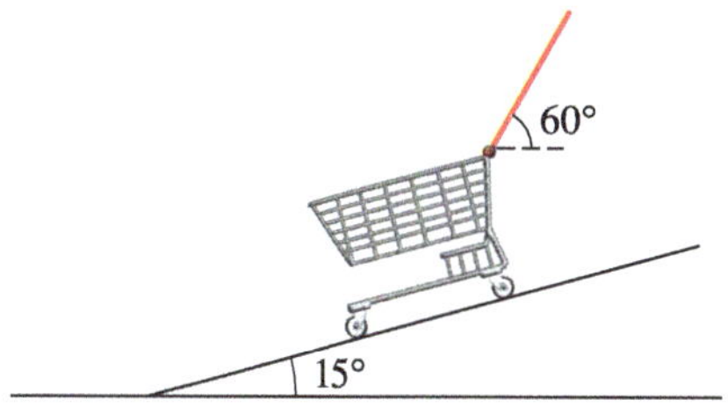

DISCUSS ■ DISCOVER ■ PROVE ■ WRITE

53. DISCUSS ■ DISCOVER ■ WRITE: Distance from a Point to a Line Let L be the line $2x + 4y = 8$, and let P be the point $(3, 4)$.

(a) Show that the points $Q(0, 2)$ and $R(2, 1)$ lie on L.

(b) Let $\mathbf{u} = \overrightarrow{QP}$ and $\mathbf{v} = \overrightarrow{QR}$, as shown in the figure. Find $\mathbf{w} = \text{proj}_{\mathbf{v}}\, \mathbf{u}$.

(c) Sketch a graph that explains why $|\mathbf{u} - \mathbf{w}|$ is the distance from P to L. Find this distance.

(d) Write a short paragraph describing the steps you would take to find the distance from a given point to a given line.

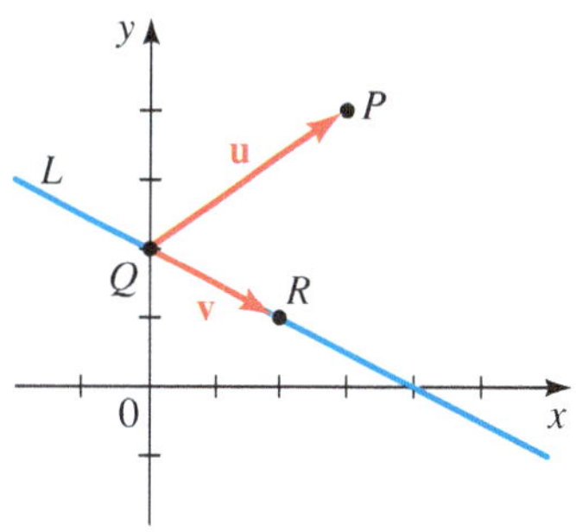

9.3 THREE-DIMENSIONAL COORDINATE GEOMETRY

■ The Three-Dimensional Rectangular Coordinate System ■ Distance Formula in Three Dimensions ■ The Equation of a Sphere

To locate a point in a plane, two numbers are necessary. We know that any point in the Cartesian plane can be represented as an ordered pair (a, b) of real numbers, where a is the x-coordinate and b is the y-coordinate. In three-dimensional space, a third dimension is added, so any point in space is represented by an ordered triple (a, b, c) of real numbers.

The Three-Dimensional Rectangular Coordinate System

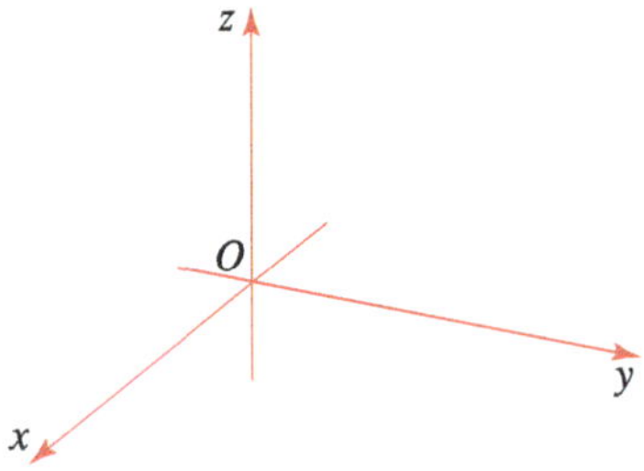

FIGURE 1 Coordinate axes

To represent points in space, we first choose a fixed point O (the origin) and three directed lines through O that are perpendicular to each other, called the **coordinate axes** and labeled the x-axis, y-axis, and z-axis. Usually we think of the x- and y-axes as being horizontal and the z-axis as being vertical, and we draw the orientation of the axes as in Figure 1.

The three coordinate axes determine the three **coordinate planes** illustrated in Figure 2(a). The xy-plane is the plane that contains the x- and y-axes; the yz-plane is the plane that contains the y- and z-axes; the xz-plane is the plane that contains the x- and z-axes. These three coordinate planes divide space into eight parts, called **octants**.

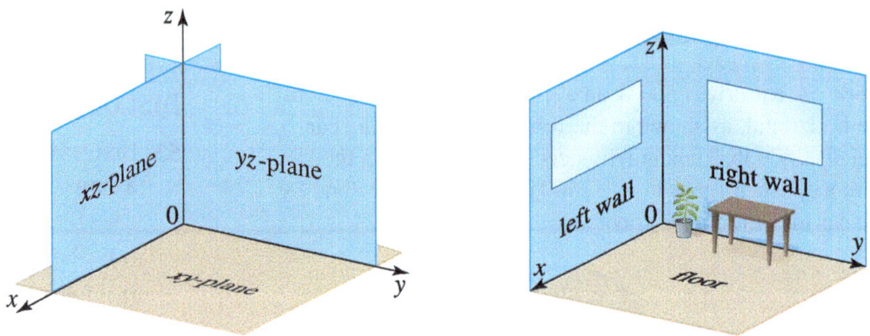

FIGURE 2 (a) Coordinate planes (b) Coordinate "walls"

Because people often have difficulty visualizing diagrams of three-dimensional figures, you may find it helpful to do the following (see Figure 2(b)). Look at any bottom corner of a room and call the corner the origin. The wall on your left is in the xz-plane, the wall on your right is in the yz-plane, and the floor is in the xy-plane. The x-axis runs along the intersection of the floor and the left wall; the y-axis runs along the intersection of the floor and the right wall. The z-axis runs up from the floor toward the ceiling along the intersection of the two walls.

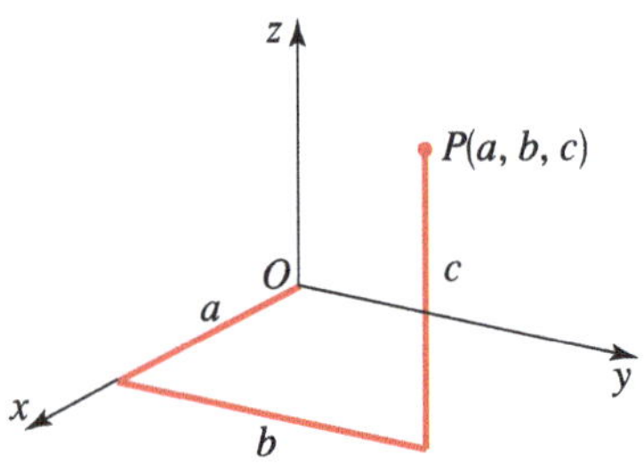

FIGURE 3 Point $P(a, b, c)$

Now any point P in space can be located by a unique **ordered triple** of real numbers (a, b, c), as shown in Figure 3. The first number a is the x-coordinate of P, the second number b is the y-coordinate of P, and the third number c is the z-coordinate of P. The set of all ordered triples $\{(x, y, z) \mid x, y, z \in \mathbb{R}\}$ forms the **three-dimensional rectangular coordinate system**.

EXAMPLE 1 ■ Plotting Points in Three Dimensions

Plot the points $(2, 4, 7)$ and $(-4, 3, -5)$.

SOLUTION The points are plotted in Figure 4.

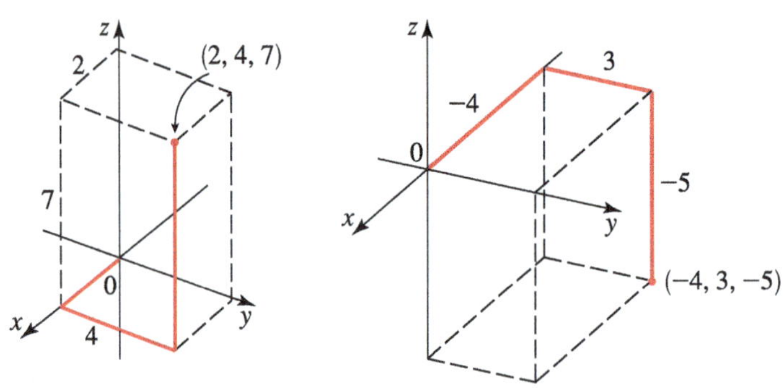

FIGURE 4

Now Try Exercise 3(a) ■

In two-dimensional geometry the graph of an equation involving x and y is a *curve* in the plane. In three-dimensional geometry an equation in x, y, and z represents a *surface* in space.

EXAMPLE 2 ■ Surfaces in Three-Dimensional Space

Describe and sketch the surfaces represented by the following equations:

(a) $z = 3$ **(b)** $y = 5$

SOLUTION

(a) The surface consists of the points $P(x, y, z)$ where the z-coordinate is 3. This is the horizontal plane that is parallel to the xy-plane and three units above it, as in Figure 5.

(b) The surface consists of the points $P(x, y, z)$ where the y-coordinate is 5. This is the vertical plane that is parallel to the xz-plane and five units to the right of it, as in Figure 6.

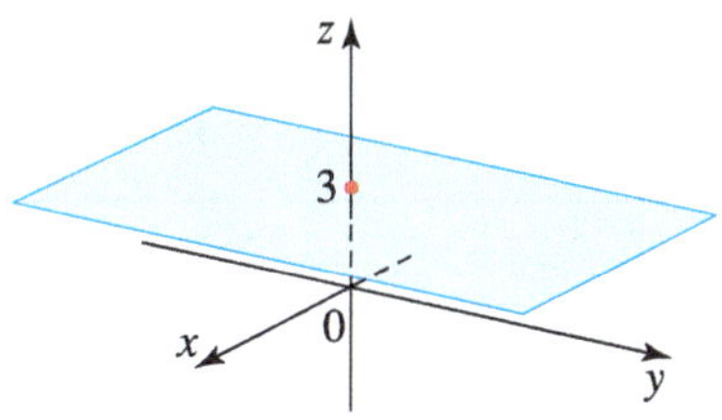

FIGURE 5 The plane $z = 3$

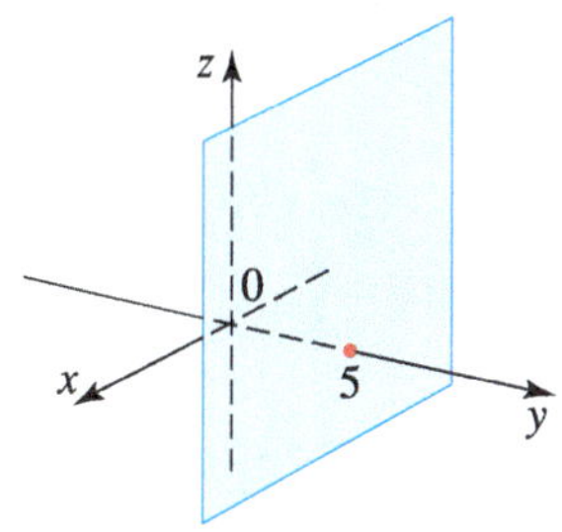

FIGURE 6 The plane $y = 5$

Now Try Exercise 7 ■

■ Distance Formula in Three Dimensions

The familiar formula for the distance between two points in a plane is easily extended to the following three-dimensional formula.

> **DISTANCE FORMULA IN THREE DIMENSIONS**
>
> The distance between the points $P(x_1, y_1, z_1)$ and $Q(x_2, y_2, z_2)$ is
>
> $$d(P, Q) = \sqrt{(x_2 - x_1)^2 + (y_2 - y_1)^2 + (z_2 - z_1)^2}$$

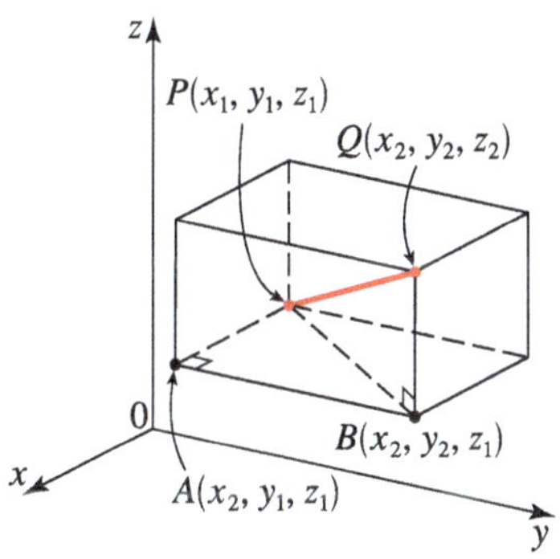

FIGURE 7

Proof To prove this formula, we construct a rectangular box as in Figure 7, where $P(x_1, y_1, z_1)$ and $Q(x_2, y_2, z_2)$ are diagonally opposite vertices and the faces of the box are parallel to the coordinate planes. If A and B are the vertices of the box that are indicated in the figure, then

$$d(P, A) = |x_2 - x_1| \qquad d(A, B) = |y_2 - y_1| \qquad d(Q, B) = |z_2 - z_1|$$

Triangles PAB and PBQ are right triangles, so by the Pythagorean Theorem we have

$$(d(P, Q))^2 = (d(P, B))^2 + (d(Q, B))^2$$

$$(d(P, B))^2 = (d(P, A))^2 + (d(A, B))^2$$

Combining these equations, we get

$$\begin{aligned}(d(P, Q))^2 &= (d(P, A))^2 + (d(A, B))^2 + (d(Q, B))^2 \\ &= |x_2 - x_1|^2 + |y_2 - y_1|^2 + |z_2 - z_1|^2\end{aligned}$$

Therefore

$$d(P, Q) = \sqrt{(x_2 - x_1)^2 + (y_2 - y_1)^2 + (z_2 - z_1)^2}$$ ■

EXAMPLE 3 ■ Using the Distance Formula

Find the distance between the points $P(2, -1, 7)$ and $Q(1, -3, 5)$.

SOLUTION We use the Distance Formula:

$$d(P, Q) = \sqrt{(1-2)^2 + (-3-(-1))^2 + (5-7)^2} = \sqrt{1+4+4} = 3$$

Now Try Exercise 3(b) ■

The Equation of a Sphere

We can use the Distance Formula to find an equation for a sphere in a three-dimensional coordinate space.

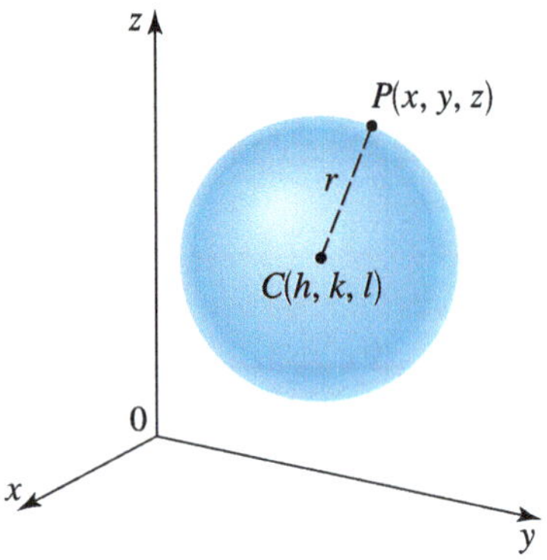

FIGURE 8 Sphere with radius r and center $C(h, k, l)$

EQUATION OF A SPHERE

An equation of a sphere with center $C(h, k, l)$ and radius r is

$$(x-h)^2 + (y-k)^2 + (z-l)^2 = r^2$$

Proof A sphere with radius r is the set of all points $P(x, y, z)$ whose distance from the center C is the constant r (see Figure 8). By the Distance Formula we have

$$[d(P, C)]^2 = (x-h)^2 + (y-k)^2 + (z-l)^2$$

Since the distance $d(P, C)$ is equal to r, we get the desired formula. ■

EXAMPLE 4 ■ Finding the Equation of a Sphere

Find an equation of a sphere with radius 5 and center $C(-2, 1, 3)$.

SOLUTION We use the general equation of a sphere, with $r = 5$, $h = -2$, $k = 1$, and $l = 3$:

$$(x+2)^2 + (y-1)^2 + (z-3)^2 = 25$$

Now Try Exercise 11 ■

EXAMPLE 5 ■ Finding the Center and Radius of a Sphere

Show that $x^2 + y^2 + z^2 + 4x - 6y + 2z + 6 = 0$ is the equation of a sphere, and find its center and radius.

SOLUTION We complete the squares in the x-, y-, and z-terms to rewrite the given equation in the form of an equation of a sphere.

$$x^2 + y^2 + z^2 + 4x - 6y + 2z + 6 = 0 \quad \text{Given equation}$$

$$(x^2 + 4x + 4) + (y^2 - 6y + 9) + (z^2 + 2z + 1) = -6 + 4 + 9 + 1 \quad \text{Complete squares}$$

$$(x+2)^2 + (y-3)^2 + (z+1)^2 = 8 \quad \text{Factor into squares}$$

Comparing this with the standard equation of a sphere, we can see that the center is $(-2, 3, -1)$ and the radius is $\sqrt{8} = 2\sqrt{2}$.

Now Try Exercise 15 ■

The intersection of a sphere with a plane is called the **trace** of the sphere in the plane.

EXAMPLE 6 ■ Finding the Trace of a Sphere

Describe the trace of the sphere $(x - 2)^2 + (y - 4)^2 + (z - 5)^2 = 36$ in **(a)** the xy-plane and **(b)** the plane $z = 9$.

SOLUTION

(a) In the xy-plane the z-coordinate is 0. So the trace of the sphere in the xy-plane consists of all the points on the sphere whose z-coordinate is 0. We replace z by 0 in the equation of the sphere and get

$$(x - 2)^2 + (y - 4)^2 + (0 - 5)^2 = 36 \qquad \text{Replace } z \text{ by } 0$$
$$(x - 2)^2 + (y - 4)^2 + 25 = 36 \qquad \text{Calculate}$$
$$(x - 2)^2 + (y - 4)^2 = 11 \qquad \text{Subtract 25}$$

Thus the trace of the sphere is the circle

$$(x - 2)^2 + (y - 4)^2 = 11 \qquad z = 0$$

which is a circle of radius $\sqrt{11}$ that is in the xy-plane, centered at $(2, 4, 0)$ (see Figure 9(a)).

(b) The trace of the sphere in the plane $z = 9$ consists of all the points on the sphere whose z-coordinate is 9. So we replace z by 9 in the equation of the sphere and get

$$(x - 2)^2 + (y - 4)^2 + (9 - 5)^2 = 36 \qquad \text{Replace } z \text{ by } 0$$
$$(x - 2)^2 + (y - 4)^2 + 16 = 36 \qquad \text{Calculate}$$
$$(x - 2)^2 + (y - 4)^2 = 20 \qquad \text{Subtract 16}$$

Thus the trace of the sphere is the circle

$$(x - 2)^2 + (y - 4)^2 = 20 \qquad z = 9$$

which is a circle of radius $\sqrt{20}$ that is 9 units above the xy-plane, centered at $(2, 4, 9)$ (see Figure 9(b)).

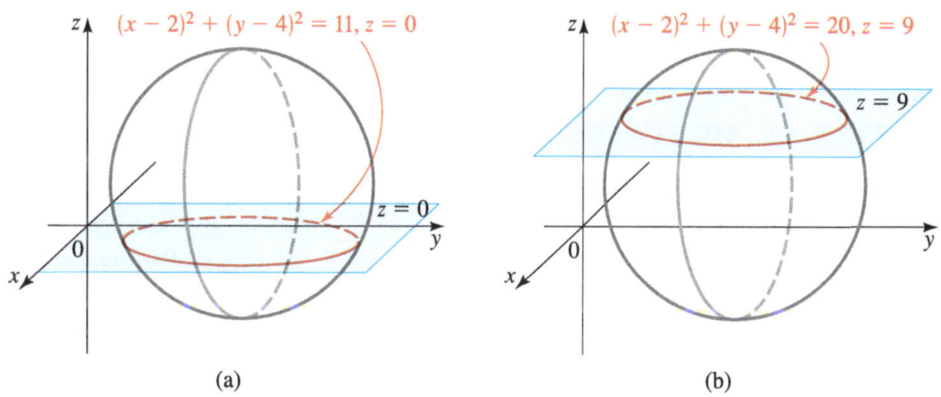

FIGURE 9 The trace of a sphere in the planes $z = 0$ and $z = 9$

Now Try Exercise 19

9.3 EXERCISES

CONCEPTS

1–2 ■ Refer to the figure.

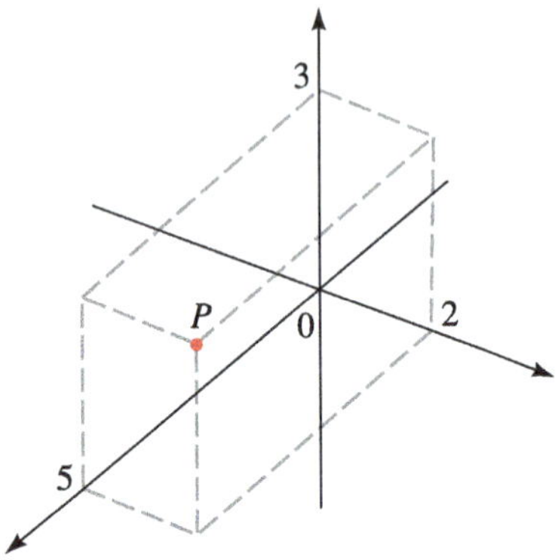

1. In a three-dimensional coordinate system the three mutually perpendicular axes are called the ___-axis, the ___-axis, and the ___-axis. Label the axes in the figure. The point P in the figure has coordinates (, ,). The equation of the plane passing through P and parallel to the xz-plane is ________.

2. The distance between the point $P(x_1, y_1, z_1)$ and $Q(x_2, y_2, z_2)$ is given by the formula $d(P, Q) =$ ________________. The distance between the point P in the figure and the origin is __________. The equation of the sphere centered at P with radius 3 is ________________.

SKILLS

3–6 ■ Plotting Points and Finding Distance in Three Dimensions Two points P and Q are given. **(a)** Plot P and Q. **(b)** Find the distance between P and Q.

3. $P(3, 1, 0)$, $Q(-1, 2, -5)$

4. $P(5, 0, 10)$, $Q(3, -6, 7)$

5. $P(-2, -1, 0)$, $Q(-12, 3, 0)$

6. $P(5, -4, -6)$, $Q(8, -7, 4)$

7–10 ■ Surfaces in Three Dimensions Describe and sketch the surface represented by the given equation.

7. $x = 4$ **8.** $y = -2$

9. $z = 8$ **10.** $y = -1$

11–14 ■ Equation of a Sphere Find an equation of a sphere with the given radius r and center C.

11. $r = 5$; $C(2, -5, 3)$

12. $r = 3$; $C(-1, 4, -7)$

13. $r = \sqrt{6}$; $C(3, -1, 0)$

14. $r = \sqrt{11}$; $C(-10, 0, 1)$

15–18 ■ Center and Radius of a Sphere Show that the equation represents a sphere, and find its center and radius.

15. $x^2 + y^2 + z^2 - 10x + 2y + 8z = 9$

16. $x^2 + y^2 + z^2 + 4x - 6y + 2z = 10$

17. $x^2 + y^2 + z^2 = 12x + 2y$

18. $x^2 + y^2 + z^2 = 14y - 6z$

19–20 ■ Trace of a Sphere In these exercises we find the trace of a sphere in a plane.

19. Describe the trace of the sphere

$$(x + 1)^2 + (y - 2)^2 + (z + 10)^2 = 100$$

(a) in the yz-plane and **(b)** in the plane $x = 4$.

20. Describe the trace of the sphere

$$x^2 + (y - 4)^2 + (z - 3)^2 = 144$$

(a) in the xz-plane and **(b)** in the plane $z = -2$.

APPLICATIONS

21. Spherical Water Tank A water tank is in the shape of a sphere of radius 5 ft. The tank is supported on a metal circle 4 ft below the center of the sphere, as shown in the figure. Find the radius of the metal circle.

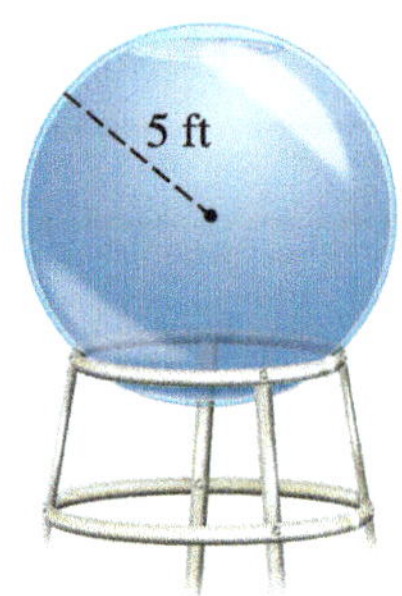

22. A Spherical Buoy A spherical buoy of radius 2 ft floats in a calm lake. Six inches of the buoy are submerged. Place a coordinate system with the origin at the center of the sphere.

(a) Find an equation of the sphere.

(b) Find an equation of the circle formed at the waterline of the buoy.

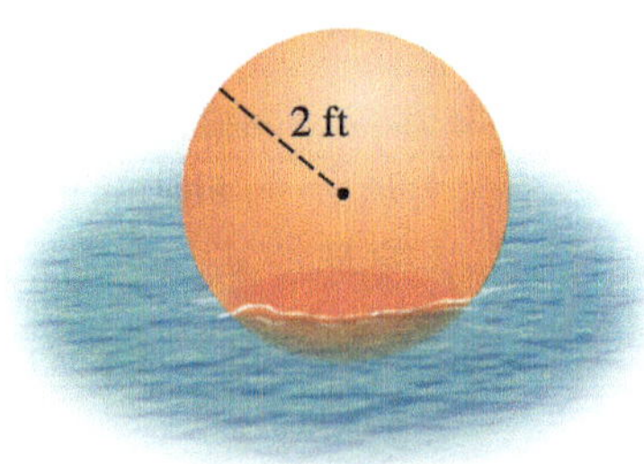

DISCUSS ■ DISCOVER ■ PROVE ■ WRITE

23. DISCUSS: Visualizing a Set in Space Try to visualize the set of all points (x, y, z) in a coordinate space that are equidistant from the points $P(0, 0, 0)$ and $Q(0, 3, 0)$. Use the Distance Formula to find an equation for this surface, and observe that it is a plane.

24. DISCUSS: Visualizing a Set in Space Try to visualize the set of all points (x, y, z) in a coordinate space that are twice as far from the points $Q(0, 3, 0)$ as from the point $P(0, 0, 0)$. Use the Distance Formula to show that the set is a sphere, and find its center and radius.

9.4 VECTORS IN THREE DIMENSIONS

■ Vectors in Space ■ Combining Vectors in Space ■ The Dot Product for Vectors in Space ■ Direction Angles of a Vector

Recall that vectors are used to indicate a quantity that has both magnitude and direction. In Section 9.1 we studied vectors in the coordinate plane, where the direction is restricted to two dimensions. Vectors in space have a direction that is in three-dimensional space. The properties that hold for vectors in the plane hold for vectors in space as well.

■ Vectors in Space

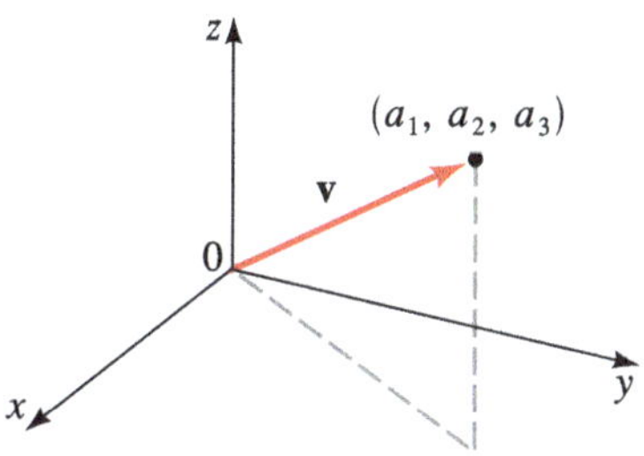

FIGURE 1 $\mathbf{v} = \langle a_1, a_2, a_3 \rangle$

Recall from Section 9.1 that a vector can be described geometrically by its initial point and terminal point. When we place a vector **v** in space with its initial point at the origin, we can describe it algebraically as an ordered triple:

$$\mathbf{v} = \langle a_1, a_2, a_3 \rangle$$

where a_1, a_2, and a_3 are the **components** of **v** (see Figure 1). Recall also that a vector has many different representations, depending on its initial point. The following definition gives the relationship between the algebraic and geometric representations of a vector.

> **COMPONENT FORM OF A VECTOR IN SPACE**
>
> If a vector **v** is represented in space with initial point $P(x_1, y_1, z_1)$ and terminal point $Q(x_2, y_2, z_2)$, then
>
> $$\mathbf{v} = \langle x_2 - x_1, y_2 - y_1, z_2 - z_1 \rangle$$

EXAMPLE 1 ■ Describing Vectors in Component Form

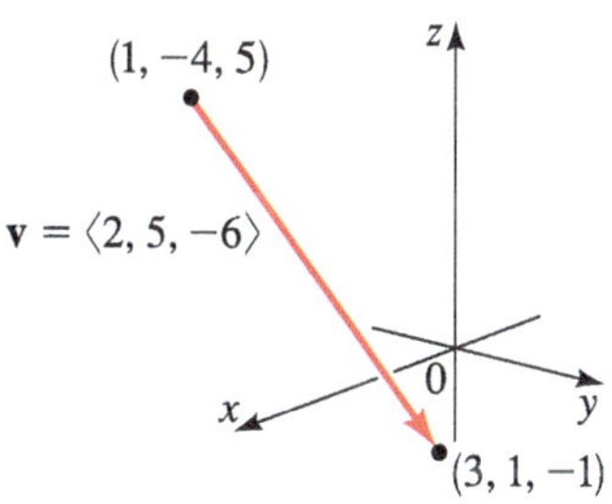

FIGURE 2 $\mathbf{v} = \langle 2, 5, -6 \rangle$

(a) Find the components of the vector **v** with initial point $P(1, -4, 5)$ and terminal point $Q(3, 1, -1)$.

(b) If the vector $\mathbf{w} = \langle -2, 1, 3 \rangle$ has initial point $(2, 1, -1)$, what is its terminal point?

SOLUTION

(a) The desired vector is

$$\mathbf{v} = \langle 3 - 1, 1 - (-4), -1 - 5 \rangle = \langle 2, 5, -6 \rangle$$

See Figure 2.

(b) Let the terminal point of **w** be (x, y, z). Then

$$\mathbf{w} = \langle x - 2, y - 1, z - (-1) \rangle$$

Since $\mathbf{w} = \langle -2, 1, 3 \rangle$, we have $x - 2 = -2$, $y - 1 = 1$, and $z + 1 = 3$. So $x = 0$, $y = 2$, and $z = 2$, and the terminal point is $(0, 2, 2)$.

Now Try Exercises 3 and 7 ■

The following formula is a consequence of the Distance Formula, since the vector $\mathbf{v} = \langle a_1, a_2, a_3 \rangle$ in standard position has initial point $(0, 0, 0)$ and terminal point (a_1, a_2, a_3).

MAGNITUDE OF A VECTOR IN THREE DIMENSIONS

The magnitude of the vector $\mathbf{v} = \langle a_1, a_2, a_3 \rangle$ is

$$|\mathbf{v}| = \sqrt{a_1^2 + a_2^2 + a_3^2}$$

EXAMPLE 2 ■ Magnitude of Vectors in Three Dimensions

Find the magnitude of the given vector.

(a) $\mathbf{u} = \langle 3, 2, 5 \rangle$ **(b)** $\mathbf{v} = \langle 0, 3, -1 \rangle$ **(c)** $\mathbf{w} = \langle 0, 0, -1 \rangle$

SOLUTION

(a) $|\mathbf{u}| = \sqrt{3^2 + 2^2 + 5^2} = \sqrt{38}$

(b) $|\mathbf{v}| = \sqrt{0^2 + 3^2 + (-1)^2} = \sqrt{10}$

(c) $|\mathbf{w}| = \sqrt{0^2 + 0^2 + (-1)^2} = 1$

Now Try Exercise 11 ■

Combining Vectors in Space

We now give definitions of the algebraic operations involving vectors in three dimensions.

ALGEBRAIC OPERATIONS ON VECTORS IN THREE DIMENSIONS

If $\mathbf{u} = \langle a_1, a_2, a_3 \rangle$, $\mathbf{v} = \langle b_1, b_2, b_3 \rangle$, and c is a scalar, then

$$\mathbf{u} + \mathbf{v} = \langle a_1 + b_1, a_2 + b_2, a_3 + b_3 \rangle$$

$$\mathbf{u} - \mathbf{v} = \langle a_1 - b_1, a_2 - b_2, a_3 - b_3 \rangle$$

$$c\mathbf{u} = \langle ca_1, ca_2, ca_3 \rangle$$

EXAMPLE 3 ■ Operations with Three-Dimensional Vectors

If $\mathbf{u} = \langle 1, -2, 4 \rangle$ and $\mathbf{v} = \langle 6, -1, 1 \rangle$ find $\mathbf{u} + \mathbf{v}$, $\mathbf{u} - \mathbf{v}$, and $5\mathbf{u} - 3\mathbf{v}$.

SOLUTION Using the definitions of algebraic operations, we have

$$\mathbf{u} + \mathbf{v} = \langle 1 + 6, -2 - 1, 4 + 1 \rangle = \langle 7, -3, 5 \rangle$$

$$\mathbf{u} - \mathbf{v} = \langle 1 - 6, -2 - (-1), 4 - 1 \rangle = \langle -5, -1, 3 \rangle$$

$$5\mathbf{u} - 3\mathbf{v} = 5\langle 1, -2, 4 \rangle - 3\langle 6, -1, 1 \rangle = \langle 5, -10, 20 \rangle - \langle 18, -3, 3 \rangle = \langle -13, -7, 17 \rangle$$

Now Try Exercise 15 ■

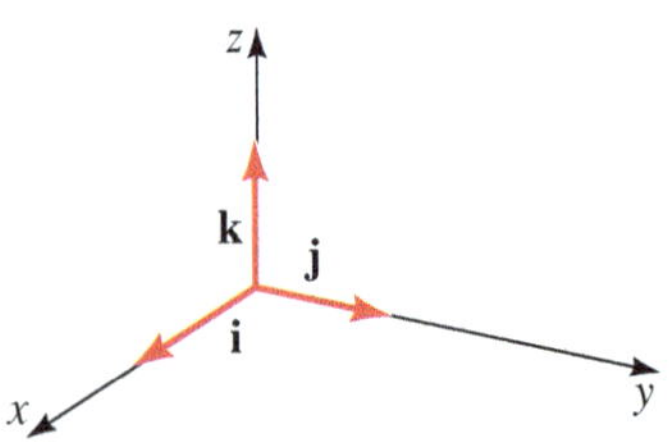

FIGURE 3

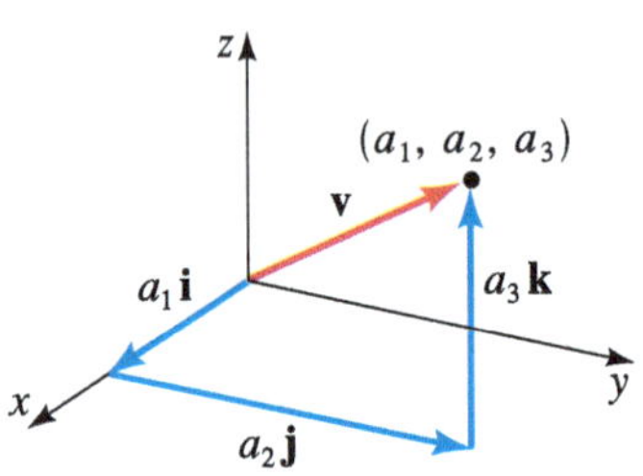

FIGURE 4

Recall that a unit vector is a vector of length 1. The vector $\mathbf{w}$ in Example 2(c) is an example of a unit vector. Some other unit vectors in three dimensions are

$$\mathbf{i} = \langle 1, 0, 0 \rangle \qquad \mathbf{j} = \langle 0, 1, 0 \rangle \qquad \mathbf{k} = \langle 0, 0, 1 \rangle$$

as shown in Figure 3. Any vector in three dimensions can be written in terms of these three vectors (see Figure 4).

EXPRESSING VECTORS IN TERMS OF i, j, AND k

The vector $\mathbf{v} = \langle a_1, a_2, a_3 \rangle$ can be expressed in terms of **i**, **j**, and **k** by

$$\mathbf{v} = \langle a_1, a_2, a_3 \rangle = a_1\mathbf{i} + a_2\mathbf{j} + a_3\mathbf{k}$$

All the properties of vectors on page 633 in Section 9.1 hold for vectors in three dimensions as well. We use these properties in the next example.

EXAMPLE 4 ■ Vectors in Terms of i, j, and k

(a) Write the vector $\mathbf{u} = \langle 5, -3, 6 \rangle$ in terms of **i**, **j**, and **k**.

(b) If $\mathbf{u} = \mathbf{i} + 2\mathbf{j} - 3\mathbf{k}$ and $\mathbf{v} = 4\mathbf{i} + 7\mathbf{k}$, express the vector $2\mathbf{u} + 3\mathbf{v}$ in terms of **i**, **j**, and **k**.

SOLUTION

(a) $\mathbf{u} = 5\mathbf{i} + (-3)\mathbf{j} + 6\mathbf{k} = 5\mathbf{i} - 3\mathbf{j} + 6\mathbf{k}$

(b) We use the properties of vectors to get the following:

$$\begin{aligned} 2\mathbf{u} + 3\mathbf{v} &= 2(2\mathbf{i} + 2\mathbf{j} - 3\mathbf{k}) + 3(4\mathbf{i} + 7\mathbf{k}) \\ &= 4\mathbf{i} + 4\mathbf{j} - 6\mathbf{k} + 12\mathbf{i} + 21\mathbf{k} \\ &= 16\mathbf{i} + 4\mathbf{j} + 15\mathbf{k} \end{aligned}$$

Now Try Exercises 19 and 23 ■

■ The Dot Product for Vectors in Space

We define the dot product for vectors in three dimensions. All the properties of the dot product, including the Dot Product Theorem (page 640), hold for vectors in three dimensions.

DEFINITION OF THE DOT PRODUCT FOR VECTORS IN THREE DIMENSIONS

If $\mathbf{u} = \langle a_1, a_2, a_3 \rangle$ and $\mathbf{v} = \langle b_1, b_2, b_3 \rangle$ are vectors in three dimensions, then their **dot product** is defined by

$$\mathbf{u} \cdot \mathbf{v} = a_1b_1 + a_2b_2 + a_3b_3$$

EXAMPLE 5 ■ Calculating Dot Products for Vectors in Three Dimensions

Find the given dot product.

(a) $\langle -1, 2, 3 \rangle \cdot \langle 6, 5, -1 \rangle$

(b) $(2\mathbf{i} - 3\mathbf{j} - \mathbf{k}) \cdot (-\mathbf{i} + 2\mathbf{j} + 8\mathbf{k})$

SOLUTION

(a) $\langle -1, 2, 3 \rangle \cdot \langle 6, 5, -1 \rangle = (-1)(6) + (2)(5) + (3)(-1) = 1$

(b) $$\begin{aligned} (2\mathbf{i} - 3\mathbf{j} - \mathbf{k}) \cdot (-\mathbf{i} + 2\mathbf{j} + 8\mathbf{k}) &= \langle 2, -3, -1 \rangle \cdot \langle -1, 2, 8 \rangle \\ &= (2)(-1) + (-3)(2) + (-1)(8) = -16 \end{aligned}$$

Now Try Exercises 25 and 27 ■

Recall that the cosine of the angle between two vectors can be calculated by using the dot product (page 641). The same property holds for vectors in three dimensions. We restate this property here for emphasis.

ANGLE BETWEEN TWO VECTORS

Let **u** and **v** be vectors in space, and let θ be the angle between them. Then

$$\cos\theta = \frac{\mathbf{u}\cdot\mathbf{v}}{|\mathbf{u}||\mathbf{v}|}$$

In particular, **u** and **v** are **perpendicular** (or **orthogonal**) if and only if $\mathbf{u}\cdot\mathbf{v} = 0$.

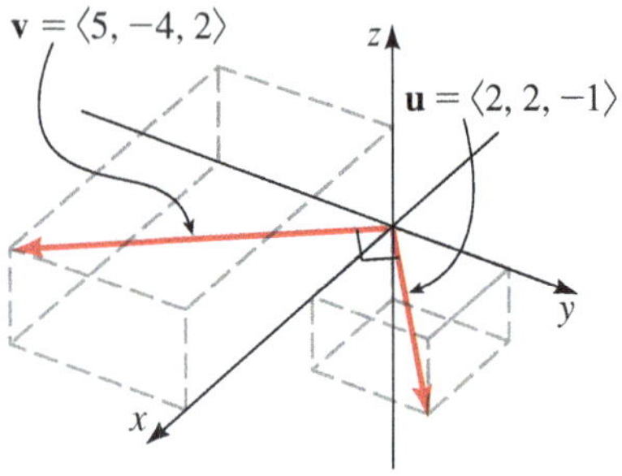

FIGURE 5 The vectors **u** and **v** are perpendicular.

EXAMPLE 6 ■ Checking Whether Two Vectors Are Perpendicular

Show that the vector $\mathbf{u} = 2\mathbf{i} + 2\mathbf{j} - \mathbf{k}$ is perpendicular to $5\mathbf{i} - 4\mathbf{j} + 2\mathbf{k}$.

SOLUTION We find the dot product.

$$(2\mathbf{i} + 2\mathbf{j} - \mathbf{k})\cdot(5\mathbf{i} - 4\mathbf{j} + 2\mathbf{k}) = (2)(5) + (2)(-4) + (-1)(2) = 0$$

Since the dot product is 0, the vectors are perpendicular. See Figure 5.

Now Try Exercise 29 ■

Direction Angles of a Vector

The **direction angles** of a nonzero vector $\mathbf{v} = a_1\mathbf{i} + a_2\mathbf{j} + a_3\mathbf{k}$ are the angles α, β, and γ in the interval $[0, \pi]$ that the vector **v** makes with the positive x-, y-, and z-axes (see Figure 6). The cosines of these angles, $\cos\alpha$, $\cos\beta$, and $\cos\gamma$, are called the **direction cosines** of the vector **v**. By using the formula for the angle between two vectors, we can find the direction cosines of **v**:

$$\cos\alpha = \frac{\mathbf{v}\cdot\mathbf{i}}{|\mathbf{v}||\mathbf{i}|} = \frac{a_1}{|\mathbf{v}|} \qquad \cos\beta = \frac{\mathbf{v}\cdot\mathbf{j}}{|\mathbf{v}||\mathbf{j}|} = \frac{a_2}{|\mathbf{v}|} \qquad \cos\gamma = \frac{\mathbf{v}\cdot\mathbf{k}}{|\mathbf{v}||\mathbf{k}|} = \frac{a_3}{|\mathbf{v}|}$$

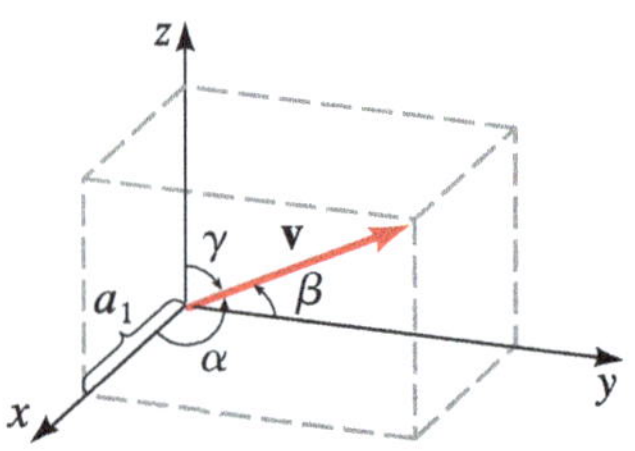

FIGURE 6 Direction angles of the vector **v**

DIRECTION ANGLES OF A VECTOR

If $\mathbf{v} = a_1\mathbf{i} + a_2\mathbf{j} + a_3\mathbf{k}$ is a nonzero vector in space, the direction angles α, β, and γ satisfy

$$\cos\alpha = \frac{a_1}{|\mathbf{v}|} \qquad \cos\beta = \frac{a_2}{|\mathbf{v}|} \qquad \cos\gamma = \frac{a_3}{|\mathbf{v}|}$$

In particular, if $|\mathbf{v}| = 1$, then the direction cosines of **v** are simply the components of **v**.

EXAMPLE 7 ■ Finding the Direction Angles of a Vector

Find the direction angles of the vector $\mathbf{v} = \mathbf{i} + 2\mathbf{j} + 3\mathbf{k}$.

SOLUTION The length of the vector **v** is $|\mathbf{v}| = \sqrt{1^2 + 2^2 + 3^2} = \sqrt{14}$. From the above box we get

$$\cos\alpha = \frac{1}{\sqrt{14}} \qquad \cos\beta = \frac{2}{\sqrt{14}} \qquad \cos\gamma = \frac{3}{\sqrt{14}}$$

Since the direction angles are in the interval $[0, \pi]$ and since $\cos^{-1}$ gives angles in that same interval, we get α, β, and γ by simply taking $\cos^{-1}$ of the above equations.

$$\alpha = \cos^{-1}\frac{1}{\sqrt{14}} \approx 74^\circ \qquad \beta = \cos^{-1}\frac{2}{\sqrt{14}} \approx 58^\circ \qquad \gamma = \cos^{-1}\frac{3}{\sqrt{14}} \approx 37^\circ$$

Now Try Exercise 37 ■

The direction angles of a vector uniquely determine its direction but not its length. If we also know the length of the vector $\mathbf{v}$, the expressions for the direction cosines of $\mathbf{v}$ allow us to express the vector as

$$\mathbf{v} = \langle |\mathbf{v}|\cos\alpha, |\mathbf{v}|\cos\beta, |\mathbf{v}|\cos\gamma \rangle$$

From this we get

$$\mathbf{v} = |\mathbf{v}|\langle \cos\alpha, \cos\beta, \cos\gamma \rangle$$

$$\frac{\mathbf{v}}{|\mathbf{v}|} = \langle \cos\alpha, \cos\beta, \cos\gamma \rangle$$

Since $\mathbf{v}/|\mathbf{v}|$ is a unit vector, we get the following.

PROPERTY OF DIRECTION COSINES

The direction angles α, β, and γ of a nonzero vector $\mathbf{v}$ in space satisfy the following equation:

$$\cos^2\alpha + \cos^2\beta + \cos^2\gamma = 1$$

This property indicates that if we know two of the direction cosines of a vector, we can find the third up to its sign.

EXAMPLE 8 ■ Finding the Direction Angles of a Vector

An angle θ is **acute** if $0 \le \theta < \pi/2$ and is **obtuse** if $\pi/2 < \theta \le \pi$.

A vector makes an angle $\alpha = \pi/3$ with the positive x-axis and an angle $\beta = 3\pi/4$ with the positive y-axis. Find the angle γ that the vector makes with the positive z-axis, given that γ is an obtuse angle.

SOLUTION By the property of the direction angles we have

$$\cos^2\alpha + \cos^2\beta + \cos^2\gamma = 1$$

$$\cos^2\frac{\pi}{3} + \cos^2\frac{3\pi}{4} + \cos^2\gamma = 1$$

$$\left(\frac{1}{2}\right)^2 + \left(-\frac{1}{\sqrt{2}}\right)^2 + \cos^2\gamma = 1$$

$$\cos^2\gamma = \frac{1}{4}$$

$$\cos\gamma = \frac{1}{2} \quad \text{or} \quad \cos\gamma = -\frac{1}{2}$$

$$\gamma = \frac{\pi}{3} \quad \text{or} \quad \gamma = \frac{2\pi}{3}$$

Since we require γ to be an obtuse angle, we conclude that $\gamma = 2\pi/3$.

Now Try Exercise 41 ■

9.4 EXERCISES

CONCEPTS

1. A vector in three dimensions can be written in either of two forms: in coordinate form as $\mathbf{v} = \langle a_1, a_2, a_3 \rangle$ and in terms of the ________ vectors **i**, **j**, and **k** as $\mathbf{v}$ = ____________.
The magnitude of the vector **v** is $|\mathbf{v}|$ = ____________.
So $\langle 4, -2, 4 \rangle$ = ▢ **i** + ▢ **j** + ▢ **k** and
$7\mathbf{j} - 24\mathbf{k} = \langle$ ▢, ▢, ▢ $\rangle$.

2. The angle θ between the vectors **u** and **v** satisfies
$\cos\theta = \dfrac{\square}{\square}$. So if **u** and **v** are perpendicular, then
$\mathbf{u}\cdot\mathbf{v}$ = ________. If $\mathbf{u} = \langle 4, 5, 6 \rangle$ and $\mathbf{v} = \langle 3, 0, -2 \rangle$ then
$\mathbf{u}\cdot\mathbf{v}$ = ________, so **u** and **v** are ____________.

SKILLS

3–6 ■ Vectors in Component Form Find the vector **v** with initial point P and terminal point Q.

3. $P(1, -1, 0)$, $Q(0, -2, 5)$
4. $P(1, 2, -1)$, $Q(3, -1, 2)$
5. $P(6, -1, 0)$, $Q(0, -3, 0)$
6. $P(1, -1, -1)$, $Q(0, 0, -1)$

7–10 ■ Terminal Point of a Vector If the vector **v** has initial point P, what is its terminal point?

7. $\mathbf{v} = \langle 3, 4, -2 \rangle$, $P(2, 0, 1)$
8. $\mathbf{v} = \langle 0, 0, 1 \rangle$, $P(0, 1, -1)$
9. $\mathbf{v} = \langle -2, 0, 2 \rangle$, $P(3, 0, -3)$
10. $\mathbf{v} = \langle 23, -5, 12 \rangle$, $P(-6, 4, 2)$

11–14 ■ Magnitude of a Vector Find the magnitude of the given vector.

11. $\langle -2, 1, 2 \rangle$ **12.** $\langle 5, 0, -12 \rangle$
13. $\langle 3, 5, -4 \rangle$ **14.** $\langle 1, -6, 2\sqrt{2} \rangle$

15–18 ■ Operations with Vectors Find the vectors $\mathbf{u} + \mathbf{v}$, $\mathbf{u} - \mathbf{v}$, and $3\mathbf{u} - \frac{1}{2}\mathbf{v}$.

15. $\mathbf{u} = \langle 2, -7, 3 \rangle$, $\mathbf{v} = \langle 0, 4, -1 \rangle$
16. $\mathbf{u} = \langle 0, 1, -3 \rangle$, $\mathbf{v} = \langle 4, 2, 0 \rangle$
17. $\mathbf{u} = \mathbf{i} + \mathbf{j}$, $\mathbf{v} = -\mathbf{j} - 2\mathbf{k}$
18. $\mathbf{u} = \langle a, 2b, 3c \rangle$, $\mathbf{v} = \langle -4a, b, -2c \rangle$

19–22 ■ Writing Vectors in Terms of i, j, and k Express the given vector in terms of the unit vectors **i**, **j**, and **k**.

19. $\langle 12, 0, 2 \rangle$ **20.** $\langle 0, -3, 5 \rangle$
21. $\langle 3, -3, 0 \rangle$ **22.** $\langle -a, \frac{1}{3}a, 4 \rangle$

23–24 ■ Operations with Vectors Two vectors **u** and **v** are given. Express the vector $-2\mathbf{u} + 3\mathbf{v}$ **(a)** in component form $\langle a_1, a_2, a_3 \rangle$ and **(b)** in terms of the unit vectors **i**, **j**, and **k**.

23. $\mathbf{u} = \langle 0, -2, 1 \rangle$, $\mathbf{v} = \langle 1, -1, 0 \rangle$
24. $\mathbf{u} = \langle 3, 1, 0 \rangle$, $\mathbf{v} = \langle 3, 0, -5 \rangle$

25–28 ■ Dot Products Two vectors **u** and **v** are given. Find their dot product $\mathbf{u}\cdot\mathbf{v}$.

25. $\mathbf{u} = \langle 2, 5, 0 \rangle$, $\mathbf{v} = \langle \frac{1}{2}, -1, 10 \rangle$
26. $\mathbf{u} = \langle -3, 0, 4 \rangle$, $\mathbf{v} = \langle 2, 4, \frac{1}{2} \rangle$
27. $\mathbf{u} = 6\mathbf{i} - 4\mathbf{j} - 2\mathbf{k}$, $\mathbf{v} = \frac{5}{6}\mathbf{i} + \frac{3}{2}\mathbf{j} - \mathbf{k}$
28. $\mathbf{u} = 3\mathbf{j} - 2\mathbf{k}$, $\mathbf{v} = \frac{5}{6}\mathbf{i} - \frac{5}{3}\mathbf{j}$

29–32 ■ Perpendicular Vectors? Determine whether or not the given vectors are perpendicular.

29. $\langle 4, -2, -4 \rangle$, $\langle 1, -2, 2 \rangle$ **30.** $4\mathbf{j} - \mathbf{k}$, $\mathbf{i} + 2\mathbf{j} + 9\mathbf{k}$
31. $\langle 0.3, 1.2, -0.9 \rangle$, $\langle 10, -5, 10 \rangle$
32. $\langle x, -2x, 3x \rangle$, $\langle 5, 7, 3 \rangle$

33–36 ■ Angle Between Two Vectors Find the angle between **u** and **v**, rounded to the nearest tenth degree.

33. $\mathbf{u} = \langle 2, -2, -1 \rangle$, $\mathbf{v} = \langle 1, 2, 2 \rangle$
34. $\mathbf{u} = \langle 4, 0, 2 \rangle$, $\mathbf{v} = \langle 2, -1, 0 \rangle$
35. $\mathbf{u} = \mathbf{j} + \mathbf{k}$, $\mathbf{v} = \mathbf{i} + 2\mathbf{j} - 3\mathbf{k}$
36. $\mathbf{u} = \mathbf{i} + 2\mathbf{j} - 2\mathbf{k}$, $\mathbf{v} = 4\mathbf{i} - 3\mathbf{k}$

37–40 ■ Direction Angles of a Vector Find the direction angles of the given vector, rounded to the nearest degree.

37. $3\mathbf{i} + 4\mathbf{j} + 5\mathbf{k}$ **38.** $\mathbf{i} - 2\mathbf{j} - \mathbf{k}$
39. $\langle 2, 3, -6 \rangle$ **40.** $\langle 2, -1, 2 \rangle$

41–44 ■ Direction Angles of a Vector Two direction angles of a vector are given. Find the third direction angle, given that it is either obtuse or acute as indicated. (In Exercises 43 and 44, round your answers to the nearest degree.)

41. $\alpha = \dfrac{\pi}{3}$, $\gamma = \dfrac{2\pi}{3}$; β is acute
42. $\beta = \dfrac{2\pi}{3}$, $\gamma = \dfrac{\pi}{4}$; α is acute
43. $\alpha = 60°$, $\beta = 50°$; γ is obtuse
44. $\alpha = 75°$, $\gamma = 15°$

SKILLS Plus

45–46 ■ Impossible Direction Angles Explain why it is impossible for a vector to have the given direction angles.

45. $\alpha = 20°$, $\beta = 45°$ **46.** $\alpha = 150°$, $\gamma = 25°$

47. Parallel Vectors Two nonzero vectors are *parallel* if they point in the same direction or in opposite directions. This means that if two vectors are parallel, one must be a scalar multiple of the other. Determine whether the given vectors **u** and **v** are parallel. If they are, express **v** as a scalar multiple of **u**.

(a) $\mathbf{u} = \langle 3, -2, 4 \rangle$, $\mathbf{v} = \langle -6, 4, -8 \rangle$

(b) $\mathbf{u} = \langle -9, -6, 12 \rangle$, $\mathbf{v} = \langle 12, 8, -16 \rangle$

(c) $\mathbf{u} = \mathbf{i} + \mathbf{j} + \mathbf{k}$, $\mathbf{v} = 2\mathbf{i} + 2\mathbf{j} - 2\mathbf{k}$

48. Unit Vectors A *unit vector* is a vector of magnitude 1. Multiplying a vector by a scalar changes its magnitude but not its direction.

(a) If a vector **v** has magnitude m, what scalar multiple of **v** has magnitude 1 (that is, is a unit vector)?

(b) Multiply each of the following vectors by an appropriate scalar to change them into unit vectors:

$$\langle 1, -2, 2 \rangle \quad \langle -6, 8, -10 \rangle \quad \langle 6, 5, 9 \rangle$$

APPLICATIONS

49. Resultant of Four Forces An object located at the origin in a three-dimensional coordinate system is held in equilibrium by four forces. One has magnitude 7 lb and points in the direction of the positive x-axis, so it is represented by the vector $7\mathbf{i}$. The second has magnitude 24 lb and points in the direction of the positive y-axis. The third has magnitude 25 lb and points in the direction of the *negative* z-axis.

(a) Use the fact that the four forces are in equilibrium (that is, their sum is **0**) to find the fourth force. Express it in terms of the unit vectors **i**, **j**, and **k**.

(b) What is the magnitude of the fourth force?

50. Central Angle of a Tetrahedron A *tetrahedron* is a solid with four triangular faces, four vertices, and six edges, as shown in the figure. In a *regular* tetrahedron the edges are all of the same length. Consider the tetrahedron with vertices $A(1, 0, 0)$, $B(0, 1, 0)$, $C(0, 0, 1)$, and $D(1, 1, 1)$.

(a) Show that the tetrahedron is regular.

(b) The center of the tetrahedron is the point $E(\frac{1}{2}, \frac{1}{2}, \frac{1}{2})$ (the "average" of the vertices). Find the angle between the vectors that join the center to any two of the vertices (for instance, $\angle AEB$). This angle is called the *central angle* of the tetrahedron.

[*Note:* In a molecule of methane (CH_4) the four hydrogen atoms form the vertices of a regular tetrahedron with the carbon atom at the center. In this case chemists refer to the central angle as the *bond angle*. In the figure, the tetrahedron in the exercise is shown, with the vertices labeled H for hydrogen and the center labeled C for carbon.]

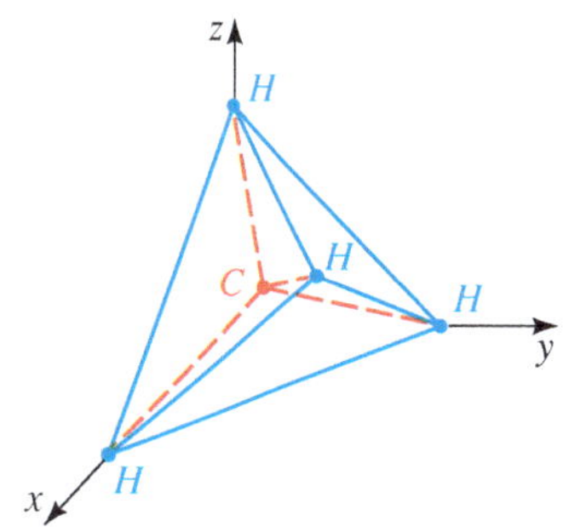

DISCUSS ■ DISCOVER ■ PROVE ■ WRITE

51. DISCUSS ■ PROVE: Vector Equation of a Sphere Let $\mathbf{u} = \langle 2, 2, 2 \rangle$, $\mathbf{v} = \langle -2, -2, 0 \rangle$, and $\mathbf{r} = \langle x, y, z \rangle$.

(a) Show that the vector equation $(\mathbf{r} - \mathbf{u}) \cdot (\mathbf{r} - \mathbf{v}) = 0$ represents a sphere, by expanding the dot product and simplifying the resulting algebraic equation.

(b) Find the center and radius of the sphere.

(c) Interpret the result of part (a) geometrically, using the fact that the dot product of two vectors is 0 only if the vectors are perpendicular. [*Hint:* Draw a diagram showing the endpoints of the vectors **u**, **v**, and **r**, noting that the endpoints of **u** and **v** are the endpoints of a diameter and the endpoint of **r** is an arbitrary point on the sphere.]

(d) Using your observations from part (a), find a vector equation for the sphere in which the points $(0, 1, 3)$ and $(2, -1, 4)$ form the endpoints of a diameter. Simplify the vector equation to obtain an algebraic equation for the sphere. What are its center and radius?

9.5 THE CROSS PRODUCT

■ The Cross Product ■ Properties of the Cross Product ■ Area of a Parallelogram ■ Volume of a Parallelepiped

In this section we define an operation on vectors that allows us to find a vector which is perpendicular to two given vectors.

The Cross Product

Given two vectors $\mathbf{u} = \langle a_1, a_2, a_3 \rangle$ and $\mathbf{v} = \langle b_1, b_2, b_3 \rangle$, we often need to find a vector **w** perpendicular to both **u** and **v**. If we write $\mathbf{w} = \langle c_1, c_2, c_3 \rangle$, then $\mathbf{u} \cdot \mathbf{w} = 0$ and $\mathbf{v} \cdot \mathbf{w} = 0$, so

$$a_1c_1 + a_2c_2 + a_3c_3 = 0$$

$$b_1c_1 + b_2c_2 + b_3c_3 = 0$$

You can check that one of the solutions of this system of equations is the vector $\mathbf{w} = \langle a_2b_3 - a_3b_2, a_3b_1 - a_1b_3, a_1b_2 - a_2b_1\rangle$. This vector is called the *cross product* of $\mathbf{u}$ and $\mathbf{v}$ and is denoted by $\mathbf{u} \times \mathbf{v}$.

THE CROSS PRODUCT

If $\mathbf{u} = \langle a_1, a_2, a_3\rangle$ and $\mathbf{v} = \langle b_1, b_2, b_3\rangle$ are three-dimensional vectors, then the **cross product** of $\mathbf{u}$ and $\mathbf{v}$ is the vector

$$\mathbf{u} \times \mathbf{v} = \langle a_2b_3 - a_3b_2,\ a_3b_1 - a_1b_3,\ a_1b_2 - a_2b_1\rangle$$

The *cross product* $\mathbf{u} \times \mathbf{v}$ of two vectors $\mathbf{u}$ and $\mathbf{v}$, unlike the dot product, is a vector (not a scalar). For this reason it is also called the *vector product.* Note that $\mathbf{u} \times \mathbf{v}$ is defined only when $\mathbf{u}$ and $\mathbf{v}$ are vectors in *three dimensions.*

Determinants and their properties are studied in Section 10.7.

To help us remember the definition of the cross product, we use the notation of determinants. A **determinant of order two** is defined by

$$\begin{vmatrix} a & b \\ c & d \end{vmatrix} = ad - bc$$

For example,

$$\begin{vmatrix} 2 & 1 \\ -6 & 4 \end{vmatrix} = 2(4) - 1(-6) = 14$$

A **determinant of order three** is defined in terms of second-order determinants as

$$\begin{vmatrix} a_1 & a_2 & a_3 \\ b_1 & b_2 & b_3 \\ c_1 & c_2 & c_3 \end{vmatrix} = a_1\begin{vmatrix} b_2 & b_3 \\ c_2 & c_3 \end{vmatrix} - a_2\begin{vmatrix} b_1 & b_3 \\ c_1 & c_3 \end{vmatrix} + a_3\begin{vmatrix} b_1 & b_2 \\ c_1 & c_2 \end{vmatrix}$$

Observe that each term on the right side of the above equation involves a number a_i in the first row of the determinant, and a_i is multiplied by the second-order determinant obtained from the left side by deleting the row and column in which a_i appears. Notice also the minus sign in the second term. For example,

$$\begin{vmatrix} 1 & 2 & -1 \\ 3 & 0 & 1 \\ -5 & 4 & 2 \end{vmatrix} = 1\begin{vmatrix} 0 & 1 \\ 4 & 2 \end{vmatrix} - 2\begin{vmatrix} 3 & 1 \\ -5 & 2 \end{vmatrix} + (-1)\begin{vmatrix} 3 & 0 \\ -5 & 4 \end{vmatrix}$$

$$= 1(0 - 4) - 2(6 - (-5)) + (-1)(12 - 0) = -38$$

We can write the definition of the cross product using determinants as

$$\begin{vmatrix} \mathbf{i} & \mathbf{j} & \mathbf{k} \\ a_1 & a_2 & a_3 \\ b_1 & b_2 & b_3 \end{vmatrix} = \begin{vmatrix} a_2 & a_3 \\ b_2 & b_3 \end{vmatrix}\mathbf{i} - \begin{vmatrix} a_1 & a_3 \\ b_1 & b_3 \end{vmatrix}\mathbf{j} + \begin{vmatrix} a_1 & a_2 \\ b_1 & b_2 \end{vmatrix}\mathbf{k}$$

$$= (a_2b_3 - a_3b_2)\mathbf{i} - (a_1b_3 - a_3b_1)\mathbf{j} + (a_1b_2 - a_2b_1)\mathbf{k}$$

Although the first row of the above determinant consists of vectors, we expand it as if it were an ordinary determinant of order 3. The symbolic formula given by the above determinant is probably the easiest way to remember and compute cross products.

EXAMPLE 1 ■ Finding a Cross Product

If $\mathbf{u} = \langle 0, -1, 3\rangle$ and $\mathbf{v} = \langle 2, 0, -1\rangle$, find $\mathbf{u} \times \mathbf{v}$.

SOLUTION We use the formula above to find the cross product of $\mathbf{u}$ and $\mathbf{v}$:

$$\begin{aligned}\mathbf{u} \times \mathbf{v} &= \begin{vmatrix} \mathbf{i} & \mathbf{j} & \mathbf{k} \\ 0 & -1 & 3 \\ 2 & 0 & -1 \end{vmatrix} \\ &= \begin{vmatrix} -1 & 3 \\ 0 & -1 \end{vmatrix}\mathbf{i} - \begin{vmatrix} 0 & 3 \\ 2 & -1 \end{vmatrix}\mathbf{j} + \begin{vmatrix} 0 & -1 \\ 2 & 0 \end{vmatrix}\mathbf{k} \\ &= (1 - 0)\mathbf{i} - (0 - 6)\mathbf{j} + (0 - (-2))\mathbf{k} \\ &= \mathbf{i} + 6\mathbf{j} + 2\mathbf{k}\end{aligned}$$

So the desired vector is $\mathbf{i} + 6\mathbf{j} + 2\mathbf{k}$.

Now Try Exercise 3 ■

Properties of the Cross Product

One of the most important properties of the cross product is the following theorem.

CROSS PRODUCT THEOREM

The vector $\mathbf{u} \times \mathbf{v}$ is orthogonal (perpendicular) to both $\mathbf{u}$ and $\mathbf{v}$.

Proof To show that $\mathbf{u} \times \mathbf{v}$ is orthogonal to $\mathbf{u}$, we compute their dot product and show that it is 0.

$$\begin{aligned}(\mathbf{u} \times \mathbf{v}) \cdot \mathbf{u} &= \begin{vmatrix} a_2 & a_3 \\ b_2 & b_3 \end{vmatrix} a_1 - \begin{vmatrix} a_1 & a_3 \\ b_1 & b_3 \end{vmatrix} a_2 + \begin{vmatrix} a_1 & a_2 \\ b_1 & b_2 \end{vmatrix} a_3 \\ &= a_1(a_2b_3 - a_3b_2) - a_2(a_1b_3 - a_3b_1) + a_3(a_1b_2 - a_2b_1) \\ &= a_1a_2b_3 - a_1a_3b_2 - a_1a_2b_3 + a_2a_3b_1 + a_1a_3b_2 - a_2a_3b_1 \\ &= 0\end{aligned}$$

A similar computation shows that $(\mathbf{u} \times \mathbf{v}) \cdot \mathbf{v} = 0$. Therefore $\mathbf{u} \times \mathbf{v}$ is orthogonal to $\mathbf{u}$ and to $\mathbf{v}$. ■

EXAMPLE 2 ■ Finding an Orthogonal Vector

If $\mathbf{u} = -\mathbf{j} + 3\mathbf{k}$ and $\mathbf{v} = 2\mathbf{i} - \mathbf{k}$, find a unit vector that is orthogonal to the plane containing the vectors $\mathbf{u}$ and $\mathbf{v}$.

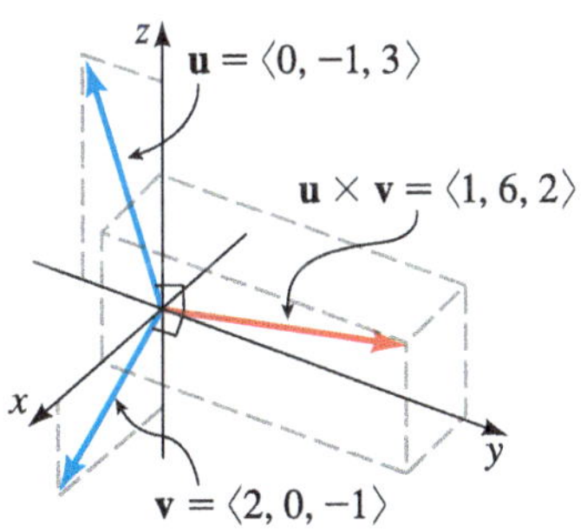

FIGURE 1 The vector $\mathbf{u} \times \mathbf{v}$ is perpendicular to $\mathbf{u}$ and $\mathbf{v}$.

SOLUTION By the Cross Product Theorem the vector $\mathbf{u} \times \mathbf{v}$ is orthogonal to the plane containing the vectors $\mathbf{u}$ and $\mathbf{v}$. (See Figure 1.) In Example 1 we found $\mathbf{u} \times \mathbf{v} = \mathbf{i} + 6\mathbf{j} + 2\mathbf{k}$. To obtain an orthogonal unit vector, we multiply $\mathbf{u} \times \mathbf{v}$ by the scalar $1/|\mathbf{u} \times \mathbf{v}|$:

$$\frac{\mathbf{u} \times \mathbf{v}}{|\mathbf{u} \times \mathbf{v}|} = \frac{\mathbf{i} + 6\mathbf{j} + 2\mathbf{k}}{\sqrt{1^2 + 6^2 + 2^2}} = \frac{\mathbf{i} + 6\mathbf{j} + 2\mathbf{k}}{\sqrt{41}}$$

So the desired vector is $\dfrac{1}{\sqrt{41}}(\mathbf{i} + 6\mathbf{j} + 2\mathbf{k})$.

Now Try Exercise 9 ■

EXAMPLE 3 ■ Finding a Vector Perpendicular to a Plane

Find a vector perpendicular to the plane that passes through the points $P(1, 4, 6)$, $Q(-2, 5, -1)$, and $R(1, -1, 1)$.

SOLUTION By the Cross Product Theorem the vector $\overrightarrow{PQ} \times \overrightarrow{PR}$ is perpendicular to both $\overrightarrow{PQ}$ and $\overrightarrow{PR}$ and is therefore perpendicular to the plane through P, Q, and R. We know that

$$\overrightarrow{PQ} = (-2 - 1)\mathbf{i} + (5 - 4)\mathbf{j} + (-1 - 6)\mathbf{k} = -3\mathbf{i} + \mathbf{j} - 7\mathbf{k}$$
$$\overrightarrow{PR} = (1 - 1)\mathbf{i} + (-1 - 4)\mathbf{j} + (1 - 6)\mathbf{k} = -5\mathbf{j} - 5\mathbf{k}$$

We compute the cross product of these vectors:

$$\overrightarrow{PQ} \times \overrightarrow{PR} = \begin{vmatrix} \mathbf{i} & \mathbf{j} & \mathbf{k} \\ -3 & 1 & -7 \\ 0 & -5 & -5 \end{vmatrix}$$
$$= (-5 - 35)\mathbf{i} - (15 - 0)\mathbf{j} + (15 - 0)\mathbf{k} = -40\mathbf{i} - 15\mathbf{j} + 15\mathbf{k}$$

So the vector $\langle -40, -15, 15\rangle$ is perpendicular to the given plane. Notice that any nonzero scalar multiple of this vector, such as $\langle -8, -3, 3\rangle$, is also perpendicular to the plane.

Now Try Exercise 17 ■

Mary Evans Picture Library/Alamy

WILLIAM ROWAN HAMILTON (1805–1865) was an Irish mathematician and physicist. He was raised by his uncle (a linguist) who noticed that Hamilton had a remarkable ability to learn languages. When he was five years old, he could read Latin, Greek, and Hebrew. At age eight he added French and Italian, and by age ten he had mastered Arabic and Sanskrit.

Hamilton was also a calculating prodigy and competed in contests of mental arithmetic. He entered Trinity College in Dublin, Ireland, where he studied science; he was appointed Professor of Astronomy there while still an undergraduate.

Hamilton made many contributions to mathematics and physics, but he is best known for his invention of quaternions. Hamilton knew that we can multiply vectors in the plane by considering them as complex numbers. He was looking for a similar multiplication for points in space. After thinking about this problem for over 20 years, he discovered the solution in a flash of insight while walking near Brougham Bridge in Dublin: He realized that a fourth dimension is needed to make the multiplication work. He carved the formula for his quaternions into the bridge, where it still stands. Later, the American mathematician Josiah Willard Gibbs extracted the dot product and cross product of vectors from the properties of quaternion multiplication. Quaternions are used today in computer graphics because of their ability to easily describe special rotations.

If $\mathbf{u}$ and $\mathbf{v}$ are represented by directed line segments with the same initial point (as in Figure 2), then the Cross Product Theorem says that the cross product $\mathbf{u} \times \mathbf{v}$ points in a direction perpendicular to the plane through $\mathbf{u}$ and $\mathbf{v}$. It turns out that the direction of $\mathbf{u} \times \mathbf{v}$ is given by the *right-hand rule:* If the fingers of your right hand curl in the direction of a rotation (through an angle less than 180°) from $\mathbf{u}$ to $\mathbf{v}$, then your thumb points in the direction of $\mathbf{u} \times \mathbf{v}$ (as in Figure 2). You can check that the vector $\mathbf{u} \times \mathbf{v}$ in Figure 1 satisfies the right-hand rule.

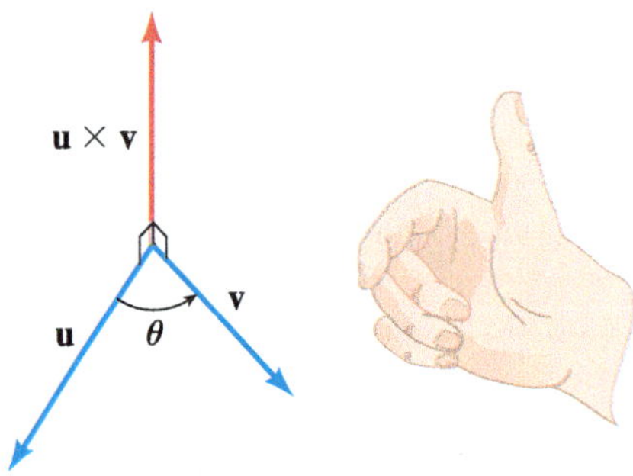

FIGURE 2 Right-hand rule

Now that we know the direction of the vector $\mathbf{u} \times \mathbf{v}$, the remaining thing we need is the length $|\mathbf{u} \times \mathbf{v}|$.

LENGTH OF THE CROSS PRODUCT

If θ is the angle between $\mathbf{u}$ and $\mathbf{v}$ (so $0 \le \theta \le \pi$), then

$$|\mathbf{u} \times \mathbf{v}| = |\mathbf{u}||\mathbf{v}|\sin\theta$$

In particular, two nonzero vectors $\mathbf{u}$ and $\mathbf{v}$ are parallel if and only if

$$\mathbf{u} \times \mathbf{v} = \mathbf{0}$$

Proof We apply the definitions of the cross product and length of a vector. You can verify the algebra in the first step by expanding the right-hand sides of the first and second lines and then comparing the results.

$$\begin{aligned} |\mathbf{u} \times \mathbf{v}|^2 &= (a_2b_3 - a_3b_2)^2 + (a_1b_3 - a_3b_1)^2 + (a_1b_2 - a_2b_1)^2 && \text{Definitions} \\ &= (a_1^2 + a_2^2 + a_3^2)(b_1^2 + b_2^2 + b_3^2) - (a_1b_1 + a_2b_2 + a_3b_3)^2 && \text{Verify algebra} \\ &= |\mathbf{u}|^2|\mathbf{v}|^2 - (\mathbf{u}\cdot\mathbf{v})^2 && \text{Definitions} \\ &= |\mathbf{u}|^2|\mathbf{v}|^2 - |\mathbf{u}|^2|\mathbf{v}|^2\cos^2\theta && \text{Property of Dot Product} \\ &= |\mathbf{u}|^2|\mathbf{v}|^2(1 - \cos^2\theta) && \text{Factor} \\ &= |\mathbf{u}|^2|\mathbf{v}|^2\sin^2\theta && \text{Pythagorean Identity} \end{aligned}$$

The result follows by taking square roots and observing that $\sqrt{\sin^2\theta} = \sin\theta$ because $\sin\theta \geq 0$ when $0 \leq \theta \leq \pi$. ■

We have now completely determined the vector $\mathbf{u} \times \mathbf{v}$ geometrically. The vector $\mathbf{u} \times \mathbf{v}$ is perpendicular to both $\mathbf{u}$ and $\mathbf{v}$, and its orientation is determined by the right-hand rule. The length of $\mathbf{u} \times \mathbf{v}$ is $|\mathbf{u}|\,|\mathbf{v}|\sin\theta$.

Area of a Parallelogram

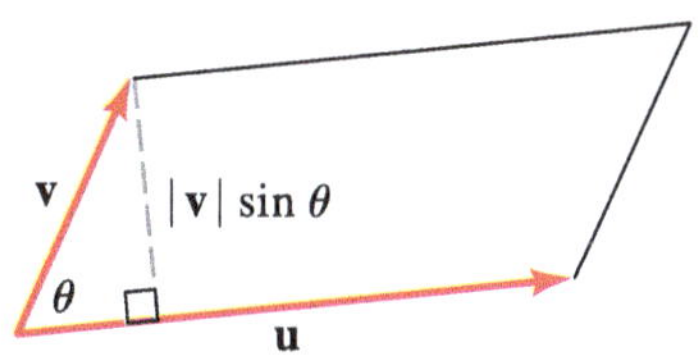

FIGURE 3 Parallelogram determined by **u** and **v**.

We can use the cross product to find the area of a parallelogram. If **u** and **v** are represented by directed line segments with the same initial point, then they determine a parallelogram with base $|\mathbf{u}|$, altitude $|\mathbf{v}|\sin\theta$, and area

$$A = |\mathbf{u}|(|\mathbf{v}|\sin\theta) = |\mathbf{u} \times \mathbf{v}|$$

(See Figure 3.) Thus we have the following way of interpreting the magnitude of a cross product.

AREA OF A PARALLELOGRAM

The length of the cross product $\mathbf{u} \times \mathbf{v}$ is the area of the parallelogram determined by **u** and **v**.

EXAMPLE 4 ■ Finding the Area of a Triangle

Find the area of the triangle with vertices $P(1, 4, 6)$, $Q(-2, 5, -1)$, and $R(1, -1, 1)$.

SOLUTION In Example 3 we computed that $\overrightarrow{PQ} \times \overrightarrow{PR} = \langle -40, -15, 15 \rangle$. The area of the parallelogram with adjacent sides PQ and PR is the length of this cross product:

$$|\overrightarrow{PQ} \times \overrightarrow{PR}| = \sqrt{(-40)^2 + (-15)^2 + 15^2} = 5\sqrt{82}$$

The area A of the triangle PQR is half the area of this parallelogram, that is, $\frac{5}{2}\sqrt{82}$.

Now Try Exercises 21 and 25 ■

Volume of a Parallelepiped

The product $\mathbf{u}\cdot(\mathbf{v} \times \mathbf{w})$ is called the **scalar triple product** of the vectors **u**, **v**, and **w**. You can check that the scalar triple product can be written as the following determinant:

$$\mathbf{u}\cdot(\mathbf{v} \times \mathbf{w}) = \begin{vmatrix} a_1 & a_2 & a_3 \\ b_1 & b_2 & b_3 \\ c_1 & c_2 & c_3 \end{vmatrix}$$

The geometric significance of the scalar triple product can be seen by considering the parallelepiped* determined by the vectors **u**, **v**, and **w** (see Figure 4). The area of the base parallelogram is $A = |\mathbf{v} \times \mathbf{w}|$. If θ is the angle between **u** and $\mathbf{v} \times \mathbf{w}$, then the height h of the parallelepiped is $h = |\mathbf{u}||\cos\theta|$. (We must use $|\cos\theta|$ instead of $\cos\theta$ in case $\theta > \pi/2$.) Therefore the volume of the parallelepiped is

$$V = Ah = |\mathbf{v} \times \mathbf{w}||\mathbf{u}||\cos\theta| = |\mathbf{u}\cdot(\mathbf{v} \times \mathbf{w})|$$

The last equality follows from the Dot Product Theorem on page 640.

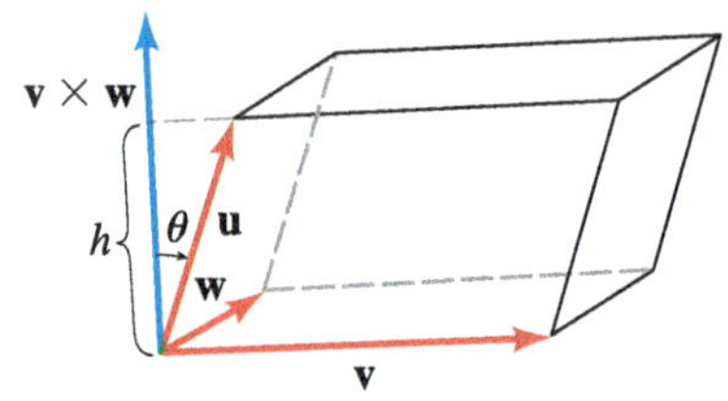

FIGURE 4 Parallelepiped determined by **u**, **v**, and **w**

We have proved the following formula.

VOLUME OF A PARALLELEPIPED

The volume of the parallelepiped determined by the vectors **u**, **v**, and **w** is the magnitude of their scalar triple product:

$$V = |\mathbf{u}\cdot(\mathbf{v} \times \mathbf{w})|$$

In particular, if the volume of the parallelepiped is 0, then the vectors **u**, **v**, and **w** are **coplanar**, that is, they lie in the same plane.

EXAMPLE 5 ■ Coplanar Vectors

Use the scalar triple product to show that the vectors $\mathbf{u} = \langle 1, 4, -7\rangle$, $\mathbf{v} = \langle 2, -1, 4\rangle$, and $\mathbf{w} = \langle 0, -9, 18\rangle$ are coplanar.

SOLUTION We compute the scalar triple product:

$$\begin{aligned}\mathbf{u}\cdot(\mathbf{v} \times \mathbf{w}) &= \begin{vmatrix} 1 & 4 & -7 \\ 2 & -1 & 4 \\ 0 & -9 & 18 \end{vmatrix} \\ &= 1\begin{vmatrix} -1 & 4 \\ -9 & 18 \end{vmatrix} - 4\begin{vmatrix} 2 & 4 \\ 0 & 18 \end{vmatrix} + (-7)\begin{vmatrix} 2 & -1 \\ 0 & -9 \end{vmatrix} \\ &= 1(18) - 4(36) - 7(-18) = 0\end{aligned}$$

So the volume of the parallelepiped is 0, and hence the vectors **u**, **v**, and **w** are coplanar.

Now Try Exercise 29 ■

*The word *parallelepiped* is derived from Greek roots which together mean, roughly, "parallel faces." While the word is often pronounced "par-al-lel-uh-PIE-ped," the more etymologically correct pronunciation is "par-al-lel-EP-uh-ped."

9.5 EXERCISES

CONCEPTS

1. The cross product of the vectors $\mathbf{u} = \langle a_1, a_2, a_3 \rangle$ and $\mathbf{v} = \langle b_1, b_2, b_3 \rangle$ is the vector

$$\mathbf{u} \times \mathbf{v} = \begin{vmatrix} \mathbf{i} & \mathbf{j} & \mathbf{k} \\ \_\_ & \_\_ & \_\_ \\ \_\_ & \_\_ & \_\_ \end{vmatrix}$$

$$= \_\_\_\_\_\_ \mathbf{i} - \_\_\_\_\_\_ \mathbf{j} + \_\_\_\_\_\_ \mathbf{k}$$

So the cross product of $\mathbf{u} = \langle 1, 0, 1 \rangle$ and $\mathbf{v} = \langle 2, 3, 0 \rangle$ is $\mathbf{u} \times \mathbf{v} =$ ______________.

2. The cross product of two vectors $\mathbf{u}$ and $\mathbf{v}$ is __________ to $\mathbf{u}$ and to $\mathbf{v}$. Thus if both vectors $\mathbf{u}$ and $\mathbf{v}$ lie in a plane, the vector $\mathbf{u} \times \mathbf{v}$ is __________ to the plane.

SKILLS

3–8 ■ Cross Products For the given vectors $\mathbf{u}$ and $\mathbf{v}$, find the cross product $\mathbf{u} \times \mathbf{v}$.

3. $\mathbf{u} = \langle 1, 0, -3 \rangle$, $\mathbf{v} = \langle 2, 3, 0 \rangle$
4. $\mathbf{u} = \langle 0, -4, 1 \rangle$, $\mathbf{v} = \langle 1, 1, -2 \rangle$
5. $\mathbf{u} = \langle 6, -2, 8 \rangle$, $\mathbf{v} = \langle -9, 3, -12 \rangle$
6. $\mathbf{u} = \langle -2, 3, 4 \rangle$, $\mathbf{v} = \langle \frac{1}{6}, -\frac{1}{4}, -\frac{1}{3} \rangle$
7. $\mathbf{u} = \mathbf{i} + \mathbf{j} + \mathbf{k}$, $\mathbf{v} = 3\mathbf{i} - 4\mathbf{k}$
8. $\mathbf{u} = 3\mathbf{i} - \mathbf{j}$, $\mathbf{v} = -3\mathbf{j} + \mathbf{k}$

9–12 ■ Orthogonal Vectors Two vectors $\mathbf{u}$ and $\mathbf{v}$ are given. **(a)** Find a vector orthogonal (perpendicular) to both $\mathbf{u}$ and $\mathbf{v}$. **(b)** Find a unit vector orthogonal (perpendicular) to both $\mathbf{u}$ and $\mathbf{v}$.

9. $\mathbf{u} = \langle 1, 1, -1 \rangle$, $\mathbf{v} = \langle -1, 1, -1 \rangle$
10. $\mathbf{u} = \langle 2, 5, 3 \rangle$, $\mathbf{v} = \langle 3, -2, -1 \rangle$
11. $\mathbf{u} = \frac{1}{2}\mathbf{i} - \mathbf{j} + \frac{2}{3}\mathbf{k}$, $\mathbf{v} = 6\mathbf{i} - 12\mathbf{j} - 6\mathbf{k}$
12. $\mathbf{u} = 3\mathbf{j} + 5\mathbf{k}$, $\mathbf{v} = -\mathbf{i} + 2\mathbf{k}$

13–16 ■ Length of a Cross Product The lengths of two vectors $\mathbf{u}$ and $\mathbf{v}$ and the angle θ between them are given. Find the length of their cross product, $|\mathbf{u} \times \mathbf{v}|$.

13. $|\mathbf{u}| = 6$, $|\mathbf{v}| = \frac{1}{2}$, $\theta = 60°$
14. $|\mathbf{u}| = 4$, $|\mathbf{v}| = 5$, $\theta = 30°$
15. $|\mathbf{u}| = 10$, $|\mathbf{v}| = 10$, $\theta = 90°$
16. $|\mathbf{u}| = 0.12$, $|\mathbf{v}| = 1.25$, $\theta = 75°$

17–20 ■ Vectors Perpendicular to a Plane Find a vector that is perpendicular to the plane passing through the three given points.

17. $P(0, 1, 0)$, $Q(1, 2, -1)$, $R(-2, 1, 0)$
18. $P(3, 4, 5)$, $Q(1, 2, 3)$, $R(4, 7, 6)$
19. $P(1, 1, -5)$, $Q(2, 2, 0)$, $R(0, 0, 0)$
20. $P(3, 0, 0)$, $Q(0, 2, -5)$, $R(-2, 0, 6)$

21–24 ■ Area of a Parallelogram Find the area of the parallelogram determined by the given vectors.

21. $\mathbf{u} = \langle 3, 2, 1 \rangle$, $\mathbf{v} = \langle 1, 2, 3 \rangle$
22. $\mathbf{u} = \langle 0, -3, 2 \rangle$, $\mathbf{v} = \langle 5, -6, 0 \rangle$
23. $\mathbf{u} = 2\mathbf{i} - \mathbf{j} + 4\mathbf{k}$, $\mathbf{v} = \frac{1}{2}\mathbf{i} + 2\mathbf{j} - \frac{3}{2}\mathbf{k}$
24. $\mathbf{u} = \mathbf{i} - \mathbf{j} + \mathbf{k}$, $\mathbf{v} = \mathbf{i} + \mathbf{j} - \mathbf{k}$

25–28 ■ Area of a Triangle Find the area of $\triangle PQR$.

25. $P(1, 0, 1)$, $Q(0, 1, 0)$, $R(2, 3, 4)$
26. $P(2, 1, 0)$, $Q(0, 0, -1)$, $R(-4, 2, 0)$
27. $P(6, 0, 0)$, $Q(0, -6, 0)$, $R(0, 0, -6)$
28. $P(3, -2, 6)$, $Q(-1, -4, -6)$, $R(3, 4, 6)$

29–34 ■ Volume of a Parallelepiped Three vectors $\mathbf{u}$, $\mathbf{v}$, and $\mathbf{w}$ are given. **(a)** Find their scalar triple product $\mathbf{u} \cdot (\mathbf{v} \times \mathbf{w})$. **(b)** Are the vectors coplanar? If not, find the volume of the parallelepiped that they determine.

29. $\mathbf{u} = \langle 1, 2, 3 \rangle$, $\mathbf{v} = \langle -3, 2, 1 \rangle$, $\mathbf{w} = \langle 0, 8, 10 \rangle$
30. $\mathbf{u} = \langle 3, 0, -4 \rangle$, $\mathbf{v} = \langle 1, 1, 1 \rangle$, $\mathbf{w} = \langle 7, 4, 0 \rangle$
31. $\mathbf{u} = \langle 2, 3, -2 \rangle$, $\mathbf{v} = \langle -1, 4, 0 \rangle$, $\mathbf{w} = \langle 3, -1, 3 \rangle$
32. $\mathbf{u} = \langle 1, -1, 0 \rangle$, $\mathbf{v} = \langle -1, 0, 1 \rangle$, $\mathbf{w} = \langle 0, -1, 1 \rangle$
33. $\mathbf{u} = \mathbf{i} - \mathbf{j} + \mathbf{k}$, $\mathbf{v} = -\mathbf{j} + \mathbf{k}$, $\mathbf{w} = \mathbf{i} + \mathbf{j} + \mathbf{k}$
34. $\mathbf{u} = 2\mathbf{i} - 2\mathbf{j} - 3\mathbf{k}$, $\mathbf{v} = 3\mathbf{i} - \mathbf{j} - \mathbf{k}$, $\mathbf{w} = 6\mathbf{i}$

APPLICATIONS

35. **Volume of a Fish Tank** A fish tank in an avant-garde restaurant is in the shape of a parallelepiped with a rectangular base that is 300 cm long and 120 cm wide. The front and back faces are vertical, but the left and right faces are slanted at 30° from the vertical and measure 120 cm by 150 cm. (See the figure.)

    (a) Let $\mathbf{u}$, $\mathbf{v}$, and $\mathbf{w}$ be the three vectors shown in the figure. Find $\mathbf{u} \cdot (\mathbf{v} \times \mathbf{w})$. [*Hint:* Recall that $\mathbf{u} \cdot \mathbf{v} = |\mathbf{u}||\mathbf{v}|\cos\theta$ and $|\mathbf{u} \times \mathbf{v}| = |\mathbf{u}||\mathbf{v}|\sin\theta$.]

    (b) What is the capacity of the tank in liters? [*Note:* 1 L = 1000 cm³.]

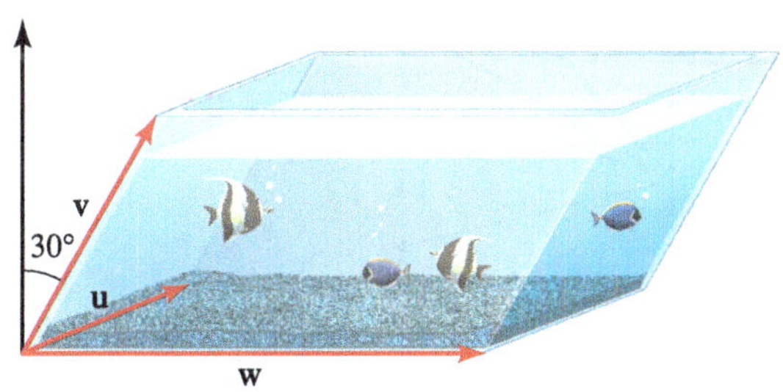

36. Rubik's Tetrahedron Rubik's Cube, a puzzle craze of the 1980s that remains popular to this day, inspired many similar puzzles. The one illustrated in the figure is called Rubik's Tetrahedron; it is in the shape of a regular tetrahedron, with each edge $\sqrt{2}$ inches long. The volume of a regular tetrahedron is one-sixth the volume of the parallelepiped determined by any three edges that meet at a corner.

(a) Use the triple product to find the volume of Rubik's Tetrahedron. [*Hint:* See Exercise 50 in Section 9.4, which gives the corners of a tetrahedron that has the same shape and size as Rubik's Tetrahedron.]

(b) Construct six identical regular tetrahedra using modeling clay. Experiment to see how they can be put together to create a parallelepiped that is determined by three edges of one of the tetrahedra (thus confirming the above statement about the volume of a regular tetrahedron).

DISCUSS ■ DISCOVER ■ PROVE ■ WRITE

37. DISCOVER ■ PROVE: Order of Operations in the Triple Product Given three vectors **u**, **v**, and **w**, their scalar triple product can be performed in six different orders:

$$\mathbf{u}\cdot(\mathbf{v}\times\mathbf{w}),\quad \mathbf{u}\cdot(\mathbf{w}\times\mathbf{v}),\quad \mathbf{v}\cdot(\mathbf{u}\times\mathbf{w}),$$
$$\mathbf{v}\cdot(\mathbf{w}\times\mathbf{u}),\quad \mathbf{w}\cdot(\mathbf{u}\times\mathbf{v}),\quad \mathbf{w}\cdot(\mathbf{v}\times\mathbf{u})$$

(a) Calculate each of these six triple products for the vectors:

$$\mathbf{u} = \langle 0, 1, 1\rangle \qquad \mathbf{v} = \langle 1, 0, 1\rangle \qquad \mathbf{w} = \langle 1, 1, 0\rangle$$

(b) On the basis of your observations in part (a), make a conjecture about the relationships between these six triple products.

(c) Prove the conjecture you made in part (b).

9.6 EQUATIONS OF LINES AND PLANES

■ Equations of Lines ■ Equations of Planes

In this section we find equations for lines and planes in a three-dimensional coordinate space. We use vectors to help us find such equations.

■ Equations of Lines

The **position vector** of a point (a_1, a_2, a_3) is the vector $\langle a_1, a_2, a_3\rangle$; that is, it is the vector from the origin to the point.

A line L in three-dimensional space is determined when we know a point $P_0(x_0, y_0, z_0)$ on L and the direction of L. In three dimensions the direction of a line is described by a vector **v** parallel to L. If we let $\mathbf{r}_0$ be the position vector of P_0 (that is, the vector $\overrightarrow{OP_0}$), then for all real numbers t the terminal points P of the position vectors $\mathbf{r}_0 + t\mathbf{v}$ trace out a line parallel to **v** and passing through P_0 (see Figure 1). Each value of the parameter t gives a point P on L. So the line L is given by the position vector **r**, where

$$\mathbf{r} = \mathbf{r}_0 + t\mathbf{v}$$

for $t \in \mathbb{R}$. This is the **vector equation of a line**.

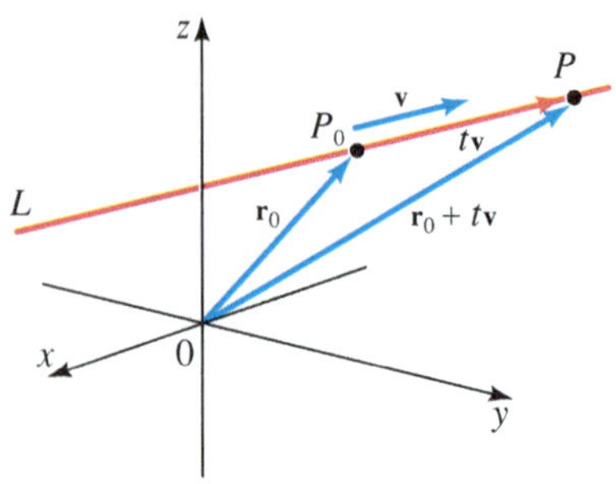

FIGURE 1

Let's write the vector **v** in component form $\mathbf{v} = \langle a, b, c\rangle$ and let $\mathbf{r}_0 = \langle x_0, y_0, z_0\rangle$ and $\mathbf{r} = \langle x, y, z\rangle$. Then the vector equation of the line becomes

$$\begin{aligned}\langle x, y, z\rangle &= \langle x_0, y_0, z_0\rangle + t\langle a, b, c\rangle \\ &= \langle x_0 + ta, y_0 + tb, z_0 + tc\rangle\end{aligned}$$

Since two vectors are equal if and only if their corresponding components are equal, we have the following result.

PARAMETRIC EQUATIONS FOR A LINE

A line passing through the point $P(x_0, y_0, z_0)$ and parallel to the vector $\mathbf{v} = \langle a, b, c\rangle$ is described by the parametric equations

$$x = x_0 + at$$
$$y = y_0 + bt$$
$$z = z_0 + ct$$

where t is any real number.

EXAMPLE 1 ■ Equations of a Line

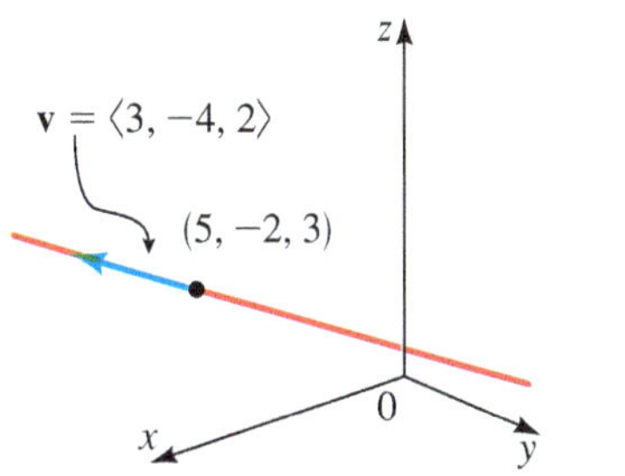

FIGURE 2 Line through $(5, -2, 3)$ with direction $\mathbf{v} = \langle 3, -4, 2\rangle$

Find parametric equations for the line that passes through the point $(5, -2, 3)$ and is parallel to the vector $\mathbf{v} = \langle 3, -4, 2\rangle$.

SOLUTION We use the above formula to find the parametric equations:

$$x = 5 + 3t$$
$$y = -2 - 4t$$
$$z = 3 + 2t$$

where t is any real number. (See Figure 2.)

Now Try Exercise 3 ■

EXAMPLE 2 ■ Equations of a Line

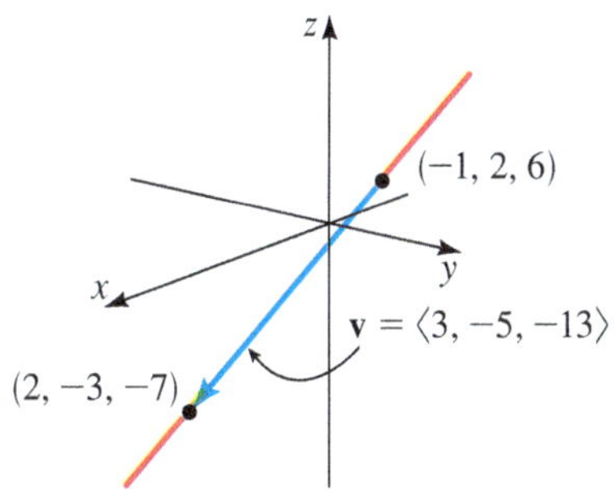

FIGURE 3 Line through $(-1, 2, 6)$ and $(2, -3, -7)$

Find parametric equations for the line that passes through the points $(-1, 2, 6)$ and $(2, -3, -7)$.

SOLUTION We first find a vector determined by the two points:

$$\mathbf{v} = \langle 2 - (-1), -3 - 2, -7 - 6\rangle = \langle 3, -5, -13\rangle$$

Now we use $\mathbf{v}$ and the point $(-1, 2, 6)$ to find the parametric equations:

$$x = -1 + 3t$$
$$y = 2 - 5t$$
$$z = 6 - 13t$$

where t is any real number. A graph of the line is shown in Figure 3.

Now Try Exercise 9 ■

In Example 2 we used the point $(-1, 2, 6)$ to get the parametric equations of the line. We could instead use the point $(2, -3, -7)$. The resulting parametric equations would look different but would still describe the same line (see Exercise 37).

■ Equations of Planes

Although a line in space is determined by a point and a direction, the "direction" of a plane cannot be described by a vector in the plane. In fact, different vectors in a plane can have different directions. But a vector perpendicular to a plane *does* completely specify the direction of the plane. Thus a plane in space is determined by a point

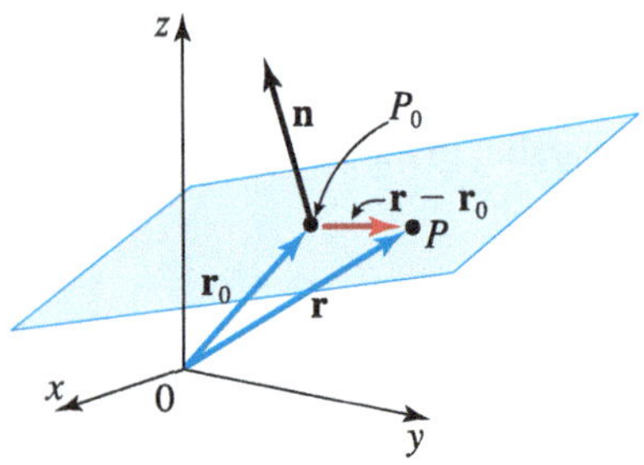

FIGURE 4

$P_0(x_0, y_0, z_0)$ in the plane and a vector $\mathbf{n}$ that is orthogonal to the plane. This orthogonal vector $\mathbf{n}$ is called a **normal vector**. To determine whether a point $P(x, y, z)$ is in the plane, we check whether the vector $\overrightarrow{P_0P}$ with initial point P_0 and terminal point P is orthogonal to the normal vector. Let $\mathbf{r}_0$ and $\mathbf{r}$ be the position vectors of P_0 and P, respectively. Then the vector $\overrightarrow{P_0P}$ is represented by $\mathbf{r} - \mathbf{r}_0$ (see Figure 4). So the plane is described by the tips of the vectors $\mathbf{r}$ satisfying

$$\mathbf{n} \cdot (\mathbf{r} - \mathbf{r}_0) = 0$$

This is the **vector equation of the plane.**

Let's write the normal vector $\mathbf{n}$ in component form $\mathbf{n} = \langle a, b, c \rangle$ and let $\mathbf{r}_0 = \langle x_0, y_0, z_0 \rangle$ and $\mathbf{r} = \langle x, y, z \rangle$. Then the vector equation of the plane becomes

$$\langle a, b, c \rangle \cdot \langle x - x_0, y - y_0, z - z_0 \rangle = 0$$

Performing the dot product, we arrive at the following equation of the plane in the variables x, y, and z.

EQUATION OF A PLANE

The plane containing the point $P(x_0, y_0, z_0)$ and having the normal vector $\mathbf{n} = \langle a, b, c \rangle$ is described by the equation

$$a(x - x_0) + b(y - y_0) + c(z - z_0) = 0$$

EXAMPLE 3 ■ Finding an Equation for a Plane

A plane has normal vector $\mathbf{n} = \langle 4, -6, 3 \rangle$ and passes through the point $P(3, -1, -2)$.

(a) Find an equation of the plane.

(b) Find the intercepts, and sketch a graph of the plane.

SOLUTION

(a) By the above formula for the equation of a plane we have

$$\begin{aligned} 4(x - 3) - 6(y - (-1)) + 3(z - (-2)) &= 0 && \text{Formula} \\ 4x - 12 - 6y - 6 + 3z + 6 &= 0 && \text{Expand} \\ 4x - 6y + 3z &= 12 && \text{Simplify} \end{aligned}$$

Thus an equation of the plane is $4x - 6y + 3z = 12$.

(b) To find the x-intercept, we set $y = 0$ and $z = 0$ in the equation of the plane and solve for x. Similarly, we find the y- and z-intercepts.

x-intercept: Setting $y = 0, z = 0$, we get $x = 3$.

y-intercept: Setting $x = 0, z = 0$, we get $y = -2$.

z-intercept: Setting $x = 0, y = 0$, we get $z = 4$.

So the graph of the plane intersects the coordinate axes at the points $(3, 0, 0)$, $(0, -2, 0)$, and $(0, 0, 4)$. This enables us to sketch the portion of the plane shown in Figure 5.

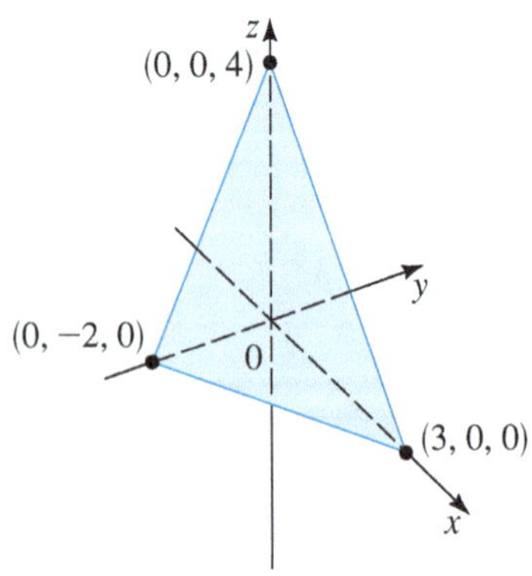

FIGURE 5 The plane $4x - 6y + 3z = 32$

Notice that in Figure 5 the axes have been rotated so that we get a better view.

Now Try Exercise 15 ■

EXAMPLE 4 ■ Finding an Equation for a Plane

Find an equation of the plane that passes through the points $P(1, 4, 6)$, $Q(-2, 5, -1)$, and $R(1, -1, 1)$.

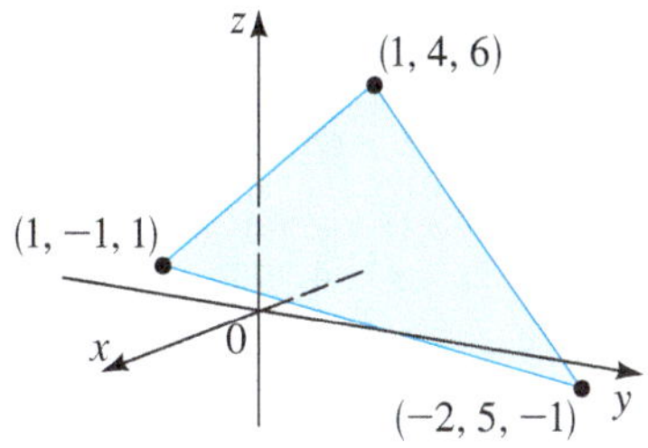

FIGURE 6 A plane through three points

SOLUTION The vector $\mathbf{n} = \overrightarrow{PQ} \times \overrightarrow{PR}$ is perpendicular to both $\overrightarrow{PQ}$ and $\overrightarrow{PR}$ and is therefore perpendicular to the plane through P, Q, and R. In Example 3 of Section 9.5 we found $\overrightarrow{PQ} \times \overrightarrow{PR} = \langle -40, -15, 15 \rangle$. Using the formula for an equation of a plane, we have

$$\begin{aligned} -40(x-1) - 15(y-4) + 15(z-6) &= 0 && \text{Formula} \\ -40x + 40 - 15y + 60 + 15z - 90 &= 0 && \text{Expand} \\ -40x - 15y + 15z &= -10 && \text{Simplify} \\ 8x + 3y - 3z &= 2 && \text{Divide by } -5 \end{aligned}$$

So an equation of the plane is $8x + 3y - 3z = 2$. A graph of this plane is shown in Figure 6.

Now Try Exercise 21 ■

In Example 4 we used the point P to obtain the equation of the plane. You can check that using Q or R gives the same equation.

9.6 EXERCISES

CONCEPTS

1. A line in space is described algebraically by using _________ equations. The line that passes through the point $P(x_0, y_0, z_0)$ and is parallel to the vector $\mathbf{v} = \langle a, b, c \rangle$ is described by the equations $x =$ _________, $y =$ _________, $z =$ _________.

2. The plane containing the point $P(x_0, y_0, z_0)$ and having the normal vector $\mathbf{n} = \langle a, b, c \rangle$ is described algebraically by the equation _________.

SKILLS

3–8 ■ Equations of Lines Find parametric equations for the line that passes through the point P and is parallel to the vector $\mathbf{v}$.

3. $P(1, 0, -2)$, $\mathbf{v} = \langle 3, 2, -3 \rangle$
4. $P(0, -5, 3)$, $\mathbf{v} = \langle 2, 0, -4 \rangle$
5. $P(3, 2, 1)$, $\mathbf{v} = \langle 0, -4, 2 \rangle$
6. $P(0, 0, 0)$, $\mathbf{v} = \langle -4, 3, 5 \rangle$
7. $P(1, 0, -2)$, $\mathbf{v} = 2\mathbf{i} - 5\mathbf{k}$
8. $P(1, 1, 1)$, $\mathbf{v} = \mathbf{i} - \mathbf{j} + \mathbf{k}$

9–14 ■ Equations of Lines Find parametric equations for the line that passes through the points P and Q.

9. $P(1, -3, 2)$, $Q(2, 1, -1)$
10. $P(2, -1, -2)$, $Q(0, 1, -3)$
11. $P(1, 1, 0)$, $Q(0, 2, 2)$
12. $P(3, 3, 3)$, $Q(7, 0, 0)$
13. $P(3, 7, -5)$, $Q(7, 3, -5)$
14. $P(12, 16, 18)$, $Q(12, -6, 0)$

15–20 ■ Equations of Planes A plane has normal vector $\mathbf{n}$ and passes through the point P. **(a)** Find an equation for the plane. **(b)** Find the intercepts, and sketch a graph of the plane.

15. $\mathbf{n} = \langle 1, 1, -1 \rangle$, $P(0, 2, -3)$
16. $\mathbf{n} = \langle 3, 2, 0 \rangle$, $P(1, 2, 7)$
17. $\mathbf{n} = \langle 3, 0, -\frac{1}{2} \rangle$, $P(2, 4, 8)$
18. $\mathbf{n} = \langle \frac{2}{3}, -\frac{1}{3}, -1 \rangle$, $P(6, 0, 3)$
19. $\mathbf{n} = 3\mathbf{i} - \mathbf{j} + 2\mathbf{k}$, $P(0, 2, -3)$
20. $\mathbf{n} = \mathbf{i} + 4\mathbf{j}$, $P(1, 0, -9)$

21–26 ■ Equations of Planes Find an equation of the plane that passes through the points P, Q, and R.

21. $P(6, -2, 1)$, $Q(5, -3, -1)$, $R(7, 0, 0)$
22. $P(3, 4, 5)$, $Q(1, 2, 3)$, $R(4, 7, 6)$
23. $P(3, \frac{1}{3}, -5)$, $Q(4, \frac{2}{3}, -3)$, $R(2, 0, 1)$
24. $P(\frac{3}{2}, 4, -2)$, $Q(-\frac{1}{2}, 2, 0)$, $R(-\frac{1}{2}, 0, 2)$
25. $P(6, 1, 1)$, $Q(3, 2, 0)$, $R(0, 0, 0)$
26. $P(2, 0, 0)$, $Q(0, 2, -2)$, $R(0, 0, 4)$

SKILLS Plus

27–30 ■ Equations of Lines A description of a line is given. Find parametric equations for the line.

27. The line crosses the z-axis where $z = 4$ and crosses the xy-plane where $x = 2$ and $y = 5$.
28. The line crosses the x-axis where $x = -2$ and crosses the z-axis where $z = 10$.
29. The line perpendicular to the xz-plane that contains the point $(2, -1, 5)$.
30. The line parallel to the y-axis that crosses the xz-plane where $x = -3$ and $z = 2$.

31–32 ■ Equations of Planes A description of a plane is given. Find an equation for the plane.

31. The plane that crosses the x-axis where $x = 1$, the y-axis where $y = 3$, and the z-axis where $z = 4$.

32. The plane that is parallel to the plane $x - 2y + 4z = 6$ and contains the origin.

33–34 ■ More Equations of Planes A description of a plane is given. Find an equation for the plane.

33. The plane that contains all the points that are equidistant from the points $P(-3, 2, 5)$ and $Q(1, -1, 4)$.

34. The plane that contains the line $x = 1 - t, y = 2 + t, z = -3t$ and the point $P(2, 0, -6)$. [*Hint:* A vector from any point on the line to P will lie in the plane.]

DISCUSS ■ DISCOVER ■ PROVE ■ WRITE

35. DISCOVER: Intersection of a Line and a Plane A line has parametric equations

$$x = 2 + t \qquad y = 3t \qquad z = 5 - t$$

and a plane has equation $5x - 2y - 2z = 1$.

(a) For what value of t does the corresponding point on the line intersect the plane?

(b) At what point do the line and the plane intersect?

36. DISCUSS ■ DISCOVER: Lines and Planes A line is parallel to the vector $\mathbf{v}$, and a plane has normal vector $\mathbf{n}$.

(a) If the line is perpendicular to the plane, what is the relationship between $\mathbf{v}$ and $\mathbf{n}$ (parallel or perpendicular)?

(b) If the line is parallel to the plane (that is, the line and the plane do not intersect), what is the relationship between $\mathbf{v}$ and $\mathbf{n}$ (parallel or perpendicular)?

(c) Parametric equations for two lines are given. Which line is parallel to the plane $x - y + 4z = 6$? Which line is perpendicular to this plane?

Line 1: $x = 2t, \quad y = 3 - 2t, \quad z = 4 + 8t$

Line 2: $x = -2t, \quad y = 5 + 2t, \quad z = 3 + t$

37. DISCUSS: Same Line: Different Parametric Equations Every line can be described by infinitely many different sets of parametric equations, since *any* point on the line and *any* vector parallel to the line can be used to construct the equations. But how can we tell whether two sets of parametric equations represent the same line? Consider the following two sets of parametric equations:

Line 1: $x = 1 - t, \quad y = 3t, \quad z = -6 + 5t$

Line 2: $x = -1 + 2t, \quad y = 6 - 6t, \quad z = 4 - 10t$

(a) Find two points that lie on Line 1 by setting $t = 0$ and $t = 1$ in its parametric equations. Then show that these points also lie on Line 2 by finding two values of the parameter that give these points when substituted into the parametric equations for Line 2.

(b) Show that the following two lines are not the same by finding a point on Line 3 and then showing that it does not lie on Line 4.

Line 3: $x = 4t, \quad y = 3 - 6t, \quad z = -5 + 2t$

Line 4: $x = 8 - 2t, \quad y = -9 + 3t, \quad z = 6 - t$

CHAPTER 9 ■ REVIEW

PROPERTIES AND FORMULAS

Vectors in Two Dimensions (p. 631)

A **vector** is a quantity with both magnitude and direction. A vector in the coordinate plane is expressed in terms of two coordinates or components

$$\mathbf{v} = \langle a_1, a_2 \rangle$$

If a vector $\mathbf{v}$ has its initial point at $P(x_1, y_1)$ and its terminal point at $Q(x_2, y_2)$, then

$$\mathbf{v} = \langle x_2 - x_1, y_2 - y_1 \rangle$$

Let $\mathbf{u} = \langle a_1, a_2 \rangle$, $\mathbf{v} = \langle b_1, b_2 \rangle$, and $c \in \mathbb{R}$. The operations on vectors are defined as follows.

$\mathbf{u} + \mathbf{v} = \langle a_1 + b_1, a_2 + b_2 \rangle$	Addition
$\mathbf{u} - \mathbf{v} = \langle a_1 - b_1, a_2 - b_2 \rangle$	Subtraction
$c\mathbf{u} = \langle ca_1, ca_2 \rangle$	Scalar multiplication

The vectors $\mathbf{i}$ and $\mathbf{j}$ are defined by

$$\mathbf{i} = \langle 1, 0 \rangle \qquad \mathbf{j} = \langle 0, 1 \rangle$$

Any vector $\mathbf{v} = \langle a_1, a_2 \rangle$ can be expressed as

$$\mathbf{v} = a_1\mathbf{i} + a_2\mathbf{j}$$

Let $\mathbf{v} = \langle a_1, a_2 \rangle$. The **magnitude** (or **length**) of $\mathbf{v}$ is

$$|\mathbf{v}| = \sqrt{a_1^2 + a_2^2}$$

The **direction** of $\mathbf{v}$ is the smallest positive angle θ in standard position formed by the positive x-axis and $\mathbf{v}$ (see the figure below).

If $\mathbf{v} = \langle a_1, a_2 \rangle$, then the components of $\mathbf{v}$ satisfy

$$a_1 = |\mathbf{v}| \cos\theta \qquad a_2 = |\mathbf{v}| \sin\theta$$

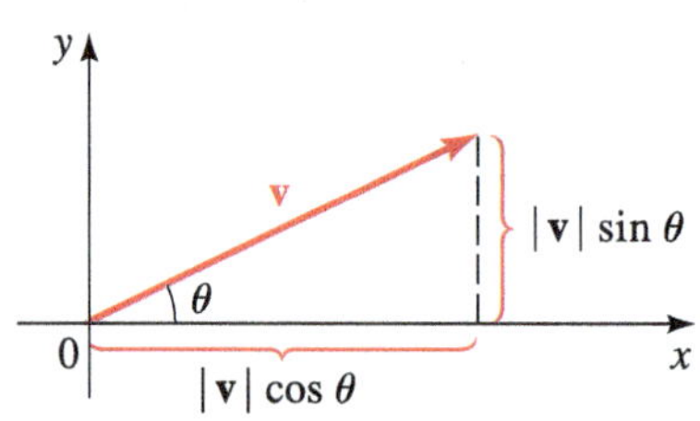

The Dot Product of Vectors (p. 640)

If $\mathbf{u} = \langle a_1, a_2 \rangle$ and $\mathbf{v} = \langle b_1, b_2 \rangle$, then their **dot product** is

$$\mathbf{u} \cdot \mathbf{v} = a_1 b_1 + a_2 b_2$$

If θ is the angle between **u** and **v**, then

$$\mathbf{u} \cdot \mathbf{v} = |\mathbf{u}|\,|\mathbf{v}|\cos\theta$$

The angle θ between **u** and **v** satisfies

$$\cos\theta = \frac{\mathbf{u} \cdot \mathbf{v}}{|\mathbf{u}|\,|\mathbf{v}|}$$

The vectors **u** and **v** are perpendicular if and only if

$$\mathbf{u} \cdot \mathbf{v} = 0$$

The **component of u along v** (a scalar) and the **projection of u onto v** (a vector) are given by

$$\text{comp}_{\mathbf{v}}\,\mathbf{u} = \frac{\mathbf{u} \cdot \mathbf{v}}{|\mathbf{v}|} \qquad \text{proj}_{\mathbf{v}}\,\mathbf{u} = \left(\frac{\mathbf{u} \cdot \mathbf{v}}{|\mathbf{v}|^2}\right)\mathbf{v}$$

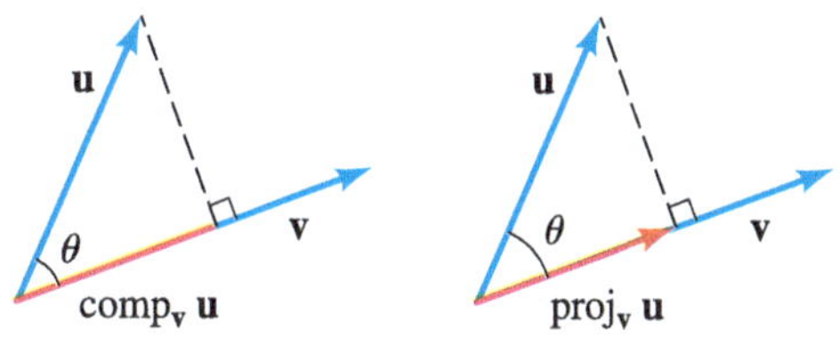

The **work** W done by a force **F** in moving along a vector **D** is

$$W = \mathbf{F} \cdot \mathbf{D}$$

Three-Dimensional Coordinate Geometry (p. 648)

A coordinate system in space consists of a fixed point O (the **origin**) and three directed lines through O that are perpendicular to each other, called the **coordinate axes** and labeled the x-axis, y-axis, and z-axis. The coordinates of a point $P(a, b, c)$ determine its location relative to the coordinate axes.

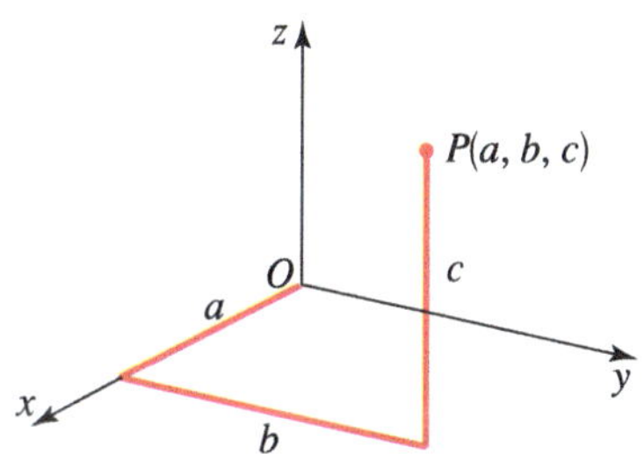

The distance between the points $P(x_1, y_1, z_1)$ and $Q(x_2, y_2, z_2)$ is given by the **Distance Formula:**

$$d(P, Q) = \sqrt{(x_2 - x_1)^2 + (y_2 - y_1)^2 + (z_2 - z_1)^2}$$

The **equation of a sphere** with center $C(h, k, l)$ and radius r is

$$(x - h)^2 + (y - k)^2 + (z - l)^2 = r^2$$

Vectors in Three Dimensions (p. 653)

A **vector in space** is a line segment with a direction. We sketch a vector as an arrow to indicate the direction. A **vector in the three dimensional coordinate system** is expressed in terms of three coordinates or components

$$\mathbf{v} = \langle a_1, a_2, a_3 \rangle$$

If a vector **v** has its initial point at $P(x_1, y_1, z_1)$ and its terminal point at $Q(x_2, y_2, z_2)$, then

$$\mathbf{v} = \langle x_2 - x_1, y_2 - y_1, z_2 - z_1 \rangle$$

Let $\mathbf{u} = \langle a_1, a_2, a_3 \rangle$, $\mathbf{v} = \langle b_1, b_2, b_3 \rangle$, and $c \in \mathbb{R}$. The operations of vector addition, vector subtraction, scalar multiplication are defined as follows:

$$\mathbf{u} + \mathbf{v} = \langle a_1 + b_1, a_2 + b_2, a_3 + b_3 \rangle$$

$$\mathbf{u} - \mathbf{v} = \langle a_1 - b_1, a_2 - b_2, a_3 - b_3 \rangle$$

$$c\mathbf{u} = \langle ca_1, ca_2, ca_3 \rangle$$

The vectors **i**, **j**, and **k** are defined by

$$\mathbf{i} = \langle 1, 0, 0 \rangle \qquad \mathbf{j} = \langle 0, 1, 0 \rangle \qquad \mathbf{k} = \langle 0, 0, 1 \rangle$$

Any vector $\mathbf{v} = \langle a_1, a_2, a_3 \rangle$ can be expressed as

$$\mathbf{v} = a_1\mathbf{i} + a_2\mathbf{j} + a_3\mathbf{k}$$

Let $\mathbf{v} = \langle a_1, a_2, a_3 \rangle$. The **magnitude** (or **length**) of **v** is

$$|\mathbf{v}| = \sqrt{a_1^2 + a_2^2 + a_3^2}$$

The **direction angles** of a nonzero vector $\mathbf{v} = \langle a_1, a_2, a_3 \rangle$ are the angles α, β, and γ in the interval $[0, \pi]$ that the vector **v** makes with the positive x-, y-, and z-axes. They are given by

$$\cos\alpha = \frac{a_1}{|\mathbf{v}|} \qquad \cos\beta = \frac{a_2}{|\mathbf{v}|} \qquad \cos\gamma = \frac{a_3}{|\mathbf{v}|}$$

The direction angles satisfy the equation

$$\cos^2\alpha + \cos^2\beta + \cos^2\gamma = 1$$

The Dot Product of Vectors in Space (p. 655)

If $\mathbf{u} = \langle a_1, a_2, a_3 \rangle$ and $\mathbf{v} = \langle b_1, b_2, b_3 \rangle$ are vectors in space, then their **dot product** is

$$\mathbf{u} \cdot \mathbf{v} = a_1 b_1 + a_2 b_2 + a_3 b_3$$

If θ is the angle between **u** and **v**, then

$$\mathbf{u} \cdot \mathbf{v} = |\mathbf{u}|\,|\mathbf{v}|\cos\theta$$

The angle θ between **u** and **v** satisfies

$$\cos\theta = \frac{\mathbf{u} \cdot \mathbf{v}}{|\mathbf{u}|\,|\mathbf{v}|}$$

The vectors **u** and **v** are perpendicular if and only if

$$\mathbf{u} \cdot \mathbf{v} = 0$$

The Cross Product of Vectors in Space (p. 659)

If $\mathbf{u} = \langle a_1, a_2, a_3 \rangle$ and $\mathbf{v} = \langle b_1, b_2, b_3 \rangle$ are vectors in space, then their **cross product** is the vector

$$\mathbf{u} \times \mathbf{v} = (a_2 b_3 - a_3 b_2)\mathbf{i} - (a_1 b_3 - a_3 b_1)\mathbf{j} + (a_1 b_2 - a_2 b_1)\mathbf{k}$$

We can calculate the cross product using determinants.

$$\mathbf{u} \times \mathbf{v} = \begin{vmatrix} \mathbf{i} & \mathbf{j} & \mathbf{k} \\ a_1 & a_2 & a_3 \\ b_1 & b_2 & b_3 \end{vmatrix}$$

The vector $\mathbf{u} \times \mathbf{v}$ is orthogonal (or perpendicular) to both **u** and **v**.

The cross product satisfies

$$|\mathbf{u} \times \mathbf{v}| = |\mathbf{u}||\mathbf{v}| \sin \theta$$

The vectors **u** and **v** are parallel if and only if

$$\mathbf{u} \times \mathbf{v} = \mathbf{0}$$

The **area of the parallelogram** determined by the vectors **u** and **v** is

$$A = |\mathbf{u} \times \mathbf{v}|$$

The **volume of the parallelepiped** determined by the vectors **u**, **v**, and **w** is

$$V = |\mathbf{u} \cdot (\mathbf{v} \times \mathbf{w})|$$

Equations of Lines and Planes (p. 666)

A **line** passing through the point $P(x_0, y_0, z_0)$ and parallel to the vector $\mathbf{v} = \langle a, b, c \rangle$ is described by the parametric equations

$$x = x_0 + at$$
$$y = y_0 + bt$$
$$z = z_0 + ct$$

where t is any real number.

A **plane** containing the point $P(x_0, y_0, z_0)$ and having the normal vector $\mathbf{n} = \langle a, b, c \rangle$ is described by the equation

$$a(x - x_0) + b(y - y_0) + c(z - z_0) = 0$$

CONCEPT CHECK

1. **(a)** What is a vector in the plane? How do we represent a vector in the coordinate plane?
   (b) Find the vector with initial point $(2, 3)$ and terminal point $(4, 10)$.
   (c) Let $\mathbf{v} = \langle 2, 1 \rangle$. If the initial point of **v** is placed at $P(1, 1)$, where is its terminal point? Sketch several representations of **v**.
   (d) How is the magnitude of $\mathbf{v} = \langle a_1, a_2 \rangle$ defined? Find the magnitude of $\mathbf{w} = \langle 3, 4 \rangle$.
   (e) What are the vectors **i** and **j**? Express the vector $\mathbf{v} = \langle 5, 9 \rangle$ in terms of **i** and **j**.
   (f) Let $\mathbf{v} = \langle a_1, a_2 \rangle$ be a vector in the coordinate plane. What is meant by the direction θ of **v**? What are the coordinates of **v** in terms of its length and direction? Sketch a figure to illustrate your answer.
   (g) Suppose that **v** has length $|\mathbf{v}| = 5$ and direction $\theta = \pi/6$. Express **v** in terms of its coordinates.
2. **(a)** Define addition and scalar multiplication for vectors.
   (b) If $\mathbf{u} = \langle 2, 3 \rangle$ and $\mathbf{v} = \langle 5, 9 \rangle$, find $\mathbf{u} + \mathbf{v}$ and $4\mathbf{u}$.
3. **(a)** Define the dot product of the vectors $\mathbf{u} = \langle a_1, a_2 \rangle$ and $\mathbf{v} = \langle b_1, b_2 \rangle$, and state the formula for the angle θ between **u** and **v**.
   (b) If $\mathbf{u} = \langle 2, 3 \rangle$ and $\mathbf{v} = \langle 1, 4 \rangle$, find $\mathbf{u} \cdot \mathbf{v}$ and find the angle between **u** and **v**.
4. **(a)** Describe the three-dimensional coordinate system. What are the coordinate planes?
   (b) What is the distance from the point $(3, -2, 5)$ to each of the coordinate planes?
   (c) State the formula for the distance between the points $P(x_1, y_1, z_1)$ and $Q(x_2, y_2, z_2)$.
   (d) Find the distance between the points $P(1, 2, 3)$ and $Q(3, -1, 4)$.
   (e) State the equation of a sphere with center $C(h, k, l)$ and radius r.
   (f) Find an equation for the sphere of radius 5 centered at the point $(1, 2, -3)$.
5. **(a)** What is a vector in space? How do we represent a vector in a three-dimensional coordinate system?
   (b) Find the vector with initial point $(2, 3, -1)$ and terminal point $(4, 10, 5)$.
   (c) How is the magnitude of $\mathbf{v} = \langle a_1, a_2, a_3 \rangle$ defined? Find the magnitude of $\mathbf{w} = \langle 3, 4, 1 \rangle$.
   (d) What are the vectors **i**, **j**, and **k**? Express the vector $\mathbf{v} = \langle 5, 9, -1 \rangle$ in terms of **i**, **j**, and **k**.
6. **(a)** Define addition and scalar multiplication for vectors.
   (b) If $\mathbf{u} = \langle 2, 3, -1 \rangle$ and $\mathbf{v} = \langle 5, 9, 2 \rangle$, find $\mathbf{u} + \mathbf{v}$ and $4\mathbf{u}$.
7. **(a)** Define the dot product of the vectors $\mathbf{u} = \langle a_1, a_2, a_3 \rangle$ and $\mathbf{v} = \langle b_1, b_2, b_3 \rangle$, and state the formula for the angle θ between **u** and **v**.
   (b) If $\mathbf{u} = \langle 2, 3, -1 \rangle$ and $\mathbf{v} = \langle 1, 4, 5 \rangle$, find $\mathbf{u} \cdot \mathbf{v}$.
8. **(a)** Define the cross product of the vectors $\mathbf{u} = \langle a_1, a_2, a_3 \rangle$ and $\mathbf{v} = \langle b_1, b_2, b_3 \rangle$.
   (b) *True* or *False*? The vector $\mathbf{u} \times \mathbf{v}$ is perpendicular to both **u** and **v**.
   (c) Let **u** and **v** be vectors in space. State the formula that relates the magnitude of $\mathbf{u} \times \mathbf{v}$ and the angle θ between **u** and **v**.
   (d) How can we use the cross product to determine whether two vectors are parallel?
9. **(a)** What are the two properties that determine a line in space? Give parametric equations for a line in space.
   (b) Find parametric equations for the line through the point $(-2, 4, 1)$ and parallel to the vector $\mathbf{v} = \langle 7, 5, 3 \rangle$.
10. **(a)** What are the two properties that determine a plane in space? State the equation of a plane.
    (b) Find an equation for the plane passing through the point $(6, -4, 3)$ and with normal vector $\mathbf{n} = \langle 5, -3, 2 \rangle$.

ANSWERS TO THE CONCEPT CHECK CAN BE FOUND AT THE BACK OF THE BOOK.

EXERCISES

Exercises 1–24 deal with vectors in two dimensions.

1–4 ■ Operations with Vectors Find $|\mathbf{u}|$, $\mathbf{u} + \mathbf{v}$, $\mathbf{u} - \mathbf{v}$, $2\mathbf{u}$, and $3\mathbf{u} - 2\mathbf{v}$.

1. $\mathbf{u} = \langle -2, 3 \rangle$, $\mathbf{v} = \langle 8, 1 \rangle$ **2.** $\mathbf{u} = \langle 5, -2 \rangle$, $\mathbf{v} = \langle -3, 0 \rangle$

3. $\mathbf{u} = 2\mathbf{i} + \mathbf{j}$, $\mathbf{v} = \mathbf{i} - 2\mathbf{j}$ **4.** $\mathbf{u} = 3\mathbf{j}$, $\mathbf{v} = -\mathbf{i} + 2\mathbf{j}$

5–6 ■ Component Form of a Vector A description of a vector is given. Express the vector in component form.

5. Find the vector with initial point $P(0, 3)$ and terminal point $Q(3, -1)$.

6. If the vector $5\mathbf{i} - 8\mathbf{j}$ is placed in the plane with its initial point at $P(5, 6)$, find its terminal point.

7–8 ■ Length and Direction of Vectors Find the length and direction of the given vector.

7. $\mathbf{u} = \langle -2, 2\sqrt{3} \rangle$ **8.** $\mathbf{v} = 2\mathbf{i} - 5\mathbf{j}$

9–10 ■ Component Form of a Vector The length $|\mathbf{u}|$ and direction θ of a vector $\mathbf{u}$ are given. Express $\mathbf{u}$ in component form.

9. $|\mathbf{u}| = 20$, $\theta = 60°$ **10.** $|\mathbf{u}| = 13.5$, $\theta = 125°$

11. Resultant Force Two tugboats are pulling a barge as shown in the figure. One pulls with a force of 2.0×10^4 lb in the direction N 50° E, and the other pulls with a force of 3.4×10^4 lb in the direction S 75° E.

(a) Find the resultant force on the barge as a vector.

(b) Find the magnitude and direction of the resultant force.

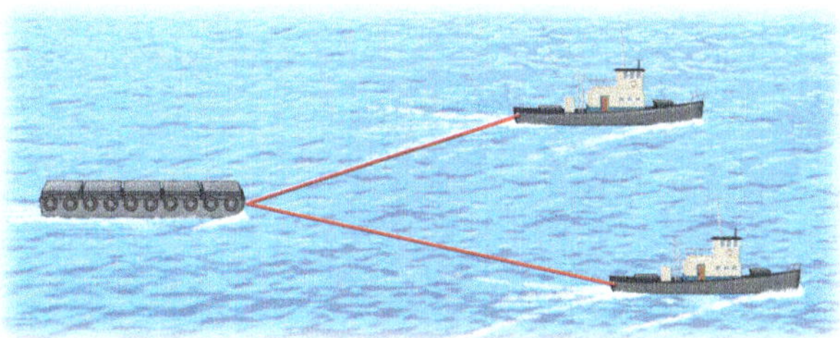

12. True Velocity of a Plane An airplane heads N 60° E at a speed of 600 mi/h relative to the air. A wind begins to blow in the direction N 30° W at 50 mi/h. (See the figure.)

(a) Find the velocity of the airplane as a vector.

(b) Find the true speed and direction of the airplane.

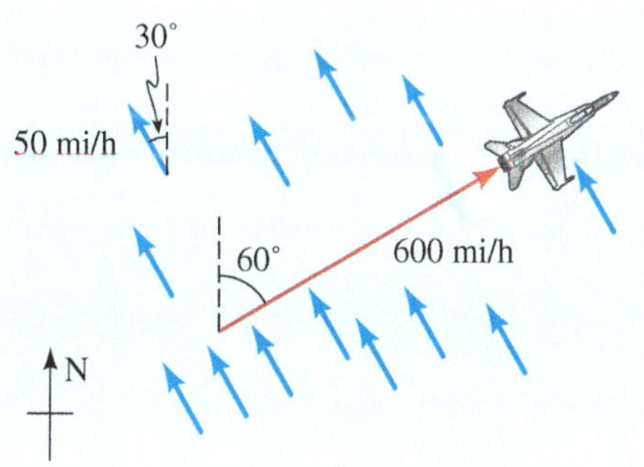

13–16 ■ Dot Products Find the vectors $|\mathbf{u}|$, $\mathbf{u} \cdot \mathbf{u}$, and $\mathbf{u} \cdot \mathbf{v}$.

13. $\mathbf{u} = \langle 4, -3 \rangle$, $\mathbf{v} = \langle 9, -8 \rangle$

14. $\mathbf{u} = \langle 5, 12 \rangle$, $\mathbf{v} = \langle 10, -4 \rangle$

15. $\mathbf{u} = -2\mathbf{i} + 2\mathbf{j}$, $\mathbf{v} = \mathbf{i} + \mathbf{j}$

16. $\mathbf{u} = 10\mathbf{j}$, $\mathbf{v} = 5\mathbf{i} - 3\mathbf{j}$

17–20 ■ Orthogonal Vectors Are $\mathbf{u}$ and $\mathbf{v}$ orthogonal? If not, find the angle between them.

17. $\mathbf{u} = \langle -4, 2 \rangle$, $\mathbf{v} = \langle 3, 6 \rangle$

18. $\mathbf{u} = \langle 5, 3 \rangle$, $\mathbf{v} = \langle -2, 6 \rangle$

19. $\mathbf{u} = 2\mathbf{i} + \mathbf{j}$, $\mathbf{v} = \mathbf{i} + 3\mathbf{j}$

20. $\mathbf{u} = \mathbf{i} - \mathbf{j}$, $\mathbf{v} = \mathbf{i} + \mathbf{j}$

21–24 ■ Scalar and Vector Projections Two vectors $\mathbf{u}$ and $\mathbf{v}$ are given. **(a)** Find the component of $\mathbf{u}$ along $\mathbf{v}$. **(b)** Find $\text{proj}_{\mathbf{v}}\mathbf{u}$. **(c)** Resolve $\mathbf{u}$ into the vectors $\mathbf{u}_1$ and $\mathbf{u}_2$, where $\mathbf{u}_1$ is parallel to $\mathbf{v}$ and $\mathbf{u}_2$ is perpendicular to $\mathbf{v}$.

21. $\mathbf{u} = \langle 3, 1 \rangle$, $\mathbf{v} = \langle 6, -1 \rangle$

22. $\mathbf{u} = \langle -8, 6 \rangle$, $\mathbf{v} = \langle 20, 20 \rangle$

23. $\mathbf{u} = \mathbf{i} + 2\mathbf{j}$, $\mathbf{v} = 4\mathbf{i} - 9\mathbf{j}$

24. $\mathbf{u} = 2\mathbf{i} + 4\mathbf{j}$, $\mathbf{v} = 10\mathbf{j}$

Exercises 25–54 deal with three-dimensional coordinate geometry.

25–26 ■ Distance Between Points Plot the given points, and find the distance between them.

25. $P(1, 0, 2)$, $Q(3, -2, 3)$ **26.** $P(0, 2, 4)$, $Q(1, 3, 0)$

27–28 ■ Finding an Equation of a Sphere Find an equation of the sphere with the given radius r and center C.

27. $r = 6$, $C(0, 0, 0)$ **28.** $r = 2$, $C(1, -2, 4)$

29–30 ■ Equations of Spheres Show that the equation represents a sphere, and find its center and radius.

29. $x^2 + y^2 + z^2 - 2x - 6y + 4z = 2$

30. $x^2 + y^2 + z^2 = 4y + 4z$

31–32 ■ Operations with Vectors Find $|\mathbf{u}|$, $\mathbf{u} + \mathbf{v}$, $\mathbf{u} - \mathbf{v}$, and $\frac{3}{4}\mathbf{u} - 2\mathbf{v}$.

31. $\mathbf{u} = \langle 4, -2, 4 \rangle$, $\mathbf{v} = \langle 2, 3, -1 \rangle$

32. $\mathbf{u} = 6\mathbf{i} - 8\mathbf{k}$, $\mathbf{v} = \mathbf{i} - \mathbf{j} + \mathbf{k}$

33–36 ■ Angle Between Vectors Two vectors $\mathbf{u}$ and $\mathbf{v}$ are given. **(a)** Find their dot product $\mathbf{u} \cdot \mathbf{v}$. **(b)** Are $\mathbf{u}$ and $\mathbf{v}$ perpendicular? If not, find the angle between them.

33. $\mathbf{u} = \langle 3, -2, 4 \rangle$, $\mathbf{v} = \langle 3, 1, -2 \rangle$

34. $\mathbf{u} = \langle 2, -6, 5 \rangle$, $\mathbf{v} = \langle 1, -\frac{1}{2}, -1 \rangle$

35. $\mathbf{u} = 2\mathbf{i} - \mathbf{j} + 4\mathbf{k}$, $\mathbf{v} = 3\mathbf{i} + 2\mathbf{j} - \mathbf{k}$

36. $\mathbf{u} = \mathbf{j} - \mathbf{k}$, $\mathbf{v} = \mathbf{i} + \mathbf{j}$

37–40 ■ Cross Products and Orthogonal Vectors Two vectors **u** and **v** are given. **(a)** Find their cross product $\mathbf{u} \times \mathbf{v}$. **(b)** Find a unit vector that is perpendicular to both **u** and **v**.

37. $\mathbf{u} = \langle 1, 1, 3 \rangle$, $\mathbf{v} = \langle 5, 0, -2 \rangle$

38. $\mathbf{u} = \langle 2, 3, 0 \rangle$, $\mathbf{v} = \langle 0, 4, -1 \rangle$

39. $\mathbf{u} = \mathbf{i} - \mathbf{j}$, $\mathbf{v} = 2\mathbf{j} - \mathbf{k}$

40. $\mathbf{u} = \mathbf{i} + \mathbf{j} - \mathbf{k}$, $\mathbf{v} = \mathbf{i} - \mathbf{j} + \mathbf{k}$

41. Area of a Triangle Find the area of the triangle with vertices $P(2, 1, 1)$, $Q(0, 0, 3)$, and $R(-2, 4, 0)$.

42. Area of a Parallelogram Find the area of the parallelogram determined by the vectors $\mathbf{u} = \langle 4, 1, 1 \rangle$ and $\mathbf{v} = \langle -1, 2, 2 \rangle$.

43. Volume of a Parallelepiped Find the volume of the parallelepiped determined by the vectors $\mathbf{u} = 2\mathbf{i} - \mathbf{j}$, $\mathbf{v} = 2\mathbf{j} + \mathbf{k}$, and $\mathbf{w} = 3\mathbf{i} + \mathbf{j} - \mathbf{k}$.

44. Volume of a Parallelepiped A parallelepiped has one vertex at the origin; the three edges that have the origin as one endpoint extend to the points $P(0, 2, 2)$, $Q(3, 1, -1)$, and $R(1, 4, 1)$. Find the volume of the parallelepiped.

45–46 ■ Equations of Lines Find parametric equations for the line that passes through P and is parallel to **v**.

45. $P(2, 0, -6)$, $\mathbf{v} = \langle 3, 1, 0 \rangle$

46. $P(5, 2, 8)$, $\mathbf{v} = 2\mathbf{i} - \mathbf{j} + 5\mathbf{k}$

47–48 ■ Equations of Lines Find parametric equations for the line that passes through the points P and Q.

47. $P(6, -2, -3)$, $Q(4, 1, -2)$

48. $P(1, 0, 0)$, $Q(3, -4, 2)$

49–50 ■ Equations of Planes Find an equation for the plane with normal vector **n** and passing through the point P.

49. $\mathbf{n} = \langle 2, 3, -5 \rangle$, $P(2, 1, 1)$

50. $\mathbf{n} = -\mathbf{i} - 2\mathbf{j} + 7\mathbf{k}$, $P(-2, 5, 2)$

51–52 ■ Equations of Planes Find an equation of the plane that passes through the points P, Q, and R.

51. $P(1, 1, 1)$, $Q(3, -4, 2)$, $R(6, -1, 0)$

52. $P(4, 0, 0)$, $Q(0, -3, 0)$, $R(0, 0, -5)$

53. Equation of a Line Find parametric equations for the line that crosses the x-axis where $x = 2$ and the z-axis where $z = -4$.

54. Equation of a Plane Find an equation of the plane that contains the line $x = 2 + 2t$, $y = 4t$, $z = -6$ and the point $P(5, 3, 0)$.

CHAPTER 9 TEST

1. Let $\mathbf{u}$ be the vector with initial point $P(3, -1)$ and terminal point $Q(-3, 9)$.

(a) Graph $\mathbf{u}$ in the coordinate plane.

(b) Express $\mathbf{u}$ in terms of $\mathbf{i}$ and $\mathbf{j}$.

(c) Find the length of $\mathbf{u}$.

2. Let $\mathbf{u} = \langle 1, 3 \rangle$, and let $\mathbf{v} = \langle -6, 2 \rangle$.

(a) Find $\mathbf{u} - 3\mathbf{v}$.

(b) Find $|\mathbf{u} + \mathbf{v}|$.

(c) Find $\mathbf{u} \cdot \mathbf{v}$.

(d) Are $\mathbf{u}$ and $\mathbf{v}$ perpendicular?

3. Let $\mathbf{u} = \langle -4\sqrt{3}, 4 \rangle$.

(a) Graph $\mathbf{u}$ in the coordinate plane, with initial point $(0, 0)$.

(b) Find the length and direction of $\mathbf{u}$.

4. A river is flowing due east at 8 mi/h. A man heads his motorboat in the direction N 30° E in the river. The speed of the motorboat relative to the water is 12 mi/h.

(a) Express the true velocity of the motorboat as a vector.

(b) Find the true speed and direction of the motorboat.

5. Let $\mathbf{u} = 3\mathbf{i} + 2\mathbf{j}$ and $\mathbf{v} = 5\mathbf{i} - \mathbf{j}$.

(a) Find the angle between $\mathbf{u}$ and $\mathbf{v}$.

(b) Find the component of $\mathbf{u}$ along $\mathbf{v}$.

(c) Find $\text{proj}_{\mathbf{v}}\,\mathbf{u}$.

6. Find the work done by the force $\mathbf{F} = 3\mathbf{i} - 5\mathbf{j}$ in moving an object from the point $(2, 2)$ to the point $(7, -13)$.

7. Let $P(4, 3, -1)$ and $Q(6, -1, 3)$ be two points in three-dimensional space.

(a) Find the distance between P and Q.

(b) Find an equation for the sphere whose center is P and for which the segment $\overrightarrow{PQ}$ is a radius of the sphere.

(c) The vector $\mathbf{u}$ has initial point P and terminal point Q. Express $\mathbf{u}$ both in component form and using the vectors $\mathbf{i}$, $\mathbf{j}$, and $\mathbf{k}$.

8. Calculate the given quantity if

$$\mathbf{u} = \mathbf{i} + \mathbf{j} - 2\mathbf{k} \qquad \mathbf{v} = 3\mathbf{i} - 2\mathbf{j} + \mathbf{k} \qquad \mathbf{w} = \mathbf{j} - 5\mathbf{k}$$

(a) $2\mathbf{u} + 3\mathbf{v}$

(b) $|\mathbf{u}|$

(c) $\mathbf{u} \cdot \mathbf{v}$

(d) $\mathbf{u} \times \mathbf{v}$

(e) $|\mathbf{v} \times \mathbf{w}|$

(f) $\mathbf{u} \cdot (\mathbf{v} \times \mathbf{w})$

(g) The angle between $\mathbf{u}$ and $\mathbf{v}$ (rounded to the nearest degree)

9. Find two unit vectors that are perpendicular to both $\mathbf{j} + 2\mathbf{k}$ and $\mathbf{i} - 2\mathbf{j} + 3\mathbf{k}$.

10. (a) Find a vector perpendicular to the plane that contains the points $P(1, 0, 0)$, $Q(2, 0, -1)$, and $R(1, 4, 3)$.

(b) Find an equation for the plane that contains P, Q, and R.

(c) Find the area of triangle PQR.

11. Find parametric equations for the line that contains the points $P(2, -4, 7)$ and $Q(0, -3, 5)$.

A CUMULATIVE REVIEW TEST FOR CHAPTERS 8 AND 9 CAN BE FOUND AT THE BOOK COMPANION WEBSITE: **www.stewartmath.com**.

Vector Fields

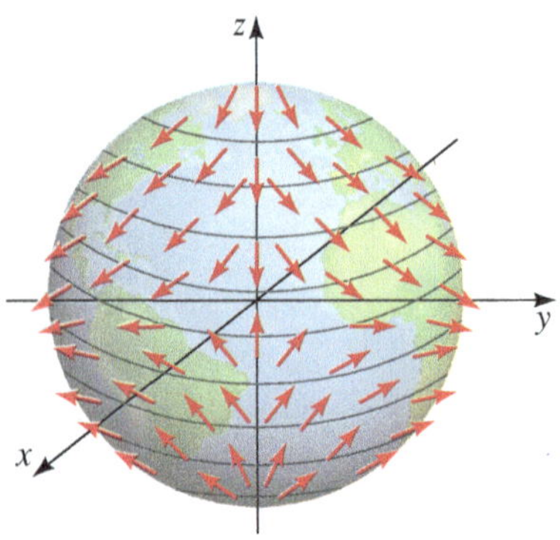

FIGURE 1 Wind represented by a vector field

To model the gravitational force near the earth or the flow of wind on a surface of the earth, we use vectors. For example, at each point on the surface of the earth air flows with a certain speed and direction. We represent the air currents by vectors. If we graph many of these vectors, we get a "picture" or a graph of the flow of the air. (See Figure 1.)

Vector Fields in the Plane

A **vector field** in the coordinate plane is a function that assigns a vector to each point in the plane (or to each point in some subset of the plane). For example,

$$\mathbf{F}(x, y) = x\mathbf{i} + y\mathbf{j}$$

is a vector field that assigns the vector $x\mathbf{i} + y\mathbf{j}$ to the point (x, y). We graph this vector field in the next example.

EXAMPLE 1 ■ Graphing a Vector Field in the Plane

Graph the vector field $\mathbf{F}(x, y) = x\mathbf{i} + y\mathbf{j}$. What does the graph indicate?

SOLUTION The table gives the vector field at several points. In Figure 2 we sketch the vectors in the table together with several other vectors in the vector field.

(x, y)	$\mathbf{F} = x\mathbf{i} + y\mathbf{j}$
$(1, 3)$	$\mathbf{i} + 3\mathbf{j}$
$(3, 3)$	$3\mathbf{i} + 3\mathbf{j}$
$(-4, 6)$	$-4\mathbf{i} + 6\mathbf{j}$
$(-6, -1)$	$-6\mathbf{i} - \mathbf{j}$
$(6, -6)$	$6\mathbf{i} - 6\mathbf{j}$

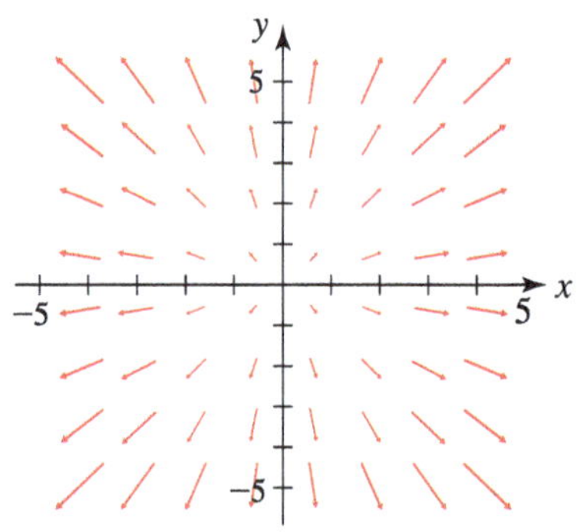

FIGURE 2

We see from the graph that the vectors in the field point away from the origin, and the farther from the origin, the greater the magnitude of the vector. ■

EXAMPLE 2 ■ Graphing a Vector Field in the Plane

A potter's wheel has a radius of 5 in. The velocity of each point on the wheel is given by the vector field $\mathbf{F}(x, y) = -y\mathbf{i} + x\mathbf{j}$. What does the graph indicate?

SOLUTION The table gives the vector field at several points. In Figure 3 we sketch the vectors in the table.

(x, y)	$\mathbf{F}(x, y)$	(x, y)	$\mathbf{F}(x, y)$
$(1, 0)$	$\langle 0, 1\rangle$	$(-1, 0)$	$\langle 0, -1\rangle$
$(2, 2)$	$\langle -2, 2\rangle$	$(-2, -2)$	$\langle 2, -2\rangle$
$(3, 0)$	$\langle 0, 3\rangle$	$(-3, 0)$	$\langle 0, -3\rangle$
$(0, 1)$	$\langle -1, 0\rangle$	$(0, -1)$	$\langle 1, 0\rangle$
$(-2, 2)$	$\langle -2, -2\rangle$	$(2, -2)$	$\langle 2, 2\rangle$
$(0, 3)$	$\langle -3, 0\rangle$	$(0, -3)$	$\langle 3, 0\rangle$

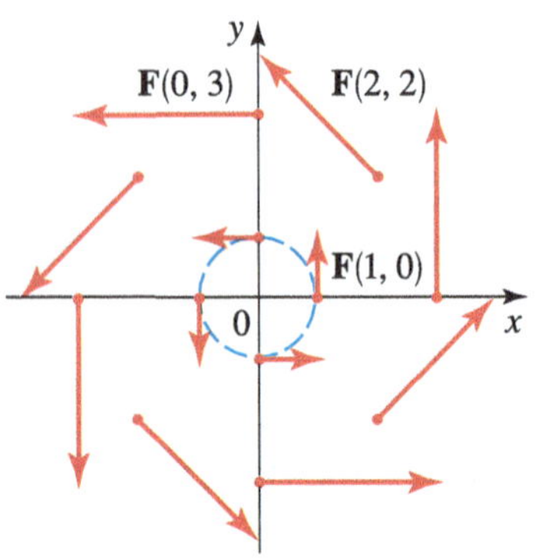

FIGURE 3

We see from the graph that the wheel is rotating counterclockwise and that the points at the edge of the wheel have a higher velocity than do the points near the center of the wheel. ■

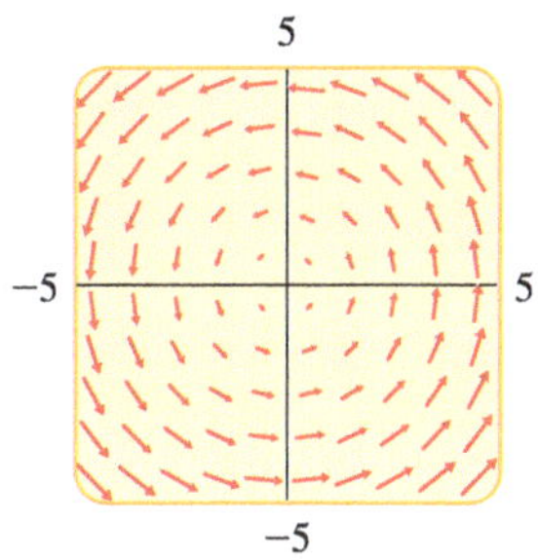

FIGURE 4

Graphing vector fields requires graphing a lot of vectors. Some graphing calculators and computer programs are capable of graphing vector fields. You can also find many Internet sites that have applets for graphing vector fields. The vector field in Example 2 is graphed with a computer program in Figure 4. Notice how the computer scales the lengths of the vectors so that they are not too long yet are proportional to their true lengths.

Vector Fields in Space

A **vector field** in three-dimensional space is a function that assigns a vector to each point in space (or to each point in some subset of space). For example,

$$\mathbf{F}(x, y, z) = 2x\mathbf{i} - y\mathbf{j} + z^2\mathbf{k}$$

is a vector field that assigns the vector $2x\mathbf{i} - y\mathbf{j} + z^3\mathbf{k}$ to the point (x, y, z). In general, it is difficult to draw a vector field in space by hand, since we must draw many vectors with the proper perspective. The vector field in the next example is particularly simple, so we'll sketch it by hand.

EXAMPLE 3 ■ Graphing a Vector Field in Space

Graph the vector field $\mathbf{F}(x, y, z) = z\mathbf{k}$. What does the graph indicate?

SOLUTION A graph is shown in Figure 5. Notice that all vectors are vertical and point upward above the xy-plane and downward below it. The magnitude of each vector increases with the distance from the xy-plane. ■

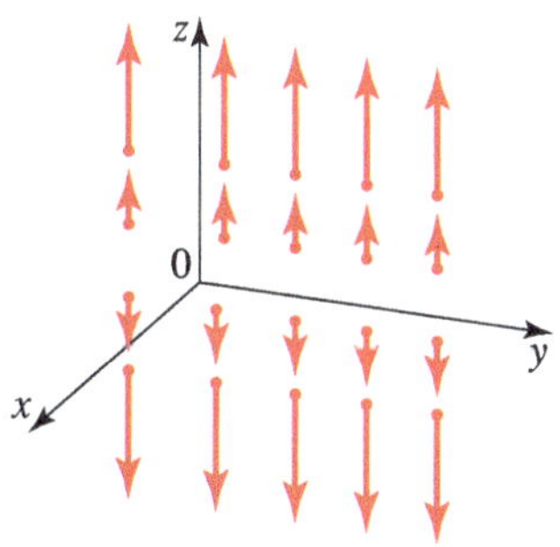

FIGURE 5

The gravitational pull of the earth in the space surrounding it is mathematically modeled by a vector field. According to Newton's Law of Gravity, the gravitational force $\mathbf{F}$ is directed toward the center of the earth and is inversely proportional to the distance from the center of the earth. The magnitude of the force is

$$F = G\frac{Mm}{r^2}$$

where M is the mass of the earth, m is mass of an object in proximity to the earth, r is the distance from the object to the center of the earth, and G is the universal gravitational constant.

To model the gravitational force, let's place a three-dimensional coordinate system with the origin at the center of the earth. The gravitational force at the point (x, y, z) is directed toward the origin. A unit vector pointing toward the origin is

$$\mathbf{u} = -\frac{x\mathbf{i} + y\mathbf{j} + z\mathbf{k}}{\sqrt{x^2 + y^2 + z^2}}$$

To obtain the gravitational vector field, we multiply this unit vector by the appropriate magnitude, namely, GMm/r^2. Since the distance r from the point (x, y, z) to the origin is $r = \sqrt{x^2 + y^2 + z^2}$, it follows that $r^2 = x^2 + y^2 + z^2$. So we can express the gravitational vector field as

$$\mathbf{F}(x, y, z) = -GMm\frac{x\mathbf{i} + y\mathbf{j} + z\mathbf{k}}{(x^2 + y^2 + z^2)^{3/2}}$$

FIGURE 6 The gravitational field

Some of the vectors in the gravitational field $\mathbf{F}$ are pictured in Figure 6.

PROBLEMS

1–6 ■ Sketch the vector field **F** by drawing a diagram as in Figure 3.

1. $\mathbf{F}(x, y) = \frac{1}{2}\mathbf{i} + \frac{1}{2}\mathbf{j}$

2. $\mathbf{F}(x, y) = \mathbf{i} + x\mathbf{j}$

3. $\mathbf{F}(x, y) = y\mathbf{i} + \frac{1}{2}\mathbf{j}$

4. $\mathbf{F}(x, y) = (x - y)\mathbf{i} + x\mathbf{j}$

5. $\mathbf{F}(x, y) = \dfrac{y\mathbf{i} + x\mathbf{j}}{\sqrt{x^2 + y^2}}$

6. $\mathbf{F}(x, y) = \dfrac{y\mathbf{i} - x\mathbf{j}}{\sqrt{x^2 + y^2}}$

7–10 ■ Sketch the vector field **F** by drawing a diagram as in Figure 5.

7. $\mathbf{F}(x, y, z) = \mathbf{j}$

8. $\mathbf{F}(x, y, z) = \mathbf{j} - \mathbf{k}$

9. $\mathbf{F}(x, y, z) = z\mathbf{j}$

10. $\mathbf{F}(x, y, z) = y\mathbf{k}$

11–14 ■ Match the vector field **F** with the graphs labeled I–IV.

11. $\mathbf{F}(x, y) = \langle y, x\rangle$

12. $\mathbf{F}(x, y) = \langle 1, \sin y\rangle$

13. $\mathbf{F}(x, y) = \langle x - 2, x + 1\rangle$

14. $\mathbf{F}(x, y) = \langle y, 1/x\rangle$

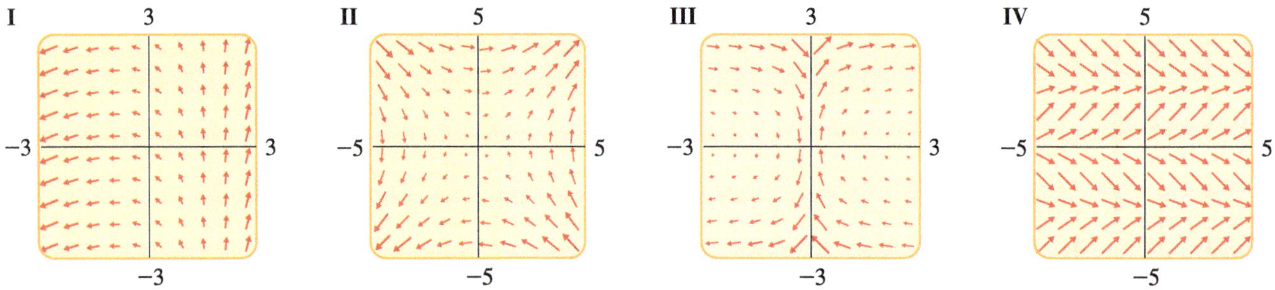

15–18 ■ Match the vector field **F** with the graphs labeled I–IV.

15. $\mathbf{F}(x, y, z) = \mathbf{i} + 2\mathbf{j} + 3\mathbf{k}$

16. $\mathbf{F}(x, y, z) = \mathbf{i} + 2\mathbf{j} + z\mathbf{k}$

17. $\mathbf{F}(x, y, z) = x\mathbf{i} + y\mathbf{j} + 3\mathbf{k}$

18. $\mathbf{F}(x, y, z) = x\mathbf{i} + y\mathbf{j} + z\mathbf{k}$

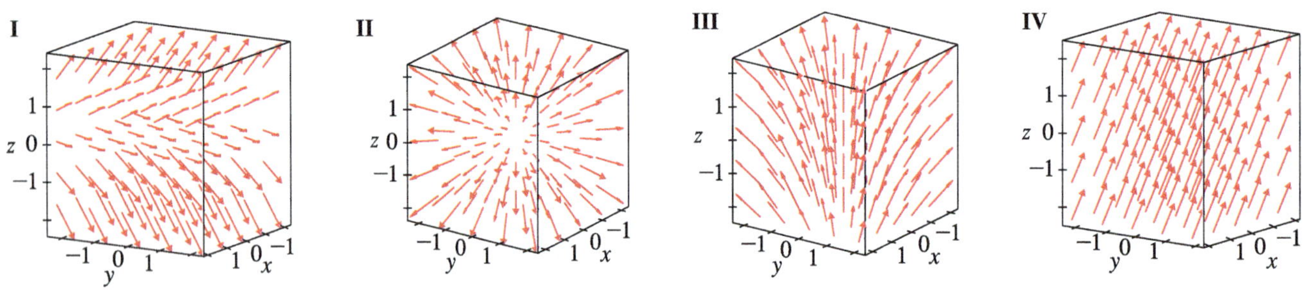

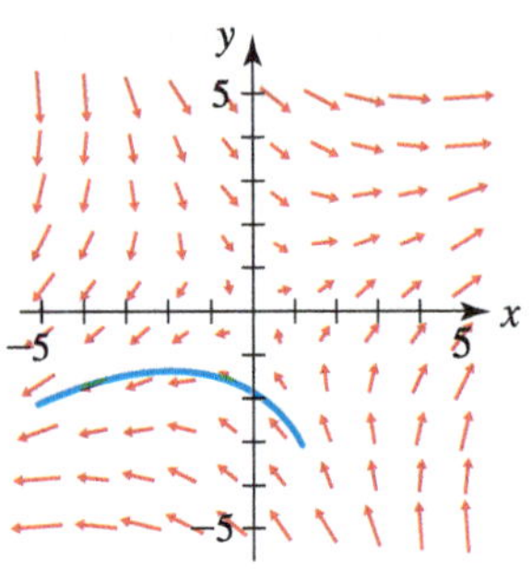

19. Flow Lines in a Current The current in a turbulent bay is described by the velocity vector field

$$\mathbf{F}(x, y) = (x + y)\mathbf{i} + (x - y)\mathbf{j}$$

A graph of the vector field **F** is shown. If a small toy boat is put in this bay, we can tell from the graph of the vector field what path the boat would follow. Such paths are called **flow lines** (or **streamlines**) of the vector field. A streamline starting at $(1, -3)$ is shown in blue in the figure. Sketch streamlines starting at the given point.

(a) $(1, 4)$ **(b)** $(-2, 1)$ **(c)** $(-1, -2)$

© Dainis Derics/Shutterstock.com

10 Systems of Equations and Inequalities

Throughout the preceding chapters we modeled real-world situations by equations. But many real-world situations involve too many variables to be modeled by a single equation. For example, weather depends on the relationships among many variables, including temperature, wind speed, air pressure, and humidity. So to model the weather (and forecast a snowstorm like the one pictured above), scientists use many equations, each having many variables. Such collections of equations, called systems of equations, *work together* to describe the weather. Systems of equations with hundreds of variables are used by airlines to establish consistent flight schedules and by telecommunications companies to find efficient routings for telephone calls. In this chapter we learn how to solve systems of equations that consist of several equations in several variables.

10.1 SYSTEMS OF LINEAR EQUATIONS IN TWO VARIABLES

■ Systems of Linear Equations and Their Solutions ■ Substitution Method ■ Elimination Method ■ Graphical Method ■ The Number of Solutions of a Linear System in Two Variables ■ Modeling with Linear Systems

■ Systems of Linear Equations and Their Solutions

A linear equation in two variables is an equation of the form

$$ax + by = c$$

The graph of a linear equation is a line (see Section 1.10).

A **system of equations** is a set of equations that involve the same variables. A **system of linear equations** is a system of equations in which each equation is linear. A **solution** of a system is an assignment of values for the variables that makes *each* equation in the system true. To **solve** a system means to find all solutions of the system.

Here is an example of a system of linear equations in two variables:

$$\begin{cases} 2x - y = 5 & \text{Equation 1} \\ x + 4y = 7 & \text{Equation 2} \end{cases}$$

We can check that $x = 3$ and $y = 1$ is a solution of this system.

Equation 1	**Equation 2**
$2x - y = 5$	$x + 4y = 7$
$2(3) - 1 = 5$ ✓	$3 + 4(1) = 7$ ✓

The solution can also be written as the ordered pair $(3, 1)$.

Note that the graphs of Equations 1 and 2 are lines (see Figure 1). Since the solution $(3, 1)$ satisfies each equation, the point $(3, 1)$ lies on each line. So it is the point of intersection of the two lines.

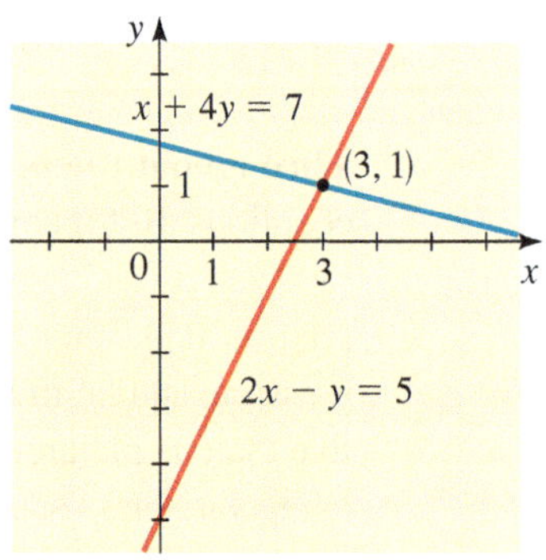

FIGURE 1

■ Substitution Method

To solve a system using the **substitution method**, we start with one equation in the system and solve for one variable in terms of the other variable.

SUBSTITUTION METHOD

1. **Solve for One Variable.** Choose one equation, and solve for one variable in terms of the other variable.
2. **Substitute.** Substitute the expression you found in Step 1 into the other equation to get an equation in one variable, then solve for that variable.
3. **Back-Substitute.** Substitute the value you found in Step 2 back into the expression found in Step 1 to solve for the remaining variable.

EXAMPLE 1 ■ Substitution Method

Find all solutions of the system.

$$\begin{cases} 2x + y = 1 \\ 3x + 4y = 14 \end{cases}$$ Equation 1, Equation 2

SOLUTION **Solve for one variable.** We solve for y in the first equation.

$$y = 1 - 2x \qquad \text{Solve for } y \text{ in Equation 1}$$

Substitute. Now we substitute for y in the second equation and solve for x.

$$\begin{aligned} 3x + 4(1 - 2x) &= 14 && \text{Substitute } y = 1 - 2x \text{ into Equation 2} \\ 3x + 4 - 8x &= 14 && \text{Expand} \\ -5x + 4 &= 14 && \text{Simplify} \\ -5x &= 10 && \text{Subtract 4} \\ x &= -2 && \text{Solve for } x \end{aligned}$$

Back-substitute. Next we back-substitute $x = -2$ into the equation $y = 1 - 2x$.

$$y = 1 - 2(-2) = 5 \qquad \text{Back-substitute}$$

Thus $x = -2$ and $y = 5$, so the solution is the ordered pair $(-2, 5)$. Figure 2 shows that the graphs of the two equations intersect at the point $(-2, 5)$.

CHECK YOUR ANSWER

$x = -2, y = 5$:

$$\begin{cases} 2(-2) + 5 = 1 \\ 3(-2) + 4(5) = 14 \end{cases} \checkmark$$

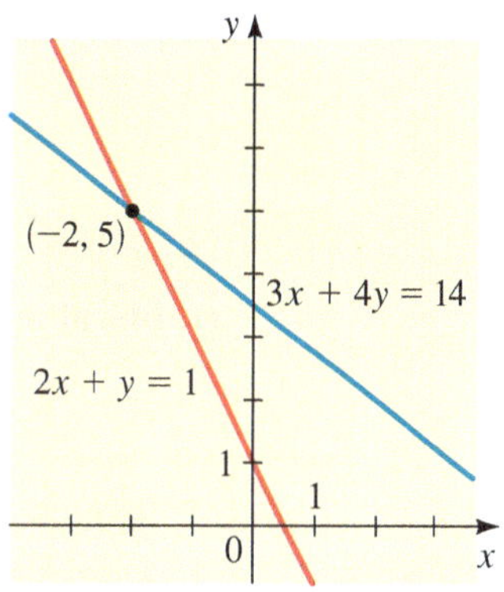

FIGURE 2

Now Try Exercise 5 ■

■ Elimination Method

To solve a system using the **elimination method**, we try to combine the equations using sums or differences so as to eliminate one of the variables.

ELIMINATION METHOD

1. **Adjust the Coefficients.** Multiply one or more of the equations by appropriate numbers so that the coefficient of one variable in one equation is the negative of its coefficient in the other equation.
2. **Add the Equations.** Add the two equations to eliminate one variable, then solve for the remaining variable.
3. **Back-Substitute.** Substitute the value that you found in Step 2 back into one of the original equations, and solve for the remaining variable.

EXAMPLE 2 ■ Elimination Method

Find all solutions of the system.

$$\begin{cases} 3x + 2y = 14 & \text{Equation 1} \\ x - 2y = 2 & \text{Equation 2} \end{cases}$$

SOLUTION Since the coefficients of the y-terms are negatives of each other, we can add the equations to eliminate y.

$$\begin{cases} 3x + 2y = 14 \\ x - 2y = 2 \end{cases} \quad \text{System}$$

$$4x \qquad = 16 \quad \text{Add}$$

$$x = 4 \quad \text{Solve for } x$$

Now we back-substitute $x = 4$ into one of the original equations and solve for y. Let's choose the second equation because it looks simpler.

$$\begin{aligned} x - 2y &= 2 && \text{Equation 2} \\ 4 - 2y &= 2 && \text{Back-substitute } x = 4 \text{ into Equation 2} \\ -2y &= -2 && \text{Subtract 4} \\ y &= 1 && \text{Solve for } y \end{aligned}$$

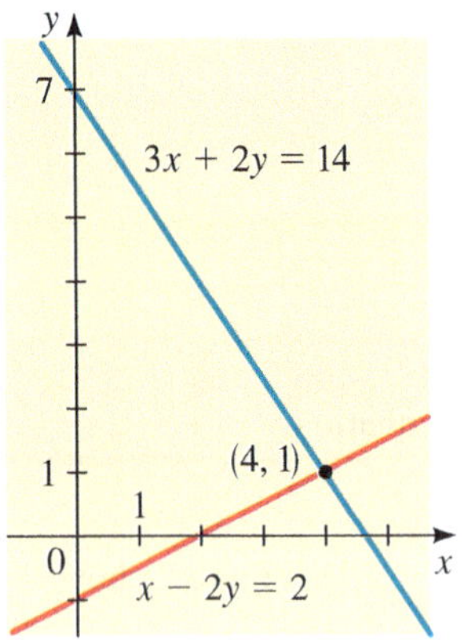

FIGURE 3

The solution is $(4, 1)$. Figure 3 shows that the graphs of the equations in the system intersect at the point $(4, 1)$.

Now Try Exercise 9 ■

Graphical Method

In the **graphical method** we use a graphing device to solve the system of equations.

GRAPHICAL METHOD

1. **Graph Each Equation.** Express each equation in a form suitable for the graphing calculator by solving for y as a function of x. Graph the equations on the same screen.
2. **Find the Intersection Point(s).** The solutions are the x- and y-coordinates of the point(s) of intersection.

EXAMPLE 3 ■ Graphical Method

See Appendix C, *Graphing with a Graphing Calculator*, for guidelines on using a graphing calculator. See Appendix D, *Using the TI-83/84 Graphing Calculator*, for specific graphing instructions. Go to **www.stewartmath.com**.

Find all solutions of the system

$$\begin{cases} 1.35x - 2.13y = -2.36 \\ 2.16x + 0.32y = 1.06 \end{cases}$$

SOLUTION Solving for y in terms of x, we get the equivalent system

$$\begin{cases} y = 0.63x + 1.11 \\ y = -6.75x + 3.31 \end{cases}$$

where we have rounded the coefficients to two decimals. Figure 4 shows that the two lines intersect. Zooming in, we see that the solution is approximately $(0.30, 1.30)$.

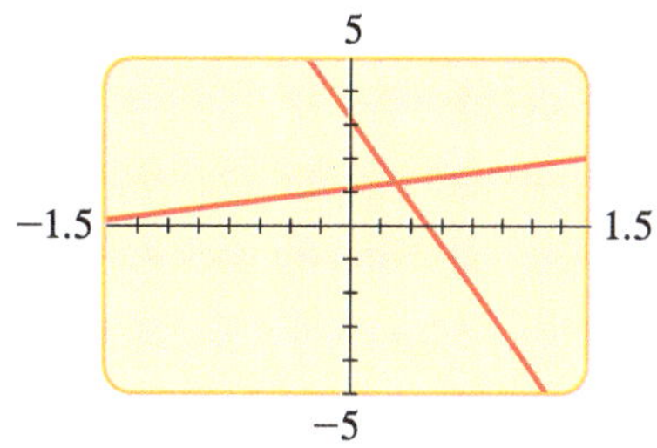

FIGURE 4

Now Try Exercises 13 and 51 ■

The Number of Solutions of a Linear System in Two Variables

The graph of a linear system in two variables is a pair of lines, so to solve the system graphically, we must find the intersection point(s) of the lines. Two lines may intersect in a single point, they may be parallel, or they may coincide, as shown in Figure 5. So there are three possible outcomes in solving such a system.

NUMBER OF SOLUTIONS OF A LINEAR SYSTEM IN TWO VARIABLES

For a system of linear equations in two variables, exactly one of the following is true. (See Figure 5.)

1. The system has exactly one solution.
2. The system has no solution.
3. The system has infinitely many solutions.

A system that has no solution is said to be **inconsistent**. A system with infinitely many solutions is called **dependent**.

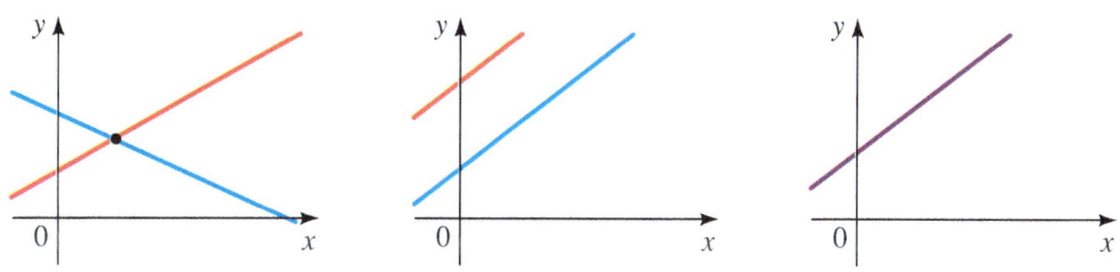

FIGURE 5

(a) Lines intersect at a single point. The system has one solution.

(b) Lines are parallel and do not intersect. The system has no solution.

(c) Lines coincide—equations are for the same line. The system has infinitely many solutions.

EXAMPLE 4 ■ A Linear System with One Solution

Solve the system and graph the lines.

$$\begin{cases} 3x - y = 0 & \text{Equation 1} \\ 5x + 2y = 22 & \text{Equation 2} \end{cases}$$

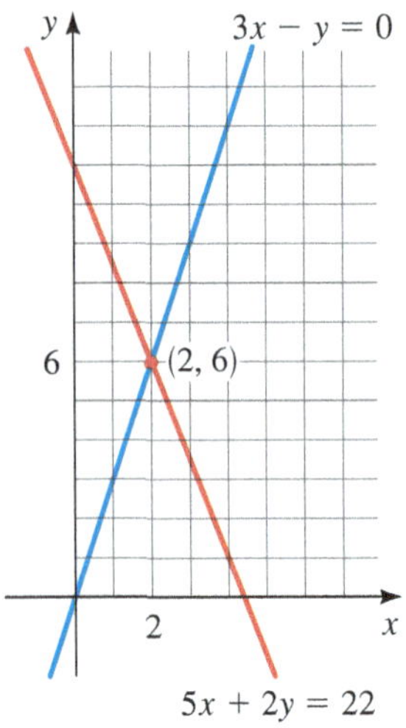

FIGURE 6

CHECK YOUR ANSWER

$x = 2, y = 6$:

$$\begin{cases} 3(2) - (6) = 0 \\ 5(2) + 2(6) = 22 \end{cases} \checkmark$$

SOLUTION We eliminate y from the equations and solve for x.

$$\begin{cases} 6x - 2y = 0 & \text{2 × Equation 1} \\ 5x + 2y = 22 \end{cases}$$

$$11x = 22 \quad \text{Add}$$

$$x = 2 \quad \text{Solve for } x$$

Now we back-substitute into the first equation and solve for y:

$$6(2) - 2y = 0 \quad \text{Back-substitute } x = 2$$

$$-2y = -12 \quad \text{Subtract 12}$$

$$y = 6 \quad \text{Solve for } y$$

The solution of the system is the ordered pair (2, 6), that is,

$$x = 2 \qquad y = 6$$

The graph in Figure 6 shows that the lines in the system intersect at the point (2, 6).

Now Try Exercise 23 ■

EXAMPLE 5 ■ A Linear System with No Solution

Solve the system.

$$\begin{cases} 8x - 2y = 5 & \text{Equation 1} \\ -12x + 3y = 7 & \text{Equation 2} \end{cases}$$

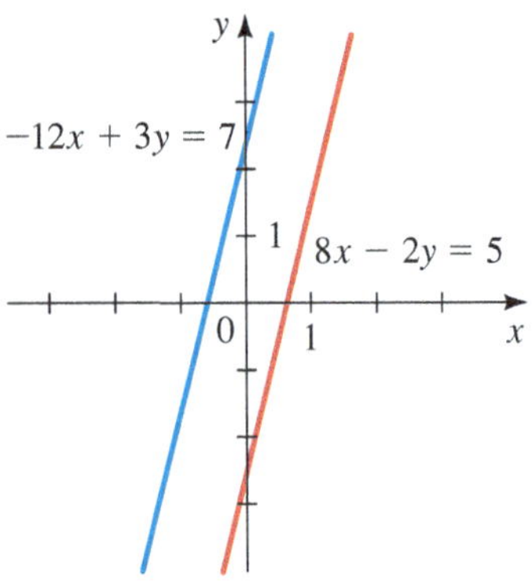

FIGURE 7

SOLUTION This time we try to find a suitable combination of the two equations to eliminate the variable y. Multiplying the first equation by 3 and the second equation by 2 gives

$$\begin{cases} 24x - 6y = 15 & \text{3 × Equation 1} \\ -24x + 6y = 14 & \text{2 × Equation 2} \end{cases}$$

$$0 = 29 \quad \text{Add}$$

Adding the two equations eliminates *both x and y* in this case, and we end up with $0 = 29$, which is obviously false. No matter what values we assign to x and y, we cannot make this statement true, so the system has *no solution*. Figure 7 shows that the lines in the system are parallel so do not intersect. The system is inconsistent.

Now Try Exercise 37 ■

EXAMPLE 6 ■ A Linear System with Infinitely Many Solutions

Solve the system.

$$\begin{cases} 3x - 6y = 12 & \text{Equation 1} \\ 4x - 8y = 16 & \text{Equation 2} \end{cases}$$

SOLUTION We multiply the first equation by 4 and the second equation by 3 to prepare for subtracting the equations to eliminate x. The new equations are

$$\begin{cases} 12x - 24y = 48 & \text{4 × Equation 1} \\ 12x - 24y = 48 & \text{3 × Equation 2} \end{cases}$$

We see that the two equations in the original system are simply different ways of expressing the equation of one single line. The coordinates of any point on this line

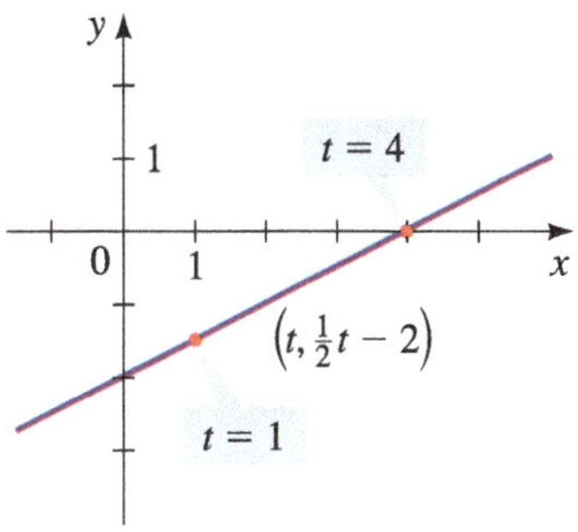

FIGURE 8

give a solution of the system. Writing the equation in slope-intercept form, we have $y = \frac{1}{2}x - 2$. So if we let t represent any real number, we can write the solution as

$$x = t$$
$$y = \tfrac{1}{2}t - 2$$

We can also write the solution in ordered-pair form as

$$\left(t, \tfrac{1}{2}t - 2\right)$$

where t is any real number. The system has infinitely many solutions (see Figure 8).

Now Try Exercise 39 ■

In Example 3, to get specific solutions we have to assign values to t. For instance, if $t = 1$, we get the solution $\left(1, -\frac{3}{2}\right)$. If $t = 4$, we get the solution $(4, 0)$. For every value of t we get a different solution. (See Figure 8.)

Modeling with Linear Systems

Frequently, when we use equations to solve problems in the sciences or in other areas, we obtain systems like the ones we've been considering. When modeling with systems of equations, we use the following guidelines, which are similar to those in Section 1.7.

GUIDELINES FOR MODELING WITH SYSTEMS OF EQUATIONS

1. **Identify the Variables.** Identify the quantities that the problem asks you to find. These are usually determined by a careful reading of the question posed at the end of the problem. Introduce notation for the variables (call them x and y or some other letters).
2. **Express All Unknown Quantities in Terms of the Variables.** Read the problem again, and express all the quantities mentioned in the problem in terms of the variables you defined in Step 1.
3. **Set Up a System of Equations.** Find the crucial facts in the problem that give the relationships between the expressions you found in Step 2. Set up a system of equations (or a model) that expresses these relationships.
4. **Solve the System and Interpret the Results.** Solve the system you found in Step 3, check your solutions, and state your final answer as a sentence that answers the question posed in the problem.

The next two examples illustrate how to model with systems of equations.

EXAMPLE 7 ■ A Distance-Speed-Time Problem

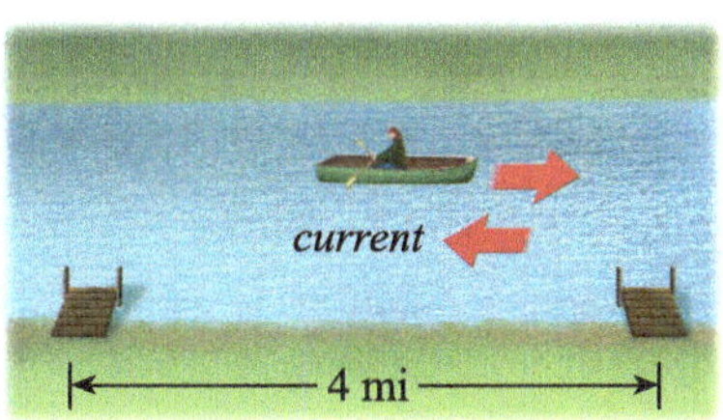

A woman rows a boat upstream from one point on a river to another point 4 mi away in $1\frac{1}{2}$ hours. The return trip, traveling with the current, takes only 45 min. How fast does she row relative to the water, and at what speed is the current flowing?

SOLUTION **Identify the variables.** We are asked to find the rowing speed and the speed of the current, so we let

$$x = \text{rowing speed (mi/h)}$$
$$y = \text{current speed (mi/h)}$$

Mathematics in the Modern World

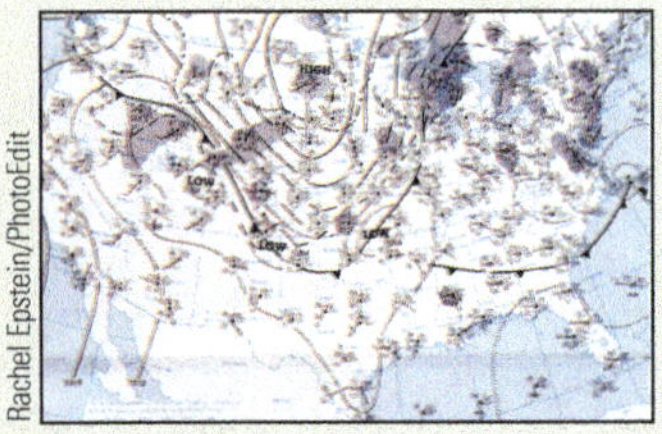

Rachel Epstein/PhotoEdit

Weather Prediction

Modern meteorologists do much more than predict tomorrow's weather. They research long-term weather patterns, depletion of the ozone layer, global warming, and other effects of human activity on the weather. But daily weather prediction is still a major part of meteorology; its value is measured by the innumerable human lives that are saved each year through accurate prediction of hurricanes, blizzards, and other catastrophic weather phenomena. Early in the 20th century mathematicians proposed to model weather with equations that used the current values of hundreds of atmospheric variables. Although this model worked in principle, it was impossible to predict future weather patterns with it because of the difficulty of measuring all the variables accurately and solving all the equations. Today, new mathematical models combined with high-speed computer simulations and better data have vastly improved weather prediction. As a result, many human as well as economic disasters have been averted. Mathematicians at the National Oceanographic and Atmospheric Administration (NOAA) are continually researching better methods of weather prediction.

Express unknown quantities in terms of the variable. The woman's speed when she rows upstream is her rowing speed minus the speed of the current; her speed downstream is her rowing speed plus the speed of the current. Now we translate this information into the language of algebra.

In Words	In Algebra
Rowing speed	x
Current speed	y
Speed upstream	$x - y$
Speed downstream	$x + y$

Set up a system of equations. The distance upstream and downstream is 4 mi, so using the fact that speed × time = distance for both legs of the trip, we get

speed upstream × time upstream = distance traveled

speed downstream × time downstream = distance traveled

In algebraic notation this translates into the following equations.

$$(x - y)\tfrac{3}{2} = 4 \qquad \text{Equation 1}$$

$$(x + y)\tfrac{3}{4} = 4 \qquad \text{Equation 2}$$

(The times have been converted to hours, since we are expressing the speeds in miles per *hour*.)

Solve the system. We multiply the equations by 2 and 4, respectively, to clear the denominators.

$$\begin{cases} 3x - 3y = 8 & 2 \times \text{Equation 1} \\ 3x + 3y = 16 & 4 \times \text{Equation 2} \end{cases}$$

$$6x = 24 \qquad \text{Add}$$

$$x = 4 \qquad \text{Solve for } x$$

Back-substituting this value of x into the first equation (the second works just as well) and solving for y, we get

$$3(4) - 3y = 8 \qquad \text{Back-substitute } x = 4$$

$$-3y = 8 - 12 \qquad \text{Subtract 12}$$

$$y = \tfrac{4}{3} \qquad \text{Solve for } y$$

The woman rows at 4 mi/h, and the current flows at $1\frac{1}{3}$ mi/h.

CHECK YOUR ANSWER

Speed upstream is

$$\frac{\text{distance}}{\text{time}} = \frac{4 \text{ mi}}{1\frac{1}{2}\text{ h}} = 2\tfrac{2}{3} \text{ mi/h}$$

and this should equal

rowing speed − current flow

$$= 4 \text{ mi/h} - \tfrac{4}{3} \text{ mi/h} = 2\tfrac{2}{3} \text{ mi/h}$$

Speed downstream is

$$\frac{\text{distance}}{\text{time}} = \frac{4 \text{ mi}}{\frac{3}{4}\text{ h}} = 5\tfrac{1}{3} \text{ mi/h}$$

and this should equal

rowing speed + current flow

$$= 4 \text{ mi/h} + \tfrac{4}{3} \text{ mi/h} = 5\tfrac{1}{3} \text{ mi/h} \checkmark$$

Now Try Exercise 65 ■

EXAMPLE 8 ■ A Mixture Problem

A vintner fortifies wine that contains 10% alcohol by adding a 70% alcohol solution to it. The resulting mixture has an alcoholic strength of 16% and fills 1000 one-liter bottles. How many liters (L) of the wine and of the alcohol solution does the vintner use?

SOLUTION **Identify the variables.** Since we are asked for the amounts of wine and alcohol, we let

$$x = \text{amount of wine used (L)}$$

$$y = \text{amount of alcohol solution used (L)}$$

Express all unknown quantities in terms of the variable. From the fact that the wine contains 10% alcohol and the solution contains 70% alcohol, we get the following.

In Words	In Algebra
Amount of wine used (L)	x
Amount of alcohol solution used (L)	y
Amount of alcohol in wine (L)	$0.10x$
Amount of alcohol in solution (L)	$0.70y$

Set up a system of equations. The volume of the mixture must be the total of the two volumes the vintner is adding together, so

$$x + y = 1000$$

Also, the amount of alcohol in the mixture must be the total of the alcohol contributed by the wine and by the alcohol solution, that is,

$$0.10x + 0.70y = (0.16)1000$$

$$0.10x + 0.70y = 160 \qquad \text{Simplify}$$

$$x + 7y = 1600 \qquad \text{Multiply by 10 to clear decimals}$$

Thus we get the system

$$\begin{cases} x + y = 1000 & \text{Equation 1} \\ x + 7y = 1600 & \text{Equation 2} \end{cases}$$

Solve the system. Subtracting the first equation from the second eliminates the variable x, and we get

$$6y = 600 \qquad \text{Subtract Equation 1 from Equation 2}$$

$$y = 100 \qquad \text{Solve for } y$$

We now back-substitute $y = 100$ into the first equation and solve for x.

$$x + 100 = 1000 \qquad \text{Back-substitute } y = 100$$

$$x = 900 \qquad \text{Solve for } x$$

The vintner uses 900 L of wine and 100 L of the alcohol solution.

Now Try Exercise 67 ■

10.1 EXERCISES

CONCEPTS

1. The system of equations

$$\begin{cases} 2x + 3y = 7 \\ 5x - y = 9 \end{cases}$$

is a system of two equations in the two variables ________ and ________. To determine whether $(5, -1)$ is a solution of this system, we check whether $x = 5$ and $y = -1$ satisfy each ________ in the system. Which of the following are solutions of this system?

$$(5, -1), \quad (-1, 3), \quad (2, 1)$$

2. A system of equations in two variables can be solved by the ______________ method, the ______________ method, or the ______________ method.

3. A system of two linear equations in two variables can have one solution, __________ solution, or __________ __________ solutions.

4. The following is a system of two linear equations in two variables.

$$\begin{cases} x + y = 1 \\ 2x + 2y = 2 \end{cases}$$

The graph of the first equation is the same as the graph of the second equation, so the system has _________ _________ solutions. We express these solutions by writing

$$x = t$$
$$y = \_\_\_\_\_\_\_\_$$

where t is any real number. Some of the solutions of this system are (1, ___), (−3, ___), and (5, ___).

SKILLS

5–8 ■ Substitution Method Use the substitution method to find all solutions of the system of equations.

5. $\begin{cases} x - y = 1 \\ 4x + 3y = 18 \end{cases}$ **6.** $\begin{cases} 3x + y = 1 \\ 5x + 2y = 1 \end{cases}$

7. $\begin{cases} x - y = 2 \\ 2x + 3y = 9 \end{cases}$ **8.** $\begin{cases} 2x + y = 7 \\ x + 2y = 2 \end{cases}$

9–12 ■ Elimination Method Use the elimination method to find all solutions of the system of equations.

9. $\begin{cases} 3x + 4y = 10 \\ x - 4y = -2 \end{cases}$ **10.** $\begin{cases} 2x + 5y = 15 \\ 4x + y = 21 \end{cases}$

11. $\begin{cases} 3x - 2y = -13 \\ -6x + 5y = 28 \end{cases}$ **12.** $\begin{cases} 2x - 5y = -18 \\ 3x + 4y = 19 \end{cases}$

13–14 ■ Graphical Method Two equations and their graphs are given. Find the intersection point(s) of the graphs by solving the system.

13. $\begin{cases} 2x + y = -1 \\ x - 2y = -8 \end{cases}$ **14.** $\begin{cases} x + y = 2 \\ 2x + y = 5 \end{cases}$

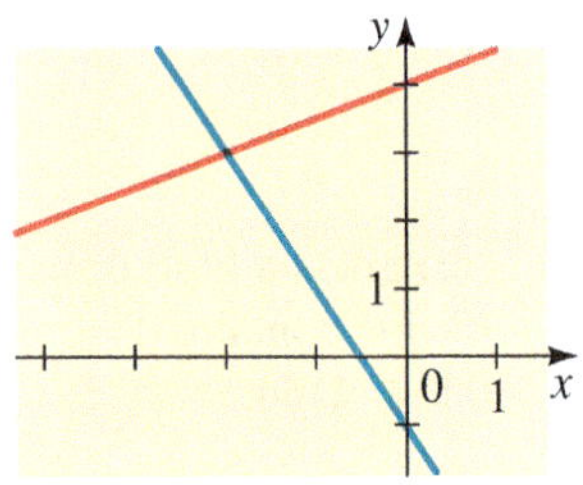

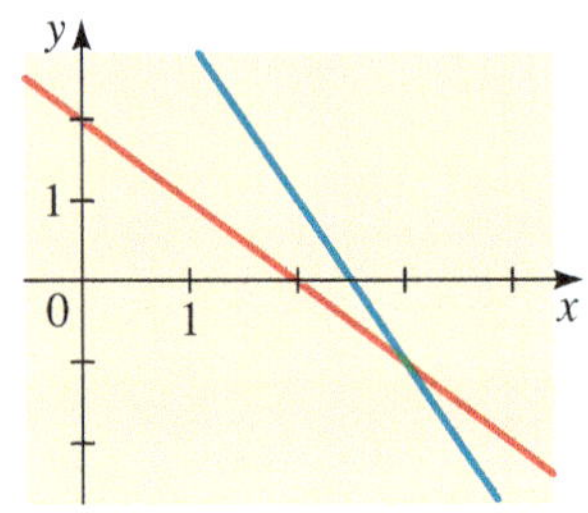

15–20 ■ Number of Solutions Determined Graphically Graph each linear system, either by hand or using a graphing device. Use the graph to determine whether the system has one solution, no solution, or infinitely many solutions. If there is exactly one solution, use the graph to find it.

15. $\begin{cases} x - y = 4 \\ 2x + y = 2 \end{cases}$ **16.** $\begin{cases} 2x - y = 4 \\ 3x + y = 6 \end{cases}$

17. $\begin{cases} 2x - 3y = 12 \\ -x + \frac{3}{2}y = 4 \end{cases}$ **18.** $\begin{cases} 2x + 6y = 0 \\ -3x - 9y = 18 \end{cases}$

19. $\begin{cases} -x + \frac{1}{2}y = -5 \\ 2x - y = 10 \end{cases}$ **20.** $\begin{cases} 12x + 15y = -18 \\ 2x + \frac{5}{2}y = -3 \end{cases}$

21–50 ■ Solving a System of Equations Solve the system, or show that it has no solution. If the system has infinitely many solutions, express them in the ordered-pair form given in Example 6.

21. $\begin{cases} x + y = 4 \\ -x + y = 0 \end{cases}$ **22.** $\begin{cases} x - y = 3 \\ x + 3y = 7 \end{cases}$

23. $\begin{cases} 2x - 3y = 9 \\ 4x + 3y = 9 \end{cases}$ **24.** $\begin{cases} 3x + 2y = 0 \\ -x - 2y = 8 \end{cases}$

25. $\begin{cases} x + 3y = 5 \\ 2x - y = 3 \end{cases}$ **26.** $\begin{cases} x + y = 7 \\ 2x - 3y = -1 \end{cases}$

27. $\begin{cases} -x + y = 2 \\ 4x - 3y = -3 \end{cases}$ **28.** $\begin{cases} 4x - 3y = 28 \\ 9x - y = -6 \end{cases}$

29. $\begin{cases} x + 2y = 7 \\ 5x - y = 2 \end{cases}$ **30.** $\begin{cases} -4x + 12y = 0 \\ 12x + 4y = 160 \end{cases}$

31. $\begin{cases} -\frac{1}{3}x - \frac{1}{6}y = -1 \\ \frac{2}{3}x + \frac{1}{6}y = 3 \end{cases}$ **32.** $\begin{cases} \frac{3}{4}x + \frac{1}{2}y = 5 \\ -\frac{1}{4}x - \frac{3}{2}y = 1 \end{cases}$

33. $\begin{cases} \frac{1}{2}x + \frac{1}{3}y = 2 \\ \frac{1}{5}x - \frac{2}{3}y = 8 \end{cases}$ **34.** $\begin{cases} 0.2x - 0.2y = -1.8 \\ -0.3x + 0.5y = 3.3 \end{cases}$

35. $\begin{cases} 3x + 2y = 8 \\ x - 2y = 0 \end{cases}$ **36.** $\begin{cases} 4x + 2y = 16 \\ x - 5y = 70 \end{cases}$

37. $\begin{cases} x + 4y = 8 \\ 3x + 12y = 2 \end{cases}$ **38.** $\begin{cases} -3x + 5y = 2 \\ 9x - 15y = 6 \end{cases}$

39. $\begin{cases} 2x - 6y = 10 \\ -3x + 9y = -15 \end{cases}$

40. $\begin{cases} 2x - 3y = -8 \\ 14x - 21y = 3 \end{cases}$

41. $\begin{cases} 6x + 4y = 12 \\ 9x + 6y = 18 \end{cases}$

42. $\begin{cases} 25x - 75y = 100 \\ -10x + 30y = -40 \end{cases}$

43. $\begin{cases} 8s - 3t = -3 \\ 5s - 2t = -1 \end{cases}$

44. $\begin{cases} u - 30v = -5 \\ -3u + 80v = 5 \end{cases}$

45. $\begin{cases} \frac{1}{2}x + \frac{3}{5}y = 3 \\ \frac{5}{3}x + 2y = 10 \end{cases}$

46. $\begin{cases} \frac{3}{2}x - \frac{1}{3}y = \frac{1}{2} \\ 2x - \frac{1}{2}y = -\frac{1}{2} \end{cases}$

47. $\begin{cases} 0.4x + 1.2y = 14 \\ 12x - 5y = 10 \end{cases}$

48. $\begin{cases} 26x - 10y = -4 \\ -0.6x + 1.2y = 3 \end{cases}$

49. $\begin{cases} \frac{1}{3}x - \frac{1}{4}y = 2 \\ -8x + 6y = 10 \end{cases}$

50. $\begin{cases} -\frac{1}{10}x + \frac{1}{2}y = 4 \\ 2x - 10y = -80 \end{cases}$

51–54 ■ Solving a System of Equations Graphically Use a graphing device to graph both lines in the same viewing rectangle. (Note that you must solve for y in terms of x before graphing if you are using a graphing calculator.) Solve the system either by zooming in and using TRACE or by using `Intersect`. Round your answers to two decimals.

51. $\begin{cases} 0.21x + 3.17y = 9.51 \\ 2.35x - 1.17y = 5.89 \end{cases}$

52. $\begin{cases} 18.72x - 14.91y = 12.33 \\ 6.21x - 12.92y = 17.82 \end{cases}$

53. $\begin{cases} 2371x - 6552y = 13{,}591 \\ 9815x + 992y = 618{,}555 \end{cases}$

54. $\begin{cases} -435x + 912y = 0 \\ 132x + 455y = 994 \end{cases}$

SKILLS Plus

55–58 ■ Solving a General System of Equations Find x and y in terms of a and b.

55. $\begin{cases} x + y = 0 \\ x + ay = 1 \end{cases} \quad (a \neq 1)$

56. $\begin{cases} ax + by = 0 \\ x + y = 1 \end{cases} \quad (a \neq b)$

57. $\begin{cases} ax + by = 1 \\ bx + ay = 1 \end{cases} \quad (a^2 - b^2 \neq 0)$

58. $\begin{cases} ax + by = 0 \\ a^2x + b^2y = 1 \end{cases} \quad (a \neq 0, b \neq 0, a \neq b)$

APPLICATIONS

59. **Number Problem** Find two numbers whose sum is 34 and whose difference is 10.

60. **Number Problem** The sum of two numbers is twice their difference. The larger number is 6 more than twice the smaller. Find the numbers.

61. **Value of Coins** A man has 14 coins in his pocket, all of which are dimes and quarters. If the total value of his change is \$2.75, how many dimes and how many quarters does he have?

62. **Admission Fees** The admission fee at an amusement park is \$1.50 for children and \$4.00 for adults. On a certain day, 2200 people entered the park, and the admission fees that were collected totaled \$5050. How many children and how many adults were admitted?

63. **Gas Station** A gas station sells regular gas for \$2.20 per gallon and premium gas for \$3.00 a gallon. At the end of a business day 280 gallons of gas had been sold, and receipts totaled \$680. How many gallons of each type of gas had been sold?

64. **Fruit Stand** A fruit stand sells two varieties of strawberries: standard and deluxe. A box of standard strawberries sells for \$7, and a box of deluxe strawberries sells for \$10. In one day the stand sold 135 boxes of strawberries for a total of \$1110. How many boxes of each type were sold?

65. **Airplane Speed** A man flies a small airplane from Fargo to Bismarck, North Dakota—a distance of 180 mi. Because he is flying into a headwind, the trip takes him 2 h. On the way back, the wind is still blowing at the same speed, so the return trip takes only 1 h 12 min. What is his speed in still air, and how fast is the wind blowing?

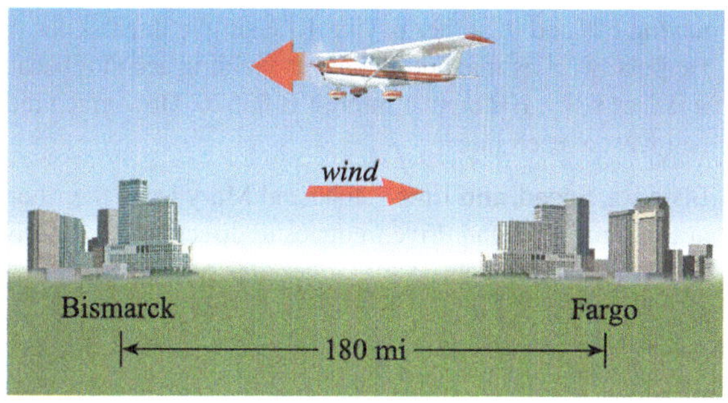

66. **Boat Speed** A boat on a river travels downstream between two points, 20 mi apart, in 1 h. The return trip against the current takes $2\frac{1}{2}$ h. What is the boat's speed, and how fast does the current in the river flow?

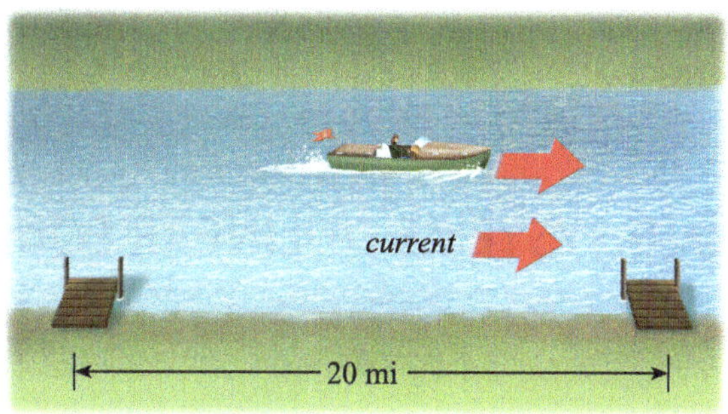

67. **Nutrition** A researcher performs an experiment to test a hypothesis that involves the nutrients niacin and retinol. She feeds one group of laboratory rats a daily diet of precisely 32 units of niacin and 22,000 units of retinol. She uses two types of commercial pellet foods. Food A contains 0.12 unit of niacin and 100 units of retinol per gram. Food B contains 0.20 unit of niacin and 50 units of retinol per gram. How many grams of each food does she feed this group of rats each day?

68. Coffee Blends A customer in a coffee shop purchases a blend of two coffees: Kenyan, costing \$3.50 a pound, and Sri Lankan, costing \$5.60 a pound. He buys 3 lb of the blend, which costs him \$11.55. How many pounds of each kind went into the mixture?

69. Mixture Problem A chemist has two large containers of sulfuric acid solution, with different concentrations of acid in each container. Blending 300 mL of the first solution and 600 mL of the second gives a mixture that is 15% acid, whereas blending 100 mL of the first with 500 mL of the second gives a $12\frac{1}{2}\%$ acid mixture. What are the concentrations of sulfuric acid in the original containers?

70. Mixture Problem A biologist has two brine solutions, one containing 5% salt and another containing 20% salt. How many milliliters of each solution should she mix to obtain 1 L of a solution that contains 14% salt?

71. Investments A woman invests a total of \$20,000 in two accounts, one paying 5% and the other paying 8% simple interest per year. Her annual interest is \$1180. How much did she invest at each rate?

72. Investments A man invests his savings in two accounts, one paying 6% and the other paying 10% simple interest per year. He puts twice as much in the lower-yielding account because it is less risky. His annual interest is \$3520. How much did he invest at each rate?

73. Distance, Speed, and Time John and Mary leave their house at the same time and drive in opposite directions. John drives at 60 mi/h and travels 35 mi farther than Mary, who drives at 40 mi/h. Mary's trip takes 15 min longer than John's. For what length of time does each of them drive?

74. Aerobic Exercise A woman keeps fit by bicycling and running every day. On Monday she spends $\frac{1}{2}$ h at each activity, covering a total of $12\frac{1}{2}$ mi. On Tuesday she runs for 12 min and cycles for 45 min, covering a total of 16 mi. Assuming that her running and cycling speeds don't change from day to day, find these speeds.

75. Number Problem The sum of the digits of a two-digit number is 7. When the digits are reversed, the number is increased by 27. Find the number.

76. Area of a Triangle Find the area of the triangle that lies in the first quadrant (with its base on the x-axis) and that is bounded by the lines $y = 2x - 4$ and $y = -4x + 20$.

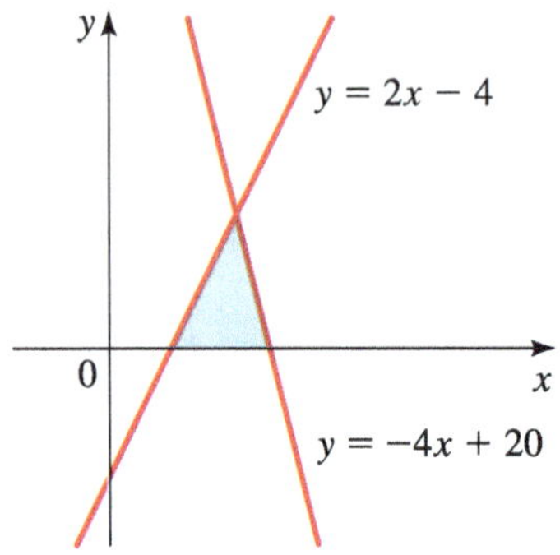

DISCUSS ■ DISCOVER ■ PROVE ■ WRITE

77. DISCUSS: The Least Squares Line The *least squares* line or *regression* line is the line that best fits a set of points in the plane. We studied this line in the *Focus on Modeling* that follows Chapter 1 (see page 139). By using calculus, it can be shown that the line that best fits the n data points $(x_1, y_1), (x_2, y_2), \ldots, (x_n, y_n)$ is the line $y = ax + b$, where the coefficients a and b satisfy the following pair of linear equations. (The notation $\sum_{k=1}^{n} x_k$ stands for the sum of all the x's. See Section 12.1 for a complete description of sigma (Σ) notation.)

$$\left(\sum_{k=1}^{n} x_k\right)a + nb = \sum_{k=1}^{n} y_k$$

$$\left(\sum_{k=1}^{n} x_k^2\right)a + \left(\sum_{k=1}^{n} x_k\right)b = \sum_{k=1}^{n} x_k y_k$$

Use these equations to find the least squares line for the following data points.

$$(1, 3), \quad (2, 5), \quad (3, 6), \quad (5, 6), \quad (7, 9)$$

Sketch the points and your line to confirm that the line fits these points well. If your calculator computes regression lines, see whether it gives you the same line as the formulas.

10.2 SYSTEMS OF LINEAR EQUATIONS IN SEVERAL VARIABLES

■ Solving a Linear System ■ The Number of Solutions of a Linear System ■ Modeling Using Linear Systems

A **linear equation in n variables** is an equation that can be put in the form

$$a_1x_1 + a_2x_2 + \cdots + a_nx_n = c$$

where $a_1, a_2, \ldots, a_n$ and c are real numbers, and $x_1, x_2, \ldots, x_n$ are the variables. If we have only three or four variables, we generally use x, y, z, and w instead of x_1, x_2, x_3, and x_4. Such equations are called *linear* because if we have just two variables, the equation is $a_1x + a_2y = c$, which is the equation of a line. Here are some examples of equations in three variables that illustrate the difference between linear and nonlinear equations.

Linear equations	Nonlinear equations	
$6x_1 - 3x_2 + \sqrt{5}x_3 = 10$	$x^2 + 3y - \sqrt{z} = 5$	Not linear because it contains the square and the square root of a variable
$x + y + z = 2w - \frac{1}{2}$	$x_1x_2 + 6x_3 = -6$	Not linear because it contains a product of variables

In this section we study systems of linear equations in three or more variables.

Solving a Linear System

The following are two examples of systems of linear equations in three variables. The second system is in **triangular form**; that is, the variable x doesn't appear in the second equation, and the variables x and y do not appear in the third equation.

A system of linear equations

$$\begin{cases} x - 2y - z = 1 \\ -x + 3y + 3z = 4 \\ 2x - 3y + z = 10 \end{cases}$$

A system in triangular form

$$\begin{cases} x - 2y - z = 1 \\ y + 2z = 5 \\ z = 3 \end{cases}$$

It's easy to solve a system that is in triangular form by using back-substitution. So our goal in this section is to start with a system of linear equations and change it to a system in triangular form that has the same solutions as the original system. We begin by showing how to use back-substitution to solve a system that is already in triangular form.

EXAMPLE 1 ■ Solving a Triangular System Using Back-Substitution

Solve the following system using back-substitution:

$$\begin{cases} x - 2y - z = 1 & \text{Equation 1} \\ y + 2z = 5 & \text{Equation 2} \\ z = 3 & \text{Equation 3} \end{cases}$$

SOLUTION From the last equation we know that $z = 3$. We back-substitute this into the second equation and solve for y.

$$y + 2(3) = 5 \qquad \text{Back-substitute } z = 3 \text{ into Equation 2}$$

$$y = -1 \qquad \text{Solve for } y$$

Then we back-substitute $y = -1$ and $z = 3$ into the first equation and solve for x.

$$x - 2(-1) - (3) = 1 \qquad \text{Back-substitute } y = -1 \text{ and } z = 3 \text{ into Equation 1}$$

$$x = 2 \qquad \text{Solve for } x$$

The solution of the system is $x = 2$, $y = -1$, $z = 3$. We can also write the solution as the ordered triple $(2, -1, 3)$.

Now Try Exercise 7 ■

To change a system of linear equations to an **equivalent system** (that is, a system with the same solutions as the original system), we use the elimination method. This means that we can use the following operations.

OPERATIONS THAT YIELD AN EQUIVALENT SYSTEM

1. Add a nonzero multiple of one equation to another.
2. Multiply an equation by a nonzero constant.
3. Interchange the positions of two equations.

To solve a linear system, we use these operations to change the system to an equivalent triangular system. Then we use back-substitution as in Example 1. This process is called **Gaussian elimination**.

EXAMPLE 2 ■ Solving a System of Three Equations in Three Variables

Solve the following system using Gaussian elimination:

$$\begin{cases} x - 2y + 3z = 1 & \text{Equation 1} \\ x + 2y - z = 13 & \text{Equation 2} \\ 3x + 2y - 5z = 3 & \text{Equation 3} \end{cases}$$

SOLUTION We need to change this to a triangular system, so we begin by eliminating the x-term from the second equation.

$$\begin{array}{rl} x + 2y - z = 13 & \text{Equation 2} \\ x - 2y + 3z = 1 & \text{Equation 1} \\ \hline 4y - 4z = 12 & \text{Equation 2} + (-1) \times \text{Equation 1} = \text{new Equation 2} \end{array}$$

This gives us a new, equivalent system that is one step closer to triangular form.

$$\begin{cases} x - 2y + 3z = 1 & \text{Equation 1} \\ 4y - 4z = 12 & \text{Equation 2} \\ 3x + 2y - 5z = 3 & \text{Equation 3} \end{cases}$$

Now we eliminate the x-term from the third equation.

$$\begin{array}{rl} 3x + 2y - 5z = 3 \\ -3x + 6y - 9z = -3 \\ \hline 8y - 14z = 0 \end{array}$$

$$\begin{cases} x - 2y + 3z = 1 \\ 4y - 4z = 12 \\ 8y - 14z = 0 & \text{Equation 3} + (-3) \times \text{Equation 1} = \text{new Equation 3} \end{cases}$$

Then we eliminate the y-term from the third equation.

$$\begin{array}{rl} 8y - 14z = 0 \\ -8y + 8z = -24 \\ \hline -6z = -24 \end{array}$$

$$\begin{cases} x - 2y + 3z = 1 \\ 4y - 4z = 12 \\ -6z = -24 & \text{Equation 3} + (-2) \times \text{Equation 2} = \text{new Equation 3} \end{cases}$$

The system is now in triangular form, but it will be easier to work with if we divide the second and third equations by the common factors of each term.

$$\begin{cases} x - 2y + 3z = 1 \\ y - z = 3 & \frac{1}{4} \times \text{Equation 2} = \text{new Equation 2} \\ z = 4 & -\frac{1}{6} \times \text{Equation 3} = \text{new Equation 3} \end{cases}$$

Now we use back-substitution to solve the system. From the third equation we get $z = 4$. We back-substitute this into the second equation and solve for y.

$$\begin{array}{rl} y - (4) = 3 & \text{Back-substitute } z = 4 \text{ into Equation 2} \\ y = 7 & \text{Solve for } y \end{array}$$

Now we back-substitute $y = 7$ and $z = 4$ into the first equation and solve for x.

$$\begin{array}{rl} x - 2(7) + 3(4) = 1 & \text{Back-substitute } y = 7 \text{ and } z = 4 \text{ into Equation 1} \\ x = 3 & \text{Solve for } x \end{array}$$

The solution of the system is $x = 3$, $y = 7$, $z = 4$, which we can write as the ordered triple $(3, 7, 4)$.

CHECK YOUR ANSWER

$x = 3, y = 7, z = 4$:

$$\begin{aligned} (3) - 2(7) + 3(4) &= 1 \\ (3) + 2(7) - (4) &= 13 \\ 3(3) + 2(7) - 5(4) &= 3 \quad ✓ \end{aligned}$$

Now Try Exercise 17 ■

The Number of Solutions of a Linear System

The graph of a linear equation in three variables is a plane in three-dimensional space (see Section 9.6). A system of three equations in three variables represents three planes in space. The solutions of the system are the points where all three planes intersect. Three planes may intersect in a point, in a line, or not at all, or all three planes may coincide. Figure 1 illustrates some of these possibilities. Checking these possibilities we see that there are three possible outcomes when solving such a system.

NUMBER OF SOLUTIONS OF A LINEAR SYSTEM

For a system of linear equations, exactly one of the following is true.

1. The system has exactly one solution.
2. The system has no solution.
3. The system has infinitely many solutions.

A system with no solution is said to be **inconsistent**, and a system with infinitely many solutions is said to be **dependent**. As we see in the next example, a linear system has no solution if we end up with a *false equation* after applying Gaussian elimination to the system.

(a) The three planes intersect at a single point. The system has one solution.

(b) The three planes intersect at more than one point. The system has infinitely many solutions.

(c) The three planes have no point in common. The system has no solution.

FIGURE 1

EXAMPLE 3 ■ A System with No Solution

Solve the following system:

$$\begin{cases} x + 2y - 2z = 1 & \text{Equation 1} \\ 2x + 2y - z = 6 & \text{Equation 2} \\ 3x + 4y - 3z = 5 & \text{Equation 3} \end{cases}$$

SOLUTION To put this in triangular form, we begin by eliminating the x-terms from the second equation and the third equation.

$$\begin{cases} x + 2y - 2z = 1 \\ -2y + 3z = 4 & \text{Equation 2} + (-2) \times \text{Equation 1} = \text{new Equation 2} \\ 3x + 4y + 3z = 5 \end{cases}$$

$$\begin{cases} x + 2y - 2z = 1 \\ -2y + 3z = 4 \\ -2y + 3z = 2 & \text{Equation 3} + (-3) \times \text{Equation 1} = \text{new Equation 3} \end{cases}$$

Now we eliminate the y-term from the third equation.

$$\begin{cases} x + 2y - 2z = 1 \\ \quad -2y + 3z = 4 \\ \qquad\quad 0 = -2 & \text{Equation 3} + (-1) \times \text{Equation 2} = \text{new Equation 3} \end{cases}$$

The system is now in triangular form, but the third equation says $0 = -2$, which is false. No matter what values we assign to x, y, and z, the third equation will never be true. This means that the system has *no solution*.

Now Try Exercise 29

EXAMPLE 4 ■ A System with Infinitely Many Solutions

Solve the following system:

$$\begin{cases} x - y + 5z = -2 & \text{Equation 1} \\ 2x + y + 4z = 2 & \text{Equation 2} \\ 2x + 4y - 2z = 8 & \text{Equation 3} \end{cases}$$

SOLUTION To put this in triangular form, we begin by eliminating the x-terms from the second equation and the third equation.

$$\begin{cases} x - y + 5z = -2 \\ \quad 3y - 6z = 6 & \text{Equation 2} + (-2) \times \text{Equation 1} = \text{new Equation 2} \\ 2x + 4y - 2z = 8 \end{cases}$$

$$\begin{cases} x - y + 5z = -2 \\ \quad 3y - 6z = 6 \\ \quad 6y - 12z = 12 & \text{Equation 3} + (-2) \times \text{Equation 1} = \text{new Equation 3} \end{cases}$$

Now we eliminate the y-term from the third equation.

$$\begin{cases} x - y + 5z = -2 \\ \quad 3y - 6z = 6 \\ \qquad\quad 0 = 0 & \text{Equation 3} + (-2) \times \text{Equation 2} = \text{new Equation 3} \end{cases}$$

The new third equation is true, but it gives us no new information, so we can drop it from the system. Only two equations are left. We can use them to solve for x and y in terms of z, but z can take on any value, so there are infinitely many solutions.

© Vitalii Nesterchuk/Shutterstock.com

DISCOVERY PROJECT

Best Fit Versus Exact Fit

The law of gravity is precise. But when we obtain data on the distance an object falls in a given time, our measurements are not exact. We can, however, find the line (or parabola) that *best* fits our data. Not all of the data points will lie on the line (or parabola). But if we are given just two points, we can find a line of *exact* fit—that is, a line that passes through the two points. Similarly, we can find a parabola through three points. In this project we compare exact data with models of real-world data. You can find the project at **www.stewartmath.com**.

To find the complete solution of the system, we begin by solving for y in terms of z, using the new second equation.

$$3y - 6z = 6 \quad \text{Equation 2}$$
$$y - 2z = 2 \quad \text{Multiply by } \tfrac{1}{3}$$
$$y = 2z + 2 \quad \text{Solve for } y$$

Then we solve for x in terms of z, using the first equation.

$$x - (2z + 2) + 5z = -2 \quad \text{Substitute } y = 2z + 2 \text{ into Equation 1}$$
$$x + 3z - 2 = -2 \quad \text{Simplify}$$
$$x = -3z \quad \text{Solve for } x$$

To describe the complete solution, we let z be any real number t. The solution is

$$x = -3t$$
$$y = 2t + 2$$
$$z = t$$

We can also write this as the ordered triple $(-3t, 2t + 2, t)$.

Now Try Exercise 33 ■

In the solution of Example 4 the variable t is called a **parameter**. To get a specific solution, we give a specific value to the parameter t. For instance, if we set $t = 2$, we get

$$x = -3(2) = -6$$
$$y = 2(2) + 2 = 6$$
$$z = 2$$

Thus $(-6, 6, 2)$ is a solution of the system. Here are some other solutions of the system obtained by substituting other values for the parameter t.

Parameter t	Solution $(-3t, 2t + 2, t)$
-1	$(3, 0, -1)$
0	$(0, 2, 0)$
3	$(-9, 8, 3)$
10	$(-30, 22, 10)$

You should check that these points satisfy the original equations. There are infinitely many choices for the parameter t, so the system has infinitely many solutions.

Modeling Using Linear Systems

Linear systems are used to model situations that involve several varying quantities. In the next example we consider an application of linear systems to finance.

EXAMPLE 5 ■ Modeling a Financial Problem Using a Linear System

Jason receives an inheritance of \$50,000. His financial advisor suggests that he invest this in three mutual funds: a money-market fund, a blue-chip stock fund, and a high-tech stock fund. The advisor estimates that the money-market fund will return 5% over the next year, the blue-chip fund 9%, and the high-tech fund 16%. Jason wants a total first-year return of \$4000. To avoid excessive risk, he decides to invest three times as much in the money-market fund as in the high-tech stock fund. How much should he invest in each fund?

SOLUTION

Let x = amount invested in the money-market fund

y = amount invested in the blue-chip stock fund

z = amount invested in the high-tech stock fund

We convert each fact given in the problem into an equation.

$$x + y + z = 50{,}000 \qquad \text{Total amount invested is \$50,000}$$
$$0.05x + 0.09y + 0.16z = 4000 \qquad \text{Total investment return is \$4000}$$
$$x = 3z \qquad \text{Money-market amount is } 3 \times \text{high-tech amount}$$

Multiplying the second equation by 100 and rewriting the third, we get the following system, which we solve using Gaussian elimination.

$$\begin{cases} x + y + z = 50{,}000 \\ 5x + 9y + 16z = 400{,}000 & 100 \times \text{Equation 2} \\ x - 3z = 0 & \text{Subtract } 3z \end{cases}$$

$$\begin{cases} x + y + z = 50{,}000 \\ 4y + 11z = 150{,}000 & \text{Equation 2} + (-5) \times \text{Equation 1} = \text{new Equation 2} \\ -y - 4z = -50{,}000 & \text{Equation 3} + (-1) \times \text{Equation 1} = \text{new Equation 3} \end{cases}$$

$$\begin{cases} x + y + z = 50{,}000 \\ -5z = -50{,}000 & \text{Equation 2} + 4 \times \text{Equation 3} = \text{new Equation 2} \\ -y - 4z = -50{,}000 \end{cases}$$

$$\begin{cases} x + y + z = 50{,}000 \\ z = 10{,}000 & \left(-\tfrac{1}{5}\right) \times \text{Equation 2} \\ y + 4z = 50{,}000 & (-1) \times \text{Equation 3} \end{cases}$$

$$\begin{cases} x + y + z = 50{,}000 \\ y + 4z = 50{,}000 & \text{Interchange Equations 2 and 3} \\ z = 10{,}000 \end{cases}$$

Now that the system is in triangular form, we use back-substitution to find that $x = 30{,}000$, $y = 10{,}000$, and $z = 10{,}000$. This means that Jason should invest

$30,000 in the money-market fund
$10,000 in the blue-chip stock fund
$10,000 in the high-tech stock fund

Now Try Exercise 39 ■

10.2 EXERCISES

CONCEPTS

1–2 ■ These exercises refer to the following system:

$$\begin{cases} x - y + z = 2 \\ -x + 2y + z = -3 \\ 3x + y - 2z = 2 \end{cases}$$

1. If we add 2 times the first equation to the second equation, the second equation becomes ________ = ____.

2. To eliminate x from the third equation, we add ________ times the first equation to the third equation. The third equation becomes ________ = ____.

SKILLS

3–6 ■ Is the System of Equations Linear? State whether the equation or system of equations is linear.

3. $6x - \sqrt{3}y + \frac{1}{2}z = 0$

4. $x^2 + y^2 + z^2 = 4$

5. $\begin{cases} xy - 3y + z = 5 \\ x - y^2 + 5z = 0 \\ 2x + yz = 3 \end{cases}$

6. $\begin{cases} x - 2y + 3z = 10 \\ 2x + 5y = 2 \\ y + 2z = 4 \end{cases}$

7–12 ■ Triangular Systems Use back-substitution to solve the triangular system.

7. $\begin{cases} x - 3y + z = 0 \\ y - z = 3 \\ z = -2 \end{cases}$

8. $\begin{cases} 3x - 3y + z = 0 \\ y + 4z = 10 \\ z = 3 \end{cases}$

9. $\begin{cases} x + 2y + z = 7 \\ -y + 3z = 9 \\ 2z = 6 \end{cases}$

10. $\begin{cases} x - 2y + 3z = 10 \\ 2y - z = 2 \\ 3z = 12 \end{cases}$

11. $\begin{cases} 2x - y + 6z = 5 \\ y + 4z = 0 \\ -2z = 1 \end{cases}$

12. $\begin{cases} 4x + 3z = 10 \\ 2y - z = -6 \\ \frac{1}{2}z = 4 \end{cases}$

13–16 ■ Eliminating a Variable Perform an operation on the given system that eliminates the indicated variable. Write the new equivalent system.

13. $\begin{cases} 3x + y + z = 4 \\ -x + y + 2z = 0 \\ x - 2y - z = -1 \end{cases}$

Eliminate the x-term from the second equation.

14. $\begin{cases} -5x + 2y - 3z = 3 \\ 10x - 3y + z = -20 \\ -x + 3y + z = 8 \end{cases}$

Eliminate the x-term from the second equation.

15. $\begin{cases} 2x + y - 3z = 5 \\ 2x + 3y + z = 13 \\ 6x - 5y - z = 7 \end{cases}$

Eliminate the x-term from the third equation.

16. $\begin{cases} x - 3y + 2z = -1 \\ y + z = -1 \\ 2y - z = 1 \end{cases}$

Eliminate the y-term from the third equation.

17–38 ■ Solving a System of Equations in Three Variables Find the complete solution of the linear system, or show that it is inconsistent.

17. $\begin{cases} x - y - z = 4 \\ 2y + z = -1 \\ -x + y - 2z = 5 \end{cases}$

18. $\begin{cases} x - y + z = 0 \\ y + 2z = -2 \\ x + y - z = 2 \end{cases}$

19. $\begin{cases} x + 2y - z = -6 \\ y - 3z = -16 \\ x - 3y + 2z = 14 \end{cases}$

20. $\begin{cases} x - 2y + 3z = -10 \\ 3y + z = 7 \\ x + y - z = 7 \end{cases}$

21. $\begin{cases} x + y + z = 4 \\ x + 3y + 3z = 10 \\ 2x + y - z = 3 \end{cases}$

22. $\begin{cases} x + y + z = 0 \\ -x + 2y + 5z = 3 \\ 3x - y = 6 \end{cases}$

23. $\begin{cases} x - 4z = 1 \\ 2x - y - 6z = 4 \\ 2x + 3y - 2z = 8 \end{cases}$

24. $\begin{cases} x - y + 2z = 2 \\ 3x + y + 5z = 8 \\ 2x - y - 2z = -7 \end{cases}$

25. $\begin{cases} 2x + 4y - z = 2 \\ x + 2y - 3z = -4 \\ 3x - y + z = 1 \end{cases}$

26. $\begin{cases} 2x + y - z = -8 \\ -x + y + z = 3 \\ -2x + 4z = 18 \end{cases}$

27. $\begin{cases} 2y + 4z = -1 \\ -2x + y + 2z = -1 \\ 4x - 2y = 0 \end{cases}$

28. $\begin{cases} y - z = -1 \\ 6x + 2y + z = 2 \\ -x - y - 3z = -2 \end{cases}$

29. $\begin{cases} x + 2y - z = 1 \\ 2x + 3y - 4z = -3 \\ 3x + 6y - 3z = 4 \end{cases}$

30. $\begin{cases} -x + 2y + 5z = 4 \\ x - 2z = 0 \\ 4x - 2y - 11z = 2 \end{cases}$

31. $\begin{cases} 2x + 3y - z = 1 \\ x + 2y = 3 \\ x + 3y + z = 4 \end{cases}$

32. $\begin{cases} x - 2y - 3z = 5 \\ 2x + y - z = 5 \\ 4x - 3y - 7z = 5 \end{cases}$

33. $\begin{cases} x + y - z = 0 \\ x + 2y - 3z = -3 \\ 2x + 3y - 4z = -3 \end{cases}$

34. $\begin{cases} x - 2y + z = 3 \\ 2x - 5y + 6z = 7 \\ 2x - 3y - 2z = 5 \end{cases}$

35. $\begin{cases} x + 3y - 2z = 0 \\ 2x + 4z = 4 \\ 4x + 6y = 4 \end{cases}$

36. $\begin{cases} 2x + 4y - z = 3 \\ x + 2y + 4z = 6 \\ x + 2y - 2z = 0 \end{cases}$

37. $\begin{cases} x + z + 2w = 6 \\ y - 2z = -3 \\ x + 2y - z = -2 \\ 2x + y + 3z - 2w = 0 \end{cases}$

38. $\begin{cases} x + y + z + w = 0 \\ x + y + 2z + 2w = 0 \\ 2x + 2y + 3z + 4w = 1 \\ 2x + 3y + 4z + 5w = 2 \end{cases}$

APPLICATIONS

39. Financial Planning Mark has \$100,000 to invest. His financial consultant advises him to diversify his investment in three types of bonds: short-term, intermediate-term, and long-term. The short-term bonds pay 4%, the intermediate-term bonds pay 5%, and the long-term bonds pay 6% simple interest per year. Mark wishes to realize a total annual income of 5.1%, with equal amounts invested in short- and intermediate-term bonds. How much should he invest in each type of bond?

40. Financial Planning Cyndee wants to invest \$50,000. Her financial planner advises her to invest in three types of accounts: one paying 3%, one paying $5\frac{1}{2}$%, and one paying 9% simple interest per year. Cyndee wants to put twice as much in the lowest-yielding, least-risky account as in the highest-yielding account. How much should she invest in each account to achieve a total annual return of \$2540?

41. Agriculture A farmer has 1200 acres of land on which he grows corn, wheat, and soybeans. It costs \$45 per acre to grow corn, \$60 to grow wheat, and \$50 to grow soybeans. Because of market demand, the farmer will grow twice as many acres of wheat as of corn. He has allocated \$63,750 for the cost of growing his crops. How many acres of each crop should he plant?

42. Gas Station A gas station sells three types of gas: Regular for \$3.00 a gallon, Performance Plus for \$3.20 a gallon, and Premium for \$3.30 a gallon. On a particular day 6500 gallons of gas were sold for a total of \$20,050. Three times as many gallons of Regular as Premium gas were sold. How many gallons of each type of gas were sold that day?

43. Nutrition A biologist is performing an experiment on the effects of various combinations of vitamins. She wishes to feed each of her laboratory rabbits a diet that contains exactly 9 mg of niacin, 14 mg of thiamin, and 32 mg of riboflavin. She has available three different types of commercial rabbit pellets; their vitamin content (per ounce) is given in the table. How many ounces of each type of food should each rabbit be given daily to satisfy the experiment requirements?

	Type A	Type B	Type C
Niacin (mg/oz)	2	3	1
Thiamin (mg/oz)	3	1	3
Riboflavin (mg/oz)	8	5	7

44. Diet Program Nicole started a new diet that requires each meal to have 460 calories, 6 g of fiber, and 11 g of fat. The table shows the fiber, fat, and calorie content of one serving of each of three breakfast foods. How many servings of each food should Nicole eat to follow her diet?

Food	Fiber (g)	Fat (g)	Calories
Toast	2	1	100
Cottage cheese	0	5	120
Fruit	2	0	60

45. Juice Blends The Juice Company offers three kinds of smoothies: Midnight Mango, Tropical Torrent, and Pineapple Power. Each smoothie contains the amounts of juices shown in the table.

Smoothie	Mango juice (oz)	Pineapple juice (oz)	Orange juice (oz)
Midnight Mango	8	3	3
Tropical Torrent	6	5	3
Pineapple Power	2	8	4

On a particular day the Juice Company used 820 oz of mango juice, 690 oz of pineapple juice, and 450 oz of orange juice. How many smoothies of each kind were sold that day?

46. Appliance Manufacturing Kitchen Korner produces refrigerators, dishwashers, and stoves at three different factories. The table gives the number of each product produced at each factory per day. Kitchen Korner receives an order for 110 refrigerators, 150 dishwashers, and 114 ovens. How many days should each plant be scheduled to fill this order?

Appliance	Factory A	Factory B	Factory C
Refrigerators	8	10	14
Dishwashers	16	12	10
Stoves	10	18	6

47. Stock Portfolio An investor owns three stocks: A, B, and C. The closing prices of the stocks on three successive trading days are given in the table.

	Stock A	Stock B	Stock C
Monday	\$10	\$25	\$29
Tuesday	\$12	\$20	\$32
Wednesday	\$16	\$15	\$32

Despite the volatility in the stock prices, the total value of the investor's stocks remained unchanged at \$74,000 at the end of each of these three days. How many shares of each stock does the investor own?

48. Electricity By using Kirchhoff's Laws, it can be shown that the currents I_1, I_2, and I_3 that pass through the three branches of the circuit in the figure satisfy the given linear system. Solve the system to find I_1, I_2, and I_3.

$$\begin{cases} I_1 + I_2 - I_3 = 0 \\ 16I_1 - 8I_2 = 4 \\ 8I_2 + 4I_3 = 5 \end{cases}$$

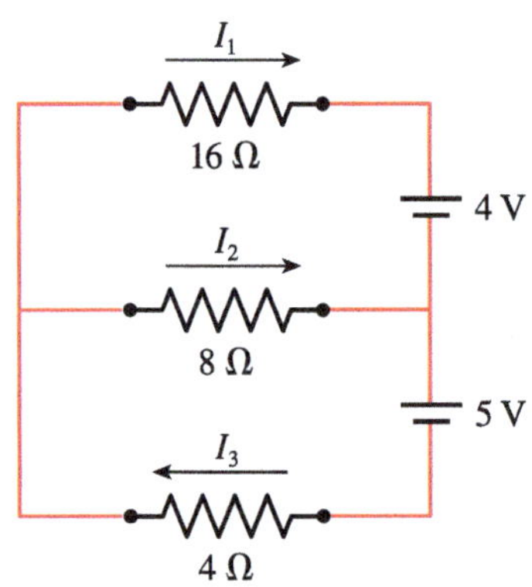

DISCUSS ■ DISCOVER ■ PROVE ■ WRITE

49. PROVE: Can a Linear System Have Exactly Two Solutions?

(a) Suppose that (x_0, y_0, z_0) and (x_1, y_1, z_1) are solutions of the system

$$\begin{cases} a_1x + b_1y + c_1z = d_1 \\ a_2x + b_2y + c_2z = d_2 \\ a_3x + b_3y + c_3z = d_3 \end{cases}$$

Show that $\left(\frac{x_0 + x_1}{2}, \frac{y_0 + y_1}{2}, \frac{z_0 + z_1}{2}\right)$ is also a solution.

(b) Use the result of part (a) to prove that if the system has two different solutions, then it has infinitely many solutions.

10.3 MATRICES AND SYSTEMS OF LINEAR EQUATIONS

■ Matrices ■ The Augmented Matrix of a Linear System ■ Elementary Row Operations ■ Gaussian Elimination ■ Gauss-Jordan Elimination ■ Inconsistent and Dependent Systems ■ Modeling with Linear Systems

A *matrix* is simply a rectangular array of numbers. Matrices* are used to organize information into categories that correspond to the rows and columns of the matrix. For example, a scientist might organize information on a population of endangered whales as follows:

	Immature	Juvenile	Adult
Male	12	52	18
Female	15	42	11

This is a compact way of saying that there are 12 immature males, 15 immature females, 18 adult males, and so on.

In this section we represent a linear system by a matrix, called the *augmented matrix* of the system.

Linear system **Augmented matrix**

$$\begin{cases} 2x - y = 5 & \text{Equation 1} \\ x + 4y = 7 & \text{Equation 2} \end{cases} \qquad \begin{bmatrix} 2 & -1 & 5 \\ 1 & 4 & 7 \end{bmatrix}$$

$$\hphantom{xxxxxxxxxxxxxxxx} x \quad y$$

The augmented matrix contains the same information as the system but in a simpler form. The operations we learned for solving systems of equations can now be performed on the augmented matrix.

■ Matrices

We begin by defining the various elements that make up a matrix.

DEFINITION OF MATRIX

An $m \times n$ **matrix** is a rectangular array of numbers with m **rows** and n **columns**.

$$\begin{bmatrix} a_{11} & a_{12} & a_{13} & \cdots & a_{1n} \\ a_{21} & a_{22} & a_{23} & \cdots & a_{2n} \\ a_{31} & a_{32} & a_{33} & \cdots & a_{3n} \\ \vdots & \vdots & \vdots & \ddots & \vdots \\ a_{m1} & a_{m2} & a_{m3} & \cdots & a_{mn} \end{bmatrix} \leftarrow m \text{ rows}$$

$$\uparrow \; n \text{ columns}$$

We say that the matrix has **dimension** $m \times n$. The numbers a_{ij} are the **entries** of the matrix. The subscript on the entry a_{ij} indicates that it is in the ith row and the jth column.

*The plural of *matrix* is *matrices*.

Here are some examples of matrices.

Matrix	Dimension	
$\begin{bmatrix} 1 & 3 & 0 \\ 2 & 4 & -1 \end{bmatrix}$	2×3	2 rows by 3 columns
$\begin{bmatrix} 6 & -5 & 0 & 1 \end{bmatrix}$	1×4	1 row by 4 columns

The Augmented Matrix of a Linear System

We can write a system of linear equations as a matrix, called the **augmented matrix** of the system, by writing only the coefficients and constants that appear in the equations. Here is an example.

Linear system	Augmented matrix
$\begin{cases} 3x - 2y + z = 5 \\ x + 3y - z = 0 \\ -x + 4z = 11 \end{cases}$	$\begin{bmatrix} 3 & -2 & 1 & 5 \\ 1 & 3 & -1 & 0 \\ -1 & 0 & 4 & 11 \end{bmatrix}$

Notice that a missing variable in an equation corresponds to a 0 entry in the augmented matrix.

EXAMPLE 1 ■ Finding the Augmented Matrix of a Linear System

Write the augmented matrix of the following system of equations:

$$\begin{cases} 6x - 2y - z = 4 \\ x + 3z = 1 \\ 7y + z = 5 \end{cases}$$

SOLUTION First we write the linear system with the variables lined up in columns.

$$\begin{cases} 6x - 2y - z = 4 \\ x \qquad\; + 3z = 1 \\ \qquad 7y + z = 5 \end{cases}$$

The augmented matrix is the matrix whose entries are the coefficients and the constants in this system.

$$\begin{bmatrix} 6 & -2 & -1 & 4 \\ 1 & 0 & 3 & 1 \\ 0 & 7 & 1 & 5 \end{bmatrix}$$

Now Try Exercise 11 ■

Elementary Row Operations

The operations that we used in Section 10.2 to solve linear systems correspond to operations on the rows of the augmented matrix of the system. For example, adding a multiple of one equation to another corresponds to adding a multiple of one row to another.

ELEMENTARY ROW OPERATIONS

1. Add a multiple of one row to another.
2. Multiply a row by a nonzero constant.
3. Interchange two rows.

Note that performing any of these operations on the augmented matrix of a system does not change its solution. We use the following notation to describe the elementary row operations:

Symbol	Description
$R_i + kR_j \to R_i$	Change the ith row by adding k times row j to it, and then put the result back in row i.
kR_i	Multiply the ith row by k.
$R_i \leftrightarrow R_j$	Interchange the ith and jth rows.

In the next example we compare the two ways of writing systems of linear equations.

EXAMPLE 2 ■ Using Elementary Row Operations to Solve a Linear System

Solve the following system of linear equations:

$$\begin{cases} x - y + 3z = 4 \\ x + 2y - 2z = 10 \\ 3x - y + 5z = 14 \end{cases}$$

SOLUTION Our goal is to eliminate the x-term from the second equation and the x- and y-terms from the third equation. For comparison we write both the system of equations and its augmented matrix.

System | **Augmented matrix**

$$\begin{cases} x - y + 3z = 4 \\ x + 2y - 2z = 10 \\ 3x - y + 5z = 14 \end{cases} \qquad \begin{bmatrix} 1 & -1 & 3 & 4 \\ 1 & 2 & -2 & 10 \\ 3 & -1 & 5 & 14 \end{bmatrix}$$

Add $(-1) \times$ Equation 1 to Equation 2.
Add $(-3) \times$ Equation 1 to Equation 3.

$$\begin{cases} x - y + 3z = 4 \\ 3y - 5z = 6 \\ 2y - 4z = 2 \end{cases} \qquad \xrightarrow[R_3 - 3R_1 \to R_3]{R_2 - R_1 \to R_2} \begin{bmatrix} 1 & -1 & 3 & 4 \\ 0 & 3 & -5 & 6 \\ 0 & 2 & -4 & 2 \end{bmatrix}$$

Multiply Equation 3 by $\frac{1}{2}$.

$$\begin{cases} x - y + 3z = 4 \\ 3y - 5z = 6 \\ y - 2z = 1 \end{cases} \qquad \xrightarrow{\frac{1}{2}R_3} \begin{bmatrix} 1 & -1 & 3 & 4 \\ 0 & 3 & -5 & 6 \\ 0 & 1 & -2 & 1 \end{bmatrix}$$

Add $(-3) \times$ Equation 3 to Equation 2 (to eliminate y from Equation 2).

$$\begin{cases} x - y + 3z = 4 \\ z = 3 \\ y - 2z = 1 \end{cases} \qquad \xrightarrow{R_2 - 3R_3 \to R_2} \begin{bmatrix} 1 & -1 & 3 & 4 \\ 0 & 0 & 1 & 3 \\ 0 & 1 & -2 & 1 \end{bmatrix}$$

Interchange Equations 2 and 3.

$$\begin{cases} x - y + 3z = 4 \\ y - 2z = 1 \\ z = 3 \end{cases} \qquad \xrightarrow{R_2 \leftrightarrow R_3} \begin{bmatrix} 1 & -1 & 3 & 4 \\ 0 & 1 & -2 & 1 \\ 0 & 0 & 1 & 3 \end{bmatrix}$$

Now we use back-substitution to find that $x = 2$, $y = 7$, and $z = 3$. The solution is $(2, 7, 3)$.

Now Try Exercise 29 ■

■ Gaussian Elimination

In general, to solve a system of linear equations using its augmented matrix, we use elementary row operations to arrive at a matrix in a certain form. This form is described in the following box.

ROW-ECHELON FORM AND REDUCED ROW-ECHELON FORM OF A MATRIX

A matrix is in **row-echelon form** if it satisfies the following conditions.

1. The first nonzero number in each row (reading from left to right) is 1. This is called the **leading entry**.
2. The leading entry in each row is to the right of the leading entry in the row immediately above it.
3. All rows consisting entirely of zeros are at the bottom of the matrix.

A matrix is in **reduced row-echelon** form if it is in row-echelon form and also satisfies the following condition.

4. Every number above and below each leading entry is a 0.

In the following matrices the first one is not in row-echelon form. The second one *is* in row-echelon form, and the third one is in reduced row-echelon form. The entries in red are the leading entries.

Not in row-echelon form

$$\begin{bmatrix} 0 & 1 & -\frac{1}{2} & 0 & 6 \\ 1 & 0 & 3 & 4 & -5 \\ 0 & 0 & 0 & 1 & 0.4 \\ 0 & 1 & 1 & 0 & 0 \end{bmatrix}$$

Leading 1's do *not* shift to the right in successive rows

Row-echelon form

$$\begin{bmatrix} 1 & 3 & -6 & 10 & 0 \\ 0 & 0 & 1 & 4 & -3 \\ 0 & 0 & 0 & 1 & \frac{1}{2} \\ 0 & 0 & 0 & 0 & 0 \end{bmatrix}$$

Leading 1's shift to the right in successive rows

Reduced row-echelon form

$$\begin{bmatrix} 1 & 3 & 0 & 0 & 0 \\ 0 & 0 & 1 & 0 & -3 \\ 0 & 0 & 0 & 1 & \frac{1}{2} \\ 0 & 0 & 0 & 0 & 0 \end{bmatrix}$$

Leading 1's have 0's above and below them

Here is a systematic way to put a matrix in row-echelon form using elementary row operations:

- Start by obtaining 1 in the top left corner. Then obtain zeros below that 1 by adding appropriate multiples of the first row to the rows below it.
- Next, obtain a leading 1 in the next row, and then obtain zeros below that 1.
- At each stage make sure that every leading entry is to the right of the leading entry in the row above it—rearrange the rows if necessary.
- Continue this process until you arrive at a matrix in row-echelon form.

This is how the process might work for a 3×4 matrix:

$$\begin{bmatrix} 1 & \blacksquare & \blacksquare & \blacksquare \\ 0 & \blacksquare & \blacksquare & \blacksquare \\ 0 & \blacksquare & \blacksquare & \blacksquare \end{bmatrix} \qquad \begin{bmatrix} 1 & \blacksquare & \blacksquare & \blacksquare \\ 0 & 1 & \blacksquare & \blacksquare \\ 0 & 0 & \blacksquare & \blacksquare \end{bmatrix} \qquad \begin{bmatrix} 1 & \blacksquare & \blacksquare & \blacksquare \\ 0 & 1 & \blacksquare & \blacksquare \\ 0 & 0 & 1 & \blacksquare \end{bmatrix}$$

Once an augmented matrix is in row-echelon form, we can solve the corresponding linear system using back-substitution. This technique is called **Gaussian elimination**, in honor of its inventor, the German mathematician C. F. Gauss (see page 290).

SOLVING A SYSTEM USING GAUSSIAN ELIMINATION

1. **Augmented Matrix.** Write the augmented matrix of the system.
2. **Row-Echelon Form.** Use elementary row operations to change the augmented matrix to row-echelon form.
3. **Back-Substitution.** Write the new system of equations that corresponds to the row-echelon form of the augmented matrix and solve by back-substitution.

EXAMPLE 3 ■ Solving a System Using Row-Echelon Form

Solve the following system of linear equations using Gaussian elimination:

$$\begin{cases} 4x + 8y - 4z = 4 \\ 3x + 8y + 5z = -11 \\ -2x + y + 12z = -17 \end{cases}$$

SOLUTION We first write the augmented matrix of the system, and then we use elementary row operations to put it in row-echelon form.

Augmented matrix:

$$\left[\begin{array}{ccc|c} 4 & 8 & -4 & 4 \\ 3 & 8 & 5 & -11 \\ -2 & 1 & 12 & -17 \end{array}\right] \quad \text{Need a 1 here}$$

$$\xrightarrow{\frac{1}{4}R_1} \left[\begin{array}{ccc|c} 1 & 2 & -1 & 1 \\ 3 & 8 & 5 & -11 \\ -2 & 1 & 12 & -17 \end{array}\right] \quad \text{Need 0's here}$$

$$\xrightarrow[R_3 + 2R_1 \to R_3]{R_2 - 3R_1 \to R_2} \left[\begin{array}{ccc|c} 1 & 2 & -1 & 1 \\ 0 & 2 & 8 & -14 \\ 0 & 5 & 10 & -15 \end{array}\right] \quad \text{Need a 1 here}$$

$$\xrightarrow{\frac{1}{2}R_2} \left[\begin{array}{ccc|c} 1 & 2 & -1 & 1 \\ 0 & 1 & 4 & -7 \\ 0 & 5 & 10 & -15 \end{array}\right] \quad \text{Need a 0 here}$$

$$\xrightarrow{R_3 - 5R_2 \to R_3} \left[\begin{array}{ccc|c} 1 & 2 & -1 & 1 \\ 0 & 1 & 4 & -7 \\ 0 & 0 & -10 & 20 \end{array}\right] \quad \text{Need a 1 here}$$

Row-echelon form:

$$\xrightarrow{-\frac{1}{10}R_3} \left[\begin{array}{ccc|c} 1 & 2 & -1 & 1 \\ 0 & 1 & 4 & -7 \\ 0 & 0 & 1 & -2 \end{array}\right]$$

We now have an equivalent matrix in row-echelon form, and the corresponding system of equations is

$$\begin{cases} x + 2y - z = 1 \\ y + 4z = -7 \\ z = -2 \end{cases}$$

Back-substitute: We use back-substitution to solve the system.

$$y + 4(-2) = -7 \qquad \text{Back-substitute } z = -2 \text{ into Equation 2}$$

$$y = 1 \qquad \text{Solve for } y$$

$$x + 2(1) - (-2) = 1 \qquad \text{Back-substitute } y = 1 \text{ and } z = -2 \text{ into Equation 1}$$

$$x = -3 \qquad \text{Solve for } x$$

So the solution of the system is $(-3, 1, -2)$.

Now Try Exercise 31 ■

```
ref([A])
  [[1 2 -1 1 ]
   [0 1 2  -3]
   [0 0 1  -2]]
```

FIGURE 1

Graphing calculators have a "row-echelon form" command that puts a matrix in row-echelon form. (On the TI-83/84 this command is `ref`.) For the augmented matrix in Example 3 the `ref` command gives the output shown in Figure 1. Notice that the

See Appendix D, *Using the TI-83/84 Graphing Calculator*, for specific instructions on working with matrices. Go to **www.stewartmath.com**.

row-echelon form that is obtained by the calculator differs from the one we got in Example 3. This is because the calculator used different row operations than we did. You should check that your calculator's row-echelon form leads to the same solution as ours.

Gauss-Jordan Elimination

If we put the augmented matrix of a linear system in *reduced* row-echelon form, then we don't need to back-substitute to solve the system. To put a matrix in reduced row-echelon form, we use the following steps.

- Use the elementary row operations to put the matrix in row-echelon form.
- Obtain zeros above each leading entry by adding multiples of the row containing that entry to the rows above it. Begin with the last leading entry and work up.

Here is how the process works for a 3×4 matrix:

$$\begin{bmatrix} 1 & \blacksquare & \blacksquare & \blacksquare \\ 0 & 1 & \blacksquare & \blacksquare \\ 0 & 0 & 1 & \blacksquare \end{bmatrix} \rightarrow \begin{bmatrix} 1 & \blacksquare & 0 & \blacksquare \\ 0 & 1 & 0 & \blacksquare \\ 0 & 0 & 1 & \blacksquare \end{bmatrix} \rightarrow \begin{bmatrix} 1 & 0 & 0 & \blacksquare \\ 0 & 1 & 0 & \blacksquare \\ 0 & 0 & 1 & \blacksquare \end{bmatrix}$$

Using the reduced row-echelon form to solve a system is called **Gauss-Jordan elimination**. The process is illustrated in the next example.

EXAMPLE 4 ■ Solving a System Using Reduced Row-Echelon Form

Solve the following system of linear equations, using Gauss-Jordan elimination:

$$\begin{cases} 4x + 8y - 4z = 4 \\ 3x + 8y + 5z = -11 \\ -2x + y + 12z = -17 \end{cases}$$

SOLUTION In Example 3 we used Gaussian elimination on the augmented matrix of this system to arrive at an equivalent matrix in row-echelon form. We continue using elementary row operations on the last matrix in Example 3 to arrive at an equivalent matrix in reduced row-echelon form.

$$\begin{bmatrix} 1 & 2 & -1 & 1 \\ 0 & 1 & 4 & -7 \\ 0 & 0 & 1 & -2 \end{bmatrix}$$

Need 0's here

Need a 0 here

$$\xrightarrow[R_1 + R_3 \to R_1]{R_2 - 4R_3 \to R_2} \begin{bmatrix} 1 & 2 & 0 & -1 \\ 0 & 1 & 0 & 1 \\ 0 & 0 & 1 & -2 \end{bmatrix}$$

$$\xrightarrow{R_1 - 2R_2 \to R_1} \begin{bmatrix} 1 & 0 & 0 & -3 \\ 0 & 1 & 0 & 1 \\ 0 & 0 & 1 & -2 \end{bmatrix}$$

We now have an equivalent matrix in reduced row-echelon form, and the corresponding system of equations is

$$\begin{cases} x = -3 \\ y = 1 \\ z = -2 \end{cases}$$

Since the system is in reduced row-echelon form, back-subsitution is not required to get the solution.

Hence we immediately arrive at the solution $(-3, 1, -2)$.

Now Try Exercise 33 ■

```
rref([A])
  [[1 0 0 -3]
   [0 1 0 1 ]
   [0 0 1 -2]]
```

FIGURE 2

Graphing calculators also have a command that puts a matrix in reduced row-echelon form. (On the TI-83/84 this command is `rref`.) For the augmented matrix in Example 4 the `rref` command gives the output shown in Figure 2. The calculator gives the same reduced row-echelon form as the one we got in Example 4. This is because every matrix has a *unique* reduced row-echelon form.

Inconsistent and Dependent Systems

The systems of linear equations that we considered in Examples 1–4 had exactly one solution. But as we know from Section 10.2, a linear system may have one solution, no solution, or infinitely many solutions. Fortunately, the row-echelon form of a system allows us to determine which of these cases applies, as described in the following box.

First we need some terminology. A **leading variable** in a linear system is one that corresponds to a leading entry in the row-echelon form of the augmented matrix of the system.

THE SOLUTIONS OF A LINEAR SYSTEM IN ROW-ECHELON FORM

Suppose the augmented matrix of a system of linear equations has been transformed by Gaussian elimination into row-echelon form. Then exactly one of the following is true.

1. **No solution.** If the row-echelon form contains a row that represents the equation $0 = c$, where c is not zero, then the system has no solution. A system with no solution is called **inconsistent**.
2. **One solution.** If each variable in the row-echelon form is a leading variable, then the system has exactly one solution, which we find using back-substitution or Gauss-Jordan elimination.
3. **Infinitely many solutions.** If the variables in the row-echelon form are not all leading variables and if the system is not inconsistent, then it has infinitely many solutions. In this case the system is called **dependent**. We solve the system by putting the matrix in reduced row-echelon form and then expressing the leading variables in terms of the nonleading variables. The nonleading variables may take on any real numbers as their values.

The matrices below, all in row-echelon form, illustrate the three cases described above.

No solution

$$\begin{bmatrix} 1 & 2 & 5 & 7 \\ 0 & 1 & 3 & 4 \\ 0 & 0 & 0 & 1 \end{bmatrix}$$

Last equation says $0 = 1$

One solution

$$\begin{bmatrix} 1 & 6 & -1 & 3 \\ 0 & 1 & 2 & -2 \\ 0 & 0 & 1 & 8 \end{bmatrix}$$

Each variable is a leading variable

Infinitely many solutions

$$\begin{bmatrix} 1 & 2 & -3 & 1 \\ 0 & 1 & 5 & -2 \\ 0 & 0 & 0 & 0 \end{bmatrix}$$

z is not a leading variable

EXAMPLE 5 ■ A System with No Solution

Solve the following system:

$$\begin{cases} x - 3y + 2z = 12 \\ 2x - 5y + 5z = 14 \\ x - 2y + 3z = 20 \end{cases}$$

SOLUTION We transform the system into row-echelon form.

$$\left[\begin{array}{rrr|r} 1 & -3 & 2 & 12 \\ 2 & -5 & 5 & 14 \\ 1 & -2 & 3 & 20 \end{array}\right] \xrightarrow[\mathrm{R}_3 - \mathrm{R}_1 \to \mathrm{R}_3]{\mathrm{R}_2 - 2\mathrm{R}_1 \to \mathrm{R}_2} \left[\begin{array}{rrr|r} 1 & -3 & 2 & 12 \\ 0 & 1 & 1 & -10 \\ 0 & 1 & 1 & 8 \end{array}\right]$$

$$\xrightarrow{\mathrm{R}_3 - \mathrm{R}_2 \to \mathrm{R}_3} \left[\begin{array}{rrr|r} 1 & -3 & 2 & 12 \\ 0 & 1 & 1 & -10 \\ 0 & 0 & 0 & 18 \end{array}\right] \xrightarrow{\frac{1}{18}\mathrm{R}_3} \left[\begin{array}{rrr|r} 1 & -3 & 2 & 12 \\ 0 & 1 & 1 & -10 \\ 0 & 0 & 0 & 1 \end{array}\right]$$

This last matrix is in row-echelon form, so we can stop the Gaussian elimination process. Now if we translate the last row back into equation form, we get $0x + 0y + 0z = 1$, or $0 = 1$, which is false. No matter what values we pick for x, y, and z, the last equation will never be a true statement. This means that the system *has no solution.*

Now Try Exercise 39 ■

```
ref([A])
[[1 -2.5 2.5 7  ]
 [0 1     1   -10]
 [0 0     0   1  ]]
```

FIGURE 3

Figure 3 shows the row-echelon form produced by a TI-83/84 calculator for the augmented matrix in Example 5. You should check that this gives the same result.

EXAMPLE 6 ■ A System with Infinitely Many Solutions

Find the complete solution of the following system:

$$\begin{cases} -3x - 5y + 36z = 10 \\ -x \qquad\quad + 7z = 5 \\ x + y - 10z = -4 \end{cases}$$

SOLUTION We transform the system into reduced row-echelon form. (The `rref` command on a TI-83 calculator gives the same result, as shown in Figure 4.)

$$\left[\begin{array}{rrr|r} -3 & -5 & 36 & 10 \\ -1 & 0 & 7 & 5 \\ 1 & 1 & -10 & -4 \end{array}\right] \xrightarrow{\mathrm{R}_1 \leftrightarrow \mathrm{R}_3} \left[\begin{array}{rrr|r} 1 & 1 & -10 & -4 \\ -1 & 0 & 7 & 5 \\ -3 & -5 & 36 & 10 \end{array}\right]$$

$$\xrightarrow[\mathrm{R}_3 + 3\mathrm{R}_1 \to \mathrm{R}_3]{\mathrm{R}_2 + \mathrm{R}_1 \to \mathrm{R}_2} \left[\begin{array}{rrr|r} 1 & 1 & -10 & -4 \\ 0 & 1 & -3 & 1 \\ 0 & -2 & 6 & -2 \end{array}\right] \xrightarrow{\mathrm{R}_3 + 2\mathrm{R}_2 \to \mathrm{R}_3} \left[\begin{array}{rrr|r} 1 & 1 & -10 & -4 \\ 0 & 1 & -3 & 1 \\ 0 & 0 & 0 & 0 \end{array}\right]$$

$$\xrightarrow{\mathrm{R}_1 - \mathrm{R}_2 \to \mathrm{R}_1} \left[\begin{array}{rrr|r} 1 & 0 & -7 & -5 \\ 0 & 1 & -3 & 1 \\ 0 & 0 & 0 & 0 \end{array}\right]$$

Reduced row-echelon form on the TI-83 calculator:

```
rref([A])
   [[1 0 -7 -5]
    [0 1 -3 1 ]
    [0 0 0  0 ]]
```

FIGURE 4

The third row corresponds to the equation $0 = 0$. This equation is always true, no matter what values are used for x, y, and z. Since the equation adds no new information about the variables, we can drop it from the system. So the last matrix corresponds to the system

$$\begin{cases} x \qquad - 7z = -5 & \text{Equation 1} \\ \qquad y - 3z = 1 & \text{Equation 2} \end{cases}$$

Leading variables

Now we solve for the leading variables x and y in terms of the nonleading variable z.

$$x = 7z - 5 \qquad \text{Solve for } x \text{ in Equation 1}$$

$$y = 3z + 1 \qquad \text{Solve for } y \text{ in Equation 2}$$

To obtain the complete solution, we let z be any real number t, and we express x, y, and z in terms of t.

$$x = 7t - 5$$
$$y = 3t + 1$$
$$z = t$$

We can also write the solution as the ordered triple $(7t - 5, 3t + 1, t)$, where t is any real number.

Now Try Exercise 41 ■

In Example 6, to get specific solutions, we give a specific value to t. For example, if $t = 1$, then

$$x = 7(1) - 5 = 2$$
$$y = 3(1) + 1 = 4$$
$$z = 1$$

Here are some other solutions of the system obtained by substituting other values for the parameter t.

Parameter t	Solution $(7t - 5, 3t + 1, t)$
-1	$(-12, -2, -1)$
0	$(-5, 1, 0)$
2	$(9, 7, 2)$
5	$(30, 16, 5)$

EXAMPLE 7 ■ A System with Infinitely Many Solutions

Find the complete solution of the following system:

$$\begin{cases} x + 2y - 3z - 4w = 10 \\ x + 3y - 3z - 4w = 15 \\ 2x + 2y - 6z - 8w = 10 \end{cases}$$

SOLUTION We transform the system into reduced row-echelon form.

$$\begin{bmatrix} 1 & 2 & -3 & -4 & 10 \\ 1 & 3 & -3 & -4 & 15 \\ 2 & 2 & -6 & -8 & 10 \end{bmatrix} \xrightarrow[\text{R}_3 - 2\text{R}_1 \to \text{R}_3]{\text{R}_2 - \text{R}_1 \to \text{R}_2} \begin{bmatrix} 1 & 2 & -3 & -4 & 10 \\ 0 & 1 & 0 & 0 & 5 \\ 0 & -2 & 0 & 0 & -10 \end{bmatrix}$$

$$\xrightarrow{\text{R}_3 + 2\text{R}_2 \to \text{R}_3} \begin{bmatrix} 1 & 2 & -3 & -4 & 10 \\ 0 & 1 & 0 & 0 & 5 \\ 0 & 0 & 0 & 0 & 0 \end{bmatrix} \xrightarrow{\text{R}_1 - 2\text{R}_2 \to \text{R}_1} \begin{bmatrix} 1 & 0 & -3 & -4 & 0 \\ 0 & 1 & 0 & 0 & 5 \\ 0 & 0 & 0 & 0 & 0 \end{bmatrix}$$

This is in reduced row-echelon form. Since the last row represents the equation $0 = 0$, we may discard it. So the last matrix corresponds to the system

$$\begin{cases} x \qquad - 3z - 4w = 0 \\ \qquad y \qquad\qquad\quad = 5 \end{cases}$$

Leading variables

To obtain the complete solution, we solve for the leading variables x and y in terms of the nonleading variables z and w, and we let z and w be any real numbers s and t, respectively. Thus the complete solution is

$$\begin{aligned} x &= 3s + 4t \\ y &= 5 \\ z &= s \\ w &= t \end{aligned}$$

where s and t are any real numbers.

Now Try Exercise 61 ■

Note that s and t do *not* have to be the *same* real number in the solution for Example 7. We can choose arbitrary values for each if we wish to construct a specific solution to the system. For example, if we let $s = 1$ and $t = 2$, then we get the solution $(11, 5, 1, 2)$. You should check that this does indeed satisfy all three of the original equations in Example 7.

Examples 6 and 7 illustrate this general fact: If a system in row-echelon form has n nonzero equations in m variables $(m > n)$, then the complete solution will have $m - n$ nonleading variables. For instance, in Example 6 we arrived at *two* nonzero equations in the *three* variables x, y, and z, which gave us $3 - 2 = 1$ nonleading variable.

■ Modeling with Linear Systems

Linear equations, often containing hundreds or even thousands of variables, occur frequently in the applications of algebra to the sciences and to other fields. For now, let's consider an example that involves only three variables.

EXAMPLE 8 ■ Nutritional Analysis Using a System of Linear Equations

A nutritionist is performing an experiment on student volunteers. He wishes to feed one of his subjects a daily diet that consists of a combination of three commercial diet foods: MiniCal, LiquiFast, and SlimQuick. For the experiment it is important that the subject consume exactly 500 mg of potassium, 75 g of protein, and 1150 units of vitamin D every day. The amounts of these nutrients in 1 oz of each food are given in the table. How many ounces of each food should the subject eat every day to satisfy the nutrient requirements exactly?

	MiniCal	LiquiFast	SlimQuick
Potassium (mg)	50	75	10
Protein (g)	5	10	3
Vitamin D (units)	90	100	50

SOLUTION Let x, y, and z represent the number of ounces of MiniCal, LiquiFast, and SlimQuick, respectively, that the subject should eat every day. This means that he will get $50x$ mg of potassium from MiniCal, $75y$ mg from LiquiFast, and $10z$ mg from SlimQuick, for a total of $50x + 75y + 10z$ mg potassium in all. Since the potassium requirement is 500 mg, we get the first equation below. Similar reasoning for the protein and vitamin D requirements leads to the system

$$\begin{cases} 50x + 75y + 10z = 500 & \text{Potassium} \\ 5x + 10y + 3z = 75 & \text{Protein} \\ 90x + 100y + 50z = 1150 & \text{Vitamin D} \end{cases}$$

Dividing the first equation by 5 and the third one by 10 gives the system

$$\begin{cases} 10x + 15y + 2z = 100 \\ 5x + 10y + 3z = 75 \\ 9x + 10y + 5z = 115 \end{cases}$$

```
rref([A])
  [[1 0 0 5 ]
   [0 1 0 2 ]
   [0 0 1 10]]
```

FIGURE 5

We can solve this system using Gaussian elimination, or we can use a graphing calculator to find the reduced row-echelon form of the augmented matrix of the system. Using the `rref` command on the TI-83/84, we get the output in Figure 5. From the reduced row-echelon form we see that $x = 5$, $y = 2$, $z = 10$. The subject should be fed 5 oz of MiniCal, 2 oz of LiquiFast, and 10 oz of SlimQuick every day.

CHECK YOUR ANSWER

$x = 5, y = 2, z = 10$:

$$\begin{cases} 10(5) + 15(2) + 2(10) = 100 \\ 5(5) + 10(2) + 3(10) = 75 \\ 9(5) + 10(2) + 5(10) = 115 \end{cases} \checkmark$$

Now Try Exercise 69 ■

A more practical application might involve dozens of foods and nutrients rather than just three. Such problems lead to systems with large numbers of variables and equations. Computers or graphing calculators are essential for solving such large systems.

10.3 EXERCISES

CONCEPTS

1. If a system of linear equations has infinitely many solutions, then the system is called ______. If a system of linear equations has no solution, then the system is called ______.

2. Write the augmented matrix of the following system of equations.

System	Augmented matrix
$\begin{cases} x + y - z = 1 \\ x + 2z = -3 \\ 2y - z = 3 \end{cases}$	$\begin{bmatrix} \square & \square & \square & \square \\ \square & \square & \square & \square \\ \square & \square & \square & \square \end{bmatrix}$

3. The following matrix is the augmented matrix of a system of linear equations in the variables x, y, and z. (It is given in reduced row-echelon form.)

$$\begin{bmatrix} 1 & 0 & -1 & 3 \\ 0 & 1 & 2 & 5 \\ 0 & 0 & 0 & 0 \end{bmatrix}$$

(a) The leading variables are ______.

(b) Is the system inconsistent or dependent? ______

(c) The solution of the system is:

$x =$ ______, $y =$ ______, $z =$ ______

4. The augmented matrix of a system of linear equations is given in reduced row-echelon form. Find the solution of the system.

(a) $\begin{bmatrix} 1 & 0 & 0 & 2 \\ 0 & 1 & 0 & 1 \\ 0 & 0 & 1 & 3 \end{bmatrix}$ **(b)** $\begin{bmatrix} 1 & 0 & 1 & 2 \\ 0 & 1 & 1 & 1 \\ 0 & 0 & 0 & 0 \end{bmatrix}$ **(c)** $\begin{bmatrix} 1 & 0 & 0 & 2 \\ 0 & 1 & 0 & 1 \\ 0 & 0 & 0 & 3 \end{bmatrix}$

$x =$ ______ $x =$ ______ $x =$ ______

$y =$ ______ $y =$ ______ $y =$ ______

$z =$ ______ $z =$ ______ $z =$ ______

SKILLS

5–10 ■ Dimension of a Matrix State the dimension of the matrix.

5. $\begin{bmatrix} 2 & 7 \\ 0 & -1 \\ 5 & -3 \end{bmatrix}$ **6.** $\begin{bmatrix} -1 & 5 & 4 & 0 \\ 0 & 2 & 11 & 3 \end{bmatrix}$

7. $\begin{bmatrix} 12 \\ 35 \end{bmatrix}$ **8.** $\begin{bmatrix} -3 \\ 0 \\ 1 \end{bmatrix}$

9. $\begin{bmatrix} 1 & 4 & 7 \end{bmatrix}$ **10.** $\begin{bmatrix} 1 & 0 \\ 0 & 1 \end{bmatrix}$

11–12 ■ The Augmented Matrix Write the augmented matrix for the system of linear equations.

11. $\begin{cases} 3x + y - z = 2 \\ 2x - y = 1 \\ x - z = 3 \end{cases}$ **12.** $\begin{cases} -x + z = -1 \\ 3y - 2z = 7 \\ x - y + 3z = 3 \end{cases}$

13–20 ■ Form of a Matrix A matrix is given. **(a)** Determine whether the matrix is in row-echelon form. **(b)** Determine whether the matrix is in reduced row-echelon form. **(c)** Write the system of equations for which the given matrix is the augmented matrix.

13. $\begin{bmatrix} 1 & 0 & -3 \\ 0 & 1 & 5 \end{bmatrix}$ **14.** $\begin{bmatrix} 1 & 3 & -3 \\ 0 & 1 & 5 \end{bmatrix}$

15. $\begin{bmatrix} 1 & 2 & 8 & 0 \\ 0 & 1 & 3 & 2 \\ 0 & 0 & 0 & 0 \end{bmatrix}$ **16.** $\begin{bmatrix} 1 & 0 & -7 & 0 \\ 0 & 1 & 3 & 0 \\ 0 & 0 & 0 & 1 \end{bmatrix}$

17. $\begin{bmatrix} 1 & 0 & 0 & 0 \\ 0 & 0 & 0 & 0 \\ 0 & 1 & 5 & 1 \end{bmatrix}$ **18.** $\begin{bmatrix} 1 & 0 & 0 & 1 \\ 0 & 1 & 0 & 2 \\ 0 & 0 & 1 & 3 \end{bmatrix}$

19. $$\begin{bmatrix} 1 & 3 & 0 & -1 & 0 \\ 0 & 0 & 1 & 2 & 0 \\ 0 & 0 & 0 & 0 & 1 \\ 0 & 0 & 0 & 0 & 0 \end{bmatrix}$$

20. $$\begin{bmatrix} 1 & 3 & 0 & 1 & 0 & 0 \\ 0 & 1 & 0 & 4 & 0 & 0 \\ 0 & 0 & 0 & 1 & 1 & 2 \\ 0 & 0 & 0 & 1 & 0 & 0 \end{bmatrix}$$

21–24 ■ Elementary Row Operations Perform the indicated elementary row operation.

21. $$\begin{bmatrix} -1 & 1 & 2 & 0 \\ 3 & 1 & 1 & 4 \\ 1 & -2 & -1 & -1 \end{bmatrix}$$
Add 3 times Row 1 to Row 2.

22. $$\begin{bmatrix} -5 & 2 & -3 & 3 \\ 10 & -3 & 1 & -20 \\ -1 & 3 & 1 & 8 \end{bmatrix}$$
Add 2 times Row 1 to Row 2.

23. $$\begin{bmatrix} 2 & 1 & -3 & 5 \\ 2 & 3 & 1 & 13 \\ 6 & -5 & -1 & 7 \end{bmatrix}$$
Add -3 times Row 1 to Row 3.

24. $$\begin{bmatrix} 1 & -3 & 2 & -1 \\ 0 & 1 & 1 & -1 \\ 0 & 2 & -1 & 1 \end{bmatrix}$$
Add -2 times Row 2 to Row 3.

25–28 ■ Back-Substitution A matrix is given in row-echelon form. **(a)** Write the system of equations for which the given matrix is the augmented matrix. **(b)** Use back-substitution to solve the system.

25. $$\begin{bmatrix} 1 & -2 & 4 & 3 \\ 0 & 1 & 2 & 7 \\ 0 & 0 & 1 & 2 \end{bmatrix}$$

26. $$\begin{bmatrix} 1 & 1 & -3 & 8 \\ 0 & 1 & -3 & 5 \\ 0 & 0 & 1 & -1 \end{bmatrix}$$

27. $$\begin{bmatrix} 1 & 2 & 3 & -1 & 7 \\ 0 & 1 & -2 & 0 & 5 \\ 0 & 0 & 1 & 2 & 5 \\ 0 & 0 & 0 & 1 & 3 \end{bmatrix}$$

28. $$\begin{bmatrix} 1 & 0 & -2 & 2 & 5 \\ 0 & 1 & 3 & 0 & -1 \\ 0 & 0 & 1 & -1 & 0 \\ 0 & 0 & 0 & 1 & -1 \end{bmatrix}$$

29–38 ■ Linear Systems with One Solution The system of linear equations has a unique solution. Find the solution using Gaussian elimination or Gauss-Jordan elimination.

29. $$\begin{cases} x - 2y + z = 1 \\ y + 2z = 5 \\ x + y + 3z = 8 \end{cases}$$

30. $$\begin{cases} x + y + 6z = 3 \\ x + y + 3z = 3 \\ x + 2y + 4z = 7 \end{cases}$$

31. $$\begin{cases} x + y + z = 2 \\ 2x - 3y + 2z = 4 \\ 4x + y - 3z = 1 \end{cases}$$

32. $$\begin{cases} x + y + z = 4 \\ -x + 2y + 3z = 17 \\ 2x - y = -7 \end{cases}$$

33. $$\begin{cases} x + 2y - z = -2 \\ x + z = 0 \\ 2x - y - z = -3 \end{cases}$$

34. $$\begin{cases} 2y + z = 4 \\ x + y = 4 \\ 3x + 3y - z = 10 \end{cases}$$

35. $$\begin{cases} x_1 + 2x_2 - x_3 = 9 \\ 2x_1 - x_3 = -2 \\ 3x_1 + 5x_2 + 2x_3 = 22 \end{cases}$$

36. $$\begin{cases} 2x_1 + x_2 = 7 \\ 2x_1 - x_2 + x_3 = 6 \\ 3x_1 - 2x_2 + 4x_3 = 11 \end{cases}$$

37. $$\begin{cases} 2x - 3y - z = 13 \\ -x + 2y - 5z = 6 \\ 5x - y - z = 49 \end{cases}$$

38. $$\begin{cases} 10x + 10y - 20z = 60 \\ 15x + 20y + 30z = -25 \\ -5x + 30y - 10z = 45 \end{cases}$$

39–48 ■ Dependent or Inconsistent Linear Systems Determine whether the system of linear equations is inconsistent or dependent. If it is dependent, find the complete solution.

39. $$\begin{cases} x + y + z = 2 \\ y - 3z = 1 \\ 2x + y + 5z = 0 \end{cases}$$

40. $$\begin{cases} x + 3z = 3 \\ 2x + y - 2z = 5 \\ -y + 8z = 8 \end{cases}$$

41. $$\begin{cases} 2x - 3y - 9z = -5 \\ x + 3z = 2 \\ -3x + y - 4z = -3 \end{cases}$$

42. $$\begin{cases} x - 2y + 5z = 3 \\ -2x + 6y - 11z = 1 \\ 3x - 16y + 20z = -26 \end{cases}$$

43. $$\begin{cases} x - y + 3z = 3 \\ 4x - 8y + 32z = 24 \\ 2x - 3y + 11z = 4 \end{cases}$$

44. $$\begin{cases} -2x + 6y - 2z = -12 \\ x - 3y + 2z = 10 \\ -x + 3y + 2z = 6 \end{cases}$$

45. $$\begin{cases} x + 4y - 2z = -3 \\ 2x - y + 5z = 12 \\ 8x + 5y + 11z = 30 \end{cases}$$

46. $$\begin{cases} 3r + 2s - 3t = 10 \\ r - s - t = -5 \\ r + 4s - t = 20 \end{cases}$$

47. $$\begin{cases} 2x + y - 2z = 12 \\ -x - \frac{1}{2}y + z = -6 \\ 3x + \frac{3}{2}y - 3z = 18 \end{cases}$$

48. $$\begin{cases} y - 5z = 7 \\ 3x + 2y = 12 \\ 3x + 10z = 80 \end{cases}$$

49–64 ■ Solving a Linear System Solve the system of linear equations.

49. $$\begin{cases} 4x - 3y + z = -8 \\ -2x + y - 3z = -4 \\ x - y + 2z = 3 \end{cases}$$

50. $$\begin{cases} 2x - 3y + 5z = 14 \\ 4x - y - 2z = -17 \\ -x - y + z = 3 \end{cases}$$

51. $$\begin{cases} 2x + y + 3z = 9 \\ -x - 7z = 10 \\ 3x + 2y - z = 4 \end{cases}$$

52. $$\begin{cases} -4x - y + 36z = 24 \\ x - 2y + 9z = 3 \\ -2x + y + 6z = 6 \end{cases}$$

53. $$\begin{cases} x + 2y - 3z = -5 \\ -2x - 4y - 6z = 10 \\ 3x + 7y - 2z = -13 \end{cases}$$

54. $$\begin{cases} 3x + y = 2 \\ -4x + 3y + z = 4 \\ 2x + 5y + z = 0 \end{cases}$$

55. $$\begin{cases} x - y + 6z = 8 \\ x + z = 5 \\ x + 3y - 14z = -4 \end{cases}$$

56. $$\begin{cases} 3x - y + 2z = -1 \\ 4x - 2y + z = -7 \\ -x + 3y - 2z = -1 \end{cases}$$

57. $$\begin{cases} -x + 2y + z - 3w = 3 \\ 3x - 4y + z + w = 9 \\ -x - y + z + w = 0 \\ 2x + y + 4z - 2w = 3 \end{cases}$$

58. $$\begin{cases} x + y - z - w = 6 \\ 2x + z - 3w = 8 \\ x - y + 4w = -10 \\ 3x + 5y - z - w = 20 \end{cases}$$

59. $$\begin{cases} x + y + 2z - w = -2 \\ 3y + z + 2w = 2 \\ x + y + 3w = 2 \\ -3x + z + 2w = 5 \end{cases}$$

60. $$\begin{cases} x - 3y + 2z + w = -2 \\ x - 2y - 2w = -10 \\ z + 5w = 15 \\ 3x + 2z + w = -3 \end{cases}$$

61. $\begin{cases} x - y + w = 0 \\ 3x - z + 2w = 0 \\ x - 4y + z + 2w = 0 \end{cases}$ **62.** $\begin{cases} 2x - y + 2z + w = 5 \\ -x + y + 4z - w = 3 \\ 3x - 2y - z = 0 \end{cases}$

63. $\begin{cases} x + z + w = 4 \\ y - z = -4 \\ x - 2y + 3z + w = 12 \\ 2x - 2z + 5w = -1 \end{cases}$

64. $\begin{cases} y - z + 2w = 0 \\ 3x + 2y + w = 0 \\ 2x + 4w = 12 \\ -2x - 2z + 5w = 6 \end{cases}$

65–68 ■ Solving a Linear System Using a Graphing Calculator Solve the system of linear equations by using the `rref` command on a graphing calculator. State your answer rounded to two decimal places.

65. $\begin{cases} 0.75x - 3.75y + 2.95z = 4.0875 \\ 0.95x - 8.75y = 3.375 \\ 1.25x - 0.15y + 2.75z = 3.6625 \end{cases}$

66. $\begin{cases} 1.31x + 2.72y - 3.71z = -13.9534 \\ -0.21x + 3.73z = 13.4322 \\ 2.34y - 4.56z = -21.3984 \end{cases}$

67. $\begin{cases} 42x - 31y - 42w = -0.4 \\ -6x - 9w = 4.5 \\ 35x - 67z + 32w = 348.8 \\ 31y + 48z - 52w = -76.6 \end{cases}$

68. $\begin{cases} 49x - 27y + 52z = -145.0 \\ 27y + 43w = -118.7 \\ -31y + 42z = -72.1 \\ 73x - 54y = -132.7 \end{cases}$

APPLICATIONS

69. Nutrition A doctor recommends that a patient take 50 mg each of niacin, riboflavin, and thiamin daily to alleviate a vitamin deficiency. In his medicine chest at home the patient finds three brands of vitamin pills. The amounts of the relevant vitamins per pill are given in the table. How many pills of each type should he take every day to get 50 mg of each vitamin?

	VitaMax	Vitron	VitaPlus
Niacin (mg)	5	10	15
Riboflavin (mg)	15	20	0
Thiamin (mg)	10	10	10

70. Mixtures A chemist has three acid solutions at various concentrations. The first is 10% acid, the second is 20%, and the third is 40%. How many milliliters of each should she use to make 100 mL of 18% solution, if she has to use four times as much of the 10% solution as the 40% solution?

71. Distance, Speed, and Time Amanda, Bryce, and Corey enter a race in which they have to run, swim, and cycle over a marked course. Their average speeds are given in the table. Corey finishes first with a total time of 1 h 45 min. Amanda comes in second with a time of 2 h 30 min. Bryce finishes last with a time of 3 h. Find the distance (in mi) for each part of the race.

	Average speed (mi/h)		
	Running	Swimming	Cycling
Amanda	10	4	20
Bryce	$7\frac{1}{2}$	6	15
Corey	15	3	40

72. Classroom Use A small school has 100 students who occupy three classrooms: A, B, and C. After the first period of the school day, half the students in room A move to room B, one-fifth of the students in room B move to room C, and one-third of the students in room C move to room A. Nevertheless, the total number of students in each room is the same for both periods. How many students occupy each room?

73. Manufacturing Furniture A furniture factory makes wooden tables, chairs, and armoires. Each piece of furniture requires three operations: cutting the wood, assembling, and finishing. Each operation requires the number of hours given in the table. The workers in the factory can provide 300 h of cutting, 400 h of assembling, and 590 h of finishing each work week. How many tables, chairs, and armoires should be produced so that all available labor-hours are used? Or is this impossible?

	Table	Chair	Armoire
Cutting (h)	$\frac{1}{2}$	1	1
Assembling (h)	$\frac{1}{2}$	$1\frac{1}{2}$	1
Finishing (h)	1	$1\frac{1}{2}$	2

74. Traffic Flow A section of a city's street network is shown in the figure. The arrows indicate one-way streets, and the numbers show how many cars enter or leave this section of the city via the indicated street in a certain one-hour period. The variables x, y, z, and w represent the number of cars that travel along the portions of First, Second, Avocado, and Birch Streets during this period. Find x, y, z, and w, assuming that none of the cars stop or park on any of the streets shown.

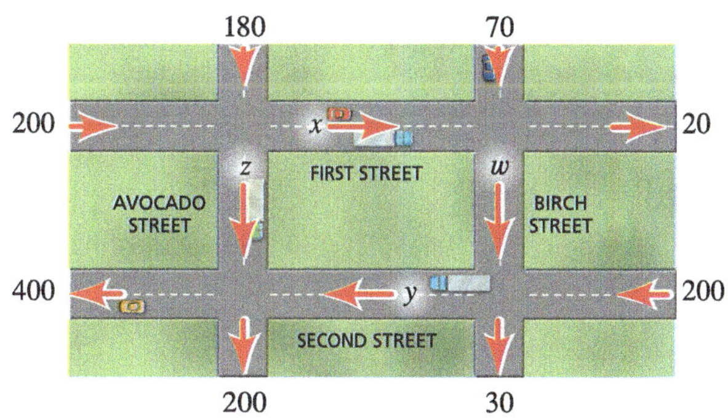

DISCUSS ■ DISCOVER ■ PROVE ■ WRITE

75. DISCUSS: Polynomials Determined by a Set of Points We all know that two points uniquely determine a line $y = ax + b$ in the coordinate plane. Similarly, three points uniquely determine a quadratic (second-degree) polynomial

$$y = ax^2 + bx + c$$

four points uniquely determine a cubic (third-degree) polynomial

$$y = ax^3 + bx^2 + cx + d$$

and so on. (Some exceptions to this rule are if the three points actually lie on a line, or the four points lie on a quadratic or line, and so on.) For the following set of five points, find the line that contains the first two points, the quadratic that contains the first three points, the cubic that contains the first four points, and the fourth-degree polynomial that contains all five points.

$$(0, 0),\quad (1, 12),\quad (2, 40),\quad (3, 6),\quad (-1, -14)$$

Graph the points and functions in the same viewing rectangle using a graphing device.

10.4 THE ALGEBRA OF MATRICES

■ Equality of Matrices ■ Addition, Subtraction, and Scalar Multiplication of Matrices ■ Multiplication of Matrices ■ Properties of Matrix Multiplication ■ Applications of Matrix Multiplication ■ Computer Graphics

Thus far, we have used matrices simply for notational convenience when solving linear systems. Matrices have many other uses in mathematics and the sciences, and for most of these applications a knowledge of matrix algebra is essential. Like numbers, matrices can be added, subtracted, multiplied, and divided. In this section we learn how to perform these algebraic operations on matrices.

■ Equality of Matrices

Two matrices are equal if they have the same entries in the same positions.

Equal matrices

$$\begin{bmatrix} \sqrt{4} & 2^2 & e^0 \\ 0.5 & 1 & 1-1 \end{bmatrix} = \begin{bmatrix} 2 & 4 & 1 \\ \frac{1}{2} & \frac{2}{2} & 0 \end{bmatrix}$$

Unequal matrices

$$\begin{bmatrix} 1 & 2 \\ 3 & 4 \\ 5 & 6 \end{bmatrix} \neq \begin{bmatrix} 1 & 3 & 5 \\ 2 & 4 & 6 \end{bmatrix}$$

EQUALITY OF MATRICES

The matrices $A = [a_{ij}]$ and $B = [b_{ij}]$ are **equal** if and only if they have the same dimension $m \times n$, and corresponding entries are equal, that is,

$$a_{ij} = b_{ij}$$

for $i = 1, 2, \ldots, m$ and $j = 1, 2, \ldots, n$.

EXAMPLE 1 ■ Equal Matrices

Find a, b, c, and d, if

$$\begin{bmatrix} a & b \\ c & d \end{bmatrix} = \begin{bmatrix} 1 & 3 \\ 5 & 2 \end{bmatrix}$$

SOLUTION Since the two matrices are equal, corresponding entries must be the same. So we must have $a = 1$, $b = 3$, $c = 5$, and $d = 2$.

Now Try Exercises 5 and 7

Courtesy UC Berkeley Office of Media Relations

JULIA ROBINSON (1919–1985) was born in St. Louis, Missouri, and grew up at Point Loma, California. Because of an illness, Robinson missed two years of school, but later, with the aid of a tutor, she completed fifth, sixth, seventh, and eighth grades, all in one year. Later, at San Diego State University, reading biographies of mathematicians in E. T. Bell's *Men of Mathematics* awakened in her what became a lifelong passion for mathematics. She said, "I cannot overemphasize the importance of such books . . . in the intellectual life of a student." Robinson is famous for her work on Hilbert's tenth problem (page 735), which asks for a general procedure for determining whether an equation has integer solutions. Her ideas led to a complete answer to the problem. Interestingly, the answer involved certain properties of the Fibonacci numbers (page 846) discovered by the then 22-year-old Russian mathematician Yuri Matijasevič. As a result of her brilliant work on Hilbert's tenth problem, Robinson was offered a professorship at the University of California, Berkeley, and became the first woman mathematician elected to the National Academy of Sciences. She also served as president of the American Mathematical Society.

Addition, Subtraction, and Scalar Multiplication of Matrices

Two matrices can be added or subtracted if they have the same dimension. (Otherwise, their sum or difference is undefined.) We add or subtract the matrices by adding or subtracting corresponding entries. To multiply a matrix by a number, we multiply every element of the matrix by that number. This is called the *scalar product.*

SUM, DIFFERENCE, AND SCALAR PRODUCT OF MATRICES

Let $A = [a_{ij}]$ and $B = [b_{ij}]$ be matrices of the same dimension $m \times n$, and let c be any real number.

1. The **sum** $A + B$ is the $m \times n$ matrix obtained by adding corresponding entries of A and B.
$$A + B = [a_{ij} + b_{ij}]$$
2. The **difference** $A - B$ is the $m \times n$ matrix obtained by subtracting corresponding entries of A and B.
$$A - B = [a_{ij} - b_{ij}]$$
3. The **scalar product** cA is the $m \times n$ matrix obtained by multiplying each entry of A by c.
$$cA = [ca_{ij}]$$

EXAMPLE 2 ■ Performing Algebraic Operations on Matrices

Let
$$A = \begin{bmatrix} 2 & -3 \\ 0 & 5 \\ 7 & -\frac{1}{2} \end{bmatrix} \qquad B = \begin{bmatrix} 1 & 0 \\ -3 & 1 \\ 2 & 2 \end{bmatrix}$$
$$C = \begin{bmatrix} 7 & -3 & 0 \\ 0 & 1 & 5 \end{bmatrix} \qquad D = \begin{bmatrix} 6 & 0 & -6 \\ 8 & 1 & 9 \end{bmatrix}$$

Carry out each indicated operation, or explain why it cannot be performed.

(a) $A + B$ **(b)** $C - D$ **(c)** $C + A$ **(d)** $5A$

SOLUTION

(a) $$A + B = \begin{bmatrix} 2 & -3 \\ 0 & 5 \\ 7 & -\frac{1}{2} \end{bmatrix} + \begin{bmatrix} 1 & 0 \\ -3 & 1 \\ 2 & 2 \end{bmatrix} = \begin{bmatrix} 3 & -3 \\ -3 & 6 \\ 9 & \frac{3}{2} \end{bmatrix}$$

(b) $$C - D = \begin{bmatrix} 7 & -3 & 0 \\ 0 & 1 & 5 \end{bmatrix} - \begin{bmatrix} 6 & 0 & -6 \\ 8 & 1 & 9 \end{bmatrix} = \begin{bmatrix} 1 & -3 & 6 \\ -8 & 0 & -4 \end{bmatrix}$$

(c) $C + A$ is undefined because we can't add matrices of different dimensions.

(d) $$5A = 5\begin{bmatrix} 2 & -3 \\ 0 & 5 \\ 7 & -\frac{1}{2} \end{bmatrix} = \begin{bmatrix} 10 & -15 \\ 0 & 25 \\ 35 & -\frac{5}{2} \end{bmatrix}$$

Now Try Exercises 23 and 25 ■

The properties in the box follow from the definitions of matrix addition and scalar multiplication and the corresponding properties of real numbers.

PROPERTIES OF ADDITION AND SCALAR MULTIPLICATION OF MATRICES

Let A, B, and C be $m \times n$ matrices and let c and d be scalars.

$A + B = B + A$	Commutative Property of Matrix Addition
$(A + B) + C = A + (B + C)$	Associative Property of Matrix Addition
$c(dA) = cdA$	Associative Property of Scalar Multiplication
$(c + d)A = cA + dA$ $c(A + B) = cA + cB$	Distributive Properties of Scalar Multiplication

EXAMPLE 3 ■ Solving a Matrix Equation

Solve the matrix equation

$$2X - A = B$$

for the unknown matrix X, where

$$A = \begin{bmatrix} 2 & 3 \\ -5 & 1 \end{bmatrix} \qquad B = \begin{bmatrix} 4 & -1 \\ 1 & 3 \end{bmatrix}$$

SOLUTION We use the properties of matrices to solve for X.

$$2X - A = B \qquad \text{Given equation}$$

$$2X = B + A \qquad \text{Add the matrix } A \text{ to each side}$$

$$X = \tfrac{1}{2}(B + A) \qquad \text{Multiply each side by the scalar } \tfrac{1}{2}$$

So $$X = \frac{1}{2}\left(\begin{bmatrix} 4 & -1 \\ 1 & 3 \end{bmatrix} + \begin{bmatrix} 2 & 3 \\ -5 & 1 \end{bmatrix}\right) \qquad \text{Substitute the matrices } A \text{ and } B$$

$$= \frac{1}{2}\begin{bmatrix} 6 & 2 \\ -4 & 4 \end{bmatrix} \qquad \text{Add matrices}$$

$$= \begin{bmatrix} 3 & 1 \\ -2 & 2 \end{bmatrix} \qquad \text{Multiply by the scalar } \tfrac{1}{2}$$

Now Try Exercise 17 ■

■ Multiplication of Matrices

Multiplying two matrices is more difficult to describe than other matrix operations. In later examples we will see why multiplying matrices involves a rather complex procedure, which we now describe.

First, the product AB (or $A \cdot B$) of two matrices A and B is defined only when the number of columns in A is equal to the number of rows in B. This means that if we write their dimensions side by side, the two inner numbers must match:

Matrices	A	B
Dimensions	$m \times n$	$n \times k$
	Columns in A	Rows in B

If we think of the row of A and the column of B as vectors, then their inner product is the same as their dot product (see Sections 9.2 and 9.4).

If the dimensions of A and B match in this fashion, then the product AB is a matrix of dimension $m \times k$. Before describing the procedure for obtaining the elements of AB, we define the *inner product* of a row of A and a column of B.

If $\begin{bmatrix} a_1 & a_2 & \cdots & a_n \end{bmatrix}$ is a row of A, and if $\begin{bmatrix} b_1 \\ b_2 \\ \vdots \\ b_n \end{bmatrix}$ is a column of B, then their **inner product**

is the number $a_1b_1 + a_2b_2 + \cdots + a_nb_n$. For example, taking the inner product of

$$\begin{bmatrix} 2 & -1 & 0 & 4 \end{bmatrix} \text{ and } \begin{bmatrix} 5 \\ 4 \\ -3 \\ \frac{1}{2} \end{bmatrix} \text{ gives}$$

$$2 \cdot 5 + (-1) \cdot 4 + 0 \cdot (-3) + 4 \cdot \tfrac{1}{2} = 8$$

We now define the **product** AB of two matrices.

MATRIX MULTIPLICATION

If $A = [a_{ij}]$ is an $m \times n$ matrix and $B = [b_{ij}]$ an $n \times k$ matrix, then their product is the $m \times k$ matrix

$$C = [c_{ij}]$$

where c_{ij} is the inner product of the ith row of A and the jth column of B. We write the product as

$$C = AB$$

This definition of matrix product says that each entry in the matrix AB is obtained from a *row* of A and a *column* of B as follows: The entry c_{ij} in the ith row and jth column of the matrix AB is obtained by multiplying the entries in the ith row of A with the corresponding entries in the jth column of B and adding the results.

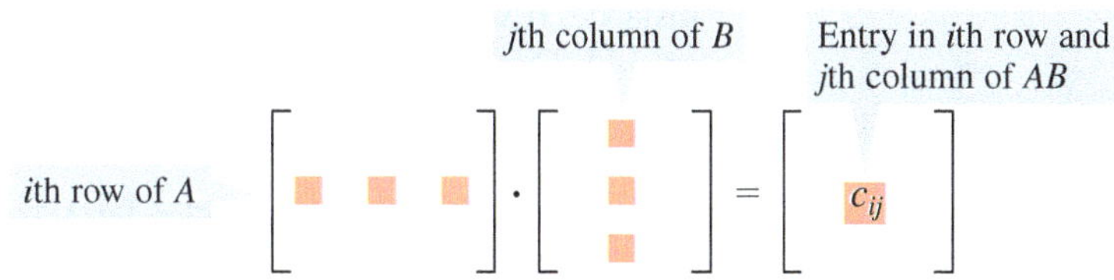

EXAMPLE 4 ■ Multiplying Matrices

Let

$$A = \begin{bmatrix} 1 & 3 \\ -1 & 0 \end{bmatrix} \quad \text{and} \quad B = \begin{bmatrix} -1 & 5 & 2 \\ 0 & 4 & 7 \end{bmatrix}$$

Calculate, if possible, the products AB and BA.

SOLUTION Since A has dimension 2×2 and B has dimension 2×3, the product AB is defined and has dimension 2×3. We can therefore write

Inner numbers match, so product is defined

$2 \times 2 \quad 2 \times 3$

Outer numbers give dimension of product: 2×3

$$AB = \begin{bmatrix} 1 & 3 \\ -1 & 0 \end{bmatrix} \begin{bmatrix} -1 & 5 & 2 \\ 0 & 4 & 7 \end{bmatrix} = \begin{bmatrix} ? & ? & ? \\ ? & ? & ? \end{bmatrix}$$

where the question marks must be filled in using the rule defining the product of two matrices. If we define $C = AB = [c_{ij}]$, then the entry c_{11} is the inner product of the first row of A and the first column of B:

$$\begin{bmatrix} 1 & 3 \\ -1 & 0 \end{bmatrix}\begin{bmatrix} -1 & 5 & 2 \\ 0 & 4 & 7 \end{bmatrix} \qquad 1\cdot(-1) + 3\cdot 0 = -1$$

Similarly, we calculate the remaining entries of the product as follows.

Entry	**Inner product of:**	**Value**	**Product matrix**
c_{12}	$\begin{bmatrix} 1 & 3 \\ -1 & 0 \end{bmatrix}\begin{bmatrix} -1 & 5 & 2 \\ 0 & 4 & 7 \end{bmatrix}$	$1\cdot 5 + 3\cdot 4 = 17$	$\begin{bmatrix} -1 & 17 & \\ & & \end{bmatrix}$
c_{13}	$\begin{bmatrix} 1 & 3 \\ -1 & 0 \end{bmatrix}\begin{bmatrix} -1 & 5 & 2 \\ 0 & 4 & 7 \end{bmatrix}$	$1\cdot 2 + 3\cdot 7 = 23$	$\begin{bmatrix} -1 & 17 & 23 \\ & & \end{bmatrix}$
c_{21}	$\begin{bmatrix} 1 & 3 \\ -1 & 0 \end{bmatrix}\begin{bmatrix} -1 & 5 & 2 \\ 0 & 4 & 7 \end{bmatrix}$	$(-1)\cdot(-1) + 0\cdot 0 = 1$	$\begin{bmatrix} -1 & 17 & 23 \\ 1 & & \end{bmatrix}$
c_{22}	$\begin{bmatrix} 1 & 3 \\ -1 & 0 \end{bmatrix}\begin{bmatrix} -1 & 5 & 2 \\ 0 & 4 & 7 \end{bmatrix}$	$(-1)\cdot 5 + 0\cdot 4 = -5$	$\begin{bmatrix} -1 & 17 & 23 \\ 1 & -5 & \end{bmatrix}$
c_{23}	$\begin{bmatrix} 1 & 3 \\ -1 & 0 \end{bmatrix}\begin{bmatrix} -1 & 5 & 2 \\ 0 & 4 & 7 \end{bmatrix}$	$(-1)\cdot 2 + 0\cdot 7 = -2$	$\begin{bmatrix} -1 & 17 & 23 \\ 1 & -5 & -2 \end{bmatrix}$

Thus we have $$AB = \begin{bmatrix} -1 & 17 & 23 \\ 1 & -5 & -2 \end{bmatrix}$$

Not equal, so product is not defined

$2 \times 3 \quad 2 \times 2$

The product BA is not defined, however, because the dimensions of B and A are

$$2 \times 3 \qquad \text{and} \qquad 2 \times 2$$

The inner two numbers are not the same, so the rows and columns won't match up when we try to calculate the product.

Now Try Exercise 27 ■

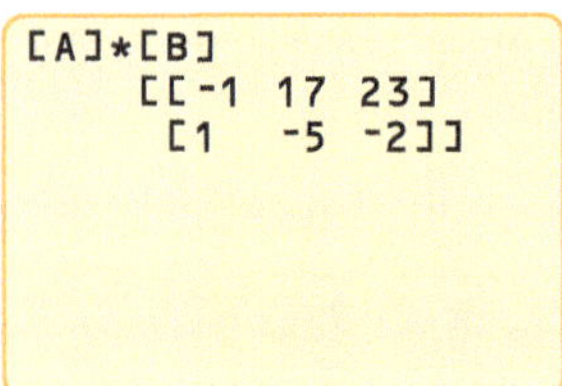

FIGURE 1

Graphing calculators and computers are capable of performing matrix algebra. For instance, if we enter the matrices in Example 4 into the matrix variables [A] and [B] on a TI-83 calculator, then the calculator finds their product as shown in Figure 1.

See Appendix D, *Using the TI-83/84 Graphing Calculator*, for specific instructions on working with matrices. Go to **www.stewartmath.com**.

Properties of Matrix Multiplication

Although matrix multiplication is not commutative, it does obey the Associative and Distributive Properties.

PROPERTIES OF MATRIX MULTIPLICATION

Let A, B, and C be matrices for which the following products are defined. Then

$A(BC) = (AB)C$	Associative Property
$A(B + C) = AB + AC$ $(B + C)A = BA + CA$	Distributive Property

The next example shows that even when both AB and BA are defined, they aren't necessarily equal. This proves that matrix multiplication is *not* commutative.

EXAMPLE 5 ■ Matrix Multiplication Is Not Commutative

Let $$A = \begin{bmatrix} 5 & 7 \\ -3 & 0 \end{bmatrix} \quad \text{and} \quad B = \begin{bmatrix} 1 & 2 \\ 9 & -1 \end{bmatrix}$$

Calculate the products AB and BA.

SOLUTION Since both matrices A and B have dimension 2×2, both products AB and BA are defined, and each product is also a 2×2 matrix.

$$AB = \begin{bmatrix} 5 & 7 \\ -3 & 0 \end{bmatrix}\begin{bmatrix} 1 & 2 \\ 9 & -1 \end{bmatrix} = \begin{bmatrix} 5\cdot 1 + 7\cdot 9 & 5\cdot 2 + 7\cdot(-1) \\ (-3)\cdot 1 + 0\cdot 9 & (-3)\cdot 2 + 0\cdot(-1) \end{bmatrix}$$

$$= \begin{bmatrix} 68 & 3 \\ -3 & -6 \end{bmatrix}$$

$$BA = \begin{bmatrix} 1 & 2 \\ 9 & -1 \end{bmatrix}\begin{bmatrix} 5 & 7 \\ -3 & 0 \end{bmatrix} = \begin{bmatrix} 1\cdot 5 + 2\cdot(-3) & 1\cdot 7 + 2\cdot 0 \\ 9\cdot 5 + (-1)\cdot(-3) & 9\cdot 7 + (-1)\cdot 0 \end{bmatrix}$$

$$= \begin{bmatrix} -1 & 7 \\ 48 & 63 \end{bmatrix}$$

This shows that, in general, $AB \neq BA$. In fact, in this example AB and BA don't even have an entry in common.

Now Try Exercise 29 ■

■ Applications of Matrix Multiplication

We now consider some applied examples that give some indication of why mathematicians chose to define the matrix product in such an apparently bizarre fashion. Example 6 shows how our definition of matrix product allows us to express a system of linear equations as a single matrix equation.

EXAMPLE 6 ■ Writing a Linear System as a Matrix Equation

Show that the following matrix equation is equivalent to the system of equations in Example 2 of Section 10.3.

Matrix equations like this one are described in more detail on page 728.

$$\begin{bmatrix} 1 & -1 & 3 \\ 1 & 2 & -2 \\ 3 & -1 & 5 \end{bmatrix}\begin{bmatrix} x \\ y \\ z \end{bmatrix} = \begin{bmatrix} 4 \\ 10 \\ 14 \end{bmatrix}$$

SOLUTION If we perform matrix multiplication on the left-hand side of the equation, we get

$$\begin{bmatrix} x - y + 3z \\ x + 2y - 2z \\ 3x - y + 5z \end{bmatrix} = \begin{bmatrix} 4 \\ 10 \\ 14 \end{bmatrix}$$

Because two matrices are equal only if their corresponding entries are equal, we equate entries to get

$$\begin{cases} x - y + 3z = 4 \\ x + 2y - 2z = 10 \\ 3x - y + 5z = 14 \end{cases}$$

This is exactly the system of equations in Example 2 of Section 10.3.

Now Try Exercise 47 ■

Courtesy of the Archives, California Institute of Technology

OLGA TAUSSKY-TODD (1906–1995) was instrumental in developing applications of matrix theory. Described as "in love with anything matrices can do," she successfully applied matrices to aerodynamics, a field used in the design of airplanes and rockets. Taussky-Todd was also famous for her work in number theory, which deals with prime numbers and divisibility. Although number theory has often been called the least applicable branch of mathematics, it is now used in significant ways throughout the computer industry.

Taussky-Todd studied mathematics at a time when young women rarely aspired to be mathematicians. She said, "When I entered university I had no idea what it meant to study mathematics." One of the most respected mathematicians of her day, she was for many years a professor of mathematics at Caltech in Pasadena.

EXAMPLE 7 ■ Representing Demographic Data by Matrices

In a certain city the proportions of voters in each age group who are registered as Democrats, Republicans, or Independents are given by the following matrix.

$$\begin{array}{l} \\ \\ \textbf{Democrat} \\ \textbf{Republican} \\ \textbf{Independent} \end{array} \begin{array}{c} \textbf{Age} \\ \begin{array}{ccc} \textbf{18–30} & \textbf{31–50} & \textbf{Over 50} \end{array} \\ \begin{bmatrix} 0.30 & 0.60 & 0.50 \\ 0.50 & 0.35 & 0.25 \\ 0.20 & 0.05 & 0.25 \end{bmatrix} \end{array} = A$$

The next matrix gives the distribution, by age and sex, of the voting population of this city.

$$\textbf{Age} \begin{array}{l} \\ \textbf{18–30} \\ \textbf{31–50} \\ \textbf{Over 50} \end{array} \begin{array}{c} \begin{array}{cc} \textbf{Male} & \textbf{Female} \end{array} \\ \begin{bmatrix} 5{,}000 & 6{,}000 \\ 10{,}000 & 12{,}000 \\ 12{,}000 & 15{,}000 \end{bmatrix} \end{array} = B$$

For this problem, let's make the (highly unrealistic) assumption that within each age group, political preference is not related to gender. That is, the percentage of Democrat males in the 18–30 group, for example, is the same as the percentage of Democrat females in this group.

(a) Calculate the product AB.

(b) How many males are registered as Democrats in this city?

(c) How many females are registered as Republicans?

SOLUTION

(a) $$AB = \begin{bmatrix} 0.30 & 0.60 & 0.50 \\ 0.50 & 0.35 & 0.25 \\ 0.20 & 0.05 & 0.25 \end{bmatrix} \begin{bmatrix} 5{,}000 & 6{,}000 \\ 10{,}000 & 12{,}000 \\ 12{,}000 & 15{,}000 \end{bmatrix} = \begin{bmatrix} 13{,}500 & 16{,}500 \\ 9{,}000 & 10{,}950 \\ 4{,}500 & 5{,}550 \end{bmatrix}$$

(b) When we take the inner product of a row in A with a column in B, we are adding the number of people in each age group who belong to the category in question. For example, the entry c_{21} of AB (the 9000) is obtained by taking the inner product of the Republican row in A with the Male column in B. This number is therefore the total number of male Republicans in this city. We can label the rows and columns of AB as follows.

$$\begin{array}{l} \\ \textbf{Democrat} \\ \textbf{Republican} \\ \textbf{Independent} \end{array} \begin{array}{c} \begin{array}{cc} \textbf{Male} & \textbf{Female} \end{array} \\ \begin{bmatrix} 13{,}500 & 16{,}500 \\ 9{,}000 & 10{,}950 \\ 4{,}500 & 5{,}550 \end{bmatrix} \end{array} = AB$$

Thus 13,500 males are registered as Democrats in this city.

(c) There are 10,950 females registered as Republicans.

Now Try Exercise 53 ■

In Example 7 the entries in each column of A add up to 1. (Can you see why this has to be true, given what the matrix describes?) A matrix with this property is called **stochastic**. Stochastic matrices are used extensively in statistics, where they arise frequently in situations like the one described here.

Computer Graphics

One important use of matrices is in the digital representation of images. A digital camera or a scanner converts an image into a matrix by dividing the image into a rectangular array of elements called pixels. Each pixel is assigned a value that represents the color, brightness, or some other feature of that location. For example, in a 256-level gray-scale image each pixel is assigned a value between 0 and 255, where 0 represents white, 255 represents black, and the numbers in between represent increasing gradations of gray. The gradations of a much simpler eight-level gray scale are shown in Figure 2. We use this eight-level gray scale to illustrate the process.

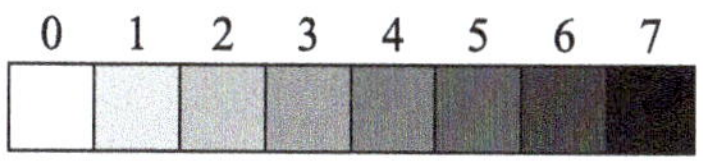

FIGURE 2

To digitize the black and white image in Figure 3(a), we place a grid over the picture as shown in Figure 3(b). Each cell in the grid is compared to the gray scale and then assigned a value between 0 and 7 depending on which gray square in the scale most closely matches the "darkness" of the cell. (If the cell is not uniformly gray, an average value is assigned.) The values are stored in the matrix shown in Figure 3(c). The digital image corresponding to this matrix is shown in Figure 3(d). Obviously, the grid that we have used is far too coarse to provide good image resolution. In practice, currently available high-resolution digital cameras use matrices with dimension as large as 2048×2048.

(a) Original image (b) 10×10 grid

$$\begin{bmatrix} 1&1&1&1&1&1&1&2&2&1\\ 1&1&1&1&1&1&4&6&5&2\\ 1&1&1&1&2&3&3&5&5&3\\ 1&1&1&1&3&5&4&6&3&2\\ 1&1&1&1&1&2&3&2&2&1\\ 1&1&1&1&1&3&3&2&1&1\\ 1&1&1&1&1&1&4&1&1&1\\ 1&1&1&1&2&2&4&2&2&2\\ 2&2&3&5&5&2&2&3&4&4\\ 3&3&3&4&3&2&3&3&3&4 \end{bmatrix}$$

(c) Matrix representation

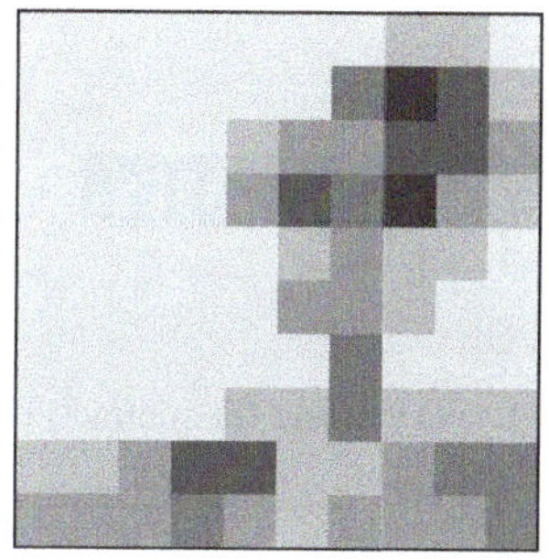

(d) Digital image

FIGURE 3

Once the image is stored as a matrix, it can be manipulated by using matrix operations. For example, to darken the image, we add a constant to each entry in the matrix; to lighten the image, we subtract a constant. To increase the contrast, we darken the darker areas and lighten the lighter areas, so we could add 1 to each entry that is 4, 5, or 6 and subtract 1 from each entry that is 1, 2, or 3. (Note that we can-

DISCOVERY PROJECT

Will the Species Survive?

To study how a species survives, scientists observe the stages in the life cycle of the species—for example, young, juvenile, adult. The proportion of the population at each stage and the proportion that survives to the next stage in each season are modeled by matrices. In this project we explore how matrix multiplication is used to predict the population proportions for the next season, the season after that, and so on, ultimately predicting the long-term prospects for the survival of the species. You can find the project at **www.stewartmath.com**.

not darken an entry of 7 or lighten a 0.) Applying this process to the matrix in Figure 3(c) produces the new matrix in Figure 4(a). This generates the high-contrast image shown in Figure 4(b).

$$\begin{bmatrix} 0&0&0&0&0&0&0&1&1&0\\ 0&0&0&0&0&0&5&7&6&1\\ 0&0&0&0&1&2&2&6&6&2\\ 0&0&0&0&2&6&5&7&2&1\\ 0&0&0&0&0&1&2&1&1&0\\ 0&0&0&0&0&2&2&1&0&0\\ 0&0&0&0&0&0&5&0&0&0\\ 0&0&0&0&1&1&5&1&1&1\\ 1&1&2&6&6&1&1&2&5&5\\ 2&2&2&5&2&1&2&2&2&5 \end{bmatrix}$$

(a) Matrix modified to increase contrast

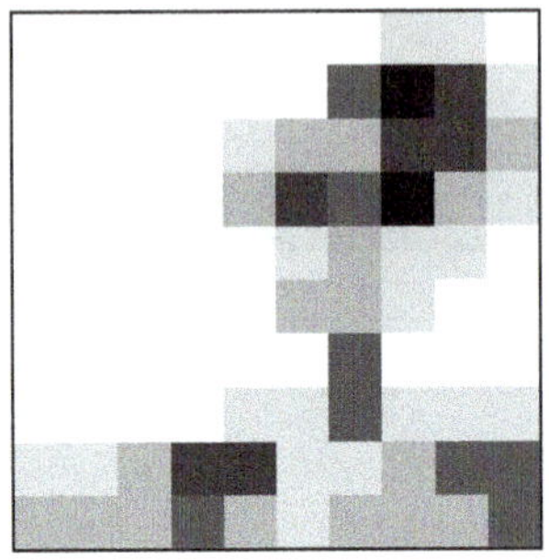

(b) High contrast image

FIGURE 4

Other ways of representing and manipulating images using matrices are discussed in the Discovery Projects *Computer Graphics I* and *II* at the book companion website, **www.stewartmath.com**.

10.4 EXERCISES

CONCEPTS

1. We can add (or subtract) two matrices only if they have the same ________________.

2. (a) We can multiply two matrices only if the number of ________ in the first matrix is the same as the number of ________ in the second matrix.
   (b) If A is a 3×3 matrix and B is a 4×3 matrix, which of the following matrix multiplications are possible?
   (i) AB (ii) BA (iii) AA (iv) BB

3. Which of the following operations can we perform for a matrix A of any dimension?
   (i) $A + A$ (ii) $2A$ (iii) $A \cdot A$

4. Fill in the missing entries in the product matrix.

$$\begin{bmatrix} 3 & 1 & 2 \\ -1 & 2 & 0 \\ 1 & 3 & -2 \end{bmatrix} \begin{bmatrix} -1 & 3 & -2 \\ 3 & -2 & -1 \\ 2 & 1 & 0 \end{bmatrix} = \begin{bmatrix} 4 & \blacksquare & -7 \\ 7 & -7 & \blacksquare \\ \blacksquare & -5 & -5 \end{bmatrix}$$

SKILLS

5–6 ■ Equality of Matrices Determine whether the matrices A and B are equal.

5. $A = \begin{bmatrix} 1 & -2 & 0 \\ \frac{1}{2} & 6 & 0 \end{bmatrix} \quad B = \begin{bmatrix} 1 & -2 \\ \frac{1}{2} & 6 \end{bmatrix}$

6. $A = \begin{bmatrix} \frac{1}{4} & \ln 1 \\ 2 & 3 \end{bmatrix} \quad B = \begin{bmatrix} 0.25 & 0 \\ \sqrt{4} & \frac{6}{2} \end{bmatrix}$

7–8 ■ Equality of Matrices Find the values of a and b that make the matrices A and B equal.

7. $A = \begin{bmatrix} 3 & 4 \\ -1 & a \end{bmatrix} \quad B = \begin{bmatrix} b & 4 \\ -1 & -5 \end{bmatrix}$

8. $A = \begin{bmatrix} 3 & 5 & 7 \\ -4 & a & 2 \end{bmatrix} \quad B = \begin{bmatrix} 3 & 5 & b \\ -4 & -5 & 2 \end{bmatrix}$

9–16 ■ Matrix Operations Perform the matrix operation, or if it is impossible, explain why.

9. $\begin{bmatrix} 2 & 6 \\ -5 & 3 \end{bmatrix} + \begin{bmatrix} -1 & -3 \\ 6 & 2 \end{bmatrix}$

10. $\begin{bmatrix} 0 & 1 & 1 \\ 1 & 1 & 0 \end{bmatrix} - \begin{bmatrix} 2 & 1 & -1 \\ 1 & 3 & -2 \end{bmatrix}$

11. $3\begin{bmatrix} 1 & 2 \\ 4 & -1 \\ 1 & 0 \end{bmatrix}$

12. $2\begin{bmatrix} 1 & 1 & 0 \\ 1 & 0 & 1 \\ 0 & 1 & 1 \end{bmatrix} + \begin{bmatrix} 1 & 1 \\ 2 & 1 \\ 3 & 1 \end{bmatrix}$

13. $\begin{bmatrix} 2 & 6 \\ 1 & 3 \\ 2 & 4 \end{bmatrix} \begin{bmatrix} 1 & -2 \\ 3 & 6 \\ -2 & 0 \end{bmatrix}$

14. $\begin{bmatrix} 2 & 1 & 2 \\ 6 & 3 & 4 \end{bmatrix} \begin{bmatrix} 1 & -2 \\ 3 & 6 \\ -2 & 0 \end{bmatrix}$

15. $\begin{bmatrix} 1 & 2 \\ -1 & 4 \end{bmatrix} \begin{bmatrix} 1 & -2 & 3 \\ 2 & 2 & -1 \end{bmatrix}$

16. $\begin{bmatrix} 2 & -3 \\ 0 & 1 \\ 1 & 2 \end{bmatrix} \begin{bmatrix} 5 \\ 1 \end{bmatrix}$

17–22 ■ Matrix Equations Solve the matrix equation for the unknown matrix X, or explain why no solution exists.

$$A = \begin{bmatrix} 4 & 6 \\ 1 & 3 \end{bmatrix} \qquad B = \begin{bmatrix} 2 & 5 \\ 3 & 7 \end{bmatrix}$$

$$C = \begin{bmatrix} 2 & 3 \\ 1 & 0 \\ 0 & 2 \end{bmatrix} \qquad D = \begin{bmatrix} 10 & 20 \\ 30 & 20 \\ 10 & 0 \end{bmatrix}$$

17. $2X + A = B$ **18.** $3X - B = C$

19. $2(B - X) = D$ **20.** $5(X - C) = D$

21. $\frac{1}{5}(X + D) = C$ **22.** $2A = B - 3X$

23–36 ■ Matrix Operations The matrices A, B, C, D, E, F, G and H are defined as follows.

$$A = \begin{bmatrix} 2 & -5 \\ 0 & 7 \end{bmatrix} \qquad B = \begin{bmatrix} 3 & \frac{1}{2} & 5 \\ 1 & -1 & 3 \end{bmatrix} \qquad C = \begin{bmatrix} 2 & -\frac{5}{2} & 0 \\ 0 & 2 & -3 \end{bmatrix}$$

$$D = \begin{bmatrix} 7 & 3 \end{bmatrix} \qquad E = \begin{bmatrix} 1 \\ 2 \\ 0 \end{bmatrix} \qquad F = \begin{bmatrix} 1 & 0 & 0 \\ 0 & 1 & 0 \\ 0 & 0 & 1 \end{bmatrix}$$

$$G = \begin{bmatrix} 5 & -3 & 10 \\ 6 & 1 & 0 \\ -5 & 2 & 2 \end{bmatrix} \qquad H = \begin{bmatrix} 3 & 1 \\ 2 & -1 \end{bmatrix}$$

Carry out the indicated algebraic operation, or explain why it cannot be performed.

23. **(a)** $B + C$ **(b)** $B + F$

24. **(a)** $C - B$ **(b)** $2C - 6B$

25. **(a)** $5A$ **(b)** $C - 5A$

26. **(a)** $3B + 2C$ **(b)** $2H + D$

27. **(a)** AD **(b)** DA

28. **(a)** DH **(b)** HD

29. **(a)** AH **(b)** HA

30. **(a)** BC **(b)** BF

31. **(a)** GF **(b)** GE

32. **(a)** B^2 **(b)** F^2

33. **(a)** A^2 **(b)** A^3

34. **(a)** $(DA)B$ **(b)** $D(AB)$

35. **(a)** ABE **(b)** AHE

36. **(a)** $DB + DC$ **(b)** $BF + FE$

37–42 ■ Matrix Operations The matrices A, B, and C are defined as follows.

$$A = \begin{bmatrix} 0.3 & 1.1 & 2.4 \\ 0.9 & -0.1 & 0.4 \\ -0.7 & 0.3 & -0.5 \end{bmatrix} \qquad B = \begin{bmatrix} 1.2 & -0.1 \\ 0 & -0.5 \\ 0.5 & -2.1 \end{bmatrix}$$

$$C = \begin{bmatrix} -0.2 & 0.2 & 0.1 \\ 1.1 & 2.1 & -2.1 \end{bmatrix}$$

Use a graphing calculator to carry out the indicated algebraic operation, or explain why it cannot be performed.

37. AB **38.** BA **39.** BC

40. CB **41.** $B + C$ **42.** A^2

43–46 ■ Equality of Matrices Solve for x and y.

43. $\begin{bmatrix} x & 2y \\ 4 & 6 \end{bmatrix} = \begin{bmatrix} 2 & -2 \\ 2x & -6y \end{bmatrix}$

44. $3\begin{bmatrix} x & y \\ y & x \end{bmatrix} = \begin{bmatrix} 6 & -9 \\ -9 & 6 \end{bmatrix}$

45. $2\begin{bmatrix} x & y \\ x + y & x - y \end{bmatrix} = \begin{bmatrix} 2 & -4 \\ -2 & 6 \end{bmatrix}$

46. $\begin{bmatrix} x & y \\ -y & x \end{bmatrix} - \begin{bmatrix} y & x \\ x & -y \end{bmatrix} = \begin{bmatrix} 4 & -4 \\ -6 & 6 \end{bmatrix}$

47–50 ■ Linear Systems as Matrix Equations Write the system of equations as a matrix equation (see Example 6).

47. $\begin{cases} 2x - 5y = 7 \\ 3x + 2y = 4 \end{cases}$

48. $\begin{cases} 6x - y + z = 12 \\ 2x + z = 7 \\ y - 2z = 4 \end{cases}$

49. $\begin{cases} 3x_1 + 2x_2 - x_3 + x_4 = 0 \\ x_1 - x_3 = 5 \\ 3x_2 + x_3 - x_4 = 4 \end{cases}$

50. $\begin{cases} x - y + z = 2 \\ 4x - 2y - z = 2 \\ x + y + 5z = 2 \\ -x - y - z = 2 \end{cases}$

SKILLS Plus

51. Products of Matrices The matrices A, B, and C are defined as follows.

$$A = \begin{bmatrix} 1 & 0 & 6 & -1 \\ 2 & \frac{1}{2} & 4 & 0 \end{bmatrix}$$

$$B = \begin{bmatrix} 1 & 7 & -9 & 2 \end{bmatrix} \qquad C = \begin{bmatrix} 1 \\ 0 \\ -1 \\ -2 \end{bmatrix}$$

Determine which of the following products are defined, and calculate the ones that are.

ABC ACB BAC

BCA CAB CBA

52. Expanding Matrix Bionomials

(a) Prove that if A and B are 2×2 matrices, then

$$(A + B)^2 = A^2 + AB + BA + B^2$$

(b) If A and B are 2×2 matrices, is it necessarily true that

$$(A + B)^2 \stackrel{?}{=} A^2 + 2AB + B^2$$

APPLICATIONS

53. Education and Income A women's group takes a survey to determine the education and income of its members. Matrix A summarizes the proportions of members in various categories of years of postsecondary education and income. Matrix B shows the total number of members in each income category.

(a) Calculate the product matrix AB.

(b) Interpret the entries of the matrix AB.

	Income level: Less than \$50,000	\$50,000 to 100,000	\$100,000 or more
None	0.75	0.10	0
1 to 4	0.25	0.70	0.70
More than 4	0	0.20	0.30

$= A$

	Total
Less than \$50,000	4
\$50,000 to 100,000	20
\$100,000 or more	10

$= B$

54. Exam Scores A large physics class takes a survey of the number of hours the students slept before an exam and their exam scores. Matrix A summarizes the proportions of students in different categories of exam scores and hours of sleep. Matrix B shows the total number of students in three exam score categories.

(a) Calculate the product matrix AB.

(b) Interpret the entries of the matrix AB.

	Exam Score: Below 60	60 to 80	Above 80
Less than 4	0.75	0.20	0.05
4 to 7	0.60	0.30	0.10
More than 7	0.40	0.30	0.30

$= A$

	Total
Below 60	80
60 to 80	170
Above 80	40

$= B$

55. Frozen-Food Revenue Some of the frozen foods that Joe's Specialty Foods sells are pesto pizza, spinach ravioli, and macaroni and cheese. The sales distribution for these products is tabulated in matrix A. The retail price (in dollars) for each item is tabulated in matrix B.

(a) Calculate the product matrix AB.

(b) What is the total revenue for Monday?

(c) What is the total revenue from all three days?

	Specialty Food: Pizza	Ravioli	Mac & Cheese
Monday	50	20	15
Tuesday	40	75	20
Wednesday	35	60	100

$= A$

	Price (\$)
Pizza	3.50
Ravioli	5.75
Mac & Cheese	4.25

$= B$

56. Fast-Food Sales A small fast-food chain with restaurants in Santa Monica, Long Beach, and Anaheim sells only hamburgers, hot dogs, and milk shakes. On a certain day, sales were distributed according to the following matrix.

	Number of items sold: Santa Monica	Long Beach	Anaheim
Hamburgers	4000	1000	3500
Hot dogs	400	300	200
Milk shakes	700	500	9000

$= A$

The price of each item is given by the following matrix.

Hamburger	Hot dog	Milk shake
\$0.90	\$0.80	\$1.10

$= B$

(a) Calculate the product BA.

(b) Interpret the entries in the product matrix BA.

57. Car-Manufacturing Profits A specialty-car manufacturer has plants in Auburn, Biloxi, and Chattanooga. Three models are produced, with daily production given in the following matrix.

	Cars produced each day: Model K	Model R	Model W
Auburn	12	10	0
Biloxi	4	4	20
Chattanooga	8	9	12

$= A$

Because of a wage increase, February profits are lower than January profits. The profit per car is tabulated by model in the following matrix.

	January	February
Model K	\$1000	\$500
Model R	\$2000	\$1200
Model W	\$1500	\$1000

$= B$

(a) Calculate AB.

(b) Assuming that all cars produced were sold, what was the daily profit in January from the Biloxi plant?

(c) What was the total daily profit (from all three plants) in February?

58. Canning Tomato Products Jaeger Foods produces tomato sauce and tomato paste, canned in small, medium, large, and giant-sized cans. The matrix A gives the size (in ounces) of each container.

	Small	Medium	Large	Giant
Ounces	6	10	14	28

$= A$

The matrix B tabulates one day's production of tomato sauce and tomato paste.

	Cans of sauce	Cans of paste
Small	2000	2500
Medium	3000	1500
Large	2500	1000
Giant	1000	500

$= B$

(a) Calculate the product AB.

(b) Interpret the entries in the product matrix AB.

59. Produce Sales A farmer's three children, Amy, Beth, and Chad, run three roadside produce stands during the summer months. One weekend they all sell watermelons, yellow squash, and tomatoes. The matrices A and B tabulate the number of pounds of each product sold by each sibling on Saturday and Sunday.

Saturday

	Melons	Squash	Tomatoes
Amy	120	50	60
Beth	40	25	30
Chad	60	30	20

$$\begin{bmatrix} 120 & 50 & 60 \\ 40 & 25 & 30 \\ 60 & 30 & 20 \end{bmatrix} = A$$

Sunday

	Melons	Squash	Tomatoes
Amy	100	60	30
Beth	35	20	20
Chad	60	25	30

$$\begin{bmatrix} 100 & 60 & 30 \\ 35 & 20 & 20 \\ 60 & 25 & 30 \end{bmatrix} = B$$

The matrix C gives the price per pound (in dollars) for each type of produce that they sell.

Price per pound

Melons	0.10
Squash	0.50
Tomatoes	1.00

$$\begin{bmatrix} 0.10 \\ 0.50 \\ 1.00 \end{bmatrix} = C$$

Perform each of the following matrix operations, and interpret the entries in each result.

(a) AC **(b)** BC **(c)** $A + B$ **(d)** $(A + B)C$

60. Digital Images A four-level gray scale is shown below.

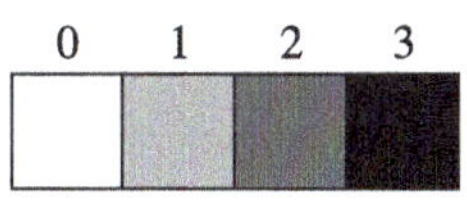

(a) Use the gray scale to find a 6×6 matrix that digitally represents the image in the figure.

(b) Find a matrix that represents a darker version of the image in the figure.

(c) The **negative** of an image is obtained by reversing light and dark, as in the negative of a photograph. Find the matrix that represents the negative of the image in the figure. How do you change the elements of the matrix to create the negative?

(d) Increase the contrast of the image by changing each 1 to a 0 and each 2 to a 3 in the matrix you found in part (a). Draw the image represented by the resulting matrix. Does this clarify the image?

(e) Draw the image represented by the matrix I. Can you recognize what this is? If you don't, try increasing the contrast.

$$I = \begin{bmatrix} 1 & 2 & 3 & 3 & 2 & 0 \\ 0 & 3 & 0 & 1 & 0 & 1 \\ 1 & 3 & 2 & 3 & 0 & 0 \\ 0 & 3 & 0 & 1 & 0 & 1 \\ 1 & 3 & 3 & 2 & 3 & 0 \\ 0 & 1 & 0 & 1 & 0 & 1 \end{bmatrix}$$

DISCUSS ■ DISCOVER ■ PROVE ■ WRITE

61. DISCUSS: When Are Both Products Defined? What must be true about the dimensions of the matrices A and B if both products AB and BA are defined?

62. DISCOVER: Powers of a Matrix Let

$$A = \begin{bmatrix} 1 & 1 \\ 0 & 1 \end{bmatrix}$$

Calculate $A^2, A^3, A^4, \ldots$ until you detect a pattern. Write a general formula for A^n.

63. DISCOVER: Powers of a Matrix Let

$$A = \begin{bmatrix} 1 & 1 \\ 1 & 1 \end{bmatrix}$$

Calculate $A^2, A^3, A^4, \ldots$ until you detect a pattern. Write a general formula for A^n.

64. DISCUSS: Square Roots of Matrices A **square root** of a matrix B is a matrix A with the property that $A^2 = B$. (This is the same definition as for a square root of a number.) Find as many square roots as you can of each matrix:

$$\begin{bmatrix} 4 & 0 \\ 0 & 9 \end{bmatrix} \qquad \begin{bmatrix} 1 & 5 \\ 0 & 9 \end{bmatrix}$$

[*Hint:* If $A = \begin{bmatrix} a & b \\ c & d \end{bmatrix}$, write the equations that a, b, c, and d would have to satisfy if A is the square root of the given matrix.]

10.5 INVERSES OF MATRICES AND MATRIX EQUATIONS

■ The Inverse of a Matrix ■ Finding the Inverse of a 2×2 Matrix ■ Finding the Inverse of an $n \times n$ Matrix ■ Matrix Equations ■ Modeling with Matrix Equations

In Section 10.4 we saw that when the dimensions are appropriate, matrices can be added, subtracted, and multiplied. In this section we investigate division of matrices. With this operation we can solve equations that involve matrices.

The Inverse of a Matrix

First, we define *identity matrices*, which play the same role for matrix multiplication as the number 1 does for ordinary multiplication of numbers; that is, $1 \cdot a = a \cdot 1 = a$ for all numbers a. A **square matrix** is one that has the same number of rows as columns. The **main diagonal** of a square matrix consists of the entries whose row and column numbers are the same. These entries stretch diagonally down the matrix, from top left to bottom right.

IDENTITY MATRIX

The **identity matrix** I_n is the $n \times n$ matrix for which each main diagonal entry is a 1 and for which all other entries are 0.

Thus the 2×2, 3×3, and 4×4 identity matrices are

$$I_2 = \begin{bmatrix} 1 & 0 \\ 0 & 1 \end{bmatrix} \qquad I_3 = \begin{bmatrix} 1 & 0 & 0 \\ 0 & 1 & 0 \\ 0 & 0 & 1 \end{bmatrix} \qquad I_4 = \begin{bmatrix} 1 & 0 & 0 & 0 \\ 0 & 1 & 0 & 0 \\ 0 & 0 & 1 & 0 \\ 0 & 0 & 0 & 1 \end{bmatrix}$$

Identity matrices behave like the number 1 in the sense that

$$A \cdot I_n = A \qquad \text{and} \qquad I_n \cdot B = B$$

whenever these products are defined.

EXAMPLE 1 ■ Identity Matrices

The following matrix products show how multiplying a matrix by an identity matrix of the appropriate dimension leaves the matrix unchanged.

$$\begin{bmatrix} 1 & 0 \\ 0 & 1 \end{bmatrix} \begin{bmatrix} 3 & 5 & 6 \\ -1 & 2 & 7 \end{bmatrix} = \begin{bmatrix} 3 & 5 & 6 \\ -1 & 2 & 7 \end{bmatrix}$$

$$\begin{bmatrix} -1 & 7 & \frac{1}{2} \\ 12 & 1 & 3 \\ -2 & 0 & 7 \end{bmatrix} \begin{bmatrix} 1 & 0 & 0 \\ 0 & 1 & 0 \\ 0 & 0 & 1 \end{bmatrix} = \begin{bmatrix} -1 & 7 & \frac{1}{2} \\ 12 & 1 & 3 \\ -2 & 0 & 7 \end{bmatrix}$$

Now Try Exercise 1(a), (b) ■

If A and B are $n \times n$ matrices, and if $AB = BA = I_n$, then we say that B is the *inverse* of A, and we write $B = A^{-1}$. The concept of the inverse of a matrix is analogous to that of the reciprocal of a real number.

INVERSE OF A MATRIX

Let A be a square $n \times n$ matrix. If there exists an $n \times n$ matrix A^{-1} with the property that

$$AA^{-1} = A^{-1}A = I_n$$

then we say that A^{-1} is the **inverse** of A.

EXAMPLE 2 ■ Verifying That a Matrix Is an Inverse

Verify that B is the inverse of A, where

$$A = \begin{bmatrix} 2 & 1 \\ 5 & 3 \end{bmatrix} \quad \text{and} \quad B = \begin{bmatrix} 3 & -1 \\ -5 & 2 \end{bmatrix}$$

SOLUTION We perform the matrix multiplications to show that $AB = I$ and $BA = I$.

$$\begin{bmatrix} 2 & 1 \\ 5 & 3 \end{bmatrix}\begin{bmatrix} 3 & -1 \\ -5 & 2 \end{bmatrix} = \begin{bmatrix} 2 \cdot 3 + 1(-5) & 2(-1) + 1 \cdot 2 \\ 5 \cdot 3 + 3(-5) & 5(-1) + 3 \cdot 2 \end{bmatrix} = \begin{bmatrix} 1 & 0 \\ 0 & 1 \end{bmatrix}$$

$$\begin{bmatrix} 3 & -1 \\ -5 & 2 \end{bmatrix}\begin{bmatrix} 2 & 1 \\ 5 & 3 \end{bmatrix} = \begin{bmatrix} 3 \cdot 2 + (-1)5 & 3 \cdot 1 + (-1)3 \\ (-5)2 + 2 \cdot 5 & (-5)1 + 2 \cdot 3 \end{bmatrix} = \begin{bmatrix} 1 & 0 \\ 0 & 1 \end{bmatrix}$$

Now Try Exercise 3 ■

Finding the Inverse of a 2 × 2 Matrix

The following rule provides a simple way for finding the inverse of a 2×2 matrix, when it exists. For larger matrices there is a more general procedure for finding inverses, which we consider later in this section.

INVERSE OF A 2 × 2 MATRIX

If $A = \begin{bmatrix} a & b \\ c & d \end{bmatrix}$, then

$$A^{-1} = \frac{1}{ad - bc}\begin{bmatrix} d & -b \\ -c & a \end{bmatrix}$$

If $ad - bc = 0$, then A has no inverse.

Bygone Collection/Alamy

ARTHUR CAYLEY (1821–1895) was an English mathematician who was instrumental in developing the theory of matrices. He was the first to use a single symbol such as A to represent a matrix, thereby introducing the idea that a matrix is a single entity rather than just a collection of numbers. Cayley practiced law until the age of 42, but his primary interest from adolescence was mathematics, and he published almost 200 articles on the subject in his spare time. In 1863 he accepted a professorship in mathematics at Cambridge, where he taught until his death. Cayley's work on matrices was of purely theoretical interest in his day, but in the 20th century many of his results found application in physics, the social sciences, business, and other fields. One of the most common uses of matrices today is in computers, where matrices are employed for data storage, error correction, image manipulation, and many other purposes. These applications have made matrix algebra more useful than ever.

EXAMPLE 3 ■ Finding the Inverse of a 2 × 2 Matrix

Let

$$A = \begin{bmatrix} 4 & 5 \\ 2 & 3 \end{bmatrix}$$

Find A^{-1}, and verify that $AA^{-1} = A^{-1}A = I_2$.

SOLUTION Using the rule for the inverse of a 2 × 2 matrix, we get

$$A^{-1} = \frac{1}{4 \cdot 3 - 5 \cdot 2}\begin{bmatrix} 3 & -5 \\ -2 & 4 \end{bmatrix} = \frac{1}{2}\begin{bmatrix} 3 & -5 \\ -2 & 4 \end{bmatrix} = \begin{bmatrix} \frac{3}{2} & -\frac{5}{2} \\ -1 & 2 \end{bmatrix}$$

To verify that this is indeed the inverse of A, we calculate AA^{-1} and $A^{-1}A$:

$$AA^{-1} = \begin{bmatrix} 4 & 5 \\ 2 & 3 \end{bmatrix}\begin{bmatrix} \frac{3}{2} & -\frac{5}{2} \\ -1 & 2 \end{bmatrix} = \begin{bmatrix} 4 \cdot \frac{3}{2} + 5(-1) & 4(-\frac{5}{2}) + 5 \cdot 2 \\ 2 \cdot \frac{3}{2} + 3(-1) & 2(-\frac{5}{2}) + 3 \cdot 2 \end{bmatrix} = \begin{bmatrix} 1 & 0 \\ 0 & 1 \end{bmatrix}$$

$$A^{-1}A = \begin{bmatrix} \frac{3}{2} & -\frac{5}{2} \\ -1 & 2 \end{bmatrix}\begin{bmatrix} 4 & 5 \\ 2 & 3 \end{bmatrix} = \begin{bmatrix} \frac{3}{2} \cdot 4 + (-\frac{5}{2})2 & \frac{3}{2} \cdot 5 + (-\frac{5}{2})3 \\ (-1)4 + 2 \cdot 2 & (-1)5 + 2 \cdot 3 \end{bmatrix} = \begin{bmatrix} 1 & 0 \\ 0 & 1 \end{bmatrix}$$

Now Try Exercise 7 ■

The quantity $ad - bc$ that appears in the rule for calculating the inverse of a 2 × 2 matrix is called the **determinant** of the matrix. If the determinant is 0, then the matrix does not have an inverse (since we cannot divide by 0).

Finding the Inverse of an *n* × *n* Matrix

For 3 × 3 and larger square matrices the following technique provides the most efficient way to calculate their inverses. If A is an $n \times n$ matrix, we first construct the $n \times 2n$ matrix that has the entries of A on the left and of the identity matrix I_n on the right:

$$\left[\begin{array}{cccc|cccc} a_{11} & a_{12} & \cdots & a_{1n} & 1 & 0 & \cdots & 0 \\ a_{21} & a_{22} & \cdots & a_{2n} & 0 & 1 & \cdots & 0 \\ \vdots & \vdots & \ddots & \vdots & \vdots & \vdots & \ddots & \vdots \\ a_{n1} & a_{n2} & \cdots & a_{nn} & 0 & 0 & \cdots & 1 \end{array}\right]$$

We then use the elementary row operations on this new large matrix to change the left side into the identity matrix. (This means that we are changing the large matrix to reduced row-echelon form.) The right side is transformed automatically into A^{-1}. (We omit the proof of this fact.)

EXAMPLE 4 ■ Finding the Inverse of a 3 × 3 Matrix

Let A be the matrix

$$A = \begin{bmatrix} 1 & -2 & -4 \\ 2 & -3 & -6 \\ -3 & 6 & 15 \end{bmatrix}$$

(a) Find A^{-1}.

(b) Verify that $AA^{-1} = A^{-1}A = I_3$.

SOLUTION

(a) We begin with the 3 × 6 matrix whose left half is A and whose right half is the identity matrix.

$$\left[\begin{array}{ccc|ccc} 1 & -2 & -4 & 1 & 0 & 0 \\ 2 & -3 & -6 & 0 & 1 & 0 \\ -3 & 6 & 15 & 0 & 0 & 1 \end{array}\right]$$

We then transform the left half of this new matrix into the identity matrix by performing the following sequence of elementary row operations on the *entire* new matrix.

$$\xrightarrow[\text{R}_3 + 3\text{R}_1 \to \text{R}_3]{\text{R}_2 - 2\text{R}_1 \to \text{R}_2} \left[\begin{array}{ccc|ccc} 1 & -2 & -4 & 1 & 0 & 0 \\ 0 & 1 & 2 & -2 & 1 & 0 \\ 0 & 0 & 3 & 3 & 0 & 1 \end{array}\right]$$

$$\xrightarrow{\frac{1}{3}\text{R}_3} \left[\begin{array}{ccc|ccc} 1 & -2 & -4 & 1 & 0 & 0 \\ 0 & 1 & 2 & -2 & 1 & 0 \\ 0 & 0 & 1 & 1 & 0 & \frac{1}{3} \end{array}\right]$$

$$\xrightarrow{\text{R}_1 + 2\text{R}_2 \to \text{R}_1} \left[\begin{array}{ccc|ccc} 1 & 0 & 0 & -3 & 2 & 0 \\ 0 & 1 & 2 & -2 & 1 & 0 \\ 0 & 0 & 1 & 1 & 0 & \frac{1}{3} \end{array}\right]$$

$$\xrightarrow{\text{R}_2 - 2\text{R}_3 \to \text{R}_2} \left[\begin{array}{ccc|ccc} 1 & 0 & 0 & -3 & 2 & 0 \\ 0 & 1 & 0 & -4 & 1 & -\frac{2}{3} \\ 0 & 0 & 1 & 1 & 0 & \frac{1}{3} \end{array}\right]$$

We have now transformed the left half of this matrix into an identity matrix. (This means that we have put the entire matrix in reduced row-echelon form.) Note that to do this in as systematic a fashion as possible, we first changed the elements below the main diagonal to zeros, just as we would if we were using Gaussian elimination. We then changed each main diagonal element to a 1 by multiplying by the appropriate constant(s). Finally, we completed the process by changing the remaining entries on the left side to zeros.

The right half is now A^{-1}.

$$A^{-1} = \begin{bmatrix} -3 & 2 & 0 \\ -4 & 1 & -\frac{2}{3} \\ 1 & 0 & \frac{1}{3} \end{bmatrix}$$

(b) We calculate AA^{-1} and $A^{-1}A$ and verify that both products give the identity matrix I_3.

$$AA^{-1} = \begin{bmatrix} 1 & -2 & -4 \\ 2 & -3 & -6 \\ -3 & 6 & 15 \end{bmatrix} \begin{bmatrix} -3 & 2 & 0 \\ -4 & 1 & -\frac{2}{3} \\ 1 & 0 & \frac{1}{3} \end{bmatrix} = \begin{bmatrix} 1 & 0 & 0 \\ 0 & 1 & 0 \\ 0 & 0 & 1 \end{bmatrix}$$

$$A^{-1}A = \begin{bmatrix} -3 & 2 & 0 \\ -4 & 1 & -\frac{2}{3} \\ 1 & 0 & \frac{1}{3} \end{bmatrix} \begin{bmatrix} 1 & -2 & -4 \\ 2 & -3 & -6 \\ -3 & 6 & 15 \end{bmatrix} = \begin{bmatrix} 1 & 0 & 0 \\ 0 & 1 & 0 \\ 0 & 0 & 1 \end{bmatrix}$$

Now Try Exercises 9 and 19 ■

See Appendix D, *Using the TI-83/84 Graphing Calculator*, for specific instructions on working with matrices. Go to **www.stewartmath.com**.

Graphing calculators are also able to calculate matrix inverses. On the TI-83 and TI-84 calculators, matrices are stored in memory using names such as `[A]`, `[B]`, `[C]`, ... To find the inverse of `[A]`, we key in

`[A]` `X⁻¹` `ENTER`

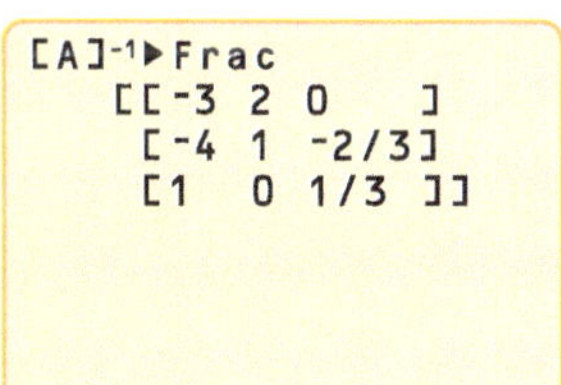

FIGURE 1

For the matrix of Example 4 this results in the output shown in Figure 1 (where we have also used the `▶Frac` command to display the output in fraction form rather than in decimal form).

The next example shows that not every square matrix has an inverse.

EXAMPLE 5 ■ A Matrix That Does Not Have an Inverse

Find the inverse of the matrix

$$\begin{bmatrix} 2 & -3 & -7 \\ 1 & 2 & 7 \\ 1 & 1 & 4 \end{bmatrix}$$

SOLUTION We proceed as follows.

$$\left[\begin{array}{ccc|ccc} 2 & -3 & -7 & 1 & 0 & 0 \\ 1 & 2 & 7 & 0 & 1 & 0 \\ 1 & 1 & 4 & 0 & 0 & 1 \end{array}\right] \xrightarrow{R_1 \leftrightarrow R_2} \left[\begin{array}{ccc|ccc} 1 & 2 & 7 & 0 & 1 & 0 \\ 2 & -3 & -7 & 1 & 0 & 0 \\ 1 & 1 & 4 & 0 & 0 & 1 \end{array}\right]$$

$$\xrightarrow[R_3 - R_1 \to R_3]{R_2 - 2R_1 \to R_2} \left[\begin{array}{ccc|ccc} 1 & 2 & 7 & 0 & 1 & 0 \\ 0 & -7 & -21 & 1 & -2 & 0 \\ 0 & -1 & -3 & 0 & -1 & 1 \end{array}\right]$$

$$\xrightarrow{-\frac{1}{7}R_2} \left[\begin{array}{ccc|ccc} 1 & 2 & 7 & 0 & 1 & 0 \\ 0 & 1 & 3 & -\frac{1}{7} & \frac{2}{7} & 0 \\ 0 & -1 & -3 & 0 & -1 & 1 \end{array}\right]$$

$$\xrightarrow[R_1 - 2R_2 \to R_1]{R_3 + R_2 \to R_3} \left[\begin{array}{ccc|ccc} 1 & 0 & 1 & \frac{2}{7} & \frac{3}{7} & 0 \\ 0 & 1 & 3 & -\frac{1}{7} & \frac{2}{7} & 0 \\ 0 & 0 & 0 & -\frac{1}{7} & -\frac{5}{7} & 1 \end{array}\right]$$

At this point we would like to change the 0 in the (3, 3) position of this matrix to a 1 without changing the zeros in the (3, 1) and (3, 2) positions. But there is no way to accomplish this, because no matter what multiple of rows 1 and/or 2 we add to row 3, we can't change the third zero in row 3 without changing the first or second zero as well. Thus we cannot change the left half to the identity matrix, so the original matrix doesn't have an inverse.

Now Try Exercise 21 ■

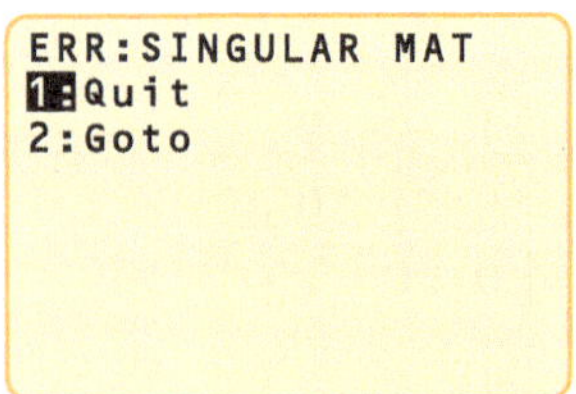

FIGURE 2

If we encounter a row of zeros on the left when trying to find an inverse, as in Example 5, then the original matrix does not have an inverse. If we try to calculate the inverse of the matrix from Example 5 on a TI-83 calculator, we get the error message shown in Figure 2. (A matrix that has no inverse is called *singular*.)

■ Matrix Equations

We saw in Example 6 in Section 10.4 that a system of linear equations can be written as a single matrix equation. For example, the system

$$\begin{cases} x - 2y - 4z = 7 \\ 2x - 3y - 6z = 5 \\ -3x + 6y + 15z = 0 \end{cases}$$

is equivalent to the matrix equation

$$\underbrace{\begin{bmatrix} 1 & -2 & -4 \\ 2 & -3 & -6 \\ -3 & 6 & 15 \end{bmatrix}}_{A} \underbrace{\begin{bmatrix} x \\ y \\ z \end{bmatrix}}_{X} = \underbrace{\begin{bmatrix} 7 \\ 5 \\ 0 \end{bmatrix}}_{B}$$

If we let

$$A = \begin{bmatrix} 1 & -2 & -4 \\ 2 & -3 & -6 \\ -3 & 6 & 15 \end{bmatrix} \qquad X = \begin{bmatrix} x \\ y \\ z \end{bmatrix} \qquad B = \begin{bmatrix} 7 \\ 5 \\ 0 \end{bmatrix}$$

then this matrix equation can be written as

$$AX = B$$

The matrix A is called the **coefficient matrix**.

We solve this matrix equation by multiplying each side by the inverse of A (provided that this inverse exists).

Solving the matrix equation $AX = B$ is very similar to solving the simple real-number equation

$$3x = 12$$

which we do by multiplying each side by the reciprocal (or inverse) of 3.

$$\tfrac{1}{3}(3x) = \tfrac{1}{3}(12)$$

$$x = 4$$

$AX = B$	
$A^{-1}(AX) = A^{-1}B$	Multiply on left by A^{-1}
$(A^{-1}A)X = A^{-1}B$	Associative Property
$I_3X = A^{-1}B$	Property of inverses
$X = A^{-1}B$	Property of identity matrix

In Example 4 we showed that

$$A^{-1} = \begin{bmatrix} -3 & 2 & 0 \\ -4 & 1 & -\frac{2}{3} \\ 1 & 0 & \frac{1}{3} \end{bmatrix}$$

So from $X = A^{-1}B$ we have

$$\underbrace{\begin{bmatrix} x \\ y \\ z \end{bmatrix}}_{X} = \underbrace{\begin{bmatrix} -3 & 2 & 0 \\ -4 & 1 & -\frac{2}{3} \\ 1 & 0 & \frac{1}{3} \end{bmatrix}}_{A^{-1}} \underbrace{\begin{bmatrix} 7 \\ 5 \\ 0 \end{bmatrix}}_{B} = \begin{bmatrix} -11 \\ -23 \\ 7 \end{bmatrix}$$

Thus $x = -11$, $y = -23$, $z = 7$ is the solution of the original system.

We have proved that the matrix equation $AX = B$ can be solved by the following method.

SOLVING A MATRIX EQUATION

If A is a square $n \times n$ matrix that has an inverse A^{-1} and if X is a variable matrix and B a known matrix, both with n rows, then the solution of the matrix equation

$$AX = B$$

is given by

$$X = A^{-1}B$$

EXAMPLE 6 ■ Solving a System Using a Matrix Inverse

A system of equations is given.

(a) Write the system of equations as a matrix equation.

(b) Solve the system by solving the matrix equation.

$$\begin{cases} 2x - 5y = 15 \\ 3x - 6y = 36 \end{cases}$$

SOLUTION

(a) We write the system as a matrix equation of the form $AX = B$.

$$\underbrace{\begin{bmatrix} 2 & -5 \\ 3 & -6 \end{bmatrix}}_{A}\underbrace{\begin{bmatrix} x \\ y \end{bmatrix}}_{X} = \underbrace{\begin{bmatrix} 15 \\ 36 \end{bmatrix}}_{B}$$

(b) Using the rule for finding the inverse of a 2×2 matrix, we get

$$A^{-1} = \begin{bmatrix} 2 & -5 \\ 3 & -6 \end{bmatrix}^{-1} = \frac{1}{2(-6) - (-5)3}\begin{bmatrix} -6 & -(-5) \\ -3 & 2 \end{bmatrix} = \frac{1}{3}\begin{bmatrix} -6 & 5 \\ -3 & 2 \end{bmatrix}$$

Multiplying each side of the matrix equation by this inverse matrix, we get

$$\underbrace{\begin{bmatrix} x \\ y \end{bmatrix}}_{X} = \underbrace{\frac{1}{3}\begin{bmatrix} -6 & 5 \\ -3 & 2 \end{bmatrix}}_{A^{-1}}\underbrace{\begin{bmatrix} 15 \\ 36 \end{bmatrix}}_{B} = \begin{bmatrix} 30 \\ 9 \end{bmatrix}$$

So $x = 30$ and $y = 9$.

Now Try Exercise 39 ■

Modeling with Matrix Equations

Suppose we need to solve several systems of equations with the same coefficient matrix. Then converting the systems to matrix equations provides an efficient way to obtain the solutions, because we need to find the inverse of the coefficient matrix only once. This procedure is particularly convenient if we use a graphing calculator to perform the matrix operations, as in the next example.

EXAMPLE 7 ■ Modeling Nutritional Requirements Using Matrix Equations

A pet-store owner feeds his hamsters and gerbils different mixtures of three types of rodent food: KayDee Food, Pet Pellets, and Rodent Chow. He wishes to feed his animals the correct amount of each brand to satisfy their daily requirements for protein, fat, and carbohydrates exactly. Suppose that hamsters require 340 mg of protein, 280 mg of fat, and 440 mg of carbohydrates, and gerbils need 480 mg of protein, 360 mg of fat, and 680 mg of carbohydrates each day. The amount of each nutrient (in mg) in 1 g of each brand is given in the following table. How many grams of each food should the storekeeper feed his hamsters and gerbils daily to satisfy their nutrient requirements?

	KayDee Food	Pet Pellets	Rodent Chow
Protein (mg)	10	0	20
Fat (mg)	10	20	10
Carbohydrates (mg)	5	10	30

SOLUTION We let x_1, x_2, and x_3 be the respective amounts (in grams) of KayDee Food, Pet Pellets, and Rodent Chow that the hamsters should eat, and we let y_1, y_2,

Mathematics in the Modern World

Mathematical Ecology

In the 1970s humpback whales became a center of controversy. Environmentalists believed that whaling threatened the whales with imminent extinction; whalers saw their livelihood threatened by any attempt to stop whaling. Are whales really threatened to extinction by whaling? What level of whaling is safe to guarantee survival of the whales? These questions motivated mathematicians to study population patterns of whales and other species more closely.

As early as the 1920s Lotka and Volterra had founded the field of mathematical biology by creating predator-prey models. Their models, which draw on a branch of mathematics called differential equations, take into account the rates at which predator eats prey and the rates of growth of each population. Note that as predator eats prey, the prey population decreases; this means less food supply for the predators, so their population begins to decrease; with fewer predators the prey population begins to increase, and so on. Normally, a state of equilibrium develops, and the two populations alternate between a minimum and a maximum. Notice that if the predators eat the prey too fast, they will be left without food and will thus ensure their own extinction.

Since Lotka and Volterra's time, more detailed mathematical models of animal populations have been developed. For many species the population is divided into several stages: immature, juvenile, adult, and so on. The proportion of each stage that survives or reproduces in a given time period is entered into a matrix (called a transition matrix); matrix multiplication is then used to predict the population in succeeding time periods. (See *Discovery Project: Will the Species Survive?* at the book companion website: **www.stewartmath.com**.)

As you can see, the power of mathematics to model and predict is an invaluable tool in the ongoing debate over the environment.

and y_3 be the corresponding amounts for the gerbils. Then we want to solve the matrix equations

$$\begin{bmatrix} 10 & 0 & 20 \\ 10 & 20 & 10 \\ 5 & 10 & 30 \end{bmatrix}\begin{bmatrix} x_1 \\ x_2 \\ x_3 \end{bmatrix} = \begin{bmatrix} 340 \\ 280 \\ 440 \end{bmatrix} \qquad \text{Hamster equation}$$

$$\begin{bmatrix} 10 & 0 & 20 \\ 10 & 20 & 10 \\ 5 & 10 & 30 \end{bmatrix}\begin{bmatrix} y_1 \\ y_2 \\ y_3 \end{bmatrix} = \begin{bmatrix} 480 \\ 360 \\ 680 \end{bmatrix} \qquad \text{Gerbil equation}$$

Let

$$A = \begin{bmatrix} 10 & 0 & 20 \\ 10 & 20 & 10 \\ 5 & 10 & 30 \end{bmatrix} \quad B = \begin{bmatrix} 340 \\ 280 \\ 440 \end{bmatrix} \quad C = \begin{bmatrix} 480 \\ 360 \\ 680 \end{bmatrix} \quad X = \begin{bmatrix} x_1 \\ x_2 \\ x_3 \end{bmatrix} \quad Y = \begin{bmatrix} y_1 \\ y_2 \\ y_3 \end{bmatrix}$$

Then we can write these matrix equations as

$$AX = B \qquad \text{Hamster equation}$$

$$AY = C \qquad \text{Gerbil equation}$$

We want to solve for X and Y, so we multiply both sides of each equation by A^{-1}, the inverse of the coefficient matrix. We could find A^{-1} by hand, but it is more convenient to use a graphing calculator as shown in Figure 3.

FIGURE 3 (a) (b)

So

$$X = A^{-1}B = \begin{bmatrix} 10 \\ 3 \\ 12 \end{bmatrix} \qquad Y = A^{-1}C = \begin{bmatrix} 8 \\ 4 \\ 20 \end{bmatrix}$$

Thus each hamster should be fed 10 g of KayDee Food, 3 g of Pet Pellets, and 12 g of Rodent Chow; and each gerbil should be fed 8 g of KayDee Food, 4 g of Pet Pellets, and 20 g of Rodent Chow daily.

Now Try Exercise 61

10.5 EXERCISES

CONCEPTS

1. **(a)** The matrix $I = \begin{bmatrix} 1 & 0 \\ 0 & 1 \end{bmatrix}$ is called an ________ matrix.

(b) If A is a 2×2 matrix, then $A \times I =$ ________ and $I \times A =$ ________.

(c) If A and B are 2×2 matrices with $AB = I$, then B is the ________ of A.

2. **(a)** Write the following system as a matrix equation $AX = B$.

System	Matrix equation
$5x + 3y = 4$ $3x + 2y = 3$	$A \cdot X = B$ $\begin{bmatrix} \square & \square \\ \square & \square \end{bmatrix}\begin{bmatrix} \square \\ \square \end{bmatrix} = \begin{bmatrix} \square \\ \square \end{bmatrix}$

(b) The inverse of A is $A^{-1} = \begin{bmatrix} \square & \square \\ \square & \square \end{bmatrix}$.

(c) The solution of the matrix equation is $X = A^{-1}B$.

$$\underset{X}{\begin{bmatrix} x \\ y \end{bmatrix}} = \underset{A^{-1}}{\begin{bmatrix} \square & \square \\ \square & \square \end{bmatrix}} \underset{B}{\begin{bmatrix} \square \\ \square \end{bmatrix}} = \begin{bmatrix} \square \\ \square \end{bmatrix}$$

(d) The solution of the system is $x =$ ________, $y =$ ________.

SKILLS

3–6 ■ Verifying the Inverse of a Matrix Calculate the products AB and BA to verify that B is the inverse of A.

3. $A = \begin{bmatrix} 4 & 1 \\ 7 & 2 \end{bmatrix}$ $B = \begin{bmatrix} 2 & -1 \\ -7 & 4 \end{bmatrix}$

4. $A = \begin{bmatrix} 2 & -3 \\ 4 & -7 \end{bmatrix}$ $B = \begin{bmatrix} \frac{7}{2} & -\frac{3}{2} \\ 2 & -1 \end{bmatrix}$

5. $A = \begin{bmatrix} 1 & 3 & -1 \\ 1 & 4 & 0 \\ -1 & -3 & 2 \end{bmatrix}$ $B = \begin{bmatrix} 8 & -3 & 4 \\ -2 & 1 & -1 \\ 1 & 0 & 1 \end{bmatrix}$

6. $A = \begin{bmatrix} 3 & 2 & 4 \\ 1 & 1 & -6 \\ 2 & 1 & 12 \end{bmatrix}$ $B = \begin{bmatrix} 9 & -10 & -8 \\ -12 & 14 & 11 \\ -\frac{1}{2} & \frac{1}{2} & \frac{1}{2} \end{bmatrix}$

7–8 ■ The Inverse of a 2 × 2 Matrix Find the inverse of the matrix and verify that $A^{-1}A = AA^{-1} = I_2$ and $B^{-1}B = BB^{-1} = I_3$.

7. $A = \begin{bmatrix} 7 & 4 \\ 3 & 2 \end{bmatrix}$

8. $B = \begin{bmatrix} 1 & 3 & 2 \\ 0 & 2 & 2 \\ -2 & -1 & 0 \end{bmatrix}$

9–10 ■ The Inverse of a 2 × 2 Matrix Use a graphing calculator to find the inverse of the matrix and to verify that $A^{-1}A = AA^{-1} = I_2$ and $B^{-1}B = BB^{-1} = I_3$. (On a TI-83, use the ▶Frac command to obtain the answer in fractions.)

9. $A = \begin{bmatrix} 1.2 & 0.3 \\ -1.2 & 0.2 \end{bmatrix}$

10. $B = \begin{bmatrix} 5 & -1 & 3 \\ 6 & -1 & 3 \\ 7 & 1 & -2 \end{bmatrix}$

11–26 ■ Finding the Inverse of a Matrix Find the inverse of the matrix if it exists.

11. $\begin{bmatrix} -3 & -5 \\ 2 & 3 \end{bmatrix}$

12. $\begin{bmatrix} 3 & 4 \\ 7 & 9 \end{bmatrix}$

13. $\begin{bmatrix} 2 & 5 \\ -5 & -13 \end{bmatrix}$

14. $\begin{bmatrix} -7 & 4 \\ 8 & -5 \end{bmatrix}$

15. $\begin{bmatrix} 6 & -3 \\ -8 & 4 \end{bmatrix}$

16. $\begin{bmatrix} \frac{1}{2} & \frac{1}{3} \\ 5 & 4 \end{bmatrix}$

17. $\begin{bmatrix} 0.4 & -1.2 \\ 0.3 & 0.6 \end{bmatrix}$

18. $\begin{bmatrix} 4 & 2 & 3 \\ 3 & 3 & 2 \\ 1 & 0 & 1 \end{bmatrix}$

19. $\begin{bmatrix} 2 & 4 & 1 \\ -1 & 1 & -1 \\ 1 & 4 & 0 \end{bmatrix}$

20. $\begin{bmatrix} 5 & 7 & 4 \\ 3 & -1 & 3 \\ 6 & 7 & 5 \end{bmatrix}$

21. $\begin{bmatrix} 1 & 2 & 3 \\ 4 & 5 & -1 \\ 1 & -1 & -10 \end{bmatrix}$

22. $\begin{bmatrix} 2 & 1 & 0 \\ 1 & 1 & 4 \\ 2 & 1 & 2 \end{bmatrix}$

23. $\begin{bmatrix} 0 & -2 & 2 \\ 3 & 1 & 3 \\ 1 & -2 & 3 \end{bmatrix}$

24. $\begin{bmatrix} 3 & -2 & 0 \\ 5 & 1 & 1 \\ 2 & -2 & 0 \end{bmatrix}$

25. $\begin{bmatrix} 1 & 2 & 0 & 3 \\ 0 & 1 & 1 & 1 \\ 0 & 1 & 0 & 1 \\ 1 & 2 & 0 & 2 \end{bmatrix}$

26. $\begin{bmatrix} 1 & 0 & 1 & 0 \\ 0 & 1 & 0 & 1 \\ 1 & 1 & 1 & 0 \\ 1 & 1 & 1 & 1 \end{bmatrix}$

27–34 ■ Finding the Inverse of a Matrix Use a graphing calculator to find the inverse of the matrix, if it exists. (On a TI-83, use the ▶Frac command to obtain the answer in fractions.)

27. $\begin{bmatrix} -3 & 2 & 3 \\ 0 & -1 & 3 \\ 1 & 0 & -2 \end{bmatrix}$

28. $\begin{bmatrix} -5 & 2 & 1 \\ 5 & 1 & 0 \\ 0 & -1 & -2 \end{bmatrix}$

29. $\begin{bmatrix} -1 & -4 & 0 & 1 \\ 1 & 0 & -1 & 0 \\ 0 & 4 & 1 & -2 \\ 2 & 2 & -2 & 0 \end{bmatrix}$

30. $\begin{bmatrix} -3 & 0 & -1 & 1 \\ 3 & -1 & 1 & -1 \\ 1 & 3 & 0 & 1 \\ -2 & -3 & 1 & 0 \end{bmatrix}$

31. $\begin{bmatrix} 1 & 7 & 3 \\ 0 & 2 & 1 \\ 0 & 0 & 3 \end{bmatrix}$

32. $\begin{bmatrix} 1 & 0 & 0 & 0 \\ 2 & 5 & 0 & 0 \\ 4 & 2 & 3 & 0 \\ 5 & 1 & 2 & 1 \end{bmatrix}$

33. $\begin{bmatrix} 1 & 0 & 0 & 0 \\ 0 & 2 & 0 & 0 \\ 0 & 0 & 4 & 0 \\ 0 & 0 & 0 & 7 \end{bmatrix}$

34. $\begin{bmatrix} -1 & 0 & 0 \\ 0 & 2 & 0 \\ 0 & 0 & -3 \end{bmatrix}$

35–38 ■ Products Involving Matrices and Inverses The matrices A and B are defined as follows.

$$A = \begin{bmatrix} -1 & 0 & 2 \\ 0 & -2 & -1 \\ 4 & 2 & 1 \end{bmatrix} \qquad B = \begin{bmatrix} 2 & -1 & -2 \\ 0 & 3 & 1 \\ -1 & 0 & 2 \end{bmatrix}$$

Use a graphing calculator to carry out the indicated algebraic operations, or explain why they cannot be performed. State the answer using fractions. (On a TI-83, use the ▶Frac command to obtain the answer in fractions.)

35. $A^{-1}B$ **36.** AB^{-1} **37.** BAB^{-1} **38.** $B^{-1}AB$

39–46 ■ Solving a Linear System as a Matrix Equation Solve the system of equations by converting to a matrix equation and using the inverse of the coefficient matrix, as in Example 6. Use the inverses from Exercises 11–14, 19, 20, 23, and 25.

39. $\begin{cases} -3x - 5y = 4 \\ 2x + 3y = 0 \end{cases}$

40. $\begin{cases} 3x + 4y = 10 \\ 7x + 9y = 20 \end{cases}$

41. $\begin{cases} 2x + 5y = 2 \\ -5x - 13y = 20 \end{cases}$

42. $\begin{cases} -7x + 4y = 0 \\ 8x - 5y = 100 \end{cases}$

43. $\begin{cases} 2x + 4y + z = 7 \\ -x + y - z = 0 \\ x + 4y = -2 \end{cases}$

44. $\begin{cases} 5x + 7y + 4z = 1 \\ 3x - y + 3z = 1 \\ 6x + 7y + 5z = 1 \end{cases}$

45. $\begin{cases} -2y + 2z = 12 \\ 3x + y + 3z = -2 \\ x - 2y + 3z = 8 \end{cases}$

46. $\begin{cases} x + 2y + 3w = 0 \\ y + z + w = 1 \\ y + w = 2 \\ x + 2y + 2w = 3 \end{cases}$

47–52 ■ Solving a Linear System Solve the system of equations by converting to a matrix equation. Use a graphing calculator to perform the necessary matrix operations, as in Example 7.

47. $\begin{cases} x + y - 2z = 3 \\ 2x + 5z = 11 \\ 2x + 3y = 12 \end{cases}$

48. $\begin{cases} 3x + 4y - z = 2 \\ 2x - 3y + z = -5 \\ 5x - 2y + 2z = -3 \end{cases}$

49. $\begin{cases} 12x + \frac{1}{2}y - 7z = 21 \\ 11x - 2y + 3z = 43 \\ 13x + y - 4z = 29 \end{cases}$

50. $\begin{cases} x + \frac{1}{2}y - \frac{1}{3}z = 4 \\ x - \frac{1}{4}y + \frac{1}{6}z = 7 \\ x + y - z = -6 \end{cases}$

51. $\begin{cases} x + y - 3w = 0 \\ x - 2z = 8 \\ 2y - z + w = 5 \\ 2x + 3y - 2w = 13 \end{cases}$

52. $\begin{cases} x + y + z + w = 15 \\ x - y + z - w = 5 \\ x + 2y + 3z + 4w = 26 \\ x - 2y + 3z - 4w = 2 \end{cases}$

SKILLS Plus

53–54 ■ Solving a Matrix Equation Solve the matrix equation by multiplying each side by the appropriate inverse matrix.

53. $\begin{bmatrix} 3 & -2 \\ -4 & 3 \end{bmatrix} \begin{bmatrix} x & y & z \\ u & v & w \end{bmatrix} = \begin{bmatrix} 1 & 0 & -1 \\ 2 & 1 & 3 \end{bmatrix}$

54. $\begin{bmatrix} 0 & -2 & 2 \\ 3 & 1 & 3 \\ 1 & -2 & 3 \end{bmatrix} \begin{bmatrix} x & u \\ y & v \\ z & w \end{bmatrix} = \begin{bmatrix} 3 & 6 \\ 6 & 12 \\ 0 & 0 \end{bmatrix}$

55–56 ■ Inverses of Special Matrices Find the inverse of the matrix.

55. $\begin{bmatrix} a & -a \\ a & a \end{bmatrix}$ $(a \neq 0)$

56. $\begin{bmatrix} a & 0 & 0 & 0 \\ 0 & b & 0 & 0 \\ 0 & 0 & c & 0 \\ 0 & 0 & 0 & d \end{bmatrix}$ $(abcd \neq 0)$

57–60 ■ When Do Matrices Have Inverses? Find the inverse of the matrix. For what value(s) of x, if any, does the matrix have no inverse?

57. $\begin{bmatrix} 2 & x \\ x & x^2 \end{bmatrix}$

58. $\begin{bmatrix} e^x & -e^{2x} \\ e^{2x} & e^{3x} \end{bmatrix}$

59. $\begin{bmatrix} 1 & e^x & 0 \\ e^x & -e^{2x} & 0 \\ 0 & 0 & 2 \end{bmatrix}$

60. $\begin{bmatrix} x & 1 \\ -x & \frac{1}{x-1} \end{bmatrix}$

APPLICATIONS

61. Nutrition A nutritionist is studying the effects of the nutrients folic acid, choline, and inositol. He has three types of food available, and each type contains the following amounts of these nutrients per ounce.

	Type A	Type B	Type C
Folic acid (mg)	3	1	3
Choline (mg)	4	2	4
Inositol (mg)	3	2	4

(a) Find the inverse of the matrix

$$\begin{bmatrix} 3 & 1 & 3 \\ 4 & 2 & 4 \\ 3 & 2 & 4 \end{bmatrix}$$

and use it to solve the remaining parts of this problem.

(b) How many ounces of each food should the nutritionist feed his laboratory rats if he wants their daily diet to contain 10 mg of folic acid, 14 mg of choline, and 13 mg of inositol?

(c) How much of each food is needed to supply 9 mg of folic acid, 12 mg of choline, and 10 mg of inositol?

(d) Will any combination of these foods supply 2 mg of folic acid, 4 mg of choline, and 11 mg of inositol?

62. Nutrition Refer to Exercise 61. Suppose food type C has been improperly labeled, and it actually contains 4 mg of folic acid, 6 mg of choline, and 5 mg of inositol per ounce. Would it still be possible to use matrix inversion to solve parts (b), (c), and (d) of Exercise 61? Why or why not?

63. Sales Commissions A saleswoman works at a kiosk that offers three different models of cell phones: standard with 16 GB capacity, deluxe with 32 GB capacity, and super-deluxe with 64 GB capacity. For each phone that she sells, she earns a commission based on the cell phone model. One week she sells 9 standard, 11 deluxe, and 8 super-deluxe and makes \$740 in commission. The next week she sells 13 standard, 15 deluxe, and 16 super-deluxe for a \$1204 commission. The third week she sells 8 standard, 7 deluxe, and 14 super-deluxe, earning \$828 in commission.

(a) Let x, y, and z represent the commission she earns on standard, deluxe, and super-deluxe, respectively. Translate the given information into a system of equations in x, y, and z.

(b) Express the system of equations you found in part (a) as a matrix equation of the form $AX = B$.

(c) Find the inverse of the coefficient matrix A and use it to solve the matrix equation in part (b). How much commission does the saleswoman earn on each model of cell phone?

DISCUSS ■ DISCOVER ■ PROVE ■ WRITE

64. DISCUSS: No Zero-Product Property for Matrices We have used the Zero-Product Property to solve algebraic equations. Matrices do *not* have this property. Let O represent the **2 × 2 zero matrix**

$$O = \begin{bmatrix} 0 & 0 \\ 0 & 0 \end{bmatrix}$$

Find 2×2 matrices $A \neq O$ and $B \neq O$ such that $AB = O$. Can you find a matrix $A \neq O$ such that $A^2 = O$?

10.6 DETERMINANTS AND CRAMER'S RULE

■ Determinant of a 2 × 2 Matrix ■ Determinant of an *n* × *n* Matrix ■ Row and Column Transformations ■ Cramer's Rule ■ Areas of Triangles Using Determinants

If a matrix is **square** (that is, if it has the same number of rows as columns), then we can assign to it a number called its *determinant*. Determinants can be used to solve systems of linear equations, as we will see later in this section. They are also useful in determining whether a matrix has an inverse.

Determinant of a 2 × 2 Matrix

We denote the determinant of a square matrix A by the symbol $\det(A)$ or $|A|$. We first define $\det(A)$ for the simplest cases. If $A = [a]$ is a 1×1 matrix, then $\det(A) = a$. The following box gives the definition of a 2×2 determinant.

We will use both notations, $\det(A)$ and $|A|$, for the determinant of A. Although the symbol $|A|$ looks like the absolute value symbol, it will be clear from the context which meaning is intended.

DETERMINANT OF A 2 × 2 MATRIX

The **determinant** of the 2×2 matrix $A = \begin{bmatrix} a & b \\ c & d \end{bmatrix}$ is

$$\det(A) = |A| = \begin{vmatrix} a & b \\ c & d \end{vmatrix} = ad - bc$$

EXAMPLE 1 ■ Determinant of a 2 × 2 Matrix

Evaluate $|A|$ for $A = \begin{bmatrix} 6 & -3 \\ 2 & 3 \end{bmatrix}$.

SOLUTION

$$\begin{vmatrix} 6 & -3 \\ 2 & 3 \end{vmatrix} = 6 \cdot 3 - (-3)2 = 18 - (-6) = 24$$

To evaluate a 2×2 determinant, we take the product of the diagonal from top left to bottom right and subtract the product from top right to bottom left, as indicated by the arrows.

Now Try Exercise 5 ■

Determinant of an *n* × *n* Matrix

To define the concept of determinant for an arbitrary $n \times n$ matrix, we need the following terminology.

MINORS AND COFACTORS

Let A be an $n \times n$ matrix.

1. The **minor** M_{ij} of the element a_{ij} is the determinant of the matrix obtained by deleting the ith row and jth column of A.
2. The **cofactor** A_{ij} of the element a_{ij} is

$$A_{ij} = (-1)^{i+j} M_{ij}$$

For example, if A is the matrix

$$\begin{bmatrix} 2 & 3 & -1 \\ 0 & 2 & 4 \\ -2 & 5 & 6 \end{bmatrix}$$

DAVID HILBERT (1862–1943) was born in Königsberg, Germany, and became a professor at Göttingen University. He is considered by many to be the greatest mathematician of the 20th century. At the International Congress of Mathematicians held in Paris in 1900, Hilbert set the direction of mathematics for the about-to-dawn 20th century by posing 23 problems that he believed to be of crucial importance. He said that "these are problems whose solutions we expect from the future." Most of Hilbert's problems have now been solved (see Julia Robinson, page 713, and Alan Turing, page 118), and their solutions have led to important new areas of mathematical research. Yet as we proceed into the new millennium, some of Hilbert's problems remain unsolved. In his work, Hilbert emphasized structure, logic, and the foundations of mathematics. Part of his genius lay in his ability to see the most general possible statement of a problem. For instance, Euler proved that every whole number is the sum of four squares; Hilbert proved a similar statement for all powers of positive integers.

then the minor M_{12} is the determinant of the matrix obtained by deleting the first row and second column from A. Thus

$$M_{12} = \begin{vmatrix} 2 & 3 & -1 \\ 0 & 2 & 4 \\ -2 & 5 & 6 \end{vmatrix} = \begin{vmatrix} 0 & 4 \\ -2 & 6 \end{vmatrix} = 0(6) - 4(-2) = 8$$

So the cofactor $A_{12} = (-1)^{1+2}M_{12} = -8$. Similarly,

$$M_{33} = \begin{vmatrix} 2 & 3 & -1 \\ 0 & 2 & 4 \\ -2 & 5 & 6 \end{vmatrix} = \begin{vmatrix} 2 & 3 \\ 0 & 2 \end{vmatrix} = 2 \cdot 2 - 3 \cdot 0 = 4$$

So $A_{33} = (-1)^{3+3}M_{33} = 4$.

Note that the cofactor of a_{ij} is simply the minor of a_{ij} multiplied by either 1 or -1, depending on whether $i + j$ is even or odd. Thus in a 3×3 matrix we obtain the cofactor of any element by prefixing its minor with the sign obtained from the following checkerboard pattern.

$$\begin{bmatrix} + & - & + \\ - & + & - \\ + & - & + \end{bmatrix}$$

We are now ready to define the determinant of any square matrix.

THE DETERMINANT OF A SQUARE MATRIX

If A is an $n \times n$ matrix, then the **determinant** of A is obtained by multiplying each element of the first row by its cofactor and then adding the results. In symbols,

$$\det(A) = |A| = \begin{vmatrix} a_{11} & a_{12} & \cdots & a_{1n} \\ a_{21} & a_{22} & \cdots & a_{2n} \\ \vdots & \vdots & \ddots & \vdots \\ a_{n1} & a_{n2} & \cdots & a_{nn} \end{vmatrix} = a_{11}A_{11} + a_{12}A_{12} + \cdots + a_{1n}A_{1n}$$

EXAMPLE 2 ■ Determinant of a 3 × 3 Matrix

Evaluate the determinant of the matrix

$$A = \begin{bmatrix} 2 & 3 & -1 \\ 0 & 2 & 4 \\ -2 & 5 & 6 \end{bmatrix}$$

SOLUTION

$$\begin{aligned} \det(A) = \begin{vmatrix} 2 & 3 & -1 \\ 0 & 2 & 4 \\ -2 & 5 & 6 \end{vmatrix} &= 2\begin{vmatrix} 2 & 4 \\ 5 & 6 \end{vmatrix} - 3\begin{vmatrix} 0 & 4 \\ -2 & 6 \end{vmatrix} + (-1)\begin{vmatrix} 0 & 2 \\ -2 & 5 \end{vmatrix} \\ &= 2(2 \cdot 6 - 4 \cdot 5) - 3[0 \cdot 6 - 4(-2)] - [0 \cdot 5 - 2(-2)] \\ &= -16 - 24 - 4 \\ &= -44 \end{aligned}$$

Now Try Exercises 21 and 29 ■

In our definition of the determinant we used the cofactors of elements in the first row only. This is called **expanding the determinant by the first row**. In fact, *we can expand the determinant by any row or column in the same way and obtain the same result in each case* (although we won't prove this). The next example illustrates this principle.

EXAMPLE 3 ■ Expanding a Determinant About a Row and a Column

Let A be the matrix of Example 2. Evaluate the determinant of A by expanding

(a) by the second row

(b) by the third column

Verify that each expansion gives the same value.

SOLUTION

(a) Expanding by the second row, we get

$$\det(A) = \begin{vmatrix} 2 & 3 & -1 \\ 0 & 2 & 4 \\ -2 & 5 & 6 \end{vmatrix} = -0\begin{vmatrix} 3 & -1 \\ 5 & 6 \end{vmatrix} + 2\begin{vmatrix} 2 & -1 \\ -2 & 6 \end{vmatrix} - 4\begin{vmatrix} 2 & 3 \\ -2 & 5 \end{vmatrix}$$

$$= 0 + 2[2\cdot 6 - (-1)(-2)] - 4[2\cdot 5 - 3(-2)]$$

$$= 0 + 20 - 64 = -44$$

(b) Expanding by the third column gives

$$\det(A) = \begin{vmatrix} 2 & 3 & -1 \\ 0 & 2 & 4 \\ -2 & 5 & 6 \end{vmatrix}$$

$$= -1\begin{vmatrix} 0 & 2 \\ -2 & 5 \end{vmatrix} - 4\begin{vmatrix} 2 & 3 \\ -2 & 5 \end{vmatrix} + 6\begin{vmatrix} 2 & 3 \\ 0 & 2 \end{vmatrix}$$

$$= -[0\cdot 5 - 2(-2)] - 4[2\cdot 5 - 3(-2)] + 6(2\cdot 2 - 3\cdot 0)$$

$$= -4 - 64 + 24 = -44$$

In both cases we obtain the same value for the determinant as when we expanded by the first row in Example 2.

We can also use a graphing calculator to compute determinants, as shown in Figure 1.

See Appendix D, *Using the TI-83/84 Graphing Calculator*, for specific instructions on calculating determinants. Go to **www.stewartmath.com**.

Here is the output when the TI-83 is used to calculate the determinant in Example 3:

```
[A]
        [[2  3 -1]
         [0  2  4 ]
         [-2 5  6 ]]
det([A])
                -44
```

FIGURE 1

Now Try Exercise 39 ■

The following criterion allows us to determine whether a square matrix has an inverse without actually calculating the inverse. This is one of the most important uses of the determinant in matrix algebra, and it is the reason for the name *determinant*.

INVERTIBILITY CRITERION

If A is a square matrix, then A has an inverse if and only if $\det(A) \neq 0$.

We will not prove this fact, but from the formula for the inverse of a 2×2 matrix (page 725) you can see why it is true in the 2×2 case.

Pictorial Parade/Hulton Archive/Getty Images

EMMY NOETHER (1882–1935) was one of the foremost mathematicians of the early 20th century. Her groundbreaking work in abstract algebra provided much of the foundation for this field, and her work in invariant theory was essential in the development of Einstein's theory of general relativity. Although women weren't allowed to study at German universities at that time, she audited courses unofficially and went on to receive a doctorate at Erlangen *summa cum laude*, despite the opposition of the academic senate, which declared that women students would "overthrow all academic order." She subsequently taught mathematics at Göttingen, Moscow, and Frankfurt. In 1933 she left Germany to escape Nazi persecution, accepting a position at Bryn Mawr College in suburban Philadelphia. She lectured there and at the Institute for Advanced Study in Princeton, New Jersey, until her untimely death in 1935.

EXAMPLE 4 ■ Using the Determinant to Show That a Matrix Is Not Invertible

Show that the matrix A has no inverse.

$$A = \begin{bmatrix} 1 & 2 & 0 & 4 \\ 0 & 0 & 0 & 3 \\ 5 & 6 & 2 & 6 \\ 2 & 4 & 0 & 9 \end{bmatrix}$$

SOLUTION We begin by calculating the determinant of A. Since all but one of the elements of the second row is zero, we expand the determinant by the second row. If we do this, we see from the following equation that only the cofactor A_{24} will have to be calculated.

$$\begin{aligned} \det(A) &= \begin{vmatrix} 1 & 2 & 0 & 4 \\ 0 & 0 & 0 & 3 \\ 5 & 6 & 2 & 6 \\ 2 & 4 & 0 & 9 \end{vmatrix} \\ &= -0 \cdot A_{21} + 0 \cdot A_{22} - 0 \cdot A_{23} + 3 \cdot A_{24} = 3A_{24} \\ &= 3\begin{vmatrix} 1 & 2 & 0 \\ 5 & 6 & 2 \\ 2 & 4 & 0 \end{vmatrix} \qquad \text{Expand this by column 3} \\ &= 3(-2)\begin{vmatrix} 1 & 2 \\ 2 & 4 \end{vmatrix} \\ &= 3(-2)(1 \cdot 4 - 2 \cdot 2) = 0 \end{aligned}$$

Since the determinant of A is zero, A cannot have an inverse, by the Invertibility Criterion.

Now Try Exercise 25 ■

Row and Column Transformations

The preceding example shows that if we expand a determinant about a row or column that contains many zeros, our work is reduced considerably because we don't have to evaluate the cofactors of the elements that are zero. The following principle often simplifies the process of finding a determinant by introducing zeros into the matrix without changing the value of the determinant.

ROW AND COLUMN TRANSFORMATIONS OF A DETERMINANT

If A is a square matrix and if the matrix B is obtained from A by adding a multiple of one row to another or a multiple of one column to another, then $\det(A) = \det(B)$.

EXAMPLE 5 ■ Using Row and Column Transformations to Calculate a Determinant

Find the determinant of the matrix A. Does it have an inverse?

$$A = \begin{bmatrix} 8 & 2 & -1 & -4 \\ 3 & 5 & -3 & 11 \\ 24 & 6 & 1 & -12 \\ 2 & 2 & 7 & -1 \end{bmatrix}$$

SOLUTION If we add -3 times row 1 to row 3, we change all but one element of row 3 to zeros.

$$\begin{bmatrix} 8 & 2 & -1 & -4 \\ 3 & 5 & -3 & 11 \\ 0 & 0 & 4 & 0 \\ 2 & 2 & 7 & -1 \end{bmatrix}$$

This new matrix has the same determinant as A, and if we expand its determinant by the third row, we get

$$\det(A) = 4\begin{vmatrix} 8 & 2 & -4 \\ 3 & 5 & 11 \\ 2 & 2 & -1 \end{vmatrix}$$

Now, adding 2 times column 3 to column 1 in this determinant gives us

$$\det(A) = 4\begin{vmatrix} 0 & 2 & -4 \\ 25 & 5 & 11 \\ 0 & 2 & -1 \end{vmatrix} \qquad \text{Expand this by column 1}$$

$$= 4(-25)\begin{vmatrix} 2 & -4 \\ 2 & -1 \end{vmatrix}$$

$$= 4(-25)[2(-1) - (-4)2] = -600$$

Since the determinant of A is not zero, A does have an inverse.

Now Try Exercise 35 ■

Cramer's Rule

The solutions of linear equations can sometimes be expressed by using determinants. To illustrate, let's solve the following pair of linear equations for the variable x.

$$\begin{cases} ax + by = r \\ cx + dy = s \end{cases}$$

To eliminate the variable y, we multiply the first equation by d and the second by b and subtract.

$$\begin{array}{r} adx + bdy = rd \\ \underline{bcx + bdy = bs} \\ adx - bcx = rd - bs \end{array}$$

DISCOVERY PROJECT

Computer Graphics I

Matrix algebra is the basic tool used in computer graphics. Properties of each pixel in an image are stored in a large matrix in the computer memory. In this project we discover how matrix multiplication can be used to "move" a point in the plane to a prescribed location. Combining such moves for each pixel in an image enables us to stretch, compress, translate, and otherwise transform an image on a computer screen by using matrix algebra. You can find the project at **www.stewartmath.com**.

Factoring the left-hand side, we get $(ad - bc)x = rd - bs$. Assuming that $ad - bc \neq 0$, we can now solve this equation for x:

$$x = \frac{rd - bs}{ad - bc}$$

Similarly, we find

$$y = \frac{as - cr}{ad - bc}$$

The numerator and denominator of the fractions for x and y are determinants of 2×2 matrices. So we can express the solution of the system using determinants as follows.

CRAMER'S RULE FOR SYSTEMS IN TWO VARIABLES

The linear system

$$\begin{cases} ax + by = r \\ cx + dy = s \end{cases}$$

has the solution

$$x = \frac{\begin{vmatrix} r & b \\ s & d \end{vmatrix}}{\begin{vmatrix} a & b \\ c & d \end{vmatrix}} \qquad y = \frac{\begin{vmatrix} a & r \\ c & s \end{vmatrix}}{\begin{vmatrix} a & b \\ c & d \end{vmatrix}}$$

provided that $\begin{vmatrix} a & b \\ c & d \end{vmatrix} \neq 0$.

Using the notation

$$D = \begin{bmatrix} a & b \\ c & d \end{bmatrix} \qquad D_x = \begin{bmatrix} r & b \\ s & d \end{bmatrix} \qquad D_y = \begin{bmatrix} a & r \\ c & s \end{bmatrix}$$

Coefficient matrix

Replace first column of D by r and s

Replace second column of D by r and s

we can write the solution of the system as

$$x = \frac{|D_x|}{|D|} \quad \text{and} \quad y = \frac{|D_y|}{|D|}$$

EXAMPLE 6 ■ Using Cramer's Rule to Solve a System with Two Variables

Use Cramer's Rule to solve the system.

$$\begin{cases} 2x + 6y = -1 \\ x + 8y = 2 \end{cases}$$

SOLUTION For this system we have

$$|D| = \begin{vmatrix} 2 & 6 \\ 1 & 8 \end{vmatrix} = 2 \cdot 8 - 6 \cdot 1 = 10$$

$$|D_x| = \begin{vmatrix} -1 & 6 \\ 2 & 8 \end{vmatrix} = (-1)8 - 6 \cdot 2 = -20$$

$$|D_y| = \begin{vmatrix} 2 & -1 \\ 1 & 2 \end{vmatrix} = 2 \cdot 2 - (-1)1 = 5$$

The solution is

$$x = \frac{|D_x|}{|D|} = \frac{-20}{10} = -2$$

$$y = \frac{|D_y|}{|D|} = \frac{5}{10} = \frac{1}{2}$$

Now Try Exercise 41 ■

Cramer's Rule can be extended to apply to any system of n linear equations in n variables in which the determinant of the coefficient matrix is not zero. As we saw in the preceding section, any such system can be written in matrix form as

$$\begin{bmatrix} a_{11} & a_{12} & \cdots & a_{1n} \\ a_{21} & a_{22} & \cdots & a_{2n} \\ \vdots & \vdots & \ddots & \vdots \\ a_{n1} & a_{n2} & \cdots & a_{nn} \end{bmatrix} \begin{bmatrix} x_1 \\ x_2 \\ \vdots \\ x_n \end{bmatrix} = \begin{bmatrix} b_1 \\ b_2 \\ \vdots \\ b_n \end{bmatrix}$$

By analogy with our derivation of Cramer's Rule in the case of two equations in two unknowns, we let D be the coefficient matrix in this system, and D_{x_i} be the matrix obtained by replacing the ith column of D by the numbers $b_1, b_2, \ldots, b_n$ that appear to the right of the equal sign. The solution of the system is then given by the following rule.

CRAMER'S RULE

If a system of n linear equations in the n variables $x_1, x_2, \ldots, x_n$ is equivalent to the matrix equation $DX = B$, and if $|D| \neq 0$, then its solutions are

$$x_1 = \frac{|D_{x_1}|}{|D|} \qquad x_2 = \frac{|D_{x_2}|}{|D|} \qquad \cdots \qquad x_n = \frac{|D_{x_n}|}{|D|}$$

where D_{x_i} is the matrix obtained by replacing the ith column of D by the $n \times 1$ matrix B.

EXAMPLE 7 ■ Using Cramer's Rule to Solve a System with Three Variables

Use Cramer's Rule to solve the system.

$$\begin{cases} 2x - 3y + 4z = 1 \\ x \qquad\quad + 6z = 0 \\ 3x - 2y \qquad\quad = 5 \end{cases}$$

SOLUTION First, we evaluate the determinants that appear in Cramer's Rule. Note that D is the coefficient matrix and that D_x, D_y, and D_z are obtained by replacing the first, second, and third columns of D by the constant terms.

$$|D| = \begin{vmatrix} 2 & -3 & 4 \\ 1 & 0 & 6 \\ 3 & -2 & 0 \end{vmatrix} = -38 \qquad |D_x| = \begin{vmatrix} 1 & -3 & 4 \\ 0 & 0 & 6 \\ 5 & -2 & 0 \end{vmatrix} = -78$$

$$|D_y| = \begin{vmatrix} 2 & 1 & 4 \\ 1 & 0 & 6 \\ 3 & 5 & 0 \end{vmatrix} = -22 \qquad |D_z| = \begin{vmatrix} 2 & -3 & 1 \\ 1 & 0 & 0 \\ 3 & -2 & 5 \end{vmatrix} = 13$$

Now we use Cramer's Rule to get the solution:

$$x = \frac{|D_x|}{|D|} = \frac{-78}{-38} = \frac{39}{19} \qquad y = \frac{|D_y|}{|D|} = \frac{-22}{-38} = \frac{11}{19}$$

$$z = \frac{|D_z|}{|D|} = \frac{13}{-38} = -\frac{13}{38}$$

Now Try Exercise 47 ■

Solving the system in Example 7 using Gaussian elimination would involve matrices whose elements are fractions with fairly large denominators. Thus in cases like Examples 6 and 7, Cramer's Rule gives us an efficient way to solve systems of linear equations. But in systems with more than three equations, evaluating the various determinants that are involved is usually a long and tedious task (unless you are using a graphing calculator). Moreover, the rule doesn't apply if $|D| = 0$ or if D is not a square matrix. So Cramer's Rule is a useful alternative to Gaussian elimination, but only in some situations.

Areas of Triangles Using Determinants

Determinants provide a simple way to calculate the area of a triangle in the coordinate plane.

AREA OF A TRIANGLE

If a triangle in the coordinate plane has vertices (a_1, b_1), (a_2, b_2), and (a_3, b_3), then its area is

$$\mathcal{A} = \pm\frac{1}{2}\begin{vmatrix} a_1 & b_1 & 1 \\ a_2 & b_2 & 1 \\ a_3 & b_3 & 1 \end{vmatrix}$$

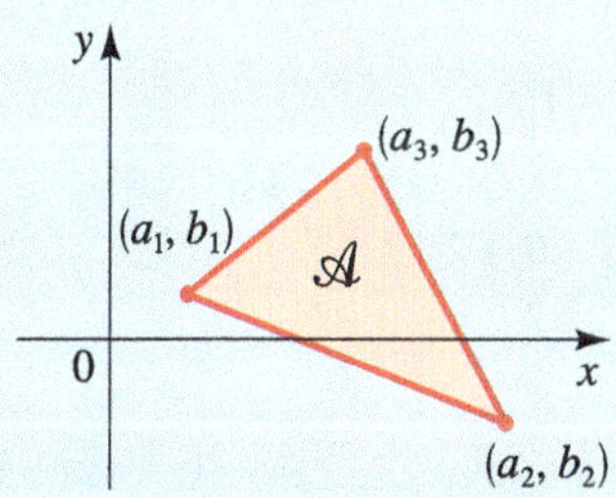

where the sign is chosen to make the area positive.

You are asked to prove this formula in Exercise 74.

EXAMPLE 8 ■ Area of a Triangle

Find the area of the triangle shown in Figure 2.

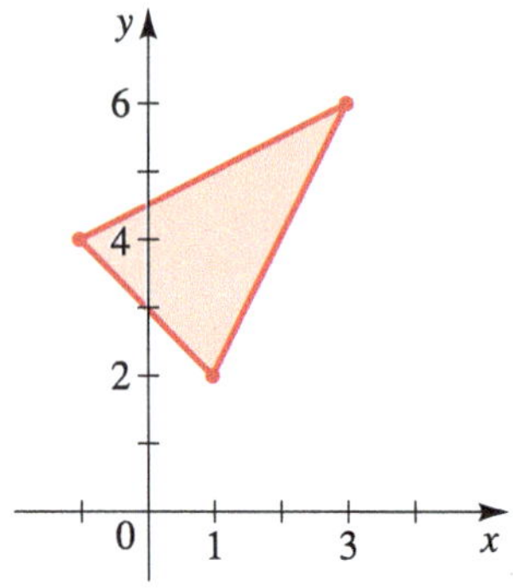

FIGURE 2

We can calculate the determinant by hand or by using a graphing calculator.

```
[A]
        [[-1  4  1]
         [3   6  1]
         [1   2  1]]
det([A])
                -12
```

SOLUTION The vertices are (1, 2), (3, 6), and (−1, 4). Using the formula in the preceding box, we get

$$\mathscr{A} = \pm\tfrac{1}{2}\begin{vmatrix} -1 & 4 & 1 \\ 3 & 6 & 1 \\ 1 & 2 & 1 \end{vmatrix} = \pm\tfrac{1}{2}(-12)$$

To make the area positive, we choose the negative sign in the formula. Thus the area of the triangle is

$$\mathscr{A} = -\tfrac{1}{2}(-12) = 6$$

Now Try Exercise 57 ■

10.6 EXERCISES

CONCEPTS

1. *True or false?* det(A) is defined only for a square matrix A.
2. *True or false?* det(A) is a number, not a matrix.
3. *True or false?* If det(A) = 0, then A is not invertible.
4. Fill in the blanks with appropriate numbers to calculate the determinant. Where there is "±", choose the appropriate sign (+ or −).

(a) $\begin{vmatrix} 2 & 1 \\ -3 & 4 \end{vmatrix} = \square\square - \square\square = $ ______

(b) $\begin{vmatrix} 1 & 0 & 2 \\ 3 & 2 & 1 \\ 0 & -3 & 4 \end{vmatrix} = \pm\square(\square\square - \square\square) \pm \square(\square\square - \square\square)$

$\pm\square(\square\square - \square\square) = $ ______

SKILLS

5–14 ■ Finding Determinants Find the determinant of the matrix, if it exists.

5. $\begin{bmatrix} 2 & 0 \\ 0 & 3 \end{bmatrix}$

6. $\begin{bmatrix} 0 & -1 \\ 2 & 0 \end{bmatrix}$

7. $\begin{bmatrix} \frac{3}{2} & 1 \\ -1 & -\frac{2}{3} \end{bmatrix}$

8. $\begin{bmatrix} 0.2 & 0.4 \\ -0.4 & -0.8 \end{bmatrix}$

9. $\begin{bmatrix} 4 & 5 \\ 0 & -1 \end{bmatrix}$

10. $\begin{bmatrix} -2 & 1 \\ 3 & -2 \end{bmatrix}$

11. $\begin{bmatrix} 2 & 5 \end{bmatrix}$

12. $\begin{bmatrix} 3 \\ 0 \end{bmatrix}$

13. $\begin{bmatrix} \frac{1}{2} & \frac{1}{8} \\ 1 & \frac{1}{2} \end{bmatrix}$

14. $\begin{bmatrix} 2.2 & -1.4 \\ 0.5 & 1.0 \end{bmatrix}$

15–20 ■ Minors and Cofactors Evaluate the minor and cofactor using the matrix A.

$$A = \begin{bmatrix} 1 & 0 & \frac{1}{2} \\ -3 & 5 & 2 \\ 0 & 0 & 4 \end{bmatrix}$$

15. M_{11}, A_{11}

16. M_{33}, A_{33}

17. M_{12}, A_{12}

18. M_{13}, A_{13}

19. M_{23}, A_{23}

20. M_{32}, A_{32}

21–28 ■ Finding Determinants Find the determinant of the matrix. Determine whether the matrix has an inverse, but don't calculate the inverse.

21. $\begin{bmatrix} 2 & 1 & 0 \\ 0 & -2 & 4 \\ 0 & 1 & -3 \end{bmatrix}$ **22.** $\begin{bmatrix} 1 & 2 & 5 \\ -2 & -3 & 2 \\ 3 & 5 & 3 \end{bmatrix}$

23. $\begin{bmatrix} 30 & 0 & 20 \\ 0 & -10 & -20 \\ 40 & 0 & 10 \end{bmatrix}$ **24.** $\begin{bmatrix} -2 & -\frac{3}{2} & \frac{1}{2} \\ 2 & 4 & 0 \\ \frac{1}{2} & 2 & 1 \end{bmatrix}$

25. $\begin{bmatrix} 1 & 3 & 7 \\ 2 & 0 & 8 \\ 0 & 2 & 2 \end{bmatrix}$ **26.** $\begin{bmatrix} 0 & -1 & 0 \\ 2 & 6 & 4 \\ 1 & 0 & 3 \end{bmatrix}$

27. $\begin{bmatrix} 1 & 3 & 3 & 0 \\ 0 & 2 & 0 & 1 \\ -1 & 0 & 0 & 2 \\ 1 & 6 & 4 & 1 \end{bmatrix}$ **28.** $\begin{bmatrix} 1 & 2 & 0 & 2 \\ 3 & -4 & 0 & 4 \\ 0 & 1 & 6 & 0 \\ 1 & 0 & 2 & 0 \end{bmatrix}$

29–34 ■ Finding Determinants Use a graphing calculator to find the determinant of the matrix. Determine whether the matrix has an inverse, but don't calculate the inverse.

29. $\begin{bmatrix} 1 & 2 & -1 \\ 2 & 2 & 1 \\ 1 & 2 & 2 \end{bmatrix}$

30. $\begin{bmatrix} 10 & -20 & 31 \\ 10 & -11 & 45 \\ -20 & 40 & -50 \end{bmatrix}$

31. $\begin{bmatrix} 1 & 10 & 2 & 7 \\ 2 & 18 & 18 & 13 \\ -3 & -30 & -4 & -24 \\ 1 & 10 & 2 & 10 \end{bmatrix}$

32. $\begin{bmatrix} 1 & 3 & -2 & 5 \\ -3 & -9 & 11 & 5 \\ 2 & 6 & 0 & 31 \\ 5 & 15 & -10 & 39 \end{bmatrix}$

33. $\begin{bmatrix} 4 & 3 & -2 & 10 \\ -8 & -6 & 24 & -1 \\ 20 & 15 & 3 & 27 \\ 12 & 9 & -6 & -1 \end{bmatrix}$

34. $\begin{bmatrix} 2 & 3 & -5 & 10 \\ -2 & -2 & 26 & 3 \\ 6 & 9 & -16 & 45 \\ -8 & -12 & 20 & -36 \end{bmatrix}$

35–38 ■ Determinants Using Row and Column Operations Evaluate the determinant, using row or column operations whenever possible to simplify your work.

35. $\begin{vmatrix} 0 & 0 & 4 & 6 \\ 2 & 1 & 1 & 3 \\ 2 & 1 & 2 & 3 \\ 3 & 0 & 1 & 7 \end{vmatrix}$ **36.** $\begin{vmatrix} -2 & 3 & -1 & 7 \\ 4 & 6 & -2 & 3 \\ 7 & 7 & 0 & 5 \\ 3 & -12 & 4 & 0 \end{vmatrix}$

37. $\begin{vmatrix} 1 & 2 & 3 & 4 & 5 \\ 0 & 2 & 4 & 6 & 8 \\ 0 & 0 & 3 & 6 & 9 \\ 0 & 0 & 0 & 4 & 8 \\ 0 & 0 & 0 & 0 & 5 \end{vmatrix}$ **38.** $\begin{vmatrix} 2 & -1 & 6 & 4 \\ 7 & 2 & -2 & 5 \\ 4 & -2 & 10 & 8 \\ 6 & 1 & 1 & 4 \end{vmatrix}$

39. Calculating a Determinant in Different Ways Consider the matrix

$$B = \begin{bmatrix} 4 & 1 & 0 \\ -2 & -1 & 1 \\ 4 & 0 & 3 \end{bmatrix}$$

(a) Evaluate $\det(B)$ by expanding by the second row.

(b) Evaluate $\det(B)$ by expanding by the third column.

(c) Do your results in parts (a) and (b) agree?

40. Determinant of a Special Matrix Find the determinant of a 10×10 matrix which has a 2 in each main diagonal entry and zeros everywhere else.

41–56 ■ Cramer's Rule Use Cramer's Rule to solve the system.

41. $\begin{cases} 2x - y = -9 \\ x + 2y = 8 \end{cases}$ **42.** $\begin{cases} 6x + 12y = 33 \\ 4x + 7y = 20 \end{cases}$

43. $\begin{cases} x \quad 6y = 3 \\ 3x + 2y = 1 \end{cases}$ **44.** $\begin{cases} \frac{1}{2}x + \frac{1}{3}y = 1 \\ \frac{1}{4}x - \frac{1}{6}y = -\frac{3}{2} \end{cases}$

45. $\begin{cases} 0.4x + 1.2y = 0.4 \\ 1.2x + 1.6y = 3.2 \end{cases}$ **46.** $\begin{cases} 10x - 17y = 21 \\ 20x - 31y = 39 \end{cases}$

47. $\begin{cases} x - y + 2z = 0 \\ 3x + z = 11 \\ -x + 2y = 0 \end{cases}$ **48.** $\begin{cases} 5x - 3y + z = 6 \\ 4y - 6z = 22 \\ 7x + 10y = -13 \end{cases}$

49. $\begin{cases} 2x_1 + 3x_2 - 5x_3 = 1 \\ x_1 + x_2 - x_3 = 2 \\ 2x_2 + x_3 = 8 \end{cases}$ **50.** $\begin{cases} -2a + c = 2 \\ a + 2b - c = 9 \\ 3a + 5b + 2c = 22 \end{cases}$

51. $\begin{cases} \frac{1}{3}x - \frac{1}{5}y + \frac{1}{2}z = \frac{7}{10} \\ -\frac{2}{3}x + \frac{2}{5}y + \frac{3}{2}z = \frac{11}{10} \\ x - \frac{4}{5}y + z = \frac{9}{5} \end{cases}$ **52.** $\begin{cases} 2x - y = 5 \\ 5x + 3z = 19 \\ 4y + 7z = 17 \end{cases}$

53. $\begin{cases} 3y + 5z = 4 \\ 2x - z = 10 \\ 4x + 7y = 0 \end{cases}$ **54.** $\begin{cases} 2x - 5y = 4 \\ x + y - z = 8 \\ 3x + 5z = 0 \end{cases}$

55. $\begin{cases} x + y + z + w = 0 \\ 2x + w = 0 \\ y - z = 0 \\ x + 2z = 1 \end{cases}$ **56.** $\begin{cases} x + y = 1 \\ y + z = 2 \\ z + w = 3 \\ w - x = 4 \end{cases}$

57–60 ■ Area of a Triangle Sketch the triangle with the given vertices, and use a determinant to find its area.

57. $(0, 0), (6, 2), (3, 8)$ **58.** $(1, 0), (3, 5), (-2, 2)$

59. $(-1, 3), (2, 9), (5, -6)$ **60.** $(-2, 5), (7, 2), (3, -4)$

SKILLS Plus

61–62 ■ Determinants of Special Matrices Evaluate the determinants.

61. $\begin{vmatrix} a & 0 & 0 & 0 & 0 \\ 0 & b & 0 & 0 & 0 \\ 0 & 0 & c & 0 & 0 \\ 0 & 0 & 0 & d & 0 \\ 0 & 0 & 0 & 0 & e \end{vmatrix}$ **62.** $\begin{vmatrix} a & a & a & a & a \\ 0 & a & a & a & a \\ 0 & 0 & a & a & a \\ 0 & 0 & 0 & a & a \\ 0 & 0 & 0 & 0 & a \end{vmatrix}$

63–66 ■ Determinant Equations Solve for x.

63. $\begin{vmatrix} x & 12 & 13 \\ 0 & x-1 & 23 \\ 0 & 0 & x-2 \end{vmatrix} = 0$ **64.** $\begin{vmatrix} x & 1 & 1 \\ 1 & 1 & x \\ x & 1 & x \end{vmatrix} = 0$

65. $\begin{vmatrix} 1 & 0 & x \\ x^2 & 1 & 0 \\ x & 0 & 1 \end{vmatrix} = 0$ **66.** $\begin{vmatrix} a & b & x-a \\ x & x+b & x \\ 0 & 1 & 1 \end{vmatrix} = 0$

67. Using Determinants Show that

$$\begin{vmatrix} 1 & x & x^2 \\ 1 & y & y^2 \\ 1 & z & z^2 \end{vmatrix} = (x-y)(y-z)(z-x)$$

68. Number of Solutions of a Linear System Consider the system

$$\begin{cases} x + 2y + 6z = 5 \\ -3x - 6y + 5z = 8 \\ 2x + 6y + 9z = 7 \end{cases}$$

(a) Verify that $x = -1$, $y = 0$, $z = 1$ is a solution of the system.

(b) Find the determinant of the coefficient matrix.

(c) Without solving the system, determine whether there are any other solutions.

(d) Can Cramer's Rule be used to solve this system? Why or why not?

69. Collinear Points and Determinants

(a) If three points lie on a line, what is the area of the "triangle" that they determine? Use the answer to this question, together with the determinant formula for the area of a triangle, to explain why the points (a_1, b_1), (a_2, b_2), and (a_3, b_3) are collinear if and only if

$$\begin{vmatrix} a_1 & b_1 & 1 \\ a_2 & b_2 & 1 \\ a_3 & b_3 & 1 \end{vmatrix} = 0$$

(b) Use a determinant to check whether each set of points is collinear. Graph them to verify your answer.

(i) $(-6, 4), (2, 10), (6, 13)$

(ii) $(-5, 10), (2, 6), (15, -2)$

70. Determinant Form for the Equation of a Line

(a) Use the result of Exercise 69(a) to show that the equation of the line containing the points (x_1, y_1) and (x_2, y_2) is

$$\begin{vmatrix} x & y & 1 \\ x_1 & y_1 & 1 \\ x_2 & y_2 & 1 \end{vmatrix} = 0$$

(b) Use the result of part (a) to find an equation for the line containing the points $(20, 50)$ and $(-10, 25)$.

APPLICATIONS

71. Buying Fruit A roadside fruit stand sells apples at 75¢ a pound, peaches at 90¢ a pound, and pears at 60¢ a pound. Muriel buys 18 lb of fruit at a total cost of \$13.80. Her peaches and pears together cost \$1.80 more than her apples.

(a) Set up a linear system for the number of pounds of apples, peaches, and pears that she bought.

(b) Solve the system using Cramer's Rule.

72. The Arch of a Bridge The opening of a railway bridge over a roadway is in the shape of a parabola. A surveyor measures the heights of three points on the bridge, as shown in the figure. He wishes to find an equation of the form

$$y = ax^2 + bx + c$$

to model the shape of the arch.

(a) Use the surveyed points to set up a system of linear equations for the unknown coefficients a, b, and c.

(b) Solve the system using Cramer's Rule.

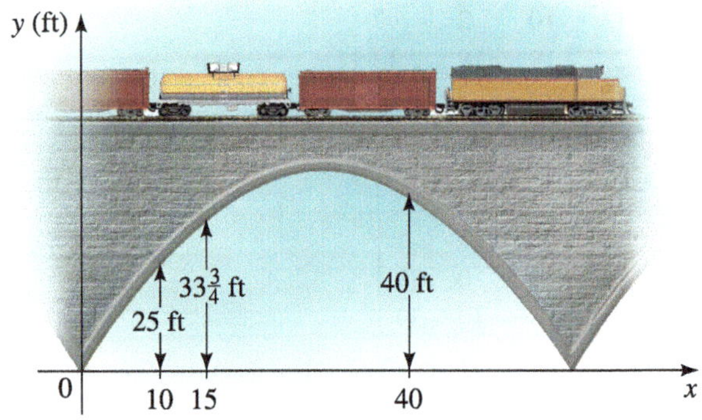

73. A Triangular Plot of Land An outdoors club is purchasing land to set up a conservation area. The last remaining piece they need to buy is the triangular plot shown in the figure. Use the determinant formula for the area of a triangle to find the area of the plot.

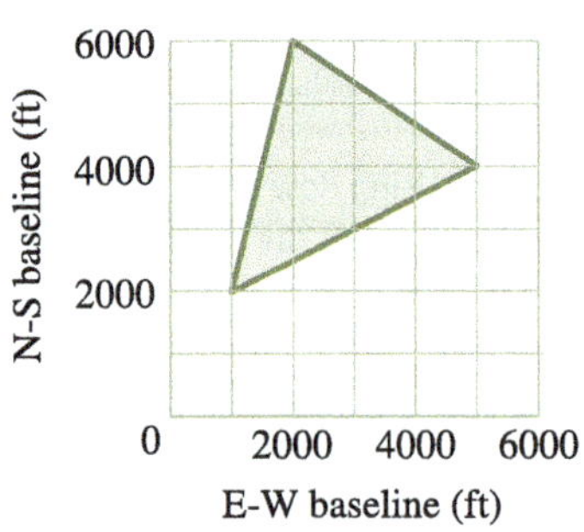

DISCUSS ■ DISCOVER ■ PROVE ■ WRITE

74. DISCOVER ■ PROVE: Determinant Formula for the Area of a Triangle The figure shows a triangle in the plane with vertices (a_1, b_1), (a_2, b_2), and (a_3, b_3).

(a) Find the coordinates of the vertices of the surrounding rectangle, and find its area.

(b) Find the area of the red triangle by subtracting the areas of the three blue triangles from the area of the rectangle.

(c) Use your answer to part (b) to show that the area $\mathcal{A}$ of the red triangle is given by

$$\mathcal{A} = \pm\frac{1}{2}\begin{vmatrix} a_1 & b_1 & 1 \\ a_2 & b_2 & 1 \\ a_3 & b_3 & 1 \end{vmatrix}$$

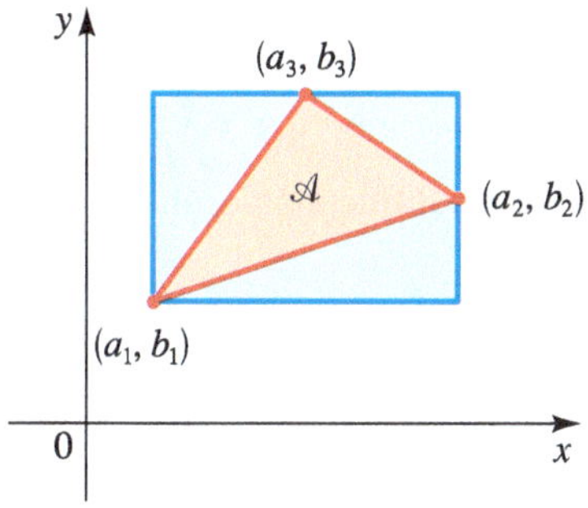

75. DISCUSS: Matrices with Determinant Zero Use the definition of determinant and the elementary row and column operations to explain why matrices of the following types have determinant 0.

(a) A matrix with a row or column consisting entirely of zeros

(b) A matrix with two rows the same or two columns the same

(c) A matrix in which one row is a multiple of another row, or one column is a multiple of another column

76. DISCUSS ■ WRITE: Solving Linear Systems Suppose you have to solve a linear system with five equations and five variables without the assistance of a calculator or computer. Which method would you prefer: Cramer's Rule or Gaussian elimination? Write a short paragraph explaining the reasons for your answer.

10.7 PARTIAL FRACTIONS

■ Distinct Linear Factors ■ Repeated Linear Factors ■ Irreducible Quadratic Factors ■ Repeated Irreducible Quadratic Factors

Common denominator →

$$\frac{1}{x-1} + \frac{1}{2x+1} = \frac{3x}{2x^2 - x - 1}$$

← Partial fractions

To write a sum or difference of fractional expressions as a single fraction, we bring them to a common denominator. For example,

$$\frac{1}{x-1} + \frac{1}{2x+1} = \frac{(2x+1)+(x-1)}{(x-1)(2x+1)} = \frac{3x}{2x^2 - x - 1}$$

But for some applications of algebra to calculus we must reverse this process—that is, we must express a fraction such as $3x/(2x^2 - x - 1)$ as the sum of the simpler fractions $1/(x-1)$ and $1/(2x+1)$. These simpler fractions are called *partial fractions*; we learn how to find them in this section.

Let r be the rational function

$$r(x) = \frac{P(x)}{Q(x)}$$

where the degree of P is less than the degree of Q. By the Linear and Quadratic Factors Theorem in Section 3.5, every polynomial with real coefficients can be factored completely into linear and irreducible quadratic factors, that is, factors of the form $ax + b$ and $ax^2 + bx + c$, where a, b, and c are real numbers. For instance,

$$x^4 - 1 = (x^2 - 1)(x^2 + 1) = (x-1)(x+1)(x^2+1)$$

After we have completely factored the denominator Q of r, we can express $r(x)$ as a sum of **partial fractions** of the form

$$\frac{A}{(ax+b)^i} \quad \text{and} \quad \frac{Ax+B}{(ax^2+bx+c)^j}$$

This sum is called the **partial fraction decomposition** of r. Let's examine the details of the four possible cases.

Distinct Linear Factors

We first consider the case in which the denominator factors into distinct linear factors.

> **CASE 1: THE DENOMINATOR IS A PRODUCT OF DISTINCT LINEAR FACTORS**
>
> Suppose that we can factor $Q(x)$ as
>
> $$Q(x) = (a_1x+b_1)(a_2x+b_2)\cdots(a_nx+b_n)$$
>
> with no factor repeated. In this case the partial fraction decomposition of $P(x)/Q(x)$ takes the form
>
> $$\frac{P(x)}{Q(x)} = \frac{A_1}{a_1x+b_1} + \frac{A_2}{a_2x+b_2} + \cdots + \frac{A_n}{a_nx+b_n}$$

The constants $A_1, A_2, \ldots, A_n$ are determined as in the following example.

EXAMPLE 1 ■ Distinct Linear Factors

Find the partial fraction decomposition of $\dfrac{5x+7}{x^3+2x^2-x-2}$.

SOLUTION The denominator factors as follows.

$$\begin{aligned} x^3+2x^2-x-2 &= x^2(x+2)-(x+2) = (x^2-1)(x+2) \\ &= (x-1)(x+1)(x+2) \end{aligned}$$

This gives us the partial fraction decomposition

$$\frac{5x+7}{x^3+2x^2-x-2} = \frac{A}{x-1} + \frac{B}{x+1} + \frac{C}{x+2}$$

Multiplying each side by the common denominator, $(x-1)(x+1)(x+2)$, we get

$$\begin{aligned} 5x+7 &= A(x+1)(x+2) + B(x-1)(x+2) + C(x-1)(x+1) \\ &= A(x^2+3x+2) + B(x^2+x-2) + C(x^2-1) && \text{Expand} \\ &= (A+B+C)x^2 + (3A+B)x + (2A-2B-C) && \text{Combine like terms} \end{aligned}$$

If two polynomials are equal, then their coefficients are equal. Thus since $5x+7$ has no x^2-term, we have $A+B+C=0$. Similarly, by comparing the coefficients of x, we see that $3A+B=5$, and by comparing constant terms, we get $2A-2B-C=7$. This leads to the following system of linear equations for A, B, and C.

$$\begin{cases} A + B + C = 0 & \text{Equation 1: Coefficients of } x^2 \\ 3A + B = 5 & \text{Equation 2: Coefficients of } x \\ 2A - 2B - C = 7 & \text{Equation 3: Constant coefficients} \end{cases}$$

We use Gaussian elimination to solve this system.

$$\begin{cases} A + B + C = 0 \\ \quad - 2B - 3C = 5 & \text{Equation 2} + (-3) \times \text{Equation 1} \\ \quad - 4B - 3C = 7 & \text{Equation 3} + (-2) \times \text{Equation 1} \end{cases}$$

$$\begin{cases} A + B + C = 0 \\ \quad - 2B - 3C = 5 \\ \quad\quad 3C = -3 & \text{Equation 3} + (-2) \times \text{Equation 2} \end{cases}$$

From the third equation we get $C = -1$. Back-substituting, we find that $B = -1$ and $A = 2$. So the partial fraction decomposition is

$$\frac{5x + 7}{x^3 + 2x^2 - x - 2} = \frac{2}{x - 1} + \frac{-1}{x + 1} + \frac{-1}{x + 2}$$

Now Try Exercises 3 and 13 ■

THE RHIND PAPYRUS is the oldest known mathematical document. It is an Egyptian scroll written in 1650 B.C. by the scribe Ahmes, who explains that it is an exact copy of a scroll written 200 years earlier. Ahmes claims that his papyrus contains "a thorough study of all things, insight into all that exists, knowledge of all obscure secrets." Actually, the document contains rules for doing arithmetic, including multiplication and division of fractions and several exercises with solutions. The exercise shown below reads: "A heap and its seventh make nineteen; how large is the heap?" In solving problems of this sort, the Egyptians used partial fractions because their number system required all fractions to be written as sums of reciprocals of whole numbers. For example, $\frac{7}{12}$ would be written as $\frac{1}{3} + \frac{1}{4}$.

The papyrus gives a correct formula for the volume of a truncated pyramid, which the ancient Egyptians used when building the pyramids at Giza. It also gives the formula $A = \left(\frac{8}{9}d\right)^2$ for the area of a circle with diameter d. How close is this to the actual area?

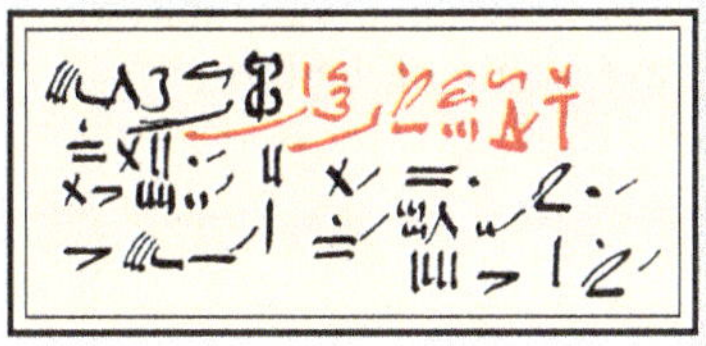

The same approach works in the remaining cases. We set up the partial fraction decomposition with the unknown constants $A, B, C, \ldots$. Then we multiply each side of the resulting equation by the common denominator, combine like terms on the right-hand side of the equation, and equate coefficients. This gives a set of linear equations that will always have a unique solution (provided that the partial fraction decomposition has been set up correctly).

Repeated Linear Factors

We now consider the case in which the denominator factors into linear factors, some of which are repeated.

CASE 2: THE DENOMINATOR IS A PRODUCT OF LINEAR FACTORS, SOME OF WHICH ARE REPEATED

Suppose the complete factorization of $Q(x)$ contains the linear factor $ax + b$ repeated k times; that is, $(ax + b)^k$ is a factor of $Q(x)$. Then, corresponding to each such factor, the partial fraction decomposition for $P(x)/Q(x)$ contains

$$\frac{A_1}{ax + b} + \frac{A_2}{(ax + b)^2} + \cdots + \frac{A_k}{(ax + b)^k}$$

EXAMPLE 2 ■ Repeated Linear Factors

Find the partial fraction decomposition of $\dfrac{x^2 + 1}{x(x - 1)^3}$.

SOLUTION Because the factor $x - 1$ is repeated three times in the denominator, the partial fraction decomposition has the form

$$\frac{x^2 + 1}{x(x - 1)^3} = \frac{A}{x} + \frac{B}{x - 1} + \frac{C}{(x - 1)^2} + \frac{D}{(x - 1)^3}$$

Multiplying each side by the common denominator, $x(x - 1)^3$, gives

$$\begin{aligned} x^2 + 1 &= A(x - 1)^3 + Bx(x - 1)^2 + Cx(x - 1) + Dx \\ &= A(x^3 - 3x^2 + 3x - 1) + B(x^3 - 2x^2 + x) + C(x^2 - x) + Dx && \text{Expand} \\ &= (A + B)x^3 + (-3A - 2B + C)x^2 + (3A + B - C + D)x - A && \text{Combine like terms} \end{aligned}$$

Equating coefficients, we get the following equations.

$$\begin{cases} A + B = 0 & \text{Coefficients of } x^3 \\ -3A - 2B + C = 1 & \text{Coefficients of } x^2 \\ 3A + B - C + D = 0 & \text{Coefficients of } x \\ -A = 1 & \text{Constant coefficients} \end{cases}$$

If we rearrange these equations by putting the last one in the first position, we can easily see (using substitution) that the solution to the system is $A = -1$, $B = 1$, $C = 0$, $D = 2$, so the partial fraction decomposition is

$$\frac{x^2 + 1}{x(x-1)^3} = \frac{-1}{x} + \frac{1}{x-1} + \frac{2}{(x-1)^3}$$

Now Try Exercises 5 and 29 ■

Irreducible Quadratic Factors

We now consider the case in which the denominator has distinct irreducible quadratic factors.

CASE 3: THE DENOMINATOR HAS IRREDUCIBLE QUADRATIC FACTORS, NONE OF WHICH IS REPEATED

Suppose the complete factorization of $Q(x)$ contains the quadratic factor $ax^2 + bx + c$ (which can't be factored further). Then, corresponding to this, the partial fraction decomposition of $P(x)/Q(x)$ will have a term of the form

$$\frac{Ax + B}{ax^2 + bx + c}$$

EXAMPLE 3 ■ Distinct Quadratic Factors

Find the partial fraction decomposition of $\dfrac{2x^2 - x + 4}{x^3 + 4x}$.

SOLUTION Since $x^3 + 4x = x(x^2 + 4)$, which can't be factored further, we write

$$\frac{2x^2 - x + 4}{x^3 + 4x} = \frac{A}{x} + \frac{Bx + C}{x^2 + 4}$$

Multiplying by $x(x^2 + 4)$, we get

$$\begin{aligned} 2x^2 - x + 4 &= A(x^2 + 4) + (Bx + C)x \\ &= (A + B)x^2 + Cx + 4A \end{aligned}$$

Equating coefficients gives us the equations

$$\begin{cases} A + B = 2 & \text{Coefficients of } x^2 \\ C = -1 & \text{Coefficients of } x \\ 4A = 4 & \text{Constant coefficients} \end{cases}$$

so $A = 1$, $B = 1$, and $C = -1$. The required partial fraction decomposition is

$$\frac{2x^2 - x + 4}{x^3 + 4x} = \frac{1}{x} + \frac{x - 1}{x^2 + 4}$$

Now Try Exercises 7 and 37 ■

Repeated Irreducible Quadratic Factors

We now consider the case in which the denominator has irreducible quadratic factors, some of which are repeated.

CASE 4: THE DENOMINATOR HAS A REPEATED IRREDUCIBLE QUADRATIC FACTOR

Suppose the complete factorization of $Q(x)$ contains the factor $(ax^2 + bx + c)^k$, where $ax^2 + bx + c$ can't be factored further. Then the partial fraction decomposition of $P(x)/Q(x)$ will have the terms

$$\frac{A_1x + B_1}{ax^2 + bx + c} + \frac{A_2x + B_2}{(ax^2 + bx + c)^2} + \cdots + \frac{A_kx + B_k}{(ax^2 + bx + c)^k}$$

EXAMPLE 4 ■ Repeated Quadratic Factors

Write the form of the partial fraction decomposition of

$$\frac{x^5 - 3x^2 + 12x - 1}{x^3(x^2 + x + 1)(x^2 + 2)^3}$$

SOLUTION

$$\frac{x^5 - 3x^2 + 12x - 1}{x^3(x^2 + x + 1)(x^2 + 2)^3}$$

$$= \frac{A}{x} + \frac{B}{x^2} + \frac{C}{x^3} + \frac{Dx + E}{x^2 + x + 1} + \frac{Fx + G}{x^2 + 2} + \frac{Hx + I}{(x^2 + 2)^2} + \frac{Jx + K}{(x^2 + 2)^3}$$

Now Try Exercises 11 and 41 ■

To find the values of $A, B, C, D, E, F, G, H, I, J$, and K in Example 4, we would have to solve a system of 11 linear equations. Although possible, this would certainly involve a great deal of work!

The techniques that we have described in this section apply only to rational functions $P(x)/Q(x)$ in which the degree of P is less than the degree of Q. If this isn't the case, we must first use long division to divide Q into P.

EXAMPLE 5 ■ Using Long Division to Prepare for Partial Fractions

Find the partial fraction decomposition of

$$\frac{2x^4 + 4x^3 - 2x^2 + x + 7}{x^3 + 2x^2 - x - 2}$$

SOLUTION Since the degree of the numerator is larger than the degree of the denominator, we use long division to obtain

$$\begin{array}{r} 2x \\ x^3 + 2x^2 - x - 2 \overline{) 2x^4 + 4x^3 - 2x^2 + x + 7} \\ 2x^4 + 4x^3 - 2x^2 - 4x \\ \hline 5x + 7 \end{array}$$

$$\frac{2x^4 + 4x^3 - 2x^2 + x + 7}{x^3 + 2x^2 - x - 2} = 2x + \frac{5x + 7}{x^3 + 2x^2 - x - 2}$$

The remainder term now satisfies the requirement that the degree of the numerator is less than the degree of the denominator. At this point we proceed as in Example 1 to obtain the decomposition

$$\frac{2x^4 + 4x^3 - 2x^2 + x + 7}{x^3 + 2x^2 - x - 2} = 2x + \frac{2}{x - 1} + \frac{-1}{x + 1} + \frac{-1}{x + 2}$$

Now Try Exercise 43 ■

10.7 EXERCISES

CONCEPTS

1–2 ■ For each rational function r, choose from (i)–(iv) the appropriate form for its partial fraction decomposition.

1. $r(x) = \dfrac{4}{x(x-2)^2}$

(i) $\dfrac{A}{x} + \dfrac{B}{x-2}$ (ii) $\dfrac{A}{x} + \dfrac{B}{(x-2)^2}$

(iii) $\dfrac{A}{x} + \dfrac{B}{x-2} + \dfrac{C}{(x-2)^2}$ (iv) $\dfrac{A}{x} + \dfrac{B}{x-2} + \dfrac{Cx+D}{(x-2)^2}$

2. $r(x) = \dfrac{2x+8}{(x-1)(x^2+4)}$

(i) $\dfrac{A}{x-1} + \dfrac{B}{x^2+4}$

(ii) $\dfrac{A}{x-1} + \dfrac{Bx+C}{x^2+4}$

(iii) $\dfrac{A}{x-1} + \dfrac{B}{x+2} + \dfrac{C}{x^2+4}$

(iv) $\dfrac{Ax+B}{x-1} + \dfrac{Cx+D}{x^2+4}$

SKILLS

3–12 ■ Form of the Partial Fraction Decomposition Write the form of the partial fraction decomposition of the function (as in Example 4). Do not determine the numerical values of the coefficients.

3. $\dfrac{1}{(x-1)(x+2)}$ **4.** $\dfrac{x}{x^2+3x-4}$

5. $\dfrac{x^2-3x+5}{(x-2)^2(x+4)}$ **6.** $\dfrac{1}{x^4-x^3}$

7. $\dfrac{x^2}{(x-3)(x^2+4)}$ **8.** $\dfrac{1}{x^4-1}$

9. $\dfrac{x^3-4x^2+2}{(x^2+1)(x^2+2)}$ **10.** $\dfrac{x^4+x^2+1}{x^2(x^2+4)^2}$

11. $\dfrac{x^3+x+1}{x(2x-5)^3(x^2+2x+5)^2}$ **12.** $\dfrac{1}{(x^3-1)(x^2-1)}$

13–44 ■ Partial Fraction Decomposition Find the partial fraction decomposition of the rational function.

13. $\dfrac{2}{(x-1)(x+1)}$ **14.** $\dfrac{2x}{(x-1)(x+1)}$

15. $\dfrac{5}{(x-1)(x+4)}$ **16.** $\dfrac{x+6}{x(x+3)}$

17. $\dfrac{12}{x^2-9}$ **18.** $\dfrac{x-12}{x^2-4x}$

19. $\dfrac{4}{x^2-4}$ **20.** $\dfrac{2x+1}{x^2+x-2}$

21. $\dfrac{x+14}{x^2-2x-8}$ **22.** $\dfrac{8x-3}{2x^2-x}$

23. $\dfrac{x}{8x^2-10x+3}$ **24.** $\dfrac{7x-3}{x^3+2x^2-3x}$

25. $\dfrac{9x^2-9x+6}{2x^3-x^2-8x+4}$ **26.** $\dfrac{-3x^2-3x+27}{(x+2)(2x^2+3x-9)}$

27. $\dfrac{x^2+1}{x^3+x^2}$ **28.** $\dfrac{3x^2+5x-13}{(3x+2)(x^2-4x+4)}$

29. $\dfrac{2x}{4x^2+12x+9}$ **30.** $\dfrac{x-4}{(2x-5)^2}$

31. $\dfrac{4x^2-x-2}{x^4+2x^3}$ **32.** $\dfrac{x^3-2x^2-4x+3}{x^4}$

33. $\dfrac{-10x^2+27x-14}{(x-1)^3(x+2)}$ **34.** $\dfrac{-2x^2+5x-1}{x^4-2x^3+2x-1}$

35. $\dfrac{3x^3+22x^2+53x+41}{(x+2)^2(x+3)^2}$ **36.** $\dfrac{3x^2+12x-20}{x^4-8x^2+16}$

37. $\dfrac{x-3}{x^3+3x}$ **38.** $\dfrac{3x^2-2x+8}{x^3-x^2+2x-2}$

39. $\dfrac{2x^3+7x+5}{(x^2+x+2)(x^2+1)}$ **40.** $\dfrac{x^2+x+1}{2x^4+3x^2+1}$

41. $\dfrac{x^4+x^3+x^2-x+1}{x(x^2+1)^2}$ **42.** $\dfrac{2x^2-x+8}{(x^2+4)^2}$

43. $\dfrac{x^5-2x^4+x^3+x+5}{x^3-2x^2+x-2}$

44. $\dfrac{x^5-3x^4+3x^3-4x^2+4x+12}{(x-2)^2(x^2+2)}$

SKILLS Plus

45. Partial Fractions Determine A and B in terms of a and b.

$$\frac{ax+b}{x^2-1} = \frac{A}{x-1} + \frac{B}{x+1}$$

46. Partial Fractions Determine A, B, C, and D in terms of a and b.

$$\frac{ax^3+bx^2}{(x^2+1)^2} = \frac{Ax+B}{x^2+1} + \frac{Cx+D}{(x^2+1)^2}$$

DISCUSS ■ DISCOVER ■ PROVE ■ WRITE

47. DISCUSS: Recognizing Partial Fraction Decompositions For each expression, determine whether it is already a partial fraction decomposition or whether it can be decomposed further.

(a) $\dfrac{x}{x^2+1} + \dfrac{1}{x+1}$ **(b)** $\dfrac{x}{(x+1)^2}$

(c) $\dfrac{1}{x+1} + \dfrac{2}{(x+1)^2}$ **(d)** $\dfrac{x+2}{(x^2+1)^2}$

48. DISCUSS: Assembling and Disassembling Partial Fractions The following expression is a partial fraction decomposition.

$$\frac{2}{x-1} + \frac{1}{(x-1)^2} + \frac{1}{x+1}$$

Use a common denominator to combine the terms into one fraction. Then use the techniques of this section to find its partial fraction decomposition. Did you get back the original expression?

10.8 SYSTEMS OF NONLINEAR EQUATIONS

■ Substitution and Elimination Methods ■ Graphical Method

In this section we solve systems of equations in which the equations are not all linear. The methods we learned in Section 10.1 can also be used to solve nonlinear systems.

■ Substitution and Elimination Methods

To solve a system of nonlinear equations, we can use the substitution or elimination method, as illustrated in the next examples.

EXAMPLE 1 ■ Substitution Method

Find all solutions of the system.

$$\begin{cases} x^2 + y^2 = 100 & \text{Equation 1} \\ 3x - y = 10 & \text{Equation 2} \end{cases}$$

SOLUTION **Solve for one variable.** We start by solving for y in the second equation.

$$y = 3x - 10 \qquad \text{Solve for } y \text{ in Equation 2}$$

Substitute. Next we substitute for y in the first equation and solve for x.

$$\begin{aligned} x^2 + (3x - 10)^2 &= 100 && \text{Substitute } y = 3x - 10 \text{ into Equation 1} \\ x^2 + (9x^2 - 60x + 100) &= 100 && \text{Expand} \\ 10x^2 - 60x &= 0 && \text{Simplify} \\ 10x(x - 6) &= 0 && \text{Factor} \\ x = 0 \quad \text{or} \quad x &= 6 && \text{Solve for } x \end{aligned}$$

Back-substitute. Now we back-substitute these values of x into the equation $y = 3x - 10$.

$$\begin{aligned} &\text{For } x = 0: && y = 3(0) - 10 = -10 && \text{Back-substitute} \\ &\text{For } x = 6: && y = 3(6) - 10 = 8 && \text{Back-substitute} \end{aligned}$$

So we have two solutions: $(0, -10)$ and $(6, 8)$.

The graph of the first equation is a circle, and the graph of the second equation is a line. Figure 1 shows that the graphs intersect at the two points $(0, -10)$ and $(6, 8)$.

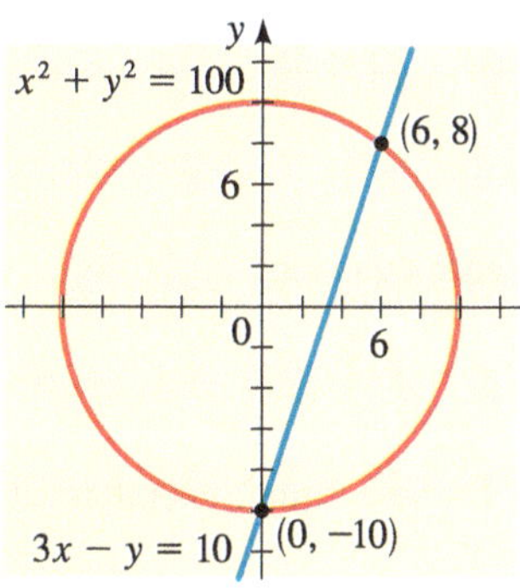

FIGURE 1

CHECK YOUR ANSWERS

$x = 0, y = -10$:

$$\begin{cases} (0)^2 + (-10)^2 = 100 \\ 3(0) - (-10) = 10 \end{cases} \checkmark$$

$x = 6, y = 8$:

$$\begin{cases} (6)^2 + (8)^2 = 36 + 64 = 100 \\ 3(6) - (8) = 18 - 8 = 10 \end{cases} \checkmark$$

Now Try Exercise 5 ■

EXAMPLE 2 ■ Elimination Method

Find all solutions of the system.

$$\begin{cases} 3x^2 + 2y = 26 & \text{Equation 1} \\ 5x^2 + 7y = 3 & \text{Equation 2} \end{cases}$$

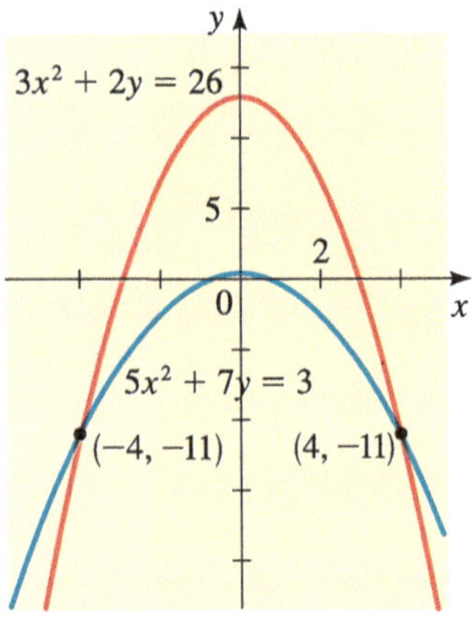

FIGURE 2

CHECK YOUR ANSWERS

$x = -4, y = -11$:

$$\begin{cases} 3(-4)^2 + 2(-11) = 26 \\ 5(-4)^2 + 7(-11) = 3 \end{cases} \checkmark$$

$x = 4, y = -11$:

$$\begin{cases} 3(4)^2 + 2(-11) = 26 \\ 5(4)^2 + 7(-11) = 3 \end{cases} \checkmark$$

SOLUTION We choose to eliminate the x-term, so we multiply the first equation by 5 and the second equation by -3. Then we add the two equations and solve for y.

$$\begin{cases} 15x^2 + 10y = 130 & 5 \times \text{Equation 1} \\ -15x^2 - 21y = -9 & (-3) \times \text{Equation 2} \end{cases}$$

$$-11y = 121 \quad \text{Add}$$

$$y = -11 \quad \text{Solve for } y$$

Now we back-substitute $y = -11$ into one of the original equations, say $3x^2 + 2y = 26$, and solve for x.

$$3x^2 + 2(-11) = 26 \quad \text{Back-substitute } y = -11 \text{ into Equation 1}$$

$$3x^2 = 48 \quad \text{Add 22}$$

$$x^2 = 16 \quad \text{Divide by 3}$$

$$x = -4 \quad \text{or} \quad x = 4 \quad \text{Solve for } x$$

So we have two solutions: $(-4, -11)$ and $(4, -11)$.

The graphs of both equations are parabolas (see Section 3.1). Figure 2 shows that the graphs intersect at the two points $(-4, -11)$ and $(4, -11)$.

Now Try Exercise 11 ■

Graphical Method

The graphical method is particularly useful in solving systems of nonlinear equations.

EXAMPLE 3 ■ Graphical Method

Find all solutions of the system

$$\begin{cases} x^2 - y = 2 \\ 2x - y = -1 \end{cases}$$

SOLUTION **Graph each equation.** To graph, we solve for y in each equation.

$$\begin{cases} y = x^2 - 2 \\ y = 2x + 1 \end{cases}$$

Find intersection points. Figure 3 shows that the graphs of these equations intersect at two points. Zooming in, we see that the solutions are

$$(-1, -1) \quad \text{and} \quad (3, 7)$$

See Appendix C, *Graphing with a Graphing Calculator*, for guidelines on using a graphing calculator. See Appendix D, *Using the TI-83/84 Graphing Calculator*, for specific graphing instructions. Go to **www.stewartmath.com**.

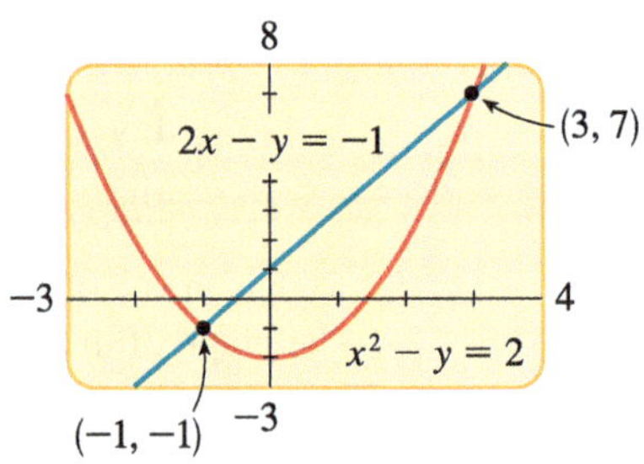

FIGURE 3

CHECK YOUR ANSWERS

$x = -1, y = -1$:

$$\begin{cases} (-1)^2 - (-1) = 2 \\ 2(-1) - (-1) = -1 \end{cases} \checkmark$$

$x = 3, y = 7$:

$$\begin{cases} 3^2 - 7 = 2 \\ 2(3) - 7 = -1 \end{cases} \checkmark$$

Now Try Exercise 33 ■

EXAMPLE 4 ■ Solving a System of Equations Graphically

Find all solutions of the system, rounded to one decimal place.

$$\begin{cases} x^2 + y^2 = 12 & \text{Equation 1} \\ y = 2x^2 - 5x & \text{Equation 2} \end{cases}$$

SOLUTION The graph of the first equation is a circle, and the graph of the second is a parabola. To graph the circle on a graphing calculator, we must first solve for y in terms of x.

$$\begin{aligned} x^2 + y^2 &= 12 \\ y^2 &= 12 - x^2 && \text{Isolate } y^2 \text{ on LHS} \\ y &= \pm\sqrt{12 - x^2} && \text{Take square roots} \end{aligned}$$

To graph the circle, we must graph both functions.

$$y = \sqrt{12 - x^2} \quad \text{and} \quad y = -\sqrt{12 - x^2}$$

In Figure 4 the graph of the circle is shown in red, and the parabola is shown in blue. The graphs intersect in Quadrants I and II. Zooming in, or using the `Intersect` command, we see that the intersection points are $(-0.559, 3.419)$ and $(2.847, 1.974)$. There also appears to be an intersection point in Quadrant IV. However, when we zoom in, we see that the curves come close to each other but don't intersect (see Figure 5). Thus the system has two solutions; rounded to the nearest tenth, they are

$$(-0.6, 3.4) \quad \text{and} \quad (2.8, 2.0)$$

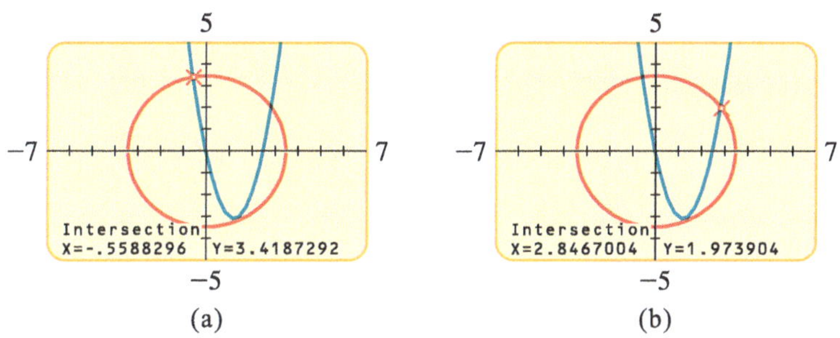

FIGURE 4 $x^2 + y^2 = 12,\ y = 2x^2 - 5x$

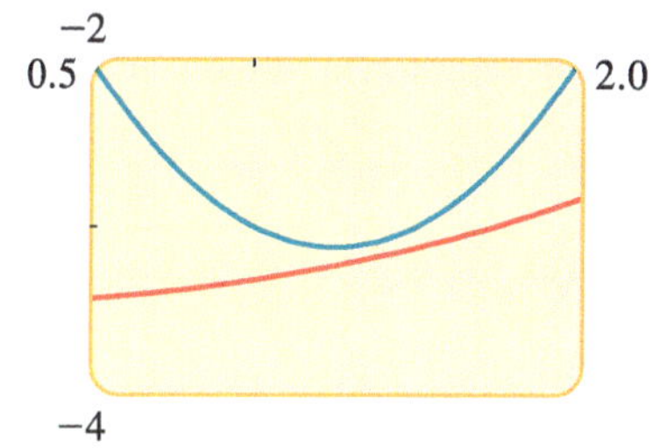

FIGURE 5 Zooming in

Now Try Exercise 37 ■

Mathematics in the Modern World

Courtesy of NASA

Global Positioning System (GPS)

On a cold, foggy day in 1707 a British naval fleet was sailing home at a fast clip. The fleet's navigators didn't know it, but the fleet was only a few yards from the rocky shores of England. In the ensuing disaster the fleet was totally destroyed. This tragedy could have been avoided had the navigators known their positions. In those days latitude was determined by the position of the North Star (and this could be done only at night in good weather), and longitude was determined by the position of the sun relative to where it would be in England *at that same time*. So navigation required an accurate method of telling time on ships. (The invention of the spring-loaded clock brought about the eventual solution.)

Since then, several different methods have been developed to determine position, and all rely heavily on mathematics (see LORAN, page 804). The latest method, called the Global Positioning System (GPS), uses triangulation. In this system, 24 satellites are strategically located above the surface of the earth. A handheld GPS device measures distance from a satellite, using the travel time of radio signals emitted from the satellite. Knowing the distances to three different satellites tells us that we are at the point of intersection of three different spheres. This uniquely determines our position (see Exercise 51, page 755).

10.8 EXERCISES

CONCEPTS

1–2 ■ The system of equations

$$\begin{cases} 2y - x^2 = 0 \\ y - x = 4 \end{cases}$$

is graphed below.

1. Use the graph to find the solution(s) of the system.
2. Check that the solutions you found in Exercise 1 satisfy the system.

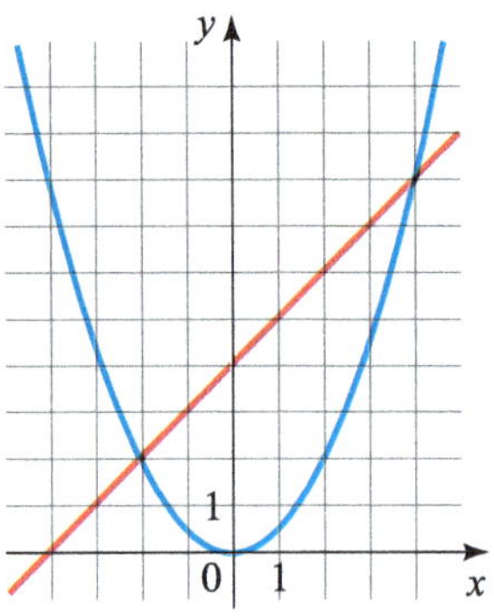

SKILLS

3–8 ■ Substitution Method Use the substitution method to find all solutions of the system of equations.

3. $\begin{cases} y = x^2 \\ y = x + 12 \end{cases}$

4. $\begin{cases} x^2 + y^2 = 25 \\ y = 2x \end{cases}$

5. $\begin{cases} x^2 + y^2 = 8 \\ x + y = 0 \end{cases}$

6. $\begin{cases} x^2 + y = 9 \\ x - y + 3 = 0 \end{cases}$

7. $\begin{cases} x + y^2 = 0 \\ 2x + 5y^2 = 75 \end{cases}$

8. $\begin{cases} x^2 - y = 1 \\ 2x^2 + 3y = 17 \end{cases}$

9–14 ■ Elimination Method Use the elimination method to find all solutions of the system of equations.

9. $\begin{cases} x^2 - 2y = 1 \\ x^2 + 5y = 29 \end{cases}$

10. $\begin{cases} 3x^2 + 4y = 17 \\ 2x^2 + 5y = 2 \end{cases}$

11. $\begin{cases} 3x^2 - y^2 = 11 \\ x^2 + 4y^2 = 8 \end{cases}$

12. $\begin{cases} 2x^2 + 4y = 13 \\ x^2 - y^2 = \frac{7}{2} \end{cases}$

13. $\begin{cases} x - y^2 + 3 = 0 \\ 2x^2 + y^2 - 4 = 0 \end{cases}$

14. $\begin{cases} x^2 - y^2 = 1 \\ 2x^2 - y^2 = x + 3 \end{cases}$

15–18 ■ Finding Intersection Points Graphically Two equations and their graphs are given. Find the intersection point(s) of the graphs by solving the system.

15. $\begin{cases} x^2 + y = 8 \\ x - 2y = -6 \end{cases}$

16. $\begin{cases} x - y^2 = -4 \\ x - y = 2 \end{cases}$

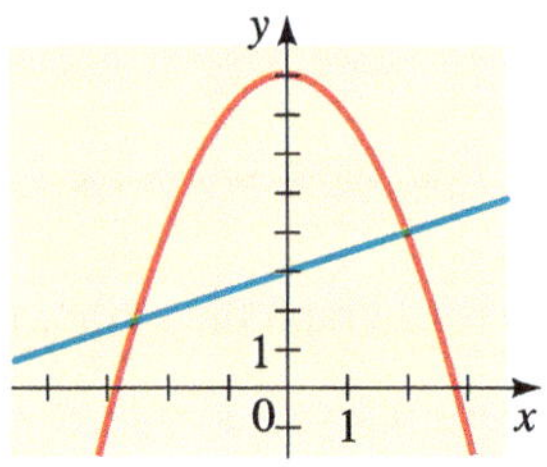

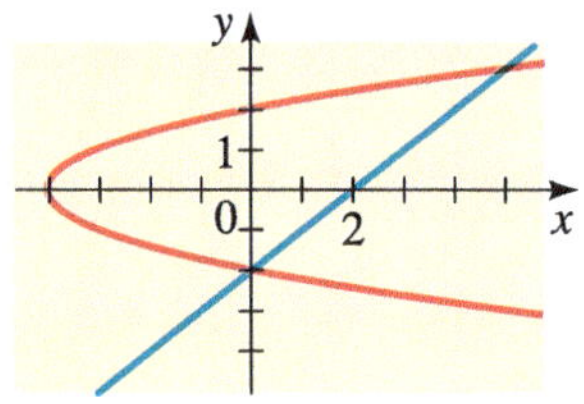

17. $\begin{cases} x^2 + y = 0 \\ x^3 - 2x - y = 0 \end{cases}$

18. $\begin{cases} x^2 + y^2 = 4x \\ x = y^2 \end{cases}$

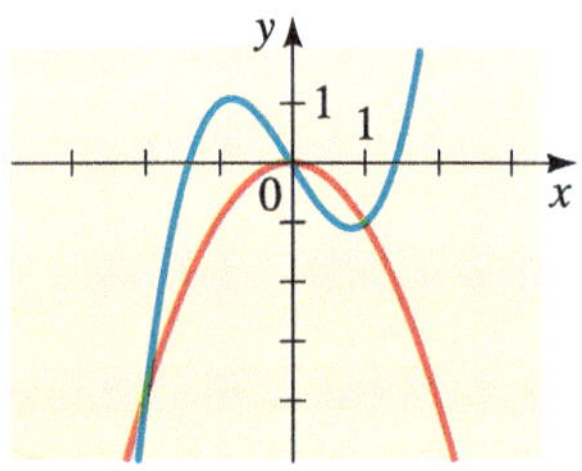

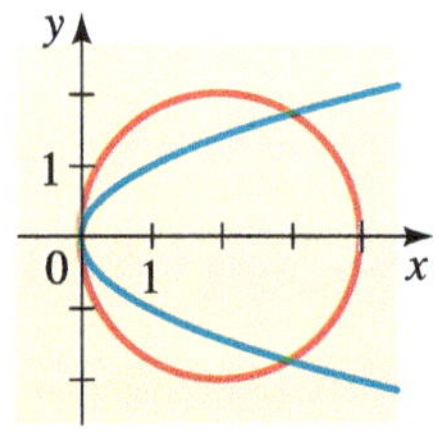

19–32 ■ Solving Nonlinear Systems Find all solutions of the system of equations.

19. $\begin{cases} y + x^2 = 4x \\ y + 4x = 16 \end{cases}$

20. $\begin{cases} x - y^2 = 0 \\ y - x^2 = 0 \end{cases}$

21. $\begin{cases} x - 2y = 2 \\ y^2 - x^2 = 2x + 4 \end{cases}$

22. $\begin{cases} y = 4 - x^2 \\ y = x^2 - 4 \end{cases}$

23. $\begin{cases} x - y = 4 \\ xy = 12 \end{cases}$

24. $\begin{cases} xy = 24 \\ 2x^2 - y^2 + 4 = 0 \end{cases}$

25. $\begin{cases} x^2y = 16 \\ x^2 + 4y + 16 = 0 \end{cases}$

26. $\begin{cases} x + \sqrt{y} = 0 \\ y^2 - 4x^2 = 12 \end{cases}$

27. $\begin{cases} x^2 + y^2 = 9 \\ x^2 - y^2 = 1 \end{cases}$

28. $\begin{cases} x^2 + 2y^2 = 2 \\ 2x^2 - 3y = 15 \end{cases}$

29. $\begin{cases} 2x^2 - 8y^3 = 19 \\ 4x^2 + 16y^3 = 34 \end{cases}$

30. $\begin{cases} x^4 + y^3 = 17 \\ 3x^4 + 5y^3 = 53 \end{cases}$

31. $\begin{cases} \dfrac{2}{x} - \dfrac{3}{y} = 1 \\ -\dfrac{4}{x} + \dfrac{7}{y} = 1 \end{cases}$

32. $\begin{cases} \dfrac{4}{x^2} + \dfrac{6}{y^4} = \dfrac{7}{2} \\ \dfrac{1}{x^2} - \dfrac{2}{y^4} = 0 \end{cases}$

33–40 ■ Graphical Method Use the graphical method to find all solutions of the system of equations, rounded to two decimal places.

33. $\begin{cases} y = x^2 + 8x \\ y = 2x + 16 \end{cases}$ **34.** $\begin{cases} y = x^2 - 4x \\ 2x - y = 2 \end{cases}$

35. $\begin{cases} x^2 + y^2 = 25 \\ x + 3y = 2 \end{cases}$ **36.** $\begin{cases} x^2 + y^2 = 17 \\ x^2 - 2x + y^2 = 13 \end{cases}$

37. $\begin{cases} \dfrac{x^2}{9} + \dfrac{y^2}{18} = 1 \\ y = -x^2 + 6x - 2 \end{cases}$ **38.** $\begin{cases} x^2 - y^2 = 3 \\ y = x^2 - 2x - 8 \end{cases}$

39. $\begin{cases} x^4 + 16y^4 = 32 \\ x^2 + 2x + y = 0 \end{cases}$ **40.** $\begin{cases} y = e^x + e^{-x} \\ y = 5 - x^2 \end{cases}$

SKILLS Plus

41–44 ■ Some Trickier Systems Follow the hints and solve the systems.

41. $\begin{cases} \log x + \log y = \frac{3}{2} \\ 2\log x - \log y = 0 \end{cases}$ [*Hint:* Add the equations.]

42. $\begin{cases} 2^x + 2^y = 10 \\ 4^x + 4^y = 68 \end{cases}$ [*Hint:* Note that $4^x = 2^{2x} = (2^x)^2$.]

43. $\begin{cases} x - y = 3 \\ x^3 - y^3 = 387 \end{cases}$ [*Hint:* Factor the left-hand side of the second equation.]

44. $\begin{cases} x^2 + xy = 1 \\ xy + y^2 = 3 \end{cases}$ [*Hint:* Add the equations, and factor the result.]

APPLICATIONS

45. Dimensions of a Rectangle A rectangle has an area of 180 cm^2 and a perimeter of 54 cm. What are its dimensions?

46. Legs of a Right Triangle A right triangle has an area of 84 ft^2 and a hypotenuse 25 ft long. What are the lengths of its other two sides?

47. Dimensions of a Rectangle The perimeter of a rectangle is 70, and its diagonal is 25. Find its length and width.

48. Dimensions of a Rectangle A circular piece of sheet metal has a diameter of 20 in. The edges are to be cut off to form a rectangle of area 160 in^2 (see the figure). What are the dimensions of the rectangle?

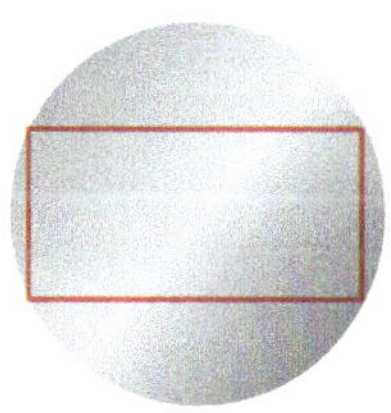

49. Flight of a Rocket A hill is inclined so that its "slope" is $\frac{1}{2}$, as shown in the figure. We introduce a coordinate system with the origin at the base of the hill and with the scales on the axes measured in meters. A rocket is fired from the base of the hill in such a way that its trajectory is the parabola $y = -x^2 + 401x$. At what point does the rocket strike the hillside? How far is this point from the base of the hill (to the nearest centimeter)?

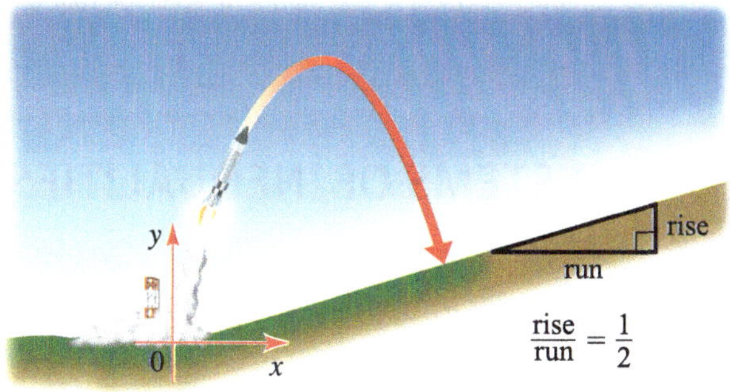

50. Making a Stovepipe A rectangular piece of sheet metal with an area of 1200 in^2 is to be bent into a cylindrical length of stovepipe having a volume of 600 in^3. What are the dimensions of the sheet metal?

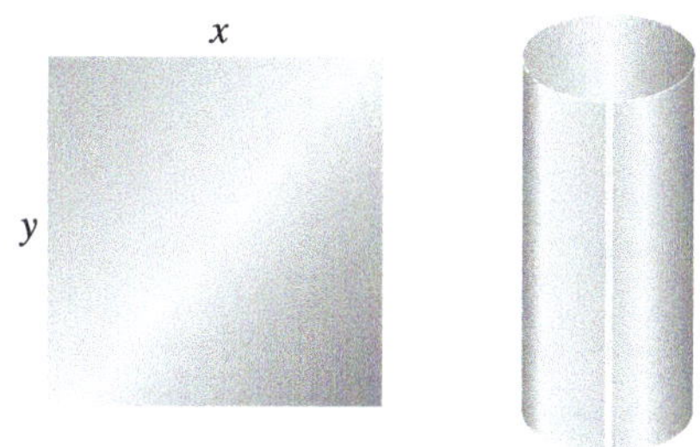

51. Global Positioning System (GPS) The Global Positioning System determines the location of an object from its distances to satellites in orbit around the earth. In the simplified, two-dimensional situation shown in the following figure, determine the coordinates of P from the fact that P is 26 units from satellite A and 20 units from satellite B.

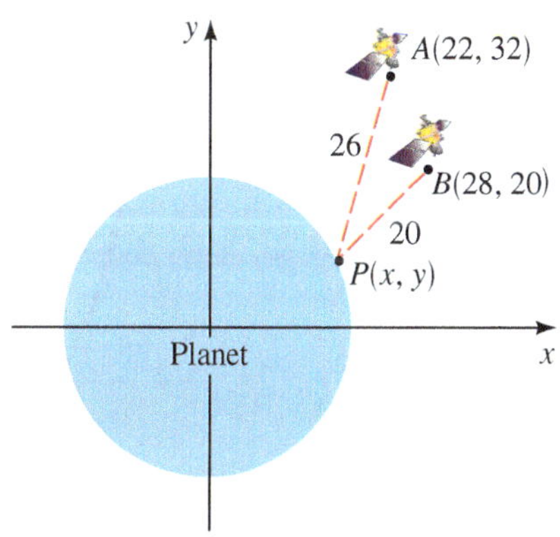

DISCUSS ■ DISCOVER ■ PROVE ■ WRITE

52. DISCOVER ■ PROVE: Intersection of a Parabola and a Line On a sheet of graph paper or using a graphing calculator, draw the parabola $y = x^2$. Then draw the graphs of the linear equation $y = x + k$ on the same coordinate plane for various values of k. Try to choose values of k so that the line and the parabola intersect at two points for some of your k's and not for others. For what value of k is there exactly one intersection point? Use the results of your experiment to make a conjecture about the values of k for which the following system has two solutions, one solution, and no solution. Prove your conjecture.

$$\begin{cases} y = x^2 \\ y = x + k \end{cases}$$

10.9 SYSTEMS OF INEQUALITIES

■ Graphing an Inequality ■ Systems of Inequalities ■ Systems of Linear Inequalities ■ Application: Feasible Regions

In this section we study systems of inequalities in two variables from a graphical point of view.

■ Graphing an Inequality

We begin by considering the graph of a single inequality. We already know that the graph of $y = x^2$, for example, is the *parabola* in Figure 1. If we replace the equal sign by the symbol $\geq$, we obtain the *inequality*

$$y \geq x^2$$

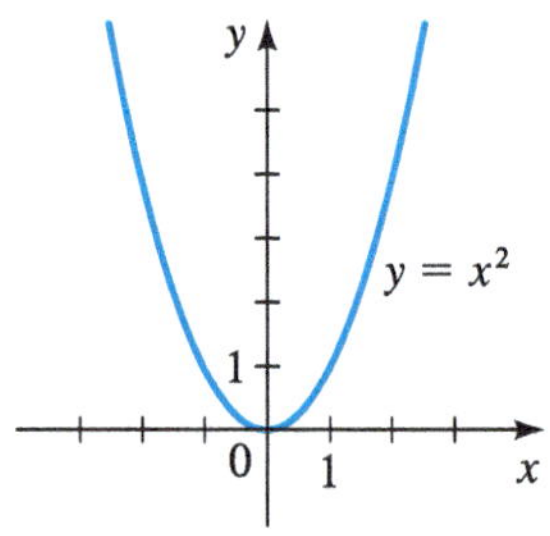

FIGURE 1

Its graph consists of not just the parabola in Figure 1, but also every point whose y-coordinate is *larger* than x^2. We indicate the solution in Figure 2(a) by shading the points *above* the parabola.

Similarly, the graph of $y \leq x^2$ in Figure 2(b) consists of all points on and *below* the parabola. However, the graphs of $y > x^2$ and $y < x^2$ do not include the points on the parabola itself, as indicated by the dashed curves in Figures 2(c) and 2(d).

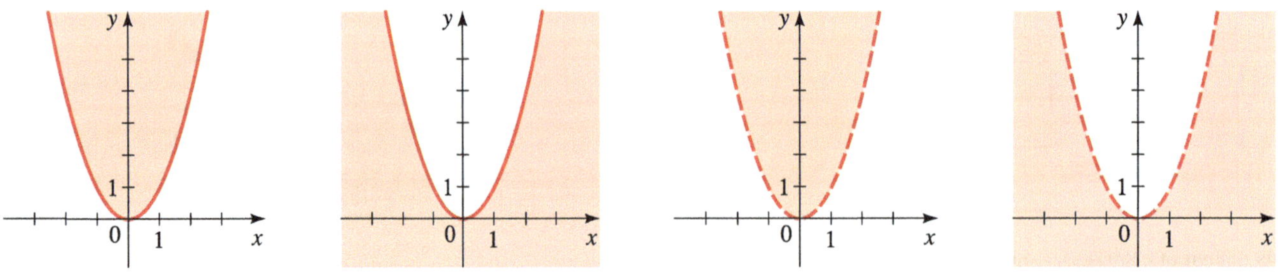

(a) $y \geq x^2$ (b) $y \leq x^2$ (c) $y > x^2$ (d) $y < x^2$

FIGURE 2

The graph of an inequality, in general, consists of a region in the plane whose boundary is the graph of the equation obtained by replacing the inequality sign ($\geq$, $\leq$, $>$, or $<$) with an equal sign. To determine which side of the graph gives the solution set of the inequality, we need only check **test points**.

GRAPHING AN INEQUALITY

To graph an inequality, we carry out the following steps.

1. **Graph the Equation.** Graph the equation that corresponds to the inequality. Use a dashed curve for $>$ or $<$ and a solid curve for $\leq$ or $\geq$.
2. **Graph the Inequality.** The graph of the inequality consists of all the points on one side of the curve that we graphed in Step 1. We use **test points** on either side of the curve to determine whether the points on that side satisfy the inequality. If the point satisfies the inequality, then all the points on that side of the curve satisfy the inequality. In that case, **shade that side of the curve** to indicate that it is part of the graph. If the test point does not satisfy the inequality, then the region isn't part of the graph.

EXAMPLE 1 ■ Graphs of Inequalities

Graph each inequality.

(a) $x^2 + y^2 < 25$ **(b)** $x + 2y \geq 5$

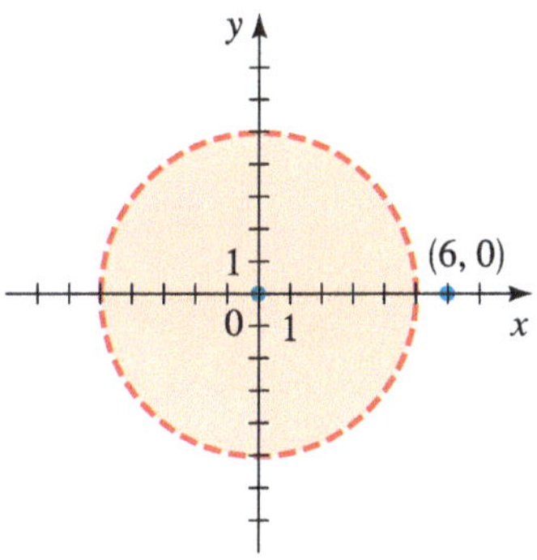

FIGURE 3 Graph of $x^2 + y^2 < 25$

Note that *any* point inside or outside the circle can serve as a test point. We have chosen these points for simplicity.

SOLUTION We follow the guidelines given above.

(a) **Graph the equation.** The graph of the equation $x^2 + y^2 = 25$ is a circle of radius 5 centered at the origin. The points on the circle itself do not satisfy the inequality because it is of the form $<$, so we graph the circle with a dashed curve, as shown in Figure 3.

Graph the inequality. To determine whether the inside or the outside of the circle satisfies the inequality, we use the test points $(0, 0)$ on the inside and $(6, 0)$ on the outside. To do this, we substitute the coordinates of each point into the inequality and check whether the result satisfies the inequality.

Test point	Inequality $x^2 + y^2 < 25$	Conclusion
$(0, 0)$	$0^2 + 0^2 \stackrel{?}{<} 25$ ✓	Part of graph
$(6, 0)$	$6^2 + 0^2 \stackrel{?}{<} 25$ ✗	Not part of graph

Our check shows that the points *inside* the circle satisfy the inequality. A graph of the inequality is shown in Figure 3.

(b) **Graph the equation.** We first graph the equation $x + 2y = 5$. The graph is the line shown in Figure 4.

Graph the inequality. Let's use the test points $(0, 0)$ and $(5, 5)$ on either side of the line.

Test point	Inequality $x + 2y \geq 5$	Conclusion
$(0, 0)$	$0 + 2(0) \stackrel{?}{\geq} 5$ ✗	Not part of graph
$(5, 5)$	$5 + 2(5) \stackrel{?}{\geq} 5$ ✓	Part of graph

We can write the inequality in Example 1 as

$$y \geq -\tfrac{1}{2}x + \tfrac{5}{2}$$

From this form of the inequality we see that the solution consists of the points with y-values *on* or *above* the line $y = -\frac{1}{2}x + \frac{5}{2}$. So the graph of the inequality is the region *above* the line.

Our check shows that the points *above* the line satisfy the inequality. A graph of the inequality is shown in Figure 4.

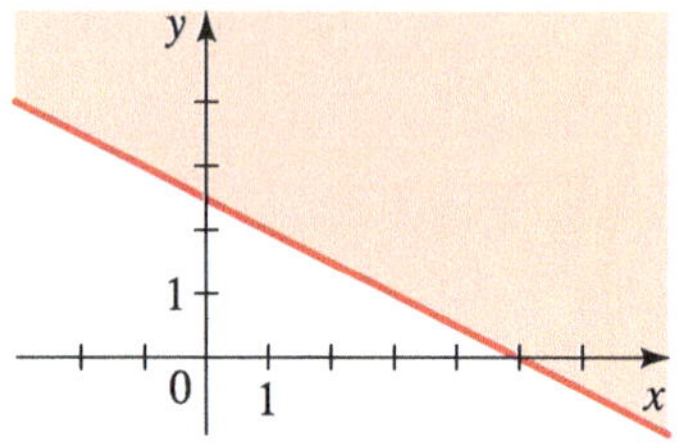

FIGURE 4 Graph of $x + 2y \geq 5$

Now Try Exercises 15 and 21 ■

Systems of Inequalities

We now consider *systems* of inequalities. The **solution set of a system of inequalities** in two variables is the set of all points in the coordinate plane that satisfy every inequality in the system. The **graph of a system of inequalities** is the graph of the solution set.

To find the solution of a system of inequalities, we first graph each inequality in the system. The solution of the system consists of those points in the coordinate plane that belong to the solution of each inequality in the system. In other words, the solution of the system is the intersection of the solutions of the individual inequalities in the system. So to solve a system of inequalities, we use the following guidelines.

THE SOLUTION OF A SYSTEM OF INEQUALITIES

To graph the solution of a system of inequalities, we carry out the following steps.

1. **Graph Each Inequality.** Graph each inequality in the system on the same graph.
2. **Graph the Solution of the System.** Shade the region where the graphs of all the inequalities intersect. All the points in this region satisfy each inequality, so they belong to the solution of the system.
3. **Find the Vertices.** Label the vertices of the region that you shaded in Step 2.

EXAMPLE 2 ■ A System of Two Inequalities

Graph the solution of the system of inequalities, and label its vertices.

$$\begin{cases} x^2 + y^2 < 25 \\ x + 2y \geq 5 \end{cases}$$

SOLUTION These are the two inequalities of Example 1. Here we want to graph only those points that simultaneously satisfy both inequalities.

Graph each inequality. In Figure 5(a) we graph the solutions of the two inequalities on the same axes (in different colors).

Graph the solution of the system. The solution of the system of inequalities is the intersection of the two graphs. This is the region where the two regions overlap, which is the purple region graphed in Figure 5(b).

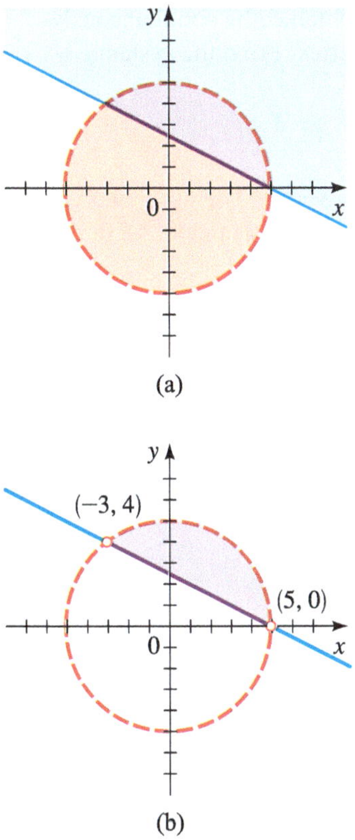

FIGURE 5 $\begin{cases} x^2 + y^2 < 25 \\ x + 2y \geq 5 \end{cases}$

Find the Vertices. The points $(-3, 4)$ and $(5, 0)$ in Figure 5(b) are the **vertices** of the solution set. They are obtained by solving the system of *equations*

$$\begin{cases} x^2 + y^2 = 25 \\ x + 2y = 5 \end{cases}$$

We solve this system of equations by substitution. Solving for x in the second equation gives $x = 5 - 2y$, and substituting this into the first equation gives

$$\begin{aligned} (5 - 2y)^2 + y^2 &= 25 && \text{Substitute } x = 5 - 2y \\ (25 - 20y + 4y^2) + y^2 &= 25 && \text{Expand} \\ -20y + 5y^2 &= 0 && \text{Simplify} \\ -5y(4 - y) &= 0 && \text{Factor} \end{aligned}$$

Thus $y = 0$ or $y = 4$. When $y = 0$, we have $x = 5 - 2(0) = 5$, and when $y = 4$, we have $x = 5 - 2(4) = -3$. So the points of intersection of these curves are $(5, 0)$ and $(-3, 4)$.

Note that in this case the vertices are not part of the solution set, since they don't satisfy the inequality $x^2 + y^2 < 25$ (so they are graphed as open circles in the figure). They simply show where the "corners" of the solution set lie.

Now Try Exercise 43 ■

Systems of Linear Inequalities

An inequality is **linear** if it can be put into one of the following forms:

$$ax + by \geq c \qquad ax + by \leq c \qquad ax + by > c \qquad ax + by < c$$

In the next example we graph the solution set of a system of linear inequalities.

EXAMPLE 3 ■ A System of Four Linear Inequalities

Graph the solution set of the system, and label its vertices.

$$\begin{cases} x + 3y \leq 12 \\ x + y \leq 8 \\ x \geq 0 \\ y \geq 0 \end{cases}$$

SOLUTION **Graph each inequality.** In Figure 6 we first graph the lines given by the equations that correspond to each inequality. To determine the graphs of the first two inequalities, we need to check only one test point. For simplicity let's use the point $(0, 0)$.

Inequality	Test point $(0, 0)$	Conclusion
$x + 3y \leq 12$	$0 + 3(0) \overset{?}{\leq} 12$ ✓	Satisfies inequality
$x + y \leq 8$	$0 + 0 \overset{?}{\leq} 8$ ✓	Satisfies inequality

Since $(0, 0)$ is below the line $x + 3y = 12$, our check shows that the region on or below the line must satisfy the inequality. Likewise, since $(0, 0)$ is below the line $x + y = 8$, our check shows that the region on or below this line must satisfy the inequality. The inequalities $x \geq 0$ and $y \geq 0$ say that x and y are nonnegative. These regions are sketched in Figure 6(a).

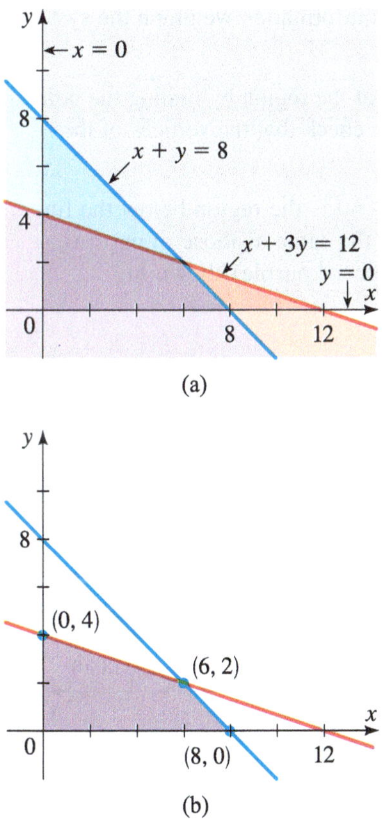

FIGURE 6

Graph the solution of the system. The solution of the system of inequalities is the intersection of the graphs. This is the purple region graphed in Figure 6(b).

Find the Vertices. The coordinates of each vertex are obtained by simultaneously solving the equations of the lines that intersect at that vertex. From the system

$$\begin{cases} x + 3y = 12 \\ x + y = 8 \end{cases}$$

we get the vertex (6, 2). The origin (0, 0) is also clearly a vertex. The other two vertices are at the x- and y-intercepts of the corresponding lines: (8, 0) and (0, 4). In this case all the vertices *are* part of the solution set.

Now Try Exercise 51 ■

EXAMPLE 4 ■ A System of Linear Inequalities

Graph the solution set of the system of inequalities, and label the vertices.

(a) $\begin{cases} 10x + 20y \geq 60 \\ 30x + 20y \geq 100 \\ 10x + 40y \geq 80 \\ x \geq 0, \quad y \geq 0 \end{cases}$ **(b)** $\begin{cases} 10x + 20y \leq 60 \\ 30x + 20y \geq 100 \\ 10x + 40y \geq 80 \\ x \geq 0, \quad y \geq 0 \end{cases}$

SOLUTION

(a) **Graph each inequality.** We must graph the lines that correspond to these inequalities and then shade the appropriate regions. The graph of $10x + 20y \geq 60$ is the region above the line $y = 3 - \frac{1}{2}x$. The graph of $30x + 20y \geq 100$ is the region above the line $y = 5 - \frac{3}{2}x$, and the graph of $10x + 40y \geq 80$ is the region above the line $y = 2 - \frac{1}{4}x$.

Graph the solution of the system. The inequalities $x \geq 0$ and $y \geq 0$ indicate that the region is in the first quadrant. With this information we graph the system of inequalities in Figure 7.

Find the vertices. We determine the vertices of the region by finding the points of intersection of the appropriate lines. You can check that the vertices of the region are the ones indicated in Figure 7.

(b) The graph of the first inequality $10x + 20y \leq 60$ is the region below the line $y = 3 - \frac{1}{2}x$, and all the other inequalities are the same as those in part (a), so the solution to the system is the region (colored purple) shown in Figure 8.

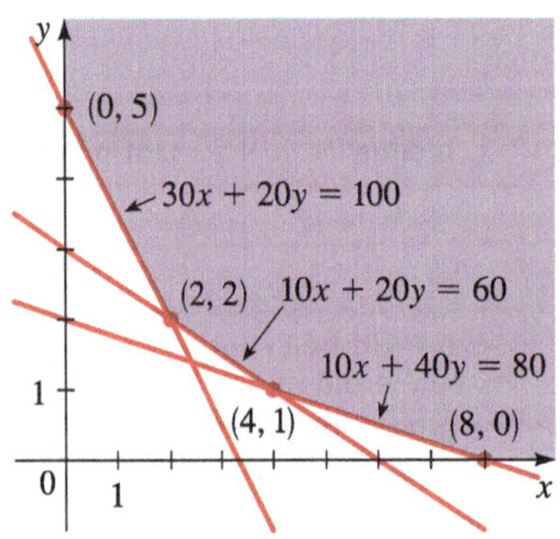

FIGURE 7

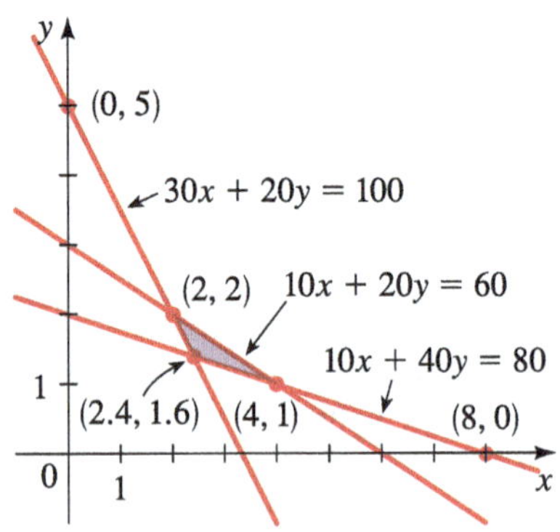

FIGURE 8

Now Try Exercises 59 and 63 ■

EXAMPLE 5 ■ A System of Linear Inequalities

Graph the solution set of the system of inequalities, and label the vertices.

$$\begin{cases} x + 2y \geq 8 \\ -x + 2y \leq 4 \\ 3x - 2y \leq 8 \end{cases}$$

SOLUTION **Graph each inequality.** We must graph the lines that correspond to these inequalities and then shade the appropriate regions, as in Example 2. We will use a graphing calculator, so we must first isolate y on the left-hand side of each inequality.

$$\begin{cases} y \geq -\frac{1}{2}x + 4 \\ y \leq \frac{1}{2}x + 2 \\ y \geq \frac{3}{2}x - 4 \end{cases}$$

Using the shading feature of the calculator, we obtain the graph in Figure 9(a). Note that the calculator shades each region in a different pattern.

Graph the solution of the system. The solution set is the triangular region that is shaded in all three patterns. The solution set is graphed in Figure 9(b).

Find the vertices. We use the TRACE or the `Intersect` command to find the vertices of the region. The vertices are labeled in Figure 9(b).

See Appendix D, *Using the TI-83/84 Graphing Calculator*, for specific instructions on graphing inequalities. Go to **www.stewartmath.com**.

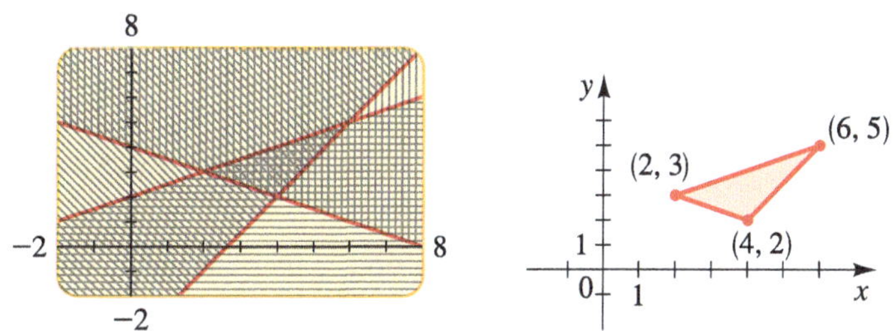

(a) Graphing calculator output

(b) Graph of solution set

FIGURE 9

Now Try Exercise 65 ■

A region in the plane is called **bounded** if it can be enclosed in a (sufficiently large) circle. A region that is not bounded is called **unbounded**. For example, the regions graphed in Figures 3, 5(b), 6(b), 8, and 9 are bounded because they can be enclosed in a circle, as illustrated in Figure 10(a). But the regions graphed in Figures 2, 4, and 7 are unbounded, because we cannot enclose them in a circle as illustrated in Figure 10(b).

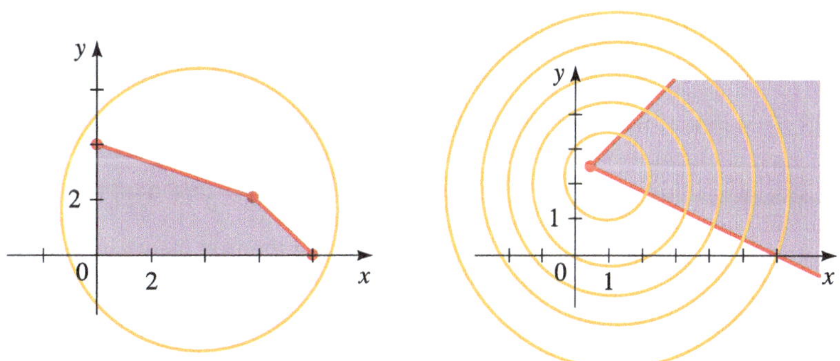

(a) A bounded region can be enclosed in a circle.

(b) An unbounded region cannot be enclosed in a circle.

FIGURE 10

Application: Feasible Regions

Many applied problems involve **constraints** on the variables. For instance, a factory manager has only a certain number of workers who can be assigned to perform jobs on the factory floor. A farmer deciding what crops to cultivate has only a certain amount of land that can be seeded. Such constraints or limitations can usually be expressed as systems of inequalities. When dealing with applied inequalities, we usually refer to the solution set of a system as a **feasible region**, because the points in the solution set represent feasible (or possible) values for the quantities being studied.

EXAMPLE 6 ■ Restricting Pollutant Outputs

A factory produces two agricultural pesticides, A and B. For every barrel of pesticide A, the factory emits 0.25 kg of carbon monoxide (CO) and 0.60 kg of sulfur dioxide (SO_2); and for every barrel of pesticide B, it emits 0.50 kg of CO and 0.20 kg of SO_2. Pollution laws restrict the factory's output of CO to a maximum of 75 kg per day and its output of SO_2 to a maximum of 90 kg per day.

(a) Find a system of inequalities that describes the number of barrels of each pesticide the factory can produce per day and still satisfy the pollution laws. Graph the feasible region.

(b) Would it be legal for the factory to produce 100 barrels of pesticide A and 80 barrels of pesticide B per day?

(c) Would it be legal for the factory to produce 60 barrels of pesticide A and 160 barrels of pesticide B per day?

SOLUTION

(a) We state the constraints as a system of inequalities and then graph the solution of the system.

Set up the inequalities. We first identify and name the variables, and we then express each statement in the problem in terms of the variables. We let the variable x represent the number of barrels of A produced per day and let y be the number of barrels of B produced per day. We can organize the information in the problem as follows.

In Words	In Algebra
Barrels of A produced	x
Barrels of B produced	y
Total CO produced	$0.25x + 0.50y$
Total SO_2 produced	$0.60x + 0.20y$

From the information in the problem and the fact that x and y can't be negative we obtain the following inequalities.

$$\begin{cases} 0.25x + 0.50y \le 75 & \text{At most 75 kg of CO can be produced} \\ 0.60x + 0.20y \le 90 & \text{At most 90 kg of SO}_2\text{ can be produced} \\ x \ge 0, \quad y \ge 0 \end{cases}$$

Multiplying the first inequality by 4 and the second by 5 simplifies the system to the following:

$$\begin{cases} x + 2y \le 300 \\ 3x + y \le 450 \\ x \ge 0, \quad y \ge 0 \end{cases}$$

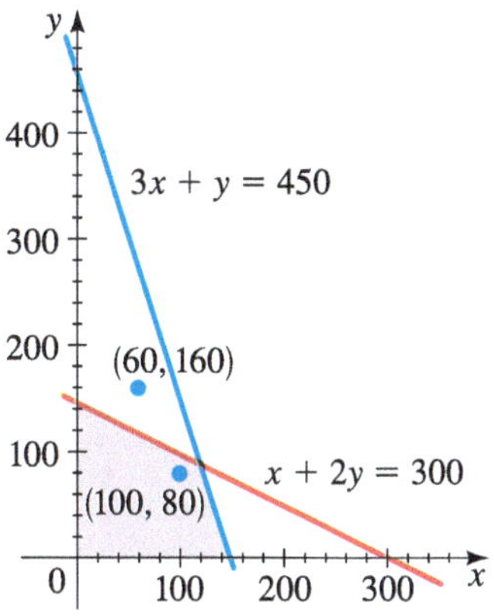

FIGURE 11

Graph the solution set. We first graph the equations

$$x + 2y = 300$$
$$3x + y = 450$$

The graphs are the two lines shown in Figure 11. Using the test point $(0, 0)$, we see that the solution set of each of these inequalities is the region below the corresponding line. So the solution to the system is the intersection of these sets as shown in Figure 11.

(b) Since the point $(100, 80)$ lies inside the feasible region, this production plan is legal (see Figure 11).

(c) Since the point $(60, 160)$ lies outside the feasible region, this production plan is not legal. It violates the CO restriction, although it does not violate the SO_2 restriction (see Figure 11).

Now Try Exercise 69 ■

10.9 EXERCISES

CONCEPTS

1. If the point $(2, 3)$ is a solution of an inequality in x and y, then the inequality is satisfied when we replace x by ________ and y by ________. Is the point $(2, 3)$ a solution of the inequality $4x - 2y \geq 1$?

2. To graph an inequality, we first graph the corresponding ________. So to graph the inequality $y \leq x + 1$, we first graph the equation ________. To decide which side of the graph of the equation is the graph of the inequality, we use ________ points. Complete the table, and sketch a graph of the inequality by shading the appropriate region.

Test point	Inequality $y \leq x + 1$	Conclusion
(0, 0) (0, 2)		

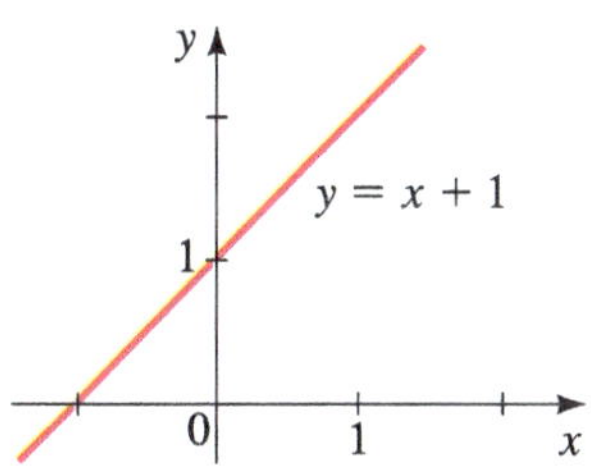

3. If the point $(2, 3)$ is a solution of a *system* of inequalities in x and y, then *each* inequality is satisfied when we replace x by ________ and y by ________. Is the point $(2, 3)$ a solution of the following system?

$$\begin{cases} 2x + 4y \leq 17 \\ 6x + 5y \leq 29 \end{cases}$$

4. Shade the solution of each system of inequalities on the given graph.

(a) $\begin{cases} x - y \geq 0 \\ x + y \geq 2 \end{cases}$ **(b)** $\begin{cases} x - y \leq 0 \\ x + y \leq 2 \end{cases}$

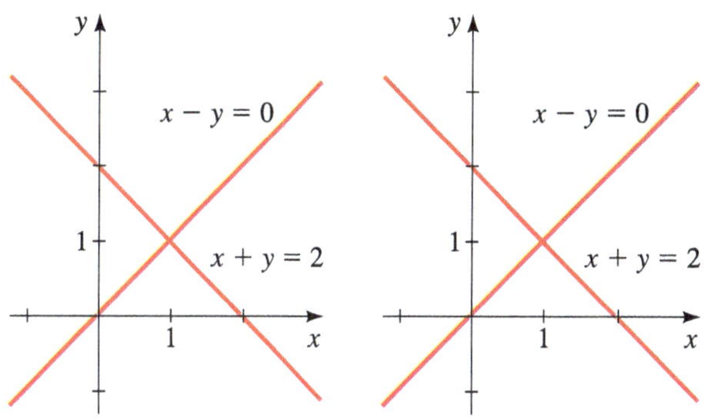

(c) $\begin{cases} x - y \geq 0 \\ x + y \leq 2 \end{cases}$ **(d)** $\begin{cases} x - y \leq 0 \\ x + y \geq 2 \end{cases}$

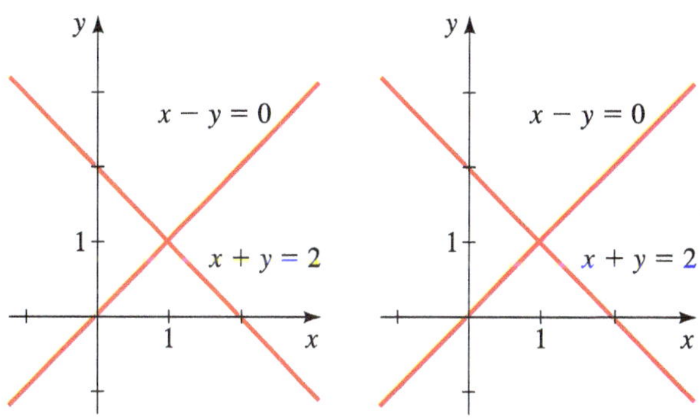

SKILLS

5–6 ■ Solutions of Inequalities An inequality and several points are given. For each point determine whether it is a solution of the inequality.

5. $x - 5y > 3$; $(-1, -2), (1, -2), (1, 2), (8, 1)$

6. $3x + 2y \le 2$; $(-2, 1), (1, 3), (1, -3), (0, 1)$

7–8 ■ Solutions of Systems of Inequalities A system of inequalities and several points are given. Determine which points are solutions of the system.

7. $\begin{cases} 3x - 2y \le 5 \\ 2x + y \ge 3 \end{cases}$; $(0, 0), (1, 2), (1, 1), (3, 1)$

8. $\begin{cases} x + 2y \ge 4 \\ 4x + 3y \ge 11 \end{cases}$; $(0, 0), (1, 3), (3, 0), (1, 2)$

9–22 ■ Graphing Inequalities Graph the inequality.

9. $y < -2x$ **10.** $y \ge 3x$

11. $y \ge 2$ **12.** $x \le -1$

13. $x < 2$ **14.** $y > 1$

15. $y > x - 3$ **16.** $y \le 1 - x$

17. $2x - y \ge -4$ **18.** $3x - y - 9 < 0$

19. $-x^2 + y \ge 5$ **20.** $y > x^2 + 1$

21. $x^2 + y^2 > 9$ **22.** $x^2 + (y - 2)^2 \le 4$

23–26 ■ Graphing Inequalities Use a graphing calculator to graph the linear inequality.

23. $3x - 2y \ge 18$ **24.** $4x + 3y \le 9$

25. $5x + 2y > 8$ **26.** $5x - 3y \ge 15$

27–30 ■ Finding Inequalities from a Graph An equation and its graph are given. Find an inequality whose solution is the shaded region.

27. $y = \frac{1}{2}x - 1$

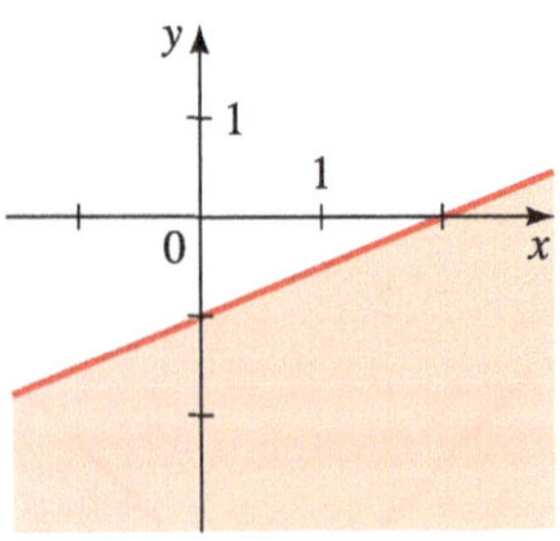

28. $y = x^2 + 2$

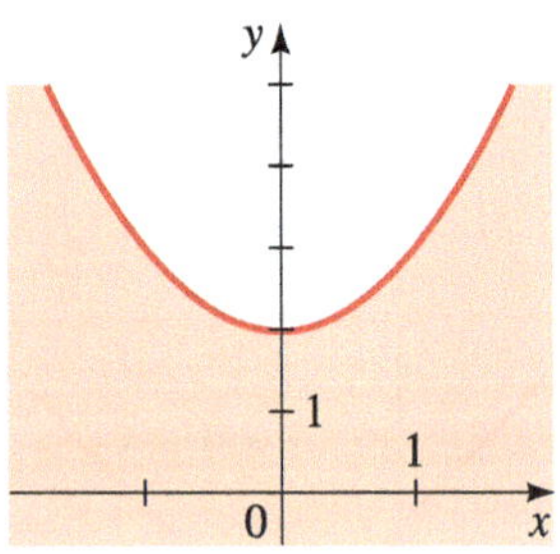

29. $x^2 + y^2 = 4$

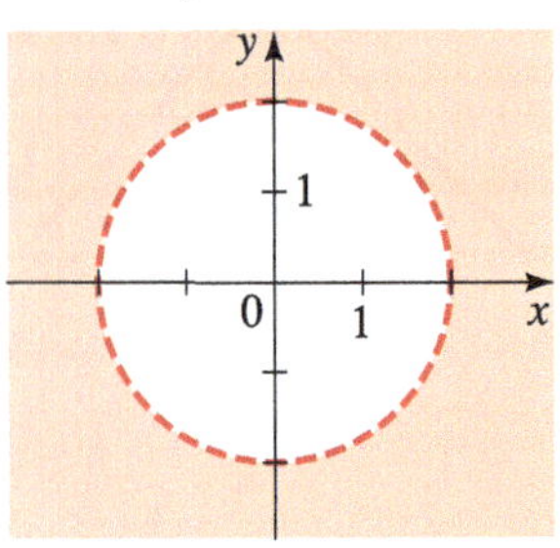

30. $y = x^3 - 4x$

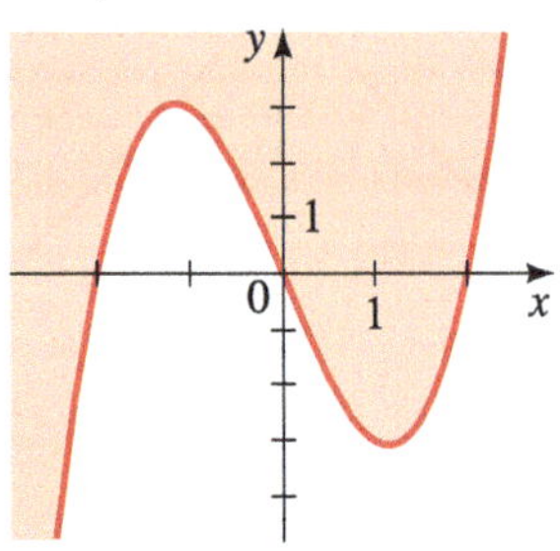

31–58 ■ Systems of Inequalities Graph the solution set of the system of inequalities. Find the coordinates of all vertices, and determine whether the solution set is bounded.

31. $\begin{cases} x + y \le 4 \\ y \ge x \end{cases}$ **32.** $\begin{cases} 2x + 3y > 12 \\ 3x - y < 21 \end{cases}$

33. $\begin{cases} y < \frac{1}{4}x + 2 \\ y \ge 2x - 5 \end{cases}$ **34.** $\begin{cases} x - y > 0 \\ 4 + y \le 2x \end{cases}$

35. $\begin{cases} y \le -2x + 8 \\ y \le -\frac{1}{2}x + 5 \\ x \ge 0, \quad y \ge 0 \end{cases}$ **36.** $\begin{cases} 4x + 3y \le 18 \\ 2x + y \le 8 \\ x \ge 0, \quad y \ge 0 \end{cases}$

37. $\begin{cases} x \ge 0 \\ y \ge 0 \\ 3x + 5y \le 15 \\ 3x + 2y \le 9 \end{cases}$ **38.** $\begin{cases} x > 2 \\ y < 12 \\ 2x - 4y > 8 \end{cases}$

39. $\begin{cases} y \le 9 - x^2 \\ x \ge 0, \quad y \ge 0 \end{cases}$ **40.** $\begin{cases} y \ge x^2 \\ y \le 4 \\ x \ge 0 \end{cases}$

41. $\begin{cases} y < 9 - x^2 \\ y \ge x + 3 \end{cases}$ **42.** $\begin{cases} y \ge x^2 \\ x + y \ge 6 \end{cases}$

43. $\begin{cases} x^2 + y^2 \le 4 \\ x - y > 0 \end{cases}$ **44.** $\begin{cases} x > 0 \\ y > 0 \\ x + y < 10 \\ x^2 + y^2 > 9 \end{cases}$

45. $\begin{cases} x^2 - y \le 0 \\ 2x^2 + y \le 12 \end{cases}$ **46.** $\begin{cases} 2x^2 + y > 4 \\ x^2 - y \le 8 \end{cases}$

47. $\begin{cases} x^2 + y^2 \le 9 \\ 2x + y^2 \le 1 \end{cases}$ **48.** $\begin{cases} x^2 + y^2 \le 4 \\ x^2 - 2y > 1 \end{cases}$

49. $\begin{cases} x + 2y \le 14 \\ 3x - y \ge 0 \\ x - y \ge 2 \end{cases}$ **50.** $\begin{cases} y < x + 6 \\ 3x + 2y \ge 12 \\ x - 2y \le 2 \end{cases}$

51. $\begin{cases} x \ge 0 \\ y \ge 0 \\ x \le 5 \\ x + y \le 7 \end{cases}$ **52.** $\begin{cases} x \ge 0 \\ y \ge 0 \\ y \le 4 \\ 2x + y \le 8 \end{cases}$

53. $\begin{cases} y > x + 1 \\ x + 2y \le 12 \\ x + 1 > 0 \end{cases}$ **54.** $\begin{cases} x + y > 12 \\ y < \frac{1}{2}x - 6 \\ 3x + y < 6 \end{cases}$

55. $\begin{cases} x^2 + y^2 \le 8 \\ x \ge 2 \\ y \ge 0 \end{cases}$ **56.** $\begin{cases} x^2 - y \ge 0 \\ x + y < 6 \\ x - y < 6 \end{cases}$

57. $\begin{cases} x^2 + y^2 < 9 \\ x + y > 0 \\ x \le 0 \end{cases}$ **58.** $\begin{cases} y \ge x^3 \\ y \le 2x + 4 \\ x + y \ge 0 \end{cases}$

59–64 ■ Systems of Inequalities Graph the system of inequalities, label the vertices, and determine whether the region is bounded or unbounded.

59. $\begin{cases} x + 2y \le 14 \\ 3x - y \ge 0 \\ x - y \le 2 \end{cases}$ **60.** $\begin{cases} x + 2y \le 14 \\ 3x - y \ge 0 \\ x - y \ge 2 \end{cases}$

61. $\begin{cases} x + y \le 12 \\ y \le \frac{1}{2}x - 6 \\ y \le 2x + 6 \end{cases}$

62. $\begin{cases} y \ge x + 1 \\ x + 2y \le 12 \\ x + 1 \ge 0 \end{cases}$

63. $\begin{cases} 30x + 10y \ge 50 \\ 10x + 20y \ge 50 \\ 10x + 60y \ge 90 \\ x \ge 0, \quad y \ge 0 \end{cases}$

64. $\begin{cases} x + y \ge 6 \\ 4x + 7y \le 39 \\ x + 5y \ge 13 \\ x \ge 0, \quad y \ge 0 \end{cases}$

65–68 ■ Graphing Systems of Inequalities Use a graphing calculator to graph the solution of the system of inequalities. Find the coordinates of all vertices, rounded to one decimal place.

65. $\begin{cases} y \ge x - 3 \\ y \ge -2x + 6 \\ y \le 8 \end{cases}$

66. $\begin{cases} x + y \ge 12 \\ 2x + y \le 24 \\ x - y \ge -6 \end{cases}$

67. $\begin{cases} y \le 6x - x^2 \\ x + y \ge 4 \end{cases}$

68. $\begin{cases} y \ge x^3 \\ 2x + y \ge 0 \\ y \le 2x + 6 \end{cases}$

APPLICATIONS

69. **Planting Crops** A farmer has 500 acres of arable land on which he wants to plant potatoes and corn. The farmer has \$40,000 available for planting and \$30,000 for fertilizer. Planting 1 acre of potatoes costs \$90, and planting 1 acre of corn costs \$50. Fertilizer costs \$30 for 1 acre of potatoes and \$80 for 1 acre of corn.

(a) Find a system of inequalities that describes the number of acres of each crop that the farmer can plant with the available resources. Graph the feasible region.

(b) Can the farmer plant 300 acres of potatoes and 180 acres of corn?

(c) Can the farmer plant 150 acres of potatoes and 325 acres of corn?

70. **Planting Crops** A farmer has 300 acres of arable land on which she wants to plant cauliflower and cabbage. The farmer has \$17,500 available for planting and \$12,000 for fertilizer. Planting 1 acre of cauliflower costs \$70, and planting 1 acre of cabbage costs \$35. Fertilizer costs \$25 for 1 acre of cauliflower and \$55 for 1 acre of cabbage.

(a) Find a system of inequalities that describes the number of acres of each crop that the farmer can plant with the available resources. Graph the feasible region.

(b) Can the farmer plant 155 acres of cauliflower and 115 acres of cabbage?

(c) Can the farmer plant 115 acres of cauliflower and 175 acres of cabbage?

71. **Publishing Books** A publishing company publishes a total of no more than 100 books every year. At least 20 of these are nonfiction, but the company always publishes at least as much fiction as nonfiction. Find a system of inequalities that describes the possible numbers of fiction and nonfiction books that the company can produce each year consistent with these policies. Graph the solution set.

72. **Furniture Manufacturing** A man and his daughter manufacture unfinished tables and chairs. Each table requires 3 h of sawing and 1 h of assembly. Each chair requires 2 h of sawing and 2 h of assembly. Between the two of them, they can put in up to 12 h of sawing and 8 h of assembly work each day. Find a system of inequalities that describes all possible combinations of tables and chairs that they can make daily. Graph the solution set.

73. **Coffee Blends** A coffee merchant sells two different coffee blends. The Standard blend uses 4 oz of arabica and 12 oz of robusta beans per package; the Deluxe blend uses 10 oz of arabica and 6 oz of robusta beans per package. The merchant has 80 lb of arabica and 90 lb of robusta beans available. Find a system of inequalities that describes the possible number of Standard and Deluxe packages the merchant can make. Graph the solution set.

74. **Nutrition** A cat food manufacturer uses fish and beef by-products. The fish contains 12 g of protein and 3 g of fat per ounce. The beef contains 6 g of protein and 9 g of fat per ounce. Each can of cat food must contain at least 60 g of protein and 45 g of fat. Find a system of inequalities that describes the possible number of ounces of fish and beef by-products that can be used in each can to satisfy these minimum requirements. Graph the solution set.

DISCUSS ■ DISCOVER ■ PROVE ■ WRITE

75. **DISCUSS: Shading Unwanted Regions** To graph the solution of a system of inequalities, we have shaded the solution of each inequality in a different color; the solution of the system is the region where all the shaded parts overlap. Here is a different method: For each inequality, shade the region that does *not* satisfy the inequality. Explain why the part of the plane that is left unshaded is the solution of the system. Solve the following system by both methods. Which do you prefer? Why?

$$\begin{cases} x + 2y > 4 \\ -x + y < 1 \\ x + 3y < 9 \\ x < 3 \end{cases}$$

CHAPTER 10 ■ REVIEW

PROPERTIES AND FORMULAS

Systems of Equations (p. 680)

A **system of equations** is a set of equations that involve the same variables. A **system of linear equations** is a system of equations in which each equation is linear. Systems of of linear equations in two variables (x and y) and three variables (x, y, and z) have the following forms:

Linear system 2 variables	Linear system 3 variables
$a_{11}x + a_{12}y = b_1$	$a_{11}x + a_{12}y + a_{13}z = b_1$
$a_{21}x + a_{22}y = b_2$	$a_{21}x + a_{22}y + a_{23}z = b_2$
	$a_{31}x + a_{32}y + a_{33}z = b_3$

A **solution** of a system of equations is an assignment of values for the variables that makes *each* equation in the system true. To **solve** a system means to find all solutions of the system.

Substitution Method (p. 680)

To solve a pair of equations in two variables by substitution:

1. **Solve for one variable** in terms of the other variable in one equation.
2. **Substitute** into the other equation to get an equation in one variable, and solve for this variable.
3. **Back-substitute** the value(s) of the variable you have found into either original equation, and solve for the remaining variable.

Elimination Method (p. 681)

To solve a pair of equations in two variables by elimination:

1. **Adjust the coefficients** by multiplying the equations by appropriate constants so that the term(s) involving one of the variables are of opposite sign in the equations.
2. **Add the equations** to eliminate that one variable; this gives an equation in the other variable. Solve for this variable.
3. **Back-substitute** the value(s) of the variable that you have found into either original equation, and solve for the remaining variable.

Graphical Method (p. 682)

To solve a pair of equations in two variables graphically, first put each equation in function form, $y = f(x)$.

1. **Graph the equations** on a common screen.
2. **Find the points of intersection** of the graphs. The solutions are the x- and y-coordinates of the points of intersection.

Gaussian Elimination (p. 691)

When we use **Gaussian elimination** to solve a system of linear equations, we use the following operations to change the system to an **equivalent** simpler system:

1. Add a nonzero multiple of one equation to another.
2. Multiply an equation by a nonzero constant.
3. Interchange the position of two equations in the system.

Number of Solutions of a Linear System (p. 693)

A system of linear equations can have:

1. A unique solution for each variable.
2. No solution, in which case the system is **inconsistent**.
3. Infinitely many solutions, in which case the system is **dependent**.

How to Determine the Number of Solutions of a Linear System (p. 693)

When we use **Gaussian elimination** to solve a system of linear equations, then we can tell that the system has:

1. **No solution** (is *inconsistent*) if we arrive at a false equation of the form $0 = c$, where c is nonzero.
2. **Infinitely many solutions** (is *dependent*) if the system is consistent but we end up with fewer equations than variables (after discarding redundant equations of the form $0 = 0$).

Matrices (p. 699)

A **matrix** A of **dimension** $m \times n$ is a rectangular array of numbers with m **rows** and n **columns**:

$$A = \begin{bmatrix} a_{11} & a_{12} & \cdots & a_{1n} \\ a_{21} & a_{22} & \cdots & a_{2n} \\ \vdots & \vdots & \ddots & \vdots \\ a_{m1} & a_{m2} & \cdots & a_{mn} \end{bmatrix}$$

Augmented Matrix of a System (p. 700)

The **augmented matrix** of a system of linear equations is the matrix consisting of the coefficients and the constant terms. For example, for the two-variable system

$$a_{11}x + a_{12}x = b_1$$
$$a_{21}x + a_{22}x = b_2$$

the augmented matrix is

$$\begin{bmatrix} a_{11} & a_{12} & b_1 \\ a_{21} & a_{22} & b_2 \end{bmatrix}$$

Elementary Row Operations (p. 700)

To solve a system of linear equations using the augmented matrix of the system, the following operations can be used to transform the rows of the matrix:

1. Add a nonzero multiple of one row to another.
2. Multiply a row by a nonzero constant.
3. Interchange two rows.

Row-Echelon Form of a Matrix (p. 702)

A matrix is in **row-echelon form** if its entries satisfy the following conditions:

1. The first nonzero entry in each row (the **leading entry**) is the number 1.

2. The leading entry of each row is to the right of the leading entry in the row above it.
3. All rows consisting entirely of zeros are at the bottom of the matrix.

If the matrix also satisfies the following condition, it is in **reduced row-echelon** form:

4. If a column contains a leading entry, then every other entry in that column is a 0.

Number of Solutions of a Linear System (p. 705)

If the augmented matrix of a system of linear equations has been reduced to row-echelon form using elementary row operations, then the system has:

1. **No solution** if the row-echelon form contains a row that represents the equation $0 = 1$. In this case the system is **inconsistent**.
2. **One solution** if each variable in the row-echelon form is a leading variable.
3. **Infinitely many solutions** if the system is not inconsistent but not every variable is a leading variable. In this case the system is **dependent**.

Operations on Matrices (p. 713)

If A and B are $m \times n$ matrices and c is a scalar (real number), then:

1. The **sum** $A + B$ is the $m \times n$ matrix that is obtained by adding corresponding entries of A and B.
2. The **difference** $A - B$ is the $m \times n$ matrix that is obtained by subtracting corresponding entries of A and B.
3. The **scalar product** cA is the $m \times n$ matrix that is obtained by multiplying each entry of A by c.

Multiplication of Matrices (p. 715)

If A is an $m \times n$ matrix and B is an $n \times k$ matrix (so the number of columns of A is the same as the number of rows of B), then the **matrix product** AB is the $m \times k$ matrix whose ij-entry is the inner product of the ith row of A and the jth column of B.

Properties of Matrix Operations (pp. 714, 716)

If A, B, and C are matrices of compatible dimensions then the following properties hold:

1. **Commutativity of addition:**

$$A + B = B + A$$

2. **Associativity:**

$$(A + B) + C = A + (B + C)$$

$$(AB)C = A(BC)$$

3. **Distributivity:**

$$A(B + C) = AB + AC$$

$$(B + C)A = BA + CA$$

(Note that matrix *multiplication* is *not* commutative.)

Identity Matrix (p. 724)

The **identity matrix** I_n is the $n \times n$ matrix whose main diagonal entries are all 1 and whose other entries are all 0:

$$I_n = \begin{bmatrix} 1 & 0 & \cdots & 0 \\ 0 & 1 & \cdots & 0 \\ \vdots & \vdots & \ddots & \vdots \\ 0 & 0 & \cdots & 1 \end{bmatrix}$$

If A is an $m \times n$ matrix, then

$$AI_n = A \qquad \text{and} \qquad I_mA = A$$

Inverse of a Matrix (p. 725)

If A is an $n \times n$ matrix, then the inverse of A is the $n \times n$ matrix A^{-1} with the following properties:

$$A^{-1}A = I_n \qquad \text{and} \qquad AA^{-1} = I_n$$

To find the inverse of a matrix, we use a procedure involving elementary row operations (explained on page 726). (Note that *some* square matrices do not have an inverse.)

Inverse of a 2 × 2 Matrix (p. 725)

For 2×2 matrices the following special rule provides a shortcut for finding the inverse:

$$A = \begin{bmatrix} a & b \\ c & d \end{bmatrix} \Rightarrow A^{-1} = \frac{1}{ad - bc}\begin{bmatrix} d & -b \\ -c & a \end{bmatrix}$$

Writing a Linear System as a Matrix Equation (p. 728)

A system of n linear equations in n variables can be written as a single matrix equation

$$AX = B$$

where A is the $n \times n$ matrix of coefficients, X is the $n \times 1$ matrix of the variables, and B is the $n \times 1$ matrix of the constants. For example, the linear system of two equations in two variables

$$a_{11}x + a_{12}x = b_1$$

$$a_{21}x + a_{22}x = b_2$$

can be expressed as

$$\begin{bmatrix} a_{11} & a_{12} \\ a_{21} & a_{22} \end{bmatrix}\begin{bmatrix} x \\ y \end{bmatrix} = \begin{bmatrix} b_1 \\ b_2 \end{bmatrix}$$

Solving Matrix Equations (p. 729)

If A is an invertible $n \times n$ matrix, X is an $n \times 1$ variable matrix, and B is an $n \times 1$ constant matrix, then the matrix equation

$$AX = B$$

has the unique solution

$$X = A^{-1}B$$

Determinant of a 2 × 2 Matrix (p. 734)

The **determinant** of the matrix

$$A = \begin{bmatrix} a & b \\ c & d \end{bmatrix}$$

is the *number*

$$\det(A) = |A| = ad - bc$$

Minors and Cofactors (p. 734)

If $A = |a_{ij}|$ is an $n \times n$ matrix, then the **minor** M_{ij} of the entry a_{ij} is the determinant of the matrix obtained by deleting the ith row and the jth column of A.

The **cofactor** A_{ij} of the entry a_{ij} is

$$A_{ij} = (-1)^{i+j} M_{ij}$$

(Thus, the minor and the cofactor of each entry either are the same or are negatives of each other.)

Determinant of an $n \times n$ Matrix (p. 735)

To find the **determinant** of the $n \times n$ matrix

$$A = \begin{bmatrix} a_{11} & a_{12} & \cdots & a_{1n} \\ a_{21} & a_{22} & \cdots & a_{2n} \\ \vdots & \vdots & \ddots & \vdots \\ a_{n1} & a_{n2} & \cdots & a_{nn} \end{bmatrix}$$

we choose a row or column to **expand**, and then we calculate the number that is obtained by multiplying each element of that row or column by its cofactor and then adding the resulting products. For example, if we choose to expand about the first row, we get

$$\det(A) = |A| = a_{11}A_{11} + a_{12}A_{12} + \cdots + a_{1n}A_{1n}$$

Invertibility Criterion (p. 736)

A square matrix has an inverse if and only if its determinant is not 0.

Row and Column Transformations (p. 737)

If we add a nonzero multiple of one row to another row in a square matrix or a nonzero multiple of one column to another column, then the determinant of the matrix is unchanged.

Cramer's Rule (pp. 738–740)

If a system of n linear equations in the n variables $x_1, x_2, \ldots, x_n$ is equivalent to the matrix equation $DX = B$ and if $|D| \neq 0$, then the solutions of the system are

$$x_1 = \frac{|D_{x_1}|}{|D|} \qquad x_2 = \frac{|D_{x_2}|}{|D|} \qquad \cdots \qquad x_n = \frac{|D_{x_n}|}{|D|}$$

where D_{x_i} is the matrix that is obtained from D by replacing its ith column by the constant matrix B.

Area of a Triangle Using Determinants (p. 741)

If a triangle in the coordinate plane has vertices (a_1, b_1), (a_2, b_2), and (a_3, b_3), then the area of the triangle is given by

$$\mathcal{A} = \pm\frac{1}{2}\begin{vmatrix} a_1 & b_1 & 1 \\ a_2 & b_2 & 1 \\ a_3 & b_3 & 1 \end{vmatrix}$$

where the sign is chosen to make the area positive.

Partial Fractions (pp. 745–749)

The *partial fraction decomposition* of a rational function

$$r(x) = \frac{P(x)}{Q(x)}$$

(where the degree of P is less than the degree of Q) is a sum of simpler fractional expressions that equal $r(x)$ when brought to a common denominator. The denominator of each simpler fraction is either a linear or quadratic factor of $Q(x)$ or a power of such a linear or quadratic factor. So to find the terms of the partial fraction decomposition, we first factor $Q(x)$ into linear and irreducible quadratic factors. The terms then have the following forms, depending on the factors of $Q(x)$.

1. For every **distinct linear factor** $ax + b$ there is a term of the form
$$\frac{A}{ax + b}$$
2. For every **repeated linear factor** $(ax + b)^m$ there are terms of the form
$$\frac{A_1}{ax + b} + \frac{A_2}{(ax + b)^2} + \cdots + \frac{A_m}{(ax + b)^m}$$
3. For every **distinct quadratic factor** $ax^2 + bx + c$ there is a term of the form
$$\frac{Ax + B}{ax^2 + bx + c}$$
4. For every **repeated quadratic factor** $(ax^2 + bx + c)^m$ there are terms of the form
$$\frac{A_1x + B_1}{ax^2 + bx + c} + \frac{A_2x + B_2}{(ax^2 + bx + c)^2} + \cdots + \frac{A_mx + B_m}{(ax^2 + bx + c)^m}$$

Graphing Inequalities (pp. 756–757)

To graph an inequality:

1. Graph the equation that corresponds to the inequality. This "boundary curve" divides the coordinate plane into separate regions.
2. Use **test points** to determine which region(s) satisfy the inequality.
3. Shade the region(s) that satisfy the inequality, and use a solid line for the boundary curve if it satisfies the inequality ($\leq$ or $\geq$) and a dashed line if it does not ($<$ or $>$).

Graphing Systems of Inequalities (p. 758)

To graph the solution of a system of inequalities (or **feasible region** determined by the inequalities):

1. Graph all the inequalities on the same coordinate plane.
2. The solution is the intersection of the solutions of all the inequalities, so shade the region that satisfies all the inequalities.
3. Determine the coordinates of the intersection points of all the boundary curves that touch the solution set of the system. These points are the **vertices** of the solution.

CONCEPT CHECK

1. **(a)** What are the three methods we use to solve a system of equations?
   (b) Solve the system by the elimination method and by the graphical method.
$$\begin{cases} x + y = 3 \\ 3x - y = 1 \end{cases}$$
2. For a system of two linear equations in two variables:
   (a) How many solutions are possible?
   (b) What is meant by an inconsistent system? a dependent system?
3. What operations can be performed on a linear system so as to arrive at an equivalent system?
4. **(a)** Explain how Gaussian elimination works.
   (b) Use Gaussian elimination to put the following system in triangular form, and then solve the system.

System	**Triangular form**
$\begin{cases} x + y - 2z = 3 \\ x + 2y + z = 5 \\ 3x - y + 5z = 1 \end{cases}$	

5. What does it mean to say that A is a matrix with dimension $m \times n$?
6. What is the row-echelon form of a matrix? What is a leading entry?
7. **(a)** What is the augmented matrix of a system? What are leading variables?
   (b) What are the elementary row operations on an augmented matrix?
   (c) How do we solve a system using the augmented matrix?
   (d) Write the augmented matrix of the following system of linear equations.
$$\begin{cases} x + y - 2z = 3 \\ x + 2y + z = 5 \\ 3x - y + 5z = 1 \end{cases}$$
   (e) Solve the system in part (d).
8. Suppose you have used Gaussian elimination to transform the augmented matrix of a linear system into row-echelon form. How can you tell whether the system has exactly one solution? no solution? infinitely many solutions?
9. What is the reduced row echelon form of a matrix?
10. **(a)** How do Gaussian elimination and Gauss-Jordan elimination differ?
    (b) Use Gauss-Jordan elimination to solve the linear system in part 7(d).
11. If A and B are matrices with the same dimension and k is a real number, how do you find $A + B$ and kA?
12. **(a)** What must be true of the dimensions of A and B for the product AB to be defined?
    (b) If A has dimension 2×3 and if B has dimension 3×2, is the product AB defined? If so, what is the dimension of AB?
    (c) Find the matrix product.
$$\begin{bmatrix} 2 & 1 \\ 4 & 0 \end{bmatrix}\begin{bmatrix} 3 & 4 & 1 \\ 5 & 1 & 2 \end{bmatrix}$$
13. **(a)** What is an identity matrix I_n? If A is an $n \times n$ matrix, what are the products AI_n and I_nA?
    (b) If A is an $n \times n$ matrix, what is its inverse matrix?
    (c) Complete the formula for the inverse of a 2×2 matrix
$$A = \begin{bmatrix} a & b \\ c & d \end{bmatrix}$$
    (d) Find the inverse of A.
$$A = \begin{bmatrix} 1 & 1 \\ 3 & -1 \end{bmatrix}$$
14. **(a)** Express the system in 1(b) as a matrix equation $AX = B$.
    (b) If a linear system is expressed as a matrix equation $AX = B$, how do we solve the system? Solve the system in part (a).
15. **(a)** Is it true that the determinant det A of a matrix A is defined only if A is a square matrix?
    (b) Find the determinant of the matrix A in part 13(d).
    (c) Use Cramer's Rule to solve the system in 1(b).
16. **(a)** How do we express a rational function r as a partial fraction decomposition?
    (b) Give the form of the partial fraction decomposition.
    (i) $\dfrac{2x}{(x-5)(x-2)^2}$ (ii) $\dfrac{2x}{(x-5)(x^2+1)}$
17. **(a)** How do we graph an inequality in two variables?
    (b) Graph the solution set of the inequality $x + y \geq 3$.
    (c) Graph the solution set of the system of inequalities: $x + y \geq 3$, $3x - y \geq 1$.

ANSWERS TO THE CONCEPT CHECK CAN BE FOUND AT THE BACK OF THE BOOK.

EXERCISES

1–6 ■ Systems of Linear Equations in Two Variables Solve the system of equations, and graph the lines.

1. $\begin{cases} 3x - y = 5 \\ 2x + y = 5 \end{cases}$

2. $\begin{cases} y = 2x + 6 \\ y = -x + 3 \end{cases}$

3. $\begin{cases} 2x - 7y = 28 \\ y = \frac{2}{7}x - 4 \end{cases}$

4. $\begin{cases} 6x - 8y = 15 \\ -\frac{3}{2}x + 2y = -4 \end{cases}$

5. $\begin{cases} 2x - y = 1 \\ x + 3y = 10 \\ 3x + 4y = 15 \end{cases}$

6. $\begin{cases} 2x + 5y = 9 \\ -x + 3y = 1 \\ 7x - 2y = 14 \end{cases}$

7–10 ■ Systems of Nonlinear Equations Solve the system of equations.

7. $\begin{cases} y = x^2 + 2x \\ y = 6 + x \end{cases}$

8. $\begin{cases} x^2 + y^2 = 8 \\ y = x + 2 \end{cases}$

9. $\begin{cases} 3x + \dfrac{4}{y} = 6 \\ x - \dfrac{8}{y} = 4 \end{cases}$

10. $\begin{cases} x^2 + y^2 = 10 \\ x^2 + 2y^2 - 7y = 0 \end{cases}$

11–14 ■ Systems of Nonlinear Equations Use a graphing device to solve the system. Round answers to the nearest hundredth.

11. $\begin{cases} 0.32x + 0.43y = 0 \\ 7x - 12y = 341 \end{cases}$

12. $\begin{cases} \sqrt{12}x - 3\sqrt{2}y = 660 \\ 7137x + 3931y = 20{,}000 \end{cases}$

13. $\begin{cases} x - y^2 = 10 \\ x = \frac{1}{22}y + 12 \end{cases}$

14. $\begin{cases} y = 5^x + x \\ y = x^5 + 5 \end{cases}$

15–20 ■ Matrices A matrix is given.

(a) State the dimension of the matrix.
(b) Is the matrix in row-echelon form?
(c) Is the matrix in reduced row-echelon form?
(d) Write the system of equations for which the given matrix is the augmented matrix.

15. $\begin{bmatrix} 1 & 2 & -5 \\ 0 & 1 & 3 \end{bmatrix}$

16. $\begin{bmatrix} 1 & 0 & 6 \\ 0 & 1 & 0 \end{bmatrix}$

17. $\begin{bmatrix} 1 & 0 & 8 & 0 \\ 0 & 1 & 5 & -1 \\ 0 & 0 & 0 & 0 \end{bmatrix}$

18. $\begin{bmatrix} 1 & 3 & 6 & 2 \\ 2 & 1 & 0 & 5 \\ 0 & 0 & 1 & 0 \end{bmatrix}$

19. $\begin{bmatrix} 0 & 1 & -3 & 4 \\ 1 & 1 & 0 & 7 \\ 1 & 2 & 1 & 2 \end{bmatrix}$

20. $\begin{bmatrix} 1 & 8 & 6 & -4 \\ 0 & 1 & -3 & 5 \\ 0 & 0 & 2 & -7 \\ 1 & 1 & 1 & 0 \end{bmatrix}$

21–42 ■ Systems of Linear Equations in Several Variables Find the complete solution of the system, or show that the system has no solution.

21. $\begin{cases} x + y + 2z = 6 \\ 2x + 5z = 12 \\ x + 2y + 3z = 9 \end{cases}$

22. $\begin{cases} x - 2y + 3z = 1 \\ x - 3y - z = 0 \\ 2x - 6z = 6 \end{cases}$

23. $\begin{cases} x - 2y + 3z = 1 \\ 2x - y + z = 3 \\ 2x - 7y + 11z = 2 \end{cases}$

24. $\begin{cases} x + y + z + w = 2 \\ 2x - 3z = 5 \\ x - 2y + 4w = 9 \\ x + y + 2z + 3w = 5 \end{cases}$

25. $\begin{cases} x + 2y + 2z = 6 \\ x - y = -1 \\ 2x + y + 3z = 7 \end{cases}$

26. $\begin{cases} x - y + z = 2 \\ x + y + 3z = 6 \\ 2y + 3z = 5 \end{cases}$

27. $\begin{cases} x - 2y + 3z = -2 \\ 2x - y + z = 2 \\ 2x - 7y + 11z = -9 \end{cases}$

28. $\begin{cases} x - y + z = 2 \\ x + y + 3z = 6 \\ 3x - y + 5z = 10 \end{cases}$

29. $\begin{cases} x + y + z + w = 0 \\ x - y - 4z - w = -1 \\ x - 2y + 4w = -7 \\ 2x + 2y + 3z + 4w = -3 \end{cases}$

30. $\begin{cases} x + 3z = -1 \\ y - 4w = 5 \\ 2y + z + w = 0 \\ 2x + y + 5z - 4w = 4 \end{cases}$

31. $\begin{cases} x - 3y + z = 4 \\ 4x - y + 15z = 5 \end{cases}$

32. $\begin{cases} 2x - 3y + 4z = 3 \\ 4x - 5y + 9z = 13 \\ 2x + 7z = 0 \end{cases}$

33. $\begin{cases} -x + 4y + z = 8 \\ 2x - 6y + z = -9 \\ x - 6y - 4z = -15 \end{cases}$

34. $\begin{cases} x - z + w = 2 \\ 2x + y - 2w = 12 \\ 3y + z + w = 4 \\ x + y - z = 10 \end{cases}$

35. $\begin{cases} x - y + 3z = 2 \\ 2x + y + z = 2 \\ 3x + 4z = 4 \end{cases}$

36. $\begin{cases} x - y = 1 \\ x + y + 2z = 3 \\ x - 3y - 2z = -1 \end{cases}$

37. $\begin{cases} x - y + z - w = 0 \\ 3x - y - z - w = 2 \end{cases}$

38. $\begin{cases} x - y = 3 \\ 2x + y = 6 \\ x - 2y = 9 \end{cases}$

39. $\begin{cases} x - y + z = 0 \\ 3x + 2y - z = 6 \\ x + 4y - 3z = 3 \end{cases}$

40. $\begin{cases} x + 2y + 3z = 2 \\ 2x - y - 5z = 1 \\ 4x + 3y + z = 6 \end{cases}$

41. $\begin{cases} x + y - z - w = 2 \\ x - y + z - w = 0 \\ 2x + 2w = 2 \\ 2x + 4y - 4z - 2w = 6 \end{cases}$

42. $\begin{cases} x - y - 2z + 3w = 0 \\ y - z + w = 1 \\ 3x - 2y - 7z + 10w = 2 \end{cases}$

43. Investments A man invests his savings in two accounts, one paying 6% interest per year and the other paying 7%. He has twice as much invested in the 7% account as in the 6% account, and his annual interest income is \$600. How much is invested in each account?

44. Number of Coins A piggy bank contains 50 coins, all of them nickels, dimes, or quarters. The total value of the coins is \$5.60, and the value of the dimes is five times the value of the nickels. How many coins of each type are there?

45. Investments Clarisse invests \$60,000 in money-market accounts at three different banks. Bank A pays 2% interest per year, bank B pays 2.5%, and bank C pays 3%. She decides to invest twice as much in bank B as in the other two banks. After 1 year, Clarisse has earned \$1575 in interest. How much did she invest in each bank?

46. Number of Fish Caught A commercial fisherman fishes for haddock, sea bass, and red snapper. He is paid \$1.25/lb for haddock, \$0.75/lb for sea bass, and \$2.00/lb for red snapper. Yesterday he caught 560 lb of fish worth \$575. The haddock and red snapper together are worth \$320. How many pounds of each fish did he catch?

47–58 ■ Matrix Operations Let

$$A = \begin{bmatrix} 2 & 0 & -1 \end{bmatrix} \qquad B = \begin{bmatrix} 1 & 2 & 4 \\ -2 & 1 & 0 \end{bmatrix}$$

$$C = \begin{bmatrix} \frac{1}{2} & 3 \\ 2 & \frac{3}{2} \\ -2 & 1 \end{bmatrix} \qquad D = \begin{bmatrix} 1 & 4 \\ 0 & -1 \\ 2 & 0 \end{bmatrix}$$

$$E = \begin{bmatrix} 2 & -1 \\ -\frac{1}{2} & 1 \end{bmatrix} \qquad F = \begin{bmatrix} 4 & 0 & 2 \\ -1 & 1 & 0 \\ 7 & 5 & 0 \end{bmatrix} \qquad G = [5]$$

Carry out the indicated operation, or explain why it cannot be performed.

47. $A + B$ **48.** $C - D$

49. $2C + 3D$ **50.** $5B - 2C$

51. GA **52.** AG

53. BC **54.** CB

55. BF **56.** FC

57. $(C + D)E$ **58.** $F(2C - D)$

59–60 ■ Inverse Matrices Verify that the matrices A and B are inverses of each other by calculating the products AB and BA.

59. $A = \begin{bmatrix} 2 & -5 \\ -2 & 6 \end{bmatrix}, \quad B = \begin{bmatrix} 3 & \frac{5}{2} \\ 1 & 1 \end{bmatrix}$

60. $A = \begin{bmatrix} 2 & -1 & 3 \\ 2 & -2 & 1 \\ 0 & 1 & 1 \end{bmatrix}, \quad B = \begin{bmatrix} -\frac{3}{2} & 2 & \frac{5}{2} \\ -1 & 1 & 2 \\ 1 & -1 & -1 \end{bmatrix}$

61–66 ■ Matrix Equations Solve the matrix equation for the unknown matrix X, or show that no solution exists, where

$$A = \begin{bmatrix} 2 & 1 \\ 3 & 2 \end{bmatrix} \qquad B = \begin{bmatrix} 1 & -2 \\ -2 & 4 \end{bmatrix} \qquad C = \begin{bmatrix} 0 & 1 & 3 \\ -2 & 4 & 0 \end{bmatrix}$$

61. $A + 3X = B$ **62.** $\frac{1}{2}(X - 2B) = A$

63. $2(X - A) = 3B$ **64.** $2X + C = 5A$

65. $AX = C$ **66.** $AX = B$

67–74 ■ Determinants and Inverse Matrices Find the determinant and, if possible, the inverse of the matrix.

67. $\begin{bmatrix} 1 & 4 \\ 2 & 9 \end{bmatrix}$ **68.** $\begin{bmatrix} 2 & 2 \\ 1 & -3 \end{bmatrix}$

69. $\begin{bmatrix} 4 & -12 \\ -2 & 6 \end{bmatrix}$ **70.** $\begin{bmatrix} 2 & 4 & 0 \\ -1 & 1 & 2 \\ 0 & 3 & 2 \end{bmatrix}$

71. $\begin{bmatrix} 3 & 0 & 1 \\ 2 & -3 & 0 \\ 4 & -2 & 1 \end{bmatrix}$ **72.** $\begin{bmatrix} 1 & 2 & 3 \\ 2 & 4 & 5 \\ 2 & 5 & 6 \end{bmatrix}$

73. $\begin{bmatrix} 1 & 0 & 0 & 1 \\ 0 & 2 & 0 & 2 \\ 0 & 0 & 3 & 3 \\ 0 & 0 & 0 & 4 \end{bmatrix}$ **74.** $\begin{bmatrix} 1 & 0 & 1 & 0 \\ 0 & 1 & 0 & 1 \\ 1 & 1 & 1 & 2 \\ 1 & 2 & 1 & 2 \end{bmatrix}$

75–78 ■ Using Inverse Matrices to Solve a System Express the system of linear equations as a matrix equation. Then solve the matrix equation by multiplying each side by the inverse of the coefficient matrix.

75. $\begin{cases} 12x - 5y = 10 \\ 5x - 2y = 17 \end{cases}$ **76.** $\begin{cases} 6x - 5y = 1 \\ 8x - 7y = -1 \end{cases}$

77. $\begin{cases} 2x + y + 5z = \frac{1}{3} \\ x + 2y + 2z = \frac{1}{4} \\ x + 3z = \frac{1}{6} \end{cases}$ **78.** $\begin{cases} 2x + 3z = 5 \\ x + y + 6z = 0 \\ 3x - y + z = 5 \end{cases}$

79–82 ■ Using Cramer's Rule to Solve a System Solve the system using Cramer's Rule.

79. $\begin{cases} 2x + 7y = 13 \\ 6x + 16y = 30 \end{cases}$ **80.** $\begin{cases} 12x - 11y = 140 \\ 7x + 9y = 20 \end{cases}$

81. $\begin{cases} 2x - y + 5z = 0 \\ -x + 7y = 9 \\ 5x + 4y + 3z = -9 \end{cases}$ **82.** $\begin{cases} 3x + 4y - z = 10 \\ x - 4z = 20 \\ 2x + y + 5z = 30 \end{cases}$

83–84 ■ Area of a Triangle Use the determinant formula for the area of a triangle to find the area of the triangle in the figure.

83.

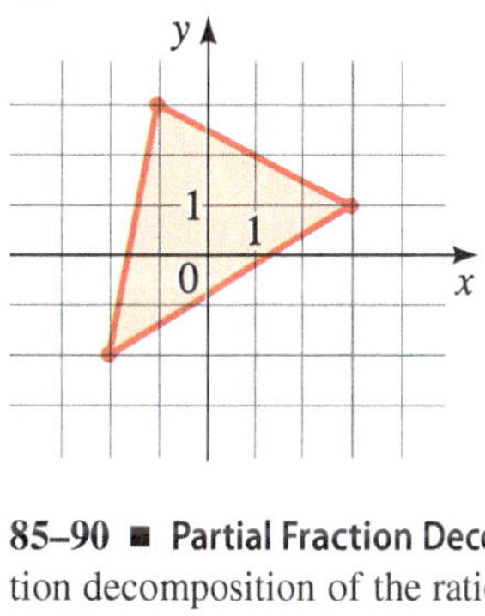

84.

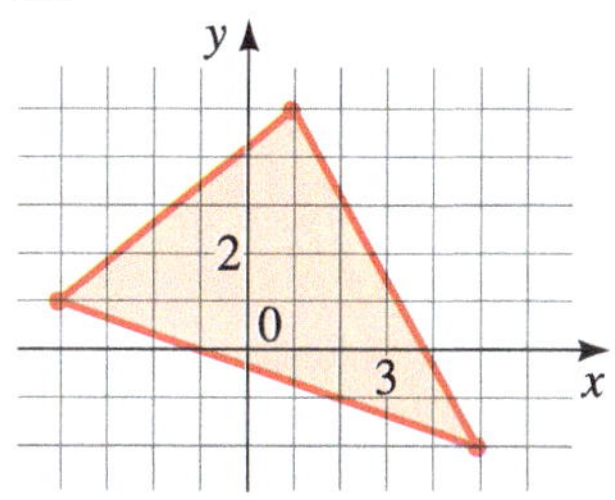

85–90 ■ Partial Fraction Decomposition Find the partial fraction decomposition of the rational expression.

85. $\dfrac{3x + 1}{x^2 - 2x - 15}$ **86.** $\dfrac{8}{x^3 - 4x}$

87. $\dfrac{2x - 4}{x(x - 1)^2}$ **88.** $\dfrac{x + 6}{x^3 - 2x^2 + 4x - 8}$

89. $\dfrac{2x - 1}{x^3 + x}$ **90.** $\dfrac{5x^2 - 3x + 10}{x^4 + x^2 - 2}$

91–94 ■ Intersection Points Two equations and their graphs are given. Find the intersection point(s) of the graphs by solving the system.

91. $\begin{cases} 2x + 3y = 7 \\ x - 2y = 0 \end{cases}$

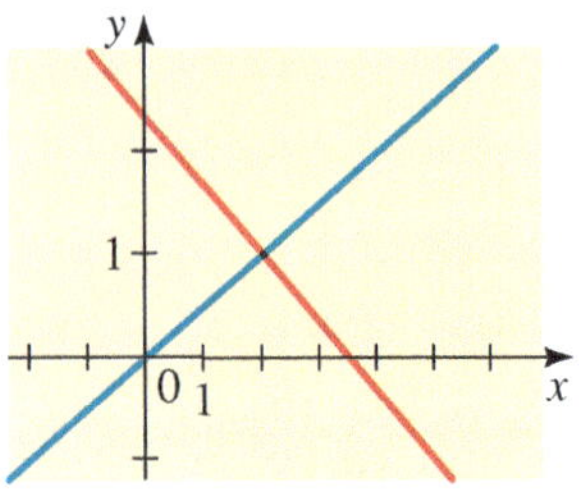

92. $\begin{cases} 3x + y = 8 \\ y = x^2 - 5x \end{cases}$

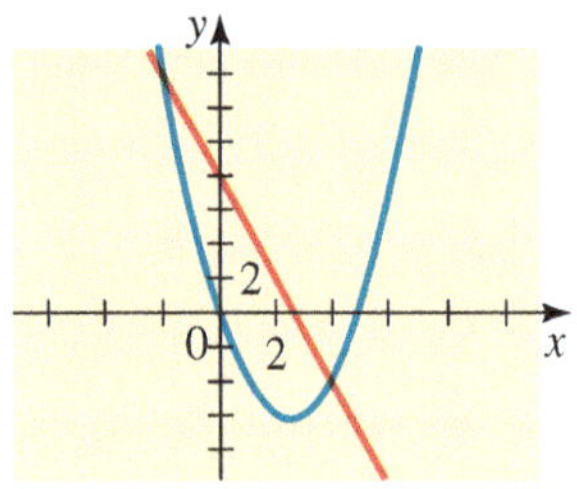

93. $\begin{cases} x^2 + y = 2 \\ x^2 - 3x - y = 0 \end{cases}$

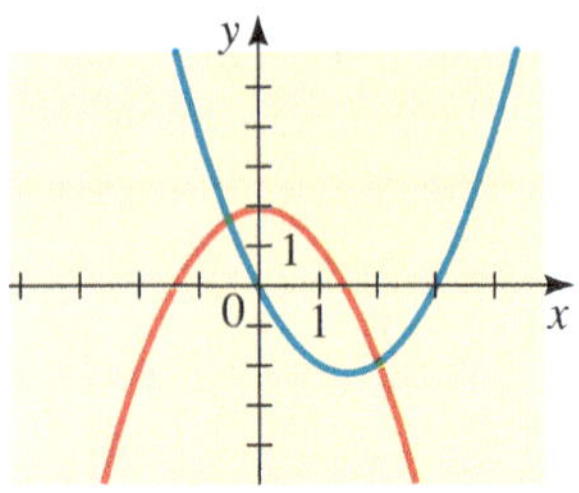

94. $\begin{cases} x - y = -2 \\ x^2 + y^2 - 4y = 4 \end{cases}$

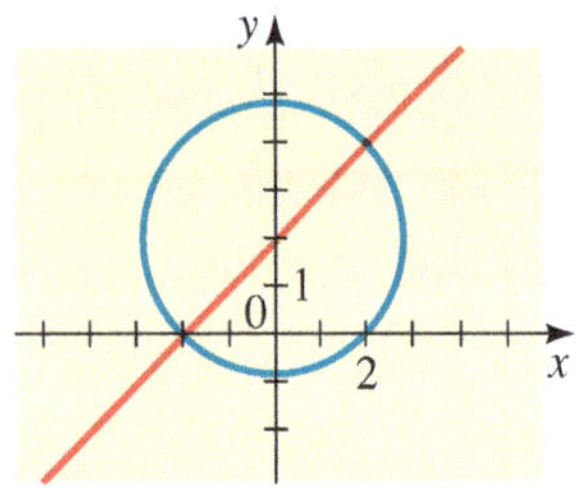

95–96 ■ Finding an Inequality from a Graph An equation and its graph are given. Find an inequality whose solution is the shaded region.

95. $x + y^2 = 4$

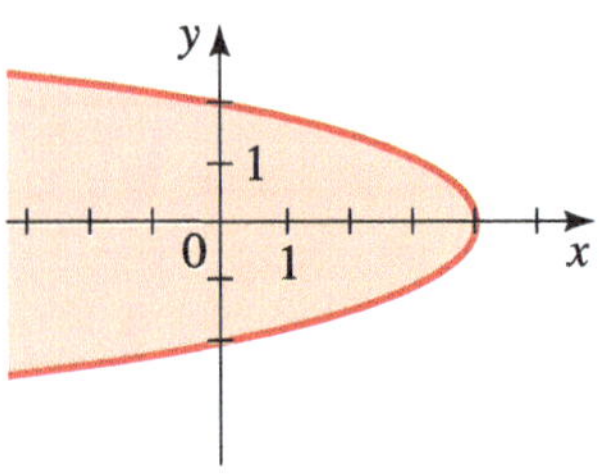

96. $x^2 + y^2 = 8$

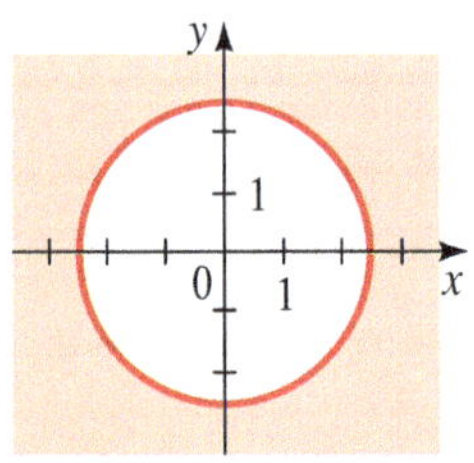

97–100 ■ Graphing Inequalities Graph the inequality.

97. $3x + y \le 6$

98. $y \ge x^2 - 3$

99. $x^2 + y^2 > 9$

100. $x - y^2 < 4$

101–104 ■ Solution Set of a System of Inequalities The figure shows the graphs of the equations corresponding to the given inequalities. Shade the solution set of the system of inequalities.

101. $\begin{cases} y \ge x^2 - 3x \\ y \le \frac{1}{3}x - 1 \end{cases}$

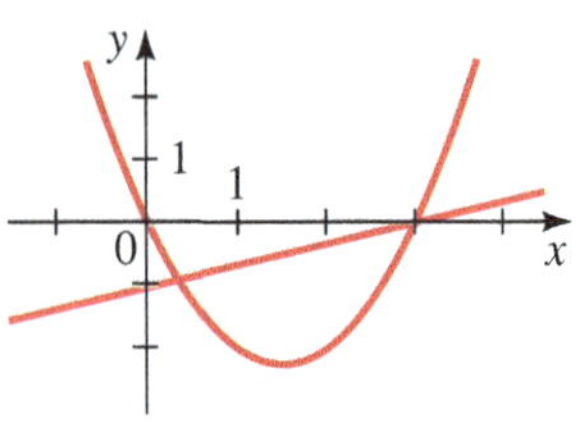

102. $\begin{cases} y \ge x - 1 \\ x^2 + y^2 \le 1 \end{cases}$

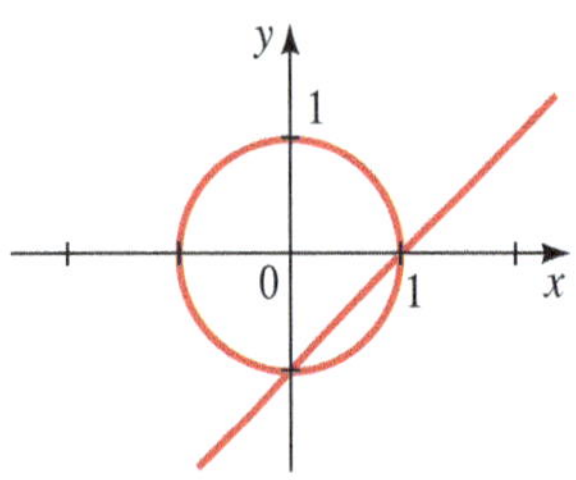

103. $\begin{cases} x + y \ge 2 \\ y - x \le 2 \\ x \le 3 \end{cases}$

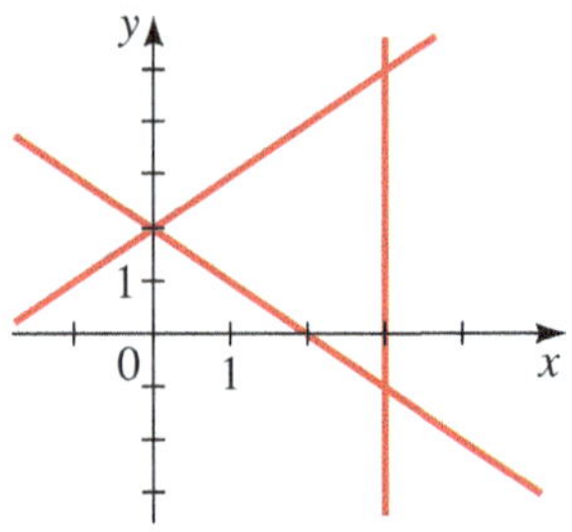

104. $\begin{cases} y \ge -2x \\ y \le 2x \\ y \le -\frac{1}{2}x + 2 \end{cases}$

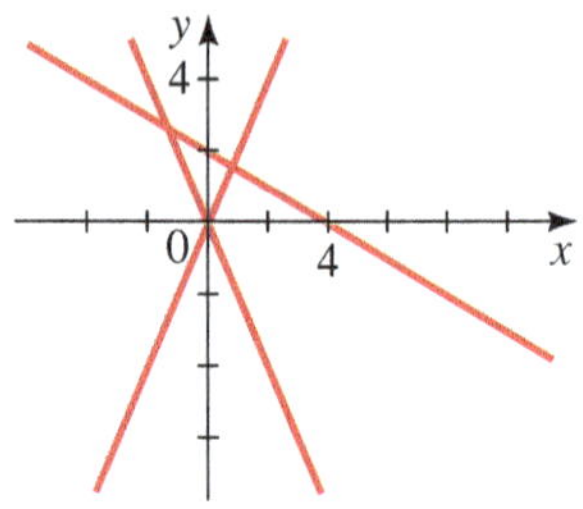

105–108 ■ Systems of Inequalities Graph the solution set of the system of inequalities. Find the coordinates of all vertices, and determine whether the solution set is bounded or unbounded.

105. $\begin{cases} x^2 + y^2 < 9 \\ x + y < 0 \end{cases}$

106. $\begin{cases} y - x^2 \ge 4 \\ y < 20 \end{cases}$

107. $\begin{cases} x \ge 0, \quad y \ge 0 \\ x + 2y \le 12 \\ y \le x + 4 \end{cases}$

108. $\begin{cases} x \ge 4 \\ x + y \ge 24 \\ x \le 2y + 12 \end{cases}$

109–110 ■ General Systems of Equations Solve for x, y, and z in terms of a, b, and c.

109. $\begin{cases} -x + y + z = a \\ x - y + z = b \\ x + y - z = c \end{cases}$

110. $\begin{cases} ax + by + cz = a - b + c \\ bx + by + cz = c \\ cx + cy + cz = c \end{cases} \qquad (a \ne b, b \ne c, c \ne 0)$

111. General Systems of Equations For what values of k do the following three lines have a common point of intersection?

$$x + y = 12$$
$$kx - y = 0$$
$$y - x = 2k$$

112. General Systems of Equations For what value of k does the following system have infinitely many solutions?

$$\begin{cases} kx + y + z = 0 \\ x + 2y + kz = 0 \\ -x + 3z = 0 \end{cases}$$

CHAPTER 10 TEST

1–2 ■ A system of equations is given. **(a)** Determine whether the system is linear or nonlinear. **(b)** Find all solutions of the system.

1. $\begin{cases} x + 3y = 7 \\ 5x + 2y = -4 \end{cases}$ **2.** $\begin{cases} 6x + y^2 = 10 \\ 3x - y = 5 \end{cases}$

3. Use a graphing device to find all solutions of the system rounded to two decimal places.

$$\begin{cases} x - 2y = 1 \\ y = x^3 - 2x^2 \end{cases}$$

4. In $2\frac{1}{2}$ h an airplane travels 600 km against the wind. It takes 50 min to travel 300 km with the wind. Find the speed of the wind and the speed of the airplane in still air.

5. Determine whether each matrix is in reduced row-echelon form, row-echelon form, or neither.

(a) $\begin{bmatrix} 1 & 2 & 4 & -6 \\ 0 & 1 & -3 & 0 \end{bmatrix}$ **(b)** $\begin{bmatrix} 1 & 0 & -1 & 0 & 0 \\ 0 & 1 & 3 & 0 & 0 \\ 0 & 0 & 0 & 1 & 0 \\ 0 & 0 & 0 & 0 & 1 \end{bmatrix}$ **(c)** $\begin{bmatrix} 1 & 1 & 0 \\ 0 & 0 & 1 \\ 0 & 1 & 3 \end{bmatrix}$

6. Use Gaussian elimination to find the complete solution of the system, or show that no solution exists.

(a) $\begin{cases} x - y + 2z = 0 \\ 2x - 4y + 5z = -5 \\ 2y - 3z = 5 \end{cases}$ **(b)** $\begin{cases} 2x - 3y + z = 3 \\ x + 2y + 2z = -1 \\ 4x + y + 5z = 4 \end{cases}$

7. Use Gauss-Jordan elimination to find the complete solution of the system.

$$\begin{cases} x + 3y - z = 0 \\ 3x + 4y - 2z = -1 \\ -x + 2y = 1 \end{cases}$$

8. Anne, Barry, and Cathy enter a coffee shop. Anne orders two coffees, one juice, and two doughnuts and pays \$6.25. Barry orders one coffee and three doughnuts and pays \$3.75. Cathy orders three coffees, one juice, and four doughnuts and pays \$9.25. Find the price of coffee, juice, and doughnuts at this coffee shop.

9. Let

$$A = \begin{bmatrix} 2 & 3 \\ 2 & 4 \end{bmatrix} \qquad B = \begin{bmatrix} 2 & 4 \\ -1 & 1 \\ 3 & 0 \end{bmatrix} \qquad C = \begin{bmatrix} 1 & 0 & 4 \\ -1 & 1 & 2 \\ 0 & 1 & 3 \end{bmatrix}$$

Carry out the indicated operation, or explain why it cannot be performed.

(a) $A + B$ **(b)** AB **(c)** $BA - 3B$ **(d)** CBA
(e) A^{-1} **(f)** B^{-1} **(g)** $\det(B)$ **(h)** $\det(C)$

10. **(a)** Write a matrix equation equivalent to the following system.

$$\begin{cases} 4x - 3y = 10 \\ 3x - 2y = 30 \end{cases}$$

(b) Find the inverse of the coefficient matrix, and use it to solve the system.

11. Only one of the following matrices has an inverse. Find the determinant of each matrix, and use the determinants to identify the one that has an inverse. Then find the inverse.

$$A = \begin{bmatrix} 1 & 4 & 1 \\ 0 & 2 & 0 \\ 1 & 0 & 1 \end{bmatrix} \qquad B = \begin{bmatrix} 1 & 4 & 0 \\ 0 & 2 & 0 \\ -3 & 0 & 1 \end{bmatrix}$$

12. Solve using Cramer's Rule:

$$\begin{cases} 2x \phantom{{}-y} - z = 14 \\ 3x - y + 5z = 0 \\ 4x + 2y + 3z = -2 \end{cases}$$

13. Find the partial fraction decomposition of the rational expression.

(a) $\dfrac{4x - 1}{(x - 1)^2(x + 2)}$ **(b)** $\dfrac{2x - 3}{x^3 + 3x}$

14. Graph the solution set of the system of inequalities. Label the vertices with their coordinates.

(a) $\begin{cases} 2x + y \le 8 \\ x - y \ge -2 \\ x + 2y \ge 4 \end{cases}$ **(b)** $\begin{cases} x^2 + y \le 5 \\ y \ge 2x + 5 \end{cases}$

Linear Programming

Linear programming is a modeling technique that is used to determine the optimal allocation of resources in business, the military, and other areas of human endeavor. For example, a manufacturer who makes several different products from the same raw materials can use linear programming to determine how much of each product should be produced to maximize the profit. This modeling technique is probably the most important practical application of systems of linear inequalities. In 1975 Leonid Kantorovich and T. C. Koopmans won the Nobel Prize in economics for their work in the development of this technique.

Although linear programming can be applied to very complex problems with hundreds or even thousands of variables, we consider only a few simple examples to which the graphical methods of Section 10.9 can be applied. (For large numbers of variables a linear programming method based on matrices is used.) Let's examine a typical problem.

EXAMPLE 1 ■ Manufacturing for Maximum Profit

A small shoe manufacturer makes two styles of shoes: oxfords and loafers. Two machines are used in the process: a cutting machine and a sewing machine. Each type of shoe requires 15 min per pair on the cutting machine. Oxfords require 10 min of sewing per pair, and loafers require 20 min of sewing per pair. Because the manufacturer can hire only one operator for each machine, each process is available for just 8 h per day. If the profit is \$15 on each pair of oxfords and \$20 on each pair of loafers, how many pairs of each type should be produced per day for maximum profit?

Because loafers produce more profit, it would seem best to manufacture only loafers. Surprisingly, this does not turn out to be the most profitable solution.

SOLUTION First we organize the given information into a table. To be consistent, let's convert all times to hours.

	Oxfords	Loafers	Time available
Time on cutting machine (h)	$\frac{1}{4}$	$\frac{1}{4}$	8
Time on sewing machine (h)	$\frac{1}{6}$	$\frac{1}{3}$	8
Profit	\$15	\$20	

We describe the model and solve the problem in four steps.

■ Choose the Variables

To make a mathematical model, we first give names to the variable quantities. For this problem we let

$$x = \text{number of pairs of oxfords made daily}$$

$$y = \text{number of pairs of loafers made daily}$$

■ Find the Objective Function

Our goal is to determine which values for x and y give maximum profit. Since each pair of oxfords provides \$15 profit and each pair of loafers provides \$20, the total profit is given by

$$P = 15x + 20y$$

This function is called the *objective function*.

Graph the Feasible Region

The larger x and y are, the greater is the profit. But we cannot choose arbitrarily large values for these variables because of the restrictions, or *constraints*, in the problem. Each restriction is an inequality in the variables.

In this problem the total number of cutting hours needed is $\frac{1}{4}x + \frac{1}{4}y$. Since only 8 h are available on the cutting machine, we have

$$\tfrac{1}{4}x + \tfrac{1}{4}y \le 8$$

Similarly, by considering the amount of time needed and available on the sewing machine, we get

$$\tfrac{1}{6}x + \tfrac{1}{3}y \le 8$$

We cannot produce a negative number of shoes, so we also have

$$x \ge 0 \qquad \text{and} \qquad y \ge 0$$

Thus x and y must satisfy the constraints

$$\begin{cases} \frac{1}{4}x + \frac{1}{4}y \le 8 \\ \frac{1}{6}x + \frac{1}{3}y \le 8 \\ x \ge 0, \quad y \ge 0 \end{cases}$$

If we multiply the first inequality by 4 and the second by 6, we obtain the simplified system

$$\begin{cases} x + y \le 32 \\ x + 2y \le 48 \\ x \ge 0, \quad y \ge 0 \end{cases}$$

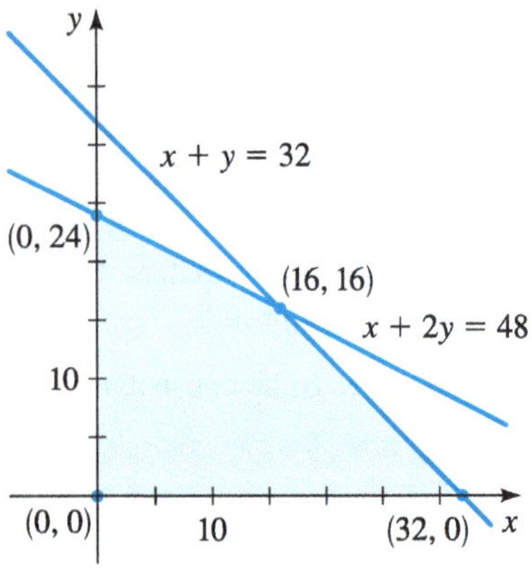

FIGURE 1

The solution of this system (with vertices labeled) is sketched in Figure 1. The only values that satisfy the restrictions of the problem are the ones that correspond to points of the shaded region in Figure 1. This is called the *feasible region* for the problem.

Find the Maximum Profit

As x or y increases, profit increases as well. Thus it seems reasonable that the maximum profit will occur at a point on one of the outside edges of the feasible region, where it is impossible to increase x or y without going outside the region. In fact, it can be shown that the maximum value occurs at a vertex. This means that we need to check the profit only at the vertices. The largest value of P occurs at the point $(16, 16)$, where $P = \$560$. Thus the manufacturer should make 16 pairs of oxfords and 16 pairs of loafers, for a maximum daily profit of \$560.

Vertex	$P = 15x + 20y$
(0, 0)	0
(0, 24)	$15(0) + 20(24) = \$480$
(16, 16)	$15(16) + 20(16) = \$560$ ← Maximum profit
(32, 0)	$15(32) + 20(0) = \$480$

Linear Programming helps the telephone industry to determine the most efficient way to route telephone calls. The computerized routing decisions must be made very rapidly so that callers are not kept waiting for connections. Since the database of customers and routes is huge, an extremely fast method for solving linear programming problems is essential. In 1984 the 28-year-old mathematician **Narendra Karmarkar**, working at Bell Labs in Murray Hill, New Jersey, discovered just such a method. His idea is so ingenious and his method so fast that the discovery caused a sensation in the mathematical world. Although mathematical discoveries rarely make the news, this one was reported in *Time*, on December 3, 1984. Today airlines routinely use Karmarkar's technique to minimize costs in scheduling passengers, flight personnel, fuel, baggage, and maintenance workers.

The linear programming problems that we consider all follow the pattern of Example 1. Each problem involves two variables. The problem describes restrictions, called **constraints**, that lead to a system of linear inequalities whose solution is called the **feasible region**. The function that we wish to maximize or minimize is called the **objective function**. This function always attains its largest and smallest values at the **vertices** of the feasible region. This modeling technique involves four steps, summarized in the following box.

GUIDELINES FOR LINEAR PROGRAMMING

1. **Choose the Variables.** Decide what variable quantities in the problem should be named x and y.
2. **Find the Objective Function.** Write an expression for the function we want to maximize or minimize.
3. **Graph the Feasible Region.** Express the constraints as a system of inequalities, and graph the solution of this system (the feasible region).
4. **Find the Maximum or Minimum.** Evaluate the objective function at the vertices of the feasible region to determine its maximum or minimum value.

EXAMPLE 2 ■ A Shipping Problem

A car dealer has warehouses in Millville and Trenton and dealerships in Camden and Atlantic City. Every car that is sold at the dealerships must be delivered from one of the warehouses. On a certain day the Camden dealers sell 10 cars, and the Atlantic City dealers sell 12. The Millville warehouse has 15 cars available, and the Trenton warehouse has 10. The cost of shipping one car is \$50 from Millville to Camden, \$40 from Millville to Atlantic City, \$60 from Trenton to Camden, and \$55 from Trenton to Atlantic City. How many cars should be moved from each warehouse to each dealership to fill the orders at minimum cost?

SOLUTION Our first step is to organize the given information. Rather than construct a table, we draw a diagram to show the flow of cars from the warehouses to the dealerships (see Figure 2 below). The diagram shows the number of cars available at each warehouse or required at each dealership and the cost of shipping between these locations.

■ Choose the Variables

The arrows in Figure 2 indicate four possible routes, so the problem seems to involve four variables. But we let

$$x = \text{number of cars to be shipped from Millville to Camden}$$

$$y = \text{number of cars to be shipped from Millville to Atlantic City}$$

To fill the orders, we must have

$$10 - x = \text{number of cars shipped from Trenton to Camden}$$

$$12 - y = \text{number of cars shipped from Trenton to Atlantic City}$$

So the only variables in the problem are x and y.

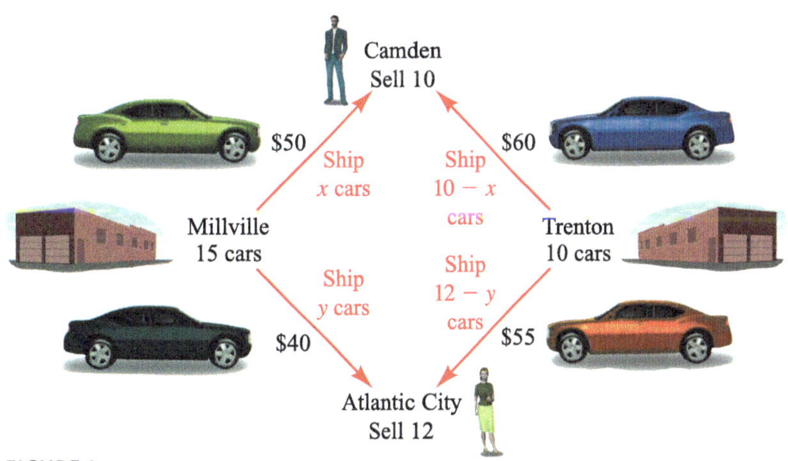

FIGURE 2

Find the Objective Function

The objective of this problem is to minimize cost. From Figure 2 we see that the total cost C of shipping the cars is

$$\begin{aligned} C &= 50x + 40y + 60(10 - x) + 55(12 - y) \\ &= 50x + 40y + 600 - 60x + 660 - 55y \\ &= 1260 - 10x - 15y \end{aligned}$$

This is the objective function.

Graph the Feasible Region

Now we derive the constraint inequalities that define the feasible region. First, the number of cars shipped on each route can't be negative, so we have

$$\begin{aligned} x \geq 0 \qquad & \qquad y \geq 0 \\ 10 - x \geq 0 \qquad & \qquad 12 - y \geq 0 \end{aligned}$$

Second, the total number of cars shipped from each warehouse can't exceed the number of cars available there, so

$$\begin{aligned} x + y &\leq 15 \\ (10 - x) + (12 - y) &\leq 10 \end{aligned}$$

Simplifying the latter inequality, we get

$$\begin{aligned} 22 - x - y &\leq 10 \\ -x - y &\leq -12 \\ x + y &\geq 12 \end{aligned}$$

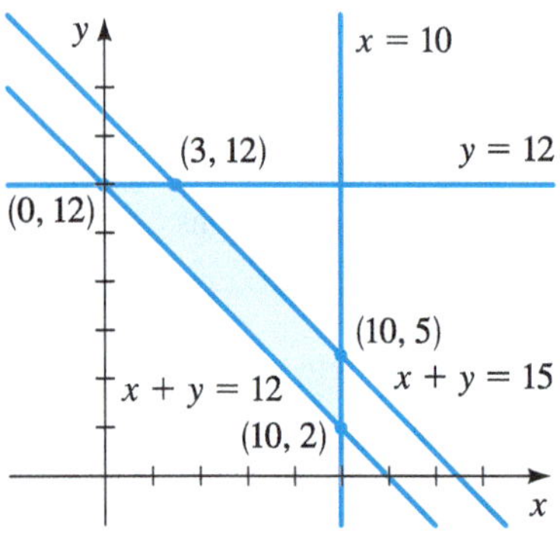

FIGURE 3

The inequalities $10 - x \geq 0$ and $12 - y \geq 0$ can be rewritten as $x \leq 10$ and $y \leq 12$. Thus the feasible region is described by the constraints

$$\begin{cases} x + y \leq 15 \\ x + y \geq 12 \\ 0 \leq x \leq 10 \\ 0 \leq y \leq 12 \end{cases}$$

The feasible region is graphed in Figure 3.

Find the Minimum Cost

We check the value of the objective function at each vertex of the feasible region.

Vertex	$C = 1260 - 10x - 15y$
(0, 12)	$1260 - 10(0) - 15(12) = \$1080$
(3, 12)	$1260 - 10(3) - 15(12) = \$1050$ ← Minimum cost
(10, 5)	$1260 - 10(10) - 15(5) = \$1085$
(10, 2)	$1260 - 10(10) - 15(2) = \$1130$

The lowest cost is incurred at the point (3, 12). Thus the dealer should ship

3 cars from Millville to Camden
12 cars from Millville to Atlantic City
7 cars from Trenton to Camden
0 cars from Trenton to Atlantic City ■

In the 1940s mathematicians developed matrix methods for solving linear programming problems that involve more than two variables. These methods were first used by the Allies in World War II to solve supply problems similar to (but, of course, much more complicated than) Example 2. Improving such matrix methods is an active and exciting area of current mathematical research.

PROBLEMS

1–4 ■ Find the maximum and minimum values of the given objective function on the indicated feasible region.

1. $M = 200 - x - y$

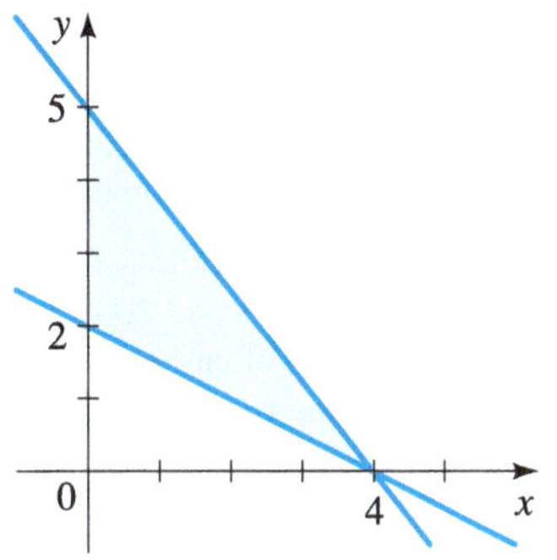

2. $N = \frac{1}{2}x + \frac{1}{4}y + 40$

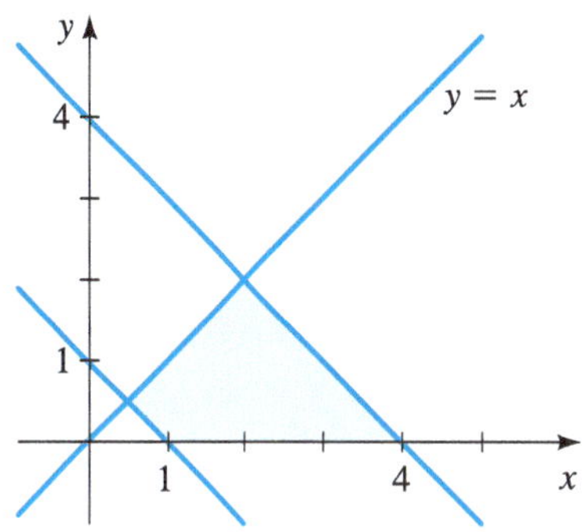

3. $P = 140 - x + 3y$

$$\begin{cases} x \geq 0, y \geq 0 \\ 2x + y \leq 10 \\ 2x + 4y \leq 28 \end{cases}$$

4. $Q = 70x + 82y$

$$\begin{cases} x \geq 0, y \geq 0 \\ x \leq 10, y \leq 20 \\ x + y \geq 5 \\ x + 2y \leq 18 \end{cases}$$

5. Making Furniture A furniture manufacturer makes wooden tables and chairs. The production process involves two basic types of labor: carpentry and finishing. A table requires 2 h of carpentry and 1 h of finishing, and a chair requires 3 h of carpentry and $\frac{1}{2}$ h of finishing. The profit is \$35 per table and \$20 per chair. The manufacturer's employees can supply a maximum of 108 h of carpentry work and 20 h of finishing work per day. How many tables and chairs should be made each day to maximize profit?

6. A Housing Development A housing contractor has subdivided a farm into 100 building lots. She has designed two types of homes for these lots: colonial and ranch style. A colonial requires \$30,000 of capital and produces a profit of \$4000 when sold. A ranch-style house requires \$40,000 of capital and provides an \$8000 profit. If the contractor has \$3.6 million of capital on hand, how many houses of each type should she build for maximum profit? Will any of the lots be left vacant?

7. Hauling Fruit A trucker hauls citrus fruit from Florida to Montreal. Each crate of oranges is 4 ft^3 in volume and weighs 80 lb. Each crate of grapefruit has a volume of 6 ft^3 and weighs 100 lb. His truck has a maximum capacity of 300 ft^3 and can carry no more than 5600 lb. Moreover, he is not permitted to carry more crates of grapefruit than crates of oranges. If his profit is \$2.50 on each crate of oranges and \$4 on each crate of grapefruit, how many crates of each fruit should he carry for maximum profit?

8. Manufacturing Calculators A manufacturer of calculators produces two models: standard and scientific. Long-term demand for the two models mandates that the company manufacture at least 100 standard and 80 scientific calculators each day. However, because of limitations on production capacity, no more than 200 standard and 170 scientific calculators can be made daily. To satisfy a shipping contract, a total of at least 200 calculators must be shipped every day.

(a) If the production cost is \$5 for a standard calculator and \$7 for a scientific one, how many of each model should be produced daily to minimize this cost?

(b) If each standard calculator results in a \$2 loss but each scientific one produces a \$5 profit, how many of each model should be made daily to maximize profit?

9. Shipping Televisions An electronics discount chain has a sale on a certain brand of 60-in. high-definition television set. The chain has stores in Santa Monica and El Toro and warehouses in Long Beach and Pasadena. To satisfy rush orders, 15 sets must be shipped from the warehouses to the Santa Monica store, and 19 must be shipped to the El Toro store. The cost of shipping a set is \$5 from Long Beach to Santa Monica, \$6 from Long Beach to El Toro, \$4 from Pasadena to Santa Monica, and \$5.50 from Pasadena to El Toro.

If the Long Beach warehouse has 24 sets and the Pasadena warehouse has 18 sets in stock, how many sets should be shipped from each warehouse to each store to fill the orders at a minimum shipping cost?

10. **Delivering Plywood** A man owns two building supply stores, one on the east side and one on the west side of a city. Two customers order some $\frac{1}{2}$-inch plywood. Customer A needs 50 sheets, and customer B needs 70 sheets. The east-side store has 80 sheets, and the west-side store has 45 sheets of this plywood in stock. The east-side store's delivery costs per sheet are \$0.50 to customer A and \$0.60 to customer B. The west-side store's delivery costs per sheet are \$0.40 to customer A and \$0.55 to customer B. How many sheets should be shipped from each store to each customer to minimize delivery costs?

11. **Packaging Nuts** A confectioner sells two types of nut mixtures. The standard-mixture package contains 100 g of cashews and 200 g of peanuts and sells for \$1.95. The deluxe-mixture package contains 150 g of cashews and 50 g of peanuts and sells for \$2.25. The confectioner has 15 kg of cashews and 20 kg of peanuts available. On the basis of past sales, the confectioner needs to have at least as many standard as deluxe packages available. How many bags of each mixture should he package to maximize his revenue?

12. **Feeding Lab Rabbits** A biologist wishes to feed laboratory rabbits a mixture of two types of foods. Type I contains 8 g of fat, 12 g of carbohydrate, and 2 g of protein per ounce. Type II contains 12 g of fat, 12 g of carbohydrate, and 1 g of protein per ounce. Type I costs \$0.20 per ounce and type II costs \$0.30 per ounce. Each rabbit receives a daily minimum of 24 g of fat, 36 g of carbohydrate, and 4 g of protein, but get no more than 5 oz of food per day. How many ounces of each food type should be fed to each rabbit daily to satisfy the dietary requirements at minimum cost?

13. **Investing in Bonds** A woman wishes to invest \$12,000 in three types of bonds: municipal bonds paying 7% interest per year, bank certificates paying 8%, and high-risk bonds paying 12%. For tax reasons she wants the amount invested in municipal bonds to be at least three times the amount invested in bank certificates. To keep her level of risk manageable, she will invest no more than \$2000 in high-risk bonds. How much should she invest in each type of bond to maximize her annual interest yield? [*Hint:* Let x = amount in municipal bonds and y = amount in bank certificates. Then the amount in high-risk bonds will be $12{,}000 - x - y$.]

14. **Annual Interest Yield** Refer to Problem 13. Suppose the investor decides to increase the maximum invested in high-risk bonds to \$3000 but leaves the other conditions unchanged. By how much will her maximum possible interest yield increase?

15. **Business Strategy** A small software company publishes computer games, educational software, and utility software. Their business strategy is to market a total of 36 new programs each year, at least four of these being games. The number of utility programs published is never more than twice the number of educational programs. On average, the company makes an annual profit of \$5000 on each computer game, \$8000 on each educational program, and \$6000 on each utility program. How many of each type of software should the company publish annually for maximum profit?

16. **Feasible Region** All parts of this problem refer to the following feasible region and objective function.

$$\begin{cases} x \geq 0 \\ x \geq y \\ x + 2y \geq 12 \\ x + y \leq 10 \end{cases}$$

$$P = x + 4y$$

(a) Graph the feasible region.

(b) On your graph from part (a), sketch the graphs of the linear equations obtained by setting P equal to 40, 36, 32, and 28.

(c) If you continue to decrease the value of P, at which vertex of the feasible region will these lines first touch the feasible region?

(d) Verify that the maximum value of P on the feasible region occurs at the vertex you chose in part (c).

Harald Sund/The Image Bank/Getty Images

11 Conic Sections

Conic sections are the curves that are formed when a plane cuts a cone, as shown in the figure. For example, if a cone is cut horizontally, the cross section is a circle. So a circle is a conic section. Other ways of cutting a cone produce ellipses, parabolas, and hyperbolas.

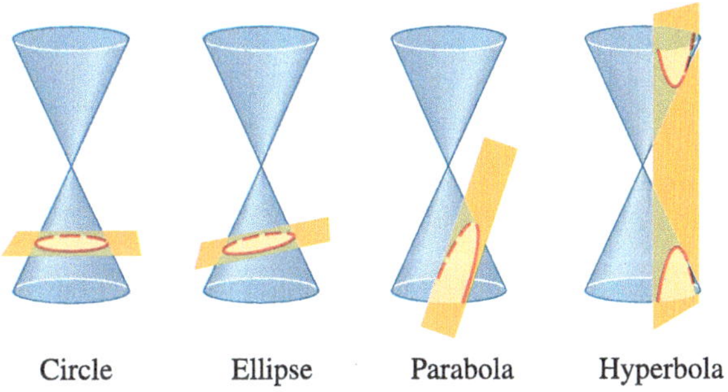

Our goal in this chapter is to find equations whose graphs are conic sections. We will find such equations by analyzing the geometric properties of conic sections. These properties make conic sections useful for many real-world applications. For instance, a reflecting surface with parabolic cross sections concentrates light at a single point. This property of a parabola is used in the construction of solar power plants, like the one in California pictured above. In the *Focus on Modeling* at the end of the chapter we explore how these curves are used in architecture.

11.1 PARABOLAS

■ Geometric Definition of a Parabola ■ Equations and Graphs of Parabolas
■ Applications

■ Geometric Definition of a Parabola

We saw in Section 3.1 that the graph of the equation

$$y = ax^2 + bx + c$$

is a U-shaped curve called a *parabola* that opens either upward or downward, depending on whether the number a is positive or negative.

In this section we study parabolas from a geometric, rather than an algebraic, point of view. We begin with the geometric definition of a parabola and show how this leads to the algebraic formula that we are already familiar with.

> **GEOMETRIC DEFINITION OF A PARABOLA**
>
> A **parabola** is the set of all points in the plane that are equidistant from a fixed point F (called the **focus**) and a fixed line l (called the **directrix**).

This definition is illustrated in Figure 1. The **vertex** V of the parabola lies halfway between the focus and the directrix, and the **axis of symmetry** is the line that runs through the focus perpendicular to the directrix.

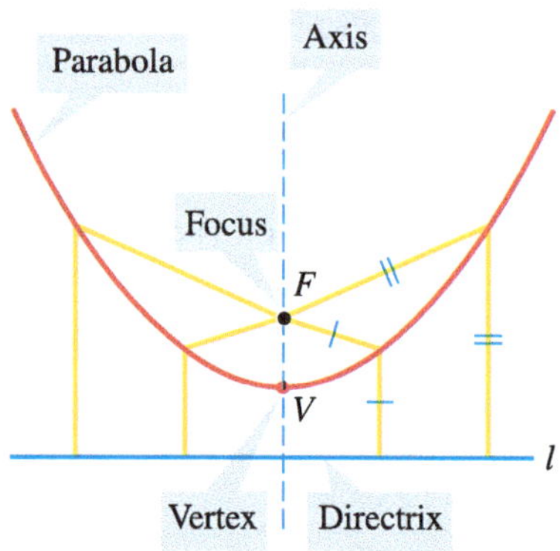

FIGURE 1

In this section we restrict our attention to parabolas that are situated with the vertex at the origin and that have a vertical or horizontal axis of symmetry. (Parabolas in more general positions will be considered in Section 11.4.) If the focus of such a parabola is the point $F(0, p)$, then the axis of symmetry must be vertical, and the directrix has the equation $y = -p$. Figure 2 illustrates the case $p > 0$.

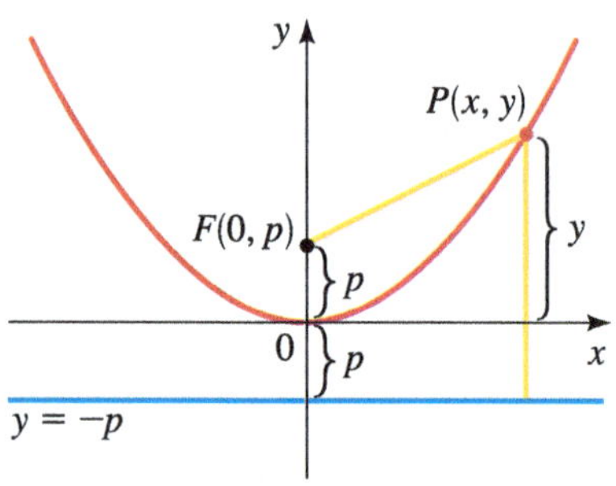

FIGURE 2

Deriving the Equation of a Parabola If $P(x, y)$ is any point on the parabola, then the distance from P to the focus F (using the Distance Formula) is

$$\sqrt{x^2 + (y - p)^2}$$

The distance from P to the directrix is

$$|y - (-p)| = |y + p|$$

By the definition of a parabola these two distances must be equal.

$$\begin{aligned}
\sqrt{x^2 + (y - p)^2} &= |y + p| \\
x^2 + (y - p)^2 &= |y + p|^2 = (y + p)^2 && \text{Square both sides} \\
x^2 + y^2 - 2py + p^2 &= y^2 + 2py + p^2 && \text{Expand} \\
x^2 - 2py &= 2py && \text{Simplify} \\
x^2 &= 4py
\end{aligned}$$

If $p > 0$, then the parabola opens upward; but if $p < 0$, it opens downward. When x is replaced by $-x$, the equation remains unchanged, so the graph is symmetric about the y-axis.

Equations and Graphs of Parabolas

The following box summarizes what we have just proved about the equation and features of a parabola with a vertical axis.

PARABOLA WITH VERTICAL AXIS

The graph of the equation

$$x^2 = 4py$$

is a parabola with the following properties.

VERTEX	$V(0, 0)$
FOCUS	$F(0, p)$
DIRECTRIX	$y = -p$

The parabola opens upward if $p > 0$ or downward if $p < 0$.

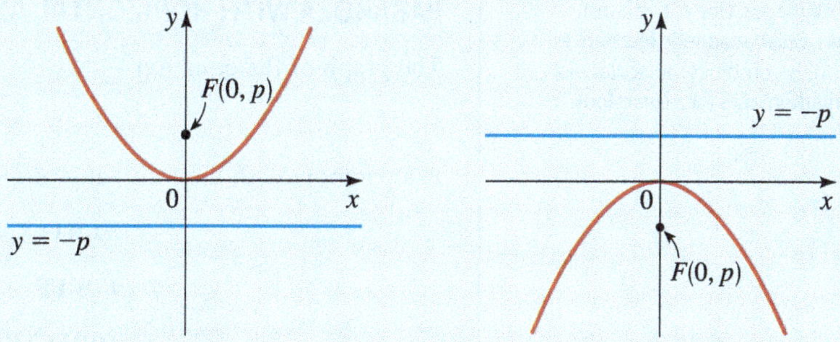

$x^2 = 4py$ with $p > 0$ $x^2 = 4py$ with $p < 0$

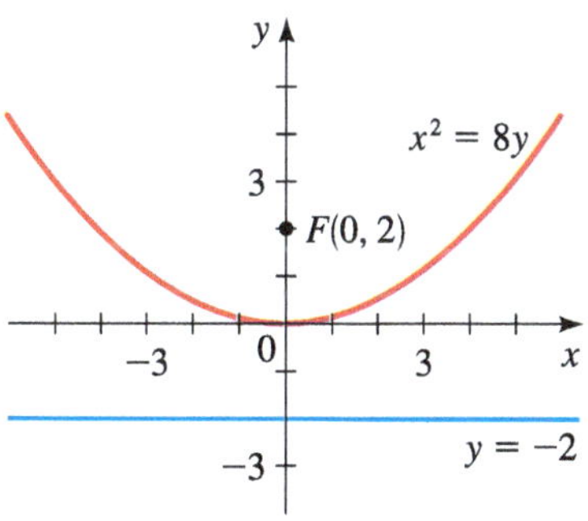

FIGURE 3

EXAMPLE 1 ■ Finding the Equation of a Parabola

Find an equation for the parabola with vertex $V(0, 0)$ and focus $F(0, 2)$, and sketch its graph.

SOLUTION Since the focus is $F(0, 2)$, we conclude that $p = 2$ (so the directrix is $y = -2$). Thus the equation of the parabola is

$$\begin{aligned}
x^2 &= 4(2)y && x^2 = 4py \text{ with } p = 2 \\
x^2 &= 8y
\end{aligned}$$

Since $p = 2 > 0$, the parabola opens upward. See Figure 3.

Now Try Exercises 31 and 49 ■

Mathematics in the Modern World

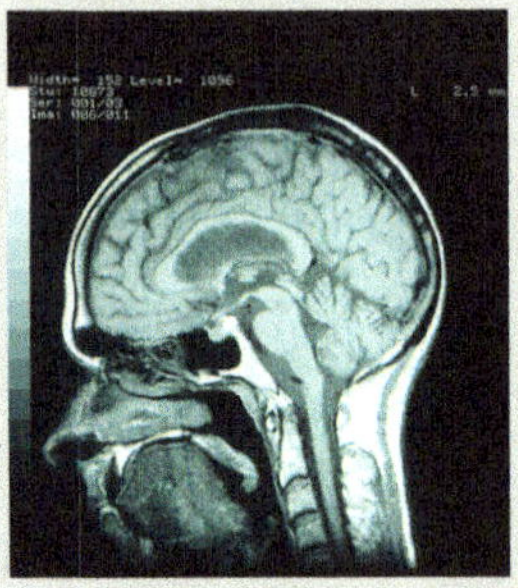

Looking Inside Your Head

Would you like to look inside your head? The idea isn't particularly appealing to most of us, but doctors often need to do just that. If they can look without invasive surgery, all the better. An X-ray doesn't really give a look inside, it simply gives a "graph" of the density of tissue the X-rays must pass through. So an X-ray is a "flattened" view in one direction. Suppose you get an X-ray view from many different directions. Can these "graphs" be used to reconstruct the three-dimensional inside view? This is a purely mathematical problem and was solved by mathematicians a long time ago. However, reconstructing the inside view requires thousands of tedious computations. Today, mathematics and high-speed computers make it possible to "look inside" by a process called computer-aided tomography (CAT scan). Mathematicians continue to search for better ways of using mathematics to reconstruct images. One of the latest techniques, called magnetic resonance imaging (MRI), combines molecular biology and mathematics for a clear "look inside."

EXAMPLE 2 ■ Finding the Focus and Directrix of a Parabola from Its Equation

Find the focus and directrix of the parabola $y = -x^2$, and sketch the graph.

SOLUTION To find the focus and directrix, we put the given equation in the standard form $x^2 = -y$. Comparing this to the general equation $x^2 = 4py$, we see that $4p = -1$, so $p = -\frac{1}{4}$. Thus the focus is $F(0, -\frac{1}{4})$, and the directrix is $y = \frac{1}{4}$. The graph of the parabola, together with the focus and the directrix, is shown in Figure 4(a). We can also draw the graph using a graphing calculator as shown in Figure 4(b).

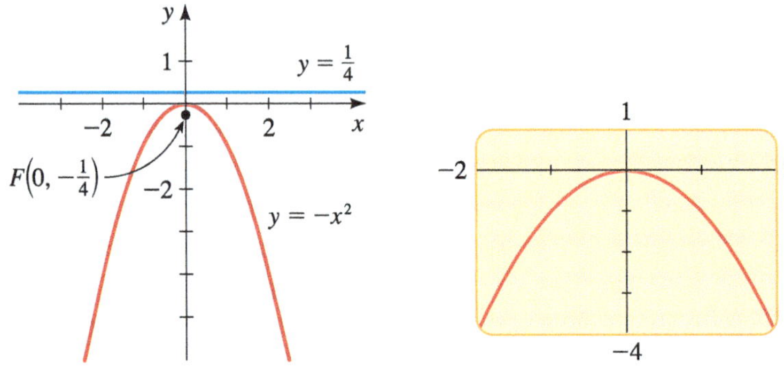

FIGURE 4

Now Try Exercise 11 ■

Reflecting the graph in Figure 2 about the diagonal line $y = x$ has the effect of interchanging the roles of x and y. This results in a parabola with horizontal axis. By the same method as before, we can prove the following properties.

PARABOLA WITH HORIZONTAL AXIS

The graph of the equation

$$y^2 = 4px$$

is a parabola with the following properties.

VERTEX	$V(0, 0)$
FOCUS	$F(p, 0)$
DIRECTRIX	$x = -p$

The parabola opens to the right if $p > 0$ or to the left if $p < 0$.

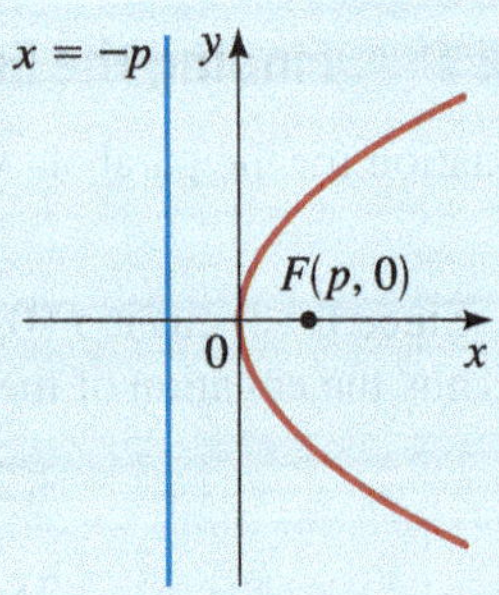

$y^2 = 4px$ with $p > 0$

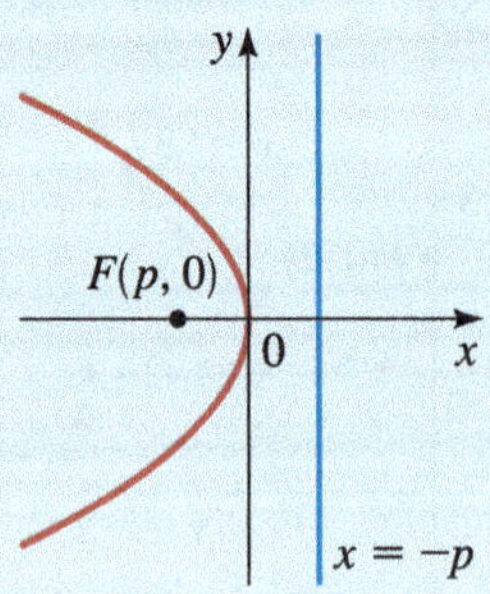

$y^2 = 4px$ with $p < 0$

EXAMPLE 3 ■ A Parabola with Horizontal Axis

A parabola has the equation $6x + y^2 = 0$.

(a) Find the focus and directrix of the parabola, and sketch the graph.

(b) Use a graphing calculator to draw the graph.

SOLUTION

(a) To find the focus and directrix, we put the given equation in the standard form $y^2 = -6x$. Comparing this to the general equation $y^2 = 4px$, we see that $4p = -6$, so $p = -\frac{3}{2}$. Thus the focus is $F\left(-\frac{3}{2}, 0\right)$, and the directrix is $x = \frac{3}{2}$. Since $p < 0$, the parabola opens to the left. The graph of the parabola, together with the focus and the directrix, is shown in Figure 5(a).

(b) To draw the graph using a graphing calculator, we need to solve for y.

$$6x + y^2 = 0$$

$$y^2 = -6x \qquad \text{Subtract } 6x$$

$$y = \pm\sqrt{-6x} \qquad \text{Take square roots}$$

To obtain the graph of the parabola, we graph both functions

$$y = \sqrt{-6x} \qquad \text{and} \qquad y = -\sqrt{-6x}$$

as shown in Figure 5(b).

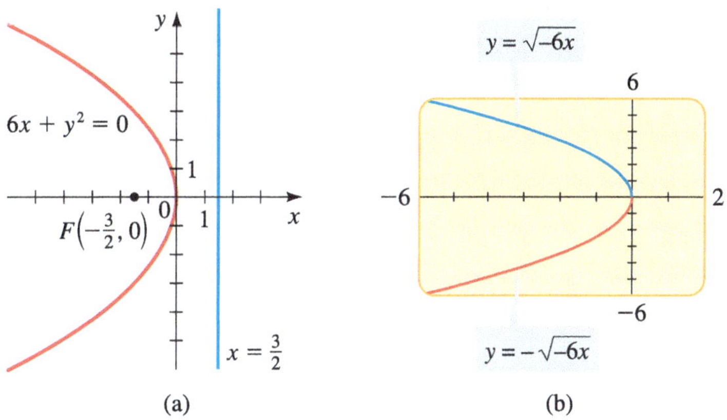

FIGURE 5

Now Try Exercises 13 and 25 ■

Graphing Calculator Note The equation $y^2 = 4px$ does not define y as a function of x (see page 165). So to use a graphing calculator to graph a parabola with a horizontal axis, we must first solve for y. This leads to two functions: $y = \sqrt{4px}$ and $y = -\sqrt{4px}$. We need to graph both functions to get the complete graph of the parabola. For example, in Figure 5(b) we had to graph both $y = \sqrt{-6x}$ and $y = -\sqrt{-6x}$ to graph the parabola $y^2 = -6x$.

We can use the coordinates of the focus to estimate the "width" of a parabola when sketching its graph. The line segment that runs through the focus perpendicular to the axis, with endpoints on the parabola, is called the **latus rectum**, and its length is the **focal diameter** of the parabola. From Figure 6 we can see that the distance from an endpoint Q of the latus rectum to the directrix is $|2p|$. Thus the distance from Q to the focus must be $|2p|$ as well (by the definition of a parabola), so the focal diameter is $|4p|$. In the next example we use the focal diameter to determine the "width" of a parabola when graphing it.

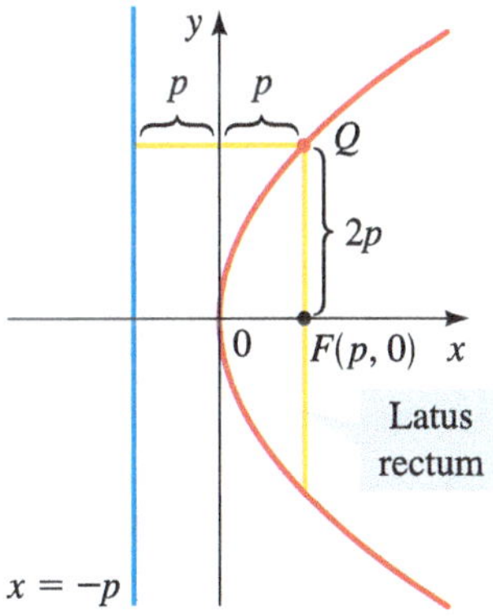

FIGURE 6

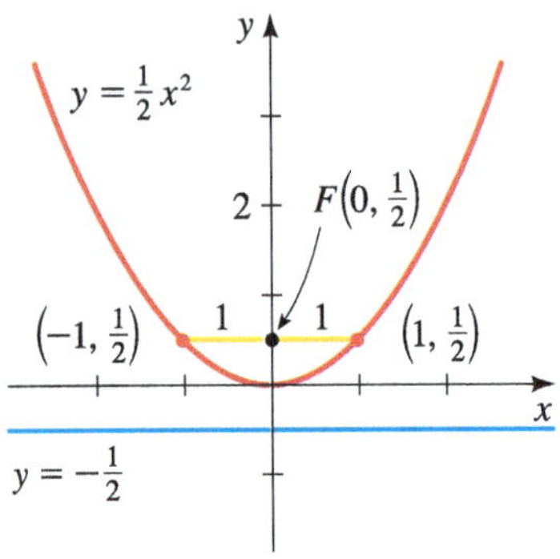

FIGURE 7

EXAMPLE 4 ■ The Focal Diameter of a Parabola

Find the focus, directrix, and focal diameter of the parabola $y = \frac{1}{2}x^2$, and sketch its graph.

SOLUTION We first put the equation in the form $x^2 = 4py$.

$$y = \tfrac{1}{2}x^2$$

$$x^2 = 2y \qquad \text{Multiply by 2, switch sides}$$

From this equation we see that $4p = 2$, so the focal diameter is 2. Solving for p gives $p = \frac{1}{2}$, so the focus is $\left(0, \frac{1}{2}\right)$, and the directrix is $y = -\frac{1}{2}$. Since the focal diameter is 2, the latus rectum extends 1 unit to the left and 1 unit to the right of the focus. The graph is sketched in Figure 7.

Now Try Exercise 15 ■

In the next example we graph a family of parabolas to show how changing the distance between the focus and the vertex affects the "width" of a parabola.

EXAMPLE 5 ■ A Family of Parabolas

(a) Find equations for the parabolas with vertex at the origin and foci $F_1\left(0, \frac{1}{8}\right)$, $F_2\left(0, \frac{1}{2}\right)$, $F_3(0, 1)$, and $F_4(0, 4)$.

(b) Draw the graphs of the parabolas in part (a). What do you conclude?

SOLUTION

(a) Since the foci are on the positive y-axis, the parabolas open upward and have equations of the form $x^2 = 4py$. This leads to the following equations.

Focus	p	Equation $x^2 = 4py$	Form of the equation for graphing calculator
$F_1\left(0, \frac{1}{8}\right)$	$p = \frac{1}{8}$	$x^2 = \frac{1}{2}y$	$y = 2x^2$
$F_2\left(0, \frac{1}{2}\right)$	$p = \frac{1}{2}$	$x^2 = 2y$	$y = 0.5x^2$
$F_3(0, 1)$	$p = 1$	$x^2 = 4y$	$y = 0.25x^2$
$F_4(0, 4)$	$p = 4$	$x^2 = 16y$	$y = 0.0625x^2$

(b) The graphs are drawn in Figure 8. We see that the closer the focus is to the vertex, the narrower the parabola.

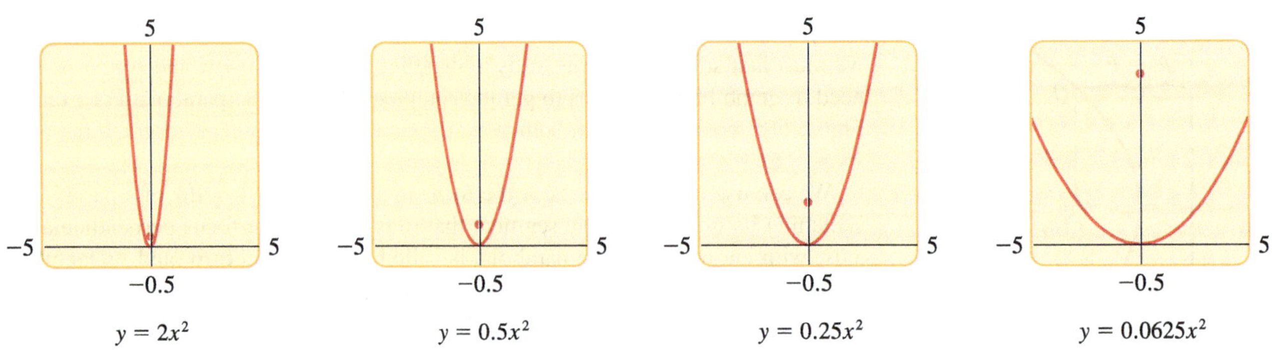

FIGURE 8 A family of parabolas

Now Try Exercise 59 ■

Applications

Parabolas have an important property that makes them useful as reflectors for lamps and telescopes. Light from a source placed at the focus of a surface with parabolic cross section will be reflected in such a way that it travels parallel to the axis of the parabola (see Figure 9). Thus a parabolic mirror reflects the light into a beam of parallel rays. Conversely, light approaching the reflector in rays parallel to its axis of symmetry is concentrated to the focus. This *reflection property*, which can be proved by using calculus, is used in the construction of reflecting telescopes.

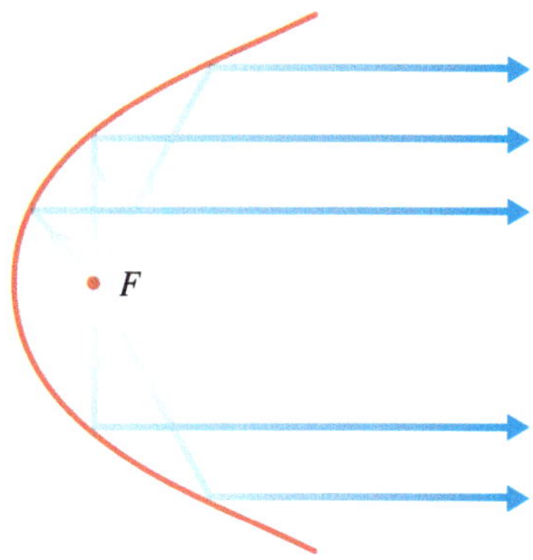

FIGURE 9 Parabolic reflector

EXAMPLE 6 ■ Finding the Focal Point of a Searchlight Reflector

A searchlight has a parabolic reflector that forms a "bowl," which is 12 in. wide from rim to rim and 8 in. deep, as shown in Figure 10. If the filament of the light bulb is located at the focus, how far from the vertex of the reflector is it?

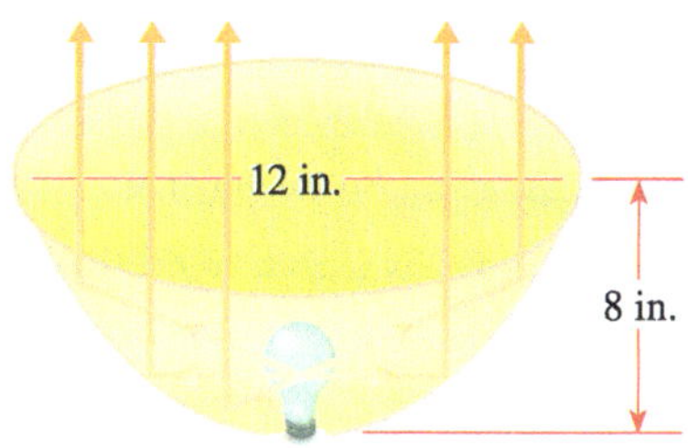

FIGURE 10 A parabolic reflector

ARCHIMEDES (287–212 B.C.) was the greatest mathematician of the ancient world. He was born in Syracuse, a Greek colony on Sicily, a generation after Euclid (see page 542). One of his many discoveries is the Law of the Lever (see page 79). He famously said, "Give me a place to stand and a fulcrum for my lever, and I can lift the earth."

Renowned as a mechanical genius for his many engineering inventions, he designed pulleys for lifting heavy ships and the spiral screw for transporting water to higher levels. He is said to have used parabolic mirrors to concentrate the rays of the sun to set fire to Roman ships attacking Syracuse.

King Hieron II of Syracuse once suspected a goldsmith of keeping part of the gold intended for the king's crown and replacing it with an equal amount of silver. The king asked Archimedes for advice. While in deep thought at a public bath, Archimedes discovered the solution to the king's problem when he noticed that his body's volume was the same as the volume of water it displaced from the tub. Using this insight, he was able to measure the volume of each crown and so determine which was the denser, all-gold crown. As the story is told, he ran home naked, shouting, "Eureka, eureka!" ("I have found it, I have found it!") This incident attests to his enormous powers of concentration.

In spite of his engineering prowess, Archimedes was most proud of his mathematical discoveries. These include the formulas for the volume of a sphere, $\left(V = \frac{4}{3}\pi r^3\right)$ and the surface area of a sphere $\left(S = 4\pi r^2\right)$ and a careful analysis of the properties of parabolas and other conics.

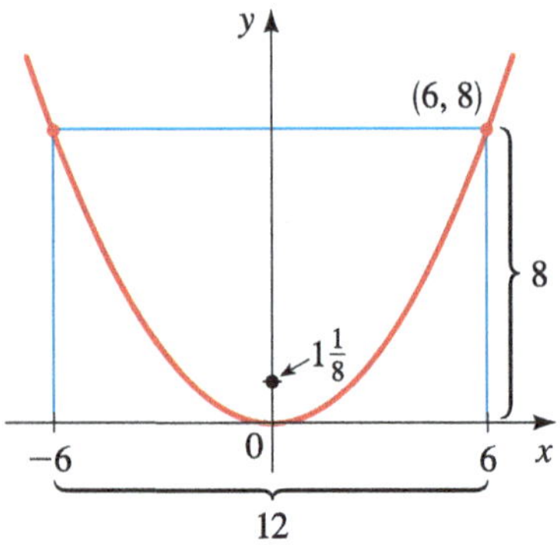

FIGURE 11

SOLUTION We introduce a coordinate system and place a parabolic cross section of the reflector so that its vertex is at the origin and its axis is vertical (see Figure 11). Then the equation of this parabola has the form $x^2 = 4py$. From Figure 11 we see that the point $(6, 8)$ lies on the parabola. We use this to find p.

$$6^2 = 4p(8) \qquad \text{The point (6, 8) satisfies the equation } x^2 = 4py$$
$$36 = 32p$$
$$p = \tfrac{9}{8}$$

The focus is $F\left(0, \frac{9}{8}\right)$, so the distance between the vertex and the focus is $\frac{9}{8} = 1\frac{1}{8}$ in. Because the filament is positioned at the focus, it is located $1\frac{1}{8}$ in. from the vertex of the reflector.

Now Try Exercise 61

11.1 EXERCISES

CONCEPTS

1. A parabola is the set of all points in the plane that are equidistant from a fixed point called the ________ and a fixed line called the ________ of the parabola.

2. The graph of the equation $x^2 = 4py$ is a parabola with focus $F($__, __$)$ and directrix $y =$ ________. So the graph of $x^2 = 12y$ is a parabola with focus $F($__, __$)$ and directrix $y =$ ________.

3. The graph of the equation $y^2 = 4px$ is a parabola with focus $F($__, __$)$ and directrix $x =$ ________. So the graph of $y^2 = 12x$ is a parabola with focus $F($__, __$)$ and directrix $x =$ ________.

4. Label the focus, directrix, and vertex on the graphs given for the parabolas in Exercises 2 and 3.

(a) $x^2 = 12y$ **(b)** $y^2 = 12x$

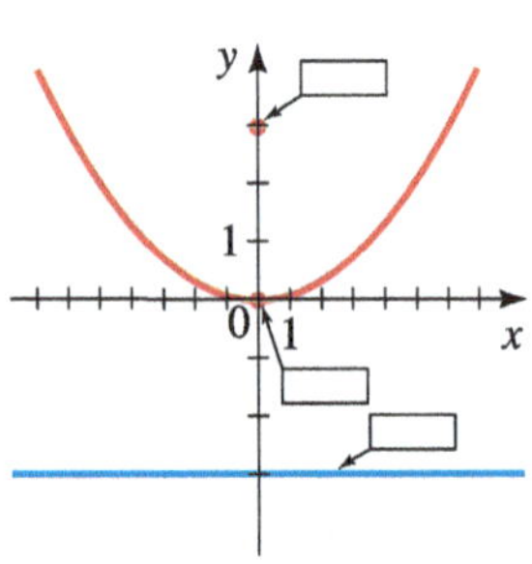

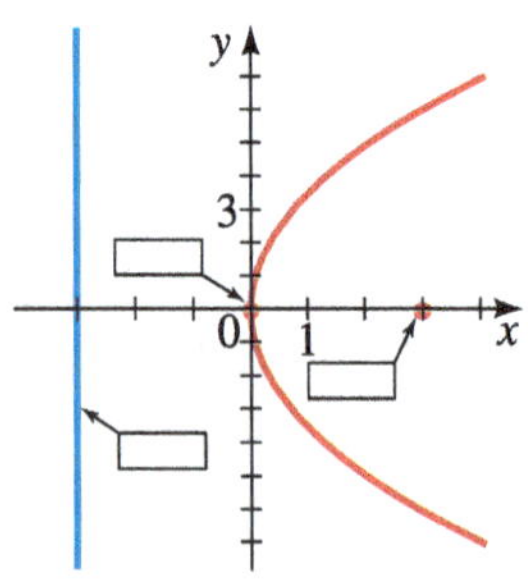

SKILLS

5–10 ■ Graphs of Parabolas Match the equation with the graphs labeled I–VI. Give reasons for your answers.

5. $y^2 = 2x$ **6.** $y^2 = -\frac{1}{4}x$

7. $x^2 = -6y$ **8.** $2x^2 = y$

9. $y^2 - 8x = 0$ **10.** $12y + x^2 = 0$

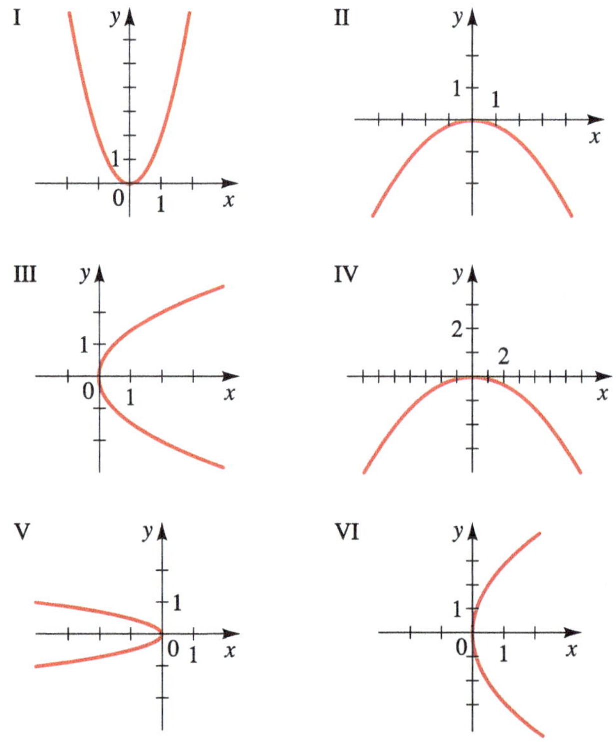

11–24 ■ Graphing Parabolas An equation of a parabola is given. **(a)** Find the focus, directrix, and focal diameter of the parabola. **(b)** Sketch a graph of the parabola and its directrix.

11. $x^2 = 8y$ **12.** $x^2 = -4y$

13. $y^2 = -24x$ **14.** $y^2 = 16x$

15. $y = -\frac{1}{8}x^2$ **16.** $x = 2y^2$

17. $x = -2y^2$ **18.** $y = \frac{1}{4}x^2$

19. $5y = x^2$ **20.** $9x = y^2$

21. $x^2 + 12y = 0$ **22.** $x + \frac{1}{5}y^2 = 0$

23. $5x + 3y^2 = 0$ **24.** $8x^2 + 12y = 0$

25–30 ■ Graphing Parabolas Use a graphing device to graph the parabola.

25. $x^2 = 16y$
26. $x^2 = -8y$
27. $y^2 = -\frac{1}{3}x$
28. $8y^2 = x$
29. $4x + y^2 = 0$
30. $x - 2y^2 = 0$

31–48 ■ Finding the Equation of a Parabola Find an equation for the parabola that has its vertex at the origin and satisfies the given condition(s).

31. Focus: $F(0, 6)$
32. Focus: $F\left(0, -\frac{1}{4}\right)$
33. Focus: $F(-8, 0)$
34. Focus: $F(5, 0)$
35. Focus: $F\left(0, -\frac{3}{4}\right)$
36. Focus: $F\left(-\frac{1}{12}, 0\right)$
37. Directrix: $x = -4$
38. Directrix: $y = \frac{1}{2}$
39. Directrix: $y = \frac{1}{10}$
40. Directrix: $x = -\frac{1}{8}$
41. Directrix: $x = \frac{1}{20}$
42. Directrix: $y = -5$
43. Focus on the positive x-axis, 2 units away from the directrix
44. Focus on the negative y-axis, 6 units away from the directrix
45. Opens downward with focus 10 units away from the vertex
46. Opens upward with focus 5 units away from the vertex
47. Directrix has y-intercept 6
48. Focal diameter 8 and focus on the negative y-axis

49–58 ■ Finding the Equation of a Parabola Find an equation of the parabola whose graph is shown.

49.

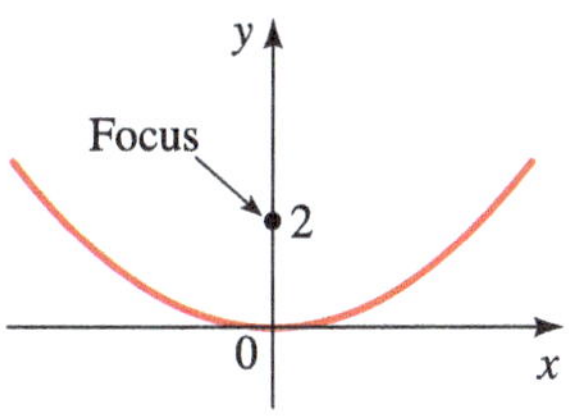

50.

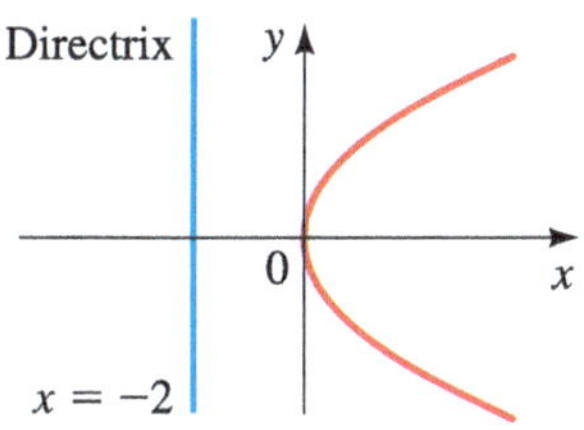

51.

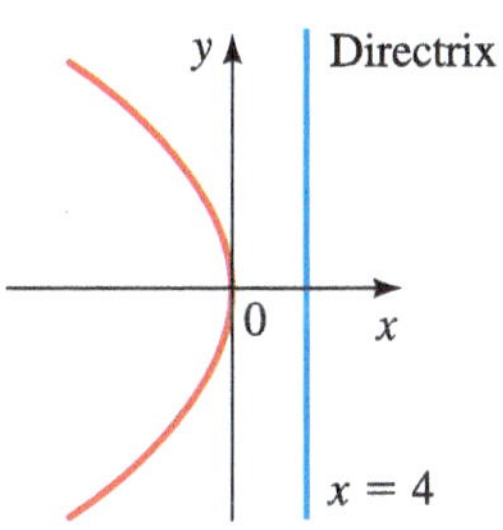

52.

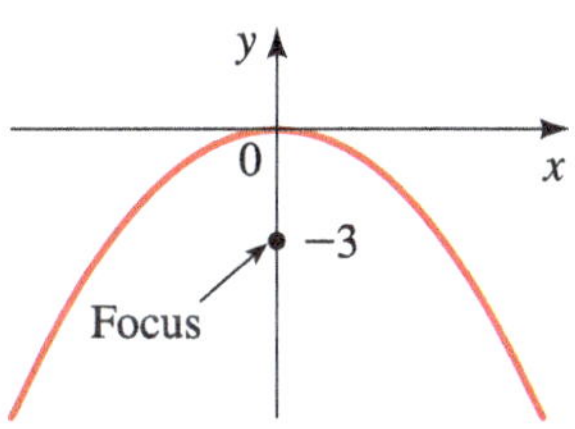

53.

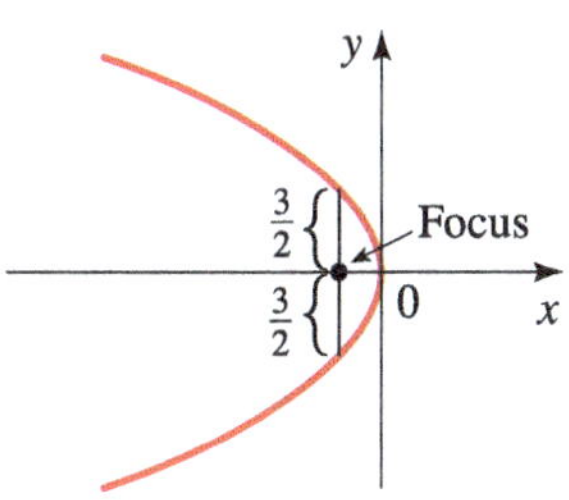

54.

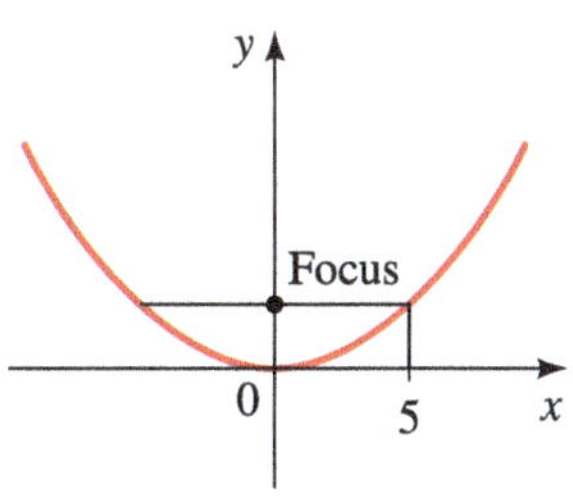

55.

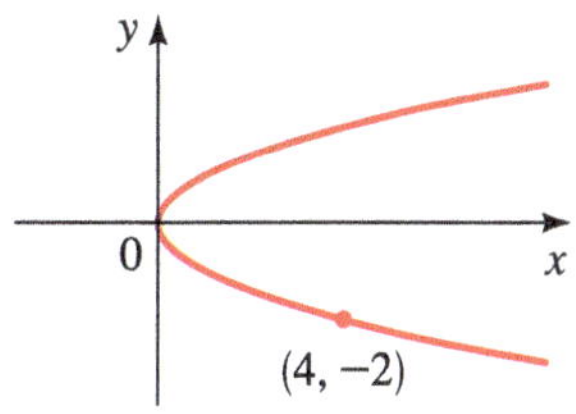

56.

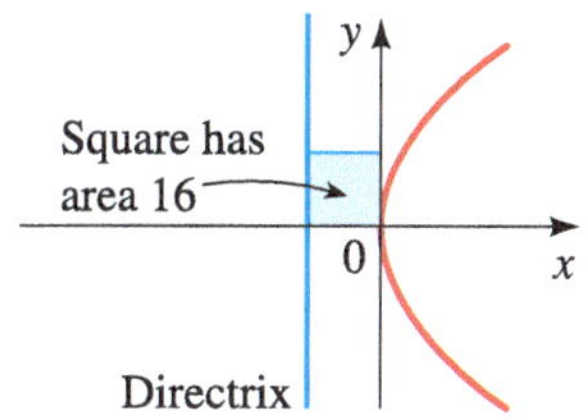

57.

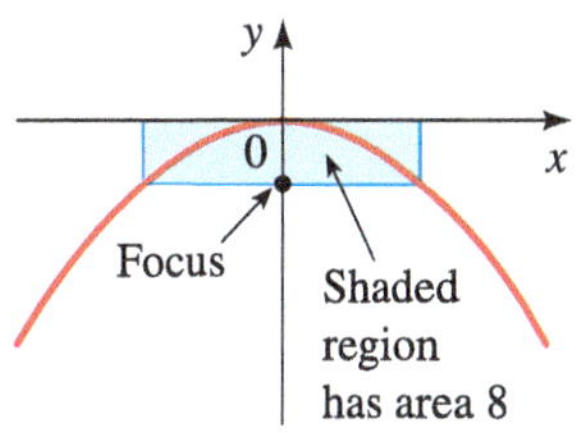

58.

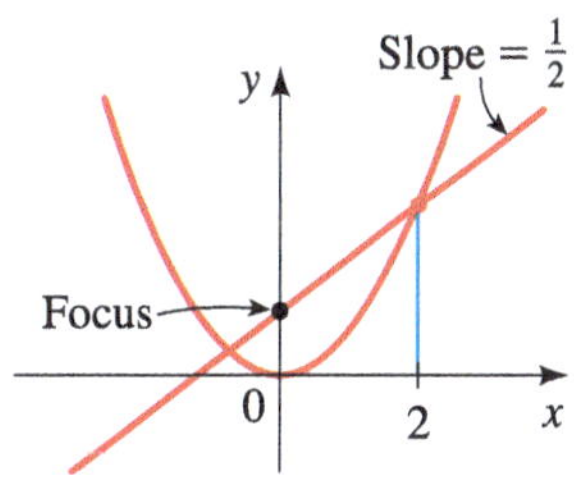

59–60 ■ Families of Parabolas **(a)** Find equations for the family of parabolas with the given description. **(b)** Draw the graphs. What do you conclude?

59. The family of parabolas with vertex at the origin and with directrixes $y = \frac{1}{2}$, $y = 1$, $y = 4$, and $y = 8$

60. The family of parabolas with vertex at the origin, focus on the positive y-axis, and with focal diameters 1, 2, 4, and 8

APPLICATIONS

61. Parabolic Reflector A lamp with a parabolic reflector is shown in the figure. The bulb is placed at the focus, and the focal diameter is 12 cm.

(a) Find an equation of the parabola.

(b) Find the diameter $d(C, D)$ of the opening, 20 cm from the vertex.

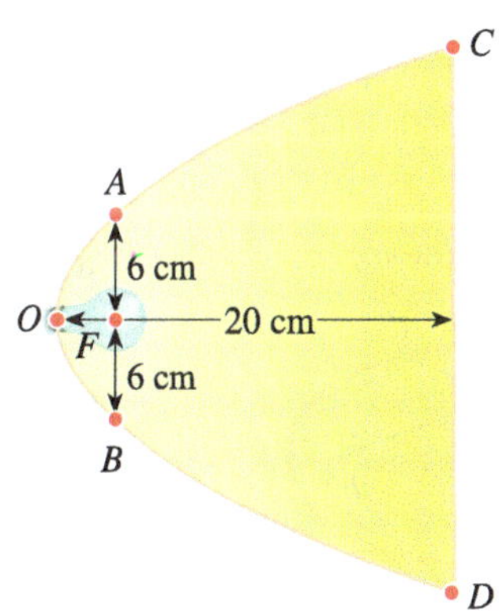

62. Satellite Dish A reflector for a satellite dish is parabolic in cross section, with the receiver at the focus F. The reflector is 1 ft deep and 20 ft wide from rim to rim (see the figure). How far is the receiver from the vertex of the parabolic reflector?

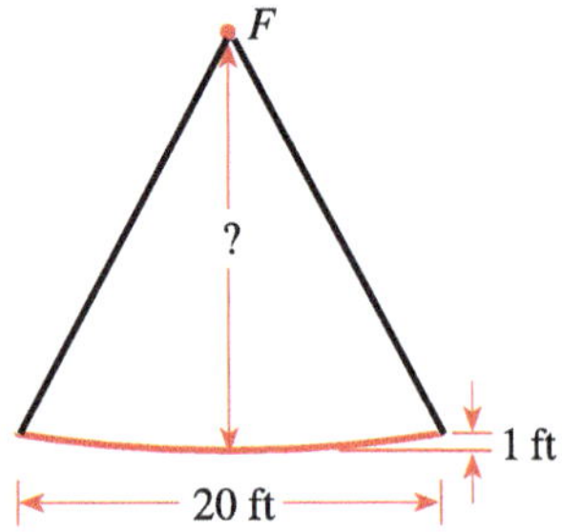

63. Suspension Bridge In a suspension bridge the shape of the suspension cables is parabolic. The bridge shown in the figure has towers that are 600 m apart, and the lowest point of the suspension cables is 150 m below the top of the towers. Find the equation of the parabolic part of the cables, placing the origin of the coordinate system at the vertex. [*Note*: This equation is used to find the length of cable needed in the construction of the bridge.]

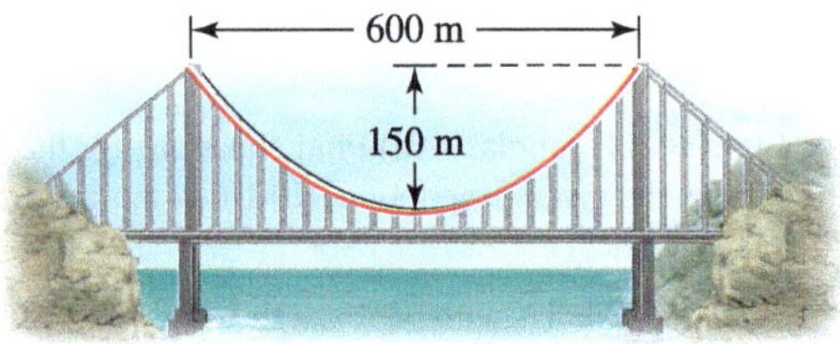

64. Reflecting Telescope The Hale telescope at the Mount Palomar Observatory has a 200-in. mirror, as shown in the figure. The mirror is constructed in a parabolic shape that collects light from the stars and focuses it at the **prime focus**, that is, the focus of the parabola. The mirror is 3.79 in. deep at its center. Find the **focal length** of this parabolic mirror, that is, the distance from the vertex to the focus.

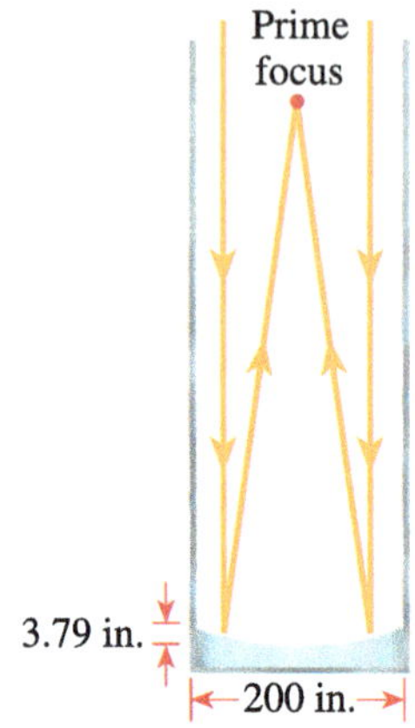

DISCUSS ■ DISCOVER ■ PROVE ■ WRITE

65. DISCUSS ■ WRITE: Parabolas in the Real World Several examples of the uses of parabolas are given in the text. Find other situations in real life in which parabolas occur. Consult a scientific encyclopedia in the reference section of your library, or search the Internet.

66. DISCUSS: Light Cone from a Flashlight A flashlight is held to form a lighted area on the ground, as shown in the figure. Is it possible to angle the flashlight in such a way that the boundary of the lighted area is a parabola? Explain your answer.

11.2 ELLIPSES

■ Geometric Definition of an Ellipse ■ Equations and Graphs of Ellipses
■ Eccentricity of an Ellipse

■ Geometric Definition of an Ellipse

An ellipse is an oval curve that looks like an elongated circle. More precisely, we have the following definition.

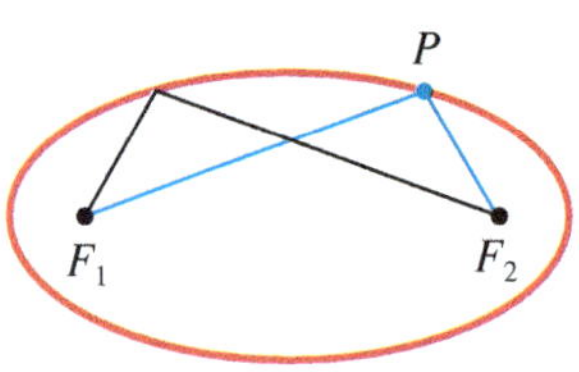

FIGURE 1

> **GEOMETRIC DEFINITION OF AN ELLIPSE**
>
> An **ellipse** is the set of all points in the plane the sum of whose distances from two fixed points F_1 and F_2 is a constant. (See Figure 1.) These two fixed points are the **foci** (plural of **focus**) of the ellipse.

The geometric definition suggests a simple method for drawing an ellipse. Place a sheet of paper on a drawing board, and insert thumbtacks at the two points that are to be the foci of the ellipse. Attach the ends of a string to the tacks, as shown in Figure 2(a). With the point of a pencil, hold the string taut. Then carefully move the pencil around the foci, keeping the string taut at all times. The pencil will trace out an ellipse, because the sum of the distances from the point of the pencil to the foci will always equal the length of the string, which is constant.

If the string is only slightly longer than the distance between the foci, then the ellipse that is traced out will be elongated in shape, as in Figure 2(a), but if the foci are close together relative to the length of the string, the ellipse will be almost circular, as shown in Figure 2(b).

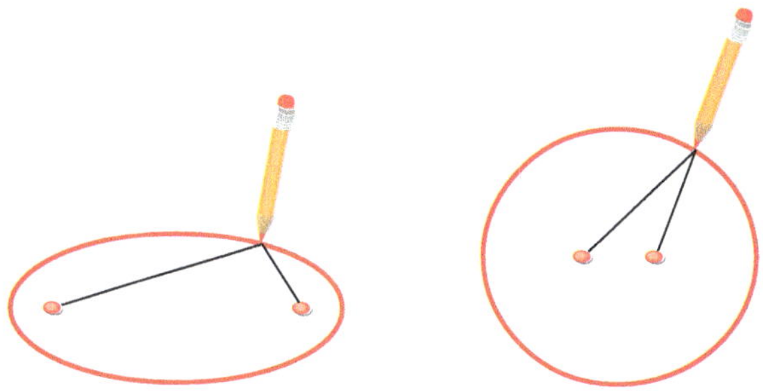

FIGURE 2 (a) (b)

Deriving the Equation of an Ellipse To obtain the simplest equation for an ellipse, we place the foci on the x-axis at $F_1(-c, 0)$ and $F_2(c, 0)$ so that the origin is halfway between them (see Figure 3).

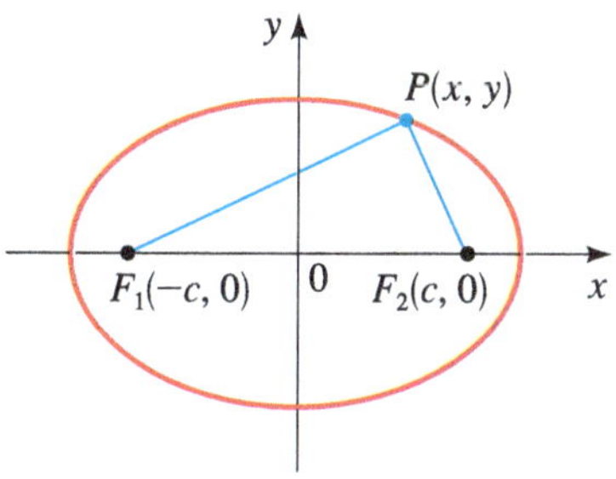

FIGURE 3

For later convenience we let the sum of the distances from a point on the ellipse to the foci be $2a$. Then if $P(x, y)$ is any point on the ellipse, we have

$$d(P, F_1) + d(P, F_2) = 2a$$

So from the Distance Formula we have

$$\sqrt{(x + c)^2 + y^2} + \sqrt{(x - c)^2 + y^2} = 2a$$

or

$$\sqrt{(x - c)^2 + y^2} = 2a - \sqrt{(x + c)^2 + y^2}$$

Squaring each side and expanding, we get

$$x^2 - 2cx + c^2 + y^2 = 4a^2 - 4a\sqrt{(x + c)^2 + y^2} + (x^2 + 2cx + c^2 + y^2)$$

which simplifies to

$$4a\sqrt{(x + c)^2 + y^2} = 4a^2 + 4cx$$

Dividing each side by 4 and squaring again, we get

$$a^2[(x + c)^2 + y^2] = (a^2 + cx)^2$$

$$a^2x^2 + 2a^2cx + a^2c^2 + a^2y^2 = a^4 + 2a^2cx + c^2x^2$$

$$(a^2 - c^2)x^2 + a^2y^2 = a^2(a^2 - c^2)$$

Since the sum of the distances from P to the foci must be larger than the distance between the foci, we have that $2a > 2c$, or $a > c$. Thus $a^2 - c^2 > 0$, and we can divide each side of the preceding equation by $a^2(a^2 - c^2)$ to get

$$\frac{x^2}{a^2} + \frac{y^2}{a^2 - c^2} = 1$$

For convenience let $b^2 = a^2 - c^2$ (with $b > 0$). Since $b^2 < a^2$, it follows that $b < a$. The preceding equation then becomes

$$\frac{x^2}{a^2} + \frac{y^2}{b^2} = 1 \qquad a > b$$

This is the equation of the ellipse. To graph it, we need to know the x- and y-intercepts. Setting $y = 0$, we get

$$\frac{x^2}{a^2} = 1$$

so $x^2 = a^2$, or $x = \pm a$. Thus the ellipse crosses the x-axis at $(a, 0)$ and $(-a, 0)$, as in Figure 4. These points are called the **vertices** of the ellipse, and the segment that joins them is called the **major axis**. Its length is $2a$.

If $a = b$ in the equation of an ellipse, then

$$\frac{x^2}{a^2} + \frac{y^2}{a^2} = 1$$

so $x^2 + y^2 = a^2$. This shows that in this case the "ellipse" is a circle with radius a.

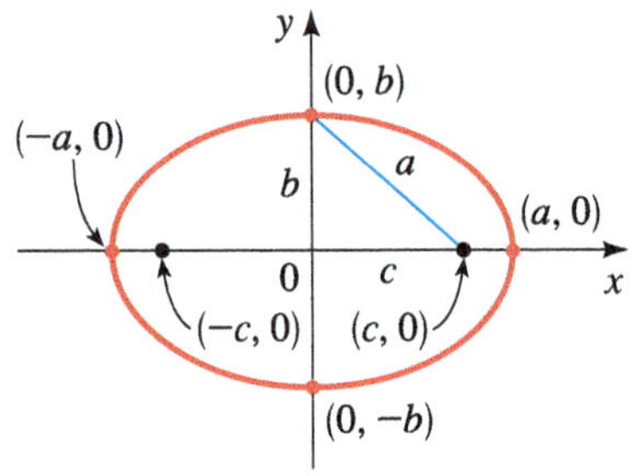

FIGURE 4
$\frac{x^2}{a^2} + \frac{y^2}{b^2} = 1$ with $a > b$

Similarly, if we set $x = 0$, we get $y = \pm b$, so the ellipse crosses the y-axis at $(0, b)$ and $(0, -b)$. The segment that joins these points is called the **minor axis**, and it has length $2b$. Note that $2a > 2b$, so the major axis is longer than the minor axis. The origin is the **center** of the ellipse.

If the foci of the ellipse are placed on the y-axis at $(0, \pm c)$ rather than on the x-axis, then the roles of x and y are reversed in the preceding discussion, and we get a vertical ellipse.

Equations and Graphs of Ellipses

The following box summarizes what we have just proved about ellipses centered at the origin.

ELLIPSE WITH CENTER AT THE ORIGIN

The graph of each of the following equations is an ellipse with center at the origin and having the given properties.

EQUATION	$\frac{x^2}{a^2} + \frac{y^2}{b^2} = 1$ $a > b > 0$	$\frac{x^2}{b^2} + \frac{y^2}{a^2} = 1$ $a > b > 0$
VERTICES	$(\pm a, 0)$	$(0, \pm a)$
MAJOR AXIS	Horizontal, length $2a$	Vertical, length $2a$
MINOR AXIS	Vertical, length $2b$	Horizontal, length $2b$
FOCI	$(\pm c, 0)$, $c^2 = a^2 - b^2$	$(0, \pm c)$, $c^2 = a^2 - b^2$
GRAPH	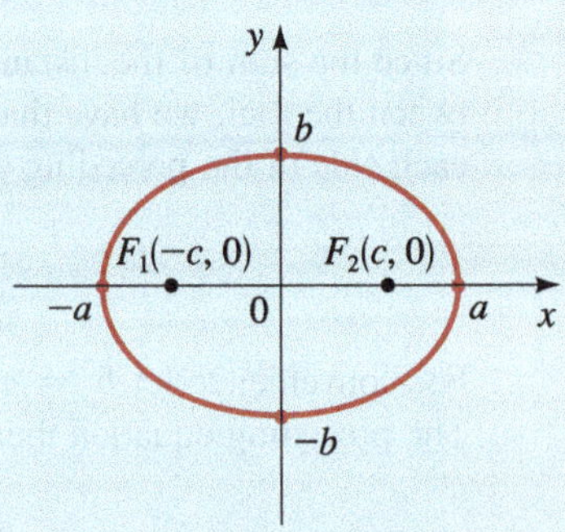	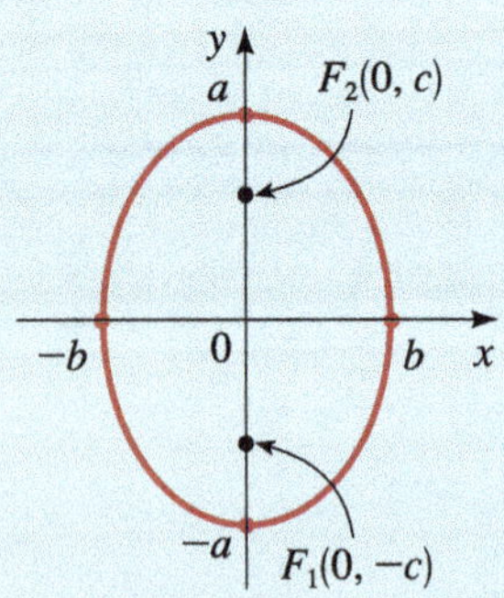

In the standard equation for an ellipse, a^2 is the *larger* denominator, and b^2 is the *smaller*. To find c^2, we subtract: larger denominator minus smaller denominator.

EXAMPLE 1 ■ Sketching an Ellipse

An ellipse has the equation

$$\frac{x^2}{9} + \frac{y^2}{4} = 1$$

(a) Find the foci, the vertices, and the lengths of the major and minor axes, and sketch the graph.

(b) Draw the graph using a graphing calculator.

SOLUTION

(a) Since the denominator of x^2 is larger, the ellipse has a horizontal major axis. This gives $a^2 = 9$ and $b^2 = 4$, so $c^2 = a^2 - b^2 = 9 - 4 = 5$. Thus $a = 3$, $b = 2$, and $c = \sqrt{5}$.

FOCI	$(\pm\sqrt{5}, 0)$
VERTICES	$(\pm 3, 0)$
LENGTH OF MAJOR AXIS	6
LENGTH OF MINOR AXIS	4

The graph is shown in Figure 5(a).

(b) To draw the graph using a graphing calculator, we need to solve for y.

$$\frac{x^2}{9} + \frac{y^2}{4} = 1$$

$$\frac{y^2}{4} = 1 - \frac{x^2}{9} \qquad \text{Subtract } \frac{x^2}{9}$$

$$y^2 = 4\left(1 - \frac{x^2}{9}\right) \qquad \text{Multiply by 4}$$

$$y = \pm 2\sqrt{1 - \frac{x^2}{9}} \qquad \text{Take square roots}$$

The orbits of the planets are ellipses, with the sun at one focus.

To obtain the graph of the ellipse, we graph both functions

$$y = 2\sqrt{1 - x^2/9} \qquad \text{and} \qquad y = -2\sqrt{1 - x^2/9}$$

as shown in Figure 5(b).

Note that the equation of an ellipse does not define y as a function of x (see page 165). That's why we need to graph two functions to graph an ellipse.

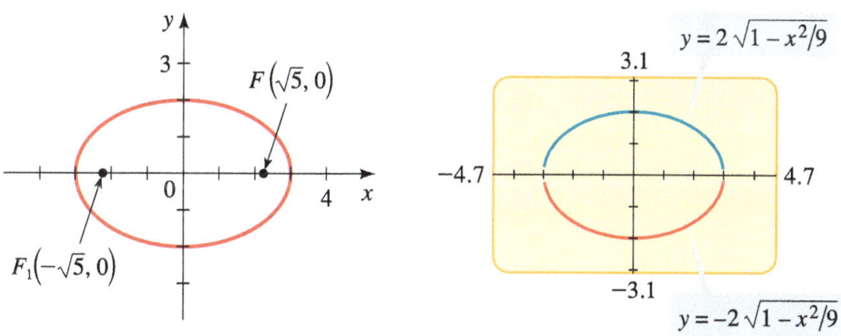

(a) (b)

FIGURE 5 $\frac{x^2}{9} + \frac{y^2}{4} = 1$

Now Try Exercises 9 and 35 ■

EXAMPLE 2 ■ Finding the Foci of an Ellipse

Find the foci of the ellipse $16x^2 + 9y^2 = 144$, and sketch its graph.

SOLUTION First we put the equation in standard form. Dividing by 144, we get

$$\frac{x^2}{9} + \frac{y^2}{16} = 1$$

Since $16 > 9$, this is an ellipse with its foci on the y-axis and with $a = 4$ and $b = 3$. We have

$$c^2 = a^2 - b^2 = 16 - 9 = 7$$
$$c = \sqrt{7}$$

Thus the foci are $(0, \pm\sqrt{7})$. The graph is shown in Figure 6(a).

We can also draw the graph using a graphing calculator as shown in Figure 6(b).

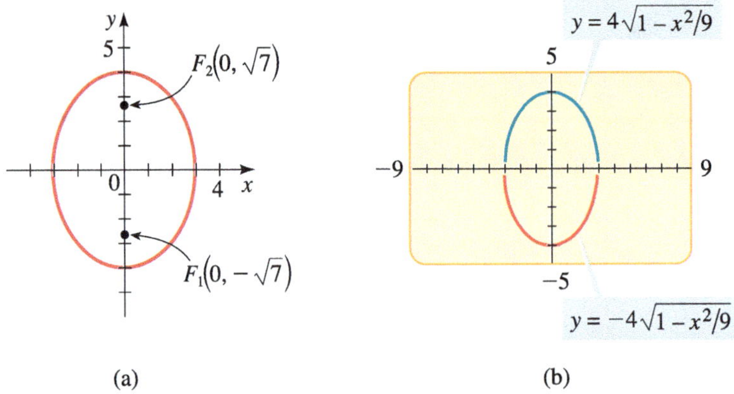

FIGURE 6
$16x^2 + 9y^2 = 144$

Now Try Exercise 15 ■

EXAMPLE 3 ■ Finding the Equation of an Ellipse

The vertices of an ellipse are $(\pm 4, 0)$, and the foci are $(\pm 2, 0)$. Find its equation, and sketch the graph.

SOLUTION Since the vertices are $(\pm 4, 0)$, we have $a = 4$, and the major axis is horizontal. The foci are $(\pm 2, 0)$, so $c = 2$. To write the equation, we need to find b. Since $c^2 = a^2 - b^2$, we have

$$2^2 = 4^2 - b^2$$
$$b^2 = 16 - 4 = 12$$

Thus the equation of the ellipse is

$$\frac{x^2}{16} + \frac{y^2}{12} = 1$$

The graph is shown in Figure 7.

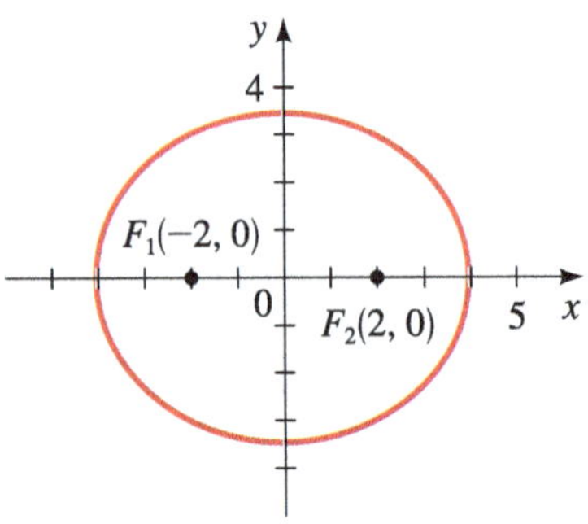

FIGURE 7
$\frac{x^2}{16} + \frac{y^2}{12} = 1$

Now Try Exercises 31 and 39 ■

■ Eccentricity of an Ellipse

We saw earlier in this section (Figure 2) that if $2a$ is only slightly greater than $2c$, the ellipse is long and thin, whereas if $2a$ is much greater than $2c$, the ellipse is almost circular. We measure the deviation of an ellipse from being circular by the ratio of a and c.

DEFINITION OF ECCENTRICITY

For the ellipse $\dfrac{x^2}{a^2} + \dfrac{y^2}{b^2} = 1$ or $\dfrac{x^2}{b^2} + \dfrac{y^2}{a^2} = 1$ (with $a > b > 0$), the **eccentricity *e*** is the number

$$e = \frac{c}{a}$$

where $c = \sqrt{a^2 - b^2}$. The eccentricity of every ellipse satisfies $0 < e < 1$.

Thus if e is close to 1, then c is almost equal to a, and the ellipse is elongated in shape, but if e is close to 0, then the ellipse is close to a circle in shape. The eccentricity is a measure of how "stretched" the ellipse is.

In Figure 8 we show a number of ellipses to demonstrate the effect of varying the eccentricity e.

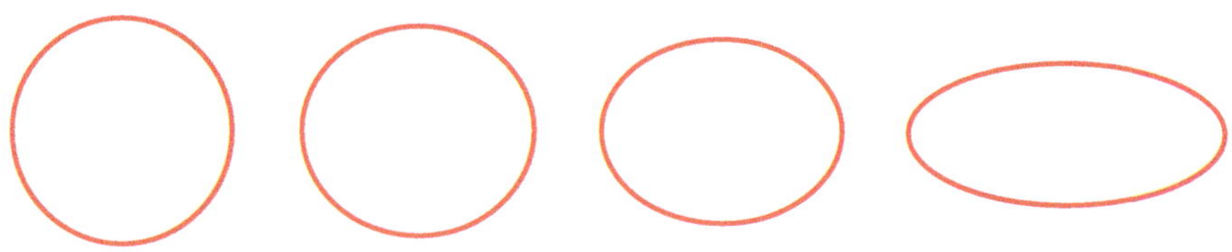

FIGURE 8 Ellipses with various eccentricities

EXAMPLE 4 ■ Finding the Equation of an Ellipse from Its Eccentricity and Foci

Find the equation of the ellipse with foci $(0, \pm 8)$ and eccentricity $e = \frac{4}{5}$, and sketch its graph.

SOLUTION We are given $e = \frac{4}{5}$ and $c = 8$. Thus

$$\frac{4}{5} = \frac{8}{a} \qquad \text{Eccentricity } e = \frac{c}{a}$$

$$4a = 40 \qquad \text{Cross-multiply}$$

$$a = 10$$

To find b, we use the fact that $c^2 = a^2 - b^2$.

$$8^2 = 10^2 - b^2$$

$$b^2 = 10^2 - 8^2 = 36$$

$$b = 6$$

Thus the equation of the ellipse is

$$\frac{x^2}{36} + \frac{y^2}{100} = 1$$

Because the foci are on the y-axis, the ellipse is oriented vertically. To sketch the ellipse, we find the intercepts. The x-intercepts are ± 6, and the y-intercepts are ± 10. The graph is sketched in Figure 9.

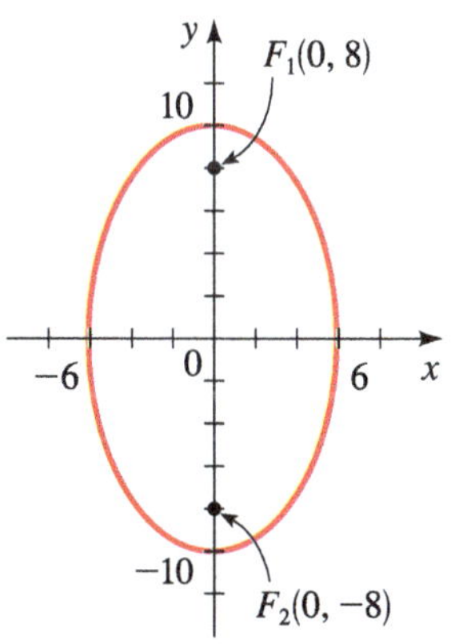

FIGURE 9
$\dfrac{x^2}{36} + \dfrac{y^2}{100} = 1$

Now Try Exercise 53 ■

Eccentricities of the Orbits of the Planets

The orbits of the planets are ellipses with the sun at one focus. For most planets these ellipses have very small eccentricity, so they are nearly circular. However, Mercury and Pluto, the innermost and outermost known planets, respectively, have visibly elliptical orbits.

Planet	Eccentricity
Mercury	0.206
Venus	0.007
Earth	0.017
Mars	0.093
Jupiter	0.048
Saturn	0.056
Uranus	0.046
Neptune	0.010
Pluto*	0.248

*Pluto is a "dwarf planet."

Gravitational attraction causes the planets to move in elliptical orbits around the sun with the sun at one focus. This remarkable property was first observed by Johannes Kepler and was later deduced by Isaac Newton from his inverse square Law of Gravity, using calculus. The orbits of the planets have different eccentricities, but most are nearly circular (see the margin).

Ellipses, like parabolas, have an interesting *reflection property* that leads to a number of practical applications. If a light source is placed at one focus of a reflecting surface with elliptical cross sections, then all the light will be reflected off the surface to the other focus, as shown in Figure 10. This principle, which works for sound waves as well as for light, is used in *lithotripsy*, a treatment for kidney stones. The patient is placed in a tub of water with elliptical cross sections in such a way that the kidney stone is accurately located at one focus. High-intensity sound waves generated at the other focus are reflected to the stone and destroy it with minimal damage to surrounding tissue. The patient is spared the trauma of surgery and recovers within days instead of weeks.

The reflection property of ellipses is also used in the construction of *whispering galleries*. Sound coming from one focus bounces off the walls and ceiling of an elliptical room and passes through the other focus. In these rooms even quiet whispers spoken at one focus can be heard clearly at the other. Famous whispering galleries include the National Statuary Hall of the U.S. Capitol in Washington, D.C. (see page 836), and the Mormon Tabernacle in Salt Lake City, Utah.

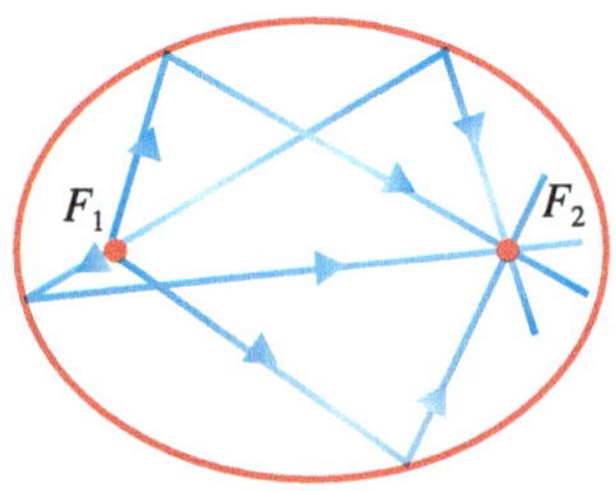

FIGURE 10

11.2 EXERCISES

CONCEPTS

1. An ellipse is the set of all points in the plane for which the ________ of the distances from two fixed points F_1 and F_2 is constant. The points F_1 and F_2 are called the ________ of the ellipse.

2. The graph of the equation $\dfrac{x^2}{a^2} + \dfrac{y^2}{b^2} = 1$ with $a > b > 0$ is an ellipse with vertices (__, __) and (__, __) and foci $(\pm c, 0)$, where $c =$ ________. So the graph of $\dfrac{x^2}{5^2} + \dfrac{y^2}{4^2} = 1$ is an ellipse with vertices (__, __) and (__, __) and foci (__, __) and (__, __).

3. The graph of the equation $\dfrac{x^2}{b^2} + \dfrac{y^2}{a^2} = 1$ with $a > b > 0$ is an ellipse with vertices (__, __) and (__, __) and foci $(0, \pm c)$, where $c =$ ________. So the graph of $\dfrac{x^2}{4^2} + \dfrac{y^2}{5^2} = 1$ is an ellipse with vertices (__, __) and (__, __) and foci (__, __) and (__, __).

4. Label the vertices and foci on the graphs given for the ellipses in Exercises 2 and 3.

(a) $\dfrac{x^2}{5^2} + \dfrac{y^2}{4^2} = 1$ **(b)** $\dfrac{x^2}{4^2} + \dfrac{y^2}{5^2} = 1$

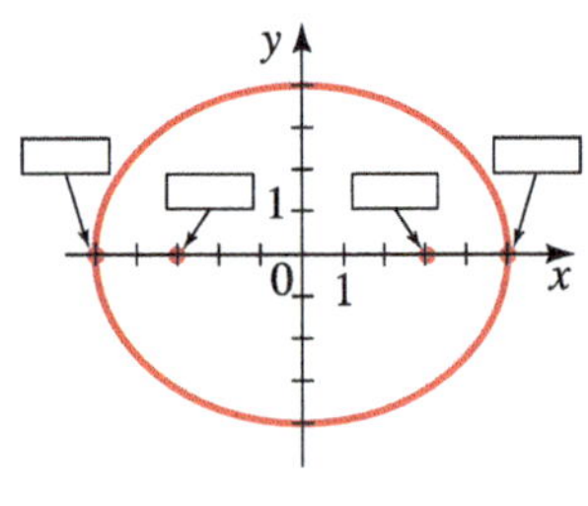

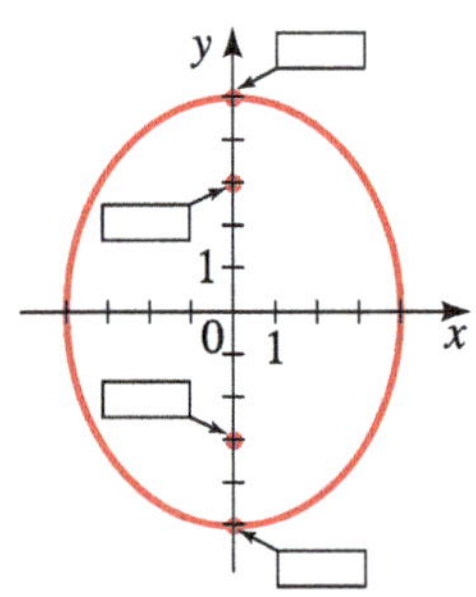

SKILLS

5–8 ■ Graphs of Ellipses Match the equation with the graphs labeled I–IV. Give reasons for your answers.

5. $\dfrac{x^2}{16} + \dfrac{y^2}{4} = 1$

6. $x^2 + \dfrac{y^2}{9} = 1$

7. $4x^2 + y^2 = 4$

8. $16x^2 + 25y^2 = 400$

I

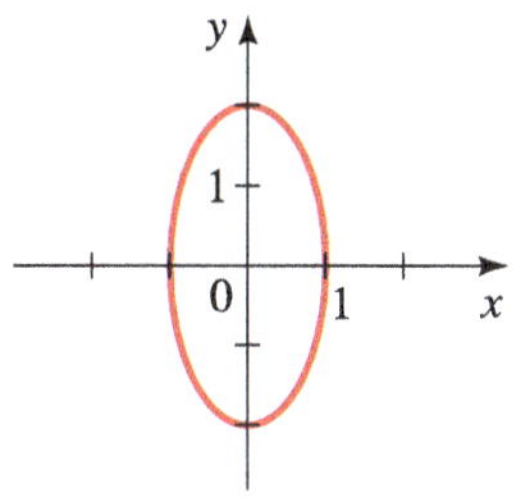

II

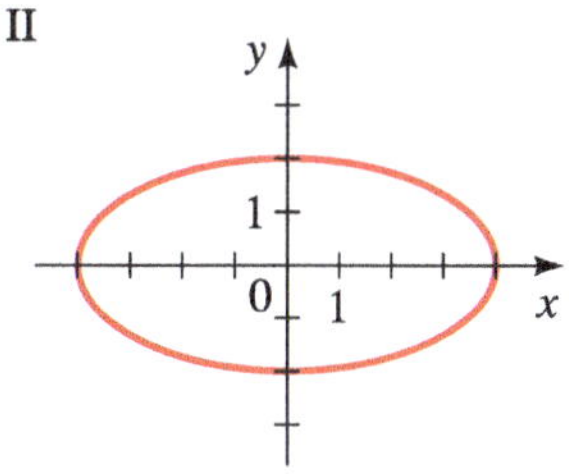

III

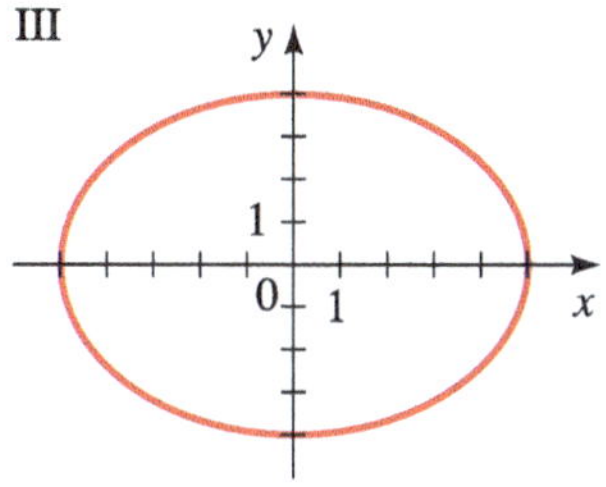

IV

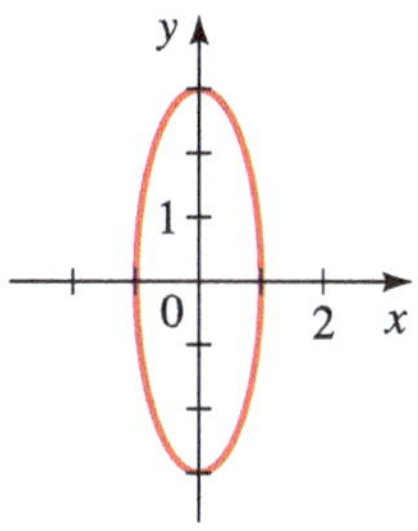

9–28 ■ Graphing Ellipses An equation of an ellipse is given. **(a)** Find the vertices, foci, and eccentricity of the ellipse. **(b)** Determine the lengths of the major and minor axes. **(c)** Sketch a graph of the ellipse.

9. $\dfrac{x^2}{25} + \dfrac{y^2}{9} = 1$

10. $\dfrac{x^2}{16} + \dfrac{y^2}{25} = 1$

11. $\dfrac{x^2}{36} + \dfrac{y^2}{81} = 1$

12. $\dfrac{x^2}{4} + y^2 = 1$

13. $\dfrac{x^2}{49} + \dfrac{y^2}{25} = 1$

14. $\dfrac{x^2}{9} + \dfrac{y^2}{64} = 1$

15. $9x^2 + 4y^2 = 36$

16. $4x^2 + 25y^2 = 100$

17. $x^2 + 4y^2 = 16$

18. $4x^2 + y^2 = 16$

19. $16x^2 + 25y^2 = 1600$

20. $2x^2 + 49y^2 = 98$

21. $3x^2 + y^2 = 9$

22. $x^2 + 3y^2 = 9$

23. $2x^2 + y^2 = 4$

24. $3x^2 + 4y^2 = 12$

25. $x^2 + 4y^2 = 1$

26. $9x^2 + 4y^2 = 1$

27. $x^2 = 4 - 2y^2$

28. $y^2 = 1 - 2x^2$

29–34 ■ Finding the Equation of an Ellipse Find an equation for the ellipse whose graph is shown.

29.

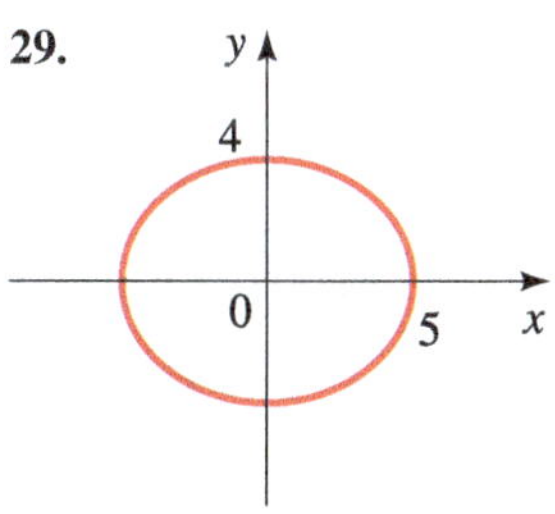

30.

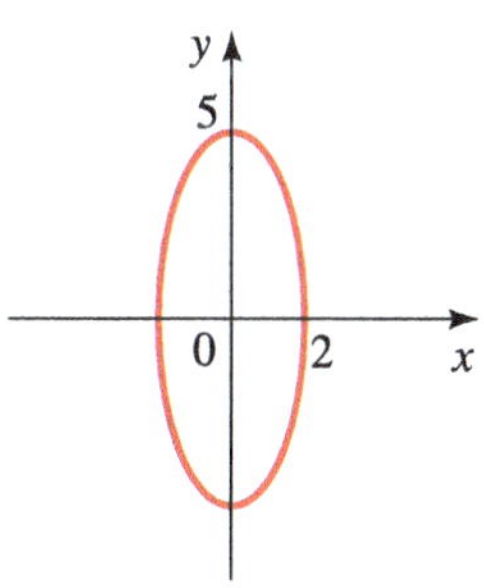

31.

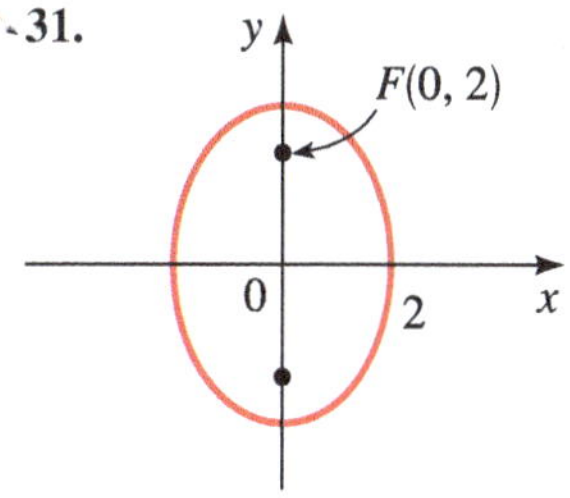

32.

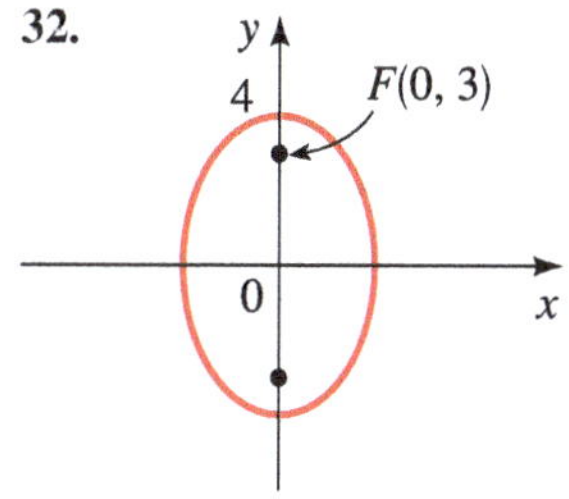

33.

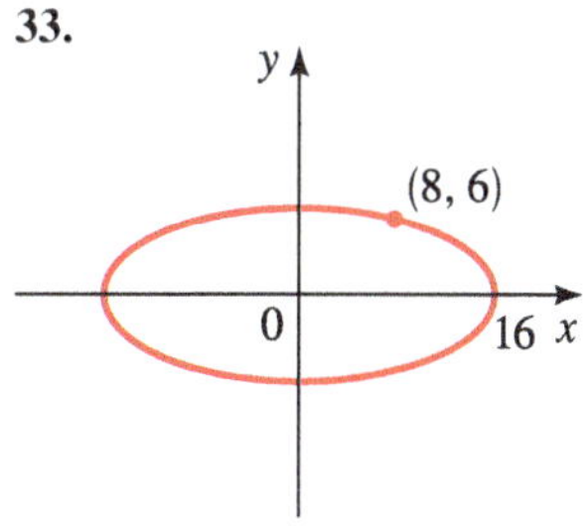

34.

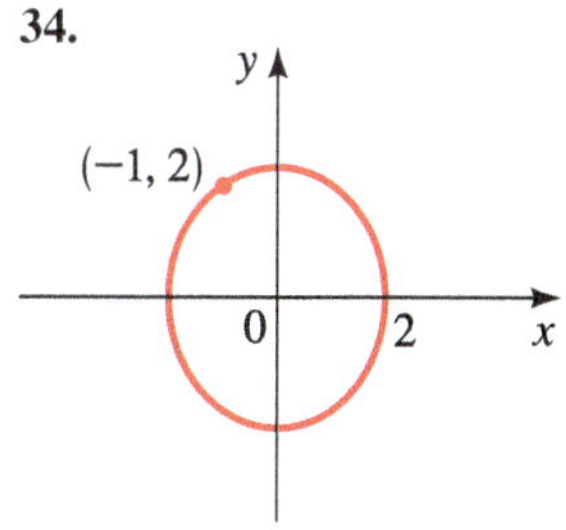

35–38 ■ Graphing Ellipses Use a graphing device to graph the ellipse.

35. $\dfrac{x^2}{25} + \dfrac{y^2}{20} = 1$

36. $x^2 + \dfrac{y^2}{12} = 1$

37. $6x^2 + y^2 = 36$

38. $x^2 + 2y^2 = 8$

39–56 ■ Finding the Equation of an Ellipse Find an equation for the ellipse that satisfies the given conditions.

39. Foci: $(\pm 4, 0)$, vertices: $(\pm 5, 0)$

40. Foci: $(0, \pm 3)$, vertices: $(0, \pm 5)$

41. Foci: $F(\pm 1, 0)$, vertices: $(\pm 2, 0)$

42. Foci: $F(0, \pm 2)$, vertices: $(0, \pm 3)$

43. Foci: $F(0, \pm\sqrt{10})$, vertices: $(0, \pm 7)$

44. Foci: $F(\pm\sqrt{15}, 0)$, vertices: $(\pm 6, 0)$

45. Length of major axis: 4, length of minor axis: 2, foci on y-axis

46. Length of major axis: 6, length of minor axis: 4, foci on x-axis

47. Foci: $(0, \pm 2)$, length of minor axis: 6

48. Foci: $(\pm 5, 0)$, length of major axis: 12

49. Endpoints of major axis: $(\pm 10, 0)$, distance between foci: 6

50. Endpoints of minor axis: $(0, \pm 3)$, distance between foci: 8

51. Length of major axis: 10, foci on x-axis, ellipse passes through the point $(\sqrt{5}, 2)$

52. Length of minor axis: 10, foci on y-axis, ellipse passes through the point $(\sqrt{5}, \sqrt{40})$

53. Eccentricity: $\frac{1}{3}$, foci: $(0, \pm 2)$

54. Eccentricity: 0.75, foci: $(\pm 1.5, 0)$

55. Eccentricity: $\sqrt{3}/2$, foci on y-axis, length of major axis: 4

56. Eccentricity: $\sqrt{5}/3$, foci on x-axis, length of major axis: 12

SKILLS Plus

57–60 ■ Intersecting Ellipses Find the intersection points of the pair of ellipses. Sketch the graphs of each pair of equations on the same coordinate axes, and label the points of intersection.

57. $\begin{cases} 4x^2 + y^2 = 4 \\ 4x^2 + 9y^2 = 36 \end{cases}$

58. $\begin{cases} \dfrac{x^2}{16} + \dfrac{y^2}{9} = 1 \\ \dfrac{x^2}{9} + \dfrac{y^2}{16} = 1 \end{cases}$

59. $\begin{cases} 100x^2 + 25y^2 = 100 \\ x^2 + \dfrac{y^2}{9} = 1 \end{cases}$

60. $\begin{cases} 25x^2 + 144y^2 = 3600 \\ 144x^2 + 25y^2 = 3600 \end{cases}$

61. Ancillary Circle The **ancillary circle** of an ellipse is the circle with radius equal to half the length of the minor axis and center the same as the ellipse (see the figure). The ancillary circle is thus the largest circle that can fit within an ellipse.

(a) Find an equation for the ancillary circle of the ellipse $x^2 + 4y^2 = 16$.

(b) For the ellipse and ancillary circle of part (a), show that if (s, t) is a point on the ancillary circle, then $(2s, t)$ is a point on the ellipse.

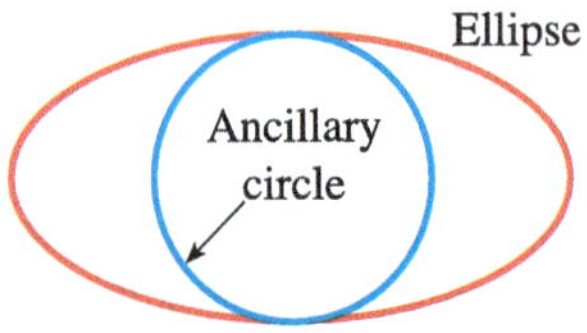

62. Family of Ellipses

(a) Use a graphing device to sketch the top half (the portion in the first and second quadrants) of the family of ellipses $x^2 + ky^2 = 100$ for $k = 4, 10, 25$, and 50.

(b) What do the members of this family of ellipses have in common? How do they differ?

63. Family of Ellipses If $k > 0$, the following equation represents an ellipse:

$$\frac{x^2}{k} + \frac{y^2}{4 + k} = 1$$

Show that all the ellipses represented by this equation have the same foci, no matter what the value of k.

64. How Wide Is an Ellipse at a Focus? A *latus rectum* for an ellipse is a line segment perpendicular to the major axis at a focus, with endpoints on the ellipse, as shown in the figure. Show that the length of a latus rectum is $2b^2/a$ for the ellipse

$$\frac{x^2}{a^2} + \frac{y^2}{b^2} = 1 \qquad a > b$$

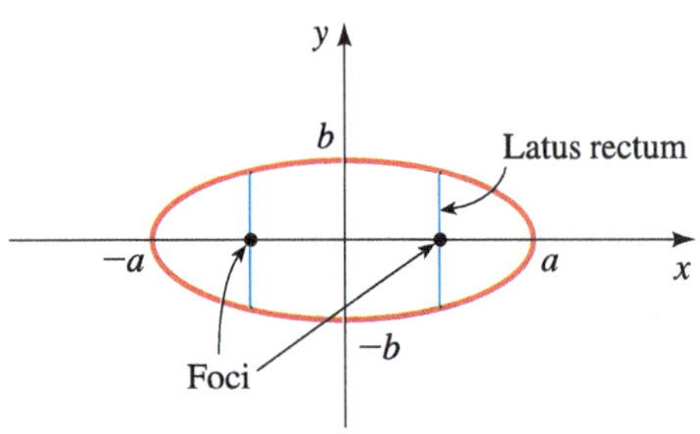

APPLICATIONS

65. Perihelion and Aphelion The planets move around the sun in elliptical orbits with the sun at one focus. The point in the orbit at which the planet is closest to the sun is called **perihelion**, and the point at which it is farthest is called **aphelion**. These points are the vertices of the orbit. The earth's distance from the sun is 147,000,000 km at perihelion and 153,000,000 km at aphelion. Find an equation for the earth's orbit. (Place the origin at the center of the orbit with the sun on the x-axis.)

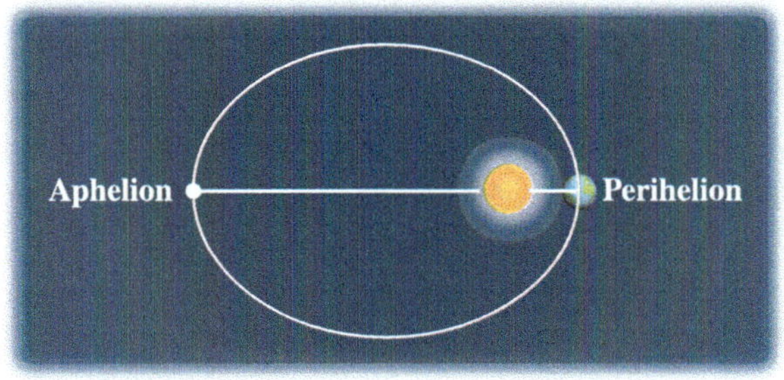

66. The Orbit of Pluto With an eccentricity of 0.25, Pluto's orbit is the most eccentric in the solar system. The length of the minor axis of its orbit is approximately 10,000,000,000 km. Find the distance between Pluto and the sun at perihelion and at aphelion. (See Exercise 65.)

67. Lunar Orbit For an object in an elliptical orbit around the moon, the points in the orbit that are closest to and farthest from the center of the moon are called **perilune** and **apolune**, respectively. These are the vertices of the orbit. The center of the moon is at one focus of the orbit. The *Apollo 11* spacecraft was placed in a lunar orbit with perilune at 68 mi and apolune at 195 mi above the surface of the moon. Assuming that the moon is a sphere of radius 1075 mi, find an equation for the orbit of *Apollo 11*. (Place the coordinate axes so that the origin is at the center of the orbit and the foci are located on the x-axis.)

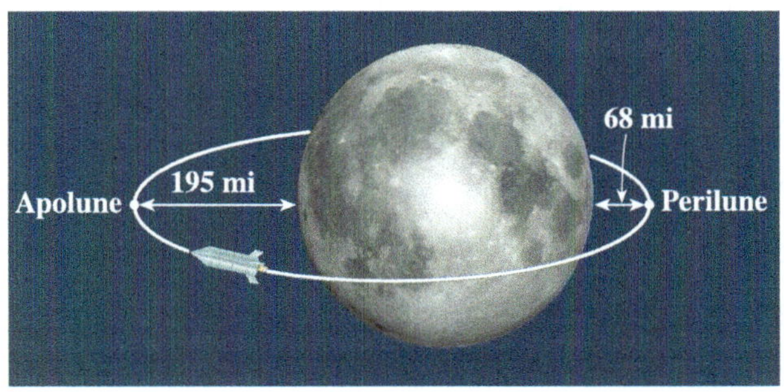

68. Plywood Ellipse A carpenter wishes to construct an elliptical table top from a 4 ft by 8 ft sheet of plywood. He will trace out the ellipse using the "thumbtack and string" method illustrated in Figures 2 and 3. What length of string should he use, and how far apart should the tacks be located, if the ellipse is to be the largest possible that can be cut out of the plywood sheet?

69. Sunburst Window A "sunburst" window above a doorway is constructed in the shape of the top half of an ellipse, as shown in the figure. The window is 20 in. tall at its highest point and 80 in. wide at the bottom. Find the height of the window 25 in. from the center of the base.

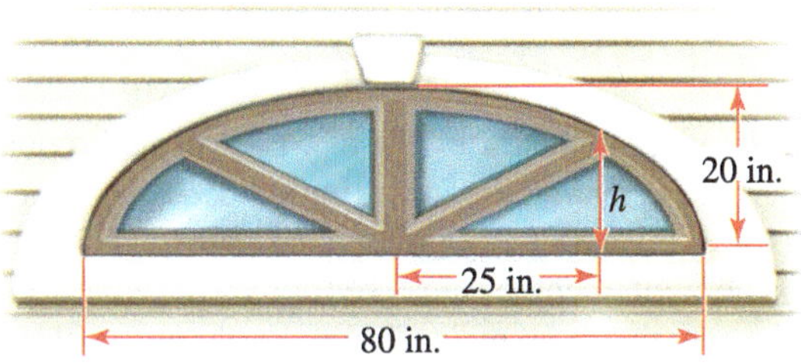

DISCUSS ■ DISCOVER ■ PROVE ■ WRITE

70. DISCUSS: Drawing an Ellipse on a Blackboard Try drawing an ellipse as accurately as possible on a blackboard. How would a piece of string and two friends help this process?

71. DISCUSS: Light Cone from a Flashlight A flashlight shines on a wall, as shown in the figure. What is the shape of the boundary of the lighted area? Explain your answer.

72. DISCUSS: Is It an Ellipse? A piece of paper is wrapped around a cylindrical bottle, and then a compass is used to draw a circle on the paper, as shown in the figure. When the paper is laid flat, is the shape drawn on the paper an ellipse? (You don't need to prove your answer, but you might want to do the experiment and see what you get.)

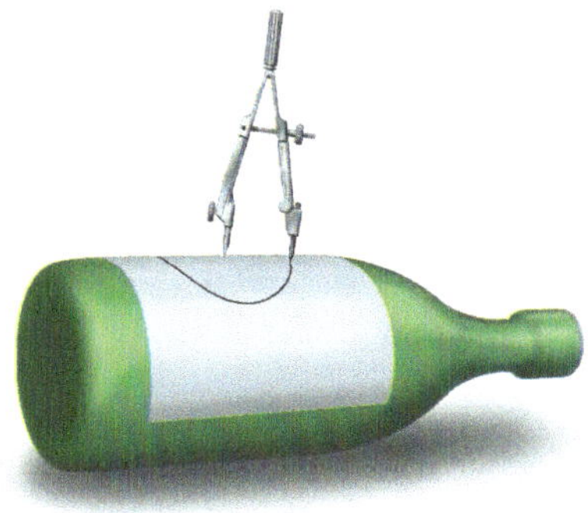

11.3 HYPERBOLAS

■ Geometric Definition of a Hyperbola ■ Equations and Graphs of Hyperbolas

■ Geometric Definition of a Hyperbola

Although ellipses and hyperbolas have completely different shapes, their definitions and equations are similar. Instead of using the *sum* of distances from two fixed foci, as in the case of an ellipse, we use the *difference* to define a hyperbola.

> **GEOMETRIC DEFINITION OF A HYPERBOLA**
>
> A **hyperbola** is the set of all points in the plane, the difference of whose distances from two fixed points F_1 and F_2 is a constant. (See Figure 1.) These two fixed points are the **foci** of the hyperbola.

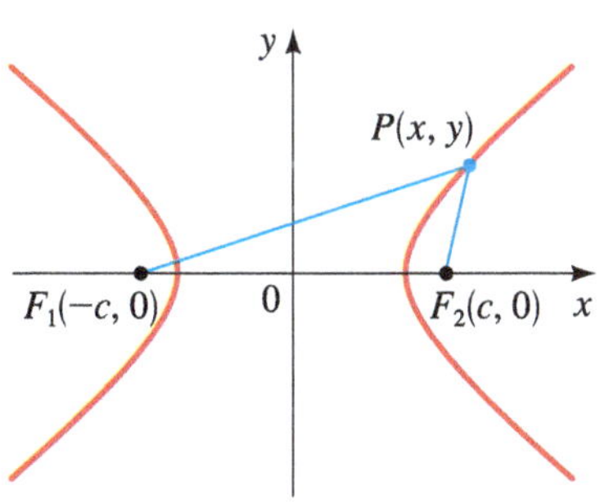

FIGURE 1 P is on the hyperbola if $|d(P, F_1) - d(P, F_2)| = 2a$.

Deriving the Equation of a Hyperbola As in the case of the ellipse, we get the simplest equation for the hyperbola by placing the foci on the x-axis at $(\pm c, 0)$, as shown in Figure 1. By definition, if $P(x, y)$ lies on the hyperbola, then either $d(P, F_1) - d(P, F_2)$ or $d(P, F_2) - d(P, F_1)$ must equal some positive constant, which we call $2a$. Thus we have

$$d(P, F_1) - d(P, F_2) = \pm 2a$$

or

$$\sqrt{(x+c)^2 + y^2} - \sqrt{(x-c)^2 + y^2} = \pm 2a$$

Proceeding as we did in the case of the ellipse (Section 11.2), we simplify this to

$$(c^2 - a^2)x^2 - a^2y^2 = a^2(c^2 - a^2)$$

From triangle PF_1F_2 in Figure 1 we see that $|d(P, F_1) - d(P, F_2)| < 2c$. It follows that $2a < 2c$, or $a < c$. Thus $c^2 - a^2 > 0$, so we can set $b^2 = c^2 - a^2$. We then simplify the last displayed equation to get

$$\frac{x^2}{a^2} - \frac{y^2}{b^2} = 1$$

This is the *equation of the hyperbola*. If we replace x by $-x$ or y by $-y$ in this equation, it remains unchanged, so the hyperbola is symmetric about both the x- and y-axes and about the origin. The x-intercepts are $\pm a$, and the points $(a, 0)$ and $(-a, 0)$ are the **vertices** of the hyperbola. There is no y-intercept, because setting $x = 0$ in the equation of the hyperbola leads to $-y^2 = b^2$, which has no real solution. Furthermore, the equation of the hyperbola implies that

$$\frac{x^2}{a^2} = \frac{y^2}{b^2} + 1 \geq 1$$

so $x^2/a^2 \geq 1$; thus $x^2 \geq a^2$, and hence $x \geq a$ or $x \leq -a$. This means that the hyperbola consists of two parts, called its **branches**. The segment joining the two vertices on the separate branches is the **transverse axis** of the hyperbola, and the origin is called its **center**.

If we place the foci of the hyperbola on the y-axis rather than on the x-axis, this has the effect of reversing the roles of x and y in the derivation of the equation of the hyperbola. This leads to a hyperbola with a vertical transverse axis.

Equations and Graphs of Hyperbolas

The main properties of hyperbolas are listed in the following box.

HYPERBOLA WITH CENTER AT THE ORIGIN

The graph of each of the following equations is a hyperbola with center at the origin and having the given properties.

EQUATION	$\dfrac{x^2}{a^2} - \dfrac{y^2}{b^2} = 1 \quad a > 0, b > 0$	$\dfrac{y^2}{a^2} - \dfrac{x^2}{b^2} = 1 \quad a > 0, b > 0$
VERTICES	$(\pm a, 0)$	$(0, \pm a)$
TRANSVERSE AXIS	Horizontal, length $2a$	Vertical, length $2a$
ASYMPTOTES	$y = \pm \dfrac{b}{a}x$	$y = \pm \dfrac{a}{b}x$
FOCI	$(\pm c, 0), \quad c^2 = a^2 + b^2$	$(0, \pm c), \quad c^2 = a^2 + b^2$
GRAPH	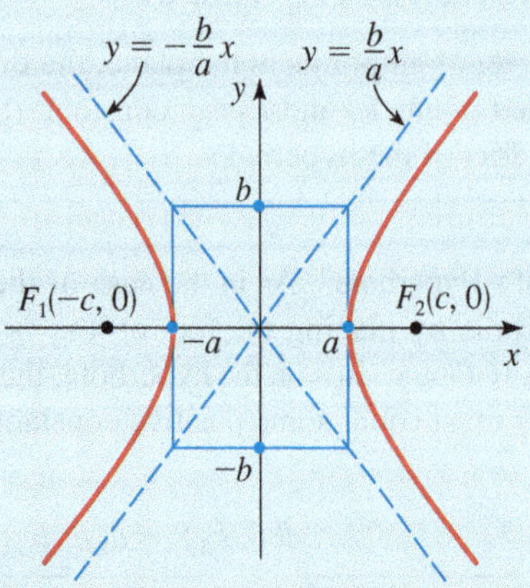	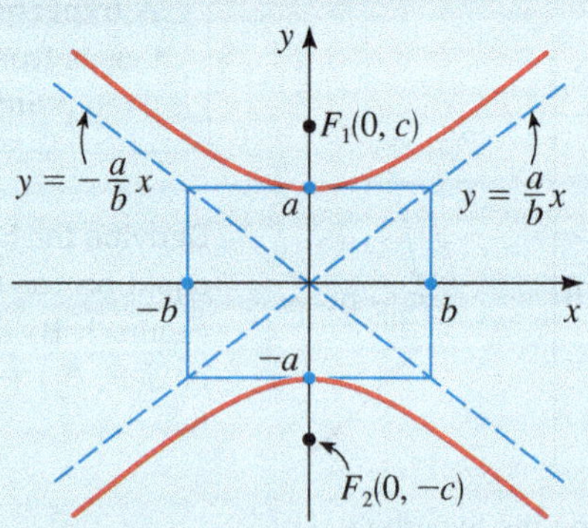

Asymptotes of rational functions are discussed in Section 3.6.

The *asymptotes* mentioned in this box are lines that the hyperbola approaches for large values of x and y. To find the asymptotes in the first case in the box, we solve the equation for y to get

$$\begin{aligned} y &= \pm\frac{b}{a}\sqrt{x^2 - a^2} \\ &= \pm\frac{b}{a}x\sqrt{1 - \frac{a^2}{x^2}} \end{aligned}$$

As x gets large, a^2/x^2 gets closer to zero. In other words, as $x \to \infty$, we have $a^2/x^2 \to 0$. So for large x the value of y can be approximated as $y = \pm(b/a)x$. This shows that these lines are asymptotes of the hyperbola.

Asymptotes are an essential aid for graphing a hyperbola; they help us to determine its shape. A convenient way to find the asymptotes, for a hyperbola with horizontal transverse axis, is to first plot the points $(a, 0)$, $(-a, 0)$, $(0, b)$, and $(0, -b)$. Then sketch horizontal and vertical segments through these points to construct a rectangle, as shown in Figure 2(a). We call this rectangle the **central box** of the hyperbola. The slopes of the diagonals of the central box are $\pm b/a$, so by extending them, we obtain the asymptotes $y = \pm(b/a)x$, as sketched in Figure 2(b). Finally, we plot the vertices and use the asymptotes as a guide in sketching the hyperbola shown in Figure 2(c). (A similar procedure applies to graphing a hyperbola that has a vertical transverse axis.)

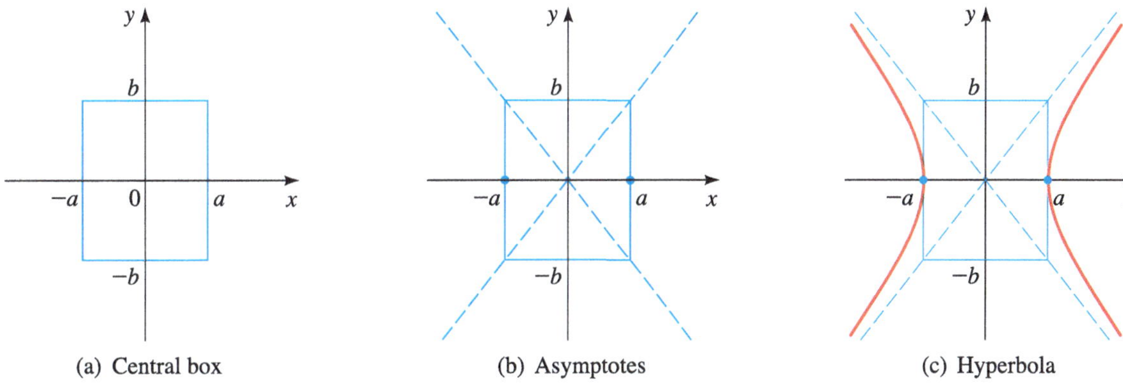

FIGURE 2 Steps in graphing the hyperbola $\dfrac{x^2}{a^2} - \dfrac{y^2}{b^2} = 1$

HOW TO SKETCH A HYPERBOLA

1. **Sketch the Central Box.** This is the rectangle centered at the origin, with sides parallel to the axes, that crosses one axis at $\pm a$ and the other at $\pm b$.
2. **Sketch the Asymptotes.** These are the lines obtained by extending the diagonals of the central box.
3. **Plot the Vertices.** These are the two x-intercepts or the two y-intercepts.
4. **Sketch the Hyperbola.** Start at a vertex, and sketch a branch of the hyperbola, approaching the asymptotes. Sketch the other branch in the same way.

EXAMPLE 1 ■ A Hyperbola with Horizontal Transverse Axis

A hyperbola has the equation

$$9x^2 - 16y^2 = 144$$

(a) Find the vertices, foci, length of the transverse axis, and asymptotes, and sketch the graph.

(b) Draw the graph using a graphing calculator.

SOLUTION

(a) First we divide both sides of the equation by 144 to put it into standard form:

$$\frac{x^2}{16} - \frac{y^2}{9} = 1$$

Because the x^2-term is positive, the hyperbola has a horizontal transverse axis; its vertices and foci are on the x-axis. Since $a^2 = 16$ and $b^2 = 9$, we get $a = 4$, $b = 3$, and $c = \sqrt{16 + 9} = 5$. Thus we have

VERTICES	$(\pm 4, 0)$
FOCI	$(\pm 5, 0)$
ASYMPTOTES	$y = \pm\frac{3}{4}x$

The length of the transverse axis is $2a = 8$. After sketching the central box and asymptotes, we complete the sketch of the hyperbola as in Figure 3(a).

Note that the equation of a hyperbola does not define y as a function of x (see page 165). That's why we need to graph two functions to graph a hyperbola.

(b) To draw the graph using a graphing calculator, we need to solve for y.

$$\begin{aligned} 9x^2 - 16y^2 &= 144 \\ -16y^2 &= -9x^2 + 144 && \text{Subtract } 9x^2 \\ y^2 &= 9\left(\frac{x^2}{16} - 1\right) && \text{Divide by } -16 \text{ and factor } 9 \\ y &= \pm 3\sqrt{\frac{x^2}{16} - 1} && \text{Take square roots} \end{aligned}$$

To obtain the graph of the hyperbola, we graph the functions

$$y = 3\sqrt{(x^2/16) - 1} \quad \text{and} \quad y = -3\sqrt{(x^2/16) - 1}$$

as shown in Figure 3(b).

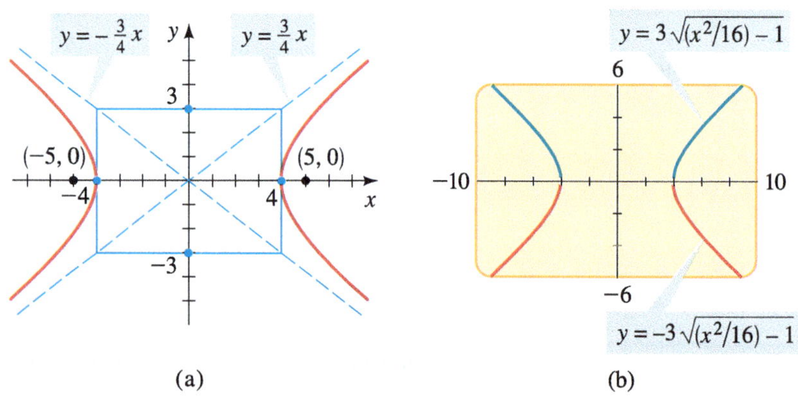

FIGURE 3
$9x^2 - 16y^2 = 144$

Now Try Exercises 9 and 33 ■

EXAMPLE 2 ■ A Hyperbola with Vertical Transverse Axis

Find the vertices, foci, length of the transverse axis, and asymptotes of the hyperbola, and sketch its graph.

$$x^2 - 9y^2 + 9 = 0$$

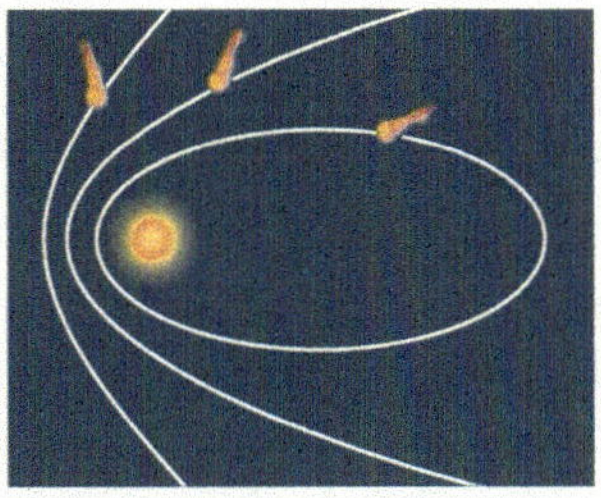

Paths of Comets

The path of a comet is an ellipse, a parabola, or a hyperbola with the sun at a focus. This fact can be proved by using calculus and Newton's Laws of Motion.* If the path is a parabola or a hyperbola, the comet will never return. If the path is an ellipse, it can be determined precisely when and where the comet can be seen again. Halley's comet has an elliptical path and returns every 75 years; it was last seen in 1987. The brightest comet of the 20th century was comet Hale-Bopp, seen in 1997. Its orbit is a very eccentric ellipse; it is expected to return to the inner solar system around the year 4377.

---

*James Stewart, *Calculus*, 7th ed. (Belmont, CA: Brooks/Cole, 2012), pages 892 and 896.

SOLUTION We begin by writing the equation in the standard form for a hyperbola:

$$x^2 - 9y^2 = -9$$

$$y^2 - \frac{x^2}{9} = 1 \qquad \text{Divide by } -9$$

Because the y^2-term is positive, the hyperbola has a vertical transverse axis; its foci and vertices are on the y-axis. Since $a^2 = 1$ and $b^2 = 9$, we get $a = 1$, $b = 3$, and $c = \sqrt{1 + 9} = \sqrt{10}$. Thus we have

VERTICES	$(0, \pm 1)$
FOCI	$(0, \pm\sqrt{10})$
ASYMPTOTES	$y = \pm\frac{1}{3}x$

The length of the transverse axis is $2a = 2$. We sketch the central box and asymptotes, then complete the graph, as shown in Figure 4(a). We can also draw the graph using a graphing calculator, as shown in Figure 4(b).

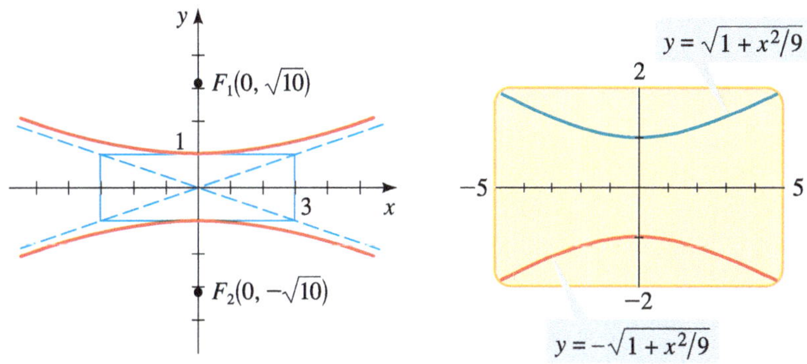

(a) (b)

FIGURE 4 $x^2 - 9y^2 + 9 = 0$

Now Try Exercises 21 and 35 ■

EXAMPLE 3 ■ Finding the Equation of a Hyperbola from Its Vertices and Foci

Find the equation of the hyperbola with vertices $(\pm 3, 0)$ and foci $(\pm 4, 0)$. Sketch the graph.

SOLUTION Since the vertices are on the x-axis, the hyperbola has a horizontal transverse axis. Its equation is of the form

$$\frac{x^2}{3^2} - \frac{y^2}{b^2} = 1$$

We have $a = 3$ and $c = 4$. To find b, we use the relation $a^2 + b^2 = c^2$.

$$3^2 + b^2 = 4^2$$

$$b^2 = 4^2 - 3^2 = 7$$

$$b = \sqrt{7}$$

Thus the equation of the hyperbola is

$$\frac{x^2}{9} - \frac{y^2}{7} = 1$$

The graph is shown in Figure 5.

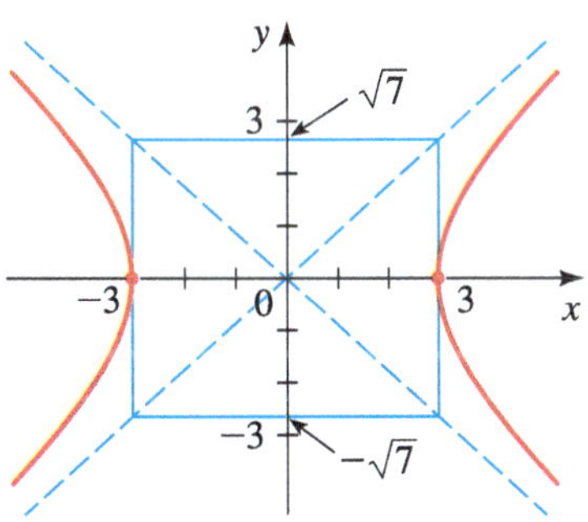

FIGURE 5 $\dfrac{x^2}{9} - \dfrac{y^2}{7} = 1$

Now Try Exercises 27 and 37 ■

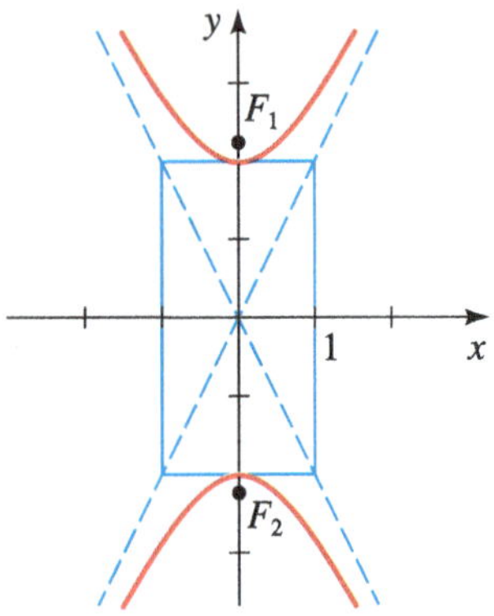

FIGURE 6
$\dfrac{y^2}{4} - x^2 = 1$

EXAMPLE 4 ■ Finding the Equation of a Hyperbola from Its Vertices and Asymptotes

Find the equation and the foci of the hyperbola with vertices $(0, \pm 2)$ and asymptotes $y = \pm 2x$. Sketch the graph.

SOLUTION Since the vertices are on the y-axis, the hyperbola has a vertical transverse axis with $a = 2$. From the asymptote equation we see that $a/b = 2$. Since $a = 2$, we get $2/b = 2$, so $b = 1$. Thus the equation of the hyperbola is

$$\frac{y^2}{4} - x^2 = 1$$

To find the foci, we calculate $c^2 = a^2 + b^2 = 2^2 + 1^2 = 5$, so $c = \sqrt{5}$. Thus the foci are $(0, \pm\sqrt{5})$. The graph is shown in Figure 6.

Now Try Exercises 31 and 41 ■

Like parabolas and ellipses, hyperbolas have an interesting *reflection property*. Light aimed at one focus of a hyperbolic mirror is reflected toward the other focus, as shown in Figure 7. This property is used in the construction of Cassegrain-type telescopes. A hyperbolic mirror is placed in the telescope tube so that light reflected from the primary parabolic reflector is aimed at one focus of the hyperbolic mirror. The light is then refocused at a more accessible point below the primary reflector (Figure 8).

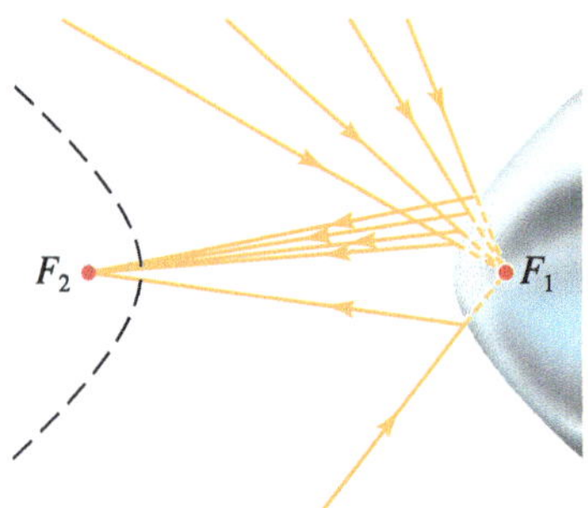

FIGURE 7 Reflection property of hyperbolas

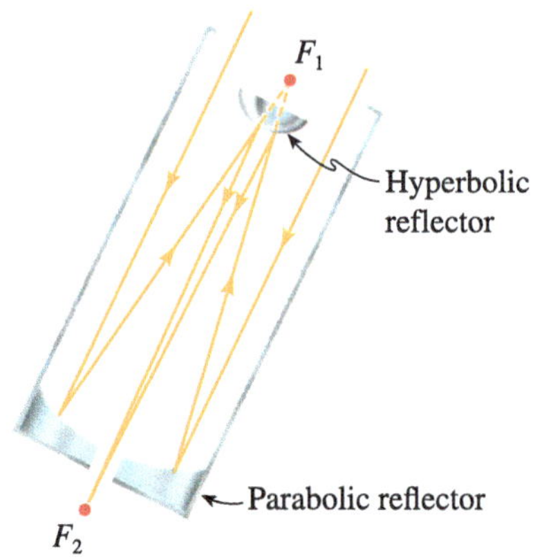

FIGURE 8 Cassegrain-type telescope

The LORAN (LOng RAnge Navigation) system was used until the early 1990s; it has now been superseded by the GPS system (see page 753). In the LORAN system, hyperbolas are used onboard a ship to determine its location. In Figure 9 radio stations at A and B transmit signals simultaneously for reception by the ship at P. The onboard computer converts the time difference in reception of these signals into a distance difference $d(P, A) - d(P, B)$. From the definition of a hyperbola this locates the ship on one branch of a hyperbola with foci at A and B (sketched in black in the figure). The same procedure is carried out with two other radio stations at C and D, and this locates the ship on a second hyperbola (shown in red in the figure). (In practice, only three stations are needed because one station can be used as a focus for both hyperbolas.) The coordinates of the intersection point of these two hyperbolas, which can be calculated precisely by the computer, give the location of P.

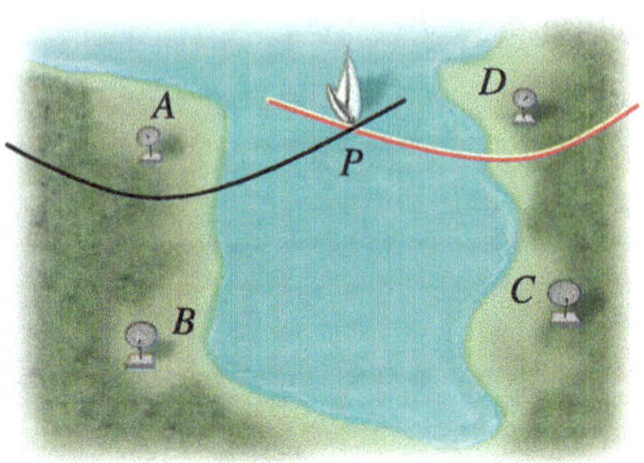

FIGURE 9 LORAN system for finding the location of a ship

11.3 EXERCISES

CONCEPTS

1. A hyperbola is the set of all points in the plane for which the ________ of the distances from two fixed points F_1 and F_2 is constant. The points F_1 and F_2 are called the ________ of the hyperbola.

2. The graph of the equation $\dfrac{x^2}{a^2} - \dfrac{y^2}{b^2} = 1$ with $a > 0, b > 0$ is a hyperbola with ________ (horizontal/vertical) transverse axis, vertices (___, ___) and (___, ___) and foci $(\pm c, 0)$, where $c =$ ________. So the graph of $\dfrac{x^2}{4^2} - \dfrac{y^2}{3^2} = 1$ is a hyperbola with vertices (___, ___) and (___, ___) and foci (___, ___) and (___, ___).

3. The graph of the equation $\dfrac{y^2}{a^2} - \dfrac{x^2}{b^2} = 1$ with $a > 0, b > 0$ is a hyperbola with ________ (horizontal/vertical) transverse axis, vertices (___, ___) and (___, ___) and foci $(0, \pm c)$, where $c =$ ________. So the graph of $\dfrac{y^2}{4^2} - \dfrac{x^2}{3^2} = 1$ is a hyperbola with vertices (___, ___) and (___, ___) and foci (___, ___) and (___, ___).

4. Label the vertices, foci, and asymptotes on the graphs given for the hyperbolas in Exercises 2 and 3.

(a) $\dfrac{x^2}{4^2} - \dfrac{y^2}{3^2} = 1$

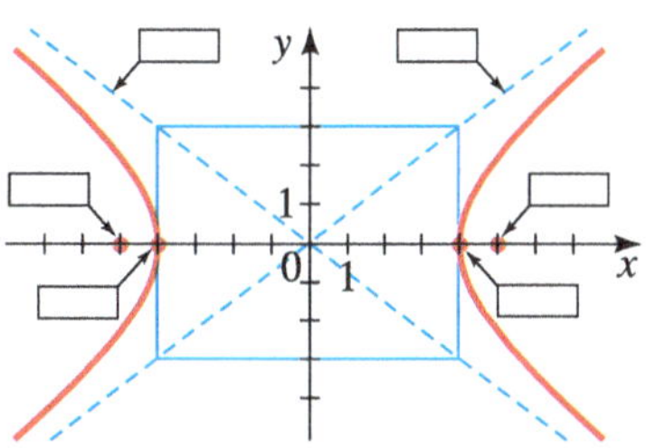

(b) $\dfrac{y^2}{4^2} - \dfrac{x^2}{3^2} = 1$

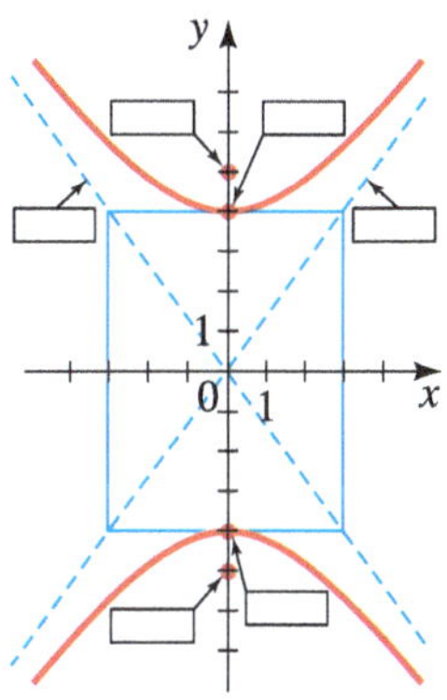

SKILLS

5–8 ■ Graphs of Hyperbolas Match the equation with the graphs labeled I–IV. Give reasons for your answers.

5. $\dfrac{x^2}{4} - y^2 = 1$

6. $y^2 - \dfrac{x^2}{9} = 1$

7. $16y^2 - x^2 = 144$

8. $9x^2 - 25y^2 = 225$

I

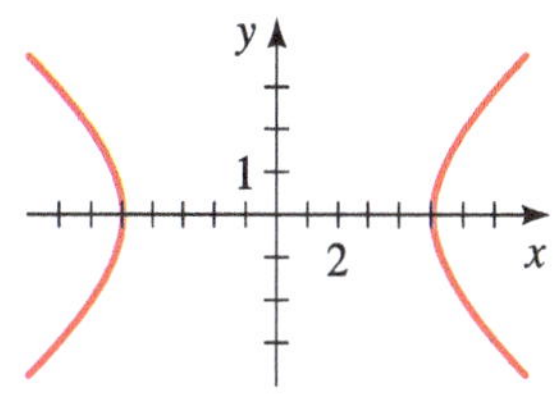

II

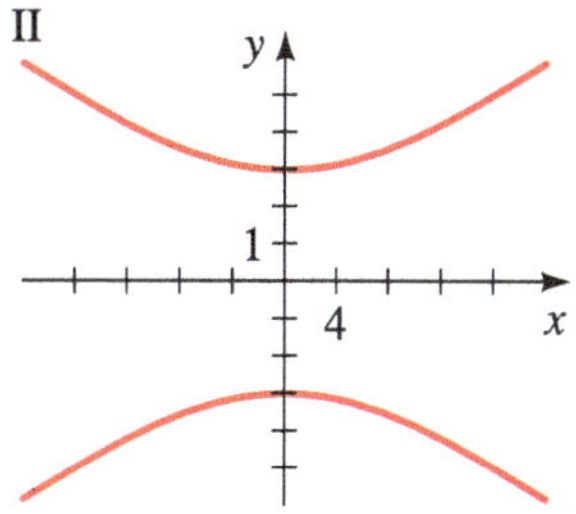

III

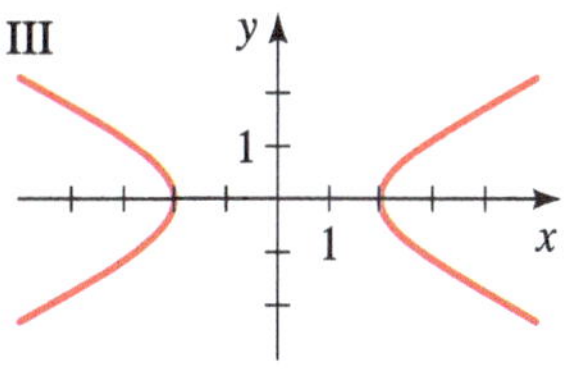

IV

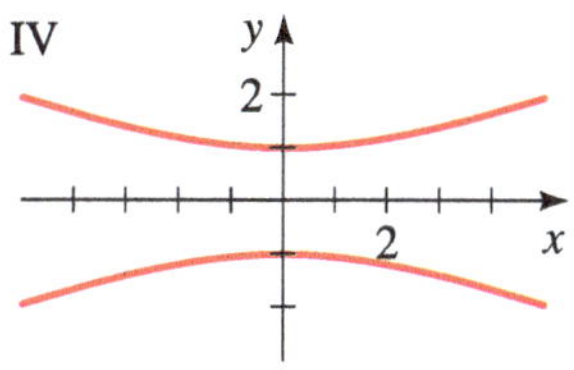

9–26 ■ Graphing Hyperbolas An equation of a hyperbola is given. **(a)** Find the vertices, foci, and asymptotes of the hyperbola. **(b)** Determine the length of the transverse axis. **(c)** Sketch a graph of the hyperbola.

9. $\dfrac{x^2}{4} - \dfrac{y^2}{16} = 1$

10. $\dfrac{y^2}{9} - \dfrac{x^2}{16} = 1$

11. $\dfrac{y^2}{36} - \dfrac{x^2}{4} = 1$

12. $\dfrac{x^2}{9} - \dfrac{y^2}{64} = 1$

13. $y^2 - \dfrac{x^2}{25} = 1$

14. $\dfrac{x^2}{2} - y^2 = 1$

15. $x^2 - y^2 = 1$

16. $\dfrac{x^2}{16} - \dfrac{y^2}{12} = 1$

17. $9x^2 - 4y^2 = 36$

18. $25y^2 - 9x^2 = 225$

19. $4y^2 - 9x^2 = 144$

20. $y^2 - 25x^2 = 100$

21. $x^2 - 4y^2 - 8 = 0$

22. $3y^2 - x^2 - 9 = 0$

23. $x^2 - y^2 + 4 = 0$

24. $x^2 - 3y^2 + 12 = 0$

25. $4y^2 - x^2 = 1$

26. $9x^2 - 16y^2 = 1$

27–32 ■ Finding an Equation of a Hyperbola Find the equation for the hyperbola whose graph is shown.

27.

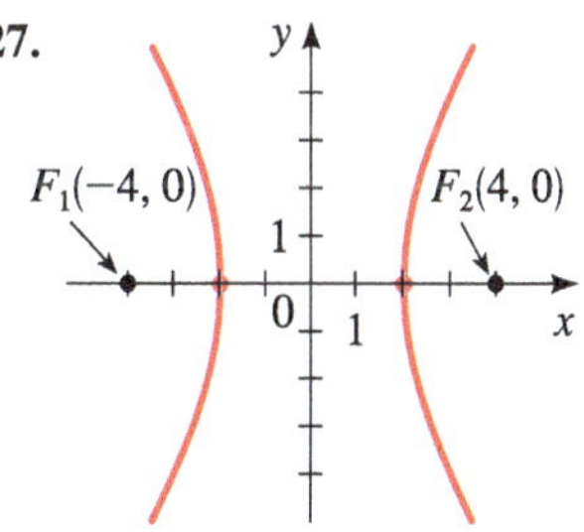

28.

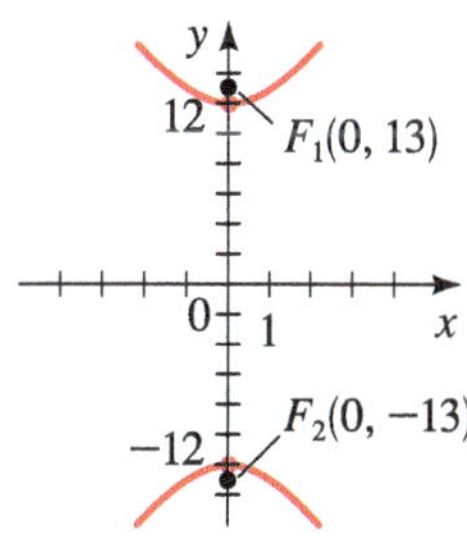

29. 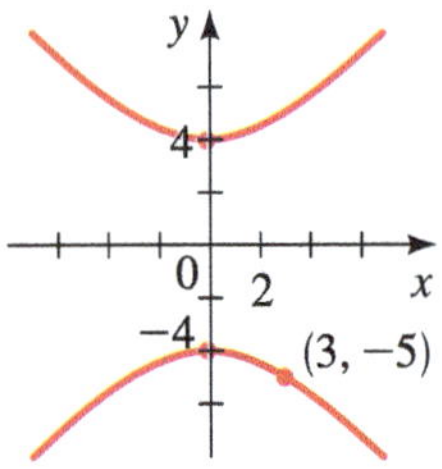

30. 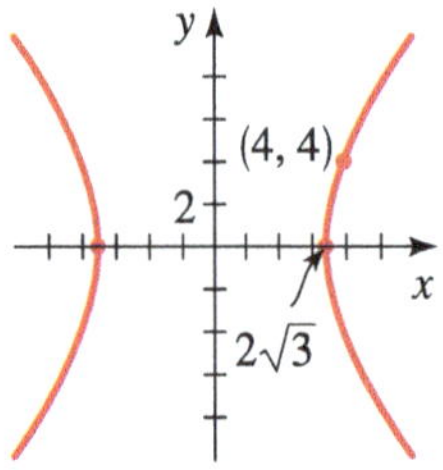

31. 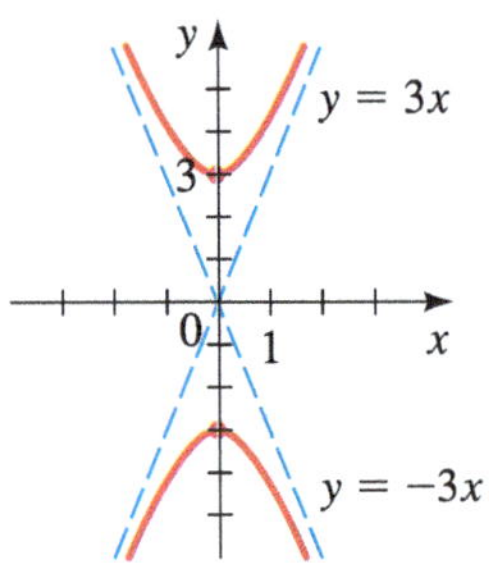

32. 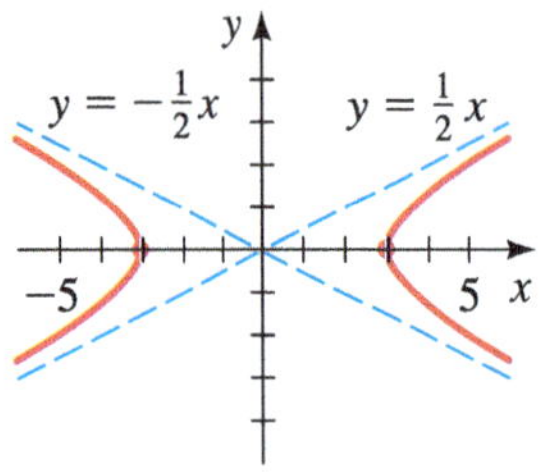

33–36 ■ Graphing Hyperbolas Use a graphing device to graph the hyperbola.

33. $x^2 - 2y^2 = 8$

34. $3y^2 - 4x^2 = 24$

35. $\dfrac{y^2}{2} - \dfrac{x^2}{6} = 1$

36. $\dfrac{x^2}{100} - \dfrac{y^2}{64} = 1$

37–50 ■ Finding the Equation of a Hyperbola Find an equation for the hyperbola that satisfies the given conditions.

37. Foci: $(\pm 5, 0)$, vertices: $(\pm 3, 0)$

38. Foci: $(0, \pm 10)$, vertices: $(0, \pm 8)$

39. Foci: $(0, \pm 2)$, vertices: $(0, \pm 1)$

40. Foci: $(\pm 6, 0)$, vertices: $(\pm 2, 0)$

41. Vertices: $(\pm 1, 0)$, asymptotes: $y = \pm 5x$

42. Vertices: $(0, \pm 6)$, asymptotes: $y = \pm \frac{1}{3}x$

43. Vertices: $(0, \pm 6)$, hyperbola passes through $(-5, 9)$

44. Vertices: $(\pm 2, 0)$, hyperbola passes through $(3, \sqrt{30})$

45. Asymptotes: $y = \pm x$, hyperbola passes through $(5, 3)$

46. Asymptotes: $y = \pm x$, hyperbola passes through $(1, 2)$

47. Foci: $(0, \pm 3)$, hyperbola passes through $(1, 4)$

48. Foci: $(\pm\sqrt{10}, 0)$, hyperbola passes through $(4, \sqrt{18})$

49. Foci: $(\pm 5, 0)$, length of transverse axis: 6

50. Foci: $(0, \pm 1)$, length of transverse axis: 1

SKILLS Plus

51. Perpendicular Asymptotes

(a) Show that the asymptotes of the hyperbola $x^2 - y^2 = 5$ are perpendicular to each other.

(b) Find an equation for the hyperbola with foci $(\pm c, 0)$ and with asymptotes perpendicular to each other.

52. Conjugate Hyperbolas The hyperbolas

$$\frac{x^2}{a^2} - \frac{y^2}{b^2} = 1 \qquad \text{and} \qquad \frac{x^2}{a^2} - \frac{y^2}{b^2} = -1$$

are said to be *conjugate* to each other.

(a) Show that the hyperbolas

$$x^2 - 4y^2 + 16 = 0 \qquad \text{and} \qquad 4y^2 - x^2 + 16 = 0$$

are conjugate to each other, and sketch their graphs on the same coordinate axes.

(b) What do the hyperbolas of part (a) have in common?

(c) Show that any pair of conjugate hyperbolas have the relationship you discovered in part (b).

53. Equation of a Hyperbola In the derivation of the equation of the hyperbola at the beginning of this section we said that the equation

$$\sqrt{(x+c)^2 + y^2} - \sqrt{(x-c)^2 + y^2} = \pm 2a$$

simplifies to

$$(c^2 - a^2)x^2 - a^2y^2 = a^2(c^2 - a^2)$$

Supply the steps needed to show this.

54. Verifying a Geometric Property of a Hyperbola

(a) For the hyperbola

$$\frac{x^2}{9} - \frac{y^2}{16} = 1$$

determine the values of a, b, and c, and find the coordinates of the foci F_1 and F_2.

(b) Show that the point $P(5, \frac{16}{3})$ lies on this hyperbola.

(c) Find $d(P, F_1)$ and $d(P, F_2)$.

(d) Verify that the difference between $d(P, F_1)$ and $d(P, F_2)$ is $2a$.

55. Confocal Hyperbolas Hyperbolas are called *confocal* if they have the same foci.

(a) Show that the hyperbolas

$$\frac{y^2}{k} - \frac{x^2}{16 - k} = 1 \qquad 0 < k < 16$$

are confocal.

(b) Use a graphing device to draw the top branches of the family of hyperbolas in part (a) for $k = 1, 4, 8$, and 12. How does the shape of the graph change as k increases?

APPLICATIONS

56. Navigation In the figure on the next page, the LORAN stations at A and B are 500 mi apart, and the ship at P receives station A's signal 2640 microseconds (μs) before it receives the signal from station B.

(a) Assuming that radio signals travel at 980 ft/μs, find $d(P, A) - d(P, B)$.

(b) Find an equation for the branch of the hyperbola indicated in red in the figure. (Use miles as the unit of distance.)

(c) If A is due north of B and if P is due east of A, how far is P from A?

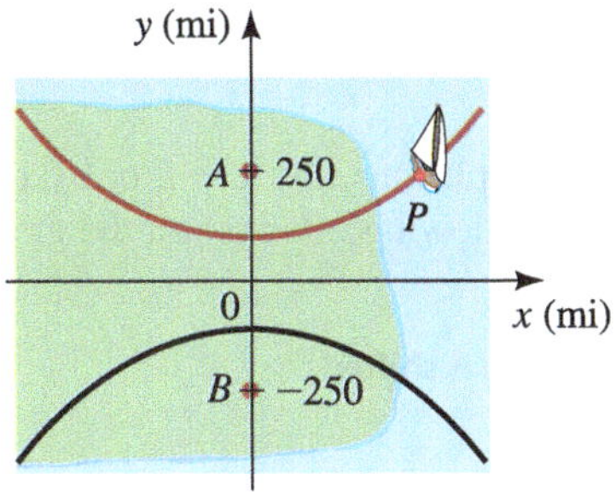

57. Comet Trajectories Some comets, such as Halley's comet, are a permanent part of the solar system, traveling in elliptical orbits around the sun. Other comets pass through the solar system only once, following a hyperbolic path with the sun at a focus. The figure below shows the path of such a comet. Find an equation for the path, assuming that the closest the comet comes to the sun is 2×10^9 mi and that the path the comet was taking before it neared the solar system is at a right angle to the path it continues on after leaving the solar system.

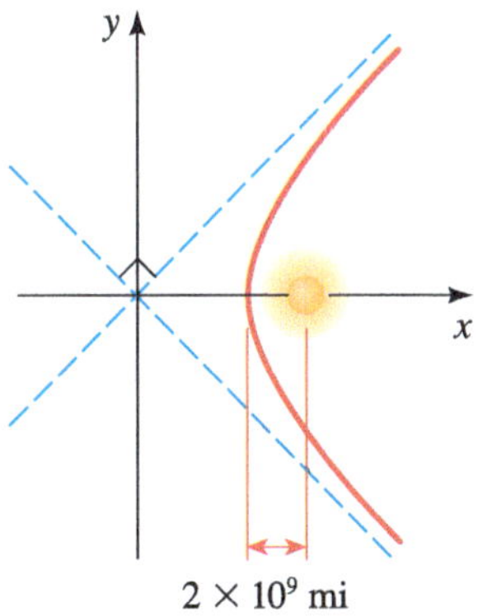

58. Ripples in Pool Two stones are dropped simultaneously into a calm pool of water. The crests of the resulting waves form equally spaced concentric circles, as shown in the figures. The waves interact with each other to create certain interference patterns.

(a) Explain why the red dots lie on an ellipse.

(b) Explain why the blue dots lie on a hyperbola.

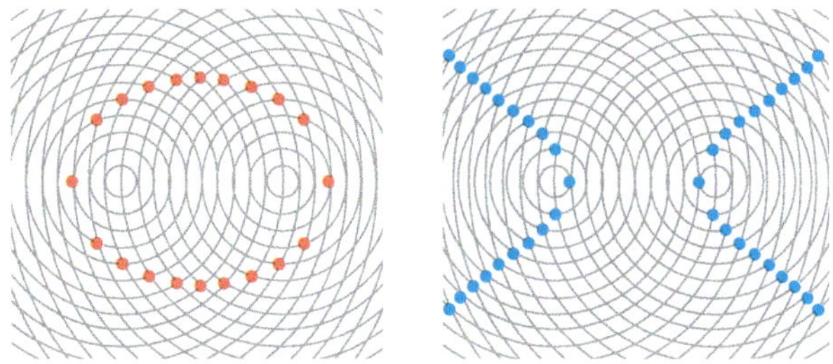

DISCUSS ■ DISCOVER ■ PROVE ■ WRITE

59. DISCUSS ■ WRITE: Hyperbolas in the Real World Several examples of the uses of hyperbolas are given in the text. Find other situations in real life in which hyperbolas occur. Consult a scientific encyclopedia in the reference section of your library, or search the Internet.

60. DISCUSS: Light from a Lamp The light from a lamp forms a lighted area on a wall, as shown in the figure. Why is the boundary of this lighted area a hyperbola? How can one hold a flashlight so that its beam forms a hyperbola on the ground?

11.4 SHIFTED CONICS

■ Shifting Graphs of Equations ■ Shifted Ellipses ■ Shifted Parabolas ■ Shifted Hyperbolas ■ The General Equation of a Shifted Conic

In the preceding sections we studied parabolas with vertices at the origin and ellipses and hyperbolas with centers at the origin. We restricted ourselves to these cases because these equations have the simplest form. In this section we consider conics whose vertices and centers are not necessarily at the origin, and we determine how this affects their equations.

■ Shifting Graphs of Equations

In Section 2.6 we studied transformations of functions that have the effect of shifting their graphs. In general, for any equation in x and y, if we replace x by $x - h$ or by $x + h$, the graph of the new equation is simply the old graph shifted horizontally; if y is

JOHANNES KEPLER (1571–1630) was the first to give a correct description of the motion of the planets. The cosmology of his time postulated complicated systems of circles moving on circles to describe these motions. Kepler sought a simpler and more harmonious description. As the official astronomer at the imperial court in Prague, he studied the astronomical observations of the Danish astronomer Tycho Brahe, whose data were the most accurate available at the time. After numerous attempts to find a theory, Kepler made the momentous discovery that the orbits of the planets are elliptical. His three great laws of planetary motion are

1. The orbit of each planet is an ellipse with the sun at one focus.
2. The line segment that joins the sun to a planet sweeps out equal areas in equal time (see the figure).
3. The square of the period of revolution of a planet is proportional to the cube of the length of the major axis of its orbit.

Kepler's formulation of these laws is perhaps the most impressive deduction from empirical data in the history of science.

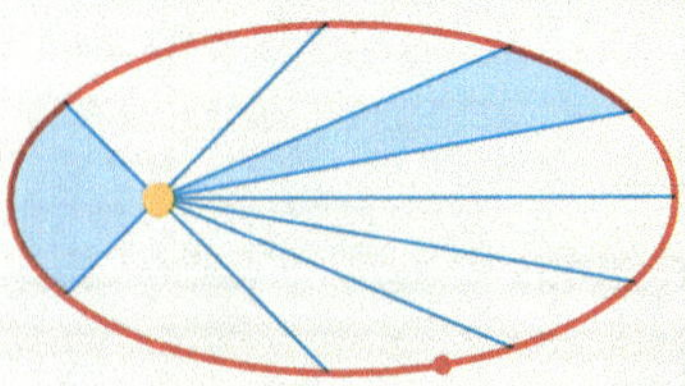

replaced by $y - k$ or by $y + k$, the graph is shifted vertically. The following box gives the details.

SHIFTING GRAPHS OF EQUATIONS

If h and k are positive real numbers, then replacing x by $x - h$ or by $x + h$ and replacing y by $y - k$ or by $y + k$ has the following effect(s) on the graph of any equation in x and y.

Replacement	How the graph is shifted
1. x replaced by $x - h$	Right h units
2. x replaced by $x + h$	Left h units
3. y replaced by $y - k$	Upward k units
4. y replaced by $y + k$	Downward k units

Shifted Ellipses

Let's apply horizontal and vertical shifting to the ellipse with equation

$$\frac{x^2}{a^2} + \frac{y^2}{b^2} = 1$$

whose graph is shown in Figure 1. If we shift it so that its center is at the point (h, k) instead of at the origin, then its equation becomes

$$\frac{(x - h)^2}{a^2} + \frac{(y - k)^2}{b^2} = 1$$

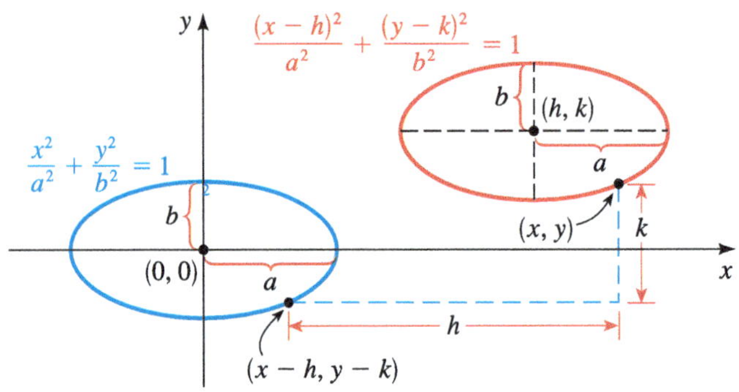

FIGURE 1 Shifted ellipse

EXAMPLE 1 ■ Sketching the Graph of a Shifted Ellipse

Sketch a graph of the ellipse

$$\frac{(x + 1)^2}{4} + \frac{(y - 2)^2}{9} = 1$$

and determine the coordinates of the foci.

SOLUTION The ellipse

$$\frac{(x + 1)^2}{4} + \frac{(y - 2)^2}{9} = 1 \qquad \text{Shifted ellipse}$$

is shifted so that its center is at $(-1, 2)$. It is obtained from the ellipse

$$\frac{x^2}{4} + \frac{y^2}{9} = 1 \qquad \text{Ellipse with center at origin}$$

by shifting it left 1 unit and upward 2 units. The endpoints of the minor and major axes of the ellipse with center at the origin are $(2, 0)$, $(-2, 0)$, $(0, 3)$, $(0, -3)$. We apply the required shifts to these points to obtain the corresponding points on the shifted ellipse.

$$(2, 0) \to (2 - 1, 0 + 2) = (1, 2)$$
$$(-2, 0) \to (-2 - 1, 0 + 2) = (-3, 2)$$
$$(0, 3) \to (0 - 1, 3 + 2) = (-1, 5)$$
$$(0, -3) \to (0 - 1, -3 + 2) = (-1, -1)$$

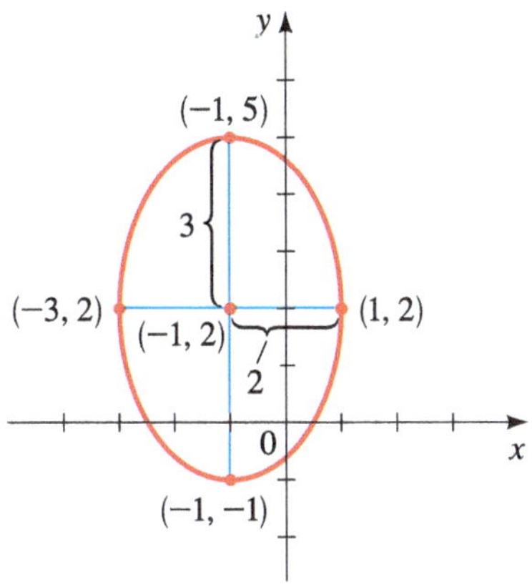

FIGURE 2
$$\frac{(x+1)^2}{4} + \frac{(y-2)^2}{9} = 1$$

This helps us sketch the graph in Figure 2.

To find the foci of the shifted ellipse, we first find the foci of the ellipse with center at the origin. Since $a^2 = 9$ and $b^2 = 4$, we have $c^2 = 9 - 4 = 5$, so $c = \sqrt{5}$. So the foci are $(0, \pm\sqrt{5})$. Shifting left 1 unit and upward 2 units, we get

$$(0, \sqrt{5}) \to (0 - 1, \sqrt{5} + 2) = (-1, 2 + \sqrt{5})$$
$$(0, -\sqrt{5}) \to (0 - 1, -\sqrt{5} + 2) = (-1, 2 - \sqrt{5})$$

Thus the foci of the shifted ellipse are

$$(-1, 2 + \sqrt{5}) \quad \text{and} \quad (-1, 2 - \sqrt{5})$$

Now Try Exercise 7 ■

EXAMPLE 2 ■ Finding the Equation of a Shifted Ellipse

The vertices of an ellipse are $(-7, 3)$ and $(3, 3)$, and the foci are $(-6, 3)$ and $(2, 3)$. Find the equation for the ellipse, and sketch its graph.

SOLUTION The center of the ellipse is the midpoint of the line segment between the vertices. By the Midpoint Formula the center is

The Midpoint Formula is given on page 94.

$$\left(\frac{-7+3}{2}, \frac{3+3}{2}\right) = (-2, 3) \qquad \text{Center}$$

Since the vertices lie on a horizontal line, the major axis is horizontal. The length of the major axis is $3 - (-7) = 10$, so $a = 5$. The distance between the foci is $2 - (-6) = 8$, so $c = 4$. Since $c^2 = a^2 - b^2$, we have

$$4^2 = 5^2 - b^2 \qquad c = 4, a = 5$$
$$b^2 = 25 - 16 = 9 \qquad \text{Solve for } b^2$$

Thus the equation of the ellipse is

$$\frac{(x+2)^2}{25} + \frac{(y-3)^2}{9} = 1 \qquad \text{Equation of shifted ellipse}$$

The graph is shown in Figure 3.

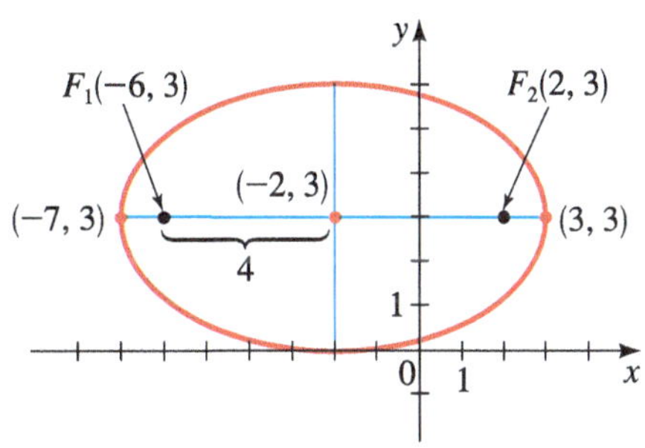

FIGURE 3 Graph of $\frac{(x+2)^2}{25} + \frac{(y-3)^2}{9} = 1$

Now Try Exercise 35 ■

Shifted Parabolas

Applying shifts to parabolas leads to the equations and graphs shown in Figure 4.

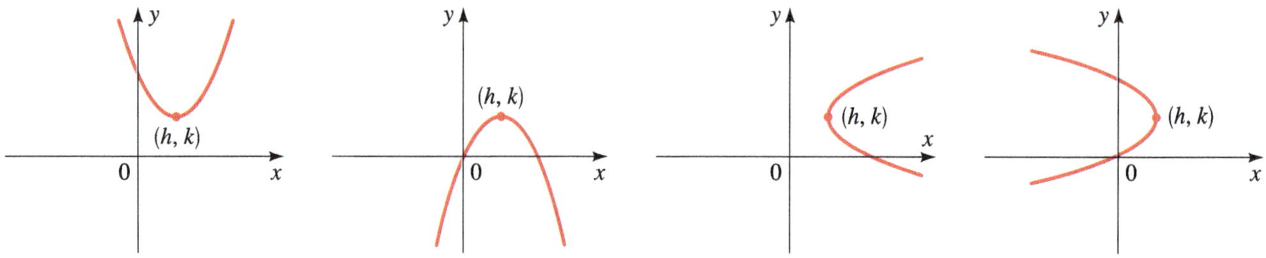

(a) $(x-h)^2 = 4p(y-k)$ $p > 0$

(b) $(x-h)^2 = 4p(y-k)$ $p < 0$

(c) $(y-k)^2 = 4p(x-h)$ $p > 0$

(d) $(y-k)^2 = 4p(x-h)$ $p < 0$

FIGURE 4 Shifted parabolas

EXAMPLE 3 ■ Graphing a Shifted Parabola

Determine the vertex, focus, and directrix, and sketch a graph of the parabola.

$$x^2 - 4x = 8y - 28$$

SOLUTION We complete the square in x to put this equation into one of the forms in Figure 4.

$$x^2 - 4x + 4 = 8y - 28 + 4 \qquad \text{Add 4 to complete the square}$$
$$(x-2)^2 = 8y - 24 \qquad \text{Perfect square}$$
$$(x-2)^2 = 8(y-3) \qquad \text{Shifted parabola}$$

This parabola opens upward with vertex at $(2, 3)$. It is obtained from the parabola

$$x^2 = 8y \qquad \text{Parabola with vertex at origin}$$

by shifting right 2 units and upward 3 units. Since $4p = 8$, we have $p = 2$, so the focus is 2 units above the vertex and the directrix is 2 units below the vertex. Thus the focus is $(2, 5)$, and the directrix is $y = 1$. The graph is shown in Figure 5.

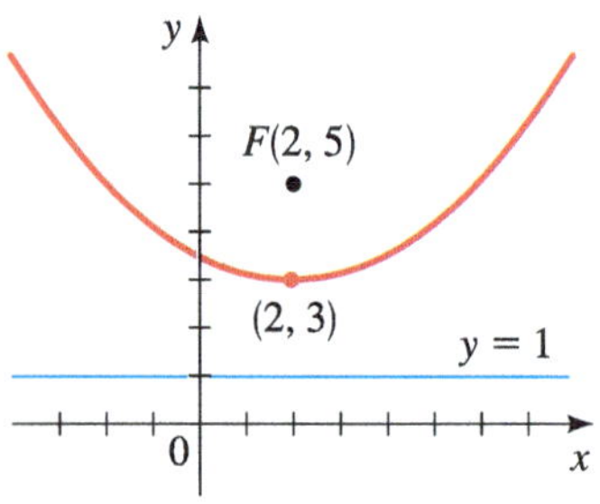

FIGURE 5
$x^2 - 4x = 8y - 28$

Now Try Exercises 13 and 19 ■

Shifted Hyperbolas

Applying shifts to hyperbolas leads to the equations and graphs shown in Figure 6.

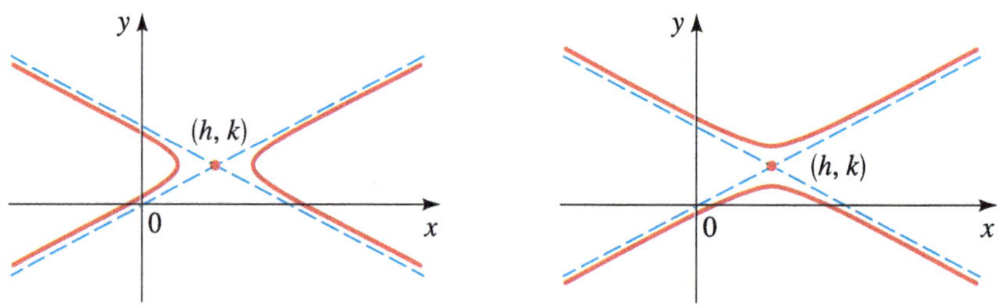

(a) $\dfrac{(x-h)^2}{a^2} - \dfrac{(y-k)^2}{b^2} = 1$

(b) $-\dfrac{(x-h)^2}{b^2} + \dfrac{(y-k)^2}{a^2} = 1$

FIGURE 6 Shifted hyperbolas

EXAMPLE 4 ■ Graphing a Shifted Hyperbola

A shifted conic has the equation

$$9x^2 - 72x - 16y^2 - 32y = 16$$

(a) Complete the square in x and y to show that the equation represents a hyperbola.

(b) Find the center, vertices, foci, and asymptotes of the hyperbola, and sketch its graph.

(c) Draw the graph using a graphing calculator.

SOLUTION

(a) We complete the squares in both x and y.

$$9(x^2 - 8x \qquad) - 16(y^2 + 2y \qquad) = 16 \qquad \text{Group terms and factor}$$

$$9(x^2 - 8x + 16) - 16(y^2 + 2y + 1) = 16 + 9 \cdot 16 - 16 \cdot 1 \qquad \text{Complete the squares}$$

$$9(x-4)^2 - 16(y+1)^2 = 144 \qquad \text{Divide this by 144}$$

$$\frac{(x-4)^2}{16} - \frac{(y+1)^2}{9} = 1 \qquad \text{Shifted hyperbola}$$

Comparing this to Figure 6(a), we see that this is the equation of a shifted hyperbola.

(b) The shifted hyperbola has center $(4, -1)$ and a horizontal transverse axis.

CENTER $(4, -1)$

Its graph will have the same shape as the unshifted hyperbola

$$\frac{x^2}{16} - \frac{y^2}{9} = 1 \qquad \text{Hyperbola with center at origin}$$

Since $a^2 = 16$ and $b^2 = 9$, we have $a = 4$, $b = 3$, and $c = \sqrt{a^2 + b^2} = \sqrt{16 + 9} = 5$. Thus the foci lie 5 units to the left and to the right of the center, and the vertices lie 4 units to either side of the center.

FOCI $(-1, -1)$ and $(9, -1)$

VERTICES $(0, -1)$ and $(8, -1)$

The asymptotes of the unshifted hyperbola are $y = \pm\frac{3}{4}x$, so the asymptotes of the shifted hyperbola are found as follows.

ASYMPTOTES

$$y + 1 = \pm\tfrac{3}{4}(x - 4)$$

$$y + 1 = \pm\tfrac{3}{4}x \mp 3$$

$$y = \tfrac{3}{4}x - 4 \qquad \text{and} \qquad y = -\tfrac{3}{4}x + 2$$

To help us sketch the hyperbola, we draw the central box; it extends 4 units left and right from the center and 3 units upward and downward from the center. We then draw the asymptotes and complete the graph of the shifted hyperbola as shown in Figure 7(a).

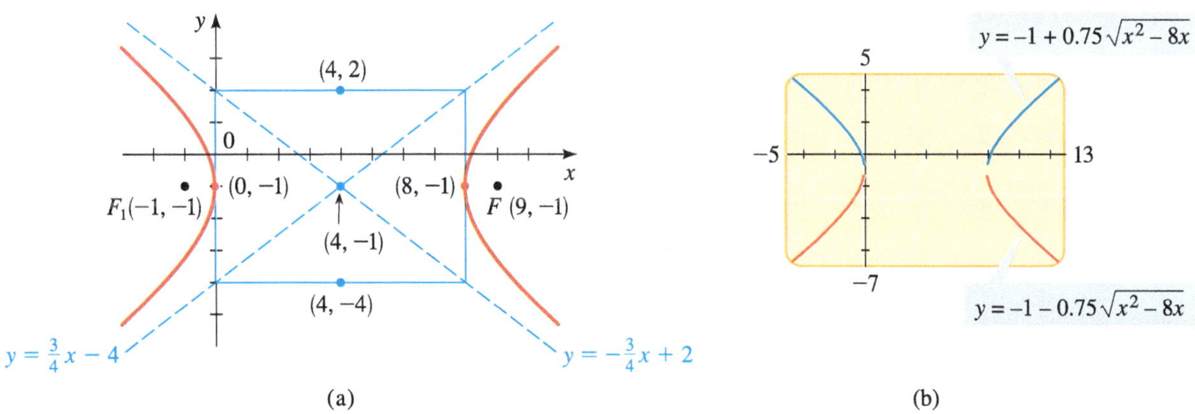

FIGURE 7 $9x^2 - 72x - 16y^2 - 32y = 16$

(c) To draw the graph using a graphing calculator, we need to solve for y. The given equation is a quadratic equation in y, so we use the Quadratic Formula to solve for y. Writing the equation in the form

$$16y^2 + 32y - 9x^2 + 72x + 16 = 0$$

Note that the equation of a hyperbola does not define y as a function of x (see page 165). That's why we need to graph two functions to graph a hyperbola.

we get

$$\begin{aligned} y &= \frac{-32 \pm \sqrt{32^2 - 4(16)(-9x^2 + 72x + 16)}}{2(16)} && \text{Quadratic Formula} \\ &= \frac{-32 \pm \sqrt{576x^2 - 4608x}}{32} && \text{Expand} \\ &= \frac{-32 \pm 24\sqrt{x^2 - 8x}}{32} && \text{Factor 576 from under the radical} \\ &= -1 \pm \tfrac{3}{4}\sqrt{x^2 - 8x} && \text{Simplify} \end{aligned}$$

To obtain the graph of the hyperbola, we graph the functions

$$y = -1 + 0.75\sqrt{x^2 - 8x}$$

and

$$y = -1 - 0.75\sqrt{x^2 - 8x}$$

as shown in Figure 7(b).

Now Try Exercises 21, 27 and 61 ■

The General Equation of a Shifted Conic

If we expand and simplify the equations of any of the shifted conics illustrated in Figures 1, 4, and 6, then we will always obtain an equation of the form

$$Ax^2 + Cy^2 + Dx + Ey + F = 0$$

where A and C are not both 0. Conversely, if we begin with an equation of this form, then we can complete the square in x and y to see which type of conic section the equation represents. In some cases the graph of the equation turns out to be just a pair of lines or a single point, or there might be no graph at all. These cases are called **degenerate conics**. If the equation is not degenerate, then we can tell whether it represents a parabola, an ellipse, or a hyperbola simply by examining the signs of A and C, as described in the following box.

GENERAL EQUATION OF A SHIFTED CONIC

The graph of the equation

$$Ax^2 + Cy^2 + Dx + Ey + F = 0$$

where A and C are not both 0, is a conic or a degenerate conic. In the nondegenerate cases the graph is

1. a parabola if A or C is 0,
2. an ellipse if A and C have the same sign (or a circle if $A = C$),
3. a hyperbola if A and C have opposite signs.

EXAMPLE 5 ■ An Equation That Leads to a Degenerate Conic

Sketch the graph of the equation

$$9x^2 - y^2 + 18x + 6y = 0$$

SOLUTION Because the coefficients of x^2 and y^2 are of opposite sign, this equation looks as if it should represent a hyperbola (like the equation of Example 4). To see whether this is in fact the case, we complete the squares.

$$9(x^2 + 2x \quad) - (y^2 - 6y \quad) = 0 \qquad \text{Group terms and factor 9}$$

$$9(x^2 + 2x + 1) - (y^2 - 6y + 9) = 0 + 9 \cdot 1 - 9 \qquad \text{Complete the squares}$$

$$9(x + 1)^2 - (y - 3)^2 = 0 \qquad \text{Factor}$$

$$(x + 1)^2 - \frac{(y - 3)^2}{9} = 0 \qquad \text{Divide by 9}$$

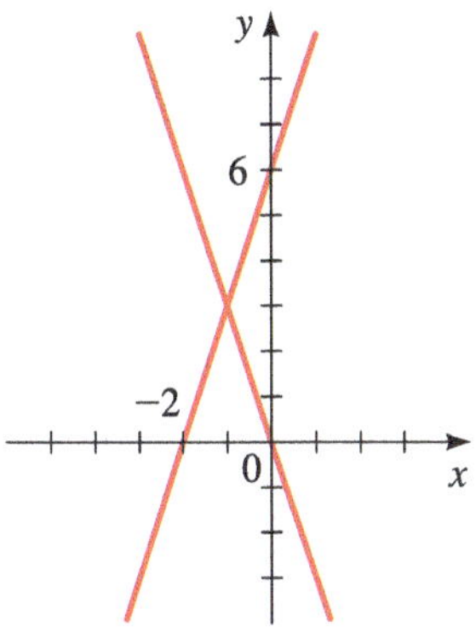

FIGURE 8
$9x^2 - y^2 + 18x + 6y = 0$

For this to fit the form of the equation of a hyperbola, we would need a nonzero constant to the right of the equal sign. In fact, further analysis shows that this is the equation of a pair of intersecting lines.

$$(y - 3)^2 = 9(x + 1)^2$$

$$y - 3 = \pm 3(x + 1) \qquad \text{Take square roots}$$

$$y = 3(x + 1) + 3 \qquad \text{or} \qquad y = -3(x + 1) + 3$$

$$y = 3x + 6 \qquad\qquad y = -3x$$

These lines are graphed in Figure 8.

Now Try Exercise 55

Because the equation in Example 5 looked at first glance like the equation of a hyperbola but, in fact, turned out to represent simply a pair of lines, we refer to its graph as a **degenerate hyperbola**. Degenerate ellipses and parabolas can also arise when we complete the square(s) in an equation that seems to represent a conic. For example, the equation

$$4x^2 + y^2 - 8x + 2y + 6 = 0$$

looks as if it should represent an ellipse, because the coefficients of x^2 and y^2 have the same sign. But completing the squares leads to

$$(x - 1)^2 + \frac{(y + 1)^2}{4} = -\frac{1}{4}$$

which has no solution at all (since the sum of two squares cannot be negative). This equation is therefore degenerate.

11.4 EXERCISES

CONCEPTS

1. Suppose we want to graph an equation in x and y.
   (a) If we replace x by $x - 3$, the graph of the equation is shifted to the ________ by 3 units. If we replace x by $x + 3$, the graph of the equation is shifted to the ________ by 3 units.
   (b) If we replace y by $y - 1$, the graph of the equation is shifted ________ by 1 unit. If we replace y by $y + 1$, the graph of the equation is shifted ________ by 1 unit.

2. The graphs of $x^2 = 12y$ and $(x - 3)^2 = 12(y - 1)$ are given. Label the focus, directrix, and vertex on each parabola.

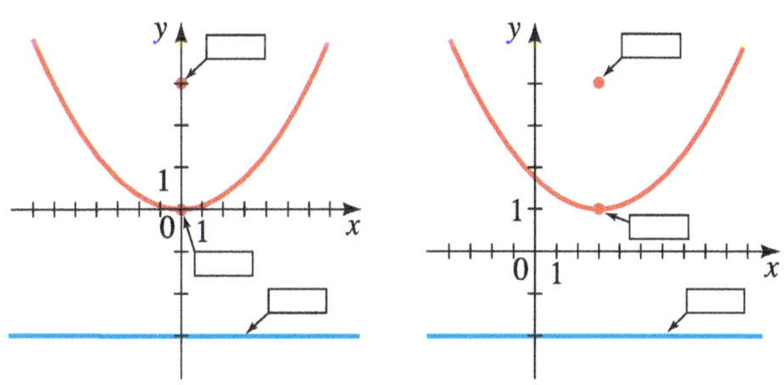

3. The graphs of $\dfrac{x^2}{5^2} + \dfrac{y^2}{4^2} = 1$ and $\dfrac{(x-3)^2}{5^2} + \dfrac{(y-1)^2}{4^2} = 1$ are given. Label the vertices and foci on each ellipse.

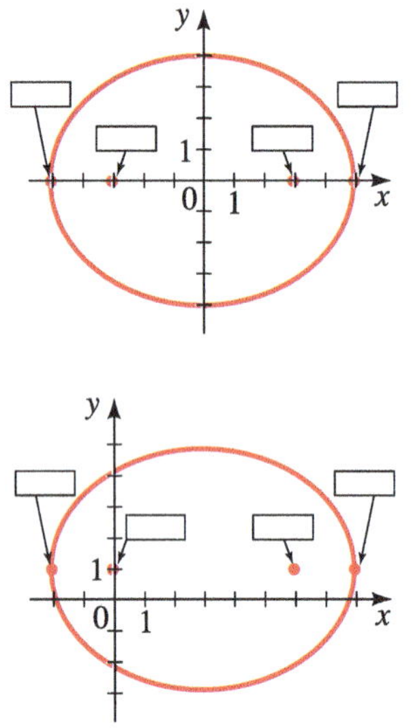

4. The graphs of $\dfrac{x^2}{4^2} - \dfrac{y^2}{3^2} = 1$ and $\dfrac{(x-3)^2}{4^2} - \dfrac{(y-1)^2}{3^2} = 1$ are given. Label the vertices, foci, and asymptotes on each hyperbola.

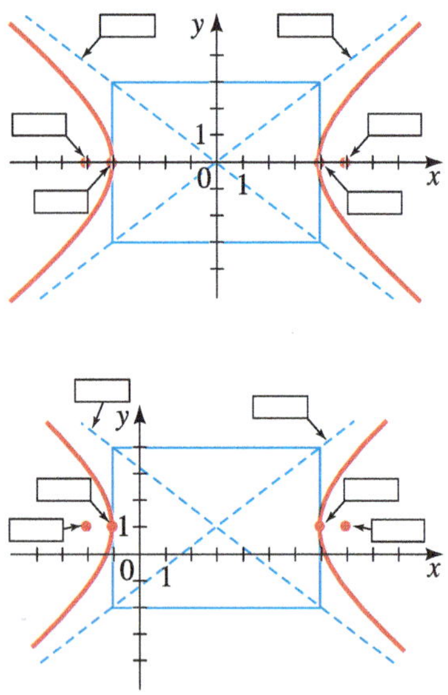

SKILLS

5–12 ■ Graphing Shifted Ellipses An equation of an ellipse is given. **(a)** Find the center, vertices, and foci of the ellipse. **(b)** Determine the lengths of the major and minor axes. **(c)** Sketch a graph of the ellipse.

5. $\dfrac{(x-2)^2}{9} + \dfrac{(y-1)^2}{4} = 1$ **6.** $\dfrac{(x-3)^2}{16} + (y+3)^2 = 1$

7. $\dfrac{x^2}{9} + \dfrac{(y+5)^2}{25} = 1$ **8.** $x^2 + \dfrac{(y+2)^2}{4} = 1$

9. $\dfrac{(x+5)^2}{16} + \dfrac{(y-1)^2}{4} = 1$

10. $\dfrac{(x+1)^2}{36} + \dfrac{(y+1)^2}{64} = 1$

11. $4x^2 + 25y^2 - 50y = 75$

12. $9x^2 - 54x + y^2 + 2y + 46 = 0$

13–20 ■ Graphing Shifted Parabolas An equation of a parabola is given. **(a)** Find the vertex, focus, and directrix of the parabola. **(b)** Sketch a graph showing the parabola and its directrix.

13. $(x-3)^2 = 8(y+1)$ **14.** $(y+1)^2 = 16(x-3)$

15. $(y+5)^2 = -6x + 12$ **16.** $y^2 = 16x - 8$

17. $2(x-1)^2 = y$ **18.** $-4\left(x + \frac{1}{2}\right)^2 = y$

19. $y^2 - 6y - 12x + 33 = 0$

20. $x^2 + 2x - 20y + 41 = 0$

21–28 ■ Graphing Shifted Hyperbolas An equation of a hyperbola is given. **(a)** Find the center, vertices, foci, and asymptotes of the hyperbola. **(b)** Sketch a graph showing the hyperbola and its asymptotes.

21. $\dfrac{(x+1)^2}{9} - \dfrac{(y-3)^2}{16} = 1$ **22.** $(x-8)^2 - (y+6)^2 = 1$

23. $y^2 - \dfrac{(x+1)^2}{4} = 1$

24. $\dfrac{(y-1)^2}{25} - (x+3)^2 = 1$

25. $\dfrac{(x+1)^2}{9} - \dfrac{(y+1)^2}{4} = 1$ **26.** $\dfrac{(y+2)^2}{36} - \dfrac{x^2}{64} = 1$

27. $36x^2 + 72x - 4y^2 + 32y + 116 = 0$

28. $25x^2 - 9y^2 - 54y = 306$

29–34 ■ Finding the Equation of a Shifted Conic Find an equation for the conic whose graph is shown.

29.

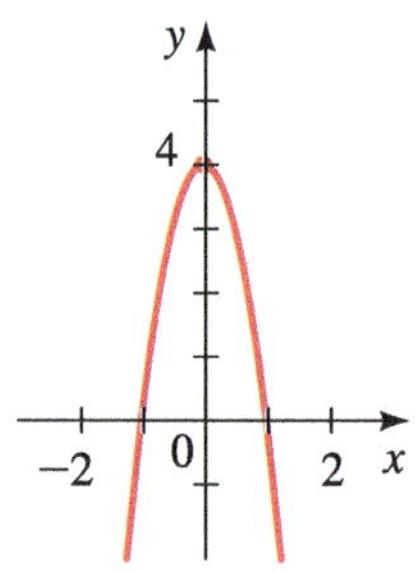

30.

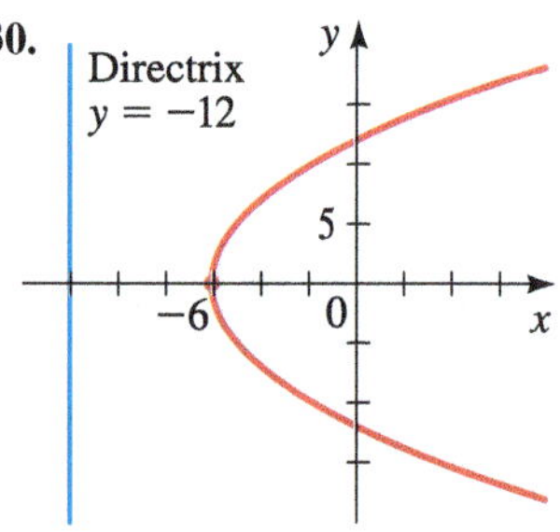

31.

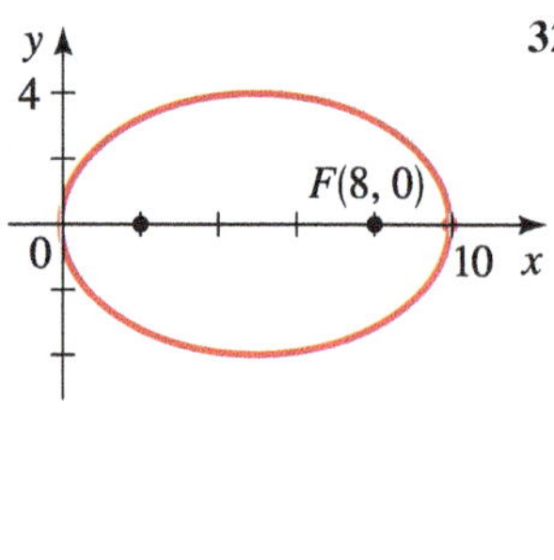

32.

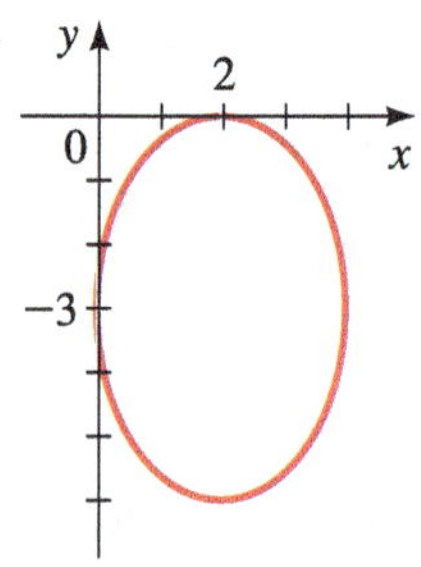

33.

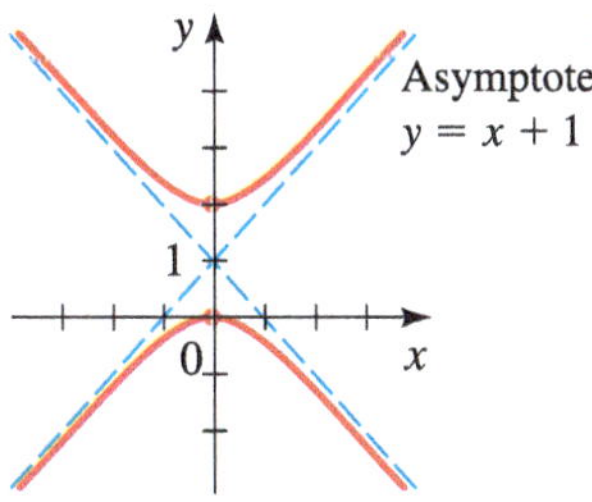

34.

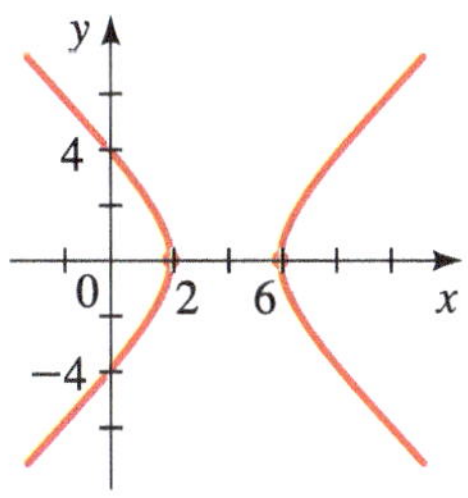

35–46 ■ Finding the Equation of a Shifted Conic Find an equation for the conic section with the given properties.

35. The ellipse with center $C(2, -3)$, vertices $V_1(-8, -3)$ and $V_2(12, -3)$, and foci $F_1(-4, -3)$ and $F_2(8, -3)$

36. The ellipse with vertices $V_1(-1, -4)$ and $V_2(-1, 6)$ and foci $F_1(-1, -3)$ and $F_2(-1, 5)$

37. The hyperbola with center $C(-1, 4)$, vertices $V_1(-1, -3)$ and $V_2(-1, 11)$, and foci $F_1(-1, -5)$ and $F_2(-1, 13)$

38. The hyperbola with vertices $V_1(-1, -1)$ and $V_2(5, -1)$ and foci $F_1(-4, -1)$ and $F_2(8, -1)$

39. The parabola with vertex $V(-3, 5)$ and directrix $y = 2$

40. The parabola with focus $F(1, 3)$ and directrix $x = 3$

41. The hyperbola with foci $F_1(1, -5)$ and $F_2(1, 5)$ that passes through the point $(1, 4)$

42. The hyperbola with foci $F_1(-2, 2)$ and $F_2(4, 2)$ that passes through the point $(3, 2)$.

43. The ellipse with foci $F_1(1, -4)$ and $F_2(5, -4)$ that passes through the point $(3, 1)$

44. The ellipse with foci $F_1(3, -4)$ and $F_2(3, 4)$, and x-intercepts 0 and 6

45. The parabola that passes through the point $(6, 1)$, with vertex $V(-1, 2)$ and horizontal axis of symmetry

46. The parabola that passes through the point $(6, -2)$, with vertex $V(4, -1)$ and vertical axis of symmetry

47–58 ■ Graphing Shifted Conics Complete the square to determine whether the graph of the equation is an ellipse, a parabola, a hyperbola, or a degenerate conic. If the graph is an ellipse, find the center, foci, vertices, and lengths of the major and minor axes. If it is a parabola, find the vertex, focus, and directrix. If it is a hyperbola, find the center, foci, vertices, and asymptotes. Then sketch the graph of the equation. If the equation has no graph, explain why.

47. $y^2 = 4(x + 2y)$

48. $9x^2 - 36x + 4y^2 = 0$

49. $x^2 - 5y^2 - 2x + 20y = 44$

50. $x^2 + 6x + 12y + 9 = 0$

51. $4x^2 + 25y^2 - 24x + 250y + 561 = 0$

52. $2x^2 + y^2 = 2y + 1$

53. $16x^2 - 9y^2 - 96x + 288 = 0$

54. $4x^2 - 4x - 8y + 9 = 0$

55. $x^2 + 16 = 4(y^2 + 2x)$

56. $x^2 - y^2 = 10(x - y) + 1$

57. $3x^2 + 4y^2 - 6x - 24y + 39 = 0$

58. $x^2 + 4y^2 + 20x - 40y + 300 = 0$

59–62 ■ Graphing Shifted Conics Use a graphing device to graph the conic.

59. $2x^2 - 4x + y + 5 = 0$

60. $4x^2 + 9y^2 - 36y = 0$

61. $9x^2 + 36 = y^2 + 36x + 6y$

62. $x^2 - 4y^2 + 4x + 8y = 0$

SKILLS Plus

63. Degenerate Conic Determine what the value of F must be if the graph of the equation

$$4x^2 + y^2 + 4(x - 2y) + F = 0$$

is **(a)** an ellipse, **(b)** a single point, or **(c)** the empty set.

64. Common Focus and Vertex Find an equation for the ellipse that shares a vertex and a focus with the parabola $x^2 + y = 100$ and has its other focus at the origin.

65. Confocal Parabolas This exercise deals with *confocal* parabolas, that is, families of parabolas that have the same focus.

(a) Draw graphs of the family of parabolas

$$x^2 = 4p(y + p)$$

for $p = -2, -\frac{3}{2}, -1, -\frac{1}{2}, \frac{1}{2}, 1, \frac{3}{2}, 2$.

(b) Show that each parabola in this family has its focus at the origin.

(c) Describe the effect on the graph of moving the vertex closer to the origin.

APPLICATIONS

66. Path of a Cannonball A cannon fires a cannonball as shown in the figure. The path of the cannonball is a parabola with vertex at the highest point of the path. If the cannonball lands 1600 ft from the cannon and the highest point it reaches is 3200 ft above the ground, find an equation for the path of the cannonball. Place the origin at the location of the cannon.

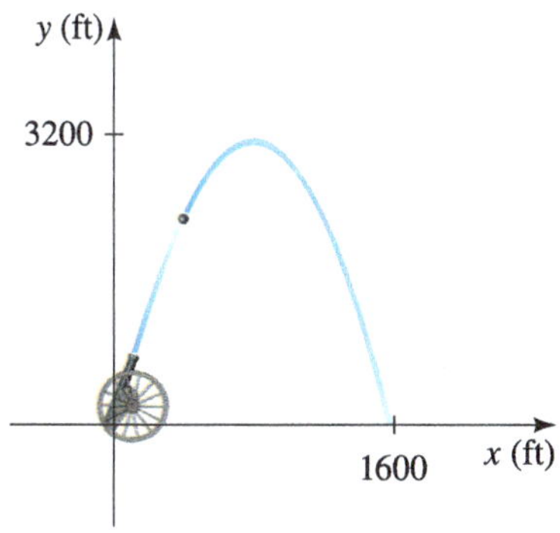

67. Orbit of a Satellite A satellite is in an elliptical orbit around the earth with the center of the earth at one focus, as shown in the figure. The height of the satellite above the earth varies between 140 mi and 440 mi. Assume that the earth is a sphere with radius 3960 mi. Find an equation for the path of the satellite with the origin at the center of the earth.

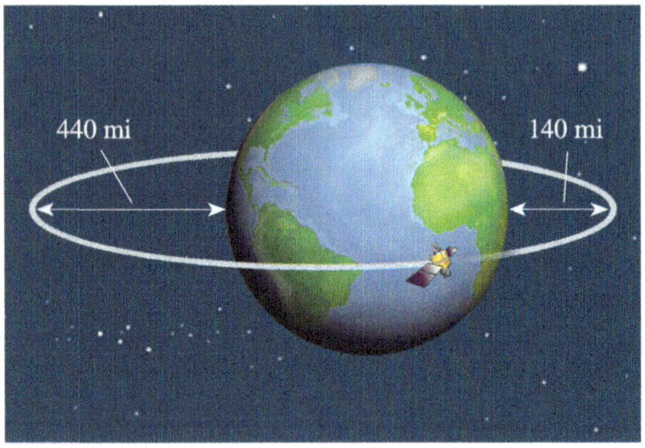

DISCUSS ■ DISCOVER ■ PROVE ■ WRITE

68. DISCUSS: A Family of Confocal Conics Conics that share a focus are called **confocal**. Consider the family of conics that have a focus at $(0, 1)$ and a vertex at the origin, as shown in the figure.

(a) Find equations of two different ellipses that have these properties.

(b) Find equations of two different hyperbolas that have these properties.

(c) Explain why only one parabola satisfies these properties. Find its equation.

(d) Sketch the conics you found in parts (a), (b), and (c) on the same coordinate axes (for the hyperbolas, sketch the top branches only).

(e) How are the ellipses and hyperbolas related to the parabola?

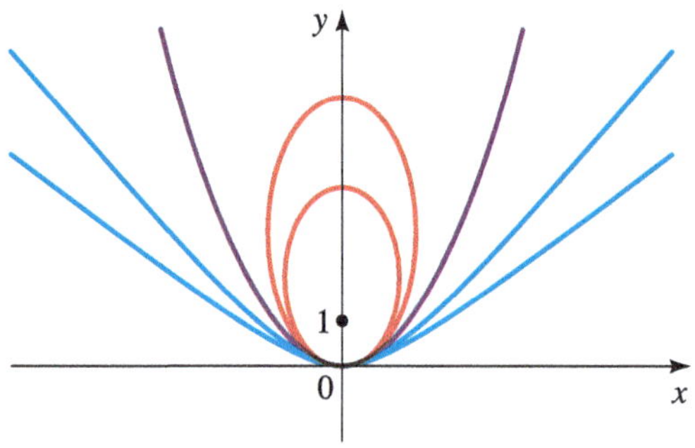

11.5 ROTATION OF AXES

■ Rotation of Axes ■ General Equation of a Conic ■ The Discriminant

In Section 11.4 we studied conics with equations of the form

$$Ax^2 + Cy^2 + Dx + Ey + F = 0$$

We saw that the graph is always an ellipse, parabola, or hyperbola with horizontal or vertical axes (except in the degenerate cases). In this section we study the most general second-degree equation

$$Ax^2 + Bxy + Cy^2 + Dx + Ey + F = 0$$

We will see that the graph of an equation of this form is also a conic. In fact, by rotating the coordinate axes through an appropriate angle, we can eliminate the term Bxy and then use our knowledge of conic sections to analyze the graph.

■ Rotation of Axes

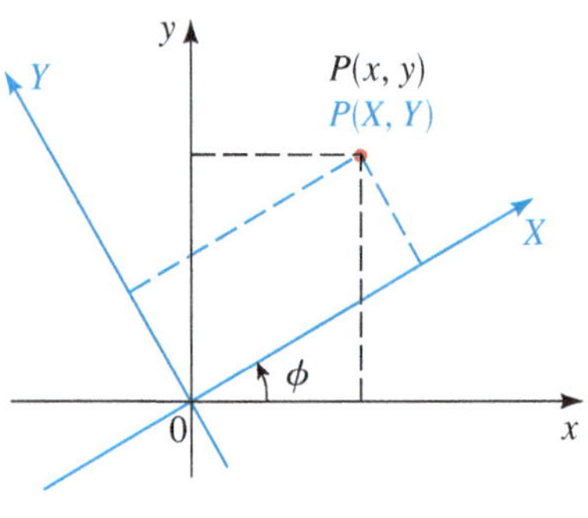

FIGURE 1

In Figure 1 the x- and y-axes have been rotated through an acute angle ϕ about the origin to produce a new pair of axes, which we call the X- and Y-axes. A point P that has coordinates (x, y) in the old system has coordinates (X, Y) in the new system. If we let r denote the distance of P from the origin and let θ be the angle that the segment OP

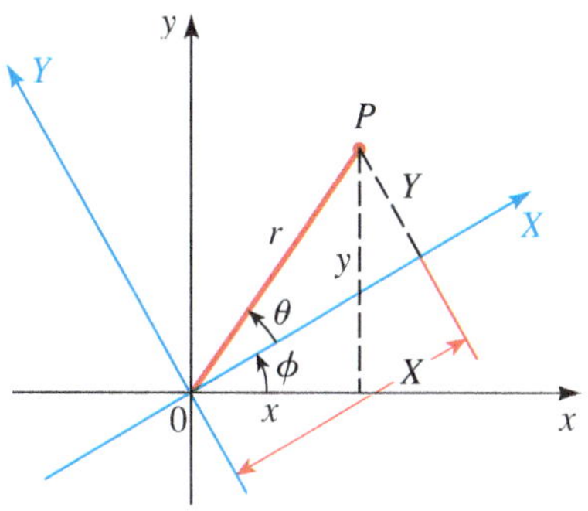

FIGURE 2

makes with the new X-axis, then we can see from Figure 2 (by considering the two right triangles in the figure) that

$$X = r\cos\theta \qquad Y = r\sin\theta$$

$$x = r\cos(\theta + \phi) \qquad y = r\sin(\theta + \phi)$$

Using the Addition Formula for Cosine, we see that

$$\begin{aligned} x &= r\cos(\theta + \phi) \\ &= r(\cos\theta\cos\phi - \sin\theta\sin\phi) \\ &= (r\cos\theta)\cos\phi - (r\sin\theta)\sin\phi \\ &= X\cos\phi - Y\sin\phi \end{aligned}$$

Similarly, we can apply the Addition Formula for Sine to the expression for y to obtain $y = X\sin\phi + Y\cos\phi$. By treating these equations for x and y as a system of linear equations in the variables X and Y (see Exercise 35), we obtain expressions for X and Y in terms of x and y, as detailed in the following box.

ROTATION OF AXES FORMULAS

Suppose the x- and y-axes in a coordinate plane are rotated through the acute angle ϕ to produce the X- and Y-axes, as shown in Figure 1. Then the coordinates (x, y) and (X, Y) of a point in the xy- and the XY-planes are related as follows.

$$x = X\cos\phi - Y\sin\phi \qquad X = x\cos\phi + y\sin\phi$$

$$y = X\sin\phi + Y\cos\phi \qquad Y = -x\sin\phi + y\cos\phi$$

EXAMPLE 1 ■ Rotation of Axes

If the coordinate axes are rotated through 30°, find the XY-coordinates of the point with xy-coordinates $(2, -4)$.

SOLUTION Using the Rotation of Axes Formulas with $x = 2$, $y = -4$, and $\phi = 30°$, we get

$$X = 2\cos 30° + (-4)\sin 30° = 2\left(\frac{\sqrt{3}}{2}\right) - 4\left(\frac{1}{2}\right) = \sqrt{3} - 2$$

$$Y = -2\sin 30° + (-4)\cos 30° = -2\left(\frac{1}{2}\right) - 4\left(\frac{\sqrt{3}}{2}\right) = -1 - 2\sqrt{3}$$

The XY-coordinates are $(-2 + \sqrt{3}, -1 - 2\sqrt{3})$.

Now Try Exercise 3 ■

EXAMPLE 2 ■ Rotating a Hyperbola

Rotate the coordinate axes through 45° to show that the graph of the equation $xy = 2$ is a hyperbola.

SOLUTION We use the Rotation of Axes Formulas with $\phi = 45°$ to obtain

$$x = X\cos 45° - Y\sin 45° = \frac{X}{\sqrt{2}} - \frac{Y}{\sqrt{2}}$$

$$y = X\sin 45° + Y\cos 45° = \frac{X}{\sqrt{2}} + \frac{Y}{\sqrt{2}}$$

Substituting these expressions into the original equation gives

$$\left(\frac{X}{\sqrt{2}} - \frac{Y}{\sqrt{2}}\right)\left(\frac{X}{\sqrt{2}} + \frac{Y}{\sqrt{2}}\right) = 2$$

$$\frac{X^2}{2} - \frac{Y^2}{2} = 2$$

$$\frac{X^2}{4} - \frac{Y^2}{4} = 1$$

We recognize this as a hyperbola with vertices $(\pm 2, 0)$ in the XY-coordinate system. Its asymptotes are $Y = \pm X$, which correspond to the coordinate axes in the xy-system (see Figure 3).

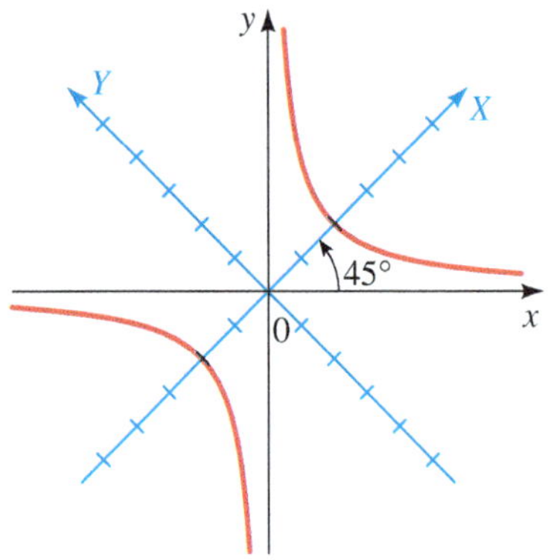

FIGURE 3
$xy = 2$

Now Try Exercise 11

General Equation of a Conic

The method of Example 2 can be used to transform any equation of the form

$$Ax^2 + Bxy + Cy^2 + Dx + Ey + F = 0$$

into an equation in X and Y that doesn't contain an XY-term by choosing an appropriate angle of rotation. To find the angle that works, we rotate the axes through an angle ϕ and substitute for x and y using the Rotation of Axes Formulas.

$$\begin{aligned} A(X\cos\phi - Y\sin\phi)^2 &+ B(X\cos\phi - Y\sin\phi)(X\sin\phi + Y\cos\phi) \\ &+ C(X\sin\phi + Y\cos\phi)^2 + D(X\cos\phi - Y\sin\phi) \\ &+ E(X\sin\phi + Y\cos\phi) + F = 0 \end{aligned}$$

If we expand this and collect like terms, we obtain an equation of the form

$$A'X^2 + B'XY + C'Y^2 + D'X + E'Y + F' = 0$$

where

$$\begin{aligned} A' &= A\cos^2\phi + B\sin\phi\cos\phi + C\sin^2\phi \\ B' &= 2(C - A)\sin\phi\cos\phi + B(\cos^2\phi - \sin^2\phi) \\ C' &= A\sin^2\phi - B\sin\phi\cos\phi + C\cos^2\phi \\ D' &= D\cos\phi + E\sin\phi \\ E' &= -D\sin\phi + E\cos\phi \\ F' &= F \end{aligned}$$

To eliminate the XY-term, we would like to choose ϕ so that $B' = 0$, that is,

Double-Angle Formulas

$\sin 2\phi = 2 \sin\phi \cos\phi$

$\cos 2\phi = \cos^2\phi - \sin^2\phi$

$$2(C - A)\sin\phi\cos\phi + B(\cos^2\phi - \sin^2\phi) = 0$$

$$(C - A)\sin 2\phi + B\cos 2\phi = 0 \qquad \text{Double-Angle Formulas for Sine and Cosine}$$

$$B\cos 2\phi = (A - C)\sin 2\phi$$

$$\cot 2\phi = \frac{A - C}{B} \qquad \text{Divide by } B \sin 2\phi$$

The preceding calculation proves the following theorem.

SIMPLIFYING THE GENERAL CONIC EQUATION

To eliminate the xy-term in the general conic equation

$$Ax^2 + Bxy + Cy^2 + Dx + Ey + F = 0$$

rotate the axes through the acute angle ϕ that satisfies

$$\cot 2\phi = \frac{A - C}{B}$$

EXAMPLE 3 ■ Eliminating the *xy*-Term

Use a rotation of axes to eliminate the xy-term in the equation

$$6\sqrt{3}x^2 + 6xy + 4\sqrt{3}y^2 = 21\sqrt{3}$$

Identify and sketch the curve.

SOLUTION To eliminate the xy-term, we rotate the axes through an angle ϕ that satisfies

$$\cot 2\phi = \frac{A - C}{B} = \frac{6\sqrt{3} - 4\sqrt{3}}{6} = \frac{\sqrt{3}}{3}$$

Thus $2\phi = 60°$, and hence $\phi = 30°$. With this value of ϕ we get

$$x = X\left(\frac{\sqrt{3}}{2}\right) - Y\left(\frac{1}{2}\right) \qquad \text{Rotation of Axes Formulas}$$

$$y = X\left(\frac{1}{2}\right) + Y\left(\frac{\sqrt{3}}{2}\right) \qquad \cos\phi = \frac{\sqrt{3}}{2},\ \sin\phi = \frac{1}{2}$$

Substituting these values for x and y into the given equation leads to

$$6\sqrt{3}\left(\frac{X\sqrt{3}}{2} - \frac{Y}{2}\right)^2 + 6\left(\frac{X\sqrt{3}}{2} - \frac{Y}{2}\right)\left(\frac{X}{2} + \frac{Y\sqrt{3}}{2}\right) + 4\sqrt{3}\left(\frac{X}{2} + \frac{Y\sqrt{3}}{2}\right)^2 = 21\sqrt{3}$$

Expanding and collecting like terms, we get

$$7\sqrt{3}X^2 + 3\sqrt{3}Y^2 = 21\sqrt{3}$$

$$\frac{X^2}{3} + \frac{Y^2}{7} = 1 \qquad \text{Divide by } 21\sqrt{3}$$

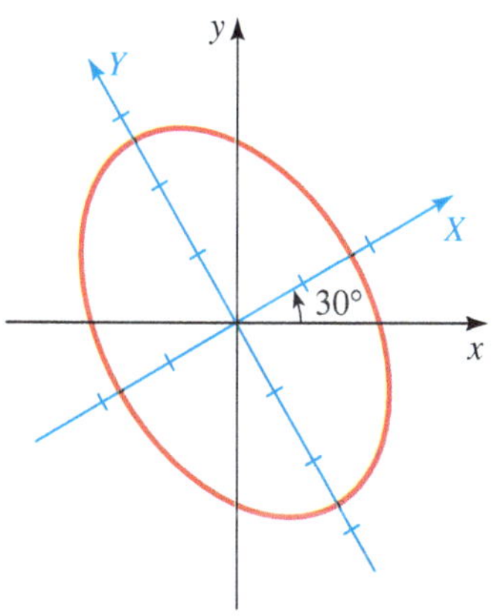

FIGURE 4
$6\sqrt{3}x^2 + 6xy + 4\sqrt{3}y^2 = 21\sqrt{3}$

This is the equation of an ellipse in the XY-coordinate system. The foci lie on the Y-axis. Because $a^2 = 7$ and $b^2 = 3$, the length of the major axis is $2\sqrt{7}$, and the length of the minor axis is $2\sqrt{3}$. The ellipse is sketched in Figure 4.

Now Try Exercise 17 ■

In the preceding example we were able to determine ϕ without difficulty, since we remembered that $\cot 60° = \sqrt{3}/3$. In general, finding ϕ is not quite so easy. The next example illustrates how the following Half-Angle Formulas, which are valid for $0 < \phi < \pi/2$, are useful in determining ϕ (see Section 7.3).

$$\cos \phi = \sqrt{\frac{1 + \cos 2\phi}{2}} \qquad \sin \phi = \sqrt{\frac{1 - \cos 2\phi}{2}}$$

EXAMPLE 4 ■ Graphing a Rotated Conic

A conic has the equation

$$64x^2 + 96xy + 36y^2 - 15x + 20y - 25 = 0$$

(a) Use a rotation of axes to eliminate the xy-term.

(b) Identify and sketch the graph.

(c) Draw the graph using a graphing calculator.

SOLUTION

(a) To eliminate the xy-term, we rotate the axes through an angle ϕ that satisfies

$$\cot 2\phi = \frac{A - C}{B} = \frac{64 - 36}{96} = \frac{7}{24}$$

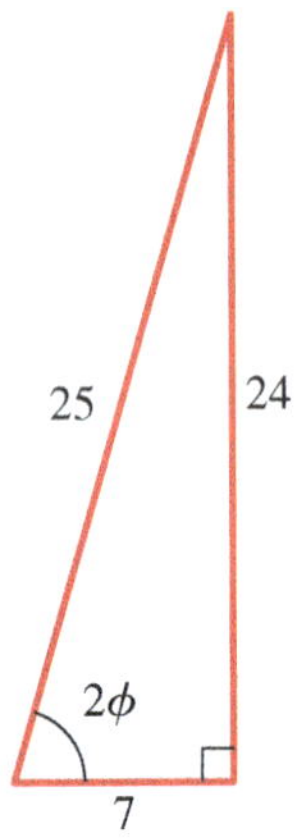

FIGURE 5

In Figure 5 we sketch a triangle with $\cot 2\phi = \frac{7}{24}$. We see that

$$\cos 2\phi = \tfrac{7}{25}$$

so, using the Half-Angle Formulas, we get

$$\cos \phi = \sqrt{\frac{1 + \frac{7}{25}}{2}} = \sqrt{\frac{16}{25}} = \frac{4}{5}$$

$$\sin \phi = \sqrt{\frac{1 - \frac{7}{25}}{2}} = \sqrt{\frac{9}{25}} = \frac{3}{5}$$

The Rotation of Axes Formulas then give

$$x = \tfrac{4}{5}X - \tfrac{3}{5}Y \qquad \text{and} \qquad y = \tfrac{3}{5}X + \tfrac{4}{5}Y$$

Substituting into the given equation, we have

$$64\left(\tfrac{4}{5}X - \tfrac{3}{5}Y\right)^2 + 96\left(\tfrac{4}{5}X - \tfrac{3}{5}Y\right)\left(\tfrac{3}{5}X + \tfrac{4}{5}Y\right)$$
$$+ 36\left(\tfrac{3}{5}X + \tfrac{4}{5}Y\right)^2 - 15\left(\tfrac{4}{5}X - \tfrac{3}{5}Y\right) + 20\left(\tfrac{3}{5}X + \tfrac{4}{5}Y\right) - 25 = 0$$

DISCOVERY PROJECT

Computer Graphics II

An image on a computer screen is stored in the computer memory as a large matrix. Each matrix entry contain information about one pixel in the image. In *Discovery Project: Computer Graphics I* we experimented with using matrix operations to transform an image—stretch, shrink, reflect, or shear. But rotating an image requires knowledge of the rotation formulas we study in this section. In this project we experiment with using rotation matrices to rotate an image. You can find the project at **www.stewartmath.com**.

Expanding and collecting like terms, we get

$$100X^2 + 25Y - 25 = 0$$

$$4X^2 = -Y + 1 \qquad \text{Simplify}$$

$$X^2 = -\tfrac{1}{4}(Y - 1) \qquad \text{Divide by 4}$$

(b) We recognize this as the equation of a parabola that opens along the negative Y-axis and has vertex $(0, 1)$ in XY-coordinates. Since $4p = -\frac{1}{4}$, we have $p = -\frac{1}{16}$, so the focus is $\left(0, \frac{15}{16}\right)$ and the directrix is $Y = \frac{17}{16}$. Using

$$\phi = \cos^{-1}\tfrac{4}{5} \approx 37°$$

we sketch the graph in Figure 6(a).

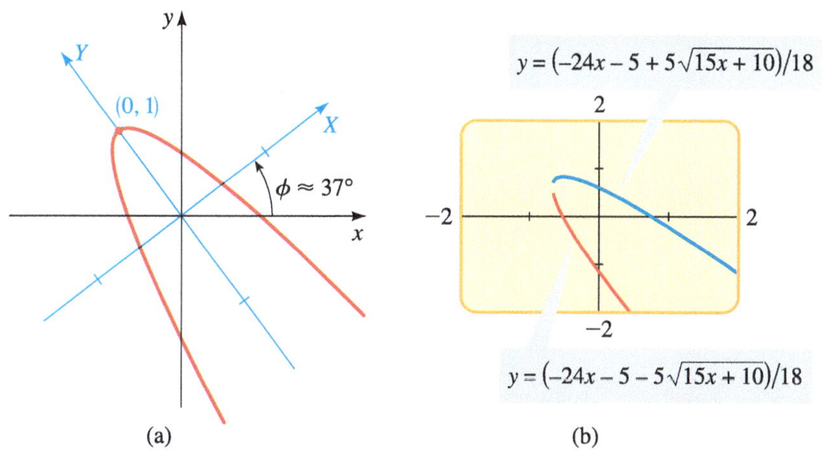

FIGURE 6

$64x^2 + 96xy + 36y^2 - 15x + 20y - 25 = 0$

(c) To draw the graph using a graphing calculator, we need to solve for y. The given equation is a quadratic equation in y, so we can use the Quadratic Formula to solve for y. Writing the equation in the form

$$36y^2 + (96x + 20)y + (64x^2 - 15x - 25) = 0$$

we get

$$y = \frac{-(96x + 20) \pm \sqrt{(96x + 20)^2 - 4(36)(64x^2 - 15x - 25)}}{2(36)} \qquad \text{Quadratic Formula}$$

$$= \frac{-(96x + 20) \pm \sqrt{6000x + 4000}}{72} \qquad \text{Expand}$$

$$= \frac{-96x - 20 \pm 20\sqrt{15x + 10}}{72} \qquad \text{Simplify}$$

$$= \frac{-24x - 5 \pm 5\sqrt{15x + 10}}{18} \qquad \text{Simplify}$$

To obtain the graph of the parabola, we graph the functions

$$y = (-24x - 5 + 5\sqrt{15x + 10})/18 \qquad \text{and} \qquad y = (-24x - 5 - 5\sqrt{15x + 10})/18$$

as shown in Figure 6(b).

Now Try Exercise 23 ■

The Discriminant

In Examples 3 and 4 we were able to identify the type of conic by rotating the axes. The next theorem gives rules for identifying the type of conic directly from the equation, without rotating axes.

IDENTIFYING CONICS BY THE DISCRIMINANT

The graph of the equation

$$Ax^2 + Bxy + Cy^2 + Dx + Ey + F = 0$$

is either a conic or a degenerate conic. In the nondegenerate cases the graph is

1. a parabola if $B^2 - 4AC = 0$,
2. an ellipse if $B^2 - 4AC < 0$,
3. a hyperbola if $B^2 - 4AC > 0$.

The quantity $B^2 - 4AC$ is called the **discriminant** of the equation.

Proof If we rotate the axes through an angle ϕ, we get an equation of the form

$$A'X^2 + B'XY + C'Y^2 + D'X + E'Y + F' = 0$$

where $A', B', C', \ldots$ are given by the formulas on page 818. A straightforward calculation shows that

$$(B')^2 - 4A'C' = B^2 - 4AC$$

Thus the expression $B^2 - 4AC$ remains unchanged for any rotation. In particular, if we choose a rotation that eliminates the xy-term $(B' = 0)$, we get

$$A'X^2 + C'Y^2 + D'X + E'Y + F' = 0$$

In this case $B^2 - 4AC = -4A'C'$. So $B^2 - 4AC = 0$ if either A' or C' is zero; $B^2 - 4AC < 0$ if A' and C' have the same sign; and $B^2 - 4AC > 0$ if A' and C' have opposite signs. According to the box on page 812, these cases correspond to the graph of the last displayed equation being a parabola, an ellipse, or a hyperbola, respectively. ■

In the proof we indicated that the discriminant is unchanged by any rotation; for this reason the discriminant is said to be **invariant** under rotation.

EXAMPLE 5 ■ Identifying a Conic by the Discriminant

A conic has the equation

$$3x^2 + 5xy - 2y^2 + x - y + 4 = 0$$

(a) Use the discriminant to identify the conic.

(b) Confirm your answer to part (a) by graphing the conic with a graphing calculator.

SOLUTION

(a) Since $A = 3$, $B = 5$, and $C = -2$, the discriminant is

$$B^2 - 4AC = 5^2 - 4(3)(-2) = 49 > 0$$

So the conic is a hyperbola.

(b) Using the Quadratic Formula, we solve for y to get

$$y = \frac{5x - 1 \pm \sqrt{49x^2 - 2x + 33}}{4}$$

We graph these functions in Figure 7. The graph confirms that this is a hyperbola.

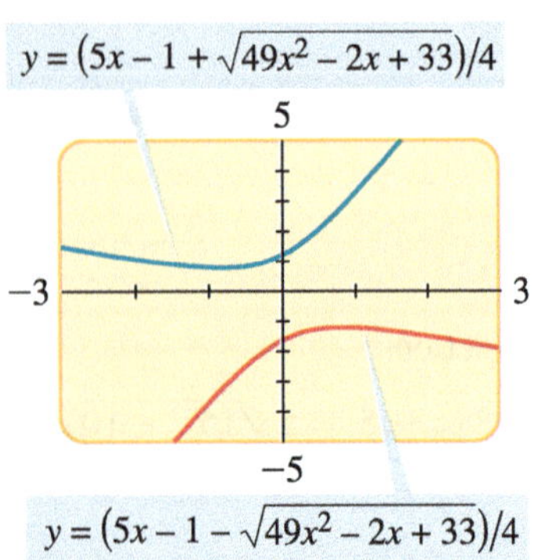

FIGURE 7

Now Try Exercise 29 ■

11.5 EXERCISES

CONCEPTS

1. Suppose the x- and y-axes are rotated through an acute angle ϕ to produce the new X- and Y-axes. A point P in the plane can be described by its xy-coordinates (x, y) or its XY-coordinates (X, Y). These coordinates are related by the following formulas.

   $x =$ ________ $X =$ ________

   $y =$ ________ $Y =$ ________

2. Consider the equation

$$Ax^2 + Bxy + Cy^2 + Dx + Ey + F = 0$$

   (a) In general, the graph of this equation is a ________.

   (b) To eliminate the xy-term from this equation, we rotate the axes through an angle ϕ that satisfies

   $\cot 2\phi =$ ________.

   (c) The discriminant of this equation is ________.

   If the discriminant is 0, the graph is a ________;

   if it is negative, the graph is ________; and

   if it is positive, the graph is ________.

SKILLS

3–8 ■ Rotation of Axes Determine the XY-coordinates of the given point if the coordinate axes are rotated through the indicated angle.

3. $(1, 1)$, $\phi = 45°$
4. $(-2, 1)$, $\phi = 30°$
5. $(3, -\sqrt{3})$, $\phi = 60°$
6. $(2, 0)$, $\phi = 15°$
7. $(0, 2)$, $\phi = 55°$
8. $(\sqrt{2}, 4\sqrt{2})$, $\phi = 45°$

9–14 ■ Finding the Equation for a Rotated Conic Determine the equation of the given conic in XY-coordinates when the coordinate axes are rotated through the indicated angle.

9. $x^2 - 3y^2 = 4$, $\phi = 60°$
10. $y = (x - 1)^2$, $\phi = 45°$
11. $x^2 - y^2 = 2y$, $\phi = \cos^{-1}\frac{3}{5}$
12. $x^2 + 2y^2 = 16$, $\phi = \sin^{-1}\frac{3}{5}$
13. $x^2 + 2\sqrt{3}xy - y^2 = 4$, $\phi = 30°$
14. $xy = x + y$, $\phi = \pi/4$

15–28 ■ Graphing a Rotated Conic **(a)** Use the discriminant to determine whether the graph of the equation is a parabola, an ellipse, or a hyperbola. **(b)** Use a rotation of axes to eliminate the xy-term. **(c)** Sketch the graph.

15. $xy = 8$
16. $xy + 4 = 0$
17. $x^2 + 2\sqrt{3}xy - y^2 + 2 = 0$
18. $13x^2 + 6\sqrt{3}xy + 7y^2 = 16$
19. $11x^2 - 24xy + 4y^2 + 20 = 0$
20. $21x^2 + 10\sqrt{3}xy + 31y^2 = 144$
21. $\sqrt{3}x^2 + 3xy = 3$
22. $153x^2 + 192xy + 97y^2 = 225$
23. $x^2 + 2xy + y^2 + x - y = 0$
24. $25x^2 - 120xy + 144y^2 - 156x - 65y = 0$
25. $2\sqrt{3}x^2 - 6xy + \sqrt{3}x + 3y = 0$
26. $9x^2 - 24xy + 16y^2 = 100(x - y - 1)$
27. $52x^2 + 72xy + 73y^2 = 40x - 30y + 75$
28. $(7x + 24y)^2 = 600x - 175y + 25$

29–32 ■ Identifying a Conic from Its Discriminant **(a)** Use the discriminant to identify the conic. **(b)** Confirm your answer by graphing the conic using a graphing device.

29. $2x^2 - 4xy + 2y^2 - 5x - 5 = 0$
30. $x^2 - 2xy + 3y^2 = 8$
31. $6x^2 + 10xy + 3y^2 - 6y = 36$
32. $9x^2 - 6xy + y^2 + 6x - 2y = 0$

SKILLS Plus

33. **Identifying a Hyperbola Using Rotation of Axes**

   (a) Use rotation of axes to show that the following equation represents a hyperbola.

$$7x^2 + 48xy - 7y^2 - 200x - 150y + 600 = 0$$

   (b) Find the XY- and xy-coordinates of the center, vertices, and foci.

   (c) Find the equations of the asymptotes in XY- and xy-coordinates.

34. **Identifying a Parabola Using Rotation of Axes**

   (a) Use rotation of axes to show that the following equation represents a parabola.

$$2\sqrt{2}(x + y)^2 = 7x + 9y$$

   (b) Find the XY- and xy-coordinates of the vertex and focus.

   (c) Find the equation of the directrix in XY- and xy-coordinates.

35. **Rotation of Axes Formulas** Solve the equations

$$x = X\cos\phi - Y\sin\phi$$
$$y = X\sin\phi + Y\cos\phi$$

   for X and Y in terms of x and y. [*Hint:* To begin, multiply the first equation by $\cos\phi$ and the second by $\sin\phi$, and then add the two equations to solve for X.]

36. **Graphing an Equation Using Rotation of Axes** Show that the graph of the equation

$$\sqrt{x} + \sqrt{y} = 1$$

   is part of a parabola by rotating the axes through an angle of 45°. [*Hint:* First convert the equation to one that does not involve radicals.]

DISCUSS ■ DISCOVER ■ PROVE ■ WRITE

37. PROVE: Matrix Form of Rotation of Axes Formulas Let Z, Z', and R be the matrices

$$Z = \begin{bmatrix} x \\ y \end{bmatrix} \qquad Z' = \begin{bmatrix} X \\ Y \end{bmatrix}$$

$$R = \begin{bmatrix} \cos\phi & -\sin\phi \\ \sin\phi & \cos\phi \end{bmatrix}$$

(a) Show that the Rotation of Axes Formulas can be written as

$$Z = RZ' \qquad \text{and} \qquad Z' = R^{-1}Z$$

(b) Let R_1 and R_2 be matrices that represent rotations through the angles ϕ_1 and ϕ_2, respectively. Show that the product matrix R_1R_2 represents a rotation through an angle $\phi_1 + \phi_2$. [*Hint:* Use the Addition Formulas for Sine and Cosine to simplify the entries of the matrix R_1R_2.]

38. PROVE: Algebraic Invariants A quantity is invariant under rotation if it does not change when the axes are rotated. It was stated in the text that for the general equation of a conic the quantity $B^2 - 4AC$ is invariant under rotation.

(a) Use the formulas for A', B', and C' on page 818 to prove that the quantity $B^2 - 4AC$ is invariant under rotation; that is, show that

$$B^2 - 4AC = B'^2 - 4A'C'$$

(b) Prove that $A + C$ is invariant under rotation.

(c) Is the quantity F invariant under rotation?

39. DISCOVER ■ PROVE: Geometric Invariants Do you expect that the distance between two points is invariant under rotation? Prove your answer by comparing the distance $d(P, Q)$ and $d(P', Q')$ where P' and Q' are the images of P and Q under a rotation of axes.

11.6 POLAR EQUATIONS OF CONICS

■ A Unified Geometric Description of Conics ■ Polar Equations of Conics

■ A Unified Geometric Description of Conics

Earlier in this chapter, we defined a parabola in terms of a focus and directrix, but we defined the ellipse and hyperbola in terms of two foci. In this section we give a more unified treatment of all three types of conics in terms of a focus and directrix. If we place one focus at the origin, then a conic section has a simple polar equation. Moreover, in polar form, rotation of conics becomes a simple matter. Polar equations of ellipses are crucial in the derivation of Kepler's Laws (see page 808).

EQUIVALENT DESCRIPTION OF CONICS

Let F be a fixed point (the **focus**), ℓ a fixed line (the **directrix**), and let e be a fixed positive number (the **eccentricity**). The set of all points P such that the ratio of the distance from P to F to the distance from P to ℓ is the constant e is a conic. That is, the set of all points P such that

$$\frac{d(P, F)}{d(P, \ell)} = e$$

is a conic. The conic is a parabola if $e = 1$, an ellipse if $e < 1$, or a hyperbola if $e > 1$.

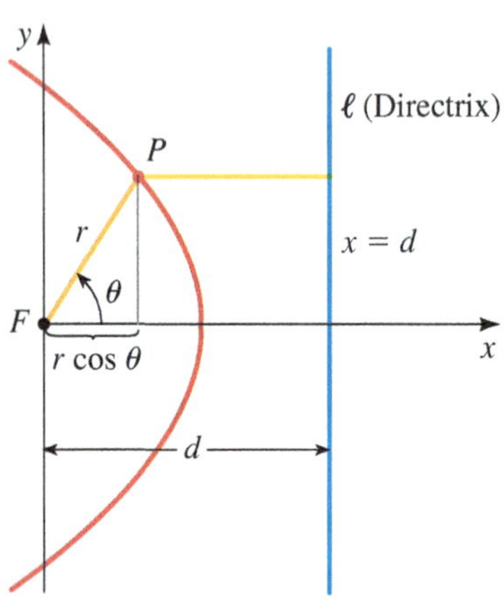

FIGURE 1

Proof If $e = 1$, then $d(P, F) = d(P, \ell)$, and so the given condition becomes the definition of a parabola as given in Section 11.1.

Now, suppose $e \neq 1$. Let's place the focus F at the origin and the directrix parallel to the y-axis and d units to the right. In this case the directrix has equation $x = d$ and is perpendicular to the polar axis. If the point P has polar coordinates (r, θ), we see from Figure 1 that $d(P, F) = r$ and $d(P, \ell) = d - r\cos\theta$. Thus the condition $d(P, F)/d(P, \ell) = e$, or $d(P, F) = e \cdot d(P, \ell)$, becomes

$$r = e(d - r\cos\theta)$$

If we square both sides of this polar equation and convert to rectangular coordinates, we get

$$x^2 + y^2 = e^2(d - x)^2$$

$$(1 - e^2)x^2 + 2de^2x + y^2 = e^2d^2 \qquad \text{Expand and simplify}$$

$$\left(x + \frac{e^2d}{1 - e^2}\right)^2 + \frac{y^2}{1 - e^2} = \frac{e^2d^2}{(1 - e^2)^2} \qquad \text{Divide by } 1 - e^2 \text{ and complete the square}$$

If $e < 1$, then dividing both sides of this equation by $e^2d^2/(1 - e^2)^2$ gives an equation of the form

$$\frac{(x - h)^2}{a^2} + \frac{y^2}{b^2} = 1$$

where

$$h = \frac{-e^2d}{1 - e^2} \qquad a^2 = \frac{e^2d^2}{(1 - e^2)^2} \qquad b^2 = \frac{e^2d^2}{1 - e^2}$$

This is the equation of an ellipse with center $(h, 0)$. In Section 11.2 we found that the foci of an ellipse are a distance c from the center, where $c^2 = a^2 - b^2$. In our case

$$c^2 = a^2 - b^2 = \frac{e^4d^2}{(1 - e^2)^2}$$

Thus $c = e^2d/(1 - e^2) = -h$, which confirms that the focus defined in the theorem (namely the origin) is the same as the focus defined in Section 11.2. It also follows that

$$\boxed{e = \frac{c}{a}}$$

If $e > 1$, a similar proof shows that the conic is a hyperbola with $e = c/a$, where $c^2 = a^2 + b^2$. ■

Polar Equations of Conics

In the proof we saw that the polar equation of the conic in Figure 1 is $r = e(d - r\cos\theta)$. Solving for r, we get

$$r = \frac{ed}{1 + e\cos\theta}$$

If the directrix is chosen to be to the *left* of the focus $(x = -d)$, then we get the equation $r = ed/(1 - e\cos\theta)$. If the directrix is *parallel* to the polar axis $(y = d$ or $y = -d)$, then we get $\sin\theta$ instead of $\cos\theta$ in the equation. These observations are summarized in the following box and in Figure 2.

POLAR EQUATIONS OF CONICS

A polar equation of the form

$$r = \frac{ed}{1 \pm e\cos\theta} \qquad \text{or} \qquad r = \frac{ed}{1 \pm e\sin\theta}$$

represents a conic with one focus at the origin and with eccentricity e. The conic is

1. a parabola if $e = 1$,

2. an ellipse if $0 < e < 1$,

3. a hyperbola if $e > 1$.

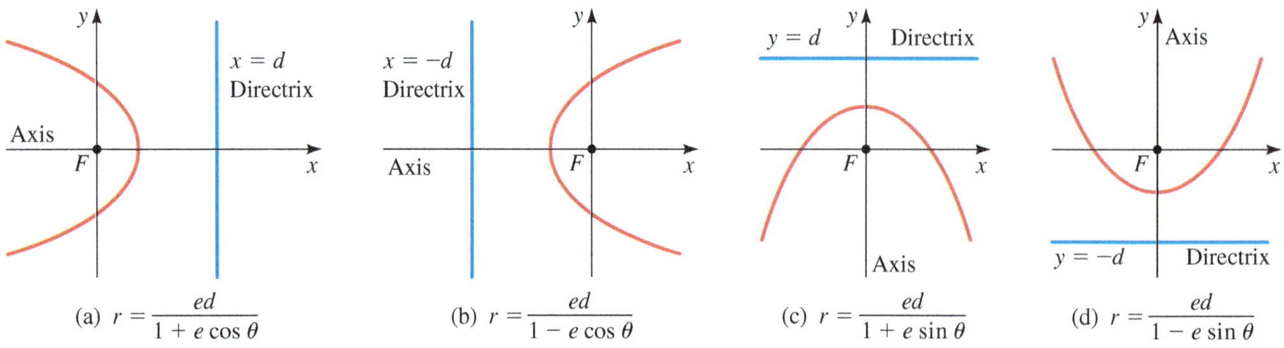

FIGURE 2 The form of the polar equation of a conic indicates the location of the directrix.

To graph the polar equation of a conic, we first determine the location of the directrix from the form of the equation. The four cases that arise are shown in Figure 2. (The figure shows only the parts of the graphs that are close to the focus at the origin. The shape of the rest of the graph depends on whether the equation represents a parabola, an ellipse, or a hyperbola.) The axis of a conic is perpendicular to the directrix—specifically we have the following:

1. For a parabola the axis of symmetry is perpendicular to the directrix.
2. For an ellipse the major axis is perpendicular to the directrix.
3. For a hyperbola the transverse axis is perpendicular to the directrix.

EXAMPLE 1 ■ Finding a Polar Equation for a Conic

Find a polar equation for the parabola that has its focus at the origin and whose directrix is the line $y = -6$.

SOLUTION Using $e = 1$ and $d = 6$ and using part (d) of Figure 2, we see that the polar equation of the parabola is

$$r = \frac{6}{1 - \sin\theta}$$

Now Try Exercise 3 ■

To graph a polar conic, it is helpful to plot the points for which $\theta = 0, \pi/2, \pi$, and $3\pi/2$. Using these points and a knowledge of the type of conic (which we obtain from the eccentricity), we can easily get a rough idea of the shape and location of the graph.

EXAMPLE 2 ■ Identifying and Sketching a Conic

A conic is given by the polar equation

$$r = \frac{10}{3 - 2\cos\theta}$$

(a) Show that the conic is an ellipse, and sketch its graph.

(b) Find the center of the ellipse and the lengths of the major and minor axes.

SOLUTION

(a) Dividing the numerator and denominator by 3, we have

$$r = \frac{\frac{10}{3}}{1 - \frac{2}{3}\cos\theta}$$

Since $e = \frac{2}{3} < 1$, the equation represents an ellipse. For a rough graph we plot the points for which $\theta = 0, \pi/2, \pi, 3\pi/2$ (see Figure 3).

θ	r
0	10
$\frac{\pi}{2}$	$\frac{10}{3}$
π	2
$\frac{3\pi}{2}$	$\frac{10}{3}$

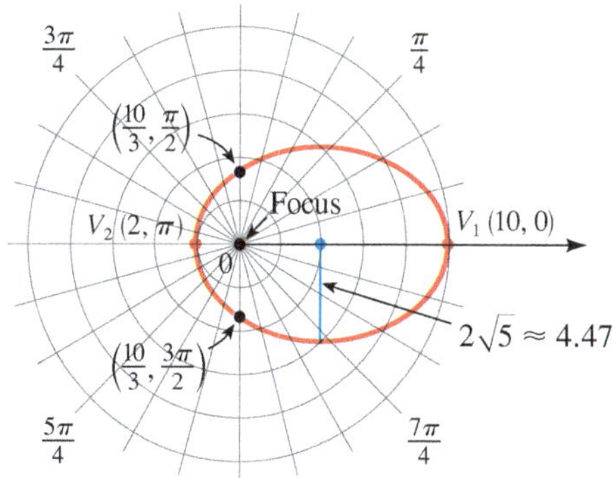

FIGURE 3 $r = \dfrac{10}{3 - 2\cos\theta}$

(b) Comparing the equation to those in Figure 2, we see that the major axis is horizontal. Thus the endpoints of the major axis are $V_1(10, 0)$ and $V_2(2, \pi)$. So the center of the ellipse is at $C(4, 0)$, the midpoint of V_1V_2.

The distance between the vertices V_1 and V_2 is 12; thus the length of the major axis is $2a = 12$, so $a = 6$. To determine the length of the minor axis, we need to find b. From page 825 we have $c = ae = 6(\frac{2}{3}) = 4$, so

$$b^2 = a^2 - c^2 = 6^2 - 4^2 = 20$$

Thus $b = \sqrt{20} = 2\sqrt{5} \approx 4.47$, and the length of the minor axis is $2b = 4\sqrt{5} \approx 8.94$.

Now Try Exercises 17 and 21 ■

EXAMPLE 3 ■ Identifying and Sketching a Conic

A conic is given by the polar equation

$$r = \frac{12}{2 + 4\sin\theta}$$

(a) Show that the conic is a hyperbola, and sketch its graph.

(b) Find the center of the hyperbola, and sketch the asymptotes.

SOLUTION

(a) Dividing the numerator and denominator by 2, we have

$$r = \frac{6}{1 + 2\sin\theta}$$

Since $e = 2 > 1$, the equation represents a hyperbola. For a rough graph we plot the points for which $\theta = 0, \pi/2, \pi, 3\pi/2$ (see Figure 4).

(b) Comparing the equation to those in Figure 2, we see that the transverse axis is vertical. Thus the endpoints of the transverse axis (the vertices of the hyperbola) are $V_1(2, \pi/2)$ and $V_2(-6, 3\pi/2) = V_2(6, \pi/2)$. So the center of the hyperbola is $C(4, \pi/2)$, the midpoint of V_1V_2.

To sketch the asymptotes, we need to find a and b. The distance between V_1 and V_2 is 4; thus the length of the transverse axis is $2a = 4$, so $a = 2$. To find b, we first find c. From page 825 we have $c = ae = 2 \cdot 2 = 4$, so

$$b^2 = c^2 - a^2 = 4^2 - 2^2 = 12$$

Thus $b = \sqrt{12} = 2\sqrt{3} \approx 3.46$. Knowing a and b allows us to sketch the central box, from which we obtain the asymptotes shown in Figure 4.

θ	r
0	6
$\frac{\pi}{2}$	2
π	6
$\frac{3\pi}{2}$	-6

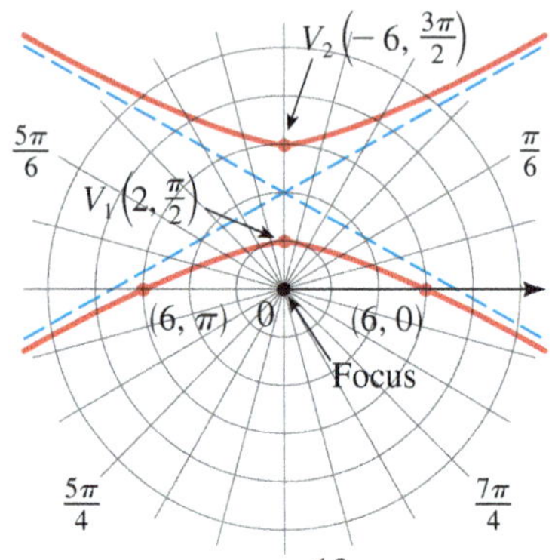

FIGURE 4 $r = \dfrac{12}{2 + 4\sin\theta}$

Now Try Exercise 25 ■

When we rotate conic sections, it is much more convenient to use polar equations than Cartesian equations. We use the fact that the graph of $r = f(\theta - \alpha)$ is the graph of $r = f(\theta)$ rotated counterclockwise about the origin through an angle α (see Exercise 65 in Section 8.2).

EXAMPLE 4 ■ Rotating an Ellipse

Suppose the ellipse of Example 2 is rotated through an angle $\pi/4$ about the origin. Find a polar equation for the resulting ellipse, and draw its graph.

SOLUTION We get the equation of the rotated ellipse by replacing θ with $\theta - \pi/4$ in the equation given in Example 2. So the new equation is

$$r = \frac{10}{3 - 2\cos(\theta - \pi/4)}$$

We use this equation to graph the rotated ellipse in Figure 5. Notice that the ellipse has been rotated about the focus at the origin.

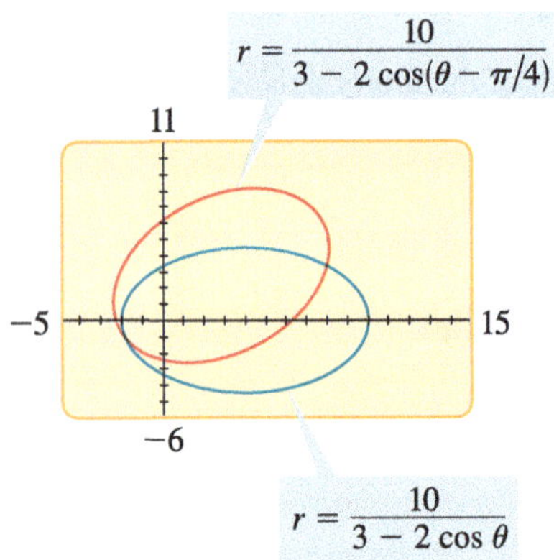

FIGURE 5

Now Try Exercise 37 ■

In Figure 6 we use a computer to sketch a number of conics to demonstrate the effect of varying the eccentricity e. Notice that when e is close to 0, the ellipse is nearly circular, and it becomes more elongated as e increases. When $e = 1$, of course, the conic is a parabola. As e increases beyond 1, the conic is an ever steeper hyperbola.

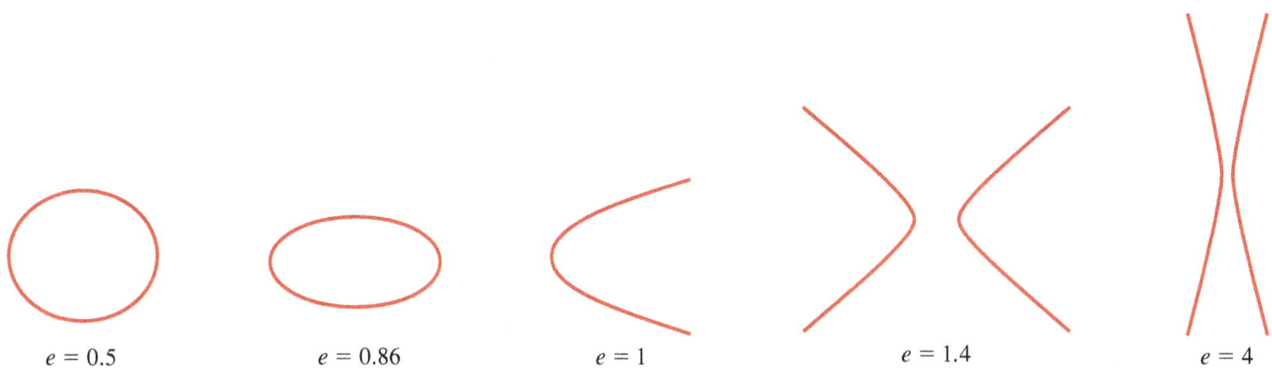

FIGURE 6

11.6 EXERCISES

CONCEPTS

1. All conics can be described geometrically by using a fixed point F called the ______________ and a fixed line ℓ called the ______________. For a fixed positive number e the set of all points P satisfying

$$\frac{\qquad}{\qquad} = e$$

is a ______________________. If $e = 1$, the conic is a(n) ______________; if $e < 1$, the conic is a(n) ______________; and if $e > 1$, the conic is a(n) ______________. The number e is called the ______________ of the conic.

2. The polar equation of a conic with eccentricity e has one of the following forms:

$r =$ ______________ or $r =$ ______________

SKILLS

3–10 ■ Finding a Polar Equation for a Conic Write a polar equation of a conic that has its focus at the origin and satisfies the given conditions.

3. Ellipse, eccentricity $\frac{2}{3}$, directrix $x = 3$

4. Hyperbola, eccentricity $\frac{4}{3}$, directrix $x = -3$

5. Parabola, directrix $y = 2$

6. Ellipse, eccentricity $\frac{1}{2}$, directrix $y = -4$

7. Hyperbola, eccentricity 4, directrix $r = 5 \sec \theta$

8. Ellipse, eccentricity 0.6, directrix $r = 2 \csc \theta$

9. Parabola, vertex at $(5, \pi/2)$

10. Ellipse, eccentricity 0.4, vertex at $(2, 0)$

11–16 ■ Graphs of Polar Equations of Conics Match the polar equations with the graphs labeled I–VI. Give reasons for your answer.

11. $r = \dfrac{6}{1 + \cos \theta}$

12. $r = \dfrac{2}{2 - \cos \theta}$

13. $r = \dfrac{3}{1 - 2 \sin \theta}$

14. $r = \dfrac{5}{3 - 3 \sin \theta}$

15. $r = \dfrac{12}{3 + 2 \sin \theta}$

16. $r = \dfrac{12}{2 + 3 \cos \theta}$

I

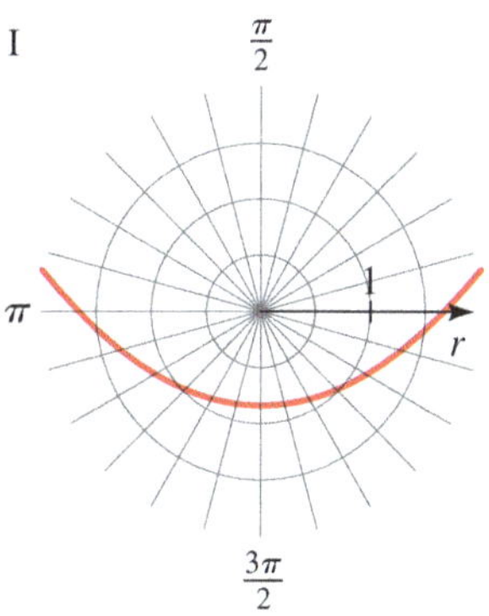

II

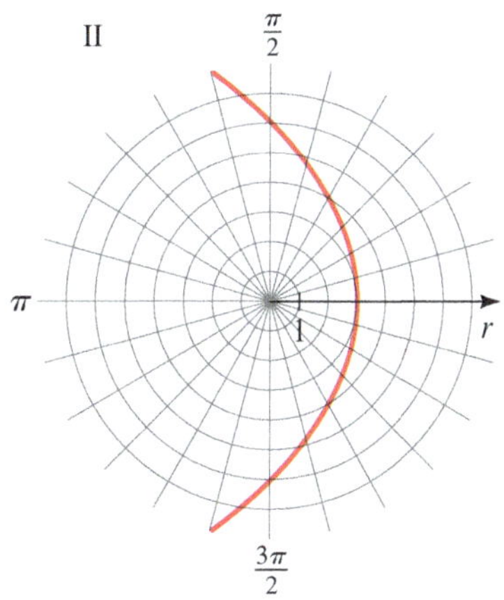

III

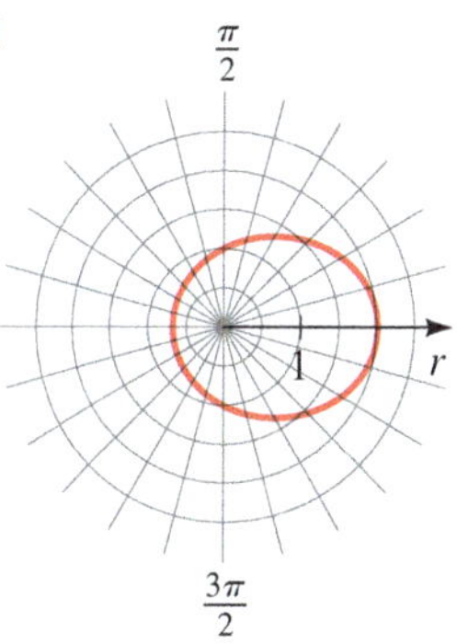

IV

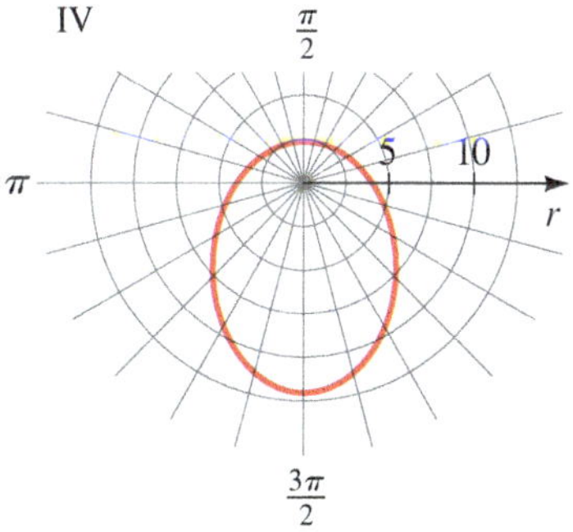

V

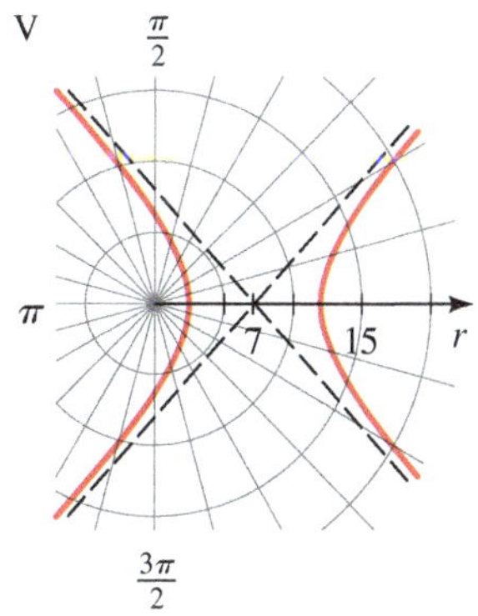

VI

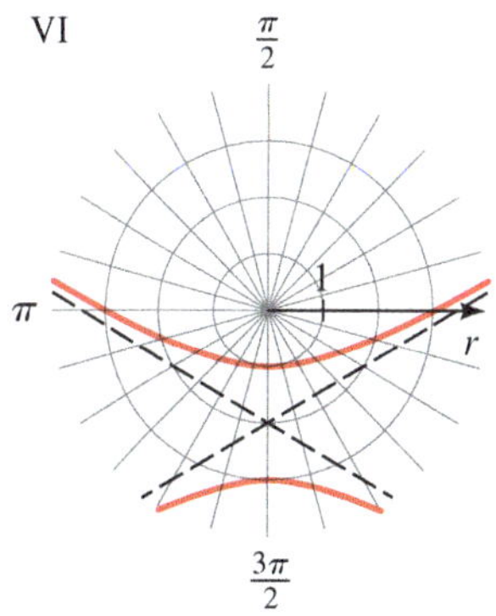

17–20 ■ Polar Equation for a Parabola A polar equation of a conic is given. **(a)** Show that the conic is a parabola, and sketch its graph. **(b)** Find the vertex and directrix, and indicate them on the graph.

17. $r = \dfrac{4}{1 - \sin\theta}$ **18.** $r = \dfrac{3}{2 + 2\sin\theta}$

19. $r = \dfrac{5}{3 + 3\cos\theta}$ **20.** $r = \dfrac{2}{5 - 5\cos\theta}$

21–24 ■ Polar Equation for an Ellipse A polar equation of a conic is given. **(a)** Show that the conic is an ellipse, and sketch its graph. **(b)** Find the vertices and directrix, and indicate them on the graph. **(c)** Find the center of the ellipse and the lengths of the major and minor axes.

21. $r = \dfrac{4}{2 - \cos\theta}$ **22.** $r = \dfrac{6}{3 - 2\sin\theta}$

23. $r = \dfrac{12}{4 + 3\sin\theta}$ **24.** $r = \dfrac{18}{4 + 3\cos\theta}$

25–28 ■ Polar Equation for a Hyperbola A polar equation of a conic is given. **(a)** Show that the conic is a hyperbola, and sketch its graph. **(b)** Find the vertices and directrix, and indicate them on the graph. **(c)** Find the center of the hyperbola, and sketch the asymptotes.

25. $r = \dfrac{8}{1 + 2\cos\theta}$ **26.** $r = \dfrac{10}{1 - 4\sin\theta}$

27. $r = \dfrac{20}{2 - 3\sin\theta}$ **28.** $r = \dfrac{6}{2 + 7\cos\theta}$

29–36 ■ Identifying and Graphing a Conic **(a)** Find the eccentricity, and identify the conic. **(b)** Sketch the conic, and label the vertices.

29. $r = \dfrac{4}{1 + 3\cos\theta}$ **30.** $r = \dfrac{8}{3 + 3\cos\theta}$

31. $r = \dfrac{2}{1 - \cos\theta}$ **32.** $r = \dfrac{10}{3 - 2\sin\theta}$

33. $r = \dfrac{6}{2 + \sin\theta}$ **34.** $r = \dfrac{5}{2 - 3\sin\theta}$

35. $r = \dfrac{7}{2 - 5\sin\theta}$ **36.** $r = \dfrac{8}{3 + \cos\theta}$

37–40 ■ Rotating a Conic A polar equation of a conic is given. **(a)** Find the eccentricity and the directrix of the conic. **(b)** If this conic is rotated about the origin through the given angle θ, write the resulting equation. **(c)** Draw graphs of the original conic and the rotated conic on the same screen.

37. $r = \dfrac{1}{4 - 3\cos\theta}$; $\theta = \dfrac{\pi}{3}$ **38.** $r = \dfrac{2}{5 - 3\sin\theta}$; $\theta = \dfrac{2\pi}{3}$

39. $r = \dfrac{2}{1 + \sin\theta}$; $\theta = -\dfrac{\pi}{4}$

40. $r = \dfrac{9}{2 + 2\cos\theta}$; $\theta = -\dfrac{5\pi}{6}$

SKILLS Plus

41. Families of Conics Graph the conics $r = e/(1 - e\cos\theta)$ with $e = 0.4, 0.6, 0.8,$ and 1.0 on a common screen. How does the value of e affect the shape of the curve?

42. Families of Conics

(a) Graph the conics

$$r = \frac{ed}{(1 + e\sin\theta)}$$

for $e = 1$ and various values of d. How does the value of d affect the shape of the conic?

(b) Graph these conics for $d = 1$ and various values of e. How does the value of e affect the shape of the conic?

APPLICATIONS

43. Orbit of the Earth The polar equation of an ellipse can be expressed in terms of its eccentricity e and the length a of its major axis.

(a) Show that the polar equation of an ellipse with directrix $x = -d$ can be written in the form

$$r = \frac{a(1 - e^2)}{1 - e\cos\theta}$$

[*Hint:* Use the relation $a^2 = e^2d^2/(1 - e^2)^2$ given in the proof on page 825.]

(b) Find an approximate polar equation for the elliptical orbit of the earth around the sun (at one focus) given that the eccentricity is about 0.017 and the length of the major axis is about 2.99×10^8 km.

44. Perihelion and Aphelion The planets move around the sun in elliptical orbits with the sun at one focus. The positions of a planet that are closest to, and farthest from, the sun are called its **perihelion** and **aphelion**, respectively.

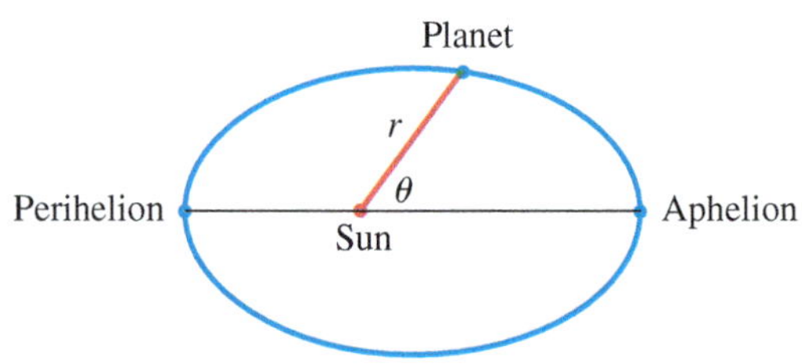

(a) Use Exercise 43(a) to show that the perihelion distance from a planet to the sun is $a(1 - e)$ and the aphelion distance is $a(1 + e)$.

(b) Use the data of Exercise 43(b) to find the distances from the earth to the sun at perihelion and at aphelion.

45. Orbit of Pluto The distance from Pluto to the sun is 4.43×10^9 km at perihelion and 7.37×10^9 km at aphelion. Use Exercise 44 to find the eccentricity of Pluto's orbit.

DISCUSS ■ DISCOVER ■ PROVE ■ WRITE

46. DISCUSS: Distance to a Focus When we found polar equations for the conics, we placed one focus at the pole. It's easy to find the distance from that focus to any point on the conic. Explain how the polar equation gives us this distance.

47. DISCUSS: Polar Equations of Orbits When a satellite orbits the earth, its path is an ellipse with one focus at the center of the earth. Why do scientists use polar (rather than rectangular) coordinates to track the position of satellites? [*Hint:* Your answer to Exercise 46 is relevant here.]

CHAPTER 11 ■ REVIEW

PROPERTIES AND FORMULAS

Geometric Definition of a Parabola (p. 782)

A **parabola** is the set of points in the plane that are equidistant from a fixed point F (the **focus**) and a fixed line l (the **directrix**).

Graphs of Parabolas with Vertex at the Origin (pp. 783, 784)

A parabola with vertex at the origin has an equation of the form $x^2 = 4py$ if its axis is vertical and an equation of the form $y^2 = 4px$ if its axis is horizontal.

$x^2 = 4py$

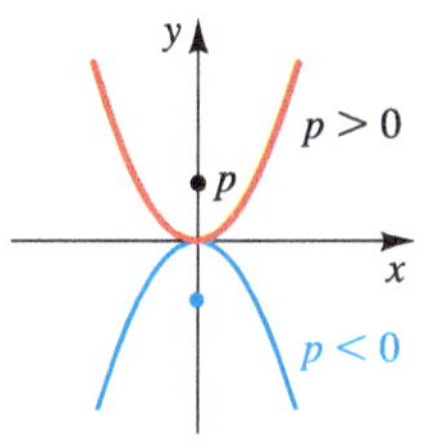

Focus $(0, p)$, directrix $y = -p$

$y^2 = 4px$

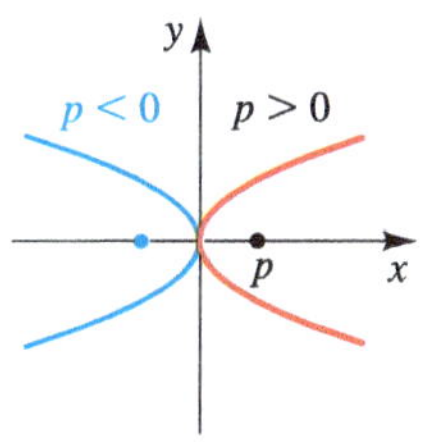

Focus $(p, 0)$, directrix $x = -p$

Geometric Definition of an Ellipse (p. 790)

An **ellipse** is the set of all points in the plane for which the sum of the distances to each of two given points F_1 and F_2 (the **foci**) is a fixed constant.

Graphs of Ellipses with Center at the Origin (p. 792)

An ellipse with center at the origin has an equation of the form

$\frac{x^2}{a^2} + \frac{y^2}{b^2} = 1$ if its axis is horizontal and an equation of the form

$\frac{x^2}{b^2} + \frac{y^2}{a^2} = 1$ if its axis is vertical (where in each case $a > b > 0$).

$\frac{x^2}{a^2} + \frac{y^2}{b^2} = 1$

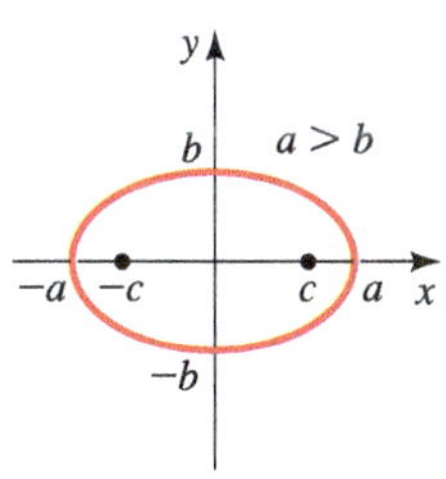

Foci $(\pm c, 0)$, $c^2 = a^2 - b^2$

$\frac{x^2}{b^2} + \frac{y^2}{a^2} = 1$

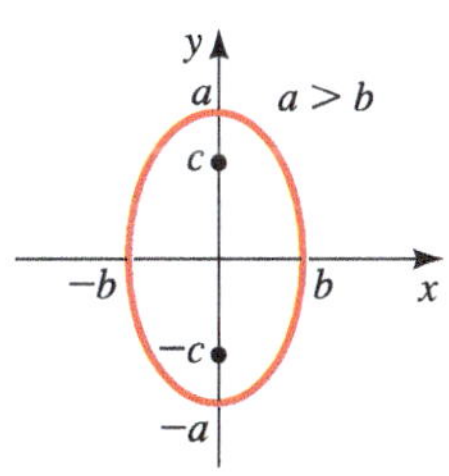

Foci $(0, \pm c)$, $c^2 = a^2 - b^2$

Eccentricity of an Ellipse (p. 795)

The **eccentricity** of an ellipse with equation $\frac{x^2}{a^2} + \frac{y^2}{b^2} = 1$ or $\frac{x^2}{b^2} + \frac{y^2}{a^2} = 1$ (where $a > b > 0$) is the number

$$e = \frac{c}{a}$$

where $c = \sqrt{a^2 - b^2}$. The eccentricity e of any ellipse is a number between 0 and 1. If e is close to 0, then the ellipse is nearly circular; the closer e gets to 1, the more elongated it becomes.

Geometric Definition of a Hyperbola (p. 799)

A **hyperbola** is the set of all points in the plane for which the absolute value of the difference of the distances to each of two given points F_1 and F_2 (the **foci**) is a fixed constant.

Graphs of Hyperbolas with Center at the Origin (p. 800)

A **hyperbola** with center at the origin has an equation of the form

$\frac{x^2}{a^2} - \frac{y^2}{b^2} = 1$ if its axis is horizontal and an equation of the form

$-\frac{x^2}{b^2} + \frac{y^2}{a^2} = 1$ if its axis is vertical.

$\frac{x^2}{a^2} - \frac{y^2}{b^2} = 1$

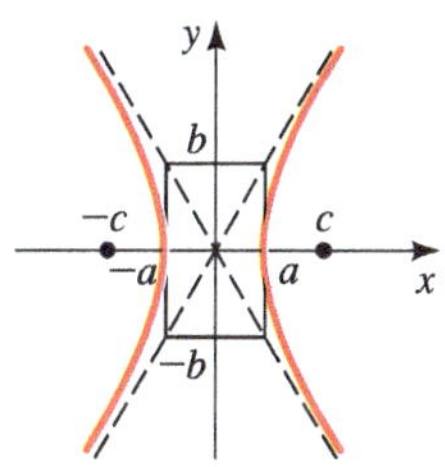

Foci $(\pm c, 0)$, $c^2 = a^2 + b^2$

Asymptotes: $y = \pm \frac{b}{a} x$

$-\frac{x^2}{b^2} + \frac{y^2}{a^2} = 1$

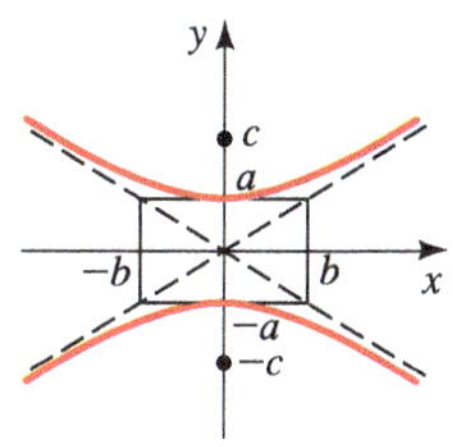

Foci $(0, \pm c)$, $c^2 = a^2 + b^2$

Asymptotes: $y = \pm \frac{a}{b} x$

Shifted Conics (p. 808)

If the vertex of a parabola or the center of an ellipse or a hyperbola does not lie at the origin but rather at the point (h, k), then we refer to the curve as a **shifted conic**. To find the equation of the shifted conic, we use the "unshifted" form for the appropriate curve and replace x by $x - h$ and y by $y - k$.

General Equation of a Shifted Conic (p. 812)

The graph of the equation

$$Ax^2 + Cy^2 + Dx + Ey + F = 0$$

(where A and C are not both 0) is either a conic or a degenerate conic. In the nondegenerate cases the graph is

1. a **parabola** if $A = 0$ or $C = 0$,
2. an **ellipse** if A and C have the same sign (or a circle if $A = C$),
3. a **hyperbola** if A and C have opposite sign.

To graph a conic whose equation is given in general form, complete the squares in x and y to put the equation in standard form for a parabola, an ellipse, or a hyperbola.

Rotation of Axes (p. 817)

Suppose the x- and y-axes in a coordinate plane are rotated through the acute angle ϕ to produce the X- and Y-axes, as shown in the figure below. Then the coordinates of a point in the xy- and the XY-planes are related as follows:

$$x = X\cos\phi - Y\sin\phi \qquad X = x\cos\phi + y\sin\phi$$

$$y = X\sin\phi + Y\cos\phi \qquad Y = -x\sin\phi + y\cos\phi$$

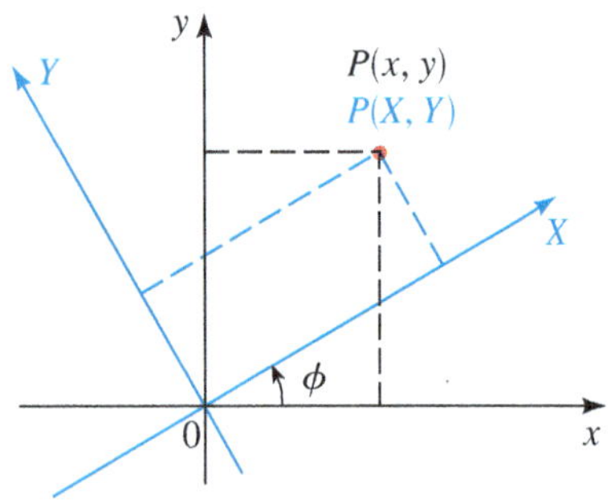

The General Conic Equation (pp. 819, 822)

The general equation of a conic is of the form

$$Ax^2 + Bxy + Cy^2 + Dx + Ey + F = 0$$

The quantity $B^2 - 4AC$ is called the **discriminant** of the equation. The graph is

1. a parabola if $B^2 - 4AC = 0$,
2. an ellipse if $B^2 - 4AC < 0$,
3. a hyperbola if $B^2 - 4AC > 0$.

To eliminate the xy-term in the general equation of a conic, rotate the axes through an angle ϕ that satisfies

$$\cot 2\phi = \frac{A - C}{B}$$

Polar Equations of Conics (p. 825)

A polar equation of the form

$$r = \frac{ed}{1 \pm e\cos\theta} \quad \text{or} \quad r = \frac{ed}{1 \pm e\sin\theta}$$

represents a conic with one focus at the origin and with eccentricity e. The conic is

1. a parabola if $e = 1$,
2. an ellipse if $0 < e < 1$,
3. a hyperbola if $e > 1$.

CONCEPT CHECK

1. **(a)** Give the geometric definition of a parabola.
   (b) Give the equation of a parabola with vertex at the origin and with vertical axis. Where is the focus? What is the directrix?
   (c) Graph the equation $x^2 = 8y$. Indicate the focus on the graph.
2. **(a)** Give the geometric definition of an ellipse.
   (b) Give the equation of an ellipse with center at the origin and with major axis along the x-axis. How long is the major axis? How long is the minor axis? Where are the foci? What is the eccentricity of the ellipse?
   (c) Graph the equation $\dfrac{x^2}{16} + \dfrac{y^2}{9} = 1$. What are the lengths of the major and minor axes? Where are the foci?
3. **(a)** Give the geometric definition of a hyperbola.
   (b) Give the equation of a hyperbola with center at the origin and with transverse axis along the x-axis. How long is the transverse axis? Where are the vertices? What are the asymptotes? Where are the foci?
   (c) What is a good first step in graphing the hyperbola that is described in part (b)?
   (d) Graph the equation $\dfrac{x^2}{16} - \dfrac{y^2}{9} = 1$. What are the asymptotes? Where are the vertices? Where are the foci? What is the length of the transverse axis?
4. **(a)** Suppose we are given an equation in x and y. Let h and k be positive numbers. What is the effect on the graph of the equation if x is replaced by $x - h$ or $x + h$ and if y is replaced by $y - k$ or $y + k$?
   (b) Sketch a graph of $\dfrac{(x+2)^2}{16} + \dfrac{(y-4)^2}{9} = 1$
5. **(a)** How can you tell whether the following nondegenerate conic is a parabola, an ellipse, or a hyperbola?

   $$Ax^2 + Cy^2 + Dx + Ey + F = 0$$

   (b) What conic does $3x^2 - 5y^2 + 4x + 5y - 8 = 0$ represent?

6. (a) Suppose that the x- and y-axes are rotated through an acute angle ϕ to produce the X- and Y-axes. What are the equations that relate the coordinates (x,y) and (X,Y) of a point in the xy-plane and XY-plane, respectively?

(b) In the equation below, how do you eliminate the xy-term?

$$Ax^2 + Bxy + Cy^2 + Dx + Ey + F = 0$$

(c) Use a rotation of axes to eliminate the xy-term in the equation

$$25x^2 - 14xy + 25y^2 = 288$$

Graph the equation.

7. (a) What is the discriminant of the equation in 6(b)? How can you use the discriminant to determine the type of conic that the equation represents?

(b) Use the discriminant to identify the equation in 6(c).

8. (a) Write polar equations that represent a conic with eccentricity e. For what values of e is the conic an ellipse? a hyperbola? a parabola?

(b) What conic does the polar equation $r = 2/(1 - \cos\theta)$ represent? Graph the conic.

ANSWERS TO THE CONCEPT CHECK CAN BE FOUND AT THE BACK OF THE BOOK.

EXERCISES

1–12 ■ Graphing Parabolas An equation of a parabola is given. **(a)** Find the vertex, focus, and directrix of the parabola. **(b)** Sketch a graph of the parabola and its directrix.

1. $y^2 = 4x$

2. $x = \frac{1}{12}y^2$

3. $\frac{1}{8}x^2 = y$

4. $x^2 = -8y$

5. $x^2 + 8y = 0$

6. $2x - y^2 = 0$

7. $(y - 2)^2 = 4(x + 2)$

8. $(x + 3)^2 = -20(y + 2)$

9. $\frac{1}{2}(y - 3)^2 + x = 0$

10. $2(x + 1)^2 = y$

11. $\frac{1}{2}x^2 + 2x = 2y + 4$

12. $x^2 = 3(x + y)$

13–24 ■ Graphing Ellipses An equation of an ellipse is given. **(a)** Find the center, vertices, and foci of the ellipse. **(b)** Determine the lengths of the major and minor axes. **(c)** Sketch a graph of the ellipse.

13. $\dfrac{x^2}{9} + \dfrac{y^2}{25} = 1$

14. $\dfrac{x^2}{49} + \dfrac{y^2}{9} = 1$

15. $\dfrac{x^2}{49} + \dfrac{y^2}{4} = 1$

16. $\dfrac{x^2}{4} + \dfrac{y^2}{36} = 1$

17. $x^2 + 4y^2 = 16$

18. $9x^2 + 4y^2 = 1$

19. $\dfrac{(x - 3)^2}{9} + \dfrac{y^2}{16} = 1$

20. $\dfrac{(x - 2)^2}{25} + \dfrac{(y + 3)^2}{16} = 1$

21. $\dfrac{(x - 2)^2}{9} + \dfrac{(y + 3)^2}{36} = 1$

22. $\dfrac{x^2}{3} + \dfrac{(y + 5)^2}{25} = 1$

23. $4x^2 + 9y^2 = 36y$

24. $2x^2 + y^2 = 2 + 4(x - y)$

25–36 ■ Graphing Hyperbolas An equation of a hyperbola is given. **(a)** Find the center, vertices, foci, and asymptotes of the hyperbola. **(b)** Sketch a graph of the hyperbola.

25. $-\dfrac{x^2}{9} + \dfrac{y^2}{16} = 1$

26. $\dfrac{x^2}{49} - \dfrac{y^2}{32} = 1$

27. $\dfrac{x^2}{4} - \dfrac{y^2}{49} = 1$

28. $\dfrac{y^2}{25} - \dfrac{x^2}{4} = 1$

29. $x^2 - 2y^2 = 16$

30. $x^2 - 4y^2 + 16 = 0$

31. $\dfrac{(x + 4)^2}{16} - \dfrac{y^2}{16} = 1$

32. $\dfrac{(x - 2)^2}{8} - \dfrac{(y + 2)^2}{8} = 1$

33. $\dfrac{(y - 3)^2}{4} - \dfrac{(x + 1)^2}{36} = 1$

34. $\dfrac{(y - 3)^2}{3} - \dfrac{x^2}{16} = 1$

35. $9y^2 + 18y = x^2 + 6x + 18$

36. $y^2 = x^2 + 6y$

37–42 ■ Finding the Equation of a Conic Find an equation for the conic whose graph is shown.

37.

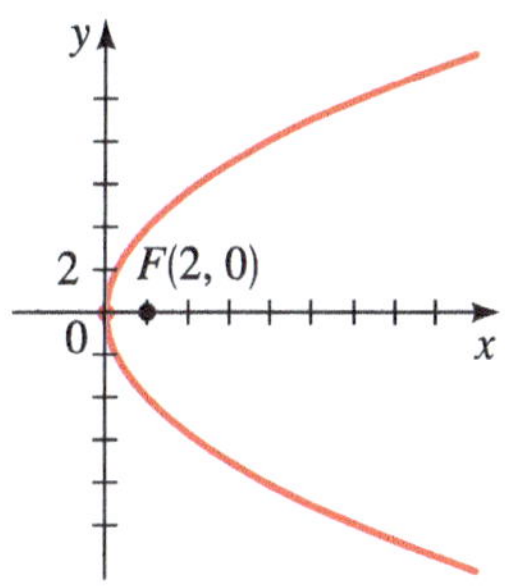

38.

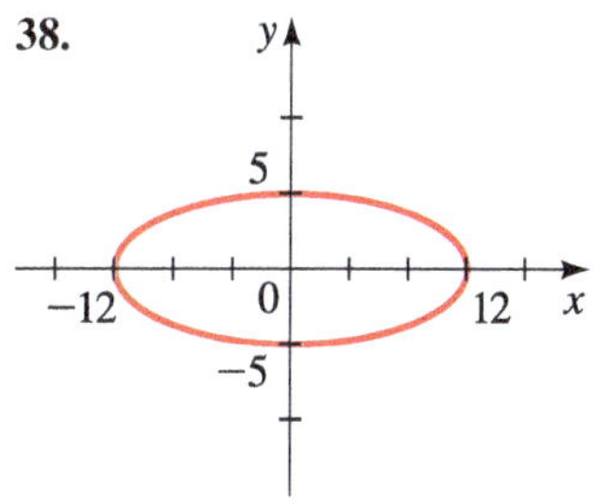

39.

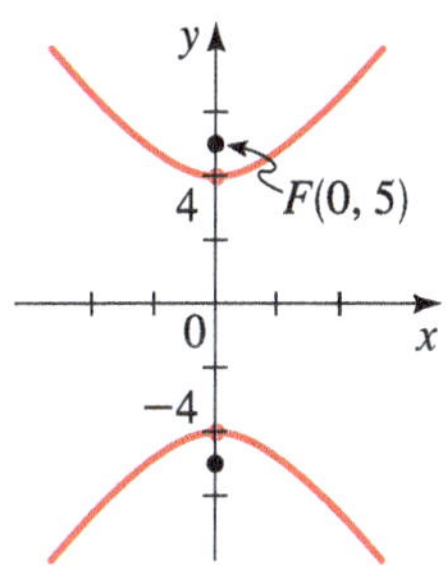

40.

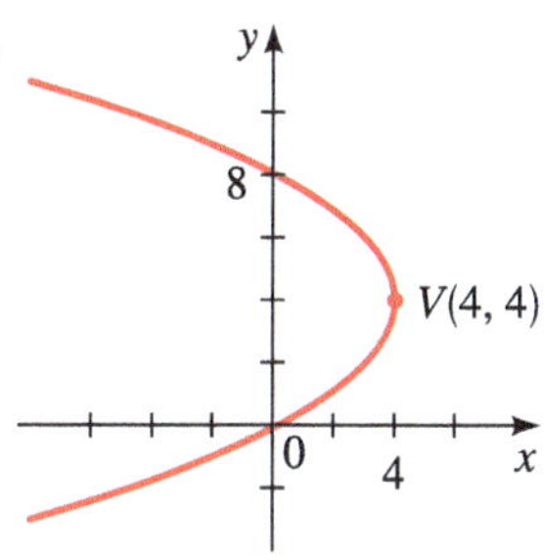

41.

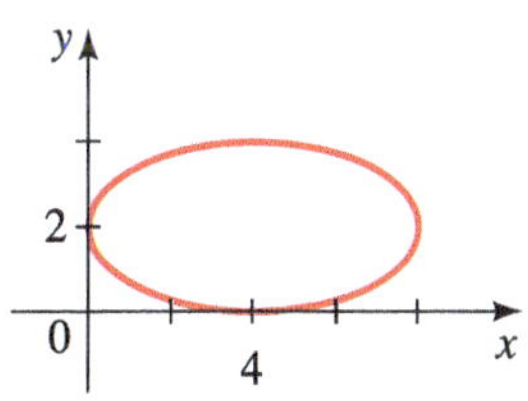

42.

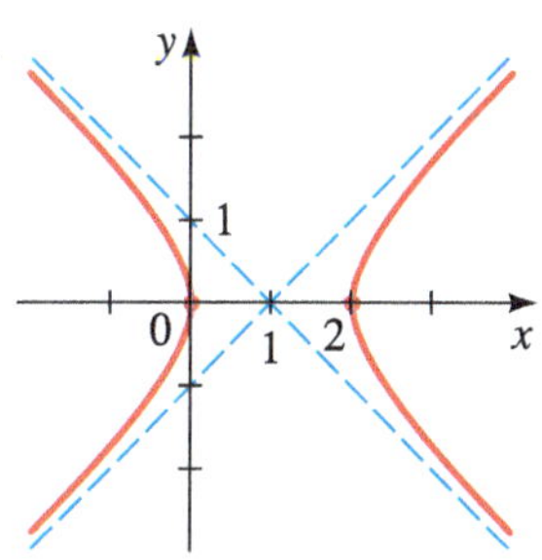

43–54 ■ Identifying and Graphing a Conic Determine whether the equation represents an ellipse, a parabola, a hyperbola, or a degenerate conic. If the graph is an ellipse, find the center, foci, and vertices. If it is a parabola, find the vertex, focus, and directrix. If it is a hyperbola, find the center, foci, vertices, and asymptotes. Then sketch the graph of the equation. If the equation has no graph, explain why.

43. $\dfrac{x^2}{12} + y = 1$

44. $\dfrac{x^2}{12} + \dfrac{y^2}{144} = \dfrac{y}{12}$

45. $x^2 - y^2 + 144 = 0$

46. $x^2 + 6x = 9y^2$

47. $4x^2 + y^2 = 8(x + y)$

48. $3x^2 - 6(x + y) = 10$

49. $x = y^2 - 16y$

50. $2x^2 + 4 = 4x + y^2$

51. $2x^2 - 12x + y^2 + 6y + 26 = 0$

52. $36x^2 - 4y^2 - 36x - 8y = 31$

53. $9x^2 + 8y^2 - 15x + 8y + 27 = 0$

54. $x^2 + 4y^2 = 4x + 8$

55–64 ■ Finding the Equation of a Conic Find an equation for the conic section with the given properties.

55. The parabola with focus $F(0, 1)$ and directrix $y = -1$

56. The parabola with vertex at the origin and focus $F(5, 0)$

57. The ellipse with center at the origin and with x-intercepts ± 2 and y-intercepts ± 5

58. The hyperbola with vertices $V(0, \pm 2)$ and asymptotes $y = \pm \frac{1}{2}x$

59. The ellipse with center $C(0, 4)$, foci $F_1(0, 0)$ and $F_2(0, 8)$, and major axis of length 10

60. The hyperbola with center $C(2, 4)$, foci $F_1(2, 1)$ and $F_2(2, 7)$, and vertices $V_1(2, 6)$ and $V_2(2, 2)$

61. The ellipse with foci $F_1(1, 1)$ and $F_2(1, 3)$ and with one vertex on the x-axis

62. The parabola with vertex $V(5, 5)$ and directrix the y-axis

63. The ellipse with vertices $V_1(7, 12)$ and $V_2(7, -8)$ and passing through the point $P(1, 8)$

64. The parabola with vertex $V(-1, 0)$ and horizontal axis of symmetry and crossing the y-axis at $y = 2$

65. Path of the Earth The path of the earth around the sun is an ellipse with the sun at one focus. The ellipse has major axis of length 186,000,000 mi and eccentricity 0.017. Find the distance between the earth and the sun when the earth is **(a)** closest to the sun and **(b)** farthest from the sun.

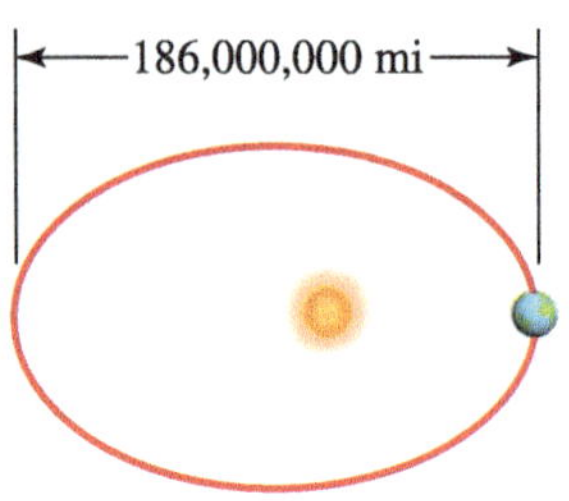

66. LORAN A ship is located 40 mi from a straight shoreline. LORAN stations are located at points A and B on the shoreline, 300 mi apart. From the LORAN signals, the captain determines that the ship is 80 mi closer to A than to B. Find the location of the ship. (Place A and B on the y-axis with the x-axis halfway between them. Find the x- and y-coordinates of the ship.)

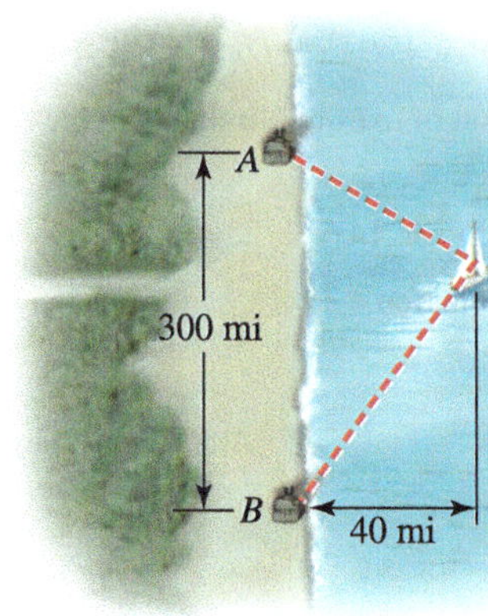

67. Families of Ellipses

(a) Draw graphs of the following family of ellipses for $k = 1, 2, 4$, and 8.

$$\frac{x^2}{16 + k^2} + \frac{y^2}{k^2} = 1$$

(b) Prove that all the ellipses in part (a) have the same foci.

68. Families of Parabolas

(a) Draw graphs of the following family of parabolas for $k = \frac{1}{2}, 1, 2$, and 4.

$$y = kx^2$$

(b) Find the foci of the parabolas in part (a).

(c) How does the location of the focus change as k increases?

69–72 ■ Identifying a Conic An equation of a conic is given. **(a)** Use the discriminant to determine whether the graph of the equation is a parabola, an ellipse, or a hyperbola. **(b)** Use a rotation of axes to eliminate the xy-term. **(c)** Sketch the graph.

69. $x^2 + 4xy + y^2 = 1$

70. $5x^2 - 6xy + 5y^2 - 8\sqrt{2}x + 8\sqrt{2}y - 4 = 0$

71. $7x^2 - 6\sqrt{3}xy + 13y^2 - 4\sqrt{3}x - 4y = 0$

72. $9x^2 + 24xy + 16y^2 = 25$

73–76 ■ Identify a Conic from Its Graph Use a graphing device to graph the conic. Identify the type of conic from the graph.

73. $5x^2 + 3y^2 = 60$

74. $9x^2 - 12y^2 + 36 = 0$

75. $6x + y^2 - 12y = 30$

76. $52x^2 - 72xy + 73y^2 = 100$

77–80 ■ Polar Equations of Conics A polar equation of a conic is given. **(a)** Find the eccentricity, and identify the conic. **(b)** Sketch the conic, and label the vertices.

77. $r = \dfrac{1}{1 - \cos\theta}$

78. $r = \dfrac{2}{3 + 2\sin\theta}$

79. $r = \dfrac{4}{1 + 2\sin\theta}$

80. $r = \dfrac{12}{1 - 4\cos\theta}$

CHAPTER 11 TEST

1. Find the focus and directrix of the parabola $x^2 = -12y$, and sketch its graph.

2. Find the vertices, foci, and the lengths of the major and minor axes for the ellipse $\frac{x^2}{16} + \frac{y^2}{4} = 1$. Then sketch its graph.

3. Find the vertices, foci, and asymptotes of the hyperbola $\frac{y^2}{9} - \frac{x^2}{16} = 1$. Then sketch its graph.

4. Find an equation for the parabola with vertex $(0, 0)$ and focus $(4, 0)$.

5. Find an equation for the ellipse with foci $(\pm 3, 0)$ and vertices $(\pm 4, 0)$.

6. Find an equation for the hyperbola with foci $(0, \pm 5)$ and with asymptotes $y = \pm \frac{3}{4}x$.

7–9 ■ Find an equation for the conic whose graph is shown.

7.

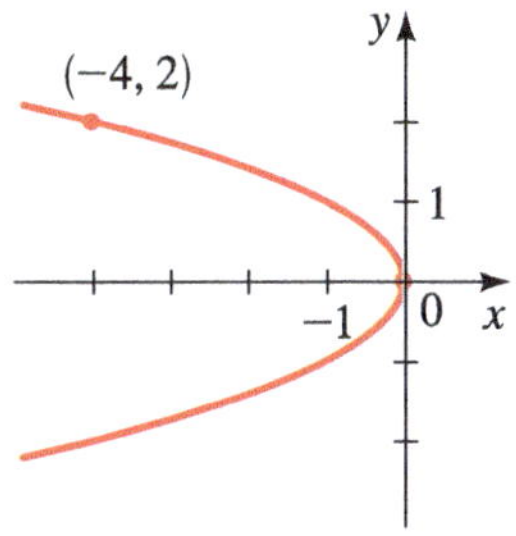

8.

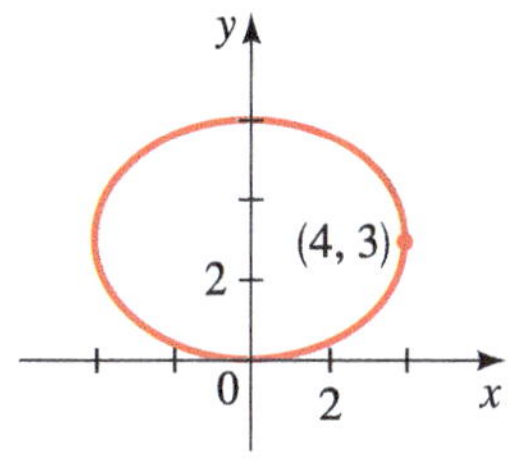

9.

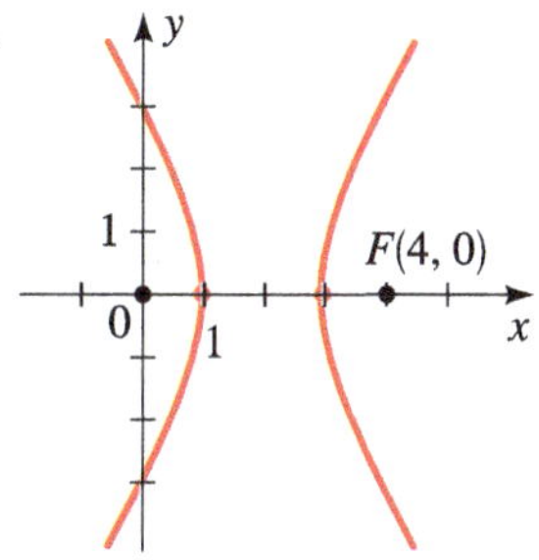

10–12 ■ Determine whether the equation represents an ellipse, a parabola, or a hyperbola. If the graph is an ellipse, find the center, foci, and vertices. If it is a parabola, find the vertex, focus, and directrix. If it is a hyperbola, find the center, foci, vertices, and asymptotes. Then sketch the graph of the equation.

10. $16x^2 + 36y^2 - 96x + 36y + 9 = 0$

11. $9x^2 - 8y^2 + 36x + 64y = 164$

12. $2x + y^2 + 8y + 8 = 0$

13. Find an equation for the ellipse with center $(2, 0)$, foci $(2, \pm 3)$ and major axis of length 8.

14. Find an equation for the parabola with focus $(2, 4)$ and directrix the x-axis.

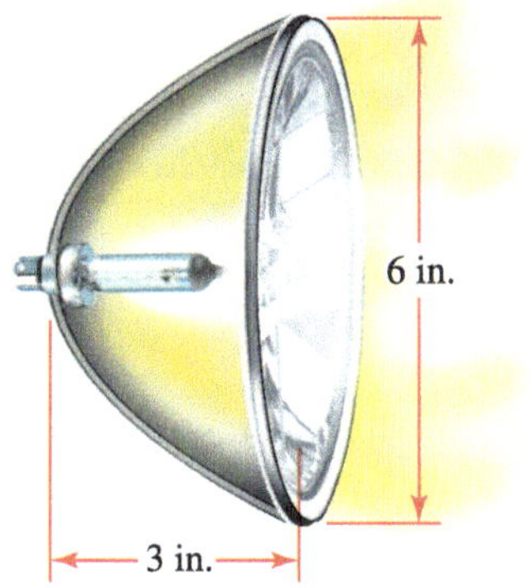

15. A parabolic reflector for a car headlight forms a bowl shape that is 6 in. wide at its opening and 3 in. deep, as shown in the figure at the left. How far from the vertex should the filament of the bulb be placed if it is to be located at the focus?

16. **(a)** Use the discriminant to determine whether the graph of the following equation is a parabola, an ellipse, or a hyperbola:

$$5x^2 + 4xy + 2y^2 = 18$$

(b) Use rotation of axes to eliminate the xy-term in the equation.

(c) Sketch a graph of the equation.

(d) Find the coordinates of the vertices of this conic (in the xy-coordinate system).

17. **(a)** Find the polar equation of the conic that has a focus at the origin, eccentricity $e = \frac{1}{2}$, and directrix $x = 2$. Sketch a graph of the conic.

(b) What type of conic is represented by the following equation? Sketch its graph.

$$r = \frac{3}{2 - \sin\theta}$$

A CUMULATIVE REVIEW TEST FOR CHAPTERS 10 AND 11 CAN BE FOUND AT THE BOOK COMPANION WEBSITE: **www.stewartmath.com**.

Many buildings employ conic sections in their design. Architects have various reasons for using these curves, ranging from structural stability to simple beauty. But how can a huge parabola, ellipse, or hyperbola be accurately constructed in concrete and steel? In this *Focus on Modeling,* we will see how the geometric properties of the conics can be used to construct these shapes.

Conics in Buildings

In ancient times architecture was part of mathematics, so architects had to be mathematicians. Many of the structures they built—pyramids, temples, amphitheaters, and irrigation projects—still stand. In modern times architects employ even more sophisticated mathematical principles. The photographs below show some structures that employ conic sections in their design.

Roman Amphitheater in Alexandria, Egypt (circle)
Nik Wheeler/Encyclopedia/Corbis

Ceiling of Statuary Hall in the U.S. Capitol (ellipse)
Architect of the Capitol

Roof of the Skydome in Toronto, Canada (parabola)
Walter Schmid/The Image Bank/Getty Images

Roof of Washington Dulles Airport (hyperbola and parabola)
Andrew Holt/Photographer's Choice/Getty Images

McDonnell Planetarium, St. Louis, MO (hyperbola)
Joe Sohm/VisionsofAmerica/Photodisc/Getty Images

Attic in La Pedrera, Barcelona, Spain (parabola)
O. Alamany & E. Vicens/Terra/Corbis

Architects have different reasons for using conics in their designs. For example, the Spanish architect Antoni Gaudí used parabolas in the attic of La Pedrera (see photo above). He reasoned that since a rope suspended between two points with an equally distributed load (as in a suspension bridge) has the shape of a parabola, an inverted parabola would provide the best support for a flat roof.

Constructing Conics

The equations of the conics are helpful in manufacturing small objects, because a computer-controlled cutting tool can accurately trace a curve given by an equation. But in a building project, how can we construct a portion of a parabola, ellipse, or hyperbola that spans the ceiling or walls of a building? The geometric properties of the conics provide practical ways of constructing them. For example, if you were building a circular tower, you would choose a center point, then make sure that the walls of the tower

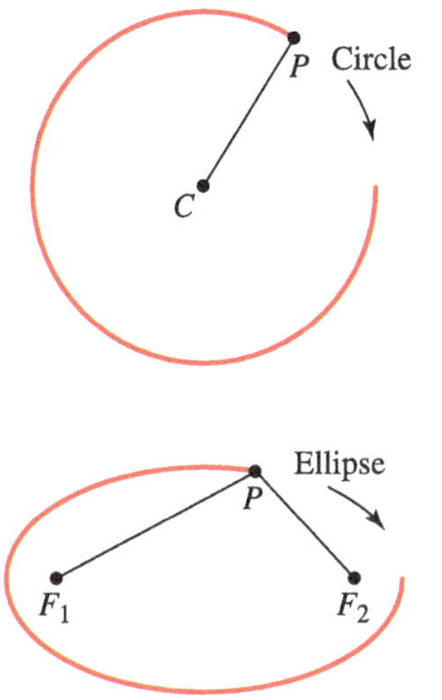

FIGURE 1 Constructing a circle and an ellipse

were a fixed distance from that point. Elliptical walls can be constructed by using a string anchored at two points, as shown in Figure 1.

To construct a parabola, we can use the apparatus shown in Figure 2. A piece of string of length a is anchored at F and A. The T-square, also of length a, slides along the straight bar L. A pencil at P holds the string taut against the T-square. As the T-square slides to the right, the pencil traces out a curve.

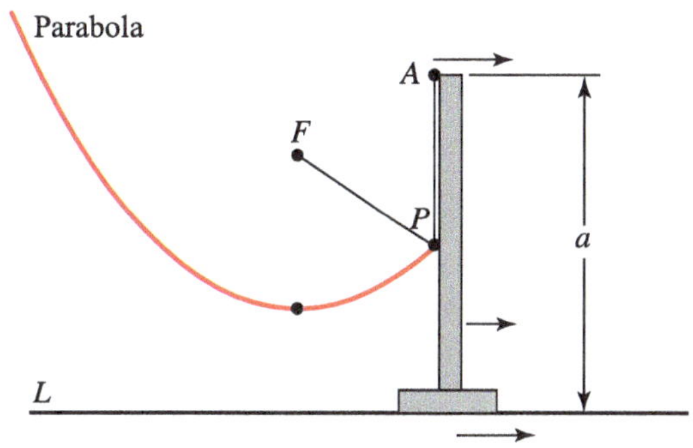

FIGURE 2 Constructing a parabola

From the figure we see that

$$d(F, P) + d(P, A) = a \qquad \text{The string is of length } a$$

$$d(L, P) + d(P, A) = a \qquad \text{The T-square is of length } a$$

It follows that $d(F, P) + d(P, A) = d(L, P) + d(P, A)$. Subtracting $d(P, A)$ from each side, we get

$$d(F, P) = d(L, P)$$

The last equation says that the distance from F to P is equal to the distance from P to the line L. Thus the curve is a parabola with focus F and directrix L.

In building projects, it is easier to construct a straight line than a curve. So in some buildings, such as in the Kobe Tower (see Problem 4), a curved surface is produced by using many straight lines. We can also produce a curve using straight lines, such as the parabola shown in Figure 3.

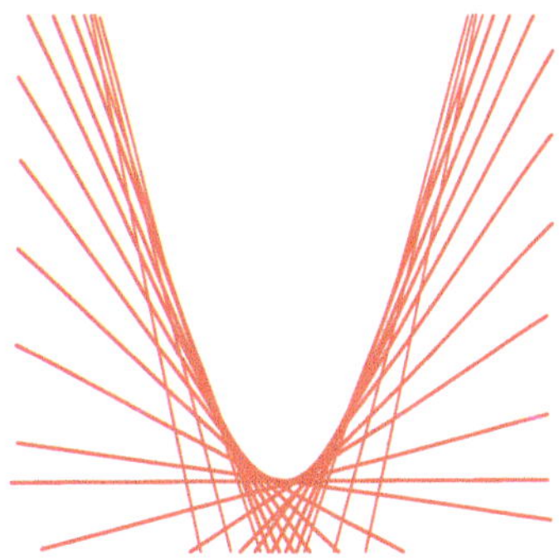

FIGURE 3 Tangent lines to a parabola

Each line is **tangent** to the parabola; that is, the line meets the parabola at exactly one point and does not cross the parabola. The line tangent to the parabola $y = x^2$ at the point (a, a^2) is

$$y = 2ax - a^2$$

You are asked to show this in Problem 6. The parabola is called the **envelope** of all such lines.

PROBLEMS

1. **Conics in Architecture** The photographs on page 836 show six examples of buildings that contain conic sections. Search the Internet to find other examples of structures that employ parabolas, ellipses, or hyperbolas in their design. Find at least one example for each type of conic.

2. **Constructing a Hyperbola** In this problem we construct a hyperbola. The wooden bar in the figure can pivot at F_1. A string that is shorter than the bar is anchored at F_2 and at A, the other end of the bar. A pencil at P holds the string taut against the bar as it moves counterclockwise around F_1.

   (a) Show that the curve traced out by the pencil is one branch of a hyperbola with foci at F_1 and F_2.

   (b) How should the apparatus be reconfigured to draw the other branch of the hyperbola?

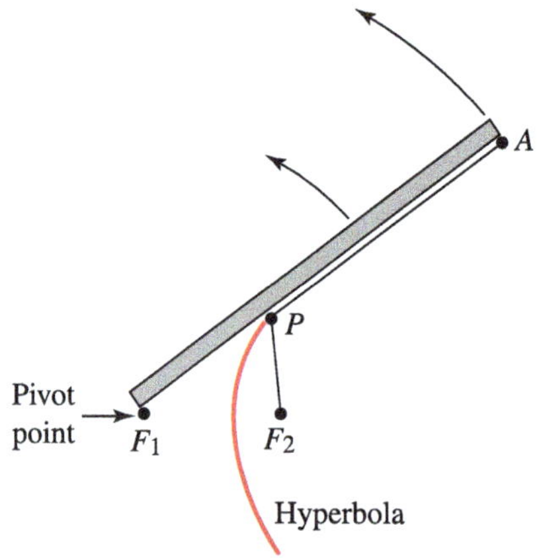

3. **A Parabola in a Rectangle** The following method can be used to construct a parabola that fits in a given rectangle. The parabola will be approximated by many short line segments.

   First, draw a rectangle. Divide the rectangle in half by a vertical line segment, and label the top endpoint V. Next, divide the length and width of each half rectangle into an equal number of parts to form grid lines, as shown in the figure below. Draw lines from V to the endpoints of horizontal grid line 1, and mark the points where these lines cross the vertical grid lines labeled 1. Next, draw lines from V to the endpoints of horizontal grid line 2, and mark the points where these lines cross the vertical grid lines labeled 2. Continue in this way until you have used all the horizontal grid lines. Now use line segments to connect the points you have marked to obtain an approximation to the desired parabola. Apply this procedure to draw a parabola that fits into a 6 ft by 10 ft rectangle on a lawn.

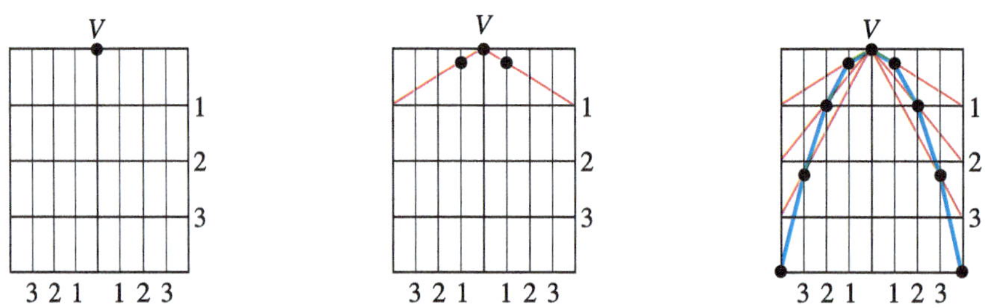

4. **Hyperbolas from Straight Lines** In this problem we construct hyperbolic shapes using straight lines. Punch equally spaced holes into the edges of two large plastic lids. Connect corresponding holes with strings of equal lengths as shown in the figure on the next page. Holding the strings taut, twist one lid against the other. An imaginary surface passing through the strings has hyperbolic cross sections. (An architectural example of this is the

Kobe Tower in Japan, shown in the photograph.) What happens to the vertices of the hyperbolic cross sections as the lids are twisted more?

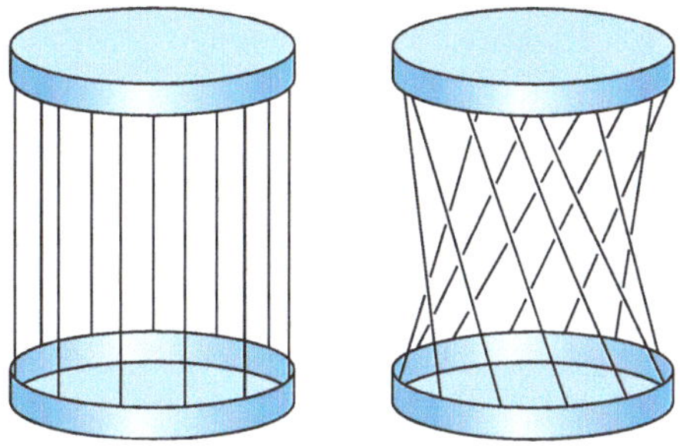

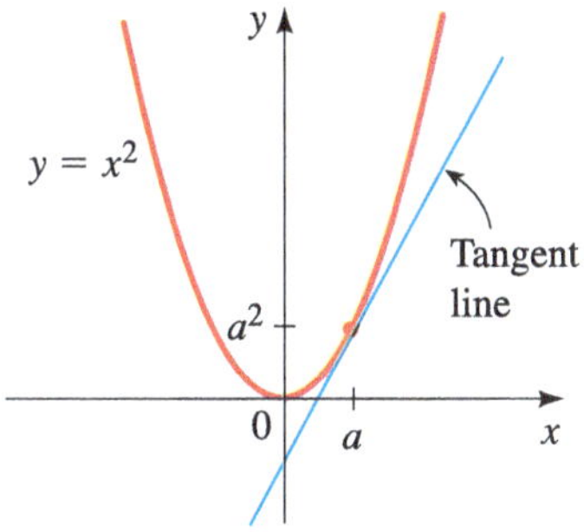

5. **Tangent Lines to a Parabola** In this problem we show that the line tangent to the parabola $y = x^2$ at the point (a, a^2) has the equation $y = 2ax - a^2$.

   (a) Let m be the slope of the tangent line at (a, a^2). Show that the equation of the tangent line is $y - a^2 = m(x - a)$.

   (b) Use the fact that the tangent line intersects the parabola at only one point to show that (a, a^2) is the only solution of the system.

$$\begin{cases} y - a^2 = m(x - a) \\ y = x^2 \end{cases}$$

   (c) Eliminate y from the system in part (b) to get a quadratic equation in x. Show that the discriminant of this quadratic is $(m - 2a)^2$. Since the system in part (b) has exactly one solution, the discriminant must equal 0. Find m.

   (d) Substitute the value for m you found in part (c) into the equation in part (a), and simplify to get the equation of the tangent line.

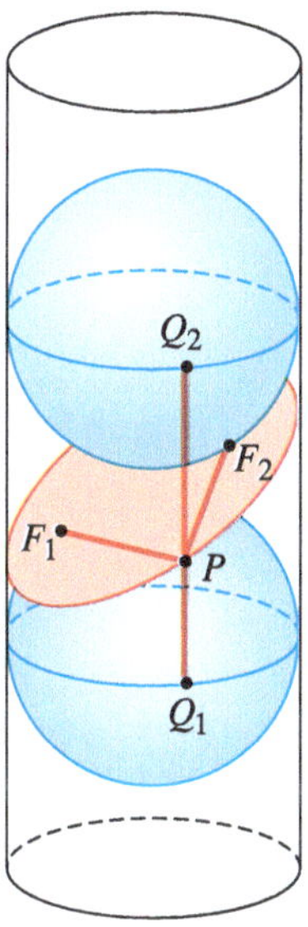

6. **A Cut Cylinder** In this problem we prove that when a cylinder is cut by a plane, an ellipse is formed. An architectural example of this is the Tycho Brahe Planetarium in Copenhagen (see the photograph). In the figure, a cylinder is cut by a plane, resulting in the red curve. Two spheres with the same radius as the cylinder slide inside the cylinder so that they just touch the plane at F_1 and F_2. Choose an arbitrary point P on the curve, and let Q_1 and Q_2 be the two points on the cylinder where a vertical line through P touches the "equator" of each sphere.

   (a) Show that $PF_1 = PQ_1$ and $PF_2 = PQ_2$. [*Hint:* Use the fact that all tangents to a sphere from a given point outside the sphere are of the same length.]

   (b) Explain why $PQ_1 + PQ_2$ is the same for all points P on the curve.

   (c) Show that $PF_1 + PF_2$ is the same for all points P on the curve.

   (d) Conclude that the curve is an ellipse with foci F_1 and F_2.

© Mircea BEZERGHEANU/Shutterstock.com

12 Sequences and Series

Throughout this book we have used functions to model real-world situations. The functions we've used have always had real numbers as inputs. But many real-world situations occur in stages: stage 1, 2, 3, ... To model such situations, we need functions whose inputs are the natural numbers 1, 2, 3, . . . (representing the stages). For example, the peaks of a bouncing ball are represented by the natural numbers 1, 2, 3, . . . (representing peak 1, 2, 3, . . .). A function f that models the height of the ball at each peak has natural numbers 1, 2, 3, . . . as inputs and gives the heights as $f(1)$, $f(2)$, $f(3)$, ... In general a function whose inputs are the natural numbers is called a *sequence*. We can think of a sequence as simply a list of numbers written in a specific order.

The amount in a bank account at the end of each month, mortgage payments, and the amount of an annuity are sequences. The formulas that generate these sequences drive our economy—they allow us to borrow money to buy our dream home closer to graduation than to retirement.

Many patterns in nature can be modeled by sequences. For example, the Fibonacci sequence describes such varied natural patterns as the growth of a rabbit population, the arrangements of leaves on a plant, the arrangement of scales on a pineapple, and the intricate pattern in a nautilus (pictured above).

12.1 SEQUENCES AND SUMMATION NOTATION

■ Sequences ■ Recursively Defined Sequences ■ The Partial Sums of a Sequence
■ Sigma Notation

Roughly speaking, a sequence is an infinite list of numbers. The numbers in the sequence are often written as $a_1, a_2, a_3, \ldots$. The dots mean that the list continues forever. A simple example is the sequence

$$\begin{array}{ccccc} 5, & 10, & 15, & 20, & 25, \ldots \\ \uparrow & \uparrow & \uparrow & \uparrow & \uparrow \\ a_1 & a_2 & a_3 & a_4 & a_5 \ldots \end{array}$$

We can describe the pattern of the sequence displayed above by the following *formula*:

$$a_n = 5n$$

You may have already thought of a different way to describe the pattern—namely, "you go from one number to the next by adding 5." This natural way of describing the sequence is expressed by the *recursive formula*:

$$a_n = a_{n-1} + 5$$

starting with $a_1 = 5$. Try substituting $n = 1, 2, 3, \ldots$ in each of these formulas to see how they produce the numbers in the sequence. In this section we see how these different ways are used to describe specific sequences.

■ Sequences

Any ordered list of numbers can be viewed as a function whose input values are 1, 2, 3, . . . and whose output values are the numbers in the list. So we define a sequence as follows.

> **DEFINITION OF A SEQUENCE**
>
> A **sequence** is a function a whose domain is the set of natural numbers. The **terms of the sequence** are the function values
>
> $$a(1), a(2), a(3), \ldots, a(n), \ldots$$
>
> We usually write a_n instead of the function notation $a(n)$. So the terms of the sequence are written as
>
> $$a_1, a_2, a_3, \ldots, a_n, \ldots$$
>
> The number a_1 is called the **first term**, a_2 is called the **second term**, and in general, a_n is called the ***n*th term**.

Here is a simple example of a sequence:

$$2, 4, 6, 8, 10, \ldots$$

We can write a sequence in this way when it's clear what the subsequent terms of the sequence are. This sequence consists of even numbers. To be more accurate, however, we need to specify a procedure for finding *all* the terms of the sequence. This can be done by giving a formula for the nth term a_n of the sequence. In this case,

$$a_n = 2n$$

Another way to write this sequence is to use function notation:

$$a(n) = 2n$$

so $a(1) = 2, a(2) = 4, a(3) = 6, \ldots$

and the sequence can be written as

$$2, \quad 4, \quad 6, \quad 8, \quad \ldots, \quad 2n, \quad \ldots$$

1st term, 2nd term, 3rd term, 4th term, *n*th term

Notice how the formula $a_n = 2n$ gives all the terms of the sequence. For instance, substituting 1, 2, 3, and 4 for n gives the first four terms:

$$a_1 = 2 \cdot 1 = 2 \qquad a_2 = 2 \cdot 2 = 4$$

$$a_3 = 2 \cdot 3 = 6 \qquad a_4 = 2 \cdot 4 = 8$$

To find the 103rd term of this sequence, we use $n = 103$ to get

$$a_{103} = 2 \cdot 103 = 206$$

EXAMPLE 1 ■ Finding the Terms of a Sequence

Find the first five terms and the 100th term of the sequence defined by each formula.

(a) $a_n = 2n - 1$ **(b)** $c_n = n^2 - 1$

(c) $t_n = \dfrac{n}{n+1}$ **(d)** $r_n = \dfrac{(-1)^n}{2^n}$

SOLUTION To find the first five terms, we substitute $n = 1, 2, 3, 4$, and 5 in the formula for the nth term. To find the 100th term, we substitute $n = 100$. This gives the following.

*n*th term	First five terms	100th term
(a) $2n - 1$	1, 3, 5, 7, 9	199
(b) $n^2 - 1$	0, 3, 8, 15, 24	9999
(c) $\dfrac{n}{n+1}$	$\dfrac{1}{2}, \dfrac{2}{3}, \dfrac{3}{4}, \dfrac{4}{5}, \dfrac{5}{6}$	$\dfrac{100}{101}$
(d) $\dfrac{(-1)^n}{2^n}$	$-\dfrac{1}{2}, \dfrac{1}{4}, -\dfrac{1}{8}, \dfrac{1}{16}, -\dfrac{1}{32}$	$\dfrac{1}{2^{100}}$

Now Try Exercises 3, 5, 7, and 9 ■

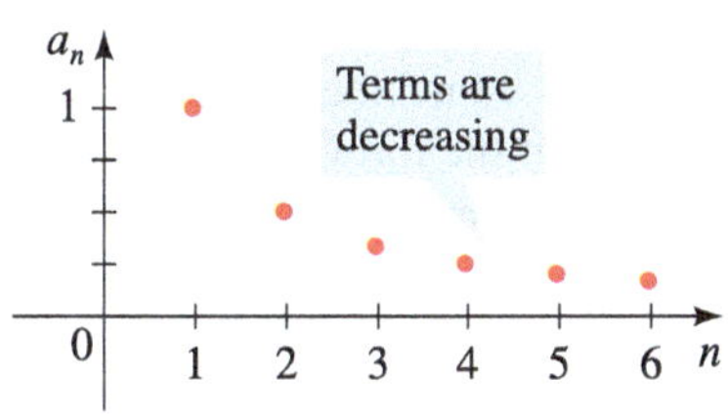

FIGURE 1

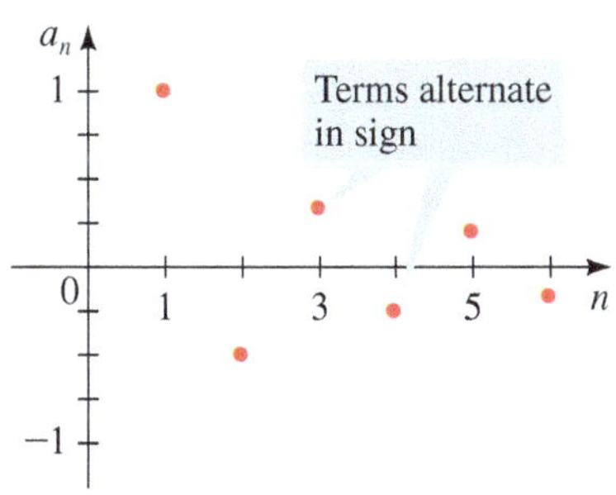

FIGURE 2

In Example 1(d) the presence of $(-1)^n$ in the sequence has the effect of making successive terms alternately negative and positive.

It is often useful to picture a sequence by sketching its graph. Since a sequence is a function whose domain is the natural numbers, we can draw its graph in the Cartesian plane. For instance, the graph of the sequence

$$1, \frac{1}{2}, \frac{1}{3}, \frac{1}{4}, \frac{1}{5}, \frac{1}{6}, \ldots, \frac{1}{n}, \ldots$$

is shown in Figure 1.

Compare the graph of the sequence shown in Figure 1 to the graph of

$$1, -\frac{1}{2}, \frac{1}{3}, -\frac{1}{4}, \frac{1}{5}, -\frac{1}{6}, \ldots, \frac{(-1)^{n+1}}{n}, \ldots$$

shown in Figure 2. The graph of every sequence consists of isolated points that are *not* connected.

See Appendix D, *Using the TI-83/84 Graphing Calculator*, for additional instructions on working with sequences. Go to **www.stewartmath.com**.

Graphing calculators are useful in analyzing sequences. To work with sequences on a TI-83, we put the calculator in `Seq` mode ("sequence" mode) as in Figure 3(a). If we enter the sequence $u(n) = n/(n + 1)$ of Example 1(c), we can display the terms using the `TABLE` command as shown in Figure 3(b). We can also graph the sequence as shown in Figure 3(c).

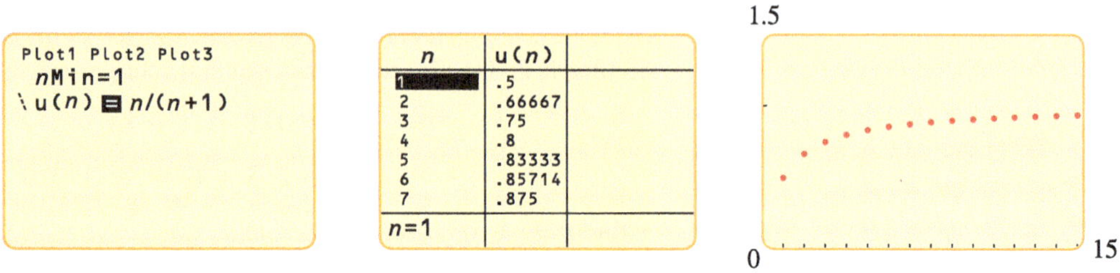

(a) (b) (c)

FIGURE 3
$u(n) = n/(n + 1)$

Not all sequences can be defined by a formula. For example, there is no known formula for the sequence of prime numbers:*

2, 3, 5, 7, 11, 13, 17, 19, 23, . . .

Finding patterns is an important part of mathematics. Consider a sequence that begins

$$1, 4, 9, 16, \ldots$$

Can you detect a pattern in these numbers? In other words, can you define a sequence whose first four terms are these numbers? The answer to this question seems easy; these numbers are the squares of the numbers 1, 2, 3, 4. Thus the sequence we are looking for is defined by $a_n = n^2$. However, this is not the *only* sequence whose first four terms are 1, 4, 9, 16. In other words, the answer to our problem is not unique (see Exercise 86). In the next example we are interested in finding an *obvious* sequence whose first few terms agree with the given ones.

ERATOSTHENES (circa 276–195 B.C.) was a renowned Greek geographer, mathematician, and astronomer. He accurately calculated the circumference of the earth by an ingenious method. He is most famous, however, for his method for finding primes, now called the ***sieve of Eratosthenes***. The method consists of listing the integers, beginning with 2 (the first prime), and then crossing out all the multiples of 2, which are not prime. The next number remaining on the list is 3 (the second prime), so we again cross out all multiples of it. The next remaining number is 5 (the third prime number), and we cross out all multiples of it, and so on. In this way all numbers that are not prime are crossed out, and the remaining numbers are the primes.

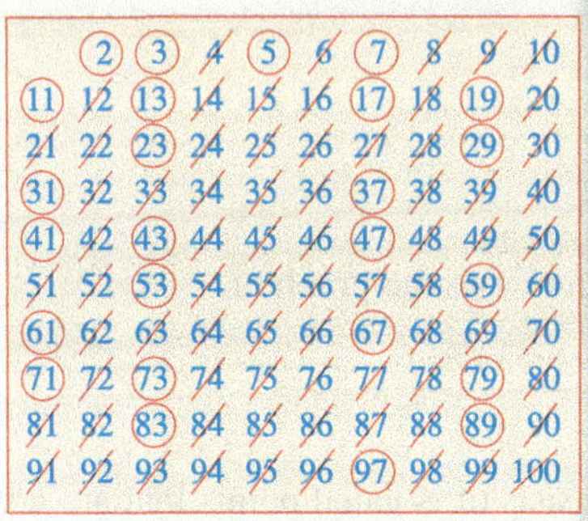

EXAMPLE 2 ■ Finding the *n*th Term of a Sequence

Find the nth term of a sequence whose first several terms are given.

(a) $\frac{1}{2}, \frac{3}{4}, \frac{5}{6}, \frac{7}{8}, \ldots$ **(b)** $-2, 4, -8, 16, -32, \ldots$

SOLUTION

(a) We notice that the numerators of these fractions are the odd numbers and the denominators are the even numbers. Even numbers are of the form $2n$, and odd numbers are of the form $2n - 1$ (an odd number differs from an even number by 1). So a sequence that has these numbers for its first four terms is given by

$$a_n = \frac{2n - 1}{2n}$$

(b) These numbers are powers of 2, and they alternate in sign, so a sequence that agrees with these terms is given by

$$a_n = (-1)^n 2^n$$

You should check that these formulas do indeed generate the given terms.

Now Try Exercises 29 and 35 ■

---

* A *prime number* is a whole number p whose only divisors are p and 1. (By convention the number 1 is not considered prime.)

Large Prime Numbers
The search for large primes fascinates many people. As of this writing, the largest known prime number is

$$2^{57,885,161} - 1$$

It was discovered by Dr. Curtis Cooper of the University of Central Missouri in January 2013. In decimal notation this number contains 17,425,170 digits. If it were written in full, it would occupy more than four times as many pages as this book contains. Cooper was working with a large Internet group known as GIMPS (the Great Internet Mersenne Prime Search). Numbers of the form $2^p - 1$, where p is prime, are called Mersenne numbers and are named for the French monk who first studied them in the 1600s. Such numbers are more easily checked for primality than others. That is why the largest known primes are of this form.

Recursively Defined Sequences

Some sequences do not have simple defining formulas like those of the preceding example. The nth term of a sequence may depend on some or all of the terms preceding it. A sequence defined in this way is called **recursive**. Here are two examples.

EXAMPLE 3 ■ Finding the Terms of a Recursively Defined Sequence

A sequence is defined recursively by $a_1 = 1$ and

$$a_n = 3(a_{n-1} + 2)$$

(a) Find the first five terms of the sequence.

(b) Use a graphing calculator to find the 20th term of the sequence.

SOLUTION

(a) The defining formula for this sequence is recursive. It allows us to find the nth term a_n if we know the preceding term a_{n-1}. Thus we can find the second term from the first term, the third term from the second term, the fourth term from the third term, and so on. Since we are given the first term $a_1 = 1$, we can proceed as follows.

$$a_2 = 3(a_1 + 2) = 3(1 + 2) = 9$$
$$a_3 = 3(a_2 + 2) = 3(9 + 2) = 33$$
$$a_4 = 3(a_3 + 2) = 3(33 + 2) = 105$$
$$a_5 = 3(a_4 + 2) = 3(105 + 2) = 321$$

Thus the first five terms of this sequence are

$$1, 9, 33, 105, 321, \ldots$$

(b) Note that to find the 20th term of the recursive sequence, we must first find all 19 preceding terms. This is most easily done by using a graphing calculator. Figure 4(a) shows how to enter this sequence on the TI-83 calculator. From Figure 4(b) we see that the 20th term of the sequence is

$$a_{20} = 4{,}649{,}045{,}865$$

See Appendix D, *Using the TI-83/84 Graphing Calculator*, for additional instructions on working with sequences. Go to **www.stewartmath.com**.

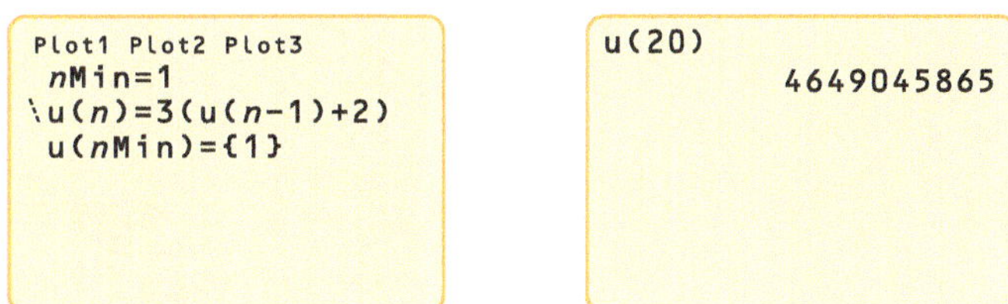

(a) (b)

FIGURE 4
$u(n) = 3(u(n - 1) + 2), u(1) = 1$

Now Try Exercises 15 and 25 ■

EXAMPLE 4 ■ The Fibonacci Sequence

Find the first 11 terms of the sequence defined recursively by $F_1 = 1$, $F_2 = 1$ and

$$F_n = F_{n-1} + F_{n-2}$$

FIBONACCI (1175–1250) was born in Pisa, Italy, and was educated in North Africa. He traveled widely in the Mediterranean area and learned the various methods then in use for writing numbers. On returning to Pisa in 1202, Fibonacci advocated the use of the Hindu-Arabic decimal system, the one we use today, over the Roman numeral system that was used in Europe in his time. His most famous book, *Liber Abaci*, expounds on the advantages of the Hindu-Arabic numerals. In fact, multiplication and division were so complicated using Roman numerals that a college degree was necessary to master these skills. Interestingly, in 1299 the city of Florence outlawed the use of the decimal system for merchants and businesses, requiring numbers to be written in Roman numerals or words. One can only speculate about the reasons for this law.

SOLUTION To find F_n, we need to find the two preceding terms, F_{n-1} and F_{n-2}. Since we are given F_1 and F_2, we proceed as follows.

$$F_3 = F_2 + F_1 = 1 + 1 = 2$$
$$F_4 = F_3 + F_2 = 2 + 1 = 3$$
$$F_5 = F_4 + F_3 = 3 + 2 = 5$$

It's clear what is happening here. Each term is simply the sum of the two terms that precede it, so we can easily write down as many terms as we please. Here are the first 11 terms. (You can also find these using a graphing calculator.)

$$1, 1, 2, 3, 5, 8, 13, 21, 34, 55, 89, \ldots$$

Now Try Exercise 19 ■

The sequence in Example 4 is called the **Fibonacci sequence**, named after the 13th century Italian mathematician who used it to solve a problem about the breeding of rabbits (see Exercise 85). The sequence also occurs in numerous other applications in nature. (See Figures 5 and 6.) In fact, so many phenomena behave like the Fibonacci sequence that one mathematical journal, the *Fibonacci Quarterly*, is devoted entirely to its properties.

FIGURE 5 The Fibonacci sequence in the branching of a tree

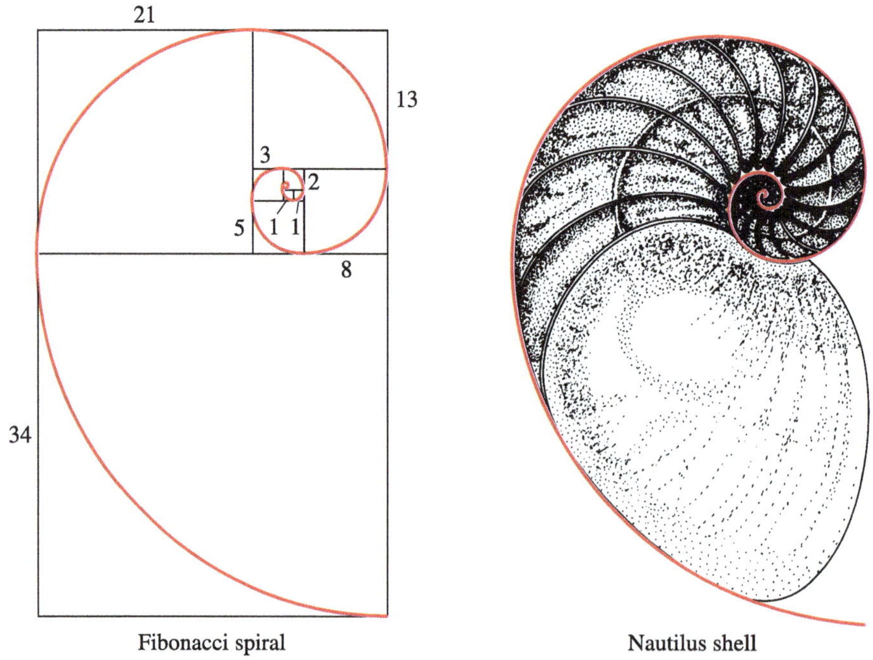

FIGURE 6

The Partial Sums of a Sequence

In calculus we are often interested in adding the terms of a sequence. This leads to the following definition.

THE PARTIAL SUMS OF A SEQUENCE

For the sequence

$$a_1, a_2, a_3, a_4, \ldots, a_n, \ldots$$

the **partial sums** are

$$\begin{aligned} S_1 &= a_1 \\ S_2 &= a_1 + a_2 \\ S_3 &= a_1 + a_2 + a_3 \\ S_4 &= a_1 + a_2 + a_3 + a_4 \\ &\vdots \\ S_n &= a_1 + a_2 + a_3 + \cdots + a_n \\ &\vdots \end{aligned}$$

S_1 is called the **first partial sum**, S_2 is the **second partial sum**, and so on. S_n is called the ***n*th partial sum**. The sequence $S_1, S_2, S_3, \ldots, S_n, \ldots$ is called the **sequence of partial sums**.

EXAMPLE 5 ■ Finding the Partial Sums of a Sequence

Find the first four partial sums and the nth partial sum of the sequence given by $a_n = 1/2^n$.

SOLUTION The terms of the sequence are

$$\frac{1}{2}, \frac{1}{4}, \frac{1}{8}, \ldots$$

The first four partial sums are

$$\begin{aligned} S_1 &= \frac{1}{2} && = \frac{1}{2} \\ S_2 &= \frac{1}{2} + \frac{1}{4} && = \frac{3}{4} \\ S_3 &= \frac{1}{2} + \frac{1}{4} + \frac{1}{8} && = \frac{7}{8} \\ S_4 &= \frac{1}{2} + \frac{1}{4} + \frac{1}{8} + \frac{1}{16} && = \frac{15}{16} \end{aligned}$$

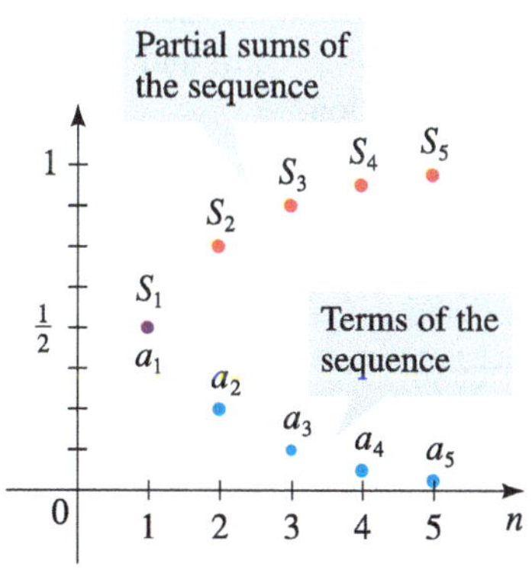

FIGURE 7 Graph of the sequence a_n and the sequence of partial sums S_n

Notice that in the value of each partial sum, the denominator is a power of 2 and the numerator is one less than the denominator. In general, the nth partial sum is

$$S_n = \frac{2^n - 1}{2^n} = 1 - \frac{1}{2^n}$$

The first five terms of a_n and S_n are graphed in Figure 7.

Now Try Exercise 43 ■

EXAMPLE 6 ■ Finding the Partial Sums of a Sequence

Find the first four partial sums and the nth partial sum of the sequence given by

$$a_n = \frac{1}{n} - \frac{1}{n+1}$$

SOLUTION The first four partial sums are

$$S_1 = \left(1 - \frac{1}{2}\right) = 1 - \frac{1}{2}$$

$$S_2 = \left(1 - \frac{1}{2}\right) + \left(\frac{1}{2} - \frac{1}{3}\right) = 1 - \frac{1}{3}$$

$$S_3 = \left(1 - \frac{1}{2}\right) + \left(\frac{1}{2} - \frac{1}{3}\right) + \left(\frac{1}{3} - \frac{1}{4}\right) = 1 - \frac{1}{4}$$

$$S_4 = \left(1 - \frac{1}{2}\right) + \left(\frac{1}{2} - \frac{1}{3}\right) + \left(\frac{1}{3} - \frac{1}{4}\right) + \left(\frac{1}{4} - \frac{1}{5}\right) = 1 - \frac{1}{5}$$

Do you detect a pattern here? Of course. The nth partial sum is

$$S_n = 1 - \frac{1}{n+1}$$

Now Try Exercise 45 ■

■ Sigma Notation

Given a sequence

$$a_1, a_2, a_3, a_4, \ldots$$

we can write the sum of the first n terms using **summation notation**, or **sigma notation**. This notation derives its name from the Greek letter Σ (capital sigma, corresponding to our S for "sum"). Sigma notation is used as follows:

This tells us to end with $k = n$

This tells us to add $\sum_{k=1}^{n} a_k$

This tells us to start with $k = 1$

$$\sum_{k=1}^{n} a_k = a_1 + a_2 + a_3 + a_4 + \cdots + a_n$$

The left side of this expression is read, "The sum of a_k from $k = 1$ to $k = n$." The letter k is called the **index of summation**, or the **summation variable**, and the idea is to replace k in the expression after the sigma by the integers 1, 2, 3, . . . , n, and add the resulting expressions, arriving at the right-hand side of the equation.

EXAMPLE 7 ■ Sigma Notation

Find each sum.

(a) $\sum_{k=1}^{5} k^2$ **(b)** $\sum_{j=3}^{5} \frac{1}{j}$ **(c)** $\sum_{k=5}^{10} k$ **(d)** $\sum_{i=1}^{6} 2$

SOLUTION

(a) $\sum_{k=1}^{5} k^2 = 1^2 + 2^2 + 3^2 + 4^2 + 5^2 = 55$

(b) $\sum_{j=3}^{5} \frac{1}{j} = \frac{1}{3} + \frac{1}{4} + \frac{1}{5} = \frac{47}{60}$

(c) $\displaystyle\sum_{k=5}^{10} k = 5 + 6 + 7 + 8 + 9 + 10 = 45$

(d) $\displaystyle\sum_{i=1}^{6} 2 = 2 + 2 + 2 + 2 + 2 + 2 = 12$

Now Try Exercises 47 and 49 ■

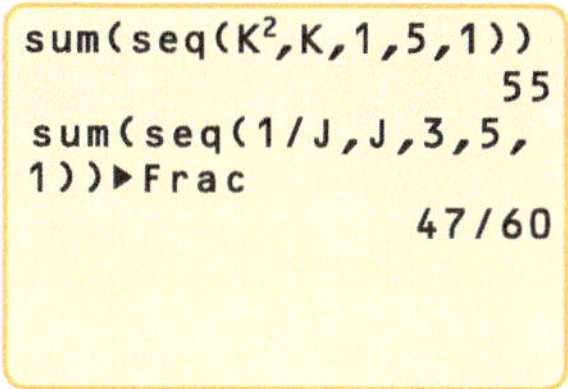

FIGURE 8

We can use a graphing calculator to evaluate sums. For instance, Figure 8 shows how the TI-83 can be used to evaluate the sums in parts (a) and (b) of Example 7.

EXAMPLE 8 ■ Writing Sums in Sigma Notation

Write each sum using sigma notation.

(a) $1^3 + 2^3 + 3^3 + 4^3 + 5^3 + 6^3 + 7^3$

(b) $\sqrt{3} + \sqrt{4} + \sqrt{5} + \cdots + \sqrt{77}$

SOLUTION

(a) We can write

$$1^3 + 2^3 + 3^3 + 4^3 + 5^3 + 6^3 + 7^3 = \sum_{k=1}^{7} k^3$$

(b) A natural way to write this sum is

$$\sqrt{3} + \sqrt{4} + \sqrt{5} + \cdots + \sqrt{77} = \sum_{k=3}^{77} \sqrt{k}$$

However, there is no unique way of writing a sum in sigma notation. We could also write this sum as

$$\sqrt{3} + \sqrt{4} + \sqrt{5} + \cdots + \sqrt{77} = \sum_{k=0}^{74} \sqrt{k+3}$$

or

$$\sqrt{3} + \sqrt{4} + \sqrt{5} + \cdots + \sqrt{77} = \sum_{k=1}^{75} \sqrt{k+2}$$

Now Try Exercises 67 and 69 ■

The Golden Ratio

The ancient Greeks considered a line segment to be divided into the **golden ratio** if the ratio of the shorter part to the longer part is the same as the ratio of the longer part to the whole segment.

Thus the segment shown is divided into the golden ratio if

$$\frac{1}{x} = \frac{x}{1+x}$$

This leads to a quadratic equation whose positive solution is

$$x = \frac{1+\sqrt{5}}{2} \approx 1.618$$

This ratio occurs naturally in many places. For instance, psychology experiments show that the most pleasing shape of rectangle is one whose sides are in golden ratio. The ancient Greeks agreed with this and built their temples in this ratio.

The golden ratio is related to the Fibonacci sequence. In fact, it can be shown by using calculus* that the ratio of two successive Fibonacci numbers

$$\frac{F_{n+1}}{F_n}$$

gets closer to the golden ratio the larger the value of n. Try finding this ratio for $n = 10$.

Clark Dunbar/Flame/Corbis

*See *Principles of Problem Solving* 13 at the book companion website: **www.stewartmath.com**.

The following properties of sums are natural consequences of properties of the real numbers.

PROPERTIES OF SUMS

Let $a_1, a_2, a_3, a_4, \ldots$ and $b_1, b_2, b_3, b_4, \ldots$ be sequences. Then for every positive integer n and any real number c the following properties hold.

1. $\sum_{k=1}^{n} (a_k + b_k) = \sum_{k=1}^{n} a_k + \sum_{k=1}^{n} b_k$
2. $\sum_{k=1}^{n} (a_k - b_k) = \sum_{k=1}^{n} a_k - \sum_{k=1}^{n} b_k$
3. $\sum_{k=1}^{n} ca_k = c\left(\sum_{k=1}^{n} a_k\right)$

Proof To prove Property 1, we write out the left side of the equation to get

$$\sum_{k=1}^{n} (a_k + b_k) = (a_1 + b_1) + (a_2 + b_2) + (a_3 + b_3) + \cdots + (a_n + b_n)$$

Because addition is commutative and associative, we can rearrange the terms on the right-hand side to read

$$\sum_{k=1}^{n} (a_k + b_k) = (a_1 + a_2 + a_3 + \cdots + a_n) + (b_1 + b_2 + b_3 + \cdots + b_n)$$

Rewriting the right side using sigma notation gives Property 1. Property 2 is proved in a similar manner. To prove Property 3, we use the Distributive Property:

$$\begin{aligned}\sum_{k=1}^{n} ca_k &= ca_1 + ca_2 + ca_3 + \cdots + ca_n \\ &= c(a_1 + a_2 + a_3 + \cdots + a_n) = c\left(\sum_{k=1}^{n} a_k\right)\end{aligned}$$

■

12.1 EXERCISES

CONCEPTS

1. A sequence is a function whose domain is ________.

2. The nth partial sum of a sequence is the sum of the first ________ terms of the sequence. So for the sequence $a_n = n^2$ the fourth partial sum is S_4 = ____ + ____ + ____ + ____ = ____.

SKILLS

3–14 ■ Terms of a Sequence Find the first four terms and the 100th term of the sequence whose nth term is given.

3. $a_n = n - 3$
4. $a_n = 2n - 1$
5. $a_n = \dfrac{1}{2n + 1}$
6. $a_n = n^2 - 1$
7. $a_n = 5^n$
8. $a_n = \left(\dfrac{-1}{3}\right)^n$
9. $a_n = \dfrac{(-1)^n}{n^2}$
10. $a_n = \dfrac{1}{n^2}$
11. $a_n = 1 + (-1)^n$
12. $a_n = (-1)^{n+1}\dfrac{n}{n + 1}$
13. $a_n = n^n$
14. $a_n = 3$

15–20 ■ Recursive Sequences A sequence is defined recursively by the given formulas. Find the first five terms of the sequence.

15. $a_n = 2(a_{n-1} + 3)$ and $a_1 = 4$
16. $a_n = \dfrac{a_{n-1}}{6}$ and $a_1 = -24$
17. $a_n = 2a_{n-1} + 1$ and $a_1 = 1$
18. $a_n = \dfrac{1}{1 + a_{n-1}}$ and $a_1 = 1$
19. $a_n = a_{n-1} + a_{n-2}$ and $a_1 = 1, a_2 = 2$
20. $a_n = a_{n-1} + a_{n-2} + a_{n-3}$ and $a_1 = a_2 = a_3 = 1$

21–26 ■ Terms of a Sequence Use a graphing calculator to do the following. **(a)** Find the first ten terms of the sequence. **(b)** Graph the first ten terms of the sequence.

21. $a_n = 4n + 3$

22. $a_n = n^2 + n$

23. $a_n = \dfrac{12}{n}$

24. $a_n = 4 - 2(-1)^n$

25. $a_n = \dfrac{1}{a_{n-1}}$ and $a_1 = 2$

26. $a_n = a_{n-1} - a_{n-2}$ and $a_1 = 1, a_2 = 3$

27–38 ■ *n*th term of a Sequence Find the nth term of a sequence whose first several terms are given.

27. 2, 4, 6, 8, . . .

28. 1, 3, 5, 7, . . .

29. 2, 4, 8, 16, . . .

30. $-\frac{1}{3}, \frac{1}{9}, -\frac{1}{27}, \frac{1}{81}, \ldots$

31. $-2, 3, 8, 13, \ldots$

32. $7, 4, 1, -2, \ldots$

33. $5, -25, 125, -625, \ldots$

34. 3, 0.3, 0.03, 0.003, . . .

35. $1, \frac{3}{4}, \frac{5}{9}, \frac{7}{16}, \frac{9}{25}, \ldots$

36. $\frac{3}{4}, \frac{4}{5}, \frac{5}{6}, \frac{6}{7}, \ldots$

37. 0, 2, 0, 2, 0, 2, . . .

38. $1, \frac{1}{2}, 3, \frac{1}{4}, 5, \frac{1}{6}, \ldots$

39–42 ■ Partial Sums Find the first six partial sums $S_1, S_2, S_3, S_4, S_5, S_6$ of the sequence whose nth term is given.

39. 1, 3, 5, 7, . . .

40. $1^2, 2^2, 3^2, 4^2, \ldots$

41. $\dfrac{1}{3}, \dfrac{1}{3^2}, \dfrac{1}{3^3}, \dfrac{1}{3^4}, \ldots$

42. $-1, 1, -1, 1, \ldots$

43–46 ■ *n*th Partial Sum Find the first four partial sums and the nth partial sum of the sequence a_n.

43. $a_n = \dfrac{2}{3^n}$

44. $a_n = \dfrac{1}{n+1} - \dfrac{1}{n+2}$

45. $a_n = \sqrt{n} - \sqrt{n+1}$

46. $a_n = \log\left(\dfrac{n}{n+1}\right)$ [*Hint:* Use a property of logarithms to write the nth term as a difference.]

47–54 ■ Evaluating a Sum Find the sum.

47. $\displaystyle\sum_{k=1}^{4} k$

48. $\displaystyle\sum_{k=1}^{4} k^2$

49. $\displaystyle\sum_{k=1}^{3} \frac{1}{k}$

50. $\displaystyle\sum_{j=1}^{100} (-1)^j$

51. $\displaystyle\sum_{i=1}^{8} [1 + (-1)^i]$

52. $\displaystyle\sum_{i=4}^{12} 10$

53. $\displaystyle\sum_{k=1}^{5} 2^{k-1}$

54. $\displaystyle\sum_{i=1}^{3} i2^i$

55–60 ■ Evaluating a Sum Use a graphing calculator to evaluate the sum.

55. $\displaystyle\sum_{k=1}^{10} k^2$

56. $\displaystyle\sum_{k=1}^{100} (3k + 4)$

57. $\displaystyle\sum_{j=7}^{20} j^2(1 + j)$

58. $\displaystyle\sum_{j=5}^{15} \frac{1}{j^2 + 1}$

59. $\displaystyle\sum_{n=0}^{22} (-1)^n 2n$

60. $\displaystyle\sum_{n=1}^{100} \frac{(-1)^n}{n}$

61–66 ■ Sigma Notation Write the sum without using sigma notation.

61. $\displaystyle\sum_{k=1}^{4} k^3$

62. $\displaystyle\sum_{j=1}^{4} \sqrt{\frac{j-1}{j+1}}$

63. $\displaystyle\sum_{k=0}^{6} \sqrt{k+4}$

64. $\displaystyle\sum_{k=6}^{9} k(k+3)$

65. $\displaystyle\sum_{k=3}^{100} x^k$

66. $\displaystyle\sum_{j=1}^{n} (-1)^{j+1} x^j$

67–74 ■ Sigma Notation Write the sum using sigma notation.

67. $2 + 4 + 6 + \cdots + 50$

68. $2 + 5 + 8 + \cdots + 29$

69. $1^2 + 2^2 + 3^2 + \cdots + 10^2$

70. $\dfrac{1}{2\ln 2} - \dfrac{1}{3\ln 3} + \dfrac{1}{4\ln 4} - \dfrac{1}{5\ln 5} + \cdots + \dfrac{1}{100\ln 100}$

71. $\dfrac{1}{1\cdot 2} + \dfrac{1}{2\cdot 3} + \dfrac{1}{3\cdot 4} + \cdots + \dfrac{1}{999\cdot 1000}$

72. $\dfrac{\sqrt{1}}{1^2} + \dfrac{\sqrt{2}}{2^2} + \dfrac{\sqrt{3}}{3^2} + \cdots + \dfrac{\sqrt{n}}{n^2}$

73. $1 + x + x^2 + x^3 + \cdots + x^{100}$

74. $1 - 2x + 3x^2 - 4x^3 + 5x^4 + \cdots - 100x^{99}$

SKILLS Plus

75. *n*th Term of a Sequence Find a formula for the nth term of the sequence

$$\sqrt{2},\ \sqrt{2\sqrt{2}},\ \sqrt{2\sqrt{2\sqrt{2}}},\ \sqrt{2\sqrt{2\sqrt{2\sqrt{2}}}}, \ldots$$

[*Hint:* Write each term as a power of 2.]

76. Comparing a Sequence to the Fibonacci Sequence Define the sequence

$$G_n = \frac{1}{\sqrt{5}}\left(\frac{(1+\sqrt{5})^n - (1-\sqrt{5})^n}{2^n}\right)$$

Use the TABLE command on a graphing calculator to find the first ten terms of this sequence. Compare to the Fibonacci sequence F_n.

APPLICATIONS

77. Compound Interest Julio deposits \$2000 in a savings account that pays 2.4% interest per year compounded monthly. The amount in the account after n months is given by

$$A_n = 2000\left(1 + \frac{0.024}{12}\right)^n$$

(a) Find the first six terms of the sequence.

(b) Find the amount in the account after 3 years.

78. Compound Interest Helen deposits \$100 at the end of each month into an account that pays 6% interest per year compounded monthly. The amount of interest she has accumulated after n months is given by

$$I_n = 100\left(\frac{1.005^n - 1}{0.005} - n\right)$$

(a) Find the first six terms of the sequence.

(b) Find the interest she has accumulated after 5 years.

79. Population of a City A city was incorporated in 2004 with a population of 35,000. It is expected that the population will increase at a rate of 2% per year. The population n years after 2004 is given by

$$P_n = 35{,}000(1.02)^n$$

(a) Find the first five terms of the sequence.

(b) Find the population in 2014.

80. Paying off a Debt Margarita borrows \$10,000 from her uncle and agrees to repay it in monthly installments of \$200. Her uncle charges 0.5% interest per month on the balance.

(a) Show that her balance A_n in the nth month is given recursively by $A_0 = 10{,}000$ and

$$A_n = 1.005A_{n-1} - 200$$

(b) Find her balance after 6 months.

81. Fish Farming A fish farmer has 5000 catfish in his pond. The number of catfish increases by 8% per month, and the farmer harvests 300 catfish per month.

(a) Show that the catfish population P_n after n months is given recursively by $P_0 = 5000$ and

$$P_n = 1.08P_{n-1} - 300$$

(b) How many fish are in the pond after 12 months?

82. Price of a House The median price of a house in Orange County increases by about 6% per year. In 2002 the median price was \$240,000. Let P_n be the median price n years after 2002.

(a) Find a formula for the sequence P_n.

(b) Find the expected median price in 2010.

83. Salary Increases A newly hired salesman is promised a beginning salary of \$30,000 a year with a \$2000 raise every year. Let S_n be his salary in his nth year of employment.

(a) Find a recursive definition of S_n.

(b) Find his salary in his fifth year of employment.

84. Concentration of a Solution A biologist is trying to find the optimal salt concentration for the growth of a certain species of mollusk. She begins with a brine solution that has 4 g/L of salt and increases the concentration by 10% every day. Let C_0 denote the initial concentration, and let C_n be the concentration after n days.

(a) Find a recursive definition of C_n.

(b) Find the salt concentration after 8 days.

85. Fibonacci's Rabbits Fibonacci posed the following problem: Suppose that rabbits live forever and that every month each pair produces a new pair that becomes productive at age 2 months. If we start with one newborn pair, how many pairs of rabbits will we have in the nth month? Show that the answer is F_n, where F_n is the nth term of the Fibonacci sequence.

DISCUSS ■ DISCOVER ■ PROVE ■ WRITE

86. DISCOVER ■ PROVE: Different Sequences That Start the Same

(a) Show that the first four terms of the sequence defined by $a_n = n^2$ are

$$1, 4, 9, 16, \ldots$$

(b) Show that the first four terms of the sequence defined by $a_n = n^2 + (n-1)(n-2)(n-3)(n-4)$ are also

$$1, 4, 9, 16, \ldots$$

(c) Find a sequence whose first six terms are the same as those of $a_n = n^2$ but whose succeeding terms differ from this sequence.

(d) Find two different sequences that begin

$$2, 4, 8, 16, \ldots$$

87. DISCUSS: A Recursively Defined Sequence Find the first 40 terms of the sequence defined by

$$a_{n+1} = \begin{cases} \dfrac{a_n}{2} & \text{if } a_n \text{ is an even number} \\ 3a_n + 1 & \text{if } a_n \text{ is an odd number} \end{cases}$$

and $a_1 = 11$. Do the same if $a_1 = 25$. Make a conjecture about this type of sequence. Try several other values for a_1, to test your conjecture.

88. DISCUSS: A Different Type of Recursion Find the first ten terms of the sequence defined by

$$a_n = a_{n-a_{n-1}} + a_{n-a_{n-2}}$$

with

$$a_1 = 1 \quad \text{and} \quad a_2 = 1$$

How is this recursive sequence different from the others in this section?

12.2 ARITHMETIC SEQUENCES

■ Arithmetic Sequences ■ Partial Sums of Arithmetic Sequences

In this section we study a special type of sequence, called an arithmetic sequence.

■ Arithmetic Sequences

Perhaps the simplest way to generate a sequence is to start with a number a and add to it a fixed constant d, over and over again.

DEFINITION OF AN ARITHMETIC SEQUENCE

An **arithmetic sequence** is a sequence of the form

$$a, a + d, a + 2d, a + 3d, a + 4d, \ldots$$

The number a is the **first term**, and d is the **common difference** of the sequence. The ***n*th term** of an arithmetic sequence is given by

$$a_n = a + (n - 1)d$$

The number d is called the common difference because any two consecutive terms of an arithmetic sequence differ by d.

EXAMPLE 1 ■ Arithmetic Sequences

(a) If $a = 2$ and $d = 3$, then we have the arithmetic sequence

$$2, 2 + 3, 2 + 6, 2 + 9, \ldots$$

or $$2, 5, 8, 11, \ldots$$

Any two consecutive terms of this sequence differ by $d = 3$. The nth term is $a_n = 2 + 3(n - 1)$.

(b) Consider the arithmetic sequence

$$9, 4, -1, -6, -11, \ldots$$

Here the common difference is $d = -5$. The terms of an arithmetic sequence decrease if the common difference is negative. The nth term is $a_n = 9 - 5(n - 1)$.

(c) The graph of the arithmetic sequence $a_n = 1 + 2(n - 1)$ is shown in Figure 1. Notice that the points in the graph lie on the straight line $y = 2x - 1$, which has slope $d = 2$.

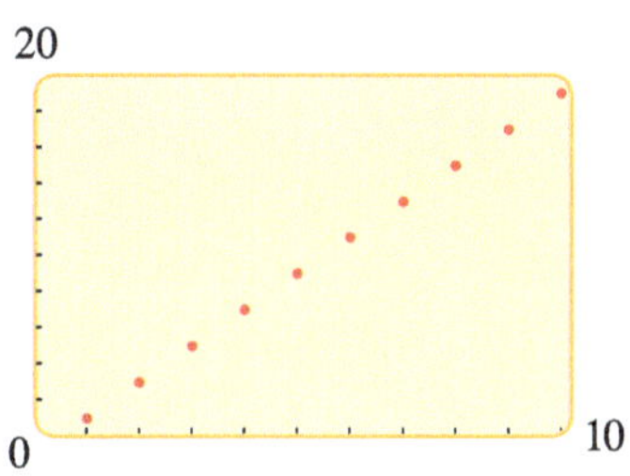

FIGURE 1

Now Try Exercises 5, 11, and 17 ■

See Appendix D, *Using the TI-83/84 Graphing Calculator*, for instructions on how to graph sequences. Go to **www.stewartmath.com**.

An arithmetic sequence is determined completely by the first term a and the common difference d. Thus if we know the first two terms of an arithmetic sequence, then we can find a formula for the nth term, as the next example shows.

EXAMPLE 2 ■ Finding Terms of an Arithmetic Sequence

Find the common difference, the first six terms, the nth term, and the 300th term of the arithmetic sequence

$$13, 7, 1, -5, \ldots$$

Mathematics in the Modern World

Fair Division of Assets

Dividing an asset fairly among a number of people is of great interest to mathematicians. Problems of this nature include dividing the national budget, disputed land, or assets in divorce cases. In 1994 Brams and Taylor found a mathematical way of dividing things fairly. Their solution has been applied to division problems in political science, legal proceedings, and other areas. To understand the problem, consider the following example. Suppose persons A and B want to divide a property fairly between them. To divide it *fairly* means that both A and B must be satisfied with the outcome of the division. Solution: A gets to divide the property into two pieces, then B gets to choose the piece he or she wants. Since both A and B had a part in the division process, each should be satisfied. The situation becomes much more complicated if three or more people are involved (and that's where mathematics comes in).

Dividing things fairly involves much more than simply cutting things in half; it must take into account the *relative worth* each person attaches to the thing being divided. A story from the Bible illustrates this clearly. Two women appear before King Solomon, each claiming to be the mother of the same newborn baby. To discover which of these two women is the real mother, King Solomon ordered his swordsman to cut the baby in half! The real mother, who attaches far more worth to the baby than anyone else does, immediately gives up her claim to the baby to save the baby's life.

Mathematical solutions to fair-division problems have recently been applied in an international treaty, the Convention on the Law of the Sea. If a country wants to develop a portion of the sea floor, it is required to divide the portion into two parts, one part to be used by itself and the other by a consortium that will preserve it for later use by a less developed country. The consortium gets first pick.

SOLUTION Since the first term is 13, we have $a = 13$. The common difference is $d = 7 - 13 = -6$. Thus the nth term of this sequence is

$$a_n = 13 - 6(n - 1)$$

From this we find the first six terms:

$$13, 7, 1, -5, -11, -17, \ldots$$

The 300th term is $a_{300} = 13 - 6(300 - 1) = -1781$.

Now Try Exercise 33 ■

The next example shows that an arithmetic sequence is determined completely by *any* two of its terms.

EXAMPLE 3 ■ Finding Terms of an Arithmetic Sequence

The 11th term of an arithmetic sequence is 52, and the 19th term is 92. Find the 1000th term.

SOLUTION To find the nth term of this sequence, we need to find a and d in the formula

$$a_n = a + (n - 1)d$$

From this formula we get

$$a_{11} = a + (11 - 1)d = a + 10d$$

$$a_{19} = a + (19 - 1)d = a + 18d$$

Since $a_{11} = 52$ and $a_{19} = 92$, we get the following two equations:

$$\begin{cases} 52 = a + 10d \\ 92 = a + 18d \end{cases}$$

Solving this system for a and d, we get $a = 2$ and $d = 5$. (Verify this.) Thus the nth term of this sequence is

$$a_n = 2 + 5(n - 1)$$

The 1000th term is $a_{1000} = 2 + 5(1000 - 1) = 4997$.

Now Try Exercise 47 ■

Partial Sums of Arithmetic Sequences

Suppose we want to find the sum of the numbers 1, 2, 3, 4, . . . , 100, that is,

$$\sum_{k=1}^{100} k$$

When the famous mathematician C. F. Gauss (see page 290) was a schoolboy, his teacher posed this problem to the class and expected that it would keep the students busy for a long time. But Gauss answered the question almost immediately. His idea was this: Since we are adding numbers produced according to a fixed pattern, there must also be a pattern (or formula) for finding the sum. He started by writing the numbers from 1 to 100 and then below them wrote the same numbers in reverse order. Writing S for the sum and adding corresponding terms give

$$\begin{array}{rcl} S &=& 1 + 2 + 3 + \cdots + 98 + 99 + 100 \\ S &=& 100 + 99 + 98 + \cdots + 3 + 2 + 1 \\ \hline 2S &=& 101 + 101 + 101 + \cdots + 101 + 101 + 101 \end{array}$$

It follows that $2S = 100(101) = 10{,}100$, so $S = 5050$.

Of course, the sequence of natural numbers 1, 2, 3, . . . is an arithmetic sequence (with $a = 1$ and $d = 1$), and the method for summing the first 100 terms of this sequence can be used to find a formula for the nth partial sum of any arithmetic sequence. We want to find the sum of the first n terms of the arithmetic sequence whose terms are $a_k = a + (k - 1)d$; that is, we want to find

$$\begin{aligned} S_n &= \sum_{k=1}^{n} [a + (k - 1)d] \\ &= a + (a + d) + (a + 2d) + (a + 3d) + \cdots + [a + (n - 1)d] \end{aligned}$$

Using Gauss's method, we write

$$\begin{array}{rcccccccc} S_n = & a & + & (a+d) & + \cdots + & [a+(n-2)d] & + & [a+(n-1)d] \\ S_n = & [a+(n-1)d] & + & [a+(n-2)d] & + \cdots + & (a+d) & + & a \\ \hline 2S_n = & [2a+(n-1)d] & + & [2a+(n-1)d] & + \cdots + & [2a+(n-1)d] & + & [2a+(n-1)d] \end{array}$$

There are n identical terms on the right side of this equation, so

$$\begin{aligned} 2S_n &= n[2a + (n - 1)d] \\ S_n &= \frac{n}{2}[2a + (n - 1)d] \end{aligned}$$

Notice that $a_n = a + (n - 1)d$ is the nth term of this sequence. So we can write

$$S_n = \frac{n}{2}[a + a + (n - 1)d] = n\left(\frac{a + a_n}{2}\right)$$

This last formula says that the sum of the first n terms of an arithmetic sequence is the average of the first and nth terms multiplied by n, the number of terms in the sum. We now summarize this result.

PARTIAL SUMS OF AN ARITHMETIC SEQUENCE

For the arithmetic sequence given by $a_n = a + (n - 1)d$, the ***n*th partial sum**

$$S_n = a + (a + d) + (a + 2d) + (a + 3d) + \cdots + [a + (n - 1)d]$$

is given by either of the following formulas.

1. $S_n = \dfrac{n}{2}[2a + (n - 1)d]$ **2.** $S_n = n\left(\dfrac{a + a_n}{2}\right)$

EXAMPLE 4 ■ Finding a Partial Sum of an Arithmetic Sequence

Find the sum of the first 50 odd numbers.

SOLUTION The odd numbers form an arithmetic sequence with $a = 1$ and $d = 2$. The nth term is $a_n = 1 + 2(n - 1) = 2n - 1$, so the 50th odd number is $a_{50} = 2(50) - 1 = 99$. Substituting in Formula 2 for the partial sum of an arithmetic sequence, we get

$$S_{50} = 50\left(\frac{a + a_{50}}{2}\right) = 50\left(\frac{1 + 99}{2}\right) = 50 \cdot 50 = 2500$$

Now Try Exercise 51 ■

EXAMPLE 5 ■ Finding a Partial Sum of an Arithmetic Sequence

Find the following partial sum of an arithmetic sequence:

$$3 + 7 + 11 + 15 + \cdots + 159$$

SOLUTION For this sequence $a = 3$ and $d = 4$, so $a_n = 3 + 4(n - 1)$. To find which term of the sequence is the last term 159, we use the formula for the nth term and solve for n.

$$159 = 3 + 4(n - 1) \quad \text{Set } a_n = 159$$

$$39 = n - 1 \quad \text{Subtract 3; divide by 4}$$

$$n = 40 \quad \text{Add 1}$$

To find the partial sum of the first 40 terms, we use Formula 1 for the nth partial sum of an arithmetic sequence:

$$S_{40} = \tfrac{40}{2}[2(3) + 4(40 - 1)] = 3240$$

Now Try Exercise 57 ■

EXAMPLE 6 ■ Finding the Seating Capacity of an Amphitheater

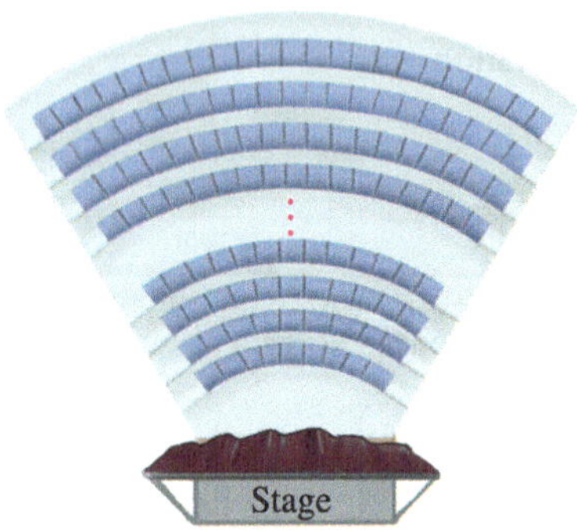

An amphitheater has 50 rows of seats with 30 seats in the first row, 32 in the second, 34 in the third, and so on. Find the total number of seats.

SOLUTION The numbers of seats in the rows form an arithmetic sequence with $a = 30$ and $d = 2$. Since there are 50 rows, the total number of seats is the sum

$$S_{50} = \tfrac{50}{2}[2(30) + 49(2)] \quad S_n = \frac{n}{2}[2a + (n - 1)d]$$

$$= 3950$$

Thus the amphitheater has 3950 seats.

Now Try Exercise 75 ■

EXAMPLE 7 ■ Finding the Number of Terms in a Partial Sum

How many terms of the arithmetic sequence 5, 7, 9, . . . must be added to get 572?

SOLUTION We are asked to find n when $S_n = 572$. Substituting $a = 5$, $d = 2$, and $S_n = 572$ in Formula 1 for the partial sum of an arithmetic sequence, we get

$$572 = \frac{n}{2}[2 \cdot 5 + (n - 1)2] \quad S_n = \frac{n}{2}[2a + (n - 1)d]$$

$$572 = 5n + n(n - 1) \quad \text{Distributive Property}$$

$$0 = n^2 + 4n - 572 \quad \text{Expand}$$

$$0 = (n - 22)(n + 26) \quad \text{Factor}$$

This gives $n = 22$ or $n = -26$. But since n is the *number* of terms in this partial sum, we must have $n = 22$.

Now Try Exercise 65 ■

12.2 EXERCISES

CONCEPTS

1. An arithmetic sequence is a sequence in which the ________ between successive terms is constant.

2. The sequence given by $a_n = a + (n - 1)d$ is an arithmetic sequence in which a is the first term and d is the ________ ________. So for the arithmetic sequence $a_n = 2 + 5(n - 1)$ the first term is ________, and the common difference is ________.

3–4 ■ *True or False*? If *False*, give a reason.

3. The nth partial sum of an arithmetic sequence is the average of the first and last terms times n.

4. If we know the first and second terms of an arithmetic sequence, then we can find any other term.

SKILLS

5–10 ■ Terms of an Arithmetic Sequence The nth term of an arithmetic sequence is given. **(a)** Find the first five terms of the sequence. **(b)** What is the common difference d? **(c)** Graph the terms you found in part (a).

5. $a_n = 7 + 3(n - 1)$ **6.** $a_n = -10 + 20(n - 1)$

7. $a_n = -6 - 4(n - 1)$ **8.** $a_n = -10 + 4(n - 1)$

9. $a_n = \frac{5}{2} - (n - 1)$ **10.** $a_n = \frac{1}{2}(n - 1)$

11–16 ■ *n*th Term of an Arithmetic Sequence Find the nth term of the arithmetic sequence with given first term a and common difference d. What is the 10th term?

11. $a = 9, \quad d = 4$ **12.** $a = -5, \quad d = 4$

13. $a = -0.7, \quad d = -0.2$ **14.** $a = 14, \quad d = -\frac{3}{2}$

15. $a = \frac{5}{2}, \quad d = -\frac{1}{2}$ **16.** $a = \sqrt{3}, \quad d = \sqrt{3}$

17–26 ■ Arithmetic Sequence? The first four terms of a sequence are given. Can these terms be the terms of an arithmetic sequence? If so, find the common difference.

17. $11, 17, 23, 29, \ldots$ **18.** $-31, -19, -7, 5, \ldots$

19. $16, 9, 2, -4, \ldots$ **20.** $100, 68, 36, 4, \ldots$

21. $2, 4, 8, 16, \ldots$ **22.** $2, 4, 6, 8, \ldots$

23. $3, \frac{3}{2}, 0, -\frac{3}{2}, \ldots$ **24.** $\ln 2, \ln 4, \ln 8, \ln 16, \ldots$

25. $2.6, 4.3, 6.0, 7.7, \ldots$ **26.** $\frac{1}{2}, \frac{1}{3}, \frac{1}{4}, \frac{1}{5}, \ldots$

27–32 ■ Arithmetic Sequence? Find the first five terms of the sequence, and determine whether it is arithmetic. If it is arithmetic, find the common difference, and express the nth term of the sequence in the standard form $a_n = a + (n - 1)d$.

27. $a_n = 4 + 7n$ **28.** $a_n = 4 + 2^n$

29. $a_n = \dfrac{1}{1 + 2n}$ **30.** $a_n = 1 + \dfrac{n}{2}$

31. $a_n = 6n - 10$ **32.** $a_n = 3 + (-1)^n n$

33–44 ■ Terms of an Arithmetic Sequence Determine the common difference, the fifth term, the nth term, and the 100th term of the arithmetic sequence.

33. $4, 10, 16, 22, \ldots$ **34.** $-1, 11, 23, 35, \ldots$

35. $29, 11, -7, -25 \ldots$ **36.** $64, 49, 34, 19, \ldots$

37. $4, 9, 14, 19, \ldots$ **38.** $11, 8, 5, 2, \ldots$

39. $-12, -8, -4, 0, \ldots$ **40.** $\frac{7}{6}, \frac{5}{3}, \frac{13}{6}, \frac{8}{3}, \ldots$

41. $25, 26.5, 28, 29.5, \ldots$ **42.** $15, 12.3, 9.6, 6.9, \ldots$

43. $2, 2 + s, 2 + 2s, 2 + 3s, \ldots$

44. $-t, -t + 3, -t + 6, -t + 9, \ldots$

45–50 ■ Finding Terms of an Arithmetic Sequence Find the indicated term of the arithmetic sequence with the given description.

45. The 50th term is 1000, and the common difference is 6. Find the first and second terms.

46. The 100th term is -750, and the common difference is -20. Find the fifth term.

47. The fourteenth term is $\frac{2}{3}$, and the ninth term is $\frac{1}{4}$. Find the first term and the nth term.

48. The twelfth term is 118, and the eighth term is 146. Find the first term and the nth term.

49. The first term is 25, and the common difference is 18. Which term of the sequence is 601?

50. The first term is 3500, and the common difference is -15. Which term of the sequence is 2795?

51–56 ■ Partial Sums of an Arithmetic Sequence Find the partial sum S_n of the arithmetic sequence that satisfies the given conditions.

51. $a = 3, \quad d = 5, \quad n = 20$

52. $a = 10, \quad d = -8, \quad n = 30$

53. $a = -40, \quad d = 14, \quad n = 15$

54. $a = -2, \quad d = 23, \quad n = 25$

55. $a_1 = 55, \quad d = 12, \quad n = 10$

56. $a_2 = 8, \quad a_5 = 9.5, \quad n = 15$

57–64 ■ Partial Sums of an Arithmetic Sequence A partial sum of an arithmetic sequence is given. Find the sum.

57. $1 + 5 + 9 + \cdots + 401$

58. $-3 + \left(-\frac{3}{2}\right) + 0 + \frac{3}{2} + 3 + \cdots + 30$

59. $250 + 233 + 216 + \cdots + 97$

60. $89 + 85 + 81 + \cdots + 13$

61. $0.7 + 2.7 + 4.7 + \cdots + 56.7$

62. $-10 - 9.9 - 9.8 - \cdots - 0.1$

63. $\displaystyle\sum_{k=0}^{10} (3 + 0.25k)$ **64.** $\displaystyle\sum_{n=0}^{20} (1 - 2n)$

65–66 ■ Adding Terms of an Arithmetic Sequence Find the number of terms of the arithmetic sequence with the given description that must be added to get a value of 2700.

65. The first term is 5, and the common difference is 2.

66. The first term is 12, and the common difference is 8.

SKILLS Plus

67. Special Triangle Show that a right triangle whose sides are in arithmetic progression is similar to a 3–4–5 triangle.

68. Product of Numbers Find the product of the numbers

$$10^{1/10}, 10^{2/10}, 10^{3/10}, 10^{4/10}, \ldots, 10^{19/10}$$

69. Harmonic Sequence A sequence is **harmonic** if the reciprocals of the terms of the sequence form an arithmetic sequence. Determine whether the following sequence is harmonic:

$$1, \tfrac{3}{5}, \tfrac{3}{7}, \tfrac{1}{3}, \ldots$$

70. Harmonic Mean The **harmonic mean** of two numbers is the reciprocal of the average of the reciprocals of the two numbers. Find the harmonic mean of 3 and 5.

APPLICATIONS

71. Depreciation The purchase value of an office computer is \$12,500. Its annual depreciation is \$1875. Find the value of the computer after 6 years.

72. Poles in a Pile Telephone poles are being stored in a pile with 25 poles in the first layer, 24 in the second, and so on. If there are 12 layers, how many telephone poles does the pile contain?

73. Salary Increases A man gets a job with a salary of \$30,000 a year. He is promised a \$2300 raise each subsequent year. Find his total earnings for a 10-year period.

74. Drive-In Theater A drive-in theater has spaces for 20 cars in the first parking row, 22 in the second, 24 in the third, and so on. If there are 21 rows in the theater, find the number of cars that can be parked.

75. Theater Seating An architect designs a theater with 15 seats in the first row, 18 in the second, 21 in the third, and so on. If the theater is to have a seating capacity of 870, how many rows must the architect use in his design?

76. Falling Ball When an object is allowed to fall freely near the surface of the earth, the gravitational pull is such that the object falls 16 ft in the first second, 48 ft in the next second, 80 ft in the next second, and so on.

(a) Find the total distance a ball falls in 6 s.

(b) Find a formula for the total distance a ball falls in n seconds.

77. The Twelve Days of Christmas In the well-known song "The Twelve Days of Christmas," a person gives his sweetheart k gifts on the kth day for each of the 12 days of Christmas. The person also repeats each gift identically on each subsequent day. Thus on the 12th day the sweetheart receives a gift for the first day, 2 gifts for the second, 3 gifts for the third, and so on. Show that the number of gifts received on the 12th day is a partial sum of an arithmetic sequence. Find this sum.

DISCUSS ■ DISCOVER ■ PROVE ■ WRITE

78. DISCUSS: Arithmetic Means The **arithmetic mean** (or average) of two numbers a and b is

$$m = \frac{a + b}{2}$$

Note that m is the same distance from a as from b, so a, m, b is an arithmetic sequence. In general, if $m_1, m_2, \ldots, m_k$ are equally spaced between a and b so that

$$a, m_1, m_2, \ldots, m_k, b$$

is an arithmetic sequence, then $m_1, m_2, \ldots, m_k$ are called k arithmetic means between a and b.

(a) Insert two arithmetic means between 10 and 18.

(b) Insert three arithmetic means between 10 and 18.

(c) Suppose a doctor needs to increase a patient's dosage of a certain medicine from 100 mg to 300 mg per day in five equal steps. How many arithmetic means must be inserted between 100 and 300 to give the progression of daily doses, and what are these means?

12.3 GEOMETRIC SEQUENCES

■ Geometric Sequences ■ Partial Sums of Geometric Sequences ■ What Is an Infinite Series? ■ Infinite Geometric Series

In this section we study geometric sequences. This type of sequence occurs frequently in applications to finance, population growth, and other fields.

■ Geometric Sequences

Recall that an arithmetic sequence is generated when we repeatedly add a number d to an initial term a. A *geometric* sequence is generated when we start with a number a and repeatedly *multiply* by a fixed nonzero constant r.

DEFINITION OF A GEOMETRIC SEQUENCE

A **geometric sequence** is a sequence of the form

$$a, ar, ar^2, ar^3, ar^4, \ldots$$

The number a is the **first term**, and r is the **common ratio** of the sequence. The ***n*th term** of a geometric sequence is given by

$$a_n = ar^{n-1}$$

The number r is called the common ratio because the ratio of any two consecutive terms of the sequence is r.

EXAMPLE 1 ■ Geometric Sequences

(a) If $a = 3$ and $r = 2$, then we have the geometric sequence

$$3, \quad 3 \cdot 2, \quad 3 \cdot 2^2, \quad 3 \cdot 2^3, \quad 3 \cdot 2^4, \quad \ldots$$

or

$$3, 6, 12, 24, 48, \ldots$$

Notice that the ratio of any two consecutive terms is $r = 2$. The nth term is $a_n = 3(2)^{n-1}$.

(b) The sequence

$$2, -10, 50, -250, 1250, \ldots$$

is a geometric sequence with $a = 2$ and $r = -5$. When r is negative, the terms of the sequence alternate in sign. The nth term is $a_n = 2(-5)^{n-1}$.

(c) The sequence

$$1, \frac{1}{3}, \frac{1}{9}, \frac{1}{27}, \frac{1}{81}, \ldots$$

is a geometric sequence with $a = 1$ and $r = \frac{1}{3}$. The nth term is $a_n = 1\left(\frac{1}{3}\right)^{n-1}$.

(d) The graph of the geometric sequence defined by $a_n = \frac{1}{5} \cdot 2^{n-1}$ is shown in Figure 1. Notice that the points in the graph lie on the graph of the exponential function $y = \frac{1}{5} \cdot 2^{x-1}$.

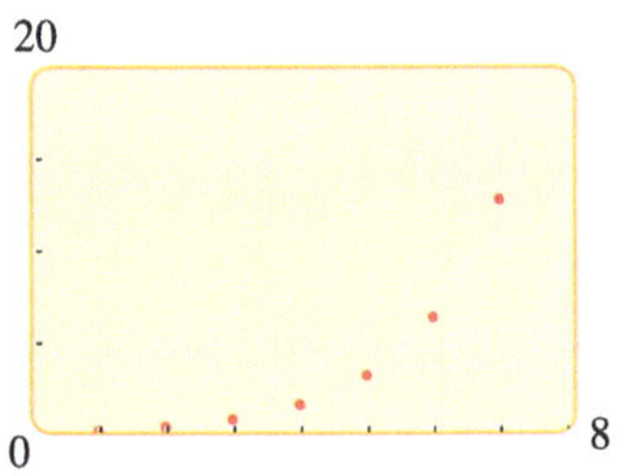

FIGURE 1

If $0 < r < 1$, then the terms of the geometric sequence ar^{n-1} decrease, but if $r > 1$, then the terms increase. (What happens if $r = 1$?)

Now Try Exercises 5, 9, and 13 ■

Geometric sequences occur naturally. Here is a simple example. Suppose a ball has elasticity such that when it is dropped, it bounces up one-third of the distance it has fallen. If this ball is dropped from a height of 2 m, then it bounces up to a height of $2\left(\frac{1}{3}\right) = \frac{2}{3}$ m. On its second bounce, it returns to a height of $\left(\frac{2}{3}\right)\left(\frac{1}{3}\right) = \frac{2}{9}$ m, and so on (see Figure 2). Thus the height h_n that the ball reaches on its nth bounce is given by the geometric sequence

$$h_n = \tfrac{2}{3}\left(\tfrac{1}{3}\right)^{n-1} = 2\left(\tfrac{1}{3}\right)^n$$

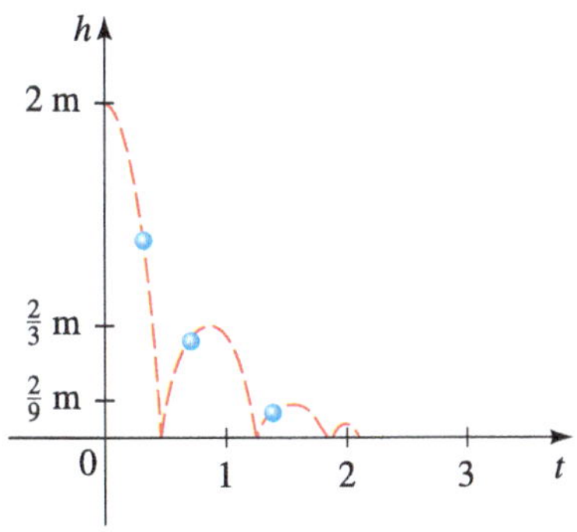

FIGURE 2

We can find the nth term of a geometric sequence if we know any two terms, as the following examples show.

EXAMPLE 2 ■ Finding Terms of a Geometric Sequence

Find the common ratio, the first term, the nth term, and the eighth term of the geometric sequence

$$5, 15, 45, 135, \ldots$$

SOLUTION To find a formula for the nth term of this sequence, we need to find the first term a and the common ratio r. Clearly, $a = 5$. To find r, we find the ratio of any two consecutive terms. For instance, $r = \frac{45}{15} = 3$. Thus

$$a_n = 5(3)^{n-1} \qquad a_n = ar^{n-1}$$

The eighth term is $a_8 = 5(3)^{8-1} = 5(3)^7 = 10{,}935$.

Now Try Exercise 29 ■

Science Source

SRINIVASA RAMANUJAN (1887–1920) was born into a poor family in the small town of Kumbakonam in India. Self-taught in mathematics, he worked in virtual isolation from other mathematicians. At the age of 25 he wrote a letter to G. H. Hardy, the leading British mathematician at the time, listing some of his discoveries. His discoveries included the following series for calculating π:

$$\frac{1}{\pi} = \frac{2\sqrt{2}}{9801} \sum_{k=0}^{\infty} \frac{(4k)!(1103 + 26390k)}{(k!)^4 396^{4k}}$$

Hardy immediately recognized Ramanujan's genius, and for the next six years the two worked together in London until Ramanujan fell ill and returned to his hometown in India, where he died a year later. Ramanujan was a genius with a phenomenal ability to see hidden patterns in the properties of numbers. Most of his discoveries were written as complicated infinite series, the importance of which was not recognized until many years after his death. In the last year of his life he wrote 130 pages of mysterious formulas, many of which still defy proof. Hardy tells the story that when he visited Ramanujan in a hospital and arrived in a taxi, he remarked to Ramanujan that the cab's number, 1729, was uninteresting. Ramanujan replied "No, it is a very interesting number. It is the smallest number expressible as the sum of two cubes in two different ways."

EXAMPLE 3 ■ Finding Terms of a Geometric Sequence

The third term of a geometric sequence is $\frac{63}{4}$, and the sixth term is $\frac{1701}{32}$. Find the fifth term.

SOLUTION Since this sequence is geometric, its nth term is given by the formula $a_n = ar^{n-1}$. Thus

$$a_3 = ar^{3-1} = ar^2$$
$$a_6 = ar^{6-1} = ar^5$$

From the values we are given for these two terms, we get the following system of equations:

$$\begin{cases} \frac{63}{4} = ar^2 \\ \frac{1701}{32} = ar^5 \end{cases}$$

We solve this system by dividing.

$$\frac{ar^5}{ar^2} = \frac{\frac{1701}{32}}{\frac{63}{4}}$$

$$r^3 = \frac{27}{8} \qquad \text{Simplify}$$

$$r = \frac{3}{2} \qquad \text{Take cube root of each side}$$

Substituting for r in the first equation gives

$$\frac{63}{4} = a\left(\frac{3}{2}\right)^2 \qquad \text{Substitute } r = \tfrac{3}{2} \text{ in } \tfrac{63}{4} = ar^2$$

$$a = 7 \qquad \text{Solve for } a$$

It follows that the nth term of this sequence is

$$a_n = 7\left(\frac{3}{2}\right)^{n-1}$$

Thus the fifth term is

$$a_5 = 7\left(\frac{3}{2}\right)^{5-1} = 7\left(\frac{3}{2}\right)^4 = \frac{567}{16}$$

Now Try Exercise 41 ■

■ Partial Sums of Geometric Sequences

For the geometric sequence $a, ar, ar^2, ar^3, ar^4, \ldots, ar^{n-1}, \ldots$, the nth partial sum is

$$S_n = \sum_{k=1}^{n} ar^{k-1} = a + ar + ar^2 + ar^3 + ar^4 + \cdots + ar^{n-1}$$

To find a formula for S_n, we multiply S_n by r and subtract from S_n.

$$S_n = a + ar + ar^2 + ar^3 + ar^4 + \cdots + ar^{n-1}$$
$$rS_n = \qquad ar + ar^2 + ar^3 + ar^4 + \cdots + ar^{n-1} + ar^n$$
$$S_n - rS_n = a - ar^n$$

So

$$S_n(1 - r) = a(1 - r^n)$$

$$S_n = \frac{a(1 - r^n)}{1 - r} \qquad r \neq 1$$

We summarize this result.

PARTIAL SUMS OF A GEOMETRIC SEQUENCE

For the geometric sequence defined by $a_n = ar^{n-1}$, the **nth partial sum**

$$S_n = a + ar + ar^2 + ar^3 + ar^4 + \cdots + ar^{n-1} \qquad r \neq 1$$

is given by

$$S_n = a\frac{1 - r^n}{1 - r}$$

EXAMPLE 4 ■ Finding a Partial Sum of a Geometric Sequence

Find the following partial sum of a geometric sequence:

$$1 + 4 + 16 + \cdots + 4096$$

SOLUTION For this sequence $a = 1$ and $r = 4$, so $a_n = 4^{n-1}$. Since $4^6 = 4096$, we use the formula for S_n with $n = 7$, and we have

$$S_7 = 1 \cdot \frac{1 - 4^7}{1 - 4} = 5461$$

Thus this partial sum is 5461.

Now Try Exercises 49 and 53 ■

EXAMPLE 5 ■ Finding a Partial Sum of a Geometric Sequence

Find the sum $\sum_{k=1}^{6} 7\left(-\frac{2}{3}\right)^{k-1}$.

SOLUTION The given sum is the sixth partial sum of a geometric sequence with first term $a = 7\left(-\frac{2}{3}\right)^0 = 7$ and $r = -\frac{2}{3}$. Thus by the formula for S_n with $n = 6$ we have

$$S_6 = 7 \cdot \frac{1 - \left(-\frac{2}{3}\right)^6}{1 - \left(-\frac{2}{3}\right)} = 7 \cdot \frac{1 - \frac{64}{729}}{\frac{5}{3}} = \frac{931}{243} \approx 3.83$$

Now Try Exercise 59 ■

■ What Is an Infinite Series?

An expression of the form

$$\sum_{k=1}^{\infty} a_k = a_1 + a_2 + a_3 + a_4 + \cdots$$

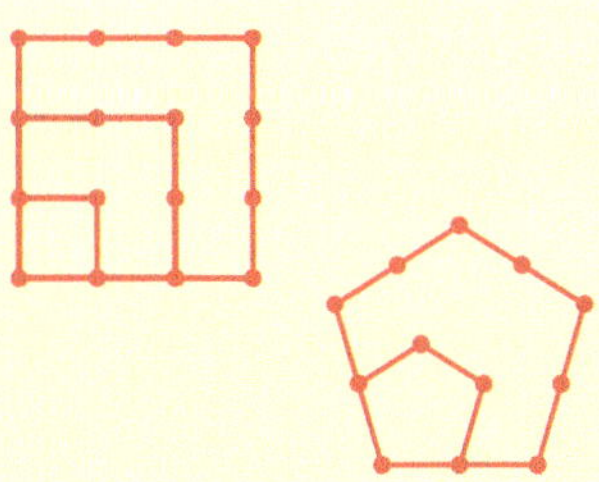

DISCOVERY PROJECT

Finding Patterns

Finding patterns in nature is an important part of mathematical modeling. If we can find a pattern (or a formula) that describes the terms of a sequence, then we can use the pattern to predict subsequent terms of the sequence. In this project we investigate difference sequences and how they help us find patterns in triangular, square, pentagonal, and other polygonal numbers. You can find the project at **www.stewartmath.com**.

is called an **infinite series**. The dots mean that we are to continue the addition indefinitely. What meaning can we attach to the sum of infinitely many numbers? It seems at first that it is not possible to add infinitely many numbers and arrive at a finite number. But consider the following problem. You have a cake, and you want to eat it by first eating half the cake, then eating half of what remains, then again eating half of what remains. This process can continue indefinitely because at each stage, some of the cake remains. (See Figure 3.)

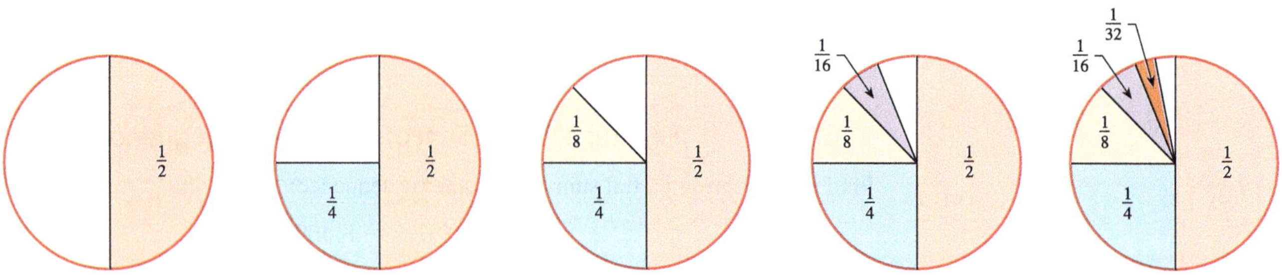

FIGURE 3

Does this mean that it's impossible to eat all of the cake? Of course not. Let's write down what you have eaten from this cake:

$$\sum_{k=1}^{\infty} \frac{1}{2^k} = \frac{1}{2} + \frac{1}{4} + \frac{1}{8} + \frac{1}{16} + \cdots$$

This is an infinite series, and we note two things about it: First, from Figure 3 it's clear that no matter how many terms of this series we add, the total will never exceed 1. Second, the more terms of this series we add, the closer the sum is to 1 (see Figure 3). This suggests that the number 1 can be written as the sum of infinitely many smaller numbers:

$$1 = \frac{1}{2} + \frac{1}{4} + \frac{1}{8} + \frac{1}{16} + \cdots + \frac{1}{2^n} + \cdots$$

To make this more precise, let's look at the partial sums of this series:

$$\begin{aligned} S_1 &= \frac{1}{2} &&= \frac{1}{2} \\ S_2 &= \frac{1}{2} + \frac{1}{4} &&= \frac{3}{4} \\ S_3 &= \frac{1}{2} + \frac{1}{4} + \frac{1}{8} &&= \frac{7}{8} \\ S_4 &= \frac{1}{2} + \frac{1}{4} + \frac{1}{8} + \frac{1}{16} &&= \frac{15}{16} \end{aligned}$$

and, in general (see Example 5 of Section 12.1),

$$S_n = 1 - \frac{1}{2^n}$$

As n gets larger and larger, we are adding more and more of the terms of this series. Intuitively, as n gets larger, S_n gets closer to the sum of the series. Now notice that as n gets large, $1/2^n$ gets closer and closer to 0. Thus S_n gets close to $1 - 0 = 1$. Using the notation of Section 3.6, we can write

$$S_n \to 1 \quad \text{as} \quad n \to \infty$$

In general, if S_n gets close to a finite number S as n gets large, we say that the infinite series **converges** (or is **convergent**). The number S is called the **sum of the infinite series**. If an infinite series does not converge, we say that the series **diverges** (or is **divergent**).

Infinite Geometric Series

An **infinite geometric series** is a series of the form

$$a + ar + ar^2 + ar^3 + ar^4 + \cdots + ar^{n-1} + \cdots$$

We can apply the reasoning used earlier to find the sum of an infinite geometric series. The nth partial sum of such a series is given by the formula

$$S_n = a\frac{1 - r^n}{1 - r} \qquad r \neq 1$$

It can be shown that if $|r| < 1$, then r^n gets close to 0 as n gets large (you can easily convince yourself of this using a calculator). It follows that S_n gets close to $a/(1 - r)$ as n gets large, or

$$S_n \to \frac{a}{1 - r} \quad \text{as} \quad n \to \infty$$

Thus the sum of this infinite geometric series is $a/(1 - r)$.

Here is another way to arrive at the formula for the sum of an infinite geometric series:

$$\begin{aligned} S &= a + ar + ar^2 + ar^3 + \cdots \\ &= a + r(a + ar + ar^2 + \cdots) \\ &= a + rS \end{aligned}$$

Solve the equation $S = a + rS$ for S to get

$$\begin{aligned} S - rS &= a \\ (1 - r)S &= a \\ S &= \frac{a}{1 - r} \end{aligned}$$

SUM OF AN INFINITE GEOMETRIC SERIES

If $|r| < 1$, then the infinite geometric series

$$\sum_{k=1}^{\infty} ar^{k-1} = a + ar + ar^2 + ar^3 + \cdots$$

converges and has the sum

$$S = \frac{a}{1 - r}$$

If $|r| \geq 1$, the series diverges.

EXAMPLE 6 ■ Infinite Series

Determine whether the infinite geometric series is convergent or divergent. If it is convergent, find its sum.

(a) $2 + \frac{2}{5} + \frac{2}{25} + \frac{2}{125} + \cdots$ **(b)** $1 + \frac{7}{5} + \left(\frac{7}{5}\right)^2 + \left(\frac{7}{5}\right)^3 + \cdots$

Mathematics in the Modern World

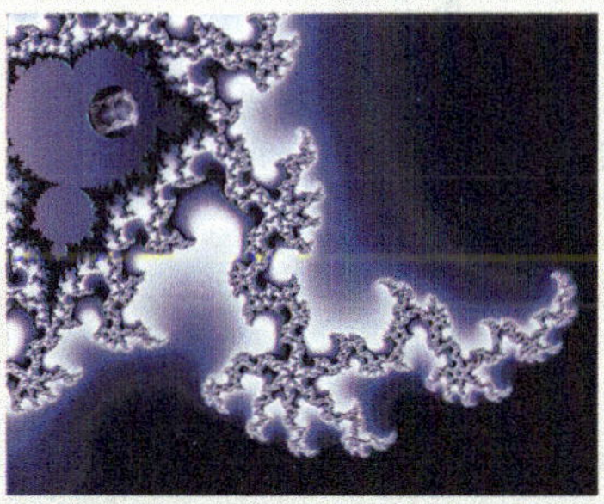

Bill Ross/Cusp/Corbis

Fractals

Many of the things we model in this book have regular predictable shapes. But recent advances in mathematics have made it possible to model such seemingly random or even chaotic shapes as those of a cloud, a flickering flame, a mountain, or a jagged coastline. The basic tools in this type of modeling are the fractals invented by the mathematician Benoit Mandelbrot. A *fractal* is a geometric shape built up from a simple basic shape by scaling and repeating the shape indefinitely according to a given rule. Fractals have infinite detail; this means the closer you look, the more you see. They are also *self-similar*; that is, zooming in on a portion of the fractal yields the same detail as the original shape. Because of their beautiful shapes, fractals are used by movie makers to create fictional landscapes and exotic backgrounds.

Although a fractal is a complex shape, it is produced according to very simple rules. This property of fractals is exploited in a process of storing pictures on a computer called *fractal image compression*. In this process a picture is stored as a simple basic shape and a rule; repeating the shape according to the rule produces the original picture. This is an extremely efficient method of storage; that's how thousands of color pictures can be put on a single flash drive.

SOLUTION

(a) This is an infinite geometric series with $a = 2$ and $r = \frac{1}{5}$. Since $|r| = \left|\frac{1}{5}\right| < 1$, the series converges. By the formula for the sum of an infinite geometric series we have

$$S = \frac{2}{1 - \frac{1}{5}} = \frac{5}{2}$$

(b) This is an infinite geometric series with $a = 1$ and $r = \frac{7}{5}$. Since $|r| = \left|\frac{7}{5}\right| > 1$, the series diverges.

Now Try Exercises 65 and 69 ■

EXAMPLE 7 ■ Writing a Repeated Decimal as a Fraction

Find the fraction that represents the rational number $2.3\overline{51}$.

SOLUTION This repeating decimal can be written as a series:

$$\frac{23}{10} + \frac{51}{1000} + \frac{51}{100{,}000} + \frac{51}{10{,}000{,}000} + \frac{51}{1{,}000{,}000{,}000} + \cdots$$

After the first term, the terms of this series form an infinite geometric series with

$$a = \frac{51}{1000} \qquad \text{and} \qquad r = \frac{1}{100}$$

Thus the sum of this part of the series is

$$S = \frac{\frac{51}{1000}}{1 - \frac{1}{100}} = \frac{\frac{51}{1000}}{\frac{99}{100}} = \frac{51}{1000} \cdot \frac{100}{99} = \frac{51}{990}$$

So

$$2.3\overline{51} = \frac{23}{10} + \frac{51}{990} = \frac{2328}{990} = \frac{388}{165}$$

Now Try Exercise 77 ■

12.3 EXERCISES

CONCEPTS

1. A geometric sequence is a sequence in which the ________ of successive terms is constant.

2. The sequence given by $a_n = ar^{n-1}$ is a geometric sequence in which a is the first term and r is the ________ ________. So for the geometric sequence $a_n = 2(5)^{n-1}$ the first term is ________, and the common ratio is ________.

3. *True or False*? If we know the first and second terms of a geometric sequence, then we can find any other term.

4. (a) The nth partial sum of a geometric sequence $a_n = ar^{n-1}$ is given by $S_n =$ ________.

(b) The series $\sum_{k=1}^{\infty} ar^{k-1} = a + ar + ar^2 + ar^3 + \cdots$ is an infinite ________ series. If $|r| < 1$, then this series ________, and its sum is $S =$ ________. If $|r| \geq 1$, the series ________.

SKILLS

5–8 ■ *n*th Term of a Geometric Sequence The nth term of a sequence is given. **(a)** Find the first five terms of the sequence. **(b)** What is the common ratio r? **(c)** Graph the terms you found in (a).

5. $a_n = 7(3)^{n-1}$

6. $a_n = 6(-0.5)^{n-1}$

7. $a_n = \frac{5}{2}\left(-\frac{1}{2}\right)^{n-1}$

8. $a_n = 3^{n-1}$

9–12 ■ *n*th Term of a Geometric Sequence Find the nth term of the geometric sequence with given first term a and common ratio r. What is the fourth term?

9. $a = 7, \quad r = 4$

10. $a = -3, \quad r = -2$

11. $a = \frac{5}{2}, \quad r = -\frac{1}{2}$

12. $a = \sqrt{3}, \quad r = \sqrt{3}$

13–22 ■ Geometric Sequence? The first four terms of a sequence are given. Determine whether these terms can be the terms of a geometric sequence. If the sequence is geometric, find the common ratio.

13. 3, 6, 12, 24, . . .

14. 3, 48, 93, 138, . . .

15. 3072, 1536, 768, 384, . . .

16. 432, −144, 48, −16, . . .

17. $3, \frac{3}{2}, \frac{3}{4}, \frac{3}{8}, \ldots$

18. 27, −9, 3, −1, . . .

19. $\frac{1}{2}, \frac{1}{3}, \frac{1}{4}, \frac{1}{5}, \ldots$

20. $e^2, e^4, e^6, e^8, \ldots$

21. 1.0, 1.1, 1.21, 1.331, . . .

22. $\frac{1}{2}, \frac{1}{4}, \frac{1}{6}, \frac{1}{8}, \ldots$

23–28 ■ Geometric Sequence? Find the first five terms of the sequence, and determine whether it is geometric. If it is geometric, find the common ratio, and express the nth term of the sequence in the standard form $a_n = ar^{n-1}$.

23. $a_n = 2(3)^n$

24. $a_n = 4 + 3^n$

25. $a_n = \dfrac{1}{4^n}$

26. $a_n = (-1)^n 2^n$

27. $a_n = \ln(5^{n-1})$

28. $a_n = n^n$

29–38 ■ Terms of a Geometric Sequence Determine the common ratio, the fifth term, and the nth term of the geometric sequence.

29. 2, 6, 18, 54, . . .

30. $7, \frac{14}{3}, \frac{28}{9}, \frac{56}{27}, \ldots$

31. 0.3, −0.09, 0.027, −0.0081, . . .

32. $1, \sqrt{2}, 2, 2\sqrt{2}, \ldots$

33. $144, -12, 1, -\frac{1}{12}, \ldots$

34. $-8, -2, -\frac{1}{2}, -\frac{1}{8}, \ldots$

35. $3, 3^{5/3}, 3^{7/3}, 27, \ldots$

36. $t, \dfrac{t^2}{2}, \dfrac{t^3}{4}, \dfrac{t^4}{8}, \ldots$

37. $1, s^{2/7}, s^{4/7}, s^{6/7}, \ldots$

38. $5, 5^{c+1}, 5^{2c+1}, 5^{3c+1}, \ldots$

39–46 ■ Finding Terms of a Geometric Sequence Find the indicated term(s) of the geometric sequence with the given description.

39. The first term is 15 and the second term is 6. Find the fourth term.

40. The first term is $\frac{1}{12}$ and the second term is $-\frac{1}{2}$. Find the sixth term.

41. The third term is $-\frac{1}{3}$ and the sixth term is 9. Find the first and second terms.

42. The fourth term is 12 and the seventh term is $\frac{32}{9}$. Find the first and nth terms.

43. The third term is −18 and the sixth term is 9216. Find the first and nth terms.

44. The third term is −54 and the sixth term is $\frac{729}{256}$. Find the first and second terms.

45. The common ratio is 0.75 and the fourth term is 729. Find the first three terms.

46. The common ratio is $\frac{1}{6}$ and the third term is 18. Find the first and seventh terms.

47. Which Term? The first term of a geometric sequence is 1536 and the common ratio is $\frac{1}{2}$. Which term of the sequence is 6?

48. Which Term? The second and fifth terms of a geometric sequence are 30 and 3750, respectively. Which term of the sequence is 468,750?

49–52 ■ Partial Sums of a Geometric Sequence Find the partial sum S_n of the geometric sequence that satisfies the given conditions.

49. $a = 5, \; r = 2, \; n = 6$

50. $a = \frac{2}{3}, \; r = \frac{1}{3}, \; n = 4$

51. $a_3 = 28, \; a_6 = 224, \; n = 6$

52. $a_2 = 0.12, \; a_5 = 0.00096, \; n = 4$

53–58 ■ Partial Sums of a Geometric Sequence Find the sum.

53. $1 + 3 + 9 + \cdots + 2187$

54. $1 - \frac{1}{2} + \frac{1}{4} - \frac{1}{8} + \cdots - \frac{1}{512}$

55. $-15 + 30 - 60 + \cdots - 960$

56. $5120 + 2560 + 1280 + \cdots + 20$

57. $1.25 + 12.5 + 125 + \cdots + 12{,}500{,}000$

58. $10800 + 1080 + 108 + \cdots + 0.000108$

59–64 ■ Partial Sums of a Geometric Sequence Find the sum.

59. $\sum_{k=1}^{5} 3\left(\frac{1}{2}\right)^{k-1}$

60. $\sum_{k=1}^{5} 8\left(-\frac{3}{2}\right)^{k-1}$

61. $\sum_{k=1}^{6} 5(-2)^{k-1}$

62. $\sum_{k=1}^{6} 10(5)^{k-1}$

63. $\sum_{k=1}^{5} 3\left(\frac{2}{3}\right)^{k-1}$

64. $\sum_{k=1}^{6} 64\left(\frac{3}{2}\right)^{k-1}$

65–76 ■ Infinite Geometric Series Determine whether the infinite geometric series is convergent or divergent. If it is convergent, find its sum.

65. $1 + \dfrac{1}{3} + \dfrac{1}{9} + \dfrac{1}{27} + \cdots$

66. $1 - \dfrac{1}{2} + \dfrac{1}{4} - \dfrac{1}{8} + \cdots$

67. $1 - \dfrac{1}{3} + \dfrac{1}{9} - \dfrac{1}{27} + \cdots$

68. $\dfrac{2}{5} + \dfrac{4}{25} + \dfrac{8}{125} + \cdots$

69. $1 + \dfrac{3}{2} + \left(\dfrac{3}{2}\right)^2 + \left(\dfrac{3}{2}\right)^3 + \cdots$

70. $\dfrac{1}{3^6} + \dfrac{1}{3^8} + \dfrac{1}{3^{10}} + \dfrac{1}{3^{12}} + \cdots$

71. $3 - \dfrac{3}{2} + \dfrac{3}{4} - \dfrac{3}{8} + \cdots$

72. $1 - 1 + 1 - 1 + \cdots$

73. $3 - 3(1.1) + 3(1.1)^2 - 3(1.1)^3 + \cdots$

74. $-\dfrac{100}{9} + \dfrac{10}{3} - 1 + \dfrac{3}{10} - \cdots$

75. $\dfrac{1}{\sqrt{2}} + \dfrac{1}{2} + \dfrac{1}{2\sqrt{2}} + \dfrac{1}{4} + \cdots$

76. $1 - \sqrt{2} + 2 - 2\sqrt{2} + 4 - \cdots$

77–82 ■ Repeated Decimal Express the repeating decimal as a fraction.

77. 0.777 . . .

78. $0.\overline{253}$

79. 0.030303 . . .

80. $2.11\overline{25}$

81. $0.\overline{112}$

82. 0.123123123 . . .

SKILLS Plus

83. Geometric Means If the numbers $a_1, a_2, \ldots, a_n$ form a geometric sequence, then $a_2, a_3, \ldots, a_{n-1}$ are **geometric means** between a_1 and a_n. Insert three geometric means between 5 and 80.

84. Partial Sum of a Geometric Sequence Find the sum of the first ten terms of the sequence

$$a + b, a^2 + 2b, a^3 + 3b, a^4 + 4b, \ldots$$

85–86 ■ Arithmetic or Geometric? The first four terms of a sequence are given. Determine whether these terms can be the terms of an arithmetic sequence, a geometric sequence, or neither. If the sequence is arithmetic or geometric, find the next term.

85. **(a)** $5, -3, 5, -3, \ldots$ **(b)** $\frac{1}{3}, 1, \frac{5}{3}, \frac{7}{3}, \ldots$

(c) $\sqrt{3}, 3, 3\sqrt{3}, 9, \ldots$ **(d)** $-3, -\frac{3}{2}, 0, \frac{3}{2}, \ldots$

86. **(a)** $1, -1, 1, -1, \ldots$ **(b)** $\sqrt{5}, \sqrt[3]{5}, \sqrt[6]{5}, 1, \ldots$

(c) $2, -1, \frac{1}{2}, 2, \ldots$ **(d)** $x - 1, x, x + 1, x + 2, \ldots$

APPLICATIONS

87. Depreciation A construction company purchases a bulldozer for \$160,000. Each year the value of the bulldozer depreciates by 20% of its value in the preceding year. Let V_n be the value of the bulldozer in the nth year. (Let $n = 1$ be the year the bulldozer is purchased.)

(a) Find a formula for V_n.

(b) In what year will the value of the bulldozer be less than \$100,000?

88. Family Tree A person has two parents, four grandparents, eight great-grandparents, and so on. How many ancestors does a person have 15 generations back?

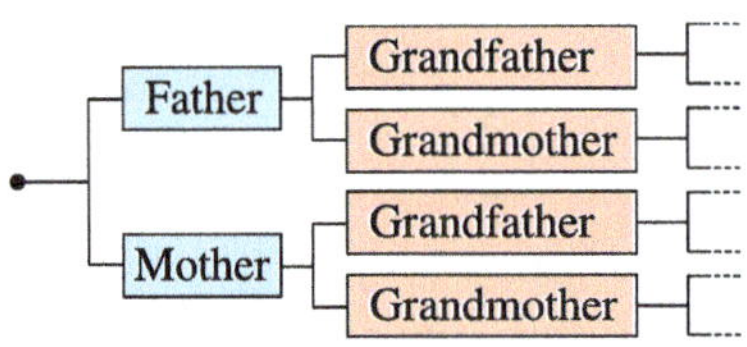

89. Bouncing Ball A ball is dropped from a height of 80 ft. The elasticity of this ball is such that it rebounds three-fourths of the distance it has fallen. How high does the ball rebound on the fifth bounce? Find a formula for how high the ball rebounds on the nth bounce.

90. Bacteria Culture A culture initially has 5000 bacteria, and its size increases by 8% every hour. How many bacteria are present at the end of 5 hours? Find a formula for the number of bacteria present after n hours.

91. Mixing Coolant A truck radiator holds 5 gal and is filled with water. A gallon of water is removed from the radiator and replaced with a gallon of antifreeze; then a gallon of the mixture is removed from the radiator and again replaced by a gallon of antifreeze. This process is repeated indefinitely. How much water remains in the tank after this process is repeated 3 times? 5 times? n times?

92. Musical Frequencies The frequencies of musical notes (measured in cycles per second) form a geometric sequence. Middle C has a frequency of 256, and the C that is an octave higher has a frequency of 512. Find the frequency of C two octaves below middle C.

93. Bouncing Ball A ball is dropped from a height of 9 ft. The elasticity of the ball is such that it always bounces up one-third the distance it has fallen.

(a) Find the total distance the ball has traveled at the instant it hits the ground the fifth time.

(b) Find a formula for the total distance the ball has traveled at the instant it hits the ground the nth time.

94. Geometric Savings Plan A very patient woman wishes to become a billionaire. She decides to follow a simple scheme: She puts aside 1 cent the first day, 2 cents the second day, 4 cents the third day, and so on, doubling the number of cents each day. How much money will she have at the end of 30 days? How many days will it take this woman to realize her wish?

95. St. Ives The following is a well-known children's rhyme:

As I was going to St. Ives,
I met a man with seven wives;
Every wife had seven sacks;
Every sack had seven cats;
Every cat had seven kits;
Kits, cats, sacks, and wives,
How many were going to St. Ives?

Assuming that the entire group is actually going to St. Ives, show that the answer to the question in the rhyme is a partial sum of a geometric sequence, and find the sum.

96. Drug Concentration A certain drug is administered once a day. The concentration of the drug in the patient's bloodstream increases rapidly at first, but each successive dose has less effect than the preceding one. The total amount of the drug (in mg) in the bloodstream after the nth dose is given by

$$\sum_{k=1}^{n} 50\left(\tfrac{1}{2}\right)^{k-1}$$

(a) Find the amount of the drug in the bloodstream after $n = 10$ days.

(b) If the drug is taken on a long-term basis, the amount in the bloodstream is approximated by the infinite series $\sum_{k=1}^{\infty} 50\left(\tfrac{1}{2}\right)^{k-1}$. Find the sum of this series.

97. Bouncing Ball A certain ball rebounds to half the height from which it is dropped. Use an infinite geometric series to approximate the total distance the ball travels after being dropped from 1 m above the ground until it comes to rest.

98. Bouncing Ball If the ball in Exercise 97 is dropped from a height of 8 ft, then 1 s is required for its first complete bounce—from the instant it first touches the ground until it next touches the ground. Each subsequent complete bounce requires $1/\sqrt{2}$ as long as the preceding complete bounce. Use an infinite geometric series to estimate the time interval from the instant the ball first touches the ground until it stops bouncing.

99. Geometry The midpoints of the sides of a square of side 1 are joined to form a new square. This procedure is repeated for each new square. (See the figure.)

(a) Find the sum of the areas of all the squares.

(b) Find the sum of the perimeters of all the squares.

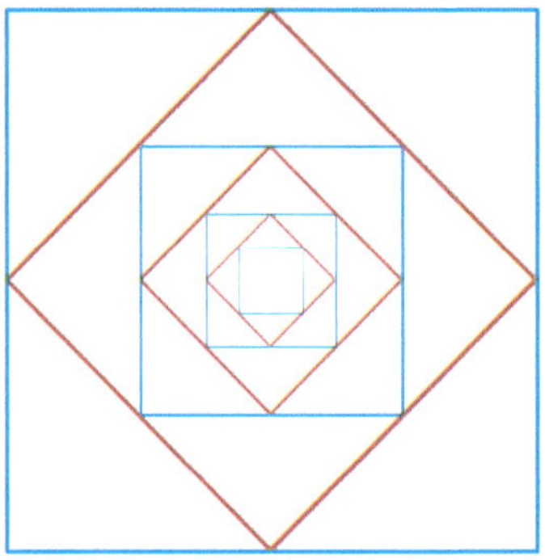

100. Geometry A circular disk of radius R is cut out of paper, as shown in figure (a). Two disks of radius $\frac{1}{2}R$ are cut out of paper and placed on top of the first disk, as in figure (b), and then four disks of radius $\frac{1}{4}R$ are placed on these two disks, as in figure (c). Assuming that this process can be repeated indefinitely, find the total area of all the disks.

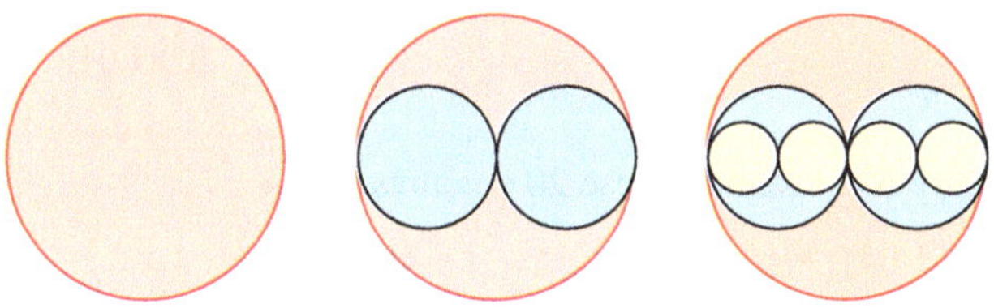

(a) (b) (c)

101. Geometry A yellow square of side 1 is divided into nine smaller squares, and the middle square is colored blue as shown in the figure. Each of the smaller yellow squares is in turn divided into nine squares, and each middle square is colored blue. If this process is continued indefinitely, what is the total area that is colored blue?

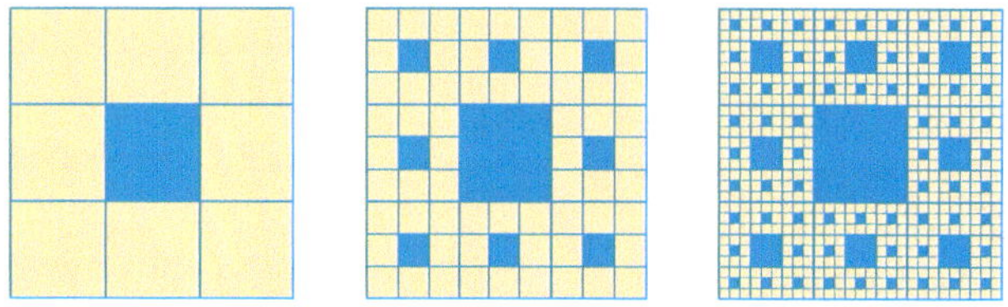

DISCUSS ■ DISCOVER ■ PROVE ■ WRITE

102. PROVE: Reciprocals of a Geometric Sequence If $a_1, a_2, a_3, \ldots$ is a geometric sequence with common ratio r, show that the sequence

$$\frac{1}{a_1}, \frac{1}{a_2}, \frac{1}{a_3}, \ldots$$

is also a geometric sequence, and find the common ratio.

103. PROVE: Logarithms of a Geometric Sequence If $a_1, a_2, a_3, \ldots$ is a geometric sequence with a common ratio $r > 0$ and $a_1 > 0$, show that the sequence

$$\log a_1, \log a_2, \log a_3, \ldots$$

is an arithmetic sequence, and find the common difference.

104. PROVE: Exponentials of an Arithmetic Sequence If $a_1, a_2, a_3, \ldots$ is an arithmetic sequence with common difference d, show that the sequence

$$10^{a_1}, 10^{a_2}, 10^{a_3}, \ldots$$

is a geometric sequence, and find the common ratio.

12.4 MATHEMATICS OF FINANCE

■ The Amount of an Annuity ■ The Present Value of an Annuity ■ Installment Buying

Many financial transactions involve payments that are made at regular intervals. For example, if you deposit \$100 each month in an interest-bearing account, what will the value of your account be at the end of 5 years? If you borrow \$100,000 to buy a house, how much must your monthly payments be in order to pay off the loan in 30 years? Each of these questions involves the sum of a sequence of numbers; we use the results of the preceding section to answer them here.

■ The Amount of an Annuity

An **annuity** is a sum of money that is paid in regular equal payments. Although the word *annuity* suggests annual (or yearly) payments, they can be made semiannually, quarterly, monthly, or at some other regular interval. Payments are usually made at the end of the payment interval. The **amount of an annuity** is the sum of all the individual payments from the time of the first payment until the last payment is made, together with all the interest. We denote this sum by A_f (the subscript f here is used to denote *final* amount).

EXAMPLE 1 ■ Calculating the Amount of an Annuity

An investor deposits \$400 every December 15 and June 15 for 10 years in an account that earns interest at the rate of 8% per year, compounded semiannually. How much will be in the account immediately after the last payment?

When using interest rates in calculators, remember to convert percentages to decimals. For example, 8% is 0.08.

SOLUTION We need to find the amount of an annuity consisting of 20 semiannual payments of \$400 each. Since the interest rate is 8% per year, compounded semiannually, the interest rate per time period is $i = 0.08/2 = 0.04$. The first payment is in the account for 19 time periods, the second for 18 time periods, and so on.

The last payment receives no interest. The situation can be illustrated by the time line in Figure 1.

FIGURE 1

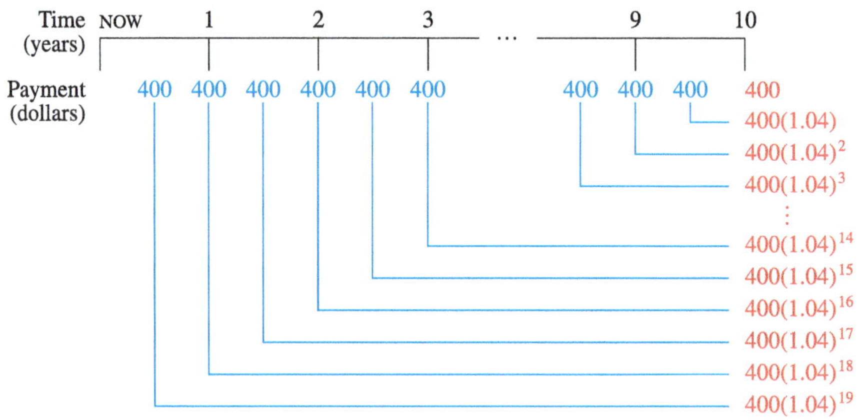

The amount A_f of the annuity is the sum of these 20 amounts. Thus

$$A_f = 400 + 400(1.04) + 400(1.04)^2 + \cdots + 400(1.04)^{19}$$

But this is a geometric series with $a = 400$, $r = 1.04$, and $n = 20$, so

$$A_f = 400\frac{1-(1.04)^{20}}{1-1.04} \approx 11{,}911.23$$

Thus the amount in the account after the last payment is \$11,911.23.

Now Try Exercise 3 ■

In general, the regular annuity payment is called the **periodic rent** and is denoted by R. We also let i denote the interest rate per time period and let n denote the number of payments. *We always assume that the time period in which interest is compounded is equal to the time between payments.* By the same reasoning as in Example 1, we see that the amount A_f of an annuity is

$$A_f = R + R(1+i) + R(1+i)^2 + \cdots + R(1+i)^{n-1}$$

Since this is the nth partial sum of a geometric sequence with $a = R$ and $r = 1 + i$, the formula for the partial sum gives

$$A_f = R\frac{1-(1+i)^n}{1-(1+i)} = R\frac{1-(1+i)^n}{-i} = R\frac{(1+i)^n-1}{i}$$

Mathematics in the Modern World

Mathematical Economics

The health of the global economy is determined by such interrelated factors as supply, demand, production, consumption, pricing, distribution, and thousands of other factors. These factors are in turn determined by economic decisions (for example, whether or not you buy a certain brand of toothpaste) made by billions of different individuals each day. How will today's creation and distribution of goods affect tomorrow's economy? Such questions are tackled by mathematicians who work on mathematical models of the economy. In the 1940s Wassily Leontief, a pioneer in this area, created a model consisting of thousands of equations that describe how different sectors of the economy, such as the oil industry, transportation, and communication, interact with each other. A different approach to economic models, one dealing with individuals in the economy as opposed to large sectors, was pioneered by John Nash in the 1950s. In his model, which uses ***game theory,*** the economy is a game where individual players make decisions that often lead to mutual gain. Leontief and Nash were awarded the Nobel Prize in Economics in 1973 and 1994, respectively. Economic theory continues to be a major area of mathematical research.

AMOUNT OF AN ANNUITY

The amount A_f of an annuity consisting of n regular equal payments of size R with interest rate i per time period is given by

$$A_f = R\,\frac{(1+i)^n - 1}{i}$$

EXAMPLE 2 ■ Calculating the Amount of an Annuity

How much money should be invested every month at 12% per year, compounded monthly, in order to have \$4000 in 18 months?

SOLUTION In this problem $i = 0.12/12 = 0.01$, $A_f = 4000$, and $n = 18$. We need to find the amount R of each payment. By the formula for the amount of an annuity,

$$4000 = R\,\frac{(1+0.01)^{18} - 1}{0.01}$$

Solving for R, we get

$$R = \frac{4000(0.01)}{(1+0.01)^{18} - 1} \approx 203.928$$

Thus the monthly investment should be \$203.93.

Now Try Exercise 9 ■

■ The Present Value of an Annuity

If you were to receive \$10,000 five years from now, it would be worth much less than if you got \$10,000 right now. This is because of the interest you could accumulate during the next 5 years if you invested the money now. What smaller amount would you be willing to accept *now* instead of receiving \$10,000 in 5 years? This is the amount of money that, together with interest, would be worth \$10,000 in 5 years. The amount that we are looking for here is called the *discounted value* or *present value*. If the interest rate is 8% per year, compounded quarterly, then the interest per time period is $i = 0.08/4 = 0.02$, and there are $4 \times 5 = 20$ time periods. If we let PV denote the present value, then by the formula for compound interest (Section 4.1) we have

$$10{,}000 = PV(1+i)^n = PV(1+0.02)^{20}$$

so

$$PV = 10{,}000(1+0.02)^{-20} \approx 6729.713$$

Thus in this situation the present value of \$10,000 is \$6729.71. This reasoning leads to a general formula for present value. If an amount A_f is to be paid in a lump sum n time periods from now and the interest rate per time period is i, then its **present value** A_p is given by

$$A_p = A_f(1+i)^{-n}$$

Similarly, the **present value of an annuity** is the amount A_p that must be invested now at the interest rate i per time period to provide n payments, each of amount R. Clearly, A_p is the sum of the present values of each individual payment (see Exercise 29). Another way of finding A_p is to note that A_p is the present value of A_f:

$$A_p = A_f(1+i)^{-n} = R\,\frac{(1+i)^n - 1}{i}(1+i)^{-n} = R\,\frac{1-(1+i)^{-n}}{i}$$

THE PRESENT VALUE OF AN ANNUITY

The **present value** A_p of an annuity consisting of n regular equal payments of size R and interest rate i per time period is given by

$$A_p = R\frac{1-(1+i)^{-n}}{i}$$

EXAMPLE 3 ■ Calculating the Present Value of an Annuity

A person wins \$10,000,000 in the California lottery, and the amount is paid in yearly installments of half a million dollars each for 20 years. What is the present value of his winnings? Assume that he can earn 10% interest, compounded annually.

SOLUTION Since the amount won is paid as an annuity, we need to find its present value. Here $i = 0.1$, $R = \$500{,}000$, and $n = 20$. Thus

$$A_p = 500{,}000\frac{1-(1+0.1)^{-20}}{0.1} \approx 4{,}256{,}781.859$$

This means that the winner really won only \$4,256,781.86 if it were paid immediately.

Now Try Exercise 11 ■

■ Installment Buying

When you buy a house or a car by installment, the payments that you make are an annuity whose present value is the amount of the loan.

EXAMPLE 4 ■ The Amount of a Loan

A student wishes to buy a car. She can afford to pay \$200 per month but has no money for a down payment. If she can make these payments for 4 years and the interest rate is 12%, what purchase price can she afford?

SOLUTION The payments that the student makes constitute an annuity whose present value is the price of the car (which is also the amount of the loan, in this case). Here we have $i = 0.12/12 = 0.01$, $R = 200$, and $n = 12 \times 4 = 48$, so

$$A_p = R\frac{1-(1+i)^{-n}}{i} = 200\frac{1-(1+0.01)^{-48}}{0.01} \approx 7594.792$$

Thus the student can buy a car priced at \$7594.79.

Now Try Exercise 19 ■

When a bank makes a loan that is to be repaid with regular equal payments R, then the payments form an annuity whose present value A_p is the amount of the loan. So to find the size of the payments, we solve for R in the formula for the amount of an annuity. This gives the following formula for R.

INSTALLMENT BUYING

If a loan A_p is to be repaid in n regular equal payments with interest rate i per time period, then the size R of each payment is given by

$$R = \frac{iA_p}{1-(1+i)^{-n}}$$

EXAMPLE 5 ■ Calculating Monthly Mortgage Payments

A couple borrows \$100,000 at 9% interest as a mortage loan on a house. They expect to make monthly payments for 30 years to repay the loan. What is the size of each payment?

SOLUTION The mortgage payments form an annuity whose present value is $A_p = \$100{,}000$. Also, $i = 0.09/12 = 0.0075$, and $n = 12 \times 30 = 360$. We are looking for the amount R of each payment.

From the formula for installment buying we get

$$R = \frac{iA_p}{1 - (1 + i)^{-n}} = \frac{(0.0075)(100{,}000)}{1 - (1 + 0.0075)^{-360}} \approx 804.623$$

Thus the monthly payments are \$804.62.

Now Try Exercise 15 ■

We now illustrate the use of graphing devices in solving problems related to installment buying.

EXAMPLE 6 ■ Calculating the Interest Rate from the Size of Monthly Payments

A car dealer sells a new car for \$18,000. He offers the buyer payments of \$405 per month for 5 years. What interest rate is this car dealer charging?

SOLUTION The payments form an annuity with present value $A_p = \$18{,}000$, $R = 405$, and $n = 12 \times 5 = 60$. To find the interest rate, we must solve for i in the equation

$$R = \frac{iA_p}{1 - (1 + i)^{-n}}$$

A little experimentation will convince you that it is not possible to solve this equation for i algebraically. So to find i, we use a graphing device to graph R as a function of the interest rate x, and we then use the graph to find the interest rate corresponding to the value of R we want (\$405 in this case). Since $i = x/12$, we graph the function

$$R(x) = \frac{\frac{x}{12}(18{,}000)}{1 - \left(1 + \frac{x}{12}\right)^{-60}}$$

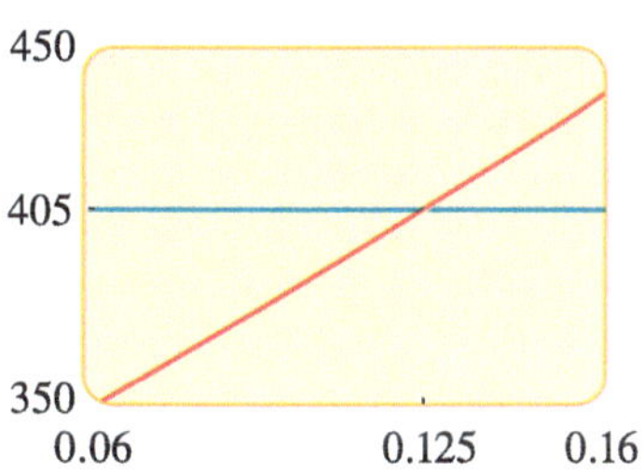

FIGURE 2

in the viewing rectangle $[0.06, 0.16] \times [350, 450]$, as shown in Figure 2. We also graph the horizontal line $R(x) = 405$ in the same viewing rectangle. Then, by moving the cursor to the point of intersection of the two graphs, we find that the corresponding x-value is approximately 0.125. Thus the interest rate is about $12\frac{1}{2}\%$.

Now Try Exercise 25 ■

12.4 EXERCISES

CONCEPTS

1. An annuity is a sum of money that is paid in regular equal payments. The ________ of an annuity is the sum of all the individual payments together with all the interest.

2. The ________ ________ of an annuity is the amount that must be invested now at interest rate i per time period to provide n payments each of amount R.

APPLICATIONS

3. **Annuity** Find the amount of an annuity that consists of ten annual payments of $1000 each into an account that pays 6% interest per year.

4. **Annuity** Find the amount of an annuity that consists of 24 monthly payments of $500 each into an account that pays 8% interest per year, compounded monthly.

5. **Annuity** Find the amount of an annuity that consists of 20 annual payments of $5000 each into an account that pays interest of 12% per year.

6. **Annuity** Find the amount of an annuity that consists of 20 semiannual payments of $500 each into an account that pays 6% interest per year, compounded semiannually.

7. **Annuity** Find the amount of an annuity that consists of 16 quarterly payments of $300 each into an account that pays 8% interest per year, compounded quarterly.

8. **Annuity** Find the amount of an annuity that consists of 40 annual payments of $2000 each into an account that pays interest of 5% per year.

9. **Saving** How much money should be invested every quarter at 10% per year, compounded quarterly, to have $5000 in 2 years?

10. **Saving** How much money should be invested monthly at 6% per year, compounded monthly, to have $2000 in 8 months?

11. **Annuity** What is the present value of an annuity that consists of 20 semiannual payments of $1000 at an interest rate of 9% per year, compounded semiannually?

12. **Annuity** What is the present value of an annuity that consists of 30 monthly payments of $300 at an interest rate of 8% per year, compounded monthly?

13. **Funding an Annuity** How much money must be invested now at 9% per year, compounded semiannually, to fund an annuity of 20 payments of $200 each, paid every 6 months, the first payment being 6 months from now?

14. **Funding an Annuity** A 55-year-old man deposits $50,000 to fund an annuity with an insurance company. The money will be invested at 8% per year, compounded semiannually. He is to draw semiannual payments until he reaches age 65. What is the amount of each payment?

15. **Financing a Car** A woman wants to borrow $12,000 to buy a car. She wants to repay the loan by monthly installments for 4 years. If the interest rate on this loan is $10\frac{1}{2}$% per year, compounded monthly, what is the amount of each payment?

16. **Mortgage** What is the monthly payment on a 30-year mortgage of $80,000 at 9% interest? What is the monthly payment on this same mortgage if it is to be repaid over a 15-year period?

17. **Mortgage** What is the monthly payment on a 30-year mortgage of $100,000 at 8% interest per year, compounded monthly? What is the total amount paid on this loan over the 30-year period?

18. **Mortgage** What is the monthly payment on a 15-year mortgage of $200,000 at 6% interest? What is the total amount paid on this loan over the 15-year period?

19. **Mortgage** Dr. Gupta is considering a 30-year mortgage at 6% interest. She can make payments of $3500 a month. What size loan can she afford?

20. **Mortgage** A couple can afford to make a monthly mortgage payment of $650. If the mortgage rate is 9% and the couple intends to secure a 30-year mortgage, how much can they borrow?

21. **Financing a Car** Jane agrees to buy a car for a down payment of $2000 and payments of $220 per month for 3 years. If the interest rate is 8% per year, compounded monthly, what is the actual purchase price of her car?

22. **Financing a Ring** Mike buys a ring for his fiancee by paying $30 a month for one year. If the interest rate is 10% per year, compounded monthly, what is the price of the ring?

23. **Mortgage** A couple secures a 30-year loan of $100,000 at $9\frac{3}{4}$% per year, compounded monthly, to buy a house.
   (a) What is the amount of their monthly payment?
   (b) What total amount will they pay over the 30-year period?
   (c) If, instead of taking the loan, the couple deposits the monthly payments in an account that pays $9\frac{3}{4}$% interest per year, compounded monthly, how much will be in the account at the end of the 30-year period?

24. **Mortgage** A couple needs a mortgage of $300,000. Their mortgage broker presents them with two options: a 30-year mortgage at $6\frac{1}{2}$% interest or a 15-year mortgage at $5\frac{3}{4}$% interest.
   (a) Find the monthly payment on the 30-year mortgage and on the 15-year mortgage. Which mortgage has the larger monthly payment?
   (b) Find the total amount to be paid over the life of each loan. Which mortgage has the lower total payment over its lifetime?

25. **Interest Rate** John buys a stereo system for $640. He agrees to pay $32 a month for 2 years. Assuming that interest is compounded monthly, what interest rate is he paying?

26. **Interest Rate** Janet's payments on her $12,500 car are $420 a month for 3 years. Assuming that interest is compounded monthly, what interest rate is she paying on the car loan?

27. **Interest Rate** An item at a department store is priced at $189.99 and can be bought by making 20 payments of $10.50. Find the interest rate, assuming that interest is compounded monthly.

28. **Interest Rate** A man purchases a $2000 diamond ring for a down payment of $200 and monthly installments of $88 for 2 years. Assuming that interest is compounded monthly, what interest rate is he paying?

DISCUSS ■ DISCOVER ■ PROVE ■ WRITE

29. **DISCOVER: Present Value of an Annuity**
   (a) Draw a time line as in Example 1 to show that the present value of an annuity is the sum of the present values of each payment, that is,

$$A_p = \frac{R}{1+i} + \frac{R}{(1+i)^2} + \frac{R}{(1+i)^3} + \cdots + \frac{R}{(1+i)^n}$$

   (b) Use part (a) to derive the formula for A_p given in the text.

30. DISCOVER: An Annuity That Lasts Forever An **annuity in perpetuity** is one that continues forever. Such annuities are useful in setting up scholarship funds to ensure that the award continues.

(a) Draw a time line (as in Example 1) to show that to set up an annuity in perpetuity of amount R per time period, the amount that must be invested now is

$$A_p = \frac{R}{1+i} + \frac{R}{(1+i)^2} + \frac{R}{(1+i)^3} + \cdots + \frac{R}{(1+i)^n} + \cdots$$

where i is the interest rate per time period.

(b) Find the sum of the infinite series in part (a) to show that

$$A_p = \frac{R}{i}$$

(c) How much money must be invested now at 10% per year, compounded annually, to provide an annuity in perpetuity of \$5000 per year? The first payment is due in 1 year.

(d) How much money must be invested now at 8% per year, compounded quarterly, to provide an annuity in perpetuity of \$3000 per year? The first payment is due in 1 year.

31. DISCOVER: Amortizing a Mortgage When they bought their house, John and Mary took out a \$90,000 mortgage at 9% interest, repayable monthly over 30 years. Their payment is \$724.17 per month (check this, using the formula in the text). The bank gave them an **amortization schedule**, which is a table showing how much of each payment is interest, how much goes toward the principal, and the remaining principal after each payment. The table below shows the first few entries in the amortization schedule.

Payment number	Total payment	Interest payment	Principal payment	Remaining principal
1	724.17	675.00	49.17	89,950.83
2	724.17	674.63	49.54	89,901.29
3	724.17	674.26	49.91	89,851.38
4	724.17	673.89	50.28	89,801.10

After 10 years they have made 120 payments and are wondering how much they still owe, but they have lost the amortization schedule.

(a) How much do John and Mary still owe on their mortgage? [*Hint:* The remaining balance is the present value of the 240 remaining payments.]

(b) How much of their next payment is interest, and how much goes toward the principal? [*Hint:* Since $9\% \div 12 = 0.75\%$, they must pay 0.75% of the remaining principal in interest each month.]

12.5 MATHEMATICAL INDUCTION

■ Conjecture and Proof ■ Mathematical Induction

There are two aspects to mathematics—discovery and proof—and they are of equal importance. We must discover something before we can attempt to prove it, and we cannot be certain of its truth until it has been proved. In this section we examine the relationship between these two key components of mathematics more closely.

■ Conjecture and Proof

Let's try a simple experiment. We add more and more of the odd numbers as follows:

$$1 = 1$$
$$1 + 3 = 4$$
$$1 + 3 + 5 = 9$$
$$1 + 3 + 5 + 7 = 16$$
$$1 + 3 + 5 + 7 + 9 = 25$$

What do you notice about the numbers on the right-hand side of these equations? They are, in fact, all perfect squares. These equations say the following:

The sum of the first 1 odd number is 1^2.

The sum of the first 2 odd numbers is 2^2.

The sum of the first 3 odd numbers is 3^2.

The sum of the first 4 odd numbers is 4^2.

The sum of the first 5 odd numbers is 5^2.

Consider the polynomial

$$p(n) = n^2 - n + 41$$

Here are some values of $p(n)$:

$$p(1) = 41 \quad p(2) = 43$$
$$p(3) = 47 \quad p(4) = 53$$
$$p(5) = 61 \quad p(6) = 71$$
$$p(7) = 83 \quad p(8) = 97$$

All the values so far are prime numbers. In fact, if you keep going, you will find that $p(n)$ is prime for all natural numbers up to $n = 40$. It might seem reasonable at this point to conjecture that $p(n)$ is prime for *every* natural number n. But that conjecture would be too hasty, because it is easily seen that $p(41)$ is *not* prime. This illustrates that we cannot be certain of the truth of a statement no matter how many special cases we check. We need a convincing argument—a *proof*—to determine the truth of a statement.

This leads naturally to the following question: Is it true that for every natural number n, the sum of the first n odd numbers is n^2? Could this remarkable property be true? We could try a few more numbers and find that the pattern persists for the first 6, 7, 8, 9, and 10 odd numbers. At this point we feel fairly confident that this is always true, so we make a *conjecture*:

The sum of the first n odd numbers is n^2.

Since we know that the nth odd number is $2n - 1$, we can write this statement more precisely as

$$1 + 3 + 5 + \cdots + (2n - 1) = n^2$$

It is important to realize that this is still a conjecture. We cannot conclude by checking a finite number of cases that a property is true for all numbers (there are infinitely many). To see this more clearly, suppose someone tells us that he has added up the first trillion odd numbers and found that they do *not* add up to 1 trillion squared. What would you tell this person? It would be silly to say that you're sure it's true because you have already checked the first five cases. You could, however, take out paper and pencil and start checking it yourself, but this task would probably take the rest of your life. The tragedy would be that after completing this task, you would still not be sure of the truth of the conjecture! Do you see why?

Herein lies the power of mathematical proof. A **proof** is a clear argument that demonstrates the truth of a statement beyond doubt.

Mathematical Induction

Let's consider a special kind of proof called **mathematical induction**. Here is how it works: Suppose we have a statement that says something about all natural numbers n. For example, for any natural number n, let $P(n)$ be the following statement:

$P(n)$: The sum of the first n odd numbers is n^2

Since this statement is about all natural numbers, it contains infinitely many statements; we will call them $P(1)$, $P(2)$, ...

$P(1)$: The sum of the first 1 odd number is 1^2.

$P(2)$: The sum of the first 2 odd numbers is 2^2.

$P(3)$: The sum of the first 3 odd numbers is 3^2.

$\vdots \qquad\qquad \vdots$

How can we prove all of these statements at once? Mathematical induction is a clever way of doing just that.

The crux of the idea is this: Suppose we can prove that whenever one of these statements is true, then the one following it in the list is also true. In other words,

For every k, if $P(k)$ is true, then $P(k + 1)$ is true.

This is called the **induction step** because it leads us from the truth of one statement to the truth of the next. Now suppose that we can also prove that

$P(1)$ is true.

The induction step now leads us through the following chain of statements:

$P(1)$ is true, so $P(2)$ is true.

$P(2)$ is true, so $P(3)$ is true.

$P(3)$ is true, so $P(4)$ is true.

$\vdots \qquad\qquad \vdots$

BLAISE PASCAL (1623–1662) is considered one of the most versatile minds in modern history. He was a writer and philosopher as well as a gifted mathematician and physicist. Among his contributions that appear in this book are Pascal's triangle and the Principle of Mathematical Induction.

Pascal's father, himself a mathematician, believed that his son should not study mathematics until he was 15 or 16. But at age 12, Blaise insisted on learning geometry and proved most of its elementary theorems himself. At 19 he invented the first mechanical adding machine. In 1647, after writing a major treatise on the conic sections, he abruptly abandoned mathematics because he felt that his intense studies were contributing to his ill health. He devoted himself instead to frivolous recreations such as gambling, but this only served to pique his interest in probability. In 1654 he miraculously survived a carriage accident in which his horses ran off a bridge. Taking this to be a sign from God, Pascal entered a monastery, where he pursued theology and philosophy, writing his famous *Pensées*. He also continued his mathematical research. He valued faith and intuition more than reason as the source of truth, declaring that "the heart has its own reasons, which reason cannot know."

So we see that if both the induction step and $P(1)$ are proved, then statement $P(n)$ is proved for all n. Here is a summary of this important method of proof.

PRINCIPLE OF MATHEMATICAL INDUCTION

For each natural number n, let $P(n)$ be a statement depending on n. Suppose that the following two conditions are satisfied.

1. $P(1)$ is true.
2. For every natural number k, if $P(k)$ is true then $P(k+1)$ is true.

Then $P(n)$ is true for all natural numbers n.

To apply this principle, there are two steps:

Step 1 Prove that $P(1)$ is true.

Step 2 Assume that $P(k)$ is true, and use this assumption to prove that $P(k+1)$ is true.

Notice that in Step 2 we do not prove that $P(k)$ is true. We only show that *if* $P(k)$ is true, *then* $P(k+1)$ is also true. The assumption that $P(k)$ is true is called the **induction hypothesis**.

We now use mathematical induction to prove that the conjecture that we made at the beginning of this section is true.

EXAMPLE 1 ■ A Proof by Mathematical Induction

Prove that for all natural numbers n,

$$1 + 3 + 5 + \cdots + (2n - 1) = n^2$$

SOLUTION Let $P(n)$ denote the statement $1 + 3 + 5 + \cdots + (2n-1) = n^2$.

Step 1 We need to show that $P(1)$ is true. But $P(1)$ is simply the statement that $1 = 1^2$, which is of course true.

Step 2 We assume that $P(k)$ is true. Thus our induction hypothesis is

$$1 + 3 + 5 + \cdots + (2k - 1) = k^2$$

We want to use this to show that $P(k+1)$ is true, that is,

$$1 + 3 + 5 + \cdots + (2k - 1) + [2(k + 1) - 1] = (k + 1)^2$$

[Note that we get $P(k+1)$ by substituting $k+1$ for each n in the statement $P(n)$.] We start with the left-hand side and use the induction hypothesis to obtain the right-hand side of the equation.

This equals k^2 by the induction hypothesis

$$\begin{aligned}
&1 + 3 + 5 + \cdots + (2k-1) + [2(k+1) - 1] \\
&\quad = [1 + 3 + 5 + \cdots + (2k-1)] + [2(k+1) - 1] && \text{Group the first } k \text{ terms} \\
&\quad = k^2 + [2(k+1) - 1] && \text{Induction hypothesis} \\
&\quad = k^2 + [2k + 2 - 1] && \text{Distributive Property} \\
&\quad = k^2 + 2k + 1 && \text{Simplify} \\
&\quad = (k+1)^2 && \text{Factor}
\end{aligned}$$

Thus $P(k+1)$ follows from $P(k)$, and this completes the induction step.

Having proved Steps 1 and 2, we conclude by the Principle of Mathematical Induction that $P(n)$ is true for all natural numbers n.

Now Try Exercise 3 ■

EXAMPLE 2 ■ A Proof by Mathematical Induction

Prove that for every natural number n,

$$1 + 2 + 3 + \cdots + n = \frac{n(n+1)}{2}$$

SOLUTION Let $P(n)$ be the statement $1 + 2 + 3 + \cdots + n = n(n+1)/2$. We want to show that $P(n)$ is true for all natural numbers n.

Step 1 We need to show that $P(1)$ is true. But $P(1)$ says that

$$1 = \frac{1(1+1)}{2}$$

and this statement is clearly true.

Step 2 Assume that $P(k)$ is true. Thus our induction hypothesis is

$$1 + 2 + 3 + \cdots + k = \frac{k(k+1)}{2}$$

We want to use this to show that $P(k+1)$ is true, that is,

$$1 + 2 + 3 + \cdots + k + (k+1) = \frac{(k+1)[(k+1)+1]}{2}$$

So we start with the left-hand side and use the induction hypothesis to obtain the right side.

This equals $\frac{k(k+1)}{2}$ by the induction hypothesis

$$\begin{aligned}
&1 + 2 + 3 + \cdots + k + (k+1) \\
&\quad = [1 + 2 + 3 + \cdots + k] + (k+1) && \text{Group the first } k \text{ terms} \\
&\quad = \frac{k(k+1)}{2} + (k+1) && \text{Induction hypothesis} \\
&\quad = (k+1)\left(\frac{k}{2} + 1\right) && \text{Factor } k+1 \\
&\quad = (k+1)\left(\frac{k+2}{2}\right) && \text{Common denominator} \\
&\quad = \frac{(k+1)[(k+1)+1]}{2} && \text{Write } k+2 \text{ as } k+1+1
\end{aligned}$$

Thus $P(k+1)$ follows from $P(k)$, and this completes the induction step.

Having proved Steps 1 and 2, we conclude by thc Principle of Mathematical Induction that $P(n)$ is true for all natural numbers n.

Now Try Exercise 5 ■

The following box gives formulas for the sums of powers of the first n natural numbers. These formulas are important in calculus. Formula 1 is proved in Example 2. The other formulas are also proved by using mathematical induction (see Exercises 6 and 9).

SUMS OF POWERS

0. $\sum_{k=1}^{n} 1 = n$

1. $\sum_{k=1}^{n} k = \dfrac{n(n+1)}{2}$

2. $\sum_{k=1}^{n} k^2 = \dfrac{n(n+1)(2n+1)}{6}$

3. $\sum_{k=1}^{n} k^3 = \dfrac{n^2(n+1)^2}{4}$

It might happen that a statement $P(n)$ is false for the first few natural numbers but true from some number on. For example, we might want to prove that $P(n)$ is true for $n \geq 5$. Notice that if we prove that $P(5)$ is true, then this fact, together with the induction step, would imply the truth of $P(5)$, $P(6)$, $P(7)$, ... The next example illustrates this point.

EXAMPLE 3 ■ Proving an Inequality by Mathematical Induction

Prove that $4n < 2^n$ for all $n \geq 5$.

SOLUTION Let $P(n)$ denote the statement $4n < 2^n$.

Step 1 $P(5)$ is the statement that $4 \cdot 5 < 2^5$, or $20 < 32$, which is true.

Step 2 Assume that $P(k)$ is true. Thus our induction hypothesis is

$$4k < 2^k$$

We want to use this to show that $P(k+1)$ is true, that is,

We get $P(k+1)$ by replacing n by $k+1$ in the statement $P(n)$.

$$4(k+1) < 2^{k+1}$$

So we start with the left-hand side of the inequality and use the induction hypothesis to show that it is less than the right-hand side. For $k \geq 5$ we have

$$\begin{aligned} 4(k+1) &= 4k + 4 && \text{Distributive Property} \\ &< 2^k + 4 && \text{Induction hypothesis} \\ &< 2^k + 4k && \text{Because } 4 < 4k \\ &< 2^k + 2^k && \text{Induction hypothesis} \\ &= 2 \cdot 2^k \\ &= 2^{k+1} && \text{Property of exponents} \end{aligned}$$

Thus $P(k+1)$ follows from $P(k)$, and this completes the induction step.

Having proved Steps 1 and 2, we conclude by the Principle of Mathematical Induction that $P(n)$ is true for all natural numbers $n \geq 5$.

Now Try Exercise 21 ■

12.5 EXERCISES

CONCEPTS

1. Mathematical induction is a method of proving that a statement $P(n)$ is true for all ________ numbers n. In Step 1 we prove that ________ is true.

2. Which of the following is true about Step 2 in a proof by mathematical induction?

(i) We prove "$P(k + 1)$ is true."

(ii) We prove "If $P(k)$ is true, then $P(k + 1)$ is true."

SKILLS

3–14 ■ Proving a Formula Use mathematical induction to prove that the formula is true for all natural numbers n.

3. $2 + 4 + 6 + \cdots + 2n = n(n + 1)$

4. $1 + 4 + 7 + \cdots + (3n - 2) = \dfrac{n(3n - 1)}{2}$

5. $5 + 8 + 11 + \cdots + (3n + 2) = \dfrac{n(3n + 7)}{2}$

6. $1^2 + 2^2 + 3^2 + \cdots + n^2 = \dfrac{n(n + 1)(2n + 1)}{6}$

7. $1 \cdot 2 + 2 \cdot 3 + 3 \cdot 4 + \cdots + n(n + 1) = \dfrac{n(n + 1)(n + 2)}{3}$

8. $1 \cdot 3 + 2 \cdot 4 + 3 \cdot 5 + \cdots + n(n + 2) = \dfrac{n(n + 1)(2n + 7)}{6}$

9. $1^3 + 2^3 + 3^3 + \cdots + n^3 = \dfrac{n^2(n + 1)^2}{4}$

10. $1^3 + 3^3 + 5^3 + \cdots + (2n - 1)^3 = n^2(2n^2 - 1)$

11. $2^3 + 4^3 + 6^3 + \cdots + (2n)^3 = 2n^2(n + 1)^2$

12. $\dfrac{1}{1 \cdot 2} + \dfrac{1}{2 \cdot 3} + \dfrac{1}{3 \cdot 4} + \cdots + \dfrac{1}{n(n + 1)} = \dfrac{n}{(n + 1)}$

13. $1 \cdot 2 + 2 \cdot 2^2 + 3 \cdot 2^3 + 4 \cdot 2^4 + \cdots + n \cdot 2^n$
$= 2[1 + (n - 1)2^n]$

14. $1 + 2 + 2^2 + \cdots + 2^{n-1} = 2^n - 1$

15–24 ■ Proving a Statement Use mathematical induction to show that the given statement is true.

15. $n^2 + n$ is divisible by 2 for all natural numbers n.

16. $5^n - 1$ is divisible by 4 for all natural numbers n.

17. $n^2 - n + 41$ is odd for all natural numbers n.

18. $n^3 - n + 3$ is divisible by 3 for all natural numbers n.

19. $8^n - 3^n$ is divisible by 5 for all natural numbers n.

20. $3^{2n} - 1$ is divisible by 8 for all natural numbers n.

21. $n < 2^n$ for all natural numbers n.

22. $(n + 1)^2 < 2n^2$ for all natural numbers $n \geq 3$.

23. If $x > -1$, then $(1 + x)^n \geq 1 + nx$ for all natural numbers n.

24. $100n \leq n^2$ for all $n \geq 100$.

25. Formula for a Recursive Sequence A sequence is defined recursively by $a_{n+1} = 3a_n$ and $a_1 = 5$. Show that $a_n = 5 \cdot 3^{n-1}$ for all natural numbers n.

26. Formula for a Recursive Sequence A sequence is defined recursively by $a_{n+1} = 3a_n - 8$ and $a_1 = 4$. Find an explicit formula for a_n, and then use mathematical induction to prove that the formula you found is true.

27. Proving a Factorization Show that $x - y$ is a factor of $x^n - y^n$ for all natural numbers n.
[*Hint:* $x^{k+1} - y^{k+1} = x^k(x - y) + (x^k - y^k)y$.]

28. Proving a Factorization Show that $x + y$ is a factor of $x^{2n-1} + y^{2n-1}$ for all natural numbers n.

SKILLS Plus

29–33 ■ Fibonacci Sequence F_n denotes the nth term of the Fibonacci sequence discussed in Section 12.1. Use mathematical induction to prove the statement.

29. F_{3n} is even for all natural numbers n.

30. $F_1 + F_2 + F_3 + \cdots + F_n = F_{n+2} - 1$

31. $F_1^2 + F_2^2 + F_3^2 + \cdots + F_n^2 = F_n F_{n+1}$

32. $F_1 + F_3 + \cdots + F_{2n-1} = F_{2n}$

33. For all $n \geq 2$,

$$\begin{bmatrix} 1 & 1 \\ 1 & 0 \end{bmatrix}^n = \begin{bmatrix} F_{n+1} & F_n \\ F_n & F_{n-1} \end{bmatrix}$$

34. Formula Using Fibonacci Numbers Let a_n be the nth term of the sequence defined recursively by

$$a_{n+1} = \frac{1}{1 + a_n}$$

and let $a_1 = 1$. Find a formula for a_n in terms of the Fibonacci numbers F_n. Prove that the formula you found is valid for all natural numbers n.

35. Discover and Prove an Inequality Let F_n be the nth term of the Fibonacci sequence. Find and prove an inequality relating n and F_n for natural numbers n.

36. Discover and Prove an Inequality Find and prove an inequality relating $100n$ and n^3.

DISCUSS ■ DISCOVER ■ PROVE ■ WRITE

37. DISCUSS: True or False? Determine whether each statement is true or false. If you think the statement is true, prove it. If you think it is false, give an example in which it fails.

(a) $p(n) = n^2 - n + 11$ is prime for all n.

(b) $n^2 > n$ for all $n \geq 2$.

(c) $2^{2n+1} + 1$ is divisible by 3 for all $n \geq 1$.

(d) $n^3 \geq (n + 1)^2$ for all $n \geq 2$.

(e) $n^3 - n$ is divisible by 3 for all $n \geq 2$.

(f) $n^3 - 6n^2 + 11n$ is divisible by 6 for all $n \geq 1$.

38. DISCUSS: All Cats Are Black? What is wrong with the following "proof" by mathematical induction that all cats are black? Let $P(n)$ denote the statement "In any group of n cats, if one cat is black, then they are all black."

Step 1 The statement is clearly true for $n = 1$.

Step 2 Suppose that $P(k)$ is true. We show that $P(k + 1)$ is true.

Suppose we have a group of $k + 1$ cats, one of whom is black; call this cat "Tadpole." Remove some other cat (call it "Sparky") from the group. We are left with k cats, one of whom (Tadpole) is black, so by the induction hypothesis, all k of these are black. Now put Sparky back in the group and take out Tadpole. We again have a group of k cats, all of whom—except possibly Sparky—are black. Then by the induction hypothesis, Sparky must be black too. So all $k + 1$ cats in the original group are black.

Thus by induction $P(n)$ is true for all n. Since everyone has seen at least one black cat, it follows that all cats are black.

12.6 THE BINOMIAL THEOREM

■ Expanding $(a + b)^n$ ■ The Binomial Coefficients ■ The Binomial Theorem ■ Proof of the Binomial Theorem

An expression of the form $a + b$ is called a **binomial**. Although in principle it's easy to raise $a + b$ to any power, raising it to a very high power would be tedious. In this section we find a formula that gives the expansion of $(a + b)^n$ for any natural number n and then prove it using mathematical induction.

■ Expanding $(a + b)^n$

To find a pattern in the expansion of $(a + b)^n$, we first look at some special cases.

$$
\begin{aligned}
(a + b)^1 &= a + b \\
(a + b)^2 &= a^2 + 2ab + b^2 \\
(a + b)^3 &= a^3 + 3a^2b + 3ab^2 + b^3 \\
(a + b)^4 &= a^4 + 4a^3b + 6a^2b^2 + 4ab^3 + b^4 \\
(a + b)^5 &= a^5 + 5a^4b + 10a^3b^2 + 10a^2b^3 + 5ab^4 + b^5 \\
&\vdots
\end{aligned}
$$

The following simple patterns emerge for the expansion of $(a + b)^n$.

1. There are $n + 1$ terms, the first being a^n and the last being b^n.
2. The exponents of a decrease by 1 from term to term, while the exponents of b increase by 1.
3. The sum of the exponents of a and b in each term is n.

For instance, notice how the exponents of a and b behave in the expansion of $(a + b)^5$.

The exponents of *a* decrease:

$$(a + b)^5 = a^{5} + 5a^{4}b^1 + 10a^{3}b^2 + 10a^{2}b^3 + 5a^{1}b^4 + b^5$$

The exponents of *b* increase:

$$(a + b)^5 = a^5 + 5a^4b^{1} + 10a^3b^{2} + 10a^2b^{3} + 5a^1b^{4} + b^{5}$$

What we now call **Pascal's triangle** appears in this Chinese document by Chu Shikie, dated 1303. The title reads "The Old Method Chart of the Seven Multiplying Squares." The triangle was rediscovered by Pascal (see page 875).

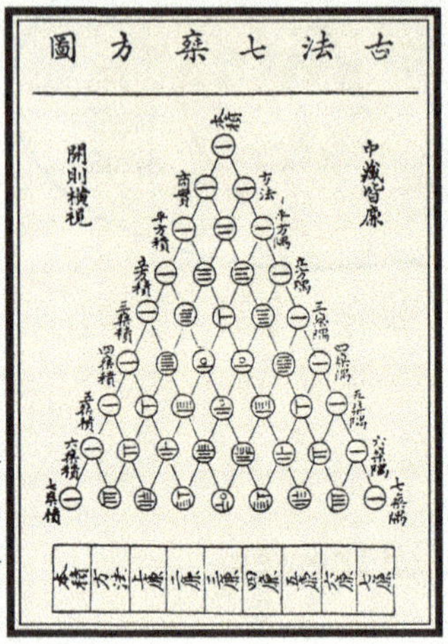

University of York, Department of Mathematics

With these observations we can write the form of the expansion of $(a + b)^n$ for any natural number n. For example, writing a question mark for the missing coefficients, we have

$$(a + b)^8 = a^8 + ?a^7b + ?a^6b^2 + ?a^5b^3 + ?a^4b^4 + ?a^3b^5 + ?a^2b^6 + ?ab^7 + b^8$$

To complete the expansion, we need to determine these coefficients. To find a pattern, let's write the coefficients in the expansion of $(a + b)^n$ for the first few values of n in a triangular array as shown in the following array, which is called **Pascal's triangle**.

$$\begin{array}{lccccccccccc}
(a+b)^0 & & & & & & 1 & & & & & \\
(a+b)^1 & & & & & 1 & & 1 & & & & \\
(a+b)^2 & & & & 1 & & 2 & & 1 & & & \\
(a+b)^3 & & & 1 & & 3 & & 3 & & 1 & & \\
(a+b)^4 & & 1 & & 4 & & 6 & & 4 & & 1 & \\
(a+b)^5 & 1 & & 5 & & 10 & & 10 & & 5 & & 1
\end{array}$$

The row corresponding to $(a + b)^0$ is called the zeroth row and is included to show the symmetry of the array. The key observation about Pascal's triangle is the following property.

KEY PROPERTY OF PASCAL'S TRIANGLE

Every entry (other than a 1) is the sum of the two entries diagonally above it.

From this property it is easy to find any row of Pascal's triangle from the row above it. For instance, we find the sixth and seventh rows, starting with the fifth row:

$$\begin{array}{lccccccccccccccc}
(a+b)^5 & & & 1 & & 5 & & 10 & & 10 & & 5 & & 1 & & \\
(a+b)^6 & & 1 & & 6 & & 15 & & 20 & & 15 & & 6 & & 1 & \\
(a+b)^7 & 1 & & 7 & & 21 & & 35 & & 35 & & 21 & & 7 & & 1
\end{array}$$

To see why this property holds, let's consider the following expansions:

$$(a + b)^5 = a^5 + 5a^4b + 10a^3b^2 + 10a^2b^3 + 5ab^4 + b^5$$

$$(a + b)^6 = a^6 + 6a^5b + 15a^4b^2 + 20a^3b^3 + 15a^2b^4 + 6ab^5 + b^6$$

We arrive at the expansion of $(a + b)^6$ by multiplying $(a + b)^5$ by $(a + b)$. Notice, for instance, that the circled term in the expansion of $(a + b)^6$ is obtained via this multiplication from the two circled terms above it. We get this term when the two terms above it are multiplied by b and a, respectively. Thus its coefficient is the sum of the coefficients of these two terms. We will use this observation at the end of this section when we prove the Binomial Theorem.

Having found these patterns, we can now easily obtain the expansion of any binomial, at least to relatively small powers.

EXAMPLE 1 ■ Expanding a Binomial Using Pascal's Triangle

Find the expansion of $(a + b)^7$ using Pascal's triangle.

SOLUTION The first term in the expansion is a^7, and the last term is b^7. Using the fact that the exponent of a decreases by 1 from term to term and that of b increases by 1 from term to term, we have

$$(a + b)^7 = a^7 + ?a^6b + ?a^5b^2 + ?a^4b^3 + ?a^3b^4 + ?a^2b^5 + ?ab^6 + b^7$$

The appropriate coefficients appear in the seventh row of Pascal's triangle. Thus

$$(a + b)^7 = a^7 + 7a^6b + 21a^5b^2 + 35a^4b^3 + 35a^3b^4 + 21a^2b^5 + 7ab^6 + b^7$$

Now Try Exercise 5 ■

EXAMPLE 2 ■ Expanding a Binomial Using Pascal's Triangle

Use Pascal's triangle to expand $(2 - 3x)^5$.

SOLUTION We find the expansion of $(a + b)^5$ and then substitute 2 for a and $-3x$ for b. Using Pascal's triangle for the coefficients, we get

$$(a + b)^5 = a^5 + 5a^4b + 10a^3b^2 + 10a^2b^3 + 5ab^4 + b^5$$

Substituting $a = 2$ and $b = -3x$ gives

$$\begin{aligned}(2 - 3x)^5 &= (2)^5 + 5(2)^4(-3x) + 10(2)^3(-3x)^2 + 10(2)^2(-3x)^3 + 5(2)(-3x)^4 + (-3x)^5 \\ &= 32 - 240x + 720x^2 - 1080x^3 + 810x^4 - 243x^5\end{aligned}$$

Now Try Exercise 13 ■

■ The Binomial Coefficients

Although Pascal's triangle is useful in finding the binomial expansion for reasonably small values of n, it isn't practical for finding $(a + b)^n$ for large values of n. The reason is that the method we use for finding the successive rows of Pascal's triangle is recursive. Thus to find the 100th row of this triangle, we must first find the preceding 99 rows.

We need to examine the pattern in the coefficients more carefully to develop a formula that allows us to calculate directly any coefficient in the binomial expansion. Such a formula exists, and the rest of this section is devoted to finding and proving it. However, to state this formula, we need some notation.

The product of the first n natural numbers is denoted by $\boldsymbol{n!}$ and is called $\boldsymbol{n}$ **factorial**.

$4! = 1 \cdot 2 \cdot 3 \cdot 4 = 24$

$7! = 1 \cdot 2 \cdot 3 \cdot 4 \cdot 5 \cdot 6 \cdot 7 = 5040$

$10! = 1 \cdot 2 \cdot 3 \cdot 4 \cdot 5 \cdot 6 \cdot 7 \cdot 8 \cdot 9 \cdot 10$
$= 3,628,800$

$$\boxed{n! = 1 \cdot 2 \cdot 3 \cdot \cdots \cdot (n - 1) \cdot n}$$

We also define 0! as follows:

$$\boxed{0! = 1}$$

This definition of 0! makes many formulas involving factorials shorter and easier to write.

THE BINOMIAL COEFFICIENT

Let n and r be nonnegative integers with $r \le n$. The **binomial coefficient** is denoted by $\binom{n}{r}$ and is defined by

$$\binom{n}{r} = \frac{n!}{r!(n - r)!}$$

EXAMPLE 3 ■ Calculating Binomial Coefficients

(a) $$\binom{9}{4} = \frac{9!}{4!(9-4)!} = \frac{9!}{4!\,5!} = \frac{\cancel{1\cdot 2\cdot 3\cdot 4\cdot 5}\cdot 6\cdot 7\cdot 8\cdot 9}{(1\cdot 2\cdot 3\cdot 4)(\cancel{1\cdot 2\cdot 3\cdot 4\cdot 5})}$$
$$= \frac{6\cdot 7\cdot 8\cdot 9}{1\cdot 2\cdot 3\cdot 4} = 126$$

(b) $$\binom{100}{3} = \frac{100!}{3!(100-3)!} = \frac{\cancel{1\cdot 2\cdot 3\cdot \cdots \cdot 97}\cdot 98\cdot 99\cdot 100}{(1\cdot 2\cdot 3)(\cancel{1\cdot 2\cdot 3\cdot \cdots \cdot 97})}$$
$$= \frac{98\cdot 99\cdot 100}{1\cdot 2\cdot 3} = 161{,}700$$

(c) $$\binom{100}{97} = \frac{100!}{97!(100-97)!} = \frac{\cancel{1\cdot 2\cdot 3\cdot \cdots \cdot 97}\cdot 98\cdot 99\cdot 100}{(\cancel{1\cdot 2\cdot 3\cdot \cdots \cdot 97})(1\cdot 2\cdot 3)}$$
$$= \frac{98\cdot 99\cdot 100}{1\cdot 2\cdot 3} = 161{,}700$$

Now Try Exercises 17 and 19 ■

Although the binomial coefficient $\binom{n}{r}$ is defined in terms of a fraction, all the results of Example 3 are natural numbers. In fact, $\binom{n}{r}$ is always a natural number (see Exercise 54). Notice that the binomial coefficients in parts (b) and (c) of Example 3 are equal. This is a special case of the following relation, which you are asked to prove in Exercise 52.

$$\binom{n}{r} = \binom{n}{n-r}$$

To see the connection between the binomial coefficients and the binomial expansion of $(a + b)^n$, let's calculate the following binomial coefficients:

$$\binom{5}{2} = \frac{5!}{2!(5-2)!} = 10$$

$$\binom{5}{0} = 1 \quad \binom{5}{1} = 5 \quad \binom{5}{2} = 10 \quad \binom{5}{3} = 10 \quad \binom{5}{4} = 5 \quad \binom{5}{5} = 1$$

These are precisely the entries in the fifth row of Pascal's triangle. In fact, we can write Pascal's triangle as follows.

$$\begin{array}{ccccccccccc}
 & & & & & \binom{0}{0} & & & & & \\
 & & & & \binom{1}{0} & & \binom{1}{1} & & & & \\
 & & & \binom{2}{0} & & \binom{2}{1} & & \binom{2}{2} & & & \\
 & & \binom{3}{0} & & \binom{3}{1} & & \binom{3}{2} & & \binom{3}{3} & & \\
 & \binom{4}{0} & & \binom{4}{1} & & \binom{4}{2} & & \binom{4}{3} & & \binom{4}{4} & \\
\binom{5}{0} & & \binom{5}{1} & & \binom{5}{2} & & \binom{5}{3} & & \binom{5}{4} & & \binom{5}{5} \\
\cdot & & \cdot & & \cdot & \cdot & \cdot & & \cdot & & \cdot \\
\binom{n}{0} & & \binom{n}{1} & & \binom{n}{2} & \cdot\ \cdot\ \cdot & & & \binom{n}{n-1} & & \binom{n}{n}
\end{array}$$

To demonstrate that this pattern holds, we need to show that any entry in this version of Pascal's triangle is the sum of the two entries diagonally above it. In other words, we must show that each entry satisfies the key property of Pascal's triangle. We now state this property in terms of the binomial coefficients.

KEY PROPERTY OF THE BINOMIAL COEFFICIENTS

For any nonnegative integers r and k with $r \le k$,

$$\binom{k}{r-1} + \binom{k}{r} = \binom{k+1}{r}$$

Notice that the two terms on the left-hand side of this equation are adjacent entries in the kth row of Pascal's triangle and the term on the right-hand side is the entry diagonally below them, in the $(k+1)$st row. Thus this equation is a restatement of the key property of Pascal's triangle in terms of the binomial coefficients. A proof of this formula is outlined in Exercise 53.

The Binomial Theorem

We are now ready to state the Binomial Theorem.

THE BINOMIAL THEOREM

$$(a+b)^n = \binom{n}{0}a^n + \binom{n}{1}a^{n-1}b + \binom{n}{2}a^{n-2}b^2 + \cdots + \binom{n}{n-1}ab^{n-1} + \binom{n}{n}b^n$$

We prove this theorem at the end of this section. First, let's look at some of its applications.

EXAMPLE 4 ■ Expanding a Binomial Using the Binomial Theorem

Use the Binomial Theorem to expand $(x+y)^4$.

SOLUTION By the Binomial Theorem,

$$(x+y)^4 = \binom{4}{0}x^4 + \binom{4}{1}x^3y + \binom{4}{2}x^2y^2 + \binom{4}{3}xy^3 + \binom{4}{4}y^4$$

Verify that

$$\binom{4}{0} = 1 \quad \binom{4}{1} = 4 \quad \binom{4}{2} = 6 \quad \binom{4}{3} = 4 \quad \binom{4}{4} = 1$$

It follows that

$$(x+y)^4 = x^4 + 4x^3y + 6x^2y^2 + 4xy^3 + y^4$$

Now Try Exercise 25 ■

EXAMPLE 5 ■ Expanding a Binomial Using the Binomial Theorem

Use the Binomial Theorem to expand $(\sqrt{x} - 1)^8$.

SOLUTION We first find the expansion of $(a + b)^8$ and then substitute $\sqrt{x}$ for a and -1 for b. Using the Binomial Theorem, we have

$$(a + b)^8 = \binom{8}{0}a^8 + \binom{8}{1}a^7b + \binom{8}{2}a^6b^2 + \binom{8}{3}a^5b^3 + \binom{8}{4}a^4b^4 + \binom{8}{5}a^3b^5 + \binom{8}{6}a^2b^6 + \binom{8}{7}ab^7 + \binom{8}{8}b^8$$

Verify that

$$\binom{8}{0} = 1 \quad \binom{8}{1} = 8 \quad \binom{8}{2} = 28 \quad \binom{8}{3} = 56 \quad \binom{8}{4} = 70$$

$$\binom{8}{5} = 56 \quad \binom{8}{6} = 28 \quad \binom{8}{7} = 8 \quad \binom{8}{8} = 1$$

So

$$(a + b)^8 = a^8 + 8a^7b + 28a^6b^2 + 56a^5b^3 + 70a^4b^4 + 56a^3b^5 + 28a^2b^6 + 8ab^7 + b^8$$

Performing the substitutions $a = x^{1/2}$ and $b = -1$ gives

$$(\sqrt{x} - 1)^8 = (x^{1/2})^8 + 8(x^{1/2})^7(-1) + 28(x^{1/2})^6(-1)^2 + 56(x^{1/2})^5(-1)^3 + 70(x^{1/2})^4(-1)^4 + 56(x^{1/2})^3(-1)^5 + 28(x^{1/2})^2(-1)^6 + 8(x^{1/2})(-1)^7 + (-1)^8$$

This simplifies to

$$(\sqrt{x} - 1)^8 = x^4 - 8x^{7/2} + 28x^3 - 56x^{5/2} + 70x^2 - 56x^{3/2} + 28x - 8x^{1/2} + 1$$

Now Try Exercise 27 ■

The Binomial Theorem can be used to find a particular term of a binomial expansion without having to find the entire expansion.

Recall that

$$\binom{n}{r} = \binom{n}{n-r}$$

(See page 882.)

GENERAL TERM OF THE BINOMIAL EXPANSION

The term that contains a^r in the expansion of $(a + b)^n$ is

$$\binom{n}{r}a^rb^{n-r}$$

EXAMPLE 6 ■ Finding a Particular Term in a Binomial Expansion

Find the term that contains x^5 in the expansion of $(2x + y)^{20}$.

SOLUTION The term that contains x^5 is given by the formula for the general term with $a = 2x$, $b = y$, $n = 20$, and $r = 5$. So this term is

$$\binom{20}{5}a^5b^{15} = \frac{20!}{5!(20-5)!}(2x)^5y^{15} = \frac{20!}{5!\,15!}32x^5y^{15} = 496{,}128x^5y^{15}$$

Now Try Exercise 39 ■

EXAMPLE 7 ■ Finding a Particular Term in a Binomial Expansion

Find the coefficient of x^8 in the expansion of $\left(x^2 + \frac{1}{x}\right)^{10}$.

SOLUTION Both x^2 and $1/x$ are powers of x, so the power of x in each term of the expansion is determined by both terms of the binomial. To find the required coefficient, we first find the general term in the expansion. By the formula we have $a = x^2$, $b = 1/x$, and $n = 10$, so the general term is

$$\binom{10}{r}(x^2)^r\left(\frac{1}{x}\right)^{10-r} = \binom{10}{r}x^{2r}(x^{-1})^{10-r} = \binom{10}{r}x^{3r-10}$$

Thus the term that contains x^8 is the term in which

$$3r - 10 = 8$$
$$r = 6$$

So the required coefficient is

$$\binom{10}{6} = 210$$

Now Try Exercise 41 ■

Proof of the Binomial Theorem

We now give a proof of the Binomial Theorem using mathematical induction.

Proof Let $P(n)$ denote the statement

$$(a+b)^n = \binom{n}{0}a^n + \binom{n}{1}a^{n-1}b + \binom{n}{2}a^{n-2}b^2 + \cdots + \binom{n}{n-1}ab^{n-1} + \binom{n}{n}b^n$$

Step 1 We show that $P(1)$ is true. But $P(1)$ is just the statement

$$(a+b)^1 = \binom{1}{0}a^1 + \binom{1}{1}b^1 = 1a + 1b = a + b$$

which is certainly true.

Step 2 We assume that $P(k)$ is true. Thus our induction hypothesis is

$$(a+b)^k = \binom{k}{0}a^k + \binom{k}{1}a^{k-1}b + \binom{k}{2}a^{k-2}b^2 + \cdots + \binom{k}{k-1}ab^{k-1} + \binom{k}{k}b^k$$

We use this to show that $P(k+1)$ is true.

$$(a+b)^{k+1} = (a+b)[(a+b)^k]$$

$$= (a+b)\left[\binom{k}{0}a^k + \binom{k}{1}a^{k-1}b + \binom{k}{2}a^{k-2}b^2 + \cdots + \binom{k}{k-1}ab^{k-1} + \binom{k}{k}b^k\right]$$ Induction hypothesis

$$= a\left[\binom{k}{0}a^k + \binom{k}{1}a^{k-1}b + \binom{k}{2}a^{k-2}b^2 + \cdots + \binom{k}{k-1}ab^{k-1} + \binom{k}{k}b^k\right]$$
$$+ b\left[\binom{k}{0}a^k + \binom{k}{1}a^{k-1}b + \binom{k}{2}a^{k-2}b^2 + \cdots + \binom{k}{k-1}ab^{k-1} + \binom{k}{k}b^k\right]$$ Distributive Property

$$= \binom{k}{0}a^{k+1} + \binom{k}{1}a^kb + \binom{k}{2}a^{k-1}b^2 + \cdots + \binom{k}{k-1}a^2b^{k-1} + \binom{k}{k}ab^k$$
$$+ \binom{k}{0}a^kb + \binom{k}{1}a^{k-1}b^2 + \binom{k}{2}a^{k-2}b^3 + \cdots + \binom{k}{k-1}ab^k + \binom{k}{k}b^{k+1}$$ Distributive Property

$$= \binom{k}{0}a^{k+1} + \left[\binom{k}{0} + \binom{k}{1}\right]a^kb + \left[\binom{k}{1} + \binom{k}{2}\right]a^{k-1}b^2$$
$$+ \cdots + \left[\binom{k}{k-1} + \binom{k}{k}\right]ab^k + \binom{k}{k}b^{k+1}$$ Group like terms

Using the key property of the binomial coefficients, we can write each of the expressions in square brackets as a single binomial coefficient. Also, writing the first and last coefficients as $\binom{k+1}{0}$ and $\binom{k+1}{k+1}$ (these are equal to 1 by Exercise 50) gives

$$(a+b)^{k+1} = \binom{k+1}{0}a^{k+1} + \binom{k+1}{1}a^k b + \binom{k+1}{2}a^{k-1}b^2 + \cdots + \binom{k+1}{k}ab^k + \binom{k+1}{k+1}b^{k+1}$$

But this last equation is precisely $P(k+1)$, and this completes the induction step.

Having proved Steps 1 and 2, we conclude by the Principle of Mathematical Induction that the theorem is true for all natural numbers n. ■

12.6 EXERCISES

CONCEPTS

1. An algebraic expression of the form $a + b$, which consists of a sum of two terms, is called a ________.

2. We can find the coefficients in the expansion of $(a+b)^n$ from the nth row of ________ triangle. So

$(a+b)^4 = \square a^4 + \square a^3b + \square a^2b^2 + \square ab^3 + \square b^4$

3. The binomial coefficients can be calculated directly by using the formula $\binom{n}{k} =$ ________. So $\binom{4}{3} =$ ________.

4. To expand $(a+b)^n$, we can use the ________ Theorem. Using this theorem, we find the expansion $(a+b)^4 =$

$\binom{\square}{\square}a^4 + \binom{\square}{\square}a^3b + \binom{\square}{\square}a^2b^2 + \binom{\square}{\square}ab^3 + \binom{\square}{\square}b^4$

SKILLS

5–16 ■ Pascal's Triangle Use Pascal's triangle to expand the expression.

5. $(x+y)^6$
6. $(2x+1)^4$
7. $\left(x + \frac{1}{x}\right)^4$
8. $(x-y)^5$
9. $(x-1)^5$
10. $(\sqrt{a} + \sqrt{b})^6$
11. $(x^2y - 1)^5$
12. $(1 + \sqrt{2})^6$
13. $(2x - 3y)^3$
14. $(1 + x^3)^3$
15. $\left(\frac{1}{x} - \sqrt{x}\right)^5$
16. $\left(2 + \frac{x}{2}\right)^5$

17–24 ■ Calculating Binomial Coefficients Evaluate the expression.

17. $\binom{6}{4}$
18. $\binom{8}{3}$
19. $\binom{100}{98}$
20. $\binom{10}{5}$
21. $\binom{3}{1}\binom{4}{2}$
22. $\binom{5}{2}\binom{5}{3}$
23. $\binom{5}{0} + \binom{5}{1} + \binom{5}{2} + \binom{5}{3} + \binom{5}{4} + \binom{5}{5}$
24. $\binom{5}{0} - \binom{5}{1} + \binom{5}{2} - \binom{5}{3} + \binom{5}{4} - \binom{5}{5}$

25–28 ■ Binomial Theorem Use the Binomial Theorem to expand the expression.

25. $(x+2y)^4$
26. $(1-x)^5$
27. $\left(1 + \frac{1}{x}\right)^6$
28. $(2A + B^2)^4$

29–42 ■ Terms of a Binomial Expansion Find the indicated terms in the expansion of the given binomial.

29. The first three terms in the expansion of $(x+2y)^{20}$
30. The first four terms in the expansion of $(x^{1/2}+1)^{30}$
31. The last two terms in the expansion of $(a^{2/3} + a^{1/3})^{25}$
32. The first three terms in the expansion of

$$\left(x + \frac{1}{x}\right)^{40}$$

33. The middle term in the expansion of $(x^2+1)^{18}$
34. The fifth term in the expansion of $(ab-1)^{20}$
35. The 24th term in the expansion of $(a+b)^{25}$
36. The 28th term in the expansion of $(A-B)^{30}$
37. The 100th term in the expansion of $(1+y)^{100}$
38. The second term in the expansion of

$$\left(x^2 - \frac{1}{x}\right)^{25}$$

39. The term containing x^4 in the expansion of $(x+2y)^{10}$
40. The term containing y^3 in the expansion of $(\sqrt{2}+y)^{12}$
41. The term containing b^8 in the expansion of $(a+b^2)^{12}$
42. The term that does not contain x in the expansion of

$$\left(8x + \frac{1}{2x}\right)^8$$

43–46 ■ Factoring Factor using the Binomial Theorem.

43. $x^4 + 4x^3y + 6x^2y^2 + 4xy^3 + y^4$
44. $(x-1)^5 + 5(x-1)^4 + 10(x-1)^3 + 10(x-1)^2 + 5(x-1) + 1$

45. $8a^3 + 12a^2b + 6ab^2 + b^3$

46. $x^8 + 4x^6y + 6x^4y^2 + 4x^2y^3 + y^4$

47–48 ■ Simplifying a Difference Quotient Simplify using the Binomial Theorem.

47. $\dfrac{(x+h)^3 - x^3}{h}$ **48.** $\dfrac{(x+h)^4 - x^4}{h}$

SKILLS Plus

49–52 ■ Proving a Statement Show that the given statement is true.

49. $(1.01)^{100} > 2$. [*Hint:* Note that $(1.01)^{100} = (1 + 0.01)^{100}$, and use the Binomial Theorem to show that the sum of the first three terms of the expansion is greater than 2.]

50. $\binom{n}{0} = 1$ and $\binom{n}{n} = 1$

51. $\binom{n}{1} = \binom{n}{n-1} = n$

52. $\binom{n}{r} = \binom{n}{n-r}$ for $0 \le r \le n$

53. Proving an Identity In this exercise we prove the identity

$$\binom{n}{r-1} + \binom{n}{r} = \binom{n+1}{r}$$

(a) Write the left-hand side of this equation as the sum of two fractions.

(b) Show that a common denominator of the expression that you found in part (a) is $r!(n - r + 1)!$.

(c) Add the two fractions using the common denominator in part (b), simplify the numerator, and note that the resulting expression is equal to the right-hand side of the equation.

54. Proof Using Induction Prove that $\binom{n}{r}$ is an integer for all n and for $0 \le r \le n$. [*Suggestion:* Use induction to show that the statement is true for all n, and use Exercise 53 for the induction step.]

APPLICATIONS

55. Difference in Volumes of Cubes The volume of a cube of side x inches is given by $V(x) = x^3$, so the volume of a cube of side $x + 2$ inches is given by $V(x + 2) = (x + 2)^3$. Use the Binomial Theorem to show that the difference in volume between the larger and smaller cubes is $6x^2 + 12x + 8$ cubic inches.

56. Probability of Hitting a Target The probability that an archer hits the target is $p = 0.9$, so the probability that he misses the target is $q = 0.1$. It is known that in this situation the probability that the archer hits the target exactly r times in n attempts is given by the term containing p^r in the binomial expansion of $(p + q)^n$. Find the probability that the archer hits the target exactly three times in five attempts.

DISCUSS ■ DISCOVER ■ PROVE ■ WRITE

57. DISCUSS: Powers of Factorials Which is larger, $(100!)^{101}$ or $(101!)^{100}$? [*Hint:* Try factoring the expressions. Do they have any common factors?]

58. DISCOVER ■ PROVE: Sums of Binomial Coefficients Add each of the first five rows of Pascal's triangle, as indicated. Do you see a pattern?

$$1 + 1 = ?$$
$$1 + 2 + 1 = ?$$
$$1 + 3 + 3 + 1 = ?$$
$$1 + 4 + 6 + 4 + 1 = ?$$
$$1 + 5 + 10 + 10 + 5 + 1 = ?$$

On the basis of the pattern you have found, find the sum of the nth row:

$$\binom{n}{0} + \binom{n}{1} + \binom{n}{2} + \cdots + \binom{n}{n}$$

Prove your result by expanding $(1 + 1)^n$ using the Binomial Theorem.

59. DISCOVER ■ PROVE: Alternating Sums of Binomial Coefficients Find the sum

$$\binom{n}{0} - \binom{n}{1} + \binom{n}{2} - \cdots + (-1)^n\binom{n}{n}$$

by finding a pattern as in Exercise 58. Prove your result by expanding $(1 - 1)^n$ using the Binomial Theorem.

CHAPTER 12 ■ REVIEW

PROPERTIES AND FORMULAS

Sequences (p. 842)

A **sequence** is a function whose domain is the set of natural numbers. Instead of writing $a(n)$ for the value of the sequence at n, we generally write a_n, and we refer to this value as the ***n*th term** of the sequence. Sequences are often described in list form:

$$a_1, a_2, a_3, \ldots$$

Partial Sums of a Sequence (pp. 847–848)

For the sequence $a_1, a_2, a_3, \ldots$ the ***n*th partial sum** S_n is the sum of the first n terms of the sequence:

$$S_n = a_1 + a_2 + a_3 + \cdots + a_n$$

The nth partial sum of a sequence can also be expressed by using **sigma notation**:

$$S_n = \sum_{k=1}^{n} a_k$$

Arithmetic Sequences (p. 853)

An **arithmetic sequence** is a sequence whose terms are obtained by adding the same fixed constant d to each term to get the next term. Thus an arithmetic sequence has the form

$$a, a + d, a + 2d, a + 3d, \ldots$$

The number a is the **first term** of the sequence, and the number d is the **common difference**. The nth term of the sequence is

$$a_n = a + (n - 1)d$$

Partial Sums of an Arithmetic Sequence (p. 855)

For the arithmetic sequence $a_n = a + (n - 1)d$ the nth partial sum $S_n = \sum_{k=1}^{n} [a + (k - 1)d]$ is given by either of the following equivalent formulas:

1. $S_n = \frac{n}{2}[2a + (n - 1)d]$ **2.** $S_n = n\left(\frac{a + a_n}{2}\right)$

Geometric Sequences (p. 858)

A **geometric sequence** is a sequence whose terms are obtained by multiplying each term by the same fixed constant r to get the next term. Thus a geometric sequence has the form

$$a, ar, ar^2, ar^3, \ldots$$

The number a is the **first term** of the sequence, and the number r is the **common ratio**. The nth term of the sequence is

$$a_n = ar^{n-1}$$

Partial Sums of a Geometric Sequence (p. 861)

For the geometric sequence $a_n = ar^{n-1}$ the nth partial sum $S_n = \sum_{k=1}^{n} ar^{k-1}$ (where $r \neq 1$) is given by

$$S_n = a\frac{1 - r^n}{1 - r}$$

Infinite Geometric Series (p. 863)

An **infinite geometric series** is a series of the form

$$a + ar + ar^2 + ar^3 + \cdots + ar^{n-1} + \cdots$$

An infinite geometric series for which $|r| < 1$ has the sum

$$S = \frac{a}{1 - r}$$

Amount of an Annuity (p. 869)

The amount A_f of an **annuity** consisting of n regular equal payments of size R with interest rate i per time period is given by

$$A_f = R\frac{(1 + i)^n - 1}{i}$$

Present Value of an Annuity (p. 870)

The **present value** A_p of an annuity consisting of n regular equal payments of size R with interest rate i per time period is given by

$$A_p = R\frac{1 - (1 + i)^{-n}}{i}$$

Present Value of a Future Amount (p. 869)

If an amount A_f is to be paid in one lump sum, n time periods from now, and the interest rate per time period is i, then its **present value** A_p is given by

$$A_p = A_f(1 + i)^{-n}$$

Installment Buying (p. 870)

If a loan A_p is to be repaid in n regular equal payments with interest rate i per time period, then the size R of each payment is given by

$$R = \frac{iA_p}{1 - (1 + i)^{-n}}$$

Principle of Mathematical Induction (p. 875)

For each natural number n, let $P(n)$ be a statement that depends on n. Suppose that each of the following conditions is satisfied.

1. $P(1)$ is true.

2. For every natural number k, if $P(k)$ is true, then $P(k + 1)$ is true.

Then $P(n)$ is true for all natural numbers n.

Sums of Powers (p. 877)

0. $\sum_{k=1}^{n} 1 = n$

1. $\sum_{k=1}^{n} k = \frac{n(n + 1)}{2}$

2. $\sum_{k=1}^{n} k^2 = \frac{n(n + 1)(2n + 1)}{6}$

3. $\sum_{k=1}^{n} k^3 = \frac{n^2(n + 1)^2}{4}$

Binomial Coefficients (pp. 881–883)

If n and r are positive integers with $n \geq r$, then the **binomial coefficient** $\binom{n}{r}$ is defined by

$$\binom{n}{r} = \frac{n!}{r!(n - r)!}$$

Binomial coefficients satisfy the following properties:

$$\binom{n}{r} = \binom{n}{n - r}$$

$$\binom{k}{r - 1} + \binom{k}{r} = \binom{k + 1}{r}$$

The Binomial Theorem (pp. 883–884)

$$(a + b)^n = \binom{n}{0}a^n + \binom{n}{1}a^{n-1}b + \binom{n}{2}a^{n-2}b^2 + \cdots + \binom{n}{n}b^n$$

The term that contains a^r in the expansion of $(a + b)^n$ is $\binom{n}{r}a^r b^{n-r}$.

CONCEPT CHECK

1. (a) What is a sequence? What notation do we use to denote the terms of a sequence?

(b) Find a formula for the sequence of even numbers and a formula for the sequence of odd numbers.

(c) Find the first three terms and the 10th term of the sequence given by $a_n = n/(n+1)$.

2. (a) What is a recursively defined sequence?

(b) Find the first four terms of the sequence recursively defined by $a_1 = 3$ and $a_n = n + 2a_{n-1}$.

3. (a) What is meant by the partial sums of a sequence?

(b) Find the first three partial sums of the sequence given by $a_n = 1/n$.

4. (a) What is an arithmetic sequence? Write a formula for the nth term of an arithmetic sequence.

(b) Write a formula for the arithmetic sequence that starts as follows: 3, 8, . . . Write the first five terms of this sequence.

(c) Write two different formulas for the sum of the first n terms of an arithmetic sequence.

(d) Find the sum of the first 20 terms of the sequence in part (b).

5. (a) What is a geometric sequence? Write an expression for the nth term of a geometric sequence that has first term a and common ratio r.

(b) Write an expression for the geometric sequence with first term $a = 3$ and common ratio $r = \frac{1}{2}$. Give the first five terms of this sequence.

(c) Write an expression for the sum of the first n terms of a geometric sequence.

(d) Find the sum of the first five terms of the sequence in part (b).

6. (a) What is an infinite geometric series?

(b) What does it mean for an infinite series to converge? For what values of r does an infinite geometric series converge? If an infinite geometric series converges, then what is its sum?

(c) Write the first four terms of the infinite geometric series with first term $a = 5$ and common ratio $r = 0.4$. Does the series converge? If so, find its sum.

7. (a) Write $1^3 + 2^3 + 3^3 + 4^3 + 5^3$ using sigma notation.

(b) Write $\sum_{k=3}^{5} 2k^2$ without using sigma notation.

8. (a) What is an annuity? Write an expression for the amount A_f of an annuity consisting of n regular equal payments of size R with interest rate i per time period.

(b) An investor deposits \$200 each month into an account that pays 6% compounded monthly. How much is in the account at the end of 3 years?

(c) What is the formula for calculating the present value of the annuity in part (b)?

(d) What is the present value of the annuity in part (b)?

(e) When buying on installment, what is the formula for calculating the periodic payments?

(f) If you take out a 5-year loan for \$10,000 at 3% interest compounded monthly, what is the size of each monthly payment?

9. (a) State the Principle of Mathematical Induction.

(b) Use mathematical induction to prove that for all natural numbers n, $3^n - 1$ is an even number.

10. (a) Write Pascal's triangle. How are the entries in the triangle related to each other?

Row 0
Row 1
Row 2
Row 3

(b) Use Pascal's triangle to expand $(x + c)^3$.

11. (a) What does the symbol $n!$ mean? Find $5!$.

(b) Define $\binom{n}{r}$, and find $\binom{5}{2}$.

12. (a) State the Binomial Theorem.

(b) Use the Binomial Theorem to expand $(x + 2)^3$.

(c) Use the Binomial Theorem to find the term containing x^4 in the expansion of $(x + 2)^{10}$.

ANSWERS TO THE CONCEPT CHECK CAN BE FOUND AT THE BACK OF THE BOOK.

EXERCISES

1–6 ■ Terms of a Sequence Find the first four terms as well as the tenth term of the sequence with the given nth term.

1. $a_n = \dfrac{n^2}{n+1}$

2. $a_n = (-1)^n \dfrac{2^n}{n}$

3. $a_n = \dfrac{(-1)^n + 1}{n^3}$

4. $a_n = \dfrac{n(n+1)}{2}$

5. $a_n = \dfrac{(2n)!}{2^n n!}$

6. $a_n = \dbinom{n+1}{2}$

7–10 ■ Recursive Sequences A sequence is defined recursively. Find the first seven terms of the sequence.

7. $a_n = a_{n-1} + 2n - 1, \quad a_1 = 1$

8. $a_n = \dfrac{a_{n-1}}{n}, \quad a_1 = 1$

9. $a_n = a_{n-1} + 2a_{n-2}, \quad a_1 = 1, a_2 = 3$

10. $a_n = \sqrt{3a_{n-1}}, \quad a_1 = \sqrt{3}$

11–14 ■ Arithmetic or Geometric? The nth term of a sequence is given. **(a)** Find the first five terms of the sequence. **(b)** Graph the terms you found in part (a). **(c)** Find the fifth partial sum of the sequence. **(d)** Determine whether the sequence is arithmetic or geometric. Find the common difference or the common ratio.

11. $a_n = 2n + 5$

12. $a_n = \dfrac{5}{2^n}$

13. $a_n = \dfrac{3^n}{2^{n+1}}$

14. $a_n = 4 - \dfrac{n}{2}$

15–22 ■ Arithmetic or Geometric? The first four terms of a sequence are given. Determine whether they can be the terms of an arithmetic sequence, a geometric sequence, or neither. If the sequence is arithmetic or geometric, find the fifth term.

15. $5, 5.5, 6, 6.5, \ldots$

16. $\sqrt{2}, 2\sqrt{2}, 3\sqrt{2}, 4\sqrt{2}, \ldots$

17. $t - 3, t - 2, t - 1, t, \ldots$

18. $\sqrt{2}, 2, 2\sqrt{2}, 4, \ldots$

19. $t^3, t^2, t, 1, \ldots$

20. $1, -\frac{3}{2}, 2, -\frac{5}{2}, \ldots$

21. $\frac{3}{4}, \frac{1}{2}, \frac{1}{3}, \frac{2}{9}, \ldots$

22. $a, 1, \dfrac{1}{a}, \dfrac{1}{a^2}, \ldots$

23. Proving a Sequence Is Geometric Show that $3, 6i, -12, -24i, \ldots$ is a geometric sequence, and find the common ratio. (Here $i = \sqrt{-1}$.)

24. nth Term of a Geometric Sequence Find the nth term of the geometric sequence $2, 2 + 2i, 4i, -4 + 4i, -8, \ldots$ (Here $i = \sqrt{-1}$.)

25–28 ■ Finding Terms of Arithmetic and Geometric Sequences Find the indicated term of the arithmetic or geometric sequence with the given description.

25. The fourth term of an arithmetic sequence is 11, and the sixth term is 17. Find the second term.

26. The 20th term of an arithmetic sequence is 96, and the common difference is 5. Find the nth term.

27. The third term of a geometric sequence is 9, and the common ratio is $\frac{3}{2}$. Find the fifth term.

28. The second term of a geometric sequence is 10, and the fifth term is $\frac{1250}{27}$. Find the nth term.

29. Salary A teacher makes \$32,000 in his first year at Lakeside School and gets a 5% raise each year.

(a) Find a formula for his salary A_n in his nth year at this school.

(b) List his salaries for his first 8 years at this school.

30. Salary A colleague of the teacher in Exercise 29, hired at the same time, makes \$35,000 in her first year and gets a \$1200 raise each year.

(a) What is her salary A_n in her nth year at this school?

(b) Find her salary in her eighth year at this school, and compare it to the salary of the teacher in Exercise 29 in his eighth year.

31. Bacteria Culture A certain type of bacteria divides every 5 s. If three of these bacteria are put into a petri dish, how many bacteria are in the dish at the end of 1 min?

32. Arithmetic Sequences If $a_1, a_2, a_3, \ldots$ and $b_1, b_2, b_3, \ldots$ are arithmetic sequences, show that $a_1 + b_1, a_2 + b_2, a_3 + b_3, \ldots$ is also an arithmetic sequence.

33. Geometric Sequences If $a_1, a_2, a_3, \ldots$ and $b_1, b_2, b_3, \ldots$ are geometric sequences, show that $a_1b_1, a_2b_2, a_3b_3, \ldots$ is also a geometric sequence.

34. Arithmetic or Geometric?

(a) If $a_1, a_2, a_3, \ldots$ is an arithmetic sequence, is the sequence $a_1 + 2, a_2 + 2, a_3 + 2, \ldots$ arithmetic?

(b) If $a_1, a_2, a_3, \ldots$ is a geometric sequence, is the sequence $5a_1, 5a_2, 5a_3, \ldots$ geometric?

35. Arithmetic and Geometric Sequences Find the values of x for which the sequence $6, x, 12, \ldots$ is

(a) arithmetic **(b)** geometric

36. Arithmetic and Geometric Sequences Find the values of x and y for which the sequence $2, x, y, 17, \ldots$ is

(a) arithmetic **(b)** geometric

37–40 ■ Partial Sums Find the sum.

37. $\displaystyle\sum_{k=3}^{6} (k + 1)^2$

38. $\displaystyle\sum_{i=1}^{4} \frac{2i}{2i - 1}$

39. $\displaystyle\sum_{k=1}^{6} (k + 1)2^{k-1}$

40. $\displaystyle\sum_{m=1}^{5} 3^{m-2}$

41–44 ■ Sigma Notation Write the sum without using sigma notation. Do not evaluate.

41. $\displaystyle\sum_{k=1}^{10} (k - 1)^2$

42. $\displaystyle\sum_{j=2}^{100} \frac{1}{j - 1}$

43. $\displaystyle\sum_{k=1}^{50} \frac{3^k}{2^{k+1}}$

44. $\displaystyle\sum_{n=1}^{10} n^2 2^n$

45–48 ■ Sigma Notation Write the sum using sigma notation. Do not evaluate.

45. $3 + 6 + 9 + 12 + \cdots + 99$

46. $1^2 + 2^2 + 3^2 + \cdots + 100^2$

47. $1 \cdot 2^3 + 2 \cdot 2^4 + 3 \cdot 2^5 + 4 \cdot 2^6 + \cdots + 100 \cdot 2^{102}$

48. $\dfrac{1}{1 \cdot 2} + \dfrac{1}{2 \cdot 3} + \dfrac{1}{3 \cdot 4} + \cdots + \dfrac{1}{999 \cdot 1000}$

49–54 ■ Sums of Arithmetic and Geometric Sequences Determine whether the expression is a partial sum of an arithmetic or geometric sequence. Then find the sum.

49. $1 + 0.9 + (0.9)^2 + \cdots + (0.9)^5$

50. $3 + 3.7 + 4.4 + \cdots + 10$

51. $\sqrt{5} + 2\sqrt{5} + 3\sqrt{5} + \cdots + 100\sqrt{5}$

52. $\frac{1}{3} + \frac{2}{3} + 1 + \frac{4}{3} + \cdots + 33$

53. $\displaystyle\sum_{n=0}^{6} 3(-4)^n$

54. $\displaystyle\sum_{k=0}^{8} 7(5)^{k/2}$

55–60 ■ Infinite Geometric Series Determine whether the infinite geometric series is convergent or divergent. If it is convergent, find its sum.

55. $1 - \frac{2}{5} + \frac{4}{25} - \frac{8}{125} + \cdots$

56. $0.1 + 0.01 + 0.001 + 0.0001 + \cdots$

57. $5 - 5(1.01) + 5(1.01)^2 - 5(1.01)^3 + \cdots$

58. $1 + \frac{1}{3^{1/2}} + \frac{1}{3} + \frac{1}{3^{3/2}} + \cdots$

59. $-1 + \frac{9}{8} - \left(\frac{9}{8}\right)^2 + \left(\frac{9}{8}\right)^3 - \cdots$

60. $a + ab^2 + ab^4 + ab^6 + \cdots, \quad |b| < 1$

61. Terms of an Arithmetic Sequence The first term of an arithmetic sequence is $a = 7$, and the common difference is $d = 3$. How many terms of this sequence must be added to obtain 325?

62. Terms of an Geometric Sequence The sum of the first three terms of a geometric series is 52, and the common ratio is $r = 3$. Find the first term.

63. Ancestors A person has two parents, four grandparents, eight great-grandparents, and so on. What is the total number of a person's ancestors in 15 generations?

64. Annuity Find the amount of an annuity consisting of 16 annual payments of $1000 each into an account that pays 8% interest per year, compounded annually.

65. Investment How much money should be invested every quarter at 12% per year, compounded quarterly, in order to have $10,000 in one year?

66. Mortgage What are the monthly payments on a mortgage of $60,000 at 9% interest if the loan is to be repaid in

(a) 30 years? **(b)** 15 years?

67–69 ■ Mathematical Induction Use mathematical induction to prove that the formula is true for all natural numbers n.

67. $1 + 4 + 7 + \cdots + (3n - 2) = \dfrac{n(3n - 1)}{2}$

68. $\dfrac{1}{1 \cdot 3} + \dfrac{1}{3 \cdot 5} + \dfrac{1}{5 \cdot 7} + \cdots + \dfrac{1}{(2n - 1)(2n + 1)} = \dfrac{n}{2n + 1}$

69. $\left(1 + \frac{1}{1}\right)\left(1 + \frac{1}{2}\right)\left(1 + \frac{1}{3}\right)\cdots\left(1 + \frac{1}{n}\right) = n + 1$

70–72 ■ Proof by Induction Use mathematical induction to show that the given statement is true.

70. $7^n - 1$ is divisible by 6 for all natural numbers n.

71. The Fibonacci number F_{4n} is divisible by 3 for all natural numbers n.

72. Formula for a Recursive Sequence A sequence is defined recursively by $a_{n+1} = 3a_n + 4$ and $a_1 = 4$. Show that $a_n = 2 \cdot 3^n - 2$ for all natural numbers n.

73–76 ■ Binomial Coefficients Evaluate the expression.

73. $\binom{5}{2}\binom{5}{3}$

74. $\binom{10}{2} + \binom{10}{6}$

75. $\sum_{k=0}^{5} \binom{5}{k}$

76. $\sum_{k=0}^{8} \binom{8}{k}\binom{8}{8 - k}$

77–80 ■ Binomial Expansion Expand the expression.

77. $(A - B)^3$

78. $(x + 2)^5$

79. $(1 - x^2)^6$

80. $(2x + y)^4$

81–83 ■ Terms in a Binomial Expansion Find the indicated terms in the given binomial expansion.

81. Find the 20th term in the expansion of $(a + b)^{22}$.

82. Find the first three terms in the expansion of $(b^{-2/3} + b^{1/3})^{20}$.

83. Find the term containing A^6 in the expansion of $(A + 3B)^{10}$.

CHAPTER 12 TEST

1. Find the first six terms and the sixth partial sum of the sequence whose nth term is $a_n = 2n^2 - n$.

2. A sequence is defined recursively by $a_{n+1} = 3a_n - n$, $a_1 = 2$. Find the first six terms of the sequence.

3. An arithmetic sequence begins 2, 5, 8, 11, 14, ...
   (a) Find the common difference d for this sequence.
   (b) Find a formula for the nth term a_n of the sequence.
   (c) Find the 35th term of the sequence.

4. A geometric sequence begins 12, 3, $\frac{3}{4}$, $\frac{3}{16}$, $\frac{3}{64}$, ...
   (a) Find the common ratio r for this sequence.
   (b) Find a formula for the nth term a_n of the sequence.
   (c) Find the tenth term of the sequence.

5. The first term of a geometric sequence is 25, and the fourth term is $\frac{1}{5}$.
   (a) Find the common ratio r and the fifth term.
   (b) Find the partial sum of the first eight terms.

6. The first term of an arithmetic sequence is 10, and the tenth term is 2.
   (a) Find the common difference and the 100th term of the sequence.
   (b) Find the partial sum of the first ten terms.

7. Let $a_1, a_2, a_3, \ldots$ be a geometric sequence with initial term a and common ratio r. Show that $a_1^2, a_2^2, a_3^2, \ldots$ is also a geometric sequence by finding its common ratio.

8. Write the expression without using sigma notation, and then find the sum.
   (a) $\displaystyle\sum_{n=1}^{5} (1 - n^2)$ **(b)** $\displaystyle\sum_{n=3}^{6} (-1)^n 2^{n-2}$

9. Find the sum.
   (a) $\dfrac{1}{3} + \dfrac{2}{3^2} + \dfrac{2^2}{3^3} + \dfrac{2^3}{3^4} + \cdots + \dfrac{2^9}{3^{10}}$
   (b) $1 + \dfrac{1}{2^{1/2}} + \dfrac{1}{2} + \dfrac{1}{2^{3/2}} + \cdots$

10. Use mathematical induction to prove that for all natural numbers n,
$$1^2 + 2^2 + 3^2 + \cdots + n^2 = \frac{n(n+1)(2n+1)}{6}$$

11. Expand $(2x + y^2)^5$.

12. Find the term containing x^3 in the binomial expansion of $(3x - 2)^{10}$.

13. A puppy weighs 0.85 lb at birth, and each week he gains 24% in weight. Let a_n be his weight in pounds at the end of his nth week of life.
   (a) Find a formula for a_n.
   (b) How much does the puppy weigh when he is 6 weeks old?
   (c) Is the sequence $a_1, a_2, a_3, \ldots$ arithmetic, geometric, or neither?

FOCUS ON MODELING Modeling with Recursive Sequences

Many real-world processes occur in stages. Population growth can be viewed in stages—each new generation represents a new stage in population growth. Compound interest is paid in stages—each interest payment creates a new account balance. Many things that change continuously are more easily measured in discrete stages. For example, we can measure the temperature of a continuously cooling object in one-hour intervals. In this *Focus on Modeling* we learn how recursive sequences are used to model such situations. In some cases we can get an explicit formula for a sequence from the recursion relation that defines it by finding a pattern in the terms of the sequence.

Recursive Sequences as Models

Suppose you deposit some money in an account that pays 6% interest compounded monthly. The bank has a definite rule for paying interest: At the end of each month the bank adds to your account $\frac{1}{2}$% (or 0.005) of the amount in your account at that time. Let's express this rule as follows:

$$\text{amount at the end of this month} = \text{amount at the end of last month} + 0.005 \times \text{amount at the end of last month}$$

Using the Distributive Property, we can write this as

$$\text{amount at the end of this month} = 1.005 \times \text{amount at the end of last month}$$

To model this statement using algebra, let A_0 be the amount of the original deposit, let A_1 be the amount at the end of the first month, let A_2 be the amount at the end of the second month, and so on. So A_n is the amount at the end of the nth month. Thus

$$A_n = 1.005A_{n-1}$$

We recognize this as a recursively defined sequence—it gives us the amount at each stage in terms of the amount at the preceding stage.

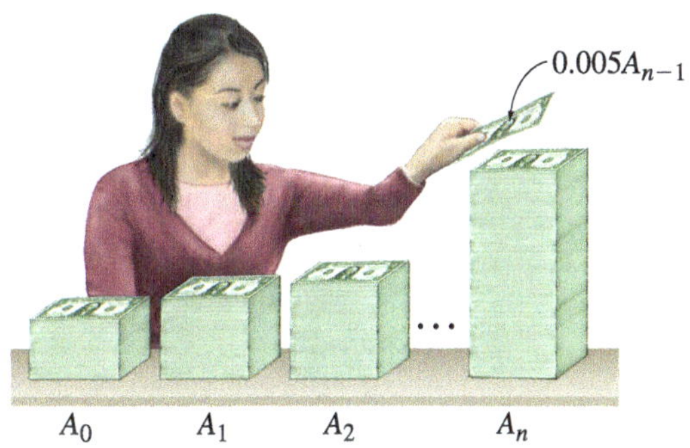

To find a formula for A_n, let's find the first few terms of the sequence and look for a pattern.

$$\begin{aligned}
A_1 &= 1.005A_0 \\
A_2 &= 1.005A_1 = (1.005)^2A_0 \\
A_3 &= 1.005A_2 = (1.005)^3A_0 \\
A_4 &= 1.005A_3 = (1.005)^4A_0
\end{aligned}$$

We can use mathematical induction to prove that the formula we found for A_n is valid for all natural numbers n.

We see that in general, $A_n = (1.005)^nA_0$.

EXAMPLE 1 ■ Population Growth

A certain animal population grows by 2% each year. The initial population is 5000.

(a) Find a recursive sequence that models the population P_n at the end of the nth year.

(b) Find the first five terms of the sequence P_n.

(c) Find a formula for P_n.

SOLUTION

(a) We can model the population using the following rule:

population at the end of this year = 1.02 × population at the end of last year

Algebraically, we can write this as the recursion relation

$$P_n = 1.02P_{n-1}$$

(b) Since the initial population is 5000, we have

$$\begin{aligned} P_0 &= 5000 \\ P_1 &= 1.02P_0 = (1.02)5000 \\ P_2 &= 1.02P_1 = (1.02)^2 5000 \\ P_3 &= 1.02P_2 = (1.02)^3 5000 \\ P_4 &= 1.02P_3 = (1.02)^4 5000 \end{aligned}$$

(c) We see from the pattern exhibited in part (b) that $P_n = (1.02)^n 5000$. (Note that P_n is a geometric sequence, with common ratio $r = 1.02$.) ■

EXAMPLE 2 ■ Daily Drug Dose

A patient is to take a 50-mg pill of a certain drug every morning. It is known that the body eliminates 40% of the drug every 24 h.

(a) Find a recursive sequence that models the amount A_n of the drug in the patient's body after each pill is taken.

(b) Find the first four terms of the sequence A_n.

(c) Find a formula for A_n.

(d) How much of the drug remains in the patient's body after 5 days? How much will accumulate in his system after prolonged use?

SOLUTION

(a) Each morning, 60% of the drug remains in his system, plus he takes an additional 50 mg (his daily dose).

amount of drug this morning = 0.6 × amount of drug yesterday morning + 50 mg

We can express this as a recursion relation

$$A_n = 0.6A_{n-1} + 50$$

(b) Since the initial dose is 50 mg, we have

$$\begin{aligned} A_0 &= 50 \\ A_1 &= 0.6A_0 + 50 = 0.6(50) + 50 \end{aligned}$$

$$\begin{aligned} A_2 &= 0.6A_1 + 50 = 0.6[0.6(50) + 50] + 50 \\ &= 0.6^2(50) + 0.6(50) + 50 \\ &= 50(0.6^2 + 0.6 + 1) \\ A_3 &= 0.6A_2 + 50 = 0.6[0.6^2(50) + 0.6(50) + 50] + 50 \\ &= 0.6^3(50) + 0.6^2(50) + 0.6(50) + 50 \\ &= 50(0.6^3 + 0.6^2 + 0.6 + 1) \end{aligned}$$

(c) From the pattern in part (b) we see that

$$\begin{aligned} A_n &= 50(1 + 0.6 + 0.6^2 + \cdots + 0.6^n) \\ &= 50\left(\frac{1 - 0.6^{n+1}}{1 - 0.6}\right) && \text{Partial sum of a geometric sequence (page 861)} \\ &= 125(1 - 0.6^{n+1}) && \text{Simplify} \end{aligned}$$

(d) To find the amount remaining after 5 days, we substitute $n = 5$ and get $A_5 = 125(1 - 0.6^{5+1}) \approx 119$ mg.

To find the amount remaining after prolonged use, we let n become large. As n gets large, 0.6^n approaches 0. That is, $0.6^n \to 0$ as $n \to \infty$ (see Section 4.1, page 332). So as $n \to \infty$,

$$A_n = 125(1 - 0.6^{n+1}) \to 125(1 - 0) = 125$$

Thus after prolonged use, the amount of drug in the patient's system approaches 125 mg (see Figure 1, where we have used a graphing calculator to graph the sequence).

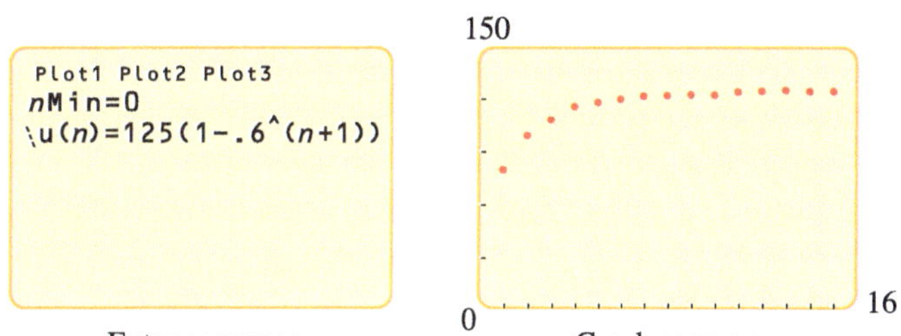

Enter sequence

Graph sequence

FIGURE 1

PROBLEMS

1. **Retirement Accounts** Many college professors keep retirement savings with TIAA, the largest annuity program in the world. Interest on these accounts is compounded and credited *daily*. Professor Brown has \$275,000 on deposit with TIAA at the start of 2015 and receives 3.65% interest per year on his account.

   (a) Find a recursive sequence that models the amount A_n in his account at the end of the nth day of 2015.

   (b) Find the first eight terms of the sequence A_n, rounded to the nearest cent.

   (c) Find a formula for A_n.

2. **Fitness Program** Sheila decides to embark on a swimming program as the best way to maintain cardiovascular health. She begins by swimming 5 min on the first day, then adds $1\frac{1}{2}$ min every day after that.

   (a) Find a recursive formula for the number of minutes T_n that she swims on the nth day of her program.

   (b) Find the first 6 terms of the sequence T_n.

   (c) Find a formula for T_n. What kind of sequence is this?

   (d) On what day does Sheila attain her goal of swimming at least 65 min a day?

   (e) What is the total amount of time she will have swum after 30 days?

3. Monthly Savings Program Alice opens a savings account that pays 3% interest per year, compounded monthly. She begins by depositing \$100 at the start of the first month and adds \$100 at the end of each month, when the interest is credited.

(a) Find a recursive formula for the amount A_n in her account at the end of the nth month. (Include the interest credited for that month and her monthly deposit.)

(b) Find the first five terms of the sequence A_n.

(c) Use the pattern you observed in (b) to find a formula for A_n. [*Hint:* To find the pattern most easily, it's best *not* to simplify the terms *too* much.]

(d) How much has she saved after 5 years?

4. Pollution A chemical plant discharges 2400 tons of pollutants every year into an adjacent lake. Through natural runoff, 70% of the pollutants contained in the lake at the beginning of the year are expelled by the end of the year.

(a) Explain why the following sequence models the amount A_n of the pollutant in the lake at the end of the nth year that the plant is operating.

$$A_n = 0.30A_{n-1} + 2400$$

(b) Find the first five terms of the sequence A_n.

(c) Find a formula for A_n.

(d) How much of the pollutant remains in the lake after 6 years? How much will remain after the plant has been operating a long time?

(e) Verify your answer to part (d) by graphing A_n with a graphing calculator for $n = 1$ to $n = 20$.

5. Annual Savings Program Ursula opens a 1-year CD that yields 5% interest per year. She begins with a deposit of \$5000. At the end of each year when the CD matures, she reinvests at the same 5% interest rate, also adding 10% to the value of the CD from her other savings. (So for example, after the first year her CD has earned 5% of \$5000 in interest, for a value of \$5250 at maturity. She then adds 10%, or \$525, bringing the total value of her renewed CD to \$5775.)

(a) Find a recursive formula for the amount U_n in Ursula's CD when she reinvests at the end of the nth year.

(b) Find the first five terms of the sequence U_n. Does this appear to be a geometric sequence?

(c) Use the pattern you observed in (b) to find a formula for U_n.

(d) How much has she saved after 10 years?

6. Annual Savings Program Victoria opens a one-year CD with a 5% annual interest yield at the same time as her friend Ursula in Problem 5. She also starts with an initial deposit of \$5000. However, Victoria decides to add \$500 to her CD when she reinvests at the end of the first year, \$1000 at the end of the second, \$1500 at the end of the third, and so on.

(a) Explain why the recursive formula displayed below gives the amount V_n in Victoria's CD when she reinvests at the end of the nth year.

$$V_n = 1.05V_{n-1} + 500n$$

(b) Using the `Seq` ("sequence") mode on your graphing calculator, enter the sequences U_n and V_n as shown in the figure. Then use the `TABLE` command to compare the two sequences. For the first few years, Victoria seems to be accumulating more savings than Ursula. Scroll down in the table to verify that Ursula eventually pulls ahead of Victoria in the savings race. In what year does this occur?

```
Plot1 Plot2 Plot3
\u(n) = 1.05 u(n - 1)
  +0.1(1.05 u(n - 1))
  u(nMin) = {5000}
\v(n) = 1.05 v(n - 1)
  +500 n
  v(nMin) = {5000}
```

Entering the sequences

n	u(n)	v(n)
0	5000	5000
1	5775	5750
2	6670.1	7037.5
3	7704	8889.4
4	8898.1	11334
5	10277	14401
6	11870	18121
n=0		

Table of values of the sequences

PF-(space1)/Alamy

13 Limits: A Preview of Calculus

In this chapter we study the central idea underlying calculus: the concept of a *limit*. Calculus is used in modeling numerous real-life phenomena, particularly situations that involve change or motion. Limits are used in finding the instantaneous rate of change of a function as well as the area of a region with curved boundary. You will learn in calculus that these two apparently different problems are closely related. In this chapter we see how limits allow us to solve both problems.

In Chapter 2 we learned how to find the average rate of change of a function. For example, to find the average speed, we divide the total distance traveled by the total time. But how can we find *instantaneous* speed—that is, the speed at a given instant? We can't divide the total distance by the total time because in an instant the total distance traveled is zero and the total time spent traveling is zero! But we can find the average rate of change on smaller and smaller intervals, zooming in on the instant we want. In other words, the instantaneous speed is a *limit* of the average speeds.

In this chapter we also learn how to find areas of regions with curved sides by using the limit process.

13.1 FINDING LIMITS NUMERICALLY AND GRAPHICALLY

■ Definition of Limit ■ Estimating Limits Numerically and Graphically ■ Limits That Fail to Exist ■ One-Sided Limits

In this section we use tables of values and graphs of functions to answer the question, What happens to the values $f(x)$ of a function f as the variable x approaches the number a?

■ Definition of Limit

We begin by investigating the behavior of the function f defined by

$$f(x) = x^2 - x + 2$$

for values of x near 2. The following tables give values of $f(x)$ for values of x close to 2 but not equal to 2.

x	$f(x)$
1.0	2.000000
1.5	2.750000
1.8	3.440000
1.9	3.710000
1.95	3.852500
1.99	3.970100
1.995	3.985025
1.999	3.997001

x	$f(x)$
3.0	8.000000
2.5	5.750000
2.2	4.640000
2.1	4.310000
2.05	4.152500
2.01	4.030100
2.005	4.015025
2.001	4.003001

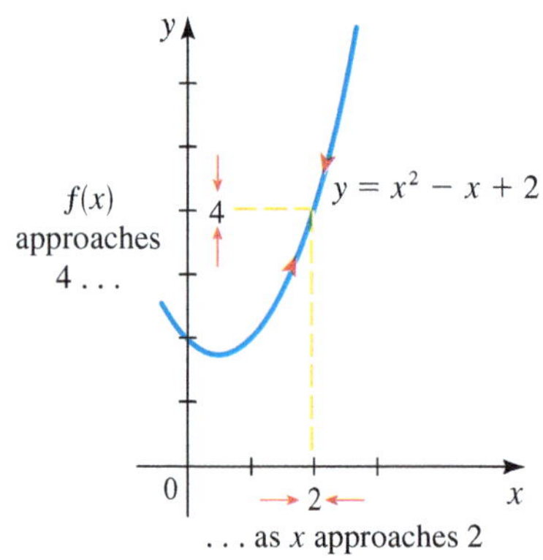

FIGURE 1

From the table and the graph of f (a parabola) shown in Figure 1 we see that when x is close to 2 (on either side of 2), $f(x)$ is close to 4. In fact, it appears that we can make the values of $f(x)$ as close as we like to 4 by taking x sufficiently close to 2. We express this by saying "the limit of the function $f(x) = x^2 - x + 2$ as x approaches 2 is equal to 4." The notation for this is

$$\lim_{x \to 2} (x^2 - x + 2) = 4$$

In general, we use the following notation.

DEFINITION OF THE LIMIT OF A FUNCTION

We write

$$\lim_{x \to a} f(x) = L$$

and say

"the limit of $f(x)$, as x approaches a, equals L"

if we can make the values of $f(x)$ arbitrarily close to L (as close to L as we like) by taking x to be sufficiently close to a, but not equal to a.

Roughly speaking, this says that the values of $f(x)$ get closer and closer to the number L as x gets closer and closer to the number a (from either side of a) but $x \neq a$.

An alternative notation for $\lim_{x\to a} f(x) = L$ is

$$f(x) \to L \qquad \text{as} \qquad x \to a$$

which is usually read "$f(x)$ approaches L as x approaches a." This is the notation we used in Section 3.6 when discussing asymptotes of rational functions.

Notice the phrase "but $x \neq a$" in the definition of limit. This means that in finding the limit of $f(x)$ as x approaches a, we never consider $x = a$. In fact, $f(x)$ need not even be defined when $x = a$. The only thing that matters is how f is defined *near* a.

Figure 2 shows the graphs of three functions. Note that in part (c), $f(a)$ is not defined, and in part (b), $f(a) \neq L$. But in each case, regardless of what happens at a, $\lim_{x\to a} f(x) = L$.

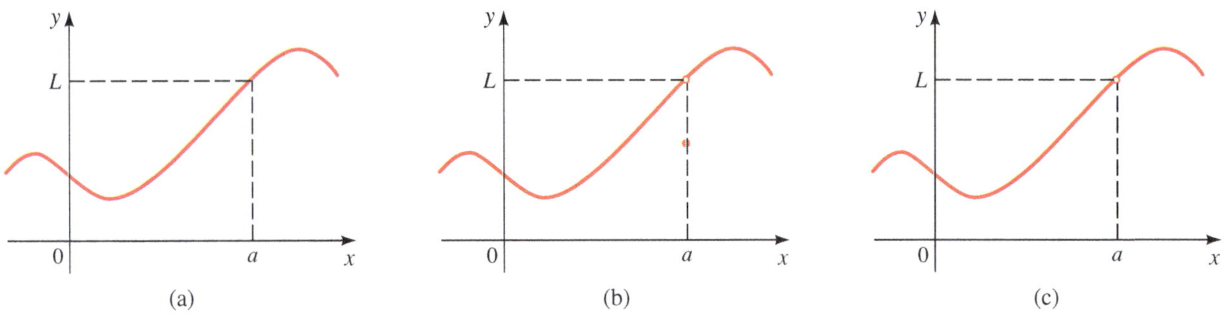

FIGURE 2 $\lim_{x\to a} f(x) = L$ in all three cases

Estimating Limits Numerically and Graphically

In Section 13.2 we will develop techniques for finding exact values of limits. For now, we use tables and graphs to estimate limits of functions.

EXAMPLE 1 ■ Estimating a Limit Numerically and Graphically

Estimate the value of the following limit by making a table of values. Check your work with a graph.

$$\lim_{x\to 1} \frac{x-1}{x^2-1}$$

SOLUTION Notice that the function $f(x) = (x-1)/(x^2-1)$ is not defined when $x = 1$, but this doesn't matter because the definition of $\lim_{x\to a} f(x)$ says that we consider values of x that are close to a but not equal to a. The following tables give values of $f(x)$ (rounded to six decimal places) for values of x that approach 1 (but are not equal to 1).

$x < 1$	$f(x)$
0.5	0.666667
0.9	0.526316
0.99	0.502513
0.999	0.500250
0.9999	0.500025

$x > 1$	$f(x)$
1.5	0.400000
1.1	0.476190
1.01	0.497512
1.001	0.499750
1.0001	0.499975

On the basis of the values in the two tables we make the guess that

$$\lim_{x\to 1} \frac{x-1}{x^2-1} = 0.5$$

As a graphical verification we use a graphing device to produce Figure 3. We see that when x is close to 1, y is close to 0.5. If we use the ZOOM and TRACE features to get a closer look, as in Figure 4, we notice that as x gets closer and closer to 1, y becomes closer and closer to 0.5. This reinforces our conclusion.

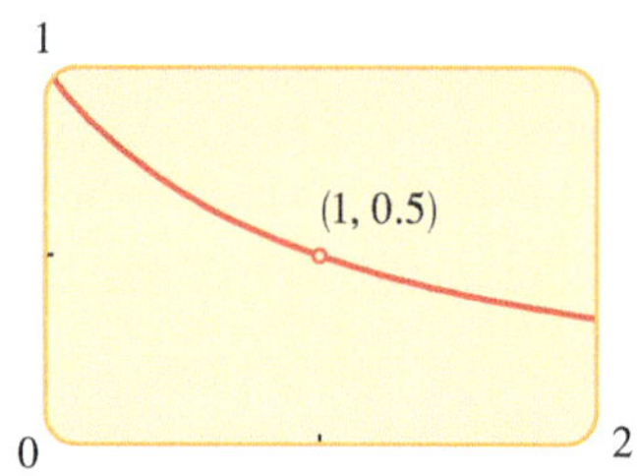

FIGURE 3

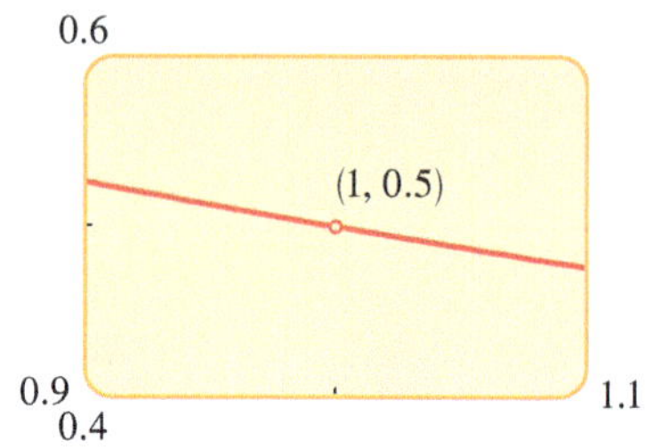

FIGURE 4

Now Try Exercise 3 ■

EXAMPLE 2 ■ Finding a Limit from a Table

Find $\displaystyle\lim_{t\to 0} \frac{\sqrt{t^2+9}-3}{t^2}$.

t	$\dfrac{\sqrt{t^2+9}-3}{t^2}$
±1.0	0.16228
±0.5	0.16553
±0.1	0.16662
±0.05	0.16666
±0.01	0.16667

SOLUTION The upper table in the margin lists values of the function for several values of t near 0. As t approaches 0, the values of the function seem to approach 0.1666666 . . . , so we guess that

$$\lim_{t\to 0} \frac{\sqrt{t^2+9}-3}{t^2} = \frac{1}{6}$$

Now Try Exercise 5 ■

t	$\dfrac{\sqrt{t^2+9}-3}{t^2}$
±0.0005	0.16800
±0.0001	0.20000
±0.00005	0.00000
±0.00001	0.00000

What would have happened in Example 2 if we had taken even smaller values of t? The lower table in the margin shows the results from one calculator; you can see that something strange seems to be happening.

If you try these calculations on your own calculator, you might get different values, but eventually, you will get the value 0 if you make t sufficiently small. Does this mean that the answer is really 0 instead of $\frac{1}{6}$? No, the value of the limit is $\frac{1}{6}$, as we will show in the next section. The problem is that the calculator gave false values because $\sqrt{t^2+9}$ is very close to 3 when t is small. (In fact, when t is sufficiently small, a calculator's value for $\sqrt{t^2+9}$ is 3.000 . . . to as many digits as the calculator is capable of carrying.)

Something similar happens when we try to graph the function of Example 2 on a graphing device. Parts (a) and (b) of Figure 5 show quite accurate graphs of this function, and when we use the TRACE feature, we can easily estimate that the limit is about $\frac{1}{6}$. But if we zoom in too far, as in parts (c) and (d), then we get inaccurate graphs, again because of problems with subtraction.

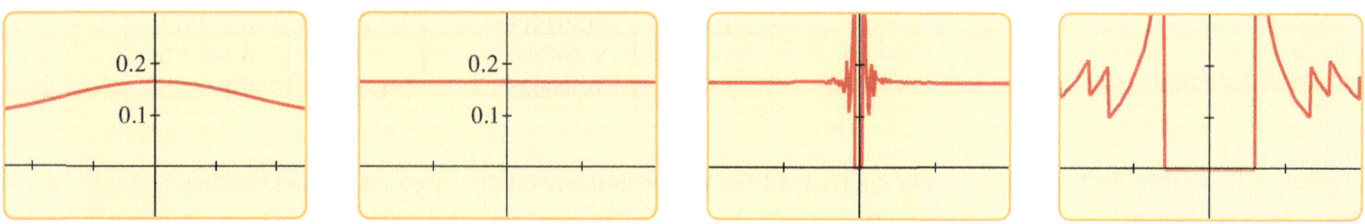

(a) $[-5, 5]$ by $[-0.1, 0.3]$ (b) $[-0.1, 0.1]$ by $[-0.1, 0.3]$ (c) $[-10^{-6}, 10^{-6}]$ by $[-0.1, 0.3]$ (d) $[-10^{-7}, 10^{-7}]$ by $[-0.1, 0.3]$

FIGURE 5

Limits That Fail to Exist

Functions do not necessarily approach a finite value at every point. In other words, it's possible for a limit not to exist. The next three examples illustrate ways in which this can happen.

EXAMPLE 3 ■ A Limit That Fails to Exist (A Function with a Jump)

The Heaviside function H is defined by

$$H(t) = \begin{cases} 0 & \text{if } t < 0 \\ 1 & \text{if } t \geq 0 \end{cases}$$

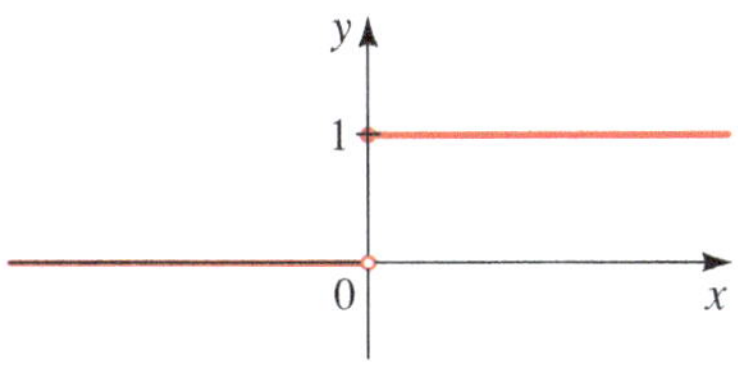

FIGURE 6

[This function, named after the electrical engineer Oliver Heaviside (1850–1925), can be used to describe an electric current that is switched on at time $t = 0$.] Its graph is shown in Figure 6. Notice the "jump" in the graph at $x = 0$.

As t approaches 0 from the left, $H(t)$ approaches 0. As t approaches 0 from the right, $H(t)$ approaches 1. There is no single number that $H(t)$ approaches as t approaches 0. Therefore $\lim_{t\to 0} H(t)$ does not exist.

Now Try Exercise 27 ■

EXAMPLE 4 ■ A Limit That Fails to Exist (A Function That Oscillates)

Find $\displaystyle\lim_{x\to 0} \sin\frac{\pi}{x}$.

SOLUTION The function $f(x) = \sin(\pi/x)$ is undefined at 0. Evaluating the function for some small values of x, we get

$$f(1) = \sin \pi = 0 \qquad f(\tfrac{1}{2}) = \sin 2\pi = 0$$
$$f(\tfrac{1}{3}) = \sin 3\pi = 0 \qquad f(\tfrac{1}{4}) = \sin 4\pi = 0$$
$$f(0.1) = \sin 10\pi = 0 \qquad f(0.01) = \sin 100\pi = 0$$

Similarly, $f(0.001) = f(0.0001) = 0$. On the basis of this information we might be tempted to guess that

$$\lim_{x\to 0} \sin\frac{\pi}{x} \stackrel{?}{=} 0$$

but this time our guess is wrong. Note that although $f(1/n) = \sin n\pi = 0$ for any integer n, it is also true that $f(x) = 1$ for infinitely many values of x that approach 0. (See the graph in Figure 7.)

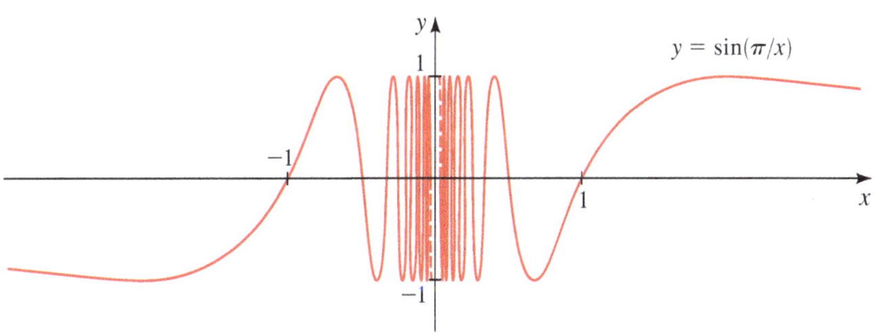

FIGURE 7

The dashed lines indicate that the values of $\sin(\pi/x)$ oscillate between 1 and -1 infinitely often as x approaches 0. Since the values of $f(x)$ do not approach a fixed number as x approaches 0,

$$\lim_{x\to 0} \sin\frac{\pi}{x} \quad \text{does not exist}$$

Now Try Exercise 25 ■

Example 4 illustrates some of the pitfalls in guessing the value of a limit. It is easy to guess the wrong value if we use inappropriate values of x, but it is difficult to know when to stop calculating values. And as the discussion after Example 2 shows, sometimes calculators and computers give incorrect values. In the next two sections, however, we will develop foolproof methods for calculating limits.

EXAMPLE 5 ■ A Limit That Fails to Exist (A Function with a Vertical Asymptote)

Find $\lim_{x \to 0} \dfrac{1}{x^2}$ if it exists.

x	$\dfrac{1}{x^2}$
± 1	1
± 0.5	4
± 0.2	25
± 0.1	100
± 0.05	400
± 0.01	10,000
± 0.001	1,000,000

SOLUTION As x becomes close to 0, x^2 also becomes close to 0, and $1/x^2$ becomes very large. (See the table in the margin.) In fact, it appears from the graph of the function $f(x) = 1/x^2$ shown in Figure 8 that the values of $f(x)$ can be made arbitrarily large by taking x close enough to 0. Thus the values of $f(x)$ do not approach a number, so $\lim_{x \to 0}(1/x^2)$ does not exist.

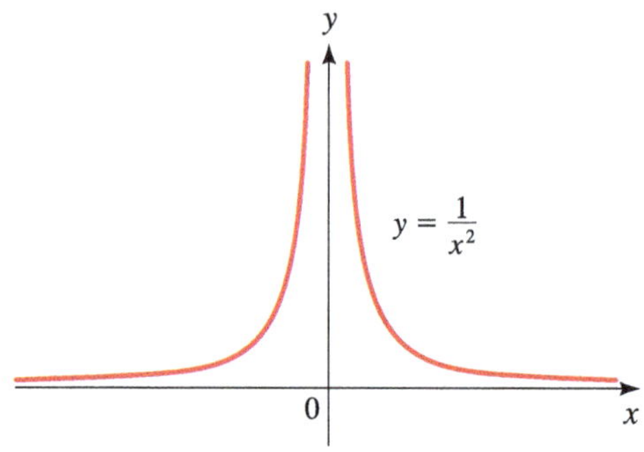

FIGURE 8

Now Try Exercise 23 ■

To indicate the kind of behavior exhibited in Example 5, we use the notation

$$\lim_{x \to 0} \frac{1}{x^2} = \infty$$

This does not mean that we are regarding ∞ as a number. Nor does it mean that the limit exists. It simply expresses the particular way in which the limit does not exist: $1/x^2$ can be made as large as we like by taking x close enough to 0. Notice that the line $x = 0$ (the y-axis) is a vertical asymptote in the sense that we described in Section 3.6.

One-Sided Limits

We noticed in Example 3 that $H(t)$ approaches 0 as t approaches 0 from the left and $H(t)$ approaches 1 as t approaches 0 from the right. We indicate this situation symbolically by writing

$$\lim_{t \to 0^-} H(t) = 0 \qquad \text{and} \qquad \lim_{t \to 0^+} H(t) = 1$$

The symbol "$t \to 0^-$" indicates that we consider only values of t that are less than 0. Likewise, "$t \to 0^+$" indicates that we consider only values of t that are greater than 0.

DEFINITION OF A ONE-SIDED LIMIT

We write

$$\lim_{x \to a^-} f(x) = L$$

and say that the "left-hand limit of $f(x)$ as x approaches a" [or the "limit of $f(x)$ as x approaches a from the left"] is equal to L if we can make the values of $f(x)$ arbitrarily close to L by taking x to be sufficiently close to a and x less than a.

Notice that this definition differs from the definition of a two-sided limit only in that we require x to be less than a. Similarly, if we require that x be greater than a, we get "the **right-hand limit of $f(x)$ as x approaches a** is equal to L," and we write

$$\lim_{x \to a^+} f(x) = L$$

Thus the symbol "$x \to a^+$" means that we consider only $x > a$. These definitions are illustrated in Figure 9.

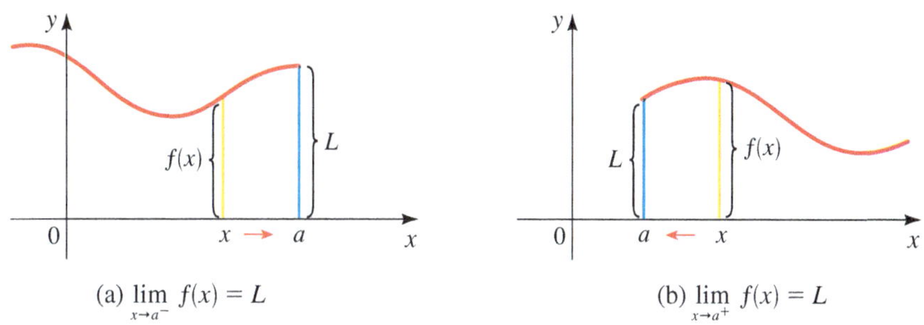

FIGURE 9

By comparing the definitions of two-sided and one-sided limits, we see that the following is true.

$$\lim_{x \to a} f(x) = L \quad \text{if and only if} \quad \lim_{x \to a^-} f(x) = L \quad \text{and} \quad \lim_{x \to a^+} f(x) = L$$

Thus if the left-hand and right-hand limits are different, the (two-sided) limit does not exist. We use this fact in the next two examples.

EXAMPLE 6 ■ Limits from a Graph

The graph of a function g is shown in Figure 10. Use it to state the values (if they exist) of the following:

(a) $\lim_{x \to 2^-} g(x)$, $\lim_{x \to 2^+} g(x)$, $\lim_{x \to 2} g(x)$

(b) $\lim_{x \to 5^-} g(x)$, $\lim_{x \to 5^+} g(x)$, $\lim_{x \to 5} g(x)$

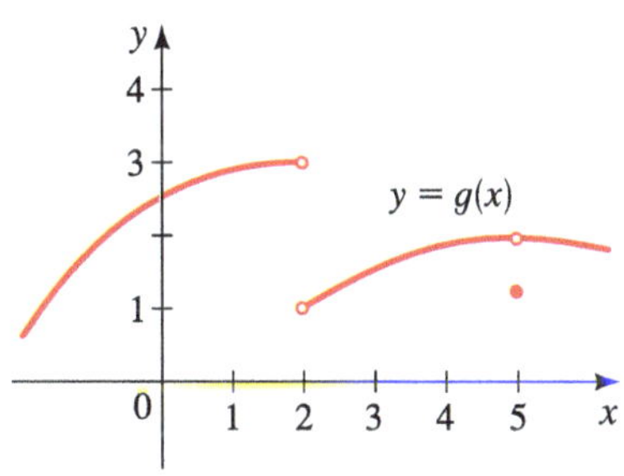

FIGURE 10

SOLUTION

(a) From the graph we see that the values of $g(x)$ approach 3 as x approaches 2 from the left, but they approach 1 as x approaches 2 from the right. Therefore

$$\lim_{x \to 2^-} g(x) = 3 \qquad \text{and} \qquad \lim_{x \to 2^+} g(x) = 1$$

Since the left- and right-hand limits are different, we conclude that $\lim_{x \to 2} g(x)$ does not exist.

(b) The graph also shows that

$$\lim_{x \to 5^-} g(x) = 2 \qquad \text{and} \qquad \lim_{x \to 5^+} g(x) = 2$$

This time the left- and right-hand limits are the same, so we have

$$\lim_{x \to 5} g(x) = 2$$

Despite this fact, notice that $g(5) \neq 2$.

Now Try Exercise 19

EXAMPLE 7 ■ A Piecewise-Defined Function

Let f be the function defined by

$$f(x) = \begin{cases} 2x^2 & \text{if } x < 1 \\ 4 - x & \text{if } x \geq 1 \end{cases}$$

Graph f, and use the graph to find the following:

(a) $\lim_{x \to 1^-} f(x)$ **(b)** $\lim_{x \to 1^+} f(x)$ **(c)** $\lim_{x \to 1} f(x)$

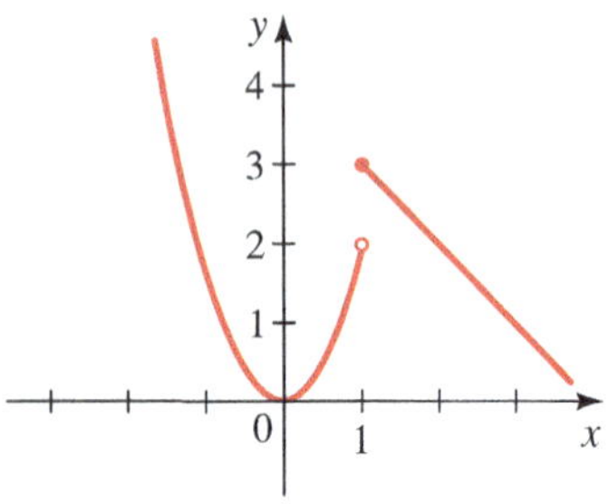

FIGURE 11

SOLUTION The graph of f is shown in Figure 11. From the graph we see that the values of $f(x)$ approach 2 as x approaches 1 from the left, but they approach 3 as x approaches 1 from the right. Thus the left- and right-hand limits are not equal. So we have

(a) $\lim_{x \to 1^-} f(x) = 2$ **(b)** $\lim_{x \to 1^+} f(x) = 3$ **(c)** $\lim_{x \to 1} f(x)$ does not exist.

Now Try Exercise 29

13.1 EXERCISES

CONCEPTS

1. When we write $\lim_{x \to a} f(x) = L$ then, roughly speaking, the values of $f(x)$ get closer and closer to the number ________ as the values of x get closer and closer to ________. To determine $\lim_{x \to 5} \frac{x - 5}{x - 5}$, we try values for x closer and closer to ________ and find that the limit is ________.

2. We write $\lim_{x \to a^-} f(x) = L$ and say that the ________ of $f(x)$ as x approaches a from the ________ (left/right) is equal to ________. To find the left-hand limit, we try values for x that are ________ (less/greater) than a. A limit exists if and only if both the ________-hand and ________-hand limits exist and are ________.

SKILLS

3–4 ■ Estimating Limits Numerically and Graphically Estimate the value of the limit by making a table of values. Check your work with a graph.

3. $\lim_{x \to 5} \frac{x^2 - 25}{x - 5}$ **4.** $\lim_{x \to 3} \frac{x^2 - x - 6}{x - 3}$

5–10 ■ Estimating Limits Numerically Complete the table of values (to five decimal places), and use the table to estimate the value of the limit.

5. $\lim_{x \to 4} \frac{\sqrt{x} - 2}{x - 4}$

x	3.9	3.99	3.999	4.001	4.01	4.1
$f(x)$						

6. $\lim_{x \to 2} \frac{x - 2}{x^2 + x - 6}$

x	1.9	1.99	1.999	2.001	2.01	2.1
$f(x)$						

7. $\lim_{x \to 1} \frac{x - 1}{x^3 - 1}$

x	0.9	0.99	0.999	1.001	1.01	1.1
$f(x)$						

8. $\lim_{x \to 0} \dfrac{e^x - 1}{x}$

x	-0.1	-0.01	-0.001	0.001	0.01	0.1
$f(x)$						

9. $\lim_{x \to 0} \dfrac{\sin x}{x}$

x	± 1	± 0.5	± 0.1	± 0.05	± 0.01
$f(x)$					

10. $\lim_{x \to 0^+} x \ln x$

x	0.1	0.01	0.001	0.0001	0.00001
$f(x)$					

11–16 ■ Estimating Limits Numerically and Graphically Use a table of values to estimate the value of the limit. Then use a graphing device to confirm your result graphically.

11. $\lim_{x \to -4} \dfrac{x + 4}{x^2 + 7x + 12}$ **12.** $\lim_{x \to 1} \dfrac{x^3 - 1}{x^2 - 1}$

13. $\lim_{x \to 0} \dfrac{5^x - 3^x}{x}$ **14.** $\lim_{x \to 0} \dfrac{\sqrt{x + 9} - 3}{x}$

15. $\lim_{x \to 1} \left(\dfrac{1}{\ln x} - \dfrac{1}{x - 1} \right)$ **16.** $\lim_{x \to 0} \dfrac{\tan 2x}{\tan 3x}$

17–20 ■ Limits from a Graph For the function f whose graph is given, state the value of the given quantity if it exists. If it does not exist, explain why.

17. **(a)** $\lim_{x \to 1^-} f(x)$ **(b)** $\lim_{x \to 1^+} f(x)$ **(c)** $\lim_{x \to 1} f(x)$
(d) $\lim_{x \to 5} f(x)$ **(e)** $f(5)$

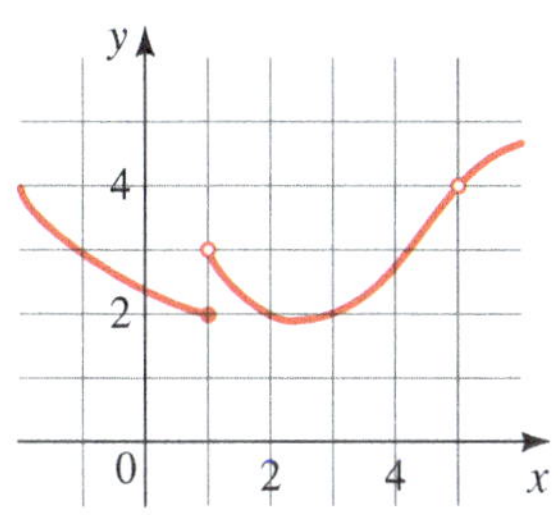

18. **(a)** $\lim_{x \to 0} f(x)$ **(b)** $\lim_{x \to 3^-} f(x)$ **(c)** $\lim_{x \to 3^+} f(x)$
(d) $\lim_{x \to 3} f(x)$ **(e)** $f(3)$

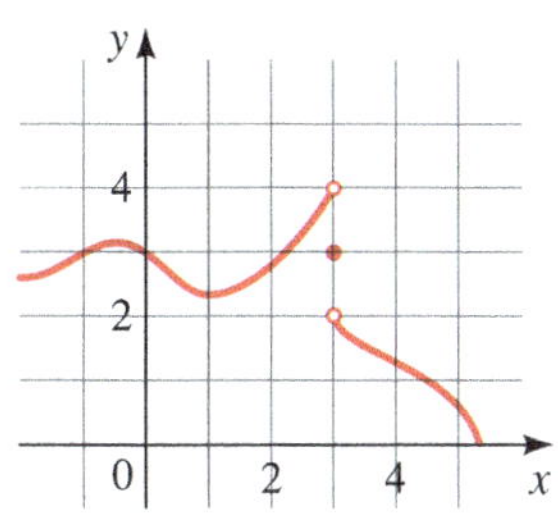

19. **(a)** $\lim_{t \to 0^-} f(t)$ **(b)** $\lim_{t \to 0^+} f(t)$ **(c)** $\lim_{t \to 0} f(t)$
(d) $\lim_{t \to 2^-} f(t)$ **(e)** $\lim_{t \to 2^+} f(t)$ **(f)** $\lim_{t \to 2} f(t)$
(g) $f(2)$ **(h)** $\lim_{t \to 4} f(t)$

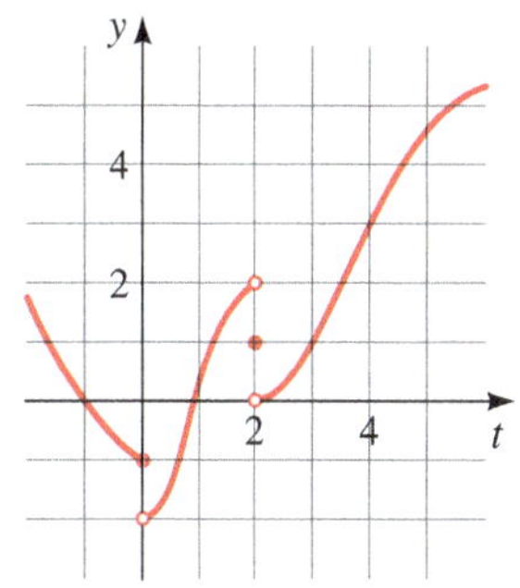

20. **(a)** $\lim_{x \to 3} f(x)$ **(b)** $\lim_{x \to 1} f(x)$ **(c)** $\lim_{x \to -3} f(x)$
(d) $\lim_{x \to 2^-} f(x)$ **(e)** $\lim_{x \to 2^+} f(x)$ **(f)** $\lim_{x \to 2} f(x)$

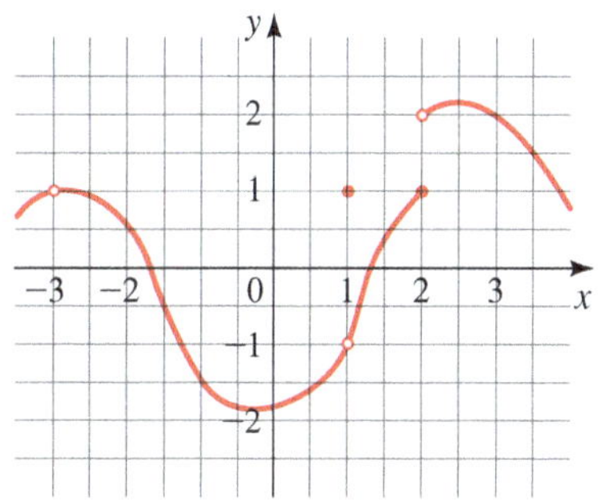

21–28 ■ Estimating Limits Graphically Use a graphing device to determine whether the limit exists. If the limit exists, estimate its value to two decimal places.

21. $\lim_{x \to 1} \dfrac{x^3 + x^2 + 3x - 5}{2x^2 - 5x + 3}$ **22.** $\lim_{x \to 0} \dfrac{x^2}{\cos 5x - \cos 4x}$

23. $\lim_{x \to 0} \ln(\sin^2 x)$ **24.** $\lim_{x \to 2} \dfrac{x^3 + 6x^2 - 5x + 1}{x^3 - x^2 - 8x + 12}$

25. $\lim_{x \to 0} \cos \dfrac{1}{x}$ **26.** $\lim_{x \to 0} \sin \dfrac{2}{x}$

27. $\lim_{x \to 3} \dfrac{|x - 3|}{x - 3}$ **28.** $\lim_{x \to 0} \dfrac{1}{1 + e^{1/x}}$

29–32 ■ One-Sided Limits Graph the piecewise-defined function and use your graph to find the values of the limits, if they exist.

29. $f(x) = \begin{cases} x^2 & \text{if } x \le 2 \\ 6 - x & \text{if } x > 2 \end{cases}$

(a) $\lim_{x \to 2^-} f(x)$ **(b)** $\lim_{x \to 2^+} f(x)$ **(c)** $\lim_{x \to 2} f(x)$

30. $f(x) = \begin{cases} 2 & \text{if } x < 0 \\ x + 1 & \text{if } x \ge 0 \end{cases}$

(a) $\lim_{x \to 0^-} f(x)$ **(b)** $\lim_{x \to 0^+} f(x)$ **(c)** $\lim_{x \to 0} f(x)$

31. $f(x) = \begin{cases} -x + 3 & \text{if } x < -1 \\ 3 & \text{if } x \geq -1 \end{cases}$

(a) $\lim_{x \to -1^-} f(x)$ **(b)** $\lim_{x \to -1^+} f(x)$ **(c)** $\lim_{x \to -1} f(x)$

32. $f(x) = \begin{cases} 2x + 10 & \text{if } x \leq -2 \\ -x + 4 & \text{if } x > -2 \end{cases}$

(a) $\lim_{x \to -2^-} f(x)$ **(b)** $\lim_{x \to -2^+} f(x)$ **(c)** $\lim_{x \to -2} f(x)$

DISCUSS ■ DISCOVER ■ PROVE ■ WRITE

33. DISCUSS: A Function with Specified Limits Sketch the graph of an example of a function f that satisfies all of the following conditions.

$$\lim_{x \to 0^-} f(x) = 2 \qquad \lim_{x \to 0^+} f(x) = 0$$

$$\lim_{x \to 2} f(x) = 1 \qquad f(0) = 2 \qquad f(2) = 3$$

How many such functions are there?

34. DISCUSS: Graphing Calculator Pitfalls

(a) Evaluate

$$h(x) = \frac{\tan x - x}{x^3}$$

for $x = 1, 0.5, 0.1, 0.05, 0.01$, and 0.005.

(b) Guess the value of $\lim_{x \to 0} \dfrac{\tan x - x}{x^3}$.

(c) Evaluate $h(x)$ for successively smaller values of x until you finally get a value of 0 for $h(x)$. Are you still confident that your guess in part (b) is correct? Explain why you eventually got a value of 0 for $h(x)$.

(d) Graph the function h in the viewing rectangle $[-1, 1]$ by $[0, 1]$. Then zoom in toward the point where the graph crosses the y-axis to estimate the limit of $h(x)$ as x approaches 0. Continue to zoom in until you observe distortions in the graph of h. Compare with your results in part (c).

13.2 FINDING LIMITS ALGEBRAICALLY

■ Limit Laws ■ Applying the Limit Laws ■ Finding Limits Using Algebra and the Limit Laws ■ Using Left- and Right-Hand Limits

In Section 13.1 we used calculators and graphs to guess the values of limits, but we saw that such methods don't always lead to the correct answer. In this section we use algebraic methods to find limits exactly.

■ Limit Laws

We use the following properties of limits, called the *Limit Laws*, to calculate limits.

LIMIT LAWS

Suppose that c is a constant and that the following limits exist:

$$\lim_{x \to a} f(x) \quad \text{and} \quad \lim_{x \to a} g(x)$$

Then

1. $\lim_{x \to a} [f(x) + g(x)] = \lim_{x \to a} f(x) + \lim_{x \to a} g(x)$ — Limit of a Sum
2. $\lim_{x \to a} [f(x) - g(x)] = \lim_{x \to a} f(x) - \lim_{x \to a} g(x)$ — Limit of a Difference
3. $\lim_{x \to a} [cf(x)] = c \lim_{x \to a} f(x)$ — Limit of a Constant Multiple
4. $\lim_{x \to a} [f(x)g(x)] = \lim_{x \to a} f(x) \cdot \lim_{x \to a} g(x)$ — Limit of a Product
5. $\lim_{x \to a} \dfrac{f(x)}{g(x)} = \dfrac{\lim_{x \to a} f(x)}{\lim_{x \to a} g(x)}$ if $\lim_{x \to a} g(x) \neq 0$ — Limit of a Quotient

These five laws can be stated verbally as follows:

Limit of a Sum
1. The limit of a sum is the sum of the limits.

Limit of a Difference
2. The limit of a difference is the difference of the limits.

Limit of a Constant Multiple
3. The limit of a constant times a function is the constant times the limit of the function.

Limit of a Product
4. The limit of a product is the product of the limits.

Limit of a Quotient
5. The limit of a quotient is the quotient of the limits (provided that the limit of the denominator is not 0).

It's easy to believe that these properties are true. For instance, if $f(x)$ is close to L and $g(x)$ is close to M, it is reasonable to conclude that $f(x) + g(x)$ is close to $L + M$. This gives us an intuitive basis for believing that Law 1 is true.

If we use Law 4 (Limit of a Product) repeatedly with $g(x) = f(x)$, we obtain the following Law 6 for the limit of a power. A similar law holds for roots.

LIMIT LAWS

6. $\lim_{x\to a} [f(x)]^n = [\lim_{x\to a} f(x)]^n$ where n is a positive integer — Limit of a Power

7. $\lim_{x\to a} \sqrt[n]{f(x)} = \sqrt[n]{\lim_{x\to a} f(x)}$ where n is a positive integer — Limit of a Root

[If n is even, we assume that $\lim_{x\to a} f(x) > 0$.]

In words, these laws say the following:

Limit of a Power
6. The limit of a power is the power of the limit.

Limit of a Root
7. The limit of a root is the root of the limit.

EXAMPLE 1 ■ Using the Limit Laws

Use the Limit Laws and the graphs of f and g in Figure 1 to evaluate the following limits if they exist.

(a) $\lim_{x\to -2} [f(x) + 5g(x)]$ **(b)** $\lim_{x\to 1} [f(x)g(x)]$

(c) $\lim_{x\to 2} \dfrac{f(x)}{g(x)}$ **(d)** $\lim_{x\to 1} [f(x)]^3$

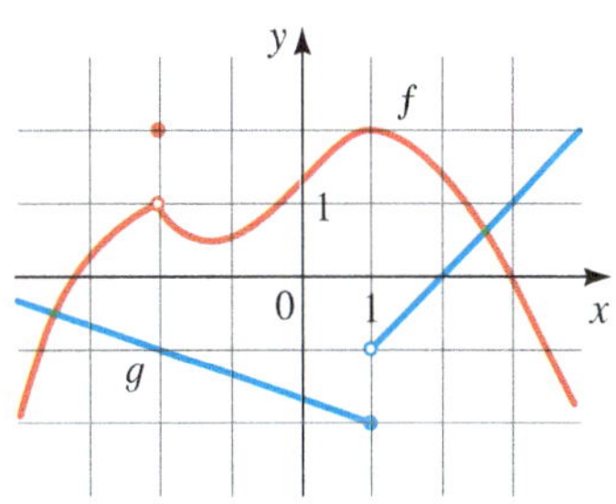

FIGURE 1

SOLUTION

(a) From the graphs of f and g we see that

$$\lim_{x\to -2} f(x) = 1 \quad \text{and} \quad \lim_{x\to -2} g(x) = -1$$

Therefore we have

$$\begin{aligned}\lim_{x\to-2}[f(x)+5g(x)] &= \lim_{x\to-2}f(x)+\lim_{x\to-2}[5g(x)] && \text{Limit of a Sum}\\ &= \lim_{x\to-2}f(x)+5\lim_{x\to-2}g(x) && \text{Limit of a Constant Multiple}\\ &= 1+5(-1)=-4\end{aligned}$$

(b) We see that $\lim_{x\to1}f(x)=2$. But $\lim_{x\to1}g(x)$ does not exist because the left- and right-hand limits are different:

$$\lim_{x\to1^-}g(x)=-2 \qquad \lim_{x\to1^+}g(x)=-1$$

So we can't use Law 4 (Limit of a Product). The given limit does not exist, since the left-hand limit is not equal to the right-hand limit.

(c) The graphs show that

$$\lim_{x\to2}f(x)\approx1.4 \qquad \text{and} \qquad \lim_{x\to2}g(x)=0$$

Because the limit of the denominator is 0, we can't use Law 5 (Limit of a Quotient). The given limit does not exist because the denominator approaches 0 while the numerator approaches a nonzero number.

(d) Since $\lim_{x\to1}f(x)=2$, we use Law 6 to get

$$\begin{aligned}\lim_{x\to1}[f(x)]^3 &= [\lim_{x\to1}f(x)]^3 && \text{Limit of a Power}\\ &= 2^3=8\end{aligned}$$

Now Try Exercise 3 ■

Applying the Limit Laws

In applying the Limit Laws, we need to use four special limits.

SOME SPECIAL LIMITS

1. $\lim_{x\to a}c=c$
2. $\lim_{x\to a}x=a$
3. $\lim_{x\to a}x^n=a^n$ where n is a positive integer
4. $\lim_{x\to a}\sqrt[n]{x}=\sqrt[n]{a}$ where n is a positive integer and $a>0$

Special Limits 1 and 2 are intuitively obvious—looking at the graphs of $y=c$ and $y=x$ will convince you of their validity. Limits 3 and 4 are special cases of Limit Laws 6 and 7 (Limits of a Power and of a Root).

EXAMPLE 2 ■ Using the Limit Laws

Evaluate the following limits, and justify each step.

(a) $\lim_{x\to5}(2x^2-3x+4)$ **(b)** $\lim_{x\to-2}\dfrac{x^3+2x^2-1}{5-3x}$

SOLUTION

(a)
$$\begin{aligned}\lim_{x\to 5}(2x^2 - 3x + 4) &= \lim_{x\to 5}(2x^2) - \lim_{x\to 5}(3x) + \lim_{x\to 5} 4 && \text{Limits of a Difference and Sum}\\ &= 2\lim_{x\to 5} x^2 - 3\lim_{x\to 5} x + \lim_{x\to 5} 4 && \text{Limit of a Constant Multiple}\\ &= 2(5^2) - 3(5) + 4 && \text{Special Limits 3, 2, and 1}\\ &= 39\end{aligned}$$

(b) We start by using Law 5, but its use is fully justified only at the final stage when we see that the limits of the numerator and denominator exist and the limit of the denominator is not 0.

$$\begin{aligned}\lim_{x\to -2}\frac{x^3 + 2x^2 - 1}{5 - 3x} &= \frac{\lim_{x\to -2}(x^3 + 2x^2 - 1)}{\lim_{x\to -2}(5 - 3x)} && \text{Limit of a Quotient}\\ &= \frac{\lim_{x\to -2} x^3 + 2\lim_{x\to -2} x^2 - \lim_{x\to -2} 1}{\lim_{x\to -2} 5 - 3\lim_{x\to -2} x} && \text{Limits of Sums, Differences, and Constant Multiples}\\ &= \frac{(-2)^3 + 2(-2)^2 - 1}{5 - 3(-2)} && \text{Special Limits 3, 2, and 1}\\ &= -\frac{1}{11}\end{aligned}$$

Now Try Exercises 9 and 11 ■

If we let $f(x) = 2x^2 - 3x + 4$, then $f(5) = 39$. In Example 2(a) we found that $\lim_{x\to 5} f(x) = 39$. In other words, we would have gotten the correct answer by substituting 5 for x. Similarly, direct substitution provides the correct answer in part (b). The functions in Example 2 are a polynomial and a rational function, respectively, and similar use of the Limit Laws proves that direct substitution always works for such functions. We state this fact as follows.

LIMITS BY DIRECT SUBSTITUTION

If f is a polynomial or a rational function and a is in the domain of f, then

$$\lim_{x\to a} f(x) = f(a)$$

Functions with this direct substitution property are called **continuous at** a. You will learn more about continuous functions when you study calculus.

EXAMPLE 3 ■ Finding Limits by Direct Substitution

Evaluate the following limits.

(a) $\lim_{x\to 3}(2x^3 - 10x - 8)$

(b) $\lim_{x\to -1}\dfrac{x^2 + 5x}{x^4 + 2}$

SOLUTION

(a) The function $f(x) = 2x^3 - 10x - 12$ is a polynomial, so we can find the limit by direct substitution:

$$\lim_{x \to 3} (2x^3 - 10x - 12) = 2(3)^3 - 10(3) - 8 = 16$$

(b) The function $f(x) = (x^2 + 5x)/(x^4 + 2)$ is a rational function, and $x = -1$ is in its domain (because the denominator is not zero for $x = -1$). Thus we can find the limit by direct substitution:

$$\lim_{x \to -1} \frac{x^2 + 5x}{x^4 + 2} = \frac{(-1)^2 + 5(-1)}{(-1)^4 + 2} = -\frac{4}{3}$$

Now Try Exercise 13 ■

Finding Limits Using Algebra and the Limit Laws

As we saw in Example 3, evaluating limits by direct substitution is easy. But not all limits can be evaluated this way. In fact, most of the situations in which limits are useful require us to work harder to evaluate the limit. The next three examples illustrate how we can use algebra to find limits.

EXAMPLE 4 ■ Finding a Limit by Canceling a Common Factor

Find $\lim_{x \to 1} \frac{x - 1}{x^2 - 1}$.

SOLUTION Let $f(x) = (x - 1)/(x^2 - 1)$. We can't find the limit by substituting $x = 1$ because $f(1)$ isn't defined. Nor can we apply Law 5 (Limit of a Quotient) because the limit of the denominator is 0. Instead, we need to do some preliminary algebra. We factor the denominator as a difference of squares:

$$\frac{x - 1}{x^2 - 1} = \frac{x - 1}{(x - 1)(x + 1)}$$

The numerator and denominator have a common factor of $x - 1$. When we take the limit as x approaches 1, we have $x \neq 1$, and so $x - 1 \neq 0$. Therefore we can cancel the common factor and compute the limit as follows.

$$\begin{aligned} \lim_{x \to 1} \frac{x - 1}{x^2 - 1} &= \lim_{x \to 1} \frac{x - 1}{(x - 1)(x + 1)} && \text{Factor} \\ &= \lim_{x \to 1} \frac{1}{x + 1} && \text{Cancel} \\ &= \frac{1}{1 + 1} = \frac{1}{2} && \text{Let } x \to 1 \end{aligned}$$

This calculation confirms algebraically the answer we got numerically and graphically in Example 1 in Section 13.1.

Now Try Exercise 19 ■

EXAMPLE 5 ■ Finding a Limit by Simplifying

Evaluate $\lim_{h \to 0} \frac{(3 + h)^2 - 9}{h}$.

B. Sanerson/Science Source

SIR ISAAC NEWTON (1642–1727) is universally regarded as one of the giants of physics and mathematics. He is well known for discovering the laws of motion and gravity and for inventing calculus, but he also proved the Binomial Theorem and the laws of optics, and he developed methods for solving polynomial equations to any desired accuracy. He was born on Christmas Day, a few months after the death of his father. After an unhappy childhood, he entered Cambridge University, where he learned mathematics by studying the writings of Euclid and Descartes.

During the plague years of 1665 and 1666, when the university was closed, Newton thought and wrote about ideas that, once published, instantly revolutionized the sciences. Imbued with a pathological fear of criticism, he published these writings only after many years of encouragement from Edmund Halley (who discovered the now-famous comet) and other colleagues.

Newton's works brought him enormous fame and prestige. Even poets were moved to praise; Alexander Pope wrote:

> Nature and Nature's Laws
> lay hid in Night.
> God said, "Let Newton be"
> and all was Light.

Newton was far more modest about his accomplishments. He said, "I seem to have been only like a boy playing on the seashore . . . while the great ocean of truth lay all undiscovered before me." Newton was knighted by Queen Anne in 1705 and was buried with great honor in Westminster Abbey.

SOLUTION We can't use direct substitution to evaluate this limit, because the limit of the denominator is 0. So we first simplify the limit algebraically.

$$\begin{aligned}\lim_{h\to 0}\frac{(3+h)^2-9}{h} &= \lim_{h\to 0}\frac{(9+6h+h^2)-9}{h} && \text{Expand}\\ &= \lim_{h\to 0}\frac{6h+h^2}{h} && \text{Simplify}\\ &= \lim_{h\to 0}(6+h) && \text{Cancel } h\\ &= 6 && \text{Let } h\to 0\end{aligned}$$

Now Try Exercise 25 ■

EXAMPLE 6 ■ Finding a Limit by Rationalizing

Find $\displaystyle\lim_{t\to 0}\frac{\sqrt{t^2+9}-3}{t^2}$.

SOLUTION We can't apply Law 5 (Limit of a Quotient) immediately, since the limit of the denominator is 0. Here the preliminary algebra consists of rationalizing the numerator:

$$\begin{aligned}\lim_{t\to 0}\frac{\sqrt{t^2+9}-3}{t^2} &= \lim_{t\to 0}\frac{\sqrt{t^2+9}-3}{t^2}\cdot\frac{\sqrt{t^2+9}+3}{\sqrt{t^2+9}+3} && \text{Rationalize numerator}\\ &= \lim_{t\to 0}\frac{(t^2+9)-9}{t^2\left(\sqrt{t^2+9}+3\right)} = \lim_{t\to 0}\frac{t^2}{t^2\left(\sqrt{t^2+9}+3\right)}\\ &= \lim_{t\to 0}\frac{1}{\sqrt{t^2+9}+3} = \frac{1}{\sqrt{\lim_{t\to 0}(t^2+9)}+3} = \frac{1}{3+3} = \frac{1}{6}\end{aligned}$$

This calculation confirms the guess that we made in Example 2 in Section 13.1.

Now Try Exercise 27 ■

■ Using Left- and Right-Hand Limits

Some limits are best calculated by first finding the left- and right-hand limits. The following theorem is a reminder of what we discovered in Section 13.1. It says that *a two-sided limit exists if and only if both of the one-sided limits exist and are equal.*

$$\lim_{x\to a} f(x) = L \qquad \text{if and only if} \qquad \lim_{x\to a^-} f(x) = L = \lim_{x\to a^+} f(x)$$

When computing one-sided limits, we use the fact that the Limit Laws also hold for one-sided limits.

EXAMPLE 7 ■ Comparing Right and Left Limits

Show that $\lim_{x\to 0}|x| = 0$.

The result of Example 7 looks plausible from Figure 2.

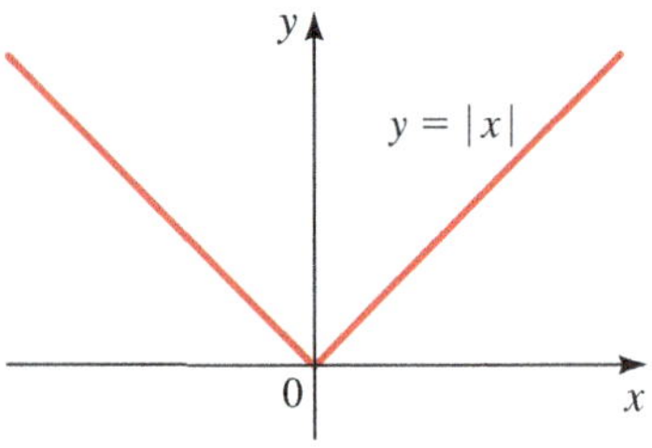

FIGURE 2

SOLUTION Recall that

$$|x| = \begin{cases} x & \text{if } x \geq 0 \\ -x & \text{if } x < 0 \end{cases}$$

Since $|x| = x$ for $x > 0$, we have

$$\lim_{x \to 0^+} |x| = \lim_{x \to 0^+} x = 0$$

For $x < 0$ we have $|x| = -x$, so

$$\lim_{x \to 0^-} |x| = \lim_{x \to 0^-} (-x) = 0$$

Therefore

$$\lim_{x \to 0} |x| = 0$$

Now Try Exercise 37 ■

EXAMPLE 8 ■ Comparing Right and Left Limits

Prove that $\lim_{x \to 0} \dfrac{|x|}{x}$ does not exist.

SOLUTION Since $|x| = x$ for $x > 0$ and $|x| = -x$ for $x < 0$, we have

$$\lim_{x \to 0^+} \frac{|x|}{x} = \lim_{x \to 0^+} \frac{x}{x} = \lim_{x \to 0^+} 1 = 1$$

$$\lim_{x \to 0^-} \frac{|x|}{x} = \lim_{x \to 0^-} \frac{-x}{x} = \lim_{x \to 0^-} (-1) = -1$$

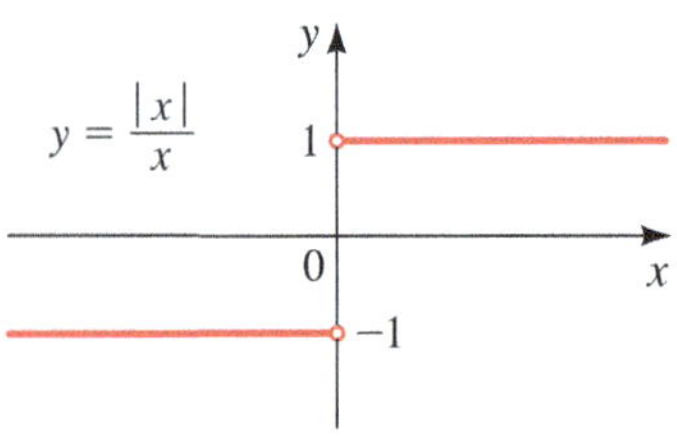

FIGURE 3

Since the right-hand and left-hand limits exist and are different, it follows that $\lim_{x \to 0} |x|/x$ does not exist. The graph of the function $f(x) = |x|/x$ is shown in Figure 3 and supports the limits that we found.

Now Try Exercise 39 ■

EXAMPLE 9 ■ The Limit of a Piecewise Defined Function

Let

$$f(x) = \begin{cases} \sqrt{x - 4} & \text{if } x > 4 \\ 8 - 2x & \text{if } x < 4 \end{cases}$$

Determine whether $\lim_{x \to 4} f(x)$ exists.

SOLUTION Since $f(x) = \sqrt{x - 4}$ for $x > 4$, we have

$$\lim_{x \to 4^+} f(x) = \lim_{x \to 4^+} \sqrt{x - 4} = \sqrt{4 - 4} = 0$$

Since $f(x) = 8 - 2x$ for $x < 4$, we have

$$\lim_{x \to 4^-} f(x) = \lim_{x \to 4^-} (8 - 2x) = 8 - 2 \cdot 4 = 0$$

The right- and left-hand limits are equal. Thus the limit exists, and

$$\lim_{x \to 4} f(x) = 0$$

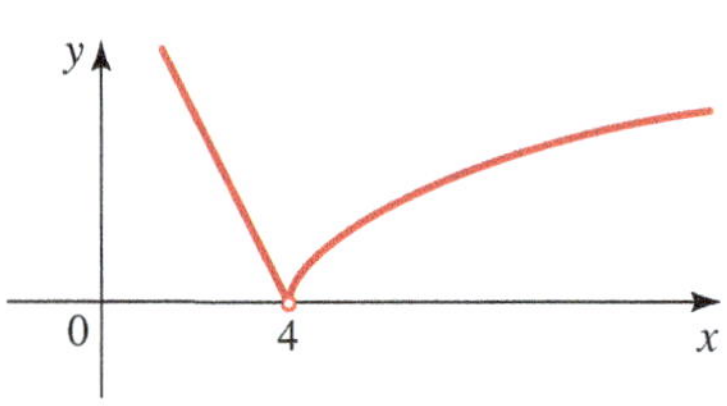

FIGURE 4

The graph of f is shown in Figure 4.

Now Try Exercise 43 ■

13.2 EXERCISES

CONCEPTS

1. Suppose the following limits exist:

$$\lim_{x\to a} f(x) \qquad \text{and} \qquad \lim_{x\to a} g(x)$$

Then $\lim_{x\to a}[f(x) + g(x)] =$ ________ $+$ ________, and

$\lim_{x\to a}[f(x)g(x)] =$ ________ $\cdot$ ________.

These formulas can be stated verbally as follows: The limit of a sum is the ________ of the limits, and the limit of a product is the ________ of the limits.

2. If f is a polynomial or a rational function and a is in the domain of f, then $\lim_{x\to a} f(x) =$ ________.

SKILLS

3. Limits from a Graph The graphs of f and g are given. Use them to evaluate each limit if it exists. If the limit does not exist, explain why.

(a) $\lim_{x\to 2}[f(x) + g(x)]$ **(b)** $\lim_{x\to 1}[f(x) + g(x)]$

(c) $\lim_{x\to 0}[f(x)g(x)]$ **(d)** $\lim_{x\to -1}\dfrac{f(x)}{g(x)}$

(e) $\lim_{x\to 2} x^3 f(x)$ **(f)** $\lim_{x\to 1}\sqrt{3 + f(x)}$

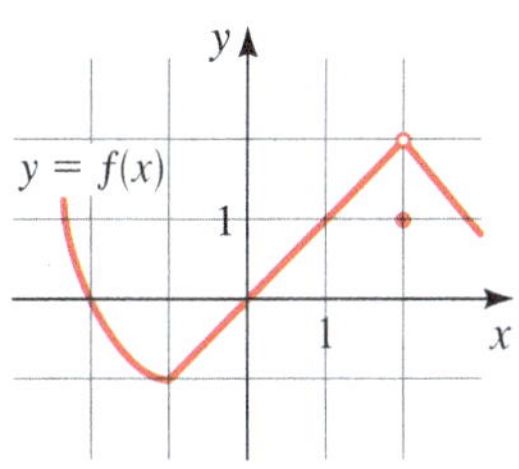

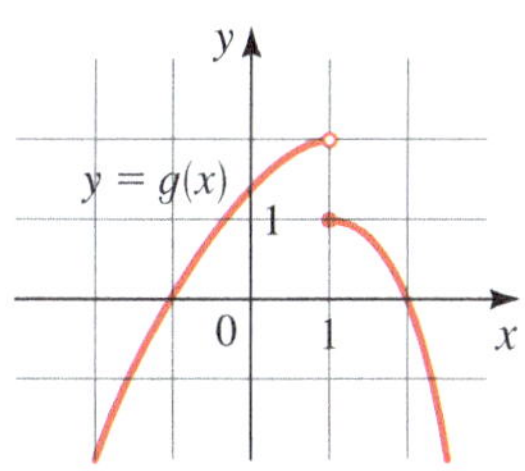

4. Using Limit Laws Suppose that

$$\lim_{x\to a} f(x) = -3 \qquad \lim_{x\to a} g(x) = 0 \qquad \lim_{x\to a} h(x) = 8$$

Find the value of the given limit. If the limit does not exist, explain why.

(a) $\lim_{x\to a}[f(x) + h(x)]$ **(b)** $\lim_{x\to a}[f(x)]^2$

(c) $\lim_{x\to a}\sqrt[3]{h(x)}$ **(d)** $\lim_{x\to a}\dfrac{1}{f(x)}$

(e) $\lim_{x\to a}\dfrac{f(x)}{h(x)}$ **(f)** $\lim_{x\to a}\dfrac{g(x)}{f(x)}$

(g) $\lim_{x\to a}\dfrac{f(x)}{g(x)}$ **(h)** $\lim_{x\to a}\dfrac{2f(x)}{h(x) - f(x)}$

5–18 ■ Using Limit Laws Evaluate the limit and justify each step by indicating the appropriate Limit Law(s).

5. $\lim_{x\to 5} x$ **6.** $\lim_{x\to 0} 3$

7. $\lim_{t\to 3} 4t$ **8.** $\lim_{t\to 2}(1 - 3t)$

9. $\lim_{x\to 4}(5x^2 - 2x + 3)$ **10.** $\lim_{x\to 0}(3x^3 - 2x^2 + 5)$

11. $\lim_{x\to -1}\dfrac{x - 2}{x^2 + 4x - 3}$ **12.** $\lim_{x\to 2}\dfrac{2 - x}{x^2 + 1}$

13. $\lim_{x\to 3}(x^3 + 2)(x^2 - 5x)$ **14.** $\lim_{t\to -2}(t + 1)^9(t^2 - 1)$

15. $\lim_{x\to 1}\left(\dfrac{x^4 + x^2 - 6}{x^4 + 2x + 3}\right)^2$

16. $\lim_{x\to 0}\left(\dfrac{-5x^{20} - 2x^2 + 3000}{x^2 - 1}\right)^{1/3}$

17. $\lim_{x\to 12}(\sqrt{x^2 + 25} - \sqrt{3x})$ **18.** $\lim_{u\to -2}\sqrt{u^4 + 3u + 6}$

19–32 ■ Finding Limits Evaluate the limit if it exists.

19. $\lim_{x\to 2}\dfrac{x^2 + x - 6}{x - 2}$ **20.** $\lim_{x\to -4}\dfrac{x^2 + 5x + 4}{x^2 + 3x - 4}$

21. $\lim_{x\to -2}\dfrac{x^2 - x + 6}{x + 2}$ **22.** $\lim_{x\to 1}\dfrac{x^3 - 1}{x^2 - 1}$

23. $\lim_{t\to -3}\dfrac{t^2 - 9}{2t^2 + 7t + 3}$ **24.** $\lim_{x\to 2}\dfrac{x^4 - 16}{x - 2}$

25. $\lim_{h\to 0}\dfrac{(2 + h)^2 - 4}{h}$ **26.** $\lim_{h\to 0}\dfrac{(2 + h)^3 - 8}{h}$

27. $\lim_{x\to 7}\dfrac{\sqrt{x + 2} - 3}{x - 7}$ **28.** $\lim_{h\to 0}\dfrac{\sqrt{1 + h} - 1}{h}$

29. $\lim_{x\to -4}\dfrac{\frac{1}{4} + \frac{1}{x}}{4 + x}$ **30.** $\lim_{t\to 0}\left(\dfrac{1}{t} - \dfrac{1}{t^2 + t}\right)$

31. $\lim_{h\to 0}\dfrac{(3 + h)^{-1} - 3^{-1}}{h}$ **32.** $\lim_{t\to 4}\dfrac{\frac{1}{\sqrt{t}} - \frac{1}{2}}{t - 4}$

33–36 ■ Finding Limits Find the limit, and use a graphing device to confirm your result graphically.

33. $\lim_{x\to 1}\dfrac{x^2 - 1}{\sqrt{x} - 1}$ **34.** $\lim_{x\to 0}\dfrac{(4 + x)^3 - 64}{x}$

35. $\lim_{x\to -1}\dfrac{x^2 - x - 2}{x^3 - x}$ **36.** $\lim_{x\to 1}\dfrac{x^8 - 1}{x^5 - 1}$

37–42 ■ Does the Limit Exist? Find the limit, if it exists. If the limit does not exist, explain why.

37. $\lim_{x\to -4}|x + 4|$ **38.** $\lim_{x\to -4^-}\dfrac{|x + 4|}{x + 4}$

39. $\lim_{x\to 2}\dfrac{|x - 2|}{x - 2}$ **40.** $\lim_{x\to 1.5}\dfrac{2x^2 - 3x}{|2x - 3|}$

41. $\lim_{x\to 0^-}\left(\dfrac{1}{x} - \dfrac{1}{|x|}\right)$ **42.** $\lim_{x\to 0^+}\left(\dfrac{1}{x} - \dfrac{1}{|x|}\right)$

43. Does the Limit Exist? Let

$$f(x) = \begin{cases} x - 1 & \text{if } x < 2 \\ x^2 - 4x + 6 & \text{if } x \geq 2 \end{cases}$$

(a) Find $\lim_{x\to 2^-} f(x)$ and $\lim_{x\to 2^+} f(x)$.
(b) Does $\lim_{x\to 2} f(x)$ exist?
(c) Sketch the graph of f.

44. Does the Limit Exist? Let

$$h(x) = \begin{cases} x & \text{if } x < 0 \\ x^2 & \text{if } 0 < x \leq 2 \\ 8 - x & \text{if } x > 2 \end{cases}$$

(a) Evaluate each limit if it exists.
(i) $\lim_{x\to 0^+} h(x)$
(ii) $\lim_{x\to 0} h(x)$
(iii) $\lim_{x\to 1} h(x)$
(iv) $\lim_{x\to 2^-} h(x)$
(v) $\lim_{x\to 2^+} h(x)$
(vi) $\lim_{x\to 2} h(x)$
(b) Sketch the graph of h.

SKILLS Plus

45. Finding Limits Numerically and Graphically
(a) Estimate the value of

$$\lim_{x\to 0} \frac{x}{\sqrt{1 + 3x} - 1}$$

by graphing the function $f(x) = x/(\sqrt{1 + 3x} - 1)$.
(b) Make a table of values of $f(x)$ for x close to 0, and guess the value of the limit.
(c) Use the Limit Laws to prove that your guess is correct.

46. Finding Limits Numerically and Graphically
(a) Use a graph of

$$f(x) = \frac{\sqrt{3 + x} - \sqrt{3}}{x}$$

to estimate the value of $\lim_{x\to 0} f(x)$ to two decimal places.
(b) Use a table of values of $f(x)$ to estimate the limit to four decimal places.
(c) Use the Limit Laws to find the exact value of the limit.

DISCUSS ■ DISCOVER ■ PROVE ■ WRITE

47. DISCUSS: Cancellation and Limits
(a) What is wrong with the following equation?

$$\frac{x^2 + x - 6}{x - 2} = x + 3$$

(b) In view of part (a), explain why the equation

$$\lim_{x\to 2} \frac{x^2 + x - 6}{x - 2} = \lim_{x\to 2} (x + 3)$$

is correct.

48. DISCUSS: The Lorentz Contraction In the theory of relativity the Lorentz contraction formula

$$L = L_0\sqrt{1 - v^2/c^2}$$

expresses the length L of an object as a function of its velocity v with respect to an observer, where L_0 is the length of the object at rest and c is the speed of light. Find $\lim_{v\to c^-} L$, and interpret the result. Why is a left-hand limit necessary?

49. DISCUSS ■ PROVE: Limits of Sums and Products
(a) Show by means of an example that $\lim_{x\to a} [f(x) + g(x)]$ may exist even though neither $\lim_{x\to a} f(x)$ nor $\lim_{x\to a} g(x)$ exists.
(b) Show by means of an example that $\lim_{x\to a} [f(x)g(x)]$ may exist even though neither $\lim_{x\to a} f(x)$ nor $\lim_{x\to a} g(x)$ exists.

13.3 TANGENT LINES AND DERIVATIVES

■ Tangent Lines ■ Derivatives ■ Instantaneous Rates of Change

In this section we see how limits arise when we attempt to find the tangent line to a curve or the instantaneous rate of change of a function.

■ Tangent Lines

A *tangent line* is a line that *just* touches a curve. For instance, Figure 1 shows the parabola $y = x^2$ and the tangent line t that touches the parabola at the point $P(1, 1)$. We will be able to find an equation of the tangent line t as soon as we know its slope m. The difficulty is that we know only one point, P, on t, whereas we need two points to compute the slope. But observe that we can compute an approximation to m by choosing a nearby point $Q(x, x^2)$ on the parabola (as in Figure 2) and computing the slope m_{PQ} of the secant line PQ.

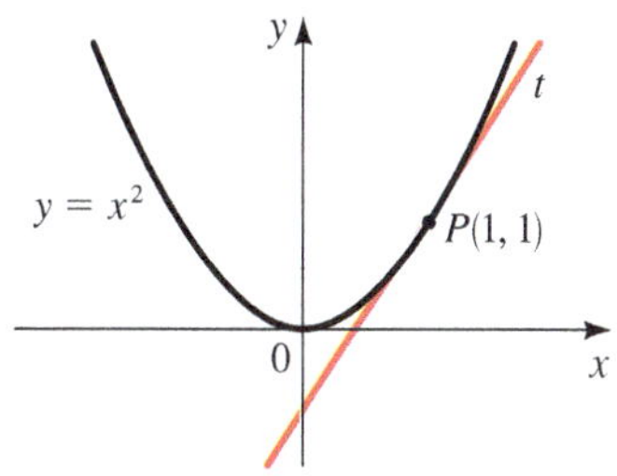

FIGURE 1

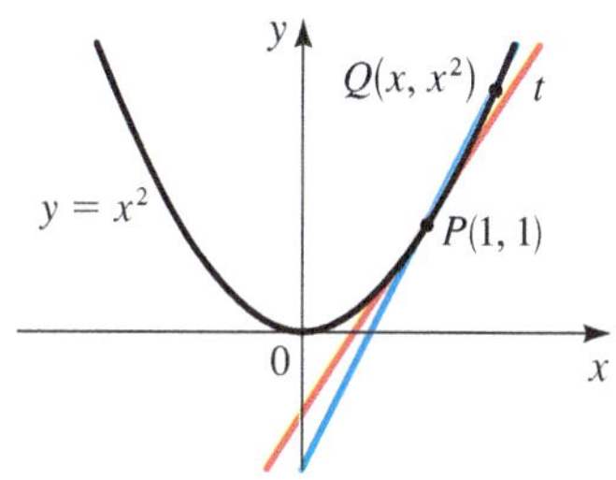

FIGURE 2

We choose $x \neq 1$ so that $Q \neq P$. Then

$$m_{PQ} = \frac{x^2 - 1}{x - 1}$$

Now we let x approach 1, so Q approaches P along the parabola. Figure 3 shows how the corresponding secant lines rotate about P and approach the tangent line t.

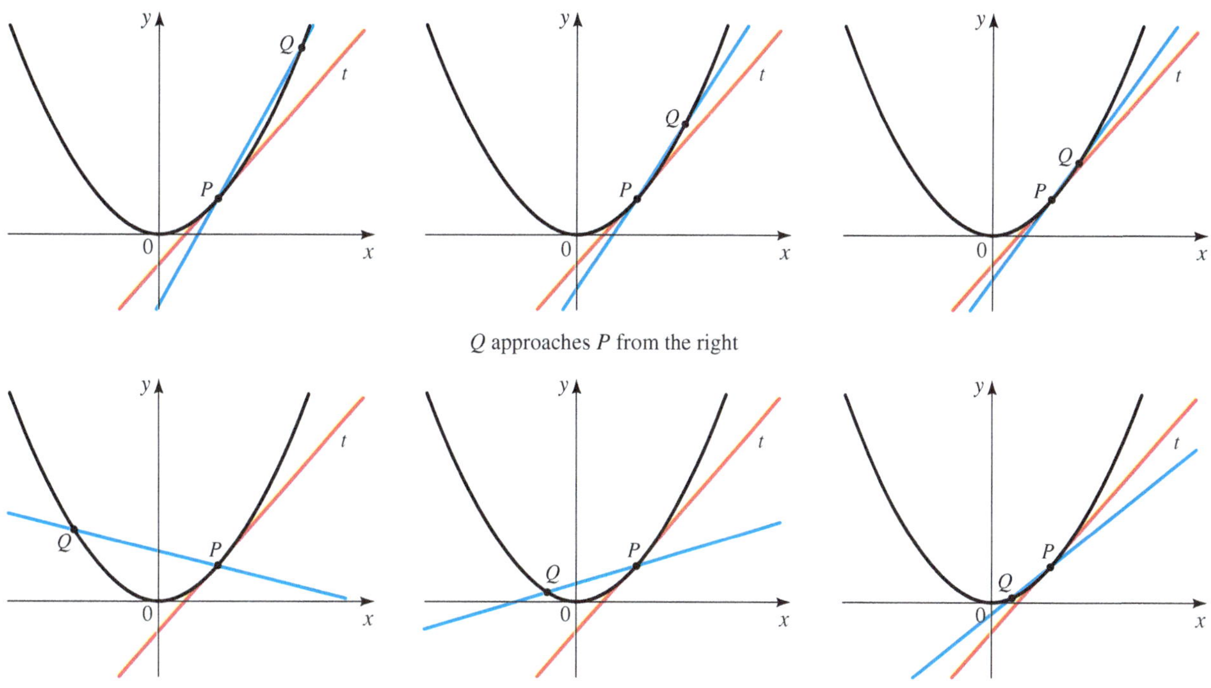

FIGURE 3

The slope of the tangent line is the limit of the slopes of the secant lines:

$$m = \lim_{Q \to P} m_{PQ}$$

So using the method of Section 13.2, we have

$$m = \lim_{x \to 1} \frac{x^2 - 1}{x - 1} = \lim_{x \to 1} \frac{(x - 1)(x + 1)}{x - 1}$$

$$= \lim_{x \to 1} (x + 1) = 1 + 1 = 2$$

The point-slope form for the equation of a line through the point (x_1, y_1) with slope m is

$$y - y_1 = m(x - x_1)$$

(See Section 1.10.)

Now that we know the slope of the tangent line is $m = 2$, we can use the point-slope form of the equation of a line to find its equation.

$$y - 1 = 2(x - 1) \qquad \text{or} \qquad y = 2x - 1$$

We sometimes refer to the slope of the tangent line to a curve at a point as the **slope of the curve** at the point. The idea is that if we zoom in far enough toward the point, the curve looks almost like a straight line. Figure 4 illustrates this procedure for the curve $y = x^2$. The more we zoom in, the more the parabola looks like a line. In other words, the curve becomes almost indistinguishable from its tangent line.

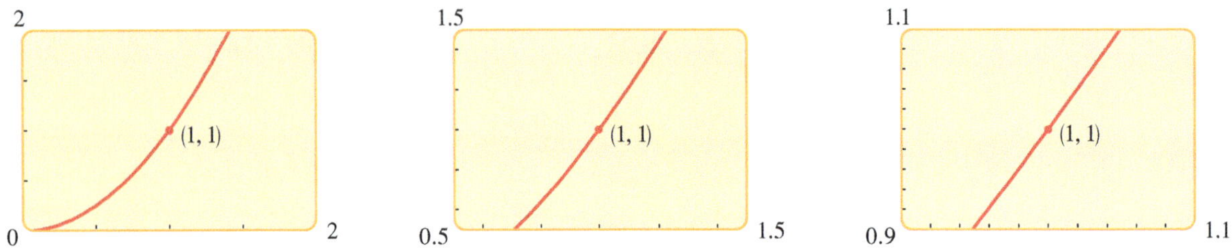

FIGURE 4 Zooming in toward the point $(1, 1)$ on the parabola $y = x^2$

If we have a general curve C with equation $y = f(x)$ and we want to find the tangent line to C at the point $P(a, f(a))$, then we consider a nearby point $Q(x, f(x))$, where $x \neq a$, and compute the slope of the secant line PQ.

$$m_{PQ} = \frac{f(x) - f(a)}{x - a}$$

Then we let Q approach P along the curve C by letting x approach a. If m_{PQ} approaches a number m, then we define the *tangent* t to be the line through P with slope m. (This amounts to saying that the tangent line is the limiting position of the secant line PQ as Q approaches P. See Figure 5.)

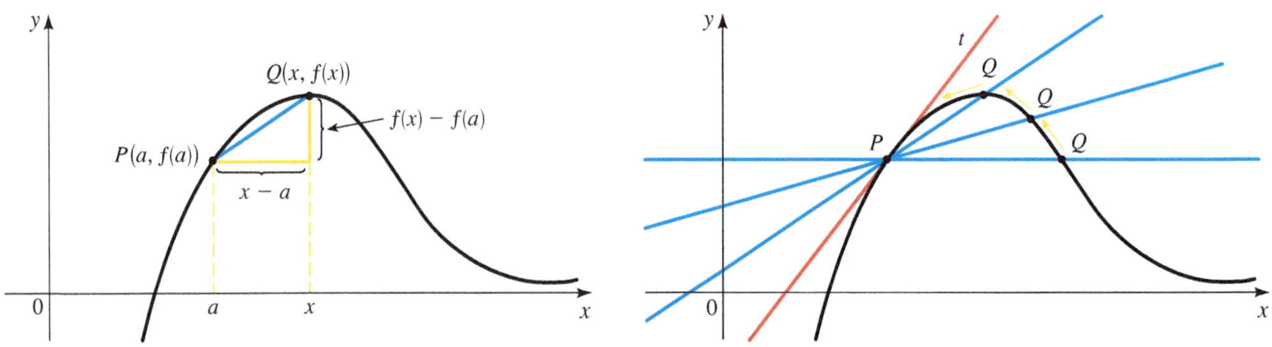

FIGURE 5

DEFINITION OF A TANGENT LINE

The **tangent line** to the curve $y = f(x)$ at the point $P(a, f(a))$ is the line through P with slope

$$m = \lim_{x \to a} \frac{f(x) - f(a)}{x - a}$$

provided that this limit exists.

EXAMPLE 1 ■ Finding a Tangent Line to a Hyperbola

Find an equation of the tangent line to the hyperbola $y = 3/x$ at the point $(3, 1)$.

SOLUTION Let $f(x) = 3/x$. Then the slope of the tangent line at $(3, 1)$ is

$$\begin{aligned} m &= \lim_{x\to 3} \frac{f(x) - f(3)}{x - 3} && \text{Definition of } m \\ &= \lim_{x\to 3} \frac{\dfrac{3}{x} - 1}{x - 3} && f(x) = \frac{3}{x} \\ &= \lim_{x\to 3} \frac{3 - x}{x(x - 3)} && \text{Multiply numerator and denominator by } x \\ &= \lim_{x\to 3}\left(-\frac{1}{x}\right) && \text{Cancel } x - 3 \\ &= -\frac{1}{3} && \text{Let } x \to 3 \end{aligned}$$

Therefore an equation of the tangent line at the point $(3, 1)$ is

$$y - 1 = -\tfrac{1}{3}(x - 3)$$

which simplifies to

$$x + 3y - 6 = 0$$

The hyperbola and its tangent are shown in Figure 6.

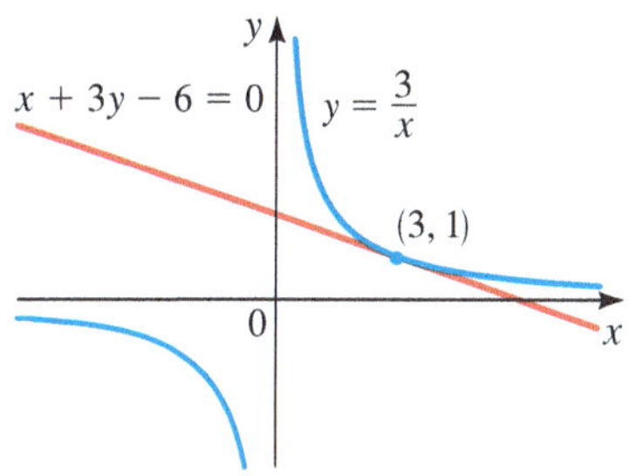

FIGURE 6

Now Try Exercises 3 and 11 ■

There is another expression for the slope of a tangent line that is sometimes easier to use. Let $h = x - a$. Then $x = a + h$, so the slope of the secant line PQ is

$$m_{PQ} = \frac{f(a + h) - f(a)}{h}$$

See Figure 7, in which the case $h > 0$ is illustrated and Q is to the right of P. If it happened that $h < 0$, however, Q would be to the left of P.

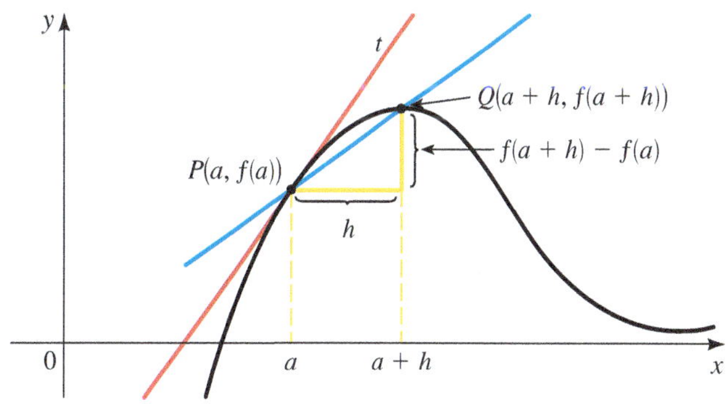

FIGURE 7

Newton and Limits
In 1687 Isaac Newton (see page 911) published his masterpiece *Principia Mathematica*. In this work, the greatest scientific treatise ever written, Newton set forth his version of calculus and used it to investigate mechanics, fluid dynamics, and wave motion and to explain the motion of planets and comets.

The beginnings of calculus are found in the calculations of areas and volumes by ancient Greek scholars such as Eudoxus and Archimedes. Although aspects of the idea of a limit are implicit in their "method of exhaustion," Eudoxus and Archimedes never explicitly formulated the concept of a limit. Likewise, mathematicians such as Cavalieri, Fermat, and Barrow, the immediate precursors of Newton in the development of calculus, did not actually use limits. It was Isaac Newton who first talked explicitly about limits. He explained that the main idea behind limits is that quantities "approach nearer than by any given difference." Newton stated that the limit was the basic concept in calculus, but it was left to later mathematicians like Cauchy and Weierstrass to clarify these ideas.

Notice that as x approaches a, h approaches 0 (because $h = x - a$), so the expression for the slope of the tangent line becomes

$$m = \lim_{h \to 0} \frac{f(a+h) - f(a)}{h}$$

EXAMPLE 2 ■ Finding a Tangent Line

Find an equation of the tangent line to the curve $y = x^3 - 2x + 3$ at the point $(1, 2)$.

SOLUTION If $f(x) = x^3 - 2x + 3$, then the slope of the tangent line where $a = 1$ is

$$\begin{aligned} m &= \lim_{h \to 0} \frac{f(1+h) - f(1)}{h} && \text{Definition of } m \\ &= \lim_{h \to 0} \frac{[(1+h)^3 - 2(1+h) + 3] - [1^3 - 2(1) + 3]}{h} && f(x) = x^3 - 2x + 3 \\ &= \lim_{h \to 0} \frac{1 + 3h + 3h^2 + h^3 - 2 - 2h + 3 - 2}{h} && \text{Expand numerator} \\ &= \lim_{h \to 0} \frac{h + 3h^2 + h^3}{h} && \text{Simplify} \\ &= \lim_{h \to 0} (1 + 3h + h^2) && \text{Cancel } h \\ &= 1 && \text{Let } h \to 0 \end{aligned}$$

So an equation of the tangent line at $(1, 2)$ is

$$y - 2 = 1(x - 1) \qquad \text{or} \qquad y = x + 1$$

Now Try Exercise 13 ■

■ Derivatives

We have seen that the slope of the tangent line to the curve $y = f(x)$ at the point $(a, f(a))$ can be written as

$$\lim_{h \to 0} \frac{f(a+h) - f(a)}{h}$$

It turns out that this expression arises in many other contexts as well, such as finding velocities and other rates of change. Because this type of limit occurs so widely, it is given a special name and notation.

DEFINITION OF A DERIVATIVE

The **derivative of a function f at a number a**, denoted by $f'(a)$, is

$$f'(a) = \lim_{h \to 0} \frac{f(a+h) - f(a)}{h}$$

if this limit exists.

EXAMPLE 3 ■ Finding a Derivative at a Point

Find the derivative of the function $f(x) = 5x^2 + 3x - 1$ at the number 2.

SOLUTION According to the definition of a derivative, with $a = 2$, we have

$$\begin{aligned}
f'(2) &= \lim_{h \to 0} \frac{f(2+h) - f(2)}{h} && \text{Definition of } f'(2) \\
&= \lim_{h \to 0} \frac{[5(2+h)^2 + 3(2+h) - 1] - [5(2)^2 + 3(2) - 1]}{h} && f(x) = 5x^2 + 3x - 1 \\
&= \lim_{h \to 0} \frac{20 + 20h + 5h^2 + 6 + 3h - 1 - 25}{h} && \text{Expand} \\
&= \lim_{h \to 0} \frac{23h + 5h^2}{h} && \text{Simplify} \\
&= \lim_{h \to 0} (23 + 5h) && \text{Cancel } h \\
&= 23 && \text{Let } h \to 0
\end{aligned}$$

Now Try Exercise 19 ■

We see from the definition of a derivative that the number $f'(a)$ is the same as the slope of the tangent line to the curve $y = f(x)$ at the point $(a, f(a))$. So the result of Example 3 shows that the slope of the tangent line to the parabola $y = 5x^2 + 3x - 1$ at the point $(2, 25)$ is $f'(2) = 23$.

EXAMPLE 4 ■ Finding a Derivative

Let $f(x) = \sqrt{x}$.

(a) Find $f'(a)$. **(b)** Find $f'(1)$, $f'(4)$, and $f'(9)$.

SOLUTION

(a) We use the definition of the derivative at a:

$$\begin{aligned}
f'(a) &= \lim_{h \to 0} \frac{f(a+h) - f(a)}{h} && \text{Definition of derivative} \\
&= \lim_{h \to 0} \frac{\sqrt{a+h} - \sqrt{a}}{h} && f(x) = \sqrt{x} \\
&= \lim_{h \to 0} \frac{\sqrt{a+h} - \sqrt{a}}{h} \cdot \frac{\sqrt{a+h} + \sqrt{a}}{\sqrt{a+h} + \sqrt{a}} && \text{Rationalize numerator} \\
&= \lim_{h \to 0} \frac{(a+h) - a}{h\left(\sqrt{a+h} + \sqrt{a}\right)} && \text{Difference of squares} \\
&= \lim_{h \to 0} \frac{h}{h\left(\sqrt{a+h} + \sqrt{a}\right)} && \text{Simplify numerator} \\
&= \lim_{h \to 0} \frac{1}{\sqrt{a+h} + \sqrt{a}} && \text{Cancel } h \\
&= \frac{1}{\sqrt{a} + \sqrt{a}} = \frac{1}{2\sqrt{a}} && \text{Let } h \to 0
\end{aligned}$$

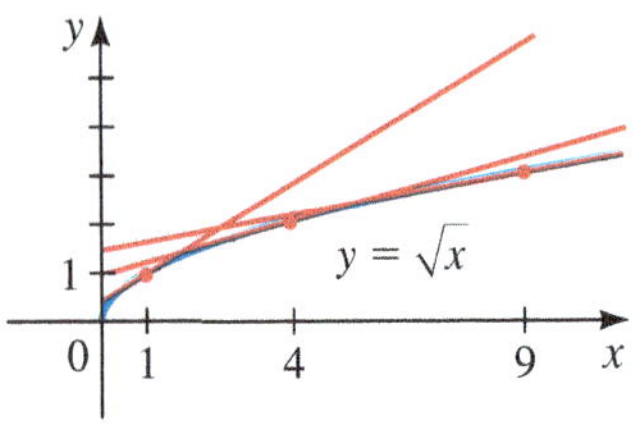

FIGURE 8

(b) Substituting $a = 1$, $a = 4$, and $a = 9$ into the result of part (a), we get

$$f'(1) = \frac{1}{2\sqrt{1}} = \frac{1}{2} \qquad f'(4) = \frac{1}{2\sqrt{4}} = \frac{1}{4} \qquad f'(9) = \frac{1}{2\sqrt{9}} = \frac{1}{6}$$

These values of the derivative are the slopes of the tangent lines shown in Figure 8.

Now Try Exercises 25 and 27 ■

Instantaneous Rates of Change

In Section 2.4 we defined the average rate of change of a function f between the numbers a and x as

$$\text{average rate of change} = \frac{\text{change in } y}{\text{change in } x} = \frac{f(x) - f(a)}{x - a}$$

Suppose we consider the average rate of change over smaller and smaller intervals by letting x approach a. The limit of these average rates of change is called the instantaneous rate of change.

INSTANTANEOUS RATE OF CHANGE

If $y = f(x)$, the **instantaneous rate of change of y with respect to x** at $x = a$ is the limit of the average rates of change as x approaches a:

$$\text{instantaneous rate of change} = \lim_{x \to a} \frac{f(x) - f(a)}{x - a} = f'(a)$$

Notice that we now have two ways of interpreting the derivative:

- $f'(a)$ is the slope of the tangent line to $y = f(x)$ at $x = a$
- $f'(a)$ is the instantaneous rate of change of y with respect to x at $x = a$

In the special case in which $x = t =$ time and $s = f(t) =$ displacement (directed distance) at time t of an object traveling in a straight line, the instantaneous rate of change is called the **instantaneous velocity**.

EXAMPLE 5 ■ Instantaneous Velocity of a Falling Object

If an object is dropped from a height of 3000 ft, its distance above the ground (in feet) after t seconds is given by $h(t) = 3000 - 16t^2$. Find the object's instantaneous velocity after 4 seconds.

© Racheal Grazias/Shutterstock.com

DISCOVERY PROJECT

Designing a Roller Coaster

To ensure an exhilarating ride, a roller coaster ought to consist of steep rises and drops joined by thrilling curves. For a safe ride, these curves must fit together "smoothly." In designing a roller coaster, you can choose where to locate the ascents and drops. We'll explore how the derivative can help us join these ascents and drops smoothly. You can find the project at **www.stewartmath.com**.

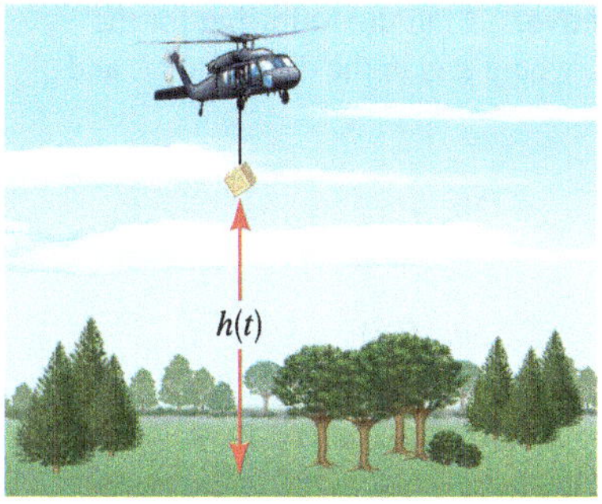

SOLUTION After 4 s have elapsed, the height is $h(4) = 2744$ ft. The instantaneous velocity is

$$h'(4) = \lim_{t\to 4} \frac{h(t) - h(4)}{t - 4} \qquad \text{Definition of } h'(4)$$

$$= \lim_{t\to 4} \frac{3000 - 16t^2 - 2744}{t - 4} \qquad h(t) = 3000 - 16t^2$$

$$= \lim_{t\to 4} \frac{256 - 16t^2}{t - 4} \qquad \text{Simplify}$$

$$= \lim_{t\to 4} \frac{16(4 - t)(4 + t)}{t - 4} \qquad \text{Factor numerator}$$

$$= \lim_{t\to 4} -16(4 + t) \qquad \text{Cancel } t - 4$$

$$= -16(4 + 4) = -128 \text{ ft/s} \qquad \text{Let } t \to 4$$

The negative sign indicates that the height is *decreasing* at a rate of 128 ft/s.

Now Try Exercise 37 ■

U.S. Population

t	$P(t)$
2004	292,805,298
2006	298,379,912
2008	304,093,966
2010	309,349,689
2012	313,914,040

Source: U.S. Census Bureau

t	$\dfrac{P(t) - P(2008)}{t - 2008}$
2004	2,822,167
2006	2,857,027
2010	2,627,862
2012	2,455,019

Here, we have estimated the derivative by averaging the slopes of two secant lines. Another method is to plot the population function and estimate the slope of the tangent line when $t = 2008$.

EXAMPLE 6 ■ Estimating an Instantaneous Rate of Change

Let $P(t)$ be the population of the United States at time t. The first table in the margin gives approximate values of this function by providing midyear population estimates from 2004 to 2012. Interpret and estimate the value of $P'(2008)$.

SOLUTION The derivative $P'(2008)$ means the rate of change of P with respect to t when $t = 2008$, that is, the rate of increase of the population in 2008.

According to the definition of a derivative, we have

$$P'(2008) = \lim_{t\to 2008} \frac{P(t) - P(2008)}{t - 2008}$$

So we compute and tabulate values of the difference quotient (the average rates of change) as shown in the second table in the margin. We see that $P'(2008)$ lies somewhere between 2,857,027 and 2,627,862. (Here we are making the reasonable assumption that the population didn't fluctuate wildly between 2004 and 2012.) We estimate that the rate of increase of the U.S. population in 2008 was the average of these two numbers, namely,

$$P'(2008) \approx 2.74 \text{ million people/year}$$

Now Try Exercise 43 ■

13.3 EXERCISES

CONCEPTS

1. The derivative of a function f at a number a is

$$f'(a) = \lim_{h\to 0} \frac{\quad - \quad}{\quad}$$

if the limit exists. The derivative $f'(a)$ is the ________ of the tangent line to the curve $y = f(x)$ at the point (,).

2. If $y = f(x)$, the average rate of change of f between the numbers x and a is $\dfrac{\quad - \quad}{\quad - \quad}$. The limit of the average rates of change as x approaches a is the ________ rate of change of y with respect to x at $x = a$; this is also the derivative $f'(\quad)$.

SKILLS

3–10 ■ Slope of a Tangent Line Find the slope of the tangent line to the graph of f at the given point.

3. $f(x) = 3x + 4$, at $(1, 7)$

4. $f(x) = 5 - 2x$, at $(-3, 11)$

5. $f(x) = 4x^2 - 3x$, at $(-1, 7)$

6. $f(x) = 1 + 2x - 3x^2$, at $(1, 0)$

7. $f(x) = 2x^3$, at $(2, 16)$

8. $f(x) = x^3 + 1$, at $(2, 9)$

9. $f(x) = \dfrac{5}{x + 2}$, at $(3, 1)$

10. $f(x) = \dfrac{6}{x + 1}$, at $(2, 2)$

11–18 ■ Equation of a Tangent Line Find an equation of the tangent line to the curve at the given point. Graph the curve and the tangent line.

11. $f(x) = -2x^2 + 1$, at $(2, -7)$

12. $f(x) = 4x^2 - 3$, at $(-1, 1)$

13. $y = x + x^2$, at $(-1, 0)$

14. $y = 2x - x^3$, at $(1, 1)$

15. $y = \dfrac{x}{x - 1}$, at $(2, 2)$

16. $y = \dfrac{1}{x^2}$, at $(-1, 1)$

17. $y = \sqrt{x + 3}$, at $(1, 2)$

18. $y = \sqrt{1 + 2x}$, at $(4, 3)$

19–26 ■ The Derivative at a Number Find the derivative of the function at the given number.

19. $f(x) = 1 - 3x^2$, at 2

20. $f(x) = 2 - 3x + x^2$, at -1

21. $f(x) = x - 3x^2$, at -1

22. $f(x) = x + x^3$, at 1

23. $f(x) = \dfrac{1}{x + 1}$, at 2

24. $f(x) = \dfrac{x}{2 - x}$, at -3

25. $F(x) = \dfrac{1}{\sqrt{x}}$, at 4

26. $G(x) = 1 + 2\sqrt{x}$, at 4

27–30 ■ Evaluating Derivatives Find the following for the given function f: **(a)** $f'(a)$, where a is in the domain of f, and **(b)** $f'(3)$ and $f'(4)$.

27. $f(x) = x^2 + 2x$

28. $f(x) = -\dfrac{1}{x^2}$

29. $f(x) = \dfrac{x}{x + 1}$

30. $f(x) = \sqrt{x - 2}$

SKILLS Plus

31. Tangent Lines

(a) If $f(x) = x^3 - 2x + 4$, find $f'(a)$.

(b) Find equations of the tangent lines to the graph of f at the points whose x-coordinates are 0, 1, and 2.

(c) Graph f and the three tangent lines.

32. Tangent Lines

(a) If $g(x) = 1/(2x - 1)$, find $g'(a)$.

(b) Find equations of the tangent lines to the graph of g at the points whose x-coordinates are -1, 0, and 1.

(c) Graph g and the three tangent lines.

33–36 ■ Which Derivative Does the Limit Represent? The given limit represents the derivative of a function f at a number a. Find f and a.

33. $\displaystyle\lim_{h \to 0} \frac{(1 + h)^{10} - 1}{h}$

34. $\displaystyle\lim_{x \to 5} \frac{2^x - 32}{x - 5}$

35. $\displaystyle\lim_{t \to 1} \frac{\sqrt{t + 1} - \sqrt{2}}{t - 1}$

36. $\displaystyle\lim_{h \to 0} \frac{\cos(\pi + h) + 1}{h}$

APPLICATIONS

37. Velocity of a Ball If a ball is thrown straight up with a velocity of 40 ft/s, its height (in ft) after t seconds is given by $y = 40t - 16t^2$. Find the instantaneous velocity when $t = 2$.

38. Velocity on the Moon If an arrow is shot upward on the moon with a velocity of 58 m/s, its height (in meters) after t seconds is given by $H = 58t - 0.83t^2$.

(a) Find the instantaneous velocity of the arrow after 1 second.

(b) Find the instantaneous velocity of the arrow when $t = a$.

(c) At what time t will the arrow hit the moon?

(d) With what velocity will the arrow hit the moon?

39. Velocity of a Particle The displacement s (in meters) of a particle moving in a straight line is given by the equation of motion $s = 4t^3 + 6t + 2$, where t is measured in seconds. Find the instantaneous velocity of the particle s at times $t = a$, $t = 1$, $t = 2$, $t = 3$.

40. Inflating a Balloon A spherical balloon is being inflated. Find the rate of change of the surface area $(S = 4\pi r^2)$ with respect to the radius r when $r = 2$ ft.

41. Temperature Change A roast turkey is taken from an oven when its temperature has reached 185°F and is placed on a table in a room where the temperature is 75°F. The graph shows how the temperature of the turkey decreases and eventually approaches room temperature. By measuring the slope of the tangent, estimate the rate of change of the temperature after 1 hour.

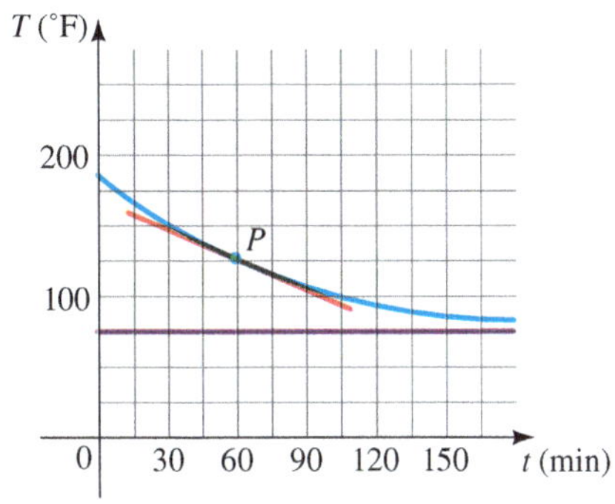

42. Heart Rate A cardiac monitor is used to measure the heart rate of a patient after surgery. It compiles the number of heartbeats after t min. When the data in the table are graphed, the slope of the tangent line represents the heart rate in beats per minute.

t (min)	Heartbeats
36	2530
38	2661
40	2806
42	2948
44	3080

(a) Find the average heart rates (slopes of the secant lines) over the time intervals $[40, 42]$ and $[42, 44]$.

(b) Estimate the patient's heart rate after 42 min by averaging the slopes of these two secant lines.

43. Water Flow A tank holds 1000 gal of water, which drains from the bottom of the tank in half an hour. The values in the table show the volume V of water remaining in the tank (in gal) after t minutes.

t (min)	V (gal)
5	694
10	444
15	250
20	111
25	28
30	0

(a) Find the average rates at which water flows from the tank (slopes of secant lines) for the time intervals $[10, 15]$ and $[15, 20]$.

(b) The slope of the tangent line at the point $(15, 250)$ represents the rate at which water is flowing from the tank after 15 min. Estimate this rate by averaging the slopes of the secant lines in part (a).

44. World Population Growth The table gives approximate values for the world population by providing midyear population estimates for the years 1900–2010

Year	Population (millions)	Year	Population (millions)
1900	1650	1960	3040
1910	1750	1970	3710
1920	1860	1980	4450
1930	2070	1990	5290
1940	2300	2000	6090
1950	2560	2010	6870

Source: U.S. Census Bureau

Estimate the rate of population growth in 1920 and in 2000 by averaging the slopes of two secant lines.

DISCUSS ■ DISCOVER ■ PROVE ■ WRITE

45. DISCUSS: Estimating Derivatives from a Graph For the function g whose graph is given, arrange the following numbers in increasing order, and explain your reasoning..

$$0 \qquad g'(-2) \qquad g'(0) \qquad g'(2) \qquad g'(4)$$

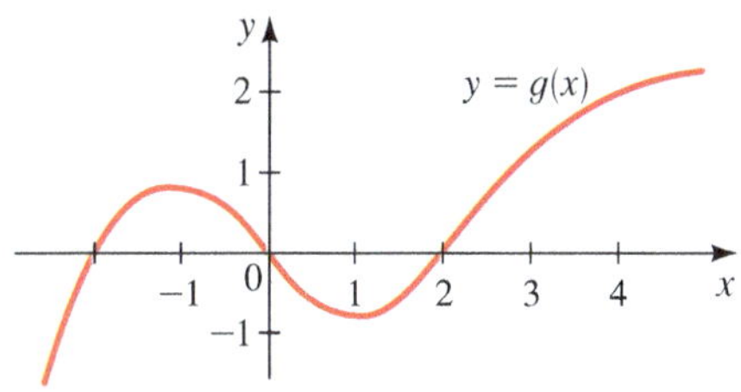

46. DISCUSS: Estimating Velocities from a Graph The graph shows the position function of a car. Use the shape of the graph to explain your answers to the following questions.

(a) What was the initial velocity of the car?

(b) Was the car going faster at B or at C?

(c) Was the car slowing down or speeding up at A, B, and C?

(d) What happened between D and E?

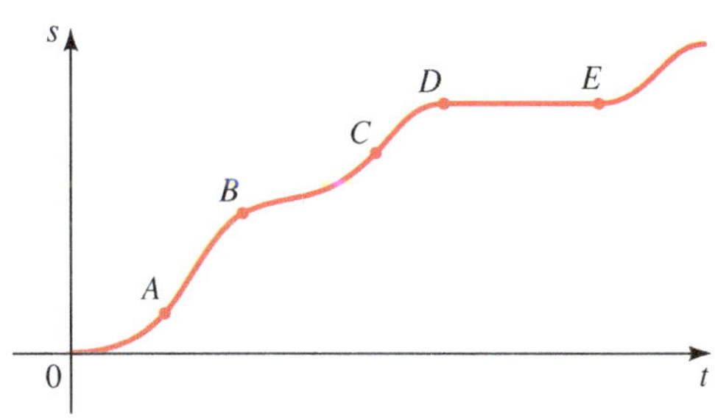

13.4 LIMITS AT INFINITY; LIMITS OF SEQUENCES

■ Limits at Infinity ■ Limits of Sequences

In this section we study a special kind of limit called a *limit at infinity*. We examine the limit of a function $f(x)$ as x becomes large. We also examine the limit of a sequence a_n as n becomes large. Limits of sequences will be used in Section 13.5 to help us find the area under the graph of a function.

■ Limits at Infinity

Let's investigate the behavior of the function f defined by

$$f(x) = \frac{x^2 - 1}{x^2 + 1}$$

as x becomes large. The table in the margin gives values of this function rounded to six decimal places, and the graph of f has been drawn by a computer in Figure 1.

x	$f(x)$
0	−1.000000
±1	0.000000
±2	0.600000
±3	0.800000
±4	0.882353
±5	0.923077
±10	0.980198
±50	0.999200
±100	0.999800
±1000	0.999998

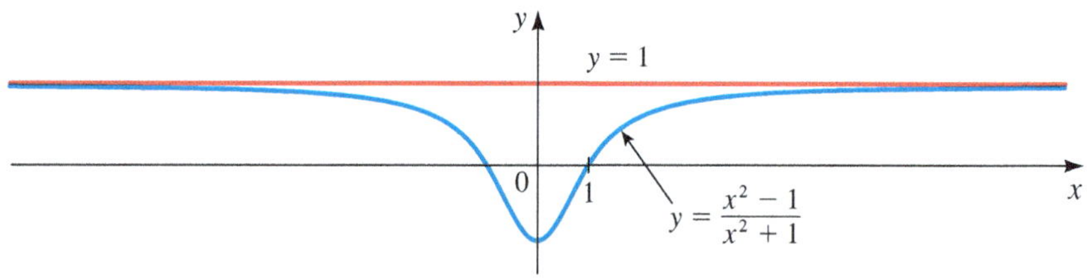

FIGURE 1

As x grows larger and larger, you can see that the values of $f(x)$ get closer and closer to 1. In fact, it seems that we can make the values of $f(x)$ as close as we like to 1 by taking x sufficiently large. This situation is expressed symbolically by writing

$$\lim_{x \to \infty} \frac{x^2 - 1}{x^2 + 1} = 1$$

In general, we use the notation

$$\lim_{x \to \infty} f(x) = L$$

to indicate that the values of $f(x)$ become closer and closer to L as x becomes larger and larger.

LIMIT AT INFINITY

Let f be a function defined on some interval (a, ∞). Then

$$\lim_{x \to \infty} f(x) = L$$

means that the values of $f(x)$ can be made arbitrarily close to L by taking x sufficiently large.

Another notation for $\lim_{x \to \infty} f(x) = L$ is

$$f(x) \to L \qquad \text{as} \qquad x \to \infty$$

Limits at infinity are also discussed in Section 3.6.

The symbol ∞ does not represent a number. Nevertheless, we often read the expression $\lim_{x\to\infty} f(x) = L$ as

"the limit of $f(x)$, as x approaches infinity, is L"

or "the limit of $f(x)$, as x becomes infinite, is L"

or "the limit of $f(x)$, as x increases without bound, is L"

Geometric illustrations are shown in Figure 2. Notice that there are many ways for the graph of f to approach the line $y = L$ (which is called a *horizontal asymptote*) as we look to the far right.

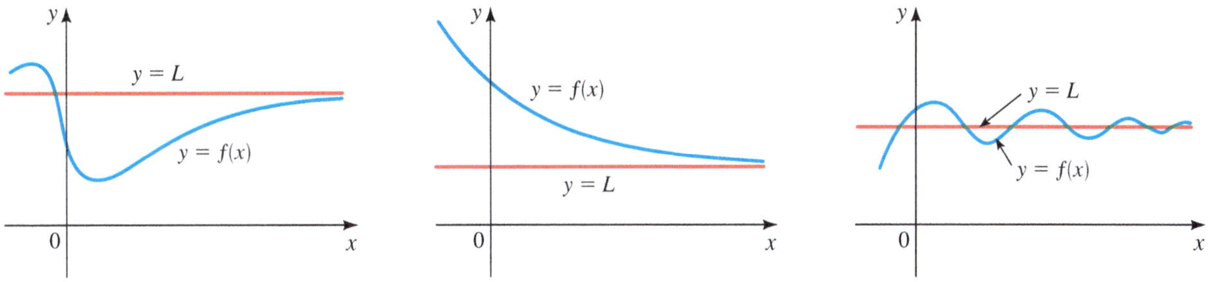

FIGURE 2 Examples illustrating $\lim_{x\to\infty} f(x) = L$

Referring back to Figure 1, we see that for numerically large negative values of x, the values of $f(x)$ are close to 1. By letting x decrease through negative values without bound, we can make $f(x)$ as close as we like to 1. This is expressed by writing

$$\lim_{x\to-\infty} \frac{x^2 - 1}{x^2 + 1} = 1$$

The general definition is as follows.

LIMIT AT NEGATIVE INFINITY

Let f be a function defined on some interval $(-\infty, a)$. Then

$$\lim_{x\to-\infty} f(x) = L$$

means that the values of $f(x)$ can be made arbitrarily close to L by taking x sufficiently large negative.

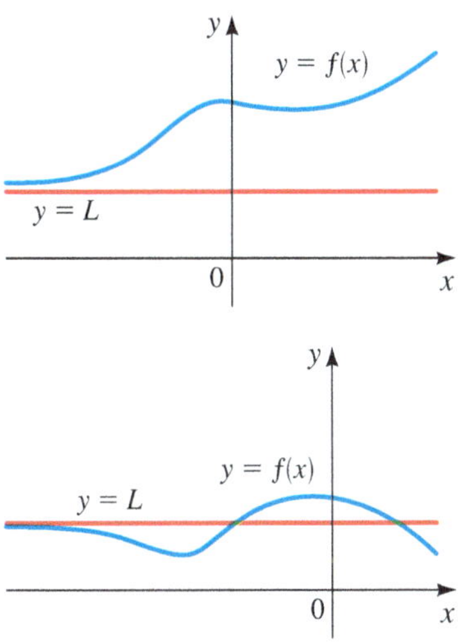

FIGURE 3 Examples illustrating $\lim_{x\to-\infty} f(x) = L$

Again, the symbol $-\infty$ does not represent a number, but the expression $\lim_{x\to-\infty} f(x) = L$ is often read as

"the limit of $f(x)$, as x approaches negative infinity, is L"

The definition is illustrated in Figure 3. Notice that the graph approaches the line $y = L$ as we look to the far left.

HORIZONTAL ASYMPTOTE

The line $y = L$ is called a **horizontal asymptote** of the curve $y = f(x)$ if either

$$\lim_{x\to\infty} f(x) = L \qquad \text{or} \qquad \lim_{x\to-\infty} f(x) = L$$

For instance, the curve illustrated in Figure 1 has the line $y = 1$ as a horizontal asymptote because

$$\lim_{x\to\infty} \frac{x^2 - 1}{x^2 + 1} = 1$$

As we discovered in Section 5.5, an example of a curve with two horizontal asymptotes is $y = \tan^{-1} x$ (see Figure 4). In fact,

$$\lim_{x\to-\infty} \tan^{-1} x = -\frac{\pi}{2} \qquad \text{and} \qquad \lim_{x\to\infty} \tan^{-1} x = \frac{\pi}{2}$$

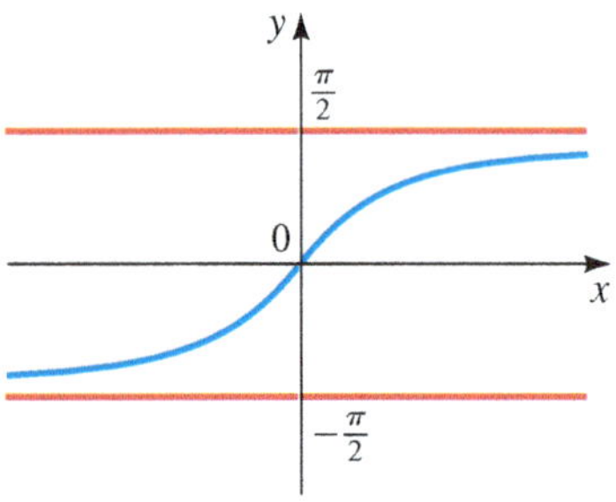

FIGURE 4 $y = \tan^{-1} x$

so both of the lines $y = -\pi/2$ and $y = \pi/2$ are horizontal asymptotes. (This follows from the fact that the lines $x = \pm\pi/2$ are vertical asymptotes of the graph of tan.)

EXAMPLE 1 ■ Limits at Infinity

Find $\lim_{x\to\infty} \frac{1}{x}$ and $\lim_{x\to-\infty} \frac{1}{x}$.

We first investigated horizontal asymptotes and limits at infinity for rational functions in Section 3.6.

SOLUTION Observe that when x is large, $1/x$ is small. For instance,

$$\frac{1}{100} = 0.01 \qquad \frac{1}{10{,}000} = 0.0001 \qquad \frac{1}{1{,}000{,}000} = 0.000001$$

In fact, by taking x large enough, we can make $1/x$ as close to 0 as we please. Therefore

$$\lim_{x\to\infty} \frac{1}{x} = 0$$

Similar reasoning shows that when x is large negative, $1/x$ is small negative, so we also have

$$\lim_{x\to-\infty} \frac{1}{x} = 0$$

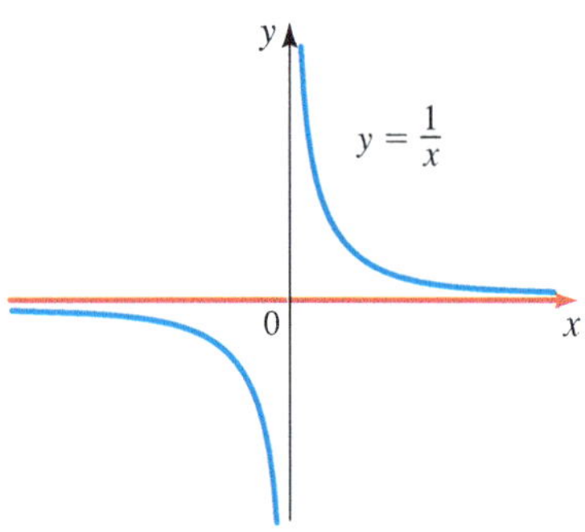

FIGURE 5 $\lim_{x\to\infty} \frac{1}{x} = 0, \ \lim_{x\to-\infty} \frac{1}{x} = 0$

It follows that the line $y = 0$ (the x-axis) is a horizontal asymptote of the curve $y = 1/x$. (This is a hyperbola; see Figure 5.)

Now Try Exercise 5 ■

The Limit Laws that we studied in Section 13.2 also hold for limits at infinity. In particular, if we combine Law 6 (Limit of a Power) with the results of Example 1, we obtain the following important rule for calculating limits.

> If k is any positive integer, then
>
> $$\lim_{x\to\infty} \frac{1}{x^k} = 0 \qquad \text{and} \qquad \lim_{x\to-\infty} \frac{1}{x^k} = 0$$

EXAMPLE 2 ■ Finding a Limit at Infinity

Evaluate $\lim_{x\to\infty} \frac{3x^2 - x - 2}{5x^2 + 4x + 1}$.

SOLUTION To evaluate the limit at infinity of a rational function, we first divide both the numerator and denominator by the highest power of x that occurs in the denominator. (We may assume that $x \neq 0$, since we are interested only in large values of x.) In this case the highest power of x in the denominator is x^2, so we have

$$\lim_{x\to\infty} \frac{3x^2 - x - 2}{5x^2 + 4x + 1} = \lim_{x\to\infty} \frac{3 - \dfrac{1}{x} - \dfrac{2}{x^2}}{5 + \dfrac{4}{x} + \dfrac{1}{x^2}} \quad \text{Divide numerator and denominator by } x^2$$

$$= \frac{\lim_{x\to\infty}\left(3 - \dfrac{1}{x} - \dfrac{2}{x^2}\right)}{\lim_{x\to\infty}\left(5 + \dfrac{4}{x} + \dfrac{1}{x^2}\right)} \quad \text{Limit of a Quotient}$$

$$= \frac{\lim_{x\to\infty} 3 - \lim_{x\to\infty} \dfrac{1}{x} - 2\lim_{x\to\infty} \dfrac{1}{x^2}}{\lim_{x\to\infty} 5 + 4\lim_{x\to\infty} \dfrac{1}{x} + \lim_{x\to\infty} \dfrac{1}{x^2}} \quad \text{Limits of Sums, Differences, and Constant Multiples}$$

$$= \frac{3 - 0 - 0}{5 + 0 + 0} = \frac{3}{5} \quad \text{Let } x \to \infty$$

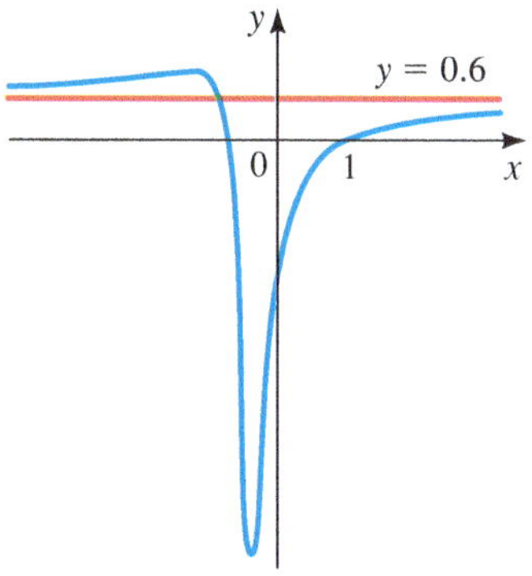

FIGURE 6

A similar calculation shows that the limit as $x \to -\infty$ is also $\frac{3}{5}$. Figure 6 illustrates the results of these calculations by showing how the graph of the given rational function approaches the horizontal asymptote $y = \frac{3}{5}$.

Now Try Exercise 9 ■

EXAMPLE 3 ■ A Limit at Negative Infinity

Use numerical and graphical methods to find $\lim_{x\to -\infty} e^x$.

SOLUTION From the graph of the natural exponential function $y = e^x$ in Figure 7 and the corresponding table of values we see that

$$\lim_{x\to -\infty} e^x = 0$$

It follows that the line $y = 0$ (the x-axis) is a horizontal asymptote.

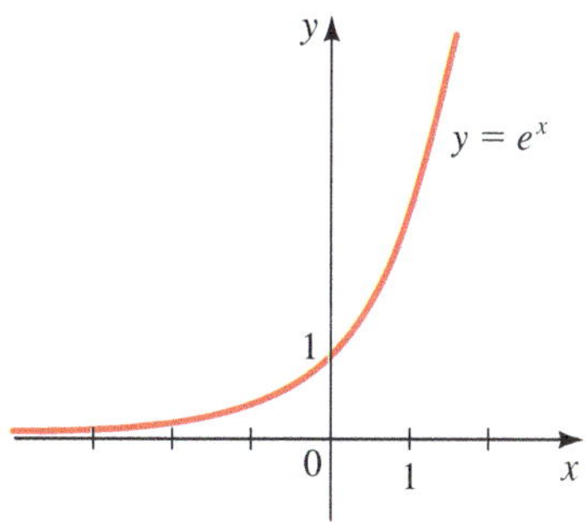

x	e^x
0	1.00000
−1	0.36788
−2	0.13534
−3	0.04979
−5	0.00674
−8	0.00034
−10	0.00005

FIGURE 7

Now Try Exercise 19 ■

EXAMPLE 4 ■ A Function with No Limit at Infinity

Evaluate $\lim_{x\to\infty} \sin x$.

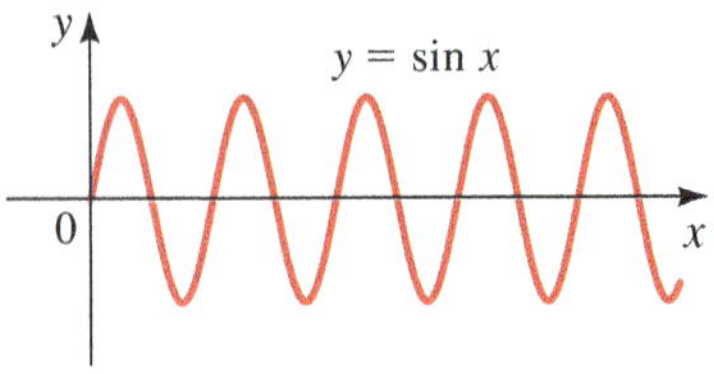

FIGURE 8

SOLUTION From the graph in Figure 8 and the periodic nature of the sine function we see that as x increases, the values of $\sin x$ oscillate between 1 and -1 infinitely often, so they don't approach any definite number. Therefore $\lim_{x\to\infty} \sin x$ does not exist.

Now Try Exercise 17 ■

Limits of Sequences

In Section 12.1 we introduced the idea of a sequence of numbers $a_1, a_2, a_3, \ldots$. Here we are interested in their behavior as n becomes large. For instance, the sequence defined by

$$a_n = \frac{n}{n+1}$$

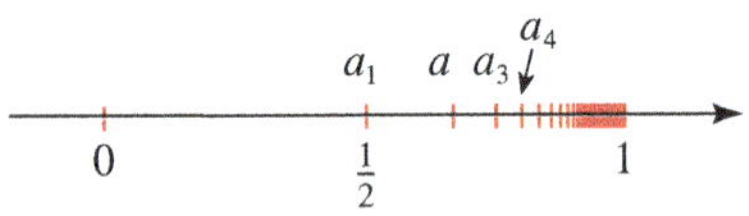

FIGURE 9

is pictured in Figure 9 by plotting its terms on a number line and in Figure 10 by plotting its graph. From Figure 9 or 10 it appears that the terms of the sequence $a_n = n/(n+1)$ are approaching 1 as n becomes large. We indicate this by writing

$$\lim_{n\to\infty} \frac{n}{n+1} = 1$$

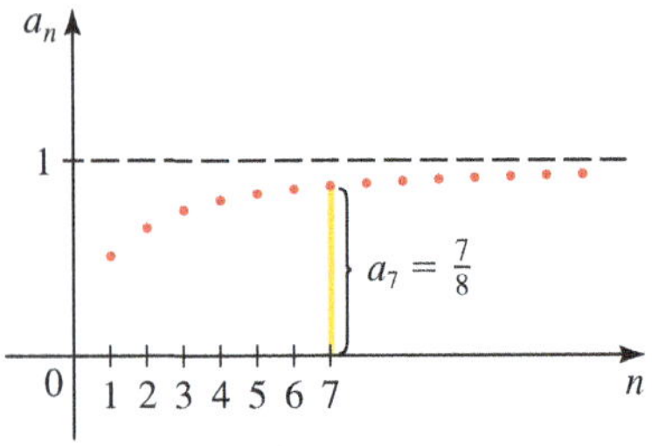

FIGURE 10

> **DEFINITION OF THE LIMIT OF A SEQUENCE**
>
> A sequence $a_1, a_2, a_3, \ldots$ has the **limit** L and we write
>
> $$\lim_{n\to\infty} a_n = L \qquad \text{or} \qquad a_n \to L \text{ as } n \to \infty$$
>
> if the nth term a_n of the sequence can be made arbitrarily close to L by taking n sufficiently large. If $\lim_{n\to\infty} a_n$ exists, we say the sequence **converges** (or is **convergent**). Otherwise, we say the sequence **diverges** (or is **divergent**).

This definition is illustrated by Figure 11.

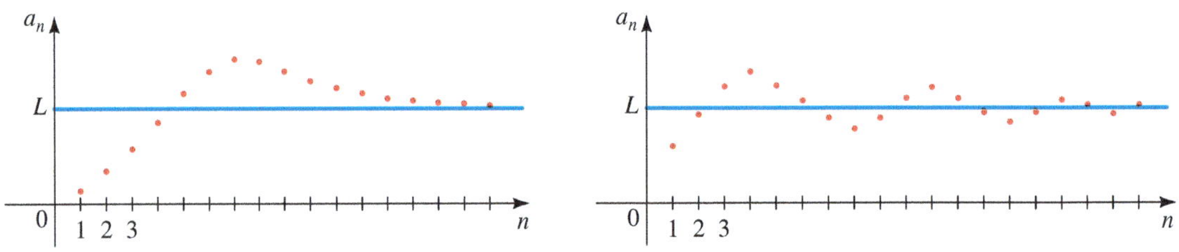

FIGURE 11 Graphs of two sequences with $\lim_{n\to\infty} a_n = L$

If we compare the definitions of $\lim_{n\to\infty} a_n = L$ and $\lim_{x\to\infty} f(x) = L$, we see that the only difference is that n is required to be an integer. Thus the following is true.

> If $\lim_{x\to\infty} f(x) = L$ and $f(n) = a_n$ when n is an integer, then $\lim_{n\to\infty} a_n = L$.

In particular, since we know that $\lim_{x\to\infty}(1/x^k) = 0$ when k is a positive integer, we have

$$\lim_{n\to\infty} \frac{1}{n^k} = 0 \qquad \text{if } k \text{ is a positive integer}$$

Note that the Limit Laws given in Section 13.2 also hold for limits of sequences.

EXAMPLE 5 ■ Finding the Limit of a Sequence

Find $\lim_{n\to\infty} \dfrac{n}{n+1}$.

SOLUTION The method is similar to the one we used in Example 2: Divide the numerator and denominator by the highest power of n, and then use the Limit Laws.

$$\lim_{n\to\infty} \frac{n}{n+1} = \lim_{n\to\infty} \frac{1}{1+\dfrac{1}{n}} \qquad \text{Divide numerator and denominator by } n$$

$$= \frac{\lim_{n\to\infty} 1}{\lim_{n\to\infty} 1 + \lim_{n\to\infty} \dfrac{1}{n}} \qquad \text{Limits of a Quotient and a Sum}$$

$$= \frac{1}{1+0} = 1 \qquad \text{Let } n \to \infty$$

This result shows that the guess we made earlier from Figures 9 and 10 was correct.

Therefore the sequence $a_n = n/(n+1)$ is convergent.

Now Try Exercise 23 ■

EXAMPLE 6 ■ A Sequence That Diverges

Determine whether the sequence $a_n = (-1)^n$ is convergent or divergent.

SOLUTION If we write out the terms of the sequence, we obtain

$$-1, 1, -1, 1, -1, 1, -1, \ldots$$

The graph of this sequence is shown in Figure 12. Since the terms oscillate between 1 and -1 infinitely often, a_n does not approach any number. Thus $\lim_{n\to\infty}(-1)^n$ does not exist; that is, the sequence $a_n = (-1)^n$ is divergent.

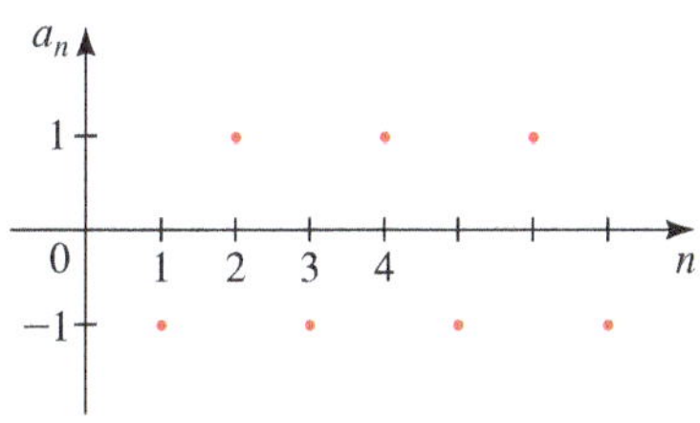

FIGURE 12

Now Try Exercise 29 ■

EXAMPLE 7 ■ Finding the Limit of a Sequence

Find the limit of the sequence given by

$$a_n = \frac{15}{n^3}\left[\frac{n(n+1)(2n+1)}{6}\right]$$

SOLUTION Before calculating the limit, let's first simplify the expression for a_n. Because $n^3 = n \cdot n \cdot n$, we place a factor of n beneath each factor in the numerator that contains an n:

$$a_n = \frac{15}{6} \cdot \frac{n}{n} \cdot \frac{n+1}{n} \cdot \frac{2n+1}{n} = \frac{5}{2} \cdot 1 \cdot \left(1 + \frac{1}{n}\right)\left(2 + \frac{1}{n}\right)$$

Now we can compute the limit.

$$\lim_{n\to\infty} a_n = \lim_{n\to\infty} \frac{5}{2}\left(1+\frac{1}{n}\right)\left(2+\frac{1}{n}\right) \qquad \text{Definition of } a_n$$

$$= \frac{5}{2} \lim_{n\to\infty}\left(1+\frac{1}{n}\right) \lim_{n\to\infty}\left(2+\frac{1}{n}\right) \qquad \text{Limit of a Product}$$

$$= \frac{5}{2}(1)(2) = 5 \qquad \text{Let } n \to \infty$$

Now Try Exercise 31 ■

13.4 EXERCISES

CONCEPTS

1. Let f be a function defined on some interval (a, ∞). Then

$$\lim_{x\to\infty} f(x) = L$$

means that the values of $f(x)$ can be made arbitrarily close to ________ by taking ________ sufficiently large. In this case the line $y = L$ is called a ________ ________ of the function $y = f(x)$. For example, $\lim_{x\to\infty} \frac{1}{x} =$ ________, and the line $y =$ ________ is a horizontal asymptote.

2. A sequence $a_1, a_2, a_3, \ldots$ has the limit L if the nth term a_n of the sequence can be made arbitrarily close to ________ by taking n to be sufficiently ________. If the limit exists, we say that the sequence ________________; otherwise, the sequence ________________.

SKILLS

3–4 ■ Limits from a Graph

(a) Use the graph of f to find the following limits.

(i) $\lim_{x\to\infty} f(x)$ (ii) $\lim_{x\to-\infty} f(x)$

(b) State the equations of the horizontal asymptotes.

3. 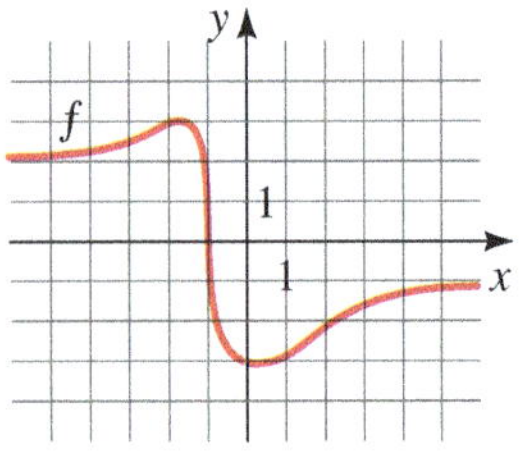

4. 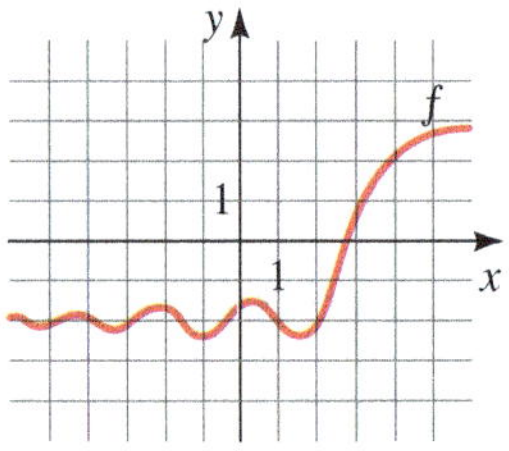

5–18 ■ Limits at Infinity Find the limit.

5. $\lim_{x\to\infty} \frac{6}{x}$

6. $\lim_{x\to\infty} \frac{3}{x^4}$

7. $\lim_{x\to\infty} \frac{2x + 1}{5x - 1}$

8. $\lim_{x\to\infty} \frac{2 - 3x}{4x + 5}$

9. $\lim_{x\to-\infty} \frac{4x^2 + 1}{2 + 3x^2}$

10. $\lim_{x\to-\infty} \frac{x^2 + 2}{x^3 + x + 1}$

11. $\lim_{t\to\infty} \frac{8t^3 + t}{(2t - 1)(2t^2 + 1)}$

12. $\lim_{r\to\infty} \frac{4r^3 - r^2}{(r + 1)^3}$

13. $\lim_{x\to\infty} \frac{x^4}{1 - x^2 + x^3}$

14. $\lim_{t\to\infty} \left(\frac{1}{t} - \frac{2t}{t - 1}\right)$

15. $\lim_{x\to-\infty} \left(\frac{x - 1}{x + 1} + 6\right)$

16. $\lim_{x\to-\infty} \left(\frac{3 - x}{3 + x} - 2\right)$

17. $\lim_{x\to\infty} \cos x$

18. $\lim_{x\to\infty} \sin^2 x$

19–22 ■ Estimating Limits Numerically and Graphically Use a table of values to estimate the limit. Then use a graphing device to confirm your result graphically.

19. $\lim_{x\to-\infty} \frac{\sqrt{x^2 + 4x}}{4x + 1}$

20. $\lim_{x\to\infty} \left(\sqrt{9x^2 + x} - 3x\right)$

21. $\lim_{x\to\infty} \frac{x^5}{e^x}$

22. $\lim_{x\to\infty} \left(1 + \frac{2}{x}\right)^{3x}$

23–34 ■ Limits of Sequences If the sequence with the given nth term is convergent, find its limit. If it is divergent, explain why.

23. $a_n = \frac{1 + n}{n + n^2}$

24. $a_n = \frac{5n}{n + 5}$

25. $a_n = \frac{n^2}{n + 1}$

26. $a_n = \frac{n - 1}{n^3 + 1}$

27. $a_n = \frac{1}{3^n}$

28. $a_n = \frac{(-1)^n}{n}$

29. $a_n = \sin(n\pi/2)$

30. $a_n = \cos n\pi$

31. $a_n = \frac{3}{n^2}\left[\frac{n(n + 1)}{2}\right]$

32. $a_n = \frac{5}{n}\left(n + \frac{4}{n}\left[\frac{n(n + 1)}{2}\right]\right)$

33. $a_n = \frac{24}{n^3}\left[\frac{n(n + 1)(2n + 1)}{6}\right]$

34. $a_n = \frac{12}{n^4}\left[\frac{n(n + 1)}{2}\right]^2$

SKILLS Plus

35–36 ■ A Function from a Description Find a formula from a function f that satisfies the following conditions.

35. Vertical asymptotes $x = 1$ and $x = 3$ and horizontal asymptote $y = 1$.

36. $\lim_{x\to\infty} f(x) = 0$, $\lim_{x\to 0} f(x) = -\infty$, $f(2) = 0$, $\lim_{x\to 3^-} f(x) = \infty$, $\lim_{x\to 3^+} f(x) = -\infty$

37. **Asymptote Behavior** How close to -3 do we have to take x so that

$$\frac{1}{(x + 3)^2} > 10{,}000$$

38. **Equivalent Limits** Show that

$$\lim_{x\to\infty} f(x) = \lim_{t\to 0^+} f\left(\frac{1}{t}\right)$$

and

$$\lim_{x\to-\infty} f(x) = \lim_{t\to 0^-} f\left(\frac{1}{t}\right)$$

if these limits exist.

APPLICATIONS

39. Salt Concentration

(a) A tank contains 5000 L of pure water. Brine that contains 30 g of salt per liter of water is pumped into the tank at a rate of 25 L/min. Show that the concentration of salt after t minutes (in g/L) is

$$C(t) = \frac{30t}{200 + t}$$

(b) What happens to the concentration as $t \to \infty$?

40. Velocity of a Raindrop The downward velocity of a falling raindrop at time t is modeled by the function

$$v(t) = 1.2(1 - e^{-8.2t})$$

(a) Find the terminal velocity of the raindrop by evaluating $\lim_{t\to\infty} v(t)$. (Use the result of Example 3.)

(b) Graph $v(t)$, and use the graph to estimate how long it takes for the velocity of the raindrop to reach 99% of its terminal velocity.

DISCUSS ■ DISCOVER ■ PROVE ■ WRITE

41. DISCUSS: The Limit of a Recursive Sequence

(a) A sequence is defined recursively by $a_1 = 0$ and

$$a_{n+1} = \sqrt{2 + a_n}$$

Find the first ten terms of this sequence rounded to eight decimal places. Does this sequence appear to be convergent? If so, guess the value of the limit.

(b) Assuming that the sequence in part (a) is convergent, let $\lim_{n\to\infty} a_n = L$. Explain why $\lim_{n\to\infty} a_{n+1} = L$ also and therefore

$$L = \sqrt{2 + L}$$

Solve this equation to find the exact value of L.

13.5 AREAS

■ The Area Problem ■ Definition of Area

We have seen that limits are needed to compute the slope of a tangent line or an instantaneous rate of change. Here we will see that they are also needed to find the area of a region with a curved boundary. The problem of finding such areas has consequences far beyond simply finding area. (See the *Focus on Modeling* on page 944.)

■ The Area Problem

One of the central problems in calculus is the *area problem*: Find the area of the region S that lies under the curve $y = f(x)$ from a to b. This means that S, illustrated in Figure 1, is bounded by the graph of a function f (where $f(x) \geq 0$), the vertical lines $x = a$ and $x = b$, and the x-axis.

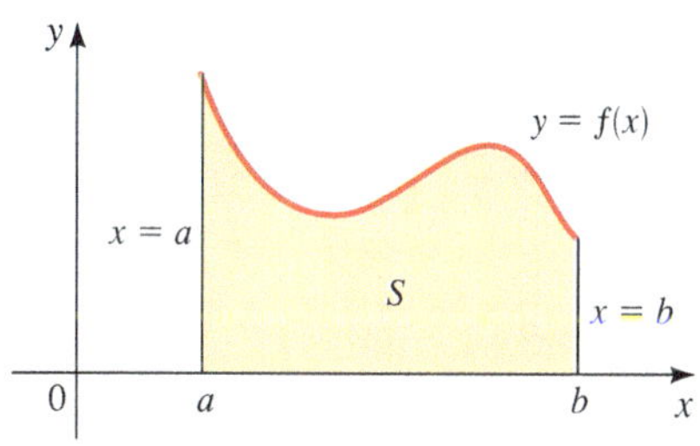

FIGURE 1

In trying to solve the area problem, we have to ask ourselves: What is the meaning of the word *area*? This question is easy to answer for regions with straight sides. For a rectangle, the area is defined as the product of the length and the width. The area of a triangle is half the base times the height. The area of a polygon is found by dividing it into triangles (as in Figure 2) and adding the areas of the triangles.

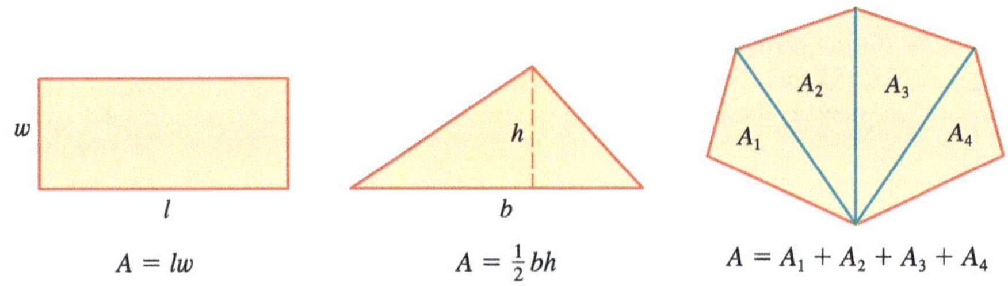

FIGURE 2

However, it is not so easy to find the area of a region with curved sides. We all have an intuitive idea of what the area of a region is. But part of the area problem is to make this intuitive idea precise by giving an exact definition of area.

Recall that in defining a tangent, we first approximated the slope of the tangent line by slopes of secant lines, and then we took the limit of these approximations. We pursue a similar idea for areas. We first approximate the region S by rectangles, and then we take the limit of the areas of these rectangles as we increase the number of rectangles. The following example illustrates the procedure.

EXAMPLE 1 ■ Estimating an Area Using Rectangles

Use rectangles to estimate the area under the parabola $y = x^2$ from 0 to 1 (the parabolic region S illustrated in Figure 3).

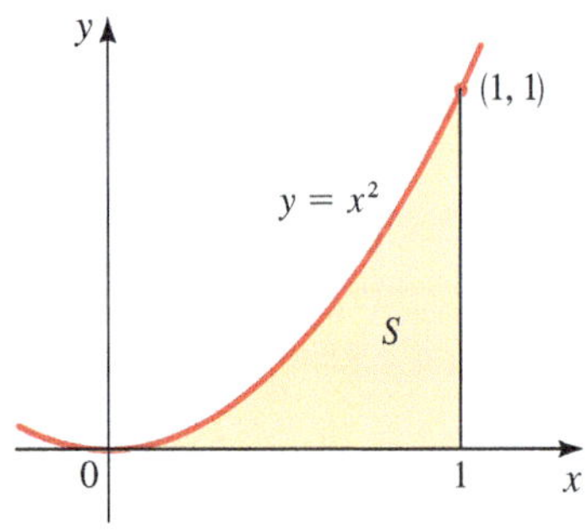

FIGURE 3

SOLUTION We first notice that the area of S must be somewhere between 0 and 1 because S is contained in a square with side length 1, but we can certainly do better than that. Suppose we divide S into four strips S_1, S_2, S_3, and S_4 by drawing the vertical lines $x = \frac{1}{4}$, $x = \frac{1}{2}$, and $x = \frac{3}{4}$ as in Figure 4(a). We can approximate each strip by a rectangle whose base is the same as the strip and whose height is the same as the right edge of the strip (see Figure 4(b)). In other words, the heights of these rectangles are the values of the function $f(x) = x^2$ at the right endpoints of the subintervals $\left[0, \frac{1}{4}\right]$, $\left[\frac{1}{4}, \frac{1}{2}\right]$, $\left[\frac{1}{2}, \frac{3}{4}\right]$, and $\left[\frac{3}{4}, 1\right]$.

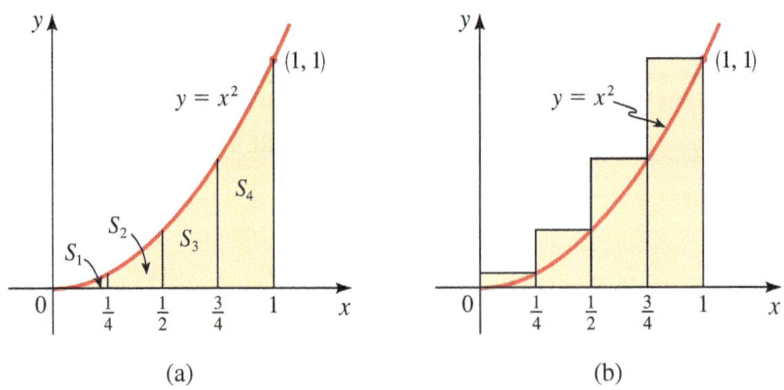

FIGURE 4

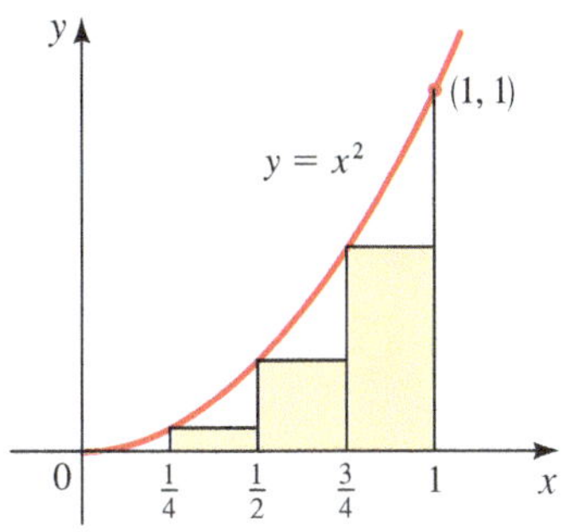

FIGURE 5

Each rectangle has width $\frac{1}{4}$, and the heights are $\left(\frac{1}{4}\right)^2$, $\left(\frac{1}{2}\right)^2$, $\left(\frac{3}{4}\right)^2$, and 1^2. If we let R_4 be the sum of the areas of these approximating rectangles, we get

$$R_4 = \tfrac{1}{4}\cdot\left(\tfrac{1}{4}\right)^2 + \tfrac{1}{4}\cdot\left(\tfrac{1}{2}\right)^2 + \tfrac{1}{4}\cdot\left(\tfrac{3}{4}\right)^2 + \tfrac{1}{4}\cdot 1^2 = \tfrac{15}{32} = 0.46875$$

From Figure 4(b) we see that the area A of S is less than R_4, so

$$A < 0.46875$$

Instead of using the rectangles in Figure 4(b), we could use the smaller rectangles in Figure 5 whose heights are the values of f at the left endpoints of the subintervals. (The leftmost rectangle has collapsed because its height is 0.) The sum of the areas of these approximating rectangles is

$$L_4 = \tfrac{1}{4}\cdot 0^2 + \tfrac{1}{4}\cdot\left(\tfrac{1}{4}\right)^2 + \tfrac{1}{4}\cdot\left(\tfrac{1}{2}\right)^2 + \tfrac{1}{4}\cdot\left(\tfrac{3}{4}\right)^2 = \tfrac{7}{32} = 0.21875$$

We see that the area of S is larger than L_4, so we have lower and upper estimates for A:

$$0.21875 < A < 0.46875$$

We can repeat this procedure with a larger number of strips. Figure 6 shows what happens when we divide the region S into eight strips of equal width. By computing the sum of the areas of the smaller rectangles (L_8) and the sum of the areas of the larger rectangles (R_8), we obtain better lower and upper estimates for A:

$$0.2734375 < A < 0.3984375$$

So one possible answer to the question is to say that the true area of S lies somewhere between 0.2734375 and 0.3984375.

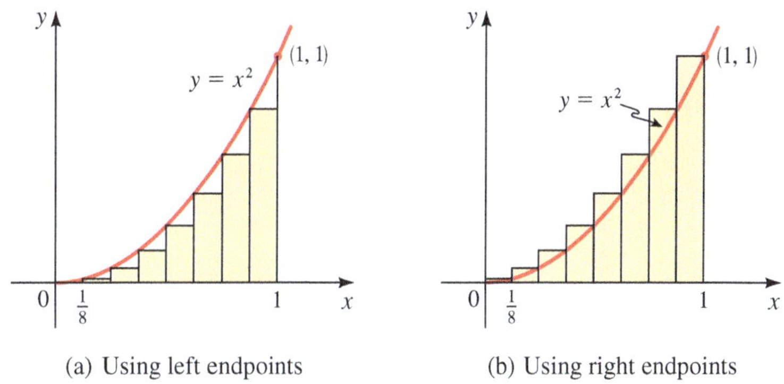

FIGURE 6 Approximating S with eight rectangles

n	L_n	R_n
10	0.2850000	0.3850000
20	0.3087500	0.3587500
30	0.3168519	0.3501852
50	0.3234000	0.3434000
100	0.3283500	0.3383500
1000	0.3328335	0.3338335

We could obtain better estimates by increasing the number of strips. The table in the margin shows the results of similar calculations (with a computer) using n rectangles whose heights are found with left endpoints (L_n) or right endpoints (R_n). In particular, we see by using 50 strips that the area lies between 0.3234 and 0.3434. With 1000 strips we narrow it down even more: A lies between 0.3328335 and 0.3338335. A good estimate is obtained by averaging these numbers: $A \approx 0.3333335$.

Now Try Exercise 3 ■

From the values in the table it looks as if R_n is approaching $\frac{1}{3}$ as n increases. We confirm this in the next example.

EXAMPLE 2 ■ The Limit of Approximating Sums

For the region S in Example 1, show that the sum of the areas of the upper approximating rectangles approaches $\frac{1}{3}$, that is,

$$\lim_{n\to\infty} R_n = \tfrac{1}{3}$$

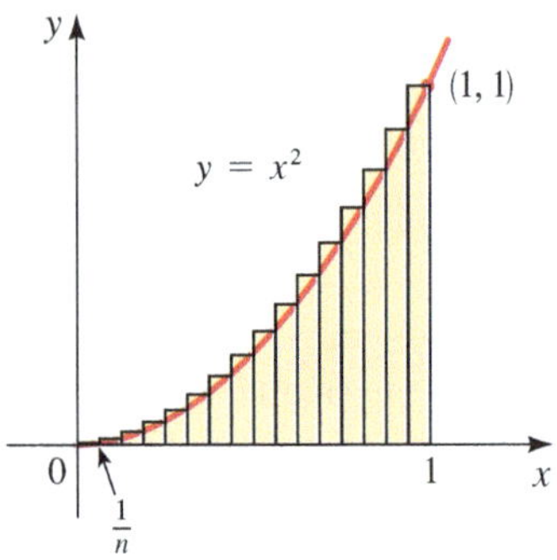

FIGURE 7

SOLUTION Let R_n be the sum of the areas of the n rectangles shown in Figure 7. Each rectangle has width $1/n$, and the heights are the values of the function $f(x) = x^2$ at the points $1/n, 2/n, 3/n, \ldots, n/n$. That is, the heights are $(1/n)^2, (2/n)^2, (3/n)^2, \ldots, (n/n)^2$. Thus

$$\begin{aligned} R_n &= \frac{1}{n}\left(\frac{1}{n}\right)^2 + \frac{1}{n}\left(\frac{2}{n}\right)^2 + \frac{1}{n}\left(\frac{3}{n}\right)^2 + \cdots + \frac{1}{n}\left(\frac{n}{n}\right)^2 \\ &= \frac{1}{n}\cdot\frac{1}{n^2}(1^2 + 2^2 + 3^2 + \cdots + n^2) \\ &= \frac{1}{n^3}(1^2 + 2^2 + 3^2 + \cdots + n^2) \end{aligned}$$

Here we need the formula for the sum of the squares of the first n positive integers:

$$1^2 + 2^2 + 3^2 + \cdots + n^2 = \frac{n(n+1)(2n+1)}{6}$$

This formula was discussed in Section 12.5.

Putting the preceding formula into our expression for R_n, we get

$$R_n = \frac{1}{n^3}\cdot\frac{n(n+1)(2n+1)}{6} = \frac{(n+1)(2n+1)}{6n^2}$$

Thus we have

$$\begin{aligned} \lim_{n\to\infty} R_n &= \lim_{n\to\infty} \frac{(n+1)(2n+1)}{6n^2} \\ &= \lim_{n\to\infty} \frac{1}{6}\left(\frac{n+1}{n}\right)\left(\frac{2n+1}{n}\right) \\ &= \lim_{n\to\infty} \frac{1}{6}\left(1 + \frac{1}{n}\right)\left(2 + \frac{1}{n}\right) \\ &= \tfrac{1}{6}\cdot 1\cdot 2 = \tfrac{1}{3} \end{aligned}$$

Now Try Exercise 13 ■

It can be shown that the lower approximating sums also approach $\frac{1}{3}$, that is,

$$\lim_{n\to\infty} L_n = \tfrac{1}{3}$$

From Figures 8 and 9 it appears that as n increases, both R_n and L_n become better and better approximations to the area of S. Therefore we *define* the area A to be the limit of the sums of the areas of the approximating rectangles, that is,

$$A = \lim_{n\to\infty} R_n = \lim_{n\to\infty} L_n = \tfrac{1}{3}$$

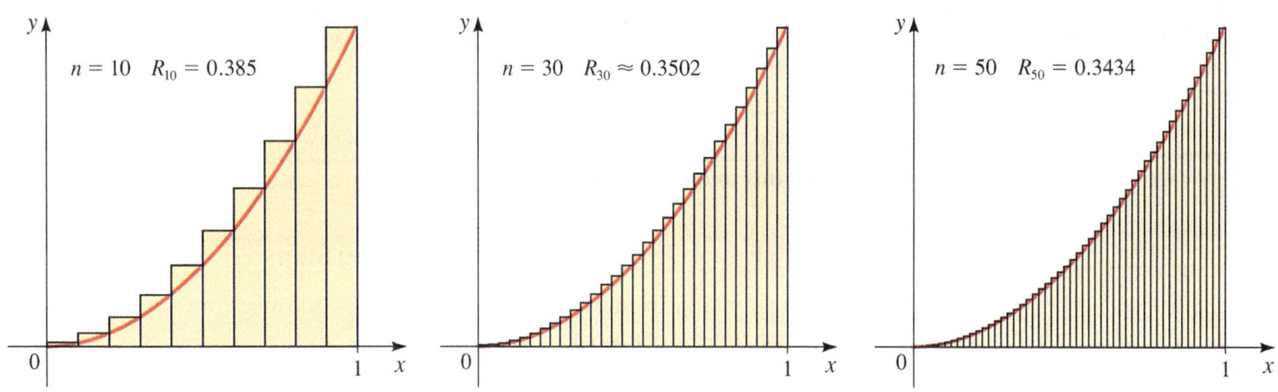

FIGURE 8

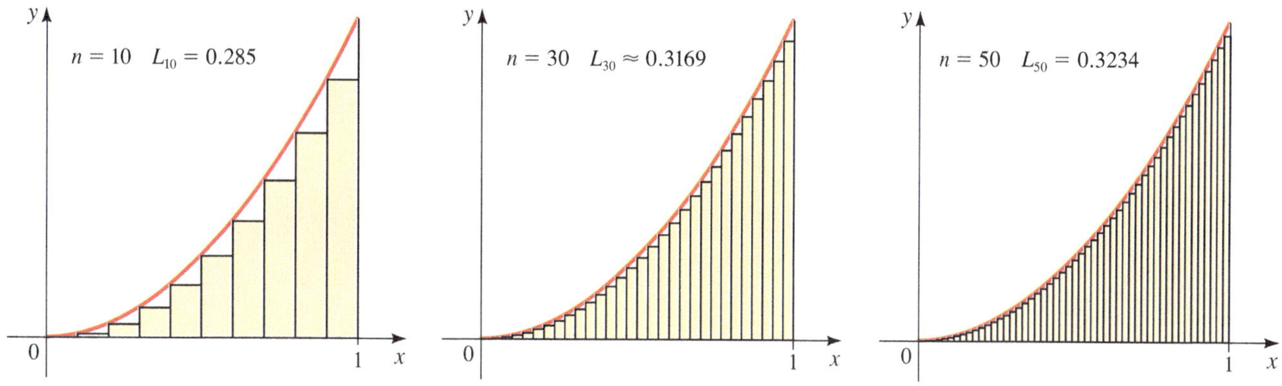

FIGURE 9

Definition of Area

Let's apply the idea of Examples 1 and 2 to the more general region S of Figure 1. We start by subdividing S into n strips $S_1, S_2, \ldots, S_n$ of equal width as in Figure 10.

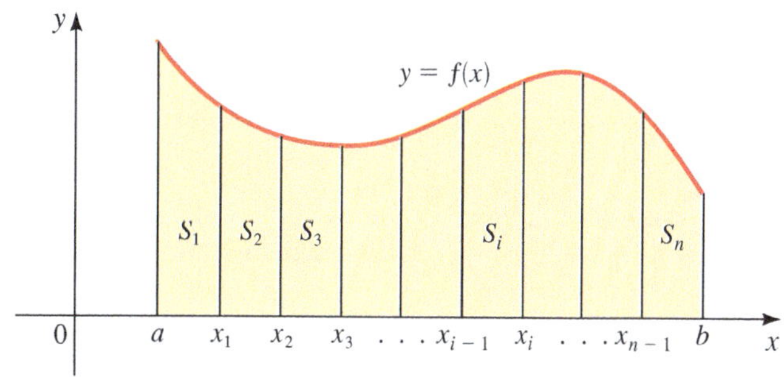

FIGURE 10

The width of the interval $[a, b]$ is $b - a$, so the width of each of the n strips is

$$\Delta x = \frac{b - a}{n}$$

These strips divide the interval $[a, b]$ into n subintervals

$$[x_0, x_1], \quad [x_1, x_2], \quad [x_2, x_3], \quad \ldots, \quad [x_{n-1}, x_n]$$

where $x_0 = a$ and $x_n = b$. The right endpoints of the subintervals are

$$x_1 = a + \Delta x, \quad x_2 = a + 2\,\Delta x, \quad x_3 = a + 3\,\Delta x, \quad \ldots, \quad x_k = a + k\,\Delta x, \quad \ldots$$

Let's approximate the kth strip S_k by a rectangle with width Δx and height $f(x_k)$, which is the value of f at the right endpoint (see Figure 11). Then the area of the kth rectangle is $f(x_k)\,\Delta x$. What we think of intuitively as the area of S is approximated by the sum of the areas of these rectangles, which is

$$R_n = f(x_1)\,\Delta x + f(x_2)\,\Delta x + \cdots + f(x_n)\,\Delta x$$

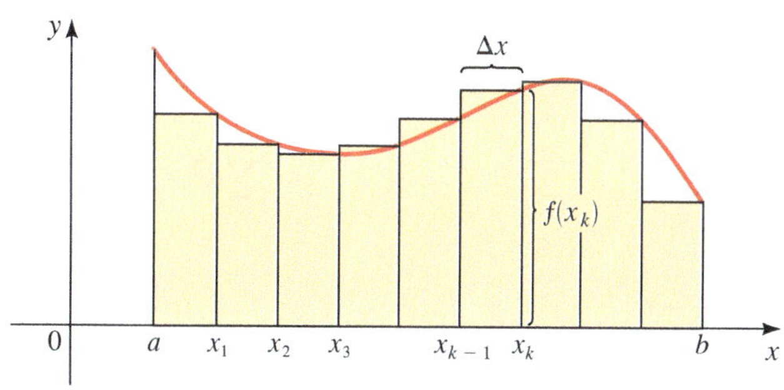

FIGURE 11

Figure 12 shows this approximation for $n = 2, 4, 8$, and 12.

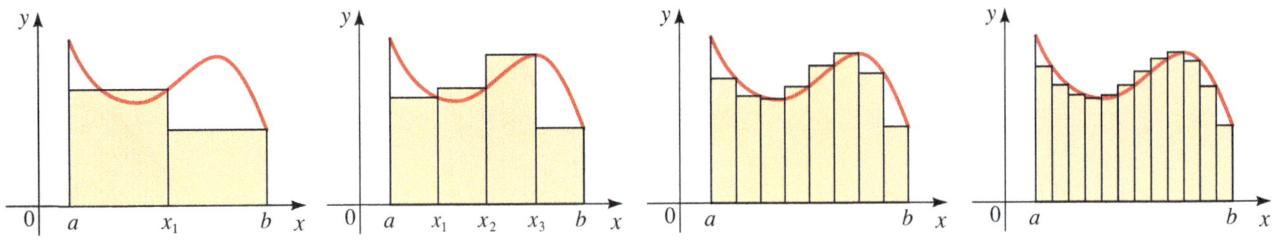

(a) $n = 2$ (b) $n = 4$ (c) $n = 8$ (d) $n = 12$

FIGURE 12

Notice that this approximation appears to become better and better as the number of strips increases, that is, as $n \to \infty$. Therefore we define the area A of the region S in the following way.

DEFINITION OF AREA

The **area** A of the region S that lies under the graph of the continuous function f is the limit of the sum of the areas of approximating rectangles:

$$A = \lim_{n\to\infty} R_n = \lim_{n\to\infty}[f(x_1)\,\Delta x + f(x_2)\,\Delta x + \cdots + f(x_n)\,\Delta x]$$

Using sigma notation, we write this as follows:

$$A = \lim_{n\to\infty} \sum_{k=1}^{n} f(x_k)\,\Delta x$$

In using this formula for area, remember that Δx is the width of an approximating rectangle, x_k is the right endpoint of the kth rectangle, and $f(x_k)$ is its height. So

Width: $\Delta x = \dfrac{b-a}{n}$

Right endpoint: $x_k = a + k\,\Delta x$

Height: $f(x_k) = f(a + k\,\Delta x)$

When working with sums, we will need the following properties from Section 12.1:

$$\sum_{k=1}^{n}(a_k \pm b_k) = \sum_{k=1}^{n} a_k \pm \sum_{k=1}^{n} b_k \qquad \sum_{k=1}^{n} ca_k = c\sum_{k=1}^{n} a_k$$

We will also need the following formulas for the sums of the powers of the first n natural numbers from Section 12.5.

$$\sum_{k=1}^{n} c = nc \qquad \sum_{k=1}^{n} k = \frac{n(n+1)}{2}$$

$$\sum_{k=1}^{n} k^2 = \frac{n(n+1)(2n+1)}{6} \qquad \sum_{k=1}^{n} k^3 = \frac{n^2(n+1)^2}{4}$$

EXAMPLE 3 ■ Finding the Area Under a Curve

Find the area of the region that lies under the parabola $y = x^2$, where $0 \le x \le 5$.

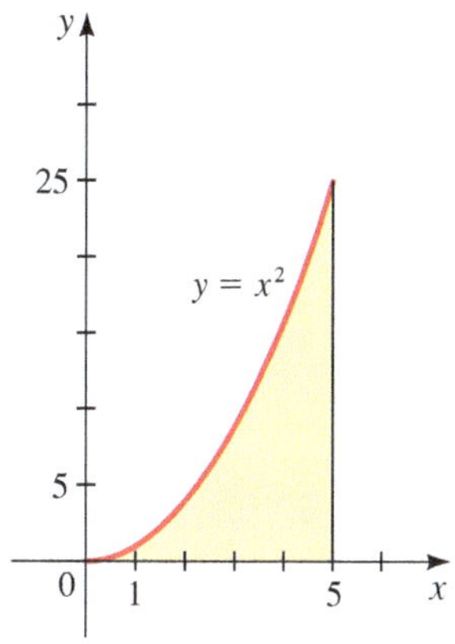

FIGURE 13

SOLUTION The region is graphed in Figure 13. To find the area, we first find the dimensions of the approximating rectangles at the nth stage.

$$\text{Width:} \qquad \Delta x = \frac{b-a}{n} = \frac{5-0}{n} = \frac{5}{n}$$

$$\text{Right endpoint:} \qquad x_k = a + k\,\Delta x = 0 + k\left(\frac{5}{n}\right) = \frac{5k}{n}$$

$$\text{Height:} \qquad f(x_k) = f\left(\frac{5k}{n}\right) = \left(\frac{5k}{n}\right)^2 = \frac{25k^2}{n^2}$$

Now we substitute these values into the definition of area.

$$\begin{aligned}
A &= \lim_{n\to\infty} \sum_{k=1}^{n} f(x_k)\,\Delta x && \text{Definition of area} \\
&= \lim_{n\to\infty} \sum_{k=1}^{n} \frac{25k^2}{n^2} \cdot \frac{5}{n} && f(x_k) = \frac{25k^2}{n^2},\ \Delta x = \frac{5}{n} \\
&= \lim_{n\to\infty} \sum_{k=1}^{n} \frac{125k^2}{n^3} && \text{Simplify} \\
&= \lim_{n\to\infty} \frac{125}{n^3} \sum_{k=1}^{n} k^2 && \text{Factor } \frac{125}{n^3} \\
&= \lim_{n\to\infty} \frac{125}{n^3} \cdot \frac{n(n+1)(2n+1)}{6} && \text{Sum of Squares Formula} \\
&= \lim_{n\to\infty} \frac{125(2n^2+3n+1)}{6n^2} && \text{Cancel } n\text{, and expand the numerator} \\
&= \lim_{n\to\infty} \frac{125}{6}\left(2 + \frac{3}{n} + \frac{1}{n^2}\right) && \text{Divide the numerator and denominator by } n^2 \\
&= \frac{125}{6}(2+0+0) = \frac{125}{3} && \text{Let } n \to \infty
\end{aligned}$$

We can also calculate the limit by writing

$$\frac{125}{n^3} \cdot \frac{n(n+1)(2n+1)}{6} = \frac{125}{6}\left(\frac{n}{n}\right)\left(\frac{n+1}{n}\right)\left(\frac{2n+1}{n}\right)$$

as in Example 2.

Thus the area of the region is $\frac{125}{3} \approx 41.7$.

Now Try Exercise 15 ■

EXAMPLE 4 ■ Finding the Area Under a Curve

Find the area of the region that lies under the parabola $y = 4x - x^2$, where $1 \le x \le 3$.

The figure below shows the region whose area is computed in Example 4.

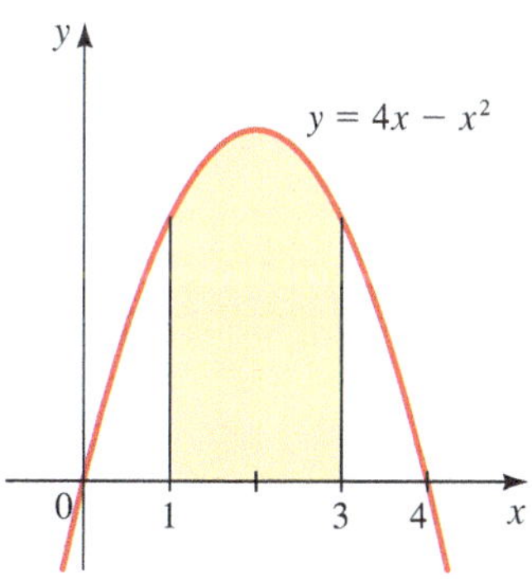

SOLUTION We start by finding the dimensions of the approximating rectangles at the nth stage.

$$\text{Width:} \qquad \Delta x = \frac{b-a}{n} = \frac{3-1}{n} = \frac{2}{n}$$

$$\text{Right endpoint:} \qquad x_k = a + k\,\Delta x = 1 + k\left(\frac{2}{n}\right) = 1 + \frac{2k}{n}$$

$$\begin{aligned}
\text{Height:} \qquad f(x_k) &= f\left(1 + \frac{2k}{n}\right) = 4\left(1 + \frac{2k}{n}\right) - \left(1 + \frac{2k}{n}\right)^2 \\
&= 4 + \frac{8k}{n} - 1 - \frac{4k}{n} - \frac{4k^2}{n^2} \\
&= 3 + \frac{4k}{n} - \frac{4k^2}{n^2}
\end{aligned}$$

Thus according to the definition of area, we get

$$\begin{aligned} A &= \lim_{n\to\infty} \sum_{k=1}^{n} f(x_k)\,\Delta x = \lim_{n\to\infty} \sum_{k=1}^{n} \left(3 + \frac{4k}{n} - \frac{4k^2}{n^2}\right)\left(\frac{2}{n}\right) \\ &= \lim_{n\to\infty} \left(\sum_{k=1}^{n} 3 + \frac{4}{n}\sum_{k=1}^{n} k - \frac{4}{n^2}\sum_{k=1}^{n} k^2\right)\left(\frac{2}{n}\right) \\ &= \lim_{n\to\infty} \left(\frac{2}{n}\sum_{k=1}^{n} 3 + \frac{8}{n^2}\sum_{k=1}^{n} k - \frac{8}{n^3}\sum_{k=1}^{n} k^2\right) \\ &= \lim_{n\to\infty} \left(\frac{2}{n}(3n) + \frac{8}{n^2}\left[\frac{n(n+1)}{2}\right] - \frac{8}{n^3}\left[\frac{n(n+1)(2n+1)}{6}\right]\right) \\ &= \lim_{n\to\infty} \left(6 + 4\cdot\frac{n}{n}\cdot\frac{n+1}{n} - \frac{4}{3}\cdot\frac{n}{n}\cdot\frac{n+1}{n}\cdot\frac{2n+1}{n}\right) \\ &= \lim_{n\to\infty} \left[6 + 4\left(1 + \frac{1}{n}\right) - \frac{4}{3}\left(1 + \frac{1}{n}\right)\left(2 + \frac{1}{n}\right)\right] \\ &= 6 + 4\cdot 1 - \frac{4}{3}\cdot 1\cdot 2 = \frac{22}{3} \end{aligned}$$

Now Try Exercise 17 ■

13.5 EXERCISES

CONCEPTS

1–2 ■ The graph of a function f is shown below.

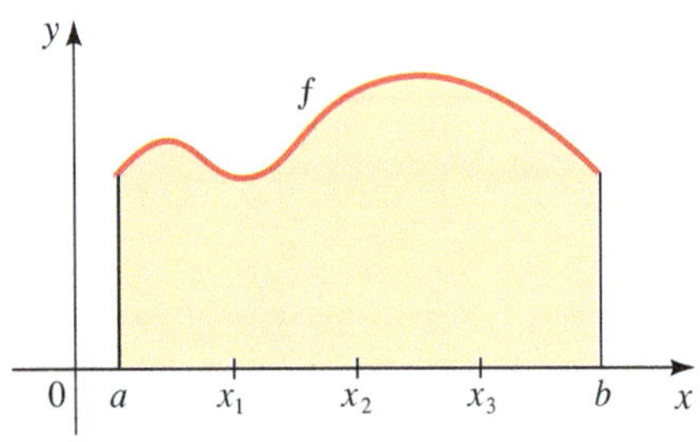

1. To find the area under the graph of f, we first approximate the area by ________. Approximate the area by drawing four rectangles. The area R_4 of this approximation is

$$R_4 = \quad + \quad + \quad + $$

2. Let R_n be the approximation obtained by using n rectangles of equal width. The exact area under the graph of f is

$$A = \lim_{n\to\infty}$$

SKILLS

3. Estimating an Area Using Rectangles

(a) By reading values from the given graph of f, use five rectangles to find a lower estimate and an upper estimate for the area under the given graph of f from $x = 0$ to $x = 10$. In each case, sketch the rectangles that you use.

(b) Find new estimates using ten rectangles in each case.

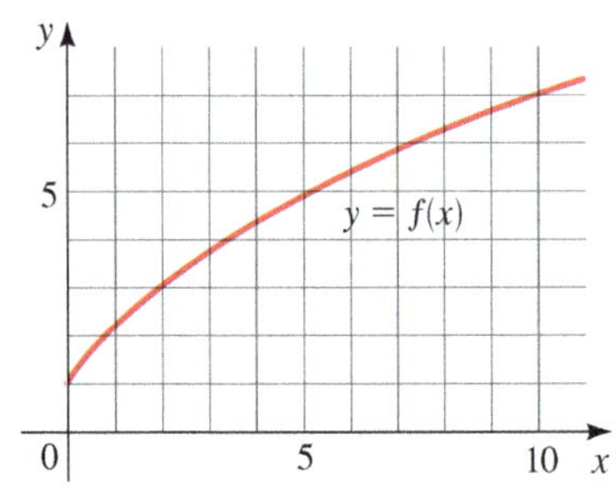

4. Estimating an Area Using Rectangles

(a) Use six rectangles to find estimates of each type for the area under the given graph of f from $x = 0$ to $x = 12$.
(i) L_6 (using left endpoints)
(ii) R_6 (using right endpoints)

(b) Is L_6 an underestimate or an overestimate of the true area?

(c) Is R_6 an underestimate or an overestimate of the true area?

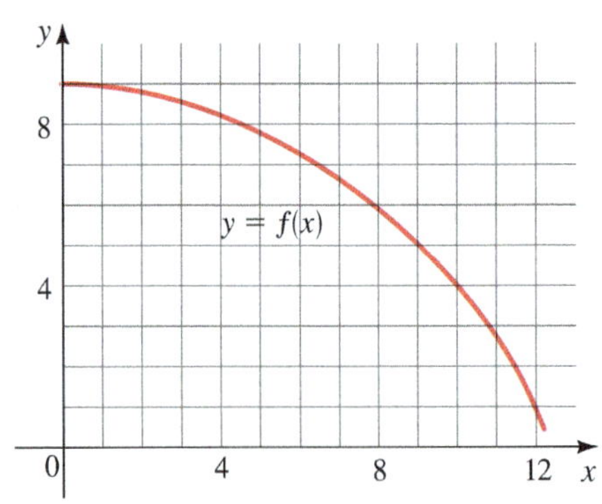

5–8 ■ Estimating Areas Using Rectangles Approximate the area of the shaded region under the graph of the given function by using the indicated rectangles. (The rectangles have equal width.)

5. $f(x) = \frac{1}{2}x + 2$

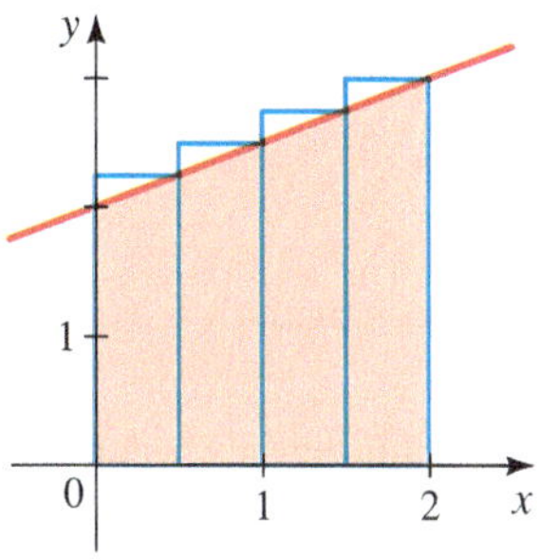

6. $f(x) = 4 - x^2$

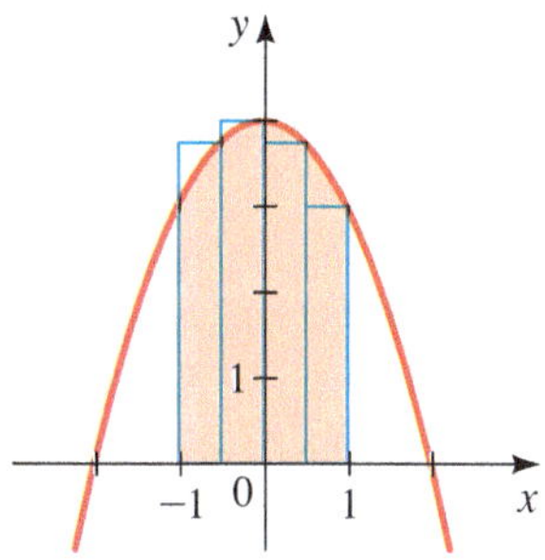

7. $f(x) = \dfrac{4}{x}$

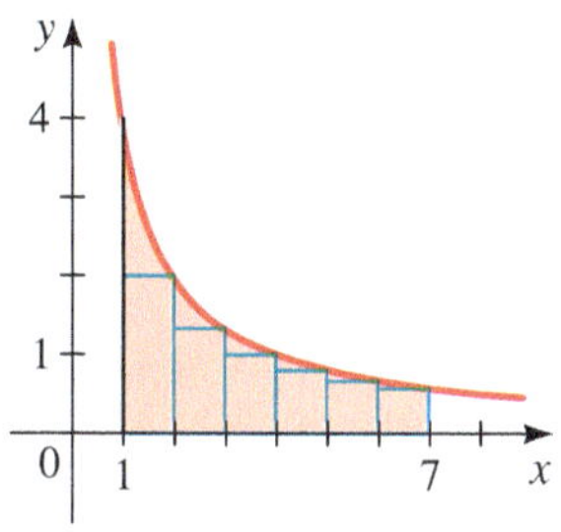

8. $f(x) = 9x - x^3$

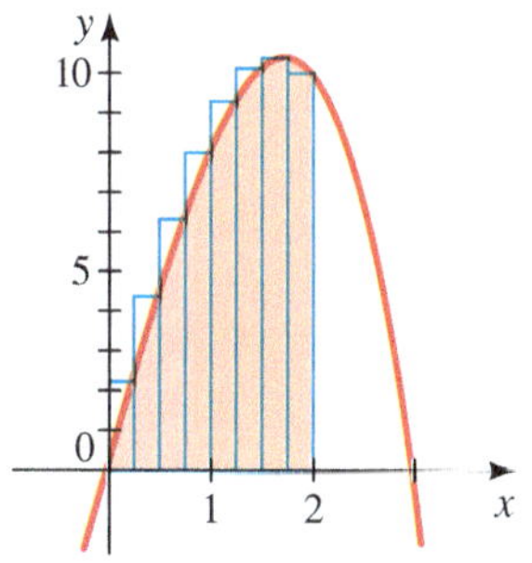

9–12 ■ Estimating Areas Using Rectangles In these exercises we estimate the area under the graph of a function by using rectangles.

9. **(a)** Estimate the area under the graph of $f(x) = 1/x$ from $x = 1$ to $x = 5$ using four approximating rectangles and right endpoints. Sketch the graph and the rectangles. Is your estimate an underestimate or an overestimate?

(b) Repeat part (a), using left endpoints.

10. **(a)** Estimate the area under the graph of $f(x) = 25 - x^2$ from $x = 0$ to $x = 5$ using five approximating rectangles and right endpoints. Sketch the graph and the rectangles. Is your estimate an underestimate or an overestimate?

(b) Repeat part (a) using left endpoints.

11. **(a)** Estimate the area under the graph of $f(x) = 1 + x^2$ from $x = -1$ to $x = 2$ using three rectangles and right endpoints. Then improve your estimate by using six rectangles. Sketch the curve and the approximating rectangles.

(b) Repeat part (a) using left endpoints.

12. **(a)** Estimate the area under the graph of $f(x) = e^{-x}$, $0 \le x \le 4$, using four approximating rectangles and taking the sample points to be

(i) right endpoints

(ii) left endpoints

In each case, sketch the curve and the rectangles.

(b) Improve your estimates in part (a) by using eight rectangles.

13–14 ■ Finding the Area Under A Curve Use the definition of area as a limit to find the area of the region that lies under the curve. Check your answer by sketching the region and using geometry.

13. $y = 3x, \quad 0 \le x \le 5$

14. $y = 2x + 1, \quad 1 \le x \le 3$

15–20 ■ Finding the Area Under a Curve Find the area of the region that lies under the graph of f over the given interval.

15. $f(x) = 3x^2, \quad 0 \le x \le 2$

16. $f(x) = x + x^2, \quad 0 \le x \le 1$

17. $f(x) = x^3 + 2, \quad 0 \le x \le 5$

18. $f(x) = 4x^3, \quad 2 \le x \le 5$

19. $f(x) = x + 6x^2, \quad 1 \le x \le 4$

20. $f(x) = 20 - 2x^2, \quad 2 \le x \le 3$

DISCUSS ■ DISCOVER ■ PROVE ■ WRITE

21. DISCUSS: Approximating Area with a Calculator When we approximate areas using rectangles as in Example 1, then the more rectangles we use, the more accurate the answer. The following TI-83 program finds the approximate area under the graph of f on the interval $[a, b]$ using n rectangles. To use the program, first store the function f in Y_1. The program prompts you to enter N, the number of rectangles, and A and B, the endpoints of the interval.

(a) Approximate the area under the graph of $f(x) = x^5 + 2x + 3$ on $[1, 3]$, using 10, 20, and 100 rectangles.

(b) Approximate the area under the graph of f on the given interval, using 100 rectangles.

(i) $f(x) = \sin x$, on $[0, \pi]$

(ii) $f(x) = e^{-x^2}$, on $[-1, 1]$

```
PROGRAM:AREA
:Prompt N
:Prompt A
:Prompt B
:(B-A)/N→D
:0→S
:A→X
:For (K,1,N)
:X+D→X
:S+Y1→S
:End
:D*S→S
:Disp "AREA IS"
:Disp S
```

22. WRITE: Regions with Straight Versus Curved Boundaries Write a short essay that explains how you would find the area of a polygon, that is, a region bounded by straight line segments. Then explain how you would find the area of a region whose boundary is curved, as we did in this section. What is the fundamental difference between these two processes?

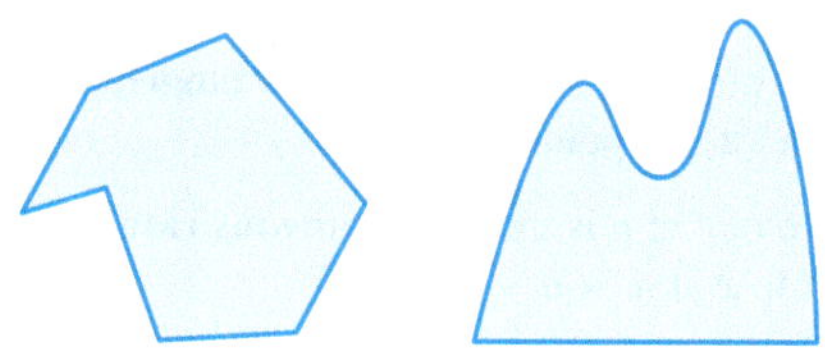

CHAPTER 13 ■ REVIEW

PROPERTIES AND FORMULAS

Limits (pp. 898, 903)

We say that the **limit of a function** f, as x approaches a, equals L, and we write

$$\lim_{x \to a} f(x) = L$$

provided that the values of $f(x)$ can be made arbitrarily close to L by taking x to be sufficiently close to a.

The **left-hand** and **right-hand** limits of f, as x approaches a, are defined similarly:

$$\lim_{x \to a^-} f(x) = L \qquad \lim_{x \to a^+} f(x) = L$$

The limit of f, as x approaches a, exists if and only if both left- and right-hand limits exist: $\lim_{x \to a} f(x) = L$ if and only if $\lim_{x \to a^-} f(x) = L$ and $\lim_{x \to a^+} f(x) = L$.

Algebraic Properties of Limits (pp. 906–908)

The following **Limit Laws** hold:

1. $\lim_{x \to a} [f(x) + g(x)] = \lim_{x \to a} f(x) + \lim_{x \to a} g(x)$
2. $\lim_{x \to a} [f(x) - g(x)] = \lim_{x \to a} f(x) - \lim_{x \to a} g(x)$
3. $\lim_{x \to a} cf(x) = c \lim_{x \to a} f(x)$
4. $\lim_{x \to a} [f(x)g(x)] = \lim_{x \to a} f(x) \cdot \lim_{x \to a} g(x)$
5. $\lim_{x \to a} \dfrac{f(x)}{g(x)} = \dfrac{\lim_{x \to a} f(x)}{\lim_{x \to a} g(x)}$, if $\lim_{x \to a} g(x) \neq 0$
6. $\lim_{x \to a} [f(x)]^n = [\lim_{x \to a} f(x)]^n$
7. $\lim_{x \to a} \sqrt[n]{f(x)} = \sqrt[n]{\lim_{x \to a} f(x)}$

The following **special limits** hold:

1. $\lim_{x \to a} c = c$
2. $\lim_{x \to a} x = a$
3. $\lim_{x \to a} x^n = a^n$
4. $\lim_{x \to a} \sqrt[n]{x} = \sqrt[n]{a}$

If f is a polynomial or a rational function and a is in the domain of f, then $\lim_{x \to a} f(x) = f(a)$.

Derivatives (p. 918)

Let $y = f(x)$ be a function. The **derivative of f at a**, denoted by $f'(a)$, is

$$f'(a) = \lim_{h \to 0} \frac{f(x+h) - f(x)}{h}$$

Equivalently, the derivative $f'(a)$ is

$$f'(a) = \lim_{x \to a} \frac{f(x) - f(a)}{x - a}$$

The derivative of f at a is the **slope of the tangent line** to the curve $y = f(x)$ at the point $P(a, f(a))$.

The derivative of f at a is the **instantaneous rate of change of y with respect to x** at $x = a$.

Limits at Infinity (pp. 924–926)

We say that **the limit of a function** f, as x **approaches infinity**, is L, and write

$$\lim_{x \to \infty} f(x) = L$$

provided that the values of $f(x)$ can be made arbitrarily close to L by taking x sufficiently large.

We say that **the limit of a function** f, as x **approaches negative infinity**, is L, and we write

$$\lim_{x \to -\infty} f(x) = L$$

provided that the values of $f(x)$ can be made arbitrarily close to L by taking x sufficiently large negative.

The line $y = L$ is a horizontal asymptote of the curve $y = f(x)$ if either

$$\lim_{x \to \infty} f(x) = L \quad \text{or} \quad \lim_{x \to -\infty} f(x) = L$$

The following special limits hold, where $k > 0$:

$$\lim_{x \to \infty} \frac{1}{x^k} = 0 \quad \text{and} \quad \lim_{x \to -\infty} \frac{1}{x^k} = 0$$

Limits of Sequences (p. 928)

We say that a sequence $a_1, a_2, a_3, \ldots$ has the limit L, and we write

$$\lim_{n \to \infty} a_n = L$$

provided that the nth term a_n of the sequence can be made arbitrarily close to L by taking n sufficiently large.

If $\lim_{x \to \infty} f(x) = L$ and if $f(n) = a_n$ when n is an integer, then $\lim_{n \to \infty} a_n = L$.

Area (pp. 935–936)

Let f be a continuous function defined on the interval $[a, b]$. The area A of the region that lies under the graph of f is the limit of the sum of the areas of approximating rectangles:

$$A = \lim_{n \to \infty} [f(x_1)\,\Delta x + f(x_2)\,\Delta x + \cdots + f(x_n)\,\Delta x]$$
$$= \lim_{n \to \infty} \sum_{k=1}^{n} f(x_k)\,\Delta x$$

where

$$\Delta x = \frac{b-a}{n} \quad \text{and} \quad x_k = a + k\,\Delta x$$

Summation Formulas (p. 936)

The following summation formulas are useful for calculating areas:

$$\sum_{k=1}^{n} c = nc \qquad \sum_{k=1}^{n} k = \frac{n(n+1)}{2}$$

$$\sum_{k=1}^{n} k^2 = \frac{n(n+1)(2n+1)}{6} \qquad \sum_{k=1}^{n} k^3 = \frac{n^2(n+1)^2}{4}$$

CONCEPT CHECK

1. **(a)** Explain what is meant by $\lim_{x\to a} f(x) = L$.
   (b) If $\lim_{x\to 2} f(x) = 5$, is it possible that $f(2) = 3$?
   (c) Find $\lim_{x\to 2} x^2$.

2. To evaluate the limit of a function, we often need to first rewrite the function using the rules of algebra. What is the logical first step in evaluating each of the following limits?
   (a) $\displaystyle\lim_{x\to 2} \frac{x^2 - 4}{x - 2}$ **(b)** $\displaystyle\lim_{h\to 0} \frac{(5 + h)^2 - 25}{h}$
   (c) $\displaystyle\lim_{x\to 3} \frac{\sqrt{x + 1} - 2}{x - 3}$ **(d)** $\displaystyle\lim_{x\to 7} \frac{\left(\frac{1}{7} - \frac{1}{x}\right)}{x - 7}$

3. **(a)** Explain what it means to say:
   $$\lim_{x\to 3^-} f(x) = 5 \qquad \lim_{x\to 3^+} f(x) = 10$$
   (b) If the two equations in part (a) are true, is it possible that $\lim_{x\to 3} f(x) = 5$?
   (c) Find $\lim_{x\to 2^-} f(x)$ and $\lim_{x\to 2^+} f(x)$, where f is defined as follows:
   $$f(x) = \begin{cases} 1 & \text{if } x \le 2 \\ x & \text{if } x > 2 \end{cases}$$
   (d) For f as in (c), does $\lim_{x\to 2} f(x)$ exist?

4. **(a)** Define the derivative $f'(a)$ of a function f at $x = a$.
   (b) State an equivalent formulation for $f'(a)$.
   (c) Find the derivative of $f(x) = x^2$ at $x = 3$.

5. **(a)** Give two different interpretations of the derivative of the function $y = f(x)$ at $x = a$.
   (b) For the function $f(x) = x^2$, find the slope of the tangent line to the graph of f at the point $(3, 9)$ on the graph.
   (c) For the function $y = x^2$, find the instantaneous rate of change of y with respect to x when $x = 3$.
   (d) Write expressions for the average rate of change of y with respect to x between a and x and for the instantaneous rate of change of y with respect to x at $x = a$.

6. **(a)** Explain what is meant by $\lim_{x\to\infty} f(x) = L$. Draw sketches to illustrate different ways in which this can happen.
   (b) Find $\displaystyle\lim_{x\to\infty} \frac{3x^2 + x}{x^2 + 1}$.
   (c) Explain why $\lim_{x\to\infty} \sin x$ does not exist.

7. **(a)** If $a_1, a_2, a_3, \ldots$ is a sequence, what is meant by $\lim_{n\to\infty} a_n = L$? What is a convergent sequence?
   (b) Find $\lim_{n\to\infty} (-1)^n/n$.

8. **(a)** Suppose S is the region under the graph of the function $y = f(x)$ and above the x-axis, where $a \le x \le b$. Explain how this area is approximated by rectangles, and write an expression for the area of S as a limit of sums.
   (b) Find the area under the graph of $f(x) = x^2$ and above the x-axis, between $x = 0$ and $x = 3$.

ANSWERS TO THE CONCEPT CHECK CAN BE FOUND AT THE BACK OF THE BOOK.

EXERCISES

1–6 ■ Estimating Limits Numerically and Graphically Use a table of values to estimate the value of the limit. Then use a graphing device to confirm your result graphically.

1. $\displaystyle\lim_{x\to 2} \frac{x - 2}{x^2 - 3x + 2}$
2. $\displaystyle\lim_{t\to -1} \frac{t + 1}{t^3 - t}$
3. $\displaystyle\lim_{x\to 0} \frac{2^x - 1}{x}$
4. $\displaystyle\lim_{x\to 0} \frac{\sin 2x}{x}$
5. $\displaystyle\lim_{x\to 1^+} \ln\sqrt{x - 1}$
6. $\displaystyle\lim_{x\to 0^-} \frac{\tan x}{|x|}$

7. **Limits from a Graph** The graph of f is shown in the figure. Find each limit, or explain why it does not exist.
   (a) $\displaystyle\lim_{x\to 2^+} f(x)$ **(b)** $\displaystyle\lim_{x\to -3^+} f(x)$
   (c) $\displaystyle\lim_{x\to -3^-} f(x)$ **(d)** $\displaystyle\lim_{x\to -3} f(x)$
   (e) $\displaystyle\lim_{x\to 4} f(x)$ **(f)** $\displaystyle\lim_{x\to\infty} f(x)$
   (g) $\displaystyle\lim_{x\to -\infty} f(x)$ **(h)** $\displaystyle\lim_{x\to 0} f(x)$

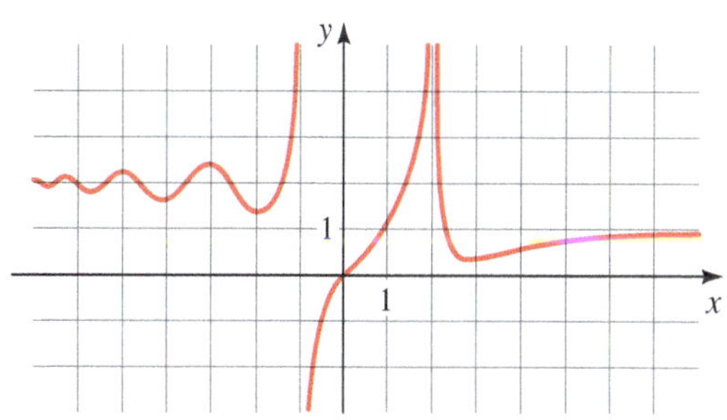

8. One-Sided Limits Let

$$f(x) = \begin{cases} 2 & \text{if } x < 1 \\ x^2 & \text{if } -1 \le x \le 2 \\ x + 2 & \text{if } x > 2 \end{cases}$$

Find each limit, or explain why it does not exist.

(a) $\lim_{x \to -1^-} f(x)$ **(b)** $\lim_{x \to -1^+} f(x)$

(c) $\lim_{x \to -1} f(x)$ **(d)** $\lim_{x \to 2^-} f(x)$

(e) $\lim_{x \to 2^+} f(x)$ **(f)** $\lim_{x \to 2} f(x)$

(g) $\lim_{x \to 0} f(x)$ **(h)** $\lim_{x \to 3} (f(x))^2$

9–20 ■ Finding Limits Evaluate the limit, if it exists. Use the Limit Laws when possible.

9. $\lim_{x \to 2} \dfrac{x + 1}{x - 3}$ **10.** $\lim_{t \to 1} (t^3 - 3t + 6)$

11. $\lim_{x \to 3} \dfrac{x^2 + x - 12}{x - 3}$ **12.** $\lim_{x \to -2} \dfrac{x^2 - 4}{x^2 + x - 2}$

13. $\lim_{u \to 0} \dfrac{(u + 1)^2 - 1}{u}$ **14.** $\lim_{z \to 9} \dfrac{\sqrt{z} - 3}{z - 9}$

15. $\lim_{x \to 3^-} \dfrac{x - 3}{|x - 3|}$ **16.** $\lim_{x \to 0} \left(\dfrac{1}{x} + \dfrac{2}{x^2 - 2x} \right)$

17. $\lim_{x \to \infty} \dfrac{2x}{x - 4}$ **18.** $\lim_{x \to \infty} \dfrac{x^2 + 1}{x^4 - 3x + 6}$

19. $\lim_{x \to \infty} \cos^2 x$ **20.** $\lim_{t \to -\infty} \dfrac{t^4}{t^3 - 1}$

21–24 ■ Derivative of a Function Find the derivative of the function at the given number.

21. $f(x) = 3x - 5$, at 4 **22.** $g(x) = 2x^2 - 1$, at -1

23. $f(x) = \sqrt{x}$, at 16 **24.** $f(x) = \dfrac{x}{x + 1}$, at 1

25–28 ■ Evaluating Derivatives **(a)** Find $f'(a)$. **(b)** Find $f'(2)$ and $f'(-2)$.

25. $f(x) = 6 - 2x$ **26.** $f(x) = x^2 - 3x$

27. $f(x) = \sqrt{x + 6}$ **28.** $f(x) = \dfrac{4}{x}$

29–30 ■ Equation of a Tangent Line Find an equation of the tangent line shown in the figure.

29.

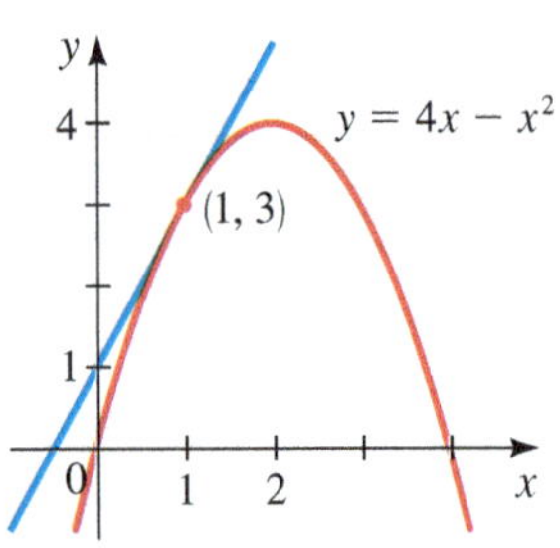

30.

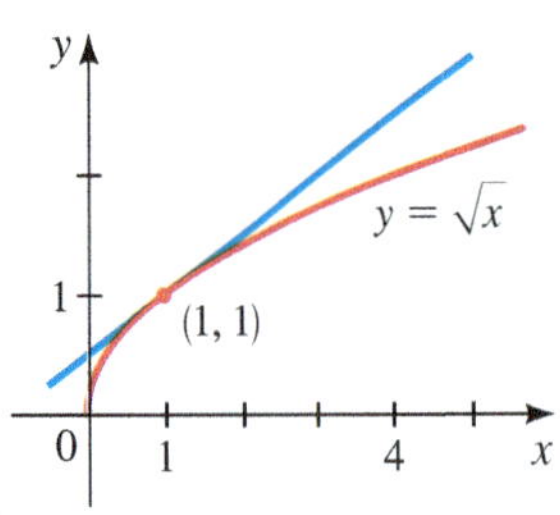

31–34 ■ Equation of a Tangent Line Find an equation of the line tangent to the graph of f at the given point.

31. $f(x) = 2x$, at $(3, 6)$ **32.** $f(x) = x^2 - 3$, at $(2, 1)$

33. $f(x) = \dfrac{1}{x}$, at $\left(2, \dfrac{1}{2}\right)$ **34.** $f(x) = \sqrt{x + 1}$, at $(3, 2)$

35. Velocity of a Dropped Stone A stone is dropped from the roof of a building 640 ft above the ground. The height of the stone (in ft) after t seconds is given by $h(t) = 640 - 16t^2$.

(a) Find the velocity of the stone when $t = 2$.

(b) Find the velocity of the stone when $t = a$.

(c) At what time t will the stone hit the ground?

(d) With what velocity will the stone hit the ground?

36. Instantaneous Rate of Change If a gas is confined in a fixed volume, then according to Boyle's Law the product of the pressure P and the temperature T is a constant. For a certain gas, $PT = 100$, where P is measured in lb/in^2 and T is measured in kelvins (K).

(a) Express P as a function of T.

(b) Find the instantaneous rate of change of P with respect to T when $T = 300$ K.

37–42 ■ Limit of a Sequence If the sequence is convergent, find its limit. If it is divergent, explain why.

37. $a_n = \dfrac{n}{5n + 1}$ **38.** $a_n = \dfrac{n^3}{n^3 + 1}$

39. $a_n = \dfrac{n(n + 1)}{2n^2}$ **40.** $a_n = \dfrac{n^3}{2n + 6}$

41. $a_n = \cos\left(\dfrac{n\pi}{2}\right)$ **42.** $a_n = \dfrac{10}{3^n}$

43–44 ■ Estimating Areas Using Rectangles Approximate the area of the shaded region under the graph of the given function by using the indicated rectangles. (The rectangles have equal width.)

43. $f(x) = \sqrt{x}$ **44.** $f(x) = 4x - x^2$

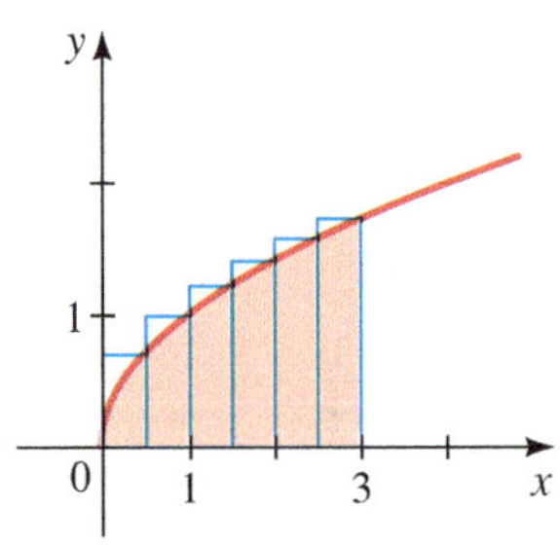

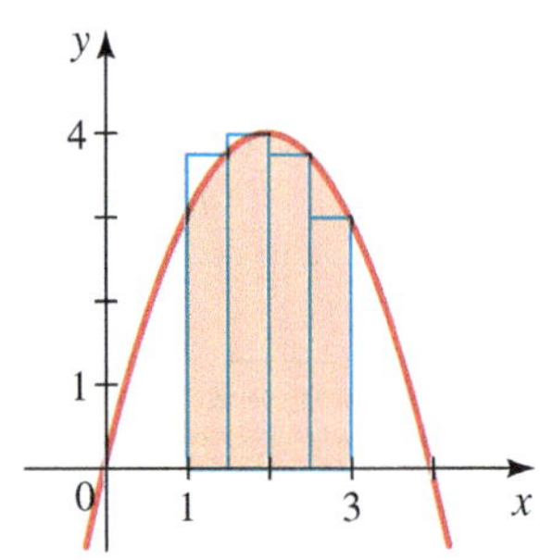

45–48 ■ Area Under a Curve Use the limit definition of area to find the area of the region that lies under the graph of f over the given interval.

45. $f(x) = 2x + 3$, $0 \le x \le 2$

46. $f(x) = x^2 + 1$, $0 \le x \le 3$

47. $f(x) = x^2 - x$, $1 \le x \le 2$

48. $f(x) = x^3$, $1 \le x \le 2$

1. **(a)** Use a table of values to estimate the limit

$$\lim_{x\to 0} \frac{x}{\sin 2x}$$

(b) Use a graphing calculator to confirm your answer graphically.

2. For the piecewise-defined function f whose graph is shown, find:

(a) $\lim_{x\to -1^-} f(x)$ **(b)** $\lim_{x\to -1^+} f(x)$ **(c)** $\lim_{x\to -1} f(x)$

(d) $\lim_{x\to 0^-} f(x)$ **(e)** $\lim_{x\to 0^+} f(x)$ **(f)** $\lim_{x\to 0} f(x)$

(g) $\lim_{x\to 2^-} f(x)$ **(h)** $\lim_{x\to 2^+} f(x)$ **(i)** $\lim_{x\to 2} f(x)$

$$f(x) = \begin{cases} 1 & \text{if } x < -1 \\ 0 & \text{if } x = -1 \\ x^2 & \text{if } -1 < x \le 2 \\ 4 - x & \text{if } 2 < x \end{cases}$$

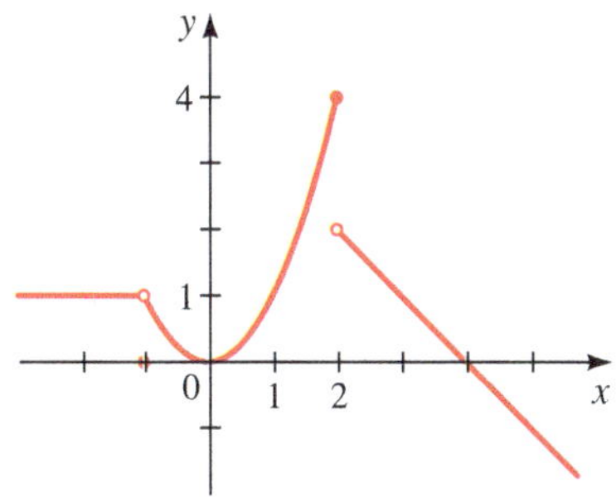

3. Evaluate the limit if it exists.

(a) $\lim_{x\to 2} \frac{x^2 + 2x - 8}{x - 2}$ **(b)** $\lim_{x\to 2} \frac{x^2 - 2x - 8}{x + 2}$ **(c)** $\lim_{x\to 2} \frac{1}{x - 2}$

(d) $\lim_{x\to 2} \frac{x - 2}{|x - 2|}$ **(e)** $\lim_{x\to 4} \frac{\sqrt{x} - 2}{x - 4}$ **(f)** $\lim_{x\to \infty} \frac{2x^2 - 4}{x^2 + x}$

4. Let $f(x) = x^2 - 2x$. Find:

(a) $f'(x)$ **(b)** $f'(-1), f'(1), f'(2)$

5. Find the equation of the line tangent to the graph of $f(x) = \sqrt{x}$ at the point where $x = 9$.

6. Find the limit of the sequence.

(a) $a_n = \frac{n}{n^2 + 4}$ **(b)** $a_n = \sec n\pi$

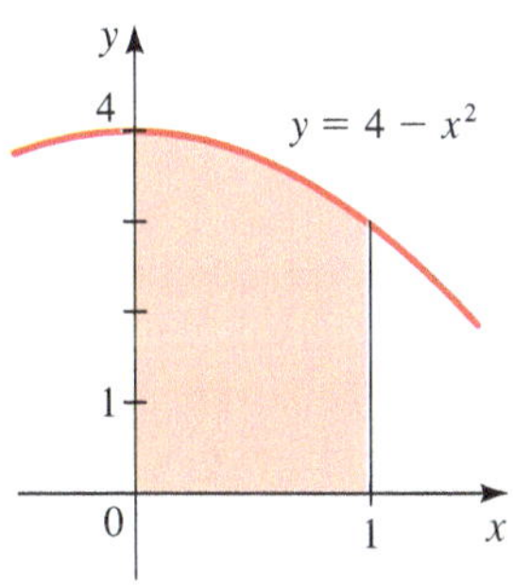

7. The region sketched in the figure in the margin lies under the graph of $f(x) = 4 - x^2$, above the interval $0 \le x \le 1$.

(a) Approximate the area of the region with five rectangles, equally spaced along the x-axis, using right endpoints to determine the heights of the rectangles.

(b) Use the limit definition of area to find the exact value of the area of the region.

A CUMULATIVE REVIEW TEST FOR CHAPTERS 12 AND 13 CAN BE FOUND AT THE BOOK COMPANION WEBSITE: **www.stewartmath.com.**

Interpretations of Area

The area under the graph of a function is used to model many quantities in physics, economics, engineering, and other fields. That is why the area problem is so important. Here, we will show how the concept of work (Section 9.2) is modeled by area. Several other applications are explored in the problems.

Recall that the work W done in moving an object is the product of the force F applied to the object and the distance d that the object moves:

$$W = Fd \qquad \text{work} = \text{force} \times \text{distance}$$

This formula is used if the force is *constant*. For example, suppose you are pushing a crate across a floor, moving along the positive x-axis from $x = a$ to $x = b$, and you apply a constant force $F = k$. The graph of F as a function of the distance x is shown in Figure 1(a). Notice that the work done is $W = Fd = k(b - a)$, which is the area under the graph of F (see Figure 1(b)).

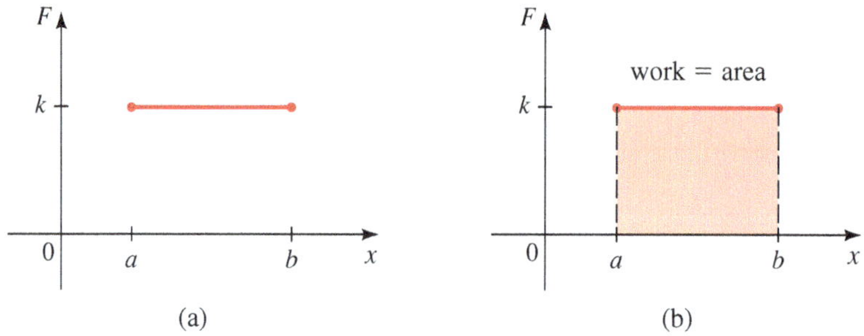

FIGURE 1 A constant force F

But what if the force is *not* constant? For example, suppose the force you apply to the crate varies with distance (you push harder at certain places than you do at others). More precisely, suppose that you push the crate along the x-axis in the positive direction, from $x = a$ to $x = b$, and at each point x between a and b you apply a force $f(x)$ to the crate. Figure 2 shows a graph of the force f as a function of the distance x.

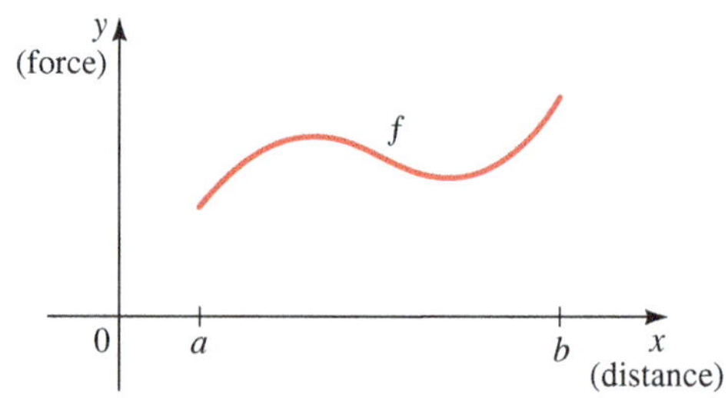

FIGURE 2 A variable force

How much work was done? We can't apply the formula for work directly because the force is not constant. So let's divide the interval $[a, b]$ into n subintervals with endpoints $x_0, x_1, \ldots, x_n$ and equal width Δx, as shown in Figure 3(a) on the next page. The force at the right endpoint of the interval $[x_{k-1}, x_k]$ is $f(x_k)$. If n is large, then Δx is small, so the values of f don't change very much over the interval $[x_{k-1}, x_k]$. In other words f is almost constant on the interval, so the work W_k that is done in moving the crate from x_{k-1} to x_k is approximately

$$W_k \approx f(x_k)\,\Delta x$$

Thus we can approximate the work done in moving the crate from $x = a$ to $x = b$ by

$$W \approx \sum_{k=1}^{n} f(x_k)\,\Delta x$$

It seems that this approximation becomes better as we make n larger (and so make the interval $[x_{k-1}, x_k]$ smaller). Therefore we define the work done in moving an object from a to b as the limit of this quantity as $n \to \infty$:

$$W = \lim_{n\to\infty} \sum_{k=1}^{n} f(x_k)\,\Delta x$$

Notice that this is precisely the area under the graph of f between $x = a$ and $x = b$ as defined in Section 13.5. See Figure 3(b).

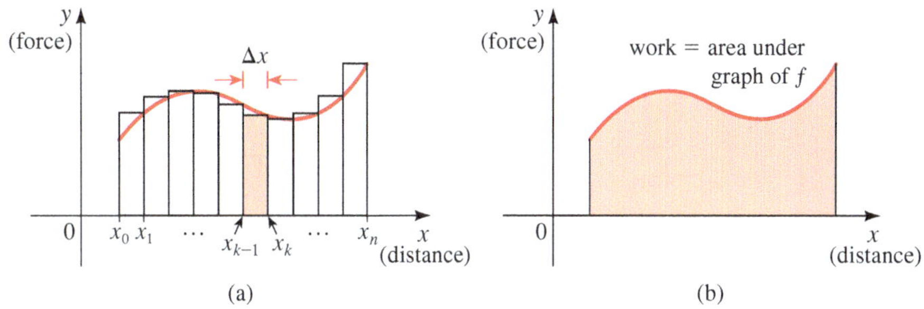

FIGURE 3 Approximating work

EXAMPLE ■ The Work Done by a Variable Force

A man pushes a crate along a straight path a distance of 18 ft. At a distance x from his starting point, he applies a force given by $f(x) = 340 - x^2$. Find the work done by the man.

SOLUTION The graph of f between $x = 0$ and $x = 18$ is shown in Figure 4. Notice how the force the man applies varies: He starts by pushing with a force of 340 lb but steadily applies less force.

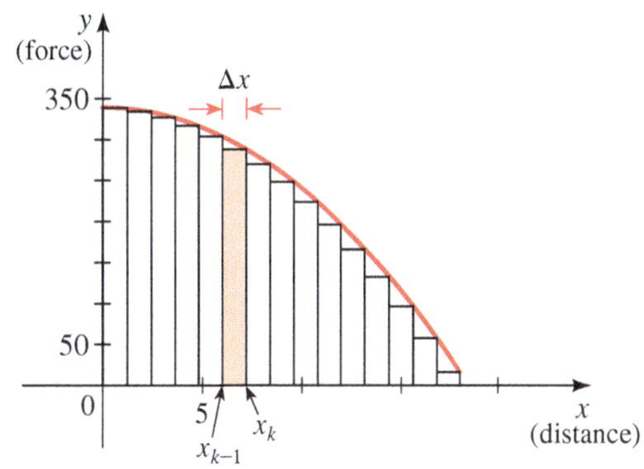

FIGURE 4

The work done is the area under the graph of f on the interval $[0, 18]$. To find this area, we start by finding the dimensions of the approximating rectangles at the nth stage.

Width: $\quad \Delta x = \dfrac{b-a}{n} = \dfrac{18-0}{n} = \dfrac{18}{n}$

Right endpoint: $\quad x_k = a + k\,\Delta x = 0 + k\left(\dfrac{18}{n}\right) = \dfrac{18k}{n}$

Height: $\quad f(x_k) = f\left(\dfrac{18k}{n}\right) = 340 - \left(\dfrac{18k}{n}\right)^2$

$$= 340 - \frac{324k^2}{n^2}$$

Thus according to the definition of work, we get

$$\begin{aligned} W &= \lim_{n\to\infty} \sum_{k=1}^{n} f(x_k)\,\Delta x = \lim_{n\to\infty} \sum_{k=1}^{n} \left(340 - \frac{324k^2}{n^2}\right)\left(\frac{18}{n}\right) \\ &= \lim_{n\to\infty} \left(\frac{18}{n}\sum_{k=1}^{n} 340 - \frac{(18)(324)}{n^3}\sum_{k=1}^{n} k^2\right) \\ &= \lim_{n\to\infty} \left(\frac{18}{n}340n - \frac{5832}{n^3}\left[\frac{n(n+1)(2n+1)}{6}\right]\right) \\ &= \lim_{n\to\infty} \left(6120 - 972\cdot\frac{n}{n}\cdot\frac{n+1}{n}\cdot\frac{2n+1}{n}\right) \\ &= 6120 - 972\cdot 1\cdot 1\cdot 2 = 4176 \end{aligned}$$

So the work done by the man in moving the crate is 4176 ft-lb. ■

PROBLEMS

1. Work Done by a Winch A motorized winch is being used to pull a felled tree to a logging truck. The motor exerts a force of $f(x) = 1500 + 10x - \frac{1}{2}x^2$ lb on the tree at the instant when the tree has moved x ft. The tree must be moved a distance of 40 ft, from $x = 0$ to $x = 40$. How much work is done by the winch in moving the tree?

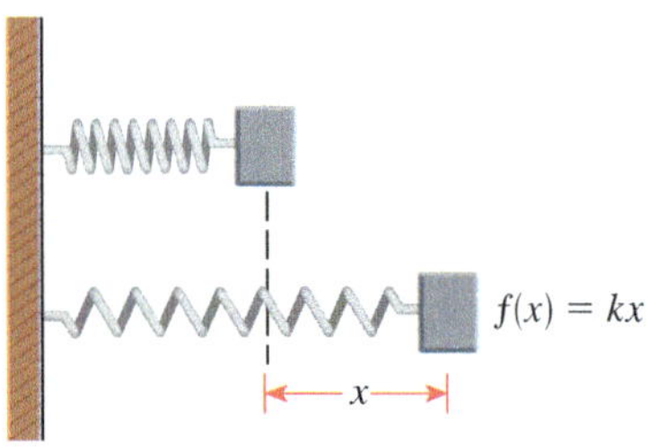

2. Work Done by a Spring Hooke's law states that when a spring is stretched, it pulls back with a force proportional to the amount of the stretch. The constant of proportionality is a characteristic of the spring known as the **spring constant**. Thus a spring with spring constant k exerts a force $f(x) = kx$ when it is stretched a distance x.

A certain spring has spring constant $k = 20$ lb/ft. Find the work done when the spring is pulled so that the amount by which it is stretched increases from $x = 0$ to $x = 2$ ft.

3. Force of Water As any diver knows, an object submerged in water experiences pressure, and as depth increases, so does the water pressure. At a depth of x ft, the water pressure is $p(x) = 62.5x$ lb/ft^2. To find the force exerted by the water on a surface, we multiply the pressure by the area of the surface:

$$\text{force} = \text{pressure} \times \text{area}$$

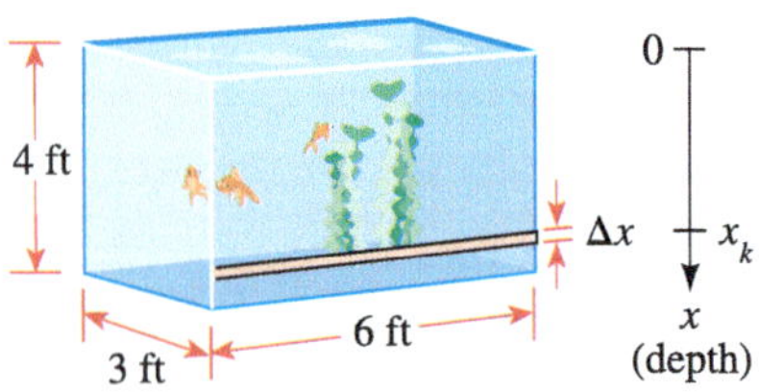

Suppose an aquarium that is 3 ft wide, 6 ft long, and 4 ft high is full of water. The bottom of the aquarium has area $3 \times 6 = 18$ ft^2, and it experiences water pressure of $p(4) = 62.5 \times 4 = 250$ lb/ft^2. Thus the total force exerted by the water on the bottom is $250 \times 18 = 4500$ lb.

The water also exerts a force on the sides of the aquarium, but this is not as easy to calculate because the pressure increases from top to bottom. To calculate the force on one of the 4 ft by 6 ft sides, we divide its area into n thin horizontal strips of width Δx, as shown in the figure. The area of each strip is

$$\text{length} \times \text{width} = 6\,\Delta x$$

If the bottom of the kth strip is at the depth x_k, then it experiences water pressure of approximately $p(x_k) = 62.5x_k$ lb/ft^2—the thinner the strip, the more accurate the approximation. Thus on each strip, the water exerts a force of

$$\text{pressure} \times \text{area} = 62.5x_k \times 6\,\Delta x = 375x_k\,\Delta x \text{ lb}$$

(a) Explain why the total force exerted by the water on the 4 ft by 6 ft sides of the aquarium is

$$\lim_{n\to\infty} \sum_{k=1}^{n} 375x_k\,\Delta x$$

where $\Delta x = 4/n$ and $x_k = 4k/n$.

(b) What area does the limit in part (a) represent?

(c) Evaluate the limit in part (a) to find the force exerted by the water on one of the 4 ft by 6 ft sides of the aquarium.

(d) Use the same technique to find the force exerted by the water on one of the 4 ft by 3 ft sides of the aquarium.

[*Note:* Engineers use the technique outlined in this problem to find the total force exerted on a dam by the water in the reservoir behind the dam.]

4. **Distance Traveled by a Car** Since distance = speed × time, it is easy to see that a car moving, say, at 70 mi/h for 5 h will travel a distance of 350 mi. But what if the speed varies, as it usually does in practice?

(a) Suppose the speed of a moving object at time t is $v(t)$. Explain why the distance traveled by the object between times $t = a$ and $t = b$ is the area under the graph of v between $t = a$ and $t = b$.

(b) The speed of a car t seconds after it starts moving is given by the function $v(t) = 6t + 0.1t^3$ ft/s. Find the distance traveled by the car from $t = 0$ to $t = 5$ s.

5. **Heating Capacity** If the outdoor temperature reaches a maximum of 90°F one day and only 80°F the next, then we would probably say that the first day was hotter than the second. Suppose, however, that on the first day the temperature was below 60°F for most of the day, reaching the high only briefly, whereas on the second day the temperature stayed above 75°F all the time. Now which day is the hotter one? To better measure how hot a particular day is, scientists use the concept of **heating degree-hour**. If the temperature is a constant D degrees for t hours, then the "heating capacity" generated over this period is Dt heating degree-hours.

$$\text{heating degree-hours} = \text{temperature} \times \text{time}$$

If the temperature is not constant, then the number of heating degree-hours equals the area under the graph of the temperature function over the time period in question.

(a) On a particular day the temperature (in °F) was modeled by the function $D(t) = 61 + \frac{6}{5}t - \frac{1}{25}t^2$, where t was measured in hours since midnight. How many heating degree-hours were experienced on this day, from $t = 0$ to $t = 24$?

(b) What was the maximum temperature on the day described in part (a)?

(c) On another day the temperature (in °F) was modeled by the function $E(t) = 50 + 5t - \frac{1}{4}t^2$. How many heating degree-hours were experienced on this day?

(d) What was the maximum temperature on the day described in part (c)?

(e) Which day was "hotter"?

APPENDIX A Geometry Review

In this appendix we review the concepts of similarity and congruence as well as the Pythagorean Theorem.

Congruent Triangles

In general, two geometric figures are congruent if they have the same shape and size. In particular, two line segments are congruent if they have the same length, and two angles are congruent if they have the same measure. For triangles we have the following definition.

CONGRUENT TRIANGLES

Two triangles are **congruent** if their vertices can be matched up so that corresponding sides and angles are congruent.

We write $\triangle ABC \cong \triangle PQR$ to mean that triangle ABC is congruent to triangle PQR and that the sides and angles correspond as follows.

$AB = PQ$	$\angle A = \angle P$
$BC = QR$	$\angle B = \angle Q$
$AC = PR$	$\angle C = \angle R$

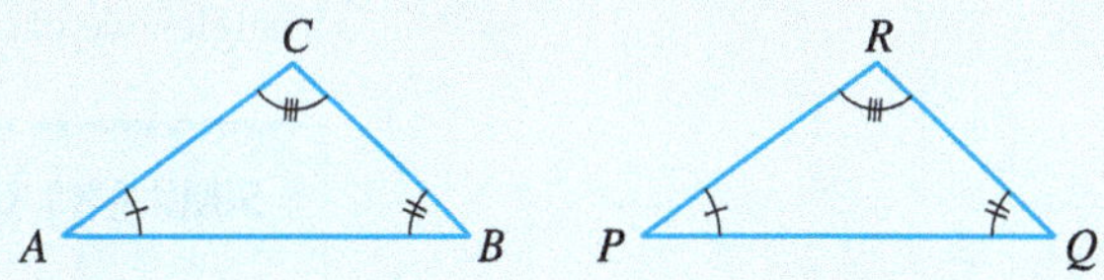

To prove that two triangles are congruent, we don't need to show that all six corresponding parts (side and angles) are congruent. For instance, if all three sides are congruent, then all three angles must also be congruent. You can easily see why the following properties lead to congruent triangles.

- **Side-Side-Side (SSS).** If each side of one triangle is congruent to the corresponding side of another triangle, then the two triangles are congruent. See Figure 1(a).
- **Side-Angle-Side (SAS).** If two sides and the included angle in one triangle are congruent to the corresponding sides and angle in another triangle, then the two triangles are congruent. See Figure 1(b).
- **Angle-Side-Angle (ASA).** If two angles and the included side in one triangle are congruent to the corresponding angles and side in another triangle, then the triangles are congruent. See Figure 1(c).

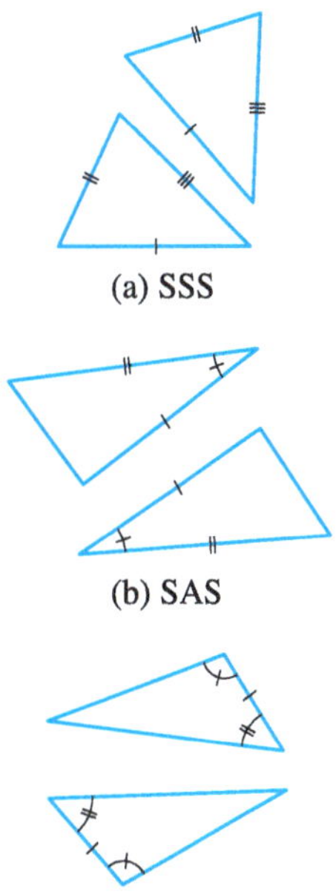

(a) SSS

(b) SAS

(c) ASA

FIGURE 1

EXAMPLE 1 ■ Congruent Triangles

(a) $\triangle ADB \cong \triangle CBD$ by SSS. **(b)** $\triangle ABE \cong \triangle CBD$ by SAS.

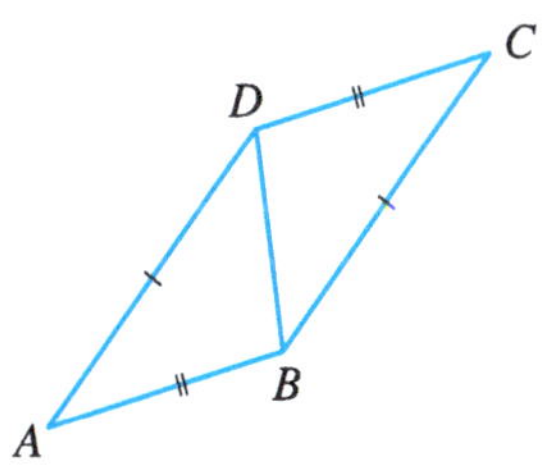

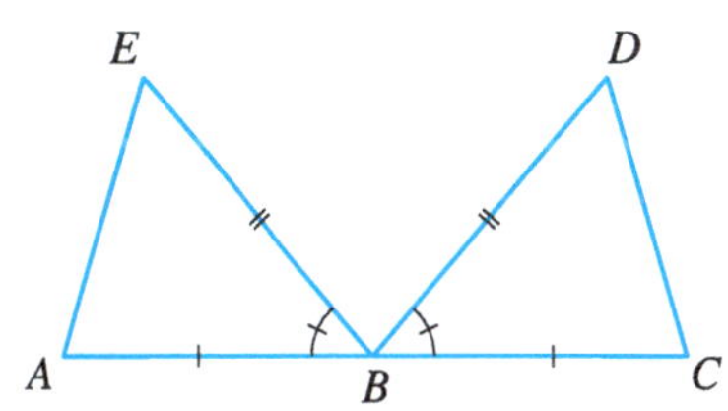

(c) $\triangle ABD \cong \triangle CBD$ by ASA.

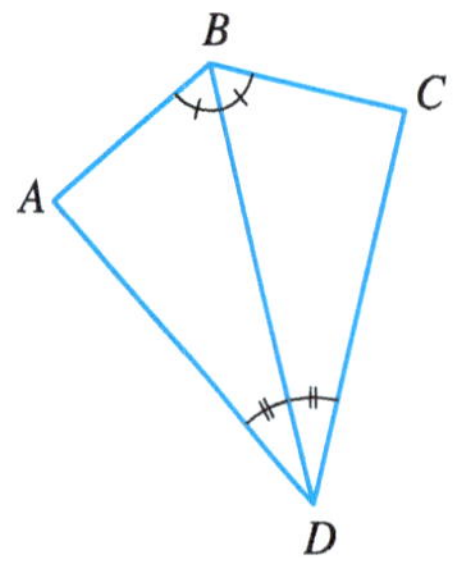

(d) These triangles are not necessarily congruent. "Side-side-angle" does *not* determine congruence.

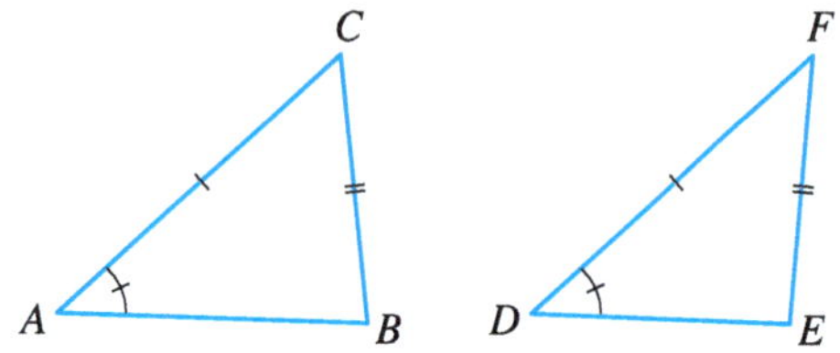

Similar Triangles

Two geometric figures are similar if they have the same shape, but not necessarily the same size. (See *Discovery Project: Similarity* referenced on page 484.) In the case of triangles we can define similarity as follows.

SIMILAR TRIANGLES

Two triangles are **similar** if their vertices can be matched up so that corresponding angles are congruent. In this case corresponding sides are proportional.

We write $\triangle ABC \sim \triangle PQR$ to mean that triangle ABC is similar to triangle PQR and that the following conditions hold.

The angles correspond as follows:

$$\angle A = \angle P, \quad \angle B = \angle Q, \quad \angle C = \angle R$$

The sides are proportional as follows:

$$\frac{AB}{PQ} = \frac{BC}{QR} = \frac{AC}{PR}$$

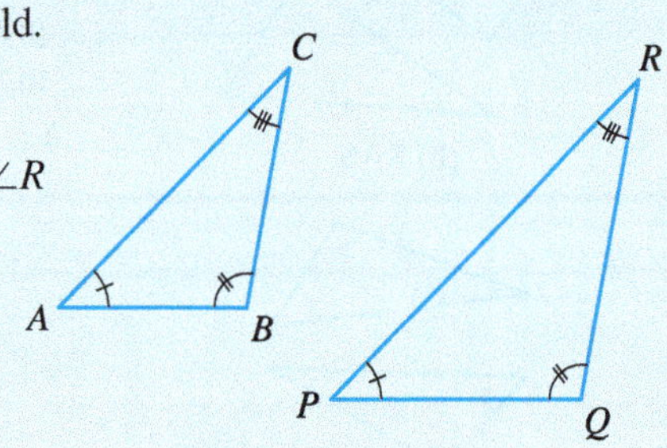

The sum of the angles in any triangle is 180°. So if we know two angles in a triangle, the third is determined. Thus to prove that two triangles are similar, we need only show that two angles in one are congruent to two angles in the other.

EXAMPLE 2 ■ Similar Triangles

Find all pairs of similar triangles in the figures.

(a)

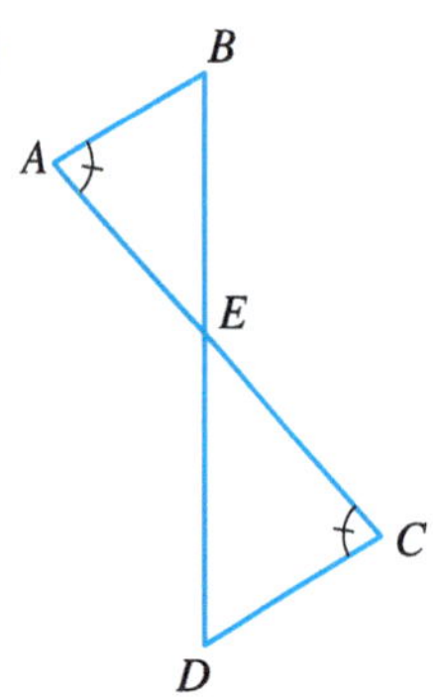

(b)

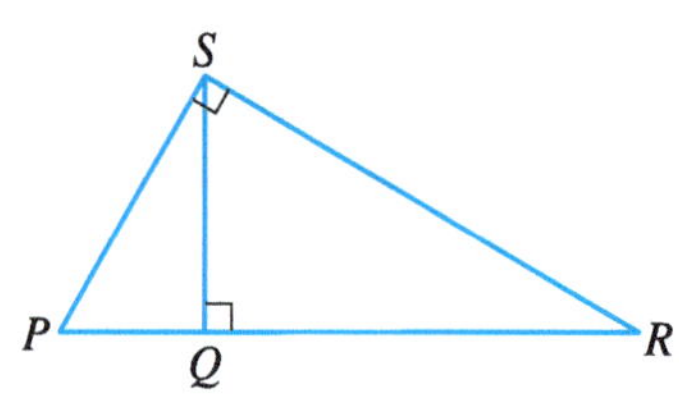

SOLUTION

(a) Since $\angle AEB$ and $\angle CED$ are opposite angles, they are equal. Thus

$$\triangle AEB \sim \triangle CED$$

(b) Since all triangles in the figure are right triangles, we have

$$\angle QSR + \angle QRS = 90°$$

$$\angle QSR + \angle QSP = 90°$$

Subtracting these equations we find that $\angle QSP = \angle QRS$. Thus

$$\triangle PQS \sim \triangle SQR \sim \triangle PSR$$

EXAMPLE 3 ■ Proportional Sides in Similar Triangles

Given that the triangles in the figure are similar, find the lengths x and y.

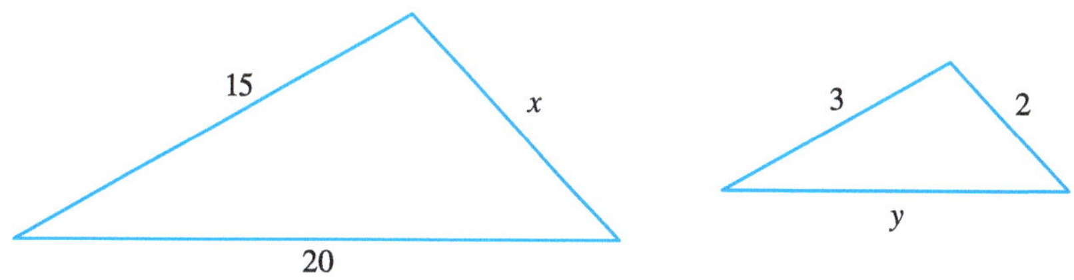

SOLUTION By similarity, we know that the lengths of corresponding sides in the triangles are proportional. First we find x.

$$\frac{x}{2} = \frac{15}{3} \qquad \text{Corresponding sides are proportional}$$

$$x = \frac{2 \cdot 15}{3} = 10 \qquad \text{Solve for } x$$

Now we find y.

$$\frac{15}{3} = \frac{20}{y} \qquad \text{Corresponding sides are proportional}$$

$$y = \frac{20 \cdot 3}{15} = 4 \qquad \text{Solve for } y$$

The Pythagorean Theorem

In a right triangle the side opposite the right angle is called the **hypotenuse**, and the other two sides are called the **legs**.

THE PYTHAGOREAN THEOREM

In a right triangle the square of the hypotenuse is equal to the sum of the squares of the legs. That is, in triangle ABC in the figure

$$a^2 + b^2 = c^2$$

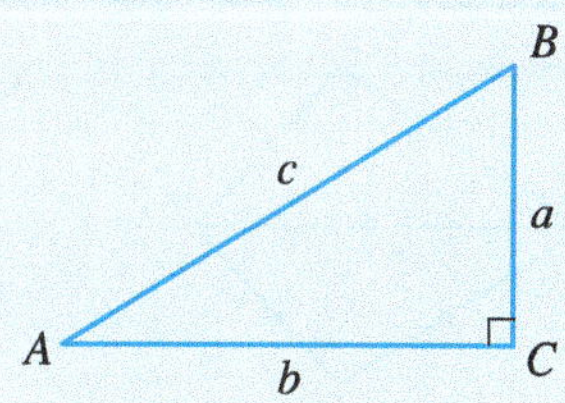

EXAMPLE 4 ■ Using the Pythagorean Theorem

Find the lengths x and y in the right triangles shown.

(a)

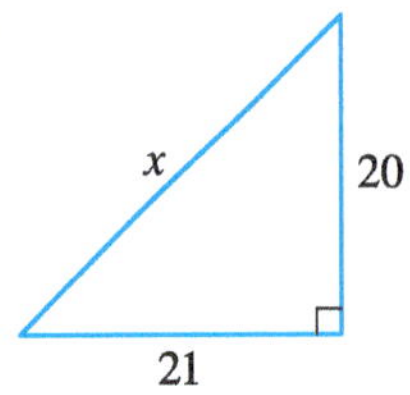

(b)

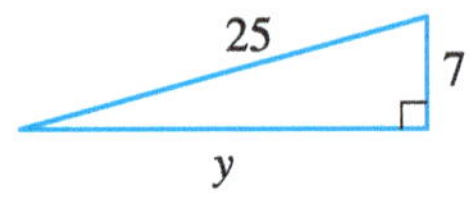

SOLUTION

(a) We use the Pythagorean Theorem with $a = 20$, and $b = 21$, and $c = x$. Then $x^2 = 20^2 + 21^2 = 841$. So $x = \sqrt{841} = 29$.

(b) We use the Pythagorean Theorem with $c = 25$, $a = 7$, and $b = y$. Then $25^2 = 7^2 + y^2$, so $y^2 = 25^2 - 7^2 = 576$. Thus $y = \sqrt{576} = 24$. ■

The converse of the Pythagorean Theorem is also true.

CONVERSE OF THE PYTHAGOREAN THEOREM

If the square of one side of a triangle is equal to the sum of the squares of the other two sides, then the triangle is a right triangle.

EXAMPLE 5 ■ Proving That a Triangle Is a Right Triangle

Prove that the triangle with sides of length 8, 15, and 17 is a right triangle.

SOLUTION You can check that $8^2 + 15^2 = 17^2$. So the triangle must be a right triangle by the converse of the Pythagorean Theorem. ■

A EXERCISES

1–4 ■ Congruent Triangles? Determine whether the pair of triangles is congruent. If so, state the congruence principle you are using.

1.

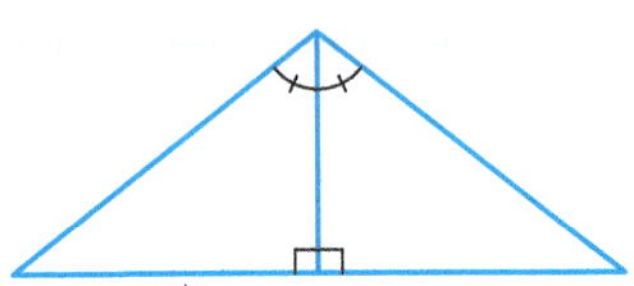

2.

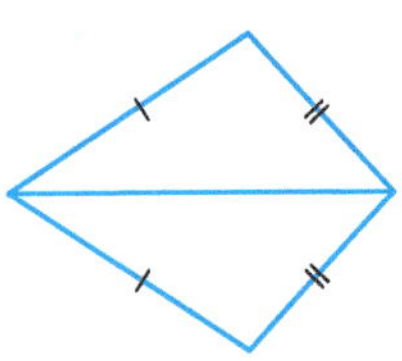

3.

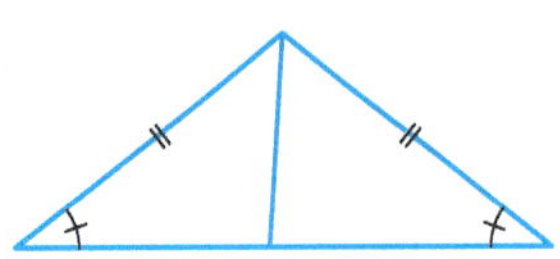

4.

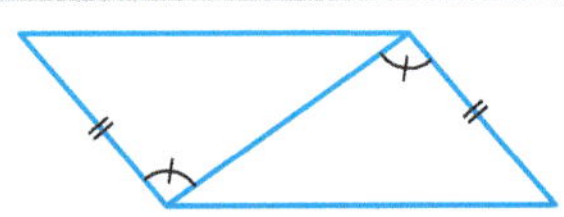

5–8 ■ Similar Triangles? Determine whether the pair of triangles is similar.

5.

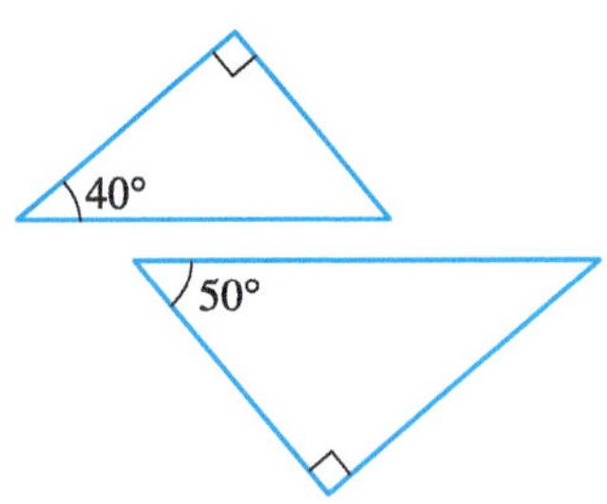

6.

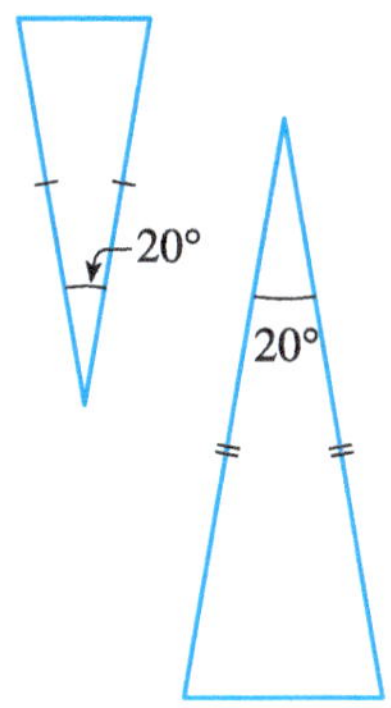

7.

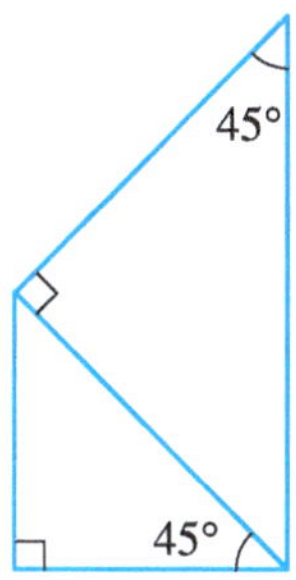

8.

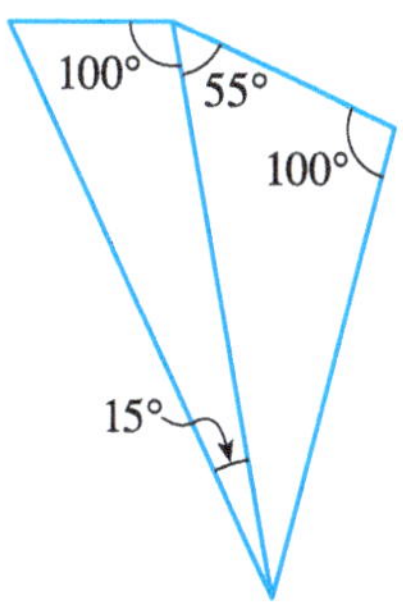

9–12 ■ Similar Triangles Given that the pair of triangles is similar, find the length(s) x and/or y.

9.

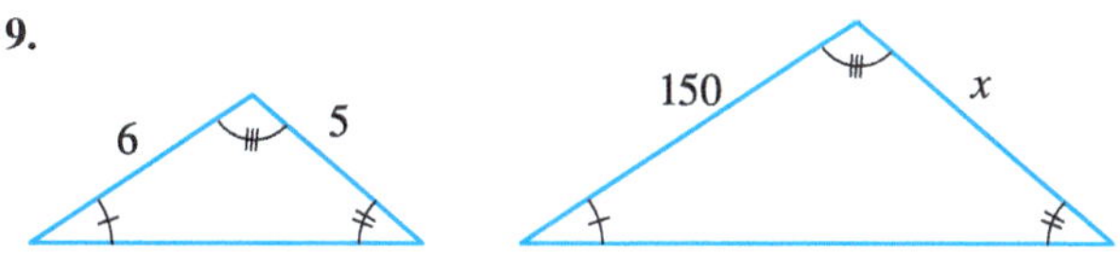

10.

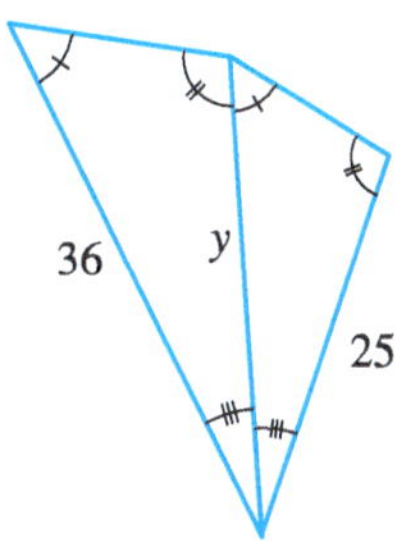

11.

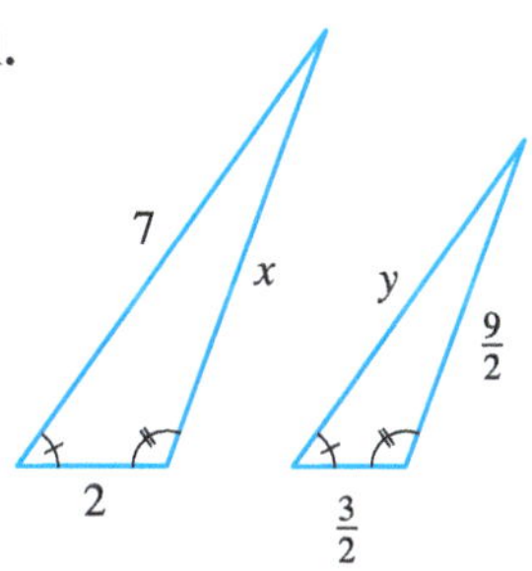

12.

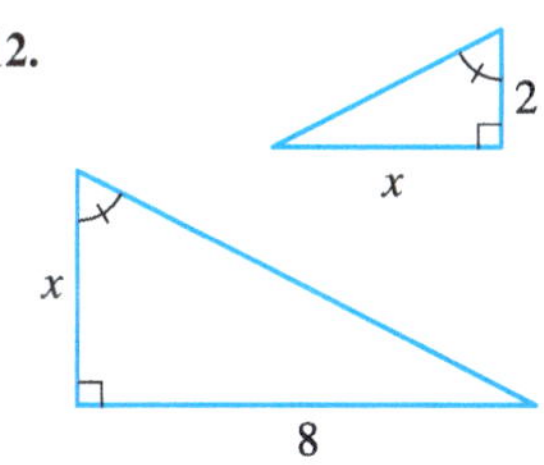

13–14 ■ Using Similarity Express x in terms of a, b, and c.

13.

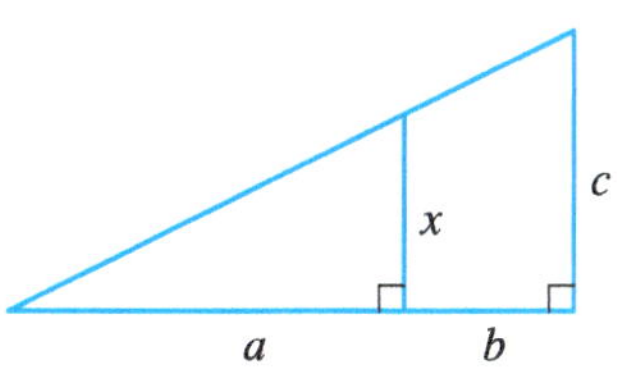

14.

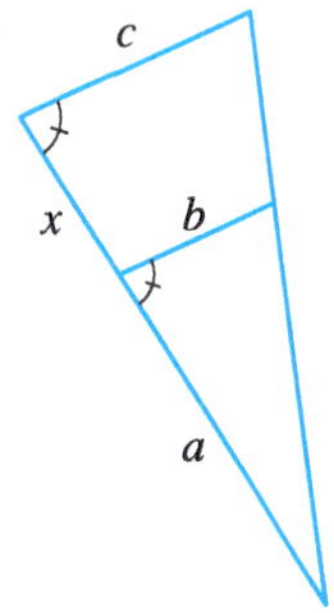

15. **Proving Similarity** In the figure CDEF is a rectangle. Prove that $\triangle ABC \sim \triangle AED \sim \triangle EBF$.

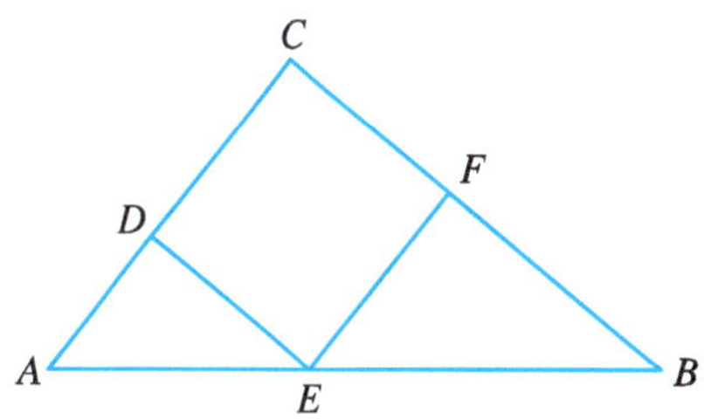

16. **Proving Similarity** In the figure $DEFG$ is a square. Prove the following:

(a) $\triangle ADG \sim \triangle GCF$

(b) $\triangle ADG \sim \triangle FEB$

(c) $AD \cdot EB = DG \cdot FE$

(d) $DE = \sqrt{AD \cdot EB}$

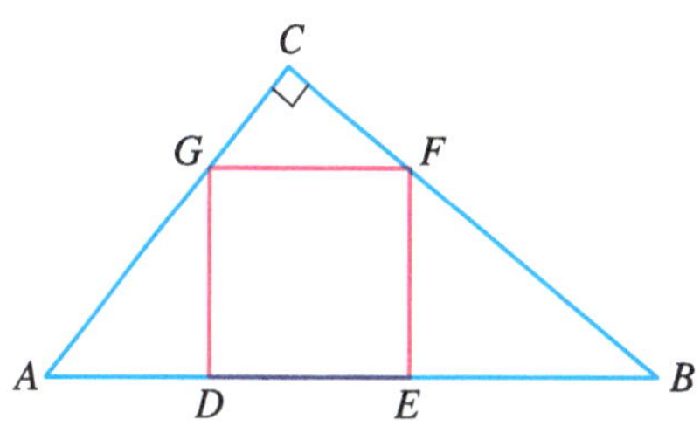

17–22 ■ Pythagorean Theorem In the given right triangle, find the side labeled x.

17.

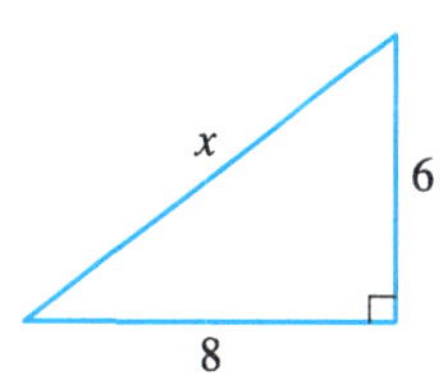

18.

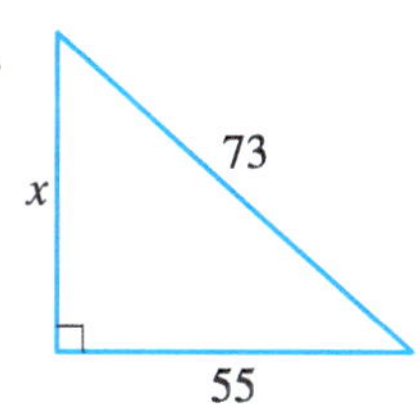

19.

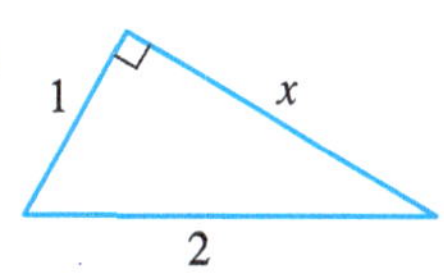

20.

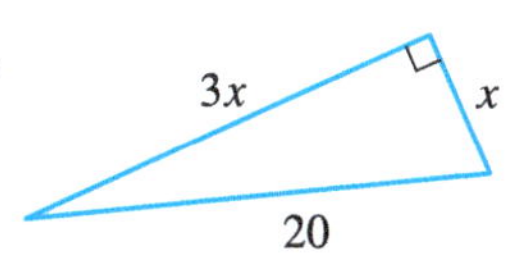

21. 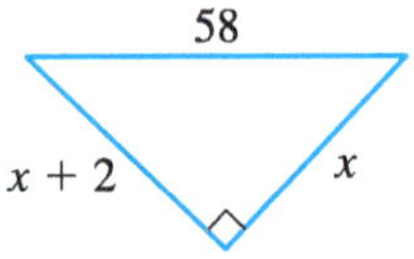

22. 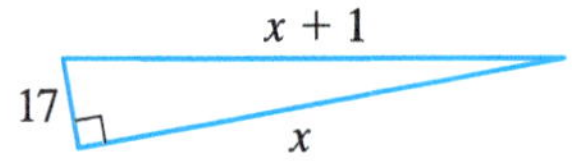

23–28 ■ Right Triangle? The lengths of the sides of a triangle are given. Determine whether the triangle is a right triangle.

23. 5, 12, 13

24. 15, 20, 25

25. 8, 10, 12

26. 6, 17, 18

27. 48, 55, 73

28. 13, 84, 85

29–32 ■ Pythagorean Theorem These exercises require the use of the Pythagorean Theorem.

29. One leg of a right triangle measures 11 cm. The hypotenuse is 1 cm longer than the other leg. Find the length of the hypotenuse.

30. The length of a rectangle is 1 ft greater than its width. Each diagonal is 169 ft long. Find the dimensions of the rectangle.

31. Each of the diagonals of a quadrilateral is 27 cm long. Two adjacent sides measure 17 cm and 21 cm. Is the quadrilateral a rectangle?

32. Find the height h of the right triangle ABC shown in the figure. [*Hint:* Find the area of triangle ABC in two different ways.]

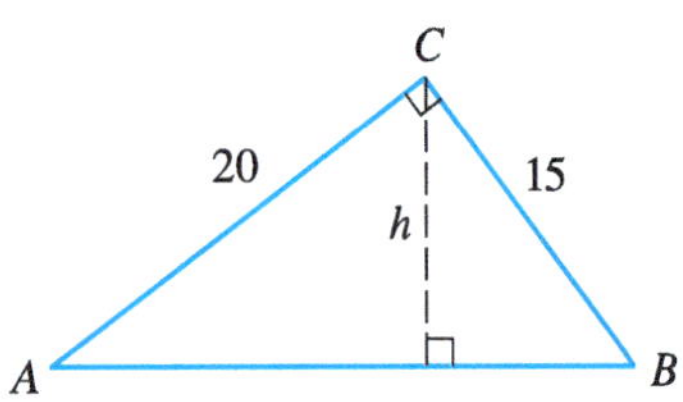

33. Diagonal of a Box Find the length of the diagonal of the rectangular box shown in the figure.

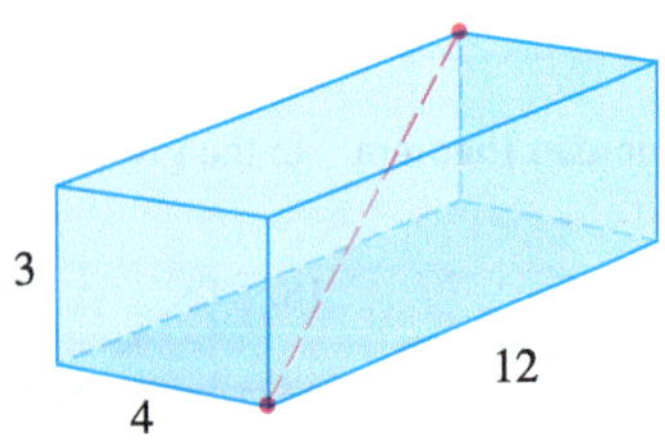

34. Pythagorean Triples If a, b, c are positive integers such that $a^2 + b^2 = c^2$, then (a, b, c) is called a **Pythagorean triple**.

(a) Let m and n be positive integers with $m > n$. Let $a = m^2 - n^2$, $b = 2mn$, and $c = m^2 + n^2$. Show that (a, b, c) is a Pythagorean triple.

(b) Use part (a) to find the rest of the Pythagorean triples in the table.

m	n	(a, b, c)
2	1	(3, 4, 5)
3	1	(8, 6, 10)
3	2	
4	1	
4	2	
4	3	
5	1	
5	2	
5	3	
5	4	

35. Finding a Length Two vertical poles, one 8 ft tall and the other 24 ft tall, have ropes stretched from the top of each to the base of the other (see the figure). How high above the ground is the point where the ropes cross? [*Hint:* Use similarity.]

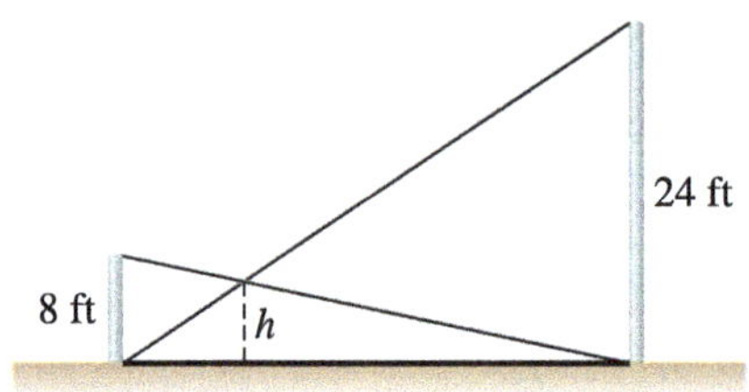

The following appendices can be found at www.stewartmath.com.

APPENDIX B: *Calculations and Significant Figures*
APPENDIX C: *Graphing with a Graphing Calculator*
APPENDIX D: *Using the TI-83/84 Graphing Calculator*

ANSWERS to Selected Exercises and Chapter Tests

PROLOGUE ■ PAGE P4

1. It can't go fast enough. **2.** 40% discount
3. 427, $3n + 1$ **4.** 57 min **5.** No, not necessarily
6. The same amount **7.** 2π
8. The North Pole is one such point; there are infinitely many others near the South Pole.

CHAPTER 1

SECTION 1.1 ■ PAGE 10

1. Answers may vary. Examples: **(a)** 2 **(b)** -3 **(c)** $\frac{3}{2}$ **(d)** $\sqrt{2}$ **2. (a)** ba; Commutative **(b)** $(a + b) + c$; Associative **(c)** $ab + ac$; Distributive **3. (a)** $\{x \mid 2 < x < 7\}$ **(b)** $(2, 7)$ **4.** absolute value; positive **5.** $|b - a|$; 7
6. (a) Yes **(b)** No **7. (a)** No **(b)** No
8. (a) Yes **(b)** Yes **9. (a)** 100 **(b)** 0, 100, -8 **(c)** $-1.5, 0, \frac{5}{2}, 2.71, 3.1\overline{4}, 100, -8$ **(d)** $\sqrt{7}, -\pi$
11. Commutative Property of Addition **13.** Associative Property of Addition **15.** Distributive Property **17.** Commutative Property of Multiplication **19.** $3 + x$ **21.** $4A + 4B$
23. $3x + 3y$ **25.** $8m$ **27.** $-5x + 10y$ **29. (a)** $\frac{17}{30}$ **(b)** $\frac{9}{20}$
31. (a) 3 **(b)** $\frac{13}{20}$ **33. (a)** $<$ **(b)** $>$ **(c)** $=$ **35. (a)** False **(b)** True **37. (a)** True **(b)** False **39. (a)** $x > 0$ **(b)** $t < 4$ **(c)** $a \geq \pi$ **(d)** $-5 < x < \frac{1}{3}$ **(e)** $|3 - p| \leq 5$
41. (a) $\{1, 2, 3, 4, 5, 6, 7, 8\}$ **(b)** $\{2, 4, 6\}$
43. (a) $\{1, 2, 3, 4, 5, 6, 7, 8, 9, 10\}$ **(b)** $\{7\}$
45. (a) $\{x \mid x \leq 5\}$ **(b)** $\{x \mid -1 < x < 4\}$

47. $-3 < x < 0$

49. $2 \leq x < 8$

51. $x \geq 2$

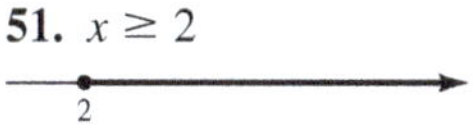

53. $(-\infty, 1]$

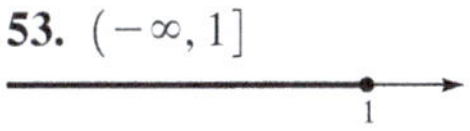

55. $(-2, 1]$ 

57. $(-1, \infty)$ 

59. (a) $[-3, 5]$ **(b)** $(-3, 5]$

61. 

63. 

65. 

67. (a) 100 **(b)** 73 **69. (a)** 2 **(b)** -1 **71. (a)** 12 **(b)** 5
73. 5 **75. (a)** 15 **(b)** 24 **(c)** $\frac{67}{40}$ **77. (a)** $\frac{7}{9}$ **(b)** $\frac{13}{45}$ **(c)** $\frac{19}{33}$
79. $\pi - 3$ **81.** $b - a$ **83. (a)** $-$ **(b)** $+$ **(c)** $+$ **(d)** $-$
85. Distributive Property
87. (a) Yes, no **(b)** 6 ft

SECTION 1.2 ■ PAGE 21

1. (a) 5^6 **(b)** base, exponent **2. (a)** add, 3^9 **(b)** subtract, 3^3
3. (a) $5^{1/3}$ **(b)** $\sqrt{5}$ **(c)** No **4.** $(4^{1/2})^3 = 8, (4^3)^{1/2} = 8$
5. $\frac{1}{\sqrt{3}} = \frac{1}{\sqrt{3}} \cdot \frac{\sqrt{3}}{\sqrt{3}} = \frac{\sqrt{3}}{3}$ **6.** $\frac{2}{3}$ **7. (a)** No **(b)** Yes **8. (a)** No **(b)** No **(c)** No **(d)** No **9.** $3^{-1/2}$ **11.** $\sqrt[3]{4^2}$
13. $5^{3/5}$ **15.** $\sqrt[5]{a^2}$ **17. (a)** -64 **(b)** 64 **(c)** $-\frac{27}{25}$
19. (a) $\frac{1}{2}$ **(b)** $\frac{1}{8}$ **(c)** $\frac{9}{4}$ **21. (a)** 625 **(b)** 25 **(c)** 64
23. (a) $6\sqrt[3]{2}$ **(b)** $\dfrac{\sqrt{2}}{3}$ **(c)** $\dfrac{3\sqrt{3}}{2}$
25. (a) $3\sqrt{5}$ **(b)** 4 **(c)** $6\sqrt[3]{2}$ **27. (a)** $2\sqrt{11}$ **(b)** 4 **(c)** $\frac{1}{4}$ **29. (a)** x^7 **(b)** $8y^6$ **(c)** y^5 **31. (a)** $\dfrac{1}{x^2}$ **(b)** $\dfrac{1}{w}$ **(c)** x^6 **33. (a)** a^6 **(b)** a^{18} **(c)** $\dfrac{5x^9}{8}$ **35. (a)** $6x^3y^5$ **(b)** $\dfrac{25w^4}{z}$ **37. (a)** $\dfrac{x^7}{y}$ **(b)** $\dfrac{a^9}{8b^6}$ **39. (a)** $\dfrac{a^{19}b}{c^9}$ **(b)** $\dfrac{v^{10}}{u^{11}}$
41. (a) $\dfrac{4a^8}{b^9}$ **(b)** $\dfrac{125}{x^6y^3}$ **43. (a)** $\dfrac{b^3}{3a}$ **(b)** $\dfrac{s^3}{q^7r^4}$
45. (a) $|x|$ **(b)** $2x^2$ **47. (a)** $2ab\sqrt[6]{b}$ **(b)** $4a^2\sqrt[3]{b^2}$
49. (a) $7\sqrt{2}$ **(b)** $9\sqrt{3}$ **51. (a)** $(3a + 1)\sqrt{a}$ **(b)** $(4 + x^2)\sqrt{x}$ **53. (a)** $9\sqrt{x^2 + 1}$ **(b)** $6\sqrt{x^2 + y^2}$
55. (a) 2 **(b)** -2 **(c)** $\frac{1}{3}$ **57. (a)** 4 **(b)** $\frac{3}{2}$ **(c)** $\frac{8}{27}$
59. (a) 5 **(b)** $\sqrt[5]{3}$ **(c)** 4 **61. (a)** x^2 **(b)** y^2
63. (a) $w^{5/3}$ **(b)** $8a^{13/4}$ **65. (a)** $4a^4b$ **(b)** $8a^9b^{12}$
67. (a) $4st^4$ **(b)** $\dfrac{4x}{y}$ **69. (a)** $\dfrac{x^4}{y}$ **(b)** $\dfrac{8y^8}{x^2}$
71. (a) $x^{3/2}$ **(b)** $x^{6/5}$ **73. (a)** $y^{3/2}$ **(b)** $10x^{7/12}$
75. (a) $2st^{11/6}$ **(b)** x **77. (a)** $y^{1/2}$ **(b)** $\dfrac{4u}{v^2}$ **79. (a)** $\dfrac{\sqrt{6}}{6}$ **(b)** $\dfrac{\sqrt{6}}{2}$ **(c)** $\dfrac{9\sqrt[4]{8}}{2}$ **81. (a)** $\dfrac{\sqrt{5x}}{5x}$ **(b)** $\dfrac{\sqrt{5x}}{5}$ **(c)** $\dfrac{\sqrt[5]{x^2}}{x}$
83. (a) 6.93×10^7 **(b)** 7.2×10^{12} **(c)** 2.8536×10^{-5} **(d)** 1.213×10^{-4} **85. (a)** 319,000 **(b)** 272,100,000 **(c)** 0.00000002670 **(d)** 0.000000009999
87. (a) 5.9×10^{12} mi **(b)** 4×10^{-13} cm **(c)** 3.3×10^{19} molecules **89.** 1.3×10^{-20}
91. 1.429×10^{19} **93.** 7.4×10^{-14} **95. (a)** Negative **(b)** Positive **(c)** Negative **(d)** Negative **(e)** Positive **(f)** Negative **97.** 2.5×10^{13} mi **99.** 1.3×10^{21} L
101. 4.03×10^{27} molecules **103. (a)** 28 mi/h **(b)** 167 ft

SECTION 1.3 ■ PAGE 33

1. (a) 3; $2x^5, 6x^4, 4x^3$ **(b)** $2x^3$; $2x^3(x^2 + 3x + 2)$
2. 10, 7; 2, 5; $(x + 2)(x + 5)$ **3.** x^2; $x^2(3x + 1)$
4. $A^2 + 2AB + B^2$; $4x^2 + 12x + 9$ **5.** $A^2 - B^2$; $25 - x^2$
6. $(A + B)(A - B)$; $(2x - 5)(2x + 5)$ **7.** $(A + B)^2$; $(x + 5)^2$
8. (a) No **(b)** Yes **(c)** Yes **(d)** Yes **9.** Binomial; $5x^3$, 6; 3
11. Monomial; -8; 0 **13.** Four terms; $-x^4, x^3, -x^2, x$; 4
15. $7x + 5$ **17.** $x^2 + 2x - 3$ **19.** $5x^3 + 3x^2 - 10x - 2$
21. $9x + 103$ **23.** $-t^4 + t^3 - t^2 - 10t + 5$
25. $21t^2 - 26t + 8$ **27.** $6x^2 + 7x - 5$ **29.** $2x^2 + 5xy - 3y^2$
31. $25x^2 + 10x + 1$ **33.** $4u^2 + 4uv + v^2$

35. $4x^2 + 12xy + 9y^2$ **37.** $x^2 - 36$ **39.** $9x^2 - 16$
41. $x - 4$ **43.** $y^3 + 6y^2 + 12y + 8$
45. $-8r^3 + 12r^2 - 6r + 1$ **47.** $x^3 + 4x^2 + 7x + 6$
49. $2x^3 - 7x^2 + 7x - 5$ **51.** $x\sqrt{x} - x$ **53.** $y^2 + y$
55. $x^4 - a^4$ **57.** $a - b^2$ **59.** $-x^4 + x^2 - 2x + 1$
61. $4x^2 + 4xy + y^2 - 9$ **63.** $x(-2x^2 + 1)$
65. $(y - 6)(y + 9)$ **67.** $xy(2x - 6y + 3)$
69. $(x + 7)(x + 1)$ **71.** $(2x - 5)(4x + 3)$
73. $(3x - 1)(x - 5)$ **75.** $(3x + 4)(3x + 8)$
77. $(3a - 4)(3a + 4)$ **79.** $(3x + y)(9x^2 - 3xy + y^2)$
81. $(2s - 5t)(4s^2 + 10st + 25t^2)$ **83.** $(x + 6)^2$
85. $(x + 4)(x^2 + 1)$ **87.** $(x^2 + 1)(5x + 1)$
89. $(x + 1)(x^2 + 1)$ **91.** $\sqrt{x}(x - 1)(x + 1)$
93. $x^{-3/2}(1 + x)^2$ **95.** $(x^2 + 1)^{-1/2}(x^2 + 3)$
97. $6x(2x^2 + 3)$ **99.** $(x - 4)(x + 2)$ **101.** $(2x + 3)(x + 1)$
103. $9(x - 5)(x + 1)$ **105.** $(7 - 2y)(7 + 2y)$
107. $(t - 3)^2$ **109.** $(2x + y)^2$ **111.** $4ab$
113. $(x - 1)(x + 1)(x - 3)(x + 3)$
115. $(2x - 5)(4x^2 + 10x + 25)$ **117.** $x(x + 1)^2$
119. $x^2y^3(x + y)(x - y)$ **121.** $(x - 2)(x + 2)(3x - 1)$
123. $3(x - 1)(x + 2)$ **125.** $(a - 1)(a + 1)(a - 2)(a + 2)$
127. $2(x^2 + 4)^4(x - 2)^3(7x^2 - 10x + 8)$
129. $(x^2 + 3)^{-4/3}(\frac{1}{3}x^2 + 3)$
133. $(a + b + c)(a + b - c)(a - b + c)(-a + b + c)$

SECTION 1.4 ■ PAGE 42

1. (a), (c) **2.** numerator; denominator; $\dfrac{x + 1}{x + 3}$

3. numerators; denominators; $\dfrac{2x}{x^2 + 4x + 3}$

4. (a) 3 **(b)** $x(x + 1)^2$ **(c)** $\dfrac{-2x^2 + 1}{x(x + 1)^2}$

5. (a) Yes **(b)** No **6. (a)** Yes **(b)** No
7. $\mathbb{R}$ **9.** $\{x \mid x \neq 3\}$ **11.** $\{x \mid x \geq -3\}$

13. $\{x \mid x \neq -1, 2\}$ **15.** $\dfrac{2x + 1}{2(x - 3)}$ **17.** $\dfrac{1}{x + 2}$

19. $\dfrac{x + 2}{x + 5}$ **21.** $\dfrac{y}{y - 1}$ **23.** $\dfrac{x(2x + 3)}{2x - 3}$ **25.** $\dfrac{1}{4(x - 2)}$

27. $\dfrac{x - 3}{x + 2}$ **29.** $\dfrac{1}{t^2 + 9}$ **31.** $\dfrac{x + 4}{x + 1}$ **33.** $\dfrac{x + 5}{(2x + 3)(x + 4)}$

35. $x^2(x + 1)$ **37.** $\dfrac{x}{yz}$ **39.** $\dfrac{x + 4}{x + 3}$ **41.** $\dfrac{3x + 7}{(x - 3)(x + 5)}$

43. $\dfrac{2x + 5}{(x + 1)(x + 2)}$ **45.** $\dfrac{2(5x - 9)}{(2x - 3)^2}$ **47.** $\dfrac{u^2 + 3u + 1}{u + 1}$

49. $\dfrac{2x + 1}{x^2(x + 1)}$ **51.** $\dfrac{2x + 7}{(x + 3)(x + 4)}$ **53.** $\dfrac{x - 2}{(x + 3)(x - 3)}$

55. $\dfrac{5x - 6}{x(x - 1)}$ **57.** $\dfrac{-5}{(x + 1)(x + 2)(x - 3)}$ **59.** $\dfrac{x + 1}{1 - 2x}$

61. $\dfrac{x + 3}{x + 1}$ **63.** $\dfrac{2}{(x - 1)(x + 3)}$ **65.** $\dfrac{x^2(y - 1)}{y^2(x - 1)}$

67. $-xy$ **69.** $\dfrac{y - x}{xy}$ **71.** $\dfrac{1}{1 - x}$ **73.** $-\dfrac{1}{(1 + x)(1 + x + h)}$

75. $-\dfrac{2x + h}{x^2(x + h)^2}$ **77.** $\dfrac{1}{\sqrt{1 - x^2}}$ **79.** $\dfrac{(x + 2)^2(x - 13)}{(x - 3)^3}$

81. $\dfrac{x + 2}{(x + 1)^{3/2}}$ **83.** $\dfrac{2x + 3}{(x + 1)^{4/3}}$ **85.** $\dfrac{\sqrt{3} + 5}{22}$

87. $\dfrac{2(\sqrt{7} - \sqrt{2})}{5}$ **89.** $\dfrac{y\sqrt{3} - y\sqrt{y}}{3 - y}$ **91.** $\dfrac{-4}{3(1 + \sqrt{5})}$

93. $\dfrac{r - 2}{5(\sqrt{r} - \sqrt{2})}$ **95.** $\dfrac{1}{\sqrt{x^2 + 1} + x}$ **97. (a)** $\dfrac{R_1R_2}{R_1 + R_2}$
(b) $\frac{20}{3} \approx 6.7$ ohms

SECTION 1.5 ■ PAGE 55

1. (a) Yes **(b)** Yes **(c)** No **2. (a)** Take (positive and negative) square roots of both sides. **(b)** Subtract 5 from both sides. **(c)** Subtract 2 from both sides.
3. (a) Factor into $(x + 1)(x - 5)$, and use the Zero-Product Property. **(b)** Add 5 to each side, then complete the square by adding 4 to both sides. **(c)** Insert coefficients into the Quadratic Formula. **4. (a)** 0, 4 **(b)** factor **5. (a)** $\sqrt{2x} = -x$
(b) $2x = x^2$ **(c)** 0, 2 **(d)** 0
6. quadratic; $x + 1$; $W^2 - 5W + 6 = 0$
7. $x(x + 2)$; $3(x + 2) + 5x = 2x(x + 2)$
8. square; $(2x + 1)^2 = x + 1$ **9. (a)** No **(b)** Yes
11. (a) Yes **(b)** No **13.** 4 **15.** 18 **17.** $\frac{3}{5}$ **19.** $-\frac{27}{4}$
21. $-\frac{3}{4}$ **23.** 30 **25.** $-\frac{1}{3}$ **27.** $\frac{13}{3}$ **29.** -2

31. $R = \dfrac{PV}{nT}$ **33.** $w = \dfrac{P - 2l}{2}$ **35.** $x = \dfrac{2d - b}{a - 2c}$

37. $x = \dfrac{1 - a}{a^2 - a - 1}$ **39.** $r = \pm\sqrt{\dfrac{3V}{\pi h}}$

41. $b = \pm\sqrt{c^2 - a^2}$ **43.** $t = \dfrac{-v_0 \pm \sqrt{v_0^2 + 2gh}}{g}$

45. $-4, 3$ **47.** 3, 4 **49.** $-\frac{3}{2}, \frac{5}{2}$ **51.** $-2, \frac{1}{3}$ **53.** ± 2
55. $-2, 7$ **57.** $-1 \pm \sqrt{6}$ **59.** $3 \pm 2\sqrt{5}$ **61.** $-2 \pm \frac{\sqrt{14}}{2}$
63. $0, \frac{1}{4}$ **65.** $-3, 5$ **67.** 6, 7 **69.** $-\frac{3}{2}, 1$ **71.** $-1 \pm \frac{2\sqrt{6}}{3}$

73. $-\frac{2}{3}$ **75.** $-\frac{9}{2}, \frac{1}{2}$ **77.** No solution **79.** $\dfrac{8 \pm \sqrt{14}}{10}$ **81.** 2

83. 1 **85.** No real solution **87.** $-50, 100$ **89.** $-\frac{7}{5}, 2$
91. $-\frac{3}{2}, 5$ **93.** 7 **95.** 4 **97.** 4 **99.** 3 **101.** 8
103. $\pm 2\sqrt{2}, \pm\sqrt{5}$ **105.** No real solution
107. $\pm 3\sqrt{3}, \pm 2\sqrt{2}$ **109.** $-1, 0, 3$ **111.** 27, 729
113. $-2, -\frac{4}{3}$ **115.** 3.99, 4.01 **117.** $-\frac{1}{2}$ **119.** 20

121. $-3, \dfrac{1 \pm \sqrt{13}}{2}$ **123.** $\pm\sqrt{a}, \pm 2\sqrt{a}$ **125.** $\sqrt{a^2 + 36}$

127. 4.24 s **129. (a)** After 1 s and $1\frac{1}{2}$ s **(b)** Never **(c)** 25 ft
(d) After $1\frac{1}{4}$ s **(e)** After $2\frac{1}{2}$ s **131. (a)** 0.00055, 12.018 m
(b) 234.375 kg/m^3 **133. (a)** After 17 years, on Jan. 1, 2019
(b) After 18.612 years, on Aug. 12, 2020 **135.** 50
137. 132.6 ft

SECTION 1.6 ■ PAGE 63

1. -1 **2.** 3, 4 **3. (a)** $3 - 4i$ **(b)** $9 + 16 = 25$ **4.** $3 - 4i$
5. Yes **6.** Yes **7.** Real part 5, imaginary part -7
9. Real part $-\frac{2}{3}$, imaginary part $-\frac{5}{3}$ **11.** Real part 3, imaginary part 0 **13.** Real part 0, imaginary part $-\frac{2}{3}$ **15.** Real part $\sqrt{3}$, imaginary part 2 **17.** $3 + 7i$ **19.** $1 - 10i$ **21.** $3 + 5i$
23. $2 - 2i$ **25.** $-19 + 4i$ **27.** $-4 + 8i$ **29.** $30 + 10i$
31. $27 - 8i$ **33.** 29 **35.** $-21 + 20i$ **37.** $-i$ **39.** $\frac{8}{5} + \frac{1}{5}i$
41. $-4 + 2i$ **43.** $2 - \frac{4}{3}i$ **45.** $-i$ **47.** $-i$ **49.** $243i$

51. 1 **53.** $7i$ **55.** -6 **57.** $(3+\sqrt{5})+(3-\sqrt{5})i$
59. 2 **61.** $\pm 7i$ **63.** $\frac{1}{2} \pm \frac{\sqrt{7}}{2}i$ **65.** $-\frac{3}{2} \pm \frac{\sqrt{19}}{2}i$
67. $-\frac{1}{2} \pm \frac{\sqrt{3}}{2}i$ **69.** $\frac{1}{2} \pm \frac{1}{2}i$ **71.** $-1 \pm \frac{\sqrt{6}}{6}i$
73. $8+2i$ **75.** 25

SECTION 1.7 ■ PAGE 75

2. principal; interest rate; time in years
3. (a) x^2 **(b)** lw **(c)** πr^2 **4.** 1.6
5. $\frac{1}{x}$ **6.** $r = \frac{d}{t}, t = \frac{d}{r}$ **7.** $3n+3$ **9.** $3n+6$ **11.** $\frac{160+s}{3}$
13. $0.025x$ **15.** $4w^2$ **17.** $\frac{d}{55}$ **19.** $\frac{25}{3+x}$ **21.** 400 mi
23. 86 **25.** \$9000 at $4\frac{1}{2}\%$ and \$3000 at 4% **27.** 7.5%
29. \$7400 **31.** 8 h **33.** 40 years old **35.** 9 pennies, 9 nickels, 9 dimes **37.** 45 ft **39.** 120 ft by 120 ft
41. 25 ft by 35 ft **43.** 60 ft by 40 ft **45.** 120 ft
47. (a) 9 cm **(b)** 5 in. **49.** 4 in. **51.** 18 ft **53.** 5 m
55. 200 mL **57.** 18 g **59.** 0.6 L **61.** 35% **63.** 37 min 20 s
65. 3 h **67.** Irene 3 h, Henry $4\frac{1}{2}$ h **69.** 4 h **71.** 500 mi/h
73. 50 mi/h (or 240 mi/h) **75.** 6 km/h
77. 6.4 ft from the fulcrum **79.** 2 ft by 6 ft by 15 ft
81. 13 in. by 13 in. **83.** 2.88 ft **85.** 16 mi; no **87.** 7.52 ft
89. 18 ft **91.** 4.55 ft

SECTION 1.8 ■ PAGE 88

1. (a) $<$ **(b)** $\le$ **(c)** $\le$ **(d)** $>$
2. $-1, 2$

Interval	$(-\infty, -1)$	$(-1, 2)$	$(2, \infty)$
Sign of $x+1$ Sign of $x-2$	$-$ $-$	$+$ $-$	$+$ $+$
Sign of $(x+1)/(x-2)$	$+$	$-$	$+$

yes, 2; $[-1, 2)$
3. (a) $[-3, 3]$ **(b)** $(-\infty, -3], [3, \infty)$ **4. (a)** <3 **(b)** >3
5. (a) No **(b)** No **6. (a)** Divide by 3 **(b)** Add 2
(c) Rewrite as $-8 \le 3x + 2 \le 8$ **7.** $\{\frac{5}{6}, 1, \sqrt{5}, 3, 5\}$
9. $\{3, 5\}$ **11.** $\{-5, -1, \sqrt{5}, 3, 5\}$

13. $(-\infty, \frac{7}{2}]$
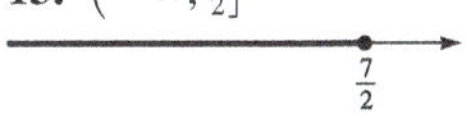

15. $(4, \infty)$
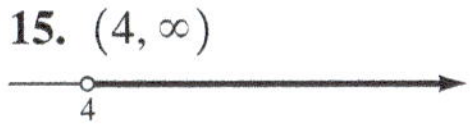

17. $(-\infty, 2]$
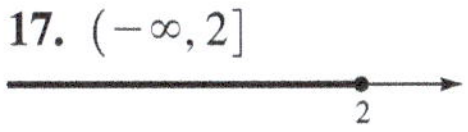

19. $(-\infty, -\frac{1}{2})$
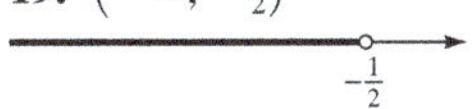

21. $(-3, \infty)$
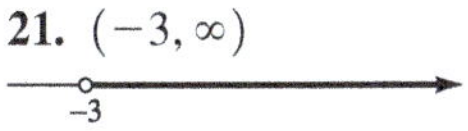

23. $(\frac{16}{3}, \infty)$
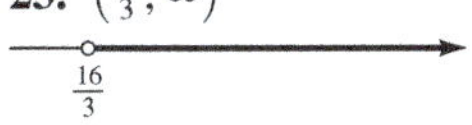

25. $(-\infty, -18)$
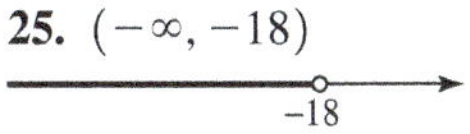

27. $(-\infty, -1]$
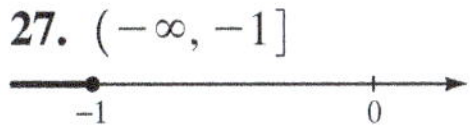

29. $[-3, -1)$
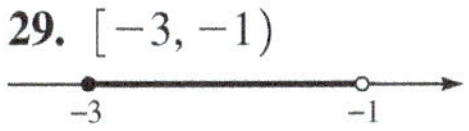

31. $(2, 6)$
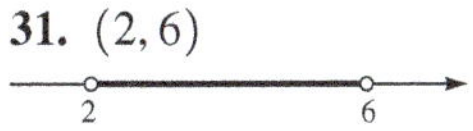

33. $[\frac{9}{2}, 5)$

35. $(\frac{15}{2}, \frac{21}{2}]$

37. $(-2, 3)$

39. $(-\infty, -\frac{7}{2}] \cup [0, \infty)$
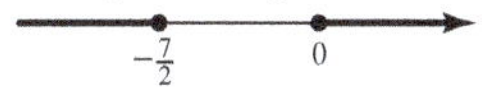

41. $[-3, 6]$

43. $(-\infty, -1] \cup [\frac{1}{2}, \infty)$

45. $(-1, 4)$

47. $(-\infty, -3) \cup (6, \infty)$

49. $(-2, 2)$

51. $(-\infty, -2] \cup [1, 3]$

53. $(-\infty, -2) \cup (-2, 4)$

55. $(-\infty, -5] \cup \{-3\} \cup [2, \infty)$

57. $(-2, 0) \cup (2, \infty)$
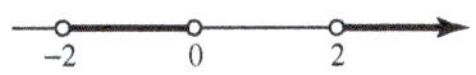

59. $(-\infty, -1) \cup [3, \infty)$

61. $(-\frac{3}{2}, -1)$
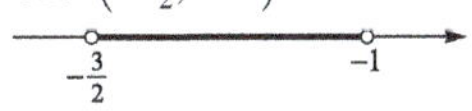

63. $(-\infty, 5) \cup [16, \infty)$
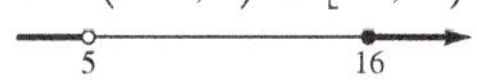

65. $(-2, 0) \cup (2, \infty)$

67. $[-2, -1) \cup (0, 1]$
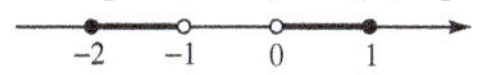

69. $[-2, 0) \cup (1, 3]$
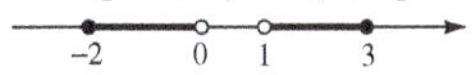

71. $(-3, -\frac{1}{2}) \cup (2, \infty)$
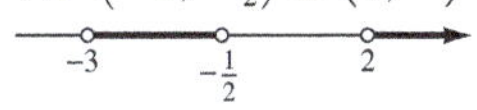

73. $(-\infty, -1) \cup (1, \infty)$

75. $(-4, 4)$

77. $(-\infty, -\frac{7}{2}) \cup (\frac{7}{2}, \infty)$
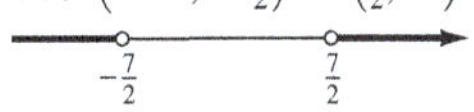

79. $[2, 8]$

81. $(-2, \frac{2}{3})$
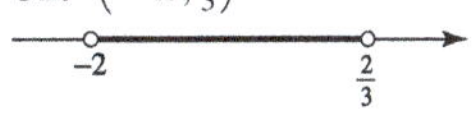

83. $(-\infty, -1] \cup [\frac{7}{3}, \infty)$
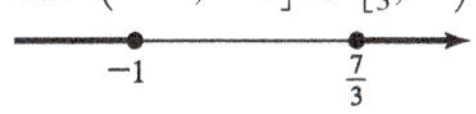

85. $(-4, 8)$

87. $(-6.001, -5.999)$
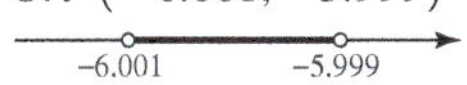

89. $[-\frac{1}{2}, \frac{3}{2}]$
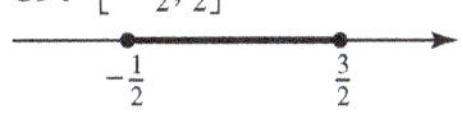

91. $|x| < 3$ **93.** $|x - 7| \ge 5$
95. $|x| \le 2$ **97.** $|x| > 3$
99. $|x - 1| \le 3$ **101.** $x \le -3$ or $x \ge 3$
103. $x < -2$ or $x > 5$ **105.** $x \ge \frac{(a+b)c}{ab}$
107. $x \le \frac{ac - 4a + d}{ab}$ or $x \ge \frac{ac + 4a - d}{ab}$

109. $68 \le F \le 86$ **111.** More than 200 mi
113. Between 12,000 mi and 14,000 mi
115. **(a)** $-\frac{1}{3}P + \frac{560}{3}$ **(b)** From \$215 to \$290
117. Distances between 20,000 km and 100,000 km
119. From 0 s to 3 s **121.** Between 0 and 60 mi/h
123. Between 20 and 40 ft
125. Between 62.4 and 74.0 in.

SECTION 1.9 ■ PAGE 101

1. $(3, -5)$ **2.** $\sqrt{(c-a)^2 + (d-b)^2}$; 10
3. $\left(\frac{a+c}{2}, \frac{b+d}{2}\right)$; $(4, 6)$
4. 2; 3; No

x	y	(x, y)
-2	$-\frac{1}{2}$	$(-2, -\frac{1}{2})$
-1	0	$(-1, 0)$
0	$\frac{1}{2}$	$(0, \frac{1}{2})$
1	1	$(1, 1)$
2	$\frac{3}{2}$	$(2, \frac{3}{2})$

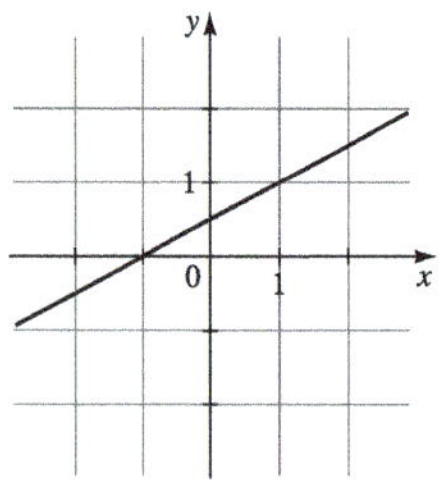

5. **(a)** y; x; -1 **(b)** x; y; $\frac{1}{2}$ **6.** $(1, 2)$; 3
7. **(a)** $(a, -b)$ **(b)** $(-a, b)$ **(c)** $(-a, -b)$
8. **(a)** -3 and 3; -1 and 2 **(b)** y-axis
9. Yes **10.** No
11. $A(5, 1)$, $B(1, 2)$, $C(-2, 6)$, $D(-6, 2)$, $E(-4, -1)$, $F(-2, 0)$, $G(-1, -3)$, $H(2, -2)$

13.

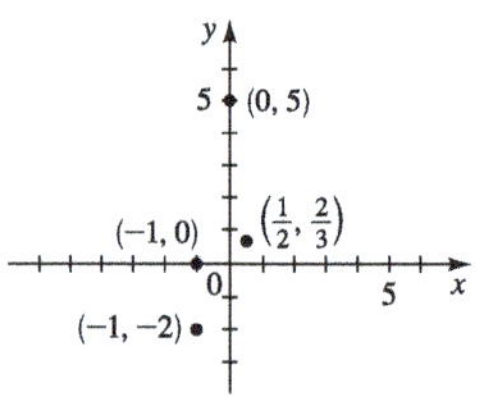

15. **(a)**

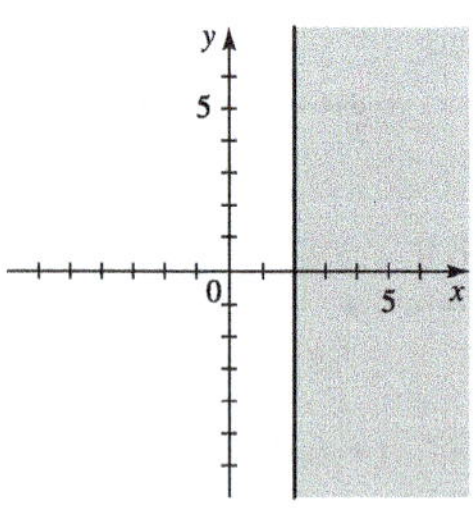

(b)

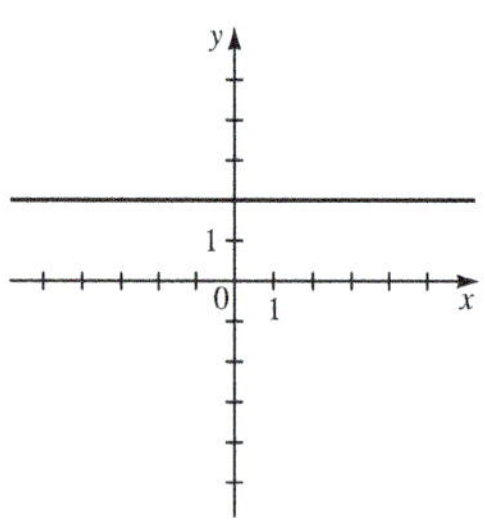

17. **(a)**

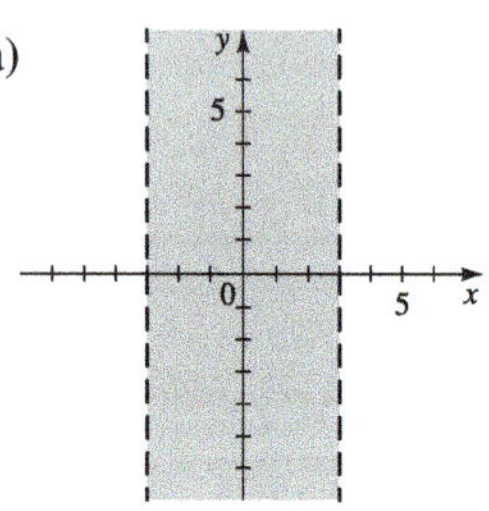

(b)

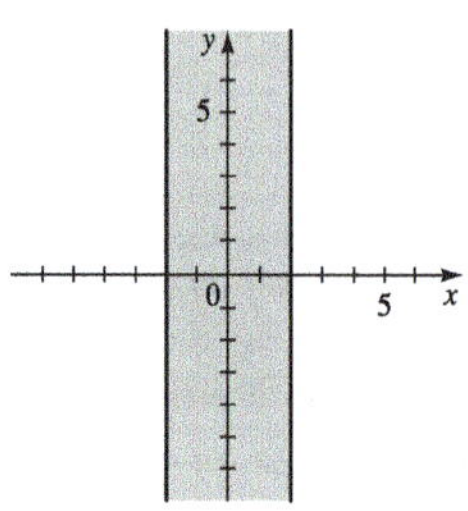

19. **(a)**

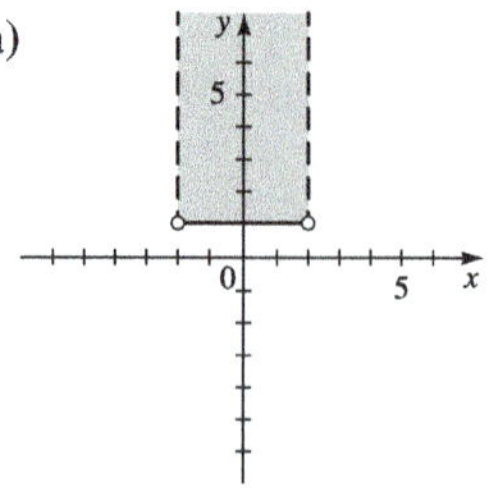

(b)

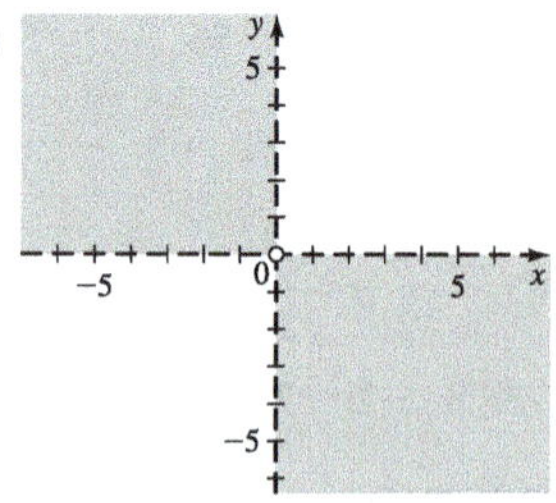

21. **(a)** $\sqrt{13}$ **(b)** $(\frac{3}{2}, 1)$ **23.** **(a)** 10 **(b)** $(1, 0)$

25. **(a)**

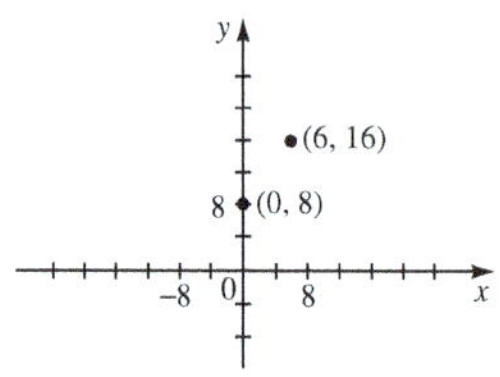

(b) 10 **(c)** $(3, 12)$

27. **(a)**

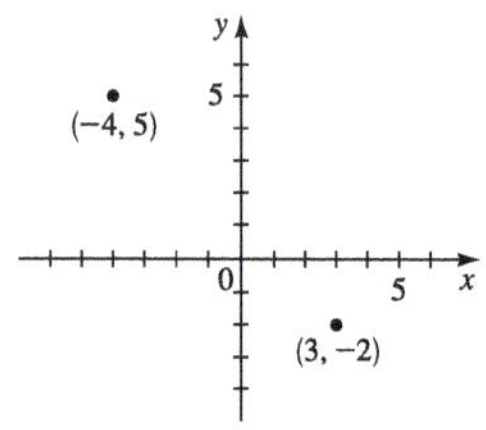

(b) $7\sqrt{2}$ **(c)** $(-\frac{1}{2}, \frac{3}{2})$

29. **(a)**

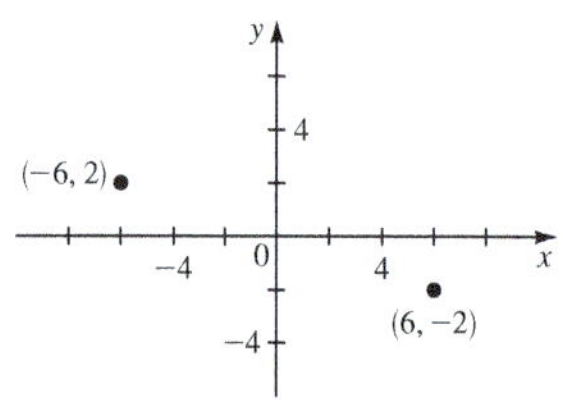

(b) $4\sqrt{10}$ **(c)** $(0, 0)$

31. 24

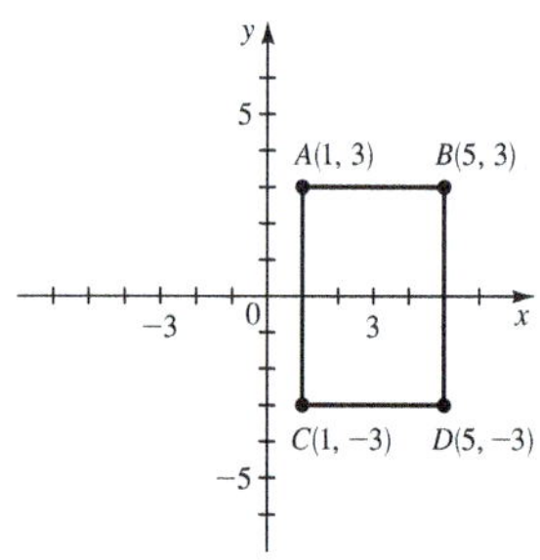

33. Trapezoid, area = 9

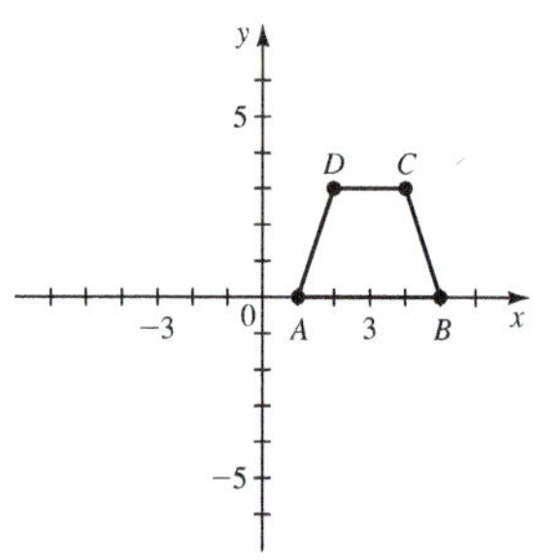

35. $A(6, 7)$ **37.** $Q(-1, 3)$ **41.** **(b)** 10 **45.** $(0, -4)$
47. $(2, -3)$

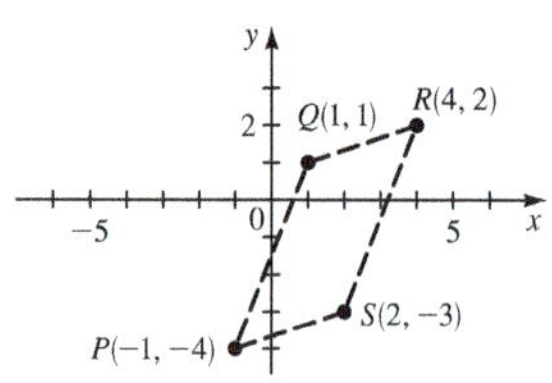

49. **(a)**

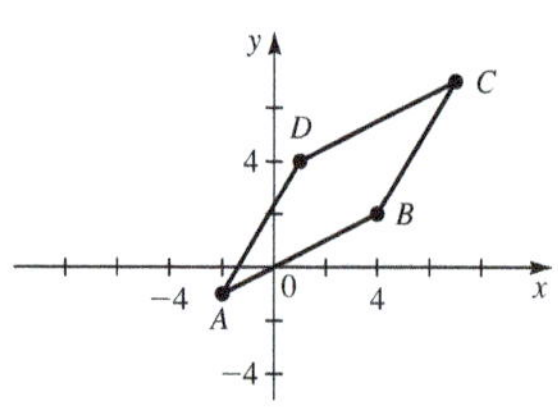

(b) $(\frac{5}{2}, 3)$, $(\frac{5}{2}, 3)$

51. No, yes, yes **53.** Yes, no, yes

55.

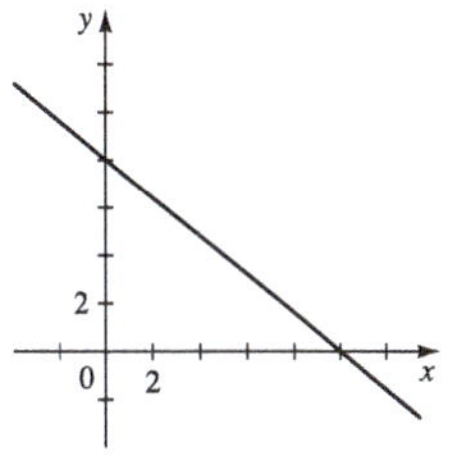

57.

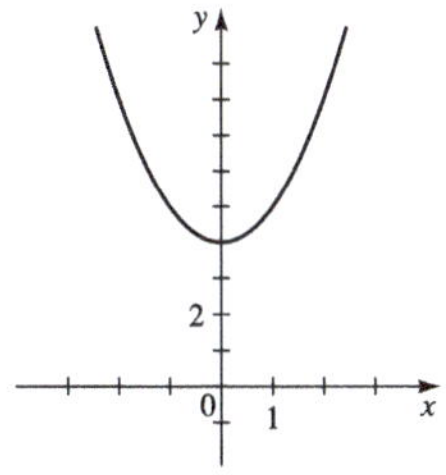

59.

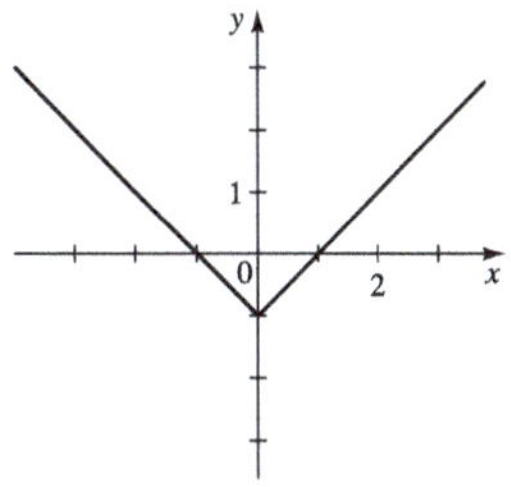

61.

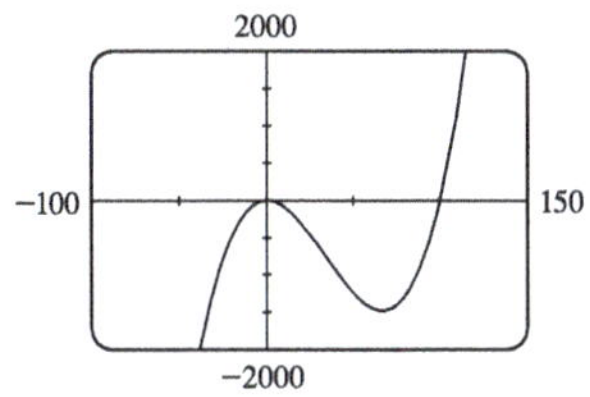

63.

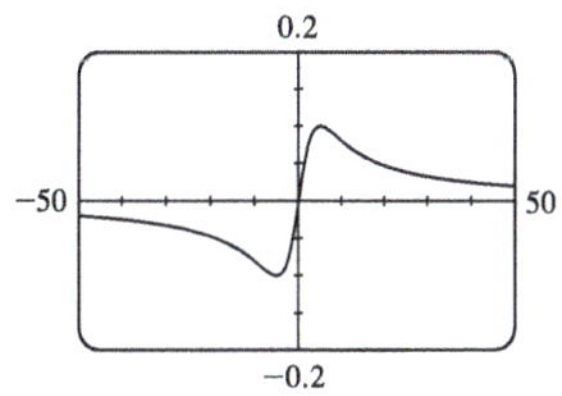

65. **(a)** x-intercept 3, y-intercept -6, no symmetry

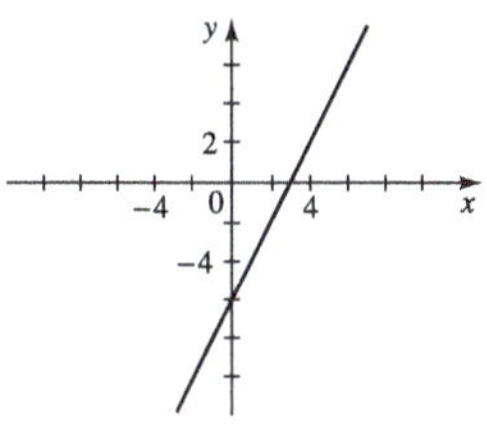

(b) x-intercept -1, y-intercept -1, no symmetry

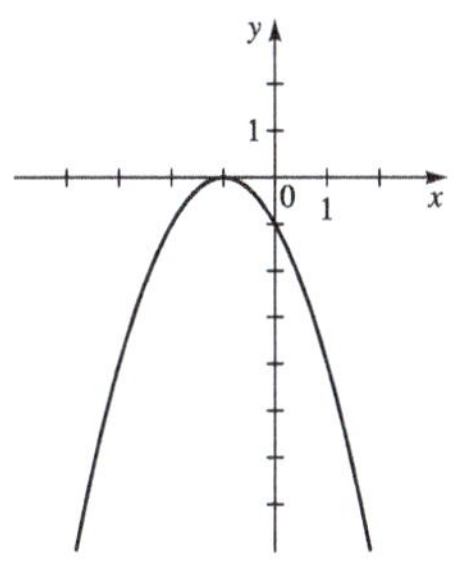

67. **(a)** x-intercept -1, y-intercept 1, no symmetry

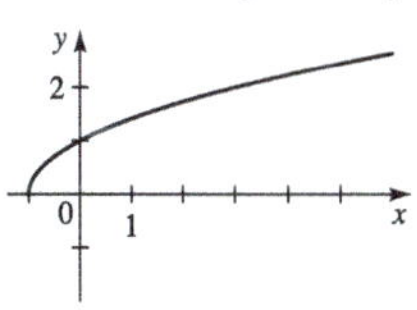

(b) x-intercept 0, y-intercept 0, symmetry about y-axis

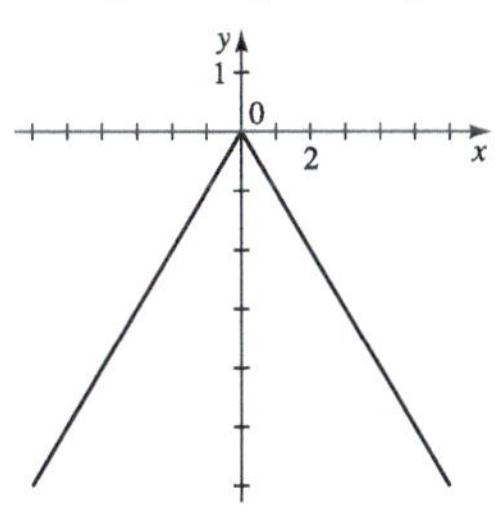

69. **(a)** x-intercepts ± 2, y-intercept 2, symmetry about y-axis

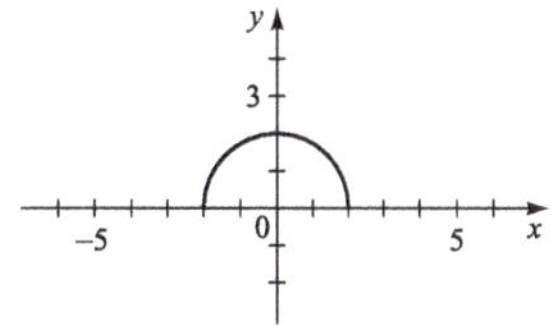

(b) x-intercept 0, y-intercept 0, symmetry about origin

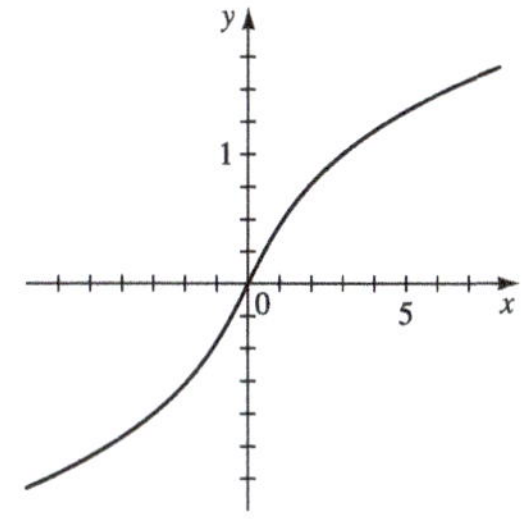

71. **(a)** x-intercept -6; y-intercept 6
(b) x-intercepts $\pm\sqrt{5}$; y-intercept -5
73. **(a)** x-intercepts ± 2; no y-intercept
(b) x-intercept $\frac{1}{4}$; y-intercept 1
75. x-intercepts 0, 4; y-intercept 0
77. x-intercepts -2, 2; y-intercepts -4, 4
79. **(a)**

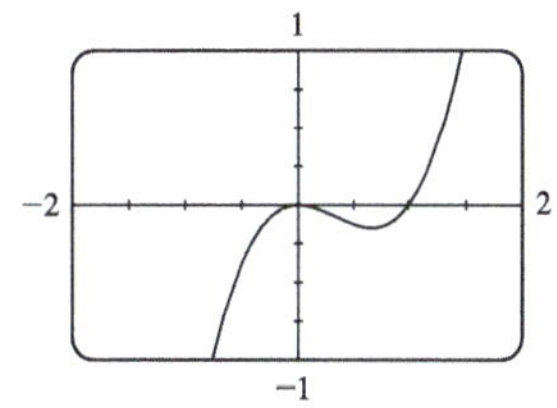

(b) x-intercepts 0, 1; y-intercept 0

81. **(a)**

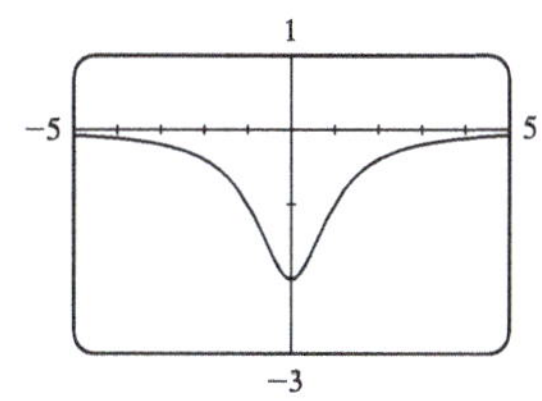

(b) No x-intercept; y-intercept -2

83. $(0, 0)$, 3

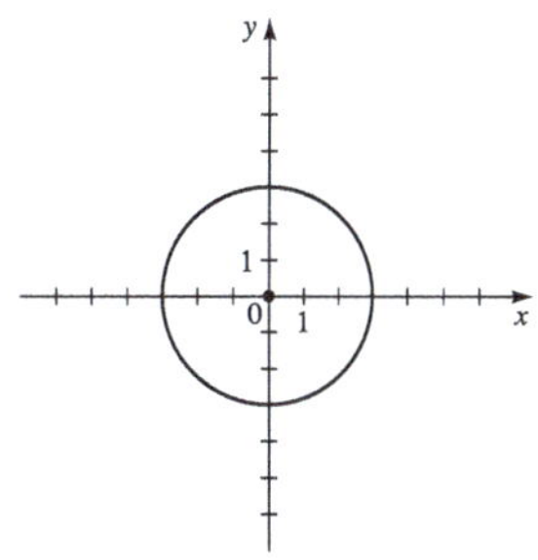

85. $(0, 4)$, 1

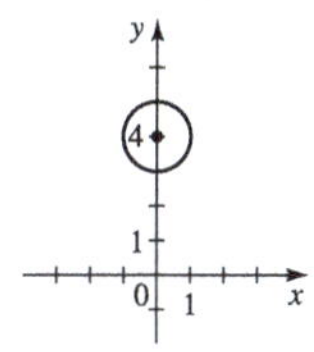

87. $(-3, 4)$, 5

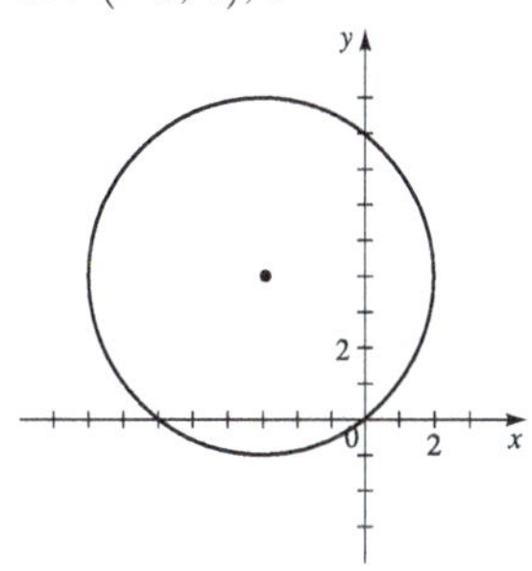

89. $(x - 2)^2 + (y + 1)^2 = 9$ **91.** $x^2 + y^2 = 65$
93. $(x - 2)^2 + (y - 5)^2 = 25$ **95.** $(x - 7)^2 + (y + 3)^2 = 9$
97. $(x + 2)^2 + (y - 2)^2 = 4$ **99.** $(-2, 3), 1$ **101.** $\left(\frac{1}{4}, -\frac{1}{4}\right), \frac{1}{2}$
103. $\left(\frac{3}{4}, 0\right), \frac{3}{4}$ **105.** Symmetry about y-axis
107. Symmetry about origin **109.** Symmetry about origin

111.

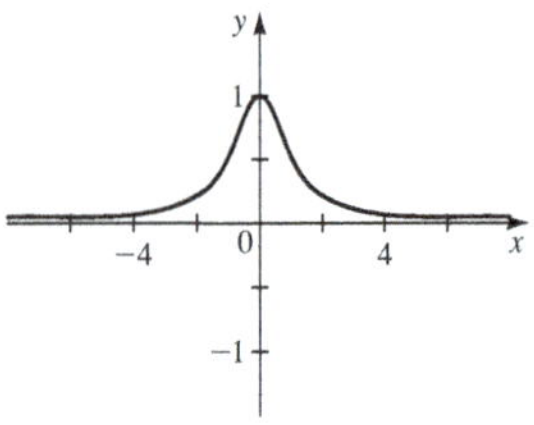

113.

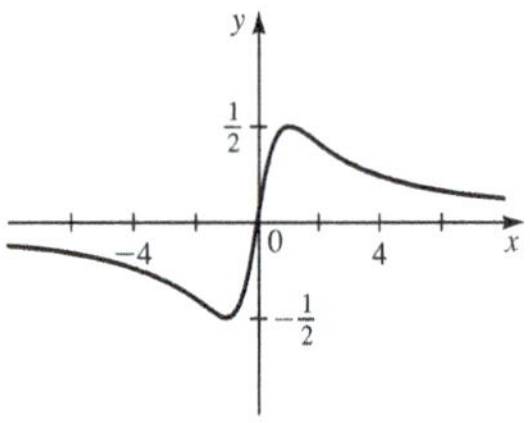

115.

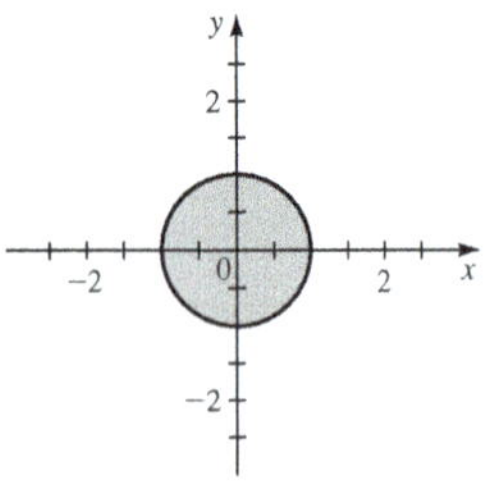

117. 12π **119. (a)** $(8, 5)$ **(b)** $(a + 3, b + 2)$ **(c)** $(0, 2)$ **(d)** $A'(-2, 1), B'(0, 4), C'(5, 3)$

121. (a)

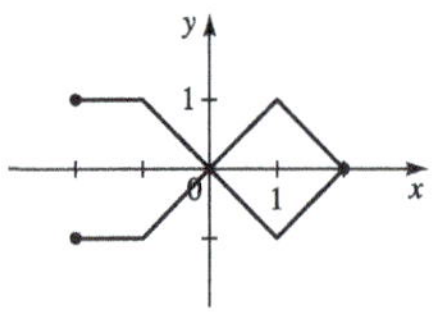

(b)

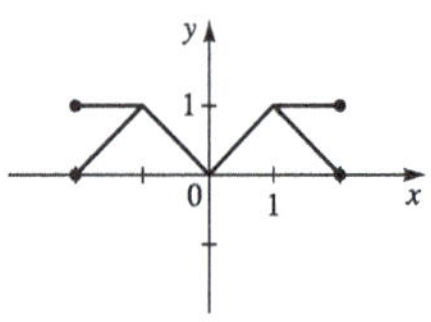

(c)

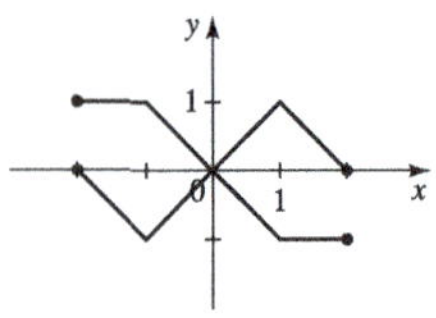

123. (a) 15th Street and 12th Avenue **(b)** 17 blocks

SECTION 1.10 ■ PAGE 113

1. y; x; 2 **2. (a)** 3 **(b)** 3 **(c)** $-\frac{1}{3}$ **3.** $y - 2 = 3(x - 1)$
4. 6, 4; $-\frac{2}{3}x + 4$; $-\frac{2}{3}$ **5.** 0; $y = 3$ **6.** Undefined; $x = 2$
7. (a) Yes **(b)** Yes **(c)** No **(d)** Yes

8.

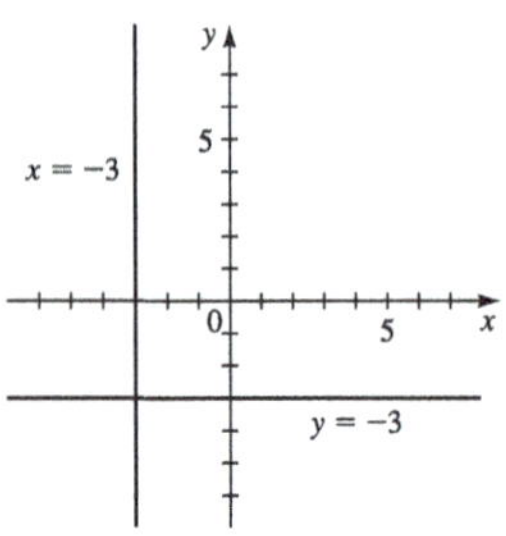

Yes

9. -2 **11.** $\frac{1}{5}$ **13.** 0 **15.** $\frac{3}{4}$ **17.** $-2, \frac{1}{2}, 3, -\frac{1}{4}$
19. $x + y - 4 = 0$ **21.** $3x - 2y - 6 = 0$ **23.** $3x - y - 2 = 0$
25. $5x - y - 7 = 0$ **27.** $2x - 3y + 19 = 0$
29. $5x + y - 11 = 0$ **31.** $8x + y + 11 = 0$
33. $3x - y - 3 = 0$ **35.** $y = 3$ **37.** $x = 2$
39. $3x - y - 1 = 0$ **41.** $y = 5$ **43.** $x + 2y + 11 = 0$
45. $x = -1$ **47.** $5x - 2y + 1 = 0$ **49.** $x - y + 6 = 0$

51. (a)

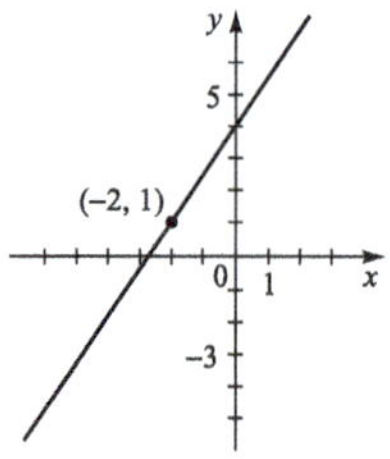

(b) $3x - 2y + 8 = 0$

53. They all have the same slope.

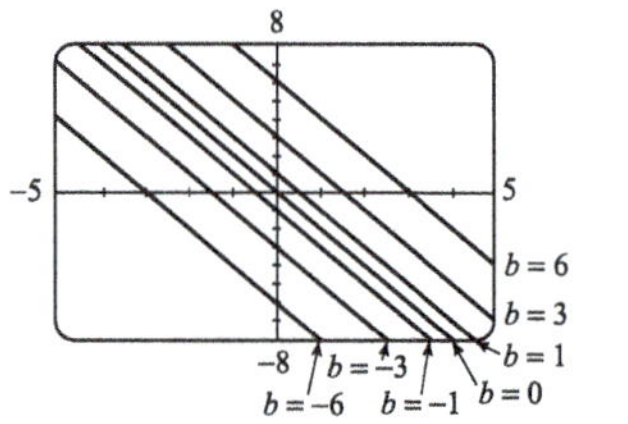

55. They all have the same x-intercept.

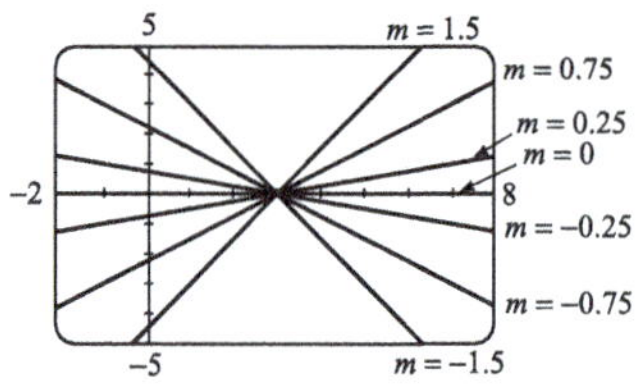

57. $-1, 3$

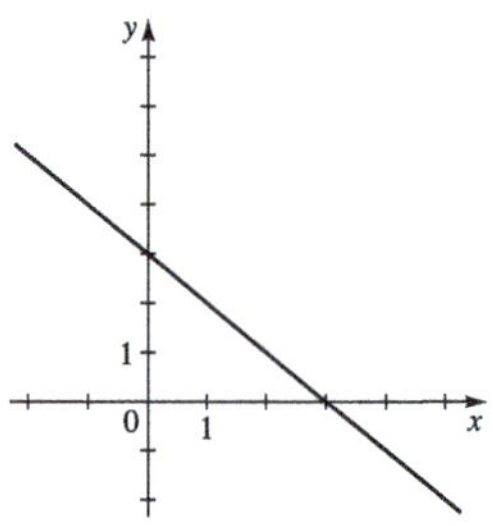

59. 2, 7

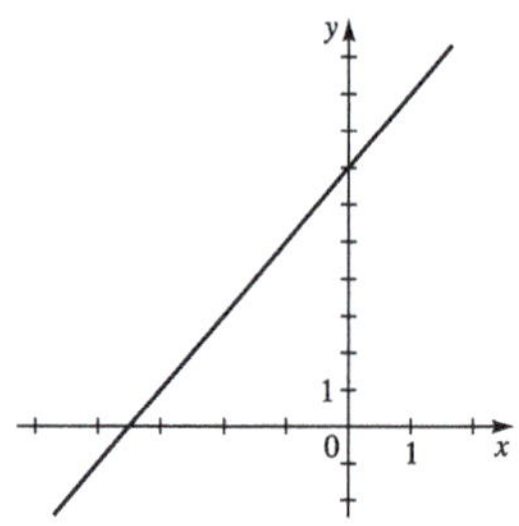

61. $-\frac{4}{5}, 2$

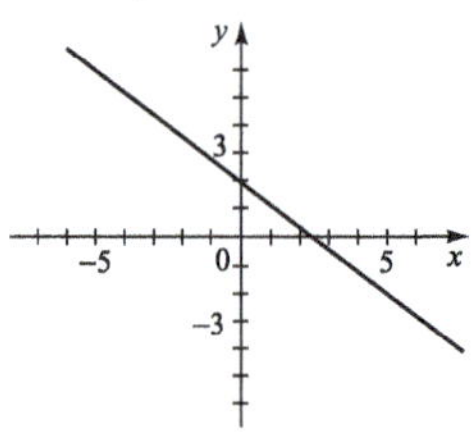

63. 0, 4

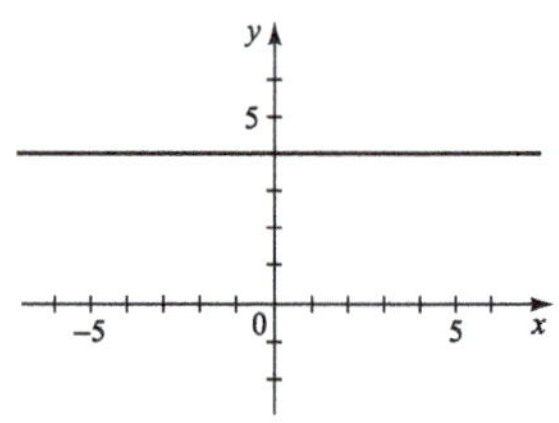

65. Undefined, none

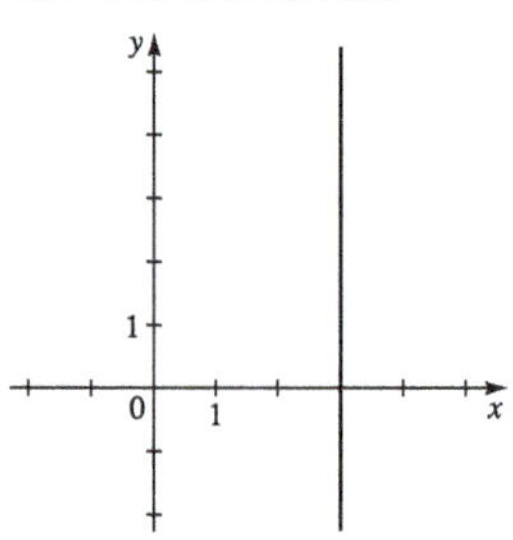

67. 2, 5

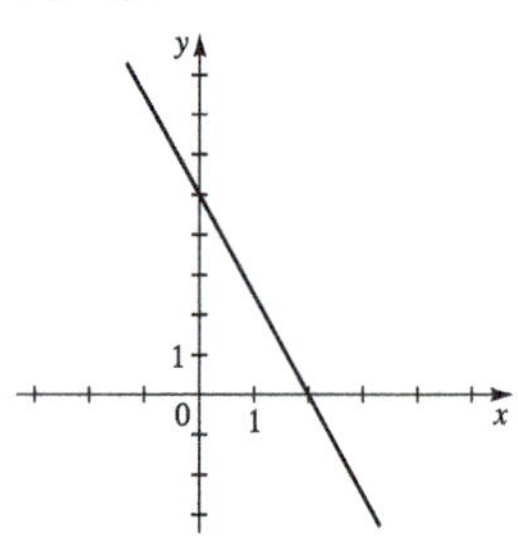

69. $-2, 3$

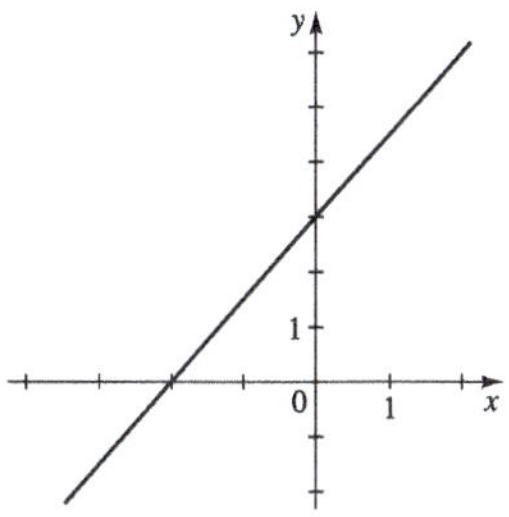

71. $-\frac{2}{3}, 4$

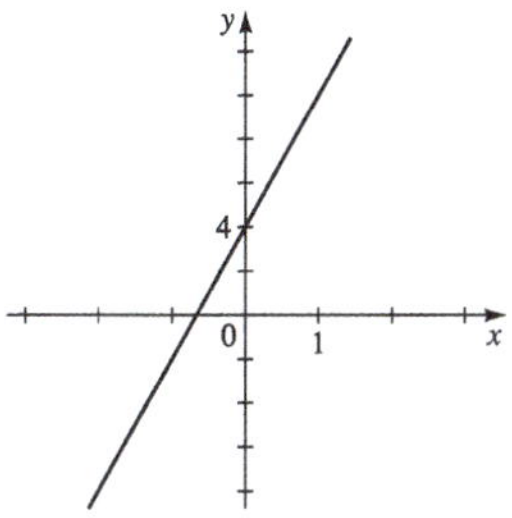

73. Parallel **75.** Perpendicular **77.** Neither
83. $x - y - 3 = 0$ **85. (b)** $4x - 3y - 24 = 0$
87. (a) The slope represents an increase of 0.02°C every year, and the T-intercept is the average surface temperature in 1950. **(b)** 17.0°C
89. (a)

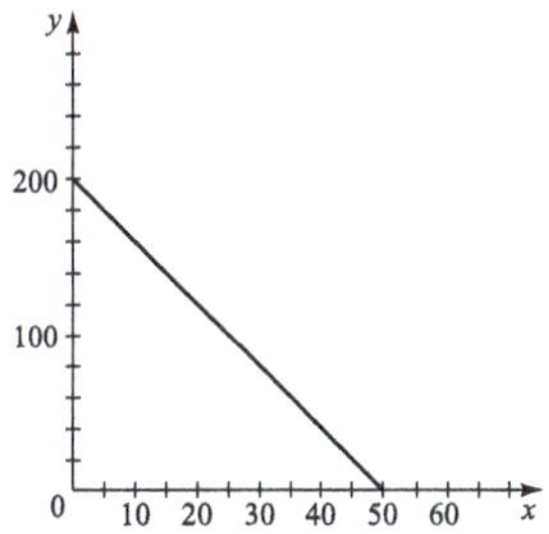

(b) The slope represents a decrease of 4 spaces rented for each one dollar increase in rental price, the y-intercept indicates that 200 spaces are rented if there is no increase in price, and the x-intercept indicates that no spaces are rented with an increase of \$50 in rental price.

91. (a)

C	−30°	−20°	−10°	0°	10°	20°	30°
F	−22°	−4°	14°	32°	50°	68°	86°

(b) −40°
93. (a) $V = -950t + 4000$
(b)

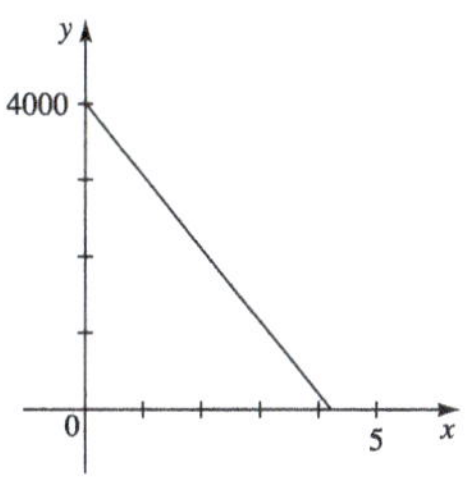

(c) The slope represents a decrease of \$950 each year in the value of the computer, and the V-intercept is the original price of the computer. **(d)** \$1150

SECTION 1.11 ■ PAGE 121

1. x **2.** above **3. (a)** $x = -1, 0, 1, 3$ **(b)** $[-1, 0] \cup [1, 3]$
4. (a) $x = 1, 4$ **(b)** $(1, 4)$ **5.** -4 **7.** $\frac{5}{14}$
9. $\pm 4\sqrt{2} \approx \pm 5.7$ **11.** No solution **13.** 2.5, −2.5
15. $5 + 2\sqrt[4]{5} \approx 7.99, 5 - 2\sqrt[4]{5} \approx 2.01$ **17.** 3.00, 4.00
19. 1.00, 2.00, 3.00 **21.** 1.62 **23.** −1.00, 0.00, 1.00 **25.** 4
27. No solution **29.** 2.55 **31.** −2.05, 0, 1.05
33. $[-2.00, 5.00]$ **35.** $(-\infty, 1.00] \cup [2.00, 3.00]$
37. $(-1.00, 0) \cup (1.00, \infty)$ **39.** $(-\infty, 0)$ **41.** $(-1, 4)$
43. $(-\infty, -5] \cup \{-3\} \cup [2, \infty)$ **45.** 2.27
47. (a)

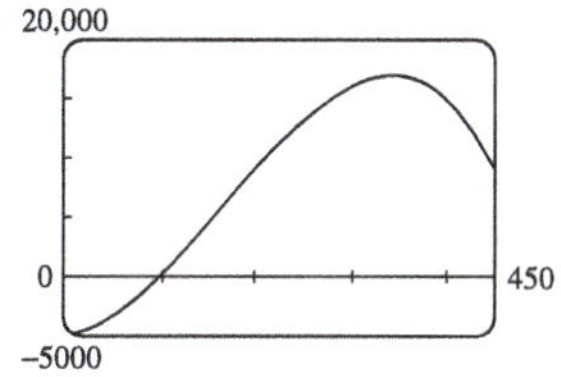

(b) 101 cooktops
(c) $279 < x < 400$

SECTION 1.12 ■ PAGE 127

1. directly proportional; proportionality **2.** inversely proportional; proportionality **3.** directly proportional; inversely proportional **4.** $\frac{1}{2}xy$
5. (a) Directly proportional **(b)** Not proportional
6. (a) Not proportional **(b)** Inversely proportional
7. $T = kx$ **9.** $v = k/z$ **11.** $y = ks/t$ **13.** $z = k\sqrt{y}$
15. $V = klwh$ **17.** $R = \dfrac{kP^2t^2}{b^3}$ **19.** $y = 7x$ **21.** $A = \dfrac{21}{r}$
23. $A = \dfrac{18x}{t}$ **25.** $W = 360/r^2$ **27.** $C = 16lwh$
29. $R = \dfrac{27.5}{\sqrt{x}}$ **31. (a)** $z = k\dfrac{x^3}{y^2}$ **(b)** $\frac{27}{4}$
33. (a) $z = kx^3y^5$ **(b)** 864 **35. (a)** $F = kx$ **(b)** 7.5
(c) 45 N **37. (a)** $P = ks^3$ **(b)** 0.012 **(c)** 324 **39.** 46 mi/h
41. 5.3 mi/h **43. (a)** $P = kT/V$ **(b)** 8.3 **(c)** 51.9 kPa
45. (a) $L = k/d^2$ **(b)** 7000 **(c)** $\frac{1}{4}$ **(d)** 4
47. (a) $R = kL/d^2$ **(b)** $0.00291\overline{6}$ **(c)** $R \approx 137\ \Omega$ **(d)** $\frac{3}{4}$
49. (a) 160,000 **(b)** 1,930,670,340
51. (a) $T = k\sqrt{l}$ **(b)** quadruple the length l
53. (a) $f = k/L$ **(b)** Halves it **55.** 3.47×10^{-14} W/m^2

CHAPTER 1 REVIEW ■ PAGE 133

1. Commutative Property of Addition
3. Distributive Property
5. $-2 \le x < 6$

7. $[5, \infty)$

9. 3 **11.** 4 **13.** $\frac{1}{6}$ **15.** 11 **17. (a)** b^{14} **(b)** $12xy^8$
19. (a) x^2y^2 **(b)** $|x|y^2$ **21.** 7.825×10^{10} **23.** 1.65×10^{-32}
25. $(x + 7)(x - 2)$ **27.** $(x - 1)^2(x + 1)^2$
29. $-4(t - 2)(t + 2)$
31. $(x - 1)(x^2 + x + 1)(x + 1)(x^2 - x + 1)$
33. $x^{-1/2}(5x - 3)(x + 1)$ **35.** $(x - 2)(4x^2 + 3)$
37. $(a + b - 5)(a + b + 2)$ **39.** $4y^2 - 49$
41. $2x^3 - 6x^2 + 4x$ **43.** $\dfrac{3(x + 3)}{x + 4}$ **45.** $\dfrac{3x^2 - 7x + 8}{x(x - 2)^2}$
47. $-\dfrac{1}{2x}$ **49.** $3\sqrt{2} - 2\sqrt{3}$ **51.** $\dfrac{\sqrt{11}}{11}$ **53.** $10\sqrt{2} + 10$
55. 5 **57.** No solution **59.** 2, 7 **61.** $-1, \frac{1}{2}$ **63.** $0, \pm\frac{5}{2}$
65. $\dfrac{-2 \pm \sqrt{7}}{3}$ **67.** −5 **69.** 3, 11 **71. (a)** $3 + i$
(b) $8 - i$ **73. (a)** $\frac{6}{5} + \frac{8}{5}i$ **(b)** 2 **75.** $\pm 4i$ **77.** $-3 \pm i$
79. $\pm 4, \pm 4i$ **81.** 20 lb raisins, 30 lb nuts

83. $\frac{1}{4}(\sqrt{329} - 3) \approx 3.78$ mi/h **85.** 1 h 50 min

87. $(-3, \infty)$

89. $(-\infty, -6) \cup (2, \infty)$

91. $(-\infty, -2) \cup (2, 4]$

93. $[2, 8]$

95. (a)

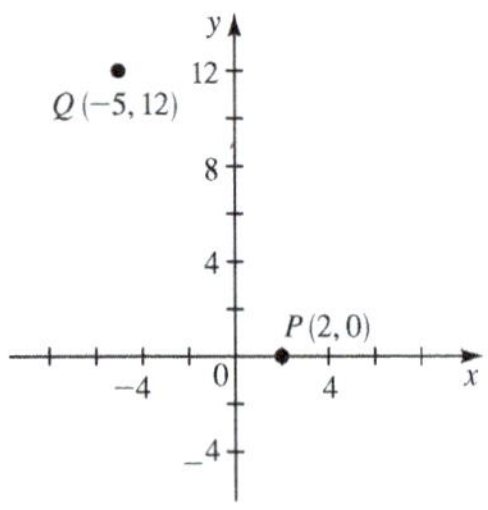

(b) $\sqrt{193}$ **(c)** $\left(-\frac{3}{2}, 6\right)$

(d) $y = -\frac{12}{7}x + \frac{24}{7}$

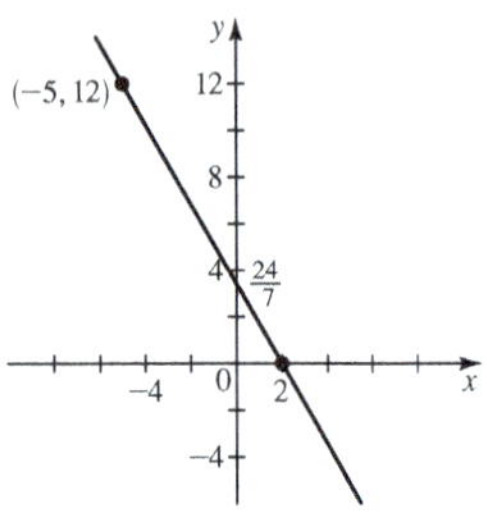

(e) $(x - 2)^2 + y^2 = 193$

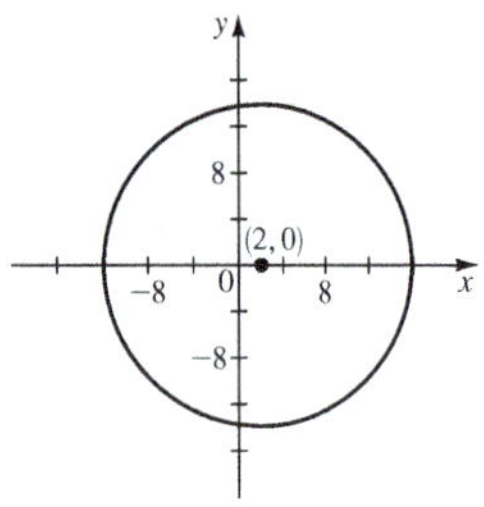

97.

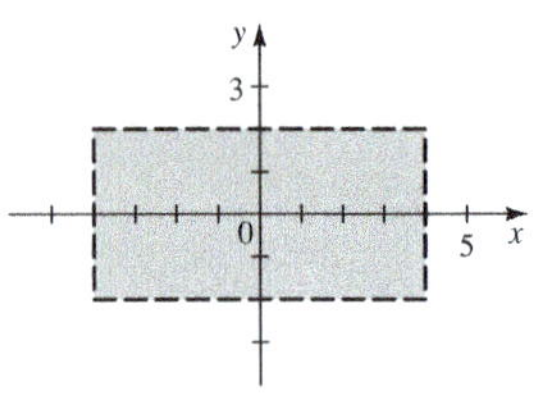

99. B **101.** $(x + 5)^2 + (y + 1)^2 = 26$

103. (a) Circle

(b) Center $(-1, 3)$, radius 1

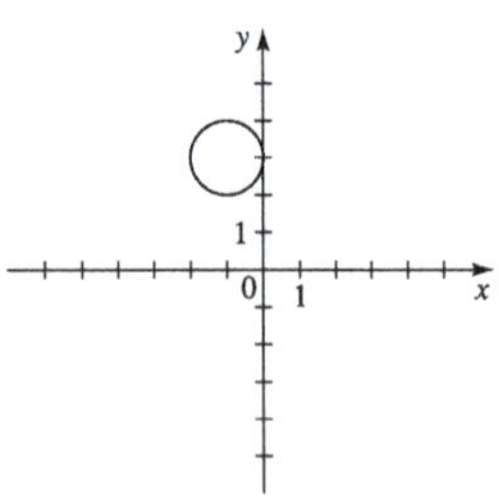

105. (a) No graph

107.

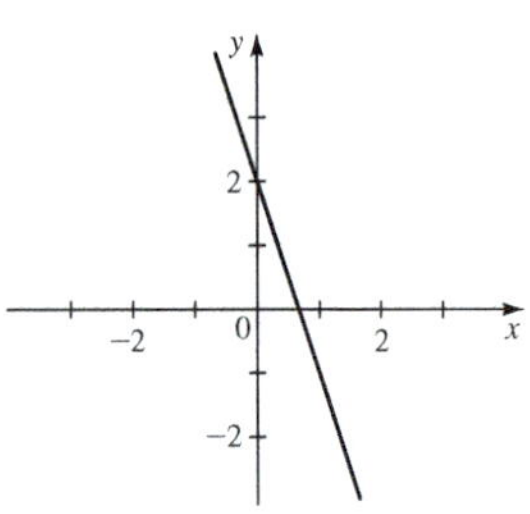

109.

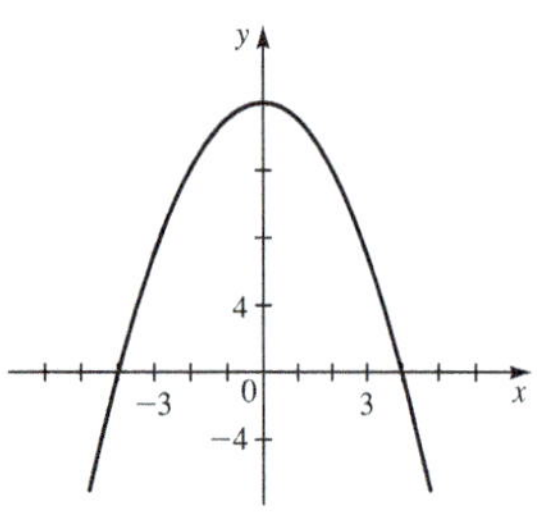

111.

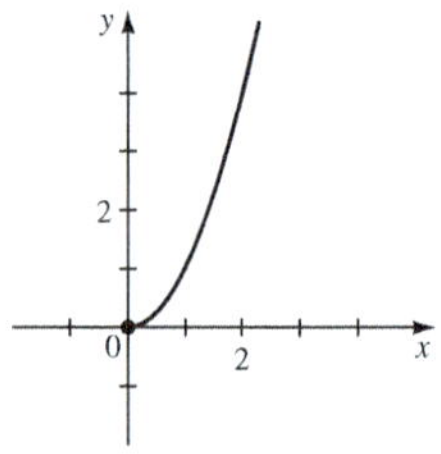

113. (a) Symmetry about y-axis

(b) x-intercepts $-3, 3$; y-intercept 9

115. (a) Symmetry about y-axis

(b) x-intercept 0; y-intercepts 0, 2

117. (a) Symmetric about origin

(b) x-intercepts $-1, 1$; y-intercepts $-1, 1$

119. (a)

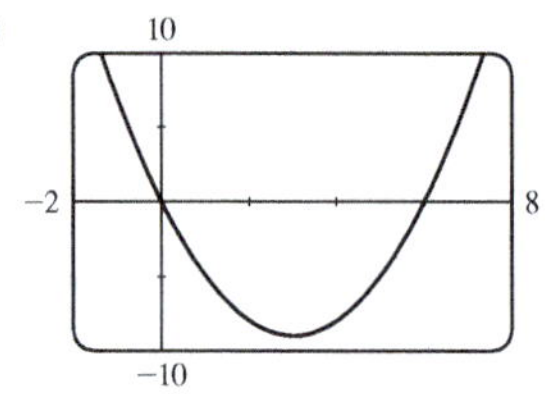

(b) x-intercepts 0, 6; y-intercept 0

121. (a)

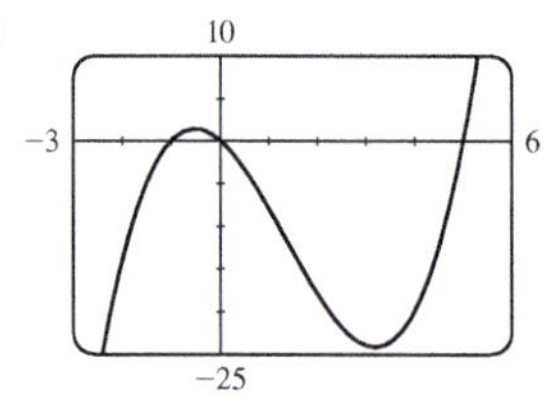

(b) x-intercepts $-1, 0, 5$; y-intercept 0

123. (a) $y = 2x + 6$

(b) $2x - y + 6 = 0$

(c)

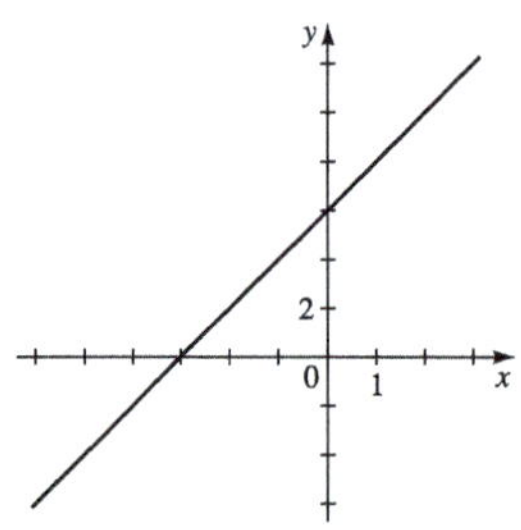

125. (a) $y = \frac{2}{3}x - \frac{16}{3}$

(b) $2x - 3y - 16 = 0$

(c)

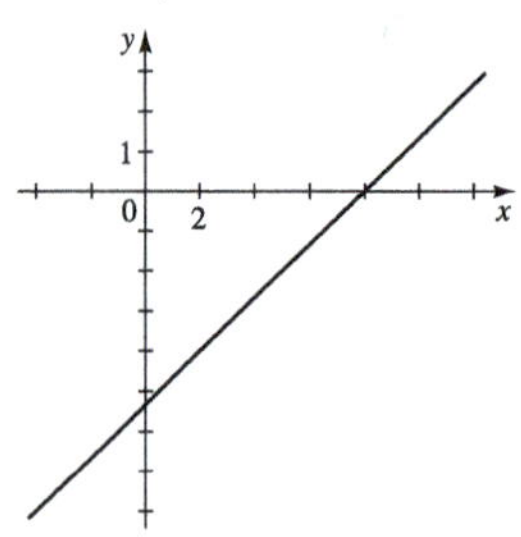

127. **(a)** $x = 3$ **(b)** $x - 3 = 0$

(c)

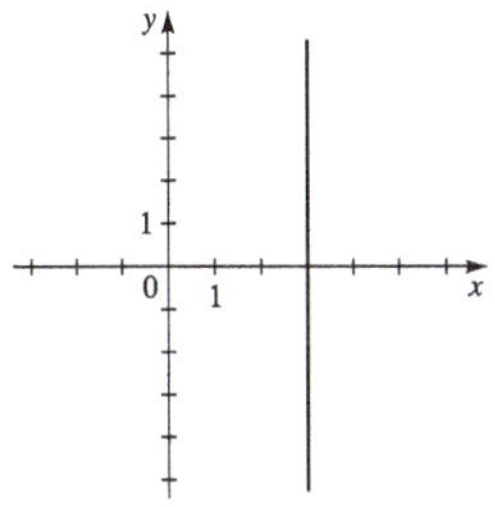

129. **(a)** $y = -4x$ **(b)** $4x + y = 0$

(c)

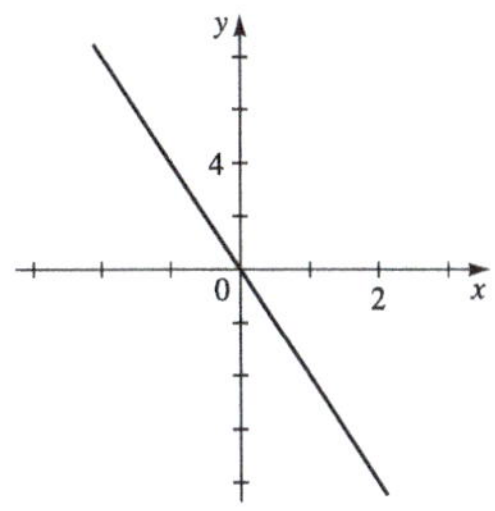

131. **(a)** The slope represents a stretch of 0.3 in. for each one-pound increase in weight. The *S*-intercept represents the unstretched length of the spring. **(b)** 4 in.
133. $-1, 6$ **135.** $[-1, 6]$ **137.** $(-\infty, 0] \cup [4, \infty)$
139. $-1, 7$ **141.** $-2.72, -1.15, 1.00, 2.87$ **143.** $[1, 3]$
145. $(-1.85, -0.60) \cup (0.45, 2.00)$
147. $x^2 + y^2 = 169$, $5x - 12y + 169 = 0$
149. $M = 8z$ **151.** **(a)** $I = k/d^2$ **(b)** 64,000
(c) 160 candles **153.** 11.0 mi/h

CHAPTER 1 TEST ■ PAGE 137

1. **(a)**

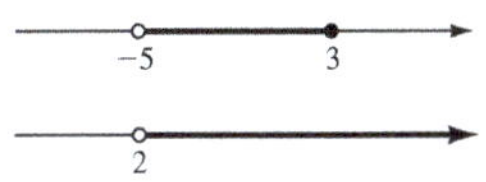

(b) $(-\infty, 3], [-1, 4)$ **(c)** 16 **2.** **(a)** 81 **(b)** -81 **(c)** $\frac{1}{81}$
(d) 25 **(e)** $\frac{9}{4}$ **(f)** $\frac{1}{8}$ **3.** **(a)** 1.86×10^{11} **(b)** 3.965×10^{-7}
4. **(a)** $6\sqrt{2}$ **(b)** $48a^5b^7$ **(c)** $\dfrac{x}{9y^7}$ **(d)** $\dfrac{x+2}{x-2}$ **(e)** $\dfrac{1}{x-2}$
(f) $-(x + y)$ **5.** $5\sqrt{2} + 2\sqrt{10}$
6. **(a)** $11x - 2$ **(b)** $4x^2 + 7x - 15$ **(c)** $a - b$
(d) $4x^2 + 12x + 9$ **(e)** $x^3 + 6x^2 + 12x + 8$
7. **(a)** $(2x - 5)(2x + 5)$ **(b)** $(2x - 3)(x + 4)$
(c) $(x - 3)(x - 2)(x + 2)$ **(d)** $x(x + 3)(x^2 - 3x + 9)$
(e) $3x^{-1/2}(x - 1)(x - 2)$ **(f)** $xy(x - 2)(x + 2)$
8. **(a)** 6 **(b)** 1 **(c)** $-3, 4$ **(d)** $-1 \pm \dfrac{\sqrt{2}}{2}$
(e) No real solution **(f)** $\pm 1, \pm\sqrt{2}$ **(g)** $\frac{2}{3}, \frac{22}{3}$ **9.** **(a)** $7 + i$
(b) $-1 - 5i$ **(c)** $18 + i$ **(d)** $\frac{6}{25} - \frac{17}{25}i$ **(e)** 1 **(f)** $6 - 2i$
10. $-1 \pm \dfrac{\sqrt{2}}{2}i$ **11.** 120 mi **12.** 50 ft by 120 ft
13. **(a)** $[-4, 3)$

(b) $(-2, 0) \cup (1, \infty)$

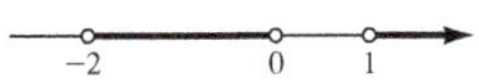

(c) $(1, 7)$

(d) $(-1, 4]$

14. Between 41°F and 50°F **15.** $0 \le x \le 6$
16. **(a)** $S(3, 6)$

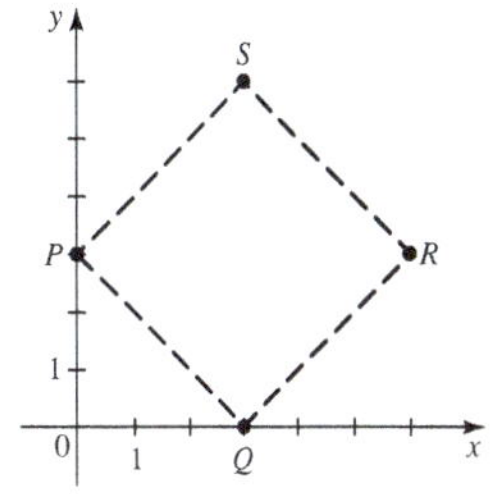

(b) 18

17. **(a)**

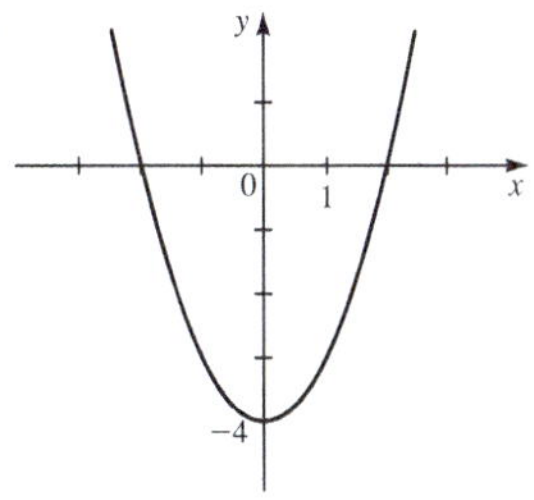

(b) *x*-intercepts $-2, 2$
y-intercept -4
(c) Symmetric about *y*-axis

18. **(a)**

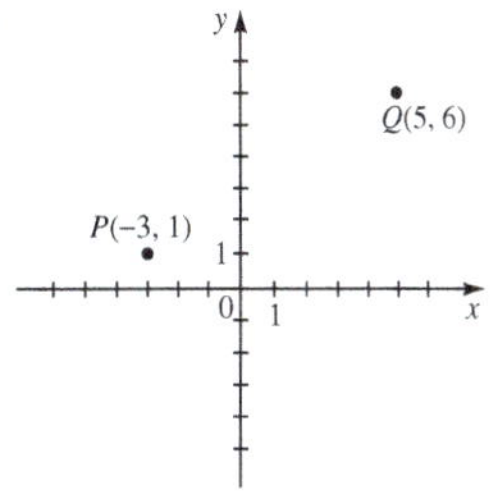

(b) $\sqrt{89}$ **(c)** $\left(1, \frac{7}{2}\right)$ **(d)** $\frac{5}{8}$ **(e)** $y = -\frac{8}{5}x + \frac{51}{10}$
(f) $(x - 1)^2 + \left(y - \frac{7}{2}\right)^2 = \frac{89}{4}$
19. **(a)** $(0, 0), 5$

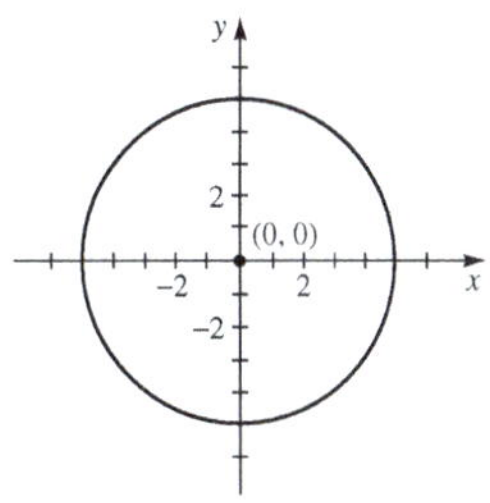

(b) $(2, -1), 3$

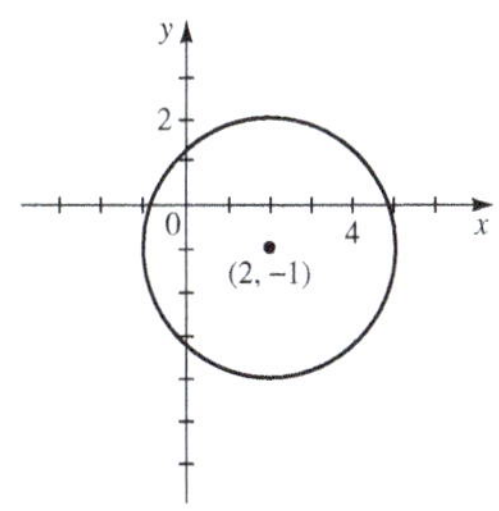

(c) $(-3, 1), 2$

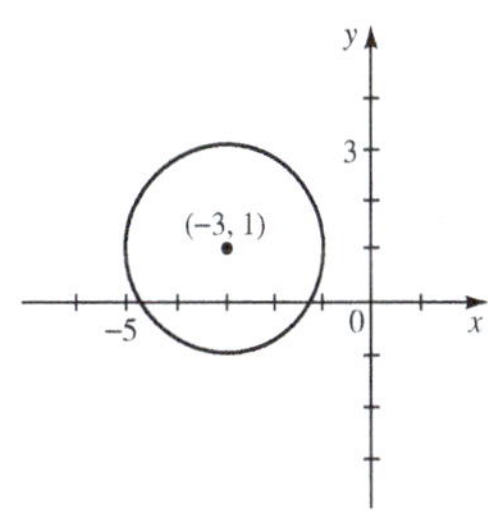

20. $y = \frac{2}{3}x - 5$

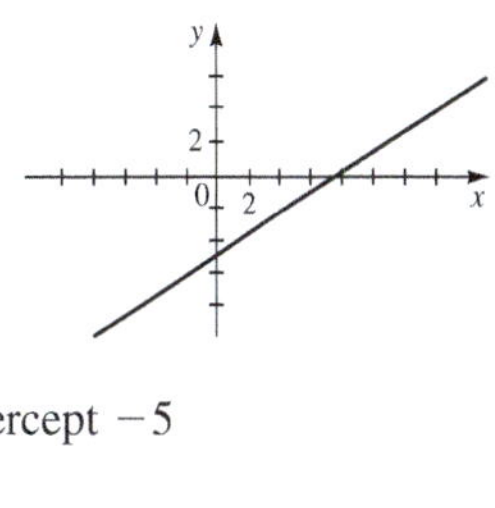

slope $\frac{2}{3}$; *y*-intercept -5

21. **(a)** $3x + y - 3 = 0$ **(b)** $2x + 3y - 12 = 0$

22. **(a)** $4°C$ **(b)**

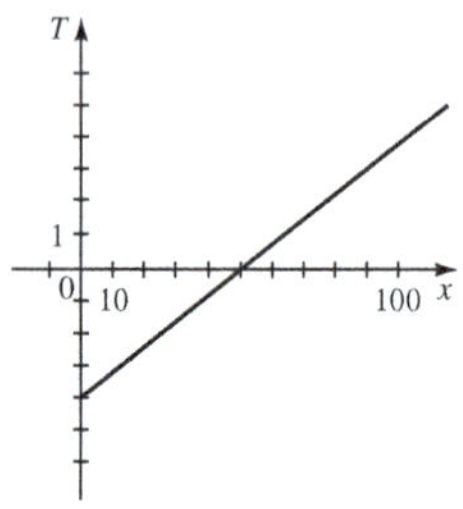

(c) The slope represents an increase of $0.08°C$ for each one-centimeter increase in depth, the x-intercept is the depth at which the temperature is $0°C$, and the T-intercept is the temperature at ground level.

23. **(a)** $-2.94, -0.11, 3.05$ **(b)** $[-1, 2]$

24. **(a)** $M = kwh^2/L$ **(b)** 400 **(c)** 12,000 lb

FOCUS ON MODELING ■ PAGE 144

1. **(a)**

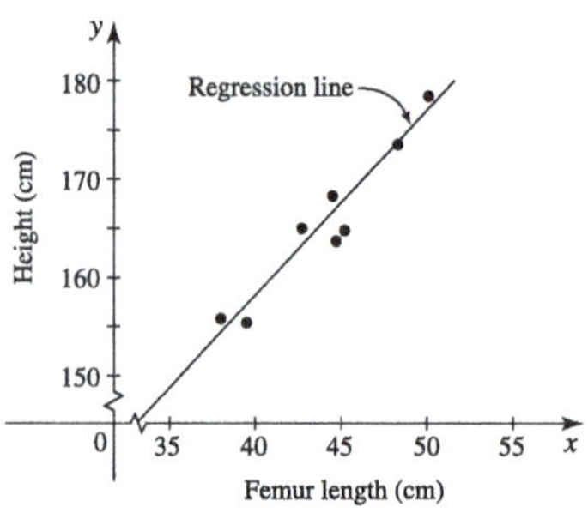

(b) $y = 1.8807x + 82.65$ **(c)** 191.7 cm

3. **(a)**

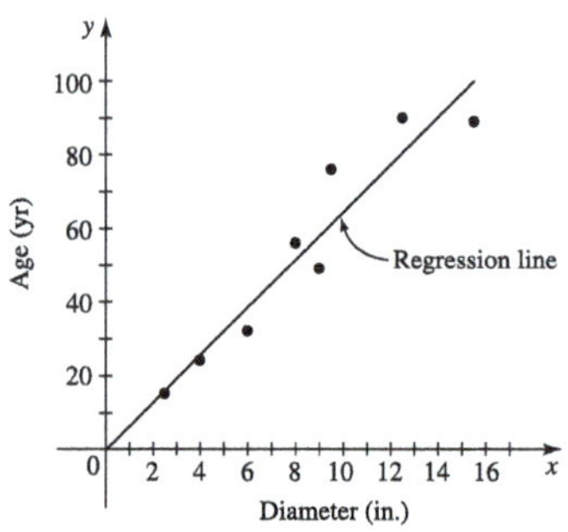

(b) $y = 6.451x - 0.1523$ **(c)** 116 years

5. **(a)**

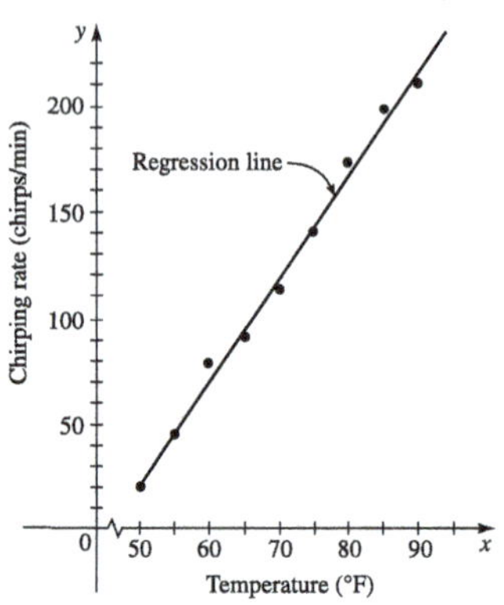

(b) $y = 4.857x - 220.97$ **(c)** 265 chirps/min

7. **(a)**

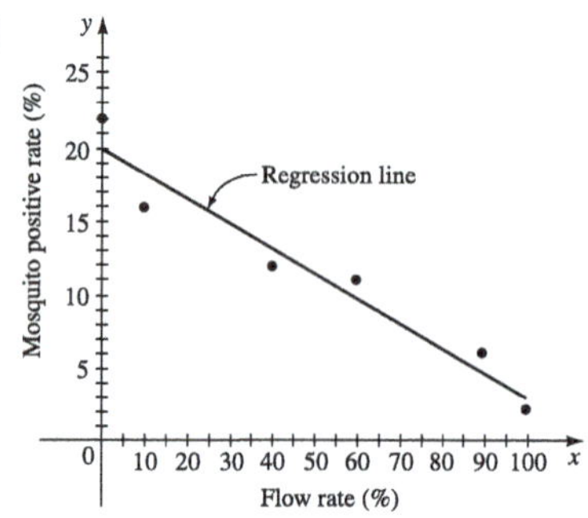

(b) $y = -0.168x + 19.89$ **(c)** 8.13%

9. **(a)**

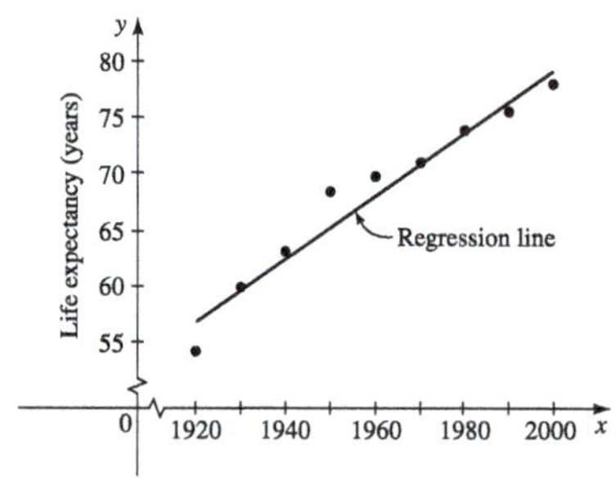

(b) $y = 0.2708x - 462.9$ **(c)** 80.4 years

CHAPTER 2

SECTION 2.1 ■ PAGE 155

1. **(a)** $f(-1) = 0$ **(b)** $f(2) = 9$ **(c)** $f(2) - f(-1) = 9$

2. domain, range **3.** **(a)** f and g **(b)** $f(5) = 10, g(5) = 0$

4. **(a)** square, add 3

(b)

x	0	2	4	6
$f(x)$	19	7	3	7

5. one; (i) **6.** **(a)** Yes **(b)** No **7.** $f(x) = 3x - 5$

9. $f(x) = (x - 1)^2$ **11.** Multiply by 2, then add 3

13. Add 1, then multiply by 5

15.

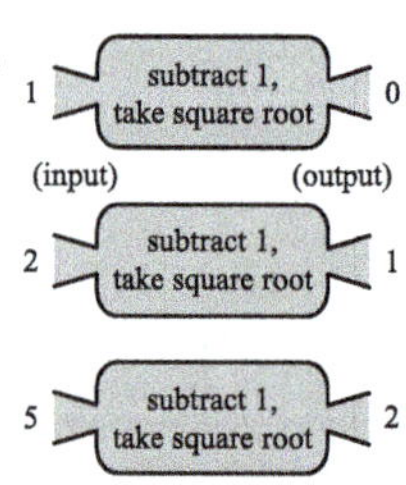

17.

x	$f(x)$
-1	8
0	2
1	0
2	2
3	8

19. $3, 3, -6, -\frac{23}{4}$ **21.** $-1, \frac{5}{3}, 0, \frac{1 - 2a}{3}, \frac{1 + 2a}{3}, \frac{3 - 2a}{3}$

23. $0, 15, 3, a^2 + 2a, x^2 - 2x, \frac{1}{a^2} + \frac{2}{a}$

25. $-\frac{1}{3}$, undefined, $\frac{1}{3}, \frac{1 - a}{1 + a}, \frac{2 - a}{a}, \frac{2 - x^2}{x^2}$

27. $3, -5, 3, 1 - 2\sqrt{2}, -a^2 - 6a - 5, -x^2 + 2x + 3, -x^4 - 2x^2 + 3$

29. $6, 2, 1, 2, 2|x|, 2(x^2 + 1)$ **31.** $4, 1, 1, 2, 3$

33. $8, -\frac{3}{4}, -1, 0, -1$ **35.** $x^2 + 4x + 5, x^2 + 6$

37. $x^2 + 4, x^2 + 8x + 16$ **39.** 12 **41.** -21

43. $5 - 2a, 5 - 2a - 2h, -2$ **45.** $5, 5, 0$

47. $\dfrac{a}{a+1}, \dfrac{a+h}{a+h+1}, \dfrac{1}{(a+h+1)(a+1)}$

49. $3 - 5a + 4a^2, 3 - 5a - 5h + 4a^2 + 8ah + 4h^2,$ $-5 + 8a + 4h$ **51.** $(-\infty, \infty), (-\infty, \infty)$

53. $[-2, 6], [-6, 18]$ **55.** $\{x \mid x \neq 3\}$ **57.** $\{x \mid x \neq \pm 1\}$

59. $[-1, \infty)$ **61.** $(-\infty, \infty)$ **63.** $(-\infty, \frac{1}{2}]$

65. $[-2, 3) \cup (3, \infty)$ **67.** $(-\infty, 0] \cup [6, \infty)$ **69.** $(4, \infty)$

71. $(\frac{1}{2}, \infty)$ **73. (a)** $f(x) = \dfrac{x}{3} + \dfrac{2}{3}$

(b)

x	$f(x)$
2	$\frac{4}{3}$
4	2
6	$\frac{8}{3}$
8	$\frac{10}{3}$

(c)

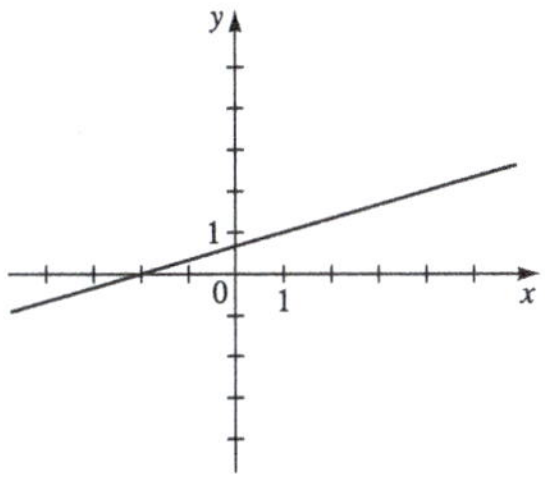

75. (a) $T(x) = 0.08x$

(b)

x	$T(x)$
2	0.16
4	0.32
6	0.48
8	0.64

(c)

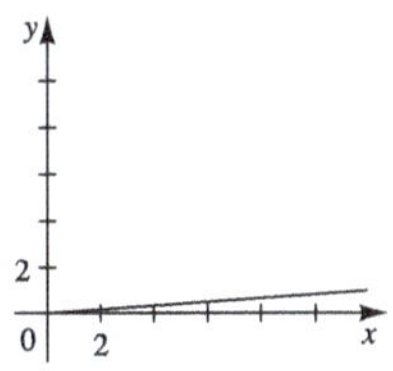

77. $(-\infty, \infty), \{1, 5\}$

79. (a) 50, 0 **(b)** $V(0)$ is the volume of the full tank, and $V(20)$ is the volume of the empty tank, 20 min later.

(c)

x	$V(x)$
0	50
5	28.125
10	12.5
15	3.125
20	0

(d) -50 gal

81. (a) 8.66 m, 6.61 m, 4.36 m **(b)** It will appear to get shorter.

83. (a) $v(0.1) = 4440, v(0.4) = 1665$

(b) Flow is faster near central axis.

(c)

r	$v(r)$
0	4625
0.1	4440
0.2	3885
0.3	2960
0.4	1665
0.5	0

(d) -4440 cm/s

85. (a) $T(5000) = 0, T(12{,}000) = 960, T(25{,}000) = 5350$

(b) The amount of tax paid on incomes of 5000, 12,000, and 25,000

87. (a) $T(x) = \begin{cases} 75x & \text{if } 0 \le x \le 2 \\ 150 + 50(x - 2) & \text{if } x > 2 \end{cases}$

(b) \$150, \$200, \$300 **(c)** Total cost of staying at the hotel

89.

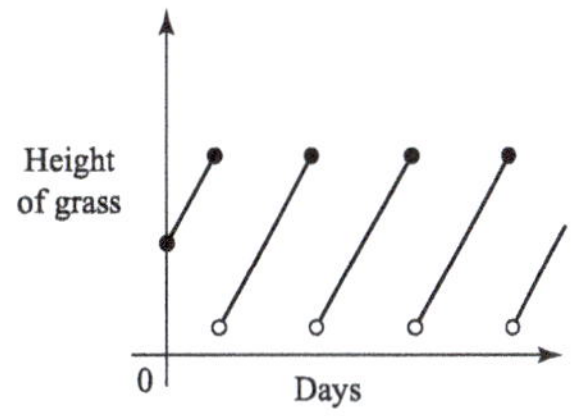

91.

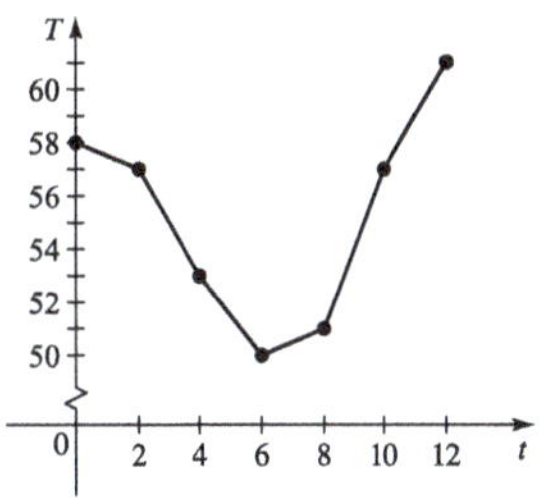

SECTION 2.2 ■ PAGE 166

1. $f(x), x^2 - 2, 7, 7$

x	$f(x)$	(x, y)
-2	2	$(-2, 2)$
-1	-1	$(-1, -1)$
0	-2	$(0, -2)$
1	-1	$(1, -1)$
2	2	$(2, 2)$

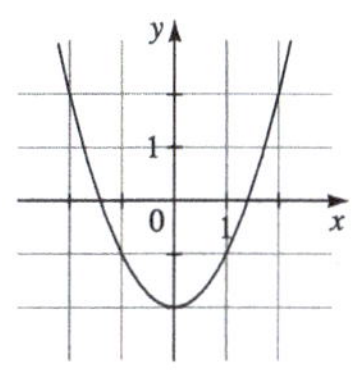

2. 10 **3.** 7 **4. (a)** IV **(b)** II **(c)** I **(d)** III

5.

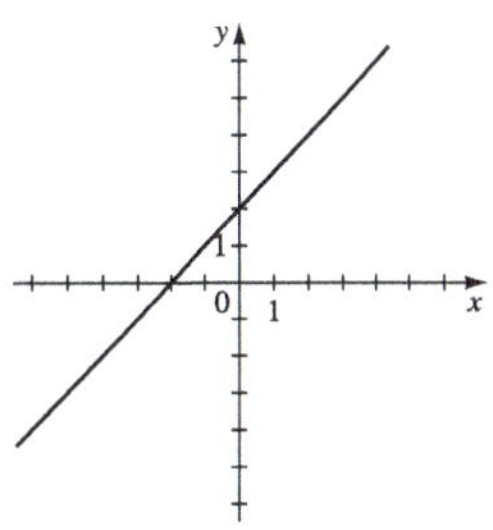

7.

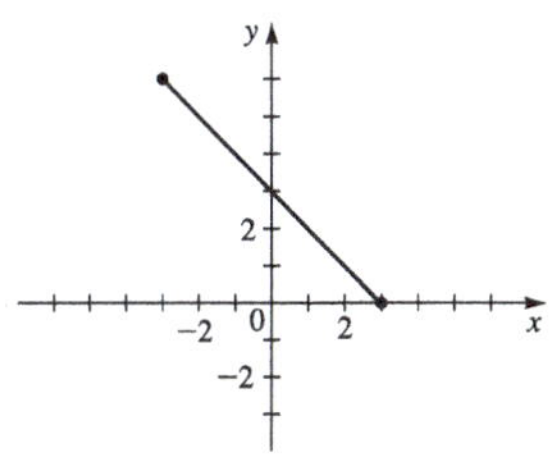

9.

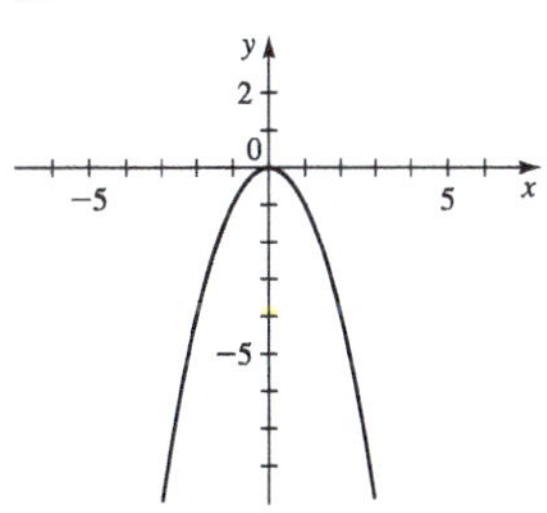

11.

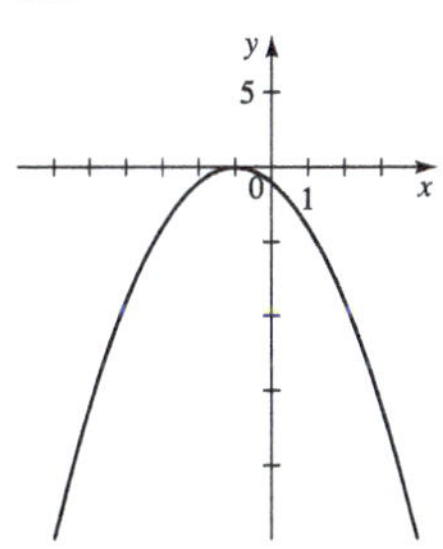

13.

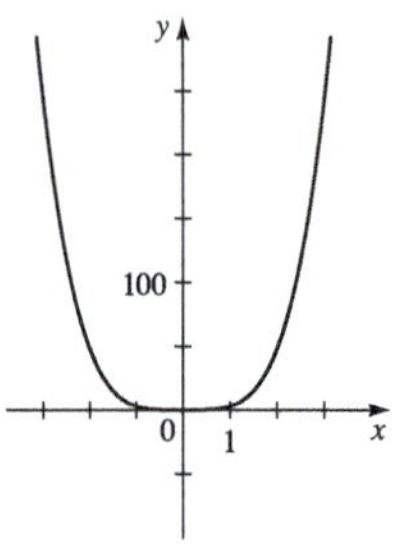

15.

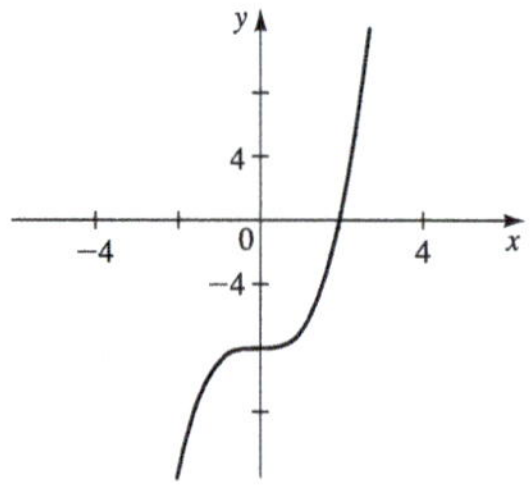

17.

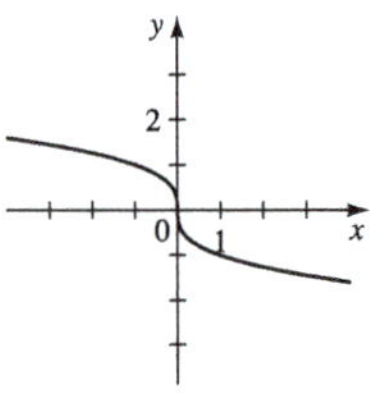

19.

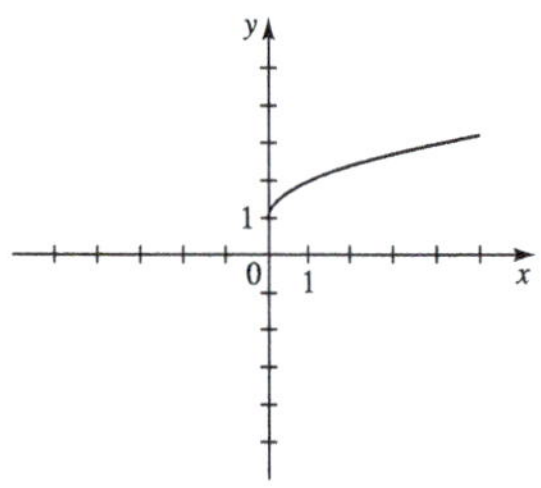

21.

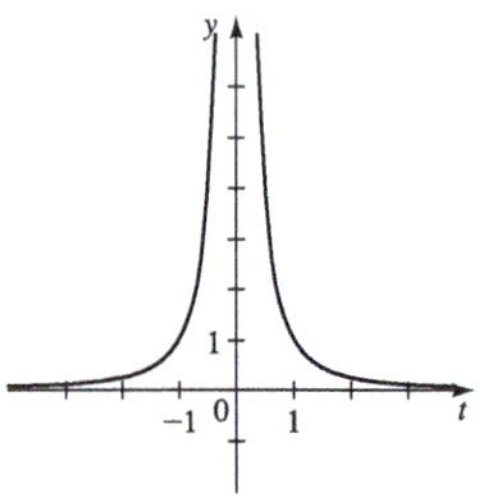

23.

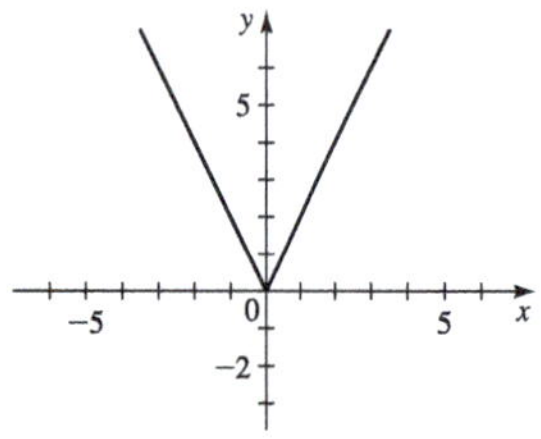

25.

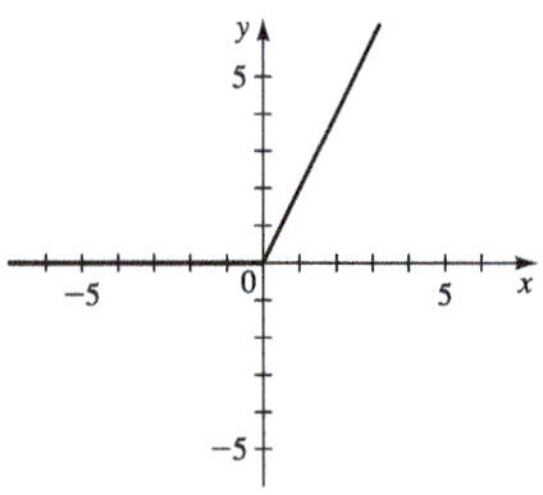

27.

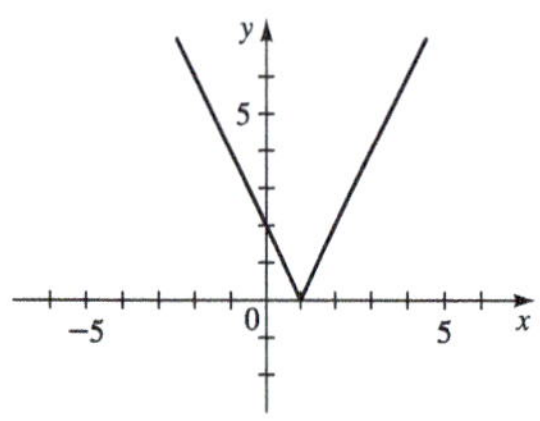

29. (a)

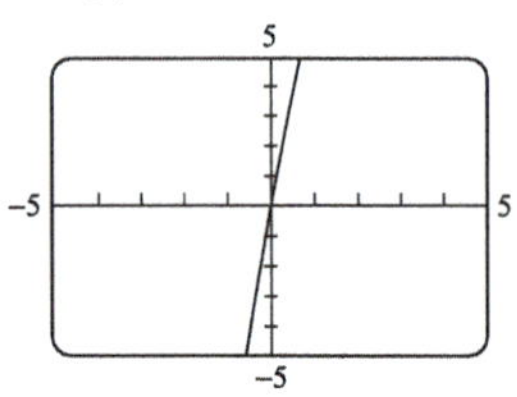

(b)

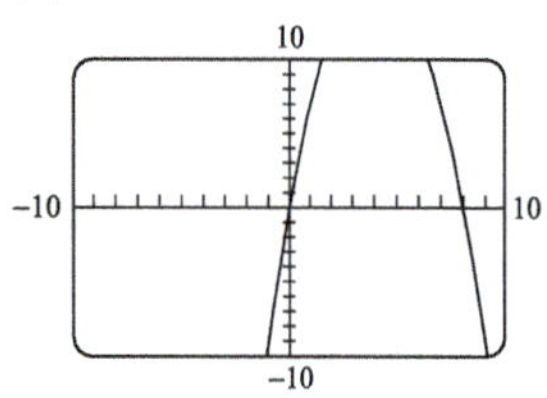

(c)

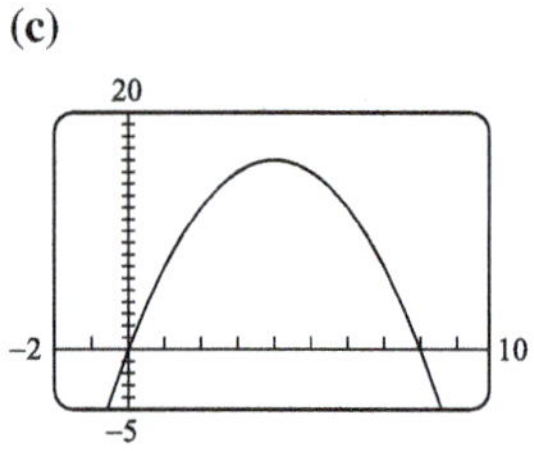

(d)

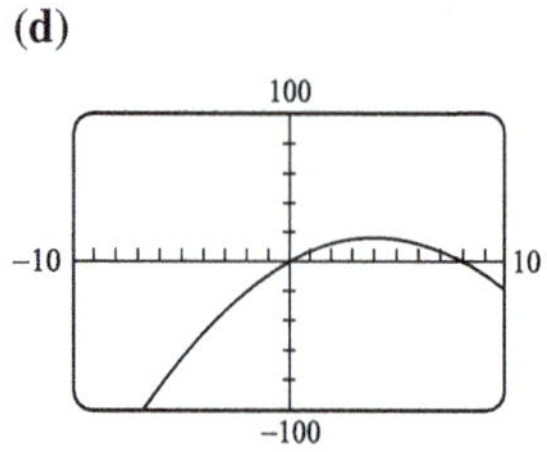

Graph (c) is the most appropriate.

31. (a)

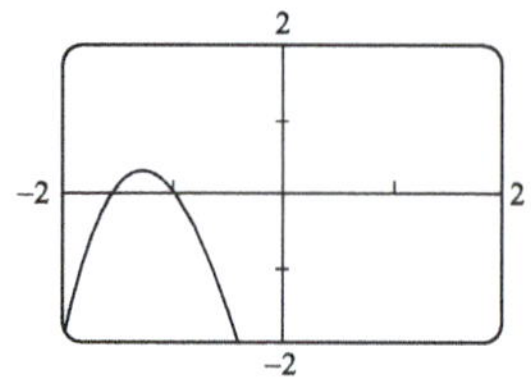

(b)

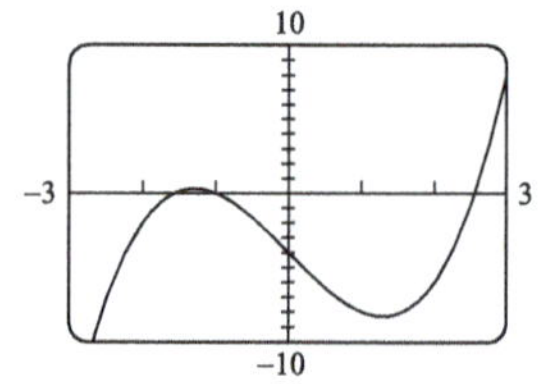

(c)

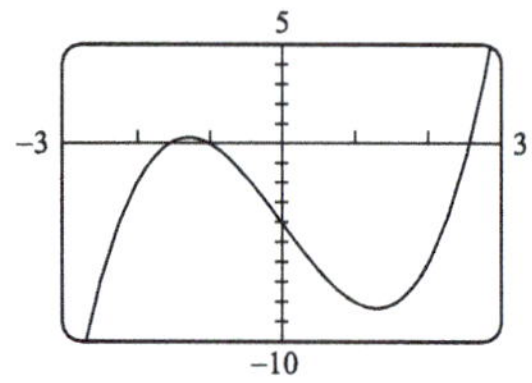

(d)

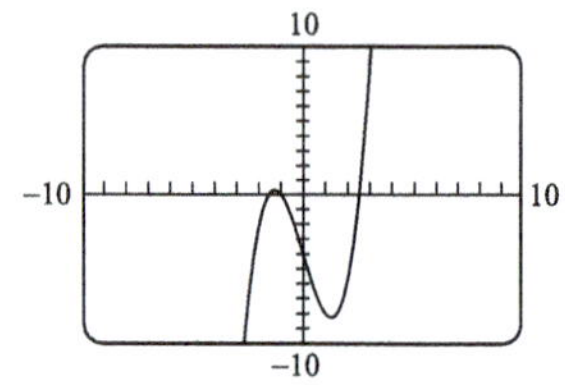

Graph (c) is the most appropriate.

33.

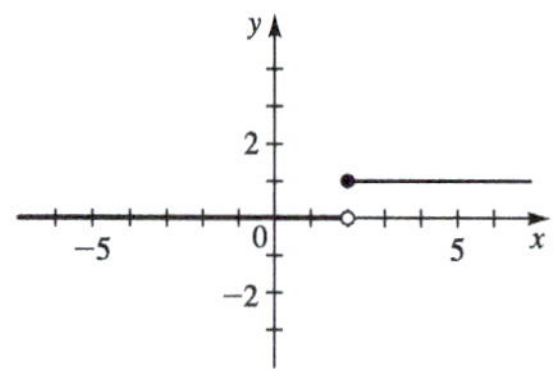

35.

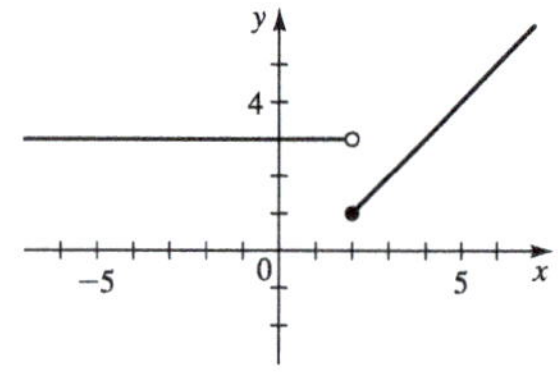

37.

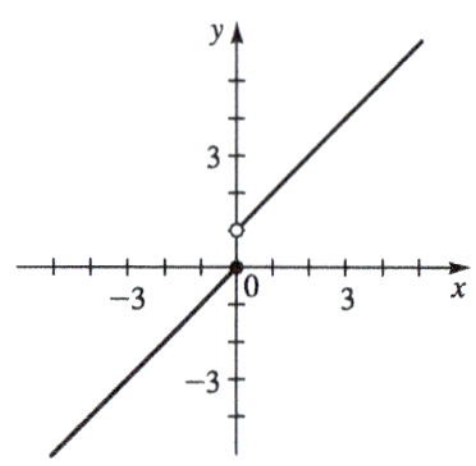

39.

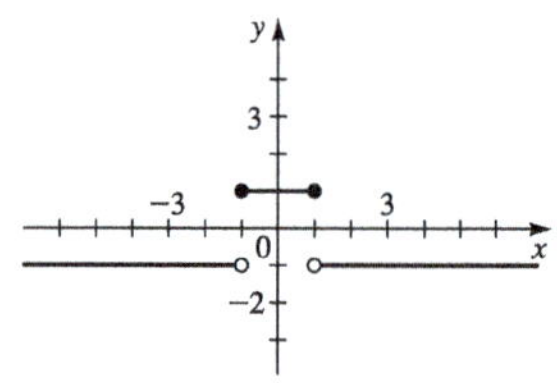

41.

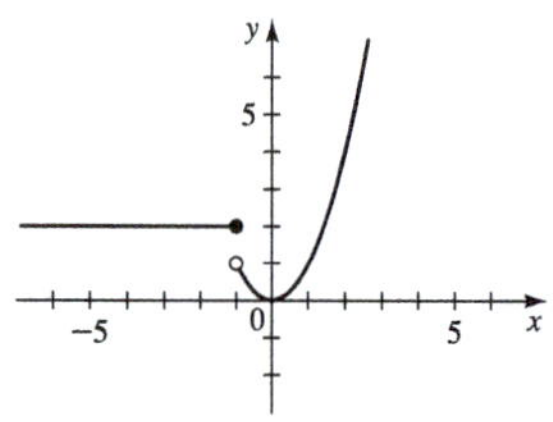

43.

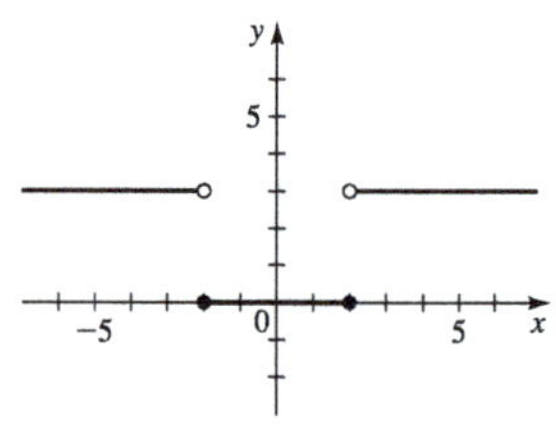

45.

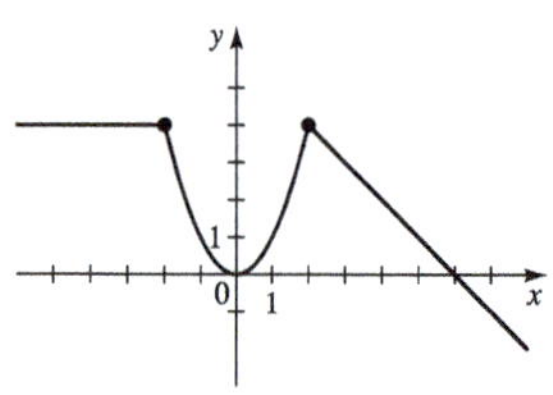

47.

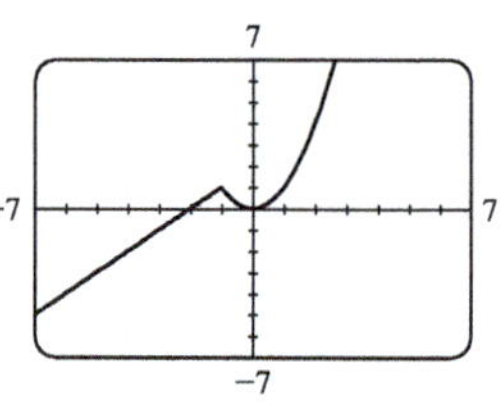

49. $f(x) = \begin{cases} -2 & \text{if } x < -2 \\ x & \text{if } -2 \le x \le 2 \\ 2 & \text{if } x > 2 \end{cases}$

51. (a) Yes **(b)** No **(c)** Yes **(d)** No
53. Function, domain $[-3, 2]$, range $[-2, 2]$
55. Not a function **57.** Yes **59.** No **61.** No **63.** No
65. Yes **67.** Yes

69. **(a)**

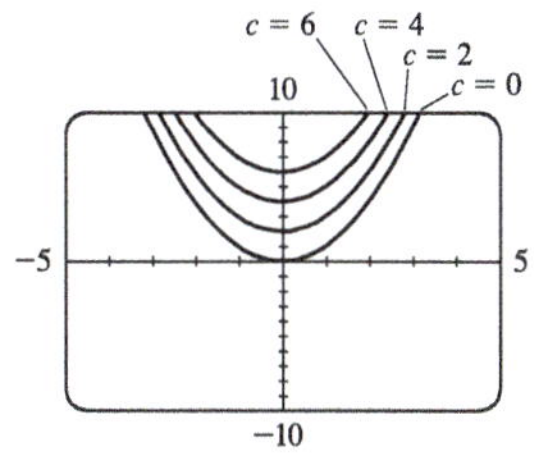

(b)

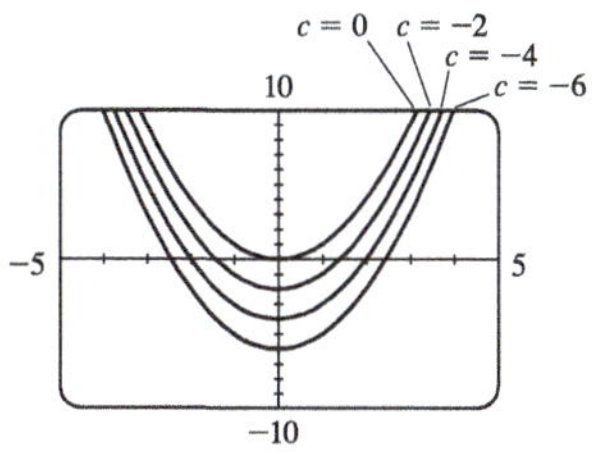

(c) If $c > 0$, then the graph of $f(x) = x^2 + c$ is the same as the graph of $y = x^2$ shifted upward c units. If $c < 0$, then the graph of $f(x) = x^2 + c$ is the same as the graph of $y = x^2$ shifted downward c units.

71. **(a)**

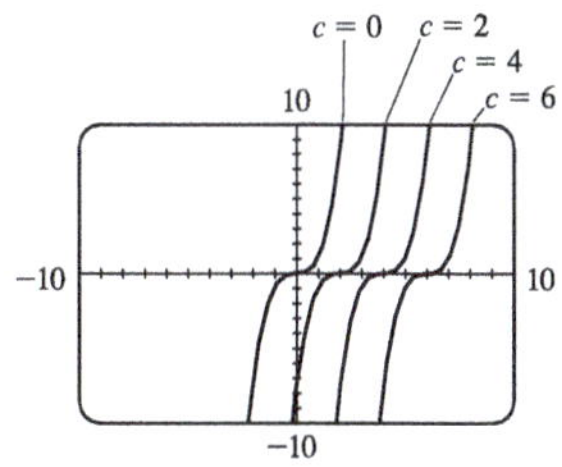

(b)

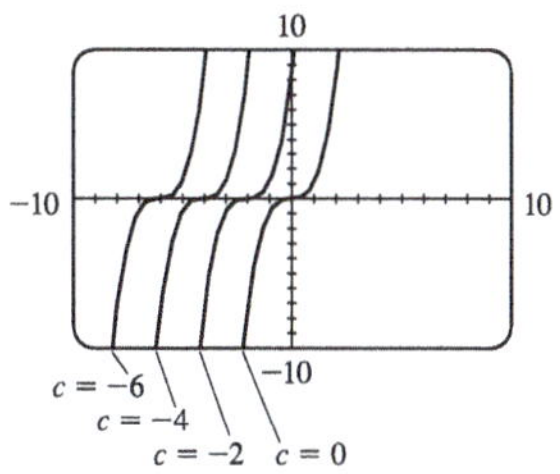

(c) If $c > 0$, then the graph of $f(x) = (x - c)^3$ is the same as the graph of $y = x^3$ shifted to the right c units. If $c < 0$, then the graph of $f(x) = (x - c)^3$ is the same as the graph of $y = x^3$ shifted to the left $|c|$ units.

73. **(a)**

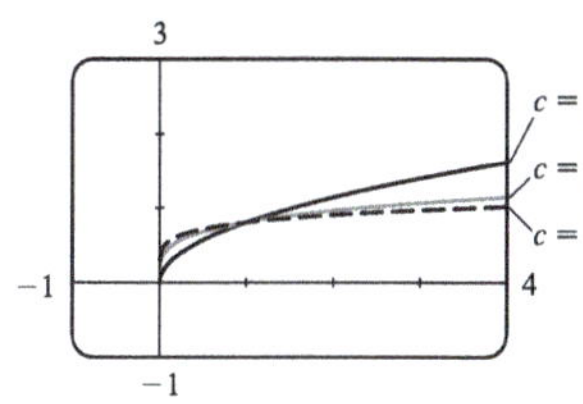

(b)

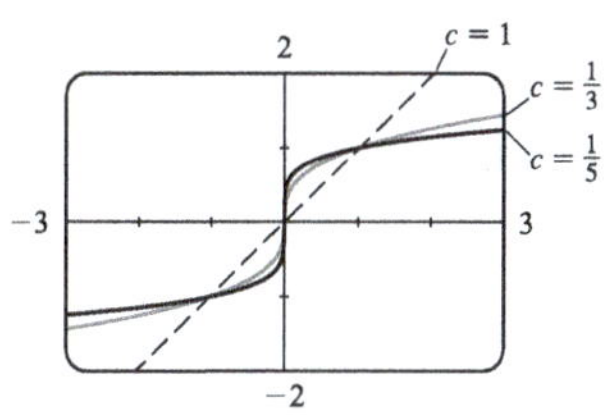

(c) Graphs of even roots are similar to $\sqrt{x}$; graphs of odd roots are similar to $\sqrt[3]{x}$. As c increases, the graph of $y = \sqrt[c]{x}$ becomes steeper near 0 and flatter when $x > 1$.

75. $f(x) = -\frac{7}{6}x - \frac{4}{3},\ -2 \le x \le 4$

77. $f(x) = \sqrt{9 - x^2},\ -3 \le x \le 3$

79.

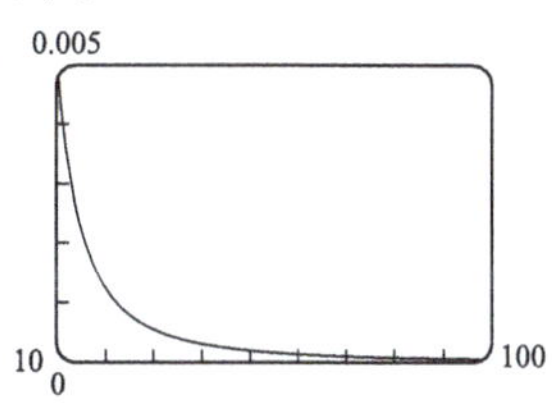

81. **(a)** $E(x) = \begin{cases} 6 + 0.10x & \text{if } 0 \le x \le 300 \\ 36 + 0.06(x - 300) & \text{if } x > 300 \end{cases}$

(b)

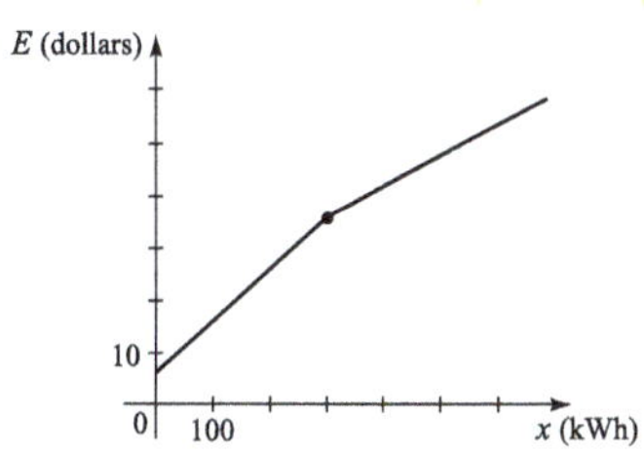

83. $P(x) = \begin{cases} 0.49 & \text{if } 0 < x \le 1 \\ 0.70 & \text{if } 1 < x \le 2 \\ 0.91 & \text{if } 2 < x \le 3 \\ 1.12 & \text{if } 3 < x \le 3.5 \end{cases}$

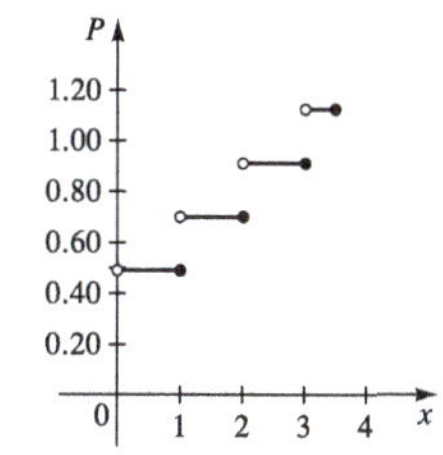

SECTION 2.3 ■ PAGE 178

1. $a, 4, 0, f(3) - f(1) = 4$ **2.** $x, y, (-\infty, \infty)(-\infty, 7]$
3. **(a)** increase, $(-\infty, 2), (4, 5)$ **(b)** decrease, $(2, 4), (5, \infty)$
4. **(a)** largest, 7, 6, 5 **(b)** smallest, 2, 4 **5.** x; x; 1, 7, $[1, 7]$
6. **(a)** $2x + 1, -x + 4$; 1 **(b)** $2x + 1, -x + 4$, higher; $(-\infty, 1)$ **7.** **(a)** $1, -1, 3, 4$ **(b)** Domain $[-3, 4]$, range $[-1, 4]$ **(c)** $-3, 2, 4$ **(d)** $-3 \le x \le 2$ and $x = 4$ **(e)** 1
9. **(a)** $f(0)$ **(b)** $g(-3)$ **(c)** $-2, 2$
(d) $\{x \mid -4 \le x \le -2 \text{ or } 2 \le x \le 3\}$ **(e)** $\{x \mid -2 < x < 2\}$

11. **(a)**

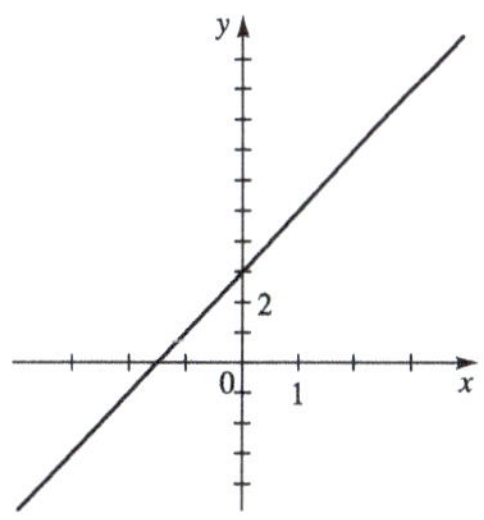

(b) $(-\infty, \infty), (-\infty, \infty)$

13. **(a)**

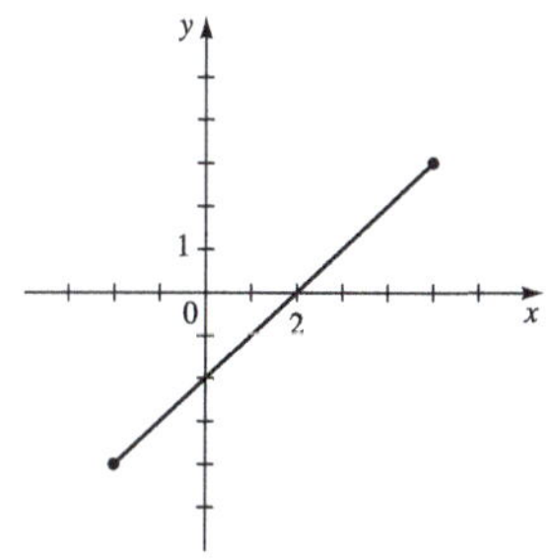

(b) $[-2, 5], [-4, 3]$

15. **(a)**

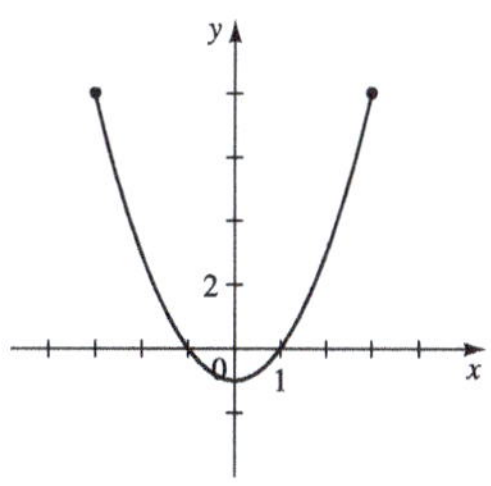

(b) $[-3, 3], [-1, 8]$

17. **(a)**

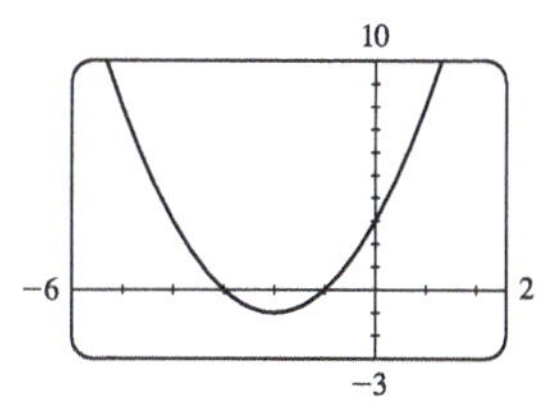

(b) Domain $(-\infty, \infty)$, range $[-1, \infty)$

19. **(a)**

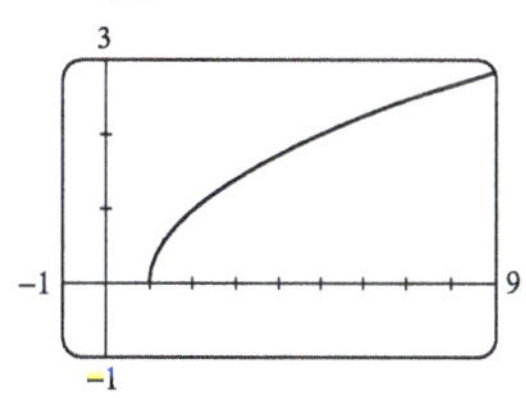

(b) Domain $[1, \infty)$, range $[0, \infty)$

21. **(a)**

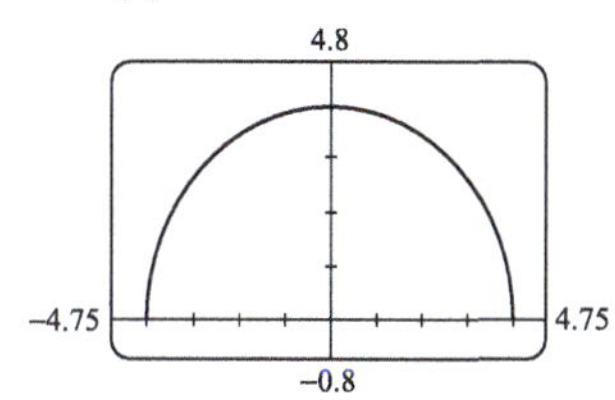

(b) Domain $[-4, 4]$, range $[0, 4]$

23. **(a)** $x = 3$ **(b)** $x > 3$
25. **(a)** $x = -2, 1$ **(b)** $-2 \le x \le 1$
27. **(a)** $x \approx -4.32, -1.12, 1.44$
(b) $-4.32 \le x \le -1.12$ or $x \ge 1.44$

29. (a) $x = -1, -0.25, 0.25$
(b) $-1 \le x \le -0.25$ or $x \ge 0.25$
31. (a) Domain $[-1, 4]$, range $[-1, 3]$ **(b)** Increasing on $(-1, 1)$ and $(2, 4)$, decreasing on $(1, 2)$
33. (a) Domain $[-3, 3]$, range $[-2, 2]$ **(b)** Increasing on $(-2, -1)$ and $(1, 2)$, decreasing on $(-3, -2)$, $(-1, 1)$, and $(2, 3)$

35. (a)

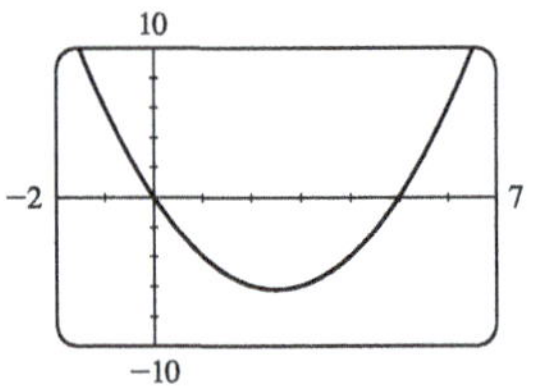

(b) Domain $(-\infty, \infty)$, range $[-6.25, \infty)$
(c) Increasing on $(2.5, \infty)$; decreasing on $(-\infty, 2.5)$

37. (a)

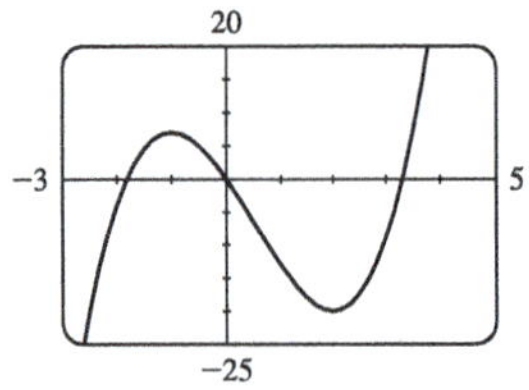

(b) Domain $(-\infty, \infty)$, range $(-\infty, \infty)$
(c) Increasing on $(-\infty, -1)$, $(2, \infty)$; decreasing on $(-1, 2)$

39. (a)

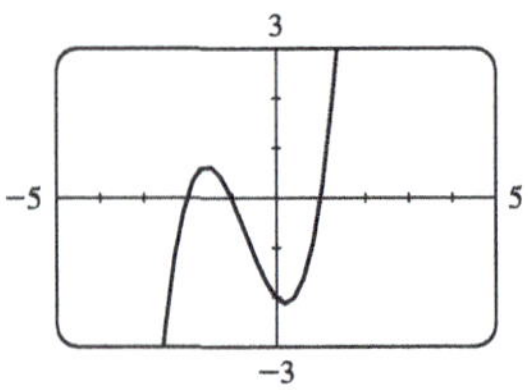

(b) Domain $(-\infty, \infty)$, range $(-\infty, \infty)$
(c) Increasing on $(-\infty, -1.55)$, $(0.22, \infty)$; decreasing on $(-1.55, 0.22)$

41. (a)

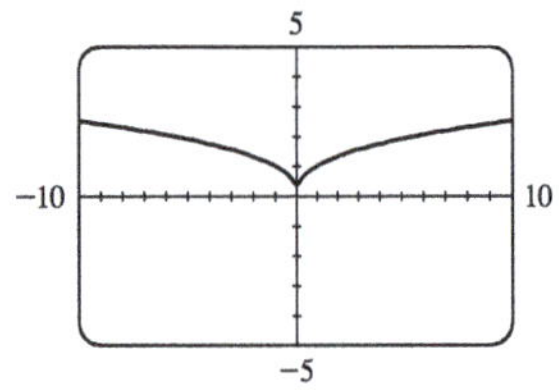

(b) Domain $(-\infty, \infty)$, range $[0, \infty)$
(c) Increasing on $(0, \infty)$; decreasing on $(-\infty, 0)$

43. (a) Local maximum 2 when $x = 0$; local minimum -1 when $x = -2$, local minimum 0 when $x = 2$ **(b)** Increasing on $(-2, 0) \cup (2, \infty)$; decreasing on $(-\infty, -2) \cup (0, 2)$
45. (a) Local maximum 0 when $x = 0$; local maximum 1 when $x = 3$, local minimum -2 when $x = -2$, local minimum -1 when $x = 1$ **(b)** Increasing on $(-2, 0) \cup (1, 3)$; decreasing on $(-\infty, -2) \cup (0, 1) \cup (3, \infty)$ **47. (a)** Local maximum ≈ 0.38 when $x \approx -0.58$; local minimum ≈ -0.38 when $x \approx 0.58$
(b) Increasing on $(-\infty, -0.58) \cup (0.58, \infty)$; decreasing on $(-0.58, 0.58)$ **49. (a)** Local maximum ≈ 0 when $x = 0$; local minimum ≈ -13.61 when $x \approx -1.71$, local minimum ≈ -73.32 when $x \approx 3.21$
(b) Increasing on $(-1.71, 0) \cup (3.21, \infty)$; decreasing on $(-\infty, -1.71) \cup (0, 3.21)$ **51. (a)** Local maximum ≈ 5.66 when $x \approx 4.00$ **(b)** Increasing on $(-\infty, 4.00)$; decreasing on $(4.00, 6.00)$ **53. (a)** Local maximum ≈ 0.38 when $x \approx -1.73$; local minimum ≈ -0.38 when $x \approx 1.73$ **(b)** Increasing on $(-\infty, -1.73) \cup (1.73, \infty)$; decreasing on $(-1.73, 0) \cup (0, 1.73)$
55. (a) 500 MW, 725 MW **(b)** Between 3:00 A.M. and 4:00 A.M. **(c)** Just before noon **(d)** -100 MW
57. (a) Increasing on $(0, 30) \cup (32, 68)$; decreasing on $(30, 32)$ **(b)** He went on a crash diet and lost weight, only to regain it again later. **(c)** 100 lb **59. (a)** Increasing on $(0, 150) \cup (300, \infty)$; decreasing on $(150, 300)$ **(b)** Local maximum when $x = 150$; local minimum when $x = 300$
(c) -50 ft **61.** Runner A won the race. All runners finished. Runner B fell but got up again to finish second.

63. (a)

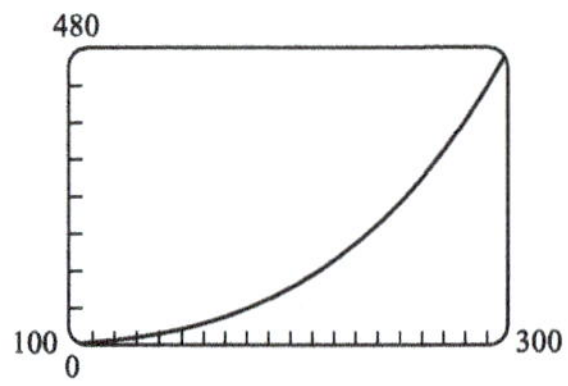

(b) Increases **65.** 7.5 mi/h

SECTION 2.4 ■ PAGE 187

1. $\dfrac{100 \text{ miles}}{2 \text{ hours}} = 50$ mi/h **2.** $\dfrac{f(b) - f(a)}{b - a}$ **3.** $\dfrac{25 - 1}{5 - 1} = 6$
4. (a) secant **(b)** 3 **5. (a)** Yes **(b)** Yes **6. (a)** No **(b)** No **7. (a)** 2 **(b)** $\frac{2}{3}$ **9. (a)** -4 **(b)** $-\frac{4}{5}$ **11. (a)** 3 **(b)** 3 **13. (a)** -5 **(b)** -1 **15. (a)** 51 **(b)** 17
17. (a) 600 **(b)** 60 **19. (a)** $5h^2 + 30h$ **(b)** $5h + 30$
21. (a) $\dfrac{1 - a}{a}$ **(b)** $-\dfrac{1}{a}$ **23. (a)** $\dfrac{-2h}{a(a + h)}$ **(b)** $\dfrac{-2}{a(a + h)}$
25. (a) $\frac{1}{2}$ **27.** f; g; 0, 1.5 **29.** -0.25 ft/day
31. (a) 245 persons/year **(b)** -328.5 persons/year **(c)** 1997–2001 **(d)** 2001–2006 **33. (a)** 14 players/year **(b)** 18 players/year **(c)** -103 players/year **(d)** 2006–2007, 2004–2005 **35.** First 20 minutes: -4.05°F/min, next 20 minutes: -1.5°F/min; first interval **37. (a)** All 10 m/s **(b)** Skier A started quickly and slowed down, skier B maintained a constant speed, and skier C started slowly and sped up.

SECTION 2.5 ■ PAGE 195

1. (a) linear, a, b **(b)** line **2. (a)** -5 **(b)** line, -5, 7 **3.** 15 **4.** 15 gal/min **5.** Upward **6.** Yes, 0, 0 **7.** Yes, $f(x) = \frac{1}{3}x + 3$ **9.** No **11.** Yes, $f(x) = \frac{1}{5}x + \frac{1}{5}$ **13.** No

15. 2

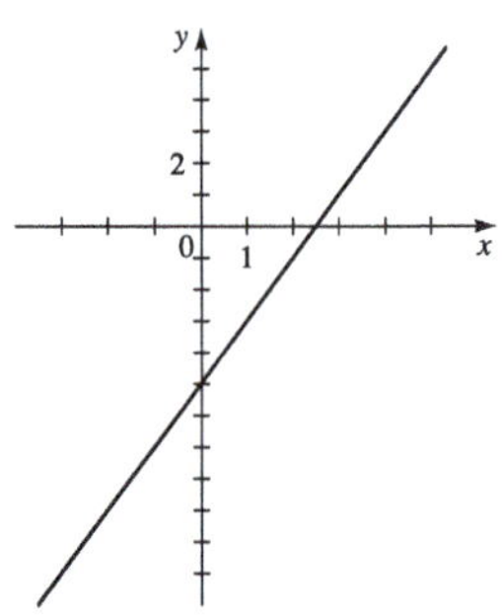

17. $-\frac{2}{3}$

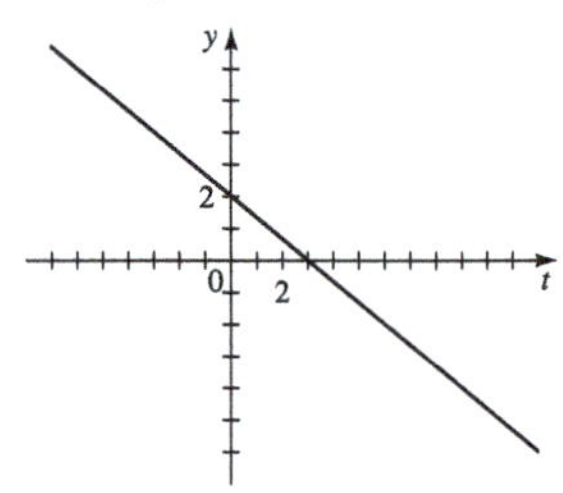

19. (a)

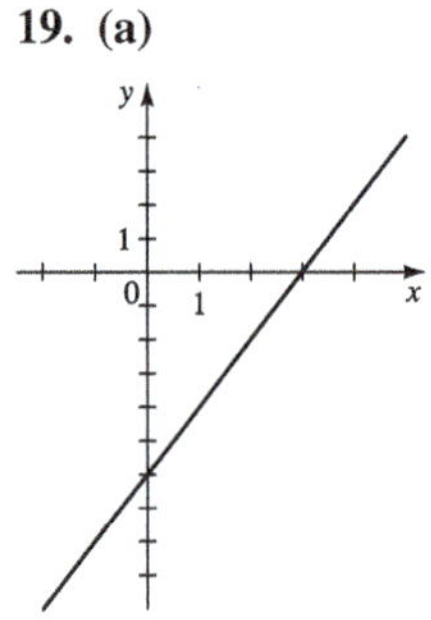

(b) 2 **(c)** 2

21. (a)

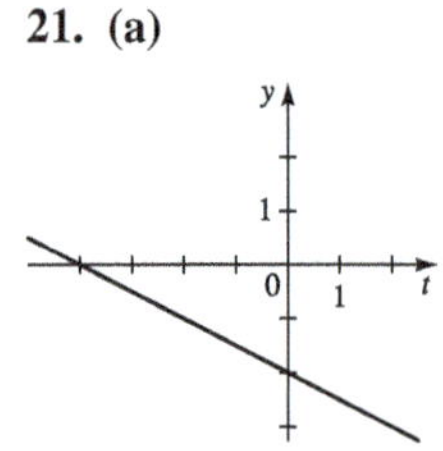

(b) -0.5 **(c)** -0.5

23. (a)

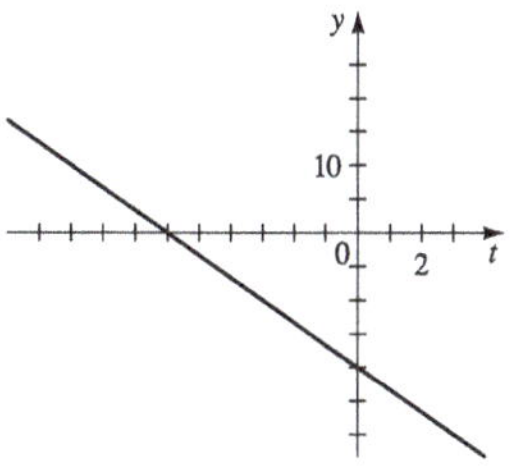

(b) $-\frac{10}{3}$ **(c)** $-\frac{10}{3}$

25. (a)

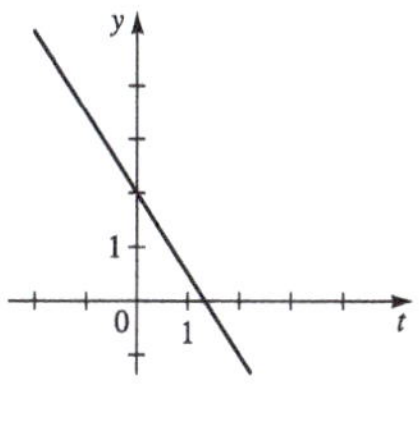

(b) $-\frac{3}{2}$ **(c)** $-\frac{3}{2}$

27. $f(x) = 3x - 1$ **29.** $h(x) = \frac{1}{2}x + 3$
31. (a) $\frac{3}{2}$ **(b)** $f(x) = \frac{3}{2}x + 7$
33. (a) 1 **(b)** $f(x) = x + 3$
35. (a) $-\frac{1}{2}$ **(b)** $f(x) = -\frac{1}{2}x + 2$
37.

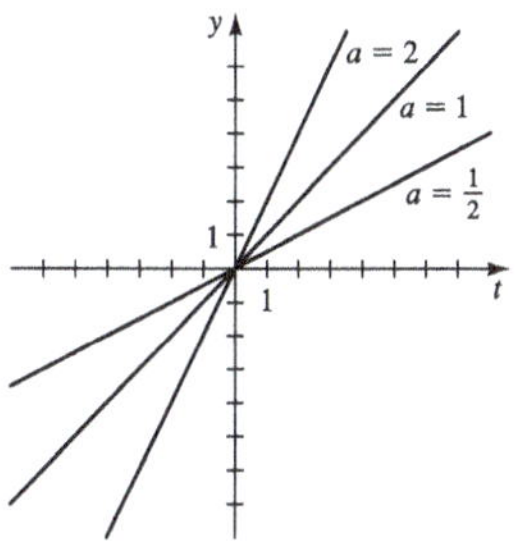

As a increases, the graph of f becomes steeper and the rate of change increases.

39. (a)

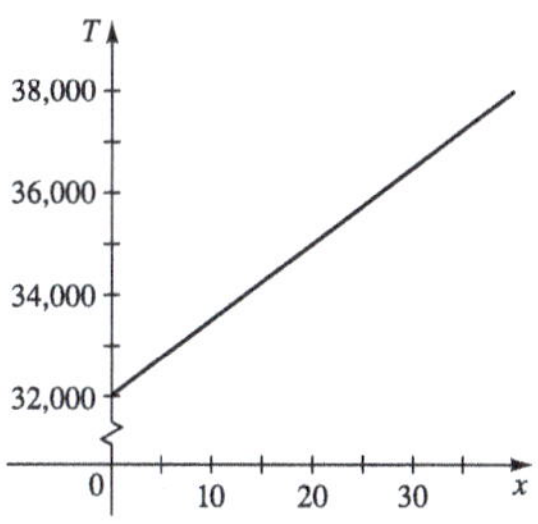

(b) 150
(c) 150,000 tons/year

41. (a) $V(t) = 0.5t + 2$ **(b)** 26 s
43. (a) $\frac{1}{12}$, $H(x) = \frac{1}{12}x$ **(b)** 12.5 in.
45. (a) Jari **(b)** Jade: 60 mi/h; Jari: 70 mi/h
(c) Jade: $f(t) = t + 10$; Jari $g(t) = \frac{7}{6}t$ **47.** 3.16 mi
49. (a) $C(x) = \frac{1}{4}x + 260$
(b) $\frac{1}{4}$ **(c)** \$0.25/mi

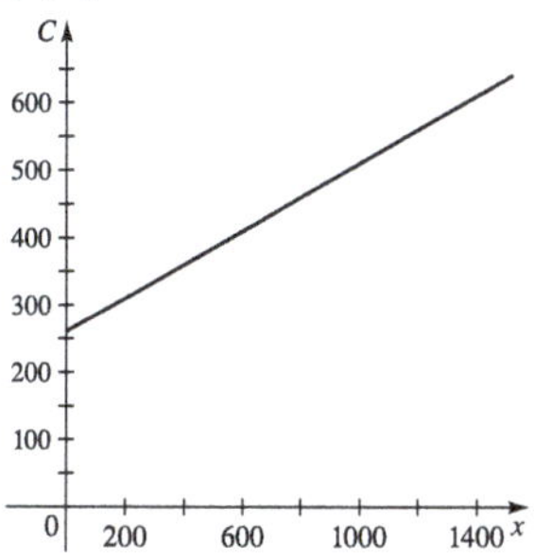

SECTION 2.6 ■ PAGE 206

1. (a) up **(b)** left **2. (a)** down **(b)** right **3. (a)** x-axis **(b)** y-axis **4. (a)** II **(b)** I **(c)** III **(d)** IV **5.** Symmetric about the y-axis **6.** Symmetric about the origin **7. (a)** Shift downward 1 unit **(b)** Shift to the right 2 units **9. (a)** Reflect about the y-axis **(b)** Stretch vertically by a factor of 3
11. (a) Shift to the right 5 units, then upward 2 units
(b) Shift to the left 1 unit, then downward 1 unit
13. (a) Reflect in the x-axis, then shift upward 5 units
(b) Stretch vertically by a factor of 3, then shift downward 5 units **15. (a)** Shift to the left 5 units, stretch vertically by a factor of 2, then shift downward 1 unit **(b)** Shift to the right 3 units, shrink vertically by a factor of $\frac{1}{4}$, then shift upward 5 units

17. (a) Shrink horizontally by a factor of $\frac{1}{4}$
(b) Stretch horizontally by a factor of 4 **19. (a)** Shift to the left 2 units **(b)** Shift upward 2 units **21. (a)** Shift to the left 2 units, then shift downward 2 units **(b)** Shift to the right 2 units, then shift upward 2 units
23. (a)

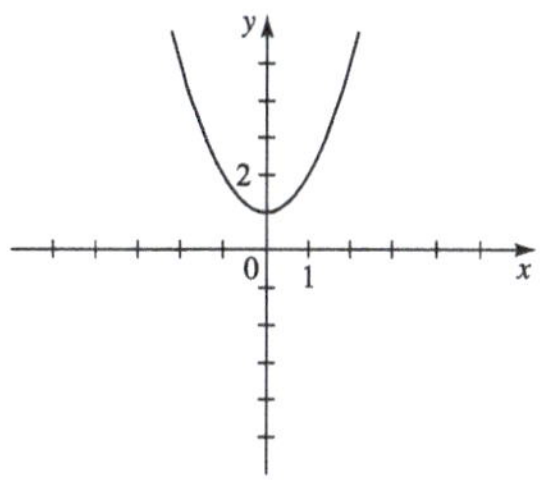

(b)

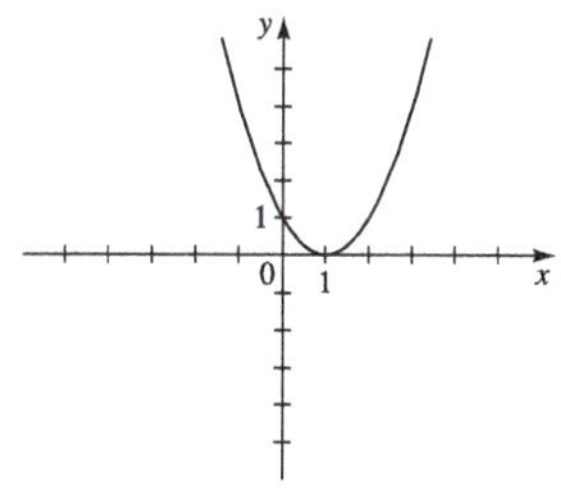

(c)

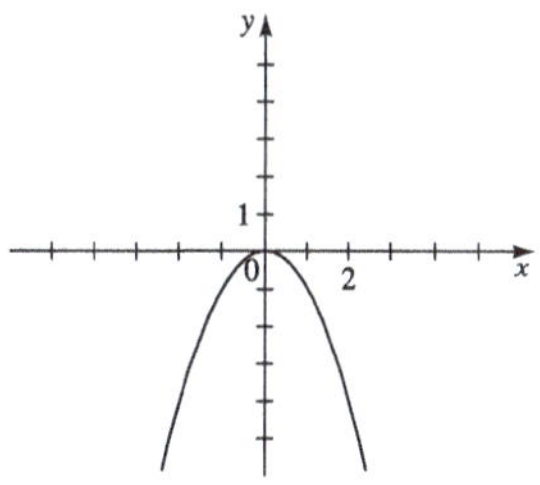

(d)

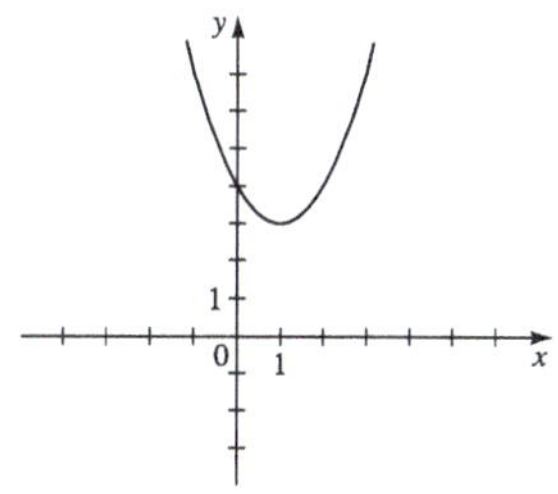

25. II **27.** I
29.

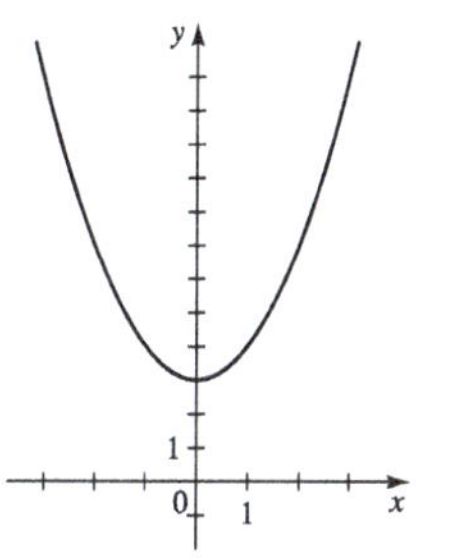

31.

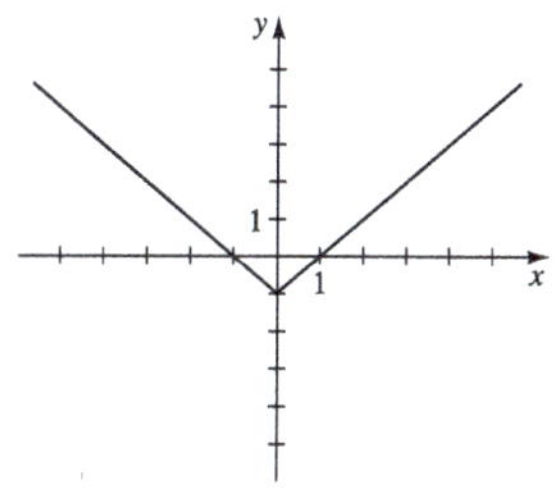

33.

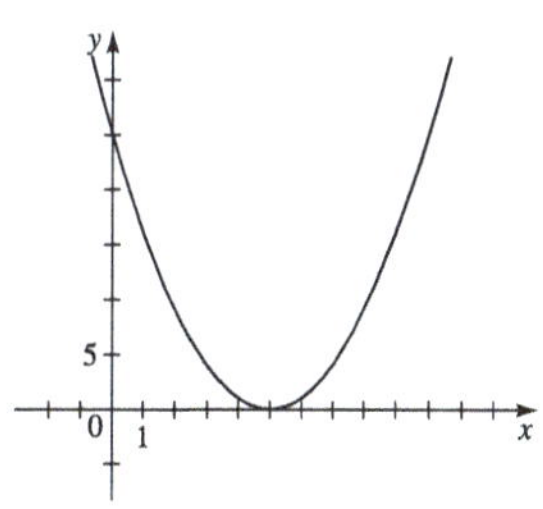

35.

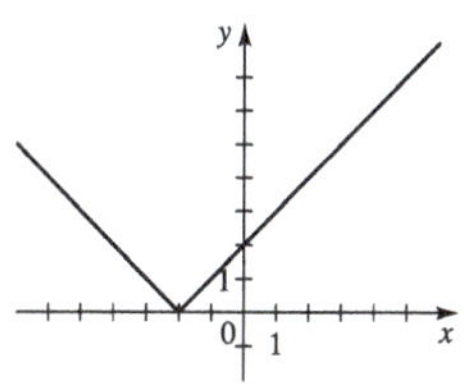

37.

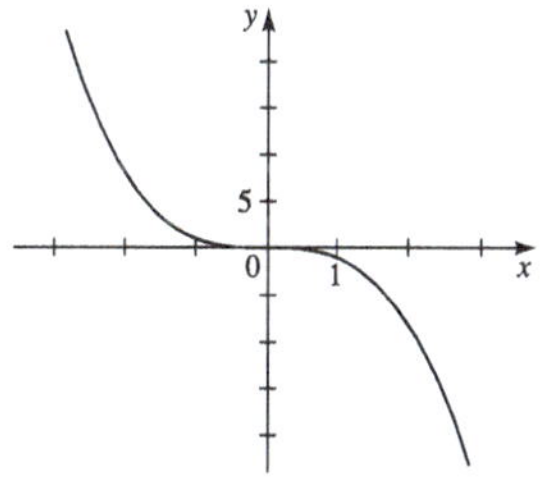

39.

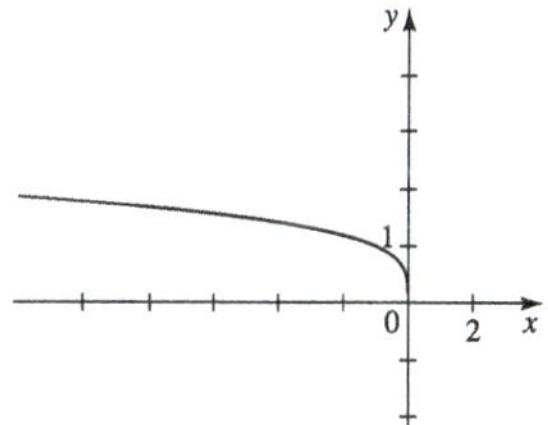

41.

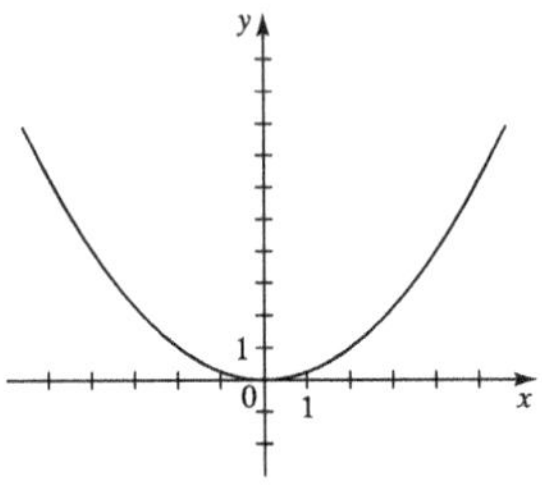

43.

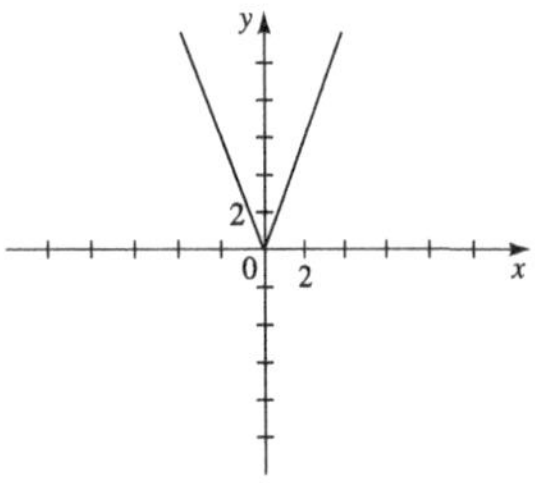

45.

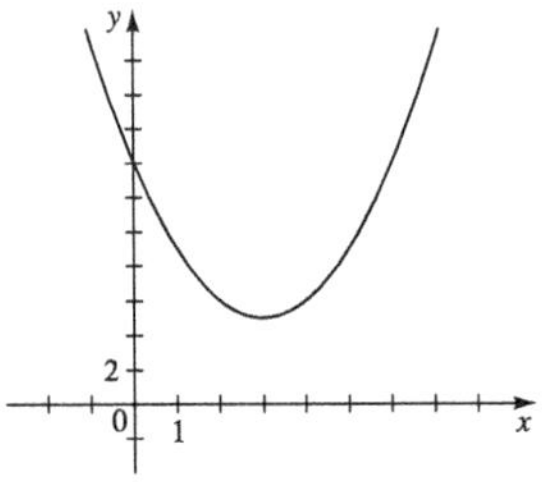

47.

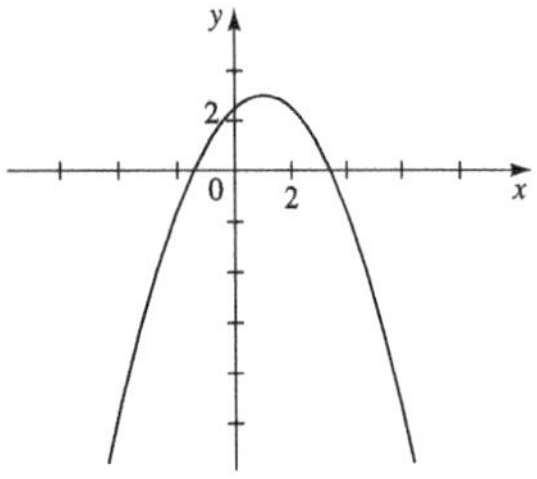

49.

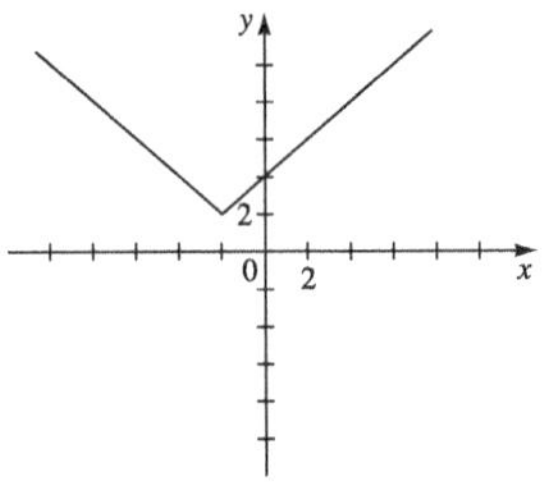

51.

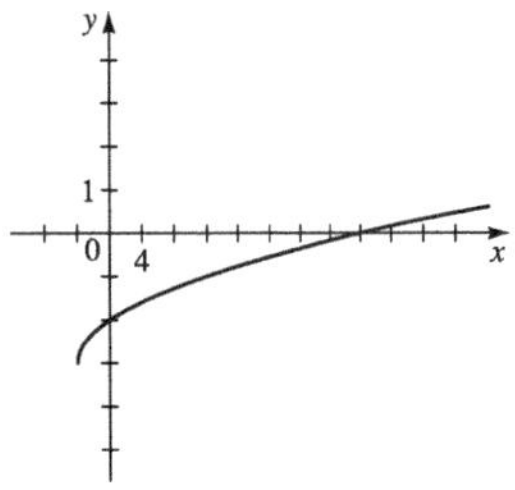

53. $y = x^2 - 3$ **55.** $y = \sqrt{x + 2}$ **57.** $y = |x + 2| - 5$
59. $y = \sqrt[4]{-x} + 1$ **61.** $y = 2(x - 3)^2 - 2$
63. $g(x) = (x - 2)^2$ **65.** $g(x) = |x + 1| + 2$
67. $g(x) = -\sqrt{x + 2}$ **69. (a)** 3 **(b)** 1 **(c)** 2 **(d)** 4

71. (a)

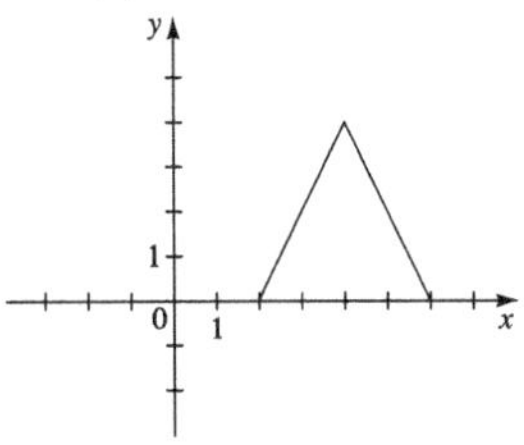

(b)

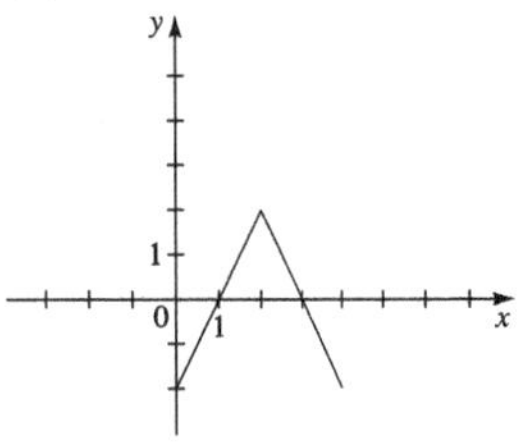

(c)

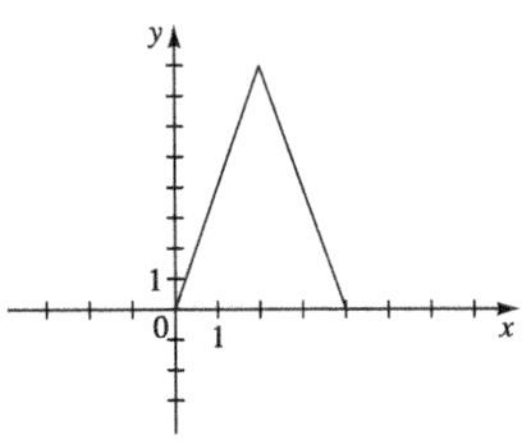

(d)

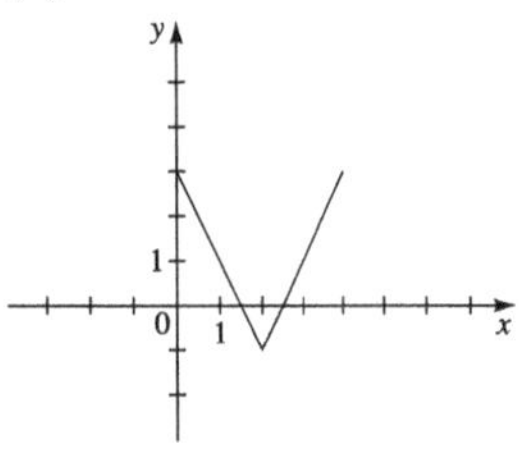

(e)

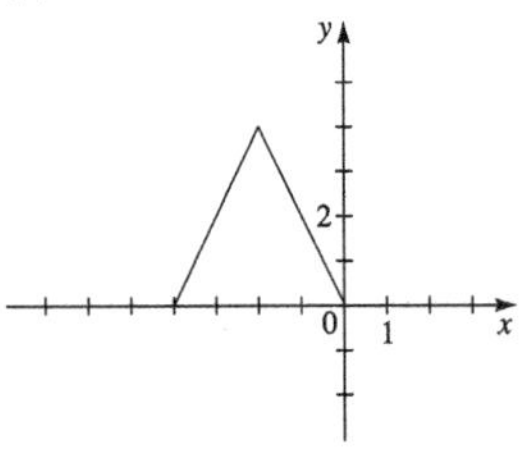

(f)

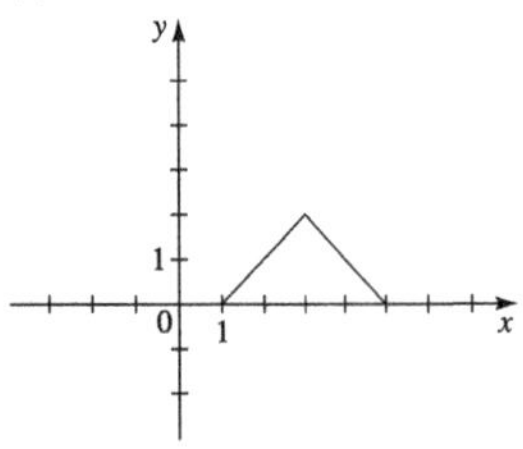

73. (a)

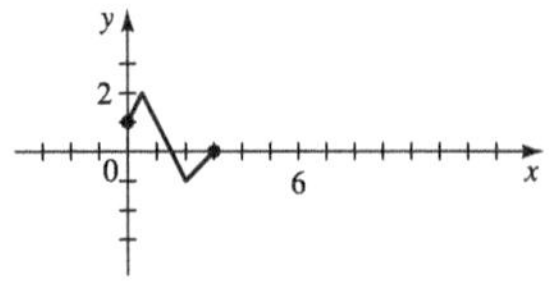

(b)

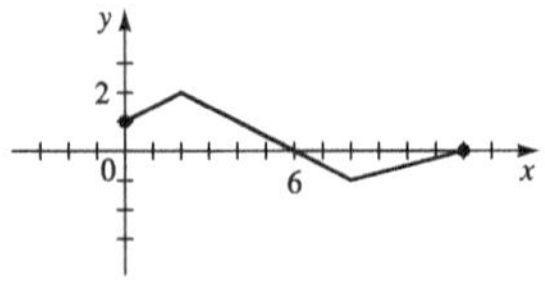

75.

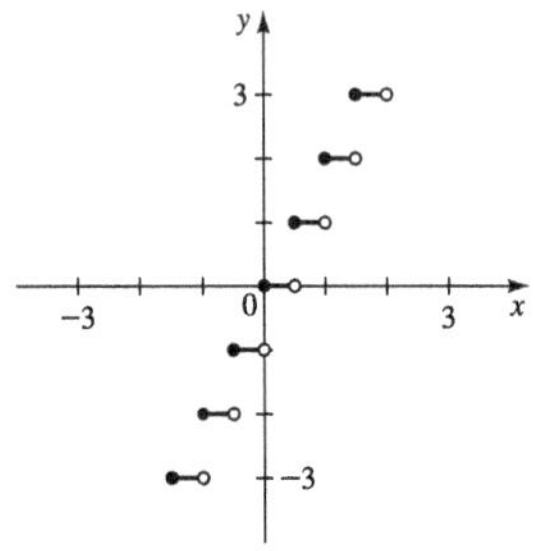

77.

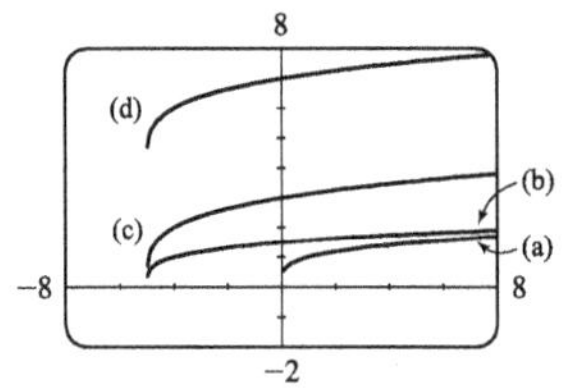

For part (b) shift the graph in (a) to the left 5 units; for part (c) shift the graph in (a) to the left 5 units and stretch vertically by a factor of 2; for part (d) shift the graph in (a) to the left 5 units, stretch vertically by a factor of 2, and then shift upward 4 units.

79.

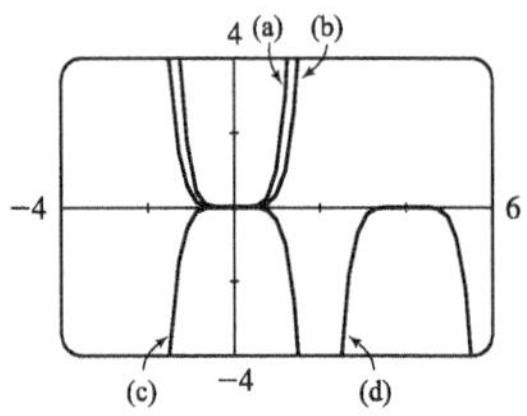

For part (b) shrink the graph in (a) vertically by a factor of $\frac{1}{3}$; for part (c) shrink the graph in (a) vertically by a factor of $\frac{1}{3}$ and reflect in the x-axis; for part (d) shift the graph in (a) to the right 4 units, shrink vertically by a factor of $\frac{1}{3}$, and then reflect in the x-axis.

81.

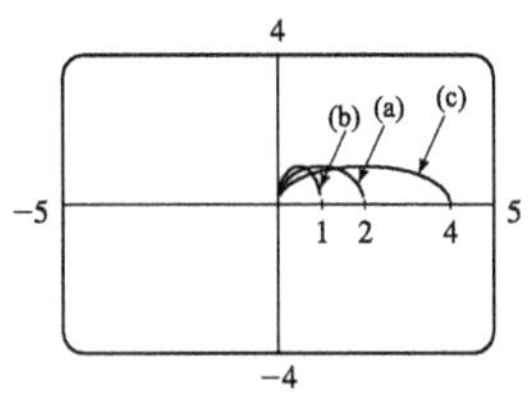

The graph in part (b) is shrunk horizontally by a factor of $\frac{1}{2}$ and the graph in part (c) is stretched by a factor of 2.

83. Even

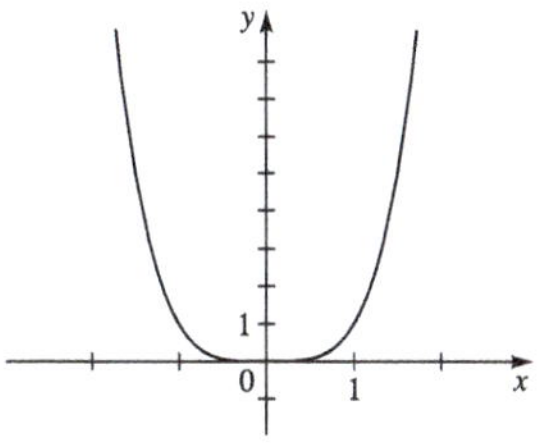

85. Neither
87. Odd

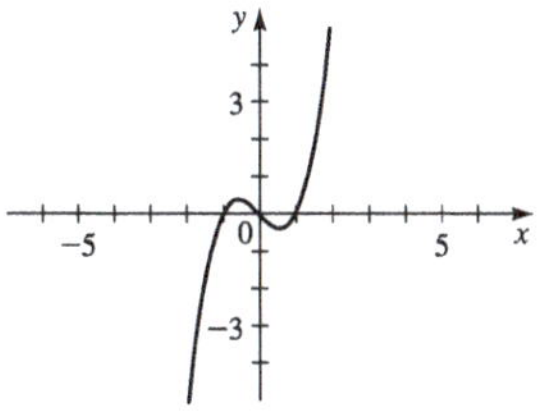

89. Neither
91. **(a)**

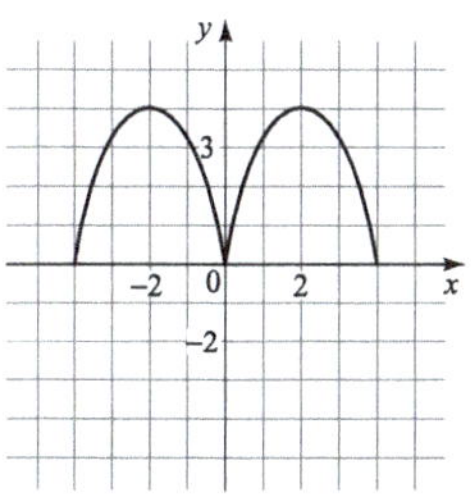

(b)

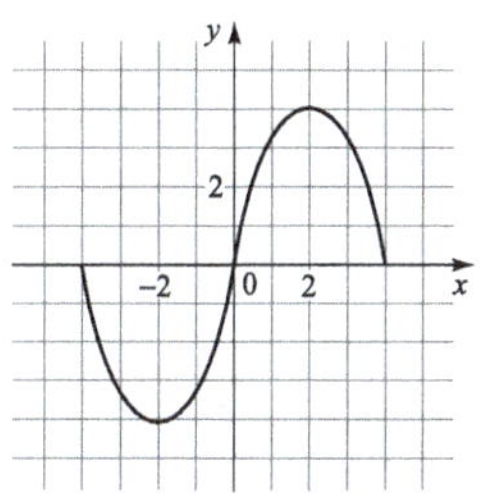

93. To obtain the graph of g, reflect in the x-axis the part of the graph of f that is below the x-axis.
95. **(a)**

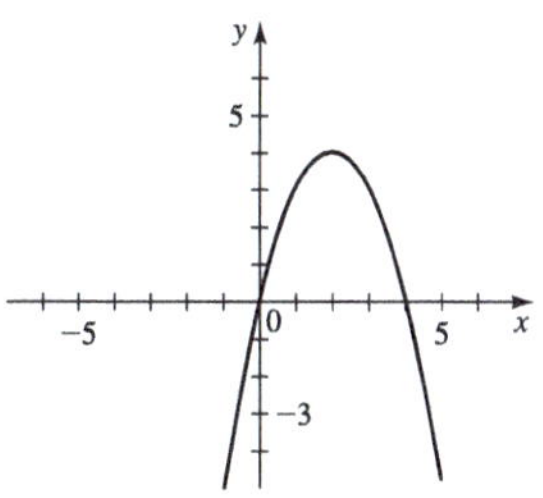

(b)

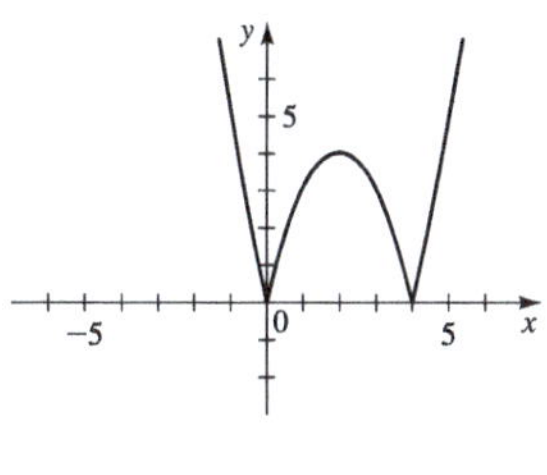

97. **(a)** She drops to 200 ft, bounces up and down, then settles at 350 ft.

(b)

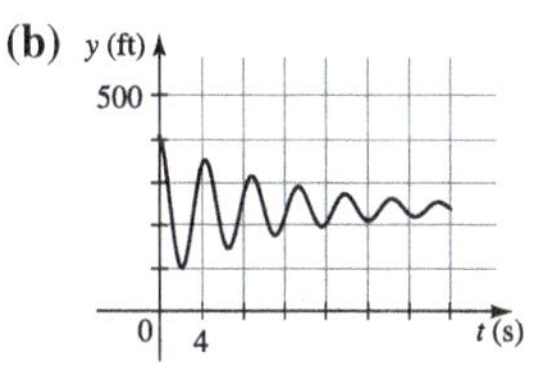

(c) Shift downward 100 ft; $H(t) = h(t) - 100$
99. **(a)** 80 ft/min; 20 min; 800 ft

(b)

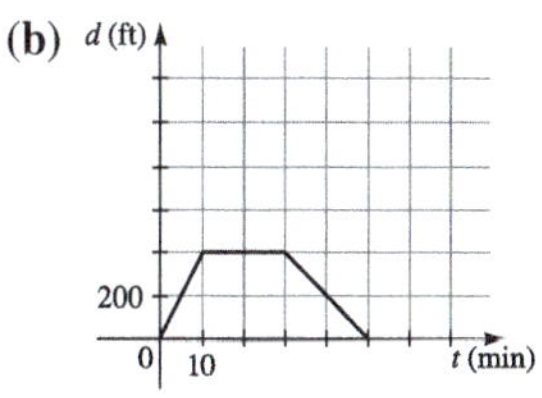

Shrunk vertically by a factor of 0.50; 40 ft/min; 400 ft

(c)

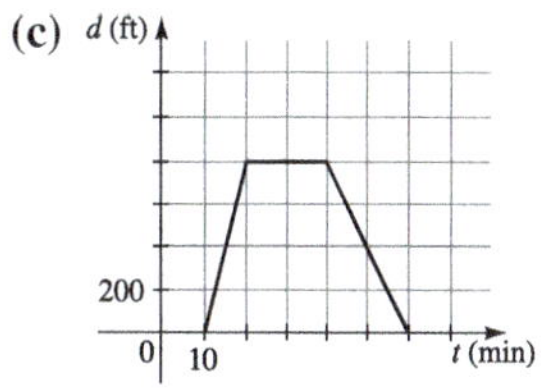

Shifted to the right 10 min; the class left 10 min later

SECTION 2.7 ■ PAGE 216

1. 8, −2, 15, $\frac{3}{5}$ **2.** $f(g(x))$, 12 **3.** Multiply by 2, then add 1; Add 1, then multiply by 2 **4.** $x + 1, 2x, 2x + 1, 2(x + 1)$
5. **(a)** f, g **(b)** f, g **(c)** $f, g, 0$ **6.** g, f
7. $(f + g)(x) = 3x, (-\infty, \infty)$; $(f - g)(x) = -x, (-\infty, \infty)$; $(fg)(x) = 2x^2, (-\infty, \infty)$; $\left(\dfrac{f}{g}\right)(x) = \dfrac{1}{2}, (-\infty, 0) \cup (0, \infty)$
9. $(f + g)(x) = 2x^2 + x, (-\infty, \infty)$; $(f - g)(x) = x, (-\infty, \infty)$; $(fg)(x) = x^4 + x^3, (-\infty, \infty)$; $\left(\dfrac{f}{g}\right)(x) = 1 + \dfrac{1}{x}$, $(-\infty, 0) \cup (0, \infty)$
11. $(f + g)(x) = x^2 - 4x + 5, (-\infty, \infty)$; $(f - g)(x) = -x^2 + 2x + 5, (-\infty, \infty)$; $(fg)(x) = -x^3 + 8x^2 - 15x, (-\infty, \infty)$; $\left(\dfrac{f}{g}\right)(x) = \dfrac{5 - x}{x^2 - 3x}, (-\infty, 0) \cup (0, 3) \cup (3, \infty)$
13. $(f + g)(x) = \sqrt{25 - x^2} + \sqrt{x + 3}, [-3, 5]$; $(f - g)(x) = \sqrt{25 - x^2} - \sqrt{x + 3}, [-3, 5]$; $(fg)(x) = \sqrt{(25 - x^2)(x + 3)}, [-3, 5]$; $\left(\dfrac{f}{g}\right)(x) = \sqrt{\dfrac{25 - x^2}{x + 3}}, (-3, 5]$
15. $(f + g)(x) = \dfrac{6x + 8}{x^2 + 4x}, x \neq -4, x \neq 0$; $(f - g)(x) = \dfrac{-2x + 8}{x^2 + 4x}, x \neq -4, x \neq 0$; $(fg)(x) = \dfrac{8}{x^2 + 4x}, x \neq -4, x \neq 0$; $\left(\dfrac{f}{g}\right)(x) = \dfrac{x + 4}{2x}, x \neq -4, x \neq 0$
17. $[0, 3]$ **19.** $(3, \infty)$
21.

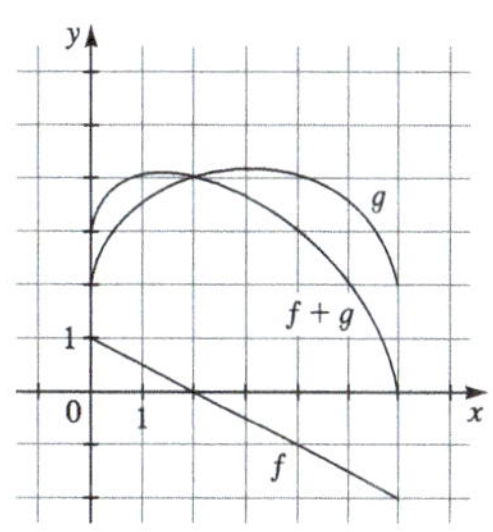

23.

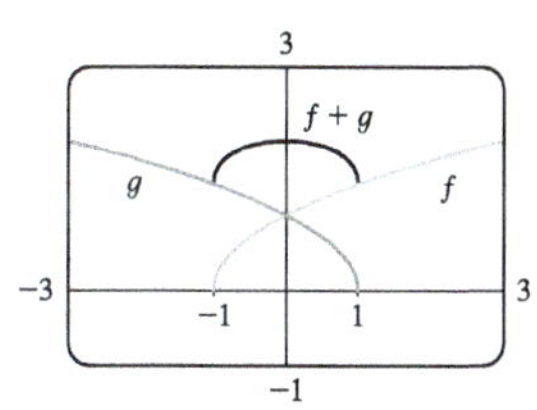

25.

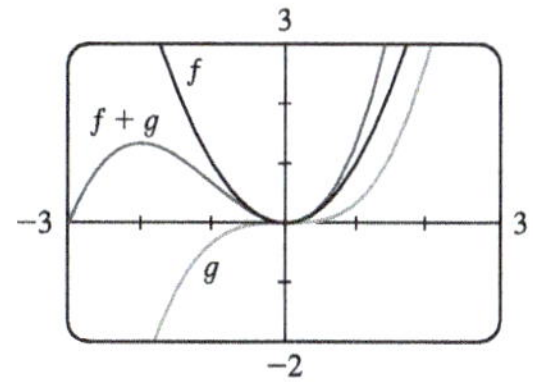

27. **(a)** 5 **(b)** -5 **29.** **(a)** -3 **(b)** -45
31. **(a)** $-2x^2+5$ **(b)** $-4x^2+12x-5$ **33.** 4
35. 5 **37.** 4 **39.** 6 **41.** 3 **43.** 1 **45.** 3
47. $(f\circ g)(x)=8x+1, (-\infty,\infty)$;
$(g\circ f)(x)=8x+11, (-\infty,\infty)$; $(f\circ f)(x)=4x+9, (-\infty,\infty)$;
$(g\circ g)(x)=16x-5, (-\infty,\infty)$
49. $(f\circ g)(x)=(x+1)^2, (-\infty,\infty)$;
$(g\circ f)(x)=x^2+1, (-\infty,\infty)$; $(f\circ f)(x)=x^4, (-\infty,\infty)$;
$(g\circ g)(x)=x+2, (-\infty,\infty)$
51. $(f\circ g)(x)=\dfrac{1}{2x+4}, x\neq -2$; $(g\circ f)(x)=\dfrac{2}{x}+4, x\neq 0$;
$(f\circ f)(x)=x, x\neq 0$, $(g\circ g)(x)=4x+12, (-\infty,\infty)$
53. $(f\circ g)(x)=|2x+3|, (-\infty,\infty)$;
$(g\circ f)(x)=2|x|+3, (-\infty,\infty)$; $(f\circ f)(x)=|x|, (-\infty,\infty)$;
$(g\circ g)(x)=4x+9, (-\infty,\infty)$
55. $(f\circ g)(x)=\dfrac{2x-1}{2x}, x\neq 0$;
$(g\circ f)(x)=\dfrac{2x}{x+1}-1, x\neq -1$;
$(f\circ f)(x)=\dfrac{x}{2x+1}, x\neq -1, x\neq -\frac{1}{2}$;
$(g\circ g)(x)=4x-3, (-\infty,\infty)$
57. $(f\circ g)(x)=\dfrac{1}{x+1}, x\neq -1, x\neq 0$; $(g\circ f)(x)=\dfrac{x+1}{x}$,
$x\neq -1, x\neq 0$; $(f\circ f)(x)=\dfrac{x}{2x+1}, x\neq -1, x\neq -\frac{1}{2}$;
$(g\circ g)(x)=x, x\neq 0$
59. $(f\circ g\circ h)(x)=\sqrt{x-1}-1$
61. $(f\circ g\circ h)(x)=(\sqrt{x}-5)^4+1$
63. $g(x)=x-9, f(x)=x^5$ **65.** $g(x)=x^2, f(x)=x/(x+4)$
67. $g(x)=1-x^3, f(x)=|x|$
69. $h(x)=x^2, g(x)=x+1, f(x)=1/x$
71. $h(x)=\sqrt[3]{x}, g(x)=4+x, f(x)=x^9$
73. Yes; m_1m_2 **75.** $R(x)=0.15x-0.000002x^2$
77. **(a)** $g(t)=60t$ **(b)** $f(r)=\pi r^2$ **(c)** $(f\circ g)(t)=3600\pi t^2$
79. $A(t)=16\pi t^2$ **81.** **(a)** $f(x)=0.9x$
(b) $g(x)=x-100$ **(c)** $(f\circ g)(x)=0.9x-90$,
$(g\circ f)(x)=0.9x-100$, $(f\circ g)$: first rebate, then discount,
$(g\circ f)$: first discount, then rebate, $g\circ f$ is the better deal

SECTION 2.8 ■ PAGE 225

1. different, Horizontal Line **2.** **(a)** one-to-one, $g(x)=x^3$
(b) $g^{-1}(x)=x^{1/3}$ **3.** **(a)** Take the cube root, subtract 5, then divide the result by 3. **(b)** $f(x)=(3x+5)^3, f^{-1}(x)=\dfrac{x^{1/3}-5}{3}$
4. Yes, 4, 5 **5.** (4, 3) **6.** **(a)** False **(b)** True **7.** No
9. Yes **11.** No **13.** Yes **15.** Yes **17.** No **19.** No **21.** Yes
23. No **25.** **(a)** 2 **(b)** 3 **27.** 1 **29.** **(a)** 6 **(b)** 2 **(c)** 0
31. 4 **33.** 1 **35.** 2 **49.** $f^{-1}(x)=\frac{1}{3}x-\frac{5}{3}$
51. $f^{-1}(x)=\sqrt[3]{\frac{1}{4}(5-x)}$ **53.** $f^{-1}(x)=(1/x)-2$
55. $f^{-1}(x)=\dfrac{4x}{1-x}$ **57.** $f^{-1}(x)=\dfrac{7x+5}{x-2}$
59. $f^{-1}(x)=\dfrac{x-3}{5x+2}$ **61.** $f^{-1}(x)=\sqrt{4-x}, x\leq 4$
63. $f^{-1}(x)=\sqrt[6]{x}, x\geq 0$ **65.** $f^{-1}(x)=\sqrt[3]{2-5x}$

67. $f^{-1}(x)=\dfrac{x^2-5}{8}, x\geq 0$ **69.** $f^{-1}(x)=(x-2)^3$
71. **(a)**

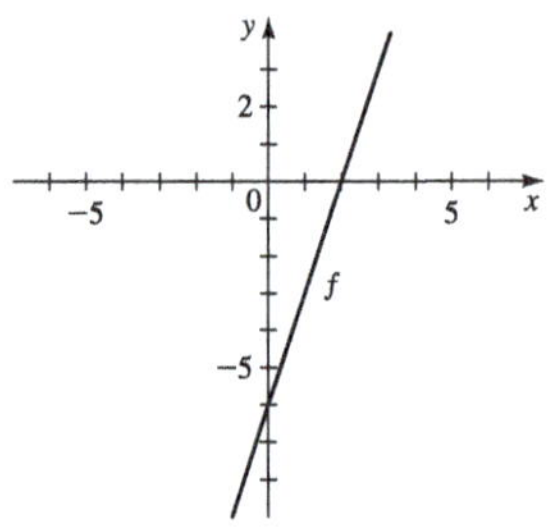

(b)

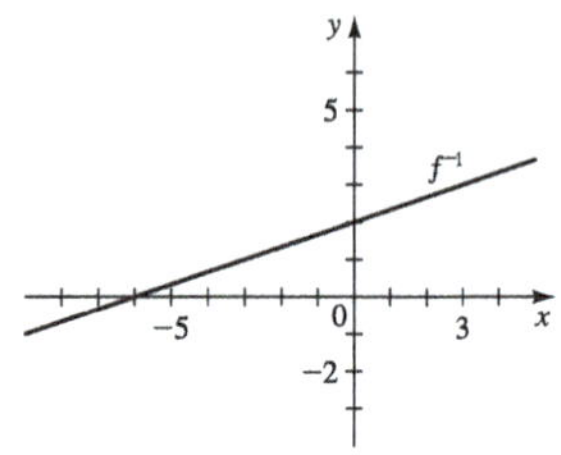

(c) $f^{-1}(x)=\frac{1}{3}(x+6)$
73. **(a)**

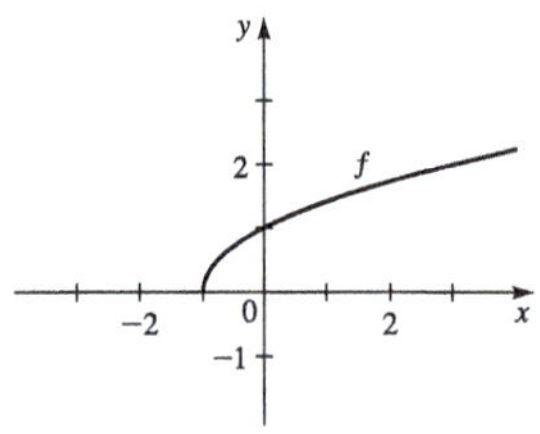

(b)

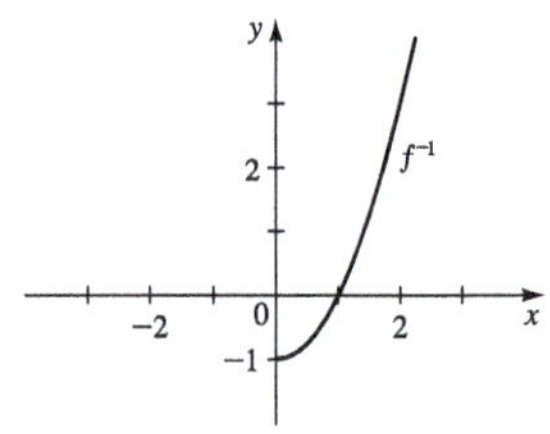

(c) $f^{-1}(x)=x^2-1, x\geq 0$
75. Not one-to-one

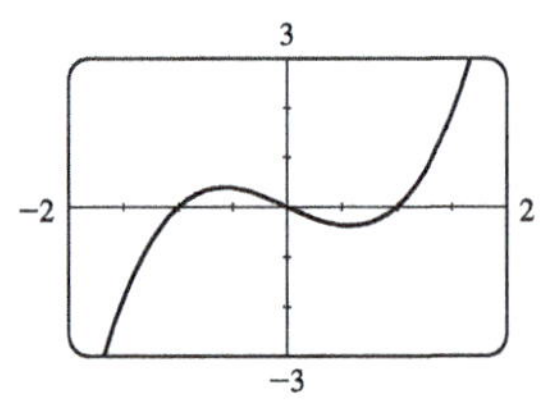

77. One-to-one

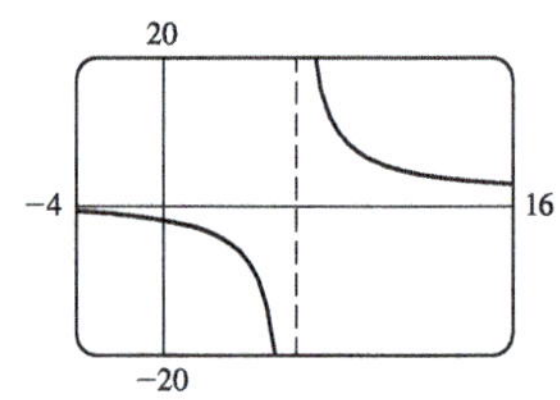

79. Not one-to-one

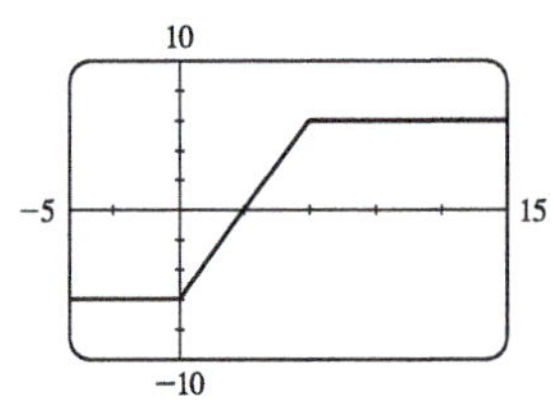

81. **(a)** $f^{-1}(x)=x-2$
(b)

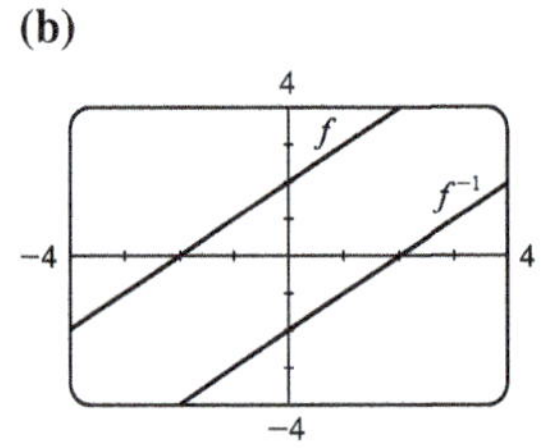

83. **(a)** $g^{-1}(x)=x^2-3, x\geq 0$
(b)

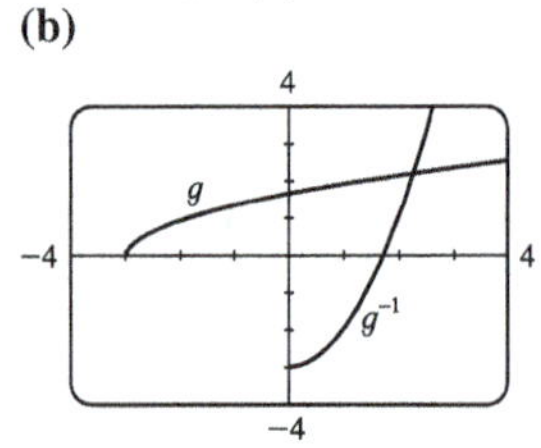

85. $x \geq 0, f^{-1}(x) = \sqrt{4 - x}$ **87.** $x \geq -2, h^{-1}(x) = \sqrt{x} - 2$

89.

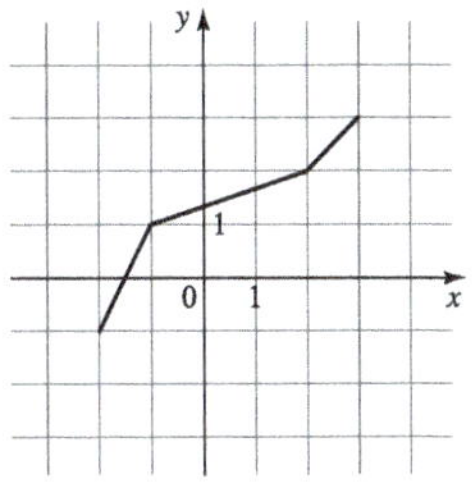

91. (a)

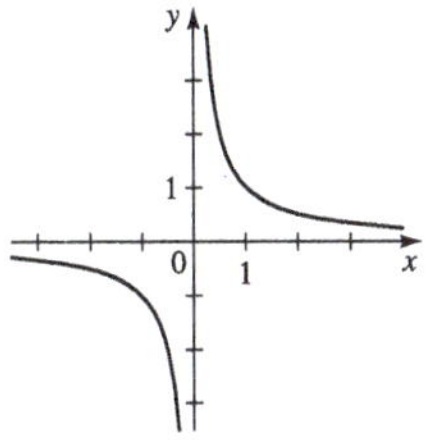

(b) Yes **(c)** $f^{-1}(x) = \dfrac{1}{x}$

93. (a) $f(n) = 16 + 1.5n$ **(b)** $f^{-1}(x) = \frac{2}{3}(x - 16)$; the number of toppings on a pizza that costs x dollars **(c)** 6 **95. (a)** $f^{-1}(V) = 40 - 4\sqrt{V}$, time elapsed when V gal of water remain **(b)** 24.5 min; in 24.5 min the tank has 15 gal of water remaining **97. (a)** $f^{-1}(D) = 50 - \frac{1}{3}D$; the price associated with the demand D **(b)** \$40; when the demand is 30 units, the price is \$40 **99. (a)** $f(x) = 0.9766x$ **(b)** $f^{-1}(x) = 1.02396x$; the exchange rate from U.S. dollars to Canadian dollars **(c)** \$12,543.52 **101. (a)** $f(x) = 0.85x$ **(b)** $g(x) = x - 1000$ **(c)** $H = 0.85x - 850$ **(d)** $H^{-1}(x) = 1.176x + 1000$, the original sticker price for a given discounted price **(e)** \$16,288, the original price of the car when the discounted price (\$1000 rebate, then 15% off) is \$13,000

CHAPTER 2 REVIEW ■ PAGE 231

1. $f(x) = x^2 - 5$ **3.** Add 10, then multiply the result by 3.

5.

x	$g(x)$
−1	5
0	0
1	−3
2	−4
3	−3

7. (a) $C(1000) = 34{,}000$, $C(10{,}000) = 205{,}000$ **(b)** The costs of printing 1000 and 10,000 copies of the book **(c)** $C(0) = 5000$; fixed costs **(d)** \$171,000; \$19/copy **9.** $6, 2, 18, a^2 - 4a + 6, a^2 + 4a + 6, x^2 - 2x + 3, 4x^2 - 8x + 6$ **11. (a)** Not a function **(b)** Function **(c)** Function, one-to-one **(d)** Not a function **13.** Domain $[-3, \infty)$, range $[0, \infty)$ **15.** $(-\infty, \infty)$ **17.** $[-4, \infty)$ **19.** $\{x \mid x \neq -2, -1, 0\}$ **21.** $(-\infty, -1] \cup [1, 4]$

23.

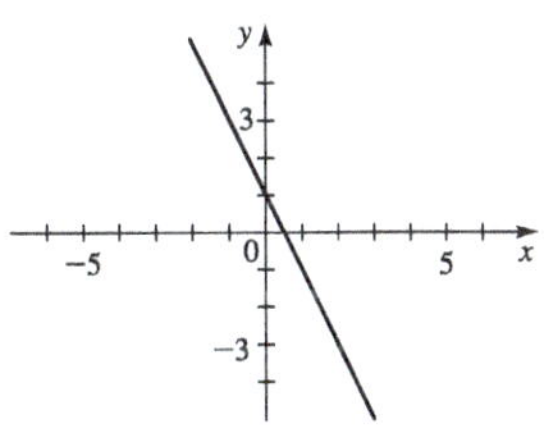

25.

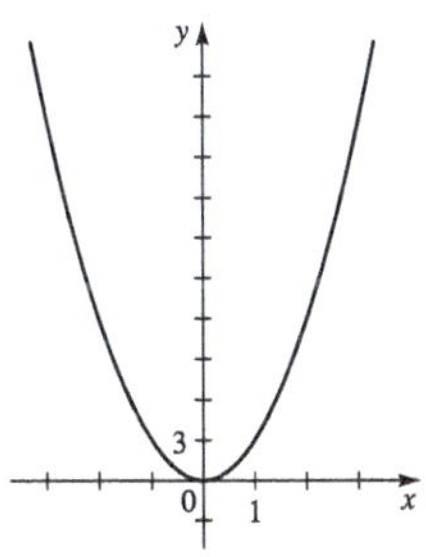

27.

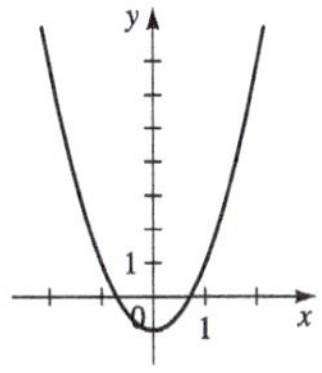

29.

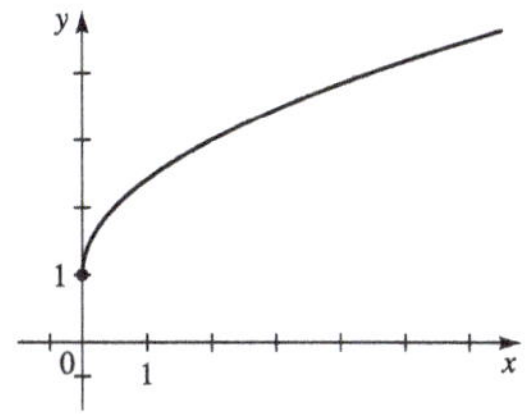

31.

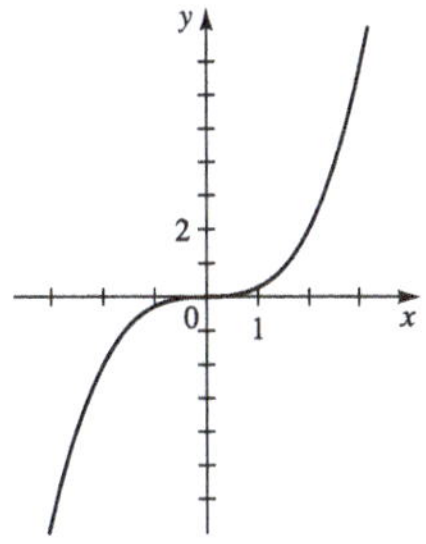

33.

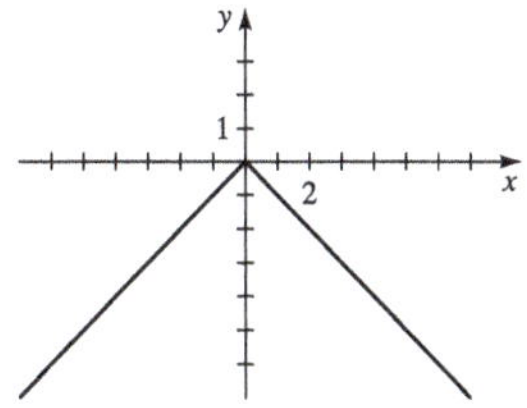

35.

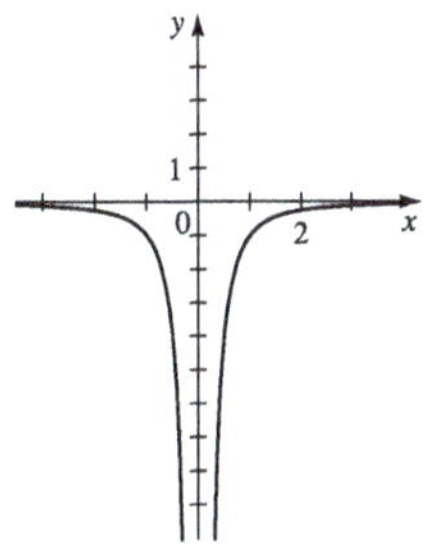

37.

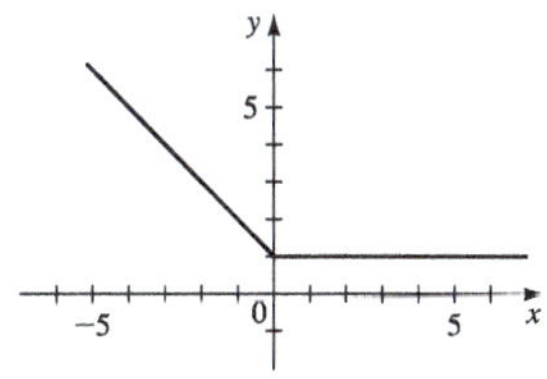

39. No **41.** Yes **43.** (iii)

45. (a)

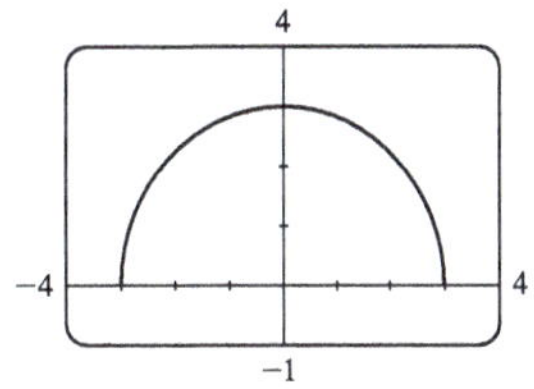

(b) Domain $[-3, 3]$, range $[0, 3]$

47. (a)

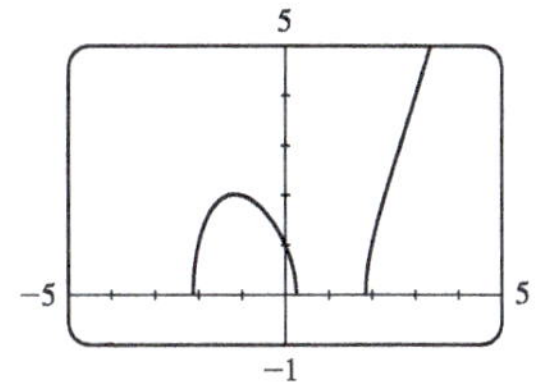

(b) Domain $[-2.11, 0.25] \cup [1.86, \infty)$, range $[0, \infty)$

49.

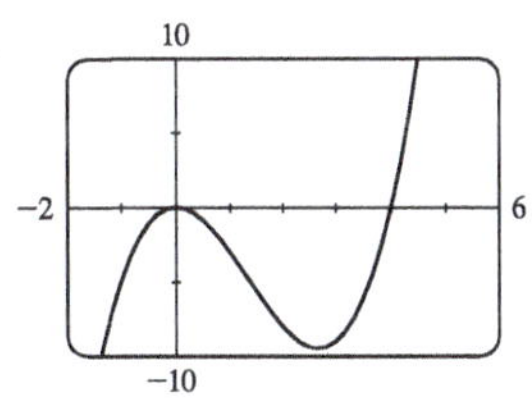

Increasing on $(-\infty, 0)$, $(2.67, \infty)$; decreasing on $(0, 2.67)$

51. $-4, -1$ **53.** $4, \frac{4}{3}$ **55.** $9, 3$ **57.** No

59. (a)

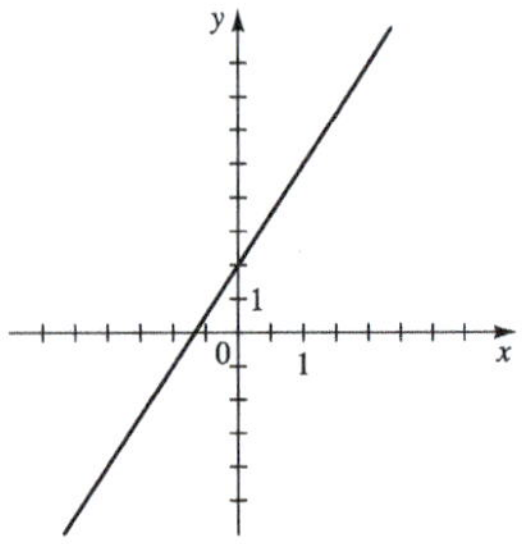

(b) 3 **(c)** 3

61. $f(x) = -2x + 3$ **63.** $f(x) = 2x + 3$

65. $f(x) = -\frac{1}{2}x + 4$ **67. (a)** $P(10) = 5010$, $P(20) = 7040$; the populations in 1995 and 2005 **(b)** 203 people/year; average annual population increase **69. (a)** $\frac{1}{2}, \frac{1}{2}$ **(b)** Yes **(c)** Yes, $\frac{1}{2}$

71. (a) Shift upward 8 units **(b)** Shift to the left 8 units **(c)** Stretch vertically by a factor of 2, then shift upward 1 unit **(d)** Shift to the right 2 units and downward 2 units **(e)** Reflect in y-axis **(f)** Reflect in y-axis, then in x-axis **(g)** Reflect in x-axis **(h)** Reflect in line $y = x$

73. (a) Neither **(b)** Odd **(c)** Even **(d)** Neither

75. Local minimum $= -7$ when $x = -1$

77. Local maximum ≈ 3.79 when $x \approx 0.46$; local minimum ≈ 2.81 when $x \approx -0.46$ **79.** 68 ft

81.

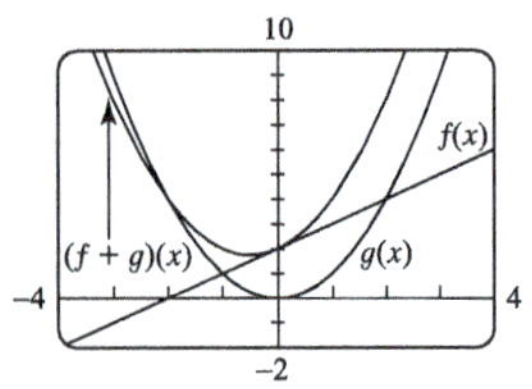

83. (a) $(f + g)(x) = x^2 - 6x + 6$ **(b)** $(f - g)(x) = x^2 - 2$ **(c)** $(fg)(x) = -3x^3 + 13x^2 - 18x + 8$ **(d)** $(f/g)(x) = (x^2 - 3x + 2)/(4 - 3x)$ **(e)** $(f \circ g)(x) = 9x^2 - 15x + 6$ **(f)** $(g \circ f)(x) = -3x^2 + 9x - 2$

85. $(f \circ g)(x) = -3x^2 + 6x - 1, (-\infty, \infty)$; $(g \circ f)(x) = -9x^2 + 12x - 3, (-\infty, \infty)$; $(f \circ f)(x) = 9x - 4, (-\infty, \infty)$; $(g \circ g)(x) = -x^4 + 4x^3 - 6x^2 + 4x, (-\infty, \infty)$

87. $(f \circ g \circ h)(x) = 1 + \sqrt{x}$ **89.** Yes **91.** No **93.** No

95. $f^{-1}(x) = \dfrac{x + 2}{3}$ **97.** $f^{-1}(x) = \sqrt[3]{x} - 1$ **99.** Yes, 1, 3

101. (a), (b)

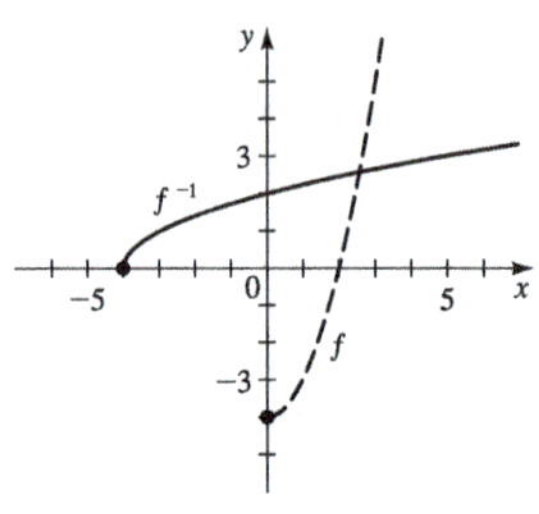

(c) $f^{-1}(x) = \sqrt{x + 4}$

CHAPTER 2 TEST ■ PAGE 235

1. (a) and (b) are graphs of functions, (a) is one-to-one

2. (a) $0, \dfrac{\sqrt{2}}{3}, \dfrac{\sqrt{a + 2}}{a + 3}$ **(b)** $[0, \infty)$

(c) $\dfrac{3\sqrt{10} - 11\sqrt{2}}{264} \approx -0.023$

3. (a) $f(x) = (x - 2)^3$

(b)

x	$f(x)$
−1	−27
0	−8
1	−1
2	0
3	1
4	8

(c)

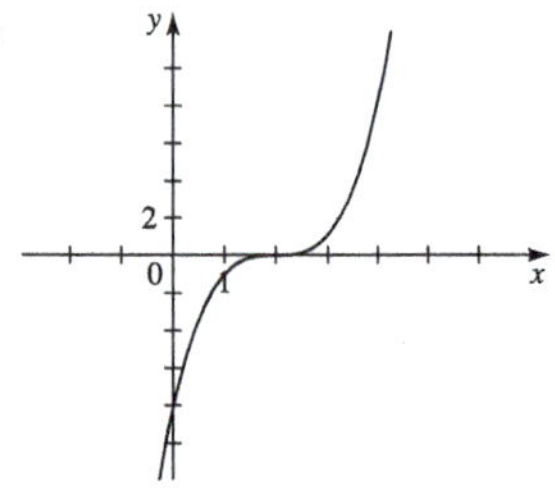

(d) By the Horizontal Line Test; take the cube root, then add 2 **(e)** $f^{-1}(x) = x^{1/3} + 2$ **4. (a)** Local minimum $f(-1) = -4$, local maxima $f(-4) = -1$ and $f(3) = 4$ **(b)** Increasing on $(-\infty, -4)$ and $(-1, 3)$, decreasing on $(-4, -1)$ and $(3, \infty)$

5. (a) $R(2) = 4000$, $R(4) = 4000$; total sales revenue with prices of \$2 and \$4

(b)

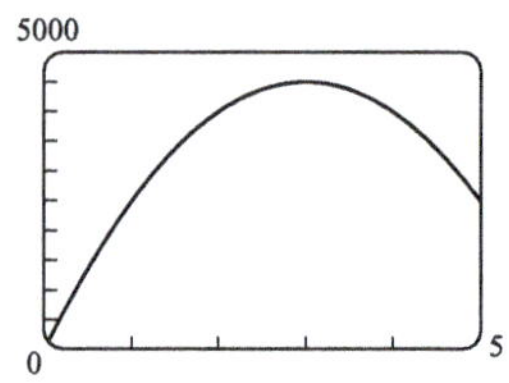

Revenue increases until price reaches \$3, then decreases

(c) \$4500; \$3 **6.** $2h + h^2, 2 + h$

7. (a) g; f is not linear because it has a square term

(b)

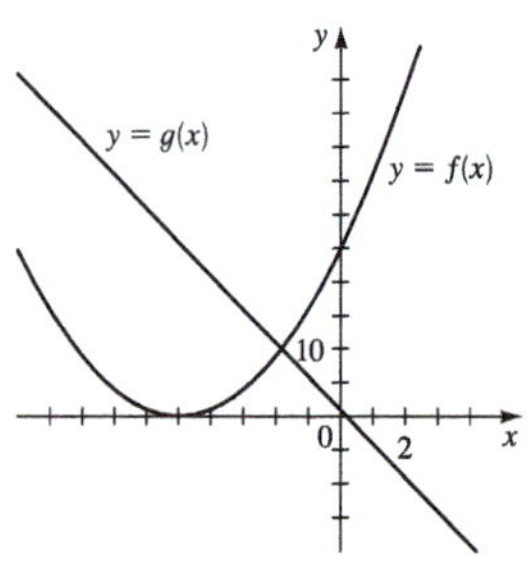

(c) −5

8. (a)

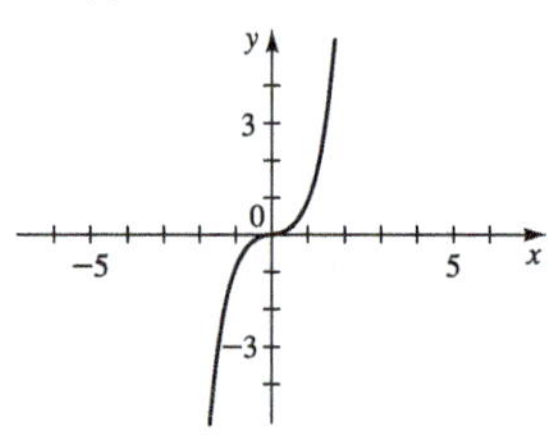

(b)

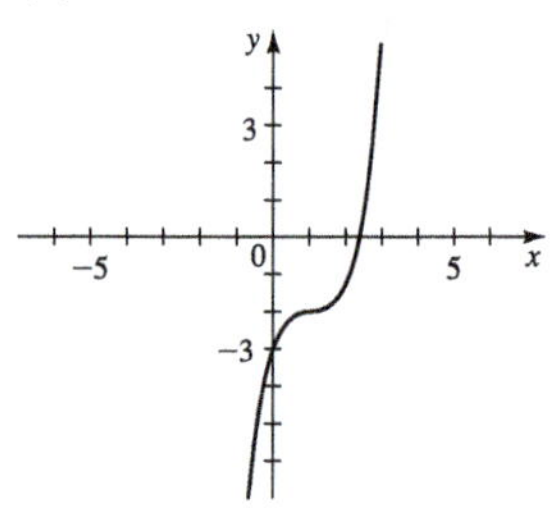

9. **(a)** Shift to the right 3 units, then shift upward 2 units
(b) Reflect in y-axis
10. **(a)** 3, 0
(b)

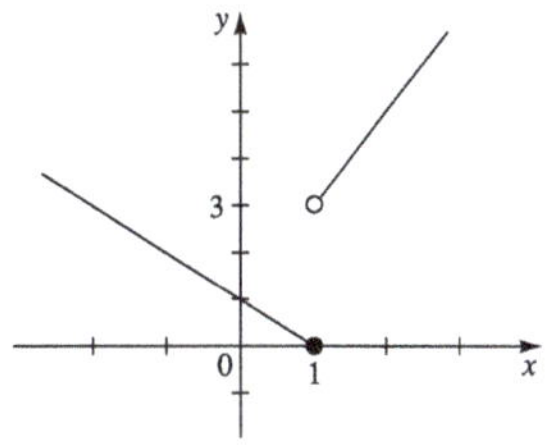

11. **(a)** $x^2 + 2x - 2$ **(b)** $x^2 + 4$ **(c)** $x^2 - 5x + 7$
(d) $x^2 + x - 2$ **(e)** 1 **(f)** 4 **(g)** $x - 9$

12. **(a)** Yes **(b)** No **14.** $f^{-1}(x) = -\dfrac{5x + 3}{2x - 1}$

15. **(a)** $f^{-1}(x) = 3 - x^2, x \geq 0$
(b)

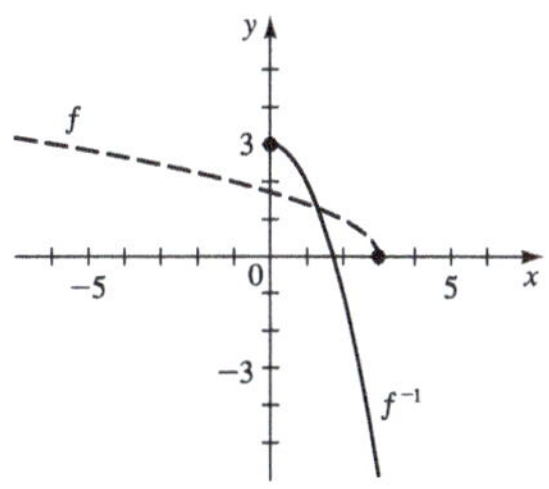

16. Domain $[0, 6]$, range $[1, 7]$ **17.** 1, 3

18.

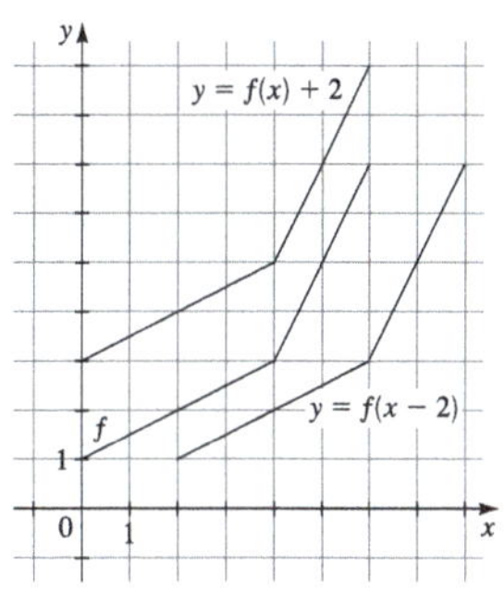

19. 5, $\frac{5}{4}$ **20.** 0, 4

21.

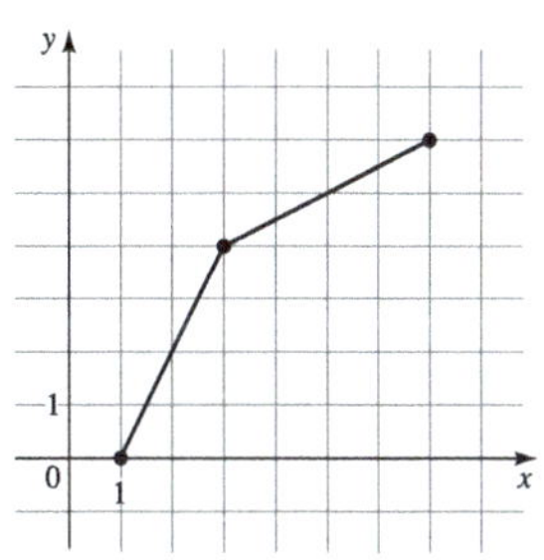

22. **(a)**

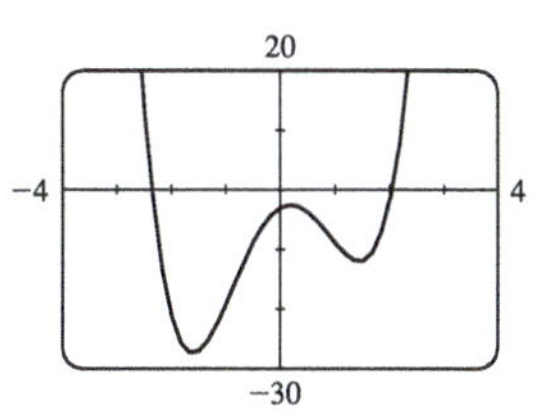

(b) No

(c) Local minimum ≈ -27.18 when $x \approx -1.61$;
local maximum ≈ -2.55 when $x \approx 0.18$;
local minimum ≈ -11.93 when $x \approx 1.43$
(d) $[-27.18, \infty)$ **(e)** Increasing on $(-1.61, 0.18) \cup (1.43, \infty)$;
decreasing on $(-\infty, -1.61) \cup (0.18, 1.43)$

FOCUS ON MODELING ■ PAGE 240

1. $A(w) = 3w^2, w > 0$ **3.** $V(w) = \frac{1}{2}w^3, w > 0$
5. $A(x) = 10x - x^2, 0 < x < 10$ **7.** $A(x) = (\sqrt{3}/4)x^2, x > 0$

9. $r(A) = \sqrt{A/\pi}, A > 0$ **11.** $S(x) = 2x^2 + \dfrac{240}{x}, x > 0$

13. $D(t) = 25t, t \geq 0$ **15.** $A(b) = b\sqrt{4 - b}, 0 < b < 4$
17. $A(h) = 2h\sqrt{100 - h^2}, 0 < h < 10$
19. **(b)** $p(x) = x(19 - x)$ **(c)** 9.5, 9.5
21. **(b)** $A(x) = x(2400 - 2x)$ **(c)** 600 ft by 1200 ft
23. **(a)** $f(w) = 8w + (7200/w)$ **(b)** Width along road is 30 ft, length is 40 ft **(c)** 15 ft to 60 ft

25. **(a)** $A(x) = 15x - \left(\dfrac{\pi + 4}{8}\right)x^2$

(b) Width $\approx$ 8.40 ft, height of rectangular part $\approx$ 4.20 ft

27. **(a)** $A(x) = x^2 + \dfrac{48}{x}$

(b) Height $\approx$ 1.44 ft, width $\approx$ 2.88 ft

29. **(a)** $A(x) = 2x + \dfrac{200}{x}$ **(b)** 10 m by 10 m

31. **(b)** To point C, 5.1 mi from B

CHAPTER 3

SECTION 3.1 ■ PAGE 251

1. square **2.** **(a)** (h, k) **(b)** upward, minimum
(c) downward, maximum **3.** upward, $(2, -6)$, -6, minimum
4. downward, $(2, -6)$, -6, maximum
5. **(a)** $(3, 4)$; x-intercepts 1, 5; y-intercept -5
(b) maximum 4 **(c)** $\mathbb{R}, (-\infty, 4]$

7. **(a)** $(1, -3)$; x-intercepts $\dfrac{2 \pm \sqrt{6}}{2}$; y-intercept -1

(b) minimum -3 **(c)** $\mathbb{R}, [-3, \infty)$

9. **(a)** $f(x) = (x - 1)^2 + 2$
(b) Vertex $(1, 2)$
no x-intercepts
y-intercept 3
(c)

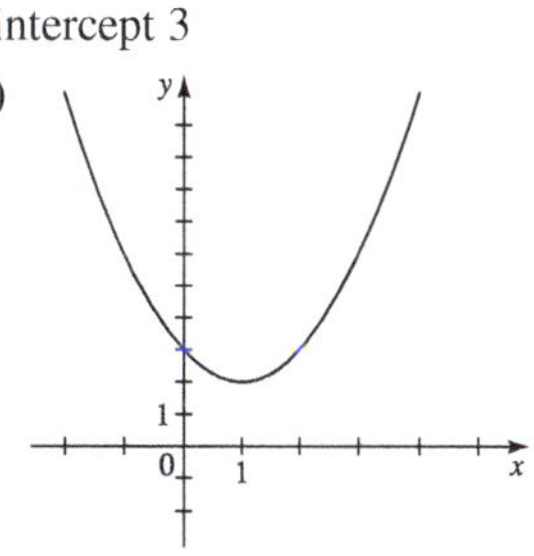

(d) $\mathbb{R}, [2, \infty)$

11. **(a)** $f(x) = (x - 3)^2 - 9$
(b) Vertex $(3, -9)$
x-intercepts 0, 6
y-intercept 0
(c)

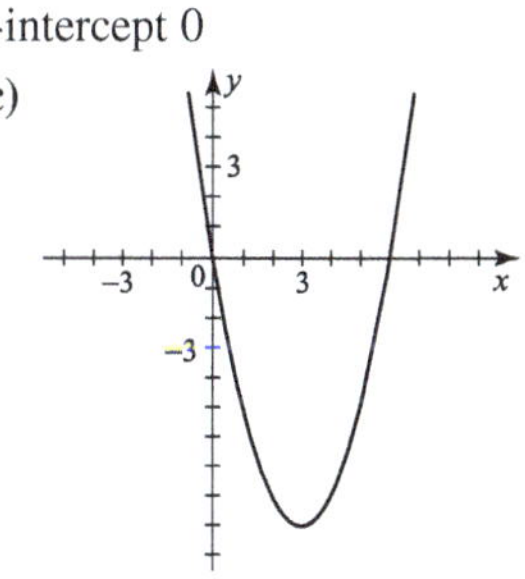

(d) $\mathbb{R}, [-9, \infty)$

13. (a) $f(x) = 3(x + 1)^2 - 3$
(b) Vertex $(-1, -3)$
x-intercepts $-2, 0$
y-intercept 0
(c)

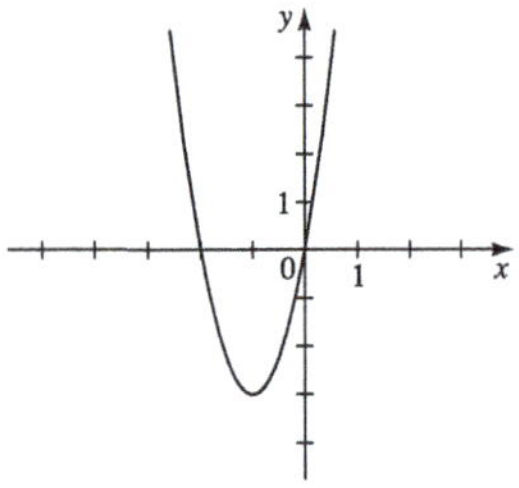

(d) $\mathbb{R}, [-3, \infty)$

15. (a) $f(x) = (x + 2)^2 - 1$
(b) Vertex $(-2, -1)$
x-intercepts $-1, -3$
y-intercept 3
(c)

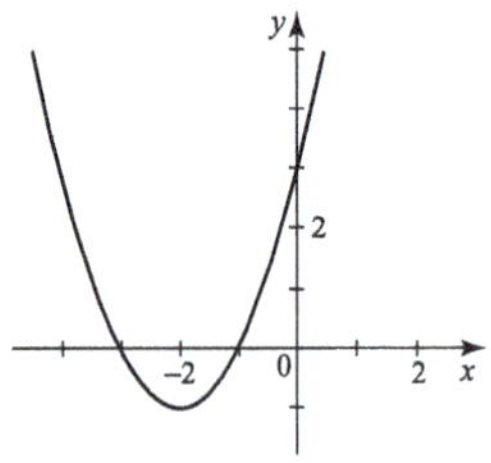

(d) $\mathbb{R}, [-1, \infty)$

17. (a) $f(x) = -(x - 3)^2 + 13$
(b) Vertex $(3, 13)$; x-intercepts $3 \pm \sqrt{13}$; y-intercept 4
(c)

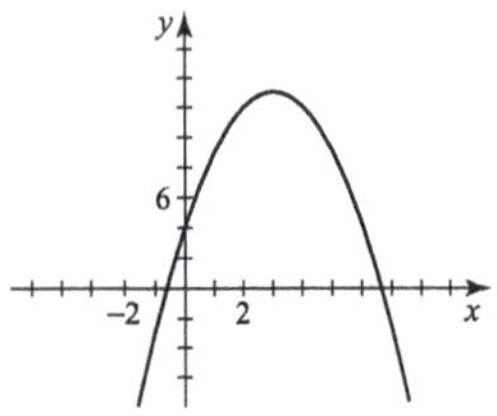

(d) $\mathbb{R}, (-\infty, 13]$

19. (a) $f(x) = 2(x + 1)^2 + 1$
(b) Vertex $(-1, 1)$; no x-intercept; y-intercept 3
(c)

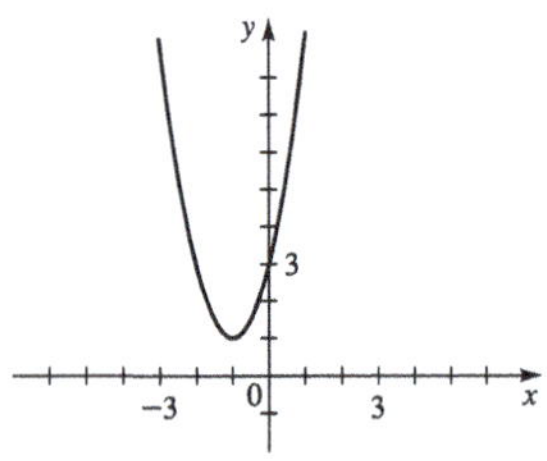

(d) $\mathbb{R}, [1, \infty)$

21. (a) $f(x) = 2(x - 5)^2 + 7$
(b) Vertex $(5, 7)$; no x-intercept; y-intercept 57
(c)

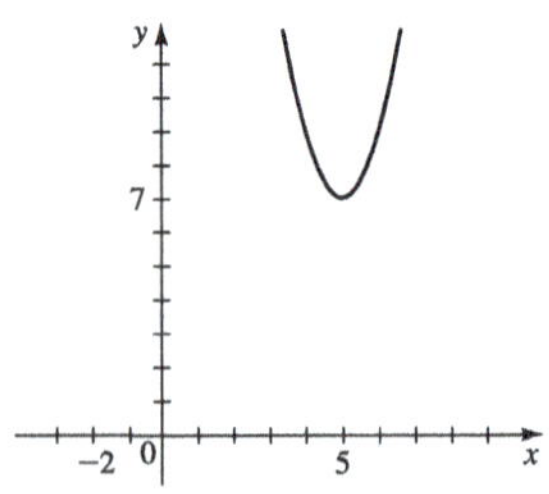

(d) $\mathbb{R}, [7, \infty)$

23. (a) $f(x) = -4\left(x + \frac{3}{2}\right)^2 + 10$
(b) Vertex $\left(-\frac{3}{2}, 10\right)$; x-intercepts $-\frac{3}{2} - \frac{\sqrt{10}}{2}, -\frac{3}{2} + \frac{\sqrt{10}}{2}$; y-intercept 1
(c)

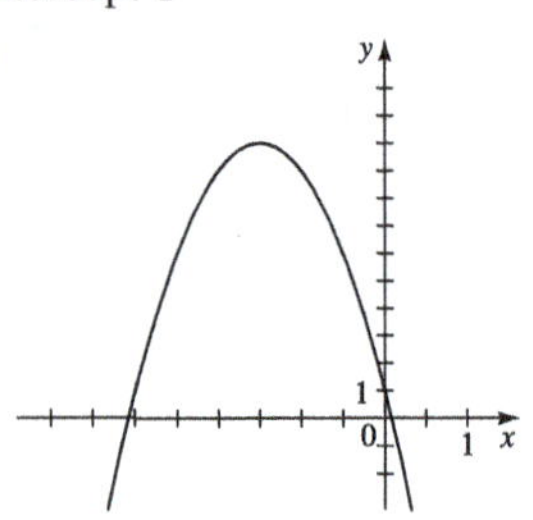

(d) $\mathbb{R}, (-\infty, 10]$

25. (a) $f(x) = (x + 1)^2 - 2$
(b)

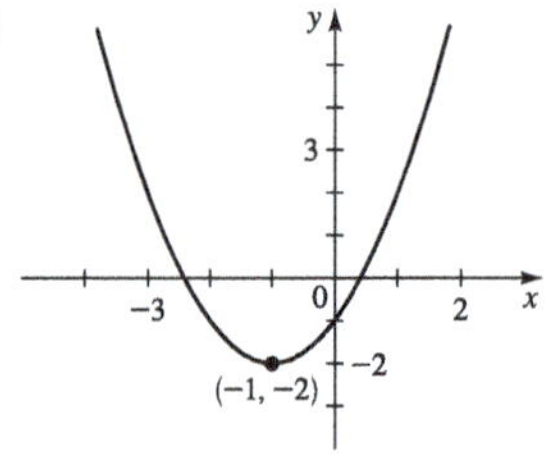

(c) Minimum $f(-1) = -2$

27. (a) $f(x) = 3(x - 1)^2 - 2$
(b)

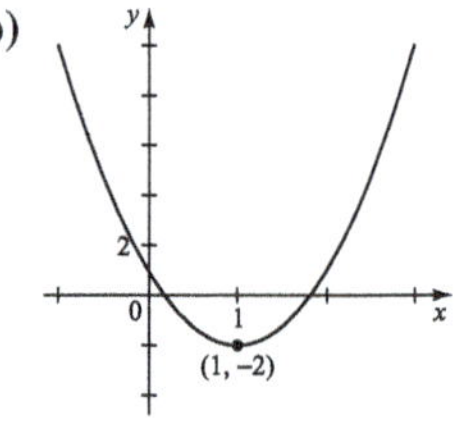

(c) Minimum $f(1) = -2$

29. (a) $f(x) = -\left(x + \frac{3}{2}\right)^2 + \frac{21}{4}$
(b)

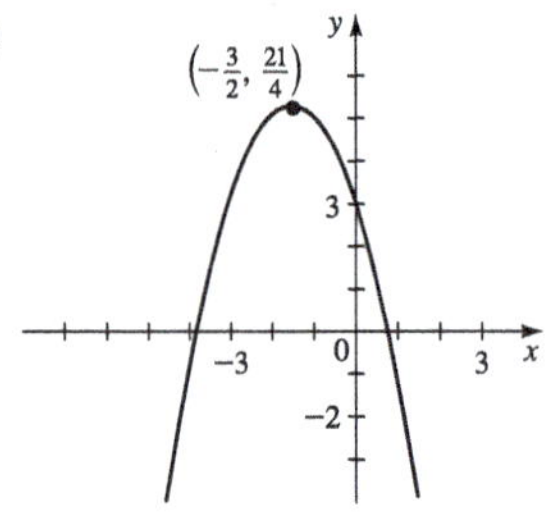

(c) Maximum $f\left(-\frac{3}{2}\right) = \frac{21}{4}$

31. (a) $g(x) = 3(x - 2)^2 + 1$
(b)

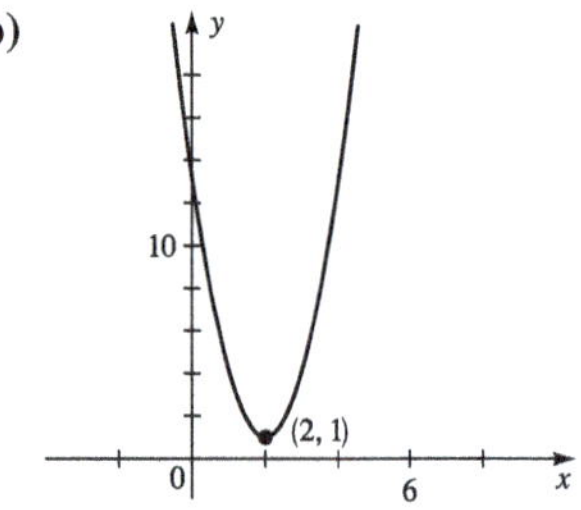

(c) Minimum $g(2) = 1$

33. (a) $h(x) = -\left(x + \frac{1}{2}\right)^2 + \frac{5}{4}$
(b)

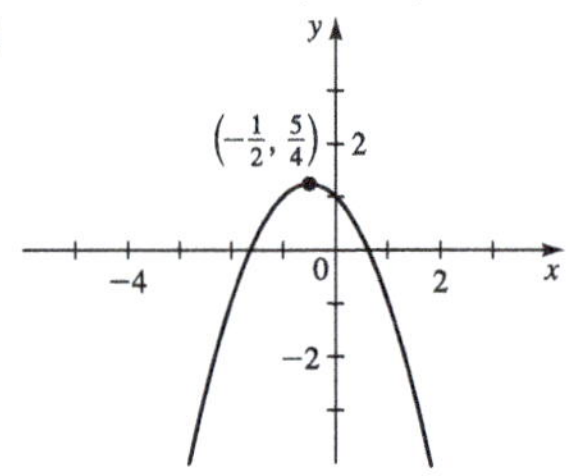

(c) Maximum $h\left(-\frac{1}{2}\right) = \frac{5}{4}$
35. Minimum $f(-1) = -3$
37. Maximum $f(2) = 77$
39. Minimum $f(0.6) = 15.64$
41. Minimum $h(-2) = -8$
43. Maximum $f(-1) = \frac{7}{2}$
45. (a) -4.01 **(b)** -4.011025
47. $f(x) = 4(x - 2)^2 - 3$
49. 7 **51.** 25 ft **53.** \$4000, 100 units **55.** 30 times
57. 50 trees/acre **59.** 600 ft by 1200 ft
61. Width 8.40 ft, height of rectangular part 4.20 ft
63. (a) $f(x) = x(1200 - x)$ **(b)** 600 ft by 600 ft
65. (a) $R(x) = x(57{,}000 - 3000x)$ **(b)** \$9.50 **(c)** \$19.00

SECTION 3.2 ■ PAGE 265

1. II **2. (a)** $\infty, -\infty$ **(b)** $-\infty, -\infty$
3. (a) 0 **(b)** factor **(c)** x **4.** (a)
5. (a)

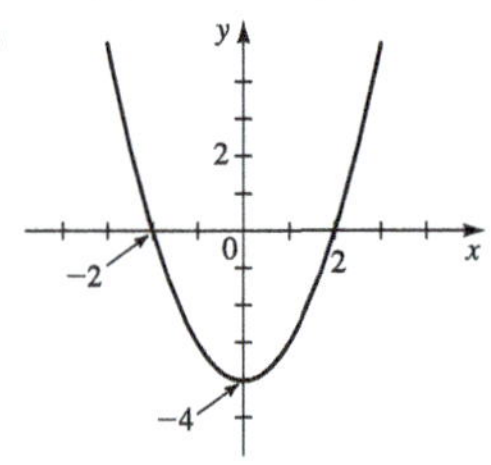

(b)

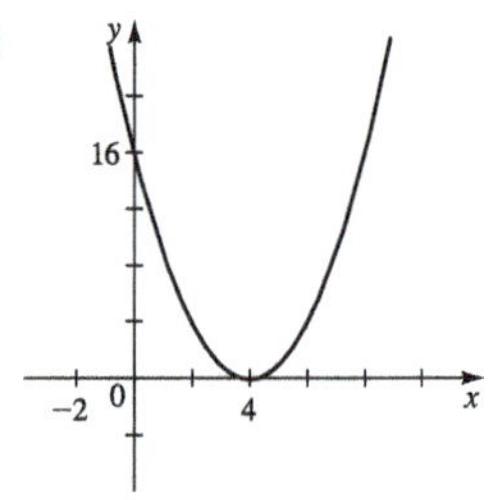

(c)

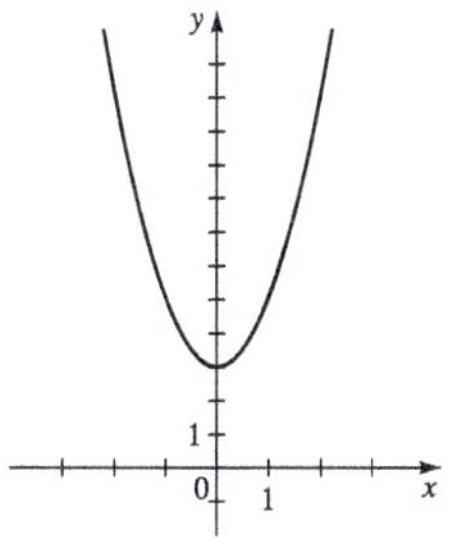

(d)

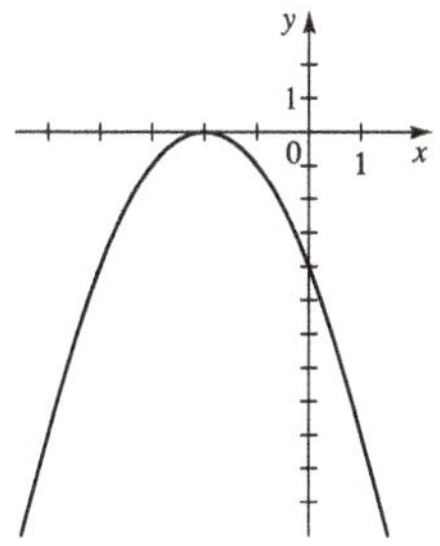

7. (a)

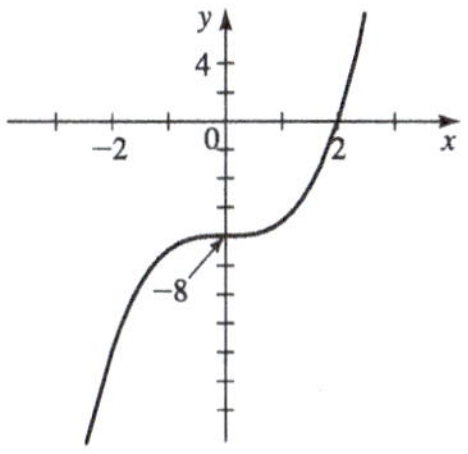

(b)

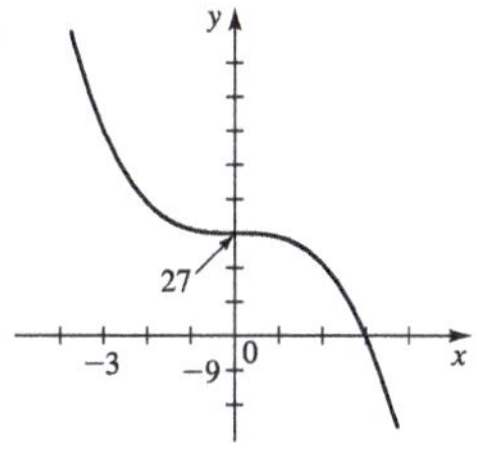

(c)

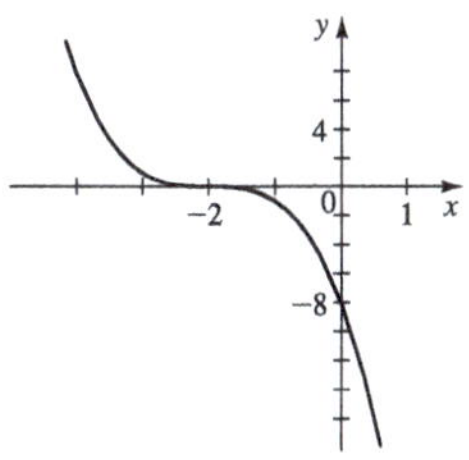

(d)

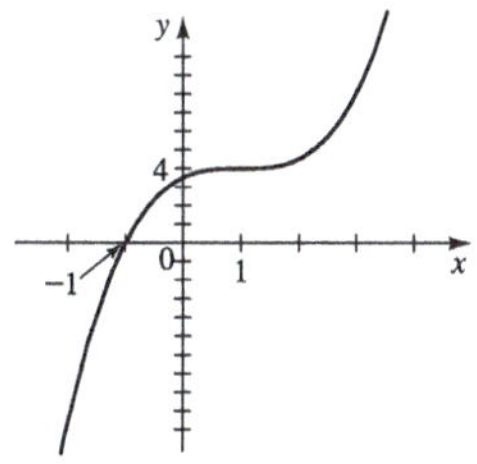

9. (a) $y \to \infty$ as $x \to \infty$, $y \to -\infty$ as $x \to -\infty$ **(b)** III

11. (a) $y \to -\infty$ as $x \to \infty$, $y \to \infty$ as $x \to -\infty$ **(b)** V

13. (a) $y \to \infty$ as $x \to \infty$, $y \to \infty$ as $x \to -\infty$ **(b)** VI

15.

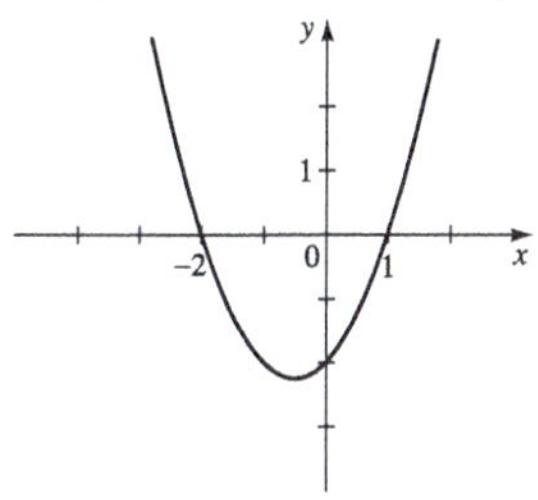

17.

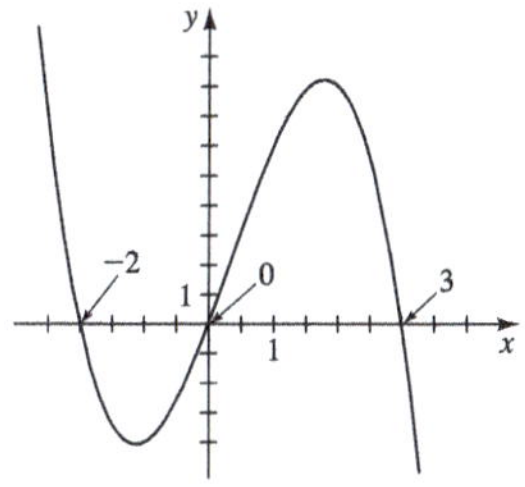

19.

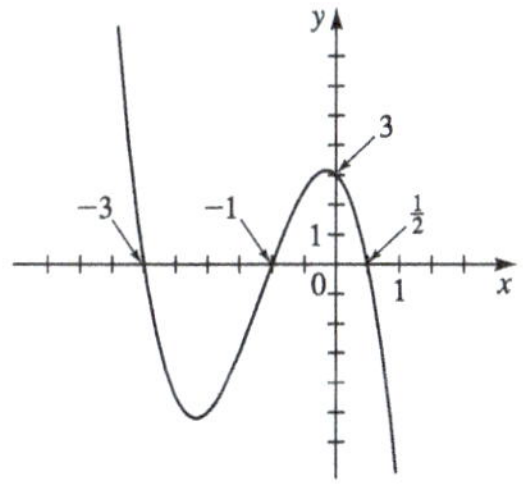

21.

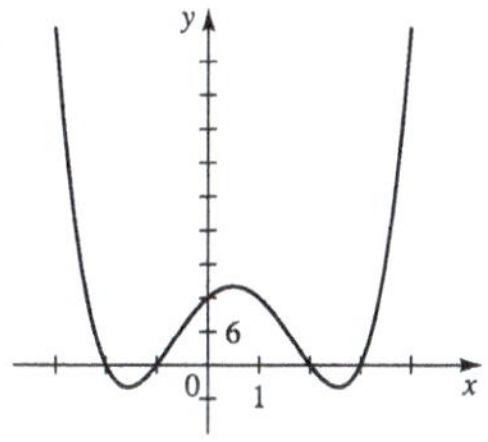

23.

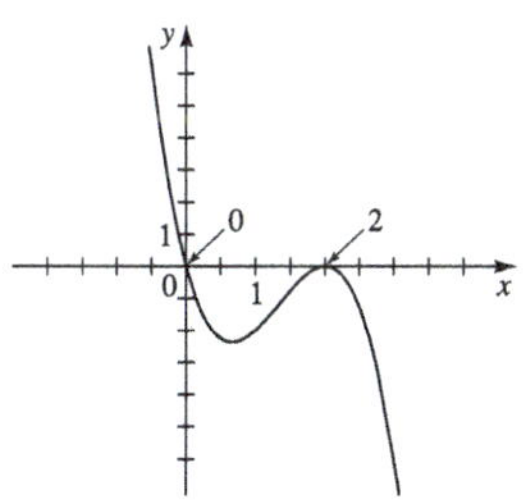

25.

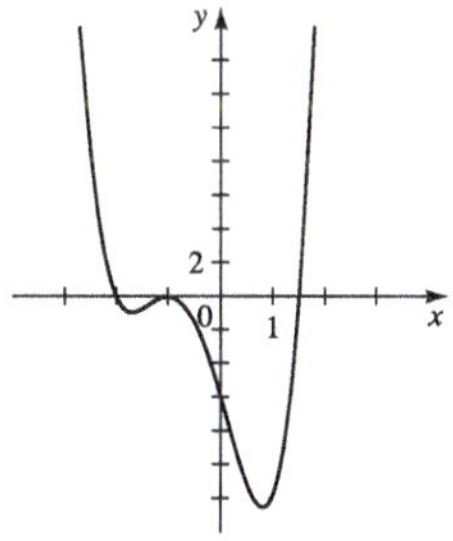

27.

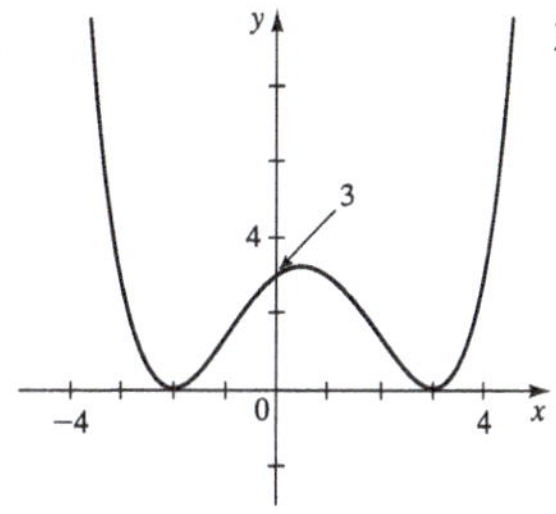

29.

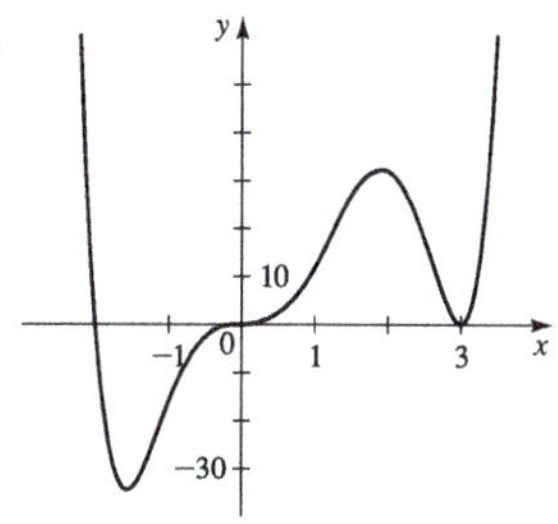

31. $P(x) = x(x + 2)(x - 3)$

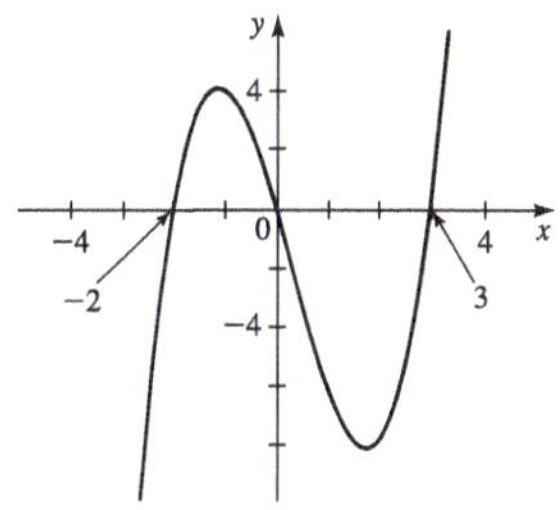

33. $P(x) = -x(x + 3)(x - 4)$

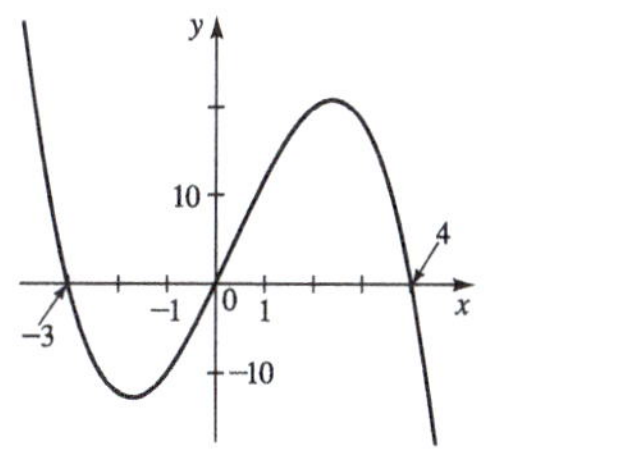

35. $P(x) = x^2(x - 1)(x - 2)$

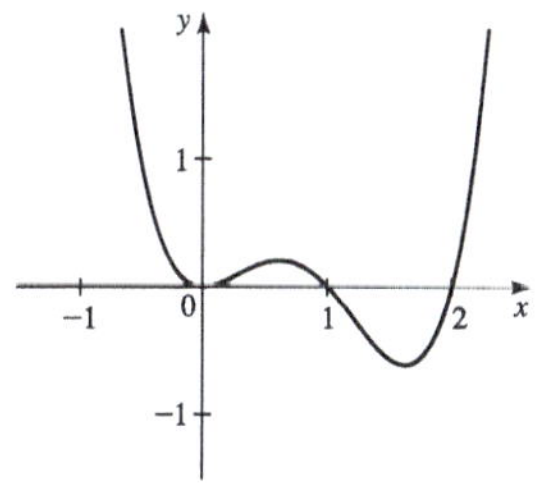

37. $P(x) = (x + 1)^2(x - 1)$

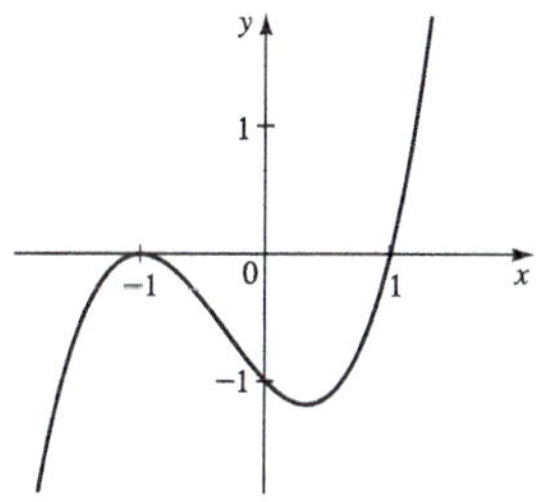

39. $P(x) = (2x - 1)(x + 3)(x - 3)$

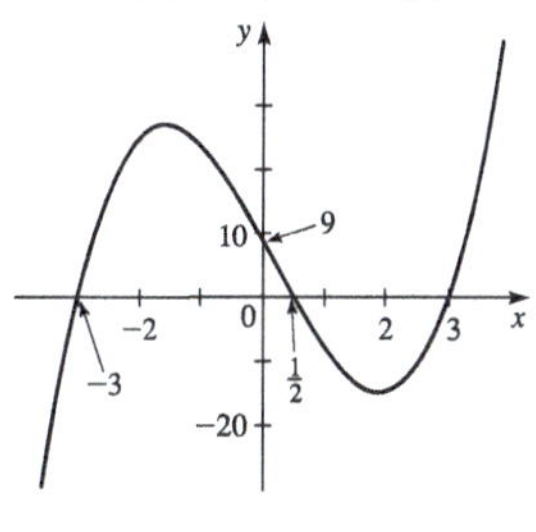

41. $P(x) = (x - 2)^2(x^2 + 2x + 4)$

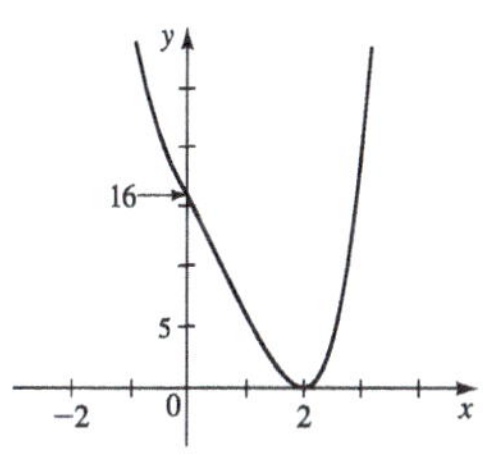

43. $P(x) = (x^2 + 1)(x + 2)(x - 2)$

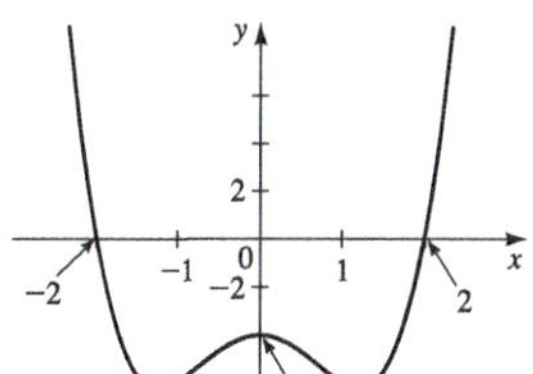

45. $y \to \infty$ as $x \to \infty$, $y \to -\infty$ as $x \to -\infty$
47. $y \to \infty$ as $x \to \pm\infty$ **49.** $y \to \infty$ as $x \to \infty$, $y \to -\infty$ as $x \to -\infty$ **51.** **(a)** x-intercepts 0, 4; y-intercept 0 **(b)** Maximum $(2, 4)$ **53.** **(a)** x-intercepts $-2, 1$; y-intercept -1 **(b)** Minimum $(-1, -2)$, maximum $(1, 0)$

55.

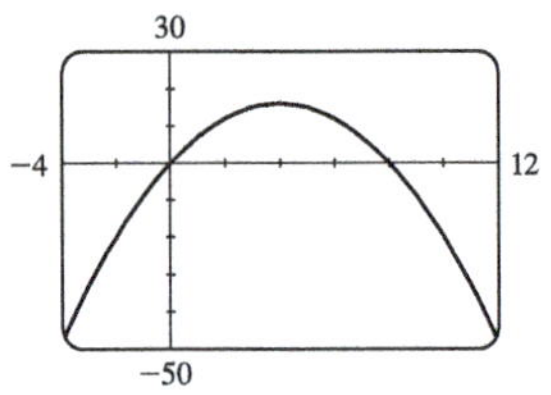

local maximum $(4, 16)$, domain $(-\infty, \infty)$, range $(-\infty, 16]$

57.

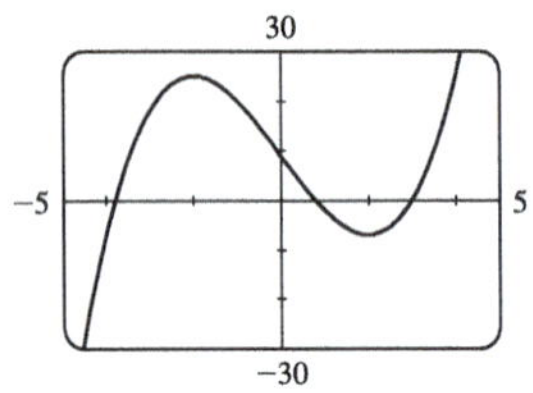

local maximum $(-2, 25)$, local minimum $(2, -7)$, domain $(-\infty, \infty)$, range $(-\infty, \infty)$

59.

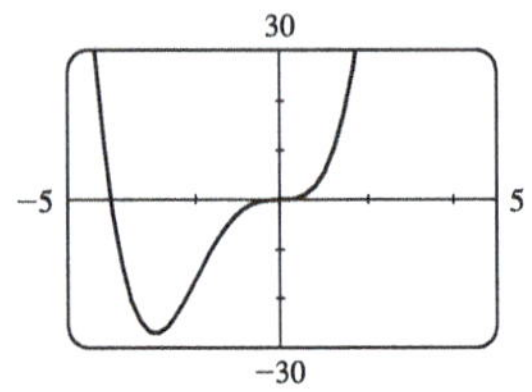

local minimum $(-3, -27)$, domain $(-\infty, \infty)$, range $[-27, \infty)$

61.

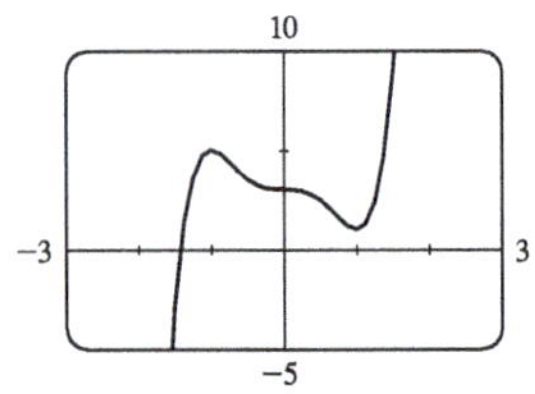

local maximum $(-1, 5)$, local minimum $(1, 1)$, domain $(-\infty, \infty)$, range $(-\infty, \infty)$

63. One local maximum, no local minimum **65.** One local maximum, one local minimum **67.** One local maximum, two local minima **69.** No local extrema **71.** One local maximum, two local minima

73.

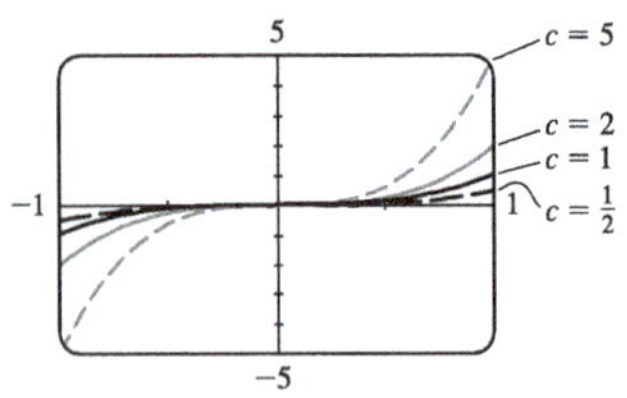

Increasing the value of c stretches the graph vertically.

75.

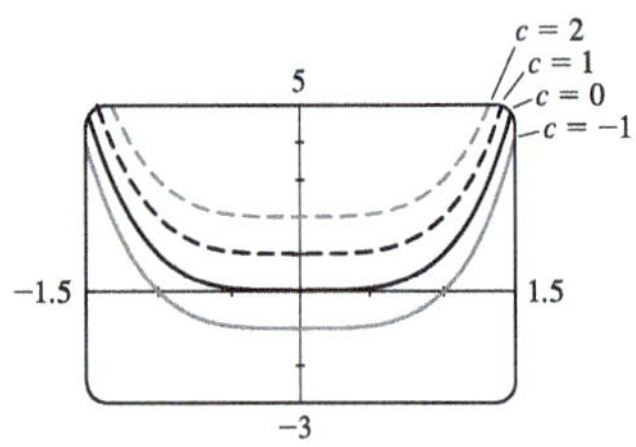

Increasing the value of c moves the graph up.

77.

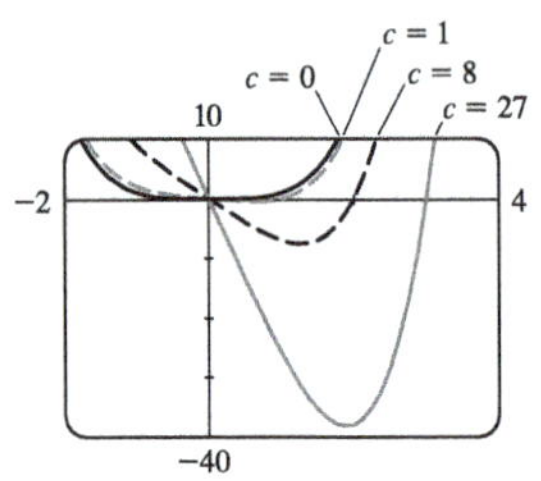

Increasing the value of c causes a deeper dip in the graph in the fourth quadrant and moves the positive x-intercept to the right.

79. **(a)**

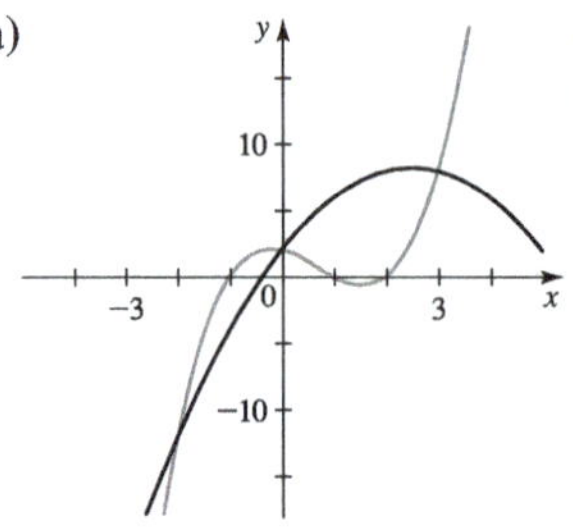

(b) Three
(c) $(0, 2)$, $(3, 8)$, $(-2, -12)$

81. **(d)** $P(x) = P_O(x) + P_E(x)$, where $P_O(x) = x^5 + 6x^3 - 2x$ and $P_E(x) = -x^2 + 5$

83. **(a)**

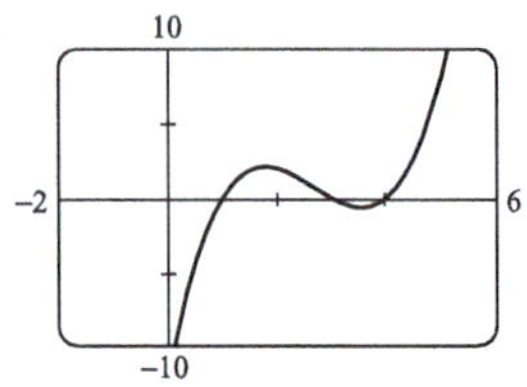

local maximum $(1.8, 2.1)$
local minimum $(3.6, -0.6)$

(b)

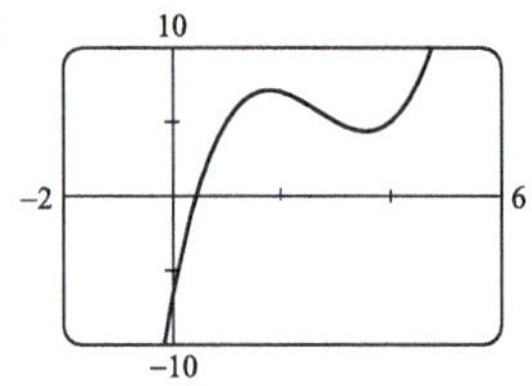

local maximum $(1.8, 7.1)$
local minimum $(3.5, 4.4)$

85. 5; there are four local extrema
87. **(a)** 26 blenders **(b)** No; \$3276.22
89. **(a)** $V(x) = 4x^3 - 120x^2 + 800x$ **(b)** $0 < x < 10$
(c) Maximum volume ≈ 1539.6 cm^3

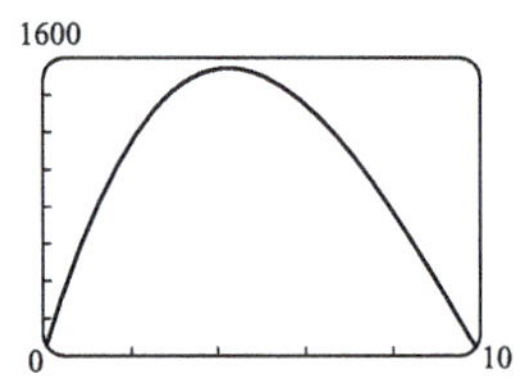

SECTION 3.3 ■ PAGE 273

1. quotient, remainder **2.** **(a)** factor **(b)** k

3. $2x - 1 + \dfrac{-9}{x - 2}$ **5.** $2x - \dfrac{1}{2} + \dfrac{-\frac{15}{2}}{2x - 1}$

7. $2x^2 - x + 1 + \dfrac{4x - 4}{x^2 + 4}$ **9.** $(x + 1)(-x^2 + x - 3) + 9$

11. $(2x - 3)(x^2 - 1) - 3$
13. $(2x^2 + 1)(4x^2 + 2x + 1) + (-2x - 1)$

In answers 15–37 the first polynomial given is the quotient, and the second is the remainder.

15. $x - 1, 5$ **17.** $2x^2 - 1, -2$ **19.** $x + 1, -2$
21. $3x + 1, 7x - 5$ **23.** $x^4 + 1, 0$ **25.** $2x + 1, 6$
27. $3x - 2, 2$ **29.** $x^2 + 2, -3$ **31.** $x^2 - 3x + 1, -1$
33. $x^4 + x^3 + 4x^2 + 4x + 4, -2$ **35.** $2x^2 + 4x, 1$
37. $x^2 + 3x + 9, 0$ **39.** -3 **41.** 12 **43.** -7 **45.** -483
47. 2159 **49.** $\frac{7}{3}$ **51.** -8.279 **57.** $-3, 3$ **59.** $-1 \pm \sqrt{6}$

61. $\dfrac{5 \pm \sqrt{37}}{6}$ **63.** $x^3 - 3x^2 - x + 3$

65. $x^4 - 8x^3 + 14x^2 + 8x - 15$

67. $-2x^4 + 4x^3 + 10x^2 - 12x$ **69.** $3x^4 - 9x^2 + 6$
71. $(x + 1)(x - 1)(x - 2)$ **73.** $(x + 2)^2(x - 1)^2$

SECTION 3.4 ■ PAGE 283

1. $a_0, a_n, \pm 1, \pm\frac{1}{2}, \pm\frac{1}{3}, \pm\frac{1}{6}, \pm 2, \pm\frac{2}{3}, \pm 5, \pm\frac{5}{2}, \pm\frac{5}{3}, \pm\frac{5}{6}, \pm 10, \pm\frac{10}{3}$
2. 1, 3, 5; 0 **3.** True **4.** False **5.** $\pm 1, \pm 3$
7. $\pm 1, \pm 2, \pm 4, \pm 8, \pm\frac{1}{2}$ **9.** $\pm 1, \pm 7, \pm\frac{1}{2}, \pm\frac{7}{2}, \pm\frac{1}{4}, \pm\frac{7}{4}$
11. (a) $\pm 1, \pm\frac{1}{5}$ **(b)** $-1, 1, \frac{1}{5}$ **13. (a)** $\pm 1, \pm 3, \pm\frac{1}{2}, \pm\frac{3}{2}$
(b) $-\frac{1}{2}, 1, 3$ **15.** $-5, 1, 2; P(x) = (x + 5)(x - 1)(x - 2)$
17. $-2, 1; P(x) = (x + 2)^2(x - 1)$
19. $2; P(x) = (x - 2)^3$
21. $-3, -2, 5; P(x) = (x + 3)(x + 2)(x - 5)$
23. $-3, -1, 1; P(x) = (x + 3)(x + 1)(x - 1)$
25. $\pm 1, \pm 2; P(x) = (x - 2)(x + 2)(x - 1)(x + 1)$
27. $-4, -2, -1, 1; P(x) = (x + 4)(x + 2)(x - 1)(x + 1)$
29. $-3, -\frac{1}{2}, \frac{1}{2}, 3; P(x) = (x + 3)(2x + 1)(2x - 1)(x - 3)$
31. $\pm 2, \frac{1}{3}, 3; P(x) = (x - 2)(x + 2)(x - 3)(3x - 1)$
33. $-1, \pm\frac{1}{2}; P(x) = (x + 1)(2x - 1)(2x + 1)$
35. $-\frac{3}{2}, \frac{1}{2}, 1; P(x) = (x - 1)(2x + 3)(2x - 1)$
37. $-\frac{2}{3}, -\frac{1}{2}, \frac{3}{4}; P(x) = (3x + 2)(2x + 1)(4x - 3)$
39. $-1, \frac{1}{2}, 2; P(x) = (x + 1)(x - 2)^2(2x - 1)$
41. $-3, -2, 1, 3; P(x) = (x + 3)(x + 2)^2(x - 1)(x - 3)$
43. $-1, -\frac{1}{3}, 2, 5; P(x) = (x + 1)^2(x - 2)(x - 5)(3x + 1)$
45. $-1, -\dfrac{1 \pm \sqrt{13}}{3}$ **47.** $-1, 4, \dfrac{3 \pm \sqrt{13}}{2}$
49. $3, \dfrac{1 \pm \sqrt{5}}{2}$ **51.** $\frac{1}{2}, \dfrac{1 \pm \sqrt{3}}{2}$ **53.** $-1, -\frac{1}{2}, -3 \pm \sqrt{10}$
55. (a) $-2, 2, 3$ **(b)**

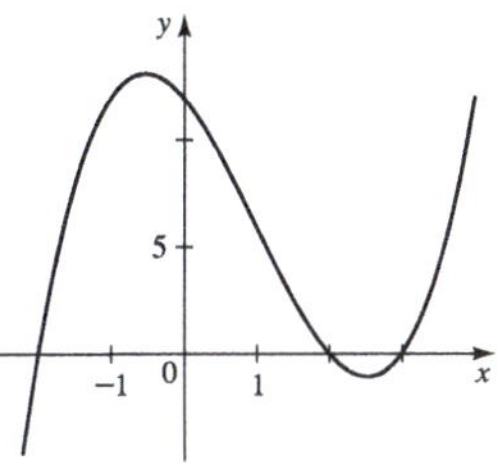

57. (a) $-\frac{1}{2}, 2$ **(b)**

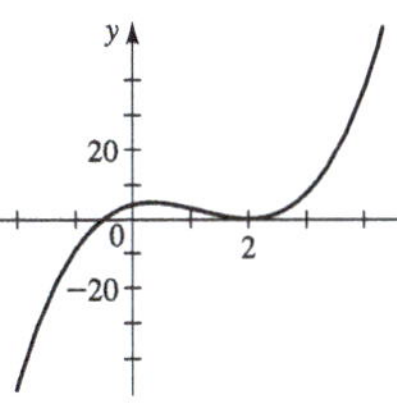

59. (a) $-1, 2$ **(b)**

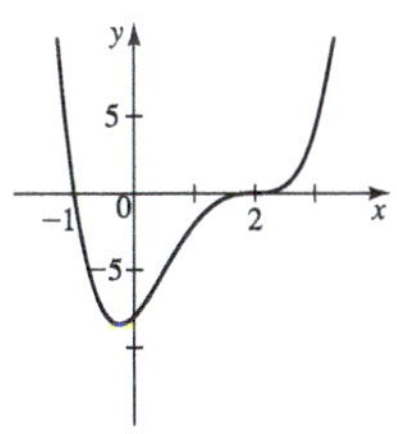

61. (a) $-1, 2$ **(b)**

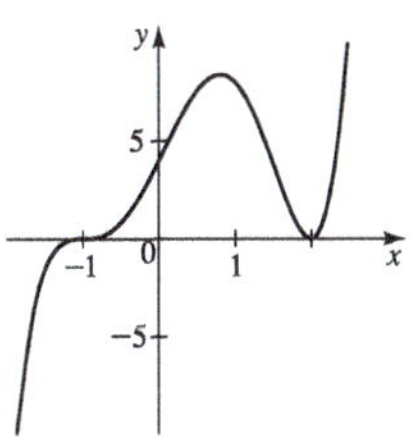

63. 1 positive, 2 or 0 negative; 3 or 1 real **65.** 1 positive, 1 negative; 2 real **67.** 2 or 0 positive, 0 negative; 3 or 1 real (since 0 is a zero but is neither positive nor negative) **77.** $3, -2$
79. $3, -1$ **81.** $-2, \frac{1}{2}, \pm 1$ **83.** $\pm\frac{1}{2}, \pm\sqrt{5}$ **85.** $-2, 1, 3, 4$
91. $-2, 2, 3$ **93.** $-\frac{3}{2}, -1, 1, 4$ **95.** $-1.28, 1.53$ **97.** -1.50
99. 11.3 ft **101. (a)** It began to snow again. **(b)** No
(c) Just before midnight on Saturday night **103.** 2.76 m
105. 88 in. (or 3.21 in.)

SECTION 3.5 ■ PAGE 293

1. 6; −7; 2, 3 **2. (a)** $x - a$ **(b)** $(x - a)^m$ **3.** n **4.** $a - bi$; $3 - i$ **5. (a)** True **(b)** True **(c)** False, $x^4 + 1 > 0$ for all real x **6. (a)** False, $x^2 + 1$ has no real zeros
(b) True **(c)** False, $x^2 + 1$ factors into linear factors with complex coefficients **7. (a)** $0, \pm 2i$ **(b)** $x^2(x - 2i)(x + 2i)$
9. (a) $0, 1 \pm i$ **(b)** $x(x - 1 - i)(x - 1 + i)$
11. (a) $\pm i$ **(b)** $(x - i)^2(x + i)^2$
13. (a) $\pm 2, \pm 2i$ **(b)** $(x - 2)(x + 2)(x - 2i)(x + 2i)$
15. (a) $-2, 1 \pm i\sqrt{3}$
(b) $(x + 2)(x - 1 - i\sqrt{3})(x - 1 + i\sqrt{3})$
17. (a) $\pm 1, \frac{1}{2} \pm \frac{1}{2}i\sqrt{3}, -\frac{1}{2} \pm \frac{1}{2}i\sqrt{3}$
(b) $(x - 1)(x + 1)(x - \frac{1}{2} - \frac{1}{2}i\sqrt{3})(x - \frac{1}{2} + \frac{1}{2}i\sqrt{3}) \times$
$(x + \frac{1}{2} - \frac{1}{2}i\sqrt{3})(x + \frac{1}{2} + \frac{1}{2}i\sqrt{3})$

In answers 19–35 the factored form is given first, then the zeros are listed with the multiplicity of each in parentheses.
19. $(x - 5i)(x + 5i)$; $\pm 5i\ (1)$
21. $[x - (-1 + i)][x - (-1 - i)]$; $-1 + i\ (1), -1 - i\ (1)$
23. $x(x - 2i)(x + 2i)$; $0\ (1), 2i\ (1), -2i\ (1)$
25. $(x - 1)(x + 1)(x - i)(x + i)$; $1\ (1), -1\ (1), i\ (1), -i\ (1)$
27. $16(x - \frac{3}{2})(x + \frac{3}{2})(x - \frac{3}{2}i)(x + \frac{3}{2}i)$; $\frac{3}{2}\ (1), -\frac{3}{2}\ (1), \frac{3}{2}i\ (1)$, $-\frac{3}{2}i\ (1)$ **29.** $(x + 1)(x - 3i)(x + 3i)$; $-1\ (1), 3i\ (1), -3i\ (1)$
31. $(x - i)^2(x + i)^2$; $i(2), -i(2)$
33. $(x - 1)(x + 1)(x - 2i)(x + 2i)$; $1\ (1), -1\ (1)$, $2i(1), -2i(1)$
35. $x(x - i\sqrt{3})^2(x + i\sqrt{3})^2$; $0\ (1), i\sqrt{3}\ (2), -i\sqrt{3}\ (2)$
37. $P(x) = x^2 - 2x + 2$ **39.** $Q(x) = x^3 - 3x^2 + 4x - 12$
41. $P(x) = x^3 - 2x^2 + x - 2$
43. $R(x) = x^4 - 4x^3 + 10x^2 - 12x + 5$
45. $T(x) = 6x^4 - 12x^3 + 18x^2 - 12x + 12$ **47.** $-2, \pm 2i$
49. $1, \dfrac{1 \pm i\sqrt{3}}{2}$ **51.** $2, \dfrac{1 \pm i\sqrt{3}}{2}$ **53.** $-\frac{3}{2}, -1 \pm i\sqrt{2}$
55. $-2, 1, \pm 3i$ **57.** $1, \pm 2i, \pm i\sqrt{3}$ **59.** 3 (multiplicity 2), $\pm 2i$
61. $-\frac{1}{2}$ (multiplicity 2), $\pm i$ **63.** 1 (multiplicity 3), $\pm 3i$
65. (a) $(x - 5)(x^2 + 4)$ **(b)** $(x - 5)(x - 2i)(x + 2i)$
67. (a) $(x - 1)(x + 1)(x^2 + 9)$
(b) $(x - 1)(x + 1)(x - 3i)(x + 3i)$
69. (a) $(x - 2)(x + 2)(x^2 - 2x + 4)(x^2 + 2x + 4)$
(b) $(x - 2)(x + 2)[x - (1 + i\sqrt{3})][x - (1 - i\sqrt{3})] \times$
$[x + (1 + i\sqrt{3})][x + (1 - i\sqrt{3})]$
71. (a) 4 real **(b)** 2 real, 2 non-real **(c)** 4 non-real

SECTION 3.6 ■ PAGE 308

1. $-\infty, \infty$ **2.** 2 **3.** $-1, 2$ **4.** $\frac{1}{3}$ **5.** $-2, 3$ **6.** 1
7. (a) False **(b)** True **(c)** False **(d)** True **8.** True

9. (a) $-3, -19, -199, -1999$; $5, 21, 201, 2001$; $1.2500, 1.0417, 1.0204, 1.0020$; $0.8333, 0.9615, 0.9804, 0.9980$
(b) $r(x) \to -\infty$ as $x \to 2^-$; $r(x) \to \infty$ as $x \to 2^+$
(c) Horizontal asymptote $y = 1$
11. (a) $-22, -430, -40{,}300, -4{,}003{,}000$; $-10, -370, -39{,}700, -3{,}997{,}000$; $0.3125, 0.0608, 0.0302, 0.0030$; $-0.2778, -0.0592, -0.0298, -0.0030$
(b) $r(x) \to -\infty$ as $x \to 2^-$; $r(x) \to -\infty$ as $x \to 2^+$
(c) Horizontal asymptote $y = 0$

13.
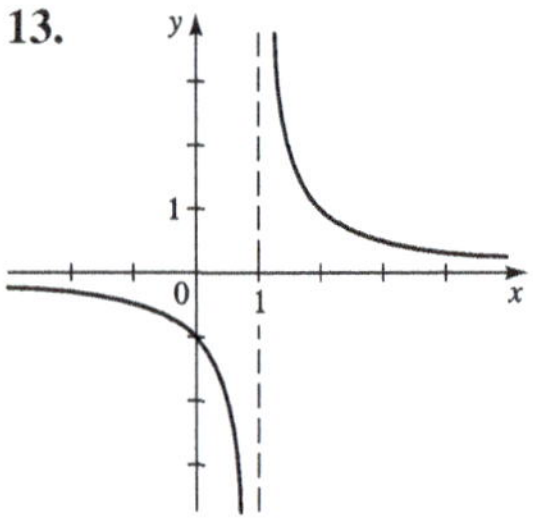

domain $\{x \mid x \neq 1\}$
range $\{y \mid y \neq 0\}$

15.
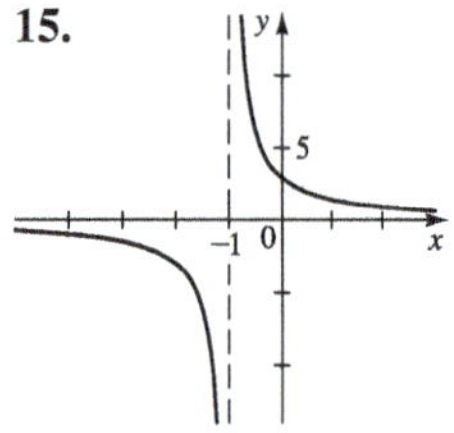

domain $\{x \mid x \neq -1\}$
range $\{y \mid y \neq 0\}$

17.
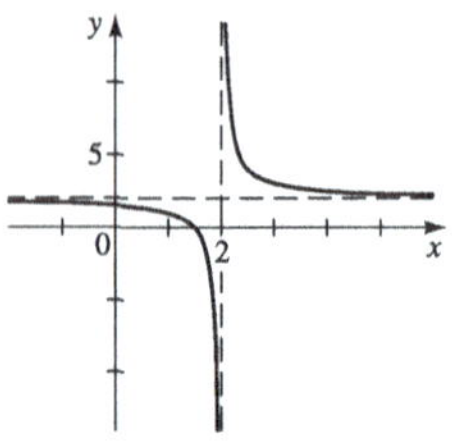

domain $\{x \mid x \neq 2\}$
range $\{y \mid y \neq 2\}$

19.
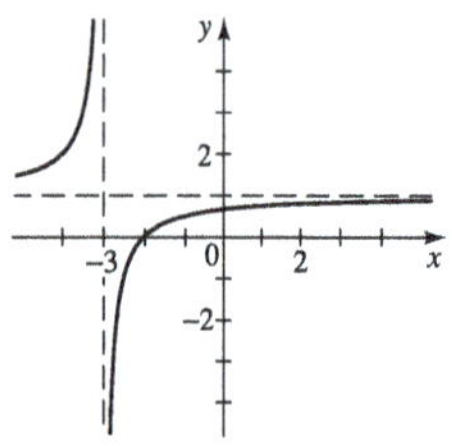

domain $\{x \mid x \neq -3\}$
range $\{y \mid y \neq 1\}$

21. x-intercept 1, y-intercept $-\frac{1}{4}$ **23.** x-intercepts $-1, 2$; y-intercept $\frac{1}{3}$ **25.** x-intercepts $-3, 3$; no y-intercept
27. x-intercept 3, y-intercept 3, vertical $x = 2$; horizontal $y = 2$
29. x-intercepts $-1, 1$; y-intercept $\frac{1}{4}$; vertical $x = -2, x = 2$; horizontal $y = 1$ **31.** Vertical $x = 2$; horizontal $y = 0$
33. Horizontal $y = 0$ **35.** Vertical $x = \frac{1}{2}, x = -1$; horizontal $y = 3$ **37.** Vertical $x = -\frac{7}{4}, x = 2$; horizontal $y = \frac{1}{2}$
39. Vertical $x = 0$; horizontal $y = 3$ **41.** Vertical $x = 1$

43.
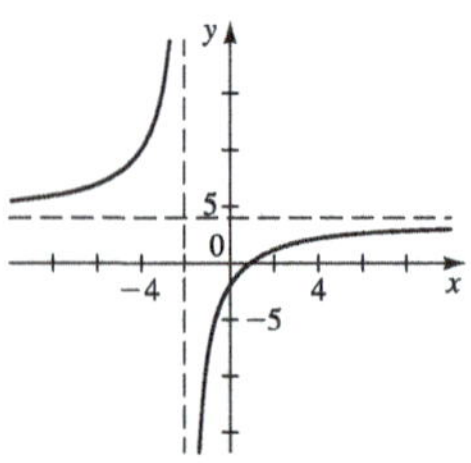

x-intercept 1
y-intercept -2
vertical $x = -2$
horizontal $y = 4$
domain $\{x \mid x \neq -2\}$
range $\{y \mid y \neq 4\}$

45.
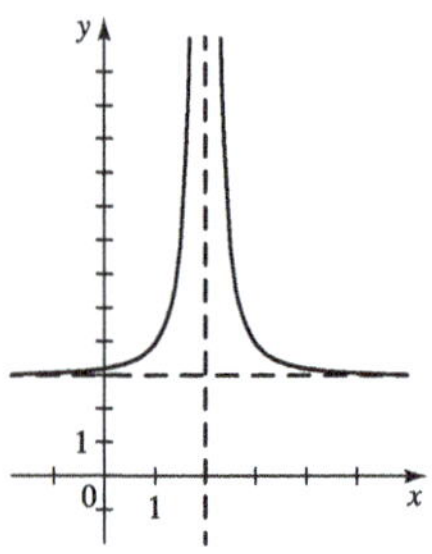

No x-intercept
y-intercept $\frac{13}{4}$
vertical $x = 2$
horizontal $y = 3$
domain $\{x \mid x \neq 2\}$
range $\{y \mid y > 3\}$

47.
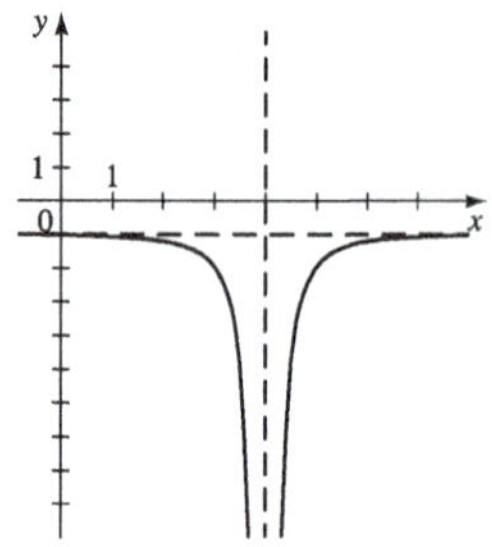

No x-intercept
y-intercept $-\frac{9}{8}$
vertical $x = 4$
horizontal $y = -1$
domain $\{x \mid x \neq 4\}$
range $\{y \mid y < -1\}$

49.
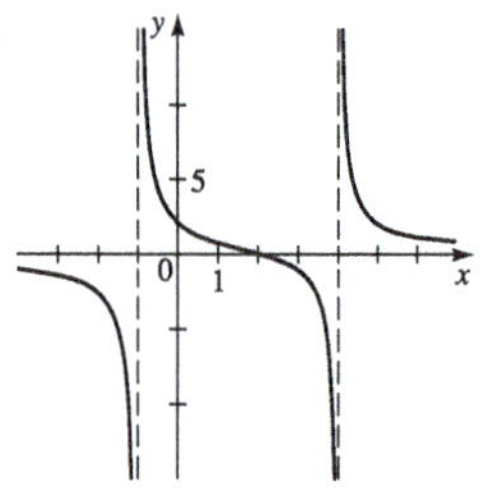

x-intercept 2
y-intercept 2
vertical $x = -1, x = 4$
horizontal $y = 0$
domain $\{x \mid x \neq -1, 4\}$
range $\mathbb{R}$

51.
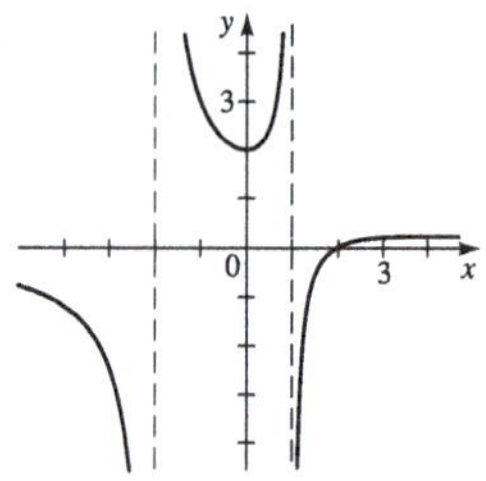

x-intercept 2
y-intercept 2
vertical $x = -2, x = 1$
horizontal $y = 0$
domain $\{x \mid x \neq -2, 1\}$
range $\{y \mid y \leq 0.2 \text{ or } y \geq 2\}$

53.
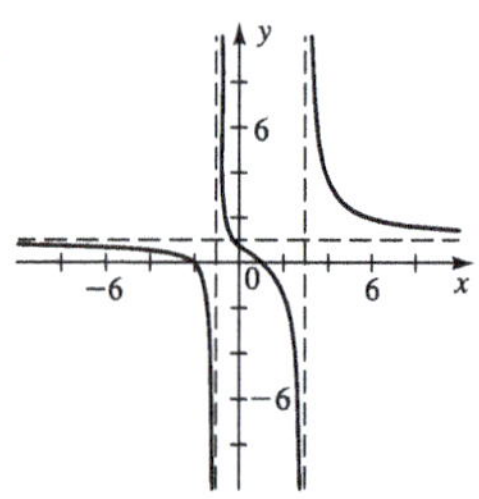

x-intercepts $-2, 1$
y-intercept $\frac{2}{3}$
vertical $x = -1, x = 3$
horizontal $y = 1$
domain $\{x \mid x \neq -1, 3\}$
range $\mathbb{R}$

55.
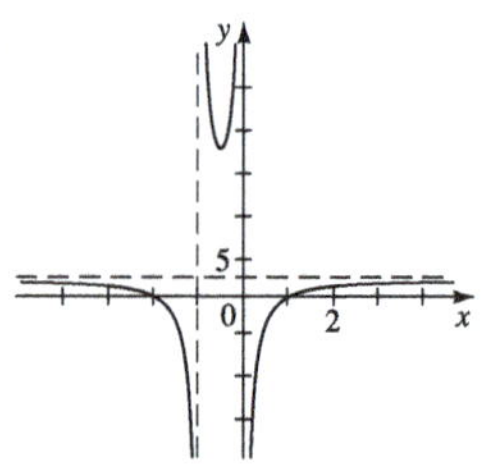

x-intercepts $1, -2$
vertical $x = -1, x = 0$
horizontal $y = 2$
domain $\{x \mid x \neq -1, 0\}$
range $\{y \mid y < 2 \text{ or } y \geq 18.4\}$

57.
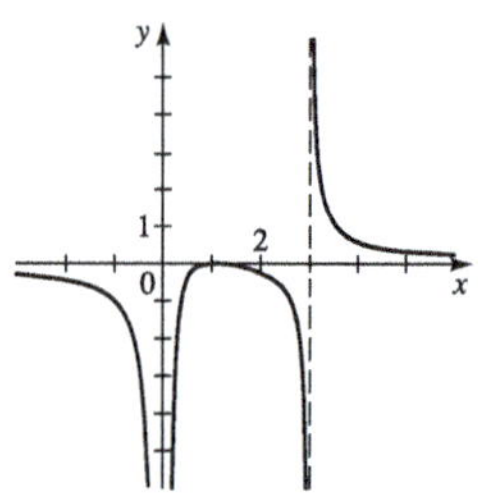

x-intercept 1
vertical $x = 0, x = 3$
horizontal $y = 0$
domain $\{x \mid x \neq 0, 3\}$
range $\mathbb{R}$

59.

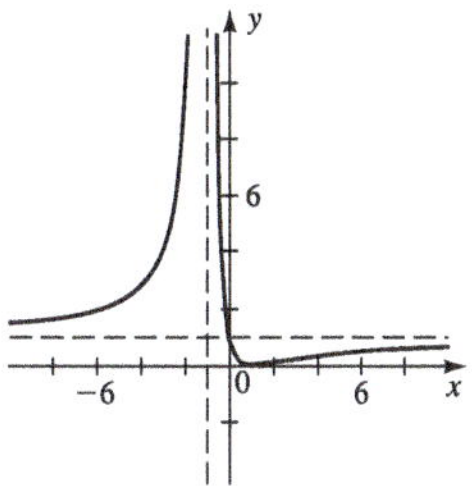

x-intercept 1
y-intercept 1
vertical $x = -1$
horizontal $y = 1$
domain $\{x \mid x \neq -1\}$
range $\{y \mid y \geq 0\}$

61.

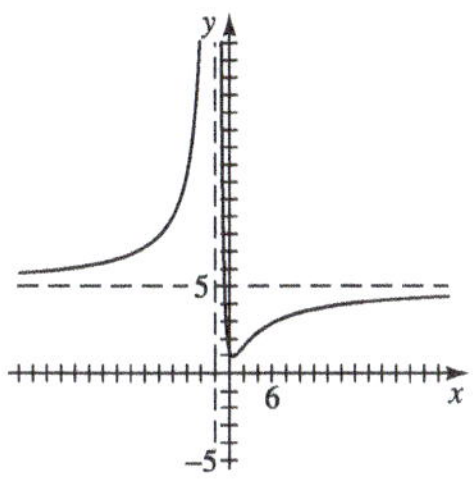

y-intercept $\frac{5}{4}$
vertical $x = -2$
horizontal $y = 5$
domain $\{x \mid x \neq -2\}$
range $\{y \mid y \geq 1.0\}$

63.

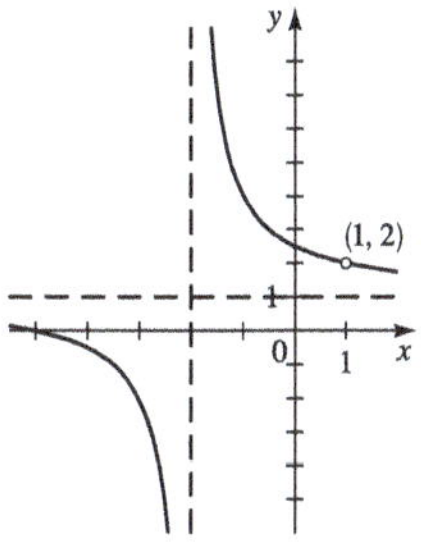

x-intercept -5
y-intercept $\frac{5}{2}$
vertical $x = -2$
horizontal $y = 1$
domain $\{x \mid x \neq -2, 1\}$
range $\{y \mid y \neq 1, 2\}$

65.

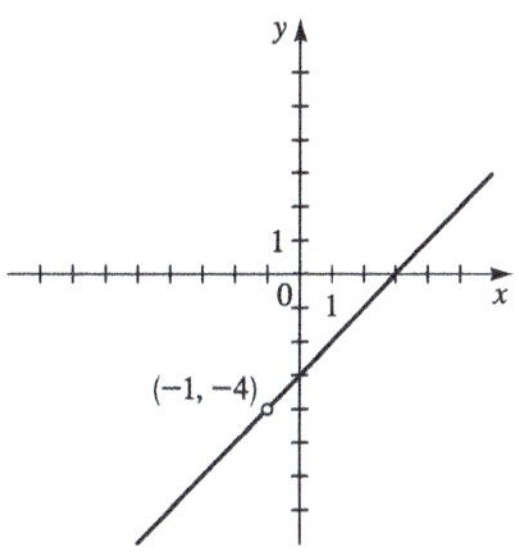

x-intercept 3
y-intercept -3
no asymptote
domain $\{x \mid x \neq -1\}$
range $\{y \mid y \neq -4\}$

67.

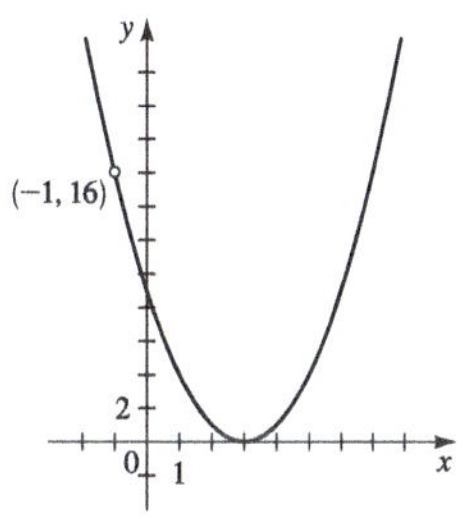

x-intercept 3
y-intercept 9
no asymptote
domain $\{x \mid x \neq -1\}$
range $\{y \mid y \geq 0\}$

69.

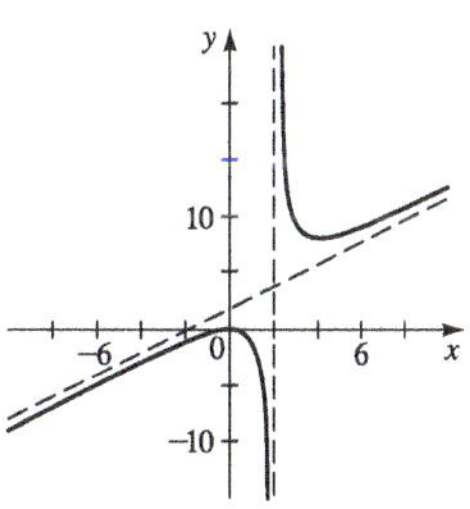

slant $y = x + 2$
vertical $x = 2$

71.

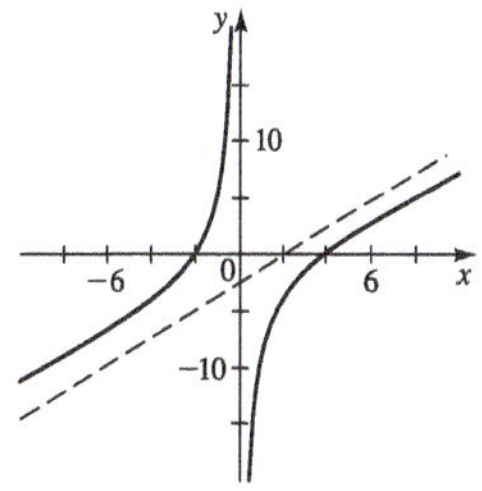

slant $y = x - 2$
vertical $x = 0$

73.

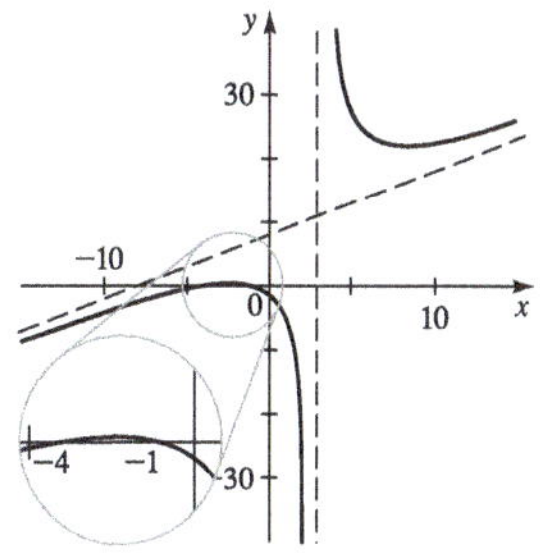

slant $y = x + 8$
vertical $x = 3$

75.

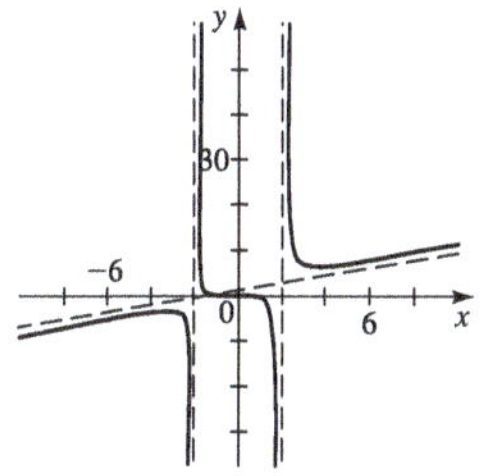

slant $y = x + 1$
vertical $x = 2, x = -2$

77.

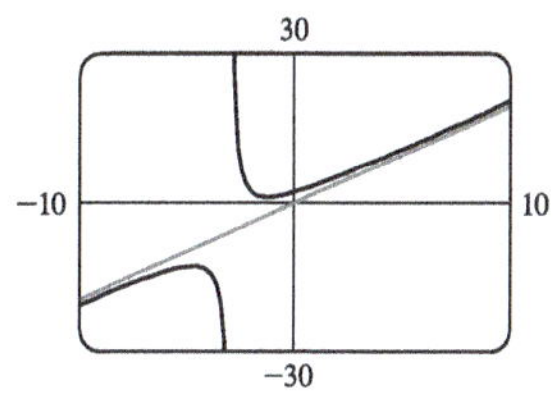

vertical $x = -3$

79.

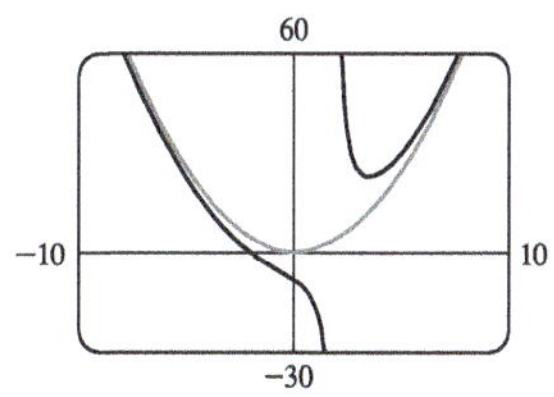

vertical $x = 2$

81.

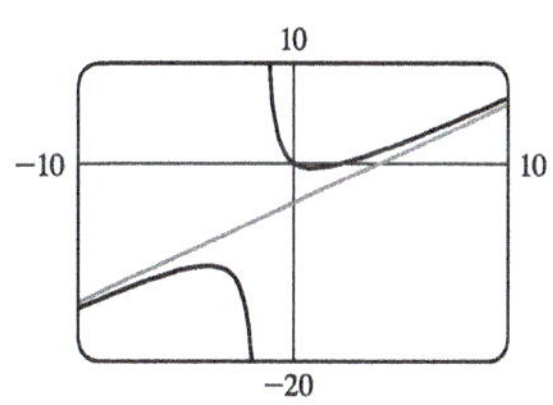

vertical $x = -1.5$
x-intercepts 0, 2.5
y-intercept 0, local maximum $(-3.9, -10.4)$
local minimum $(0.9, -0.6)$
end behavior $y = x - 4$

83.

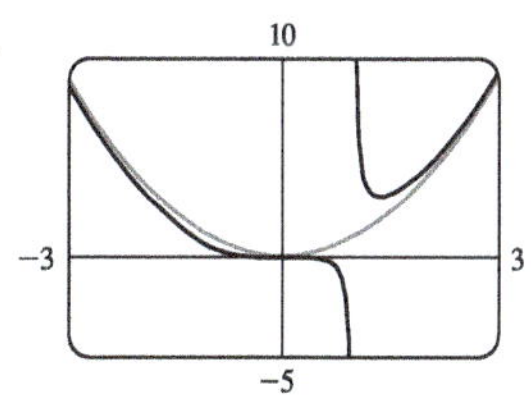

vertical $x = 1$
x-intercept 0
y-intercept 0
local minimum $(1.4, 3.1)$
end behavior $y = x^2$

85.

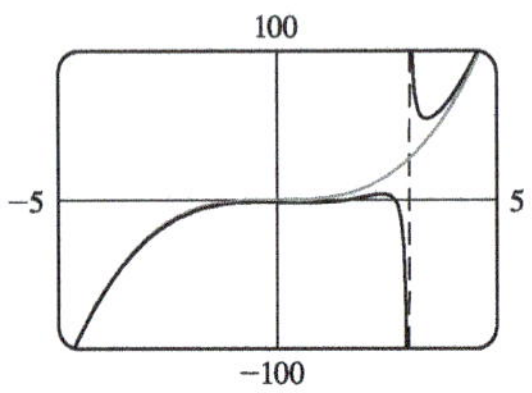

vertical $x = 3$
x-intercepts 1.6, 2.7
y-intercept -2
local maxima $(-0.4, -1.8)$, $(2.4, 3.8)$,
local minima $(0.6, -2.3)$, $(3.4, 54.3)$
end behavior $y = x^3$

87. (a)

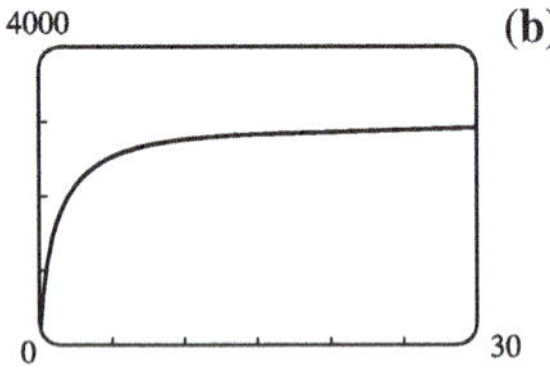

(b) It levels off at 3000.

89. (a) 2.50 mg/L **(b)** It decreases to 0. **(c)** 16.61 h

91.

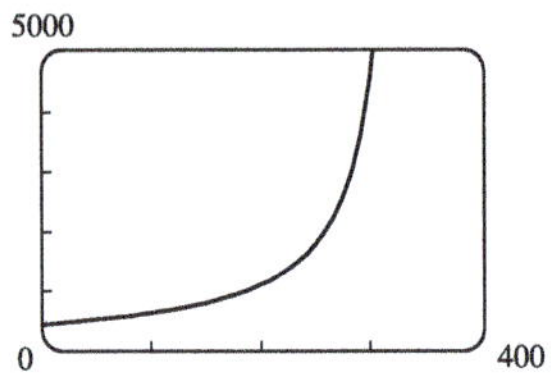

If the speed of the train approaches the speed of sound, then the pitch increases indefinitely (a sonic boom).

SECTION 3.7 ▪ PAGE 316

1. zeros; zeros; $[-2, 0]$, $[1, \infty)$

Sign of		-2		0		1	
x	−		−		+		+
$x + 2$	−		+		+		+
$x - 1$	−		−		−		+
$x(x + 2)(x - 1)$	−		+		−		+

2. zeros; zeros; cut points; $(-\infty, -4,)$, $[-2, 1]$, $(3, \infty)$

Sign of		-4		-2		1		3	
$x + 2$	−		−		+		+		+
$x - 1$	−		−		−		+		+
$x - 3$	−		−		−		−		+
$x + 4$	−		+		+		+		+
$\dfrac{(x + 2)(x - 1)}{(x - 3)(x + 4)}$	+		−		+		−		+

3. $(-\infty, -5) \cup (-\frac{5}{2}, 3)$ **5.** $(-\infty, -5) \cup (-5, -3) \cup (1, \infty)$
7. $[-4, -2] \cup [2, \infty)$ **9.** $(-\infty, \frac{1}{2})$ **11.** $(-3, 3)$
13. $[-5, 1] \cup [3, \infty)$ **15.** $(-\infty, -1) \cup (1, 7)$ **17.** $(1, 10)$
19. $(-7, -\frac{5}{2}] \cup (5, \infty)$ **21.** $(-\infty, -1 - \sqrt{3}) \cup [0, \sqrt{3} - 1)$
23. $(-\infty, -3) \cup (-\frac{2}{3}, 1) \cup (3, \infty)$ **25.** $(-4, 3]$
27. $[-8, -\frac{5}{2})$ **29.** $(0, \frac{3-\sqrt{3}}{2}] \cup (1, \frac{3+\sqrt{3}}{2}]$
31. $(-\infty, -2) \cup (-1, 1) \cup (1, \infty)$ **33.** $[-2, 0) \cup (1, 3]$
35. $(-3, -\frac{1}{2}) \cup (2, \infty)$ **37.** $(-\infty, -2) \cup (5, \infty)$
39. $(-\frac{1}{2}, 0) \cup (\frac{1}{2}, \infty)$ **41.** $[-2, 3]$ **43.** $(-\infty, -1] \cup [1, \infty)$

45. $[-2, 1] \cup [3, \infty)$ **47.** $(-\infty, -1.37) \cup (0.37, 1)$
49. $(0, 1.60)$ **51.** $(0, 1]$ **53.** $(-\infty, a] \cup [b, c] \cup [d, \infty)$
55. More than 35.6 m
57.

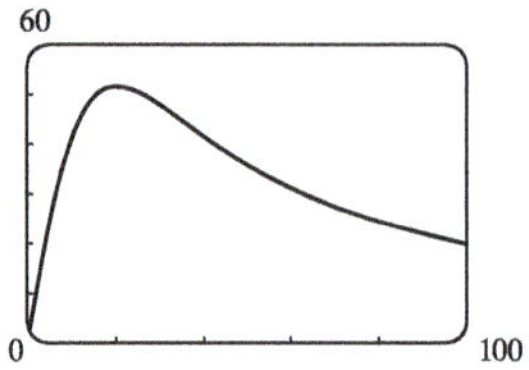

Between 9.5 and 42.3 mi/h

CHAPTER 3 REVIEW ▪ PAGE 320

1. (a) $f(x) = (x + 3)^2 - 7$ **3. (a)** $f(x) = -(x + 5)^2 + 26$

(b)

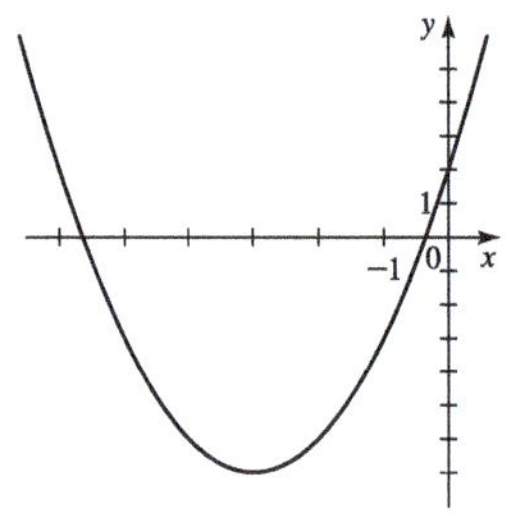

(b)

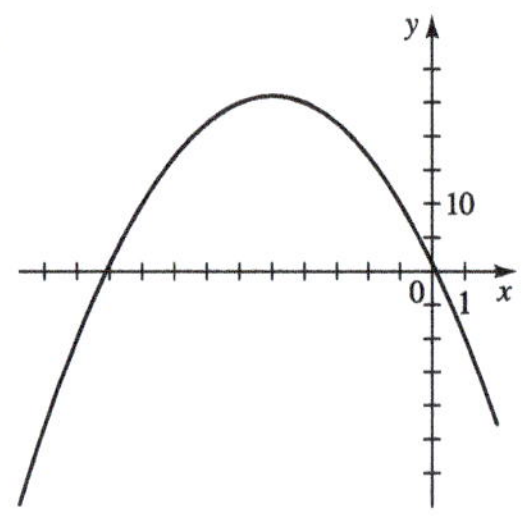

5. Maximum $f(\frac{3}{2}) = \frac{5}{4}$ **7.** 68 ft

9.

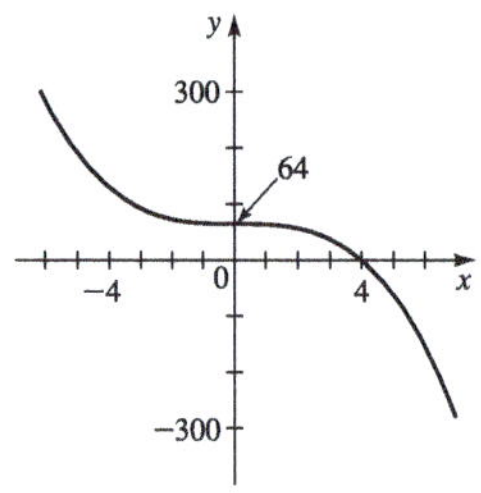

11.

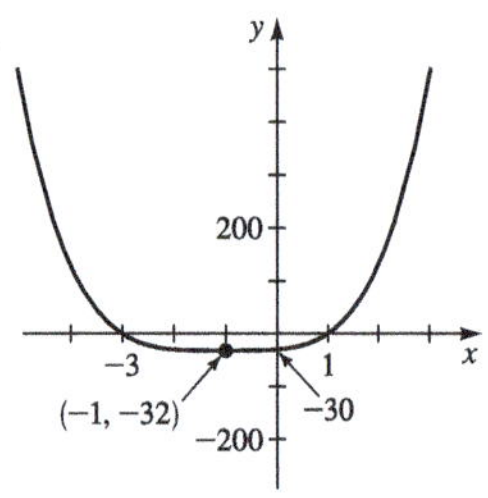

13.

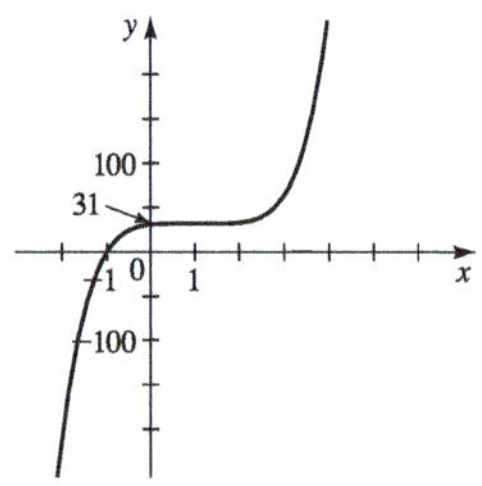

15. (a) $y \to \infty$ as $x \to \infty$, $y \to -\infty$ as $x \to -\infty$

(b)

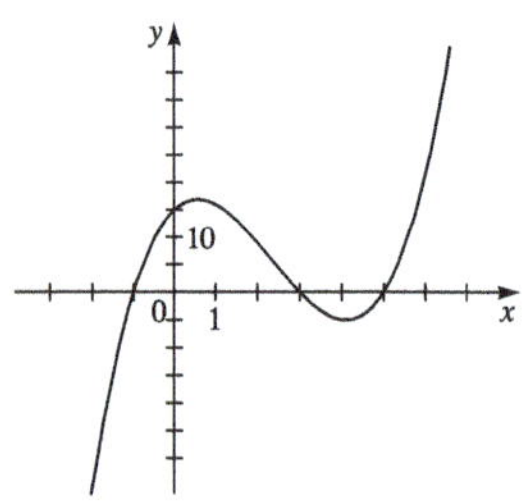

17. (a) $y \to -\infty$ as $x \to \infty$, $y \to \infty$ as $x \to -\infty$

(b)

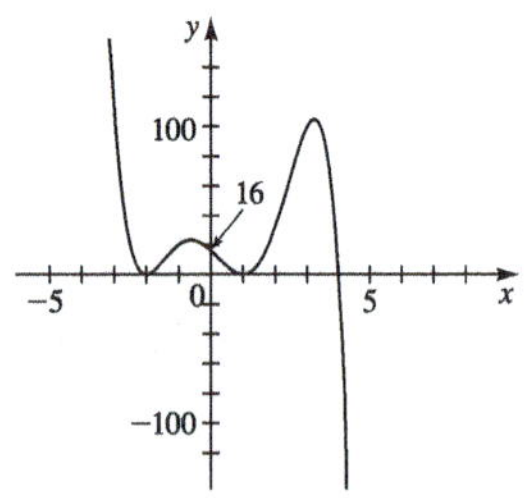

19. (a) 0 (multiplicity 3), 2 (multiplicity 2)

(b)

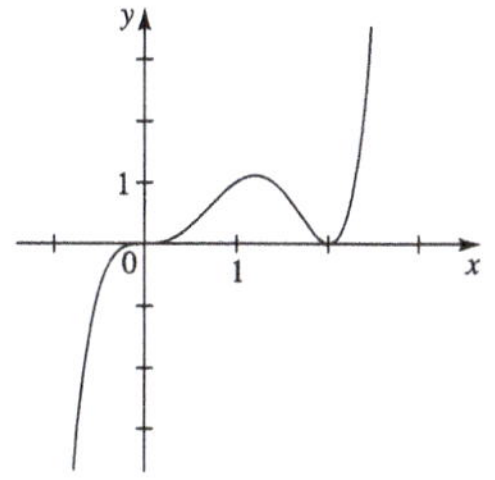

21.

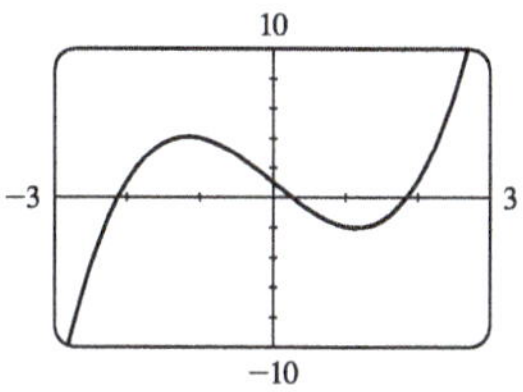

x-intercepts −2.1, 0.3, 1.9
y-intercept 1
local maximum (−1.2, 4.1)
local minimum (1.2, −2.1)
$y \to \infty$ as $x \to \infty$ and
$y \to -\infty$ as $x \to -\infty$

23.

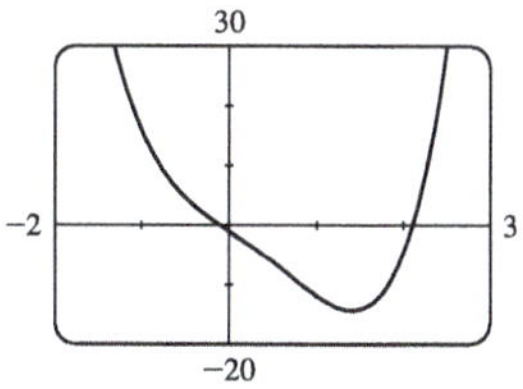

x-intercepts −0.1, 2.1
y-intercept −1
local minimum (1.4, −14.5)
$y \to \infty$ as $x \to \infty$ and
$y \to \infty$ as $x \to -\infty$

25. (a) $S = 13.8x(100 - x^2)$ **(b)** $0 \le x \le 10$

(c)

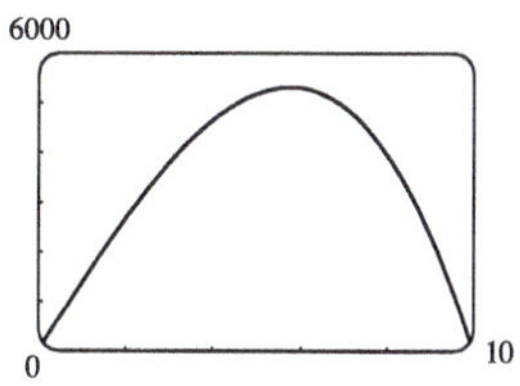

(d) 5.8 in.

In answers 27–33 the first polynomial given is the quotient, and the second is the remainder.

27. $x - 2, -4$ **29.** $2x^2 - 11x + 58, -294$
31. $x^3 - 5x^2 + 17x - 83, 422$ **33.** $2x - 3, 12$ **35.** 3 **37.** 8
41. (a) $\pm1, \pm2, \pm3, \pm6, \pm9, \pm18$ **(b)** 2 or 0 positive, 3 or 1 negative **43. (a)** $\pm1, \pm2, \pm4, \pm8, \pm\frac{1}{3}, \pm\frac{2}{3}, \pm\frac{4}{3}, \pm\frac{8}{3}$
(b) 0 or 2 positive, 1 or 3 negative
45. (a) −4, 0, 4 **(b)**

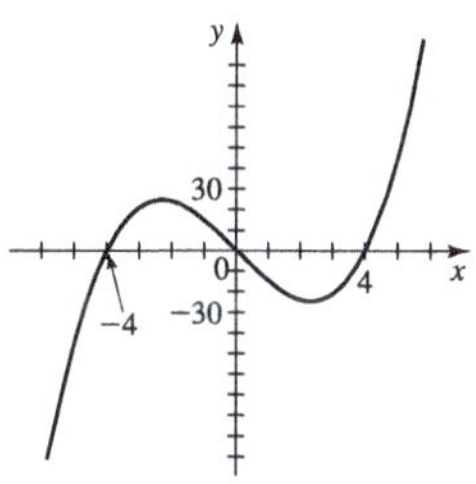

47. (a) −2, 0 (multiplicity 2), 1 **(b)**

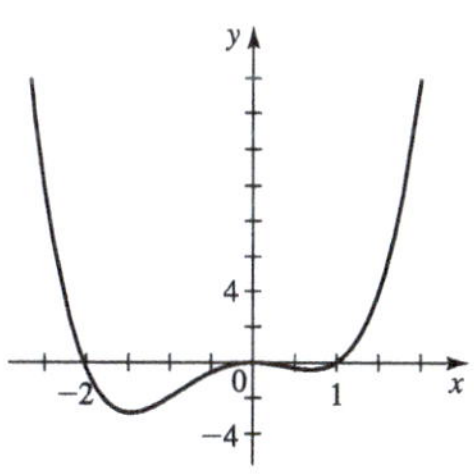

49. (a) −2, −1, 2, 3

(b)

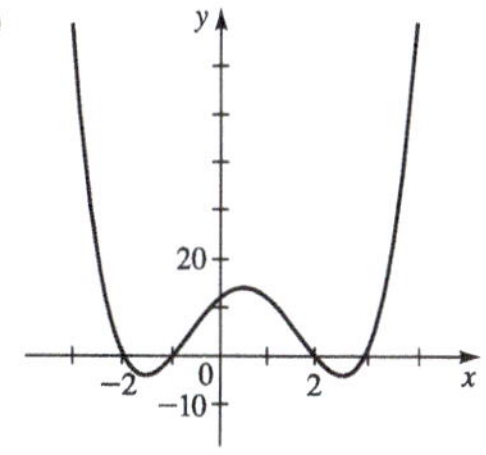

51. (a) $-\frac{1}{2}$, 1

(b)

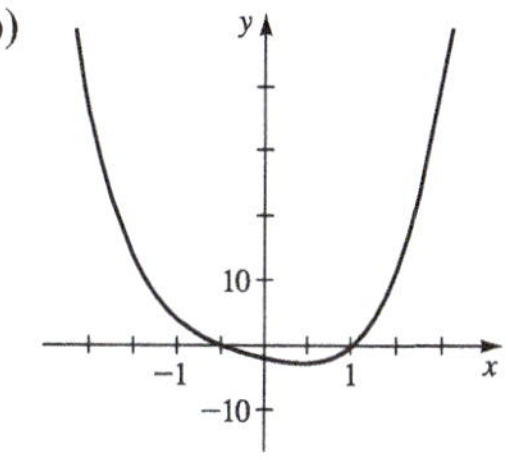

53. $4x^3 - 18x^2 + 14x + 12$
55. No; since the complex conjugates of imaginary zeros will also be zeros, the polynomial would have 8 zeros, contradicting the requirement that it have degree 4.
57. $1, \pm i$ **59.** −3, 1, 5 **61.** $-1 \pm 2i$, −2 (multiplicity 2)
63. ±2, 1 (multiplicity 3) **65.** $\pm2, \pm1 \pm i\sqrt{3}$
67. $1, 3, \dfrac{-1 \pm i\sqrt{7}}{2}$ **69.** $x = -0.5, 3$ **71.** $x \approx -0.24, 4.24$
73. $2, P(x) = (x - 2)(x^2 + 2x + 2)$

75. (a) Vertical asymptote $x = -4$, horizontal asymptote $y = 0$, no x-intercept, y-intercept $\frac{3}{4}$, domain $\{x \mid x \ne -4\}$ range $\{y \mid y \ne 0\}$

(b)

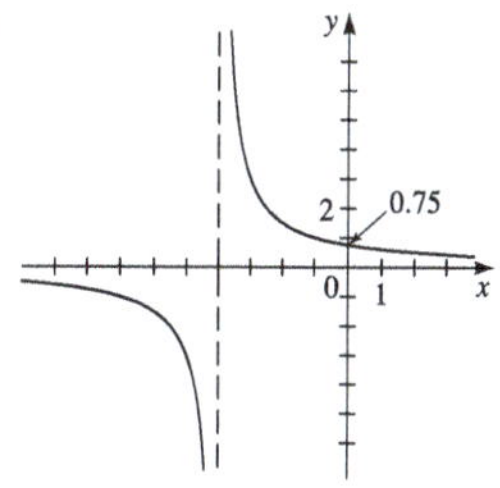

77. (a) Vertical asymptote $x = 1$, horizontal asymptote $y = 3$, x-intercept $\frac{4}{3}$, y-intercept 4, domain $\{x \mid x \ne 1\}$ range $\{y \mid y \ne 3\}$

(b)

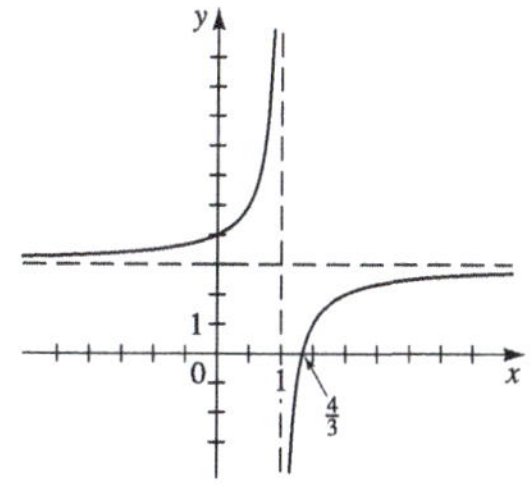

79.

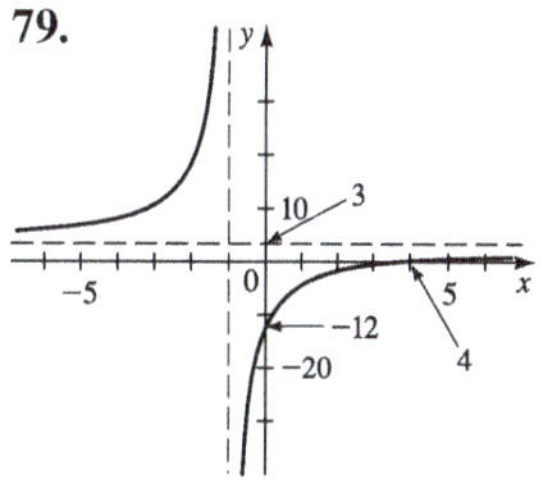

Domain $\{x \mid x \ne -1\}$, range $\{y \mid y \ne 3\}$

81.

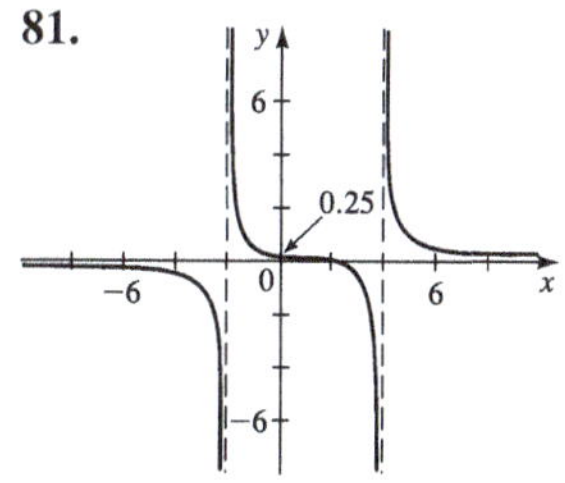

Domain $\{x \mid x \ne -2, 4\}$, range $(-\infty, \infty)$

83.

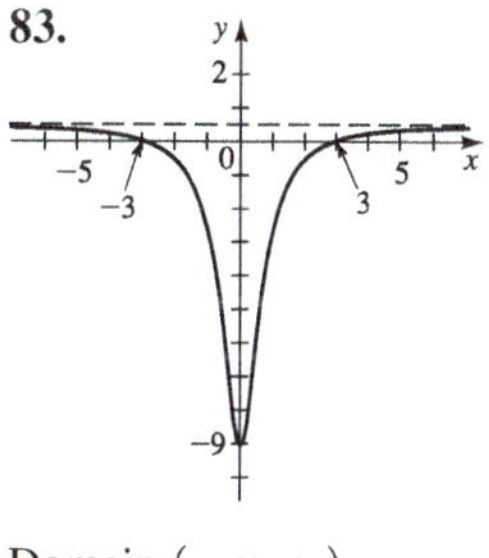

Domain $(-\infty, \infty)$, range $\{y \mid -9 \le y < \frac{1}{2}\}$

85. 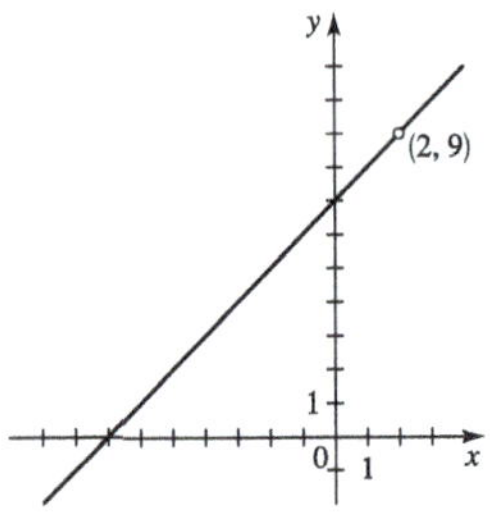

x-intercept -7
y-intercept 7
no asymptote
domain $\{x \mid x \neq 2\}$
range $\{y \mid y \neq 9\}$

87. 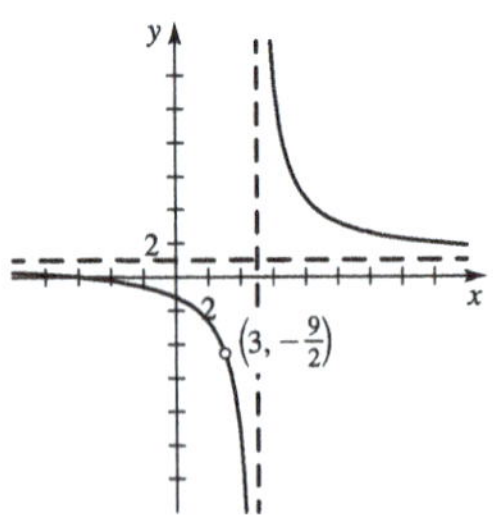

x-intercept -6
y-intercept $-\frac{6}{5}$
vertical $x = 5$
horizontal $y = 1$
domain $\{x \mid x \neq 3, 5\}$
range $\{y \mid y \neq 1, -\frac{9}{2}\}$

89. 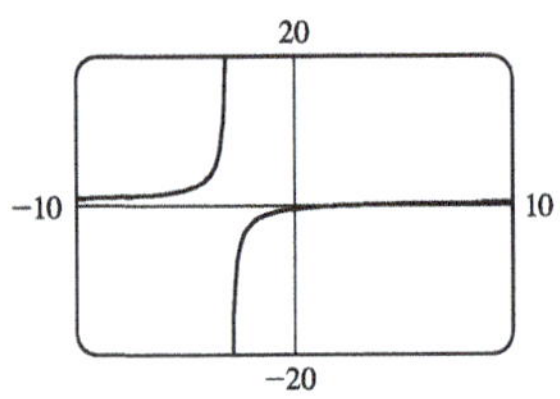

x-intercept 3
y-intercept -0.5
vertical $x = -3$
horizontal $y = 0.5$
no local extrema

91. 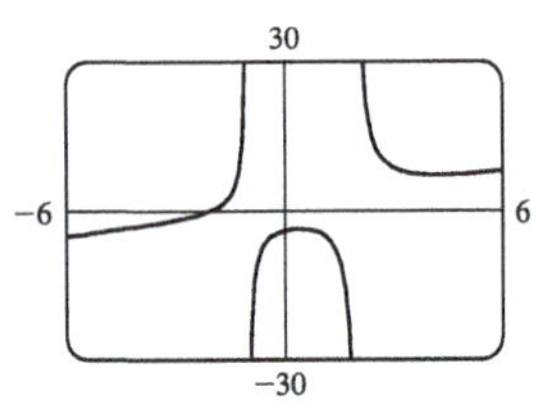

x-intercept -2
y-intercept -4
vertical $x = -1, x = 2$
slant $y = x + 1$
local maximum
$(0.425, -3.599)$
local minimum $(4.216, 7.175)$

93. $(-\infty, -1] \cup [\frac{3}{2}, \infty)$ **95.** $(-3, 3)$
97. $(-\infty, -2) \cup (1, 2)$ **99.** $(-3, 0) \cup (2, \frac{9}{2}]$
101. $[-3, \frac{8}{3}]$ **103.** $[0.74, 1.95]$

CHAPTER 3 TEST ■ PAGE 323

1. $f(x) = (x - \frac{1}{2})^2 - \frac{25}{4}$

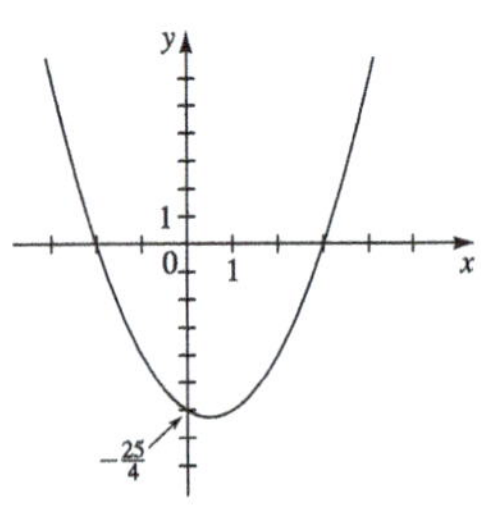

2. Minimum $g(-\frac{3}{2}) = -\frac{3}{2}$ **3. (a)** 2500 ft **(b)** 1000 ft

4.

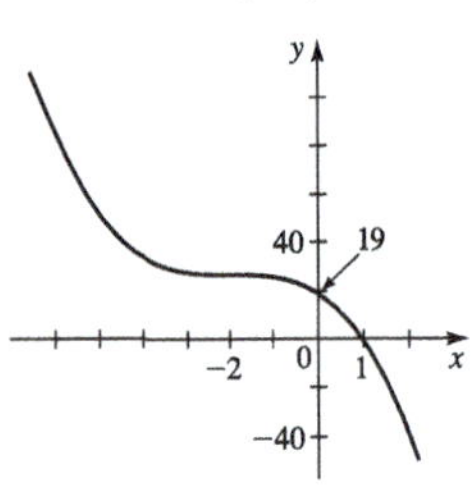

5. (a) $x^3 + 2x^2 + 2, 9$ **(b)** $x^3 + 2x^2 + \frac{1}{2}, \frac{15}{2}$
6. (a) $\pm 1, \pm 3, \pm \frac{1}{2}, \pm \frac{3}{2}$ **(b)** $2(x - 3)(x - \frac{1}{2})(x + 1)$
(c) $-1, \frac{1}{2}, 3$ **(d)**

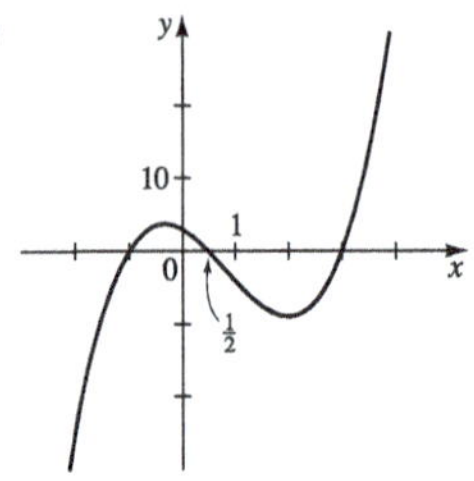

7. $3, -1 \pm i$ **8.** $(x - 1)^2(x - 2i)(x + 2i)$
9. $x^4 + 2x^3 + 10x^2 + 18x + 9$
10. (a) 4, 2, or 0 positive; 0 negative
(c) 0.17, 3.93

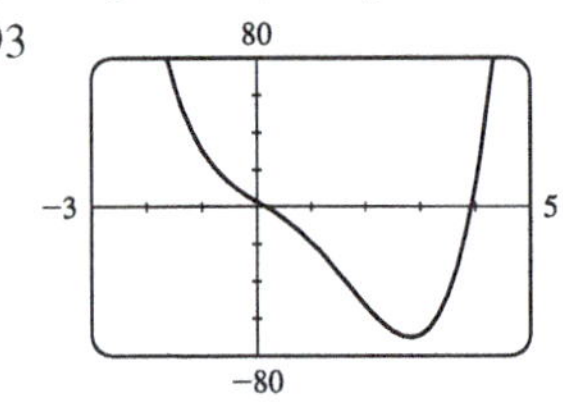

(d) Local minimum $(2.82, -70.31)$
11. (a) r, u **(b)** s **(c)** s, w **(d)** w
(e) Vertical $x = -1, x = 2$; horizontal $y = 0$
(f)

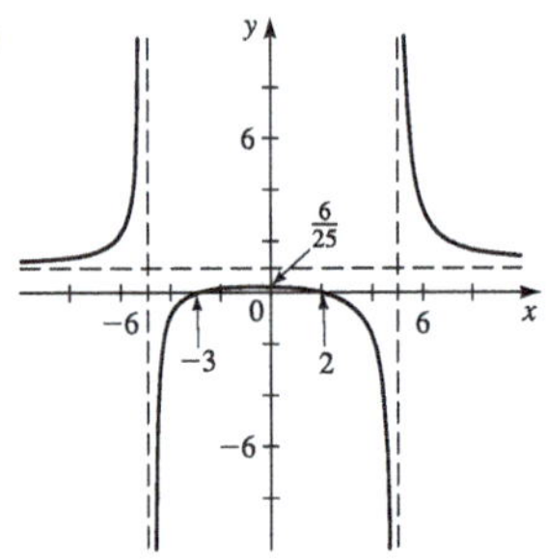

(g) $x^2 - 2x - 5$

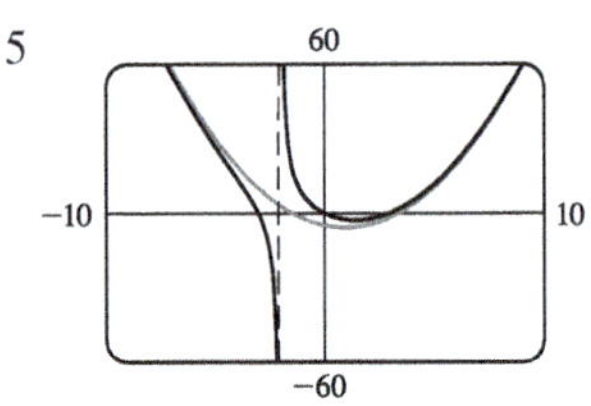

12. $\{x \mid x \le -1 \text{ or } \frac{5}{2} < x \le 3\}$
13. $\{x \mid -1 - \sqrt{5} < x < -1 + \sqrt{5}\}$
14. (a)

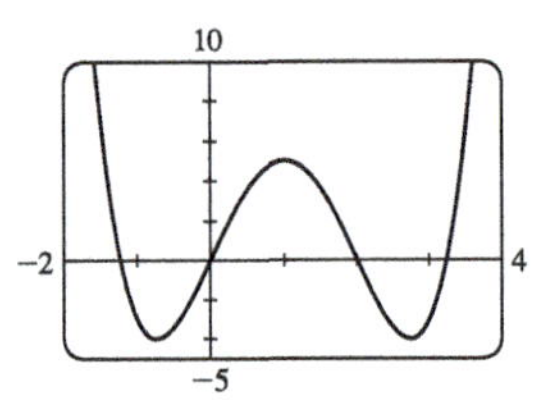

x-intercepts $-1.24, 0, 2, 3.24$, local maximum $P(1) = 5$,
local minima $P(-0.73) = P(2.73) = -4$
(b) $(-\infty, -1.24] \cup [0, 2] \cup [3.24, \infty)$

FOCUS ON MODELING ■ PAGE 327

1. (a) $y = -0.275428x^2 + 19.7485x - 273.5523$

(b)

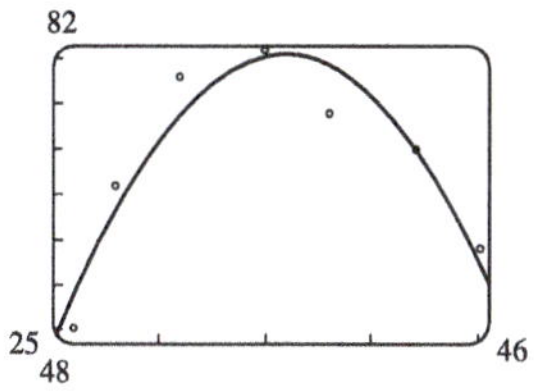

(c) 35.85 lb/in^2

3. (a) $y = 0.00203708x^3 - 0.104521x^2 + 1.966206x + 1.45576$

(b)

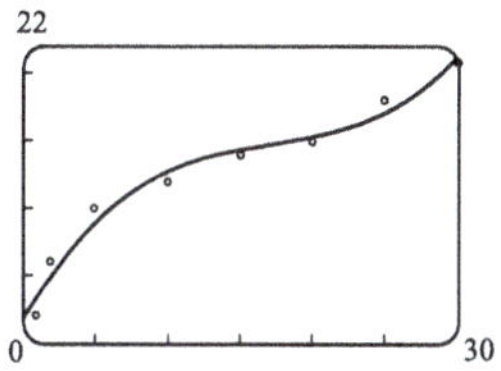

(c) 43 vegetables **(d)** 2.0 s

5. (a) $y = 0.0120536x^2 - 0.490357x + 4.96571$

(b)

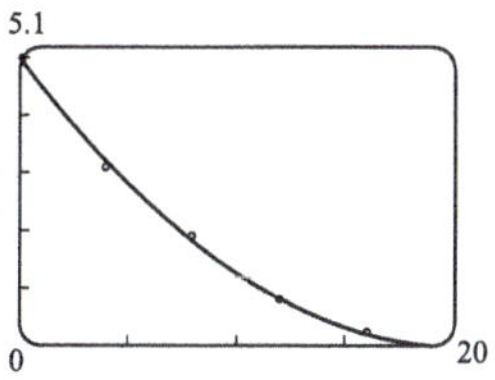

(c) 19.0 min

CHAPTER 4

SECTION 4.1 ■ PAGE 336

1. 5; $\frac{1}{25}$; 1; 25; 15,625 **2. (a)** III **(b)** I **(c)** II **(d)** IV
3. (a) downward **(b)** right **4.** principal, interest rate per year, number of times interest is compounded per year, number of years, amount after t years; \$112.65 **5.** horizontal, 0; 0
6. horizontal, 3; 3 **7.** 2.000, 22.195, 0.063, 1.516
9. 0.192, 0.070, 15.588, 1.552

11.

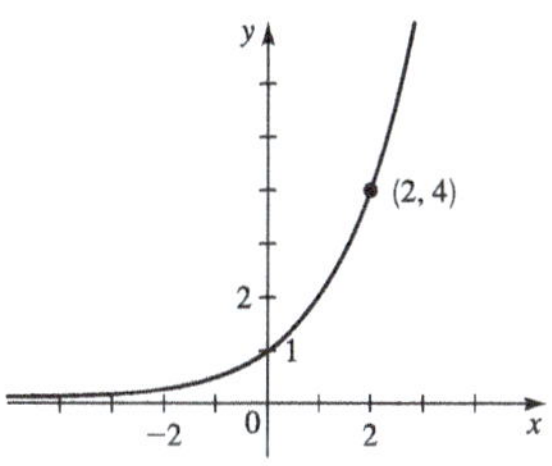

13.

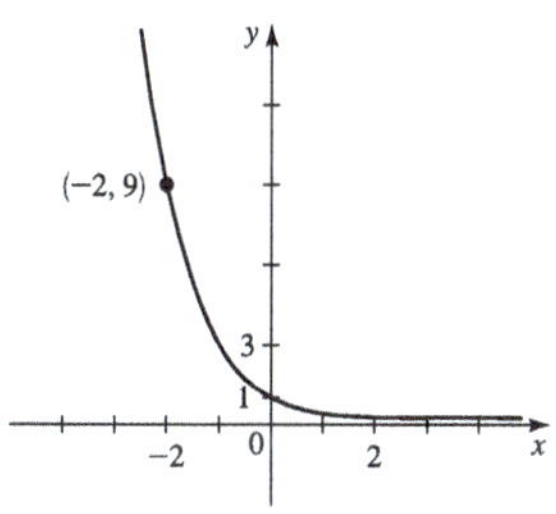

15.

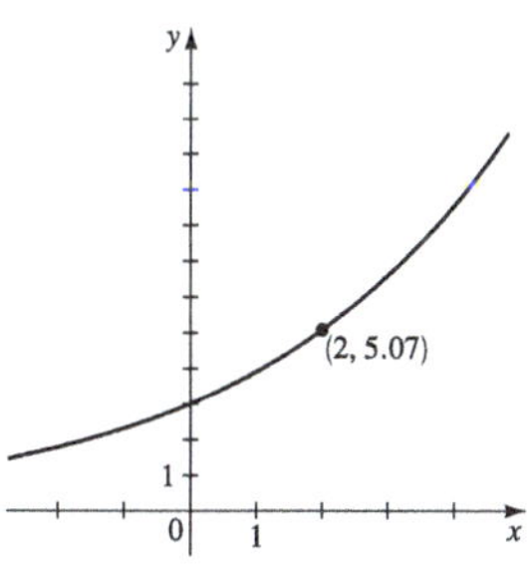

17.

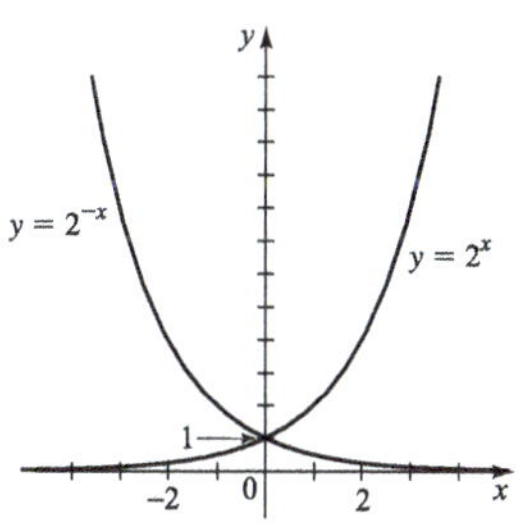

19.

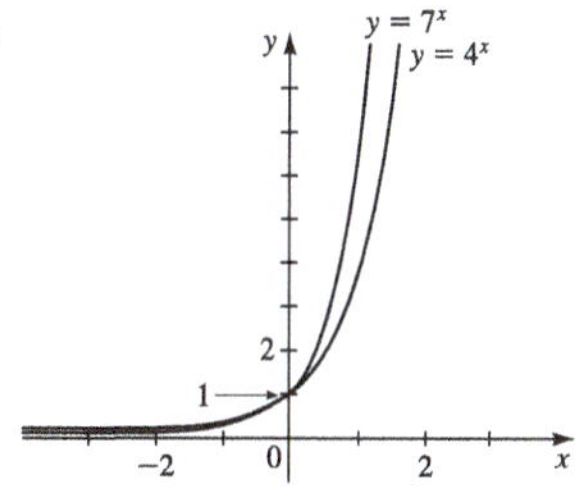

21. $f(x) = 3^x$ **23.** $f(x) = \left(\frac{1}{4}\right)^x$ **25.** II

27. $\mathbb{R}, (-3, \infty), y = -3$

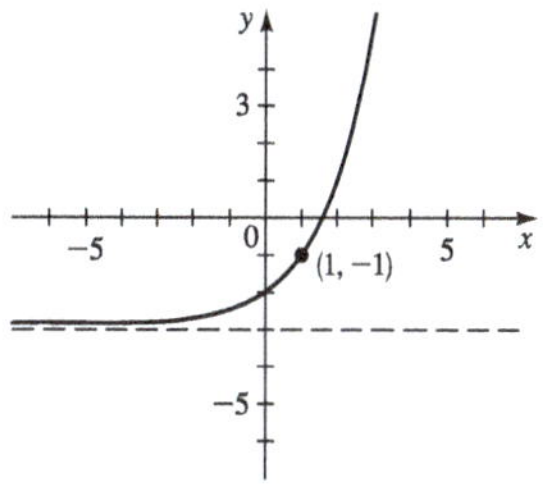

29. $\mathbb{R}, (-\infty, 0), y = 0$

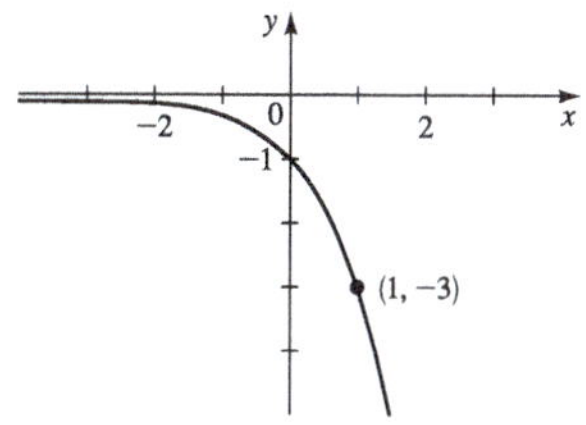

31. $\mathbb{R}, (0, \infty), y = 0$

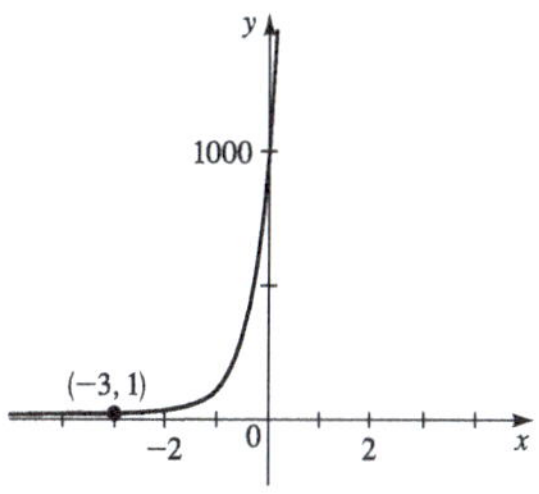

33. $\mathbb{R}, (1, \infty), y = 1$

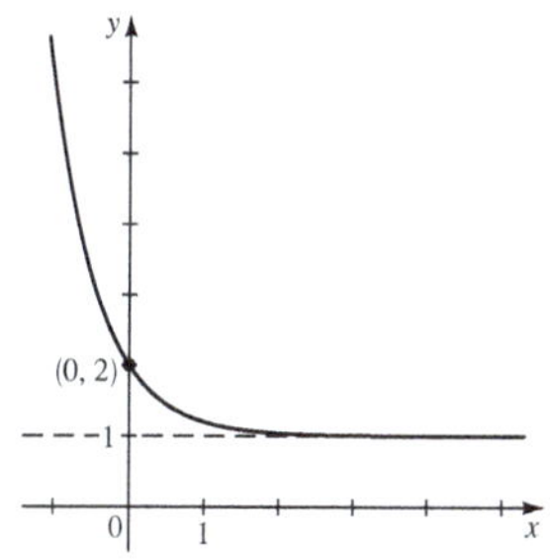

35. $\mathbb{R}, (-\infty, 2), y = 2$

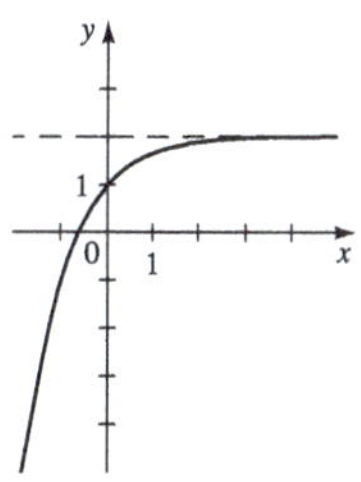

37. $\mathbb{R}, (1, \infty), y = 1$

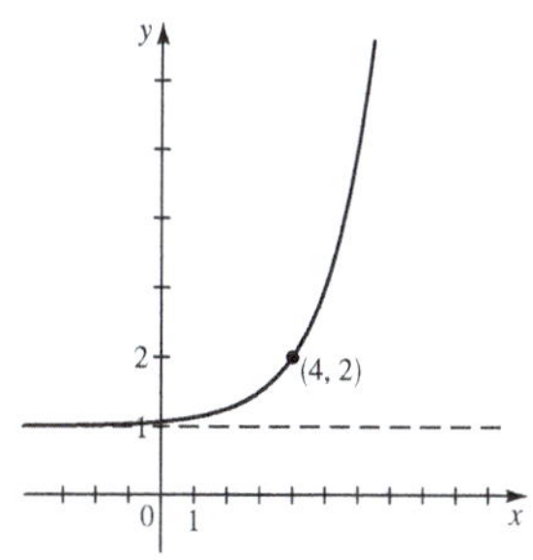

39. $\mathbb{R}, (-\infty, 1), y = 1$

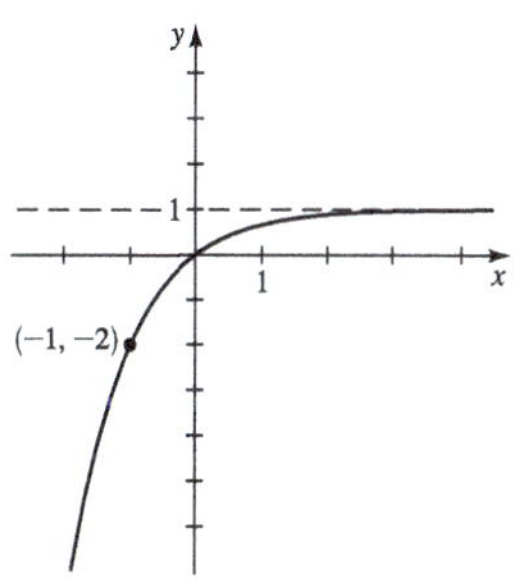

41. (a)

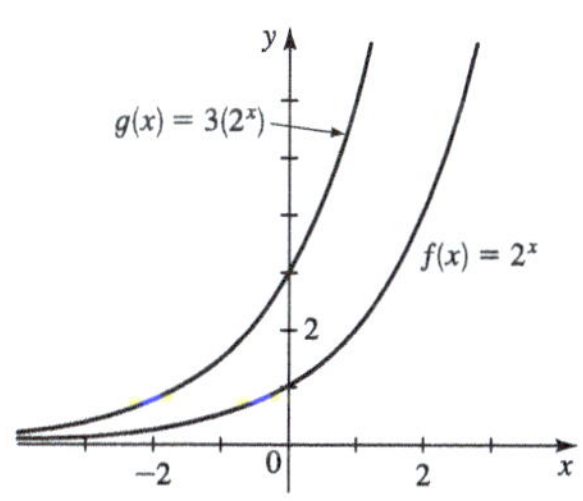

(b) The graph of g is steeper than that of f.

43.

x	0	1	2	3	4	6	8	10
$f(x)$	0	1	8	27	64	216	512	1000
$g(x)$	1	3	9	27	81	729	6561	59,049

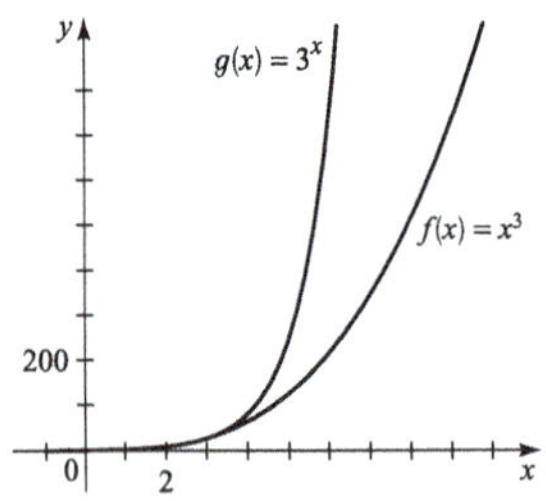

45. (a)

(i)

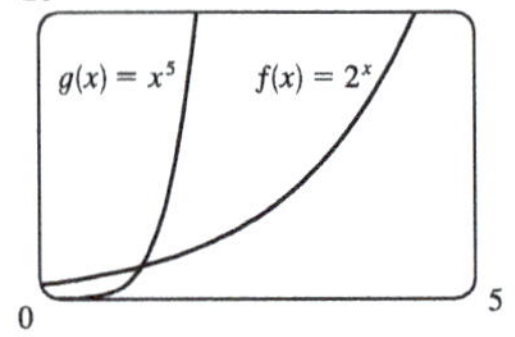

(ii)

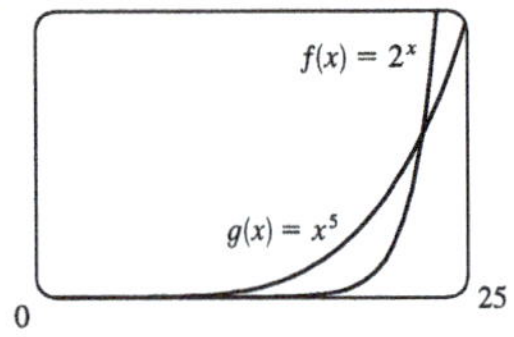

(iii)

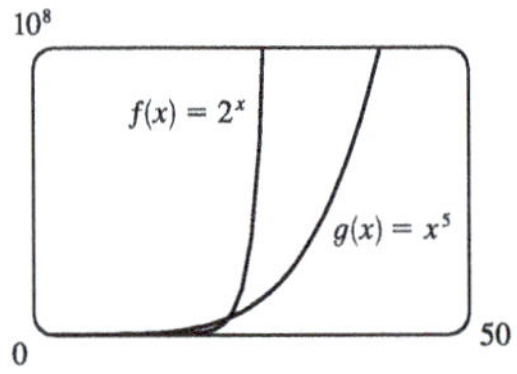

The graph of f ultimately increases much more quickly than that of g.

(b) 1.2, 22.4

47.

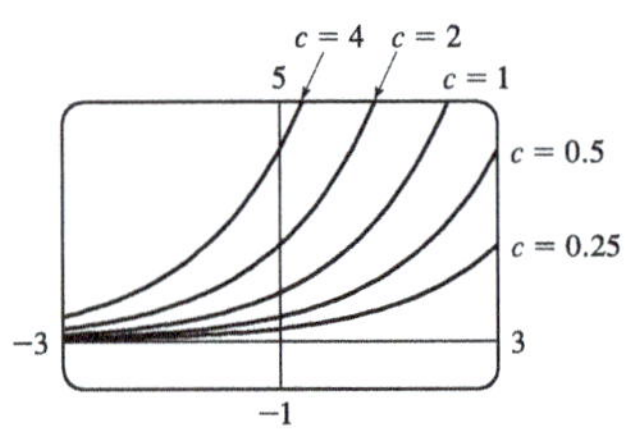

The larger the value of c, the more rapidly the graph increases.

49. (a) Increasing on $(-\infty, 0.50)$; decreasing on $(0.50, \infty)$ **(b)** $(0, 1.78)$ **53. (a)** $N(t) = 1500 \cdot 2^t$ **(b)** 25,165,824,000 **55.** \$5203.71, \$5415.71, \$5636.36, \$5865.99, \$6104.98, \$6353.71 **57. (a)** \$11,605.41 **(b)** \$13,468.55 **(c)** \$15,630.80 **59. (a)** \$519.02 **(b)** \$538.75 **(c)** \$726.23 **61.** \$7678.96 **63.** 8.30%

SECTION 4.2 ■ PAGE 341

1. natural; 2.71828 **2.** principal, interest rate per year, number of years; amount after t years; \$112.75 **3.** 2.718, 23.141, 0.050, 4.113

5.

x	y
−2	0.20
−1	0.55
−0.5	0.91
0	1.5
0.5	2.47
1	4.08
2	11.08

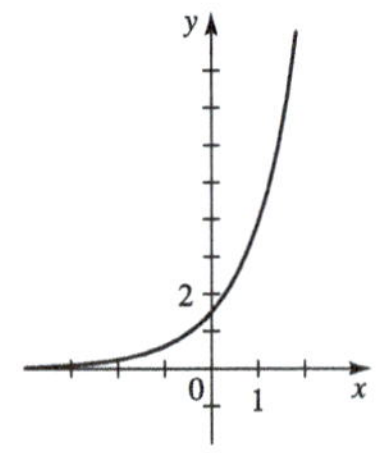

7. $\mathbb{R}$, $(2, \infty)$, $y = 2$

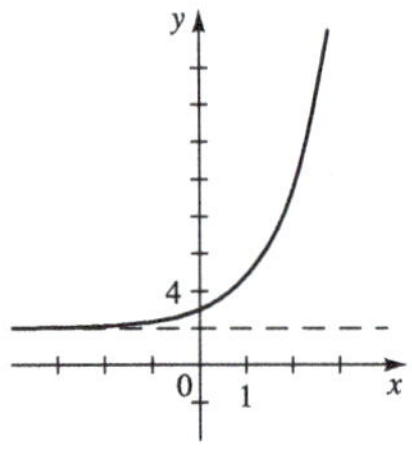

9. $\mathbb{R}$, $(-\infty, 0)$, $y = 0$

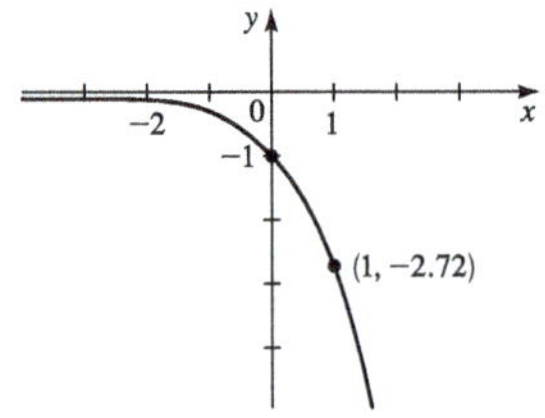

11. $\mathbb{R}$, $(-1, \infty)$, $y = -1$

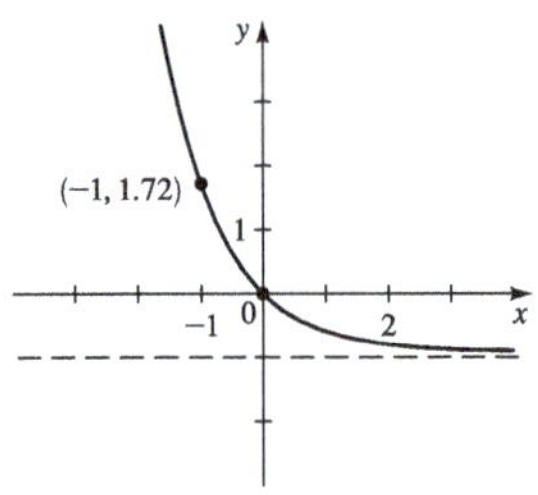

13. $\mathbb{R}$, $(0, \infty)$, $y = 0$

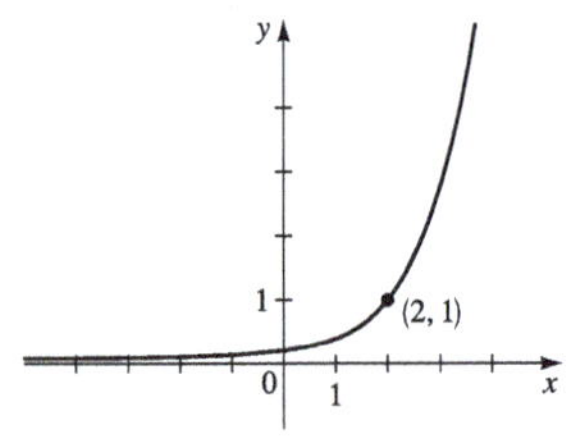

15. $\mathbb{R}$, $(-3, \infty)$, $y = -3$

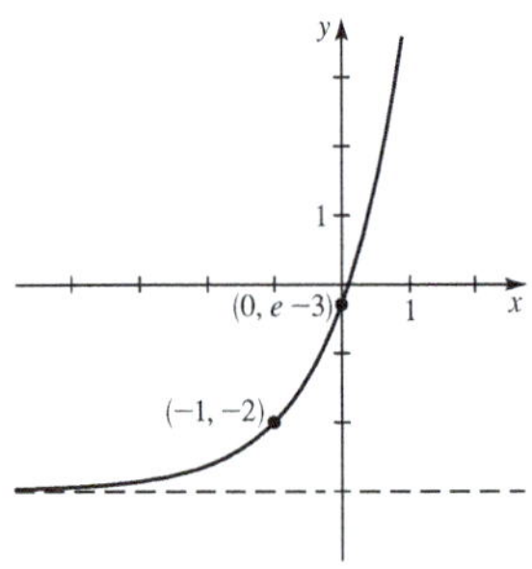

17. (a)

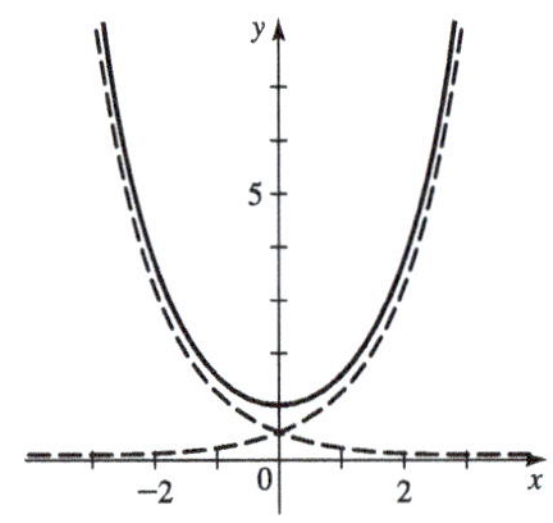

19. (a)

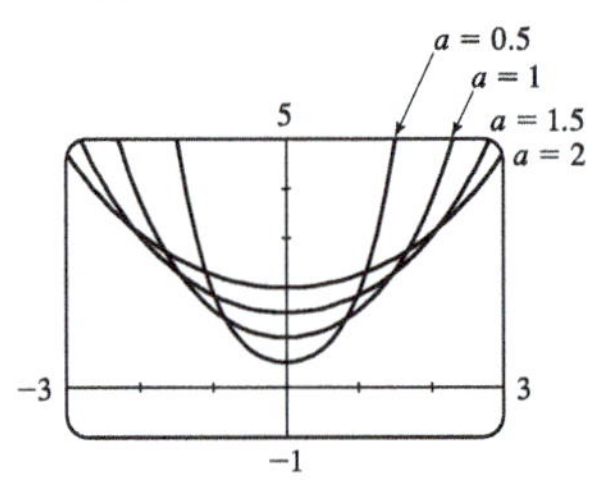

(b) The larger the value of a, the wider the graph.
21. Local minimum $(0.37, 0.69)$ **23.** 27.4 mg

25. **(a)** 0 **(b)** 113.8 ft/s, 155.6 ft/s
(c)

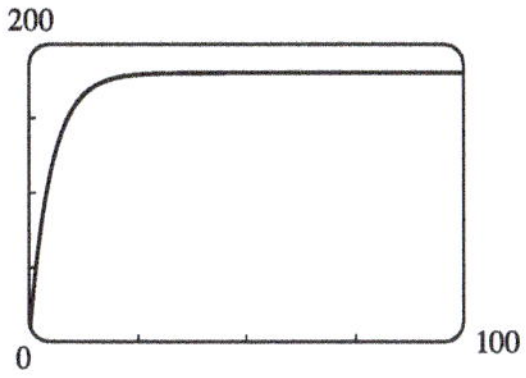

(d) 180 ft/s
27. **(a)** 100 **(b)** 482, 999, 1168 **(c)** 1200
29. **(a)** 11.79 billion, 11.97 billion
(b)

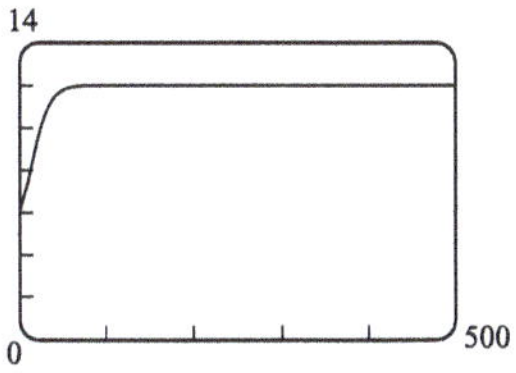

(c) 12 billion

31. \$7213.18, \$7432.86, \$7659.22, \$7892.48, \$8132.84, \$8380.52
33. **(a)** \$2145.02 **(b)** \$2300.55 **(c)** \$3043.92
35. **(a)** \$768.05 **(b)** \$769.22 **(c)** \$769.82 **(d)** \$770.42
37. (a) is best.
39. **(a)** $A(t) = 5000e^{0.09t}$ **(b)**

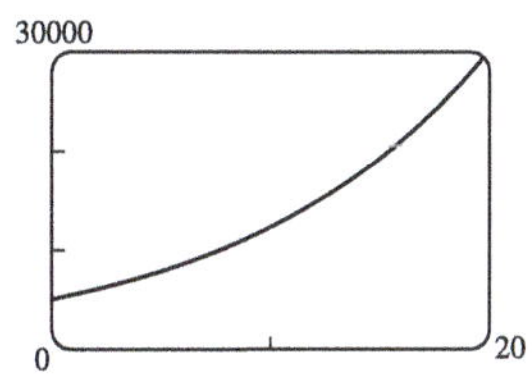

(c) After 17.88 years

SECTION 4.3 ■ PAGE 351

1. x

x	10^3	10^2	10^1	10^0	10^{-1}	10^{-2}	10^{-3}	$10^{1/2}$
$\log x$	3	2	1	0	−1	−2	−3	$\frac{1}{2}$

2. 9; 1, 0, −1, 2, $\frac{1}{2}$ **3.** **(a)** $\log_5 125 = 3$ **(b)** $5^2 = 25$
4. **(a)** III **(b)** II **(c)** I **(d)** IV **5.** vertical, 0 **6.** vertical, 1

7.

Logarithmic form	Exponential form
$\log_8 8 = 1$	$8^1 = 8$
$\log_8 64 = 2$	$8^2 = 64$
$\log_8 4 = \frac{2}{3}$	$8^{2/3} = 4$
$\log_8 512 = 3$	$8^3 = 512$
$\log_8\left(\frac{1}{8}\right) = -1$	$8^{-1} = \frac{1}{8}$
$\log_8\left(\frac{1}{64}\right) = -2$	$8^{-2} = \frac{1}{64}$

9. **(a)** $3^4 = 81$ **(b)** $3^0 = 1$ **11.** **(a)** $8^{1/3} = 2$ **(b)** $10^{-2} = 0.01$
13. **(a)** $3^x = 5$ **(b)** $7^2 = 3y$ **15.** **(a)** $e^{3y} = 5$ **(b)** $e^{-1} = t + 1$
17. **(a)** $\log_{10} 10{,}000 = 4$ **(b)** $\log_5\left(\frac{1}{25}\right) = -2$
19. **(a)** $\log_8\left(\frac{1}{8}\right) = -1$ **(b)** $\log_2\left(\frac{1}{8}\right) = -3$

21. **(a)** $\log_4 70 = x$ **(b)** $\log_3 w = 5$
23. **(a)** $\ln 2 = x$ **(b)** $\ln y = 3$
25. **(a)** 1 **(b)** 0 **(c)** 5 **27.** **(a)** 2 **(b)** 2 **(c)** 10
29. **(a)** −3 **(b)** $\frac{1}{2}$ **(c)** −1 **31.** **(a)** 5 **(b)** 27 **(c)** 10
33. **(a)** $-\frac{2}{3}$ **(b)** 4 **(c)** −1 **35.** **(a)** 64 **(b)** −2
37. **(a)** e^3 **(b)** 2 **39.** **(a)** −2 **(b)** 32
41. **(a)** −1 **(b)** $\frac{1}{1000}$ **43.** **(a)** 2 **(b)** 4
45. **(a)** 0.3010 **(b)** 1.5465 **(c)** −0.1761
47. **(a)** 1.6094 **(b)** 3.2308 **(c)** 1.0051
49.

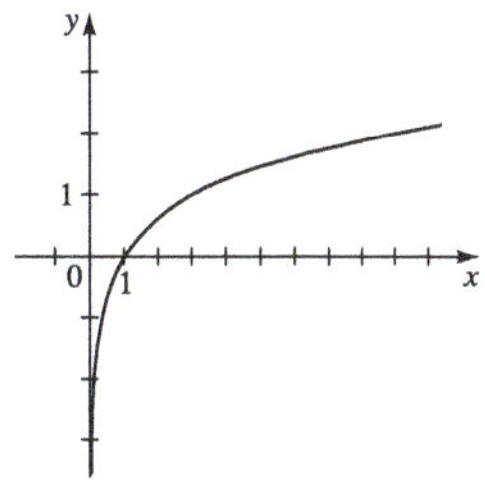

51.

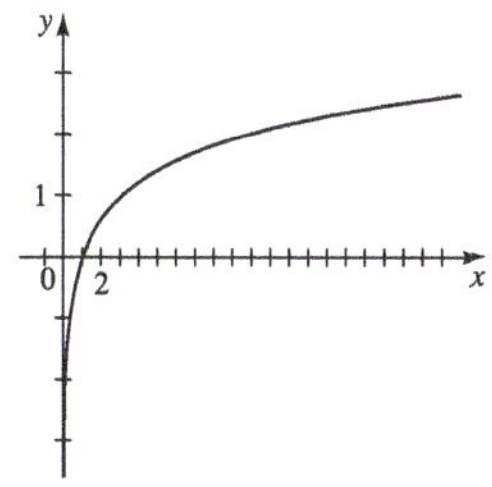

53. $y = \log_5 x$ **55.** $y = \log_9 x$ **57.** I
59.

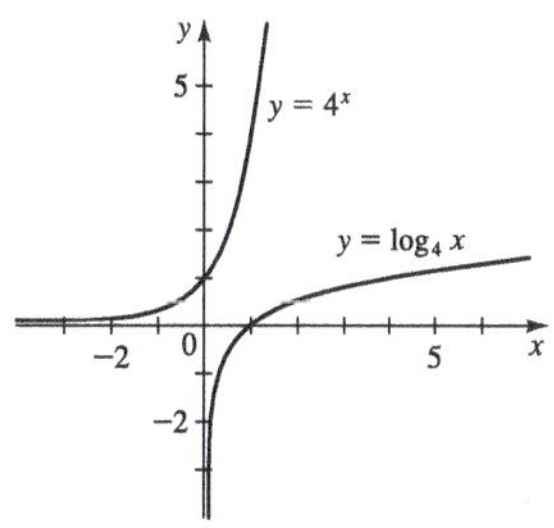

61. $(-\infty, 0), \mathbb{R}, x = 0$

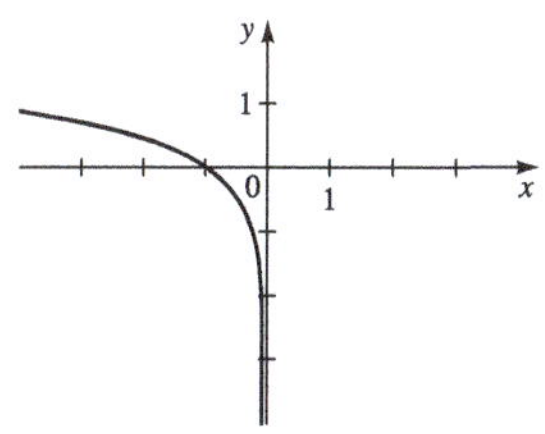

63. $(4, \infty), \mathbb{R}, x = 4$

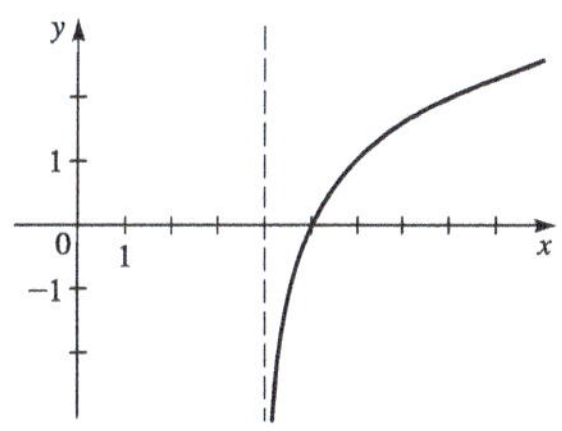

65. $(-5, \infty), \mathbb{R}, x = -5$

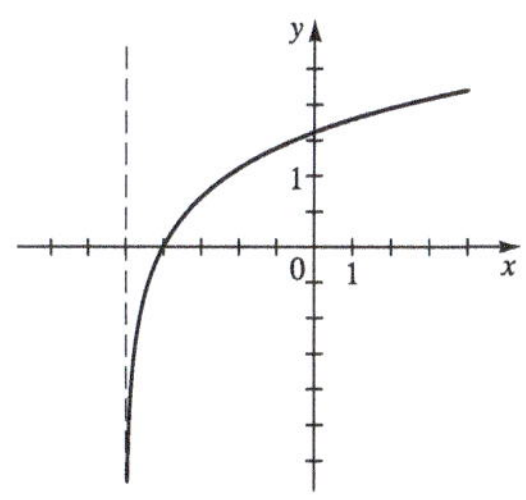

67. $(0, \infty), \mathbb{R}, x = 0$

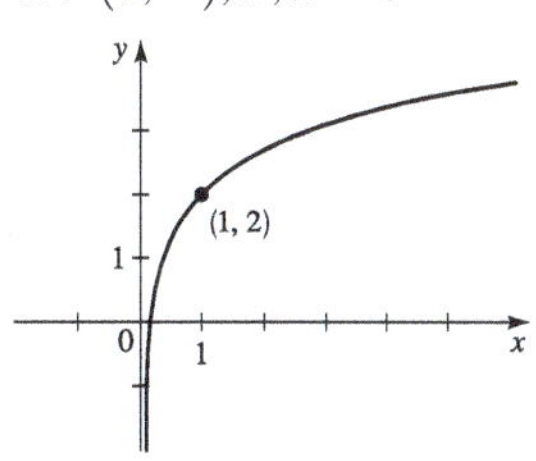

69. $(1, \infty), \mathbb{R}, x = 1$

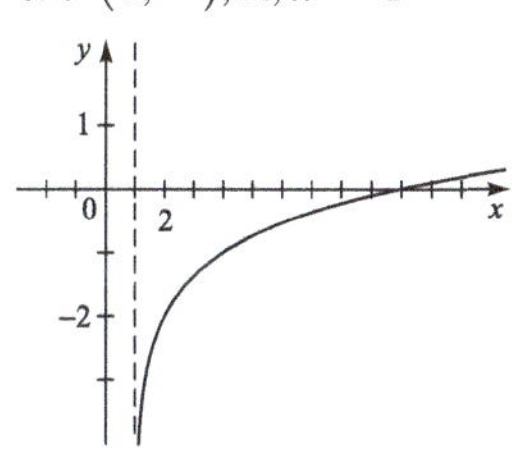

71. $(0, \infty), [0, \infty), x = 0$

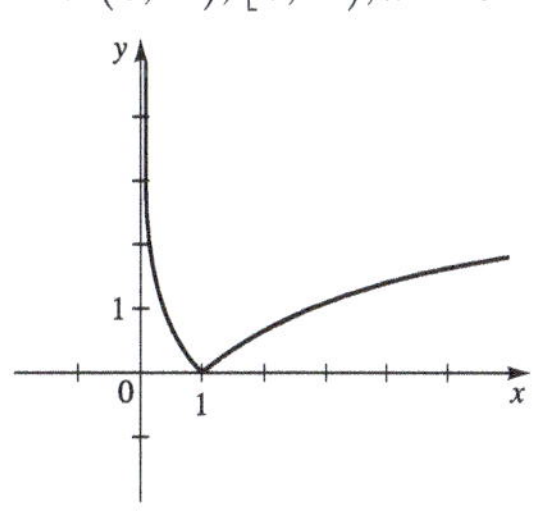

73. $(-3, \infty)$ **75.** $(-\infty, -1) \cup (1, \infty)$ **77.** $(0, 2)$

79.

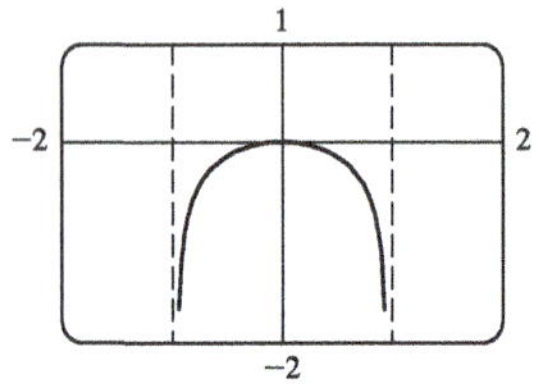

domain $(-1, 1)$
vertical asymptotes $x = 1$, $x = -1$
local maximum $(0, 0)$

81.

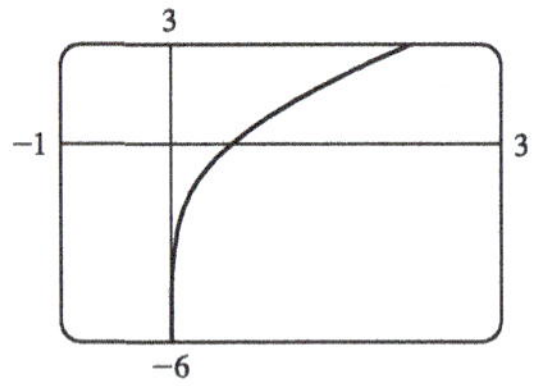

domain $(0, \infty)$
vertical asymptote $x = 0$
no maximum or minimum

83.

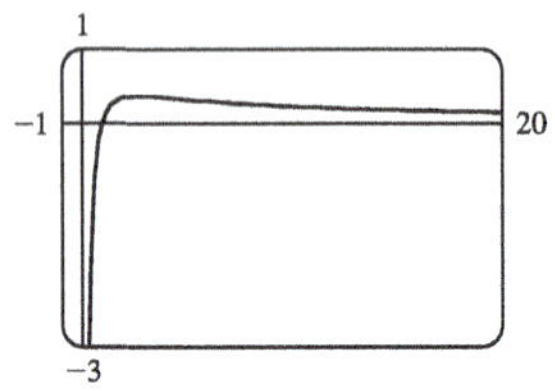

domain $(0, \infty)$
vertical asymptote $x = 0$
horizontal asymptote $y = 0$
local maximum $\approx (2.72, 0.37)$

85. $(f \circ g)(x) = 2^{x+1}, (-\infty, \infty)$; $(g \circ f)(x) = 2^x + 1, (-\infty, \infty)$
87. $(f \circ g)(x) = \log_2(x - 2), (2, \infty)$; $(g \circ f)(x) = \log_2 x - 2, (0, \infty)$
89. The graph of f grows more slowly than g.
91. (a)

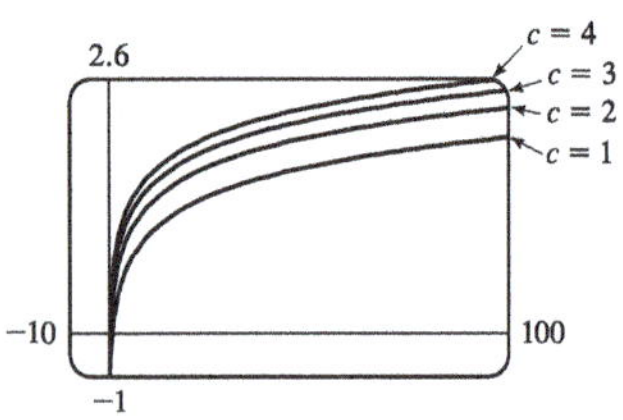

(b) The graph of $f(x) = \log(cx)$ is the graph of $f(x) = \log(x)$ shifted upward $\log c$ units.

93. (a) $(1, \infty)$ **(b)** $f^{-1}(x) = 10^{2^x}$
95. (a) $f^{-1}(x) = \log_2\left(\dfrac{x}{1 - x}\right)$ **(b)** $(0, 1)$
97. 2602 years **99.** 11.6 years, 9.9 years, 8.7 years
101. 5.32, 4.32

SECTION 4.4 ▪ PAGE 358

1. sum; $\log_5 25 + \log_5 125 = 2 + 3$
2. difference; $\log_5 25 - \log_5 125 = 2 - 3$
3. power; $10 \cdot \log_5 25 = 10 \cdot 2$ **4.** $2 \log x + \log y - \log z$
5. $\log\left(\dfrac{x^2 y}{z}\right)$ **6. (a)** 10, e; Change of Base; $\log_7 12 = \dfrac{\log 12}{\log 7} \approx 1.277$ **(b)** Yes **7. (a)** False **(b)** True **8. (a)** True **(b)** False **9.** 4 **11.** 2 **13.** 1
15. $\frac{1}{2}$ **17.** 3 **19.** 200 **21.** 4 **23.** $\log_3 8 + \log_3 x$
25. $\log_3 2 + \log_3 x + \log_3 y$ **27.** $3 \ln a$
29. $10(\log_2 x + \log_2 y)$ **31.** $\log_2 A + 2 \log_2 B$
33. $\log_3 2 + \log_3 x - \log_3 y$ **35.** $\log_5 3 + 2 \log_5 x - 3 \log_5 y$
37. $\frac{1}{2} + \frac{5}{2} \log_3 x - \log_3 y$ **39.** $3 \log x + 4 \log y - 6 \log z$
41. $\frac{1}{2} \ln(x^4 + 2)$ **43.** $\ln x + \frac{1}{2}(\ln y - \ln z)$
45. $\frac{1}{4} \log(x^2 + y^2)$
47. $\frac{1}{2}[\log(x^2 + 4) - \log(x^2 + 1) - 2 \log(x^3 - 7)]$
49. $\log_4 294$ **51.** $\log \dfrac{x^2}{(x + 1)^3}$ **53.** $\log\left(\dfrac{x^4(x - 1)^2}{\sqrt[3]{x^2 + 1}}\right)$
55. $\ln \dfrac{a^2 - b^2}{c^2}$ **57.** $\log\left(\dfrac{x^2}{x - 3}\right)$ **59.** 2.321928
61. 2.523719 **63.** 0.493008 **65.** 3.482892
67.

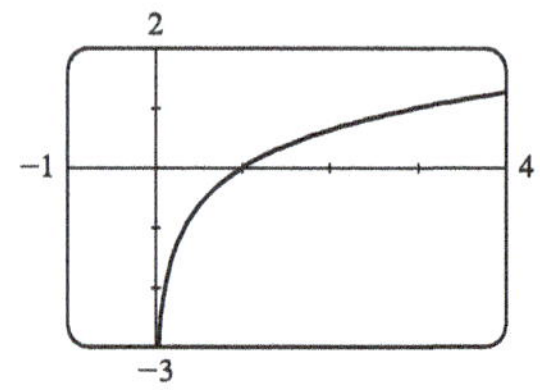

73. (a) $P = c/W^k$ **(b)** 1866, 64
75. (a) $M = -2.5 \log B + 2.5 \log B_0$

SECTION 4.5 ▪ PAGE 368

1. (a) $e^x = 25$ **(b)** $x = \ln 25$ **(c)** 3.219
2. (a) $\log 3(x - 2) = \log x$ **(b)** $3(x - 2) = x$ **(c)** 3 **3.** 4
5. $\frac{3}{2}$ **7.** −3 **9.** −1, 1 **11. (a)** $2 \log 5$ **(b)** 1.397940
13. (a) $-\frac{1}{5} \ln 10$ **(b)** −0.460517 **15. (a)** $1 - \dfrac{\log 3}{\log 2}$
(b) −0.584963 **17. (a)** $\ln(\frac{10}{3})$ **(b)** 1.203973
19. (a) $\dfrac{\ln(10/3)}{12 \ln(41/40)}$ **(b)** 4.063202 **21. (a)** $\dfrac{1 - \ln 2}{4}$
(b) 0.076713 **23. (a)** $\dfrac{5}{7} - \dfrac{\ln 15}{7 \ln 2}$ **(b)** 0.156158
25. (a) $\dfrac{14 \log 0.1}{\log 3}$ **(b)** −29.342646
27. (a) $\frac{1}{5} \log(\frac{5}{4})$ **(b)** 0.019382
29. (a) $\dfrac{1 - \ln 12}{4}$ **(b)** −0.371227
31. (a) $\dfrac{\ln(50/3)}{2 \ln 2}$ **(b)** 2.029447
33. (a) $\dfrac{\log 4}{\log(5/4)}$ **(b)** 6.212567
35. (a) $-\dfrac{\log 18}{\log(8/3)}$ **(b)** −2.946865
37. (a) $-\ln 11.5$ **(b)** −2.442347 **39.** $\ln 2 \approx 0.6931$, 0
41. $\frac{1}{2} \ln 3 \approx 0.5493$ **43.** 1 **45.** ±1 **47.** 0, $\frac{4}{3}$ **49.** 5
51. 2, 4 **53.** 5 **55.** $e^{10} \approx 22{,}026$ **57.** 0.01 **59.** $\frac{95}{3}$
61. −7 **63.** 4 **65.** 6 **67.** $\frac{13}{12}$ **69.** 2.21 **71.** 0.00, 1.14
73. −0.57 **75.** 0.36 **77.** $1/\sqrt{5} \approx 0.4472$
79. $2 < x < 4$ or $7 < x < 9$ **81.** $\log 2 < x < \log 5$
83. $f^{-1}(x) = \dfrac{\ln x}{2 \ln 2}$ **85.** $f^{-1}(x) = 2^x + 1$ **87.** $\frac{3}{2}$
89. (a) \$6435.09 **(b)** 8.24 years
91. 6.33 years **93.** 8.15 years **95.** 13 days
97. (a) 7337 **(b)** 1.73 years
99. (a) $P = P_0 e^{-h/k}$ **(b)** 56.47 kPa
101. (a) $t = -\frac{5}{13} \ln(1 - \frac{13}{60} I)$ **(b)** 0.218 s

SECTION 4.6 ▪ PAGE 378

1. (a) $n(t) = 10 \cdot 2^{2t/3}$ **(b)** 1.06×10^8 **(c)** 14.9

3. (a) 3125 **(b)** 317,480

(c) 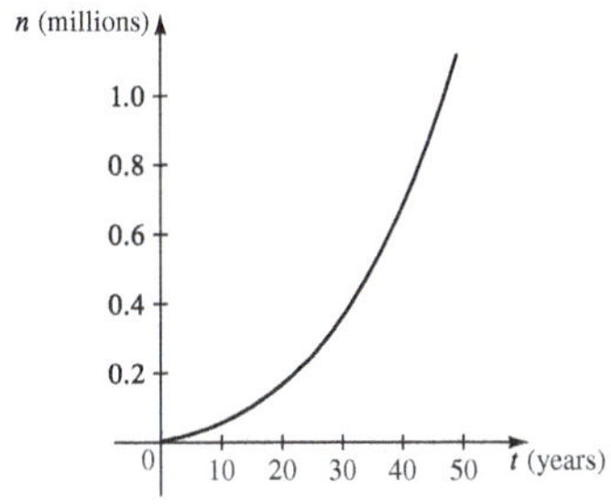

5. (a) $n(t) = 18{,}000e^{0.08t}$ **(b)** 34,100 **(c)** 4.1

(d) 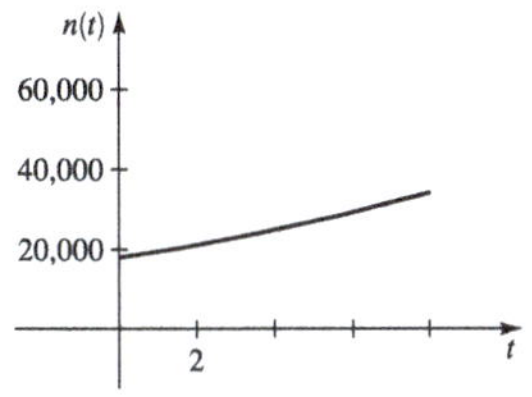

7. (a) 233 million **(b)** 181 million

9. (a) $n(t) = 112{,}000 \cdot 2^{t/18}$ **(b)** $n(t) = 112{,}000e^{0.0385t}$

(c) 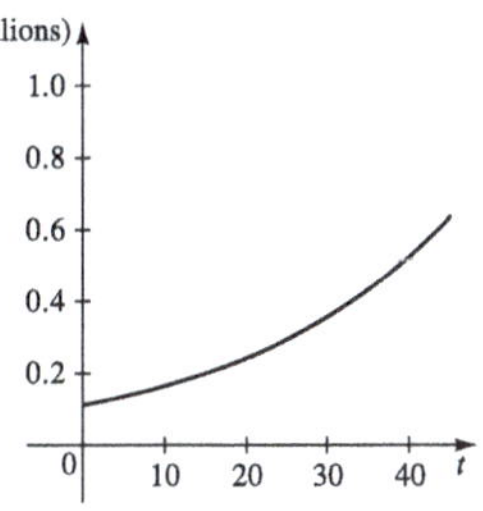

(d) 38.9 years

11. (a) 20,000 **(b)** $n(t) = 20{,}000e^{0.1096t}$ **(c)** About 48,000 **(d)** 14.7 years

13. (a) $n(t) = 8600e^{0.1508t}$ **(b)** About 11,600 **(c)** 4.6 h

15. (a) $n(t) = 29.76e^{0.012936t}$ million **(b)** 53.6 years **(c)** 38.55 million **17. (a)** $m(t) = 22 \cdot 2^{-t/1600}$ **(b)** $m(t) = 22e^{-0.000433t}$ **(c)** 3.9 mg **(d)** 463.4

19. 18 years **21.** 149 h **23.** 3560 years **25. (a)** 210°F **(b)** 153°F **(c)** 28 min **27. (a)** 137°F **(b)** About 2 h

SECTION 4.7 ■ PAGE 385

1. (a) 2.3 **(b)** 3.5 **(c)** 8.3 **3. (a)** 10^{-3} M **(b)** 3.2×10^{-7} M

5. $4.8 \le \text{pH} \le 6.4$ **7. (a)** 6.31×10^{-4} M, 1.26×10^{-3} M **(b)** California **9. (a)** 5.49 **(b)** 6.3 cm **11.** $\log 20 \approx 1.3$

13. Six times as intense **15.** 73 dB **17.** 10^{-5} W/m^2

19. (a) 75 dB **(b)** 10^{-3} W/m^2 **(c)** 32.3

CHAPTER 4 REVIEW ■ PAGE 388

1. 0.089, 9.739, 55.902 **3.** 0.269, 1.472, 12.527

5. $\mathbb{R}, (0, \infty), y = 0$

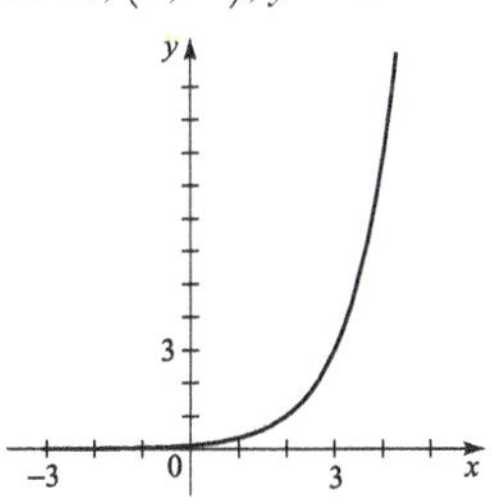

7. $\mathbb{R}, (3, \infty), y = 3$

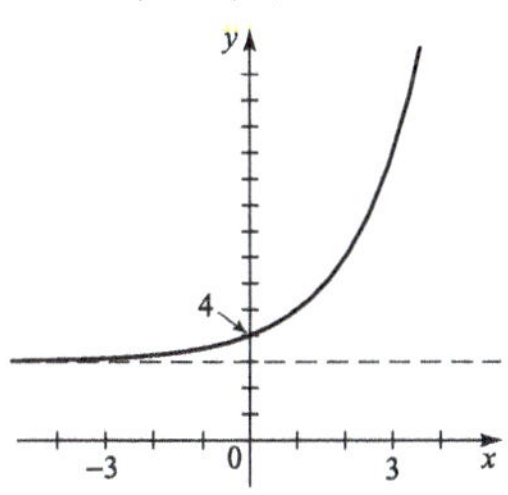

9. $\mathbb{R}, (1, \infty), y = 1$

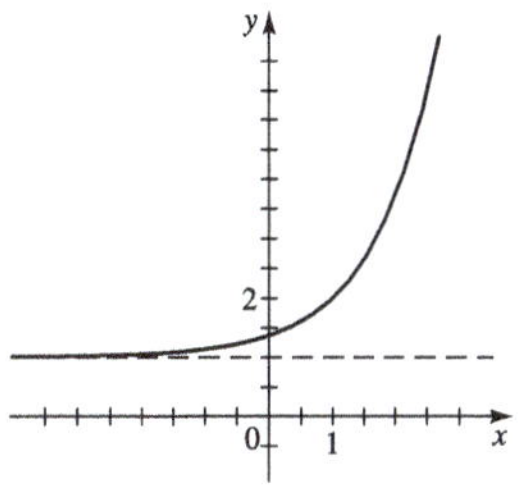

11. $(1, \infty), \mathbb{R}, x = 1$

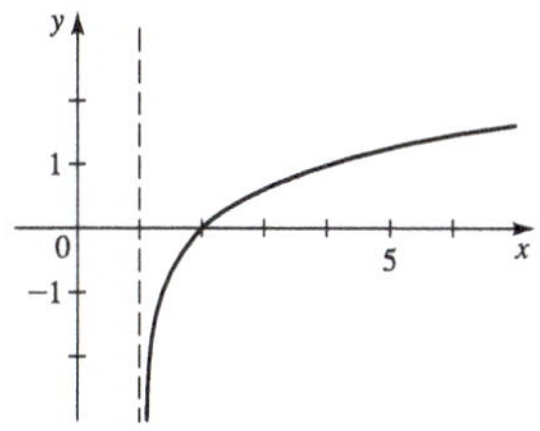

13. $(0, \infty), \mathbb{R}, x = 0$

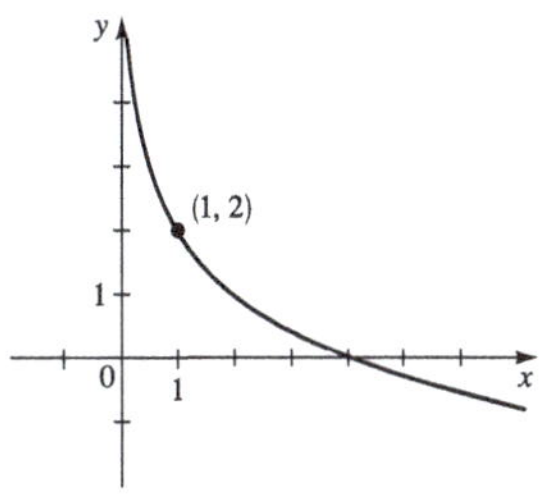

15. $(0, \infty), \mathbb{R}, x = 0$

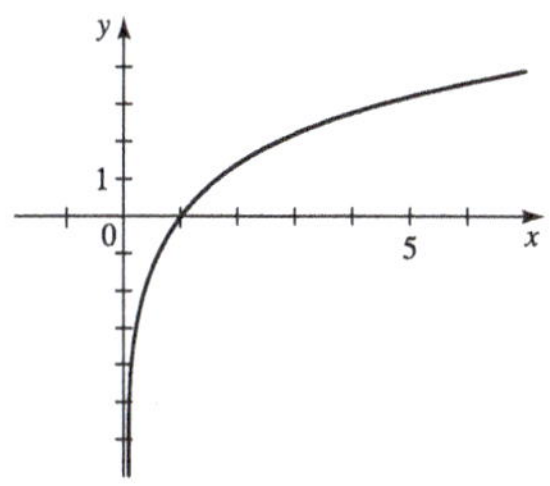

17. $\left(-\infty, \frac{1}{2}\right)$ **19.** $(-\infty, -2) \cup (2, \infty)$ **21.** $2^{10} = 1024$

23. $10^y = x$ **25.** $\log_2 64 = 6$ **27.** $\log 74 = x$ **29.** 7 **31.** 45

33. 6 **35.** -3 **37.** $\frac{1}{2}$ **39.** 2 **41.** 92 **43.** $\frac{2}{3}$

45. $\log A + 2 \log B + 3 \log C$

47. $\frac{1}{2}[\ln(x - 1) + \ln(x + 1) - \ln(x^2 + 1)]$

49. $2 \log_5 x + \frac{3}{2} \log_5(1 - 5x) - \frac{1}{2}[\log_5 x + \log_5(x - 1) + \log_5(x + 1)]$

51. $\log 96$ **53.** $\log_2\left(\dfrac{(x - y)^{3/2}}{(x^2 + y^2)^2}\right)$ **55.** $\log\left(\dfrac{x^2 - 4}{\sqrt{x^2 + 4}}\right)$

57. 5 **59.** $\dfrac{1}{3}\left[\dfrac{\log 7}{\log 2} + 5\right] \approx 2.60$ **61.** $\dfrac{\log(4/243)}{\log 36} \approx -1.15$

63. $-4, 2$ **65.** 3 **67.** -15 **69.** 9 **71.** 0.430618

73. 2.303600

75. 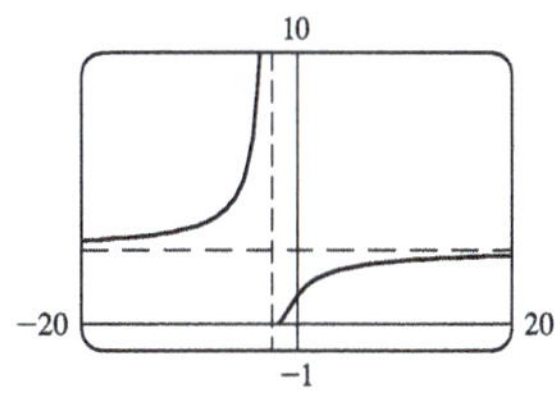

vertical asymptote $x = -2$
horizontal asymptote $y \approx 2.72$
no maximum or minimum

77. 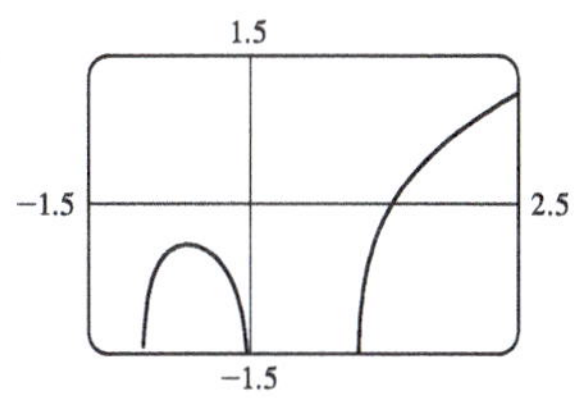

vertical asymptotes $x = -1, x = 0, x = 1$
local maximum $\approx (-0.58, -0.41)$

79. 2.42 **81.** $0.16 < x < 3.15$

83. Increasing on $(-\infty, 0)$ and $(1.10, \infty)$, decreasing on $(0, 1.10)$

85. 1.953445 **87.** -0.579352 **89.** $\log_4 258$

91. (a) \$16,081.15 **(b)** \$16,178.18 **(c)** \$16,197.64 **(d)** \$16,198.31 **93.** 1.83 years **95.** 4.341%

97. (a) $n(t) = 30e^{0.15t}$ **(b)** 55 **(c)** 19 years

99. (a) 9.97 mg **(b)** 1.39×10^5 years

101. (a) $n(t) = 150e^{-0.0004359t}$ **(b)** 97.0 mg **(c)** 2520 years

103. (a) $n(t) = 1500e^{0.1515t}$ **(b)** 7940 **105.** 7.9, basic

107. 8.0

CHAPTER 4 TEST ■ PAGE 391

1. (a) $\mathbb{R}, (4, \infty), y = 4$ **(b)** $(-3, \infty), \mathbb{R}, x = -3$

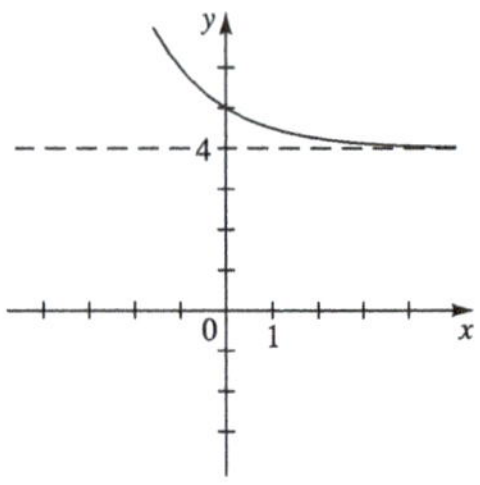
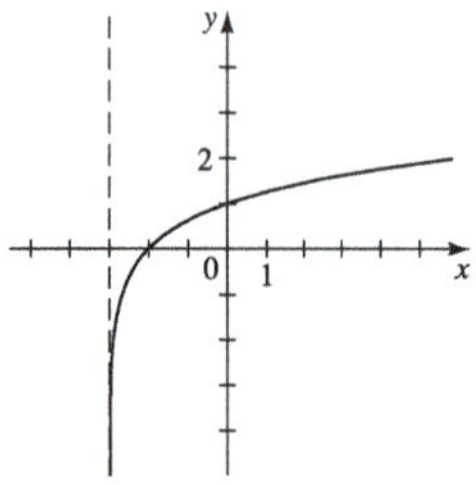

2. (a) $(\frac{3}{2}, \infty)$ **(b)** $(-\infty, -1) \cup (1, \infty)$
3. (a) $\log_6 25 = 2x$ **(b)** $e^3 = A$
4. (a) 36 **(b)** 3 **(c)** $\frac{3}{2}$ **(d)** 3 **(e)** $\frac{2}{3}$ **(f)** 2
5. (a) $\log x + 3 \log y - 2 \log z$ **(b)** $\frac{1}{2} \ln x - \frac{1}{2} \ln y$
(c) $\frac{1}{3}[\log(x + 2) - 4 \log x - \log(x^2 + 4)]$
6. (a) $\log(ab^2)$ **(b)** $\ln(x - 5)$ **(c)** $\log_2 \dfrac{3\sqrt{x + 1}}{x^3}$
7. (a) 25 **(b)** 1, 2 **(c)** 11.13 **(d)** 5.39
8. (a) 500 **(b)** $\frac{2}{3}$ **(c)** $3 - e^{4/5} \approx 0.774$ **(d)** 2 **9.** 1.326
10. (a) $n(t) = 1000e^{2.07944t}$ **(b)** 22,600 **(c)** 1.3
(d)

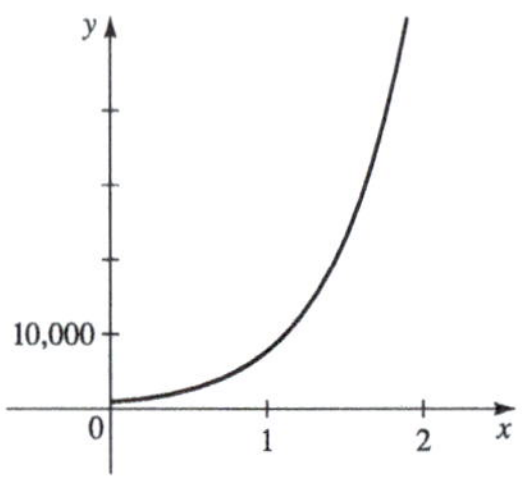

11. (a) $A(t) = 12{,}000\left(1 + \dfrac{0.056}{12}\right)^{12t}$ **(b)** \$14,195.06
(c) 9.12 years **12. (a)** $m(t) = 3 \cdot 2^{-t/10}$ **(b)** $m(t) = 3e^{-0.0693t}$
(c) 0.047 g **(d)** after 3.6 min **13.** 1995 times more intense

FOCUS ON MODELING ■ PAGE 398

1. (a)

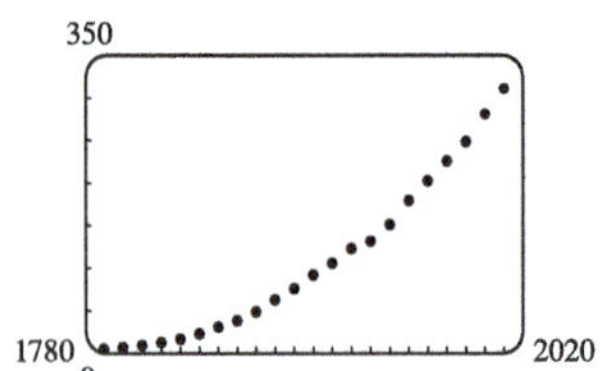

(b) $y = ab^t$, where $a = 3.334926 \times 10^{-15}$, $b = 1.019844$, and y is the population in millions in the year t **(c)** 577 million **(d)** 196 million

3. (a)

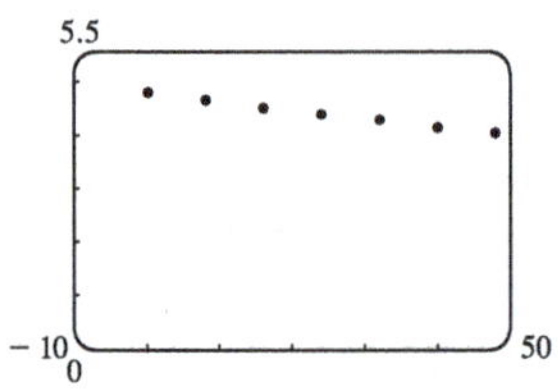

(b) $y = ab^t$, where $a = 4.79246$ and $b = 0.99642$ **(c)** 192.8 h

5. (a) $y = at^b$, where $a = 49.70030$ and $b = -0.15437$; $y = ab^t$, where $a = 44.82418$ and $b = 0.99317$
(b)

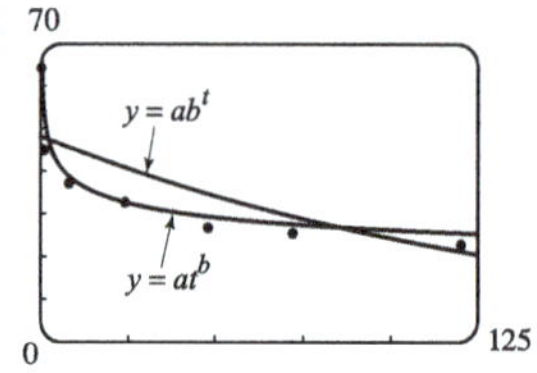

(c) The power function
7. $y = ab^x$, where $a = 2.414$ and $b = 1.05452$
9. (a)

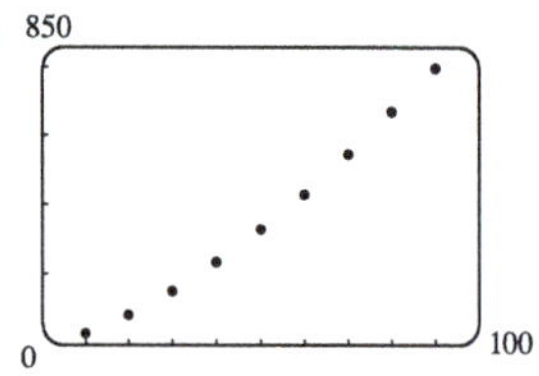

(b)

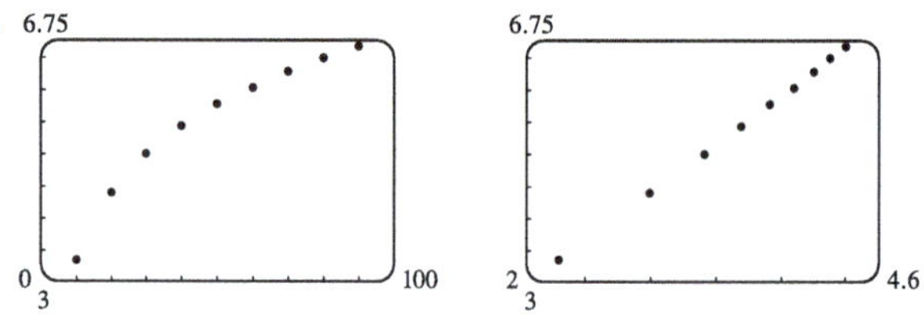

(c) The power function
(d) $y = ax^b$, where $a = 0.893421326$ and $b = 1.50983$

CHAPTER 5

SECTION 5.1 ■ PAGE 407

1. (a) $(0, 0), 1$ **(b)** $x^2 + y^2 = 1$ **(c)** (i) 0 (ii) 0 (iii) 0 (iv) 0 **2. (a)** terminal **(b)** $(0, 1), (-1, 0), (0, -1), (1, 0)$
9. $-\frac{4}{5}$ **11.** $-2\sqrt{2}/3$ **13.** $3\sqrt{5}/7$ **15.** $P(\frac{5}{13}, -\frac{12}{13})$
17. $P(-\sqrt{5}/3, \frac{2}{3})$ **19.** $P(-\sqrt{2}/3, -\sqrt{7}/3)$
21. $t = \pi/4, (\sqrt{2}/2, \sqrt{2}/2)$; $t = \pi/2, (0, 1)$; $t = 3\pi/4, (-\sqrt{2}/2, \sqrt{2}/2)$; $t = \pi, (-1, 0)$; $t = 5\pi/4, (-\sqrt{2}/2, -\sqrt{2}/2)$; $t = 3\pi/2, (0, -1)$; $t = 7\pi/4, (\sqrt{2}/2, -\sqrt{2}/2)$; $t = 2\pi, (1, 0)$
23. $(1, 0)$ **25.** $(0, -1)$ **27.** $(\sqrt{3}/2, -\frac{1}{2})$
29. $(-\sqrt{2}/2, -\sqrt{2}/2)$ **31.** $(-\sqrt{3}/2, \frac{1}{2})$
33. $(\sqrt{2}/2, \sqrt{2}/2)$ **35.** $(-\sqrt{2}/2, -\sqrt{2}/2)$
37. (a) $\pi/3$ **(b)** $\pi/3$ **(c)** $\pi/6$ **(d)** $3.5 - \pi \approx 0.36$
39. (a) $2\pi/7$ **(b)** $2\pi/9$ **(c)** $\pi - 3 \approx 0.14$
(d) $2\pi - 5 \approx 1.28$ **41. (a)** $\pi/6$ **(b)** $(\sqrt{3}/2, -\frac{1}{2})$
43. (a) $\pi/3$ **(b)** $(-\frac{1}{2}, \sqrt{3}/2)$
45. (a) $\pi/3$ **(b)** $(-\frac{1}{2}, -\sqrt{3}/2)$
47. (a) $\pi/4$ **(b)** $(-\sqrt{2}/2, -\sqrt{2}/2)$
49. (a) $\pi/6$ **(b)** $(-\sqrt{3}/2, \frac{1}{2})$
51. (a) $\pi/3$ **(b)** $(\frac{1}{2}, \sqrt{3}/2)$
53. (a) $\pi/3$ **(b)** $(-\frac{1}{2}, -\sqrt{3}/2)$
55. $(0.5, 0.8)$ **57.** $(0.5, -0.9)$
59. (a) $(-\frac{3}{5}, \frac{4}{5})$ **(b)** $(\frac{3}{5}, -\frac{4}{5})$ **(c)** $(-\frac{3}{5}, -\frac{4}{5})$ **(d)** $(\frac{3}{5}, \frac{4}{5})$

SECTION 5.2 ■ PAGE 416

1. $y, x, y/x$ **2.** 1; 1 **3.** $t = \pi/4$, $\sin t = \sqrt{2}/2$, $\cos t = \sqrt{2}/2$; $t = \pi/2$, $\sin t = 1$, $\cos t = 0$; $t = 3\pi/4$, $\sin t = \sqrt{2}/2$, $\cos t = -\sqrt{2}/2$; $t = \pi$, $\sin t = 0$, $\cos t = -1$; $t = 5\pi/4$, $\sin t = -\sqrt{2}/2$, $\cos t = -\sqrt{2}/2$; $t = 3\pi/2$, $\sin t = -1$, $\cos t = 0$; $t = 7\pi/4$, $\sin t = -\sqrt{2}/2$, $\cos t = \sqrt{2}/2$; $t = 2\pi$, $\sin t = 0$, $\cos t = 1$
5. **(a)** $-\frac{1}{2}$ **(b)** $-\sqrt{3}/2$ **(c)** $\sqrt{3}/3$
7. **(a)** $\sqrt{2}/2$ **(b)** $-\sqrt{2}/2$ **(c)** $-\sqrt{2}/2$
9. **(a)** $-\sqrt{2}/2$ **(b)** $-\sqrt{2}/2$ **(c)** $\sqrt{2}/2$
11. **(a)** $\sqrt{3}/2$ **(b)** $2\sqrt{3}/3$ **(c)** $\sqrt{3}/3$
13. **(a)** $\frac{1}{2}$ **(b)** 2 **(c)** $-\sqrt{3}/2$
15. **(a)** $\sqrt{3}/2$ **(b)** $-2\sqrt{3}/3$ **(c)** $-\sqrt{3}/3$
17. **(a)** -2 **(b)** $2\sqrt{3}/3$ **(c)** $\sqrt{3}$
19. **(a)** $-\sqrt{3}/2$ **(b)** $2\sqrt{3}/3$ **(c)** $-\sqrt{3}/3$
21. **(a)** 0 **(b)** 1 **(c)** 0
23. $\sin 0 = 0$, $\cos 0 = 1$, $\tan 0 = 0$, $\sec 0 = 1$, others undefined
25. $\sin \pi = 0$, $\cos \pi = -1$, $\tan \pi = 0$, $\sec \pi = -1$, others undefined
27. $-\frac{4}{5}, -\frac{3}{5}, \frac{4}{3}$ **29.** $2\sqrt{2}/3, -\frac{1}{3}, -2\sqrt{2}$
31. $\sqrt{13}/7, -6/7, -\sqrt{13}/6$ **33.** $-\frac{12}{13}, -\frac{5}{13}, \frac{12}{5}$ **35.** $\frac{21}{29}, -\frac{20}{29}, -\frac{21}{20}$
37. **(a)** 0.8 **(b)** 0.84147 **39.** **(a)** 0.9 **(b)** 0.93204
41. **(a)** 1 **(b)** 1.02964 **43.** **(a)** -0.6 **(b)** -0.57482
45. Negative **47.** Negative **49.** II **51.** II
53. $\sin t = \sqrt{1 - \cos^2 t}$ **55.** $\tan t = \dfrac{\sin t}{\sqrt{1 - \sin^2 t}}$
57. $\sec t = -\sqrt{1 + \tan^2 t}$ **59.** $\tan t = \sqrt{\sec^2 t - 1}$
61. $\tan^2 t = \dfrac{\sin^2 t}{1 - \sin^2 t}$
63. $\cos t = \frac{3}{5}$, $\tan t = -\frac{4}{3}$, $\csc t = -\frac{5}{4}$, $\sec t = \frac{5}{3}$, $\cot t = -\frac{3}{4}$
65. $\sin t = -2\sqrt{2}/3$, $\cos t = \frac{1}{3}$, $\tan t = -2\sqrt{2}$, $\csc t = -\frac{3}{4}\sqrt{2}$, $\cot t = -\sqrt{2}/4$
67. $\sin t = \frac{12}{13}$, $\cos t = -\frac{5}{13}$, $\csc t = \frac{13}{12}$, $\sec t = -\frac{13}{5}$, $\cot t = -\frac{5}{12}$
69. $\cos t = -\sqrt{15}/4$, $\tan t = \sqrt{15}/15$, $\csc t = -4$, $\sec t = -4\sqrt{15}/15$, $\cot t = \sqrt{15}$
71. Odd **73.** Odd **75.** Even **77.** Neither
79. $y(0) = 4$, $y(0.25) = -2.828$, $y(0.50) = 0$, $y(0.75) = 2.828$, $y(1.00) = -4$, $y(1.25) = 2.828$
81. **(a)** 0.49870 amp **(b)** -0.17117 amp

SECTION 5.3 ■ PAGE 429

1. $f(t)$; 2π, 1

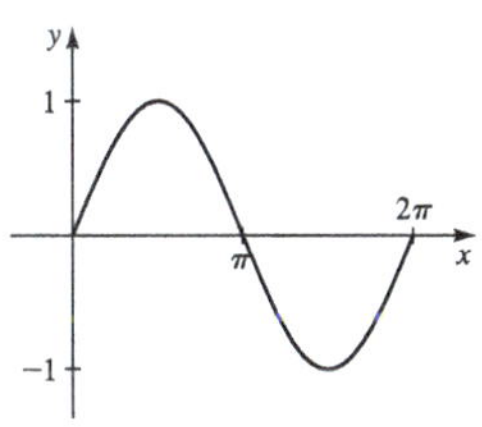

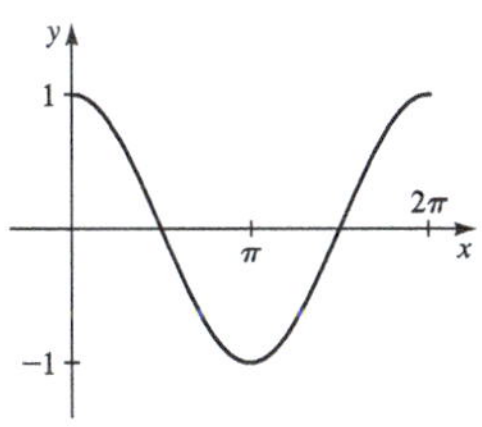

2. upward; x **3.** $|a|, 2\pi/k, 3, \pi$
4. $|a|, 2\pi/k, b; 4, 2\pi/3, \pi/6$

5.

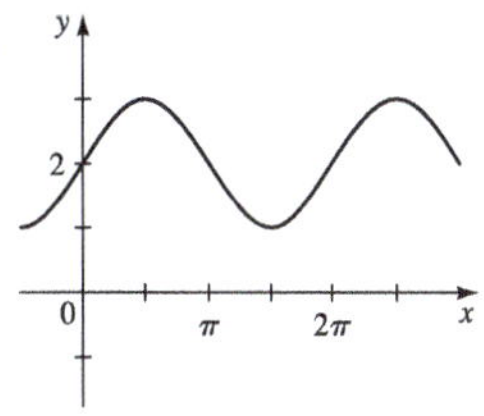

7.

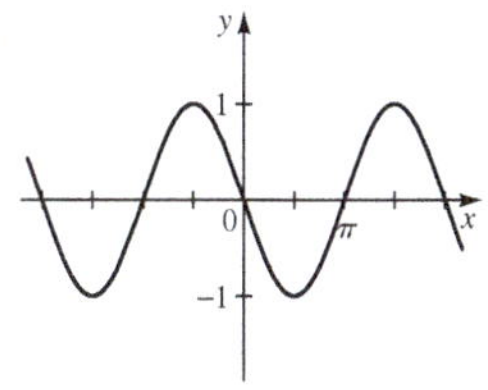

9.

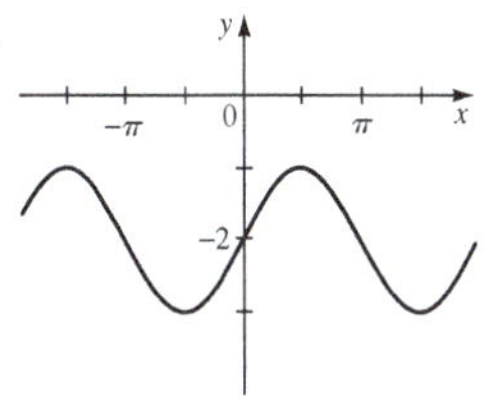

11.

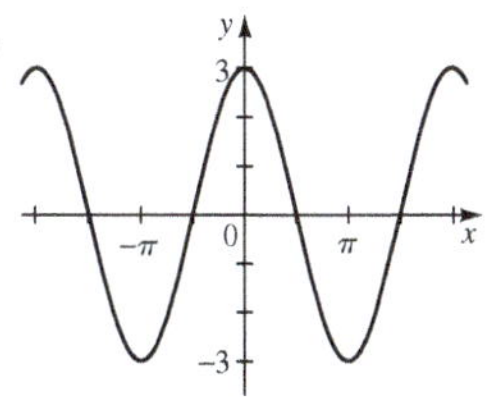

13.

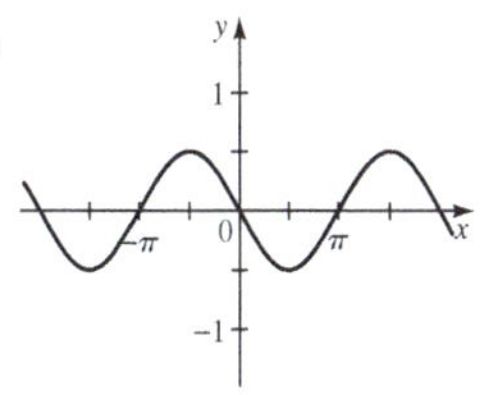

15.

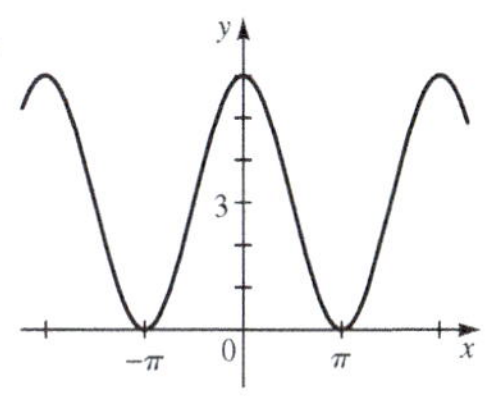

17.

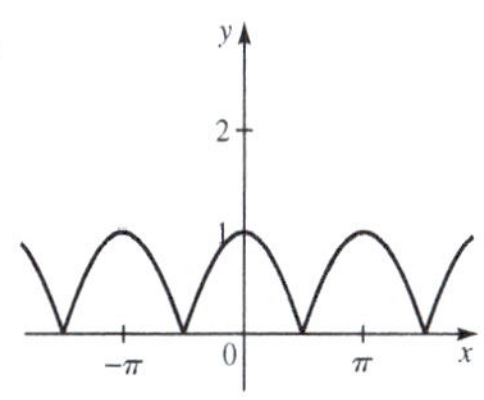

19. 1, π

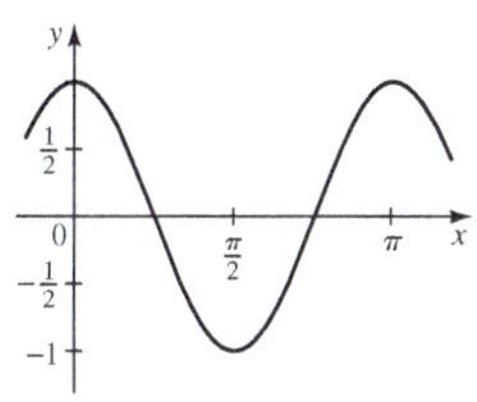

21. 1, $2\pi/3$

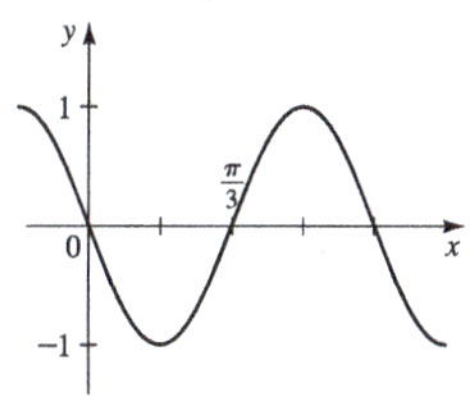

23. 2, $\frac{2}{3}$

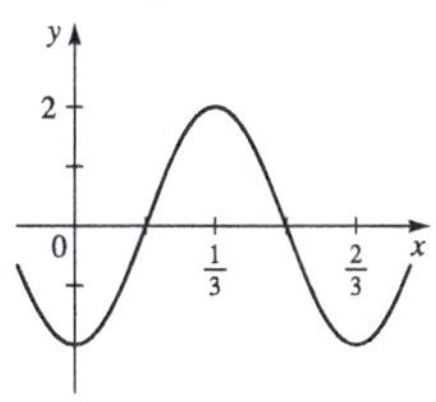

25. 10, 4π

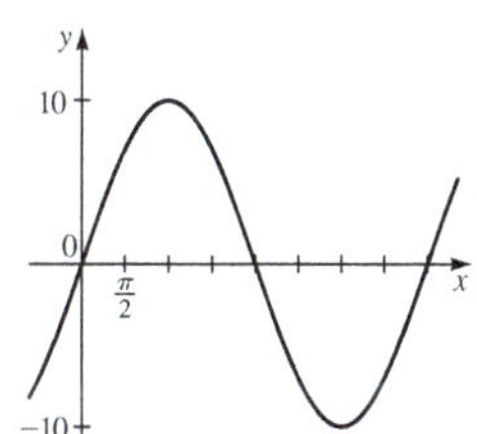

27. $\frac{1}{3}$, 6π

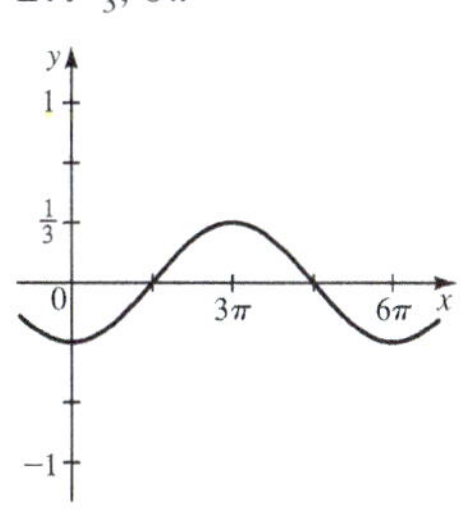

29. 2, 1

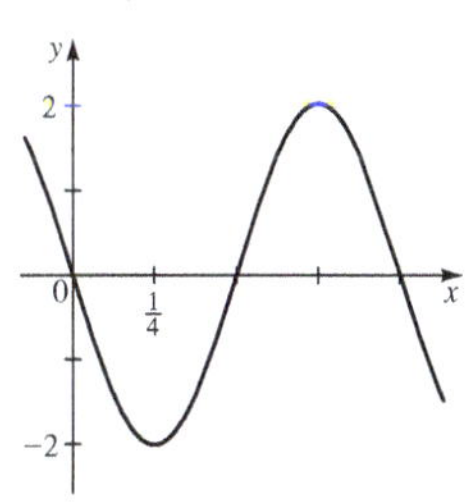

31. $\frac{1}{2}$, 2

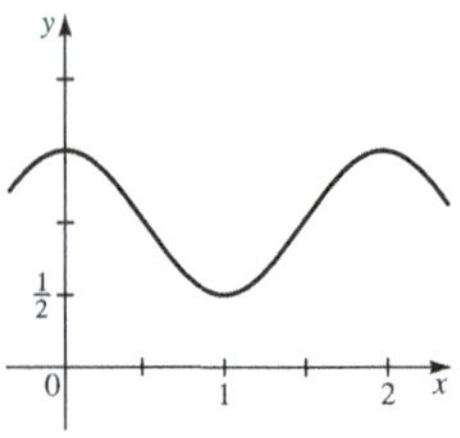

33. 1, 2π, $\pi/2$

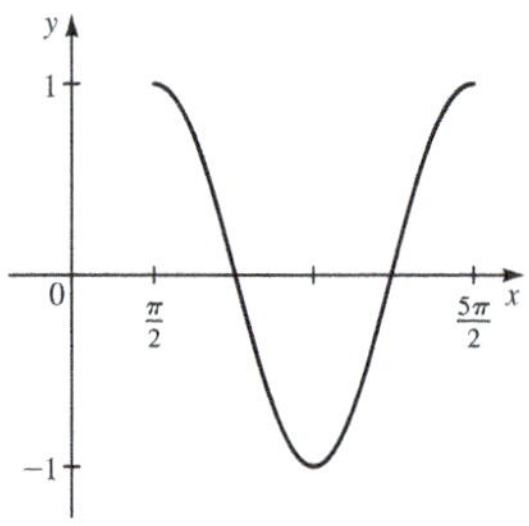

35. 2, 2π, $\pi/6$

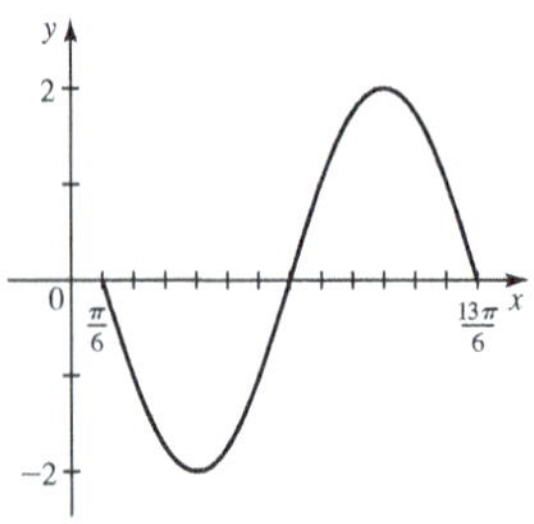

37. 4, π, $-\pi/2$

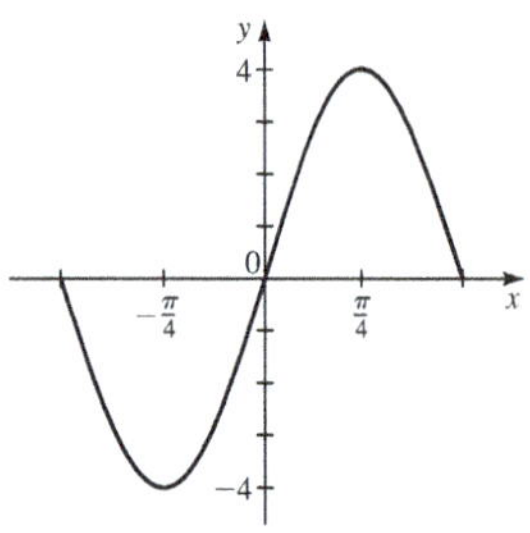

39. 5, $2\pi/3$, $\pi/12$

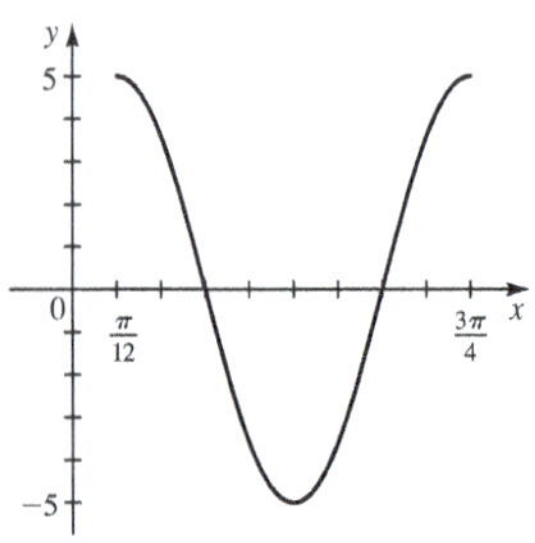

41. $\frac{1}{2}$, π, $\pi/6$

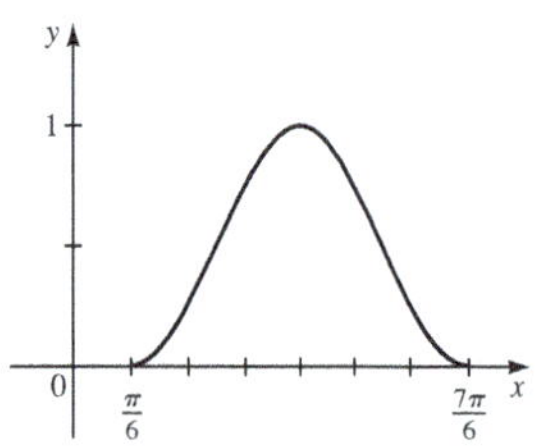

43. 3, 2, $-\frac{1}{2}$

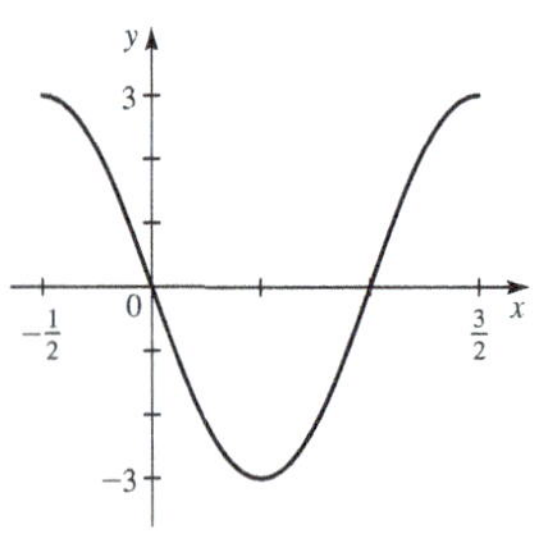

45. 1, $2\pi/3$, $-\pi/3$

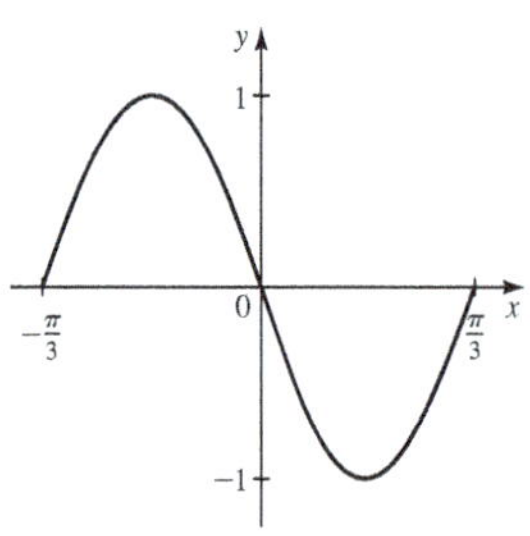

47. **(a)** 4, 2π, 0 **(b)** $y = 4 \sin x$
49. **(a)** $\frac{3}{2}$, $2\pi/3$, 0 **(b)** $y = \frac{3}{2} \cos 3x$
51. **(a)** $\frac{1}{2}$, π, $-\pi/3$ **(b)** $y = -\frac{1}{2} \cos 2(x + \pi/3)$
53. **(a)** 4, $\frac{3}{2}$, $-\frac{1}{2}$ **(b)** $y = 4 \sin 4\pi/3 \left(x + \frac{1}{2}\right)$

55.

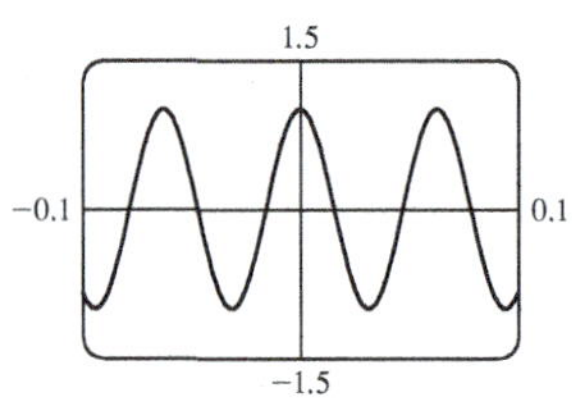

57.

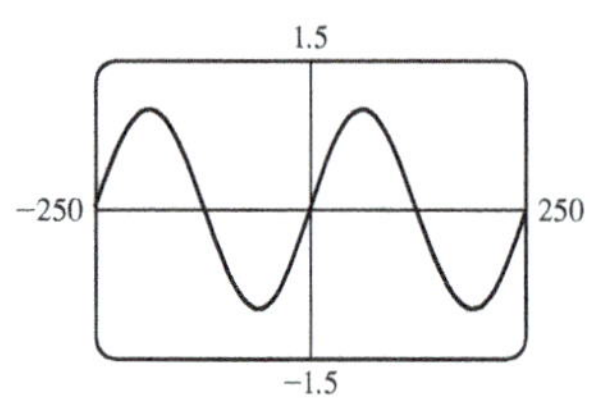

59.

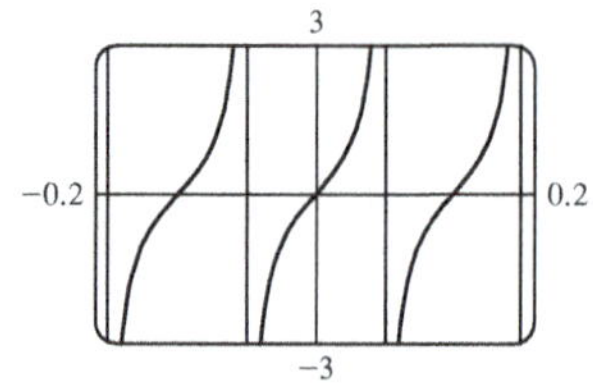

61.

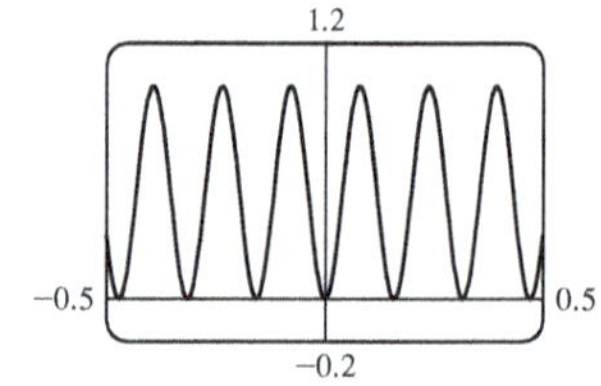

63.

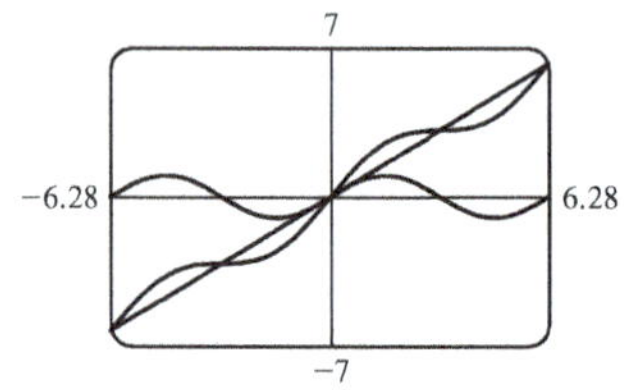

65.

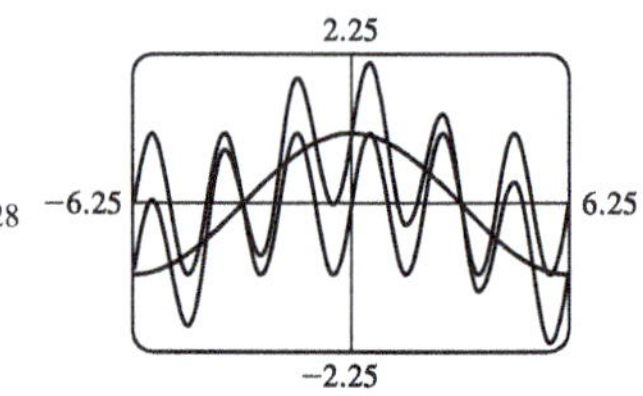

67.

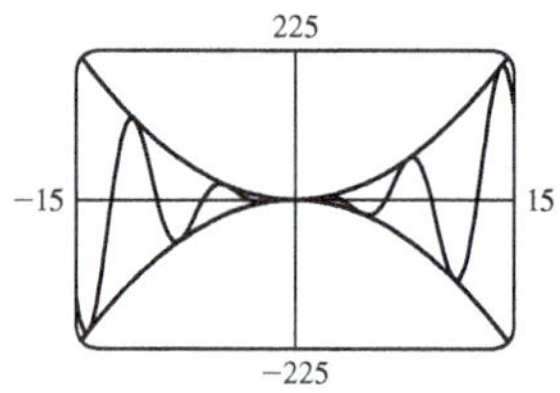

$y = x^2 \sin x$ is a sine curve that lies between the graphs of $y = x^2$ and $y = -x^2$

69.

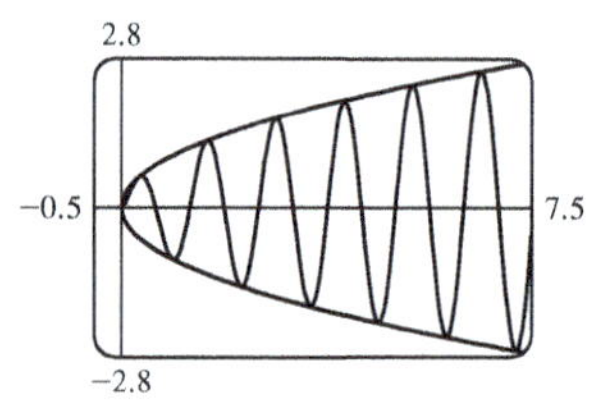

$y = \sqrt{x} \sin 5\pi x$ is a sine curve that lies between the graphs of $y = \sqrt{x}$ and $y = -\sqrt{x}$

71.

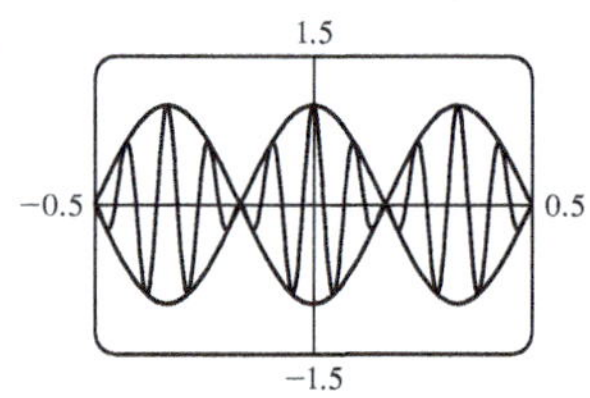

$y = \cos 3\pi x \cos 21\pi x$ is a cosine curve that lies between the graphs of $y = \cos 3\pi x$ and $y = -\cos 3\pi x$

73. Maximum value 1.76 when $x \approx 0.94$, minimum value -1.76 when $x \approx -0.94$ (The same maximum and minimum values occur at infinitely many other values of x.)
75. Maximum value 3.00 when $x \approx 1.57$, minimum value -1.00 when $x \approx -1.57$ (The same maximum and minimum values occur at infinitely many other values of x.)
77. 1.16 **79.** 0.34, 2.80
81. **(a)** Odd **(b)** $\pm 2\pi, \pm 4\pi, \pm 6\pi, \ldots$
(c)

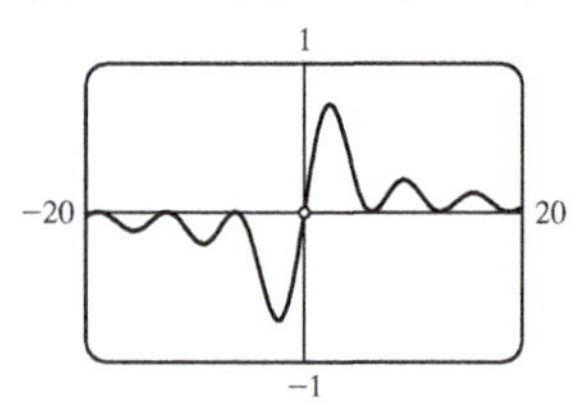

(d) $f(x)$ approaches 0
(e) $f(x)$ approaches 0

83. **(a)** 20 s **(b)** 6 ft

85. **(a)** $\frac{1}{80}$ min **(b)** 80

(c)

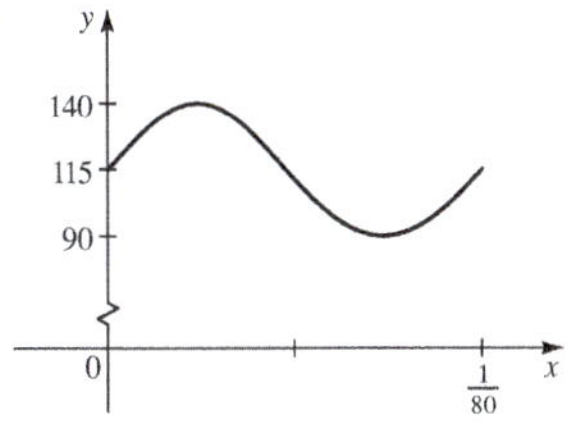

(d) $\frac{140}{90}$; higher than normal

SECTION 5.4 ■ PAGE 438

1. π; $\frac{\pi}{2} + n\pi$, n an integer

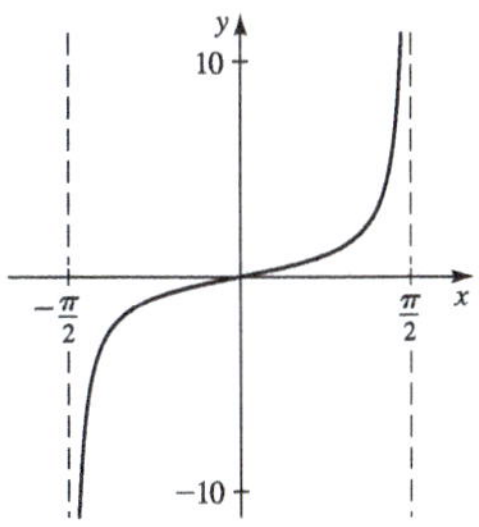

2. 2π; $n\pi$, n an integer

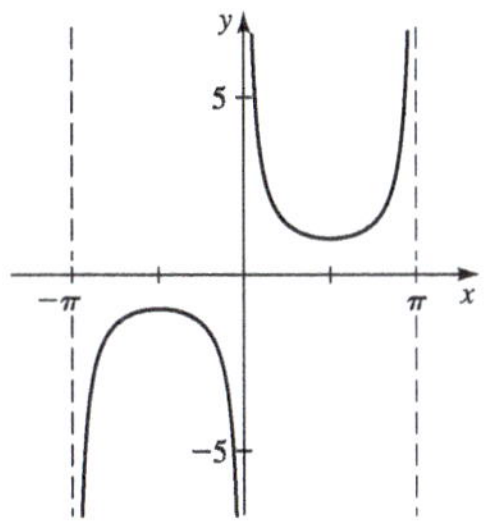

3. II **5.** VI **7.** IV

9. π

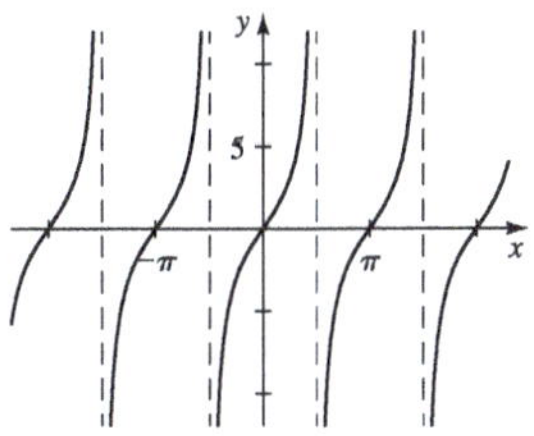

11. π

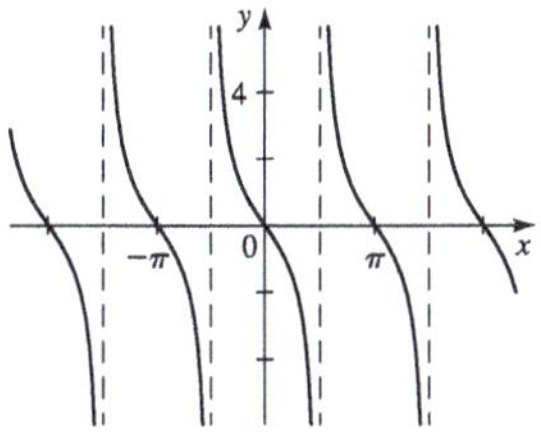

13. π

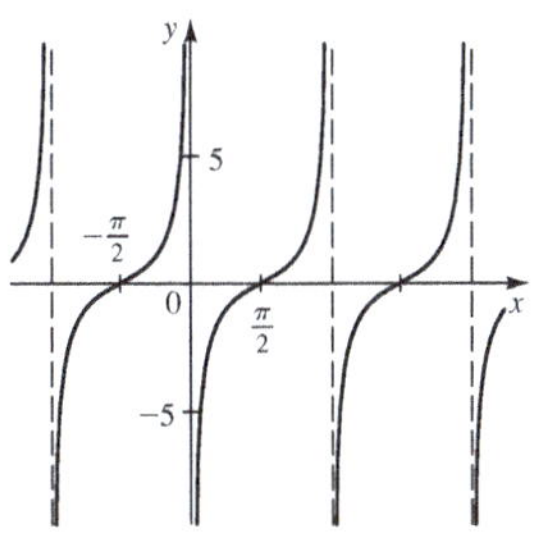

15. 2π

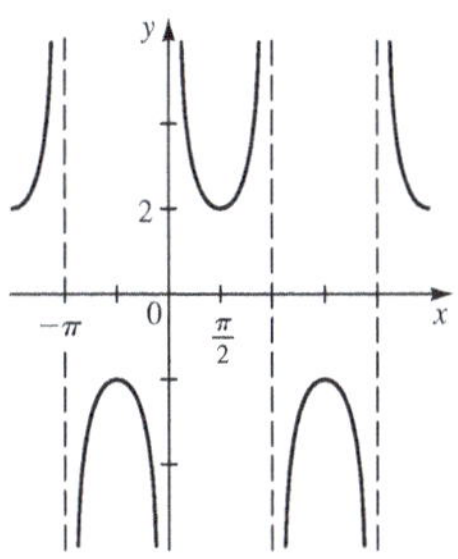

17. 2π

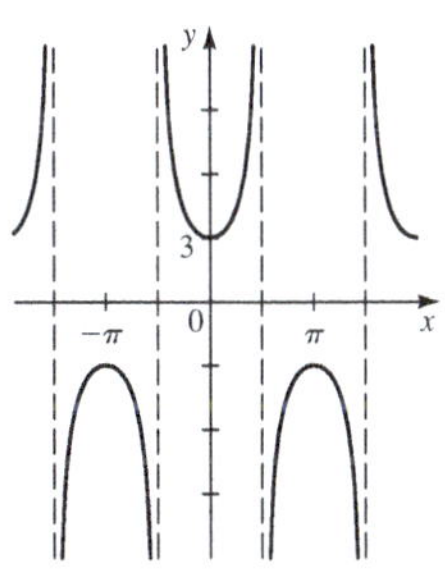

19. $\pi/3$

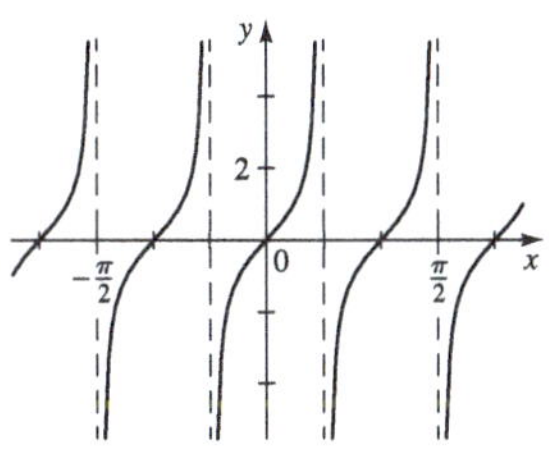

21. 1

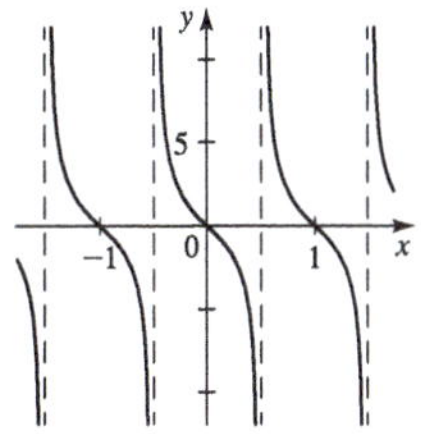

23. $\frac{1}{3}$

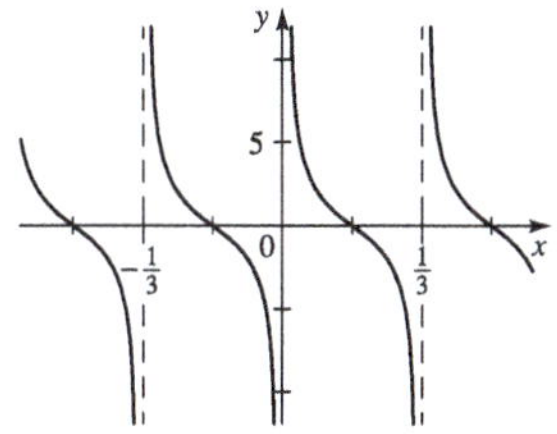

25. 4

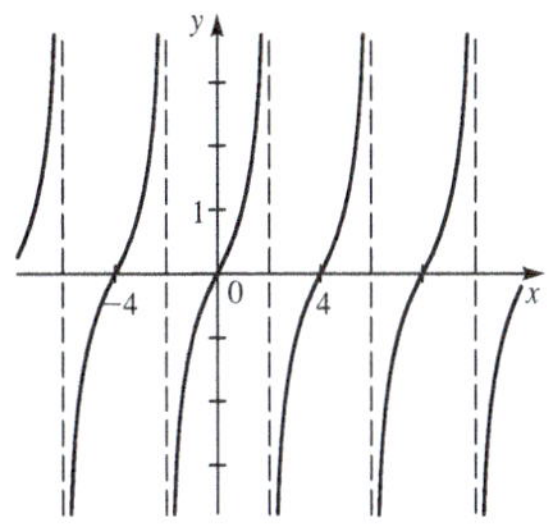

27. $\frac{1}{3}$

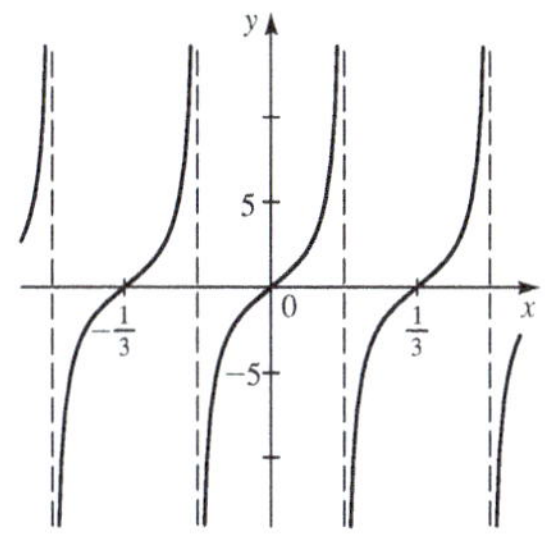

29. $\pi/2$

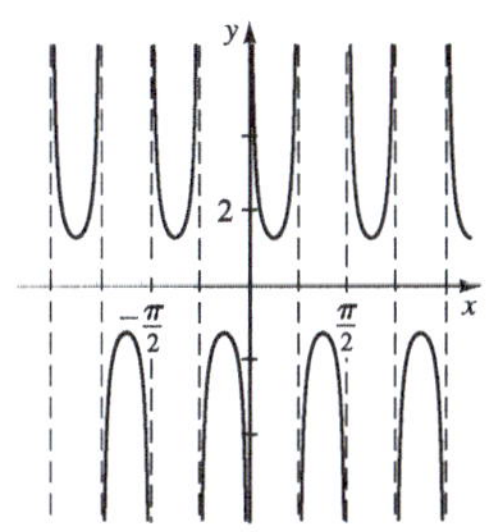

31. π

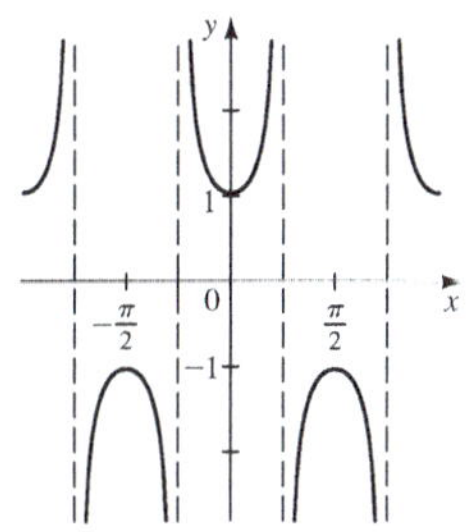

33. $\frac{4}{3}$

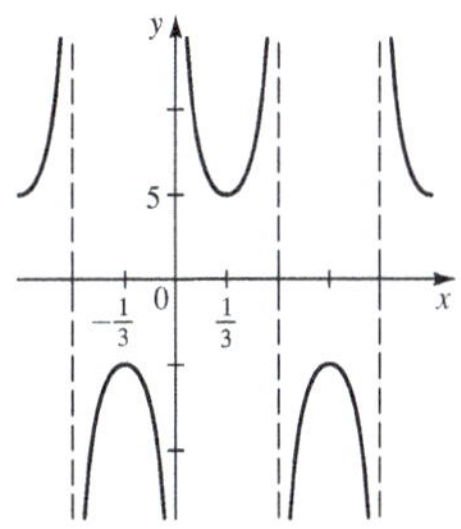

35. π

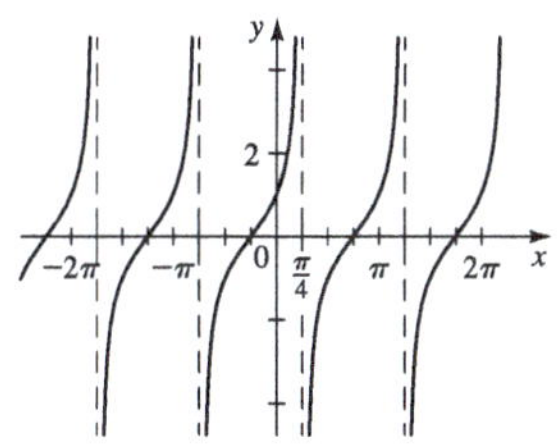

37. π

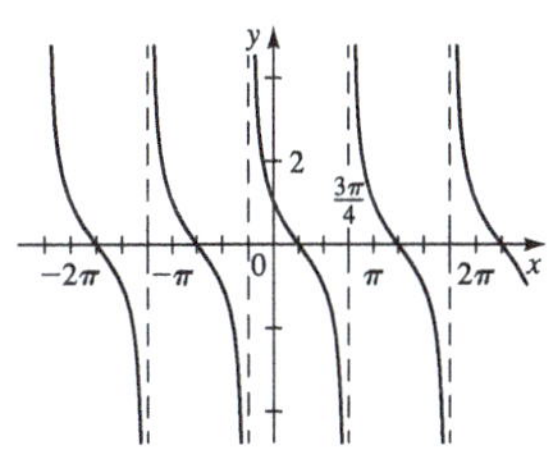

39. 2π

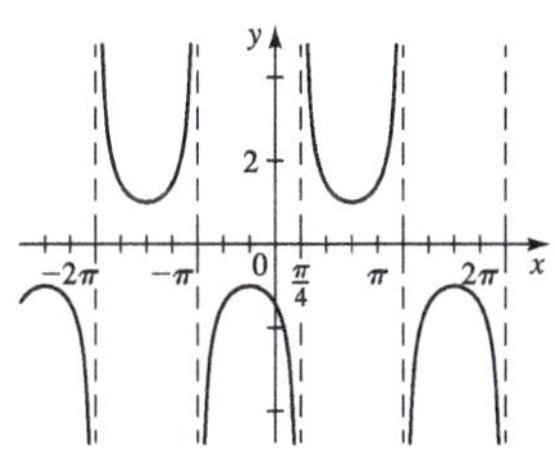

41. 2π

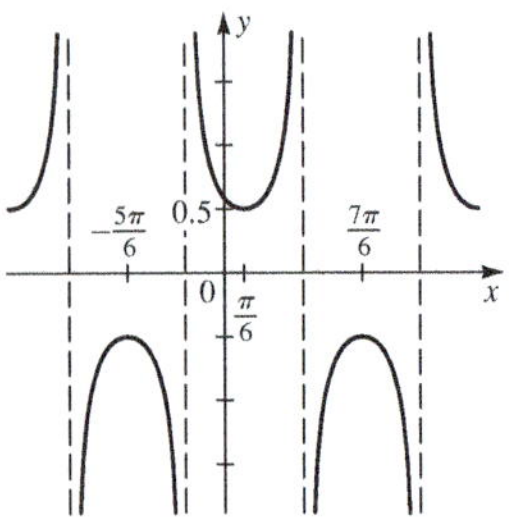

43. $\pi/2$

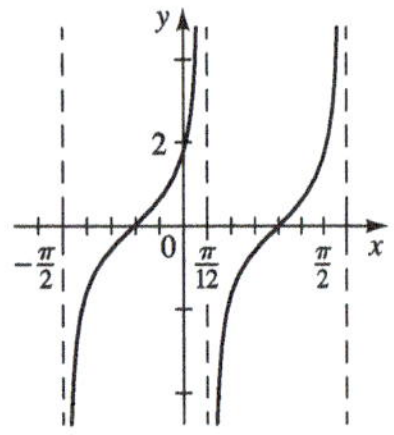

45. $\pi/3$

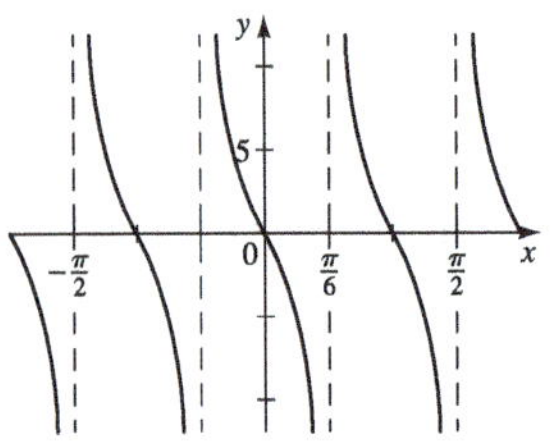

47. $\pi/2$

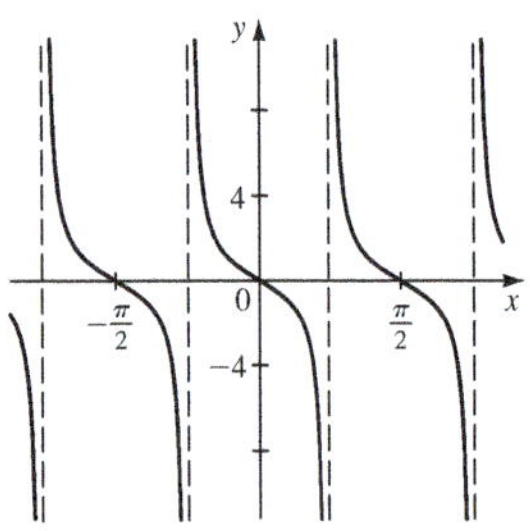

49. 2

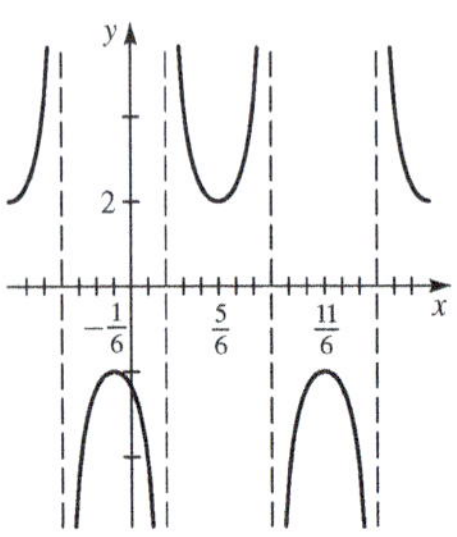

51. π

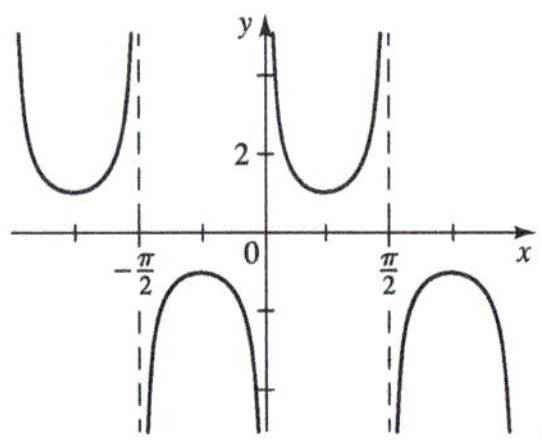

53. $2\pi/3$

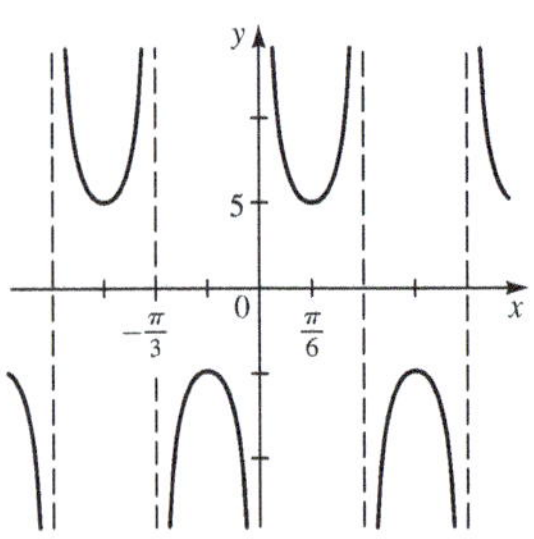

55. $3\pi/2$

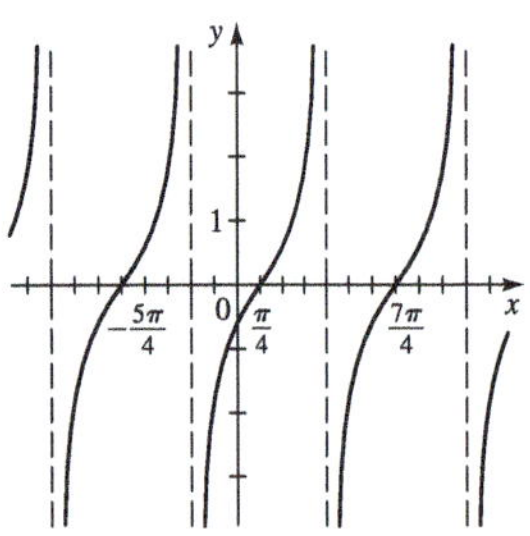

57. 2

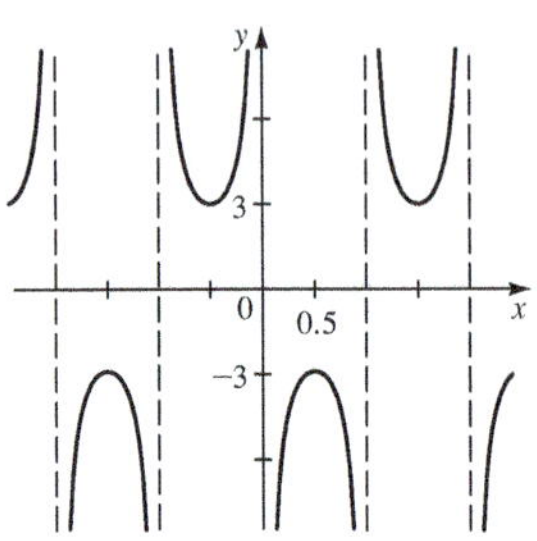

59. $\pi/2$

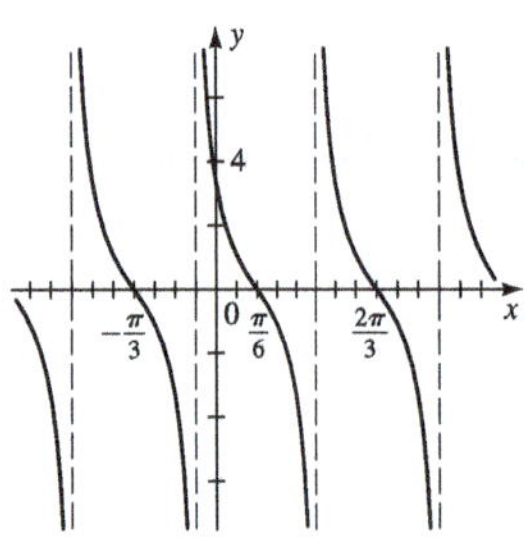

61. (a) 1.53 mi, 3.00 mi, 18.94 mi
(b)

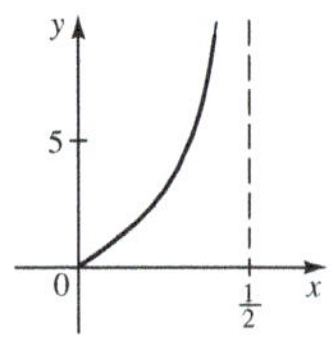

(c) $d(t)$ approaches ∞

SECTION 5.5 ■ PAGE 444

1. (a) $[-\pi/2, \pi/2]$, y, x, $\pi/6$, $\pi/6$, $\frac{1}{2}$
(b) $[0, \pi]$; y, x, $\pi/3$, $\pi/3$, $\frac{1}{2}$ **2.** $[-\pi/2, \pi/2]$; (ii)
3. (a) $\pi/2$ **(b)** $\pi/3$ **(c)** Undefined **5. (a)** π **(b)** $\pi/3$
(c) $5\pi/6$ **7. (a)** $-\pi/4$ **(b)** $\pi/3$ **(c)** $\pi/6$ **9. (a)** $2\pi/3$
(b) $-\pi/4$ **(c)** $\pi/4$ **11.** 0.72973 **13.** 2.01371
15. 2.75876 **17.** 1.47113 **19.** 0.88998 **21.** -0.26005
23. $\frac{1}{4}$ **25.** 5 **27.** Undefined **29.** $-\frac{1}{5}$ **31.** $\pi/4$ **33.** $\pi/4$
35. $5\pi/6$ **37.** $5\pi/6$ **39.** $\pi/4$ **41.** $-\pi/3$ **43.** $\sqrt{3}/3$
45. $\frac{1}{2}$ **47.** $-\sqrt{2}/2$

SECTION 5.6 ■ PAGE 456

1. (a) $a \sin \omega t$ **(b)** $a \cos \omega t$
2. (a) $ke^{-ct} \sin \omega t$ **(b)** $ke^{-ct} \cos \omega t$
3. (a) $|A|$, $2\pi/k$, b; $A \sin k\left(t - \frac{b}{k}\right)$; b/k **(b)** 5, $\pi/2$, π, $\pi/4$
4. π, $\pi/2$; $\pi/2$, out of phase
5. (a) 2, $2\pi/3$, $3/(2\pi)$
(b)

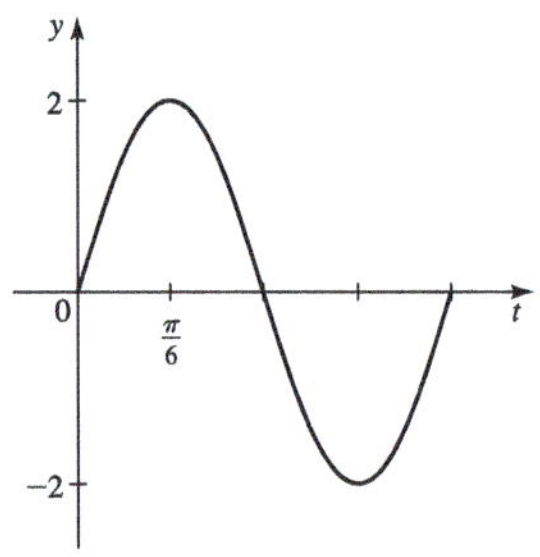

7. (a) 1, $20\pi/3$, $3/(20\pi)$
(b)

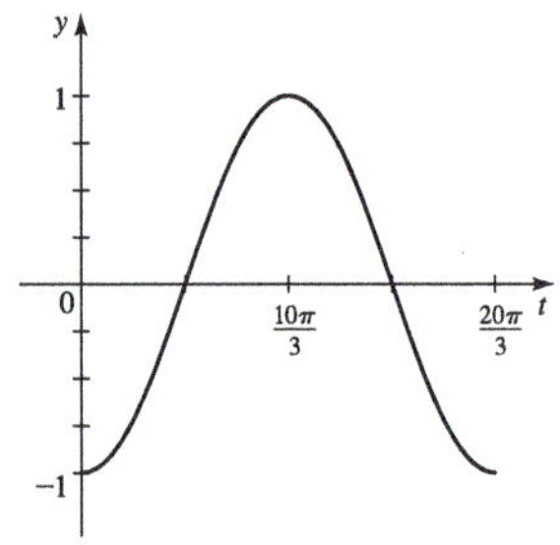

9. (a) $\frac{1}{4}$, $4\pi/3$, $3/(4\pi)$ **(b)**

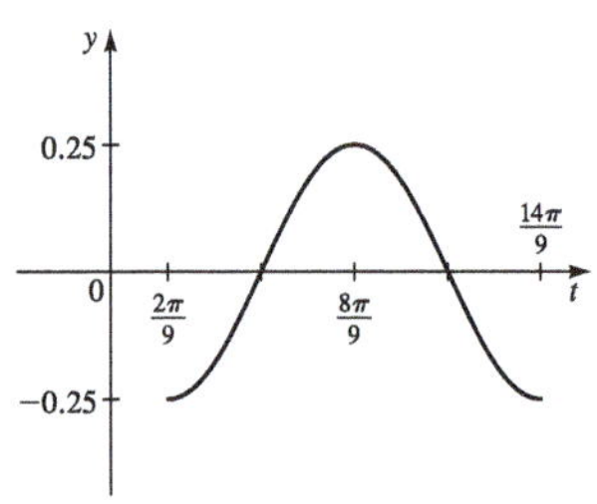

11. (a) 5, 3π, $1/(3\pi)$ **(b)**

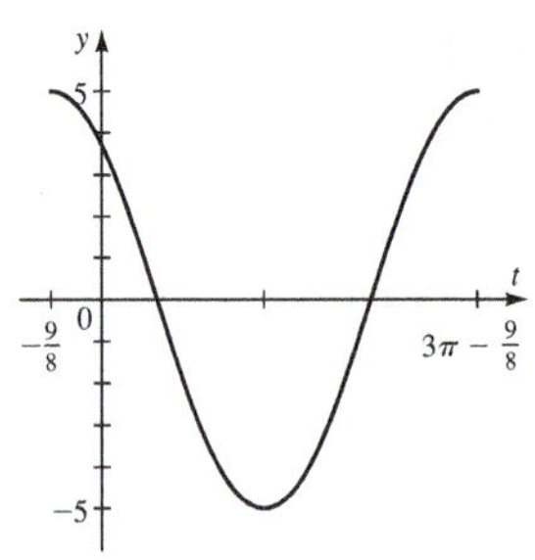

13. $y = 10\sin\left(\frac{2\pi}{3}t\right)$ **15.** $y = 6\sin(10t)$

17. $y = 60\cos(4\pi t)$ **19.** $y = 2.4\cos(1500\pi t)$

21. **(a)** $y = 2e^{-1.5t}\cos 6\pi t$ **(b)**

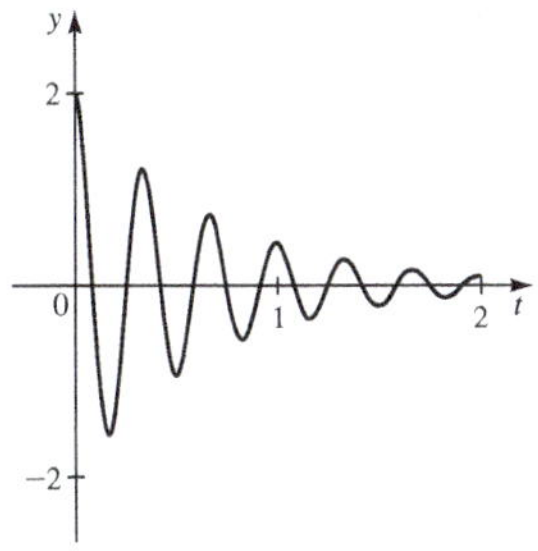

23. **(a)** $y = 100e^{-0.05t}\cos\frac{\pi}{2}t$

(b)

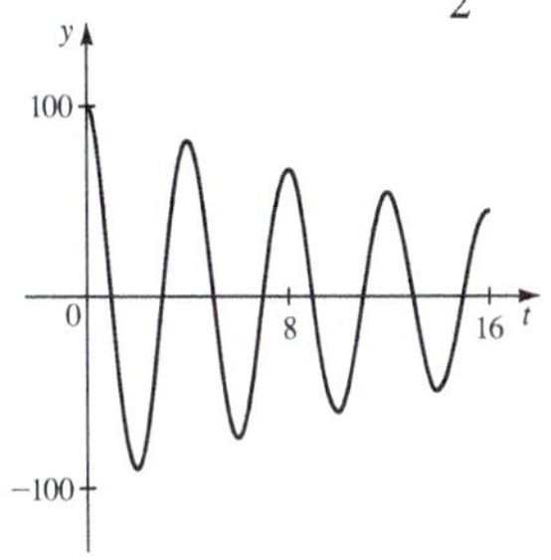

25. **(a)** $y = 7e^{-10t}\sin 12t$ **(b)**

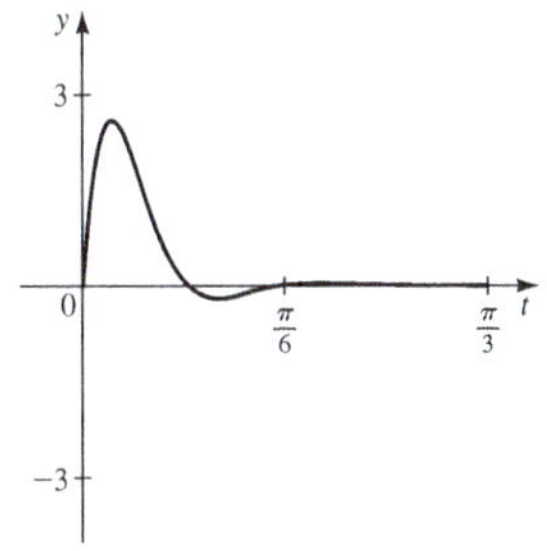

27. **(a)** $y = 0.3e^{-0.2t}\sin(40\pi t)$

(b)

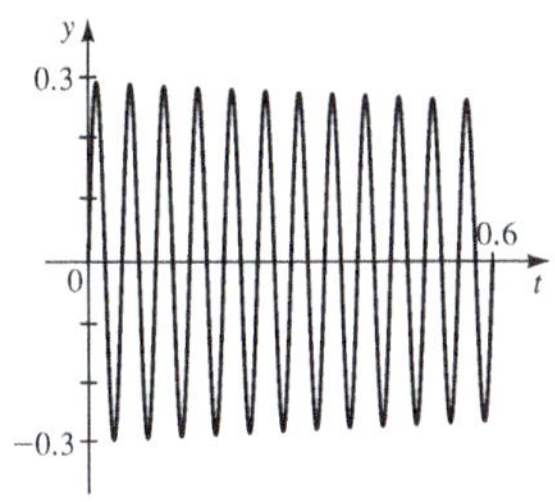

29. $5, \pi, \pi/2, \pi/4$ **31.** $100, 2\pi/5, -\pi, -\pi/5$

33. $20, \pi, \pi/2, \pi/4$

35. **(a)** $\pi/2, 5\pi/2$

(b) -2π

(c) In phase

(d)

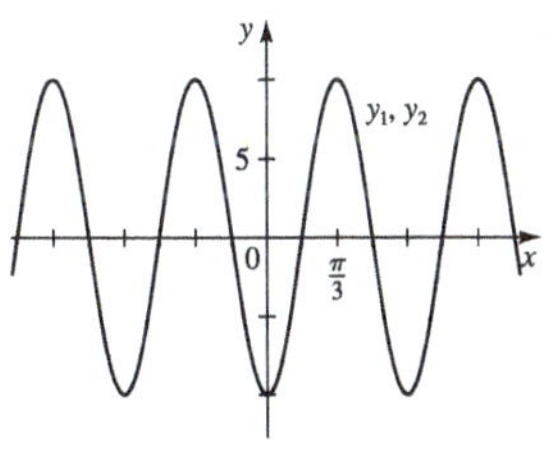

37. **(a)** $\pi/2, \pi/3$

(b) $\pi/6$

(c) Out of phase

(d)

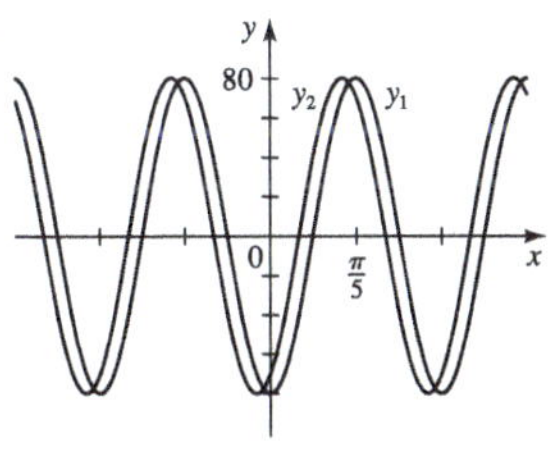

39. **(a)** 10 cycles per minute

(b)

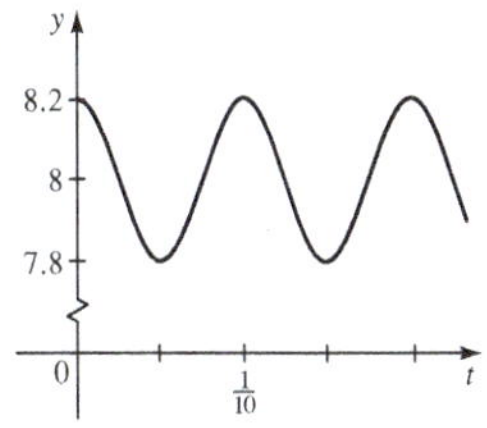

(c) 8.2 m

41. **(a)** 25, 0.0125, 80 **(b)**

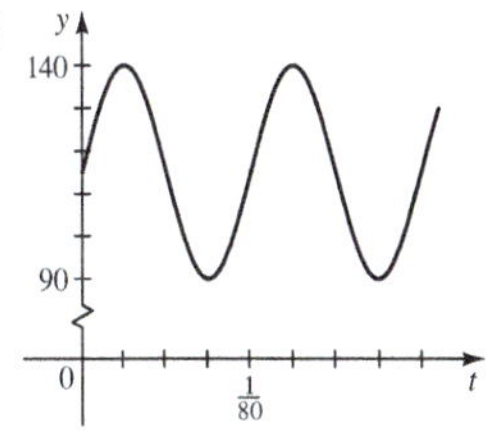

(c) The period decreases and the frequency increases.

43. $d(t) = 5\sin(5\pi t)$

45. $y = 21\sin\left(\frac{\pi}{6}t\right)$

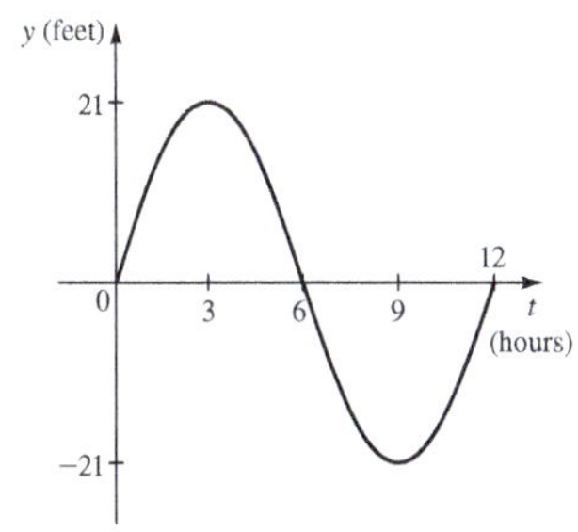

47. $y = 5\cos(2\pi t)$ **49.** $y = 11 + 10\sin\left(\frac{\pi t}{10}\right)$

51. $y = 3.8 + 0.2\sin\left(\frac{\pi}{5}t\right)$

53. $f(t) = 10\sin\left(\frac{\pi}{12}(t - 8)\right) + 90$

55. **(a)** 45 V **(b)** 40 **(c)** 40 **(d)** $E(t) = 45\cos(80\pi t)$

57. $f(t) = e^{-0.9t}\sin\pi t$ **59.** $c = \frac{1}{3}\ln 4 \approx 0.46$

61. **(a)** $y = \sin(200\pi t)$, $y = \sin\left(200\pi t + \frac{3\pi}{4}\right)$

(b) No; $3\pi/4$

CHAPTER 5 REVIEW ■ PAGE 463

1. **(b)** $\frac{1}{2}, -\sqrt{3}/2, -\sqrt{3}/3$ **3.** **(a)** $\pi/3$ **(b)** $\left(-\frac{1}{2}, \sqrt{3}/2\right)$

(c) $\sin t = \sqrt{3}/2$, $\cos t = -\frac{1}{2}$, $\tan t = -\sqrt{3}$, $\csc t = 2\sqrt{3}/3$, $\sec t = -2$, $\cot t = -\sqrt{3}/3$

5. **(a)** $\pi/4$ **(b)** $(-\sqrt{2}/2, -\sqrt{2}/2)$

(c) $\sin t = -\sqrt{2}/2$, $\cos t = -\sqrt{2}/2$, $\tan t = 1$, $\csc t = -\sqrt{2}$, $\sec t = -\sqrt{2}$, $\cot t = 1$

7. **(a)** $\sqrt{2}/2$ **(b)** $-\sqrt{2}/2$ **9.** **(a)** 0.89121 **(b)** 0.45360

11. **(a)** 0 **(b)** Undefined **13.** **(a)** Undefined **(b)** 0

15. **(a)** $-\sqrt{3}/3$ **(b)** $-\sqrt{3}$ **17.** $\dfrac{\sin t}{1 - \sin^2 t}$ **19.** $\dfrac{\sin t}{\sqrt{1 - \sin^2 t}}$

21. $\tan t = -\frac{5}{12}$, $\csc t = \frac{13}{5}$, $\sec t = -\frac{13}{12}$, $\cot t = -\frac{12}{5}$

23. $\sin t = 2\sqrt{5}/5$, $\cos t = -\sqrt{5}/5$, $\tan t = -2$, $\sec t = -\sqrt{5}$

25. $-\dfrac{\sqrt{17}}{4} + 4$ **27.** 3

29. **(a)** $10, 4\pi, 0$
(b)

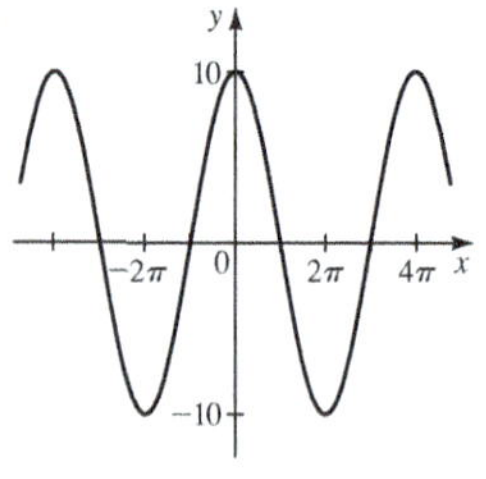

31. **(a)** $1, 4\pi, 0$
(b)

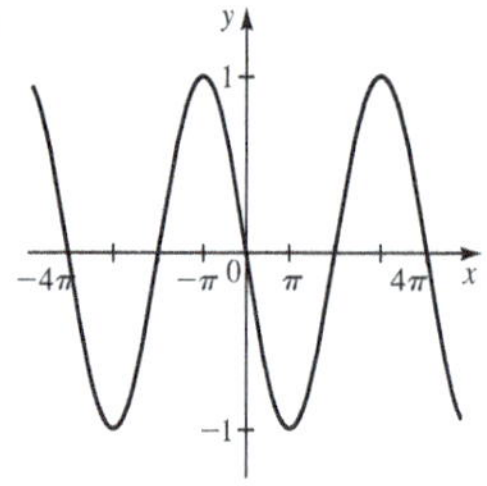

33. **(a)** $3, \pi, 1$
(b)

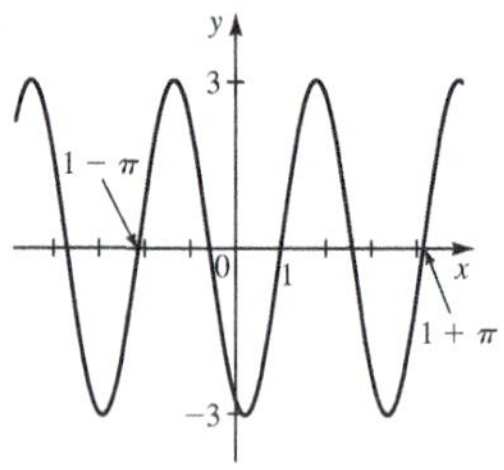

35. **(a)** $1, 4, -\frac{1}{3}$
(b)

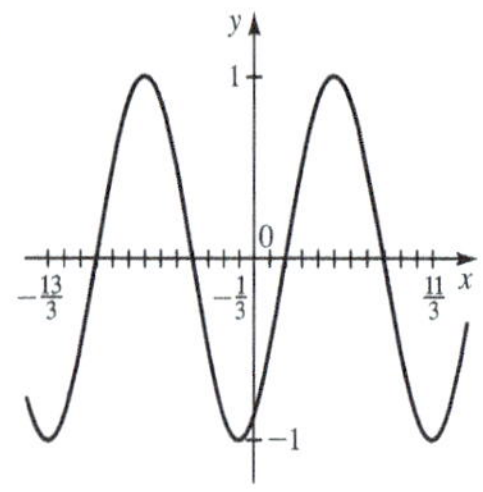

37. $y = 5 \sin 4x$ **39.** $y = \frac{1}{2} \sin 2\pi(x + \frac{1}{3})$

41. π

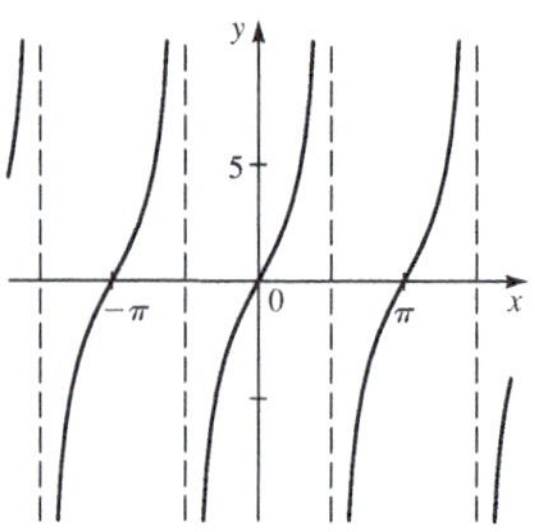

43. π

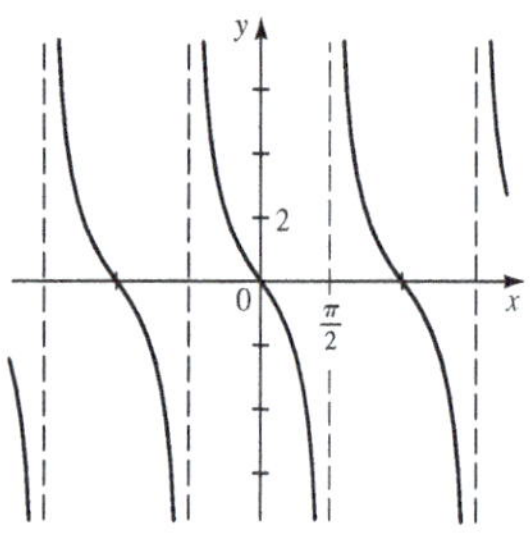

45. π

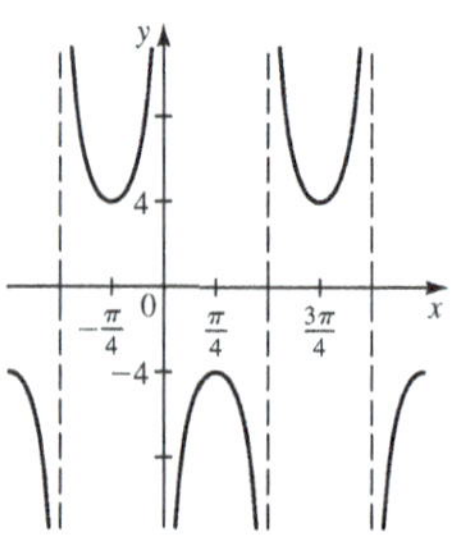

47. 2π

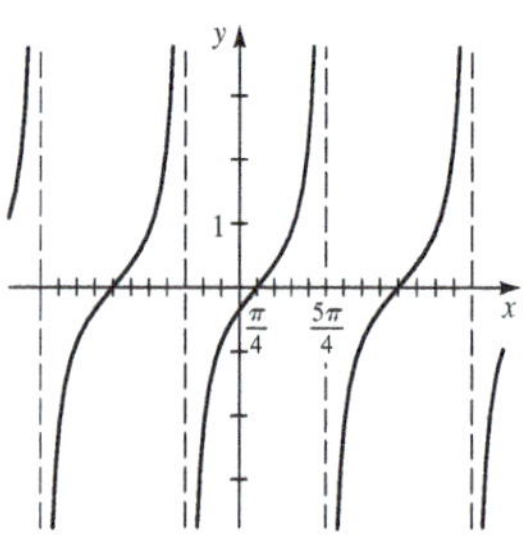

49. $\pi/2$ **51.** $\pi/6$ **53.** $100, \pi/4, -\pi/2, -\pi/16$

55. **(a)** $3\pi/2, 5\pi/2$ **(b)** $-\pi$ **(c)** Out of phase
(d)

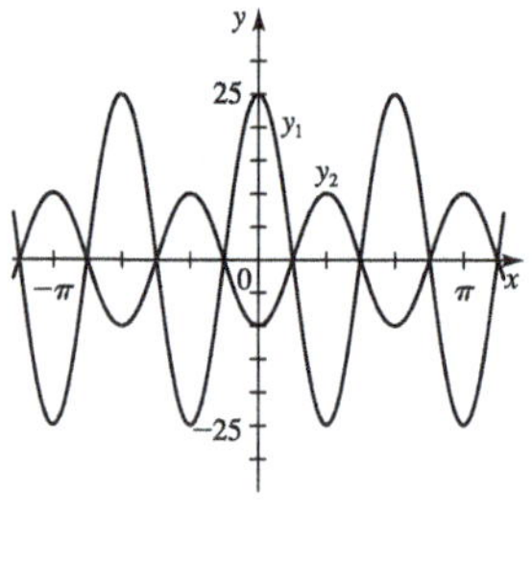

57. **(a)**

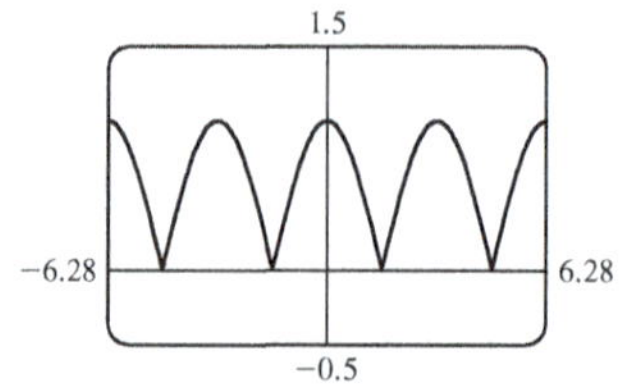

(b) Period π
(c) Even

59. **(a)**

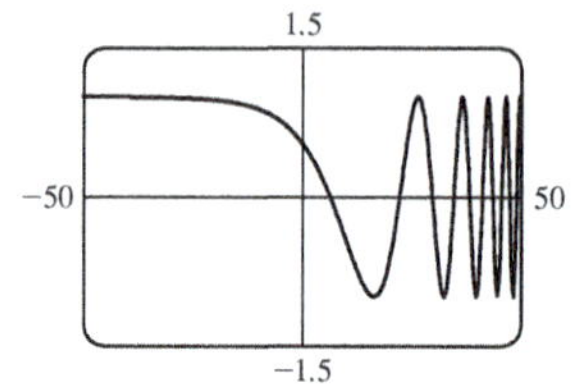

(b) Not periodic
(c) Neither

61. **(a)**

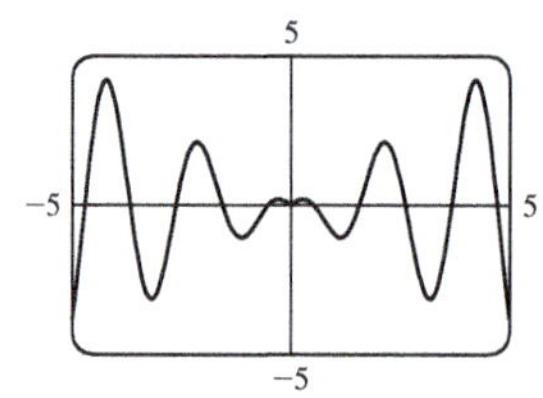

(b) Not periodic
(c) Even

63.

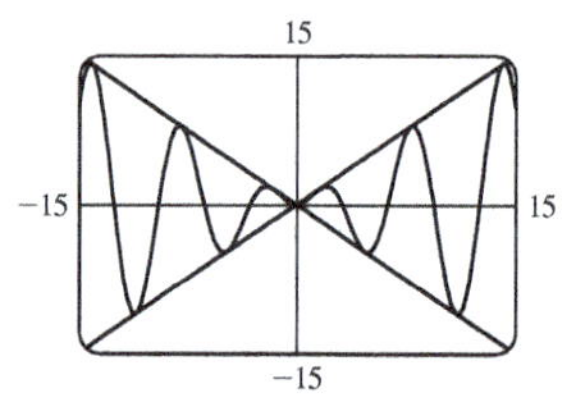

$y = x \sin x$ is a sine function whose graph lies between those of $y = x$ and $y = -x$

65.

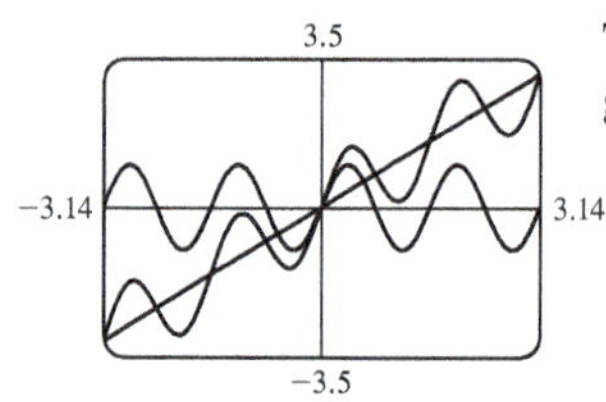

The graphs are related by graphical addition.

67. 1.76, −1.76 **69.** 0.30, 2.84

71. **(a)**

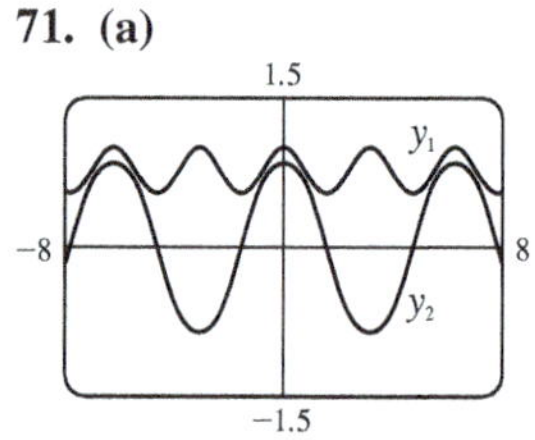

(b) y_1 has period π, y_2 has period 2π
(c) $\sin(\cos x) < \cos(\sin x)$, for all x

73. $y = -50 \cos(8\pi t)$

CHAPTER 5 TEST ■ PAGE 465

1. $y = -\frac{5}{6}$ **2.** **(a)** $\frac{4}{5}$ **(b)** $-\frac{3}{5}$ **(c)** $-\frac{4}{3}$ **(d)** $-\frac{5}{3}$

3. **(a)** $-\frac{1}{2}$ **(b)** $-\sqrt{2}/2$ **(c)** $\sqrt{3}$ **(d)** -1

4. $\tan t = -\dfrac{\sin t}{\sqrt{1 - \sin^2 t}}$ **5.** $-\frac{2}{15}$

6. (a) $5, \pi/2, 0, 0$

(b)

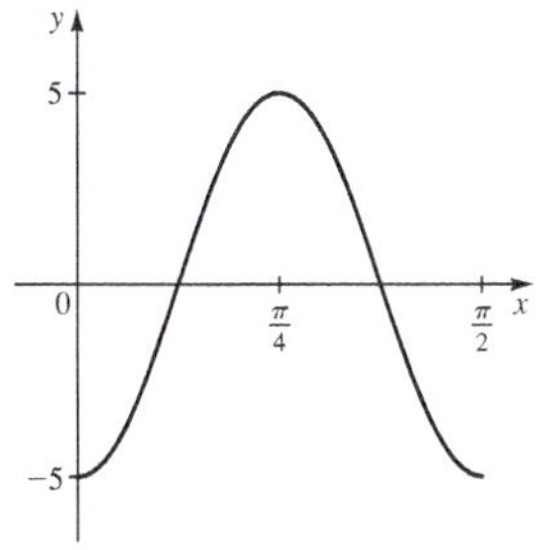

7. (a) $2, 4\pi, \pi/6, \pi/3$

(b)

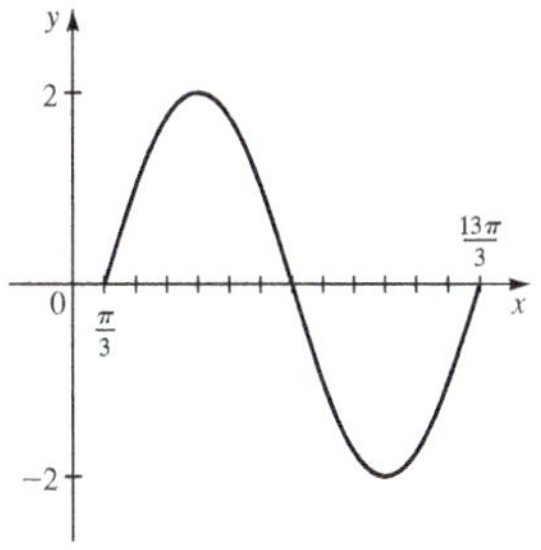

8. π

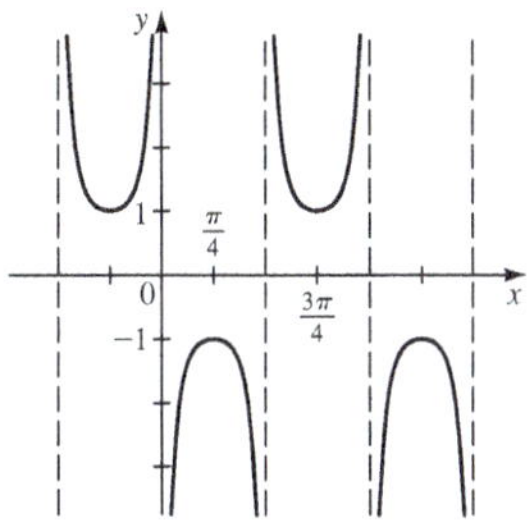

9. $\pi/2$

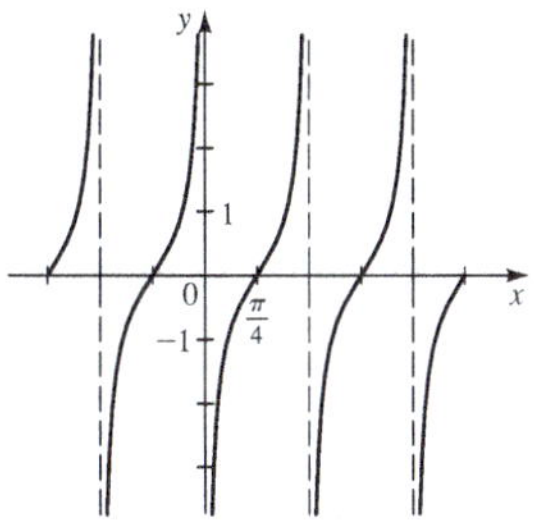

10. (a) $\pi/4$ **(b)** $5\pi/6$ **(c)** 0 **(d)** $\frac{1}{2}$

11. $y = 2 \sin 2(x + \pi/3)$

12. (a) $\pi/2, \pi/3$ **(b)** $\pi/6$

(c) Out of phase

(d)

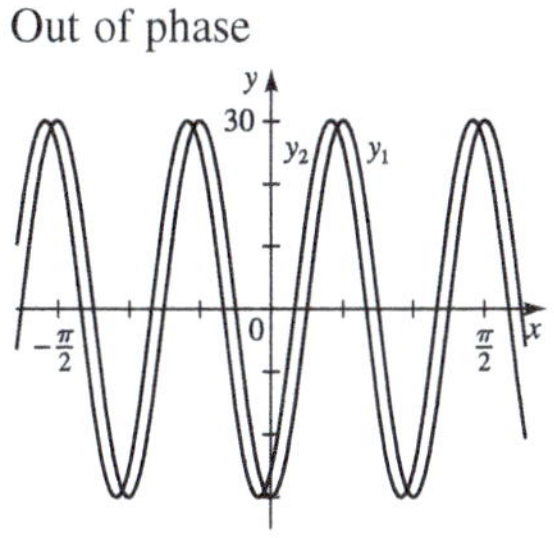

13. (a)

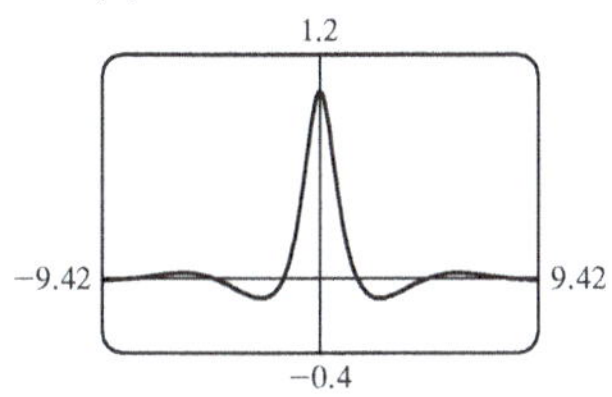

(b) Even

(c) Minimum value -0.11 when $x \approx \pm 2.54$, maximum value 1 when $x = 0$

14. $y = 5 \sin(4\pi t)$

15. $y = 16e^{-0.1t} \cos 24\pi t$

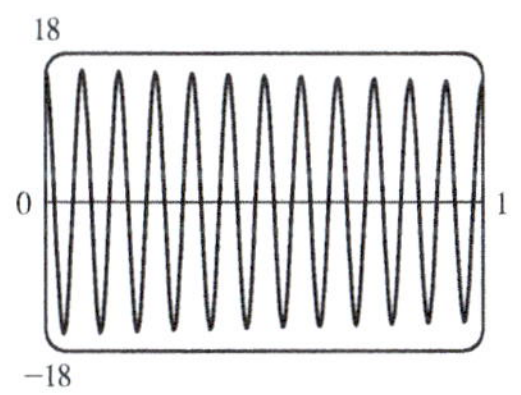

FOCUS ON MODELING ■ PAGE 469

1. (a) and **(c)**

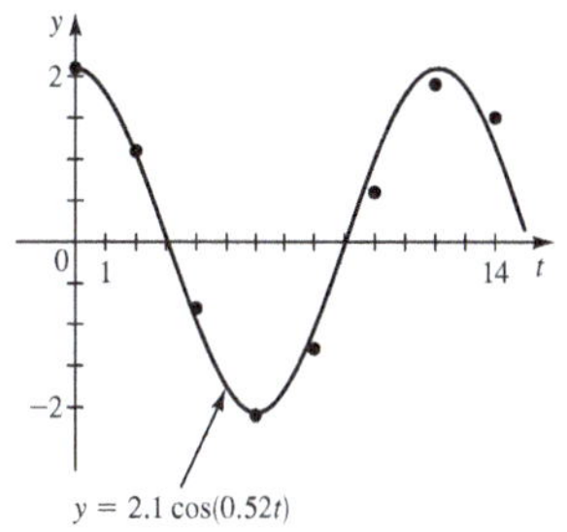

(b) $y = 2.1 \cos(0.52t)$

(d) $y = 2.05 \sin(0.50t + 1.55) - 0.01$ **(e)** The formula of (d) reduces to $y = 2.05 \cos(0.50t - 0.02) - 0.01$. Same as (b), rounded to one decimal.

3. (a) and **(c)**

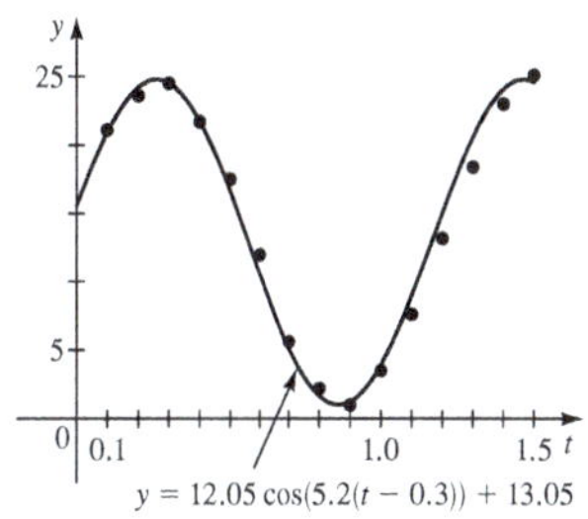

(b) $y = 12.05 \cos(5.2(t - 0.3)) + 13.05$

(d) $y = 11.72 \sin(5.05t + 0.24) + 12.96$ **(e)** The formula of (d) reduces to $y = 11.72 \cos(5.05(t - 0.26)) + 12.96$. Close, but not identical, to (b).

5. (a) and **(c)**

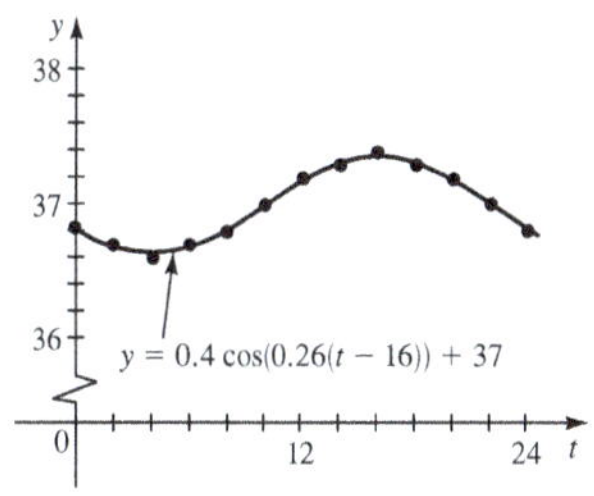

(b) $y = 0.4 \cos(0.26(t - 16)) + 37$, where y is the body temperature (°C) and t is hours since midnight

(d) $y = 0.37 \sin(0.26t - 2.62) + 37.0$

7. (a) and **(c)**

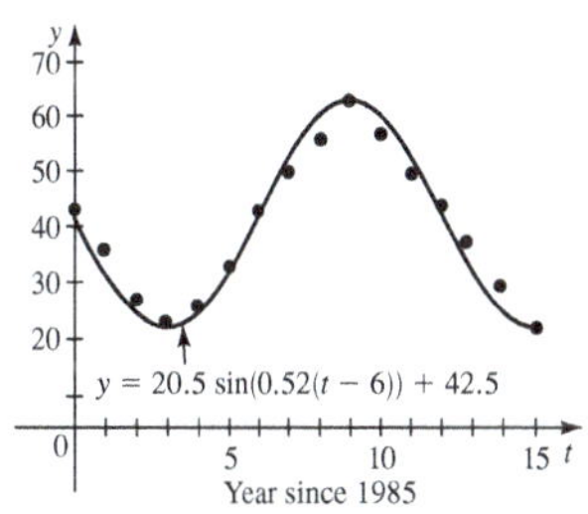

(b) $y = 20.5 \sin(0.52(t - 6)) + 42.5$, where y is the salmon population ($\times$ 1000), and t is years since 1985

(d) $y = 17.8 \sin(0.52t + 3.11) + 42.4$

CHAPTER 6

SECTION 6.1 ■ PAGE 478

1. (a) arc, 1 (b) $\pi/180$ (c) $180/\pi$ **2.** (a) $r\theta$ (b) $\frac{1}{2}r^2\theta$
3. (a) θ/t (b) s/t (c) $r\omega$ **4.** No, B **5.** $\pi/12 \approx 0.262$ rad
7. $3\pi/10 \approx 0.942$ rad **9.** $-\pi/4 \approx -0.785$ rad
11. $5\pi/9 \approx 1.745$ rad **13.** $50\pi/9 \approx 17.453$ rad
15. $-7\pi/18 \approx -1.222$ rad **17.** 300° **19.** 150°
21. $540/\pi \approx 171.9°$ **23.** $-216/\pi \approx -68.8°$
25. 18° **27.** −24° **29.** 410°, 770°, −310°, −670°
31. $11\pi/4, 19\pi/4, -5\pi/4, -13\pi/4$
33. $7\pi/4, 15\pi/4, -9\pi/4, -17\pi/4$ **35.** Yes **37.** Yes
39. Yes **41.** 40° **43.** 60° **45.** 280° **47.** $7\pi/6$
49. π **51.** $\pi/4$ **53.** $15\pi/2 \approx 23.6$
55. $360/\pi \approx 114.6°$ **57.** 15 cm **59.** $\frac{14}{9}$ rad, 89.1°
61. $18/\pi \approx 5.73$ m **63.** (a) $128\pi/9 \approx 44.68$ (b) 25
65. $100\pi/3 \approx 104.7\ \text{m}^2$ **67.** $6\sqrt{5\pi}/\pi \approx 7.6$ m
69. $\frac{1}{2}$ rad **71.** $\pi/4\ \text{ft}^2$ **73.** $3\pi/2$ rad, $\pi/8$ rad
75. 13.9 mi **77.** 330π mi ≈ 1037 mi **79.** 1.6 million mi
81. 1.15 mi **83.** $360\pi\ \text{in}^2 \approx 1130.97\ \text{in}^2$
85. (a) 90π rad/min (b) 1440π in./min ≈ 4523.9 in./min
87. $32\pi/15$ ft/s ≈ 6.7 ft/s **89.** 1039.6 mi/h **91.** 2.1 m/s
93. (a) 10π cm ≈ 31.4 cm (b) 5 cm (c) 3.32 cm
(d) $86.8\ \text{cm}^3$

SECTION 6.2 ■ PAGE 487

1. (a)

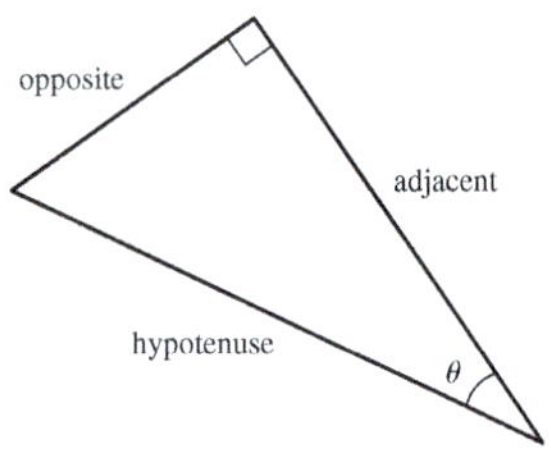

(b) $\dfrac{\text{opposite}}{\text{hypotenuse}}, \dfrac{\text{adjacent}}{\text{hypotenuse}}, \dfrac{\text{opposite}}{\text{adjacent}}$ (c) similar
2. $\sin\theta, \cos\theta, \tan\theta$
3. $\sin\theta = \frac{4}{5}, \cos\theta = \frac{3}{5}, \tan\theta = \frac{4}{3}, \csc\theta = \frac{5}{4},$
$\sec\theta = \frac{5}{3}, \cot\theta = \frac{3}{4}$
5. $\sin\theta = \frac{40}{41}, \cos\theta = \frac{9}{41}, \tan\theta = \frac{40}{9},$
$\csc\theta = \frac{41}{40}, \sec\theta = \frac{41}{9}, \cot\theta = \frac{9}{40}$
7. $\sin\theta = 2\sqrt{13}/13, \cos\theta = 3\sqrt{13}/13, \tan\theta = \frac{2}{3},$
$\csc\theta = \sqrt{13}/2, \sec\theta = \sqrt{13}/3, \cot\theta = \frac{3}{2}$
9. (a) $3\sqrt{34}/34, 3\sqrt{34}/34$ (b) $\frac{3}{5}, \frac{3}{5}$ (c) $\sqrt{34}/5, \sqrt{34}/5$
11. (a) 0.37461 (b) 2.35585 **13.** (a) 1.02630 (b) 1.23490
15. $\frac{25}{2}$ **17.** $13\sqrt{3}/2$ **19.** 16.51658
21. $x = 28\cos\theta, y = 28\sin\theta$
23. $\sin\theta = 5\sqrt{61}/61, \cos\theta = 6\sqrt{61}/61, \csc\theta = \sqrt{61}/5,$
$\sec\theta = \sqrt{61}/6, \cot\theta = \frac{6}{5}$

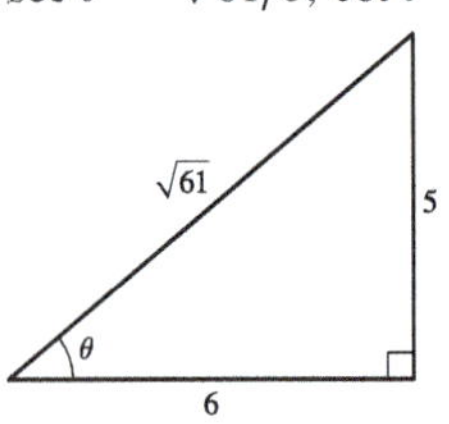

25. $\sin\theta = \sqrt{2}/2, \cos\theta = \sqrt{2}/2, \tan\theta = 1,$
$\csc\theta = \sqrt{2}, \sec\theta = \sqrt{2}$

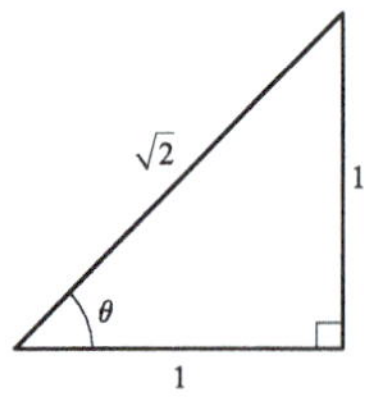

27. $\sin\theta = \frac{6}{11}, \cos\theta = \sqrt{85}/11, \tan\theta = 6\sqrt{85}/85,$
$\sec\theta = 11\sqrt{85}/85, \cot\theta = \sqrt{85}/6$

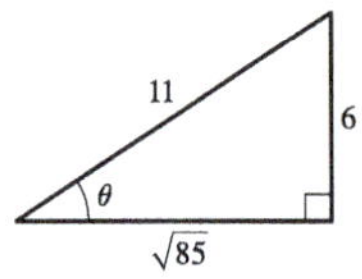

29. $(1 + \sqrt{3})/2$ **31.** 1 **33.** $\frac{1}{2}$ **35.** $\frac{3}{4} + (\sqrt{2}/2)$

37.

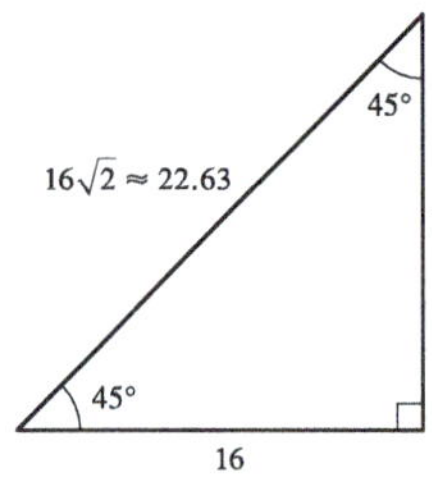

39.

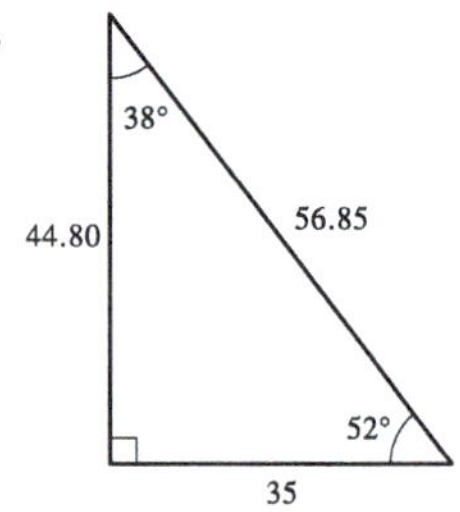

41.

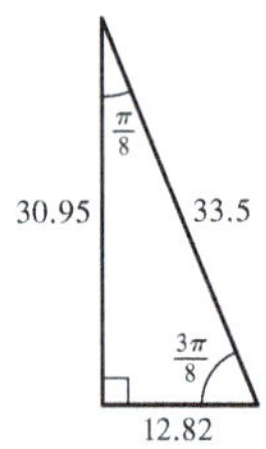

43.

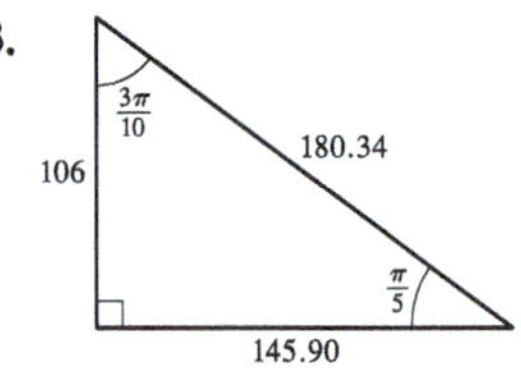

45. $\sin\theta \approx 0.44, \cos\theta \approx 0.89, \tan\theta = 0.50, \csc\theta \approx 2.25,$
$\sec\theta \approx 1.125, \cot\theta = 2.00$ **47.** 230.9 **49.** 63.7
51. $x = 10\tan\theta\sin\theta$ **53.** 1026 ft
55. (a) 2100 mi (b) No **57.** 19 ft **59.** 345 ft
61. 415 ft, 152 ft **63.** 2570 ft **65.** 5808 ft
67. 91.7 million mi **69.** 3960 mi **71.** 0.723 AU

SECTION 6.3 ■ PAGE 498

1. $y/r, x/r, y/x$ **2.** quadrant; positive; negative; negative
3. (a) x-axis; 80°, 10° (b) 80°; 10° **4.** $\frac{1}{2}ab\sin\theta$; 7
5. (a) 60° (b) 20° (c) 75° **7.** (a) 45° (b) 90° (c) 75°
9. (a) $3\pi/10$ (b) $\pi/8$ (c) $\pi/3$
11. (a) $2\pi/7$ (b) 0.4π (c) 1.4 **13.** $-\sqrt{3}/2$ **15.** $-\sqrt{3}/3$
17. $\sqrt{3}/3$ **19.** 1 **21.** $-\sqrt{3}/2$ **23.** $\sqrt{3}/3$ **25.** −1
27. $-\sqrt{3}$ **29.** −2 **31.** 2 **33.** −1
35. Undefined **37.** III **39.** IV
41. $\tan\theta = -\sqrt{1 - \cos^2\theta}/\cos\theta$
43. $\cos\theta = \sqrt{1 - \sin^2\theta}$ **45.** $\sec\theta = -\sqrt{1 + \tan^2\theta}$
47. $\cos\theta = \frac{3}{5}, \tan\theta = -\frac{4}{3}, \csc\theta = -\frac{5}{4}, \sec\theta = \frac{5}{3}, \cot\theta = -\frac{3}{4}$
49. $\sin\theta = -\sqrt{95}/12, \tan\theta = -\sqrt{95}/7, \csc\theta = -12\sqrt{95}/95,$
$\sec\theta = \frac{12}{7}, \cot\theta = -7\sqrt{95}/95$

51. $\sin\theta = \frac{1}{2}$, $\cos\theta = \sqrt{3}/2$, $\tan\theta = \sqrt{3}/3$, $\sec\theta = 2\sqrt{3}/3$, $\cot\theta = \sqrt{3}$
53. $\sin\theta = 3\sqrt{5}/7$, $\tan\theta = -3\sqrt{5}/2$, $\csc\theta = 7\sqrt{5}/15$, $\sec\theta = -\frac{7}{2}$, $\cot\theta = -2\sqrt{5}/15$
55. $\sqrt{3}/2, \sqrt{3}$ **57.** 30.0 **59.** $25\sqrt{3} \approx 43.3$
61. 66.1° **63.** $(4\pi/3) - \sqrt{3} \approx 2.46$
65. **(b)**

θ	20°	60°	80°	85°
h	1922	9145	29,944	60,351

67. **(a)** $A(\theta) = 400 \sin\theta \cos\theta$
(b)

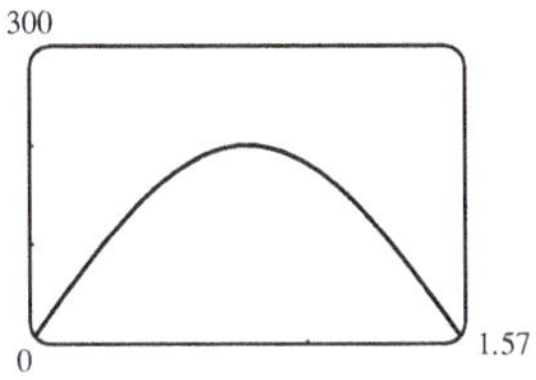

(c) width = depth ≈ 14.14 in.
69. **(a)** $9\sqrt{3}/4$ ft ≈ 3.897 ft, $\frac{9}{16}$ ft = 0.5625 ft
(b) 23.982 ft, 3.462 ft
71. **(a)**

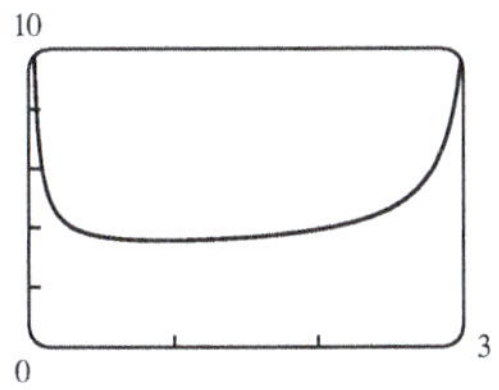

(b) 0.946 rad or 54°
73. 42°

SECTION 6.4 ■ PAGE 506

1. one-to-one; domain, $[-\pi/2, \pi/2]$
2. **(a)** $[-1, 1], [-\pi/2, \pi/2]$ **(b)** $[-1, 1], [0, \pi]$
(c) $\mathbb{R}, (-\pi/2, \pi/2)$ **3.** **(a)** $\frac{8}{10}$ **(b)** $\frac{6}{10}$ **(c)** $\frac{8}{6}$
4. $\frac{12}{13}$

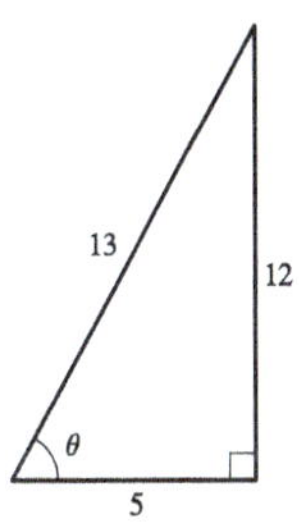

5. **(a)** $\pi/2$ **(b)** $\pi/2$ **(c)** $\pi/3$
7. **(a)** $-\pi/4$ **(b)** $3\pi/4$ **(c)** $-\pi/4$ **9.** 0.30469
11. 1.23096 **13.** 1.24905 **15.** Undefined **17.** 36.9°
19. 34.7° **21.** 34.8° **23.** 41.8°, 138.2° **25.** 113.6°
27. 78.7° **29.** $\frac{3}{5}$ **31.** $\frac{13}{5}$ **33.** $\frac{12}{5}$ **35.** $\sqrt{1 - x^2}$
37. $x/\sqrt{1 - x^2}$ **39.** 72.5°, 19 ft
41. **(a)** $h = 2\tan\theta$ **(b)** $\theta = \tan^{-1}(h/2)$
43. **(a)** $\theta = \sin^{-1}(h/680)$ **(b)** $\theta = 47.3°$
45. **(a)** 54.1° **(b)** 48.3°, 32.2°, 24.5°. The function $\sin^{-1}$ is undefined for values outside the interval $[-1, 1]$.

SECTION 6.5 ■ PAGE 513

1. $\dfrac{\sin A}{a} = \dfrac{\sin B}{b} = \dfrac{\sin C}{c}$ **2.** **(a)** ASA, SSA **(b)** SSA
3. 318.8 **5.** 24.8 **7.** 44° **9.** $\angle C = 114°$, $a \approx 51$, $b \approx 24$
11. $\angle A = 44°$, $\angle B = 68°$, $a \approx 8.99$
13. $\angle C = 62°$, $a \approx 200$, $b \approx 242$

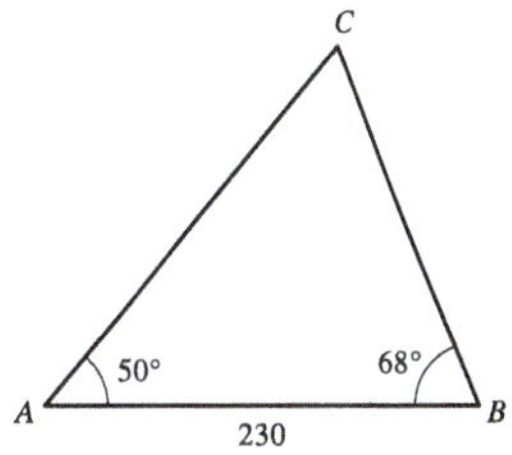

15. $\angle B = 85°$, $a \approx 5$, $c \approx 9$

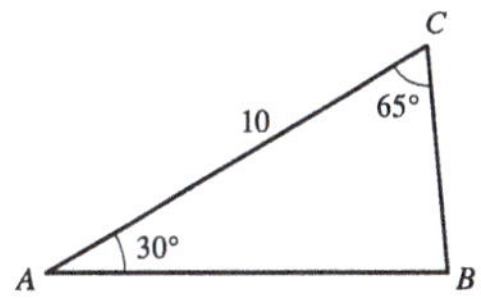

17. $\angle A = 100°$, $a \approx 89$, $c \approx 71$

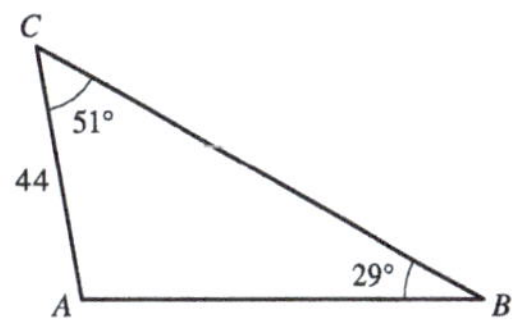

19. $\angle B \approx 30°$, $\angle C \approx 40°$, $c \approx 19$ **21.** No solution
23. $\angle A_1 \approx 125°$, $\angle C_1 \approx 30°$, $a_1 \approx 49$; $\angle A_2 \approx 5°$, $\angle C_2 \approx 150°$, $a_2 \approx 5.6$
25. No solution **27.** $\angle A_1 \approx 57.2°$, $\angle B_1 \approx 93.8°$, $b_1 \approx 30.9$; $\angle A_2 \approx 122.8°$, $\angle B_2 \approx 28.2°$, $b_2 \approx 14.6$
29. **(a)** 91.146° **(b)** 14.427° **31.** **(a)** 1018 mi **(b)** 1017 mi
33. 219 ft **35.** 55.9 m **37.** 175 ft **39.** 192 m
41. 0.427 AU, 1.119 AU

SECTION 6.6 ■ PAGE 520

1. $a^2 + b^2 - 2ab\cos C$ **2.** SSS, SAS **3.** 28.9 **5.** 47
7. 29.89° **9.** 15 **11.** $\angle A \approx 39.4°$, $\angle B \approx 20.6°$, $c \approx 24.6$
13. $\angle A \approx 48°$, $\angle B \approx 79°$, $c \approx 3.2$
15. $\angle A \approx 50°$, $\angle B \approx 73°$, $\angle C \approx 57°$
17. $\angle A_1 \approx 83.6°$, $\angle C_1 \approx 56.4°$, $a_1 \approx 193$; $\angle A_2 \approx 16.4°$, $\angle C_2 \approx 123.6$, $a_2 \approx 54.9$
19. No such triangle **21.** 2 **23.** 25.4 **25.** 89.2°
27. 24.3 **29.** 54 **31.** 26.83 **33.** 5.33 **35.** 40.77
37. 3.85 cm^2 **39.** 2.30 mi **41.** 23.1 mi **43.** 2179 mi
45. **(a)** 62.6 mi **(b)** S 18.2° E **47.** 96° **49.** 211 ft
51. 3835 ft **53.** $165,554

CHAPTER 6 REVIEW ■ PAGE 527

1. **(a)** $\pi/6$ **(b)** $5\pi/6$ **(c)** $-\pi/9$ **(d)** $-5\pi/4$
3. **(a)** 150° **(b)** −20° **(c)** −240° **(d)** 229.2°
5. $4\pi \approx 12.6$ m **7.** $90/\pi \approx 28.6$ ft **9.** 21,609 **11.** 25 m^2

13. 0.4 rad ≈ 22.9° **15.** 300π rad/min ≈ 942.5 rad/min, 7539.8 in./min = 628.3 ft/min
17. $\sin\theta = 5/\sqrt{74}$, $\cos\theta = 7/\sqrt{74}$, $\tan\theta = \frac{5}{7}$, $\csc\theta = \sqrt{74}/5$, $\sec\theta = \sqrt{74}/7$, $\cot\theta = \frac{7}{5}$
19. $x \approx 3.83$, $y \approx 3.21$ **21.** $x \approx 2.92$, $y \approx 3.11$
23. $A = 70°$, $a \approx 2.819$, $b \approx 1.026$
25. $A \approx 16.3°$, $C \approx 73.7°$, $c = 24$
27. $a = \cot\theta$, $b = \csc\theta$ **29.** 48 m **31.** 1076 mi
33. $-\sqrt{2}/2$ **35.** 1 **37.** $-\sqrt{3}/3$ **39.** $-\sqrt{2}/2$
41. $2\sqrt{3}/3$ **43.** $-\sqrt{3}$
45. $\sin\theta = \frac{12}{13}$, $\cos\theta = -\frac{5}{13}$, $\tan\theta = -\frac{12}{5}$, $\csc\theta = \frac{13}{12}$, $\sec\theta = -\frac{13}{5}$, $\cot\theta = -\frac{5}{12}$ **47.** 60°
49. $\tan\theta = \sqrt{1 - \cos^2\theta}/\cos\theta$
51. $\tan^2\theta = \sin^2\theta/(1 - \sin^2\theta)$
53. $\sin\theta = \sqrt{7}/4$, $\cos\theta = \frac{3}{4}$, $\csc\theta = 4\sqrt{7}/7$, $\cot\theta = 3\sqrt{7}/7$
55. $\cos\theta = -\frac{4}{5}$, $\tan\theta = -\frac{3}{4}$, $\csc\theta = \frac{5}{3}$, $\sec\theta = -\frac{5}{4}$, $\cot\theta = -\frac{4}{3}$ **57.** $-\sqrt{5}/5$ **59.** 1 **61.** $\pi/3$ **63.** $2/\sqrt{21}$
65. $x/\sqrt{1 + x^2}$ **67.** $\theta = \cos^{-1}(x/3)$ **69.** 5.32 **71.** 148.07
73. 9.17 **75.** 54.1° or 125.9° **77.** 80.4° **79.** 77.3 mi
81. 3.9 mi **83.** 32.12

CHAPTER 6 TEST ■ PAGE 531

1. $11\pi/6$, $-3\pi/4$ **2.** 240°, −74.5°
3. (a) 240π rad/min ≈ 753.98 rad/min
(b) 12,063.7 ft/min = 137 mi/h **4. (a)** $\sqrt{2}/2$
(b) $\sqrt{3}/3$ **(c)** 2 **(d)** 1 **5.** $(26 + 6\sqrt{13})/39$
6. $a = 24\sin\theta$, $b = 24\cos\theta$ **7.** $(4 - 3\sqrt{2})/4$
8. $-\frac{13}{12}$ **9.** $\tan\theta = -\sqrt{\sec^2\theta - 1}$ **10.** 19.6 ft
11. (a) $\theta = \tan^{-1}(x/4)$ **(b)** $\theta = \cos^{-1}(3/x)$ **12.** $\frac{40}{41}$
13. 9.1 **14.** 250.5 **15.** 8.4 **16.** 19.5 **17.** 78.6° **18.** 40.2°
19. (a) 15.3 m² **(b)** 24.3 m **20. (a)** 129.9° **(b)** 44.9
21. 554 ft

FOCUS ON MODELING ■ PAGE 534

1. 1.41 mi **3.** 14.3 m **5. (c)** 2350 ft
7.

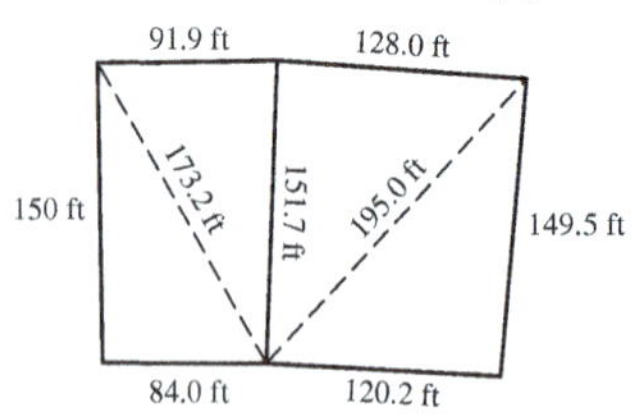

CHAPTER 7

SECTION 7.1 ■ PAGE 542

1. all; 1 **2.** $\cos(-x) = \cos x$ **3.** $\sin t$ **5.** $\tan\theta$ **7.** −1
9. $\csc u$ **11.** $\tan\theta$ **13.** 1 **15.** $\cos t + 1$ **17.** $\cos x$
19. $\sin^2 x$ **21.** $\cos y$ **23.** $2\sec u$ **25.** $1 - \sin x$ **27.** $2\sec^2\alpha$

29. (a) $\text{LHS} = \dfrac{1 - \sin^2 x}{\sin x} = \text{RHS}$

31. $\text{LHS} = \sin\theta\dfrac{\cos\theta}{\sin\theta} = \text{RHS}$

33. $\text{LHS} = \cos u\dfrac{1}{\cos u}\cot u = \text{RHS}$

35. $\text{LHS} = \dfrac{\frac{\sin y}{\cos y}}{\frac{1}{\sin y}} = \dfrac{\sin^2 y}{\cos y} = \dfrac{1 - \cos^2 y}{\cos y} = \dfrac{1}{\cos y} - \cos y = \text{RHS}$

37. $\text{LHS} = \cos x - (-\sin x) = \text{RHS}$

39. $\text{LHS} = \dfrac{\sin\theta}{\cos\theta} + \dfrac{\cos\theta}{\sin\theta} = \dfrac{\sin^2\theta + \cos^2\theta}{\cos\theta\sin\theta}$
$= \dfrac{1}{\cos\theta\sin\theta} = \text{RHS}$

41. $\text{LHS} = 1 - \cos^2\beta = \sin^2\beta = \text{RHS}$

43. $\text{LHS} = \dfrac{1}{\cos^2 y} = \sec^2 y = \text{RHS}$

45. $\text{LHS} = \tan^2 x + 2\tan x\cot x + \cot^2 x = \tan^2 x + 2 + \cot^2 x$
$= (\tan^2 x + 1) + (\cot^2 x + 1) = \text{RHS}$

47. $\text{LHS} = (2\cos^2 t)^2 + 4\sin^2 t\cos^2 t$
$= 4\cos^2 t(\cos^2 t + \sin^2 t) = \text{RHS}$

49. $\text{LHS} = \dfrac{\cos^2 x}{\sin x} + \dfrac{\sin^2 x}{\sin x} = \dfrac{1}{\sin x} = \text{RHS}$

51. $\text{LHS} = \dfrac{(\sin x + \cos x)^2}{(\sin x + \cos x)(\sin x - \cos x)} = \dfrac{\sin x + \cos x}{\sin x - \cos x}$
$= \dfrac{(\sin x + \cos x)(\sin x - \cos x)}{(\sin x - \cos x)(\sin x - \cos x)} = \text{RHS}$

53. $\text{LHS} = \dfrac{\frac{1}{\cos t} - \cos t}{\frac{1}{\cos t}} \cdot \dfrac{\cos t}{\cos t} = \dfrac{1 - \cos^2 t}{1} = \text{RHS}$

55. $\text{LHS} = \cos^2 x - (1 - \cos^2 x) = 2\cos^2 x - 1 = \text{RHS}$

57. $\text{LHS} = (\sin^2\theta)^2 - (\cos^2\theta)^2$
$= (\sin^2\theta - \cos^2\theta)(\sin^2\theta + \cos^2\theta) = \text{RHS}$

59. $\text{LHS} = \dfrac{\sin^2 t + 2\sin t\cos t + \cos^2 t}{\sin t\cos t}$
$= \dfrac{\sin^2 t + \cos^2 t}{\sin t\cos t} + \dfrac{2\sin t\cos t}{\sin t\cos t} = \dfrac{1}{\sin t\cos t} + 2$
$= \text{RHS}$

61. $\text{LHS} = \dfrac{1 + \frac{\sin^2 u}{\cos^2 u}}{1 - \frac{\sin^2 u}{\cos^2 u}} \cdot \dfrac{\cos^2 u}{\cos^2 u} = \dfrac{\cos^2 u + \sin^2 u}{\cos^2 u - \sin^2 u} = \text{RHS}$

63. $\text{LHS} = \dfrac{\frac{1}{\cos x} + \frac{1}{\sin x}}{\frac{\sin x}{\cos x} + \frac{\cos x}{\sin x}} \cdot \dfrac{\sin x\cos x}{\sin x\cos x} = \dfrac{\sin x + \cos x}{\sin^2 x + \cos^2 x} = \text{RHS}$

65. $\text{LHS} = \dfrac{1 - \cos x}{\sin x} \cdot \dfrac{1 - \cos x}{1 - \cos x} + \dfrac{\sin x}{1 - \cos x} \cdot \dfrac{\sin x}{\sin x}$
$= \dfrac{1 - 2\cos x + \cos^2 x + \sin^2 x}{\sin x(1 - \cos x)} = \dfrac{2 - 2\cos x}{\sin x(1 - \cos x)}$
$= \dfrac{2(1 - \cos x)}{\sin x(1 - \cos x)} = \text{RHS}$

67. $\text{LHS} = \dfrac{\sin^2 u}{\cos^2 u} - \dfrac{\sin^2 u\cos^2 u}{\cos^2 u} = \dfrac{\sin^2 u}{\cos^2 u}(1 - \cos^2 u) = \text{RHS}$

69. $\text{LHS} = \dfrac{1 + \frac{\sin x}{\cos x}}{1 - \frac{\sin x}{\cos x}} \cdot \dfrac{\cos x}{\cos x} = \dfrac{\cos x + \sin x}{\cos x - \sin x} = \text{RHS}$

71. $\text{LHS} = \dfrac{\sec x - \tan x + \sec x + \tan x}{(\sec x + \tan x)(\sec x - \tan x)}$
$= \dfrac{2\sec x}{\sec^2 x - \tan^2 x} = \text{RHS}$

73. $\text{LHS} = \dfrac{(1+\sin x)^2 - (1-\sin x)^2}{(1-\sin x)(1+\sin x)}$

$= \dfrac{1 + 2\sin x + \sin^2 x - 1 + 2\sin x - \sin^2 x}{1 - \sin^2 x}$

$= \dfrac{4\sin x}{\cos^2 x} = 4\dfrac{\sin x}{\cos x}\cdot\dfrac{1}{\cos x} = \text{RHS}$

75. $\text{LHS} = \dfrac{(\sin x + \cos x)(\sin^2 x - \sin x\cos x + \cos^2 x)}{\sin x + \cos x}$

$= \sin^2 x - \sin x \cos x + \cos^2 x = \text{RHS}$

77. $\text{LHS} = \dfrac{1-\cos\alpha}{\sin\alpha}\cdot\dfrac{1+\cos\alpha}{1+\cos\alpha}$

$= \dfrac{1-\cos^2\alpha}{\sin\alpha(1+\cos\alpha)} = \dfrac{\sin^2\alpha}{\sin\alpha(1+\cos\alpha)} = \text{RHS}$

79. $\text{LHS} = \dfrac{\sin w}{\sin w + \cos w}\cdot\dfrac{\frac{1}{\cos w}}{\frac{1}{\cos w}} = \dfrac{\frac{\sin w}{\cos w}}{\frac{\sin w}{\cos w} + \frac{\cos w}{\cos w}} = \text{RHS}$

81. $\text{LHS} = \dfrac{\sec x}{\sec x - \tan x}\cdot\dfrac{\sec x + \tan x}{\sec x + \tan x}$

$= \dfrac{\sec x(\sec x + \tan x)}{\sec^2 x - \tan^2 x} = \text{RHS}$

83. $\text{LHS} = \dfrac{\cos\theta}{1-\sin\theta}\cdot\dfrac{1+\sin\theta}{1+\sin\theta} = \dfrac{\cos\theta(1+\sin\theta)}{1-\sin^2\theta}$

$= \dfrac{\cos\theta(1+\sin\theta)}{\cos^2\theta} = \text{RHS}$

85. $\text{LHS} = \dfrac{1-\sin x}{1+\sin x}\cdot\dfrac{1-\sin x}{1-\sin x} = \dfrac{1 - 2\sin x + \sin^2 x}{1-\sin^2 x}$

$= \dfrac{1}{\cos^2 x} - \dfrac{2\sin x}{\cos^2 x} + \dfrac{\sin^2 x}{\cos^2 x}$

$= \sec^2 x - 2\sec x\tan x + \tan^2 x$

$= (\sec x - \tan x)^2 = \text{RHS}$

87. $\text{LHS} = \dfrac{1}{\sin x} - \dfrac{\cos x}{\sin x} = \dfrac{(1-\cos x)(1+\cos x)}{\sin x(1+\cos x)}$

$= \dfrac{\sin^2 x}{\sin x(1+\cos x)} = \dfrac{1}{\frac{1}{\sin x} + \frac{\cos x}{\sin x}} = \text{RHS}$

89. $\tan\theta$ **91.** $\tan\theta$ **93.** $3\cos\theta$

95.

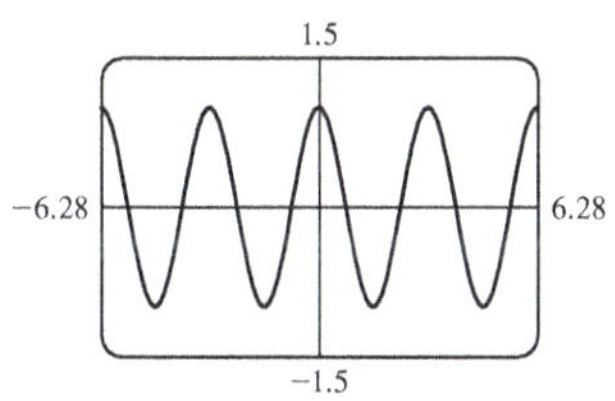

Yes

97.

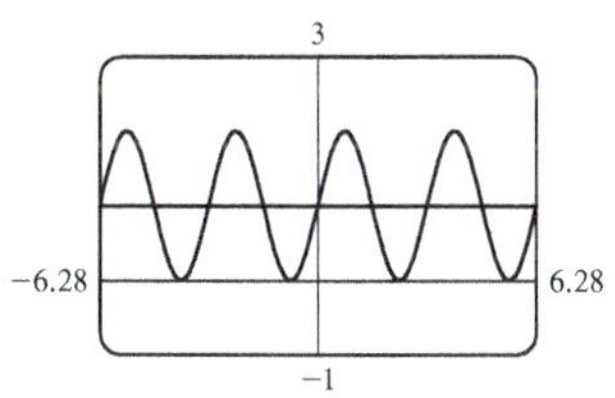

No

99. $\text{LHS} = \sin^2 x \sin^2 y - \cos^2 x\cos^2 y$

$= (1-\cos^2 x)\sin^2 y - \cos^2 x(1-\sin^2 y) = \text{RHS}$

101. $\text{LHS} = \left(\dfrac{\sin x}{\cos x} + \dfrac{\cos x}{\sin x}\right)^4 = \left(\dfrac{\sin^2 x + \cos^2 x}{\sin x\cos x}\right)^4$

$= \left(\dfrac{1}{\sin x\cos x}\right)^4 = \text{RHS}$

103. $\text{LHS} = \dfrac{(\sin y - \csc y)(\sin^2 y + \sin y\csc y + \csc^2 y)}{\sin y - \csc y}$

$= \text{RHS}$

105. $\text{LHS} = \ln|\tan x| + \ln|\sin x| = \ln\left|\dfrac{\sin x}{\cos x}\right| + \ln|\sin x|$

$= \ln|\sin x| + \ln\left|\dfrac{1}{\cos x}\right| + \ln|\sin x| = \text{RHS}$

107. $\text{LHS} = e^{1-\cos^2 x}e^{\sec^2 x - 1} = e^{1-\cos^2 x + \sec^2 x - 1} = \text{RHS}$

109. Yes **111.** $x = k\pi$, k an integer

SECTION 7.2 ■ PAGE 551

1. addition; $\sin x\cos y + \cos x\sin y$

2. subtraction; $\cos x\cos y + \sin x\sin y$

3. $\dfrac{\sqrt{6}+\sqrt{2}}{4}$ **5.** $\dfrac{\sqrt{2}-\sqrt{6}}{4}$ **7.** $2-\sqrt{3}$ **9.** $-\dfrac{\sqrt{6}+\sqrt{2}}{4}$

11. $\sqrt{3}-2$ **13.** $-\dfrac{\sqrt{6}+\sqrt{2}}{4}$ **15.** $\sqrt{2}/2$ **17.** $\frac{1}{2}$ **19.** $\sqrt{3}$

21. $\text{LHS} = \dfrac{\sin(\frac{\pi}{2}-u)}{\cos(\frac{\pi}{2}-u)} = \dfrac{\sin\frac{\pi}{2}\cos u - \cos\frac{\pi}{2}\sin u}{\cos\frac{\pi}{2}\cos u + \sin\frac{\pi}{2}\sin u}$

$= \dfrac{\cos u}{\sin u} = \text{RHS}$

23. $\text{LHS} = \dfrac{1}{\cos(\frac{\pi}{2}-u)} = \dfrac{1}{\cos\frac{\pi}{2}\cos u + \sin\frac{\pi}{2}\sin u}$

$= \dfrac{1}{\sin u} = \text{RHS}$

25. $\text{LHS} = \sin x\cos\frac{\pi}{2} - \cos x\sin\frac{\pi}{2} = \text{RHS}$

27. $\text{LHS} = \sin x\cos\pi - \cos x\sin\pi = \text{RHS}$

29. $\text{LHS} = \dfrac{\tan x - \tan\pi}{1 + \tan x\tan\pi} = \text{RHS}$

31. $\text{LHS} = \sin\left(\dfrac{\pi}{2} - x\right) = \sin\dfrac{\pi}{2}\cos x - \cos\dfrac{\pi}{2}\sin x = \cos x$

$\text{RHS} = \sin\left(\dfrac{\pi}{2} + x\right) = \sin\dfrac{\pi}{2}\cos x + \cos\dfrac{\pi}{2}\sin x = \cos x$

33. $\text{LHS} = \dfrac{\tan x + \tan\frac{\pi}{3}}{1 - \tan x\tan\frac{\pi}{3}} = \text{RHS}$

35. $\text{LHS} = \sin x\cos y + \cos x\sin y$

$-(\sin x\cos y - \cos x\sin y) = \text{RHS}$

37. $\text{LHS} = \dfrac{1}{\tan(x-y)} = \dfrac{1 + \tan x\tan y}{\tan x - \tan y}$

$= \dfrac{1 + \frac{1}{\cot x}\frac{1}{\cot y}}{\frac{1}{\cot x} - \frac{1}{\cot y}}\cdot\dfrac{\cot x\cot y}{\cot x\cot y} = \text{RHS}$

39. $\text{LHS} = \dfrac{\sin x}{\cos x} - \dfrac{\sin y}{\cos y} = \dfrac{\sin x\cos y - \cos x\sin y}{\cos x\cos y} = \text{RHS}$

41. $\text{LHS} = \dfrac{(\tan x - \tan y)(\cos x\cos y)}{(1 - \tan x\tan y)(\cos x\cos y)}$

$= \dfrac{\sin x\cos y - \cos x\sin y}{\cos x\cos y - \sin x\sin y} = \text{RHS}$

43. $\text{LHS} = (\cos x \cos y - \sin x \sin y)(\cos x \cos y + \sin x \sin y)$
$= \cos^2 x \cos^2 y - \sin^2 x \sin^2 y$
$= \cos^2 x(1 - \sin^2 y) - (1 - \cos^2 x)\sin^2 y$
$= \cos^2 x - \sin^2 y \cos^2 x + \sin^2 y \cos^2 x - \sin^2 y = \text{RHS}$

45. $\text{LHS} = \sin((x + y) + z)$
$= \sin(x + y)\cos z + \cos(x + y)\sin z$
$= \cos z[\sin x \cos y + \cos x \sin y]$
$+ \sin z[\cos x \cos y - \sin x \sin y] = \text{RHS}$

47. $\dfrac{\sqrt{1 - x^2} + xy}{\sqrt{1 + y^2}}$ **49.** $\dfrac{x - y}{\sqrt{1 + x^2}\sqrt{1 + y^2}}$

51. $\frac{1}{4}(\sqrt{6} + \sqrt{2})$ **53.** $\dfrac{3 - 2\sqrt{14}}{\sqrt{7} + 6\sqrt{2}}$ **55.** $-\frac{1}{10}(3 + 4\sqrt{3})$

57. $2\sqrt{5}/65$ **59.** $2\sin\left(x + \dfrac{5\pi}{6}\right)$ **61.** $5\sqrt{2}\sin\left(2x + \dfrac{7\pi}{4}\right)$

63. (a) $g(x) = 2\sin 2\left(x + \dfrac{\pi}{12}\right)$

(b)

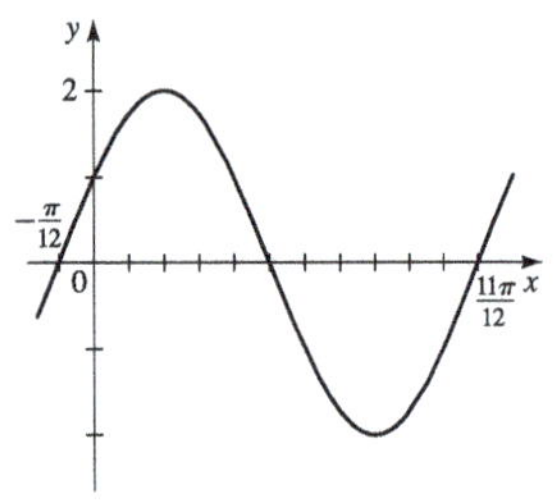

67. (a)

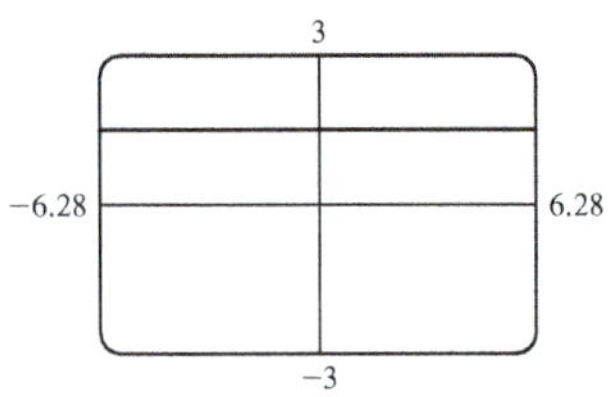

$$\sin^2\left(x + \frac{\pi}{4}\right) + \sin^2\left(x - \frac{\pi}{4}\right) = 1$$

71. $\text{LHS} = \tan^{-1}\left(\dfrac{\tan u + \tan v}{1 - \tan u \tan v}\right) = \tan^{-1}(\tan(u + v))$
$= u + v = \text{RHS}$

73. (c) $8.1°$

75. (a)

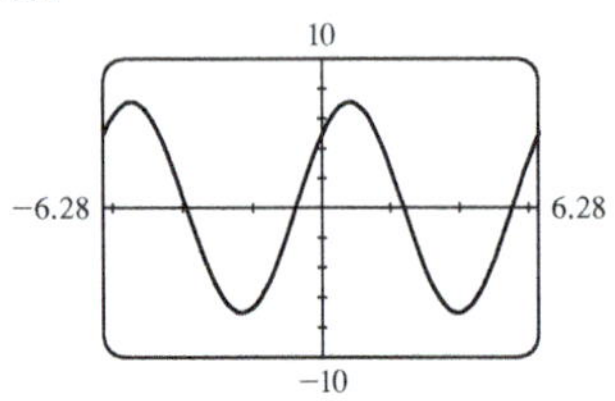

(b) $k = 5\sqrt{2}, \phi = \pi/4$

SECTION 7.3 ■ PAGE 560

1. Double-Angle; $2\sin x \cos x$

2. Half-Angle; $\pm\sqrt{(1 - \cos x)/2}$

3. $\frac{120}{169}, \frac{119}{169}, \frac{120}{119}$ **5.** $-\frac{24}{25}, \frac{7}{25}, -\frac{24}{7}$ **7.** $\frac{24}{25}, \frac{7}{25}, \frac{24}{7}$

9. $-\frac{3}{5}, \frac{4}{5}, -\frac{3}{4}$ **11.** $\frac{1}{2}(\frac{3}{4} - \cos 2x + \frac{1}{4}\cos 4x)$

13. $\frac{1}{16}(1 - \cos 2x - \cos 4x + \cos 2x \cos 4x)$

15. $\frac{1}{32}(\frac{3}{4} - \cos 4x + \frac{1}{4}\cos 8x)$

17. $\frac{1}{2}\sqrt{2 - \sqrt{3}}$ **19.** $\sqrt{2} - 1$ **21.** $-\frac{1}{2}\sqrt{2 + \sqrt{3}}$

23. $\sqrt{2} - 1$ **25.** $\frac{1}{2}\sqrt{2 + \sqrt{3}}$ **27.** $-\frac{1}{2}\sqrt{2 - \sqrt{2}}$

29. (a) $\sin 36°$ **(b)** $\sin 6\theta$ **31. (a)** $\cos 68°$ **(b)** $\cos 10\theta$

33. (a) $\tan 4°$ **(b)** $\tan 2\theta$ **37.** $\sqrt{10}/10, 3\sqrt{10}/10, \frac{1}{3}$

39. $\sqrt{(3 + 2\sqrt{2})/6}, \sqrt{(3 - 2\sqrt{2})/6}, 3 + 2\sqrt{2}$

41. $\sqrt{6}/6, -\sqrt{30}/6, -\sqrt{5}/5$ **43.** $\dfrac{2x}{1 + x^2}$ **45.** $\sqrt{\dfrac{1 - x}{2}}$

47. $\frac{336}{625}$ **49.** $\frac{8}{7}$ **51.** $\frac{7}{25}$ **53.** $-8\sqrt{3}/49$ **55.** $\frac{1}{2}(\sin 5x - \sin x)$

57. $\frac{1}{2}(\sin 5x + \sin 3x)$ **59.** $\frac{3}{2}(\cos 11x + \cos 3x)$

61. $2\sin 4x \cos x$ **63.** $2\sin 5x \sin x$ **65.** $-2\cos\frac{9}{2}x \sin\frac{5}{2}x$

67. $(\sqrt{2} + \sqrt{3})/2$ **69.** $\frac{1}{4}(\sqrt{2} - 1)$ **71.** $\sqrt{2}/2$

73. $\text{LHS} = \cos(2 \cdot 5x) = \text{RHS}$

75. $\text{LHS} = \sin^2 x + 2\sin x \cos x + \cos^2 x$
$= 1 + 2\sin x \cos x = \text{RHS}$

77. $\text{LHS} = \dfrac{2\tan x}{\sec^2 x} = 2 \cdot \dfrac{\sin x}{\cos x}\cos^2 x = 2\sin x \cos x = \text{RHS}$

79. $\text{LHS} = \dfrac{1 - \cos x}{\sin x} + \cos x\left(\dfrac{1 - \cos x}{\sin x}\right)$
$= \dfrac{1 - \cos x + \cos x - \cos^2 x}{\sin x} = \dfrac{\sin^2 x}{\sin x} = \text{RHS}$

81. $\text{LHS} = \dfrac{2\sin 2x \cos 2x}{\sin x} = \dfrac{2(2\sin x \cos x)(\cos 2x)}{\sin x} = \text{RHS}$

83. $\text{LHS} = \dfrac{2(\tan x - \cot x)}{(\tan x + \cot x)(\tan x - \cot x)} = \dfrac{2}{\tan x + \cot x}$
$= \dfrac{2}{\frac{\sin x}{\cos x} + \frac{\cos x}{\sin x}} \cdot \dfrac{\sin x \cos x}{\sin x \cos x} = \dfrac{2\sin x \cos x}{\sin^2 x + \cos^2 x}$
$= 2\sin x \cos x = \text{RHS}$

85. $\text{LHS} = \dfrac{1}{\tan 2x} = \dfrac{1}{\frac{2\tan x}{1 - \tan^2 x}} = \text{RHS}$

87. $\text{LHS} = \tan(2x + x) = \dfrac{\tan 2x + \tan x}{1 - \tan 2x \tan x}$
$= \dfrac{\frac{2\tan x}{1 - \tan^2 x} + \tan x}{1 - \frac{2\tan x}{1 - \tan^2 x}\tan x}$
$= \dfrac{2\tan x + \tan x(1 - \tan^2 x)}{1 - \tan^2 x - 2\tan x \tan x} = \text{RHS}$

89. $\text{LHS} = \dfrac{2\sin 3x \cos 2x}{2\cos 3x \cos 2x} = \dfrac{\sin 3x}{\cos 3x} = \text{RHS}$

91. $\text{LHS} = \dfrac{2\sin 5x \cos 5x}{2\sin 5x \cos 4x} = \text{RHS}$

93. $\text{LHS} = \dfrac{2\sin(\frac{x + y}{2})\cos(\frac{x - y}{2})}{2\cos(\frac{x + y}{2})\cos(\frac{x - y}{2})}$
$= \dfrac{\sin(\frac{x + y}{2})}{\cos(\frac{x + y}{2})} = \text{RHS}$

95. $\text{LHS} = \dfrac{1 - \cos 2(\frac{x}{2} + \frac{\pi}{4})}{1 + \cos 2(\frac{x}{2} + \frac{\pi}{4})} = \dfrac{1 - \cos(x + \frac{\pi}{2})}{1 + \cos(x + \frac{\pi}{2})}$
$= \dfrac{1 - (-\sin x)}{1 + (-\sin x)} = \text{RHS}$

101. $\text{LHS} = \dfrac{(\sin x + \sin 5x) + (\sin 2x + \sin 4x) + \sin 3x}{(\cos x + \cos 5x) + (\cos 2x + \cos 4x) + \cos 3x}$
$= \dfrac{2\sin 3x \cos 2x + 2\sin 3x \cos x + \sin 3x}{2\cos 3x \cos 2x + 2\cos 3x \cos x + \cos 3x}$
$= \dfrac{\sin 3x(2\cos 2x + 2\cos x + 1)}{\cos 3x(2\cos 2x + 2\cos x + 1)} = \text{RHS}$

103. $\text{RHS} = \cos^{-1}(1 - 2\sin^2 u) = \cos^{-1}(\cos 2u) = 2u = \text{LHS}$

105. **(a)**

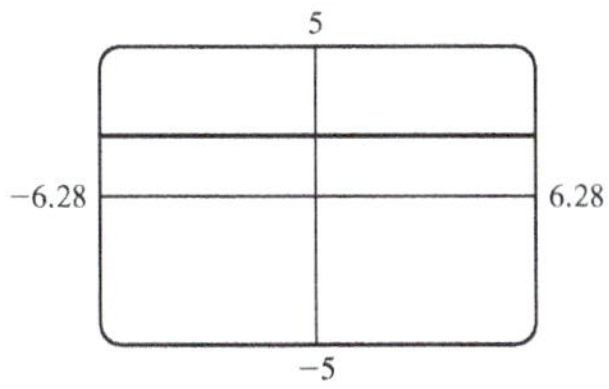

$\dfrac{\sin 3x}{\sin x} - \dfrac{\cos 3x}{\cos x} = 2$

107. **(a)**

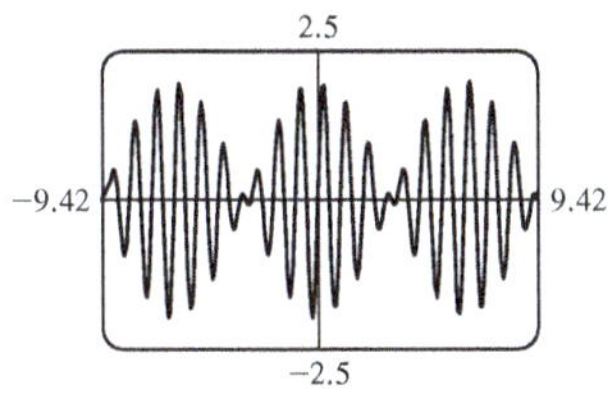

(c)

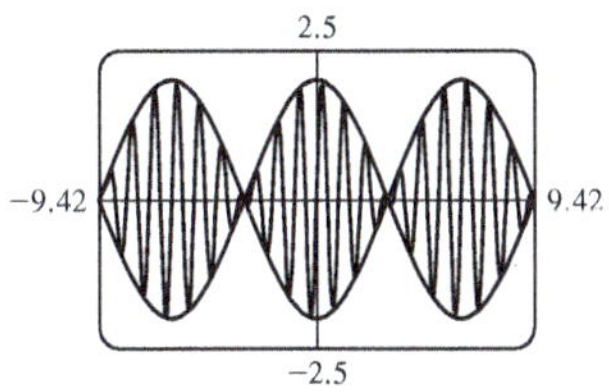

The graph of $y = f(x)$ lies between the two other graphs.

109. **(a)** $P(t) = 8t^4 - 8t^2 + 1$ **(b)** $Q(t) = 16t^5 - 20t^3 + 5t$

115. **(a)** and **(c)**

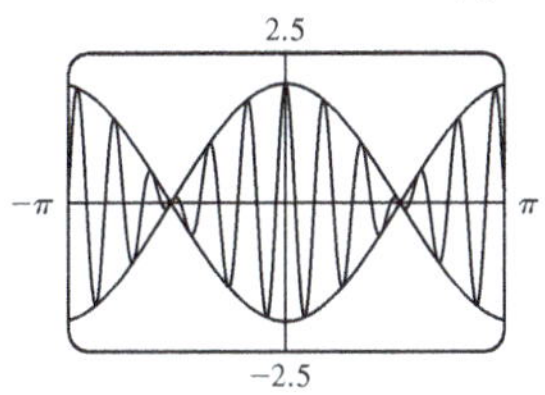

The graph of f lies between the graphs of $y = 2\cos t$ and $y = -2\cos t$. Thus, the loudness of the sound varies between $y = \pm 2\cos t$.

SECTION 7.4 ■ PAGE 568

1. infinitely many **2.** no, infinitely many
3. 0.3; $x \approx -9.7, -6.0, -3.4, 0.3, 2.8, 6.6, 9.1$
4. **(a)** 0.30, 2.84 **(b)** 2π, $0.30 + 2k\pi$, $2.84 + 2k\pi$
5. $\dfrac{\pi}{3} + 2k\pi, \dfrac{2\pi}{3} + 2k\pi$
7. $(2k+1)\pi$ **9.** $1.32 + 2k\pi, 4.97 + 2k\pi$
11. $3.61 + 2k\pi, 5.82 + 2k\pi$ **13.** $-\dfrac{\pi}{3} + k\pi$
15. $1.37 + k\pi$ **17.** $\dfrac{5\pi}{6} + 2k\pi, \dfrac{7\pi}{6} + 2k\pi$; $-7\pi/6, -5\pi/6, 5\pi/6, 7\pi/6, 17\pi/6, 19\pi/6$
19. $\dfrac{\pi}{4} + 2k\pi, \dfrac{3\pi}{4} + 2k\pi$; $-7\pi/4, -5\pi/4, \pi/4, 3\pi/4, 9\pi/4, 11\pi/4$
21. $1.29 + 2k\pi, 5.00 + 2k\pi$; $-5.00, -1.29, 1.29, 5.00, 7.57, 11.28$
23. $-1.47 + k\pi$; $-7.75, -4.61, -1.47, 1.67, 4.81, 7.95$
25. $(2k+1)\pi$ **27.** $\dfrac{5\pi}{4} + 2k\pi, \dfrac{7\pi}{4} + 2k\pi$
29. $0.20 + 2k\pi, 2.94 + 2k\pi$ **31.** $-\dfrac{\pi}{6} + k\pi, \dfrac{\pi}{6} + k\pi$
33. $\dfrac{\pi}{4} + k\pi, \dfrac{3\pi}{4} + k\pi$ **35.** $-1.11 + k\pi, 1.11 + k\pi$
37. $\dfrac{\pi}{4} + k\pi, \dfrac{3\pi}{4} + k\pi$
39. $-1.11 + k\pi, 1.11 + k\pi, \dfrac{2\pi}{3} + 2k\pi, \dfrac{4\pi}{3} + 2k\pi$
41. $\dfrac{\pi}{3} + 2k\pi, \dfrac{5\pi}{3} + 2k\pi$ **43.** $0.34 + 2k\pi, 2.80 + 2k\pi$
45. $\dfrac{\pi}{3} + 2k\pi, \dfrac{5\pi}{3} + 2k\pi$ **47.** No solution **49.** $\dfrac{3\pi}{2} + 2k\pi$
51. $\dfrac{\pi}{2} + k\pi, \dfrac{7\pi}{6} + 2k\pi, \dfrac{11\pi}{6} + 2k\pi$ **53.** $\dfrac{\pi}{2} + k\pi$
55. $k\pi, 0.73 + 2k\pi, 2.41 + 2k\pi$ **57.** 44.95°
59. **(a)** 0° **(b)** 60°, 120° **(c)** 90°, 270° **(d)** 180°

SECTION 7.5 ■ PAGE 574

1. $\sin x = 0, k\pi$ **2.** $\sin x + 2\sin x\cos x = 0$, $\sin x = 0$, $1 + 2\cos x = 0$ **3.** $\dfrac{7\pi}{6} + 2k\pi, \dfrac{11\pi}{6} + 2k\pi, \dfrac{\pi}{2} + 2k\pi$
5. $(2k+1)\pi, 1.23 + 2k\pi, 5.05 + 2k\pi$
7. $k\pi, 0.72 + 2k\pi, 5.56 + 2k\pi$ **9.** $\dfrac{\pi}{6} + 2k\pi, \dfrac{5\pi}{6} + 2k\pi$
11. $\dfrac{\pi}{3} + 2k\pi, \dfrac{5\pi}{3} + 2k\pi, (2k+1)\pi$
13. $(2k+1)\pi, \dfrac{\pi}{2} + 2k\pi$ **15.** $2k\pi$
17. **(a)** $\dfrac{\pi}{9} + \dfrac{2k\pi}{3}, \dfrac{5\pi}{9} + \dfrac{2k\pi}{3}$ **(b)** $\pi/9, 5\pi/9, 7\pi/9, 11\pi/9, 13\pi/9, 17\pi/9$
19. **(a)** $\dfrac{\pi}{3} + k\pi, \dfrac{2\pi}{3} + k\pi$ **(b)** $\pi/3, 2\pi/3, 4\pi/3, 5\pi/3$
21. **(a)** $\dfrac{5\pi}{18} + \dfrac{k\pi}{3}$ **(b)** $5\pi/18, 11\pi/18, 17\pi/18, 23\pi/18, 29\pi/18, 35\pi/18$
23. **(a)** $4k\pi$ **(b)** 0
25. **(a)** $4\pi + 6k\pi, 5\pi + 6k\pi$ **(b)** None
27. **(a)** $0.62 + \dfrac{k\pi}{2}$ **(b)** 0.62, 2.19, 3.76, 5.33
29. **(a)** $k\pi, \dfrac{\pi}{2} + 2k\pi$ **(b)** $0, \pi/2, \pi$

31. **(a)** $\frac{\pi}{6} + k\pi, \frac{\pi}{4} + k\pi, \frac{5\pi}{6} + k\pi$
(b) $\pi/6, \pi/4, 5\pi/6, 7\pi/6, 5\pi/4, 11\pi/6$

33. **(a)** $\frac{\pi}{6} + 2k\pi, \frac{5\pi}{6} + 2k\pi, \frac{3\pi}{4} + k\pi$
(b) $\pi/6, 3\pi/4, 5\pi/6, 7\pi/4$

35. **(a)**

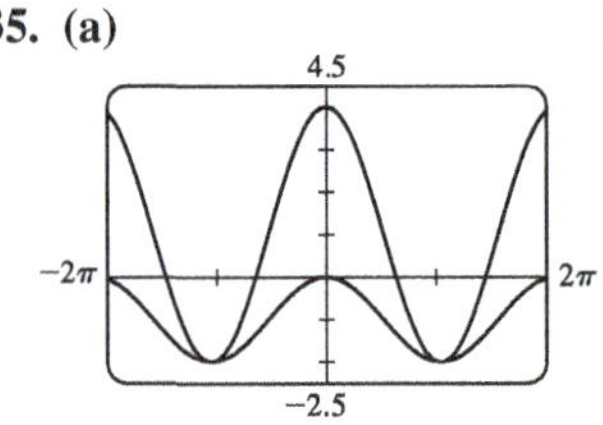

$(\pm 3.14, -2)$

(b) $((2k + 1)\pi, -2)$

37. **(a)**

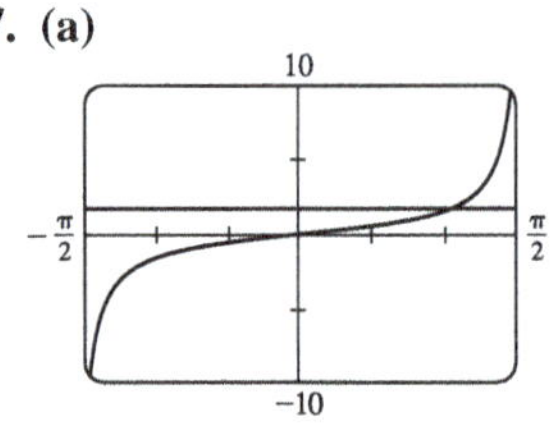

$(1.04, 1.73)$

(b) $\left(\frac{\pi}{3} + k\pi, \sqrt{3}\right)$

39. $\pi/8, 3\pi/8, 5\pi/8, 7\pi/8, 9\pi/8, 11\pi/8, 13\pi/8, 15\pi/8$
41. $\pi/3, 2\pi/3$ **43.** $\pi/2, 7\pi/6, 3\pi/2, 11\pi/6$ **45.** 0
47. $0, \pi$ **49.** $0, \pi/3, 2\pi/3, \pi, 4\pi/3, 5\pi/3$ **51.** $\pi/6, 3\pi/2$

53. $k\pi/2$ **55.** $\frac{\pi}{2} + k\pi, \frac{\pi}{9} + \frac{2k\pi}{3}, \frac{5\pi}{9} + \frac{2k\pi}{3}$

57. $0, \pm 0.95$ **59.** 1.92 **61.** ± 0.71

63. $\frac{\sqrt{17} - 3}{4}$ **65.** 0.95° or 89.1°

67. **(a)** 34th day (February 3), 308th day (November 4)
(b) 275 days

CHAPTER 7 REVIEW ■ PAGE 578

1. $\text{LHS} = \sin\theta\left(\frac{\cos\theta}{\sin\theta} + \frac{\sin\theta}{\cos\theta}\right) = \cos\theta + \frac{\sin^2\theta}{\cos\theta}$
$= \frac{\cos^2\theta + \sin^2\theta}{\cos\theta} = \text{RHS}$

3. $\text{LHS} = (1 - \sin^2 x)\csc x - \csc x$
$= \csc x - \sin^2 x \csc x - \csc x$
$= -\sin^2 x \frac{1}{\sin x} = \text{RHS}$

5. $\text{LHS} = \frac{\cos^2 x}{\sin^2 x} - \frac{\tan^2 x}{\sin^2 x} = \cot^2 x - \frac{1}{\cos^2 x} = \text{RHS}$

7. $\text{LHS} = \frac{\cos x}{\frac{1}{\cos x}(1 - \sin x)} = \frac{\cos x}{\frac{1}{\cos x} - \frac{\sin x}{\cos x}} = \text{RHS}$

9. $\text{LHS} = \sin^2 x \frac{\cos^2 x}{\sin^2 x} + \cos^2 x \frac{\sin^2 x}{\cos^2 x} = \cos^2 x + \sin^2 x = \text{RHS}$

11. $\text{LHS} = \frac{2\sin x \cos x}{1 + 2\cos^2 x - 1} = \frac{2\sin x\cos x}{2\cos^2 x} = \frac{2\sin x}{2\cos x} = \text{RHS}$

13. $\text{LHS} = \csc x - \frac{1 - \cos x}{\sin x}$
$= \csc x - (\csc x - \cot x) = \text{RHS}$

15. $\text{LHS} = \frac{2\sin x\cos x}{\sin x} - \frac{2\cos^2 x - 1}{\cos x}$
$= 2\cos x - 2\cos x + \frac{1}{\cos x} = \text{RHS}$

17. $\text{LHS} = \frac{\frac{1}{\cos x} - 1}{\sin x \frac{1}{\cos x}} = \left(\frac{1}{\cos x} - 1\right)\frac{\cos x}{\sin x}$
$= \frac{1 - \cos x}{\sin x} = \text{RHS}$

19. $\text{LHS} = \cos^2\frac{x}{2} - 2\sin\frac{x}{2}\cos\frac{x}{2} + \sin^2\frac{x}{2}$
$= 1 - \sin\left(2\cdot\frac{x}{2}\right) = \text{RHS}$

21. $\text{LHS} = \frac{2\sin\left(\frac{(x+y) + (x-y)}{2}\right)\cos\left(\frac{(x+y) - (x-y)}{2}\right)}{2\cos\left(\frac{(x+y) + (x-y)}{2}\right)\cos\left(\frac{(x+y) - (x-y)}{2}\right)}$
$= \frac{2\sin x\cos y}{2\cos x\cos y} = \text{RHS}$

23. **(a)**

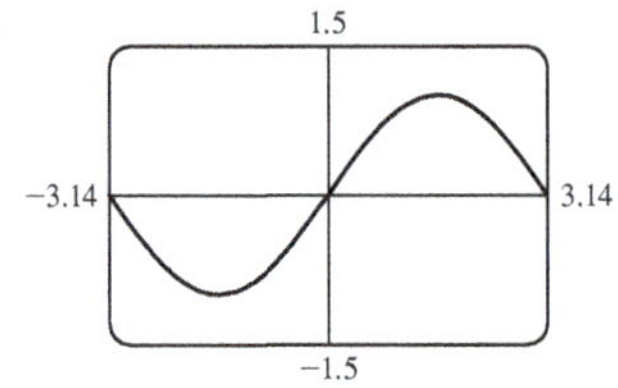

(b) Yes

25. **(a)**

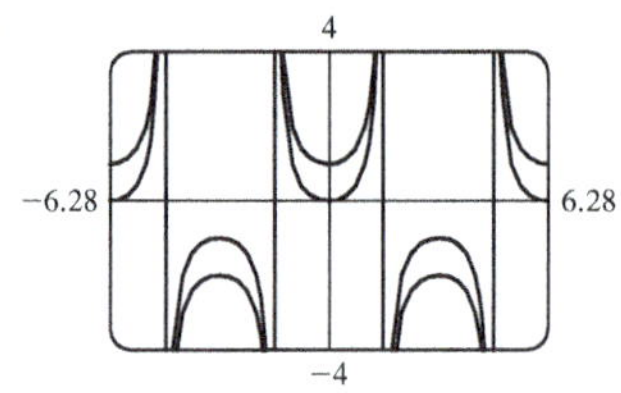

(b) No

27. **(a)**

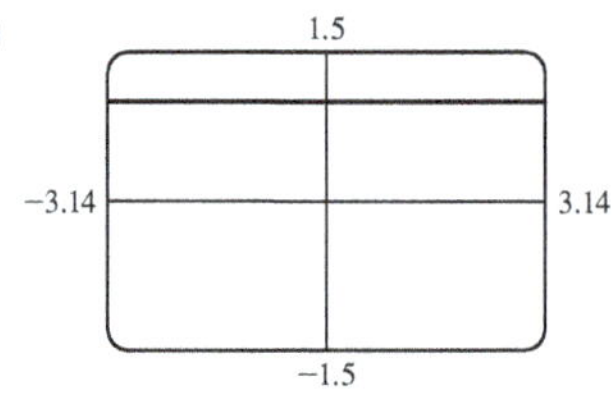

$2\sin^2 3x + \cos 6x = 1$

29. 0.85, 2.29 **31.** $0, \pi$ **33.** $\pi/6, 5\pi/6$ **35.** $\pi/3, 5\pi/3$
37. $2\pi/3, 4\pi/3$ **39.** $\pi/3, 2\pi/3, 3\pi/4, 4\pi/3, 5\pi/3, 7\pi/4$
41. $\pi/6, \pi/2, 5\pi/6, 7\pi/6, 3\pi/2, 11\pi/6$ **43.** $\pi/6$
45. 1.18 **47.** **(a)** 63.4° **(b)** No **(c)** 90°

49. $\frac{\sqrt{2} + \sqrt{6}}{4}$ or $\frac{1}{2}\sqrt{2 + \sqrt{3}}$ **51.** $\sqrt{2} - 1$ **53.** $\sqrt{2}/2$

55. $\sqrt{2}/2$ **57.** $\frac{\sqrt{2} + \sqrt{3}}{4}$ **59.** $\frac{2}{9}(\sqrt{10} + 1)$

61. $\frac{2}{3}(\sqrt{2} + \sqrt{5})$ **63.** $\sqrt{(3 + 2\sqrt{2})/6}$ **65.** $-\frac{12\sqrt{10}}{31}$

67. $\frac{2x}{1 - x^2}$ **69.** **(a)** $\theta = \tan^{-1}\left(\frac{10}{x}\right)$ **(b)** 286.4 ft

CHAPTER 7 TEST ■ PAGE 580

1. $\text{LHS} = \frac{\sin\theta}{\cos\theta}\sin\theta + \cos\theta = \frac{\sin^2\theta + \cos^2\theta}{\cos\theta} = \text{RHS}$

2. $\text{LHS} = \frac{\tan x}{1 - \cos x}\cdot\frac{1 + \cos x}{1 + \cos x} = \frac{\tan x(1 + \cos x)}{1 - \cos^2 x}$
$= \frac{\frac{\sin x}{\cos x}(1 + \cos x)}{\sin^2 x} = \frac{1}{\sin x}\cdot\frac{1 + \cos x}{\cos x} = \text{RHS}$

3. $\text{LHS} = \frac{2\tan x}{\sec^2 x} = \frac{2\sin x}{\cos x}\cdot\cos^2 x = 2\sin x\cos x = \text{RHS}$

4. $\text{LHS} = \sin x \tan\left(\dfrac{x}{2}\right) = \sin x\left(\dfrac{1-\cos x}{\sin x}\right) = \text{RHS}$

5. $\text{LHS} = 2\left(\dfrac{1-\cos 6x}{2}\right) = \text{RHS}$

6. $\text{LHS} = 1 - 2\sin^2 2x = 1 - 2(2\sin x\cos x)^2 = 1 - 8\sin^2 x(1-\sin^2 x) = \text{RHS}$

7. $\text{LHS} = \sin^2\left(\dfrac{x}{2}\right) + 2\sin\left(\dfrac{x}{2}\right)\cos\left(\dfrac{x}{2}\right) + \cos^2\left(\dfrac{x}{2}\right) = 1 + \sin 2\left(\dfrac{x}{2}\right) = \text{RHS}$

8. $\tan\theta$ **9. (a)** $\frac{1}{2}$ **(b)** $\dfrac{\sqrt{2}+\sqrt{6}}{4}$ or $\frac{1}{2}\sqrt{2+\sqrt{3}}$
(c) $\dfrac{\sqrt{6}-\sqrt{2}}{4}$ or $\frac{1}{2}\sqrt{2-\sqrt{3}}$
10. $(10-2\sqrt{5})/15$
11. $\frac{1}{2}(\sin 8x - \sin 2x)$ **12.** $-2\cos\frac{7}{2}x\sin\frac{3}{2}x$ **13.** -2
14. 0.34, 2.80 **15.** $\pi/3, \pi/2, 5\pi/3$ **16.** $2\pi/3, 4\pi/3$
17. $\pi/6, \pi/2, 5\pi/6, 3\pi/2$ **18.** 0.58, 2.56, 3.72, 5.70
19. $\pi/3, 2\pi/3, 4\pi/3, 5\pi/3$ **20.** $\pi/3, 5\pi/3$
21. $\frac{1519}{1681}$ **22.** $\dfrac{\sqrt{1-x^2}-xy}{\sqrt{1+y^2}}$

FOCUS ON MODELING ■ PAGE 584

1. (a) $y = -5\sin\left(\dfrac{\pi}{2}t\right)$

(b)

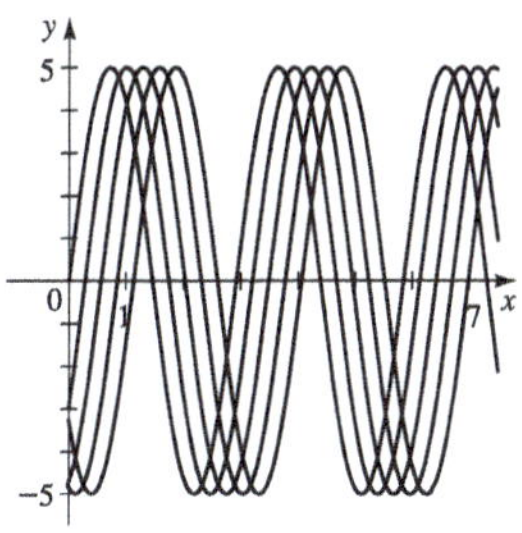

Yes, it is a traveling wave.

(c) $v = \pi/4$
3. $y(x,t) = 2.7\sin(0.68x - 4.10t)$
5. $y(x,t) = 0.6\sin(\pi x)\cos(40\pi t)$
7. (a) 1, 2, 3, 4
(b) 5:

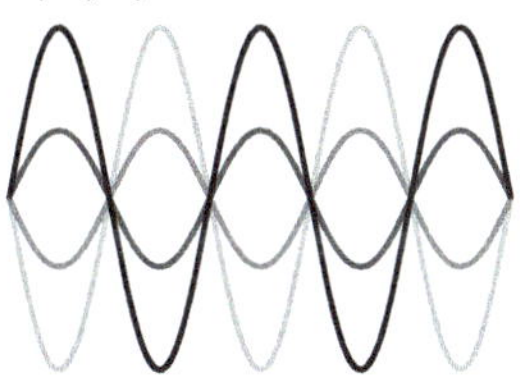

6:

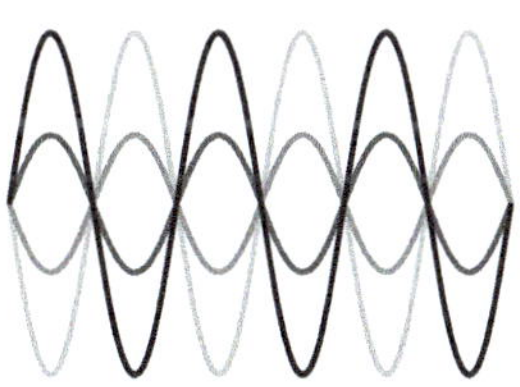

(c) 880π **(d)** $y(x,t) = \sin x\cos(880\pi t)$; $y(x,t) = \sin(2x)\cos(880\pi t)$; $y(x,t) = \sin(3x)\cos(880\pi t)$; $y(x,t) = \sin(4x)\cos(880\pi t)$

CHAPTER 8

SECTION 8.1 ■ PAGE 592

1. coordinate; $(1,1)$, $(\sqrt{2}, \pi/4)$ **2. (a)** $r\cos\theta$, $r\sin\theta$
(b) x^2+y^2, y/x **3.** Yes **4.** No; adding a multiple of 2π to θ gives the same point.

5.

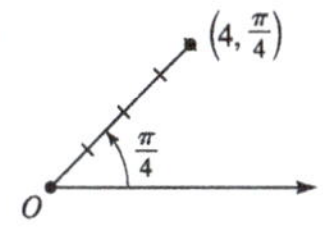

7.

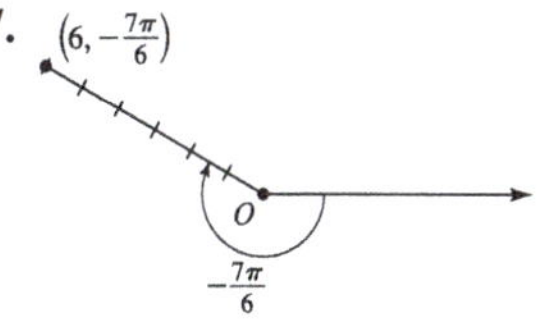

9.

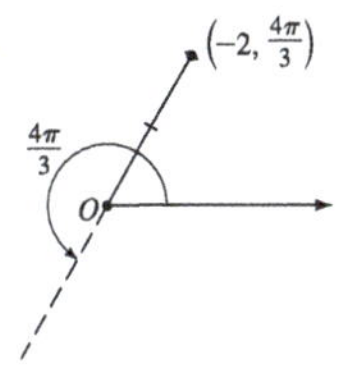

11.

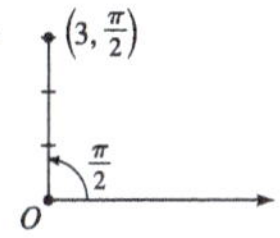

$\left(-3, \dfrac{3\pi}{2}\right), \left(3, \dfrac{5\pi}{2}\right)$

13.

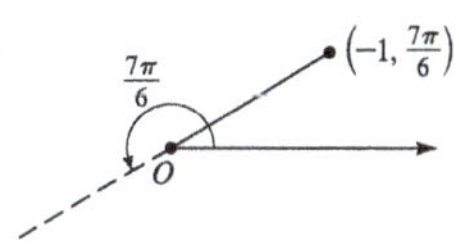

$\left(-1, -\dfrac{5\pi}{6}\right), \left(1, \dfrac{\pi}{6}\right)$

15.

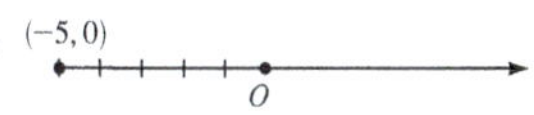

$(-5, 2\pi), (5, \pi)$

17. Q **19.** Q **21.** P **23.** P **25.** $(3\sqrt{2}, 3\pi/4)$
27. $\left(-\dfrac{5}{2}, -\dfrac{5\sqrt{3}}{2}\right)$ **29.** $(2\sqrt{3}, 2)$ **31.** $(1, -1)$
33. $(-5, 0)$ **35.** $(3\sqrt{6}, -3\sqrt{2})$ **37.** $(\sqrt{2}, 3\pi/4)$
39. $(4, \pi/4)$ **41.** $(5, \tan^{-1}\frac{4}{3})$ **43.** $(6, \pi)$ **45.** $\theta = \pi/4$
47. $r = \tan\theta\sec\theta$ **49.** $r = 4\sec\theta$ **51.** $x^2+y^2 = 49$
53. $x = 0$ **55.** $x = 6$ **57.** $x^2+y^2 = 4y$
59. $x^2+y^2 = (x^2+y^2-x)^2$ **61.** $(x^2+y^2-2y)^2 = x^2+y^2$
63. $y - x = 1$ **65.** $x^2 - 3y^2 + 16y - 16 = 0$
67. $x^2+y^2 = \dfrac{y}{x}$ **69.** $y^2 - 3x^2 = 0$

SECTION 8.2 ■ PAGE 600

1. circles, rays **2. (a)** satisfy **(b)** circle, 3, pole; line, pole, 1
3. VI **5.** II **7.** I **9.** Symmetric about $\theta = \pi/2$
11. Symmetric about the polar axis
13. Symmetric about $\theta = \pi/2$
15. All three types of symmetry

17.

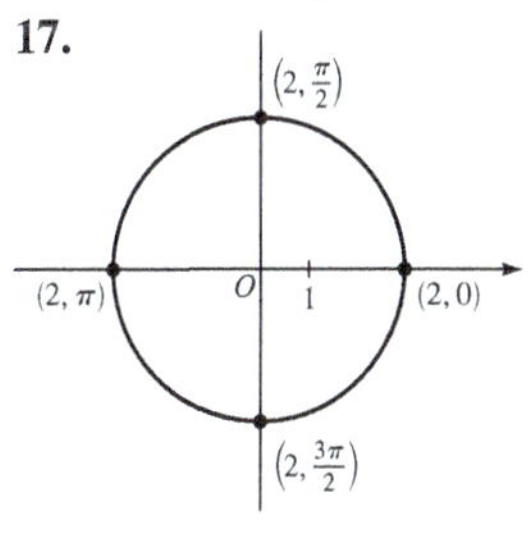

$x^2+y^2 = 4$

19.

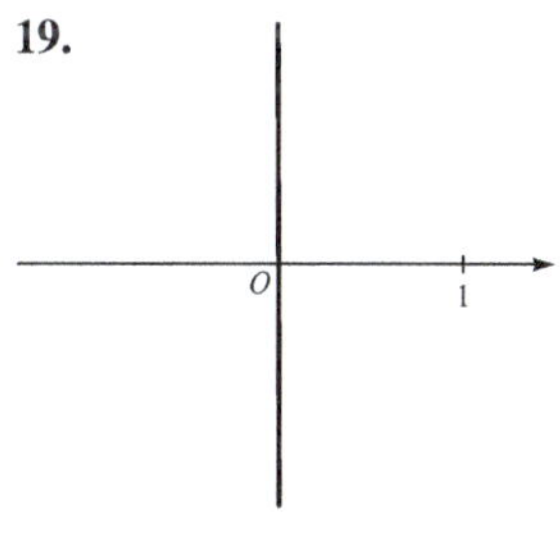

$x = 0$

21.

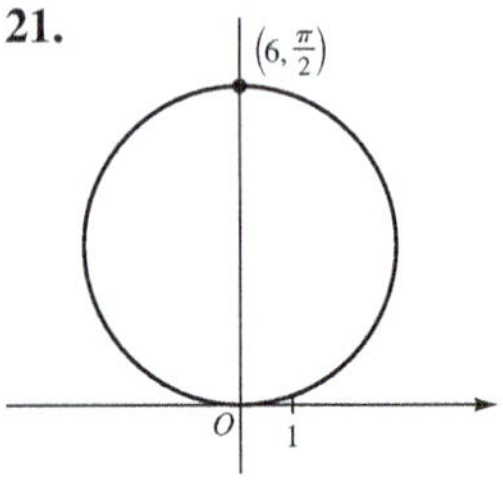

$x^2 + (y - 3)^2 = 9$

23.

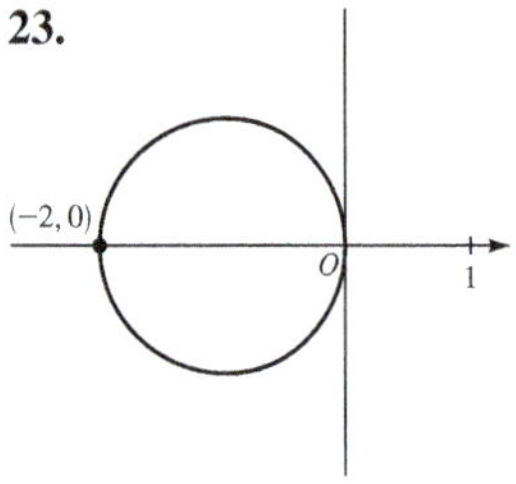

25.

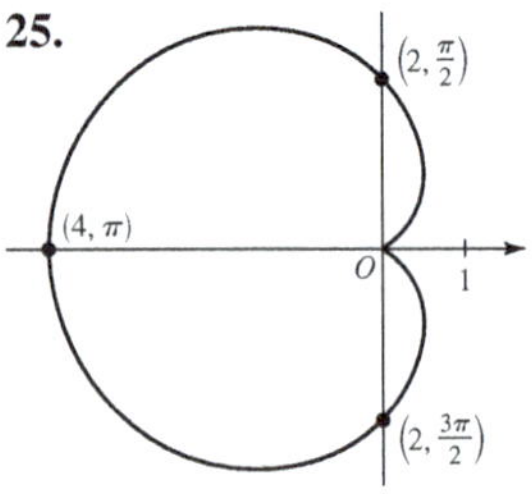

27.

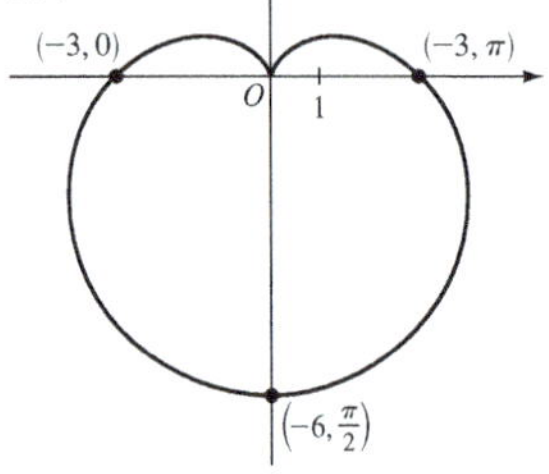

29.

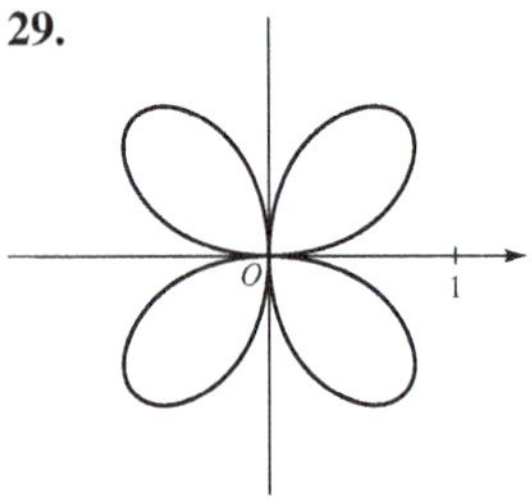

31.

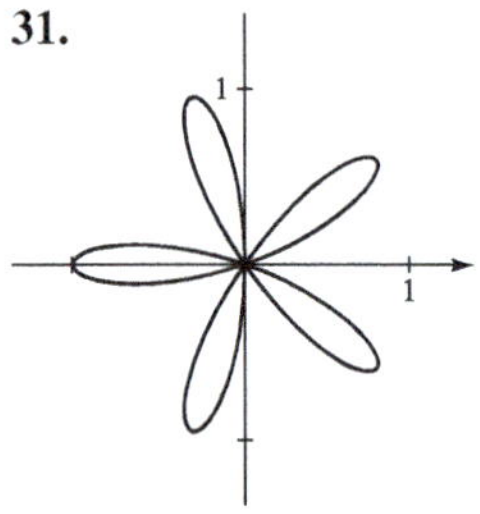

33.

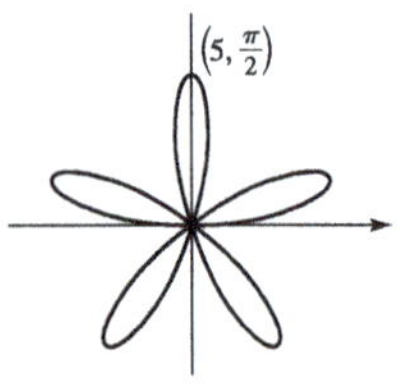

35.

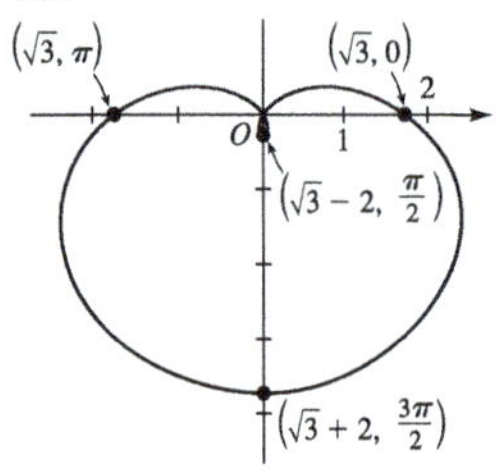

37.

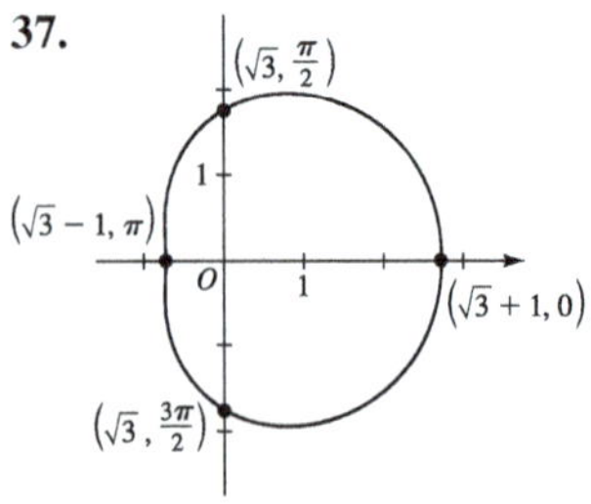

39.

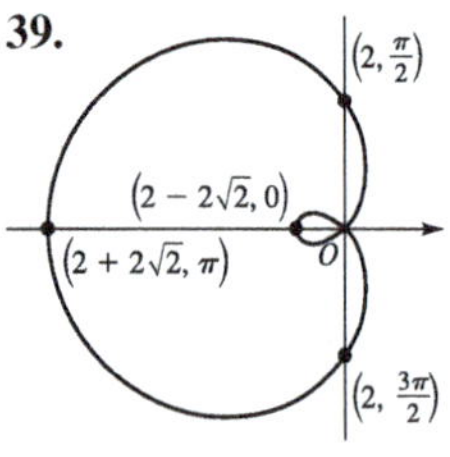

41.

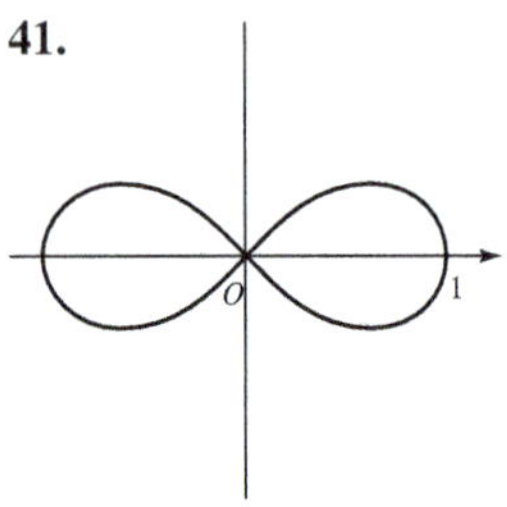

43.

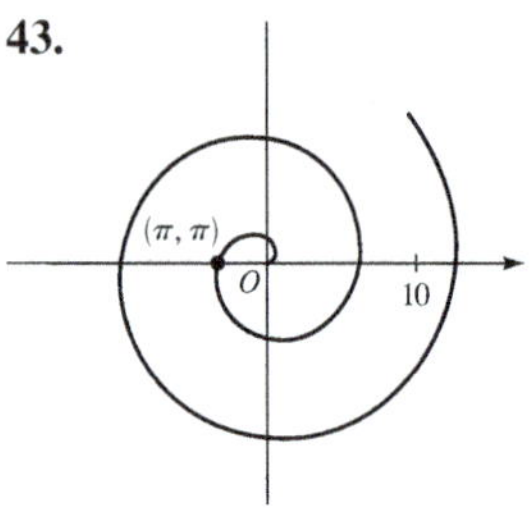

45.

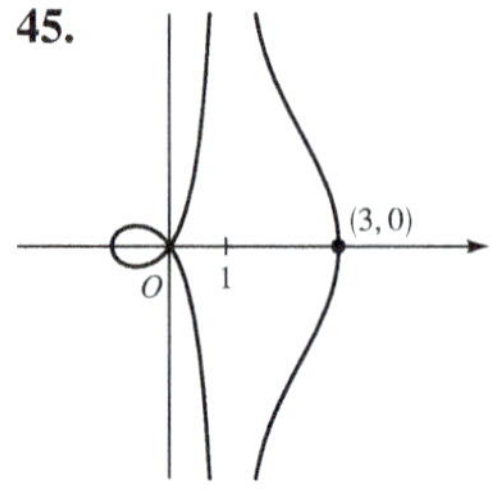

47. $0 \le \theta \le 4\pi$

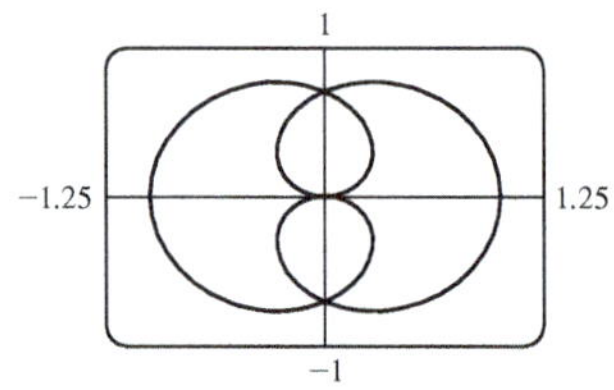

49. $0 \le \theta \le 4\pi$

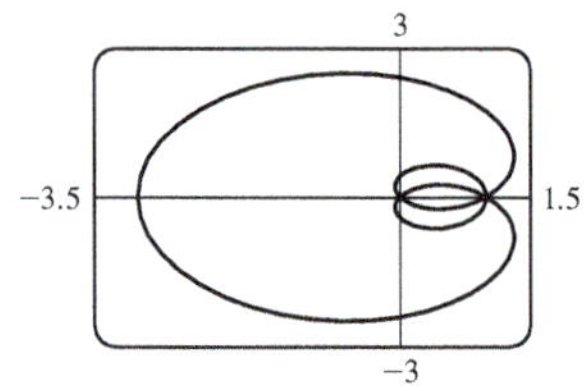

51. The graph of $r = 1 + \sin n\theta$ has n loops.
53. IV **55.** III

57.

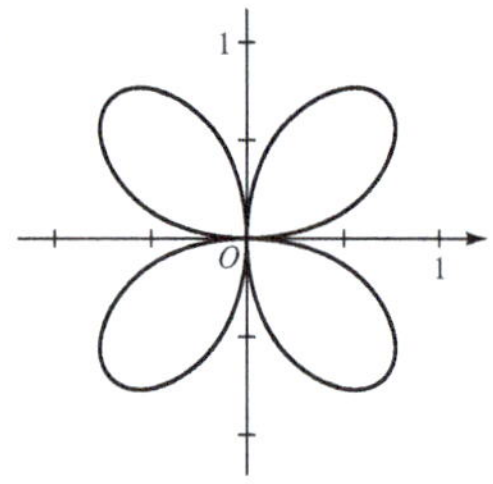

59.

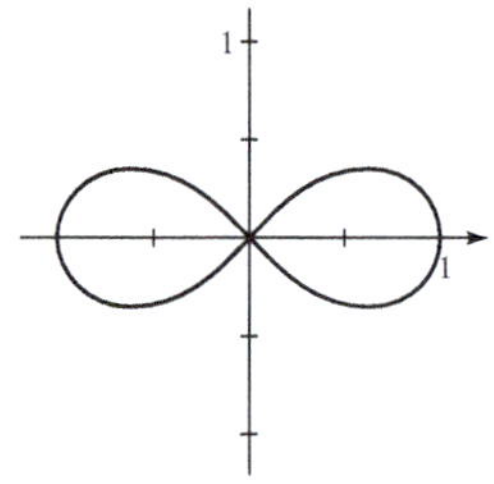

61. (a) $\left(x - \dfrac{a}{2}\right)^2 + \left(y - \dfrac{b}{2}\right)^2 = \dfrac{a^2 + b^2}{4}$,

$\left(\dfrac{a}{2}, \dfrac{b}{2}\right), \dfrac{1}{2}\sqrt{a^2 + b^2}$

(b)

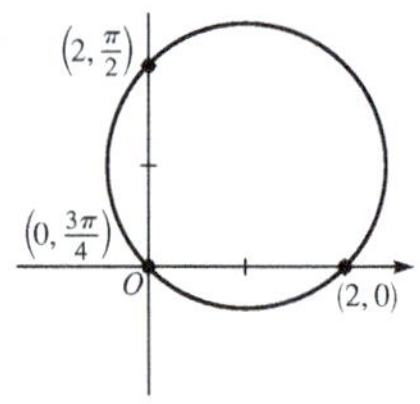

63. (a) Elliptical

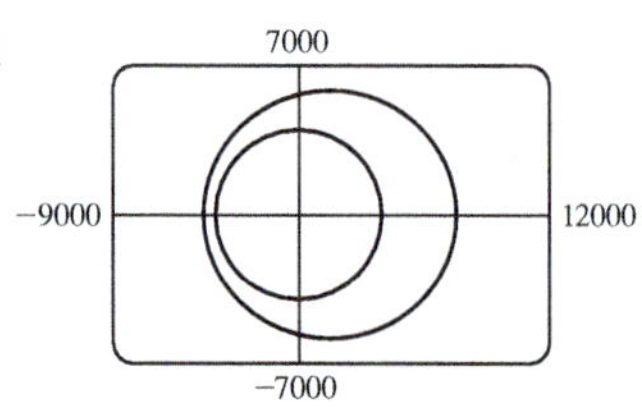

(b) π; 540 mi

SECTION 8.3 ■ PAGE 609

1. real, imaginary, (a, b) **2. (a)** $\sqrt{a^2 + b^2}, b/a$
(b) $r(\cos\theta + i\sin\theta)$

3. (a) $\sqrt{2}\left(\cos\dfrac{3\pi}{4} + i\sin\dfrac{3\pi}{4}\right)$ **(b)** $\sqrt{3} + i$

(c) $1 + i, \sqrt{2}\left(\cos\dfrac{\pi}{4} + i\sin\dfrac{\pi}{4}\right)$

4. n; four; 2, $2i$, -2, $2i$; 2

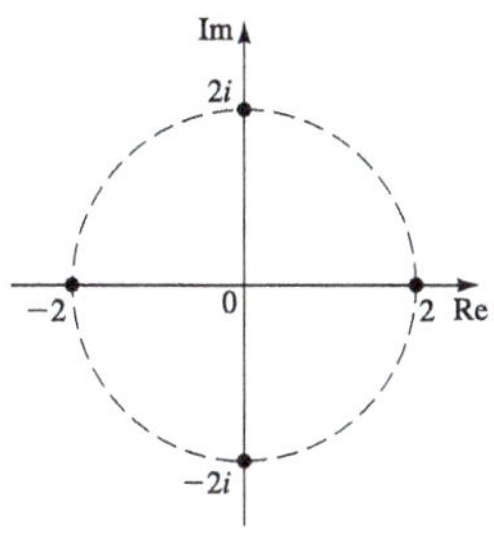

5. 4

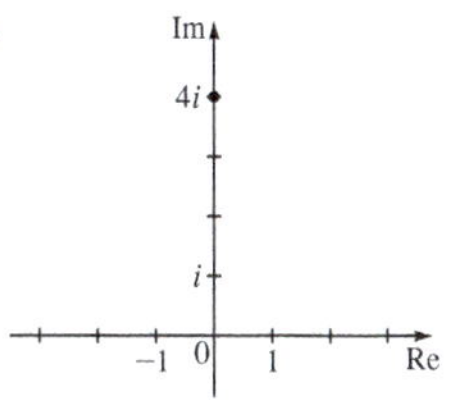

7. 2

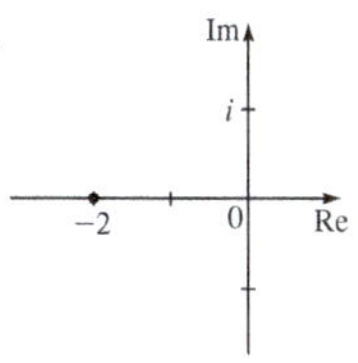

9. $\sqrt{29}$

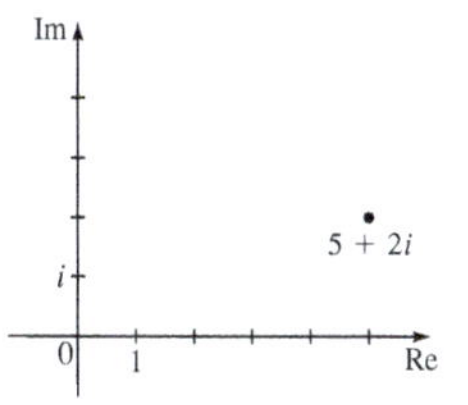

11. 2

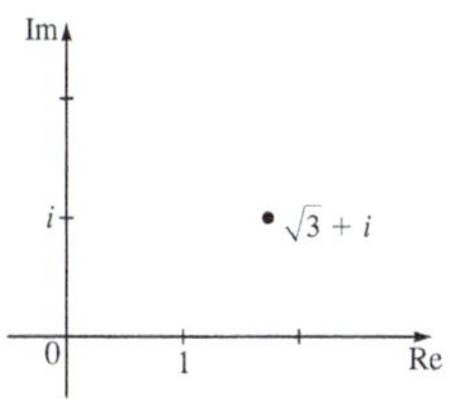

13. 1

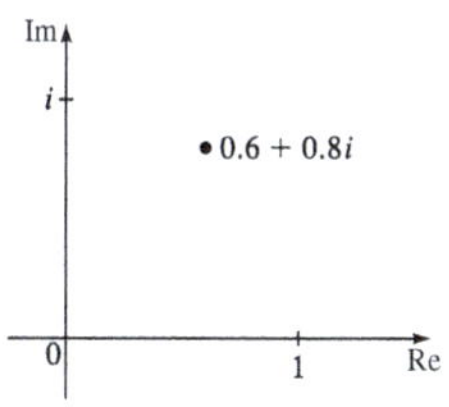

15.

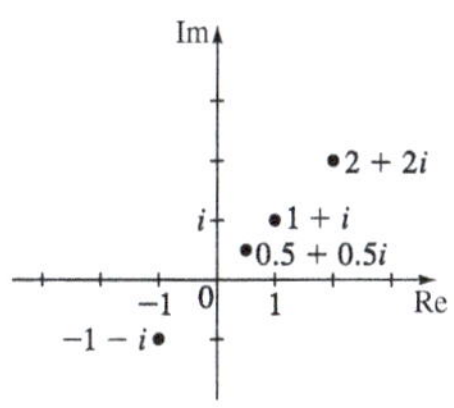

17.

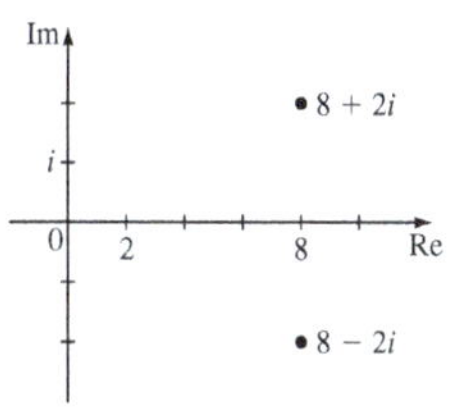

19.

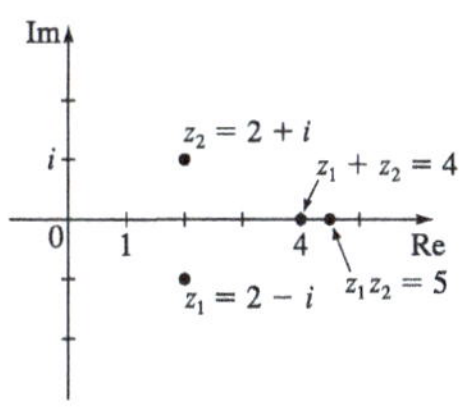

21.

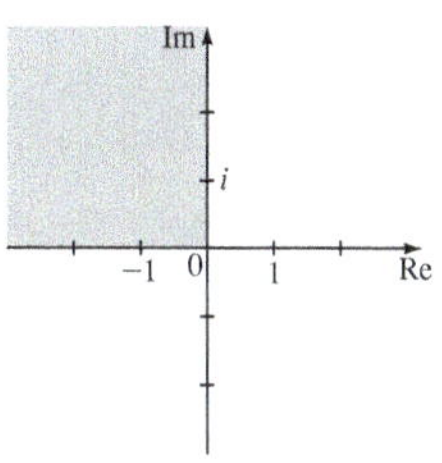

23.

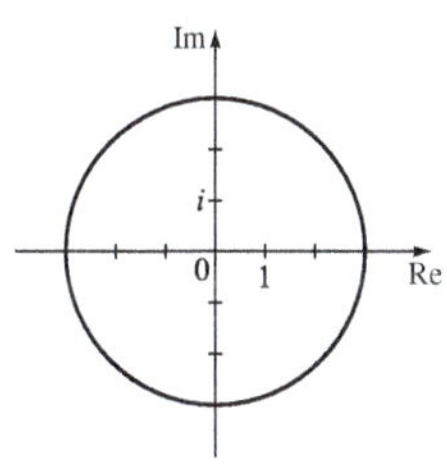

25.

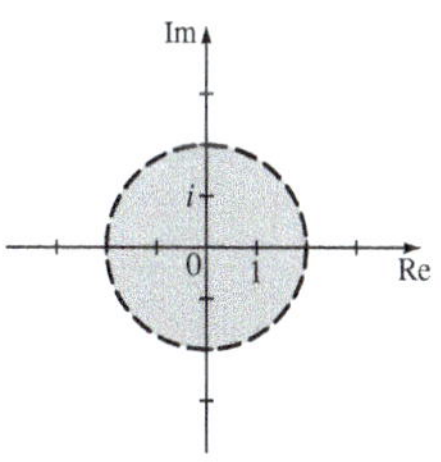

27.

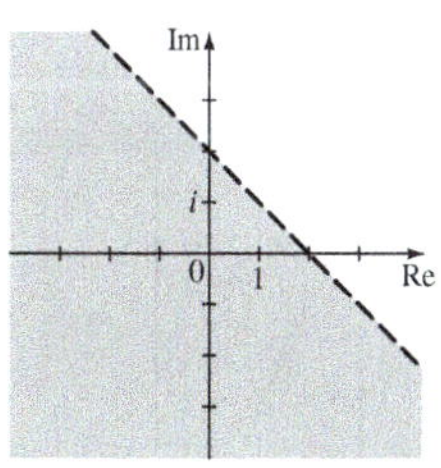

29. $\sqrt{2}\left(\cos\frac{\pi}{4} + i\sin\frac{\pi}{4}\right)$ **31.** $2\sqrt{2}\left(\cos\frac{3\pi}{4} + i\sin\frac{3\pi}{4}\right)$

33. $2\left(\cos\frac{7\pi}{6} + i\sin\frac{7\pi}{6}\right)$ **35.** $4\left(\cos\frac{11\pi}{6} + i\sin\frac{11\pi}{6}\right)$

37. $2\left(\cos\frac{\pi}{2} + i\sin\frac{\pi}{2}\right)$ **39.** $3(\cos\pi + i\sin\pi)$

41. $2\sqrt{2}\left(\cos\frac{5\pi}{6} + i\sin\frac{5\pi}{6}\right)$

43. $5\left(\cos\left(\tan^{-1}\frac{3}{4}\right) + i\sin\left(\tan^{-1}\frac{3}{4}\right)\right)$

45. $8\left(\cos\frac{11\pi}{6} + i\sin\frac{11\pi}{6}\right)$ **47.** $3\sqrt{2}\left(\cos\frac{3\pi}{4} + i\sin\frac{3\pi}{4}\right)$

49. $z_1z_2 = 6\left(\cos\frac{\pi}{2} + i\sin\frac{\pi}{2}\right), \frac{z_1}{z_2} = \frac{3}{2}\left(\cos\frac{\pi}{6} + i\sin\frac{\pi}{6}\right)$

51. $z_1z_2 = 4\left(\cos\frac{7\pi}{6} + i\sin\frac{7\pi}{6}\right)$,

$\frac{z_1}{z_2} = \frac{1}{2}\left(\cos\frac{\pi}{6} + i\sin\frac{\pi}{6}\right)$

53. $z_1z_2 = 8(\cos 150° + i\sin 150°)$
$z_1/z_2 = 2(\cos 90° + i\sin 90°)$

55. $z_1z_2 = 100(\cos 350° + i\sin 350°)$
$z_1/z_2 = \frac{4}{25}(\cos 50° + i\sin 50°)$

57. $z_1 = 2\left(\cos\frac{\pi}{6} + i\sin\frac{\pi}{6}\right)$

$z_2 = 2\left(\cos\frac{\pi}{3} + i\sin\frac{\pi}{3}\right)$

$z_1z_2 = 4\left(\cos\frac{\pi}{2} + i\sin\frac{\pi}{2}\right)$

$\frac{z_1}{z_2} = \cos\left(-\frac{\pi}{6}\right) + i\sin\left(-\frac{\pi}{6}\right)$

$\frac{1}{z_1} = \frac{1}{2}\left[\cos\left(-\frac{\pi}{6}\right) + i\sin\left(-\frac{\pi}{6}\right)\right]$

59. $z_1 = 4\left(\cos\frac{11\pi}{6} + i\sin\frac{11\pi}{6}\right)$

$z_2 = \sqrt{2}\left(\cos\frac{3\pi}{4} + i\sin\frac{3\pi}{4}\right)$

$z_1z_2 = 4\sqrt{2}\left(\cos\frac{7\pi}{12} + i\sin\frac{7\pi}{12}\right)$

$\frac{z_1}{z_2} = 2\sqrt{2}\left(\cos\frac{13\pi}{12} + i\sin\frac{13\pi}{12}\right)$

$\frac{1}{z_1} = \frac{1}{4}\left(\cos\frac{\pi}{6} + i\sin\frac{\pi}{6}\right)$

61. $z_1 = 5\sqrt{2}\left(\cos\frac{\pi}{4} + i\sin\frac{\pi}{4}\right)$

$z_2 = 4(\cos 0 + i\sin 0)$

$z_1z_2 = 20\sqrt{2}\left(\cos\frac{\pi}{4} + i\sin\frac{\pi}{4}\right)$

$\frac{z_1}{z_2} = \frac{5\sqrt{2}}{4}\left(\cos\frac{\pi}{4} + i\sin\frac{\pi}{4}\right)$

$\frac{1}{z_1} = \frac{\sqrt{2}}{10}\left[\cos\left(-\frac{\pi}{4}\right) + i\sin\left(-\frac{\pi}{4}\right)\right]$

63. $z_1 = 20(\cos \pi + i \sin \pi)$

$z_2 = 2\left(\cos \frac{\pi}{6} + i \sin \frac{\pi}{6}\right)$

$z_1 z_2 = 40\left(\cos \frac{7\pi}{6} + i \sin \frac{7\pi}{6}\right)$

$\frac{z_1}{z_2} = 10\left(\cos \frac{5\pi}{6} + i \sin \frac{5\pi}{6}\right)$

$\frac{1}{z_1} = \frac{1}{20}(\cos \pi + i \sin \pi)$

65. -64 **67.** $16\sqrt{2} + 16\sqrt{2}i$ **69.** -1 **71.** 4096

73. $8(-1 + i)$ **75.** $\frac{1}{2048}(-\sqrt{3} - i)$

77. $2\sqrt{2}\left(\cos \frac{\pi}{12} + i \sin \frac{\pi}{12}\right)$,

$2\sqrt{2}\left(\cos \frac{13\pi}{12} + i \sin \frac{13\pi}{12}\right)$

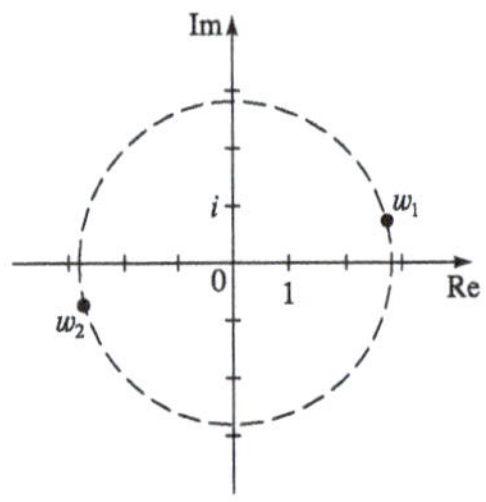

79. $3\left(\cos \frac{3\pi}{8} + i \sin \frac{3\pi}{8}\right)$,

$3\left(\cos \frac{7\pi}{8} + i \sin \frac{7\pi}{8}\right)$,

$3\left(\cos \frac{11\pi}{8} + i \sin \frac{11\pi}{8}\right)$,

$3\left(\cos \frac{15\pi}{8} + i \sin \frac{15\pi}{8}\right)$

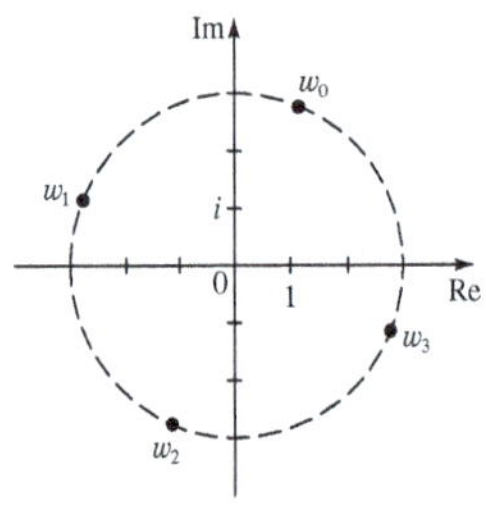

81. $\pm 1, \pm i, \frac{\sqrt{2}}{2} \pm \frac{\sqrt{2}}{2}i$,

$-\frac{\sqrt{2}}{2} \pm \frac{\sqrt{2}}{2}i$

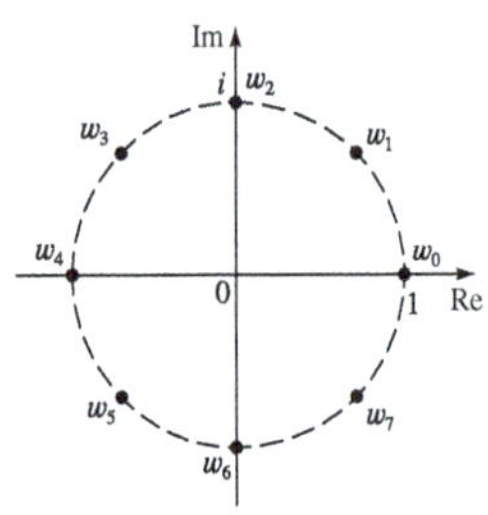

83. $\frac{\sqrt{3}}{2} + \frac{1}{2}i, -\frac{\sqrt{3}}{2} + \frac{1}{2}i, -i$

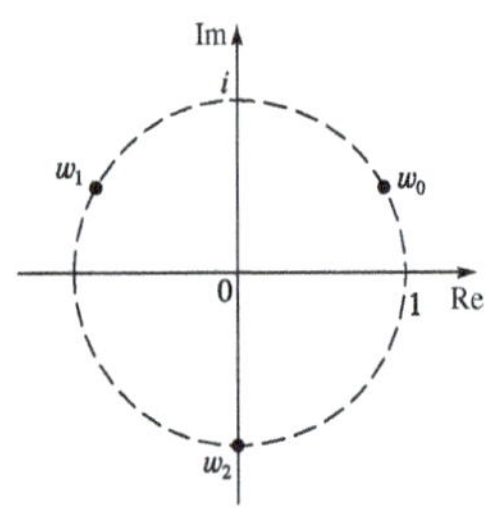

85. $\frac{\sqrt{2}}{2} \pm \frac{\sqrt{2}}{2}i$,

$-\frac{\sqrt{2}}{2} \pm \frac{\sqrt{2}}{2}i$

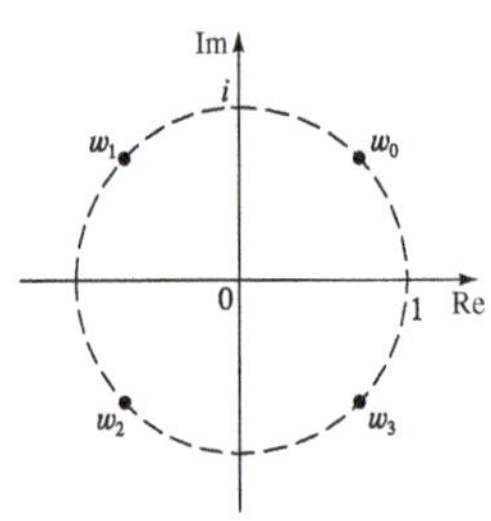

87. $\frac{\sqrt{2}}{2} \pm \frac{\sqrt{2}}{2}i, -\frac{\sqrt{2}}{2} \pm \frac{\sqrt{2}}{2}i$

89. $2\left(\cos \frac{\pi}{18} + i \sin \frac{\pi}{18}\right), 2\left(\cos \frac{13\pi}{18} + i \sin \frac{13\pi}{18}\right)$,

$2\left(\cos \frac{25\pi}{18} + i \sin \frac{25\pi}{18}\right)$

91. $2^{1/6}\left(\cos \frac{5\pi}{12} + i \sin \frac{5\pi}{12}\right), 2^{1/6}\left(\cos \frac{13\pi}{12} + i \sin \frac{13\pi}{12}\right)$,

$2^{1/6}\left(\cos \frac{21\pi}{12} + i \sin \frac{21\pi}{12}\right)$

93. $\frac{1}{2} \pm \frac{\sqrt{5}}{2}i$ **95.** $1 + i, 1 - i$

SECTION 8.4 ■ PAGE 617

1. (a) parameter **(b)** $(0, 0), (1, 1)$ **(c)** x^2; parabola

2. (a) True **(b)** $(0, 0), (2, 4)$ **(c)** x^2; path

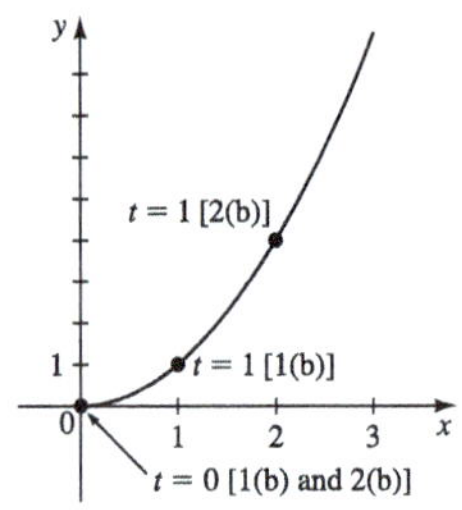

3. (a)

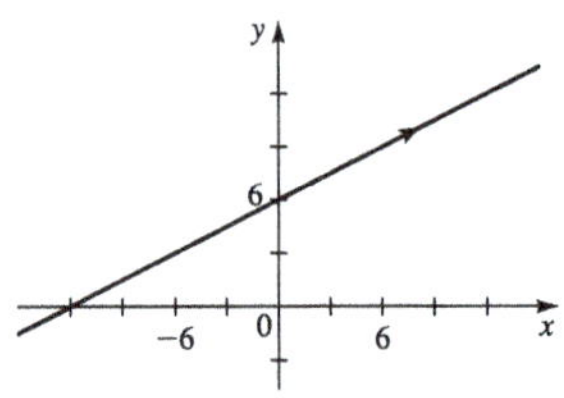

(b) $x - 2y + 12 = 0$

5. (a)

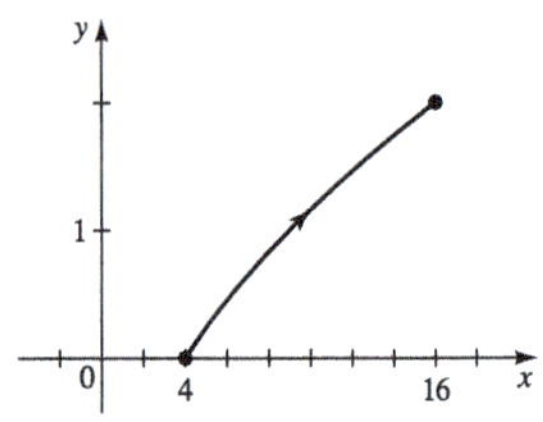

(b) $x = (y + 2)^2$

7. (a)

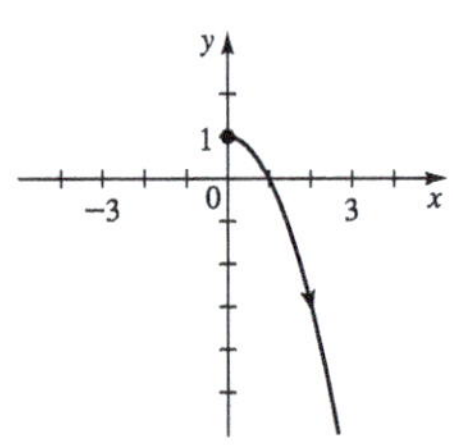

(b) $x = \sqrt{1 - y}$

9. (a)

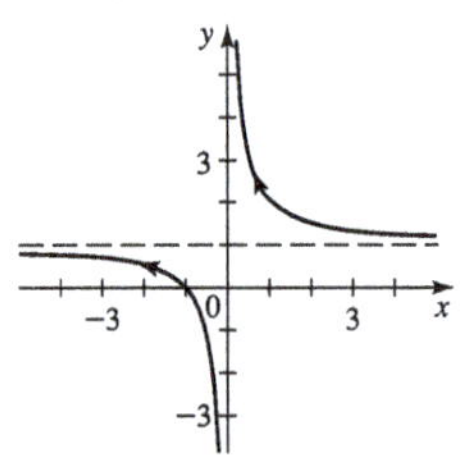

(b) $y = \frac{1}{x} + 1$

11. (a)

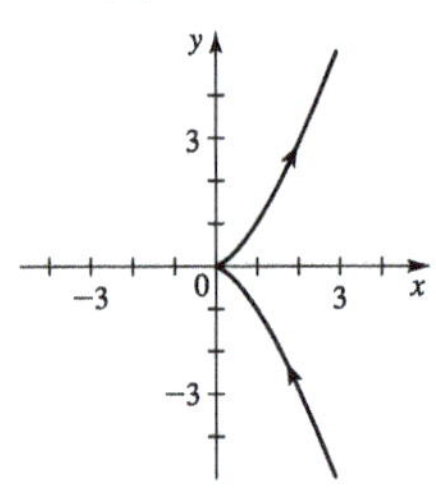

(b) $x^3 = y^2$

13. (a)

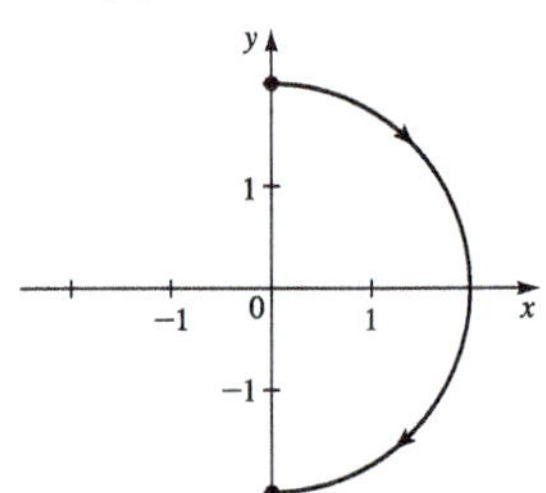

(b) $x^2 + y^2 = 4, x \geq 0$

15. (a)

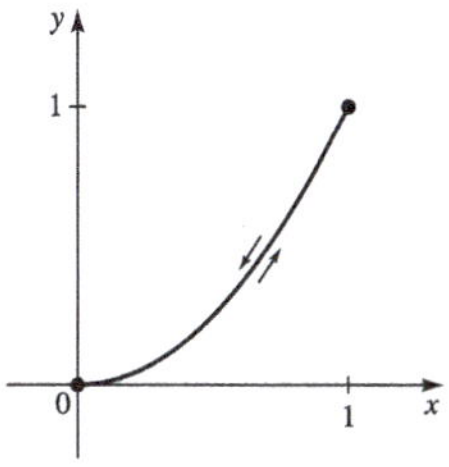

(b) $y = x^2, 0 \le x \le 1$

17. (a)

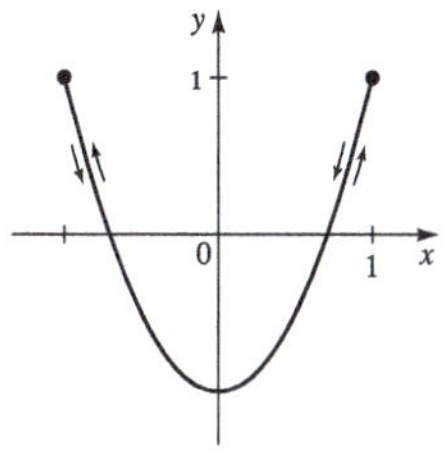

(b) $y = 2x^2 - 1, -1 \le x \le 1$

19. (a)

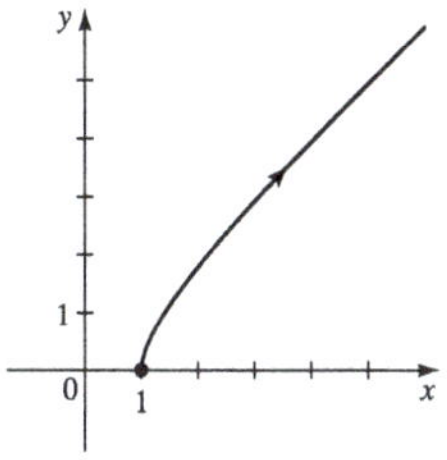

(b) $x^2 - y^2 = 1, x \ge 1, y \ge 0$

21. (a)

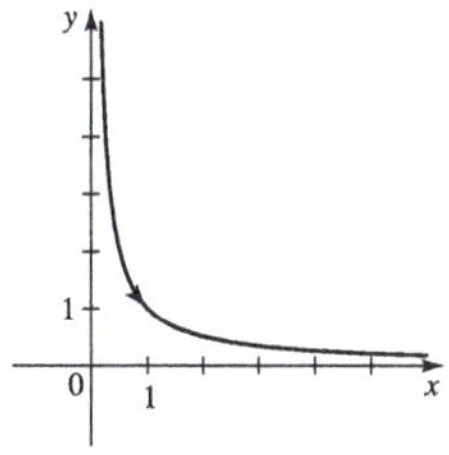

(b) $y = 1/x, x > 0$

23. (a)

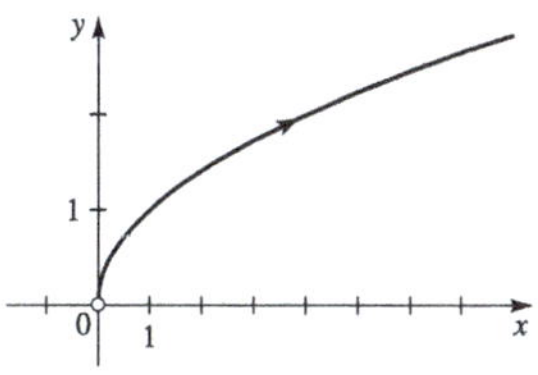

(b) $x = y^2, y > 0$

25. (a)

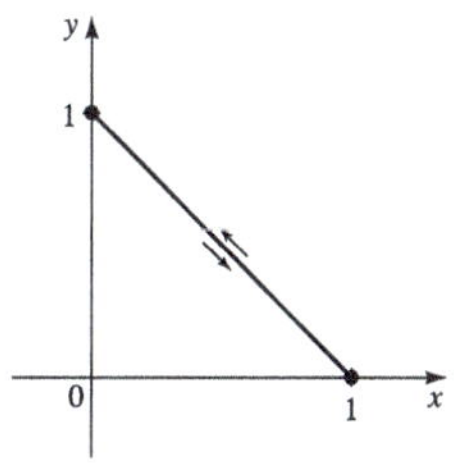

(b) $x + y = 1, 0 \le x \le 1$

27. 3, (3, 0), counterclockwise, 2π sec

29. 1, (0, 1), clockwise, π sec **31.** $x = 4 + t, y = -1 + \frac{1}{2}t$

33. $x = 6 + t, y = 7 + t$ **35.** $x = a \cos t, y = a \sin t$

39.

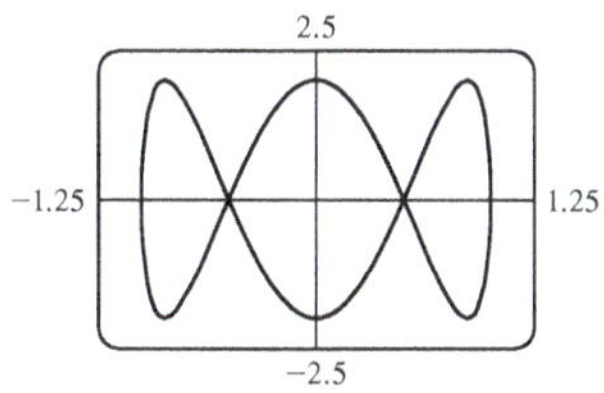

41.

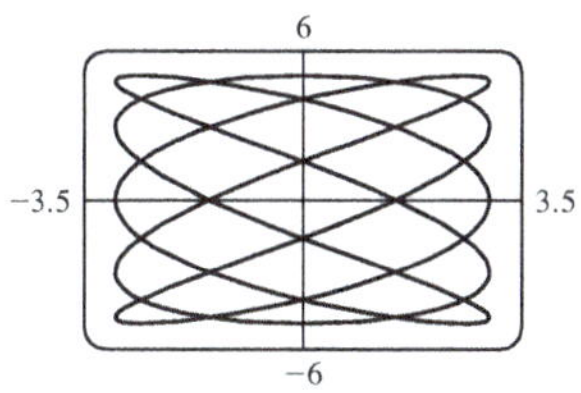

43.

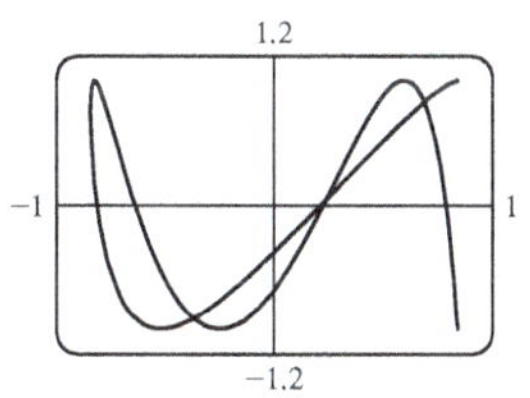

45. (a) $x = 2^{t/12} \cos t, y = 2^{t/12} \sin t$

(b)

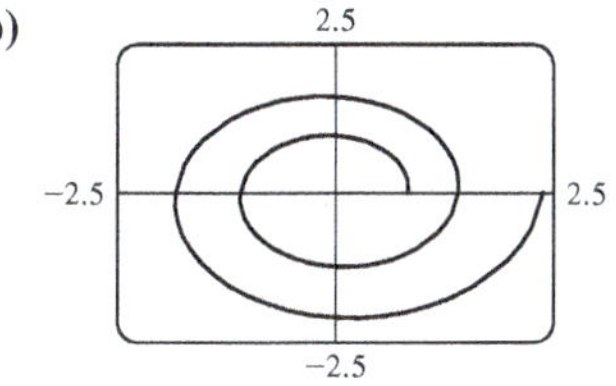

47. (a) $x = \dfrac{4 \cos t}{2 - \cos t}, y = \dfrac{4 \sin t}{2 - \cos t}$

(b)

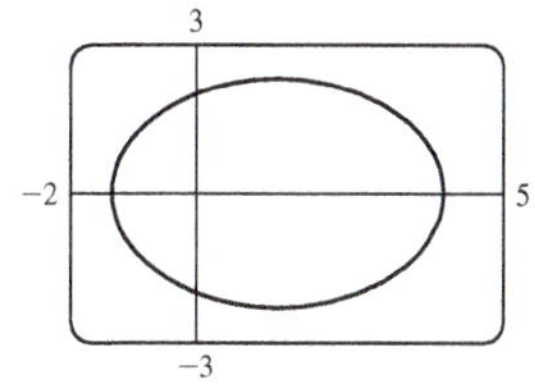

49. III **51.** II

53. (a) $x = a \cos \theta, y = b \sin \theta$

(b)

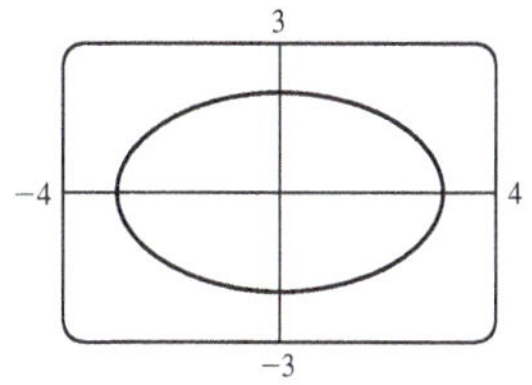

(c) $\dfrac{x^2}{a^2} + \dfrac{y^2}{b^2} = 1$

55.

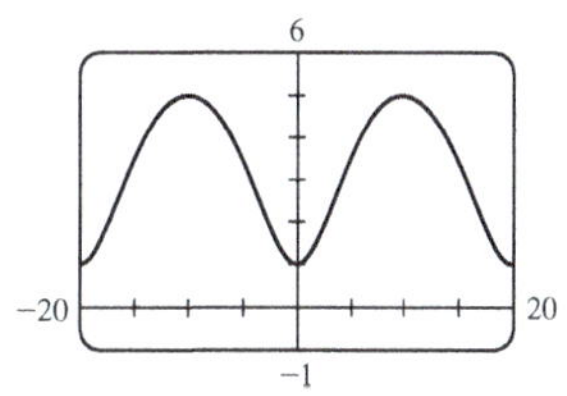

59.

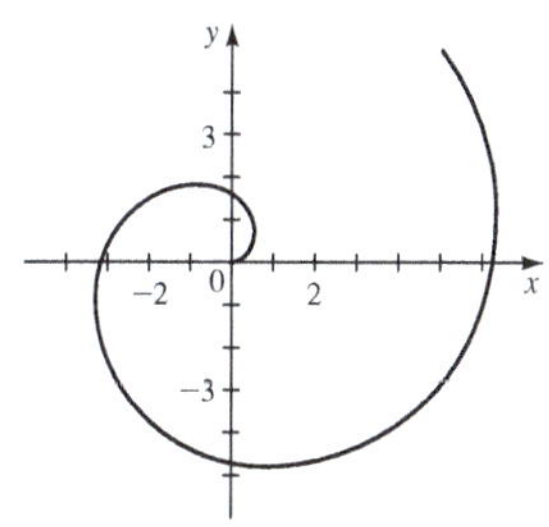

61.

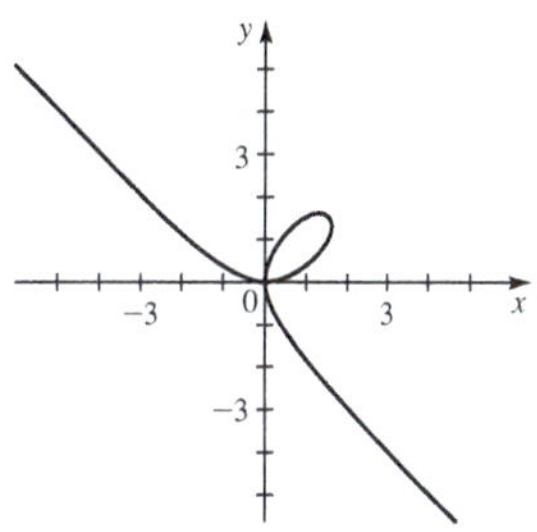

63. (b) $x^{2/3} + y^{2/3} = a^{2/3}$

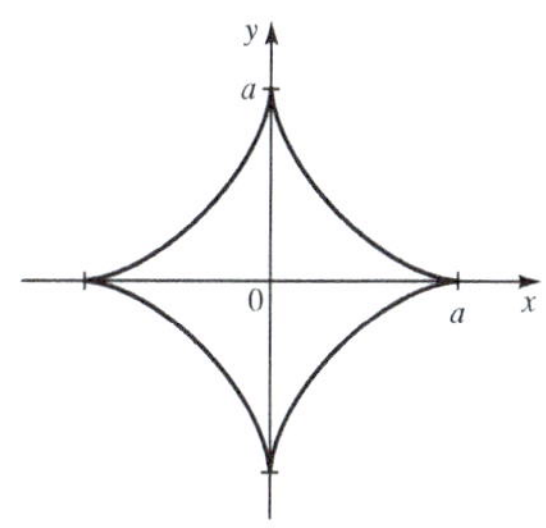

65. $x = a(\sin t \cos t + \cot t), y = a(1 + \sin^2 t)$

67. $y = a - a \cos\left(\dfrac{x + \sqrt{2ay - y^2}}{a}\right)$

69. (b)

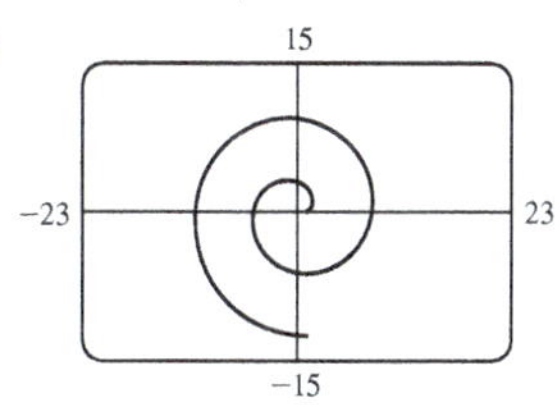

CHAPTER 8 REVIEW ■ PAGE 622

1. (a)

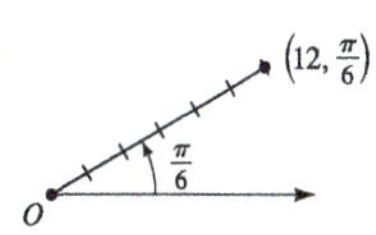

(b) $(6\sqrt{3}, 6)$

3. (a)

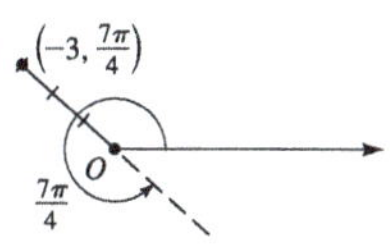

(b) $\left(\dfrac{-3\sqrt{2}}{2}, \dfrac{3\sqrt{2}}{2}\right)$

5. (a)

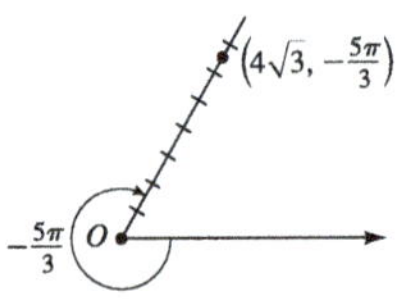

(b) $(2\sqrt{3}, 6)$

7. (a)

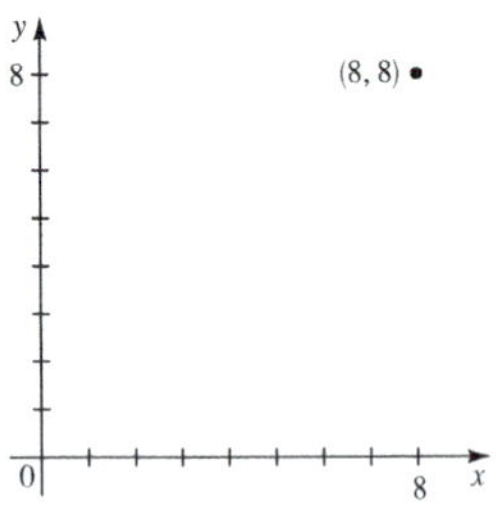

(b) $\left(8\sqrt{2}, \dfrac{\pi}{4}\right)$ **(c)** $\left(-8\sqrt{2}, \dfrac{5\pi}{4}\right)$

9. (a)

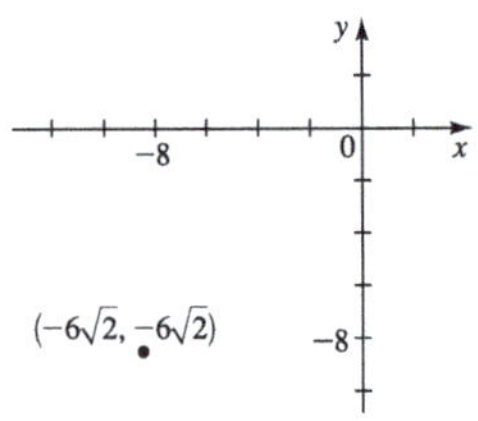

(b) $\left(12, \dfrac{5\pi}{4}\right)$ **(c)** $\left(-12, \dfrac{\pi}{4}\right)$

11. (a)

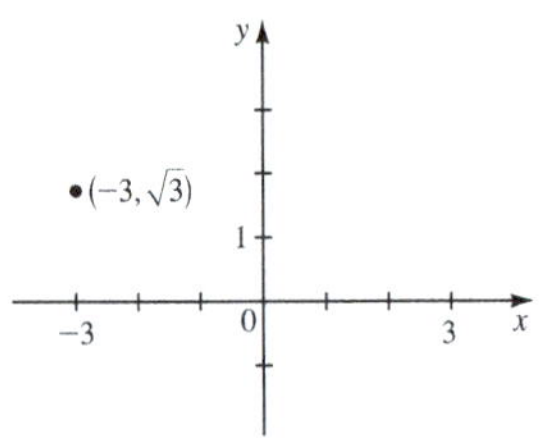

(b) $\left(2\sqrt{3}, \dfrac{5\pi}{6}\right)$ **(c)** $\left(-2\sqrt{3}, -\dfrac{\pi}{6}\right)$

13. (a) $r = \dfrac{4}{\cos\theta + \sin\theta}$ **(b)**

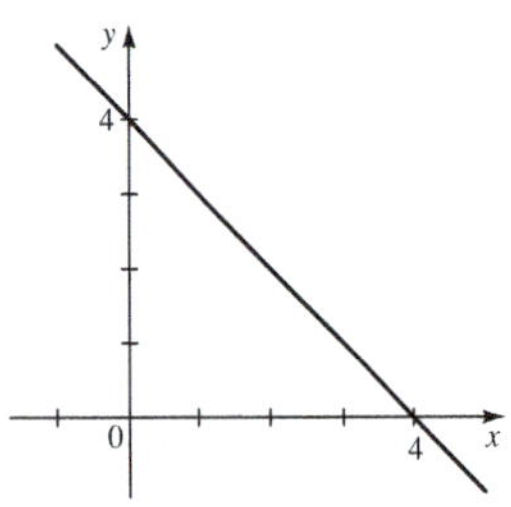

15. (a) $r = 4(\cos\theta + \sin\theta)$ **(b)**

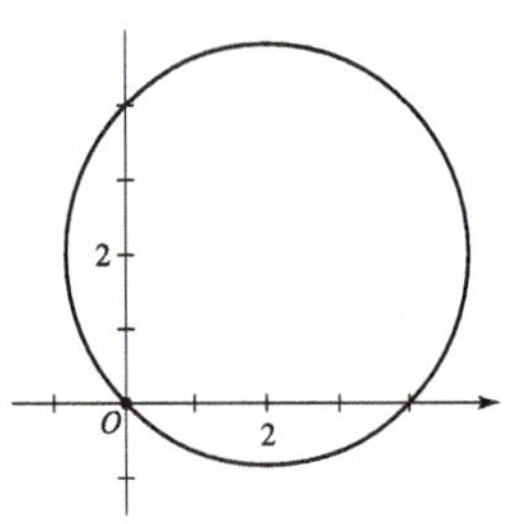

17. (a)

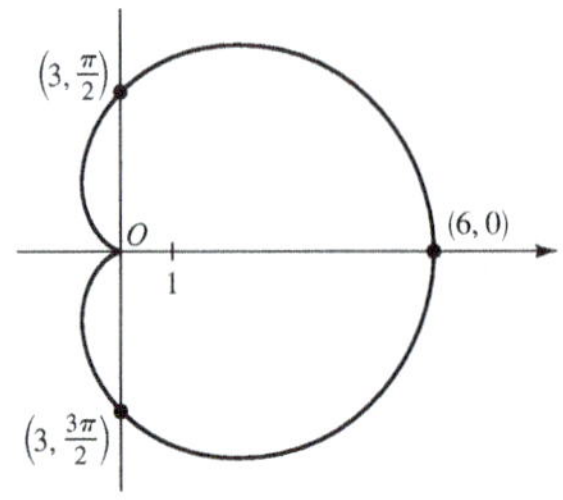

(b) $(x^2 + y^2 - 3x)^2 = 9(x^2 + y^2)$

19. (a)

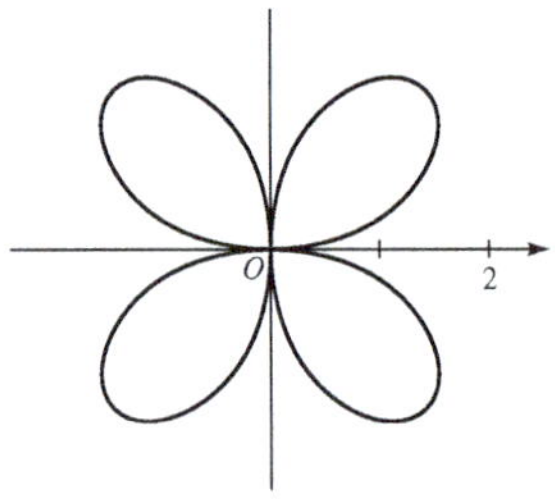

(b) $(x^2 + y^2)^3 = 16x^2y^2$

21. (a)

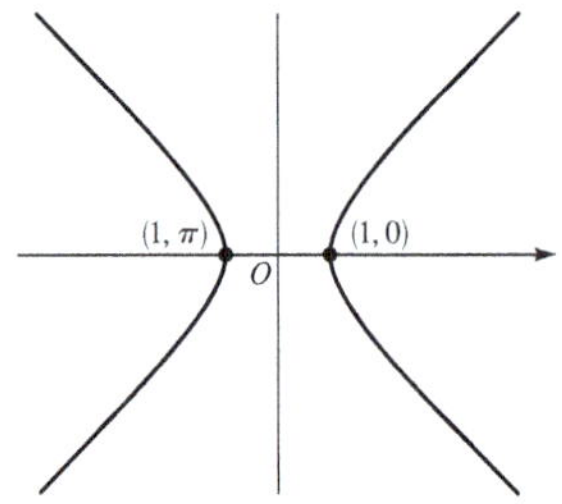

(b) $x^2 - y^2 = 1$

23. (a)

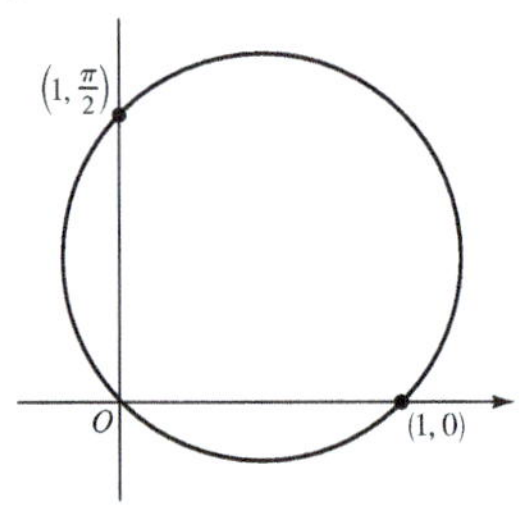

(b) $x^2 + y^2 = x + y$

25. $0 \le \theta \le 3\pi$

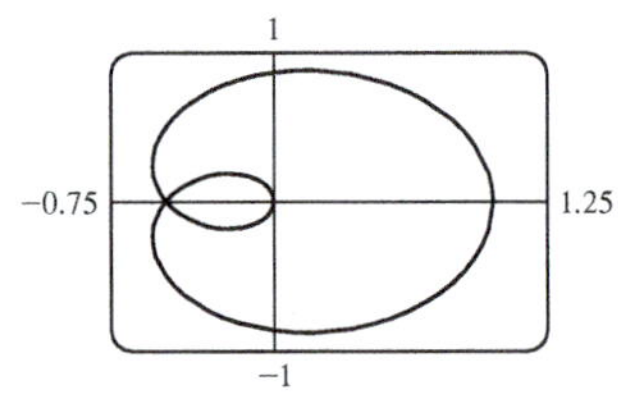

27. $0 \le \theta \le 6\pi$

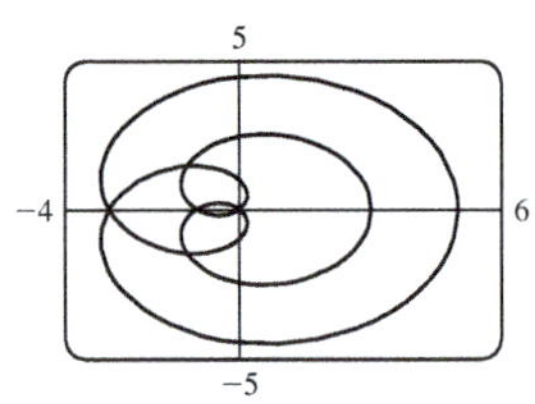

29. (a)

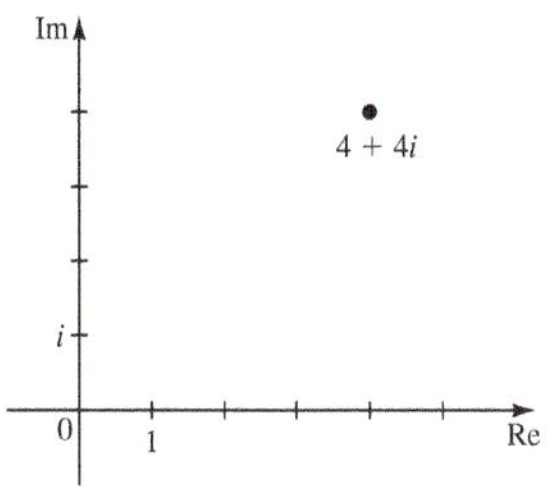

(b) $4\sqrt{2}, \dfrac{\pi}{4}$ **(c)** $4\sqrt{2}\left(\cos\dfrac{\pi}{4} + i\sin\dfrac{\pi}{4}\right)$

31. (a)

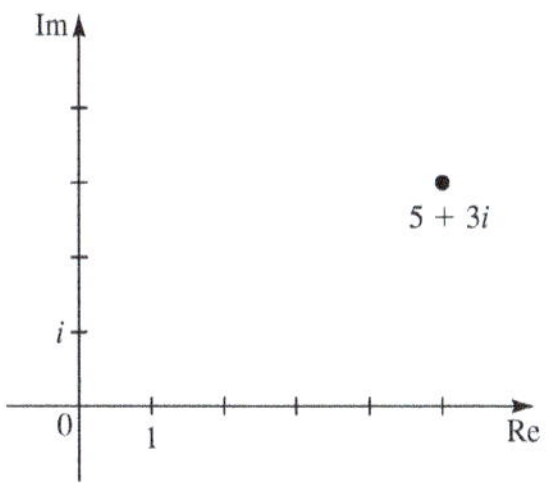

(b) $\sqrt{34}, \tan^{-1}\frac{3}{5}$ **(c)** $\sqrt{34}[\cos(\tan^{-1}\frac{3}{5}) + i\sin(\tan^{-1}\frac{3}{5})]$

33. (a)

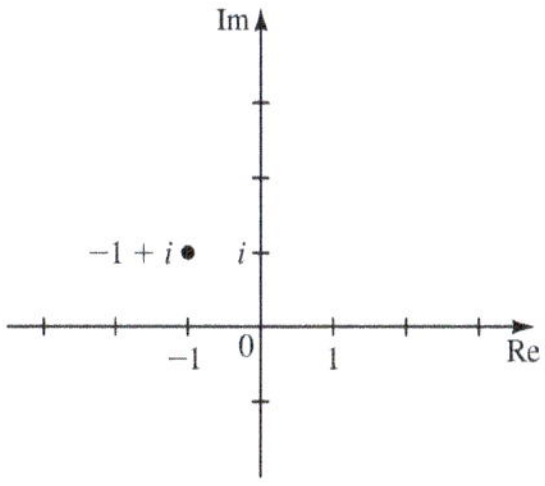

(b) $\sqrt{2}, \dfrac{3\pi}{4}$ **(c)** $\sqrt{2}\left(\cos\dfrac{3\pi}{4} + i\sin\dfrac{3\pi}{4}\right)$

35. $8(-1 + i\sqrt{3})$ **37.** $-\frac{1}{32}(1 + i\sqrt{3})$

39. $2\sqrt{2}(-1 + i), 2\sqrt{2}(1 - i)$

41. $\pm 1, \dfrac{1}{2} \pm \dfrac{\sqrt{3}}{2}i, -\dfrac{1}{2} \pm \dfrac{\sqrt{3}}{2}i$

43. (a)

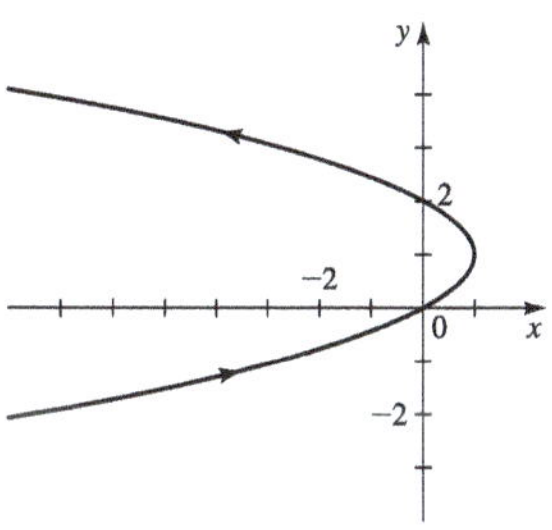

(b) $x = 2y - y^2$

45. (a)

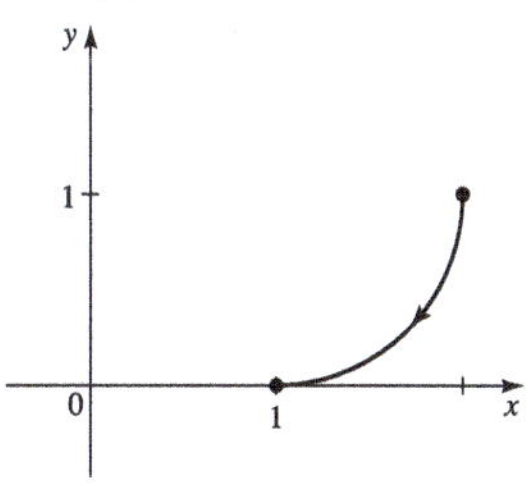

(b) $(x - 1)^2 + (y - 1)^2 = 1, 1 \le x \le 2, 0 \le y \le 1$

47.

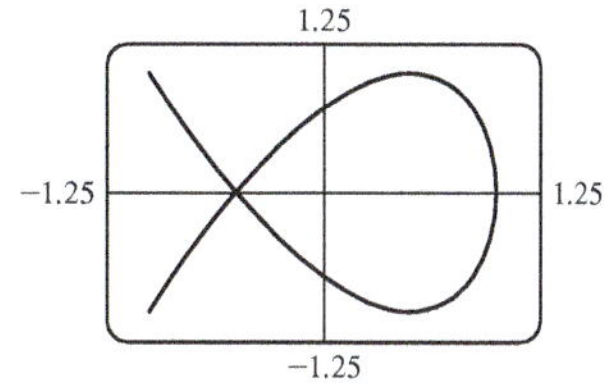

49. $x = \frac{1}{2}(1 + \cos\theta), y = \frac{1}{2}(\sin\theta + \tan\theta)$

CHAPTER 8 TEST ■ PAGE 624

1. (a) $(-4\sqrt{2}, -4\sqrt{2})$ **(b)** $(4\sqrt{3}, 5\pi/6), (-4\sqrt{3}, 11\pi/6)$

2. (a) circle

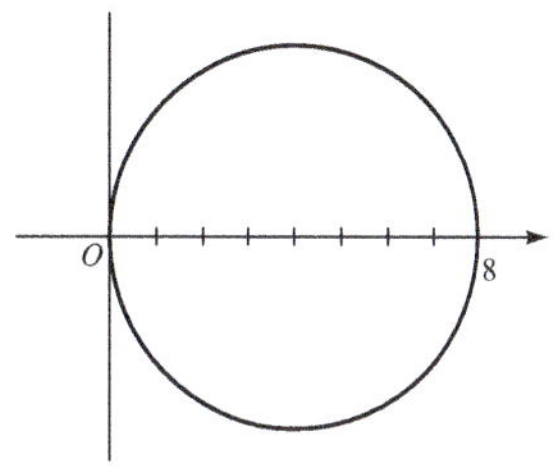

(b) $(x - 4)^2 + y^2 = 16$

3. limaçon

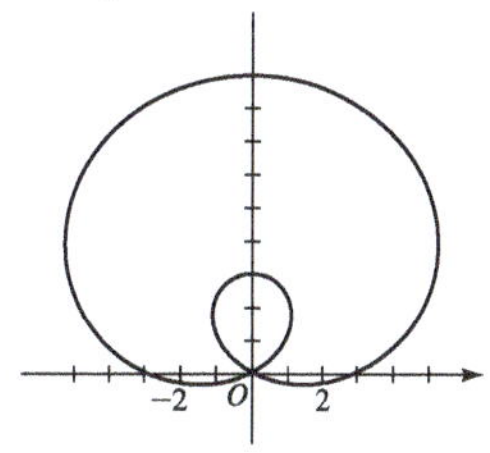

4. (a)

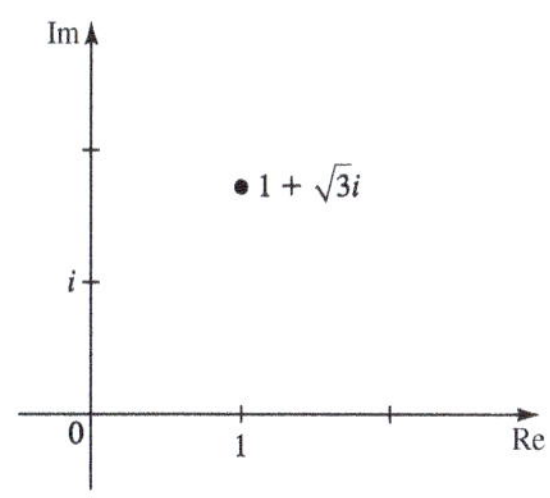

(b) $2\left(\cos\dfrac{\pi}{3} + i\sin\dfrac{\pi}{3}\right)$

(c) -512 **5.** $-8, \sqrt{3} + i$

6. $-3i, 3\left(\pm\dfrac{\sqrt{3}}{2} + \dfrac{1}{2}i\right)$

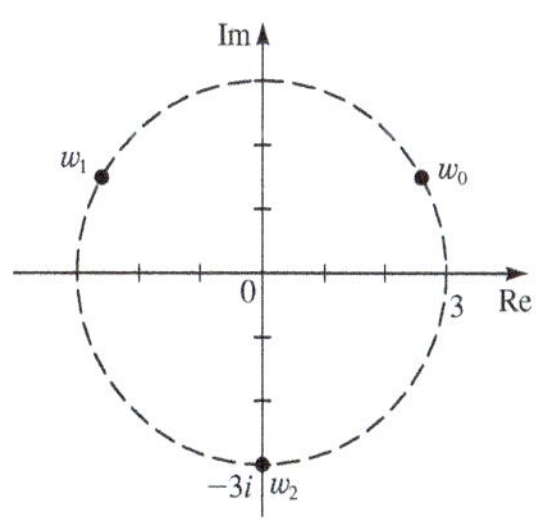

7. (a)

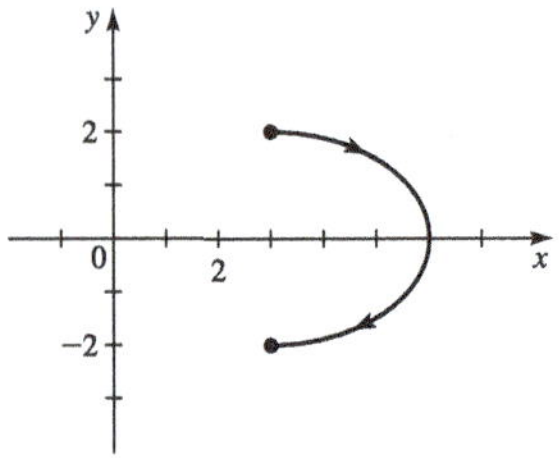

(b) $\dfrac{(x-3)^2}{9} + \dfrac{y^2}{4} = 1, x \geq 3$

8. $x = 3 + t, y = 5 + 2t$

9. (a) 3, (0, 3), clockwise, π **(b)** $x = 3 \sin 4t, y = 3 \cos 4t$ **(c)** $x^2 + y^2 = 9$ **(d)** $r = 3$

FOCUS ON MODELING ■ PAGE 627

1. $y = -\left(\dfrac{g}{2v_0^2 \cos^2\theta}\right)x^2 + (\tan\theta)x$

3. (a) 62.26 s **(b)** 15,500 ft **(c)** 5426 ft

(d)

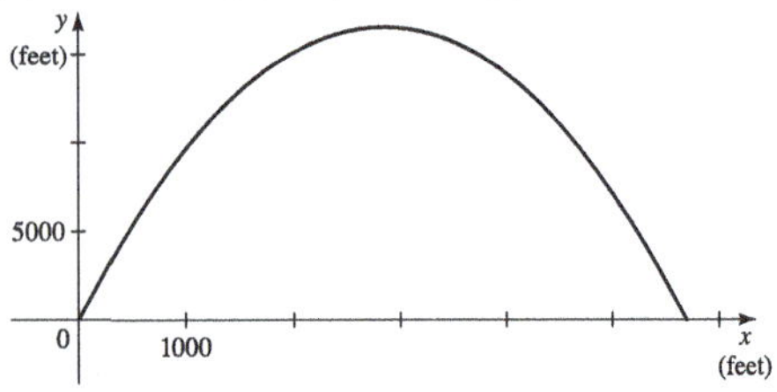

7. No, $\theta \approx 23°$

CHAPTER 9

SECTION 9.1 ■ PAGE 637

1. (a) *A, B*

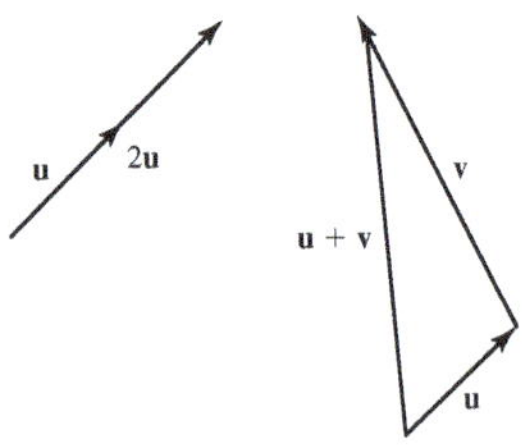

(b) (2, 1), (4, 3), ⟨2, 2⟩, ⟨−3, 6⟩, ⟨4, 4⟩, ⟨−1, 8⟩

2. (a) $\sqrt{a_1^2 + a_2^2}, 2\sqrt{2}$ **(b)** $\langle |\mathbf{w}| \cos\theta, |\mathbf{w}| \sin\theta \rangle$

3.

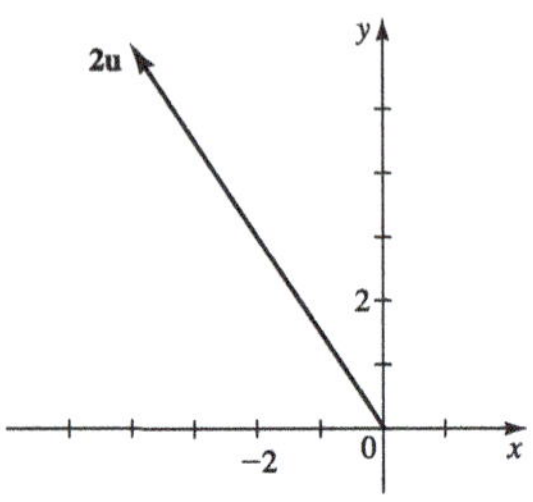

5.

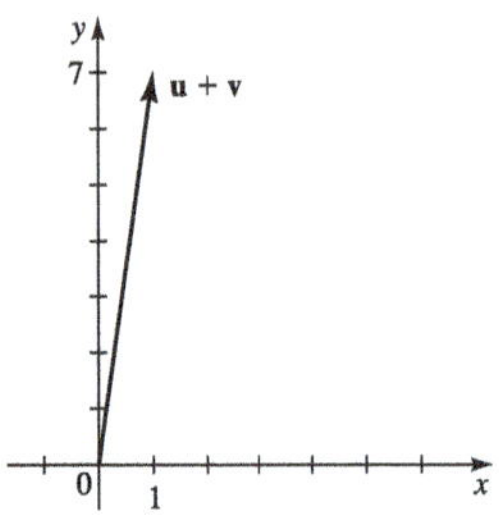

7.

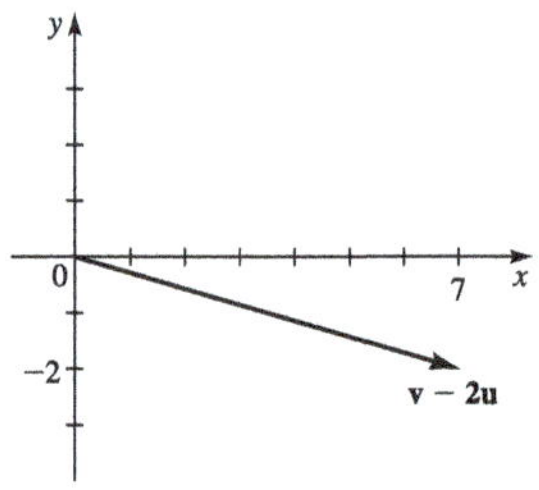

9. ⟨3, 3⟩ **11.** ⟨3, −1⟩ **13.** ⟨5, 7⟩ **15.** ⟨−4, −3⟩ **17.** ⟨0, 2⟩

19.

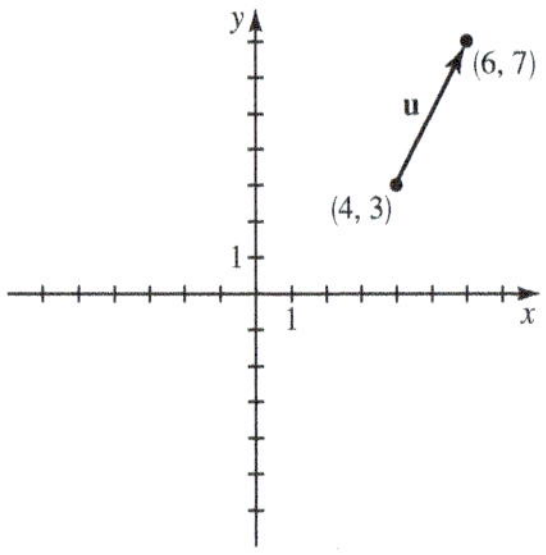

21.

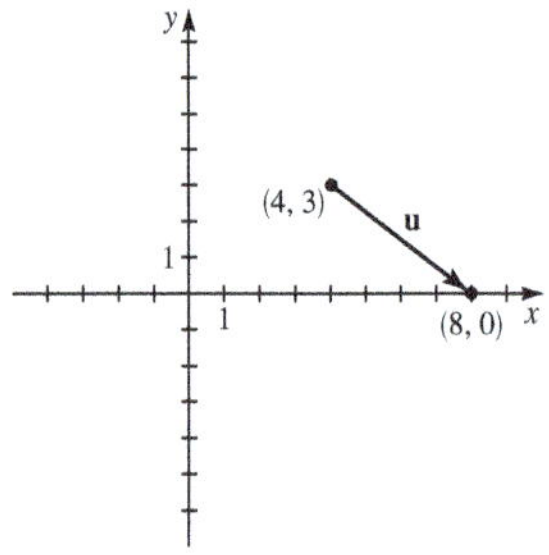

23.

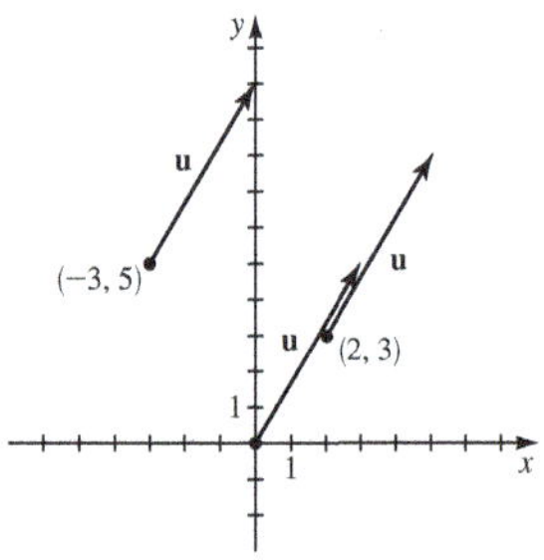

25.

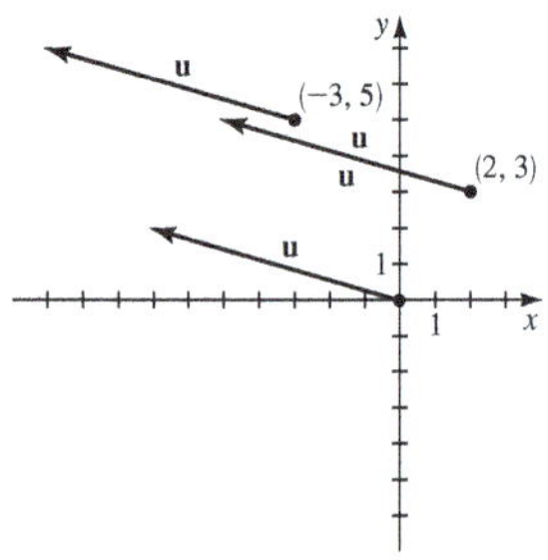

27. $\mathbf{i} + 4\mathbf{j}$ **29.** $3\mathbf{i}$

31. ⟨4, 14⟩, ⟨−9, −3⟩, ⟨5, 8⟩, ⟨−6, 17⟩

33. ⟨0, −2⟩, ⟨6, 0⟩, ⟨−2, −1⟩, ⟨8, −3⟩

35. $4\mathbf{i}, -9\mathbf{i} + 6\mathbf{j}, 5\mathbf{i} - 2\mathbf{j}, -6\mathbf{i} + 8\mathbf{j}$

37. $\sqrt{5}, \sqrt{13}, 2\sqrt{5}, \frac{1}{2}\sqrt{13}, \sqrt{26}, \sqrt{10}, \sqrt{5} - \sqrt{13}$

39. $\sqrt{101}, 2\sqrt{2}, 2\sqrt{101}, \sqrt{2}, \sqrt{73}, \sqrt{145}, \sqrt{101} - 2\sqrt{2}$

41. $20\sqrt{3}\,\mathbf{i} + 20\mathbf{j}$ **43.** $-\dfrac{\sqrt{2}}{2}\mathbf{i} - \dfrac{\sqrt{2}}{2}\mathbf{j}$

45. $4 \cos 10°\mathbf{i} + 4 \sin 10°\mathbf{j} \approx 3.94\mathbf{i} + 0.69\mathbf{j}$

47. 5, 53.13° **49.** 13, 157.38° **51.** 2, 60° **53.** $15\sqrt{3}, -15$

55. $2\mathbf{i} - 3\mathbf{j}$ **57.** S 84.26° W **59. (a)** $40\mathbf{j}$ **(b)** $425\mathbf{i}$ **(c)** $425\mathbf{i} + 40\mathbf{j}$ **(d)** 427 mi/h, N 84.6° E

61. 794 mi/h, N 26.6° W

63. (a) $10\mathbf{i}$ **(b)** $10\mathbf{i} + 10\sqrt{3}\mathbf{j}$ **(c)** $20\mathbf{i} + 10\sqrt{3}\mathbf{j}$ **(d)** 26.5 mi/h, N 49.1° E

65. (a) $22.8\mathbf{i} + 7.4\mathbf{j}$ **(b)** 7.4 mi/h, 22.8 mi/h

67. (a) ⟨5, −3⟩ **(b)** ⟨−5, 3⟩ **69. (a)** $-4\mathbf{j}$ **(b)** $4\mathbf{j}$

71. (a) ⟨−7.57, 10.61⟩ **(b)** ⟨7.57, −10.61⟩

73. $\mathbf{T}_1 \approx -56.5\mathbf{i} + 67.4\mathbf{j}, \mathbf{T}_2 \approx 56.5\mathbf{i} + 32.6\mathbf{j}$

SECTION 9.2 ■ PAGE 646

1. $a_1b_1 + a_2b_2$; real number or scalar

2. $\dfrac{\mathbf{u} \cdot \mathbf{v}}{|\mathbf{u}||\mathbf{v}|}$; perpendicular

3. (a) $\dfrac{\mathbf{u}\cdot\mathbf{v}}{|\mathbf{v}|}$ **(b)** $\left(\dfrac{\mathbf{u}\cdot\mathbf{v}}{|\mathbf{v}|^2}\right)\mathbf{v}$

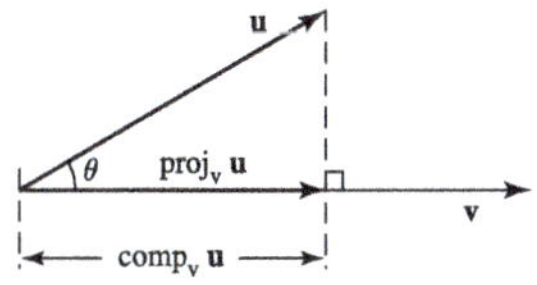

4. $\mathbf{F}\cdot\mathbf{D}$ **5. (a)** 2 **(b)** 45° **7. (a)** 13 **(b)** 56°
9. (a) −1 **(b)** 97° **11. (a)** $5\sqrt{3}$ **(b)** 30°
13. (a) 1 **(b)** 86° **15.** Yes **17.** No **19.** Yes
21. 9 **23.** −5 **25.** $-\frac{12}{5}$ **27.** −24
29. (a) $\langle 1, 1\rangle$ **(b)** $\mathbf{u}_1 = \langle 1, 1\rangle, \mathbf{u}_2 = \langle -3, 3\rangle$
31. (a) $\langle -\frac{1}{2}, \frac{3}{2}\rangle$ **(b)** $\mathbf{u}_1 = \langle -\frac{1}{2}, \frac{3}{2}\rangle, \mathbf{u}_2 = \langle \frac{3}{2}, \frac{1}{2}\rangle$
33. (a) $\langle -\frac{18}{5}, \frac{24}{5}\rangle$ **(b)** $\mathbf{u}_1 = \langle -\frac{18}{5}, \frac{24}{5}\rangle, \mathbf{u}_2 = \langle \frac{28}{5}, \frac{21}{5}\rangle$
35. −28 **37.** 25 **45.** 16 ft-lb **47.** 8660 ft-lb
49. (a) 2822 lb **(b)** 2779 lb **51.** 23.6°

SECTION 9.3 ■ PAGE 652

1. x, y, z; $(5, 2, 3)$; $y = 2$

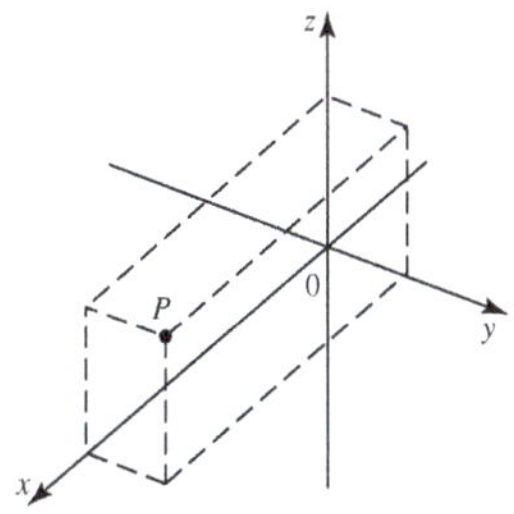

2. $\sqrt{(x_2 - x_1)^2 + (y_2 - y_1)^2 + (z_2 - z_1)^2}$;
$\sqrt{38}$; $(x - 5)^2 + (y - 2)^2 + (z - 3)^2 = 9$

3. (a)

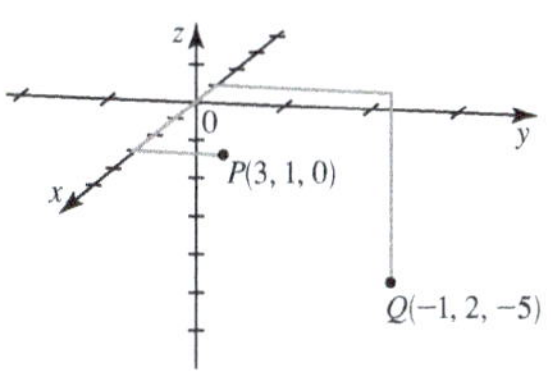

(b) $\sqrt{42}$

5. (a)

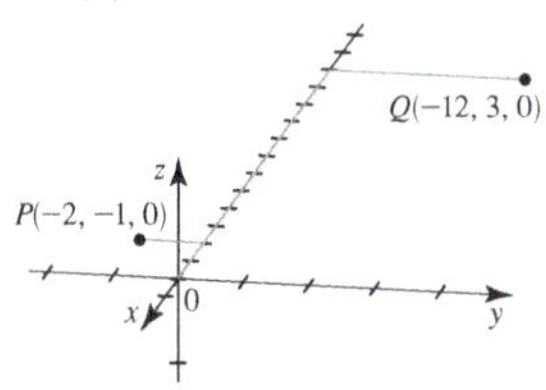

(b) $2\sqrt{29}$

7. Plane parallel to the yz-plane

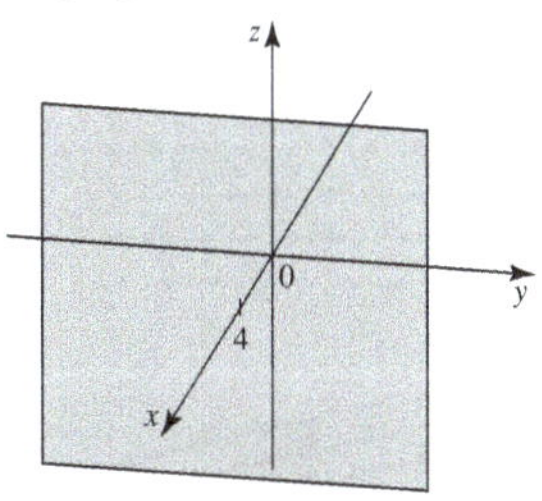

9. Plane parallel to the xy-plane

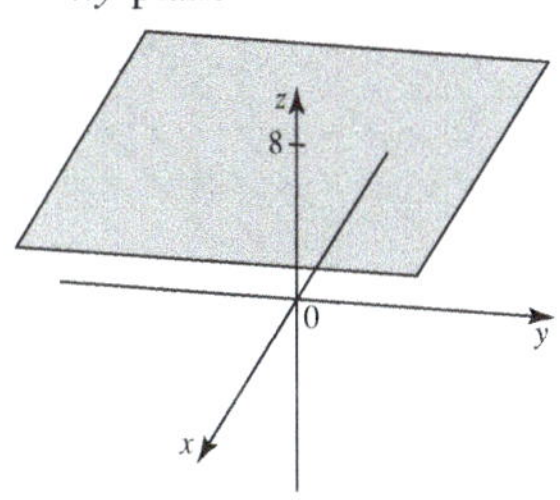

11. $(x - 2)^2 + (y + 5)^2 + (z - 3)^2 = 25$
13. $(x - 3)^2 + (y + 1)^2 + z^2 = 6$
15. Center: $(5, -1, -4)$, radius: $\sqrt{51}$
17. Center: $(6, 1, 0)$, radius: $\sqrt{37}$
19. (a) Circle, center: $(0, 2, -10)$, radius: $3\sqrt{11}$
(b) Circle, center: $(4, 2, -10)$, radius: $5\sqrt{3}$ **21. (a)** 3

SECTION 9.4 ■ PAGE 658

1. unit, $a_1\mathbf{i} + a_2\mathbf{j} + a_3\mathbf{k}$;
$\sqrt{a_1^2 + a_2^2 + a_3^2}$; $4\mathbf{i} + (-2)\mathbf{j} + \mathbf{k}$, $\langle 0, 7, -24\rangle$

2. $\dfrac{\mathbf{u}\cdot\mathbf{v}}{|\mathbf{u}||\mathbf{v}|}$; 0; 0, perpendicular **3.** $\langle -1, -1, 5\rangle$
5. $\langle -6, -2, 0\rangle$ **7.** $(5, 4, -1)$ **9.** $(1, 0, -1)$ **11.** 3
13. $5\sqrt{2}$ **15.** $\langle 2, -3, 2\rangle, \langle 2, -11, 4\rangle, \langle 6, -23, \frac{19}{2}\rangle$
17. $\mathbf{i} - 2\mathbf{k}, \mathbf{i} + 2\mathbf{j} + 2\mathbf{k}, 3\mathbf{i} + \frac{7}{2}\mathbf{j} + \mathbf{k}$ **19.** $12\mathbf{i} + 2\mathbf{k}$
21. $3\mathbf{i} - 3\mathbf{j}$ **23. (a)** $\langle 3, 1, -2\rangle$ **(b)** $3\mathbf{i} + \mathbf{j} - 2\mathbf{k}$ **25.** −4
27. 1 **29.** Yes **31.** No **33.** 116.4° **35.** 100.9°
37. $\alpha \approx 65°, \beta \approx 56°, \gamma = 45°$ **39.** $\alpha \approx 73°, \beta \approx 65°$,
$\gamma \approx 149°$ **41.** 45° **43.** 125° **47. (a)** Parallel, $\mathbf{v} = -2\mathbf{u}$
(b) Parallel, $\mathbf{v} = -\frac{4}{3}\mathbf{u}$ **(c)** Not parallel
49. (a) $-7\mathbf{i} - 24\mathbf{j} + 25\mathbf{k}$ **(b)** $25\sqrt{2}$

SECTION 9.5 ■ PAGE 665

1. $\begin{vmatrix} \mathbf{i} & \mathbf{j} & \mathbf{k} \\ a_1 & a_2 & a_3 \\ b_1 & b_2 & b_3 \end{vmatrix} = (a_2b_3 - a_3b_2)\mathbf{i} + (a_3b_1 - a_1b_3)\mathbf{j} + (a_1b_2 - a_2b_1)\mathbf{k}$, $-3\mathbf{i} + 2\mathbf{j} + 3\mathbf{k}$

2. perpendicular; perpendicular **3.** $9\mathbf{i} - 6\mathbf{j} + 3\mathbf{k}$
5. $\mathbf{0}$ **7.** $-4\mathbf{i} + 7\mathbf{j} - 3\mathbf{k}$
9. (a) $\langle 0, 2, 2\rangle$ **(b)** $\left\langle 0, \dfrac{\sqrt{2}}{2}, \dfrac{\sqrt{2}}{2}\right\rangle$
11. (a) $14\mathbf{i} + 7\mathbf{j}$ **(b)** $\dfrac{2\sqrt{5}}{5}\mathbf{i} + \dfrac{\sqrt{5}}{5}\mathbf{j}$
13. $\dfrac{3\sqrt{3}}{2}$ **15.** 100 **17.** $\langle 0, 2, 2\rangle$ **19.** $\langle 10, -10, 0\rangle$
21. $4\sqrt{6}$ **23.** $\dfrac{5\sqrt{14}}{2}$ **25.** $\sqrt{14}$ **27.** $18\sqrt{3}$
29. (a) 0 **(b)** Yes **31. (a)** 55 **(b)** No, 55
33. (a) −2 **(b)** No, 2
35. (a) $2{,}700{,}000\sqrt{3}$ **(b)** 4677 liters

SECTION 9.6 ■ PAGE 669

1. parametric; $x = x_0 + at, y = y_0 + bt, z = z_0 + ct$
2. $a(x - x_0) + b(y - y_0) + c(z - z_0) = 0$
3. $x = 1 + 3t, y = 2t, z = -2 - 3t$
5. $x = 3, y = 2 - 4t, z = 1 + 2t$
7. $x = 1 + 2t, y = 0, z = -2 - 5t$
9. $x = 1 + t, y = -3 + 4t, z = 2 - 3t$
11. $x = 1 - t, y = 1 + t, z = 2t$
13. $x = 3 + 4t, y = 7 - 4t, z = -5$
15. (a) $x + y - z = 5$ **(b)** x-intercept 5, y-intercept 5, z-intercept −5

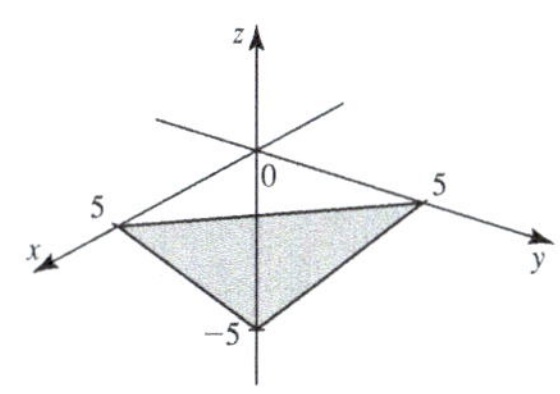

17. (a) $6x - z = 4$ **(b)** x-intercept $\frac{2}{3}$, no y-intercept, z-intercept −4

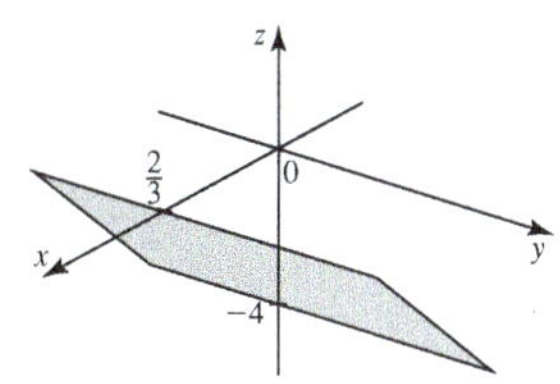

19. (a) $3x - y + 2z = -8$ **(b)** x-intercept $-\frac{8}{3}$, y-intercept 8, z-intercept -4

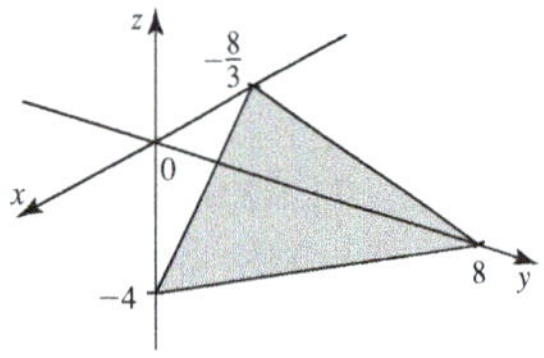

21. $5x - 3y - z = 35$ **23.** $x - 3y = 2$
25. $2x - 3y - 9z = 0$ **27.** $x = 2t, y = 5t, z = 4 - 4t$
29. $x = 2, y = -1 + t, z = 5$ **31.** $12x + 4y + 3z = 12$
33. $4x - 3y - z = -10$

CHAPTER 9 REVIEW ■ PAGE 673

1. $\sqrt{13}, \langle 6, 4\rangle, \langle -10, 2\rangle, \langle -4, 6\rangle, \langle -22, 7\rangle$
3. $\sqrt{5}, 3\mathbf{i} - \mathbf{j}, \mathbf{i} + 3\mathbf{j}, 4\mathbf{i} + 2\mathbf{j}, 4\mathbf{i} + 7\mathbf{j}$
5. $\langle 3, -4\rangle$ **7.** 4, 120° **9.** $\langle 10, 10\sqrt{3}\rangle$
11. (a) $10^4(4.8\mathbf{i} + 0.4\mathbf{j})$ **(b)** 4.8×10^4 lb, N 85.2° E
13. 5, 25, 60 **15.** $2\sqrt{2}$, 8, 0 **17.** Yes **19.** No, 45°
21. (a) $\dfrac{17\sqrt{37}}{37}$ **(b)** $\langle \frac{102}{37}, -\frac{17}{37}\rangle$
(c) $\mathbf{u}_1 = \langle \frac{102}{37}, -\frac{17}{37}\rangle, \mathbf{u}_2 = \langle \frac{9}{37}, \frac{54}{37}\rangle$
23. (a) $-\dfrac{14\sqrt{97}}{97}$ **(b)** $-\frac{56}{97}\mathbf{i} + \frac{126}{97}\mathbf{j}$
(c) $\mathbf{u}_1 = -\frac{56}{97}\mathbf{i} + \frac{126}{97}\mathbf{j}, \mathbf{u}_2 = \frac{153}{97}\mathbf{i} + \frac{68}{97}\mathbf{j}$
25. 3

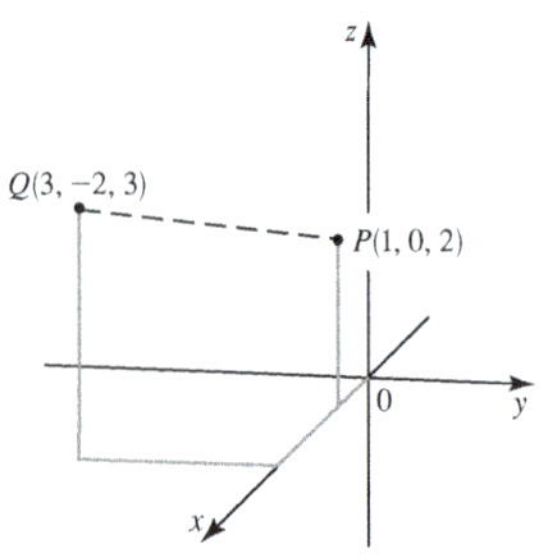

27. $x^2 + y^2 + z^2 = 36$
29. Center: $(1, 3, -2)$, radius: 4
31. $6, \langle 6, 1, 3\rangle, \langle 2, -5, 5\rangle, \langle -1, -\frac{15}{2}, 5\rangle$
33. (a) -1 **(b)** No, 92.8° **35. (a)** 0 **(b)** Yes
37. (a) $\langle -2, 17, -5\rangle$
(b) $\left\langle -\dfrac{\sqrt{318}}{159}, \dfrac{17\sqrt{318}}{318}, -\dfrac{5\sqrt{318}}{318}\right\rangle$
39. (a) $\mathbf{i} + \mathbf{j} + 2\mathbf{k}$
(b) $\dfrac{\sqrt{6}}{6}\mathbf{i} + \dfrac{\sqrt{6}}{6}\mathbf{j} + \dfrac{\sqrt{6}}{3}\mathbf{k}$
41. $\frac{15}{2}$ **43.** 9 **45.** $x = 2 + 3t, y = t, z = -6$
47. $x = 6 - 2t, y = -2 + 3t, z = -3 + t$
49. $2x + 3y - 5z = 2$ **51.** $x + y + 3z = 5$
53. $x = 2 - 2t, y = 0, z = -4t$

CHAPTER 9 TEST ■ PAGE 675

1. (a)

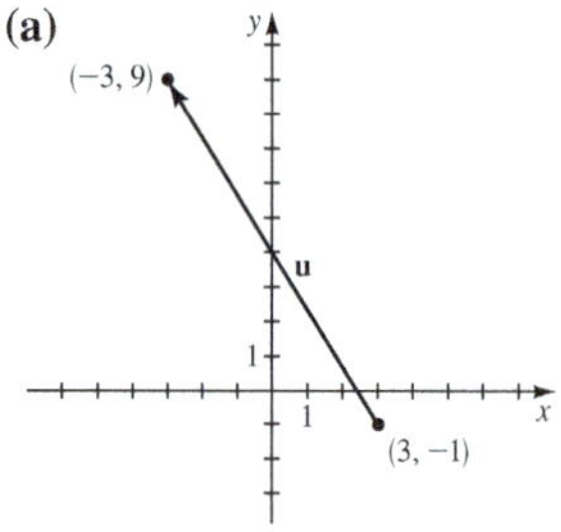

(b) $-6\mathbf{i} + 10\mathbf{j}$ **(c)** $2\sqrt{34}$

2. (a) $\langle 19, -3\rangle$ **(b)** $5\sqrt{2}$ **(c)** 0 **(d)** Yes
3. (a)

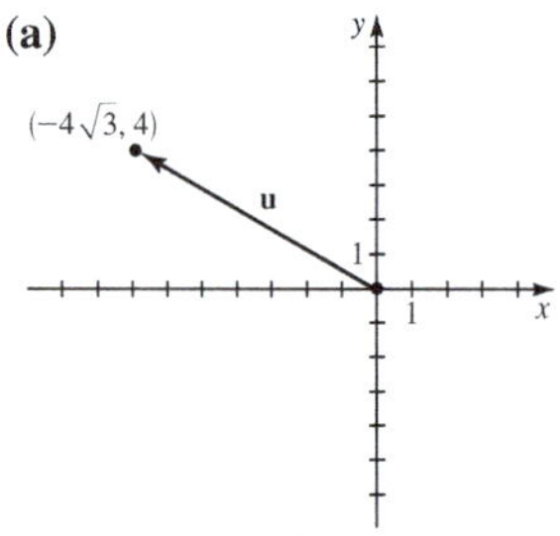

(b) 8, 150°

4. (a) $14\mathbf{i} + 6\sqrt{3}\mathbf{j}$ **(b)** 17.4 mi/h, N 53.4° E
5. (a) 45° **(b)** $\dfrac{\sqrt{26}}{2}$ **(c)** $\frac{5}{2}\mathbf{i} - \frac{1}{2}\mathbf{j}$ **6.** 90
7. (a) 6 **(b)** $(x - 4)^2 + (y - 3)^2 + (z + 1)^2 = 36$
(c) $\langle 2, -4, 4\rangle = 2\mathbf{i} - 4\mathbf{j} + 4\mathbf{k}$ **8. (a)** $11\mathbf{i} - 4\mathbf{j} - \mathbf{k}$ **(b)** $\sqrt{6}$
(c) -1 **(d)** $-3\mathbf{i} - 7\mathbf{j} - 5\mathbf{k}$ **(e)** $3\sqrt{35}$ **(f)** 18 **(g)** 96°
9. $\left\langle \dfrac{7\sqrt{6}}{18}, \dfrac{\sqrt{6}}{9}, -\dfrac{\sqrt{6}}{18}\right\rangle, \left\langle -\dfrac{7\sqrt{6}}{18}, -\dfrac{\sqrt{6}}{9}, \dfrac{\sqrt{6}}{18}\right\rangle$
10. (a) $\langle 4, -3, 4\rangle$ **(b)** $4x - 3y + 4z = 4$ **(c)** $\dfrac{\sqrt{41}}{2}$
11. $x = 2 - 2t, y = -4 + t, z = 7 - 2t$

FOCUS ON MODELING ■ PAGE 678

1.

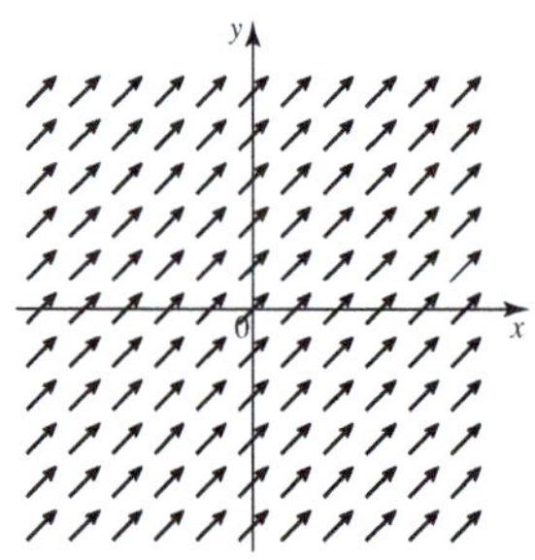

3.

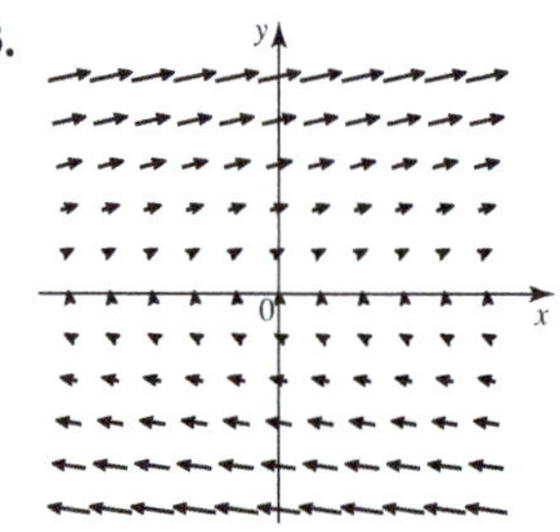

5.

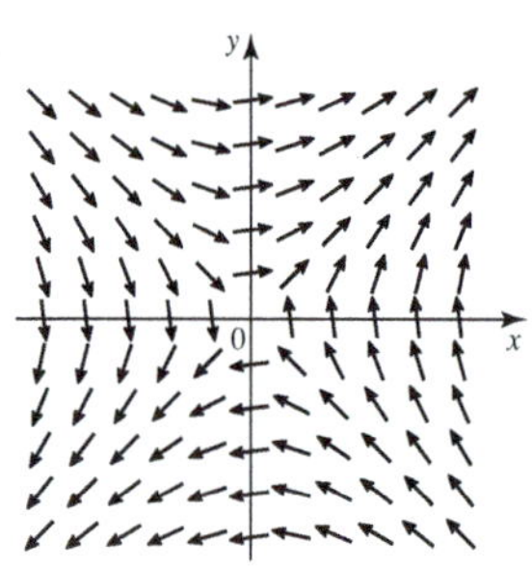

7.

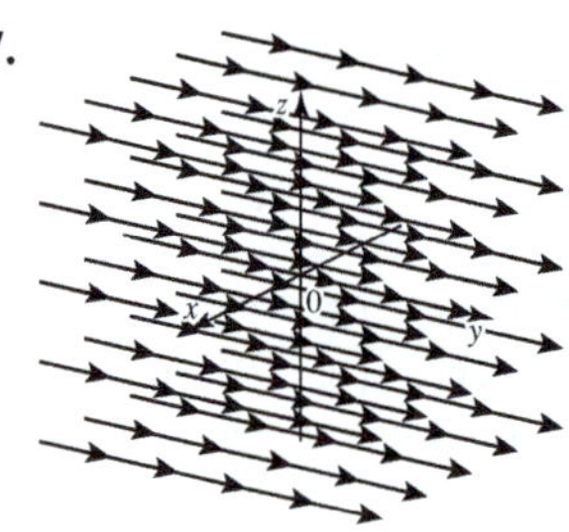

9.

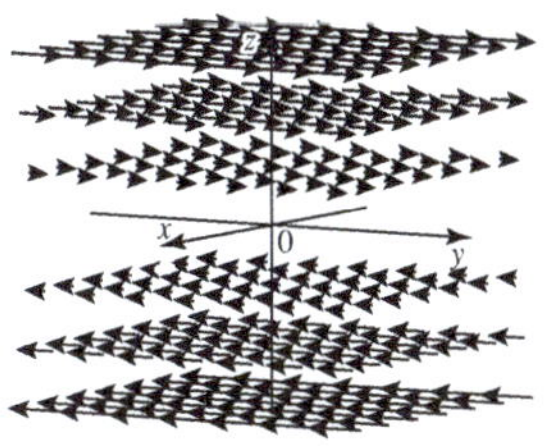

11. II **13.** I **15.** IV **17.** III

19.

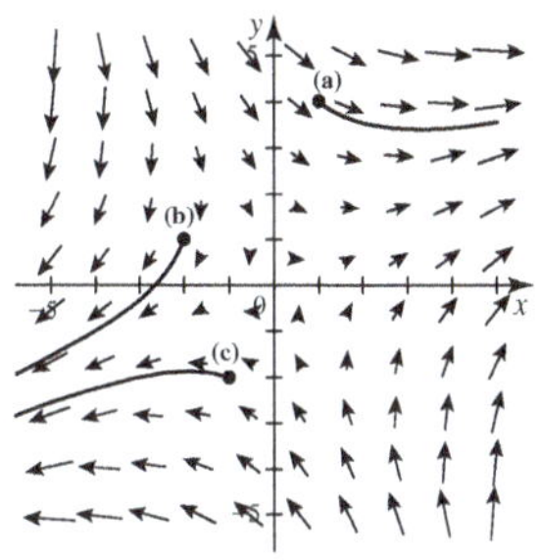

CHAPTER 10

SECTION 10.1 ■ PAGE 688

1. x, y; equation; $(2, 1)$ **2.** substitution, elimination, graphical **3.** no, infinitely many **4.** infinitely many; $1 - t$; $(1, 0), (-3, 4), (5, -4)$ **5.** $(3, 2)$ **7.** $(3, 1)$ **9.** $(2, 1)$ **11.** $(-3, 2)$ **13.** $(-2, 3)$
15. $(2, -2)$ **17.** No solution

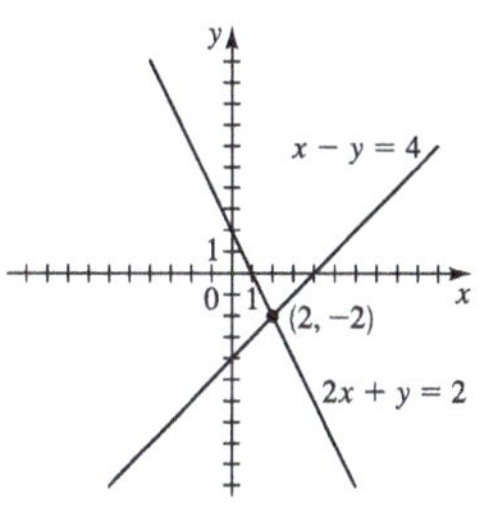

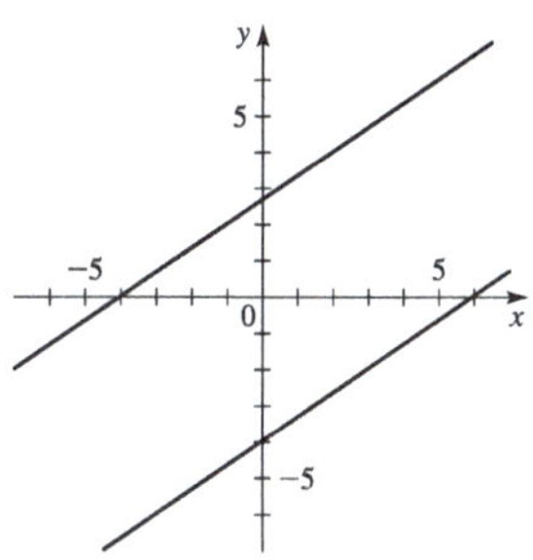

19. Infinitely many solutions

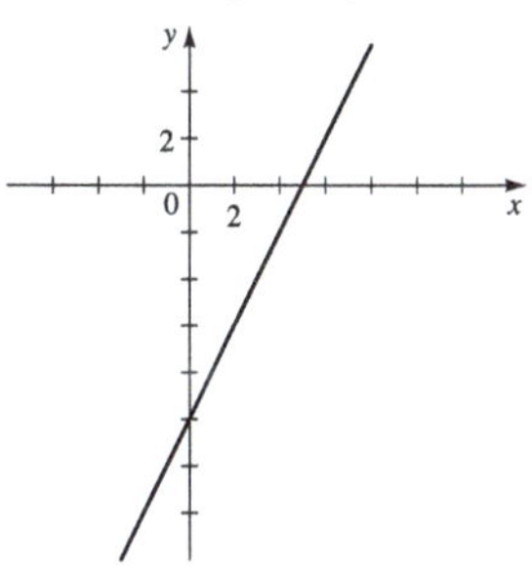

21. $(2, 2)$ **23.** $(3, -1)$ **25.** $(2, 1)$ **27.** $(3, 5)$ **29.** $(1, 3)$ **31.** $(6, -6)$ **33.** $(10, -9)$ **35.** $(2, 1)$ **37.** No solution **39.** $\left(x, \frac{1}{3}x - \frac{5}{3}\right)$ **41.** $\left(x, 3 - \frac{3}{2}x\right)$ **43.** $(-3, -7)$ **45.** $\left(x, 5 - \frac{5}{6}x\right)$ **47.** $(5, 10)$ **49.** No solution

51. $(3.87, 2.74)$ **53.** $(61.00, 20.00)$ **55.** $\left(-\dfrac{1}{a-1}, \dfrac{1}{a-1}\right)$

57. $\left(\dfrac{1}{a+b}, \dfrac{1}{a+b}\right)$ **59.** 22, 12 **61.** 5 dimes, 9 quarters

63. 200 gallons of regular gas, 80 gallons of premium gas

65. Plane's speed 120 mi/h, wind speed 30 mi/h **67.** 200 g of A, 40 g of B **69.** 25%, 10% **71.** \$14,000 at 5%, \$6,000 at 8% **73.** John $2\frac{1}{4}$ h, Mary $2\frac{1}{2}$ h **75.** 25

SECTION 10.2 ■ PAGE 696

1. $x + 3z = 1$ **2.** -3; $4y - 5z = -4$ **3.** Linear **5.** Nonlinear **7.** $(5, 1, -2)$ **9.** $(4, 0, 3)$ **11.** $\left(5, 2, -\frac{1}{2}\right)$

13. $\begin{cases} 3x + y + z = 4 \\ -y + z = -1 \\ x - 2y - z = -1 \end{cases}$ **15.** $\begin{cases} 2x + y - 3z = 5 \\ 2x + 3y + z = 13 \\ -8y + 8z = -8 \end{cases}$

17. $(2, 1, -3)$ **19.** $(1, -1, 5)$ **21.** $(1, 2, 1)$ **23.** $(5, 0, 1)$ **25.** $(0, 1, 2)$ **27.** $\left(\frac{1}{4}, \frac{1}{2}, -\frac{1}{2}\right)$ **29.** No solution **31.** No solution **33.** $(3 - t, -3 + 2t, t)$ **35.** $\left(2 - 2t, -\frac{2}{3} + \frac{4}{3}t, t\right)$ **37.** $(1, -1, 1, 2)$

39. \$30,000 in short-term, \$30,000 in intermediate-term, \$40,000 in long-term **41.** 250 acres corn, 500 acres wheat, 450 acres soybeans **43.** Impossible **45.** 50 Midnight Mango, 60 Tropical Torrent, 30 Pineapple Power **47.** 1500 shares of A, 1200 shares of B, 1000 shares of C

SECTION 10.3 ■ PAGE 709

1. dependent, inconsistent

2. $\begin{bmatrix} 1 & 1 & -1 & 1 \\ 1 & 0 & 2 & -3 \\ 0 & 2 & -1 & 3 \end{bmatrix}$

3. **(a)** x and y **(b)** dependent **(c)** $x = 3 + t, y = 5 - 2t, z = t$ **4.** **(a)** $x = 2, y = 1, z = 3$ **(b)** $x = 2 - t, y = 1 - t, z = t$ **(c)** No solution **5.** 3×2 **7.** 2×1 **9.** 1×3

11. $\begin{bmatrix} 3 & 1 & -1 & 2 \\ 2 & -1 & 0 & 1 \\ 1 & 0 & -1 & 3 \end{bmatrix}$

13. **(a)** Yes **(b)** Yes **(c)** $\begin{cases} x = -3 \\ y = 5 \end{cases}$

15. **(a)** Yes **(b)** No **(c)** $\begin{cases} x + 2y + 8z = 0 \\ y + 3z = 2 \\ 0 = 0 \end{cases}$

17. **(a)** No **(b)** No **(c)** $\begin{cases} x = 0 \\ 0 = 0 \\ y + 5z = 1 \end{cases}$

19. **(a)** Yes **(b)** Yes **(c)** $\begin{cases} x + 3y - w = 0 \\ z + 2w = 0 \\ 0 = 1 \\ 0 = 0 \end{cases}$

21. $\begin{bmatrix} -1 & 1 & 2 & 0 \\ 0 & 4 & 7 & 4 \\ 1 & -2 & -1 & -1 \end{bmatrix}$ **23.** $\begin{bmatrix} 2 & 1 & -3 & 5 \\ 2 & 3 & 1 & 13 \\ 0 & -8 & 8 & -8 \end{bmatrix}$

25. **(a)** $\begin{cases} x - 2y + 4z = 3 \\ y + 2z = 7 \\ z = 2 \end{cases}$ **(b)** $(1, 3, 2)$

27. **(a)** $\begin{cases} x + 2y + 3z - w = 7 \\ y - 2z = 5 \\ z + 2w = 5 \\ w = 3 \end{cases}$ **(b)** $(7, 3, -1, 3)$

29. $(1, 1, 2)$ **31.** $(1, 0, 1)$ **33.** $(-1, 0, 1)$ **35.** $(-1, 5, 0)$ **37.** $(10, 3, -2)$ **39.** No solution **41.** $(2 - 3t, 3 - 5t, t)$ **43.** No solution **45.** $(-2t + 5, t - 2, t)$ **47.** $\left(-\frac{1}{2}s + t + 6, s, t\right)$ **49.** $(-2, 1, 3)$ **51.** No solution **53.** $(-9, 2, 0)$ **55.** $(5 - t, -3 + 5t, t)$ **57.** $(0, -3, 0, -3)$ **59.** $(-1, 0, 0, 1)$ **61.** $\left(\frac{1}{3}s - \frac{2}{3}t, \frac{1}{3}s + \frac{1}{3}t, s, t\right)$ **63.** $\left(\frac{7}{4} - \frac{7}{4}t, -\frac{7}{4} + \frac{3}{4}t, \frac{9}{4} + \frac{3}{4}t, t\right)$ **65.** $x = 1.25, y = -0.25, z = 0.75$ **67.** $x = 1.2, y = 3.4, z = -5.2, w = -1.3$ **69.** 2 VitaMax, 1 Vitron, 2 VitaPlus **71.** 5-mile run, 2-mile swim, 30-mile cycle **73.** Impossible

SECTION 10.4 ■ PAGE 720

1. dimension **2. (a)** columns, rows **(b)** (ii), (iii) **3.** (i), (ii)

4. $\begin{bmatrix} 4 & 9 & -7 \\ 7 & -7 & 0 \\ 4 & -5 & -5 \end{bmatrix}$ **5.** No **7.** $a = -5, b = 3$

9. $\begin{bmatrix} 1 & 3 \\ 1 & 5 \end{bmatrix}$ **11.** $\begin{bmatrix} 3 & 6 \\ 12 & -3 \\ 3 & 0 \end{bmatrix}$ **13.** Impossible

15. $\begin{bmatrix} 5 & 2 & 1 \\ 7 & 10 & -7 \end{bmatrix}$ **17.** $\begin{bmatrix} -1 & -\frac{1}{2} \\ 1 & 2 \end{bmatrix}$ **19.** Impossible

21. $\begin{bmatrix} 0 & -5 \\ -25 & -20 \\ -10 & 10 \end{bmatrix}$ **23. (a)** $\begin{bmatrix} 5 & -2 & 5 \\ 1 & 1 & 0 \end{bmatrix}$ **(b)** Impossible

25. (a) $\begin{bmatrix} 10 & -25 \\ 0 & 35 \end{bmatrix}$ **(b)** Impossible

27. (a) Impossible **(b)** $\begin{bmatrix} 14 & -14 \end{bmatrix}$

29. (a) $\begin{bmatrix} -4 & 7 \\ 14 & -7 \end{bmatrix}$ **(b)** $\begin{bmatrix} 6 & -8 \\ 4 & -17 \end{bmatrix}$

31. (a) $\begin{bmatrix} 5 & -3 & 10 \\ 6 & 1 & 0 \\ -5 & 2 & 2 \end{bmatrix}$ **(b)** $\begin{bmatrix} -1 \\ 8 \\ -1 \end{bmatrix}$

33. (a) $\begin{bmatrix} 4 & -45 \\ 0 & 49 \end{bmatrix}$ **(b)** $\begin{bmatrix} 8 & -335 \\ 0 & 343 \end{bmatrix}$

35. (a) $\begin{bmatrix} 13 \\ -7 \end{bmatrix}$ **(b)** Impossible **37.** $\begin{bmatrix} 1.56 & -5.62 \\ 1.28 & -0.88 \\ -1.09 & 0.97 \end{bmatrix}$

39. $\begin{bmatrix} -0.35 & 0.03 & 0.33 \\ -0.55 & -1.05 & 1.05 \\ -2.41 & -4.31 & 4.46 \end{bmatrix}$ **41.** Impossible

43. $x = 2, y = -1$ **45.** $x = 1, y = -2$

47. $\begin{bmatrix} 2 & -5 \\ 3 & 2 \end{bmatrix}\begin{bmatrix} x \\ y \end{bmatrix} = \begin{bmatrix} 7 \\ 4 \end{bmatrix}$

49. $\begin{bmatrix} 3 & 2 & -1 & 1 \\ 1 & 0 & -1 & 0 \\ 0 & 3 & 1 & -1 \end{bmatrix}\begin{bmatrix} x_1 \\ x_2 \\ x_3 \\ x_4 \end{bmatrix} = \begin{bmatrix} 0 \\ 5 \\ 4 \end{bmatrix}$

51. Only ACB is defined. $ACB = \begin{bmatrix} -3 & -21 & 27 & -6 \\ -2 & -14 & 18 & -4 \end{bmatrix}$

53. (a) $\begin{bmatrix} 5 \\ 22 \\ 7 \end{bmatrix}$

(b) Five members have no postsecondary education, 22 have 1 to 4 years, and seven have more than 4 years.

55. (a) $\begin{bmatrix} 353.75 \\ 656.25 \\ 892.50 \end{bmatrix}$ **(b)** \$353.75 **(c)** \$1902.50

57. (a) $\begin{bmatrix} \$32{,}000 & \$18{,}000 \\ \$42{,}000 & \$26{,}800 \\ \$44{,}000 & \$26{,}800 \end{bmatrix}$ **(b)** \$42,000 **(c)** \$71,600

59. (a) $\begin{bmatrix} 97.00 \\ 46.50 \\ 41.00 \end{bmatrix}$ Amy's stand sold \$97 of produce on Saturday. Beth's stand sold \$46.50. Chad's stand sold \$41.

(b) $\begin{bmatrix} 70.00 \\ 33.50 \\ 48.50 \end{bmatrix}$ Amy's stand sold \$70 of produce on Sunday. Beth's stand sold \$33.50. Chad's stand sold \$48.50.

(c) $\begin{bmatrix} 220 & 110 & 90 \\ 75 & 45 & 50 \\ 120 & 55 & 50 \end{bmatrix}$ This represents the number of melons, squash, and tomatoes they sold during the weekend.

(d) $\begin{bmatrix} 167.00 \\ 80.00 \\ 89.50 \end{bmatrix}$ During the weekend Amy's stand sold \$167, Beth's stand sold \$80, and Chad's stand sold \$89.50 of produce.

SECTION 10.5 ■ PAGE 731

1. (a) identity **(b)** A, A **(c)** inverse

2. (a) $\overset{A}{\begin{bmatrix} 5 & 3 \\ 3 & 2 \end{bmatrix}}\overset{X}{\begin{bmatrix} x \\ y \end{bmatrix}} = \overset{B}{\begin{bmatrix} 4 \\ 3 \end{bmatrix}}$ **(b)** $\begin{bmatrix} 2 & -3 \\ -3 & 5 \end{bmatrix}$

(c) $\overset{A^{-1}}{\begin{bmatrix} 2 & -3 \\ -3 & 5 \end{bmatrix}}\overset{B}{\begin{bmatrix} 4 \\ 3 \end{bmatrix}} = \begin{bmatrix} -1 \\ 3 \end{bmatrix}$ **(d)** $x = -1, y = 3$

7. $\begin{bmatrix} 1 & -2 \\ -\frac{3}{2} & \frac{7}{2} \end{bmatrix}$ **9.** $\begin{bmatrix} \frac{1}{3} & -\frac{1}{2} \\ 2 & 2 \end{bmatrix}$ **11.** $\begin{bmatrix} 3 & 5 \\ -2 & -3 \end{bmatrix}$

13. $\begin{bmatrix} 13 & 5 \\ -5 & -2 \end{bmatrix}$ **15.** No inverse **17.** $\begin{bmatrix} 1 & 2 \\ -\frac{1}{2} & \frac{2}{3} \end{bmatrix}$

19. $\begin{bmatrix} -4 & -4 & 5 \\ 1 & 1 & -1 \\ 5 & 4 & -6 \end{bmatrix}$ **21.** No inverse

23. $\begin{bmatrix} -\frac{9}{2} & -1 & 4 \\ 3 & 1 & -3 \\ \frac{7}{2} & 1 & -3 \end{bmatrix}$ **25.** $\begin{bmatrix} 0 & 0 & -2 & 1 \\ -1 & 0 & 1 & 1 \\ 0 & 1 & -1 & 0 \\ 1 & 0 & 0 & -1 \end{bmatrix}$

27. $\begin{bmatrix} \frac{2}{3} & \frac{4}{3} & 3 \\ 1 & 1 & 3 \\ \frac{1}{3} & \frac{2}{3} & 1 \end{bmatrix}$ **29.** $\begin{bmatrix} -2 & 3 & -1 & -2 \\ 0 & -1 & 0 & \frac{1}{2} \\ -2 & 2 & -1 & -2 \\ -1 & -1 & -1 & 0 \end{bmatrix}$

31. $\begin{bmatrix} 1 & -\frac{7}{2} & \frac{1}{6} \\ 0 & \frac{1}{2} & -\frac{1}{6} \\ 0 & 0 & \frac{1}{3} \end{bmatrix}$ **33.** $\begin{bmatrix} 1 & 0 & 0 & 0 \\ 0 & \frac{1}{2} & 0 & 0 \\ 0 & 0 & \frac{1}{4} & 0 \\ 0 & 0 & 0 & \frac{1}{7} \end{bmatrix}$

35. $\begin{bmatrix} -\frac{1}{4} & \frac{3}{4} & \frac{3}{4} \\ -\frac{7}{16} & -\frac{23}{16} & -\frac{3}{16} \\ \frac{7}{8} & -\frac{1}{8} & -\frac{5}{8} \end{bmatrix}$ **37.** $\begin{bmatrix} -7 & -3 & -4 \\ \frac{22}{7} & -\frac{2}{7} & \frac{16}{7} \\ \frac{50}{7} & \frac{26}{7} & \frac{37}{7} \end{bmatrix}$

39. $x = 12, y = -8$ **41.** $x = 126, y = -50$
43. $x = -38, y = 9, z = 47$ **45.** $x = -20, y = 10, z = 16$
47. $x = 3, y = 2, z = 1$ **49.** $x = 3, y = -2, z = 2$
51. $x = 8, y = 1, z = 0, w = 3$

53. $\begin{bmatrix} 7 & 2 & 3 \\ 10 & 3 & 5 \end{bmatrix}$ **55.** $\frac{1}{2a}\begin{bmatrix} 1 & 1 \\ -1 & 1 \end{bmatrix}$

57. $\begin{bmatrix} 1 & -\frac{1}{x} \\ -\frac{1}{x} & \frac{2}{x^2} \end{bmatrix}$; inverse does not exist for $x = 0$

59. $\frac{1}{2}\begin{bmatrix} 1 & e^{-x} & 0 \\ e^{-x} & -e^{-2x} & 0 \\ 0 & 0 & 1 \end{bmatrix}$; inverse exists for all x

61. (a) $\begin{bmatrix} 0 & 1 & -1 \\ -2 & \frac{3}{2} & 0 \\ 1 & -\frac{3}{2} & 1 \end{bmatrix}$ **(b)** 1 oz A, 1 oz B, 2 oz C

(c) 2 oz A, 0 oz B, 1 oz C **(d)** No

63. (a) $\begin{cases} 9x + 11y + 8z = 740 \\ 13x + 15y + 16z = 1204 \\ 8x + 7y + 14z = 828 \end{cases}$

(b) $\begin{bmatrix} 9 & 11 & 8 \\ 13 & 15 & 16 \\ 8 & 7 & 14 \end{bmatrix}\begin{bmatrix} x \\ y \\ z \end{bmatrix} = \begin{bmatrix} 740 \\ 1204 \\ 828 \end{bmatrix}$

(c) $A^{-1} = \begin{bmatrix} \frac{7}{4} & -\frac{7}{4} & 1 \\ -\frac{27}{28} & \frac{31}{28} & -\frac{5}{7} \\ -\frac{29}{56} & \frac{25}{56} & -\frac{1}{7} \end{bmatrix}$

She earns \$16 on a standard model, \$28 on a deluxe model and \$36 on a super-deluxe model.

SECTION 10.6 ■ PAGE 742

1. True **2.** True **3.** True **4. (a)** $2 \cdot 4 - (-3) \cdot 1 = 11$
(b)
$+1(2 \cdot 4 - (-3) \cdot 1) - 0(3 \cdot 4 - 0 \cdot 1) + 2(3 \cdot (-3) - 0 \cdot 2) = -7$
5. 6 **7.** 0 **9.** −4 **11.** Does not exist **13.** $\frac{1}{8}$ **15.** 20, 20
17. −12, 12 **19.** 0, 0 **21.** 4, has an inverse
23. 5000, has an inverse **25.** 0, does not have an inverse
27. −4, has an inverse **29.** −6, has an inverse
31. −12, has an inverse **33.** 0, does not have an inverse
35. −18 **37.** 120 **39. (a)** −2 **(b)** −2 **(c)** Yes **41.** $(-2, 5)$
43. $(0.6, -0.4)$ **45.** $(4, -1)$ **47.** $(4, 2, -1)$ **49.** $(1, 3, 2)$
51. $(0, -1, 1)$ **53.** $\left(\frac{189}{29}, -\frac{108}{29}, \frac{88}{29}\right)$ **55.** $\left(\frac{1}{2}, \frac{1}{4}, \frac{1}{4}, -1\right)$
57. 21 **59.** $\frac{63}{2}$ **61.** *abcde* **63.** 0, 1, 2 **65.** 1, −1
69. (a) 0 **(b)** (i) Yes, (ii) No

71. (a) $\begin{cases} x + y + z = 18 \\ 75x + 90y + 60z = 1380 \\ -75x + 90y + 60z = 180 \end{cases}$

(b) 8 apples, 6 peaches, 4 pears
73. 7 million ft^2

SECTION 10.7 ■ PAGE 750

1. (iii) **2.** (ii) **3.** $\dfrac{A}{x-1} + \dfrac{B}{x+2}$

5. $\dfrac{A}{x-2} + \dfrac{B}{(x-2)^2} + \dfrac{C}{x+4}$

7. $\dfrac{A}{x-3} + \dfrac{Bx + C}{x^2 + 4}$ **9.** $\dfrac{Ax + B}{x^2 + 1} + \dfrac{Cx + D}{x^2 + 2}$

11. $\dfrac{A}{x} + \dfrac{B}{2x-5} + \dfrac{C}{(2x-5)^2} + \dfrac{D}{(2x-5)^3} + \dfrac{Ex + F}{x^2 + 2x + 5} + \dfrac{Gx + H}{(x^2 + 2x + 5)^2}$

13. $\dfrac{1}{x-1} - \dfrac{1}{x+1}$ **15.** $\dfrac{1}{x-1} - \dfrac{1}{x+4}$

17. $\dfrac{2}{x-3} - \dfrac{2}{x+3}$ **19.** $\dfrac{1}{x-2} - \dfrac{1}{x+2}$

21. $\dfrac{3}{x-4} - \dfrac{2}{x+2}$ **23.** $\dfrac{-\frac{1}{2}}{2x-1} + \dfrac{\frac{3}{2}}{4x-3}$

25. $\dfrac{2}{x-2} + \dfrac{3}{x+2} - \dfrac{1}{2x-1}$ **27.** $\dfrac{2}{x+1} - \dfrac{1}{x} + \dfrac{1}{x^2}$

29. $\dfrac{1}{2x+3} - \dfrac{3}{(2x+3)^2}$ **31.** $\dfrac{2}{x} - \dfrac{1}{x^3} - \dfrac{2}{x+2}$

33. $\dfrac{4}{x+2} - \dfrac{4}{x-1} + \dfrac{2}{(x-1)^2} + \dfrac{1}{(x-1)^3}$

35. $\dfrac{3}{x+2} - \dfrac{1}{(x+2)^2} - \dfrac{1}{(x+3)^2}$ **37.** $\dfrac{x+1}{x^2+3} - \dfrac{1}{x}$

39. $\dfrac{2x-5}{x^2+x+2} + \dfrac{5}{x^2+1}$ **41.** $\dfrac{1}{x^2+1} - \dfrac{x+2}{(x^2+1)^2} + \dfrac{1}{x}$

43. $x^2 + \dfrac{3}{x-2} - \dfrac{x+1}{x^2+1}$ **45.** $A = \dfrac{a+b}{2}, B = \dfrac{a-b}{2}$

SECTION 10.8 ■ PAGE 754

1. $(4, 8), (-2, 2)$ **3.** $(4, 16), (-3, 9)$ **5.** $(2, -2), (-2, 2)$
7. $(-25, 5), (-25, -5)$ **9.** $(-3, 4)\ (3, 4)$
11. $(-2, -1), (-2, 1), (2, -1), (2, 1)$
13. $(-1, \sqrt{2}), (-1, -\sqrt{2}), (\frac{1}{2}, \sqrt{\frac{7}{2}}), (\frac{1}{2}, -\sqrt{\frac{7}{2}})$
15. $(2, 4), (-\frac{5}{2}, \frac{7}{4})$ **17.** $(0, 0), (1, -1), (-2, -4)$
19. $(4, 0)$ **21.** $(-2, -2)$ **23.** $(6, 2), (-2, -6)$
25. No solution **27.** $(\sqrt{5}, 2), (\sqrt{5}, -2), (-\sqrt{5}, 2), (-\sqrt{5}, -2)$
29. $(3, -\frac{1}{2}), (-3, -\frac{1}{2})$ **31.** $(\frac{1}{5}, \frac{1}{3})$ **33.** $(2.00, 20.00), (-8.00, 0)$
35. $(-4.51, 2.17), (4.91, -0.97)$
37. $(1.23, 3.87), (-0.35, -4.21)$
39. $(-2.30, -0.70), (0.48, -1.19)$ **41.** $(\sqrt{10}, 10)$
43. $(-5, -8), (8, 5)$ **45.** 12 cm by 15 cm
47. 15, 20 **49.** $(400.50, 200.25)$, 447.77 m **51.** $(12, 8)$

SECTION 10.9 ■ PAGE 763

1. 2, 3; yes

2. equation; $y = x + 1$; test

Test point	Inequality $y \le x + 1$	Conclusion
(0, 0)	$0 \overset{?}{\le} 0 + 1$ ✓	Part of graph
(0, 2)	$2 \overset{?}{\le} 0 + 1$ ✗	Not part of graph

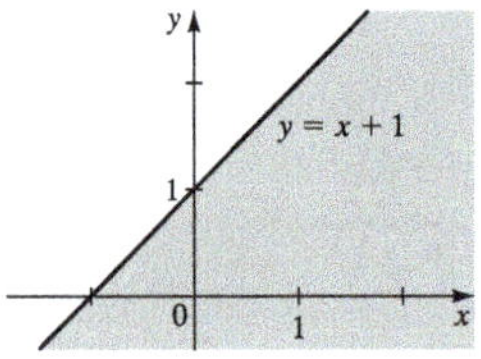

3. 2, 3; yes

4. **(a)**

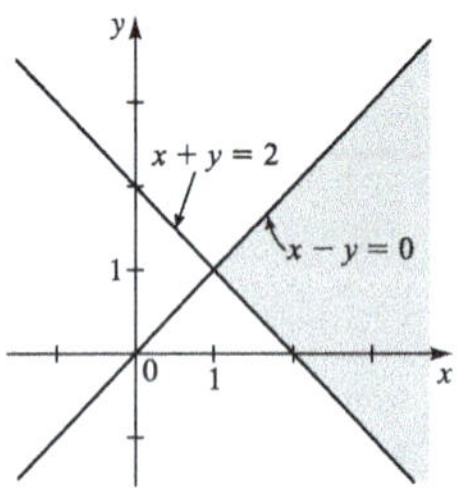

(b)

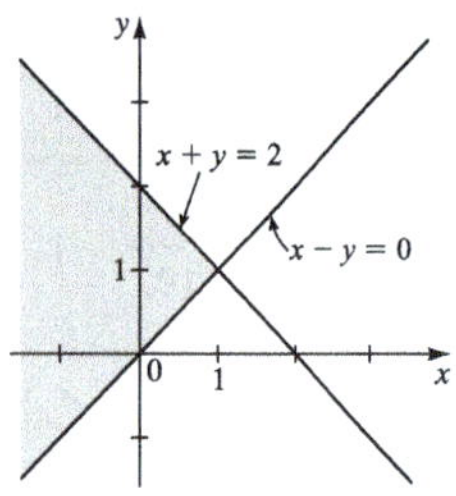

(c)

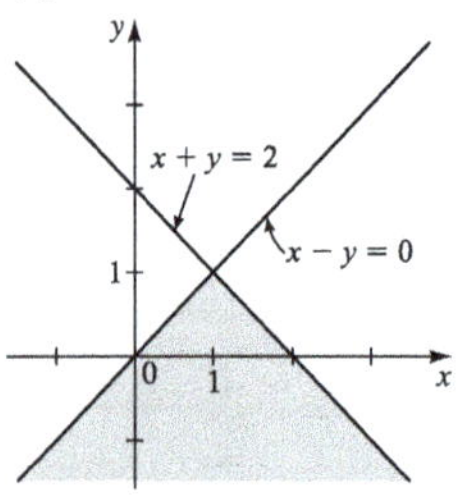

(d)

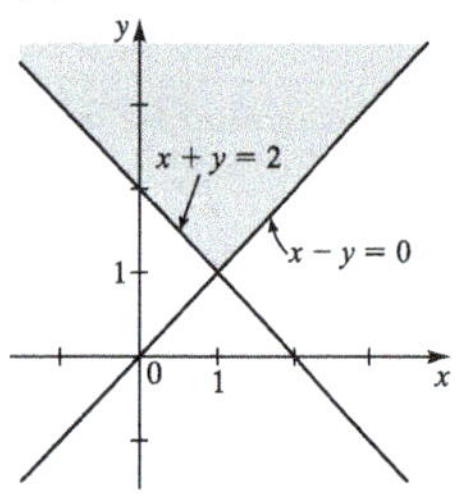

5. $(-1, -2), (1, -2)$ **7.** $(1, 2), (1, 1)$

9.

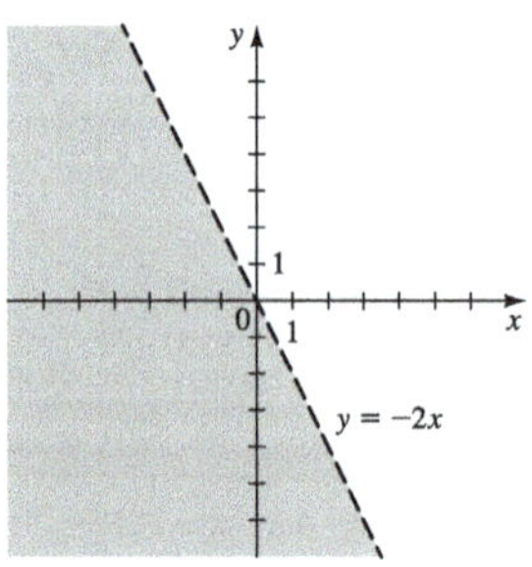

11.

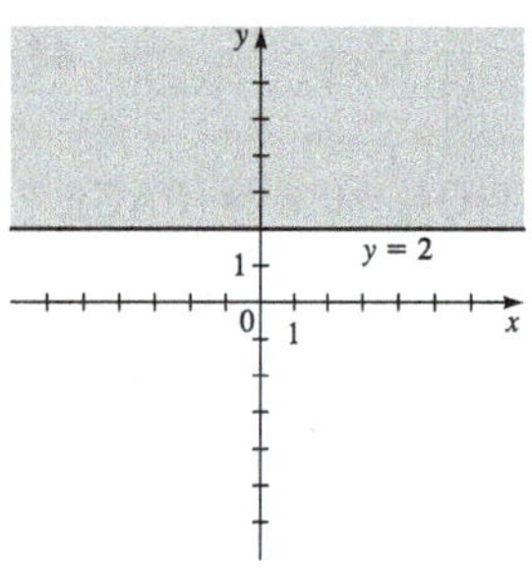

13.

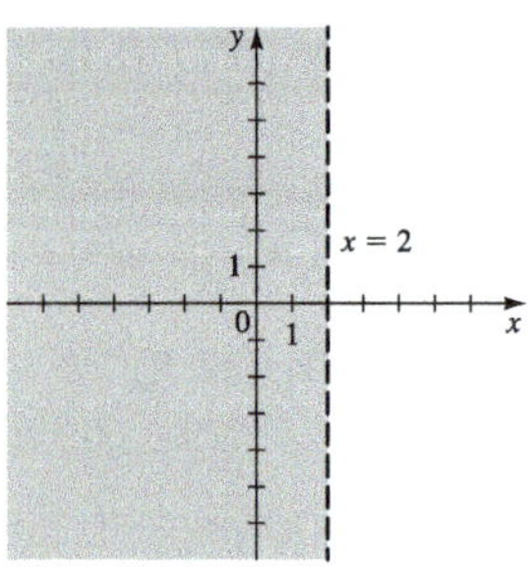

15.

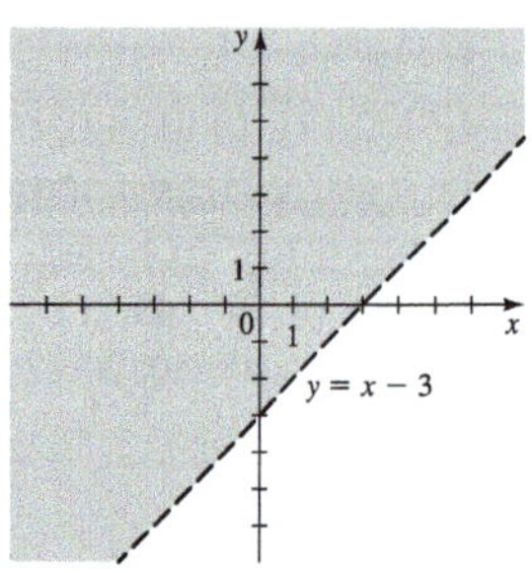

17.

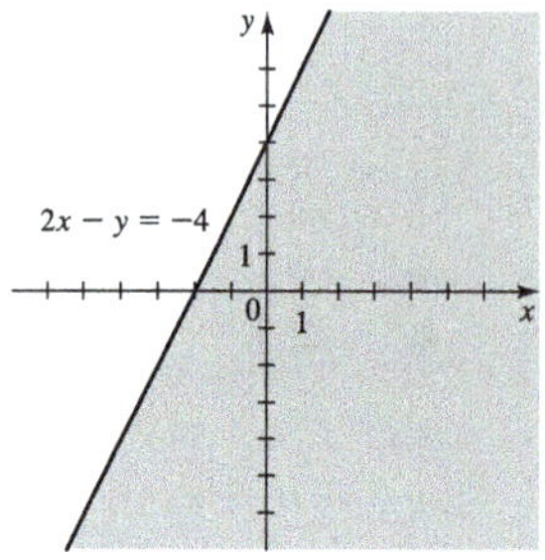

19.

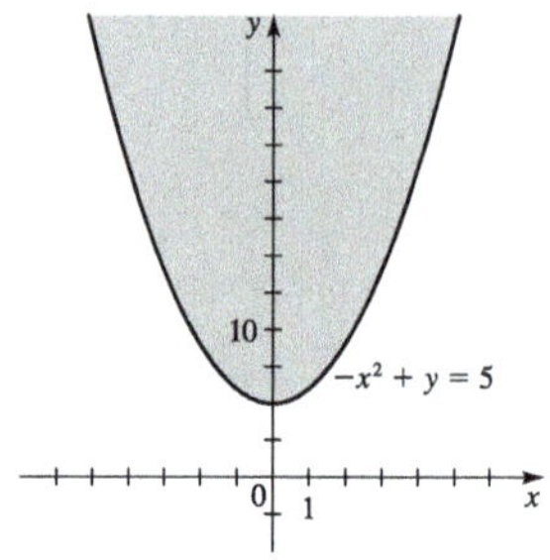

21.

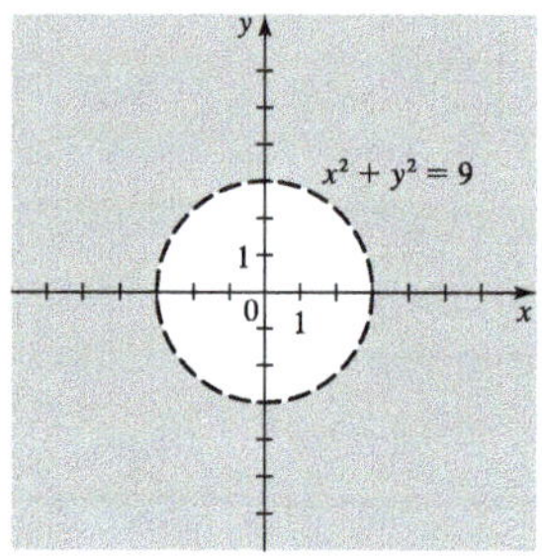

23.

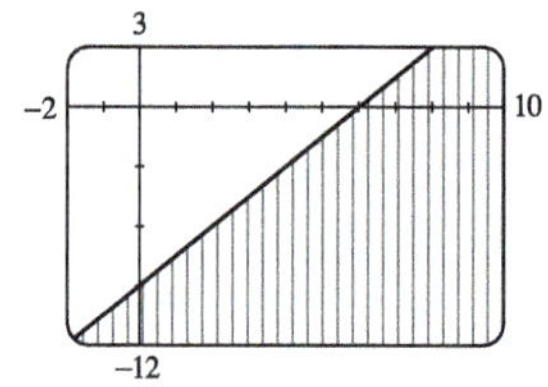

25.

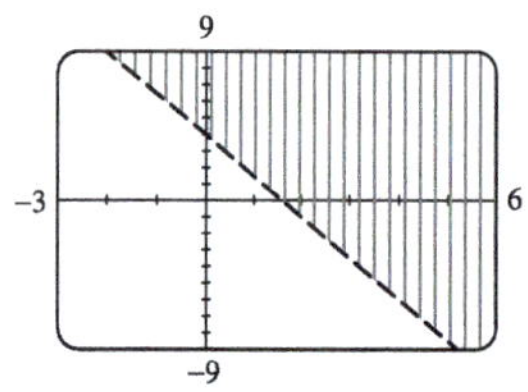

27. $y \le \frac{1}{2}x - 1$ **29.** $x^2 + y^2 > 4$

31.

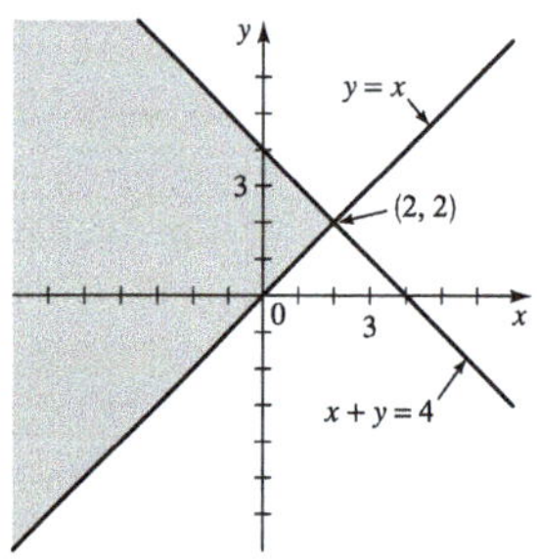

Not bounded

33.

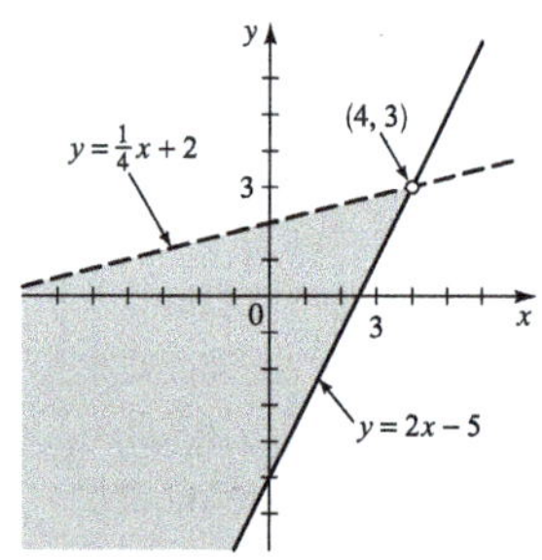

Not bounded

35.

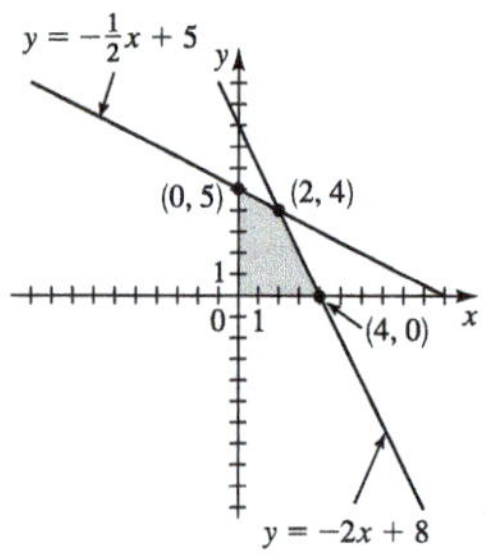

Bounded

37.

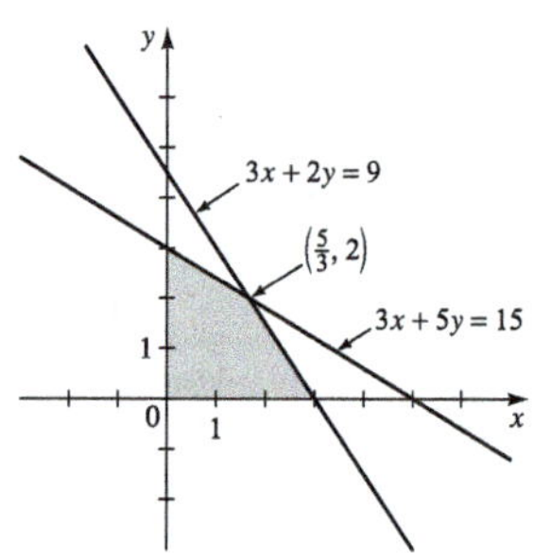

Bounded

39.

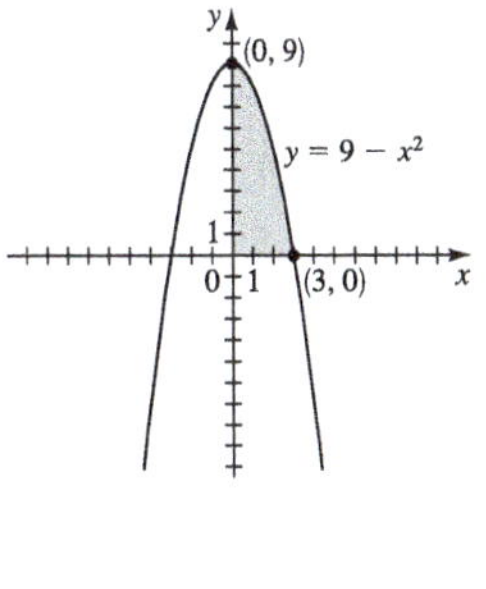

Bounded

41.

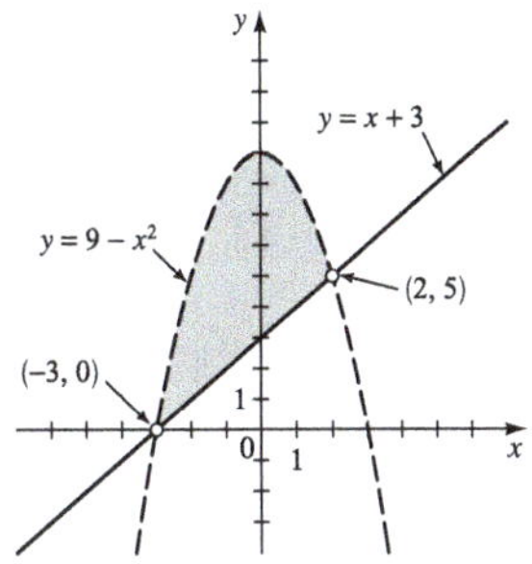

Bounded

43.

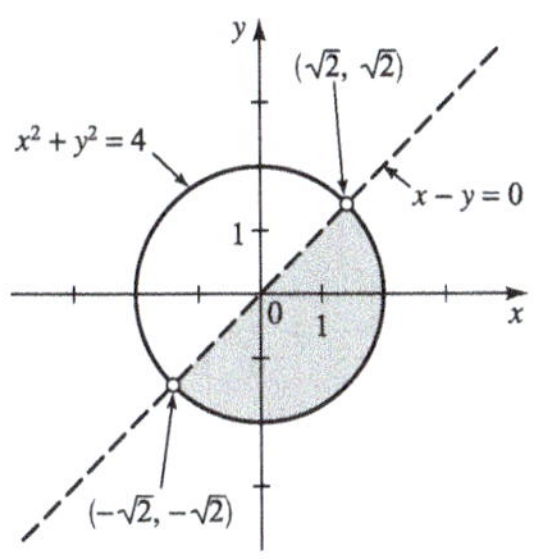

Bounded

45.

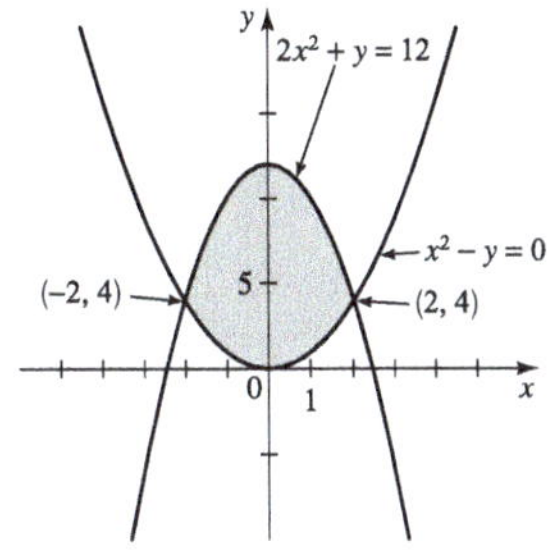

Bounded

47.

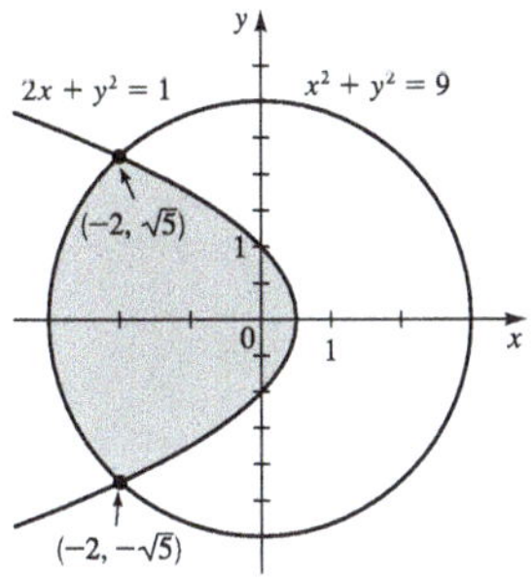

Bounded

49.

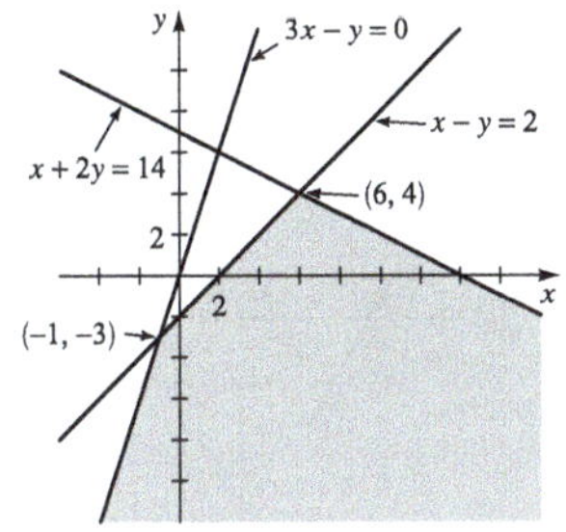

Not bounded

51.

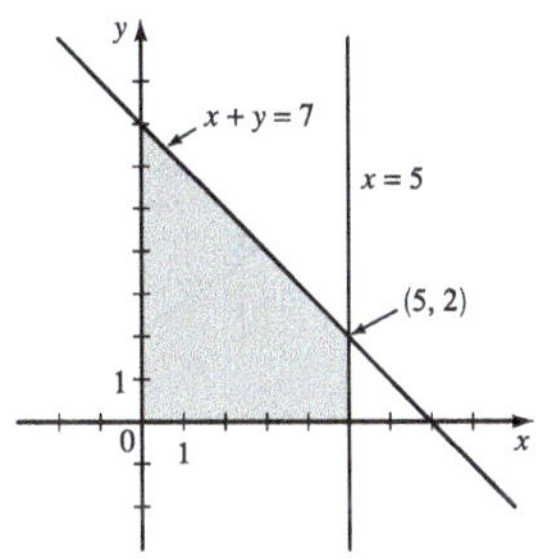

Bounded

53.

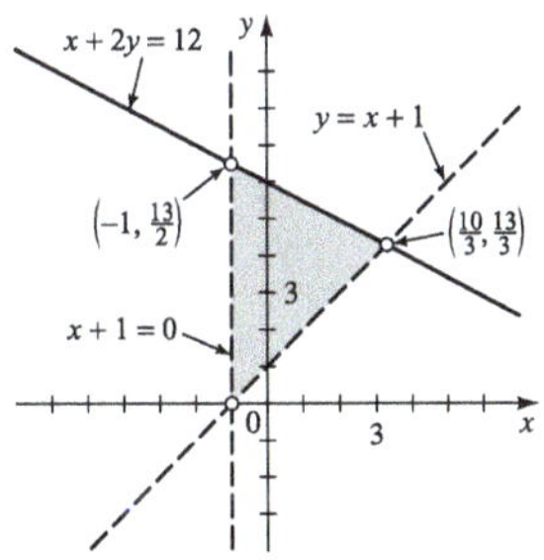

Bounded

55.

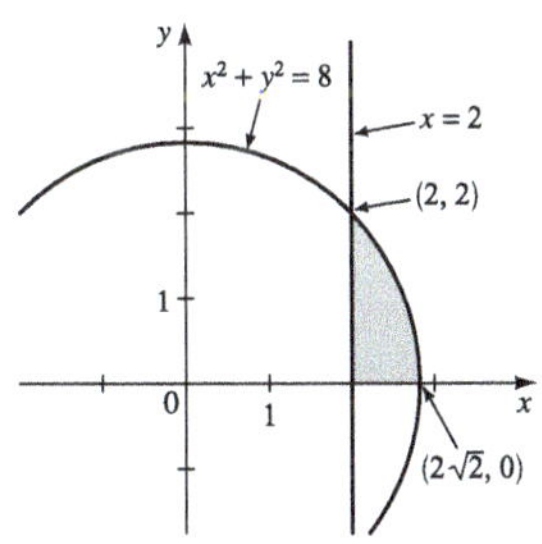

Bounded

57.

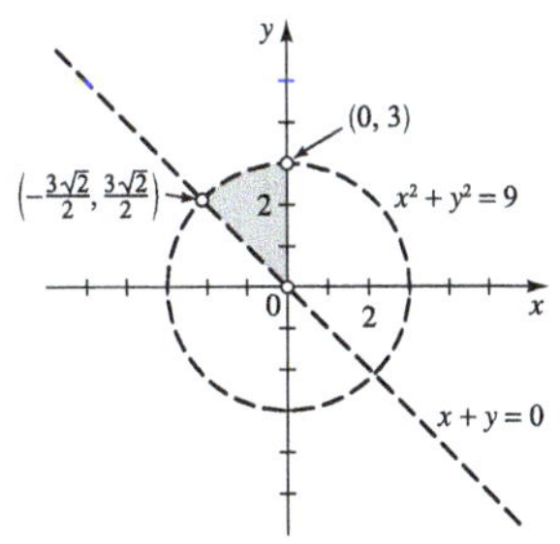

Bounded

59.

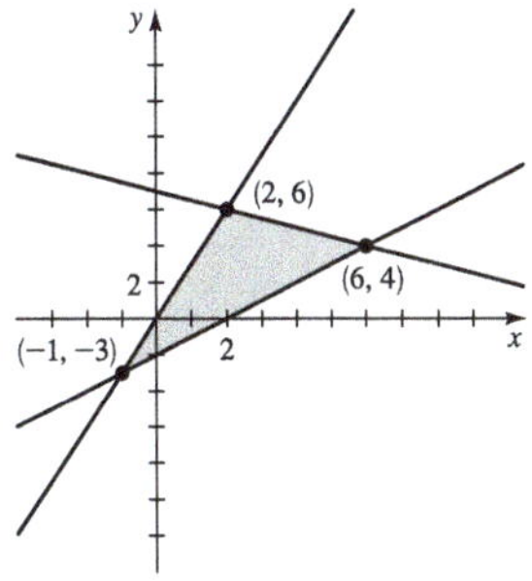

Bounded

61.

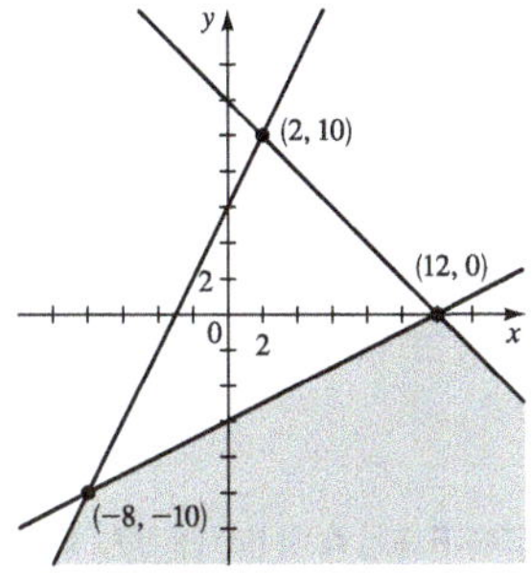

Not bounded

63.

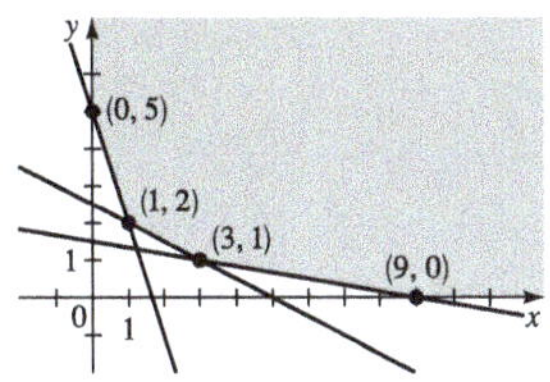

Not bounded

65.

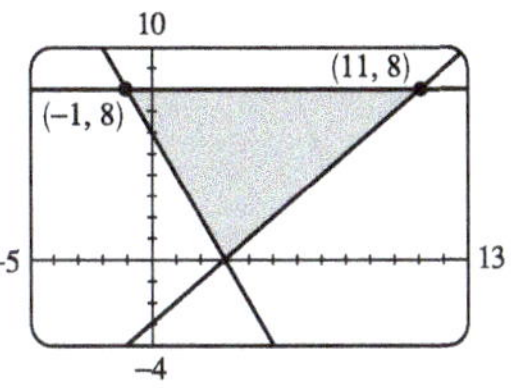

67.

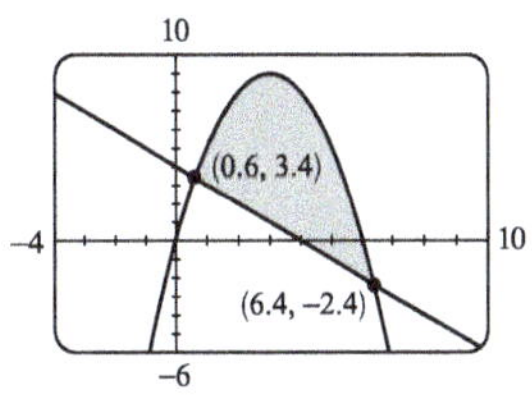

69. (a) $\begin{cases} x + y \le 500 \\ 90x + 50y \le 40{,}000 \\ 30x + 80y \le 30{,}000 \\ x \ge 0, \quad y \ge 0 \end{cases}$

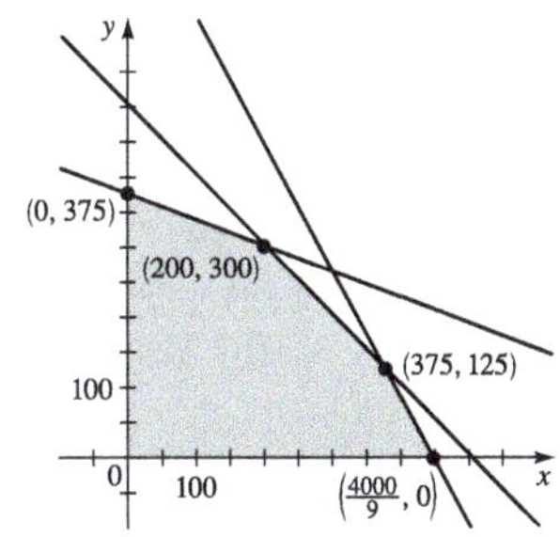

(b) Yes **(c)** No

71. x = number of fiction books
y = number of nonfiction books

$\begin{cases} x + y \le 100 \\ 20 \le y, \quad x \ge y \\ x \ge 0, \quad y \ge 0 \end{cases}$

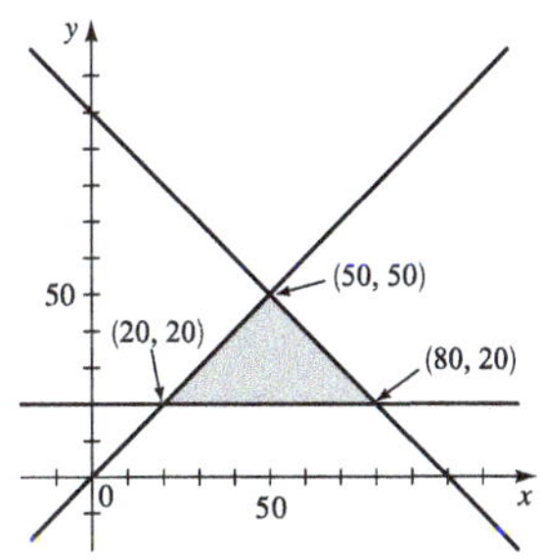

73. x = number of standard packages
y = number of deluxe packages

$$\begin{cases} \frac{1}{4}x + \frac{5}{8}y \le 80 \\ \frac{3}{4}x + \frac{3}{8}y \le 90 \\ x \ge 0, \quad y \ge 0 \end{cases}$$

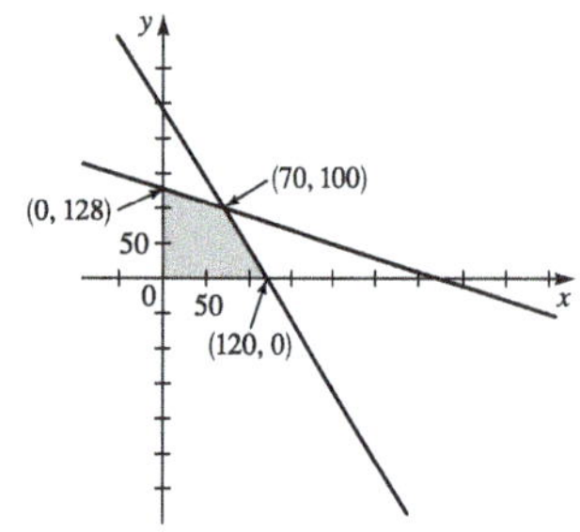

CHAPTER 10 REVIEW ■ PAGE 770

1. $(2, 1)$

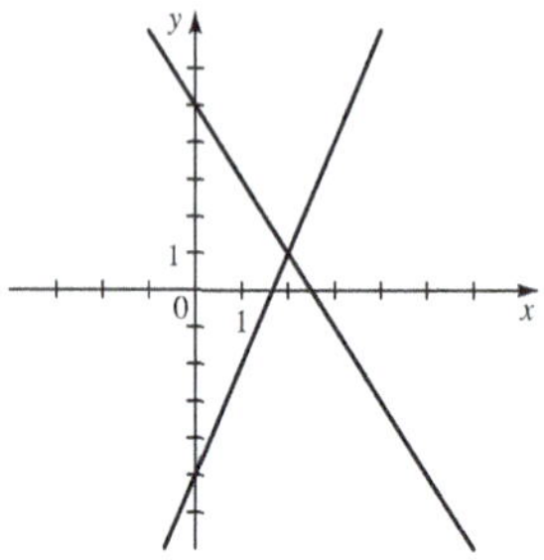

3. x = any number
$y = \frac{2}{7}x - 4$

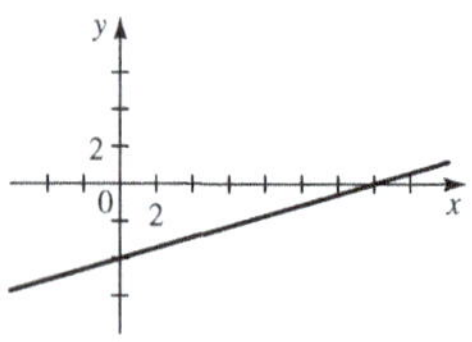

5. No solution

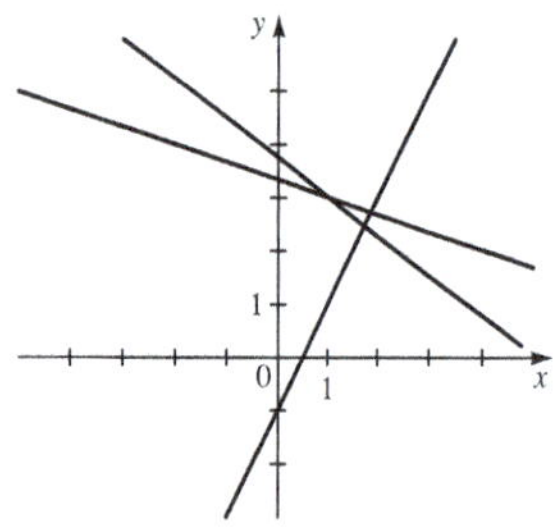

7. $(-3, 3), (2, 8)$ **9.** $\left(\frac{16}{7}, -\frac{14}{3}\right)$ **11.** $(21.41, -15.93)$
13. $(11.94, -1.39), (12.07, 1.44)$
15. **(a)** 2×3 **(b)** Yes **(c)** No

(d) $\begin{cases} x + 2y = -5 \\ y = 3 \end{cases}$

17. **(a)** 3×4 **(b)** Yes **(c)** Yes

(d) $\begin{cases} x + 8z = 0 \\ y + 5z = -1 \\ 0 = 0 \end{cases}$

19. **(a)** 3×4 **(b)** No **(c)** No

(d) $\begin{cases} y - 3z = 4 \\ x + y = 7 \\ x + 2y + z = 2 \end{cases}$

21. $(1, 1, 2)$ **23.** No solution **25.** $(0, 1, 2)$ **27.** No solution
29. $(1, 0, 1, -2)$ **31.** $(-4t + 1, -t - 1, t)$
33. $(6 - 5t, \frac{1}{2}(7 - 3t), t)$ **35.** $\left(-\frac{4}{3}t + \frac{4}{3}, \frac{5}{3}t - \frac{2}{3}, t\right)$
37. $(s + 1, 2s - t + 1, s, t)$ **39.** No solution
41. $(1, t + 1, t, 0)$ **43.** \$3000 at 6%, \$6000 at 7%
45. \$2500 in bank A, \$40,000 in bank B, \$17,500 in bank C
47. Impossible

49. $\begin{bmatrix} 4 & 18 \\ 4 & 0 \\ 2 & 2 \end{bmatrix}$ **51.** $[10 \quad 0 \quad -5]$ **53.** $\begin{bmatrix} -\frac{7}{2} & 10 \\ 1 & -\frac{9}{2} \end{bmatrix}$

55. $\begin{bmatrix} 30 & 22 & 2 \\ -9 & 1 & -4 \end{bmatrix}$ **57.** $\begin{bmatrix} -\frac{1}{2} & \frac{11}{2} \\ \frac{15}{4} & -\frac{3}{2} \\ -\frac{1}{2} & 1 \end{bmatrix}$ **61.** $\frac{1}{3}\begin{bmatrix} -1 & -3 \\ -5 & 2 \end{bmatrix}$

63. $\begin{bmatrix} \frac{7}{2} & -2 \\ 0 & 8 \end{bmatrix}$ **65.** $\begin{bmatrix} 2 & -2 & 6 \\ -4 & 5 & -9 \end{bmatrix}$ **67.** $1, \begin{bmatrix} 9 & -4 \\ -2 & 1 \end{bmatrix}$

69. 0, no inverse **71.** $-1, \begin{bmatrix} 3 & 2 & -3 \\ 2 & 1 & -2 \\ -8 & -6 & 9 \end{bmatrix}$

73. $24, \begin{bmatrix} 1 & 0 & 0 & -\frac{1}{4} \\ 0 & \frac{1}{2} & 0 & -\frac{1}{4} \\ 0 & 0 & \frac{1}{3} & -\frac{1}{4} \\ 0 & 0 & 0 & \frac{1}{4} \end{bmatrix}$ **75.** $(65, 154)$ **77.** $\left(-\frac{1}{12}, \frac{1}{12}, \frac{1}{12}\right)$

79. $\left(\frac{1}{5}, \frac{9}{5}\right)$ **81.** $\left(-\frac{87}{26}, \frac{21}{26}, \frac{3}{2}\right)$ **83.** 11 **85.** $\dfrac{2}{x - 5} + \dfrac{1}{x + 3}$

87. $\dfrac{-4}{x} + \dfrac{4}{x - 1} + \dfrac{-2}{(x - 1)^2}$ **89.** $\dfrac{-1}{x} + \dfrac{x + 2}{x^2 + 1}$

91. $(2, 1)$ **93.** $\left(-\frac{1}{2}, \frac{7}{4}\right), (2, -2)$ **95.** $x + y^2 \le 4$

97.

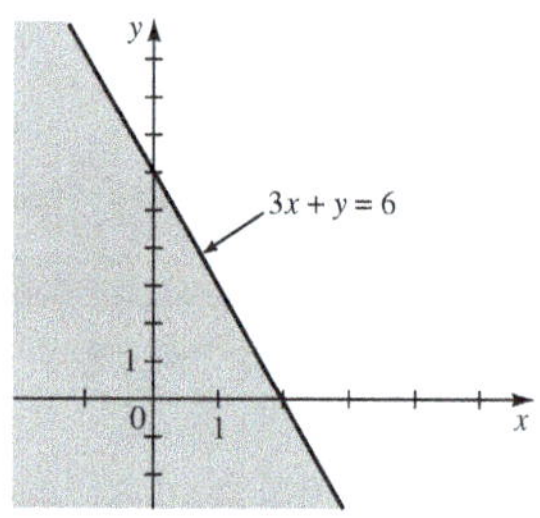

99.

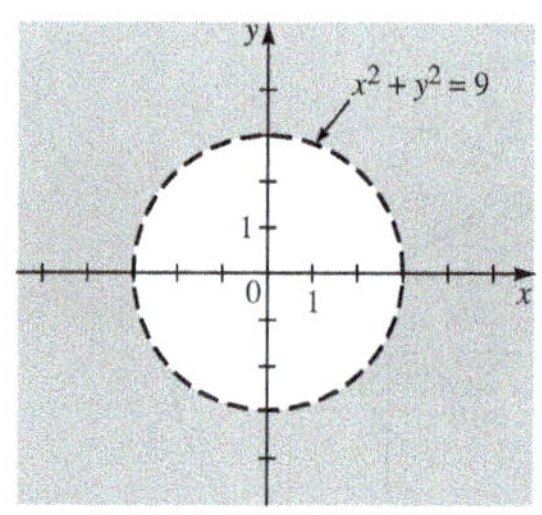

101.

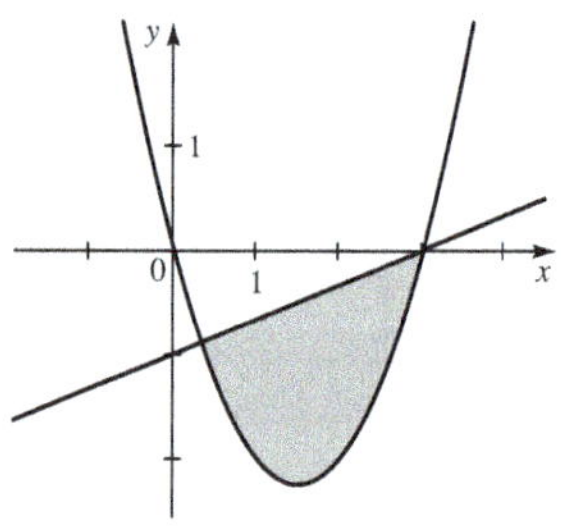

103.

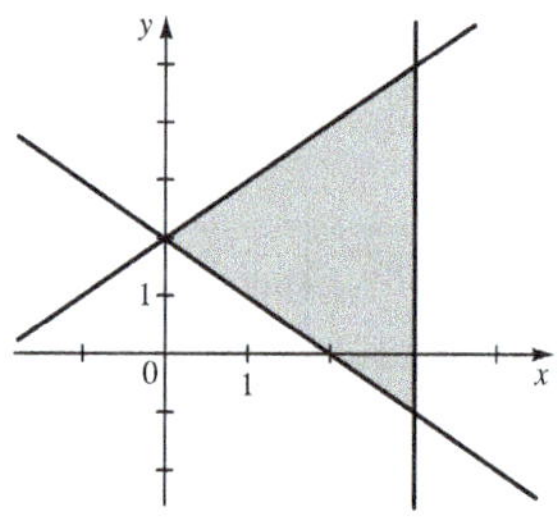

105.

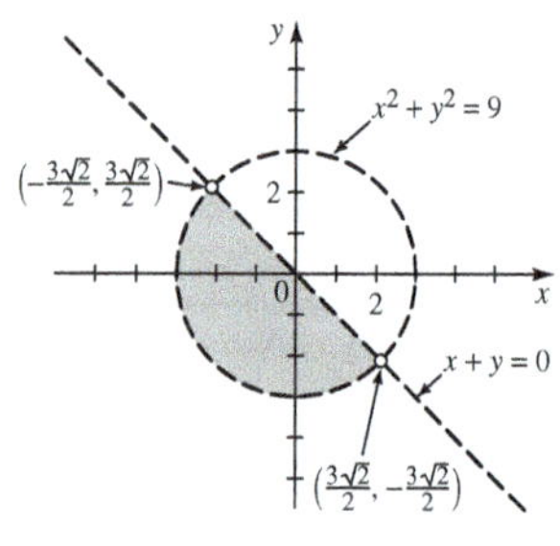

Bounded

107.

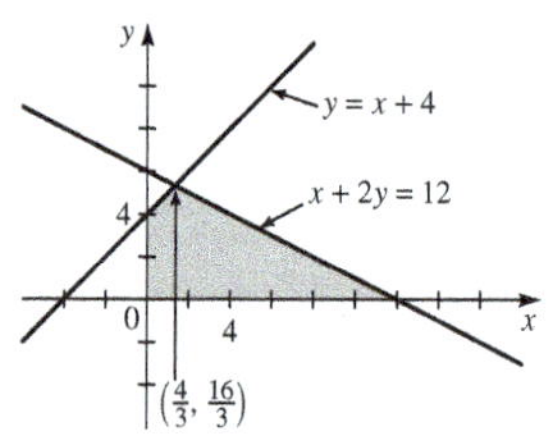

Bounded

109. $x = \dfrac{b + c}{2}, y = \dfrac{a + c}{2}, z = \dfrac{a + b}{2}$ **111.** 2, 3

CHAPTER 10 TEST ■ PAGE 773

1. **(a)** Linear **(b)** $(-2, 3)$ **2.** **(a)** Nonlinear **(b)** $(1, -2), (\frac{5}{3}, 0)$
3. $(-0.55, -0.78), (0.43, -0.29), (2.12, 0.56)$
4. Wind 60 km/h, airplane 300 km/h
5. **(a)** Row-echelon form **(b)** Reduced row-echelon form **(c)** Neither **6.** **(a)** $(\frac{5}{2}, \frac{5}{2}, 0)$ **(b)** No solution
7. $\left(-\frac{3}{5} + \frac{2}{5}t, \frac{1}{5} + \frac{1}{5}t, t\right)$
8. Coffee \$1.50, juice \$1.75, donut \$0.75
9. **(a)** Incompatible dimensions
(b) Incompatible dimensions
(c) $\begin{bmatrix} 6 & 10 \\ 3 & -2 \\ -3 & 9 \end{bmatrix}$ **(d)** $\begin{bmatrix} 36 & 58 \\ 0 & -3 \\ 18 & 28 \end{bmatrix}$ **(e)** $\begin{bmatrix} 2 & -\frac{3}{2} \\ -1 & 1 \end{bmatrix}$
(f) B is not square **(g)** B is not square **(h)** -3
10. **(a)** $\begin{bmatrix} 4 & -3 \\ 3 & -2 \end{bmatrix}\begin{bmatrix} x \\ y \end{bmatrix} = \begin{bmatrix} 10 \\ 30 \end{bmatrix}$ **(b)** $(70, 90)$
11. $|A| = 0, |B| = 2, B^{-1} = \begin{bmatrix} 1 & -2 & 0 \\ 0 & \frac{1}{2} & 0 \\ 3 & -6 & 1 \end{bmatrix}$
12. $(5, -5, -4)$
13. **(a)** $\dfrac{1}{x-1} + \dfrac{1}{(x-1)^2} - \dfrac{1}{x+2}$ **(b)** $-\dfrac{1}{x} + \dfrac{x+2}{x^2+3}$
14. **(a)**

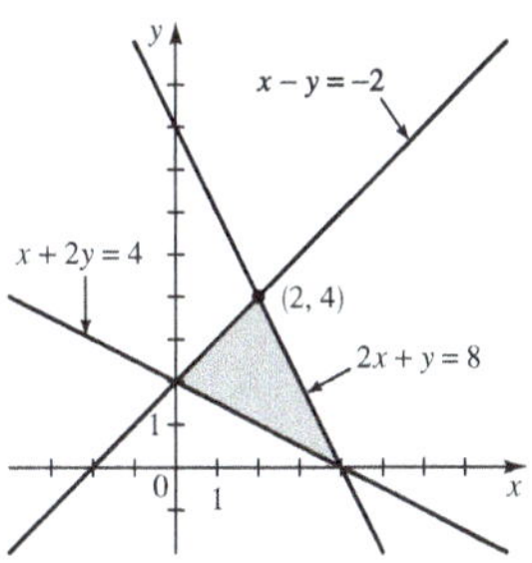

(b)

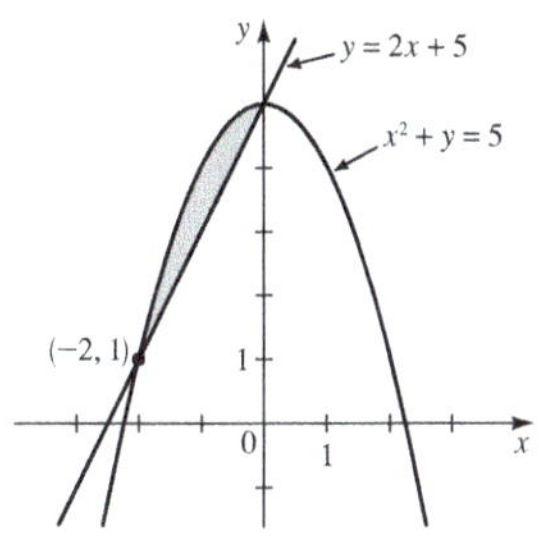

FOCUS ON MODELING ■ PAGE 779

1. 198, 195
3.

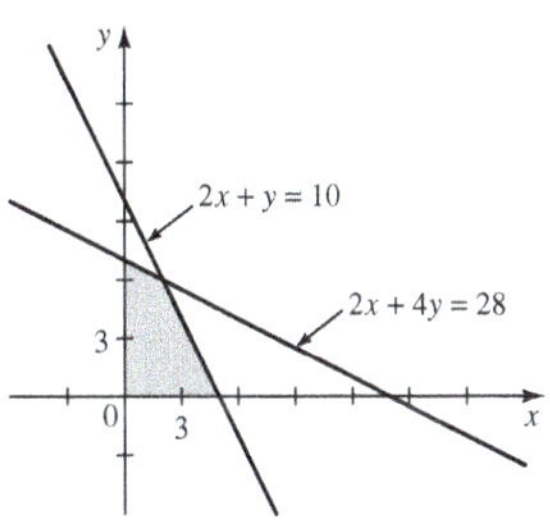

maximum 161
minimum 135

5. 3 tables, 34 chairs **7.** 30 grapefruit crates, 30 orange crates
9. 15 Pasadena to Santa Monica, 3 Pasadena to El Toro, 0 Long Beach to Santa Monica, 16 Long Beach to El Toro
11. 90 standard, 40 deluxe **13.** \$7500 in municipal bonds, \$2500 in bank certificates, \$2000 in high-risk bonds
15. 4 games, 32 educational, 0 utility

CHAPTER 11

SECTION 11.1 ■ PAGE 788

1. focus, directrix **2.** $F(0, p), y = -p, F(0, 3), y = -3$
3. $F(p, 0), x = -p, F(3, 0), x = -3$
4. **(a)**

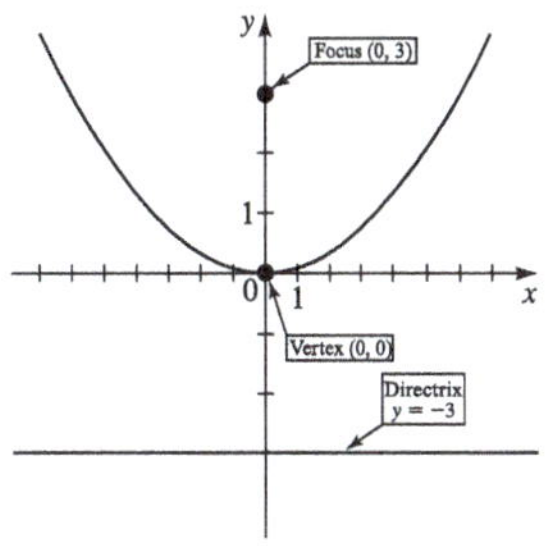

(b)

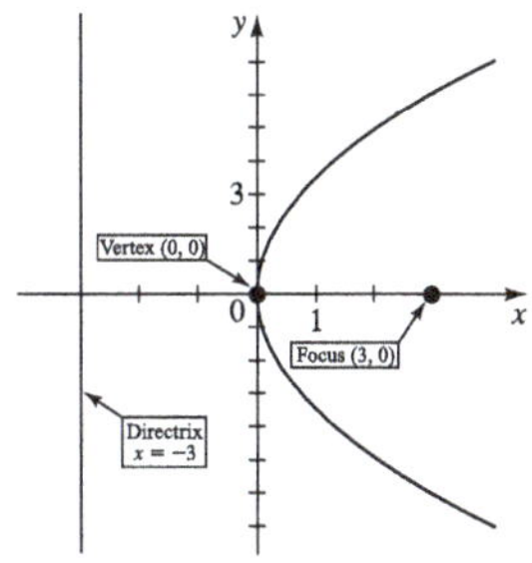

5. III **7.** II **9.** VI

Order of answers for 11–23, part (a): focus; directrix; focal diameter

11. **(a)** $F(0, 2); y = -2; 8$
(b)

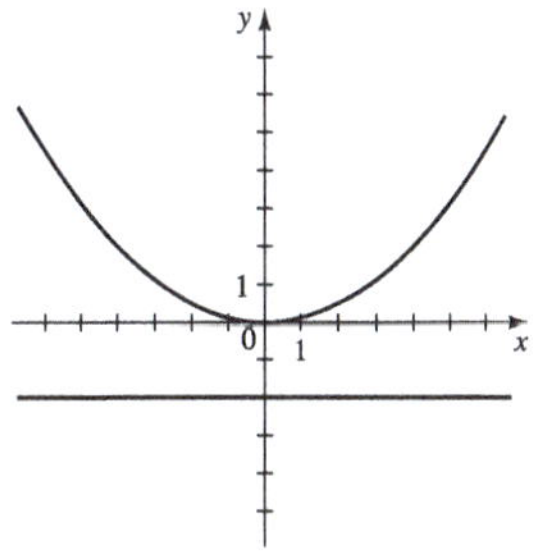

13. **(a)** $F(-6, 0); x = 6; 24$
(b)

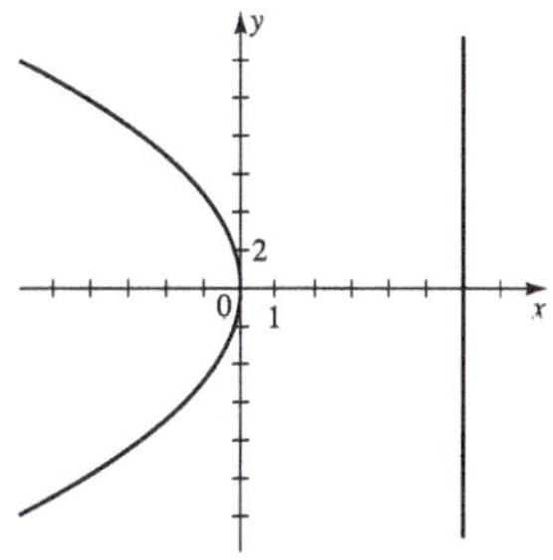

15. **(a)** $F(0, -2); y = 2; 8$
(b)

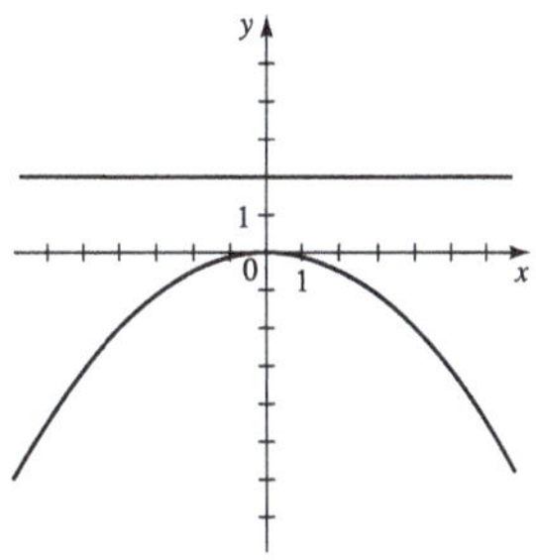

17. **(a)** $F(-\frac{1}{8}, 0); x = \frac{1}{8}; \frac{1}{2}$
(b)

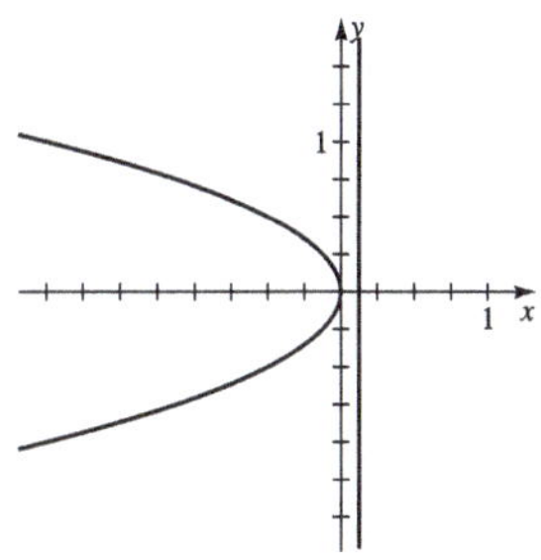

19. **(a)** $F(0, \frac{5}{4}); y = -\frac{5}{4}; 5$
(b)

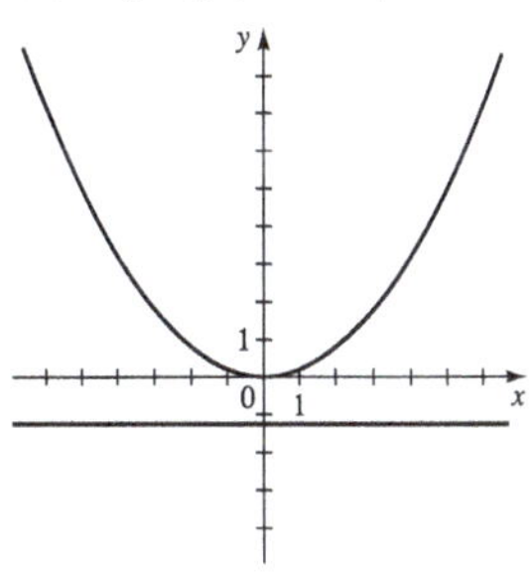

21. **(a)** $F(0, -3); y = 3; 12$
(b)

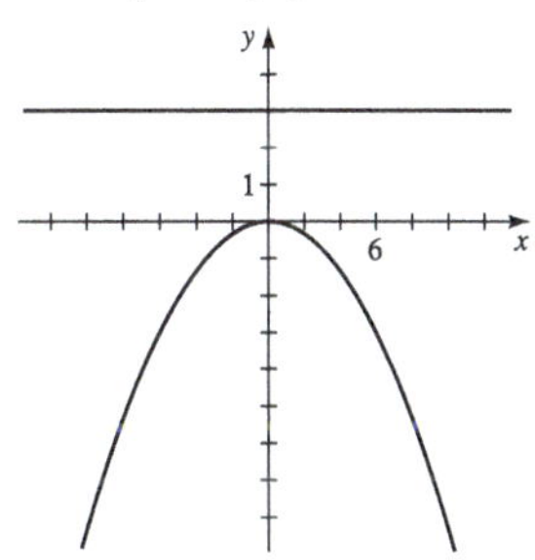

23. (a) $F\left(-\frac{5}{12}, 0\right)$; $x = \frac{5}{12}$; $\frac{5}{3}$ **(b)**

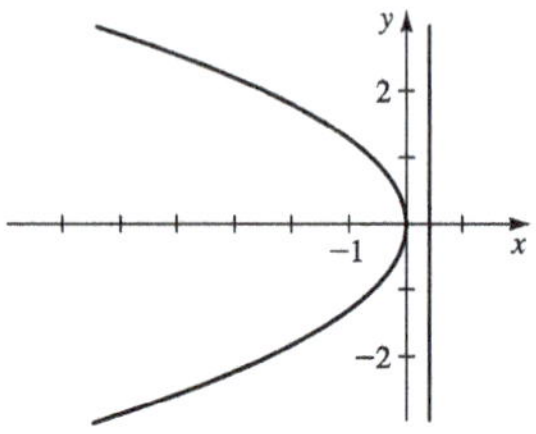

25.

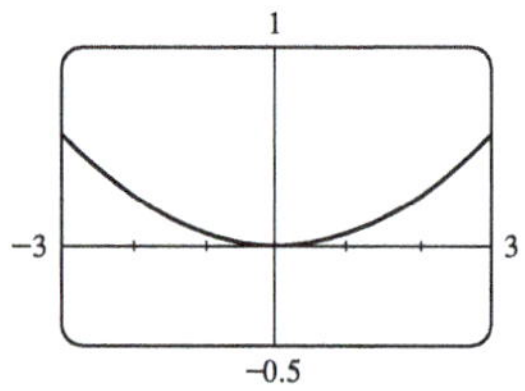

27.

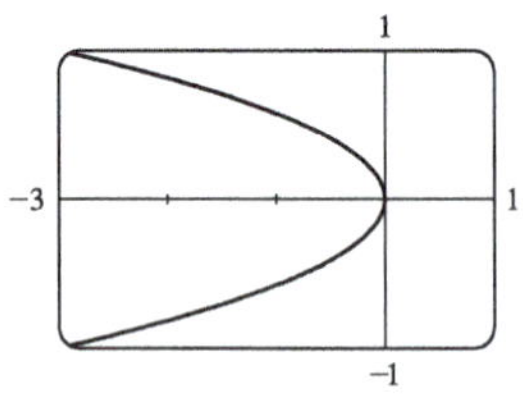

29.

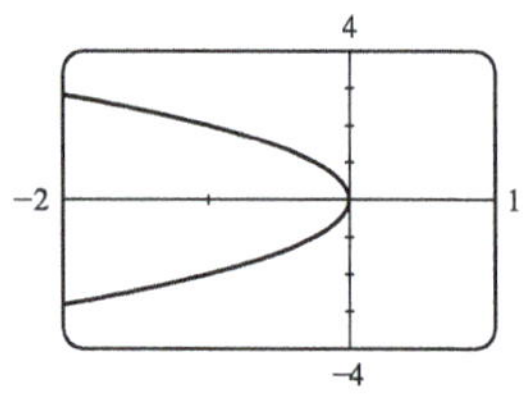

31. $x^2 = 24y$ **33.** $y^2 = -32x$ **35.** $x^2 = -3y$ **37.** $y^2 = 16x$
39. $x^2 = -\frac{2}{5}y$ **41.** $y^2 = -\frac{1}{5}x$ **43.** $y^2 = 4x$ **45.** $x^2 = -40y$
47. $x^2 = -24y$ **49.** $x^2 = 8y$ **51.** $y^2 = -16x$
53. $y^2 = -3x$ **55.** $x = y^2$ **57.** $x^2 = -4\sqrt{2}y$
59. (a) $x^2 = -4py$, $p = \frac{1}{2}$, 1, 4, and 8
(b) The closer the directrix to the vertex, the steeper the parabola.

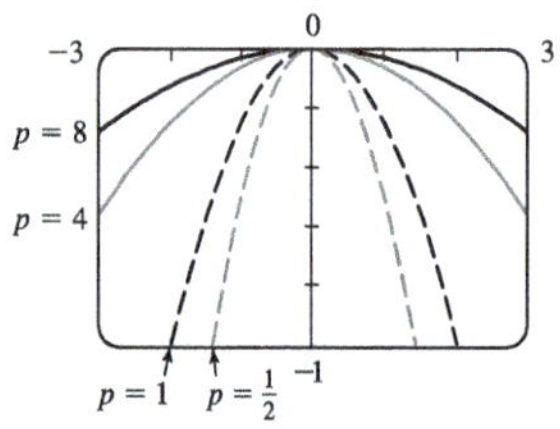

61. (a) $y^2 = 12x$ **(b)** $8\sqrt{15} \approx 31$ cm **63.** $x^2 = 600y$

SECTION 11.2 ■ PAGE 796

1. sum; foci
2. $(a, 0), (-a, 0)$; $c = \sqrt{a^2 - b^2}$; $(5, 0), (-5, 0), (3, 0), (-3, 0)$
3. $(0, a), (0, -a)$; $c = \sqrt{a^2 - b^2}$; $(0, 5), (0, -5), (0, 3), (0, -3)$
4. (a)

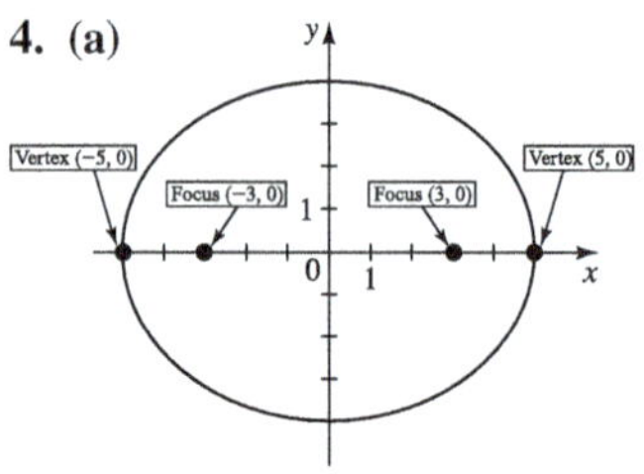

(b)

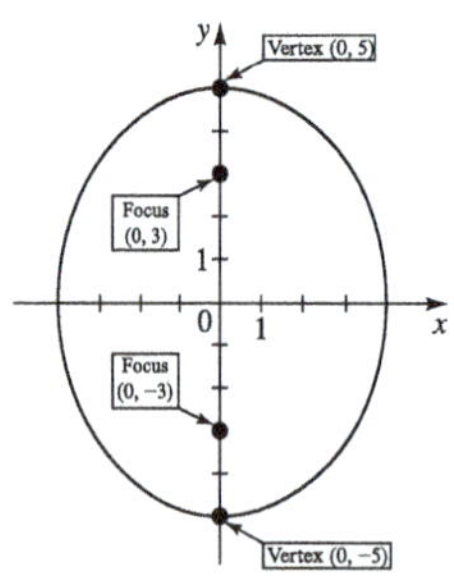

5. II **7.** I

Order of answers for 9–27 part (a): vertices; foci; eccentricity

9. (a) $V(\pm 5, 0)$; $F(\pm 4, 0)$; $\frac{4}{5}$
(b) 10, 6
(c)

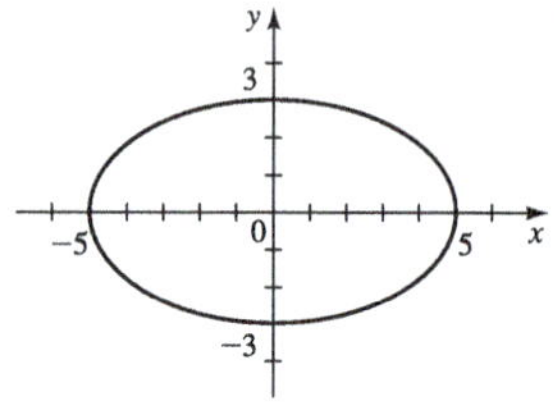

11. (a) $V(0, \pm 9)$; $F(0, \pm 3\sqrt{5})$; $\sqrt{5}/3$
(b) 18, 12
(c)

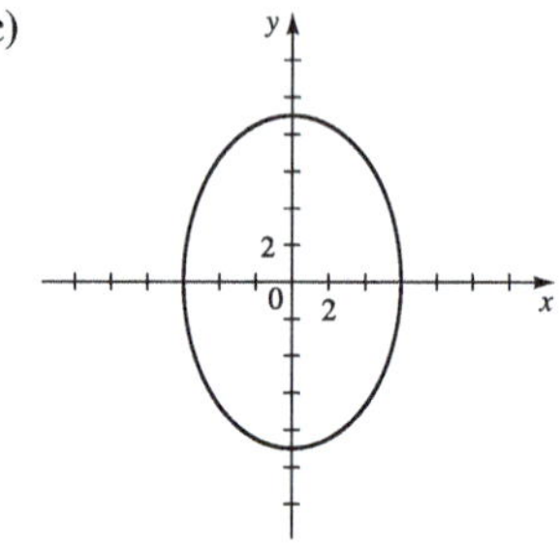

13. (a) $V(\pm 7, 0)$; $F(\pm 2\sqrt{6}, 0)$; $2\sqrt{6}/7$
(b) 14, 10
(c)

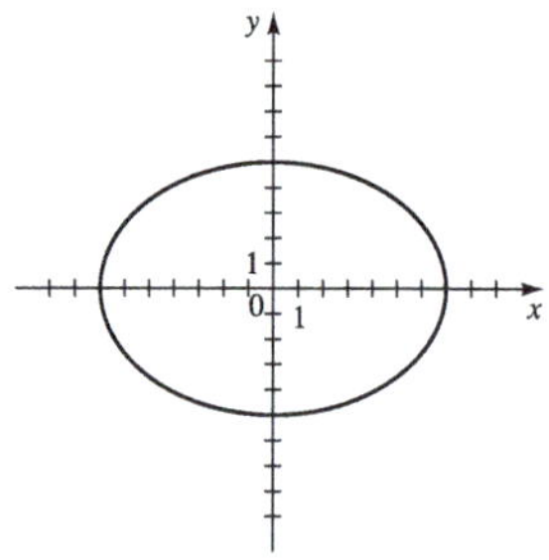

15. (a) $V(0, \pm 3)$; $F(0, \pm\sqrt{5})$; $\sqrt{5}/3$
(b) 6, 4
(c)

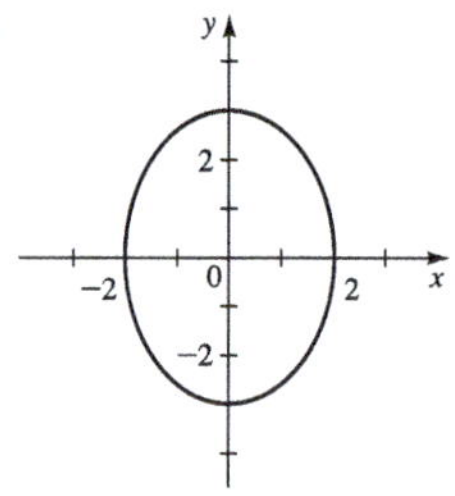

17. (a) $V(\pm 4, 0)$; $F\left(\pm 2\sqrt{3}, 0\right)$; $\sqrt{3}/2$
(b) 8, 4
(c)

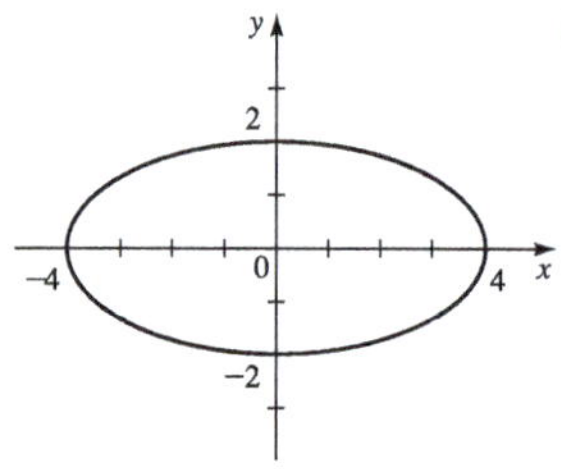

19. (a) $V(\pm 10, 0)$; $F(\pm 6, 0)$; $\frac{3}{5}$
(b) 20, 16
(c)

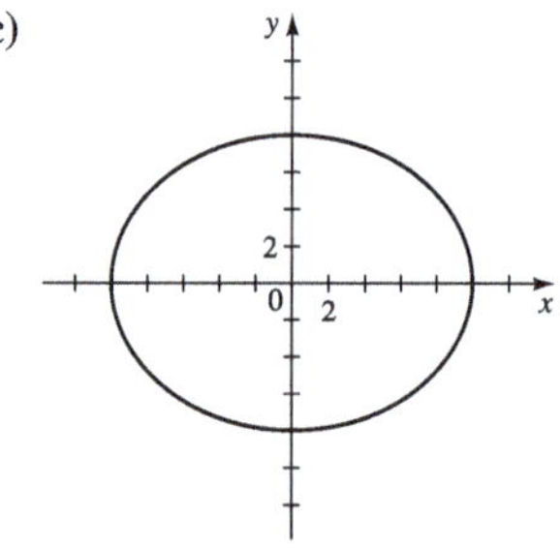

21. (a) $V(0, \pm 3)$; $F(0, \pm\sqrt{6})$; $\sqrt{6}/3$
(b) 6, $2\sqrt{3}$
(c)

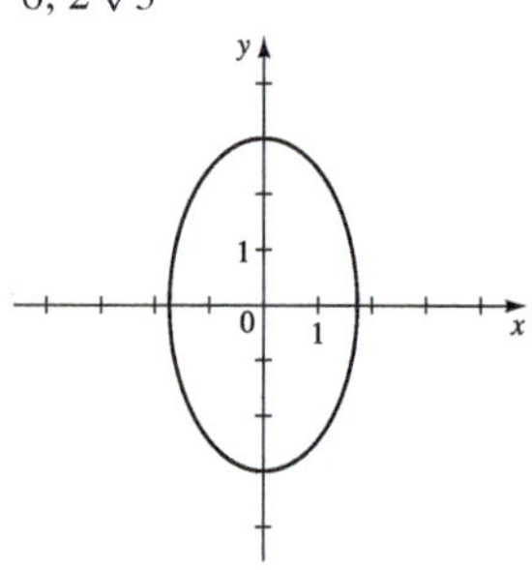

23. (a) $V(0, \pm 2)$; $F(0, \pm\sqrt{2})$; $\sqrt{2}/2$
(b) 4, $2\sqrt{2}$
(c)

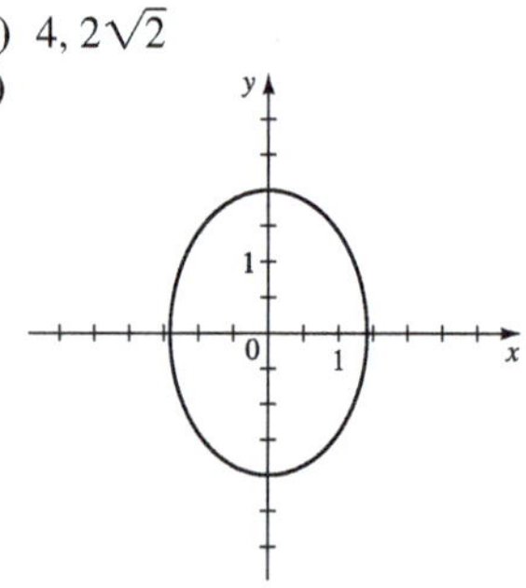

25. **(a)** $V(\pm 1, 0)$; $F(\pm\sqrt{3}/2, 0)$; $\sqrt{3}/2$
(b) 2, 1
(c)

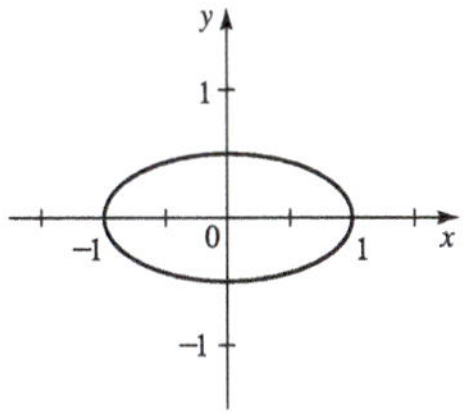

27. **(a)** $V(\pm 2, 0)$; $F(\pm\sqrt{2}, 0)$; $\sqrt{2}/2$
(b) 4, $2\sqrt{2}$
(c)

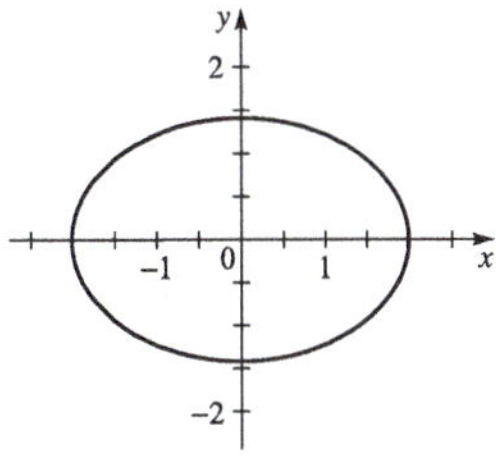

29. $\dfrac{x^2}{25} + \dfrac{y^2}{16} = 1$ **31.** $\dfrac{x^2}{4} + \dfrac{y^2}{8} = 1$ **33.** $\dfrac{x^2}{256} + \dfrac{y^2}{48} = 1$

35.

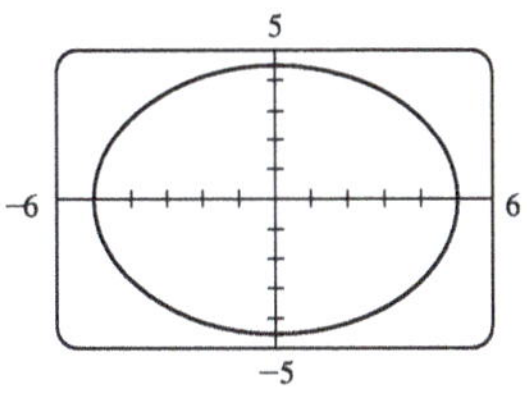

37.

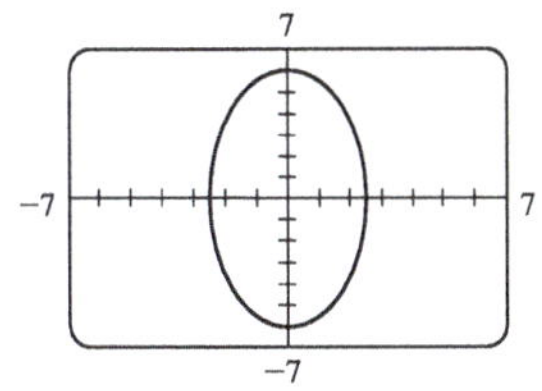

39. $\dfrac{x^2}{25} + \dfrac{y^2}{9} = 1$ **41.** $\dfrac{x^2}{4} + \dfrac{y^2}{3} = 1$ **43.** $\dfrac{x^2}{39} + \dfrac{y^2}{49} = 1$

45. $x^2 + \dfrac{y^2}{4} = 1$ **47.** $\dfrac{x^2}{9} + \dfrac{y^2}{13} = 1$ **49.** $\dfrac{x^2}{100} + \dfrac{y^2}{91} = 1$

51. $\dfrac{x^2}{25} + \dfrac{y^2}{5} = 1$ **53.** $\dfrac{x^2}{32} + \dfrac{y^2}{36} = 1$ **55.** $x^2 + \dfrac{y^2}{4} = 1$

57. $(0, \pm 2)$

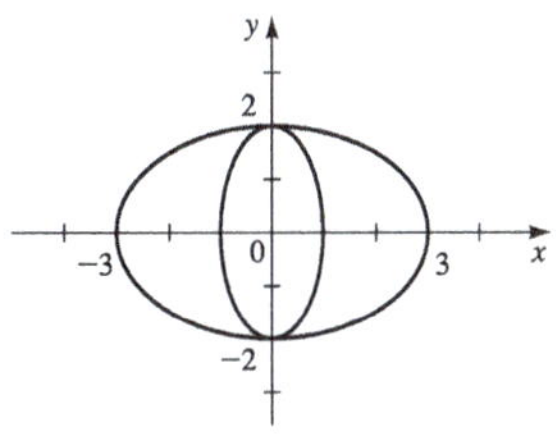

59. $(\pm 1, 0)$

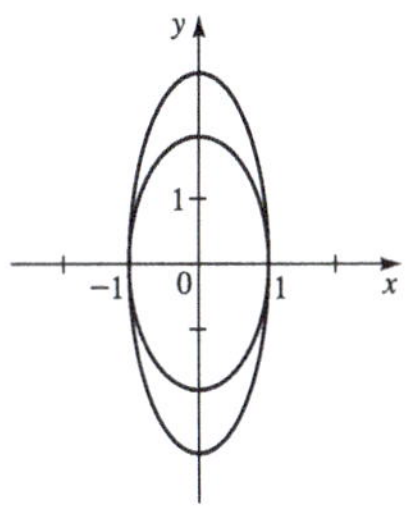

61. **(a)** $x^2 + y^2 = 4$

65. $\dfrac{x^2}{2.2500 \times 10^{16}} + \dfrac{y^2}{2.2491 \times 10^{16}} = 1$

67. $\dfrac{x^2}{1{,}455{,}642} + \dfrac{y^2}{1{,}451{,}610} = 1$ **69.** $5\sqrt{39}/2 \approx 15.6$ in.

SECTION 11.3 ■ PAGE 805

1. difference; foci
2. horizontal; $(-a, 0), (a, 0)$; $\sqrt{a^2 + b^2}$; $(-4, 0), (4, 0), (-5, 0), (5, 0)$
3. vertical; $(0, -a), (0, a)$; $\sqrt{a^2 + b^2}$; $(0, -4), (0, 4), (0, -5), (0, 5)$

4. **(a)**

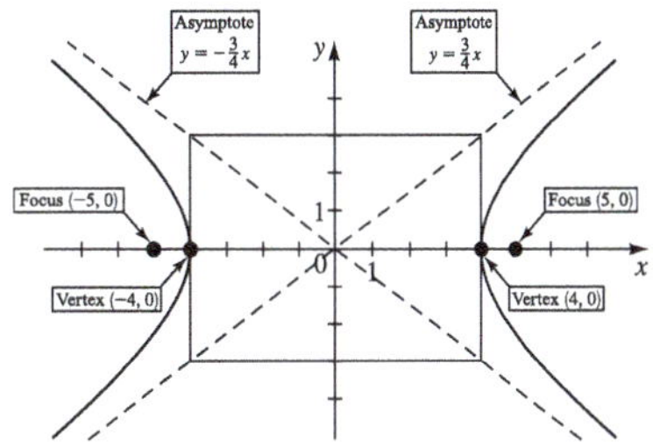

(b)

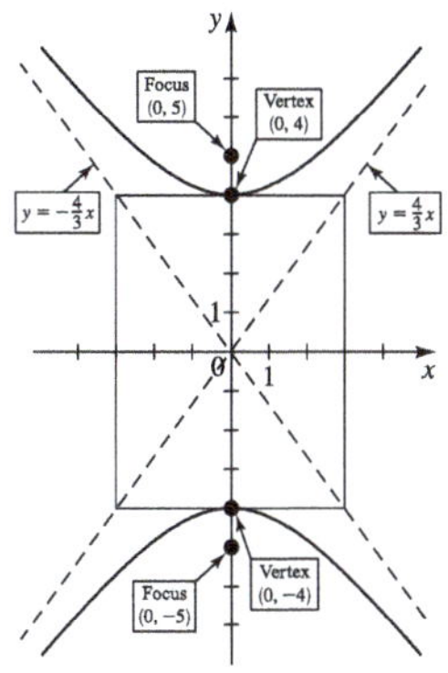

5. III **7.** II

Order of answers for 9–25, part (a): vertices; foci; asymptotes

9. **(a)** $V(\pm 2, 0)$; $F(\pm 2\sqrt{5}, 0)$; $y = \pm 2x$
(b) 4
(c)

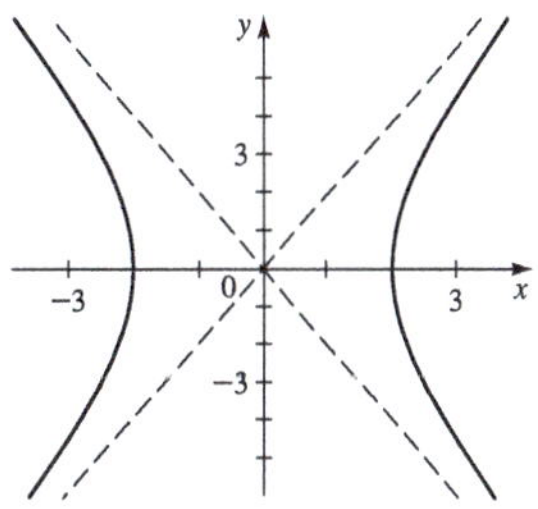

11. **(a)** $V(0, \pm 6)$; $F(0, \pm 2\sqrt{10})$; $y = \pm 3x$
(b) 12
(c)

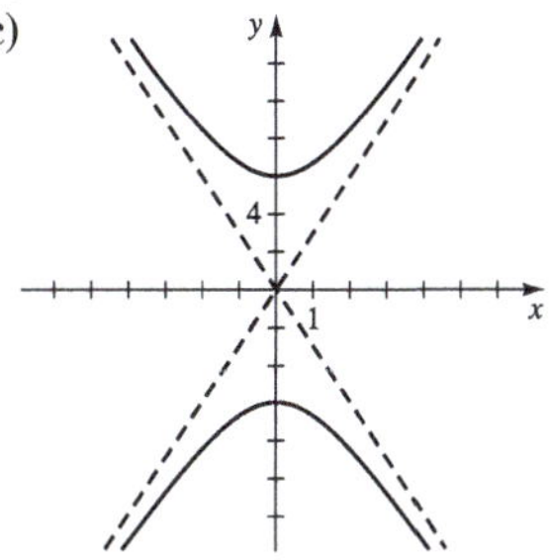

13. **(a)** $V(0, \pm 1)$; $F(0, \pm\sqrt{26})$; $y = \pm\frac{1}{5}x$
(b) 2
(c)

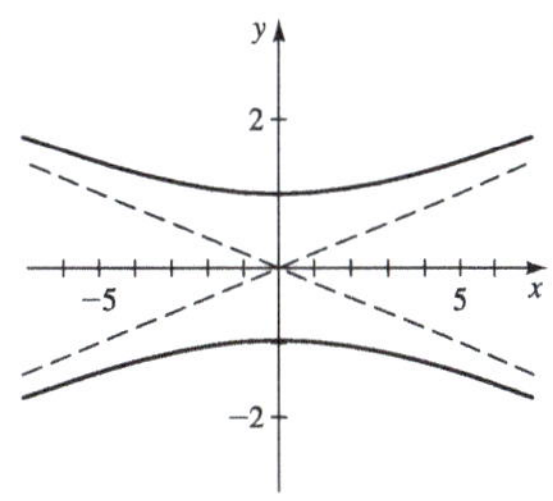

15. **(a)** $V(\pm 1, 0)$; $F(\pm\sqrt{2}, 0)$; $y = \pm x$
(b) 2
(c)

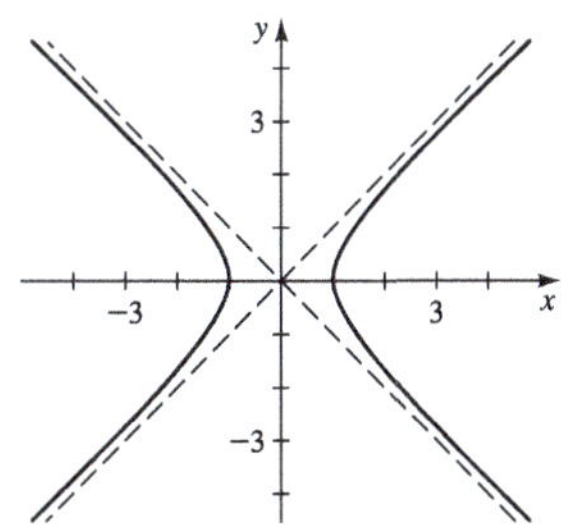

17. **(a)** $V(\pm 2, 0)$; $F(\pm\sqrt{13}, 0)$; $y = \pm\frac{3}{2}x$
(b) 4
(c)

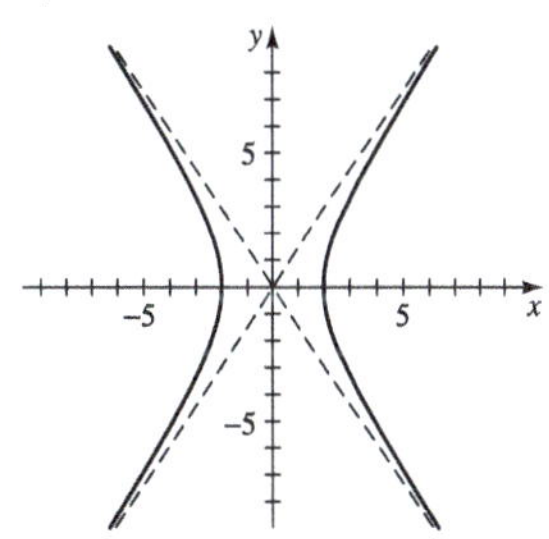

19. **(a)** $V(0, \pm 6)$; $F(0, \pm 2\sqrt{13})$; $y = \pm\frac{3}{2}x$
(b) 12
(c)

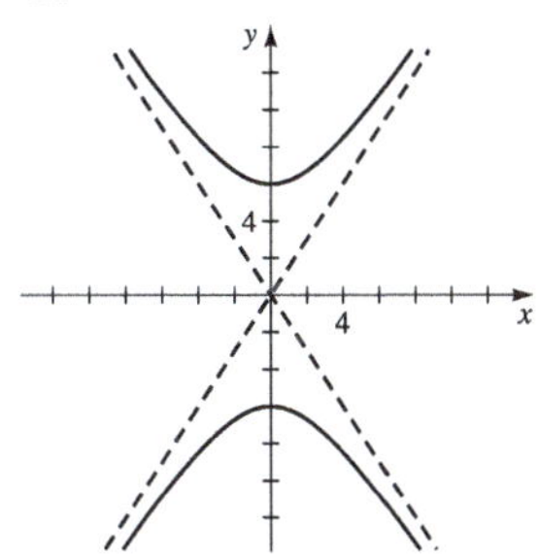

21. (a) $V(\pm 2\sqrt{2}, 0)$; $F(\pm\sqrt{10}, 0)$; $y = \pm\frac{1}{2}x$
(b) $4\sqrt{2}$
(c)

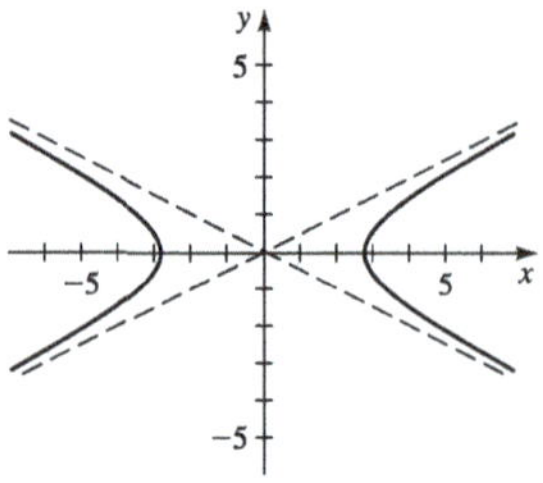

23. (a) $V(0, \pm 2)$; $F(0, \pm 2\sqrt{2})$; $y = \pm x$
(b) 4
(c)

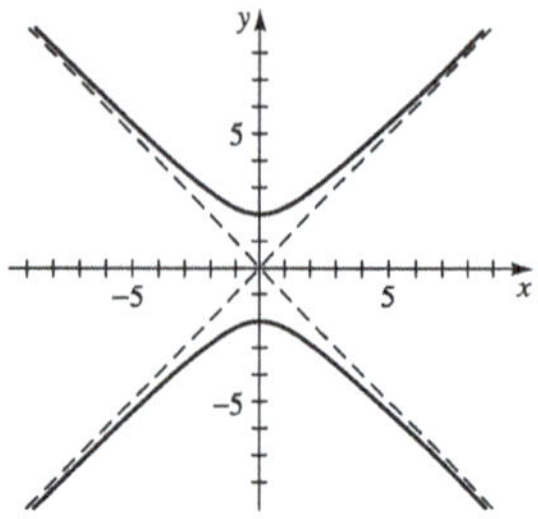

25. (a) $V(0, \pm\frac{1}{2})$; $F(0, \pm\sqrt{5}/2)$; $y = \pm\frac{1}{2}x$
(b) 1
(c)

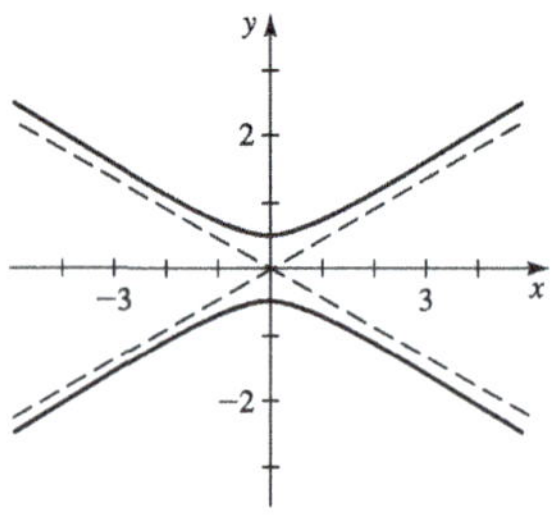

27. $\frac{x^2}{4} - \frac{y^2}{12} = 1$ **29.** $\frac{y^2}{16} - \frac{x^2}{16} = 1$ **31.** $\frac{y^2}{9} - x^2 = 1$

33.

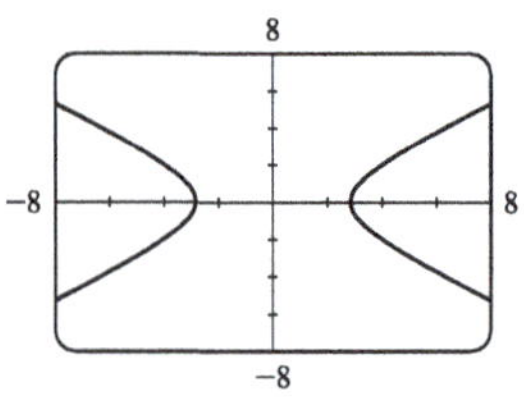

35.

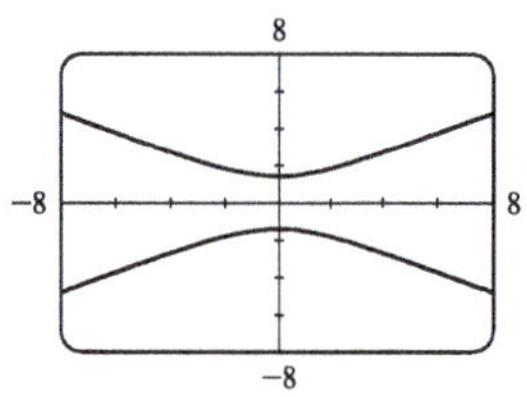

37. $\frac{x^2}{9} - \frac{y^2}{16} = 1$ **39.** $y^2 - \frac{x^2}{3} = 1$ **41.** $x^2 - \frac{y^2}{25} = 1$

43. $\frac{y^2}{36} - \frac{x^2}{20} = 1$ **45.** $\frac{x^2}{16} - \frac{y^2}{16} = 1$ **47.** $\frac{y^2}{8} - x^2 = 1$

49. $\frac{x^2}{9} - \frac{y^2}{16} = 1$ **51. (b)** $x^2 - y^2 = c^2/2$

55. (b)

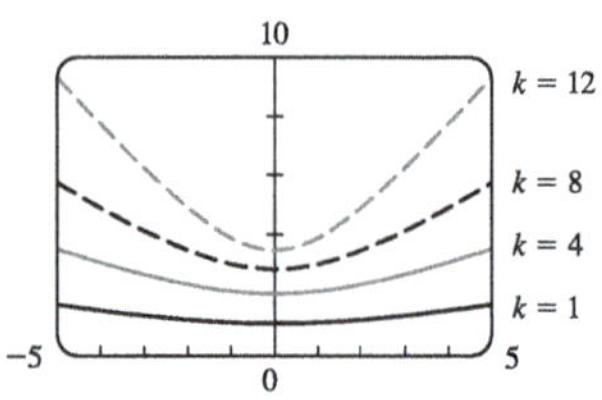

As k increases, the asymptotes get steeper.

57. $x^2 - y^2 = 2.3 \times 10^{19}$

SECTION 11.4 ▪ PAGE 813

1. (a) right; left **(b)** upward; downward

2.

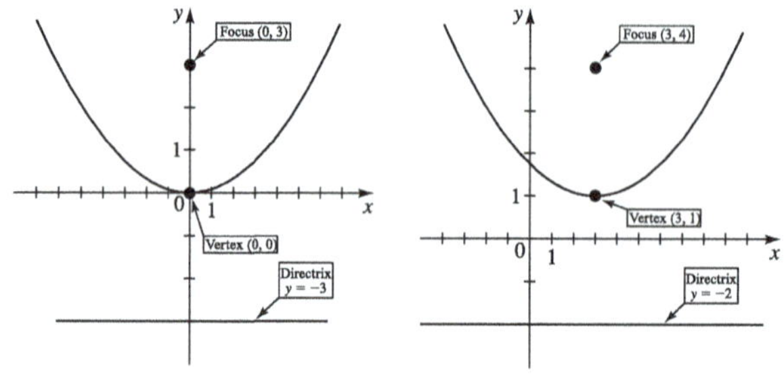

3.

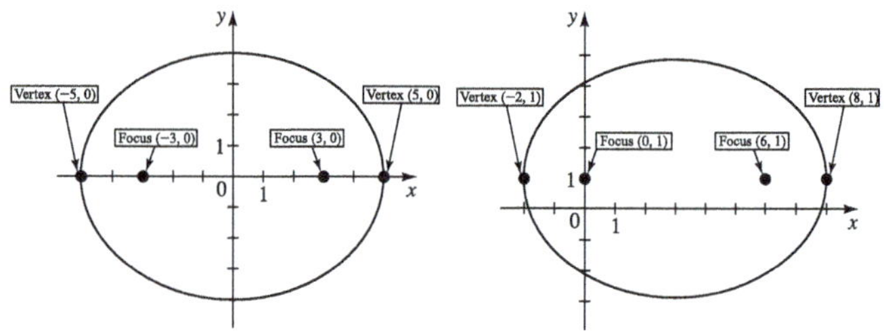

4.

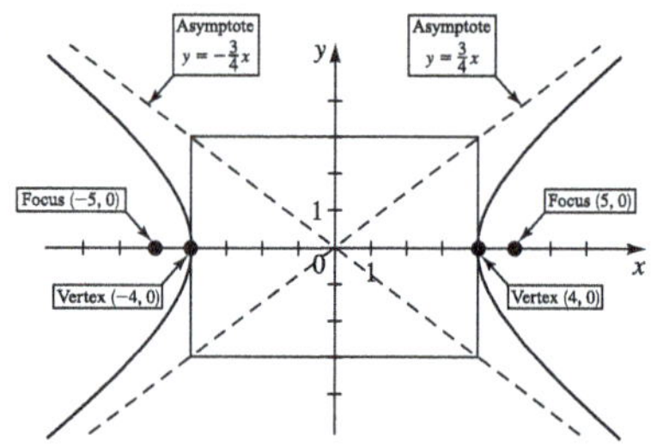

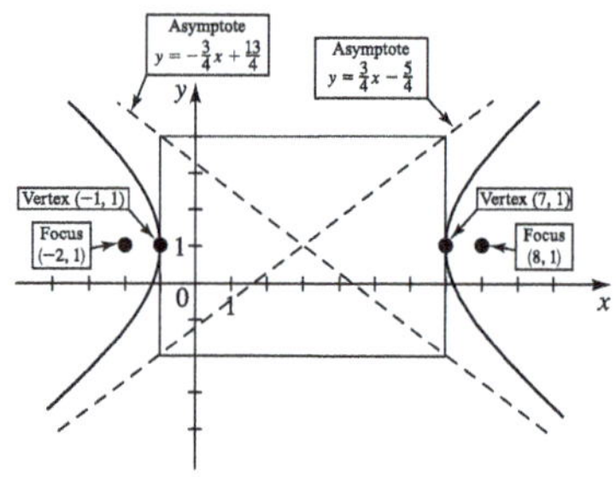

5. (a) $C(2, 1)$; $V_1(-1, 1)$, $V_2(5, 1)$; $F(2 \pm \sqrt{5}, 1)$
(b) 6, 4
(c)

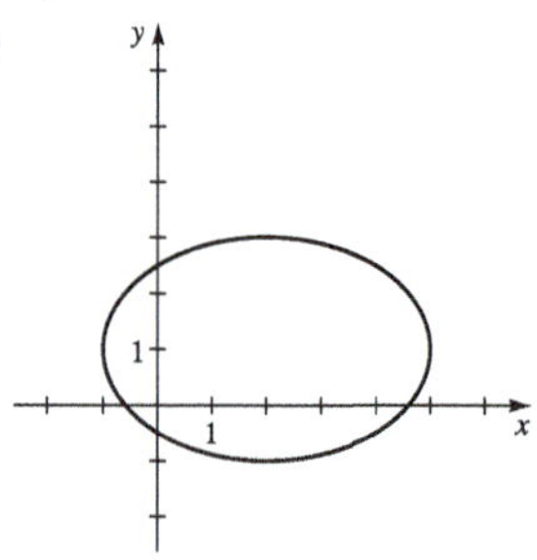

7. (a) $C(0, -5)$; $V_1(0, -10)$, $V_2(0, 0)$; $F_1(0, -9)$, $F_2(0, -1)$
(b) 10, 6
(c)

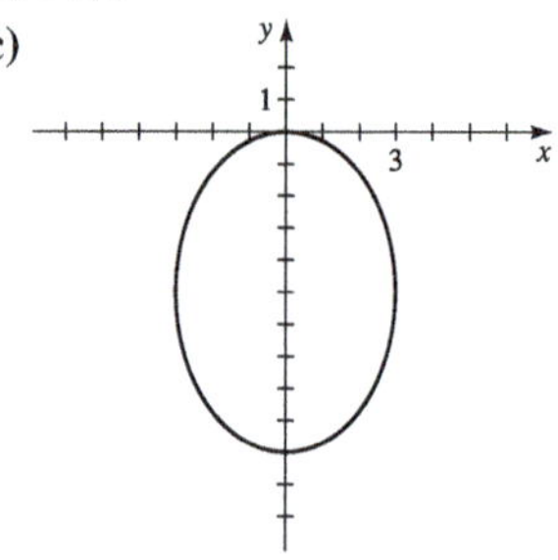

9. (a) $C(-5, 1)$; $V_1(-9, 1)$, $V_2(-1, 1)$; $F(-5 \pm 2\sqrt{3}, 1)$
(b) 8, 4
(c)

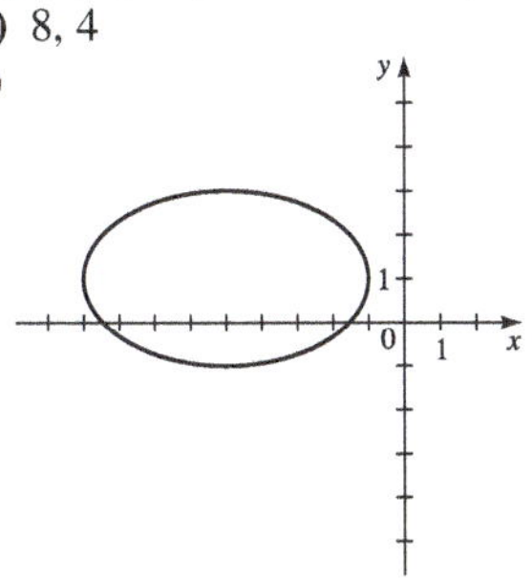

11. (a) $C(0, 1)$; $V(\pm 5, 1)$; $F(\pm\sqrt{21}, 1)$
(b) 10, 4
(c)

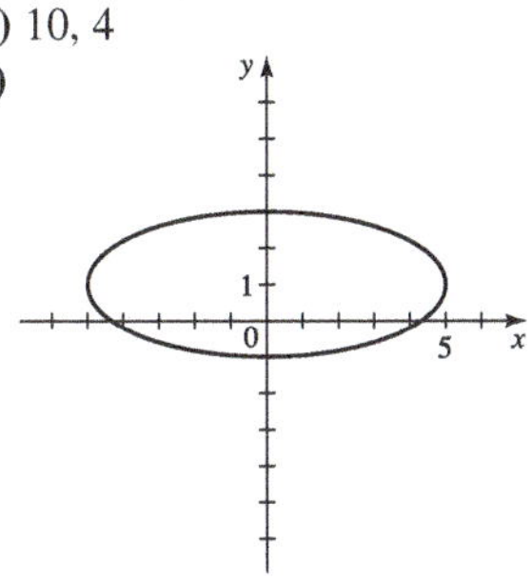

13. (a) $V(3, -1)$; $F(3, 1)$; directrix $y = -3$
(b)

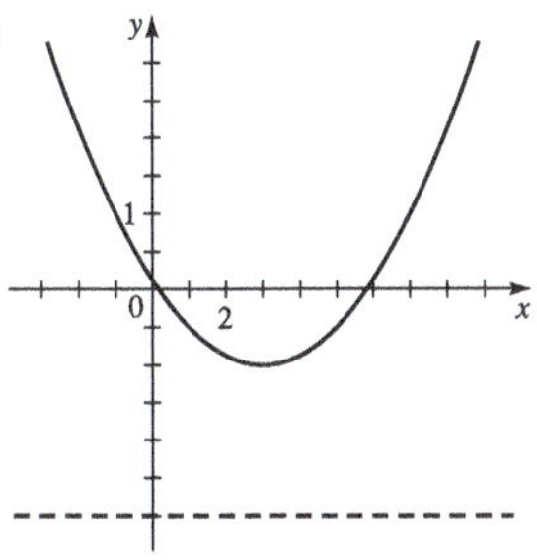

15. (a) $V(2, -5)$; $F(\frac{1}{2}, -5)$; directrix $x = \frac{7}{2}$
(b)

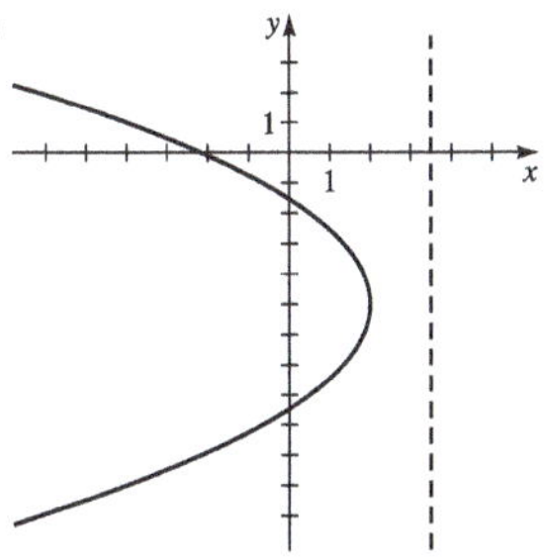

17. (a) $V(1, 0)$; $F\left(1, \frac{1}{8}\right)$; directrix $y = -\frac{1}{8}$
(b)

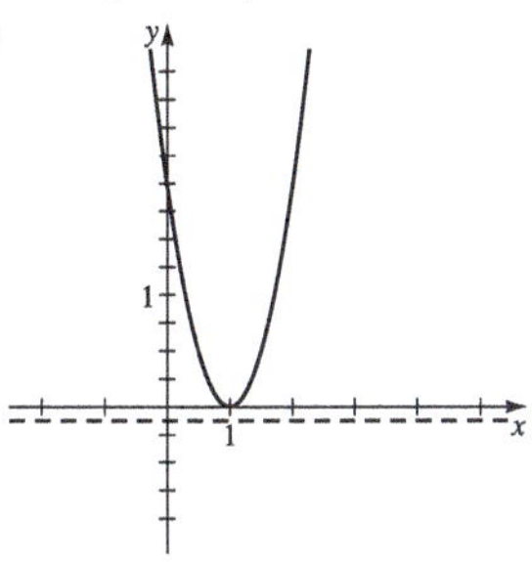

19. (a) $V(2, 3)$; $F(5, 3)$; directrix $x = -1$
(b)

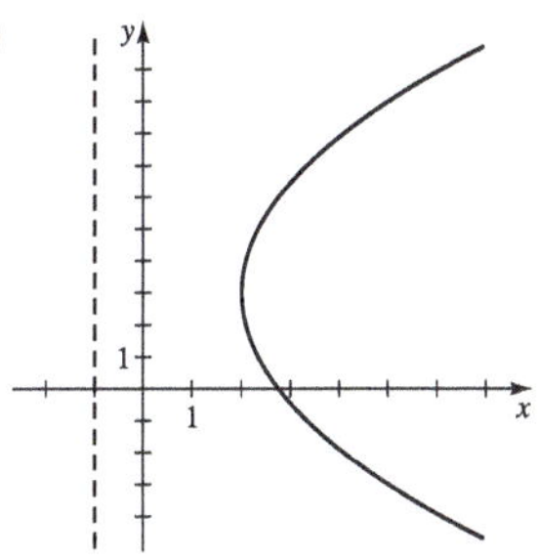

21. (a) $C(-1, 3)$; $V_1(-4, 3)$, $V_2(2, 3)$; $F_1(-6, 3)$, $F_2(4, 3)$; asymptotes $y = \frac{4}{3}x + \frac{13}{3}$ and $y = -\frac{4}{3}x + \frac{5}{3}$
(b)

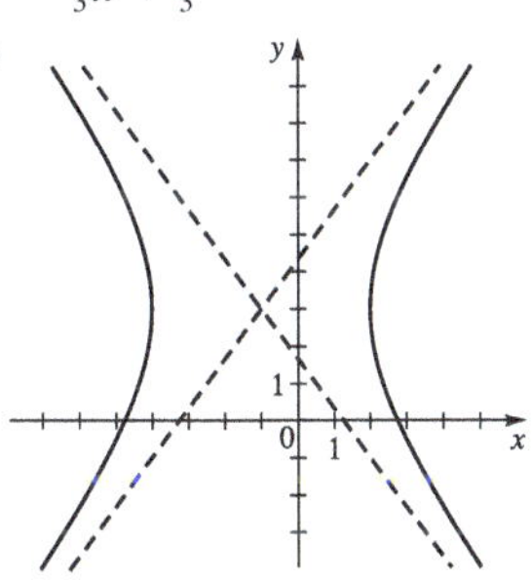

23. (a) $C(-1, 0)$; $V(-1, \pm 1)$; $F(-1, \pm\sqrt{5})$; asymptotes $y = \frac{1}{2}x + \frac{1}{2}$ and $y = -\frac{1}{2}x - \frac{1}{2}$
(b)

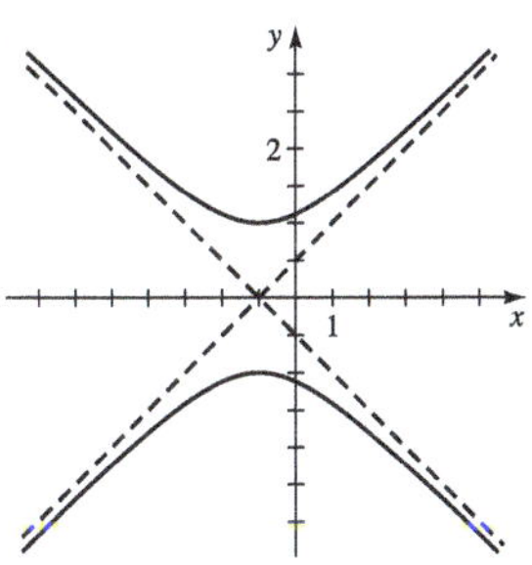

25. (a) $C(-1, -1)$; $V_1(-4, -1)$, $V_2(2, -1)$; $F(-1 \pm \sqrt{13}, -1)$; asymptotes $y = \frac{2}{3}x - \frac{1}{3}$ and $y = -\frac{2}{3}x - \frac{5}{3}$
(b)

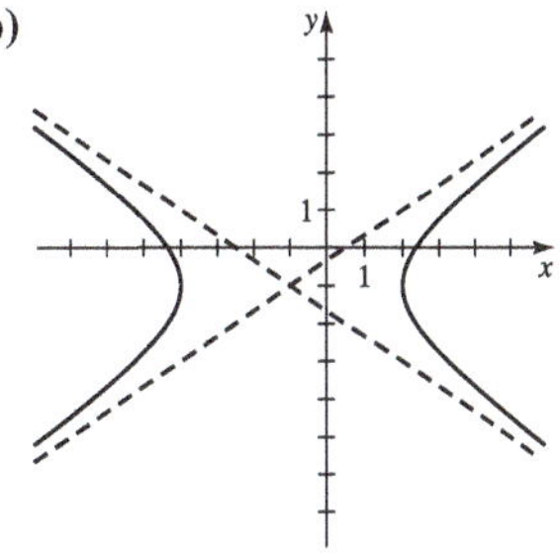

27. (a) $C(-1, 4)$; $V_1(-1, -2)$, $V_2(-1, 10)$; $F(-1, 4 \pm 2\sqrt{10})$; asymptotes $y = 3x + 7$ and $y = -3x + 1$
(b)

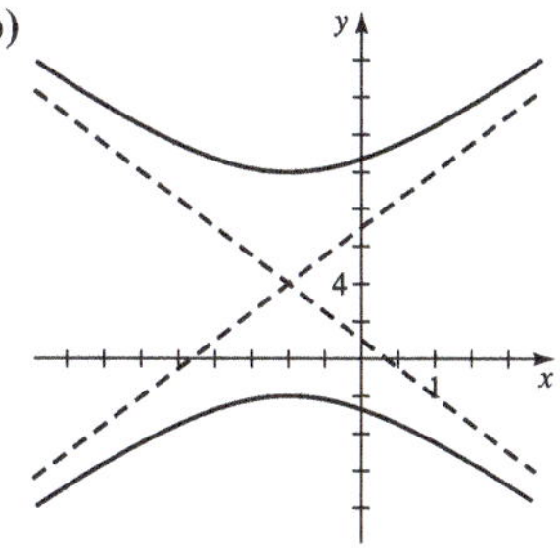

29. $x^2 = -\frac{1}{4}(y - 4)$ **31.** $\dfrac{(x-5)^2}{25} + \dfrac{y^2}{16} = 1$

33. $(y - 1)^2 - x^2 = 1$ **35.** $\dfrac{(x-2)^2}{100} + \dfrac{(y+3)^2}{64} = 1$

37. $\dfrac{(y-4)^2}{49} - \dfrac{(x+1)^2}{32} = 1$ **39.** $(x + 3)^2 = 12(y - 5)$

41. $\dfrac{y^2}{16} - \dfrac{(x-1)^2}{9} = 1$ **43.** $\dfrac{(x-3)^2}{29} + \dfrac{(y+4)^2}{25} = 1$

45. $(y - 2)^2 = \frac{1}{7}(x + 1)$

47. Parabola; $V(-4, 4)$; $F(-3, 4)$; directrix $x = -5$

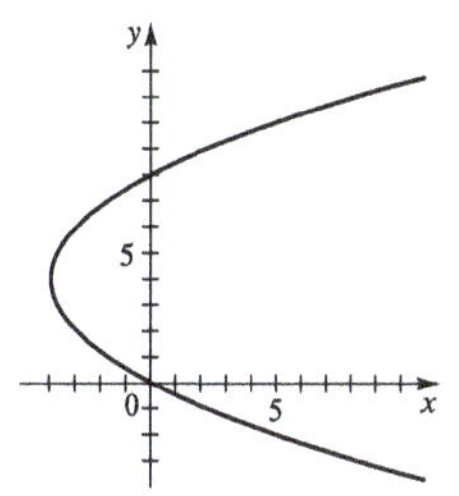

49. Hyperbola; $C(1, 2)$; $F(1 \pm \sqrt{30}, 2)$; $V_1(-4, 2)$, $V_2(6, 2)$; asymptotes $y = \pm\dfrac{\sqrt{5}}{5}(x - 1) + 2$

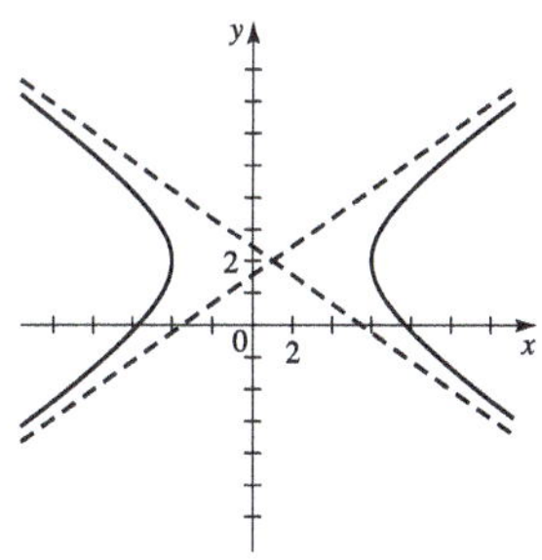

51. Ellipse; $C(3, -5)$; $F(3 \pm \sqrt{21}, -5)$; $V_1(-2, -5)$, $V_2(8, -5)$; major axis 10, minor axis 4

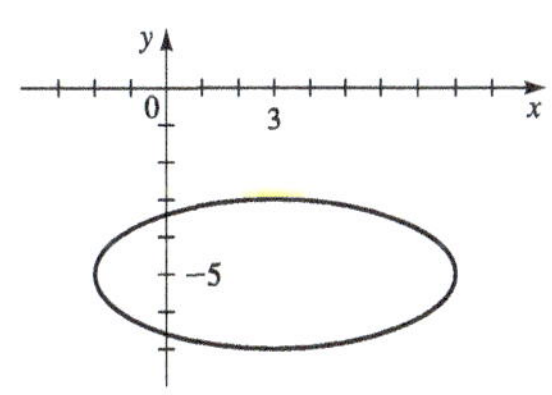

53. Hyperbola; $C(3, 0)$; $F(3, \pm 5)$; $V(3, \pm 4)$; asymptotes $y = \pm\frac{4}{3}(x - 3)$

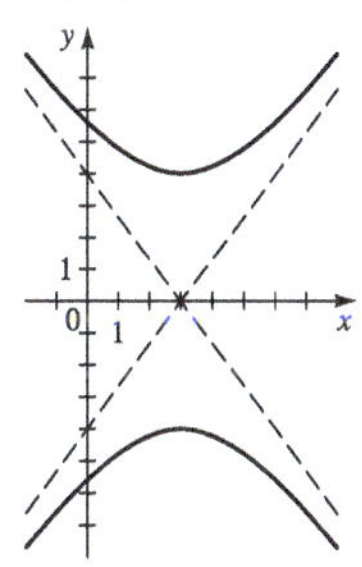

55. Degenerate conic (pair of lines), $y = \pm\frac{1}{2}(x - 4)$

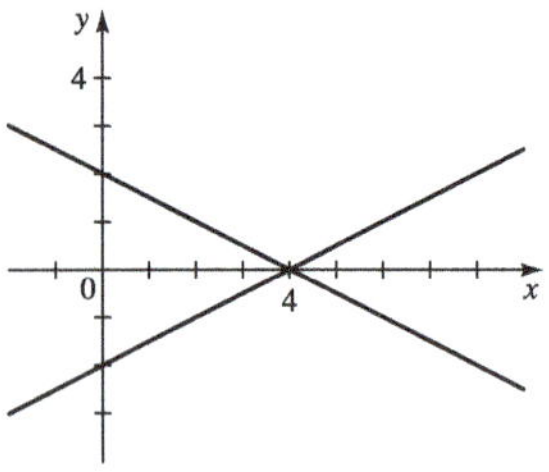

57. Point $(1, 3)$

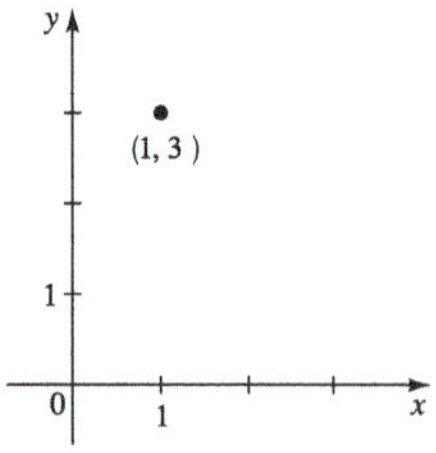

59.

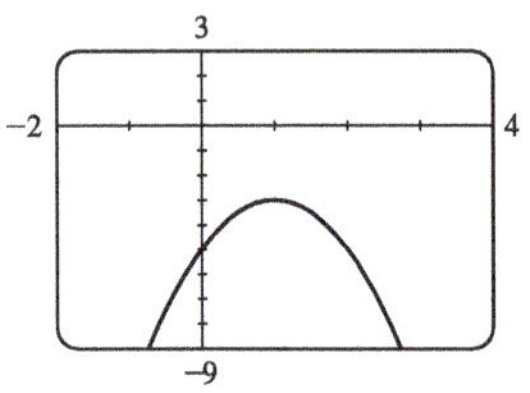

61.

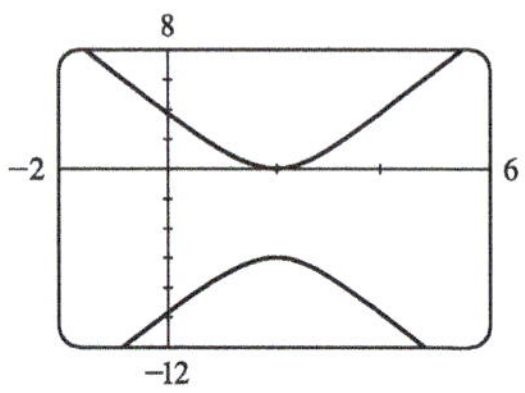

63. (a) $F < 17$ **(b)** $F = 17$ **(c)** $F > 17$

65. (a)

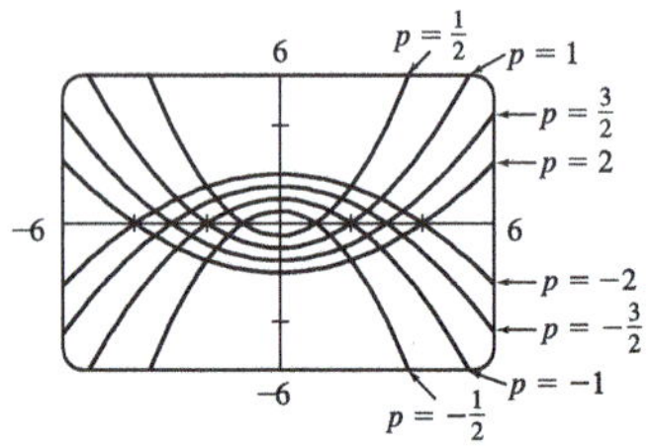

(c) The parabolas become narrower.

67. $\dfrac{(x + 150)^2}{18{,}062{,}500} + \dfrac{y^2}{18{,}040{,}000} = 1$

SECTION 11.5 ■ PAGE 823

1. $x = X\cos\phi - Y\sin\phi$, $y = X\sin\phi + Y\cos\phi$, $X = x\cos\phi + y\sin\phi$, $Y = -x\sin\phi + y\cos\phi$
2. (a) conic section **(b)** $(A - C)/B$ **(c)** $B^2 - 4AC$, parabola, ellipse, hyperbola **3.** $(\sqrt{2}, 0)$ **5.** $(0, -2\sqrt{3})$
7. $(1.6383, 1.1472)$ **9.** $X^2 + \sqrt{3}XY + 2 = 0$
11. $7Y^2 - 48XY - 7X^2 - 40X - 30Y = 0$ **13.** $X^2 - Y^2 = 2$
15. (a) Hyperbola
(b) $X^2 - Y^2 = 16$
(c) $\phi = 45°$

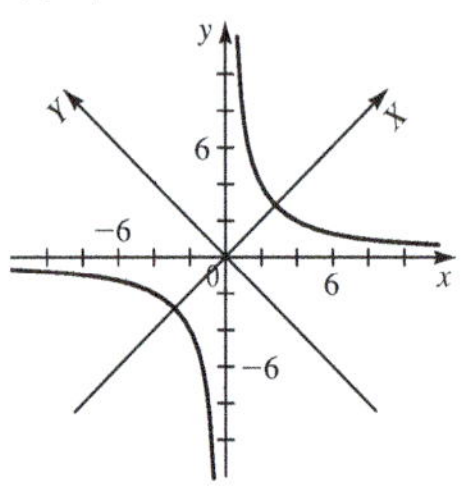

17. (a) Hyperbola
(b) $Y^2 - X^2 = 1$
(c) $\phi = 30°$

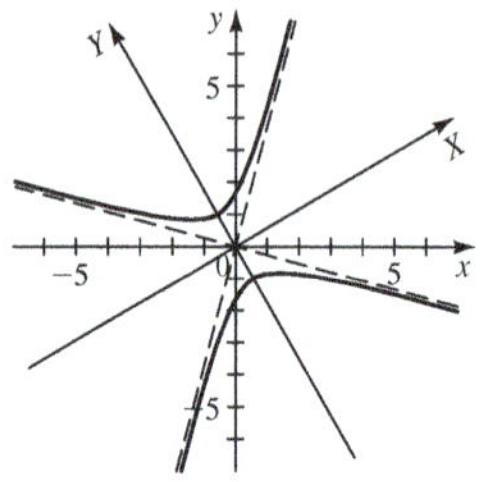

19. (a) Hyperbola
(b) $\dfrac{X^2}{4} - Y^2 = 1$
(c) $\phi \approx 53°$

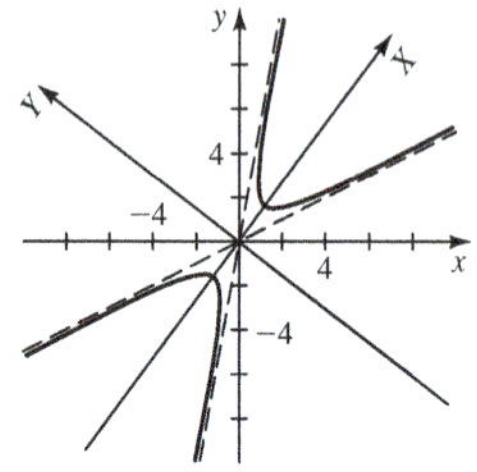

21. (a) Hyperbola
(b) $3X^2 - Y^2 = 2\sqrt{3}$
(c) $\phi = 30°$

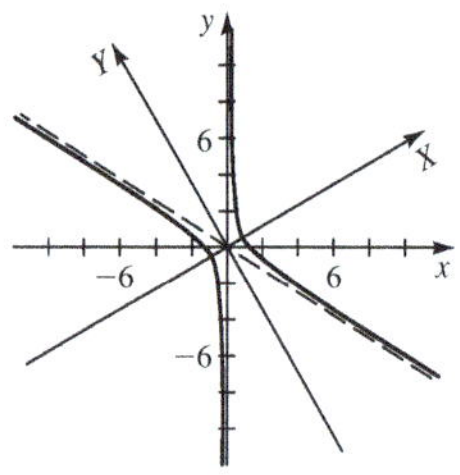

23. (a) Parabola
(b) $Y = \sqrt{2}X^2$
(c) $\phi = 45°$

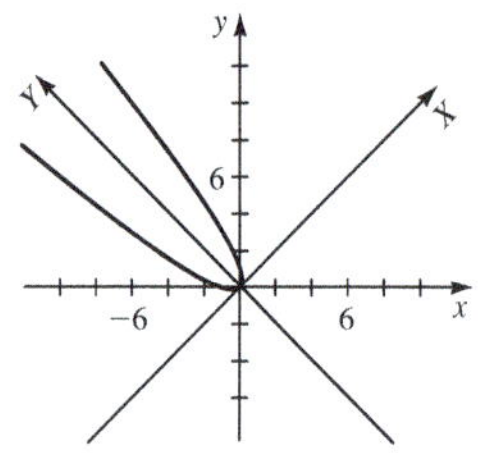

25. (a) Hyperbola
(b) $(X - 1)^2 - 3Y^2 = 1$
(c) $\phi = 60°$

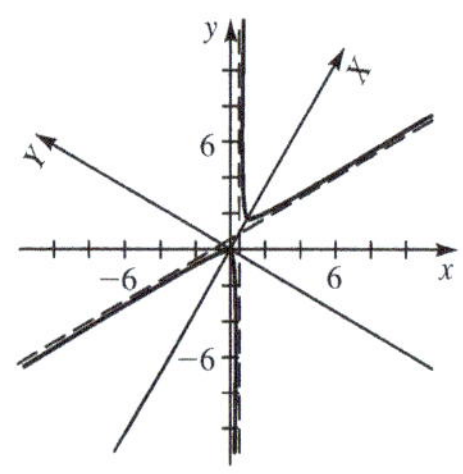

27. (a) Ellipse
(b) $X^2 + \dfrac{(Y + 1)^2}{4} = 1$
(c) $\phi \approx 53°$

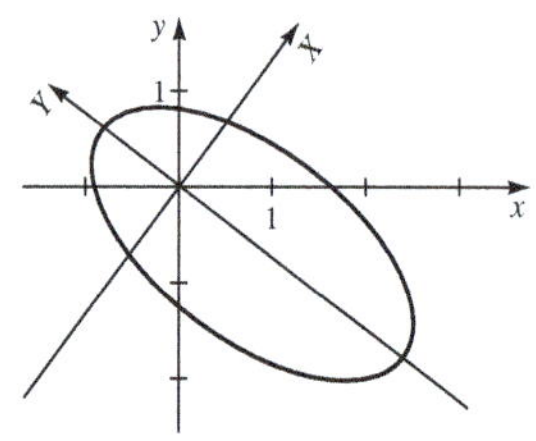

29. (a) Parabola
(b)

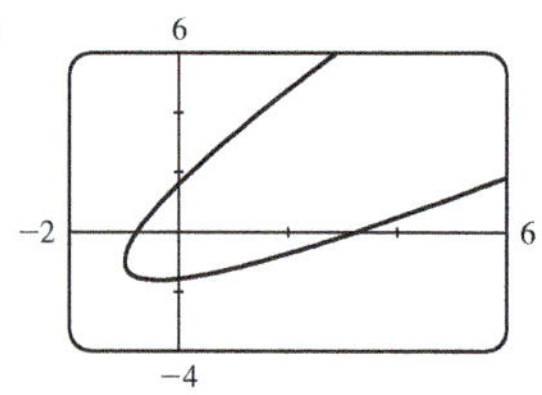

31. (a) Hyperbola
(b)

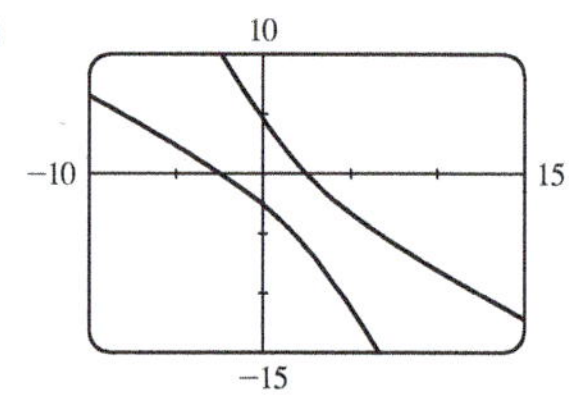

33. (a) $(X - 5)^2 - Y^2 = 1$
(b) XY-coordinates: $C(5, 0)$; $V_1(6, 0)$, $V_2(4, 0)$; $F(5 \pm \sqrt{2}, 0)$; xy-coordinates:
$C(4, 3)$; $V_1(\frac{24}{5}, \frac{18}{5})$, $V_2(\frac{16}{5}, \frac{12}{5})$; $F_1(4 + \frac{4}{5}\sqrt{2}, 3 + \frac{3}{5}\sqrt{2})$, $F_2(4 - \frac{4}{5}\sqrt{2}, 3 - \frac{3}{5}\sqrt{2})$
(c) $Y = \pm(X - 5)$; $7x - y - 25 = 0$, $x + 7y - 25 = 0$
35. $X = x\cos\phi + y\sin\phi$; $Y = -x\sin\phi + y\cos\phi$

SECTION 11.6 ■ PAGE 829

1. focus, directrix; $\dfrac{\text{distance from } P \text{ to } F}{\text{distance from } P \text{ to } \ell}$, conic section; parabola, ellipse, hyperbola, eccentricity

2. $\dfrac{ed}{1 \pm e\cos\theta}, \dfrac{ed}{1 \pm e\sin\theta}$ **3.** $r = 6/(3 + 2\cos\theta)$

5. $r = 2/(1 + \sin\theta)$ **7.** $r = 20/(1 + 4\cos\theta)$

9. $r = 10/(1 + \sin\theta)$ **11.** II **13.** VI **15.** IV

17. (a), (b)

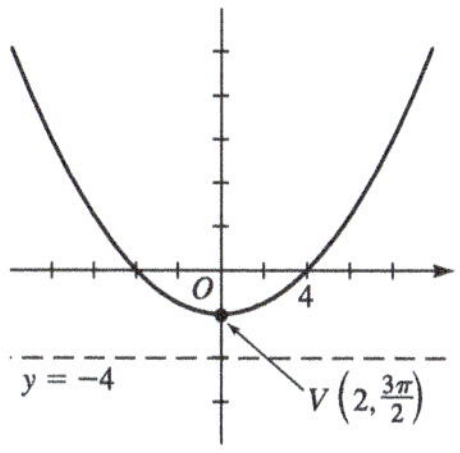

19. (a), (b)

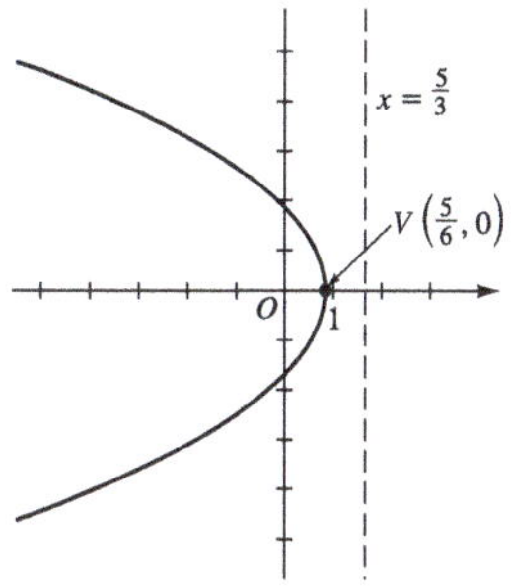

21. (a), (b)

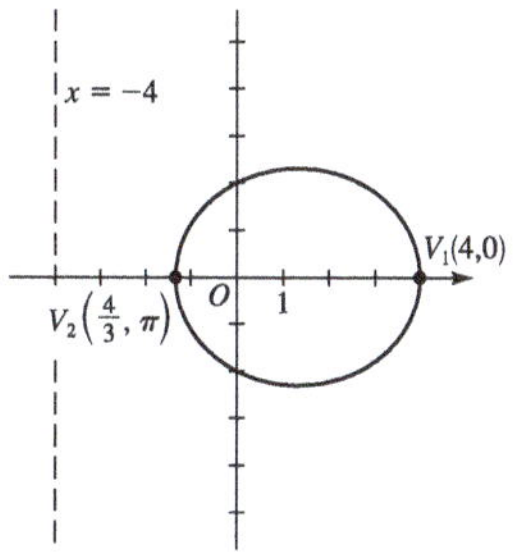

(c) $C(\frac{4}{3}, 0)$, major axis: $\frac{16}{3}$, minor axis: $\frac{8\sqrt{3}}{3}$

23. (a), (b)

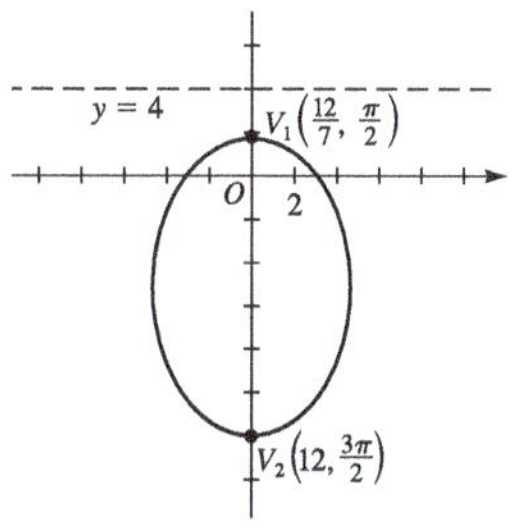

(c) $C(\frac{36}{7}, \frac{3\pi}{2})$, major axis: $\frac{96}{7}$, minor axis: $\frac{24\sqrt{7}}{7}$

25. (a), (b)

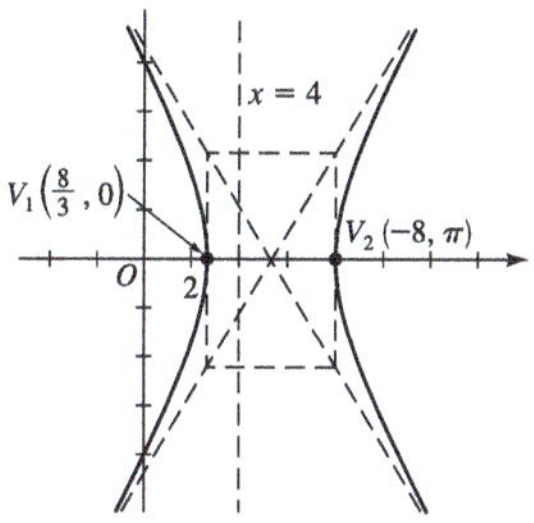

(c) $(\frac{16}{3}, 0)$

27. (a), (b)

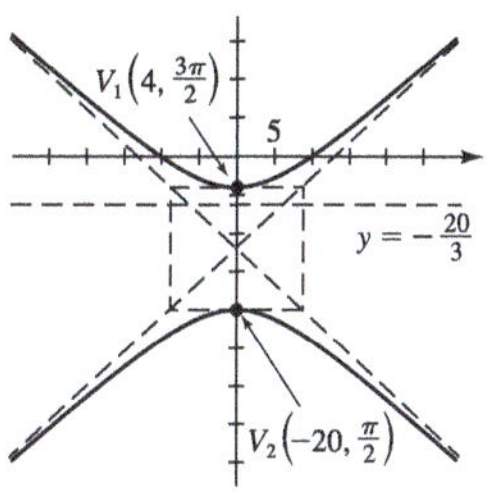

(c) $(12, \frac{3\pi}{2})$

29. (a) 3, hyperbola

(b)

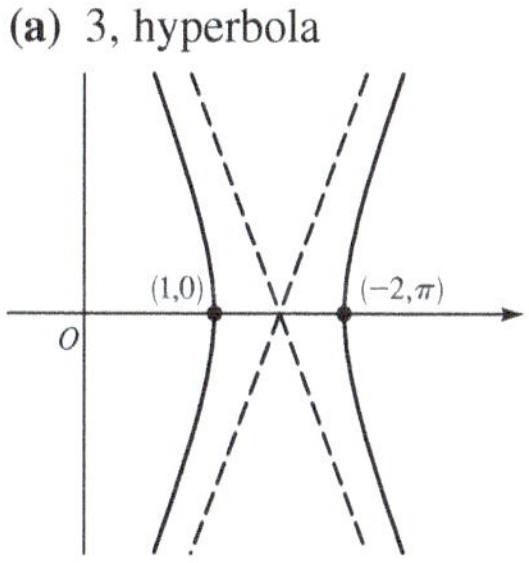

31. (a) 1, parabola

(b)

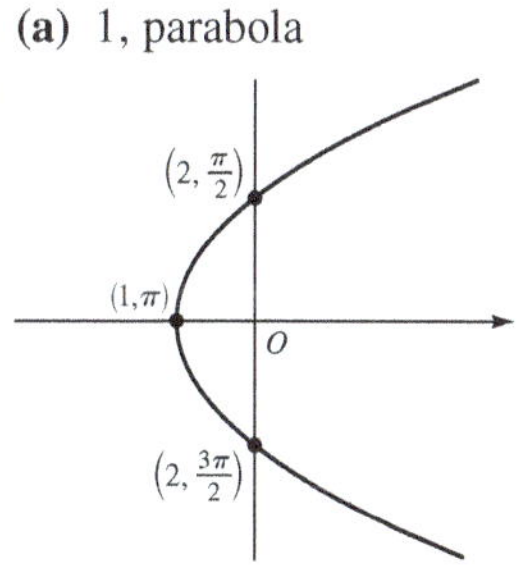

33. (a) $\frac{1}{2}$, ellipse

(b)

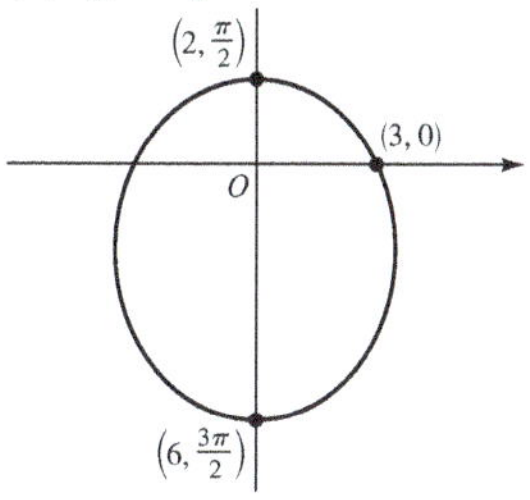

35. (a) $\frac{5}{2}$, hyperbola

(b)

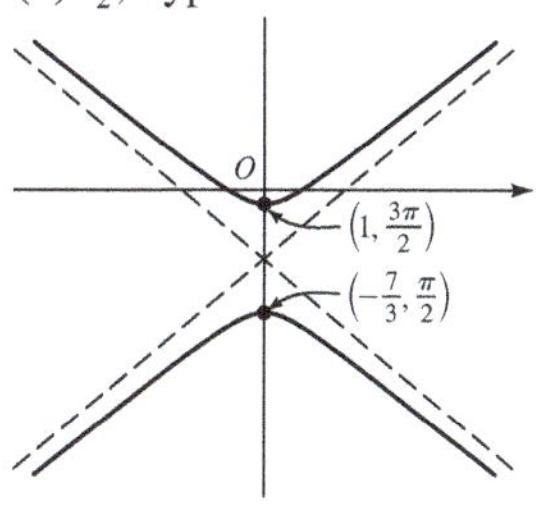

37. (a) eccentricity $\frac{3}{4}$, directrix $x = -\frac{1}{3}$

(b) $r = \dfrac{1}{4 - 3\cos(\theta - \frac{\pi}{3})}$

(c)

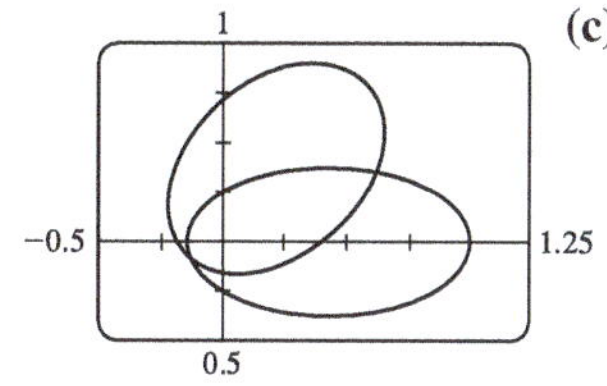

39. (a) eccentricity 1, directrix $y = 2$

(b) $r = \dfrac{2}{1 + \sin(\theta + \frac{\pi}{4})}$

(c)

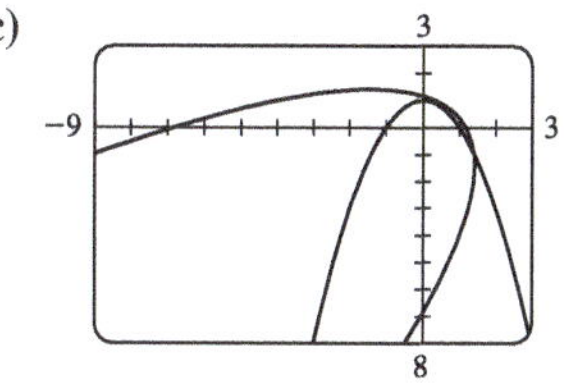

41. The ellipse is nearly circular when e is close to 0 and becomes more elongated as $e \to 1^-$. At $e = 1$ the curve becomes a parabola.

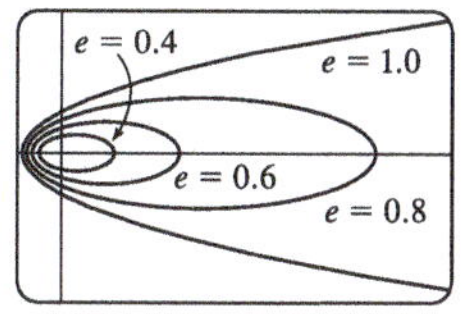

43. (b) $r = (1.49 \times 10^8)/(1 - 0.017\cos\theta)$ **45.** 0.25

CHAPTER 11 REVIEW ■ PAGE 833

1. (a) $V(0, 0)$; $F(1, 0)$; directrix $x = -1$

(b)

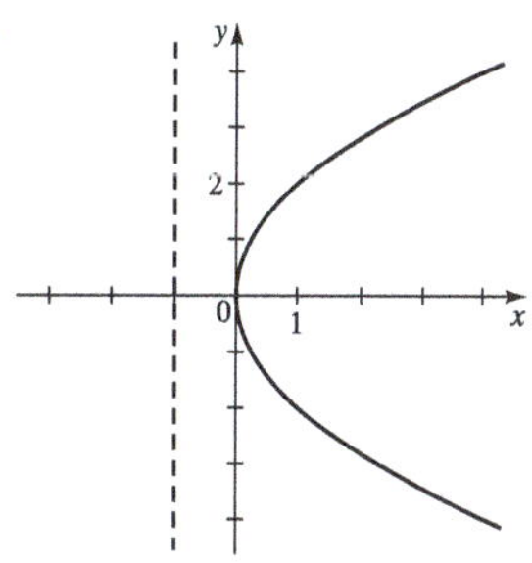

3. (a) $V(0, 0)$; $F(0, 2)$; directrix $y = -2$

(b)

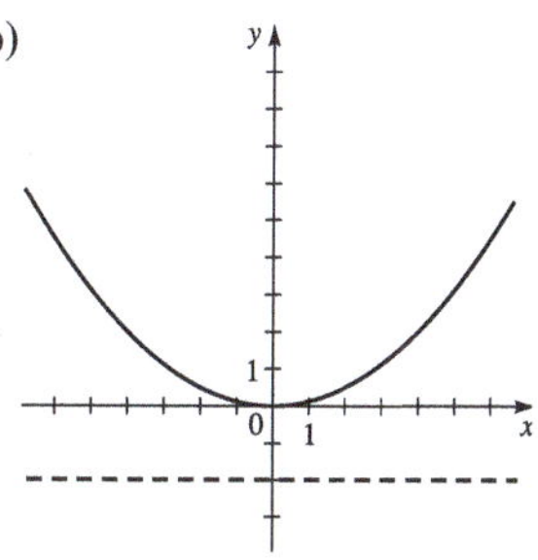

5. (a) $V(0, 0)$; $F(0, -2)$; directrix $y = 2$

(b)

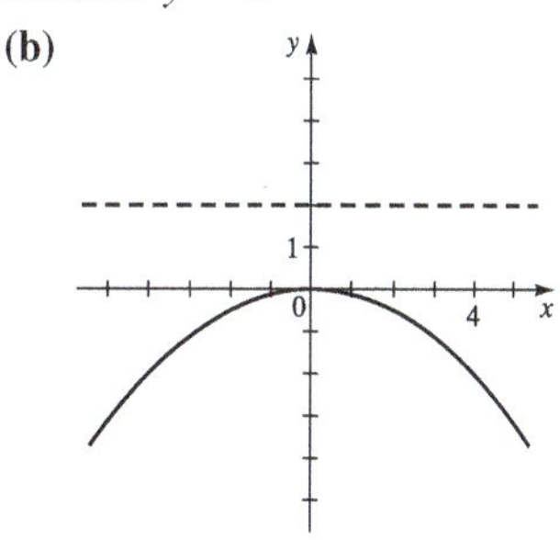

7. (a) $V(-2, 2)$; $F(-1, 2)$; directrix $x = -3$

(b)

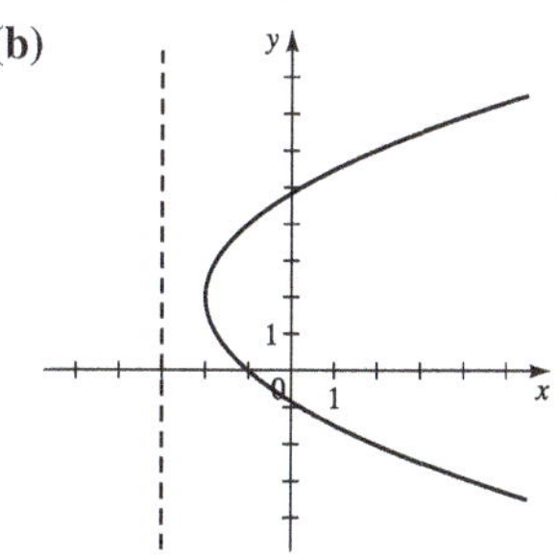

9. (a) $V(0, 3)$; $F(-\frac{1}{2}, 3)$; directrix $x = \frac{1}{2}$

(b)

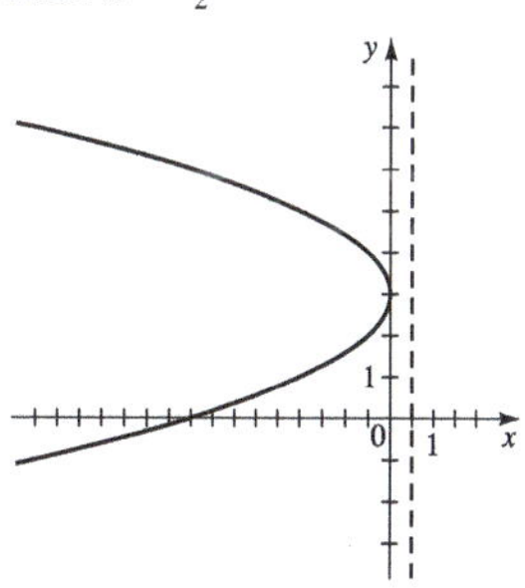

11. (a) $V(-2, -3)$; $F(-2, -2)$; directrix $y = -4$

(b)

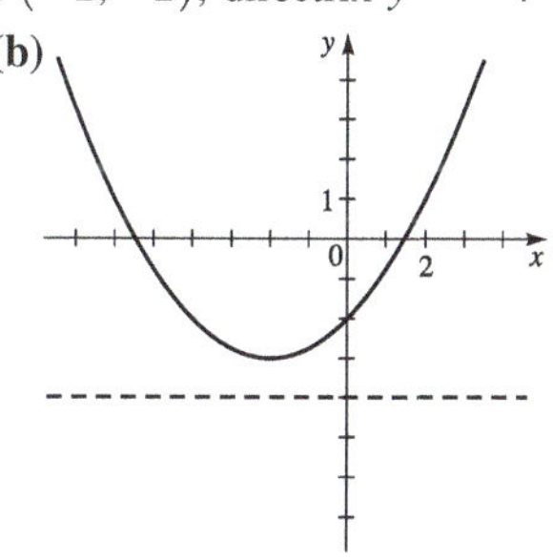

13. **(a)** $C(0, 0)$; $V(0, \pm 5)$; $F(0, \pm 4)$
(b) 10, 6
(c)

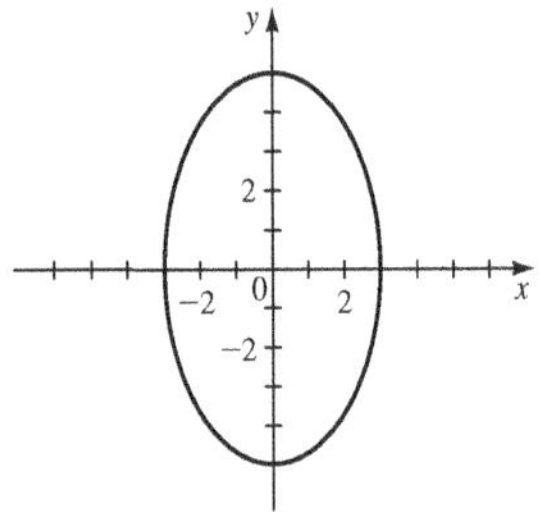

15. **(a)** $C(0, 0)$; $V(\pm 7, 0)$; $F(\pm 3\sqrt{5}, 0)$
(b) 14, 4
(c)

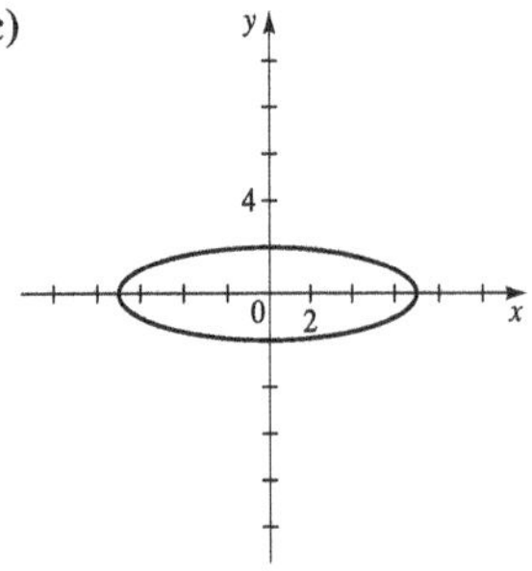

17. **(a)** $C(0, 0)$; $V(\pm 4, 0)$; $F(\pm 2\sqrt{3}, 0)$
(b) 8, 4
(c)

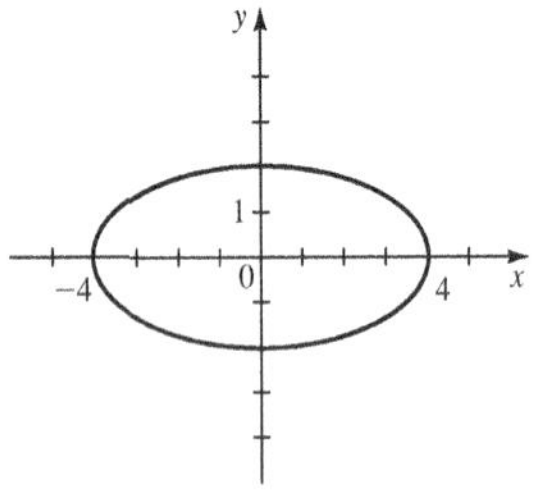

19. **(a)** $C(3, 0)$; $V(3, \pm 4)$; $F(3, \pm\sqrt{7})$
(b) 8, 6
(c)

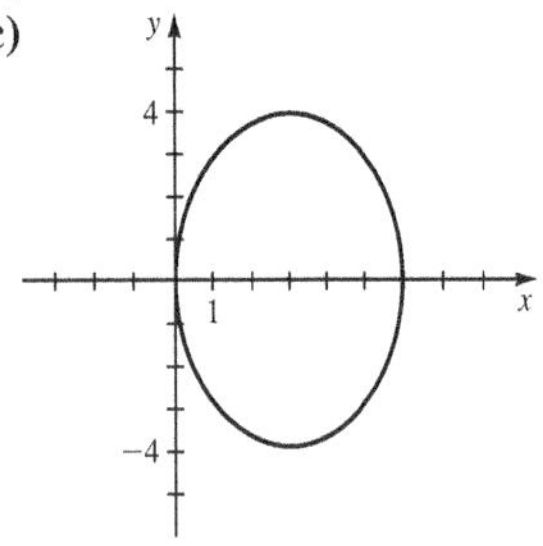

21. **(a)** $C(2, -3)$; $V_1(2, -9)$, $V_2(2, 3)$; $F(2, -3 \pm 3\sqrt{3})$
(b) 12, 6
(c)

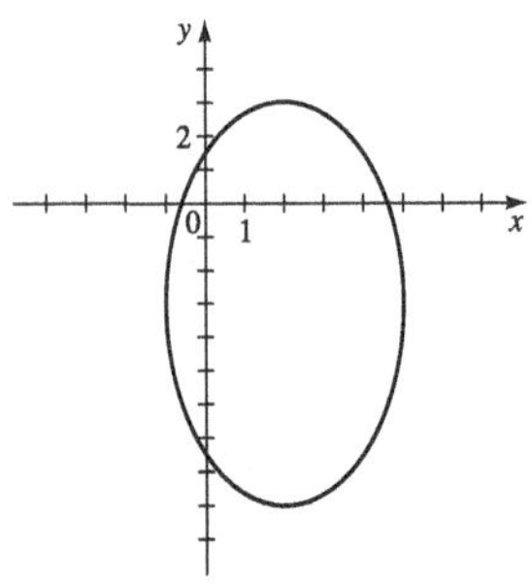

23. **(a)** $C(0, 2)$; $V(\pm 3, 2)$; $F(\pm\sqrt{5}, 2)$
(b) 6, 4
(c)

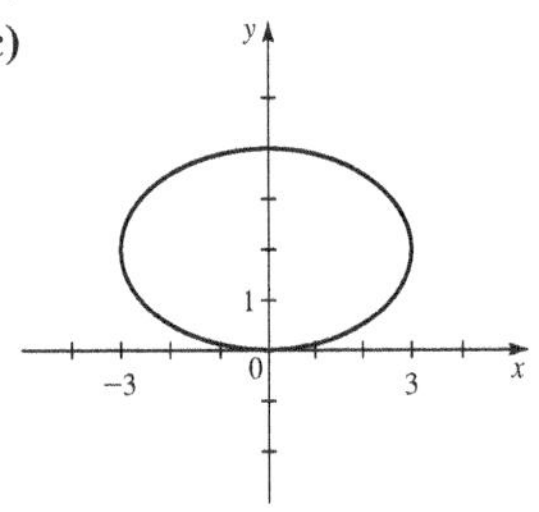

25. **(a)** $C(0, 0)$; $V(0, \pm 4)$; $F(0, \pm 5)$; asymptotes $y = \pm\frac{4}{3}x$
(b)

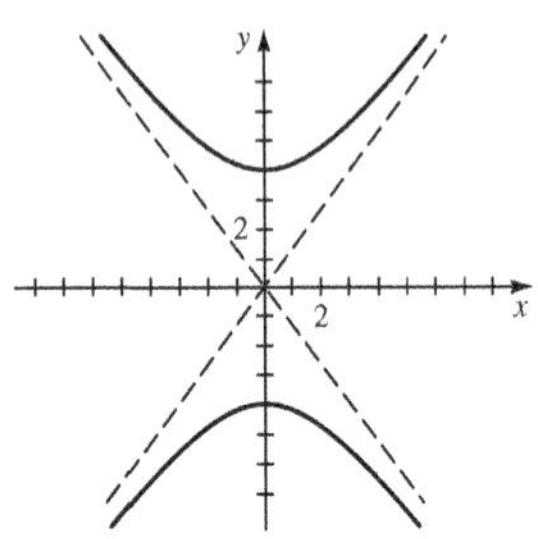

27. **(a)** $C(0, 0)$; $V(\pm 2, 0)$; $F(\pm\sqrt{53}, 0)$; asymptotes $y = \pm\frac{7}{2}x$
(b)

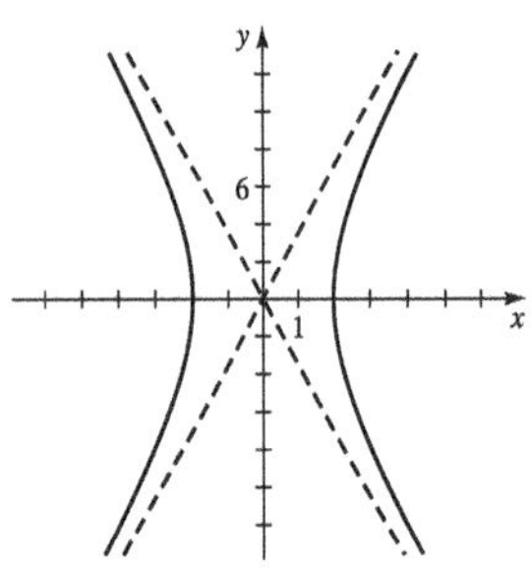

29. **(a)** $C(0, 0)$; $V(\pm 4, 0)$; $F(\pm 2\sqrt{6}, 0)$; asymptotes $y = \pm\dfrac{1}{\sqrt{2}}x$
(b)

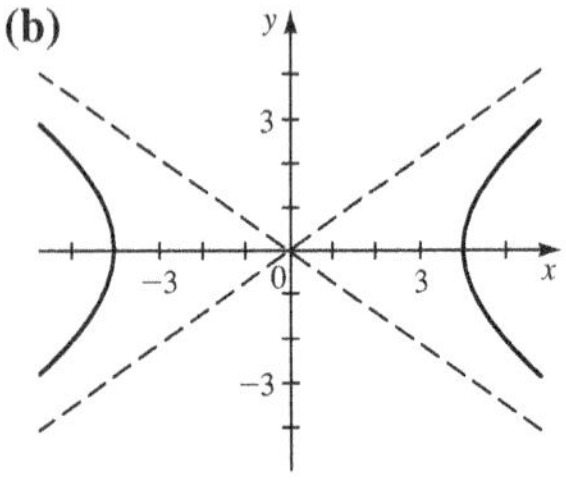

31. **(a)** $C(-4, 0)$; $V_1(-8, 0)$, $V_2(0, 0)$; $F(-4 \pm 4\sqrt{2}, 0)$; asymptotes $y = \pm(x + 4)$
(b)

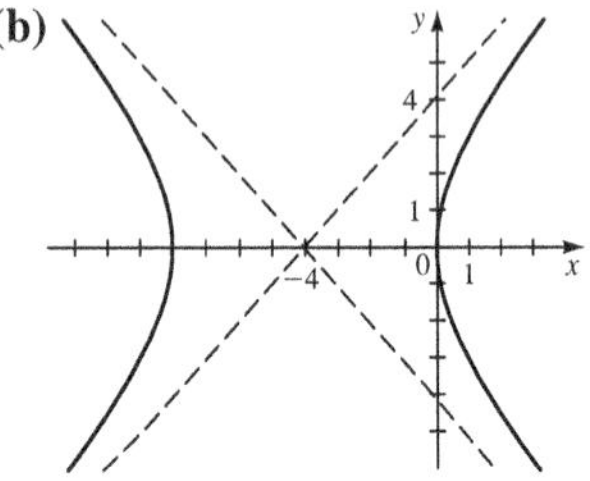

33. **(a)** $C(-1, 3)$; $V_1(-1, 1)$, $V_2(-1, 5)$; $F(-1, 3 \pm 2\sqrt{10})$; asymptotes $y = \frac{1}{3}x + \frac{10}{3}$ and $y = -\frac{1}{3}x + \frac{8}{3}$
(b)

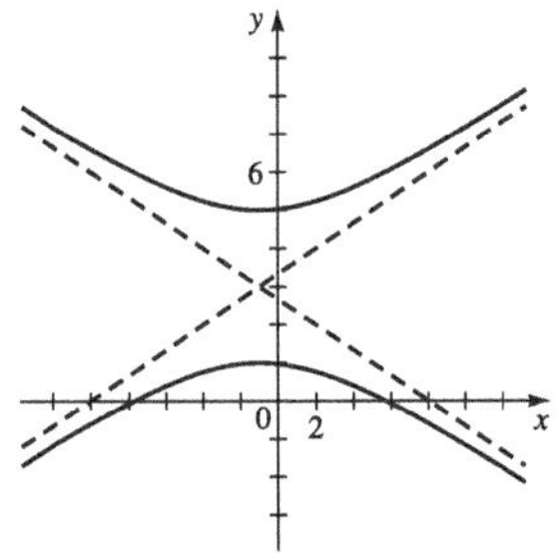

35. **(a)** $C(-3, -1)$; $V(-3, -1 \pm \sqrt{2})$; $F(-3, -1 \pm 2\sqrt{5})$; asymptotes $y = \frac{1}{3}x$, $y = -\frac{1}{3}x - 2$
(b)

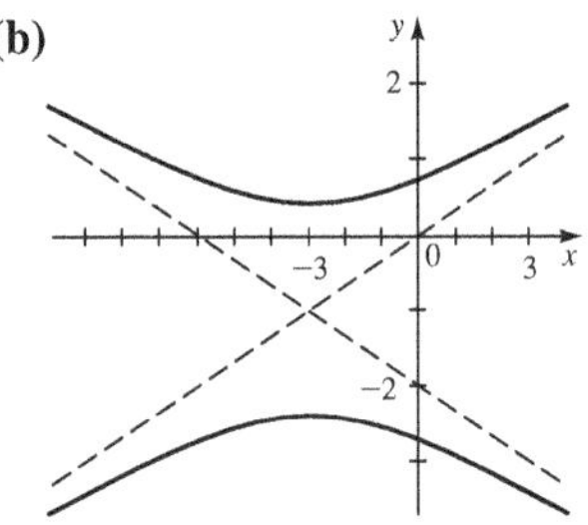

37. $y^2 = 8x$ **39.** $\dfrac{y^2}{16} - \dfrac{x^2}{9} = 1$ **41.** $\dfrac{(x-4)^2}{16} + \dfrac{(y-2)^2}{4} = 1$

43. Parabola; $V(0, 1)$; $F(0, -2)$; directrix $y = 4$

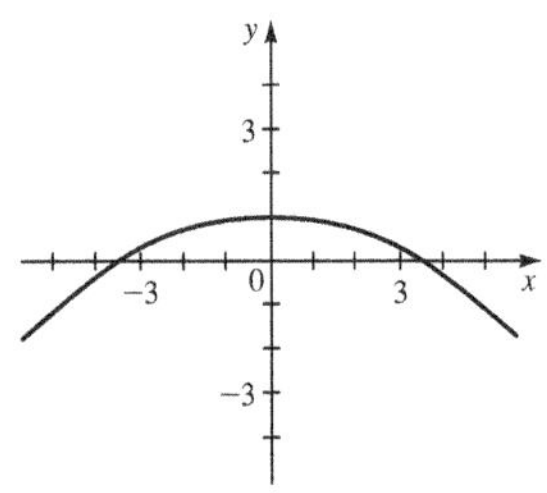

45. Hyperbola; $C(0, 0)$; $F(0, \pm 12\sqrt{2})$; $V(0, \pm 12)$; asymptotes $y = \pm x$

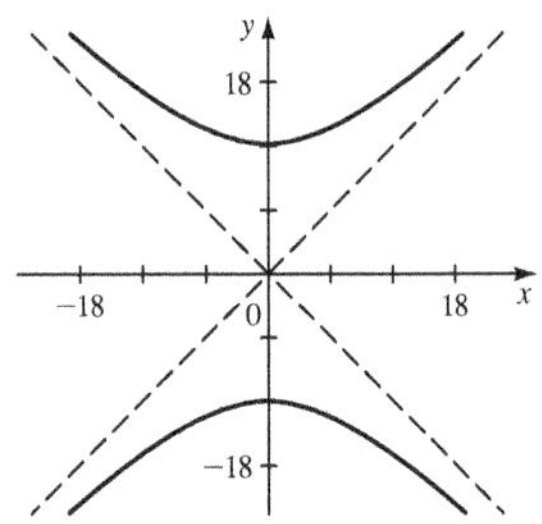

47. Ellipse; $C(1, 4)$; $F(1, 4 \pm \sqrt{15})$; $V(1, 4 \pm 2\sqrt{5})$

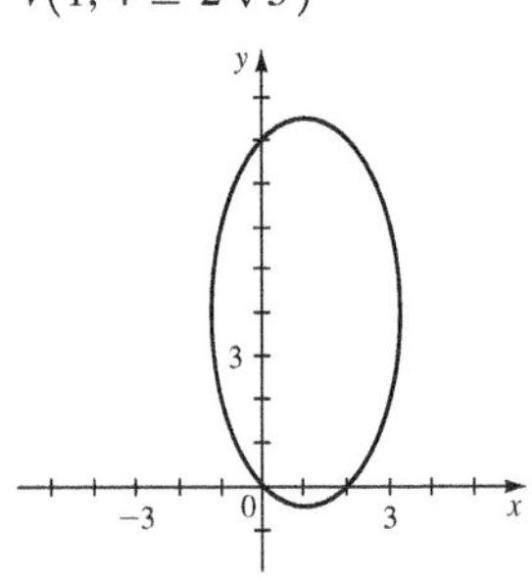

49. Parabola; $V(-64, 8)$; $F\left(-\frac{255}{4}, 8\right)$; directrix $x = -\frac{257}{4}$

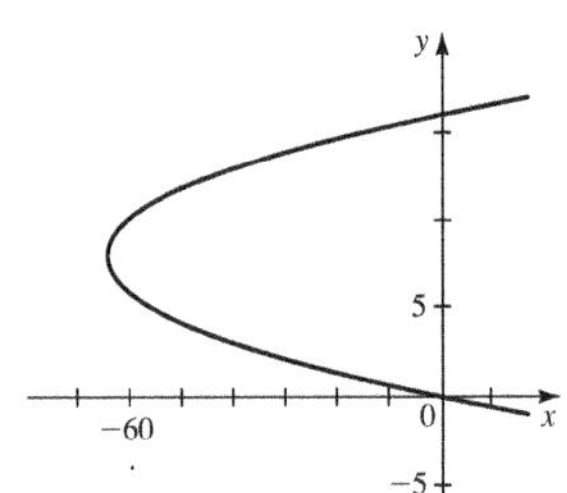

51. Ellipse; $C(3, -3)$; $F\left(3, -3 \pm \dfrac{\sqrt{2}}{2}\right)$; $V_1(3, -4)$, $V_2(3, -2)$

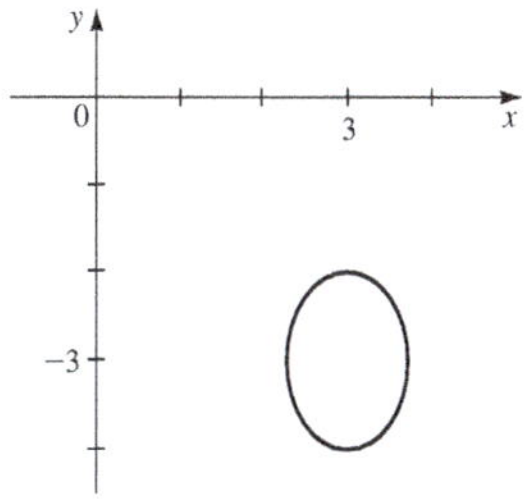

53. Has no graph

55. $x^2 = 4y$ **57.** $\dfrac{x^2}{4} + \dfrac{y^2}{25} = 1$

59. $\dfrac{x^2}{9} + \dfrac{(y-4)^2}{25} = 1$

61. $\dfrac{(x-1)^2}{3} + \dfrac{(y-2)^2}{4} = 1$

63. $\dfrac{4(x-7)^2}{225} + \dfrac{(y-2)^2}{100} = 1$

65. (a) 91,419,000 mi **(b)** 94,581,000 mi

67. (a)

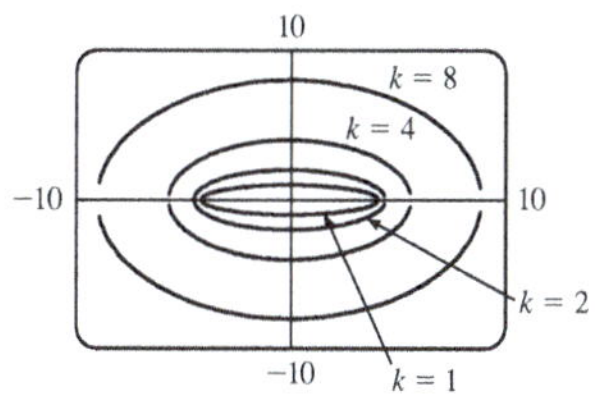

69. (a) Hyperbola **(b)** $3X^2 - Y^2 = 1$
(c) $\phi = 45°$

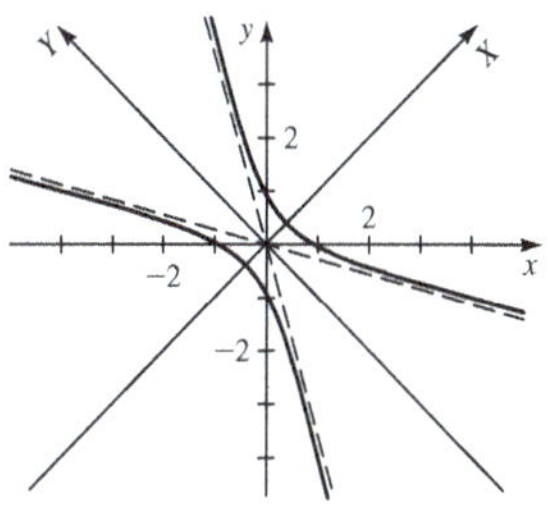

71. (a) Ellipse
(b) $(X-1)^2 + 4Y^2 = 1$
(c) $\phi = 30°$

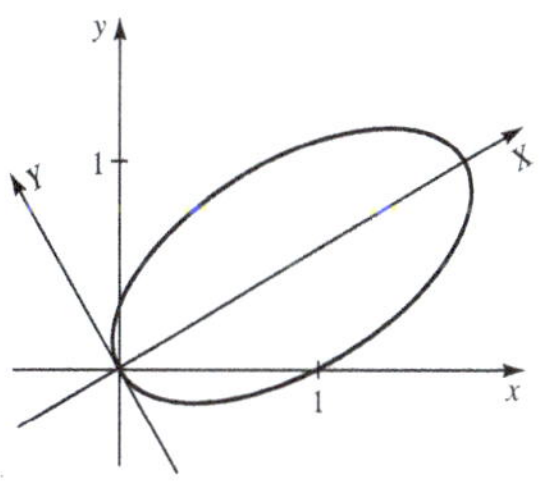

73. Ellipse

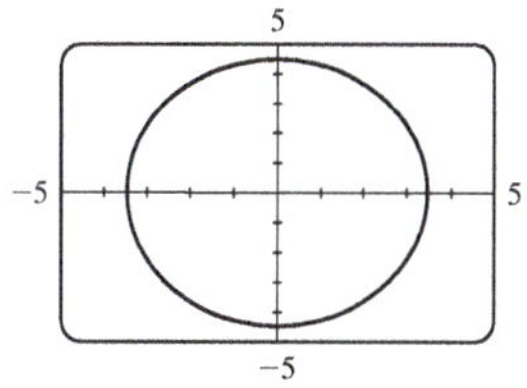

75. Parabola

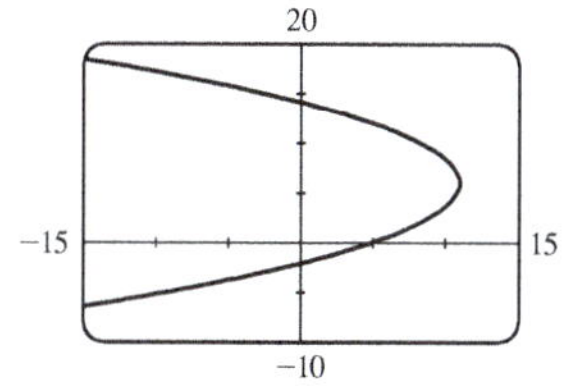

77. (a) $e = 1$, parabola
(b)

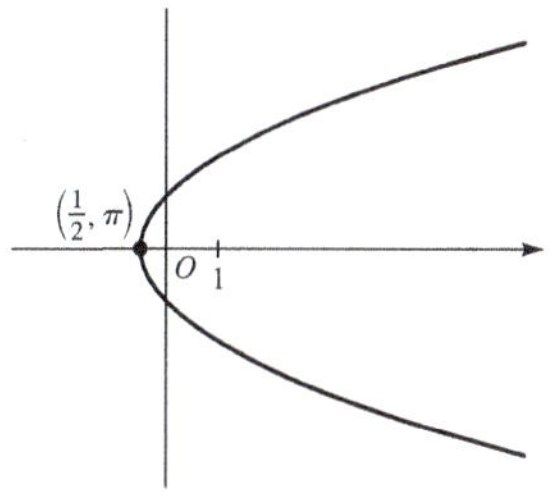

79. (a) $e = 2$, hyperbola
(b)

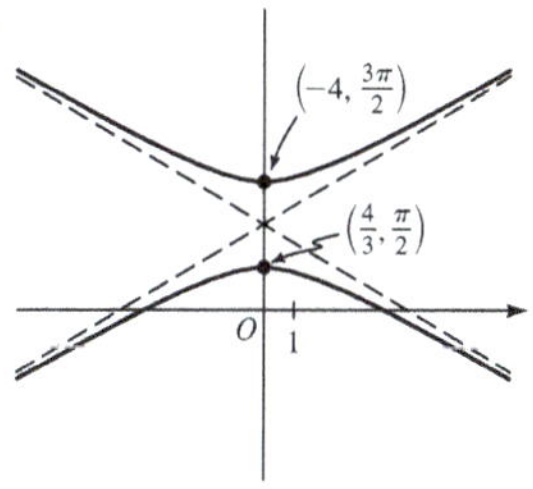

CHAPTER 11 TEST ■ PAGE 835

1. $F(0, -3)$, $y = 3$

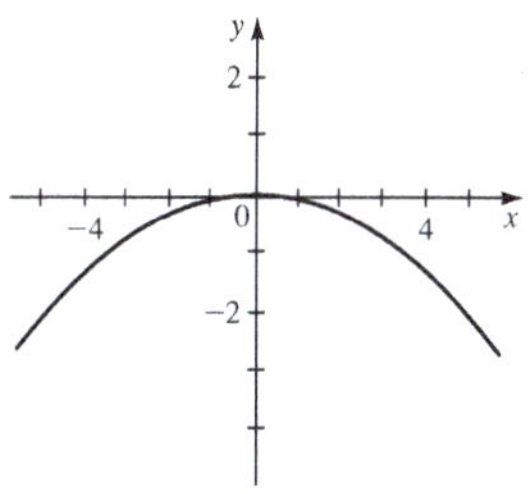

2. $V(\pm 4, 0)$; $F(\pm 2\sqrt{3}, 0)$; 8, 4

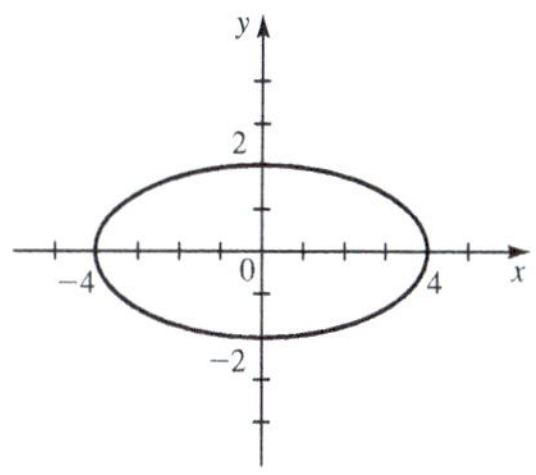

3. $V(0, \pm 3)$; $F(0, \pm 5)$; $y = \pm \frac{3}{4}x$

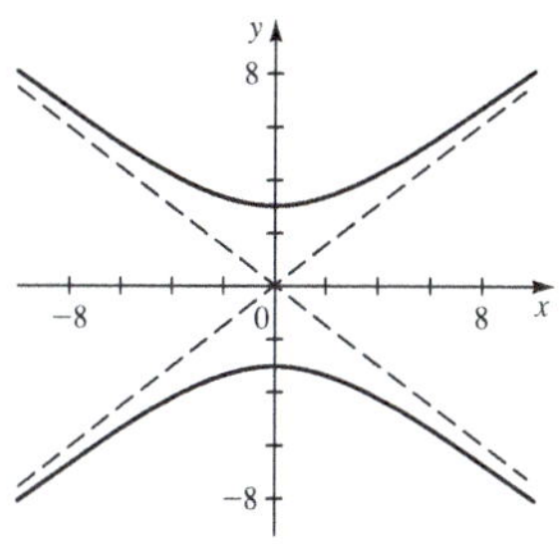

4. $y^2 = 16x$ **5.** $\dfrac{x^2}{16} + \dfrac{y^2}{7} = 1$ **6.** $\dfrac{y^2}{9} - \dfrac{x^2}{16} = 1$

7. $y^2 = -x$ **8.** $\dfrac{x^2}{16} + \dfrac{(y-3)^2}{9} = 1$ **9.** $(x-2)^2 - \dfrac{y^2}{3} = 1$

10. Ellipse; $C\left(3, -\frac{1}{2}\right)$; $F\left(3 \pm \sqrt{5}, -\frac{1}{2}\right)$; $V_1\left(0, -\frac{1}{2}\right)$, $V_2\left(6, -\frac{1}{2}\right)$

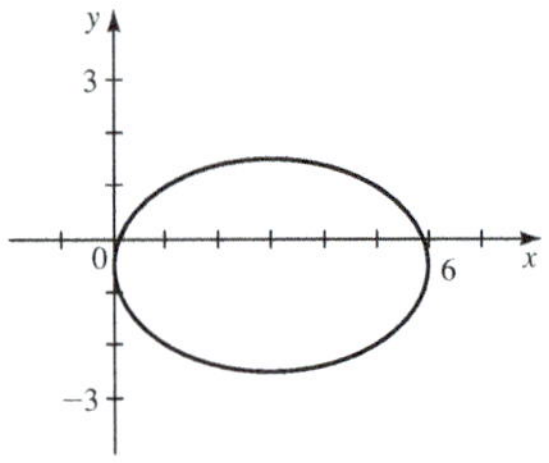

11. Hyperbola; $C(-2, 4)$, $F(-2 \pm \sqrt{17}, 4)$, $V(-2 \pm 2\sqrt{2}, 4)$, asymptotes $y - 4 = \pm\dfrac{3\sqrt{2}}{4}(x + 2)$

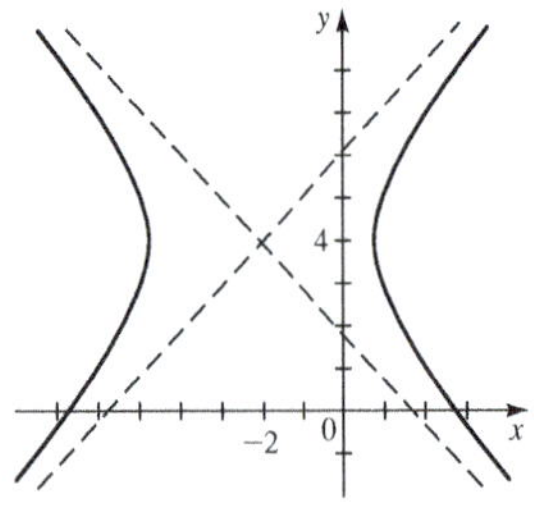

12. Parabola; $V(4, -4)$; $F\left(\frac{7}{2}, -4\right)$; directrix $x = \frac{9}{2}$

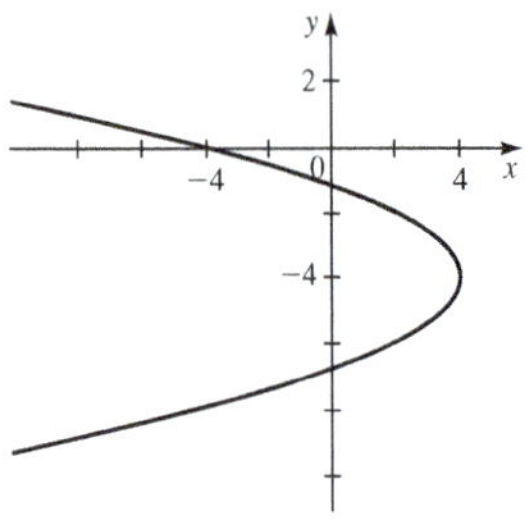

13. $\dfrac{(x-2)^2}{7} + \dfrac{y^2}{16} = 1$ **14.** $(x-2)^2 = 8(y-2)$ **15.** $\frac{3}{4}$ in.

16. (a) Ellipse **(b)** $\dfrac{X^2}{3} + \dfrac{Y^2}{18} = 1$

(c) $\phi \approx 27°$

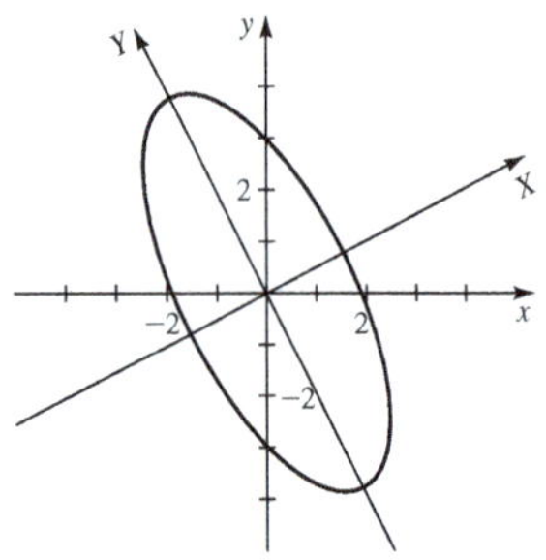

(d) $\left(-3\sqrt{2/5}, 6\sqrt{2/5}\right), \left(3\sqrt{2/5}, -6\sqrt{2/5}\right)$

17. (a) $r = \dfrac{1}{1 + 0.5\cos\theta}$ **(b)** Ellipse

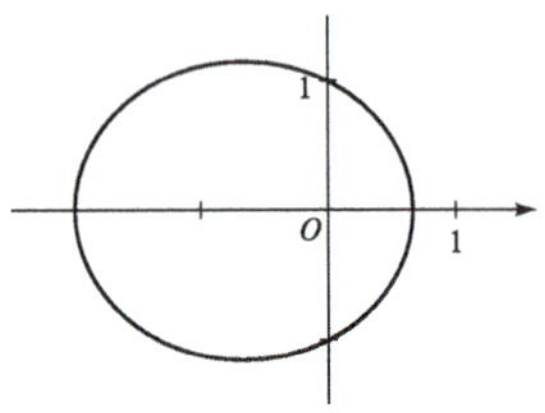

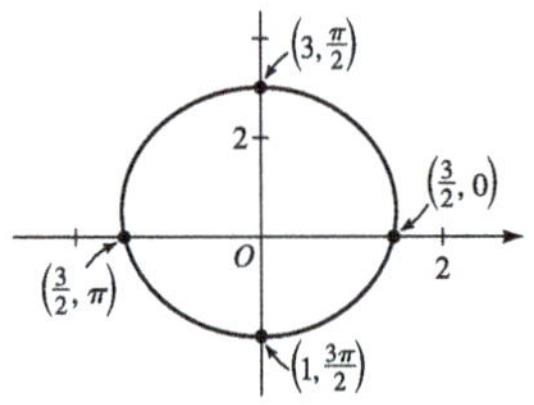

FOCUS ON MODELING ■ PAGE 838

5. (c) $x^2 - mx + (ma - a^2) = 0$, discriminant $m^2 - 4ma + 4a^2 = (m - 2a)^2$, $m = 2a$

CHAPTER 12

SECTION 12.1 ■ PAGE 850

1. the natural numbers **2.** n; $1^2 + 2^2 + 3^2 + 4^2 = 30$
3. $-2, -1, 0, 1$; 97 **5.** $\frac{1}{3}, \frac{1}{5}, \frac{1}{7}, \frac{1}{9}$; $\frac{1}{201}$ **7.** $5, 25, 125, 625$; 5^{100}
9. $-1, \frac{1}{4}, -\frac{1}{9}, \frac{1}{16}$; $\frac{1}{10{,}000}$ **11.** $0, 2, 0, 2$; 2
13. $1, 4, 27, 256$; 100^{100} **15.** $4, 14, 34, 74, 154$
17. $1, 3, 7, 15, 31$ **19.** $1, 2, 3, 5, 8$
21. (a) $7, 11, 15, 19, 23, 27, 31, 35, 39, 43$

(b)

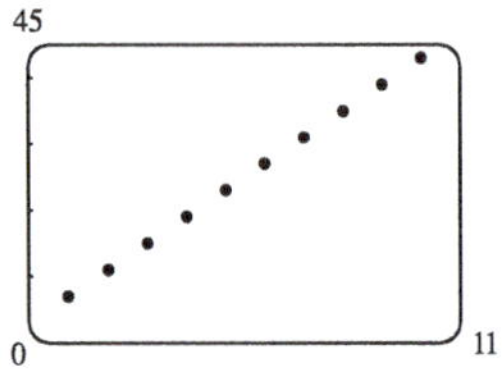

23. (a) $12, 6, 4, 3, \frac{12}{5}, 2, \frac{12}{7}, \frac{3}{2}, \frac{4}{3}, \frac{6}{5}$

(b)

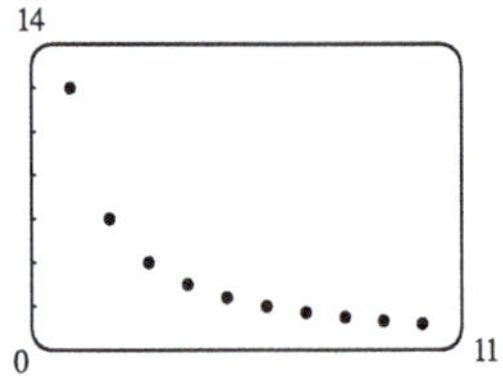

25. (a) $2, \frac{1}{2}, 2, \frac{1}{2}, 2, \frac{1}{2}, 2, \frac{1}{2}, 2, \frac{1}{2}$

(b)

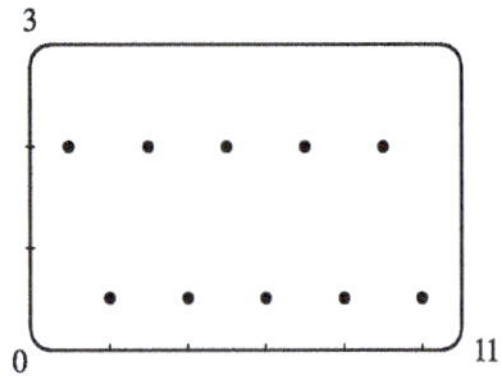

27. $2n$ **29.** 2^n **31.** $5n - 7$ **33.** $a_n = (-1)^{n+1}5^n$
35. $(2n - 1)/n^2$ **37.** $1 + (-1)^n$ **39.** $1, 4, 9, 16, 25, 36$
41. $\frac{1}{3}, \frac{4}{9}, \frac{13}{27}, \frac{40}{81}, \frac{121}{243}, \frac{364}{729}$ **43.** $\frac{2}{3}, \frac{8}{9}, \frac{26}{27}, \frac{80}{81}$; $S_n = 1 - \dfrac{1}{3^n}$
45. $1 - \sqrt{2}, 1 - \sqrt{3}, -1, 1 - \sqrt{5}$; $S_n = 1 - \sqrt{n+1}$
47. 10 **49.** $\frac{11}{6}$ **51.** 8 **53.** 31 **55.** 385 **57.** $46{,}438$
59. 22 **61.** $1^3 + 2^3 + 3^3 + 4^3$
63. $\sqrt{4} + \sqrt{5} + \sqrt{6} + \sqrt{7} + \sqrt{8} + \sqrt{9} + \sqrt{10}$
65. $x^3 + x^4 + \cdots + x^{100}$ **67.** $\displaystyle\sum_{k=1}^{25} 2k$ **69.** $\displaystyle\sum_{k=1}^{10} k^2$
71. $\displaystyle\sum_{k=1}^{999} \frac{1}{k(k+1)}$ **73.** $\displaystyle\sum_{k=0}^{100} x^k$ **75.** $2^{(2^n - 1)/2^n}$
77. (a) 2004.00, 2008.01, 2012.02, 2016.05, 2020.08, 2024.12 **(b)** \$2149.16 **79. (a)** 35,700, 36,414, 37,142, 37,885, 38,643 **(b)** 42,665 **81. (b)** 6898 **83. (a)** $S_n = S_{n-1} + 2000$ **(b)** \$38,000

SECTION 12.2 ■ PAGE 856

1. difference **2.** common difference; 2, 5 **3.** True **4.** True
5. (a) $7, 10, 13, 16, 19$ **(b)** 3

(c)

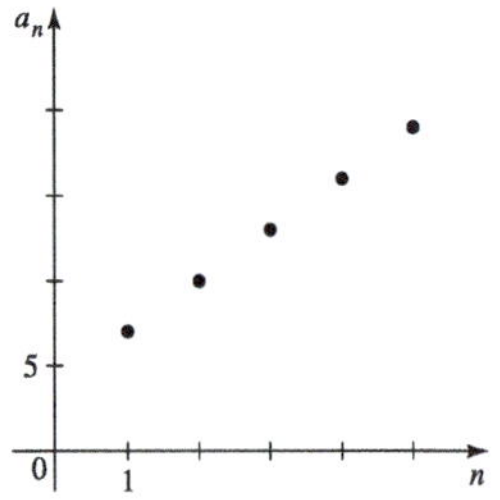

7. (a) $-6, -10, -14, -18, -22$ **(b)** -4

(c)

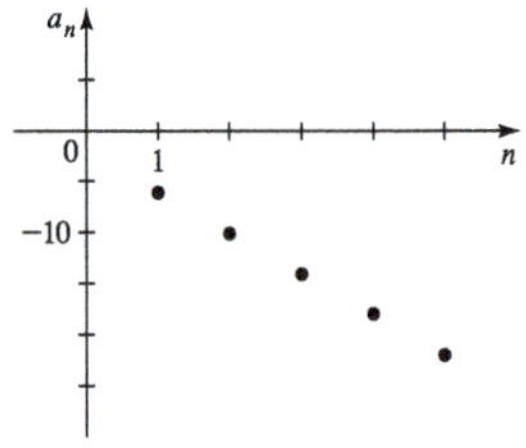

9. (a) $\frac{5}{2}, \frac{3}{2}, \frac{1}{2}, -\frac{1}{2}, -\frac{3}{2}$
(b) -1
(c)

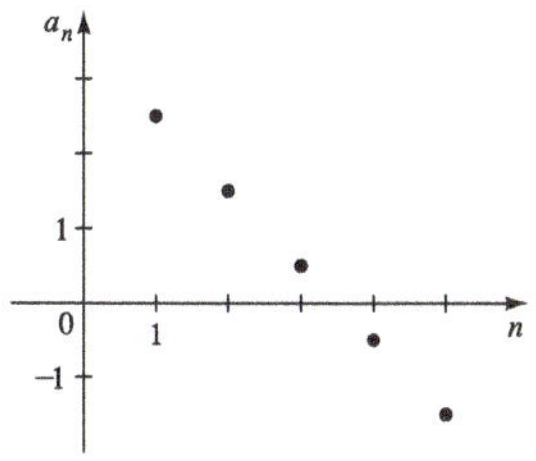

11. $a_n = 9 + 4(n - 1), a_{10} = 45$
13. $a_n = -0.7 - 0.2(n - 1), a_{10} = -2.5$
15. $a_n = \frac{5}{2} - \frac{1}{2}(n - 1), a_{10} = -2$ **17.** Yes, 6 **19.** No
21. No **23.** Yes, $-\frac{3}{2}$ **25.** Yes, 1.7
27. 11, 18, 25, 32, 39; 7; $a_n = 11 + 7(n - 1)$
29. $\frac{1}{3}, \frac{1}{5}, \frac{1}{7}, \frac{1}{9}, \frac{1}{11}$; not arithmetic
31. $-4, 2, 8, 14, 20; 6; a_n = -4 + 6(n - 1)$
33. $6, a_5 = 28, a_n = 4 + 6(n - 1), a_{100} = 598$
35. $-18, a_5 = -43, a_n = 29 - 18(n - 1), a_{100} = -1753$
37. $5, a_5 = 24, a_n = 4 + 5(n - 1), a_{100} = 499$
39. $4, a_5 = 4, a_n = -12 + 4(n - 1), a_{100} = 384$
41. $1.5, a_5 = 31, a_n = 25 + 1.5(n - 1), a_{100} = 173.5$
43. $s, a_5 = 2 + 4s, a_n = 2 + (n - 1)s, a_{100} = 2 + 99s$
45. 706, 712 **47.** $a_1 = -\frac{5}{12}, a_n = -\frac{5}{12} + \frac{1}{12}(n - 1)$
49. 33rd **51.** 1010 **53.** 870 **55.** 1090 **57.** 20,301
59. 1735 **61.** 832.3 **63.** 46.75 **65.** 50 **69.** Yes
71. \$1250 **73.** \$403,500 **75.** 20 **77.** 78

SECTION 12.3 ■ PAGE 864

1. ratio **2.** common ratio; 2, 5 **3.** True **4. (a)** $a\left(\frac{1 - r^n}{1 - r}\right)$
(b) geometric; converges, $a/(1 - r)$; diverges
5. (a) 7, 21, 63, 189, 567
(b) 3
(c)

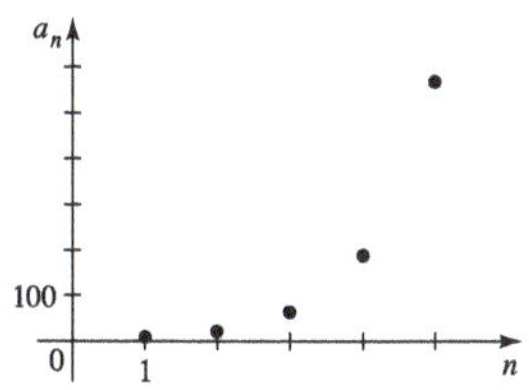

7. (a) $\frac{5}{2}, -\frac{5}{4}, \frac{5}{8}, -\frac{5}{16}, \frac{5}{32}$
(b) $-\frac{1}{2}$
(c)

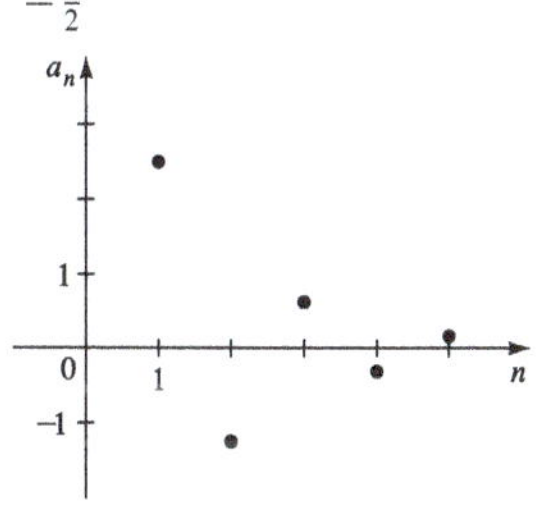

9. $a_n = 7(4)^{n-1}, a_4 = 448$ **11.** $a_n = \frac{5}{2}\left(-\frac{1}{2}\right)^{n-1}, a_4 = -\frac{5}{16}$
13. Yes, 2 **15.** Yes, $\frac{1}{2}$ **17.** Yes, $\frac{1}{2}$ **19.** No **21.** Yes, 1.1
23. 6, 18, 54, 162, 486; geometric, common ratio 3; $a_n = 6 \cdot 3^{n-1}$
25. $\frac{1}{4}, \frac{1}{16}, \frac{1}{64}, \frac{1}{256}, \frac{1}{1024}$; geometric, common ratio $\frac{1}{4}$; $a_n = \frac{1}{4}\left(\frac{1}{4}\right)^{n-1}$
27. 0, ln 5, 2 ln 5, 3 ln 5, 4 ln 5; not geometric
29. $3, a_5 = 162, a_n = 2 \cdot 3^{n-1}$
31. $-0.3, a_5 = 0.00243, a_n = (0.3)(-0.3)^{n-1}$
33. $-\frac{1}{12}, a_5 = \frac{1}{144}, a_n = 144\left(-\frac{1}{12}\right)^{n-1}$
35. $3^{2/3}, a_5 = 3^{11/3}, a_n = 3^{(2n+1)/3}$
37. $s^{2/7}, a_5 = s^{8/7}, a_n = s^{2(n-1)/7}$ **39.** $\frac{24}{25}$
41. $a_1 = -\frac{1}{27}, a_2 = \frac{1}{9}$ **43.** $a_1 = -\frac{9}{32}, a_n = -\frac{9}{32}(-8)^{n-1}$
45. $a_1 = 1728, a_2 = 1296, a_3 = 972$ **47.** Ninth **49.** 315
51. 441 **53.** 3280 **55.** -645 **57.** 13,888,888.75
59. $\frac{93}{16}$ **61.** -105 **63.** $\frac{211}{27}$ **65.** $\frac{3}{2}$ **67.** $\frac{3}{4}$
69. divergent **71.** 2 **73.** divergent **75.** $\sqrt{2} + 1$
77. $\frac{7}{9}$ **79.** $\frac{1}{33}$ **81.** $\frac{112}{999}$ **83.** 10, 20, 40 **85. (a)** Neither
(b) Arithmetic, 3 **(c)** Geometric, $9\sqrt{3}$ **(d)** Arithmetic, 3
87. (a) $V_n = 160{,}000(0.80)^{n-1}$ **(b)** 4th year **89.** 19 ft, $80\left(\frac{3}{4}\right)^n$
91. $\frac{64}{25}, \frac{1024}{625}, 5\left(\frac{4}{5}\right)^n$ **93. (a)** $17\frac{8}{9}$ ft **(b)** $18 - \left(\frac{1}{3}\right)^{n-3}$
95. 2801 **97.** 3 m **99. (a)** 2 **(b)** $8 + 4\sqrt{2}$ **101.** 1

SECTION 12.4 ■ PAGE 871

1. amount **2.** present value **3.** \$13,180.79
5. \$360,262.21 **7.** \$5,591.79 **9.** \$572.34
11. \$13,007.94 **13.** \$2,601.59 **15.** \$307.24
17. \$733.76, \$264,153.60 **19.** \$583,770.65 **21.** \$9020.60
23. (a) \$859.15 **(b)** \$309,294.00 **(c)** \$1,841,519.29
25. 18.16% **27.** 11.68%

SECTION 12.5 ■ PAGE 878

1. natural; $P(1)$ **2.** (ii)

3. Let $P(n)$ denote the statement $2 + 4 + \cdots + 2n = n(n + 1)$.

Step 1 $P(1)$ is true, since $2 = 1(1 + 1)$.
Step 2 Suppose $P(k)$ is true. Then

$$\begin{aligned} &2 + 4 + \cdots + 2k + 2(k + 1) \\ &= k(k + 1) + 2(k + 1) \quad \text{Induction hypothesis} \\ &= (k + 1)(k + 2) \end{aligned}$$

So $P(k + 1)$ follows from $P(k)$. Thus by the Principle of Mathematical Induction $P(n)$ holds for all n.

5. Let $P(n)$ denote the statement

$$5 + 8 + \cdots + (3n + 2) = \frac{n(3n + 7)}{2}.$$

Step 1 $P(1)$ is true, since $5 = \dfrac{1(3 \cdot 1 + 7)}{2}$

Step 2 Suppose $P(k)$ is true. Then

$$\begin{aligned} &5 + 8 + \cdots + (3k + 2) + [3(k + 1) + 2] \\ &= \frac{k(3k + 7)}{2} + (3k + 5) \quad \text{Induction hypothesis} \\ &= \frac{3k^2 + 13k + 10}{2} \\ &= \frac{(k + 1)[3(k + 1) + 7]}{2} \end{aligned}$$

So $P(k + 1)$ follows from $P(k)$. Thus by the Principle of Mathematical Induction $P(n)$ holds for all n.

7. Let $P(n)$ denote the statement $1\cdot 2 + 2\cdot 3 + \cdots + n(n+1) = \dfrac{n(n+1)(n+2)}{3}$.

Step 1 $P(1)$ is true, since $1\cdot 2 = \dfrac{1\cdot(1+1)\cdot(1+2)}{3}$.

Step 2 Suppose $P(k)$ is true. Then

$$\begin{aligned}&1\cdot 2 + 2\cdot 3 + \cdots + k(k+1) + (k+1)(k+2)\\ &= \frac{k(k+1)(k+2)}{3} + (k+1)(k+2) \quad \text{Induction hypothesis}\\ &= \frac{(k+1)(k+2)(k+3)}{3}\end{aligned}$$

So $P(k+1)$ follows from $P(k)$. Thus by the Principle of Mathematical Induction $P(n)$ holds for all n.

9. Let $P(n)$ denote the statement $1^3 + 2^3 + \cdots + n^3 = \dfrac{n^2(n+1)^2}{4}$.

Step 1 $P(1)$ is true, since $1^3 = \dfrac{1^2\cdot(1+1)^2}{4}$.

Step 2 Suppose $P(k)$ is true. Then

$$\begin{aligned}&1^3 + 2^3 + \cdots + k^3 + (k+1)^3\\ &= \frac{k^2(k+1)^2}{4} + (k+1)^3 \quad \text{Induction hypothesis}\\ &= \frac{(k+1)^2[k^2 + 4(k+1)]}{4}\\ &= \frac{(k+1)^2(k+2)^2}{4}\end{aligned}$$

So $P(k+1)$ follows from $P(k)$. Thus by the Principle of Mathematical Induction $P(n)$ holds for all n.

11. Let $P(n)$ denote the statement $2^3 + 4^3 + \cdots + (2n)^3 = 2n^2(n+1)^2$.

Step 1 $P(1)$ is true, since $2^3 = 2\cdot 1^2(1+1)^2$.
Step 2 Suppose $P(k)$ is true. Then

$$\begin{aligned}&2^3 + 4^3 + \cdots + (2k)^3 + [2(k+1)]^3\\ &= 2k^2(k+1)^2 + [2(k+1)]^3 \quad \text{Induction hypothesis}\\ &= (k+1)^2(2k^2 + 8k + 8)\\ &= 2(k+1)^2(k+2)^2\end{aligned}$$

So $P(k+1)$ follows from $P(k)$. Thus by the Principle of Mathematical Induction $P(n)$ holds for all n.

13. Let $P(n)$ denote the statement $1\cdot 2 + 2\cdot 2^2 + \cdots + n\cdot 2^n = 2[1 + (n-1)2^n]$.

Step 1 $P(1)$ is true, since $1\cdot 2 = 2[1 + 0]$.
Step 2 Suppose $P(k)$ is true. Then

$$\begin{aligned}&1\cdot 2 + 2\cdot 2^2 + \cdots + k\cdot 2^k + (k+1)\cdot 2^{k+1}\\ &= 2[1 + (k-1)2^k] + (k+1)\cdot 2^{k+1} \quad \text{Induction hypothesis}\\ &= 2 + (k-1)2^{k+1} + (k+1)\cdot 2^{k+1}\\ &= 2 + 2k2^{k+1} = 2(1 + k2^{k+1})\end{aligned}$$

So $P(k+1)$ follows from $P(k)$. Thus by the Principle of Mathematical Induction $P(n)$ holds for all n.

15. Let $P(n)$ denote the statement $n^2 + n$ is divisible by 2.

Step 1 $P(1)$ is true, since $1^2 + 1$ is divisible by 2.
Step 2 Suppose $P(k)$ is true. Now

$$\begin{aligned}(k+1)^2 + (k+1) &= k^2 + 2k + 1 + k + 1\\ &= (k^2 + k) + 2(k+1)\end{aligned}$$

But $k^2 + k$ is divisible by 2 (by the induction hypothesis), and $2(k+1)$ is clearly divisible by 2, so $(k+1)^2 + (k+1)$ is divisible by 2. So $P(k+1)$ follows from $P(k)$. Thus by the Principle of Mathematical Induction $P(n)$ holds for all n.

17. Let $P(n)$ denote the statement $n^2 - n + 41$ is odd.

Step 1 $P(1)$ is true, since $1^2 - 1 + 41$ is odd.
Step 2 Suppose $P(k)$ is true. Now

$$(k+1)^2 - (k+1) + 41 = (k^2 - k + 41) + 2k$$

But $k^2 - k + 41$ is odd (by the induction hypothesis), and $2k$ is clearly even, so their sum is odd. So $P(k+1)$ follows from $P(k)$. Thus by the Principle of Mathematical Induction $P(n)$ holds for all n.

19. Let $P(n)$ denote the statement $8^n - 3^n$ is divisible by 5.

Step 1 $P(1)$ is true, since $8^1 - 3^1$ is divisible by 5.
Step 2 Suppose $P(k)$ is true. Now

$$\begin{aligned}8^{k+1} - 3^{k+1} &= 8\cdot 8^k - 3\cdot 3^k\\ &= 8\cdot 8^k - (8-5)\cdot 3^k = 8\cdot(8^k - 3^k) + 5\cdot 3^k\end{aligned}$$

which is divisible by 5 because $8^k - 3^k$ is divisible by 5 (by the induction hypothesis) and $5\cdot 3^k$ is clearly divisible by 5. So $P(k+1)$ follows from $P(k)$. Thus by the Principle of Mathematical Induction $P(n)$ holds for all n.

21. Let $P(n)$ denote the statement $n < 2^n$.

Step 1 $P(1)$ is true, since $1 < 2^1$.
Step 2 Suppose $P(k)$ is true. Then

$$\begin{aligned}k + 1 &< 2^k + 1 &&\text{Induction hypothesis}\\ &< 2^k + 2^k &&\text{Because } 1 < 2^k\\ &= 2\cdot 2^k = 2^{k+1}\end{aligned}$$

So $P(k+1)$ follows from $P(k)$. Thus by the Principle of Mathematical Induction $P(n)$ holds for all n.

23. Let $P(n)$ denote the statement $(1+x)^n \ge 1 + nx$ for $x > -1$.

Step 1 $P(1)$ is true, since $(1+x)^1 \ge 1 + 1\cdot x$.
Step 2 Suppose $P(k)$ is true. Then

$$\begin{aligned}(1+x)^{k+1} &= (1+x)(1+x)^k\\ &\ge (1+x)(1+kx) \quad \text{Induction hypothesis}\\ &= 1 + (k+1)x + kx^2\\ &\ge 1 + (k+1)x\end{aligned}$$

So $P(k+1)$ follows from $P(k)$. Thus by the Principle of Mathematical Induction $P(n)$ holds for all n.

25. Let $P(n)$ denote the statement $a_n = 5 \cdot 3^{n-1}$.

Step 1 $P(1)$ is true, since $a_1 = 5 \cdot 3^0 = 5$.
Step 2 Suppose $P(k)$ is true. Then

$$\begin{aligned} a_{k+1} &= 3 \cdot a_k && \text{Definition of } a_{k+1} \\ &= 3 \cdot 5 \cdot 3^{k-1} && \text{Induction hypothesis} \\ &= 5 \cdot 3^k \end{aligned}$$

So $P(k+1)$ follows from $P(k)$. Thus by the Principle of Mathematical Induction $P(n)$ holds for all n.

27. Let $P(n)$ denote the statement $x - y$ is a factor of $x^n - y^n$.

Step 1 $P(1)$ is true, since $x - y$ is a factor of $x^1 - y^1$.
Step 2 Suppose $P(k)$ is true. Now

$$\begin{aligned} x^{k+1} - y^{k+1} &= x^{k+1} - x^k y + x^k y - y^{k+1} \\ &= x^k(x - y) + (x^k - y^k)y \end{aligned}$$

But $x^k(x - y)$ is clearly divisible by $x - y$, and $(x^k - y^k)y$ is divisible by $x - y$ (by the induction hypothesis), so their sum is divisible by $x - y$. So $P(k+1)$ follows from $P(k)$. Thus by the Principle of Mathematical Induction $P(n)$ holds for all n.

29. Let $P(n)$ denote the statement F_{3n} is even.

Step 1 $P(1)$ is true, since $F_{3 \cdot 1} = 2$, which is even.
Step 2 Suppose $P(k)$ is true. Now, by the definition of the Fibonacci sequence

$$\begin{aligned} F_{3(k+1)} &= F_{3k+3} = F_{3k+2} + F_{3k+1} \\ &= F_{3k+1} + F_{3k} + F_{3k+1} \\ &= F_{3k} + 2 \cdot F_{3k+1} \end{aligned}$$

But F_{3k} is even (by the induction hypothesis), and $2 \cdot F_{3k+1}$ is clearly even, so $F_{3(k+1)}$ is even. So $P(k+1)$ follows from $P(k)$. Thus by the Principle of Mathematical Induction $P(n)$ holds for all n.

31. Let $P(n)$ denote the statement $F_1^2 + F_2^2 + \cdots + F_n^2 = F_n \cdot F_{n+1}$.

Step 1 $P(1)$ is true, since $F_1^2 = F_1 \cdot F_2$ (because $F_1 = F_2 = 1$).
Step 2 Suppose $P(k)$ is true. Then

$$\begin{aligned} &F_1^2 + F_2^2 + \cdots + F_k^2 + F_{k+1}^2 \\ &= F_k \cdot F_{k+1} + F_{k+1}^2 && \text{Induction hypothesis} \\ &= F_{k+1}(F_k + F_{k+1}) && \text{Definition of the Fibonacci sequence} \\ &= F_{k+1} \cdot F_{k+2} \end{aligned}$$

So $P(k+1)$ follows from $P(k)$. Thus by the Principle of Mathematical Induction $P(n)$ holds for all n.

33. Let $P(n)$ denote the statement

$$\begin{bmatrix} 1 & 1 \\ 1 & 0 \end{bmatrix}^n = \begin{bmatrix} F_{n+1} & F_n \\ F_n & F_{n-1} \end{bmatrix}.$$

Step 1 $P(2)$ is true, since

$$\begin{bmatrix} 1 & 1 \\ 1 & 0 \end{bmatrix}^2 = \begin{bmatrix} 2 & 1 \\ 1 & 1 \end{bmatrix} = \begin{bmatrix} F_3 & F_2 \\ F_2 & F_1 \end{bmatrix}.$$

Step 2 Suppose $P(k)$ is true. Then

$$\begin{aligned} \begin{bmatrix} 1 & 1 \\ 1 & 0 \end{bmatrix}^{k+1} &= \begin{bmatrix} 1 & 1 \\ 1 & 0 \end{bmatrix}^k \begin{bmatrix} 1 & 1 \\ 1 & 0 \end{bmatrix} \\ &= \begin{bmatrix} F_{k+1} & F_k \\ F_k & F_{k-1} \end{bmatrix} \begin{bmatrix} 1 & 1 \\ 1 & 0 \end{bmatrix} && \text{Induction hypothesis} \\ &= \begin{bmatrix} F_{k+1} + F_k & F_{k+1} \\ F_k + F_{k-1} & F_k \end{bmatrix} \\ &= \begin{bmatrix} F_{k+2} & F_{k+1} \\ F_{k+1} & F_k \end{bmatrix} && \text{Definition of the Fibonacci sequence} \end{aligned}$$

So $P(k+1)$ follows from $P(k)$. Thus by the Principle of Mathematical Induction $P(n)$ holds for all $n \ge 2$.

35. Let $P(n)$ denote the statement $F_n \ge n$.

Step 1 $P(5)$ is true, since $F_5 \ge 5$ (because $F_5 = 5$).
Step 2 Suppose $P(k)$ is true. Now

$$\begin{aligned} F_{k+1} &= F_k + F_{k-1} && \text{Definition of the Fibonacci sequence} \\ &\ge k + F_{k-1} && \text{Induction hypothesis} \\ &\ge k + 1 && \text{Because } F_{k-1} \ge 1 \end{aligned}$$

So $P(k+1)$ follows from $P(k)$. Thus by the Principle of Mathematical Induction $P(n)$ holds for all $n \ge 5$.

SECTION 12.6 ■ PAGE 886

1. binomial **2.** Pascal's; 1, 4, 6, 4, 1

3. $\dfrac{n!}{k!(n-k)!}$; $\dfrac{4!}{3!(4-3)!} = 4$

4. Binomial; $\binom{4}{0}, \binom{4}{1}, \binom{4}{2}, \binom{4}{3}, \binom{4}{4}$

5. $x^6 + 6x^5y + 15x^4y^2 + 20x^3y^3 + 15x^2y^4 + 6xy^5 + y^6$

7. $x^4 + 4x^2 + 6 + \dfrac{4}{x^2} + \dfrac{1}{x^4}$

9. $x^5 - 5x^4 + 10x^3 - 10x^2 + 5x - 1$
11. $x^{10}y^5 - 5x^8y^4 + 10x^6y^3 - 10x^4y^2 + 5x^2y - 1$
13. $8x^3 - 36x^2y + 54xy^2 - 27y^3$

15. $\dfrac{1}{x^5} - \dfrac{5}{x^{7/2}} + \dfrac{10}{x^2} - \dfrac{10}{x^{1/2}} + 5x - x^{5/2}$

17. 15 **19.** 4950 **21.** 18 **23.** 32
25. $x^4 + 8x^3y + 24x^2y^2 + 32xy^3 + 16y^4$

27. $1 + \dfrac{6}{x} + \dfrac{15}{x^2} + \dfrac{20}{x^3} + \dfrac{15}{x^4} + \dfrac{6}{x^5} + \dfrac{1}{x^6}$

29. $x^{20} + 40x^{19}y + 760x^{18}y^2$ **31.** $25a^{26/3} + a^{25/3}$
33. $48{,}620x^{18}$ **35.** $300a^2b^{23}$ **37.** $100y^{99}$ **39.** $13{,}440x^4y^6$
41. $495a^8b^8$ **43.** $(x+y)^4$ **45.** $(2a+b)^3$
47. $3x^2 + 3xh + h^2$

CHAPTER 12 REVIEW ■ PAGE 889

1. $\frac{1}{2}, \frac{4}{3}, \frac{9}{4}, \frac{16}{5}; \frac{100}{11}$ **3.** $0, \frac{1}{4}, 0, \frac{1}{32}; \frac{1}{500}$
5. 1, 3, 15, 105; 654,729,075
7. 1, 4, 9, 16, 25, 36, 49
9. 1, 3, 5, 11, 21, 43, 85

11. (a) 7, 9, 11, 13, 15
(b)

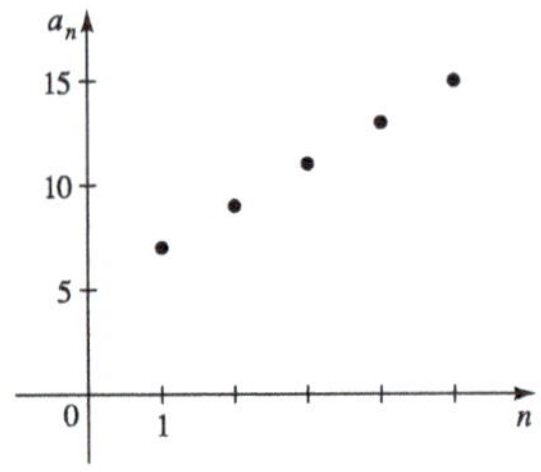

(c) 55
(d) Arithmetic, common difference 2

13. (a) $\frac{3}{4}, \frac{9}{8}, \frac{27}{16}, \frac{81}{32}, \frac{243}{64}$
(b)

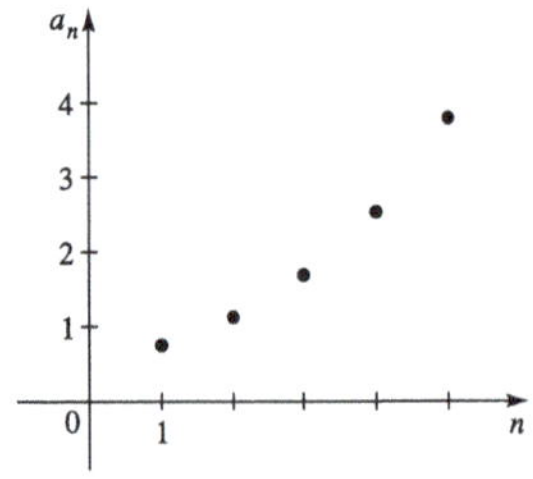

(c) $\frac{633}{64}$
(d) Geometric, common ratio $\frac{3}{2}$

15. Arithmetic, 7 **17.** Arithmetic, $t + 1$ **19.** Geometric, $\dfrac{1}{t}$
21. Geometric, $\frac{4}{27}$ **23.** $2i$ **25.** 5 **27.** $\frac{81}{4}$
29. (a) $A_n = 32{,}000(1.05)^{n-1}$ **(b)** \$32,000, \$33,600, \$35,280, \$37,044, \$38,896.20, \$40,841.01, \$42,883.06, \$45,027.21
31. 12,288 **35. (a)** 9 **(b)** $\pm 6\sqrt{2}$ **37.** 126
39. 384 **41.** $0^2 + 1^2 + 2^2 + \cdots + 9^2$
43. $\dfrac{3}{2^2} + \dfrac{3^2}{2^3} + \dfrac{3^3}{2^4} + \cdots + \dfrac{3^{50}}{2^{51}}$ **45.** $\sum_{k=1}^{33} 3k$ **47.** $\sum_{k=1}^{100} k2^{k+2}$
49. Geometric; 4.68559 **51.** Arithmetic, $5050\sqrt{5}$
53. Geometric, 9831 **55.** $\frac{5}{7}$ **57.** Divergent **59.** Divergent
61. 13 **63.** 65,534 **65.** \$2390.27

67. Let $P(n)$ denote the statement

$$1 + 4 + 7 + \cdots + (3n - 2) = \frac{n(3n-1)}{2}.$$

Step 1 $P(1)$ is true, since $1 = \dfrac{1(3 \cdot 1 - 1)}{2}$.

Step 2 Suppose $P(k)$ is true. Then

$$\begin{aligned} 1 + 4 + 7 + \cdots + (3k - 2) + [3(k+1) - 2] &= \frac{k(3k-1)}{2} + [3k + 1] \quad \text{Induction hypothesis} \\ &= \frac{3k^2 - k + 6k + 2}{2} \\ &= \frac{(k+1)(3k+2)}{2} \\ &= \frac{(k+1)[3(k+1) - 1]}{2} \end{aligned}$$

So $P(k + 1)$ follows from $P(k)$. Thus by the Principle of Mathematical Induction $P(n)$ holds for all n.

69. Let $P(n)$ denote the statement
$(1 + \frac{1}{1})(1 + \frac{1}{2}) \cdots (1 + \frac{1}{n}) = n + 1$.
Step 1 $P(1)$ is true, since $(1 + \frac{1}{1}) = 1 + 1$.
Step 2 Suppose $P(k)$ is true. Then

$$\begin{aligned} \left(1 + \frac{1}{1}\right)\left(1 + \frac{1}{2}\right) \cdots \left(1 + \frac{1}{k}\right)\left(1 + \frac{1}{k+1}\right) &= (k+1)\left(1 + \frac{1}{k+1}\right) \quad \text{Induction hypothesis} \\ &= (k+1) + 1 \end{aligned}$$

So $P(k + 1)$ follows from $P(k)$. Thus by the Principle of Mathematical Induction $P(n)$ holds for all n.

71. Let $P(n)$ denote the statement that F_{4n} is divisible by 3.

Step 1 $P(1)$ is true, since $F_4 = 3$.
Step 2 Suppose $P(k)$ is true. Then F_{4k} is divisible by 3. Using the definition of the Fibonacci sequence repeatedly, we get

$$\begin{aligned} F_{4(k+1)} &= F_{4k+4} = F_{4k+3} + F_{4k+2} \\ &= (F_{4k+2} + F_{4k+1}) + (F_{4k+1} + F_{4k}) \\ &= [(F_{4k+1} + F_{4k}) + F_{4k+1}] + (F_{4k+1} + F_{4k}) \\ &= 3F_{4k+1} + 2F_{4k} \end{aligned}$$

The first term is clearly divisible by 3, and so is the second by the induction hypothesis. So $P(k + 1)$ follows from $P(k)$. Thus by the Principle of Mathematical Induction $P(n)$ holds for all n.

73. 100 **75.** 32 **77.** $A^3 - 3A^2B + 3AB^2 - B^3$
79. $1 - 6x^2 + 15x^4 - 20x^6 + 15x^8 - 6x^{10} + x^{12}$
81. $1540a^3b^{19}$ **83.** $17{,}010A^6B^4$

CHAPTER 12 TEST ▪ PAGE 892

1. 1, 6, 15, 28, 45, 66; 161 **2.** 2, 5, 13, 36, 104, 307
3. (a) 3 **(b)** $a_n = 2 + (n - 1)3$ **(c)** 104
4. (a) $\frac{1}{4}$ **(b)** $a_n = 12(\frac{1}{4})^{n-1}$ **(c)** $3/4^8$
5. (a) $\frac{1}{5}, \frac{1}{25}$ **(b)** $\dfrac{5^8 - 1}{12{,}500}$ **6. (a)** $-\frac{8}{9}, -78$ **(b)** 60
8. (a) $(1 - 1^2) + (1 - 2^2) + (1 - 3^2) + (1 - 4^2) + (1 - 5^2) = -50$
(b) $(-1)^3 2^1 + (-1)^4 2^2 + (-1)^5 2^3 + (-1)^6 2^4 = 10$
9. (a) $\frac{58{,}025}{59{,}049}$ **(b)** $2 + \sqrt{2}$

10. Let $P(n)$ denote the statement

$$1^2 + 2^2 + \cdots + n^2 = \frac{n(n+1)(2n+1)}{6}.$$

Step 1 $P(1)$ is true, since $1^2 = \dfrac{1(1+1)(2 \cdot 1 + 1)}{6}$.

Step 2 Suppose $P(k)$ is true. Then

$$\begin{aligned} 1^2 + 2^2 + \cdots + k^2 + (k+1)^2 &= \frac{k(k+1)(2k+1)}{6} + (k+1)^2 \quad \text{Induction hypothesis} \\ &= \frac{k(k+1)(2k+1) + 6(k+1)^2}{6} \\ &= \frac{(k+1)[k(2k+1) + 6(k+1)]}{6} \\ &= \frac{(k+1)(2k^2 + 7k + 6)}{6} \\ &= \frac{(k+1)[(k+1) + 1][2(k+1) + 1]}{6} \end{aligned}$$

So $P(k + 1)$ follows from $P(k)$. Thus by the Principle of Mathematical Induction $P(n)$ holds for all n.

11. $32x^5 + 80x^4y^2 + 80x^3y^4 + 40x^2y^6 + 10xy^8 + y^{10}$
12. $\binom{10}{3}(3x)^3(-2)^7 = -414{,}720x^3$
13. (a) $a_n = (0.85)(1.24)^n$ **(b)** 3.09 lb **(c)** Geometric

FOCUS ON MODELING ■ PAGE 895

1. **(a)** $A_n = 1.0001A_{n-1}, A_0 = 275{,}000$ **(b)** $A_0 = 275{,}000$, $A_1 = 275{,}027.50$, $A_2 = 275{,}055.00$, $A_3 = 275{,}082.51$, $A_4 = 275{,}110.02$, $A_5 = 275{,}137.53$, $A_6 = 275{,}165.04$, $A_7 = 275{,}192.56$ **(c)** $A_n = 1.0001^n(275{,}000)$
3. **(a)** $A_n = 1.0025A_{n-1} + 100, A_0 = 100$ **(b)** $A_0 = 100$, $A_1 = 200.25$, $A_2 = 300.75$, $A_3 = 401.50$, $A_4 = 502.51$ **(c)** $A_n = 100[(1.0025^{n+1} - 1)/0.0025]$ **(d)** \$6580.83
5. **(a)** $U_n = U_{n-1} + 0.05U_{n-1} + 0.1(U_{n-1} + 0.05U_{n-1}) = 1.155U_{n-1}, U_0 = 5000$ **(b)** $U_0 = 5000$, $U_1 = 5775$, $U_2 = 6670.13$, $U_3 = 7703.99$, $U_4 = 8898.11$
(c) $U_n = 5000(1.155)^n$ **(d)** \$21,124.67

CHAPTER 13

SECTION 13.1 ■ PAGE 904

1. L, a; 5, 1 **2.** limit, left, L; less; left, right, equal
3. 10 **5.** $\frac{1}{4}$ **7.** $\frac{1}{3}$ **9.** 1 **11.** -1 **13.** 0.51 **15.** $\frac{1}{2}$
17. **(a)** 2 **(b)** 3 **(c)** Does not exist **(d)** 4 **(e)** Not defined
19. **(a)** -1 **(b)** -2 **(c)** Does not exist **(d)** 2 **(e)** 0 **(f)** Does not exist **(g)** 1 **(h)** 3 **21.** -8
23. Does not exist **25.** Does not exist **27.** Does not exist
29. **(a)** 4 **(b)** 4 **(c)** 4

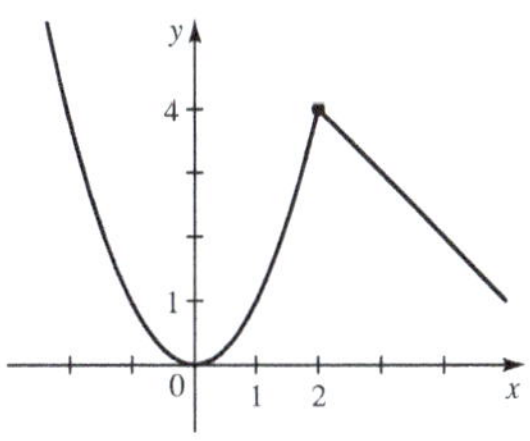

31. **(a)** 4 **(b)** 3 **(c)** Does not exist

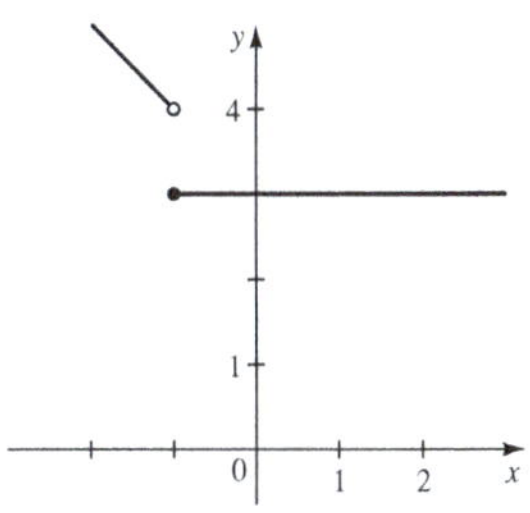

SECTION 13.2 ■ PAGE 913

1. $\lim_{x\to a} f(x) + \lim_{x\to a} g(x), \lim_{x\to a} f(x) \cdot \lim_{x\to a} g(x)$; sum, product
2. $f(a)$ **3.** **(a)** 2 **(b)** Does not exist **(c)** 0 **(d)** Does not exist **(e)** 16 **(f)** 2
5. 5 **7.** 12 **9.** 75 **11.** $\frac{1}{2}$ **13.** -174 **15.** $\frac{4}{9}$ **17.** 7 **19.** 5
21. Does not exist **23.** $\frac{6}{5}$ **25.** 4 **27.** $\frac{1}{6}$ **29.** $-\frac{1}{16}$ **31.** $-\frac{1}{9}$
33. 4 **35.** $-\frac{3}{2}$

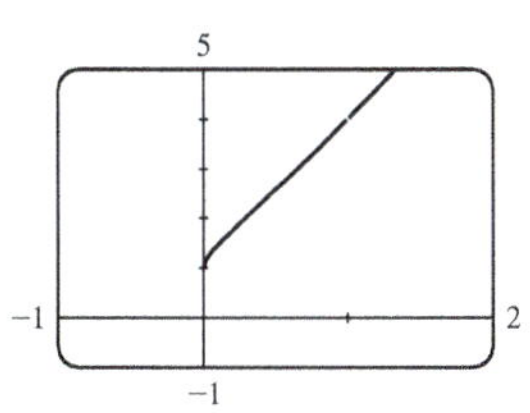

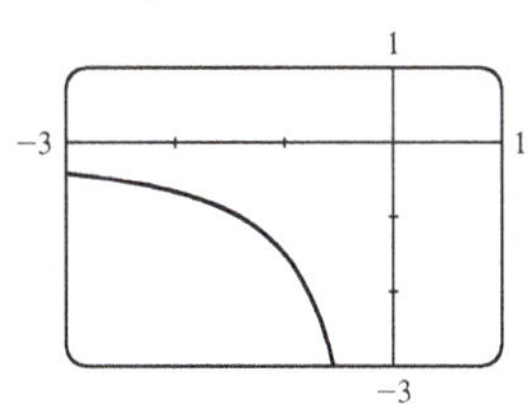

37. 0 **39.** Does not exist **41.** Does not exist
43. **(a)** 1, 2 **(b)** Does not exist
(c)

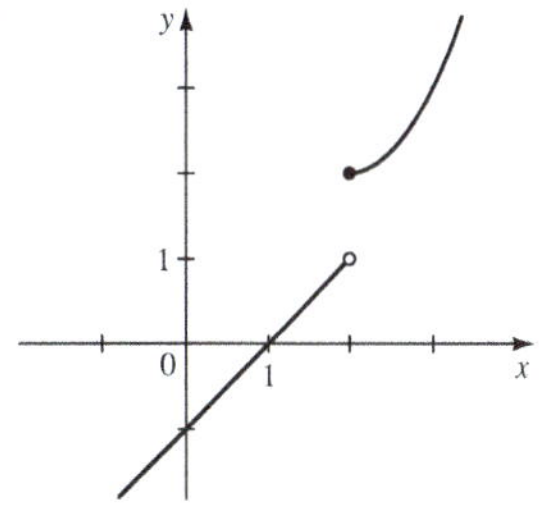

45. **(a)** 0.667

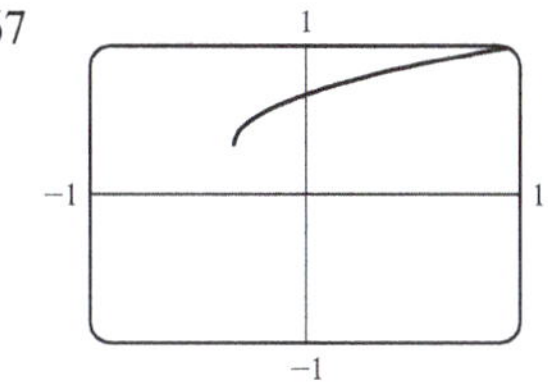

(b) 0.667

x	$f(x)$
0.1	0.71339
0.01	0.67163
0.001	0.66717
0.0001	0.66672

x	$f(x)$
-0.1	0.61222
-0.01	0.66163
-0.001	0.66617
-0.0001	0.66662

(c) $\frac{2}{3}$

SECTION 13.3 ■ PAGE 921

1. $\dfrac{f(a + h) - f(a)}{h}$; slope, $(a, f(a))$ **2.** $\dfrac{f(x) - f(a)}{x - a}$, instantaneous, a **3.** 3 **5.** -11 **7.** 24 **9.** $-\frac{1}{5}$
11. $y = -8x + 9$ **13.** $y = -x - 1$

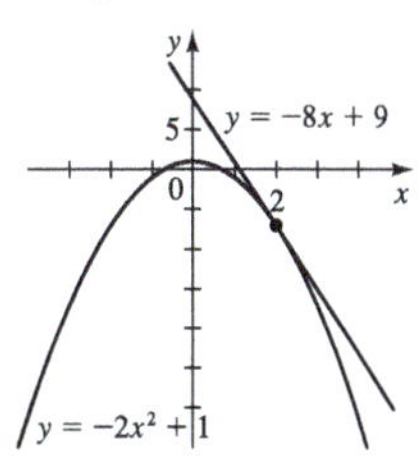

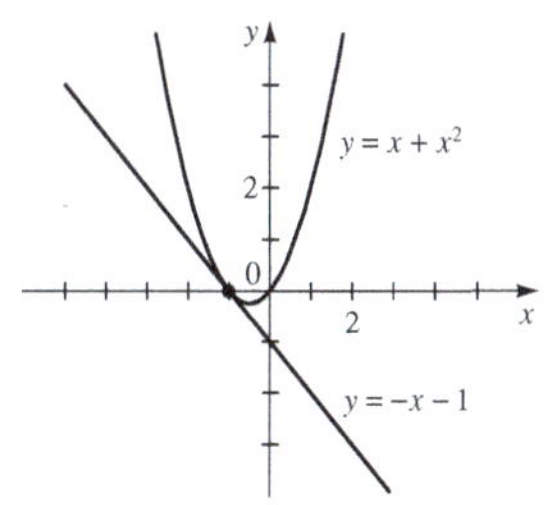

15. $y = -x + 4$ **17.** $y = \frac{1}{4}x + \frac{7}{4}$

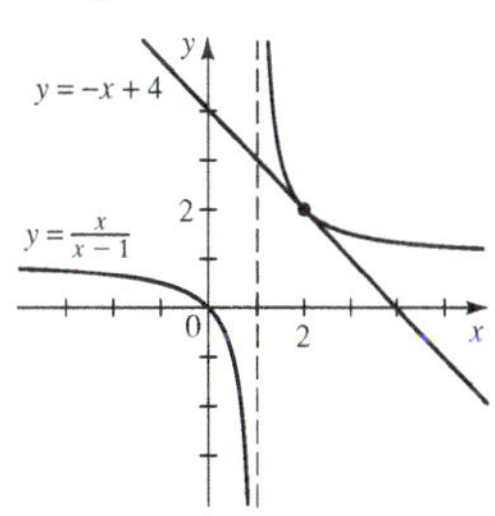

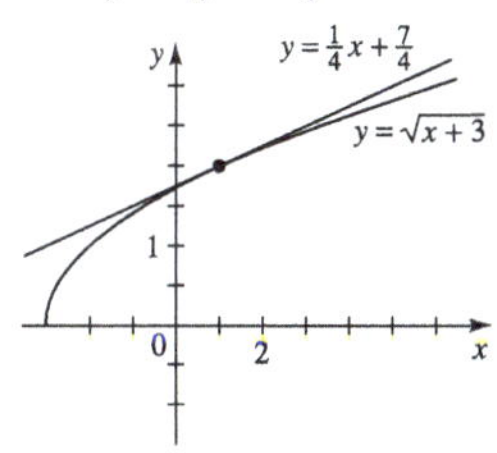

19. $f'(2) = -12$ **21.** $f'(-1) = 7$ **23.** $f'(2) = -\frac{1}{9}$
25. $F'(4) = -\frac{1}{16}$ **27.** **(a)** $2a + 2$ **(b)** 8, 10

29. **(a)** $\dfrac{1}{(a+1)^2}$ **(b)** $\frac{1}{16}, \frac{1}{25}$

31. **(a)** $f'(a) = 3a^2 - 2$
(b) $y = -2x + 4, y = x + 2, y = 10x - 12$
(c)

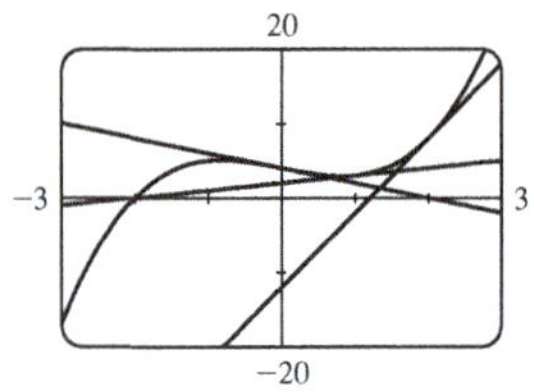

33. $f(x) = x^{10}, a = 1$ **35.** $f(t) = \sqrt{t+1}, a = 1$
37. −24 ft/s **39.** $12a^2 + 6$ m/s, 18 m/s, 54 m/s, 114 m/s
41. −0.8°/min **43.** **(a)** −38.8 gal/min, −27.8 gal/min
(b) −33.8 gal/min

SECTION 13.4 ■ PAGE 930

1. L, x; horizontal asymptote; 0, 0
2. L, large; converges, diverges
3. **(a)** −1, 2 **(b)** $y = -1, y = 2$ **5.** 0
7. $\frac{2}{5}$ **9.** $\frac{4}{3}$ **11.** 2 **13.** Does not exist **15.** 7
17. Does not exist **19.** $-\frac{1}{4}$ **21.** 0 **23.** 0
25. Divergent **27.** 0 **29.** Divergent **31.** $\frac{3}{2}$ **33.** 8
35. $f(x) = \dfrac{x^2}{(x-1)(x-3)}$ [Other answers are possible.]
37. Within 0.01 **39.** **(b)** 30 g/L

SECTION 13.5 ■ PAGE 938

1. rectangles;
$f(x_1)(x_1 - a) + f(x_2)(x_2 - x_1) + f(x_3)(x_3 - x_2) + f(b)(b - x_3)$
2. $\sum_{k=1}^{n} f(x_k)\,\Delta x$
3. **(a)** 40, 52

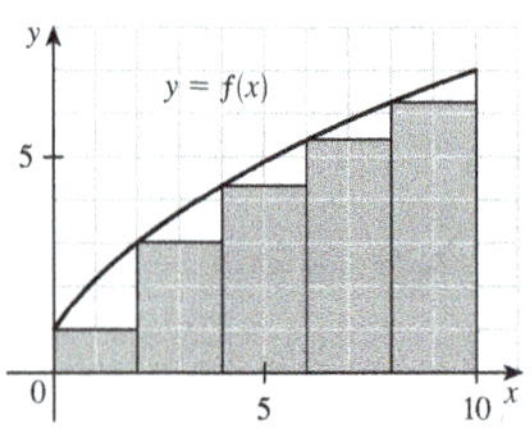

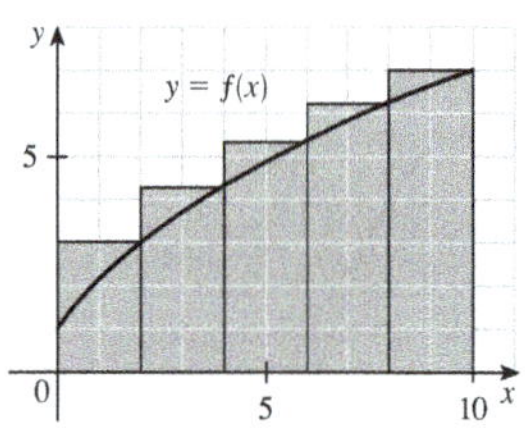

(b) 43, 49
5. 5.25 **7.** $\frac{223}{35}$
9. **(a)** $\frac{77}{60}$, underestimate **(b)** $\frac{25}{12}$, overestimate

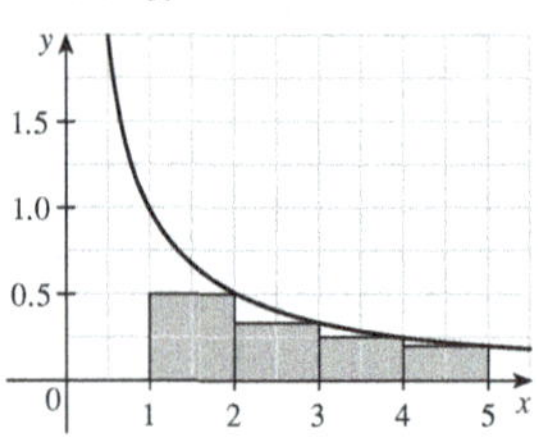

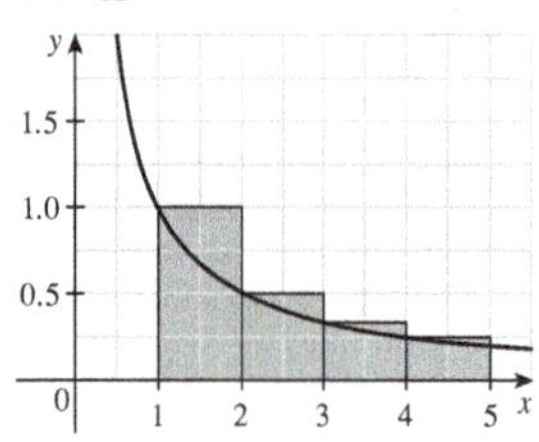

11. **(a)** 8, 6.875 **(b)** 5, 5.375

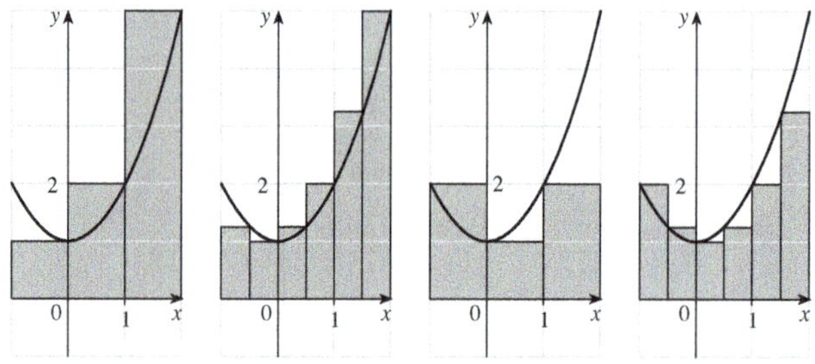

13. 37.5 **15.** 8 **17.** 166.25 **19.** 133.5

CHAPTER 13 REVIEW ■ PAGE 941

1. 1 **3.** 0.69 **5.** Does not exist
7. **(a)** Does not exist **(b)** 2.4 **(c)** 2.4 **(d)** 2.4 **(e)** 0.5
(f) 1 **(g)** 2 **(h)** 0 **9.** −3 **11.** 7 **13.** 2 **15.** −1 **17.** 2
19. Does not exist **21.** $f'(4) = 3$ **23.** $f'(16) = \frac{1}{8}$
25. **(a)** $f'(a) = -2$ **(b)** −2, −2
27. **(a)** $f'(a) = 1/(2\sqrt{a+6})$ **(b)** $1/(4\sqrt{2}), 1/4$
29. $y = 2x + 1$ **31.** $y = 2x$ **33.** $y = -\frac{1}{4}x + 1$
35. **(a)** −64 ft/s **(b)** $-32a$ ft/s **(c)** $\sqrt{40} \approx 6.32$ s
(d) −202.4 ft/s **37.** $\frac{1}{5}$ **39.** $\frac{1}{2}$ **41.** Divergent **43.** 3.83
45. 10 **47.** $\frac{5}{6}$

CHAPTER 13 TEST ■ PAGE 943

1. **(a)** $\frac{1}{2}$ **(b)**

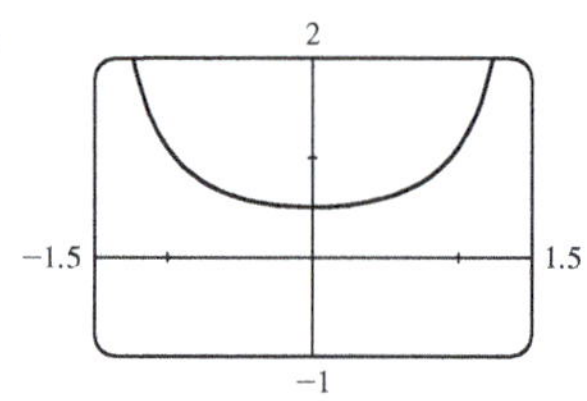

2. **(a)** 1 **(b)** 1 **(c)** 1 **(d)** 0 **(e)** 0 **(f)** 0 **(g)** 4 **(h)** 2
(i) Does not exist
3. **(a)** 6 **(b)** −2 **(c)** Does not exist
(d) Does not exist **(e)** $\frac{1}{4}$ **(f)** 2
4. **(a)** $f'(x) = 2x - 2$ **(b)** −4, 0, 2
5. $y = \frac{1}{6}x + \frac{3}{2}$ **6.** **(a)** 0 **(b)** Does not exist
7. **(a)** $\frac{89}{25}$ **(b)** $\frac{11}{3}$

FOCUS ON MODELING ■ PAGE 946

1. $57{,}333\frac{1}{3}$ ft-lb **3.** **(b)** Area under the graph of $p(x) = 375x$ between $x = 0$ and $x = 4$ **(c)** 3000 lb **(d)** 1500 lb
5. **(a)** 1625.28 heating degree-hours **(b)** 70°F
(c) 1488 heating degree-hours **(d)** 75°F
(e) The day in part (a)

APPENDIX A ■ PAGE 952

1. Congruent, ASA **2.** Congruent, SSS
3. Not necessarily congruent **4.** Congruent, SAS
5. Similar **6.** Similar **7.** Similar **8.** Not similar

9. $x = 125$ **10.** $y = 30$ **11.** $x = 6, y = \frac{21}{4}$

12. $x = 4$ **13.** $x = \frac{ac}{a + b}$ **14.** $x = \frac{ac}{b} - a$

17. $x = 10$ **18.** $x = 48$ **19.** $x = \sqrt{3}$

20. $x = 2\sqrt{10}$ **21.** $x = 40$ **22.** $x = 144$ **23.** Yes

24. Yes **25.** No **26.** No **27.** Yes **28.** Yes **29.** 61 cm

30. 119 ft by 120 ft **31.** No **32.** 12 **33.** 13

34. (b)

m	n	(a, b, c)
2	1	(3, 4, 5)
3	1	(8, 6, 10)
3	2	(5, 12, 13)
4	1	(15, 8, 17)
4	2	(12, 16, 20)
4	3	(7, 24, 25)
5	1	(24, 10, 26)
5	2	(21, 20, 29)
5	3	(16, 30, 34)
5	4	(9, 40, 41)

35. $h = 6$

INDEX

Cut here and keep for reference

CHAPTER 1 Review: Concept Check Answers

1. (a) What does the set of natural numbers consist of? What does the set of integers consist of? Give an example of an integer that is not a natural number.

The set of natural numbers consists of the counting numbers $1, 2, 3, \ldots$. The set of integers consists of the natural numbers together with their negatives and 0. The number -1 is an integer that is not a natural number.

(b) What does the set of rational numbers consist of? Give an example of a rational number that is not an integer.

The set of rational numbers is constructed by taking all ratios of nonzero integers, and then adding the number 0. The number $2/3$ is a rational number that is not an integer.

(c) What does the set of irrational numbers consist of? Give an example of an irrational number.

The set of irrational numbers consists of all those numbers that cannot be expressed as a ratio of integers. The number $\sqrt{5}$ is an irrational number.

(d) What does the set of real numbers consist of?

The set of real numbers consists of all the rational numbers along with all the irrational numbers.

2. A property of real numbers is given. State the property and give an example in which the property is used.

(i) Commutative Property:

$a + b = b + a$ and $ab = ba$. For example, $5 + 8 = 8 + 5$ and $5 \cdot 8 = 8 \cdot 5$.

(ii) Associative Property:

$(a + b) + c = a + (b + c)$ and $(ab)c = a(bc)$. For example, $(2 + 5) + 3 = 2 + (5 + 3)$ and $(2 \cdot 5)3 = 2(5 \cdot 3)$.

(iii) Distributive Property:

$a(b + c) = ab + ac$ and $(b + c)a = ab + ac$. For example, $7(1 + 4) = 7 \cdot 1 + 7 \cdot 4$ and $(2 + 5)9 = 9 \cdot 2 + 9 \cdot 5$.

3. Explain the difference between the open interval (a, b) and the closed interval $[a, b]$. Give an example of an interval that is neither open nor closed.

The open interval excludes the endpoints a and b, and the closed interval includes the endpoints a and b. The interval $(0, 1]$ is neither open nor closed.

4. Give the formula for finding the distance between two real numbers a and b. Use the formula to find the distance between 103 and -52.

The distance between a and b is $|b - a|$. The distance between 103 and -52 is $|(-52) - 103| = 155$.

5. Suppose $a \neq 0$ is any real number.

(a) In the expression a^n, which is the base and which is the exponent?

The base is a and the exponent is n.

(b) What does a^n mean if n is a positive integer? What does 6^5 mean?

The expression a^n means to multiply a by itself n times. For example, $6^5 = 6 \cdot 6 \cdot 6 \cdot 6 \cdot 6$.

(c) What does a^{-n} mean if n is a positive integer? What does 3^{-2} mean?

The expression a^{-n} means the reciprocal of a^n, that is, $a^{-n} = \dfrac{1}{a^n}$. For example, $3^{-2} = \dfrac{1}{3^2}$.

(d) What does a^n mean if n is zero?

Any number raised to the 0 power is always equal to 1.

(e) If m and n are positive integers, what does $a^{m/n}$ mean? What does $4^{3/2}$ mean?

The expression $a^{m/n}$ means the nth root of the mth power of a. So $4^{3/2}$ means that you take the square root of 4 and then raise it to the third power: $4^{3/2} = 8$.

6. State the first five Laws of Exponents. Give examples in which you would use each law.

Law 1: $a^m a^n = a^{m+n}$; $\quad 5^2 \cdot 5^6 = 5^8$

Law 2: $\dfrac{a^m}{a^n} = a^{m-n}$; $\quad \dfrac{3^4}{3^2} = 3^{4-2} = 3^2$

Law 3: $(a^m)^n = a^{mn}$; $\quad (3^2)^4 = 3^{2 \cdot 4} = 3^8$

Law 4: $(ab)^n = a^n b^n$; $\quad (3 \cdot 5)^4 = 3^4 \cdot 5^4$

Law 5: $\left(\dfrac{a}{b}\right)^n = \dfrac{a^n}{b^n}$; $\quad \left(\dfrac{3}{5}\right)^2 = \dfrac{3^2}{5^2}$

7. When you multiply two powers of the same number, what should you do with the exponents? When you raise a power to a new power, what should you do with the exponents?

When you multiply two powers of the same number, you add the exponents. When you raise a power to a new power, you multiply the two exponents.

8. (a) What does $\sqrt[n]{a} = b$ mean?

The number b is the nth root of a.

(b) Is it true that $\sqrt{a^2}$ is equal to $|a|$? Try values for a that are positive and negative.

Yes, $\sqrt{a^2} = |a|$.

(c) How many real nth roots does a positive real number have if n is even? If n is odd?

There are two real nth roots if n is even and one real nth root if n is odd.

(d) Is $\sqrt[4]{-2}$ a real number? Is $\sqrt[3]{-2}$ a real number? Explain why or why not.

The expression $\sqrt[4]{-2}$ does not represent a real number because the fourth root of a negative number is undefined. The expression $\sqrt[3]{-2}$ does represent a real number because the third root of a negative number is defined.

(continued)

CHAPTER 1 Review: Concept Check Answers *(continued)*

9. Explain the steps involved in rationalizing a denominator. What is the logical first step in rationalizing the denominator of the expression $\frac{5}{\sqrt{3}}$?

The logical first step in rationalizing $\frac{5}{\sqrt{3}}$ is to multiply the numerator and denominator by $\sqrt{3}$:

$$\frac{5}{\sqrt{3}} \cdot \frac{\sqrt{3}}{\sqrt{3}} = \frac{5\sqrt{3}}{3}$$

10. Explain the difference between expanding an expression and factoring an expression.

We use the Distributive Property to expand algebraic expressions, and we reverse this process by factoring an expression as a product of simpler ones.

11. State the Special Product Formulas used for expanding the given expression. Use the appropriate formula to expand $(x + 5)^2$ and $(x + 5)(x - 5)$.

(i) $(a + b)^2 = a^2 + 2ab + b^2$
(ii) $(a - b)^2 = a^2 - 2ab + b^2$
(iii) $(a + b)^3 = a^3 + 3a^2b + 3ab^2 + b^3$
(iv) $(a - b)^3 = a^3 - 3a^2b + 3ab^2 - b^3$
(v) $(a + b)(a - b) = a^2 - b^2$

By (i) we have $(x + 5)^2 = x^2 + 10x + 25$, and by (v) we have $(x + 5)(x - 5) = x^2 - 25$.

12. State the following Special Factoring Formulas. Use the appropriate formula to factor $x^2 - 9$.

(i) Difference of Squares: $a^2 - b^2 = (a + b)(a - b)$
(ii) Perfect Square: $a^2 + 2ab + b^2 = (a + b)^2$
(iii) Sum of Cubes: $(a + b)(a^2 - ab + b^2) = (a + b)^3$

By (i) we have $x^2 - 9 = (x + 3)(x - 3)$.

13. If the numerator and the denominator of a rational expression have a common factor, how would you simplify the expression? Simplify the expression $\frac{x^2 + x}{x + 1}$.

You would simplify the expression by canceling the common factors in the numerator and the denominator. We simplify the expression as follows:

$$\frac{x^2 + x}{x + 1} = \frac{x\cancel{(x + 1)}}{\cancel{x + 1}} = x$$

14. Explain the following.

(a) How to multiply and divide rational expressions.

To multiply two rational expressions, we multiply their numerators and multiply their denominators. To divide a rational expression by another rational expression, we invert the divisor and multiply.

(b) How to add and subtract rational expressions.

To add or subtract two rational expressions, we first find the least common denominator (LCD), then rewrite the expressions using the LCD, and then add the fractions and combine the terms in the numerator.

(c) What LCD do we use to perform the addition in the expression $\frac{3}{x - 1} + \frac{5}{x + 2}$?

We use $(x - 1)(x + 2)$.

15. What is the logical first step in rationalizing the denominator of $\frac{3}{1 + \sqrt{x}}$?

Multiply both the numerator and the denominator by $(1 - \sqrt{x})$: $\frac{3}{1 + \sqrt{x}} \cdot \frac{1 - \sqrt{x}}{1 - \sqrt{x}} = \frac{3(1 - \sqrt{x})}{1 - x}$

16. What is the difference between an algebraic expression and an equation? Give examples.

An algebraic expression is a combination of variables; for example, $2x^2 + xy + 6$. An equation is a statement that two mathematical expressions are equal; for example, $3x - 2y = 9x - 1$.

17. Write the general form of each type of equation.

(i) Linear equation: $ax + b = 0$
(ii) Quadratic equation: $ax^2 + bx + c = 0$

18. What are the three ways to solve a quadratic equation?

(i) Factor the equation and use the Zero-Product property.
(ii) Complete the square and solve.
(iii) Use the Quadratic Formula.

19. State the Zero-Product Property. Use the property to solve the equation $x(x - 1) = 0$.

The Zero-Product Property states that $AB = 0$ if and only if $A = 0$ or $B = 0$.

To solve the equation $x(x - 1) = 0$, the Zero-Product Property shows that either $x = 0$ or $x = 1$.

20. What do you need to add to $ax^2 + bx$ to complete the square? Complete the square for the expression $x^2 + 6x$.

To complete the square, add $\left(\frac{b}{2}\right)^2$. To make $x^2 + 6x$ a perfect square, add $\left(\frac{6}{2}\right)^2 = 9$, and this gives the perfect square $x^2 + 6x + 9 = (x + 3)^2$.

21. State the Quadratic Formula for the quadratic equation $ax^2 + bx + c = 0$, and use it to solve the equation $x^2 + 6x - 1 = 0$.

The Quadratic Formula is $x = \frac{-b \pm \sqrt{b^2 - 4ac}}{2a}$.

Using the Quadratic Formula we get

$$x = \frac{-6 \pm \sqrt{36 - 4(1)(-1)}}{2(1)} = -3 \pm \sqrt{10}$$

(continued)

CHAPTER 1 Review: Concept Check Answers *(continued)*

22. What is the discriminant of the quadratic equation $ax^2 + bx + c = 0$? Find the discriminant of $2x^2 - 3x + 5 = 0$. How many real solutions does this equation have?

The discriminant is $b^2 - 4ac$. The discriminant of $2x^2 - 3x + 5 = 0$ is negative, so there are no real solutions.

23. What is the logical first step in solving the equation $\sqrt{x - 1} = x - 3$? Why is it important to check your answers when solving equations of this type?

The logical first step in solving this equation is to square both sides. It is important to check your answers because the operation of squaring both sides can turn a false equation into a true one. In this case $x = 5$ and $x = 2$ are potential solutions, but after checking, we see that $x = 5$ is the only solution.

24. What is a complex number? Give an example of a complex number, and identify the real and imaginary parts.

A complex number is an expression of the form $a + bi$, where a and b are real numbers and $i^2 = -1$. The complex number $2 + 3i$ has real part 2 and imaginary part 3.

25. What is the complex conjugate of a complex number $a + bi$?

The complex conjugate of $a + bi$ is $a - bi$.

26. **(a)** How do you add complex numbers?

To add complex numbers, add the real parts and the imaginary parts.

(b) How do you multiply $(3 + 5i)(2 - i)$?

Multiply complex numbers like binomials:
$(3 + 5i)(2 - i) = 6 + 10i - 3i - 5i^2 = 11 + 7i$

(c) Is $(3 - i)(3 + i)$ a real number?

Yes, $(3 - i)(3 + i) = 9 - i^2 = 10$

(d) How do you simplify the quotient $(3 + 5i)/(3 - i)$?

Multiply the numerator and the denominator by $3 + i$, the complex conjugate of the denominator.

27. State the guidelines for modeling with equations.

(i) Identify the variable.
(ii) Translate from words to algebra.
(iii) Set up the model.
(iv) Solve the equation and check your answer.

28. Explain how to solve the given type of problem.

(a) Linear inequality: $2x \geq 1$

Divide both sides by 2; the solution set is $[\frac{1}{2}, \infty)$.

(b) Nonlinear inequality: $(x - 1)(x - 4) < 0$

Find the intervals and make a table or diagram; the solution set is $(1, 4)$.

(c) Absolute value equation: $|2x - 5| = 7$

Solve the two equations $2x - 5 = 7$ and $2x - 5 = -7$; the solutions are $x = 6$ and $x = -1$.

(d) Absolute value inequality: $|2x - 5| \leq 7$

Solve the equivalent inequality $-7 \leq 2x - 5 \leq 7$; the solution set is $[-1, 6]$.

29. **(a)** In the coordinate plane, what is the horizontal axis called and what is the vertical axis called?

The horizontal axis is called the x-axis and the vertical axis is called the y-axis.

(b) To graph an ordered pair of numbers (x, y), you need the coordinate plane. For the point $(2, 3)$, which is the x-coordinate and which is the y-coordinate?

The x-coordinate is 2, and the y-coordinate is 3.

(c) For an equation in the variables x and y, how do you determine whether a given point is on the graph? Is the point $(5, 3)$ on the graph of the equation $y = 2x - 1$?

Any point (x, y) on the graph must satisfy the equation. Since $3 \neq 2(5) - 1$, the point $(5, 3)$ is *not* on the graph of the equation $y = 2x - 1$.

30. **(a)** What is the formula for finding the distance between the points (x_1, y_1) and (x_2, y_2)?

$$d = \sqrt{(x_2 - x_1)^2 + (y_2 - y_1)^2}$$

(b) What is the formula for finding the midpoint between (x_1, y_1) and (x_2, y_2)?

$$\left(\frac{x_1 + x_2}{2}, \frac{y_1 + y_2}{2}\right)$$

31. How do you find x-intercepts and y-intercepts of a graph of an equation?

To find the x-intercepts, you set $y = 0$ and solve for x. To find the y-intercepts, you set $x = 0$ and solve for y.

32. **(a)** Write an equation of the circle with center (h, k) and radius r.

$$(x - h)^2 + (y - k)^2 = r^2$$

(b) Find the equation of the circle with center $(2, -1)$ and radius 3.

$$(x - 2)^2 + (y + 1)^2 = 9$$

33. **(a)** How do you test whether the graph of an equation is symmetric with respect to the (i) x-axis, (ii) y-axis, and (iii) origin?

(i) When you replace y by $-y$, the resulting equation is equivalent to the original one.
(ii) When you replace x by $-x$, the resulting equation is equivalent to the original one.
(iii) When you replace x by $-x$ and y by $-y$, the resulting equation is equivalent to the original one.

(b) What type of symmetry does the graph of the equation $xy^2 + y^2x^2 = 3x$ have?

The graph is symmetric with respect to the x-axis.

(continued)

34. (a) What is the slope of a line? How do you compute the slope of the line through the points $(-1, 4)$ and $(1, -2)$?

The slope of a line is a measure of "steepness." The slope of the line through the points $(-1, 4)$ and $(1, -2)$ is

$$m = \frac{\text{rise}}{\text{run}} = \frac{-2-4}{1-(-1)} = -3$$

(b) How do you find the slope and y-intercept of the line $6x + 3y = 12$?

You write the equation in slope-intercept form $y = mx + b$. The slope is m, and the y-intercept is b. The slope-intercept form of this line is $y = -2x + 4$, so the slope is -2 and the intercept is 4.

(c) How do you write the equation for a line that has slope 3 and passes through the point $(1, 2)$?

Use the point-slope form of the equation of a line. So the equation is $y - 2 = 3(x - 1)$.

35. Give an equation of a vertical line and of a horizontal line that passes through the point $(2, 3)$.

An equation of a vertical line that passes through $(2, 3)$ is $x = 2$. An equation of a horizontal line that passes through $(2, 3)$ is $y = 3$.

36. State the general equation of a line.

$Ax + By = C$, where A and B are not both zero

37. Given lines with slopes m_1 and m_2, explain how you can tell whether the lines are (i) parallel, (ii) perpendicular.

(i) The lines are parallel if $m_1 = m_2$.

(ii) The lines are perpendicular if $m_2 = -\dfrac{1}{m_1}$.

38. How do you solve an equation (i) algebraically? (ii) graphically?

(i) Use the rules of algebra to isolate the unknown on one side of the equation.

(ii) Move all terms to one side and set that side equal to y. Sketch a graph of the resulting equation to find the values of x at which $y = 0$.

39. How do you solve an inequality (i) algebraically? (ii) graphically?

(i) Use the rules of algebra to isolate the unknown on one side of the inequality.

(ii) Move all terms to one side, and set that side equal to y. Sketch a graph to find the values x where the graph is above (or below) the x-axis.

40. Write an equation that expresses each relationship.

(a) y is directly proportional to x: $y = kx$

(b) y is inversely proportional to x: $y = \dfrac{k}{x}$

(c) z is jointly proportional to x and y: $z = kxy$

Cut here and keep for reference

CHAPTER 2 Review: Concept Check Answers

1. Define each concept.

(a) Function

A function f is a rule that assigns to each input x in a set A exactly one output $f(x)$ in a set B.

(b) Domain and range of a function

The domain of a function is the set of all the possible input values, and the range is the set of all possible output values.

(c) Graph of a function

The graph of a function f is the set of all ordered pairs $(x, f(x))$ plotted in a coordinate plane for x in the domain of f.

(d) Independent and dependent variables

The symbol that represents any value in the domain of a function f is called an independent variable, and the symbol that represents any value in the range of f is called a dependent variable.

2. Describe the four ways of representing a function.

A function can be represented verbally (using words), algebraically (using a formula), visually (using a graph), and numerically (using a table of values).

3. Sketch graphs of the following functions by hand.

(a) $f(x) = x^2$

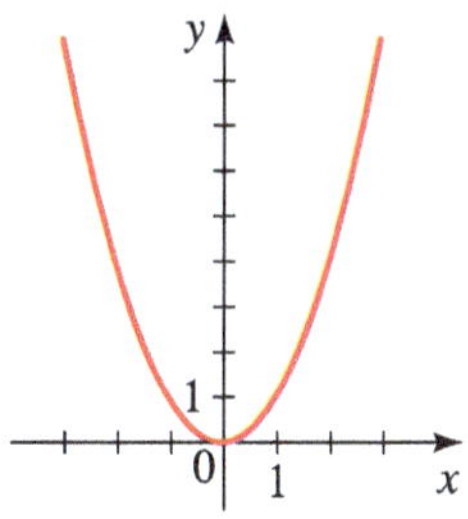

(b) $g(x) = x^3$

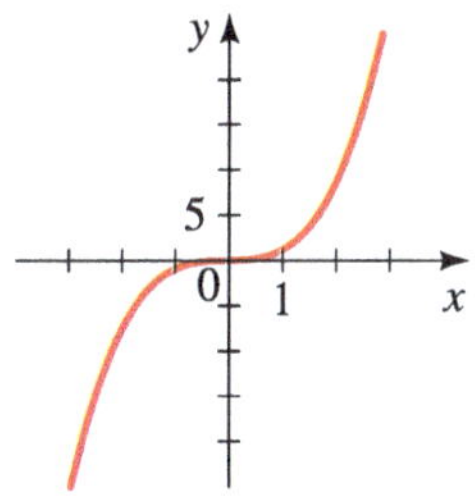

(c) $h(x) = |x|$

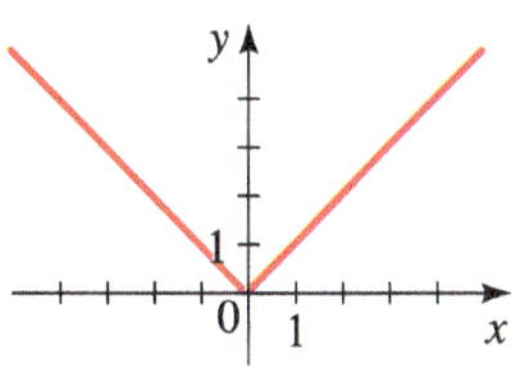

(d) $k(x) = \sqrt{x}$

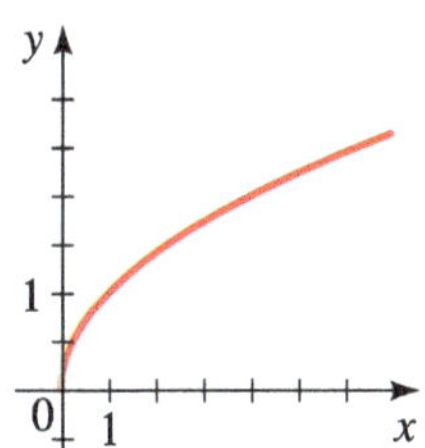

4. What is a piecewise defined function? Give an example.

A piecewise defined function is defined by different formulas on different parts of its domain. An example is

$$f(x) = \begin{cases} x^2 & \text{if } x > 0 \\ 2 & \text{if } x \leq 0 \end{cases}$$

5. (a) What is the Vertical Line Test, and what is it used for?

The Vertical Line Test states that a curve in the coordinate plane represents a function if and only if no vertical line intersects the curve more than once. It is used to determine when a given curve represents a function.

(b) What is the Horizontal Line Test, and what is it used for?

The Horizontal Line Test states that a function is one-to-one if and only if no horizontal line intersects its graph more than once. It is used to determine when a function is one-to-one.

6. Define each concept, and give an example of each.

(a) Increasing function

A function is increasing when its graph rises. More precisely, a function is increasing on an interval I if $f(x_1) < f(x_2)$ whenever $x_1 < x_2$ in I. For example, the function $f(x) = x^2$ is an increasing function on the interval $(0, \infty)$.

(b) Decreasing function

A function is decreasing when its graph falls. More precisely, a function is decreasing on an interval I if $f(x_1) > f(x_2)$ whenever $x_1 < x_2$ in I. For example, the function $f(x) = x^2$ is a decreasing function on the interval $(-\infty, 0)$.

(c) Constant function

A function f is constant if $f(x) = c$. For example, the function $f(x) = 3$ is constant.

7. Suppose we know that the point $(3, 5)$ is a point on the graph of a function f. Explain how to find $f(3)$ and $f^{-1}(5)$.

Since $(3, 5)$ is on the graph of f, the value 3 is the input and the value 5 is the output, so $f(3) = 5$ and $f^{-1}(5) = 3$.

8. What does it mean to say that $f(4)$ is a local maximum value of f?

The value $f(4)$ is a local maximum if $f(4) \geq f(x)$ for all x near 4.

9. Explain how to find the average rate of change of a function f between $x = a$ and $x = b$.

The average rate of change of f is

$$\frac{\text{change in } y}{\text{change in } x} = \frac{f(b) - f(a)}{b - a}$$

10. (a) What is the slope of a linear function? How do you find it? What is the rate of change of a linear function?

The slope of the graph of a linear function $f(x) = ax + b$ is the same as the rate of change of f, and they are both equal to a, the coefficient of x.

(b) Is the rate of change of a linear function constant? Explain.

Yes, because it is equal to the slope, and the slope is the same between any two points.

(continued)

(c) Give an example of a linear function, and sketch its graph.

An example is $f(x) = 2x + 1$, and the graph is shown below.

$f(x) = 2x + 1$

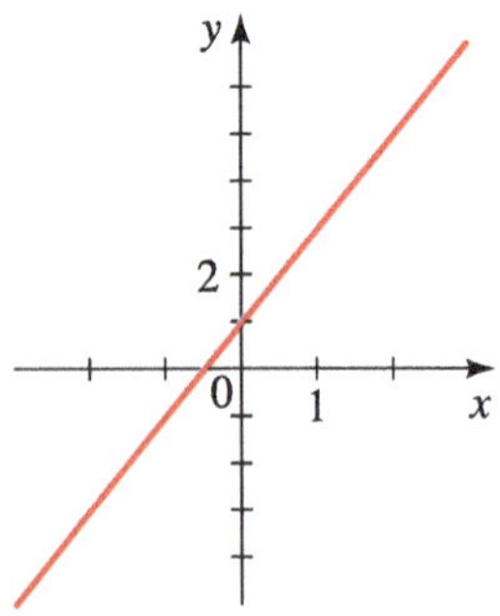

11. Suppose the graph of a function f is given. Write an equation for each of the graphs that are obtained from the graph of f as follows.

(a) Shift upward 3 units: $y = f(x) + 3$

(b) Shift downward 3 units: $y = f(x) - 3$

(c) Shift 3 units to the right: $y = f(x - 3)$

(d) Shift 3 units to the left: $y = f(x + 3)$

(e) Reflect in the x-axis: $y = -f(x)$

(f) Reflect in the y-axis: $y = f(-x)$

(g) Stretch vertically by a factor of 3: $y = 3f(x)$

(h) Shrink vertically by a factor of $\frac{1}{3}$: $y = \frac{1}{3}f(x)$

(i) Shrink horizontally by a factor of $\frac{1}{3}$: $y = f(3x)$

(j) Stretch horizontally by a factor of 3: $y = f(\frac{1}{3}x)$

12. (a) What is an even function? How can you tell that a function is even by looking at its graph? Give an example of an even function.

An even function f satisfies $f(-x) = f(x)$ for all x in its domain. If the graph of a function is symmetric with respect to the y-axis, then the function is even. Some examples are $f(x) = x^2$ and $f(x) = |x|$.

(b) What is an odd function? How can you tell that a function is odd by looking at its graph? Give an example of an odd function.

An odd function f satisfies $f(-x) = -f(x)$ for all x in its domain. If the graph of a function is symmetric with respect to the origin, then the function is odd. Some examples are $f(x) = x^3$ and $f(x) = \sqrt[3]{x}$.

13. Suppose that f has domain A and g has domain B. What are the domains of the following functions?

(a) Domain of $f + g$: $A \cap B$

(b) Domain of fg: $A \cap B$

(c) Domain of f/g: $\{x \in A \cap B \mid g(x) \neq 0\}$

14. (a) How is the composition function $f \circ g$ defined? What is its domain?

The function $f \circ g$ is defined by $f \circ g(x) = f(g(x))$. The domain is the set of all x in the domain of g such that $g(x)$ is in the domain of f.

(b) If $g(a) = b$ and $f(b) = c$, then explain how to find $(f \circ g)(a)$.

To find $f \circ g(a)$, we evaluate the following:

$$f \circ g(a) = f(g(a)) = f(b) = c$$

15. (a) What is a one-to-one function?

A function with domain A is called a one-to-one function if no two elements of A have the same image. More precisely, $f(x_1) \neq f(x_2)$ whenever $x_1 \neq x_2$.

(b) How can you tell from the graph of a function whether it is one-to-one?

We use the Horizontal Line Test, which states that a function is one-to-one if and only if no horizontal line intersects its graph more than once.

(c) Suppose that f is a one-to-one function with domain A and range B. How is the inverse function f^{-1} defined? What are the domain and range of f^{-1}?

The inverse function of f has domain B and range A and is defined by

$$f^{-1}(y) = x \iff f(x) = y$$

(d) If you are given a formula for f, how do you find a formula for f^{-1}? Find the inverse of the function $f(x) = 2x$.

We write $y = f(x)$, solve the equation for x in terms of y, and interchange x and y. The resulting equation is $y = f^{-1}(x)$. If $f(x) = 2x$, we write $y = 2x$, solve for x to get $x = \frac{1}{2}y$, interchange x and y to get $f^{-1}(x) = \frac{1}{2}x$.

(e) If you are given a graph of f, how do you find a graph of the inverse function f^{-1}?

The graph of the inverse function f^{-1} is obtained by reflecting the graph of f in the line $y = x$.

Cut here and keep for reference

CHAPTER 3 Review: Concept Check Answers

1. (a) What is the degree of a quadratic function f? What is the standard form of a quadratic function? How do you put a quadratic function into standard form?

A quadratic function f is a polynomial of degree 2. The standard form of a quadratic function f is $f(x) = a(x - h)^2 + k$. Complete the square to put a quadratic function into standard form.

(b) The quadratic function $f(x) = a(x - h)^2 + k$ is in standard form. The graph of f is a parabola. What is the vertex of the graph of f? How do you determine whether $f(h) = k$ is a minimum or a maximum value?

The vertex of the graph of f is (h, k). If the coefficient a is positive, then the graph of f opens upward and $f(h) = k$ is a minimum value. If a is negative, then the graph of f opens downward and $f(h) = k$ is a maximum value.

(c) Express $f(x) = x^2 + 4x + 1$ in standard form. Find the vertex of the graph and the maximum or minimum value of f.

We complete the square to get $f(x) = (x + 2)^2 - 3$. The graph is a parabola that opens upward with vertex $(-2, -3)$. The minimum value is $f(-2) = -3$.

2. (a) Give the general form of polynomial function P of degree n.

$$P(x) = a_n x^n + a_{n-1}x^{n-1} + \cdots + a_1 x + a_0 \qquad a_n \neq 0$$

(b) What does it mean to say that c is a zero of P? Give two equivalent conditions that tell us that c is a zero of P.

The value c is a zero of P if $P(c) = 0$. Equivalently, c is a zero of P if $x - c$ is a factor of P or if c is an x-intercept of the graph of P.

3. Sketch graphs showing the possible end behaviors of polynomials of odd degree and of even degree.

Odd degree

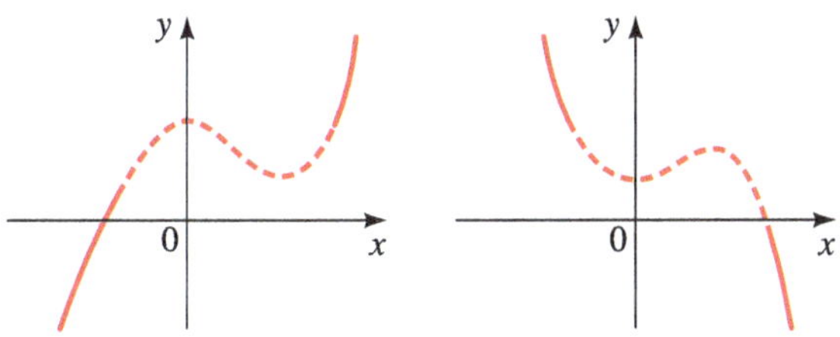

Even degree

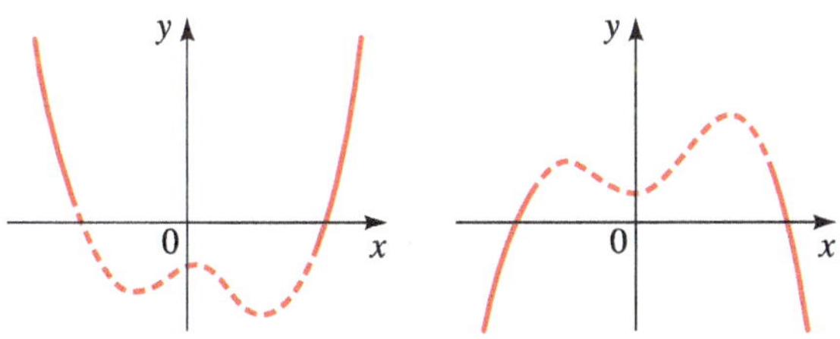

4. What steps do you follow to graph a polynomial function P?

We first find the zeros of P and then make a table using test points between successive zeros. We then determine the end behavior and use all this information to graph P.

5. (a) What is a local maximum point or local minimum point of a polynomial P?

The point $(a, P(a))$ is a local maximum if it is the highest point on the graph of P within some viewing rectangle. The point $(b, P(b))$ is a local minimum if it is the lowest point on the graph of P within some viewing rectangle.

(b) How many local extrema can a polynomial P of degree n have?

The graph of P has at most $n - 1$ local extrema.

6. When we divide a polynomial $P(x)$ by a divisor $D(x)$, the Division Algorithm tells us that we can always obtain a quotient $Q(x)$ and a remainder $R(x)$. State the two forms in which the result of this division can be written.

$$\frac{P(x)}{D(x)} = Q(x) + \frac{R(x)}{D(x)}$$

$$P(x) = D(x)Q(x) + R(x)$$

7. (a) State the Remainder Theorem.

If a polynomial $P(x)$ is divided by $x - c$, then the remainder is the value $P(c)$.

(b) State the Factor Theorem.

The number c is a zero of P if and only if $x - c$ is a factor of $P(x)$.

(c) State the Rational Zeros Theorem.

If the polynomial

$$P(x) = a_n x^n + a_{n-1}x^{n-1} + \cdots + a_1 x + a_0$$

has integer coefficients, then every rational zero of P is of the form p/q, where p is a factor of the constant coefficient a_0 and q is a factor of the leading coefficient a_n.

8. What steps would you take to find the rational zeros of a polynomial P?

First list all possible rational zeros of P given by the Rational Zeros Theorem. Evaluate P at a possible zero (using synthetic division), and note the quotient if the remainder is 0. Repeat this process on the quotient until you reach a quotient that is quadratic. Then use the quadratic formula to find the remaining zeros.

(*continued*)

9. Let $P(x) = 2x^4 - 3x^3 + x - 15$.

(a) Explain how Descartes' Rule of Signs is used to determine the possible number of positive and negative real roots of P.

Since there are three variations in sign in $P(x)$, by Descartes' Rule of Signs there are either three or one positive real zeros. Since there is one variation in sign in $P(-x)$, by Descartes' Rule of Signs there is exactly one negative real zero.

(b) What does it mean to say that a is a lower bound and b is an upper bound for the zeros of a polynomial?

We say that a is a lower bound and b is an upper bound for the zeros of a polynomial if every real zero c of the polynomial satisfies $a \le c \le b$.

(c) Explain how the Upper and Lower Bounds Theorem is used to show that all the real zeros of P lie between -3 and 3.

When we divide P by $x - 3$, the row that contains the quotient and the remainder has only nonnegative entries, so 3 is an upper bound. When we divide P by $x - (-3) = x + 3$, the row that contains the quotient and the remainder has entries that alternate in sign, so -3 is a lower bound.

10. (a) State the Fundamental Theorem of Algebra.

Every polynomial has at least one complex zero.

(b) State the Complete Factorization Theorem.

Every polynomial of degree $n \ge 1$ can be factored completely into linear factors (with complex coefficients).

(c) State the Zeros Theorem.

Every polynomial of degree $n \ge 1$ has exactly n zeros, provided that a zero of multiplicity k is counted k times.

(d) State the Conjugate Zeros Theorem.

If a polynomial has real coefficients and if the complex number z is a zero of the polynomial, then its complex conjugate $\bar{z}$ is also a zero of the polynomial.

11. (a) What is a rational function?

A rational function is a function of the form

$r(x) = \dfrac{P(x)}{Q(x)}$, where P and Q are polynomials.

(b) What does it mean to say that $x = a$ is a vertical asymptote of $y = f(x)$?

The line $x = a$ is a vertical asymptote if

$$y \to \pm\infty \quad \text{as} \quad x \to a^+ \quad \text{or} \quad x \to a^-$$

(c) What does it mean to say that $y = b$ is a horizontal asymptote of $y = f(x)$?

The line $y = b$ is a horizontal asymptote if

$$y \to b \quad \text{as} \quad x \to \infty \quad \text{or} \quad x \to -\infty$$

(d) Find the vertical and horizontal asymptotes of

$$f(x) = \frac{5x^2 + 3}{x^2 - 4}$$

The denominator factors as $(x - 2)(x + 2)$, so the vertical asymptotes are $x = 2$ and $x = -2$. The horizontal asymptote is $y = 5$.

12. (a) How do you find vertical asymptotes of rational functions?

Vertical asymptotes of a rational function are the line $x = a$, where a is a zero of the denominator.

(b) Let s be the rational function

$$s(x) = \frac{a_n x^n + a_{n-1}x^{n-1} + \cdots + a_1 x + a_0}{b_m x^m + b_{m-1}x^{m-1} + \cdots + b_1 x + b_0}$$

How do you find the horizontal asymptote of s?

If $n < m$, then the horizontal asymptote is

$$y = 0$$

If $n = m$, then the horizontal asymptote is

$$y = \frac{a_n}{b_m}$$

If $n > m$, then there is no horizontal asymptote.

13. (a) Under what circumstances does a rational function have a slant asymptote?

If $r(x) = P(x)/Q(x)$ and the degree of P is one greater than the degree of Q, then r has a slant asymptote.

(b) How do you determine the end behavior of a rational function?

Divide the numerator by the denominator; the quotient determines the end behavior of the function.

14. (a) Explain how to solve a polynomial inequality.

Move all terms to one side, factor the polynomial, find the zeros of the polynomial, use the zeros and test points to make a sign diagram, and use the diagram to solve the inequality.

(b) What are the cut points of a rational function? Explain how to solve a rational inequality.

The cut points are the zeros of the numerator and zeros of the denominator. To solve a rational inequality, move all terms to one side, factor the numerator and denominator to find all the cut points, use the cut points and test points to make a sign diagram, and use the diagram to solve the inequality.

(c) Solve the inequality $x^2 - 9 \le 8x$.

Move all terms to one side and then factor: $(x + 1)(x - 9) \le 0$. We make a sign diagram as shown.

	$(-\infty, -1)$	$(-1, 9)$	$(9, \infty)$
Sign of $x + 1$	−	+	+
Sign of $x - 9$	−	−	+
Sign of $(x + 1)(x - 9)$	+	−	+

The solution is the interval $[-1, 9]$.

Cut here and keep for reference

CHAPTER 4 Review: Concept Check Answers

1. Let f be the exponential function with base a.

(a) Write an equation that defines f.

$$f(x) = a^x$$

(b) Write an equation for the exponential function f with base 3.

$$f(x) = 3^x$$

2. Let f be the exponential function $f(x) = a^x$, where $a > 0$.

(a) What is the domain of f?

All real numbers $(-\infty, \infty)$

(b) What is the range of f?

All positive real numbers $(0, \infty)$

(c) Sketch graphs of f for the following cases.

(i) $a > 1$ (ii) $0 < a < 1$

(i) (ii)

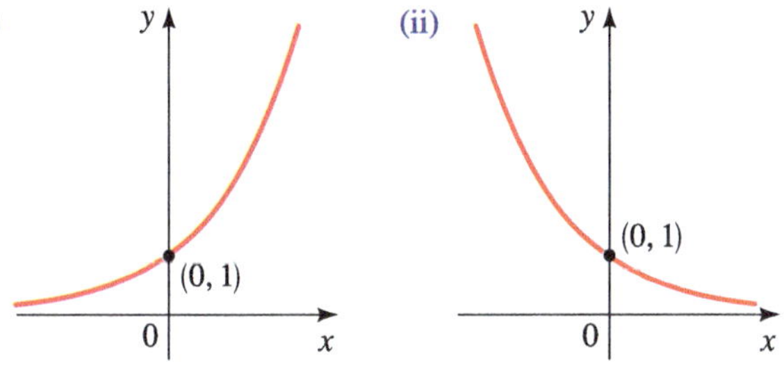

3. If x is large, which function grows faster, $f(x) = 2^x$ or $g(x) = x^2$?

The function $f(x) = 2^x$ grows faster. We can see this by graphing both functions in a sufficiently large viewing rectangle.

4. (a) How is the number e defined?

The number e is the value that $\left(1 + \frac{1}{n}\right)^n$ approaches as n becomes large.

(b) Give an approximate value of e, correct to four decimal places.

$$e \approx 2.71828$$

(c) What is the natural exponential function?

It is the exponential function with base e:

$$f(x) = e^x$$

5. (a) How is $\log_a x$ defined?

$$\log_a x = y \quad \Leftrightarrow \quad a^y = x$$

(b) Find $\log_3 9$.

$$\log_3 9 = 2 \quad \text{because} \quad 3^2 = 9$$

(c) What is the natural logarithm?

It is the logarithm with base e: $\ln x = \log_e x$

(d) What is the common logarithm?

It is the logarithm with base 10: $\log x = \log_{10} x$

(e) Write the exponential form of the equation $\log_7 49 = 2$.

$$7^2 = 49$$

6. Let f be the logarithmic function $f(x) = \log_a x$.

(a) What is the domain of f?

All positive real numbers $(0, \infty)$

(b) What is the range of f?

All real numbers $(-\infty, \infty)$

(c) Sketch a graph of the logarithmic function for the case that $a > 1$.

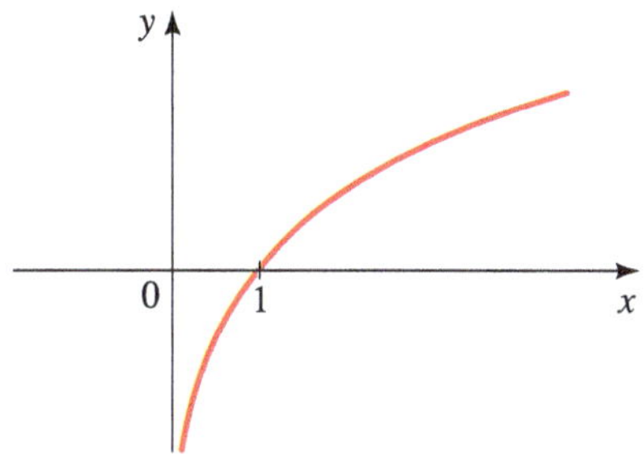

7. State the three Laws of Logarithms.

$$\log_a xy = \log_a x + \log_a y$$

$$\log_a\left(\frac{x}{y}\right) = \log_a x - \log_a y$$

$$\log_a x^N = N \log_a x$$

8. (a) State the Change of Base Formula.

$$\log_a x = \frac{\log_b x}{\log_b a}$$

(b) Find $\log_7 30$.

By the Change of Base Formula

$$\log_7 30 = \frac{\log 30}{\log 7} \approx 1.7479$$

9. (a) What is an exponential equation?

An exponential equation is one in which the unknown occurs in an exponent.

(b) How do you solve an exponential equation?

First isolate the exponential term on one side, take logarithms of each side, and then use the laws of logarithms to bring down the exponent. Then solve for the unknown.

(c) Solve for x: $2^x = 19$

$$\log 2^x = \log 19$$

$$x \log 2 = \log 19$$

$$x = \frac{\log 19}{\log 2} \approx 4.2479$$

(continued)

CHAPTER 4 Review: Concept Check Answers (*continued*)

10. (a) What is a logarithmic equation?

A logarithmic equation is one in which a logarithm of the unknown occurs.

(b) How do you solve a logarithmic equation?

First combine the logarithmic terms on one side of the equation, write the resulting equation in exponential form, and then solve for the unknown.

(c) Solve for x: $4 \log_3 x = 7$

$$4 \log_3 x = 7$$
$$\log_3 x = 1.75$$
$$x = 3^{1.75} \approx 6.84$$

11. Suppose that an amount P is invested at an interest rate r and $A(t)$ is the amount of the investment after t years. Write a formula for $A(t)$ in the following cases.

(a) Interest is compounded n times per year.

$$A(t) = P\left(1 + \frac{r}{n}\right)^{nt}$$

(b) Interest is compounded continuously.

$$A(t) = Pe^{rt}$$

12. Suppose that the initial size of a population is n_0 and the population grows exponentially. Let $n(t)$ be the size of the population at time t.

(a) Write a formula for $n(t)$ in terms of the doubling time a.

$$n(t) = n_0 2^{t/a}$$

(b) Write a formula for $n(t)$ in terms of the relative growth rate r.

$$n(t) = n_0 e^{rt}$$

13. Suppose that the initial mass of a radioactive substance is m_0 and the half-life of the substance is h. Let $m(t)$ be the mass remaining at time t.

(a) What is meant by the half-life h?

The time it takes for a mass to decay to half its amount

(b) Write a formula for $m(t)$ in terms of the half-life h.

$$m(t) = m_0 2^{-t/h}$$

(c) Write a formula for the relative decay rate r in terms of the half-life h.

$$r = \frac{\ln 2}{h}$$

(d) Write a formula for $m(t)$ in terms of the relative decay rate r.

$$m(t) = m_0 e^{-rt}$$

14. Suppose that the initial temperature difference between an object and its surroundings is D_0 and the surroundings have temperature T_s. Let $T(t)$ be the temperature at time t. State Newton's Law of Cooling for $T(t)$.

$$T(t) = T_s + D_0 e^{-kt}$$

where k is a constant that depends on the type of object.

15. What is a logarithmic scale? If we use a logarithmic scale with base 10, what do the following numbers correspond to on the logarithmic scale?

(i) 100 (ii) 100,000 (iii) 0.0001

On a logarithmic scale, numbers are represented by their logarithms.

(i) 2 (ii) 5 (iii) -4

16. (a) What does the pH scale measure?

The acidity (or alkalinity) of a substance

(b) Define the pH of a substance with hydrogen ion concentration of $[H^+]$.

$$\text{pH} = -\log [H^+]$$

17. (a) What does the Richter scale measure?

The magnitude of earthquakes

(b) Define the magnitude M of an earthquake in terms of the intensity I of the earthquake and the intensity S of a standard earthquake.

$$M = -\log \frac{I}{S}$$

18. (a) What does the decibel scale measure?

The loudness of sound

(b) Define the decibel level B of a sound in terms of the intensity I of the sound and the intensity I_0 of a barely audible sound.

$$B = 10 \log \frac{I}{I_0}$$

Cut here and keep for reference

CHAPTER 5 Review: Concept Check Answers

1. (a) What is the unit circle, and what is the equation of the unit circle?

The unit circle is the circle of radius 1 centered at $(0, 0)$. The equation of the unit circle is $x^2 + y^2 = 1$.

(b) Use a diagram to explain what is meant by the terminal point $P(x, y)$ determined by t.

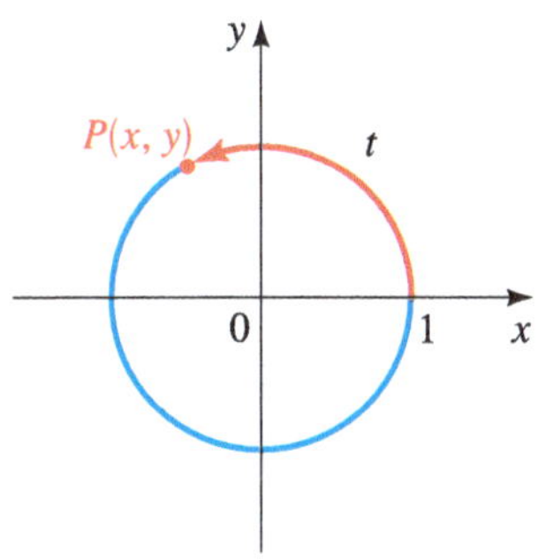

(c) Find the terminal point for $t = \dfrac{\pi}{2}$.

$$P(x, y) = (0, 1)$$

(d) What is the reference number associated with t?

The reference number is the shortest distance along the unit circle between the terminal point determined by t and the x-axis.

(e) Find the reference number and terminal point for $t = \dfrac{7\pi}{4}$.

The reference number is $\frac{\pi}{4}$. The terminal point is in Quadrant IV, so $P(x, y) = \left(\frac{\sqrt{2}}{2}, -\frac{\sqrt{2}}{2}\right)$.

2. Let t be a real number, and let $P(x, y)$ be the terminal point determined by t.

(a) Write equations that define $\sin t$, $\cos t$, $\tan t$, $\csc t$, $\sec t$, and $\cot t$.

$\sin t = y \qquad \cos t = x \qquad \tan t = \frac{y}{x}$

$\csc t = \frac{1}{y} \qquad \sec t = \frac{1}{x} \qquad \cot t = \frac{x}{y}$

(b) In each of the four quadrants, identify the trigonometric functions that are positive.

In Quadrant I all functions are positive; in Quadrant II the sine and cosecant functions are positive; in Quadrant III the tangent and cotangent functions are positive; and in Quadrant IV the cosine and secant functions are positive.

(c) List the special values of sine, cosine, and tangent.

$\sin 0 = 0$, $\sin \frac{\pi}{6} = \frac{1}{2}$, $\sin \frac{\pi}{4} = \frac{\sqrt{2}}{2}$, $\sin \frac{\pi}{3} = \frac{\sqrt{3}}{2}$, $\sin \frac{\pi}{2} = 1$

$\cos 0 = 1$, $\cos \frac{\pi}{6} = \frac{\sqrt{3}}{2}$, $\cos \frac{\pi}{4} = \frac{\sqrt{2}}{2}$, $\cos \frac{\pi}{3} = \frac{1}{2}$, $\cos \frac{\pi}{2} = 0$

$\tan 0 = 0$, $\tan \frac{\pi}{6} = \frac{\sqrt{3}}{3}$, $\tan \frac{\pi}{4} = 1$, $\tan \frac{\pi}{3} = \sqrt{3}$

3. (a) Describe the steps we use to find the value of a trigonometric function at a real number t.

We find the reference number for t, the quadrant where the terminal point lies, and the sign of the function in that quadrant, and we use all these to find the value of the function at t.

(b) Find $\sin \dfrac{5\pi}{6}$.

The terminal point of $\frac{5\pi}{6}$ is in Quadrant II. Since sine is positive in Quadrant II, $\sin \frac{5\pi}{6} = \sin \frac{\pi}{6} = \frac{1}{2}$.

4. (a) What is a periodic function?

A function f is periodic if there is a positive number p such that $f(x + p) = f(x)$ for every x. The least such p is called the period of f.

(b) What are the periods of the six trigonometric functions?

The sine, cosine, cosecant, and secant functions have period 2π, and the tangent and cotangent functions have period π.

(c) Find $\sin \dfrac{19\pi}{4}$.

$$\sin \tfrac{19\pi}{4} = \sin\left(\tfrac{3\pi}{4} + 4\pi\right) = \sin \tfrac{3\pi}{4} = \tfrac{\sqrt{2}}{2}$$

5. (a) What is an even function, and what is an odd function?

An even function satisfies $f(-x) = f(x)$.

An odd function satisfies $f(-x) = -f(x)$.

(b) Which trigonometric functions are even? Which are odd?

The cosine and secant functions are even; the sine, cosecant, tangent, and cotangent functions are odd.

(c) If $\sin t = 0.4$, find $\sin(-t)$.

Since the sine function is odd, $\sin(-t) = -0.4$.

(d) If $\cos s = 0.7$, find $\cos(-s)$.

Since the cosine function is even, $\cos(-s) = 0.7$.

6. (a) State the reciprocal identities.

$\csc t = \frac{1}{\sin t}$, $\sec t = \frac{1}{\cos t}$, $\cot t = \frac{1}{\tan t}$,
$\tan t = \frac{\sin t}{\cos t}$, $\cot t = \frac{\cos t}{\sin t}$

(b) State the Pythagorean identities.

$\sin^2 t + \cos^2 t = 1$, $\tan^2 t + 1 = \sec^2 t$,
$1 + \cot^2 t = \csc^2 t$

7. (a) Graph the sine and cosine functions.

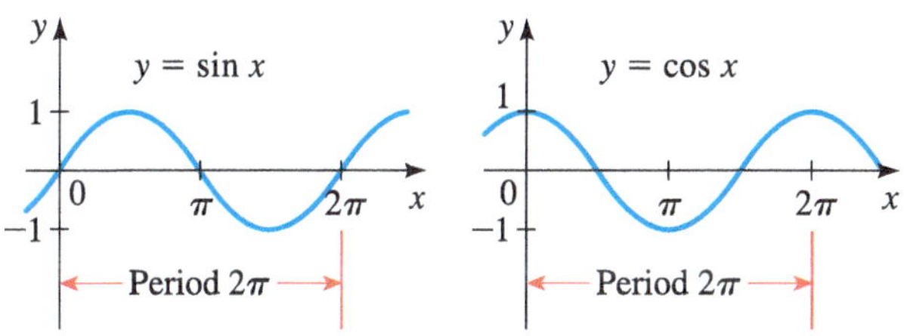

(b) What are the amplitude, period, and horizontal shift for the sine curve $y = a \sin k(x - b)$ and for the cosine curve $y = a \cos k(x - b)$?

Amplitude a; period $\frac{2\pi}{k}$; horizontal shift b

(c) Find the amplitude, period, and horizontal shift of $y = 3 \sin\left(2x - \dfrac{\pi}{6}\right)$.

We factor to get $y = 3 \sin 2\left(x - \frac{\pi}{12}\right)$.

Amplitude 3; period π; horizontal shift $\frac{\pi}{12}$

(continued)

8. (a) Graph the tangent and cotangent functions.

$y = \tan x$

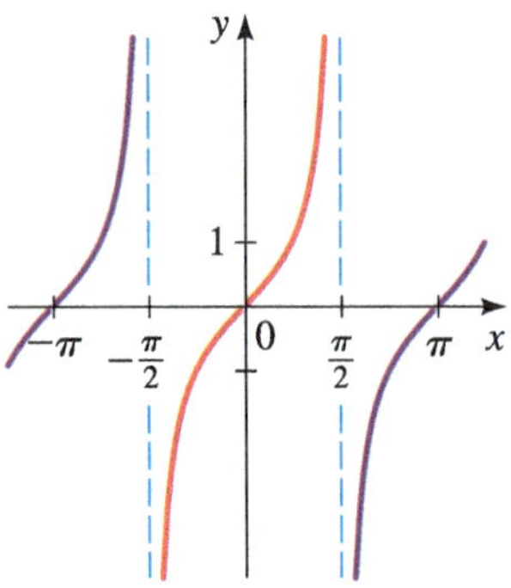

$y = \cot x$

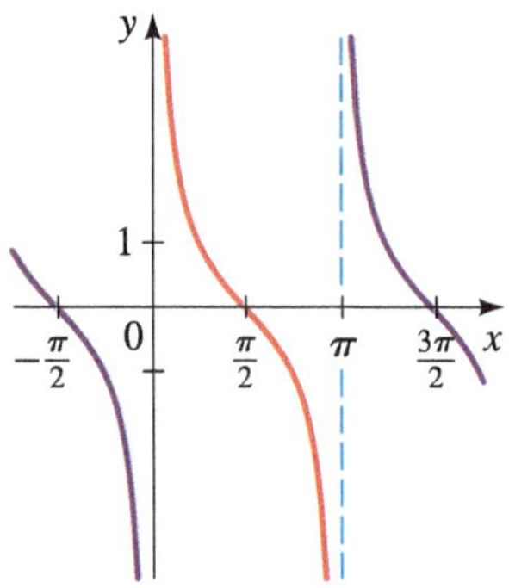

(b) For the curves $y = a \tan kx$ and $y = a \cot kx$, state appropriate intervals to graph one complete period of each curve.

An appropriate interval for $y = a \tan kx$ is $(-\pi/2k, \pi/2k)$.

An appropriate interval for $y = a \cot kx$ is $(0, \pi/k)$.

(c) Find an appropriate interval to graph one complete period of $y = 5 \tan 3x$.

An appropriate interval for $y = 5 \tan 3x$ is $(-\pi/6, \pi/6)$.

9. (a) Graph the cosecant and secant functions.

$y = \csc x$

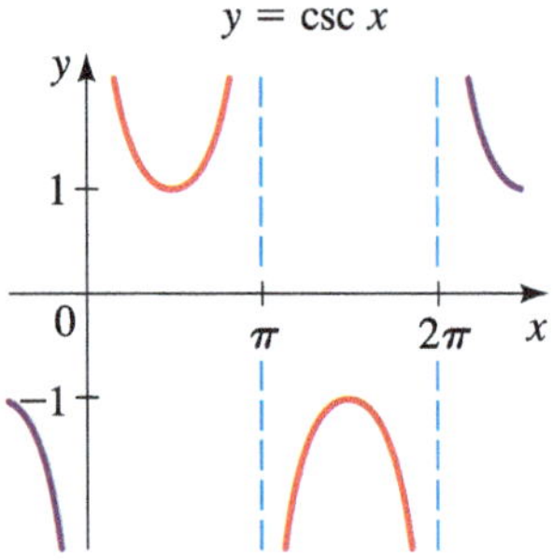

$y = \sec x$

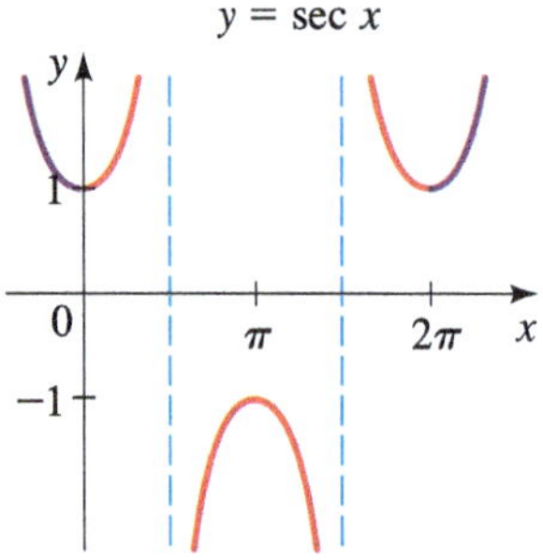

(b) For the curves $y = a \csc kx$ and $y = a \sec kx$, state appropriate intervals to graph one complete period of each curve.

An appropriate interval for $y = a \csc kx$ is $(0, 2\pi/k)$.

An appropriate interval for $y = a \sec kx$ is $(0, 2\pi/k)$.

(c) Find an appropriate interval to graph one period of $y = 3 \csc 6x$.

An appropriate interval for $y = 3 \csc 6x$ is $(0, \pi/3)$.

10. (a) Define the inverse sine function, the inverse cosine function, and the inverse tangent function.

$$\sin^{-1}x = y \quad \Leftrightarrow \quad \sin y = x$$
$$\cos^{-1}x = y \quad \Leftrightarrow \quad \cos y = x$$
$$\tan^{-1}x = y \quad \Leftrightarrow \quad \tan y = x$$

(b) Find $\sin^{-1}\frac{1}{2}$, $\cos^{-1}\frac{\sqrt{2}}{2}$, and $\tan^{-1} 1$.

From 2(c) and the definitions in part (a) we get $\sin^{-1}\frac{1}{2} = \frac{\pi}{6}$, $\cos^{-1}\frac{\sqrt{2}}{2} = \frac{\pi}{4}$, and $\tan^{-1} 1 = \frac{\pi}{4}$.

(c) For what values of x is the equation $\sin(\sin^{-1} x) = x$ true? For what values of x is the equation $\sin^{-1}(\sin x) = x$ true?

$$\sin(\sin^{-1} x) = x \quad \text{for} \quad -1 \le x \le 1$$
$$\sin^{-1}(\sin x) = x \quad \text{for} \quad -\tfrac{\pi}{2} \le x \le \tfrac{\pi}{2}$$

11. (a) What is simple harmonic motion?

An object is in simple harmonic motion if its displacement y at time t is modeled by $y = a \sin \omega t$ or $y = a \cos \omega t$.

(b) What is damped harmonic motion?

An object is in damped harmonic motion if its displacement y at time t is modeled by $y = ke^{-ct} \sin \omega t$ or $y = ke^{-ct} \cos \omega t$, $c > 0$.

(c) Give real-world examples of harmonic motion.

The motion of a vibrating mass on a spring, the vibrations of a violin string, the brightness of a variable star, and many more

12. Suppose that an object is in simple harmonic motion given by $y = 5 \sin\left(2t - \frac{\pi}{3}\right)$.

(a) Find the amplitude, period, and frequency.

Amplitude 5; period $\frac{2\pi}{2} = \pi$; frequency $\frac{2}{2\pi} = \frac{1}{\pi}$

(b) Find the phase and the horizontal shift.

The phase is $\frac{\pi}{3}$, and the horizontal shift (or lag time) is $\frac{\pi}{6}$.

13. Consider the following models of harmonic motion.

$$y_1 = 5 \sin(2t - 1) \qquad y_2 = 5 \sin(2t - 3)$$

Do both motions have the same frequency? What is the phase for each equation? What is the phase difference? Are the objects moving in phase or out of phase?

Both motions have the same frequency: $1/\pi$. The phase of the first is 1, and the phase of the second is 3. The phase difference is $3 - 1 = 2$, which is not a multiple of 2π, so the objects are moving out of phase.

Cut here and keep for reference

CHAPTER 6 Review: Concept Check Answers

1. (a) How is the degree measure of an angle defined?

An angle of $1°$ is $\frac{1}{360}$ of a complete revolution.

(b) How is the radian measure of an angle defined?

The radian measure of an angle is the length of the arc that the angle subtends in a circle of radius 1.

(c) How do you convert from degrees to radians? Convert $45°$ to radians.

To convert from degrees to radians, we multiply by $\pi/180$. So

$$45° = 45\left(\frac{\pi}{180}\right)\text{rad} = \frac{\pi}{4}$$

(d) How do you convert from radians to degrees? Convert 2 rad to degrees.

To convert from radians to degrees, we multiply by $180/\pi$. So

$$2 \text{ rad} = 2\left(\frac{180}{\pi}\right) \approx 114.6°$$

2. (a) When is an angle in standard position? Illustrate with a graph.

An angle is in standard position if it is drawn in the xy-plane with its vertex at the origin and its initial side on the positive x-axis.

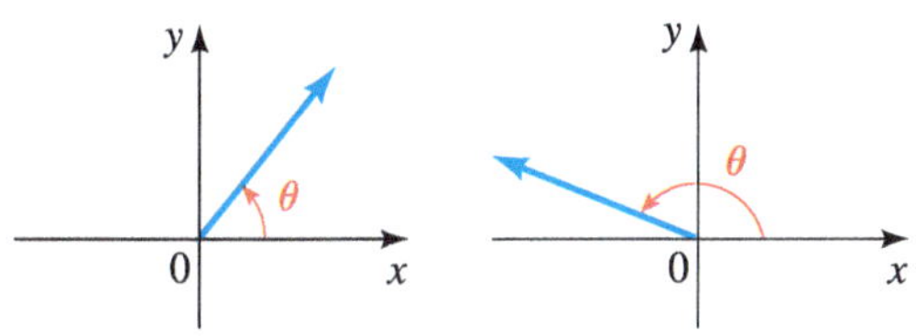

(b) When are two angles in standard position coterminal? Illustrate with a graph.

Two angles are coterminal if their sides coincide. Angles that differ by a multiple of 2π rad (or a multiple of $360°$) are coterminal.

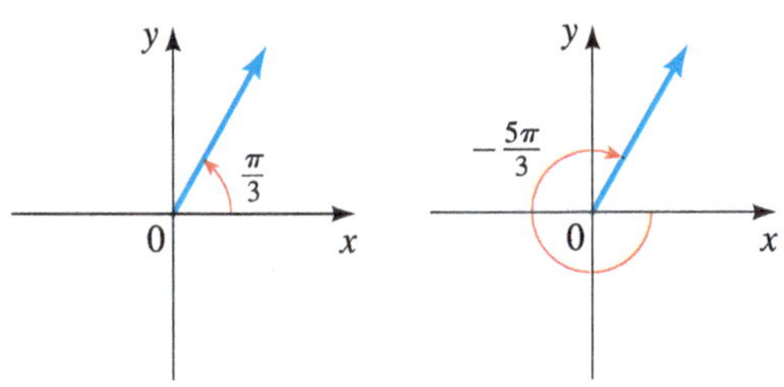

(c) Are the angles $25°$ and $745°$ coterminal?

Yes, because $745° - 25° = 720°$, which is a multiple of $360°$.

(d) How is the reference angle for an angle θ defined?

The reference angle $\bar{\theta}$ is the acute angle formed by the terminal side of θ and the x-axis.

(e) Find the reference angle for $150°$.

The reference angle is $\bar{\theta} = 180° - 150° = 30°$.

3. (a) In a circle of radius r, what is the length s of an arc that subtends a central angle of θ radians?

$$s = r\theta$$

(b) In a circle of radius r, what is the area A of a sector with central angle θ radians?

$$A = \tfrac{1}{2}r^2\theta$$

4. (a) Let θ be an acute angle in a right triangle. Identify the opposite side, the adjacent side, and the hypotenuse in the figure.

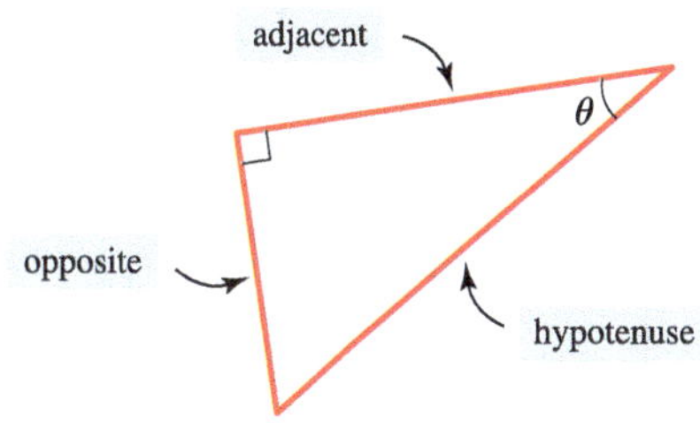

(b) Define the six trigonometric ratios in terms of the adjacent and opposite sides and the hypotenuse.

$$\sin\theta = \frac{\text{opp}}{\text{hyp}} \qquad \cos\theta = \frac{\text{adj}}{\text{hyp}} \qquad \tan\theta = \frac{\text{opp}}{\text{adj}}$$

$$\csc\theta = \frac{\text{hyp}}{\text{opp}} \qquad \sec\theta = \frac{\text{hyp}}{\text{adj}} \qquad \cot\theta = \frac{\text{adj}}{\text{opp}}$$

(c) Find the six trigonometric ratios for the angle θ shown in the figure.

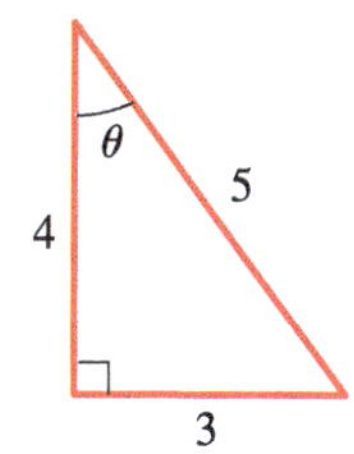

$$\sin\theta = \frac{3}{5} \qquad \cos\theta = \frac{4}{5} \qquad \tan\theta = \frac{3}{4}$$

$$\csc\theta = \frac{5}{3} \qquad \sec\theta = \frac{5}{4} \qquad \cot\theta = \frac{4}{3}$$

(d) List the special values of sine, cosine, and tangent.

$$\sin\tfrac{\pi}{6} = \tfrac{1}{2} \qquad \sin\tfrac{\pi}{4} = \tfrac{\sqrt{2}}{2} \qquad \sin\tfrac{\pi}{3} = \tfrac{\sqrt{3}}{2}$$

$$\cos\tfrac{\pi}{6} = \tfrac{\sqrt{3}}{2} \qquad \cos\tfrac{\pi}{4} = \tfrac{\sqrt{2}}{2} \qquad \cos\tfrac{\pi}{3} = \tfrac{1}{2}$$

$$\tan\tfrac{\pi}{6} = \tfrac{\sqrt{3}}{3} \qquad \tan\tfrac{\pi}{4} = 1 \qquad \tan\tfrac{\pi}{3} = \sqrt{3}$$

5. (a) What does it mean to solve a triangle?

To solve a triangle means to find all three angles and all three sides.

(b) Solve the triangle shown.

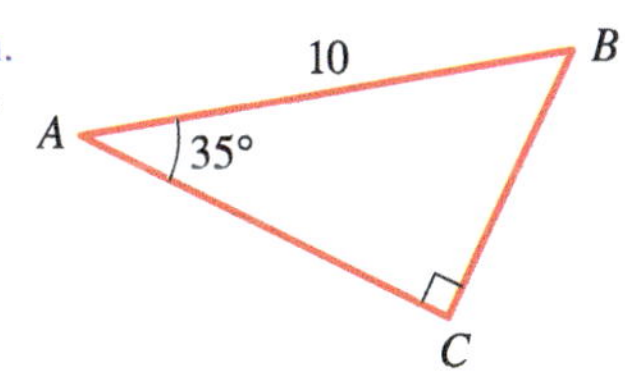

$\angle B = 90° - 35° = 55°$

$a = 10 \sin 35° \approx 5.74$

$b = 10 \cos 35° \approx 8.19$

6. (a) Let θ be an angle in standard position, let $P(x, y)$ be a point on the terminal side, and let r be the distance from

(continued)

the origin to P, as shown in the figure. Write expressions for the six trigonometric functions of θ.

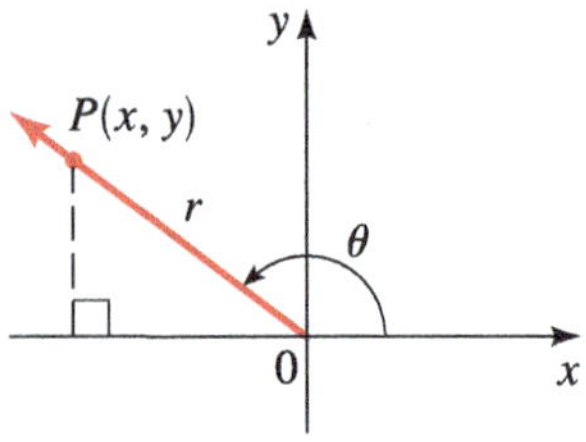

$\sin\theta = \frac{y}{r}$ $\quad\csc\theta = \frac{r}{y}$

$\cos\theta = \frac{x}{r}$ $\quad\sec\theta = \frac{r}{x}$

$\tan\theta = \frac{y}{x}$ $\quad\cot\theta = \frac{x}{y}$

(b) Find the sine, cosine, and tangent for the angle θ shown in the figure.

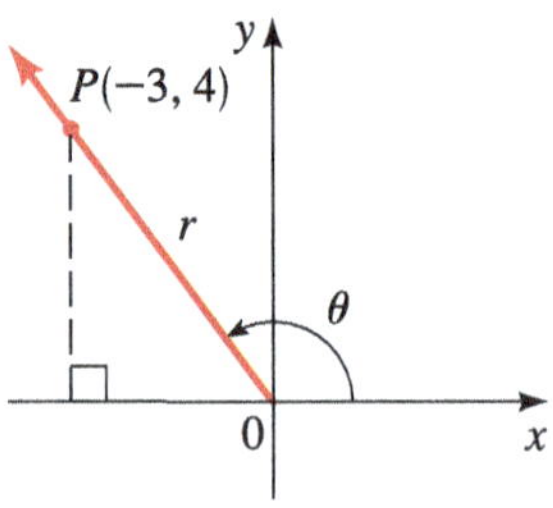

Here $x = -3$, $y = 4$, and $r = \sqrt{(-3)^2 + 4^2} = 5$. So $\sin\theta = \frac{4}{5}$, $\cos\theta = \frac{-3}{5}$, and $\tan\theta = \frac{4}{-3}$.

7. In each of the four quadrants, identify the trigonometric functions that are positive.

In Quadrant I all the trigonometric functions are positive; in Quadrant II the sine and cosecant functions are positive; in Quadrant III the tangent and cotangent functions are positive; and in Quadrant IV the cosine and secant functions are positive.

8. (a) Describe the steps we use to find the value of a trigonometric function of an angle θ.

We find the reference angle for θ, the quadrant where the terminal side lies, and the sign of the function in that quadrant, and we use all these to find the value of the function at θ.

(b) Find $\sin 5\pi/6$.

The terminal side of the angle $\frac{5\pi}{6}$ is in Quadrant II, and the reference angle is $\pi - \frac{5\pi}{6} = \frac{\pi}{6}$. Since sine is positive in Quadrant II, $\sin\frac{5\pi}{6} = \sin\frac{\pi}{6} = \frac{1}{2}$.

9. (a) State the reciprocal identities.

$$\csc\theta = \frac{1}{\sin\theta} \qquad \sec\theta = \frac{1}{\cos\theta} \qquad \cot\theta = \frac{1}{\tan\theta}$$

(b) State the Pythagorean identities.

$$\sin^2\theta + \cos^2\theta = 1 \qquad \tan^2\theta + 1 = \sec^2\theta \qquad 1 + \cot^2\theta = \csc^2\theta$$

10. (a) What is the area of a triangle with sides of length a and b and with included angle θ?

The area is $\mathcal{A} = \dfrac{1}{2}ab\sin\theta$.

(b) What is the area of a triangle with sides of length a, b, and c?

The area is given by Heron's Formula

$$\mathcal{A} = \sqrt{s(s-a)(s-b)(s-c)}$$

where $s = \frac{1}{2}(a + b + c)$ is the semiperimeter.

11. (a) Define the inverse sine function, the inverse cosine function, and the inverse tangent function.

$$\sin^{-1}x = y \quad\Leftrightarrow\quad \sin y = x$$
$$\cos^{-1}x = y \quad\Leftrightarrow\quad \cos y = x$$
$$\tan^{-1}x = y \quad\Leftrightarrow\quad \tan y = x$$

(b) Find $\sin^{-1}\frac{1}{2}$, $\cos^{-1}(\sqrt{2}/2)$, and $\tan^{-1}1$.

From 2(c) and the definitions in part (a) we get

$$\sin^{-1}\tfrac{1}{2} = \tfrac{\pi}{6} \qquad \cos^{-1}\tfrac{\sqrt{2}}{2} = \tfrac{\pi}{4} \qquad \tan^{-1}1 = \tfrac{\pi}{4}$$

(c) For what values of x is the equation $\sin(\sin^{-1}x) = x$ true? For what values of x is the equation $\sin^{-1}(\sin x) = x$ true?

$$\sin(\sin^{-1}x) = x \quad\text{for}\quad -1 \le x \le 1$$
$$\sin^{-1}(\sin x) = x \quad\text{for}\quad -\tfrac{\pi}{2} \le x \le \tfrac{\pi}{2}$$

12. (a) State the Law of Sines.

In triangle ABC we have $\dfrac{\sin A}{a} = \dfrac{\sin B}{b} = \dfrac{\sin C}{c}$.

(b) Find side a in the figure.

Note that $\angle C = 180° - (85° + 40°) = 55°$.

By the Law of Sines

$$\frac{\sin 85°}{a} = \frac{\sin 55°}{100}, \text{ so}$$

$$a = \frac{100\sin 85°}{\sin 55°} \approx 121.6.$$

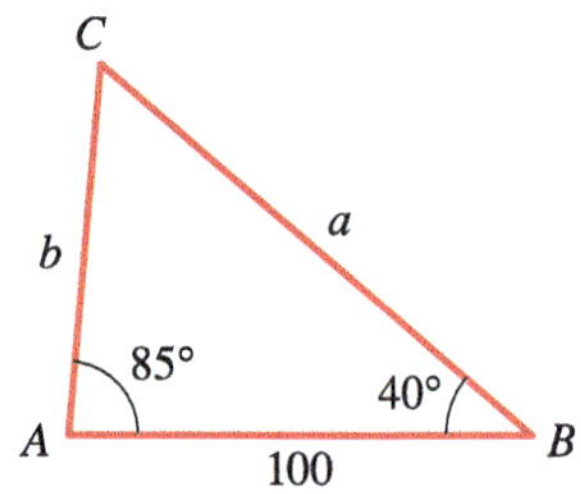

(c) Explain the ambiguous case in the Law of Sines.

In the case SSA there may be two triangles, one triangle, or no triangle with the given sides and angles.

13. (a) State the Law of Cosines.

In triangle ABC we have

$$a^2 = b^2 + c^2 - 2bc\cos A$$
$$b^2 = a^2 + c^2 - 2ac\cos B$$
$$c^2 = a^2 + b^2 - 2ab\cos C$$

(b) Find side a in the figure.

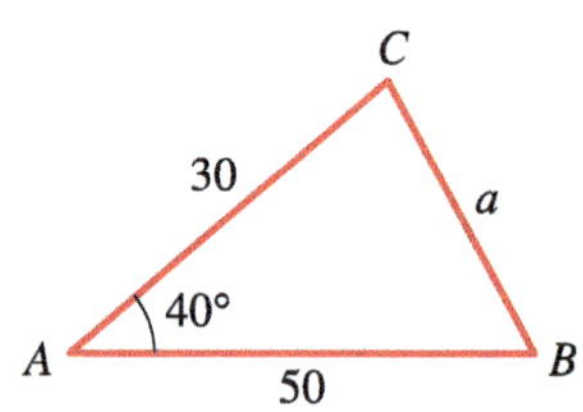

By the Law of Cosines we have

$$\begin{aligned} a &= \sqrt{b^2 + c^2 - 2bc\cos A} \\ &= \sqrt{50^2 + 30^2 - 2(50)(30)\cos 40°} \\ &\approx 33.2 \end{aligned}$$

CHAPTER 7 Review: Concept Check Answers

1. What is an identity? What is a trigonometric identity?

An identity is an equation that is true for all values of the variable(s). A trigonometric identity is an identity that involves trigonometric functions.

2. (a) State the Pythagorean identities.

$$\sin^2 x + \cos^2 x = 1$$
$$\tan^2 x + 1 = \sec^2 x$$
$$1 + \cot^2 x = \csc^2 x$$

(b) Use a Pythagorean identity to express cosine in terms of sine.

By the first Pythagorean identity we have $\cos x = \pm\sqrt{1 - \sin^2 x}$

3. (a) State the reciprocal identities for cosecant, secant, and cotangent.

$$\csc x = \frac{1}{\sin x} \qquad \sec x = \frac{1}{\cos x} \qquad \cot x = \frac{1}{\tan x}$$

(b) State the even-odd identities for sine and cosine.

$$\sin(-x) = -\sin x \qquad \cos(-x) = \cos x$$

(c) State the cofunction identities for sine, tangent, and secant.

$$\sin\left(\tfrac{\pi}{2} - x\right) = \cos x \qquad \tan\left(\tfrac{\pi}{2} - x\right) = \cot x \qquad \sec\left(\tfrac{\pi}{2} - x\right) = \csc x$$

(d) Suppose that $\cos(-x) = 0.4$; use the identities in parts (a) and (b) to find $\sec x$.

$$\sec x = \frac{1}{\cos x} = \frac{1}{\cos(-x)} = \frac{1}{0.4} = 2.5$$

(e) Suppose that $\sin 10° = a$; use the identities in part (c) to find $\cos 80°$.

Since 10° and 80° are complementary angles, we have $\cos 80° = \sin 10° = a$.

4. (a) How do you prove an identity?

Start with one side of the equation, and then use known identities to transform it to the other side.

(b) Prove the identity $\sin x(\csc x - \sin x) = \cos^2 x$

$$\begin{aligned} \text{LHS} &= \sin x(\csc x - \sin x) & \\ &= \sin x\left(\frac{1}{\sin x} - \sin x\right) & \text{Reciprocal identity} \\ &= 1 - \sin^2 x & \text{Distributive Property} \\ &= \cos^2 x = \text{RHS} & \text{Pythagorean identity} \end{aligned}$$

5. (a) State the Addition and Subtraction Formulas for Sine and Cosine.

$$\sin(s + t) = \sin s \cos t + \cos s \sin t$$
$$\cos(s + t) = \cos s \cos t - \sin s \sin t$$

(b) Use a formula from part (a) to find $\sin 75°$.

$$\begin{aligned} \sin 75° &= \sin(45° + 30°) \\ &= \sin 45° \cos 30° + \cos 45° \sin 30° \\ &= \frac{\sqrt{2}}{2}\frac{\sqrt{3}}{2} + \frac{\sqrt{2}}{2}\frac{1}{2} = \frac{\sqrt{6} + \sqrt{2}}{4} \end{aligned}$$

6. (a) State the formula for $A \sin x + B \cos x$.

Let $k = \sqrt{A^2 + B^2}$; then

$$A \sin x + B \cos x = k \sin(x + \phi)$$

where ϕ satisfies $\cos\phi = A/\sqrt{A^2 + B^2}$ and $\sin\phi = B/\sqrt{A^2 + B^2}$.

(b) Express $3 \sin x + 4 \cos x$ as a function of sine only.

We have $k = \sqrt{3^2 + 4^2} = 5$. The angle ϕ satisfies $\cos\phi = \frac{3}{5}$ and $\sin\phi = \frac{4}{5}$, so ϕ is in Quadrant I. We find $\phi = \sin^{-1}\left(\frac{4}{5}\right) \approx 53.1°$. Thus

$$3 \sin x + 4 \cos x = 5 \sin(x + 53.1°)$$

7. (a) State the Double-Angle Formula for Sine and the Double-Angle Formulas for Cosine.

$$\begin{aligned} \sin 2x &= 2 \sin x \cos x \\ \cos 2x &= \cos^2 x - \sin^2 x \\ &= 1 - 2\sin^2 x \\ &= 2\cos^2 x - 1 \end{aligned}$$

(b) Prove the identity $\sec x \sin 2x = 2 \sin x$.

$$\begin{aligned} \text{LHS} &= \sec x \sin 2x & \\ &= \sec x(2 \sin x \cos x) & \text{Double-Angle Formula} \\ &= \frac{1}{\cos x}(2 \sin x \cos x) & \text{Reciprocal identity} \\ &= 2 \sin x = \text{RHS} & \text{Pythagorean identity} \end{aligned}$$

8. (a) State the formulas for lowering powers of sine and cosine.

$$\sin^2 x = \frac{1 - \cos 2x}{2} \qquad \cos^2 x = \frac{1 + \cos 2x}{2}$$

(b) Prove the identity $4 \sin^2 x \cos^2 x = \sin^2 2x$.

$$\begin{aligned} \text{LHS} &= 4 \sin^2 x \cos^2 x & \\ &= 4\left(\frac{1 - \cos 2x}{2}\right)\left(\frac{1 + \cos 2x}{2}\right) & \text{Lower powers} \\ &= 1 - \cos^2 2x & \text{Simplify} \\ &= \sin^2 2x = \text{RHS} & \text{Pythagorean identity} \end{aligned}$$

9. (a) State the Half-Angle Formulas for Sine and Cosine.

$$\sin\frac{u}{2} = \pm\sqrt{\frac{1 - \cos u}{2}} \qquad \cos\frac{u}{2} = \pm\sqrt{\frac{1 + \cos u}{2}}$$

(b) Find $\cos 15°$.

$$\begin{aligned} \cos 15° &= \cos\left(\frac{30°}{2}\right) \\ &= \pm\sqrt{\frac{1 + \cos 30°}{2}} = \pm\sqrt{\frac{1 + \sqrt{3}/2}{2}} \\ &= \pm\sqrt{\frac{2 + \sqrt{3}}{4}} = \pm\frac{1}{2}\sqrt{2 + \sqrt{3}} \end{aligned}$$

Since 15° is in Quadrant I and since cosine is positive in Quadrant I, we conclude that $\cos 15° = \frac{1}{2}\sqrt{2 + \sqrt{3}}$.

(continued)

10. (a) State the Product-to-Sum Formula for the product $\sin u \cos v$.

$$\sin u \cos v = \tfrac{1}{2}[\sin(u + v) + \sin(u - v)]$$

(b) Express $\sin 5x \cos 3x$ as a sum of trigonometric functions.

By the formula in part (a) we have

$$\begin{aligned}\sin 5x \cos 3x &= \tfrac{1}{2}[\sin(5x + 3x) + \sin(5x - 3x)]\\ &= \tfrac{1}{2}\sin 8x + \tfrac{1}{2}\sin 2x\end{aligned}$$

11. (a) State the Sum-to-Product Formula for the sum $\sin x + \sin y$.

$$\sin x + \sin y = 2\sin\frac{x+y}{2}\cos\frac{x-y}{2}$$

(b) Express $\sin 5x + \sin 7x$ as a product of trigonometric functions.

By the formula in part (a) we have

$$\begin{aligned}\sin 5x + \sin 7x &= 2\sin\frac{5x+7x}{2}\cos\frac{5x-7x}{2}\\ &= 2\sin 6x\cos(-x)\\ &= 2\sin 6x\cos x\end{aligned}$$

12. What is a trigonometric equation? How do we solve a trigonometric equation?

A trigonometric equation is an equation involving trigonometric functions. To solve a trigonometric equation, we first find all solutions for one period of the function involved and then add integer multiples of the period to obtain all solutions.

(a) Solve the equation $\cos x = \frac{1}{2}$.

The solutions of this equation in the interval $[0, 2\pi)$ are

$$x = \frac{\pi}{3} \qquad \text{and} \qquad x = \frac{5\pi}{3}$$

To obtain all solutions, we add multiples of 2π (because $\cos x$ is periodic with period 2π). The solutions are

$$x = \frac{\pi}{3} + 2k\pi \qquad \text{and} \qquad x = \frac{5\pi}{3} + 2k\pi$$

where k is any integer.

(b) Solve the equation $2\sin x\cos x = \frac{1}{2}$.

First we use a double-angle formula to express the left-hand side as a single trigonometric function.

$$2\sin x\cos x = \tfrac{1}{2} \qquad \text{Given equation}$$

$$\sin 2x = \tfrac{1}{2} \qquad \text{Double-Angle Formula}$$

The solutions of this equation in the interval $[0, 2\pi)$ are

$$2x = \frac{\pi}{6} \qquad \text{and} \qquad 2x = \frac{5\pi}{6}$$

To obtain all solutions, we add multiples of 2π. The solutions are

$$2x = \frac{\pi}{6} + 2k\pi \qquad \text{and} \qquad 2x = \frac{5\pi}{6} + 2k\pi$$

and dividing by 2, we get the solutions

$$x = \frac{\pi}{12} + k\pi \qquad \text{and} \qquad x = \frac{5\pi}{12} + k\pi$$

where k is any integer.

Cut here and keep for reference

CHAPTER 8 Review: Concept Check Answers

1. (a) Explain the polar coordinate system.

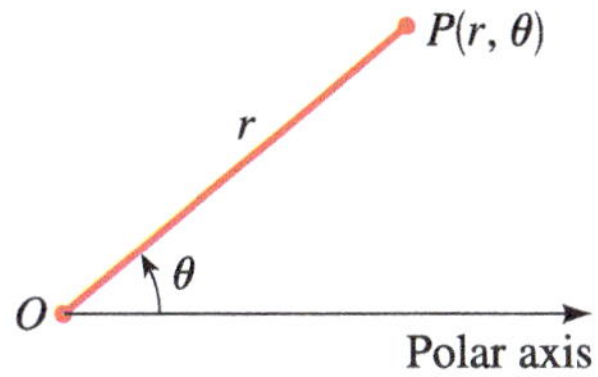

In the **polar coordinate** system the location of a point P in the plane is determined by an ordered pair (r, θ), where r is the distance from the pole O to P and θ is the angle formed by the polar axis and the ray $\overrightarrow{OP}$, as shown in the figure.

(b) Graph the points with polar coordinates $(2, \pi/3)$ and $(-1, 3\pi/4)$.

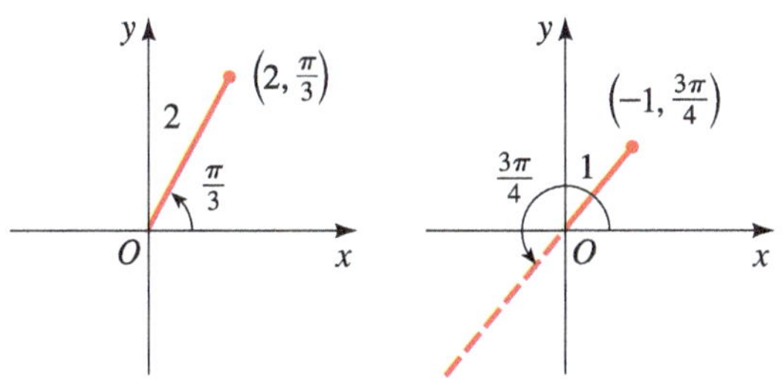

(c) State the equations that relate the rectangular coordinates of a point to its polar coordinates.

To change from polar to rectangular:

$$x = r\cos\theta \qquad \text{and} \qquad y = r\sin\theta$$

To change from rectangular to polar:

$$r^2 = x^2 + y^2 \qquad \text{and} \qquad \tan\theta = \frac{y}{x}$$

(d) Find rectangular coordinates for $(2, \pi/3)$.

$$x = 2\cos\frac{\pi}{3} = 1 \qquad \text{and} \qquad y = 2\sin\frac{\pi}{3} = \sqrt{3}$$

So in rectangular coordinates the point is $(1, \sqrt{3})$.

(e) Find polar coordinates for $P(-2, 2)$.

$r^2 = (-2)^2 + 2^2 = 8$, so $r = \sqrt{8} = 2\sqrt{2}$. $\tan\theta = 2/(-2) = -1$ and P is in Quadrant II, so $\theta = 3\pi/4$. So in polar coordinates the point is $(2\sqrt{2}, 3\pi/4)$.

2. (a) What is a polar equation?

A polar equation is an equation in the variables r and θ, where these variables are the polar coordinates of the point (r, θ).

(b) Convert the polar equation $r = \sin\theta$ to an equivalent rectangular equation.

$r = \sin\theta$	Polar equation
$r^2 = r\sin\theta$	Multiply by r
$x^2 + y^2 = y$	Convert

In the last step we substituted $r^2 = x^2 + y^2$ and $r\sin\theta = y$. So an equivalent rectangular equation is $x^2 + y^2 = y$.

3. (a) How do we graph a polar equation?

We plot all the points with polar coordinates (r, θ) that satisfy the equation.

(b) Sketch a graph of the polar equation $r = 4 + 4\cos\theta$. What is the graph called?

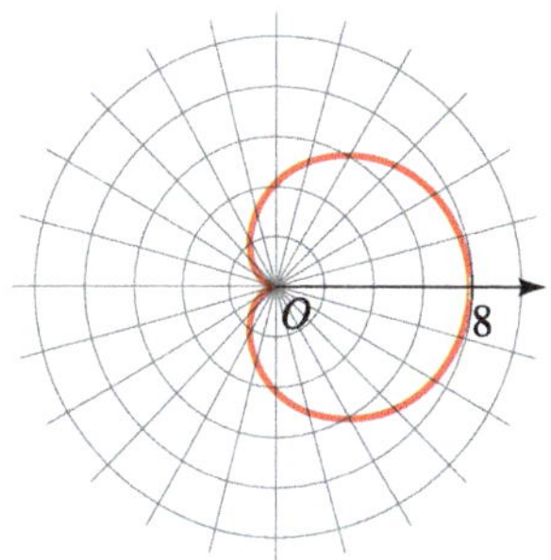

This graph is called a cardioid.

4. (a) What is the complex plane? How do we graph a complex number $z = a + bi$ in the complex plane?

The complex plane is a plane determined by two axes: the real axis and the imaginary axis. To graph the complex number $z = a + bi$, we plot the ordered pair (a, b) in this plane as shown.

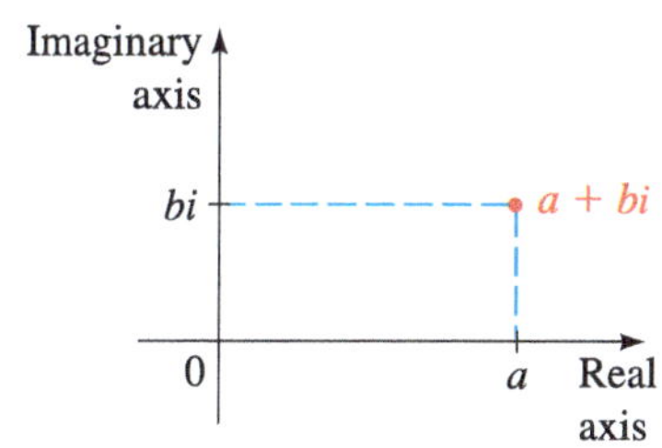

(b) What are the modulus and argument of the complex number $z = a + bi$?

The modulus of z, written $|z|$, is the distance of the point z to the origin in the complex plane. So

$$|z| = \sqrt{a^2 + b^2}$$

The argument of z is the angle θ formed by the line segment connecting the origin to the point z and the positive real axis. So $\tan\theta = b/a$.

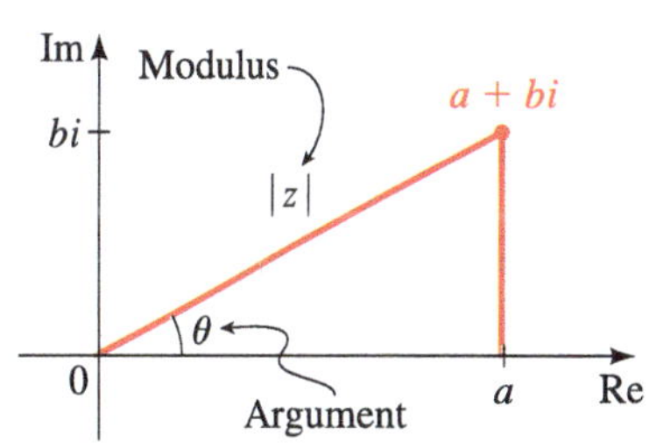

(continued)

(c) Graph the point $z = \sqrt{3} - i$, and find the modulus and argument of z.

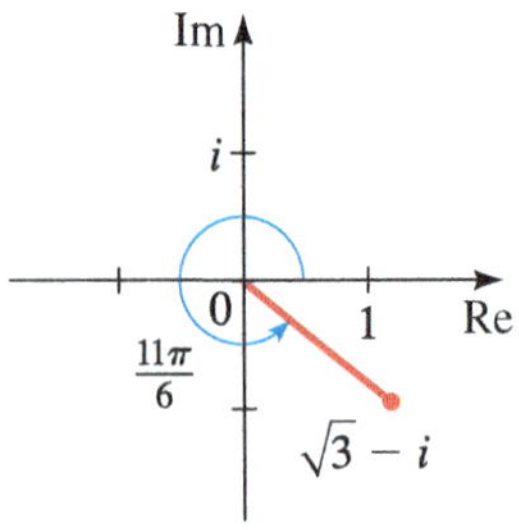

Modulus: $|z| = \sqrt{(\sqrt{3})^2 + 1^2} = \sqrt{4} = 2$

Argument: $\theta = \dfrac{11\pi}{6}$

5. (a) How do we express the complex number z in polar form?

The polar form is $z = r(\cos\theta + i\sin\theta)$, where r is the modulus and θ is the argument of z.

(b) Express $z = \sqrt{3} - i$ in polar form.

Using the moduli and arguments from Question 4(c) above, we get

$$z = 2\left(\cos\frac{11\pi}{6} + i\sin\frac{11\pi}{6}\right)$$

6. Let $\quad z_1 = 2\left(\cos\dfrac{\pi}{3} + i\sin\dfrac{\pi}{3}\right)$

and $\quad z_2 = 5\left(\cos\dfrac{\pi}{4} + i\sin\dfrac{\pi}{4}\right)$

(a) Find the product z_1z_2.

To find the product z_1z_2, we multiply the moduli and add the arguments, so

$$z_1z_2 = 10\left[\cos\left(\frac{\pi}{3} + \frac{\pi}{4}\right) + i\sin\left(\frac{\pi}{3} + \frac{\pi}{4}\right)\right]$$
$$= 10\left(\cos\frac{7\pi}{12} + i\sin\frac{7\pi}{12}\right)$$

(b) Find the quotient z_1/z_2.

To find the quotient z_1/z_2, we divide the moduli and subtract the arguments, so

$$\frac{z_1}{z_2} = \frac{2}{5}\left[\cos\left(\frac{\pi}{3} - \frac{\pi}{4}\right) + i\sin\left(\frac{\pi}{3} - \frac{\pi}{4}\right)\right]$$
$$= \frac{2}{5}\left(\cos\frac{\pi}{12} + i\sin\frac{\pi}{12}\right)$$

7. (a) State De Moivre's Theorem.

If $z = r(\cos\theta + i\sin\theta)$ then

$$z^n = r^n(\cos n\theta + i\sin n\theta)$$

(b) Use De Moivre's Theorem to find the fifth power of $z = 2\left(\cos\dfrac{\pi}{3} + i\sin\dfrac{\pi}{3}\right)$.

$$z^5 = 32\left(\cos\frac{5\pi}{3} + i\sin\frac{5\pi}{3}\right)$$

8. (a) State the formula for the nth roots of a complex number $z = r(\cos\theta + i\sin\theta)$.

The n nth roots are

$$w_k = r^{1/n}\left[\cos\left(\frac{\theta + 2k\pi}{n}\right) + i\sin\left(\frac{\theta + 2k\pi}{n}\right)\right]$$

for $k = 0, 1, 2, \ldots, n - 1$.

(b) How do we find the nth roots of a complex number?

We use the following guidelines.

1. The modulus of each nth root is $r^{1/n}$.
2. The argument of the first root is θ/n.
3. Add $2\pi/n$ to get the argument of each successive root.

(c) Find the three third roots of $z = -8$.

First we express z in polar form:

$$z = 8(\cos\pi + i\sin\pi)$$

So the modulus of each root is $8^{1/3} = 2$. The argument of the first root is $\pi/3$. We add $2\pi/3$ to get the argument of each successive root. So the three roots are

$$w_0 = 2\left[\cos\left(\frac{\pi}{3}\right) + i\sin\left(\frac{\pi}{3}\right)\right] = 1 + \sqrt{3}i$$
$$w_1 = 2\left[\cos\left(\frac{\pi + 2\pi}{3}\right) + i\sin\left(\frac{\pi + 2\pi}{3}\right)\right] = -2$$
$$w_2 = 2\left[\cos\left(\frac{\pi + 4\pi}{3}\right) + i\sin\left(\frac{\pi + 4\pi}{3}\right)\right] = 1 - \sqrt{3}i$$

9. (a) What are parametric equations?

Parametric equations are equations of the form

$$x = f(t) \qquad y = g(t)$$

where f and g are functions of the parameter t.

(b) Sketch a graph of the following parametric equations, using arrows to indicate the direction of the curve.

$$x = t + 1 \qquad y = t^2 \qquad -2 \le t \le 2$$

t	(x, y)
-2	$(-1, 4)$
-1	$(0, 1)$
0	$(0, 0)$
1	$(2, 1)$
2	$(3, 4)$

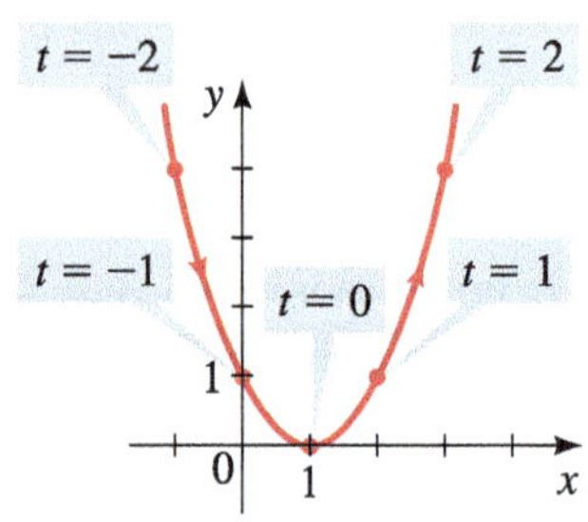

(c) Eliminate the parameter to obtain an equation in x and y.

From the first equation $t = x - 1$, so from the second equation we have $y = (x - 1)^2$.

CHAPTER 9 Review: Concept Check Answers

1. (a) What is a vector in the plane? How do we represent a vector in the coordinate plane?

A vector is a quantity that has both length (or magnitude) and direction. A vector $\mathbf{v}$ in the coordinate plane is expressed in terms of components as

$$\mathbf{v} = \langle a_1, a_2 \rangle$$

where a_1 is the horizontal component and a_2 is the vertical component.

(b) Find the vector with initial point $(2, 3)$ and terminal point $(4, 10)$.

$$\mathbf{v} = \langle 4 - 2, 10 - 3 \rangle = \langle 2, 7 \rangle$$

(c) Let $\mathbf{v} = \langle 2, 1 \rangle$. If the initial point of $\mathbf{v}$ is placed at $P(1, 1)$, where is its terminal point? Sketch several representations of $\mathbf{v}$.

The terminal point is $Q(1 + 2, 1 + 1) = Q(3, 2)$.

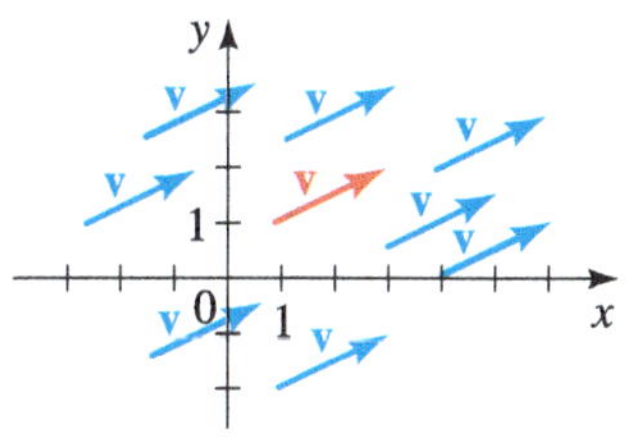

(d) How is the magnitude of $\mathbf{v} = \langle a_1, a_2 \rangle$ defined? Find the magnitude of $\mathbf{w} = \langle 3, 4 \rangle$.

The magnitude of $\mathbf{v}$ is $|\mathbf{v}| = \sqrt{a_1^2 + a_2^2}$. We have $|\mathbf{w}| = \sqrt{3^2 + 4^2} = 5$.

(e) What are the vectors $\mathbf{i}$ and $\mathbf{j}$? Express the vector $\mathbf{v} = \langle 5, 9 \rangle$ in terms of $\mathbf{i}$ and $\mathbf{j}$.

The vector $\mathbf{i} = \langle 1, 0 \rangle$ and $\mathbf{j} = \langle 0, 1 \rangle$. So $\mathbf{v} = 5\mathbf{i} + 9\mathbf{j}$.

(f) Let $\mathbf{v} = \langle a_1, a_2 \rangle$ be a vector in the coordinate plane. What is meant by the direction θ of $\mathbf{v}$? What are the coordinates of $\mathbf{v}$ in terms of its length and direction? Sketch a figure to illustrate your answer.

The **direction** of $\mathbf{v}$ is the smallest positive angle θ in standard position formed by the positive x-axis and $\mathbf{v}$. So $\mathbf{v} = |\mathbf{v}| \cos\theta\, \mathbf{i} + |\mathbf{v}| \sin\theta\, \mathbf{j}$.

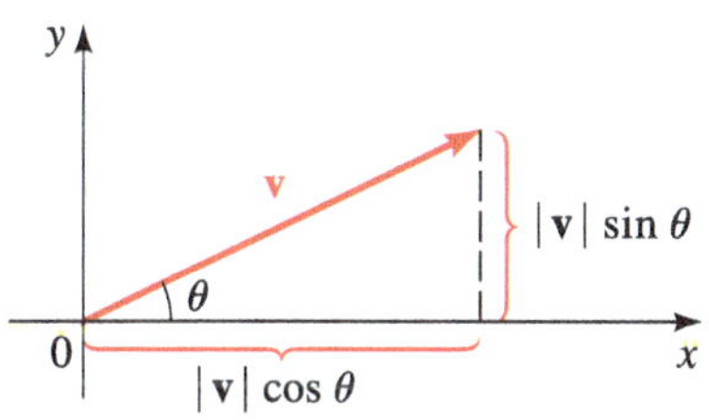

(g) Suppose that $\mathbf{v}$ has length $|\mathbf{v}| = 5$ and direction $\theta = \pi/6$. Express $\mathbf{v}$ in terms of its coordinates.

$$\mathbf{v} = 5\cos\frac{\pi}{6}\mathbf{i} + 5\sin\frac{\pi}{6}\mathbf{i} = \frac{5\sqrt{3}}{2}\mathbf{i} + \frac{5}{2}\mathbf{j}$$

2. (a) Define addition and scalar multiplication for vectors.

Let $\mathbf{u} = \langle a_1, a_2 \rangle$ and $\mathbf{v} = \langle b_1, b_2 \rangle$ and let $c \in \mathbb{R}$. Then $\mathbf{u} + \mathbf{v} = \langle a_1 + b_1, a_2 + b_2 \rangle$, and $c\mathbf{u} = \langle ca_1, ca_2 \rangle$.

(b) If $\mathbf{u} = \langle 2, 3 \rangle$ and $\mathbf{v} = \langle 5, 9 \rangle$, find $\mathbf{u} + \mathbf{v}$ and $4\mathbf{u}$.

$\mathbf{u} + \mathbf{v} = \langle 2 + 5, 3 + 9 \rangle = \langle 7, 12 \rangle$, and
$4\mathbf{u} = \langle 4 \cdot 2, 4 \cdot 3 \rangle = \langle 8, 12 \rangle$.

3. (a) Define the dot product of the vectors $\mathbf{u} = \langle a_1, a_2 \rangle$ and $\mathbf{v} = \langle b_1, b_2 \rangle$, and state the formula for the angle θ between $\mathbf{u}$ and $\mathbf{v}$.

The dot product is $\mathbf{u} \cdot \mathbf{v} = a_1b_1 + a_2b_2$. The angle θ satisfies $\cos\theta = \dfrac{\mathbf{u} \cdot \mathbf{v}}{|\mathbf{u}||\mathbf{v}|}$.

(b) If $\mathbf{u} = \langle 2, 3 \rangle$ and $\mathbf{v} = \langle 1, 4 \rangle$, find $\mathbf{u} \cdot \mathbf{v}$ and find the angle between $\mathbf{u}$ and $\mathbf{v}$.

The dot product is $\mathbf{u} \cdot \mathbf{v} = 2 \cdot 1 + 3 \cdot 4 = 14$. The angle θ between $\mathbf{u}$ and $\mathbf{v}$ satisfies

$$\cos\theta = \frac{14}{\sqrt{13}\sqrt{17}} \approx 0.942$$

So $\theta = \cos^{-1}(0.942) \approx 19.7°$.

4. (a) Describe the three-dimensional coordinate system. What are the coordinate planes?

The coordinate system consists of a point O, called the origin, and three mutually perpendicular lines called the coordinate axes, labeled as the x-, y-, and z-axes. The point $P(a, b, c)$ is plotted in this system as shown.

The coordinate planes are the xy-plane, the xz-plane, and the yz-plane as shown.

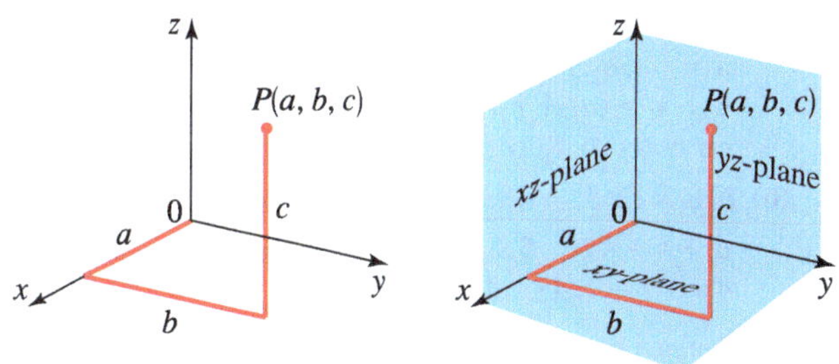

(b) What is the distance from the point $(3, -2, 5)$ to each of the coordinate planes?

The distance to the xy-plane is 5, to the xz-plane is 2, and to the yz-plane is 3.

(c) State the formula for the distance between the points $P(x_1, y_1, z_1)$ and $Q(x_2, y_2, z_2)$.

$$d(P, Q) = \sqrt{(x_2 - x_1)^2 + (y_2 - y_1)^2 + (z_2 - z_1)^2}$$

(d) Find the distance between the points $P(1, 2, 3)$ and $Q(3, -1, 4)$.

$$d(P, Q) = \sqrt{(3 - 1)^2 + (-1 - 2)^2 + (4 - 3)^2} = \sqrt{14}$$

(continued)

(e) State the equation of a sphere with center $C(h, k, l)$ and radius r.

$$(x - h)^2 + (y - k)^2 + (z - l)^2 = r^2$$

(f) Find an equation for the sphere of radius 5 centered at the point $(1, 2, -3)$.

$$(x - 1)^2 + (y - 2)^2 + (z + 3)^2 = 5^2$$

5. (a) What is a vector in space? How do we represent a vector in a three-dimensional coordinate system?

A vector is a quantity that has both length (or magnitude) and direction. A vector $\mathbf{v}$ in a three-dimensional coordinate system is expressed in terms of components as

$$\mathbf{v} = \langle a_1, a_2, a_3 \rangle$$

(b) Find the vector with initial point $(2, 3, -1)$ and terminal point $(4, 10, 5)$.

$$\mathbf{v} = \langle 4 - 2, 10 - 3, 5 - (-1) \rangle = \langle 2, 7, 6 \rangle$$

(c) How is the magnitude of $\mathbf{v} = \langle a_1, a_2, a_3 \rangle$ defined? Find the magnitude of $\mathbf{w} = \langle 3, 4, 1 \rangle$.

The magnitude of $\mathbf{v}$ is $|\mathbf{v}| = \sqrt{a_1^2 + a_2^2 + a_3^2}$.

We have $|\mathbf{w}| = \sqrt{3^2 + 4^2 + 1^2} = \sqrt{26}$.

(d) What are the vectors $\mathbf{i}$, $\mathbf{j}$, and $\mathbf{k}$? Express the vector $\mathbf{v} = \langle 5, 9, -1 \rangle$ in terms of $\mathbf{i}$, $\mathbf{j}$, and $\mathbf{k}$.

$$\mathbf{i} = \langle 1, 0, 0 \rangle$$

$$\mathbf{j} = \langle 0, 1, 0 \rangle$$

$$\mathbf{k} = \langle 0, 0, 1 \rangle$$

So $\mathbf{v} = 5\mathbf{i} + 9\mathbf{j} - \mathbf{k}$.

6. (a) Define addition and scalar multiplication for vectors.

$\mathbf{u} = \langle a_1, a_2, a_3 \rangle$ and $\mathbf{v} = \langle b_1, b_2, b_3 \rangle$, and let $c \in \mathbb{R}$. Then $\mathbf{u} + \mathbf{v} = \langle a_1 + b_1, a_2 + b_2, a_3 + b_3 \rangle$, and $c\mathbf{u} = \langle ca_1, ca_2, ca_3 \rangle$.

(b) If $\mathbf{u} = \langle 2, 3, -1 \rangle$ and $\mathbf{v} = \langle 5, 9, 2 \rangle$, find $\mathbf{u} + \mathbf{v}$ and $4\mathbf{u}$.

$\mathbf{u} + \mathbf{v} = \langle 2 + 5, 3 + 9, -1 + 2 \rangle = \langle 7, 12, 1 \rangle$ and $4\mathbf{u} = \langle 4 \cdot 2, 4 \cdot 3, 4(-1) \rangle = \langle 8, 12, -4 \rangle$.

7. (a) Define the dot product of the vectors $\mathbf{u} = \langle a_1, a_2, a_3 \rangle$ and $\mathbf{v} = \langle b_1, b_2, b_3 \rangle$, and state the formula for the angle θ between $\mathbf{u}$ and $\mathbf{v}$.

The dot product is $\mathbf{u} \cdot \mathbf{v} = a_1b_1 + a_2b_2 + a_3b_3$. The angle θ satisfies $\cos\theta = \dfrac{\mathbf{u} \cdot \mathbf{v}}{|\mathbf{u}||\mathbf{v}|}$.

(b) If $\mathbf{u} = \langle 2, 3, -1 \rangle$ and $\mathbf{v} = \langle 1, 4, 5 \rangle$, find $\mathbf{u} \cdot \mathbf{v}$.

$$\mathbf{u} \cdot \mathbf{v} = 2 \cdot 1 + 3 \cdot 4 + (-1) \cdot 5 = 9$$

8. (a) Define the cross product of the vectors $\mathbf{u} = \langle a_1, a_2, a_3 \rangle$ and $\mathbf{v} = \langle b_1, b_2, b_3 \rangle$.

$$\mathbf{u} \times \mathbf{v} = \begin{vmatrix} \mathbf{i} & \mathbf{j} & \mathbf{k} \\ a_1 & a_2 & a_3 \\ b_1 & b_2 & b_3 \end{vmatrix}$$

$$= (a_2b_3 - a_3b_2)\mathbf{i} - (a_1b_3 - a_3b_1)\mathbf{j} + (a_1b_2 - a_2b_1)\mathbf{k}$$

(b) *True* or *False*? The vector $\mathbf{u} \times \mathbf{v}$ is perpendicular to both $\mathbf{u}$ and $\mathbf{v}$.

True.

(c) Let $\mathbf{u}$ and $\mathbf{v}$ be vectors in space. State the formula that relates the magnitude of $\mathbf{u} \times \mathbf{v}$ and the angle θ between $\mathbf{u}$ and $\mathbf{v}$.

$$|\mathbf{u} \times \mathbf{v}| = |\mathbf{u}||\mathbf{v}|\sin\theta$$

(d) How can we use the cross product to determine whether two vectors are parallel?

The vectors $\mathbf{u}$ and $\mathbf{v}$ are parallel if and only if $\mathbf{u} \times \mathbf{v} = \mathbf{0}$.

9. (a) What are the two properties that determine a line in space? Give parametric equations for a line in space.

A line is determined by a point (x_0, y_0, z_0) on the line and a vector $\mathbf{v} = \langle a, b, c \rangle$ parallel to the line. Parametric equations for the line are $x = x_0 + at$, $y = y_0 + bt$, and $z = z_0 + ct$.

(b) Find parametric equations for the line through the point $(-2, 4, 1)$ and parallel to the vector $\mathbf{v} = \langle 7, 5, 3 \rangle$.

$$x = -2 + 7t \qquad y = 4 + 5t \qquad z = 1 + 3t$$

10. (a) What are the two properties that determine a plane in space? State the equation of a plane.

A plane is determined by a point (x_0, y_0, z_0) on the plane and a vector $\mathbf{n} = \langle a, b, c \rangle$ that is normal (or perpendicular) to the plane. An equation for the plane is $a(x - x_0) + b(y - y_0) + c(z - z_0) = 0$.

(b) Find an equation for the plane passing through the point $(6, -4, 3)$and with normal vector $\mathbf{n} = \langle 5, -3, 2 \rangle$.

$$5(x - 6) - 3(y + 4) + 2(z - 3) = 0$$

Cut here and keep for reference

CHAPTER 10 Review: Concept Check Answers

1. (a) What are the three methods we use to solve a system of equations?

The substitution method, the elimination method, and the graphical method

(b) Solve the system by the elimination method and by the graphical method.

$$\begin{cases} x + y = 3 \\ 3x - y = 1 \end{cases}$$

Elimination method: To eliminate y, we add the two equations to get $4x = 4$, so $x = 1$. Substituting 1 for x in the first equation, we get $1 + y = 3$, so $y = 2$. The solution is $(1, 2)$

Graphical method: We graph the two equations as shown. The point of intersection is $(1, 2)$. So the solution of the system is $(1, 2)$.

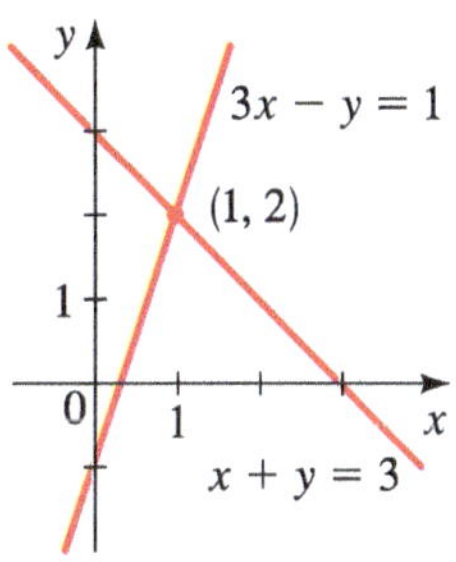

2. For a system of two linear equations in two variables:

(a) How many solutions are possible?

Such a system can have one solution, no solution, or infinitely many solutions.

(b) What is meant by an inconsistent system? a dependent system?

A system is inconsistent if it has no solution. A system is dependent if it has infinitely many solutions.

3. What operations can be performed on a linear system so as to arrive at an equivalent system?

1. Add a nonzero multiple of one equation to another.
2. Multiply an equation by a nonzero constant.
3. Interchange the position of two equations.

4. (a) Explain how Gaussian elimination works.

We use the operations in Question 3 above to obtain a system in triangular form and then use back-substitution to solve for the variables.

(b) Use Gaussian elimination to put the following system in triangular form, and then solve the system.

System

$$\begin{cases} x + y - 2z = 3 \\ x + 2y + z = 5 \\ 3x - y + 5z = 1 \end{cases}$$

Triangular form

$$\begin{cases} x + y - 2z = 3 \\ y + 3z = 2 \\ 23z = 0 \end{cases}$$

Using back-substitution, we get the solution $(1, 2, 0)$.

5. What does it mean to say that A is a matrix with dimension $m \times n$?

An $m \times n$ matrix A has m rows and n columns.

6. What is the row-echelon form of a matrix? What is a leading entry?

The first nonzero entry in a row (reading from left to right) is called a *leading entry*. A matrix is in row-echelon form if it satisfies the following:

1. The leading entry in each row is 1.
2. The leading entry in each row is to the right of the leading entry in the row above it.
3. All rows consisting entirely of 0's are at the bottom of the matrix.

7. (a) What is the augmented matrix of a system? What are leading variables?

The augmented matrix of a linear system is the matrix that contains the coefficients and the constant terms. A *leading* variable is one that corresponds to a leading entry in the augmented matrix.

(b) What are the elementary row operations on an augmented matrix?

The elementary row operations on a matrix correspond to the operations in Question 3.

1. Add a nonzero multiple of one row to another.
2. Multiply a row by a nonzero constant.
3. Interchange the position of two rows.

(c) How do we solve a system using the augmented matrix?

We perform elementary row operations to put the matrix in row-echelon form (as in Question 6). The equations that correspond to the row-echelon form can be solved using back-substitution.

(d) Write the augmented matrix of the following system of linear equations.

$$\begin{cases} x + y - 2z = 3 \\ x + 2y + z = 5 \\ 3x - y + 5z = 1 \end{cases} \qquad \begin{bmatrix} 1 & 1 & -2 & 3 \\ 1 & 2 & 1 & 5 \\ 3 & -1 & 5 & 1 \end{bmatrix}$$

(e) Solve the system in part (d).

We use elementary row operations to put the augmented matrix into row-echelon form.

$$\begin{bmatrix} 1 & 1 & -2 & 3 \\ 0 & 1 & 3 & 2 \\ 0 & 0 & 23 & 0 \end{bmatrix} \qquad \begin{cases} x + y - 2z = 3 \\ y + 3z = 2 \\ 23z = 0 \end{cases}$$

The system is solved by back-substitution as in Question 4(b). The solution is $(1, 2, 0)$.

8. Suppose you have used Gaussian elimination to transform the augmented matrix of a linear system into row-echelon form. How can you tell whether the system has exactly one solution? no solution? infinitely many solutions?

No solution: The row-echelon form has a row that represents the equation $0 = c$, where c is not zero.

Exactly one solution: Each variable in the row-echelon form is a leading variable.

Infinitely many solutions: The variables in the row-echelon form are not all leading.

(continued)

9. What is the reduced row echelon form of a matrix?

A matrix is in reduced **row-echelon form** if it is in row-echelon form and also satisfies the following:

Every number above and below each leading entry is a 0.

10. (a) How do Gaussian elimination and Gauss-Jordan elimination differ?

In each method we start with the augmented matrix of a linear system and perform row operations. In Gaussian elimination we put the matrix in row-echelon form. In Gauss-Jordan elimination we put the matrix in reduced row-echelon form.

(b) Use Gauss-Jordan elimination to solve the linear system in part 7(d).

We start with the matrix in 7(d) and continue to use row operations to obtain the following reduced-row echelon form and the corresponding system of equations:

$$\begin{bmatrix} 1 & 0 & 0 & 1 \\ 0 & 1 & 0 & 2 \\ 0 & 0 & 1 & 0 \end{bmatrix} \qquad \begin{cases} x = 1 \\ y = 2 \\ z = 0 \end{cases}$$

The solution is $(1, 2, 0)$.

11. If A and B are matrices with the same dimension and k is a real number, how do you find $A + B$ and kA?

To find $A + B$, we add corresponding entries. To find kA, we multiply each entry in A by k.

12. (a) What must be true of the dimensions of A and B for the product AB to be defined?

The number of columns of A must be the same as the number of rows of B.

(b) If A has dimension 2×3 and if B has dimension 3×2, is the product AB defined? If so, what is the dimension of AB?

The product AB is defined and has dimension 2×2.

(c) Find the matrix product.

$$\begin{bmatrix} 2 & 1 \\ 4 & 0 \end{bmatrix}\begin{bmatrix} 3 & 4 & 1 \\ 5 & 1 & 2 \end{bmatrix} = \begin{bmatrix} 11 & 9 & 4 \\ 12 & 16 & 4 \end{bmatrix}$$

13. (a) What is an identity matrix I_n? If A is an $n \times n$ matrix, what are the products AI_n and I_nA?

The identity matrix I_n is an $n \times n$ matrix with 1's on the main diagonal and 0's elsewhere: $AI_n = A$ and $I_nA = A$.

(b) If A is an $n \times n$ matrix, what is its inverse matrix?

The inverse is a matrix A^{-1} with the property that $AA^{-1} = I_n$ and $A^{-1}A = I_n$.

(c) Complete the formula for the inverse of a 2×2matrix

$$A = \begin{bmatrix} a & b \\ c & d \end{bmatrix} \qquad A^{-1} = \frac{1}{ad - bc}\begin{bmatrix} d & -b \\ -c & a \end{bmatrix}$$

(d) Find the inverse of A.

$$A = \begin{bmatrix} 1 & 1 \\ 3 & -1 \end{bmatrix} \qquad A^{-1} = \begin{bmatrix} \frac{1}{4} & \frac{1}{4} \\ \frac{3}{4} & -\frac{1}{4} \end{bmatrix}$$

14. (a) Express the system in 1(b) as a matrix equation $AX = B$.

$$\underbrace{\begin{bmatrix} 1 & 1 \\ 3 & -1 \end{bmatrix}}_{A}\underbrace{\begin{bmatrix} x \\ y \end{bmatrix}}_{X} = \underbrace{\begin{bmatrix} 3 \\ 1 \end{bmatrix}}_{B}$$

(b) If a linear system is expressed as a matrix equation $AX = B$, how do we solve the system? Solve the system in part (a).

The solution is given by the matrix $X = A^{-1}B$. We found A^{-1} in 13(d). So

$$\underbrace{\begin{bmatrix} x \\ y \end{bmatrix}}_{X} = \underbrace{\begin{bmatrix} \frac{1}{4} & \frac{1}{4} \\ \frac{3}{4} & -\frac{1}{4} \end{bmatrix}}_{A^{-1}}\underbrace{\begin{bmatrix} 3 \\ 1 \end{bmatrix}}_{B} = \begin{bmatrix} 1 \\ 2 \end{bmatrix}$$

15. (a) Is it true that the determinant det A of a matrix A is defined only if A is a square matrix?

Yes

(b) Find the determinant of the matrix A in part 13(d).

$$\det\begin{bmatrix} 1 & 1 \\ 3 & -1 \end{bmatrix} = (1)(-1) - (1)(3) = -4$$

(c) Use Cramer's Rule to solve the system in 1(b).

$$x = \frac{\begin{vmatrix} 3 & 1 \\ 1 & -1 \end{vmatrix}}{\begin{vmatrix} 1 & 1 \\ 3 & -1 \end{vmatrix}} = 1 \qquad y = \frac{\begin{vmatrix} 1 & 3 \\ 3 & 1 \end{vmatrix}}{\begin{vmatrix} 1 & 1 \\ 3 & -1 \end{vmatrix}} = 2$$

16. (a) How do we express a rational function r as a partial fraction decomposition?

We express r as a sum of fractions whose denominators consist of linear or irreducible quadratic factors.

(b) Give the form of the partial fraction decomposition.

(i) $$\frac{2x}{(x-5)(x-2)^2} = \frac{A}{x-5} + \frac{B}{x-2} + \frac{C}{(x-2)^2}$$

(ii) $$\frac{2x}{(x-5)(x^2+1)} = \frac{A}{x-5} + \frac{Bx + C}{x^2+1}$$

17. (a) How do we graph an inequality in two variables?

We first graph the corresponding equation and then use test points to determine the solution set.

(b) Graph the solution set of the inequality $x + y \geq 3$.

(c) Graph the solution set of the system of inequalities:

$$x + y \geq 3$$
$$3x - y \geq 1$$

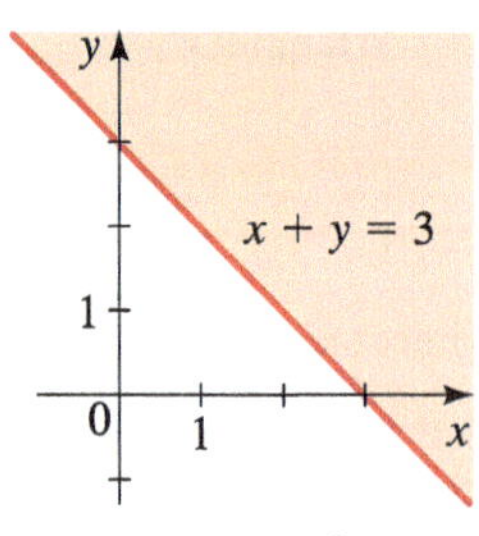

$x + y \geq 3$

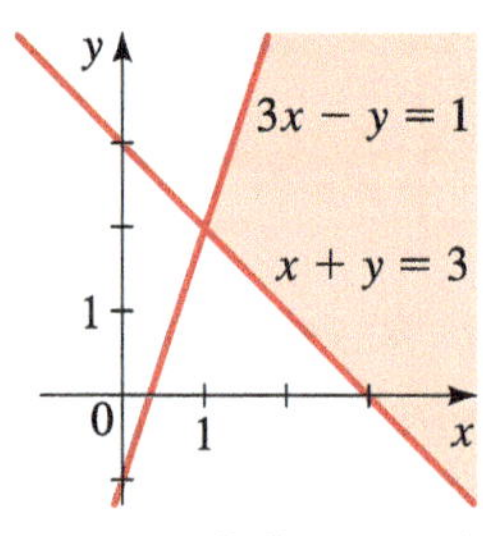

$x + y \geq 3,\ 3x - y \geq 1$

CHAPTER 11 Review: Concept Check Answers

1. (a) Give the geometric definition of a parabola.

A **parabola** is the set of points in the plane that are equidistant from a fixed point F (called the **focus**) and a fixed line l (called the **directrix**).

(b) Give the equation of a parabola with vertex at the origin and with vertical axis. Where is the focus? What is the directrix?

The equation of a parabola with vertical axis and vertex at the origin has the form

$$x^2 = 4py$$

where the focus is $F(0, p)$ and the directrix is the horizontal line $y = -p$.

(c) Graph the equation $x^2 = 8y$. Indicate the focus on the graph.

Writing the equation as $x^2 = 4(2)y$, we see that $p = 2$. So the focus is $F(0, 2)$, and the directrix is the line $y = -2$.

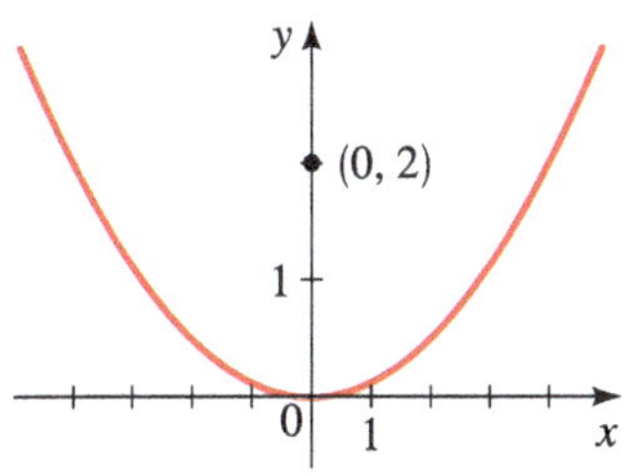

2. (a) Give the geometric definition of an ellipse.

An **ellipse** is the set of all points in the plane, the sum of whose distances from two fixed points F_1 and F_2 is a constant. These two fixed points are the **foci** (plural of **focus**) of the ellipse.

(b) Give the equation of an ellipse with center at the origin and with major axis along the x-axis. How long is the major axis? How long is the minor axis? Where are the foci? What is the eccentricity of the ellipse?

The equation of an ellipse with center at the origin and with major axis along the x-axis has the form

$$\frac{x^2}{a^2} + \frac{y^2}{b^2} = 1$$

The major axis is along the x-axis, provided that $a > b$. In this case its length is $2a$. The minor axis is along the y-axis, and its length is $2b$.

The foci are $(\pm c, 0)$, where $c^2 = a^2 - b^2$.

The eccentricity is $e = c/a$.

(c) Graph the equation $\frac{x^2}{16} + \frac{y^2}{9} = 1$. What are the lengths of the major and minor axes? Where are the foci?

Comparing this equation with the general equation of an ellipse, we see that $a = 4$ and $b = 3$. Since $4 > 3$, the major axis is along the x-axis and has length $2 \cdot 4 = 8$. The minor axis is along the y-axis, and it length is $2 \cdot 3 = 6$. We find $c^2 = 16 - 9 = 7$, so the foci are at $(\pm\sqrt{7}, 0)$.

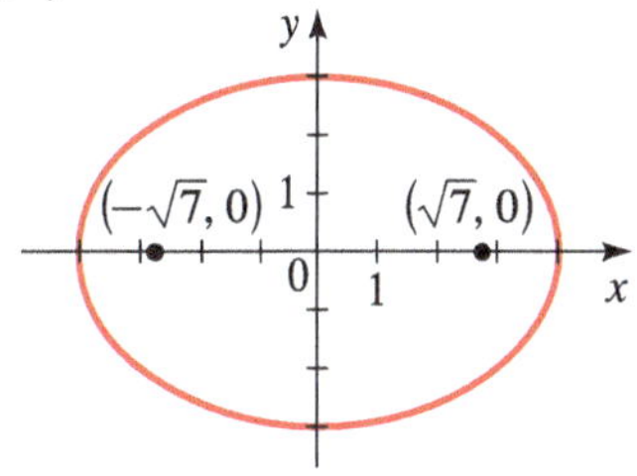

3. (a) Give the geometric definition of a hyperbola.

A **hyperbola** is the set of all points in the plane, the difference of whose distances from two fixed points F_1 and F_2 is a constant. These two fixed points are the **foci** of the hyperbola.

(b) Give the equation of a hyperbola with center at the origin and with transverse axis along the x-axis. How long is the transverse axis? Where are the vertices? What are the asymptotes? Where are the foci?

The equation of a hyperbola with center at the origin and with transverse axis along the x-axis has the form

$$\frac{x^2}{a^2} - \frac{y^2}{b^2} = 1$$

The transverse axis is along the x-axis and its length $2a$.

The vertices are at $(\pm a, 0)$.

The asymptotes are the lines $y = \pm(b/a)x$.

The foci are at $(\pm c, 0)$, where $c^2 = a^2 + b^2$.

(c) What is a good first step in graphing the hyperbola that is described in part (b)?

A good first step is to sketch the central box. This is the rectangle centered at the origin, with sides parallel to the axes, that crosses the x-axis at $\pm a$ and the y-axis at $\pm b$.

(d) Graph the equation $\frac{x^2}{16} - \frac{y^2}{9} = 1$. What are the asymptotes? Where are the vertices? Where are the foci? What is the length of the transverse axis?

This is an equation of a hyperbola with transverse axis along the x-axis. The central box crosses the axes at $(\pm 4, 0)$ and $(0, \pm 3)$. The asymptotes are $y = \pm\frac{3}{4}x$. The vertices are $(\pm 4, 0)$. We find $c^2 = 16 + 9 = 25$, so the foci are $(\pm 5, 0)$. Using the vertices and the asymptotes as guides, we graph the hyperbola as shown.

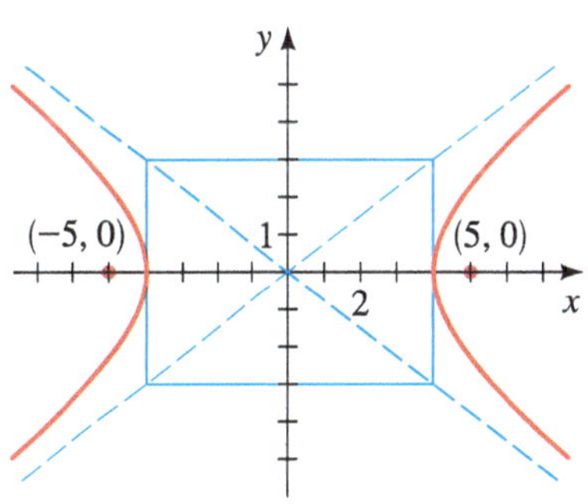

(continued)

CHAPTER 11 Review: Concept Check Answers *(continued)*

4. (a) Suppose we are given an equation in x and y. Let h and k be positive numbers. What is the effect on the graph of the equation if x is replaced by $x - h$ or $x + h$ and if y is replaced by $y - k$ or $y + k$?

Replacing x by $x - h$ or $x + h$ shifts the graph to the right or left, respectively, by h units. Replacing y by $y - k$ or $y + k$ shifts the graph upward or downward respectively, by k units.

(b) Sketch a graph of $\dfrac{(x+2)^2}{16} + \dfrac{(y-4)^2}{9} = 1$.

The graph is the same as the ellipse in 2(c) but shifted left 2 units and upward 4 units. So the center of the ellipse is at $(-2, 4)$.

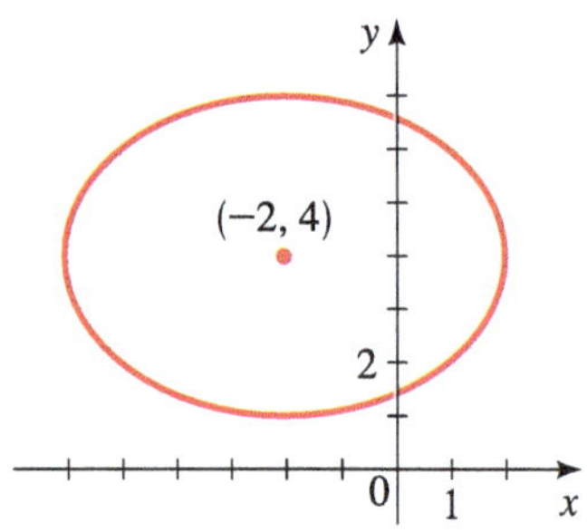

5. (a) How can you tell whether the following nondegenerate conic is a parabola, an ellipse, or a hyperbola?

$$Ax^2 + Cy^2 + Dx + Ey + F = 0$$

The graph is a parabola if either A or C is 0, an ellipse if A and C have the same sign (a circle if $A = C$), or a hyperbola if A and C have opposite signs.

(b) What conic does $3x^2 - 5y^2 + 4x + 5y - 8 = 0$ represent?

The graph is a hyperbola because the coefficients of x^2 and y^2 (3 and -5) have opposite signs.

6. (a) Suppose that the x- and y-axes are rotated through an acute angle ϕ to produce the X- and Y-axes. What are the equations that relate the coordinates (x, y) and (X, Y) of a point in the xy-plane and XY-plane, respectively?

$$x = X\cos\phi - Y\sin\phi \qquad X = x\cos\phi + y\sin\phi$$

$$y = X\sin\phi + Y\cos\phi \qquad Y = -x\sin\phi + y\cos\phi$$

(b) In the equation below, how do you eliminate the xy-term?

$$Ax^2 + Bxy + Cy^2 + Dx + Ey + F = 0$$

Rotate the axes through an angle ϕ that satisfies

$$\cot 2\phi = \frac{A - C}{B}$$

(c) Use a rotation of axes to eliminate the xy-term in the equation $25x^2 - 14xy + 25y^2 = 288$. Graph the equation.

The angle ϕ satisfies $\cot 2\phi = 0$, so $\phi = 45°$. By part (a) we have

$$x = X\cos 45° - Y\sin 45° = \frac{X - Y}{\sqrt{2}}$$

$$y = X\sin 45° + Y\cos 45° = \frac{X + Y}{\sqrt{2}}$$

Substituting into the given equation and simplifying, we get

$$\frac{X^2}{16} + \frac{Y^2}{9} = 1$$

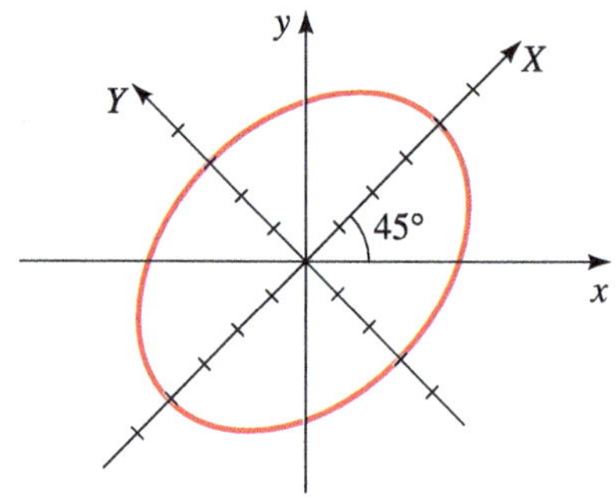

7. (a) What is the discriminant of the equation in 6(b)? How can you use the discriminant to determine the type of conic that the equation represents?

The discriminant is $B^2 - 4AC$. The conic is a parabola, ellipse, or hyperbola provided that the discriminant is zero, negative, or positive, respectively.

(b) Use the discriminant to identify the equation in 6(c).

The discriminant is $14^2 - 4(25)(25) < 0$, which confirms that the equation represents an ellipse.

8. (a) Write polar equations that represent a conic with eccentricity e. For what values of e is the conic an ellipse? a hyperbola? a parabola?

Polar equations of conics have the form

$$r = \frac{ed}{1 \pm e\cos\theta} \qquad \text{or} \qquad r = \frac{ed}{1 \pm e\sin\theta}$$

The equation represents a parabola if $e = 1$, an ellipse if $0 < e < 1$, and a hyperbola if $e > 1$.

(b) What conic does the polar equation $r = 2/(1 - \cos\theta)$ represent? Graph the conic.

This is a polar equation of a parabola.

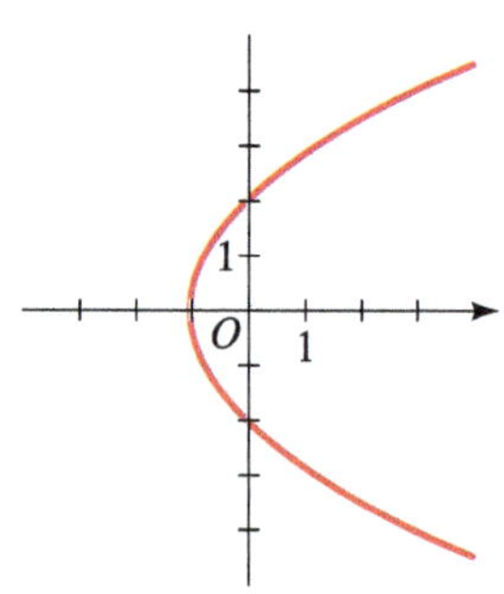

Cut here and keep for reference

CHAPTER 12 Review: Concept Check Answers

1. (a) What is a sequence? What notation do we use to denote the terms of a sequence?

A sequence is a list of numbers written in a specific order. Each number is called a term of the sequence. We denote the terms of a sequence by $a_1, a_2, a_3, \ldots$.

(b) Find a formula for the sequence of even numbers and a formula for the sequence of odd numbers.

Even numbers: $a_n = 2n$
Odd numbers: $a_n = 2n + 1$

(c) Find the first three terms and the 10th term of the sequence given by $a_n = n/(n + 1)$.

$$a_1 = \tfrac{1}{2}, \quad a_2 = \tfrac{2}{3}, \quad a_3 = \tfrac{3}{4}, \quad \text{and} \quad a_{10} = \tfrac{10}{11}$$

2. (a) What is a recursively defined sequence?

A recursively defined sequence is a sequence in which each term depends on some or all of the preceding terms.

(b) Find the first four terms of the sequence recursively defined by $a_1 = 3$ and $a_n = n + 2a_{n-1}$.

$$a_1 = 3, \quad a_2 = 8, \quad a_3 = 19, \quad a_4 = 42$$

3. (a) What is meant by the partial sums of a sequence?

The nth partial sum S_n of a sequence $a_1, a_2, a_3, \ldots$ is obtained by adding the first n terms of the sequence $S_1 = a_1$, $S_2 = a_1 + a_2, \ldots$, and in general $S_n = a_1 + a_2 + \cdots + a_n$.

(b) Find the first three partial sums of the sequence given by $a_n = 1/n$.

$$S_1 = \tfrac{1}{1} = 1, \quad S_2 = \tfrac{1}{1} + \tfrac{1}{2} = \tfrac{3}{2}, \quad S_3 = \tfrac{1}{1} + \tfrac{1}{2} + \tfrac{1}{3} = \tfrac{11}{6}$$

4. (a) What is an arithmetic sequence? Write a formula for the nth term of an arithmetic sequence.

An arithmetic sequence a_n is obtained when we start with a number a and add to it a fixed constant d over and over again. So

$$a_n = a + (n - 1)d$$

(b) Write a formula for the arithmetic sequence that starts as follows: 3, 8, . . . Write the first five terms of this sequence.

The first term is $a = 3$, and the common difference is 5. So the nth term is

$$a_n = 3 + (n - 1)5$$

which simplifies to $a_n = -2 + 5n$. So the first five terms are 3, 8, 13, 18, and 23.

(c) Write two different formulas for the sum of the first n terms of an arithmetic sequence.

$$S_n = \frac{n}{2}[2a + (n - 1)d] \quad \text{and} \quad S_n = n\left[\frac{a + a_n}{2}\right]$$

(d) Find the sum of the first 20 terms of the sequence in part (b).

Using the first formula in part (c), we get

$$S_{20} = \frac{20}{2}[2\cdot 3 + 19\cdot 5] = 1010$$

5. (a) What is a geometric sequence? Write an expression for the nth term of a geometric sequence that has first term a and common ratio r.

A geometric sequence a_n is obtained when we start with a number a and multiply it by a fixed constant r over and over again. So

$$a_n = ar^{n-1}$$

(b) Write an expression for the geometric sequence with first term $a = 3$ and common ratio $r = \frac{1}{2}$. Give the first five terms of this sequence.

The nth term is $a_n = 3\left(\frac{1}{2}\right)^{n-1}$.

The first five terms are 3, $\frac{3}{2}$, $\frac{3}{4}$, $\frac{3}{8}$, and $\frac{3}{16}$.

(c) Write an expression for the sum of the first n terms of a geometric sequence.

$$S_n = a\frac{1 - r^n}{1 - r}$$

(d) Find the sum of the first five terms of the sequence in part (b).

Using the formula in part (c), we get

$$S_5 = 3\frac{1 - \left(\frac{1}{2}\right)^5}{1 - \left(\frac{1}{2}\right)} = \frac{93}{16}$$

6. (a) What is an infinite geometric series?

An infinite geometric series is a series with infinitely many terms of the form

$$a + ar + ar^2 + \cdots + ar^n + \cdots$$

(b) What does it mean for an infinite series to converge? For what values of r does an infinite geometric series converge? If an infinite geometric series converges, then what is its sum?

An infinite series converges if its sequence of partial sums S_n approaches a finite number as $n \to \infty$. An infinite geometric series converges if $|r| < 1$; in this case its sum is $S = a/(1 - r)$.

(c) Write the first four terms of the infinite geometric series with first term $a = 5$ and common ratio $r = 0.4$. Does the series converge? If so, find its sum.

$$5 + 5(0.4) + 5(0.4)^2 + 5(0.4)^3 + \cdots + 5(0.4)^n + \cdots$$

The series converges because $|0.4| < 1$. By the formula in part (b) the sum of the series is

$$S = \frac{5}{1 - 0.4} = \frac{25}{3}$$

7. (a) Write $1^3 + 2^3 + 3^3 + 4^3 + 5^3$ using sigma notation.

$$1^3 + 2^3 + 3^3 + 4^3 + 5^3 = \sum_{k=1}^{5} k^3$$

(b) Write $\sum_{k=3}^{5} 2k^2$ without using sigma notation.

$$\sum_{k=3}^{5} 2k^2 = 2\cdot 3^2 + 2\cdot 4^2 + 2\cdot 5^2$$

(continued)

8. (a) What is an annuity? Write an expression for the amount A_f of an annuity consisting of n regular equal payments of size R with interest rate i per time period.

An annuity is a sum of money that is paid in regular equal payments. The amount is

$$A_f = R\frac{(1+i)^n - 1}{i}$$

(b) An investor deposits $200 each month into an account that pays 6% compounded monthly. How much is in the account at the end of 3 years?

The interest per time period is $i = 0.06/12 = 0.005$; the number of time periods is $n = 3 \times 12 = 36$.

$$A_f = 200\frac{(1+0.005)^{36} - 1}{0.005} = 7867.22$$

(c) What is the formula for calculating the present value of the annuity in part (b)?

$$A_p = R\frac{1 - (1+i)^{-n}}{i}$$

(d) What is the present value of the annuity in part (b)?

$$A_p = 200\frac{1 - (1+0.005)^{-36}}{0.005} = 6574.20$$

(e) When buying on installment, what is the formula for calculating the periodic payments?

If a loan A_p is to be repaid in n regular equal payments with interest rate i per time period, then the size R of each payment is

$$R = \frac{iA_p}{1 - (1+i)^{-n}}$$

(f) If you take out a 5-year loan for $10,000 at 3% interest compounded monthly, what is the size of each monthly payment?

The interest per time period is $i = 0.03/12 = 0.0025$; the number of time periods is $n = 5 \times 12 = 60$.

$$R = \frac{0.0025(10{,}000)}{1 - (1+0.0025)^{-60}} = 179.69$$

9. (a) State the Principle of Mathematical Induction.

A statement $P(n)$ about a natural number n is true for all n provided that the following hold.

1. $P(1)$ is true.
2. If $P(k)$ is true, then $P(k+1)$ is true.

(b) Use mathematical induction to prove that for all natural numbers n, $3^n - 1$ is an even number.

Let $P(n)$ be the statement that $3^n - 1$ is even.

1. $P(1)$ is true because $3^1 - 1$ is even.
2. Suppose $P(k)$ is true. Now

$$3^{k+1} - 1 = 3 \cdot 3^k - 1 = \underbrace{2 \cdot 3^k}_{\text{even}} + \underbrace{3^k - 1}_{\text{even}}$$

The first term is clearly even, and the second is even by the induction hypothesis. So $P(k+1)$ is true.

It follows that the statement is true for all n.

10. (a) Write Pascal's triangle. How are the entries in the triangle related to each other?

Row 0	1
Row 1	1 1
Row 2	1 2 1
Row 3	1 3 3 1

Each entry is the sum of the two entries above it.

(b) Use Pascal's triangle to expand $(x + c)^3$.

$$(x+c)^3 = x^3 + 3x^2c + 3xc^2 + c^3$$

11. (a) What does the symbol $n!$ mean? Find $5!$.

$$n! = 1 \cdot 2 \cdot 3 \cdot \cdots \cdot n; \quad 5! = 1 \cdot 2 \cdot 3 \cdot 4 \cdot 5 = 120$$

(b) Define $\binom{n}{r}$, and find $\binom{5}{2}$.

$$\binom{n}{r} = \frac{n!}{r!(n-r)!}, \quad \binom{5}{2} = \frac{5!}{2!\,3!} = 10$$

12. (a) State the Binomial Theorem.

$$(a+b)^n = a^n + \binom{n}{1}a^{n-1}b + \binom{n}{2}a^{n-2}b^2 + \cdots + b^n$$

(b) Use the Binomial Theorem to expand $(x + 2)^3$.

$$(x+2)^3 = x^3 + \binom{3}{1}x^2 2^1 + \binom{3}{2}x^1 \cdot 2^2 + 2^3$$

$$= x^3 + 6x^2 + 12x + 8$$

(c) Use the Binomial Theorem to find the term containing x^4 in the expansion of $(x + 2)^{10}$.

The term is $\binom{10}{4}x^4 2^6 = 13440x^4$.

Cut here and keep for reference

CHAPTER 13 Review: Concept Check Answers

1. (a) Explain what is meant by $\lim_{x\to a} f(x) = L$.

As x approaches a, the values of $f(x)$ approach the number L.

(b) If $\lim_{x\to 2} f(x) = 5$, is it possible that $f(2) = 3$?

Yes. For x close to 2, $f(x)$ is close to 5. But the value of f at 2 does not affect the limit.

(c) Find $\lim_{x\to 2} x^2$.

If x is close to 2, then x^2 is close to 4, so $\lim_{x\to 2} x^2 = 4$.

2. To evaluate the limit of a function, we often need to first rewrite the function using the rules of algebra. What is the logical first step in evaluating each of the following limits?

(a) $\displaystyle\lim_{x\to 2} \frac{x^2 - 4}{x - 2}$

Factor the numerator, and simplify.

(b) $\displaystyle\lim_{h\to 0} \frac{(5 + h)^2 - 25}{h}$

Expand the numerator, and simplify.

(c) $\displaystyle\lim_{x\to 3} \frac{\sqrt{x + 1} - 2}{x - 3}$

Rationalize the numerator by multiplying the numerator and denominator by $\sqrt{x + 1} + 2$, and then simplify.

(d) $\displaystyle\lim_{x\to 7} \frac{\left(\frac{1}{7} - \frac{1}{x}\right)}{x - 7}$

Combine the fractions in the numerator, using the common denominator $7x$, and simplify.

3. (a) Explain what it means to say:

$$\lim_{x\to 3^-} f(x) = 5 \qquad \lim_{x\to 3^+} f(x) = 10$$

The first equation says that the limit, as x approaches 3 from the left (through values to the left of 3, or less than 3), is 5.

The second equation says that the limit, as x approaches 3 from the right (through values to the right of 3, or greater than 3), is 10.

(b) If the two equations in part (a) are true, is it possible that $\lim_{x\to 3} f(x) = 5$?

No. For the limit to exist, the left- and right-hand limits must be equal.

(c) Find $\lim_{x\to 2^-} f(x)$ and $\lim_{x\to 2^+} f(x)$, where f is defined as follows:

$$f(x) = \begin{cases} 1 & \text{if } x \le 2 \\ x & \text{if } x > 2 \end{cases}$$

As x gets close to 2 from the left (through values less than 2), the function is constantly 1, so the limit from the left is 1. Similarly, as x gets close to 2 from the right (through values greater than 2), the function is equal to x, so the values of f approach 2. Thus the limit from the right is 2.

$$\lim_{x\to 2^-} f(x) = 1 \qquad \text{and} \qquad \lim_{x\to 2^+} f(x) = 2$$

(d) For f as in (c), does $\lim_{x\to 2} f(x)$ exist?

No, because the left- and right-hand limits are not equal.

4. (a) Define the derivative $f'(a)$ of a function f at $x = a$.

$$f'(a) = \lim_{h\to 0} \frac{f(a + h) - f(a)}{h}$$

(b) State an equivalent formulation for $f'(a)$.

$$f'(a) = \lim_{x\to a} \frac{f(x) - f(a)}{x - a}$$

(c) Find the derivative of $f(x) = x^2$ at $x = 3$.

Using the definition in (a), we have

$$\begin{aligned} f'(3) &= \lim_{h\to 0} \frac{f(3 + h) - f(3)}{h} \\ &= \lim_{h\to 0} \frac{(3 + h)^2 - 3^2}{h} \\ &= \lim_{h\to 0} \frac{6h + h^2}{h} \\ &= \lim_{h\to 0} (6 + h) = 6 \end{aligned}$$

So $f'(3) = 6$.

5. (a) Give two different interpretations of the derivative of the function $y = f(x)$ at $x = a$.

The derivative has the following interpretations:

$f'(a)$ is the slope of the tangent line to the graph of f at the point $P(a, f(a))$.

$f'(a)$ is the instantaneous rate of change of y with respect to x at $x = a$.

(b) For the function $f(x) = x^2$, find the slope of the tangent line to the graph of f at the point $(3, 9)$ on the graph.

The slope is $f'(3) = 6$ (from 4(c)).

(c) For the function $y = x^2$, find the instantaneous rate of change of y with respect to x when $x = 3$.

The instantaneous rate of change is $f'(3) = 6$ (from 4(c)).

(d) Write expressions for the average rate of change of y with respect to x between a and x and for the instantaneous rate of change of y with respect to x at $x = a$.

Average rate of change	Instantaneous rate of change
$\dfrac{f(x) - f(a)}{x - a}$	$\displaystyle\lim_{x\to a} \frac{f(x) - f(a)}{x - a}$

(continued)

CHAPTER 13 Review: Concept Check Answers *(continued)*

6. (a) Explain what is meant by $\lim_{x \to \infty} f(x) = L$. Draw sketches to illustrate different ways in which this can happen.

It means that the values of $f(x)$ can be made arbitrarily close to L by taking x sufficiently large.

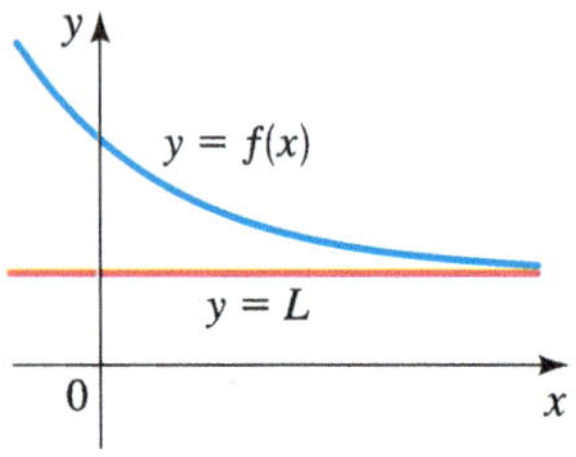

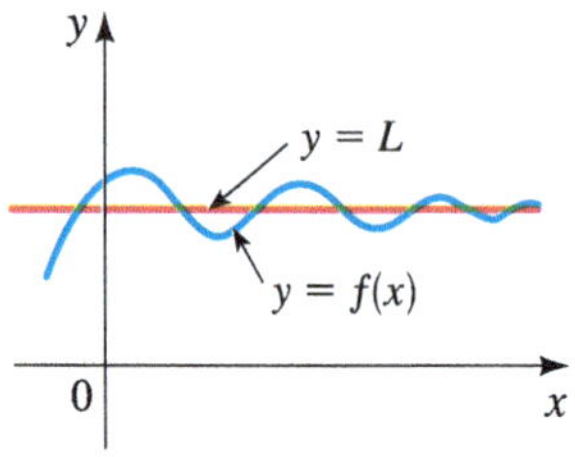

(b) Find $\lim_{x \to \infty} \dfrac{3x^2 + x}{x^2 + 1}$.

We divide numerator and denominator by x^2.

$$\lim_{x \to \infty} \frac{3x^2 + x}{x^2 + 1} = \lim_{x \to \infty} \frac{3 + \dfrac{1}{x}}{1 + \dfrac{1}{x^2}} = \frac{3 + 0}{1 + 0} = 3$$

(c) Explain why $\lim_{x \to \infty} \sin x$ does not exist.

From the graph below we see that as x increases, the values of the sine function oscillate between 1 and -1, so they don't approach a definite number.

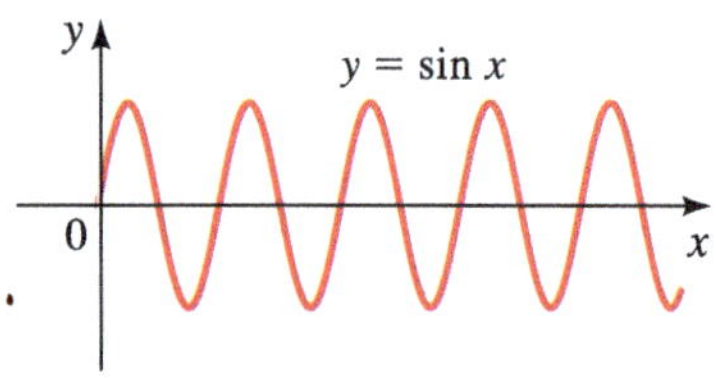

7. (a) If $a_1, a_2, a_3, \ldots$ is a sequence, what is meant by $\lim_{n \to \infty} a_n = L$? What is a convergent sequence?

It means that the nth term a_n of the sequence can be made arbitrarily close to L by taking n sufficiently large. The sequence is convergent if $\lim_{n \to \infty} a_n$ exists.

(b) Find $\lim_{n \to \infty} (-1)^n/n$.

As n gets large, the denominator gets large but the numerator is always 1 or -1, so the limit is 0.

8. (a) Suppose S is the region under the graph of the function $y = f(x)$ and above the x-axis, where $a \le x \le b$. Explain how this area is approximated by rectangles, and write an expression for the area of S as a limit of sums.

We first divide the interval $[a, b]$ into n equal intervals, each of length $\Delta x = (b - a)/n$, and then erect a rectangle on each of these intervals, where its height is the value of f at the right-hand endpoint. The area is the limit of the sum of these approximating rectangles:

$$A = \lim_{n \to \infty} \sum_{k=1}^{n} f(x_k)\, \Delta x$$

where

$$\Delta x = \frac{b - a}{n} \qquad \text{and} \qquad x_k = a + k\,\Delta x$$

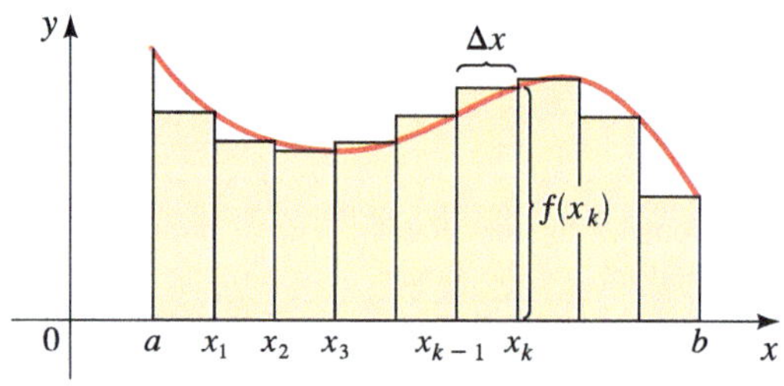

(b) Find the area under the graph of $f(x) = x^2$ and above the x-axis, between $x = 0$ and $x = 3$.

We first divide the interval $[0, 3]$ into n equal intervals, each of length

$$\Delta x = \frac{3 - 0}{n} = \frac{3}{n}$$

The right-hand endpoint of the kth interval is

$$x_k = 0 + k\frac{3}{n} = \frac{3k}{n}$$

So the area is

$$\begin{aligned} A &= \lim_{n \to \infty} \sum_{k=1}^{n} f(x_k)\, \Delta x = \lim_{n \to \infty} \sum_{k=1}^{n} f\!\left(\frac{3k}{n}\right) \cdot \frac{3}{n} \\ &= \lim_{n \to \infty} \sum_{k=1}^{n} \left(\frac{3k}{n}\right)^2 \cdot \frac{3}{n} = \lim_{n \to \infty} \frac{3^3}{n^3} \sum_{k=1}^{n} k^2 \\ &= \lim_{n \to \infty} \frac{27}{n^3} \cdot \frac{n(n + 1)(2n + 1)}{6} \\ &= \lim_{n \to \infty} \frac{54n^3 + 81n^2 + 27n}{6n^3} = 9 \end{aligned}$$

WebAssign.

WebAssign Student Quick Start Guide

You can use WebAssign to access your homework, quizzes, and tests — whatever your instructor chooses — at any time of day or night, from any computer with a connection to the Internet and a Web browser. Your instructor creates your assignments, schedules them, and decides how many submissions you get. Your instructor also determines if you can have an extension, if you can save your work without submitting it at the time, and how much feedback you get after you submit an assignment.

The WebAssign support staff cannot change your username or password, give extensions, change your score, give you extra submissions, or help you with the content of your assignments.

Logging In

You can log in to WebAssign using any Web browser connected to the Internet. There are two different ways to log in to WebAssign. Each requires information from your teacher. If you are unsure about how to log in, please check with your teacher or another student in your class.

Go to the login page at http://webassign.net/login.html or the web address provided by your teacher. The way you log in depends on how your instructor set up the class:

- If your teacher created a WebAssign account for you, they will provide you with a **Username**, an **Institution** code and a **Password**. Simply enter this information in the boxes provided and click the **Log In** button.

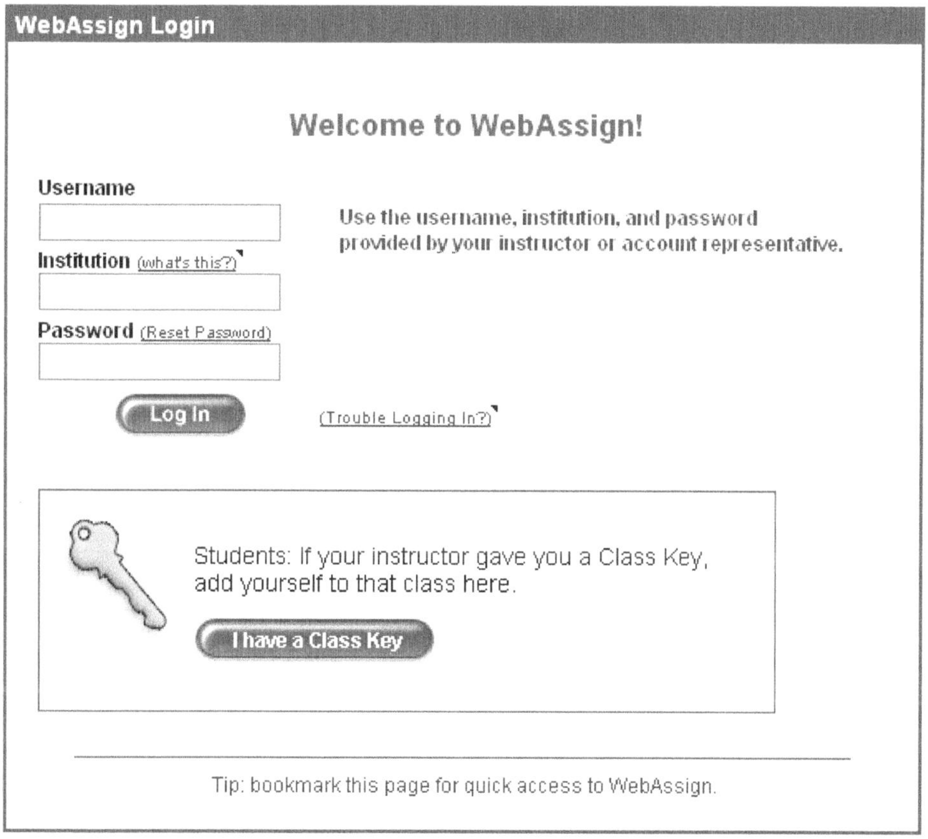

WebAssign.

- If your teacher wants you to **Self-Enroll** in the WebAssign course they will provide you with a **Class Key**. You will create your own username and password. It is important that you remember this information so you can log in for the remainder of the class. In this case, just click the **I have a Class Key** button. You don't need to enter any other information on this page.

  Then, enter the **Class Key** your instructor provided and click **Submit.** Verify you are enrolling in the correct class on the next page.

Class Key

Enter the Class Key that you received from your instructor. You will only need to complete this once. After you have created your account, you can log in on the main page.

Class Key

Class Keys generally start with an institution code, followed by two sets of four digits.

Submit

- Enter your preferred Login and Student information.

- Click the **Create My Account** button to complete the enrollment process.

- A review screen will display, showing your username, institution code, and password. **Retain a copy of this information.** You will need it to log into WebAssign.

Log In Information

Required fields are marked with an asterisk (*).

Preferred Username * Check Availability

Your username may contain letters, numbers, and the following characters: underscore (_), hyphen (-), period (.)

Institution Code **webassign**

Password *

Re-Enter Password *

Passwords are case-sensitive.

Student Information

Required fields are marked with an asterisk (*).

First Name *

Last Name *

Email Address *

Student ID Number

Create My Account

WebAssign.

Access Codes

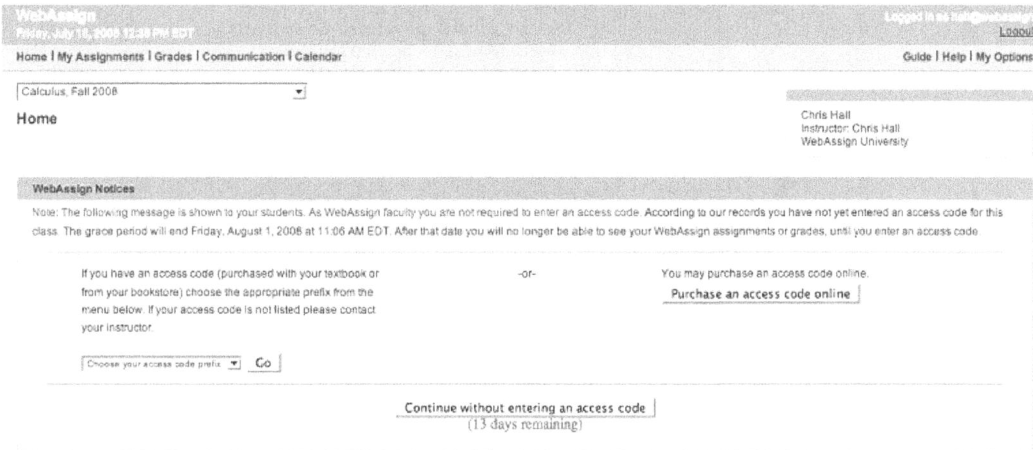

Once you log in, you may see a WebAssign Notice about entering an access code for your class. You can get an Access Code from any of the following places if you need to use one:

- A new textbook you purchased for the class.
- Your bookstore, which may sell Access Code cards.
- Online, where you can purchase an access code with a credit card.

You have a 14 day grace period to use WebAssign, starting with the WebAssign class start date. During this time you can work on and view your WebAssign assignments without registering a code.

After the grace period is over you will only see the code registration message until you submit or purchase a code.

There are two types of WebAssign access code cards. The small card requires you to scratch off the silver surface in order to reveal the complete access code.

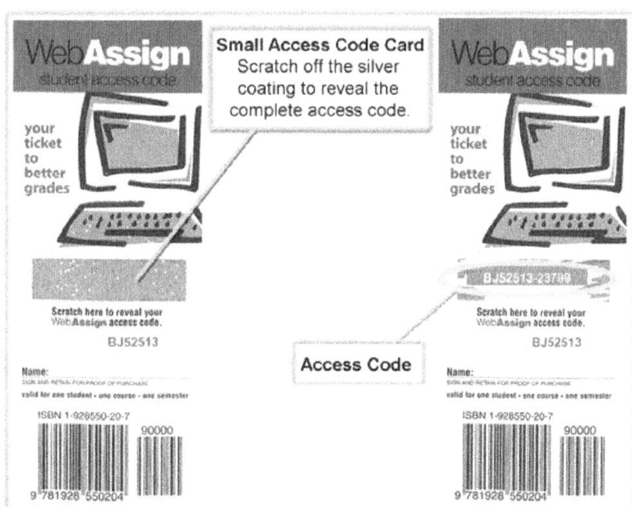

WebAssign.

The larger security envelope card requires you to open the card to reveal the access code number.

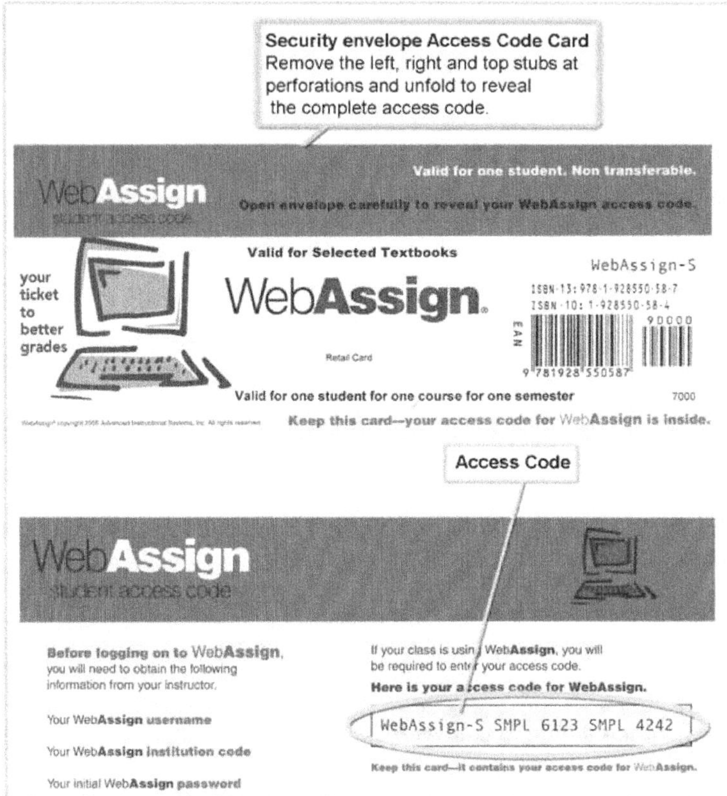

If you would like to purchase an access code directly from WebAssign online, you may do so with a credit card. Your code will be automatically registered to your WebAssign account as soon as the transaction is complete. You will receive an email confirmation. Please keep a copy for your records.

WebAssign.

Your WebAssign Home Page

Once you have successfully logged in you will see your WebAssign homepage. If you are taking more than one WebAssign class, you will need to select which class you wish to view first.

The upper right corner features links to a complete Student **Guide**, as well as a link to WebAssign Technical Support under **Help**. If you want to change your password or add or update your email address, simply click **My Options** in the upper right hand corner.

You will see your assignments and due dates listed, as well as any Communications, Grades, and Announcements posted by your teacher.

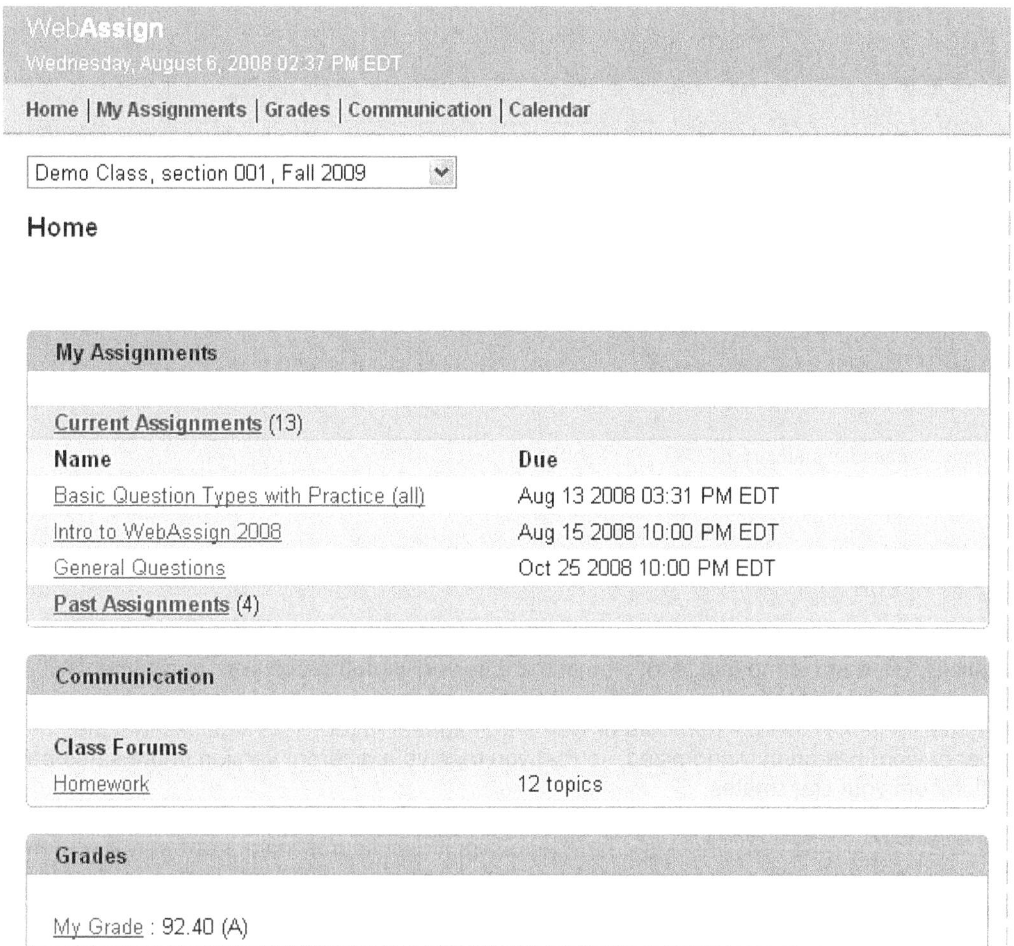

WebAssign.

Answering Questions

WebAssign has a variety of different question types, ranging from multiple choice to fill-in-the-blank to symbolic questions. Here are some things to keep in mind as you work through your assignments:

- Some questions may include numbers or words that appear in red. This signifies that the number or word has been randomized, so that you receive a different version of the same basic question from your classmates.

- Some WebAssign questions check the number of significant figures in your answer. If you enter the correct value with the wrong number of significant figures, you will not receive credit, but you will receive a hint that your number does not have the correct number of significant figures.

- Some questions require entering symbolic notation. Answer symbolic questions by using calculator notation. You must use the exact variables specified in the questions. The order is not important as long as it is mathematically correct. Clicking on the eye button previews the expression you enter in proper mathematical notation. Clicking on the symbolic formatting help button provides tips for using the correct keystrokes.

WebAssign®

- When you click on some WebAssign chemistry or math questions an input palette will open. These palettes, called chemPad and mathPad, will help you enter your answer in proper notation.

- Some questions may require the use of an Active Figure simulation. Active Figures require the free Macromedia Flash Player plug-in, downloadable from www.macromedia.com.

- If your instructor allows it, you can save your work without grading by selecting the Save Work button at the end of the question. After you save your work, it will be available to you the next time you click the assignment.

- Please note that WebAssign will **not** automatically submit your answers for scoring if you only **Save** your work. Your teacher will not be able to see your submissions. Please be sure to **Submit** prior to the due date and time.

- If your instructor allows it, you can submit answers by question part or for the entire assignment. To submit an individual question answer for grading, click the **Submit New Answers to Question __** button at the bottom of each question. To submit the entire assignment for grading, click the **Submit All New Answers** button at the end of the assignment.

Technical Support

If you are having difficulty logging in, please be sure to check with your teacher and verify whether an account has been created for you or whether you need to self-enroll. In either case your teacher needs to provide the appropriate information (username, institution code and password OR Class Key).

To email WebAssign Support go to http://www.webassign.net/info/support/report.html. This page also lists answers to **Common Problems**, and provides links to the **Student Guide**.

August 7, 2008

REVIEW OF ALGEBRA

Here we review the basic rules and procedures of algebra that you need to know in order to be successful in calculus.

ARITHMETIC OPERATIONS

The real numbers have the following properties:

$a + b = b + a$	$ab = ba$	(Commutative Law)
$(a + b) + c = a + (b + c)$	$(ab)c = a(bc)$	(Associative Law)
$a(b + c) = ab + ac$		(Distributive law)

In particular, putting $a = -1$ in the Distributive Law, we get

$$-(b + c) = (-1)(b + c) = (-1)b + (-1)c$$

and so

$$-(b + c) = -b - c$$

EXAMPLE 1

(a) $(3xy)(-4x) = 3(-4)x^2y = -12x^2y$

(b) $2t(7x + 2tx - 11) = 14tx + 4t^2x - 22t$

(c) $4 - 3(x - 2) = 4 - 3x + 6 = 10 - 3x$ ■

If we use the Distributive Law three times, we get

$$(a + b)(c + d) = (a + b)c + (a + b)d = ac + bc + ad + bd$$

This says that we multiply two factors by multiplying each term in one factor by each term in the other factor and adding the products. Schematically, we have

$$(a + b)(c + d)$$

In the case where $c = a$ and $d = b$, we have

$$(a + b)^2 = a^2 + ba + ab + b^2$$

or

1 $$(a + b)^2 = a^2 + 2ab + b^2$$

Similarly, we obtain

2 $$(a - b)^2 = a^2 - 2ab + b^2$$

EXAMPLE 2

(a) $(2x + 1)(3x - 5) = 6x^2 + 3x - 10x - 5 = 6x^2 - 7x - 5$

(b) $(x + 6)^2 = x^2 + 12x + 36$

(c) $3(x - 1)(4x + 3) - 2(x + 6) = 3(4x^2 - x - 3) - 2x - 12$

$= 12x^2 - 3x - 9 - 2x - 12 = 12x^2 - 5x - 21$ ■

FRACTIONS

To add two fractions with the same denominator, we use the Distributive Law:

$$\frac{a}{b} + \frac{c}{b} = \frac{1}{b} \times a + \frac{1}{b} \times c = \frac{1}{b}(a + c) = \frac{a + c}{b}$$

Thus, it is true that

$$\boxed{\frac{a + c}{b} = \frac{a}{b} + \frac{c}{b}}$$

But remember to avoid the following common error:

$$\frac{a}{b + c} \not= \frac{a}{b} + \frac{a}{c}$$

(For instance, take $a = b = c = 1$ to see the error.)

To add two fractions with different denominators, we use a common denominator:

$$\boxed{\frac{a}{b} + \frac{c}{d} = \frac{ad + bc}{bd}}$$

We multiply such fractions as follows:

$$\boxed{\frac{a}{b} \cdot \frac{c}{d} = \frac{ac}{bd}}$$

In particular, it is true that

$$\boxed{\frac{-a}{b} = -\frac{a}{b} = \frac{a}{-b}}$$

To divide two fractions, we invert and multiply:

$$\boxed{\frac{\frac{a}{b}}{\frac{c}{d}} = \frac{a}{b} \times \frac{d}{c} = \frac{ad}{bc}}$$

EXAMPLE 3

(a) $\dfrac{x + 3}{x} = \dfrac{x}{x} + \dfrac{3}{x} = 1 + \dfrac{3}{x}$

(b) $\dfrac{3}{x - 1} + \dfrac{x}{x + 2} = \dfrac{3(x + 2) + x(x - 1)}{(x - 1)(x + 2)} = \dfrac{3x + 6 + x^2 - x}{x^2 + x - 2} = \dfrac{x^2 + 2x + 6}{x^2 + x - 2}$

(c) $\dfrac{s^2t}{u} \cdot \dfrac{ut}{-2} = \dfrac{s^2t^2u}{-2u} = -\dfrac{s^2t^2}{2}$

(d) $$\frac{\dfrac{x}{y}+1}{1-\dfrac{y}{x}} = \frac{\dfrac{x+y}{y}}{\dfrac{x-y}{x}} = \frac{x+y}{y} \times \frac{x}{x-y} = \frac{x(x+y)}{y(x-y)} = \frac{x^2+xy}{xy-y^2}$$ ■

FACTORING

We have used the Distributive Law to expand certain algebraic expressions. We sometimes need to reverse this process (again using the Distributive Law) by factoring an expression as a product of simpler ones. The easiest situation occurs when the expression has a common factor as follows:

$$\xrightarrow{\text{Expanding}}$$
$$3x(x-2) = 3x^2 - 6x$$
$$\xleftarrow{\text{Factoring}}$$

To factor a quadratic of the form $x^2 + bx + c$ we note that

$$(x+r)(x+s) = x^2 + (r+s)x + rs$$

so we need to choose numbers r and s so that $r + s = b$ and $rs = c$.

EXAMPLE 4 Factor $x^2 + 5x - 24$.

SOLUTION The two integers that add to give 5 and multiply to give -24 are -3 and 8. Therefore

$$x^2 + 5x - 24 = (x-3)(x+8)$$ ■

EXAMPLE 5 Factor $2x^2 - 7x - 4$.

SOLUTION Even though the coefficient of x^2 is not 1, we can still look for factors of the form $2x + r$ and $x + s$, where $rs = -4$. Experimentation reveals that

$$2x^2 - 7x - 4 = (2x+1)(x-4)$$ ■

Some special quadratics can be factored by using Equations 1 or 2 (from right to left) or by using the formula for a difference of squares:

3 $$a^2 - b^2 = (a-b)(a+b)$$

The analogous formula for a difference of cubes is

4 $$a^3 - b^3 = (a-b)(a^2+ab+b^2)$$

which you can verify by expanding the right side. For a sum of cubes we have

5 $$a^3 + b^3 = (a+b)(a^2-ab+b^2)$$

EXAMPLE 6

(a) $x^2 - 6x + 9 = (x-3)^2$ (Equation 2; $a = x, b = 3$)

(b) $4x^2 - 25 = (2x-5)(2x+5)$ (Equation 3; $a = 2x, b = 5$)

(c) $x^3 + 8 = (x+2)(x^2 - 2x + 4)$ (Equation 5; $a = x, b = 2$) ■

EXAMPLE 7 Simplify $\dfrac{x^2 - 16}{x^2 - 2x - 8}$.

SOLUTION Factoring numerator and denominator, we have

$$\frac{x^2 - 16}{x^2 - 2x - 8} = \frac{(x-4)(x+4)}{(x-4)(x+2)} = \frac{x+4}{x+2}$$

To factor polynomials of degree 3 or more, we sometimes use the following fact.

6 The Factor Theorem If P is a polynomial and $P(b) = 0$, then $x - b$ is a factor of $P(x)$.

EXAMPLE 8 Factor $x^3 - 3x^2 - 10x + 24$.

SOLUTION Let $P(x) = x^3 - 3x^2 - 10x + 24$. If $P(b) = 0$, where b is an integer, then b is a factor of 24. Thus, the possibilities for b are $\pm1, \pm2, \pm3, \pm4, \pm6, \pm8, \pm12$, and ±24. We find that $P(1) = 12$, $P(-1) = 30$, $P(2) = 0$. By the Factor Theorem, $x - 2$ is a factor. Instead of substituting further, we use long division as follows:

$$\begin{array}{r l}
 & x^2 - x - 12 \\
x - 2 \,\big) & x^3 - 3x^2 - 10x + 24 \\
 & \underline{x^3 - 2x^2} \\
 & -x^2 - 10x \\
 & \underline{-x^2 + 2x} \\
 & -12x + 24 \\
 & \underline{-12x + 24}
\end{array}$$

Therefore
$$\begin{aligned} x^3 - 3x^2 - 10x + 24 &= (x-2)(x^2 - x - 12) \\ &= (x-2)(x+3)(x-4) \end{aligned}$$

COMPLETING THE SQUARE

Completing the square is a useful technique for graphing parabolas or integrating rational functions. Completing the square means rewriting a quadratic $ax^2 + bx + c$ in the form $a(x + p)^2 + q$ and can be accomplished by:

1. Factoring the number a from the terms involving x.
2. Adding and subtracting the square of half the coefficient of x.

In general, we have

$$\begin{aligned} ax^2 + bx + c &= a\left[x^2 + \frac{b}{a}x\right] + c \\ &= a\left[x^2 + \frac{b}{a}x + \left(\frac{b}{2a}\right)^2 - \left(\frac{b}{2a}\right)^2\right] + c \\ &= a\left(x + \frac{b}{2a}\right)^2 + \left(c - \frac{b^2}{4a}\right) \end{aligned}$$

EXAMPLE 9 Rewrite $x^2 + x + 1$ by completing the square.

SOLUTION The square of half the coefficient of x is $\frac{1}{4}$. Thus

$$x^2 + x + 1 = x^2 + x + \tfrac{1}{4} - \tfrac{1}{4} + 1 = \left(x + \tfrac{1}{2}\right)^2 + \tfrac{3}{4}$$

EXAMPLE 10

$$2x^2 - 12x + 11 = 2[x^2 - 6x] + 11 = 2[x^2 - 6x + 9 - 9] + 11$$
$$= 2[(x - 3)^2 - 9] + 11 = 2(x - 3)^2 - 7$$ ■

QUADRATIC FORMULA

By completing the square as above we can obtain the following formula for the roots of a quadratic equation.

7 The Quadratic Formula The roots of the quadratic equation $ax^2 + bx + c = 0$ are

$$x = \frac{-b \pm \sqrt{b^2 - 4ac}}{2a}$$

EXAMPLE 11 Solve the equation $5x^2 + 3x - 3 = 0$.

SOLUTION With $a = 5$, $b = 3$, $c = -3$, the quadratic formula gives the solutions

$$x = \frac{-3 \pm \sqrt{3^2 - 4(5)(-3)}}{2(5)} = \frac{-3 \pm \sqrt{69}}{10}$$ ■

The quantity $b^2 - 4ac$ that appears in the quadratic formula is called the **discriminant**. There are three possibilities:

1. If $b^2 - 4ac > 0$, the equation has two real roots.
2. If $b^2 - 4ac = 0$, the roots are equal.
3. If $b^2 - 4ac < 0$, the equation has no real root. (The roots are complex.)

These three cases correspond to the fact that the number of times the parabola $y = ax^2 + bx + c$ crosses the x-axis is 2, 1, or 0 (see Figure 1). In case (3) the quadratic $ax^2 + bx + c$ can't be factored and is called **irreducible.**

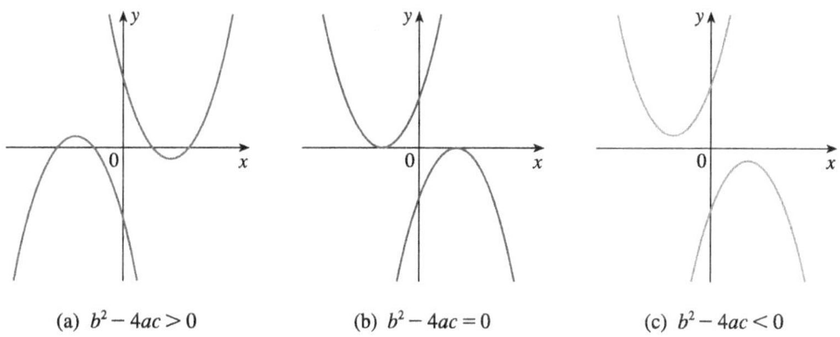

FIGURE 1
Possible graphs of $y = ax^2 + bx + c$

EXAMPLE 12 The quadratic $x^2 + x + 2$ is irreducible because its discriminant is negative:

$$b^2 - 4ac = 1^2 - 4(1)(2) = -7 < 0$$

Therefore, it is impossible to factor $x^2 + x + 2$. ■

THE BINOMIAL THEOREM

Recall the binomial expression from Equation 1:

$$(a + b)^2 = a^2 + 2ab + b^2$$

If we multiply both sides by $(a + b)$ and simplify, we get the binomial expansion

8
$$(a + b)^3 = a^3 + 3a^2b + 3ab^2 + b^3$$

Repeating this procedure, we get

$$(a + b)^4 = a^4 + 4a^3b + 6a^2b^2 + 4ab^3 + b^4$$

In general, we have the following formula.

9 The Binomial Theorem If k is a positive integer, then

$$\begin{aligned}(a + b)^k = a^k + ka^{k-1}b &+ \frac{k(k-1)}{1 \cdot 2} a^{k-2}b^2 \\ &+ \frac{k(k-1)(k-2)}{1 \cdot 2 \cdot 3} a^{k-3}b^3 \\ &+ \cdots + \frac{k(k-1)\cdots(k-n+1)}{1 \cdot 2 \cdot 3 \cdot \cdots \cdot n} a^{k-n}b^n \\ &+ \cdots + kab^{k-1} + b^k\end{aligned}$$

EXAMPLE 13 Expand $(x - 2)^5$.

SOLUTION Using the Binomial Theorem with $a = x$, $b = -2$, $k = 5$, we have

$$\begin{aligned}(x - 2)^5 &= x^5 + 5x^4(-2) + \frac{5 \cdot 4}{1 \cdot 2} x^3(-2)^2 + \frac{5 \cdot 4 \cdot 3}{1 \cdot 2 \cdot 3} x^2(-2)^3 + 5x(-2)^4 + (-2)^5 \\ &= x^5 - 10x^4 + 40x^3 - 80x^2 + 80x - 32\end{aligned}$$ ■

RADICALS

The most commonly occurring radicals are square roots. The symbol $\sqrt{}$ means "the positive square root of." Thus

$$x = \sqrt{a} \qquad \text{means} \qquad x^2 = a \quad \text{and} \quad x \geqslant 0$$

Since $a = x^2 \geqslant 0$, the symbol $\sqrt{a}$ makes sense only when $a \geqslant 0$. Here are two rules for working with square roots:

10
$$\sqrt{ab} = \sqrt{a}\,\sqrt{b} \qquad \sqrt{\frac{a}{b}} = \frac{\sqrt{a}}{\sqrt{b}}$$

However, there is no similar rule for the square root of a sum. In fact, you should remember to avoid the following common error:

⊘
$$\sqrt{a + b} = \sqrt{a} + \sqrt{b}$$

(For instance, take $a = 9$ and $b = 16$ to see the error.)

EXAMPLE 14

(a) $\dfrac{\sqrt{18}}{\sqrt{2}} = \sqrt{\dfrac{18}{2}} = \sqrt{9} = 3$ (b) $\sqrt{x^2y} = \sqrt{x^2}\sqrt{y} = |x|\sqrt{y}$

Notice that $\sqrt{x^2} = |x|$ because $\sqrt{\ }$ indicates the positive square root. (See **Absolute Value**.) ■

In general, if n is a positive integer,

$$x = \sqrt[n]{a} \quad \text{means} \quad x^n = a$$

If n is even, then $a \geqslant 0$ and $x \geqslant 0$.

Thus $\sqrt[3]{-8} = -2$ because $(-2)^3 = -8$, but $\sqrt[4]{-8}$ and $\sqrt[6]{-8}$ are not defined. The following rules are valid:

$$\sqrt[n]{ab} = \sqrt[n]{a}\,\sqrt[n]{b} \qquad \sqrt[n]{\frac{a}{b}} = \frac{\sqrt[n]{a}}{\sqrt[n]{b}}$$

EXAMPLE 15 $\sqrt[3]{x^4} = \sqrt[3]{x^3x} = \sqrt[3]{x^3}\,\sqrt[3]{x} = x\sqrt[3]{x}$ ■

To **rationalize** a numerator or denominator that contains an expression such as $\sqrt{a} - \sqrt{b}$, we multiply both the numerator and the denominator by the conjugate radical $\sqrt{a} + \sqrt{b}$. Then we can take advantage of the formula for a difference of squares:

$$(\sqrt{a} - \sqrt{b})(\sqrt{a} + \sqrt{b}) = (\sqrt{a})^2 - (\sqrt{b})^2 = a - b$$

EXAMPLE 16 Rationalize the numerator in the expression $\dfrac{\sqrt{x+4} - 2}{x}$.

SOLUTION We multiply the numerator and the denominator by the conjugate radical $\sqrt{x+4} + 2$:

$$\frac{\sqrt{x+4} - 2}{x} = \left(\frac{\sqrt{x+4} - 2}{x}\right)\left(\frac{\sqrt{x+4} + 2}{\sqrt{x+4} + 2}\right) = \frac{(x+4) - 4}{x(\sqrt{x+4} + 2)}$$

$$= \frac{x}{x(\sqrt{x+4} + 2)} = \frac{1}{\sqrt{x+4} + 2}$$ ■

EXPONENTS

Let a be any positive number and let n be a positive integer. Then, by definition,

1. $a^n = \underbrace{a \cdot a \cdot \cdots \cdot a}_{n \text{ factors}}$
2. $a^0 = 1$
3. $a^{-n} = \dfrac{1}{a^n}$
4. $a^{1/n} = \sqrt[n]{a}$

   $a^{m/n} = \sqrt[n]{a^m} = (\sqrt[n]{a})^m \qquad m$ is any integer

11 Laws of Exponents Let a and b be positive numbers and let r and s be any rational numbers (that is, ratios of integers). Then

1. $a^r \times a^s = a^{r+s}$ **2.** $\dfrac{a^r}{a^s} = a^{r-s}$ **3.** $(a^r)^s = a^{rs}$

4. $(ab)^r = a^r b^r$ **5.** $\left(\dfrac{a}{b}\right)^r = \dfrac{a^r}{b^r} \quad b \neq 0$

In words, these five laws can be stated as follows:

1. To multiply two powers of the same number, we add the exponents.
2. To divide two powers of the same number, we subtract the exponents.
3. To raise a power to a new power, we multiply the exponents.
4. To raise a product to a power, we raise each factor to the power.
5. To raise a quotient to a power, we raise both numerator and denominator to the power.

EXAMPLE 17

(a) $2^8 \times 8^2 = 2^8 \times (2^3)^2 = 2^8 \times 2^6 = 2^{14}$

(b) $$\frac{x^{-2} - y^{-2}}{x^{-1} + y^{-1}} = \frac{\dfrac{1}{x^2} - \dfrac{1}{y^2}}{\dfrac{1}{x} + \dfrac{1}{y}} = \frac{\dfrac{y^2 - x^2}{x^2y^2}}{\dfrac{y + x}{xy}} = \frac{y^2 - x^2}{x^2y^2} \cdot \frac{xy}{y + x}$$
$$= \frac{(y - x)(y + x)}{xy(y + x)} = \frac{y - x}{xy}$$

(c) $4^{3/2} = \sqrt{4^3} = \sqrt{64} = 8$ Alternative solution: $4^{3/2} = \left(\sqrt{4}\right)^3 = 2^3 = 8$

(d) $\dfrac{1}{\sqrt[3]{x^4}} = \dfrac{1}{x^{4/3}} = x^{-4/3}$

(e) $\left(\dfrac{x}{y}\right)^3 \left(\dfrac{y^2x}{z}\right)^4 = \dfrac{x^3}{y^3} \cdot \dfrac{y^8x^4}{z^4} = x^7y^5z^{-4}$ ■

INEQUALITIES

When working with inequalities, note the following rules.

Rules for Inequalities

1. If $a < b$, then $a + c < b + c$.
2. If $a < b$ and $c < d$, then $a + c < b + d$.
3. If $a < b$ and $c > 0$, then $ac < bc$.
4. If $a < b$ and $c < 0$, then $ac > bc$.
5. If $0 < a < b$, then $1/a > 1/b$.

Rule 1 says that we can add any number to both sides of an inequality, and Rule 2 says that two inequalities can be added. However, we have to be careful with multiplication. Rule 3 says that we can multiply both sides of an inequality by a *positive* number, but Rule 4 says that *if we multiply both sides of an inequality by a negative number, then we reverse the direction of the inequality.* For example, if we take the inequality

$3 < 5$ and multiply by 2, we get $6 < 10$, but if we multiply by -2, we get $-6 > -10$. Finally, Rule 5 says that if we take reciprocals, then we reverse the direction of an inequality (provided the numbers are positive).

EXAMPLE 18 Solve the inequality $1 + x < 7x + 5$.

SOLUTION The given inequality is satisfied by some values of x but not by others. To *solve* an inequality means to determine the set of numbers x for which the inequality is true. This is called the *solution set.*

First we subtract 1 from each side of the inequality (using Rule 1 with $c = -1$):

$$x < 7x + 4$$

Then we subtract $7x$ from both sides (Rule 1 with $c = -7x$):

$$-6x < 4$$

Now we divide both sides by -6 (Rule 4 with $c = -\frac{1}{6}$):

$$x > -\tfrac{4}{6} = -\tfrac{2}{3}$$

These steps can all be reversed, so the solution set consists of all numbers greater than $-\frac{2}{3}$. In other words, the solution of the inequality is the interval $\left(-\frac{2}{3}, \infty\right)$. ■

EXAMPLE 19 Solve the inequality $x^2 - 5x + 6 \leqslant 0$.

SOLUTION First we factor the left side:

$$(x - 2)(x - 3) \leqslant 0$$

We know that the corresponding equation $(x - 2)(x - 3) = 0$ has the solutions 2 and 3. The numbers 2 and 3 divide the real line into three intervals:

$$(-\infty, 2) \qquad (2, 3) \qquad (3, \infty)$$

On each of these intervals we determine the signs of the factors. For instance,

$$x \in (-\infty, 2) \quad \Rightarrow \quad x < 2 \quad \Rightarrow \quad x - 2 < 0$$

Then we record these signs in the following chart:

Interval	$x - 2$	$x - 3$	$(x - 2)(x - 3)$
$x < 2$	$-$	$-$	$+$
$2 < x < 3$	$+$	$-$	$-$
$x > 3$	$+$	$+$	$+$

■■ A visual method for solving Example 19 is to use a graphing device to graph the parabola $y = x^2 - 5x + 6$ (as in Figure 2) and observe that the curve lies on or below the x-axis when $2 \leqslant x \leqslant 3$.

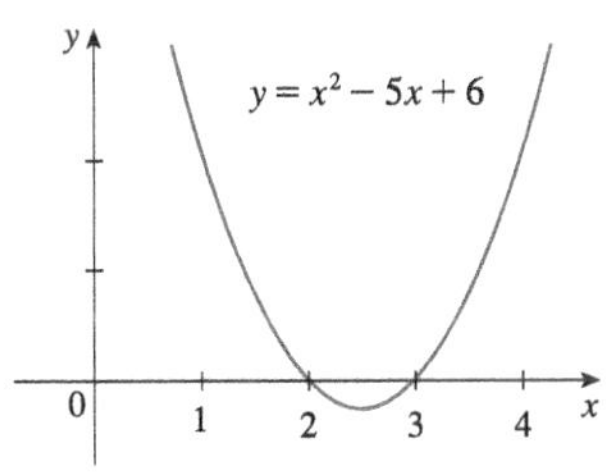

FIGURE 2

Another method for obtaining the information in the chart is to use *test values.* For instance, if we use the test value $x = 1$ for the interval $(-\infty, 2)$, then substitution in $x^2 - 5x + 6$ gives

$$1^2 - 5(1) + 6 = 2$$

The polynomial $x^2 - 5x + 6$ doesn't change sign inside any of the three intervals, so we conclude that it is positive on $(-\infty, 2)$.

Then we read from the chart that $(x - 2)(x - 3)$ is negative when $2 < x < 3$. Thus, the solution of the inequality $(x - 2)(x - 3) \leqslant 0$ is

$$\{x \mid 2 \leqslant x \leqslant 3\} = [2, 3]$$

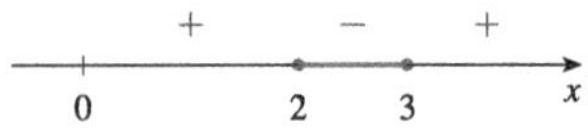

FIGURE 3

Notice that we have included the endpoints 2 and 3 because we are looking for values of x such that the product is either negative or zero. The solution is illustrated in Figure 3. ■

EXAMPLE 20 Solve $x^3 + 3x^2 > 4x$.

SOLUTION First we take all nonzero terms to one side of the inequality sign and factor the resulting expression:

$$x^3 + 3x^2 - 4x > 0 \qquad \text{or} \qquad x(x - 1)(x + 4) > 0$$

As in Example 19 we solve the corresponding equation $x(x - 1)(x + 4) = 0$ and use the solutions $x = -4$, $x = 0$, and $x = 1$ to divide the real line into four intervals $(-\infty, -4)$, $(-4, 0)$, $(0, 1)$, and $(1, \infty)$. On each interval the product keeps a constant sign as shown in the following chart.

Interval	x	$x - 1$	$x + 4$	$x(x - 1)(x + 4)$
$x < -4$	−	−	−	−
$-4 < x < 0$	−	−	+	+
$0 < x < 1$	+	−	+	−
$x > 1$	+	+	+	+

Then we read from the chart that the solution set is

$$\{x \mid -4 < x < 0 \text{ or } x > 1\} = (-4, 0) \cup (1, \infty)$$

FIGURE 4

The solution is illustrated in Figure 4. ■

ABSOLUTE VALUE

The **absolute value** of a number a, denoted by $|a|$, is the distance from a to 0 on the real number line. Distances are always positive or 0, so we have

$$|a| \geqslant 0 \qquad \text{for every number } a$$

For example,

$$|3| = 3 \qquad |-3| = 3 \qquad |0| = 0$$

$$|\sqrt{2} - 1| = \sqrt{2} - 1 \qquad |3 - \pi| = \pi - 3$$

In general, we have

■■ Remember that if a is negative, then $-a$ is positive.

12

$$|a| = a \qquad \text{if } a \geqslant 0$$
$$|a| = -a \qquad \text{if } a < 0$$

EXAMPLE 21 Express $|3x - 2|$ without using the absolute-value symbol.

SOLUTION

$$|3x - 2| = \begin{cases} 3x - 2 & \text{if } 3x - 2 \geqslant 0 \\ -(3x - 2) & \text{if } 3x - 2 < 0 \end{cases}$$

$$= \begin{cases} 3x - 2 & \text{if } x \geqslant \frac{2}{3} \\ 2 - 3x & \text{if } x < \frac{2}{3} \end{cases}$$ ■

Recall that the symbol $\sqrt{}$ means "the positive square root of." Thus, $\sqrt{r} = s$ means $s^2 = r$ and $s \geqslant 0$. Therefore, the equation $\sqrt{a^2} = a$ is not always true. It is true only when $a \geqslant 0$. If $a < 0$, then $-a > 0$, so we have $\sqrt{a^2} = -a$. In view of (12), we then have the equation

13 $$\sqrt{a^2} = |a|$$

which is true for all values of a.

Hints for the proofs of the following properties are given in the exercises.

Properties of Absolute Values Suppose a and b are any real numbers and n is an integer. Then

1. $|ab| = |a||b|$ **2.** $\left|\dfrac{a}{b}\right| = \dfrac{|a|}{|b|} \quad (b \neq 0)$ **3.** $|a^n| = |a|^n$

For solving equations or inequalities involving absolute values, it's often very helpful to use the following statements.

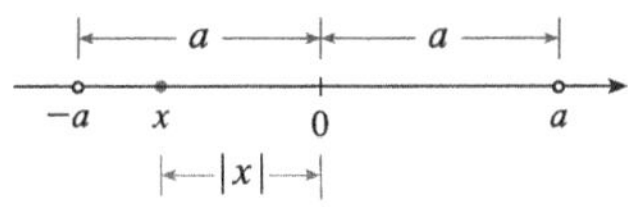

FIGURE 5

Suppose $a > 0$. Then

4. $|x| = a$ if and only if $x = \pm a$

5. $|x| < a$ if and only if $-a < x < a$

6. $|x| > a$ if and only if $x > a$ or $x < -a$

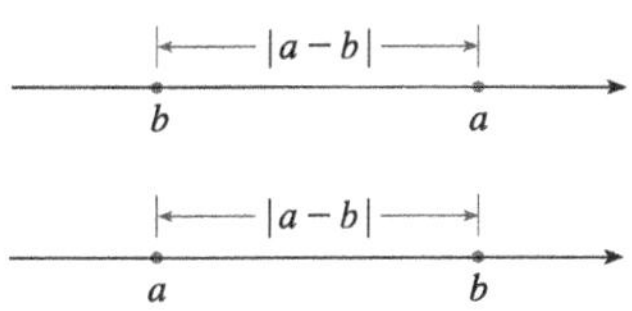

FIGURE 6
Length of a line segment $= |a - b|$

For instance, the inequality $|x| < a$ says that the distance from x to the origin is less than a, and you can see from Figure 5 that this is true if and only if x lies between $-a$ and a.

If a and b are any real numbers, then the distance between a and b is the absolute value of the difference, namely, $|a - b|$, which is also equal to $|b - a|$. (See Figure 6.)

EXAMPLE 22 Solve $|2x - 5| = 3$.

SOLUTION By Property 4 of absolute values, $|2x - 5| = 3$ is equivalent to

$$2x - 5 = 3 \qquad \text{or} \qquad 2x - 5 = -3$$

So $2x = 8$ or $2x = 2$. Thus, $x = 4$ or $x = 1$. ■

EXAMPLE 23 Solve $|x - 5| < 2$.

SOLUTION 1 By Property 5 of absolute values, $|x - 5| < 2$ is equivalent to

$$-2 < x - 5 < 2$$

Therefore, adding 5 to each side, we have

$$3 < x < 7$$

and the solution set is the open interval $(3, 7)$.

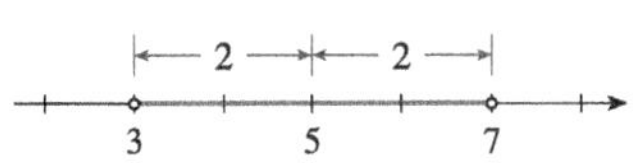

FIGURE 7

SOLUTION 2 Geometrically, the solution set consists of all numbers x whose distance from 5 is less than 2. From Figure 7 we see that this is the interval $(3, 7)$. ■

EXAMPLE 24 Solve $|3x + 2| \geqslant 4$.

SOLUTION By Properties 4 and 6 of absolute values, $|3x + 2| \geqslant 4$ is equivalent to

$$3x + 2 \geqslant 4 \qquad \text{or} \qquad 3x + 2 \leqslant -4$$

In the first case, $3x \geqslant 2$, which gives $x \geqslant \frac{2}{3}$. In the second case, $3x \leqslant -6$, which gives $x \leqslant -2$. So the solution set is

$$\left\{x \mid x \leqslant -2 \text{ or } x \geqslant \tfrac{2}{3}\right\} = (-\infty, -2] \cup \left[\tfrac{2}{3}, \infty\right)$$ ■

EXERCISES

A Click here for answers. S Click here for solutions.

1–16 ■ Expand and simplify.

1. $(-6ab)(0.5ac)$
2. $-(2x^2y)(-xy^4)$
3. $2x(x - 5)$
4. $(4 - 3x)x$
5. $-2(4 - 3a)$
6. $8 - (4 + x)$
7. $4(x^2 - x + 2) - 5(x^2 - 2x + 1)$
8. $5(3t - 4) - (t^2 + 2) - 2t(t - 3)$
9. $(4x - 1)(3x + 7)$
10. $x(x - 1)(x + 2)$
11. $(2x - 1)^2$
12. $(2 + 3x)^2$
13. $y^4(6 - y)(5 + y)$
14. $(t - 5)^2 - 2(t + 3)(8t - 1)$
15. $(1 + 2x)(x^2 - 3x + 1)$
16. $(1 + x - x^2)^2$

17–28 ■ Perform the indicated operations and simplify.

17. $\dfrac{2 + 8x}{2}$
18. $\dfrac{9b - 6}{3b}$
19. $\dfrac{1}{x + 5} + \dfrac{2}{x - 3}$
20. $\dfrac{1}{x + 1} + \dfrac{1}{x - 1}$
21. $u + 1 + \dfrac{u}{u + 1}$
22. $\dfrac{2}{a^2} - \dfrac{3}{ab} + \dfrac{4}{b^2}$
23. $\dfrac{x/y}{z}$
24. $\dfrac{x}{y/z}$
25. $\left(\dfrac{-2r}{s}\right)\left(\dfrac{s^2}{-6t}\right)$
26. $\dfrac{a}{bc} \div \dfrac{b}{ac}$
27. $\dfrac{1 + \dfrac{1}{c - 1}}{1 - \dfrac{1}{c - 1}}$
28. $1 + \dfrac{1}{1 + \dfrac{1}{1 + x}}$

29–48 ■ Factor the expression.

29. $2x + 12x^3$
30. $5ab - 8abc$
31. $x^2 + 7x + 6$
32. $x^2 - x - 6$
33. $x^2 - 2x - 8$
34. $2x^2 + 7x - 4$
35. $9x^2 - 36$
36. $8x^2 + 10x + 3$
37. $6x^2 - 5x - 6$
38. $x^2 + 10x + 25$
39. $t^3 + 1$
40. $4t^2 - 9s^2$
41. $4t^2 - 12t + 9$
42. $x^3 - 27$
43. $x^3 + 2x^2 + x$
44. $x^3 - 4x^2 + 5x - 2$
45. $x^3 + 3x^2 - x - 3$
46. $x^3 - 2x^2 - 23x + 60$
47. $x^3 + 5x^2 - 2x - 24$
48. $x^3 - 3x^2 - 4x + 12$

49–54 ■ Simplify the expression.

49. $\dfrac{x^2 + x - 2}{x^2 - 3x + 2}$
50. $\dfrac{2x^2 - 3x - 2}{x^2 - 4}$
51. $\dfrac{x^2 - 1}{x^2 - 9x + 8}$
52. $\dfrac{x^3 + 5x^2 + 6x}{x^2 - x - 12}$
53. $\dfrac{1}{x + 3} + \dfrac{1}{x^2 - 9}$
54. $\dfrac{x}{x^2 + x - 2} - \dfrac{2}{x^2 - 5x + 4}$

55–60 ■ Complete the square.

55. $x^2 + 2x + 5$
56. $x^2 - 16x + 80$
57. $x^2 - 5x + 10$
58. $x^2 + 3x + 1$
59. $4x^2 + 4x - 2$
60. $3x^2 - 24x + 50$

61–68 ■ Solve the equation.

61. $x^2 + 9x - 10 = 0$
62. $x^2 - 2x - 8 = 0$
63. $x^2 + 9x - 1 = 0$
64. $x^2 - 2x - 7 = 0$
65. $3x^2 + 5x + 1 = 0$
66. $2x^2 + 7x + 2 = 0$
67. $x^3 - 2x + 1 = 0$
68. $x^3 + 3x^2 + x - 1 = 0$

69–72 ■ Which of the quadratics are irreducible?

69. $2x^2 + 3x + 4$
70. $2x^2 + 9x + 4$
71. $3x^2 + x - 6$
72. $x^2 + 3x + 6$

73–76 ■ Use the Binomial Theorem to expand the expression.

73. $(a + b)^6$
74. $(a + b)^7$
75. $(x^2 - 1)^4$
76. $(3 + x^2)^5$

77–82 ■ Simplify the radicals.

77. $\sqrt{32}\sqrt{2}$ **78.** $\dfrac{\sqrt[3]{-2}}{\sqrt[3]{54}}$ **79.** $\dfrac{\sqrt[4]{32x^4}}{\sqrt[4]{2}}$

80. $\sqrt{xy}\sqrt{x^3y}$ **81.** $\sqrt{16a^4b^3}$ **82.** $\dfrac{\sqrt[5]{96a^6}}{\sqrt[5]{3a}}$

83–100 ■ Use the Laws of Exponents to rewrite and simplify the expression.

83. $3^{10} \times 9^8$ **84.** $2^{16} \times 4^{10} \times 16^6$

85. $\dfrac{x^9(2x)^4}{x^3}$ **86.** $\dfrac{a^n \times a^{2n+1}}{a^{n-2}}$

87. $\dfrac{a^{-3}b^4}{a^{-5}b^5}$ **88.** $\dfrac{x^{-1} + y^{-1}}{(x + y)^{-1}}$

89. $3^{-1/2}$ **90.** $96^{1/5}$

91. $125^{2/3}$ **92.** $64^{-4/3}$

93. $(2x^2y^4)^{3/2}$ **94.** $(x^{-5}y^3z^{10})^{-3/5}$

95. $\sqrt[5]{y^6}$ **96.** $(\sqrt[4]{a})^3$

97. $\dfrac{1}{(\sqrt{t})^5}$ **98.** $\dfrac{\sqrt[8]{x^5}}{\sqrt[4]{x^3}}$

99. $\sqrt[4]{\dfrac{t^{1/2}\sqrt{st}}{s^{2/3}}}$ **100.** $\sqrt[4]{r^{2n+1}} \times \sqrt[4]{r^{-1}}$

101–108 ■ Rationalize the expression.

101. $\dfrac{\sqrt{x} - 3}{x - 9}$ **102.** $\dfrac{(1/\sqrt{x}) - 1}{x - 1}$

103. $\dfrac{x\sqrt{x} - 8}{x - 4}$ **104.** $\dfrac{\sqrt{2 + h} + \sqrt{2 - h}}{h}$

105. $\dfrac{2}{3 - \sqrt{5}}$ **106.** $\dfrac{1}{\sqrt{x} - \sqrt{y}}$

107. $\sqrt{x^2 + 3x + 4} - x$ **108.** $\sqrt{x^2 + x} - \sqrt{x^2 - x}$

109–116 ■ State whether or not the equation is true for all values of the variable.

109. $\sqrt{x^2} = x$ **110.** $\sqrt{x^2 + 4} = |x| + 2$

111. $\dfrac{16 + a}{16} = 1 + \dfrac{a}{16}$ **112.** $\dfrac{1}{x^{-1} + y^{-1}} = x + y$

113. $\dfrac{x}{x + y} = \dfrac{1}{1 + y}$ **114.** $\dfrac{2}{4 + x} = \dfrac{1}{2} + \dfrac{2}{x}$

115. $(x^3)^4 = x^7$

116. $6 - 4(x + a) = 6 - 4x - 4a$

117–126 ■ Rewrite the expression without using the absolute value symbol.

117. $|5 - 23|$ **118.** $|\pi - 2|$

119. $|\sqrt{5} - 5|$ **120.** $||-2| - |-3||$

121. $|x - 2|$ if $x < 2$ **122.** $|x - 2|$ if $x > 2$

123. $|x + 1|$ **124.** $|2x - 1|$

125. $|x^2 + 1|$ **126.** $|1 - 2x^2|$

127–142 ■ Solve the inequality in terms of intervals and illustrate the solution set on the real number line.

127. $2x + 7 > 3$ **128.** $4 - 3x \geqslant 6$

129. $1 - x \leqslant 2$ **130.** $1 + 5x > 5 - 3x$

131. $0 \leqslant 1 - x < 1$ **132.** $1 < 3x + 4 \leqslant 16$

133. $(x - 1)(x - 2) > 0$ **134.** $x^2 < 2x + 8$

135. $x^2 < 3$ **136.** $x^2 \geqslant 5$

137. $x^3 - x^2 \leqslant 0$

138 $(x + 1)(x - 2)(x + 3) \geqslant 0$

139. $x^3 > x$ **140.** $x^3 + 3x < 4x^2$

141. $\dfrac{1}{x} < 4$ **142.** $-3 < \dfrac{1}{x} \leqslant 1$

143. The relationship between the Celsius and Fahrenheit temperature scales is given by $C = \frac{5}{9}(F - 32)$, where C is the temperature in degrees Celsius and F is the temperature in degrees Fahrenheit. What interval on the Celsius scale corresponds to the temperature range $50 \leqslant F \leqslant 95$?

144. Use the relationship between C and F given in Exercise 143 to find the interval on the Fahrenheit scale corresponding to the temperature range $20 \leqslant C \leqslant 30$.

145. As dry air moves upward, it expands and in so doing cools at a rate of about 1°C for each 100-m rise, up to about 12 km.
(a) If the ground temperature is 20°C, write a formula for the temperature at height h.
(b) What range of temperature can be expected if a plane takes off and reaches a maximum height of 5 km?

146. If a ball is thrown upward from the top of a building 128 ft high with an initial velocity of 16 ft/s, then the height h above the ground t seconds later will be

$$h = 128 + 16t - 16t^2$$

During what time interval will the ball be at least 32 ft above the ground?

147–148 ■ Solve the equation for x.

147. $|x + 3| = |2x + 1|$ **148.** $|3x + 5| = 1$

149–156 ■ Solve the inequality.

149. $|x| < 3$ **150.** $|x| \geqslant 3$

151. $|x - 4| < 1$ **152.** $|x - 6| < 0.1$

153. $|x + 5| \geqslant 2$ **154.** $|x + 1| \geqslant 3$

155. $|2x - 3| \leqslant 0.4$ **156.** $|5x - 2| < 6$

157. Solve the inequality $a(bx - c) \geqslant bc$ for x, assuming that a, b, and c are positive constants.

158. Solve the inequality $ax + b < c$ for x, assuming that a, b, and c are negative constants.

159 Prove that $|ab| = |a||b|$. [*Hint:* Use Equation 3.]

160. Show that if $0 < a < b$, then $a^2 < b^2$.

ANSWERS

[S] Click here for solutions.

1. $-3a^2bc$ **2.** $2x^3y^5$ **3.** $2x^2 - 10x$ **4.** $4x - 3x^2$
5. $-8 + 6a$ **6.** $4 - x$ **7.** $-x^2 + 6x + 3$
8. $-3t^2 + 21t - 22$ **9.** $12x^2 + 25x - 7$
10. $x^3 + x^2 - 2x$ **11.** $4x^2 - 4x + 1$
12. $9x^2 + 12x + 4$ **13.** $30y^4 + y^5 - y^6$
14. $-15t^2 - 56t + 31$ **15.** $2x^3 - 5x^2 - x + 1$
16. $x^4 - 2x^3 - x^2 + 2x + 1$ **17.** $1 + 4x$ **18.** $3 - 2/b$
19. $\dfrac{3x + 7}{x^2 + 2x - 15}$ **20.** $\dfrac{2x}{x^2 - 1}$ **21.** $\dfrac{u^2 + 3u + 1}{u + 1}$
22. $\dfrac{2b^2 - 3ab + 4a^2}{a^2b^2}$ **23.** $\dfrac{x}{yz}$ **24.** $\dfrac{zx}{y}$ **25.** $\dfrac{rs}{3t}$
26. $\dfrac{a^2}{b^2}$ **27.** $\dfrac{c}{c - 2}$ **28.** $\dfrac{3 + 2x}{2 + x}$ **29.** $2x(1 + 6x^2)$
30. $ab(5 - 8c)$ **31.** $(x + 6)(x + 1)$ **32.** $(x - 3)(x + 2)$
33. $(x - 4)(x + 2)$ **34.** $(2x - 1)(x + 4)$
35. $9(x - 2)(x + 2)$ **36.** $(4x + 3)(2x + 1)$
37. $(3x + 2)(2x - 3)$ **38.** $(x + 5)^2$
39. $(t + 1)(t^2 - t + 1)$ **40.** $(2t - 3s)(2t + 3s)$
41. $(2t - 3)^2$ **42.** $(x - 3)(x^2 + 3x + 9)$
43. $x(x + 1)^2$ **44.** $(x - 1)^2(x - 2)$
45. $(x - 1)(x + 1)(x + 3)$ **46.** $(x - 3)(x + 5)(x - 4)$
47. $(x - 2)(x + 3)(x + 4)$ **48.** $(x - 2)(x - 3)(x + 2)$
49. $\dfrac{x + 2}{x - 2}$ **50.** $\dfrac{2x + 1}{x + 2}$ **51.** $\dfrac{x + 1}{x - 8}$ **52.** $\dfrac{x(x + 2)}{x - 4}$
53. $\dfrac{x - 2}{x^2 - 9}$ **54.** $\dfrac{x^2 - 6x - 4}{(x - 1)(x + 2)(x - 4)}$
55. $(x + 1)^2 + 4$ **56.** $(x - 8)^2 + 16$ **57.** $\left(x - \frac{5}{2}\right)^2 + \frac{15}{4}$
58. $\left(x + \frac{3}{2}\right)^2 - \frac{5}{4}$ **59.** $(2x + 1)^2 - 3$
60. $3(x - 4)^2 + 2$ **61.** $1, -10$ **62.** $-2, 4$
63. $\dfrac{-9 \pm \sqrt{85}}{2}$ **64.** $1 \pm 2\sqrt{2}$ **65.** $\dfrac{-5 \pm \sqrt{13}}{6}$
66. $\dfrac{-7 \pm \sqrt{33}}{4}$ **67.** $1, \dfrac{-1 \pm \sqrt{5}}{2}$ **68.** $-1, -1 \pm \sqrt{2}$
69. Irreducible **70.** Not irreducible
71. Not irreducible (two real roots) **72.** Irreducible
73. $a^6 + 6a^5b + 15a^4b^2 + 20a^3b^3 + 15a^2b^4 + 6ab^5 + b^6$
74. $a^7 + 7a^6b + 21a^5b^2 + 35a^4b^3 + 35a^3b^4 + 21a^2b^5 + 7ab^6 + b^7$
75. $x^8 - 4x^6 + 6x^4 - 4x^2 + 1$
76. $243 + 405x^2 + 270x^4 + 90x^6 + 15x^8 + x^{10}$
77. 8 **78.** $-\frac{1}{3}$ **79.** $2|x|$ **80.** $x^2|y|$
81. $4a^2b\sqrt{b}$ **82.** $2a$ **83.** 3^{26} **84.** 2^{60} **85.** $16x^{10}$
86. a^{2n+3} **87.** $\dfrac{a^2}{b}$ **88.** $\dfrac{(x + y)^2}{xy}$ **89.** $\dfrac{1}{\sqrt{3}}$
90. $2^5\sqrt{3}$ **91.** 25 **92.** $\frac{1}{256}$ **93.** $2\sqrt{2}|x|^3y^6$
94. $\dfrac{x^3}{y^{9/5}z^6}$ **95.** $y^{6/5}$ **96.** $a^{3/4}$ **97.** $t^{-5/2}$ **98.** $\dfrac{1}{x^{1/8}}$
99. $\dfrac{t^{1/4}}{s^{1/24}}$ **100.** $r^{n/2}$ **101.** $\dfrac{1}{\sqrt{x} + 3}$ **102.** $\dfrac{-1}{\sqrt{x} + x}$
103. $\dfrac{x^2 + 4x + 16}{x\sqrt{x} + 8}$ **104.** $\dfrac{2}{\sqrt{2 + h} - \sqrt{2 - h}}$
105. $\dfrac{3 + \sqrt{5}}{2}$ **106.** $\dfrac{\sqrt{x} + \sqrt{y}}{x - y}$ **107.** $\dfrac{3x + 4}{\sqrt{x^2 + 3x + 4} + x}$
108. $\dfrac{2x}{\sqrt{x^2 + x} + \sqrt{x^2 - x}}$ **109.** False **110.** False
111. True **112.** False **113.** False **114.** False
115. False **116.** True **117.** 18 **118.** $\pi - 2$
119. $5 - \sqrt{5}$ **120.** 1 **121.** $2 - x$ **122.** $x - 2$
123. $|x + 1| = \begin{cases} x + 1 & \text{if } x \geqslant -1 \\ -x - 1 & \text{if } x < -1 \end{cases}$
124. $|2x - 1| = \begin{cases} 2x - 1 & \text{if } x \geqslant \frac{1}{2} \\ 1 - 2x & \text{if } x < \frac{1}{2} \end{cases}$
125. $x^2 + 1$
126. $|1 - 2x^2| = \begin{cases} 1 - 2x^2 & \text{if } -1/\sqrt{2} \leqslant x \leqslant 1/\sqrt{2} \\ 2x^2 - 1 & \text{if } x < -1/\sqrt{2} \text{ or } x > 1/\sqrt{2} \end{cases}$

127. $(-2, \infty)$
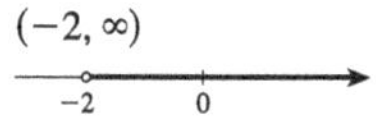
128. $\left(-\infty, -\frac{2}{3}\right]$
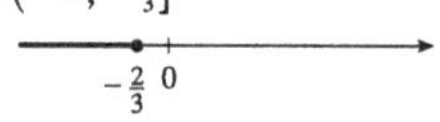
129. $[-1, \infty)$
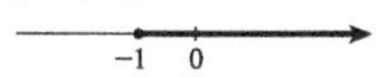
130. $\left(\frac{1}{2}, \infty\right)$
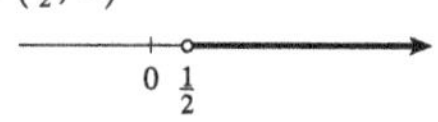
131. $(0, 1]$
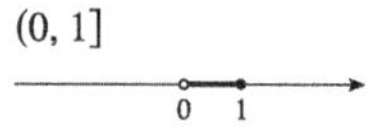
132. $(-1, 4]$
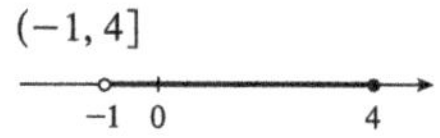
133. $(-\infty, 1) \cup (2, \infty)$
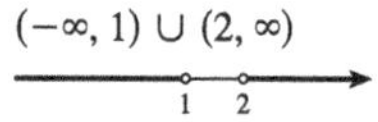
134. $(-2, 4)$
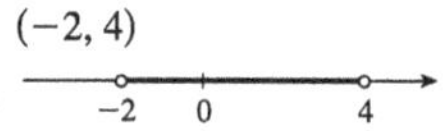
135. $(-\sqrt{3}, \sqrt{3})$
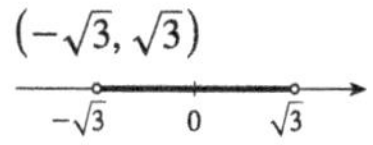
136. $\left(-\infty, -\sqrt{5}\right] \cup \left[\sqrt{5}, \infty\right)$
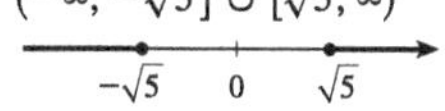
137. $(-\infty, 1]$
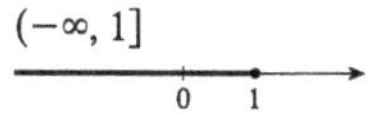
138. $[-3, -1] \cup [2, \infty)$
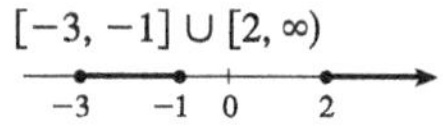
139. $(-1, 0) \cup (1, \infty)$
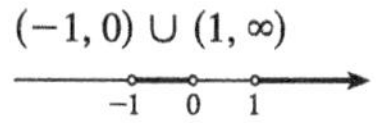
140. $(-\infty, 0) \cup (1, 3)$
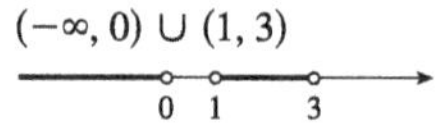
141. $(-\infty, 0) \cup \left(\frac{1}{4}, \infty\right)$
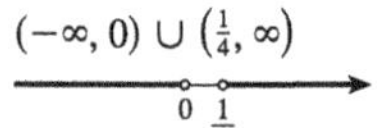
142. $\left(-\infty, -\frac{1}{3}\right) \cup [1, \infty)$
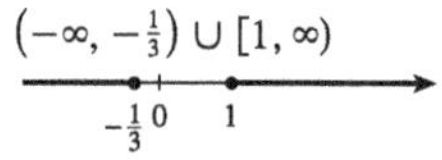

143. $[10, 35]$ **144.** $[68, 86]$
145. (a) $T = 20 - 10h, 0 \leqslant h \leqslant 12$ (b) $-30°\text{C} \leqslant T \leqslant 20°\text{C}$
146. $[0, 3]$ **147.** $2, -\frac{4}{3}$ **148.** $-\frac{4}{3}, -2$ **149.** $(-3, 3)$
150. $(-\infty, -3] \cup [3, \infty)$ **151.** $(3, 5)$ **152.** $(5.9, 6.1)$
153. $(-\infty, -7] \cup [-3, \infty)$ **154.** $(-\infty, -4] \cup [2, \infty)$
155. $[1.3, 1.7]$ **156.** $\left(-\frac{4}{5}, \frac{8}{5}\right)$
157. $x \geqslant \dfrac{(a + b)c}{ab}$ **158.** $x > \dfrac{c - b}{a}$

SOLUTIONS

1. $(-6ab)(0.5ac) = (-6)(0.5)(a \cdot abc) = -3a^2bc$
2. $-(2x^2y)(-xy^4) = 2x^2xyy^4 = 2x^3y^5$
3. $2x(x-5) = 2x \cdot x - 2x \cdot 5 = 2x^2 - 10x$
4. $(4-3x)x = 4 \cdot x - 3x \cdot x = 4x - 3x^2$
5. $-2(4-3a) = -2 \cdot 4 + 2 \cdot 3a = -8 + 6a$
6. $8 - (4+x) = 8 - 4 - x = 4 - x$
7. $4(x^2 - x + 2) - 5(x^2 - 2x + 1) = 4x^2 - 4x + 8 - 5x^2 - 5(-2x) - 5$
   $= 4x^2 - 5x^2 - 4x + 10x + 8 - 5 = -x^2 + 6x + 3$
8. $5(3t-4) - (t^2+2) - 2t(t-3) = 15t - 20 - t^2 - 2 - 2t^2 + 6t$
   $= (-1-2)t^2 + (15+6)t - 20 - 2 = -3t^2 + 21t - 22$
9. $(4x-1)(3x+7) = 4x(3x+7) - (3x+7) = 12x^2 + 28x - 3x - 7 = 12x^2 + 25x - 7$
10. $x(x-1)(x+2) = (x^2-x)(x+2) = x^2(x+2) - x(x+2) = x^3 + 2x^2 - x^2 - 2x$
    $= x^3 + x^2 - 2x$
11. $(2x-1)^2 = (2x)^2 - 2(2x)(1) + 1^2 = 4x^2 - 4x + 1$
12. $(2+3x)^2 = 2^2 + 2(2)(3x) + (3x)^2 = 9x^2 + 12x + 4$
13. $y^4(6-y)(5+y) = y^4[6(5+y) - y(5+y)] = y^4(30 + 6y - 5y - y^2)$
    $= y^4(30 + y - y^2) = 30y^4 + y^5 - y^6$
14. $(t-5)^2 - 2(t+3)(8t-1) = t^2 - 2(5t) + 5^2 - 2(8t^2 - t + 24t - 3)$
    $= t^2 - 10t + 25 - 16t^2 + 2t - 48t + 6 = -15t^2 - 56t + 31$
15. $(1+2x)(x^2-3x+1) = 1(x^2-3x+1) + 2x(x^2-3x+1) = x^2 - 3x + 1 + 2x^3 - 6x^2 + 2x$
    $= 2x^3 - 5x^2 - x + 1$
16. $(1+x-x^2)^2 = (1+x-x^2)(1+x-x^2) = 1(1+x-x^2) + x(1+x-x^2) - x^2(1+x-x^2)$
    $= 1 + x - x^2 + x + x^2 - x^3 - x^2 - x^3 + x^4 = x^4 - 2x^3 - x^2 + 2x + 1$
17. $\dfrac{2+8x}{2} = \dfrac{2}{2} + \dfrac{8x}{2} = 1 + 4x$
18. $\dfrac{9b-6}{3b} = \dfrac{9b}{3b} - \dfrac{6}{3b} = 3 - \dfrac{2}{b}$
19. $\dfrac{1}{x+5} + \dfrac{2}{x-3} = \dfrac{(1)(x-3) + 2(x+5)}{(x+5)(x-3)} = \dfrac{x - 3 + 2x + 10}{(x+5)(x-3)} = \dfrac{3x+7}{x^2+2x-15}$
20. $\dfrac{1}{x+1} + \dfrac{1}{x-1} = \dfrac{1(x-1) + 1(x+1)}{(x+1)(x-1)} = \dfrac{x - 1 + x + 1}{x^2-1} = \dfrac{2x}{x^2-1}$
21. $u + 1 + \dfrac{u}{u+1} = \dfrac{(u+1)(u+1) + u}{u+1} = \dfrac{u^2 + 2u + 1 + u}{u+1} = \dfrac{u^2+3u+1}{u+1}$
22. $\dfrac{2}{a^2} - \dfrac{3}{ab} + \dfrac{4}{b^2} = \dfrac{2b^2}{a^2b^2} - \dfrac{3ab}{a^2b^2} + \dfrac{4a^2}{a^2b^2} = \dfrac{2b^2 - 3ab + 4a^2}{a^2b^2}$
23. $\dfrac{x/y}{z} = \dfrac{x/y}{z/1} = \dfrac{1}{z} \cdot \dfrac{x}{y} = \dfrac{x}{yz}$
24. $\dfrac{x}{y/z} = \dfrac{x/1}{y/z} = \dfrac{z}{y} \cdot \dfrac{x}{1} = \dfrac{zx}{y}$
25. $\left(\dfrac{-2r}{s}\right)\left(\dfrac{s^2}{-6t}\right) = \dfrac{-2rs^2}{-6st} = \dfrac{rs}{3t}$
26. $\dfrac{a}{bc} \div \dfrac{b}{ac} = \dfrac{a}{bc} \times \dfrac{ac}{b} = \dfrac{a^2c}{b^2c} = \dfrac{a^2}{b^2}$

27. $\dfrac{1+\dfrac{1}{c-1}}{1-\dfrac{1}{c-1}} = \dfrac{\dfrac{c-1+1}{c-1}}{\dfrac{c-1-1}{c-1}} = \dfrac{\dfrac{c}{c-1}}{\dfrac{c-2}{c-1}} = \dfrac{c-1}{c-2}\cdot\dfrac{c}{c-1} = \dfrac{c}{c-2}$

28. $1+\dfrac{1}{1+\dfrac{1}{1+x}} = 1+\dfrac{1}{\dfrac{1+x+1}{1+x}} = 1+\dfrac{1+x}{2+x} = \dfrac{2+x+1+x}{2+x} = \dfrac{3+2x}{2+x}$

29. $2x+12x^3 = 2x\cdot 1 + 2x\cdot 6x^2 = 2x(1+6x^2)$

30. $5ab-8abc = ab\cdot 5 - ab\cdot 8c = ab(5-8c)$

31. The two integers that add to give 7 and multiply to give 6 are 6 and 1. Therefore $x^2+7x+6=(x+6)(x+1)$.

32. The two integers that add to give -1 and multiply to give -6 are -3 and 2.
Therefore $x^2-2x-6=(x-3)(x+2)$.

33. The two integers that add to give -2 and multiply to give -8 are -4 and 2.
Therefore $x^2-2x-8=(x-4)(x+2)$.

34. $2x^2+7x-4=(2x-1)(x+4)$

35. $9x^2-36=9(x^2-4)=9(x-2)(x+2)$ [Equation 3 with $a=x$, $b=2$]

36. $8x^2+10x+3=(4x+3)(2x+1)$

37. $6x^2-5x-6=(3x+2)(2x-3)$

38. $x^2+10x+25=(x+5)^2$ [Equation 1 with $a-x$, $b=5$]

39. $t^3+1=(t+1)(t^2-t+1)$ [Equation 5 with $a=t$, $b=1$]

40. $4t^2-9s^2=(2t)^2-(3s)^2=(2t-3s)(2t+3s)$ [Equation 3 with $a=2t$, $b=3s$]

41. $4t^2-12t+9=(2t-3)^2$ [Equation 2 with $a=2t$, $b=3$]

42. $x^3-27=(x-3)(x^2+3x+9)$ [Equation 4 with $a=x$, $b=3$]

43. $x^3+2x^2+x=x(x^2+2x+1)=x(x+1)^2$ [Equation 1 with $a=x$, $b=1$]

44. Let $p(x)=x^3-4x^2+5x-2$, and notice that $p(1)=0$, so by the Factor Theorem, $(x-1)$ is a factor.
Use long division (as in Example 8):

$$\begin{array}{r|l}
 & x^2 - 3x + 2 \\
\hline
x-1 & x^3 - 4x^2 + 5x - 2 \\
 & \underline{x^3 - x^2} \\
 & \quad - 3x^2 + 5x \\
 & \quad \underline{- 3x^2 + 3x} \\
 & \qquad\qquad 2x - 2 \\
 & \qquad\qquad \underline{2x - 2}
\end{array}$$

Therefore $x^3-4x^2+5x-2=(x-1)(x^2-3x+2)=(x-1)(x-2)(x-1)=(x-1)^2(x-2)$.

45. Let $p(x)=x^3+3x^2-x-3$, and notice that $p(1)=0$, so by the Factor Theorem, $(x-1)$ is a factor.
Use long division (as in Example 8):

$$\begin{array}{r|l}
 & x^2 + 4x + 3 \\
\hline
x-1 & x^3 + 3x^2 - x - 3 \\
 & \underline{x^3 - x^2} \\
 & \quad 4x^2 - x \\
 & \quad \underline{4x^2 - 4x} \\
 & \qquad\qquad 3x - 3 \\
 & \qquad\qquad \underline{3x - 3}
\end{array}$$

Therefore $x^3+3x^2-x-3=(x-1)(x^2+4x+3)=(x-1)(x+1)(x+3)$.

46. Let $p(x) = x^3 - 2x^2 - 23x + 60$, and notice that $p(3) = 0$, so by the Factor Theorem, $(x - 3)$ is a factor. Use long division (as in Example 8):

$$\begin{array}{r|l} & x^2 + x - 20 \\ \hline x - 3 & x^3 - 2x^2 - 23x + 60 \\ & x^3 - 3x^2 \\ \hline & \quad x^2 - 23x \\ & \quad x^2 - 3x \\ \hline & \quad -20x + 60 \\ & \quad -20x + 60 \\ \hline \end{array}$$

Therefore $x^3 - 2x^2 - 23x + 60 = (x - 3)(x^2 + x - 20) = (x - 3)(x + 5)(x - 4)$.

47. Let $p(x) = x^3 + 5x^2 - 2x - 24$, and notice that $p(2) = 2^3 + 5(2)^2 - 2(2) - 24 = 0$, so by the Factor Theorem, $(x - 2)$ is a factor. Use long division (as in Example 8):

$$\begin{array}{r|l} & x^2 + 7x + 12 \\ \hline x - 2 & x^3 + 5x^2 - 2x - 24 \\ & x^3 - 2x^2 \\ \hline & \quad 7x^2 - 2x \\ & \quad 7x^2 - 14x \\ \hline & \quad 12x - 24 \\ & \quad 12x - 24 \\ \hline \end{array}$$

Therefore $x^3 + 5x^2 - 2x - 24 = (x - 2)(x^2 + 7x + 12) = (x - 2)(x + 3)(x + 4)$.

48. Let $p(x) = x^3 - 3x^2 - 4x + 12$, and notice that $p(2) = 0$, so by the Factor Theorem, $(x - 2)$ is a factor. Use long division (as in Example 8):

$$\begin{array}{r|l} & x^2 - x - 6 \\ \hline x - 2 & x^3 - 3x^2 - 4x + 12 \\ & x^3 - 2x^2 \\ \hline & \quad -x^2 - 4x \\ & \quad -x^2 + 2x \\ \hline & \quad -6x + 12 \\ & \quad -6x + 12 \\ \hline \end{array}$$

Therefore $x^3 - 3x^2 - 4x + 12 = (x - 2)(x^2 - x - 6) = (x - 2)(x - 3)(x + 2)$.

49. $\dfrac{x^2 + x - 2}{x^2 - 3x + 2} = \dfrac{(x + 2)(x - 1)}{(x - 2)(x - 1)} = \dfrac{x + 2}{x - 2}$

50. $\dfrac{2x^2 - 3x - 2}{x^2 - 4} = \dfrac{(2x + 1)(x - 2)}{(x - 2)(x + 2)} = \dfrac{2x + 1}{x + 2}$

51. $\dfrac{x^2 - 1}{x^2 - 9x + 8} = \dfrac{(x - 1)(x + 1)}{(x - 8)(x - 1)} = \dfrac{x + 1}{x - 8}$

52. $\dfrac{x^3 + 5x^2 + 6x}{x^2 - x - 12} = \dfrac{x(x^2 + 5x + 6)}{(x - 4)(x + 3)} = \dfrac{x(x + 3)(x + 2)}{(x - 4)(x + 3)} = \dfrac{x(x + 2)}{x - 4}$

53. $\dfrac{1}{x + 3} + \dfrac{1}{x^2 - 9} = \dfrac{1}{x + 3} + \dfrac{1}{(x - 3)(x + 3)} - \dfrac{1(x - 3) + 1}{(x - 3)(x + 3)} = \dfrac{x - 2}{x^2 - 9}$

54. $\dfrac{x}{x^2 + x - 2} - \dfrac{2}{x^2 - 5x + 4} = \dfrac{x}{(x - 1)(x + 2)} - \dfrac{2}{(x - 4)(x - 1)} = \dfrac{x(x - 4) - 2(x + 2)}{(x - 1)(x + 2)(x - 4)}$

$$= \frac{x^2 - 4x - 2x - 4}{(x - 1)(x + 2)(x - 4)} = \frac{x^2 - 6x - 4}{(x - 1)(x + 2)(x - 4)}$$

55. $x^2 + 2x + 5 = [x^2 + 2x] + 5 = [x^2 + 2x + (1)^2 - (1)^2] + 5 = (x + 1)^2 + 5 - 1 = (x + 1)^2 + 4$

56. $x^2 - 16x + 80 = [x^2 - 16x] + 80 = [x^2 - 16x + (8)^2 - (8)^2] + 80 = (x-8)^2 + 80 - 64 = (x-8)^2 + 16$

57. $x^2 - 5x + 10 = [x^2 - 5x] + 10 = \left[x^2 - 5x + \left(-\frac{5}{2}\right)^2 - \left(-\frac{5}{2}\right)^2\right] + 10 = \left(x - \frac{5}{2}\right)^2 + 10 - \frac{25}{4} = \left(x - \frac{5}{2}\right)^2 + \frac{15}{4}$

58. $x^2 + 3x + 1 = [x^2 + 3x] + 1 = \left[x^2 + 3x + \left(\frac{3}{2}\right)^2 - \left(\frac{3}{2}\right)^2\right] + 1 = \left(x + \frac{3}{2}\right)^2 + 1 - \left(\frac{3}{2}\right)^2 = \left(x + \frac{3}{2}\right)^2 - \frac{5}{4}$

59. $4x^2 + 4x - 2 = 4[x^2 + x] - 2 = 4\left[x^2 + x + \left(\frac{1}{2}\right)^2 - \left(\frac{1}{2}\right)^2\right] - 2 = 4\left(x + \frac{1}{2}\right)^2 - 2 - 4\left(\frac{1}{4}\right) = 4\left(x + \frac{1}{2}\right)^2 - 3$

60. $3x^2 - 24x + 50 = 3[x^2 - 8x] + 50 = 3[x^2 - 8x + (-4)^2 - (-4)^2] + 50 = 3(x-4)^2 + 50 - 3(-4)^2$
$= 3(x-4)^2 + 2$

61. $x^2 - 9x - 10 = 0 \;\Leftrightarrow\; (x+10)(x-1) = 0 \;\Leftrightarrow\; x + 10 = 0$ or $x - 1 = 0 \;\Leftrightarrow\; x = -10$ or $x = 1$.

62. $x^2 - 2x - 8 = 0 \;\Leftrightarrow\; (x-4)(x+2) = 0 \;\Leftrightarrow\; x - 4 = 0$ or $x + 2 = 0 \;\Leftrightarrow\; x = 4$ or $x = -2$.

63. Using the quadratic formula, $x^2 + 9x - 1 = 0 \;\Leftrightarrow\; x = \dfrac{-9 \pm \sqrt{9^2 - 4(1)(-1)}}{2(1)} = \dfrac{9 \pm \sqrt{85}}{2}$.

64. Using the quadratic formula, $x^2 - 2x - 7 = 0 \;\Leftrightarrow\; x = \dfrac{2 \pm \sqrt{4 - 4(1)(-7)}}{2} = \dfrac{2 \pm \sqrt{32}}{2} = 1 \pm 2\sqrt{2}$.

65. Using the quadratic formula, $3x^2 + 5x + 1 = 0 \;\Leftrightarrow\; x = \dfrac{-5 \pm \sqrt{5^2 - 4(3)(1)}}{2(3)} = \dfrac{-5 \pm \sqrt{13}}{6}$.

66. Using the quadratic formula, $2x^2 + 7x + 2 = 0 \;\Leftrightarrow\; x = \dfrac{-7 \pm \sqrt{49 - 4(2)(2)}}{2(2)} = \dfrac{-7 \pm \sqrt{33}}{4}$.

67. Let $p(x) = x^3 - 2x + 1$, and notice that $p(1) = 0$, so by the Factor Theorem, $(x-1)$ is a factor.
Use long division:

$$
\begin{array}{r|l}
 & x^2 + x - 1 \\
\hline
x - 1 & x^3 + 0x^2 - 2x + 1 \\
 & \underline{x^3 - x^2} \\
 & \quad x^2 - 2x \\
 & \quad \underline{x^2 - x} \\
 & \qquad -x + 1 \\
 & \qquad \underline{-x + 1}
\end{array}
$$

Therefore $x^3 - 2x + 1 = (x-1)(x^2 + x - 1) = 0 \;\Leftrightarrow\; x - 1 = 0$ or $x^2 + x - 1 = 0 \;\Leftrightarrow$
$x = 1$ or [using the quadratic formula] $x = \dfrac{-1 \pm \sqrt{1^2 - 4(1)(-1)}}{2(1)} = \dfrac{-1 \pm \sqrt{5}}{2}$.

68. Let $p(x) = x^3 + 3x^2 + x - 1$, and notice that $p(-1) = 0$, so by the Factor Theorem, $(x+1)$ is a factor.
Use long division:

$$
\begin{array}{r|l}
 & x^2 + 2x - 1 \\
\hline
x + 1 & x^3 + 3x^2 + x - 1 \\
 & \underline{x^3 + x^2} \\
 & \quad 2x^2 + x \\
 & \quad \underline{2x^2 + 2x} \\
 & \qquad -x - 1 \\
 & \qquad \underline{-x - 1}
\end{array}
$$

Therefore $x^3 + 3x^2 + x - 1 = (x+1)(x^2 + 2x - 1) = 0 \;\Leftrightarrow\; x + 1 = 0$ or $x^2 + 2x - 1 = 0 \;\Leftrightarrow$
$x = -1$ or [using the quadratic formula] $x = \dfrac{-2 \pm \sqrt{2^2 - 4(1)(-1)}}{2} = -1 \pm \sqrt{2}$.

69. $2x^2 + 3x + 4$ is irreducible because its discriminant is negative: $b^2 - 4ac = 9 - 4(2)(4) = -23 < 0$.

70. The quadratic $2x^2 + 9x + 4$ is not irreducible because $b^2 - 4ac = 9^2 - 4(2)(4) = 49 > 0$.

71. $3x^2 + x - 6$ is not irreducible because its discriminant is nonnegative: $b^2 - 4ac = 1 - 4(3)(-6) = 73 > 0$.

72. The quadratic $x^2 + 3x + 6$ is irreducible because $b^2 - 4ac = 3^2 - 4(1)(6) = -15 < 0$.

73. Using the Binomial Theorem with $k = 6$ we have

$$\begin{aligned}(a+b)^6 &= a^6 + 6a^5b + \frac{6\cdot 5}{1\cdot 2}a^4b^2 + \frac{6\cdot 5\cdot 4}{1\cdot 2\cdot 3}a^3b^3 + \frac{6\cdot 5\cdot 4\cdot 3}{1\cdot 2\cdot 3\cdot 4}a^2b^4 + 6ab^5 + b^6\\ &= a^6 + 6a^5b + 15a^4b^2 + 20a^3b^3 + 15a^2b^4 + 6ab^5 + b^6\end{aligned}$$

74. Using the Binomial Theorem with $k = 7$ we have

$$\begin{aligned}(a+b)^7 &= a^7 + 7a^6b + \frac{7\cdot 6}{1\cdot 2}a^5b^2 + \frac{7\cdot 6\cdot 5}{1\cdot 2\cdot 3}a^4b^3 + \frac{7\cdot 6\cdot 5\cdot 4}{1\cdot 2\cdot 3\cdot 4}a^3b^4 + \frac{7\cdot 6\cdot 5\cdot 4\cdot 3}{1\cdot 2\cdot 3\cdot 4\cdot 5}a^2b^5 + 7ab^6 + b^7\\ &= a^7 + 7a^6b + 21a^5b^2 + 35a^4b^3 + 35a^3b^4 + 21a^2b^5 + 7ab^6 + b^7\end{aligned}$$

75. Using the Binomial Theorem with $a = x^2$, $b = -1$, $k = 4$ we have

$$\begin{aligned}(x^2-1)^4 = [x^2 + (-1)]^4 &= (x^2)^4 + 4(x^2)^3(-1) + \frac{4\cdot 3}{1\cdot 2}(x^2)^2(-1)^2 + 4(x^2)(-1)^3 + (-1)^4\\ &= x^8 - 4x^6 + 6x^4 - 4x^2 + 1\end{aligned}$$

76. Using the Binomial Theorem with $a = 3$, $b = x^2$, $k = 5$ we have

$$\begin{aligned}(3+x^2)^5 &= 3^5 + 5(3)^4(x^2)^1 + \frac{5\cdot 4}{1\cdot 2}(3)^3(x^2)^2 + \frac{5\cdot 4\cdot 3}{1\cdot 2\cdot 3}(3)^2(x^2)^3 + 5(3)(x^2)^4 + (x^2)^5\\ &= 243 + 405x^2 + 270x^4 + 90x^6 + 15x^8 + x^{10}\end{aligned}$$

77. Using Equation 10, $\sqrt{32}\,\sqrt{2} = \sqrt{32\cdot 2} = \sqrt{64} = 8$.

78. $\dfrac{\sqrt[3]{-2}}{\sqrt[3]{54}} = \sqrt[3]{\dfrac{-2}{54}} = \sqrt[3]{\dfrac{-1}{27}} = \dfrac{\sqrt[3]{-1}}{\sqrt[3]{27}} = \dfrac{-1}{3} = -\dfrac{1}{3}$

79. Using Equation 10, $\dfrac{\sqrt[4]{32x^4}}{\sqrt[4]{2}} = \dfrac{\sqrt[4]{32}\,\sqrt[4]{x^4}}{\sqrt[4]{2}} = \sqrt[4]{\dfrac{32}{2}}\,\sqrt[4]{x^4} = \sqrt[4]{16}\,|x| = 2\,|x|$.

80. $\sqrt{xy}\sqrt{x^3y} = \sqrt{(xy)(x^3y)} = \sqrt{x^4y^2} = x^2\,|y|$

81. Using Equation 10, $\sqrt{16a^4b^3} = \sqrt{16}\sqrt{a^4}\sqrt{b^3} = 4a^2b^{3/2} = 4a^2b\,b^{1/2} = 4a^2b\sqrt{b}$.

82. $\dfrac{\sqrt[5]{96a^6}}{\sqrt[5]{3a}} = \sqrt[5]{\dfrac{96a^6}{3a}} = \sqrt[5]{32a^5} = 2a$

83. Using Laws 3 and 1 of Exponents respectively, $3^{10}\times 9^8 = 3^{10}\times(3^2)^8 = 3^{10}\times 3^{2\cdot 8} = 3^{10+16} = 3^{26}$.

84. Using Laws 3 and 1, $2^{16}\times 4^{10}\times 16^6 = 2^{16}\times(2^2)^{10}\times(2^4)^6 = 2^{16}\times 2^{20}\times 2^{24} = 2^{60}$.

85. Using Laws 4, 1, and 2 of Exponents respectively, $\dfrac{x^9(2x)^4}{x^3} = \dfrac{x^9(2^4)x^4}{x^3} = \dfrac{16x^{9+4}}{x^3} = 16x^{9+4-3} = 16x^{10}$.

86. Using Laws 1 and 2, $\dfrac{a^n\times a^{2n+1}}{a^{n-2}} = \dfrac{a^{n+2n+1}}{a^{n-2}} = \dfrac{a^{3n+1}}{a^{n-2}} = a^{3n+1-(n-2)} = a^{2n+3}$.

87. Using Law 2 of Exponents, $\dfrac{a^{-3}b^4}{a^{-5}b^5} = a^{-3-(-5)}b^{4-5} = a^2b^{-1} = \dfrac{a^2}{b}$.

88. $\dfrac{x^{-1}+y^{-1}}{(x+y)^{-1}} = (x+y)\left(\dfrac{1}{x}+\dfrac{1}{y}\right) = (x+y)\left(\dfrac{y+x}{xy}\right) = \dfrac{(y+x)^2}{xy}$

89. By definitions 3 and 4 for exponents respectively, $3^{-1/2} = \dfrac{1}{3^{1/2}} = \dfrac{1}{\sqrt{3}}$.

90. $96^{1/5} = \sqrt[5]{96} = \sqrt[5]{32\cdot 3} = \sqrt[5]{32}\,\sqrt[5]{3} = 2\,\sqrt[5]{3}$

91. Using definition 4 for exponents, $125^{2/3} = \left[\sqrt[3]{125}\right]^2 = 5^2 = 25$.

92. $64^{-4/3} = \dfrac{1}{64^{4/3}} = \dfrac{1}{\left[\sqrt[3]{64}\right]^4} = \dfrac{1}{4^4} = \dfrac{1}{256}$

93. $(2x^2y^4)^{3/2} = 2^{3/2}(x^2)^{3/2}(y^4)^{3/2} = 2 \cdot 2^{1/2}\left[\sqrt{x^2}\right]^3\left[\sqrt{y^4}\right]^3 = 2\sqrt{2}\,|x|^3\,(y^2)^3 = 2\sqrt{2}\,|x|^3\,y^6$

94. $(x^{-5}y^3z^{10})^{-3/5} = (x^{-5})^{-3/5}(y^3)^{-3/5}(z^{10})^{-3/5} = x^{15/5}y^{-9/5}z^{-30/5} = \dfrac{x^2}{y^{9/5}z^6}$

95. $\sqrt[5]{y^6} = y^{6/5}$ by definition 4 for exponents.

96. $(\sqrt[4]{a})^3 = (a^{1/4})^3 = a^{3/4}$

97. $\dfrac{1}{(\sqrt{t})^5} = \dfrac{1}{(t^{1/2})^5} = \dfrac{1}{t^{5/2}} = t^{-5/2}$

98. $\dfrac{\sqrt[8]{x^5}}{\sqrt[4]{x^3}} = \dfrac{x^{5/8}}{x^{3/4}} = x^{(5/8)-(3/4)} = x^{-1/8} = \dfrac{1}{x^{1/8}}$

99. $\sqrt[4]{\dfrac{t^{1/2}\sqrt{st}}{s^{2/3}}} = \left(\dfrac{t^{1/2}s^{1/2}t^{1/2}}{s^{2/3}}\right)^{1/4} = \left(t^{(1/2)+(1/2)}s^{(1/2)-(2/3)}\right)^{1/4} = (ts^{-1/6})^{1/4}$

$= t^{1/4}s^{(-1/6)\cdot(1/4)} = \dfrac{t^{1/4}}{s^{1/24}}$

100. $\sqrt[4]{r^{2n+1}} \times \sqrt[4]{r^{-1}} = \sqrt[4]{r^{2n+1} \times r^{-1}} = \sqrt[4]{r^{2n+1-1}} = \sqrt[4]{r^{2n}} = (r^{2n})^{1/4} = r^{2n/4} = r^{n/2}$

101. $\dfrac{\sqrt{x}-3}{x-9} = \dfrac{\sqrt{x}-3}{x-9} \cdot \dfrac{\sqrt{x}+3}{\sqrt{x}+3} = \dfrac{(x-9)}{(x-9)(\sqrt{x}+3)} = \dfrac{1}{\sqrt{x}+3}$

102. $\dfrac{\dfrac{1}{\sqrt{x}}-1}{x-1} = \dfrac{\dfrac{1}{\sqrt{x}}-1}{x-1} \cdot \dfrac{\dfrac{1}{\sqrt{x}}+1}{\dfrac{1}{\sqrt{x}}+1} = \dfrac{\dfrac{1}{x}-1}{(x-1)\left(\dfrac{1}{\sqrt{x}}+1\right)} = \dfrac{\dfrac{1-x}{x}}{(x-1)\left(\dfrac{1}{\sqrt{x}}+1\right)} = \dfrac{-1}{x\left(\dfrac{1}{\sqrt{x}}+1\right)} = \dfrac{-1}{\sqrt{x}+x}$

103. $\dfrac{x\sqrt{x}-8}{x-4} = \dfrac{x\sqrt{x}-8}{x-4} \cdot \dfrac{x\sqrt{x}+8}{x\sqrt{x}+8} = \dfrac{x^3-64}{(x-4)(x\sqrt{x}+8)}$

$= \dfrac{(x-4)(x^2+4x+16)}{(x-4)(x\sqrt{x}+8)}$ [Equation 4 with $a = x$, $b = 4$] $= \dfrac{x^2+4x+16}{x\sqrt{x}+8}$

104. $\dfrac{\sqrt{2+h}+\sqrt{2-h}}{h} = \dfrac{\sqrt{2+h}+\sqrt{2-h}}{h} \cdot \dfrac{\sqrt{2+h}-\sqrt{2-h}}{\sqrt{2+h}-\sqrt{2-h}} = \dfrac{2+h-(2-h)}{h\left(\sqrt{2+h}-\sqrt{2-h}\right)}$

$= \dfrac{2}{\sqrt{2+h}-\sqrt{2-h}}$

105. $\dfrac{2}{3-\sqrt{5}} = \dfrac{2}{3-\sqrt{5}} \cdot \dfrac{3+\sqrt{5}}{3+\sqrt{5}} = \dfrac{2\left(3+\sqrt{5}\right)}{9-5} = \dfrac{3+\sqrt{5}}{2}$

106. $\dfrac{1}{\sqrt{x}-\sqrt{y}} = \dfrac{1}{\sqrt{x}-\sqrt{y}} \cdot \dfrac{\sqrt{x}+\sqrt{y}}{\sqrt{x}+\sqrt{y}} = \dfrac{\sqrt{x}+\sqrt{y}}{x-y}$

107. $\sqrt{x^2+3x+4}-x = \left(\sqrt{x^2+3x+4}-x\right) \cdot \dfrac{\sqrt{x^2+3x+4}+x}{\sqrt{x^2+3x+4}+x} = \dfrac{x^2+3x+4-x^2}{\sqrt{x^2+3x+4}+x} = \dfrac{3x+4}{\sqrt{x^2+3x+4}+x}$

108. $\sqrt{x^2+x}-\sqrt{x^2-x} = \left(\sqrt{x^2+x}-\sqrt{x^2-x}\right) \cdot \dfrac{\sqrt{x^2+x}+\sqrt{x^2-x}}{\sqrt{x^2+x}+\sqrt{x^2-x}} = \dfrac{x^2+x-(x^2-x)}{\sqrt{x^2+x}+\sqrt{x^2-x}}$

$= \dfrac{2x}{\sqrt{x^2+x}+\sqrt{x^2-x}}$

109. False. See Example 14(b).

110. False. See the warning after Equation 10.

111. True: $\dfrac{16+a}{16} = \dfrac{16}{16} + \dfrac{a}{16} = 1 + \dfrac{a}{16}$

112. False: $\dfrac{1}{x^{-1}+y^{-1}} = \dfrac{1}{\dfrac{1}{x}+\dfrac{1}{y}} = \dfrac{1}{\dfrac{x+y}{xy}} = \dfrac{xy}{x+y} \neq x+y$

113. False.

114. False. See the warning on page 2.

115. False. Using Law 3 of Exponents, $(x^3)^4 = x^{3\cdot 4} = x^{12} \neq x^7$.

116. True.

117. $|5-23| = |-18| = 18$

118. $|\pi - 2| = \pi - 2$ because $\pi - 2 > 0$.

119. $|\sqrt{5}-5| = -(\sqrt{5}-5) = 5-\sqrt{5}$ because $\sqrt{5}-5<0$.

120. $\big||-2|-|-3|\big| = |2-3| = |-1| = 1$

121. If $x<2$, $x-2<0$, so $|x-2| = -(x-2) = 2-x$.

122. If $x>2$, $x-2>0$, so $|x-2| = x-2$.

123. $|x+1| = \begin{cases} x+1 & \text{if } x+1\geq 0 \\ -(x+1) & \text{if } x+1<0 \end{cases} = \begin{cases} x+1 & \text{if } x\geq -1 \\ -x-1 & \text{if } x<-1 \end{cases}$

124. $|2x-1| = \begin{cases} 2x-1 & \text{if } 2x-1\geq 0 \\ -(2x-1) & \text{if } 2x-1<0 \end{cases} = \begin{cases} 2x-1 & \text{if } x\geq \frac{1}{2} \\ 1-2x & \text{if } x<\frac{1}{2} \end{cases}$

125. $|x^2+1| = x^2+1$ (since $x^2+1\geq 0$ for all x).

126. Determine when $1-2x^2<0 \;\Leftrightarrow\; 1<2x^2 \;\Leftrightarrow\; x^2>\frac{1}{2} \;\Leftrightarrow\; \sqrt{x^2}>\sqrt{\frac{1}{2}} \;\Leftrightarrow\; |x|>\sqrt{\frac{1}{2}} \;\Leftrightarrow$ $x<-\frac{1}{\sqrt{2}}$ or $x>\frac{1}{\sqrt{2}}$. Thus, $|1-2x^2| = \begin{cases} 1-2x^2 & \text{if } -\frac{1}{\sqrt{2}}\leq x\leq \frac{1}{\sqrt{2}} \\ 2x^2-1 & \text{if } x<-\frac{1}{\sqrt{2}} \text{ or } x>\frac{1}{\sqrt{2}} \end{cases}$

127. $2x+7>3 \;\Leftrightarrow\; 2x>-4 \;\Leftrightarrow\; x>-2$, so $x\in(-2,\infty)$.

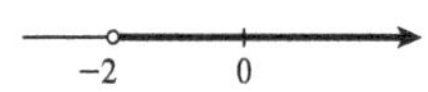

128. $4-3x\geq 6 \;\Leftrightarrow\; -3x\geq 2 \;\Leftrightarrow\; x\leq -\frac{2}{3}$, so $x\in\left(-\infty,-\frac{2}{3}\right]$.

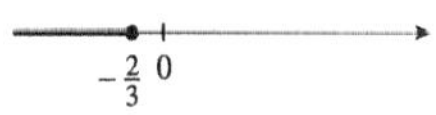

129. $1-x\leq 2 \;\Leftrightarrow\; -x\leq 1 \;\Leftrightarrow\; x\geq -1$, so $x\in[-1,\infty)$.

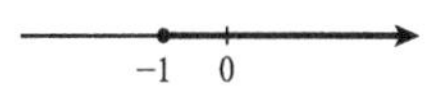

130. $1+5x>5-3x \;\Leftrightarrow\; 8x>4 \;\Leftrightarrow\; x>\frac{1}{2}$, so $x\in\left(\frac{1}{2},\infty\right)$.

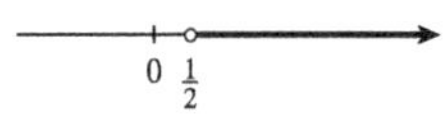

131. $0\leq 1-x<1 \;\Leftrightarrow\; -1\leq -x<0 \;\Leftrightarrow\; 1\geq x>0$, so $x\in(0,1]$.

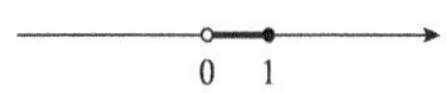

132. $1<3x+4\leq 16 \;\Leftrightarrow\; -3<3x\leq 12 \;\Leftrightarrow\; -1<x\leq 4$, so $x\in(-1,4]$.

133. $(x-1)(x-2)>0$. *Case 1:* (both factors are positive, so their product is positive)

$x-1>0 \;\Leftrightarrow\; x>1$, and $x-2>0 \;\Leftrightarrow\; x>2$, so $x\in(2,\infty)$.

Case 2: (both factors are negative, so their product is positive)

$x-1<0 \;\Leftrightarrow\; x<1$, and $x-2<0 \;\Leftrightarrow\; x<2$, so $x\in(-\infty,1)$.

Thus, the solution set is $(-\infty,1)\cup(2,\infty)$.

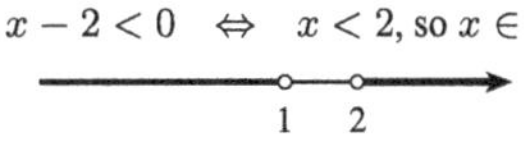

134. $x^2<2x+8 \;\Leftrightarrow\; x^2-2x-8<0 \;\Leftrightarrow\; (x-4)(x+2)<0$. *Case 1:* $x>4$ and $x<-2$, which is impossible. *Case 2:* $x<4$ and $x>-2$. Thus, the solution set is $(-2,4)$.

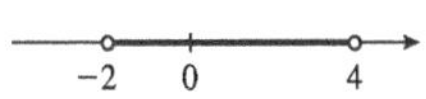

135. $x^2<3 \;\Leftrightarrow\; x^2-3<0 \;\Leftrightarrow\; (x-\sqrt{3})(x+\sqrt{3})<0$. *Case 1:* $x>\sqrt{3}$ and $x<-\sqrt{3}$, which is impossible. *Case 2:* $x<\sqrt{3}$ and $x>-\sqrt{3}$. Thus, the solution set is $(-\sqrt{3},\sqrt{3})$.

Another method: $x^2<3 \;\Leftrightarrow\; |x|<\sqrt{3} \;\Leftrightarrow\; -\sqrt{3}<x<\sqrt{3}$.

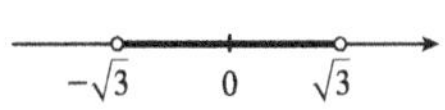

136. $x^2 \geq 5 \quad \Leftrightarrow \quad x^2 - 5 \geq 0 \quad \Leftrightarrow \quad (x - \sqrt{5})(x + \sqrt{5}) \geq 0$. *Case 1:* $x \geq \sqrt{5}$ and $x \geq -\sqrt{5}$, so $x \in [\sqrt{5}, \infty)$. *Case 2:* $x \leq \sqrt{5}$ and $x \leq -\sqrt{5}$, so $x \in (-\infty, -\sqrt{5}]$. Thus, the solution set is $(-\infty, -\sqrt{5}] \cup [\sqrt{5}, \infty)$.
Another method: $x^2 \geq 5 \quad \Leftrightarrow \quad |x| \geq \sqrt{5} \quad \Leftrightarrow \quad x \geq \sqrt{5}$ or $x \leq -\sqrt{5}$.

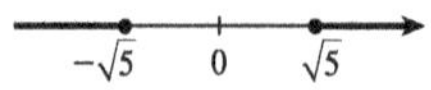

137. $x^3 - x^2 \leq 0 \quad \Leftrightarrow \quad x^2(x - 1) \leq 0$. Since $x^2 \geq 0$ for all x, the inequality is satisfied when $x - 1 \leq 0 \quad \Leftrightarrow \quad x \leq 1$. Thus, the solution set is $(-\infty, 1]$.

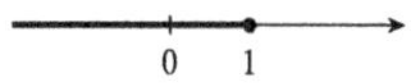

138. $(x + 1)(x - 2)(x + 3) = 0 \quad \Leftrightarrow \quad x = -1, 2$, or -3. Construct a chart:

Interval	$x+1$	$x-2$	$x+3$	$(x+1)(x-2)(x+3)$
$x < -3$	$-$	$-$	$-$	$-$
$-3 < x < -1$	$-$	$-$	$+$	$+$
$-1 < x < 2$	$+$	$-$	$+$	$-$
$x > 2$	$+$	$+$	$+$	$+$

Thus, $(x + 1)(x - 2)(x + 3) \geq 0$ on $[-3, -1]$ and $[2, \infty)$, and the solution set is $[-3, -1] \cup [2, \infty)$.

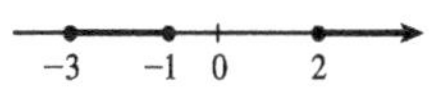

139. $x^3 > x \quad \Leftrightarrow \quad x^3 - x > 0 \quad \Leftrightarrow \quad x(x^2 - 1) > 0 \quad \Leftrightarrow \quad x(x - 1)(x + 1) > 0$. Construct a chart:

Interval	x	$x-1$	$x+1$	$x(x-1)(x+1)$
$x < -1$	$-$	$-$	$-$	$-$
$-1 < x < 0$	$-$	$-$	$+$	$+$
$0 < x < 1$	$+$	$-$	$+$	$-$
$x > 1$	$+$	$+$	$+$	$+$

Since $x^3 > x$ when the last column is positive, the solution set is $(-1, 0) \cup (1, \infty)$.

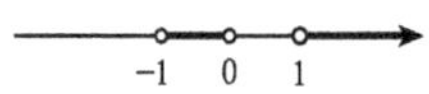

140. $x^3 + 3x < 4x^2 \quad \Leftrightarrow \quad x^3 - 4x^2 + 3x < 0 \quad \Leftrightarrow \quad x(x^2 - 4x + 3) < 0 \quad \Leftrightarrow \quad x(x - 1)(x - 3) < 0$.

Interval	x	$x-1$	$x-3$	$x(x-1)(x-3)$
$x < 0$	$-$	$-$	$-$	$-$
$0 < x < 1$	$+$	$-$	$-$	$+$
$1 < x < 3$	$+$	$+$	$-$	$-$
$x > 3$	$+$	$+$	$+$	$+$

Thus, the solution set is $(-\infty, 0) \cup (1, 3)$.

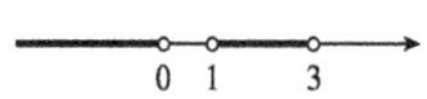

141. $1/x < 4$. This is clearly true for $x < 0$. So suppose $x > 0$. then $1/x < 4 \quad \Leftrightarrow \quad 1 < 4x \quad \Leftrightarrow \quad \frac{1}{4} < x$. Thus, the solution set is $(-\infty, 0) \cup (\frac{1}{4}, \infty)$.

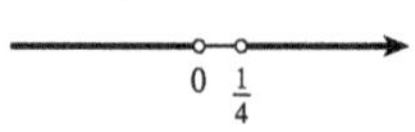

142. $-3 < 1/x \leq 1$. We solve the two inequalities separately and take the intersection of the solution sets. First, $-3 < 1/x$ is clearly true for $x > 0$. So suppose $x < 0$. Then $-3 < 1/x \quad \Leftrightarrow \quad -3x > 1 \quad \Leftrightarrow \quad x < -\frac{1}{3}$, so for this inequality, the solution set is $\left(-\infty, -\frac{1}{3}\right) \cup (0, \infty)$. Now $1/x \leq 1$ is clearly true if $x < 0$. So suppose $x > 0$. Then $1/x \leq 1 \quad \Leftrightarrow \quad 1 \leq x$, and the solution set here is $(-\infty, 0) \cup [1, \infty)$. Taking the intersection of the two solution sets gives the final solution set: $\left(-\infty, -\frac{1}{3}\right) \cup [1, \infty)$.

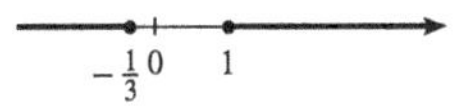

143. $C = \frac{5}{9}(F - 32) \quad \Rightarrow \quad F = \frac{9}{5}C + 32$. So $50 \leq F \leq 95 \quad \Rightarrow \quad 50 \leq \frac{9}{5}C + 32 \leq 95 \quad \Rightarrow \quad 18 \leq \frac{9}{5}C \leq 63 \quad \Rightarrow$ $10 \leq C \leq 35$. So the interval is $[10, 35]$.

144. Since $20 \leq C \leq 30$ and $C = \frac{5}{9}(F - 32)$, we have $20 \leq \frac{5}{9}(F - 32) \leq 30 \quad \Rightarrow \quad 36 \leq F - 32 \leq 54 \quad \Rightarrow$ $68 \leq F \leq 86$. So the interval is $[68, 86]$.

145. (a) Let T represent the temperature in degrees Celsius and h the height in km. $T = 20$ when $h = 0$ and T decreases by $10°\mathrm{C}$ for every km ($1°\mathrm{C}$ for each 100-m rise). Thus, $T = 20 - 10h$ when $0 \leq h \leq 12$.

(b) From part (a), $T = 20 - 10h \quad \Rightarrow \quad 10h = 20 - T \quad \Rightarrow \quad h = 2 - T/10$. So $0 \leq h \leq 5 \quad \Rightarrow$ $0 \leq 2 - T/10 \leq 5 \quad \Rightarrow \quad -2 \leq -T/10 \leq 3 \quad \Rightarrow \quad -20 \leq -T \leq 30 \quad \Rightarrow \quad 20 \geq T \geq -30 \quad \Rightarrow$ $-30 \leq T \leq 20$. Thus, the range of temperatures (in $°\mathrm{C}$) to be expected is $[-30, 20]$.

146. The ball will be at least 32 ft above the ground if $h \geq 32 \quad \Leftrightarrow \quad 128 + 16t - 16t^2 \geq 32 \quad \Leftrightarrow$ $16t^2 - 16t - 96 \leq 0 \quad \Leftrightarrow \quad 16(t - 3)(t + 2) \leq 0$. $t = 3$ and $t = -2$ are endpoints of the interval we're looking for, and constructing a table gives $-2 \leq t \leq 3$. But $t \geq 0$, so the ball will be at least 32 ft above the ground in the time interval $[0, 3]$.

147. $|x + 3| = |2x + 1| \quad \Leftrightarrow \quad$ either $x + 3 = 2x + 1$ or $x + 3 = -(2x + 1)$. In the first case, $x = 2$, and in the second case, $x + 3 = -2x - 1 \quad \Leftrightarrow \quad 3x = -4 \quad \Leftrightarrow \quad x = -\frac{4}{3}$. So the solutions are $-\frac{4}{3}$ and 2.

148. $|3x + 5| = 1 \quad \Leftrightarrow \quad$ either $3x + 5 = 1$ or -1. In the first case, $3x = -4 \quad \Leftrightarrow \quad x = -\frac{4}{3}$, and in the second case, $3x = -6 \quad \Leftrightarrow \quad x = -2$. So the solutions are -2 and $-\frac{4}{3}$.

149. By Property 5 of absolute values, $|x| < 3 \quad \Leftrightarrow \quad -3 < x < 3$, so $x \in (-3, 3)$.

150. By Properties 4 and 6 of absolute values, $|x| \geq 3 \quad \Leftrightarrow \quad x \leq -3$ or $x \geq 3$, so $x \in (-\infty, -3] \cup [3, \infty)$.

151. $|x - 4| < 1 \quad \Leftrightarrow \quad -1 < x - 4 < 1 \quad \Leftrightarrow \quad 3 < x < 5$, so $x \in (3, 5)$.

152. $|x - 6| < 0.1 \quad \Leftrightarrow \quad -0.1 < x - 6 < 0.1 \quad \Leftrightarrow \quad 5.9 < x < 6.1$, so $x \in (5.9, 6.1)$.

153. $|x + 5| \geq 2 \quad \Leftrightarrow \quad x + 5 \geq 2$ or $x + 5 \leq -2 \quad \Leftrightarrow \quad x \geq -3$ or $x \leq -7$, so $x \in (-\infty, -7] \cup [-3, \infty)$.

154. $|x + 1| \geq 3 \quad \Leftrightarrow \quad x + 1 \geq 3$ or $x + 1 \leq -3 \quad \Leftrightarrow \quad x \geq 2$ or $x \leq -4$, so $x \in (-\infty, -4] \cup [2, \infty)$.

155. $|2x - 3| \leq 0.4 \quad \Leftrightarrow \quad -0.4 \leq 2x - 3 \leq 0.4 \quad \Leftrightarrow \quad 2.6 \leq 2x \leq 3.4 \quad \Leftrightarrow \quad 1.3 \leq x \leq 1.7$, so $x \in [1.3, 1.7]$.

156. $|5x - 2| < 6 \quad \Leftrightarrow \quad -6 < 5x - 2 < 6 \quad \Leftrightarrow \quad -4 < 5x < 8 \quad \Leftrightarrow \quad -\frac{4}{5} < x < \frac{8}{5}$, so $x \in \left(-\frac{4}{5}, \frac{8}{5}\right)$.

157. $a(bx - c) \geq bc \quad \Leftrightarrow \quad bx - c \geq \dfrac{bc}{a} \quad \Leftrightarrow \quad bx \geq \dfrac{bc}{a} + c = \dfrac{bc + ac}{a} \quad \Leftrightarrow \quad x \geq \dfrac{bc + ac}{ab}$

158. $ax + b < c \quad \Leftrightarrow \quad ax < c - b \quad \Leftrightarrow \quad x > \dfrac{c - b}{a}$ (since $a < 0$)

159. $|ab| = \sqrt{(ab)^2} = \sqrt{a^2 b^2} = \sqrt{a^2}\,\sqrt{b^2} = |a|\,|b|$

160. If $0 < a < b$, then $a \cdot a < a \cdot b$ and $a \cdot b < b \cdot b$ [using Rule 3 of Inequalities]. So $a^2 < ab < b^2$ and hence $a^2 < b^2$.